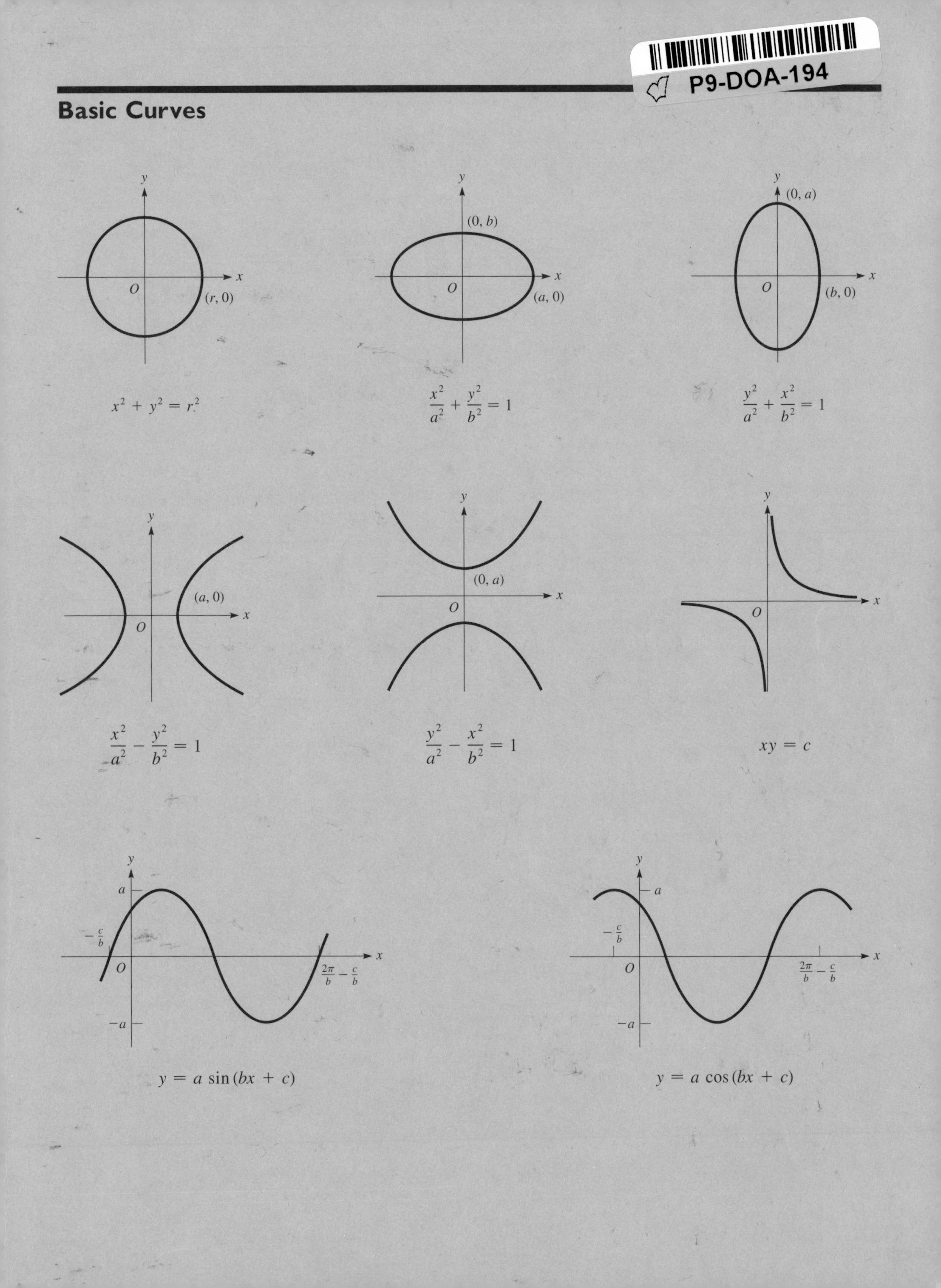
P9-DOA-194
Basic Curves
y
x
O
(r, 0)
$x^2 + y^2 = r^2$
(0, b)
(a, 0)
$\frac{x^2}{a^2} + \frac{y^2}{b^2} = 1$
(0, a)
(b, 0)
$\frac{y^2}{a^2} + \frac{x^2}{b^2} = 1$
(a, 0)
$\frac{x^2}{a^2} - \frac{y^2}{b^2} = 1$
(0, a)
$\frac{y^2}{a^2} - \frac{x^2}{b^2} = 1$
$xy = c$
a
$-\frac{c}{b}$
$\frac{2\pi}{b} - \frac{c}{b}$
−a
$y = a \sin(bx + c)$
$y = a \cos(bx + c)$

6th Edition

Basic Technical Mathematics with Calculus

Metric Version

Other Addison-Wesley Titles of Related Interest

Basic Technical Mathematics, Sixth Edition, by Allyn J. Washington

Basic Technical Mathematics with Calculus, Sixth Edition, by Allyn J. Washington

Introduction to Technical Mathematics, Fourth Edition, by Allyn J. Washington and Mario F. Triola

Technical Calculus with Analytic Geometry, Third Edition, by Allyn J. Washington

6th Edition

Basic Technical Mathematics with Calculus

Metric Version

Allyn J. Washington

Dutchess Community College

Addison-Wesley Publishing Company

Reading, Massachusetts • Menlo Park, California • New York
Don Mills, Ontario • Wokingham, England • Amsterdam • Bonn
Sydney • Singapore • Tokyo • Madrid • San Juan • Milan • Paris

Publisher: Ron Doleman
Managing Editor: Linda Scott
Sponsoring Editor: Jason A. Jordan
Developmental Editor: David Chelton
Production Services: Greg Hubit Bookworks
Copy Editor: Margaret Moore
Text Design: Darci Mehall
Illustrator: Lotus Art
Photo Research: Amy Branowicki
Prepress Buying Manager: Sarah McCracken
Manufacturing Supervisor: Roy Logan
Marketing Manager: Andy Fisher
Cover Design: Leslie Haimes
Cover Photograph: Mel Lindstrom
Composition: Beacon Graphics Company
Printer: Von Hoffmann

Photo Credits

Chapter 1 United Airlines; *Chapter 2* National Park Service; *Chapter 3* Michael Dwyer/Stock, Boston; *Chapter 4* John M. Baron; *Chapter 5* General Motors Corporation; *Chapter 6* Courtesy of International Business Machines Corporation; *Chapter 7* CraftMaster Interior Door Design by Masonite; *Chapter 8* Martin Marietta Astro Space; *Chapter 9* Courtesy of Beech Aircraft Corporation; *Chapter 10* Donald Dietz/Stock, Boston; *Chapter 11* Department of Energy; *Chapter 12* Laimute Druskis/Stock, Boston; *Chapter 13* Spencer Grant/Stock, Boston; *Chapter 14* © 1988 Blair Seitz/Photo Researchers; *Chapter 15* © 1994, Comstock, Inc.; *Chapter 16* © 1986 Ray Ellis/Photo Researchers; *Chapter 17* © David R. Frazier/Photo Researchers; *Chapter 18* NASA; *Chapter 19* Tim Barnwell/Stock, Boston; *Chapter 20* Bruce Robinson & Mildred Washington; *Chapter 21* Eric Neurath/Stock, Boston; *Chapter 22* Chris Morrow/Stock, Boston; *Chapter 23* John M. Baron; *Chapter 24* Mildred Washington; *Chapter 25* Bureau of Reclamation; *Chapter 26* Lionel Delevingne/Stock, Boston; *Chapter 27* NASA; *Chapter 28* Whirlpool Home Appliances; *Chapter 29* Mel Lindstrom; *Chapter 30* Department of Energy.

Reprinted with corrections, March 1997.

ISBN 0-201-76642-6

E F VH 9998

Preface

Scope of the Book

This book is intended primarily for students in technical and pre-engineering technology programs or other programs for which a coverage of basic mathematics is required. The distinctive feature of this metric version is that all units of measurement are metric (SI) units, or are acceptable for use with SI units. All other features are the same as those in the regular sixth edition.

Chapters 1 through 20 provide the necessary background in algebra and trigonometry for analytic geometry and calculus courses, and Chapters 1 through 21 provide the necessary background for calculus courses. There is an integrated treatment of mathematical topics, from algebra to calculus, which are necessary for a sound mathematical background for the technician. Numerous applications from many fields of technology are included, primarily to indicate where and how mathematical techniques are used. However, it is not necessary that the student have a specific knowledge of the technical area from which any given problem is taken.

Most students using this text will have a background including some algebra and geometry. However, the material is presented in sufficient detail for use by those whose background is possibly deficient to some extent in these areas. The material presented here is sufficient for three or four semesters.

One of the primary reasons for the arrangement of topics in this text is to present material in an order that allows a student to take courses in allied technical areas, such as physics and electricity, concurrently. These allied courses normally require a student to know certain mathematical topics by certain definite times; yet the traditional order of topics in mathematics makes it difficult to attain this coverage without loss of continuity. However, the material in this book can be rearranged to fit any appropriate sequence of topics if this is deemed necessary. Another feature of this text is that certain topics that are traditionally included primarily for mathematical completeness have been covered only briefly or have been omitted.

The approach used here is basically an intuitive one. It is not unduly rigorous mathematically, although all appropriate terms and concepts are introduced as needed and given an intuitive or algebraic foundation. The book's aim is to help the student develop a feeling for mathematical methods, and not simply to provide a collection of formulas. The text material emphasizes that it is essential for the student to have a sound background in algebra and trigonometry in order to understand and succeed in any subsequent work in mathematics.

New Features

The sixth edition of *Basic Technical Mathematics with Calculus, Metric Version,* includes all the basic features of the earlier editions. However, most sections have been rewritten to some degree to include additional or revised explanatory material, examples, and exercises. Some sections have been extensively revised. Specifically, among the new features of this edition are the following:

◆ **The Graphing Calculator**
The graphing calculator (or other graphing utility) is used throughout the text in examples and exercises to help develop and reinforce the coverage of many topics. The graphing calculator is noted in Section 1-3, with an introduction to its use in Section 3-5. Additional coverage on its use is found in Appendix C and in many examples throughout the book.

◆ **Geometry Chapter**
A new chapter on geometry has been included as Chapter 2. It is intended for those students who need a thorough review of the basic topics of geometry. It has been found that many students need a more complete coverage than that which was provided in earlier editions with the appendix on geometry.

◆ **Problem-Solving Techniques**
Techniques and procedures that summarize the approaches to solving many types of problems have been clearly outlined in boxes.

◆ **Revised Coverage**
In addition to the chapter on geometry and the sections that now include a coverage of the graphing calculator, other sections also have revised coverage from the fifth edition. Section 1-3 now includes the material on approximate numbers that had been in Appendix B in earlier editions. Section 6-4 is a separate section on the algebraic expansion of the sum and difference of cubes. In Section 11-5, the coverage of the multiplication and division of radicals has been reduced from two sections to one.

◆ **Writing Exercises**
In order to have the student practice written explanations of solutions, some writing exercises have been included. There is one specific writing exercise at the end of each chapter. Also, there are over 120 additional exercises throughout the book (at least two in each chapter) that require a sentence or two of explanation along with the answer. A list of these additional exercises is found in the Instructor's Guide.

◆ **Subheads and Key Terms**
Many sections now include subheads to indicate where the discussion of a new topic starts within the section. Also, other key terms are noted in the margin for emphasis and easy reference.

◆ **Special Caution and Note Margin Indicators**

CAUTION ▶ NOTE ▶

Two special margin indicators (as shown at the left) are used in this edition. The CAUTION indicator is used to clearly identify points with which students commonly make errors or which they tend to have more difficulty in handling. The NOTE indicator is used to point out text material that is of particular importance in developing or understanding the topic under discussion.

◆ **Page Layout**

Special attention has been given to the page layout in this edition. Although each page is approximately 20% larger, the amount of material on each page is about the same as in the fifth edition, and this allows for a more open appearance. Also, nearly all examples are started and completed on the same page (there are only four exceptions, and each of these is presented on facing pages). Also, all figures are shown immediately adjacent to the material in which they are discussed.

◆ **Figures**

In addition to over 150 new figures, all of the figures from the fifth edition have been redrawn. There are now over 1000 figures in the text, in addition to those in the answer section.

◆ **Exercises**

There are now over 2200 exercises illustrating technical applications, an increase of about 15% from the fifth edition. Of these new exercises, many illustrate applications related to real-world structures and places. In all, there are about 10 000 exercises in this edition.

◆ **Examples**

There are now over 280 worked examples illustrating technical applications. Of these, about 15% are new to the sixth edition. In all, there are about 1350 examples.

Additional Features

◆ **Special Explanatory Comments**

Throughout the book, special explanatory comments in color have been used in the examples to emphasize and clarify certain important points. Arrows are used to clearly indicate the part of the example to which reference is made.

◆ **Chapter Introductions**

Each chapter is introduced by identifying the topics to be developed and some of the important areas of technical applications. A particular type of application is shown in a photograph, and in Chapters 2 through 30 a problem related to this application is solved in an example later in the chapter.

◆ **Chapter Equations**

At the end of each chapter, all important chapter equations are listed together for easy reference.

◆ **Review Exercises**
Each chapter is followed by review exercises. These can be used either as additional problems or as review assignments.

◆ **Chapter Practice Tests**
At the end of each chapter there is a practice test that a student may use to check on his or her understanding of the material. Solutions to all problems in each test are given at the back of the book.

◆ **Applications**
The text material and exercises illustrate the application of mathematics to all fields of technology. Many of the exercises relate to modern technology such as computer design, computer-assisted design (CAD), electronics, solar energy, lasers, fiber optics, holography, the environment, and space technology. A special *Index of Applications* is included near the end of the book.

◆ **Supplementary Topics**
In order to respond to the needs of certain programs, eight additional sections of text material are included after Chapter 30. The topics covered are *Gaussian elimination, rotation of axes, functions of two variables, curves and surfaces in three dimensions, partial derivatives, double integrals,* and two sections on *integration by partial fractions.*

◆ **Graphical Methods**
In addition to specific topics from analytic geometry, graphical techniques and interpretations are included throughout the text. The graphing calculator has been used frequently in this regard.

◆ **Stated Problems**
Stated problems are introduced in Chapter 1, and over 120 examples and 900 exercises appear throughout the text. Including them a few at a time, but regularly, allows students to better develop techniques of solution. As it has been noted, procedures for solving these types of problems are clearly outlined.

◆ **Important Formulas**
These formulas are set off and displayed so that they are easily located and used.

◆ **Units of Measurement**
Units are in *Le Système International d' Unités* (SI). All uses of metric units are in keeping with accepted international standards.

◆ **Answers to Exercises**
The answers to all the odd-numbered exercises (except the writing exercises) are given at the back of the book. Also, the *Student Solutions Manual* contains solutions for every other odd-numbered exercise. The answers to all exercises are given in the *Instructor's Guide.*

◆ **Use of Color**
A second color is used extensively to assist in highlighting important points and features. It is used for special symbols, special explanatory comments, graphs and figures, and other display features.

◆ **Flexibility of Material Coverage**
The order of material coverage can be changed in many places, and certain sections may be omitted without loss of continuity of coverage. Users of earlier editions have indicated the successful use of numerous variations in coverage. Any changes or omissions will depend on the type of course and completeness required. Several of the possible variations in coverage are discussed in the *Instructor's Guide.*

◆ **Computer Supplement**
TECHDISK 2.0 is a software package that follows the sequence of topics in the text and allows students to have additional practice on the concepts presented. It does not require previous computer experience.

◆ **Other Supplements**
Other supplements to this text include an *Instructor's Guide,* which presents information on the use of the text as well as answers to all of the exercises. A *Student Solutions Manual* includes detailed solutions for every other odd-numbered exercise, as was previously noted. A *Graphing Utility Activities Manual,* which is correlated to the sixth edition, is also available. For test preparation, there is a *Printed Test Bank,* as well as *OmniTest*^{3}, an algorithm-driven computer testing system. *OmniTest*^{3} is available for both DOS- and Macintosh-based computers. Instructors may obtain a copy of any of these supplements by contacting their Addison-Wesley sales representative.

◆ **Questions/Comments/Information**
We welcome your comments about this text. You may write the publisher or the author at:

Addison-Wesley Publishing Company
Mathematics Marketing—Higher Ed
1 Jacob Way
Reading, MA 01867

You may also use our Internet address: TECHMATH@AW.COM to leave comments, suggestions, or questions for Addison-Wesley or the author. We will also make information relevant to Technical Mathematics available with automatic return e-mail.

Acknowledgments

The author gratefully acknowledges the contributions of the following reviewers. Their detailed comments and many valuable suggestions were of great assistance in preparing this sixth edition.

Stan Adamski
Owens Community College

Vincent Bates
Anne Arundel Community College

Deborah Bennett
State University of New York at Farmingdale

Christopher M. Burgess
Conestoga College

Edward Champy
Northern Essex Community College

Ken Chow
Mohawk College

James Corbett, Jr.
J. S. Reynolds Community College

Sandra Dashiell
Thomas Nelson Community College

David Earnshaw
Lambton College

Robert Ewen
State University of New York at Morrisville

William Ferguson
Columbus State Community College

Lionel Geller
Dawson College

Glen A. Goodale
Dawson College

Gary Helmer
Mohawk College

Tommy Hinson
Forsyth Technical Community College

Henry Hosek
Purdue University–Calumet

Fred Janusek
Northeast Wisconsin Technical College

Wendell Johnson
University of Akron
Community and Technical College

Judy Ann Jones
Madison Area Technical College

Joseph Jordan
John Tyler Community College

Maureen Kelley
Northern Essex Community College

Robert Kimball
Wake Technical Community College

Colin Lawrence
British Columbia Institute of Technology

Roger Loiseau
Naugatuck Valley Community–Technical College

Douglas MacDonald
New Brunswick Community College

S. Paul Maini
Suffolk County Community College

Robert Maynard
Tidewater Community College

Joan Page
Onondaga Community College

Peter Papadopoulos
DeVry Institute of Technology

Jorge Sarmiento
County College of Morris

J. William Schaller
Hocking College

Doris Schoonmaker
Hudson Valley Community College

Daniel Sims
Thomas Nelson Community College

Tracy Shields
Seneca College

Rod Somppi
Confederation College

Thomas Stark
Cincinnati Technical College

Frank Stoyles
Algonquin College

Richard Watkins
Tidewater Community College

Joseph Wilson
University of Pittsburgh–Johnstown

Bill Wunderlich
Cincinnati Technical College

Ben Zirkle
Virginia Western Community College

I again wish to thank the members of the Mathematics Department of Dutchess Community College for their help and suggestions for the earlier editions.

Special thanks go to Anne Zeigler of Florence Darlington Technical College and Frances Willbanks of Southern Bell for again preparing the *Student Solutions Manual,* and to Robert Seaver of Lorain County Community College and William Thomas of the University of Toledo for preparing *TECHDISK 2.0.* I thank Robert Seaver again for preparing the *Graphing Utility Activities Manual.*

My thanks and gratitude go to Carolyn Edmond for preparing and checking the answers to all of the exercises. I wish to thank Gloria Langer, Doreen Clark, and Steven Finch for assisting in the tedious task of reading proofs. In addition, I wish to thank Tracy Shields for reviewing the manuscript to check the usage of metric units.

I gratefully acknowledge the assistance, cooperation, and support of Ron Doleman of Addison-Wesley of Canada; my editor, Jason Jordan; the production supervisor, Jack Casteel; the production editor, Greg Hubit; the copy editor, Margaret Moore; and Amy Branowicki, Anju Chatani, David Chelton, and many others of the Addison-Wesley staff during the production of this edition. Finally, special mention is due my wife, Millie, who helped with reading proof and checking answers, as well as with her patience and support all through the preparation of this edition, and all earlier editions.

A. J. W.

Contents

20

Additional Topics in Trigonometry 503

21

Plane Analytic Geometry 536

22

Introduction to Statistics and Empirical Curve Fitting 584

23

The Derivative 609

24

Applications of the Derivative 656

25

Integration 692

26

Applications of Integration 722

1

Basic Algebraic Operations

Mathematics has played a most important role in the development and understanding of the great advances in technology and science. This has resulted in a continually increased use of mathematics by technicians in all fields.

With the mathematics developed in this text, we can solve many kinds of applied problems. Consider the jet as shown at the right. A few of the technical areas of application associated with the jet are

- surveying and construction (the runway)
- physics, mechanical design, machine design, and drafting (the plane)
- electronics (communications and circuitry)
- architecture (the terminal buildings)

as well as many others. In support of these technologies, many other technologies would also be used. Lasers might be used in surveying, and computers would certainly be used in many ways. To solve the applied problems in this text will require a knowledge of the mathematics presented, but will *not* require any prior knowledge of the field of application.

Although we cannot solve the more advanced types of problems that arise, we can form a foundation for the more advanced mathematics which is used to solve such problems. Therefore, a real understanding of the mathematics given in this text will be of great value to you in your future work.

A thorough understanding of algebra is essential in the study of any of the fields of mathematics. It is very important for you to learn and understand the basic concepts developed in this text; otherwise, the result will be a weak foundation in mathematics and in the various technical fields where mathematics is applied. Development of this understanding will require a serious commitment on your part. The author sincerely wishes you the best success.

We begin our study of mathematics by reviewing some of the basic concepts and operations that deal with numbers and symbols. With these we shall be able to develop the topics in algebra necessary for progress into other fields of mathematics, such as trigonometry and calculus.

1-1 Numbers

The way we represent numbers has been evolving for thousands of years. *The first numbers used were those which stand for whole quantities, and these are called the* **positive integers** (*or* **natural numbers**). The positive integers are represented by the symbols 1, 2, 3, 4, and so forth.

It is also necessary to have numbers that can represent parts of certain quantities. *The name* **positive rational number** *is given to any number that we can represent by the division of one positive integer by another. Numbers that cannot be written as the division of one integer by another are termed* **irrational.**

EXAMPLE 1 ▸▸ The numbers 5 and 19 are positive integers. They are also rational numbers since they may be written as $\frac{5}{1}$ and $\frac{19}{1}$. Normally we do not write the 1's in the denominators.

The numbers $\frac{5}{8}$, $\frac{11}{3}$, and $\frac{106}{17}$ are positive rational numbers, since the numerator and denominator of each are positive integers.

The numbers $\sqrt{2}$ and π are irrational. It is not possible to find two integers that represent these numbers if one of the integers is divided by the other. For example, $\frac{22}{7}$ is not *exactly* equal to π; it is an *approximation*.

The number $\frac{2}{\sqrt{3}}$ is irrational. The numerator is an integer, but the denominator is irrational and $\frac{2}{\sqrt{3}}$ cannot be written as one integer divided by another. ◂◂

It is also necessary to have **negative numbers** in order to have a numerical answer to problems such as $5 - 8$. *Thus,* -1, -2, -3, *and so on are the* **negative integers**. *The number* **zero** *is an integer, but it is neither positive nor negative. This means that the* **integers** *are the numbers* . . . , -3, -2, -1, 0, 1, 2, 3, *and so on.*

The Real Number System

The integers, the rational numbers, and the irrational numbers, which include all such numbers that are zero, positive, or negative, make up what we call the **real number system**. We shall use real numbers throughout this text, with one important exception. In Chapter 12 we shall be using **imaginary numbers**, *which is the name given to square roots of negative numbers.* (The symbol j is used to designate $\sqrt{-1}$, which is not part of the real number system.) However, until Chapter 12, when we discuss operations on imaginary numbers in detail, it will be necessary only to recognize them if they occur.

EXAMPLE 2 ▸▸ The number 7 is an integer. It is also rational since $7 = \frac{7}{1}$, and it is a real number since the real numbers include all the rational numbers.

The number 3π is irrational, and it is real since the real numbers include all the irrational numbers.

The numbers $\sqrt{-10}$ and $-\sqrt{-7}$ are imaginary numbers.

The number $\frac{-3}{7}$ is rational and real. The number $-\sqrt{7}$ is irrational and real.

The number $\frac{\pi}{6}$ is irrational and real. The number $\frac{\sqrt{-3}}{2}$ is imaginary. ◂◂

A **fraction** *may contain any number or symbol representing a number in its numerator or in its denominator.* Therefore, a fraction may be a number that is rational, irrational, or imaginary.

EXAMPLE 3 ▸▸ The numbers $\frac{2}{7}$ and $\frac{-3}{2}$ are fractions, and they are rational.

The numbers $\frac{\sqrt{2}}{9}$ and $\frac{6}{\pi}$ are fractions, but they are not rational numbers. It is not possible to express either as one integer divided by another integer.

The number $\frac{\sqrt{-3}}{2}$ is a fraction, and it is an imaginary number. ◂◂

The Number Line Real numbers may be represented by points on a line. We draw a horizontal line and designate some point on it by O, which we call the **origin** (see Fig. 1-1). The integer *zero* is located at this point. Equal intervals are marked to the right of the origin, and the positive integers are placed at these positions. The other positive rational numbers are located between the integers.

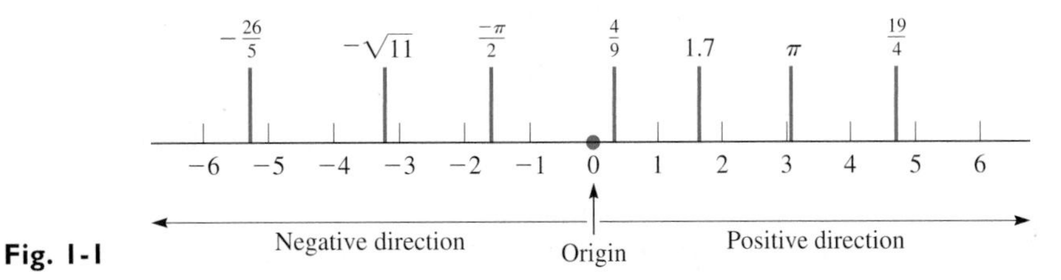

Fig. 1-1

Now we can give a meaning of direction to negative numbers. By starting at the origin and proceeding to the left, *defined as the* **negative direction**, we locate all the negative numbers. *As shown in Fig. 1-1, the positive numbers are to the right of the origin and the negative numbers are to the left of the origin.* Representing numbers in this way will be especially useful when we study graphical methods.

It will not be proved here, but the rational numbers do not take up all the positions on the line; the remaining points represent irrational numbers.

We next define another important concept of a number. *The* **absolute value** *of a positive number is the number itself, and the absolute value of a negative number is the corresponding positive number.* On the number line we may interpret the absolute value of a number as the distance between the origin and the number. The absolute value is denoted by writing the number between vertical lines, as shown in the following example.

EXAMPLE 4 ▸▸ The absolute value of 6 is 6, and the absolute value of -7 is 7. We write these as $|6| = 6$ and $|-7| = 7$. See Fig. 1-2.

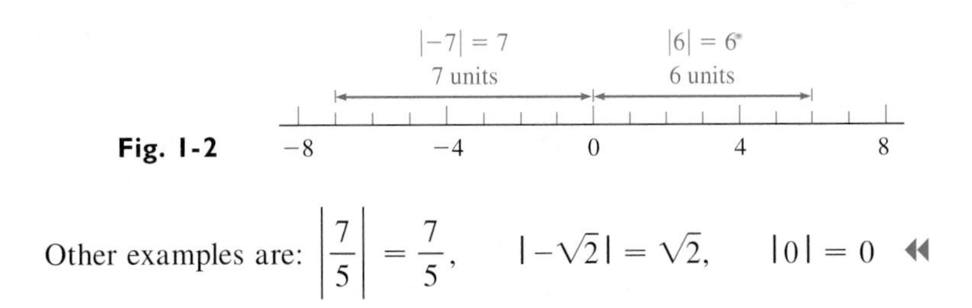

Fig. 1-2

Other examples are: $\left|\frac{7}{5}\right| = \frac{7}{5}$, $|-\sqrt{2}| = \sqrt{2}$, $|0| = 0$ ◂◂

On the number line, *if a first number is to the right of a second number, then the first number is said to be* **greater than** *the second. If the first number is to the left of the second, it is* **less than** *the second number.* The symbol $>$ is used to designate "is greater than," and the symbol $<$ is used to designate "is less than." These are called **signs of inequality**.

EXAMPLE 5 ▸▸

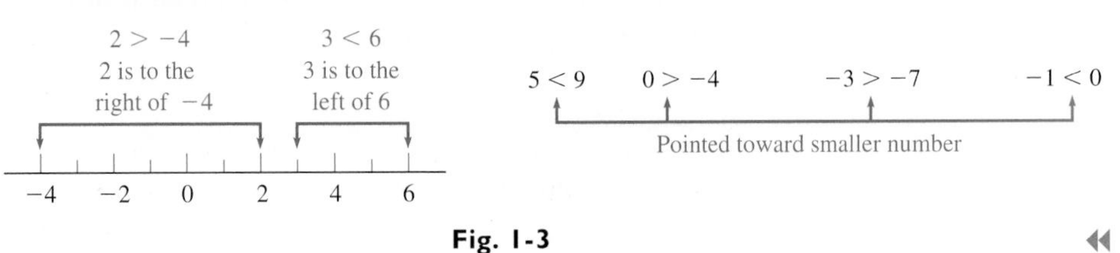

Fig. 1-3

◂◂

Every number, except zero, has a **reciprocal**. *The reciprocal of a number is 1 divided by the number.*

EXAMPLE 6 ▸▸ The reciprocal of 7 is $\frac{1}{7}$. The reciprocal of $\frac{2}{3}$ is

$$\frac{1}{\frac{2}{3}} = 1 \times \frac{3}{2} = \frac{3}{2} \qquad \text{invert and multiply (from arithmetic)}$$

The reciprocal of π is $1/\pi$. The reciprocal of -5 is $-\frac{1}{5}$. Note that the negative sign is retained in the reciprocal of a negative number. ◂◂

For reference, see Appendix B for units of measurement and the symbols used for them.

In applications, *numbers that represent a measurement and are written with units of measurement are called* **denominate numbers**. The next example illustrates the use of units and the symbols that represent them.

EXAMPLE 7 ▸▸ To show that a wire is 10 metres long, we write the length as 10 m.

To show that the speed of a rocket is 1500 metres per second, we write the speed as 1500 m/s. (Note the use of s for second. We use s rather than sec.)

To show that the area of a computer chip is 0.75 square centimetre, we write this as 0.75 cm^2. (We will not use sq cm.)

To show that the volume of a container is 500 cubic centimetres, we write the volume as 500 cm^3. (We will not use cu cm or cc.) ◂◂

Literal Numbers

Until now we have used numbers in their explicit form. However, it is usually more convenient to state definitions and operations on numbers in a general form. *To do this we represent the numbers by letters, referred to as* **literal numbers**.

For example, we can say, "If a is to the right of b on the number line, then $a > b$." This is more convenient than saying, "If a first number is to the right of a second number on the number line, then the first number is greater than the second number." The statement "the reciprocal of a number n is $1/n$" is another example of using a letter to stand for a number in general.

In an algebraic discussion, certain literal numbers may take on any allowable value, whereas other literal numbers represent the same number throughout the discussion. *Those literal numbers that may vary in a given problem are called* **variables**, *and those that are held fixed are called* **constants**.

Common usage normally designates letters near the end of the alphabet as variables and letters near the beginning of the alphabet as constants. Any exceptions to this usage would be specifically noted.

EXAMPLE 8 ▸▸ (a) The resistance of an electric resistor is R. The current I in the resistor equals the voltage V divided by R, written as $I = V/R$. For this resistor, I and V may take on various values and R is fixed. This means I and V are variables and R is a constant. For *another* resistor, the value of R may differ.

(b) The fixed cost for a calculator manufacturer to operate a certain plant is b dollars per day, and it costs a dollars to produce each calculator. The total daily cost C to produce n calculators is

$$C = an + b$$

Here C and n are variables, and a and b are constants. For *another* plant, the values of a and b may differ. ◂◂

Exercises 1–1

In Exercises 1–4, designate each of the given numbers as being an integer, rational, irrational, real, or imaginary. (More than one designation may be correct.)

1. $3, -\pi$

2. $\frac{5}{4}, \sqrt{-4}$

3. $-\sqrt{-6}, \frac{\sqrt{7}}{3}$

4. $-\frac{7}{3}, \frac{\pi}{6}$

In Exercises 5–8, find the absolute value of each number.

5. $3, \frac{7}{2}$

6. $-4, \sqrt{2}$

7. $-\frac{6}{7}, -\sqrt{3}$

8. $-\frac{\pi}{2}, -\frac{19}{4}$

In Exercises 9–16, insert the correct sign of inequality (> or <) between the given pairs of numbers.

9. $6 \quad 8$

10. $7 \quad 5$

11. $\pi \quad -1$

12. $-4 \quad 0$

13. $-4 \quad -3$

14. $-\sqrt{2} \quad -9$

15. $-\frac{1}{3} \quad -\frac{1}{2}$

16. $-0.6 \quad 0.2$

In Exercises 17–20, find the reciprocal of each number.

17. $3, -2$

18. $\frac{1}{6}, -\frac{7}{4}$

19. $-\frac{5}{\pi}, x$

20. $-\frac{8}{3}, \frac{y}{b}$

In Exercises 21–24, locate each number on a number line as in Fig. 1-1.

21. $2.5, -\frac{1}{2}$

22. $\sqrt{3}, -\frac{12}{5}$

23. $-\frac{\sqrt{2}}{2}, 2\pi$

24. $\frac{123}{19}, -\frac{\pi}{6}$

In Exercises 25–40, solve the given problems. Refer to Appendix B for units of measurement and their symbols.

25. List the following numbers in numerical order, starting with the smallest: $-1, 9, \pi, \sqrt{5}, |-8|, -|-3|, -18$.

26. List the following numbers in numerical order, starting with the smallest: $\frac{1}{5}, -\sqrt{10}, -|-6|, -4, 0.25, |-\pi|$.

27. If a and b are positive integers and $b > a$, what type of number is represented by

(a) $b - a$, (b) $a - b$, (c) $\frac{b - a}{b + a}$?

28. If a and b represent positive integers, what kind of number is represented by (a) $a + b$, (b) a/b, (c) $a \times b$?

29. For any integer: (a) Is its absolute value always an integer? (b) Is its reciprocal always an integer?

30. For any positive or negative rational number: (a) Is its absolute value always a rational number? (b) Is its reciprocal always a rational number?

31. Describe the location of a number x on the number line when (a) $x > 0$, (b) $x < -4$.

32. Describe the location of a number x on the number line when (a) $|x| < 1$, (b) $|x| > 2$.

33. For a number $x > 1$, describe the location on the number line of the reciprocal of x.

34. For a number $x < 0$, describe the location on the number line of the number with a value of $|x|$.

35. The heat loss L through a certain type of insulation of thickness t is given by $L = a/t$, where a has a fixed value for this type of insulation. Identify the variables and constants.

36. A sensitive gauge measures the total weight w of a container and the water that forms in it as vapor condenses. It is found that $w = c\sqrt{0.1t + 1}$, where c is the weight of the container and t is the time of condensation. Identify the variables and constants.

37. The memory of a certain computer has a bits in each byte. Express the number N of bits in n kilobytes in an equation. (A *bit* is a single digit, and bits are grouped in *bytes* in order to represent special characters. Commonly there are 6 or 8 bits per byte. If necessary, see Appendix B for the meaning of "kilo.")

38. A piece y centimetres long is cut from a board x metres long. Give an equation for the length L, in centimetres, of the remaining piece.

39. In writing a laboratory report, a student wrote "$-20°C > -30°C$." Is this statement correct?

40. After 5 s, the pressure on a valve is less than 600 kPa. Using t to represent time and p to represent pressure, this statement can be written "for $t > 5$ s, $p < 600$ kPa." In this way, write the statement "when the current I in a circuit is less than 4 A, the voltage V is greater than 12 V."

1-2 Fundamental Laws and Operations of Algebra

In performing operations with numbers, we know that certain basic laws are valid. *These are called the* **fundamental laws of algebra**.

The Commutative and Associative Laws

For example, if two numbers are added, it does not matter in which order they are added. (For example, $5 + 3 = 8$ and $3 + 5 = 8$, or $5 + 3 = 3 + 5$.) This statement, generalized and assumed correct for all possible combinations of numbers to be added, is called the **commutative law** for addition. The law states that *the sum of two numbers is the same, regardless of the order in which they are added*. We make no attempt to prove this law in general but accept its validity.

In the same way, we have the **associative law** for addition, which states that *the sum of three or more numbers is the same, regardless of the way in which they are grouped for addition*. For example, $3 + (5 + 6) = (3 + 5) + 6$.

The laws just stated for addition are also true for multiplication. Therefore, *the product of two numbers is the same, regardless of the order in which they are multiplied*, and *the product of three or more numbers is the same, regardless of the way in which they are grouped for multiplication*. For example, $2 \times 5 = 5 \times 2$, and $5 \times (4 \times 2) = (5 \times 4) \times 2$.

The Distributive Law

There is one more very important law, called the **distributive law**. It states that *the product of one number and the sum of two or more other numbers is equal to the sum of the products of the first number and each of the other numbers of the sum.* For example,

$$4(3 + 5) = 4 \times 3 + 4 \times 5$$

In practice these laws are used intuitively, except perhaps for the distributive law. However, it is necessary to state them and to accept them so that we may use them as a basis for later results.

Not all operations are commutative and associative. For example, division is not commutative, since the order of division of two numbers does matter. For example, $\frac{6}{5} \neq \frac{5}{6}$ ($\neq$ is read "does not equal").

Using literal numbers, the fundamental laws of algebra are as follows:

Commutative law of addition: $a + b = b + a$

Associative law of addition: $a + (b + c) = (a + b) + c$

Commutative law of multiplication: $ab = ba$

Associative law of multiplication: $a(bc) = (ab)c$

Distributive law: $a(b + c) = ab + ac$

Operations on Positive and Negative Numbers

Following are the rules used to determine whether the result of using one of the basic operations (addition, subtraction, multiplication, division) is positive or negative when we are dealing with positive and negative numbers. In stating these rules we refer to the *sign* of the number. From Section 1-1 we recall that a negative number is preceded by a negative sign and that a *positive number is preceded by no sign*. Therefore, we indicate the "sign" of a positive number by simply writing the number itself.

Addition of two numbers of the same sign. *Add their absolute values and assign the sum their common sign.*

EXAMPLE 1 ▸▸ (a) $2 + 6 = 8$ — the sum of two positive numbers is positive

(b) $-2 + (-6) = -(2 + 6) = -8$ — the sum of two negative numbers is negative

The negative number -6 is placed in parentheses since it is also preceded by a plus sign showing addition. It is not necessary to place the negative number -2 in parentheses. ◂◂

Addition of two numbers of different signs. *Subtract the number of smaller absolute value from the number of larger absolute value, and assign to the result the sign of the number of larger absolute value.*

EXAMPLE 2 ▸▸

(a) $2 + (-6) = -(6 - 2) = -4$

(b) $-6 + 2 = -(6 - 2) = -4$

(the negative 6 has the larger absolute value)

(c) $6 + (-2) = 6 - 2 = 4$

(d) $-2 + 6 = 6 - 2 = 4$

(the positive 6 has the larger absolute value)

(the subtraction of absolute values)

Note that the commutative law of addition is shown in illustrations (a) and (b), where the same numbers are added in different orders with the same result. This is also the case for illustrations (c) and (d). ◂◂

Subtraction of one number from another. *Change the sign of the number being subtracted, and change the subtraction to addition. Perform the addition.*

EXAMPLE 3 ▸▸ (a) $2 - 6 = 2 + (-6) = -(6 - 2) = -4$

Note that after changing the subtraction to addition, and changing the sign of 6 to make it -6, we have precisely the same illustration as Example 2(a).

(b) $-2 - 6 = -2 + (-6) = -(2 + 6) = -8$

Note that after changing the subtraction to addition, and changing the sign of 6 to make it -6, we have precisely the same illustration as Example 1(b).

(c) $-a - (-a) = -a + a = 0$

This shows that subtracting a number from itself results in zero, even if the number is negative. Therefore, *subtracting a negative number is equivalent to adding a positive number of the same absolute value.* ◂◂

Subtraction of a Negative Number

Multiplication and division of two numbers. *The product (or quotient) of two numbers of the same sign is positive. The product (or quotient) of two numbers of different signs is negative.*

EXAMPLE 4 ▸▸

(a) $3(12) = 3 \times 12 = 36 \qquad \dfrac{12}{3} = 4$ — result is positive if both numbers are positive

(b) $-3(-12) = 3 \times 12 = 36 \qquad \dfrac{-12}{-3} = 4$ — result is positive if both numbers are negative

(c) $3(-12) = -(3 \times 12) = -36 \qquad \dfrac{-12}{3} = -\dfrac{12}{3} = -4$ — result is negative if one number is positive and the other is negative

(d) $-3(12) = -(3 \times 12) = -36 \qquad \dfrac{12}{-3} = -\dfrac{12}{3} = -4$ ◂◂

Order of Operations

For an expression in which there is a combination of operations, we must perform them in the proper order. Often it is clear by the grouping of numbers as to the proper order of performing the operations. Numbers are grouped by symbols such as **parentheses**, ( ), and the **bar**, ____, between the numerator and the denominator of a fraction. However, if the order of some operations is not defined by specific groupings, we use the following order of operations:

> *Order of Operations*
>
> **1.** *Operations within specific groupings are done first.*
>
> **2.** *Perform multiplications and divisions (from left to right).*
>
> **3.** *Then perform additions and subtractions (from left to right).*

EXAMPLE 5 ▸▸ (a) $20 \div (2 + 3)$ is evaluated by first adding $2 + 3$ and then dividing. The grouping of $2 + 3$ is clearly shown by the parentheses. Therefore, we have $20 \div (2 + 3) = 20 \div 5 = 4$.

(b) $20 \div 2 + 3$ is evaluated by first dividing 20 by 2 and then adding. No specific grouping is shown, and therefore the division is done before the addition. This means $20 \div 2 + 3 = 10 + 3 = 13$.

CAUTION ▶ (c) $16 - 2 \times 3$ is evaluated by ***first multiplying*** 2 ***by*** 3 and then subtracting. We *do* **not** *first subtract* 2 *from* 16. Therefore, $16 - 2 \times 3 = 16 - 6 = 10$.

(d) $16 \div 2 \times 4$ is evaluated by first dividing 16 by 2 and then multiplying. From left to right, the division occurs first. Therefore, $16 \div 2 \times 4 = 8 \times 4 = 32$.

(e) $16 \div (2 \times 4)$ is evaluted by first multiplying 2 by 4 and then dividing. The grouping of 2×4 is clearly shown by parentheses. Therefore, this means that $16 \div (2 \times 4) = 16 \div 8 = 2$. ◂◂

When evaluating expressions, it is generally more convenient to change the operations and numbers so that the result is found by the addition and subtraction of positive numbers. When this is done, we must remember that

$$a + (-b) = a - b \qquad \textbf{(1-1)}$$

$$a - (-b) = a + b \qquad \textbf{(1-2)}$$

EXAMPLE 6 ▸▸ (a) $7 + (-3) - 6 = 7 - 3 - 6 = 7 - 9 = -2$ using Eq. (1-1)

(b) $\dfrac{18}{-6} + 5 - (-2) = -3 + 5 + 2 = -3 + 7 = 4$ using Eq. (1-2)

(c) $2(-3) - 2(-4) + \dfrac{25}{-5} = -6 - (-8) + (-5) = -6 + 8 - 5 = -3$

(d) $\dfrac{-12}{2 - 8} + \dfrac{5 - 1}{2(-1)} = \dfrac{-12}{-6} + \dfrac{4}{-2} = 2 + (-2) = 2 - 2 = 0$

In illustrations (b) and (c) we see that the multiplications and divisions were done before the additions and subtractions. In (d) we see that the groupings ($2 - 8$ and $5 - 1$) were evaluated first. Then we did the divisions, and finally the addition. ◂◂

EXAMPLE 7 ▸▸ A 1500-kg car going at 40 km/h (kilometres per hour) ran head-on into a 1000-kg car going at 20 km/h. An insurance investigator determined the velocity of the cars immediately following the collision from the following calculation.

$$\frac{1500(40) + (1000)(-20)}{1500 + 1000} = \frac{60\,000 + (-20\,000)}{1500 + 1000} = \frac{60\,000 - 20\,000}{2500}$$

$$= \frac{40\,000}{2500} = 16 \text{ km/h}$$

The numerator and the denominator must be evaluated before the division is performed. The multiplications in the numerator are performed first, followed by the addition in the denominator and the subtraction in the numerator. ◂◂

Operations with Zero

Since operations with zero tend to cause some difficulty, we will show them here.

If a is a real number, the operations of addition, subtraction, multiplication, and division with zero are as follows:

$$\mathbf{a + 0 = a}$$

$$\mathbf{a - 0 = a} \qquad \mathbf{0 - a = -a}$$

$$\mathbf{a \times 0 = 0}$$

$$\mathbf{0 \div a = \frac{0}{a} = 0} \quad \textbf{if} \quad \mathbf{a \neq 0} \qquad (\neq \text{ means “is not equal to”})$$

EXAMPLE 8 ▸▸ (a) $5 + 0 = 5$ (b) $-6 - 0 = -6$ (c) $0 - 4 = -4$ (d) $\frac{0}{6} = 0$ (e) $\frac{0}{-3} = 0$ (f) $\frac{5 \times 0}{7} = \frac{0}{7} = 0$ ◂◂

Note that there is no result defined for division by zero. To understand the reason for this, consider the results for $\frac{6}{2}$ and $\frac{6}{0}$.

$$\frac{6}{2} = 3 \qquad \text{since} \qquad 2 \times 3 = 6$$

If $\frac{6}{0} = b$, then $0 \times b = 6$. This cannot be true because $0 \times b = 0$ for any value of b. Thus,

division by zero is undefined

(The special case of $\frac{0}{0}$ is termed *indeterminate*. Setting $\frac{0}{0} = b$, we see that $0 = 0 \times b$, which is true for any value of b. Thus, no specific value of b can be determined.)

EXAMPLE 9 ▸▸

$\frac{2}{5} \div 0$ is undefined $\frac{8}{0}$ is undefined $\left(\frac{7 \times 0}{0 \times 6} \text{ is indeterminate}\right)$ ◂◂

The operations with zero will not cause any difficulty if we remember to

CAUTION ▸ ***never divide by 0***

Division by zero is the only undefined basic operation. All the other operations with zero may be performed as for any other number.

Exercises 1–2

In Exercises 1–32, evaluate each of the given expressions by performing the indicated operations.

1. $8 + (-4)$ **2.** $-4 + (-7)$ **3.** $-3 + 9$

4. $18 - 21$ **5.** $-19 - (-16)$ **6.** $8 - (-4)$

7. $8(-3)$ **8.** $-9(3)$ **9.** $-7(-5)$

10. $\frac{-9}{3}$ **11.** $\frac{-60}{-3}$ **12.** $\frac{28}{-7}$

13. $-2(4)(-5)$ **14.** $3(-4)(6)$ **15.** $\frac{2(-5)}{10}$

16. $\frac{-64}{2(-4)}$ **17.** $9 - 0$ **18.** $\frac{0}{-6}$

19. $\frac{17}{0}$ **20.** $\frac{2 - (-5)}{0}$

21. $8 - 3(-4)$ **22.** $20 + 8 \div 4$

23. $3 - 2(6) + \frac{8}{2}$ **24.** $0 - (-6)(-8) + (-10)$

25. $\frac{3(-6)(-2)}{0 - 4}$ **26.** $\frac{7 - (-5)}{-1(-2)}$

27. $\frac{24}{3 + (-5)} - 4(-9)$ **28.** $\frac{-18}{3} - \frac{4 - 6}{-1}$

29. $-7 - \frac{-14}{2} - 3(2)$ **30.** $-7(-3) + \frac{6}{-3} - (-9)$

31. $\frac{3(-9) - 2(-3)}{3 - 10}$ **32.** $\frac{2(-7) - 4(-2)}{-9 - (-9)}$

In Exercises 33–40, determine which of the fundamental laws of algebra is demonstrated.

33. $6(7) = 7(6)$ **34.** $6 + 8 = 8 + 6$

35. $6(3 + 1) = 6(3) + 6(1)$ **36.** $4(5 \times 7) = (4 \times 5)(7)$

37. $3 + (5 + 9) = (3 + 5) + 9$

38. $8(3 - 2) = 8(3) - 8(2)$

39. $(2 \times 3) \times 9 = 2 \times (3 \times 9)$

40. $(3 \times 6) \times 7 = 7 \times (3 \times 6)$

In Exercises 41–44, for numbers a and b determine which of the following expressions equals the given expression. (a) $a + b$, (b) $a - b$, (c) $b - a$, (d) $-a - b$

41. $-a + (-b)$ **42.** $b - (-a)$

43. $-b - (-a)$ **44.** $-a - (-b)$

In Exercises 45–52, answer the given questions.

45. What is the sign of the product of an even number of negative numbers?

46. What is the sign of the product of an odd number of negative numbers?

47. Using the division of 4 by 2 and then 2 by 4, show that division is not commutative.

48. Is subtraction commutative? Illustrate.

49. One oil well drilling rig drills 100 m deep the first day and 200 m deeper the second day. A second rig drills 200 m deep the first day and 100 m deeper the second day. In showing that the total depth drilled by each rig was the same, state what fundamental law is illustrated.

50. Eight wires, each supporting 10 N, hold up one end of a beam. Ten wires, each supporting 8 N, hold up the other end. In showing that the weight supported at one end equals the weight supported at the other end, what fundamental law of algebra is illustrated?

51. In a certain month, a computer dealer sold eight computers, which cost the dealer $2000 each. On each a profit of $1000 was made. Set up the expression to show the total amount the dealer received for these computers. What fundamental law of algebra is illustrated?

52. A jet travels 600 km/h relative to the air. The wind is blowing at 50 km/h. If the jet travels with the wind for 3 h, set up the expression to find the distance traveled. What fundamental law of algebra is illustrated?

1-3 Calculators and Approximate Numbers

In the preceding sections, a number of basic calculations have been made. In the remainder of the book it is assumed that you will use a calculator for most of the calculations to be done. A scientific calculator or a *graphing calculator* can be used for these calculations, but a graphing calculator will be needed at other times.

A discussion of calculator usage appears in Appendix C. However, if you are not familiar with the use of your calculator, you should practice using it and *review its manual*. ***You must know the order in which the keys are used*** for calculations.

The order in which operations are done by the calculator, which determines the order in which the keys are used, is called the **logic** of the calculator. Following is an example using a calculator with *algebraic logic*.

EXAMPLE 1 ▸▸ Calculate the value of $38.3 - 12.9(-3.58)$.

Recalling the order of operations, the multiplication $12.9(-3.58)$ is done first. If your calculator uses algebraic logic, the numbers are entered as shown, and the calculator will do the multiplication first. The sign of -3.58 is entered by use of the [+/−] key. Therefore, the sequence of keys to be used is

38.3 [−] 12.9 [×] 3.58 [+/−] [=] [84.482] ← final display

This means that $38.3 - 12.9(-3.58) = 84.482$. ◂◂

Looking back to Section 1-2, we see that *the minus sign is used in two different ways*: (1) to indicate subtraction and (2) to designate a negative number. This is clearly shown on a calculator since there is a key for each purpose. The [−] key is used for subtraction, and the [+/−] key is used to change the sign of a number.

The [−] and [+/−] keys also show one of the many possible differences in the logic used in different calculators. To enter the calculation of Example 1 in a graphing calculator, the negative sign of -3.58 is entered (with the [(−)] key) *before* the 3.58 is entered. Although algebraic logic is used, the keys are used in a different order. On a typical graphing calculator (where the display shows the numbers entered as well as the result), the display for Example 1 is

38.3−12.9*-3.58

84.482 ← result obtained by pressing ENTER

Note that the negative sign of -3.58 is smaller to distinguish it from the minus sign for subtraction. Also note the * shown in the display for multiplication. The asterisk (*) is the standard computer symbol for multiplication.

Approximate Numbers and Significant Digits

We must consider the accuracy of the numbers used in calculations, as the final result should not be written with any more accuracy than is proper. For example, if the numbers in Example 1 are *approximate*, the result should be written as 84.5, not 84.482. We will see the reason for this later in this section.

Most numbers in technical and scientific work are **approximate numbers**, having been determined by some measurement. Certain other numbers are **exact numbers**, having been determined by a definition or a counting process.

EXAMPLE 2 ▸▸ If a voltage shown on a voltmeter is read as 116 V, the 116 is approximate. Another voltmeter may show the voltage as 115.7 V. However, this voltage cannot be determined *exactly*.

If a computer prints out the number of names on a list as 97, this 97 is exact. We know it is not 96 or 98. Since 97 was found by precise counting, it is exact.

By definition, 60 s = 1 min, and the 60 and the 1 are exact. ◂◂

Significant Digits

An approximate number may have to include some zeros to properly locate the decimal point. *Except for these zeros, all other digits are called* **significant digits**. The next example illustrates how we determine the number of significant digits.

EXAMPLE 3 ▸▸ All numbers in this example are assumed to be approximate.

34.7 has three significant digits.

0.039 has two significant digits. The zeros properly locate the decimal point.

706.1 has four significant digits. The zero is not used for the location of the decimal point. It shows the number of tens in 706.1.

CAUTION ▶

5.90 has three significant digits. ***The zero is not necessary as a placeholder*** and should not be written unless it is significant.

8900 has two significant digits, unless information is known about the number that makes one or both zeros significant. Without such information, we assume the zeros are placeholders for proper location of the decimal point.

Other approximate numbers with the number of significant digits are:
0.0005 (one), 960 000 (two), 0.0709 (three), 1.070 (four), 700.00 (five) ◂◂

From Example 3 we see that *all nonzero digits are significant. Also, zeros not used as placeholders (for location of the decimal point) are significant.*

Accuracy and Precision

In calculations with approximate numbers, the number of significant digits and the position of the decimal point are important. *The* **accuracy** *of a number refers to the number of significant digits it has,* whereas *the* **precision** *of a number refers to the decimal position of the last significant digit.*

EXAMPLE 4 ▸▸ An electric current is measured as 0.31 A on one ammeter and as 0.312 A on another ammeter. Here 0.312 is more precise since its last digit represents thousandths and 0.31 is expressed only to hundredths. Also, 0.312 is more accurate since it has three significant digits and 0.31 has only two.

A concrete driveway is 130 m long and 0.1 m thick. Here 130 is more accurate (two significant digits) and 0.1 is more precise (expressed to tenths). ◂◂

Rounding Off

The last significant digit of an approximate number is not completely accurate. It has usually been determined by estimation or *rounding off.* However, it is in error at most by one-half of a unit in its place value.

EXAMPLE 5 ▸▸ When we write the voltage in Example 2 as 115.7 V, we are saying that the voltage is at least 115.65 V and no more than 115.75 V. Any value between these two, rounded off to tenths, would be expressed as 115.7 V.

In changing the fraction 2/3 to the approximate decimal value 0.667, we are saying that the value is between 0.6665 and 0.6675. ◂◂

To **round off** *a number to a specified number of significant digits, discard all digits to the right of the last significant digit (replace them with zeros if needed to place the decimal point). If the first digit discarded is 5 or more, increase the last significant digit by 1 (round up). If the first digit discarded is less than 5, do not change the last significant digit (round down).*

Note that if the only digit discarded is a 5, then rounding up, rounding down, or rounding to the nearest even number are all equally appropriate. Rounding up is probably most commonly used, although rounding to the nearest even number is used in many references.

EXAMPLE 6 ▸▸ 70 360 rounded off to three significant digits is 70 400. Here 3 is the third significant digit and the next digit is 6. Since $6 > 5$, we add 1 to 3 and the result, 4, becomes the third significant digit of the approximation. The 6 is then replaced with a zero in order to keep the decimal point in the proper position.

70 430 rounded off to three significant digits, or to the nearest hundred, is 70 400. Here the 3 is replaced with a zero.

187.35 rounded off to four significant digits, or to tenths, is 187.4.

187.349 rounded off to four significant digits is 187.3.

71 500 rounded off to two significant digits is 72 000.

71 499 rounded off to two significant digits is 71 000. ◂◂

Operations with Approximate Numbers

When performing operations on approximate numbers, *we must not express the result to an accuracy or precision that is not valid.* The next two examples show how results might be written with incorrect accuracy.

EXAMPLE 7 ▸▸ A pipe is made in two sections. One is measured to be 16.3 m long, and the other is measured as 0.927 m. A plumber wants to know the total length when the two sections are put together.

$$\begin{array}{r} 16.3\ \ \ \ \text{m} \\ \underline{0.927\ \text{m}} \\ 17.227\ \text{m} \end{array}$$

To obtain the result, it appears that we simply add the numbers as shown at the left. However, 16.3 m is precise only to tenths, which means it might have been as small as 16.25 m or as large as 16.35 m. If we consider only the precision of 16.3, the total length might be as small as 17.177 m or as large as 17.277 m. These values agree when rounded off to two significant digits (17). They vary by 0.1 if rounded off to tenths (17.2 and 17.3). When rounded to hundredths, they do not agree at all, since the third significant digit is different (17.18 and 17.28). Thus, there is no agreement when these numbers are rounded off to a precision beyond tenths. This is reasonable, since 16.3 is not expressed beyond tenths. The length 0.927 m does not change the precision, since it is expressed to thousandths. We conclude that the total length should be rounded off to *tenths*, the precision of 16.3. This means that the total length should be written as 17.2 m. ◂◂

EXAMPLE 8 ▸▸ We find the area of the rectangular piece of land in Fig. 1-4 by multiplying the length, 207.54 m, by the width, 81.4 m. Using a calculator, we find that $(207.54)(81.4) = 16\,893.756$. This apparently means the area is $16\,893.756\ \text{m}^2$.

However, *the area should not be expressed with this accuracy.* We know that the length and width are both approximate, and the least they could be is 207.535 m and 81.35 m. Using these values, we find the least possible value for the area is

$$(207.535\ \text{m})(81.35\ \text{m}) = 16\,882.972\,25\ \text{m}^2$$

The greatest possible value for the area is

$$(207.545\ \text{m})(81.45\ \text{m}) = 16\,904.540\,25\ \text{m}^2$$

These values agree when rounded off to three significant digits ($16\,900\ \text{m}^2$), but they do not agree when rounded off to a greater accuracy. Therefore, we conclude that the result is accurate only to *three* significant digits, which means the area is $16\,900\ \text{m}^2$. Note that the width is accurate only to *three* significant digits, and the length to five significant digits. ◂◂

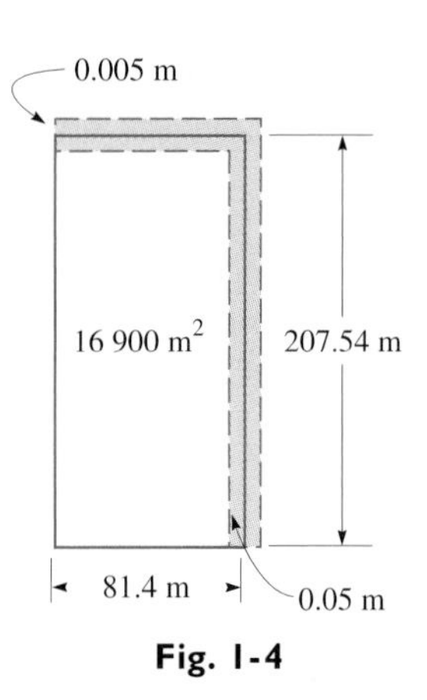

Fig. 1-4

Following are the rules used in expressing the result when we perform basic operations on approximate numbers. They are based on reasoning similar to that shown in Examples 7 and 8.

Operations with Approximate Numbers

1. *When approximate numbers are added or subtracted, the result is expressed with the precision of the least precise number.*
2. *When approximate numbers are multiplied or divided, the result is expressed with the accuracy of the least accurate number.*
3. *When the root of an approximate number is found, the result is expressed with the accuracy of the number.*

CAUTION ▶ *We should always express the result of a calculation with the proper accuracy or precision.* ***Using a calculator, it is necessary to round off the result if additional digits are displayed.*** *Therefore, it is necessary to note the accuracy or precision of the numbers being used.*

EXAMPLE 9 ▸▸ Find the sum of the approximate numbers 73.2, 8.0627, and 93.57.

Showing the addition in the standard way, and using a calculator, we have

$$\begin{array}{r l} 73.2 & \longleftarrow \text{least precise number (expressed to tenths)} \\ 8.0627 & \\ \underline{93.57} & \\ 174.8327 & \longleftarrow \text{final display must be rounded to tenths} \end{array}$$

Therefore, the sum of these approximate numbers is 174.8. ◂◂

EXAMPLE 10 ▸▸ In finding the product of the approximate numbers 2.4832 and 30.5 on a calculator, the final display shows 75.7376. However, since 30.5 has only three significant digits, the product is 75.7.

In Example 1 we calculated that $38.3 - 12.9(-3.58) = 84.482$. We know that $38.3 - 12.9(-3.58) = 38.3 + 46.182 = 84.482$. If these numbers are approximate, we must round off the result to tenths, which means the sum is 84.5. We see that *where there is a combination of operations, the final operation determines how the final result is to be rounded off.* ◂◂

NOTE ▶ ***If an exact number is used in a calculation, there is no limit to the number of decimal places it may take on.*** *The accuracy of the result is limited only by the approximate numbers involved.*

EXAMPLE 11 ▸▸ Using the exact number 600 and the approximate number 2.7, we express the result to tenths if the numbers are added or subtracted. If they are multiplied or divided, we express the result to two significant digits. Since 600 is exact, the accuracy of the result depends only on the approximate number 2.7.

$$600 + 2.7 = 602.7 \qquad 600 - 2.7 = 597.3$$

$$600 \times 2.7 = 1600 \qquad 600 \div 2.7 = 220$$ ◂◂

A note regarding the equal sign (=) is in order. We will use it for its defined meaning of "equals exactly" and when the result is an approximate number that has been properly rounded off. Although $\sqrt{27.8} \approx 5.27$, where $\approx$ means "equals approximately," we write $\sqrt{27.8} = 5.27$, since 5.27 has been properly rounded off.

Estimating Results

It is a good practice to *make a rough estimation* of the result when using a calculator. An estimation can possibly prevent accepting an incorrect result after using an incorrect calculator sequence, particularly when the value in the final display is far from the estimated value.

EXAMPLE 12 ▸▸ In Example 1 we found that

$$38.3 - 12.9(-3.58) = 84.482 \quad \text{using exact numbers}$$

When using the calculator, if we forgot to make 3.58 negative, the display would be −7.882, or if we incorrectly entered 38.3 as 83.3, the display would be 129.482.

However, if we estimate the result as

$$40 - 10(-4) = 80$$

we know that a result of −7.882 or 129.482 cannot be correct. If we obtain a calculator display that is far from the estimation, we should do the calculation again. ◂◂

In estimating a result, we can usually use one-significant-digit approximations. We should be able to do the estimation mentally most of the time.

Exercises 1–3

In Exercises 1–4, determine whether the given numbers are approximate or exact.

1. A car with 8 cylinders travels at 55 km/h.

2. A computer chip 0.002 mm thick is priced at $7.50.

3. A cube of gold 1 cm on an edge weighs 19.3 g.

4. A calculator has 50 keys and its battery lasted for 50 h.

In Exercises 5–12, determine the number of significant digits in the given approximate numbers.

5. 37.2; 6844 **6.** 3600; 730

7. 107; 3004 **8.** 0.8735; 0.0075

9. 6.80; 6.08 **10.** 90 050; 105 040

11. 30 000; 30 000.0 **12.** 1.00; 0.01

In Exercises 13–20, determine which of the pair of approximate numbers is (a) more precise and (b) more accurate.

13. 3.764; 2.81 **14.** 0.041; 7.673

15. 30.8; 0.01 **16.** 70 370; 50 400

17. 0.1; 78.0 **18.** 7040; 0.004

19. 7000; 0.004 **20.** 50.060; 8.914

In Exercises 21–28, round off the given approximate numbers (a) to three significant digits and (b) to two significant digits.

21. 4.936 **22.** 80.53 **23.** 50 893 **24.** 31 490

25. 9545 **26.** 30.96 **27.** 0.9499 **28.** 0.9999

In Exercises 29–44, perform the indicated operations on a calculator. Assume that all numbers are approximate.

29. $3.8 + 0.154 + 47.26$ **30.** $12.78 + 1.0495 - 1.633$

31. $3.64(17.06)$ **32.** $0.49 \div 827$

33. $3.168 + 53.91 \div (-17.85)$

34. $8070 - 2450 \times (-3.191)$

35. $0.0350 - \dfrac{0.0450}{1.909}$ **36.** $\dfrac{0.3275}{1.096 \times 0.500\,85}$

37. $46.7(0.923) + 39.8(-0.362)$

38. $4760 - 256(19.6) - 3.7(1667)$

39. $\dfrac{0.26(-0.4095)}{50.75(0.937)}$ **40.** $\dfrac{326.0}{2.060(3894) - 4008}$

41. $\dfrac{23.962 \times 0.015\,37}{10.965 - 8.249}$ **42.** $\dfrac{0.693\,78 + 0.049\,97}{257.4 \times 3.216}$

43. $\dfrac{3872}{503.1} - \dfrac{2.056 \times 309.6}{395.2}$ **44.** $\dfrac{1}{0.5926} + \dfrac{3.6957}{2.935 - 1.054}$

In Exercises 45–48, perform the indicated operations. The first number is approximate, and the second number is exact.

45. $0.9788 + 14.9$ **46.** $17.311 - 22.98$

47. $3.142(65)$ **48.** $8.62 \div 1728$

In Exercises 49–52, answer the given questions. Refer to Appendix B for units of measurement and their symbols.

49. A surveyor measured the road frontage of a parcel of land as 128.3 m. What are the least possible and the greatest possible frontages?

50. An automobile manufacturer states that the gas tank on a certain car holds approximately 82 L. What are the least possible and greatest possible capacities of this gas tank?

51. A flash of lightning struck a tower 5.23 km from a person. The thunder was heard 15 s later. The person calculated the speed of sound and reported it as 348.7 m/s. What is wrong with this conclusion?

52. A student reports the electric current in a certain experiment as 0.02 A and later notes that it is 0.023 A. The student states the change in current is 0.003 A. What is wrong with this conclusion?

In Exercises 53–64, perform the indicated calculations on a calculator.

53. Show that π is not exactly equal to 3.1416.

54. Show that π is not exactly equal to 22/7.

55. What is the calculator display for $2 \div 0$?

56. What is the calculator display for $0 \div 0$?

57. At some point in the decimal equivalent of a rational number, some sequence of digits will start repeating endlessly. For an irrational number there is never such an endlessly repeating sequence of digits. Find the decimal equivalents to (a) 8/33 and (b) π. Note the repetition for 8/33 and that no such repetition occurs for π.

58. Following Exercise 57, show that the decimal equivalent of the fraction 124/990 indicates that it is rational. Why is the last digit different?

59. Three adjacent lots have road frontages of 75.4 m, 39.66 m, and 81 m, respectively. What is the total frontage of these lots?

60. Two jets flew at 938 km/h and 1450 km/h, respectively. How much faster was the second jet?

61. If 1 K of computer memory has 1024 bytes, how many bytes are there in 256 K of memory? (All numbers are exact.)

62. The power (in watts) developed in an electric circuit is the product of the current (in amperes) and the voltage. What is the power developed in a circuit in which the current is 0.0125 A and the voltage is 12.68 V?

63. The percent of alcohol in a certain car engine coolant is found by performing the calculation $\frac{100(40.63 + 52.96)}{105.30 + 52.96}$. Find this percent of alcohol. The number 100 is exact.

64. The evaporation rate (in L/day) of a wastewater holding pond is found by calculating $145(1.05 + \frac{1}{236})$. Determine this rate.

1-4 Exponents

In this section we shall introduce some basic terminology and notation that are important to the algebraic expressions developed in later sections.

We often have a number multiplied by itself several times. Rather than writing the number over and over, we use the notation a^n, where a is the number and n is the number of times it appears in the product. *In the expression a^n, the number a is called the* **base** *and the number n is called the* **exponent**; *in words, a^n is read as the* "**nth power of a.**"

EXAMPLE 1 (a) $4 \times 4 \times 4 \times 4 \times 4 = 4^5$ (the fifth power of 4)

(b) $(-2)(-2)(-2)(-2) = (-2)^4$ (the fourth power of -2)

(c) $a \times a = a^2$ (the second power of a, called "a squared")

(d) $(\frac{1}{5})(\frac{1}{5})(\frac{1}{5}) = (\frac{1}{5})^3$ (the third power of $\frac{1}{5}$, called "$\frac{1}{5}$ cubed")

The basic operations with exponents will now be stated symbolically. We first state them for positive integers as exponents. Therefore, if m and n are positive integers, we have the following important operations for exponents.

$$a^m \times a^n = a^{m+n} \tag{1-3}$$

$$\frac{a^m}{a^n} = a^{m-n} \quad (m > n, a \neq 0), \qquad \frac{a^m}{a^n} = \frac{1}{a^{n-m}} \quad (m < n, a \neq 0) \tag{1-4}$$

$$(a^m)^n = a^{mn} \tag{1-5}$$

$$(ab)^n = a^n b^n, \qquad \left(\frac{a}{b}\right)^n = \frac{a^n}{b^n} \quad (b \neq 0) \tag{1-6}$$

Two forms are shown for Eqs. (1-4) in order that the resulting exponent is a positive integer. This is generally, but not always, the best form of the result. After the next three examples, we discuss zero and negative exponents. This will cover the case for $m = n$ and show that the first form of Eqs. (1-4) is its basic form.

EXAMPLE 2 ▸▸ Applying Eq. (1-3), we have

add exponents

$$a^3 \times a^5 = a^{3+5} = a^8$$

We see that this result is correct since we can also write

(3 factors of a) (5 factors of a) → 8 factors of a

$$a^3 \times a^5 = (a \times a \times a)(a \times a \times a \times a \times a) = a^8$$

Applying the first form of Eqs. (1-4), we have

5 > 3

$$\frac{a^5}{a^3} = a^{5-3} = a^2, \qquad \frac{a^5}{a^3} = \frac{\not{a} \times \not{a} \times \not{a} \times a \times a}{\not{a} \times \not{a} \times \not{a}} = a^2$$

Applying the second form of Eqs. (1-4), we have

$$\frac{a^3}{a^5} = \frac{1}{a^{5-3}} = \frac{1}{a^2}, \qquad \frac{a^3}{a^5} = \frac{\not{a} \times \not{a} \times \not{a}}{\not{a} \times \not{a} \times \not{a} \times a \times a} = \frac{1}{a^2}$$

5 > 3 ◂◂

EXAMPLE 3 ▸▸ Applying Eq. (1-5), we have

multiply exponents

$$(a^5)^3 = a^{5(3)} = a^{15}, \qquad (a^5)^3 = (a^5)(a^5)(a^5) = a^{5+5+5} = a^{15}$$

Applying the first form of Eqs. (1-6), we have

$$(ab)^3 = a^3b^3, \qquad (ab)^3 = (ab)(ab)(ab) = a^3b^3$$

Applying the second form of Eqs. (1-6), we have

$$\left(\frac{a}{b}\right)^3 = \frac{a^3}{b^3}, \qquad \left(\frac{a}{b}\right)^3 = \left(\frac{a}{b}\right)\left(\frac{a}{b}\right)\left(\frac{a}{b}\right) = \frac{a^3}{b^3}$$ ◂◂

CAUTION ▶ In applying Eqs. (1-3) and (1-4), the base a must be the same for the exponents to be added or subtracted. When a problem involves a product of different bases, ***only exponents of the same base may be combined.***

EXAMPLE 4 ▸▸ Other illustrations using Eqs. (1-3) to (1-6) are as follows:

(a) $(-x^2)^3 = [(-1)x^2]^3 = (-1)^3(x^2)^3 = -x^6$

(b) $ax^2(ax)^3 = ax^2(a^3x^3) = a^4x^5$ (exponent of 1; add exponents of a; add exponents of x)

(c) $\dfrac{(3 \times 2)^4}{(3 \times 5)^3} = \dfrac{3^4 2^4}{3^3 5^3} = \dfrac{3 \times 2^4}{5^3}$ (d) $\dfrac{(ry^3)^2}{r(y^2)^4} = \dfrac{r^2y^6}{ry^8} = \dfrac{r}{y^2}$ ◂◂

CAUTION ▶ In illustration (b), note that ***ax^2 means a times the square of x and does* not *mean a^2x^2***, whereas $(ax)^3$ *does* mean a^3x^3.

EXAMPLE 5 ▸▸ In the analysis of the deflection of a beam, the expression that follows is simplified as shown.

$$\frac{1}{2}\left(\frac{PL}{4EI}\right)\left(\frac{2}{3}\right)\left(\frac{L}{2}\right)^2 = \frac{1}{2}\left(\frac{PL}{4EI}\right)\left(\frac{2}{3}\right)\left(\frac{L^2}{2^2}\right) = \frac{\overset{1}{\cancel{2}}PL(L^2)}{\underset{1}{\cancel{2}}(3)(4)(4)EI} = \frac{PL^3}{48EI}$$

L is the length of the beam, and P is the force applied to it. E and I are constants related to the beam. In *simplifying* this expression, we combined exponents of L and divided out the 2 that was in the numerator and in the denominator. ◂◂

Zero and Negative Exponents

We developed Eqs. (1-3) to (1-6) using positive integers as exponents. We now show how zero and negative integers are used as exponents.

In Eqs. (1-4), if $n = m$, we would have $a^m/a^m = a^{m-m} = a^0$. Also $a^m/a^m = 1$, since any nonzero quantity divided by itself equals 1. Therefore, for Eqs. (1-4) to hold when $m = n$, we have

$$a^0 = 1 \quad (a \neq 0) \tag{1-7}$$

Equation (1-7) gives the definition of zero as an exponent. Since a has not been specified, this equation states that *any nonzero algebraic expression raised to the zero power is* 1. Also, the other laws of exponents are valid for this definition.

EXAMPLE 6 ▸▸ (a) $5^0 = 1$ (b) $(2x)^0 = 1$ (c) $(ax + b)^0 = 1$

(d) $(a^2b^0c)^2 = a^4b^0c^2 = a^4c^2$ ($b^0 = 1$) (e) $2t^0 = 2(1) = 2$

CAUTION ▶ We note in illustration (e) that *only t is raised to the zero power.* If the quantity $2t$ were raised to the zero power, it would be written as $(2t)^0$. ◂◂

If we apply the first form of Eqs. (1-4) to the case where $n > m$, the resulting exponent is negative. This leads to the definition of a negative exponent.

EXAMPLE 7 ▸▸ Applying both forms of Eqs. (1-4) to a^2/a^7, we have

$$\frac{a^2}{a^7} = a^{2-7} = a^{-5} \qquad \text{and} \qquad \frac{a^2}{a^7} = \frac{1}{a^{7-2}} = \frac{1}{a^5}$$

If these results are to be consistent, then $a^{-5} = \dfrac{1}{a^5}$. ◂◂

Following the reasoning of Example 7, if we define

$$a^{-n} = \frac{1}{a^n} \qquad (a \neq 0) \tag{1-8}$$

then all of the laws of exponents will hold for negative integers.

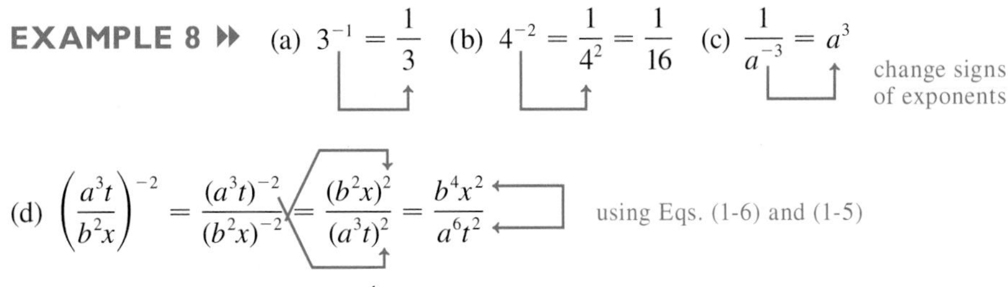

EXAMPLE 8
(a) $3^{-1} = \frac{1}{3}$
(b) $4^{-2} = \frac{1}{4^2} = \frac{1}{16}$
(c) $\frac{1}{a^{-3}} = a^3$
change signs of exponents
(d) $\left(\frac{a^3t}{b^2x}\right)^{-2} = \frac{(a^3t)^{-2}}{(b^2x)^{-2}} = \frac{(b^2x)^2}{(a^3t)^2} = \frac{b^4x^2}{a^6t^2}$
using Eqs. (1-6) and (1-5)

From illustration (a) where $3^{-1} = 1/3$, we can see in general that the reciprocal of x is x^{-1}. ◂◂

Order of Operations

In Section 1-2 we saw that it is necessary to follow a particular order of operations when performing the basic operations on numbers. Since raising a number to a power is a form of multiplication, this operation is performed before additions and subtractions. In fact, it is performed before multiplications and divisions.

> *Order of Operations*
>
> 1. *Operations within specific groupings*
> 2. *Powers*
> 3. *Multiplications and divisions (from left to right)*
> 4. *Additions and subtractions (from left to right)*

EXAMPLE 9 ▸▸ $8 - (-1)^2 - 2(-3)^2 = 8 - 1 - 2(9)$
$= 8 - 1 - 18 = -11$

CAUTION ▸ Since there were no specific groupings, we first squared -1 and -3. Next we found the product $2(9)$ in the last term. Finally, the subtractions were performed. Note carefully that *we did not change the sign of -1 before we squared it.* ◂◂

Evaluating Algebraic Expressions

It is often necessary to **evaluate** an algebraic expression for given values of the literal numbers. *The evaluation is done by* **substituting** *the given values for the literal numbers and then calculating the value of the result.* On a calculator use the $\boxed{x^2}$ key to square numbers. For other powers use the $\boxed{x^y}$ (or $\boxed{\wedge}$) key.

EXAMPLE 10 ▸▸ The distance (in m) that an object falls in 4.2 s is found by substituting 4.2 for t in the expression $4.90t^2$. We show this as

$t = 4.2$ s ← substituting

↓

$4.90(4.2)^2 = 86$ m   estimation → $5(4)^2 = 80$

The result is rounded off to two significant digits (the accuracy of t). The calculator will square the 4.2 before multiplying. ◂◂

As we stated in Section 1-3, in evaluating an expression on a calculator, we should also estimate its value as in Example 10. For the estimate we note that *a negative number raised to an even power gives a positive value, and a negative number raised to an odd power gives a negative value.* (On some calculators the $\boxed{x^y}$ key can be used only if x is positive. When using this type of calculator we must then use this to determine the sign of a negative number raised to a power.)

EXAMPLE 11 ▸▸ Using the meaning of a power of a number, we have

$$(-2)^2 = (-2)(-2) = 4 \quad \text{and} \quad (-2)^3 = (-2)(-2)(-2) = 4(-2) = -8$$

Using a calculator that can calculate powers of a negative base, we have

$$(-2)^4 = 16, \quad (-2)^5 = -32, \quad (-2)^6 = 64, \quad (-2)^7 = -128$$ ◂◂

EXAMPLE 12 ▸▸ A wire made of a special alloy has an electric resistance R (in ohms) that is given by $R = a + 0.0115T^3$, where T (in °C) is the temperature (between -4°C and 4°C). Find the value of R for $a = 0.838\ \Omega$ (Ω is the symbol for ohm) and $T = -2.87$°C.

Substituting these values, we have

$$R = 0.838 + 0.0115(-2.87)^3$$
$$= 0.566\ \Omega$$

estimation:
$0.8 + 0.01(-3)^3 = 0.8 + 0.01(-27) = 0.53$

Note in the estimation that $(-3)^3 = -27$. ◂◂

Some graphing calculators use computer symbols in the display for some of the operations to be performed. These symbols are as follows:

Multiplication: *   Division: /   Powers: ^

Therefore, to calculate the value of $20 \times 6 + 200/5 - 3^4$, we enter

$20 \times 6 + 200 \div 5 - 3$ ^ 4 and the display shows 20*6+200/5−3^4.

CAUTION ▸ The result is 79. ***Note carefully that 200 is divided only by 5.*** If it were divided by $5 - 3^4$, then we would use parentheses and show the expression to be evaluated as $20 \times 6 + 200/(5 - 3^4)$.

Exercises 1–4

In Exercises 1–48, simplify the given expressions. Express results with positive exponents only.

1. x^3x^4 **2.** y^2y^7 **3.** $2b^4b^2$

4. $3k(k^5)$ **5.** $\frac{m^5}{m^3}$ **6.** $\frac{x^6}{x}$

7. $\frac{n^5}{n^9}$ **8.** $\frac{s}{s^4}$ **9.** $(a^2)^4$

10. $(x^8)^3$ **11.** $(t^5)^4$ **12.** $(n^3)^7$

13. $(2n)^3$ **14.** $(ax)^5$ **15.** $(ax^4)^2$

16. $(3a^2)^3$ **17.** $\left(\frac{2}{b}\right)^3$ **18.** $\left(\frac{x}{y}\right)^7$

19. $\left(\frac{x^2}{2}\right)^4$ **20.** $\left(\frac{3}{n^3}\right)^3$ **21.** 7^0

22. $(8a)^0$ **23.** $-3x^0$ **24.** $6v^0$

25. 6^{-1} **26.** -10^{-3} **27.** $\frac{1}{s^{-2}}$

28. $\frac{1}{t^{-5}}$ **29.** $(-t^2)^7$ **30.** $(-y^3)^5$

31. $(2x^2)^6$ **32.** $-(-c^4)^4$ **33.** $(4xa^{-2})^0$

34. $3(ab^{-1})^0$ **35.** $-b^5b^{-3}$ **36.** $2c^4c^{-7}$

37. $\frac{2a^4}{(2a)^4}$ **38.** $\frac{x^2x^3}{(x^2)^3}$ **39.** $\frac{(n^2)^4}{(n^4)^2}$

40. $\frac{(3t)^{-1}}{3t^0}$ **41.** $(5^0x^2a^{-1})^{-1}$ **42.** $(3m^{-2}n^4)^{-2}$

43. $\left(\frac{4a}{x}\right)^{-3}$ **44.** $\left(\frac{2b^2}{y^5}\right)^{-2}$ **45.** $(-8gs^3)^2$

46. $ax^2(-a^2x)^2$ **47.** $\frac{15a^2n^5}{3an^6}$ **48.** $\frac{(ab^2)^3}{a^2b^8}$

In Exercises 49–56, evaluate the given expressions. In Exercises 51–56, all numbers are approximate.

49. $7(-4) - (-5)^2$ **50.** $6 + (-2)^5 - (-2)(8)$

51. $-(-26.5)^2 - (-9.85)^3$ **52.** $-0.711^2 - (-0.809)^6$

53. $\frac{3.07(-1.86)}{(-1.86)^4 + 1.596}$ **54.** $\frac{15.66^2 - (-4.017)^4}{1.044(-3.68)}$

55. $2.38(-60.7)^2 - 2540/1.17^3 + 0.806^5(26.1^3 - 9.88^4)$

56. $0.513(-2.778) - (-3.67)^3 + 0.889^4/(1.89 - 1.09^2)$

In Exercises 57–60, perform the indicated operations.

57. In designing a cam for a pump, the expression $\pi\left(\frac{r}{2}\right)^3\left(\frac{4}{3\pi r^2}\right)$ is used. Simplify this expression.

58. For an integrated electric circuit it might be necessary to simplify the expression $\frac{gM}{2\pi fC(2\pi fM)^2}$. Perform this simplification.

59. In order to find the electric power (in watts) consumed by an electric light, the expression i^2R must be evaluated, where i is the current in amperes and R is the resistance in ohms. Find the power consumed if $i = 0.525$ A and $R = 250\ \Omega$.

60. In designing a building, it was determined that the forces acting on an I beam would deflect the beam an amount, in centimetres, given by $\frac{x(1000 - 20x^2 + x^3)}{1850}$, where x is the distance, in metres, from one end of the beam. Find the deflection for $x = 6.85$ m. (The 1000 and 20 are exact.)

1-5 Scientific Notation

In technical and scientific work we often encounter numbers that are either very large or very small. Such numbers are illustrated in the next example.

EXAMPLE 1 Television signals travel at about 30 000 000 000 cm/s. The mass of the earth is about 6 000 000 000 000 000 000 000 000 kg. A typical individual fiber in a fiber-optic communications cable has a diameter of 0.000 005 m. Some X rays have a wavelength of about 0.000 000 095 cm.

Writing numbers like those in Example 1 is inconvenient in ordinary notation. Also, calculators and computers require a more efficient way of expressing such numbers in order to work with them. Therefore, a convenient and useful notation, called *scientific notation*, is used to represent such numbers.

A number in **scientific notation** *is expressed as the product of a number greater than or equal to 1 and less than 10, and a power of 10, and is written as*

$$P \times 10^k$$

where $1 \le P < 10$ and k is an integer. (The symbol $\le$ means "is less than or equal to.") See the following example.

EXAMPLE 2 ▸▸ (a) $340\,000 = 3.4(100\,000) = 3.4 \times 10^5$

(b) $0.000\,503 = \dfrac{5.03}{10\,000} = \dfrac{5.03}{10^4} = 5.03 \times 10^{-4}$

between 1 and 10

(c) $6.82 = 6.82(1) = 6.82 \times 10^0$ ◂◂

From Example 2 we can establish a way of changing numbers from ordinary notation to scientific notation. *The decimal point is moved so that only one nonzero digit is to its left. The number of places moved is the value of k, which is positive if the decimal point is moved to the left and negative if moved to the right.*

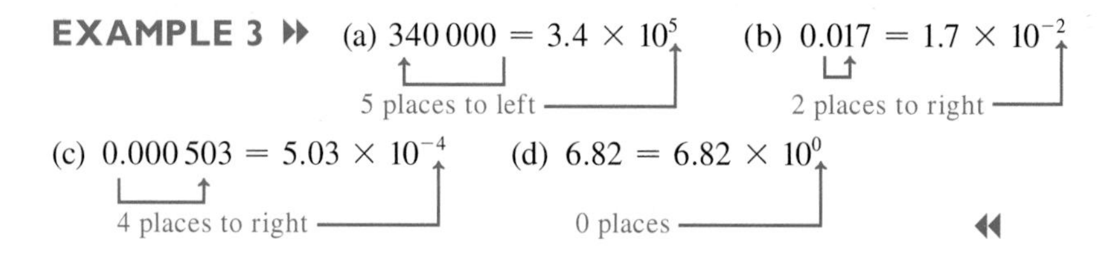

To change a number from scientific notation to ordinary notation, we reverse the procedure used in Example 3. This is shown in the next example.

EXAMPLE 4 ▸▸ To change 5.83×10^6 to ordinary notation, we move the decimal point 6 places to the right. Additional zeros must be included to properly locate the decimal point. This means we write

$$5.83 \times 10^6 = 5\,830\,000$$

6 places to right

To change 8.06×10^{-3} to ordinary notation, we must move the decimal point 3 places to the left. Again, additional zeros must be included. Therefore,

$$8.06 \times 10^{-3} = 0.008\,06$$

3 places to left ◂◂

The importance of scientific notation is demonstrated by the SI use of prefixes on units to denote certain powers of 10. These prefixes are listed in Table B-2 in Appendix B and we illustrate their use here as well as in Appendix B.

EXAMPLE 5 ▸▸ The prefix symbol k represents a multiplying factor of 10^3. Since the symbol g represents gram, we have kg represent 10^3 g or kilogram. Thus,

$$5600 \text{ g} = 5.6 \times 10^3 \text{ g} = 5.6 \text{ kg}$$

Similarly,

$$0.000\,000\,028 \text{ s} = 2.8 \times 10^{-8} \text{ s} = 28 \times 10^{-9} \text{ s} = 28 \text{ ns}$$ ◂◂

In writing the numbers with SI prefixes in Example 5, we have followed the SI practice by which *the multiple is chosen such that the numerical value is between 0.1 and 1000*. This is followed to the extent that it is practical to do so.

Scientific notation also provides a practical way to handle calculations with very large or very small numbers. First, all numbers are expressed in scientific notation. Then the calculation can be performed on numbers between 1 and 10, using the laws of exponents to find the power of 10 in the final result.

EXAMPLE 6 ▸▸ In designing a computer, it was determined that it would be able to process 803 000 bits of data in 0.000 005 25 s. (See Exercise 37 of Exercises 1-1 for a brief note on computer data.) The rate of processing the data is

$$\frac{803\,000}{0.000\,005\,25} = \frac{8.03 \times 10^5}{5.25 \times 10^{-6}} = \left(\frac{8.03}{5.25}\right) \times \overbrace{10^{11}}^{5-(-6)=11} = 1.53 \times 10^{11} \text{ bits/s}$$

As shown, it is proper to leave the result (*rounded off*) in scientific notation. This method is useful when using a calculator and then estimating the result. In this case the estimate is $(8 \times 10^5) \div (5 \times 10^{-6}) = 1.6 \times 10^{11}$. ◂◂

EXAMPLE 7 ▸▸ In evaluating the product $(7.50 \times 10^{11})(6.44 \times 10^{-3})$, we have

$$(7.50 \times 10^{11})(6.44 \times 10^{-3}) = 48.3 \times 10^8$$

However, since the answer is to be given in scientific notation, we should express it as the product of a number between 1 and 10 and a power of 10, which means we should rewrite it as

$$\underbrace{48.3}_{\text{not between 1 and 10}} \times 10^8 = (4.83 \times 10)(10^8)$$
$$= 4.83 \times 10^9$$ ◂◂

In Example 7 the product 7.50×6.44 can be found on a calculator, but it would not be possible to enter 7.50×10^{11} on a scientific calculator in the form 750 000 000 000. (It could be entered on a graphing calculator.) However, since calculators use scientific notation it is possible to enter numbers in scientific notation as well as have answers given automatically in scientific notation.

EXAMPLE 8 ▸▸ Using a calculator for the following calculation, we first write each number in scientific notation, as shown. The number 100 000 000 000 is exact.

$$\frac{69\,500\,000 \times 0.000\,004\,36}{100\,000\,000\,000} = \frac{(6.95 \times 10^7)(4.36 \times 10^{-6})}{10^{11}}$$
$$= 3.03 \times 10^{-9}$$

CAUTION ▸ The display of 3.0302 −9 (or 3.0302 E −9) means the calculator gives the result as 3.0302×10^{-9}. *Note that we must enter* 10^{11} *as* 1 [EE] 11, *not* 10 [EE] 11. ◂◂

Exercises 1–5

In Exercises 1–8, change the numbers from scientific notation to ordinary notation.

1. 4.5×10^4 **2.** 6.8×10^7 **3.** 2.01×10^{-3}

4. 9.61×10^{-5} **5.** 3.23×10^0 **6.** 8.40×10^0

7. 1.86×10 **8.** 1×10^{-1}

In Exercises 9–16, change the numbers from ordinary notation to scientific notation.

9. 40 000 **10.** 560 000 **11.** 0.0087

12. 0.7 **13.** 6 **14.** 1.09

15. 0.063 **16.** 0.000 090 8

In Exercises 17–20, perform the indicated calculations using a calculator and by first expressing all numbers in scientific notation.

17. 28 000(2 000 000 000)

18. 50 000(0.006)

19. $\dfrac{88\,000}{0.0004}$

20. $\dfrac{0.000\,03}{6\,000\,000}$

In Exercises 21–28, perform the indicated calculations using a calculator. All numbers are approximate.

21. 1280(865 000) (43.8)

22. 0.000 065 9(0.004 86)(3 190 000 000)

23. $\dfrac{0.0732(6710)}{0.001\,34(0.0231)}$

24. $\dfrac{0.004\,52}{2430(97\,100)}$

25. $(3.642 \times 10^{-8})(2.736 \times 10^{5})$

26. $\dfrac{9.368 \times 10^{-12}}{4.651 \times 10^{4}}$

27. $\dfrac{10^{7}}{(3.1075 \times 10^{-5})(1.0772 \times 10^{14})}$

28. $\dfrac{7.3009 \times 10^{-2}}{5.9843(2.5036 \times 10^{-20})}$

In Exercises 29–40, change numbers in ordinary notation to scientific notation or change numbers in scientific notation to ordinary notation. See Appendix B for an explanation of symbols used.

29. The power plant at Grand Coulee Dam produces 6 500 000 kW of power.

30. The maximum pressure exerted by the human heart is 16 000 Pa.

31. The power of a radio signal received from Galileo, the space probe to Jupiter, is 1.6×10^{-12} W.

32. Some computers can perform an addition in 4.5×10^{-15} s.

33. The signal on Channel 10 of a TV set is transmitted at a frequency of 1.92×10^{8} Hz.

34. To attain an energy density of that in some laser beams, an object would have to be heated to about 10^{30}°C.

35. A fiber-optic system requires 0.000 003 W of power.

36. Uranium is used in nuclear energy reactors to generate electricity. About 0.000 000 039% of the uranium disintegrates each day.

37. The electrical force between two electrons is about 2.4×10^{-43} times the gravitational force between them.

38. Among the stars nearest the earth, Centaurus A is about 4.07×10^{13} km away.

39. The altitude of a communications satellite is 36 000 km.

40. A reforestation machine can plant about 250 000 seedlings in one day.

In Exercises 41–44, perform the indicated calculations by first expressing all numbers in scientific notation.

41. The area of Lake Erie is 25 700 km². What is the area of Lake Erie in square centimetres?

42. The rate of energy radiation (in watts) from an object is found by evaluating the expression kT^4, where T is the thermodynamic temperature. Find this value for the human body, for which $k = 0.000\,000\,057$ W/K^4 and $T = 303$ K.

43. In a microwave receiver circuit, the resistance R of a wire 1 m long is given by $R = k/d^2$, where d is the diameter of the wire. Find R if $k = 0.000\,000\,021\,96\ \Omega \cdot \text{m}^2$ and $d = 0.000\,079\,98$ m.

44. At 0°C, the refrigerant Freon is a vapor at a pressure of $P = 1.378 \times 10^{5}$ Pa. If the volume of vapor is $V = 1.185 \times 10^{3}$ cm³, find the value of PV.

1-6 Roots and Radicals

A problem often encountered is: What number multiplied by itself n times gives another specified number? For example, what number squared is 9? We can see that either 3 or −3 is a proper answer. *We call either 3 or −3 a* **square root** *of 9,* since $3^2 = 9$ or $(-3)^2 = 9$.

In order to have a general notation for the square root, and have it represent *one* number, *we define the* **principal square root** *of a to be positive if a is positive and represent it by* $\sqrt{a}$. This means $\sqrt{9} = 3$ and not -3.

The general notation for the **principal *n*th root** *of a is* $\sqrt[n]{a}$. (When $n = 2$, we do not put the 2 for n.) *The* $\sqrt{\ }$ *sign is called a* **radical sign**. Unless we state otherwise, when we refer to the root of a number, it is the principal root.

EXAMPLE 1 ▸▸ (a) $\sqrt{2}$ (the square root of two) (b) $\sqrt[3]{2}$ (the cube root of two)

(c) $\sqrt[4]{2}$ (the fourth root of two) (d) $\sqrt[7]{6}$ (the seventh root of six) ◂◂

In order to have a single defined value for all roots (not just square roots), and to consider only real number roots, *we define the* **principal *n*th root** *of a to be positive if a is positive and to be negative if a is negative and n is odd.* (If a is negative and n is even, the roots are not real.)

EXAMPLE 2 ▸▸ (a) $\sqrt{169} = 13$ ($\sqrt{169} \neq -13$) (b) $-\sqrt{64} = -8$
(c) $\sqrt[3]{27} = 3$

(d) $-\sqrt[4]{256} = -4$ (e) $\sqrt[3]{-27} = -3$ (odd) (f) $-\sqrt[3]{27} = -(+3) = -3$ ◂◂

Another property of square roots is developed by noting illustrations such as $\sqrt{36} = \sqrt{4 \times 9} = \sqrt{4} \times \sqrt{9} = 2 \times 3 = 6$. In general, this property states that *the square root of a product of positive numbers is the product of their square roots.*

$$\sqrt{ab} = \sqrt{a}\sqrt{b} \quad (a \text{ and } b \text{ positive real numbers}) \tag{1-9}$$

This property is used in simplifying radicals. It is most useful if either a or b is a **perfect square**, *which is the square of a rational number.*

EXAMPLE 3 ▸▸ (a) $\sqrt{8} = \sqrt{(4)(2)} = \sqrt{4}\sqrt{2} = 2\sqrt{2}$

perfect squares — simplest form

(b) $\sqrt{75} = \sqrt{(25)(3)} = \sqrt{25}\sqrt{3} = 5\sqrt{3}$ ◂◂

An *exact* value of a square root is sometimes preferred, and we use Eq. (1-9) to write it in simplest form. However, a decimal *approximation* is generally acceptable, and we use the $\sqrt{x}$ key on a calculator. (Finding other roots on a calculator is discussed in Chapter 11.)

EXAMPLE 4 ▸▸ After reaching its greatest height, the time (in seconds) for a rocket to fall h metres due to gravity is found by evaluating $0.45\sqrt{h}$. Find the time for the rocket to fall 360 m.

Using the $\sqrt{x}$ key on a calculator, we find that $0.45\sqrt{360} = 8.538\,149\,7$. This means that the rocket takes 8.5 s to fall 360 m. The result is rounded off to two significant digits. ◂◂

In simplifying a radical, *all operations under a radical sign must be done before finding the root.* The horizontal bar groups the numbers under it.

EXAMPLE 5 ▸▸ (a) $\sqrt{16 + 9} = \sqrt{25} = 5$ first perform the addition 16 + 9

CAUTION ▸ However, $\sqrt{16 + 9}$ **is *not*** $\sqrt{16} + \sqrt{9} = 4 + 3 = 7$

(b) $\sqrt{2^2 + 6^2} = \sqrt{4 + 36} = \sqrt{40} = \sqrt{4}\sqrt{10} = 2\sqrt{10}$, but $\sqrt{2^2 + 6^2}$ is ***not*** $\sqrt{2^2} + \sqrt{6^2} = 2 + 6 = 8$ ◂◂

In defining the principal square root, we did not define the square root of a negative number. However, in Section 1-1 we defined the square root of a negative number to be an **imaginary number.** More generally, *the even root of a negative number is an imaginary number, and the odd root of a negative number is a negative real number.* A more detailed discussion of exponents, radicals, and imaginary numbers is found in Chapters 11 and 12.

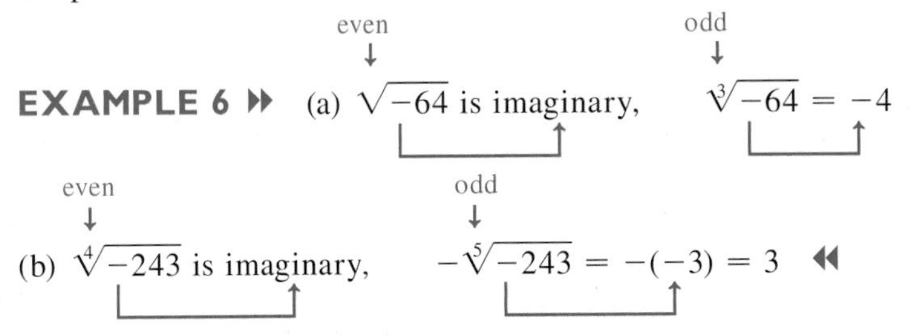

EXAMPLE 6 ▸▸ (a) $\sqrt{-64}$ is imaginary, $\sqrt[3]{-64} = -4$

(b) $\sqrt[4]{-243}$ is imaginary, $-\sqrt[5]{-243} = -(-3) = 3$ ◂◂

Exercises 1–6

In Exercises 1–28, simplify the given expressions. In each of 1–16, the result is an integer.

1. $\sqrt{25}$ **2.** $\sqrt{81}$ **3.** $-\sqrt{121}$

4. $-\sqrt{36}$ **5.** $-\sqrt{49}$ **6.** $\sqrt{225}$

7. $\sqrt{400}$ **8.** $-\sqrt{900}$ **9.** $\sqrt[3]{125}$

10. $\sqrt[4]{16}$ **11.** $\sqrt[3]{-216}$ **12.** $\sqrt[5]{-32}$

13. $(\sqrt{5})^2$ **14.** $(\sqrt{19})^2$ **15.** $(\sqrt[3]{31})^3$

16. $(\sqrt[4]{53})^4$ **17.** $\sqrt{18}$ **18.** $\sqrt{32}$

19. $\sqrt{12}$ **20.** $\sqrt{50}$ **21.** $2\sqrt{84}$

22. $4\sqrt{108}$ **23.** $\dfrac{7^2\sqrt{81}}{3^2\sqrt{49}}$ **24.** $\dfrac{2^5\sqrt[5]{243}}{3\sqrt{144}}$

25. $\sqrt{36 + 64}$ **26.** $\sqrt{25 + 144}$

27. $\sqrt{3^2 + 9^2}$ **28.** $\sqrt{8^2 - 4^2}$

In Exercises 29–36, find the value of each square root by use of a calculator. Each number is approximate.

29. $\sqrt{85.4}$ **30.** $\sqrt{3762}$

31. $\sqrt{0.4729}$ **32.** $\sqrt{0.0627}$

33. (a) $\sqrt{1296 + 2304}$ (b) $\sqrt{1296} + \sqrt{2304}$

34. (a) $\sqrt{10.6276 + 2.1609}$ (b) $\sqrt{10.6276} + \sqrt{2.1609}$

35. (a) $\sqrt{0.0429^2 - 0.0183^2}$ (b) $\sqrt{0.0429^2} - \sqrt{0.0183^2}$

36. (a) $\sqrt{3.625^2 + 0.614^2}$ (b) $\sqrt{3.625^2} + \sqrt{0.614^2}$

In Exercises 37–44, solve the given problems.

37. The period, in seconds, of a pendulum whose length is L metres can be found by evaluating $2.01\sqrt{L}$. Find the period of a pendulum for which $L = 1.75$ m.

38. The resistance in an amplifier circuit is found by evaluating $\sqrt{Z^2 - X^2}$. Find the resistance for $Z = 5.362\ \Omega$ and $X = 2.875\ \Omega$.

39. The speed (in m/s) of sound in sea water is found by evaluating $\sqrt{B/d}$ for $B = 2.18 \times 10^9$ Pa and $d = 1.03 \times 10^3$ kg/m^3. Find this speed, which is important in locating underwater objects using sonar.

40. The time (in seconds) for an object to fall h metres is given by the expression $\sqrt{h}/2.2$. How long does it take a person to fall 17 m from a fifth-floor window into a net while escaping a fire?

41. A TV screen is 38.6 cm wide and 29.0 cm high. The length of a diagonal (the dimension used to describe it—from one corner to the opposite corner) is found by evaluating $\sqrt{w^2 + h^2}$, where w is the width and h is the height. Find the diagonal of this TV screen.

42. A car costs \$22 000 new and is worth \$15 000 two years later. The annual rate of depreciation is found by evaluating $100(1 - \sqrt{V/C})$, where C is the cost and V is the value after two years. At what rate did the car depreciate? (100 and 1 are exact.)

43. Is it always true that $\sqrt{a^2} = a$? Explain.

44. If $0 < x < 1$ (x between 0 and 1), is $x > \sqrt{x}$? Explain.

1-7 Addition and Subtraction of Algebraic Expressions

Since we use letters to represent numbers, we conclude that all operations valid for numbers are also valid for literal numbers. In this section we discuss the methods for combining literal numbers.

Addition, subtraction, multiplication, division, and taking of roots are known as **algebraic operations.** *Any combination of numbers and literal symbols that results from algebraic operations is known as an* **algebraic expression.**

When an algebraic expression consists of several parts connected by plus signs and minus signs, each part (along with its sign) is known as a **term** *of the expression. If a given expression is made up of the product of a number of quantities, each of these quantities, or any product of them, is called a* **factor** *of the expression.*

CAUTION ▶ It is important to ***distinguish clearly between terms and factors*** since some operations that are valid for terms are not valid for factors, and conversely.

EXAMPLE 1 ▸▸ $3xy + 6x^2 - 7x\sqrt{y}$ is an algebraic expression with three terms: $3xy$, $6x^2$, and $-7x\sqrt{y}$.

(terms; factors)

The first term, $3xy$, has individual factors of 3, x, and y. Also, any product of these factors is also a factor of $3xy$. Thus, additional expressions that are factors of $3xy$ are $3x$, $3y$, xy, and $3xy$ itself. ◂◂

EXAMPLE 2 ▸▸ $7x(y^2 + x) - \dfrac{x + y}{6x}$ is an algebraic expression with terms $7x(y^2 + x)$ and $-\dfrac{x + y}{6x}$.

The term $7x(y^2 + x)$ has individual factors of 7, x, and $(y^2 + x)$, as well as products of these factors. The factor $y^2 + x$ has two terms, y^2 and x.

The numerator of the term $-\dfrac{x + y}{6x}$ has two terms, and the denominator has factors of 2, 3, and x. The negative sign can be treated as a factor of -1. ◂◂

An algebraic expression containing only one term is called a **monomial.** *An expression containing two terms is a* **binomial,** *and one containing three terms is a* **trinomial.** *Any expression containing two or more terms is called a* **multinomial.** Therefore, any binomial or trinomial is also a multinomial.

In any given term, the numbers and literal symbols multiplying any given factor constitute the **coefficient** *of that factor. The product of all the numbers in explicit form is known as the* **numerical coefficient** *of the term. All terms that differ at most in their numerical coefficients are known as* **similar** *or* **like** *terms.* That is, similar terms have the same variables with the same exponents.

EXAMPLE 3 ▸▸ (a) $7x^3\sqrt{y}$ is a monomial. It has a numerical coefficient of 7. The coefficient of $\sqrt{y}$ is $7x^3$, and the coefficient of x^3 is $7\sqrt{y}$.

(b) The expression $3a + 2b - 5a$ is a trinomial. The terms $3a$ and $-5a$ are similar since the literal parts are the same. ◂◂

like terms — unlike other terms due to factor of a

EXAMPLE 4 ▸▸ (a) $4 \times 2b + 81b - 6ab$
is a multinomial of three terms (a trinomial). The first term has a numerical coefficient of 8 (= 4×2), the second has a numerical coefficient of 81, and the third has a numerical coefficient of -6 (the sign is attached to the numerical coefficient). The first two terms are similar, since they differ only in numerical coefficients. The third term is not similar to the others, for it has the factor a.

(b) The commutative law tells us that $x^2y^3 = y^3x^2$. Therefore, the two terms of the expression $3x^2y^3 + 5y^3x^2$ are similar. ◂◂

In adding and subtracting algebraic expressions, we combine similar (or like) terms into a single term. The final simplified expression will contain only terms that are not similar.

EXAMPLE 5 ▸▸ (a) $3x + 2x - 5y = 5x - 5y$
Since there are two similar terms in the original expression, they are added together so that the simplified result has two unlike terms.

(b) $6a^2 - 7a + 8ax$ cannot be simplified since there are no like terms.

(c) $6a + 5c + 2a - c = 6a + 2a + 5c - c$ commutative law
$= 8a + 4c$ add like terms ◂◂

In writing algebraic expressions, it is often necessary to group certain terms together. For this purpose we use **symbols of grouping**. In this text we use **parentheses**, ( ), **brackets**, [ ], and **braces**, { }. The **bar**, which is used with radicals and fractions, also groups terms. In earlier sections we used parentheses and the bar for grouping.

CAUTION ▸ When adding and subtracting algebraic expressions, it is often necessary to remove symbols of grouping. To do so we must *change the sign of* ***every term*** *within the symbols if the grouping is preceded by a minus sign. If the symbols of grouping are preceded by a plus sign, each term within the symbols retains its original sign.* This is a result of the distributive law, $a(b + c) = ab + ac$.

EXAMPLE 6 ▸▸ (a) $2(a + 2x) = 2a + 2(2x)$ use distributive law
$= 2a + 4x$

(b) $-(+a - 3c) = (-1)(+a - 3c)$ treat − sign as −1
$= (-1)(+a) + (-1)(-3c)$
$= -a + 3c$ note change of signs

Normally, the $+a$ would be written simply as a. ◂◂

EXAMPLE 7 ▸▸ + sign before parentheses

(a) $3c + (2b - c) = 3c + 2b - c = 2b + 2c$ use distributive law

$2b = +2b$ signs retained

− sign before parentheses

(b) $3c - (2b - c) = 3c - 2b + c = -2b + 4c$ use distributive law

$2b = +2b$ signs changed

(c) $3c - (-2b + c) = 3c + 2b - c = 2b + 2c$ use distributive law

signs changed

Note in each case that the parentheses are removed and the sign before the parentheses is also removed. ◂◂

EXAMPLE 8 ▸▸ (a) $-(2x - 3c) + (c - x) = -2x + 3c + c - x = 4c - 3x$

(b) $4(t - 3 - 2t^2) - (6t + t^2 - 4) = 4t - 12 - 8t^2 - 6t - t^2 + 4$

$= -9t^2 - 2t - 8$ ◂◂

EXAMPLE 9 ▸▸ In designing a certain machine part, it is necessary to perform the following simplification.

$$16(8 - x) - 2(8x - x^2) - (64 - 16x + x^2) = 128 - 16x - 16x + 2x^2 - 64 + 16x - x^2$$
$$= 64 - 16x + x^2$$ ◂◂

It is fairly common to have expressions in which more than one symbol of grouping is to be removed in the simplification. Normally, *when several symbols of grouping are to be removed, it is more convenient to remove the innermost symbols first.* This is illustrated in the following example.

EXAMPLE 10 ▸▸ (a) $3ax - [ax - (5s - 2ax)] = 3ax - [ax - 5s + 2ax]$ ← remove parentheses

$= 3ax - ax + 5s - 2ax$ ← remove brackets

$= 5s$

(b) $3a^2b - \{[a - (2a^2b - a)] + 2b\} = 3a^2b - \{[a - 2a^2b + a] + 2b\}$ ← remove parentheses

$= 3a^2b - \{a - 2a^2b + a + 2b\}$ ← remove brackets

$= 3a^2b - a + 2a^2b - a - 2b$ ← remove braces

$= 5a^2b - 2a - 2b$ ◂◂

Calculators use only parentheses for grouping symbols, and we often need to use one set of parentheses within another set. These are called **nested parentheses**. In the next example, note that the innermost parentheses are removed first.

EXAMPLE 11 ▸▸ $2 - (3x - 2(5 - (7 - x))) = 2 - (3x - 2(5 - 7 + x))$

$= 2 - (3x - 10 + 14 - 2x)$

$= 2 - 3x + 10 - 14 + 2x$

$= -x - 2$ ◂◂

CAUTION ▶ One of the most common errors made by beginning students is changing the sign of only the first term when removing symbols of grouping preceded by a minus sign. *Remember,* ***if the symbols are preceded by a minus sign, we must change the sign of* all *terms.***

Exercises 1–7

In the following exercises, simplify the given algebraic expressions.

1. $5x + 7x - 4x$
2. $6t - 3t - 4t$
3. $2y - y + 4x$
4. $4c + d - 6c$
5. $2a - 2c - 2 + 3c - a$
6. $x - 2y + 3x - y + z$
7. $a^2b - a^2b^2 - 2a^2b$
8. $xy^2 - 3x^2y^2 + 2xy^2$
9. $s + (4 + 3s)$
10. $5 + (3 - 4n + p)$
11. $v - (4 - 5x + 2v)$
12. $2a - (b - a)$
13. $2 - 3 - (4 - 5a)$
14. $\sqrt{x} + (y - 2\sqrt{x}) - 3\sqrt{x}$
15. $(a - 3) + (5 - 6a)$
16. $(4x - y) - (-2x - 4y)$
17. $-(t - 2u) + (3u - t)$
18. $2(x - 2y) + (5x - y)$
19. $3(2r + s) - (-5s - r)$
20. $3(a - b) - 2(a - 2b)$
21. $-7(6 - 3c) - 2(c + 4)$
22. $-(5t + a^2) - 2(3a^2 - 2st)$
23. $-[(6 - n) - (2n - 3)]$
24. $-[(a - b) - (b - a)]$
25. $2[4 - (t^2 - 5)]$
26. $3[-3 - (a - 4)]$
27. $-2[-x - 2a - (a - x)]$
28. $-2[-3(x - 2y) + 4y]$
29. $a\sqrt{xy} - [3 - (a\sqrt{xy} + 4)]$
30. $9v - [6 - (v - 4) + 4v]$
31. $8c - \{5 - [2 - (3 + 4c)]\}$
32. $7y - \{y - [2y - (x - y)]\}$
33. $5p - (q - 2p) - [3q - (p - q)]$
34. $-(4 - x) - [(5x - 7) - (6x + 2)]$
35. $-2\{-(4 - x^2) - [3 + (4 - x^2)]\}$
36. $-\{-[-(x - 2a) - b] - a\}$
37. $3a - (6 - (a + 3))$
38. $-2x + 2((2x - 1) - 5)$
39. $-(3t - (7 + 2t - (5t - 6)))$
40. $a^2 - 2(x - 5 - (7 - 2(a^2 - 2x) - 3x))$

41. In determining the size of a V belt to be used with an engine, the expression $3D - (D - d)$ is used. Simplify this expression.

42. When finding the current in a transistor circuit, the expression $i_1 - (2 - 3i_2) + i_2$ is used. Simplify this expression. (The numbers below the i's are *subscripts.* Different subscripts denote different variables.)

43. Water leaked into a gasoline storage tank at an oil refinery. Finding the pressure in the tank leads to the expression $-4(b - c) - 3(a - b)$. Simplify this expression.

44. Research on a plastic building material leads to $[(B + \frac{4}{3}\alpha) + 2(B - \frac{2}{3}\alpha)] - [(B + \frac{4}{3}\alpha) - (B - \frac{2}{3}\alpha)]$. Simplify this expression.

1-8 Multiplication of Algebraic Expressions

To find the product of two or more monomials, we use the laws of exponents as given in Section 1-4 and the laws for multiplying signed numbers as stated in Section 1-2. We first multiply the numerical coefficients to find the numerical coefficient of the product. Then we multiply the literal numbers, remembering that *the exponents may be combined only if the base is the same.*

EXAMPLE 1 ▶▶ (a) $3c^5(-4c^2) = -12c^7$ — multiply numerical coefficients and add exponents of c

(b) $(-2b^2y^3)(-9aby^5) = 18ab^3y^8$ — add exponents of same base

(c) $2xy(-6cx^2)(3xcy^2) = -36c^2x^4y^3$ ◀◀

If a product contains a monomial that is raised to a power, *we must first raise it to the indicated power* before proceeding with the multiplication.

EXAMPLE 2 ▸▸ (a) $3(2a^2x)^3(-ax) = 3(8a^6x^3)(-ax) = -24a^7x^4$

(b) $2s^3(-st^4)^2(4s^2t) = 2s^3(s^2t^8)(4s^2t) = 8s^7t^9$ ◂◂

We find the product of a monomial and a multinomial by using the distributive law, which states that we *multiply each term of the multinomial by the monomial.* We must be careful to assign the correct sign to each term of the result, using the rules for multiplication of signed numbers.

EXAMPLE 3 ▸▸ (a) $2ax(3ax^2 - 4yz) = 2ax(3ax^2) + (2ax)(-4yz) = 6a^2x^3 - 8axyz$

(b) $5cy^2(-7cx - ac) = (5cy^2)(-7cx) + (5cy^2)(-ac)$

$= -35c^2xy^2 - 5ac^2y^2$ ◂◂

It is generally not necessary to write out the middle step as it appears in the preceding example. We write the answer directly. For example, illustration (a) in Example 3 would appear as $2ax(3ax^2 - 4yz) = 6a^2x^3 - 8axyz$.

We find the product of two or more multinomials by using the distributive law and the laws of exponents. The result is that *we multiply each term of one multinomial by each term of the other, and add the results.*

EXAMPLE 4 ▸▸ $(x - 2)(x + 3) = x(x) + x(3) + (-2)(x) + (-2)(3)$

$= x^2 + 3x - 2x - 6 = x^2 + x - 6$ ◂◂

Finding the power of an algebraic expression is equivalent to using the expression as a factor the number of times indicated by the exponent. It is often convenient to write the power of an algebraic expression in this form before multiplying.

EXAMPLE 5 ▸▸ (a) $(x + 5)^2 = (x + 5)(x + 5) = x^2 + 5x + 5x + 25$

$= x^2 + 10x + 25$

(b) $(2a - b)^3 = (2a - b)(2a - b)(2a - b)$

$= (2a - b)(4a^2 - 4ab + b^2)$

$= 8a^3 - 8a^2b + 2ab^2 - 4a^2b + 4ab^2 - b^3$

$= 8a^3 - 12a^2b + 6ab^2 - b^3$

We should note in illustration (a) that

CAUTION ▸ $(x + 5)^2$ ***is not equal to*** $x^2 + 25$

since the term $10x$ is not included. We must follow the proper procedure and not simply square each of the terms within the parentheses. ◂◂

EXAMPLE 6 ▸▸ An expression used with a lens of a certain telescope is simplified as shown.

$$a(a+b)^2 + a^3 - (a+b)(2a^2 - s^2)$$
$$= a(a+b)(a+b) + a^3 - (2a^3 - as^2 + 2a^2b - bs^2)$$
$$= a(a^2 + ab + ab + b^2) + a^3 - 2a^3 + as^2 - 2a^2b + bs^2$$
$$= a^3 + a^2b + a^2b + ab^2 - a^3 + as^2 - 2a^2b + bs^2$$
$$= ab^2 + as^2 + bs^2$$ ◂◂

Exercises 1–8

In the following exercises, perform the indicated multiplications.

1. $(a^2)(ax)$ **2.** $(2xy)(x^2y^3)$

3. $-ac^2(acx^3)$ **4.** $-2s^2(-4cs)^2$

5. $(2ax^2)^2(-2ax)$ **6.** $6pq^3(3pq^2)^2$

7. $a(-a^2x)^3(-2a)$ **8.** $-2m^2(-3mn)(m^2n)^2$

9. $a^2(x+y)$ **10.** $2x(p-q)$

11. $-3s(s^2-5t)$ **12.** $-3b(2b^2-b)$

13. $5m(m^2n+3mn)$ **14.** $a^2bc(2ac-3a^2b)$

15. $3x(-x-y+2)$ **16.** $b^2x^2(x^2-2x+1)$

17. $ab^2c^4(ac-bc-ab)$ **18.** $-4c^2(-9gc-2c+g^2)$

19. $ax(cx^2)(x+y^3)$ **20.** $-2(-3st^3)(3s-4t)$

21. $(x-3)(x+5)$ **22.** $(a+7)(a+1)$

23. $(x+5)(2x-1)$ **24.** $(4t+s)(2t-3s)$

25. $(2a-b)(3a-2b)$ **26.** $(4x-3)(3x-1)$

27. $(2s+7t)(3s-5t)$ **28.** $(5p-2q)(p+8q)$

29. $(x^2-1)(2x+5)$ **30.** $(3y^2+2)(2y-9)$

31. $(x^2-2x)(x+4)$ **32.** $(2ab^2-5t)(-ab^2-6t)$

33. $(x+1)(x^2-3x+2)$ **34.** $(2x+3)(x^2-x-5)$

35. $(4x-x^3)(2+x-x^2)$ **36.** $(5a-3c)(a^2+ac-c^2)$

37. $2(a+1)(a-9)$ **38.** $-5(y-3)(y+6)$

39. $2x(x-1)(x+4)$ **40.** $ax(x+4)(7-x^2)$

41. $(2x-5)^2$ **42.** $(x-3)^2$

43. $(x+3a)^2$ **44.** $(2m+1)^2$

45. $(xyz-2)^2$ **46.** $(b-2x^2)^2$

47. $2(x+8)^2$ **48.** $3(a+4)^2$

49. $(2+x)(3-x)(x-1)$ **50.** $(3x-c^2)^3$

51. $3x(x+2)^2(2x-1)$ **52.** $[(x-2)^2(x+2)]^2$

53. In a particular computer design containing n circuit elements, n^2 switches are needed. Find the expression for the number of switches needed for $n+100$ circuit elements.

54. An expression found in chemical thermodynamics is $p(c-1) + 2 - c(p-1)$. Multiply and simplify.

55. In finding the maximum power in part of a microwave transmitter circuit, the expression $(R+r)^2 - 2r(R+r)$ is used. Multiply and simplify.

56. In determining the deflection of a certain steel beam, the expression $27x^2 - 24(x-6)^2 - (x-12)^3$ is used. Multiply and simplify.

1-9 Division of Algebraic Expressions

To find the quotient of one monomial divided by another, we use the laws of exponents and the laws for dividing signed numbers. Again, the *exponents may be combined only if the base is the same.*

EXAMPLE 1 ▸▸ (a) $\dfrac{3c^7}{c^2} = 3c^{7-2} = 3c^5$ (b) $\dfrac{16x^3y^5}{4xy^2} = \dfrac{16}{4}(x^{3-1})(y^{5-2}) = 4x^2y^3$

(c) $\dfrac{-6a^2xy^2}{2axy^4} = -\left(\dfrac{6}{2}\right)\dfrac{a^{2-1}x^{1-1}}{y^{4-2}} = -\dfrac{3a}{y^2}$

As shown in illustration (c), we use only positive exponents in the final result unless there are specific instructions otherwise. ◂◂

From arithmetic we may show how a multinomial is to be divided by a monomial. When adding fractions, say $\frac{2}{7}$ and $\frac{3}{7}$, we have

$$\frac{2}{7} + \frac{3}{7} = \frac{2 + 3}{7}$$

Looking at this from *right to left*, we see that *the quotient of a multinomial divided by a monomial is found by dividing each term of the multinomial by the monomial and adding the results.* This can be shown as

$$\frac{a + b}{c} = \frac{a}{c} + \frac{b}{c}$$

CAUTION ▶ Be careful: Although $\frac{a + b}{c} = \frac{a}{c} + \frac{b}{c}$, we must note that $\frac{c}{a + b}$ ***is not*** $\frac{c}{a} + \frac{c}{b}$.

EXAMPLE 2 ▶▶ (a) $\frac{4a^2 + 8a}{2a} = \frac{4a^2}{2a} + \frac{8a}{2a} = 2a + 4$

each term of numerator divided by denominator

(b) $\frac{4x^3y - 8x^3y^2 + 2x^2y}{2x^2y} = \frac{4x^3y}{2x^2y} - \frac{8x^3y^2}{2x^2y} + \frac{2x^2y}{2x^2y}$

$= 2x - 4xy + 1$

(c) $\frac{a^3bc^4 - 6abc + 9a^2b^3c - 3}{3ab^2c^3} = \frac{a^2c}{3b} - \frac{2}{bc^2} + \frac{3ab}{c^2} - \frac{1}{ab^2c^3}$

We usually do not write the middle step shown in illustrations (a) and (b). The divisions are done by inspection, and the result appears as in illustration (c). ◀◀

EXAMPLE 3 ▶▶ The expression $\frac{2p + v^2d + 2ydg}{2dg}$ is used when analyzing the operation of an irrigation pump. Performing the indicated division, we have

$$\frac{2p + v^2d + 2ydg}{2dg} = \frac{p}{dg} + \frac{v^2}{2g} + y$$ ◀◀

If each term in an algebraic sum is a number or is of the form ax^n, where n is a nonnegative integer, we call the expression a **polynomial** *in x. The greatest value of the exponent n that appears is the* **degree** *of the polynomial.*

A multinomial and a polynomial may differ in that a polynomial cannot have terms like $\sqrt{x}$ or $1/x^2$, but a multinomial may have such terms. Also, a polynomial may have only one term, whereas a multinomial has at least two terms.

EXAMPLE 4 ▶▶ (a) $3 + 2x^2 - x^3$ is a polynomial of degree 3.

(b) $x^4 - 3x^2 - \sqrt{x}$ is not a polynomial.

(c) $4x^5$ is a polynomial of degree 5. (d) $\frac{1}{4x^5}$ is not a polynomial.

Expressions (a) and (b) are also multinomials since each has more than one term. Expression (b) is not a polynomial due to the $\sqrt{x}$ term. Expression (c) is a polynomial since the exponent is a positive integer. It is also a monomial. Expression (d) is not a polynomial since it can be written as $\frac{1}{4}x^{-5}$, and when written in the form ax^n, n is not positive. ◀◀

To divide one polynomial by another, first arrange the dividend (the polynomial to be divided) and the divisor in descending powers of the variable. Then divide the first term of the dividend by the first term of the divisor. The result is the first term of the quotient. Next, multiply the entire divisor by the first term of the quotient and subtract the product from the dividend. Divide the first term of this difference by the first term of the divisor. This gives the second term of the quotient. Multiply this term by the entire divisor, and subtract the product from the first difference. Repeat this process until the remainder is zero or a term of lower degree than the divisor. This process is similar to long division of numbers.

EXAMPLE 5 ▸▸ Perform the division $(6x^2 + x - 2) \div (2x - 1)$. $\left(\text{This division can also be indicated in the fractional form } \frac{6x^2 + x - 2}{2x - 1}.\right)$

We set up the division as we would for long division in arithmetic. Then, following the procedure outlined above, we have the following:

$$
\begin{array}{r|l}
 & 3x + 2 \qquad \frac{6x^2}{2x} \\
2x - 1 & 6x^2 + x - 2 \\
\text{subtract} & 6x^2 - 3x \leftarrow 3x(2x - 1) \\
6x^2 - 6x^2 = 0,\ x - (-3x) = 4x & \quad\ \ 4x - 2 \\
 & \quad\ \ 4x - 2 \quad \text{subtract} \\
 & \qquad\quad 0
\end{array}
$$

CAUTION ▸

The remainder is zero and the quotient is $3x + 2$. Note that when we subtracted $-3x$ from x, we obtained $4x$. ◂◂

EXAMPLE 6 ▸▸ Perform the division $(4x^3 + 6x^2 + 1) \div (2x - 1)$. Since there is no x-term in the dividend, we should leave space for any x-terms that might arise.

$$
\begin{array}{r|l}
 & 2x^2 + 4x + 2 \\
\text{divisor} \quad 2x - 1 & 4x^3 + 6x^2 \qquad\ + 1 \quad \text{dividend} \\
 & 4x^3 - 2x^2 \\
6x^2 - (-2x^2) = 8x^2 \rightarrow & \qquad\ 8x^2 \qquad\ + 1 \\
 & \qquad\ 8x^2 - 4x \\
0 - (-4x) = 4x \rightarrow & \qquad\qquad\ 4x + 1 \\
 & \qquad\qquad\ 4x - 2 \\
 & \qquad\qquad\qquad 3 \quad \text{remainder}
\end{array}
$$

The quotient in this case is written as $2x^2 + 4x + 2 + \frac{3}{2x - 1}$. ◂◂

Exercises 1–9

In the following exercises, perform the indicated divisions.

1. $\frac{8x^3y^2}{-2xy}$
2. $\frac{-18b^7c^3}{bc^2}$
3. $\frac{-16r^3t^5}{-4r^5t}$
4. $\frac{51mn^5}{17m^2n^2}$
5. $\frac{(15x^2)(4bx)(2y)}{30bxy}$
6. $\frac{(5st)(8s^2t^3)}{10s^3t^2}$
7. $\frac{6(ax)^2}{-ax^2}$
8. $\frac{12a^2b}{(3ab^2)^2}$
9. $\frac{a^2x + 4xy}{x}$
10. $\frac{2m^2n - 6mn}{2m}$
11. $\frac{3rst - 6r^2st^2}{3rs}$
12. $\frac{-5a^2n - 10an^2}{5an}$

13. $\dfrac{4pq^3 + 8p^2q^2 - 16pq^5}{4pq^2}$

14. $\dfrac{a^2xy^2 + ax^3 - ax}{ax}$

15. $\dfrac{2\pi fL - \pi fR^2}{\pi fR}$

16. $\dfrac{2(ab)^4 - a^3b^4}{3(ab)^3}$

17. $\dfrac{3ab^2 - 6ab^3 + 9a^2b^2}{9a^2b^2}$

18. $\dfrac{2x^2y^2 + 8xy - 12x^2y^4}{2x^2y^2}$

19. $\dfrac{x^{n+2} + ax^n}{x^n}$

20. $\dfrac{3a(x + y)b^2 - (x + y)}{a(x + y)}$

21. $(2x^2 + 7x + 3) \div (x + 3)$

22. $(3x^2 - 11x - 4) \div (x - 4)$

23. $\dfrac{x^2 - 3x + 2}{x - 2}$

24. $\dfrac{2x^2 - 5x - 7}{x + 1}$

25. $\dfrac{x - 14x^2 + 8x^3}{2x - 3}$

26. $\dfrac{6x^2 + 6 + 7x}{2x + 1}$

27. $(4x^2 + 23x + 15) \div (4x + 3)$

28. $(6x^2 - 20x + 16) \div (3x - 4)$

29. $\dfrac{x^3 + 3x^2 - 4x - 12}{x + 2}$

30. $\dfrac{3x^3 + 19x^2 + 16x - 20}{3x - 2}$

31. $\dfrac{2x^4 + 4x^3 + 2}{x^2 - 1}$

32. $\dfrac{2x^3 - 3x^2 + 8x - 2}{x^2 - x + 2}$

33. $\dfrac{x^3 + 8}{x + 2}$

34. $\dfrac{x^3 - 1}{x - 1}$

35. $\dfrac{x^2 - 2xy + y^2}{x - y}$

36. $\dfrac{3a^2 - 5ab + 2b^2}{a - 3b}$

37. In the optical theory dealing with lasers, the following expression arises: $\dfrac{8A^5 + 4A^3\mu^2E^2 - A\mu^4E^4}{8A^4}$. Perform the indicated division. (μ is the Greek letter mu.)

38. In finding the total resistance of the resistors shown in Fig. 1-5, the expression $\dfrac{6R_1 + 6R_2 + R_1R_2}{6R_1R_2}$ is used. Perform the indicated division.

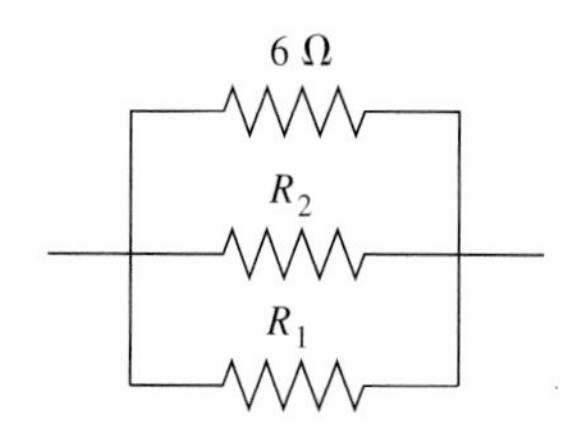

Fig. 1-5

39. A computer model shows that the temperature change T in a certain freezing unit is found by using the expression $\dfrac{3T^3 - 8T^2 + 8}{T - 2}$. Perform the indicated division.

40. In analyzing the displacement of a certain valve, the expression $\dfrac{s^2 - 2s - 2}{s^4 + 4}$ is used. Find the reciprocal of this expression, and perform the indicated division.

1-10 Solving Equations

In this section we show how algebraic operations are used in solving equations. In the following sections we show some of the important applications of equations.

An **equation** *is an algebraic statement that two algebraic expressions are equal.* Any value of the literal number representing the **unknown** that produces equality when **substituted** in the equation is said to **satisfy** the equation.

EXAMPLE 1 ▸▸ The equation $3x - 5 = x + 1$ is true only if $x = 3$. For $x = 3$ we have $4 = 4$; for $x = 2$ we have $1 = 3$, which is not correct.

This equation is valid for only one value of the unknown. *An equation valid only for certain values of the unknown is a* **conditional equation**. In this section, nearly all equations we solve will be conditional equations that are satisfied by only one value of the unknown. ◂◂

EXAMPLE 2 ▸▸ (a) The equation $x^2 - 4 = (x - 2)(x + 2)$ is true for all values of x. For example, if $x = 3$, we have $9 - 4 = (3 - 2)(3 + 2)$, or $5 = 5$. If $x = -1$, we have $-3 = -3$. *An equation valid for all values of the unknown is an* **identity**.

(b) The equation $x + 5 = x + 1$ is not true for any value of x. For any value of x we try, we find that the left side is 4 greater than the right side. *Such an equation is called a* **contradiction**. ◂◂

To **solve** *an equation we find the values of the unknown that satisfy it.* There is one basic rule to follow when solving an equation:

Perform the same operation on both sides of the equation.

We do this to isolate the unknown and thus to find its values.

By performing the same operation on both sides of an equation, the two sides remain equal. Thus,

we may add the same number to both sides, subtract the same number from both sides, multiply both sides by the same number, or divide both sides by the same number (not zero).

Although we may multiply both sides of an equation by zero, this produces $0 = 0$, which is not useful in finding the solution.

EXAMPLE 3 ▸▸ For each of the following equations, we may isolate x, and thereby solve the equation, by performing the indicated operation.

$x - 3 = 12$	$x + 3 = 12$	$\frac{x}{3} = 12$	$3x = 12$
add 3 to both sides	subtract 3 from both sides	multiply both sides by 3	divide both sides by 3
$x - 3 + 3 = 12 + 3$	$x + 3 - 3 = 12 - 3$	$3\left(\frac{x}{3}\right) = 3(12)$	$\frac{3x}{3} = \frac{12}{3}$
$x = 15$	$x = 9$	$x = 36$	$x = 4$

Each solution should be checked by substitution in the original equation. ◂◂

The solution of an equation generally requires a combination of the basic operations. The following examples illustrate the solution of such equations.

EXAMPLE 4 ▸▸ Solve the equation $2t - 7 = 9$.

We are to perform basic operations to both sides of the equation to finally isolate t on one side. The steps to be followed are suggested by the form of the equation, and in this case are as follows:

$$2t - 7 = 9 \quad \text{original equation}$$
$$2t - 7 + 7 = 9 + 7 \quad \text{add 7 to both sides}$$
$$2t = 16 \quad \text{combine like terms}$$
$$\frac{2t}{2} = \frac{16}{2} \quad \text{divide both sides by 2}$$
$$t = 8 \quad \text{simplify}$$

NOTE ▸ Therefore, we conclude that $t = 8$. Checking *in the original equation,* we have

$$2(8) - 7 \stackrel{?}{=} 9, \qquad 16 - 7 \stackrel{?}{=} 9, \quad \text{or} \quad 9 = 9$$

Therefore, the solution checks. ◂◂

When simpler numbers are involved, the step of adding or subtracting a term, or multiplying or dividing by a factor, is usually done by inspection and not actually written down. This is done in the later examples, when applicable.

EXAMPLE 5 ▶▶ Solve the equation $3n + 4 = n - 6$.

$$2n + 4 = -6 \quad \text{n subtracted from both sides—by inspection}$$
$$2n = -10 \quad \text{4 subtracted from both sides—by inspection}$$
$$n = -5 \quad \text{both sides divided by 2—by inspection}$$

Checking *in the original equation,* we have $-11 = -11$. ◀◀

EXAMPLE 6 ▶▶ Solve the equation $x - 7 = 3x - (6x - 8)$.

$$x - 7 = 3x - 6x + 8 \quad \text{parentheses removed}$$
$$x - 7 = -3x + 8 \quad \text{x-terms combined on right}$$
$$4x - 7 = 8 \quad \text{3x added to both sides}$$
$$4x = 15 \quad \text{7 added to both sides}$$
$$x = \tfrac{15}{4} \quad \text{both sides divided by 4}$$

Checking in the original equation, we obtain (after simplifying) $-\frac{13}{4} = -\frac{13}{4}$. ◀◀

NOTE ▶ Note that we ***always check in the* original *equation.*** This is done since errors may have been made in finding the later equations.

From these examples, we see that the following steps are used in solving the basic equations of this section.

Procedure for Solving Equations

1. *Remove grouping symbols (distributive law).*
2. *Combine any like terms on each side (also after step 3).*
3. *Perform the same operations on both sides until* $x =$ *result is obtained.*
4. *Check the solution in the original equation.*

If an equation contains decimals, the best procedure is to *set up the solution first, before actually performing the calculations.* Once the unknown is isolated, a calculator can be used to perform the calculations.

EXAMPLE 7 ▶▶ When finding the electric current i (in amperes) in a circuit in a radio, the following equation and solution are used.

$$0.0595 - 0.525i - 8.85(i + 0.00316) = 0$$
$$0.0595 - 0.525i - 8.85i - 8.85(0.00316) = 0 \quad \text{note how the above procedure is followed}$$
$$(-0.525 - 8.85)i = 8.85(0.00316) - 0.0595$$
$$i = \frac{8.85(0.00316) - 0.0595}{-0.525 - 8.85} \quad \text{evaluate}$$
$$= 0.00336 \text{ A}$$

In checking, we find the left side of the original equation to be 3.4×10^{-5}. If we use the unrounded calculator value, we get 0. Although the *result* should be rounded off, *no rounding off should be done before the final calculation.* An inaccurate result might otherwise be obtained. ◀◀

Except for the third illustration of Example 3, we have not solved an equation that contained a fraction. We now consider one type of equation with fractions.

The quotient a/b is also called the **ratio** *of a to b. An equation stating that two ratios are equal is called a* **proportion**. Since a proportion is an equation, if one of the numbers is unknown, we can solve for its value as with any equation.

EXAMPLE 8 ▸▸ The ratio of 2 to 3 is the quotient 2/3. The ratio of 4 to 6 is 4/6. Since these ratios are equal, we have the proportion

$$\frac{2}{3} = \frac{4}{6}$$

If the ratio of x to 8 equals the ratio of 3 to 4, we have the proportion

$$\frac{x}{8} = \frac{3}{4}$$

We can solve this equation by multiplying both sides by 8. This gives

$$8\left(\frac{x}{8}\right) = 8\left(\frac{3}{4}\right) \quad \text{or} \quad x = 6$$

Since $6/8 = 3/4$, the solution checks. ◂◂

EXAMPLE 9 ▸▸ The ratio of electric current I (in amperes) to the voltage V across a resistor is constant. If $I = 1.52$ A for $V = 60.0$ V, find I for $V = 82.0$ V.

Since the ratio of I to V is constant, we have the following solution.

$$\frac{1.52}{60.0} = \frac{I}{82.0}$$

$$82.0\left(\frac{1.52}{60.0}\right) = 82.0\left(\frac{I}{82.0}\right)$$

$$2.08 \text{ A} = I$$

$$I = 2.08 \text{ A}$$

Checking in the original equation, we have 0.0253 A/V = 0.0254 A/V. Since the value of I is rounded up, the solution checks. ◂◂

We will use the terms *ratio* and *proportion* (particularly ratio) when we study trigonometry in Chapter 4. A more detailed discussion of ratio and proportion is found in Chapter 18. Also, a general method of solving equations with fractions is given in Chapter 6.

Exercises 1–10

In Exercises 1–32, solve the given equations.

1. $x - 2 = 7$

2. $x - 4 = 1$

3. $x + 5 = 4$

4. $s + 6 = -3$

5. $\frac{t}{2} = 5$

6. $\frac{x}{4} = -2$

7. $4x = -20$

8. $2x = 12$

9. $3t + 5 = -4$

10. $5x - 2 = 13$

11. $5 - 2y = 3$

12. $8 - 5t = 18$

13. $3x + 7 = x$

14. $6 + 8y = 5 - y$

15. $2(s - 4) = s$

16. $3(n - 2) = -n$

17. $6 - (r - 4) = 2r$

18. $5 - (x + 2) = 5x$

19. $2(x - 3) - 5x = 7$

20. $4(x + 7) - x = -7$

21. $x - 5(x - 2) = 2$

22. $5x - 2(x - 5) = 4x$

23. $7 - 3(1 - 2p) = 4 + 2p$

24. $3 - 6(2 - 3t) = t - 5$

In Exercises 25–32, all numbers are approximate.

25. $5.8 - 0.3(x - 6.0) = 0.5x$

26. $1.9t = 0.5(4.0 - t) - 0.8$

27. $0.15 - 0.24(y - 0.50) = 0.63$

28. $27.5(5.17 - 1.44x) = 73.4$

29. $\frac{x}{2.0} = \frac{17}{6.0}$

30. $\frac{3.0}{7.0} = \frac{x}{42}$

31. $\frac{165}{223} = \frac{13x}{15}$

32. $\frac{276x}{17.0} = \frac{1360}{46.4}$

In Exercises 33–40, solve the given problems.

33. In finding the rate (in km/h) at which a polluted stream flows, the equation $15(5.5 + v) = 24(5.5 - v)$ is used. Find the rate v.

34. To find the voltage V in a circuit in a TV remote-control unit, the equation $1.12V - 0.67(10.5 - V) = 0$ is used. Find V.

35. In blending two gasolines of different octanes, in order to find the number n of litres of one octane needed, the equation $0.14n + 0.06(2000 - n) = 0.09(2000)$ is used. Find n, given that 0.06 and 0.09 are exact and the first zero of 2000 is significant.

36. In order to find the distance x such that the weights are balanced on the lever shown in Fig. 1-6, the equation $210(3x) = 55.3x + 38.5(8.25 - 3x)$ must be solved. Find x. (3 is exact.)

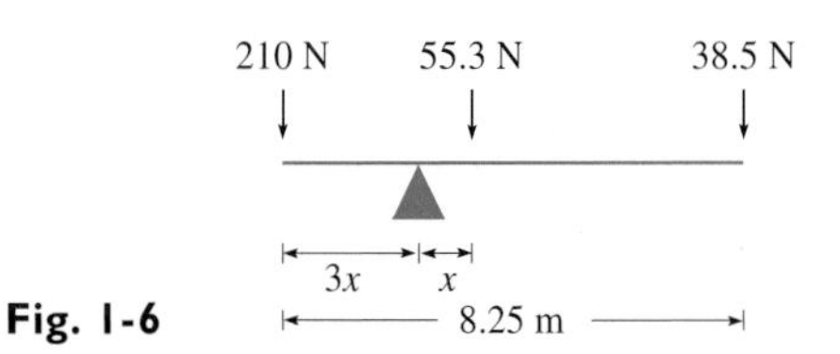

Fig. 1-6

37. A capsule contains two medications in the ratio of 5 to 2. If there are 150 mg of the first medication in the capsule, how many milligrams of the second are there?

38. A person 1.8 m tall is photographed with a 35 mm camera, and the film image is 20 mm. Under the same conditions, how tall is a person whose film image is 16 mm?

39. In solving the equation $2(x - 3) + 1 = 2x - 5$, what conclusion can be made?

40. In solving the equation $7 - (2 - x) = x + 2$, what conclusion can be made?

1-11 Formulas and Literal Equations

One of the most important applications of equations occurs in the use of formulas in geometry, physics, engineering, and many other fields. *A* **formula** *is an equation that expresses a rule and uses letters to represent certain quantities.* For example, one of the most famous formulas in the history of science is Einstein's equivalence of mass and energy, $E = mc^2$. The symbol E represents energy, as does mc^2. The formula states that the equivalent energy E of a mass m is found by multiplying m by the square of c, the speed of light.

Often it is necessary to solve a formula for a particular symbol that appears in it. We do this in the same way as we solve any equation. *We isolate the desired symbol by use of the basic algebraic operations.*

EXAMPLE 1 ▶▶ In the formula for the equivalent energy E of a mass m, $E = mc^2$, solve for m.

$$E = mc^2 \quad \text{original equation}$$

$$\frac{E}{c^2} = m \quad \text{divide both sides by } c^2$$

$$m = \frac{E}{c^2}$$

Since the symbol for which we are solving is usually shown on the left of the equal sign, the sides were switched, as shown. ◀◀

EXAMPLE 2 ▸▸ A formula relating acceleration a, velocity v, initial velocity v_0, and time t is $v = v_0 + at$. Solve for t.

$$v - v_0 = at \qquad v_0 \text{ subtracted from both sides}$$

$$t = \frac{v - v_0}{a} \qquad \text{both sides divided by } a \text{ and then sides switched}$$ ◂◂

As we can see from Examples 1 and 2, we can solve for the indicated literal number just as we solved for the unknown in the previous section. That is, we perform the basic algebraic operations on the various literal numbers that appear in the same way we perform them on explicit numbers. Other illustrations appear in the following examples.

EXAMPLE 3 ▸▸ In the study of the forces on a certain beam, the equation

$$M = \frac{L(wL + 2P)}{8}$$

is used. Solve for P.

$$8M = \frac{8L(wL + 2P)}{8} \qquad \text{multiply both sides by 8}$$

$$8M = L(wL + 2P) \qquad \text{simplify right side}$$

$$8M = wL^2 + 2LP \qquad \text{remove parentheses}$$

$$8M - wL^2 = 2LP \qquad \text{subtract } wL^2 \text{ from both sides}$$

$$P = \frac{8M - wL^2}{2L} \qquad \text{divide both sides by } 2L \text{ and switch sides}$$ ◂◂

EXAMPLE 4 ▸▸ The effect of temperature is important when measurements must be made with great accuracy. The volume V of a special precision container at temperature T in terms of the volume at temperature T_0 is given by

$$V = V_0[1 + b(T - T_0)]$$

where b depends on the material of which the container is made. Solve for T.

Since we are to solve for T, we must isolate the term containing T. This can be done by first removing the grouping symbols and then isolating the term with T.

$$V = V_0[1 + b(T - T_0)] \qquad \text{original equation}$$

$$V = V_0[1 + bT - bT_0] \qquad \text{remove parentheses}$$

$$V = V_0 + bTV_0 - bT_0V_0 \qquad \text{remove brackets}$$

$$V - V_0 + bT_0V_0 = bTV_0 \qquad \text{subtract } V_0 \text{ and add } bT_0V_0 \text{ to both sides}$$

$$T = \frac{V - V_0 + bT_0V_0}{bV_0} \qquad \text{divide both sides by } bV_0 \text{ and switch sides}$$

◂◂

If we wish to determine the value of any literal number in an expression for which we know values of the other literal numbers, we should *first solve for the required symbol and then substitute the given values.* This is illustrated in the following example.

EXAMPLE 5 ▸▸ The electric resistance R (in ohms) of a resistor changes with the temperature T (in °C) according to the formula $R = R_0 + R_0\alpha T$, where R_0 is the resistance at 0°C. For a given resistor $R_0 = 712\ \Omega$ and $\alpha = 0.00455/°\text{C}$. Find the value of T for $R = 825\ \Omega$.

We first solve for T and then substitute the given values.

$$R = R_0 + R_0\alpha T$$
$$R - R_0 = R_0\alpha T$$
$$T = \frac{R - R_0}{\alpha R_0}$$

Now substituting, we have

$$T = \frac{825 - 712}{(0.00455)(712)}$$
$$= 34.9°\text{C} \quad \text{rounded off}$$

estimation:
$$\frac{800 - 700}{0.005(700)} = \frac{1}{0.035} = 30$$ ◂◂

Exercises 1–11

In Exercises 1–8, solve for the given letter.

1. $ax = b$, for x
2. $cy + d = 0$, for y
3. $4n + 1 = 4m$, for n
4. $bt - 3 = a$, for t
5. $ax + 6 = 2ax - c$, for x
6. $s - 6n^2 = 3s + 4$, for s
7. $\frac{1}{2}t - (4 - a) = 2a$, for t
8. $7 - (p - \frac{1}{3}x) = 3p$, for x

In Exercises 9–32, each of the given formulas arises in the technical or scientific area of study shown. Solve for the indicated letter.

9. $\theta = kA + \lambda$, for λ (robotics)
10. $C = a + bx$, for a (economics)
11. $E = IR$, for R (electricity)
12. $Q = SLd^2$, for L (machine design)
13. $P = 2\pi Tf$, for T (mechanics)
14. $PV = nRT$, for T (chemistry)
15. $p = p_a + dgh$, for h (hydrodynamics)
16. $2Q = 2I + A + S$, for I (nuclear physics)
17. $A = \dfrac{Rt}{PV}$, for t (jet engine design)
18. $u = -\dfrac{eL}{2m}$, for L (spectroscopy)
19. $s = vt - 16t^2$, for v (physics: motion)
20. $FL = P_1L - P_1d + P_2L$, for d (construction)
21. $C_0^2 = C_1^2(1 + 2V)$, for V (electronics)
22. $A_1 = A(M + 1)$, for M (photography)
23. $a = V(k - PV)$, for k (biology)
24. $T = 3(T_2 - T_1)$, for T_1 (oil drilling)
25. $Q_1 = P(Q_2 - Q_1)$, for Q_2 (refrigeration)
26. $p - p_a = dg(y_2 - y_1)$, for y_2 (pressure gauges)
27. $N = N_1T - N_2(1 - T)$, for N_1 (machine design)
28. $t_a = t_c + (1 - h)t_m$, for h (computer access time)
29. $L = \pi(r_1 + r_2) + 2x_1 + x_2$, for r_1 (pulleys)
30. $r_e + r_c(1 - a) = \dfrac{1}{h}$, for r_e (electronics: transistors)
31. $P = \dfrac{V_1(V_2 - V_1)}{gJ}$, for V_2 (jet engine power)
32. $W = T(S_1 - S_2) - Q$, for S_2 (refrigeration)

In Exercises 33–36, find the indicated values.

33. The pressure p (in kPa) at a depth h (in metres) below the surface of water is given by the formula $p = p_0 + kh$, where p_0 is the atmospheric pressure. Find h for $p = 205$ kPa, $p_0 = 101$ kPa, and $k = 9.80$ kPa/m.
34. A formula used in determining the total transmitted power P_t in an AM radio signal is $P_t = P_c(1 + 0.5m^2)$. Find P_c if $P_t = 680$ W and $m = 0.925$.
35. The efficiency E of a certain type of engine operating between thermodynamic temperatures T_1 and T_2 is given by $E = 1 - T_2/T_1$. Find T_2 if $E = 0.455$ and $T_1 = 375$ K.
36. In forestry, a formula used to determine the volume V of a log is $V = \frac{1}{2}L(B + b)$, where L is the length of the log and B and b are the areas of the ends. Find b, in square feet, if $V = 1.09\text{ m}^3$, $L = 4.91$ m, and $B = 0.244\text{ m}^2$.

1-12 Applied Verbal Problems

Mathematics allows us to solve many kinds of applied problems. Some of these problems are given in equation or formula form and can be solved directly. However, in practice it is often necessary to first set up equations by using known formulas and the given conditions. Such problems are at first verbal problems, and it is necessary to translate them into mathematical terms for solution.

Usually the most difficult part in solving a verbal problem is identifying the information that leads to the equation. Finding this information requires that you carefully read the problem, being sure that you understand all the terms and expressions that are used. Following is a general approach that you should use.

Problem-Solving Procedure

1. *Read the statement of the problem.* First read it quickly for a general overview. Then *reread* slowly and carefully, *listing the information* given.
2. *Clearly identify the unknown quantities,* and then *assign an appropriate letter to represent one of them,* stating this choice clearly.
3. *Specify the other unknown quantities* in terms of the one in step 2.
4. *If possible, make a sketch* using the known and unknown quantities.
5. *Analyze the statement* of the problem and *write the necessary equation.* This is often the most difficult step because ***some of the information may be implied, and not explicitly stated.*** Again, a very careful reading of the statement is necessary.

CAUTION ▶

6. *Solve the equation,* clearly stating the solution.
7. *Check the solution with the original statement* of the problem.

Carefully read the following examples. Note how the above steps are followed.

EXAMPLE 1 ▸▸ A 17-N beam is supported at each end. The supporting force at one end is 3 N more than at the other end. Find the forces.

Since the force at each end is required, we write

let F = the smaller force step 2

as a way of establishing the unknown for the equation. Any appropriate letter could be used, and we could have let it represent the larger force.

F 17 N $F + 3$

Fig. 1-7

Also, since the other force is 3 N more, we write

$F + 3$ = the larger force step 3

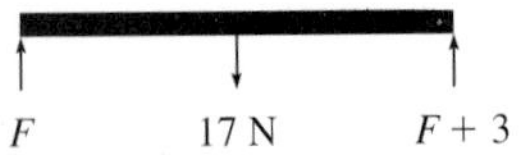

We now draw the sketch in Fig. 1-7. step 4

Since the forces at each end of the beam support the weight of the beam, we have the equation

$$F + (F + 3) = 17$$ step 5

This equation can now be solved. $2F = 14$

$$F = 7 \text{ N}$$ step 6

Thus, the smaller force is 7 N and the larger force is 10 N. This checks with the original statement of the problem. step 7 ◂◂

EXAMPLE 2 ▸▸ In designing an electric circuit, it is found that 34 resistors with a total resistance of 56 Ω are required. Two different resistances, 1.5 Ω and 2.0 Ω, are used. How many of each are in the circuit?

Since we want to find the number of each resistance, we

let x = number of 1.5-Ω resistors

Also, since there are 34 resistors in all,

$34 - x$ = number of 2.0-Ω resistors

We also know that the total resistance of all resistors is 56 Ω. This means that

1.5 Ω each, number → $1.5x$ (total resistance of 1.5-Ω resistors)

2.0 Ω each, number → $2.0(34 - x)$ (total resistance of 2.0-Ω resistors)

$$1.5x + 2.0(34 - x) = 56 \quad \leftarrow \text{total resistance of all resistors}$$

$$1.5x + 68 - 2.0x = 56$$

$$-0.5x = -12$$

$$x = 24$$

Therefore, there are 24 1.5-Ω resistors and 10 2.0-Ω resistors. The total resistance of these is $24(1.5) + 10(2.0) = 36 + 20 = 56$ Ω. We see that this checks with the statement of the problem. ◂◂

EXAMPLE 3 ▸▸ A medical researcher finds that a given sample of an experimental drug can be divided into 4 more slides with 5 mg each than with 6 mg each. How many slides with 5 mg each can be made up?

We are asked to find the number of slides with 5 mg, and therefore we

let x = number of slides with 5 mg

Since the sample may be divided into 4 more slides with 5 mg each than of 6 mg each, we know that

$x - 4$ = number of slides with 6 mg

Since *it is the same sample that is to be divided,* the total mass of the drug on each type is the same. This means

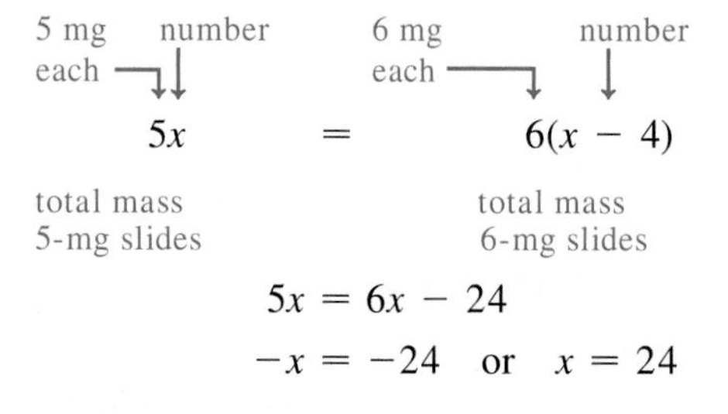

$$5x = 6x - 24$$

$$-x = -24 \quad \text{or} \quad x = 24$$

Therefore, the sample can be divided into 24 slides with 5 mg each, or 20 slides with 6 mg each. Since the total mass, 120 mg, is the same for each set of slides, the solution checks with the statement of the problem. ◂◂

Shuttle

6000 km

Satellite

Fig. 1-8

EXAMPLE 4 ▸▸ A space shuttle maneuvers so that it may "capture" an already orbiting satellite that is 6000 km ahead. If the satellite is moving at 27 000 km/h and the shuttle is moving at 29 500 km/h, how long will it take the shuttle to reach the satellite? (All digits shown are significant.)

First we let t = the time for the shuttle to reach the satellite. Then, using the fact that the shuttle must go 6000 km farther in the same time, we draw the sketch in Fig. 1-8. Next we use the formula *distance* = *rate* × *time* ($d = rt$). This leads to the following equation and solution.

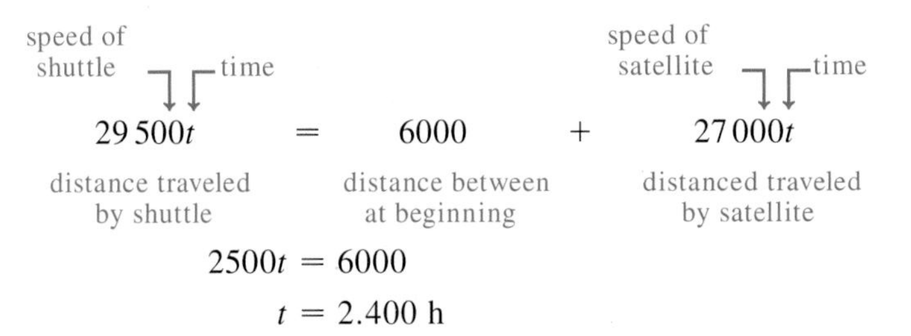

$$2500t = 6000$$

$$t = 2.400 \text{ h}$$

This means that it will take the shuttle 2.400 h to reach the satellite. In 2.400 h the shuttle will travel 70 800 km and the satellite will travel 64 800 km. We see that the solution checks with the statement of the problem. ◂◂

EXAMPLE 5 ▸▸ A refinery has 7600 L of a gasoline and methanol blend, which is 5.00% methanol. How much pure methanol must be added to this blend so that the resulting blend is 10.0% methanol? (All data have three significant digits.)

First let x = the number of liters of methanol to be added. We want the total volume of methanol to be 10.0% of the volume of the final blend, which is the volume in the original blend plus the volume that is added. See Fig. 1-9. This leads to the following equation and solution.

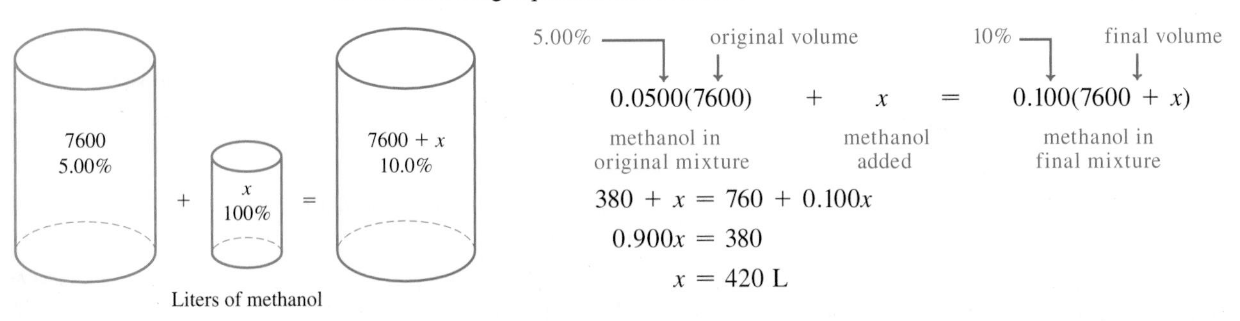

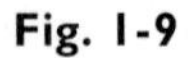

Fig. 1-9

Therefore, 420 L of methanol are to be added. The result checks (to three significant digits) since there would be 800 L of methanol of a total final volume of 8020 L. ◂◂

Exercises 1–12

Solve the following problems by first setting up an appropriate equation. Assume all data are accurate to two significant digits unless greater accuracy is given.

1. Two computer software programs cost \$390 together. If one costs \$114 more than the other, what is the cost of each?
2. Two pipes drain an oil tank. One pipe releases 50 L/min more than the other. If they release 3300 L in 10 min together, what is the drainage rate of each?
3. Approximately 4.5 million wrecked cars are recycled in two consecutive years. There were 700 000 more recycled the second year than in the first year. How many are recycled each year?
4. In order to produce equilibrium on a particular beam, the sum of two forces must equal a third force. If the second of the two forces is 6.4 N more than the first and the third force is four times the first, what are the forces?

5. A developer purchased 70 hectares of land for \$900 000. If part cost \$20 000 per hectare and the remainder cost \$10 000 per hectare, how many hectares did the developer buy at each price? (1 hectare = 1 hm^2)

6. A vial contains 2000 mg, which is to be used for two dosages. One patient is to be administered 660 mg more than another. How much should be administered to each?

7. In the design of a bridge, an engineer determines that four fewer 18-m girders are needed for the span than 15-m girders. How many 18-m girders are needed?

8. A fuel oil storage depot had an 8-week supply on hand. However, cold weather caused the supply to be used in 6 weeks when 20 000 L extra were used each week. How many litres were in the original supply?

9. The sum of three electric currents that come together at a point in an integrated circuit is zero. If the second current is double the first, and the third current is 9.2 μA more than the first, what are the currents? (The sign of a current indicates the direction of flow.)

10. A trucking firm uses one fleet of trucks on round-trip delivery routes of 8 h, and a second fleet, with five more trucks than the first, is used on round-trip delivery routes of 6 h. Budget allotments allow 198 h of delivery time in a week. How many trucks are in each fleet?

11. A primary natural gas pipeline feeds into three smaller pipelines, each of which is 2.6 km longer than the main pipeline. If the total length of the four pipelines is 35.4 km, what is the length of each section of the line?

12. Ten 6.0-V and 12-V batteries have a total voltage of 84 V. How many of each type are there?

13. If \$8000 is invested at 7.00%, how much must be invested at 9.00% to have total interest of \$1550?

14. Two stock investments cost \$15 000. One stock then had a 40% gain and the other a 10% loss. If the net profit is \$2000, how much was invested in each stock?

15. One lap at the Indianapolis Speedway is 4.02 km. In a race, a car stalls and then starts 30.0 s after a second car. The first car travels at 80.0 m/s, and the second car travels at 73.8 m/s. How long does it take the first car to overtake the second, and which car will be ahead after eight laps?

16. The supersonic jet Concorde made a trip averaging 100 km/h less than the speed of sound for 1.00 h and averaging 400 km/h more than the speed of sound for 3.00 h. If the trip covered 5740 km, what is the speed of sound?

17. Trains at each end of the 50.0-km long Eurotunnel under the English Channel start at the same time into the tunnel. Find their speeds if the train from France travels 8.0 km/h faster than the train from England and they pass in 17.0 min.

18. A corporate executive leaves the manufacturing plant and travels on a highway at 88 km/h to corporate headquarters. The executive later returns in a helicopter, which travels at 205 km/h on a route parallel to the highway. If the total travel time is 1.8 h, how far is it from the manufacturing plant to the corporate headquarters?

19. A ski lift takes a skier up a slope at 50 m/min. The skier then skis down the slope at 150 m/min. If one round-trip takes 24 min, how long is the slope?

20. A computer chip manufacturer produces two types of chips. In testing a total of 6100 chips of both types, 0.50% of one type and 0.80% of the other type were defective. If a total of 38 defective chips were found, how many of each type were tested?

21. Two gasoline distributors, A and B, are 367 km apart on a highway. A charges \$0.45 per litre and B charges \$0.42 per litre. Each charges 0.04¢ per litre per kilometre for delivery. Where on the highway is the cost to the customer the same?

22. An outboard engine uses a gasoline-oil fuel mixture in the ratio of 15 to 1. How much gasoline must be mixed with a gasoline-oil mixture, which is 75% gasoline, to make 8.0 L of the mixture for the outboard engine?

23. How many grams of solder that is 15% tin must be mixed with 90 g of solder that is 45% tin in order to have solder that is 25% tin?

24. By weight, a certain roadbed material is 75% crushed rock, and another is 30% crushed rock. How many tonnes of each must be mixed in order to have 250 tonnes of material that is 50% crushed rock? (1 tonne = 1 Mg)

Chapter Equations, Review Exercises, and Practice Test

Chapter Equations

Commutative law of addition: $a + b = b + a$

Commutative law of multiplication: $ab = ba$

Associative law of addition: $a + (b + c) = (a + b) + c$

Associative law of multiplication: $a(bc) = (ab)c$

Distributive law: $a(b + c) = ab + ac$

$$a + (-b) = a - b \quad (1\text{-}1) \qquad a - (-b) = a + b \quad (1\text{-}2)$$

$$a^m \times a^n = a^{m+n} \quad (1\text{-}3)$$

$$\frac{a^m}{a^n} = a^{m-n} \quad (m > n,\ a \neq 0), \qquad \frac{a^m}{a^n} = \frac{1}{a^{n-m}} \quad (m < n,\ a \neq 0) \quad (1\text{-}4)$$

$$(a^m)^n = a^{mn} \quad (1\text{-}5) \qquad (ab)^n = a^n b^n, \quad \left(\frac{a}{b}\right)^n = \frac{a^n}{b^n} \quad (b \neq 0) \quad (1\text{-}6)$$

$$a^0 = 1 \quad (a \neq 0) \quad (1\text{-}7) \qquad a^{-n} = \frac{1}{a^n} \quad (a \neq 0) \quad (1\text{-}8)$$

$$\sqrt{ab} = \sqrt{a}\sqrt{b} \quad (a \text{ and } b \text{ positive real numbers}) \quad (1\text{-}9)$$

Review Exercises

In Exercises 1–12, evaluate the given expressions.

1. $(-2) + (-5) - 3$ **2.** $6 - 8 - (-4)$

3. $\frac{(-5)(6)(-4)}{(-2)(3)}$ **4.** $\frac{(-9)(-12)(-4)}{24}$

5. $-5 - 2(-6) + \frac{-15}{3}$ **6.** $3 - 5(-2) - \frac{12}{-4}$

7. $\frac{18}{3 - 5} - (-4)^2$ **8.** $-(-3)^2 - \frac{-8}{(-2) - (-4)}$

9. $\sqrt{16} - \sqrt{64}$ **10.** $-\sqrt{144} + \sqrt{49}$

11. $(\sqrt{7})^2 - \sqrt[3]{8}$ **12.** $-\sqrt[4]{16} + (\sqrt{6})^2$

In Exercises 13–20, simplify the given expressions. Where appropriate, express results with positive exponents only.

13. $(-2rt^2)^2$ **14.** $(3a^0b^{-2})^3$ **15.** $\frac{18m^3n^4t}{-3mn^5t^3}$

16. $\frac{15p^4q^2r}{5pq^5r}$ **17.** $\frac{-16s^{-2}(st^2)}{-2st^{-1}}$ **18.** $\frac{-35x^{-1}y(x^2y)}{5xy^{-1}}$

19. $\sqrt{45}$ **20.** $\sqrt{9 + 36}$

In Exercises 21–24, for each number (a) determine the number of significant digits, and (b) round off each to two significant digits.

21. 8840 **22.** 21 450 **23.** 9.040 **24.** 0.700

In Exercises 25–28, evaluate the given expressions. All numbers are approximate.

25. $37.3 - 16.92(1.067)^2$ **26.** $\frac{8.896 \times 10^{-12}}{3.5954 + 6.0449}$

27. $\frac{\sqrt{0.1958 + 2.844}}{3.142(65)^2}$ **28.** $\frac{1}{0.03568} + \frac{37466}{29.63^2}$

In Exercises 29–60, perform the indicated operations.

29. $a - 3ab - 2a + ab$ **30.** $xy - y - 5y - 4xy$

31. $6xy - (xy - 3)$ **32.** $-(2x - b) - 3(-x - 5b)$

33. $(2x - 1)(x + 5)$ **34.** $(x - 4y)(2x + y)$

35. $(x + 8)^2$ **36.** $(2x + 3y)^2$

37. $\frac{2h^3k^2 - 6h^4k^5}{2h^2k}$ **38.** $\frac{4a^2x^3 - 8ax^4}{2ax^2}$

39. $4a - [2b - (3a - 4b)]$

40. $3b - [2b + 3a - (2a - 3b)] + 4a$

41. $2xy - \{3z - [5xy - (7z - 6xy)]\}$

42. $x^2 + 3b + [(b - y) - 3(2b - y + z)]$

43. $(2x + 1)(x^2 - x - 3)$ **44.** $(x - 3)(2x^2 - 3x + 1)$

45. $-3y(x - 4y)^2$ **46.** $-s(4s - 3t)^2$

47. $3p[(q - p) - 2p(1 - 3q)]$

48. $3x[2y - r - 4(s - 2r)]$

49. $\frac{12p^3q^2 - 4p^4q + 6pq^5}{2p^4q}$ **50.** $\frac{27s^3t^2 - 18s^4t + 9s^2t}{9s^2t}$

51. $(2x^2 + 7x - 30) \div (x + 6)$

52. $(4x^2 + 15x - 21) \div (2x + 7)$

53. $\frac{3x^3 - 7x^2 + 11x - 3}{3x - 1}$ **54.** $\frac{x^3 - 4x^2 + 7x - 12}{x - 3}$

55. $\frac{4x^4 + 10x^3 + 18x - 1}{x + 3}$ **56.** $\frac{8x^3 - 14x + 3}{2x + 3}$

57. $-3\{(r + s - t) - 2[(3r - 2s) - (t - 2s)]\}$

58. $(1 - 2x)(x - 3) - (x + 4)(4 - 3x)$

59. $\frac{2y^3 + 9y^2 - 7y + 5}{2y - 1}$ **60.** $\frac{6x^2 + 5xy - 4y^2}{2x - y}$

In Exercises 61–72, solve the given equations.

61. $3x + 1 = x - 8$

62. $4y - 3 = 5y + 7$

63. $\frac{5x}{7} = \frac{3}{2}$

64. $\frac{1}{4} = \frac{2x}{9}$

65. $6x - 5 = 3(x - 4)$

66. $-2(-4 - y) = 3y$

67. $2s + 4(3 - s) = 6$

68. $-(4 + v) = 2(2v - 5)$

69. $3t - 2(7 - t) = 5(2t + 1)$

70. $6 - 3x - (8 - x) = x - 2(2 - x)$

71. $2.7 + 2.0(2.1x - 3.4) = 0.1$

72. $0.250(6.721 - 2.44x) = 2.08$

In Exercises 73–80, change numbers in ordinary notation to scientific notation or change numbers in scientific notation to ordinary notation. (See Appendix B for an explanation of the symbols that are used.)

73. The escape velocity (the velocity required to leave the earth's gravitational field) is about 40 000 km/h.

74. When the first pictures of the surface of Mars (in 1976) were transmitted to earth, Mars was about 340 000 000 km from earth.

75. Police radar has a frequency of 1.02×10^9 Hz.

76. The World Trade Center has about 8.4×10^5 m^2 of office space.

77. A biological cell has a surface area of 0.000 001 2 cm^2.

78. An optical coating on glass to reduce reflections is about 0.000 000 15 m thick.

79. Helium (used in the Goodyear blimp) has a density of 1.8×10^{-1} kg/m^3.

80. A computer can execute an instruction in 6.3×10^{-10} s.

In Exercises 81–96, solve for the indicated letter. Where noted, the given formula arises in the technical or scientific area of study.

81. $3s + 2 = 5a$, for s

82. $5 - 7t = 6b$, for t

83. $3(4 - x) = 8 + 2n$, for x

84. $6 - 3b = 5(7 + 2v)$, for v

85. $R = n^2Z$, for Z (electricity)

86. $R = \frac{2GM}{c^2}$, for G (astronomy: black holes)

87. $I = P + Prt$, for t (business)

88. $V = IR + Ir$, for R (electricity)

89. $m = dV(1 - e)$, for e (solar heating)

90. $mu = (m + M)v$, for M (physics: momentum)

91. $C = \frac{m(N_1 + N_2)}{2}$, for N_1 (mechanics: gears)

92. $2(J + 1) = \frac{f}{B}$, for J (spectroscopy)

93. $E = \frac{J - K}{1 - S}$, for K (chemistry)

94. $Z^2\left(1 - \frac{\lambda}{2a}\right) = k$, for λ (radar design)

95. $d = kx^2[3(a + b) - x]$, for a (mechanics: beams)

96. $V = V_0[1 + 3a(T_2 - T_1)]$, for T_2 (thermal expansion)

In Exercises 97–100, perform the indicated calculations.

97. The CN Tower in Toronto is 0.553 km high. The Sears Tower in Chicago is 443 m high. How much higher is the CN Tower than the Sears Tower?

98. The time (in seconds) it takes a computer to check n memory cells is found by evaluating $(n/2650)^2$. Find the time to check 48 cells.

99. The combined electric resistance of two parallel resistors is found by evaluating the expression $\frac{R_1R_2}{R_1 + R_2}$. Evalute this for $R_1 = 0.0275\ \Omega$ and $R_2 = 0.0590\ \Omega$.

100. The distance (in metres) from the earth for which the gravitational force of the earth on a spacecraft equals the gravitational force of the sun on it is found by evaluating $1.5 \times 10^{11}\sqrt{m/M}$, where m and M are the masses of the earth and sun, respectively. Find this distance for $m = 5.98 \times 10^{24}$ kg and $M = 1.99 \times 10^{30}$ kg.

In Exercises 101–104, simplify the given expressions.

101. An analysis of the electric potential of a conductor includes the expression $2V(r - a) - V(b - a)$. Simplify this expression.

102. In finding the value of an annuity, the expression $(Ai - R)(1 + i)^2$ is used. Multiply out this expression.

103. A computer analysis of the velocity of a link in an industrial robot leads to the expression $4(t + h) - 2(t + h)^2$. Simplify this expression.

104. When analyzing the motion of a communications satellite, the expression $\frac{k^2r - 2h^2k + h^2rv^2}{k^2r}$ is used. Perform the indicated division.

In Exercises 105–116, solve the given problems. All data are accurate to two significant digits unless more are given.

105. One computer has four times the memory of another computer. If their combined memory is 10.5 megabytes, what is the memory of each?

106. Three chemical reactions each produce oxygen. If the first produces twice that of the second, the third produces twice that of the first, and the combined total is 560 cm^3, what volume is produced by each reaction?

107. The voltage across a resistor equals the current times the resistance. In a microprocessor circuit, one resistor is 1200 Ω greater than another. The sum of the voltages across them is 12.0 mV. Find the resistances if the current is 2.4 μA in each.

108. A air sample contains 4.0 ppm (parts per million) of two pollutants. The concentration of one is four times the other. What are the concentrations?

109. A 5.0-m piece of tubing weighs 6.8 N. What is the weight of a 27-m piece of the same tubing?

110. The fuel for a two-cycle motorboat engine is a mixture of gasoline and oil in the ratio of 15 to 1. How many liters of each are in 6.6 L of mixture?

111. Two motorboats, 55.3 km apart, start toward each other. One travels at 22.5 km/h and the other at 17.0 km/h. When will they meet?

112. A helicopter used in fighting a forest fire travels at 175 km/h from the fire to a pond and 115 km/h with water from the pond to the fire. If a round-trip takes 30 min, how long does it take from the pond to the fire?

113. One grade of oil has 0.50% of an additive, and a higher grade has 0.75% of the additive. How many liters of each must be used to have 1000 L of a mixture with 0.65% of the additive?

114. Fifty pounds of a cement-sand mixture is 40% sand. How many pounds of sand must be added for the resulting mixture to be 60% sand?

115. An architect plans to have 25% of the floor area of a house in ceramic tile. In all but the kitchen and entry, there are 205 m^2 of floor area, 15% of which is tile. What area can be planned for the kitchen and entry if each has an all-tile floor?

116. A *karat* equals 1/24 part of gold in an alloy (e.g., 9-karat gold is 9/24 gold). How many grams of 9-karat gold must be mixed with 18-karat gold to get 200 g of 14-karat gold?

Writing Exercise

117. A person was asked to explain how to evaluate the expression $\frac{1.70}{0.0246(0.0309)}$, where the numbers are approximate. The explanation was: "Divide 1.70 by 0.0246, and then multiply by 0.0309. The calculator display is the answer." Write a short paragraph stating what is wrong with the explanation. (What is the correct result?)

Practice Test

In Problems 1–5, evaluate the given expressions. In Problems 3 and 5, the numbers are approximate.

1. $\sqrt{9+16}$

2. $\frac{(7)(-3)(-2)}{(-6)(0)}$

3. $\frac{3.372 \times 10^{-3}}{7.526 \times 10^{12}}$

4. $\frac{(+6)(-2) - 3(-1)}{5 - 2}$

5. $\frac{346.4 - 23.5}{287.7} - \frac{0.944^3}{(3.46)(0.109)}$

In Problems 6–12, perform the indicated operations and simplify. Use only positive exponents when exponents are used in the result.

6. $(2a^0b^{-2}c^3)^{-3}$

7. $(2x + 3)^2$

8. $3m^2(am - 2m^3)$

9. $\frac{8a^3x^2 - 4a^2x^4}{-2ax^2}$

10. $\frac{6x^2 - 13x + 7}{2x - 1}$

11. $(2x - 3)(x + 7)$

12. $3x - [4x - (3 - 2x)]$

13. Solve for y: $5y - 2(y - 4) = 7$

14. Solve for x: $3(x - 3) = x - (2 - 3d)$

15. Express 0.000 003 6 in scientific notation.

16. List the numbers -3, $|-4|$, $-\pi$, $\sqrt{2}$, and 0.3 in numerical order.

17. What fundamental law is illustrated by $3(5 + 8) = 3(5) + 3(8)$?

18. (a) How many significant digits are in the number 3.0450? (b) Round it off to two significant digits.

19. If P dollars is deposited in a bank that compounds interest n times a year, the value of the account after t years is found by evaluating $P(1 + i/n)^{nt}$, where i is the annual interest rate. Find the value of an account for which $P = \$1000$, $i = 5\%$, $n = 2$, and $t = 3$ years (values are exact).

20. In finding the illuminance from a light source, the expression $8(100 - x)^2 + x^2$ is used. Simplify this expression.

21. The equation $L = L_0[1 + \alpha(t_2 - t_1)]$ is used when studying thermal expansion. Solve for t_2.

22. An alloy weighing 20 N is 30% copper. How many pounds of another alloy, which is 80% copper, must be added for the final alloy to be 60% copper?

2

Geometry

Many applied problems in technology and science involve the shape and size of objects. Since geometry deals with shape and size, the topics and methods of geometry are important in many areas of technology. These include architecture, construction, instrumentation, surveying, mechanical design, product design of all possible types, surveying and civil engineering, and many other areas of engineering.

The study of geometry includes the properties and measurements of angles, lines, and surfaces and the basic figures they form. In this chapter we review the more important methods and formulas for calculating the important geometric measures, such as area and volume. Numerous technical applications are shown.

Geometric figures and concepts are also basic to the development of many other areas of mathematics, such as graphing and trigonometry. We will start our study of graphs in Chapter 3 and trigonometry in Chapter 4.

In Section 2-5 we see how to find an excellent approximation of the area of an irregular geometric figure, such as the surface of a lake.

2-1 Lines and Angles

In establishing the properties of the basic geometric figures, it is not possible to define every word and prove every statement. Certain words and concepts must be used without definition. In general, *the words* **point**, **line**, *and* **plane** *are accepted in geometry without being defined.* This gives us a starting point for defining other terms.

The amount of rotation of a **ray** *(or* ***half-line****) about its endpoint is called an* **angle**. A ray is that part of a line (the word *line* means *straight line*) to one side of a fixed point on the line. *The fixed point is the* **vertex** *of the angle. One complete rotation of a ray is an angle of 360* **degrees,** *written as 360°*. Some special types of angles are as follows:

Name of angle	Measure of angle
Right angle	*90°*
Straight angle	*180°*
Acute angle	*Between 0° and 90°*
Obtuse angle	*Between 90° and 180°*

These types of angles are illustrated in the following example.

EXAMPLE 1 ▸▸ Figure 2-1(a) shows a right angle (marked as ⊾). The vertex of the angle is point B, and the ray is the half-line BA. Figure 2-1(b) shows a straight angle. Figure 2-1(c) shows an acute angle, denoted as $\angle E$ (or $\angle DEF$ or $\angle FED$). In Fig. 2-1(d), $\angle G$ is an obtuse angle.

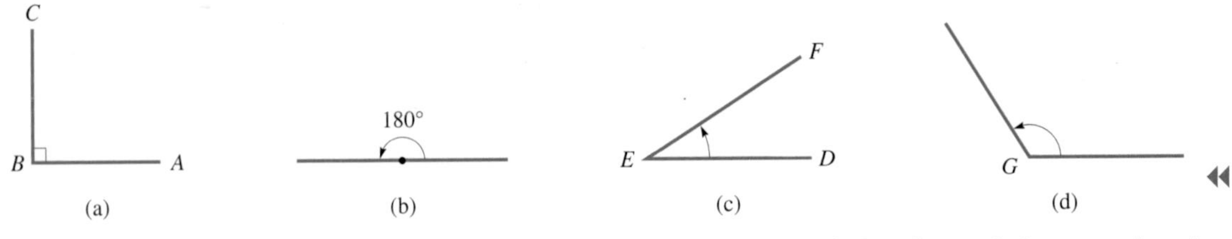

Fig. 2-1

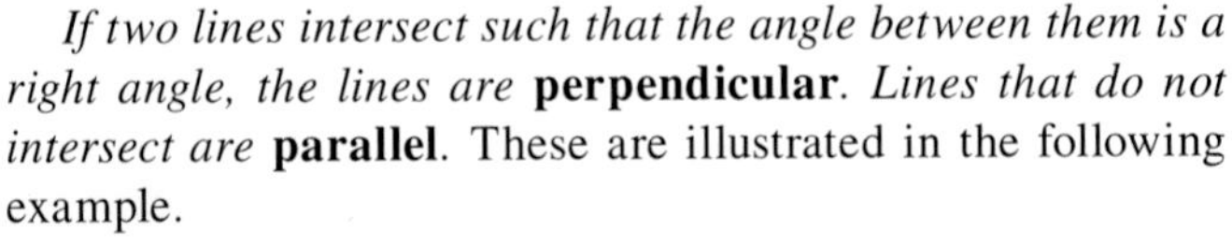

If two lines intersect such that the angle between them is a right angle, the lines are **perpendicular**. *Lines that do not intersect are* **parallel**. These are illustrated in the following example.

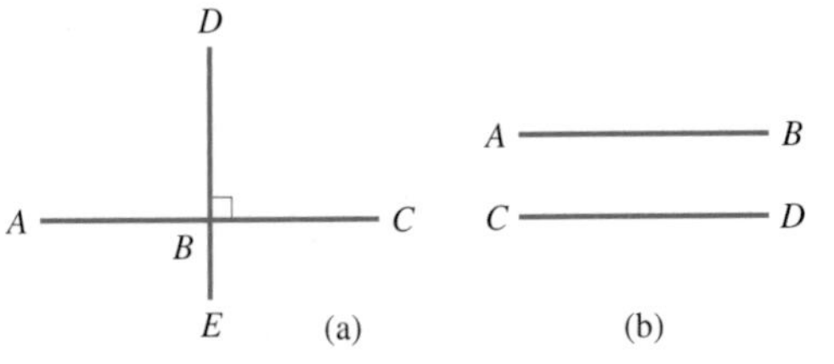

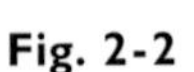

Fig. 2-2

EXAMPLE 2 ▸▸ In Fig. 2-2(a), lines AC and DE are perpendicular (which can be shown as $AC \perp DE$) since they meet in a right angle (again, shown as ⊾) at B. In Fig. 2-2(b), lines AB and CD are drawn so that they do not meet, even if extended. Therefore, these lines are parallel (which can be shown as $AB \parallel CD$). ◂◂

If the sum of two angles is 180°, the angles are called **supplementary angles**. Each angle is the **supplement** of the other. *If the sum of two angles is 90°, the angles are called* **complementary angles**. Each is the **complement** of the other.

EXAMPLE 3 ▸▸ (a) In Fig. 2-3(a), $\angle BAC = 55°$, and in Fig. 2-3(b), $\angle DEF = 125°$. Since $55° + 125° = 180°$, $\angle BAC$ and $\angle DEF$ are supplementary angles.

(b) In Fig. 2-4, we see that $\angle POQ$ is a right angle, or $\angle POQ = 90°$. Since $\angle POR + \angle ROQ = \angle POQ = 90°$, $\angle POR$ is the complement of $\angle ROQ$ (or $\angle ROQ$ is the complement of $\angle POR$).

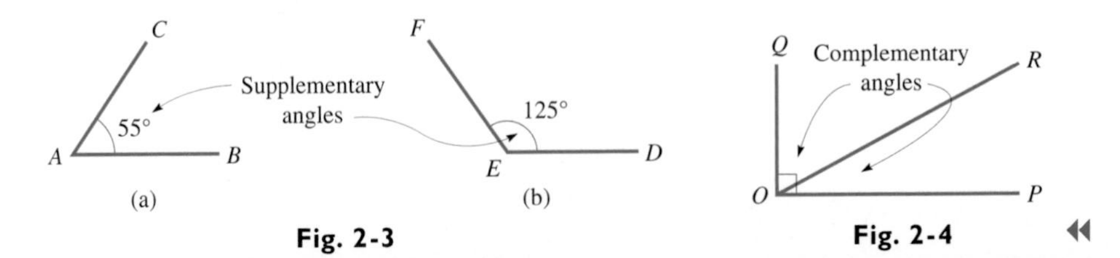

Fig. 2-3 Fig. 2-4 ◂◂

In many geometric figures, it is necessary to refer to certain specific pairs of angles. *Two angles that have a common vertex and a side common between them are known as* **adjacent angles.** *If two lines cross to form equal angles on opposite sides of the point of intersection, which is the common vertex, these equal angles are called* **vertical angles.** These are illustrated in the following example.

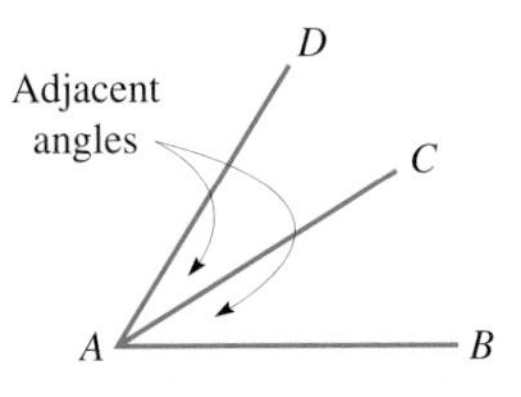

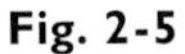
Fig. 2-5

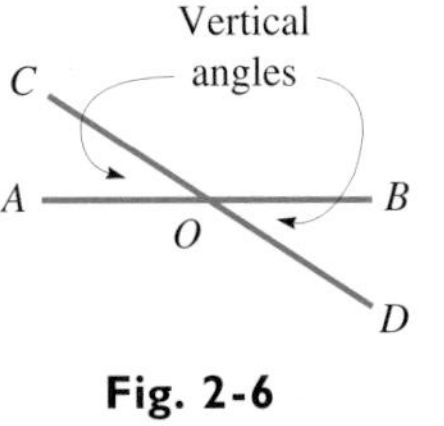

Fig. 2-6

EXAMPLE 4 ▸▸ (a) In Fig. 2-5, $\angle BAC$ and $\angle CAD$ have a common vertex at A and the common side AC between them so that $\angle BAC$ and $\angle CAD$ are adjacent angles.

(b) In Fig. 2-6, lines AB and CD intersect at point O. Here $\angle AOC$ and $\angle BOD$ are vertical angles, and they are equal. Also, $\angle BOC$ and $\angle AOD$ are vertical angles and are equal. ◂◂

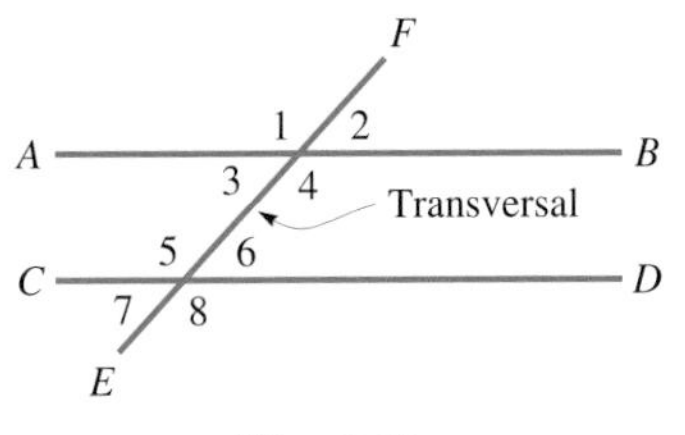

Fig. 2-7

In a plane, *if a line crosses two or more parallel or nonparallel lines, it is called a* **transversal.** In Fig. 2-7, $AB \parallel CD$, and the transversal of these parallel lines is the line EF.

When a transversal crosses a pair of parallel lines, certain pairs of equal angles result. In Fig. 2-7, the **corresponding angles** are equal (that is, $\angle 1 = \angle 5$, $\angle 2 = \angle 6$, $\angle 3 = \angle 7$, and $\angle 4 = \angle 8$). Also, the **alternate-interior angles** are equal ($\angle 3 = \angle 6$ and $\angle 4 = \angle 5$) and the **alternate-exterior angles** are equal ($\angle 1 = \angle 8$ and $\angle 2 = \angle 7$).

When more than two parallel lines are crossed by *two* transversals, such as is shown in Fig. 2-8, *the segments of the transversals between the same two parallel lines are called* **corresponding segments**. A useful theorem is that *the ratios of corresponding segments of the transversals are equal.* In Fig. 2-8, this means that

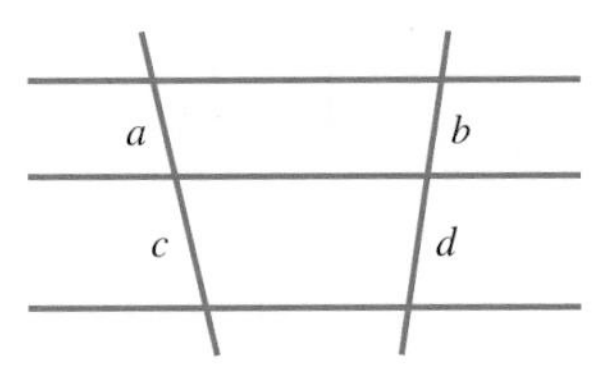

Fig. 2-8

$$\frac{a}{b} = \frac{c}{d} \tag{2-1}$$

EXAMPLE 5 ▸▸ In Fig. 2-9, part of the beam structure within a building is shown. The vertical beams are parallel. From the distances between beams that are shown, determine the distance x between the middle and right vertical beams.

Using Eq. (2-1), we have

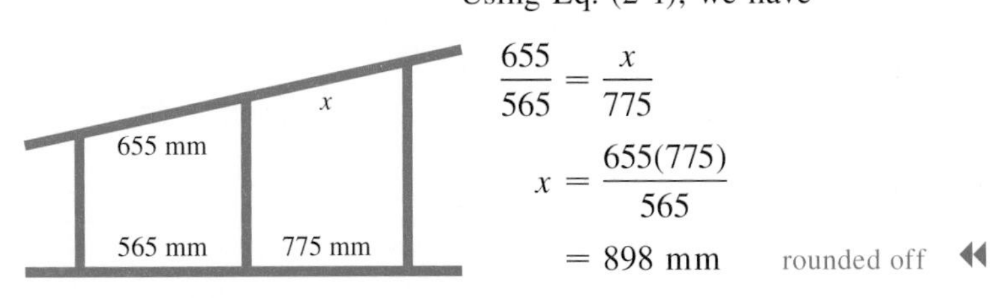

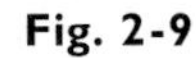
Fig. 2-9

$$\frac{655}{565} = \frac{x}{775}$$

$$x = \frac{655(775)}{565}$$

$$= 898 \text{ mm} \qquad \text{rounded off}$$ ◂◂

Exercises 2–1

In Exercises 1-4, identify the indicated angles in Fig. 2-10. In Exercises 3 and 4, also evaluate the indicated angles.

1. Two acute angles

2. The obtuse angle

3. *If* $\angle CBD = 65°$, determine its complement.

4. *If* $\angle CBD = 65°$, determine its supplement.

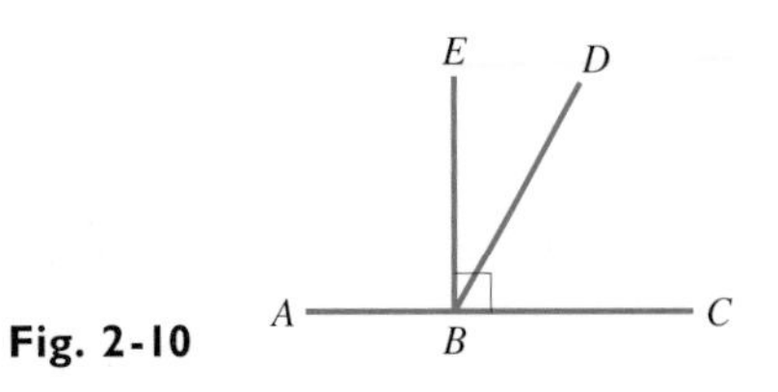

Fig. 2-10

In Exercises 5 and 6, use Fig. 2-11. In Exercises 7 and 8, use Fig. 2-12. Determine the indicated angles.

5. $\angle AOB$ **6.** $\angle AOC$ **7.** $\angle 3$ **8.** $\angle 4$

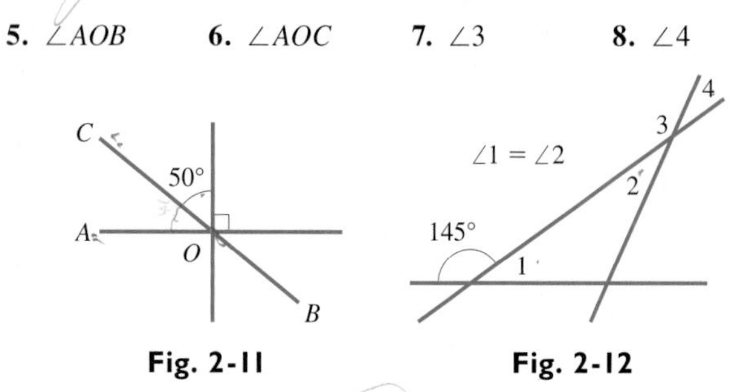

Fig. 2-11 **Fig. 2-12**

In Exercises 9–12, use Fig. 2-13. In Exercises 13–16, use Fig. 2-14. Determine the indicated angles.

9. $\angle 1$ **10.** $\angle 2$ **11.** $\angle 3$ **12.** $\angle 4$

13. $\angle FCE$ **14.** $\angle ECD$ **15.** $\angle BCE$ **16.** $\angle BFC$

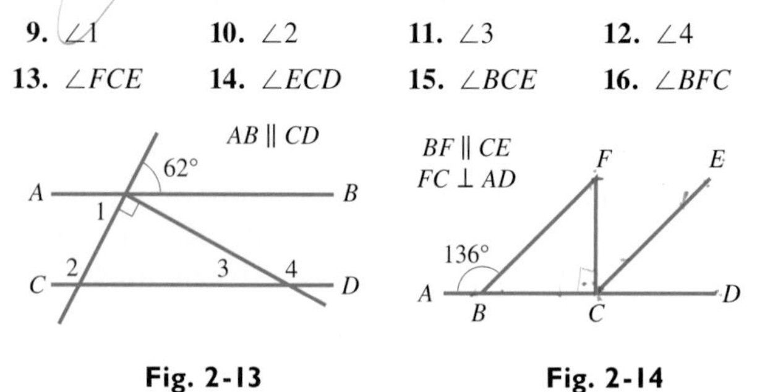

Fig. 2-13 **Fig. 2-14**

In Exercises 17–20, determine the indicated segments in Fig. 2-15.

17. a **18.** b **19.** c **20.** d

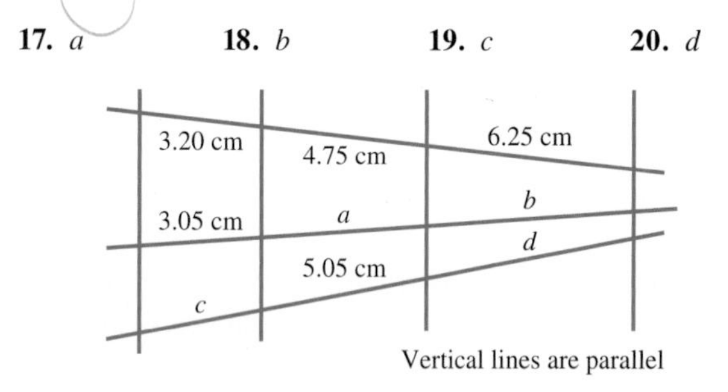

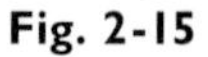

Fig. 2-15

In Exercises 21–24, solve the given problems.

21. A steam-pipe is connected in sections AB, BC, and CD, as shown in Fig. 2-16. What is the angle between sections BC and CD if $AB \parallel CD$?

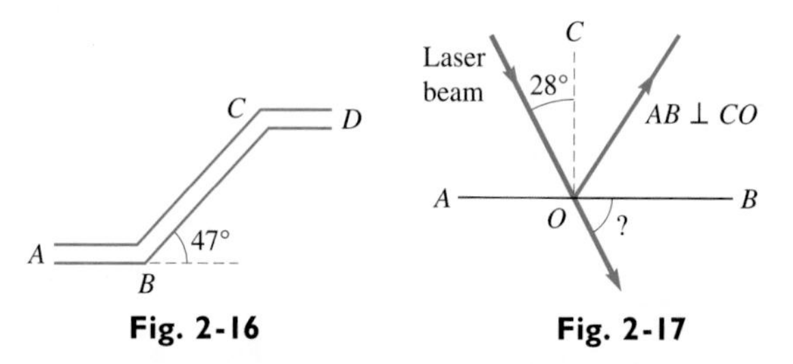

Fig. 2-16 **Fig. 2-17**

22. A laser beam striking a surface is partly reflected, and the remainder of the beam passes straight through the surface, as shown in Fig. 2-17. Find the angle between the surface and the part that passes through.

23. Find the distance on Pine St. between Second Ave. and Third Ave. in Fig. 2-18. The avenues are parallel to each other.

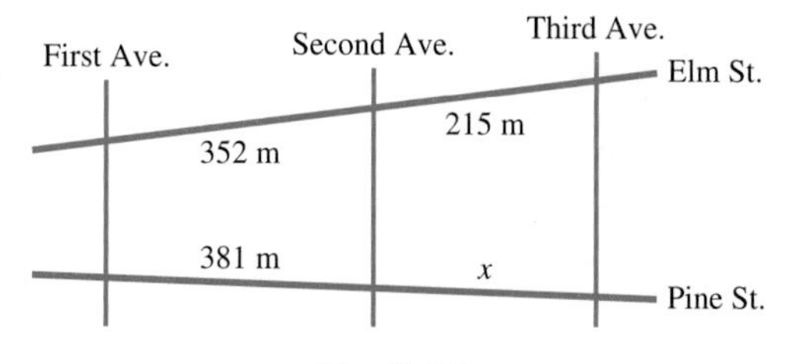

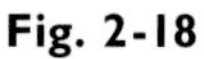

Fig. 2-18

24. An electric circuit board has equally spaced parallel wires with connections at points A, B, and C, as shown in Fig. 2-19. How far is A from C, if $BC = 2.15$ cm?

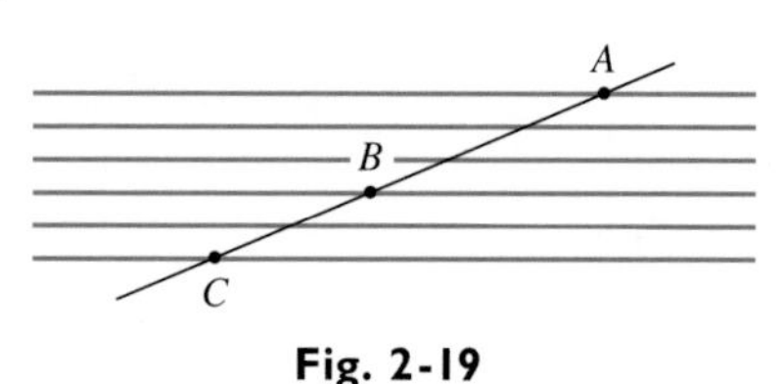

Fig. 2-19

2-2 Triangles

When part of a plane is bounded and closed by straight-line segments, it is called a **polygon**. In general, polygons are named according to the number of sides they have. *A* **triangle** *has three sides, a* **quadrilateral** *has four sides, a* **pentagon** *has five sides, a* **hexagon** *has six sides, and so on.* The polygons of greatest importance are the triangle and the quadrilateral. Therefore, in this section we review the properties of triangles, and in the following section we will consider the quadrilateral. Many of the properties of triangles are important in the study of trigonometry, which we will start in Chapter 4.

Types and Properties of Triangles

There are several important types of triangles. In a **scalene triangle,** no two sides are equal in length. In an **isosceles triangle,** two of the sides are equal in length, and the two *base angles* (the angles opposite the equal sides) are equal. In an **equilateral triangle,** the three sides are equal in length, and each of the three angles is 60°.

One of the most important triangles in scientific and technical applications is the **right triangle.** *In a right triangle, one of the angles is a right angle. The side opposite the right angle is called the* **hypotenuse,** *and the other two sides are called* **legs.** Each of these triangles is illustrated in the following example.

EXAMPLE 1 ▸▸ Figure 2-20(a) shows a scalene triangle. We see that each side is of a different length. Figure 2-20(b) shows an isosceles triangle with two equal sides of 2 cm and equal base angles of 40°. Figure 2-20(c) shows an equilateral triangle, each side of which is 5 cm and each interior angle of which is 60°. Figure 2-20(d) shows a right triangle. The hypotenuse is side AB.

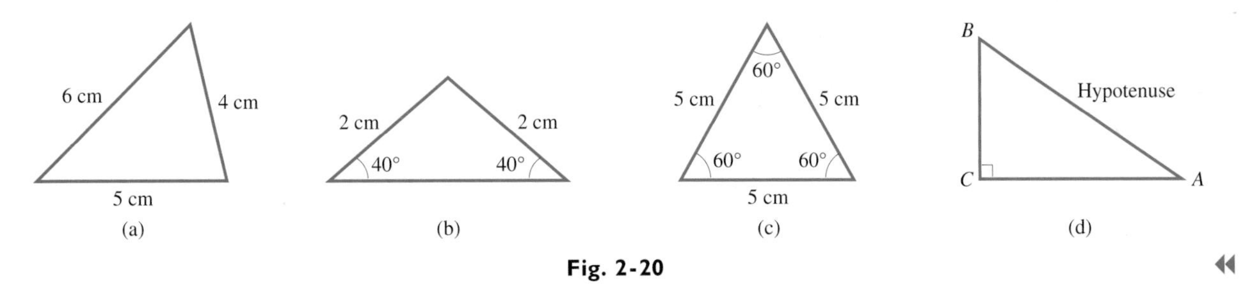

Fig. 2-20 ◂◂

One very important property of a triangle is that

the sum of the three angles of a triangle is 180°.

In the next example, we show this property by using material from Section 2-1.

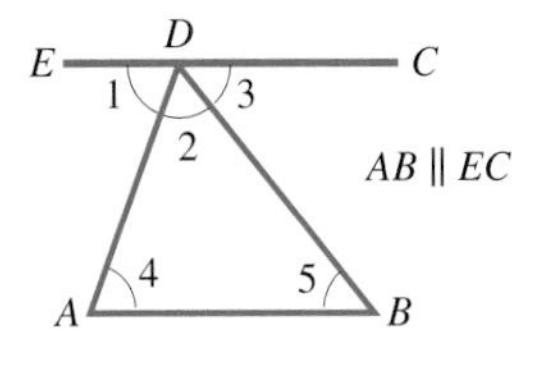

Fig. 2-21

EXAMPLE 2 ▸▸ In Fig. 2-21, since $\angle 1$, $\angle 2$, and $\angle 3$ constitute a straight angle,

$$\angle 1 + \angle 2 + \angle 3 = 180°$$

Also, by noting alternate interior angles, we see that $\angle 1 = \angle 4$ and $\angle 3 = \angle 5$. Therefore, by substitution we have

$$\angle 4 + \angle 2 + \angle 5 = 180°$$

Therefore, if two of the angles of a triangle are known, the third may be found by subtracting the sum of the first two from 180°. ◂◂

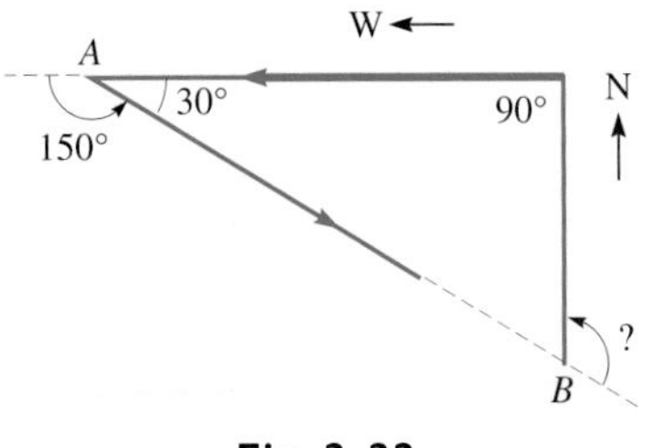

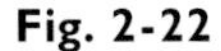
Fig. 2-22

EXAMPLE 3 ▸▸ An airplane is flying north and then makes a 90° turn to the west. Later it makes another left turn of 150°. What is the angle of a third left turn that will cause the plane to again fly north? See Fig. 2-22.

From Fig. 2-22, we see that the interior angle of the triangle at A is the supplement of 150°, or 30°. Since the sum of the interior angles of the triangle is 180°, the interior angle at B is

$$\angle B = 180° - (90° + 30°) = 60°$$

The required angle is the supplement of 60°, which is 120°. ◂◂

A line segment drawn from a vertex of a triangle to the *midpoint* of the opposite side is called a **median** of the triangle. A basic property of a triangle is that *the three medians meet at a single point, called the* **centroid** *of the triangle.* See Fig. 2-23. Also, *the three* **angle bisectors** (lines from the vertices that divide the angles in half) *meet at a common point.* See Fig. 2-24.

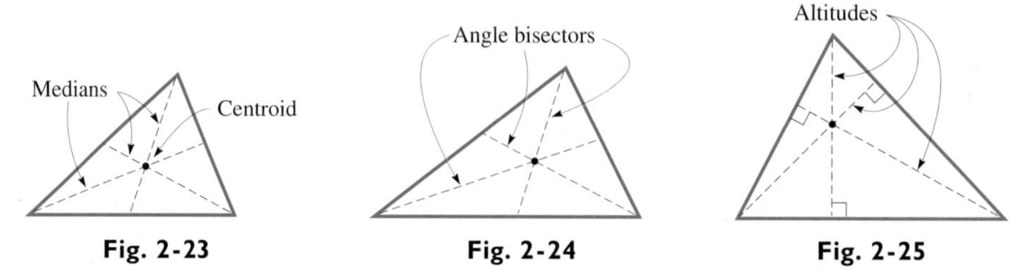

Fig. 2-23 Fig. 2-24 Fig. 2-25

An **altitude** *(or* **height***) of a triangle is the line segment drawn from a vertex perpendicular to the opposite side (or its extension), which is called the* **base** *of the triangle. The three altitudes of a triangle meet at a common point.* See Fig. 2-25. The three common points of the medians, angle bisectors, and altitudes are generally not the same point for a given triangle.

Perimeter and Area of a Triangle

We now consider two of the most basic measures of a plane geometric figure. The first of these is its **perimeter**, *which is the total distance around it.* In the following example we find the perimeter of a triangle.

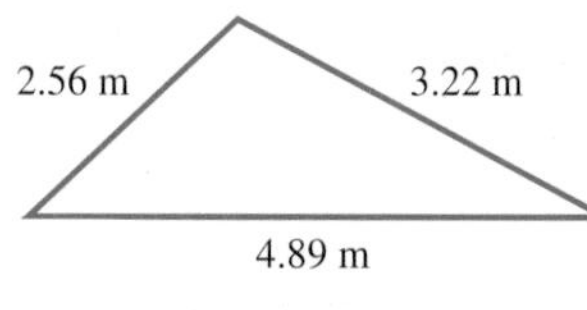

Fig. 2-26

EXAMPLE 4 ⏵⏵ Find the perimeter p of a triangle with sides 2.56 m, 3.22 m, and 4.89 m. See Fig. 2-26.

Using the definition of perimeter, for this triangle we have

$$p = 2.56 + 3.22 + 4.89 = 10.67 \text{ m}$$

Therefore, the distance around the triangle is 10.67 m. We express the results to hundredths since each side is given to hundredths. ⏴⏴

The second important measure of a geometric figure is its **area**. Although the concept of area is primarily intuitive, it is easily defined and calculated for the basic geometric figures. *Area gives a measure of the surface of the figure,* just as perimeter gives the measure of the distance around it.

The area A of a triangle of base b and altitude h is

$$A = \tfrac{1}{2}bh \qquad (2\text{-}2)$$

The following example illustrates the use of Eq. (2-2).

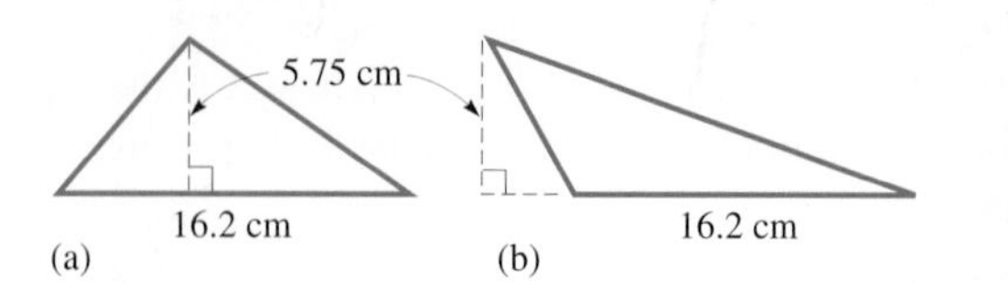

Fig. 2-27

EXAMPLE 5 ⏵⏵ Find the area of the triangles in Fig. 2-27(a) and Fig. 2-27(b).

Even though the triangles are of different shapes, we see that the base b of each triangle is 16.2 cm and that the altitude h of each is 5.75 cm. Therefore, the area of each triangle is

$$A = \tfrac{1}{2}bh = \tfrac{1}{2}(16.2)(5.75) = 46.6 \text{ cm}^2$$ ⏴⏴

Hero's formula is named for Hero (or Heron), a first-century A.D. Greek mathematician.

Another formula for the area of a triangle that is particularly useful when we have *a triangle with three known sides and no right angle* is **Hero's formula**, which is given in Eq. (2-3).

$$A = \sqrt{s(s-a)(s-b)(s-c)} \quad \text{where } s = \tfrac{1}{2}(a+b+c) \tag{2-3}$$

We note that s is one-half of the perimeter.

EXAMPLE 6 ▸▸ A surveyor measures the three sides of a triangular parcel of land between two intersecting straight roads to be 206 m, 293 m, and 187 m, as shown in Fig. 2-28. Find the area of this parcel.

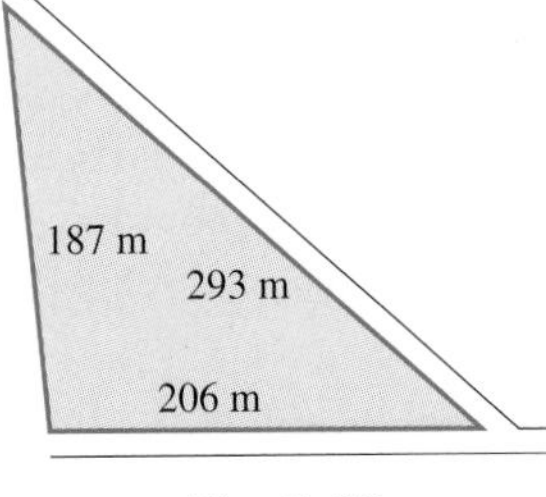

Fig. 2-28

In order to use Eq. (2-3), we first find s.

$$s = \tfrac{1}{2}(206 + 293 + 187) = \tfrac{1}{2}(686) = 343 \text{ m}$$

Now, substituting in Eq. (2-3), we have

$$A = \sqrt{343(343-206)(343-293)(343-187)} = 19\,100 \text{ m}^2$$

The result has been rounded off to three significant digits. In using a calculator, the value of s is stored in memory and then used to find A. It is not necessary to write down anything except the final result. ◂◂

The Pythagorean Theorem

Named for the Greek mathematician Pythagoras (6th-century B.C.).

As we have noted, one of the most important geometric figures in technical applications is the right triangle. A very important property of a right triangle is given by the **Pythagorean theorem,** which states that

> *in a* ***right triangle,*** *the square of the length of the hypotenuse equals the sum of the squares of the lengths of the other two sides.*

If c is the length of the hypotenuse, and a and b are the lengths of the other two sides, the Pythagorean theorem is

$$c^2 = a^2 + b^2 \tag{2-4}$$

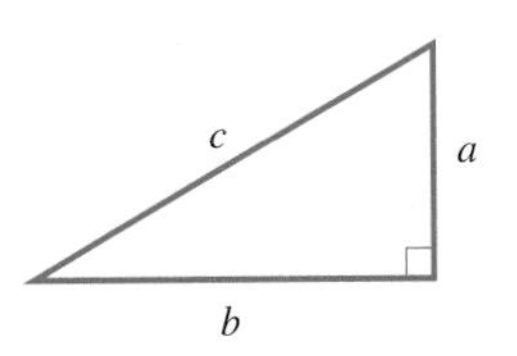

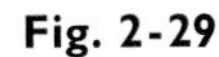
Fig. 2-29

See Fig. 2-29.

EXAMPLE 7 ▸▸ A pole is perpendicular to the level ground around it. A guy wire is attached 3.20 m up the pole and at a point on the ground, 2.65 m from the pole. How long is the guy wire?

From the given information, we sketch the pole and guy wire as shown in Fig. 2-30. Using the Pythagorean theorem, and then substituting, we have

$$AC^2 = AB^2 + BC^2$$
$$= 2.65^2 + 3.20^2$$
$$AC = \sqrt{2.65^2 + 3.20^2} = 4.15 \text{ m}$$

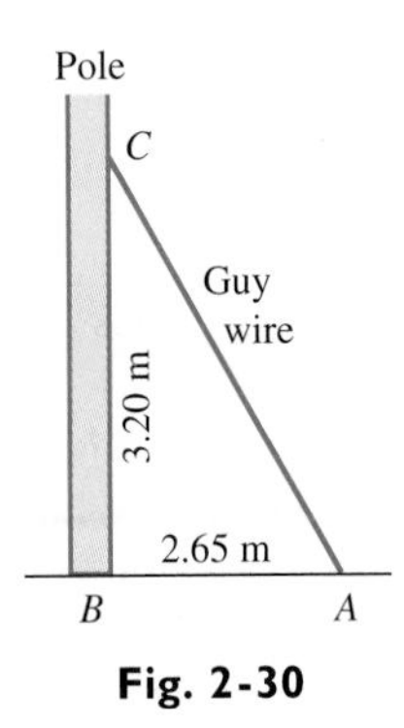

Fig. 2-30

The guy wire is 4.15 m long. In using the calculator, parentheses are used to group $2.65^2 + 3.20^2$. ◂◂

Similar Triangles

The perimeter and area of a triangle are measures of its *size*. We now consider the shape of triangles.

Two triangles are **similar** *if they have the same shape (but not necessarily the same size).* There are two very important properties of similar triangles.

Properties of Similar Triangles

1. *The corresponding angles of similar triangles are equal.*
2. *The corresponding sides of similar triangles are proportional.*

In the two triangles that are similar, *the* **corresponding sides** *are the sides, one in each triangle, which are between the same pair of equal corresponding angles.* These properties are illustrated in the following example.

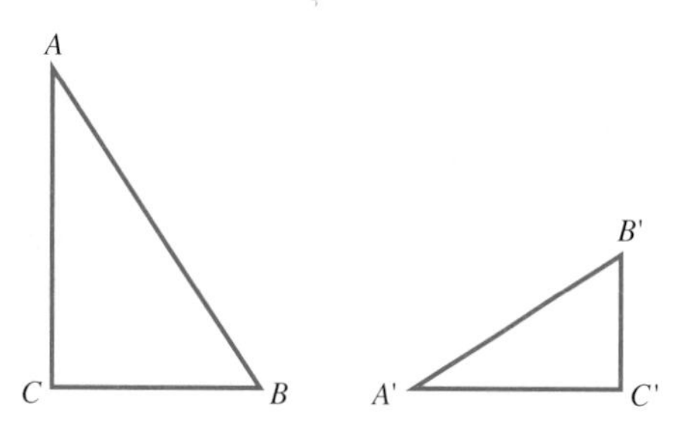

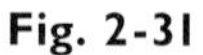
Fig. 2-31

EXAMPLE 8 ▸▸ In Fig. 2-31, a pair of similar triangles are shown. They are similar even though the corresponding parts are not in the same position relative to the page. Using standard symbols, we can write $\triangle ABC \sim \triangle A'B'C'$, where $\triangle$ means "triangle" and $\sim$ means "is similar to."

The pairs of corresponding angles are A and A', B and B', and C and C'. This means $A = A'$, $B = B'$, and $C = C'$.

The pairs of corresponding sides are AB and $A'B'$, BC and $B'C'$, and AC and $A'C'$. In order to show that these corresponding sides are proportional, we write

$$\frac{AB}{A'B'} = \frac{BC}{B'C'} = \frac{AC}{A'C'} \begin{array}{l} \longleftarrow \text{sides of } \triangle ABC \\ \longleftarrow \text{sides of } \triangle A'B'C' \end{array}$$ ◂◂

If we know that two triangles are similar, we can use the two basic properties of similar triangles to find the unknown parts of one triangle from the known parts of the other triangle. The following example illustrates how this is done in a practical application.

EXAMPLE 9 ▸▸ On level ground a silo casts a shadow 24 m long. At the same time, a pole casts a shadow 3.0 m long. How tall is the silo? See Fig. 2-32.

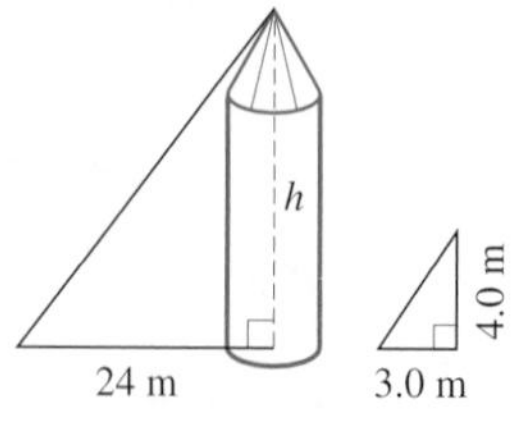

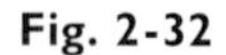
Fig. 2-32

The rays of the sun are essentially parallel. The two triangles in Fig. 2-32 are similar since *each has a right angle and the angles at the tops are equal.* The other angles must be equal since the sum of the angles is 180°. The lengths of the hypotenuses are of no importance in this problem, so we use only the other sides in stating the ratios of corresponding sides. Denoting the height of the silo as h, we have

$$\frac{h}{4.0} = \frac{24}{3.0} \qquad h = 32 \text{ m}$$

We conclude that the silo is 32 m high. ◂◂

One of the most practical uses of similar geometric figures is that of **scale drawings**. Maps, charts, blueprints, and most drawings that appear in books are familiar examples of scale drawings. Actually, there have been many scale drawings used in this book already.

In any scale drawing, all distances are drawn a certain ratio of the distances they represent and all angles equal the angles they represent. Consider the following example.

EXAMPLE 10 ▸▸ In drawing a map of the area shown in Fig. 2-33, a scale of 1 cm = 200 km is used. In measuring the distance between Chicago and Toronto on the map, we find it to be 3.5 cm. The actual distance x between Chicago and Toronto is found from the proportion

$$\underset{\text{distance on map}\longrightarrow}{\overset{\text{actual distance}\longrightarrow}{}}\ \frac{x}{3.5\text{ cm}} = \overset{\text{scale}\downarrow}{\frac{200\text{ km}}{1\text{ cm}}} \quad \text{or} \quad x = 700\text{ km}$$

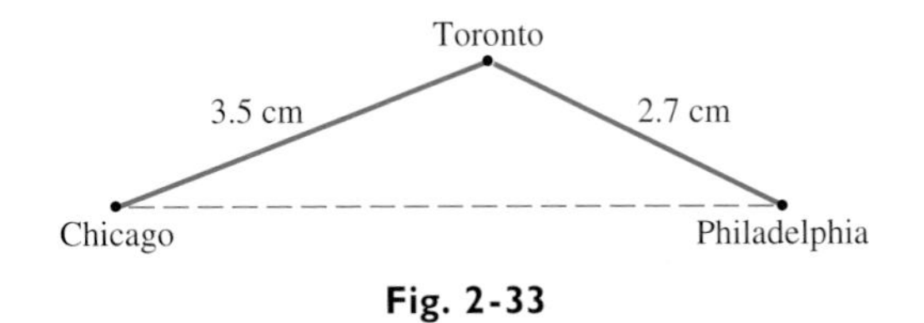

Fig. 2-33

If we did not have the scale but knew that the distance between Chicago and Toronto is 700 km, then by measuring distances on the map between Chicago and Toronto (3.5 cm) and between Toronto and Philadelphia (2.7 cm), we could find the distance between Toronto and Philadelphia. It is found from the following proportion, determined by use of similar triangles:

$$\frac{700\text{ km}}{3.5\text{ cm}} = \frac{y}{2.7\text{ cm}}$$

$$y = \frac{2.7(700)}{3.5} = 540\text{ km}$$ ◂◂

Similarity requires *equal* angles and *proportional* sides. *If the corresponding angles and the corresponding sides of two triangles are equal, the two triangles are* **congruent.** As a result of this definition, the areas and perimeters of congruent triangles are also equal. Informally, we can say that similar triangles have the same shape, whereas congruent triangles have the same shape and same size.

EXAMPLE 11 ▸▸ A right triangle with legs of 2 cm and 4 cm is congruent to any other right triangle with legs of 2 cm and 4 cm. However, it is similar to any right triangle with legs of 5 cm and 10 cm, since the corresponding sides are proportional. See Fig. 2-34.

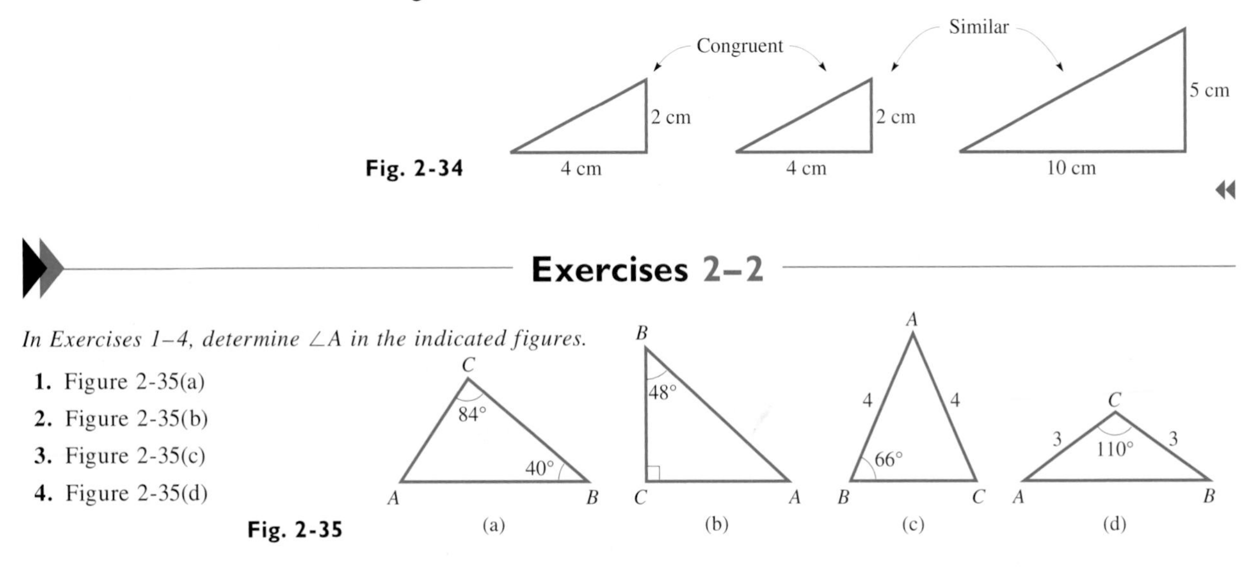

Fig. 2-34

◂◂

Exercises 2–2

In Exercises 1–4, determine ∠A in the indicated figures.

1. Figure 2-35(a)
2. Figure 2-35(b)
3. Figure 2-35(c)
4. Figure 2-35(d)

Fig. 2-35

In Exercises 5–8, find the perimeter of each triangle.

5. Fig. 2-36 **6.** Fig. 2-37

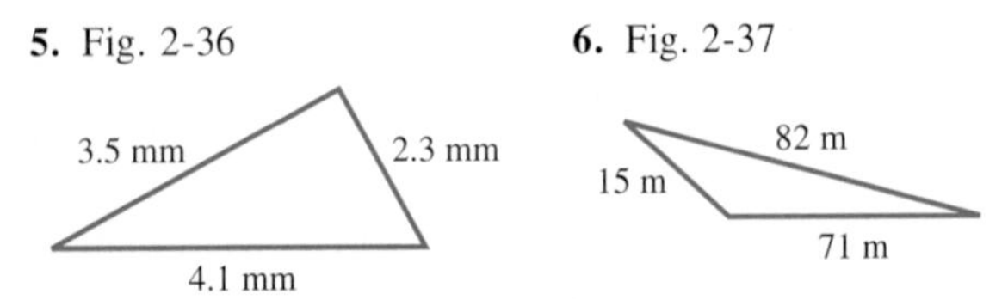

7. An equilateral triangle of side 21.5 cm

8. An isosceles triangle with equal sides of 2.45 dm and third side of 3.22 dm.

In Exercises 9–16, find the area of each triangle.

9. Fig. 2-38 **10.** Fig. 2-39

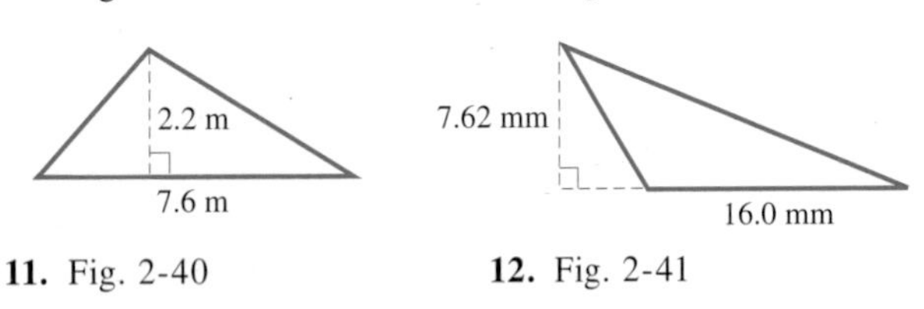

11. Fig. 2-40 **12.** Fig. 2-41

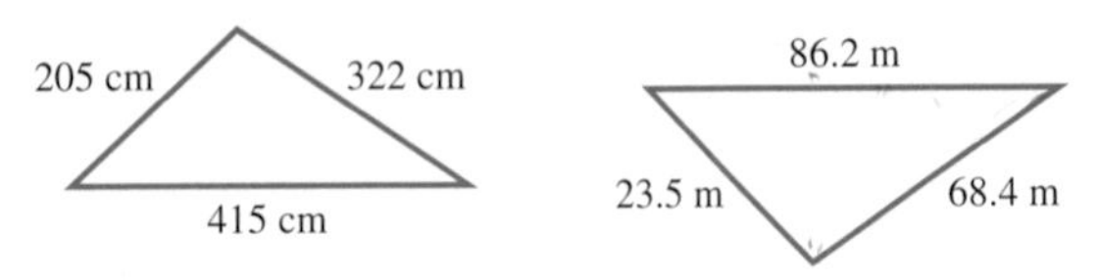

13. Right triangle with legs 3.46 cm and 2.55 cm

14. Right triangle with legs 234 mm and 342 mm

15. An isosceles triangle with equal sides of 0.986 m and third side of 0.884 m

16. An equilateral triangle of side 3.20 dm

In Exercises 17–20, find the third side of the right triangle of Fig. 2-42 for the given values.

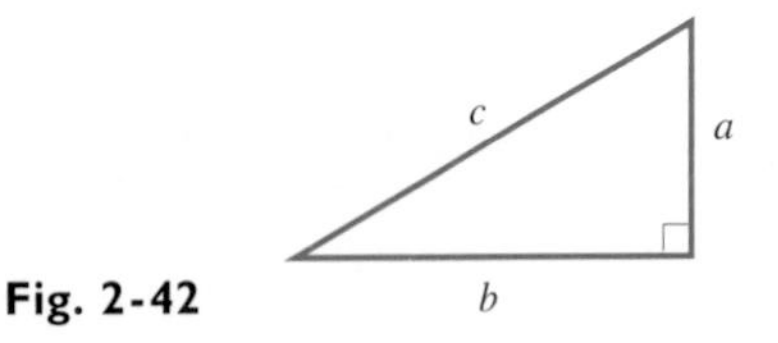

Fig. 2-42

17. $a = 13.8$ mm, $b = 22.7$ mm

18. $a = 2.48$ m, $b = 1.45$ m

19. $a = 17.5$ cm, $c = 55.1$ cm

20. $b = 0.474$ km, $c = 0.836$ km

In Exercises 21–24, use the right triangle in Fig. 2-43.

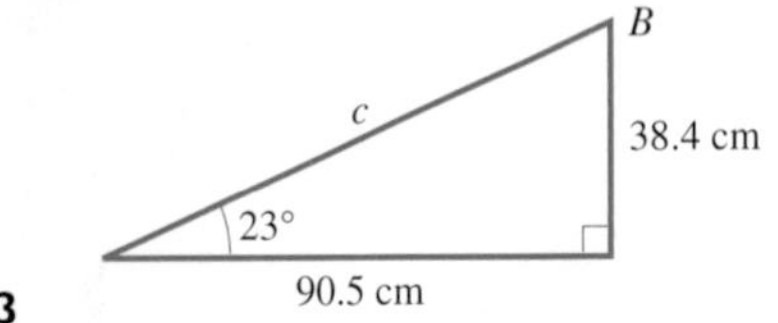

Fig. 2-43

21. Find $\angle B$. **22.** Find side c.

23. Find the perimeter. **24.** Find the area.

In Exercises 25–36, solve the given problems.

25. In Fig. 2-44, show that $\triangle MKL \sim \triangle MNO$.

26. In Fig. 2-45, show that $\triangle ACB \sim \triangle ADC$.

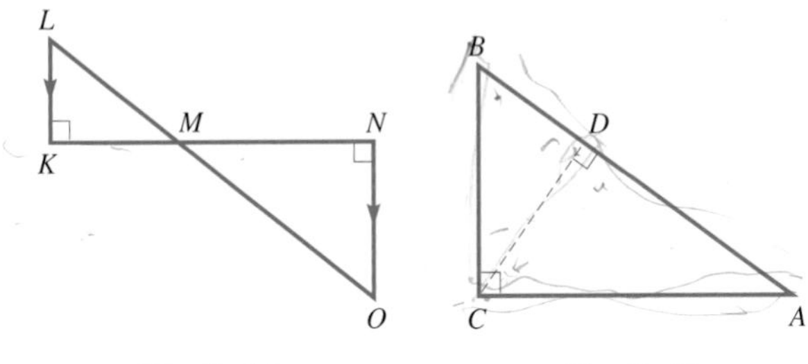

Fig. 2-44

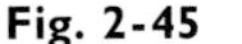

Fig. 2-45

27. In Fig. 2-44, if $KN = 15$, $MN = 9$, and $MO = 12$, find LM.

28. In Fig. 2-45, if $AD = 9$ and $AC = 12$, find AB.

29. A tooth on a saw is in the shape of an isosceles triangle. If the angle at the point is 38°, find the two base angles.

30. A transmitting tower is supported by a wire that makes an angle of 52° with the level ground. What is the angle between the tower and the wire?

31. Find the area of a triangular patio with sides of 17.5 m, 18.7 m, and 19.5 m.

32. Three straight city streets enclose a right-triangular city block. If the shorter sides are 260 m and 320 m, find (a) the area of the block and (b) the distance around the block.

33. An observer is 550 m from the launch pad of a rocket. After the rocket has ascended 750 m, how far is it from the observer?

34. The base of a 6.0-m ladder is 1.8 m from a wall. How far up on the wall does the ladder reach?

35. On a blueprint, a hallway is 45.6 cm long. The scale is 1.2 cm = 1.0 m. How long is the hallway?

36. To find the width ED of a river, a surveyor places markers at A, B, C, and D, as shown in Fig. 2-46. The markers are placed such that $AB \parallel ED$, $BC = 50.0$ m, $DC = 312$ m, and $AB = 80.0$ m. How wide is the river?

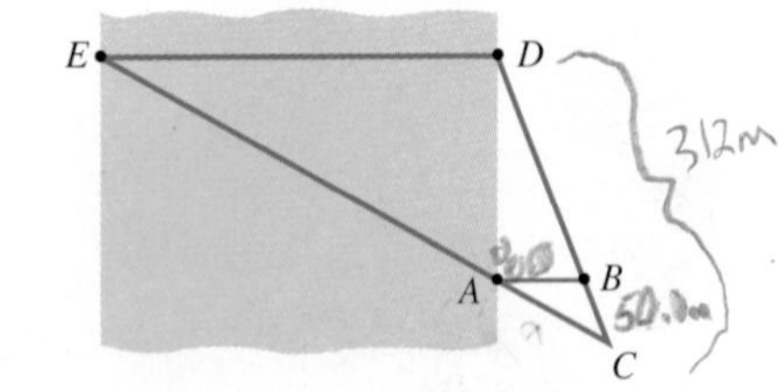

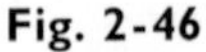

Fig. 2-46

2-3 Quadrilaterals

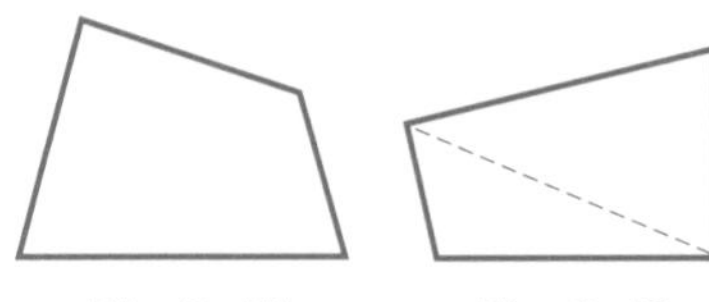

Fig. 2-47 Fig. 2-48

In this section we consider another important type of polygon. *A* **quadrilateral** *is a closed plane figure that has four sides,* and these four sides form four interior angles. A general quadrilateral is shown in Fig. 2-47.

A **diagonal** *of a polygon is a straight line segment joining any two nonadjacent vertices.* The dashed line is one of two possible diagonals of the quadrilateral shown in Fig. 2-48.

Types of Quadrilaterals

A **parallelogram** *is a quadrilateral in which opposite sides are parallel.* In a parallelogram, opposite sides are equal and opposite angles are equal. *A* **rhombus** *is a parallelogram with four equal sides.*

A **rectangle** *is a parallelogram in which intersecting sides are perpendicular,* which means that all four interior angles are right angles. In a rectangle, the longer side is usually called the **length** and the shorter side is called the **width**. *A* **square** *is a rectangle with four equal sides.*

A **trapezoid** *is a quadrilateral in which two sides are parallel.* The parallel sides are called the **bases** of the trapezoid.

EXAMPLE 1 ▸▸ A parallelogram is shown in Fig. 2-49(a). Opposite sides a are equal in length, as are opposite sides b. A rhombus with equal sides s is shown in Fig. 2-49(b). A rectangle is shown in Fig. 2-49(c). The length is labeled l, and the width is labeled w. A square with equal sides s is shown in Fig. 2-49(d). A trapezoid with bases b_1 and b_2 is shown in Fig. 2-49(e). ◂◂

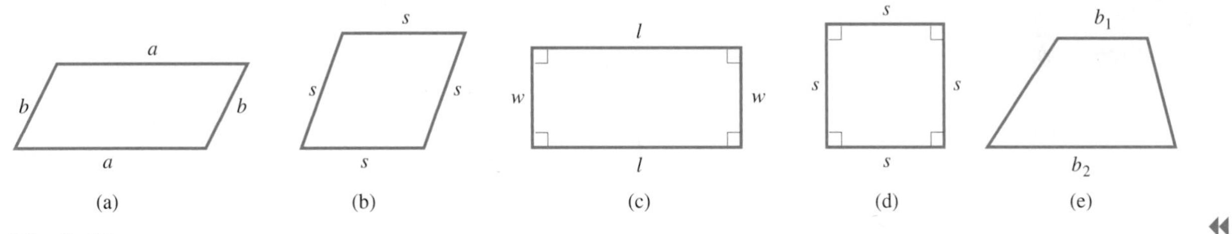

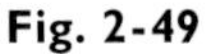

Fig. 2-49

Perimeter and Area of a Quadrilateral

Since the perimeter of a polygon is the distance around it, *the perimeter of a quadrilateral is the sum of the lengths of its four sides.* Consider the next example.

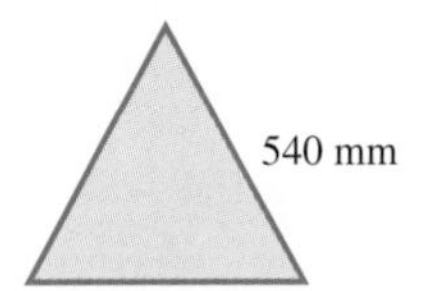

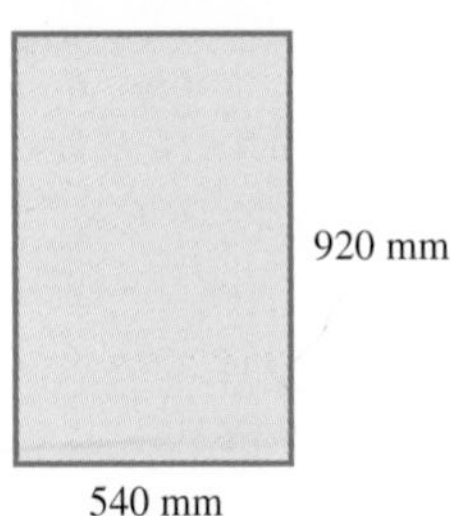

Fig. 2-50

EXAMPLE 2 ▸▸ An architect designs a room with a rectangular window 920 mm high and 540 mm wide, with another window above in the shape of an equilateral triangle, 540 mm on a side. See Fig. 2-50. How much molding is needed for these windows?

The length of molding is the sum of the perimeters of the windows. For the rectangular window, the opposite sides are equal, which means the perimeter is twice the length l plus twice the width w. For the equilateral triangle, the perimeter is three times the side s. Therefore, the length L of molding is

$$\begin{aligned} L &= 2l + 2w + 3s \\ &= 2(920) + 2(540) + 3(540) \\ &= 4540 \text{ mm} \end{aligned}$$ ◂◂

Following are the formulas for the areas of the square, rectangle, parallelogram, and trapezoid.

$A = s^2$	square of side s (Fig. 2-51)	(2-5)
$A = lw$	rectangle of length l and width w (Fig. 2-52)	(2-6)
$A = bh$	parallelogram of base b and height h (Fig. 2-53)	(2-7)
$A = \frac{1}{2}h(b_1 + b_2)$	trapezoid of bases b_1 and b_2, and height h (Fig. 2-54)	(2-8)

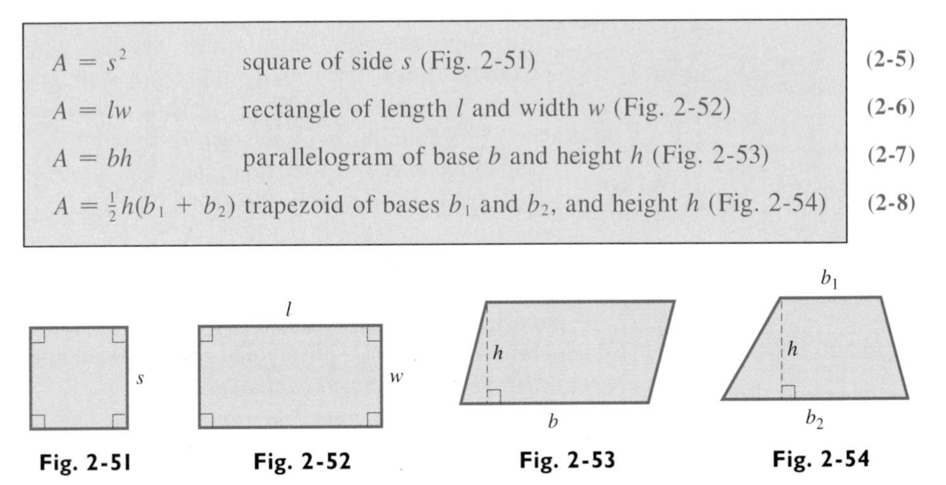

Fig. 2-51 Fig. 2-52 Fig. 2-53 Fig. 2-54

Since a rectangle, a square, and a rhombus are special types of parallelograms, the area of these figures can be found from Eq. (2-7). The area of a trapezoid is of importance when we find areas of irregular geometric figures in Section 2-5.

EXAMPLE 3 ▸▸ A city park is designed with lawn areas in the shape of a right triangle, a parallelogram, and a trapezoid, as shown in Fig. 2-55, with walkways between them. Find the area of each section of lawn and the total lawn area.

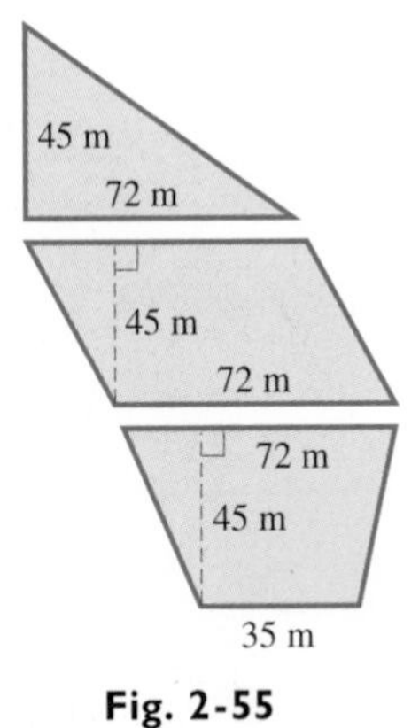

Fig. 2-55

$$A_1 = \frac{1}{2}bh = \frac{1}{2}(72)(45) = 1600 \text{ m}^2 \qquad A_2 = bh = (72)(45) = 3200 \text{ m}^2$$
$$A_3 = \frac{1}{2}h(b_1 + b_2) = \frac{1}{2}(45)(72 + 35) = 2400 \text{ m}^2$$

The total lawn area is about 7200 m^2. ◂◂

Stated Problems Involving Geometric Figures

In Chapter 1 we solved stated problems by setting up and solving an appropriate equation. Following is an example of a stated problem involving a quadrilateral.

EXAMPLE 4 ▸▸ The length of a rectangular computer chip is 2.0 mm longer than its width. Find the dimensions of the chip if its perimeter is 26.4 mm.

Since the dimensions, the length and the width, are required, we let $w =$ the width of the chip. Since the length is 2.0 mm more than the width, we know that $w + 2.0 =$ the length of the chip. See Fig. 2-56.

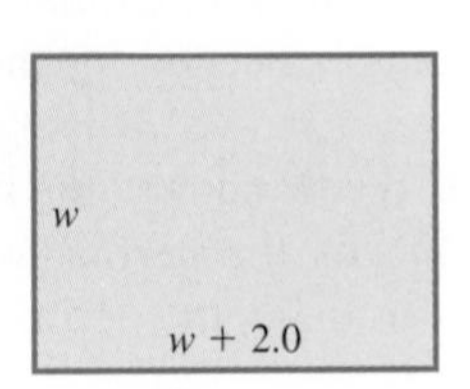

Fig. 2-56

Since the perimeter of a rectangle is twice the length plus twice the width, we have the equation

$$2(w + 2.0) + 2w = 26.4$$

since the perimeter is given as 26.4 mm. This is the equation we need.

Solving this equation, we have

$$2w + 4.0 + 2w = 26.4$$
$$4w = 22.4$$
$$w = 5.6 \text{ mm} \quad \text{and} \quad w + 2.0 = 7.6 \text{ mm}$$

Therefore, the length is 7.6 mm and the width is 5.6 mm. These values check with the statements of the original problem. ◂◂

Exercises 2–3

In Exercises 1–8, find the perimeter of each figure.

1. Square: side of 0.65 m
2. Rhombus: side of 2.46 km
3. Rectangle: $l = 46.5$ mm, $w = 37.4$ mm
4. Rectangle: $l = 14.2$ cm, $w = 12.6$ cm
5. The parallelogram in Fig. 2-57
6. The parallelogram in Fig. 2-58
7. The trapezoid in Fig. 2-59
8. The trapezoid in Fig. 2-60

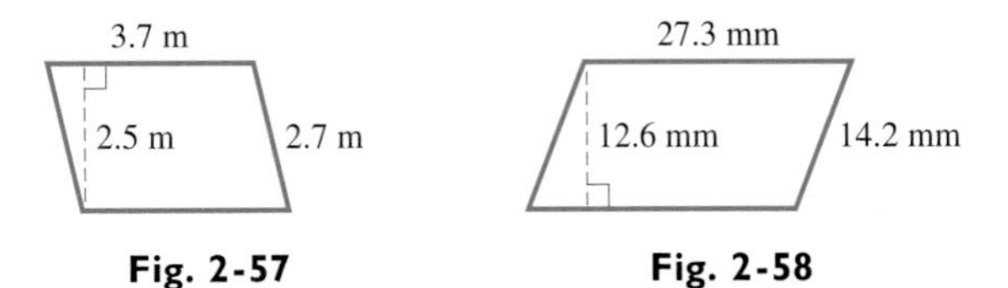

Fig. 2-57

Fig. 2-58

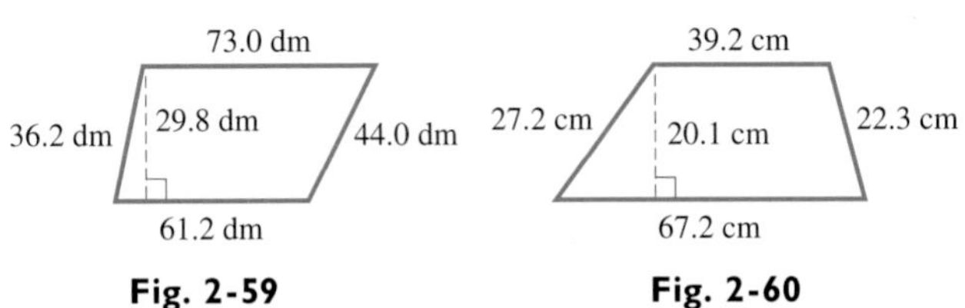

Fig. 2-59

Fig. 2-60

In Exercises 9–16, find the area of each figure.

9. Square: $s = 2.7$ mm
10. Square: $s = 15.6$ m
11. Rectangle: $l = 0.465$ km, $w = 0.374$ km
12. Rectangle: $l = 14.2$ cm, $w = 12.6$ cm
13. The parallelogram in Fig. 2-57
14. The parallelogram in Fig. 2-58
15. The trapezoid in Fig. 2-59
16. The trapezoid in Fig. 2-60

In Exercises 17–20, set up a formula for the indicated perimeter or area. (Do not include dashed lines.)

17. The perimeter of the figure in Fig. 2-61 (a parallelogram and a square attached)
18. The perimeter of the figure in Fig. 2-62 (two trapezoids attached)
19. The area of the figure in Fig. 2-61
20. The area of the figure in Fig. 2-62

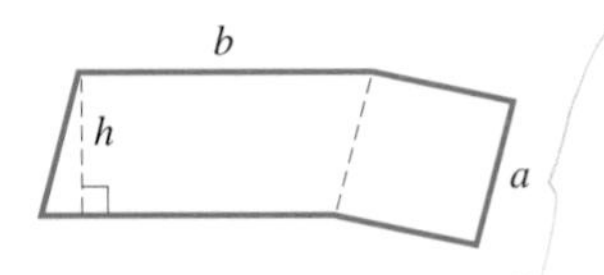

Fig. 2-61

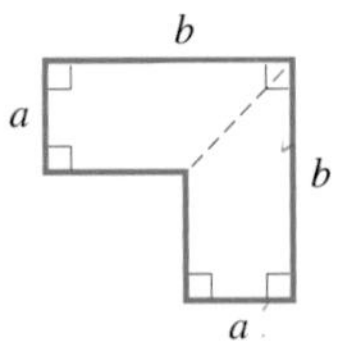

b
a
b
a

Fig. 2-62

In Exercises 21–28, solve the given problems.

21. A machine part is in the shape of a square with equilateral triangles attached to two sides (see Fig. 2-63). Find the perimeter of the machine part.

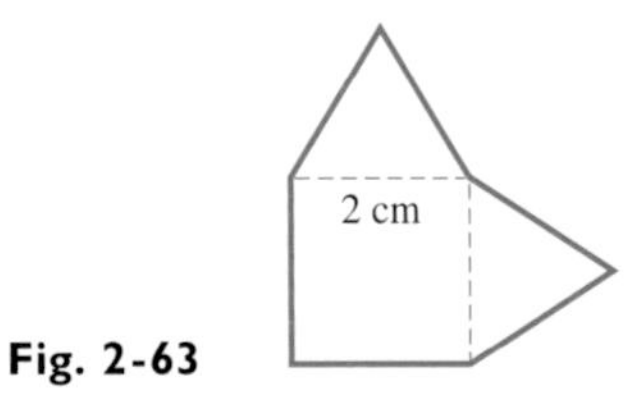

Fig. 2-63

22. The frame of the top of a square card table is made of metal tubing that costs 12¢ per centimetre. What is the cost of the frames for two tabletops that are 84 cm on a side?
23. A beam support in a building is in the shape of a parallelogram as shown in Fig. 2-64. Find the area of the side of the beam shown.

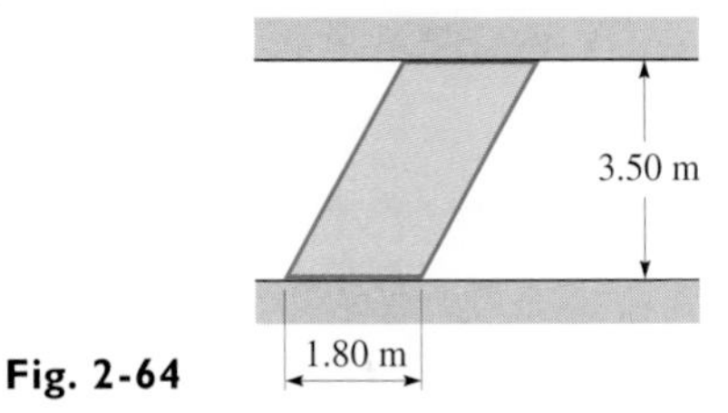

Fig. 2-64

24. A litre of paint will cover 7.5 m^2. How much paint is needed to paint the walls of a rectangular room 3600 mm by 4800 mm, if the walls are 2400 mm high?
25. A walkway 3.0 m wide is constructed along the outside edge of a square courtyard. If the perimeter of the courtyard is 320 m, what is the perimeter of the square formed by the outer edge of the walkway?
26. An architect designs a rectangular window such that the width of the window is 450 mm less than the height. If the perimeter of the window is 4500 mm, what are its dimensions?
27. A rectangular field is enclosed with a fence of four parallel strands of barbed wire. The length of the field is 20.0 m more than the width, and a total of 1360 m of wire is used for the fencing. What are the dimensions of the field?
28. A rectangular security area is enclosed on one side by a wall, and the other sides are fenced. The length of the wall is twice the width of the area. The total cost of building the wall and fence is \$13 200. If the wall costs \$50.00/m and the fence costs \$5.00/m, find the dimensions of the area.

2-4 Circles

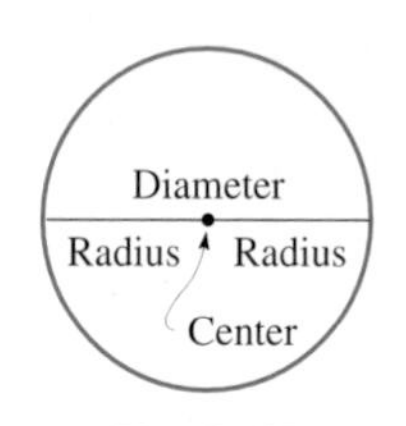

Fig. 2-65

The next geometric figure we consider is the circle. *All points on a* **circle** *are at the same distance from a fixed point, the* **center** *of the circle. The distance from the center to a point on the circle is the* **radius** *of the circle. The distance between two points on the circle on a line through the center is the* **diameter**. Therefore, the diameter d is twice the radius r, or $d = 2r$. See Fig. 2-65.

There are also certain special types of lines associated with a circle. *A* **chord** *is a line segment having its endpoints on the circle. A* **tangent** *is a line that touches (does not pass through) the circle at one point. A* **secant** *is a line that passes through two points of the circle.* See Fig. 2-66.

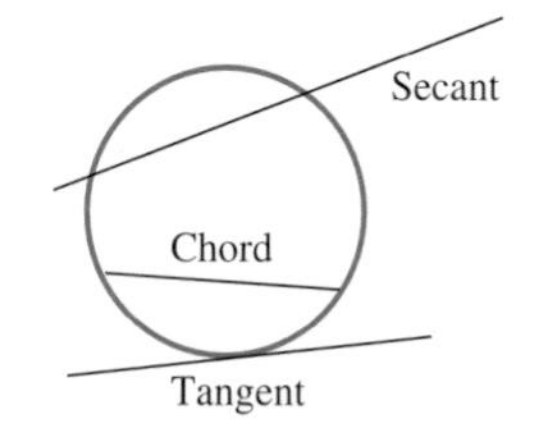

Fig. 2-66

An important property of a tangent is that *a tangent to a circle is perpendicular to the radius drawn to the point of contact.* This is illustrated in the next example.

EXAMPLE 1 ▸▸ In Fig. 2-67, O is the center of the circle and AB is tangent at B. If $\angle OAB = 25°$, find $\angle AOB$.

Since the center is O, OB is a radius of the circle. A tangent is perpendicular to a radius at the point of tangency, which means $\angle ABO = 90°$ so that

$$\angle OAB + \angle OBA = 25° + 90° = 115°$$

Since the sum of the angles of a triangle is 180°, we have

$$\angle AOB = 180° - 115° = 65°$$ ◂◂

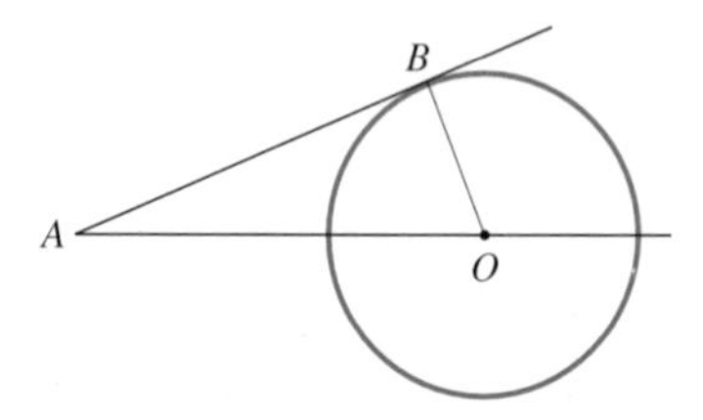

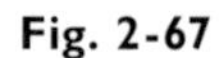

Fig. 2-67

Circumference and Area of a Circle

The perimeter of a circle is called the **circumference.** The formulas for the circumference and area of a circle are as follows:

$c = 2\pi r$	circumference of a circle of radius r	(2-9)
$A = \pi r^2$	area of a circle of radius r	(2-10)

Here π equals approximately 3.1416. In using a calculator, π can be entered by using the [π] key.

EXAMPLE 2 ▸▸ A circular oil spill has a diameter of 2.4 km. This oil spill is to be enclosed within a length of special flexible tubing. What is the area of the spill, and how long must the tubing be? See Fig. 2-68.

We find the area by using Eq. (2-10). Since $d = 2r$, $r = d/2 = 1.2$ km. Therefore, the area is

$$A = \pi r^2 = \pi(1.2)^2$$
$$= 4.5 \text{ km}^2$$

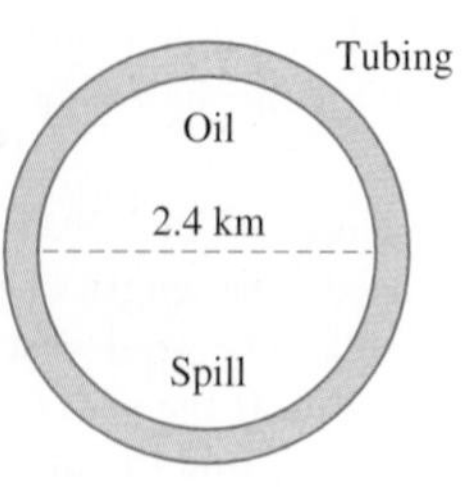

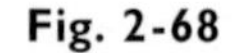

Fig. 2-68

The length of the tubing needed to enclose the oil spill is the circumference of the circle. Therefore,

$$c = 2\pi r = 2\pi(1.2) \qquad \text{note that } c = \pi d$$
$$= 7.5 \text{ km}$$

Results have been rounded off to two significant digits, the accuracy of d. ◂◂

Many applied problems involve a combination of geometric figures. The following example illustrates one such combination.

EXAMPLE 3 ▸▸ A machine part is a square of side 3.25 cm with a quarter circle removed (see Fig. 2-69). Find the perimeter and the area of one side of the part.

Setting up a formula for the perimeter, we add the two sides of length s to *one-fourth of the circumference of a circle with radius s*. For the area, we *subtract the area of one-fourth of a circle from the area of the square*. This gives

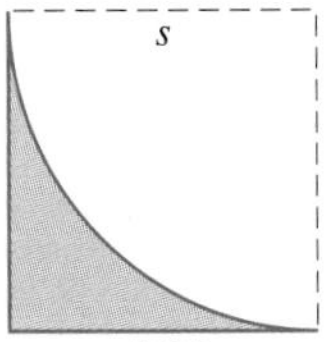

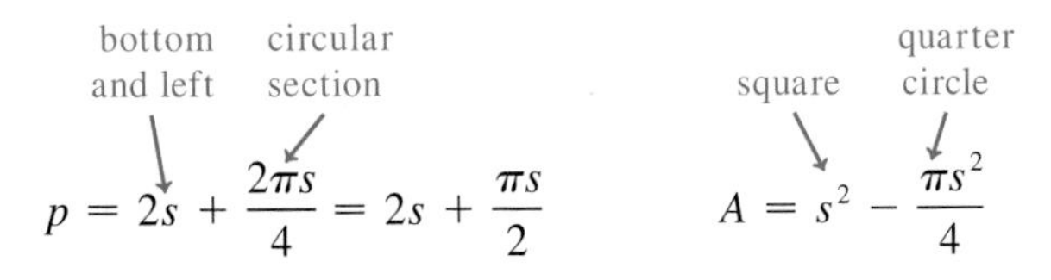

$$p = 2s + \frac{2\pi s}{4} = 2s + \frac{\pi s}{2} \qquad A = s^2 - \frac{\pi s^2}{4}$$

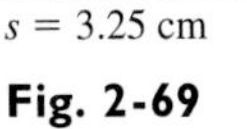

Fig. 2-69

where s is the side of the square and the radius of the circle. Evaluating, we have

$$p = 2(3.25) + \frac{\pi(3.25)}{2} = 11.6 \text{ cm}$$

$$A = 3.25^2 - \frac{\pi(3.25)^2}{4} = 2.27 \text{ cm}^2$$ ◂◂

Circular Arcs and Angles

An **arc** *is part of a circle,* and *an angle formed at the center by two radii is a* **central angle.** The measure of an arc is the same as the central angle between the ends of the radii that define the arc. *A* **sector** *of a circle is the region bounded by two radii and the arc they intercept.* These are illustrated in the following example.

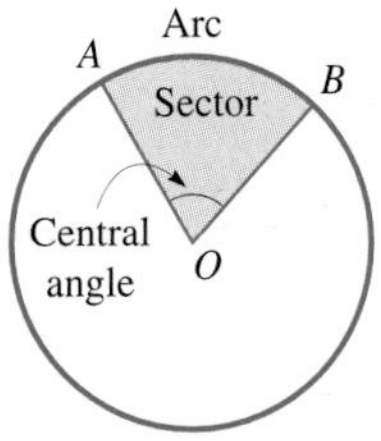

Arc, A, B, Sector, Central angle, O

Fig. 2-70

EXAMPLE 4 ▸▸ In Fig. 2-70, a sector of the circle is between radii OA and OB, and arc AB (which is denoted by $\overset{\frown}{AB}$). If the measure of the central angle at O between the radii is 70°, the measure of $\overset{\frown}{AB}$ is also 70°. ◂◂

An **inscribed angle** *of an arc is one for which the endpoints of the arc are points on the sides of the angle, and for which the vertex is a point (not an endpoint) of the arc.* An important property of a circle is that *the measure of an inscribed angle is one-half of its intercepted arc.*

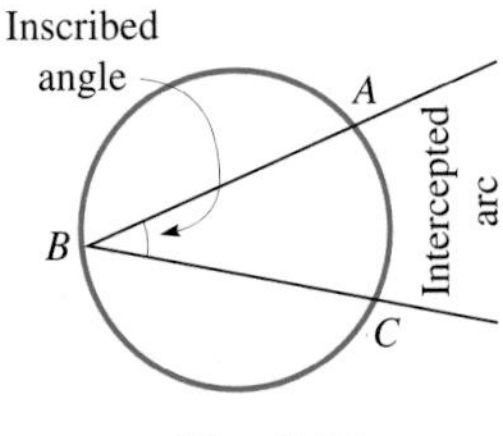

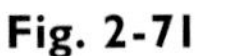

Fig. 2-71

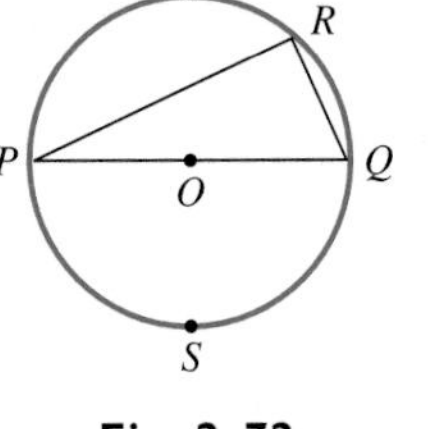

Fig. 2-72

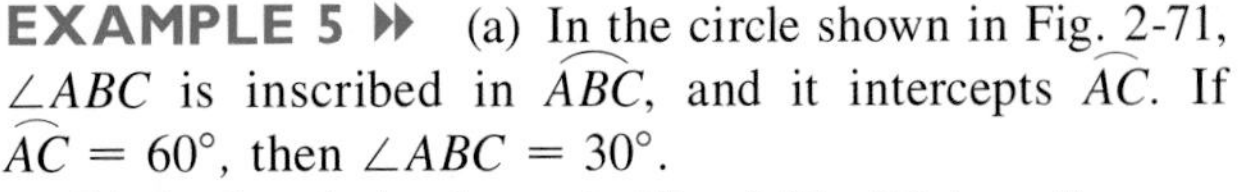

EXAMPLE 5 ▸▸ (a) In the circle shown in Fig. 2-71, $\angle ABC$ is inscribed in $\overset{\frown}{ABC}$, and it intercepts $\overset{\frown}{AC}$. If $\overset{\frown}{AC} = 60°$, then $\angle ABC = 30°$.

(b) In the circle shown in Fig. 2-72, PQ is a diameter and $\angle PRQ$ is inscribed in the semicircular $\overset{\frown}{PRQ}$. Since $\overset{\frown}{PSQ} = 180°$, $\angle PRQ = 90°$. From this we conclude that *an angle inscribed in a semicircle is a right angle.* ◂◂

Radian Measure of an Angle

To this point we have measured all angles in degrees. There is another measure of an angle, the *radian*, which is defined in terms of an arc of a circle. We will find it of importance when we study trigonometry.

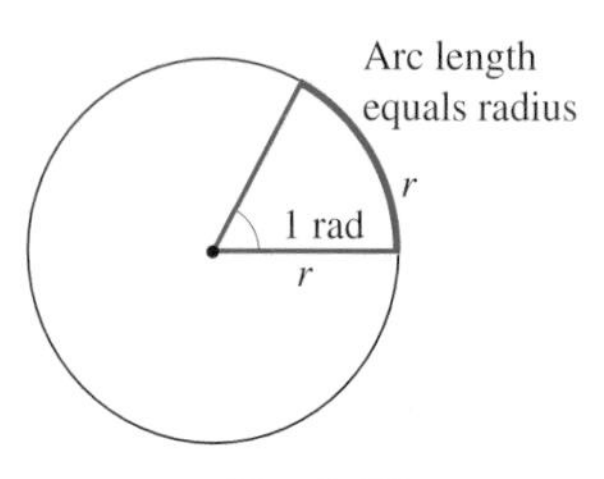

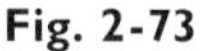
Fig. 2-73

If a central angle of a circle intercepts an arc equal in length to the radius of the circle, the central angle is defined as one **radian.** See Fig. 2-73. Since the radius can be marked off along the circumference 2π times (about 6.283 times), we see that 2π rad $= 360°$ (where rad is the symbol for radian). Therefore,

$$\pi \text{ rad} = 180° \tag{2-11}$$

is a basic relationship between radians and degrees.

EXAMPLE 6 ▸▸ (a) If we divide each side of Eq. (2-11) by π, we get

$$1 \text{ rad} = 57.3°$$

where the result has been rounded off.

(b) To change an angle of 118.2° to radian measure, we have

$$118.2° = 118.2°\left(\frac{\pi \text{ rad}}{180°}\right) = 2.06 \text{ rad}$$

By multiplying 118.2° by π rad/180°, the unit of measurement that remains is rad, since the degrees "cancel." We will review radian measure again when we study trigonometry. ◂◂

Exercises 2–4

In Exercises 1–4, find the circumference of the circle with the given radius or diameter.

1. $r = 2.75$ cm
2. $r = 0.563$ m
3. $d = 23.1$ mm
4. $d = 8.2$ dm

In Exercises 5–8, find the area of the circle with the given radius or diameter.

5. $r = 0.0952$ km
6. $r = 45.8$ cm
7. $d = 2.33$ m
8. $d = 12.56$ mm

In Exercises 9–12, refer to Fig. 2-74, where AB is a diameter, TB is a tangent line at B, and $\angle ABC = 65°$. Determine the indicated angles.

9. $\angle CBT$
10. $\angle BCT$
11. $\angle CAB$
12. $\angle BTC$

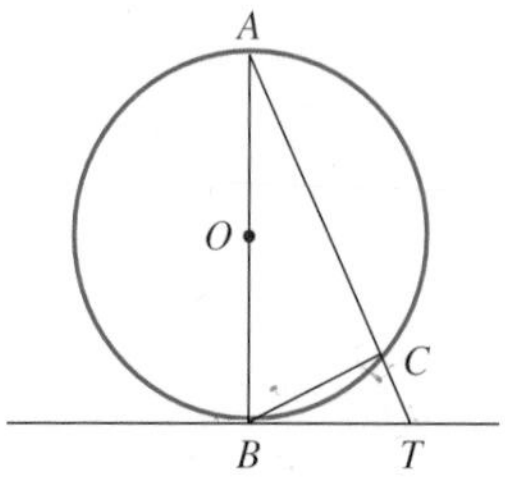

Fig. 2-74

In Exercises 13–16, refer to Fig. 2-75. Determine the indicated arcs and angles.

13. $\overset{\frown}{BC}$
14. $\overset{\frown}{AB}$
15. $\angle ABC$
16. $\angle ACB$

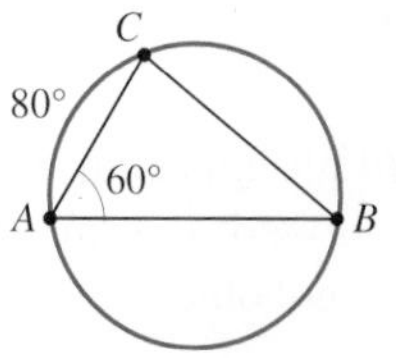

Fig. 2-75

In Exercises 17–20, change the given angles to radian measure.

17. 22.5°
18. 60.0°
19. 125.2°
20. 323.0°

In Exercises 21–24, find a formula for the indicated perimeter or area.

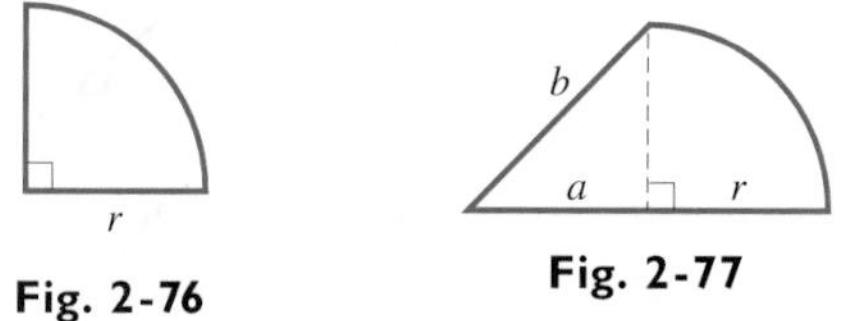

Fig. 2-76 Fig. 2-77

21. The perimeter of the quarter-circle in Fig. 2-76.
22. The perimeter of the figure in Fig. 2-77. A quarter-circle is attached to a triangle.
23. The area of the quarter-circle in Fig. 2-76.
24. The area of the figure in Fig. 2-77.

In Exercises 25–32, solve the given problems.

25. The radius of the earth's equator is 6370 km. What is the circumference?
26. As a ball bearing rolls along a straight track, it makes 11.0 revolutions while traveling a distance of 109 mm. Find its radius.
27. Using a tape measure, the circumference of a tree is found to be 112 cm. What is the diameter of the tree (assuming a circular cross section)?
28. What is the area of the largest circle that can be cut from a rectangular plate 21.2 cm by 15.8 cm?

29. A patio is designed with semicircular areas attached to a square, as shown in Fig. 2-78. Find the area of the patio.

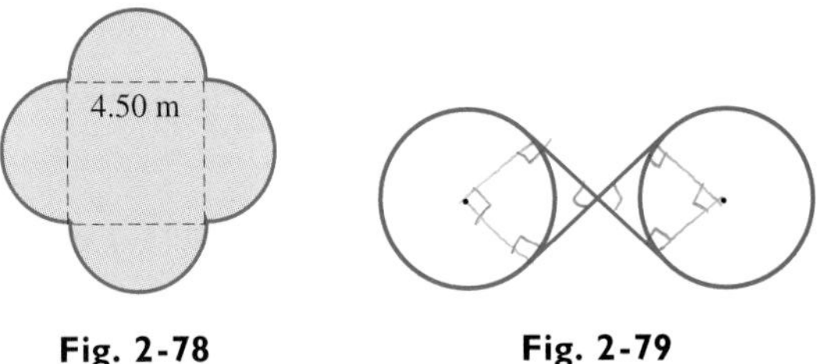

Fig. 2-78 **Fig. 2-79**

30. Find the length of the pulley belt shown in Fig. 2-79, if the belt crosses at right angles. The radius of each pulley wheel is 5.50 cm.

31. The velocity of an object moving in a circular path is directed tangent to the circle in which it is moving. A stone on a string moves in a vertical circle, and the string breaks after 5.5 revolutions. If the string was initially in a vertical position, in what direction does it move after the string breaks?

32. The gear in Fig. 2-80 has 24 teeth. Find the indicated angle.

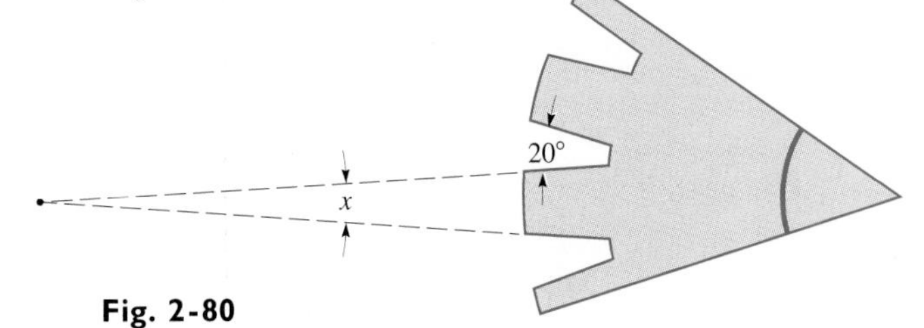

Fig. 2-80

2-5 Measurement of Irregular Areas

To this point the figures for which we have found areas are well defined, and the areas can be found by direct use of a specific formula. In practice, however, it may be necessary to find the area of a figure with an irregular perimeter or one for which there is no specific formula. In this section we show two methods of finding a very good *approximation* of such an area. These methods are particularly useful in technical areas such as surveying, architecture, and mechanical design.

The Trapezoidal Rule

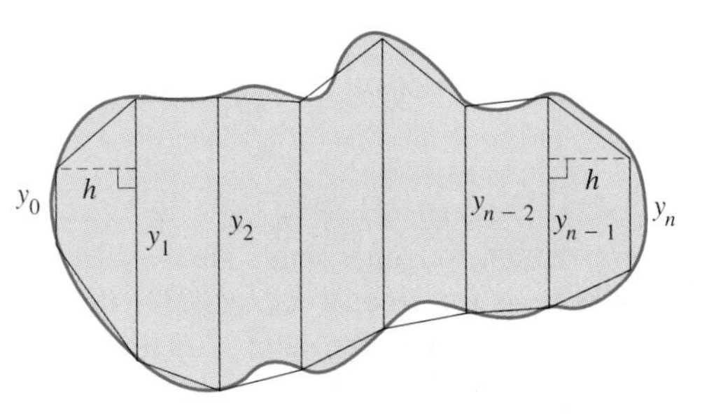

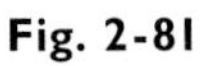
Fig. 2-81

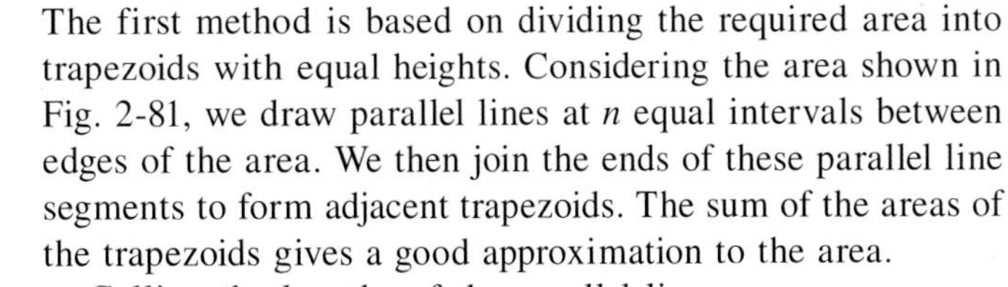

The first method is based on dividing the required area into trapezoids with equal heights. Considering the area shown in Fig. 2-81, we draw parallel lines at n equal intervals between edges of the area. We then join the ends of these parallel line segments to form adjacent trapezoids. The sum of the areas of the trapezoids gives a good approximation to the area.

Calling the lengths of the parallel lines $y_0, y_1, y_2, \ldots, y_n$, and the height of each trapezoid h (the distance between the parallel lines), the total area A is the sum of the areas of all the trapezoids. This gives us

$$A = \underbrace{\frac{h}{2}(y_0 + y_1)}_{\text{first trapezoid}} + \underbrace{\frac{h}{2}(y_1 + y_2)}_{\text{second trapezoid}} + \underbrace{\frac{h}{2}(y_2 + y_3)}_{\text{third trapezoid}} + \cdots + \underbrace{\frac{h}{2}(y_{n-2} + y_{n-1})}_{\text{next-to-last trapezoid}} + \underbrace{\frac{h}{2}(y_{n-1} + y_n)}_{\text{last trapezoid}}$$

$$= \frac{h}{2}(y_0 + y_1 + y_1 + y_2 + y_2 + y_3 + \cdots + y_{n-2} + y_{n-1} + y_{n-1} + y_n)$$

Therefore, the approximate area is

$$A = \frac{h}{2}(y_0 + 2y_1 + 2y_2 + \cdots + 2y_{n-1} + y_n) \qquad \textbf{(2-12)}$$

Equation (2-12) is known as the **trapezoidal rule.** Consider the following examples.

EXAMPLE 1 ▸▸ A plate cam for opening and closing a valve is shown in Fig. 2-82. Widths of the face of the cam are shown at 2.00-cm intervals. Find the area of the face of the cam.

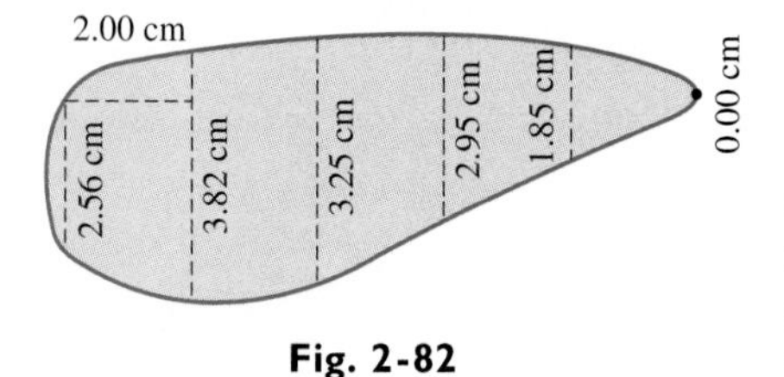

Fig. 2-82

From the figure we see that

$$y_0 = 2.56 \text{ cm}, y_1 = 3.82 \text{ cm}, y_2 = 3.25 \text{ cm}, y_3 = 2.95 \text{ cm},$$
$$y_4 = 1.85 \text{ cm}, y_5 = 0.00 \text{ cm}$$

(In making such measurements, often a y-value at one end (or both ends) is zero. The end "trapezoid" is actually a triangle.) From the given information, $h = 2.00$ cm. Therefore, using the trapezoidal rule, Eq. (2-12), we have

$$A = \frac{2.00}{2}[2.56 + 2(3.82) + 2(3.25) + 2(2.95) + 2(1.85) + 0.00]$$
$$= 26.3 \text{ cm}^2$$

The area of the face of the cam is approximately 26.3 cm^2. ◂◂

See the chapter introduction.

EXAMPLE 2 ▸▸ A surveyor measures the width of a small lake at 250-m intervals from one end of the lake, as shown in Fig. 2-83. The widths found are as follows:

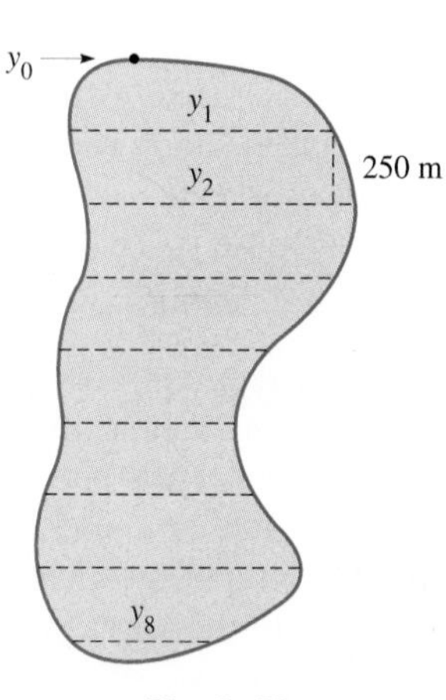

Fig. 2-83

Distance from one end (m)	0	250	500	750	1000	1250	1500	1750	2000
Width (m)	0	940	920	890	740	550	770	960	220

From the table we see that $y_0 = 0$ m, $y_1 = 940$ m, . . . , $y_8 = 220$ m, and $h = 250$ m. Therefore, using the trapezoidal rule, the approximate area of the lake is found as follows:

$$A = \frac{250}{2}[0 + 2(940) + 2(920) + 2(890) + 2(740) + 2(550) + 2(770) + 2(960) + 220]$$
$$= 1\,500\,000 \text{ m}^2$$ ◂◂

When approximating the area with trapezoids, we omit small parts of the area for some trapezoids and include small extra areas for other trapezoids. The omitted areas often approximate the extra areas, which makes the approximation better. Also, the use of smaller intervals improves the approximation since the total omitted area or total extra area is smaller.

Simpson's Rule

Named for the English mathematician Thomas Simpson (1710–1761).

For the second method of measuring an irregular area, we also draw parallel lines at equal intervals between the edges of the area. We then join the ends of these parallel lines with curved *arcs.* This takes into account the fact that the perimeters of most figures are curved. The arcs used in this method are not arcs of a circle, but arcs of a *parabola.* A parabola is shown in Fig. 2-84 and is discussed in detail in Chapter 21. (Examples of parabolas are (1) the path of a ball that has been thrown and (2) the cross section of a microwave "dish.")

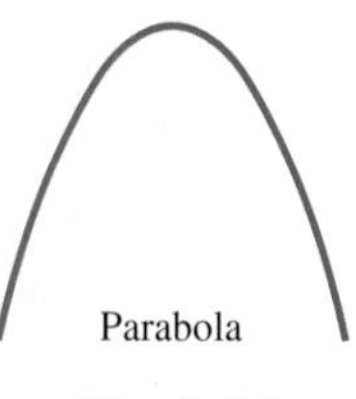
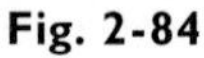

Fig. 2-84

The development of this method requires advanced mathematics. Therefore, we will simply state the formula to be used. It might be noted that the form of the equation is similar to that of the trapezoidal rule.

The approximate area of the geometric figure shown in Fig. 2-85 is given by

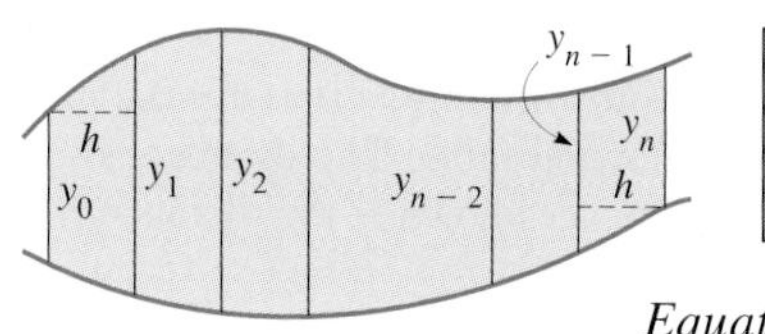

Fig. 2-85

$$A = \frac{h}{3}(y_0 + 4y_1 + 2y_2 + 4y_3 + \cdots + 2y_{n-2} + 4y_{n-1} + y_n) \qquad (2\text{-}13)$$

Equation (2-13) is known as **Simpson's rule.** In using Eq. (2-13), *the number n of intervals of width h must be even.*

EXAMPLE 3 ▸▸ A parking lot is proposed for a riverfront area in a town. The town engineer measured the widths of the area at 30.0-m intervals, as shown in Fig. 2-86. Find the area available for parking.

First we see that there are six intervals, which means Eq. (2-13) may be used. With $y_0 = 124$ m, $y_1 = 147$ m, . . . , $y_6 = 144$ m, and $h = 30.0$ m, we have

$$A = \frac{30.0}{3}[124 + 4(147) + 2(116) + 4(115) + 2(87) + 4(117) + 144]$$

$$= 21\,900 \text{ m}^2 \quad ◂◂$$

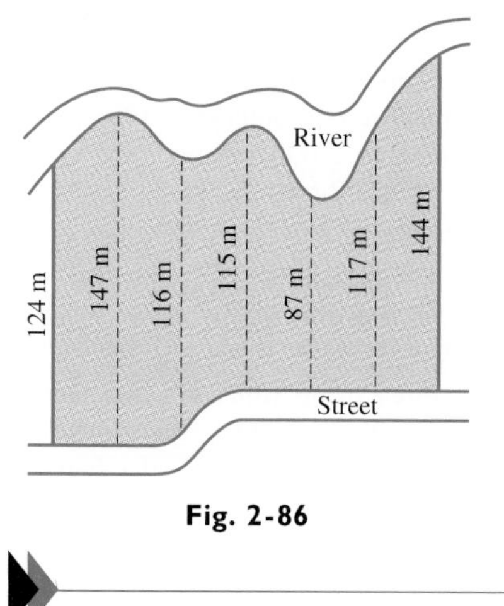

Fig. 2-86

For most areas, Simpson's rule gives a somewhat better approximation than the trapezoidal rule. The accuracy of Simpson's rule is also usually improved by using smaller intervals.

Exercises 2–5

In Exercises 1–12, calculate the indicated areas. All data is accurate to at least two significant digits.

1. The widths of a kidney-shaped swimming pool were measured at 2.0-m intervals as shown in Fig. 2-87. Calculate the surface area of the pool using the trapezoidal rule.

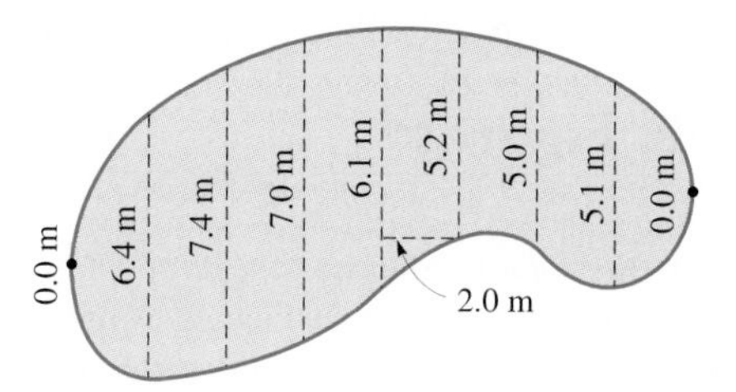

Fig. 2-87

2. Calculate the area of the swimming pool in Fig. 2-87 using Simpson's rule.
3. The widths of a cross section of an airplane wing are measured at 0.30-m intervals as shown in Fig. 2-88. Calculate the area of the cross section using Simpson's rule.

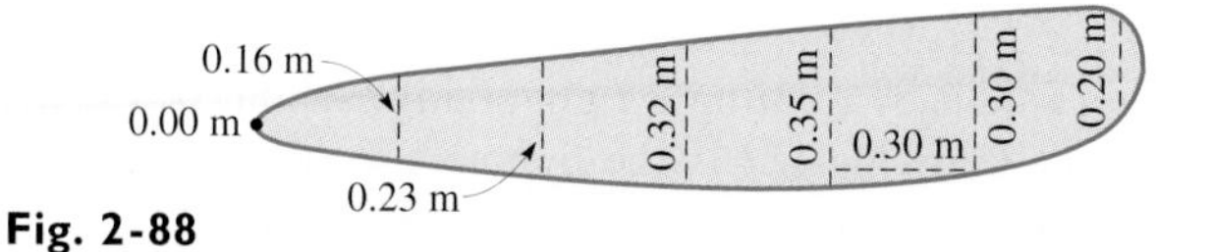

Fig. 2-88

4. Calculate the area of the cross section of the airplane wing in Fig. 2-88 using the trapezoidal rule.
5. Using aerial photography, the widths of an area burned by a forest fire were measured at 0.5-km intervals as shown in the following table.

Distance (km)	0.0	0.5	1.0	1.5	2.0	2.5	3.0	3.5	4.0
Width (km)	0.6	2.2	4.7	3.1	3.6	1.6	2.2	1.5	0.8

Determine the area burned by the fire by using the trapezoidal rule.

6. Find the area burned by the forest fire of Exercise 5 by using Simpson's rule.
7. The widths of a metal plate were measured at 5.0-cm intervals as shown in the following table.

Distance (cm)	0.0	5.0	10.0	15.0	20.0	25.0	30.0	35.0	40.0	45.0
Width (cm)	4.8	6.9	12.0	14.5	15.5	14.5	12.0	12.5	15.2	10.0

Find the area of the plate.

8. The widths of an oval-shaped floor were measured at 1.5-m intervals as shown in the following table.

Distance (m)	0.0	1.5	3.0	4.5	6.0	7.5	9.0	10.5	12.0
Width (m)	0.0	5.0	7.2	8.3	8.6	8.3	7.2	5.0	0.0

Find the area of the floor by using Simpson's rule.

9. The widths of the baseball playing area in Boston's Fenway Park at 14-m intervals are shown in Fig. 2-89. Find the playing area using the trapezoidal rule.

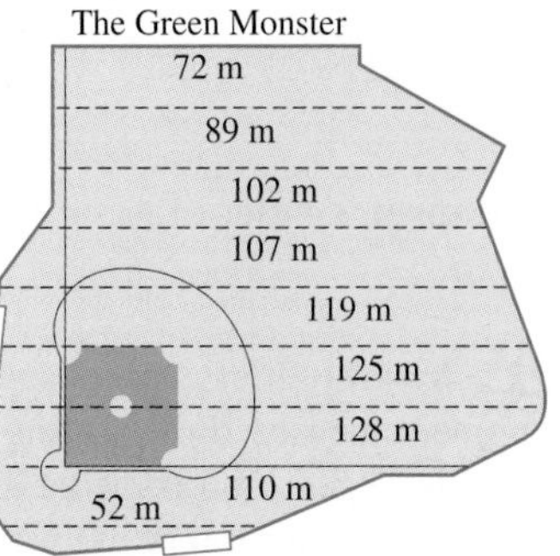

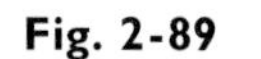

Fig. 2-89

10. Find the playing area of Fenway Park (see Exercise 9) by Simpson's rule.

11. Soundings taken across a river channel give the following values of distance from one shore with the corresponding depth of the channel.

Distance (m)	0	50	100	150	200	250	300	350	400	450	500
Depth (m)	5	12	17	21	22	25	26	16	10	8	0

Find the area of the channel using Simpson's rule.

12. The widths of a bell crank are measured at 2.0-cm intervals as shown in Fig. 2-90. Find the area of the bell crank if the two connector holes are each 2.50 cm in diameter.

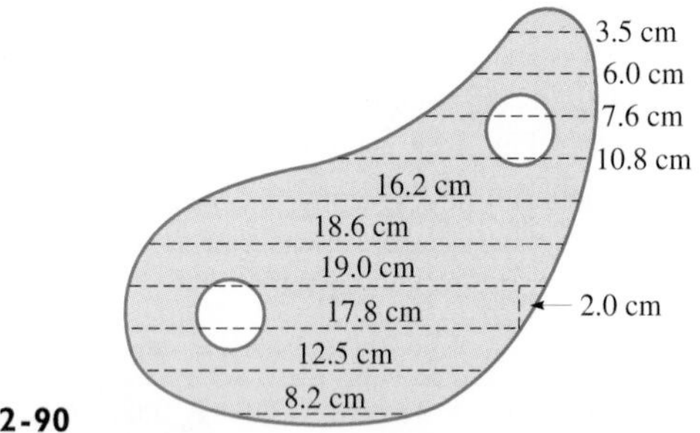

Fig. 2-90

In Exercises 13–16, calculate the area of the circle by the indicated method.

The lengths of parallel chords of a circle that are 0.250 cm apart are given in the following table. The diameter of the circle is 2.000 cm. The distance shown is the distance from one end of a diameter.

Distance (cm)	0.000	0.250	0.500	0.750	1.000	1.250	1.500	1.750	2.000
Length (cm)	0.000	1.323	1.732	1.936	2.000	1.936	1.732	1.323	0.000

Using the formula $A = \pi r^2$, the area of the circle is 3.14 cm^2.

13. Find the area of the circle using the trapezoidal rule and only the values of distance of 0.000 cm, 0.500 cm, 1.000 cm, 1.500 cm, and 2.000 cm with the corresponding values of the chord lengths. Explain why the value found is less than 3.14 cm^2.

14. Find the area of the circle using the trapezoidal rule and all values in the table. Explain why the value found is closer to 3.14 cm^2 than the value found in Exercise 13.

15. Find the area of the circle using Simpson's rule and the same table values as in Exercise 13. Explain why the value found is closer to 3.14 cm^2 than the value found in Exercise 13.

16. Find the area of the circle using Simpson's rule and all values in the table. Explain why the value found is closer to 3.14 cm^2 than the value found in Exercise 15.

2-6 Solid Geometric Figures

We now review the formulas for the *volume* and *surface area* of some basic solid geometric figures. Just as area is a measure of the surface of a plane geometric figure, **volume** is a measure of the space occupied by a solid geometric figure.

One of the most common solid figures is the **rectangular solid**. This figure has six sides (**faces**), and opposite sides are rectangles. All intersecting sides are perpendicular to each other. The **bases** of the rectangular solid are the top and bottom faces. A **cube** is a rectangular solid with all six faces being equal squares.

A **right circular cylinder** is *generated* by rotating a rectangle about one of its sides. Each **base** is a circle, and the *cylindrical surface* is perpendicular to each of the bases. The **height** is one side of the rectangle, and the **radius** of the base is the other side.

A **right circular cone** is generated by rotating a right triangle about one of its legs. The **base** is a circle, and the **slant height** is the hypotenuse of the right triangle. The **height** is one leg of the right triangle, and the **radius** of the base is the other leg.

The bases of a **right prism** are equal and parallel polygons, and the sides are rectangles. The **height** of a prism is the perpendicular distance between bases. The base of a **pyramid** is a polygon, and the other faces, the **lateral faces**, are triangles that meet at a common point, the **vertex**. A **regular pyramid** has congruent triangles for its lateral faces.

A **sphere** is generated by rotating a circle about a diameter. The **radius** is a line segment joining the center and a point on the sphere. The **diameter** is a line segment through the center and having its endpoints on the sphere.

In the following formulas, V represents the *volume*, A represents the *total surface area*, S represents the *lateral surface area* (bases not included), B represents the *area of the base*, and p represents the *perimeter of the base.*

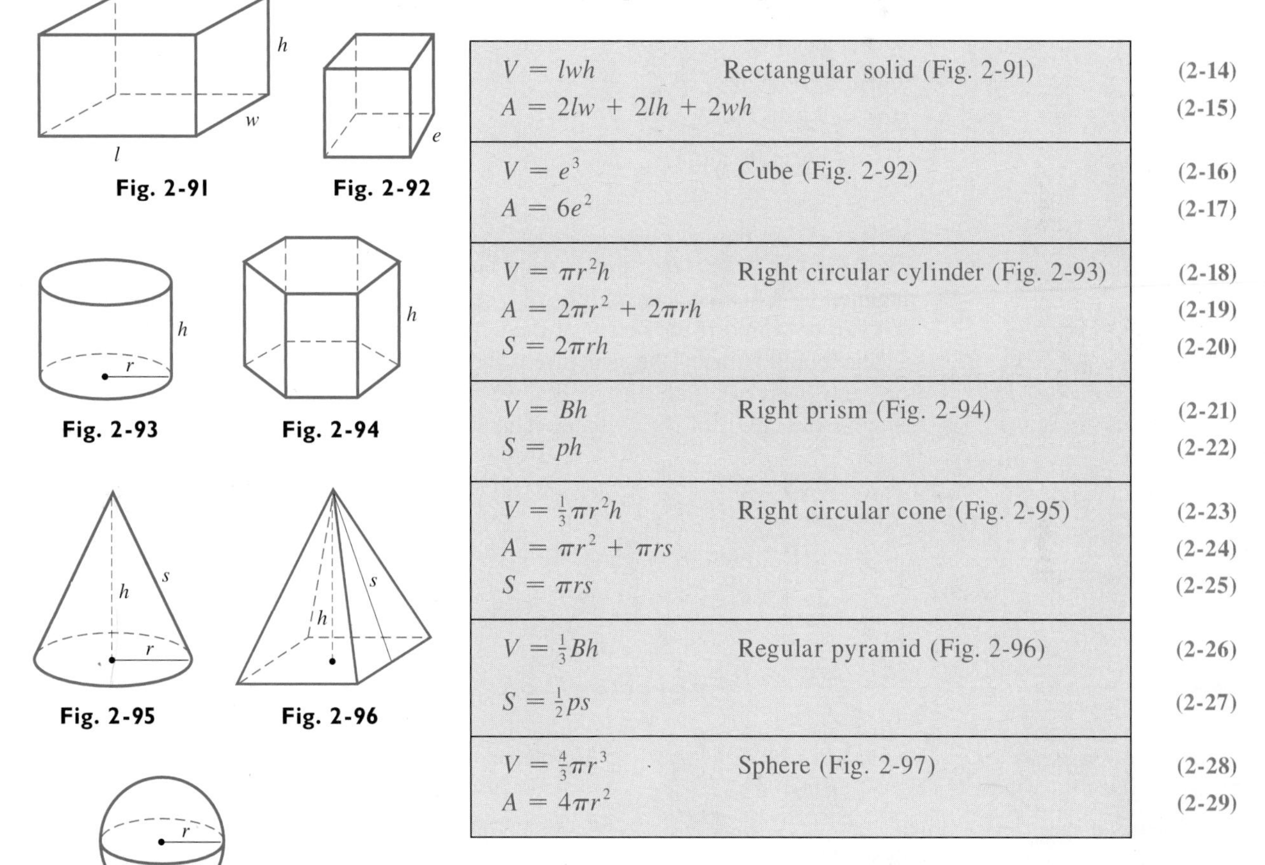

Fig. 2-91 Fig. 2-92 Fig. 2-93 Fig. 2-94 Fig. 2-95 Fig. 2-96

$V = lwh$	Rectangular solid (Fig. 2-91)	(2-14)
$A = 2lw + 2lh + 2wh$		(2-15)
$V = e^3$	Cube (Fig. 2-92)	(2-16)
$A = 6e^2$		(2-17)
$V = \pi r^2 h$	Right circular cylinder (Fig. 2-93)	(2-18)
$A = 2\pi r^2 + 2\pi rh$		(2-19)
$S = 2\pi rh$		(2-20)
$V = Bh$	Right prism (Fig. 2-94)	(2-21)
$S = ph$		(2-22)
$V = \frac{1}{3}\pi r^2 h$	Right circular cone (Fig. 2-95)	(2-23)
$A = \pi r^2 + \pi rs$		(2-24)
$S = \pi rs$		(2-25)
$V = \frac{1}{3}Bh$	Regular pyramid (Fig. 2-96)	(2-26)
$S = \frac{1}{2}ps$		(2-27)
$V = \frac{4}{3}\pi r^3$	Sphere (Fig. 2-97)	(2-28)
$A = 4\pi r^2$		(2-29)

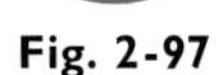

Fig. 2-97

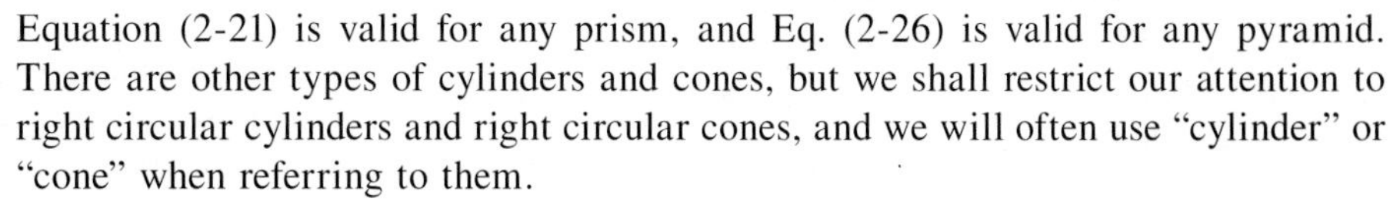

Equation (2-21) is valid for any prism, and Eq. (2-26) is valid for any pyramid. There are other types of cylinders and cones, but we shall restrict our attention to right circular cylinders and right circular cones, and we will often use "cylinder" or "cone" when referring to them.

The **frustum** of a cone or pyramid is the solid figure that remains after the top is cut off by a plane parallel to the base. Figure 2-98 shows the frustum of a cone.

Fig. 2-98

EXAMPLE 1 ▸▸ What volume of concrete is needed for a driveway 25.0 m long, 2.75 m wide, and 0.100 m thick?

The driveway is a rectangular solid for which $l = 25.0$ m, $w = 2.75$ m, and $h = 0.100$ m. Using Eq. (2-14), we have

$V = (25.0)(2.75)(0.100) = 6.88 \text{ m}^3$ ◂◂

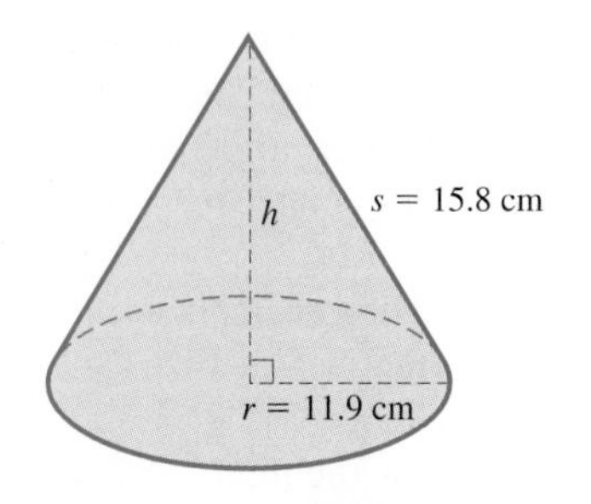

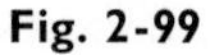
Fig. 2-99

EXAMPLE 2 ▸▸ Calculate the volume of a right circular cone for which the radius $r = 11.9$ cm and the *slant height* $s = 15.8$ cm. See Fig. 2-99.

To find the volume using Eq. (2-23), we need the radius and height of the cone. Therefore, we must first find the height. As noted, the radius and height are the legs of a right triangle. and the slant height is the hypotenuse. To find the height we *use the Pythagorean theorem.*

$$
\begin{aligned}
s^2 &= r^2 + h^2 && \text{Pythagorean theorem} \\
h^2 &= s^2 - r^2 && \text{solve for } h \\
h &= \sqrt{s^2 - r^2} \\
&= \sqrt{15.8^2 - 11.9^2} = 10.4 \text{ cm}
\end{aligned}
$$

Now calculating the volume (in using a calculator it is not necessary to record the value of h), we have

$$
\begin{aligned}
V &= \tfrac{1}{3}\pi r^2 h && \text{Eq. (2-23)} \\
&= \tfrac{1}{3}\pi(11.9^2)(10.4) && \text{substituting} \\
&= 1540 \text{ cm}^3
\end{aligned}
$$

◂◂

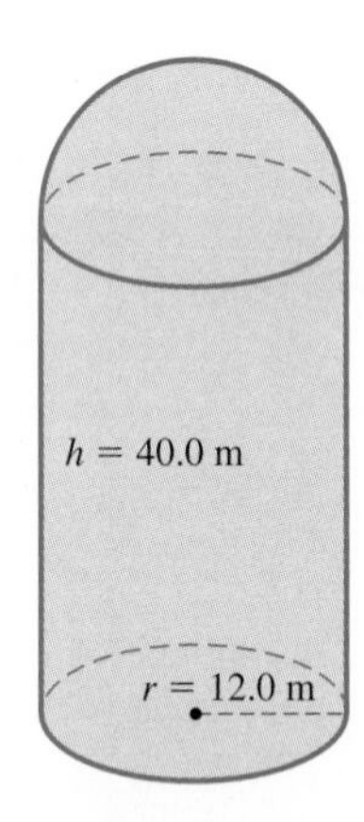

Fig. 2-100

EXAMPLE 3 ▸▸ A grain storage building is in the shape of a cylinder surmounted by a hemisphere (*half a sphere*). See Fig. 2-100. Find the volume of grain that can be stored if the height of the cylinder is 40.0 m and its radius is 12.0 m.

The total volume of the structure is the volume of the cylinder plus the volume of the hemisphere. By the construction we see that the radius of the hemisphere is the same as the radius of the cylinder. Therefore,

$$
\begin{aligned}
V &= \underbrace{\pi r^2 h}_{\text{cylinder}} + \underbrace{\tfrac{1}{2}\left(\tfrac{4}{3}\pi r^3\right)}_{\text{hemisphere}} = \pi r^2 h + \tfrac{2}{3}\pi r^3 \\
&= \pi(12.0)^2(40.0) + \tfrac{2}{3}\pi(12.0)^3 \\
&= 21\,700 \text{ m}^3
\end{aligned}
$$

◂◂

Exercises 2–6

In Exercises 1–16, find the volume or area of each solid figure for the given values. See Figs. 2-91 to 2-97.

1. Volume of cube: $e = 7.15$ dm
2. Volume of right circular cylinder: $r = 23.5$ cm, $h = 48.4$ cm
3. Total surface area of right circular cylinder: $r = 6.89$ m, $h = 2.33$ m
4. Area of sphere: $r = 67$ mm
5. Volume of sphere: $r = 0.877$ m
6. Volume of right circular cone: $r = 25.1$ mm, $h = 5.66$ mm
7. Lateral area of right circular cone: $r = 78.0$ cm, $s = 83.8$ cm
8. Lateral area of regular pyramid: $p = 3.45$ m, $s = 2.72$ m
9. Volume of regular pyramid: square base of side 16 dm, $h = 13$ dm
10. Volume of right prism: square base of side 29.0 cm, $h = 11.2$ cm
11. Lateral area of regular prism: equilateral triangle base of side 1.092 m, $h = 1.025$ m
12. Lateral area of right circular cylinder: diameter = 25.0 mm, $h = 34.7$ mm
13. Volume of hemisphere: diameter = 0.83 cm
14. Volume of regular pyramid: square base of side 22.4 m, $s = 14.2$ m
15. Total surface area of right circular cone: $r = 0.339$ cm, $h = 0.274$ cm
16. Total surface area of pyramid: all faces and base are equilateral triangles of side 3.67 dm

In Exercises 17–24, find the indicated areas and volumes.

17. A rectangular box is to be used to store radioactive materials. The inside of the box is 12.0 cm long, 9.50 cm wide, and 8.75 cm deep. What is the area of sheet lead that must be used to line the inside of the box?

18. A swimming pool is 15.0 m wide, 24.0 m long, 1.00 m deep at one end, and 2.60 m deep at the other end. How many cubic metres of water can it hold? (The slope on the bottom is constant.)

19. The Alaskan oil pipeline is 1200 km long and has a diameter of 1.2 m. What is the maximum volume of the pipeline?

20. A glass prism used in the study of optics has a right triangular base. The legs of the triangle are 3.00 cm and 4.00 cm. The prism is 8.50 cm high. What is the total surface area of the prism? See Fig. 2-101.

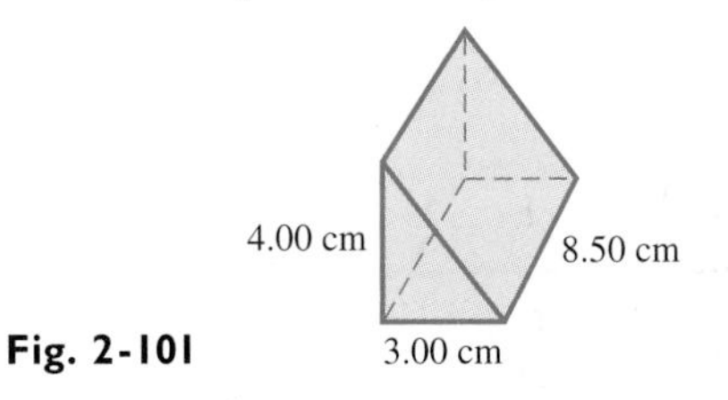

Fig. 2-101

21. The Great Pyramid of Egypt has a square base approximately 230 m on a side. The height of the pyramid is about 150 m. What is its volume?

22. A paper cup is in the shape of a cone with radius 4.60 cm and height 8.90 cm. What is the surface area of the cup?

23. *Spaceship Earth* (shown in Fig. 2-102) at Epcot Center in Florida is a sphere of 50.3 m in diameter. What is the volume of *Spaceship Earth*?

Fig. 2-102

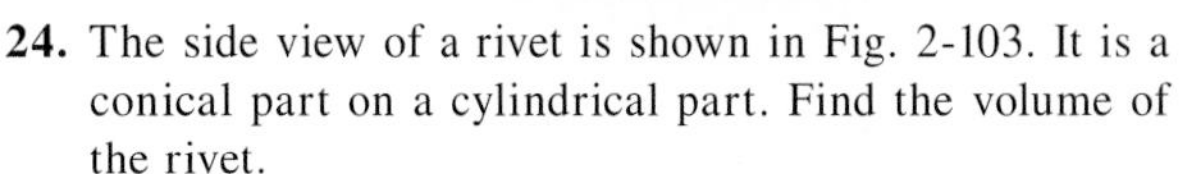

24. The side view of a rivet is shown in Fig. 2-103. It is a conical part on a cylindrical part. Find the volume of the rivet.

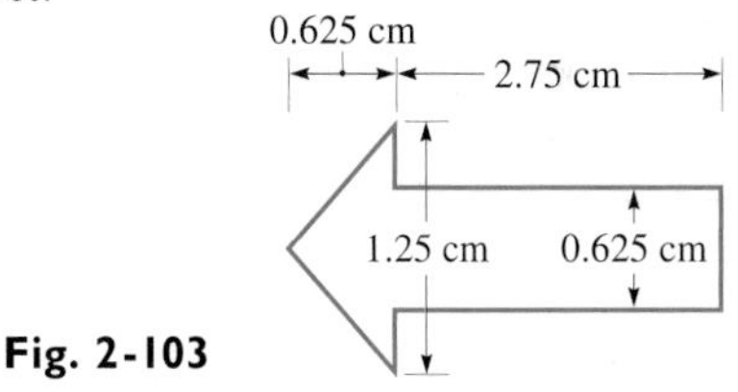

Fig. 2-103

Chapter Equations, Review Exercises, and Practice Test

Chapter Equations

Line segments	Fig. 2-8	$\frac{a}{b} = \frac{c}{d}$	(2-1)
Triangle		$A = \frac{1}{2}bh$	(2-2)
Hero's formula		$A = \sqrt{s(s-a)(s-b)(s-c)}$ where $s = \frac{1}{2}(a+b+c)$	(2-3)
Pythagorean theorem	Fig. 2-29	$c^2 = a^2 + b^2$	(2-4)
Square	Fig. 2-51	$A = s^2$	(2-5)
Rectangle	Fig. 2-52	$A = lw$	(2-6)
Parallelogram	Fig. 2-53	$A = bh$	(2-7)
Trapezoid	Fig. 2-54	$A = \frac{1}{2}h(b_1 + b_2)$	(2-8)
Circle		$c = 2\pi r$	(2-9)
		$A = \pi r^2$	(2-10)

Radians	Fig. 2-73	$\pi \text{ rad} = 180°$	(2-11)
Trapezoidal rule	Fig. 2-81	$A = \frac{h}{2}(y_0 + 2y_1 + 2y_2 + \cdots + 2y_{n-1} + y_n)$	(2-12)
Simpson's rule	Fig. 2-85	$A = \frac{h}{3}(y_0 + 4y_1 + 2y_2 + 4y_3 + \cdots + 2y_{n-2} + 4y_{n-1} + y_n)$	(2-13)
Rectangular solid	Fig. 2-91	$V = lwh$	(2-14)
		$A = 2lw + 2lh + 2wh$	(2-15)
Cube	Fig. 2-92	$V = e^3$	(2-16)
		$A = 6e^2$	(2-17)
Right circular cylinder	Fig. 2-93	$V = \pi r^2 h$	(2-18)
		$A = 2\pi r^2 + 2\pi rh$	(2-19)
		$S = 2\pi rh$	(2-20)
Right prism	Fig. 2-94	$V = Bh$	(2-21)
		$S = ph$	(2-22)
Right circular cone	Fig. 2-95	$V = \frac{1}{3}\pi r^2 h$	(2-23)
		$A = \pi r^2 + \pi rs$	(2-24)
		$S = \pi rs$	(2-25)
Regular pyramid	Fig. 2-96	$V = \frac{1}{3}Bh$	(2-26)
		$S = \frac{1}{2}ps$	(2-27)
Sphere	Fig. 2-97	$V = \frac{4}{3}\pi r^3$	(2-28)
		$A = 4\pi r^2$	(2-29)

Review Exercises

In Exercises 1–4, use Fig. 2-104. Determine the indicated angles.

1. $\angle CGE$ **2.** $\angle EGF$ 3. $\angle DGH$ 4. $\angle EGI$

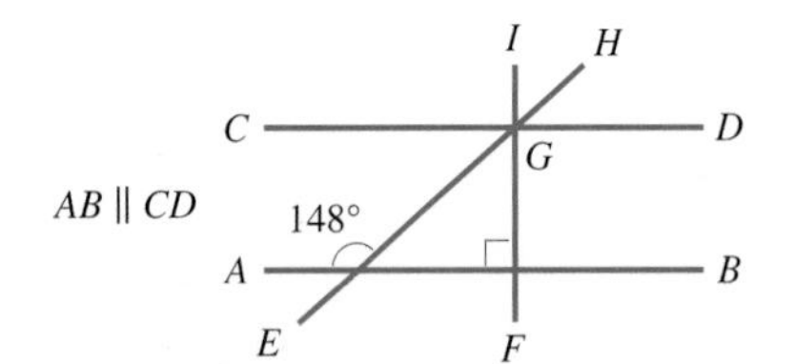

Fig. 2-104

In Exercises 5–12, find the indicated sides of the right triangle shown in Fig. 2-105.

5. $a = 9, b = 40, c = ?$

6. $a = 14, b = 48, c = ?$

7. $a = 40, c = 58, b = ?$

8. $b = 56, c = 65, a = ?$

9. $a = 6.30, b = 3.80, c = ?$

Fig. 2-105

10. $a = 126, b = 251, c = ?$

11. $b = 29.3, c = 36.1, a = ?$

12. $a = 0.782, c = 0.885, b = ?$

In Exercises 13–20, find the perimeter or area of the indicated figure.

13. Perimeter: equilateral triangle of side 8.5 mm

14. Perimeter: rhombus of side 15.2 cm

15. Area: triangle, $b = 3.25$ m, $h = 1.88$ m

16. Area: triangle of sides 17.5 cm, 13.8 cm, 11.9 cm

17. Circumference of circle: $d = 98.4$ mm

18. Perimeter: rectangle, $l = 2.98$ dm, $w = 1.86$ dm

19. Area: trapezoid, $b_1 = 67.2$ cm, $b_2 = 83.8$ cm, $h = 34.2$ cm

20. Area: circle, $d = 32.8$ m

In Exercises 21–24, find the volume of the indicated solid geometric figure.

21. Prism: base is right triangle with legs 26.0 cm and 34.0 cm, height is 14.0 cm

22. Cylinder: base radius 36.0 cm, height 24.0 cm

23. Pyramid: base area 3850 m^2, height 125 m

24. Sphere: diameter 22.1 mm

In Exercises 25–28, find the surface area of the indicated solid geometric figure.

25. Total area of cube of edge 5.20 m

26. Total area of cylinder: base diameter 1.20 cm, height 5.80 cm

27. Lateral area of cone: base radius 18.2 mm, height 11.5 mm

28. Total area of sphere: diameter 0.884 m

In Exercises 29–32, use Fig. 2-106. Line CT is tangent to the circle with center at O. Find the indicated angles.

29. $\angle BTA$

30. $\angle TAB$

31. $\angle BTC$

32. $\angle ABT$

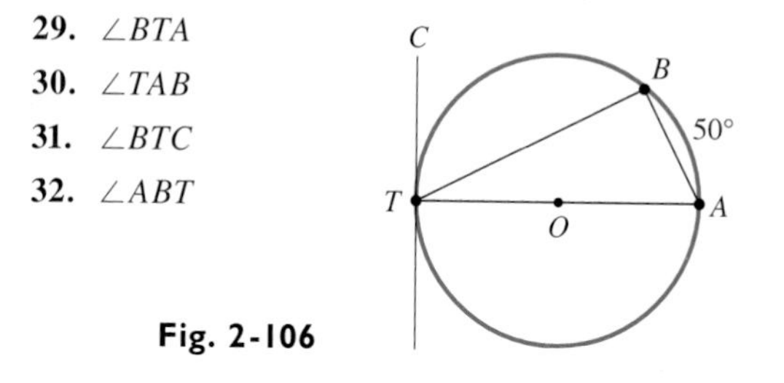

Fig. 2-106

In Exercises 33–36, use Fig. 2-107. Given that $AB = 4$, $BC = 4$, $CD = 6$, and $\angle ADC = 53°$, find the indicated angle and lengths.

33. $\angle ABE$

34. AD

35. BE

36. AE

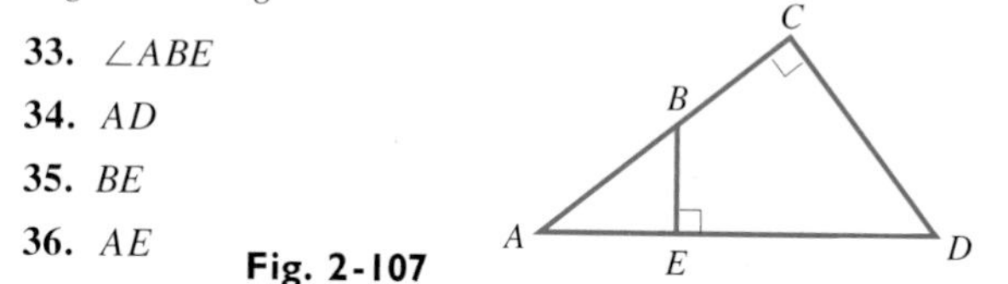

Fig. 2-107

In Exercises 37–40, find the formulas for the indicated perimeters and areas.

37. Perimeter of Fig. 2-108 (a right triangle and semicircle attached)

38. Perimeter of Fig. 2-109 (a square with a quarter circle at each end)

39. Area of Fig. 2-108

40. Area of Fig. 2-109

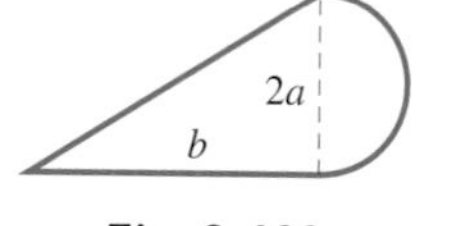

Fig. 2-108

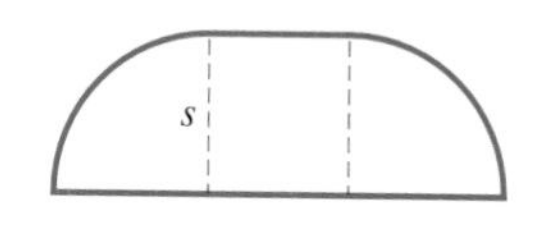

Fig. 2-109

In Exercises 41–56, solve the given problems.

41. A ramp for the disabled is designed so that it rises 1.2 m over a horizontal distance of 7.8 m. How long is the ramp?

42. An airplane is 640 m directly above one end of a 3200-m runway. How far is the plane from the glide-slope indicator on the ground at the other end of the runway?

43. A radio transmitting tower is supported by guy wires. The tower and three parallel guy wires are shown in Fig. 2-110. Find distance AB along the tower.

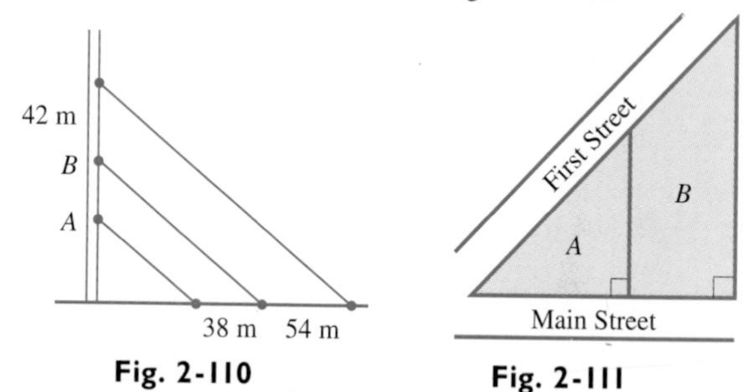

Fig. 2-110 **Fig. 2-111**

44. Find the areas of lots A and B in Fig. 2-111. A has a frontage on Main Street of 140 m, and B has a frontage on Main Street of 84 m. The boundary between lots is 120 m.

45. Find the area of the side of the building shown in Fig. 2-112 (a triangle over a rectangle).

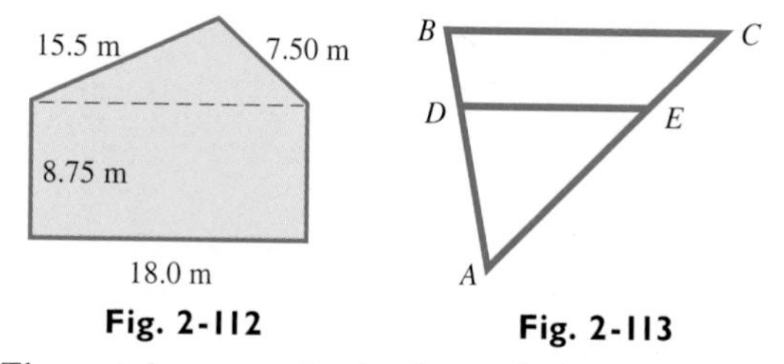

Fig. 2-112 **Fig. 2-113**

46. The metal support in the form of $\triangle ABC$ shown in Fig. 2-113 is strengthened by brace DE, which is parallel to BC. How long is the brace if $AB = 24$ cm, $AD = 16$ cm, and $BC = 33$ cm?

47. A typical scale for an aerial photograph is 1/18 450. In a 20.0-cm-by-25.0-cm photograph with this scale, what is the longest distance in kilometres between two locations in the photograph?

48. For a hydraulic press, the mechanical advantage is the ratio of the large piston area to the small piston area. Find the mechanical advantage if the pistons have diameters of 3.10 cm and 2.25 cm.

49. The diameter of the earth is 12 700 km, and a satellite is in orbit at an altitude of 340 km. How far does the satellite travel in one rotation about the earth?

50. The roof of the Louisiana Superdome in New Orleans is supported by a circular steel tension ring 651 m in circumference. Find the area covered by the roof.

51. A rectangular piece of wallboard is 2400 mm long and 1200 mm wide. Two 350-mm diameter holes are cut out for heating ducts. What is the area of the remaining piece?

52. The diameter of the sun is 1.38×10^6 km, the diameter of the earth is 1.27×10^4 km, and the distance from the earth to the sun (center to center) is 1.50×10^8 km. What is the distance from the center of the earth to the end of the shadow due to the rays of the sun?

53. Using aerial photography, the width of an oil spill is measured at 250-m intervals as shown in Fig. 2-114. Using Simpson's rule, find the area of the oil spill.

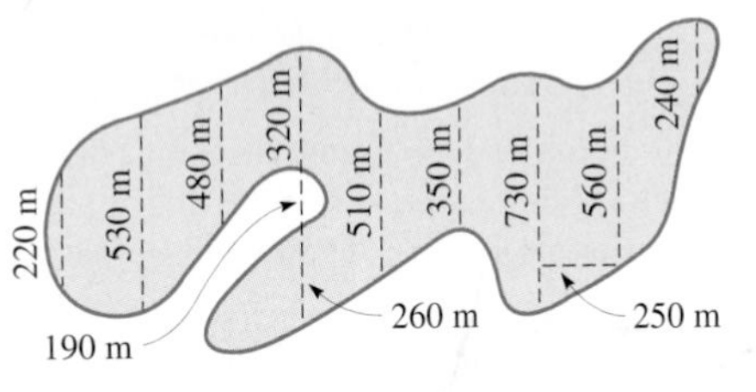

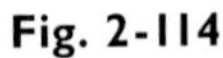
Fig. 2-114

54. To build a highway, it is necessary to cut through a hill. A surveyor measured the cross-sectional areas at 250-m intervals through the cut as shown in the following table. Using the trapezoidal rule, determine the volume of soil to be removed.

dist. (m)	0	250	500	750	1000	1250	1500	1750
area (m^2)	560	1780	4650	6730	5600	6280	2260	230

55. A hot-water tank is in the shape of a right circular cylinder surmounted by a hemisphere. The total height of the tank is 2.05 m, and the diameter of the base is 0.760 m. How many litres does the tank hold?

56. A tent is in the shape of a regular pyramid surmounted on a cube. If the edge of the cube is 2.50 m and the total height of the tent is 3.25 m, find the area of the material used in making the tent (not including any floor area).

Writing Exercise

57. The Pentagon, headquarters of the U.S. Department of Defense, is one of the world's largest office buildings. It is a regular pentagon (five sides, all equal in length, and all internal angles are equal) 281 m on a side, with a diagonal of length 454 m. Using these data, draw a sketch, and write one or two paragraphs to explain how to find the area covered within the outside perimeter of the Pentagon. (What is the area?)

Practice Test

1. In Fig. 2-115, determine $\angle 1$.

2. In Fig. 2-115, determine $\angle 2$.

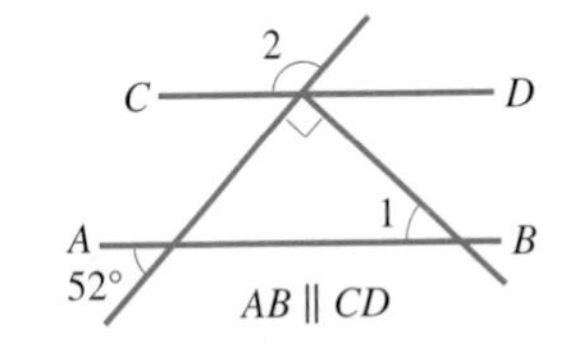

Fig. 2-115

3. A tree is 2.4 m high and casts a shadow 3.0 m long. At the same time, a telephone pole casts a shadow 7.6 m long. How tall is the pole?

4. Find the area of a triangle with sides of 2.46 cm, 3.65 cm, and 4.07 cm.

5. What is the diagonal distance along the floor between corners of a rectangular room 3810 mm wide and 5180 mm long?

6. Find the surface area of a tennis ball whose circumference is 21.0 cm.

7. Find the volume of a right circular cone of radius 2.08 m and height 1.78 m.

8. In Fig. 2-116, find $\angle 1$.

9. In Fig. 2-116, find $\angle 2$.

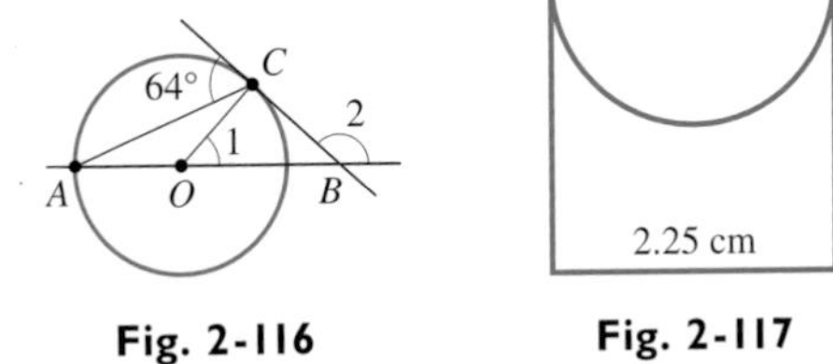

Fig. 2-116 **Fig. 2-117**

10. In Fig. 2-117, find the perimeter of the figure shown. It is a square, with a semicircle removed.

11. In Fig. 2-117, find the area of the figure shown.

12. The width of a marshy area is measured at 50-m intervals, with the results shown in the following table. Using the trapezoidal rule, find the area of the marsh. (All data accurate to two or more significant digits.)

distance (m)	0	50	100	150	200	250	300
width (m)	0	90	145	260	205	110	20

3

Functions and Graphs

In technology and science, as well as in everyday life, we see that one quantity depends on one or more other quantities. Plant growth depends on sunlight and rainfall; traffic flow depends on the roadway design; the sales tax on an item depends on the cost of the item. These are but a few of the innumerable possible examples.

Determining how one quantity depends on other quantities is one of the primary goals of science. A rule that relates such quantities is of great importance and usefulness in science and technology. In mathematics such a rule is called a *function*, and we will start this chapter with a discussion of functions.

A way of actually seeing how one quantity depends on another is by means of a *graph*. The basic method of drawing and using graphs is also taken up in this chapter.

The electric power produced in a circuit depends on the resistance R in the circuit. In Section 3-4 we draw a graph of $P = \frac{100R}{(0.50 + R)^2}$ *to see this type of relationship.*

3-1 Introduction to Functions

In most of the formulas of Chapter 1, one quantity was given in terms of one or more other quantities. It is obvious, then, that the various quantities are related by means of the formula. One important method of finding such formulas is through scientific observation and experimentation.

If we were to perform an experiment to determine whether or not a relationship exists between the distance an object drops and the time it falls, observation of the results would indicate (approximately, at least) that $s = 4.9t^2$, where s is the distance in metres and t is the time in seconds. We would therefore see that distance and time for a falling object are related.

A similar study of the pressure and the volume of a gas at constant temperature would show that as pressure increases, volume decreases according to the formula $PV = k$, where k is a constant. Electrical measurements of current and voltage with respect to a particular resistor would show that $V = kI$, where V is the voltage, I is the current, and k is a constant.

Considerations such as these lead us to one of the most important and basic concepts in mathematics.

Definition of a Function

> *Whenever a relationship exists between two variables such that for every value of the first, there is only one corresponding value of the second, we say that the second variable is a* **function** *of the first variable.*
>
> *The first variable is called the* **independent variable**, *and the second variable is called the* **dependent variable**.

The first variable is termed *independent* since permissible values can be assigned to it arbitrarily, and the second variable is termed *dependent* since its value is determined by the choice of the independent variable. *Values of the independent variable and dependent variable are to be real numbers.* Therefore, there may be restrictions on their possible values. This is discussed in the following section.

EXAMPLE 1 ▸▸ In the equation $y = 2x$, we see that y is a function of x, since for each value of x there is only one value of y. For example, if we substitute $x = 3$, we get $y = 6$ and no other value. By arbitrarily assigning values to x and then substituting, we see that the values of y we obtain *depend* on the values chosen for x. Therefore, x is the independent variable and y is the dependent variable. ◂◂

EXAMPLE 2 ▸▸ The power P developed in a certain resistor by a current I is given by $P = 4I^2$. Here P is a function of I. The dependent variable is P, and the independent variable is I. ◂◂

EXAMPLE 3 ▸▸ Figure 3-1 shows a cube of edge e. In order to express the volume V as a function of the edge e, we recall from Chapter 2 that $V = e^3$. Here V is a function of e, since for each value of e there is only one value of V. The dependent variable is V, and the independent variable is e.

If the equation relating the volume and the edge of a cube is written as $e = \sqrt[3]{V}$—that is, if the edge is expressed in terms of the volume—e is a function of V. In this case, e is the dependent variable and V is the independent variable. ◂◂

e

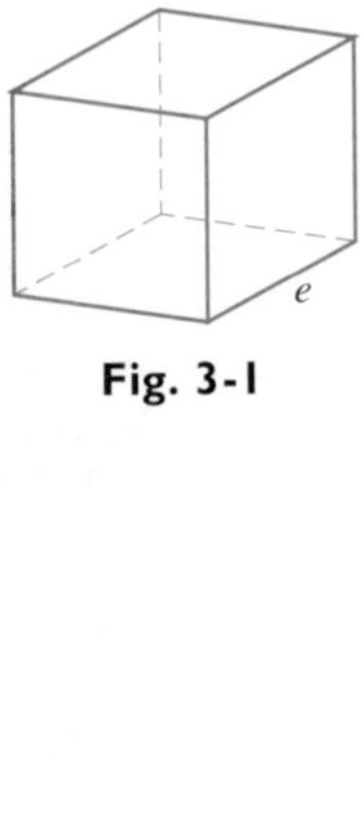

Fig. 3-1

There are many ways to express functions. Formulas such as those we have discussed define functions. Other ways to express functions are by means of tables, charts, and graphs. Functions will be of importance throughout the book, and we will use a number of different types of functions in later chapters.

Functional Notation

For convenience of notation, the phrase "*function of x*" *is written as* $f(x)$. This means that "y is a function of x" may be written as $y = f(x)$. In this form, y and x may represent quantities such as volume and edge, or power and current. However, the f denotes *dependence* and does not represent a quantity. Thus,

CAUTION ▸ ***f(x) does not mean f times x.***

EXAMPLE 4 ▸▸ If $y = 6x^3 - 5x$, we say that y is a function of x. This function is $6x^3 - 5x$. It is also common to write such a function as $f(x) = 6x^3 - 5x$. However, y and $f(x)$ represent the same expression, $6x^3 - 5x$. Using y, the quantities are shown, and using $f(x)$, the functional dependence is shown. ◂◂

One of the most important uses of functional notation is to designate the value of the function for a particular value of the independent variable. That is,

the value of the function $f(x)$ when $x = a$ is written as $f(a)$.

This is illustrated in the next example.

EXAMPLE 5 ▸▸ For the function $f(x) = 3x - 7$, the value of $f(x)$ for $x = 2$ may be expressed as $f(2)$. Thus, substituting 2 for x, we have

$$f(2) = 3(2) - 7 = -1 \qquad \text{substitute 2 for } x$$

The value of $f(x)$ for $x = -1.4$ is

$$f(-1.4) = 3(-1.4) - 7 = -11.2 \qquad \text{substitute } -1.4 \text{ for } x$$ ◂◂

At times we need to define more than one function of the independent variable. We then use different symbols to denote the different functions. For example, $f(x)$ and $g(x)$ may represent different functions, such as $f(x) = 5x - 3$ and $g(x) = ax^2 + x$ (where a is a constant).

EXAMPLE 6 ▸▸ For $f(x) = 5x - 3$ and $g(x) = ax^2 + x$, we have

$$f(-4) = 5(-4) - 3 = -23 \qquad \text{substitute } -4 \text{ for } x \text{ in } f(x)$$

and

$$g(-4) = a(-4)^2 + (-4) = 16a - 4 \qquad \text{substitute } -4 \text{ for } x \text{ in } g(x)$$ ◂◂

There are occasions when we wish to evaluate a function for a literal number value rather than an explicit number. However, whatever number a represents in $f(a)$, we substitute a for x in $f(x)$.

EXAMPLE 7 ▸▸ If $g(t) = \dfrac{t^2}{2t + 1}$, to find $g(a^3)$ we substitute a^3 for t in $g(t)$.

$$g(a^3) = \frac{(a^3)^2}{2a^3 + 1} = \frac{a^6}{2a^3 + 1}$$

For the same function, $g(3) = \dfrac{3^2}{2(3) + 1} = \dfrac{9}{7}$. In both cases, to obtain the value of $g(t)$, we simply substitute the value within the parentheses for t. ◂◂

EXAMPLE 8 ▸▸ The electric resistance R of a particular resistor as a function of the temperature T is given by $R = 10.0 + 0.10T + 0.001T^2$. If a given temperature T is increased by 10°C, what is the value of R for the increased temperature as a function of the temperature T?

We are to determine R for a temperature of $T + 10$. Since

$$f(T) = 10.0 + 0.10T + 0.001T^2$$

then

$$\begin{aligned} f(T + 10) &= 10.0 + 0.10(T + 10) + 0.001(T + 10)^2 \qquad \text{substitute } T + 10 \text{ for } T \\ &= 10.0 + 0.10T + 1.0 + 0.001T^2 + 0.02T + 0.1 \\ &= 11.1 + 0.12T + 0.001T^2 \end{aligned}$$ ◂◂

A function may be looked upon as a set of instructions. These instructions tell us how to obtain the value of the dependent variable for a particular value of the independent variable, even if the instructions are expressed in literal symbols.

EXAMPLE 9 The function $f(x) = x^2 - 3x$ tells us to "square the value of the independent variable, multiply the value of the independent variable by 3, and subtract the second result from the first." An analogy would be a computer which was programmed so that when a number was entered into the program, it would square the number, then multiply the number by 3, and finally subtract the second result from the first. This is represented in diagram form in Fig. 3-2.

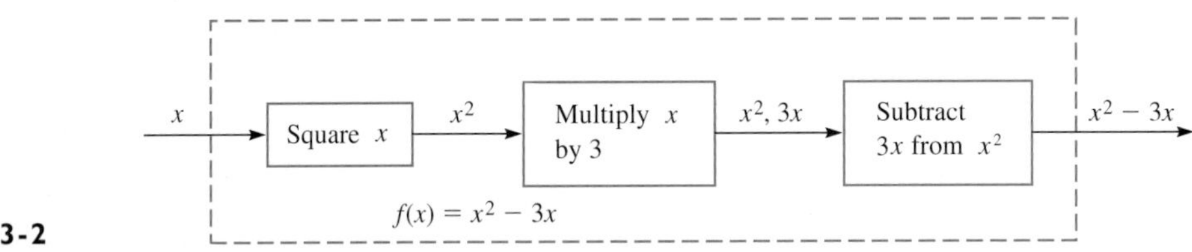

Fig. 3-2

The functions $f(t) = t^2 - 3t$ and $f(n) = n^2 - 3n$ are the same as the function $f(x) = x^2 - 3x$, since the operations performed on the independent variable are the same. **NOTE ▶** ***Although different literal symbols appear, this does not change the function.***

In later chapters we will use a number of special functions, and these special functions are represented by particular symbols. For example, in trigonometry we shall come across the "sine of the angle θ," where the sine is a function of θ. This is designated by $\sin\theta$.

Exercises 3–1

In Exercises 1–8, determine the appropriate functions.

1. Express the area A of a circle as a function of (a) its radius r, (b) its diameter d.
2. Express the circumference c of a circle as a function of (a) its radius r, (b) its diameter d.
3. Express the volume V of a sphere as a function of its diameter d.
4. Express the edge e of a cube as a function of its surface area A.
5. Express the area A of a rectangle of width 5 as a function of its length l.
6. Express the volume V of a right circular cone of height 8 as a function of the radius r of the base.
7. Express the area A of a square as a function of its side s; express the side s of a square as a function of its area A.
8. Express the perimeter p of a square as a function of its side s; express the side s of a square as a function of its perimeter p.

In Exercises 9–20, evaluate the given functions.

9. Given $f(x) = 2x + 1$, find $f(1)$ and $f(-1)$.
10. Given $f(x) = 5x - 9$, find $f(2)$ and $f(-2)$.
11. Given $f(x) = 5 - 3x$, find $f(-2)$ and $f(0.4)$.
12. Given $f(x) = 7 - 2x$, find $f(2.6)$ and $f(-4)$.
13. Given $f(n) = n^2 - 9n$, find $f(3)$ and $f(-5)$.
14. Given $f(v) = 2v^3 - 7v$, find $f(1)$ and $f(\frac{1}{2})$.
15. Given $\phi(x) = \dfrac{6 - x^2}{2x}$, find $\phi(1)$ and $\phi(-2)$.
16. Given $H(q) = \dfrac{8}{q} + 2\sqrt{q}$, find $H(4)$ and $H(0.16)$.
17. Given $g(t) = at^2 - a^2t$, find $g(-\frac{1}{2})$ and $g(a)$.
18. Given $s(y) = 6\sqrt{y+1} - 3$, find $s(8)$ and $s(a^2)$.
19. Given $K(s) = 3s^2 - s + 6$, find $K(-s)$ and $K(2s)$.
20. Given $T(t) = 5t + 7$, find $T(-2t)$ and $T(t + 1)$.

In Exercises 21–24, evaluate the given functions. The values of the independent variable are approximate.

21. Given $f(x) = 5x^2 - 3x$, find $f(3.86)$ and $f(-6.92)$.

22. Given $g(t) = \sqrt{t + 1.0604} - 6t^3$, find $g(0.9261)$.

23. Given $F(y) = \dfrac{2y^2}{y + 0.03685}$, find $F(-0.08466)$.

24. Given $f(x) = \dfrac{x^4 - 2.0965}{6x}$, find $f(1.9654)$ and $f(-2.3865)$.

In Exercises 25–28, state the instructions of the function in words as in Example 9.

25. $f(x) = x^2 + 2$

26. $f(x) = 2x - 6$

27. $g(y) = 6y - y^3$

28. $\phi(s) = 8 - 5s + s^5$

In Exercises 29–32, write the equation as given by the statement. Then write the indicated function using functional notation.

29. y is equal to the square of x.

30. s is equal to the square root of $t + 2$.

31. The net profit P made on selling 40 items, if each costs $24, is equal to the product of 40 and $p - 24$, where p is the price charged.

32. The electrical resistance R of a certain ammeter, in which the resistance of the coil is R_c, is the product of 10 and R_c divided by the sum of 10 and R_c.

In Exercises 33–36, solve the given problems.

33. A demolition ball is used to tear down a building. Its distance s, in metres, above the ground as a function of the time t, in seconds, after it is dropped is given by $s = 17.5 - 4.9t^2$. Since $s = f(t)$, find $f(1.2)$.

34. The change C (in cm) in the length of a 100-m steel bridge girder from its length at 10°C, as a function of the temperature T, is given by $C = 0.12(T - 10)$. Since $C = f(T)$, find $f(-10)$.

35. The stopping distance d, in metres, of a car going v kilometres per hour is given by the function $d = 0.2v + 0.008v^2$. Since $d = f(v)$, find $f(30)$, $f(2v)$, and $f(60)$ using both $f(v)$ and $f(2v)$.

36. The electric power P, in watts, dissipated in a resistor of resistance R, in ohms, is given by the function $P = \dfrac{200R}{(100 + R)^2}$. Since $P = f(R)$, find $f(R + 10)$.

3-2 More About Functions

As we mentioned in the previous section, using only real numbers may result in restrictions as to the permissible values of the independent and dependent variables. *The complete set of possible values of the independent variable is called the* **domain** *of the function,* and *the complete set of all possible resulting values of the dependent variable is called the* **range** *of the function.* Therefore, using real numbers in the domain and range of a function,

Domain and Range

values that lead to division by zero or
to imaginary values may not be included.

EXAMPLE 1 ▸▸ The function $f(x) = x^2 + 2$ is defined for all real values of x. This means its domain is written as *all real numbers.* However, since x^2 is never negative, $x^2 + 2$ is never less than 2. We then write the range as *all real numbers* $f(x) \geq 2$, where the symbol $\geq$ means "is greater than or equal to."

The function $f(t) = \dfrac{1}{t + 2}$ is not defined for $t = -2$, for this value would require division by zero. Also, no matter how large t becomes, $f(t)$ will never exactly equal zero. Therefore, the domain of this function is *all real numbers except* -2, and the range is *all real numbers except 0.* ◂◂

EXAMPLE 2 ▶▶ The function $g(s) = \sqrt{3 - s}$ is not defined for real numbers greater than 3, since such values make $3 - s$ negative and would result in imaginary values for $g(s)$. This means that the domain of this function is *all real numbers* $s \leq 3$, where the symbol $\leq$ means "is less than or equal to."

Also, since $\sqrt{3 - s}$ means the principal square root of $3 - s$ (see Section 1-6), we know that $g(s)$ cannot be negative. This tells us that the range of the function is *all real numbers* $g(s) \geq 0$. ◀◀

In Examples 1 and 2 we determined the domain of each function by looking for those values of the independent variable which cannot be used. The range of each was found through an inspection of the function. This is normally the procedure, except that more advanced methods are often necessary to find the range. (As we develop methods of graphing, we will also see that the graphing calculator is useful in finding the range of a function.) Even the second illustration in Example 1 required a special look at the function. For this reason, we will look only for the domain of some functions.

EXAMPLE 3 ▶▶ Find the domain of the function $f(x) = 16\sqrt{x} + \dfrac{1}{x}$.

From the term $16\sqrt{x}$ we see that x must be greater than or equal to zero in order to have real values. The term $\dfrac{1}{x}$ indicates that x cannot be zero, because of division by zero. Thus, putting these together, the domain is *all real numbers* $x > 0$.

As for the range, it is *all real numbers* $f(x) \geq 12$. More advanced methods are needed to determine this. ◀◀

We have seen that the domains of some functions are restricted to particular values. It can also happen that the domain of a function is restricted by definition or by practical considerations in an application. Consider the illustrations in the following example.

EXAMPLE 4 ▶▶ A function defined as

$$f(x) = x^2 + 4 \quad \text{for } x > 2$$

has a domain restricted to real numbers greater than 2 by definition. Thus, $f(5) = 29$, but $f(1)$ is not defined, since 1 is not in the domain. Also, the range is all real numbers greater than 8.

The height h, in meters, of a certain projectile as a function of the time t, in seconds, is

$$h = 20t - 4.9t^2$$

Generally, negative values of time do not have meaning in such an application. This leads us to state the domain as values of $t \geq 0$. Of course, the projectile will not continue in flight indefinitely, and there is some upper limit on the value of t. These restrictions are not usually stated unless there is a particular reason that affects the solution. ◀◀

The following example illustrates a function that is defined differently for different intervals of the domain.

EXAMPLE 5 ⯮ For the function $f(x) = \begin{cases} 2x - 1 & \text{for } x < 2 \\ 5 & \text{for } x \geq 2 \end{cases}$ find $f(-1)$ and $f(3)$.

We see that values of this function are found differently for values of x less than 2 than for values of x greater than or equal to 2. Since -1 is less than 2,

$$f(-1) = 2(-1) - 1 = -3$$

Since 3 is greater than 2,

$$f(3) = 5$$

We note that $f(x) = 5$ for all values of x which are 2 or greater. ⯬

Functions from Verbal Statements

In the previous section we wrote functions in mathematical form from given statements and by using geometric information. It is often necessary to determine a mathematical function from a given statement. The function is formed using methods similar to those we used in establishing equations from statements in Chapter 1. In the following examples we set up such functions.

EXAMPLE 6 ⯮ The fixed cost for a company to operate a certain plant is \$3000 per day. It also costs \$4 for each unit produced in the plant. Express the daily cost C of operating the plant as a function of the number n of units produced.

The daily total cost C equals the fixed cost of \$3000 plus the cost of producing n units. Since the cost of producing one unit is \$4, the cost of producing n units is $4n$. Thus, the total cost C, where $C = f(n)$, is

$$C = 3000 + 4n$$

Here we know that the domain is all values of $n \geq 0$, with some upper limit on n based on the production capacity of the plant. ⯬

EXAMPLE 7 ⯮ A metallurgist melts and mixes m grams of solder that is 40% tin with n grams of another solder that is 20% tin to get a final solder mixture that contains 200 g of tin. Express n as a function of m. See Fig. 3-3.

The statement leads to the following equation.

$$\underset{\substack{\text{tin in} \\ \text{first solder}}}{0.40m} + \underset{\substack{\text{tin in} \\ \text{second solder}}}{0.20n} = \underset{\substack{\text{total amount} \\ \text{of tin}}}{200}$$

Since we want $n = f(m)$, we now solve for n.

$$0.20n = 200 - 0.40m$$
$$n = 1000 - 2m$$

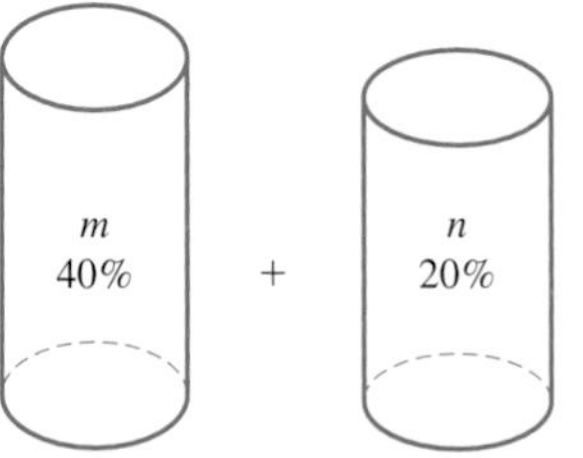

= 200

Grams of tin

Fig. 3-3

This is the required function. Since neither m nor n can be negative, the domain is all values $0 \leq m \leq 500$ g, which means that m is greater than or equal to 0 g and less than or equal to 500 g. The range is all values $0 \leq n \leq 1000$ g. ⯬

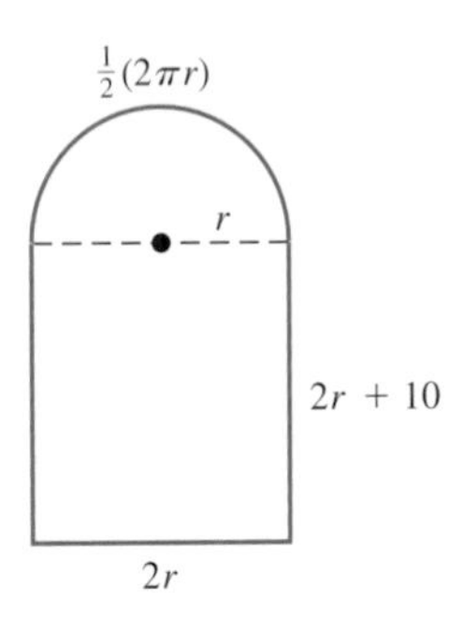

Fig. 3-4

EXAMPLE 8 ▸▸ An architect designs a window such that it has the shape of a rectangle with a semicircle on top, as shown in Fig. 3-4. The base of the window is 10 cm less than the height of the rectangular part. Express the perimeter p of the window as a function of the radius r of the circular part.

We know that the perimeter is the distance around the window. Since the top part is a semicircle, and the circumference of a circle is $2\pi r$, the length of the top circular part is $\frac{1}{2}(2\pi r)$. This also tells us that the dashed line, and therefore the base of the window, is $2r$. Finally, the fact that the base is 10 cm less than the height of the rectangular part tells us that each vertical part is $2r + 10$. This means that the perimeter p, where $p = f(r)$, is

$$\begin{aligned} p &= \tfrac{1}{2}(2\pi r) + 2r + 2(2r + 10) \\ &= \pi r + 2r + 4r + 20 \\ &= \pi r + 6r + 20 \end{aligned}$$

We see that the required function is $p = \pi r + 6r + 20$. Since the radius cannot be negative, and there would be no window if $r = 0$, the domain of the function is all values $0 < r \leq R$, where R is a maximum possible value of r determined by design considerations. ◂◂

Relation

In the definition of a function, it was stipulated that any value for the independent variable must yield only a single value of the dependent variable. This requirement is stressed in more advanced courses, and we will use it again in Chapter 20. *If a value of the independent variable yields more than one value of the dependent variable, the relationship is called a* **relation** instead of a function. A relation involves two variables related so that values of the second variable can be determined from values of the first variable. A function is a relation in which each value of the first variable yields only one value of the second. A function is therefore a special type of relation. However, there are relations that are not functions.

EXAMPLE 9 ▸▸ For $y^2 = 4x^2$, if $x = 2$, then y can be either 4 or -4. Since a value of x yields more than one single value for y, we see that $y^2 = 4x^2$ is a relation, not a function. ◂◂

Exercises 3–2

In Exercises 1–8, determine the domain and range of the given functions.

1. $f(x) = x + 5$
2. $g(u) = 3 - u^2$
3. $G(z) = \dfrac{3}{z}$
4. $F(r) = \sqrt{r + 4}$
5. $f(s) = \dfrac{2}{s^2}$
6. $f(x) = \dfrac{6}{\sqrt{2 - x}}$
7. $H(h) = 2h + \sqrt{h} + 1$
8. $T(t) = 2t^4 + t^2 - 1$

In Exercises 9–12, determine the domain of the given functions.

9. $Y(y) = \dfrac{y + 1}{\sqrt{y - 2}}$
10. $f(n) = \dfrac{n}{6 - 2n}$
11. $f(u) = \dfrac{u}{u - 2} + \dfrac{4}{u + 4}$
12. $g(x) = \dfrac{\sqrt{x - 2}}{x - 3}$

In Exercises 13–16, evaluate the indicated functions.

$F(t) = 3t - t^2 \quad \text{for } t \leq 2 \qquad h(s) = \begin{cases} 2s & \text{for } s < -1 \\ s + 1 & \text{for } s \geq -1 \end{cases}$

$f(x) = \begin{cases} x + 1 & \text{for } x < 1 \\ \sqrt{x + 3} & \text{for } x \geq 1 \end{cases} \qquad g(x) = \begin{cases} \dfrac{1}{x} & \text{for } x \neq 0 \\ 0 & \text{for } x = 0 \end{cases}$

13. Find $F(2)$ and $F(3)$.

14. Find $h(-8)$ and $h\left(-\frac{1}{2}\right)$.

15. Find $f(1)$ and $f\left(-\frac{1}{4}\right)$.

16. Find $g\left(\frac{1}{5}\right)$ and $g(0)$.

In Exercises 17–28, determine the appropriate functions.

17. A motorist travels at 40 km/h for 2 h and then at 55 km/h for t hours. Express the distance d traveled as a function of t.

18. Express the cost C of insulating a cylindrical water tank of height 2 m as a function of its radius r, if the cost of insulation is \$3 per square metre.

19. A rocket burns up at the rate of 2 Mg/min after falling out of orbit into the atmosphere. If the rocket weighed 5500 Mg before reentry, express its weight w as a function of the time t, in minutes, of reentry.

20. A computer part costs \$3 to produce and distribute. Express the profit p made by selling 100 of these parts as a function of the price of c dollars each.

21. Upon ascending, a weather balloon ices up at the rate of 0.5 kg/m after reaching an altitude of 1000 m. If the mass of the balloon below 1000 m is 110 kg, express its mass m as a function of its altitude h if $h > 1000$ m.

22. A chemist adds x litres of a solution that is 50% alcohol to 100 L of a solution that is 70% alcohol. Express the number n of litres of alcohol in the final solution as a function of x.

23. A company installs underground cable at a cost of \$500 for the first 50 m (or up to 50 m) and \$5 for each metre thereafter. Express the cost C as a function of the length l of underground cable if $l > 50$ m.

24. The *mechanical advantage* of an inclined plane is the ratio of the length of the plane to its height. Express the mechanical advantage M of a plane of length 8 m as a function of its height h.

25. The capacities (in L) of two oil-storage tanks are x and y. The tanks are initially full; 1200 L of oil is removed from them by taking 10% of the contents of the first tank and 40% of the contents of the second tank. Express y as a function of x.

26. A manufacturer finds that it earns a profit of \$2 on each box of computer disks and a profit of \$4 on each box of computer paper. If x boxes of disks and y boxes of paper are produced, the profit is \$5000. Express y as a function of x.

27. In studying the electric current that is induced in wire rotating through a magnetic field, a piece of wire 60 cm long is cut into two pieces. One of these is bent into a circle, and the other is bent into a square. Express the total area A of the two figures as a function of the perimeter p of the square.

28. The cross section of an air-conditioning duct is in the shape of a square with semicircles on each side. See Fig. 3-5. Express the area A of this cross section as a function of the diameter d (in cm) of the circular part.

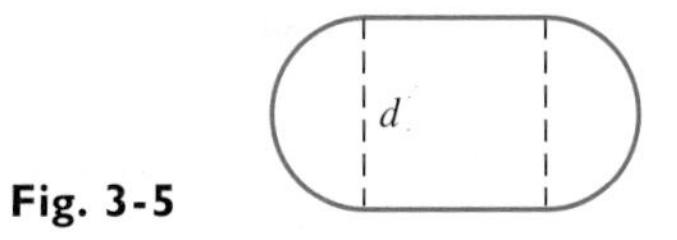

Fig. 3-5

In Exercises 29–36, solve the given problems.

29. A computer program displays a circular image of radius 6 cm. If the radius is decreased by x cm, express the area of the image as a function of x. What are the domain and range of $A = f(x)$?

30. A helicopter 120 m from a person takes off vertically. Express the distance d from the person to the helicopter as a function of the height h of the helicopter. What are the domain and the range of $d = f(h)$? See Fig. 3-6.

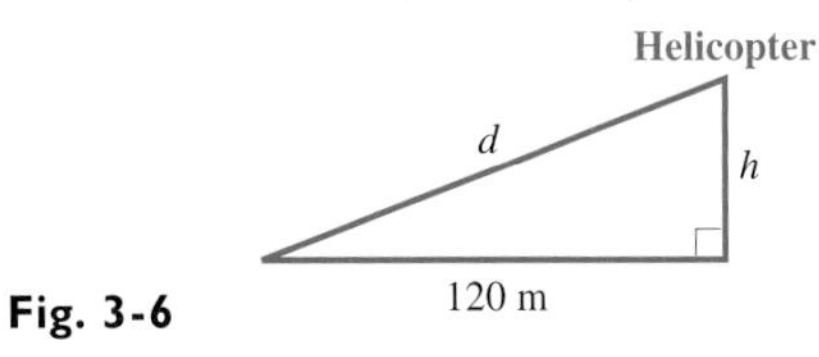

Fig. 3-6

31. A truck travels 300 km in t hours. Express the average speed s of the truck as a function of t. What are the domain and range of $s = f(t)$?

32. A rectangular grazing range with an area of 8 km^2 is to be fenced. Express the length l of the field as a function of its width w. What are the domain and range of $l = f(w)$?

33. The resonant frequency f (in Hz) in a certain electric circuit as a function of the capacitance is $f = \frac{1}{2\pi\sqrt{C}}$. Describe the domain of this function.

34. A jet is traveling directly between cities A and B, which are 2400 km apart. If the jet is x km from city A and y km from city B, find the domain of $y = f(x)$.

35. Express the mass m of the weather balloon in Exercise 21 as a function of any height h in the same manner as the function in Example 5 (and Exercises 13 through 16) was expressed.

36. Express the cost C of installing any length l of the underground cable in Exercise 23 in the same manner as the function in Example 5 (and Exercises 13 through 16) was represented.

3-3 Rectangular Coordinates

One of the most valuable ways of representing a function is by graphical representation. By using graphs we are able to obtain a "picture" of the function, and by using this picture we can learn a great deal about the function.

To make a graphical representation of a function, we recall from Chapter 1 that numbers can be represented by points on a line. For a function, we have values of the independent variable as well as the corresponding values of the dependent variable. Therefore, it is necessary to use two different lines to represent the values from each of these sets of numbers. We do this by placing the lines perpendicular to each other.

Rectangular (Cartesian) coordinates were developed by the French mathematician Descartes (1596–1650).

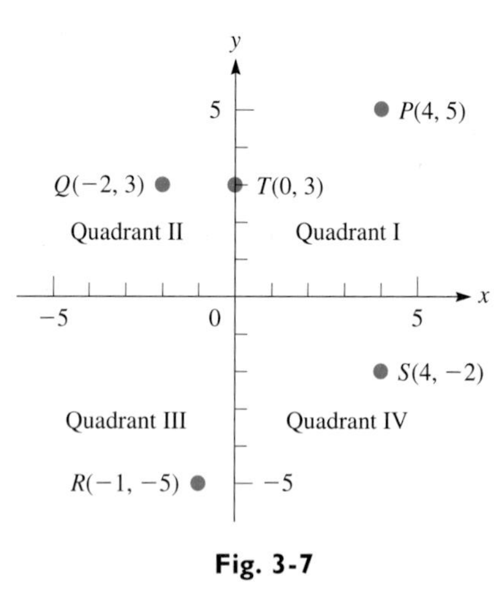

Fig. 3-7

We place one line horizontally and label it the **x-axis**. The numbers of the set for the independent variable are normally placed on this axis. *The other line we place vertically and label the* **y-axis**. Normally the y-axis is used for values of the dependent variable. *The point of intersection is called the* **origin**. This is the **rectangular coordinate system**.

On the x-axis, positive values are to the right of the origin, and negative values are to the left of the origin. On the y-axis, positive values are above the origin, and negative values are below it. *The four parts into which the plane is divided are called* **quadrants**, which are numbered as in Fig. 3-7.

A point P on the plane is designated by the pair of numbers (x, y), where x is the value of the independent variable and y is the corresponding value of the dependent variable. *The x-value, called the* **abscissa**, *is the perpendicular distance of P from the y-axis. The y-value, called the* **ordinate**, *is the perpendicular distance of P from the x-axis.* The values x and y together, written as (x, y), are the **coordinates** of the point P.

EXAMPLE 1 ▸▸ Locate the points $A(2, 1)$ and $B(-4, -3)$ on the rectangular coordinate system.

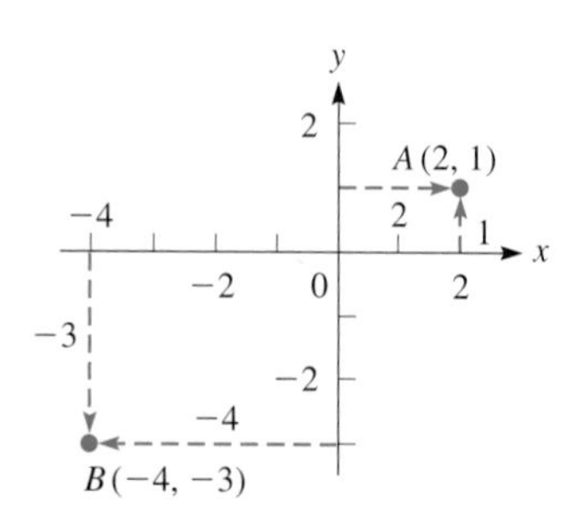

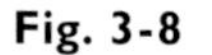
Fig. 3-8

The coordinates (2, 1) for A mean that the point is 2 units to the *right* of the y-axis and 1 unit *above* the x-axis, as shown in Fig. 3-8. The coordinates $(-4, -3)$ for B mean that the point is 4 units to the *left* of the y-axis and 3 units *below* the x-axis, as shown. ◂◂

EXAMPLE 2 ▸▸ The positions of points $P(4, 5)$, $Q(-2, 3)$, $R(-1, -5)$, $S(4, -2)$, and $T(0, 3)$ are shown in Fig. 3-7, which is located on the previous page. We see that this representation allows for *one point for any pair of values (x, y)*. Also, we note that the point $T(0, 3)$ is on the y-axis. Any such point that is on either axis is not *in* any of the quadrants. ◂◂

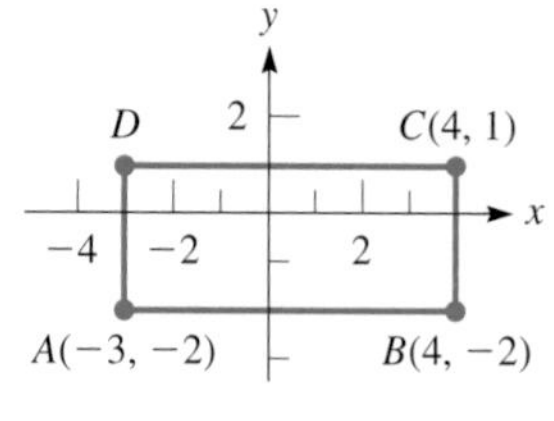

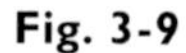
Fig. 3-9

EXAMPLE 3 ▸▸ Three vertices of the rectangle in Fig. 3-9 are $A(-3, -2)$, $B(4, -2)$, and $C(4, 1)$. What is the fourth vertex?

We use the fact that opposite sides of a rectangle are equal and parallel to find the solution. Since both vertices of the base AB of the rectangle have a y-coordinate of -2, the base is parallel to the x-axis. Therefore, the top of the rectangle must also be parallel to the x-axis. Thus, the vertices of the top must both have a y-coordinate of 1, since one of them has a y-coordinate of 1. In the same way, the x-coordinates of the left side must both be -3. Therefore, the fourth vertex is $D(-3, 1)$. ◂◂

EXAMPLE 4 ▸▸ Where are all points whose ordinates are 2?

Since the ordinate is the y-value, we can see that the question could be stated as: "Where are all points for which $y = 2$?" Since all such points are 2 units above the x-axis, the answer can be stated as "on a line 2 units above the x-axis." See Fig. 3-10. ◂◂

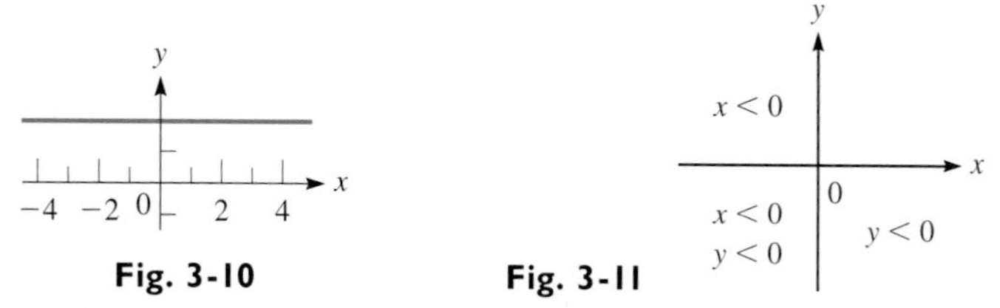

Fig. 3-10

Fig. 3-11

EXAMPLE 5 ▸▸ Where are all points (x, y) for which $x < 0$ and $y < 0$?

Noting that $x < 0$ means "x is less than zero," or "x is negative," and that $y < 0$ means the same for y, we want to determine where both x and y are negative. Our answer is "in the third quadrant," since both coordinates are negative for all points in the third quadrant, and this is the only quadrant for which this is true. See Fig. 3-11. ◂◂

Exercises 3–3

In Exercises 1 and 2, determine (at least approximately) the coordinates of the points specified in Fig. 3-12.

1. A, B, C
2. D, E, F

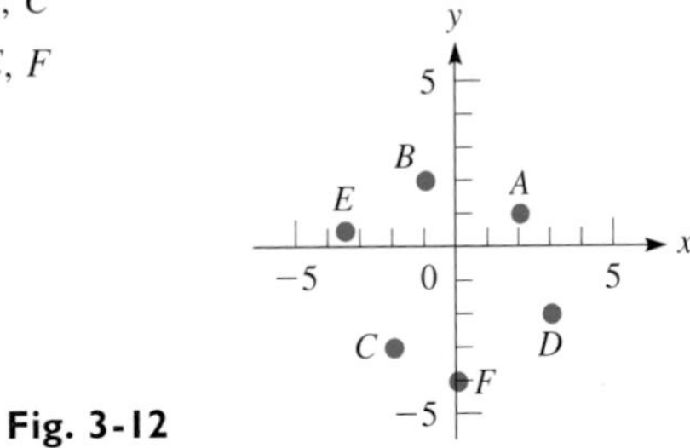

Fig. 3-12

In Exercises 3 and 4, plot the given points.

3. $A(2,7)$, $B(-1,-2)$, $C(-4,2)$
4. $A(3,\frac{1}{2})$, $B(-6,0)$, $C(-\frac{5}{2},-5)$

In Exercises 5–8, plot the given points and then join these points, in the order given, by straight-line segments. Name the geometric figure formed.

5. $A(-1,4)$, $B(3,4)$, $C(1,-2)$
6. $A(0,3)$, $B(0,-1)$, $C(4,-1)$
7. $A(-2,-1)$, $B(3,-1)$, $C(3,5)$, $D(-2,5)$
8. $A(-5,-2)$, $B(4,-2)$, $C(6,3)$, $D(-3,3)$

In Exercises 9–12, find the indicated coordinates.

9. Three vertices of a rectangle are $(5,2)$, $(-1,2)$, and $(-1,4)$. What are the coordinates of the fourth vertex?
10. Two vertices of an equilateral triangle are $(7,1)$ and $(2,1)$. What is the abscissa of the third vertex?
11. P is the point $(3,2)$. Locate point Q such that the x-axis is the perpendicular bisector of the line segment joining P and Q.
12. P is the point $(-4,1)$. Locate point Q such that the line segment joining P and Q is bisected by the origin.

In Exercises 13–28, answer the given questions.

13. Where are all the points whose abscissas are 1?
14. Where are all the points whose ordinates are -3?
15. Where are all points such that $y = 3$?
16. Where are all points such that $x = -2$?
17. Where are all the points whose abscissas equal their ordinates?
18. Where are all the points whose abscissas equal the negative of their ordinates?
19. What is the abscissa of all points on the y-axis?
20. What is the ordinate of all points on the x-axis?
21. Where are all the points for which $x > 0$?
22. Where are all the points for which $y < 0$?
23. Where are all the points for which $x < -1$?
24. Where are all the points for which $y > 4$?
25. Where are all points (x, y) for which $x = 0$ and $y < 0$?
26. Where are all points (x, y) for which $x < 0$ and $y > 1$?
27. In which quadrants is the ratio y/x positive?
28. In which quadrants is the ratio y/x negative?

3-4 The Graph of a Function

Now that we have introduced the concepts of a function and the rectangular coordinate system, we are in a position to determine the graph of a function. In this way we shall obtain a visual representation of a function.

The graph of a function is the set of all points whose coordinates (x, y) satisfy the functional relationship $y = f(x)$. Since $y = f(x)$, we can write the coordinates of the points on the graph as $(x, f(x))$. Writing the coordinates in this manner tells us exactly how to find them. *We assume a certain value for x and then find the value of the function of x. These two numbers are the coordinates of the point.*

Since there is no limit to the possible number of points that can be chosen, we normally select a few values of x, obtain the corresponding values of the function, plot these points, and then join them. Therefore, we use the following basic procedure in plotting the graph of a function.

1. *Let x take on several values and calculate the corresponding values of y.*
2. *Tabulate these values, arranging the table so that values of x are increasing.*
3. **NOTE ▶** *Plot the points and join them from left to right by a smooth curve* (not short straight-line segments).

EXAMPLE 1 ▸▸ Graph the function $f(x) = 3x - 5$.

For purposes of graphing, we let $y = f(x)$, or $y = 3x - 5$. We then let x take on various values and determine the corresponding values of y. Note that once we choose a given value of x, we have no choice about the corresponding y-value, as it is determined by evaluating the function. If $x = 0$, we find that $y = -5$. This means that the point $(0, -5)$ is on the graph of the function $3x - 5$. Choosing another value of x, for example, 1, we find that $y = -2$. This means that the point $(1, -2)$ is on the graph of the function $3x - 5$. Continuing to choose a few other values of x, we tabulate the results, as shown in Fig. 3-13. It is best to arrange the table so that the values of x increase; then there is no doubt how they are to be connected, for they are then connected in the order shown. Finally, we connect the points as shown in Fig. 3-13, and see that the graph of the function $3x - 5$ is a straight line.

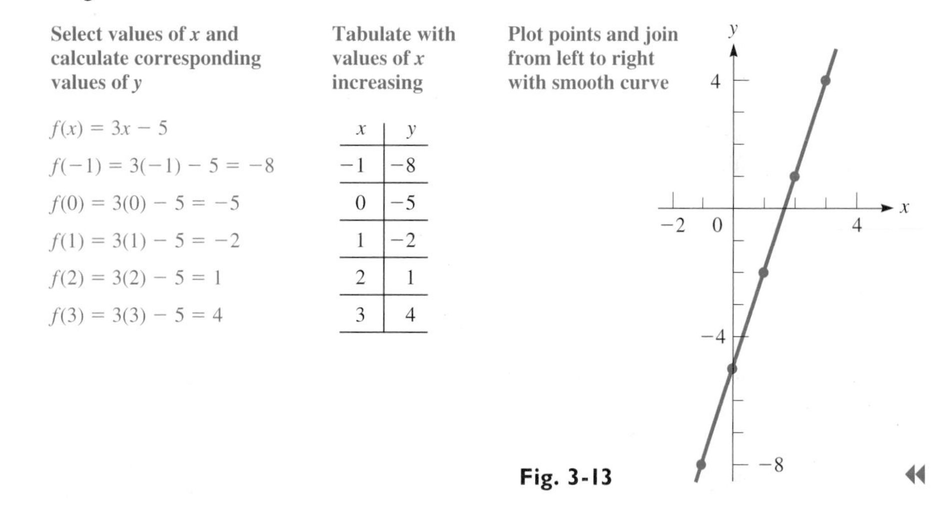

$f(x) = 3x - 5$
$f(-1) = 3(-1) - 5 = -8$
$f(0) = 3(0) - 5 = -5$
$f(1) = 3(1) - 5 = -2$
$f(2) = 3(2) - 5 = 1$
$f(3) = 3(3) - 5 = 4$

x	y
−1	−8
0	−5
1	−2
2	1
3	4

Fig. 3-13 ◂◂

EXAMPLE 2 ▸▸ Graph the function $f(x) = 2x^2 - 4$.

First, we let $y = 2x^2 - 4$ and tabulate the values as shown in Fig. 3-14. In determining the values in the table, we must take particular care to obtain the correct values of y for negative values of x. ***Mistakes are relatively common when dealing with negative numbers.*** We must carefully use the laws for signed numbers. For example, if $x = -2$, we have $y = 2(-2)^2 - 4 = 2(4) - 4 = 8 - 4 = 4$. Once the values are obtained, we plot and connect the points with a smooth curve, as shown.

CAUTION ▶

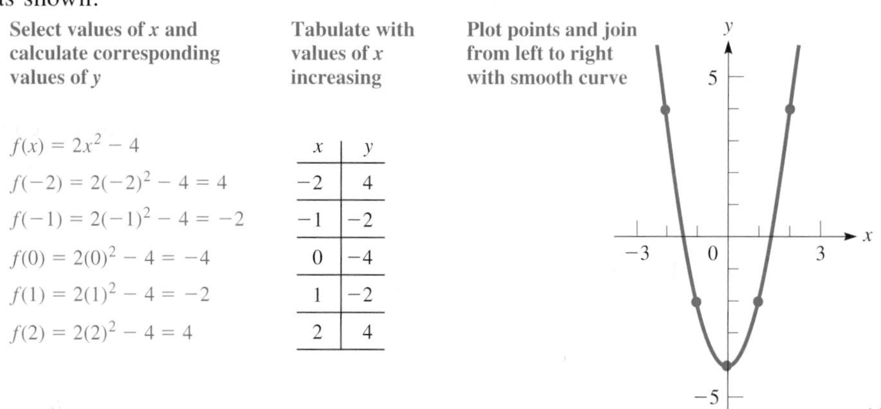

Select values of x and calculate corresponding values of y
Tabulate with values of x increasing
Plot points and join from left to right with smooth curve
y
x
5
−3
0
3
−5

$f(x) = 2x^2 - 4$
$f(-2) = 2(-2)^2 - 4 = 4$
$f(-1) = 2(-1)^2 - 4 = -2$
$f(0) = 2(0)^2 - 4 = -4$
$f(1) = 2(1)^2 - 4 = -2$
$f(2) = 2(2)^2 - 4 = 4$

x	y
−2	4
−1	−2
0	−4
1	−2
2	4

Fig. 3-14 ◂◂

When graphing a function, there are some special points about which we should be careful. These include the following:

Restrictions on a Graph

1. Since the graphs of most common functions are smooth, any irregularities in the graph should be carefully checked. In these cases it usually helps to take values of x between those values where the question arises.
2. The domain of the function may not include all values of x (remember, *division by zero is not defined, and only real values of the variables are permissible).*
3. In applications, we must be careful to use values that are meaningful. As we have seen, negative values for many quantities, such as time, are not generally meaningful.

The following examples illustrate these points.

EXAMPLE 3 ▸▸ Graph the function $y = x - x^2$.

First we determine the values in the table, as shown with Fig. 3-15. Again, we must be careful with negative values of x. For the value $x = -1$, we have $y = (-1) - (-1)^2 = -1 - (+1) = -1 - 1 = -2$. Once all the values have been

NOTE ▸ found and plotted, we note that ***y* = 0 *for both x* = 0 *and x* = 1**. The question arises—***what happens between these values?*** Trying $x = \frac{1}{2}$, we find that $y = \frac{1}{4}$. Using this point completes the necessary information. Note that in plotting these graphs we do not stop the graph with the last point determined, but indicate that the curve continues. ◂◂

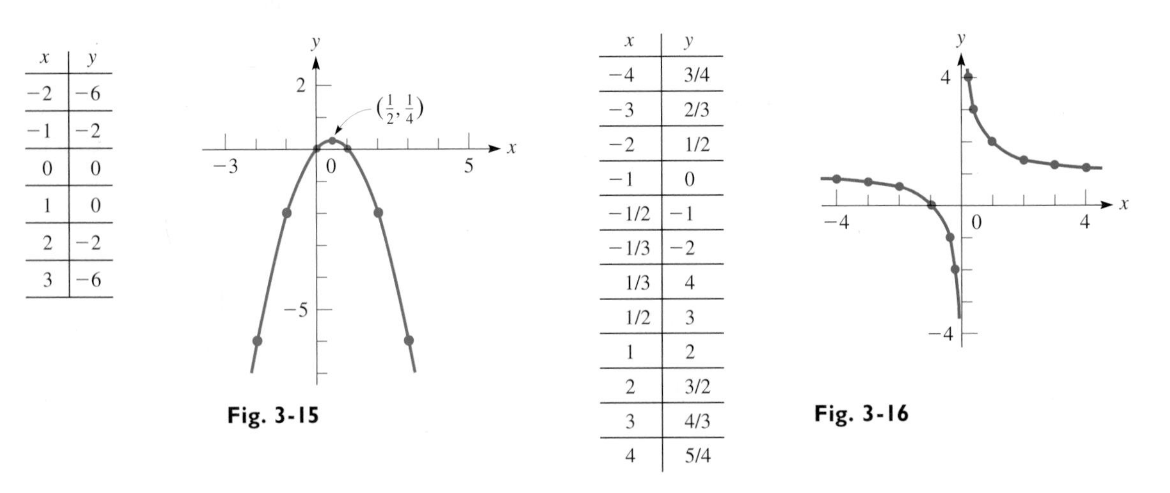

x	y
−2	−6
−1	−2
0	0
1	0
2	−2
3	−6

Fig. 3-15

x	y
−4	3/4
−3	2/3
−2	1/2
−1	0
−1/2	−1
−1/3	−2
1/3	4
1/2	3
1	2
2	3/2
3	4/3
4	5/4

Fig. 3-16

EXAMPLE 4 ▸▸ Graph the function $y = 1 + \dfrac{1}{x}$.

In finding the points on this graph, as shown in Fig. 3-16, we note that y is not

CAUTION ▸ defined for $x = 0$, due to division by zero. Thus, $x = 0$ is not in the domain and ***we must be careful not to have any part of the curve cross the y-axis*** ($x = 0$). Although we cannot let $x = 0$, we can choose other values of x between -1 and 1 that are close to zero. In doing so, we find that as x gets closer to zero, the points get closer and closer to the y-axis, although they do not reach or touch it. In this case the y-axis is called an **asymptote** of the curve. ◂◂

EXAMPLE 5 ▸▸ Graph the function $y = \sqrt{x + 1}$.

When finding the points for the graph, we may not let x take on any value less than -1, for ***all such values would lead to imaginary values for y*** and are not in the domain. Also, since we have the positive square root indicated, the range consists of all values of y which are positive or zero ($y \geq 0$). See Fig. 3-17. Note that the graph starts at $(-1, 0)$. ◂◂

NOTE ▸

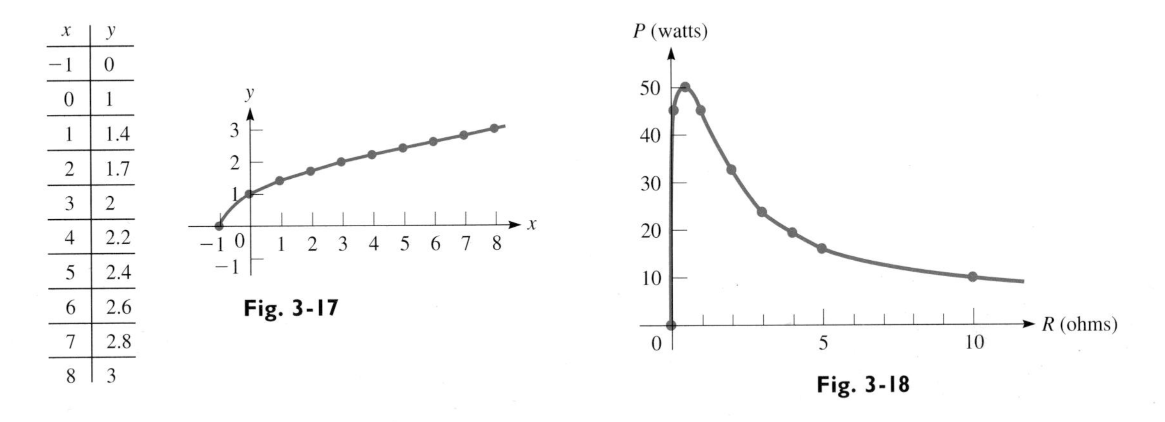

x	y
−1	0
0	1
1	1.4
2	1.7
3	2
4	2.2
5	2.4
6	2.6
7	2.8
8	3

Fig. 3-17

Fig. 3-18

See the chapter introduction.

EXAMPLE 6 ▸▸ The electric power P delivered by a certain battery as a function of the resistance R in the circuit is given by

$$P = \frac{100R}{(0.50 + R)^2}$$

where P is measured in watts and R in ohms. Plot the power as a function of the resistance.

Since negative values for the resistance have no physical significance, we should not plot any values of P for negative values of R. The following table of values is obtained.

R (ohms)	0	0.25	0.50	1.0	2.0	3.0	4.0	5.0	10.0
P (watts)	0.0	44.4	50.0	44.4	32.0	24.5	19.8	16.5	9.1

The values 0.25 and 0.50 are used for R when it is found that P is less for $R = 2$ than for $R = 1$. The sharp change in direction at $R = 1$ should be checked by using these additional points to better see how the curve changes near $R = 1$, and thereby obtain a smoother curve. See Fig. 3-18.

Also note that the scale on the P-axis is different from that on the R-axis. This reflects the different magnitudes and ranges of values used for each of the variables. Different scales are normally used in such cases.

We can make various conclusions from the graph. For example, we see that the maximum power of 50 W occurs for $R = 0.5\ \Omega$. Also, P continually decreases as R increases beyond $0.5\ \Omega$. We will consider further the information that can be read from a graph in the next section. ◂◂

The following example illustrates the graph of a function that is defined differently for different intervals of the domain.

EXAMPLE 7 ▸▸ Graph the function $f(x) = \begin{cases} 2x + 1 & \text{for } x \le 1 \\ 6 - x^2 & \text{for } x > 1 \end{cases}$

First we let $y = f(x)$ and then tabulate the necessary values. In evaluating $f(x)$ we must be careful to use the proper part of the definition. In order to see where to start the curve for $x > 1$, we evaluate $6 - x^2$ for $x = 1$, but we must realize that ***the curve does not include this point*** $(1, 5)$ and starts immediately to its right. To show that this is not part of the curve, we draw it as an open circle. See Fig. 3-19. Such a function, with a "break" in it, is called *discontinuous*. ◂◂

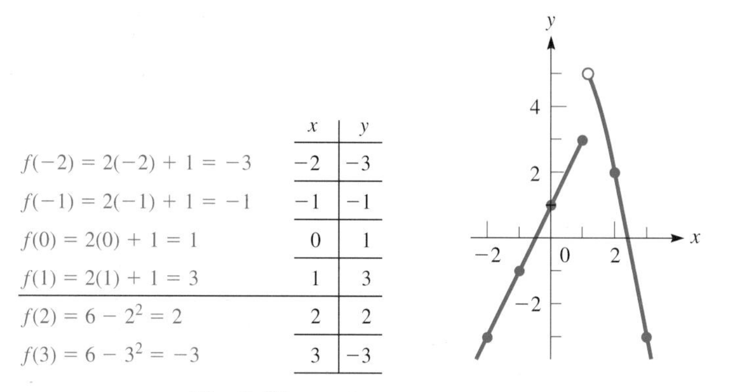

	x	y
$f(-2) = 2(-2) + 1 = -3$	-2	-3
$f(-1) = 2(-1) + 1 = -1$	-1	-1
$f(0) = 2(0) + 1 = 1$	0	1
$f(1) = 2(1) + 1 = 3$	1	3
$f(2) = 6 - 2^2 = 2$	2	2
$f(3) = 6 - 3^2 = -3$	3	-3

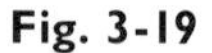
Fig. 3-19

In Section 3-1, we defined a function such that it has only one value of the dependent variable for each value of the independent variable. At the end of Section 3-2, we stated that a relation may have more than one such value of the dependent variable. The following example illustrates how to use the graph to determine whether or not a relation is a function.

EXAMPLE 8 ▸▸ In Example 9 of Section 3-2, we noted that $y^2 = 4x^2$ is a relation, but not a function. Since $y = 4$ or $y = -4$, normally written as $y = \pm 4$, for $x = 2$, we have two values of y for $x = 2$. Therefore, it is not a function. Making the table as shown at the left, we then draw the graph in Fig. 3-20. When making the table, we also note that there are two values of y for every value of x, except $x = 0$.

If any vertical line that crosses the x-axis in the domain intersects the graph in more than one point, it is the graph of a relation that is not a function. Any such vertical line in any of the previous graphs of this section would intersect the graph in only one point. This shows that they are graphs of functions. ◂◂

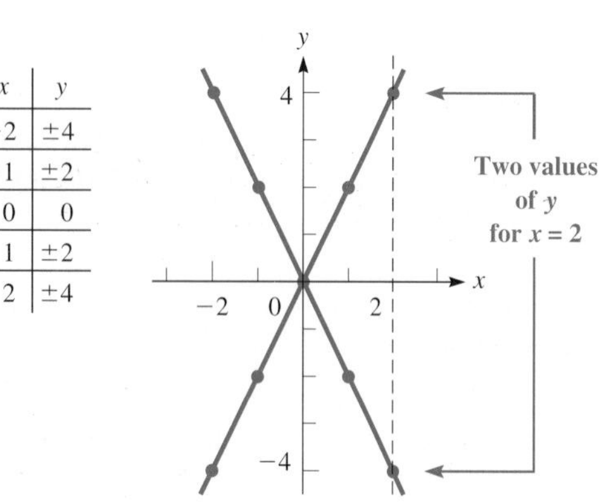

x	y
-2	± 4
-1	± 2
0	0
1	± 2
2	± 4

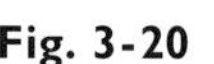
Fig. 3-20

Functions of a particular type have graphs of a specific basic shape, and many have been named. Two examples of this are the *straight line* (Example 1) and the *parabola* (Examples 2 and 3). We consider the straight line again in Chapter 5 and the parabola in Chapter 7. Other types of graphs are found in many of the later chapters, with a detailed analysis of some of them in Chapter 21.

The use of graphs is extensive in mathematics and in applications. For example, in the next section we use graphs to solve equations. Numerous applications can be noted for most types of curves. For example, the parabola has applications in microwave dish design, suspension bridge design, and the path (trajectory) of a baseball. We will note many more applications of graphs throughout the book.

Exercises 3–4

In Exercises 1–32, graph the given functions.

1. $y = 3x$
2. $y = -2x$
3. $y = 2x - 4$
4. $y = 3x + 5$
5. $y = 7 - 2x$
6. $y = 5 - 3x$
7. $y = \frac{1}{2}x - 2$
8. $y = 6 - \frac{1}{3}x$
9. $y = x^2$
10. $y = -2x^2$
11. $y = 3 - x^2$
12. $y = x^2 - 3$
13. $y = \frac{1}{2}x^2 + 2$
14. $y = 2x^2 + 1$
15. $y = x^2 + 2x$
16. $y = 2x - x^2$
17. $y = x^2 - 3x + 1$
18. $y = 2 + 3x + x^2$
19. $y = x^3$
20. $y = -2x^3$
21. $y = x^3 - x^2$
22. $y = 3x - x^3$
23. $y = x^4 - 4x^2$
24. $y = x^3 - x^4$
25. $y = \dfrac{1}{x}$
26. $y = \dfrac{2}{x + 2}$
27. $y = \dfrac{4}{x^2}$
28. $y = \dfrac{1}{x^2 + 1}$
29. $y = \sqrt{x}$
30. $y = \sqrt{4 - x}$
31. $y = \sqrt{16 - x^2}$
32. $y = \sqrt{x^2 - 16}$

In Exercises 33–52, graph the indicated functions.

33. A force F (in N) stretches a spring x cm according to the function $F = 4x$. Plot F as a function of x.

34. The length L (in cm) of a pulley belt is 12 cm longer than the circumference of one of the pulley wheels. Express L as a function of the radius r of the wheel, and plot L as a function of r.

35. For a certain model of truck, its resale value V, in dollars, as a function of the total distance m it has been driven is $V = 50\,000 - 0.2m$. Plot V as a function of m for $m \le 100\,000$ km.

36. The resistance R, in ohms, of a resistor as a function of the temperature T, in degrees Celsius, is given by $R = 250(1 + 0.0032T)$. Plot R as a function of T.

37. The consumption c of fuel, in litres per hour (L/h), of a certain engine is determined as a function of the number r, in revolutions per minute (r/min) of the engine, to be $c = 0.011r + 4.0$. This formula is valid from 500 r/min to 3000 r/min. Plot c as a function of r.

38. The profit P, in dollars, a manufacturer makes in producing x units is given by $P = 3.5x - 120$. Plot P as a function of x for $x \le 100$.

39. The rate H (in watts) at which heat is developed in the filament of an electric light bulb as a function of the electric current I (in amperes) is $H = 240I^2$. Plot H as a function of I.

40. The total annual fraction f of energy supplied by solar energy to a home as a function of the area A (in m^2) of the solar collector is $f = 0.065\sqrt{A}$. Plot f as a function of A.

41. The maximum speed v (in km/h) at which a car can safely travel around a circular turn of radius r, in metres, is given by $r = 0.55v^2$. Plot r as a function of v.

42. The height h, in metres, of a rocket as a function of the time t, in seconds, is given by the function $h = 1500t - 4.9t^2$. Plot h as a function of t, assuming level terrain.

43. A formula used to determine the amount of lumber V (in dm^3) that can be cut from a 2-m section of a log of diameter d (in mm) is $V = 0.0013d^2 - 0.043d$. Plot N as a function of d for values of d from 300 mm to 1000 mm.

44. A copper electrode with a mass of 25.0 g is placed in a solution of copper sulfate. An electric current is passed through the solution, and 1.6 g of copper is deposited on the electrode each hour. Express the total mass m of the electrode as a function of the time t and plot the graph.

45. The perimeter of a rectangular tarpaulin is 60 m. Express the area A of the tarpaulin as a function of its width w and plot the graph.

46. The distance p, in metres, from a camera with a 50-mm lens to the object being photographed is a function of the magnification m of the camera given by $p = \dfrac{0.05(1 + m)}{m}$. Plot the graph for positive values of m up to 0.50.

47. A measure of the light beam that can be passed through an optic fiber is its numerical aperture N. For a particular optic fiber, N is a function of the index of refraction n of the glass in the fiber, given by $N = \sqrt{n^2 - 1.69}$. Plot the graph for values of n up to 2.00.

48. The force F, in newtons, exerted by a cam on the arm of a robot is a function of the distance x shown in Fig. 3-21. The function is $F = x^4 - 12x^3 + 46x^2 - 60x + 25$. Plot the graph. (Note that x varies from 1 cm to 5 cm.)

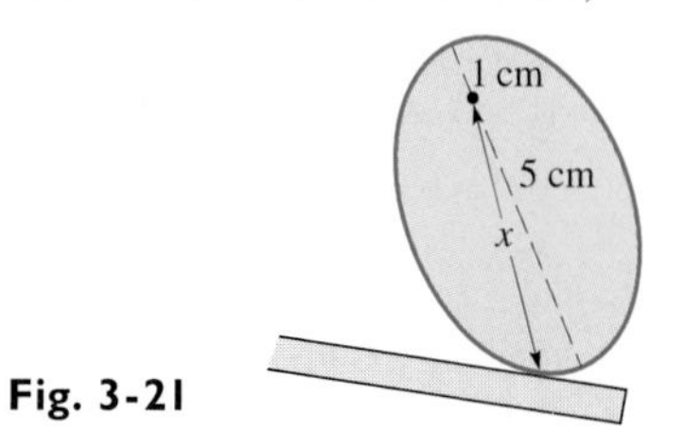

Fig. 3-21

49. Plot the graphs of $y = x$ and $y = |x|$ on the same coordinate system. Explain why the graphs differ.

50. Plot the graphs of $y = 2 - x$ and $y = |2 - x|$ on the same coordinate system. Explain why the graphs differ.

51. Plot the graph of $f(x) = \begin{cases} 3 - x & \text{for } x < 1 \\ x^2 + 1 & \text{for } x \geq 1 \end{cases}$

52. Plot the graph of $f(x) = \begin{cases} \dfrac{1}{x - 1} & \text{for } x < 0 \\ \sqrt{x + 1} & \text{for } x \geq 0 \end{cases}$

In Exercises 53–56, determine whether or not the indicated graph is that of a relation which is a function.

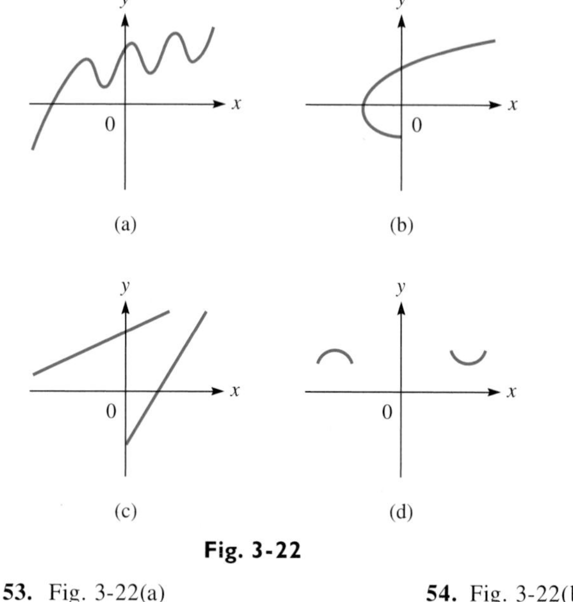

Fig. 3-22

53. Fig. 3-22(a)

54. Fig. 3-22(b)

55. Fig. 3-22(c)

56. Fig. 3-22(d)

3-5 The Graphing Calculator

The graphing calculator was first introduced in the late 1980s.

In the previous section we developed the basic method of graphing a function to see how graphs are constructed. A graphing calculator (or a computer *grapher*) can be used to show the graph of a function, and it can do it very quickly. In this section we discuss the use of a graphing calculator to display graphs.

As with any calculator, *you must be familiar with the sequence of keys to be used.* For example, on some calculators the GRAPH key is used after the function has been entered, while on others it puts the calculator into the graphing mode. A brief discussion of the features of graphing calculators is found in Appendix C. For a detailed discussion, the manual for any particular calculator should be used.

Many different views of a graph can be shown in the *viewing window* (or *viewing rectangle*) of a graphing calculator. One view of $y = 2x + 8$ is shown in Fig. 3-23(a). (The views in Fig. 3-23(b) and 3-23(c) are discussed in Example 1 on the next page.)

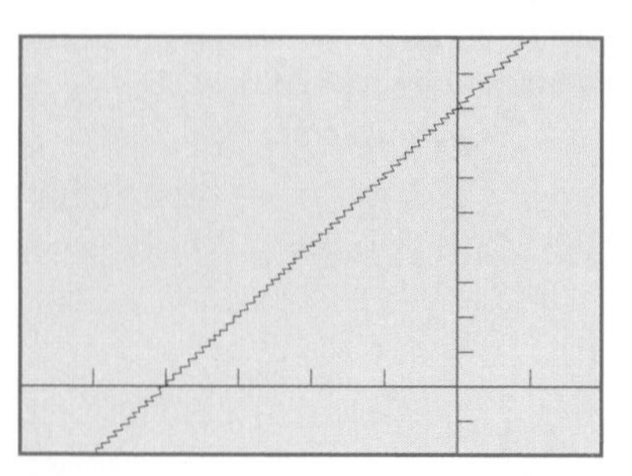

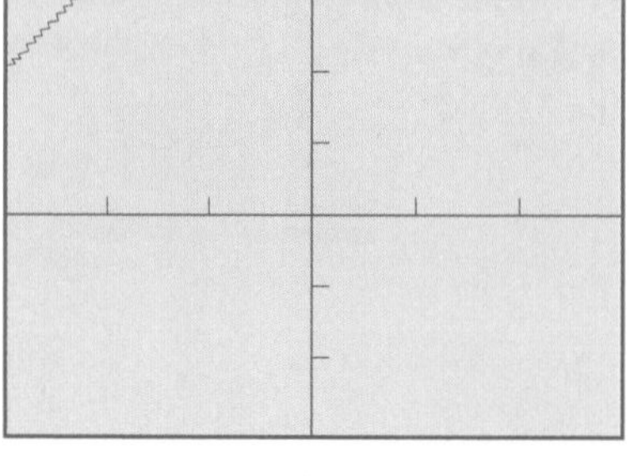

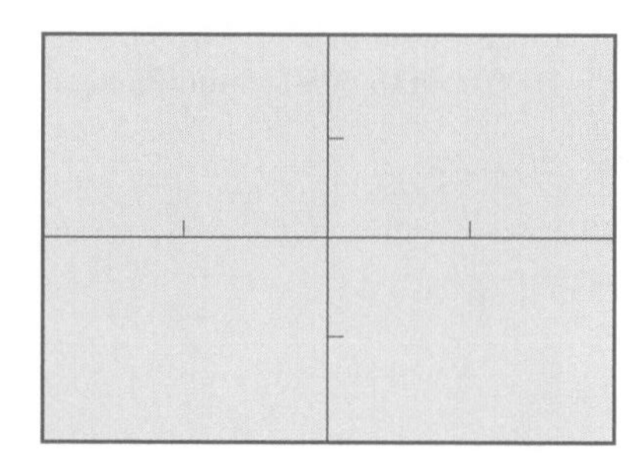

Fig. 3-23 (a) (b) (c)

In using a graphing calculator, particular attention must be paid to the RANGE key. It is used to set the intervals of the domain and range of the function, which are seen in the viewing window. The choice of these intervals affects what part of the graph is actually seen.

EXAMPLE 1 ▸▸ For the function $y = 2x + 8$, as shown in Fig. 3-23(a), the RANGE key was used to set the domain and range with

$$\text{Xmin} = -6,\ \text{Xmax} = 2,\ \text{Ymin} = -2,\ \text{Ymax} = 10$$

and we get a good view of the graph.

However, using Xmin $= -3$, Xmax $= 3$, Ymin $= -3$, and Ymax $= 3$, we get the view shown in Fig. 3-23(b). It is possible that no part of the graph will be seen, such as is shown in Fig. 3-23(c), where Xmin $= -2$, Xmax $= 2$, Ymin $= -2$, and Ymax $= 2$. As we see, care must be used in choosing these values.

Since the RANGE values are easily changed, at first it is usually best to choose intervals that are larger than is probably necessary to get the general location of the curve. These intervals can then be reduced to obtain the view needed.

Also note that the calculator constructs the graph just as we did in the previous section—by plotting points (called *pixels*). It just does it a lot faster. ◂◂

Solving Equations Graphically

An equation can be solved by use of a graph. Most of the time the solution will be approximate, but with a graphing calculator it is possible to get good accuracy in the result. The procedure used is as follows:

> *Procedure for Solving an Equation Graphically*
>
> **1.** *Collect all terms on one side of the equal sign.* This gives us the equation $f(x) = 0$.
>
> **2.** Set $y = f(x)$ *and graph this function.*
>
> **3.** *Find the points where the graph crosses the x-axis. These points are called the* **x-intercepts** *of the graph.* At these points, $y = 0$.
>
> **4.** *The values of x for which $y = 0$ are the solutions of the equation. (These values are called the* **zeros** *of the function $f(x)$.)*

The next example is done without the graphing calculator to illustrate the method.

EXAMPLE 2 ▸▸ Graphically solve the equation $x^2 - 2x = 1$.

Following the above procedure, we write the equation as $x^2 - 2x - 1 = 0$, and then set $y = x^2 - 2x - 1$. Next we plot the graph of this function as shown in Fig. 3-24.

For this graph we see that it crosses the x-axis between $x = -1$ and $x = 0$, and between $x = 2$ and $x = 3$. This means that there are *two* solutions of the equation, and they can be estimated as

$$x = -0.5 \text{ and } x = 2.5$$

For these values, $y = 0$. ◂◂

x	y
−2	7
−1	2
0	−1
1	−2
2	−1
3	2
4	7

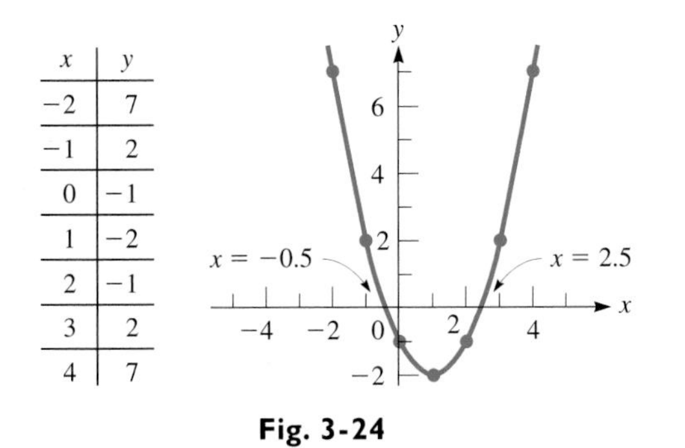

Fig. 3-24

We now show the use of the graphing calculator in solving equations graphically. In the next example we solve the equation of Example 2.

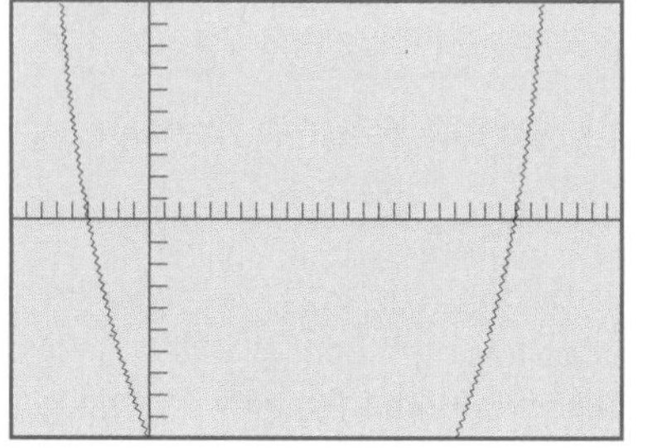

Fig. 3-25

EXAMPLE 3 ▸▸ Using the graphing calculator, solve the equation $x^2 - 2x = 1$.

Again, first write the equation as $x^2 - 2x - 1 = 0$, and then set $y = x^2 - 2x - 1$.

Knowing the approximate solution from Example 2, we use the RANGE key to set

$$\text{Xmin} = -1, \text{Xmax} = 3, \text{Ymin} = -1, \text{Ymax} = 1$$

and the scale on each axis at 0.1.

This gives us the view shown in Fig. 3-25. Then, using the TRACE feature, we can see that the solutions are approximately

$$x = -0.4 \quad \text{and} \quad x = 2.4$$

Using the ZOOM feature, the solutions can be found to a much greater accuracy.

Checking the solutions, we substitute the values into the original equation and get $0.96 = 1$ for each. This shows that the solutions check, since we know that they are approximate.

When using the TRACE feature, note that we must use the x-value which corresponds to the y-value closest to zero. Generally the pixel closest to the axis will not have a y-coordinate of *exactly* $y = 0$.

If we start the solution with the graphing calculator (assuming we did not plot it first), we then would probably need two or three settings for the domain and range (using the RANGE key) before getting a view that gives us reasonably accurate solutions. ◂◂

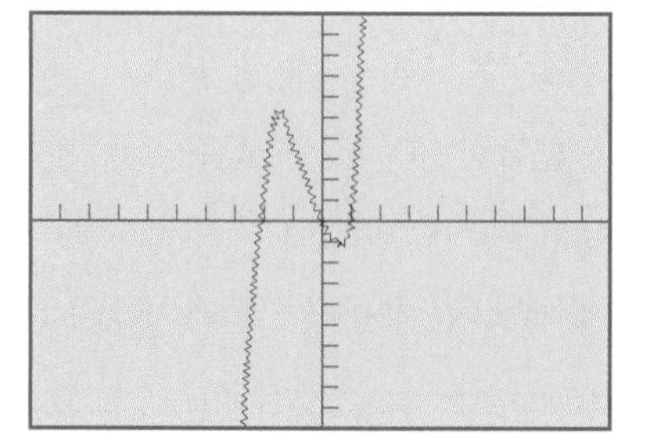

(a)

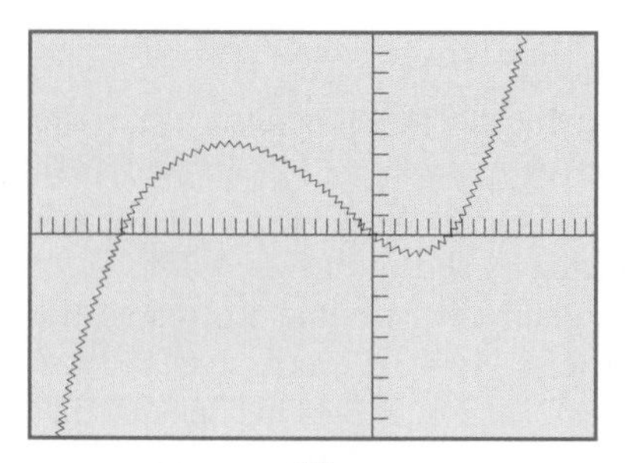

(b)

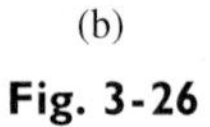

Fig. 3-26

EXAMPLE 4 ▸▸ Solve the equation $x^2(2x + 3) = 3x$ graphically.

Collecting all terms on the left leads to the equation $2x^3 + 3x^2 - 3x = 0$. We then let $y = 2x^3 + 3x^2 - 3x$.

Using a graphing calculator, we graph this function as shown in Fig. 3-26(a). Here, we have used the RANGE settings $\text{Xmin} = -10$, $\text{Xmax} = 10$, $\text{Ymin} = -10$, and $\text{Ymax} = 10$. (These are usually the *default* settings.)

This gives a general idea of the graph, but to get better accuracy we change the X settings to $\text{Xmin} = -3$ and $\text{Xmax} = 2$. We get the view shown in Fig. 3-26(b).

Now, using the TRACE feature, we get the solutions of

$$x = -2.2, x = 0, \text{and } x = 0.7$$

Here, $x = 0$ is exact since the calculator showed coordinates of $x = 0$, $y = 0$, with no other decimal positions displayed. The other solutions are approximate.

More accurate values can be obtained by using the ZOOM feature.

Checking the solutions, we find that they check (and also that $x = 0$ is exact). ◂◂

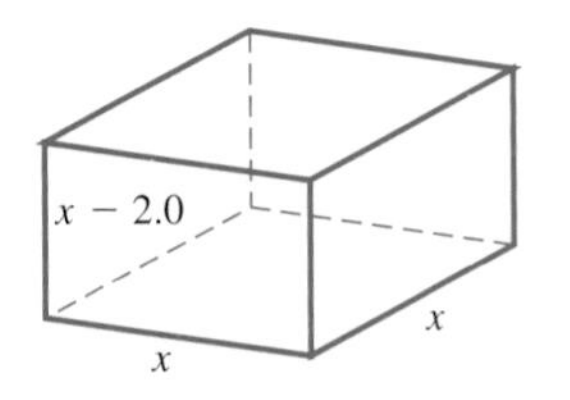

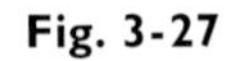

Fig. 3-27

EXAMPLE 5 ▸▸ A rectangular box whose volume is 30 cm^3 (two significant digits) is made with a square base and a height that is 2.0 cm less than the length of a side of the base. Find the dimensions of the box by first setting up the necessary equation and then solving it graphically.

Let x = the length of a side of the square base (see Fig. 3-27); the height is then $x - 2.0$. This means that the volume is $(x)(x)(x - 2.0)$, or $x^2(x - 2.0)$. Since the volume is 30 cm^3, we have the equation

$$x^2(x - 2.0) = 30$$

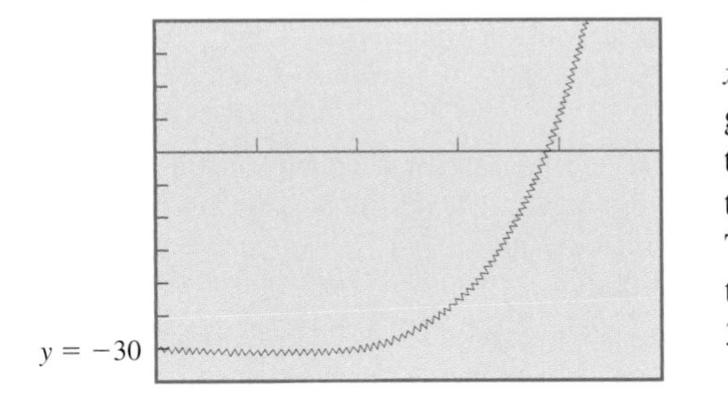

Fig. 3-28

To solve it graphically, we first write it as $x^3 - 2.0x^2 - 30 = 0$, and then set $y = x^3 - 2.0x^2 - 30$. A graphing calculator shows the view in Fig. 3-28. We use only positive values of x since negative values of x have no meaning. Using the TRACE feature, we see that $x = 3.9$ is the approximate solution. Thus, the dimensions are 3.9 cm, 3.9 cm, and 1.9 cm. Checking, these dimensions give a volume of 29 cm^3. The difference from 30 cm^3 is due to the approximate value of x. ◂◂

Finding the Range of a Function

In Section 3-2, we stated that advanced methods are often necessary to find the range of a function and that we would see that a graphing calculator is useful for this purpose. As with solving equations, we can get very good approximations of the range for most functions by using a graphing calculator.

The method is simply to graph the function and see for what values of y there is a point on the graph. These values of y give us the range of the function.

EXAMPLE 6 ▸▸ Graphically find the range of the function $y = 16\sqrt{x} + \dfrac{1}{x}$.

If we start by using the default RANGE settings of Xmin = −10, Xmax = 10, Ymin = −10, Ymax = 10, we see no part of the graph. Looking back to the function, we note that x cannot be zero because that would require dividing by zero in the $1/x$ term. Also, $x > 0$ since $\sqrt{x}$ is not defined for $x < 0$. Thus, we should try Xmin = 0 and Ymin = 0.

We also can easily see that $y = 17$ for $x = 1$. This means we should try a value more than 17 for Ymax. Using Ymax = 20, after two or three RANGE settings, we use Xmin = 0, Xmax = 2, Ymin = 10, Ymax = 20 to get the view in Fig. 3-29. It shows the range to be all real numbers greater than about 12. (Actually, the range is all real numbers $y \geq 12$.)

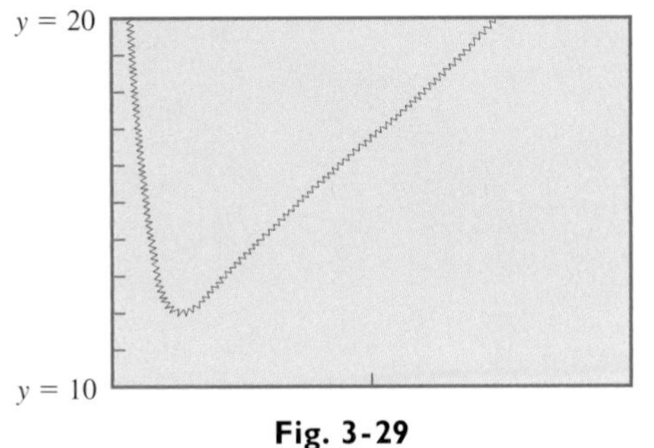

Fig. 3-29 ◂◂

Exercises 3–5

In Exercises 1–16, use a graphing calculator to solve the given equations to the nearest 0.1.

1. $7x - 5 = 0$

2. $8x + 3 = 0$

3. $x^2 - 41 = 0$

4. $2x^2 = 1$

5. $x^2 + x - 5 = 0$

6. $x^2 - 2x - 4 = 0$

7. $2x^2 - x = 7$

8. $x(x - 4) = 9$

9. $x^3 = 4x$

10. $x^3 - 3 = 3x$

11. $x^4 - 2x = 0$

12. $5x = 2x^5$

13. $\sqrt{2x + 2} = 3$

14. $\sqrt{x} + 3x = 7$

15. $\dfrac{1}{x^2 + 1} = 0$

16. $x - 2 = \dfrac{1}{x}$

In Exercises 17–24, use a graphing calculator to find the range of the given functions. (The functions of Exercises 21–24 are the same as the functions of Exercises 9–12 of Section 3-2.)

17. $y = \dfrac{4}{x^2 - 4}$

18. $y = \dfrac{x + 1}{x^2}$

19. $y = \dfrac{x^2}{x + 1}$

20. $y = \dfrac{x}{x^2 - 4}$

21. $Y(y) = \dfrac{y + 1}{\sqrt{y - 2}}$

22. $f(n) = \dfrac{n}{6 - 2n}$

23. $f(u) = \dfrac{u}{u - 2} + \dfrac{4}{u + 4}$

24. $g(x) = \dfrac{\sqrt{x - 2}}{x - 3}$

In Exercises 25–32, solve the indicated equations graphically. Assume all data are accurate to two significant digits unless greater accuracy is given.

25. In an electric circuit, the current i, in amperes, as a function of the voltage v is given by $i = 0.01v - 0.06$. Find v for $i = 0$.

26. For tax purposes, a corporation assumes that one of its computers depreciates in value according to the equation $V = 90\,000 - 12\,000t$, where V is the value, in dollars, of the computer after t years. According to this formula, when will the computer be fully depreciated (no value)?

27. A cubical cooler holds $0.026\ \text{m}^3$. Find its inside dimension.

28. The pressure loss P (in kPa/100 m) in a fire hose is given by $P = 0.000\,12Q^2 + 0.0055Q$, where Q is the rate of flow (in L/min). Find Q if $P = 15$ kPa.

29. The length of a rectangular solar panel is 12 cm more than its width. If its area is $520\ \text{cm}^2$, find its dimensions.

30. The height h, in metres, of a rocket as a function of the time t, in seconds, of flight is given by $h = 15 + 86t - 4.9t^2$. Determine when the rocket is at ground level.

31. In finding the illumination at a point x metres from one of two light sources that are 100 m apart, it is necessary to solve the equation $9x^3 - 2400x^2 + 240\,000x - 8\,000\,000 = 0$. Find x.

32. A rectangular storage bin is to be made from a rectangular piece of sheet metal 12 cm by 10 cm, by cutting out equal corners of side x and bending up the sides. See Fig. 3-30. Find x if the storage bin is to hold $90\ \text{cm}^3$.

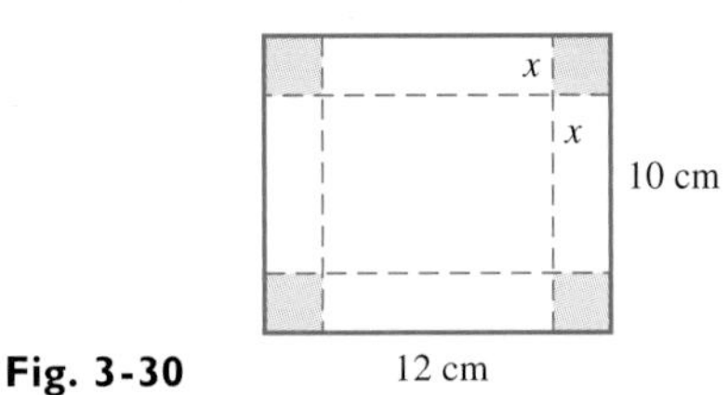

Fig. 3-30

In Exercises 33–36, solve the given problems.

33. Solve the equation of Example 3 to the nearest 0.001 by using the ZOOM feature of a graphing calculator.

34. Solve the equation of Example 4 to the nearest 0.001 by using the ZOOM feature of the graphing calculator.

35. The cutting speed s (in cm/s) of a saw in cutting a particular type of metal piece is given by $s = \sqrt{t - 4t^2}$, where t is the time in seconds. What is the maximum cutting speed in this operation (to two significant digits)? (*Hint:* Find the range.)

36. Referring to Exercise 32, explain how to determine the maximum possible capacity for a storage bin constructed in this way. What is the maximum possible capacity (to three significant digits)?

3-6 Graphs of Functions Defined by Tables of Data

In Section 3-4 we showed how to construct the graph of a function, where the function was defined by a formula or an equation. However, as we stated in Section 3-1, there are other ways to express functions. One of the most important ways to show the relationship between variables is by using a table of values obtained by observation or experimentation.

Often the data from an experiment indicate that the variables could have a formula that relates them, although the formula may not be known. Data from experiments from the various fields of science and technology generally have variables that are related in such a way. In this case, when we are plotting the graph, the points should be connected by a smooth curve.

Statistical data in tabular form often give values that are taken only for certain intervals or are averaged over specified intervals. In such cases, no real meaning can be given to the intervals between the points on the graph. On these graphs, the points should be connected by straight-line segments, where this is done only to make the points stand out better and make the graph easier to read. Example 1 illustrates this type of graph.

EXAMPLE 1 ▸▸ The electric energy usage (in MJ) for a particular all-electric house for each month of a certain year is given in the following table. Plot these data.

Month	Jan	Feb	Mar	Apr	May	Jun
Energy usage	10 504	12 363	10 168	7500	4825	3568

Month	July	Aug	Sep	Oct	Nov	Dec
Energy usage	2548	2887	3301	5748	7302	9706

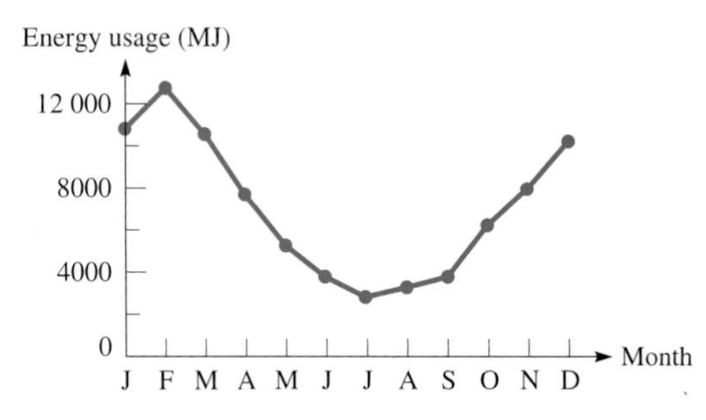

Fig. 3-31

We know that there is no meaning to the intervals between the months, since we have the total number of megajoules for each month. Therefore, we use straight-line segments, but only to make the points stand out better. ◂◂

In the next example, we illustrate data that give a graph in which the points are connected by a smooth curve and not by straight-line segments.

EXAMPLE 2 ▸▸ Steam in a boiler was heated to 150°C. Its temperature was recorded each minute, giving the values in the following table. Plot the graph.

Time (minutes)	0.0	1.0	2.0	3.0	4.0	5.0
Temperature (°C)	150.0	142.8	138.5	135.2	132.7	130.8

Since the temperature changes in a continuous way, there is meaning to the values in the intervals between points. Therefore, these points are joined by a smooth curve, as in Fig. 3-32. Also note that most of the vertical scale was used for the required values, with the indicated break in the scale between 0 and 130. ◂◂

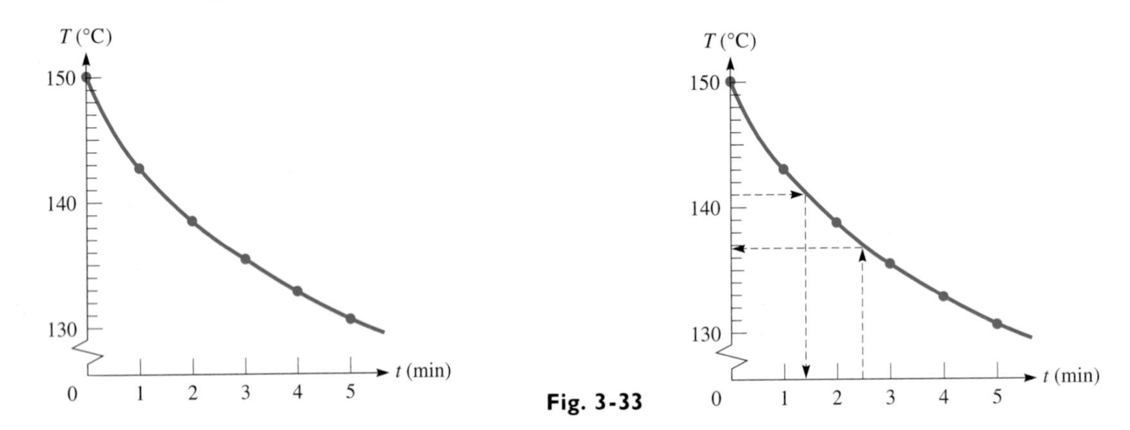

Fig. 3-32

Fig. 3-33

If the graph relating two variables is known, values can be obtained directly from the graph. In finding such a value through the inspection of the graph, we are *reading the graph.*

EXAMPLE 3 ▸▸ For the cooling steam in Example 2, we can estimate values of one variable for given values of the other.

If we want to know the temperature after 2.5 min, we estimate 0.5 of the interval between 2 and 3 on the t-axis and mark it. See Fig. 3-33. Then we draw a line vertically from this point to the graph. From the point where it intersects the graph, we draw a horizontal line to the T-axis. We now estimate that the line crosses at $T = 136.7$°C. Obviously, the number of tenths is a rough estimate.

In the same way, if we want to determine how long the steam took to cool down to 141.0°, we go from 141.0 on the T-axis to the graph and then to the t-axis. This crosses at about $t = 1.4$ min. ◂◂

Linear Interpolation

In Example 3 we see that we can estimate values from a graph. However, unless a very accurate graph is drawn with expanded scales for both variables, only very approximate values can be found. There is a method, called *linear interpolation,* which uses the table itself to get more accurate results.

Linear interpolation *assumes that if a particular value of one variable lies between two of those listed in the table, then the corresponding value of the other variable is at the same proportional distance between the listed values.* On the graph, linear interpolation assumes that two points defined in the table are connected by a straight line. Although this is generally not correct, it is a good approximation if the values in the table are sufficiently close together.

EXAMPLE 4 ▸▸ For the cooling steam in Example 2, we can use interpolation to find its temperature after 1.4 min. Since 1.4 min is $\frac{4}{10}$ of the way from 1.0 min to 2.0 min, we shall assume that the value of T we want is $\frac{4}{10}$ of the way between 142.8 and 138.5, the values of T for 1.0 min and 2.0 min, respectively. The difference between these values is 4.3, and $\frac{4}{10}$ of 4.3 is 1.7 (rounded off to tenths). Subtracting (the values of T are decreasing) 1.7 from 142.8, we obtain 141.1. Thus, the required value of T is about 141.1°C. (Note that this agrees well with the result in Example 3.)

Another method of indicating the interpolation is shown in Fig. 3-34. From the figure we have the proportion

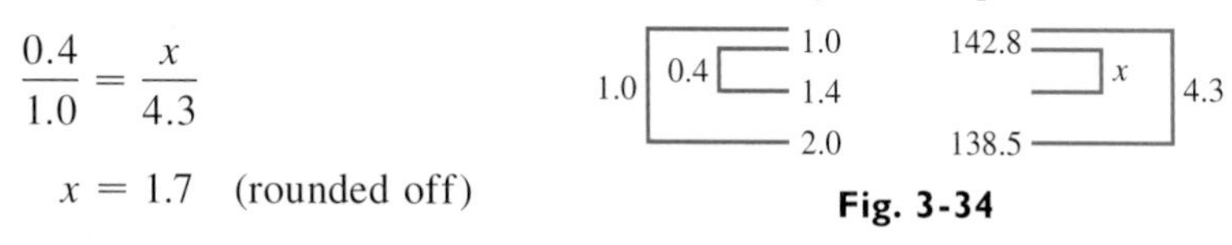

Fig. 3-34

$$\frac{0.4}{1.0} = \frac{x}{4.3}$$

$$x = 1.7 \quad \text{(rounded off)}$$

Therefore,

$$142.8 - 1.7 = 141.1°\text{C}$$

is the required value of T. If the values of T had been increasing, we would have added 1.7 to the value of T for 1.0 min. ◂◂

Exercises 3–6

In Exercises 1–8, represent the data graphically.

1. The diesel fuel production, in thousands of litres, at a certain refinery during an 8-week period was as follows.

Week	1	2	3	4	5	6	7	8
Production	765	780	840	850	880	840	760	820

2. The *exchange rate* for the number of Canadian dollars equal to one United States dollar for 1980–1994 is given in the following table.

Year	1980	1982	1984	1986	1988	1990	1992	1994
No. Can. dollars	1.17	1.23	1.30	1.39	1.23	1.17	1.21	1.38

3. The amount of material necessary to make a cylindrical gallon container depends on the diameter, as shown in the following table.

Diameter (cm)	6.0	8.0	10	12	14	16
Material (cm²)	723	601	557	560	594	652

4. An oil burner propels air that has been heated to 90°C. The temperature then drops as the distance from the burner increases, as shown in the following table.

Distance (m)	0.0	1.0	2.0	3.0	4.0	5.0	6.0
Temperature (°C)	90	84	76	66	54	46	41

5. A changing electric current in a coil of wire will induce a voltage in a nearby coil. Important in the design of transformers, the effect is called *mutual inductance*. For two coils, the mutual inductance, in henrys, as a function of the distance between them is given in this table.

Distance (cm)	0.0	2.0	4.0	6.0	8.0	10.0	12.0
M. ind. (H)	0.77	0.75	0.61	0.49	0.38	0.25	0.17

6. The temperatures felt by the body as a result of the *wind chill factor* for an outside temperature of 0°C are given in the following table.

Wind speed (km/h)	10	20	30	40	50	60	70
Temp. felt (°C)	−3	−9	−13	−15	−17	−18	−19

7. The time required for a sum of money to double in value, when compounded annually, is given as a function of the interest rate in the following table.

Rate (%)	4	5	6	7	8	9	10
Time (years)	17.7	14.2	11.9	10.2	9.0	8.0	7.3

8. The force required to break a metal rod, as a function of its diameter, is given in the following table.

Diameter (cm)	1.0	1.5	2.0	2.5	3.0	3.5	4.0
Force (N)	520	1250	2050	3100	4250	5800	8100

In Exercises 9 and 10, use the graph in Fig. 3-33, which relates the temperature of cooling steam and the time. Find the indicated values by reading the graph.

9. (a) For $t = 4.3$ min, find T. (b) For $T = 145.0$°C, find t.

10. (a) For $t = 1.8$ min, find T. (b) For $T = 133.5$°C, find t.

In Exercises 11 and 12, use the following table, which gives the voltage produced by a certain thermocouple as a function of the temperature of the thermocouple. Plot the graph. Find the indicated values by reading the graph.

Temperature (°C)	0	10	20	30	40	50
Voltage (volts)	0.0	2.9	5.9	9.0	12.3	15.8

11. (a) For $T = 26$°C, find V. (b) For $V = 13.5$ V, find T.

12. (a) For $T = 32$°C, find V. (b) For $V = 4.8$ V, find T.

In Exercises 13–16, find the indicated values by means of linear interpolation.

13. In Exercise 5, find the inductance for $d = 9.2$ cm.

14. In Exercise 6, find the temperature for $s = 23$ km/h.

15. In Exercise 7, find the rate for $t = 10.0$ years.

16. In Exercise 8, find the diameter for $F = 6500$ N.

In Exercises 17–20, use the following table, which gives the rate R of discharge from a tank of water as a function of the height H of water in the tank. For Exercises 17 and 18, plot the graph and find the values from the graph. For Exercises 19 and 20, find the indicated values by means of linear interpolation. See Fig. 3-35.

Height (cm)	0	50	100	200	300	400	600
Rate (m^3/s)	0	1.0	1.5	2.2	2.7	3.1	3.5

17. (a) For $R = 2.0\ m^3/s$, find H. (b) For $H = 240$ cm, find R.

18. (a) For $R = 3.4\ m^3/s$, find H. (b) For $H = 320$ cm, find R.

19. Find R for $H = 80$ cm. **20.** Find H for $R = 2.5\ m^3/s$.

In Exercises 21–24, use the following table, which gives the fraction (as a decimal) of the total heating load of a certain system that will be supplied by a solar collector of area A (in m^2). Find the indicated values by means of linear interpolation.

f	0.22	0.30	0.37	0.44	0.50	0.56	0.61
A (m^2)	20	30	40	50	60	70	80

21. For $A = 36\ m^2$, find f. **22.** For $A = 52\ m^2$, find f.

23. For $f = 0.59$, find A. **24.** For $f = 0.27$, find A.

In Exercises 25–28, a method of finding values beyond those given is considered. By using a straight-line segment to extend a graph beyond the last known point we can estimate values from the extension of the graph. The method is known as **linear extrapolation**. *Use this method to estimate the required values from the given graphs.*

25. Using Fig. 3-33, estimate T for $t = 5.3$ min.

26. Using the graph for Exercises 11 and 12, estimate V for $T = 55$°C.

27. Using the graph for Exercises 17–20, estimate R for $H = 700$ cm.

28. Using the graph for Exercises 17–20, estimate R for $H = 800$ cm.

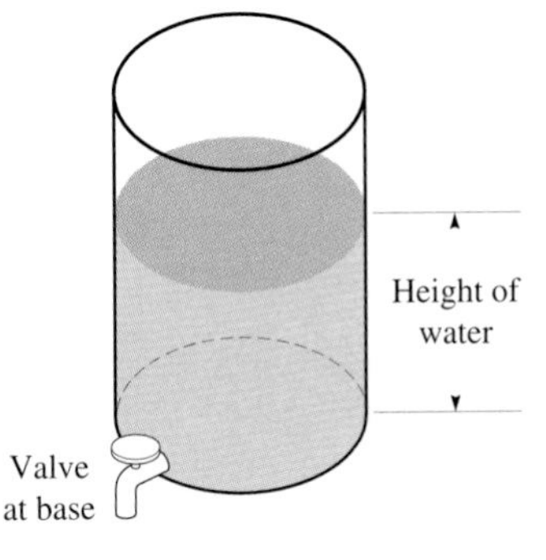

Fig. 3-35

Review Exercises and Practice Test

Review Exercises

In Exercises 1–4, determine the appropriate function.

1. The radius of a circular water wave increases at the rate of 2 m/s. Express the area of the circle as a function of the time t (in seconds).

2. A conical sheet metal hood is to cover an area 6 m in diameter. Find the total surface area of the hood as a function of its height h.

3. One computer printer prints at the rate of 2000 lines/min for x minutes, and a second printer prints at the rate of 1800 lines/min for y minutes. Together they print 50 000 lines. Find y as a function of x.

4. Fencing around a rectangular storage depot area costs twice as much along the front as along the other three sides. The back costs \$10 per metre. Express the cost C of the fencing as a function of the width if the length (along the front) is 20 m more than the width.

In Exercises 5–12, evaluate the given functions.

5. Given $f(x) = 7x - 5$, find $f(3)$ and $f(-6)$.

6. Given $g(x) = 8 - 3x$, find $g\left(\frac{1}{6}\right)$ and $g(-4)$.

7. Given $F(u) = 3u - 2u^2$, find $F(-1)$ and $F(-0.3)$.

8. Given $h(y) = 2y^2 - y - 2$, find $h(5)$ and $h(-3)$.

9. Given $H(h) = \sqrt{1 - 2h}$, find $H(-4)$ and $H(2h)$.

10. Given $\phi(v) = \dfrac{3v - 2}{v + 1}$, find $\phi(-2)$ and $\phi(v + 1)$.

11. Given $f(x) = 3x^2 - 2x + 4$, find $f(x + h) - f(x)$.

12. Given $F(x) = x^3 + 2x^2 - 3x$, find $F(3 + h) - F(3)$.

In Exercises 13–16, evaluate the given functions. Values of the independent variable are approximate.

13. Given $f(x) = 8.07 - 2x$, find $f(5.87)$ and $f(-4.29)$.

14. Given $g(x) = 7x - x^2$, find $g(45.81)$ and $g(-21.85)$.

15. Given $G(y) = \dfrac{y - 0.087\,629}{3y}$, find $G(0.174\,27)$ and $G(0.053\,206)$.

16. Given $h(t) = \dfrac{t^2 - 4t}{t^3 + 564}$, find $h(8.91)$ and $h(-4.91)$.

In Exercises 17–20, determine the domain and the range of the given functions.

17. $f(x) = x^4 + 1$

18. $G(z) = \dfrac{4}{z^3}$

19. $g(t) = \dfrac{2}{\sqrt{t + 4}}$

20. $F(y) = 1 - 2\sqrt{y}$

In Exercises 21–32, plot the graphs of the given functions. Check these graphs by using a graphing calculator.

21. $y = 4x + 2$

22. $y = 5x - 10$

23. $y = 4x - x^2$

24. $y = x^2 - 8x - 5$

25. $y = 3 - x - 2x^2$

26. $y = 6 + 4x + x^2$

27. $y = x^3 - 6x$

28. $y = 3 - x^3$

29. $y = 2 - x^4$

30. $y = x^4 - 4x$

31. $y = \dfrac{x}{x + 1}$

32. $y = \sqrt{25 - x^2}$

In Exercises 33–40, use a graphing calculator to solve the given equations to the nearest 0.1.

33. $7x - 3 = 0$

34. $3x + 11 = 0$

35. $x^2 + 1 = 6x$

36. $3x - 2 = x^2$

37. $x^3 - x^2 = 2 - x$

38. $5 - x^3 = 2x^2$

39. $\dfrac{1}{x} = 2x$

40. $\sqrt{x} = 2x - 1$

In Exercises 41–44, use a graphing calculator to find the range of the given function.

41. $y = x^4 - 5x^2$

42. $y = x\sqrt{4 - x^2}$

43. $y = x + \dfrac{2}{x}$

44. $y = 2x + \dfrac{3}{\sqrt{x}}$

In Exercises 45–52, answer the given questions.

45. Explain how $A(a, b)$ and $B(b, a)$ may be in different quadrants.

46. Determine the distance from the origin to the point (a, b).

47. Two vertices of an equilateral triangle are $(0, 0)$ and $(2, 0)$. What is the third vertex?

48. The points $(1, 2)$ and $(1, -3)$ are two adjacent vertices of a square. Find the other vertices.

49. An equation used in electronics with a transformer antenna is $I = 12.5\sqrt{1 + 0.5m^2}$. For $I = f(m)$, find $f(0.55)$.

50. The percent p of wood lost in cutting it into boards 38 mm thick due to the thickness t (in mm) of the saw blade is $p = \dfrac{100t}{t + 38}$. Find p if $t = 5.0$ mm. That is, since $p = f(t)$, find $f(5.0)$.

51. The angle A (in degrees) of a robot arm with the horizontal as a function of time t (in seconds, for 0.0 s to 6.0 s) is given by $A = 8.0 + 12t^2 - 2.0t^3$. What is the greatest value of A to the nearest 0.1°? (*Hint:* Find the range.)

52. The electric power P (in watts) produced by a battery is $P = \dfrac{24R}{R^2 + 1.40R + 0.49}$, where R is the resistance in the circuit. What is the maximum power produced? (*Hint:* Find the range.) See Example 6 on p. 89.

In Exercises 53–64, plot the graphs of the indicated functions.

53. In determining the forces that act on a certain structure, it was found that two of them satisfied the equation $F_2 = 0.75F_1 + 45$, where the forces are in newtons. Plot the graph of F_2 as a function of F_1.

54. A company which makes digital clocks determines that in order to sell x clocks, the price must be $p = 50 - 0.25x$. Plot the graph of p as a function of x, for $x = 1$ to $x = 60$. (Although only the points for integral values of x have real meaning, join the points.)

55. The thermodynamic temperature T (in kelvins) of an object is 273 more than the Celsius reading C. Plot the graph of $T = f(C)$.

56. There are 5000 L of oil in a tank that has a capacity of 100 000 L. It is filled at the rate of 7000 L/h. Determine the function relating the number of liters N and the time t while the tank is being filled. Plot N as a function of t.

57. For a certain laser device, the laser output power P, in milliwatts, is negligible if the drive current i, in milliamperes, is less than 80 mA. From 80 mA to 140 mA, $P = 1.5 \times 10^{-6}i^3 - 0.77$. Plot the graph of $P = f(i)$.

58. It is determined that a good approximation for the cost C, in cents per kilometre, of operating a certain car at a constant speed v (in km/h) is given by $C = 0.025v^2 - 1.4v + 35$. Plot C as a function of v for $v = 10$ km/h to $v = 60$ km/h.

59. A medical researcher exposed a virus culture to an experimental vaccine. It was observed that the number of live cells N in the culture as a function of the time t, in hours, after exposure was given by $N = \dfrac{1000}{\sqrt{t+1}}$. Plot the graph of $N = f(t)$.

60. The electric field E, in volts per metre, from a certain electric charge is given by the function $E = 25/r^2$, where r is the distance, in metres, from the charge. Plot the graph of $E = f(r)$ for values of r up to 10 cm.

61. To draw the approximate shape of an irregular shoreline, a surveyor measured the distances from a straight wall to the shoreline at 20-m intervals along the wall. The measurements are in the following table. Plot the graph of the distance d to the shoreline as a function of the distance D along the wall.

D (m)	0	20	40	60	80	100	120	140	160
d (m)	15	32	56	33	29	47	68	31	52

62. The percent of a computer network that is in use during a particular loading cycle as a function of the time, in seconds, is given in the following table. Plot the graph of the function.

Time (s)	0.0	0.2	0.4	0.6	0.8	1.0	1.2	1.4	1.6
(%)	0	45	85	90	85	85	60	10	0

63. In an experiment measuring the pressure p, in kilopascals, at a given depth d, in metres, of seawater, the results in the following table were found. Plot the graph of $p = f(d)$, and from the graph determine $f(10)$.

Depth (m)	0.0	3.0	6.0	9.0	12	15
Pressure (kPa)	101	131	161	193	225	256

64. The vertical sag s, in metres, at the middle of an 800-m power line, as a function of the temperature T (in °C) is given in the following table. For $s = f(T)$, find $f(14)$ by linear interpolation.

T (°C)	−10	0	10	20
s (m)	3.11	3.23	3.38	3.57

In Exercises 65–72, solve the indicated equations graphically.

65. A box is suspended by two ropes, as shown in Fig. 3-36. In analyzing the tensions in the supporting ropes, it was found that they satisfied the following equation: $0.60T_1 + 0.87T_2 = 10.0$. Find T_1 if $T_2 = 8.0$ N.

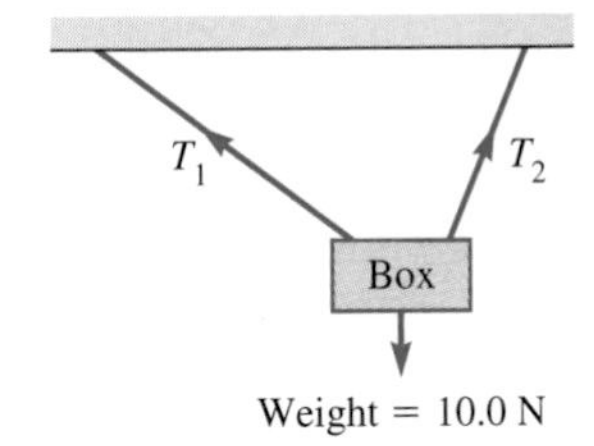

Fig. 3-36

66. One industrial cleaner contains 30% of a certain solvent, and another contains 10% of the solvent. To get a mixture containing 50 L of the solvent, 120 L of the first cleaner is used. How much of the second must be used?

67. The solubility s, in kilograms per cubic metre of water, of a certain type of fertilizer is given by $s = 135 + 4.9T + 0.19T^2$, where T is the temperature in degrees Celsius. Find T for $s = 500$ kg/m^3.

68. The rectangular cross section of a beam has a perimeter of 420 mm. Express the cross-sectional area A as a function of the width w of the beam. Then find w for $A = 9600$ mm^2.

69. In an oil pipeline, the velocity v (in m/s) of the oil as a function of the distance x (in m) from the wall of the pipe is $v = 9.6x - 7.5x^2$. Find x for $v = 2.6$ m/s. The diameter of the pipe is 1.20 m.

70. One ball bearing is 1.00 mm more in radius and has twice the volume of another ball bearing. What is the radius of each?

71. A computer, using data from a refrigeration plant, estimates that in the event of a power failure, the temperature, in degrees Celsius, in the freezers would be given by $T = \frac{4t^2}{t + 2} - 20$, where t is the number of hours after the power failure. How long would it take for the temperature to reach 0°C?

72. Two electrical resistors in parallel have a combined resistance R_T, in ohms, given by $R_T = \frac{R_1 R_2}{R_1 + R_2}$. If $R_2 = R_1 + 2.0$, express R_T as a function of R_1 and find R_1 if $R_T = 6.0\ \Omega$.

Writing Exercise

73. In one or two paragraphs, explain how you would solve the following problem using a graphing calculator: The inner surface area A of a 250.0-cm^3 cylindrical cup as a function of the radius r of the base is $A = 3.1416r^2 + \frac{500.0}{r}$. Find r if $A = 175.0$ cm^2. (What is the answer?) (See if you can derive the formula.)

Practice Test

1. Given $f(x) = 2x - x^2 + \frac{8}{x}$, find $f(-4)$ and $f(2.385)$.

2. A rocket has a mass of 2000 Mg at liftoff. If the first-stage engines burn fuel at the rate of 10 Mg/s, find the mass m of the rocket as a function of the time t while the first-stage engines operate.

3. Plot the graph of the function $f(x) = 4 - 2x$.

4. Use a graphing calculator to solve the equation $2x^2 - 3 = 3x$ to the nearest 0.1.

5. Plot the graph of the function $y = \sqrt{4 + 2x}$.

6. Locate all points (x, y) for which $x < 0$ and $y = 0$.

7. Find the domain and the range of the function $f(x) = \sqrt{6 - x}$.

8. A window has the shape of a semicircle over a square, as shown in Fig. 3-37. Express the area of the window as a function of the radius of the circular part.

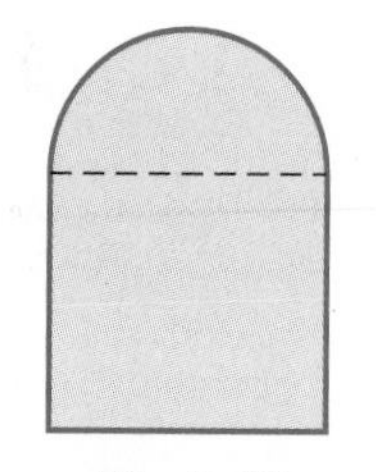

Fig. 3-37

9. The voltage V and current i, in milliamperes, for a certain electrical experiment were measured as shown in the following table. Plot the graph of $i = f(V)$, and from the graph find $f(45.0)$.

Voltage (V)	10.0	20.0	30.0	40.0	50.0	60.0
Current (mA)	145	188	220	255	285	315

10. From the table in Problem 9, find the voltage for $i = 200$ mA.

4

The Trigonometric Functions

A highway engineer must be sure that you can read the word STOP when you are approaching an intersection. In Section 4-5 we show how we can measure fhe angle through which STOP is seen by the driver.

Many applied problems in science and technology require the use of triangles, especially right triangles, for their solution. Included among these are problems in air navigation, surveying, the motion of rockets and missiles, and optics. Problems involving forces acting on objects and the measurement of distances between various parts of the universe can also be solved. Even certain types of electric circuits are analyzed by the use of triangles.

In **trigonometry** we develop methods for measuring sides and angles of triangles as well as for solving related applied problems. Because of the extensive use of these concepts, trigonometry is considered one of the most practical and relevant branches of mathematics.

In this chapter we introduce the basic trigonometric functions and show a number of applications of right triangles from many areas of science and technology. In later chapters we will use the trigonometric functions with other types of triangles. We will also see these important functions applied to periodic motion (for example, mechanical vibrations and various types of wave motion).

4-1 Angles

In the review of geometry in Chapter 2, we stated a basic definition of an *angle.* In this section we extend this definition and give other important definitions related to angles.

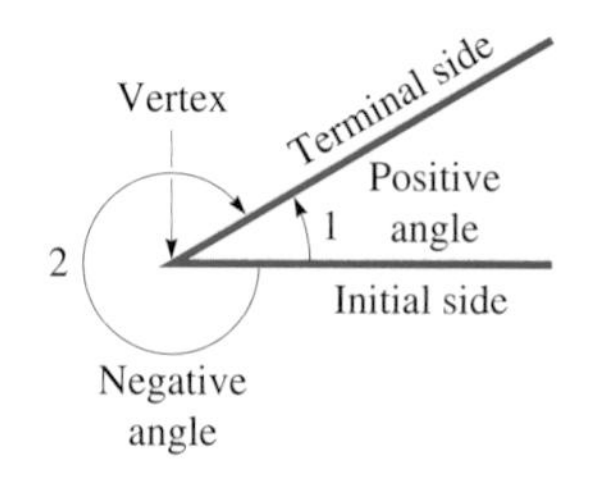

Fig. 4-1

An **angle** *is generated by rotating a ray about its fixed endpoint from an* ***initial position*** *to a* ***terminal position****. The initial position is called the* **initial side** *of the angle, the terminal position is called the* **terminal side**, *and the fixed endpoint is the* **vertex**. The angle itself is the amount of rotation from the initial side to the terminal side.

If the rotation of the terminal side from the initial side is ***counterclockwise****, the angle is defined as* ***positive****. If the rotation is* ***clockwise****, the angle is* ***negative***. In Fig. 4-1, $\angle 1$ is positive and $\angle 2$ is negative.

There are many symbols used to designate angles. Among the most widely used are certain Greek letters such as θ (theta), ϕ (phi), α (alpha), and β (beta). Capital letters representing the vertex (e.g., $\angle A$ or simply A) and other literal symbols, such as x and y, are also used commonly.

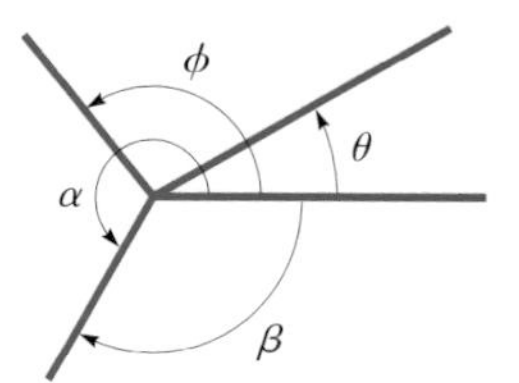

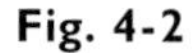
Fig. 4-2

In Chapter 2 we introduced two measurements of an angle. These are the *degree* and the *radian*. Since degrees and radians are both used on calculators and computers, we will briefly review the relationship between them in this section. However, we will not make use of radians until Chapter 8.

From Section 2-1 we recall that *a* **degree** *is* 1/360 *of one complete rotation*. In Fig. 4-2, $\angle\theta = 30°$, $\angle\phi = 140°$, $\angle\alpha = 240°$, and $\angle\beta = -120°$. Note that β is drawn in a clockwise direction to show that it is negative. The other angles are drawn in a counterclockwise direction to show that they are positive angles.

In Chapter 2 we used degrees and decimal parts of a degree. Most calculators use degrees in this decimal form. Another traditional way is to divide a degree into 60 equal parts called **minutes**; each minute is divided into 60 equal parts called **seconds.** The symbols ′ and ″ are used to designate minutes and seconds, respectively.

In Fig. 4-2 we note that angles α and β have the same initial and terminal sides. *Such angles are called* **coterminal angles**. An understanding of coterminal angles is important in certain concepts of trigonometry.

EXAMPLE 1 ⏩ Determine the values of two angles that are coterminal with an angle of 145.6°.

Since there are 360° in a complete rotation, we can find one coterminal angle by considering the angle that is 360° larger than the given angle. This gives us an angle of 505.6°. Another method of finding a coterminal angle is to subtract 145.6° from 360° and then consider the resulting angle to be negative. This means that the original angle and the derived angle would make up one complete rotation, when put together. This method leads us to the angle of −214.4° (see Fig. 4-3). These methods could be employed repeatedly to find other coterminal angles. ⏪

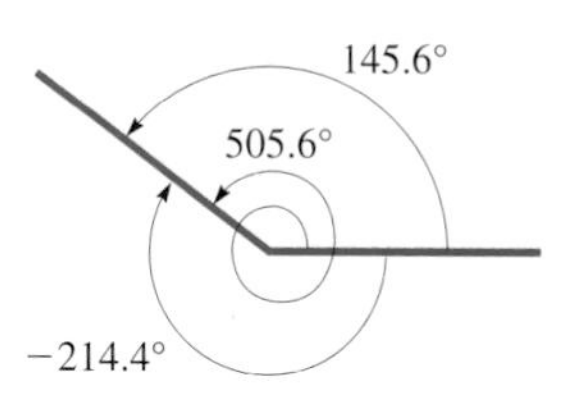

Fig. 4-3

Angle Conversions

Although we will use only degrees in this chapter, when using a calculator or computer we must be careful to be in the proper mode. Also, in using a computer it may be necessary to change an angle expressed in radians to one expressed in degrees. Therefore, at this point we briefly note that this can be done on a calculator by use of the DRG key (see Appendix C). Also, we can use the definition of a radian in Section 2-4, which leads to the fact that π rad = 180°.

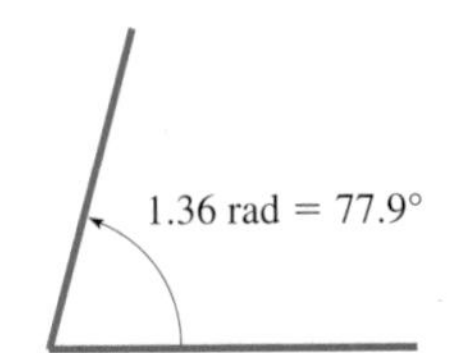

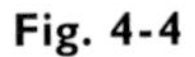
Fig. 4-4

EXAMPLE 2 ▸▸ Express 1.36 rad in degrees.

We know that π rad = 180°, which means 1 rad = $180°/\pi$. Therefore,

$$1.36 \text{ rad} = 1.36\left(\frac{180°}{\pi}\right) = 77.9° \qquad \text{to nearest } 0.1°$$

This angle is shown in Fig. 4-4. We again note that degrees and radians are simply two different ways of measuring an angle. ◂◂

As stated previously, the radian is discussed in detail in Chapter 8. It might be noted here that the "G" on the DRG key of a calculator represents *grads*, where 100 grad = 90°. We will not use grads in this text.

Traditionally it was common to use degrees and minutes in tables, and calculators use degrees and decimal parts. Changing from one form to the other can be done directly on many calculators by using the DMS (or °′″) key (see Appendix C). It can also be done by using the meanings of minutes and seconds.

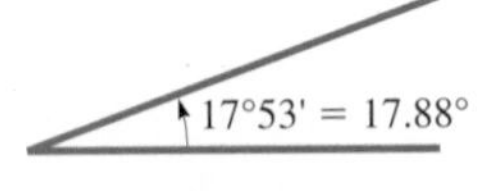

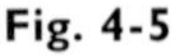
Fig. 4-5

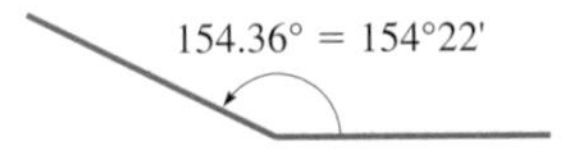

Fig. 4-6

EXAMPLE 3 ▸▸ To change 17°53′ to decimal form, we use the fact that $1° = 60'$. Therefore, $53' = \left(\frac{53}{60}\right)^{\circ} = 0.88°$ to nearest 0.01°

This means that 17°53′ = 17.88°. This angle is shown in Fig. 4-5.

To change 154.36° to an angle measured to the nearest minute, we have

$$0.36° = 0.36(60') = 22'$$

This means that 154.36° = 154°22′. See Fig. 4-6. ◂◂

Standard Position of an Angle

If the initial side of the angle is the positive x-axis and the vertex is the origin, the angle is said to be in **standard position**. The angle is then determined by the position of the terminal side. *If the terminal side is in the first quadrant, the angle is called a* **first-quadrant angle**. Similar terms are used when the terminal side is in the other quadrants. *If the terminal side coincides with one of the axes, the angle is a* **quadrantal angle**. For an angle in standard position, the terminal side can be determined if we know any point, except the origin, on the terminal side.

EXAMPLE 4 ▸▸ A standard position angle of 60° is a first-quadrant angle with its terminal side 60° from the x-axis. See Fig. 4-7(a).

A second-quadrant angle of 130° is shown in Fig. 4-7(b).

A third-quadrant angle of 225° is shown in Fig. 4-7(c).

A fourth-quadrant angle of 340° is shown in Fig. 4-7(d).

A standard position angle of −120° is shown in Fig. 4-7(e). Since the terminal side is in the third quadrant, it is a third-quadrant angle.

A standard position angle of 90° is a quadrantal angle since its terminal side is the positive y-axis. See Fig. 4-7(f).

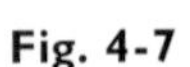
Fig. 4-7

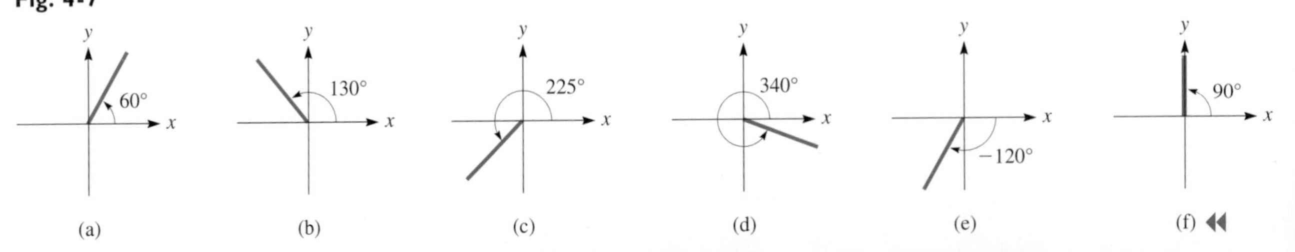

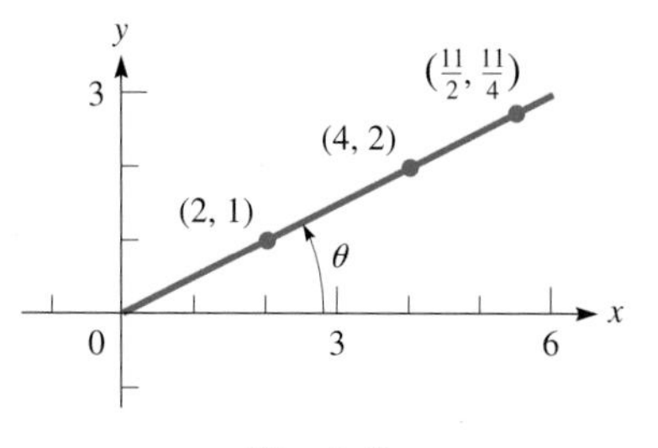

Fig. 4-8

EXAMPLE 5 ▸▸ In Fig. 4-8, θ is in standard position and the terminal side is uniquely determined by knowing that it passes through the point (2, 1). The same terminal side passes through (4, 2) and $(\frac{11}{2}, \frac{11}{4})$, among other points. Knowing that the terminal side passes through any one of these points makes it possible to determine the terminal side of the angle. ◂◂

Exercises 4–1

In Exercises 1–4, draw the given angles.

1. 60°, 120°, −90° **2.** 330°, −150°, 450°

3. 50°, −360°, −30° **4.** 45°, 245°, −250°

In Exercises 5–12, determine one positive and one negative coterminal angle for each angle given.

5. 45° **6.** 73°

7. −150° **8.** 162°

9. 70°30′ **10.** 153°47′

11. 278.1° **12.** −197.6°

In Exercises 13–16, change the given angles in radians to equal angles expressed in degrees to the nearest 0.01°.

13. 0.265 rad **14.** 0.838 rad

15. 1.447 rad **16.** 3.642 rad

In Exercises 17–24, change the given angles to equal angles expressed in decimal form to the nearest 0.01°.

17. 15°12′ **18.** 246°48′

19. 86°3′ **20.** 157°39′

21. 301°16′ **22.** −4°47′

23. −96°8′ **24.** 38°28′

In Exercises 25–32, change the given angles to equal angles expressed to the nearest minute.

25. 47.50° **26.** 315.80°

27. 19.75° **28.** −84.55°

29. −5.62° **30.** 238.21°

31. 24.92° **32.** 142.87°

In Exercises 33–40, draw angles in standard position such that the terminal side passes through the given point.

33. (4, 2) **34.** (−3, 8)

35. (−3, −5) **36.** (6, −1)

37. (−7, 5) **38.** (−4, −2)

39. (2, −5) **40.** (1, 6)

In Exercises 41 and 42, change the given angles to equal angles expressed in decimal form to the nearest 0.001°. In Exercises 43 and 44, change the given angles to equal angles expressed to the nearest second.

41. 21°42′36″ **42.** 7°16′23″

43. 86.274° **44.** 57.019°

▸ 4-2 Defining the Trigonometric Functions

In Chapter 2 we reviewed many of the basic geometric figures and their properties. Important to the definitions and development in this section are the right triangle, the Pythagorean theorem, and the properties of similar triangles. We now briefly review similar triangles and their properties.

As stated in Section 2-2, *two triangles are* **similar** *if they have the same shape (but not necessarily the same size).* Similar triangles have the following important properties.

Properties of Similar Triangles

1. *Corresponding angles are equal.*
2. *Corresponding sides are proportional.*

The **corresponding sides** *are the sides, one in each triangle, which are between the same pair of equal* **corresponding angles**.

EXAMPLE 1 ▸▸ In Fig. 4-9 the triangles are similar and are lettered so that corresponding sides and angles have the same letters. That is, angles A_1 and A_2, angles B_1 and B_2, and angles C_1 and C_2 are pairs of corresponding angles. The pairs of corresponding sides are a_1 and a_2, b_1 and b_2, and c_1 and c_2. From the properties of similar triangles we know that the corresponding angles are equal, or

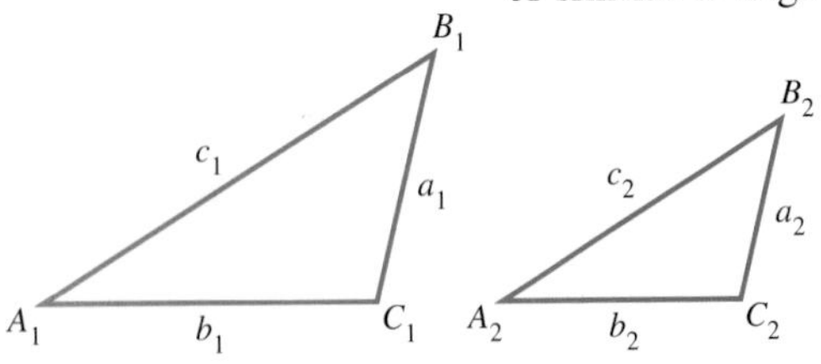

Fig. 4-9

$$\angle A_1 = \angle A_2,\ \angle B_1 = \angle B_2,\ \angle C_1 = \angle C_2$$

Also, the corresponding sides are proportional, which we can show as

$$\frac{a_1}{a_2} = \frac{b_1}{b_2},\quad \frac{a_1}{a_2} = \frac{c_1}{c_2},\quad \frac{b_1}{b_2} = \frac{c_1}{c_2}$$ ◂◂

In Example 1, if we multiply both sides of

$$\frac{a_1}{a_2} = \frac{b_1}{b_2} \quad \text{by} \quad \frac{a_2}{b_1} \quad \text{we get} \quad \frac{a_1}{a_2}\left(\frac{a_2}{b_1}\right) = \frac{b_1}{b_2}\left(\frac{a_2}{b_1}\right)$$

which when simplified gives

$$\frac{a_1}{b_1} = \frac{a_2}{b_2}$$

This shows us that *when two triangles are similar*

> *the ratio of one side to another side in one triangle is the same as the ratio of the corresponding sides in the other triangle.*

Using this we now proceed to the definitions of the trigonometric functions.

We now place an angle θ in standard position and drop perpendicular lines from points on the terminal side to the x-axis as shown in Fig. 4-10. In doing this we set up similar triangles, each with one vertex at the origin and one side along the x-axis.

EXAMPLE 2 ▸▸ In Fig. 4-10 we can see that triangles ORP and OSQ are similar since their corresponding angles are equal (each has the same angle at O, a right angle, and therefore equal angles at P and Q). This means that ratios of the lengths of corresponding sides are equal. For example,

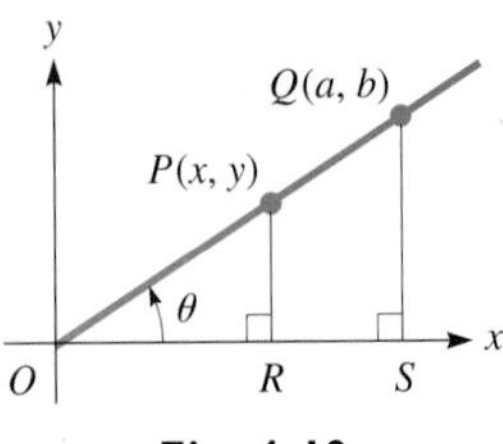

Fig. 4-10

$$\frac{RP}{OR} = \frac{SQ}{OS} \quad \text{which is the same as} \quad \frac{y}{x} = \frac{b}{a}$$

For any position (except at the origin) of Q on the terminal side of θ, the ratio b/a of its ordinate to its abscissa will equal y/x. ◂◂

For any angle θ in standard position, six different ratios may be set up. Because of the property of similar triangles, any given ratio has the same value for any point on the terminal side that is chosen. For a different angle, with a different terminal side, the ratio will have a different value. Thus the values of the ratios depend on the position of the terminal side, which means that *the values of the ratios depend on the size of the angle, and there is only one value for each ratio for a given angle.* Recalling the meaning of a function, we see that *the ratios are functions of the angle. These functions are called the* **trigonometric functions,** and they are defined as follows (see Fig. 4-11):

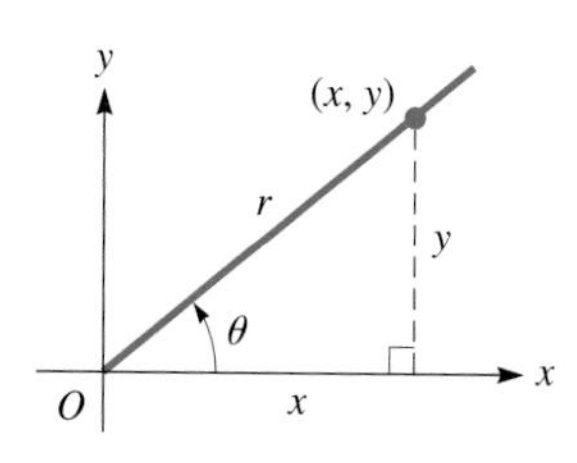

Fig. 4-11

$$\text{sine of } \theta\text{: } \mathbf{sin}\ \theta = \frac{y}{r} \qquad \text{cosine of } \theta\text{: } \mathbf{cos}\ \theta = \frac{x}{r}$$

$$\text{tangent of } \theta\text{: } \mathbf{tan}\ \theta = \frac{y}{x} \qquad \text{cotangent of } \theta\text{: } \mathbf{cot}\ \theta = \frac{x}{y} \qquad (4\text{-}1)$$

$$\text{secant of } \theta\text{: } \mathbf{sec}\ \theta = \frac{r}{x} \qquad \text{cosecant of } \theta\text{: } \mathbf{csc}\ \theta = \frac{r}{y}$$

Here, *the distance r from the origin to the point is called the* **radius vector.** Also note that we have used the abbreviations which are used most of the time when working with these functions.

In this chapter we shall restrict our attention to the trigonometric functions of acute angles (angles between 0° and 90°). However, *the definitions in Eqs. (4-1) are general and may be used for angles of any magnitude.* Discussion of the trigonometric functions of angles in general, along with other important properties, is found in Chapters 8 and 20.

We should note that a given function is not defined if the denominator is zero. The denominator is zero in tan θ and sec θ for $x = 0$, and in cot θ and csc θ for $y = 0$. In all cases we will assume that $r > 0$. If $r = 0$, there would be no terminal side and therefore no angle. These restrictions affect the domain of these functions. We will discuss the domains and ranges of the trigonometric functions in Chapter 10, when we consider the graphs of these functions.

Evaluating the Trigonometric Functions

When evaluating the trigonometric functions we use the definitions in Eqs. (4-1). We also often use the *Pythagorean theorem,* which we discussed in Section 2-2. For reference, we restate it here. For the right triangle in Fig. 4-12, with hypotenuse c and legs a and b, we have

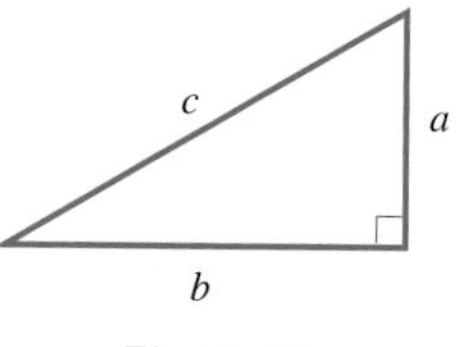

Fig. 4-12

$$c^2 = a^2 + b^2 \qquad (4\text{-}2)$$

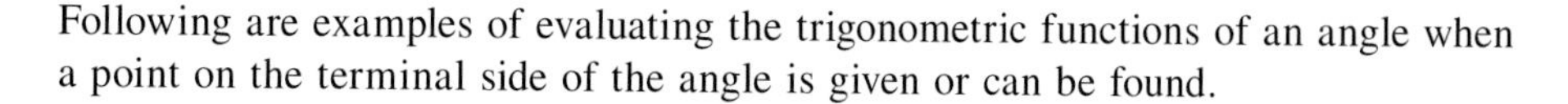
Following are examples of evaluating the trigonometric functions of an angle when a point on the terminal side of the angle is given or can be found.

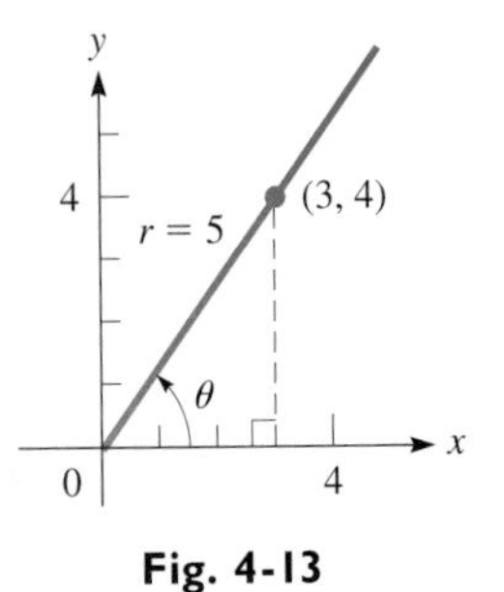

Fig. 4-13

EXAMPLE 3 ▸▸ Find the values of the trigonometric functions of the angle θ with its terminal side passing through the point (3, 4).

By placing the angle in standard position, as shown in Fig. 4-13, and drawing the terminal side through (3, 4), we find by use of the Pythagorean theorem that

$$r = \sqrt{3^2 + 4^2} = \sqrt{25} = 5$$

Using the values $x = 3$, $y = 4$, and $r = 5$, we find that

$$\sin\theta = \frac{4}{5} \qquad \cos\theta = \frac{3}{5} \qquad \tan\theta = \frac{4}{3}$$

$$\cot\theta = \frac{3}{4} \qquad \sec\theta = \frac{5}{3} \qquad \csc\theta = \frac{5}{4}$$ ◂◂

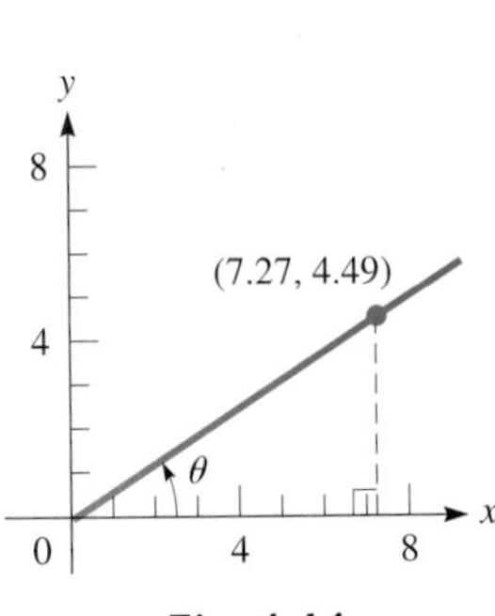

Fig. 4-14

EXAMPLE 4 ▸▸ Find the values of the trigonometric functions of the angle whose terminal side passes through (7.27, 4.49). The coordinates are approximate.

We show the angle and the given point in Fig. 4-14. From the Pythagorean theorem, we have

$$r = \sqrt{7.27^2 + 4.49^2} = 8.545$$

(Here we show a rounded-off value of r. It is not actually necessary to record the value of r since its value can be stored in the memory of a calculator. The reason for recording it here is to show the values used in the calculation of each of the trigonometric functions.) Therefore, we have the following values:

$$\sin\theta = \frac{4.49}{8.545} = 0.525 \qquad \cos\theta = \frac{7.27}{8.545} = 0.851$$

$$\tan\theta = \frac{4.49}{7.27} = 0.618 \qquad \cot\theta = \frac{7.27}{4.49} = 1.62$$

$$\sec\theta = \frac{8.545}{7.27} = 1.18 \qquad \csc\theta = \frac{8.545}{4.49} = 1.90$$

Since the coordinates are approximate, the results are rounded off.

The sequence of keys on a *scientific calculator* that can be used to find r and place its value in memory is

7.27 [x^2] [+] 4.49 [x^2] [=] [$\sqrt{x}$] [STO]

At this point the necessary divisions can be performed using the [RCL] key for r. For example, to find $\sin\theta$ we continue the calculator sequence with

4.49 [÷] [RCL] [=] 0.525467965

On a *graphing calculator* a sequence of keys that can be used is

[$\sqrt{\ }$] [(] 7.27 [x^2] [+] 4.49 [x^2] [)] [ENTER]

(Some calculators use [EXE] rather than [ENTER].) At this point the calculator stores the last answer (in this case r) under [ANS] (and it can also be stored in memory). To find $\sin\theta$ we continue with

4.49 [÷] [ANS] [ENTER]

to get the result of 0.525 467 965. (To get values of all functions, we use the value of r in memory.) Note that the sequences used on the different types of calculators are different, but they are similar. ◂◂

In Example 4, we expressed the result as $\sin \theta = 0.525$. A common error is to omit the angle and give the value as $\sin = 0.525$. This is a meaningless expression, for ***we must show the angle*** for which we have the value of a function.

CAUTION ▶

If one of the trigonometric functions is known, it is possible to find the values of the other functions. The following example illustrates the method.

EXAMPLE 5 ▸▸ If we know that $\sin \theta = 3/7$ and that θ is a first-quadrant angle, we know the ratio of the ordinate to the radius vector (y to r) is 3 to 7. Therefore, the point on the terminal side for which $y = 3$ can be found by use of the Pythagorean theorem. The x-value for this point is

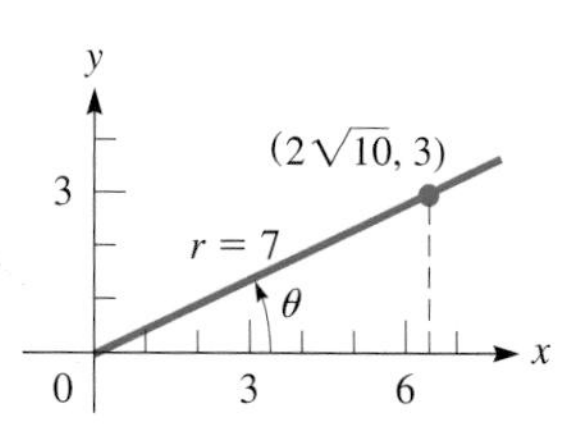

Fig. 4-15

$$x = \sqrt{7^2 - 3^2} = \sqrt{49 - 9} = \sqrt{40} = 2\sqrt{10}$$

Therefore, the point $(2\sqrt{10}, 3)$ is on the terminal side, as shown in Fig. 4-15.

Therefore, using the values $x = 2\sqrt{10}$, $y = 3$, and $r = 7$, we have the other trigonometric functions of θ. They are

$$\cos \theta = \frac{2\sqrt{10}}{7}, \quad \tan \theta = \frac{3}{2\sqrt{10}}, \quad \cot \theta = \frac{2\sqrt{10}}{3}, \quad \sec \theta = \frac{7}{2\sqrt{10}}, \quad \csc \theta = \frac{7}{3}$$

These values are *exact*. *Approximate* decimal values found on a calculator are

$$\cos \theta = 0.9035, \quad \tan \theta = 0.4743, \quad \cot \theta = 2.108,$$
$$\sec \theta = 1.107, \quad \csc \theta = 2.333$$ ◂◂

Exercises 4–2

In Exercises 1–16, find the values of the trigonometric functions of the angle (in standard position) whose terminal side passes through the given points. For Exercises 1–12, give answers in exact form. For Exercises 13–16, the coordinates are approximate.

1. $(6, 8)$ **2.** $(5, 12)$ **3.** $(15, 8)$ **4.** $(24, 7)$

5. $(9, 40)$ **6.** $(16, 30)$ **7.** $(1, \sqrt{15})$ **8.** $(\sqrt{3}, 2)$

9. $(1, 1)$ **10.** $(6, 5)$ **11.** $(5, 2)$ **12.** $(1, \frac{1}{2})$

13. $(3.25, 5.15)$ **14.** $(0.687, 0.943)$

15. $(0.08623, 0.01327)$ **16.** $(37.65, 21.87)$

In Exercises 17–24, find the indicated functions. In Exercises 17–20, give answers in exact form. In Exercises 21–24, the values are approximate.

17. Given $\cos \theta = 12/13$, find $\sin \theta$ and $\cot \theta$.

18. Given $\sin \theta = 1/2$, find $\cos \theta$ and $\csc \theta$.

19. Given $\tan \theta = 1$, find $\sin \theta$ and $\sec \theta$.

20. Given $\sec \theta = 4/3$, find $\tan \theta$ and $\cos \theta$.

21. Given $\sin \theta = 0.750$, find $\cot \theta$ and $\csc \theta$.

22. Given $\cos \theta = 0.326$, find $\sin \theta$ and $\tan \theta$.

23. Given $\cot \theta = 0.254$, find $\cos \theta$ and $\tan \theta$.

24. Given $\csc \theta = 1.20$, find $\sec \theta$ and $\cos \theta$.

In Exercises 25–28, each listed point is on the terminal side of an angle. Show that each of the indicated functions is the same for each of the points.

25. $(3, 4)$, $(6, 8)$, $(4.5, 6)$, $\sin \theta$ and $\tan \theta$

26. $(5, 12)$, $(15, 36)$, $(7.5, 18)$, $\cos \theta$ and $\cot \theta$

27. $(2, 1)$, $(4, 2)$, $(8, 4)$, $\tan \theta$ and $\sec \theta$

28. $(3, 2)$, $(6,4)$, $(9, 6)$, $\csc \theta$ and $\cos \theta$

In Exercises 29–32, answer the given questions.

29. From the definitions of the trigonometric functions, it can be seen that $\csc \theta$ is the reciprocal of $\sin \theta$. What function is the reciprocal of $\cos \theta$?

30. Following Exercise 29, what function is the reciprocal of $\tan \theta$?

31. Multiply the expression for $\cot \theta$ by that for $\sin \theta$. Is the result the expression for any of the other functions?

32. Refer to the definitions of the trigonometric functions in Eqs. (4-1). Is the quotient of one of the functions divided by $\cos \theta$ equal to $\tan \theta$? Explain.

4-3 Values of the Trigonometric Functions

We have been able to calculate the trigonometric functions if we knew one point on the terminal side of the angle. However, in practice it is more common to know the angle in degrees, for example, and to be required to find the functions of this angle. Therefore, we must be able to determine the trigonometric functions of angles in degrees.

One way to determine the functions of a given angle is to make a scale drawing. That is, we draw the angle in standard position using a protractor and then measure the lengths of the values of x, y, and r for some point on the terminal side. By using the proper ratios we may determine the functions of this angle.

We may also use certain geometric facts to determine the functions of some particular angles. The following two examples illustrate this procedure.

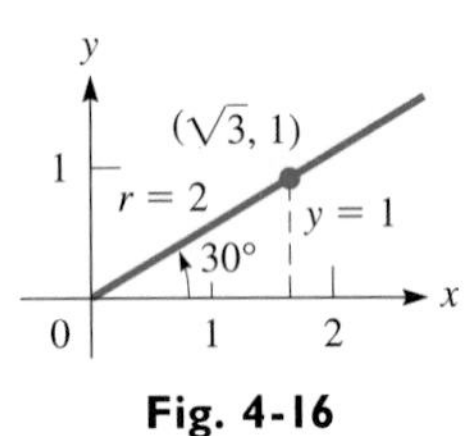

Fig. 4-16

EXAMPLE 1 ▸▸ From geometry we find that the side opposite a 30° angle in a right triangle is one-half of the hypotenuse. Using this fact and letting $y = 1$ and $r = 2$ (see Fig. 4-16), we calculate that $x = \sqrt{2^2 - 1^2} = \sqrt{3}$ from the Pythagorean theorem. Therefore, with $x = \sqrt{3}$, $y = 1$, and $r = 2$, we have

$$\sin 30° = \frac{1}{2}, \quad \cos 30° = \frac{\sqrt{3}}{2}, \quad \text{and} \quad \tan 30° = \frac{1}{\sqrt{3}}$$

In a similar way we may determine the values of the functions of 60° to be

$$\sin 60° = \frac{\sqrt{3}}{2}, \quad \cos 60° = \frac{1}{2}, \quad \text{and} \quad \tan 60° = \sqrt{3}$$ ◂◂

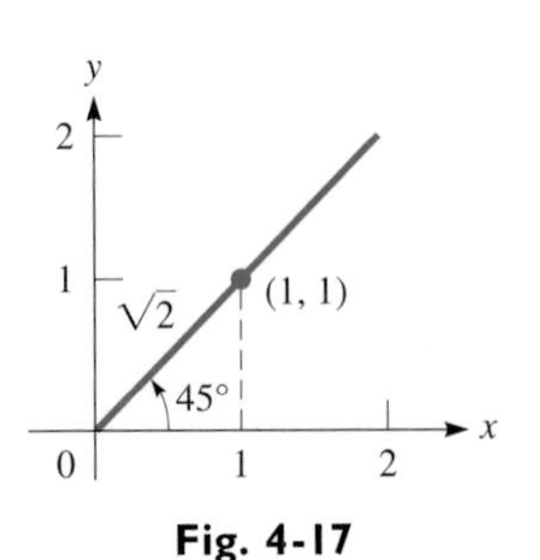

Fig. 4-17

EXAMPLE 2 ▸▸ Find sin 45°, cos 45°, and tan 45°.

From geometry we know that in an isosceles right triangle the angles are 45°, 45°, and 90°. We know that the sides are in proportion 1, 1, $\sqrt{2}$, respectively. Putting the 45° angle in standard position, we find $x = 1$, $y = 1$, and $r = \sqrt{2}$ (see Fig. 4-17). From this we determine

$$\sin 45° = \frac{1}{\sqrt{2}}, \quad \cos 45° = \frac{1}{\sqrt{2}}, \quad \text{and} \quad \tan 45° = 1$$

As in Example 1, we have given the exact values. Decimal approximations are given in the table that follows. ◂◂

Summarizing the results for 30°, 45°, and 60°, we have:

				(decimal approximations)		
θ	30°	45°	60°	30°	45°	60°
$\sin\theta$	$\frac{1}{2}$	$\frac{1}{\sqrt{2}}$	$\frac{\sqrt{3}}{2}$	0.500	0.707	0.866
$\cos\theta$	$\frac{\sqrt{3}}{2}$	$\frac{1}{\sqrt{2}}$	$\frac{1}{2}$	0.866	0.707	0.500
$\tan\theta$	$\frac{1}{\sqrt{3}}$	1	$\sqrt{3}$	0.577	1.000	1.732

NOTE ▸ It is helpful to be familiar with these values, as they are used in later sections.

The scale-drawing method is only approximate, and geometric methods work only for certain angles. However, the values of the functions may be found through more advanced methods (using calculus and what are known as *power series*).

The values of the trigonometric functions have been programmed into scientific calculators and graphing calculators (use the MODE feature). For all of our work in this chapter, *be sure that your calculator is set for degrees* (not radians).

EXAMPLE 3 ▸▸ Using a standard scientific calculator to find tan 67.36°, enter the angle and then the function. Using a graphing calculator, enter the function and then the angle. Therefore, the sequences are

67.36 [tan] [2.397626383] — standard scientific calculator

[tan] 67.36 [ENTER] [2.397626383] — graphing calculator

Thus, tan 67.36° = 2.397 626 383. A ten-digit calculator display is common. ◂◂

Inverse Trigonometric Functions

We can also find the angle if we know the value of a function. In doing this, we are actually using another important type of mathematical function, an **inverse trigonometric function.** These are discussed in detail in Chapter 20. For calculator purposes at this point, it is sufficient to recognize the notation that is used.

CAUTION ▶

The notation for "the angle whose sine is x" is $\sin^{-1} x$ (or arcsin x). Similar meanings are given to $\cos^{-1} x$ and $\tan^{-1} x$. (The -1 used with a *function* indicates an *inverse function* and is *not* a negative exponent.) On a calculator the [$\sin^{-1}$] key is used to find the angle if its sine is known. (On some calculators, the sequence [INV] [sin] is used.) We will use the [$\sin^{-1}$] key in our examples.

EXAMPLE 4 ▸▸ If $\cos\theta = 0.3527$, which means that $\theta = \cos^{-1} 0.3527$, we can find the angle θ with the following calculator sequences:

.3527 [$\cos^{-1}$] [69.34745162] — standard scientific calculator

[$\cos^{-1}$] .3527 [ENTER] [69.34745162] — graphing calculator

Therefore, $\theta = 69.35°$ (rounded off—see below). ◂◂

When using the trigonometric functions, the angle is often *approximate.* Angles of 2.3°, 92.3°, and 182.3° are angles with equal accuracy, which shows that *the accuracy of an angle does not depend on the number of digits shown.* The measurement of an angle and the accuracy of its trigonometric functions are shown in the following table.

Angles and Accuracy of Trigonometric Functions

Measurement of Angle to Nearest	*Accuracy of Trigonometric Function*
1°	2 significant digits
0.1° or 10′	3 significant digits
0.01° or 1′	4 significant digits

We rounded off the result in Example 4 according to this table.

It is generally possible to set up the solution of a problem in terms of the sine, cosine, or tangent. However, if a value of the cotangent, secant, or cosecant of an angle is needed, it can also be found on a calculator.

Reciprocal Functions

From the definitions of the trigonometric functions we see that $\csc\theta = r/y$ and $\sin\theta = y/r$. This means that *the value of* $\csc\theta$ *is the reciprocal of the value of* $\sin\theta$.

NOTE ▶ In the same way, $\sec\theta$ *is the reciprocal of* $\cos\theta$ and $\cot\theta$ *is the reciprocal of* $\tan\theta$. Thus, by use of the reciprocal key, $\boxed{x^{-1}}$ (here the -1 *is* an exponent), the values of these functions can be found. (On some calculators the $\boxed{1/x}$ key is the reciprocal key. Note that $1/x = x^{-1}$.)

EXAMPLE 5 ▸▸ To find the value of sec 27.82°, we use the fact that

$$\sec 27.82° = \frac{1}{\cos 27.82°}$$

Therefore, we are to find the reciprocal of cos 27.82°. On a scientific calculator we use the sequence 27.82 $\boxed{\cos}$ $\boxed{x^{-1}}$, and on a graphing calculator we evaluate it as $(\cos 27.82)^{-1}$. The display will show 1.130 686 918.

Therefore, sec 27.82° = 1.131, with the results rounded off according to the table following Example 4. ◂◂

EXAMPLE 6 ▸▸ To find the value of θ if $\cot\theta = 0.354$, we use the fact that

$$\tan\theta = \frac{1}{\cot\theta} = \frac{1}{0.354}$$

This means the scientific calculator sequence is .354 $\boxed{x^{-1}}$ $\boxed{\tan^{-1}}$. On a graphing calculator we evaluate $\tan^{-1}.354^{-1}$. The result is 70.506 036 93, which means that $\theta = 70.5°$ (rounded off). ◂◂

In the following example, we see how to find the value of one function if we know the value of another function of the same angle.

EXAMPLE 7 ▸▸ Find $\sin\theta$ if $\sec\theta = 2.504$.

Since we know $\sec\theta$, we find θ by use of the $\boxed{x^{-1}}$ and $\boxed{\cos^{-1}}$ keys. This leads to the scientific calculator sequence of

	to get $\cos\theta$	to get θ	to get $\sin\theta$	
2.504	$\boxed{x^{-1}}$	$\boxed{\cos^{-1}}$	$\boxed{\sin}$	0.9167937466

Using a graphing calculator we evaluate $\sin\cos^{-1}2.504^{-1}$. Thus, $\sin\theta = 0.9168$ (rounded off). ◂◂

Calculators have been used extensively since the late 1970s. Until then values of the trigonometric functions were generally obtained from tables. Many standard sources are still available with precision to at least 10′ or 0.1°. However, a calculator is easier to use than a table, and it can give values to a much greater accuracy. Therefore, we will not use tables in this text.

The following example illustrates the use of the value of a trigonometric function in an applied problem. We will consider various types of applications later in the chapter.

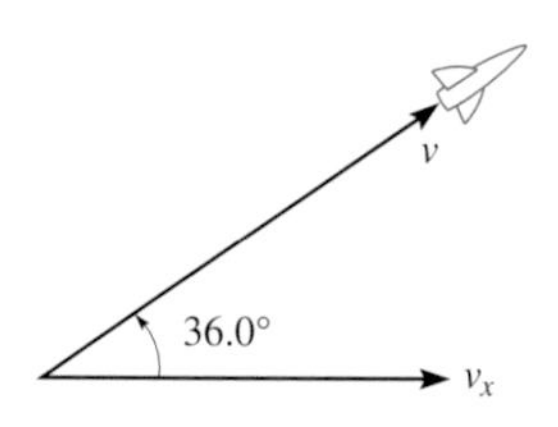

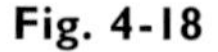
Fig. 4-18

EXAMPLE 8 When a rocket is launched, its horizontal velocity v_x is related to the velocity v with which it is fired by the equation $v_x = v \cos \theta$ (which means $v(\cos \theta)$, but does *not* mean $\cos \theta\, v$, which is the same as $\cos(\theta v)$). Here θ is the angle between the horizontal and the direction in which it is fired (see Fig. 4-18). Find v_x if $v = 1250$ m/s and $\theta = 36.0°$.

Substituting the given values of v and θ in $v_x = v \cos \theta$, we have

$$v_x = 1250 \cos 36.0°$$
$$= 1010 \text{ m/s}$$

Therefore, the horizontal velocity is 1010 m/s.

Exercises 4–3

In Exercises 1–4, use a protractor to draw the given angle. Measure off 10 units (centimetres are convenient) along the radius vector. Then measure the corresponding values of x and y. From these values determine the trigonometric functions of the angle.

1. 40° **2.** 75° **3.** 15° **4.** 53°

In Exercises 5–20, find the values of the trigonometric functions. Round off results according to the table following Example 4.

5. $\sin 22.4°$ **6.** $\cos 72.5°$
7. $\tan 57.6°$ **8.** $\sin 36.0°$
9. $\cos 15.71°$ **10.** $\tan 8.653°$
11. $\sin 84°$ **12.** $\cos 47°$
13. $\cot 67.78°$ **14.** $\csc 22.81°$
15. $\sec 50.4°$ **16.** $\cot 41.8°$
17. $\csc 49.3°$ **18.** $\sec 7.8°$
19. $\cot 85.96°$ **20.** $\csc 76.30°$

In Exercises 21–36, find θ for each of the given trigonometric functions. Round off results according to the table following Example 4.

21. $\cos \theta = 0.3261$ **22.** $\tan \theta = 2.470$
23. $\sin \theta = 0.9114$ **24.** $\cos \theta = 0.0427$
25. $\tan \theta = 0.207$ **26.** $\sin \theta = 0.109$
27. $\cos \theta = 0.650\,07$ **28.** $\tan \theta = 5.7706$
29. $\csc \theta = 1.245$ **30.** $\sec \theta = 2.045$
31. $\cot \theta = 0.1443$ **32.** $\csc \theta = 1.012$
33. $\sec \theta = 3.65$ **34.** $\cot \theta = 2.08$
35. $\csc \theta = 3.262$ **36.** $\cot \theta = 0.1519$

In Exercises 37–40, find the values of the indicated trigonometric functions.

37. Find $\sin \theta$, given $\tan \theta = 1.936$.

38. Find $\cos \theta$, given $\sin \theta = 0.6725$.

39. Find $\tan \theta$, given $\sec \theta = 1.3698$.

40. Find $\csc \theta$, given $\cos \theta = 0.1063$.

In Exercises 41–44, solve the given problems. Assume the data in Exercise 41 are accurate to two significant digits.

41. The sound produced by a jet engine was measured at a distance of 100 m in all directions. The loudness of the sound (in decibels) was found to be $d = 70 + 30 \cos \theta$, where the 0° line was directed in front of the engine. Calculate d for $\theta = 54.5°$.

42. A brace is used in the structure shown in Fig. 4-19. Its length is $l = a(\sec \theta + \csc \theta)$. Find l if $a = 28.0$ cm and $\theta = 34.5°$.

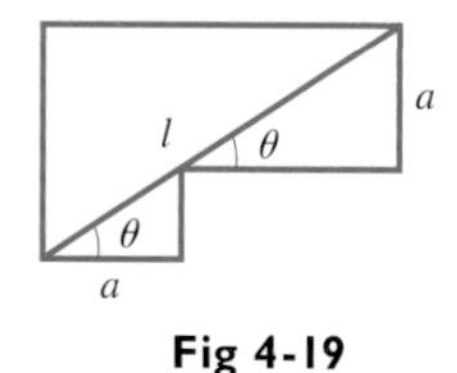

Fig 4-19

43. The voltage e at any instant in a coil of wire that is turning in a magnetic field is given by $e = E \cos \alpha$, where E is the maximum voltage and α is the angle the coil makes with the field. Find the acute angle α if $e = 56.9$ V and $E = 171$ V.

44. A submarine dives such that the horizontal distance h and vertical distance v are related by $v = h \tan \theta$. Here θ is the angle of the dive, as shown in Fig. 4-20. Find θ if $h = 2.35$ km and $v = 1.52$ km.

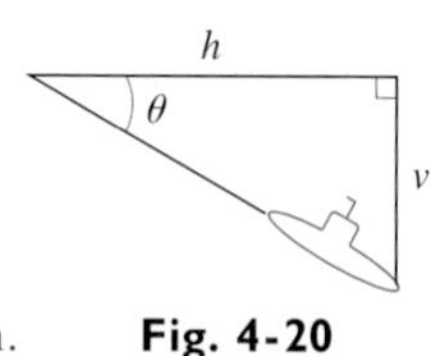

Fig. 4-20

4-4 The Right Triangle

From geometry we know that a triangle, by definition, consists of three sides and has three angles. If one side and any other two of these six parts of the triangle are known, it is possible to determine the other three parts. One of the three known parts must be a side, for if we know only the three angles, we can conclude only that an entire set of similar triangles has those particular angles.

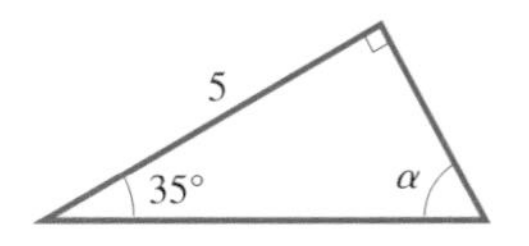

Fig. 4-21

EXAMPLE 1 ▸▸ Assume that one side and two angles are known, such as the side of 5 and the angles of 35° and 90° in the triangle in Fig. 4-21. Then we may determine the third angle α by the fact that the sum of the angles of a triangle is always 180°. Of all possible similar triangles having the three angles of 35°, 90°, and 55° (which is α), we have the one with the particular side of 5 between angles of 35° and 90°. Only one triangle with these parts is possible (in the sense that all triangles with the given parts are *congruent* and have equal corresponding angles and sides). ◂◂

To **solve a triangle** *means that, when we are given three parts of a triangle (at least one a side), we are to find the other three parts.* In this section we are going to demonstrate the method of solving a right triangle. *Since one angle of the triangle will be* 90°, *it is necessary to know one side and one other part.* Also, we know that the sum of the three angles of a triangle is 180°, and this in turn tells us that *the sum of the other two angles, both acute, is* 90°. *Any two acute angles whose sum is* 90° *are said to be* **complementary**.

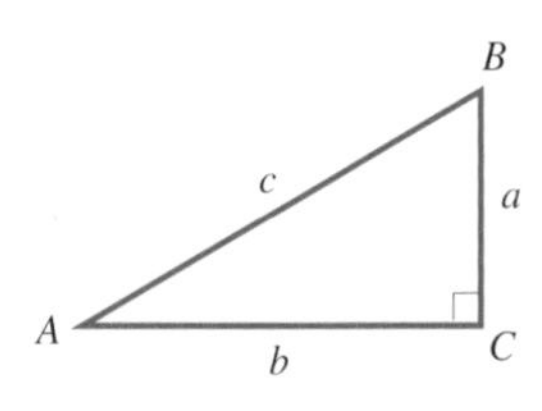

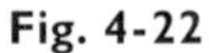
Fig. 4-22

For consistency, when we are labeling the parts of the right triangle *we shall use the letters A and B to denote the acute angles, and C to denote the right angle. The letters a, b, and c will denote the sides opposite these angles, respectively. Thus, side c is the hypotenuse of the right triangle.* See Fig. 4-22.

In solving right triangles we shall find it convenient to express the trigonometric functions of the acute angles in terms of the sides. By placing the vertex of angle A at the origin and the vertex of right angle C on the positive x-axis, as shown in Fig. 4-23, we have the following ratios for angle A in terms of the sides of the triangle.

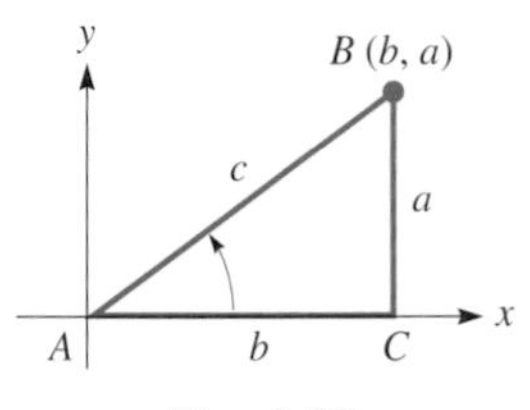

Fig. 4-23

$$\sin A = \frac{a}{c} \qquad \cos A = \frac{b}{c} \qquad \tan A = \frac{a}{b}$$
$$\cot A = \frac{b}{a} \qquad \sec A = \frac{c}{b} \qquad \csc A = \frac{c}{a} \tag{4-3}$$

If we should place the vertex of B at the origin, instead of the vertex of angle A, we would obtain the following ratios for the functions of angle B (see Fig. 4-24):

$$\sin B = \frac{b}{c} \qquad \cos B = \frac{a}{c} \qquad \tan B = \frac{b}{a}$$
$$\cot B = \frac{a}{b} \qquad \sec B = \frac{c}{a} \qquad \csc B = \frac{c}{b} \tag{4-4}$$

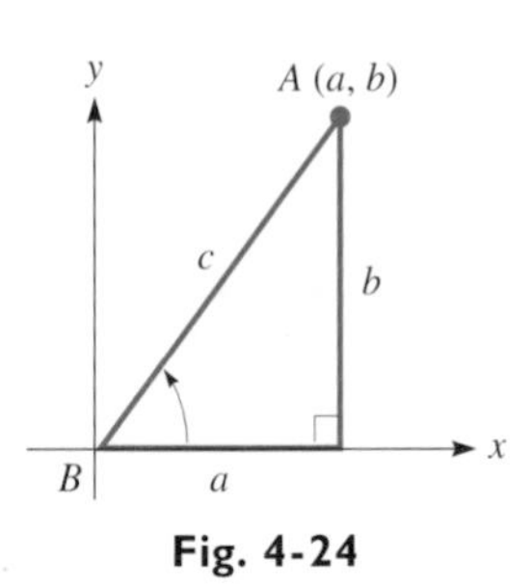

Fig. 4-24

Equations (4-3) and (4-4) show that we may generalize our definitions of the trigonometric functions of an acute angle of a right triangle (we have chosen $\angle A$ in Fig. 4-25) to be as follows:

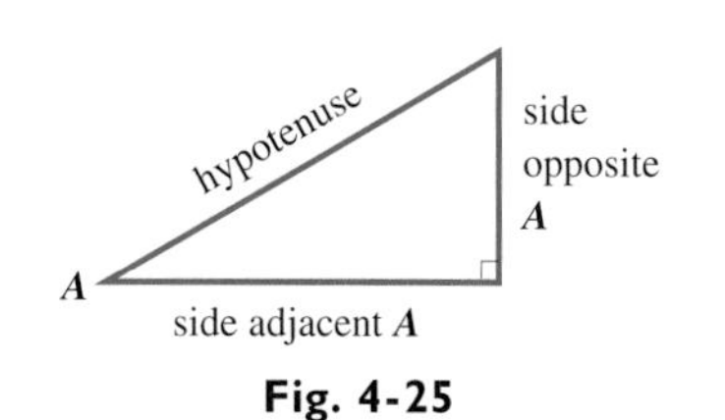

Fig. 4-25

$$\begin{aligned} \sin A &= \frac{\text{side opposite } A}{\text{hypotenuse}} & \cos A &= \frac{\text{side adjacent } A}{\text{hypotenuse}} \\ \tan A &= \frac{\text{side opposite } A}{\text{side adjacent } A} & \cot A &= \frac{\text{side adjacent } A}{\text{side opposite } A} \\ \sec A &= \frac{\text{hypotenuse}}{\text{side adjacent } A} & \csc A &= \frac{\text{hypotenuse}}{\text{side opposite } A} \end{aligned} \quad (4\text{-}5)$$

Using the definitions in this form, we can solve right triangles without placing the angle in standard position. The angle need only be a part of any right triangle.

We note from the above discussion that $\sin A = \cos B$, $\tan A = \cot B$, and $\sec A = \csc B$. From this we conclude that *cofunctions of acute complementary angles are equal.* The sine function and cosine function are cofunctions, the tangent function and cotangent function are cofunctions, and the secant function and cosecant functions are cofunctions. This property of the values of the trigonometric functions is illustrated in the following example.

EXAMPLE 2 ▸▸ Given $a = 4$, $b = 7$, and $c = \sqrt{65}$, find $\sin A$, $\cos A$, and $\tan A$ (see Fig. 4-26) in exact form and in approximate decimal form (to three significant digits).

$$\sin A = \frac{\text{side opposite angle } A}{\text{hypotenuse}} = \frac{4}{\sqrt{65}} = 0.496$$

$$\cos A = \frac{\text{side adjacent angle } A}{\text{hypotenuse}} = \frac{7}{\sqrt{65}} = 0.868$$

$$\tan A = \frac{\text{side opposite angle } A}{\text{side adjacent angle } A} = \frac{4}{7} = 0.571$$ ◂◂

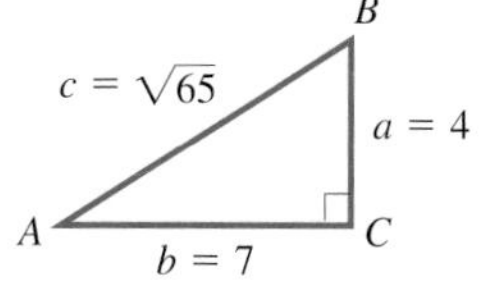

Fig. 4-26

EXAMPLE 3 ▸▸ Finding $\sin B$, $\cos B$, and $\tan B$ for the triangle in Fig. 4-26, we have

$$\sin B = \frac{\text{side opposite angle } B}{\text{hypotenuse}} = \frac{7}{\sqrt{65}} = 0.868$$

$$\cos B = \frac{\text{side adjacent angle } B}{\text{hypotenuse}} = \frac{4}{\sqrt{65}} = 0.496$$

$$\tan B = \frac{\text{side opposite angle } B}{\text{side adjacent angle } B} = \frac{7}{4} = 1.75$$

In Fig. 4-26 we note that A and B are complementary angles. Comparing with values found in Example 2, we see that $\sin A = \cos B$ and $\cos A = \sin B$. ◂◂

We are now ready to solve right triangles. The method is given and illustrated on the next page.

> ***Procedure for Solving a Right Triangle***
>
> 1. *Sketch a right triangle and label the known and unknown sides and angles.*
> 2. *Express each of the three unknown parts in terms of the known parts and solve for the unknown parts.*
> 3. *Check the results.* The sum of the angles should be 180°. If only one side is given, check the computed side with the Pythagorean theorem. If two sides are given, check the angles and computed side by using appropriate trigonometric functions.

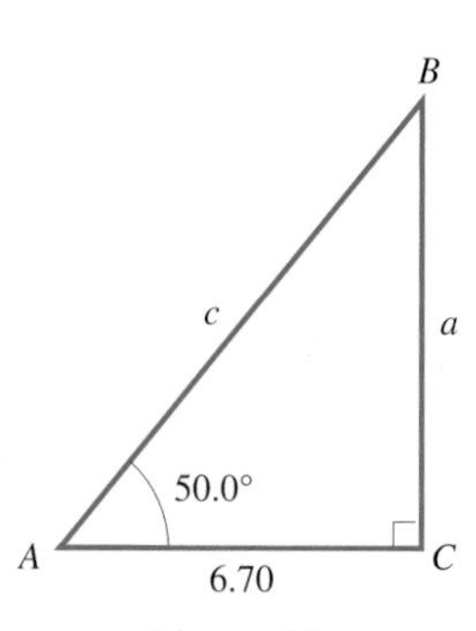

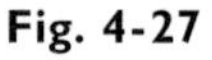
Fig. 4-27

EXAMPLE 4 ▸▸ Solve the right triangle with $A = 50.0°$ and $b = 6.70$.

We first sketch the right triangle shown in Fig. 4-27. (In making the sketch, we should be careful to follow proper labeling of the triangle as outlined on the previous page.) We then express unknown side a in terms of known side b and known angle A, and solve for a. We will then do the same for unknown side c and unknown angle B.

Finding side a, we know that $\tan A = \dfrac{a}{b}$, which means that $a = b \tan A$. Thus,

$$a = 6.70 \tan 50.0° = 7.98$$

Next, solving for side c, we have $\cos A = \dfrac{b}{c}$, which means $c = \dfrac{b}{\cos A}$.

$$c = \frac{6.70}{\cos 50.0°} = 10.4$$

Now solving for B, we know that $A + B = 90°$, or

$$B = 90° - A = 90° - 50.0° = 40.0°$$

Therefore, $a = 7.98$, $c = 10.4$, and $B = 40.0°$.

Checking the angles: $A + B + C = 50.0° + 40.0° + 90° = 180°$
Checking the sides: $10.4^2 = 108.16$
$7.98^2 + 6.70^2 = 108.57$

Since the computed values were rounded off, these values show that the sides check. ◂◂

NOTE ▸ In finding the unknown parts, we first expressed them in terms of the known parts. We do this because *it is best to use given values in calculations*. If we use one computed value to find another computed value, any error in the first would be carried to the value of the second. For example, in Example 4, if we were to find the value of c by using the value of a, any error in a would cause c to be in error as well.

We should also point out that, by inspection, we can make a rough check on the sides and angles of any triangle. *The longest side is always opposite the largest angle, and the shortest side is always opposite the smallest angle.* Also, in a right triangle, *the hypotenuse is always the longest side.* We see that this is true for the sides and angles for the triangle in Example 4, where c is the longest side (opposite the 90° angle), and b is the shortest side and is opposite the smallest angle of 40°.

EXAMPLE 5 ▸▸ Solve the right triangle with $b = 56.82$ and $c = 79.55$.

We sketch the right triangle as shown in Fig. 4-28. Since two sides are given, we will use the Pythagorean theorem to find the third side a. Also, we will use the cosine to find $\angle A$.

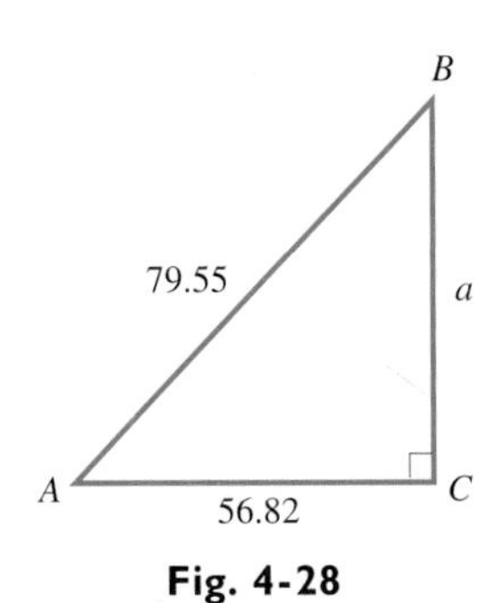

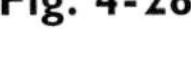

Fig. 4-28

Since $c^2 = a^2 + b^2$, $a^2 = c^2 - b^2$. Therefore

$$a = \sqrt{c^2 - b^2} = \sqrt{79.55^2 - 56.82^2}$$
$$= 55.67$$

Since $\cos A = \dfrac{b}{c}$, we have

$$\cos A = \frac{56.82}{79.55} = 0.7143$$

This means that $A = 44.42°$. We find this value by using the $\boxed{\cos^{-1}}$ key on the calculator. (It is not actually necessary to record the value of $\cos A$.)

Since $A + B = 90°$,

$$B = 90° - 44.42° = 45.58° \qquad \text{90° is exact}$$

We have now found that $a = 55.67$, $A = 44.42°$, and $B = 45.58°$.

Checking the sides and angles: $\sin 44.42° = \dfrac{55.67}{79.55}$, or $0.6999 \approx 0.6998$

$\sin 45.58° = \dfrac{56.82}{79.55}$, or $0.7142 \approx 0.7143$

This shows that the values check. (As we noted earlier, $\approx$ means "equals approximately.") We could also have used a different function in our check. ◂◂

EXAMPLE 6 ▸▸ Given that $\angle A$ and side a are known, express the unknown parts of the right triangle in terms of A and a.

We sketch a right triangle as shown in Fig. 4-29, and set up the required expressions as follows:

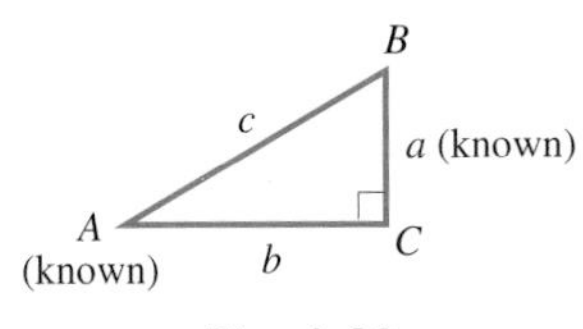

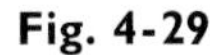

Fig. 4-29

Since $\dfrac{a}{b} = \tan A$, we have $b = \dfrac{a}{\tan A}$.

Since A is known, $B = 90° - A$.

Since $\dfrac{a}{c} = \sin A$, we have $c = \dfrac{a}{\sin A}$. ◂◂

Exercises 4–4

In Exercises 1–4, draw appropriate figures and verify through observation that only one triangle may contain the given parts (that is, any other which may be drawn will be congruent).

1. A 60° angle included between sides of 3 cm and 6 cm
2. A side of 4 cm included between angles of 40° and 50°
3. A right triangle with a hypotenuse of 5 cm and a leg of 3 cm
4. A right triangle with a 70° angle between the hypotenuse and a leg of 5 cm

In Exercises 5–24, solve the right triangles with the given parts. Round off results. Refer to Fig. 4-30.

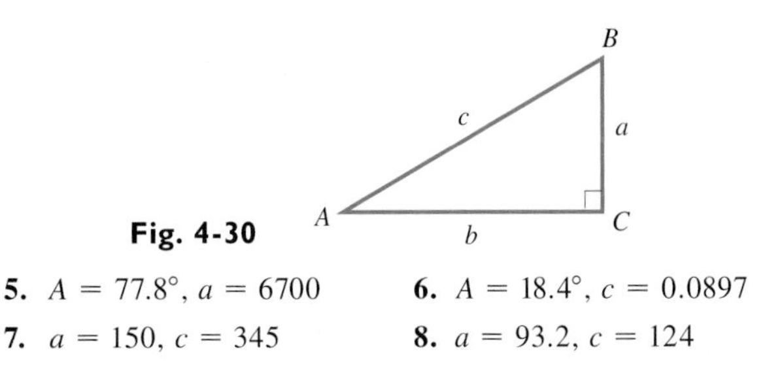

Fig. 4-30

5. $A = 77.8°$, $a = 6700$
6. $A = 18.4°$, $c = 0.0897$
7. $a = 150$, $c = 345$
8. $a = 93.2$, $c = 124$

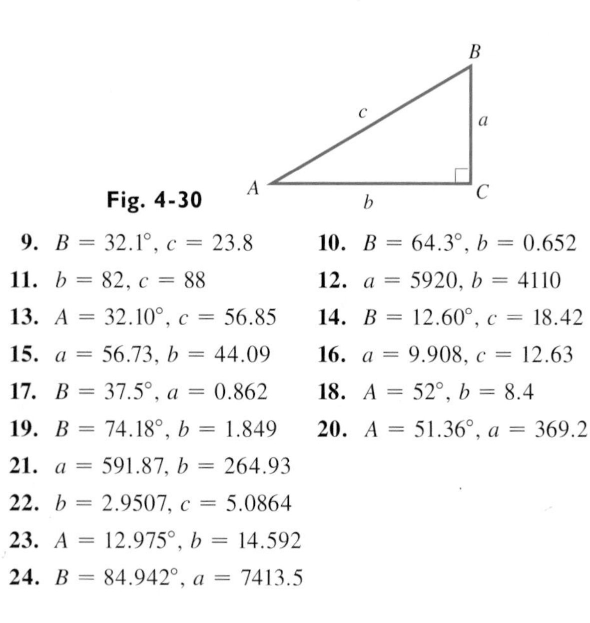

Fig. 4-30

9. $B = 32.1°, c = 23.8$
10. $B = 64.3°, b = 0.652$
11. $b = 82, c = 88$
12. $a = 5920, b = 4110$
13. $A = 32.10°, c = 56.85$
14. $B = 12.60°, c = 18.42$
15. $a = 56.73, b = 44.09$
16. $a = 9.908, c = 12.63$
17. $B = 37.5°, a = 0.862$
18. $A = 52°, b = 8.4$
19. $B = 74.18°, b = 1.849$
20. $A = 51.36°, a = 369.2$
21. $a = 591.87, b = 264.93$
22. $b = 2.9507, c = 5.0864$
23. $A = 12.975°, b = 14.592$
24. $B = 84.942°, a = 7413.5$

In Exercises 25–28, find the part of the triangle labeled either x or A in the indicated figure.

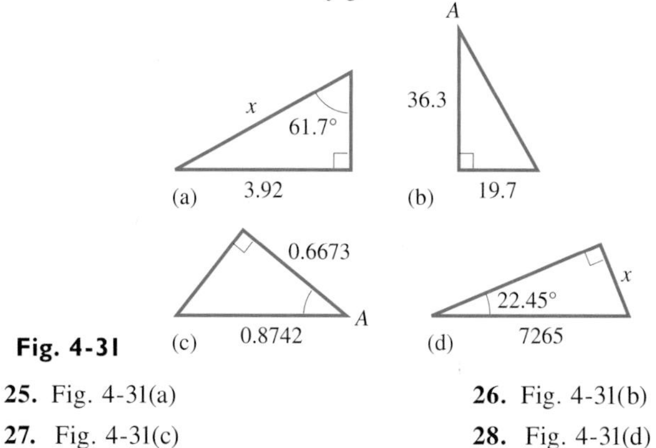

Fig. 4-31

25. Fig. 4-31(a)
26. Fig. 4-31(b)
27. Fig. 4-31(c)
28. Fig. 4-31(d)

In Exercises 29–32, refer to Fig. 4-30. In Exercises 29–31, the listed parts are assumed known. Express the other parts in terms of these known parts.

29. A, c
30. a, b
31. B, a
32. In Fig. 4-30, is there any combination of two given parts (not including $\angle C$) that does not give a unique solution of the triangle? Explain.

4-5 Applications of Right Triangles

EXAMPLE 1 ▸▸ The Sears Tower in Chicago can be seen from a point on the ground known to be 1800 m from the base of the tower. The **angle of elevation** (*the angle between the horizontal and the line of sight, when the object is* ***above*** *the horizontal*) from the observer to the top of the tower is 14°. Based on this information, how high is the Sears Tower?

By drawing an appropriate figure, as shown in Fig. 4-32, we show the given information and that which is required. Here we have let h be the height of the tower, O the position of the observer, and B the point at the base of Sears Tower. From the figure we see that

$$\frac{h}{1800} = \tan 14° \qquad \frac{\text{required opposite side}}{\text{given adjacent side}} = \text{tangent of given angle}$$

or

$$h = 1800 \tan 14°$$
$$= 450 \text{ m}$$

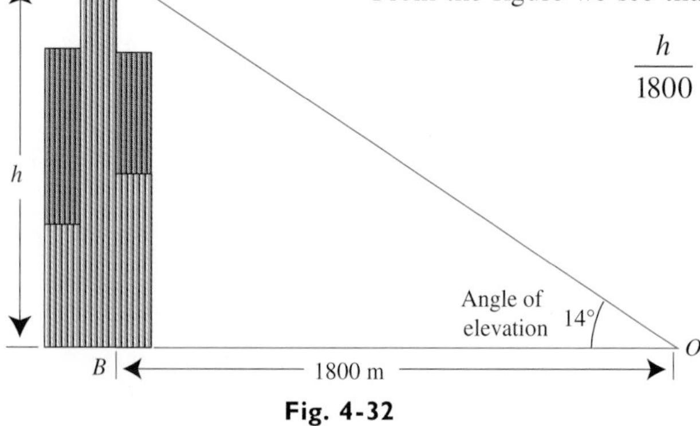

Fig. 4-32

Here we have rounded off the result since the data is good only to two significant digits. (The Sears Tower was completed in 1974 and is the tallest building in the world. Its height is actually 443 m.) ◂◂

EXAMPLE 2 ▸▸ The Goodyear blimp is 565 m above the ground and south of the Rose Bowl in California during a Super Bowl game. The **angle of depression** *(the angle between the horizontal and the line of sight, when the object is* ***below*** *the horizontal)* of the north goal line from the blimp is 58.5°. How far is the observer in the blimp from the goal line?

Again we sketch an appropriate figure as shown in Fig. 4-33. Here we let d represent the required distance. From the figure we see that

$$\frac{565}{d} = \cos 31.5° \qquad \frac{\text{given adjacent side}}{\text{required hypotenuse}} = \text{cosine of known angle}$$

or

$$d = \frac{565}{\cos 31.5°}$$
$$= 663 \text{ m}$$

Here we have rounded off the result to three significant digits, the accuracy of the given information. (The Super Bowl games in 1977, 1980, 1987, and 1993 were played in the Rose Bowl.) ◂◂

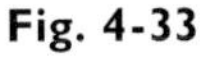

Fig. 4-33

EXAMPLE 3 ▸▸ A missile is launched at an angle of 26.55° with respect to the horizontal. If it travels in a straight line over level terrain for 2.000 min and its average speed is 6355 km/h, what is its altitude at this time?

In Fig. 4-34 we have let h represent the altitude of the missile after 2.000 min (altitude is measured on a perpendicular). Also, we determine that the missile has flown 211.8 km in a direct line from the launching site. This is found from the fact that it travels at 6355 km/h for $\frac{1}{30.00}$ h (2.000 min) and from the fact that $(6355 \text{ km/h})\left(\frac{1}{30.00}\text{ h}\right) = 211.8$ km. Therefore,

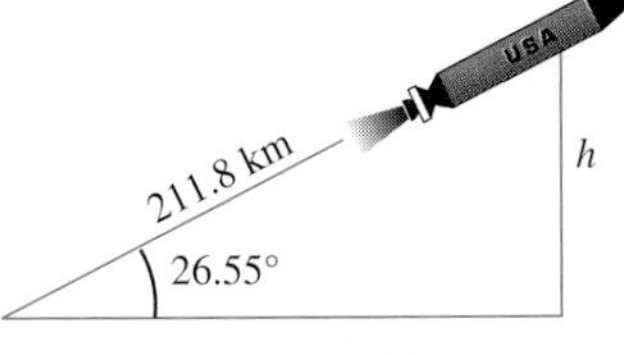

Fig. 4-34

$$\frac{h}{211.8} = \sin 26.55° \qquad \frac{\text{required opposite side}}{\text{known hypotenuse}} = \text{sine of given angle}$$
$$h = 211.8(\sin 26.55°)$$
$$= 94.67 \text{ km}$$ ◂◂

EXAMPLE 4 ▸▸ A driver coming to an intersection sees the word STOP in the roadway. From the measurements shown in Fig. 4-35, find the angle θ that the letters make at the driver's eye.

See the chapter introduction.

From the figure we know sides BS and BE in triangle BES, and sides BT and BE in triangle BET. This means we can find $\angle TEB$ and $\angle SEB$ by use of the tangent. We then find θ from the fact that $\theta = \angle TEB - \angle SEB$.

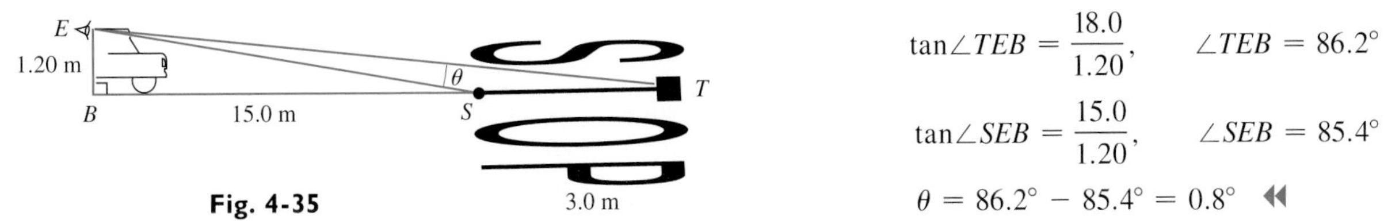

Fig. 4-35

$$\tan\angle TEB = \frac{18.0}{1.20}, \qquad \angle TEB = 86.2°$$
$$\tan\angle SEB = \frac{15.0}{1.20}, \qquad \angle SEB = 85.4°$$
$$\theta = 86.2° - 85.4° = 0.8°$$ ◂◂

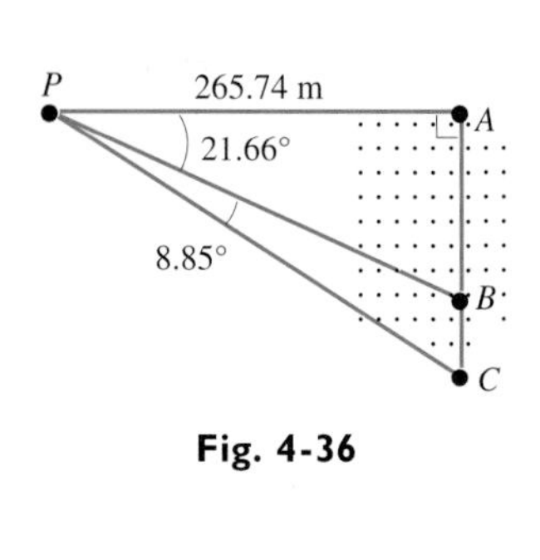

Fig. 4-36

EXAMPLE 5 ▸▸ Using lasers, a surveyor makes the measurements shown in Fig. 4-36, where points B and C are in a marsh. Find the distance between B and C.

Since the distance $BC = AC - AB$, BC is found by finding AC and AB and subtracting.

$$\frac{AB}{265.74} = \tan 21.66° \quad \text{or} \quad AB = 265.74 \tan 21.66°$$

$$\frac{AC}{265.74} = \tan(21.66° + 8.85°) \quad \text{or} \quad AC = 265.74 \tan 30.51°$$

$$BC = AC - AB = 265.74 \tan 30.51° - 265.74 \tan 21.66°$$
$$= 51.06 \text{ m} \quad ◂◂$$

Exercises 4–5

In the following exercises, solve the given problems. Sketch an appropriate figure unless the figure is given.

1. A straight 120-m culvert is built down a hillside which makes an angle of 54.0° with the horizontal. Find the height of the hill.

2. A point near the top of the Leaning Tower of Pisa is 50.5 m from a point at the base of the tower (measured along the tower). This top point is also directly above a point on the ground 4.25 m from the same base point. What angle does the tower make with the ground? See Fig. 4-37.

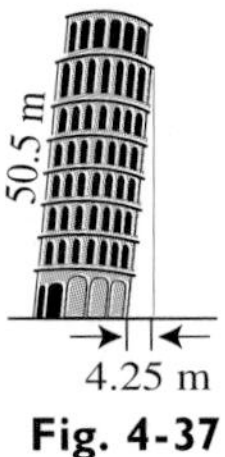

Fig. 4-37

3. A tree has a shadow 22.8 m long when the angle of elevation of the sun is 62.6°. How tall is the tree?

4. The straight arm of a robot is 1.25 m long and makes an angle of 13.0° above a horizontal conveyor belt. How high above the belt is the end of the arm? See Fig. 4-38.

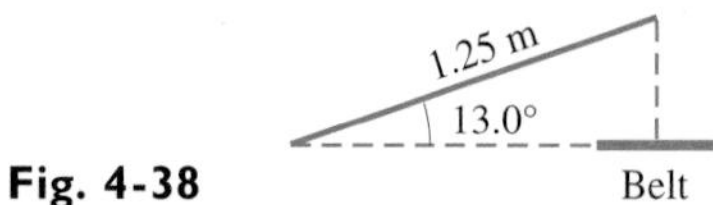

Fig. 4-38

5. The headlights of an automobile are set such that the beam drops 5.10 cm for each 7.50 m in front of the car. What is the angle between the beam and the road?

6. A bullet was fired such that it just grazed the top of a table. It entered a wall, which is 3.84 m from the graze point in the table, at a point 1.41 m above the tabletop. At what angle was the bullet fired above the horizontal? See Fig. 4-39.

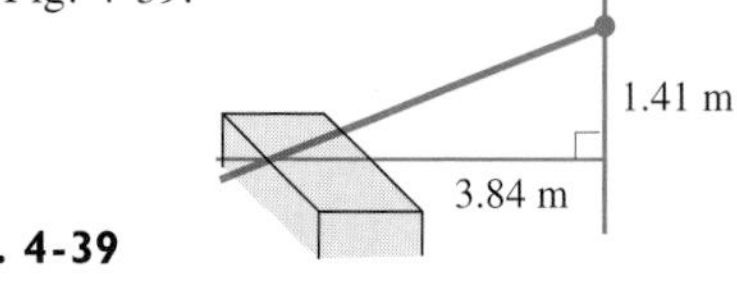

Fig. 4-39

7. A robot is on the surface of Mars. The angle of depression from a camera in the robot to a rock on the surface of Mars is 13.33°. The camera is 196.0 cm above the surface. How far is the camera from the rock?

8. An observation deck at the CN Tower in Toronto is 447 m above the level of Lake Ontario. An observer measures the angle of depression of Niagara Falls to be 0.37°. How far is Niagara Falls from Toronto? (The CN Tower, shown in Fig. 4-40, is the tallest free-standing structure in the world. It is 553 m high.)

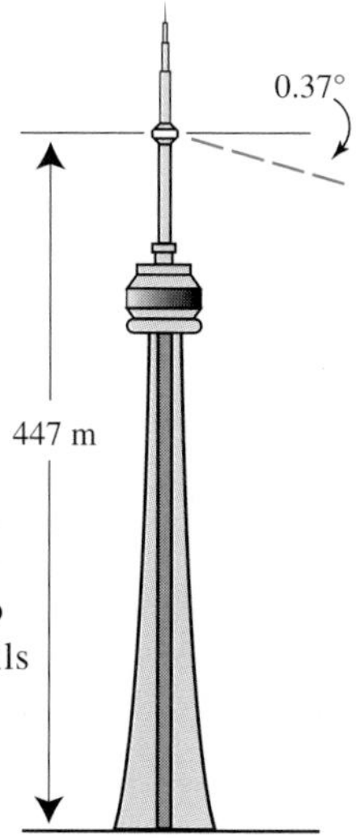

Fig. 4-40

9. In designing a new building, a doorway is 795 mm above the ground. A ramp for the disabled, at an angle of 6.0° with the ground, is to be built to the doorway. How long will the ramp be?

10. On a test flight, during the landing of the space shuttle, the ship was 105 m above the end of the landing strip. It then came in on a constant angle of 7.5° with the landing strip. How far from the end of the landing strip did it first touch?

11. A rectangular piece of plywood 1200 mm by 2400 mm is cut from one corner to an opposite corner. What are the angles between edges of the resulting pieces?

12. A guardrail is to be constructed around the top of a circular observation tower. The diameter of the observation area is 12.3 m. If the railing is constructed with 30 equal straight sections, what should be the length of each section?

13. The angle of inclination of a road is often expressed as *percent grade*, which is the vertical rise divided by the horizontal run (expressed as a percent). See Fig. 4-41. A 6.0% grade corresponds to a road that rises 6.0 m for every 100 m along the horizontal. Find the angle of inclination that corresponds to a 6.0% grade.

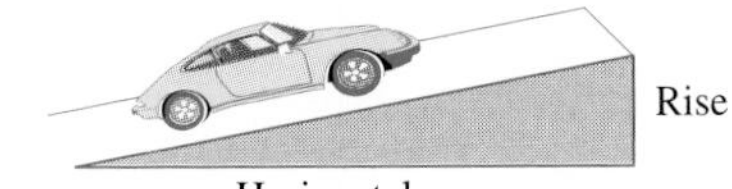

Fig. 4-41

14. A tabletop is in the shape of a regular octagon (8 sides). What is the greatest distance across the table if one edge of the octagon is 0.750 m?

15. A rectangular solar panel is 125 cm long. It is supported by a vertical rod 97.5 cm long (see Fig. 4-42). What is the angle between the panel and the horizontal?

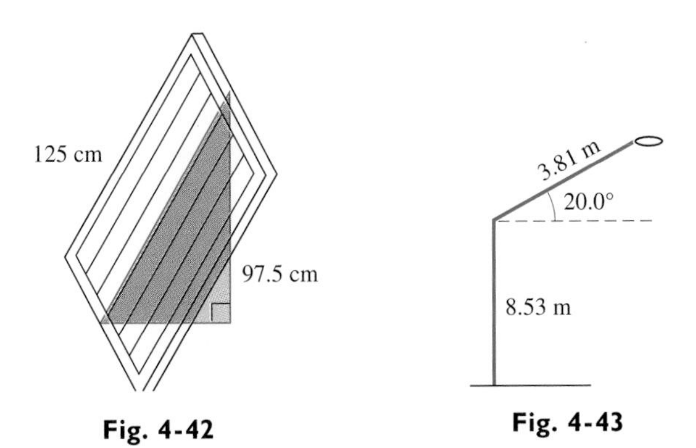

Fig. 4-42 **Fig. 4-43**

16. A street light is designed as shown in Fig. 4-43. How high above the street is the light?

17. A straight driveway is 85.0 m long, and the top is 12.0 m above the bottom. What angle does the driveway make with the horizontal?

18. A level drawbridge is 93.8 m long. When each half is raised, the distance between them is 30.0 m. What angle does each make with the horizontal?

19. A square wire loop is rotating in the magnetic field between two poles of a magnet in order to induce an electric current. The axis of rotation passes through the center of the loop and is midway between the poles, as shown in the side view in Fig. 4-44. How far is the edge of the loop from either pole if the side of the square is 7.30 cm and the poles are 7.66 cm apart when the angle between the loop and the vertical is 78.0°?

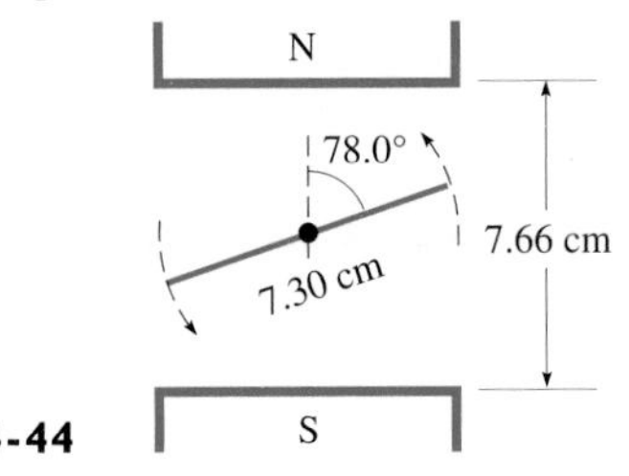

Fig. 4-44

20. From a space probe circling Io, one of Jupiter's moons, at an altitude of 552 km, it was observed that the angle of depression of the horizon was 39.7°. What is the radius of Io?

21. A manufacturing plant is designed to be in the shape of a regular pentagon with 92.5 m on each side. A security fence surrounds the building to form a circle, and each corner of the building is to be 25.0 m from the closest point on the fence. How much fencing is required?

22. A surveyor on the New York City side of the Hudson River wishes to find the height of a cliff (*palisade*) on the New Jersey side of the river. The measurements made are shown in Fig. 4-45. How high is the cliff? (In the figure, the triangle containing the height h is vertical and perpendicular to the river.)

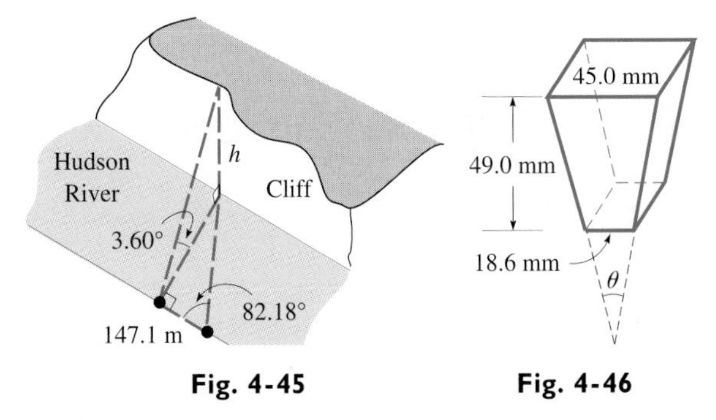

Fig. 4-45 **Fig. 4-46**

23. Find the angle θ in the taper shown in Fig. 4-46. (The front face is an isosceles trapezoid.)

24. A light beam 3.87 cm in diameter strikes the floor at an angle of 26.0° with the floor. What is the longest dimension of the area lit by the beam on the floor?

25. A stairway 1.0 m wide goes from the bottom of a cylindrical storage tank to the top at a point halfway around the tank. The handrail on the outside of the stairway makes an angle of 31.8° with the horizontal, and the radius of the tank is 11.8 m. Find the length of the handrail. See Fig. 4-47.

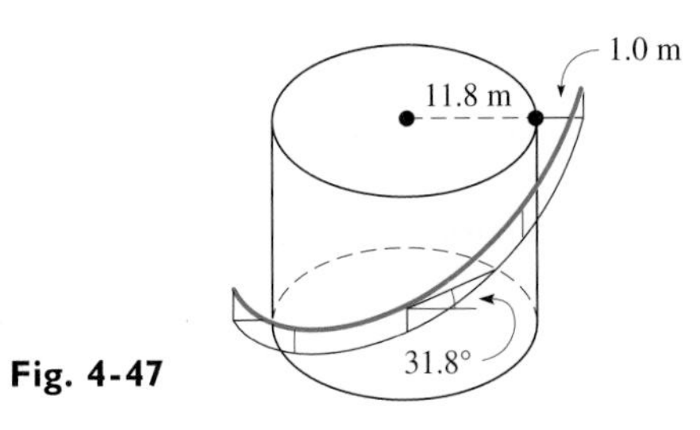

Fig. 4-47

26. An antenna is on the top of the World Trade Center. From a point on the river 2400 m from the Center, the angles of elevation of the top and bottom of the antenna are 12.1° and 9.9°, respectively. How tall is the antenna? (Disregard the small part of the antenna near the base that cannot be seen.) The World Trade Center is shown in Fig. 4-48.

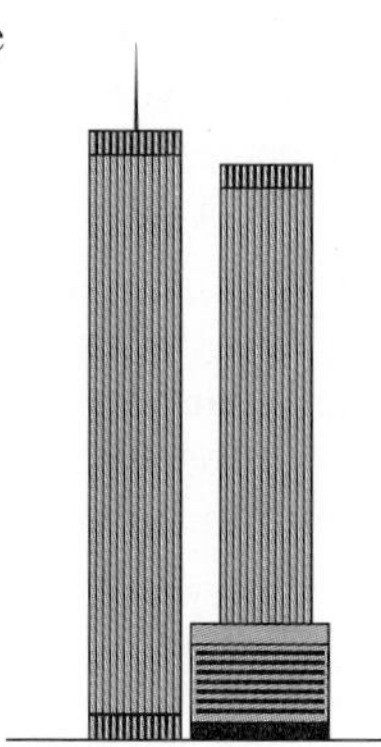

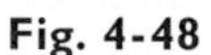

Fig. 4-48

27. Some of the streets of San Francisco are shown in Fig. 4-49. The distances between intersections A and B, A and D, and C and E are shown. How far is it between intersections C and D?

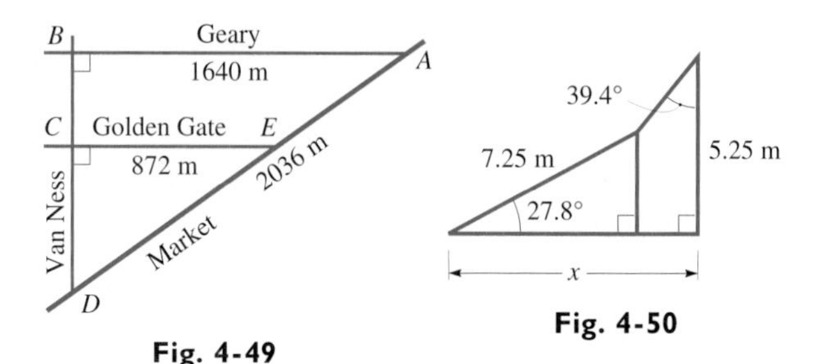

Fig. 4-49

Fig. 4-50

28. A supporting girder structure is shown in Fig. 4-50. Find the length x.

Chapter Equations, Review Exercises, and Practice Test

Chapter Equations

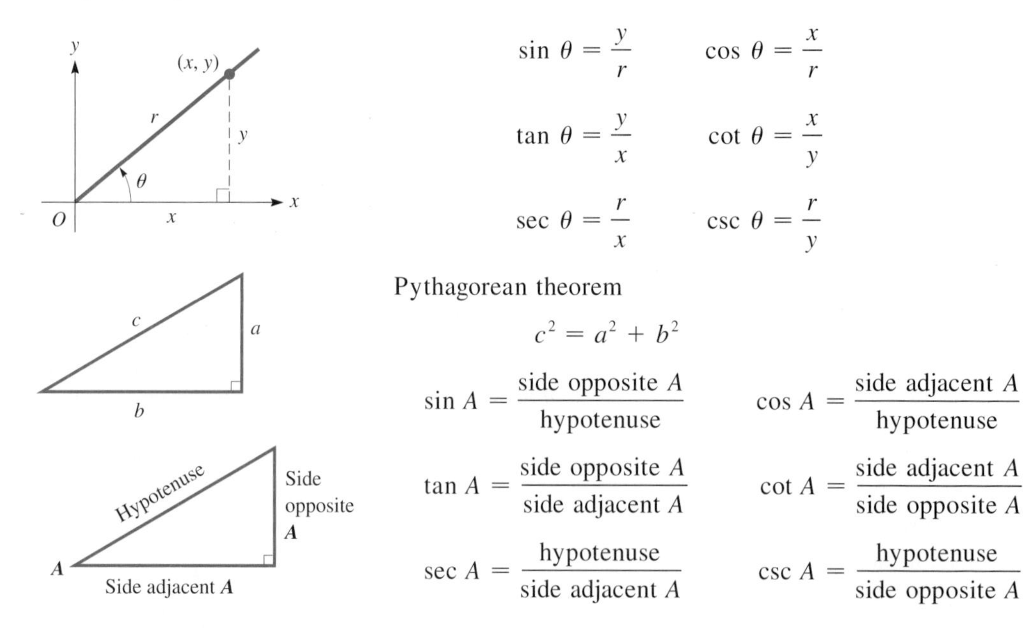

$$\sin\theta = \frac{y}{r} \qquad \cos\theta = \frac{x}{r}$$

$$\tan\theta = \frac{y}{x} \qquad \cot\theta = \frac{x}{y} \qquad \text{(4-1)}$$

$$\sec\theta = \frac{r}{x} \qquad \csc\theta = \frac{r}{y}$$

Pythagorean theorem

$$c^2 = a^2 + b^2 \qquad \text{(4-2)}$$

$$\sin A = \frac{\text{side opposite } A}{\text{hypotenuse}} \qquad \cos A = \frac{\text{side adjacent } A}{\text{hypotenuse}}$$

$$\tan A = \frac{\text{side opposite } A}{\text{side adjacent } A} \qquad \cot A = \frac{\text{side adjacent } A}{\text{side opposite } A} \qquad \text{(4-5)}$$

$$\sec A = \frac{\text{hypotenuse}}{\text{side adjacent } A} \qquad \csc A = \frac{\text{hypotenuse}}{\text{side opposite } A}$$

Review Exercises

In Exercises 1–4, find the smallest positive angle and the smallest negative angle (numerically) coterminal with, but not equal to, the given angle.

1. 17.0° **2.** 248.3° **3.** −217.5° **4.** −7.6°

In Exercises 5–8, change the given angles to equal angles expressed in decimal form.

5. 31°54′ **6.** 174°45′ **7.** 38°6′ **8.** 321°27′

In Exercises 9–12, change the given angles to equal angles expressed to the nearest minute.

9. 17.5° **10.** 65.4° **11.** 49.7° **12.** 126.25°

In Exercises 13–16, determine the trigonometric functions of the angles (in standard position) whose terminal side passes through the given points. Give answers in exact form.

13. (24, 7) **14.** (5, 4) **15.** (4, 4) **16.** (1.2, 0.5)

In Exercises 17–20, find the indicated trigonometric functions. Give answers in decimal form, rounded off to three significant digits.

17. Given $\sin\theta = \frac{5}{13}$, find $\cos\theta$ and $\cot\theta$.

18. Given $\cos\theta = \frac{3}{8}$, find $\sin\theta$ and $\tan\theta$.

19. Given $\tan\theta = 2$, find $\cos\theta$ and $\csc\theta$.

20. Given $\cot\theta = 4$, find $\sin\theta$ and $\sec\theta$.

In Exercises 21–28, find the values of the trigonometric functions. Round off results.

21. $\sin 72.1°$ **22.** $\cos 40.3°$

23. $\tan 61.64°$ **24.** $\sin 49.09°$

25. $\sec 18.4°$ **26.** $\csc 82.4°$

27. $\cot 7.06°$ **28.** $\sec 79.36°$

In Exercises 29–36, find θ for each of the given trigonometric functions. Round off results.

29. $\cos\theta = 0.950$ **30.** $\sin\theta = 0.630\,52$

31. $\tan\theta = 1.574$ **32.** $\cos\theta = 0.1345$

33. $\csc\theta = 4.713$ **34.** $\cot\theta = 0.7561$

35. $\sec\theta = 2.54$ **36.** $\csc\theta = 1.92$

In Exercises 37–48, solve the right triangles with the given parts. Refer to Fig. 4-51.

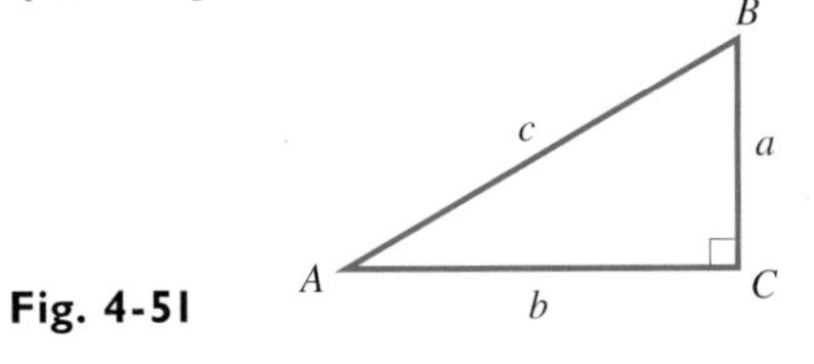

Fig. 4-51

37. $A = 17.0°$, $b = 6.00$ **38.** $B = 68.1°$, $a = 1080$

39. $a = 81.0$, $b = 64.5$ **40.** $a = 1.06$, $c = 3.82$

41. $A = 37.5°$, $a = 12.0$ **42.** $B = 15.7°$, $c = 12.6$

43. $b = 6.508$, $c = 7.642$ **44.** $a = 72.14$, $b = 14.37$

45. $A = 49.67°$, $c = 0.8253$ **46.** $B = 4.38°$, $b = 5.682$

47. $a = 11.652$, $c = 15.483$ **48.** $a = 724.39$, $b = 852.44$

In Exercises 49–76, solve the given applied problems.

49. In analyzing the forces on a certain hinge, one of the vertical forces F_y was found from the equation $F_y = F\cos\theta$, where θ is the angle between one part of the hinge and the vertical. Find the value of F_y if $F = 56.0$ N and $\theta = 37.5°$.

50. The velocity v (in m/s) of an object that slides s feet down an inclined plane is given by $v = \sqrt{2gs\sin\theta}$, where g is the acceleration due to gravity and θ is the angle the plane makes with the horizontal. Find v if $g = 9.80$ m/s^2, $s = 10.5$ m, and $\theta = 25.2°$.

51. In finding the area A of a triangular tract of land, a surveyor uses the formula $A = \frac{1}{2}ab\sin C$, where a and b are two sides of the triangle and C is the angle included between them. Find A if $a = 31.96$ m, $b = 47.25$ m, and $C = 64.09°$.

52. A water channel has the cross section of an isosceles trapezoid. See Fig. 4-52. The area of the cross section is $A = bh + h^2\cot\theta$. Find A if $b = 12.6$ m, $h = 4.75$ m, and $\theta = 37.2°$.

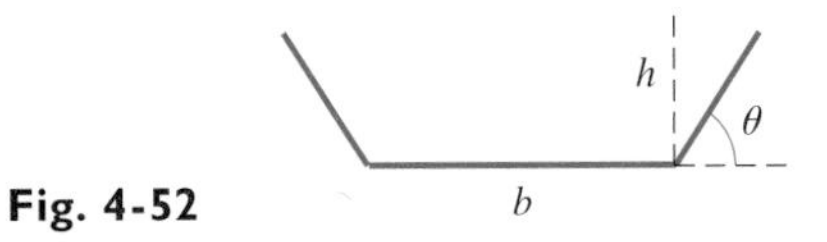

Fig. 4-52

53. For a car rounding a curve, the road should be banked at an angle θ according to the equation $\tan\theta = \dfrac{v^2}{gr}$. Here v is the speed of the car, g is the acceleration due to gravity, and r is the radius of the curve in the road. Find θ for $v = 24.2$ m/s, $g = 9.80$ m/s^2, and $r = 282$ m.

54. The *apparent power* S in an electric circuit in which the power is P and the impedance phase angle is θ is given by $S = P\sec\theta$. Given $P = 12.0$ V · A and $\theta = 29.4°$, find S.

55. In tracking an airplane on radar, it is found that the plane is 27.5 km on a direct line from the control tower, with an angle of elevation of 10.3°. What is the altitude of the plane?

56. A conveyor belt 25.0 m long is inclined at 34.7° above the horizontal. Through what height can the belt lift objects placed on it?

57. The window of a house is shaded as shown in Fig. 4-53. What percent of the window is shaded when the angle of elevation θ of the sun is 65.0°?

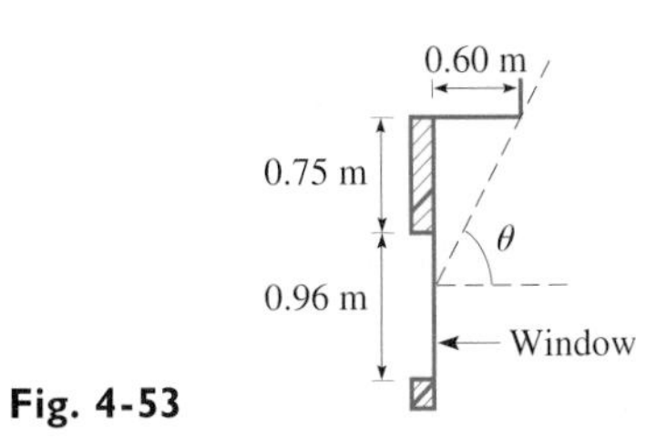

Fig. 4-53

58. The windshield on an automobile is inclined 42.5° with respect to the horizontal. Assuming that the windshield is flat and rectangular, what is its area if it is 1.50 m wide and the bottom is 0.48 m in front of the top?

59. A water slide at an amusement park is 25 m long and is inclined at an angle of 52° with the horizontal. How high is the top of the slide above the water level?

60. A theater stage inclines 3.5° upward from front to back. If it measures 12.5 m from front to back along the stage, how much higher is the back of the stage than the front?

61. The vertical cross section of an attic room in a house is shown in Fig. 4-54. Find the distance d across the floor.

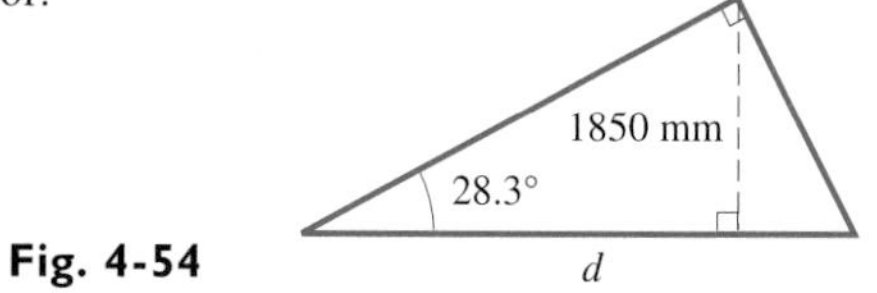

Fig. 4-54

62. The impedance Z and resistance R in an alternating-current circuit may be represented by letting the impedance be the hypotenuse of a right triangle and the resistance be the side adjacent to the phase angle θ. If $R = 1750\ \Omega$ and $\theta = 17.38°$, find Z.

63. A particular straight section of the Alaskan oil pipeline is 5.3 km long, and it rises at an angle of 6.2°. How much higher is one end of the section than the other?

64. A Coast Guard boat that is 2.75 km from a straight beach can travel at 37.5 km/h. By traveling along a line that is at 69.0° with the beach, how long will it take to reach the beach? See Fig. 4-55.

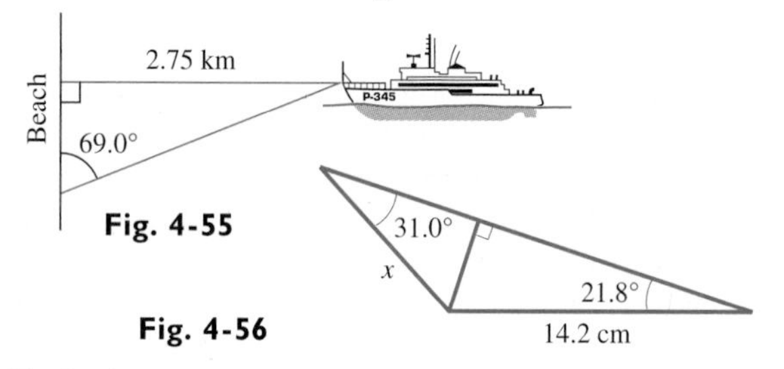

Fig. 4-55

Fig. 4-56

65. In the structural support shown in Fig. 4-56, find x.

66. A straight emergency chute for an airplane is 5.50 m long. In being tested, the end at the chute is 2.9 m above the ground. What angle does the chute make with the ground?

67. The main span of the Mackinac Bridge (see Fig. 4-57) in northern Michigan is 1160 m long. The angle *subtended* by the span at the eye of an observer in a helicopter is 2.2°. Show that the distance calculated from the helicopter to the span is about the same if the line of sight is perpendicular to the end or to the middle of the span.

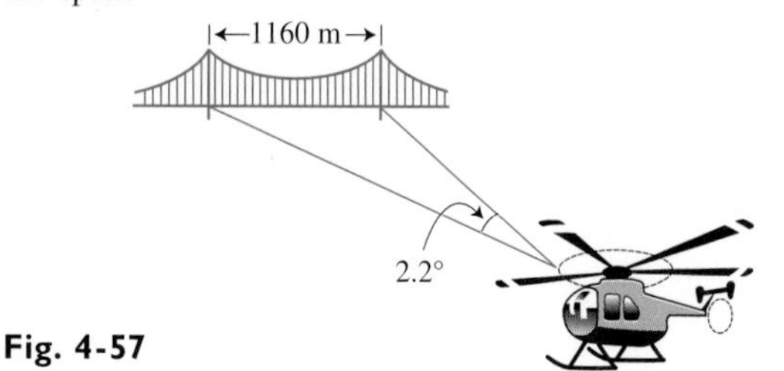

Fig. 4-57

68. The top and height of the trellis shown in Fig. 4-58 are each 2.25 m. Each side piece makes an angle of 80.0° with the ground. Find the length of each side piece and the area covered by the trellis.

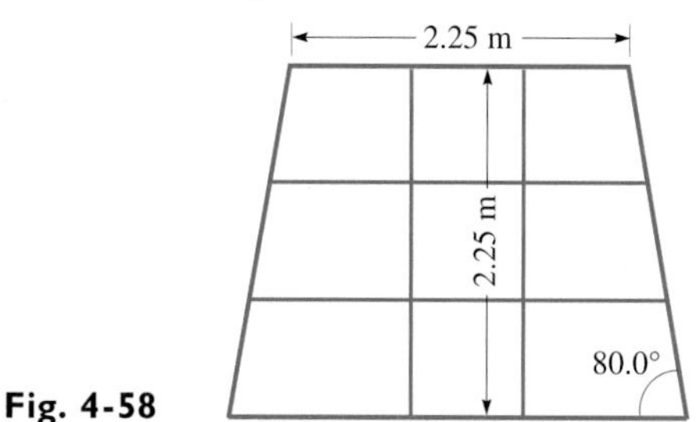

Fig. 4-58

69. The distance from ground level to the underside of a cloud is called the *ceiling*. A ground observer 950 m from a searchlight aimed vertically notes that the angle of elevation of the spot of light on a cloud is 76°. What is the ceiling?

70. Through what angle θ must the crate shown in Fig. 4-59 be tipped in order that its center C is directly above the pivot point P?

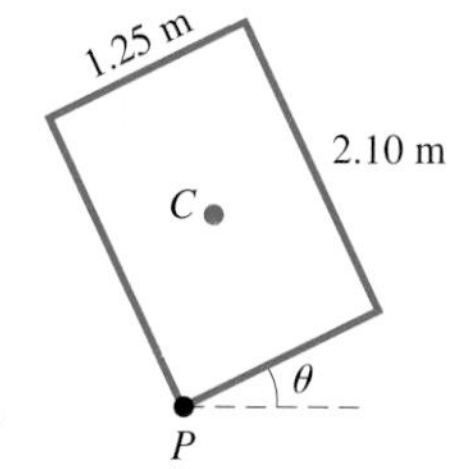

Fig. 4-59

71. A laser beam is transmitted with a "width" of 0.002 00°. What is the diameter of a spot of the beam on an object 52 500 km distant? See Fig. 4-60.

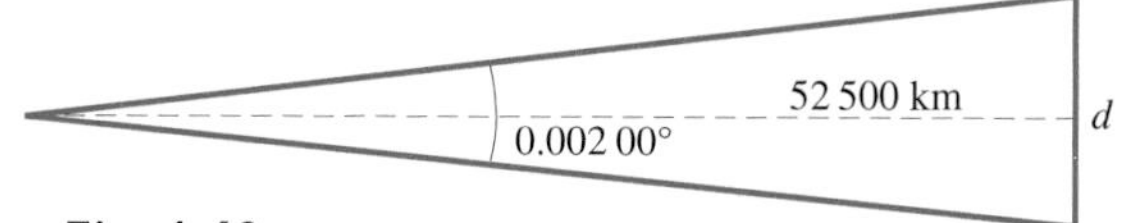

Fig. 4-60

72. Find the gear angle θ in Fig. 4-61 if $t = 0.180$ cm.

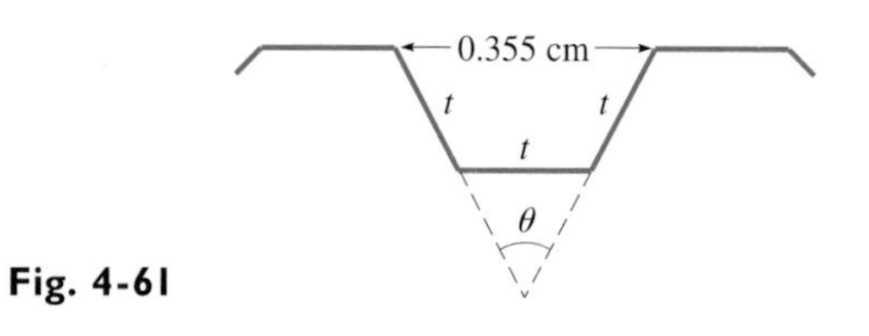

Fig. 4-61

73. A hang glider is directly above the shore of a small lake. An observer on a hill is 375 m along a straight line from the shore. From the observer, the angle of elevation of the hang glider is 42.0°, and the angle of depression of the shore is 25.0°. How far above the shore is the hang glider?

74. A ground observer sights a weather balloon to the east at an angle of elevation of 15.0°. A second observer 2.35 km to the east of the first also sights the balloon to the east at an angle of elevation of 24.0°. How high is the balloon?

75. A uniform strip of wood 50.0 mm wide frames a trapezoidal window as shown in Fig. 4-62. Find the left dimension l of the outside of the frame.

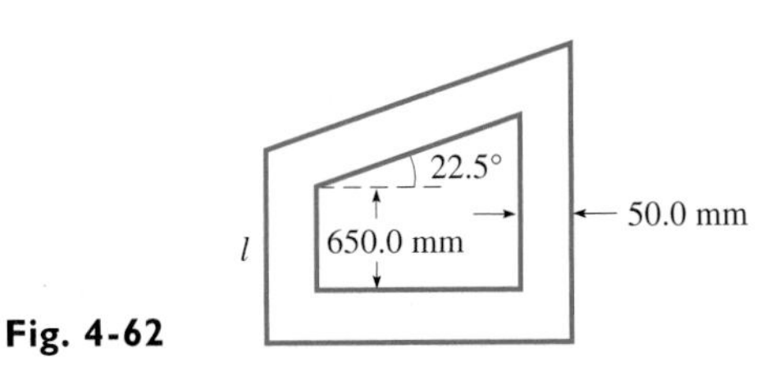

Fig. 4-62

76. A crop-dusting plane flies over a field at a height of 8.0 m. If the dust leaves the plane through a 30° angle and hits the ground after the plane travels 25 m, how wide a strip is dusted? See Fig. 4-63.

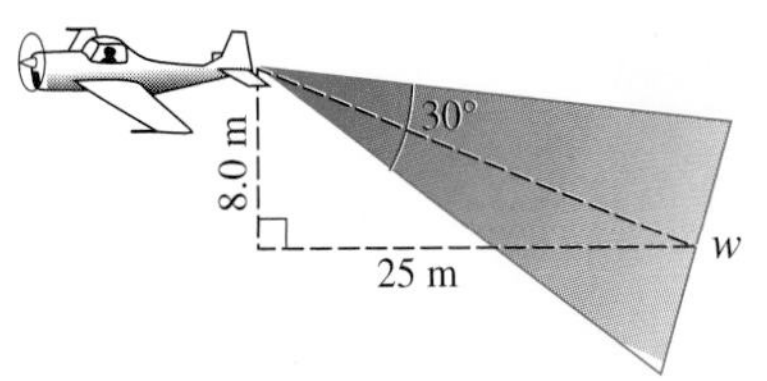

Fig. 4-63

Writing Exercise

77. Two students were discussing a problem in which a distant object was seen through an angle of 2.3°. One found the length of the object (perpendicular to the line of sight) using the tangent of the angle, and the other had the same answer using the sine of the angle. Write a paragraph explaining how this is possible.

Practice Test

1. Express 37°39′ in decimal form.
2. Find θ to the nearest 0.01° if $\cos\theta = 0.3726$.
3. A ship's captain, desiring to travel due south, discovers that due to an improperly functioning instrument, the ship has gone 22.62 km in a direction 4.05° east of south. How far from its course (to the east) is the ship?
4. Find $\tan\theta$ in fractional form if $\sin\theta = \frac{2}{3}$.
5. Find $\csc\theta$ if $\tan\theta = 1.294$.
6. Solve the right triangle in Fig. 4-64 if $A = 37.4°$ and $b = 52.8$.
7. Solve the right triangle in Fig. 4-64 if $a = 2.49$ and $c = 3.88$.

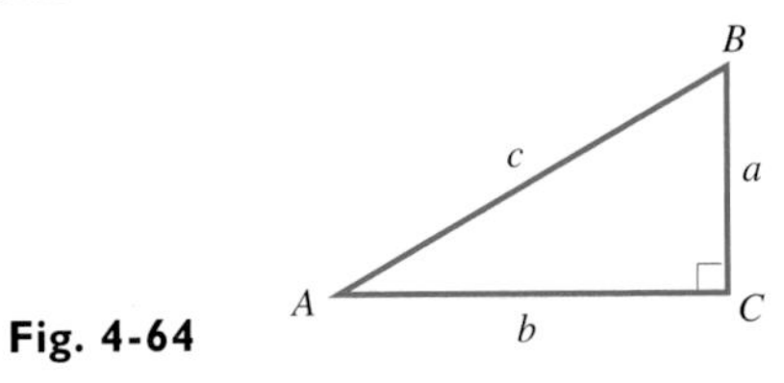

Fig. 4-64

8. In finding the wavelength λ (the Greek lambda) of light, the equation $\lambda = d\sin\theta$ is used. Find λ if $d = 30.05\ \mu$m and $\theta = 1.167°$. (μ is the prefix for 10^{-6}.)
9. Determine the trigonometric functions of an angle in standard position if its terminal side passes through (5, 2). Give answers in exact and decimal forms.
10. A surveyor sights two points directly ahead. Both are at an elevation 18.525 m lower than the observation point. How far apart are the points if the angles of depression are 13.500° and 21.375°, respectively?

5

Systems of Linear Equations; Determinants

In designing an industrial robot, the forces acting on each link must be carefully analyzed. In Section 5-6 we show how a system of equations is solved to find these forces.

The solution of most technical and scientific problems requires that we consider several quantities that may be related in a number of ways. This can lead to more than one equation relating the quantities. For example, when determining the price at which to sell a computer, we must consider the costs for research, development, and production as well as the costs of the various components.

More than one equation relating quantities is also found for the various forces acting on an object, the electric currents in different parts of a given circuit, and the different components of a chemical or of a medical dosage. Applications exist in all technical areas.

In this chapter we consider the solution of systems of two equations with two unknowns and of three equations with three unknowns.

5-1 Linear Equations

In general, *an equation is termed* **linear** *in a given set of variables if each term contains only one variable, to the first power, or is a constant.*

EXAMPLE 1 ▸▸ $5x - t + 6 = 0$ is linear in x and t, but $5x^2 - t + 6 = 0$ is not linear, due to the presence of x^2.

The equation $4x + y = 8$ is linear in x and y, but $4xy + y = 8$ is not, due to the presence of xy.

The equation $x - 6y + z - 4w = 7$ is linear in x, y, z, and w, but the equation $x - \frac{6}{y} + z - 4w = 7$ is not, due to the presence of $\frac{6}{y}$, where y appears in the denominator. ◂◂

An equation that can be written in the form

$$ax + b = 0 \quad (5\text{-}1)$$

is known as a **linear equation in one unknown**. Here a and b are constants. We have already discussed the solution to this type of equation in Section 1-10. In general, *the* **solution**, *or* **root**, *of the equation is* $x = -b/a$. Also, it is noted that the solution of the linear equation is the same as the *zero* of the **linear function** $f(x) = ax + b$.

EXAMPLE 2 ▸▸ The equation $2x + 7 = 0$ is a linear equation of the form of Eq. (5-1) with $a = 2$ and $b = 7$.

The solution to the linear equation $2x + 7 = 0$, or the zero of the linear function $f(x) = 2x + 7$, is $-7/2$. From Chapter 3 we recall that the *zero* of a function $f(x)$ is the value of x for which $f(x) = 0$. ◂◂

As we noted in the chapter introduction, there are a great many applied problems that involve more than one unknown quantity. In the next example, two specific illustrations are given.

EXAMPLE 3 ▸▸ (a) A basic law of direct-current electricity, known as *Kirchhoff's first law,* may be stated as "The algebraic sum of the currents entering any junction in a circuit is zero." If three wires are joined at a junction, this law leads to the linear equation

$$i_1 + i_2 + i_3 = 0$$

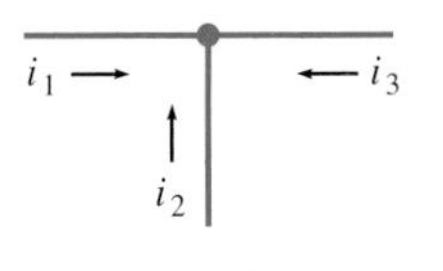

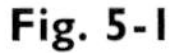
Fig. 5-1

where i_1, i_2, and i_3 are the currents in each of the wires. (Either one or two of these currents must have a negative sign, showing that it is acually leaving the junction.) See Fig. 5-1.

(b) When determining two forces F_1 and F_2 acting on a beam, we might encounter an equation such as

$$2F_1 + 4F_2 = 200$$ ◂◂

An equation that can be written in the form

$$ax + by = c \quad (5\text{-}2)$$

is known as a **linear equation in two unknowns.** In Chapter 3 we considered many equations that can be written in this form. We found that for each value of x, there is a corresponding value for y. Each of these pairs of numbers is a **solution** to the equation, although we did not call it that at the time. *A solution is any set of numbers, one for each variable, which satisfies the equation.* When we represent the solutions in the form of a graph, we see that the graph of any linear equation in two unknowns is a straight line. Also, graphs of linear equations in one unknown, those for which $a = 0$ or $b = 0$, are also straight lines. Thus we see the significance of the name *linear.* In the next section we shall further consider the graph of the linear equation.

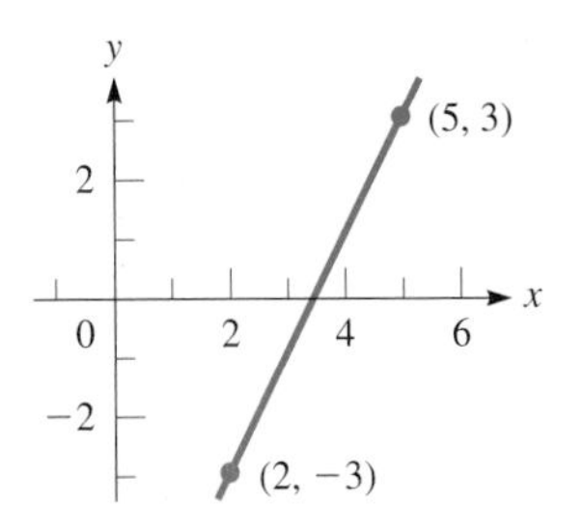

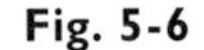
Fig. 5-6

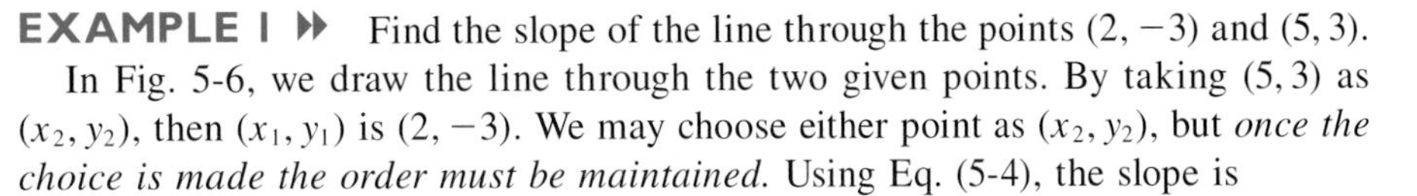

EXAMPLE 1 ▸▸ Find the slope of the line through the points $(2, -3)$ and $(5, 3)$.

In Fig. 5-6, we draw the line through the two given points. By taking $(5, 3)$ as (x_2, y_2), then (x_1, y_1) is $(2, -3)$. We may choose either point as (x_2, y_2), but *once the choice is made the order must be maintained.* Using Eq. (5-4), the slope is

$$m = \frac{3 - (-3)}{5 - 2} = \frac{6}{3} = 2$$

The rise is 2 units for each unit (of run) it moves from left to right. ◂◂

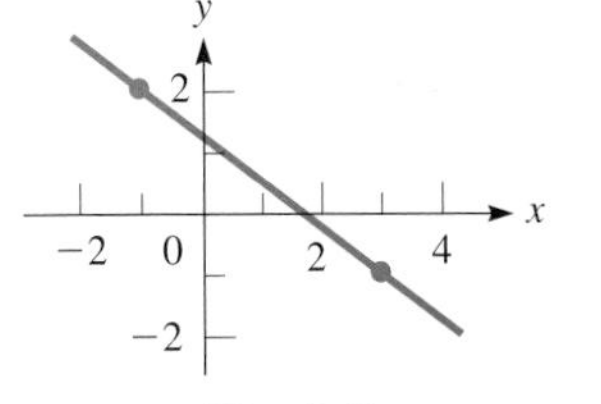

Fig. 5-7

EXAMPLE 2 ▸▸ Find the slope of the line through $(-1, 2)$ and $(3, -1)$.

In Fig. 5-7, we draw the line through these two points. By taking (x_2, y_2) as $(3, -1)$ and (x_1, y_1) as $(-1, 2)$, the slope is

$$m = \frac{-1 - 2}{3 - (-1)} = \frac{-3}{3 + 1} = -\frac{3}{4}$$

The line *falls* 3 units for each 4 units it moves from left to right. ◂◂

We note in Example 1 that *as x increases, y increases and that slope is* ***positive.*** In Example 2, *as x increases, y decreases and that slope is* ***negative.*** Also, *the larger the absolute value of the slope, the more nearly vertical is the line.*

EXAMPLE 3 ▸▸ For each of the following lines shown in Fig. 5-8, we show the difference in the y-coordinates and in the x-coordinates between two points.

In Fig. 5-8(a), a line with a slope of 5 is shown. It rises sharply.
In Fig. 5-8(b), a line with a slope of $\frac{1}{2}$ is shown. It rises slowly.
In Fig. 5-8(c), a line with a slope of -5 is shown. It falls sharply.
In Fig. 5-8(d), a line with a slope of $-\frac{1}{2}$ is shown. It falls slowly.

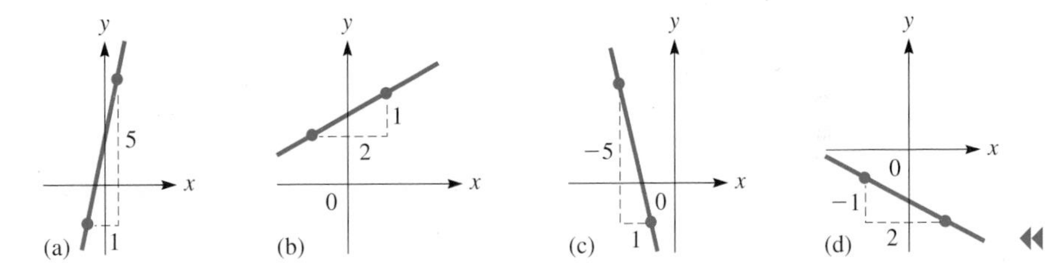

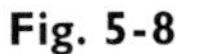
Fig. 5-8 ◂◂

Slope-Intercept Form of the Equation of a Straight Line

We now show how the slope is related to the equation of a straight line. In Fig. 5-9, if we have two points, $(0, b)$ and a general point (x, y), the slope is

$$m = \frac{y - b}{x - 0}$$

Simplifying this, we have $mx = y - b$, or

$$y = mx + b \tag{5-5}$$

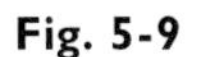
Fig. 5-9

In Eq. (5-5), m is the slope and b is the y-coordinate of the point where it crosses the y-axis. *This point is the* ***y-intercept*** *of the line,* and its coordinates are $(0, b)$. *Equation (5-5) is the* ***slope-intercept*** *form of the equation of a straight line. The coefficient of x is the slope, and the constant is the ordinate of the y-intercept.* The point $(0, b)$ and simply b are both referred to as the y-intercept. Since the x-coordinate of the y-intercept is 0, this should cause no confusion.

In Exercises 9–16, determine whether or not the given pair of values is a solution of the given system of simultaneous linear equations.

9. $x - y = 5$ $\quad x = 4, y = -1$
$2x + y = 7$

10. $2x + y = 8$ $\quad x = -1, y = 10$
$3x - y = -13$

11. $x + 5y = -7$ $\quad x = -2, y = 1$
$3x - 4y = -10$

12. $-3x + y = 1$ $\quad x = \frac{1}{3}, y = 2$
$6x - 3y = -4$

13. $2x - 5y = 0$ $\quad x = \frac{1}{2}, y = -\frac{1}{5}$
$4x + 10y = 4$

14. $6x + y = 5$ $\quad x = 1, y = -1$
$3x - 4y = -1$

15. $3x - 2y = 2.2$ $\quad x = 0.6, y = -0.2$
$5x + y = 2.8$

16. $x - 7y = -3.2$ $\quad x = -1.1, y = 0.3$
$2x + y = 2.5$

In Exercises 17–20, answer the given questions.

17. An airplane is flying at p kilometres per hour relative to a wind blowing at w kilometres per hour. Traveling with the wind, the ground speed of the plane is 300 km/h, and traveling against the wind, the ground speed is 220 km/h. This leads to the two equations

$$p + w = 300$$
$$p - w = 220$$

Are the speeds 260 km/h and 40 km/h?

18. The electric resistance R of a certain resistor is a function of the temperature T given by the equation $R = aT + b$, where a and b are constants. If $R = 1200\ \Omega$ when $T = 10.0°C$, and $R = 1280\ \Omega$ when $T = 50.0°C$, we can find the constants a and b by substituting and obtaining the equations

$$1200 = 10.0a + b$$
$$1280 = 50.0a + b$$

Are the constants $a = 4.00\ \Omega/°C$ and $b = 1160\ \Omega$?

19. The forces acting on part of a structure are shown in Fig. 5-4. An analysis of the forces leads to the equations

$$0.80F_1 + 0.50F_2 = 50$$
$$0.60F_1 - 0.87F_2 = 12$$

Are the forces 45 N and 28 N?

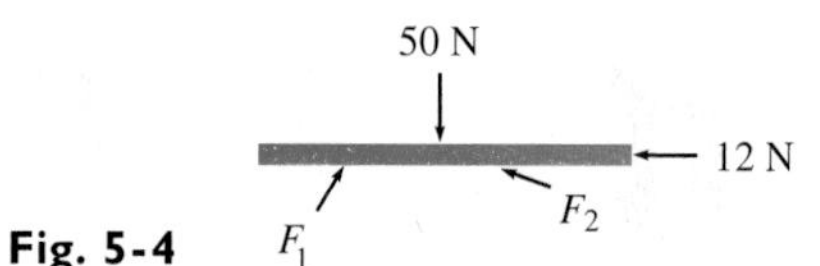

Fig. 5-4

20. Using the data that fuel consumption for transportation contributes a percent p_1 of pollution that is 16% less than the percent p_2 of all other sources combined, the equations

$$p_1 + p_2 = 100$$
$$p_2 - p_1 = 16$$

can be set up. Are the percents $p_1 = 58$ and $p_2 = 42$?

5-2 Graphs of Linear Functions

Our first method of solving a system of equations will be a graphical method. Therefore, before taking up this method, we shall develop additional ways of graphing a linear equation. We will then be able to analyze and check the graph of a linear equation quickly, often by inspection.

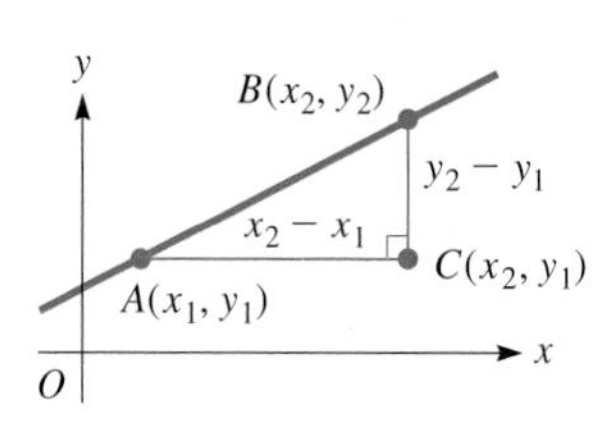

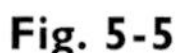

Fig. 5-5

Consider the line that passes through points A, with coordinates (x_1, y_1), and B, with coordinates (x_2, y_2), in Fig. 5-5. Point C is horizontal from A and vertical from B. Thus, C has coordinates (x_2, y_1), and there is a right angle at C. One way of measuring the steepness of this line is to find the ratio of the vertical distance to the horizontal distance between two points. Therefore, *we define the* **slope** *of the line through two points as the difference in the y-coordinates divided by the difference in the x-coordinates.* For points A and B, the slope m is

Slope

$$m = \frac{y_2 - y_1}{x_2 - x_1} \tag{5-4}$$

The slope is often referred to as the *rise* (vertical change) over the *run* (horizontal change). Note that the slope of a vertical line, for which $x_2 = x_1$, is undefined (for $x_2 = x_1$, the denominator of Eq. (5-4) is zero).

EXAMPLE 4 ▸▸ Find the slope and the y-intercept of the line $y = \frac{3}{2}x - 3$.

Since the equation is written as y as a function of x, it is in the form of Eq. (5-5). Therefore, since the coefficient of x is $\frac{3}{2}$, the slope of the line is $\frac{3}{2}$. Also, since we can write the equation as

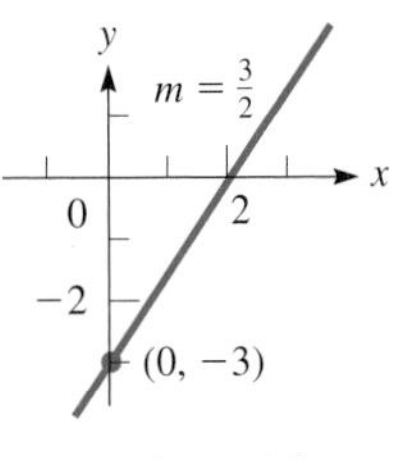

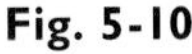
Fig. 5-10

slope ↓ y-intercept ordinate ↓

$$y = \frac{3}{2}x + (-3)$$

we see that the constant is -3, which means the y-intercept is the point $(0, -3)$. The line is shown in Fig. 5-10. ◂◂

EXAMPLE 5 ▸▸ Find the slope and the y-intercept of the line $2x + 3y = 4$.

Here, *we must first write the equation in the slope-intercept form.* Solving for y gives us

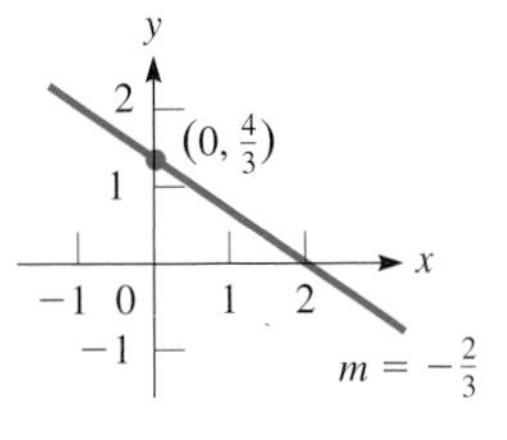

Fig. 5-11

slope ↓ y-intercept ordinate ↓

$$y = -\frac{2}{3}x + \frac{4}{3}$$

Therefore, the slope is $-\frac{2}{3}$ and the y-intercept is the point $(0, \frac{4}{3})$. The line is shown in Fig. 5-11. ◂◂

EXAMPLE 6 ▸▸ In analyzing an electric current, two of the currents, i_1 and i_2, were related by $i_2 = 2i_1 + 1$. Graph this equation using the slope-intercept form.

Since the equation is solved for i_2, we treat i_1 as the independent variable and i_2 as the dependent variable. This means that the slope of the line is 2. Also, since the b-term is 1, we can see that the intercept is $(0, 1)$.

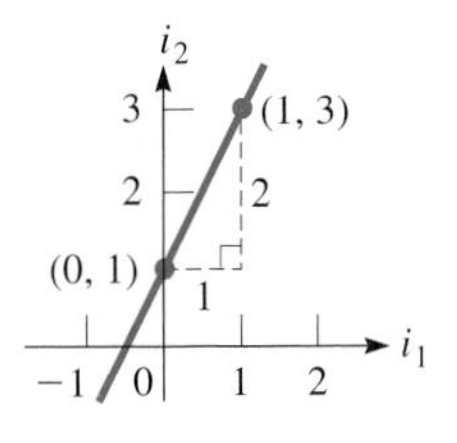

Fig. 5-12

We can use this information to sketch the line, as shown in Fig. 5-12. Since the slope is 2, we know that i_2 increases 2 units for each unit of increase of i_1. Thus, starting at the i_2-intercept $(0, 1)$, if i_1 increases by 1, i_2 increases by 2, and we are at the point $(1, 3)$. The line must pass through $(1, 3)$, as well as through $(0, 1)$. Therefore, we draw the line through these points. (Negative values may be used for electric currents since the sign shows the direction of flow.) ◂◂

Sketching Lines by Intercepts

Another way of sketching the graph of a straight line is to find two points on the line and then draw the line through these points. Two points that are easily determined are those where the line crosses the y-axis and the x-axis. We already know that the point where it crosses the y-axis is the y-intercept. In the same way, *the point where it crosses the x-axis is called the* **x-intercept**, and the coordinates of the x-intercept are $(a, 0)$. These points are easily found because in each case one of the coordinates is zero. By setting $x = 0$ and $y = 0$, in turn, and determining the corresponding value of the other unknown, we obtain the coordinates of the intercepts. A third point should be found as a check. This method is sufficient unless the line passes through the origin. Then both intercepts are at the origin and one more point must be determined, or we must use the slope-intercept method. Example 7 illustrates how a line is sketched by finding its intercepts.

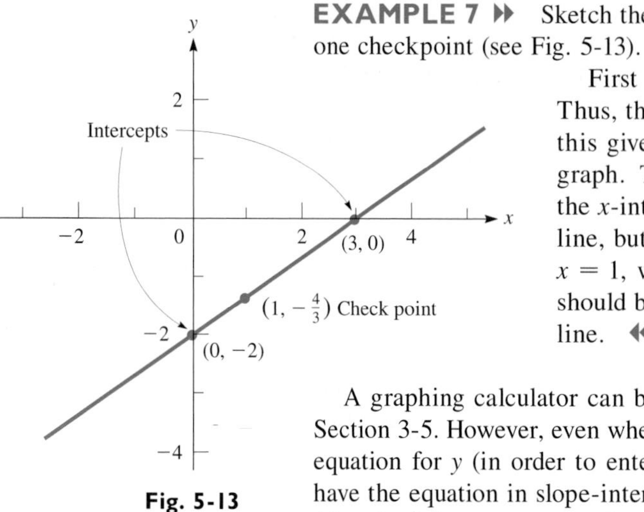

Fig. 5-13

EXAMPLE 7 ▸▸ Sketch the graph of $2x - 3y = 6$ by finding its intercepts and one checkpoint (see Fig. 5-13).

First we let $x = 0$. This gives us $-3y = 6$, or $y = -2$. Thus, the point $(0, -2)$ is on the graph. Next, let $y = 0$, and this gives $2x = 6$, or $x = 3$. Thus, the point $(3, 0)$ is on the graph. The point $(0, -2)$ is the y-intercept, and $(3, 0)$ is the x-intercept. These two points are sufficient to sketch the line, but we should find another point as a check. Choosing $x = 1$, we find that $y = -\frac{4}{3}$. This means the point $(1, -\frac{4}{3})$ should be on the line. From Fig. 5-13 we see that it is on the line. ◂◂

A graphing calculator can be used to graph a linear equation, just as we did in Section 3-5. However, even when using a graphing calculator we must first solve the equation for y (in order to enter it into the calculator). In doing so, we will often have the equation in slope-intercept form, or easily be able to write it in this form. Also, both intercepts can often be found by inspection. Making the RANGE settings can be helped by noting the slope and y-intercept, or both intercepts.

EXAMPLE 8 ▸▸ Use a graphing calculator to graph the line $4x + 5y = 100$.

Solving for y, we get

$$y = -\frac{4}{5}x + 20$$

We immediately can see that the y-intercept is $(0, 20)$, which means we must be careful in choosing the RANGE settings. We also see that the x-intercept is $(25, 0)$.

Thus, we set Xmin $= -5$, Xmax $= 30$, Ymin $= -5$, Ymax $= 25$, and the intervals at 5. This gives the graph shown in Fig. 5-14. ◂◂

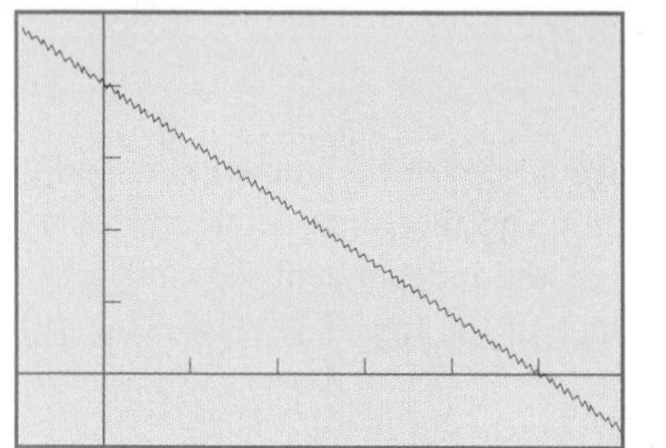
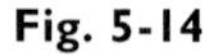
Fig. 5-14

Further details and discussion of slope and the graphs of linear equations are found in Chapter 21.

Exercises 5–2

In Exercises 1–8, find the slope of the line that passes through the given points.

1. $(1, 0), (3, 8)$
2. $(3, 1), (2, 7)$
3. $(-1, 2), (-4, 17)$
4. $(-1, -2), (2, 10)$
5. $(5, -3), (-2, -5)$
6. $(3, -4), (-7, 1)$
7. $(0.4, 0.5), (-0.2, 0.2)$
8. $(-2.8, 3.4), (1.2, 4.2)$

In Exercises 9–16, sketch the line with the given slope and y-intercept.

9. $m = 2, (0, -1)$
10. $m = 3, (0, 1)$
11. $m = -3, (0, 2)$
12. $m = -4, (0, -2)$
13. $m = \frac{1}{2}, (0, 0)$
14. $m = \frac{2}{3}, (0, -1)$
15. $m = -9, (0, 20)$
16. $m = -0.3, (0, -1.4)$

In Exercises 17–24, find the slope and the y-intercept of the line with the given equation and sketch the graph using the slope and the y-intercept. A graphing calculator can be used to check your graph.

17. $y = -2x + 1$
18. $y = -4x$
19. $y = x + 4$
20. $y = \frac{4}{5}x + 2$
21. $5x - 2y = 40$
22. $6x - 2y = 7$
23. $2x + 6y = 3$
24. $10x + 3y = 30$

In Exercises 25–32, find the x-intercept and the y-intercept of the line with the given equation. Sketch the line using the intercepts. A graphing calculator can be used to check the graph.

25. $x + 2y = 4$

26. $3x + y = 3$

27. $4x - 3y = 12$

28. $x - 5y = 5$

29. $y = 3x + 6$

30. $y = -2x - 4$

31. $y = -12x + 30$

32. $y = 0.25x + 4.5$

In Exercises 33–36, sketch the indicated lines.

33. The diameter of the large end, d (in cm), of a certain type of machine tool can be found from the equation $d = 0.2l + 1.2$, where l is the length of the tool. Sketch d as a function of l, for values of l to 10 cm.

34. In testing an anticholesterol drug, it was found that each gram of drug administered reduced a person's blood cholesterol level by 2 units. Set up the function relating the cholesterol level C as a function of the dosage d for a person whose cholesterol level is 310 before taking the drug. Sketch the graph.

35. Two alloys containing lead are being mixed to make a new alloy. An equation relating the percents of lead in the two alloys with the lead in the mixture is $0.7x + 0.4y = 50$. Sketch the graph.

36. Two electric currents, I_1 and I_2 (in amperes), in part of a circuit in a microcomputer are related by the equation $4I_1 - 5I_2 = 2$. Sketch I_2 as a function of I_1. These currents can be considered to be negative.

5-3 Solving Systems of Two Linear Equations in Two Unknowns Graphically

Since a solution of a system of simultaneous linear equations in two unknowns is any pair of values (x, y) that satisfies *both* equations, graphically *the solution would be the coordinates of the point of intersection of the two lines.* This must be the case, for the coordinates of this point constitute the only pair of values to satisfy *both* equations. (In some special cases there may be no solution; in others there may be many solutions. See Examples 5 and 6).

Therefore, when we solve two simultaneous linear equations in two unknowns graphically, *we must graph each line and determine the point of intersection.* This may, of course, lead to *approximate results* if the lines cross at points not used to determine the graph.

EXAMPLE 1 ▸▸ Solve the system of equations

$$y = x - 3$$
$$y = -2x + 1$$

Since each of the equations is in slope-intercept form, we see that $m = 1$ and $b = -3$ for the first line and that $m = -2$ and $b = 1$ for the second line. Using these values we sketch the lines, as shown in Fig. 5-15.

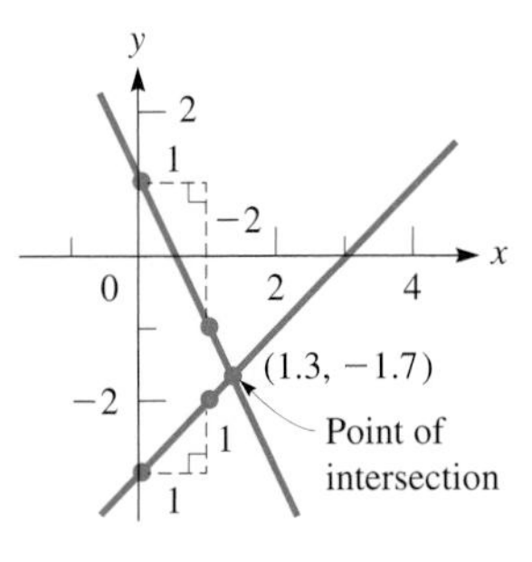

Fig. 5-15

From the figure we see that *the lines cross at about the point* $(1.3, -1.7)$. This means that the solution is approximately

$$x = 1.3, \; y = -1.7$$

(The exact solution is $x = \frac{4}{3}$, $y = -\frac{5}{3}$.)

Checking this solution in *both* equations, we have

$$-1.7 \stackrel{?}{=} 1.3 - 3 \quad \text{and} \quad -1.7 \stackrel{?}{=} -2(1.3) + 1$$
$$= -1.7 \qquad\qquad \approx -1.6$$

These values show that the solution checks. (The point $(1.3, -1.7)$ is *on* the first line and *almost on* the second line. The difference in values when checking the values for the second line is due to the fact that the solution is *approximate.*) ◂◂

EXAMPLE 2 ▸▸ Solve the system of equations

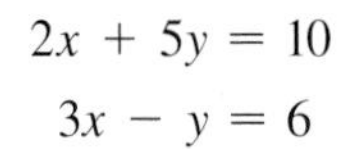

$$2x + 5y = 10$$
$$3x - y = 6$$

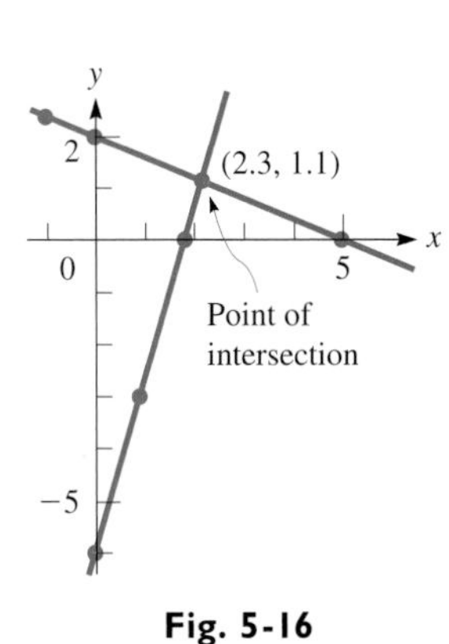

Fig. 5-16

We could write each equation in slope-intercept form in order to sketch the lines. Also, we could use the form in which they are written to find the intercepts. Choosing to find the intercepts and draw lines through them, we let $y = 0$; then $x = 0$. Therefore, we find that the intercepts of the first line are the points $(5, 0)$ and $(0, 2)$. A third point is $(-1, \frac{12}{5})$. The intercepts of the second line are $(2, 0)$ and $(0, -6)$. A third point is $(1, -3)$. Plotting these points and drawing the proper straight lines, we see that the lines cross at about $(2.3, 1.1)$. [The exact values are $(\frac{40}{17}, \frac{18}{17})$.] The solution of the system of equations is approximately $x = 2.3$, $y = 1.1$ (see Fig. 5-16).

Checking, we have

$$2(2.3) + 5(1.1) \stackrel{?}{=} 10 \qquad \text{and} \qquad 3(2.3) - 1.1 \stackrel{?}{=} 6$$
$$10.1 \approx 10 \qquad\qquad\qquad\qquad 5.8 \approx 6$$

This verifies that the solution is correct to the accuracy obtainable from the graph. ◂◂

As we have seen, it is difficult to get a good approximation of the coordinates of the point of intersection without being very careful in sketching both lines. The graphing calculator can be used to find this point with much greater accuracy than is possible by hand-sketching the lines. Once we locate the point of intersection, the TRACE and ZOOM features allow us to get the required accuracy.

EXAMPLE 3 ▸▸ Using a graphing calculator, we graph the lines of Example 1 in Fig. 5-17(a) and those of Example 2 in Fig. 5-17(b).

In Fig. 5-17(a) the RANGE values are Xmin = 0, Xmax = 3, Ymin = −3, and Ymax = 0, with intervals of 0.5. Using the TRACE and ZOOM features of the calculator, we find (to the nearest 0.001) that $x = 1.333$, $y = -1.667$.

In Fig. 5-17(b) the RANGE values are Xmin = 0, Xmax = 3, Ymin = 0, and Ymax = 2, with intervals of 0.5. We then find (to the nearest 0.001) $x = 2.353$, $y = 1.059$.

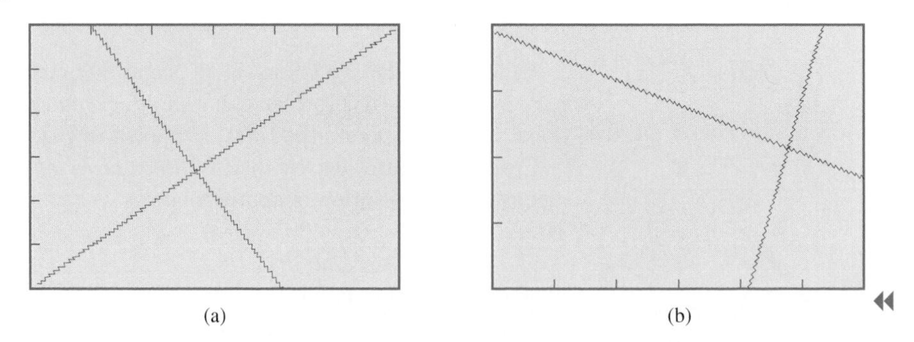

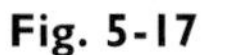

Fig. 5-17 (a) (b) ◂◂

Linear equations in two unknowns are often useful in solving verbal problems. Just as in Section 1-12, *we must read the statement carefully in order to identify the unknowns and the information for setting up the equations.* The following example illustrates the method.

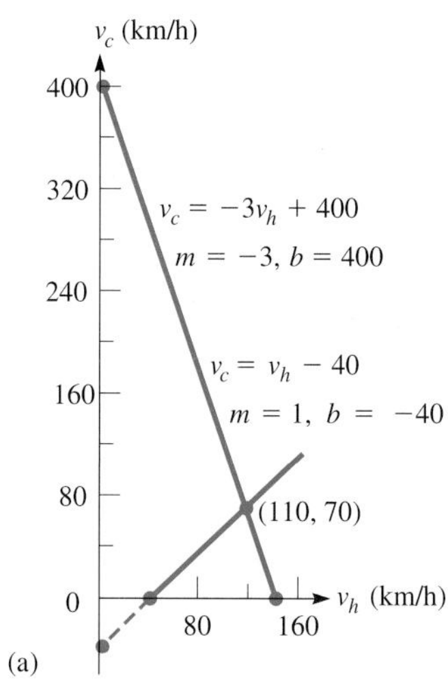

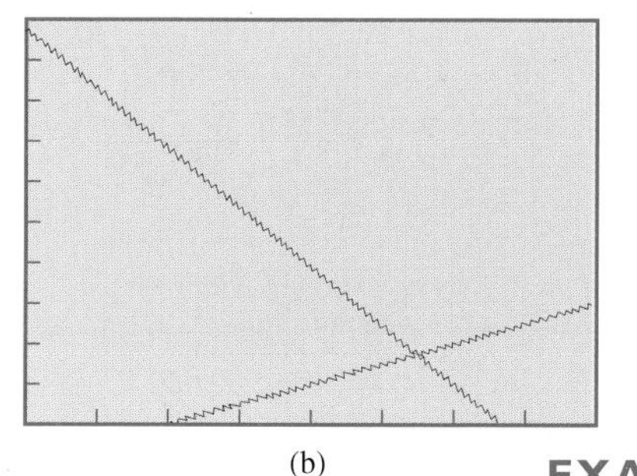

(b)

Fig. 5-18

EXAMPLE 4 ▸▸ A driver traveled for 1.5 h at a constant speed along a highway. Then, through a construction zone, the driver reduced the car's speed by 40 km/h for 30 min. If 200 km were covered in the 2.0 h, what were the two speeds?

First, we let v_h = the highway speed and v_c = the speed in the construction zone. Two equations are found by using

1. distance = rate × time (for units, km = $(\frac{\text{km}}{\text{h}})$h), and
2. the fact that "the driver reduced the car's speed by 40 km/h for 30 min."

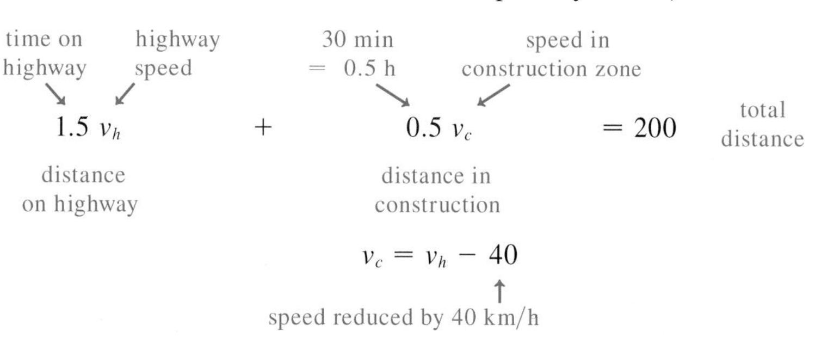

$$v_c = v_h - 40$$

speed reduced by 40 km/h

Using v_h as the independent variable and v_c as the dependent variable, the sketch is shown in Fig. 5-18(a). Figure 5-18(b) shows the calculator graph with Xmin = 0, Ymin = 0, and intervals of 20 and 40.

We see that the point of intersection is (110, 70), which means the solution is v_h = 110 km/h and v_c = 70 km/h. Checking *in the statement of the problem*, (1.5 h)(110 km/h) + (0.5 h)(70 km/h) = 200 km. ◂◂

EXAMPLE 5 ▸▸ Solve the system of equations

$$x = 2y + 6$$
$$6y = 3x - 6$$

Writing each of these equations in slope-intercept form, we have

$$y = \tfrac{1}{2}x - 3 \qquad \text{and} \qquad y = \tfrac{1}{2}x - 1$$

We note that each line has a slope of $\frac{1}{2}$ and that the y-intercepts are $(0, -3)$ and $(0, -1)$. This means that the y-intercepts are different, but the slopes are the same. Since the slope indicates that each line rises $\frac{1}{2}$ unit for y for each unit of x, *the lines are parallel and do **not** intersect* (see Fig. 5-19). Therefore, ***there are no solutions.*** *Such a system is called* **inconsistent.** ◂◂

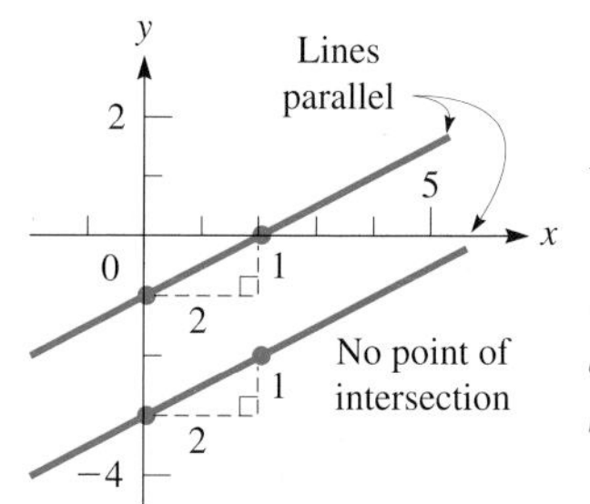

EXAMPLE 6 ▸▸ Solve the system of equations

$$x - 3y = 9$$
$$-2x + 6y = -18$$

The intercepts and a third point for the first line are $(9, 0)$, $(0, -3)$, and $(3, -2)$. In determining the intercepts for the second line, we find that they are $(9, 0)$ and $(0, -3)$, which are also the intercepts of the first line (see Fig. 5-20). As a check we note that $(3, -2)$ also satisfies the equation of the second line. This means the two lines are really the same line, and the *coordinates of any point on this common line constitute a solution of the system.* Since ***no unique solution may be determined,*** *such a system is called* **dependent.** ◂◂

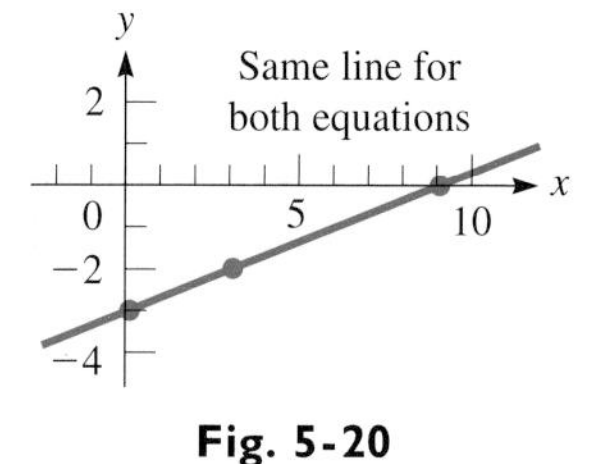

Fig. 5-20

Exercises 5–3

In Exercises 1–16, solve each system of equations by sketching the graphs. Use the slope and the y-intercept or both intercepts. Estimate each result to the nearest 0.1 if necessary.

1. $y = -x + 4$
$y = x - 2$

2. $y = \frac{1}{2}x - 1$
$y = -x + 8$

3. $y = 2x - 6$
$y = -\frac{1}{3}x + 1$

4. $y = \frac{1}{2}x - 4$
$y = 2x + 2$

5. $3x + 2y = 6$
$x - 3y = 3$

6. $4R - 3V = -8$
$6R + V = 6$

7. $2x - 5y = 10$
$3x + 4y = -12$

8. $-5x + 3y = 15$
$2x + 7y = 14$

9. $s - 4t = 8$
$2s = t + 4$

10. $y = 4x - 6$
$y = 2x + 4$

11. $y = -x + 3$
$y = -2x + 3$

12. $p - 6 = 6v$
$v = 3 - 3p$

13. $x - 4y = 6$
$2y = x + 4$

14. $x + y = 3$
$3x - 2y = 14$

15. $-2r_1 + 2r_2 = 7$
$4r_1 - 2r_2 = 1$

16. $2x - 3y = -5$
$3x + 2y = 12$

In Exercises 17–28, solve each system of equations to the nearest 0.1 for each variable by using a graphing calculator.

17. $x = 4y + 2$
$3y = 2x + 3$

18. $1.2x - 2.4y = 4.8$
$3.0x = -2.0y + 7.2$

19. $4.0x - 3.5y = 1.5$
$0.7y + 0.1x = 0.7$

20. $5F - 2T = 7$
$3F + 4T = 8$

21. $x - 5y = 10$
$2x - 10y = 20$

22. $18x - 3y = 7$
$2y = 1 + 12x$

23. $1.9v = 3.2t$
$1.2t - 2.6v = 6$

24. $4x - y = 3$
$2x + 3y = 0$

25. $5x = y + 3$
$4x = 2y - 3$

26. $0.75u + 0.67v = 5.9$
$2.1u - 3.9v = 4.8$

27. $3x = 8y + 12$
$-6x + 16y = 6$

28. $y = 6x + 2$
$12x - 2y = -4$

In Exercises 29–32, graphically solve the given problems to the stated accuracy.

29. Chains support a crate as shown in Fig. 5-21. The equations relating tensions T_1 and T_2 are given below. Determine T_1 and T_2 to the nearest 1 N from the graph.

$$0.8T_1 - 0.6T_2 = 12$$
$$0.6T_1 + 0.8T_2 = 68$$

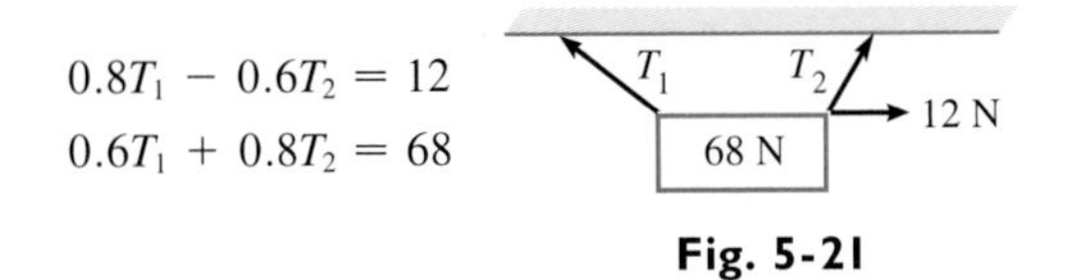

Fig. 5-21

30. The equations relating the currents i_1 and i_2 shown in Fig. 5-22 are given below. Find the currents to the nearest 0.1 A.

$$2i_1 + 6(i_1 + i_2) = 12$$
$$4i_2 + 6(i_1 + i_2) = 12$$

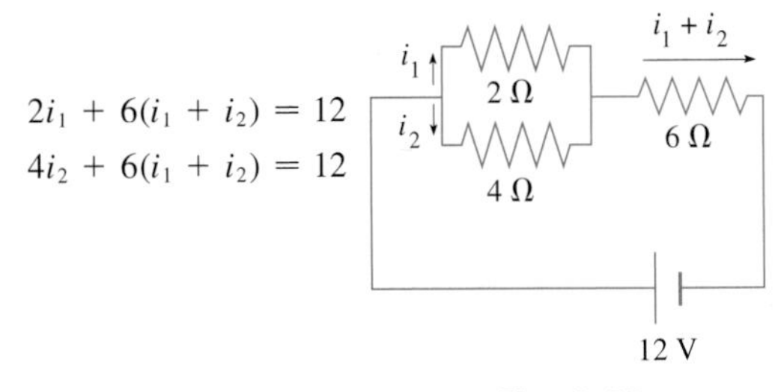

Fig. 5-22

31. An architect designing a parking lot has a row 62.0 m wide to divide into spaces for compact cars and full-size cars. The architect determines that 16 compact car spaces and 6 full-size car spaces use the width. It is also possible to have spaces for 12 compact cars and 9 full-size cars with 0.2 m not used. What are the widths of the spaces being planned? See Fig. 5-23.

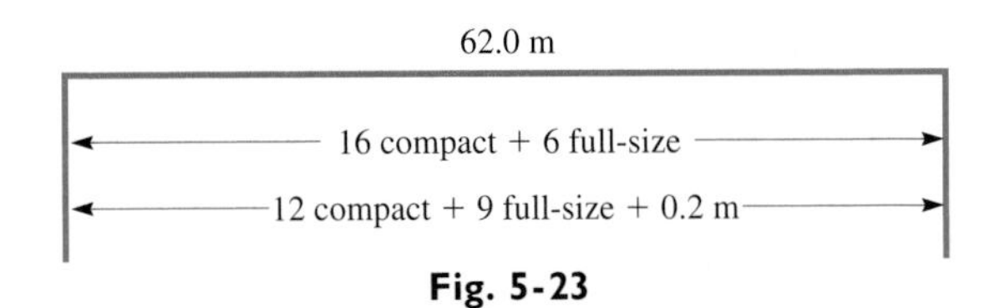

Fig. 5-23

32. A total of 42 tonnes of two types of ore is to be loaded into a smelter. The first type contains 6.0% copper, and the second contains 2.4% copper. Find the necessary amounts of each ore (to the nearest 1 tonne) to produce 2 tonnes of copper.

5-4 Solving Systems of Two Linear Equations in Two Unknowns Algebraically

We have just seen how a system of two linear equations can be solved graphically. This method is good for obtaining a "picture" of the solution. Solving systems by graphical methods has one difficulty: The results are usually approximate. If exact solutions are required, we turn to other methods. In this section we present two algebraic methods of solution.

Solution by Substitution

The first method involves the *elimination of one variable by* **substitution**. To follow this method, we

> *first solve one of the equations for one of the unknowns, and then* ***substitute*** *this solution into the other equation.*

The result is one linear equation in one unknown. This equation is then solved for the unknown it contains. By substituting this value into one of the original equations, we find the corresponding value of the other unknown. The following examples illustrate the method.

EXAMPLE 1 ▸▸ Use the method of elimination by substitution to solve the system of equations

$$x - 3y = 6$$
$$2x + 3y = 3$$

The first step is to solve one of the equations for one of the unknowns. The choice of which equation and which unknown depends on ease of algebraic manipulation. In this system, it is somewhat easier to solve the first equation for x. Therefore, performing this operation we have

$$x = 3y + 6 \qquad \text{(A1)}$$

We then substitute this expression into the second equation in place of x, giving

in second equation, x replaced by $3y + 6$

$$2(3y + 6) + 3y = 3$$

Solving this equation for y, we obtain

$$6y + 12 + 3y = 3$$
$$9y = -9$$
$$y = -1$$

We now put the value $y = -1$ into the first of the original equations. Since we have already solved this equation for x in terms of y, Eq. (A1), we obtain

$$x = 3(-1) + 6 = 3$$

Therefore, the solution of the system is $x = 3$, $y = -1$. As a check, we substitute these values into each of the original equations. We obtain $3 - 3(-1) = 6$ and $2(3) + 3(-1) = 3$, which verifies the solution. ◂◂

EXAMPLE 2 ▸▸ Use the method of elimination by substitution to solve the system of equations

$$\begin{aligned} -5x + 2y &= -4 \\ 10x + 6y &= 3 \end{aligned}$$

It makes little difference which equation or which unknown is chosen. Therefore, choosing to solve the first equation for y, we obtain

$$\begin{aligned} 2y &= 5x - 4 \\ y &= \frac{5x - 4}{2} \end{aligned} \qquad \text{(B1)}$$

Substituting this expression into the second equation, we have

$$10x + 6\left(\frac{5x - 4}{2}\right) = 3 \qquad \text{in second equation, } y \text{ replaced by } \frac{5x - 4}{2}$$

We now proceed to solve this equation for x.

$$\begin{aligned} 10x + 3(5x - 4) &= 3 \\ 10x + 15x - 12 &= 3 \\ 25x &= 15 \\ x &= \frac{3}{5} \end{aligned}$$

Substituting this value into the expression for y, Eq. (B1), we obtain

$$y = \frac{5(3/5) - 4}{2} = \frac{3 - 4}{2} = -\frac{1}{2}$$

Therefore, the solution of this system is $x = \frac{3}{5}$, $y = -\frac{1}{2}$. Substituting these values in both original equations shows that the solution checks. ◂◂

Solution by Addition or Subtraction

The method of elimination by substitution is useful if one equation can easily be solved for one of the unknowns. However, the numerical coefficients often make this method somewhat cumbersome. So we now develop another algebraic method of solving a system of linear equations.

The second method is the *elimination of a variable by means of* **addition or subtraction.** To use this method we

> *multiply each equation by a number chosen so that the coefficients for one of the unknowns will be numerically the same in both equations.*

If these numerically equal coefficients are opposite in sign, we ***add*** the two equations. If the numerically equal coefficients have the same sign, we ***subtract*** one equation from the other. That is, we *subtract* the left side of one equation from the left side of the other equation, and also do the same to the right sides. After adding or subtracting, we have a simple linear equation in one unknown, which we then solve for the unknown. We then substitute this value into one of the original equations to find the value of the other unknown.

EXAMPLE 3 ▸▸ Use the method of elimination by addition or subtraction to solve the system of equations

$$x - 3y = 6$$
$$2x + 3y = 3$$

We look at the coefficients to determine the best way to eliminate one of the unknowns. In this case, since the coefficients of the y-terms are numerically the same and are opposite in sign, we may immediately add the two equations to eliminate y. Adding the left sides together and the right sides together, we obtain

$$x + 2x - 3y + 3y = 6 + 3$$
$$3x = 9$$
$$x = 3$$

Substituting this value into the first equation, we obtain

$$3 - 3y = 6$$
$$-3y = 3$$
$$y = -1$$

The solution $x = 3$, $y = -1$ agrees with the results obtained for the same problem illustrated in Example 1 of this section. ◂◂

EXAMPLE 4 ▸▸ Use addition or subtraction to solve the system of equations

$$3x - 2y = 4$$
$$x + 3y = 2$$

Looking at the coefficients of x and y, we see that we must multiply the second equation by 3 to make the coefficients of x the same. To make the coefficients of y numerically the same, we must multiply the first equation by 3 and the second equation by 2. Thus, the best method is to multiply the second equation by 3 and eliminate x. Doing this (be careful to multiply the terms on *both* sides: *a common error is to forget to multiply the value on the right*) and then subtracting the second equation from the first, we have

CAUTION ▸

$$3x - 2y = 4$$
$$3x + 9y = 6 \quad \text{each term of second equation multiplied by 3}$$
$$-11y = -2 \quad \text{subtract}$$
$$y = \frac{2}{11}$$

$3x - 3x = 0$

$-2y - (+9y) = -11y$

$4 - 6 = -2$

In order to find the value of x, we substitute $y = \frac{2}{11}$ into one of the original equations. Choosing the second equation (its form is somewhat simpler), we have

$$x + 3\left(\frac{2}{11}\right) = 2$$
$$11x + 6 = 22 \quad \text{multiply each term by 11}$$
$$x = \frac{16}{11}$$

We arrive at the solution $x = \frac{16}{11}$, $y = \frac{2}{11}$. Substituting these values into both of the original equations shows that the solution checks. ◂◂

EXAMPLE 5 ▸▸ As we noted in Example 4, we can solve the system of equations by first multiplying the first equation by 3 and the second equation by 2, thereby eliminating y. Doing this, we have

$$\begin{aligned} 9x - 6y &= 12 && \text{each term of first equation multiplied by 3} \\ 2x + 6y &= 4 && \text{each term of second equation multiplied by 2} \\ \hline 11x \quad &= 16 && \text{add} \\ x &= \frac{16}{11} && 12 + 4 = 16 \end{aligned}$$

$9x + 2x = 11x$

$-6y + 6y = 0$

At this point we can find the value of y by substituting $x = \frac{16}{11}$ into one of the original equations, or we could eliminate x as is done in Example 4. Substitution in the first of the original equations gives us

$$\begin{aligned} 3\left(\frac{16}{11}\right) - 2y &= 4 \\ 48 - 22y &= 44 && \text{multiply each term by 11} \\ -22y &= -4 \\ y &= \frac{2}{11} \end{aligned}$$

Therefore, the solution is $x = \frac{16}{11}$, $y = \frac{2}{11}$, as we obtained in Example 4. ◂◂

The best form in which to have the equations of the system for solution by addition or subtraction is the form shown in Examples 3 and 4. That is, the x-term and then the y-term are on the left (the reverse order in *both* equations is also acceptable) and the constant is on the right. *If the equations are not written in this form, both should be written this way before proceeding with the solution.*

NOTE ▸

EXAMPLE 6 ▸▸ In solving the system of equations

$$\begin{aligned} 4x &= 2y + 3 \\ -y + 2x - 2 &= 0 \end{aligned}$$

we should first write the equations in the form noted just above. Doing this, we have

$$\begin{aligned} 4x - 2y &= 3 \\ 2x - \ \ y &= 2 \end{aligned}$$

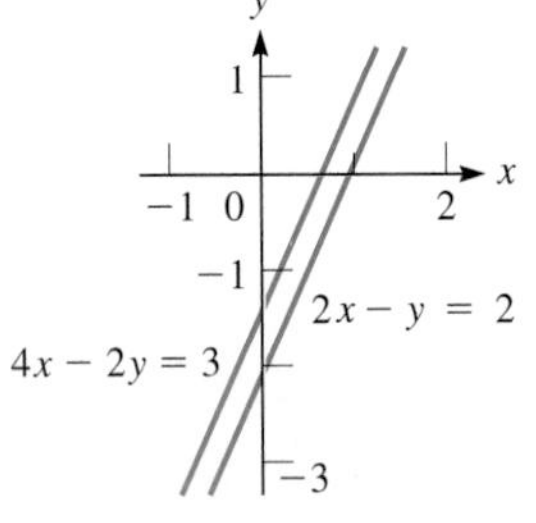

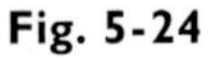

Fig. 5-24

When we multiply the second equation by 2 and subtract, we obtain

$$\begin{aligned} 4x - 2y &= 3 \\ 4x - 2y &= 4 \\ \hline 0 &= -1 \end{aligned}$$

$4x - 4x = 0 \longrightarrow 0 = -1 \longleftarrow 3 - 4 = -1$

$-2y - (-2y) = 0$

Since we know 0 does not equal -1, we conclude that there is no solution. When we obtain a result of $0 = a$ ($a \neq 0$), the system of equations is *inconsistent*. As we discussed in the previous section, the lines that represent the equations are parallel, as shown in Fig. 5-24. ◂◂

In Example 6 we showed that a result of $0 = a$ $(a \neq 0)$ indicates the system of equations is inconsistent. If we obtain the result $0 = 0$, the system is *dependent.* As shown in the previous section, this means that there is an unlimited number of solutions and the lines that represent the equations are the same line.

The following example gives another complete illustration of solving a verbal problem by first setting up the proper equations.

EXAMPLE 7 ⏩ By weight, one alloy is 70% copper and 30% zinc. Another alloy is 40% copper and 60% zinc. How many grams of each of these are required to make 300 g of an alloy that is 60% copper and 40% zinc?

Let A = the required number of grams of the first alloy, and B = the required number of grams of the second alloy. Our equations are determined from:

1. The total weight of the final alloy is 300 g: $A + B = 300$.
2. The final alloy will have 180 g of copper (60% of 300 g), and this comes from 70% of A $(0.70A)$ and 40% of B $(0.40B)$: $0.70A + 0.40B = 180$.

The second equation is based on the amount of copper. We could have used the amount of zinc, which would lead to the equation $0.30A + 0.60B = 0.40(300)$. Since we need only two equations, we may use any two of these three equations to find the solution.

These two equations can now be solved simultaneously.

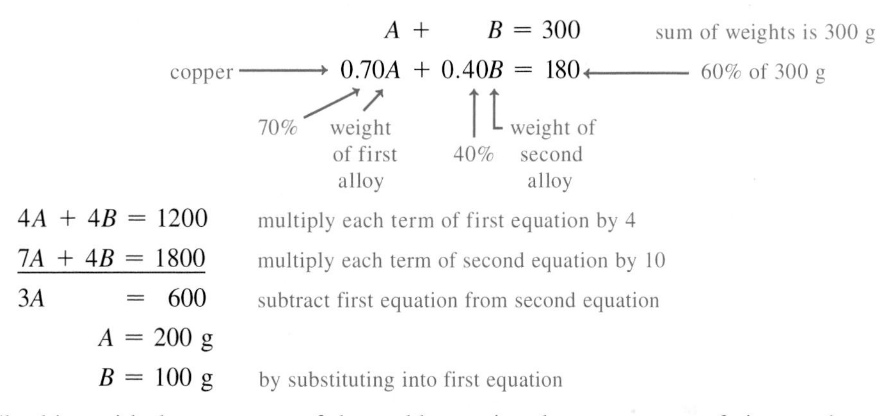

Checking with the statement of the problem, using the percentages of zinc, we have $0.30(200) + 0.60(100) = 0.40(300)$, or $60 \text{ g} + 60 \text{ g} = 120 \text{ g}$. We use zinc here since we used the equation for the percentages of copper. ⏪

Exercises 5–4

In Exercises 1–12, solve the given systems of equations by the method of elimination by substitution.

1. $x = y + 3$; $x - 2y = 5$
2. $x = 2y + 1$; $2x - 3y = 4$
3. $p = V - 4$; $V + p = 10$
4. $y = 2x + 10$; $2x + y = -2$
5. $x + y = -5$; $2x - y = 2$
6. $3x + y = 1$; $3x - 2y = 16$
7. $2x + 3y = 7$; $6x - y = 1$
8. $2s + 2t = 1$; $4s - 2t = 17$
9. $3x + 2y = 7$; $2y = 9x + 11$
10. $3x + 3y = -1$; $5x = -6y - 1$
11. $0.4p - 0.3n = 0.6$; $0.2p + 0.4n = -0.5$
12. $6.0x + 4.8y = -8.4$; $4.8x - 6.5y = -7.8$

In Exercises 13–24, solve the given systems of equations by the method of elimination by addition or subtraction.

13. $x + 2y = 5$; $x - 2y = 1$
14. $x + 3y = 7$; $2x + 3y = 5$
15. $2x - 3y = 4$; $2x + y = -4$
16. $R - 4r = 17$; $3R + 4r = 3$
17. $2x + 3y = 8$; $x = 2y - 3$
18. $3x - y = 3$; $4x = 3y + 14$
19. $v + 2t = 7$; $2v + 4t = 9$
20. $3x - y = 5$; $-9x + 3y = -15$
21. $2x - 3y - 4 = 0$; $3x + 2 = 2y$
22. $3i_1 + 5 = -4i_2$; $3i_2 = 5i_1 - 2$
23. $0.3x = 0.7y + 0.4$; $0.5y = 0.7 - 0.2x$
24. $2.50x + 2.25y = 4.00$; $3.75x - 6.75y = 3.25$

In Exercises 25–32, solve the given systems of equations by either method of this section.

25. $2x - y = 5$
$6x + 2y = -5$

26. $3x + 2y = 4$
$6x - 6y = 13$

27. $6x + 3y + 4 = 0$
$5y = -9x - 6$

28. $1 + 6q = 5p$
$3p - 4q = 7$

29. $3x - 6y = 15$
$4x - 8y = 20$

30. $2x + 6y = -3$
$-6x - 18y = 5$

31. $1.2V + 10.8 = -8.4C$
$3.6C + 4.8V + 13.2 = 0$

32. $0.66x + 0.66y = -0.77$
$0.33x - 1.32y = 1.43$

In Exercises 33–36, solve the given systems of equations by an appropriate algebraic method.

33. Find the voltages V_1 and V_2 of the batteries shown in Fig. 5-25. The terminals are aligned in the same direction in Fig. 5-25(a) and in opposite directions in Fig. 5-25(b).

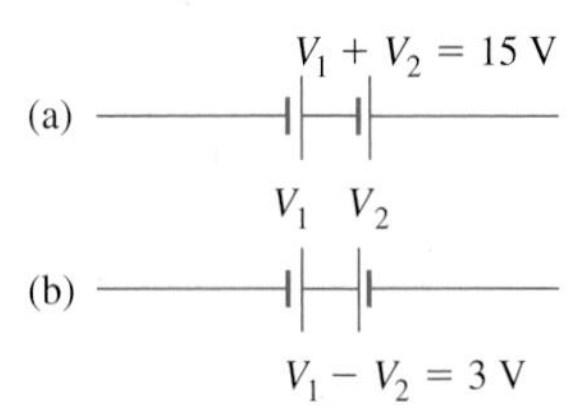

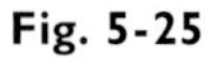
Fig. 5-25

34. In finding the dimensions of a rectangular solar panel, the following equations are used:

$2l + 2w = 4600$
$l = w + 1000$

Find the length l and width w (in mm) of the panel.

35. Two grades of gasoline are mixed to make a blend with 1.50% of a special additive. Combining x litres of a grade with 1.80% of the additive to y litres of a grade with 1.00% of the additive gives 10 000 L of the blend. The equations relating x and y are

$$x + y = 10\,000$$
$$0.0180x + 0.0100y = 0.0150(10\,000)$$

Find x and y (to three significant digits).

36. A 6.0% solution and a 15.0% solution of a drug are added to 200 mL of a 20.0% solution to make 1200 mL of a 12.0% solution for a proper dosage. The equations relating the number of millilitres of the added solutions are

$$x + y + 200 = 1200$$
$$0.060x + 0.150y + 0.200(200) = 0.120(1200)$$

Find x and y (to three significant digits).

In Exercises 37–44, set up appropriate systems of two linear equations and solve the systems algebraically. All data are accurate to at least two significant digits.

37. In a test of a heat-seeking rocket, a first rocket is launched at 600 m/s and the heat-seeking rocket is launched along the same flight path 12 s later at a speed of 960 m/s. Find the times t_1 and t_2 of flight of the rockets until the heat-seeking rocket destroys the first rocket.

38. The *torque* of a force is the product of the force and the perpendicular distance from a specified point. If a lever is supported at only one point, and is in balance, the sum of the torques (about the support) of forces acting on one side of the support must equal the sum of the torques of the forces acting on the other side. Find the forces F_1 and F_2 that are in the positions shown in Fig. 5-26(a) and then move to the positions in Fig. 5-26(b). The lever weighs 20 N and is in balance.

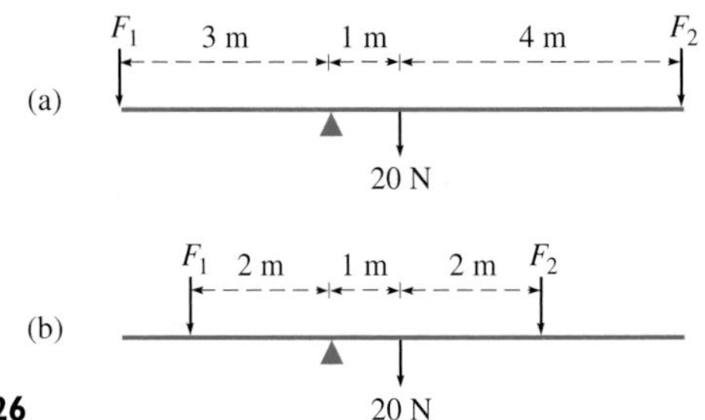

Fig. 5-26

39. One personal computer can perform a calculations per second, and a second personal computer can perform b calculations per second. If the first calculates for 3 s and the second calculates for 2 s, 17.0 million calculations are performed. If the times are reversed, 15.5 million calculations are performed. Find the rates a and b.

40. An underwater (but near the surface) explosion is detected by sonar on a ship 30 s before it is heard on the deck. If sound travels at 1500 m/s in water and 330 m/s in air, how far is the ship from the explosion?

41. As a 40-m pulley belt makes one revolution, one of the two pulley wheels makes one more revolution than the other. Another wheel of half the radius replaces the smaller wheel and makes six more revolutions than the larger wheel for one revolution of the belt. Find the circumferences of the wheels.

42. In mixing a weed-killing chemical, a 40% solution of the chemical is mixed with an 85% solution to get 20 L of a 60% solution. How much of each solution is needed?

43. What conclusion can you draw from a sales report which states that "sales this month were \$8000 more than last month, which means that total sales for both months are \$4000 more than twice the sales last month?"

44. For an electric circuit, a report stated that current i_1 is twice current i_2 and that twice the sum of the two currents less 6 times i_2 is 6. Explain your conclusion about the values of the currents found from this report.

5-5 Solving Systems of Two Linear Equations in Two Unknowns by Determinants

Consider two linear equations in two unknowns, as given in Eq. (5-3):

$$\begin{aligned} a_1x + b_1y &= c_1 \\ a_2x + b_2y &= c_2 \end{aligned} \qquad (5\text{-}3)$$

If we multiply the first of these equations by b_2 and the second by b_1, we obtain

$$\begin{aligned} a_1b_2x + b_1b_2y &= c_1b_2 \\ a_2b_1x + b_2b_1y &= c_2b_1 \end{aligned} \qquad (5\text{-}6)$$

We see that the coefficients of y are the same. Thus, subtracting the second equation from the first, we can solve for x. The solution can be shown to be

$$x = \frac{c_1b_2 - c_2b_1}{a_1b_2 - a_2b_1} \qquad (5\text{-}7)$$

In the same manner, we may show that

$$y = \frac{a_1c_2 - a_2c_1}{a_1b_2 - a_2b_1} \qquad (5\text{-}8)$$

The expression $a_1b_2 - a_2b_1$, which appears in each of the denominators of Eqs. (5-7) and (5-8), is an example of a special kind of expression called a *determinant of the second order.* The determinant $a_1b_2 - a_2b_1$ is denoted by

$$\begin{vmatrix} a_1 & b_1 \\ a_2 & b_2 \end{vmatrix}$$

Therefore, by definition, *a* **determinant of the second order** *is*

Second-Order Determinant

$$\begin{vmatrix} a_1 & b_1 \\ a_2 & b_2 \end{vmatrix} = a_1b_2 - a_2b_1 \qquad (5\text{-}9)$$

The numbers a_1 and b_1 are called the **elements** *of the first* **row** *of the determinant. The numbers a_1 and a_2 are the elements of the first* **column** *of the determinant.* In the same manner, the numbers a_2 and b_2 are the elements of the second row, and the numbers b_1 and b_2 are the elements of the second column. *The numbers a_1 and b_2 are the elements of the* **principal diagonal**, *and the numbers a_2 and b_1 are the elements of the* **secondary diagonal**. Thus, one way of stating the definition indicated in Eq. (5-9) is that *the value of a determinant of the second order is found by taking the product of the elements of the principal diagonal and subtracting the product of the elements of the secondary diagonal.*

A diagram that is often helpful for remembering the expansion of a second-order determinant is shown in Fig. 5-27. The following examples illustrate how we carry out the evaluation of determinants.

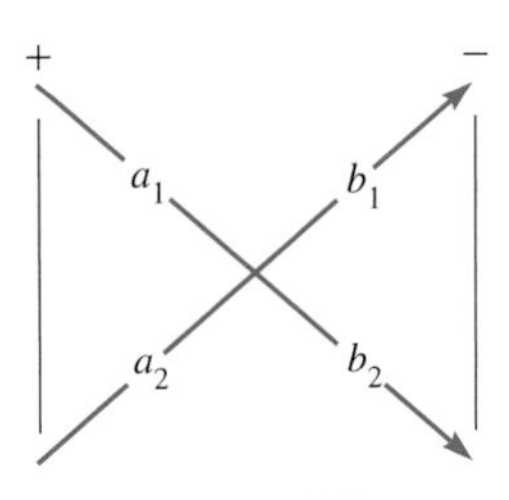

Fig. 5-27

EXAMPLE 1 ▸▸ $\begin{vmatrix} -5 & 8 \\ 3 & 7 \end{vmatrix} = (-5)(7) - 3(8) = -35 - 24 = -59$ ◂◂

EXAMPLE 2 ▸▸ (a) $\begin{vmatrix} 4 & 6 \\ 3 & 17 \end{vmatrix} = 4(17) - (3)(6) = 68 - 18 = 50$

(b) $\begin{vmatrix} 4 & 6 \\ -3 & 17 \end{vmatrix} = 4(17) - (-3)(6) = 68 + 18 = 86$

(c) $\begin{vmatrix} 3.6 & 6.1 \\ -3.2 & -17.2 \end{vmatrix} = 3.6(-17.2) - (-3.2)(6.1) = -42.4$

Note the signs of the terms being combined. ◂◂

We note that the numerators and denominators of Eqs. (5-7) and (5-8) may be written as determinants. The numerators of the equations are

$$\begin{vmatrix} c_1 & b_1 \\ c_2 & b_2 \end{vmatrix} \quad \text{and} \quad \begin{vmatrix} a_1 & c_1 \\ a_2 & c_2 \end{vmatrix}$$

Therefore, the solutions for x and y of the system of equations

$$\begin{aligned} a_1x + b_1y &= c_1 \\ a_2x + b_2y &= c_2 \end{aligned} \tag{5-3}$$

may be written directly in terms of determinants, without algebraic operations, as

Note carefully the location of c_1 and c_2 ▶

$$x = \frac{\begin{vmatrix} c_1 & b_1 \\ c_2 & b_2 \end{vmatrix}}{\begin{vmatrix} a_1 & b_1 \\ a_2 & b_2 \end{vmatrix}} \quad \text{and} \quad y = \frac{\begin{vmatrix} a_1 & c_1 \\ a_2 & c_2 \end{vmatrix}}{\begin{vmatrix} a_1 & b_1 \\ a_2 & b_2 \end{vmatrix}} \tag{5-10}$$

For this reason, determinants provide a quick and easy method of solution of systems of equations. Again, *the denominator of each of Eqs. (5-10) is the same.*

The determinant of the denominator is made up of the coefficients of x and y. Also, *the determinant of the numerator of the solution for x is obtained from the determinant of the denominator by* **replacing the column of *a*'s by the column of *c*'s**. *The determinant of the numerator of the solution for y is obtained from the determinant of the denominator by* **replacing the column of *b*'s by the column of *c*'s**.

Named for the Swiss mathematician Gabriel Cramer (1704–1752).

This result is referred to as **Cramer's rule.** In using Cramer's rule we must be sure that the equations are written in the form of Eq. (5-3) before setting up the determinants.

The following examples illustrate the method of solving systems of equations by determinants.

EXAMPLE 3 ▸▸ Solve the following system of equations by determinants:

$$2x + y = 1$$
$$5x - 2y = -11$$

We first note that the equations are in the proper form of Eq. (5-3) for solution by determinants. Next, we set up the determinant for the denominator, which consists of the four coefficients in the system, written as shown. It is

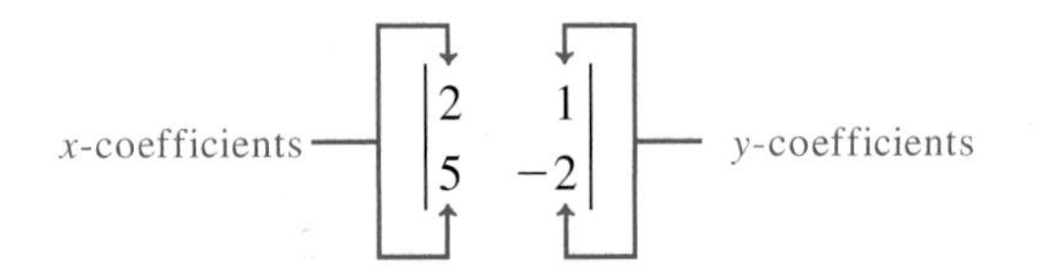

For finding x, the determinant in the numerator is obtained from this determinant by replacing the first column by the constants that appear on the right sides of the equations. Thus, *the numerator for the solution for x* is

$$\begin{vmatrix} 1 & 1 \\ -11 & -2 \end{vmatrix}$$ — replace x-coefficients with the constants

For finding y, the determinant in the numerator is obtained from the determinant of the denominator by replacing the second column by the constants that appear on the right sides of the equations. Thus, *the numerator for the solution for y* is

$$\begin{vmatrix} 2 & 1 \\ 5 & -11 \end{vmatrix}$$ — replace y-coefficients with the constants

Now we set up the solutions for x and y using the determinants above.

$$x = \frac{\begin{vmatrix} 1 & 1 \\ -11 & -2 \end{vmatrix}}{\begin{vmatrix} 2 & 1 \\ 5 & -2 \end{vmatrix}} = \frac{1(-2) - (-11)(1)}{2(-2) - (5)(1)} = \frac{-2 + 11}{-4 - 5} = \frac{9}{-9} = -1$$

$$y = \frac{\begin{vmatrix} 2 & 1 \\ 5 & -11 \end{vmatrix}}{\begin{vmatrix} 2 & 1 \\ 5 & -2 \end{vmatrix}} = \frac{2(-11) - (5)(1)}{-9} = \frac{-22 - 5}{-9} = 3$$

Therefore, the solution to the system of equations is $x = -1$, $y = 3$.

Substituting these values into the equations, we have

$$2(-1) + 3 \stackrel{?}{=} 1 \quad \text{and} \quad 5(-1) - 2(3) \stackrel{?}{=} -11$$
$$1 = 1 \qquad\qquad\qquad -11 = -11$$

which shows that they check.

Since the determinant in the denominators is the same, it needs to be evaluated only once. This means that three determinants are to be evaluated in order to solve the system. ◂◂

EXAMPLE 4 ▸▸ Solve the following system of equations by determinants. All numbers are approximate.

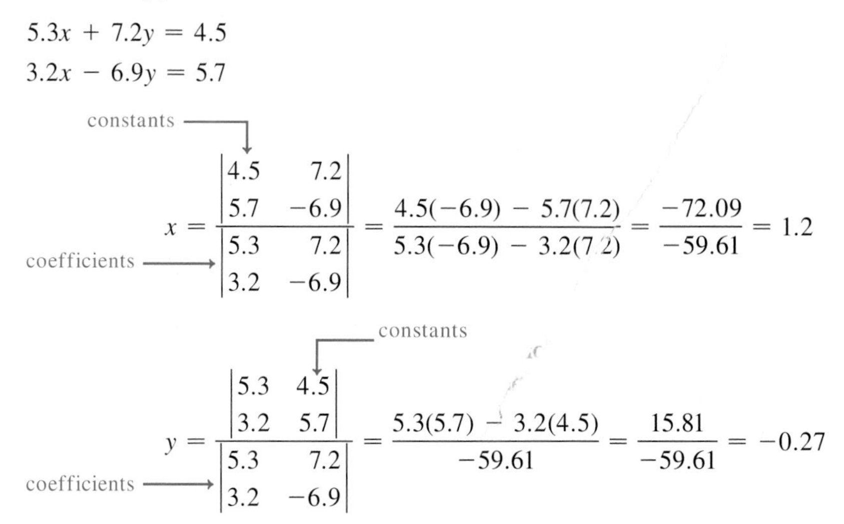

$$5.3x + 7.2y = 4.5$$
$$3.2x - 6.9y = 5.7$$

$$x = \frac{\begin{vmatrix} 4.5 & 7.2 \\ 5.7 & -6.9 \end{vmatrix}}{\begin{vmatrix} 5.3 & 7.2 \\ 3.2 & -6.9 \end{vmatrix}} = \frac{4.5(-6.9) - 5.7(7.2)}{5.3(-6.9) - 3.2(7.2)} = \frac{-72.09}{-59.61} = 1.2$$

$$y = \frac{\begin{vmatrix} 5.3 & 4.5 \\ 3.2 & 5.7 \end{vmatrix}}{\begin{vmatrix} 5.3 & 7.2 \\ 3.2 & -6.9 \end{vmatrix}} = \frac{5.3(5.7) - 3.2(4.5)}{-59.61} = \frac{15.81}{-59.61} = -0.27$$

Therefore, the solution is $x = 1.2$, $y = -0.27$. We rounded off the values to two significant digits since all numbers are approximate. Checking these values, we have

$$5.3(1.2) + 7.2(-0.27) \stackrel{?}{=} 4.5 \quad \text{and} \quad 3.2(1.2) - 6.9(-0.27) \stackrel{?}{=} 5.7$$
$$4.416 \approx 4.5 \qquad\qquad 5.703 \approx 5.7 \text{ (They check.)}$$

It is not actually necessary to calculate the numerators and denominators separately. The calculations can be done completely on the calculator using parentheses to group values in the numerator and in the denominator. Also, the values of x and y can be stored in separate memories before rounding off. This can allow us to get a closer value for checking. ◂◂

EXAMPLE 5 ▸▸ Two investments totaling \$18 000 yield an annual income of \$700. If the first investment has an interest rate of 5.5% and the second a rate of 3.0%, what is the value of each investment?

Let x = the value of the first investment and y = the value of the second investment. We know that the total of the two investments is \$18 000. This leads to the equation $x + y = 18\,000$. The first investment yields $0.055x$ dollars annually, and the second yields $0.030y$ dollars annually. This leads to the equation $0.055x + 0.030y = 700$. These two equations are then solved simultaneously.

$$x + y = 18\,000 \quad \text{sum of investments}$$
$$\underset{\text{5.5\% value}}{0.055x} + \underset{\text{3.0\% value}}{0.030y} = 700 \longleftarrow \text{income}$$

$$x = \frac{\begin{vmatrix} 18\,000 & 1 \\ 700 & 0.030 \end{vmatrix}}{\begin{vmatrix} 1 & 1 \\ 0.055 & 0.030 \end{vmatrix}} = \frac{540 - 700}{0.030 - 0.055} = \frac{-160}{-0.025} = 6400$$

The value of y can be found most easily by substituting this value of x into the first equation. $y = 18\,000 - x = 18\,000 - 6400 = 11\,600$.

Therefore, the values invested are \$6400 and \$11 600, respectively. Checking, we see that the total income is \$6400(0.055) + \$11 600(0.030) = \$700. This agrees with the statement of the problem. ◂◂

CAUTION ▶ *The equations must be in the form of Eqs. (5-3) before the determinants are set up.* The specific positions of the values in the determinants are based on that form of writing the system. If either unknown is missing from an equation, a zero must be placed in the proper position. Also, from Example 4 we see that determinants are easier to use than other algebraic methods when the coefficients are decimals.

If the determinant of the denominator is zero, we do not have a unique solution since this would require division by zero. If the determinant of the denominator is zero, and that of the numerator is not zero, the system is *inconsistent.* If the determinants of both numerator and denominator are zero, the system is *dependent.*

Exercises 5–5

In Exercises 1–12, evaluate the given determinants.

1. $\begin{vmatrix} 2 & 4 \\ 3 & 1 \end{vmatrix}$ **2.** $\begin{vmatrix} -1 & 3 \\ 2 & 6 \end{vmatrix}$ **3.** $\begin{vmatrix} 3 & -5 \\ 7 & -2 \end{vmatrix}$

4. $\begin{vmatrix} -4 & 7 \\ 1 & -3 \end{vmatrix}$ **5.** $\begin{vmatrix} 8 & -10 \\ 0 & 4 \end{vmatrix}$ **6.** $\begin{vmatrix} -4 & -3 \\ -8 & -6 \end{vmatrix}$

7. $\begin{vmatrix} -2 & 11 \\ -7 & -8 \end{vmatrix}$ **8.** $\begin{vmatrix} -6 & 12 \\ -15 & 3 \end{vmatrix}$ **9.** $\begin{vmatrix} 0.7 & -1.3 \\ 0.1 & 1.1 \end{vmatrix}$

10. $\begin{vmatrix} 0.20 & -0.05 \\ 0.28 & 0.09 \end{vmatrix}$ **11.** $\begin{vmatrix} 16 & -8 \\ 42 & -15 \end{vmatrix}$ **12.** $\begin{vmatrix} 43 & -7 \\ -81 & 16 \end{vmatrix}$

In Exercises 13–24, solve the given systems of equations by determinants. (These are the same as those for Exercises 13–24 of Section 5-4.)

13. $x + 2y = 5$
$x - 2y = 1$

14. $x + 3y = 7$
$2x + 3y = 5$

15. $2x - 3y = 4$
$2x + y = -4$

16. $R - 4r = 17$
$3R + 4r = 3$

17. $2x + 3y = 8$
$x = 2y - 3$

18. $3x - y = 3$
$4x = 3y + 14$

19. $v + 2t = 7$
$2v + 4t = 9$

20. $3x - y = 5$
$-9x + 3y = -15$

21. $2x - 3y - 4 = 0$
$3x + 2 = 2y$

22. $3i_1 + 5 = -4i_2$
$3i_2 = 5i_1 - 2$

23. $0.3x = 0.7y + 0.4$
$0.5y = 0.7 - 0.2x$

24. $2.50x + 2.25y = 4.00$
$3.75x - 6.75y = 3.25$

In Exercises 25–32, solve the given systems of equations by determinants. All numbers are approximate.

25. $2.1x - 1.0y = 5.2$
$5.8x + 1.6y = -5.4$

26. $3.3x + 1.9y = 4.2$
$5.4x - 6.4y = 13.2$

27. $5.5R = -9.0t - 6.8$
$6.0t + 3.8R + 4.0 = 0$

28. $12 + 66y = 53x$
$32x - 39y = 71$

29. $301x - 529y = 1520$
$385x - 741y = 2540$

30. $0.25d + 0.63n = -0.37$
$-0.61d - 1.80n = 0.55$

31. $1.2y + 10.8 = -8.4x$
$3.5x + 4.8y + 12.9 = 0$

32. $6541x + 4397y = -7732$
$3309x - 8755y = 7622$

In Exercises 33–36, solve the given systems of equations by determinants. All numbers are accurate to at least two significant digits.

33. The forces acting on a link of an industrial robot are shown in Fig. 5-28. The equations for finding forces F_1 and F_2 are

$$F_1 + F_2 = 21$$
$$2F_1 = 5F_2$$

Find F_1 and F_2.

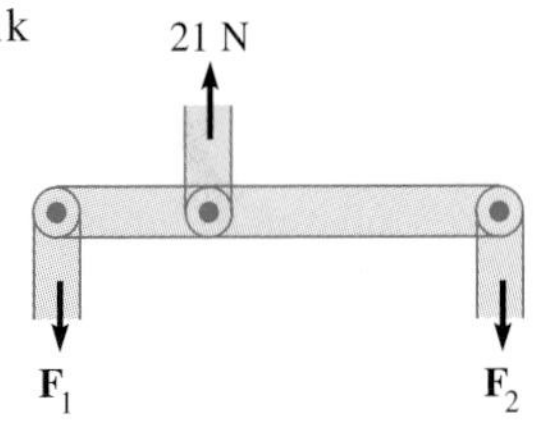

Fig. 5-28

34. The area of a quadrilateral is

$$A = \frac{1}{2}\left(\begin{vmatrix} x_0 & x_1 \\ y_0 & y_1 \end{vmatrix} + \begin{vmatrix} x_1 & x_2 \\ y_1 & y_2 \end{vmatrix} + \begin{vmatrix} x_2 & x_3 \\ y_2 & y_3 \end{vmatrix} + \begin{vmatrix} x_3 & x_0 \\ y_3 & y_0 \end{vmatrix}\right)$$

where (x_0, y_0), (x_1, y_1), (x_2, y_2), and (x_3, y_3) are the rectangular coordinates of the vertices of the quadrilateral, listed counterclockwise. (This *surveyor's formula* can be generalized to find the area of any polygon.)

A surveyor records the locations of the vertices of a quadrilateral building lot on a rectangular coordinate system as (12.79, 0.00), (67.21, 12.30), (53.05, 47.12), and (10.09, 53.11), where distances are in metres. Find the area of the lot.

35. An airplane begins a flight with a total of 144.0 L of fuel stored in two separate wing tanks. During the flight, 25.0% of the fuel in one tank is used, and in the other tank 37.5% of the fuel is used. If the total fuel used is 44.8 L, the amounts x and y used from each of the tanks can be found by solving the system of equations

$$x + y = 144.0$$
$$0.250x + 0.375y = 44.8$$

Find x and y.

36. A certain amount of fuel contains 150 MJ of potential heat. Part is burned at 80% efficiency and the remainder is burned at 70% efficiency such that the total amount of heat actually delivered is 114 MJ. To find the amounts x and y burned with each efficiency, the following equations must be solved. Find x and y.

$$x + y = 150$$
$$0.80x + 0.70y = 114$$

In Exercises 37–44, set up appropriate systems of two linear equations in two unknowns and then solve the systems by determinants. All numbers are accurate to at least two significant digits.

37. A boat carrying illegal drugs leaves a port and travels at 63 km/h. A Coast Guard cutter leaves the port 24 min later and travels at 75 km/h in pursuit of the boat. Find the times each has traveled when the cutter overtakes the boat with drugs.

38. In framing the wall of a house, each horizontal plate is 900 mm longer than each vertical stud. Find the lengths of each plate and each of the 9 studs if a total of 27 100 mm of lumber is used. See Fig. 5-29.

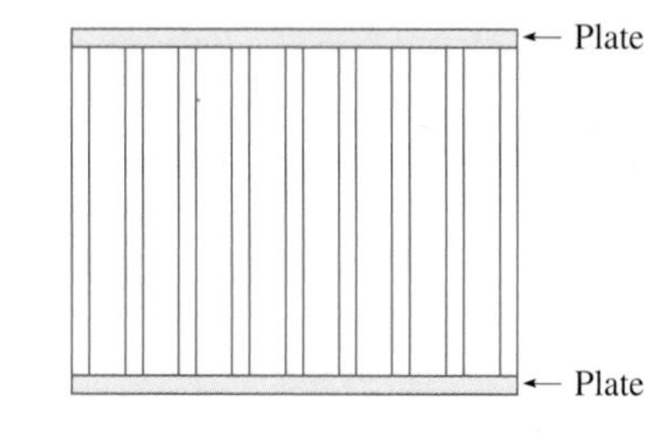

Fig. 5-29

39. A computer manufacturer receives two types of computer components in a shipment of 5600 components. Type A components cost \$0.75 each and type B components cost \$0.45 each; the total cost of the shipment was \$3780. How many of each component were in the shipment?

40. Two types of electromechanical carburetors are being assembled and tested. Each of the first type requires 15 min of assembly time and 2 min of testing time. Each of the second type requires 12 min of assembly time and 3 min of testing time. If 222 min of assembly time and 45 min of testing time are available, how many of each type can be assembled and tested if all the time is used?

41. A pharmacist is mixing a 3.0% saline solution and an 8.0% saline solution to get 2.0 L of a 6.0% solution. How much of each solution is needed?

42. The velocity of sound in steel is 4850 m/s faster than the velocity of sound in air. One end of a long steel bar is struck and an observer at the other end measures the time it takes for the sound to reach him. He finds that the sound through the bar takes 0.0120 s to reach him and that the sound through the air takes 0.180 s. What are the velocities of sound in air and in steel?

43. In an experiment, a variable voltage V is in a circuit with a fixed voltage V_0 and a resistance R. The voltage V is related to the current i in the circuit by $V = Ri - V_0$. If $V = 5.8$ V for $i = 2.0$ A, and $V = 24.7$ V for $i = 6.2$ A, find V as a function of i.

44. If \$6200 is invested, part at 5.5% and part at 2.5% with a total annual income of \$257, how much is invested at each rate?

5-6 Solving Systems of Three Linear Equations in Three Unknowns Algebraically

Many problems involve the solution of systems of linear equations that have three, four, and occasionally even more unknowns. Solving such systems algebraically or by determinants is essentially the same as solving systems of two linear equations in two unknowns. In this section we discuss the algebraic method of solving a system of three linear equations in three unknowns. Graphical solutions are not used, since a linear equation in three unknowns represents a plane in space. We shall, however, briefly show graphical interpretations of systems of three linear equations in three unknowns at the end of this section.

A system of three linear equations in three unknowns written in the form

$$\begin{aligned} a_1x + b_1y + c_1z &= d_1 \\ a_2x + b_2y + c_2z &= d_2 \\ a_3x + b_3y + c_3z &= d_3 \end{aligned} \qquad (5\text{-}11)$$

has as its solution the set of values x, y, and z that satisfy all three equations simultaneously. The method of solution involves multiplying ***two*** of the equations by the proper numbers to eliminate ***one*** of the unknowns between these equations. We then repeat this process, using a ***different pair*** of the original equations, being sure that we eliminate the same unknown as we did between the first pair of equations. At this point we have two linear equations in two unknowns that can be solved by any of the methods previously discussed.

EXAMPLE 1 ▸▸ Solve the following system of equations.

(1)	$4x + y + 3z = 1$	
(2)	$2x - 2y + 6z = 11$	
(3)	$-6x + 3y + 12z = -4$	
(4)	$8x + 2y + 6z = 2$	(1) multiplied by 2
	$2x - 2y + 6z = 11$	(2)
(5)	$10x + 12z = 13$	adding
(6)	$12x + 3y + 9z = 3$	(1) multiplied by 3
	$-6x + 3y + 12z = -4$	(3)
(7)	$18x - 3z = 7$	subtracting
	$10x + 12z = 13$	(5)
(8)	$72x - 12z = 28$	(7) multiplied by 4
(9)	$82x = 41$	adding
(10)	$x = \frac{1}{2}$	
(11)	$18(\frac{1}{2}) - 3z = 7$	substituting (10) in (7)
(12)	$-3z = -2$	
(13)	$z = \frac{2}{3}$	
(14)	$4(\frac{1}{2}) + y + 3(\frac{2}{3}) = 1$	substituting (13) and (10) in (1)
(15)	$2 + y + 2 = 1$	
(16)	$y = -3$	

Thus, the solution is $x = \frac{1}{2}$, $y = -3$, $z = \frac{2}{3}$. Substituting in the equations, we have

$$4(\tfrac{1}{2}) + (-3) + 3(\tfrac{2}{3}) \stackrel{?}{=} 1 \qquad 2(\tfrac{1}{2}) - 2(-3) + 6(\tfrac{2}{3}) \stackrel{?}{=} 11 \qquad -6(\tfrac{1}{2}) + 3(-3) + 12(\tfrac{2}{3}) \stackrel{?}{=} -4$$

$$1 = 1 \qquad 11 = 11 \qquad -4 = -4$$

We see that the solution checks. Note that we could have eliminated x first, or z first, and then completed the solution in a similar manner. ◂◂

See the chapter introduction.

EXAMPLE 2 ▸▸ An industrial robot and the forces acting on its main link are shown in Fig. 5-30. An analysis of the forces leads to the following equations relating the forces. Determine the forces.

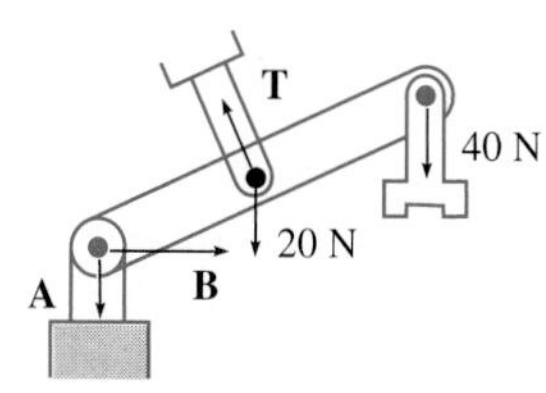

Fig. 5-30

(1)	$A + 60 = 0.8T$	
(2)	$B = 0.6T$	
(3)	$8A + 6B + 80 = 5T$	
(4)	$5A - 4T = -300$	(1) multiplied by 5 and rewritten
(5)	$5B - 3T = 0$	(2) multiplied by 5 and rewritten
(6)	$8A + 6B - 5T = -80$	(3) rewritten
(7)	$40A - 32T = -2400$	(4) multiplied by 8
(8)	$40A + 30B - 25T = -400$	(6) multiplied by 5
(9)	$-30B - 7T = -2000$	subtracting
(10)	$30B - 18T = 0$	(5) multiplied by 6
(11)	$-25T = -2000$	adding
(12)	$T = 80\text{ N}$	
(13)	$A + 60 = 64$	(12) substituted in (1)
(14)	$A = 4\text{ N}$	
(15)	$B = 48\text{ N}$	(12) substituted in (2)

Therefore, the forces are $A = 4$ N, $B = 48$ N, and $T = 80$ N. The solution checks when substituted into the original equations. Note that we could have started the solution by substituting Eq. (2) into Eq. (3), thereby first eliminating B. ◂◂

EXAMPLE 3 ▸▸ A triangular brace has a perimeter of 37 cm. The longest side is 3 cm longer than the next longest, which in turn is 8 cm longer than the shortest side. Find the length of each side.

Let a = length of the longest side, b = length of the next-longest side, and c = length of the shortest side. Since the perimeter is 37 cm, we have the equation $a + b + c = 37$. The statement of the problem also leads to the equations $a = b + 3$ and $b = c + 8$. These equations are put in standard form and solved.

(1)	$a + b + c = 37$	
(2)	$a - b = 3$	rewriting second equation
(3)	$b - c = 8$	rewriting third equation
(4)	$a + 2b = 45$	adding (1) and (3)
	$a - b = 3$	(2)
(5)	$3b = 42$	subtracting
(6)	$b = 14$	
(7)	$a - 14 = 3$	substituting (6) in (2)
(8)	$a = 17$	
(9)	$14 - c = 8$	substituting (6) in (3)
(10)	$c = 6$	

Therefore, the three sides of the triangle are 17 cm, 14 cm, and 6 cm.

Checking the solution, the sum of the lengths of the three sides of the brace is 17 cm + 14 cm + 6 cm = 37 cm, and the perimeter was given to be 37 cm. ◂◂

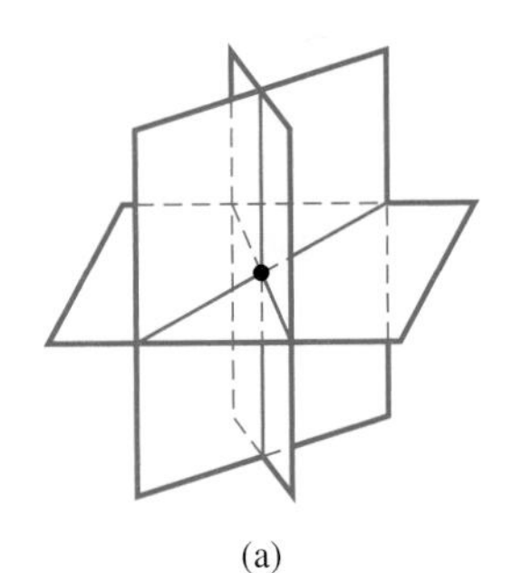
(a)

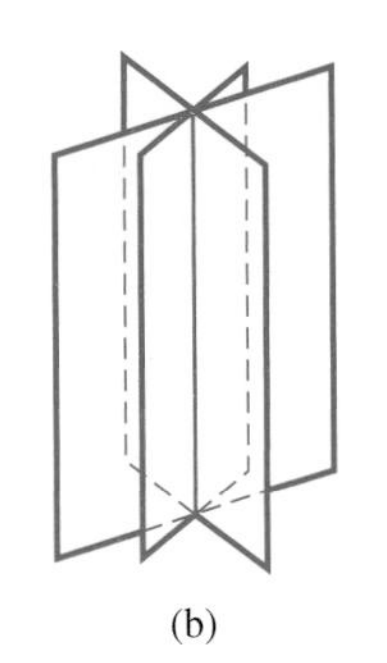
(b)

Fig. 5-31

The systems of equations we have solved in this section have had unique solutions. However, linear systems with more than two unknowns may also have an unlimited number of solutions or may be inconsistent. After eliminating unknowns, if we obtain $0 = 0$, there is an unlimited number of solutions. If we obtain $0 = a$ $(a \neq 0)$, the system is inconsistent and there is no solution. (See Exercises 25–28 of this section.)

In the introduction to this section we noted that a linear equation in three unknowns represents a plane in space. For a system of three linear equations in three unknowns, if the planes intersect at a point, there is a unique solution (Fig. 5-31(a)). If the three planes intersect in a line, there is an unlimited number of solutions (Fig. 5-31(b)). If the planes do not have a common intersection, the system is inconsistent. The planes can be parallel (Fig. 5-32(a)), two of the planes can be parallel (Fig. 5-32(b)), or they can intersect in pairs in three parallel lines (Fig. 5-32(c)). If any plane is coincident with another plane, the system is inconsistent or there is an unlimited number of solutions, depending on whether or not any of the planes are parallel.

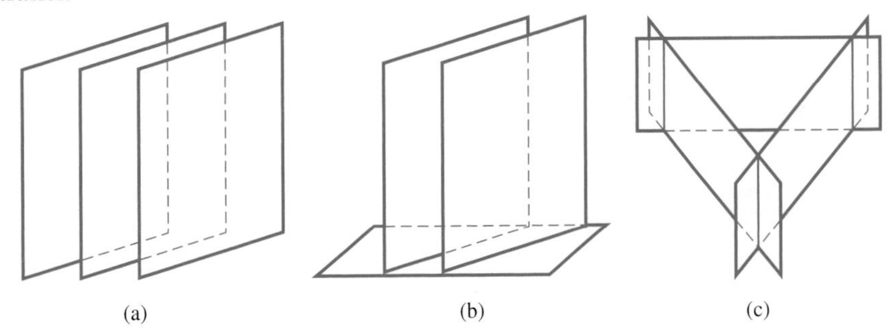
(a) (b) (c)

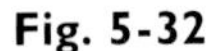
Fig. 5-32

For systems of equations with more than three unknowns, the solution is found in a manner similar to that used with three unknowns. For example, with four unknowns one of the unknowns is eliminated between three different pairs of equations. The result is three equations in the remaining three unknowns. The solution then follows the procedure used with three unknowns.

Exercises 5–6

In Exercises 1–16, solve the given systems of equations.

1. $x + y + z = 2$
$x - z = 1$
$x + y = 1$

2. $x + y - z = -3$
$x + z = 2$
$2x - y + 2z = 3$

3. $2x + 3y + z = 2$
$-x + 2y + 3z = -1$
$-3x - 3y + z = 0$

4. $2x + y - z = 4$
$4x - 3y - 2z = -2$
$8x - 2y - 3z = 3$

5. $5x + 6y - 3z = 6$
$4x - 7y - 2z = -3$
$3x + y - 7z = 1$

6. $3r + s - t = 2$
$r - 2s + t = 0$
$4r - s + t = 3$

7. $2x - 2y + 3z = 5$
$2x + y - 2z = -1$
$4x - y - 3z = 0$

8. $2u + 2v + 3w = 0$
$3u + v + 4w = 21$
$-u - 3v + 7w = 15$

9. $3x - 7y + 3z = 6$
$3x + 3y + 6z = 1$
$5x - 5y + 2z = 5$

10. $8x + y + z = 1$
$7x - 2y + 9z = -3$
$4x - 6y + 8z = -5$

11. $p + 2q + 2r = 0$
$2p + 6q - 3r = -1$
$4p - 3q + 6r = -8$

12. $3x + 3y + z = 6$
$2x + 2y - z = 9$
$4x + 2y - 3z = 16$

13. $2x + 3y - 5z = 7$
$4x - 3y - 2z = 1$
$8x - y + 4z = 3$

14. $2x - 4y - 4z = 3$
$3x + 8y + 2z = -11$
$4x + 6y - z = -8$

15. $r - s - 3t - u = 1$
$2r + 4s - 2u = 2$
$3r + 4s - 2t = 0$
$r + 2t - 3u = 3$

16. $3x + 2y - 4z + 2t = 3$
$5x - 3y - 5z + 6t = 8$
$2x - y + 3z - 2t = 1$
$-2x + 3y + 2z - 3t = -2$

In Exercises 17–20, solve the given systems of equations by determinants. All numbers are accurate to at least two significant digits.

17. A medical supply company has 1150 worker-hours for production, maintenance, and inspection. Using this and other factors, the number of hours used for each operation, P, M, and I, respectively, is found by solving the following system of equations:

$$\begin{aligned} P + M + I &= 1150 \\ P &= 4I - 100 \\ P &= 6M + 50 \end{aligned}$$

18. Three oil pumps fill three different tanks. The pumping rates of the pumps in litres per hour are r_1, r_2, and r_3, respectively. Because of malfunctions, they do not operate at capacity each time. Their rates can be found by solving the system of equations

$$\begin{aligned} r_1 + r_2 + r_3 &= 14\,000 \\ r_1 + 2r_2 \qquad &= 13\,000 \\ 3r_1 + 3r_2 + 2r_3 &= 36\,000 \end{aligned}$$

Find the pumping rates.

19. The forces acting on a certain girder, as shown in Fig. 5-33, can be found by solving the following system of equations:

$$\begin{aligned} 0.707F_1 - 0.800F_2 \qquad\quad &= 0 \\ 0.707F_1 + 0.600F_2 - F_3 &= 10.0 \\ 3.00\ F_2 - 3.00F_3 &= 20.0 \end{aligned}$$

Find the forces, in newtons.

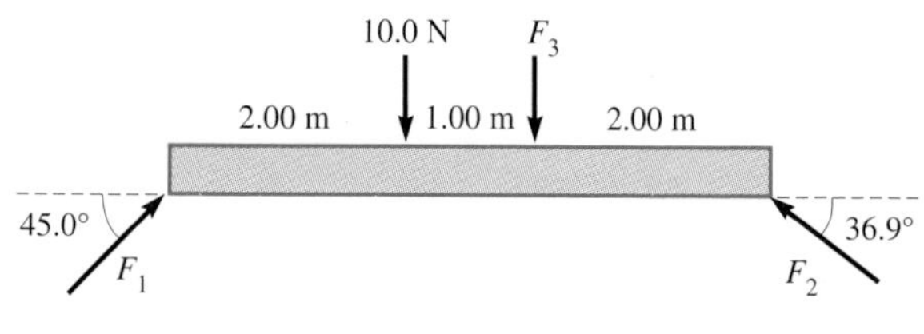

Fig. 5-33

20. In applying Kirchhoff's laws (e.g., see Beiser, *Modern Technical Physics,* 6th ed., p. 550) to the electric circuit shown in Fig. 5-34, the following equations are found. Determine the indicated currents. In Fig. 5-34, I signifies current, in amperes.

$$\begin{aligned} I_A + I_B + I_C &= 0 \\ 4I_A - 10I_B \qquad &= 3 \\ -10I_B + 5I_C &= 6 \end{aligned}$$

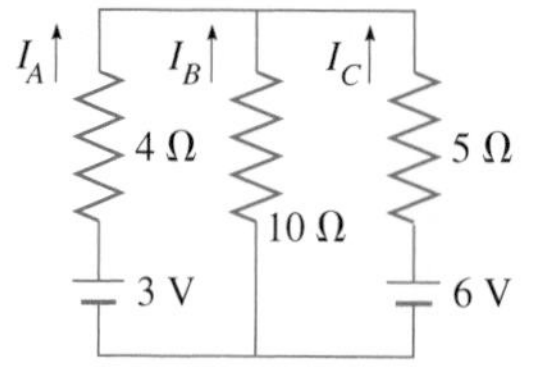

Fig. 5-34

In Exercises 21–24, set up systems of three linear equations in three unknowns and solve for the indicated quantities. All numbers are accurate to at least two significant digits.

21. A construction company orders a total of 7600 nuts, bolts, and washers for \$462.21, including \$28.21 in sales tax. In the order the total number of nuts and bolts is 400 more than the number of washers. If the nuts cost 5¢ each, the bolts cost 12¢ each, and the washers cost 3¢ each, how many of each are in the order?

22. Three computer line printers print 18 500 lines when operating together for 2 min. They print 17 500 lines if they operate for 3 min, 2 min, and 1 min, respectively. The first and third printers print 6000 lines when they operate together for 1 min. What is the rate, in lines per minute, at which each prints? (All data are accurate to three significant digits, although times are shown with only one digit.)

23. By weight, one fertilizer is 20% potassium, 30% nitrogen, and 50% phosphorus. A second fertilizer has percents of 10, 20, and 70, respectively, and a third fertilizer has percents of 0, 30, and 70, respectively. How much of each must be mixed together to get 200 kg of fertilizer with percents of 12, 25, and 63, respectively?

24. The average traffic flow (number of vehicles) from noon until 1 P.M. in a certain section of one-way streets in a city is shown in Fig. 5-35. Show that an analysis of the flow through intersections is not sufficient to obtain unique values for x, y, and z.

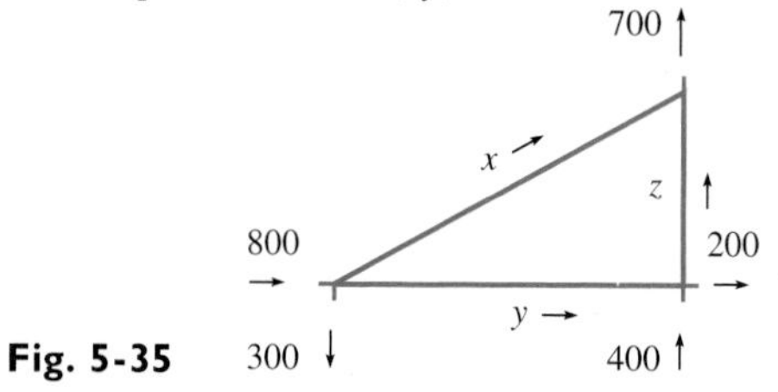

Fig. 5-35

In Exercises 25–28, show that the given systems of equations have either an unlimited number of solutions or no solution. If there is an unlimited number of solutions, find one of them.

25.
$$\begin{aligned} x - 2y - 3z &= 2 \\ x - 4y - 13z &= 14 \\ -3x + 5y + 4z &= 0 \end{aligned}$$

26.
$$\begin{aligned} x - 2y - 3z &= 2 \\ x - 4y - 13z &= 14 \\ -3x + 5y + 4z &= 2 \end{aligned}$$

27.
$$\begin{aligned} 3x + 3y - 2z &= 2 \\ 2x - y + z &= 1 \\ x - 5y + 4z &= -3 \end{aligned}$$

28.
$$\begin{aligned} 3x + y - z &= -3 \\ x + y - 3z &= -5 \\ -5x - 2y + 3z &= -7 \end{aligned}$$

5-7 Solving Systems of Three Linear Equations in Three Unknowns by Determinants

Just as systems in two linear equations in two unknowns can be solved by determinants, so can systems of three linear equations in three unknowns. The system

$$\begin{aligned} a_1x + b_1y + c_1z &= d_1 \\ a_2x + b_2y + c_2z &= d_2 \\ a_3x + b_3y + c_3z &= d_3 \end{aligned} \tag{5-11}$$

can be solved in general terms by the method of elimination by addition or subtraction. This leads to the following solutions for x, y, and z.

$$\begin{aligned} x &= \frac{d_1b_2c_3 + d_3b_1c_2 + d_2b_3c_1 - d_3b_2c_1 - d_1b_3c_2 - d_2b_1c_3}{a_1b_2c_3 + a_3b_1c_2 + a_2b_3c_1 - a_3b_2c_1 - a_1b_3c_2 - a_2b_1c_3} \\ y &= \frac{a_1d_2c_3 + a_3d_1c_2 + a_2d_3c_1 - a_3d_2c_1 - a_1d_3c_2 - a_2d_1c_3}{a_1b_2c_3 + a_3b_1c_2 + a_2b_3c_1 - a_3b_2c_1 - a_1b_3c_2 - a_2b_1c_3} \\ z &= \frac{a_1b_2d_3 + a_3b_1d_2 + a_2b_3d_1 - a_3b_2d_1 - a_1b_3d_2 - a_2b_1d_3}{a_1b_2c_3 + a_3b_1c_2 + a_2b_3c_1 - a_3b_2c_1 - a_1b_3c_2 - a_2b_1c_3} \end{aligned} \tag{5-12}$$

The expressions that appear in the numerators and denominators of Eqs. (5-12) are examples of a **determinant of the third order.** *This determinant is defined by*

$$\begin{vmatrix} a_1 & b_1 & c_1 \\ a_2 & b_2 & c_2 \\ a_3 & b_3 & c_3 \end{vmatrix} = a_1b_2c_3 + a_3b_1c_2 + a_2b_3c_1 - a_3b_2c_1 - a_1b_3c_2 - a_2b_1c_3 \tag{5-13}$$

The elements, rows, columns, and diagonals of a third-order determinant are defined just as are those of a second-order determinant. For example, the principal diagonal is made up of the elements a_1, b_2, and c_3.

Probably the easiest way of remembering the method of determining the value of a third-order determinant is as follows (this method does *not* work for determinants of order higher than three): *Rewrite the first and second columns to the right of the determinant. The products of the elements of the principal diagonal and the two parallel diagonals to the right of it are then added. The products of the elements of the secondary diagonal and the two parallel diagonals to the right of it are subtracted from the first sum. The algebraic sum of these six products gives the value of the determinant.* These products are indicated in Fig. 5-36.

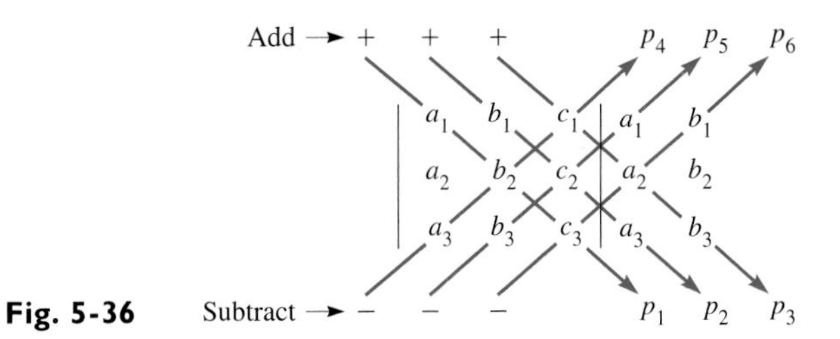

Fig. 5-36

Examples 1 and 2 illustrate this method of evaluating third-order determinants. (Other methods are shown in Chapter 16.)

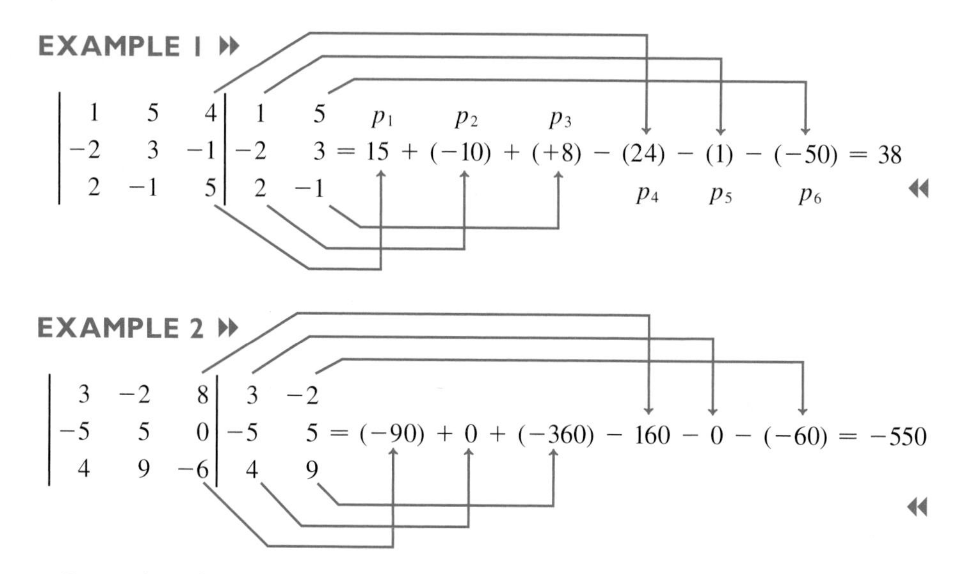

Inspection of Eqs. (5-12) reveals that the numerators of these solutions may also be written in terms of determinants. Thus, we may write the general solution to a system of three equations in three unknowns as

$$x = \frac{\begin{vmatrix} d_1 & b_1 & c_1 \\ d_2 & b_2 & c_2 \\ d_3 & b_3 & c_3 \end{vmatrix}}{\begin{vmatrix} a_1 & b_1 & c_1 \\ a_2 & b_2 & c_2 \\ a_3 & b_3 & c_3 \end{vmatrix}} \quad y = \frac{\begin{vmatrix} a_1 & d_1 & c_1 \\ a_2 & d_2 & c_2 \\ a_3 & d_3 & c_3 \end{vmatrix}}{\begin{vmatrix} a_1 & b_1 & c_1 \\ a_2 & b_2 & c_2 \\ a_3 & b_3 & c_3 \end{vmatrix}} \quad z = \frac{\begin{vmatrix} a_1 & b_1 & d_1 \\ a_2 & b_2 & d_2 \\ a_3 & b_3 & d_3 \end{vmatrix}}{\begin{vmatrix} a_1 & b_1 & c_1 \\ a_2 & b_2 & c_2 \\ a_3 & b_3 & c_3 \end{vmatrix}} \tag{5-14}$$

If the determinant of the denominator is not zero, there is a unique solution to the system of equations. (If all determinants are zero, there is an *unlimited number of solutions*. If the determinant of the denominator is zero, and any of the determinants of the numerators is not zero, the system is *inconsistent*, and there is *no solution*.)

An analysis of Eqs. (5-14) shows that the situation is precisely the same as it was when we were using determinants to solve systems of two linear equations. That is, the determinants in the denominators in the expressions for x, y, and z are the same. They consist of elements that are the coefficients of the unknowns. The determinant of the numerator of the solution for x is the same as that of the denominator, except that the column of d's replaces the column of a's. The determinant in the numerator of the solution for y is the same as that of the denominator, except that the column of d's replaces the column of b's. The determinant of the numerator of the solution for z is the same as the determinant of the denominator, except that the column of d's replaces the column of c's. To summarize:

Cramer's Rule

NOTE ▶

The determinant in the denominator of each is made up of the coefficients of x, y, and z. The determinants in the numerators are the same as that in the denominator, except that ***the column of d's replaces the column of coefficients of the unknown for which we are solving.***

This, again, is **Cramer's rule**. Remember, the equations must be written in the standard form shown in Eqs. (5-11) before the determinants are formed.

EXAMPLE 3 ▸▸ Solve the following system by determinants.

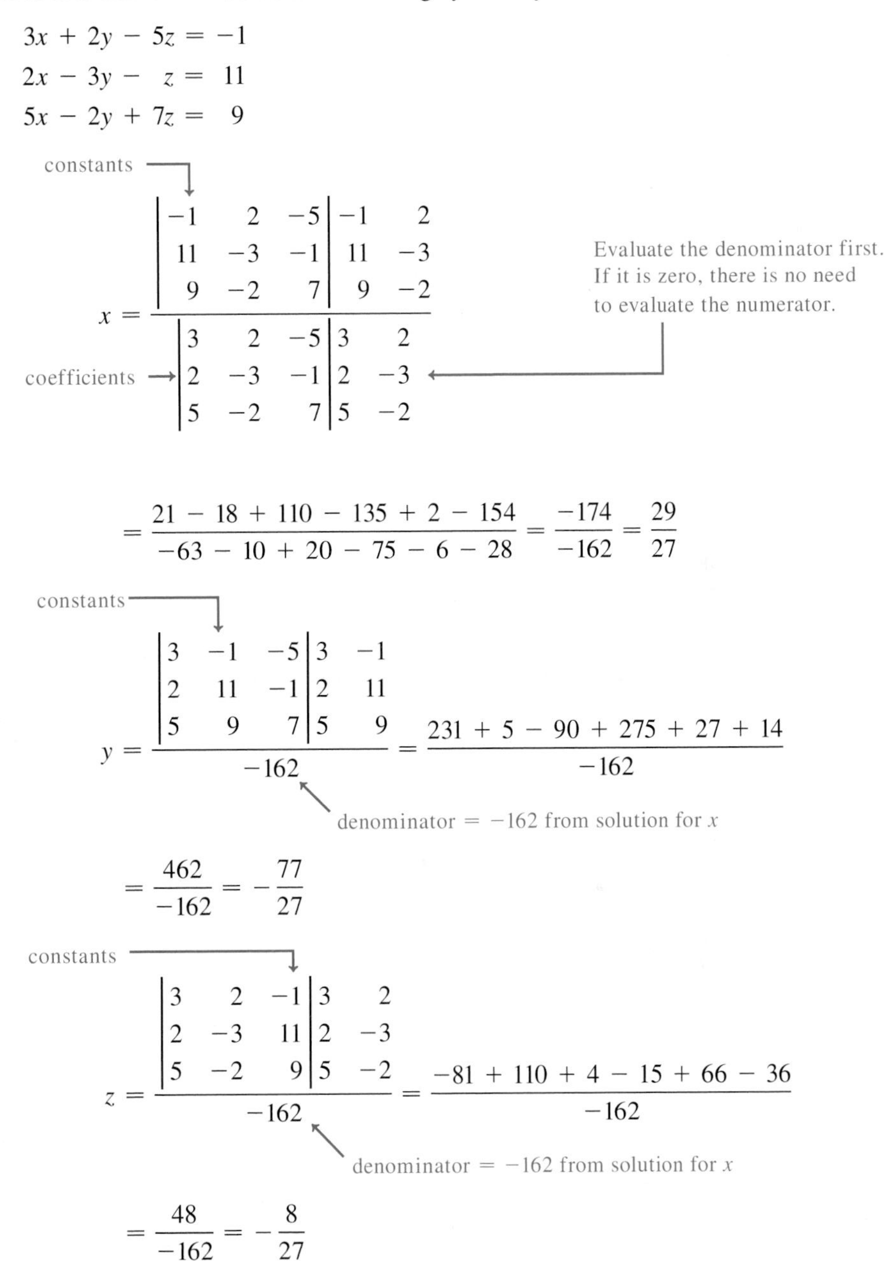

$$\begin{aligned} 3x + 2y - 5z &= -1 \\ 2x - 3y - z &= 11 \\ 5x - 2y + 7z &= 9 \end{aligned}$$

$$x = \frac{\begin{vmatrix} -1 & 2 & -5 \\ 11 & -3 & -1 \\ 9 & -2 & 7 \end{vmatrix}\begin{matrix} -1 & 2 \\ 11 & -3 \\ 9 & -2 \end{matrix}}{\begin{vmatrix} 3 & 2 & -5 \\ 2 & -3 & -1 \\ 5 & -2 & 7 \end{vmatrix}\begin{matrix} 3 & 2 \\ 2 & -3 \\ 5 & -2 \end{matrix}}$$

$$= \frac{21 - 18 + 110 - 135 + 2 - 154}{-63 - 10 + 20 - 75 - 6 - 28} = \frac{-174}{-162} = \frac{29}{27}$$

$$y = \frac{\begin{vmatrix} 3 & -1 & -5 \\ 2 & 11 & -1 \\ 5 & 9 & 7 \end{vmatrix}\begin{matrix} 3 & -1 \\ 2 & 11 \\ 5 & 9 \end{matrix}}{-162} = \frac{231 + 5 - 90 + 275 + 27 + 14}{-162}$$

$$= \frac{462}{-162} = -\frac{77}{27}$$

$$z = \frac{\begin{vmatrix} 3 & 2 & -1 \\ 2 & -3 & 11 \\ 5 & -2 & 9 \end{vmatrix}\begin{matrix} 3 & 2 \\ 2 & -3 \\ 5 & -2 \end{matrix}}{-162} = \frac{-81 + 110 + 4 - 15 + 66 - 36}{-162}$$

$$= \frac{48}{-162} = -\frac{8}{27}$$

Substituting in each of the original equations shows that the solution checks.

$$3\left(\tfrac{29}{27}\right) + 2\left(-\tfrac{77}{27}\right) - 5\left(-\tfrac{8}{27}\right) = \frac{87 - 154 + 40}{27} = \frac{-27}{27} = -1$$

$$2\left(\tfrac{29}{27}\right) - 3\left(-\tfrac{77}{27}\right) - \left(-\tfrac{8}{27}\right) = \frac{58 + 231 + 8}{27} = \frac{297}{27} = 11$$

$$5\left(\tfrac{29}{27}\right) - 2\left(-\tfrac{77}{27}\right) + 7\left(-\tfrac{8}{27}\right) = \frac{145 + 154 - 56}{27} = \frac{243}{27} = 9$$

After the values of x and y were determined, we could have evaluated z by substituting the values of x and y into one of the original equations. ◂◂

EXAMPLE 4 ▸▸ An 8.0% solution, an 11% solution, and an 18% solution of nitric acid are to be mixed to get 150 mL of a 12% solution. If the volume of acid from the 8.0% solution equals half the volume of acid from the other two solutions, how much of each is needed?

Let x = volume of 8.0% solution needed, y = volume of 11% solution needed, and z = volume of 18% solution needed.

The fact that the sum of the volumes of the three solutions is 150 mL leads to the equation $x + y + z = 150$. Since there are $0.080x$ mL of pure acid from the first solution, $0.11y$ mL from the second solution, and $0.18z$ mL from the third solution, and $0.12(150)$ mL in the final solution, we are led to the equation $0.080x + 0.11y + 0.18z = 18$. Finally, using the last stated condition, we have the equation $0.080x = 0.5(0.11y + 0.18z)$. These equations are then written in the form of Eqs. (5-13) and solved.

$$\begin{aligned}
x + y + z &= 150 && \text{sum of volumes}\\
0.080x + 0.11y + 0.18z &= 18 && \text{volumes of pure acid}\\
0.080x &= 0.055y + 0.090z && \text{one-half of acid in others}
\end{aligned}$$

acid in 8.0% solution

$$\begin{aligned}
x + y + z &= 150 && \text{standard form of}\\
0.080x + 0.11y + 0.18z &= 18 && \text{Eqs. (5-11) by}\\
0.080x - 0.055y - 0.090z &= 0 && \text{rewriting third equation}
\end{aligned}$$

$$x = \frac{\left|\begin{array}{ccc|cc} 150 & 1 & 1 & 150 & 1\\ 18 & 0.11 & 0.18 & 18 & 0.11\\ 0 & -0.055 & -0.090 & 0 & -0.055 \end{array}\right.}{\left|\begin{array}{ccc|cc} 1 & 1 & 1 & 1 & 1\\ 0.080 & 0.11 & 0.18 & 0.080 & 0.11\\ 0.080 & -0.055 & -0.090 & 0.080 & -0.055 \end{array}\right.}$$

$$= \frac{-1.485 + 0 - 0.990 - 0 + 1.485 + 1.620}{-0.0099 + 0.0144 - 0.0044 - 0.0088 + 0.0099 + 0.0072} = \frac{0.630}{0.0084} = 75$$

$$y = \frac{\left|\begin{array}{ccc|cc} 1 & 150 & 1 & 1 & 150\\ 0.080 & 18 & 0.18 & 0.080 & 18\\ 0.080 & 0 & -0.090 & 0.080 & 0 \end{array}\right.}{0.0084}$$

$$= \frac{-1.620 + 2.160 + 0 - 1.440 - 0 + 1.080}{0.0084} = \frac{0.180}{0.0084} = 21$$

$$z = \frac{\left|\begin{array}{ccc|cc} 1 & 1 & 150 & 1 & 1\\ 0.080 & 0.11 & 18 & 0.080 & 0.11\\ 0.080 & -0.055 & 0 & 0.080 & -0.055 \end{array}\right.}{0.0084}$$

The value of z can also be found by substituting $x = 75$ and $y = 21$ into the first equation.

$$= \frac{0 + 1.440 - 0.660 - 1.320 + 0.990 - 0}{0.0084} = \frac{0.450}{0.0084} = 54$$

Therefore, 75 mL of the 8.0% solution, 21 mL of the 11% solution, and 54 mL of the 18% solution are required to make the 12% solution. Results have been rounded off to two significant digits, the accuracy of the data. Checking with the statement of the problem, we see that these volumes total 150 mL.

In using the calculator, it is not necessary to record the individual products in the determinants. Each calculation can be done completely on the calculator. ◂◂

Determinants can also be evaluated directly on many models of graphing calculators by use of the MATRIX feature. Once the elements are entered, the calculator completes the evaluation. Additional methods that are useful in solving systems of equations are taken up in Chapter 16.

Exercises 5–7

In Exercises 1–12, evaluate the given third-order determinants.

1. $\begin{vmatrix} 5 & 4 & -1 \\ -2 & -6 & 8 \\ 7 & 1 & 1 \end{vmatrix}$ **2.** $\begin{vmatrix} -7 & 0 & 0 \\ 2 & 4 & 5 \\ 1 & 4 & 2 \end{vmatrix}$

3. $\begin{vmatrix} 8 & 9 & -6 \\ -3 & 7 & 2 \\ 4 & -2 & 5 \end{vmatrix}$ **4.** $\begin{vmatrix} -2 & 4 & -1 \\ 5 & -1 & 4 \\ 4 & -8 & 2 \end{vmatrix}$

5. $\begin{vmatrix} -3 & -4 & -8 \\ 5 & -1 & 0 \\ 2 & 10 & -1 \end{vmatrix}$ **6.** $\begin{vmatrix} 10 & 2 & -7 \\ -2 & -3 & 6 \\ 6 & 5 & -2 \end{vmatrix}$

7. $\begin{vmatrix} 4 & -3 & -11 \\ -9 & 2 & -2 \\ 0 & 1 & -5 \end{vmatrix}$ **8.** $\begin{vmatrix} 9 & -2 & 0 \\ -1 & 3 & -6 \\ -4 & -6 & -2 \end{vmatrix}$

9. $\begin{vmatrix} 5 & 4 & -5 \\ -3 & 2 & -1 \\ 7 & 1 & 3 \end{vmatrix}$ **10.** $\begin{vmatrix} 20 & 0 & -15 \\ -4 & 30 & 1 \\ 6 & -1 & 40 \end{vmatrix}$

11. $\begin{vmatrix} 0.1 & -0.2 & 0 \\ -0.5 & 1 & 0.4 \\ -2 & 0.8 & 2 \end{vmatrix}$ **12.** $\begin{vmatrix} 0.2 & -0.5 & -0.4 \\ 1.2 & 0.3 & 0.2 \\ -0.5 & 0.1 & -0.4 \end{vmatrix}$

In Exercises 13–28, solve the given systems of equations by use of determinants. (Exercises 15–26 are the same as Exercises 1–12 of Section 5-6.)

13. $2x + 3y + z = 4$
$3x - z = -3$
$x - 2y + 2z = -5$

14. $4x + y + z = 2$
$2x - y - z = 4$
$3y + z = 2$

15. $x + y + z = 2$
$x - z = 1$
$x + y = 1$

16. $x + y - z = -3$
$x + z = 2$
$2x - y + 2z = 3$

17. $2x + 3y + z = 2$
$-x + 2y + 3z = -1$
$-3x - 3y + z = 0$

18. $2x + y - z = 4$
$4x - 3y - 2z = -2$
$8x - 2y - 3z = 3$

19. $5x + 6y - 3z = 6$
$4x - 7y - 2z = -3$
$3x + y - 7z = 1$

20. $3r + s - t = 2$
$r - 2s + t = 0$
$4r - s + t = 3$

21. $2x - 2y + 3z = 5$
$2x + y - 2z = -1$
$4x - y - 3z = 0$

22. $2u + 2v + 3w = 0$
$3u + v + 4w = 21$
$-u - 3v + 7w = 15$

23. $3x - 7y + 3z = 6$
$3x + 3y + 6z = 1$
$5x - 5y + 2z = 5$

24. $8x + y + z = 1$
$7x - 2y + 9z = -3$
$4x - 6y + 8z = -5$

25. $p + 2q + 2r = 0$
$2p + 6q - 3r = -1$
$4p - 3q + 6r = -8$

26. $3x + 3y + z = 6$
$2x + 2y - z = 9$
$4x + 2y - 3z = 16$

27. $3.0x + 4.5y - 7.5z = 10.5$
$4.8x - 3.6y - 2.4z = 1.2$
$4.0x - 0.5y + 2.0z = 1.5$

28. $26x - 52y - 52z = 39$
$45x + 96y + 40z = -80$
$55x + 62y - 11z = -48$

In Exercises 29–32, solve the given problems by determinants. In Exercises 31 and 32, set up appropriate systems of equations and then solve them. All numbers are accurate to at least two significant digits.

29. In analyzing the forces on the bell-crank mechanism shown in Fig. 5-37, the following equations are obtained. Find the indicated forces.

$$A - 0.60F = 80$$
$$B - 0.80F = 0$$
$$6.0A - 10F = 0$$

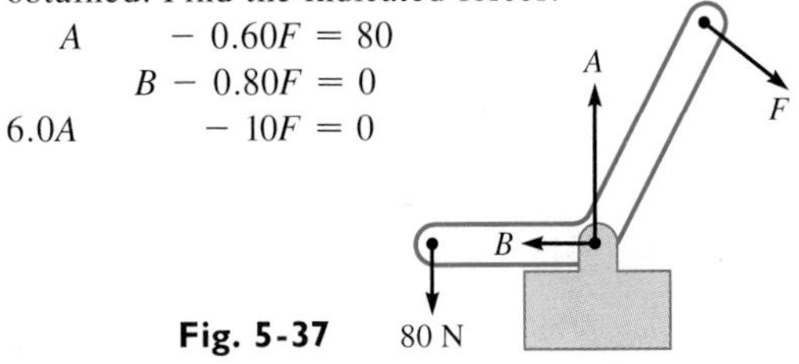

Fig. 5-37

30. Find the indicated electric currents (in amperes) in the circuit shown in Fig. 5-38 from the following equations.

$$i_A + i_B + i_C = 0$$
$$-8.2i_B + 10i_C = 0$$
$$4.3i_A - 8.2i_B = 6.5$$

Fig. 5-38

31. A person spent 1.10 h in a car going to an airport, 1.95 h flying in a jet, and 0.520 h in a taxi to reach the final destination. The jet's speed averaged 12.0 times that of the car, which averaged 15.0 km/h more than the taxi. What was the average speed of each if the trip covered 1140 km?

32. An intravenous aqueous solution is made from three mixtures to get 500 mL with 6.0% of one medication, 8.0% of a second medication, and 86% water. The percents in the mixtures are, respectively, 5.0, 20, 75 (first), 0, 5.0, 95 (second), and 10, 5.0, 85 (third). How much of each is used?

Chapter Equations, Review Exercises, and Practice Test

Chapter Equations

Linear equation in one unknown $ax + b = 0$ (5-1)

Linear equation in two unknowns $ax + by = c$ (5-2)

System of two linear equations
$$a_1x + b_1y = c_1$$
$$a_2x + b_2y = c_2 \qquad (5\text{-}3)$$

Definition of slope
$$m = \frac{y_2 - y_1}{x_2 - x_1} \qquad (5\text{-}4)$$

Slope-intercept form
$$y = mx + b \qquad (5\text{-}5)$$

Second-order determinant
$$\begin{vmatrix} a_1 & b_1 \\ a_2 & b_2 \end{vmatrix} = a_1b_2 - a_2b_1 \qquad (5\text{-}9)$$

Cramer's rule
$$x = \frac{\begin{vmatrix} c_1 & b_1 \\ c_2 & b_2 \end{vmatrix}}{\begin{vmatrix} a_1 & b_1 \\ a_2 & b_2 \end{vmatrix}} \quad \text{and} \quad y = \frac{\begin{vmatrix} a_1 & c_1 \\ a_2 & c_2 \end{vmatrix}}{\begin{vmatrix} a_1 & b_1 \\ a_2 & b_2 \end{vmatrix}} \qquad (5\text{-}10)$$

System of three linear equations
$$a_1x + b_1y + c_1z = d_1$$
$$a_2x + b_2y + c_2z = d_2$$
$$a_3x + b_3y + c_3z = d_3 \qquad (5\text{-}11)$$

Third-order determinant
$$\begin{vmatrix} a_1 & b_1 & c_1 \\ a_2 & b_2 & c_2 \\ a_3 & b_3 & c_3 \end{vmatrix} = a_1b_2c_3 + a_3b_1c_2 + a_2b_3c_1 - a_3b_2c_1 - a_1b_3c_2 - a_2b_1c_3 \qquad (5\text{-}13)$$

Cramer's rule
$$x = \frac{\begin{vmatrix} d_1 & b_1 & c_1 \\ d_2 & b_2 & c_2 \\ d_3 & b_3 & c_3 \end{vmatrix}}{\begin{vmatrix} a_1 & b_1 & c_1 \\ a_2 & b_2 & c_2 \\ a_3 & b_3 & c_3 \end{vmatrix}} \quad y = \frac{\begin{vmatrix} a_1 & d_1 & c_1 \\ a_2 & d_2 & c_2 \\ a_3 & d_3 & c_3 \end{vmatrix}}{\begin{vmatrix} a_1 & b_1 & c_1 \\ a_2 & b_2 & c_2 \\ a_3 & b_3 & c_3 \end{vmatrix}} \quad z = \frac{\begin{vmatrix} a_1 & b_1 & d_1 \\ a_2 & b_2 & d_2 \\ a_3 & b_3 & d_3 \end{vmatrix}}{\begin{vmatrix} a_1 & b_1 & c_1 \\ a_2 & b_2 & c_2 \\ a_3 & b_3 & c_3 \end{vmatrix}} \qquad (5\text{-}14)$$

Review Exercises

In Exercises 1–4, evaluate the given determinants.

1. $\begin{vmatrix} -2 & 5 \\ 3 & 1 \end{vmatrix}$

2. $\begin{vmatrix} 4 & 0 \\ -2 & -6 \end{vmatrix}$

3. $\begin{vmatrix} -18 & -33 \\ -21 & 44 \end{vmatrix}$

4. $\begin{vmatrix} 0.91 & -1.2 \\ 0.73 & -5.0 \end{vmatrix}$

In Exercises 5–8, find the slopes of the lines that pass through the given points.

5. $(2, 0), (4, -8)$

6. $(-1, -5), (-4, 4)$

7. $(4, -2), (-3, -4)$

8. $(-6, \frac{1}{2}), (1, -\frac{7}{2})$

In Exercises 9–12, find the slopes and y-intercepts of the lines with the given equations, and sketch the graphs.

9. $y = -2x + 4$

10. $y = \frac{2}{3}x - 3$

11. $8x - 2y = 5$

12. $3x = 8 + 3y$

In Exercises 13–20, solve the given systems of equations graphically.

13. $y = 2x - 4$
$y = -\frac{3}{2}x + 3$

14. $y = -3x + 3$
$y = 2x - 6$

15. $4x - y = 6$
$3x + 2y = 12$

16. $2x - 5y = 10$
$3x + y = 6$

17. $\begin{aligned} 7x &= 2y + 14 \\ y &= -4x + 4 \end{aligned}$

18. $\begin{aligned} 5x &= 15 - 3y \\ y &= 6x - 12 \end{aligned}$

19. $\begin{aligned} 3x + 4y &= 6 \\ 2x - 3y &= 2 \end{aligned}$

20. $\begin{aligned} 5x + 2y &= 5 \\ 2x - 4y &= 3 \end{aligned}$

In Exercises 21–32, solve the given systems of equations algebraically.

21. $\begin{aligned} x + 2y &= 5 \\ x + 3y &= 7 \end{aligned}$

22. $\begin{aligned} 2x - y &= 7 \\ x + y &= 2 \end{aligned}$

23. $\begin{aligned} 4x + 3y &= -4 \\ y &= 2x - 3 \end{aligned}$

24. $\begin{aligned} x &= -3y - 2 \\ -2x - 9y &= 2 \end{aligned}$

25. $\begin{aligned} 3i + 4v &= 6 \\ 9i + 8v &= 11 \end{aligned}$

26. $\begin{aligned} 3x - 6y &= 5 \\ 7x + 2y &= 4 \end{aligned}$

27. $\begin{aligned} 2x - 5y &= 8 \\ 5x - 3y &= 7 \end{aligned}$

28. $\begin{aligned} 3x + 4y &= 8 \\ 2x - 3y &= 9 \end{aligned}$

29. $\begin{aligned} 7x &= 2y - 6 \\ 7y &= 12 - 4x \end{aligned}$

30. $\begin{aligned} 3R &= 8 - 5I \\ 6I &= 8R + 11 \end{aligned}$

31. $\begin{aligned} 0.9x - 1.1y &= 0.4 \\ 0.6x - 0.3y &= 0.5 \end{aligned}$

32. $\begin{aligned} 0.42x - 0.56y &= 1.26 \\ 0.98x - 1.40y &= -0.28 \end{aligned}$

In Exercises 33–44, solve the given systems of equations by determinants. (These systems are the same as for Exercises 21–32.)

33. $\begin{aligned} x + 2y &= 5 \\ x + 3y &= 7 \end{aligned}$

34. $\begin{aligned} 2x - y &= 7 \\ x + y &= 2 \end{aligned}$

35. $\begin{aligned} 4x + 3y &= -4 \\ y &= 2x - 3 \end{aligned}$

36. $\begin{aligned} x &= -3y - 2 \\ -2x - 9y &= 2 \end{aligned}$

37. $\begin{aligned} 3i + 4v &= 6 \\ 9i + 8v &= 11 \end{aligned}$

38. $\begin{aligned} 3x - 6y &= 5 \\ 7x + 2y &= 4 \end{aligned}$

39. $\begin{aligned} 2x - 5y &= 8 \\ 5x - 3y &= 7 \end{aligned}$

40. $\begin{aligned} 3x + 4y &= 8 \\ 2x - 3y &= 9 \end{aligned}$

41. $\begin{aligned} 7x &= 2y - 6 \\ 7y &= 12 - 4x \end{aligned}$

42. $\begin{aligned} 3R &= 8 - 5I \\ 6I &= 8R + 11 \end{aligned}$

43. $\begin{aligned} 0.9x - 1.1y &= 0.4 \\ 0.6x - 0.3y &= 0.5 \end{aligned}$

44. $\begin{aligned} 0.42x - 0.56y &= 1.26 \\ 0.98x - 1.40y &= -0.28 \end{aligned}$

In Exercises 45–48, evaluate the given determinants.

45. $\begin{vmatrix} 4 & -1 & 8 \\ -1 & 6 & -2 \\ 2 & 1 & -1 \end{vmatrix}$

46. $\begin{vmatrix} -5 & 0 & -5 \\ 2 & 3 & -1 \\ -3 & 2 & 2 \end{vmatrix}$

47. $\begin{vmatrix} -2.2 & -4.1 & 7.0 \\ 1.2 & 6.4 & -3.5 \\ -7.2 & 2.4 & -1.0 \end{vmatrix}$

48. $\begin{vmatrix} 30 & 22 & -12 \\ 0 & -34 & 44 \\ 35 & -41 & -27 \end{vmatrix}$

In Exercises 49–56, solve the given systems of equations algebraically. In Exercises 55 and 56, the numbers are approximate.

49. $\begin{aligned} 2x + y + z &= 4 \\ x - 2y - z &= 3 \\ 3x + 3y - 2z &= 1 \end{aligned}$

50. $\begin{aligned} x + 2y + z &= 2 \\ 3x - 6y + 2z &= 2 \\ 2x - z &= 8 \end{aligned}$

51. $\begin{aligned} 3x + 2y + z &= 1 \\ 9x - 4y + 2z &= 8 \\ 12x - 18y &= 17 \end{aligned}$

52. $\begin{aligned} 2x + 2y - z &= 2 \\ 3x + 4y + z &= -4 \\ 5x - 2y - 3z &= 5 \end{aligned}$

53. $\begin{aligned} 2r + s + 2t &= 8 \\ 3r - 2s - 4t &= 5 \\ -2r + 3s + 4t &= -3 \end{aligned}$

54. $\begin{aligned} 2u + 2v - w &= -2 \\ 4u - 3v + 2w &= -2 \\ 8u - 4v - 3w &= 13 \end{aligned}$

55. $\begin{aligned} 3.6x + 5.2y - z &= -2.2 \\ 3.2x - 4.8y + 3.9z &= 8.1 \\ 6.4x + 4.1y + 2.3z &= 5.1 \end{aligned}$

56. $\begin{aligned} 32t + 24u + 63v &= 32 \\ 42t - 31u + 19v &= 132 \\ 48t + 12u + 11v &= 0 \end{aligned}$

In Exercises 57–64, solve the given systems of equations by determinants. In Exercises 63 and 64, the numbers are approximate. (These systems are the same as for Exercises 49–56.)

57. $\begin{aligned} 2x + y + z &= 4 \\ x - 2y - z &= 3 \\ 3x + 3y - 2z &= 1 \end{aligned}$

58. $\begin{aligned} x + 2y + z &= 2 \\ 3x - 6y + 2z &= 2 \\ 2x - z &= 8 \end{aligned}$

59. $\begin{aligned} 3x + 2y + z &= 1 \\ 9x - 4y + 2z &= 8 \\ 12x - 18y &= 17 \end{aligned}$

60. $\begin{aligned} 2x + 2y - z &= 2 \\ 3x + 4y + z &= -4 \\ 5x - 2y - 3z &= 5 \end{aligned}$

61. $\begin{aligned} 2r + s + 2t &= 8 \\ 3r - 2s - 4t &= 5 \\ -2r + 3s + 4t &= -3 \end{aligned}$

62. $\begin{aligned} 2u + 2v - w &= -2 \\ 4u - 3v + 2w &= -2 \\ 8u - 4v - 3w &= 13 \end{aligned}$

63. $\begin{aligned} 3.6x + 5.2y - z &= -2.2 \\ 3.2x - 4.8y + 3.9z &= 8.1 \\ 6.4x + 4.1y + 2.3z &= 5.1 \end{aligned}$

64. $\begin{aligned} 32t + 24u + 63v &= 32 \\ 42t - 31u + 19v &= 132 \\ 48t + 12u + 11v &= 0 \end{aligned}$

In Exercises 65–68, let $1/x = u$ and $1/y = v$. Solve for u and v, and then solve for x and y. In this way we see how to solve systems of equations involving reciprocals.

65. $\begin{aligned} \frac{1}{x} - \frac{1}{y} &= \frac{1}{2} \\ \frac{1}{x} + \frac{1}{y} &= \frac{1}{4} \end{aligned}$

66. $\begin{aligned} \frac{1}{x} + \frac{1}{y} &= 3 \\ \frac{2}{x} + \frac{1}{y} &= 1 \end{aligned}$

67. $\begin{aligned} \frac{2}{x} + \frac{3}{y} &= 3 \\ \frac{5}{x} - \frac{6}{y} &= 3 \end{aligned}$

68. $\begin{aligned} \frac{3}{x} - \frac{2}{y} &= 4 \\ \frac{2}{x} + \frac{4}{y} &= 1 \end{aligned}$

In Exercises 69 and 70, determine the value of k that makes the system dependent. In Exercises 71 and 72, determine the value of k that makes the system inconsistent.

69. $\begin{aligned} 3x - ky &= 6 \\ x + 2y &= 2 \end{aligned}$

70. $\begin{aligned} 5x + 20y &= 15 \\ 2x + ky &= 6 \end{aligned}$

71. $\begin{aligned} kx - 2y &= 5 \\ 4x + 6y &= 1 \end{aligned}$

72. $\begin{aligned} 2x - 5y &= 7 \\ kx + 10y &= 2 \end{aligned}$

In Exercises 73 and 74, solve the given systems of equations by any appropriate method. All numbers in 73 are accurate to two significant digits, and in 74 they are accurate to three significant digits.

73. A 20-m crane arm with a supporting cable and with a 9000-N box suspended from its end has forces acting on it as shown in Fig. 5-39. Find the forces (in N) from the following equations.

$$\begin{aligned} F_1 + 2.0F_2 \quad &= 26\,000 \\ 0.87F_1 \qquad - F_3 &= 0 \\ 3.0F_1 - 4.0F_2 \quad &= 54\,000 \end{aligned}$$

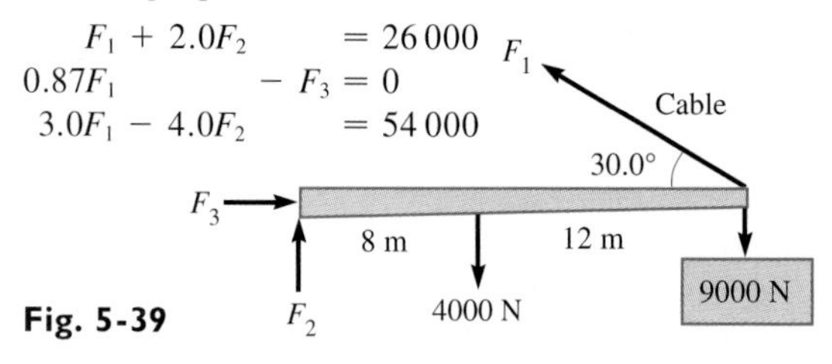

Fig. 5-39

74. In applying Kirchhoff's laws (see Exercise 20 of Section 5-6) to the electric circuit shown in Fig. 5-40, the following equations result. Find the indicated currents (in amperes).

$$\begin{aligned} i_1 + i_2 + i_3 &= 0 \\ 5.20i_1 - 3.25i_2 &= 8.33 - 6.45 \\ 3.25i_2 - 2.62i_3 &= 6.45 - 9.80 \end{aligned}$$

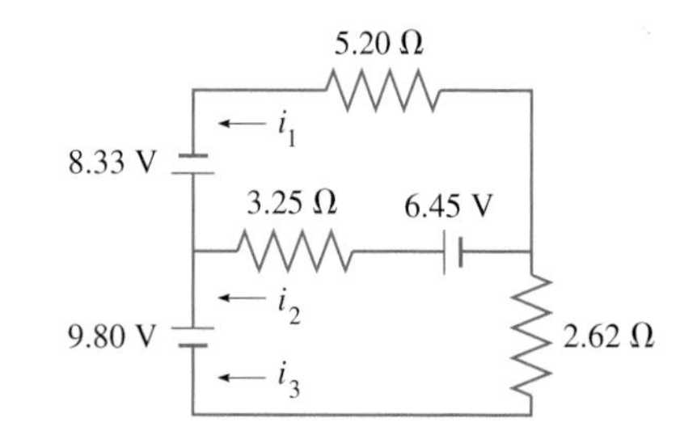

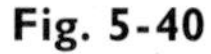
Fig. 5-40

In Exercises 75–84, set up systems of equations and solve by any appropriate method. All numbers are accurate to at least two significant digits.

75. A computer analysis showed that the temperature T of the ocean water within 1000 m of a nuclear-plant discharge pipe was given by $T = \dfrac{a}{x + 100} + b$, where x is the distance from the pipe and a and b are constants. If $T = 14°\text{C}$ for $x = 0$, and $T = 10°\text{C}$ for $x = 900$ m, find a and b.

76. A sales representative receives a fixed amount plus a percentage of sales commissions each month. If the representative received \$3260 on sales of \$22 000 in one month and \$4380 on sales of \$36 000 in the following month, what is the fixed amount and the commission percent?

77. A satellite is to be launched from a space shuttle. It is calculated that the satellite's speed will be 24 200 km/h if launched directly ahead of the shuttle, or 21 400 km/h if launched directly to the rear of the shuttle. What is the speed of the shuttle and the launching speed of the satellite relative to the shuttle?

78. The velocity v of sound is a function of the temperature T according to the function $v = aT + b$, where a and b are constants. If $v = 337.5$ m/s for $T = 10.0°\text{C}$, and $v = 346.6$ m/s for $T = 25.0°\text{C}$, find v as a function of T.

79. The power (in watts) dissipated in an electric resistance (in ohms) equals the resistance times the square of the current (in amperes). If 1.0 A flows through resistance R_1 and 3.0 A flows through resistance R_2, the total power dissipated is 14.0 W. If 3.0 A flows through R_1 and 1.0 A flows through R_2, the total power dissipated is 6.0 W. Find R_1 and R_2.

80. Twelve equal rectangular ceiling panels are used as shown in Fig. 5-41. If each panel is 150 mm longer than it is wide, and a total of 40.0 m of edge and middle strips is used, what are the dimensions of the room?

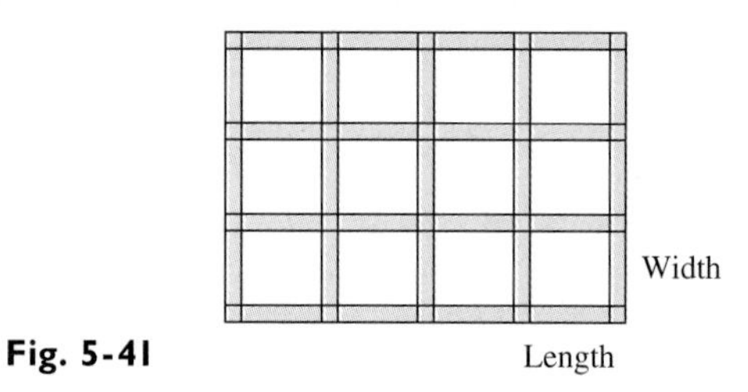

Fig. 5-41

81. The weight of a lever may be considered to be at its center. A 10-m lever of weight w is balanced on a fulcrum 4 m from one end by a load L at that end. If 4 times the load is placed at that end, it requires a 20-N weight at the other end to balance the lever. What is the initial load L and the weight w of the lever? (See Exercise 38 of Section 5-4.)

82. Two fuel mixtures, one of 2.0% oil and 98% gasoline and another of 8.0% oil and 92% gasoline, are to be mixed to make 10.0 L of a fuel that is 4.0% oil and 96% gasoline that is to be used in a chain saw. How much of each mixture is needed?

83. A manufacturer of televisions, videocassette recorders, and compact disc players makes 1750 total units each month. The number of CDs is twice that of TVs, and four times the number of CDs is 700 more than three times the number of TVs and VCRs combined. How many of each are produced?

84. One ampere of electric current is passed through a solution of sulfuric acid, silver nitrate, and cupric sulfate, releasing hydrogen gas, silver, and copper. A total mass of 1.750 g is released. The mass of silver deposited is 3.40 times the mass of copper deposited, and the mass of copper and 70 times the mass of hydrogen combined equals the mass of silver deposited less 0.037 g. How much of each is released?

Writing Exercise

85. Write one or two paragraphs giving reasons for choosing a particular method of solving the following problem. If a first pump is used for 2.2 h and a second pump is used for 2.7 h, 32 m^3 can be removed from a wastewater holding tank. If the first pump is used for 1.4 h and the second for 2.5 h, 24 m^3 can be removed. How much can each pump remove in 1.0 h? (What is the result to two significant digits?)

Practice Test

1. Find the slope of the line through $(2, -5)$ and $(-1, 4)$.
2. Solve by substitution:

$$\begin{aligned} x + 2y &= 5 \\ 4y &= 3 - 2x \end{aligned}$$

3. Solve by determinants:

$$\begin{aligned} 3x - 2y &= 4 \\ 2x + 5y &= -1 \end{aligned}$$

4. Sketch the graph of $2x + y = 4$ by finding its slope and y-intercept.
5. The perimeter of a rectangular ranch is 24 km, and the length is 6.0 km more than the width. Set up equations relating the length l and the width w, and then solve for l and w.
6. Solve the following system of equations graphically. Determine the values to the nearest 0.1.

$$\begin{aligned} 2x - 3y &= 6 \\ 4x + y &= 4 \end{aligned}$$

7. By volume, one alloy is 60% copper, 30% zinc, and 10% nickel. A second alloy has percents of 50, 30, and 20, respectively, of the three metals. A third alloy is 30% copper and 70% nickel. How much of each metal is needed to make 100 cm^3 of a resulting alloy with percents of 40, 15, and 45, respectively?
8. Solve for y by determinants:

$$\begin{aligned} 3x + 2y - z &= 4 \\ 2x - y + 3z &= -2 \\ x \qquad + 4z &= 5 \end{aligned}$$

6

Factoring and Fractions

Time for processing data by a computer system may be reduced by adding components to the system. In Section 6-8 we show how this time may be calculated.

The basic algebraic operations we introduced in Chapter 1 have been sufficient for our purposes to this point. However, material we shall encounter throughout the rest of the book will require algebraic methods beyond those we have developed to now. Therefore, in this chapter we develop additional algebraic operations with products, quotients, and fractions. In later chapters these in turn will allow us to develop other topics having technical and scientific applications.

Although the primary purpose in this chapter is to develop additional algebraic methods, we show many of the applied areas in which these methods are used. Also, certain direct applications are shown in many technical areas.

6-1 Special Products

In working with algebraic expressions, we use certain types of products so often that we should be very familiar with them. We show these here in general form.

$$a(x + y) = ax + ay \qquad (6\text{-}1)$$

$$(x + y)(x - y) = x^2 - y^2 \qquad (6\text{-}2)$$

$$(x + y)^2 = x^2 + 2xy + y^2 \qquad (6\text{-}3)$$

$$(x - y)^2 = x^2 - 2xy + y^2 \qquad (6\text{-}4)$$

$$(x + a)(x + b) = x^2 + (a + b)x + ab \qquad (6\text{-}5)$$

$$(ax + b)(cx + d) = acx^2 + (ad + bc)x + bd \qquad (6\text{-}6)$$

We recognize Eq. (6-1) as the very important *distributive law.* Equations (6-2) to (6-6) are found by using the distributive law along with the associative and commutative laws and the operations on positive and negative numbers that were developed in Chapter 1.

Equations (6-1) to (6-6) are important **special products** *and should be known* ***thoroughly.*** When these products are quickly recognized, they allow us to perform many multiplications easily, often by inspection. In using them we must realize that any of the literal numbers may represent an expression that in turn represents a number.

EXAMPLE 1 ▸▸ (a) Using Eq. (6-1) in the following product we have

$$6(3r + 2s) = 6(3r) + 6(2s) = 18r + 12s$$

(b) Using Eq. (6-2), we have

Difference of Squares

$$(3r + 2s)(3r - 2s) = (3r)^2 - (2s)^2 = 9r^2 - 4s^2$$

sum of 3r and 2s; difference of 3r and 2s; difference of squares

Using Eq. (6-1) in (a), we have $a = 6$. In (a) and (b), $3r = x$ and $2s = y$. ◂◂

EXAMPLE 2 ▸▸ Using Eqs. (6-3) and (6-4) in the following products, we have

(a) $(5a + 2)^2 = (5a)^2 + 2(5a)(2) + 2^2 = 25a^2 + 20a + 4$

square; twice product; square

(b) $(5a - 2)^2 = (5a^2) - 2(5a)(2) + 2^2 = 25a^2 - 20a + 4$

In these illustrations we let $x = 5a$ and $y = 2$. *It should be emphasized that*

CAUTION ▸ $(5a + 2)^2$ is **not** $(5a)^2 + 2^2$, or $25a^2 + 4$

We must be careful to follow the form of Eqs. (6-3) and (6-4) properly and include the middle term, $20a$. (See Example 5 of Section 1-8 on p. 31.) ◂◂

EXAMPLE 3 ▸▸ Using Eqs. (6-5) and (6-6) in the following products, we have

(a) $(x + 5)(x - 3) = x^2 + [5 + (-3)]x + (5)(-3) = x^2 + 2x - 15$

(b) $(4x + 5)(2x - 3) = (4x)(2x) + [(4)(-3) + (5)(2)]x + (5)(-3)$
$= 8x^2 - 2x - 15$ ◂◂

Generally, when we use these special products, we find the middle term mentally and write down the result directly, as shown in the next example.

EXAMPLE 4 ▸▸

(a) $(y - 5)(y + 5) = y^2 - 25$ — no middle term—Eq. (6-2)

(b) $(3x - 2)^2 = 9x^2 - 12x + 4$ — middle term $= 2(3x)(-2)$—Eq. (6-4)

(c) $(x - 4)(x + 7) = x^2 + 3x - 28$ — middle term $= (7 - 4)x$—Eq. (6-5) ◂◂

At times these special products are used in combinations. When this happens, it may be necessary to show an intermediate step.

EXAMPLE 5 ▸▸ (a) When analyzing the forces on a certain type of beam, the expression $Fa(L - a)(L + a)$ occurs. In expanding this expression, we first multiply $L - a$ by $L + a$ by use of Eq. (6-2). The expansion is completed by using Eq. (6-1), the distributive law.

$$Fa(L - a)(L + a) = Fa(L^2 - a^2) = FaL^2 - Fa^3$$

(b) In electricity, the expression $R(i_1 + i_2)^2$ is used. In expanding this expression, we first perform the square by use of Eq. (6-3) and then complete the expansion by use of Eq. (6-1).

$$R(i_1 + i_2)^2 = R(i_1^2 + 2i_1i_2 + i_2^2) = Ri_1^2 + 2Ri_1i_2 + Ri_2^2$$ ◂◂

EXAMPLE 6 ▸▸ In determining the product $(x + y - 2)^2$, we may group the quantity $(x + y)$ in an intermediate step. This leads to

$$(x + y - 2)^2 = [(x + y) - 2]^2 = (x + y)^2 - 2(x + y)(2) + 2^2$$
$$= x^2 + 2xy + y^2 - 4x - 4y + 4$$

In this example we used Eqs. (6-3) and (6-4). ◂◂

Special Products Involving Cubes

There are four other special products that occur less frequently. However, they are sufficiently important that they should be readily recognized. They are shown in Eqs. (6-7) to (6-10).

$$(x + y)^3 = x^3 + 3x^2y + 3xy^2 + y^3 \quad \textbf{(6-7)}$$
$$(x - y)^3 = x^3 - 3x^2y + 3xy^2 - y^3 \quad \textbf{(6-8)}$$
$$(x + y)(x^2 - xy + y^2) = x^3 + y^3 \quad \textbf{(6-9)}$$
$$(x - y)(x^2 + xy + y^2) = x^3 - y^3 \quad \textbf{(6-10)}$$

The following examples illustrate the use of Eqs. (6-7) to (6-10).

EXAMPLE 7 ▸▸ (a) $(x + 4)^3 = x^3 + 3(x^2)(4) + 3(x)(4^2) + 4^3$ Eq. (6-7)

$$= x^3 + 12x^2 + 48x + 64$$

(b) $(2x - 5)^3 = (2x)^3 - 3(2x)^2(5) + 3(2x)(5^2) - 5^3$ Eq. (6-8)

$$= 8x^3 - 60x^2 + 150x - 125$$ ◂◂

EXAMPLE 8 (a) $(x + 3)(x^2 - 3x + 9) = x^3 + 3^3$ Eq. (6-9)

$= x^3 + 27$

(b) $(x - 2)(x^2 + 2x + 4) = x^3 - 2^3$ Eq. (6-10)

$= x^3 - 8$

Exercises 6–1

In Exercises 1–36, find the indicated products directly by inspection. It should not be necessary to write down intermediate steps (except possibly when using Eq. (6-6)).

1. $40(x - y)$ **2.** $2x(a - 3)$

3. $2x^2(x - 4)$ **4.** $3a^2(2a + 7)$

5. $(y + 6)(y - 6)$ **6.** $(s + 2t)(s - 2t)$

7. $(3v - 2)(3v + 2)$ **8.** $(ab - c)(ab + c)$

9. $(4x - 5y)(4x + 5y)$ **10.** $(7s + 2t)(7s - 2t)$

11. $(12 + 5ab)(12 - 5ab)$ **12.** $(2xy - 11)(2xy + 11)$

13. $(5f + 4)^2$ **14.** $(i_1 + 3)^2$

15. $(2x + 7)^2$ **16.** $(5a + 2b)^2$

17. $(x - 1)^2$ **18.** $(y - 6)^2$

19. $(4a + 7xy)^2$ **20.** $(3x + 10y)^2$

21. $(4x - 2y)^2$ **22.** $(a - 5p)^2$

23. $(6s - t)^2$ **24.** $(3p - 4q)^2$

25. $(x + 1)(x + 5)$ **26.** $(y - 8)(y + 5)$

27. $(3 + c)(6 + c)$ **28.** $(1 - t)(7 - t)$

29. $(3x - 1)(2x + 5)$ **30.** $(2x - 7)(2x + 1)$

31. $(4x - 5)(5x + 1)$ **32.** $(2y - 1)(3y - 1)$

33. $(5v - 3)(4v + 5)$ **34.** $(7s + 6)(2s + 5)$

35. $(3x + 7y)(2x - 9y)$ **36.** $(8x - y)(3x + 4y)$

Use the special products of this section to determine the products of Exercises 37–60. You may need to write down one or two intermediate steps.

37. $2(x - 2)(x + 2)$ **38.** $5(n - 5)(n + 5)$

39. $2a(2a - 1)(2a + 1)$ **40.** $4c(2c - 3)(2c + 3)$

41. $6a(x + 2b)^2$ **42.** $7r(5r + 2b)^2$

43. $5n^2(2n + 5)^2$ **44.** $8p(p - 7)^2$

45. $4a(2a - 3)^2$ **46.** $6t^2(5t - 3s)^2$

47. $(x + y + 1)^2$ **48.** $(x + 2 + 3y)^2$

49. $(3 - x - y)^2$ **50.** $2(x - y + 1)^2$

51. $(5 - t)^3$ **52.** $(2s + 3)^3$

53. $(2x + 5t)^3$ **54.** $(x - 5y)^3$

55. $(x + y - 1)(x + y + 1)$

56. $(2a - c + 2)(2a - c - 2)$

57. $(x + 2)(x^2 - 2x + 4)$

58. $(a - 3)(a^2 + 3a + 9)$

59. $(4 - 3x)(16 + 12x + 9x^2)$

60. $(2x + 3a)(4x^2 - 6ax + 9a^2)$

Use the special products of this section to determine the products in Exercises 61–68. Each comes from the technical area indicated.

61. $P_1(P_0c + G)$ (computers)

62. $h^2L(L + 1)$ (spectroscopy)

63. $4(p + DA)^2$ (photography)

64. $(2J + 3)(2J - 1)$ (lasers)

65. $\frac{1}{2}\pi(R + r)(R - r)$ (architecture)

66. $w(1 - h)(4 - h^2)$ (hydrodynamics)

67. $\frac{L}{6}(x - a)^3$ (mechanics: beams)

68. $(1 - z)^2(1 + z)$ (motion: gyroscope)

In Exercises 69–72, solve the given problems.

69. The length of a piece of rectangular floor tile is 3 cm more than twice the side x of a second square piece of tile. The width of the rectangular piece is 3 cm less than twice the side of the square piece. Find the area of the rectangular piece in terms of x (in expanded form).

70. The radius of a circular oil spill is r. It then increases in radius by 40 m before being contained. Find the area of the oil spill at the time it is contained in terms of r (in expanded form).

71. Verify Eq. (6-8) by multiplication. Then show by substitution that the same value is obtained for each side for the values $x = -2$ and $y = 3$.

72. Verify Eqs. (6-9) and (6-10) by multiplication. Then explain why we may refer to Eqs. (6-1) to (6-10) as *identities*.

6-2 Factoring: Common Factor and Difference of Squares

At times we want to determine which expressions can be multiplied together to equal a given algebraic expression. We know from Section 1-7 that when an algebraic expression is the product of two or more quantities, each of these quantities is a ***factor*** of the expression. Therefore, *determining these factors, which is essentially reversing the process of finding a product, is called* **factoring.**

In our work on factoring we shall consider only the factoring of polynomials (see Section 1-9) that have integers as coefficients for all terms. Also, all factors will have integral coefficients. *A polynomial or a factor is called* **prime** *if it contains no factors other than +1 or −1 and plus or minus itself.* Also, we say that *an expression is* **factored completely** *if it is expressed as a product of its prime factors.*

EXAMPLE 1 ▸▸ When we factor the expression $12x + 6x^2$ as

$$12x + 6x^2 = 2(6x + 3x^2)$$

we see that it has not been factored completely. The factor $6x + 3x^2$ is not prime, for it may be factored as

$$6x + 3x^2 = 3x(2 + x)$$

Therefore, the expression $12x + 6x^2$ is factored completely as

$$12x + 6x^2 = 6x(2 + x)$$

The factors x and $2 + x$ are prime. We can factor the numerical coefficient, 6, as 2(3), but it is standard not to write numerical coefficients in factored form. ◂◂

NOTE ▸ To factor expressions easily, we must be familiar with algebraic multiplication, particularly the special products of the preceding section. *The solution of factoring problems is heavily dependent on the recognition of special products.* The special products also provide methods of checking answers and deciding whether or not a given factor is prime.

Common Monomial Factors

For reference, Eq. (6-1) is $a(x + y) = ax + ay$.

Often an expression contains a monomial that is common to each term of the expression. Therefore, *the first step in factoring any expression should be to factor out any* **common monomial factor** *that may exist.* To do this, we note the common factor by inspection and then use the reverse of the distributive law, Eq. (6-1), to show the factored form. The following examples illustrate factoring a common monomial factor out of an expression.

EXAMPLE 2 ▸▸ In factoring $6x - 2y$, we note each term contains a factor of 2.

$$6x - 2y = 2(3x) - 2y = 2(3x - y)$$

Here, 2 is the common monomial factor, and $2(3x - y)$ is the required factored form of $6x - 2y$. Once the common factor has been identified, it is not actually necessary to write a term like $6x$ as $2(3x)$. The result can be written directly.

We check our result by multiplication. Here, $2(3x - y) = 6x - 2y$, which is the original expression. ◂◂

In Example 2 we determined the common factor of 2 by inspection. This is normally the way in which a common factor is found. Once the common factor has been found, the other factor can be determined by dividing the original expression by the common factor.

EXAMPLE 3 ▸▸ Factor: $4ax^2 + 2ax$.

The numerical factor 2 and the literal factors a and x are common to each term. Therefore, the common monomial factor of $4ax^2 + 2ax$ is $2ax$. This means that

$$4ax^2 + 2ax = 2ax(2x + 1)$$

Note the presence of the 1 in the factored form. When we divide $4ax^2 + 2ax$ by $2ax$ we get

$$\frac{4ax^2 + 2ax}{2ax} = \frac{4ax^2}{2ax} + \frac{2ax}{2ax} = 2x + 1$$

Although it is a common error to omit the 1,

CAUTION ▸ ***we must include the 1 in the factor 2x + 1.***

Without the 1, when the factored form is multiplied out, we would not obtain the proper expression.

Usually the division shown in this example is done by inspection. However, we show it here to emphasize the actual operation that is being performed when we factor out a common factor. ◂◂

EXAMPLE 4 ▸▸ Factor: $6a^5x^2 - 9a^3x^3 + 3a^3x^2$.

After inspecting each term, we determine that each contains a factor of 3, a^3, and x^2. Thus, the common monomial factor is $3a^3x^2$. This means that

$$6a^5x^2 - 9a^3x^3 + 3a^3x^2 = 3a^3x^2(2a^2 - 3x + 1)$$ ◂◂

In these examples, we note that factoring an expression does not actually change the expression, although it does change the *form* of the expression. In equating the expression to its factored form, we write an *identity*.

It is often necessary to use factoring when solving an equation. This is illustrated in the following example.

EXAMPLE 5 ▸▸ An equation used in the analysis of FM reception is $R_F = \alpha(2R_A + R_F)$. Solve for R_F.

The steps in the solution are as follows:

$$R_F = \alpha(2R_A + R_F) \quad \text{original equation}$$

$$R_F = 2\alpha R_A + \alpha R_F \quad \text{use distributive law}$$

$$R_F - \alpha R_F = 2\alpha R_A \quad \text{subtract } \alpha R_F \text{ from both sides}$$

$$R_F(1 - \alpha) = 2\alpha R_A \quad \text{factor out } R_F \text{ on left}$$

$$R_F = \frac{2\alpha R_A}{1 - \alpha} \quad \text{divide both sides by } 1 - \alpha$$

We see that we collected both terms containing R_F on the left in order that we could factor and thereby solve for R_F. ◂◂

Factoring the Difference of Two Squares

For reference, Eq. (6-2) is $(x + y)(x - y) = x^2 - y^2$.

Another important form for factoring is based on the special product of Eq. (6-2). In Eq. (6-2) we see that the product of the sum and difference of two numbers results in the difference between the squares of the two numbers. Therefore, *factoring the difference of two squares gives factors that are the sum and the difference of the numbers.*

EXAMPLE 6 ▸▸ In factoring $x^2 - 16$, we note that x^2 is the square of x and that 16 is the square of 4. Therefore,

squares

$$x^2 - 16 = x^2 - 4^2 = (x + 4)(x - 4)$$

difference   sum   difference

Usually in factoring an expression of this type we do not actually write out the middle step as shown. ◂◂

EXAMPLE 7 ▸▸ (a) Since $4x^2$ is the square of $2x$ and 9 is the square of 3, we may factor $4x^2 - 9$ as

$$4x^2 - 9 = (2x + 3)(2x - 3)$$

(b) In the same way,

$$(y - 3)^2 - 16x^4 = (y - 3 + 4x^2)(y - 3 - 4x^2)$$

where we note that $16x^4 = (4x^2)^2$. ◂◂

Complete Factoring

NOTE ▸ As indicated previously, *if it is possible to factor out a common monomial factor, **this factoring should be done first.*** We should then inspect the resulting factors to see if more factoring can be done. It is possible, for example, that the resulting factor is a difference of squares. Thus, complete factoring often requires more than one step. Be sure to include all prime factors in writing the result.

EXAMPLE 8 ▸▸ (a) In factoring $20x^2 - 45$, we note a common factor of 5 in each term. Therefore, $20x^2 - 45 = 5(4x^2 - 9)$. However, the factor $4x^2 - 9$ itself is the difference of squares. Therefore, $20x^2 - 45$ is completely factored as

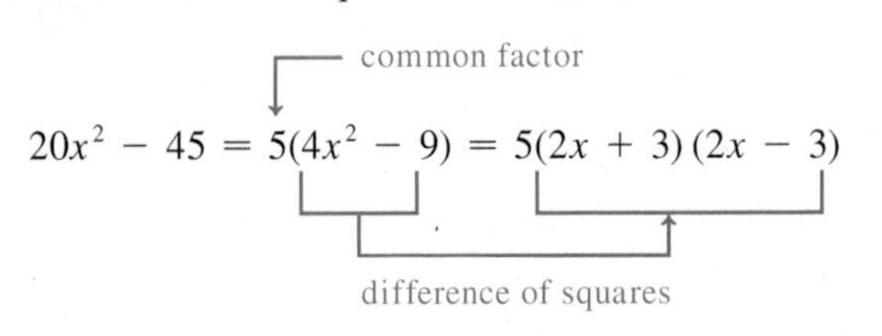

(b) In factoring $x^4 - y^4$, we note that we have the difference of two squares. Therefore, $x^4 - y^4 = (x^2 + y^2)(x^2 - y^2)$. However, the factor $x^2 - y^2$ is also the difference of squares. This means that

$$x^4 - y^4 = (x^2 + y^2)(x^2 - y^2) = (x^2 + y^2)(x + y)(x - y)$$

CAUTION ▸ ***The factor $x^2 + y^2$ is prime.*** It is **not** equal to $(x + y)^2$. (See Example 2 of Section 6-1.). ◂◂

Factoring by Grouping

The terms in a polynomial can sometimes be grouped such that the polynomial can then be factored by the methods of this section. The following example illustrates this method of *factoring by grouping.*

EXAMPLE 9 ▸▸ Factor: $2x - 2y + ax - ay$.

We note that each of the first two terms contains a factor of 2, and each of the third and fourth terms contains a factor of a. Grouping the terms in this way, and then factoring each group, we have

$$2x - 2y + ax - ay = (2x - 2y) + (ax - ay)$$
$$= 2(x - y) + a(x - y)$$

NOTE ▸ Each of the two terms we now have contains a ***common binomial factor*** of $x - y$. Since this is a common factor, we have

$$2(x - y) + a(x - y) = (x - y)(2 + a)$$

This means that

$$2x - 2y + ax - ay = (x - y)(2 + a)$$

where the expression on the right is the factored form of the polynomial. ◂◂

The general method of factoring by grouping can be used with several types of groupings. We will discuss another type in the following section.

Exercises 6–2

In Exercises 1–40, factor the given expressions completely.

1. $6x + 6y$ **2.** $3a - 3b$ **3.** $5a - 5$

4. $2x^2 + 2$ **5.** $3x^2 - 9x$ **6.** $4s^2 + 20s$

7. $7b^2y - 28b$ **8.** $5a^2 - 20ax$

9. $12n^2 + 6n$ **10.** $18p^3 - 3p^2$

11. $2x + 4y - 8z$ **12.** $10a - 5b + 15c$

13. $3ab^2 - 6ab + 12ab^3$ **14.** $4pq - 14q^2 - 16pq^2$

15. $12pq^2 - 8pq - 28pq^3$ **16.** $27a^2b - 24ab - 9a$

17. $2a^2 - 2b^2 + 4c^2 - 6d^2$

18. $5a + 10ax - 5ay + 20az$

19. $x^2 - 4$ **20.** $r^2 - 25$ **21.** $100 - y^2$

22. $49 - z^2$ **23.** $36a^2 - 1$ **24.** $81z^2 - 1$

25. $81s^2 - 25t^2$ **26.** $36s^2 - 121t^2$

27. $144n^2 - 169p^4$ **28.** $36a^2b^2 - 169c^2$

29. $(x + y)^2 - 9$ **30.** $(a - b)^2 - 1$

31. $2x^2 - 8$ **32.** $5a^2 - 125$ **33.** $3x^2 - 27z^2$

34. $4x^2 - 100y^2$ **35.** $2(a - 3)^2 - 8$

36. $a(x + 2)^2 - ay^2$ **37.** $x^4 - 16$

38. $y^4 - 81$ **39.** $x^8 - 1$ **40.** $2x^4 - 8y^4$

In Exercises 41–44, solve for the indicated letter.

41. $2a - b = ab + 3$, for a **42.** $n(x + 1) = 5 - x$, for x

43. $3 - 2s = 2(3 - st)$, for s

44. $k(2 - y) = y(2k - 1)$, for y

In Exercises 45–52, factor the given expressions by grouping as illustrated in Example 9.

45. $3x - 3y + bx - by$ **46.** $am + an + cn + cm$

47. $a^2 + ax - ab - bx$ **48.** $2y - y^2 - 6y^4 + 12y^3$

49. $x^3 + 3x^2 - 4x - 12$ **50.** $x^3 - 5x^2 - x + 5$

51. $x^2 - y^2 + x - y$ **52.** $4p^2 - q^2 + 2p + q$

In Exercises 53–58, factor the given expressions. In Exercises 59 and 60, solve for the indicated letter. Each comes from the technical area indicated.

53. $Rv + Rv^2 + Rv^3$ (business)

54. $4d^2D^2 - 4d^3D - d^4$ (machine design)

55. $aD_1^2 - aD_2^2$ (surveying)

56. $rR^2 - r^3$ (pipeline flow)

57. $PbL^2 - Pb^3$ (architecture)

58. $4RI^2 - 9Ri^2$ (electricity)

59. $ER = AtT_0 - AtT_1$ (Solve for t.) (energy conservation)

60. $R = kT_2^4 - kT_1^4$ (Solve for k and factor the resulting denominator.) (energy: radiation)

6-3 Factoring Trinomials

In the previous section we introduced the concept of factoring and considered factoring based on special products of Eqs. (6-1) and (6-2). We now note that the special products formed from Eqs. (6-3) to (6-6) all result in trinomial (three-term) polynomials. Thus, trinomials of the types formed by these products are important expressions to be factored, and this section is devoted to them.

For reference, Eq. (6-5) is $(x + a)(x + b) = x^2 + (a + b)x + ab$.

When factoring an expression based on Eq. (6-5), we start with the expression on the right and then find the factors that are at the left. Therefore, by writing Eq. (6-5) with sides reversed, we have

$$x^2 + (a + b)x + ab = (x + a)(x + b)$$

coefficient = 1 (of x^2); sum (of a and b gives coefficient of x); product (of a and b gives ab)

We are to find integers a and b, and they are found by noting that

1. *the coefficient of x^2 is 1,*
2. *the product of a and b is the final constant ab, and*
3. *the sum of a and b is the coefficient of x.*

As in Section 6-2 we shall consider only factors in which all terms have integral coefficients.

EXAMPLE 1 ▸▸ In factoring $x^2 + 3x + 2$, we set it up as

$$x^2 + 3x + 2 = (x\square)(x\square)$$

sum (3), product (2), integers (in the boxes)

The constant 2 tells us that the product of the required integers is 2. Thus, the only possibilities are 2 and 1 (or 1 and 2). The plus sign before the 2 indicates that the sign before the 1 and 2 in the factors must be the same, either plus or minus. Since the coefficient of x, 3, is the sum of the integers, the plus sign before the 3 tells us that both signs are positive. Therefore,

$$x^2 + 3x + 2 = (x + 2)(x + 1)$$

In factoring $x^2 - 3x + 2$, the analysis is the same until we note that the middle term is negative. This tells us that both integers are negative in this case. Therefore,

$$x^2 - 3x + 2 = (x - 2)(x - 1)$$

For a trinomial containing x^2 and 2 to be factorable, the middle term must be $+3x$ or $-3x$. No other combination of integers gives the proper middle term. Therefore, the expression

$$x^2 + 4x + 2$$

cannot be factored. The integers would have to be 2 and 1, but the middle term would not be $4x$. ◂◂

EXAMPLE 2 ▸▸ (a) In order to factor $x^2 + 7x - 8$, *we must find two integers whose product is* -8 *and whose sum is* $+7$. The possible factors of -8 are

NOTE ▶

$$-8 \text{ and } +1,\quad 8 \text{ and } -1,\quad -4 \text{ and } +2,\quad +4 \text{ and } -2$$

Inspecting these, we see that only $+8$ and -1 have the sum of $+7$. Therefore,

$$x^2 + 7x - 8 = (x + 8)(x - 1)$$

(b) In the same way, we have

$$x^2 - x - 12 = (x - 4)(x + 3)$$

since -4 and $+3$ is the only pair of integers whose product is -12 and sum is -1.

(c) Also,

$$x^2 - 5xy + 6y^2 = (x - 3y)(x - 2y)$$

since -3 and -2 is the only pair of integers whose product is $+6$ and whose sum is -5. Here we find second terms of each factor with a product of $6y^2$ and sum of $-5xy$, which means that each second term must have a factor of y. ◂◂

For reference, Eqs. (6-3) and (6-4) are $(x + y)^2 = x^2 + 2xy + y^2$ $(x - y)^2 = x^2 - 2xy + y^2$.

In factoring a trinomial in which the second power term is x^2, we may find that the expression fits the form of Eq. (6-3) or Eq. (6-4), as well as Eq. (6-5). The following example illustrates this case.

EXAMPLE 3 ▸▸ To factor $x^2 + 10x + 25$, we must find two integers whose product is $+25$ and whose sum is $+10$. Since $5^2 = 25$ we note that this expression may fit the form of Eq. (6-3). This can be the case only if the first and third terms are perfect squares. Since the sum of $+5$ and $+5$ is $+10$, we have

$$x^2 + 10x + 25 = (x + 5)(x + 5)$$

or

$$x^2 + 10x + 25 = (x + 5)^2$$ ◂◂

Factoring General Trinomials

For reference, Eq. (6-6) is $(ax + b)(cx + d) = acx^2 + (ad + bc)x + bd$.

Factoring expressions based on the special product of Eq. (6-6) often requires some trial and error. However, the amount of trial and error can be kept to a minimum with a careful analysis of the coefficients of x^2 and the constant. Rewriting Eq. (6-6) with sides reversed, we have

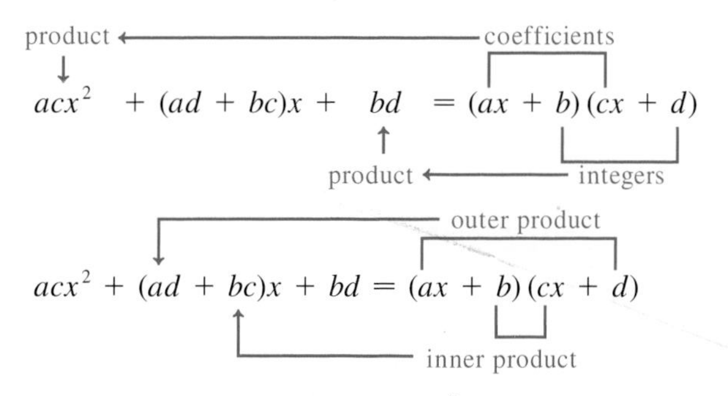

This shows that the coefficient of x^2 gives the possibilities for the coefficients a and c in the factors, and that the constant bd gives the possibilities for the integers b and d in the factors. We must try possible combinations to find which one gives the proper outer and inner products for the middle term of the given expression.

EXAMPLE 4 ▸▸ When factoring $2x^2 + 11x + 5$, we take the factors of 2 to be $+2$ and $+1$ (we will use only positive coefficients a and c when the coefficient of x^2 is positive). We now set up the factoring as

$$2x^2 + 11x + 5 = (2x\square)(x\square)$$

NOTE ▸ Since the product of the integers to be found is $+5$, only integers of the same sign need be considered. Also since ***the sum of the outer and inner products is* +11,** the integers are positive. The factors of $+5$ are $+1$ and $+5$, and -1 and -5, which means that $+1$ and $+5$ is the only possible pair. Now, trying the factors

$$(2x + 5)(x + 1)$$

inner: $+5x$, outer: $+2x$; $+2x + 5x = +7x$

we see that $7x$ is not the correct middle term.

Next, trying

$$(2x + 1)(x + 5)$$

inner: $+x$, outer: $+10x$; $+x + 10x = +11x$

we have the correct sum of $+11x$. Therefore,

$$2x^2 + 11x + 5 = (2x + 1)(x + 5)$$

According to this analysis, the expression $2x^2 + 10x + 5$ is not factorable, but the following expression is:

$$2x^2 + 7x + 5 = (2x + 5)(x + 1)$$ ◂◂

EXAMPLE 5 ▸▸ In factoring $4x^2 + 4x - 3$, the coefficient 4 in $4x^2$ shows that 4 and 1, or 2 and 2, are possible coefficients of x in the factors. The 3 shows that

NOTE ▸ only 3 and 1 are possible integers, and *the minus sign with the* 3 *shows that the integers have* ***different*** *signs. The plus sign with the* $4x$ *tells us that the larger of the inner and outer products is positive.* This gives us possible combinations of

$$(4x - 3)(x + 1), \quad (4x - 1)(x + 3), \quad \text{or} \quad (2x + 3)(2x - 1)$$

The only one of these that has a middle term of $+4x$ is $(2x + 3)(2x - 1)$. Therefore,

$$4x^2 + 4x - 3 = (2x + 3)(2x - 1)$$

inner: $+6x$, outer: $-2x$

$-2x + 6x = +4x$ ◂◂

EXAMPLE 6 ▸▸ $6s^2 + 19st - 20t^2 = (6s - 5t)(s + 4t)$

$+24st - 5st = +19st$

CAUTION ▸ There are numerous possibilities for the combinations of 6 and 20. However, we must remember to ***check carefully that the*** **middle term** *of the expression is the* ***proper result of the factors*** *we have chosen.* ◂◂

EXAMPLE 7 ▸▸ (a) In factoring $9x^2 - 6x + 1$, we note that $9x^2$ is the square of $3x$ and 1 is the square of 1. Therefore, we recognize that this expression might fit the perfect square form of Eq. (6-4). This leads us to factor it tentatively as

$$9x^2 - 6x + 1 = (3x - 1)^2$$

However, before we can be certain that this is correct, we must check to see if the middle term of the expansion of $(3x - 1)^2$ is $-6x$, which is what it must be to fit the form of Eq. (6-4). When we expand $(3x - 1)^2$, we find that the middle term is $-6x$ and therefore that the factorization is correct.

(b) In the same way, we have

$$36x^2 + 84xy + 49y^2 = (6x + 7y)^2$$ ◂◂

NOTE ▸ As pointed out in Section 6-2, we must be careful to see that we have factored an expression completely. ***We look for common monomial factors first,*** and then check each resulting factor. This check of each factor should be made each time we complete a step in factoring.

EXAMPLE 8 ▸▸ When factoring $2x^2 + 6x - 8$, we first note the common monomial factor of 2. This leads to

$$2x^2 + 6x - 8 = 2(x^2 + 3x - 4)$$

We now notice that $x^2 + 3x - 4$ is also factorable. Therefore,

$$2x^2 + 6x - 8 = 2(x + 4)(x - 1)$$

Now each factor is prime.

Having noted the common factor of 2 prevents our having to check factors of 2 and 8. If we had not noted the common factor, we might have arrived at factorizations of

$$(2x + 8)(x - 1) \quad \text{or} \quad (2x - 2)(x + 4)$$

NOTE ▸ (possibly after a number of trials). Each is correct as far as it goes, but also each of these is ***not complete.*** Since $2x + 8 = 2(x + 4)$, or $2x - 2 = 2(x - 1)$, we can arrive at the proper result shown above. In this way we would have

$$\begin{aligned} 2x^2 + 6x - 8 &= (2x + 8)(x - 1) \\ &= 2(x + 4)(x - 1) \end{aligned} \quad \text{or} \quad \begin{aligned} 2x^2 + 6x - 8 &= (2x - 2)(x + 4) \\ &= 2(x - 1)(x + 4) \end{aligned}$$

Although these factorizations are correct, *it is better to factor out the common factor first.* By doing so, the number of possible factoring combinations is greatly reduced, and the factoring can be done more easily. ◂◂

EXAMPLE 9 ▸▸ A study of the path of a certain rocket leads to the expression $16t^2 + 240t - 1600$, where t is the time of flight. Factor this expression.

An inspection shows that there is a common factor of 16. (This might be found by noting successive factors of 2 or 4.) Factoring out 16 leads to

$$\begin{aligned} 16t^2 + 240t - 1600 &= 16(t^2 + 15t - 100) \\ &= 16(t + 20)(t - 5) \end{aligned}$$

Here, factors of 100 need to be checked for sums equal to 15. This might take a little time, but it is much simpler than looking for factors of 16 and 1600 with sums equal to 240. ◂◂

Factoring by Grouping

In the previous section, we introduced the method of factoring by grouping. The following examples use this method for factoring (1) a trinomial and (2) an expression that can be written as the difference of squares.

EXAMPLE 10 ▸▸ Factor the trinomial $6x^2 + 7x - 20$ by grouping.

To factor the trinomial $ax^2 + bx + c$ by grouping, we first find two numbers whose product is ac and whose sum is b. For $6x^2 + 7x - 20$, this means we want two numbers with a product of -120 and a sum of 7. Trying products of numbers with different signs and a sum of 7, we find the numbers are -8 and 15. We write $6x^2 + 7x - 20$ with x-terms having coefficients of -8 and 15, and then complete the factorization by grouping, as follows.

$$\begin{aligned} 6x^2 + 7x - 20 &= 6x^2 - 8x + 15x - 20 \\ &= (6x^2 - 8x) + (15x - 20) && \text{group first two terms and last two terms} \\ &= 2x(3x - 4) + 5(3x - 4) && \text{find common factor of each group and} \\ &= (3x - 4)(2x + 5) && \text{note common factor of } 3x - 4 \end{aligned}$$

Multiplication verifies that these are the correct factors. ◂◂

EXAMPLE 11 ▸▸ Factor: $x^2 - 4xy + 4y^2 - 9$.

We see that the first three terms of this expression represent $(x - 2y)^2$. Thus, grouping these terms, we have the following solution:

$$\begin{aligned} x^2 - 4xy + 4y^2 - 9 &= (x^2 - 4xy + 4y^2) - 9 && \text{group terms} \\ &= (x - 2y)^2 - 9 && \text{factor grouping (note difference of squares)} \\ &= [(x - 2y) + 3][(x - 2y) - 3] && \text{factor difference of squares} \\ &= (x - 2y + 3)(x - 2y - 3) \end{aligned}$$

Note that not all groupings work. If we had seen the combination $4y^2 - 9$, which is factorable, and then grouped the first two terms and the last two terms, this would not have led to the factorization. ◂◂

Exercises 6–3

In Exercises 1–48, factor the given expressions completely.

1. $x^2 + 5x + 4$
2. $x^2 - 5x - 6$
3. $s^2 - s - 42$
4. $a^2 + 14a - 32$
5. $t^2 + 5t - 24$
6. $r^2 - 11r + 18$
7. $x^2 + 2x + 1$
8. $y^2 + 8y + 16$
9. $x^2 - 4xy + 4y^2$
10. $b^2 - 12bc + 36c^2$
11. $3x^2 - 5x - 2$
12. $2n^2 - 13n - 7$
13. $3y^2 - 8y - 3$
14. $5x^2 + 9x - 2$
15. $2s^2 + 13s + 11$
16. $7y^2 - 12y + 5$
17. $3f^2 - 16f + 5$
18. $5x^2 - 3x - 2$
19. $2t^2 + 7t - 15$
20. $3n^2 - 20n + 20$
21. $3t^2 - 7tu + 4u^2$
22. $3x^2 + xy - 14y^2$
23. $4x^2 - 3x - 7$
24. $2z^2 + 13z - 5$
25. $9x^2 + 7xy - 2y^2$
26. $4r^2 + 11rs - 3s^2$
27. $4m^2 + 20m + 25$
28. $16q^2 + 24q + 9$
29. $4x^2 - 12x + 9$
30. $a^2c^2 - 2ac + 1$
31. $9t^2 - 15t + 4$
32. $6x^2 + x - 12$
33. $8b^2 + 31b - 4$
34. $12n^2 + 8n - 15$
35. $4p^2 - 25pq + 6q^2$
36. $12x^2 + 4xy - 5y^2$
37. $12x^2 + 47xy - 4y^2$
38. $8r^2 - 14rs - 9s^2$
39. $2x^2 - 14x + 12$
40. $6y^2 - 33y - 18$
41. $4x^2 + 14x - 8$
42. $12x^2 + 22xy - 4y^2$
43. $ax^3 + 4a^2x^2 - 12a^3x$
44. $6x^4 - 13x^3 + 5x^2$
45. $a^2 + 2ab + b^2 - 4$
46. $x^2 - 6xy + 9y^2 - 4z^2$
47. $25a^2 - 25x^2 - 10xy - y^2$
48. $r^2 - s^2 + 2st - t^2$

In Exercises 49–56, factor the given expressions completely. Each is from the technical area indicated.

49. $4s^2 + 16s + 12$ (electricity)

50. $3p^2 + 9p - 54$ (business)

51. $200n^2 - 2100n - 3600$ (biology)

52. $2x^3 - 28x^2 + 98x$ (container design)

53. $wx^4 - 5wLx^3 + 6wL^2x^2$ (beam design)

54. $1 - 2r^2 + r^4$ (lasers)

55. $3Adu^2 - 4Aduv + Adv^2$ (water power)

56. $k^2A^2 + 2k\lambda A + \lambda^2 - \alpha^2$ (robotics)

6-4 The Sum and Difference of Cubes

In Section 6-2 we saw that the difference of squares is factorable, but the sum of squares is prime. By writing Eqs. (6-9) and (6-10) with sides reversed, we have

$$x^3 + y^3 = (x + y)(x^2 - xy + y^2) \qquad (6\text{-}9)$$

$$x^3 - y^3 = (x - y)(x^2 + xy + y^2) \qquad (6\text{-}10)$$

Both the sum and difference of perfect cubes are factorable. In Eqs. (6-9) and (6-10), neither $x^2 - xy + y^2$ nor $x^2 + xy + y^2$ is factorable; they are both prime.

EXAMPLE 1 ▸▸

(a) $x^3 + 8 = x^3 + 2^3$

$$= (x + 2)[(x)^2 - 2x + 2^2]$$

$$= (x + 2)(x^2 - 2x + 4)$$

(b) $x^3 - 1 = x^3 - 1^3$

$$= (x - 1)[(x)^2 + (1)(x) + 1^2]$$

$$= (x - 1)(x^2 + x + 1)$$ ◂◂

EXAMPLE 2 ▸▸ $8 - 27x^3 = 2^3 - (3x)^3$ $\qquad 8 = 2^3$ and $27x^3 = (3x)^3$

$$= (2 - 3x)[2^2 + 2(3x) + (3x)^2]$$

$$= (2 - 3x)(4 + 6x + 9x^2)$$ ◂◂

As we have stated in the previous sections on factoring, *we should start any factoring process by first checking for any common monomial factor* that might be present in each term of the expression. Also, we should check that our result has been factored ***completely.***

EXAMPLE 3 ▸▸ In factoring $ax^5 - ax^2$, we first note that each term has a common factor of ax^2. This is factored out to get $ax^2(x^3 - 1)$. However, the expression is not completely factored since $1 = 1^3$, which means that $x^3 - 1$ is the difference of cubes. We complete the factoring by the use of Eq. (6-10). Therefore,

$$ax^5 - ax^2 = ax^2(x^3 - 1)$$

$$= ax^2(x - 1)(x^2 + x + 1)$$ ◂◂

EXAMPLE 4 ▸▸ The volume of material used to make a steel bearing with a hollow core is given by $\frac{4}{3}\pi R^3 - \frac{4}{3}\pi r^3$. Factor this expression.

$$\frac{4}{3}\pi R^3 - \frac{4}{3}\pi r^3 = \frac{4}{3}\pi(R^3 - r^3) \qquad \text{common factor of } \frac{4}{3}\pi$$

$$= \frac{4}{3}\pi(R - r)(R^2 + Rr + r^2) \qquad \text{using Eq. (6-10)}$$ ◂◂

In our study of factoring, we have seen that we should

first factor out any common monomial factor

and then see if the remaining expression can be further factored as one of the following types:

1. *Difference of squares*
2. *Factorable trinomial*
3. *Sum or difference of cubes*
4. *Factorable by grouping*

Remember, an expression should be factored ***completely.***

Exercises 6–4

In Exercises 1–20, factor the given expressions completely.

1. $x^3 + 1$ **2.** $x^3 + 27$ **3.** $8 - t^3$

4. $8r^3 - 1$ **5.** $27x^3 - 8a^3$ **6.** $64x^3 + 125$

7. $2x^3 + 16$ **8.** $3y^3 - 81$ **9.** $6a^4 + 6a$

10. $8s^3 - 8$ **11.** $6x^3y - 6x^3y^4$ **12.** $12a^3 + 96a^3b^3$

13. $x^6y^3 + x^3y^6$ **14.** $16r^3 - 432$

15. $3a^6 - 3a^2$ **16.** $x^6 - 81y^2$

17. $(a + b)^3 - 64$ **18.** $125 + (2x + y)^3$

19. $64 + x^6$ **20.** $a^6 - b^6$

In Exercises 21–24, factor the given expressions completely. Each is from the technical area indicated.

21. $2x^3 + 250$ (computer image)

22. $kT^3 - kT_0^3$ (thermodynamics)

23. $D^4 - d^3D$ (machine design)

24. $pb^3 + p(8a^3)$ (business)

6-5 Equivalent Fractions

When we deal with algebraic expressions, we must be able to work effectively with fractions. Since algebraic expressions are representations of numbers, the basic operations on fractions from arithmetic form the basis of our algebraic operations. In this section we demonstrate a very important property of fractions, and in the following two sections we establish the basic algebraic operations with fractions.

Fundamental Principle of Fractions

This important property of fractions, often referred to as the **fundamental principle of fractions,** is that *the value of a fraction is unchanged if both numerator and denominator are multiplied or divided by the same number, provided this number is not zero.* Two fractions are said to be **equivalent** if one can be obtained from the other by use of the fundamental principle.

EXAMPLE 1 ►► If we multiply the numerator and the denominator of the fraction $\frac{6}{8}$ by 2, we obtain the equivalent fraction $\frac{12}{16}$. If we divide the numerator and the denominator of $\frac{6}{8}$ by 2, we obtain the equivalent fraction $\frac{3}{4}$. Therefore, the fractions $\frac{6}{8}$, $\frac{3}{4}$, and $\frac{12}{16}$ are equivalent. ◄◄

EXAMPLE 2 ►► We may write

$$\frac{ax}{2} = \frac{3a^2x}{6a}$$

since the fraction on the right is obtained from the fraction on the left by multiplying the numerator and the denominator by $3a$. Therefore, the fractions are equivalent. ◄◄

Simplest Form, or Lowest Terms, of a Fraction

One of the most important operations to be performed on a fraction is that of reducing it to its **simplest form,** or **lowest terms.** *A fraction is said to be in its simplest form if the numerator and the denominator have no common integral factors other than* $+1$ *or* -1. In reducing a fraction to its simplest form, we use the fundamental principle by dividing both the numerator and the denominator by all factors that are common to each. (It will be assumed throughout this text that if any of the literal symbols were to be evaluated, numerical values would be restricted so that none of the denominators would be zero. Thereby, we avoid the undefined operation of division by zero.)

EXAMPLE 3 ▸▸ In order to reduce the fraction

$$\frac{16ab^3c^2}{24ab^2c^5}$$

to its lowest terms, we note that both the numerator and the denominator contain the factor $8ab^2c^2$. Therefore, we may write

$$\frac{16ab^3c^2}{24ab^2c^5} = \frac{2b(8ab^2c^2)}{3c^3(8ab^2c^2)} = \frac{2b}{3c^3}$$

common factor

Here we divided out the common factor. The resulting fraction is in lowest terms, since there are no common factors in the numerator and the denominator other than $+1$ or -1. ◂◂

CAUTION ▸ We must note very carefully that in simplifying fractions, *we divide both the numerator and the denominator* ***by the common*** **factor.** *This process is called* **cancellation.** However, many students are tempted to try to remove *any* expression that appears in both the numerator and the denominator. If a ***term*** is removed in this way, it is an incorrect application of the cancellation process. The following example illustrates this common error in the simplification of fractions.

EXAMPLE 4 ▸▸ When simplifying the expression

$$\frac{x^2(x-2)}{x^2-4}$$

a term, but not a factor, of the denominator

many students would "cancel" the x^2 from the numerator and the denominator. This

CAUTION ▸ is *incorrect*, since $\boldsymbol{x^2}$ ***is a term only*** of the denominator.

In order to simplify the above fraction properly, we should factor the denominator. We obtain

$$\frac{x^2\overset{1}{\cancel{(x-2)}}}{\underset{1}{\cancel{(x-2)}}(x+2)} = \frac{x^2}{x+2}$$

Here, the common ***factor*** $x-2$ has been divided out. ◂◂

The following examples illustrate the proper simplification of fractions.

EXAMPLE 5 (a) $\frac{2a}{2ax} = \frac{1}{x}$ — $2a$ is a factor of the numerator and the denominator

We divide out the common factor of $2a$.

(b) $\frac{2a}{2a + x}$ — $2a$ is a term, but not a factor, of the denominator

CAUTION ▶ ***This cannot be reduced,*** since *there are no common **factors** in the numerator and the denominator.*

EXAMPLE 6

$$\frac{2x^2 + 8x}{x + 4} = \frac{2x(x + 4)}{x + 4} = \frac{2x}{1} = 2x$$

The numerator and the denominator were each divided by $x + 4$ after we factored the numerator. The only remaining factor in the denominator is 1, and it is generally not written in the final result. Another way of writing the denominator is $1(x + 4)$, which shows the ***factor*** of 1 more clearly.

EXAMPLE 7

$$\frac{x^2 - 4x + 4}{x^2 - 4} = \frac{(x - 2)(x - 2)}{(x + 2)(x - 2)} = \frac{x - 2}{x + 2}$$

x is a term, but not a factor

CAUTION ▶ Here the numerator and the denominator have each been *factored first and then the common factor $x - 2$ has been divided out.* In the final form, neither the x's nor the 2's may be canceled, since they are not common *factors.*

EXAMPLE 8 In the mathematical analysis of the vibrations in a certain mechanical system, the following expression and simplification are used:

$$\frac{8s + 12}{4s^2 + 26s + 30} = \frac{4(2s + 3)}{2(2s^2 + 13s + 15)} = \frac{4(2s + 3)}{2(2s + 3)(s + 5)} = \frac{2}{s + 5}$$

Factors common to the numerator and the denominator are 2 and $(2s + 3)$.

Factors That Differ Only in Sign

In simplifying fractions we must be able to distinguish between factors that differ only in ***sign.*** Since $-(y - x) = -y + x = x - y$, we have

$$x - y = -(y - x) \qquad (6\text{-}11)$$

NOTE ▶ Here we see that ***factors $x - y$ and $y - x$ differ only in sign.*** The following examples illustrate the simplification of fractions where a change of signs is necessary.

EXAMPLE 9 ▸▸ $\frac{x^2 - 1}{1 - x} = \frac{(x - 1)(x + 1)}{-(x - 1)} = \frac{x + 1}{-1} = -(x + 1)$

CAUTION ▸ In the second fraction, ***we replaced* 1 − *x with the equal expression* −(*x* − 1).** In the third fraction, the common factor $x - 1$ was divided out. Finally, we expressed the result in the more convenient form by dividing $x + 1$ by -1, which makes the quantity $x + 1$ negative. ◂◂

EXAMPLE 10 ▸▸

$$\frac{2x^4 - 128x}{20 + 7x - 3x^2} = \frac{2x(x^3 - 64)}{(4 - x)(5 + 3x)} = \frac{2x(x - 4)(x^2 + 4x + 16)}{-(x - 4)(3x + 5)}$$

$$= -\frac{2x(x^2 + 4x + 16)}{3x + 5}$$

Again, the factor $4 - x$ has been replaced by the equal expression $-(x - 4)$. This allows us to recognize the common factor of $x - 4$. Notice also that the order of the terms of the factor $5 + 3x$ has been changed to $3x + 5$. This is merely an application of the commutative law of addition. ◂◂

Exercises 6–5

In Exercises 1–8, multiply the numerator and the denominator of each fraction by the given factor and obtain an equivalent fraction.

1. $\frac{2}{3}$ (by 7)

2. $\frac{7}{5}$ (by 9)

3. $\frac{ax}{y}$ (by $2x$)

4. $\frac{2x^2y}{3n}$ (by $2xn^2$)

5. $\frac{2}{x + 3}$ (by $x - 2$)

6. $\frac{7}{a - 1}$ (by $a + 2$)

7. $\frac{a(x - y)}{x - 2y}$ (by $x + y$)

8. $\frac{x - 1}{x + 1}$ (by $x - 1$)

In Exercises 9–16, divide the numerator and the denominator of each fraction by the given factor and obtain an equivalent fraction.

9. $\frac{28}{44}$ (by 4)

10. $\frac{25}{65}$ (by 5)

11. $\frac{4x^2y}{8xy^2}$ (by $2x$)

12. $\frac{6a^3b^2}{9a^5b^4}$ (by $3a^2b^2$)

13. $\frac{2(x - 1)}{(x - 1)(x + 1)}$ (by $x - 1$)

14. $\frac{(x + 5)(x - 3)}{3(x + 5)}$ (by $x + 5$)

15. $\frac{x^2 - 3x - 10}{2x^2 + 3x - 2}$ (by $x + 2$)

16. $\frac{6x^2 + 13x - 5}{6x^3 - 2x^2}$ (by $3x - 1$)

In Exercises 17–52, reduce each fraction to simplest form.

17. $\frac{2a}{8a}$

18. $\frac{6x}{15x}$

19. $\frac{18x^2y}{24xy}$

20. $\frac{2a^2xy}{6axyz^2}$

21. $\frac{a + b}{5a^2 + 5ab}$

22. $\frac{t - a}{t^2 - a^2}$

23. $\frac{6a - 4b}{4a - 2b}$

24. $\frac{5r - 20s}{10r - 5s}$

25. $\frac{4x^2 + 1}{4x^2 - 1}$

26. $\frac{x^2 - y^2}{x^2 + y^2}$

27. $\frac{3x^2 - 6x}{x - 2}$

28. $\frac{10x^2 + 15x}{2x + 3}$

29. $\frac{2y + 3}{4y^3 + 6y^2}$

30. $\frac{3t - 6}{4t^3 - 8t^2}$

31. $\frac{x^2 - 8x + 16}{x^2 - 16}$

32. $\frac{4a^2 + 12ab + 9b^2}{4a^2 + 6ab}$

33. $\frac{2x^2 + 5x - 3}{x^2 + 11x + 24}$

34. $\frac{3y^3 + 7y^2 + 4y}{y^2 + 5y + 4}$

35. $\frac{5x^2 - 6x - 8}{x^3 + x^2 - 6x}$

36. $\frac{4r^2 - 8rs - 5s^2}{6r^2 - 17rs + 5s^2}$

37. $\frac{x^4 - 16}{x + 2}$

38. $\frac{2x^2 - 8}{4x + 8}$

39. $\frac{x^2y^4 - x^4y^2}{y^2 - 2xy + x^2}$

40. $\frac{8x^3 + 8x^2 + 2x}{4x + 2}$

41. $\frac{(x-1)(3+x)}{(3-x)(1-x)}$

42. $\frac{(2x-1)(x+6)}{(x-3)(1-2x)}$

43. $\frac{y-x}{2x-2y}$

44. $\frac{x^2-y^2}{y-x}$

45. $\frac{2x^2-9x+4}{4x-x^2}$

46. $\frac{3a^2-13a-10}{5+4a-a^2}$

47. $\frac{(x+5)(x-2)(x+2)(3-x)}{(2-x)(5-x)(3+x)(2+x)}$

48. $\frac{(2x-3)(3-x)(x-7)(3x+1)}{(3x+2)(3-2x)(x-3)(7+x)}$

49. $\frac{x^3+y^3}{2x+2y}$

50. $\frac{x^3-8}{x^2+2x+4}$

51. $\frac{6x^2+2x}{27x^3+1}$

52. $\frac{3a^3-24}{a^2-4a+4}$

In Exercises 53–56, after finding the simplest form of each fraction, explain why it cannot be simplified more.

53. (a) $\frac{x^2(x+2)}{x^2+4}$ (b) $\frac{x^4+4x^2}{x^4-16}$

54. (a) $\frac{2x+3}{2x+6}$ (b) $\frac{2(x+6)}{2x+6}$

55. (a) $\frac{x^2-x-2}{x^2-x}$ (b) $\frac{x^2-x-2}{x^2+x}$

56. (a) $\frac{x^3-x}{1-x}$ (b) $\frac{2x^2+4x}{2x^2+4}$

In Exercises 57–60, reduce each fraction to simplest form. Each is from the indicated technical area of application.

57. $\frac{mu^2-mv^2}{mu-mv}$ (nuclear energy)

58. $\frac{16(t^2-2tt_0+t_0^2)(t-t_0-3)}{3t-3t_0}$ (rocket motion)

59. $\frac{E^2R^2-E^2r^2}{(R^2+2Rr+r^2)^2}$ (electricity)

60. $\frac{r_0^3-r_i^3}{r_0^2-r_i^2}$ (machine design)

6-6 Multiplication and Division of Fractions

From arithmetic we recall that *the product of two fractions is a fraction whose numerator is the product of the numerators and whose denominator is the product of the denominators of the given fractions.* Also, we recall that *we can find the quotient of two fractions by inverting the divisor and proceeding as in multiplication.* Symbolically, multiplication of fractions is indicated by

Multiplication of Fractions

$$\frac{a}{b}\times\frac{c}{d}=\frac{ac}{bd}$$

and division is indicated by

Division of Fractions

$$\frac{a}{b}\div\frac{c}{d}=\frac{\frac{a}{b}}{\frac{c}{d}}=\frac{a}{b}\times\frac{d}{c}=\frac{ad}{bc}$$

The rule for division may be verified by use of the fundamental principle of fractions. By multiplying the numerator and the denominator of the fraction

$$\frac{\frac{a}{b}}{\frac{c}{d}} \text{ by } \frac{d}{c} \text{ we obtain } \frac{\frac{a}{b}\times\frac{d}{c}}{\frac{c}{d}\times\frac{d}{c}}=\frac{\frac{ad}{bc}}{1}=\frac{ad}{bc}$$

EXAMPLE 1 (a) $\frac{3}{5}\times\frac{2}{7}=\frac{(3)(2)}{(5)(7)}=\frac{6}{35}$ ← multiply numerators / ← multiply denominators

(b) $\frac{3a}{5b}\times\frac{15b^2}{a}=\frac{(3a)(15b^2)}{(5b)(a)}=\frac{45ab^2}{5ab}=\frac{9b}{1}=9b$

In illustration (b), we divided out the common factor of $5ab$ to reduce the resulting fraction to its lowest terms.

We shall usually want to express the result in its simplest form, which is generally its most useful form. Since all factors in the numerators and all factors in the denominators are to be multiplied, we should *first only* ***indicate*** *the multiplication, but not actually perform it, and then factor the numerator and the denominator.* In this way we can easily identify any factors common to both. If we were to multiply out the numerator and the denominator before factoring, it is very possible that we would be unable to factor the result to simplify it. The following example illustrates this point.

NOTE ▶

EXAMPLE 2 ▶▶ In performing the multiplication

$$\frac{3(x-y)}{(x-y)^2} \times \frac{(x^2-y^2)}{6x+9y}$$

if we multiplied out the numerators and the denominators before performing any factoring, we would have to simplify the fraction

$$\frac{3x^3 - 3x^2y - 3xy^2 + 3y^3}{6x^3 - 3x^2y - 12xy^2 + 9y^3}$$

It is possible to factor the resulting numerator and denominator, but finding any common factors this way is very difficult. If we first indicate the multiplications, but do not actually perform them, and then factor completely, we have

$$\begin{aligned}\frac{3(x-y)}{(x-y)^2} \times \frac{(x^2-y^2)}{6x+9y} &= \frac{3(x-y)(x^2-y^2)}{(x-y)^2(6x+9y)} = \frac{3(x-y)(x+y)(x-y)}{(x-y)^2(3)(2x+3y)} \\ &= \frac{3(x-y)^2(x+y)}{3(x-y)^2(2x+3y)} \\ &= \frac{x+y}{2x+3y}\end{aligned}$$

The common factor $3(x-y)^2$ is readily recognized using this procedure. ◀◀

EXAMPLE 3 ▶▶

$$\begin{aligned}\frac{2x-4}{4x+12} \times \frac{2x^2+x-15}{3x-1} &= \frac{2(x-2)(2x-5)(x+3)}{4(x+3)(3x-1)} \quad \text{multiplications indicated} \\ &= \frac{(x-2)(2x-5)}{2(3x-1)}\end{aligned}$$

Here the common factor is $2(x+3)$. It is permissible to multiply out the final form of the numerator and the denominator, but it is often preferable to leave the numerator and the denominator in factored form, as indicated. ◀◀

The following examples illustrate the division of fractions.

EXAMPLE 4 ▶▶

(a) $\dfrac{6x}{7} \div \dfrac{5}{3} = \dfrac{6x}{7} \times \dfrac{3}{5} = \dfrac{18x}{35}$ (multiply; invert)

(b) $\dfrac{\frac{3a^2}{5c}}{\frac{2c^2}{a}} = \dfrac{3a^2}{5c} \times \dfrac{a}{2c^2} = \dfrac{3a^3}{10c^3}$ (multiply; invert) ◀◀

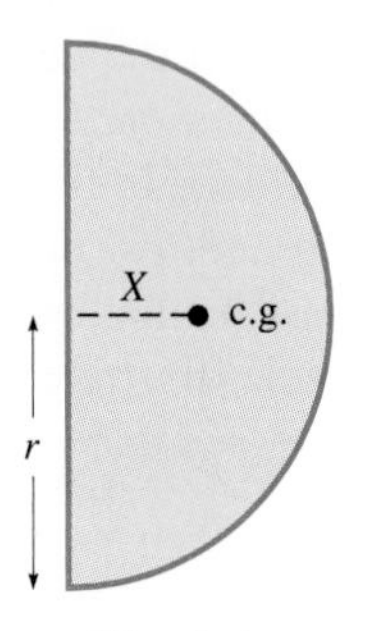

Fig. 6-1

EXAMPLE 5 ▸▸ When finding the center of gravity of a uniform flat semicircular metal plate, the equation $X = \frac{4\pi r^3}{3} \div \left(\frac{\pi r^2}{2} \times 2\pi\right)$ is derived. Simplify the right side of this equation to find X as a function of r in simplest form. See Fig. 6-1.

The parentheses indicate that we should perform the multiplication first.

$$X = \frac{4\pi r^3}{3} \div \left(\frac{\pi r^2}{2} \times 2\pi\right) = \frac{4\pi r^3}{3} \div \left(\frac{2\pi^2 r^2}{2}\right)$$

$$= \frac{4\pi r^3}{3} \div (\pi^2 r^2) = \frac{4\pi r^3}{3} \times \frac{1}{\pi^2 r^2}$$

$$= \frac{4\pi r^3}{3\pi^2 r^2} = \frac{4r}{3\pi}$$

This is the exact solution. Approximately, $X = 0.424r$. ◂◂

EXAMPLE 6 ▸▸

$$\frac{x+y}{3} \div \frac{2x+2y}{6x+15y} = \frac{x+y}{3} \times \frac{6x+15y}{2x+2y} = \frac{(x+y)(3)(2x+5y)}{3(2)(x+y)}$$ indicate multiplications

invert

$$= \frac{2x+5y}{2}$$ simplify ◂◂

EXAMPLE 7 ▸▸

$$\frac{\dfrac{4-x^2}{x^2-3x+2}}{\dfrac{x+2}{x^2-9}} = \frac{4-x^2}{x^2-3x+2} \times \frac{x^2-9}{x+2}$$ invert

$$= \frac{(2-x)(2+x)(x-3)(x+3)}{(x-2)(x-1)(x+2)}$$ factor and indicate multiplications

$$= \frac{-(x-2)(x+2)(x-3)(x+3)}{(x-2)(x-1)(x+2)}$$ replace $(2-x)$ with $-(x-2)$ and $(2+x)$ with $(x+2)$

$$= -\frac{(x-3)(x+3)}{x-1} \quad \text{or} \quad \frac{(x-3)(x+3)}{1-x}$$ simplify

Note the use of Eq. (6-11) in the simplification and in expressing an alternate form of the result. The factor $2-x$ was replaced by its equivalent $-(x-2)$, and then $x-1$ was replaced by $-(1-x)$. ◂◂

Exercises 6–6

In Exercises 1–36, simplify the given expressions involving the indicated multiplications and divisions.

1. $\frac{3}{8} \times \frac{2}{7}$ **2.** $\frac{11}{5} \times \frac{13}{33}$ **3.** $\frac{4x}{3y} \times \frac{9y^2}{2}$

4. $\frac{18sy^3}{ax^2} \times \frac{(ax)^2}{3s}$ **5.** $\frac{2}{9} \div \frac{4}{7}$ **6.** $\frac{5}{16} \div \frac{25}{13}$

7. $\frac{xy}{az} \div \frac{bz}{ay}$ **8.** $\frac{sr^2}{2t} \div \frac{st}{4}$

9. $\frac{4x+12}{5} \times \frac{15t}{3x+9}$ **10.** $\frac{y^2+2y}{6z} \times \frac{z^3}{y^2-4}$

11. $\frac{u^2 - v^2}{u + 2v}(3u + 6v)$

12. $(x - y)\frac{x + 2y}{x^2 - y^2}$

13. $\frac{2a + 8}{15} \div \frac{a^2 + 8a + 16}{25}$

14. $\frac{a^2 - a}{3a + 9} \div \frac{a^2 - 2a + 1}{a^2 - 9}$

15. $\frac{x^2 - 9}{x} \div (x + 3)^2$

16. $\frac{9x^2 - 16}{x + 1} \div (4 - 3x)$

17. $\frac{3ax^2 - 9ax}{10x^2 + 5x} \times \frac{2x^2 + x}{a^2x - 3a^2}$

18. $\frac{2x^2 - 18}{x^3 - 25x} \times \frac{3x - 15}{2x^2 + 6x}$

19. $\frac{x^4 - 1}{8x + 16} \times \frac{2x^2 - 8x}{x^3 + x}$

20. $\frac{2x^2 - 4x - 6}{x^2 - 3x} \times \frac{x^3 - 4x^2}{4x^2 - 4x - 8}$

21. $\frac{ax + x^2}{2b - cx} \div \frac{a^2 + 2ax + x^2}{2bx - cx^2}$

22. $\frac{x^2 - 11x + 28}{x + 3} \div \frac{x - 4}{2x + 3}$

23. $\frac{35a + 25}{12a + 33} \div \frac{28a + 20}{36a + 99}$

24. $\frac{2a^3 + a^2}{2b^3 + b^2} \div \frac{2ab + a}{2ab + b}$

25. $\frac{x^2 - 6x + 5}{4x^2 - 17x - 15} \times \frac{6x + 21}{2x^2 + 5x - 7}$

26. $\frac{x^2 + 5x}{3x^2 + 8x - 4} \times \frac{2x^2 - 8}{x^3 + 2x^2 - 15x}$

27. $\frac{7x^2 + 27x - 4}{6x^2 + x - 15} \div \frac{4x^2 + 17x + 4}{8x^2 - 10x - 3}$

28. $\frac{4x^3 - 9x}{8x^2 + 10x - 3} \div \frac{2x^3 - 3x^2}{8x^2 + 18x - 5}$

29. $\frac{7x^2}{3a} \div \left(\frac{a}{x} \times \frac{a^2x}{x^2}\right)$

30. $\left(\frac{3u}{8v^2} \div \frac{9u^2}{2w^2}\right) \times \frac{2u^4}{15vw}$

31. $\left(\frac{4t^2 - 1}{t - 5} \div \frac{2t + 1}{2t}\right) \times \frac{2t^2 - 50}{4t^2 + 4t + 1}$

32. $\frac{2x^2 - 5x - 3}{x - 4} \div \left(\frac{x - 3}{x^2 - 16} \times \frac{1}{3 - x}\right)$

33. $\frac{x^3 - y^3}{2x^2 - 2y^2} \times \frac{x^2 + 2xy + y^2}{x^2 + xy + y^2}$

34. $\frac{2x^2 + 4x + 2}{6x - 6} \div \frac{5x + 5}{x^2 - 1}$

35. $\frac{ax + bx + ay + by}{p - q} \times \frac{3p^2 + 4pq - 7q^2}{a + b}$

36. $\frac{x^4 + x^5 - 1 - x}{x - 1} \div \frac{x + 1}{x}$

In Exercises 37–40, simplify the given expressions. The technical application of each is indicated.

37. $\frac{n^2a^2}{v^2} \div \frac{n - an}{v}$ (chemistry)

38. $\frac{w\pi rbt}{2btg}\left(\frac{2r}{12\pi}\right)\left(\frac{144v^2}{r^2}\right)$ (stress on a rotating hoop)

39. $\frac{2\pi}{\lambda}\left(\frac{a + b}{2ab}\right)\left(\frac{ab\lambda}{2a + 2b}\right)$ (optics)

40. $(p_1 - p_2) \div \left(\frac{\pi a^4 p_1 - \pi a^4 p_2}{8lu}\right)$ (hydrodynamics)

6-7 Addition and Subtraction of Fractions

From arithmetic we recall that *the sum of a set of fractions that all have the same denominator is the sum of the numerators divided by the common denominator.* Since algebraic expressions represent numbers, this fact is also true in algebra. Addition and subtraction of such fractions are illustrated in the following example.

EXAMPLE 1 (a) $\frac{5}{9} + \frac{2}{9} - \frac{4}{9} = \frac{5 + 2 - 4}{9}$ ← sum of numerators; ← same denominators

$= \frac{3}{9} = \frac{1}{3}$ ← final result in lowest terms

use parentheses to show subtraction of both terms

(b) $\frac{b}{ax} + \frac{1}{ax} - \frac{2b - 1}{ax} = \frac{b + 1 - (2b - 1)}{ax} = \frac{b + 1 - 2b + 1}{ax}$

$= \frac{2 - b}{ax}$

Lowest Common Denominator

If the fractions to be combined do not all have the same denominator, we must first change each to an equivalent fraction so that the resulting fractions do have the same denominator. Normally the denominator that is most convenient and useful is the **lowest common denominator** (abbreviated as **LCD**). *This is the product of all the prime factors that appear in the denominators, with each factor raised to the* ***highest power*** *to which it appears* ***in any one*** *of the denominators.* This means that the lowest common denominator is the *simplest* algebraic expression into which all given denominators will divide exactly.

EXAMPLE 2 ▸▸ Find the lowest common denominator of the fractions

$$\frac{3}{4a^2b}, \quad \frac{5}{6ab^3}, \quad \text{and} \quad \frac{1}{4ab^2}$$

We now express each denominator in terms of powers of its prime factors.

highest powers (of 2^2, a^2, 3, b^3) — already seen to be highest power of 2 (of 2^2 in $4ab^2$)

$$4a^2b = 2^2a^2b, \quad 6ab^3 = 2 \times 3 \times ab^3, \quad \text{and} \quad 4ab^2 = 2^2ab^2$$

The prime factors to be considered are 2, 3, a, and b. The largest exponent of 2 that appears is 2. Therefore, 2^2 is a factor of the lowest common denominator.

CAUTION ▸ *What matters is that* ***the highest power of 2 that appears is 2,*** *not the fact that 2 appears in all three denominators with a total of five factors.*

The largest exponent of 3 that appears is 1 (understood in the second denominator). Therefore, 3 is a factor of the lowest common denominator. The largest exponent of a that appears is 2, and the largest exponent of b that appears is 3. Thus, a^2 and b^3 are factors of the lowest common denominator. Therefore, the lowest common denominator of the fractions is

$$2^2 \times 3 \times a^2b^3 = 12a^2b^3$$

This is the simplest expression into which ***each*** of the denominators above will divide exactly. ◂◂

EXAMPLE 3 ▸▸ Find the LCD of the following fractions:

$$\frac{x-4}{x^2-2x+1}, \quad \frac{1}{x^2-1}, \quad \frac{x+3}{x^2-x}$$

Factoring each of the denominators, we find that the fractions are

$$\frac{x-4}{(x-1)^2}, \quad \frac{1}{(x-1)(x+1)}, \quad \text{and} \quad \frac{x+3}{x(x-1)}$$

The factor $(x - 1)$ appears in all the denominators. It is squared in the first fraction and appears only to the first power in the other two fractions. Thus, we must have $(x - 1)^2$ as a factor in the LCD. We do not need a higher power of $(x - 1)$ since, as far as this factor is concerned, each denominator will divide into it evenly. Next, the second denominator has a factor of $(x + 1)$. Therefore, the LCD must also have a factor of $(x + 1)$; otherwise, the second denominator would not divide into it exactly. Finally, the third denominator shows that a factor of x is also needed. The LCD is, therefore, $x(x + 1)(x - 1)^2$. All three denominators will divide exactly into this expression, and there is no simpler expression for which this is true. ◂◂

Addition and Subtraction of Fractions

Once we have found the lowest common denominator for the fractions, we multiply the numerator and the denominator of each fraction by the proper quantity to make the resulting denominator in each case the lowest common denominator. After this step, it is necessary only to add the numerators, place this result over the common denominator, and simplify.

EXAMPLE 4 ▸▸ Combine $\dfrac{2}{3r^2} + \dfrac{4}{rs^3} - \dfrac{5}{3s}$.

By looking at the denominators, we see that the factors necessary in the lowest common denominator are 3, r, and s. The 3 appears only to the first power, the largest exponent of r is 2, and the largest exponent of s is 3. Therefore, the lowest common denominator is $3r^2s^3$. We now wish to write each fraction with this quantity as the denominator. Since the denominator of the first fraction already contains factors of 3 and r^2, ***it is necessary to introduce the factor of*** s^3**.** In other words, we must multiply the numerator and the denominator of this fraction by s^3. For similar reasons, we must multiply the numerators and the denominators of the second and third fractions by $3r$ and r^2s^2, respectively. This leads to

NOTE ▶

$$\frac{2}{3r^2} + \frac{4}{rs^3} - \frac{5}{3s} = \frac{2(s^3)}{(3r^2)(s^3)} + \frac{4(3r)}{(rs^3)(3r)} - \frac{5(r^2s^2)}{(3s)(r^2s^2)}$$

change to equivalent fractions with LCD

factors needed in each

$$= \frac{2s^3}{3r^2s^3} + \frac{12r}{3r^2s^3} - \frac{5r^2s^2}{3r^2s^3}$$

$$= \frac{2s^3 + 12r - 5r^2s^2}{3r^2s^3}$$

combine numerators over LCD ◂◂

EXAMPLE 5 ▸▸

$$\frac{a}{x-1} + \frac{a}{x+1} = \frac{a(x+1)}{(x-1)(x+1)} + \frac{a(x-1)}{(x+1)(x-1)}$$

change to equivalent fractions with LCD

factors needed

$$= \frac{ax + a + ax - a}{(x+1)(x-1)}$$

combine numerators over LCD

$$= \frac{2ax}{(x+1)(x-1)}$$

simplify

When we multiply each fraction by the quantity required to obtain the proper denominator, we do not actually have to write the common denominator under each numerator. Placing all the products that appear in the numerators over the common denominator is sufficient. Hence the illustration in this example would appear as

$$\frac{a}{x-1} + \frac{a}{x+1} = \frac{a(x+1) + a(x-1)}{(x-1)(x+1)} = \frac{ax + a + ax - a}{(x-1)(x+1)}$$

$$= \frac{2ax}{(x-1)(x+1)}$$ ◂◂

EXAMPLE 6 ▸▸ The following expression is found in the analysis of the dynamics of missile firing. The indicated addition is performed as shown.

$$\frac{1}{s} - \frac{1}{s+4} + \frac{8}{s^2+8s+16}$$

$$= \frac{1}{s} - \frac{1}{s+4} + \frac{8}{(s+4)^2} \quad \text{factor third denominator}$$

$$= \frac{1(s+4)^2 - 1(s)(s+4) + 8s}{s(s+4)^2} \quad \text{LCD has one factor of } s \text{ and two factors of } (s+4)$$

$$= \frac{s^2+8s+16-s^2-4s+8s}{s(s+4)^2} \quad \text{expand terms of numerator}$$

$$= \frac{12s+16}{s(s+4)^2} = \frac{4(3s+4)}{s(s+4)^2} \quad \text{simplify/factor}$$

We factored the numerator in the final result to see whether or not there were any factors common to the numerator and the denominator. Since there are none, either form of the result is acceptable. ◂◂

EXAMPLE 7 ▸▸

$$\frac{3x}{x^2-x-12} - \frac{x-1}{x^2-8x+16} - \frac{6-x}{2x-8} = \frac{3x}{(x-4)(x+3)} - \frac{x-1}{(x-4)^2} - \frac{6-x}{2(x-4)} \quad \text{factor denominators}$$

$$= \frac{3x(2)(x-4) - (x-1)(2)(x+3) - (6-x)(x-4)(x+3)}{2(x-4)^2(x+3)} \quad \text{change to equivalent fraction with LCD}$$

$$= \frac{6x^2-24x-2x^2-4x+6+x^3-7x^2-6x+72}{2(x-4)^2(x+3)} \quad \text{expand in numerator}$$

$$= \frac{x^3-3x^2-34x+78}{2(x-4)^2(x+3)} \quad \text{simplify}$$

One note of caution must be sounded here. In doing this kind of problem, many errors may arise in the use of the minus sign. Remember, if a minus sign precedes a given expression, the ***signs of* all *terms in that expression must be changed*** before they can be combined with the other terms. ◂◂

CAUTION ▸

Complex Fractions

A **complex fraction** *is one in which the numerator, the denominator, or both the numerator and the denominator contain fractions.* The following examples illustrate the simplification of complex fractions.

EXAMPLE 8 ▸▸

$$\frac{\frac{2}{x}}{1-\frac{4}{x}} = \frac{\frac{2}{x}}{\frac{x-4}{x}} \quad \text{first perform subtraction in denominator}$$

$$= \frac{2}{x} \times \frac{x}{x-4} = \frac{2x}{x(x-4)} \quad \text{invert divisor and multiply}$$

$$= \frac{2}{x-4} \quad \text{simplify} ◂◂$$

EXAMPLE 9

$$\frac{1-\frac{2}{x}}{\frac{1}{x}+\frac{2}{x^2+4x}} = \frac{1-\frac{2}{x}}{\frac{1}{x}+\frac{2}{x(x+4)}} = \frac{\frac{x-2}{x}}{\frac{x+4+2}{x(x+4)}} \quad \text{perform subtraction and addition}$$

$$= \frac{x-2}{x} \times \frac{x(x+4)}{x+6} \quad \text{invert divisor and multiply}$$

$$= \frac{(x-2)(x)(x+4)}{x(x+6)} \quad \text{indicate multiplication}$$

$$= \frac{(x-2)(x+4)}{x+6} \quad \text{simplify}$$

Exercises 6–7

In Exercises 1–40, perform the indicated operations and simplify.

1. $\frac{3}{5}+\frac{6}{5}$

2. $\frac{2}{13}+\frac{6}{13}$

3. $\frac{1}{x}+\frac{7}{x}$

4. $\frac{2}{a}+\frac{3}{a}$

5. $\frac{1}{2}+\frac{3}{4}$

6. $\frac{5}{9}-\frac{1}{3}$

7. $\frac{3}{4x}+\frac{7a}{4}$

8. $\frac{t-3}{a}-\frac{t}{2a}$

9. $\frac{a}{x}-\frac{b}{x^2}$

10. $\frac{2}{s^2}+\frac{3}{s}$

11. $\frac{6}{5x^3}+\frac{a}{25x}$

12. $\frac{a}{6y}-\frac{2b}{3y^4}$

13. $\frac{2}{5a}+\frac{1}{a}-\frac{a}{10}$

14. $\frac{1}{2a}-\frac{6}{b}-\frac{9}{4c}$

15. $\frac{x+1}{x}-\frac{x-3}{y}-\frac{2-x}{xy}$

16. $5+\frac{1-x}{2}-\frac{3+x}{4}$

17. $\frac{3}{2x-1}+\frac{1}{4x-2}$

18. $\frac{5}{6y+3}-\frac{a}{8y+4}$

19. $\frac{4}{x(x+1)}-\frac{3}{2x}$

20. $\frac{3}{ax+ay}-\frac{1}{a^2}$

21. $\frac{s}{2s-6}+\frac{1}{4}-\frac{3s}{4s-12}$

22. $\frac{2}{x+2}-\frac{3-x}{x^2+2x}+\frac{1}{x}$

23. $\frac{3x}{x^2-9}-\frac{2}{3x+9}$

24. $\frac{2}{x^2+4x+4}-\frac{3}{x+2}$

25. $\frac{3}{x^2-8x+16}-\frac{2}{4-x}$

26. $\frac{1}{a^2-1}-\frac{2}{3-3a}$

27. $\frac{3}{x^2-11x+30}-\frac{2}{x^2-25}$

28. $\frac{x-1}{2x^3-4x^2}+\frac{5}{x-2}$

29. $\frac{x-1}{3x^2-13x+4}-\frac{3x+1}{4-x}$

30. $\frac{x}{4x^2-12x+5}+\frac{2x-1}{4x^2-4x-15}$

31. $\frac{t}{t^2-t-6}-\frac{2t}{t^2+6t+9}+\frac{t}{t^2-9}$

32. $\frac{5}{2x^3-3x^2+x}-\frac{x}{x^4-x^2}+\frac{2-x}{2x^2+x-1}$

33. $\frac{1}{x^3+1}+\frac{1}{x+1}-2$

34. $\frac{2}{8-x^3}+\frac{1}{x^2-x-2}$

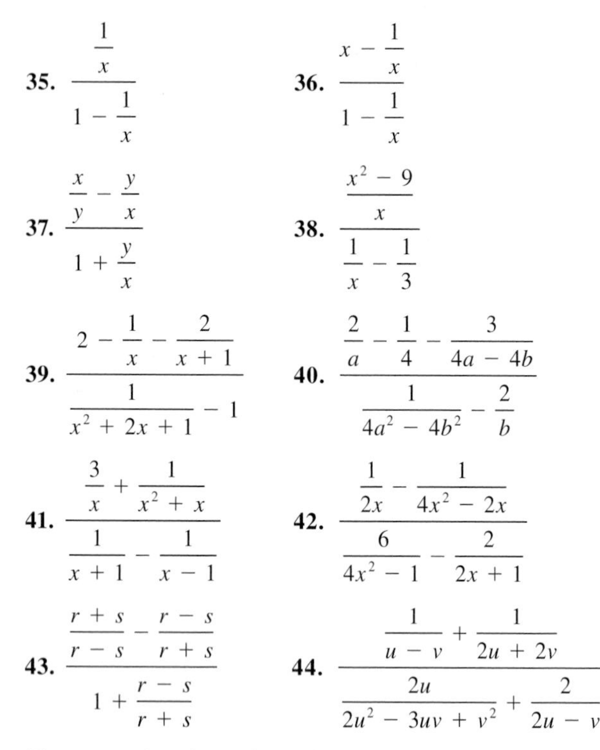

35. $\dfrac{\frac{1}{x}}{1-\frac{1}{x}}$

36. $\dfrac{x-\frac{1}{x}}{1-\frac{1}{x}}$

37. $\dfrac{\frac{x}{y}-\frac{y}{x}}{1+\frac{y}{x}}$

38. $\dfrac{\frac{x^2-9}{x}}{\frac{1}{x}-\frac{1}{3}}$

39. $\dfrac{2-\frac{1}{x}-\frac{2}{x+1}}{\frac{1}{x^2+2x+1}-1}$

40. $\dfrac{\frac{2}{a}-\frac{1}{4}-\frac{3}{4a-4b}}{\frac{1}{4a^2-4b^2}-\frac{2}{b}}$

41. $\dfrac{\frac{3}{x}+\frac{1}{x^2+x}}{\frac{1}{x+1}-\frac{1}{x-1}}$

42. $\dfrac{\frac{1}{2x}-\frac{1}{4x^2-2x}}{\frac{6}{4x^2-1}-\frac{2}{2x+1}}$

43. $\dfrac{\frac{r+s}{r-s}-\frac{r-s}{r+s}}{1+\frac{r-s}{r+s}}$

44. $\dfrac{\frac{1}{u-v}+\frac{1}{2u+2v}}{\frac{2u}{2u^2-3uv+v^2}+\frac{2}{2u-v}}$

The expression $f(x+h)-f(x)$ is frequently used in the study of calculus. (If necessary, refer to Section 3-1 for a review of functional notation.) In Exercises 45–48, determine and then simplify this expression for the given functions.

45. $f(x)=\dfrac{x}{x+1}$

46. $f(x)=\dfrac{3}{2x-1}$

47. $f(x)=\dfrac{1}{x^2}$

48. $f(x)=\dfrac{2}{x^2+4}$

In Exercises 49–56, simplify the indicated expressions.

49. Using the definitions of the trigonometric functions given in Section 4-2, find an expression that is equivalent to $(\tan\theta)(\cot\theta)+(\sin\theta)^2-\cos\theta$ in terms of x, y, and r.

50. Using the definitions of the trigonometric functions given in Section 4-2, find an expression that is equivalent to $\sec\theta-(\cot\theta)^2+\csc\theta$ in terms of x, y, and r.

51. If $f(x)=2x-x^2$, find $f\left(\dfrac{1}{a}\right)$.

52. If $f(x)=x^2+x$, find $f\left(a+\dfrac{1}{a}\right)$.

53. If $f(x)=x-\dfrac{2}{x}$, find $f(a+1)$.

54. If $f(x)=\dfrac{x+1}{2}+\dfrac{3}{x}$, find $f(2a)$.

55. The sum of two numbers a and b is divided by the sum of their reciprocals. Determine the expression for this quotient and then simplify it.

56. The difference of two numbers a and b (subtract b from a) is divided by the difference in their reciprocals. Determine the expression for this quotient and then simplify it.

In Exercises 57–64, perform the indicated operations. Each expression occurs in the indicated area of application.

57. $\dfrac{3}{4\pi}-\dfrac{3H_0}{4\pi H}$ (transistor theory)

58. $1+\dfrac{9}{128T}-\dfrac{27P}{64T^3}$ (thermodynamics)

59. $\dfrac{2n^2-n-4}{2n^2+2n-4}+\dfrac{1}{n-1}$ (optics)

60. $\dfrac{5}{3}+\dfrac{3L}{8Cr}-\dfrac{L^3}{8C^3r^3}$ (steel column safety factor)

61. $1-\dfrac{2r^2}{R^2}+\dfrac{r^4}{R^4}$ (pipeline design)

62. $\dfrac{a}{b^2h}+\dfrac{c}{bh^2}-\dfrac{1}{6bh}$ (strength of materials)

63. $\dfrac{\frac{L}{C}+\frac{R}{sC}}{sL+R+\frac{1}{sC}}$ (electricity)

64. $\dfrac{\frac{m}{c}}{1-\frac{p^2}{c^2}}$ (airfoil design)

6-8 Equations Involving Fractions

Many important equations in science and technology have fractions in them. Although the solution of these equations will still involve the use of the basic operations stated in Section 1-10, an additional procedure can be used to eliminate the fractions and thereby help lead to the solution. The method is to

NOTE ▶ ***multiply each term of the equation by the lowest common denominator.***

The resulting equation will not involve fractions and can be solved by methods previously discussed. The following examples illustrate how to solve equations involving fractions.

EXAMPLE 1 ▶▶ Solve for x: $\dfrac{x}{12} - \dfrac{1}{8} = \dfrac{x+2}{6}$.

We first note that the lowest common denominator of the terms of the equation is 24. Therefore, we multiply each term by 24. This gives

$$\frac{24(x)}{12} - \frac{24(1)}{8} = \frac{24(x+2)}{6} \qquad \text{each term multiplied by LCD}$$

We reduce each term to its lowest terms and solve the resulting equation.

$$2x - 3 = 4(x + 2) \qquad \text{each term reduced}$$
$$2x - 3 = 4x + 8$$
$$-2x = 11$$
$$x = -\frac{11}{2}$$

When we check this solution in the original equation, we obtain $-\frac{7}{12}$ on each side of the equal sign. Therefore, the solution is correct. ◀◀

EXAMPLE 2 ▶▶ Solve for x: $\dfrac{x}{2} - \dfrac{1}{b^2} = \dfrac{x}{2b}$.

We first determine that the lowest common denominator of the terms of the equation is $2b^2$. We then multiply each term by $2b^2$ and continue with the solution.

$$\frac{2b^2(x)}{2} - \frac{2b^2(1)}{b^2} = \frac{2b^2(x)}{2b} \qquad \text{each term multiplied by LCD}$$
$$b^2x - 2 = bx \qquad \text{each term reduced}$$
$$b^2x - bx = 2$$
$$x(b^2 - b) = 2 \qquad \text{factor}$$
$$x = \frac{2}{b^2 - b}$$

Note the use of factoring in arriving at the final result. Checking shows that each side of the original equation is equal to $\dfrac{1}{b^2(b-1)}$. (The check here is somewhat lengthy.) ◀◀

EXAMPLE 3 ▸▸ An equation relating the focal length f of a lens with the object distance p and the image distance q is given below. See Fig. 6-2. Solve for q.

$$f = \frac{pq}{p + q} \quad \text{given equation}$$

Since the only denominator is $p + q$, the LCD is also $p + q$. By first multiplying each term by $p + q$, the solution is completed as follows:

Fig. 6-2

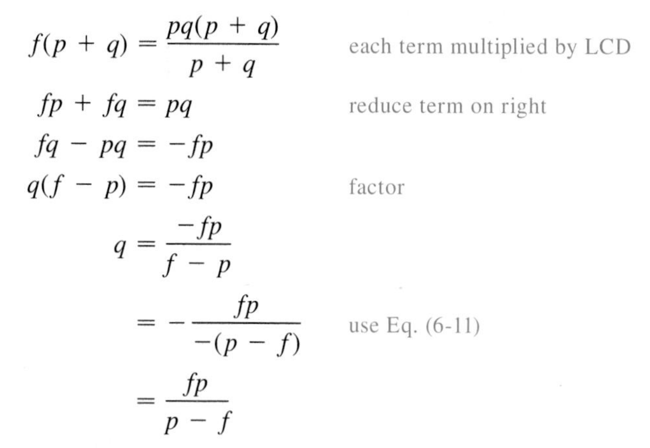

$$
\begin{aligned}
f(p + q) &= \frac{pq(p + q)}{p + q} && \text{each term multiplied by LCD} \\
fp + fq &= pq && \text{reduce term on right} \\
fq - pq &= -fp \\
q(f - p) &= -fp && \text{factor} \\
q &= \frac{-fp}{f - p} \\
&= -\frac{fp}{-(p - f)} && \text{use Eq. (6-11)} \\
&= \frac{fp}{p - f}
\end{aligned}
$$

For reference, Eq. (6-11) is $x - y = -(y - x)$.

The last form is preferred since there is no minus sign before the fraction. However, either form of the result is correct. ◂◂

EXAMPLE 4 ▸▸ When developing the equations that describe the motion of the planets, the equation

$$\frac{1}{2}v^2 - \frac{GM}{r} = -\frac{GM}{2a}$$

is found. Solve for M.

We first determine that the lowest common denominator of the terms of the equation is $2ar$. Multiplying each term by $2ar$ and proceeding, we have

$$
\begin{aligned}
\frac{2ar(v^2)}{2} - \frac{2ar(GM)}{r} &= -\frac{2ar(GM)}{2a} && \text{each term multiplied by LCD} \\
arv^2 - 2aGM &= -rGM && \text{each term reduced} \\
rGM - 2aGM &= -arv^2 \\
M(rG - 2aG) &= -arv^2 && \text{factor} \\
M &= -\frac{arv^2}{rG - 2aG} \\
&= \frac{arv^2}{2aG - rG}
\end{aligned}
$$

The second form of the result is obtained by using Eq. (6-11). Again, note the use of factoring to arrive at the final result. ◂◂

EXAMPLE 5 ▸▸ Solve for x: $\dfrac{2}{x+1} - \dfrac{1}{x} = -\dfrac{2}{x^2+x}$.

Multiplying each term by the lowest common denominator $x(x + 1)$, we have

$$\frac{2(x)(x+1)}{x+1} - \frac{x(x+1)}{x} = -\frac{2x(x+1)}{x(x+1)}$$

Now, simplifying each fraction, we have

$$2x - (x + 1) = -2$$

We now complete the solution.

$$2x - x - 1 = -2$$
$$x = -1$$

CAUTION ▸

Extraneous Solutions

Checking this solution *in the original equation,* we see that we have zero in the denominators of the first and third terms of the equation. Since division by zero is undefined (see Section 1-2), $x = -1$ cannot be a solution. *Thus there is* **no solution** *to this equation.* This example points out clearly why it is necessary to check solutions in the original equation. It also shows that *whenever we multiply each term by a common denominator which* ***contains the unknown,*** *it is possible to obtain a value which is not a solution of the original equation. Such a value is termed an* **extraneous solution.** Only certain equations will lead to extraneous solutions, but we must be careful to identify them when they occur. ◂◂

A number of stated problems give rise to equations involving fractions. The following example illustrates the solution of such a problem.

See the chapter introduction.

EXAMPLE 6 ▸▸ An industrial firm uses a computer system that processes and prints out its data for an average day in 20 h. In order to process the data more rapidly and to handle increased future computer needs, the firm plans to add new components to the system. One set of new components can process the data in 12 h, without the present system. How long would it take the new system, a combination of the present system and the new components, to process the data?

First, we let x = the number of hours for the new system to process the data. Next, we know that it takes the present system 20 h to do it. This means that it processes $\frac{1}{20}$ of the data in one hour, or $\frac{1}{20}x$ of the data in x hours. In the same way, the new components can process $\frac{1}{12}x$ of the data in x hours. When x hours have passed, the new system will have processed all of the data. Therefore,

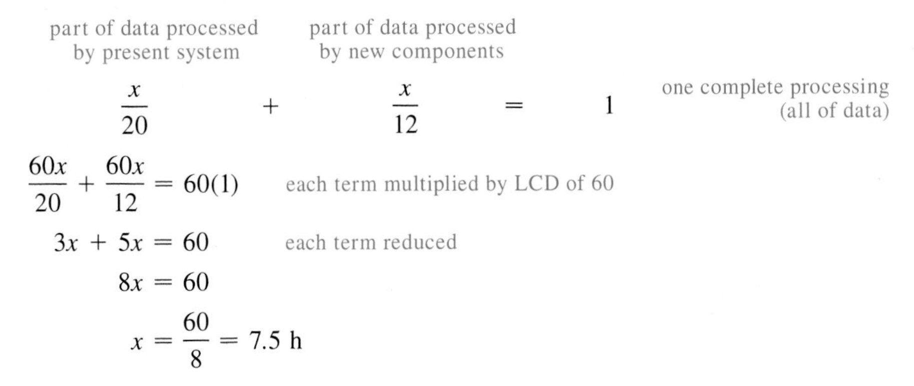

$$\frac{x}{20} + \frac{x}{12} = 1$$

$$\frac{60x}{20} + \frac{60x}{12} = 60(1)$$

$$3x + 5x = 60$$

$$8x = 60$$

$$x = \frac{60}{8} = 7.5 \text{ h}$$

Therefore, the new system should take about 7.5 h to process the data. ◂◂

Exercises 6–8

In Exercises 1–28, solve the given equations and check the results.

1. $\frac{x}{2} + 6 = 2x$
2. $\frac{x}{5} + 2 = \frac{15 + x}{10}$
3. $\frac{x}{6} - \frac{1}{2} = \frac{x}{3}$
4. $\frac{3x}{8} - \frac{3}{4} = \frac{x - 4}{2}$
5. $\frac{1}{2} - \frac{t - 5}{6} = \frac{3}{4}$
6. $\frac{2x - 7}{3} + 5 = \frac{1}{5}$
7. $\frac{3x}{7} - \frac{5}{21} = \frac{2 - x}{14}$
8. $\frac{x - 3}{12} - \frac{2}{3} = \frac{1 - 3x}{2}$
9. $\frac{3}{x} + 2 = \frac{5}{3}$
10. $\frac{1}{2y} - \frac{1}{2} = 4$
11. $3 - \frac{x - 2}{5x} = \frac{1}{5}$
12. $\frac{1}{2x} - \frac{1}{3} = \frac{2}{3x}$
13. $\frac{2y}{y - 1} = 5$
14. $\frac{x}{2x - 3} = 4$
15. $\frac{2}{s} = \frac{3}{s - 1}$
16. $\frac{5}{n + 2} = \frac{3}{2n}$
17. $\frac{5}{2x + 4} + \frac{3}{x + 2} = 2$
18. $\frac{3}{4x - 6} + \frac{1}{4} = \frac{5}{2x - 3}$
19. $\frac{2}{z - 5} - \frac{3}{10 - 2z} = 3$
20. $\frac{4}{4 - x} + 2 - \frac{2}{12 - 3x} = \frac{1}{3}$
21. $\frac{1}{x} + \frac{3}{2x} = \frac{2}{x + 1}$
22. $\frac{3}{t + 3} - \frac{1}{t} = \frac{5}{2t + 6}$
23. $\frac{7}{y} = \frac{3}{y - 4} + \frac{7}{2y^2 - 8y}$
24. $\frac{1}{2x + 3} = \frac{5}{2x} - \frac{4}{2x^2 + 3x}$
25. $\frac{1}{x^2 - x} - \frac{1}{x} = \frac{1}{x - 1}$
26. $\frac{2}{x^2 - 1} - \frac{2}{x + 1} = \frac{1}{x - 1}$
27. $\frac{2}{x^2 - 4} - \frac{1}{x - 2} = \frac{1}{2x + 4}$
28. $\frac{2}{2x^2 + 5x - 3} - \frac{1}{4x - 2} + \frac{3}{2x + 6} = 0$

In Exercises 29–32, solve for the indicated letter.

29. $2 - \frac{1}{b} + \frac{3}{c} = 0$, for c
30. $\frac{2}{3} - \frac{h}{x} = \frac{1}{6x}$, for x
31. $\frac{t - 3}{b} - \frac{t}{2b - 1} = \frac{1}{2}$, for t
32. $\frac{1}{a^2 + 2a} - \frac{y}{2a} = \frac{2y}{a + 2}$, for y

In Exercises 33–44, each of the given formulas arises in the technical or scientific area of study listed. Solve for the indicated letter.

33. $n = n_1 - \frac{n_1 v}{V}$, for v (acoustics)
34. $S = \frac{P}{A} + \frac{Mc}{I}$, for P (machine design)
35. $V_0 = \frac{V_r A}{1 + \beta A}$, for β (electricity)
36. $K = \frac{ax}{x + b}$, for x (medicine)
37. $z = \frac{1}{g_m} - \frac{jX}{g_m R}$, for R (FM transmission)
38. $A = \frac{1}{2}wp - \frac{1}{2}w^2 - \frac{\pi}{8}w^2$, for p (architecture)
39. $P = \frac{RT}{V - b} - \frac{a}{V^2}$, for T (thermodynamics)
40. $\frac{1}{x} + \frac{1}{nx} = \frac{1}{f}$, for n (photography)
41. $R = \frac{L_1}{kA_1} + \frac{L_2}{kA_2}$, for L_1 (air conditioning)
42. $D = \frac{wx^4}{24EI} - \frac{wLx^3}{6EI} + \frac{wL^2x^2}{4EI}$, for w (beam design)
43. $\frac{1}{f} = (n - 1)\left(\frac{1}{R_1} + \frac{1}{R_2}\right)$, for R_1 (optics)
44. $P = \frac{\frac{1}{1 + i}}{1 - \frac{1}{1 + i}}$, for i (business)

In Exercises 45–52, set up appropriate equations and solve the given stated problems. All numbers are accurate to at least two significant digits.

45. One pipe can fill a certain oil storage tank in 4.0 h, and a second pipe can fill it in 6.0 h. How long will it take to fill the tank if both pipes operate together?
46. One company determines that it will take its crew 450 h to clean up a chemical dump site, and a second company determines that it will take its crew 600 h to clean up the site. How long will it take the two crews working together?
47. One automatic packaging machine can package 100 boxes of machine parts in 12 min, and a second machine can do it in 10 min. A newer model machine can do it in 8.0 min. How long would it take the three machines working together?

48. A painting crew can paint a structure in 12 h and can paint the structure in 7.2 h when working with a second crew. How long would it take the second crew to do the job if working alone?

49. A jet takes the same time to travel 2580 km with the wind as it does to travel 1800 km against the wind. If its speed relative to the air is 450 km/h, what is the speed of the wind?

50. A person travels from Winnipeg to Sioux City, Iowa, on a train that averages 60 km/h. After spending 8 h in Sioux City, the person returns to Winnipeg on a jet that averages 900 km/h. If the total trip takes 24 h, how far is Winnipeg from Sioux City?

51. An industrial cleaning solution is to be $\frac{2}{5}$ acid. How many litres of pure acid must be added to 10 L of a solution that is $\frac{1}{4}$ acid to get the proper mixture?

52. How many grams of an alloy that is $\frac{3}{8}$ copper must be mixed with an alloy that is $\frac{2}{3}$ copper in order to get 100 g of alloy that is $\frac{1}{2}$ copper?

Chapter Equations, Review Exercises, and Practice Test

Chapter Equations

$$a(x + y) = ax + ay \quad (6\text{-}1)$$

$$(x + y)(x - y) = x^2 - y^2 \quad (6\text{-}2)$$

$$(x + y)^2 = x^2 + 2xy + y^2 \quad (6\text{-}3)$$

$$(x - y)^2 = x^2 - 2xy + y^2 \quad (6\text{-}4)$$

$$(x + a)(x + b) = x^2 + (a + b)x + ab \quad (6\text{-}5)$$

$$(ax + b)(cx + d) = acx^2 + (ad + bc)x + bd \quad (6\text{-}6)$$

$$(x + y)^3 = x^3 + 3x^2y + 3xy^2 + y^3 \quad (6\text{-}7)$$

$$(x - y)^3 = x^3 - 3x^2y + 3xy^2 - y^3 \quad (6\text{-}8)$$

$$(x + y)(x^2 - xy + y^2) = x^3 + y^3 \quad (6\text{-}9)$$

$$(x - y)(x^2 + xy + y^2) = x^3 - y^3 \quad (6\text{-}10)$$

$$x - y = -(y - x) \quad (6\text{-}11)$$

Review Exercises

In Exercises 1–12, find the products by inspection. No intermediate steps should be necessary.

1. $3a(4x + 5a)$

2. $-7xy(4x^2 - 7y)$

3. $(2a + 7b)(2a - 7b)$

4. $(x - 4z)(x + 4z)$

5. $(2a + 1)^2$

6. $(4x - 3y)^2$

7. $(b - 4)(b + 7)$

8. $(y - 5)(y - 7)$

9. $(2x + 5)(x - 9)$

10. $(4ax - 3)(5ax + 7)$

11. $(2c + d)(8c - d)$

12. $(3s - 2t)(8s + 3t)$

In Exercises 13–44, factor the given expressions completely.

13. $3s + 9t$

14. $7x - 28y$

15. $a^2x^2 + a^2$

16. $3ax - 6ax^4 - 9a$

17. $x^2 - 144$

18. $900 - n^2$

19. $16(x + 2)^2 - t^4$

20. $25s^4 - 36t^2$

21. $9t^2 - 6t + 1$

22. $4x^2 - 12x + 9$

23. $25t^2 + 10t + 1$

24. $4x^2 + 36xy + 81y^2$

25. $x^2 + x - 56$

26. $x^2 - 4x - 45$

27. $t^2 - 5t - 36$

28. $n^2 - 11n + 10$

29. $2x^2 - x - 36$

30. $5x^2 + 2x - 3$

31. $4x^2 - 4x - 35$

32. $9x^2 + 7x - 16$

33. $10b^2 + 23b - 5$

34. $12x^2 - 7xy - 12y^2$

35. $4x^2 - 64y^2$

36. $4a^2x^2 + 26a^2x + 36a^2$

37. $250 - 16y^6$

38. $a^4 + 64a$

39. $8x^3 + 27$

40. $x^3 - 125a^3$

41. $ab^2 - 3b^2 + a - 3$

42. $axy - ay + ax - a$

43. $nx + 5n - x^2 + 25$

44. $ty - 4t + y^2 - 16$

In Exercises 45–68, perform the indicated operations and express results in simplest form.

45. $\dfrac{48ax^3y^6}{9a^3xy^6}$

46. $\dfrac{-39r^2s^4t^8}{52rs^5t}$

47. $\dfrac{6x^2-7x-3}{4x^2-8x+3}$

48. $\dfrac{x^2-3x-4}{x^2-x-12}$

49. $\dfrac{4x+4y}{35x^2}\times\dfrac{28x}{x^2-y^2}$

50. $\left(\dfrac{6x-3}{x^2}\right)\left(\dfrac{4x^2-12x}{12x-6}\right)$

51. $\dfrac{18-6x}{x^2-6x+9}\div\dfrac{x^2-2x-15}{x^2-9}$

52. $\dfrac{6x^2-xy-y^2}{2x^2+xy-y^2}\div\dfrac{4x^2-16y^2}{x^2+3xy+2y^2}$

53. $\dfrac{\dfrac{3x}{7x^2+13x-2}}{\dfrac{6x^2}{x^2+4x+4}}$

54. $\dfrac{\dfrac{3x-3y}{2x^2+3xy-2y^2}}{\dfrac{3x^2-3y^2}{x^2+4xy+4y^2}}$

55. $\dfrac{x+\dfrac{1}{x}+1}{x^2-\dfrac{1}{x}}$

56. $\dfrac{\dfrac{4}{y}-4y}{2-\dfrac{2}{y}}$

57. $\dfrac{4}{9x}-\dfrac{5}{12x^2}$

58. $\dfrac{3}{10a^2}+\dfrac{1}{4a^3}$

59. $\dfrac{6}{x}-\dfrac{7}{2x}+\dfrac{3}{xy}$

60. $\dfrac{4}{a^2b}-\dfrac{5}{2ab}+\dfrac{1}{2b}$

61. $\dfrac{a+1}{a+2}-\dfrac{a+3}{a}$

62. $\dfrac{y}{y+2}-\dfrac{1}{y^2+2y}$

63. $\dfrac{2x}{x^2+2x-3}-\dfrac{1}{x^2+3x}$

64. $\dfrac{x}{4x^2+4x-3}-\dfrac{3}{4x^2-9}$

65. $\dfrac{3x}{2x^2-2}-\dfrac{2}{4x^2-5x+1}$

66. $\dfrac{2x-1}{4-x}+\dfrac{x+2}{5x-20}$

67. $\dfrac{3x}{x^2+2x-3}-\dfrac{2}{x^2+3x}+\dfrac{x}{x-1}$

68. $\dfrac{3}{y^4-2y^3-8y^2}+\dfrac{y-1}{y^2+2y}-\dfrac{y-3}{y^2-4y}$

In Exercises 69–76, solve the given equations.

69. $\dfrac{x}{2}-3=\dfrac{x-10}{4}$

70. $\dfrac{x}{6}-\dfrac{1}{2}=\dfrac{3-x}{12}$

71. $\dfrac{2x}{c}-\dfrac{1}{2c}=\dfrac{3}{c}-x$, for x

72. $\dfrac{x}{2a}-b+\dfrac{x}{2c}=\dfrac{a}{b}-c$, for x

73. $\dfrac{2}{t}-\dfrac{1}{at}=2+\dfrac{a}{t}$, for t

74. $\dfrac{3}{a^2y}-\dfrac{1}{ay}=\dfrac{9}{a}$, for y

75. $\dfrac{2x}{x^2-3x}-\dfrac{3}{x}=\dfrac{1}{2x-6}$

76. $\dfrac{3}{x^2+3x}-\dfrac{1}{x}=\dfrac{1}{x+3}$

In Exercises 77–92, perform the given operations. Where indicated, the expression is found in the stated technical area.

77. Show that $xy=\dfrac{1}{4}[(x+y)^2-(x-y)^2]$.

78. Show that $x^2+y^2=\dfrac{1}{2}[(x+y)^2+(x-y)^2]$.

79. Multiply: $2zS(S+1)$. (solid state physics)

80. Expand: $[2b+(n-1)\lambda]^2$. (optics)

81. Factor: $\pi r_1^2 l-\pi r_2^2 l$. (jet plane fuel supply)

82. Factor: $cT_2-cT_1+RT_2-RT_1$. (pipeline flow)

83. Factor: $256+96t-16t^2$. (projectile motion)

84. Factor: $4x^3-20x^2+25x$. (mechanics: center of mass)

85. Express in factored form: $(t+1)^2-2t(t+1)$. (velocity)

86. Express in factored form: $2R(R+r)-(R+r)^2$. (electricity: power)

87. Expand and simplify: $(n+1)^3(2n+1)^3$. (fluid flow in pipes)

88. Expand and simplify: $2(e_1-e_2)^2+2(e_2-e_3)^2$. (mechanical design)

89. Expand and simplify: $10a(T-t)+a(T-t)^2$. (instrumentation)

90. Expand the third term and then factor by grouping: $pa^2+(1-p)b^2-[pa+(1-p)b]^2$. (nuclear physics)

91. A metal cube of edge x is heated and each edge increases by 4 mm. Express the increase in volume in factored form.

92. A machine part is made from a rectangular metal plate (see Fig. 6-3) by cutting out a square piece and a rectangular piece (shaded). Express the area of the face of the machine part in factored form.

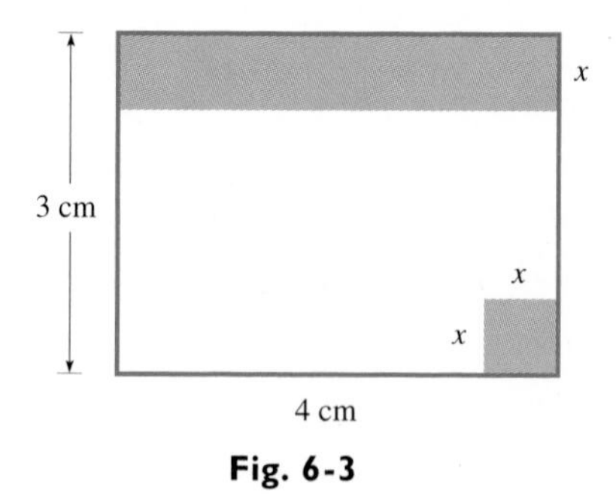

Fig. 6-3

In Exercises 93–104, perform the indicated operations and simplify the given expressions. Each expression is from the indicated technical area of application.

93. $\left(\dfrac{2wtv^2}{Dg}\right)\left(\dfrac{b\pi^2D^2}{n^2}\right)\left(\dfrac{6}{bt^2}\right)$ (machine design)

94. $\dfrac{m}{c} \div \left[1 - \left(\dfrac{p}{c}\right)^2\right]$ (airfoil design)

95. $10t + \dfrac{3}{2}t^2 + \dfrac{1}{3}t^3$ (business)

96. $\dfrac{V}{kp} - \dfrac{RT}{k^2p^2}$ (electric motors)

97. $1 - \dfrac{d^2}{2} + \dfrac{d^4}{24} - \dfrac{d^6}{120}$ (aircraft emergency locator transmitter)

98. $\dfrac{wx^2}{2T_0} + \dfrac{kx^4}{12T_0}$ (bridge design)

99. $\dfrac{N+n}{2} + \dfrac{(N-n)^2}{4\pi^2C}$ (machine design)

100. $\dfrac{Am}{k} - \dfrac{g}{2}\left(\dfrac{m}{k}\right)^2 + \dfrac{AML}{k}$ (rocket fuel)

101. $1 - \dfrac{3a}{4r} - \dfrac{a^3}{4r^3}$ (hydrodynamics)

102. $\dfrac{1}{F} + \dfrac{1}{f} - \dfrac{d}{fF}$ (optics)

103. $\dfrac{\dfrac{u^2}{2g} - x}{\dfrac{1}{2gc^2} - \dfrac{u^2}{2g} + x}$ (mechanism design)

104. $\dfrac{V}{\dfrac{1}{2R} + \dfrac{1}{2R+2}}$ (electricity)

In Exercises 105–112, solve for the indicated letter. Each equation is from the indicated technical area of application.

105. $\dfrac{q_2 - q_1}{d} = \dfrac{f + q_1}{D}$, for q_1 (photography)

106. $\dfrac{110}{R} = \dfrac{180}{R+5}$, for R (electricity)

107. $R = \dfrac{wL}{H(w+L)}$, for L (architecture)

108. $y = \dfrac{1000a - bx}{x + a}$, for x (production of medication)

109. $s^2 + \dfrac{cs}{m} + \dfrac{kL^2}{mb^2} = 0$, for c (mechanical vibrations)

110. $I = \dfrac{A}{x^2} + \dfrac{B}{(10-x)^2}$, for A (optics)

111. $X = \dfrac{H}{RT_1} - \dfrac{H}{RT}$, for T (chemistry)

112. $f = \dfrac{f_0 v_0}{v_0 - v}$, for v_0 (acoustics)

In Exercises 113–120, set up appropriate equations and solve the given stated problems. All numbers are accurate to at least two significant digits.

113. If a certain car's lights are left on, the battery will be dead in 4.0 h. If only the radio is left on, the battery will be dead in 24 h. How long will the battery last if both the lights and the radio are left on?

114. Two pumps are being used to fight a fire. One pumps 5000 L in 20 min, and the other pumps 5000 L in 25 min. How long will it take the two pumps together to pump 5000 L?

115. One computer can solve a certain problem in 30 s. With the aid of a second computer, the problem is solved in 10 s. How long would the second computer take to solve the problem alone?

116. An auto mechanic can do a certain motor job in 3.0 h, and with an assistant he can do it in 2.1 h. How long would it take the assistant to do the job alone?

117. The *relative density* of an object may be defined as its weight in air w_a divided by the difference of its weight in air and its weight when submerged in water, w_w. For a lead weight, $w_a = 1.097\ w_w$. Find the relative density of lead.

118. A car travels halfway to its destination at 80.0 km/h and the remainder of the distance at 60.0 km/h. What is the average speed of the car for the trip?

119. For electric resistors in parallel, the reciprocal of the combined resistance equals the sum of the reciprocals of the individual resistances. For three resistors of 12 Ω, R ohms, and $2R$ ohms in parallel, the combined resistance is 6.0 Ω. Find R.

120. An ambulance averaged 36 km/h going to an accident and 48 km/h on its return to the hospital. If the total time for the round-trip was 40 min, including 5 min at the accident scene, how far from the hospital was the accident?

Writing Exercise

121. An architecture student encounters the fraction

$$\frac{2r^2 + 5r - 3}{2r^2 + 7r + 3}$$

Simplify this fraction, and write a paragraph to describe your procedure. When you "cancel," explain what basic operation is being performed.

Practice Test

1. Find the product: $2x(2x - 3)^2$.
2. The following equation is used in electricity.

   Solve for R_1: $\dfrac{1}{R} = \dfrac{1}{R_1 + r} + \dfrac{1}{R_2}$.
3. Reduce to simplest form: $\dfrac{2x^2 + 5x - 3}{2x^2 + 12x + 18}$.
4. Factor: $4x^2 - 16y^2$.

In Problems 5–7, perform the indicated operations and simplify.

5. $\dfrac{3}{4x^2} - \dfrac{2}{x^2 - x} - \dfrac{x}{2x - 2}$
6. $\dfrac{x^2 + x}{2 - x} \div \dfrac{x^2}{x^2 - 4x + 4}$
7. $\dfrac{1 - \dfrac{3}{2x + 2}}{\dfrac{x}{5} - \dfrac{1}{2}}$
8. If one riveter can do a job in 12 days, and a second riveter can do it in 16 days, how long would it take for them to do it together?

7

Quadratic Equations

The solution of basic equations was introduced in Chapter 1. Then, in Chapter 5, we extended the solution of equations to systems of linear equations. There are many other types of equations that we shall discuss. Among these is the important quadratic equation. To develop the methods of solving a quadratic equation, we use factoring and the algebraic operations with fractions discussed in Chapter 6.

There are many applications of the use of quadratic equations. These include projectile motion, electric circuits, architecture, mechanical systems, and product design.

In the first three sections of this chapter we present algebraic methods of solving quadratic equations. In Section 7-4 we discuss the graphical solution.

The frame around the panels of a door affects the strength and appearance of the door. In Section 7-3 we see how the design of the frame of a particular door involves the solution of the quadratic equation $6x^2 - 6.90x + 0.69 = 0$.

7-1 Quadratic Equations; Solution by Factoring

Given that a, b, and c are constants, the equation

$$ax^2 + bx + c = 0 \tag{7-1}$$

is called the **general quadratic equation in *x*.** The left side of Eq. (7-1) is a polynomial function of degree 2. *This function, $f(x) = ax^2 + bx + c$, is known as the* **quadratic function.** Any equation that can be simplified and then written in the form of Eq. (7-1) is a quadratic equation in one unknown.

Among the applications of quadratic equations and functions we have the following examples: In finding the time t of flight of a projectile, we have the equation $s_0 + v_0t - 4.9t^2 = 0$; in analyzing the electric current i in a circuit, the function $f(i) = Ei - Ri^2$ is found; and in determining the forces at a distance x along a beam, the function $f(x) = ax^2 + bLx + cL^2$ is used.

Since it is the x^2-term in Eq. (7-1) that distinguishes the quadratic equation from other types of equations, the equation is not quadratic if $a = 0$. However, either b or c (or both) may be zero, and the equation is still quadratic. No power of x higher than the second may be present in a quadratic equation. Also, we should be able to properly identify a quadratic equation even when it does not initially appear in the form of Eq. (7-1). The following two examples illustrate how we may recognize quadratic equations.

EXAMPLE 1 ▶▶ The following are quadratic equations.

$$\underset{a=1}{x^2} \; - \; \underset{b=-4}{4x} \; - \; \underset{c=-5}{5} = 0$$

To show this equation in the form of Eq. (7-1), it can be written as $1x^2 + (-4)x + (-5) = 0$.

$$\underset{a=3}{3x^2} - \underset{c=-6}{6} = 0$$

Since there is no x-term, $b = 0$.

$$\underset{a=2}{2x^2} + \underset{b=7}{7x} = 0$$

Since no constant appears, $c = 0$.

$$(a - 3)x^2 - ax + 7 = 0$$

The constants in Eq. (7-1) may include literal expressions. In this case, $a - 3$ takes the place of a, $-a$ takes the place of b, and $c = 7$.

$$4x^2 - 2x = x^2$$

After all nonzero terms have been collected on the left side, the equation becomes $3x^2 - 2x = 0$.

$$(x + 1)^2 = 4$$

Expanding the left side and collecting all nonzero terms on the left, we have $x^2 + 2x - 3 = 0$. ◀◀

EXAMPLE 2 ▶▶ The following are not quadratic equations.

$bx - 6 = 0$ — There is no x^2-term.

$x^3 - x^2 - 5 = 0$ — There should be no term of degree higher than 2. Thus there can be no x^3-term in a quadratic equation.

$x^2 + x - 7 = x^2$ — When terms are collected, there will be no x^2-term. ◀◀

Solutions of a Quadratic Equation

From our previous work with equations, we recall that *the* **solution** *of an equation consists of all numbers* (**roots**) *which, when substituted in the equation, produce equality.* There are ***two*** such roots for a quadratic equation. Occasionally these roots turn out to be equal, as is shown in Example 3, and only one number is actually a solution. Also, due to the presence of the x^2-term, the roots may be imaginary numbers. If the roots are imaginary, at this point all we wish to do is to recognize them when they occur.

EXAMPLE 3 ▸▸ (a) The quadratic equation

$$3x^2 - 7x + 2 = 0$$

has roots $x = 1/3$ and $x = 2$. This can be seen by substituting in the equation.

$$3\left(\frac{1}{3}\right)^2 - 7\left(\frac{1}{3}\right) + 2 = 3\left(\frac{1}{9}\right) - \frac{7}{3} + 2 = \frac{1}{3} - \frac{7}{3} + 2 = \frac{0}{3} = 0$$

$$3(2)^2 - 7(2) + 2 = 3(4) - 14 + 2 = 12 - 14 + 2 = 0$$

(b) The quadratic equation

$$4x^2 - 4x + 1 = 0$$

has the **double root** *(both roots are the same)* of $x = \frac{1}{2}$. This can be seen to be a solution by substitution:

$$4\left(\frac{1}{2}\right)^2 - 4\left(\frac{1}{2}\right) + 1 = 4\left(\frac{1}{4}\right) - 2 + 1 = 1 - 2 + 1 = 0$$

(c) The quadratic equation

$$x^2 + 9 = 0$$

has the imaginary roots of $\sqrt{-3}$ and $-\sqrt{-3}$. ◂◂

In this section we shall deal only with quadratic equations whose quadratic expression is factorable. Therefore, all roots will be rational. Using the fact that

a product is zero if any of its factors is zero

we have the following steps in solving a quadratic equation.

> ***Procedure for Solving a Quadratic Equation by Factoring***
>
> 1. *Collect all terms on the left and simplify* (to the form of Eq. (7-1)).
> 2. *Factor the quadratic expression.*
> 3. *Set each factor equal to zero.*
> 4. *Solve the resulting linear equations.*
> 5. *Check the solutions in the original equation.*

The solutions of the resulting linear equations make up the solution of the quadratic equation.

EXAMPLE 4 ▸▸ $x^2 - x - 12 = 0$

$(x - 4)(x + 3) = 0$ factor

$x - 4 = 0$ $x + 3 = 0$ set each factor equal to zero

$x = 4$ $x = -3$ solve

The roots are $x = 4$ and $x = -3$. We can check them in the original equation by substitution. Therefore, we have

$$(4)^2 - (4) - 12 \stackrel{?}{=} 0 \qquad (-3)^2 - (-3) - 12 \stackrel{?}{=} 0$$

$$0 = 0 \qquad 0 = 0$$

Both roots satisfy the original equation. ◂◂

EXAMPLE 5 ▸▸ $2x^2 + 7x - 4 = 0$

$$(2x - 1)(x + 4) = 0 \qquad \text{factor}$$

$$2x - 1 = 0, \quad \text{or} \quad x = \frac{1}{2} \qquad \text{set each factor equal to zero and solve}$$

$$x + 4 = 0, \quad \text{or} \quad x = -4$$

Therefore, the roots are $x = \frac{1}{2}$ and $x = -4$. These roots can be checked by the same procedure used in Example 4. ◂◂

EXAMPLE 6 ▸▸ $x^2 + 4 = 4x$ — equation not in form of Eq. (7-1)

$$x^2 - 4x + 4 = 0 \qquad \text{subtract } 4x \text{ from both sides}$$

$$(x - 2)^2 = 0 \qquad \text{factor}$$

$$x - 2 = 0, \quad \text{or} \quad x = 2 \qquad \text{solve}$$

Since $(x - 2)^2 = (x - 2)(x - 2)$, both factors are the same. This means there is a double root of $x = 2$. Substitution shows that $x = 2$ satisfies the original equation. ◂◂

CAUTION ▸ *It is essential for the quadratic expression on the left to be equal to zero* (on the right), because if a product equals a nonzero number, it is probable that neither factor will give us a correct root. Again, the first step must be to write the equation in the form of Eq. (7-1).

A number of equations involving fractions lead to quadratic equations after the fractions are eliminated. The following two examples, the second being a stated problem, illustrate the process of solving such equations with fractions.

EXAMPLE 7 ▸▸ Solve for x: $\dfrac{1}{x} + 3 = \dfrac{2}{x + 2}$.

$$\frac{x(x + 2)}{x} + 3x(x + 2) = \frac{2x(x + 2)}{x + 2} \qquad \text{multiply each term by the LCD, } x(x + 2)$$

$$x + 2 + 3x^2 + 6x = 2x \qquad \text{reduce each term}$$

$$3x^2 + 5x + 2 = 0 \qquad \text{collect terms on left}$$

$$(3x + 2)(x + 1) = 0 \qquad \text{factor}$$

$$3x + 2 = 0, \quad \text{or} \quad x = -\tfrac{2}{3} \qquad \text{set each factor equal to zero and solve}$$

$$x + 1 = 0, \quad \text{or} \quad x = -1$$

Checking in the original equation, we have

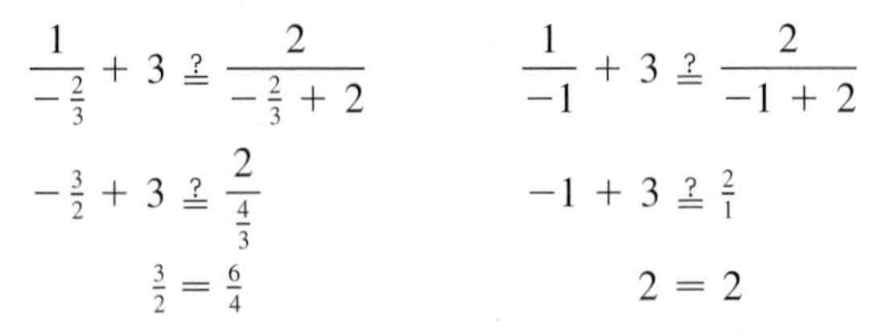

$$\frac{1}{-\frac{2}{3}} + 3 \stackrel{?}{=} \frac{2}{-\frac{2}{3} + 2} \qquad \frac{1}{-1} + 3 \stackrel{?}{=} \frac{2}{-1 + 2}$$

$$-\tfrac{3}{2} + 3 \stackrel{?}{=} \frac{2}{\frac{4}{3}} \qquad -1 + 3 \stackrel{?}{=} \tfrac{2}{1}$$

$$\tfrac{3}{2} = \tfrac{6}{4} \qquad 2 = 2$$

We see that the roots check. Remember, if either value gives division by zero, the root is extraneous. ◂◂

EXAMPLE 8 A lumber truck travels to and from a lumber camp 60 km distant in 7 h. If the average speed on the return trip is 5 km/h less than on the trip to the camp, what was the average speed of the truck going to the camp?

Let v = the average speed (in km/h) of the truck going to the camp. This means that the average speed of the return trip was $v - 5$ km/h.

We also know that $d = vt$ (distance equals speed times time), which tells us that $t = d/v$. Thus, the time for each part of the trip is the distance divided by the speed. This leads to the equation

time to camp, time from camp, total time

$$\frac{60}{v} + \frac{60}{v-5} = 7$$

$$60(v - 5) + 60v = 7v(v - 5) \quad \text{multiply each term by } v(v-5)$$

$$7v^2 - 155v + 300 = 0 \quad \text{collect terms on the left}$$

$$(7v - 15)(v - 20) = 0 \quad \text{factor}$$

$$7v - 15 = 0 \quad \text{or} \quad v = \frac{15}{7} \quad \text{set each factor equal to zero and solve}$$

$$v - 20 = 0 \quad \text{or} \quad v = 20$$

The factors lead to two possible solutions, but only one has meaning for this problem. The value $v = \frac{15}{7}$ cannot be the solution, since the return speed of 5 km/h less would be negative. Therefore, the solution is $v = 20$ km/h, which means that the return speed was 15 km/h. Thus, the trip to the camp took 3 h and the return trip took 4 h, which therefore shows that the solution checks.

Exercises 7–1

In Exercises 1–8, determine whether or not the given equations are quadratic. If the resulting form is quadratic, identify a, b, and c, with a > 0.

1. $x^2 + 5 = 8x$
2. $5x^2 = 9 - x$
3. $x(x - 2) = 4$
4. $(3x - 2)^2 = 2$
5. $x^2 = (x + 2)^2$
6. $x(2x + 5) = 7 + 2x^2$
7. $x(x^2 + x - 1) = x^3$
8. $(x - 7)^2 = (2x + 3)^2$

In Exercises 9–44, solve the given quadratic equations by factoring.

9. $x^2 - 4 = 0$
10. $x^2 - 400 = 0$
11. $4y^2 - 9 = 0$
12. $x^2 - 0.16 = 0$
13. $x^2 - 8x - 9 = 0$
14. $s^2 + s - 6 = 0$
15. $x^2 - 7x + 12 = 0$
16. $x^2 - 11x + 30 = 0$
17. $x^2 = -2x$
18. $x^2 = 7x$
19. $27m^2 = 3$
20. $5p^2 = 80$
21. $3x^2 - 13x + 4 = 0$
22. $7x^2 + 3x - 4 = 0$
23. $x^2 + 8x + 16 = 0$
24. $4x^2 - 20x + 25 = 0$
25. $6x^2 = 13x - 6$
26. $6z^2 = 6 + 5z$
27. $4x(x + 1) = 3$
28. $9t^2 = 9 - t(43 + t)$
29. $x^2 - x - 1 = 1$
30. $2x^2 - 7x + 6 = 3$
31. $x^2 - 4b^2 = 0$
32. $a^2x^2 - 1 = 0$
33. $40x - 16x^2 = 0$
34. $15x = 20x^2$
35. $8s^2 + 16s = 90$
36. $18t^2 - 48t + 32 = 0$
37. $(x + 2)^3 = x^3 + 8$
38. $x(x^2 - 4) = x^2(x - 1)$
39. $(x + a)^2 - b^2 = 0$
40. $x^2(a^2 + 2ab + b^2) - x(a + b) = 0$

41. The bending moment M of a simply supported beam of length L with a uniform load of w kg/m at a distance x from one end is $M = \frac{1}{2}wLx - \frac{1}{2}wx^2$. For what values of x is $M = 0$?

42. The mass m, in megagrams, of the fuel supply in the first-stage booster of a rocket is $m = 135 - 6t - t^2$, where t is the time, in seconds, after launch. When does the booster run out of fuel?

43. The power, in megawatts, produced between midnight and noon by a nuclear power plant is given by $P = 4h^2 - 48h + 744$, where h is the hour of the day. At what time is the power 664 MW?

44. In determining the speed s (in km/h) of a car while studying its fuel economy, the equation $s^2 - 16s = 3072$ is used. Find s.

In Exercises 45–48, solve the given equations involving fractions.

45. $\frac{1}{x-3} + \frac{4}{x} = 2$

46. $2 - \frac{1}{x} = \frac{3}{x+2}$

47. $\frac{1}{2x} - \frac{3}{4} = \frac{1}{2x+3}$

48. $\frac{x}{2} + \frac{1}{x-3} = 3$

In Exercises 49–52, set up the appropriate quadratic equations and solve.

49. The spring constant k is the force F divided by the amount x it stretches ($k = F/x$). See Fig. 7-1(a). For two springs in series (see Fig. 7-1(b)), the reciprocal of the spring constant k_c for the combination equals the sum of the reciprocals of the individual spring constants. Find the spring constants for each of two springs in series if $k_c = 2$ N/cm and one spring constant is 3 N/cm more than the other.

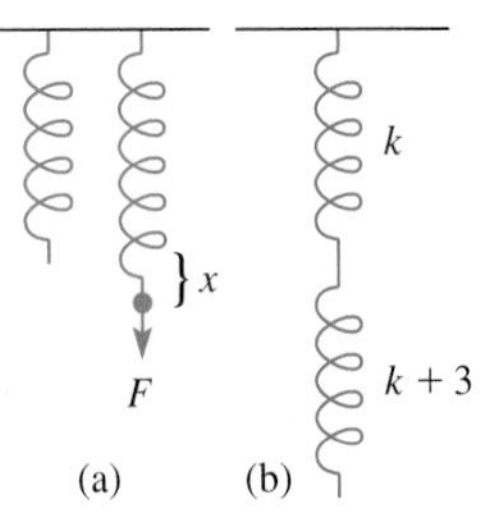

Fig. 7-1 (a) (b)

50. The reciprocal of the combined resistance R of two resistances R_1 and R_2 connected in parallel (see Fig. 7-2(a)) is equal to the sum of the reciprocals of the individual resistances. If the two resistances are connected in series (see Fig. 7-2(b)), their combined resistance is the sum of their individual resistances. If two resistances connected in parallel have a combined resistance of 3.0 Ω, and the same two resistances have a combined resistance of 16 Ω when connected in series, what are the resistances?

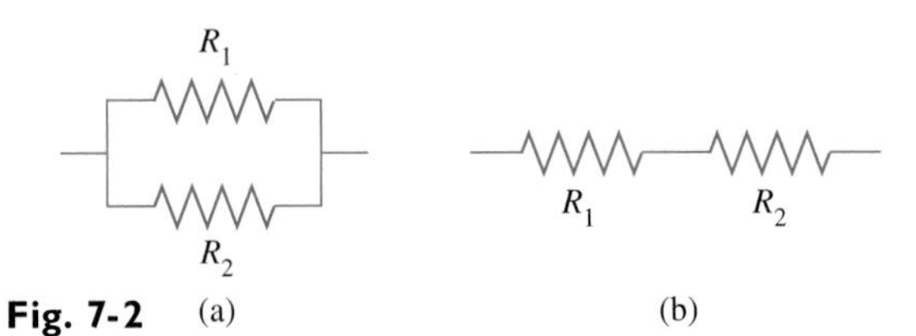

Fig. 7-2 (a) (b)

51. A jet, by increasing its speed by 200 km/h, could decrease the time needed to cover 4000 km by 1 h. What is its speed?

52. A rectangular solar panel is 20 cm by 30 cm. By adding the same amount to each dimension, the area is doubled. How much is added?

7-2 Completing the Square

Many quadratic equations cannot be solved by factoring. This is true of most quadratic equations that arise in applications. Therefore, we now develop a method that can be used to solve any quadratic equation. *The method is called* **completing the square.** In the next section we shall use completing the square to develop a general formula that can be used to solve any quadratic equation.

In the first example that follows we show the solution of a type of quadratic equation which arises while using the method of completing the square. In the examples that follow it, the method itself is used and described.

EXAMPLE 1 ▸▸ In solving $x^2 = 16$, we may write it as $x^2 - 16 = 0$ and complete the solution by factoring. This gives us $x = 4$ and $x = -4$. Therefore, we see that the principal square root of 16 and its negative both satisfy the original equation. Thus, we may solve $x^2 = 16$ by equating x to the principal square root of 16 and to its negative. The roots of $x^2 = 16$ are 4 and -4.

We may solve $(x - 3)^2 = 16$ in a similar way by equating $x - 3$ to 4 and to -4. Thus,

$$x - 3 = 4 \quad \text{or} \quad x - 3 = -4$$

Solving these equations, we obtain the roots 7 and -1.

We may solve $(x - 3)^2 = 17$ in the same way. Thus,

$$x - 3 = \sqrt{17} \quad \text{or} \quad x - 3 = -\sqrt{17}$$

The roots are therefore $3 + \sqrt{17}$ and $3 - \sqrt{17}$. Decimal approximations of these roots are 7.123 and -1.123. ◂◂

EXAMPLE 2 ▸▸ We wish to find the roots of the quadratic equation

$$x^2 - 6x - 8 = 0$$

First we note that this equation is not factorable. However, we do recognize that $x^2 - 6x$ is part of the special product $(x - 3)^2 = x^2 - 6x + 9$, and this product is a *perfect square*. By adding 9 to $x^2 - 6x$ we would have $(x - 3)^2$. Therefore, we rewrite the original equation as

$$x^2 - 6x = 8$$

NOTE ▶ and then ***add* 9 *to* both *sides*** of the equation. The result is

$$x^2 - 6x + 9 = 17$$

The left side of this equation may be rewritten, giving

$$(x - 3)^2 = 17$$

Now, as in the third illustration of Example 1, we have

$$x - 3 = \pm\sqrt{17}$$

The $\pm$ sign means that $x - 3 = \sqrt{17}$ or $x - 3 = -\sqrt{17}$.

By adding 3 to each side, we obtain

$$x = 3 \pm \sqrt{17}$$

which means that $x = 3 + \sqrt{17}$ and $x = 3 - \sqrt{17}$ are the two roots of the equation.

Therefore, we see that by creating an expression which is a perfect square and then using the principal square root and its negative, we were finally able to solve the equation as two linear equations. ◂◂

How we determine the number to be added to complete the square is based on the special products in Eqs. (6-3) and (6-4). We rewrite these as

$$(x + a)^2 = x^2 + 2ax + a^2 \qquad (7\text{-}2)$$

$$(x - a)^2 = x^2 - 2ax + a^2 \qquad (7\text{-}3)$$

We must be certain that the coefficient of the x^2-term is 1 before we start to complete the square. The coefficient of x in each case is numerically $2a$, and the number

CAUTION ▶ added to complete the square is a^2. Thus *if we **take half the coefficient of the x-term and square this result,** we have the number that completes the square.* In our example, the numerical coefficient of the x-term was 6, and 9 was added to complete the square. Therefore, the procedure for solving a quadratic equation by completing the square is as follows:

Solving a Quadratic Equation by Completing the Square

1. *Divide each side by a* (the coefficient of x^2).
2. *Rewrite the equation with the constant on the right side.*
3. *Complete the square: Add the square of one-half of the coefficient of x to both sides.*
4. *Write the left side as a square and simplify the right side.*
5. *Equate the square root of the left side to the principal square root of the right side and to its negative.*
6. *Solve the two resulting linear equations.*

The following example illustrates the step-by-step procedure for solving a quadratic equation by completing the square.

EXAMPLE 3 ▸▸ Solve the following quadratic equation by completing the square.

$2x^2 + 16x - 9 = 0$

1. Divide each side by 2 to make the coefficient of x^2 equal to 1.

$$x^2 + 8x - \frac{9}{2} = 0$$

2. Put the constant on the right by adding $\frac{9}{2}$ to both sides.

$$x^2 + 8x = \frac{9}{2}$$

CAUTION ▶

3. The coefficient of the x-term is 8. Therefore, divide 8 by 2 to get 4. Now square 4 to get 16, and add 16 to both sides.

$\frac{1}{2}(8) = 4;\ 4^2 = 16$

$$x^2 + 8x + 16 = \frac{9}{2} + 16 = \frac{41}{2}$$

4. Write the left side as $(x + 4)^2$.

$$(x + 4)^2 = \frac{41}{2}$$

5. Equate $x + 4$ to the principal square root of $\frac{41}{2}$ and its negative.

$$x + 4 = \pm\sqrt{\tfrac{41}{2}}$$

6. Solve for x.

$$x = -4 \pm \sqrt{\tfrac{41}{2}}$$

Therefore the roots are $-4 + \sqrt{\frac{41}{2}}$ and $-4 - \sqrt{\frac{41}{2}}$. Using a calculator to approximate these roots, we get 0.5277 and −8.528. ◂◂

Exercises 7–2

In Exercises 1–8, solve the given quadratic equations by finding appropriate square roots as in Example 1.

1. $x^2 = 25$
2. $x^2 = 100$
3. $x^2 = 7$
4. $x^2 = 15$
5. $(x - 2)^2 = 25$
6. $(x + 2)^2 = 100$
7. $(x + 3)^2 = 7$
8. $(x - 4)^2 = 10$

In Exercises 9–24, solve the given quadratic equations by completing the square. Exercises 9–12 and 15–18 may be checked by factoring.

9. $x^2 + 2x - 8 = 0$
10. $x^2 - x - 6 = 0$
11. $x^2 + 3x + 2 = 0$
12. $t^2 + 5t - 6 = 0$
13. $x^2 - 4x + 2 = 0$
14. $x^2 + 10x - 4 = 0$
15. $v^2 + 2v - 15 = 0$
16. $x^2 - 8x + 12 = 0$
17. $2s^2 + 5s = 3$
18. $4x^2 + x = 3$
19. $3y^2 = 3y + 2$
20. $3x^2 = 3 - 4x$
21. $2y^2 - y - 2 = 0$
22. $9v^2 - 6v - 2 = 0$
23. $x^2 + 2bx + c = 0$
24. $px^2 + qx + r = 0$

7-3 The Quadratic Formula

We shall now use the method of completing the square to derive a general formula that may be used for the solution of any quadratic equation.

Consider Eq. (7-1), the general quadratic equation:

$$ax^2 + bx + c = 0$$

When we divide through by a, we obtain

$$x^2 + \frac{b}{a}x + \frac{c}{a} = 0$$

Subtracting c/a from each side, we have

$$x^2 + \frac{b}{a}x = -\frac{c}{a}$$

Half of b/a is $b/2a$, which squared is $b^2/4a^2$. Adding $b^2/4a^2$ to each side gives us

$$x^2 + \frac{b}{a}x + \frac{b^2}{4a^2} = -\frac{c}{a} + \frac{b^2}{4a^2}$$

Writing the left side as a perfect square, and combining fractions on the right side, we have

$$\left(x + \frac{b}{2a}\right)^2 = \frac{b^2 - 4ac}{4a^2}$$

Equating $x + \dfrac{b}{2a}$ to the principal square root of the right side and its negative,

$$x + \frac{b}{2a} = \frac{\pm\sqrt{b^2 - 4ac}}{2a}$$

When we subtract $b/2a$ from each side and simplify the resulting expression, we obtain the **quadratic formula:**

Quadratic Formula

$$x = \frac{-b \pm \sqrt{b^2 - 4ac}}{2a} \qquad (7\text{-}4)$$

To solve a quadratic equation by using the quadratic formula, we need only write the equation in the standard form of Eq. (7-1), identify a, b, and c, and substitute these numbers directly into the formula. We shall use the quadratic formula to solve the quadratic equations in the following examples.

EXAMPLE 1 ▸▸ Solve: $x^2 - 5x + 6 = 0$.

$$a = 1 \quad b = -5 \quad c = 6$$

Here, using the indicated values of a, b, and c in the quadratic formula, we have

$$x = \frac{-(-5) \pm \sqrt{(-5)^2 - 4(1)(6)}}{2(1)} = \frac{5 \pm \sqrt{25 - 24}}{2} = \frac{5 \pm 1}{2}$$

$$x = \frac{5 + 1}{2} = 3, \quad \text{or} \quad x = \frac{5 - 1}{2} = 2$$

The roots $x = 3$ and $x = 2$ check when substituted in the original equation. ◂◂

EXAMPLE 2 ▸▸ Solve: $2x^2 - 7x - 5 = 0$.

$$\underset{a=2}{2x^2} \quad - \quad \underset{b=-7}{7x} \quad - \quad \underset{c=-5}{5} = 0$$

Substituting the values for a, b, and c in the quadratic formula, we have

$$x = \frac{-(-7) \pm \sqrt{(-7)^2 - 4(2)(-5)}}{2(2)} = \frac{7 \pm \sqrt{49 + 40}}{4} = \frac{7 \pm \sqrt{89}}{4}$$

$$x = \frac{7 + \sqrt{89}}{4} = 4.108, \quad \text{or} \quad x = \frac{7 - \sqrt{89}}{4} = -0.6085$$

The exact roots are $x = \frac{7 \pm \sqrt{89}}{4}$ (this form is often used when the roots are irrational). Approximate decimal values are $x = 4.108$ and $x = -0.6085$. ◂◂

EXAMPLE 3 ▸▸ Solve: $9x^2 + 24x + 16 = 0$.

In this example, $a = 9$, $b = 24$, and $c = 16$. Thus,

$$x = \frac{-24 \pm \sqrt{24^2 - 4(9)(16)}}{2(9)} = \frac{-24 \pm \sqrt{576 - 576}}{18} = \frac{-24 \pm 0}{18} = -\frac{4}{3}$$

Here both roots are $-\frac{4}{3}$, and we write the result as $x = -\frac{4}{3}$ and $x = -\frac{4}{3}$. We will get a double root when $b^2 = 4ac$, as in this case. ◂◂

EXAMPLE 4 ▸▸ Solve: $3x^2 - 5x + 4 = 0$.

In this example, $a = 3$, $b = -5$, and $c = 4$. Therefore,

$$x = \frac{-(-5) \pm \sqrt{(-5)^2 - 4(3)(4)}}{2(3)} = \frac{5 \pm \sqrt{25 - 48}}{6} = \frac{5 \pm \sqrt{-23}}{6}$$

We see that the roots contain imaginary numbers. This happens if $b^2 < 4ac$. ◂◂

We can generalize on the results of these examples as to the character of the roots of a quadratic equation. This is done by noting the value of $b^2 - 4ac$, which is called the **discriminant.** If a, b, and c are rational numbers (see Section 1-1), we have the following:

Character of the Roots of a Quadratic Equation

1. If $b^2 - 4ac$ is positive and a perfect square (see Section 1-6), the roots are real, rational, and unequal. (See Example 1, where $b^2 - 4ac = 1$.)
2. If $b^2 - 4ac$ is positive but not a perfect square, the roots are real, irrational, and unequal. (See Example 2, where $b^2 - 4ac = 89$.)
3. If $b^2 - 4ac = 0$, the roots are real, rational, and equal. (See Example 3, where $b^2 - 4ac = 0$.)
4. If $b^2 - 4ac < 0$, the roots contain imaginary numbers and are unequal. (See Example 4, where $b^2 - 4ac = -23$.)

We can use the value of $b^2 - 4ac$ to help in checking the roots or in finding the character of the roots without having to solve the equation completely.

EXAMPLE 5 ▸▸ Solve: $2x^2 = 4x + 3$.

First we must put the equation in the proper form:

$$2x^2 - 4x - 3 = 0$$

Now we identify $a = 2$, $b = -4$, and $c = -3$, which leads to the solution

$$x = \frac{-(-4) \pm \sqrt{(-4)^2 - 4(2)(-3)}}{2(2)} = \frac{4 \pm \sqrt{16 + 24}}{4}$$

$$= \frac{4 \pm \sqrt{40}}{4} = \frac{4 \pm 2\sqrt{10}}{4} = \frac{2(2 \pm \sqrt{10})}{4} = \frac{2 \pm \sqrt{10}}{2}$$

Approximate decimal results are $x = 2.581$ and $x = -0.581$. The radical form of the answer is found by simplifying radicals as shown in Section 1-6.

Here $b^2 - 4ac = 40$, and the roots are real, irrational, and unequal. ◂◂

EXAMPLE 6 ▸▸ Solve for x: $dx^2 - 3x = dx - 4$.

$$dx^2 - 3x - dx + 4 = 0$$
$$dx^2 - (3 + d)x + 4 = 0 \qquad \text{put in the form of Eq. (7-1)}$$

Note the use of factoring to get the coefficient of x. We now see that $a = d$, $b = -(3 + d)$, and $c = 4$. Using the quadratic formula, we complete the solution.

$$x = \frac{-[-(3 + d)] \pm \sqrt{[-(3 + d)]^2 - 4(d)(4)}}{2d}$$

$$= \frac{3 + d \pm \sqrt{9 + 6d + d^2 - 16d}}{2d} = \frac{3 + d \pm \sqrt{9 - 10d + d^2}}{2d}$$

We see that we can solve quadratic equations which have literal coefficients. ◂◂

See the chapter introduction.

EXAMPLE 7 ▸▸ In designing a custom door, an architect plans it as 0.90 m wide and 2.10 m high. The door is to have a frame of uniform width on the top and sides and twice this width at the bottom. The area within the frame must be 1.20 m^2 to allow for decorative panels. How wide a frame is the door to have?

First we let x be the width of the frame at the top and sides. This means the width of the frame at the bottom is $2x$. Knowing that the area within the frame is 1.20 m^2, we find x as follows:

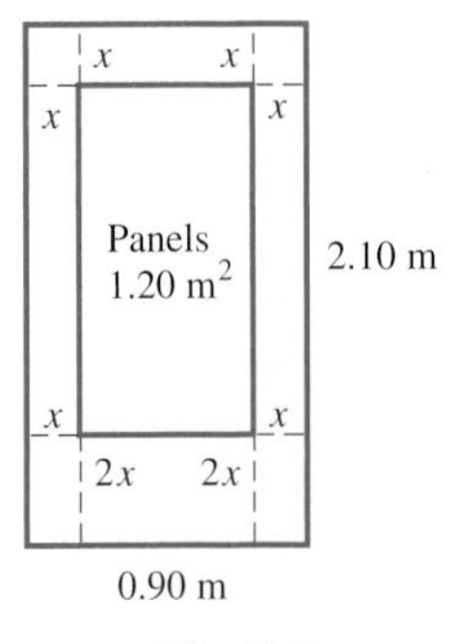

Fig. 7-3

$$\underset{\text{interior width}}{(0.90 - 2x)}\underset{\text{interior height}}{(2.10 - 3x)} = \underset{\text{interior area}}{1.20}$$

$$1.89 - 4.20x - 2.70x + 6x^2 = 1.20$$

$$6x^2 - 6.90x + 0.69 = 0$$

$$x = \frac{-(-6.90) \pm \sqrt{(-6.90)^2 - 4(6)(0.69)}}{2(6)} = \frac{6.90 \pm 5.572}{12}$$

Checking the roots, $\frac{6.90 + 5.572}{12} = 1.04$ m. This cannot be the required result, since the width of the frame would be greater than the width of the door. The other root is $\frac{6.90 - 5.572}{12} = 0.11$ m $= 11$ cm. Checking this root, the interior area is $(0.90 - 0.22)(2.10 - 0.33) = 1.20$ m^2. Thus, $x = 11$ cm, which means the frame is 11 cm wide at the top and sides and 22 cm wide at the bottom. ◂◂

CAUTION ▶ It must be emphasized that, in using the quadratic formula, the entire expression $-b \pm \sqrt{b^2 - 4ac}$ is divided by $2a$. ***It is a relatively common error to divide only the radical*** $\sqrt{b^2 - 4ac}$.

The quadratic formula provides a quick general method for solving quadratic equations. Proper recognition and substitution of the coefficients a, b, and c is all that is required to complete the solution, regardless of the nature of the roots.

Exercises 7–3

In Exercises 1–32, solve the given quadratic equations using the quadratic formula. Exercises 1–4 are the same as Exercises 9–12 of Section 7-2.

1. $x^2 + 2x - 8 = 0$ **2.** $x^2 - x - 6 = 0$

3. $x^2 + 3x + 2 = 0$ **4.** $t^2 + 5t - 6 = 0$

5. $x^2 - 4x + 2 = 0$ **6.** $x^2 + 10x - 4 = 0$

7. $v^2 + 2v - 15 = 0$ **8.** $x^2 - 8x + 12 = 0$

9. $2s^2 + 5s = 3$ **10.** $4x^2 + x = 3$

11. $3y^2 = 3y + 2$ **12.** $3x^2 = 3 - 4x$

13. $2y^2 - y - 2 = 0$ **14.** $9v^2 - 6v - 2 = 0$

15. $30y^2 + 23y - 40 = 0$ **16.** $40x^2 - 62x - 63 = 0$

17. $2t^2 + 10t = -15$ **18.** $2d(d - 2) = -7$

19. $s^2 = 9 + s(1 - 2s)$ **20.** $6r^2 = 6r + 1$

21. $4x^2 = 9$ **22.** $6x = x^2$

23. $15 + 4z = 32z^2$ **24.** $4x^2 - 12x = 7$

25. $x^2 - 0.20x - 0.40 = 0$ **26.** $3.2x^2 = 2.5x + 7.6$

27. $0.29x^2 - 0.18 = 0.63x$ **28.** $12.5x^2 + 13.2x = 15.5$

29. $x^2 + 2cx - 1 = 0$ **30.** $x^2 - 7x + (6 + a) = 0$

31. $b^2x^2 - (b + 1)x + (1 - a) = 0$

32. $c^2x^2 - x - 1 = x^2$

In Exercises 33–36, solve the given quadratic equations. All numbers are accurate to at least two significant digits.

33. In machine design, in finding the outside diameter D_0 of a hollow shaft, the equation $D_0^2 - DD_0 - 0.25D^2 = 0$ is used. Solve for D_0 if $D = 3.625$ cm.

34. A missile is fired vertically into the air. The distance (in metres) above the ground as a function of time (in seconds) is given by the formula $s = 100 + 500t - 4.9t^2$. (a) When will the missile hit the ground? (b) When will the missile be 1000 m above the ground?

35. In calculating the current in an electric circuit with an inductance L, a resistance R, and a capacitance C, it is necessary to solve the equation $Lm^2 + Rm + 1/C = 0$. Solve for m in terms of L, R, and C.

36. In finding the radius r of a circular arch of height h and span b, an architect used the formula $r = \dfrac{b^2 + 4h^2}{8h}$. Solve for h.

In Exercises 37–40, without solving the given equations, determine the character of the roots.

37. $2x^2 - 7x = -8$ **38.** $3x^2 + 19x = 14$

39. $3.6t^2 + 2.1 = 7.7t$ **40.** $0.45s^2 + 0.33 = 0.12s$

In Exercises 41–44, set up appropriate equations and solve the stated problems. All numbers are accurate to at least two significant digits.

41. The length of a tennis court is 12.8 m more than its width. If the area of a tennis court is 262 m^2, what are its dimensions? See Fig. 7-4.

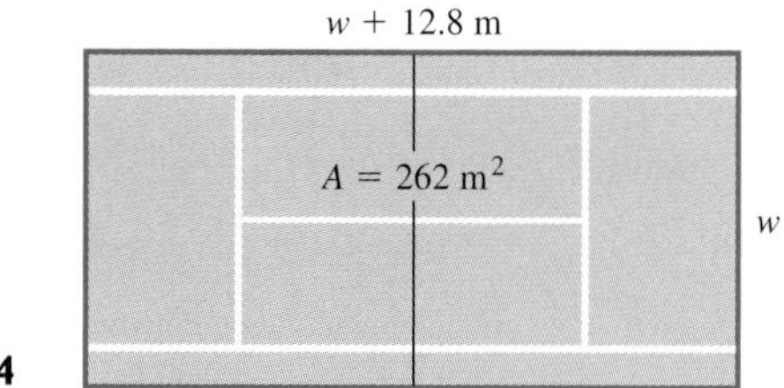

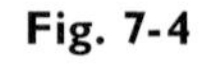
Fig. 7-4

42. Two circular oil spills are tangent to each other. If the distance between centers is 800 m, and they cover a combined area of 1.02×10^6 m^2, what is the radius of each?

43. In remodeling a house, an architect finds that by adding the same amount to each dimension of a 3.8-m by 5.0-m rectangular room the area would be increased by 11 m^2. How much must be added to each dimension?

44. Two pipes together drain a wastewater holding tank in 6.00 h. If used alone to empty the tank, one takes 2.00 h longer than the other. How long does each take to empty the tank if used alone?

7-4 The Graph of the Quadratic Function

We have developed the basic algebraic methods of solving a quadratic equation. We shall now discuss the graph of the quadratic function and show the graphical solution of a quadratic equation, using the methods developed in Chapter 3.

In Section 7-1 we noted that $ax^2 + bx + c$ is the quadratic function. As in Chapter 3, by letting $y = ax^2 + bx + c$ we can graph this function. The next example briefly reviews the graph of a quadratic function done in this way.

EXAMPLE 1 ▸▸ Graph the function $f(x) = x^2 + 2x - 3$.

First we let $y = x^2 + 2x - 3$. We can then set up a table of values and graph the function as shown in Fig. 7-5, or we can display its graph on a graphing calculator as shown in Fig. 7-6.

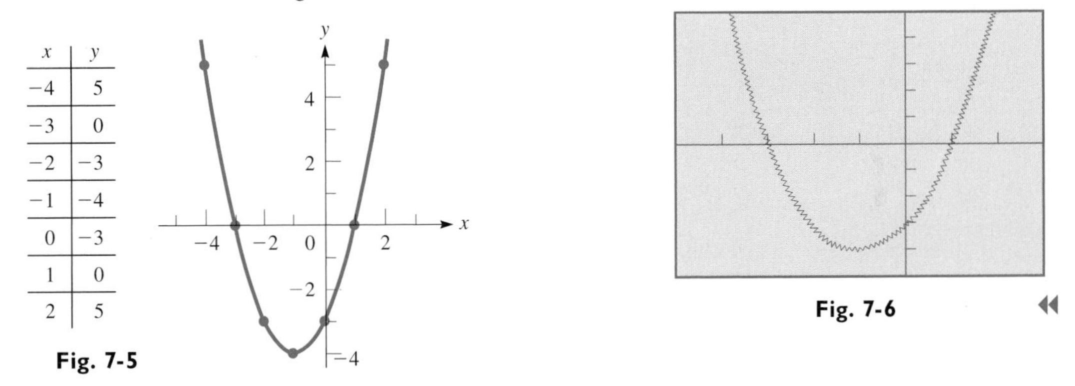

x	y
−4	5
−3	0
−2	−3
−1	−4
0	−3
1	0
2	5

Fig. 7-5

Fig. 7-6 ◂◂

The graph of any quadratic function $y = ax^2 + bx + c$ will have the same basic shape as that shown in Fig. 7-5, and it is called a **parabola.** (In Section 3-4 we briefly noted that graphs in Examples 2 and 3 were parabolas.) A parabola defined by a quadratic function can open upward (as in Example 1) or downward. The location of the parabola and how it opens depend on the values of a, b, and c.

In Example 1 the parabola has a minimum point at $(-1, -4)$ and the curve opens upward. *All parabolas have an* **extreme point** *of this type. If $a > 0$, the parabola has a* **minimum point** *and it opens* **upward.** *If $a < 0$, the parabola has a* **maximum point** *and it opens* **downward.** *The extreme point of the parabola is also known as its* **vertex.**

EXAMPLE 2 ▸▸ The graph of $y = 2x^2 - 8x + 6$ is shown in Fig. 7-7(a). For this parabola, $a = 2$ $(a > 0)$ and it opens upward. The vertex (a minimum point) is $(2, -2)$.

The graph of $y = -2x^2 + 8x - 6$ is shown in Fig. 7-7(b). For this parabola, $a = -2 (a < 0)$ and it opens downward. The vertex (a maximum point) is $(2, 2)$. ◂◂

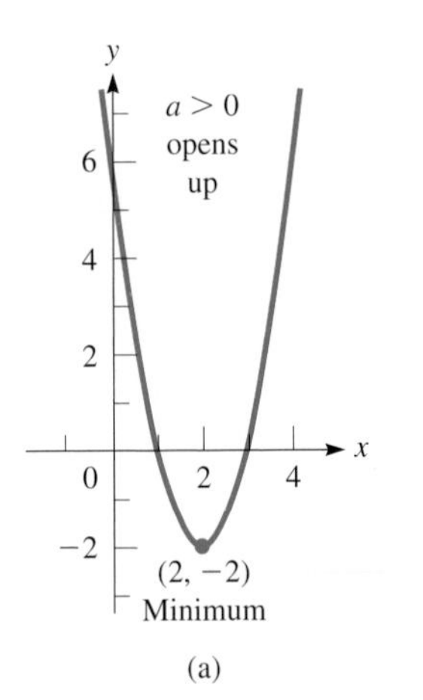

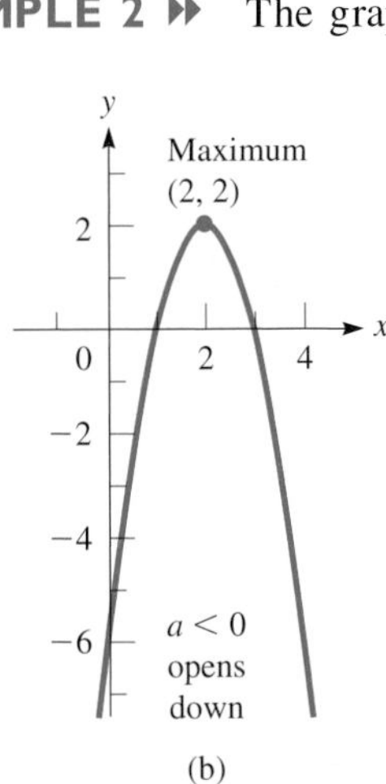

Fig. 7-7

We can sketch the graph of a parabola by using its basic shape and knowing the location of two or three points, including the vertex. Even when using a graphing calculator, we can get a check on the graph by knowing the vertex and how the parabola opens.

In order to find the coordinates of the extreme point we start with the quadratic function

$$y = ax^2 + bx + c$$

Then we factor a from the two terms containing x, obtaining

$$y = a\left(x^2 + \frac{b}{a}x\right) + c$$

Now, completing the square of the terms within parentheses, we have

$$y = a\left(x^2 + \frac{b}{a}x + \frac{b^2}{4a^2}\right) + c - \frac{b^2}{4a}$$

$$= a\left(x + \frac{b}{2a}\right)^2 + c - \frac{b^2}{4a}$$

We now look at the factor $\left(x + \frac{b}{2a}\right)^2$. If $x = -b/2a$, the term is zero. If x is any other value, $\left(x + \frac{b}{2a}\right)^2$ is positive. Thus, if $a > 0$, the value of y increases from that we have for $x = -b/2a$, and if $a < 0$, the value of y decreases from that we have for $x = -b/2a$. *This means that $x = -b/2a$ is the x-coordinate of the vertex (extreme point). The y-coordinate can be found by substituting in the function.*

Another easily found point is the y-intercept. As with a linear equation, we find the y-intercept where $x = 0$. For $y = ax^2 + bx + c$, if $x = 0$, then $y = c$. *This means that the point $(0, c)$ is the y-intercept.*

EXAMPLE 3 ▸▸ For the graph of the function $y = 2x^2 - 8x + 6$, find the vertex and y-intercept and sketch the graph. (This function is also used in Example 2.)

First, $a = 2$ and $b = -8$. This means that the x-coordinate of the vertex is

$$\frac{-b}{2a} = \frac{-(-8)}{2(2)} = \frac{8}{4} = 2$$

and the y-coordinate is

$$y = 2(2^2) - 8(2) + 6 = -2$$

Thus, the vertex is $(2, -2)$. Since $a > 0$, it is a minimum point.

Since $c = 6$, the y-intercept is $(0, 6)$.

We can use the minimum point $(2, -2)$ and the y-intercept $(0, 6)$, along with the fact that the graph is a parabola, to get an approximate sketch of the graph. Noting that a parabola increases (or decreases) away from the vertex in the same way on each side of it (it is *symmetric* to a vertical line through the vertex), we sketch the graph in Fig. 7-8. We see that it is the same graph as that shown in Fig. 7-7(a). ◂◂

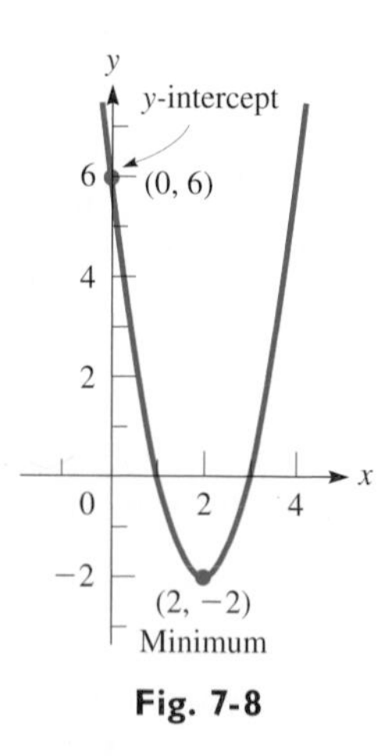

Fig. 7-8

If we are not using a graphing calculator, we may need one or two additional points to get a good sketch of a parabola. This would be true if the y-intercept is close to the vertex. Two points we can find are the x-intercepts, if the parabola crosses the x-axis (one point if the vertex is on the x-axis). They are found by setting $y = 0$ and solving the quadratic equation $ax^2 + bx + c = 0$. Also, we may simply find one or two points other than the vertex and the y-intercept. Sketching a parabola in this way is shown in the following two examples.

EXAMPLE 4 ▸▸ Sketch the graph of $y = -x^2 + x + 6$.

We first note that $a = -1$ and $b = 1$. Therefore, the x-coordinate of the maximum point ($a < 0$) is $-\frac{1}{2(-1)} = \frac{1}{2}$. The y-coordinate is $-(\frac{1}{2})^2 + \frac{1}{2} + 6 = \frac{25}{4}$. This means that the maximum point is $(\frac{1}{2}, \frac{25}{4})$.

The y-intercept is $(0, 6)$.

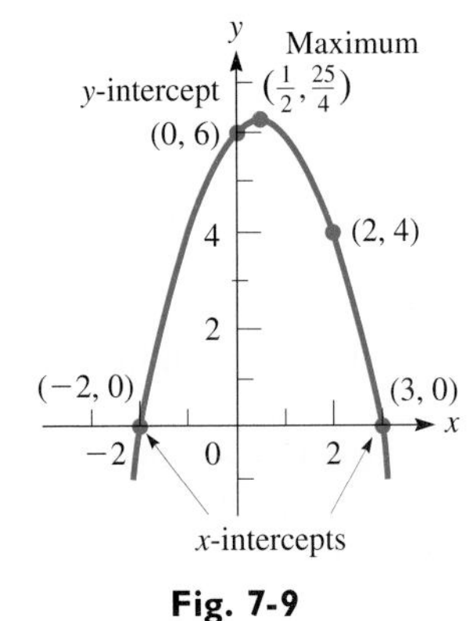

Fig. 7-9

Using these points in Fig. 7-9, we see that they are close together and do not give a good idea of how wide the parabola opens. Therefore, setting $y = 0$, we solve the equation

$$-x^2 + x + 6 = 0$$

or

$$x^2 - x - 6 = 0$$

This equation is factorable. Thus,

$$(x - 3)(x + 2) = 0$$
$$x = 3, -2$$

This means that the x-intercepts are $(3, 0)$ and $(-2, 0)$, as shown in Fig. 7-9.

Also, rather than finding the x-intercepts, we can let $x = 2$ and then use the point $(2, 4)$. ◂◂

EXAMPLE 5 ▸▸ Sketch the graph of $y = x^2 + 1$.

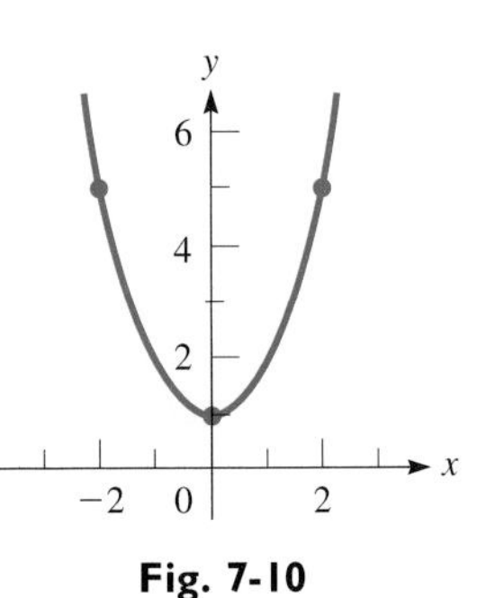

Fig. 7-10

Since there is no x-term, $b = 0$. This means that the x-coordinate of the minimum point ($a > 0$) is 0 and that the minimum point and the y-intercept are both $(0, 1)$. We know that the graph opens upward, since $a > 0$, which in turn means that it does not cross the x-axis. Now, letting $x = 2$ and $x = -2$, we find the points $(2, 5)$ and $(-2, 5)$ on the graph, which is shown in Fig. 7-10. ◂◂

We can see that the domain of the quadratic function is all x. Knowing that the graph has either a maximum point or a minimum point, we can see that the range of the quadratic function must be restricted to certain values of y. In Example 3, the range is $f(x) \geq -2$; in Example 4, the range is $f(x) \leq \frac{25}{4}$; and in Example 5, the range is $f(x) \geq 1$.

Solving Quadratic Equations Graphically

In Chapter 3 we showed how an equation can be solved graphically. Following that method, to solve the equation $ax^2 + bx + c = 0$, we let $y = ax^2 + bx + c$ and graph the function. The roots of the equation are the x-coordinates of the points for which $y = 0$ (the x-intercepts). The following examples illustrate solving quadratic equations graphically.

EXAMPLE 6 ▸▸ Solve the equation $3x = x(2 - x) + 3$ graphically.

We first collect all terms on the left of the equal sign. This gives us $x^2 + x - 3 = 0$. We then let $y = x^2 + x - 3$, and graph this function. The minimum point is $(-\frac{1}{2}, -\frac{13}{4})$ and the y-intercept is $(0, -3)$.

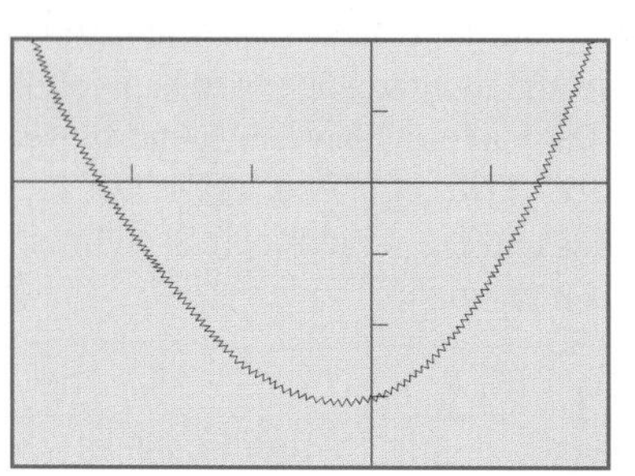

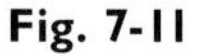

Fig. 7-11

Using the graphing calculator, these points help us choose Xmin = −3, Xmax = 2, Ymin = −4, Ymax = 2 for the RANGE feature. The graphing calculator view is shown in Fig. 7-11.

Then, using the TRACE and ZOOM features, we find the two roots to be $x = -2.30$ and $x = 1.30$. These values check when substituted in the original equation. ◂◂

EXAMPLE 7 ▸▸ A projectile is fired vertically upward from the ground with a velocity of 38 m/s. Its distance above the ground is given by $s = -4.9t^2 + 38t$, where s is the distance (in metres) and t is the time (in seconds). Graph the function, and from the graph determine (1) when the projectile will hit the ground and (2) how long it takes to reach 45 m above the ground.

Since $c = 0$, the s-intercept is at the origin. Using $-b/2a$, we find the maximum point at about (3.9, 74). These points help us choose (we use x for t, and y for s on the calculator) Xmin = 0, Xmax = 8, Ymin = 0, Ymax = 80 for the RANGE feature. The graphing calculator view is shown in Fig. 7-12.

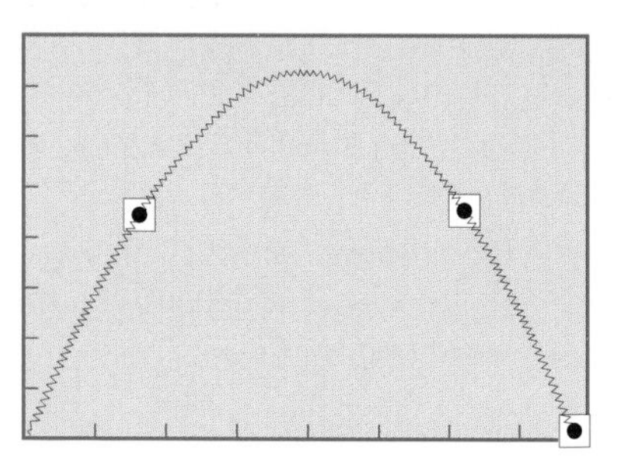

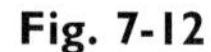

Fig. 7-12

Using the TRACE and ZOOM features, we can now determine the required answers.

1. The projectile will hit the ground for $s = 0$, which is shown at the right t-intercept. Thus, it takes about 7.8 s to hit the ground.

2. We find the time for the projectile to get 45 m above the ground by finding the values of t for $s = 45$ m. We see that there are two points for which $s = 45$ m. The values of t at these points tell us that the times are $t = 1.5$ s and $t = 6.3$ s.

From the maximum point, we can also conclude that the maximum height reached by the projectile is about 74 m. This can also be seen on the graph. ◂◂

Exercises 7–4

In Exercises 1–8, sketch the graph of each parabola by using only the vertex and the y-intercept. Check the graph using a graphing calculator.

1. $y = x^2 - 6x + 5$ **2.** $y = -x^2 - 4x - 3$

3. $y = -3x^2 + 10x - 4$ **4.** $y = 2x^2 + 8x - 5$

5. $y = x^2 - 4x$ **6.** $y = -2x^2 - 5x$

7. $y = -2x^2 - 4x - 3$ **8.** $y = x^2 - 3x + 4$

In Exercises 9–12, sketch the graph of each parabola by using the vertex, the y-intercept, and the x-intercepts. Check the graph using a graphing calculator.

9. $y = x^2 - 4$ **10.** $y = x^2 + 3x$

11. $y = -2x^2 - 6x + 8$ **12.** $y = -3x^2 + 12x - 5$

In Exercises 13–16, sketch the graph of each parabola by using the vertex, the y-intercept, and two other points, not including the x-intercepts. Check the graph using a graphing calculator.

13. $y = 2x^2 + 3$ **14.** $y = x^2 + 2x + 2$

15. $y = -2x^2 - 2x - 6$ **16.** $y = -3x^2 - x$

In Exercises 17–24, use a graphing calculator to solve the given equations. If there are no real roots, state this as the answer.

17. $2x^2 - 3 = 0$ **18.** $5 - x^2 = 0$

19. $-3x^2 + 11x - 5 = 0$ **20.** $2x^2 = 7x + 4$

21. $x(2x - 1) = -3$ **22.** $2x - 5 = x^2$

23. $6x^2 = 18 - 7x$ **24.** $3x^2 - 25 = 20x$

In Exercises 25–28, use a graphing calculator to graph all three parabolas on the same coordinate system. State the changes in the graphs in parts (b) and (c) from the graph in part (a).

25. (a) $y = x^2$ (b) $y = 3x^2$ (c) $y = \frac{1}{3}x^2$

26. (a) $y = x^2$ (b) $y = (x - 3)^2$ (c) $y = (x + 3)^2$

27. (a) $y = x^2$ (b) $y = -x^2$ (c) $y = 3x - x^2$

28. (a) $y = x^2$ (b) $y = x^2 + 3$ (c) $y = x^2 - 3$

In Exercises 29–36, solve the given applied problems.

29. An equipment company determines that the area A (in m^2) covered by a rectangular tarpaulin is given by $A = w(8 - w)$, where w is the width of the tarpaulin, and its perimeter is 16 m. Sketch the graph of A as a function of w.

30. Under specified conditions, the pressure loss L (in kPa/100 m), in the flow of water through a water line in which the flow is q litres per minute, is given by $L = 0.0001q^2 + 0.005q$. Sketch the graph of L as a function of q, for $q < 400$ L/min.

31. When analyzing the power P, in watts, dissipated in an electric circuit, the equation $P = 50i - 3i^2$ results. Here i is the current in amperes. Sketch the graph of $P = f(i)$.

32. The vertical distance d, in centimeters, of the end of a robot arm above a conveyor belt in its 8-s cycle is given by $d = 2t^2 - 16t + 47$. Sketch the graph of $d = f(t)$.

33. A missile is fired vertically upward such that its distance s, in metres, above the ground is given by $s = 50 + 90t - 4.9t^2$, where t is the time in seconds. Sketch the graph, and then determine from the graph (a) when the missile will hit the ground (to 0.1 s), (b) how high it will go (to 3 significant digits), and (c) how long it will take to reach a height of 250 m (to 0.1 s).

34. In a certain electric circuit, the resistance R, in ohms, that gives resonance is found by solving the equation $25R = 3(R^2 + 4)$. Solve this equation graphically (to 0.1 Ω).

35. A security fence is to be built around a rectangular parking area of 2000 m^2. If the front side of the fence costs \$60 per metre and the other three sides cost \$30 per metre, solve graphically for the dimensions (to the nearest metre) of the parking area if the fence is to cost \$7500.

36. An airplane pilot could decrease the time needed to travel the 1000 km from Ottawa to Milwaukee by 20 min if the plane's speed is increased by 60.0 km/h. Set up the appropriate equation and solve graphically for the plane's speed (to three significant digits).

Chapter Equations, Review Exercises, and Practice Test

Chapter Equations

Quadratic equation $\quad ax^2 + bx + c = 0 \quad$ (7-1)

Quadratic formula $\quad x = \dfrac{-b \pm \sqrt{b^2 - 4ac}}{2a} \quad$ (7-4)

Review Exercises

In Exercises 1–12, solve the given quadratic equations by factoring.

1. $x^2 + 3x - 4 = 0$ **2.** $x^2 + 3x - 10 = 0$

3. $x^2 - 10x + 16 = 0$ **4.** $x^2 - 6x - 27 = 0$

5. $3x^2 + 11x = 4$ **6.** $6y^2 = 11y - 3$

7. $6t^2 = 13t - 5$ **8.** $3x^2 + 5x + 2 = 0$

9. $6s^2 = 25s$ **10.** $6n^2 - 23n - 35 = 0$

11. $4x^2 - 8x = 21$ **12.** $6x^2 = 8 - 47x$

In Exercises 13–24, solve the given quadratic equations by using the quadratic formula.

13. $x^2 - x - 110 = 0$ **14.** $x^2 + 3x - 18 = 0$

15. $x^2 + 2x - 5 = 0$ **16.** $x^2 - 7x - 1 = 0$

17. $2x^2 - x = 36$ **18.** $3x^2 + x = 14$

19. $4x^2 - 3x - 2 = 0$ **20.** $5x^2 + 7x - 2 = 0$

21. $2.1x^2 + 2.3x + 5.5 = 0$ **22.** $0.30x^2 - 0.42x = 0.15$

23. $6x^2 = 9 - 4x$ **24.** $24x^2 = 25x + 20$

In Exercises 25–36, solve the given quadratic equations by any appropriate algebraic method.

25. $x^2 + 4x - 4 = 0$ **26.** $x^2 + 3x + 1 = 0$

27. $3x^2 + 8x + 2 = 0$ **28.** $3p^2 = 28 - 5p$

29. $4v^2 = v + 5$ **30.** $6x^2 - x + 2 = 0$

31. $2x^2 + 3x + 7 = 0$ **32.** $4y^2 - 5y = 8$

33. $a^2x^2 + 2ax + 2 = 0$ **34.** $16r^2 - 8r + 1 = 0$

35. $ax^2 = a^2 - 3x$ **36.** $2bx = x^2 - 3b$

In Exercises 37–40, solve the given quadratic equations by completing the square.

37. $x^2 - x - 30 = 0$ **38.** $x^2 - 2x - 5 = 0$

39. $2x^2 - x - 4 = 0$ **40.** $4x^2 - 8x - 3 = 0$

In Exercises 41–44, solve the given equations.

41. $\frac{x - 4}{x - 1} = \frac{2}{x}$ **42.** $\frac{x - 1}{3} = \frac{5}{x} + 1$

43. $\frac{x^2 - 3x}{x - 3} = \frac{x^2}{x + 2}$ **44.** $\frac{x - 2}{x - 5} = \frac{15}{x^2 - 5x}$

In Exercises 45–48, sketch the graphs of the given functions by using the vertex, the y-intercept, and one or two other points.

45. $y = 2x^2 - x - 1$ **46.** $y = -4x^2 - 1$

47. $y = x - 3x^2$ **48.** $y = 2x^2 + 8x - 10$

In Exercises 49–52, solve the given equations by using a graphing calculator. If there are no real roots, state this as the answer.

49. $2x^2 + x - 4 = 0$ **50.** $-4x^2 - x - 1 = 0$

51. $3x^2 = -x - 2$ **52.** $x(15x - 12) = 8$

In Exercises 53–60, solve the given quadratic equations by any appropriate method. All numbers are accurate to at least two significant digits.

53. In a natural gas pipeline, the velocity v (in m/s) of the gas as a function of the distance x (in cm) from the wall of the pipe is given by $v = 5.2x - x^2$. Determine x for $v = 4.8$ m/s.

54. At an altitude h (in metres) above sea level, the boiling point of water is lower by T °C than the boiling point at sea level, which is 100°C. The difference can be approximated by solving the equation $T^2 + 244T - h = 0$. What is the boiling point at an altitude of 1500 m?

55. The height h of an object ejected at an angle θ from a vehicle moving with velocity v is given by $h = vt \sin \theta - 4.9t^2$, where t is the time of flight. Find t (to 0.1 s) if $v = 15$ m/s, $\theta = 65°$, and $h = 6.0$ m.

56. In studying the emission of light, in order to determine the angle at which the intensity is a given value, the equation $\sin^2 A - 4 \sin A + 1 = 0$ must be solved. Find angle A (to 0.1°). ($\sin^2 A = (\sin A)^2$.)

57. A computer analysis shows that the number n of electronic components a company should produce for supply to equal demand is found by solving $\frac{n^2}{500\,000} = 144 - \frac{n}{500}$. Find n.

58. To determine the resistances of two resistors that are to be in parallel in an electric circuit, it is necessary to solve the equation $\frac{20}{R} + \frac{20}{R + 10} = \frac{1}{5}$. Find R (to nearest 1 Ω).

59. In designing a cylindrical container, the formula $A = 2\pi r^2 + 2\pi rh$ (Eq. (2-19)) is used. Solve for r.

60. In the analysis of mechanical vibrations, the equation $s^2 + \frac{c}{m}s + \frac{kl^2}{mb^2} = 0$ is found. Solve for s in terms of b, c, k, l, and m.

In Exercises 61–72, set up the necessary equation where appropriate and solve the given problems. All numbers are accurate to at least two significant digits.

61. In testing the effects of a drug, the percent of the drug in the blood was given by $p = 0.090t - 0.015t^2$, where t is the time, in hours, after the drug was administered. Sketch the graph of $p = f(t)$.

62. In an electric circuit, the voltage V as a function of the time t is given by $V = 9.8 - 9.2t + 2.3t^2$. Sketch the graph of $V = f(t)$, for $t \leq 5$ min.

63. By adding the same amount to its length and its width, a developer increased the area of a rectangular lot by 3000 m² to make it 80 m by 100 m. What were the dimensions of the lot before the change?

64. A machinery pedestal is made of two concrete cubes, one on top of the other. The pedestal is 2.0 m high and contains 3.5 m³ of concrete. Find the edge of each cube.

65. Concrete contracts as it dries. If the volume of a cubical concrete block is 29 cm³ less and each edge is 0.10 cm less after drying, what was the original length of an edge of the block?

66. A military jet flies directly over and at right angles to the straight course of a commercial jet. The military jet is flying at 200 km/h faster than four times the speed of the commercial jet. How fast is each going if they are 2050 km apart (on a direct line) after 1 h?

67. The width of a rectangular TV screen is 14.5 cm more than its height. If the diagonal is 68.6 cm, find the dimensions of the screen.

68. A given length of wire weighs 8.50 N. It is stretched to a length 1.00 m longer, and then each metre of wire weighs 0.050 N less. Find the original length of the wire.

69. An electric utility company is placing utility poles along a road. It is determined that five fewer poles per kilometre would be necessary if the distance between poles were increased by 10 m. How many poles are being placed each kilometre?

70. A rectangular duct in a building's ventilating system is made of sheet metal 2.6 m wide and has a cross-sectional area of 0.40 m^2. What are the cross-sectional dimensions of the duct?

71. A testing station found that the parts p per million of sulfur dioxide in the air as a function of the hour h of the day was given by $p = 0.00174(10 + 24h - h^2)$. Sketch the graph of $p = f(h)$, and from the graph, find the time when $p = 0.200$ part per million.

72. A compact disc (CD) is made such that it is 53.0 mm from the edge of the center hole to the edge of the disc. Find the radius of the hole if 1.36% of the disc is removed in making the hole.

Writing Exercise

73. An electronics student is asked to solve the equation $\frac{1}{R_T} = \frac{1}{R} + \frac{1}{R+1}$ for R. Write one or two paragraphs explaining your procedure for solving this equation. (What is the solution?)

Practice Test

In Problems 1–4, algebraically solve for x.

1. $x^2 - 3x - 5 = 0$

2. $2x^2 = 9x - 4$

3. $\frac{3}{x} - \frac{2}{x+2} = 1$

4. $2x^2 - x = 6 - 2x(3 - x)$

5. Sketch the graph of $y = 2x^2 + 8x + 5$ using the extreme point and the y-intercept.

6. In electricity the formula $P = EI - RI^2$ is used. Solve for I in terms of E, P, and R.

7. Solve by completing the square: $x^2 - 6x - 9 = 0$.

8. The perimeter of a rectangular window is 8.4 m, and its area is 3.8 m^2. Find its dimensions.

8

Trigonometric Functions of Any Angle

In Section 8-4 we see how the velocity v of a satellite orbiting the earth may be calculated using the equation $v = \omega r$, where ω is the angular velocity and r is its distance from earth.

When we were dealing with the trigonometric functions in Chapter 4, we restricted ourselves primarily to right triangles and functions of acute angles measured in degrees. Since we did define the functions in general, we can use these same definitions for finding the functions of any possible angle. We will find angles larger than 90° and triangles that are not right triangles important in many types of applied problems.

We shall not only find trigonometric functions of angles measured in degrees, but we shall develop the use of radian measure as well. Radian measure is also used in numerous applications, including mechanical vibrations, electric currents, and rotational motion.

8-1 Signs of the Trigonometric Functions

We recall the definitions of the trigonometric functions that were given in Section 4-2. *Here the point (x, y) is a point on the terminal side of angle θ, and r is the radius vector.* See Fig. 8-1.

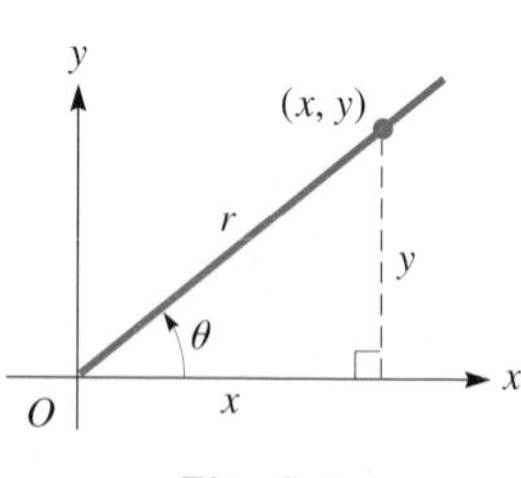

Fig. 8-1

$$\sin\theta = \frac{y}{r} \qquad \cos\theta = \frac{x}{r} \qquad \tan\theta = \frac{y}{x}$$
$$\cot\theta = \frac{x}{y} \qquad \sec\theta = \frac{r}{x} \qquad \csc\theta = \frac{r}{y} \qquad \textbf{(8-1)}$$

We see that we can find the functions if we know the values of the coordinates (x, y) on the terminal side of θ and the radius vector r. Of course, if either x or y is zero in the denominator, the function is undefined, and we will consider this further in the next section. *Remembering that r is always considered to be positive, we can see that the various functions will vary in sign in each of the quadrants, depending on the signs of x and y.*

	Quadrant I	II	III	IV
$\sin\theta$	+	+	−	−

If the terminal side of the angle is in the first or second quadrant, the value of $\sin\theta$ will be positive, but if the terminal side is in the third or fourth quadrant, $\sin\theta$ is negative. This is because *the sign of* $\sin\theta$ *depends on the sign of the y-coordinate* of the point on the terminal side, and y is positive if the point is above the x-axis, and y is negative if this point is below the x-axis. See Fig. 8-2.

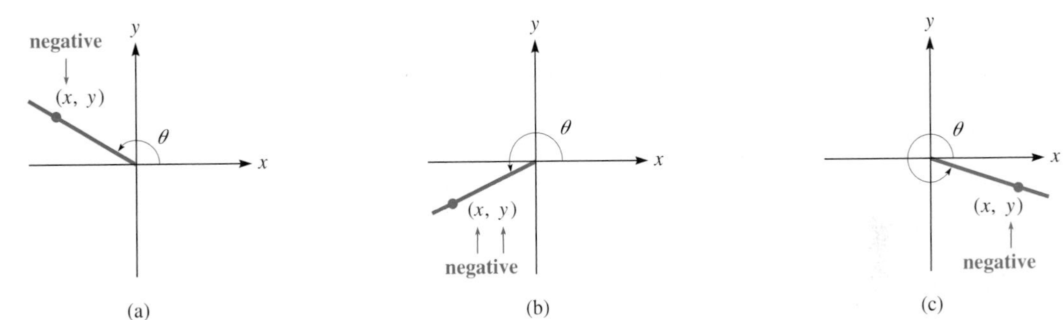

Fig. 8-2

EXAMPLE 1 ▸▸ The value of $\sin 20°$ is positive, since the terminal side of $20°$ is in the first quadrant. The value of $\sin 160°$ is positive, since the terminal side of $160°$ is in the second quadrant. The values of $\sin 200°$ and $\sin 340°$ are negative, since the terminal sides of these angles are in the third and fourth quadrants, respectively. ◂◂

	Quadrant I	II	III	IV
$\tan\theta$	+	−	+	−

The sign of $\tan\theta$ *depends on the ratio of y to x.* In the first quadrant both x and y are positive, and therefore the ratio y/x is positive. In the third quadrant both x and y are negative, and the ratio y/x is positive. In the second quadrant x is negative and y is positive, and in the fourth quadrant x is positive and y is negative. Therefore, in these quadrants, the ratio y/x is negative. See Fig. 8-2.

EXAMPLE 2 ▸▸ The values of $\tan 20°$ and $\tan 200°$ are positive, since the terminal sides of these angles are in the first and third quadrants, respectively. The values of $\tan 160°$ and $\tan 340°$ are negative, since the terminal sides of these angles are in the second and fourth quadrants, respectively. ◂◂

	Quadrant I	II	III	IV
$\cos\theta$	+	−	−	+

The sign of $\cos\theta$ *depends on the sign of x.* Since x is positive in the first and fourth quadrants, $\cos\theta$ is positive in these quadrants. Since x is negative in the second and third quadrants, $\cos\theta$ is negative in these quadrants. See Fig. 8-2.

EXAMPLE 3 ▸▸ The values of $\cos 20°$ and $\cos 340°$ are positive, since the terminal sides of these angles are in the first and fourth quadrants, respectively. The values of $\cos 160°$ and $\cos 200°$ are negative, since the terminal sides of these angles are in the second and third quadrants, respectively. ◂◂

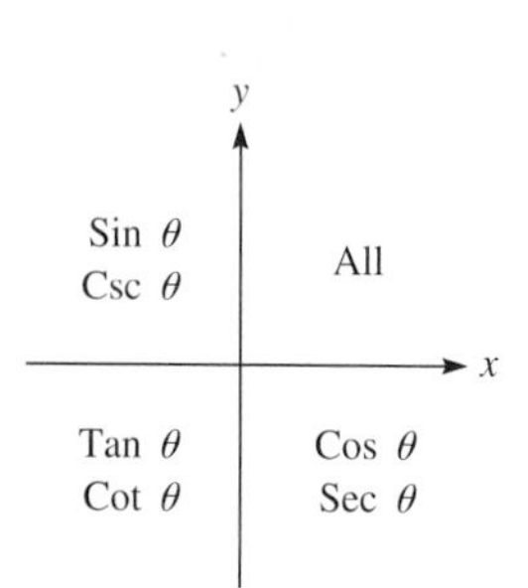

Positive functions

Fig. 8-3

Since csc θ is defined in terms of y and r, as is sin θ, the sign of csc θ is the same as that of sin θ. For the same reason, cot θ has the same sign as tan θ, and sec θ has the same sign as cos θ. Therefore:

All functions of first-quadrant angles are positive. Sin θ and csc θ are positive for second-quadrant angles. Tan θ and cot θ are positive for third-quadrant angles. Cos θ and sec θ are positive for fourth-quadrant angles. All others are negative.

This is shown in Fig. 8-3.

This discussion does not include the *quadrantal angles,* those angles with terminal sides on one of the axes. They will be discussed in the next section.

EXAMPLE 4 ▸▸ The following functions have *positive* values: sin 150°, sin (−200°), cos 8°, cos 300°, cos (−40°), tan 220°, tan (−100°), cot 260°, cot (−310°), sec 280°, sec (−37°), csc 140°, and csc (−190°). ◂◂

EXAMPLE 5 ▸▸ The following functions have *negative* values: sin 190°, sin 325°, cos 95°, cos (−120°), tan 172°, tan 295°, cot 105°, cot (−60°), sec 135°, sec (−135°), csc 240°, and csc 355°. ◂◂

EXAMPLE 6 ▸▸ Determine the trigonometric functions of θ if the terminal side of θ passes through $(-1, \sqrt{3})$. See Fig. 8-4.

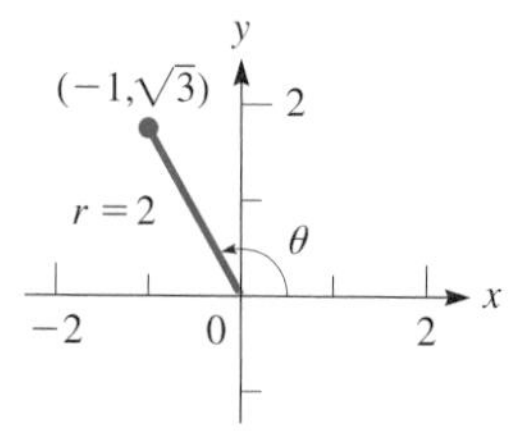

Fig. 8-4

We know that $x = -1$, $y = \sqrt{3}$, and from the Pythagorean theorem we find that $r = 2$. Therefore, the trigonometric functions of θ are

$$\sin\theta = \frac{\sqrt{3}}{2} = 0.8660 \qquad \cos\theta = -\frac{1}{2} = -0.5000 \qquad \tan\theta = -\sqrt{3} = -1.732$$

$$\cot\theta = -\frac{1}{\sqrt{3}} = -0.5774 \qquad \sec\theta = -2 = -2.000 \qquad \csc\theta = \frac{2}{\sqrt{3}} = 1.155$$

The point $(-1, \sqrt{3})$ is in the second quadrant, and the signs of the functions of θ are those of a second-quadrant angle. ◂◂

Exercises 8–1

In Exercises 1–8, determine the sign of the given trigonometric functions.

1. sin 60°, cos 120°, tan 320°
2. tan 185°, sec 115°, sin (−36°)
3. cos 300°, csc 97°, cot (−35°)
4. sin 100°, sec (−15°), cos 188°
5. cot 186°, sec 280°, sin 470°
6. tan (−91°), csc 87°, cot 103°
7. cos 700°, tan (−560°), csc 530°
8. sin 256°, tan 321°, cos (−370°)

In Exercises 9–16, find the trigonometric functions of θ if the terminal side of θ passes through the given point.

9. (2, 1)
10. (−1, 1)
11. (−2, −3)
12. (4, −3)
13. (−5, 12)
14. (−3, −4)
15. (50, −20)
16. (9, 40)

In Exercises 17–24, determine the quadrant in which the terminal side of θ lies, subject to the given conditions.

17. sin θ positive, cos θ negative
18. tan θ positive, cos θ negative
19. sec θ negative, cot θ negative
20. cos θ positive, csc θ negative
21. csc θ negative, tan θ negative
22. sec θ positive, csc θ positive
23. sin θ negative, tan θ positive
24. cot θ negative, sin θ negative

8-2 Trigonometric Functions of Any Angle

The trigonometric functions of acute angles (angles less than 90°) were discussed in Section 4-3, and in the previous section we determined the signs of the trigonometric functions in each of the four quadrants. In this section we show how we can find the trigonometric functions of an angle of any magnitude. This information will be very important in Chapter 9, when we discuss vectors and oblique triangles, and in Chapter 10, when we graph the trigonometric functions. Even *a calculator will not always give the required angle for a given value of a function.*

CAUTION ▶

Any angle in standard position is coterminal with some positive angle less than 360°. Since the terminal sides of coterminal angles are the same, the trigonometric functions of coterminal angles are the same. Therefore, we need consider only the problem of finding the values of the trigonometric functions of positive angles less than 360°.

EXAMPLE 1 ▶▶ The following pairs of angles are coterminal.

$$390° \text{ and } 30°, \qquad -60° \text{ and } 300°$$
$$900° \text{ and } 180°, \qquad -150° \text{ and } 210°$$

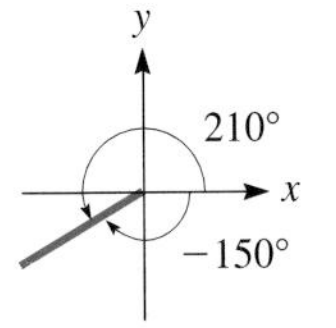

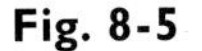
Fig. 8-5

From this we conclude that the trigonometric functions of both angles in each of these pairs are equal. That is, for example,

$$\sin 390° = \sin 30° \quad \text{and} \quad \tan(-150°) = \tan 210°$$

See Fig. 8-5. ◀◀

Considering the definitions of the functions, we see that the values of the functions depend only on the values of x, y, and r. The absolute value of a function of a second-quadrant angle is equal to the value of the same function of a first-quadrant angle. For example, considering angles θ_1 and θ_2 in Fig. 8-6, for angle θ_2 with terminal side passing through $(-3, 4)$, $\tan \theta_2 = -4/3$, or $|\tan \theta_2| = 4/3$. For angle θ_1, with terminal side passing through $(3, 4)$, $\tan \theta_1 = 4/3$, and we see that $|\tan \theta_2| = \tan \theta_1$. Triangles containing angles θ_1 and α are congruent, which means $\theta_1 = \alpha$. Knowing that the absolute value of a function of θ_2 equals the same function of θ_1 means that

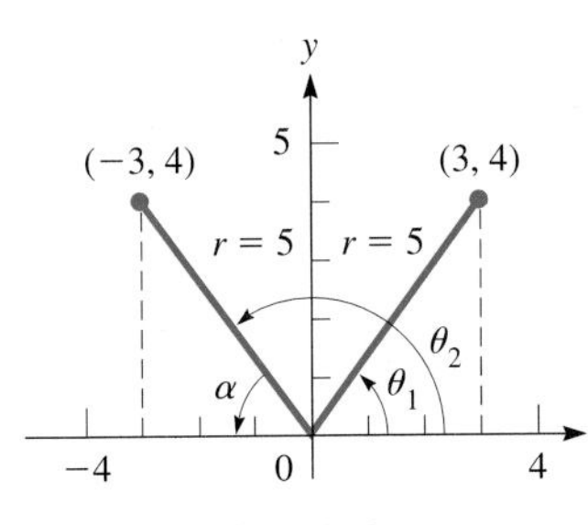

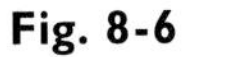
Fig. 8-6

$$|F(\theta_2)| = |F(\theta_1)| = |F(\alpha)| \tag{8-2}$$

where F represents any of the trigonometric functions.

Reference Angle

The angle labeled α is called the **reference angle.** *The reference angle of a given angle is the acute angle formed by the terminal side of the angle and the x-axis.*

Using Eq. (8-2) and the fact that $\alpha = 180° - \theta_2$, we may conclude that the value of any trigonometric function of any second-quadrant angle is found from

$$F(\theta_2) = \pm F(180° - \theta_2) = \pm F(\alpha) \tag{8-3}$$

NOTE ▶ The *sign* used depends on whether the *function* is positive or negative in the second quadrant.

EXAMPLE 2 ▸▸ In Fig. 8-6, the trigonometric functions of θ are as follows:

$$\sin\theta_2 = \sin(180^\circ - \theta_2) = \sin\alpha = \sin\theta_1 = \tfrac{4}{5} = 0.8000$$
$$\cos\theta_2 = -\cos\theta_1 = -\tfrac{3}{5} = -0.6000, \quad \tan\theta_2 = -\tfrac{4}{3} = -1.333$$
$$\cot\theta_2 = -\tfrac{3}{4} = -0.7500, \quad \sec\theta_2 = -\tfrac{5}{3} = -1.667, \quad \csc\theta_2 = \tfrac{5}{4} = 1.250$$ ◂◂

In the same way, we derive the formulas for trigonometric functions of any third- or fourth-quadrant angle. In Fig. 8-7 the reference angle α is found by subtracting 180° from θ_3 and the functions of α and θ_1 are numerically equal. In Fig. 8-8 the reference angle α is found by subtracting θ_4 from 360°.

$$F(\theta_3) = \pm F(\theta_3 - 180^\circ) = \pm F(\alpha) \tag{8-4}$$
$$F(\theta_4) = \pm F(360^\circ - \theta_4) = \pm F(\alpha) \tag{8-5}$$

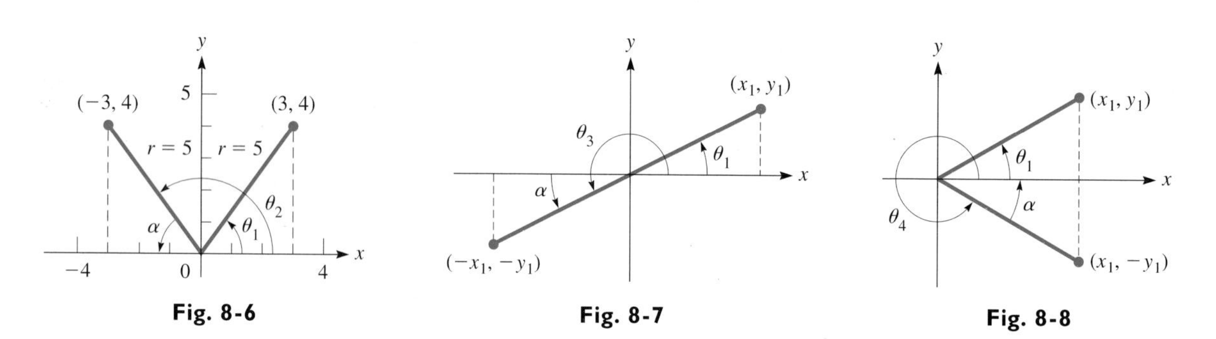

Fig. 8-6 Fig. 8-7 Fig. 8-8

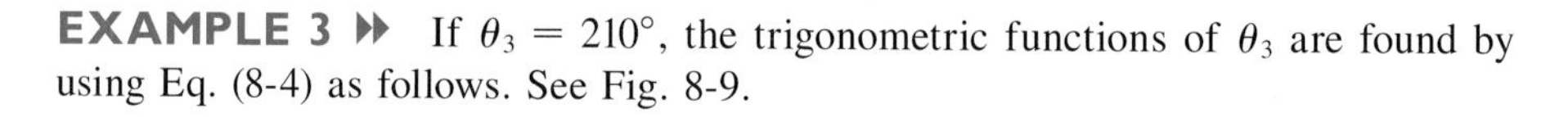

EXAMPLE 3 ▸▸ If $\theta_3 = 210^\circ$, the trigonometric functions of θ_3 are found by using Eq. (8-4) as follows. See Fig. 8-9.

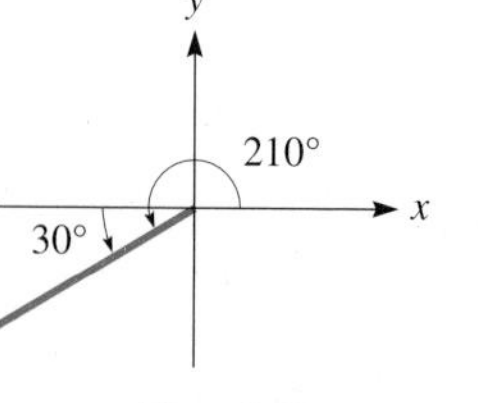

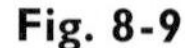
Fig. 8-9

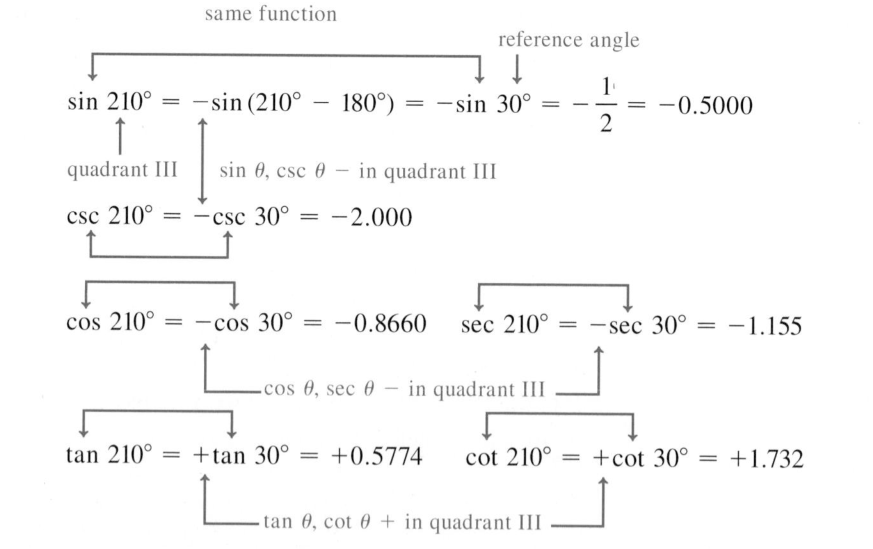

Here we see that the reference angle is $210^\circ - 180^\circ = 30^\circ$ and that we can express the value of a function of 210° in terms of the same function of 30°. We must be careful to attach the correct sign to the result. ◂◂

EXAMPLE 4 ▶▶ If $\theta_4 = 315°$, the trigonometric functions of θ_4 are found by using Eq. (8-5) as follows. See Fig. 8-10.

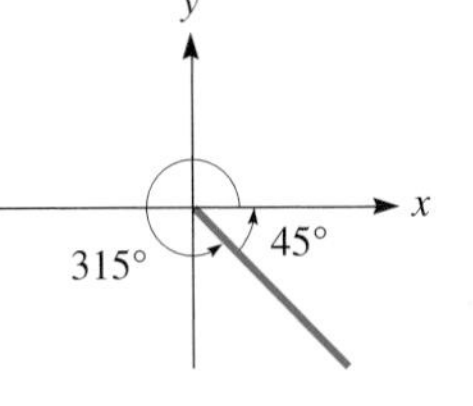

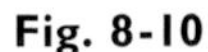
Fig. 8-10

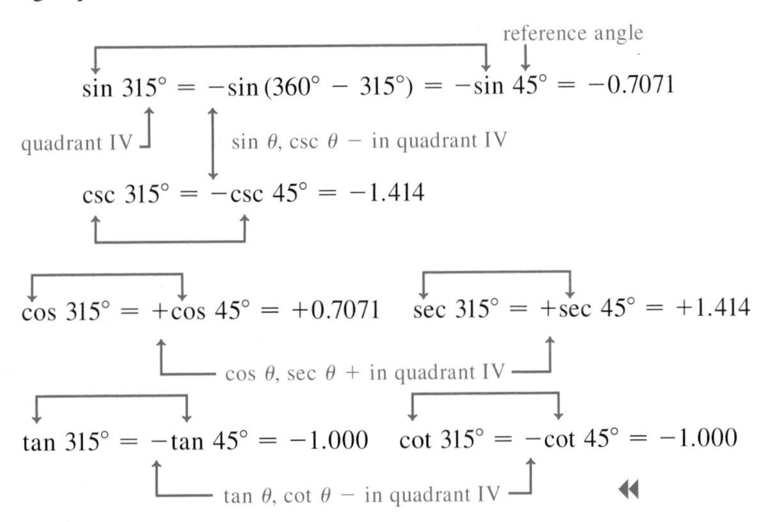

◀◀

EXAMPLE 5 ▶▶ Other illustrations of the use of Eqs. (8-3), (8-4), and (8-5) follow:

same function — reference angle

$$
\begin{aligned}
\sin 160° &= +\sin(180° - 160°) = \sin 20° = 0.3420\\
\tan 110° &= -\tan(180° - 110°) = -\tan 70° = -2.747\\
\cos 225° &= -\cos(225° - 180°) = -\cos 45° = -0.7071\\
\cot 260° &= +\cot(260° - 180°) = \cot 80° = 0.1763\\
\sec 304° &= +\sec(360° - 304°) = \sec 56° = 1.788\\
\sin 357° &= -\sin(360° - 357°) = -\sin 3° = -0.0523
\end{aligned}
$$

determines quadrant — proper sign for function in quadrant ◀◀

In Examples 3, 4, and 5, we have rounded off values to four significant digits. **NOTE** ▶ *If we know that an angle is approximate, we will use the guidelines in Section 4-3 for rounding off values of a function.*

A calculator can be used directly to find values like those in Examples 3, 4, and 5. We simply enter the function and the angle (the order in which these are entered depends on the calculator) and the calculator gives the value, with the proper sign. The reciprocal key is used to find cot θ, sec θ, and csc θ, as shown in Section 4-3.

EXAMPLE 6 ▶▶ A formula for finding the area of a triangle, knowing sides a and b and the included $\angle C$, is $A = \frac{1}{2}ab \sin C$. A surveyor uses this formula to find the area of a triangular tract of land for which $a = 173.2$ m, $b = 156.3$ m, and $C = 112.51°$. See Fig. 8-11.

To find the area, we substitute into the formula, which gives us

$$
\begin{aligned}
A &= \tfrac{1}{2}(173.2)(156.3)\sin 112.51°\\
&= 12\,500 \text{ m}^2 \quad \text{rounded to four significant digits}
\end{aligned}
$$

112.51°
b = 156.3 m
a = 173.2 m

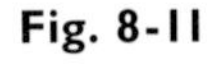
Fig. 8-11

When these numbers are entered into the calculator, the calculator automatically uses a positive value for sin 112.51°. ◀◀

Knowing how to use the reference angle is important when using a calculator, because *if we have the value of a function and want to find the angle,*

CAUTION ▶ ***the calculator will not necessarily give us directly the required angle.***

It will give us an angle we can use, but whether or not it is the required angle for the problem will depend on the problem being solved.

When a value of a trigonometric function is entered into a calculator, it is programmed to give the angle as follows: *For positive values of a function, the calculator displays positive acute angles. For negative values of* $\sin\theta$ *and* $\tan\theta$, *the calculator displays negative acute angles for* $\sin^{-1}$ and $\tan^{-1}$. *For negative values of* $\cos\theta$, *the calculator displays angles between* 90° *and* 180° *for* $\cos^{-1}$. The reason for this is shown in Chapter 20, when these functions are discussed in detail.

EXAMPLE 7 ▶▶ For $\sin\theta = 0.2250$, a calculator will display an angle of 13.00° (rounded off). Remember, depending on your calculator, use the [$\sin^{-1}$], [arcsin], or [INV] [sin] keys to find the angle θ.

This result is correct, but we must remember that

$$\sin(180° - 13.00°) = \sin 167.00° = 0.2250$$

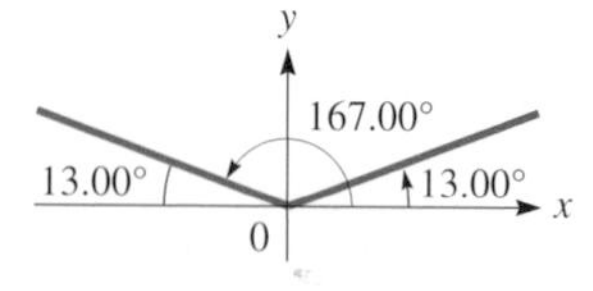

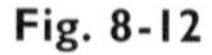
Fig. 8-12

also. If we need only an acute angle, $\theta = 13.00°$ is correct. However, if a second-quadrant angle is required, then $\theta = 167.00°$ is the angle (see Fig. 8-12). These values can be checked by finding sin 13.00° and sin 167.00°. ◀◀

EXAMPLE 8 ▶▶ For $\sec\theta = -2.722$ and $0° \leq \theta < 360°$ (this means θ may equal 0° or be between 0° and 360°), a calculator will display an angle of 111.55° (rounded off) after using the [x^{-1}] and [$\cos^{-1}$] keys.

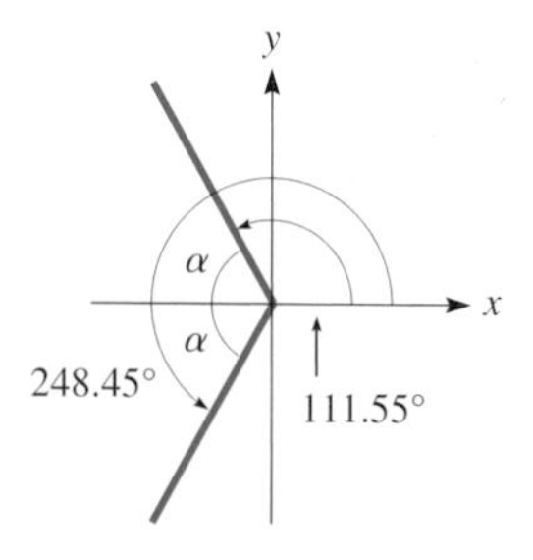

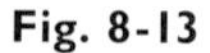
Fig. 8-13

The angle 111.55° is the second-quadrant angle, but $\sec\theta < 0$ in the third quadrant as well. The reference angle is $\alpha = 180° - 111.55° = 68.45°$, and the third-quadrant angle is $180° + 68.45° = 248.45°$. Therefore, the two angles between 0° and 360° for which $\sec\theta = -2.722$ are 111.55° and 248.45° (see Fig. 8-13). These angles can be checked by finding sec 111.55° and sec 248.45°. ◀◀

EXAMPLE 9 ▶▶ Given that $\tan\theta = 2.050$ and $\cos\theta < 0$, find the angle θ for $0° \leq \theta < 360°$.

Since $\tan\theta$ is positive and $\cos\theta$ is negative, θ must be a third-quadrant angle. A calculator will display an angle of 64.00° (rounded off) for $\tan\theta = 2.050$. However, since we need a third-quadrant angle, *we must add* 64.00° *to* 180°. Thus, the required angle is 244.00° (see Fig. 8-14). Check by finding tan 244.00°.

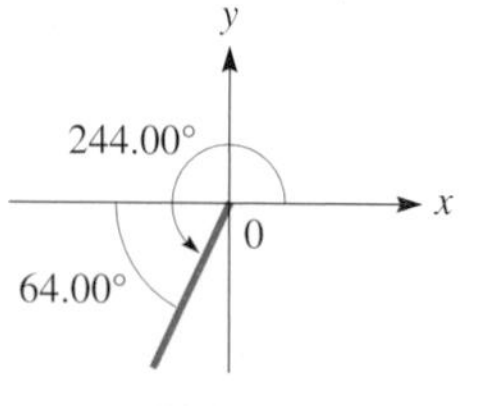

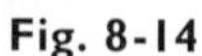
Fig. 8-14

If we are given that $\tan\theta = -2.050$ and $\cos\theta < 0$, the calculator will display an angle of $-64.00°$ for $\tan\theta = -2.050$. We would then have to *recognize that the reference angle is* 64.00° *and subtract it from* 180° *to get* 116.00°, the required second-quadrant angle. This can be checked by finding tan 116.00°. ◀◀

We see that the calculator gives the reference angle (disregarding any minus signs) in all cases except when $\cos\theta$ is negative. To avoid confusion from the angle

NOTE ▶ displayed by the calculator, *a good procedure is to find the reference angle first.* Then it can be used to determine the angle required by the problem.

We can find the reference angle by entering the absolute value of the function. The angle displayed will be the reference angle. Then the required angle θ is found by use of the reference angle α as described earlier, and as shown on the next page in Eqs. (8-6) for θ in the indicated quadrant. The angle should be checked as indicated in the examples above.

$$\begin{array}{ll} \theta = \alpha & \text{(first quadrant)} \\ \theta = 180^\circ - \alpha & \text{(second quadrant)} \\ \theta = 180^\circ + \alpha & \text{(third quadrant)} \\ \theta = 360^\circ - \alpha & \text{(fourth quadrant)} \end{array} \qquad (8\text{-}6)$$

EXAMPLE 10 ▸▸ Given that $\cos\theta = -0.1298$, find θ for $0^\circ \le \theta < 360^\circ$.

Since $\cos\theta$ is negative, θ is either a second-quadrant angle or a third-quadrant angle. Using 0.1298, the calculator tells us that the reference angle is 82.54°.

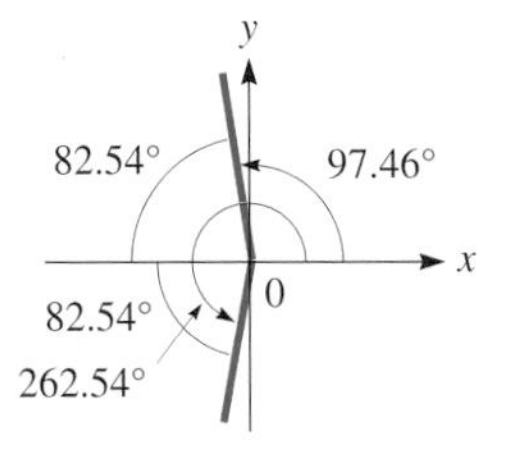

Fig. 8-15

To get the required second-quadrant angle, we subtract 82.54° from 180° and obtain 97.46°. To get the required third-quadrant angle, we add 82.54° to 180° to obtain 262.54°. See Fig. 8-15.

If we use −0.1298, the calculator displays the required second-quadrant angle of 97.46°. However, to get the third-quadrant angle we must then subtract 97.46° from 180° to get the reference angle of 82.54°. The reference angle is then added to 180° to obtain the result of 262.54°. Also note that it is better to store the reference angle in memory rather than recalculate it. ◂◂

Quadrantal Angles

Using Eqs. (8-3) through (8-5) we may find the value of any function when the terminal side of the angle lies *in* one of the quadrants. We now consider *the angle for which the terminal side is along one of the axes, a* **quadrantal angle.** Using the definitions of the functions (recalling that $r > 0$), we obtain the following values.

θ	$\sin\theta$	$\cos\theta$	$\tan\theta$	$\cot\theta$	$\sec\theta$	$\csc\theta$
0°	0.000	1.000	0.000	undef.	1.000	undef.
90°	1.000	0.000	undef.	0.000	undef.	1.000
180°	0.000	−1.000	0.000	undef.	−1.000	undef.
270°	−1.000	0.000	undef.	0.000	undef.	−1.000
360°	Same as the functions of 0° (same terminal side)					

The values in the table may be verified by referring to the figures in Fig. 8-16.

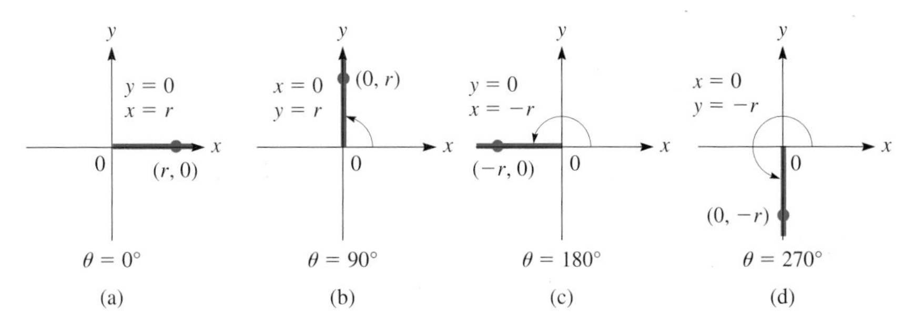

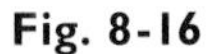
Fig. 8-16

EXAMPLE 11 ▸▸ Since $\sin\theta = y/r$, by looking at Fig. 8-16(a) we can see that $\sin 0^\circ = 0/r = 0$.

Since $\tan\theta = y/x$, from Fig. 8-16(b) we see that $\tan 90^\circ = r/0$, which is undefined due to the division by zero. If we use a calculator to find tan 90°, the display would indicate an error (due to division by zero).

Since $\cos\theta = x/r$, from Fig. 8-16(c) we see that $\cos 180^\circ = -r/r = -1$.

Since $\cot\theta = x/y$, from Fig. 8-16(d) we see that $\cot 270^\circ = 0/-r = 0$. ◂◂

Exercises 8–2

In Exercises 1–8, express the given trigonometric function in terms of the same function of a positive acute angle.

1. $\sin 160°$, $\cos 220°$ **2.** $\tan 91°$, $\sec 345°$

3. $\tan 105°$, $\csc 302°$ **4.** $\cos 190°$, $\cot 290°$

5. $\sin(-123°)$, $\cot 174°$ **6.** $\sin 98°$, $\sec(-315°)$

7. $\cos 400°$, $\tan(-400°)$ **8.** $\tan 920°$, $\csc(-550°)$

In Exercises 9–44, the given angles are approximate. In Exercises 9–16, find the values of the given trigonometric functions by finding the reference angle and attaching the proper sign.

9. $\sin 195°$ **10.** $\tan 311°$

11. $\cos 106.3°$ **12.** $\sin 103.4°$

13. $\tan 219.15°$ **14.** $\cos 198.82°$

15. $\sec 328.33°$ **16.** $\cot 136.53°$

In Exercises 17–24, find the values of the given trigonometric functions directly from a calculator.

17. $\tan 152.4°$ **18.** $\cos 341.4°$

19. $\sin 310.36°$ **20.** $\tan 242.68°$

21. $\cos 110°$ **22.** $\sin 163°$

23. $\csc 194.82°$ **24.** $\sec 261.08°$

In Exercises 25–40, find θ for $0° \le \theta < 360°$.

25. $\sin\theta = -0.8480$ **26.** $\tan\theta = -1.830$

27. $\cos\theta = 0.4003$ **28.** $\sin\theta = 0.6374$

29. $\tan\theta = 0.283$ **30.** $\cos\theta = -0.928$

31. $\cot\theta = -0.212$ **32.** $\csc\theta = -1.09$

33. $\sin\theta = 0.870$, $\cos\theta < 0$

34. $\tan\theta = 0.932$, $\sin\theta < 0$

35. $\cos\theta = -0.12$, $\tan\theta > 0$

36. $\sin\theta = -0.192$, $\tan\theta < 0$

37. $\tan\theta = -1.366$, $\cos\theta > 0$

38. $\cos\theta = 0.5726$, $\sin\theta < 0$

39. $\sec\theta = 2.047$, $\tan\theta < 0$

40. $\cot\theta = -0.3256$, $\sin\theta > 0$

In Exercises 41–44, determine the function that satisfies the given conditions.

41. Find $\tan\theta$ when $\sin\theta = -0.5736$ and $\cos\theta > 0$.

42. Find $\sin\theta$ when $\cos\theta = 0.422$ and $\tan\theta < 0$.

43. Find $\cos\theta$ when $\tan\theta = -0.809$ and $\csc\theta > 0$.

44. Find $\cot\theta$ when $\sec\theta = 1.122$ and $\sin\theta < 0$.

In Exercises 45–48, insert the proper sign, $>$ or $<$ or $=$, between the given expressions.

45. $\sin 90°$ $2\sin 45°$ **46.** $\cos 360°$ $2\cos 180°$

47. $\tan 180°$ $\tan 0°$ **48.** $\sin 270°$ $3\sin 90°$

In Exercises 49–52, evaluate the given expressions.

49. The current i in an alternating-current circuit is given by $i = i_m \sin\theta$, where i_m is the maximum current. Find i if $i_m = 0.0259$ A and $\theta = 495.2°$.

50. A force F is related to force F_x directed along the x-axis by $F = F_x \sec\theta$, where θ is the standard position angle for F. Find F if $F_x = -29.2$ N and $\theta = 127.6°$. See Fig. 8-17.

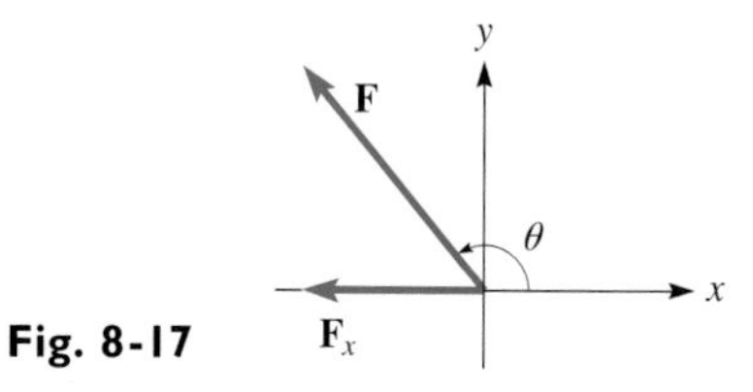

Fig. 8-17

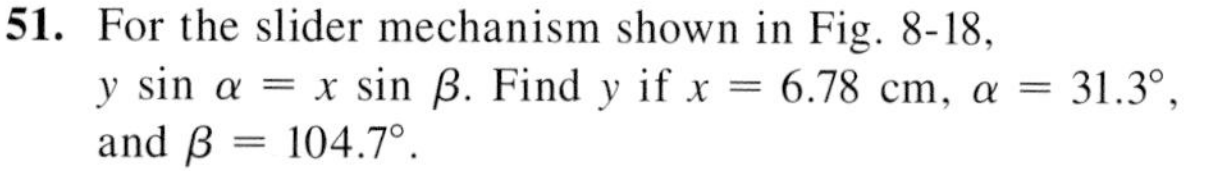

51. For the slider mechanism shown in Fig. 8-18, $y \sin\alpha = x \sin\beta$. Find y if $x = 6.78$ cm, $\alpha = 31.3°$, and $\beta = 104.7°$.

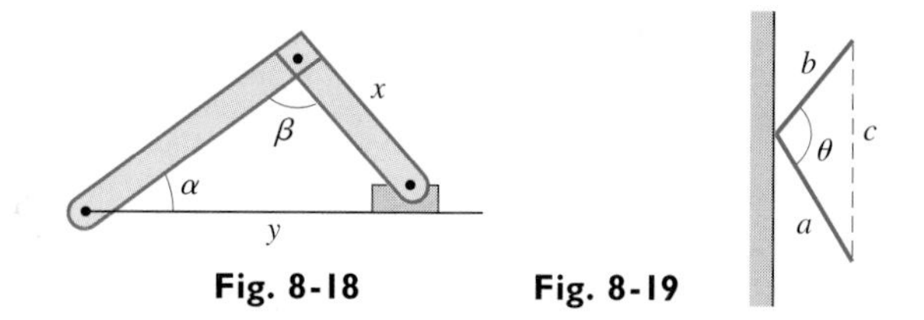

Fig. 8-18 Fig. 8-19

52. A laser follows the path shown in Fig. 8-19. The angle θ is related to the distances a, b, and c by $2ab\cos\theta = a^2 + b^2 - c^2$. Find θ if $a = 15.3$ cm, $b = 12.9$ cm, and $c = 24.5$ cm.

In Exercises 53–56, the trigonometric functions of negative angles are considered. In Exercises 54–56, use the equations derived in Exercise 53.

53. Using the definitions of the trigonometric functions, explain why $\sin(-\theta) = -\sin\theta$. See Fig. 8-20. Also, verify the remaining equations in Eqs. (8-7).

$$\begin{aligned} \sin(-\theta) &= -\sin\theta & \cos(-\theta) &= \cos\theta \\ \tan(-\theta) &= -\tan\theta & \cot(-\theta) &= -\cot\theta \\ \sec(-\theta) &= \sec\theta & \csc(-\theta) &= -\csc\theta \end{aligned} \quad (8\text{-}7)$$

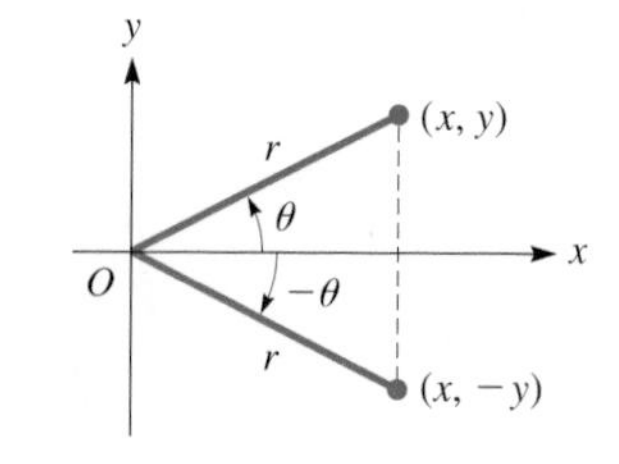

Fig. 8-20

54. Find (a) $\sin(-60°)$ and (b) $\cos(-156°)$.

55. Find (a) $\tan(-100°)$ and (b) $\cot(-215°)$.

56. Find (a) $\sec(-310°)$ and (b) $\csc(-35°)$.

8-3 Radians

For many problems in which trigonometric functions are used, particularly those involving the solution of triangles, degree measurements of angles are convenient and quite sufficient. However, division of a circle into 360 equal parts is by definition, and it is arbitrary and artificial. This definition comes from the ancient Babylonians and their use of a number system based on 60 rather than on 10, as is the system we use today. (Historians are uncertain as to the precise reason for the choice of 360° in a circle.) The *grad,* where $90° = 100$ grad, is also arbitrary and is defined to fit base 10 numbers better.

In numerous other types of applications and in more theoretical discussions, the *radian* is a more meaningful measure of an angle. We defined the radian in Chapter 2 and reviewed it briefly in Chapter 4. In this section we discuss the radian in detail and start by reviewing its definition.

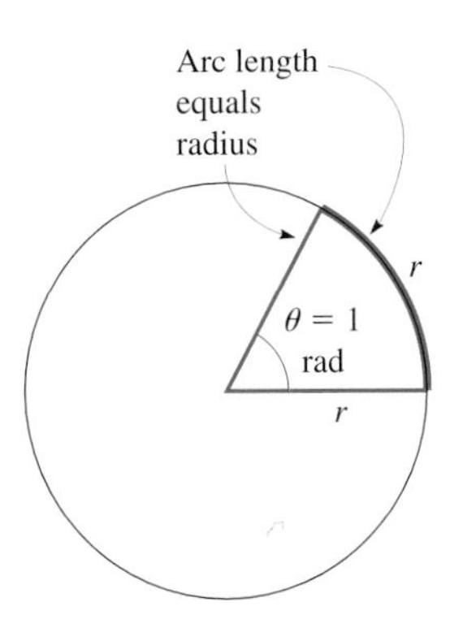

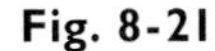
Fig. 8-21

> *A* **radian** *is the measure of an angle with its vertex at the center of a circle and with an intercepted arc on the circle equal in length to the radius of the circle.* See Fig. 8-21.

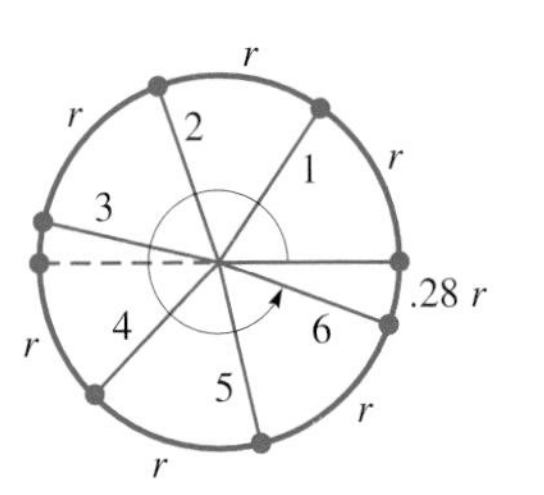

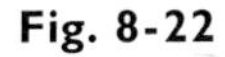
Fig. 8-22

Since the circumference of any circle in terms of its radius is given by $c = 2\pi r$, the ratio of the circumference to the radius is 2π. This means that the radius may be laid off 2π (about 6.28) times along the circumference, regardless of the length of the radius. Therefore, we see that radian measure is independent of the radius of the circle. The definition of a radian is based on an important property of a circle and is therefore a more natural measure of an angle. In Fig. 8-22 the numbers on each of the radii indicate the number of radians in the angle measured in standard position. The circular arrow shows an angle of 6 radians.

Since the radius may be laid off 2π times along the circumference, it follows that there are 2π radians in one complete rotation. Also, there are 360° in one complete rotation. Therefore, 360° is *equivalent* to 2π radians. It then follows that the relation between degrees and radians is $2\pi \text{ rad} = 360°$, or

Converting Angles

$$\pi \text{ rad} = 180° \tag{8-8}$$

Degrees to Radians

$$1° = \frac{\pi}{180} \text{ rad} = 0.017\,45 \text{ rad} \tag{8-9}$$

Radians to Degrees

$$1 \text{ rad} = \frac{180°}{\pi} = 57.30° \tag{8-10}$$

We see from Eqs. (8-8) through (8-10) that we convert angle measurements from degrees to radians or from radians to degrees as follows:

> ***Procedure for Converting Angle Measurements***
>
> **1.** *To convert an angle measured in degrees to the same angle measured in radians,* **multiply the number of degrees by $\pi/180°$.**
>
> **2.** *To convert an angle measured in radians to the same angle measured in degrees,* **multiply the number of radians by $180°/\pi$.**

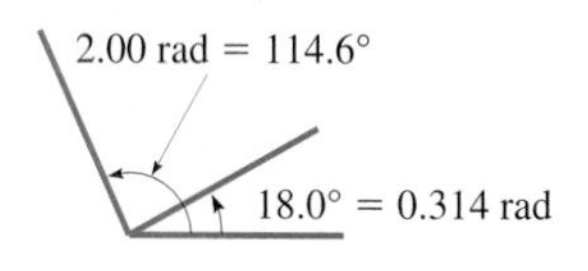

Fig. 8-23

EXAMPLE 1 ▸▸ converting degrees to radians

(a) $18.0° = \left(\frac{\pi}{180°}\right)(18.0°) = \frac{\pi}{10.0} = \frac{3.14}{10.0} = 0.314 \text{ rad}$

degrees cancel (See Fig. 8-23.)

converting radians to degrees

(b) $2.00 \text{ rad} = \left(\frac{180°}{\pi}\right)(2.00) = \frac{360°}{3.14} = 114.6°$ (See Fig. 8-23.)

In multiplying by $\pi/180°$ or $180°/\pi$, we are actually only multiplying by 1 because π rad = 180°. Although the unit of measurement is different, *the angle is the same.* See Appendix B on unit conversions. ◂◂

Due to the nature of the definition of the radian, it is very common to express radians in terms of π. Expressing angles in terms of π is illustrated in the following example.

EXAMPLE 2 ▸▸ converting degrees to radians

(a) $30° = \left(\frac{\pi}{180°}\right)(30°) = \frac{\pi}{6} \text{ rad}$

(b) $45° = \left(\frac{\pi}{180°}\right)(45°) = \frac{\pi}{4} \text{ rad}$ (See Fig. 8-24.)

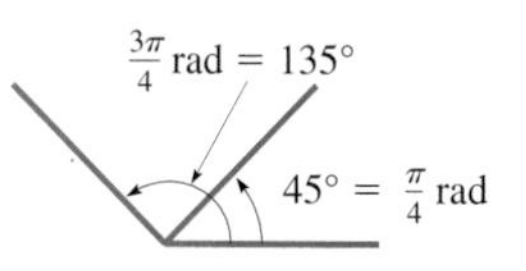

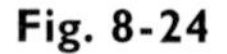

Fig. 8-24

converting radians to degrees

(c) $\frac{\pi}{2} \text{ rad} = \left(\frac{180°}{\pi}\right)\left(\frac{\pi}{2}\right) = 90°$

(d) $\frac{3\pi}{4} \text{ rad} = \left(\frac{180°}{\pi}\right)\left(\frac{3\pi}{4}\right) = 135°$ (See Fig. 8-24.) ◂◂

We wish now to make a very important point. Since π is a special way of writing the number (slightly greater than 3) that is the ratio of the circumference of a circle to its diameter, it is the ratio of one distance to another. Thus, radians really have no units and *radian measure amounts to measuring angles in terms of real numbers.* It is this property of radians that makes them useful in many situations. Therefore, when radians are being used, it is customary that no units are indicated for the angle.

CAUTION ▶ *When no units are indicated, the radian is understood to be the unit of angle measurement.*

EXAMPLE 3 ▸▸ (a) $60° = \left(\frac{\pi}{180°}\right)(60.0°) = \frac{\pi}{3.00} = 1.05$

no units indicates radian measure

(b) $3.80 = \left(\frac{180°}{\pi}\right)(3.80) = 218°$

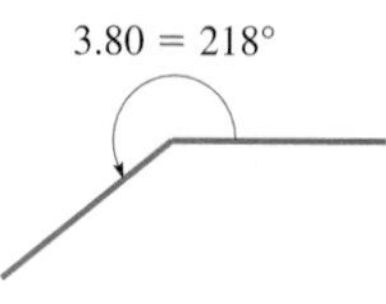

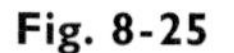

Fig. 8-25

Since no units are indicated for 1.05 and 3.80 in this example, they are known to be in radian measure. Here we must know that 3.80 is an angle measure. See Fig. 8-25. ◂◂

Many calculators have a key that can be used directly to change an angle expressed in degrees to an angle expressed in radians, or from an angle expressed in radians to an angle expressed in degrees. Generally the key is a second-function key. The angle in radians should have the same number of significant digits as the angle in degrees in decimal form.

We can use a calculator to find the value of a function of an angle in radians. If the calculator is in radian mode, it then uses values in radians directly and will *consider any angle entered to be in radians.* Some calculators have a display that always indicates that either degrees or radians are being used. Many calculators are normally in degree mode, whereas others are normally in radian mode. The mode can be changed as needed, but

CAUTION ▶ *always be careful to have your calculator in the proper mode.*

If you are working in degrees, use the degree mode, but if you are working in radians, use the radian mode. On a graphing calculator, use the MODE feature.

EXAMPLE 4 ▶▶ (a) To find the value of sin 0.7538, put the calculator in radian mode (note that no units are shown with 0.7538), and then use 0.7538 and the [sin] key. This gives us

$$\sin 0.7538 = 0.6844$$

no units indicates radian measure

(b) With the calculator in radian mode, we use the [tan] key to get

$$\tan 0.9977 = 1.550$$

(c) Using radian mode and the [cos] key we get

$$\cos 2.074 = -0.4822$$

In each case, the order in which the angle and the function are entered in the calculator depends on the calculator. ◀◀

EXAMPLE 5 ▶▶ The velocity v of an object undergoing simple harmonic motion at the end of a spring is given by

$$v = A\sqrt{\frac{k}{m}}\cos\sqrt{\frac{k}{m}}\,t$$

the angle is $\left(\sqrt{\frac{k}{m}}\right)(t)$

Here m is the mass of the object (in grams), k is a constant depending on the spring, A is the maximum distance the object moves, and t is the time (in seconds). Find the velocity (in centimeters per second) after 0.100 s of a 36.0-g object at the end of a spring for which $k = 400\ \text{g/s}^2$, if $A = 5.00$ cm.

Substituting, we have

$$v = 5.00\sqrt{\frac{400}{36.0}}\cos\sqrt{\frac{400}{36.0}}(0.100)$$

Using calculator memory for $\sqrt{\frac{400}{36.0}}$, and with the calculator in radian mode, we have

$$v = 15.7\ \text{cm/s}$$ ◀◀

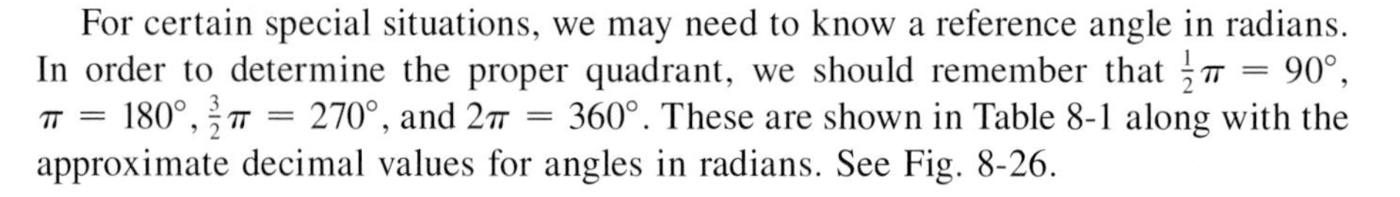

For certain special situations, we may need to know a reference angle in radians. In order to determine the proper quadrant, we should remember that $\frac{1}{2}\pi = 90°$, $\pi = 180°$, $\frac{3}{2}\pi = 270°$, and $2\pi = 360°$. These are shown in Table 8-1 along with the approximate decimal values for angles in radians. See Fig. 8-26.

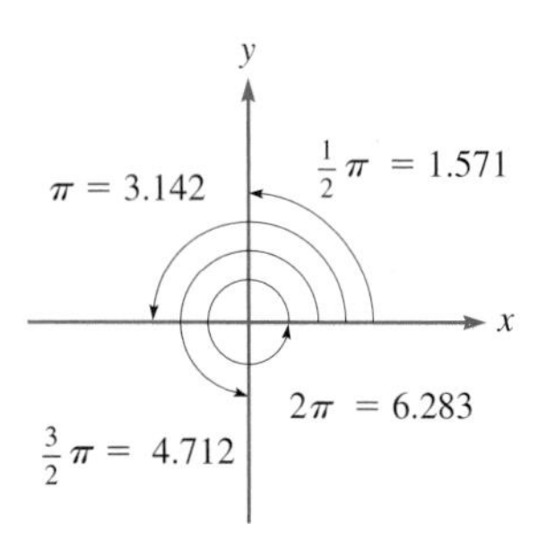

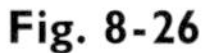
Fig. 8-26

TABLE 8-1 Quadrantal Angles

Degrees	*Radians*	*Radians (decimal)*
90°	$\frac{1}{2}\pi$	1.571
180°	π	3.142
270°	$\frac{3}{2}\pi$	4.712
360°	2π	6.283

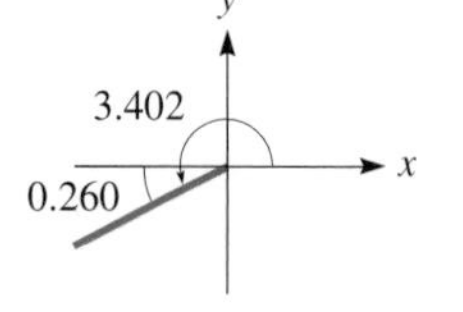

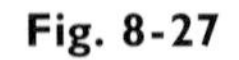
Fig. 8-27

EXAMPLE 6 ▸▸ An angle of 3.402 is greater than 3.142, but less than 4.712. Thus, it is a third-quadrant angle, and the reference angle is $3.402 - \pi = 0.260$. The π key can be used. See Fig. 8-27.

An angle of 5.210 is between 4.712 and 6.283. Therefore, it is in the fourth quadrant and the reference angle is $2\pi - 5.210 = 1.073$. ◂◂

EXAMPLE 7 ▸▸ Express θ in radians, such that $\cos\,\theta = 0.8829$ and $0 \le \theta < 2\pi$.

We are to find θ in radians for the given value of the $\cos\,\theta$. Also, since θ is restricted to values between 0 and 2π, we must find a first-quadrant angle and a fourth-quadrant angle ($\cos\,\theta$ is positive in the first and fourth quadrants). With the calculator in radian mode, we find that

$$\cos 0.4888 = 0.8829$$

Therefore, for the fourth-quadrant angle,

$$\cos\,(2\pi - 0.4888) = \cos 5.794$$

This means

$$\theta = 0.4888 \quad \text{or} \quad \theta = 5.794$$

See Fig. 8-28. ◂◂

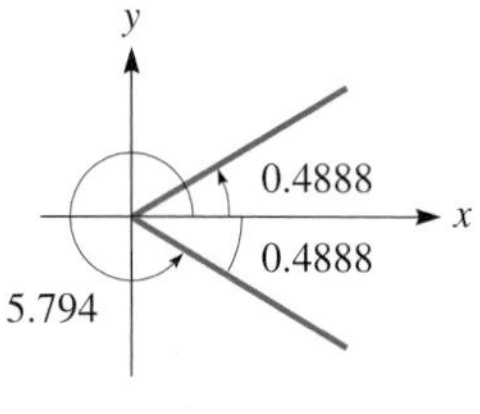

Fig. 8-28

When one first encounters radian measure,

CAUTION ▸ *expressions such as* $\sin 1$ *and* $\sin\,\theta = 1$ *are often confused.*

The first is equivalent to $\sin 57.30°$, since $57.30° = 1$ (radian). The second means that θ is the angle for which the sine is 1. Since we know that $\sin 90° = 1$, we then can say that $\theta = 90°$ or $\theta = \pi/2$. The following example gives additional illustrations of evaluating expressions involving radians.

EXAMPLE 8 ▸▸ (a) $\sin\dfrac{\pi}{3} = \dfrac{\sqrt{3}}{2}$ since $\dfrac{\pi}{3} = 60°$.

(b) $\sin 0.6050 = 0.5688$. (0.6050 rad = 34.66°.)

(c) $\tan\,\theta = 1.709$ means that $\theta = 59.67°$ (smallest positive θ).

(d) Since $59.67° = 1.041$, we can state that $\tan 1.041 = 1.708$. ◂◂

Exercises 8–3

In Exercises 1–8, express the given angle measurements in radian measure in terms of π.

1. 15°, 150°
2. 12°, 225°
3. 75°, 330°
4. 36°, 315°
5. 210°, 270°
6. 240°, 300°
7. 160°, 260°
8. 66°, 350°

In Exercises 9–16, the given numbers express angle measure. Express the measure of each angle in terms of degrees.

9. $\frac{2\pi}{5}, \frac{3\pi}{2}$
10. $\frac{3\pi}{10}, \frac{5\pi}{6}$
11. $\frac{\pi}{18}, \frac{7\pi}{4}$
12. $\frac{7\pi}{15}, \frac{4\pi}{3}$
13. $\frac{17\pi}{18}, \frac{5\pi}{3}$
14. $\frac{11\pi}{36}, \frac{5\pi}{4}$
15. $\frac{\pi}{12}, \frac{3\pi}{20}$
16. $\frac{7\pi}{30}, \frac{4\pi}{15}$

In Exercises 17–24, express the given angles in radian measure. Round off results to the number of significant digits in the given angle.

17. 23.0°
18. 54.3°
19. 252°
20. 104°
21. 333.5°
22. 168.7°
23. 178.5°
24. 86.1°

In Exercises 25–32, the given numbers express angle measure. Express the measure of each angle in terms of degrees, with the same accuracy as in the given value.

25. 0.750
26. 0.240
27. 3.407
28. 1.703
29. 2.45
30. 34.4
31. 16.42
32. 100.0

In Exercises 33–40, evaluate the given trigonometric functions by first changing the radian measure to degree measure. Round off results to four significant digits.

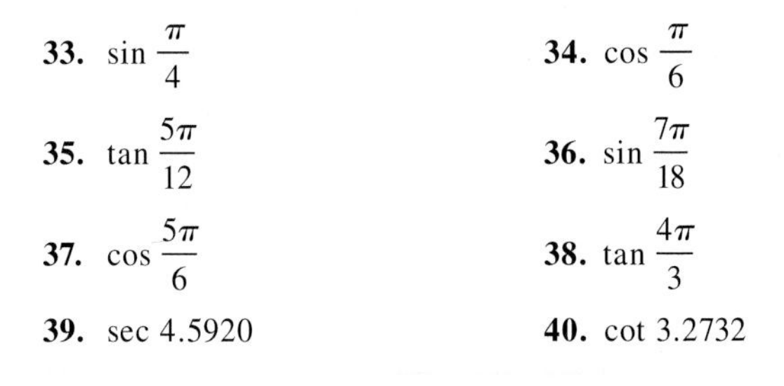

33. $\sin \frac{\pi}{4}$
34. $\cos \frac{\pi}{6}$
35. $\tan \frac{5\pi}{12}$
36. $\sin \frac{7\pi}{18}$
37. $\cos \frac{5\pi}{6}$
38. $\tan \frac{4\pi}{3}$
39. sec 4.5920
40. cot 3.2732

In Exercises 41–48, evaluate the given trigonometric functions directly, without first changing the radian measure to degree measure.

41. tan 0.7359
42. cos 0.9308
43. sin 4.24
44. tan 3.47
45. cos 2.07
46. sin 2.34
47. cot 4.86
48. csc 6.19

In Exercises 49–56, find θ to four significant digits for $0 \le \theta < 2\pi$. ($2\pi = 6.283$.)

49. $\sin \theta = 0.3090$
50. $\cos \theta = -0.9135$
51. $\tan \theta = -0.2126$
52. $\sin \theta = -0.0436$
53. $\cos \theta = 0.6742$
54. $\tan \theta = 1.860$
55. $\sec \theta = -1.307$
56. $\csc \theta = 3.940$

In Exercises 57–60, evaluate the given expressions.

57. A flat plate of weight W is attached at the top and is oscillating as shown in Fig. 8-29. Its potential energy V is given by $V = \frac{1}{2}Wb\theta^2$, where θ is measured in radians. Find V if $W = 8.75$ N, $b = 0.75$ m, and $\theta = 5.5°$.

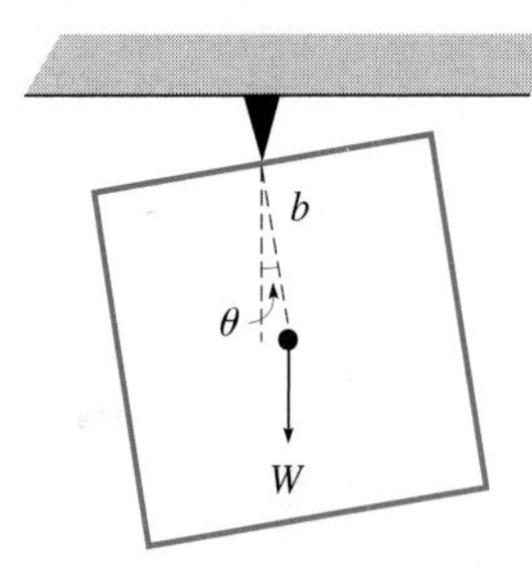

Fig. 8-29

58. The charge q (in coulombs) on a capacitor as a function of time is $q = A \sin \omega t$. If t is measured in seconds, in what units is ω measured? Explain.

59. The height h of a rocket launched 1200 m from an observer is found to be $h = 1200 \tan \frac{5t}{3t + 10}$ for $t < 10$ s, where t is the time after launch. Find h for $t = 8.0$ s.

60. The electric intensity I, in W/m², from the two radio antennas shown in Fig. 8-30 is a function of the angle θ given by $I = 0.023 \cos^2(\pi \sin \theta)$. Find I for $\theta = 40.0°$. ($\cos^2 \alpha = (\cos \alpha)^2$.)

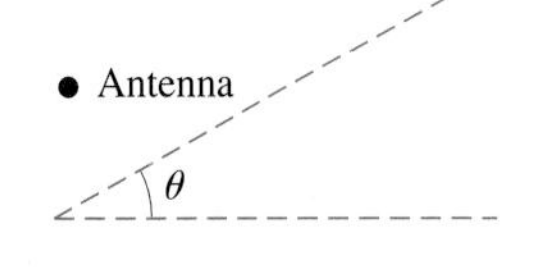

Fig. 8-30

8-4 Applications of the Use of Radian Measure

Radian measure has numerous applications in mathematics and technology, some of which were indicated in the last four exercises of the previous section. In this section we illustrate the usefulness of radian measure in certain specific geometric and physical applications.

Arc Length

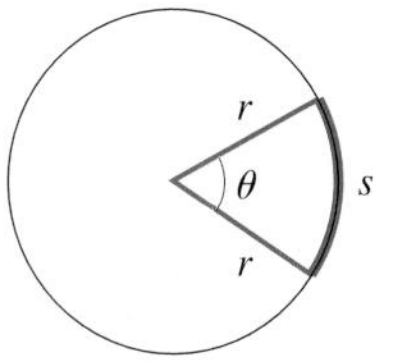

From geometry we know that *the length of an arc on a circle is proportional to the central angle* and that the length of arc of a complete circle is the circumference. Letting s stand for the length of arc, we may state that $s = 2\pi r$ for a complete circle. Since 2π is the central angle (in radians) of the complete circle, *we have for the length of arc*

Fig. 8-31

$$s = \theta r \quad (\theta \text{ in radians}) \tag{8-11}$$

for any circular arc with central angle θ. If we know the central angle in radians and the radius of a circle, we can find the length of a circular arc directly by using Eq. (8-11). See Fig. 8-31.

EXAMPLE 1 ▸▸ (a) If $\theta = \pi/6$ and $r = 3.00$ cm,

$$s = \left(\frac{\pi}{6}\right)(3.00) = \frac{\pi}{2.00} = 1.57 \text{ cm} \qquad (\theta \text{ in radians})$$

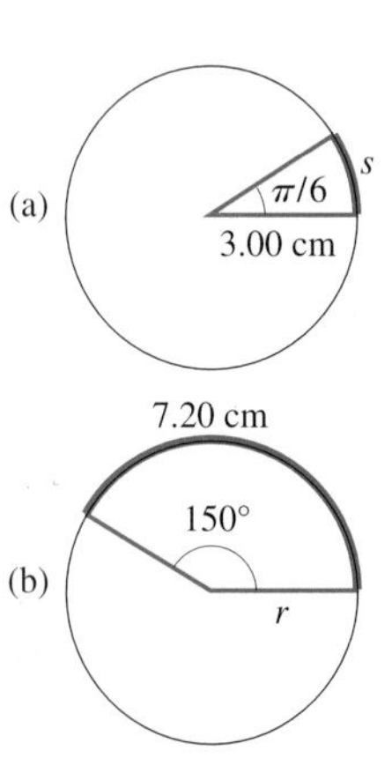

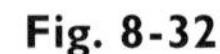
Fig. 8-32

See Fig. 8-32(a).

(b) If the arc length is 7.20 cm for a central angle of 150° on a certain circle, we may find the radius of the circle by solving $s = \theta r$ for r and then substituting. Thus, $r = s/\theta$. Substituting, we have

$$r = \frac{7.20}{150\left(\frac{\pi}{180}\right)} = \frac{(7.20)(180)}{150\pi} = 2.75 \text{ cm}$$

(θ in radians)

See Fig. 8-32(b). ◂◂

Area of a Sector of a Circle

Another geometric application of radians is in finding the area of a sector of a circle (see Fig. 8-33). We recall from geometry that areas of sectors of circles are proportional to their central angles. The area of a circle is given by $A = \pi r^2$. This can be written as $A = \frac{1}{2}(2\pi)r^2$. We now note that the angle for a complete circle is 2π, and therefore *the area of any sector of a circle in terms of the radius and the central angle (in radians) is*

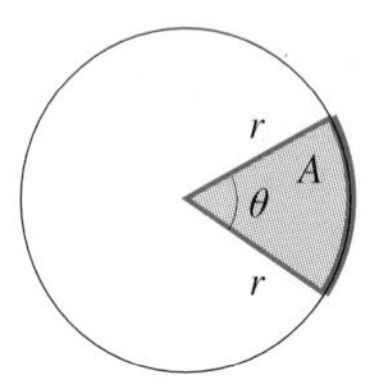

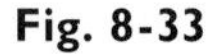
Fig. 8-33

$$A = \frac{1}{2}\theta r^2 \quad (\theta \text{ in radians}) \tag{8-12}$$

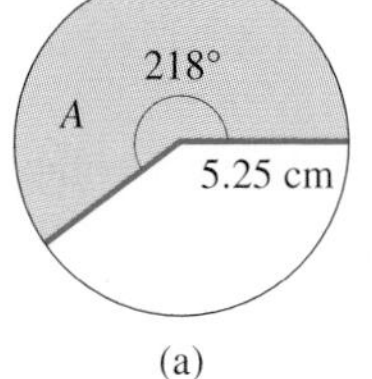

(a)

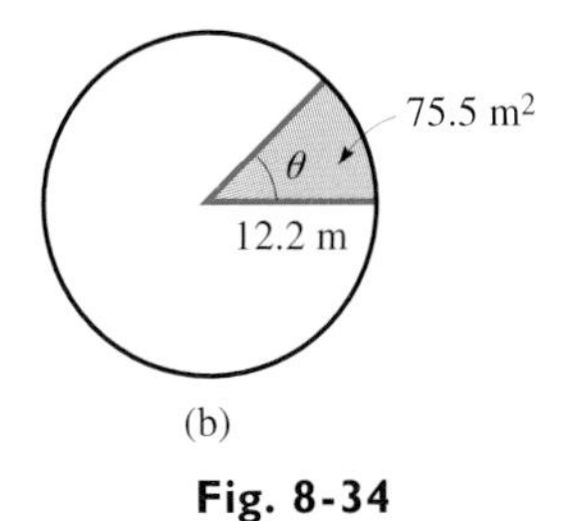

(b)

Fig. 8-34

EXAMPLE 2 ▸▸ (a) The area of a sector of a circle with central angle 218° and a radius of 5.25 cm (see Fig. 8-34(a)) is

$$A = \frac{1}{2}(218)\underbrace{\left(\frac{\pi}{180}\right)}_{\theta \text{ in radians}}(5.25)^2 = 52.4 \text{ cm}^2$$

(b) Given that the area of a sector is 75.5 m^2 and the radius is 12.2 m (see Fig. 8-34(b)), we find the central angle by solving for θ and then substituting.

$$\theta = \frac{2A}{r^2}$$

$$= \frac{2(75.5)}{(12.2)^2} = 1.01 \quad \leftarrow \text{no units indicates radian measure}$$

This means that the central angle is 1.01 rad, or 57.9°. ◂◂

CAUTION ▶ We should note again that the equations in this section require that the angle θ *is expressed in radians.* A common error is to use θ in degrees.

Angular Velocity

The next application of radians deals with velocity. We know that average velocity is defined by the equation $v = s/t$, where v is the average velocity, s is the distance traveled, and t is the elapsed time. If an object is moving around a circular path with constant speed, the actual distance traveled is the length of arc traversed. Therefore, if we divide both sides of Eq. (8-11) by t, we obtain

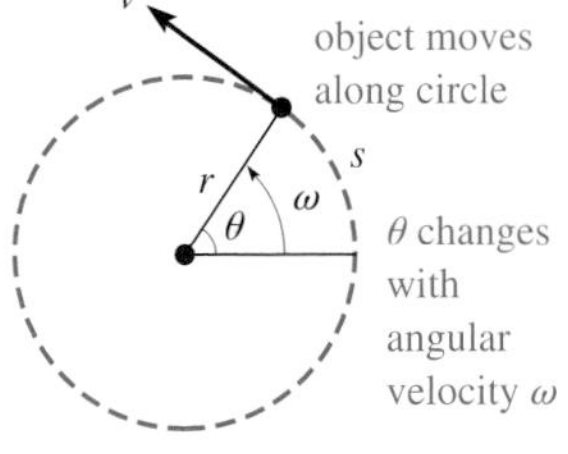

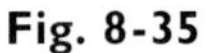

Fig. 8-35

$$\frac{s}{t} = \frac{\theta r}{t} = \frac{\theta}{t}r$$

where θ/t is called the *angular velocity* and is designated by ω. Therefore,

$$v = \omega r \tag{8-13}$$

Equation (8-13) expresses the relationship between the **linear velocity** v *and the* **angular velocity** ω *of an object moving around a circle of radius* r. See Fig. 8-35. In the figure, v is shown directed tangent to the circle, for that is its direction for the position shown. The direction of v changes constantly.

The units for ω are radians per unit of time. In this way the formula can be used directly. However, in practice, ω is often given in revolutions per minute or in some similar unit. In these cases it is necessary to convert the units of ω to radians per unit of time before substituting in Eq. (8-13).

EXAMPLE 3 ▸▸ A person on a hang glider is moving in a horizontal circular arc of radius 90.0 m with an angular velocity of 0.125 rad/s. The person's linear velocity is

$$v = (0.125)(90.0) = 11.3 \text{ m/s}$$

(Remember that radians are numbers and are not included in the final set of units.) This means that the person is moving along the circumference of the arc at 11.3 m/s (40.7 km/h). ◂◂

See the chapter introduction.

EXAMPLE 4 ▸▸ A communications satellite remains at an altitude of 35 920 km above a point on the equator. If the radius of the earth is 6370 km, what is the velocity of the satellite?

In order for the satellite to remain over a point on the equator, it must rotate exactly once each day around the center of the earth (and it must remain at an altitude of 35 920 km). Since there are 2π radians in each revolution, the angular velocity is

$$\omega = \frac{1 \text{ r}}{1 \text{ day}} = \frac{2\pi \text{ rad}}{24 \text{ h}} = 0.2618 \text{ rad/h}$$

The radius of the circle through which the satellite moves is its altitude plus the radius of the earth, or $35\,920 + 6370 = 42\,290$. Thus, the velocity is

$$v = 0.2618(42\,290) = 11\,070 \text{ km/h}$$ ◂◂

EXAMPLE 5 ▸▸ A pulley belt 4.00 m long takes 2.00 s to make one complete revolution. The radius of the pulley is 20.0 cm. What is the angular velocity (in revolutions per minute) of a point on the rim of the pulley? See Fig. 8-36.

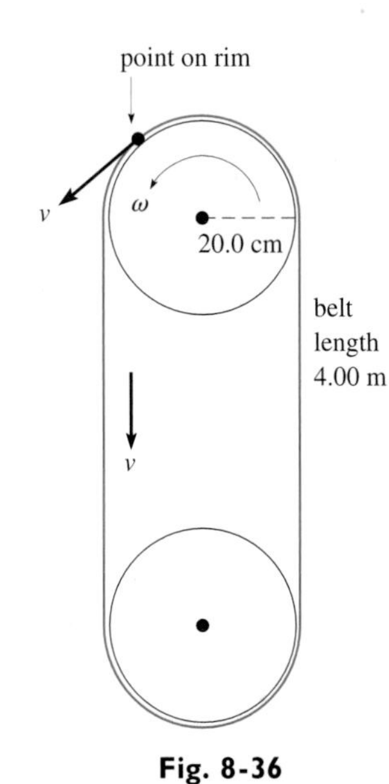

Fig. 8-36

Since the linear velocity of a point on the rim of the pulley is the same as the velocity of the belt, $v = 4.00/2.00 = 2.00$ m/s. The radius of the pulley is $r = 20.0 \text{ cm} = 0.200$ m, and we can find ω by substituting into Eq. (8-13). This gives us

$$v = \omega r$$

$$2.00 = \omega(0.200)$$

or

$$\omega = 10.0 \text{ rad/s}$$ multiply by 60 s/1 min

$$= 600 \text{ rad/min}$$ multiply by 1 r/2π rad

$$= 95.5 \text{ r/min}$$ r is the symbol for revolution

As shown in Appendix B, the change of units can be handled algebraically as

$$10.0\frac{\text{rad}}{\text{s}} \times 60\frac{\text{s}}{\text{min}} = 600\frac{\text{rad}}{\text{min}}$$

$$\frac{600 \text{ rad/min}}{2\pi \text{ rad/r}} = 600\frac{\text{rad}}{\text{min}} \times \frac{1}{2\pi}\frac{\text{r}}{\text{rad}} = 95.5 \text{ r/min}$$ ◂◂

EXAMPLE 6 ▸▸ The current at any time in a certain alternating-current electric circuit is given by $i = I \sin 120\pi t$, where I is the maximum current and t is the time in seconds. Given that $I = 0.0685$ A, find i for $t = 0.005\,00$ s.

Substituting, with the calculator in radian mode, we get

$$i = 0.0685 \sin(120\pi)(0.005\,00)$$

$$= 0.0651 \text{ A}$$ ◂◂

Exercises 8–4

In Exercises 1–12, for an arc length s, area of sector A, and central angle θ of a circle of radius r, find the indicated quantity for the given values.

1. $r = 3.30$ cm, $\theta = \pi/3$, $s = ?$
2. $r = 21.2$ cm, $\theta = 2.65$, $s = ?$
3. $r = 425$ mm, $\theta = 136°$, $s = ?$
4. $r = 0.2690$ m, $\theta = 73.61°$, $s = ?$
5. $s = 0.3913$ km, $\theta = 0.4141$, $r = ?$
6. $s = 3.19$ m, $r = 2.29$ m, $\theta = ?$
7. $r = 4.9$ cm, $\theta = 3.6$, $A = ?$
8. $r = 46.3$ dm, $\theta = 2\pi/5$, $A = ?$
9. $r = 0.0646$ m, $\theta = 326°$, $A = ?$
10. $r = 89$ mm, $\theta = 17°$, $A = ?$
11. $A = 16.5\ \text{m}^2$, $r = 4.02$ m, $\theta = ?$
12. $A = 67.8\ \text{km}^2$, $r = 67.8$ km, $\theta = ?$

In Exercises 13–48, solve the given problems.

13. Part of the center line of a highway follows a circular arc of which the radius is 320 m and the central angle is 62.0°. Find the length of this part of the center line.
14. While playing, the left spool of a VCR turns through 820°. For this part of the tape, it is 3.30 cm from the center of the spool to the tape. What length of tape is played?
15. A door 76.2 cm wide is opened through an angle of 110.0°. What floor area does the door pass over?
16. A section of sidewalk is a circular sector of radius 1.25 m and central angle 50.6°. What is the area of this section of sidewalk?
17. A section of a natural gas pipeline 3.25 km long is part of a circular arc. If the radius of the circle is 8.50 km, what is the central angle of the arc?
18. A cam is in the shape of a circular sector, as shown in Fig. 8-37. What is the perimeter of the cam?

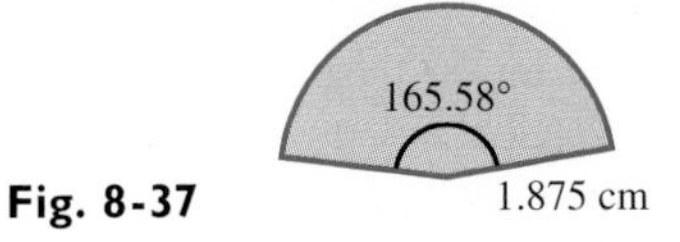

Fig. 8-37

19. A lawn sprinkler can water up to a distance of 25.0 m. It turns through an angle of 115.0°. What area can the sprinkler water?
20. A beam of light from a spotlight sweeps through a horizontal angle of 75.0°. If the range of the spotlight is 110 m, how large an area can it cover?
21. If a car makes a U-turn in 6.0 s, what is its average angular velocity in the turn?
22. The roller on a computer printer makes 2200 r/min. What is its angular velocity?
23. What is the floor area of the hallway shown in Fig. 8-38? The outside and inside of the hallway are circular arcs.

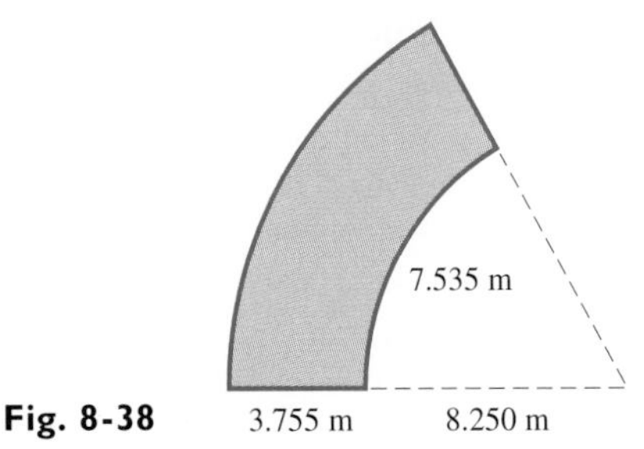

Fig. 8-38

24. The arm of a car windshield wiper is 32.4 cm long and is attached at the middle of a 38.1 cm blade. (Assume that the arm and blade are in line.) What area of the windshield is cleaned by the wiper if it swings through 110.0° arcs?
25. Part of a railroad track follows a circular arc with a central angle of 28.0°. If the radius of the arc of the inner rail is 28.55 m and the rails are 1.44 m apart, how much longer is the outer rail than the inner rail?
26. A wrecking ball is dropped as shown in Fig. 8-39. Its velocity at the bottom of its swing is $v = \sqrt{2gh}$, where g is the acceleration due to gravity. What is its angular velocity at the bottom if $g = 9.80\ \text{m/s}^2$ and $h = 4.80$ m?

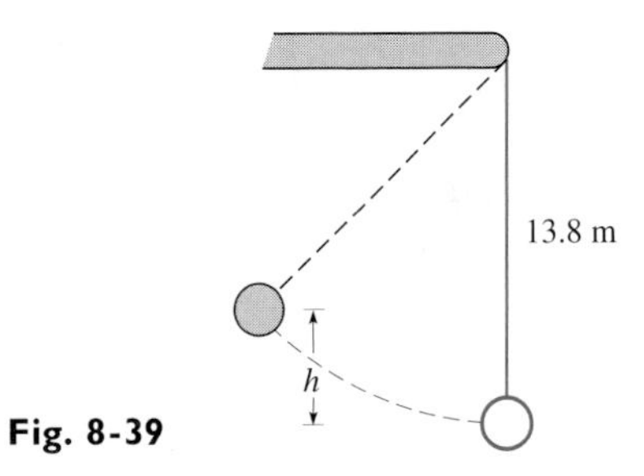

Fig. 8-39

27. Part of a security fence is built 2.50 m from a cylindrical storage tank 11.2 m in diameter. What is the area between the tank and this part of the fence if the central angle of the fence is 75.5°?
28. Through what angle does the drum in Fig. 8-40 turn in order to lower the crate 10.3 m?

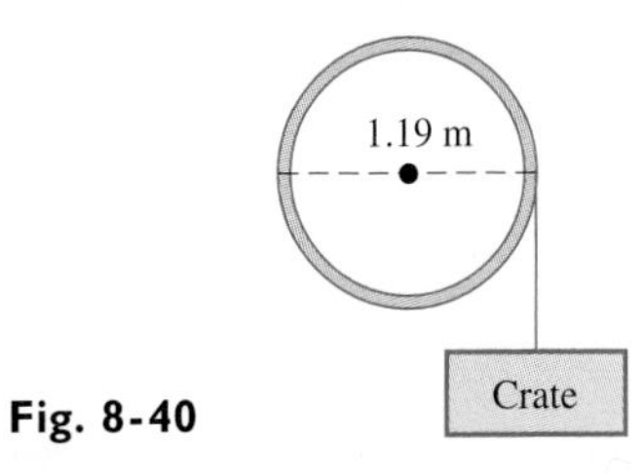

Fig. 8-40

29. A section of road follows a circular arc with a central angle of 15.6°. The radius of the inside of the curve is 285.0 m, and the road is 15.2 m wide. What is the volume of the concrete in the road if it is 0.305 m thick?

30. The propeller of the motor on a motorboat is rotating at 130 rad/s. What is the linear velocity of a point on the tip of a blade if it is 22.5 cm long?

31. A storm causes a pilot to follow a circular-arc route, with a central angle of 12.8°, from city A to city B rather than the straight-line route of 185.0 km. How much farther does the plane fly due to the storm?

32. A highway exit is a circular arc 330 m long with a central angle of 79.4°. What is the radius of curvature of the exit?

33. A special vehicle for traveling on glacial ice has tires that are 3.66 m in diameter. If the vehicle travels at 5.6 km/h, what is the angular velocity of the tire in revolutions per minute?

34. The sweep second hand of a watch is 15.0 mm long. What is the linear velocity of the tip?

35. A floppy disk used in a personal computer has a diameter of 13.3 cm and rotates at 360.0 r/min. What is the linear velocity of a point on the outer edge?

36. A rotating circular restaurant at the top of a hotel has a diameter of 25.0 m. If it completes one revolution in 30.0 min, what is the velocity of its outer surface?

37. Two streets meet at an angle of 82.0°. What is the length of the piece of curved curbing at the intersection if it is constructed along the arc of a circle 5.50 m in radius? See Fig. 8-41.

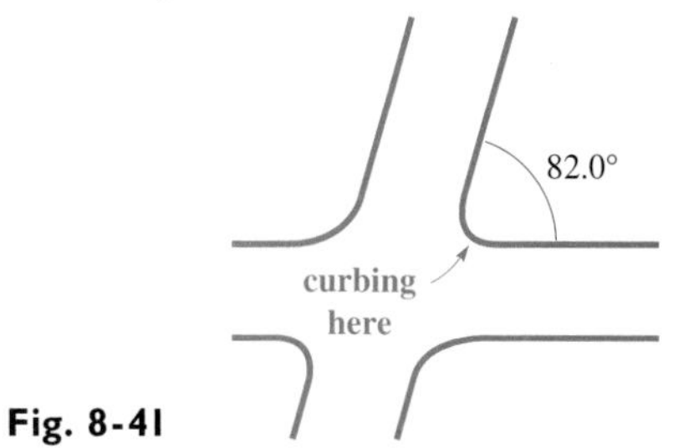

Fig. 8-41

38. An ammeter needle is deflected 52.00° by a current of 0.2500 A. The needle is 3.750 cm long, and a circular scale is used. How long is the scale for a maximum current of 1.500 A?

39. A drill bit 9.53 mm in diameter rotates at 1200 r/min. What is the linear velocity of a point on its circumference?

40. A helicopter blade is 2.75 m long and is rotating at 420 r/min. What is the linear velocity of the tip of the blade?

41. A waterwheel used to generate electricity has paddles 3.75 m long. The speed of the end of a paddle is one-fourth that of the water. If the water is flowing at the rate of 6.50 m/s, what is the angular velocity of the waterwheel?

42. A jet is traveling westward with the sun directly overhead (the jet is on a line between the sun and the center of the earth). How fast must the jet fly in order to keep the sun directly overhead? (Assume that the earth's radius is 6370 km, the altitude of the jet is low, and the earth rotates about its axis once in 24.0 h.)

43. A 1500-kW wind turbine (windmill) rotates at 40.0 r/min. What is the linear velocity of a point on the end of a blade, if the blade is 12.0 m long (from the center of rotation)?

44. What is the linear velocity of a point in Sydney, Australia, which is at a latitude of 33°46′S? The radius of the earth is 6370 km.

45. Through what total angle does the drive shaft of a car rotate in one second when the tachometer reads 2400 r/min?

46. A patio is in the shape of a circular sector with a central angle of 160.0°. It is enclosed by a railing of which the circular part is 11.6 m long. What is the area of the patio?

47. An oil storage tank 4.25 m long has a flat bottom as shown in Fig. 8-42. The radius of the circular part is 1.10 m. What volume of oil does the tank hold?

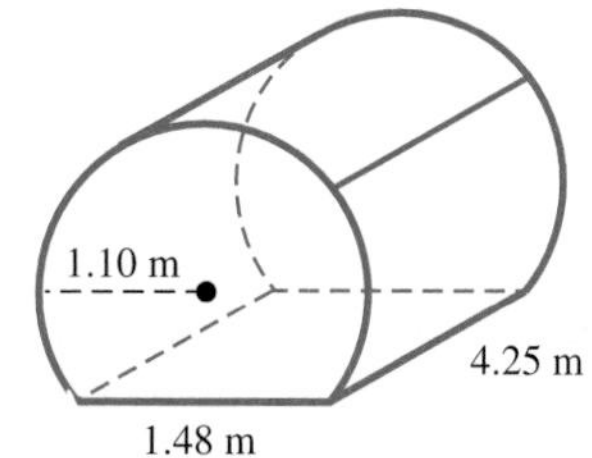

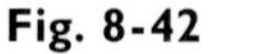
Fig. 8-42

48. Two equal beams of light illuminate the area shown in Fig. 8-43. What area is lit by both beams?

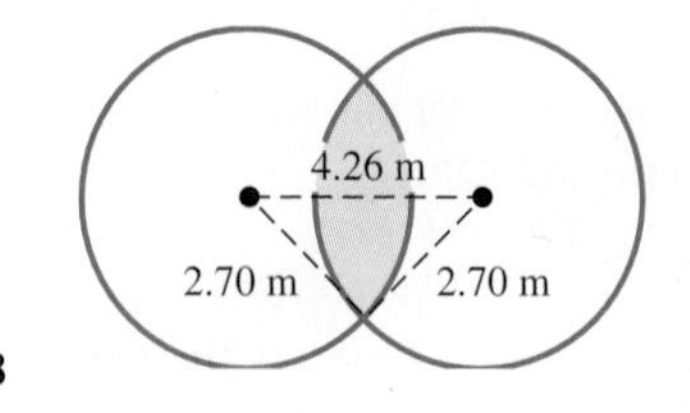

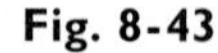
Fig. 8-43

In Exercises 49–52, another use of radians is illustrated.

49. It can be shown through advanced mathematics that an excellent approximate method of evaluating $\sin\theta$ or $\tan\theta$ is given by

$$\sin\theta = \tan\theta = \theta \qquad (8\text{-}14)$$

for small values of θ (the equivalent of a few degrees or less), if θ is expressed in radians. Equation (8-14) can be used for very small values of θ—even some calculators cannot adequately handle very small angles. Using Eq. (8-14), express $1''$ in radians and evalute $\sin 1''$ and $\tan 1''$.

50. Using Eq. (8-14), evaluate $\tan 0.001°$. Compare with a calculator value.

51. An astronomer observes that a star 12.5 light-years away moves through an angle of $0.2''$ in one year. Assuming it moved in a straight line perpendicular to the initial line of observation, how many kilometres did the star move? (One light-year = 9.46 Pm.) Use Eq. (8-14).

52. In calculating a back line of a lot, a surveyor discovers an error of 0.05° in an angle measurement. If the lot is 136.0 m deep, by how much is the back line calculation in error? See Fig. 8-44. Use Eq. (8-14).

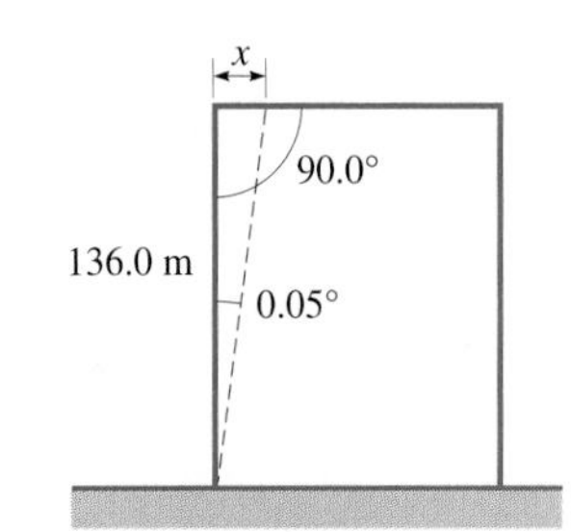

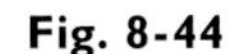
Fig. 8-44

Chapter Equations, Review Exercises, and Practice Test

Chapter Equations

α is reference angle

$$\sin\theta = \frac{y}{r} \qquad \cos\theta = \frac{x}{r} \qquad \tan\theta = \frac{y}{x}$$

$$\cot\theta = \frac{x}{y} \qquad \sec\theta = \frac{r}{x} \qquad \csc\theta = \frac{r}{y} \qquad (8\text{-}1)$$

$$F(\theta_2) = \pm F(180° - \theta_2) = \pm F(\alpha) \qquad (8\text{-}3)$$

$$F(\theta_3) = \pm F(\theta_3 - 180°) = \pm F(\alpha) \qquad (8\text{-}4)$$

$$F(\theta_4) = \pm F(360° - \theta_4) = \pm F(\alpha) \qquad (8\text{-}5)$$

$$\begin{aligned} \theta &= \alpha && \text{(first quadrant)} \\ \theta &= 180° - \alpha && \text{(second quadrant)} \\ \theta &= 180° + \alpha && \text{(third quadrant)} \\ \theta &= 360° - \alpha && \text{(fourth quadrant)} \end{aligned} \qquad (8\text{-}6)$$

$$\pi \text{ rad} = 180° \qquad (8\text{-}8)$$

Radian-degree conversions

$$1° = \frac{\pi}{180} \text{ rad} = 0.01745 \text{ rad} \qquad (8\text{-}9)$$

$$1 \text{ rad} = \frac{180°}{\pi} = 57.30° \qquad (8\text{-}10)$$

Circular arc length

$$s = \theta r \qquad (\theta \text{ in radians}) \qquad (8\text{-}11)$$

Circular sector area

$$A = \frac{1}{2}\theta r^2 \qquad (\theta \text{ in radians}) \qquad (8\text{-}12)$$

Linear and angular velocity

$$v = \omega r \qquad (8\text{-}13)$$

Review Exercises

In Exercises 1–4, find the trigonometric functions of θ. The terminal side of θ passes through the given point.

1. $(6, 8)$ **2.** $(-12, 5)$

3. $(7, -2)$ **4.** $(-2, -3)$

In Exercises 5–8, express the given trigonometric functions in terms of the same function of a positive acute angle.

5. $\cos 132°$, $\tan 194°$ **6.** $\sin 243°$, $\cot 318°$

7. $\sin 289°$, $\sec(-15°)$ **8.** $\cos 103°$, $\csc(-100°)$

In Exercises 9–12, express the given angle measurements in terms of π.

9. $40°$, $153°$ **10.** $22.5°$, $324°$

11. $48°$, $202.5°$ **12.** $27°$, $162°$

In Exercises 13–20, the given numbers represent angle measure. Express the measure of each angle in degrees.

13. $\frac{7\pi}{5}, \frac{13\pi}{18}$ **14.** $\frac{3\pi}{8}, \frac{7\pi}{20}$

15. $\frac{\pi}{15}, \frac{11\pi}{6}$ **16.** $\frac{17\pi}{10}, \frac{5\pi}{4}$

17. 0.560 **18.** 1.354

19. 3.607 **20.** 14.5

In Exercises 21–28, express the given angles in radians (not in terms of π).

21. $102°$ **22.** $305°$

23. $20.25°$ **24.** $148.38°$

25. $262.05°$ **26.** $18.72°$

27. $136.2°$ **28.** $385.4°$

In Exercises 29–48, determine the values of the given trigonometric functions directly on a calculator. Assume that the angles are approximate. Express answers to Exercises 41–44 to four significant digits.

29. $\cos 245.5°$ **30.** $\sin 141.3°$

31. $\cot 295°$ **32.** $\tan 184°$

33. $\csc 247.82°$ **34.** $\sec 96.17°$

35. $\sin 205.24°$ **36.** $\cos 326.72°$

37. $\tan 301.4°$ **38.** $\sin 103.9°$

39. $\tan 256.42°$ **40.** $\cos 162.32°$

41. $\sin \frac{9\pi}{5}$ **42.** $\sec \frac{5\pi}{8}$

43. $\cos \frac{7\pi}{6}$ **44.** $\tan \frac{23\pi}{12}$

45. $\sin 0.5906$ **46.** $\tan 0.8035$

47. $\csc 2.153$ **48.** $\cot 5.190$

In Exercises 49–52, find θ in degrees for $0° \le \theta < 360°$.

49. $\tan \theta = 0.1817$ **50.** $\sin \theta = -0.9323$

51. $\cos \theta = -0.4730$ **52.** $\cot \theta = 1.196$

In Exercises 53–56, find θ in radians for $0 \le \theta < 2\pi$.

53. $\cos \theta = 0.8387$ **54.** $\sin \theta = 0.1045$

55. $\sin \theta = -0.8650$ **56.** $\tan \theta = 2.840$

In Exercises 57–60, find θ in degrees for $0° \le \theta < 360°$.

57. $\cos \theta = -0.7222$, $\sin \theta < 0$

58. $\tan \theta = -1.683$, $\cos \theta < 0$

59. $\cot \theta = 0.4291$, $\cos \theta < 0$

60. $\sin \theta = 0.2626$, $\tan \theta < 0$

In Exercises 61–76, solve the given problems.

61. The instantaneous power p, in watts, input to a resistor in an alternating-current circuit is given by $p = p_m \sin^2 377t$, where p_m is the maximum power input and t is the time in seconds. Find p for $p_m = 0.120$ W and $t = 2.00$ ms. ($\sin^2 \theta = (\sin \theta)^2$.)

62. The horizontal distance x through which a pendulum moves is given by $x = a(\theta + \sin \theta)$, where a is a constant and θ is the angle between the vertical and the pendulum. Find x for $a = 45.0$ cm and $\theta = 0.175$.

63. A sector gear with a pitch radius of 8.25 cm and a 6.60-cm arc of contact is shown in Fig. 8-45. What is the sector angle θ?

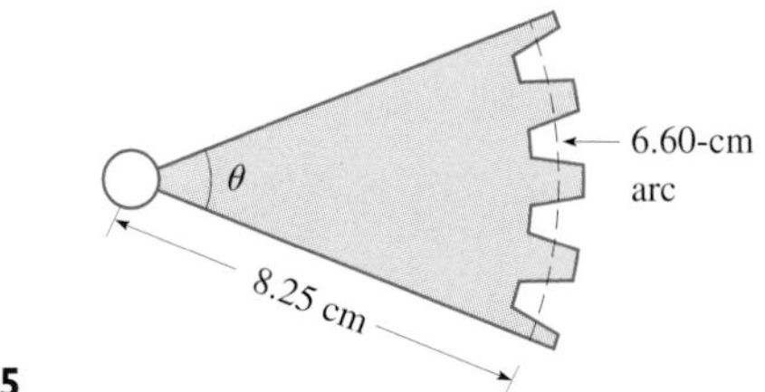

Fig. 8-45

64. Two pulleys have radii of 10.0 cm and 6.00 cm, and their centers are 40.0 cm apart. If the pulley belt is uncrossed, what must be the length of the belt?

65. A thermometer needle passes through 55.25° for a temperature change of 40.00°C. If the needle is 5.250 cm long and the scale is circular, how long must the scale be for a maximum temperature change of 150.00°C?

66. A piece of circular filter paper 15.0 cm in diameter is folded such that its effective filtering area is the same as that of a sector with central angle of 220°. What is the filtering area?

67. To produce an electric current, a circular loop of wire of diameter 25.0 cm is rotating about its diameter at 60.0 r/s in a magnetic field. What is the greatest linear velocity of any point on the loop?

68. Find the area of the decorative glass panel shown in Fig. 8-46. The panel is made of two equal circular sectors and an isosceles triangle.

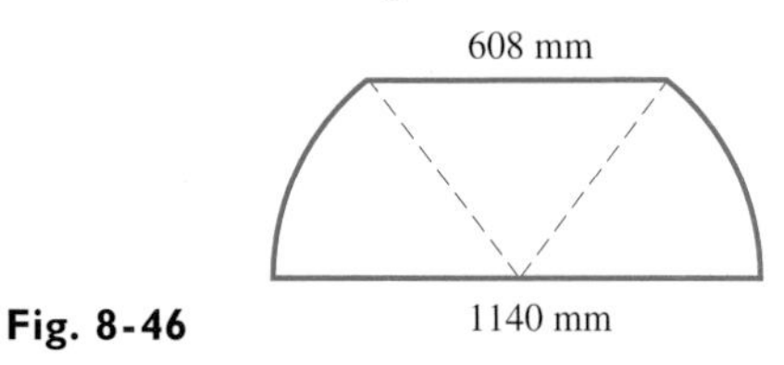

Fig. 8-46

69. A circular hood is to be used over a piece of machinery. It is to be made from a circular piece of sheet metal 1.08 m in radius. A hole 0.25 m in radius and a sector of central angle 80.0° are to be removed to make the hood. What is the area of the top of the hood?

70. The chain on a chain saw is driven by a sprocket 7.50 cm in diameter. If the chain is 108 cm long and makes one revolution in 0.250 s, what is the angular velocity, in revolutions per second, of the sprocket?

71. An *ultracentrifuge,* used to observe the sedimentation of particles such as proteins, may rotate as fast as 80 000 r/min. If it rotates at this rate and is 7.20 cm in diameter, what is the linear velocity of a particle at the outer edge?

72. A computer is programmed to shade in a sector of a pie chart 2.44 cm in radius. If the perimeter of the shaded sector is 7.32 cm, what is the central angle (in degrees) of the sector? See Fig. 8-47.

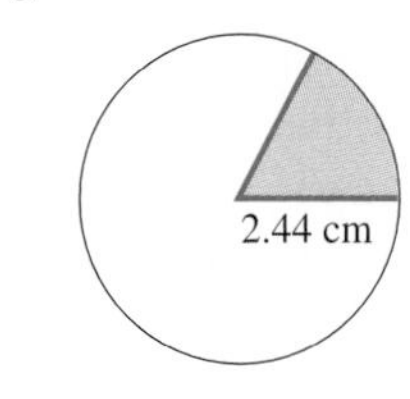

Fig. 8-47

73. Part of a parking lot is shown in Fig. 8-48. The widths of the parking spaces are equal, and the lengths of the dividing lines are equal. The circular section has a radius of 7.20 m and a central angle of 26.0°. What is the total length of the lines and circular section shown?

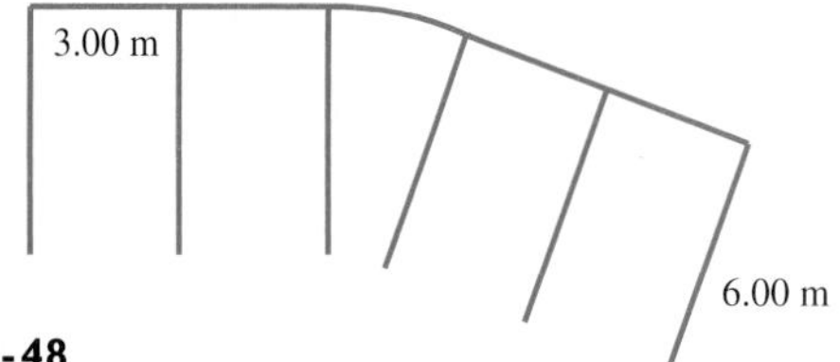

Fig. 8-48

74. The Trans-Alaska Pipeline was assembled in sections 12.2 m long and 1.22 m in diameter. If the depth of the oil in one horizontal section is 0.305 m, what is the volume of oil in this section?

75. A laser beam is transmitted with a "width" of 0.0008° and makes a circular spot of radius 2.50 km on a distant object. How far is the object from the source of the laser beam? Use Eq. (8-14).

76. The planet Venus subtends an angle of 15″ to an observer on Earth. If the distance between Venus and Earth is 167 Gm, what is the diameter of Venus? Use Eq. (8-14).

Writing Exercise

77. Write a paragraph explaining how you determine the units for the result of the following problem: An astronaut in a spacecraft circles the moon once each 1.95 h. If the altitude of the spacecraft is constant at 113 km, what is its velocity? The radius of the moon is 1740 km. (What is the answer?)

Practice Test

1. Change 150° to radians in terms of π.

2. Express sin 205° in terms of the sine of a positive acute angle. Do not evaluate.

3. Find $\sin\theta$ and $\sec\theta$ if θ is in standard position and the terminal side passes through $(-9, 12)$.

4. An airplane propeller blade is 1.40 m long and rotates at 2200 r/min. What is the linear velocity of a point on the tip of the blade?

5. Given that 3.572 is the measure of an angle, express the angle in degrees.

6. If $\tan\theta = 0.2396$, find θ, in degrees, for $0° \le \theta < 360°$.

7. If $\cos\theta = -0.8244$ and $\csc\theta < 0$, find θ in radians for $0 \le \theta < 2\pi$.

8. The floor of a sunroom is in the shape of a circular sector of arc length 16.0 m and radius 4.25 m. What is the area of the floor?

9

Vectors and Oblique Triangles

The wind must be considered to find the proper heading for an aircraft. In Section 9-5 we use vectors and an oblique triangle to show how this may be done.

In our technical applications to this point we have considered only the *magnitude* of the quantities used. In this chapter we study *vectors,* for which we must specify the *direction* as well as the *magnitude* of the quantity. In order to specify the direction we use trigonometric functions of any required angle, as developed in Chapter 8. Vectors are of great importance in many fields of science and technology, including physics, engineering, and navigation.

After developing the concept of a vector, we study methods of solving triangles that are not right triangles (*oblique* triangles). In doing so we must be able to use the trigonometric functions of oblique angles. As with right triangles, the applications of oblique triangles are numerous.

9-1 Introduction to Vectors

A great many quantities with which we deal may be described by specifying their magnitudes. Generally, one can describe lengths of objects, areas, time intervals, monetary amounts, temperatures, and numerous other quantities by specifying a number: the magnitude of the quantity. *Such quantities are known as* **scalar** *quantities.*

Scalars

Many other quantities are fully described only when both their magnitude and direction are specified. Such quantities are known as **vectors.** Examples of vectors are velocity and force. The following example illustrates a vector quantity and the difference between scalars and vectors.

EXAMPLE 1 ▸▸ A jet is traveling at 800 km/h. From this statement alone we know only the *speed* of the jet. *Speed is a scalar quantity,* and it tells us only the *magnitude* of the rate. Knowing only the speed of the jet, we know the rate at which it is moving, but we do not know where it is headed.

If we add the phrase "in a direction 10° south of west" to the sentence above about the jet, we specify the direction of travel as well as the speed. We then know the *velocity* of the jet; that is, we know the *direction* of travel as well as the *magnitude* of the rate at which it is traveling. *Velocity is a vector quantity.* Knowing the velocity of the jet, we know where it is headed and the rate at which it is moving. ◂◂

Let us analyze an example of the action of two vectors. Consider a boat moving in a river. For purposes of this example, we shall assume that the boat is driven by a motor which can move it at 8 km/h in still water. We shall assume the current is moving downstream at 6 km/h. We immediately see that the movement of the boat depends on the direction in which it is headed. If the boat heads downstream, it can travel at 14 km/h, for the current is going 6 km/h and the boat moves at 8 km/h with respect to the water. If the boat heads upstream, however, it moves at the rate of only 2 km/h, since the action of the boat and that of the river are counter to each other. If the boat heads directly across the river, the point it reaches on the other side will not be directly opposite the point from which it started. We can see that this is so because we know that as the boat heads across the river, the river is moving the boat downstream *at the same time.*

This last case should be investigated further. Assume that the river is 0.4 km wide where the boat is crossing. It then takes 0.05 h (0.4 km ÷ 8 km/h = 0.05 h) to cross. In 0.05 h the river will carry the boat 0.3 km (0.05 h × 6 km/h = 0.3 km) downstream. Therefore, when the boat reaches the other side it will be 0.3 km downstream. From the Pythagorean theorem, we find that the boat traveled 0.5 km from its starting point to its finishing point.

$$d^2 = 0.4^2 + 0.3^2 = 0.25; \; d = 0.5 \text{ km}$$

Since this 0.5 km was traveled in 0.05 h, the magnitude of the velocity of the boat was actually

$$v = \frac{d}{t} = \frac{0.5 \text{ km}}{0.05 \text{ h}} = 10 \text{ km/h}$$

Also, we see that the direction of this velocity can be represented along a line that makes an angle θ with the line directed directly across the river as shown in Fig. 9-1.

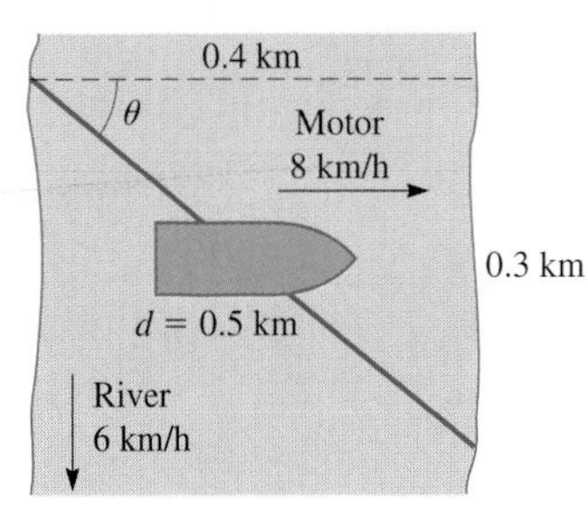

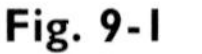
Fig. 9-1

Addition of Vectors

We have just seen two velocity vectors being *added.* Note that these vectors are not added the way numbers are added. We must take into account their directions as well as their magnitudes. Reasoning along these lines, let us now define the sum of two vectors.

We will represent a vector quantity by a letter printed in **boldface** type. The same letter in *italic* (lightface) type represents the magnitude only. Thus, **A** is a vector of magnitude A. In handwriting, one usually places an arrow over the letter to represent a vector, such as $\vec{A}$.

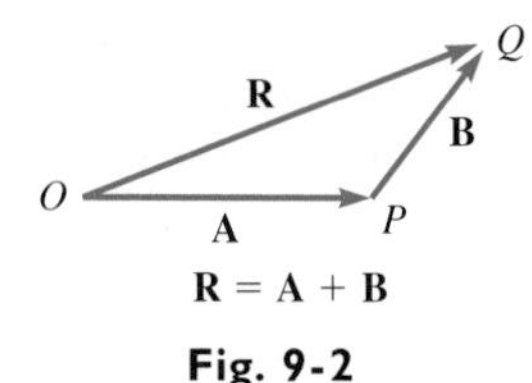

Fig. 9-2

Let **A** and **B** represent vectors directed from O to P and P to Q, respectively (see Fig. 9-2). *The vector sum* **A + B** *is the vector* **R,** *from the* **initial point** *O to the* **terminal point** *Q. Here, vector* **R** *is called the* **resultant.** *In general, a resultant is a single vector that is the vector sum of any number of other vectors.*

There are two common methods of adding vectors by means of a diagram. The first is illustrated in Fig. 9-3. To add **B** to **A,** we shift **B** parallel to itself until its tail touches the head of **A.** *The vector sum* **A + B** *is the resultant vector* **R,** *which is drawn from the tail of* **A** *to the head of* **B.** In using this method, we can move a vector for addition as long as *we keep its magnitude and direction unchanged.* (Since the magnitude and direction specify a vector, two vectors in different *locations* are considered the same if they have the same magnitude and direction.) When using a diagram to add vectors, it must be drawn with reasonable accuracy.

CAUTION ▶

Polygon Method

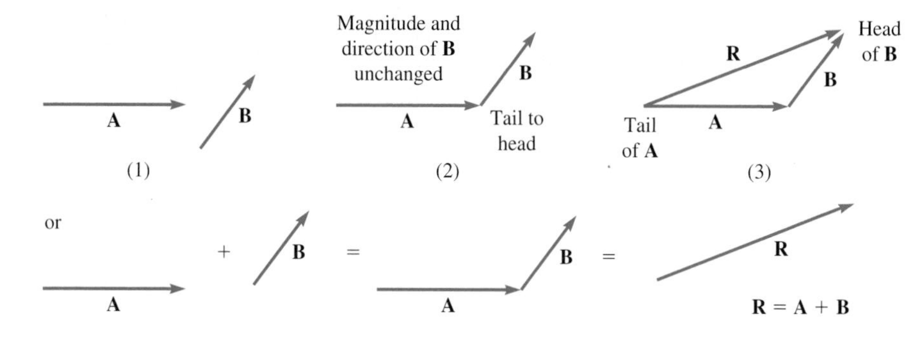

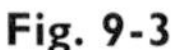
Fig. 9-3

Three or more vectors are added in the same general manner. We place the initial point of the second vector at the terminal point of the first vector, the initial point of the third vector at the terminal point of the second vector, and so on. The resultant is the vector from the initial point of the first vector to the terminal point of the last vector. The order in which they are added does not matter.

EXAMPLE 2 ▸▸ The addition of vectors **A, B,** and **C** is shown in Fig. 9-4.

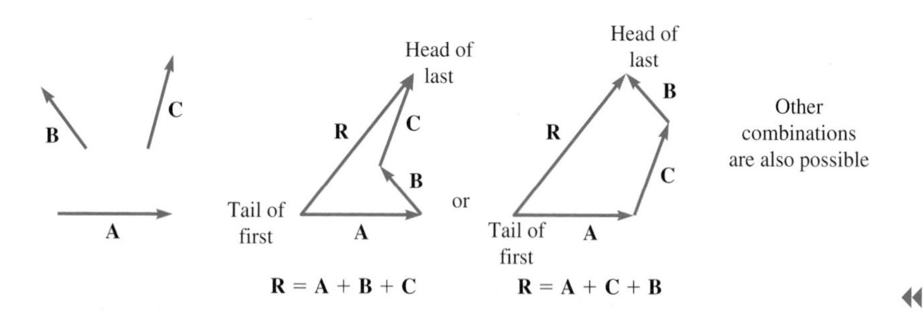

Fig. 9-4 ◂◂

Parallelogram Method

Another method that is convenient when two vectors are being added is to *let the two vectors being added be the sides of a parallelogram. The resultant is then the diagonal of the parallelogram.* The vectors are first placed tail to tail, and the initial point of the resultant is the ***common initial point*** of the vectors being added.

EXAMPLE 3 ▸▸ The addition of vectors **A** and **B** is shown in Fig. 9-5.

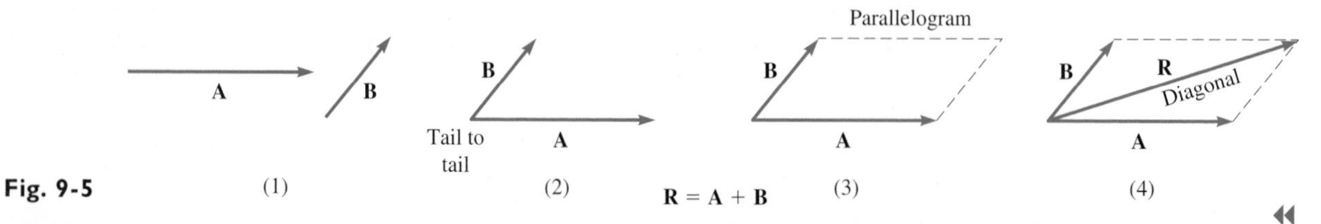

Fig. 9-5 ◂◂

Scalar Multiple of Vector

If vector **C** is in the same direction as vector **A**, and **C** has a magnitude n times that of **A**, then **C** $= n$**A**, where *the vector n***A** *is called the* **scalar multiple** *of vector* **A.** Thus, 2**A** is a vector twice as long as A, but *in the same direction.*

EXAMPLE 4 ▸▸ For vectors **A** and **B** in Fig. 9-6, find vector 3**A** + 2**B.**

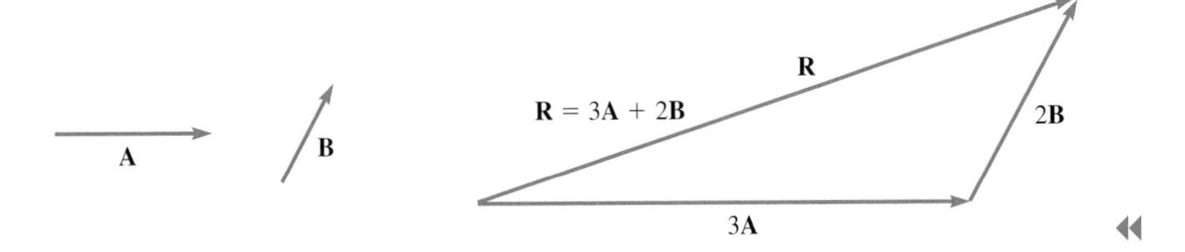

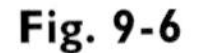
Fig. 9-6 ◂◂

Subtraction of Vectors

Vector **B** may be subtracted from vector **A** by reversing the direction of **B** and proceeding as in vector addition. Thus, **A** − **B** = **A** + (−**B**), where *the minus sign indicates that vector* −**B** *has the opposite direction of vector* **B.**

EXAMPLE 5 ▸▸ For vectors **A** and **B** in Fig. 9-7, find vector 2**A** − **B.**

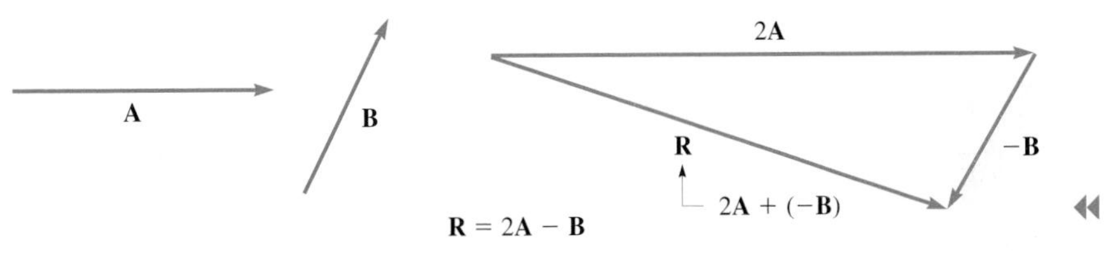

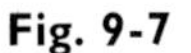
Fig. 9-7 ◂◂

One very important application of vectors is that of forces acting on an object. The next example shows the addition of forces by the parallelogram method.

EXAMPLE 6 ▸▸ Two persons pull horizontally on ropes attached to a car mired in mud. One person pulls with a force of 500 N directly to the right, and the other person pulls with a force of 350 N at 40° from the first force, as shown in Fig. 9-8(a). Find the resultant force on the car.

We make a scale drawing of the forces in Fig. 9-8(b), measuring the forces with a ruler and the angles with a protractor. (We make the scale drawing of forces larger and with a different scale than that in Fig. 9-8(a) to get reasonable accuracy.) We find that the resultant force is about 800 N and acts at an angle of about 16° from the first force. ◂◂

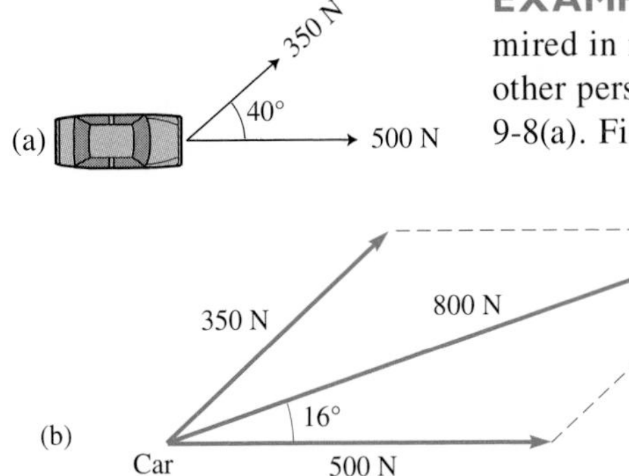

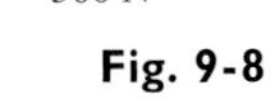
Fig. 9-8

Two other important vector quantities are *velocity* and *displacement.* Velocity as a vector is illustrated in Example 1. *The* **displacement** *of an object is given by the distance from a reference point and the angle from a reference direction.*

EXAMPLE 7 ▸▸ A jet travels due east from Indianapolis for 190 km and then turns 70° north of east and travels another 280 km to Detroit. Find the displacement of Detroit from Indianapolis.

We make a scale drawing in Fig. 9-9 to show the route taken by the jet. Measuring distances with a ruler and angles with a protractor, we find that Detroit is about 390 km from Indianapolis, at an angle of about 43° north of east. By giving both the magnitude and *the direction,* we have given the displacement.

If the jet returned directly from Detroit to Indianapolis, its *displacement* from Indianapolis would be *zero,* although it traveled a *distance* of 860 km. ◂◂

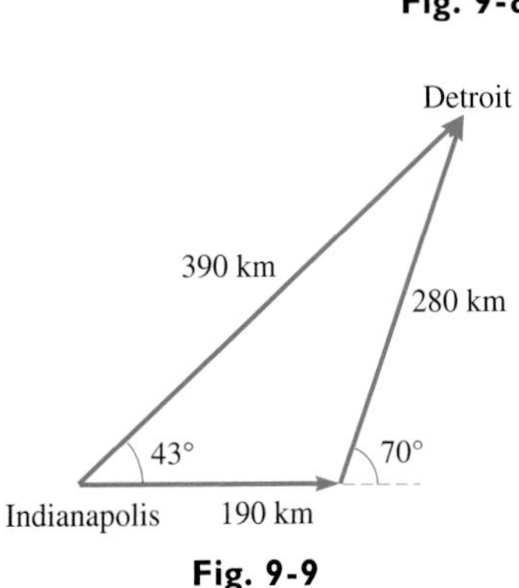

Fig. 9-9

Exercises 9–1

In Exercises 1–4, determine whether a scalar or a vector is described in (a) and (b). Explain your answers.

1. (a) A person traveled 300 km to the southwest.
(b) A person traveled 300 km.

2. (a) A small craft warning reports winds of 35 km/h.
(b) A small craft warning reports winds out of the north at 35 km/h.

3. (a) An arm of an industrial robot pushes with a 10-N force downward on a part. (b) A part is being pushed with a 10-N force by an arm of an industrial robot.

4. (a) A ballistics test shows that a bullet hit a wall at a speed of 100 m/s. (b) A ballistics test shows that a bullet hit a wall at a speed of 100 m/s perpendicular to the wall.

In Exercises 5–8, add the given vectors by drawing the appropriate resultant. Use the parallelogram method in Exercises 7 and 8.

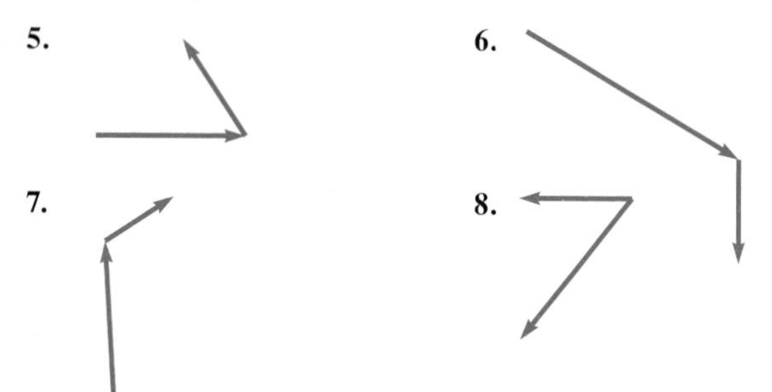

In Exercises 9–28, find the indicated vector sums and differences with the given vectors by means of diagrams. (You might find graph paper to be helpful.)

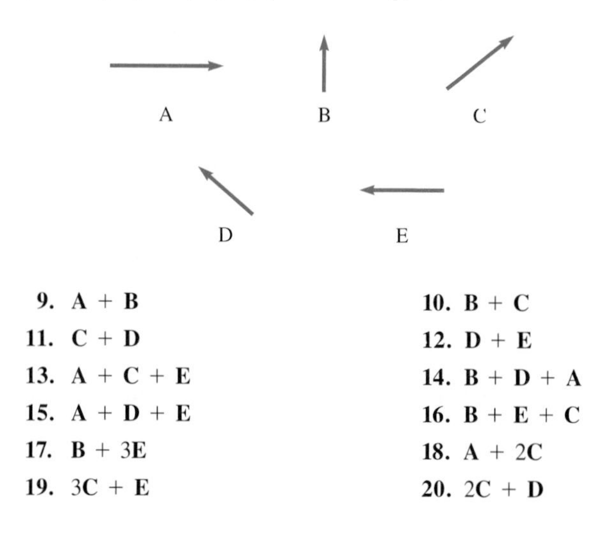

9. A + B
10. B + C
11. C + D
12. D + E
13. A + C + E
14. B + D + A
15. A + D + E
16. B + E + C
17. B + 3E
18. A + 2C
19. 3C + E
20. 2C + D
21. A − B
22. C − D
23. E − B
24. D − A
25. 3B − 2D
26. 2A − 3E
27. B + 2C − E
28. A + 4E − 3B

In Exercises 29–36, solve the given problems. Use a ruler and protractor as in Examples 6 and 7.

29. Two forces that act on an airplane wing are called the *lift* and the *drag*. Find the resultant of these forces acting on the airplane wing in Fig. 9-10.

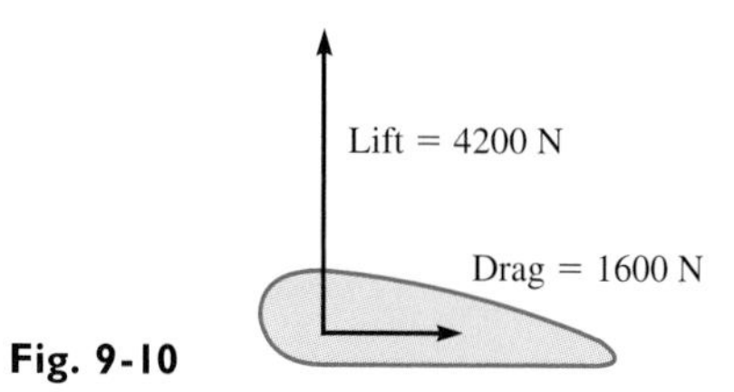

Fig. 9-10

30. Two electric charges create an electric field intensity, a vector quantity, at a given point. The field intensity is 30 kN/C to the right and 60 kN/C at an angle of 45° above the horizontal to the right. Find the resultant electric field intensity at this point.

31. A rocket takes off, moving at 640 m/s horizontally and 480 m/s vertically. Find its resultant velocity.

32. A small plane travels at 180 km/h in still air. It is headed due south in a wind of 50 km/h from the northeast. What is the resultant velocity of the plane?

33. A driver takes the wrong road at an intersection and travels 4 km north, then 6 km east, and finally 10 km to the southeast to reach the home of a friend. What is the displacement of the friend's home from the intersection?

34. A ship travels 20 km in a direction of 30° south of east and then turns due south for another 40 km. What is the ship's displacement from its initial position?

35. Three forces act on a small ring in a vertical plane. One force of 20 N acts vertically downward. A second force of 15 N acts horizontally to the right, and the third force of 25 N acts at an angle of 53° above the horizontal to the left. What is the resultant force on the ring?

36. A crate that has a weight of 100 N is suspended by two ropes. The force in one rope is 70 N and is directed to the left at an angle of 60° above the horizontal. What must be the other force in order that the resultant force (including the weight) on the crate is zero?

9-2 Components of Vectors

Adding vectors by means of diagrams is very useful in developing an understanding of vector quantities. However, unless the diagrams are drawn with great care and accuracy, the results we can obtain are quite approximate. Therefore, it is necessary to develop other methods to obtain results of sufficient accuracy.

In this section we show how a given vector can be considered to be the sum of two other vectors, with any required degree of accuracy. In the following section we show how this will allow us to add vectors in order to obtain the sum of vectors with the required accuracy in the result.

Two vectors which, when added together, have a resultant equal to the original vector are called **components** *of the original vector.* In the illustration of the boat in Section 9-1, the velocities of 8 km/h across the river and 6 km/h downstream are components of the 10 km/h vector directed at the angle θ.

Resolving a Vector into Components

In practice, there are certain components of a vector that are of particular importance. If a vector is placed with its initial point at the origin of a rectangular coordinate system and its direction is indicated by an angle in standard position, we may find its x- and y-**components.** *These components are vectors directed along the coordinate axes and which, when added together, have a resultant equal to the given vector.* The initial points of these components are at the origin, and the terminal points are located at the points where perpendicular lines from the terminal point of the given vector cross the axes. *Finding these component vectors is called* **resolving** *the vector into its components.*

EXAMPLE 1 ▸▸ Find the x- and y-components of the vector **A** shown in Fig. 9-11. The magnitude of **A** is 7.25.

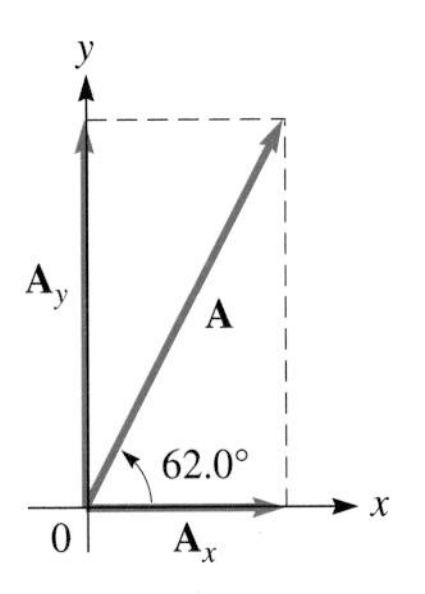

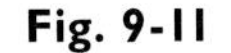
Fig. 9-11

From the figure we see that A_x, the magnitude of the x-component $\mathbf{A}_x$, is related to **A** by

$$\frac{A_x}{A} = \cos 62.0°$$

or

$$A_x = A \cos 62.0°$$

In the same way, A_y, the magnitude of the y-component $\mathbf{A}_y$, is related to **A** ($\mathbf{A}_y$ could be placed along the vertical dashed line) by

$$\frac{A_y}{A} = \sin 62.0°$$

or

$$A_y = A \sin 62.0°$$

From these relations, knowing that $A = 7.25$, we have

$$A_x = 7.25 \cos 62.0° = 3.40$$
$$A_y = 7.25 \sin 62.0° = 6.40$$

This means that the x-component is directed along the x-axis to the right and has a magnitude of 3.40. Also, the y-component is directed along the y-axis upward and its magnitude is 6.40. These two component vectors can replace vector **A,** since the effect they have is the same as **A.** ◂◂

EXAMPLE 2 ▸▸ Resolve a vector 14.4 units long and directed at an angle of 126.0° into its x- and y-components. See Fig. 9-12.

Placing the initial point of the vector at the origin and putting the angle in standard position, we see that the vector directed along the x-axis, $\mathbf{V}_x$, is related to the vector $\mathbf{V}$ of magnitude V by

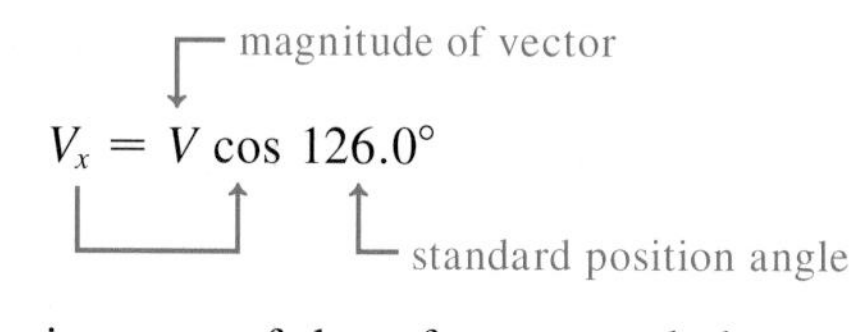

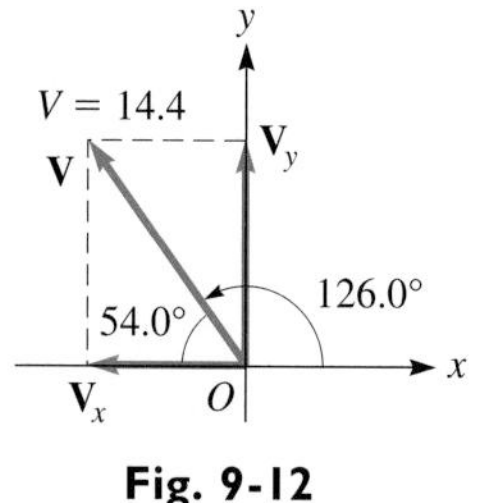

Fig. 9-12

or in terms of the reference angle by

$$V_x = -V \cos 54.0° \longleftarrow \text{reference angle}$$

(directed along negative x-axis)

since $\cos 126.0° = -\cos 54.0°$. We see that the minus sign shows that the x-component is directed in the negative direction, that is, to the left.

Since the vector directed along the y-axis, $\mathbf{V}_y$, could also be placed along the vertical dashed line, it is related to the vector $\mathbf{V}$ by

$$V_y = V \sin 126.0° = V \sin 54.0°$$

Thus, the vectors $\mathbf{V}_x$ and $\mathbf{V}_y$ have the magnitudes

$$V_x = 14.4 \cos 126.0° = -8.46, \qquad V_y = 14.4 \sin 126.0° = 11.6$$

Therefore, we have resolved the given vector into two components: one, directed along the negative x-axis, of magnitude 8.46, and the other, directed along the positive y-axis, of magnitude 11.6. ◂◂

We see that the steps used in finding the x- and y-components are as follows:

> ***Steps Used in Finding the x- and y-Components of a Vector***
>
> 1. *Place vector* **A** *such that* θ *is in standard position.*
> 2. *Calculate* A_x *and* A_y *from* $A_x = A \cos \theta$ *and* $A_y = A \sin \theta$. We may use the reference angle if we note the direction of the component.
> 3. *Check the components* to see if each is in the correct direction and has a magnitude that is proper for the reference angle.

EXAMPLE 3 ▸▸ Resolve vector **A**, of magnitude 375.4 and direction $\theta = 205.32°$, into its x- and y-components. See Fig. 9-13.

By placing **A** such that θ is in standard position, we see that

$$A_x = A \cos 205.32° = 375.4 \cos 205.32° = -339.3$$

and

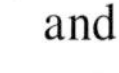

(directed along negative axis)

$$A_y = A \sin 205.32° = 375.4 \sin 205.32° = -160.5$$

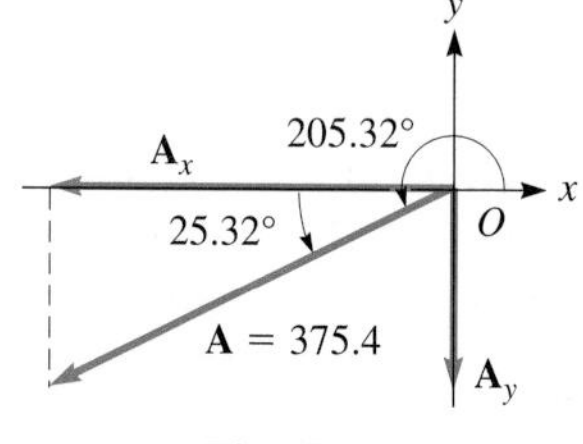

Fig. 9-13

Each component is directed along the negative axis, which must be true for a third-quadrant angle θ. Also, the magnitude of A_x is greater than the magnitude of A_y, which is the case with a reference angle less than 45°. ◂◂

EXAMPLE 4 ▸▸ The tension **T** in a cable supporting the sign shown in Fig. 9-14(a) is 85.0 N. If the cable makes an angle of 53.5° with the horizontal, find the horizontal and vertical components of the tension.

The tension in the cable is the force that the cable exerts on the sign. Showing the tension in Fig. 9-14(b), we see that

$$T_y = T \sin 53.5° = 85.0 \sin 53.5°$$
$$= 68.3 \text{ N}$$
$$T_x = T \cos 53.5° = 85.0 \cos 53.5°$$
$$= 50.6 \text{ N}$$ ◂◂

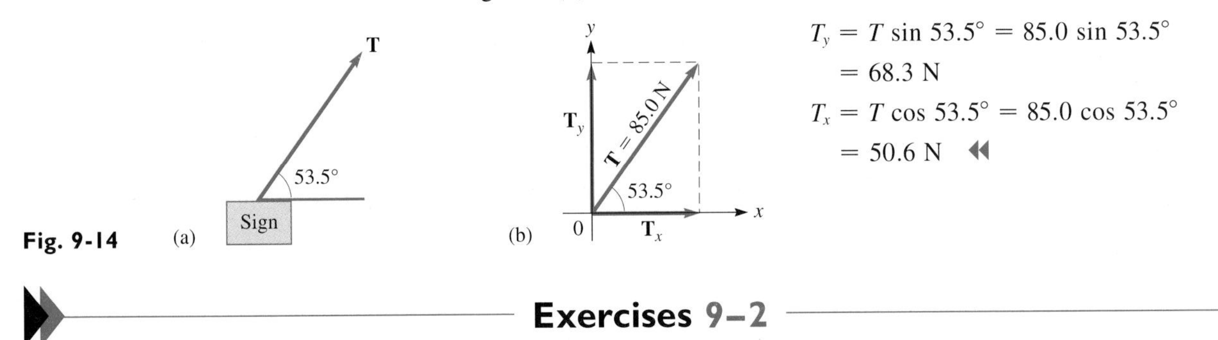

Fig. 9-14

Exercises 9–2

In Exercises 1–4, find the horizontal and vertical components of the vectors shown in the given figures. In each the magnitude of the vector is 750.

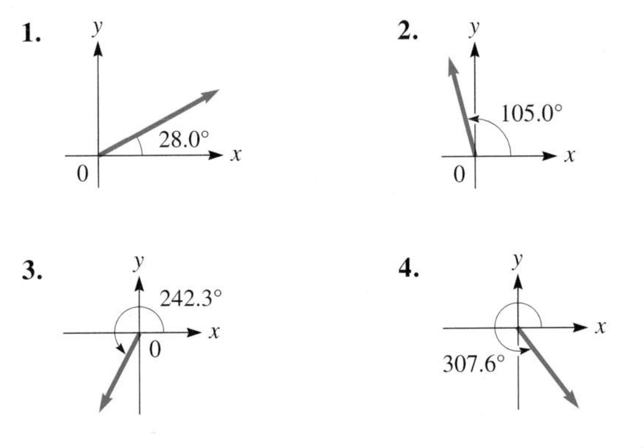

In Exercises 5–16, find the x- and y- components of the given vectors by use of the trigonometric functions.

5. Magnitude 8.60, $\theta = 68.0°$
6. Magnitude 9750, $\theta = 243.0°$
7. Magnitude 76.8, $\theta = 145.0°$
8. Magnitude 0.0998, $\theta = 296.0°$
9. Magnitude 9.04, $\theta = 283.3°$
10. Magnitude 16 400, $\theta = 156.5°$
11. Magnitude 2.65, $\theta = 197.3°$
12. Magnitude 67.8, $\theta = 22.5°$
13. Magnitude 0.8734, $\theta = 157.83°$
14. Magnitude 509.4, $\theta = 221.87°$
15. Magnitude 89 760, $\theta = 7.84°$
16. Magnitude 1.806, $\theta = 301.83°$

In Exercises 17–24, find the required horizontal and vertical components of the given vectors.

17. A nuclear submarine approaches the surface of the ocean at 25.0 km/h and at an angle of 17.3° with the surface. What are the components of its velocity?
18. Water is flowing downhill at 8.0 m/s through a pipe that is at an angle of 66.4° with the horizontal. What are the components of its velocity?
19. The tension in a rope attached to a boat is 55.0 N. The rope is attached to the boat 4.00 m below the level at which it is being drawn in. At the point where there are 12.0 m of rope out, what force tends to bring the boat toward the wharf, and what force tends to raise the boat?
20. A car is being unloaded from a ship. It is supported by a cable from a crane and guided into position by a horizontal rope. If the tension in the cable is 12 400 N and the cable makes an angle of 3.5° with the vertical, what is the weight of the car and the tension in the rope? (The weight of the cable is negligible to that of the car.)
21. A jet is 145 km at a position 37.5° north of east of a city. What are the components of the jet's displacement from the city?
22. The end of a robot arm is 1.20 m on a line 78.6° above the horizontal from the point where it does a weld. What are the components of the displacement from the end of the robot arm to the welding point?
23. At one point the *Pioneer* space probe was entering the gravitational field of Jupiter at an angle of 2.55° below the horizontal with a velocity of 29 860 km/h. What were the components of its velocity?
24. Two upward forces are acting on a bolt. One force of 60.5 N acts at an angle of 82.4° above the horizontal, and the other force of 37.2 N acts at an angle of 50.5° below the first force. What is the total upward force on the bolt?

9-3 Vector Addition by Components

Now that we have developed the meaning of the components of a vector, we are able to add vectors to any degree of required accuracy. To do this we use the components of the vector, the Pythagorean theorem, and the tangent of the standard position angle of the resultant. In the following example, two vectors at right angles are added.

EXAMPLE 1 ▸▸ Add vectors **A** and **B**, with $A = 14.5$ and $B = 9.10$. The vectors are at right angles, as shown in Fig. 9-15.

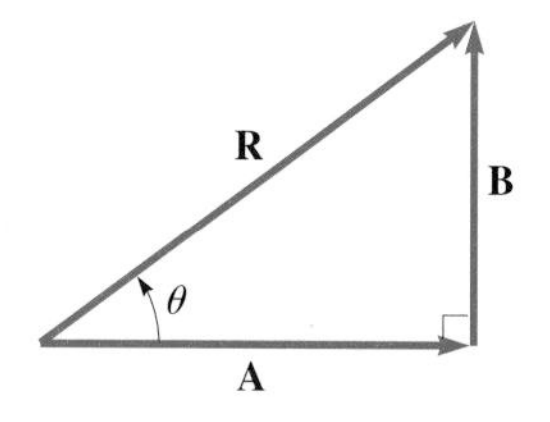

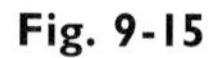
Fig. 9-15

We can find the magnitude R of the resultant vector **R** by use of the Pythagorean theorem. This leads to

$$R = \sqrt{A^2 + B^2} = \sqrt{(14.5)^2 + (9.10)^2}$$
$$= 17.1$$

We shall now determine the direction of the resultant vector **R** by specifying its direction as the angle θ in Fig. 9-15, that is, the angle that **R** makes with vector **A.** Therefore, we have

$$\tan\theta = \frac{B}{A} = \frac{9.10}{14.5}$$
$$\theta = 32.1°$$

The angle θ is the angle whose tangent is the quotient 9.10/14.5. Therefore, in using the calculator, we use the $\boxed{\tan^{-1}}$ key. Also, it should be noted that we do not have to record the value of the quotient 9.10/14.5. This means that when we find θ using the calculator, we treat θ as the angle

$$\theta = \tan^{-1}(9.10/14.5)$$

The sequence of keys is just what is shown on the right-hand side of this equation on a graphing calculator. On a scientific calculator, the $\boxed{\tan^{-1}}$ key is used after the indicated division (9.10/14.5) is entered.

Therefore, we see that **R** is a vector of magnitude $R = 17.1$ and in a direction 32.1° from vector **A.**

We note that Fig. 9-15 shows vectors **A** and **B** as horizontal and vertical, respectively. This means they are the horizontal and vertical components of the resultant vector **R.** However, we would find the resultant of any two vectors *at right angles* in the same way. ◂◂

If vectors are to be added and they are not at right angles, we first place each vector with its tail at the origin. Next, we resolve each vector into its x- and y-components. We then add all the x-components and add all the y-components to determine the x- and y-components of the resultant. Then, by using the Pythagorean theorem we find the magnitude of the resultant, and by use of the tangent we find the angle that gives us the direction of the resultant.

CAUTION ▶ *Remember, a vector is not completely specified unless both its magnitude and its direction are given.*

The following examples illustrate the addition of vectors.

EXAMPLE 2 ▸▸ Find the resultant of two vectors **A** and **B** such that $A = 1200$, $\theta_A = 270.0°$, $B = 1750$, and $\theta_B = 115.0°$.

We first place the vectors on a coordinate system with the tail of each at the origin as shown in Fig. 9-16(a). We then resolve each vector into its x- and y-components, as shown in Fig. 9-16(b) and as calculated below. (Note that **A** is vertical and has no horizontal component.) Next, the components are combined, as in Fig. 9-16(c) and as calculated. Finally, the magnitude of the resultant and the angle θ (to determine the direction), as shown in Fig. 9-16(d), are calculated.

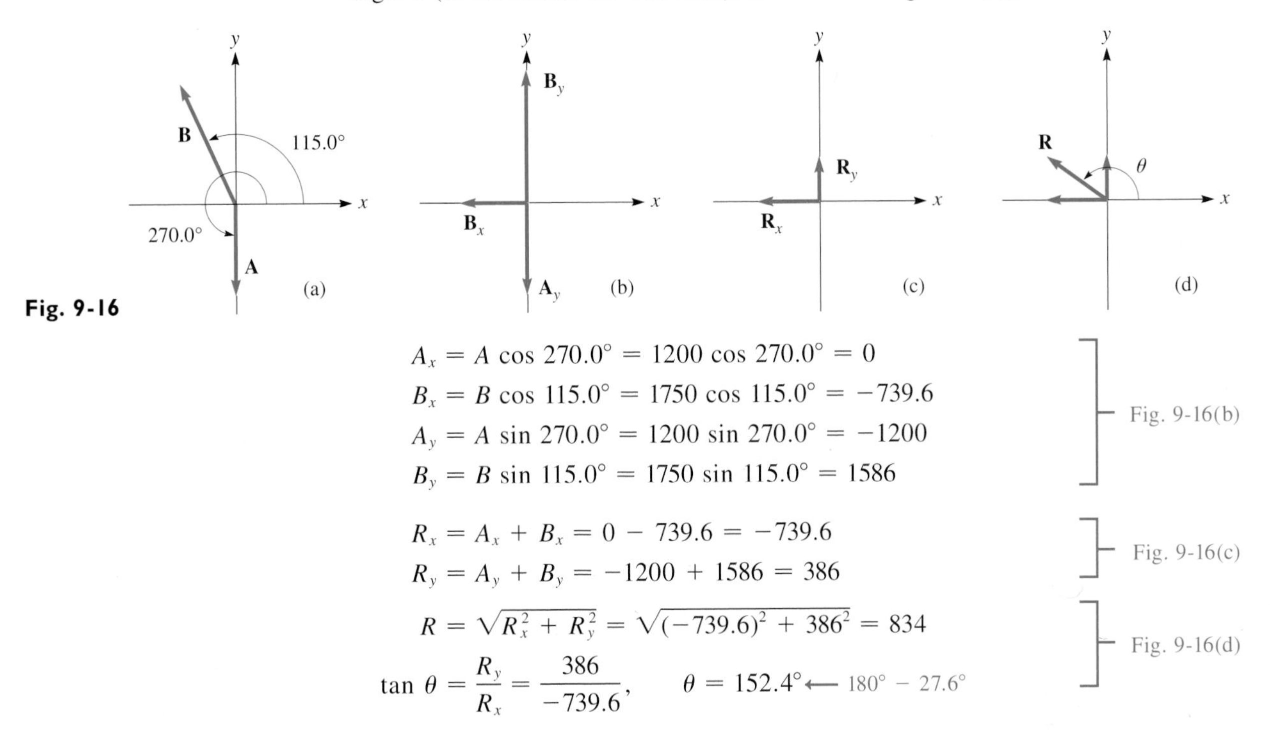

Fig. 9-16

$$A_x = A \cos 270.0° = 1200 \cos 270.0° = 0$$
$$B_x = B \cos 115.0° = 1750 \cos 115.0° = -739.6$$
$$A_y = A \sin 270.0° = 1200 \sin 270.0° = -1200$$
$$B_y = B \sin 115.0° = 1750 \sin 115.0° = 1586$$

Fig. 9-16(b)

$$R_x = A_x + B_x = 0 - 739.6 = -739.6$$
$$R_y = A_y + B_y = -1200 + 1586 = 386$$

Fig. 9-16(c)

$$R = \sqrt{R_x^2 + R_y^2} = \sqrt{(-739.6)^2 + 386^2} = 834$$
$$\tan\theta = \frac{R_y}{R_x} = \frac{386}{-739.6}, \qquad \theta = 152.4° \leftarrow 180° - 27.6°$$

Fig. 9-16(d)

Thus, the resultant has a magnitude of 834 and is directed at a standard-position angle of 152.4°. In finding θ from a calculator,

CAUTION ▶ *the calculator display shows an angle of −27.6°. However, we know θ is a second-quadrant angle, since* $\mathbf{R}_x$ *is negative and* $\mathbf{R}_y$ *is positive.*

Therefore, we must use 27.6° as a reference angle. For this reason, it is usually advisable to *find the reference angle first* by disregarding the signs of R_x and R_y when finding θ. Thus,

$$\tan\theta_{\text{ref}} = \left|\frac{R_y}{R_x}\right| = \frac{386}{739.6}, \qquad \theta_{\text{ref}} = 27.6°$$

The values in this example have been rounded off. (In order not to round off intermediate results too soon, one extra digit was carried until R and θ were found.) In using the calculator, R_x and R_y are calculated in continuous steps, but here we wished to show individual steps and results.

When we found R_y we saw that we were able to do a vector addition as a scalar addition. Also, since the magnitude of the components and the resultant are much smaller than either of the original vectors, this vector addition would have been difficult to do accurately by means of a diagram. ◂◂

EXAMPLE 3 ▶▶ Find the resultant **R** of the two vectors shown in Fig. 9-17(a), **A** of magnitude 8.075 and standard position angle of 57.26°, and **B** of magnitude 5.437 and standard-position angle of 322.15°.

In Fig. 9-17(b) we show the components of vectors **A** and **B**, and then in Fig. 9-17(c) we show the resultant and its components.

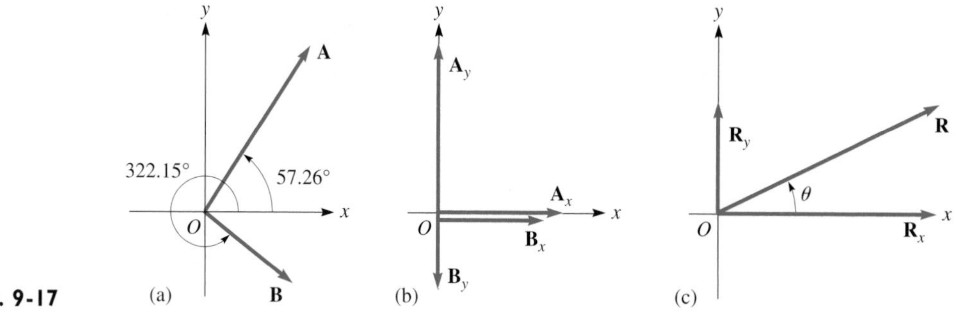

Fig. 9-17

Vector	*Magnitude*	*Angle*	*x-component*	*y-component*
A	8.075	57.26°	$A_x = 8.075 \cos 57.26° = 4.367$	$A_y = 8.075 \sin 57.26° = 6.792$
B	5.437	322.15°	$B_x = 5.437 \cos 322.15° = 4.293$	$B_y = 5.437 \sin 322.15° = -3.336$
R			$R_x = A_x + B_x = 8.660$	$R_y = A_y + B_y = 3.456$

$$R = \sqrt{R_x^2 + R_y^2} = \sqrt{(8.660)^2 + (3.456)^2} = 9.324$$

$$\tan \theta = \frac{R_y}{R_x} = \frac{3.456}{8.660} = 0.3991$$

$$\theta = 21.76° \longleftarrow \text{don't forget the direction}$$

The resultant vector is 9.324 units long and is directed at a standard-position angle of 21.76°, as shown in Fig. 9-17(c). We know that the resultant is in the first quadrant since both R_x and R_y are positive.

The above results are rounded off from the calculator values. In using a calculator we may calculate first R_x, then R_y, then R, and finally θ in a continuous sequence of steps (the specific sequence depends on the calculator). If a calculator has more than one memory, the values of R_x and R_y can be stored and need not be written down. ◀◀

Some general formulas can be derived from the previous examples. For a given vector **A**, directed at an angle θ, of magnitude A, and with components A_x and A_y, we have the following relations:

$$A_x = A \cos \theta, \quad A_y = A \sin \theta \qquad \textbf{(9-1)}$$

$$A = \sqrt{A_x^2 + A_y^2} \qquad \textbf{(9-2)}$$

$$\tan \theta = \frac{A_y}{A_x} \qquad \textbf{(9-3)}$$

From the previous examples we see that the following procedure is used for adding vectors.

> ***Procedure for Adding Vectors by Components***
>
> 1. *Resolve the given vectors into their x- and y-components.* Use Eqs. (9-1).
> 2. *Add the x-components to obtain* $\mathbf{R}_x$; *add the y-components to obtain* $\mathbf{R}_y$.
> 3. *Find the magnitude of the resultant* **R**. Use Eq. (9-2) in the form
> $R = \sqrt{R_x^2 + R_y^2}$
> 4. *Find the standard-position angle θ for the resultant* **R.** First find the reference angle θ_{ref} for the resultant **R** by using Eq. (9-3) in the form
> $$\tan\theta_{\text{ref}} = \frac{|R_y|}{|R_x|} \quad \left(\text{or } \theta_{\text{ref}} = \tan^{-1}\frac{|R_y|}{|R_x|}\right)$$

EXAMPLE 4 ▸▸ Find the resultant of the three given vectors in Fig. 9-18. The magnitudes of these vectors are $A = 422$, $B = 405$, and $C = 210$.

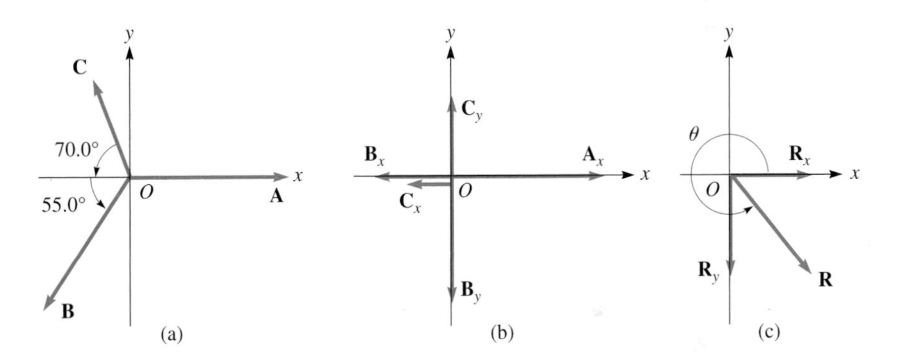

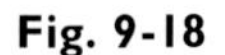
Fig. 9-18

We could change the given angles to standard-position angles. However, we will use the given angles, *being careful to give the proper sign to each component.* In the following table we show the x- and y-components of the given vectors, and the sums of these components give us the components of **R.**

Vector	*Magnitude*	*Ref. Angle*	*x-component*	*y-component*
A	422	0°	422 cos 0° = 422	422 sin 0° = 0
B	405	55.0°	−405 cos 55.0° = −232	−405 sin 55.0° = −332
C	210	70.0°	−210 cos 70.0° = −72	+210 sin 70.0° = 197
R			118	−135

Using the values of the components of **R,** we have

$$R = \sqrt{(118)^2 + (-135)^2} = 179$$

$$\tan\theta_{\text{ref}} = \frac{135}{118}, \quad \theta_{\text{ref}} = 48.8°, \qquad \theta = 311.2°$$

NOTE ▸ **Since $\mathbf{R}_x$ is positive and $\mathbf{R}_y$ is negative,** we know that θ **is a fourth-quadrant angle.** Therefore, in finding θ we subtracted θ_{ref} from 360°. ◂◂

Exercises 9–3

*In Exercises 1–4, vectors **A** and **B** are at right angles. Find the magnitude and direction (the angle from vector **A**) of the resultant.*

1. $A = 14.7$
$B = 19.2$

2. $A = 592$
$B = 195$

3. $A = 3.086$
$B = 7.143$

4. $A = 1734$
$B = 3297$

In Exercises 5–12, with the given sets of components, find R and θ.

5. $R_x = 5.18$, $R_y = 8.56$

6. $R_x = 89.6$, $R_y = -52.0$

7. $R_x = -0.982$, $R_y = 2.56$

8. $R_x = -729$, $R_y = -209$

9. $R_x = -646$, $R_y = 2030$

10. $R_x = -31.2$, $R_y = -41.2$

11. $R_x = 0.6941$, $R_y = -1.246$

12. $R_x = 7.627$, $R_y = -6.353$

In Exercises 13–28, add the given vectors by using the trigonometric functions, and the Pythagorean theorem.

13. $A = 18.0$, $\theta_A = 0.0°$
$B = 12.0$, $\theta_B = 27.0°$

14. $A = 154$, $\theta_A = 90.0°$
$B = 128$, $\theta_B = 43.0°$

15. $A = 56.0$, $\theta_A = 76.0°$
$B = 24.0$, $\theta_B = 200.0°$

16. $A = 6.89$, $\theta_A = 123.0°$
$B = 29.0$, $\theta_B = 260.0°$

17. $A = 9.821$, $\theta_A = 34.27°$
$B = 17.45$, $\theta_B = 752.50°$

18. $A = 1.653$, $\theta_A = 36.37°$
$B = 0.9807$, $\theta_B = 253.06°$

19. $A = 12.653$, $\theta_A = 98.472°$
$B = 15.147$, $\theta_B = 332.092°$

20. $A = 121.36$, $\theta_A = 292.362°$
$B = 112.98$, $\theta_B = 197.892°$

21. $A = 21.9$, $\theta_A = 236.2°$
$B = 96.7$, $\theta_B = 11.5°$
$C = 62.9$, $\theta_C = 143.4°$

22. $A = 6300$, $\theta_A = 189.6°$
$B = 1760$, $\theta_B = 320.1°$
$C = 3240$, $\theta_C = 75.4°$

23. $A = 0.364$, $\theta_A = 4.3°$
$B = 0.596$, $\theta_B = 319.5°$
$C = 0.129$, $\theta_C = 100.6°$

24. $A = 6.4$, $\theta_A = 126°$
$B = 5.9$, $\theta_B = 238°$
$C = 3.2$, $\theta_C = 72°$

25. The vectors shown in Fig. 9-19.

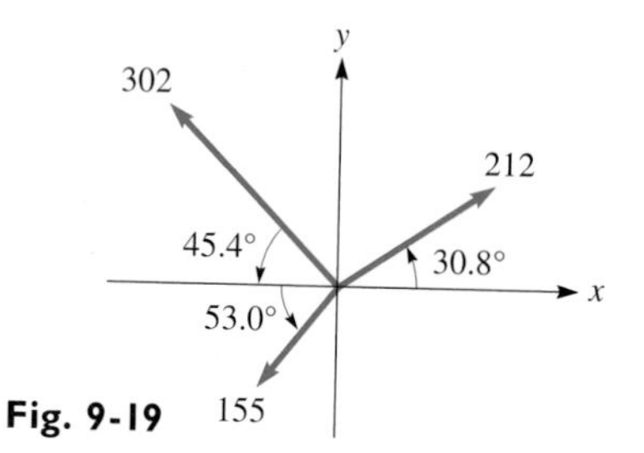

Fig. 9-19

26. The vectors shown in Fig. 9-20.

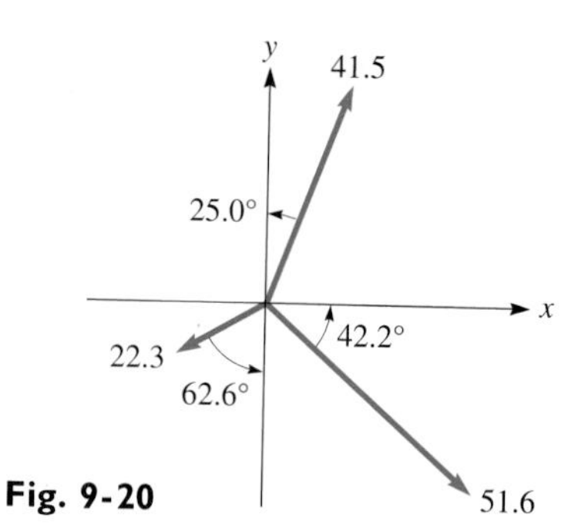

Fig. 9-20

27. The vectors shown in Fig. 9-21.

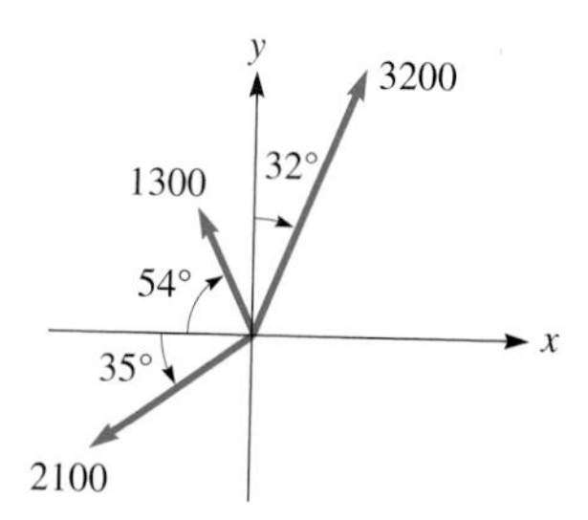

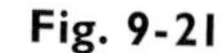

Fig. 9-21

28. The vectors shown in Fig. 9-22.

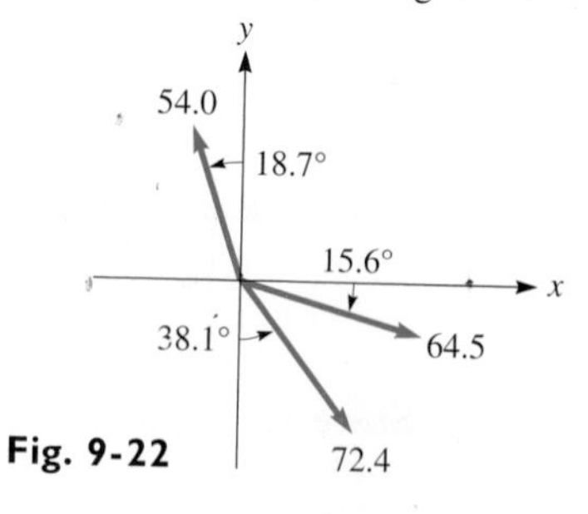

Fig. 9-22

9-4 Applications of Vectors

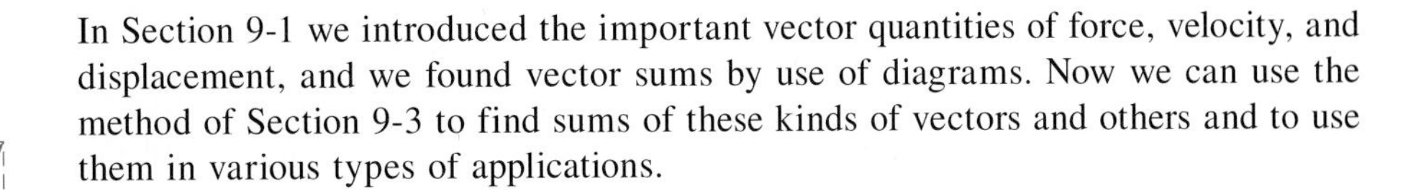

In Section 9-1 we introduced the important vector quantities of force, velocity, and displacement, and we found vector sums by use of diagrams. Now we can use the method of Section 9-3 to find sums of these kinds of vectors and others and to use them in various types of applications.

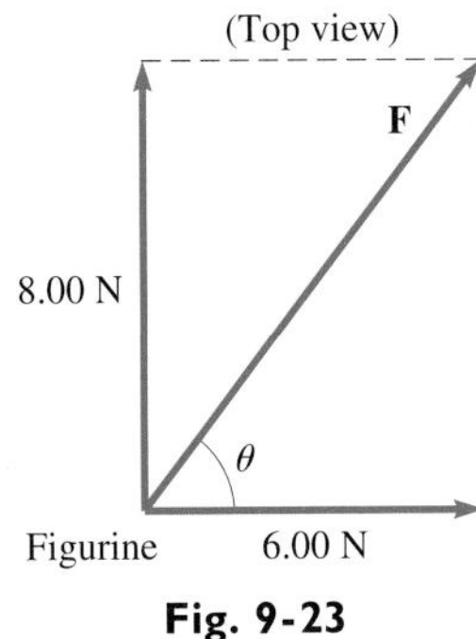

Fig. 9-23

EXAMPLE 1 ▸▸ In centering a figurine on a table, two persons apply forces on it. These forces are at right angles and have magnitudes of 6.00 N and 8.00 N. The angle between their lines of action is 90.0°. What is the resultant of these forces on the figurine?

By means of an appropriate diagram (Fig. 9-23), we may better visualize the actual situation. We note that a good choice of axes (unless specified, it is often convenient to choose the x- and y-axes to fit the problem) is to have the x-axis in the direction of the 6.00-N force and the y-axis in the direction of the 8.00-N force. (This is possible since the angle between them is 90°.) With this choice we note that the two given forces will be the x- and y-components of the resultant. Therefore, we arrive at the following results:

$$F_x = 6.00 \text{ N}, \quad F_y = 8.00 \text{ N}$$
$$F = \sqrt{(6.00)^2 + (8.00)^2} = 10.0 \text{ N}$$
$$\tan\theta = \frac{F_y}{F_x} = \frac{8.00}{6.00}, \qquad \theta = 53.1^\circ$$

We would state that the resultant has a magnitude of 10.0 N and acts at an angle of 53.1° from the 6.00-N force. ◂◂

EXAMPLE 2 ▸▸ A ship sails 32.50 km due east and then turns 41.25° north of east. After sailing another 16.18 km, where is it with reference to the starting point?

In this problem we are to find the resultant displacement of the ship from the two given displacements. The problem is diagramed in Fig. 9-24, where the first displacement is labeled vector **A** and the second as vector **B.**

Since east corresponds to the positive x-direction, we see that the x-component of the resultant is $\mathbf{A} + \mathbf{B}_x$ and the y-component of the resultant is $\mathbf{B}_y$. Therefore, we have the following results:

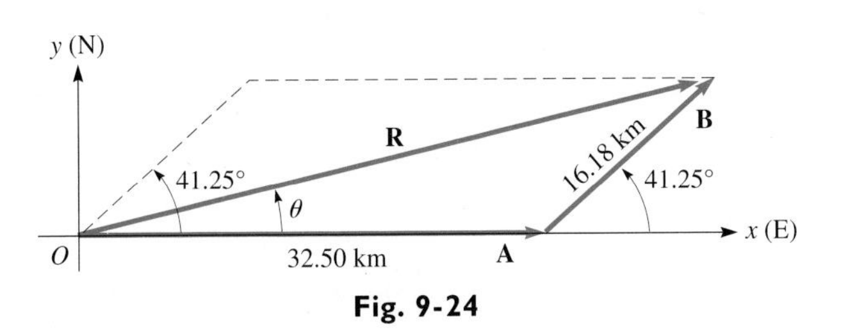

Fig. 9-24

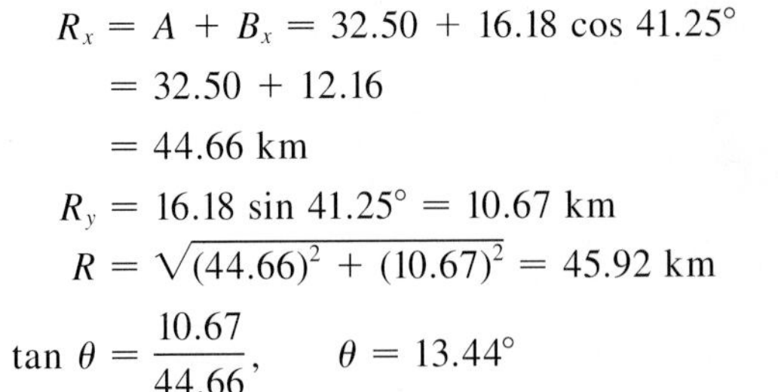

$$R_x = A + B_x = 32.50 + 16.18\cos 41.25^\circ$$
$$= 32.50 + 12.16$$
$$= 44.66 \text{ km}$$
$$R_y = 16.18\sin 41.25^\circ = 10.67 \text{ km}$$
$$R = \sqrt{(44.66)^2 + (10.67)^2} = 45.92 \text{ km}$$
$$\tan\theta = \frac{10.67}{44.66}, \qquad \theta = 13.44^\circ$$

Therefore, the ship is 45.92 km from the starting point, in a direction 13.44° north of east. ◂◂

EXAMPLE 3 ▸▸ An airplane headed due east is in a wind blowing from the southeast. What is the resultant velocity of the plane with respect to the surface of the earth if the velocity of the plane with respect to the air is 600 km/h and that of the wind is 100 km/h? See Fig. 9-25.

Let $\mathbf{v}_{px}$ be the velocity of the plane in the x-direction (east), $\mathbf{v}_{py}$ the velocity of the plane in the y-direction, $\mathbf{v}_{wx}$ the x-component of the velocity of the wind, $\mathbf{v}_{wy}$ the y-component of the velocity of the wind, and $\mathbf{v}_{pa}$ the velocity of the plane with respect to the air. Therefore,

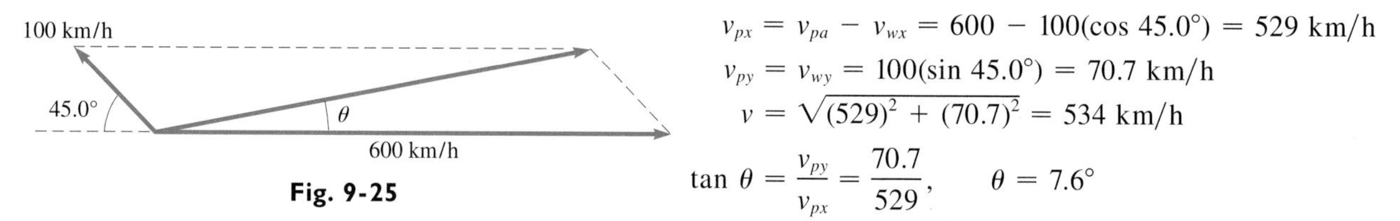

Fig. 9-25

$$v_{px} = v_{pa} - v_{wx} = 600 - 100(\cos 45.0°) = 529 \text{ km/h}$$
$$v_{py} = v_{wy} = 100(\sin 45.0°) = 70.7 \text{ km/h}$$
$$v = \sqrt{(529)^2 + (70.7)^2} = 534 \text{ km/h}$$
$$\tan\theta = \frac{v_{py}}{v_{px}} = \frac{70.7}{529}, \qquad \theta = 7.6°$$

We have determined that the plane is traveling 534 km/h and is flying in a direction 7.6° north of east. From this we observe that a plane does not necessarily head in the direction of its destination. ◂◂

Equilibrium of Forces

NOTE ▶

As we have seen, an important vector quantity is the force acting on an object. One of the most important applications of vectors involves forces that are in **equilibrium.** *For an object to be in equilibrium, the net force acting on it in any direction must be zero.* This condition is satisfied if the sum of the x-components of the force is zero and the sum of the y-components of the force is also zero. The following two examples illustrate forces in equilibrium.

EXAMPLE 4 ▸▸ A cement block is resting on a straight inclined plank that makes an angle of 30.0° with the horizontal. If the block weighs 80.0 N, what is the force of friction between the block and the plank?

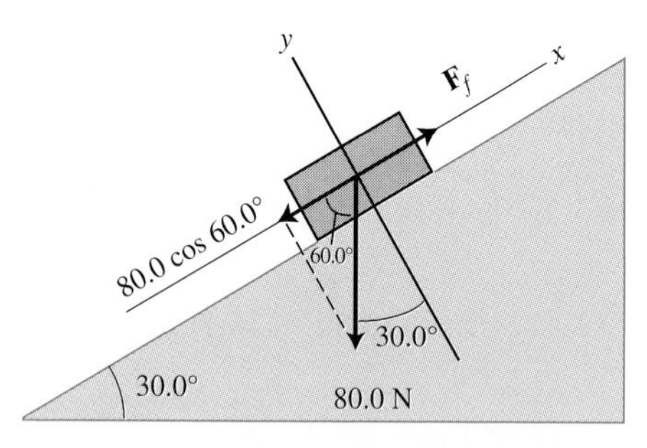

Fig. 9-26

The weight of the cement block is the force exerted on the block due to gravity. Therefore, the weight is directed vertically downward. The frictional force tends to oppose the motion of the block and is directed upward along the plank. The frictional force must be sufficient to counterbalance that component of the weight of the block that is directed down the plank for the block to be at rest (not moving). The plank itself "holds up" that component of the weight which is perpendicular to the plank. A convenient set of coordinates (see Fig. 9-26) is one with the origin at the center of the block, and with the x-axis directed up the plank and the y-axis perpendicular to the plank. The magnitude of the frictional force $\mathbf{F}_f$ is given by

$$F_f = 80.0 \cos 60.0° = 40.0 \text{ N}$$ component of weight down plank equals frictional force

We have used the 60.0° angle since it is the reference angle. We could have expressed the frictional force as $F_f = 80.0 \sin 30.0°$.

Here we have assumed that the block is small enough that we may calculate all forces as though they act at the center of the block (although we know that the frictional force acts at the surface between the block and the plank). ◂◂

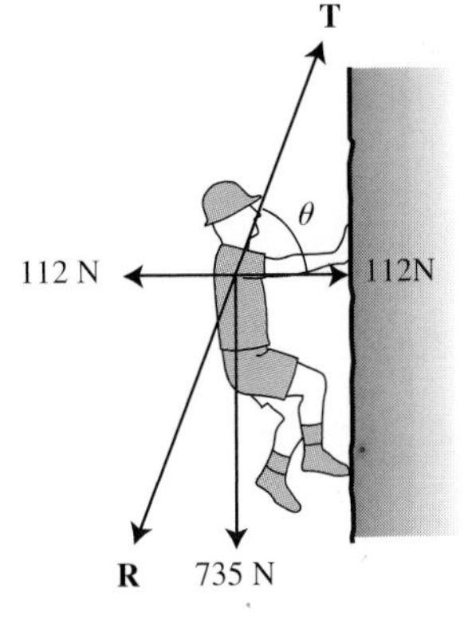

Fig. 9-27

EXAMPLE 5 ▸▸ A 735-N mountain climber suspended by a rope pushes on the side of a cliff with a horizontal force of 112 N. What is the tension **T** in the rope if the climber is in equilibrium? See Fig. 9-27.

For the climber to be in equilibrium, the tension in the rope must be equal and opposite to the resultant of the climber's weight and the force against the cliff. This means that the magnitude of the x-component of the tension is 112 N (the reaction force of the cliff—see Fig. 9-27) and the magnitude of the y-component is 735 N. Therefore,

$$T = \sqrt{112^2 + 735^2} = 743 \text{ N}$$

$$\tan\theta = \frac{735}{112}, \qquad \theta = 81.3° \quad ◂◂$$

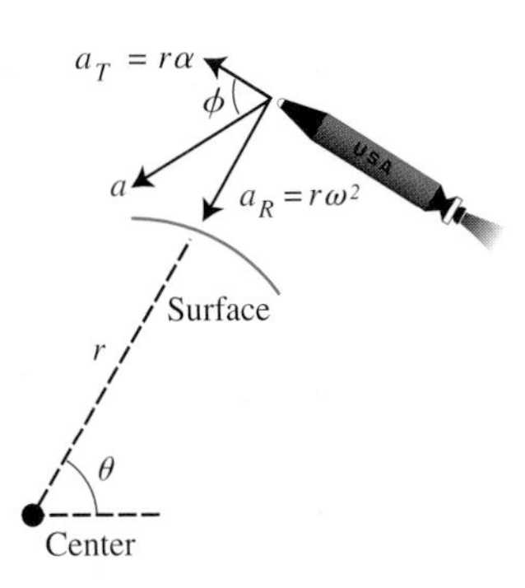

Fig. 9-28

EXAMPLE 6 ▸▸ For a spacecraft moving in a circular path around the earth, the tangential component $\mathbf{a}_T$ and the centripetal component $\mathbf{a}_R$ of its acceleration are given by the expressions shown in Fig. 9-28. The radius of the circle through which it is moving (from the center of the earth to the spacecraft) is r, its angular velocity is ω, and its angular acceleration is α (the rate at which ω is changing).

While going into orbit, at one point a spacecraft is moving in a circular path 230 km above the surface of the earth. At this point $r = 6.60 \times 10^6$ m, $\omega = 1.10 \times 10^{-3}$ rad/s, and $\alpha = 0.420 \times 10^{-6}$ rad/s^2. Calculate the magnitude of the resultant acceleration and the angle it makes with the tangential component.

$$a_R = r\omega^2 = 6.60 \times 10^6(1.10 \times 10^{-3})^2 = 7.99 \text{ m/s}^2$$

$$a_T = r\alpha = 6.60 \times 10^6(0.420 \times 10^{-6}) = 2.77 \text{ m/s}^2$$

Since a tangent line to a circle is perpendicular to the radius at the point of tangency, $\mathbf{a}_T$ is perpendicular to $\mathbf{a}_R$. Thus,

$$a = \sqrt{a_T^2 + a_R^2} = \sqrt{7.99^2 + 2.77^2} = 8.46 \text{ m/s}^2$$

$$\tan\phi = \frac{a_R}{a_T} = \frac{7.99}{2.77}, \qquad \phi = 70.9°$$

If the magnitude of the acceleration a is found directly on a calculator, without first recording the components, we get $a = 8.45$ m/s^2. The difference is due to rounding off the values of the components and then calculating a. ◂◂

Exercises 9–4

In Exercises 1–24, solve the given problems.

1. Two hockey players strike the puck at the same time, hitting it with horizontal forces of 34.5 N and 19.5 N that are perpendicular to each other. Find the resultant of these forces.

2. To straighten a small tree, two horizontal ropes perpendicular to each other are attached to the tree. If the tensions in the ropes are 82.3 N and 102 N, what is the resultant force on the tree?

3. In lifting a heavy piece of equipment from the mud, a cable from a crane exerts a vertical force of 6500 N and a cable from a truck exerts a force of 8300 N at 10.0° above the horizontal. Find the resultant of these forces.

4. At a point in the plane, two electric charges create an electric field (a vector quantity) of 25.9 kN/C at 10.8° above the horizontal to the right and 12.6 kN/C at 83.4° below the horizontal to the right. Find the resultant electric field.

5. A motorboat leaves a dock and travels 1580 m due west, then turns 35.0° to the south and travels another 1640 m to a second dock. What is the displacement of the second dock from the first dock?

6. Toronto is 650 km at 19.0° north of east from Chicago. Cincinnati is 390 km at 48.0° south of east from Chicago. What is the displacement of Cincinnati from Toronto?

7. From a fixed point, a surveyor locates a pole at 215.6 m due east and a building corner at 358.2 m at 37.72° north of east. What is the displacement of the building from the pole?

8. A rocket is launched with a vertical component of velocity of 2840 km/h and a horizontal component of velocity of 1520 km/h. What is its resultant velocity?

9. A storm front is moving east at 22.0 km/h and south at 12.5 km/h. Find the resultant velocity of the front.

10. In an accident, a truck with momentum (a vector quantity) of 22 100 kg · m/s strikes a car with momentum of 17 800 kg · m/s from the rear. The angle between their directions of motion is 25.0°. What is the resultant momentum?

11. In an automobile safety test, a shoulder and seat belt exerts a force of 425 N directly backward and a force of 368 N backward at an angle of 20.0° below the horizontal on a dummy. If the belt holds the dummy from moving farther forward, what force did the dummy exert on the belt? See Fig. 9-29.

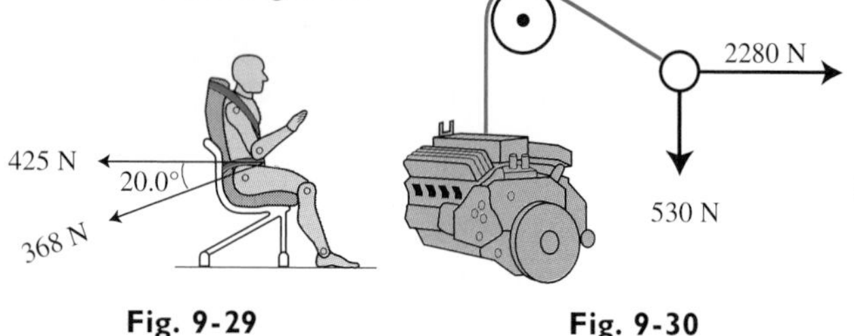

Fig. 9-29

Fig. 9-30

12. Two perpendicular forces act on a ring at the end of a chain that passes over a pulley and holds an automobile engine. If the forces have the values shown, what is the weight of the engine? See Fig. 9-30.

13. A plane flies at 550 km/h into a head wind of 60 km/h at 78° with the direction of the plane. Find the resultant velocity of the plane with respect to the ground. See Fig. 9-31.

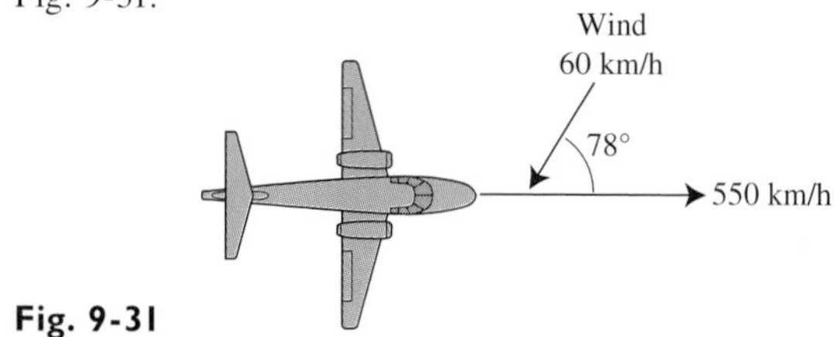

Fig. 9-31

14. A ship's navigator determines that the ship is moving through the water at 17.5 km/h with a heading of 26.3° north of east, but that the ship is actually moving at 19.3 km/h in a direction of 33.7° north of east. What is the velocity of the current?

15. A space shuttle is moving in orbit at 29 370 km/h. A satellite is launched to the rear at 190 km/h at an angle of 5.20° from the direction of the shuttle. Find the velocity of the satellite.

16. A block of ice slides down a (frictionless) ramp with an acceleration of 5.3 m/s². If the ramp makes an angle of 32.7° with the horizontal, find g, the acceleration due to gravity. See Fig. 9-32.

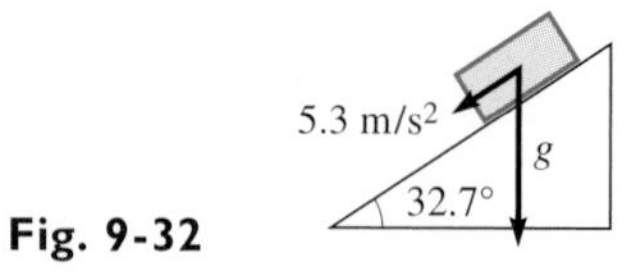

Fig. 9-32

17. While starting up, a circular saw blade 8.20 cm in diameter is rotating at 212 rad/min and has an angular acceleration of 318 rad/min². What is the acceleration of the tip of one of the teeth? (See Example 6.)

18. A boat travels across a river, reaching the opposite bank at a point directly opposite that from which it left. If the boat travels 6.00 km/h in still water, and the current of the river flows at 3.00 km/h, what was the velocity of the boat in the water?

19. In searching for a boat lost at sea, a Coast Guard cutter leaves a port and travels 75.0 km due east. It then turns 65.0° north of east and travels another 75.0 km, and finally turns another 65.0° toward the west and travels another 75.0 km. What is its displacement from the port?

20. A car is held stationary on a ramp by two forces. One is the force of 2140 N by the brakes, which hold it from rolling down the ramp. The other is a reaction force by the ramp of 9850 N, perpendicular to the ramp. This force keeps the car from going through the ramp. See Fig. 9-33. What is the weight of the car, and at what angle with the horizontal is the ramp inclined?

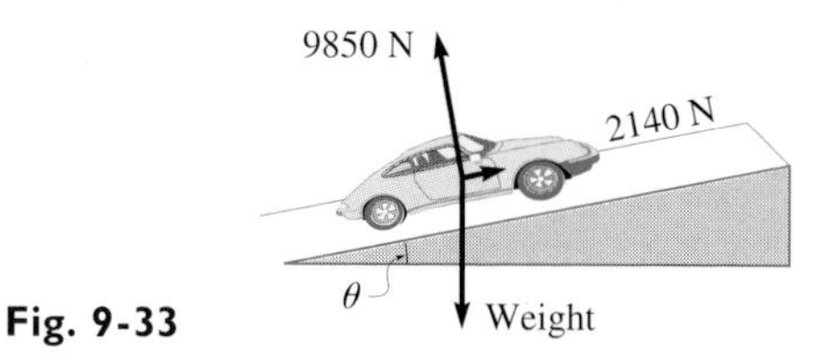

Fig. 9-33

21. A plane is moving at 75.0 m/s, and a package with weather instruments is ejected horizontally from the plane at 15.0 m/s, perpendicular to the direction of the plane. If the vertical velocity v_v (in m/s), as a function of time of fall, is given by $v_v = 9.80t$, what is the velocity of the package after 2.00 s (before its parachute opens)?

22. A flat rectangular barge, 48.0 m long and 20.0 m wide, is headed directly across a stream at 4.5 km/h. The stream flows at 3.8 km/h. What is the velocity, relative to the riverbed, of a person walking diagonally across the barge at 5.0 km/h while facing the opposite upstream bank?

23. In Fig. 9-34, a long, straight conductor perpendicular to the plane of the paper carries an electric current i. A bar magnet having poles of strength m lies in the plane of the paper. The vectors $\mathbf{H}_i$, $\mathbf{H}_N$, and $\mathbf{H}_S$ represent the components of the magnetic intensity $\mathbf{H}$ due to the current and to the N and S poles of the magnet, respectively. The magnitudes of the components of $\mathbf{H}$ are given by

$$H_i = \frac{1}{2\pi}\frac{i}{a}, \qquad H_N = \frac{1}{4\pi}\frac{m}{b^2}, \qquad \text{and} \quad H_S = \frac{1}{4\pi}\frac{m}{c^2}$$

Given that $a = 0.300$ m, $b = 0.400$ m, $c = 0.300$ m, the length of the magnet is 0.500 m, $i = 4.00$ A, and $m = 2.00$ A · m, calculate the resultant magnetic intensity $\mathbf{H}$. The component $\mathbf{H}_i$ is parallel to the magnet.

24. Solve the problem of Exercise 23 if $\mathbf{H}_i$ is directed away from the magnet, making an angle of 10.0° with the direction of the magnet.

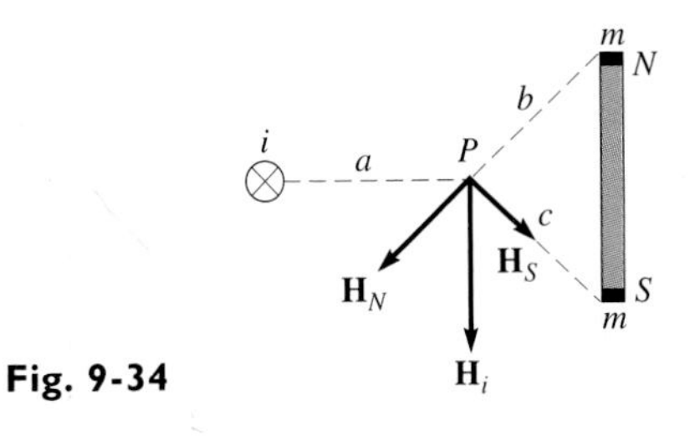

Fig. 9-34

9-5 Oblique Triangles, the Law of Sines

To this point we have limited our study of triangle solution to right triangles. However, *many triangles that require solution do not contain a right angle. Such triangles are called* **oblique triangles.** We now discuss solutions of oblique triangles.

In Section 4-4 we stated that we need to know three parts, at least one of them a side, to solve a triangle. There are four possible such combinations of parts, and these combinations are as follows:

Case 1. Two angles and one side.
Case 2. Two sides and the angle opposite one of them.
Case 3. Two sides and the included angle.
Case 4. Three sides.

There are several ways in which oblique triangles may be solved, but we shall restrict our attention to the two most useful methods, the **law of sines** and the **law of cosines.** In this section we shall discuss the law of sines and show that it may be used to solve Case 1 and Case 2.

Let ABC be an oblique triangle with sides a, b, and c opposite angles A, B, and C, respectively. By drawing a perpendicular h from B to side b, or its extension, we see from Fig. 9-35(a) that

$$h = c \sin A \quad \text{or} \quad h = a \sin C \tag{9-4}$$

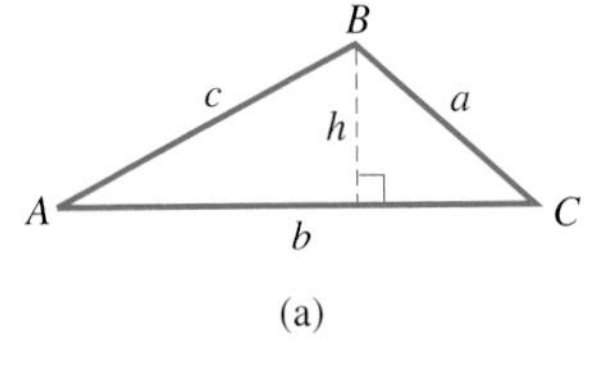

(a)

and from Fig. 9-35(b)

$$h = c \sin A \quad \text{or} \quad h = a \sin(180° - C) = a \sin C \tag{9-5}$$

We see that the results are precisely the same in Eqs. (9-4) and (9-5). Setting the results for h equal to each other, we have

$$c \sin A = a \sin C \quad \text{or} \quad \frac{a}{\sin A} = \frac{c}{\sin C} \tag{9-6}$$

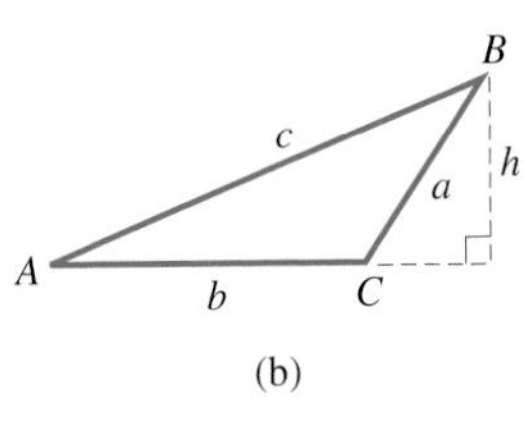

(b)

Fig. 9-35

By dropping a perpendicular from A to a, we also derive the result

$$c \sin B = b \sin C \quad \text{or} \quad \frac{b}{\sin B} = \frac{c}{\sin C} \tag{9-7}$$

Combining Eqs. (9-6) and (9-7), *for any triangle with sides a, b, and c, opposite angles A, B, and C, respectively,* such as the one shown in Fig. 9-36, *we have the* **law of sines:**

Law of Sines

$$\frac{a}{\sin A} = \frac{b}{\sin B} = \frac{c}{\sin C} \tag{9-8}$$

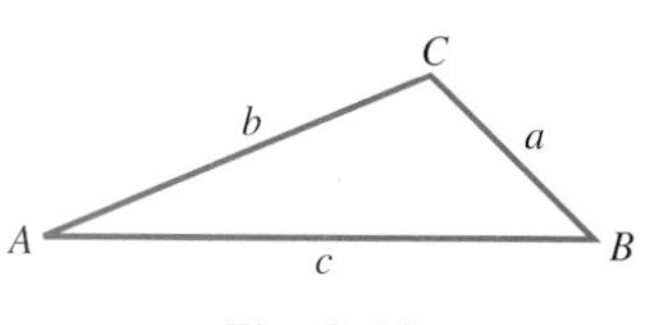

Fig. 9-36

Another form of the law of sines can be obtained by equating the reciprocals of each of the fractions in Eq. (9-8). The law of sines is a statement of proportionality between the sides of a triangle and the sines of the angles opposite them. We should note that there are actually three equations combined in Eq. (9-8). Of these, we use the one with three known parts of the triangle and we find the fourth part. In finding the complete solution of a triangle, it may be necessary to use two of the three equations.

Case I: Two Angles and One Side

Now we see how the law of sines is used in the solution of a triangle in which two angles and one side are known. If two angles are known, the third may be found from the fact that the sum of the angles in a triangle is 180°. At this point we must be able to find the ratio between the given side and the sine of the angle opposite it. Then, by use of the law of sines, we may find the other two sides.

EXAMPLE 1 ▸▸ Given $c = 6.00$, $A = 60.0°$, and $B = 40.0°$, find a, b, and C.

First we can see that

$$C = 180.0° - (60.0° + 40.0°) = 80.0°$$

We now know side c and angle C, which allows us to use Eq. (9-8). Therefore, using the equation relating a, A, c, and C, we have

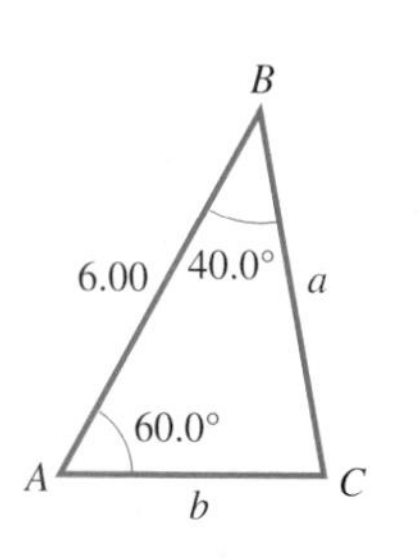

Fig. 9-37

$$\frac{a}{\underset{A}{\sin 60.0°}} = \frac{\overset{c}{6.00}}{\underset{C}{\sin 80.0°}} \quad \text{or} \quad a = \frac{6.00 \sin 60.0°}{\sin 80.0°} = 5.28$$

Now, using the equation relating b, B, c, and C, we have

$$\frac{b}{\underset{B}{\sin 40.0°}} = \frac{\overset{c}{6.00}}{\underset{C}{\sin 80.0°}} \quad \text{or} \quad b = \frac{6.00 \sin 40.0°}{\sin 80.0°} = 3.92$$

Thus, $a = 5.28$, $b = 3.92$, and $C = 80.0°$. See Fig. 9-37. We could also have used the form of Eq. (9-8) relating a, A, b, and B in order to find b, but any error in calculating a would make b in error as well. Of course, any error in calculating C would make both a and b in error.

In finding the values on a calculator, the quotient 6.00/sin 80.0° should be stored in memory since it is used in the solution for both a and b. ◂◂

EXAMPLE 2 ▸▸ Solve the triangle with the following given parts: $a = 63.71$, $A = 56.29°$, and $B = 97.06°$. See Fig. 9-38.

From the figure, we see that we are to find angle C and sides b and c. We first determine angle C.

$$\begin{aligned} C &= 180° - (A + B) = 180° - (56.29° + 97.06°) \\ &= 26.65° \end{aligned}$$

We will now use the ratio $a/\sin A$ to find sides b and c using Eq. (9-8) (the law of sines).

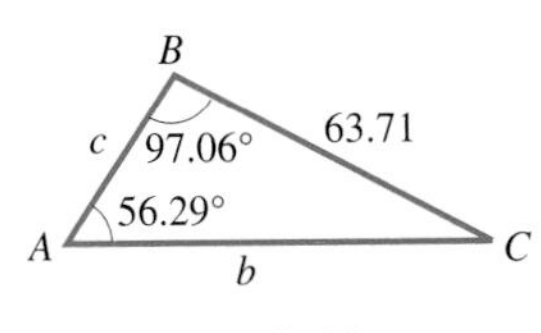

Fig. 9-38

$$\frac{b}{\underset{B}{\sin 97.06°}} = \frac{\overset{a}{63.71}}{\underset{A}{\sin 56.29°}} \quad \text{or} \quad b = \frac{63.71 \sin 97.06°}{\sin 56.29°} = 76.01$$

and

$$\frac{c}{\underset{C}{\sin 26.65°}} = \frac{\overset{a}{63.71}}{\underset{A}{\sin 56.29°}} \quad \text{or} \quad c = \frac{63.71 \sin 26.65°}{\sin 56.29°} = 34.35$$

Thus, $b = 76.01$, $c = 34.35$, and $C = 26.65°$. See Fig. 9-38. ◂◂

If the given information is appropriate, the law of sines may be used to solve applied problems. The following example illustrates the use of the law of sines in such a problem.

EXAMPLE 3 ▸▸ Two observers A and B sight a helicopter due east. The observers are 1540 m apart, and the angles of elevation they each measure to the helicopter are 32.0° and 44.0°, respectively. How far is observer A from the helicopter? See Fig. 9-39.

Letting H represent the position of the helicopter, we see that angle B within the triangle ABH is $180° - 44.0° = 136.0°$. This means that the angle at H within the triangle is

$$H = 180° - (32.0° + 136.0°) = 12.0°$$

Fig. 9-39

Now, using the law of sines to find required side AH, we have

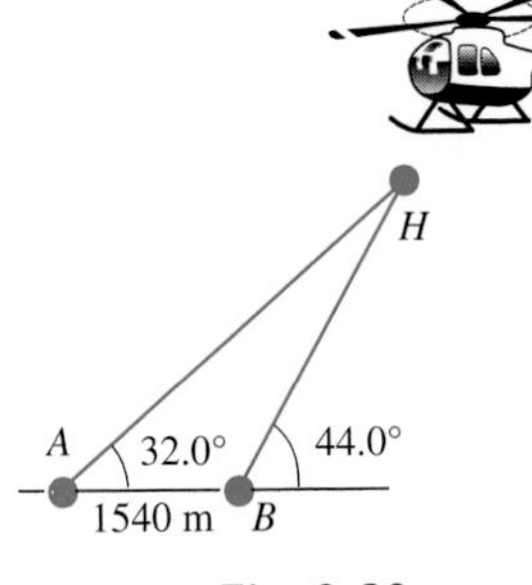

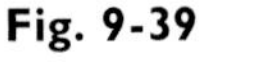

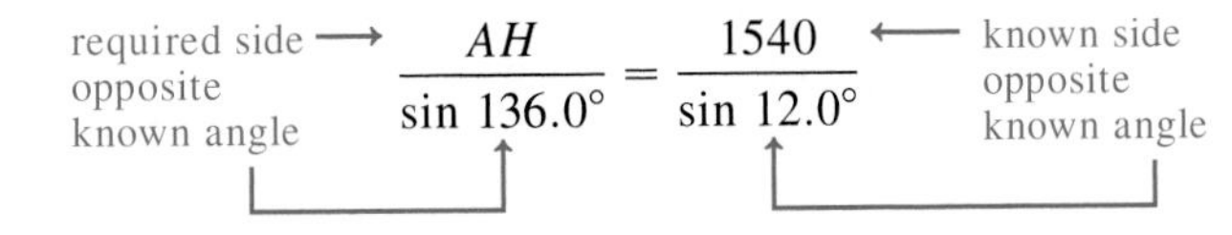

$$\text{required side opposite known angle} \longrightarrow \frac{AH}{\sin 136.0°} = \frac{1540}{\sin 12.0°} \longleftarrow \text{known side opposite known angle}$$

or

$$AH = \frac{1540 \sin 136.0°}{\sin 12.0°} = 5150 \text{ m}$$

Thus, observer A is about 5150 m from the helicopter. ◂◂

Case 2: Two Sides and the Angle Opposite One of Them

If we know two sides of a triangle and the angle that is opposite one of the given sides, *we may find that there are two triangles* which satisfy the given information. The following example illustrates this point.

EXAMPLE 4 ▸▸ Solve the triangle with the following given parts: $a = 60.0$, $b = 40.0$, and $B = 30.0°$.

By making a good scale drawing, Fig. 9-40(a), we see that the angle opposite a may be at either position A or A'. Both positions of this angle satisfy the given parts. Therefore, there are two triangles that result. Using the law of sines, we solve the case in which A, opposite side a, is an acute angle.

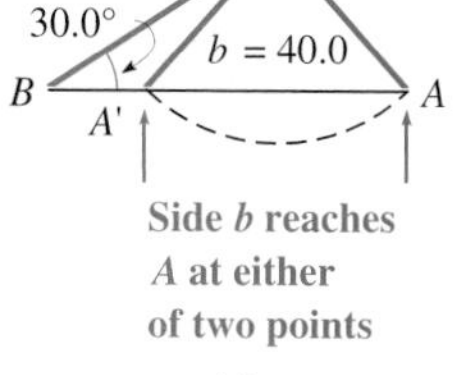

(a)

$$\frac{60.0}{\sin A} = \frac{40.0}{\sin 30.0°} \quad \text{or} \quad \sin A = \frac{60.0 \sin 30.0°}{40.0}, \qquad A = 48.6°$$

Therefore, $A = 48.6°$ and $C = 101.4°$. Using the law of sines again to find c, we have

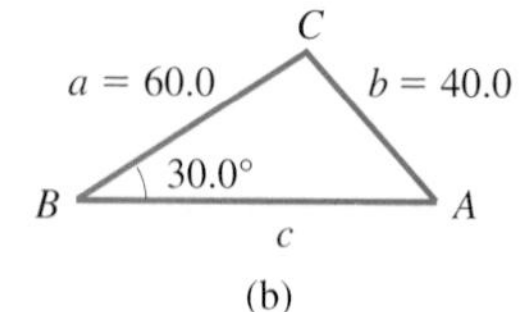

(b)

$$\frac{c}{\sin 101.4°} = \frac{40.0}{\sin 30.0°} \quad \text{or} \quad c = \frac{40.0 \sin 101.4°}{\sin 30.0°} = 78.4$$

Thus we have $A = 48.6°$, $C = 101.4°$, and $c = 78.4$. See Fig. 9-40(b).

The other solution is the case in which A', opposite side a, is an obtuse angle. Here we have

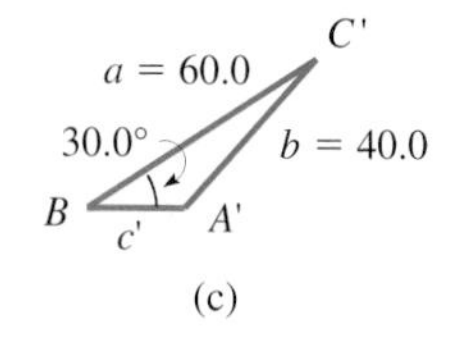

(c)

Fig. 9-40

$$\frac{60.0}{\sin A'} = \frac{40.0}{\sin 30.0°}, \qquad \sin A = 0.7500$$

which leads to $A' = 180° - 48.6° = 131.4°$. For this case we have C′ (the angle opposite c when $A' = 131.4°$) as 18.6°.

Using the law of sines to find c', we have

$$\frac{c'}{\sin 18.6°} = \frac{40.0}{\sin 30.0°} \quad \text{or} \quad c' = \frac{40.0 \sin 18.6°}{\sin 30.0°} = 25.5$$

This means that the second solution is $A' = 131.4°$, $C' = 18.6°$, and $c' = 25.5$. See Fig. 9-40(c). ◂◂

EXAMPLE 5 ▸▸ In Example 4, if $b > 60.0$, only one solution would result. In this case, side b would intercept side c at A. It also intercepts the extension of side c, but this would require that angle B not be included in the triangle (see Fig. 9-41). Thus, only one solution may result if $b > a$.

In Example 4, there would be *no solution* if side b were not at least 30.0. If this were the case, side b would not be long enough to even touch side c. It can be seen that b must at least equal $a \sin B$. If it is just equal to $a \sin B$, there is *one solution*, a right triangle. See Fig. 9-42.

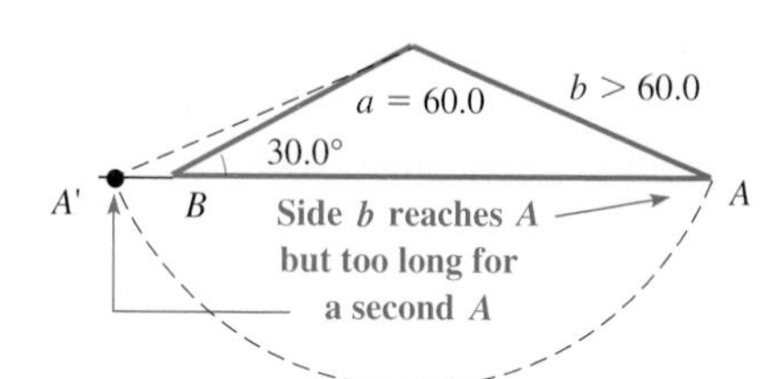

Fig. 9-41

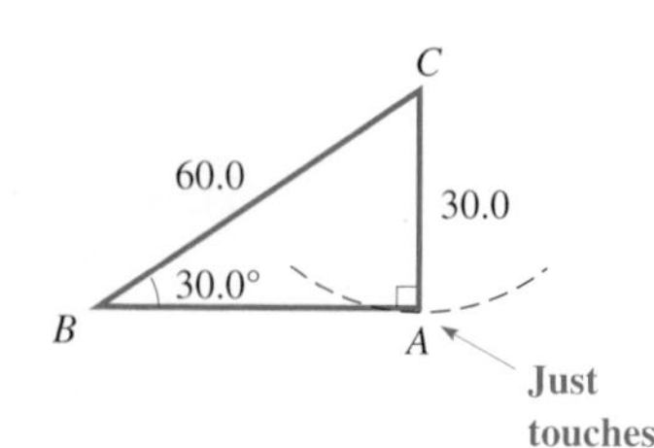

Fig. 9-42 ◂◂

Ambiguous Case

Summarizing the results for Case 2 as illustrated in Examples 4 and 5, we make the following conclusions. Given sides a and b and angle A (assuming here that a and A ($A < 90°$) are corresponding parts), we have

1. **No solution** *if* $a < b \sin A$. See Fig. 9-43(a).
2. **A right triangle solution** *if* $a = b \sin A$. See Fig. 9-43(b).
3. **Two solutions** *if* $b \sin A < a < b$. See Fig. 9-43(c). (CAUTION ▶)
4. **One solution** *if* $a > b$. See Fig. 9-43(d).

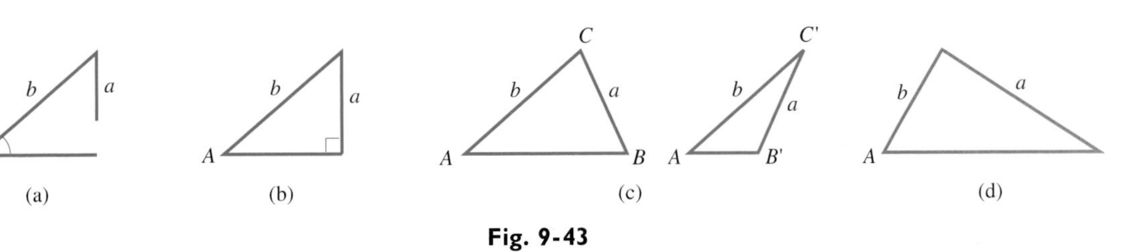

Fig. 9-43

NOTE ▶ We note that *in order to have two solutions, we must know two sides and the angle opposite one of the sides, and the shorter side must be opposite the known angle.*

If there is *no solution*, the calculator will indicate an *error.* If the solution is a *right triangle,* the calculator will show an angle of *exactly 90°* (no extra decimal digits will be displayed.)

For the reason that two solutions may result from it, Case 2 is called the **ambiguous case.** The following example illustrates Case 2 in an applied problem.

See the chapter introduction.

EXAMPLE 6 ▸▸ City B is 43.2° south of east of city A. Find the direction in which a pilot should head a plane in flying from A to B if the wind is from the west at 40.0 km/h and the plane's velocity with respect to the air is 300 km/h.

The plane's heading should be set so that the resultant of the plane's velocity with respect to the air $\mathbf{v}_{pa}$ and the velocity of the wind $\mathbf{v}_w$ will be in the direction from city A to city B. This means that the resultant velocity $\mathbf{v}_{pg}$, representing the velocity of the plane with respect to the ground, must be at an angle of 43.2° south of east from city A.

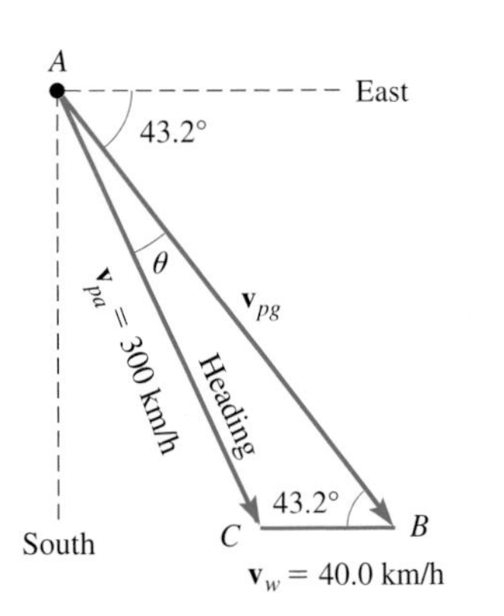

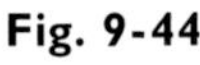

Fig. 9-44

With this analysis, we use the given information and draw the vector triangle shown in Fig. 9-44. In triangle ABC, we know that $\angle ABC = 43.2°$ by noting the alternate interior angles (see p. 51). By finding angle θ, the required heading can be determined. There can be only one solution, since $v_{pa} > v_w$. Using the law of sines, we have

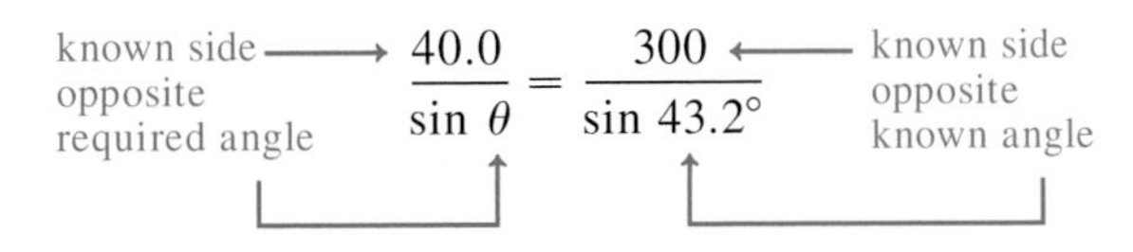

$$\frac{40.0}{\sin\theta} = \frac{300}{\sin 43.2°}$$

from which we get

$$\sin\theta = \frac{40.0 \sin 43.2°}{300}, \qquad \theta = 5.2°$$

Therefore, the heading should be $43.2° + 5.2° = 48.4°$ south of east. Compare this example with Example 3 on p. 254. ◂◂

If we attempt to use the law of sines for the solution of Case 3 or Case 4, we find that we do not have sufficient information to complete one of the ratios. These cases can, however, be solved by the law of cosines, which we shall consider in the next section.

EXAMPLE 7 ▸▸ Given the three sides of a triangle $a = 5$, $b = 6$, $c = 7$, we would set up the ratios

$$\frac{5}{\sin A} = \frac{6}{\sin B} = \frac{7}{\sin C}$$

However, since there is no way to determine a complete ratio from these equations, we cannot find the solution of the triangle in this manner. ◂◂

Exercises 9–5

In Exercises 1–20, solve the triangles with the given parts.

1. $a = 45.7$, $A = 65.0°$, $B = 49.0°$
2. $b = 3.07$, $A = 26.0°$, $C = 120.0°$
3. $c = 4380$, $A = 37.4°$, $B = 34.6°$
4. $a = 93.2$, $B = 17.9°$, $C = 82.6°$
5. $a = 4.601$, $b = 3.107$, $A = 18.23°$
6. $b = 3.625$, $c = 2.946$, $B = 69.37°$
7. $b = 77.51$, $c = 36.42$, $B = 20.73°$
8. $a = 150.4$, $c = 250.9$, $C = 76.43°$
9. $b = 0.0742$, $B = 51.0°$, $C = 3.4°$
10. $c = 729$, $B = 121.0°$, $C = 44.2°$
11. $a = 63.8$, $B = 58.4°$, $C = 22.2°$
12. $a = 13.0$, $A = 55.2°$, $B = 67.5°$
13. $b = 4384$, $B = 47.43°$, $C = 64.56°$
14. $b = 283.2$, $B = 13.79°$, $C = 76.38°$
15. $a = 5.240$, $b = 4.446$, $B = 48.13°$
16. $a = 89.45$, $c = 37.36$, $C = 15.62°$
17. $b = 2880$, $c = 3650$, $B = 31.4°$
18. $a = 0.841$, $b = 0.965$, $A = 57.1°$
19. $a = 45.0$, $b = 126$, $A = 64.8°$
20. $a = 20$, $c = 10$, $C = 30°$

In Exercises 21–32, use the law of sines to solve the given problems.

21. The Pentagon (headquarters of the U.S. Department of Defense) is one of the world's largest office buildings. It is a regular pentagon (five sides), 281 m on a side. Find the greatest straight-line distance from one point on the outside of the building to another outside point. (Find the length of a diagonal from one vertex to another.)

22. Two ropes hold a 175-N crate as shown in Fig. 9-45. Find the tensions $\mathbf{T}_1$ and $\mathbf{T}_2$ in the ropes. (*Hint:* Move vectors so that they are tail to head to form a triangle. The vector sum $\mathbf{T}_1 + \mathbf{T}_2$ must equal 175 N for equilibrium. See page 254.

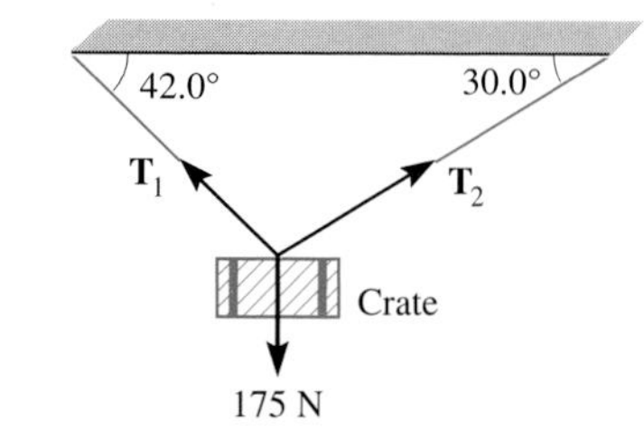

Fig. 9-45

23. Find the tension **T** in the left guy wire attached to the top of the tower shown in Fig. 9-46. (*Hint:* The horizontal components of the tensions must be equal and opposite for equilibrium. See page 254. Thus, move the tension vectors tail to head to form a triangle with a vertical resultant. This resultant equals the upward force at the top of the tower for equilibrium. This last force is not shown and does not have to be calculated.)

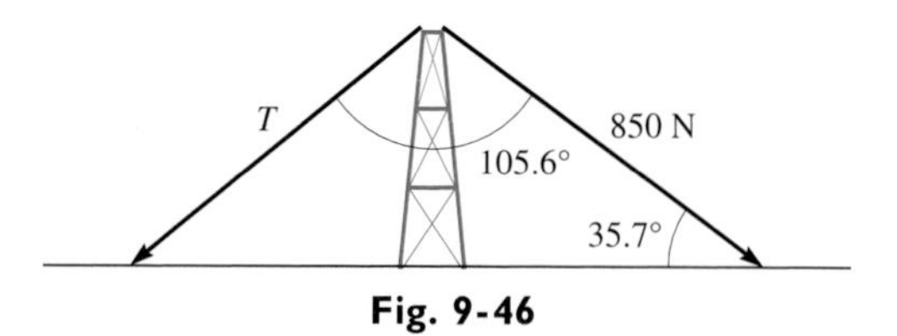

Fig. 9-46

24. Find the distance from Atlanta to Raleigh, North Carolina, from Fig. 9-47.

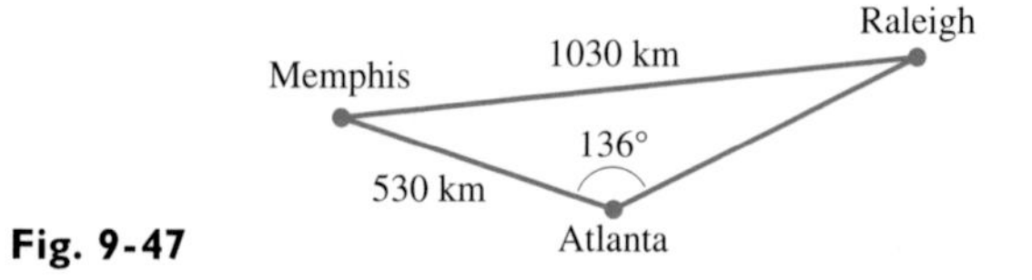

Fig. 9-47

25. Find the distance between Gravois Ave. and Jefferson Ave. along Arsenal St. in St. Louis from Fig. 9-48.

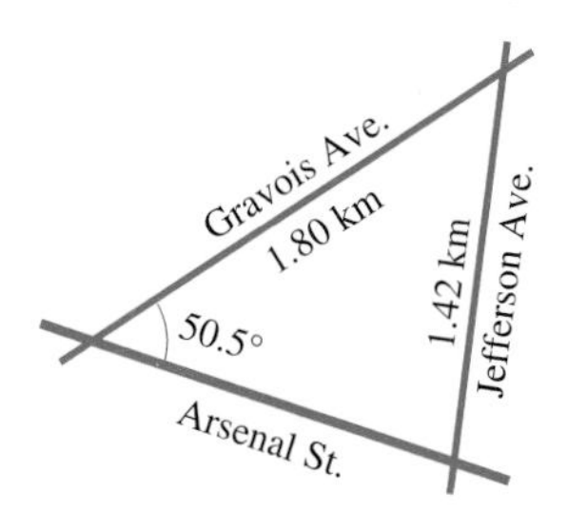

Fig. 9-48

26. When an airplane is landing at a 2510-m runway, the angles of depression to the ends of the runway are 10.0° and 13.5°. How far is the plane from the near end of the runway?

27. Find the total length of the path of the laser beam shown in Fig. 9-49.

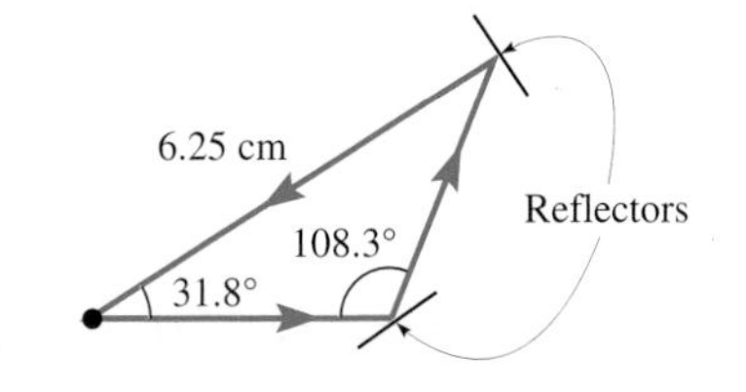

Fig. 9-49

28. In widening a highway, it is necessary for a construction crew to cut into the bank along the highway. The present angle of elevation of the straight slope of the bank is 23.0° and the new angle is to be 38.5°, leaving the top of the slope at its present position. If the slope of the present bank is 66.0 m long, how far horizontally into the bank at its base must they dig?

29. A communications satellite is directly above the extension of a line between receiving towers A and B. It is determined from radio signals that the angle of elevation of the satellite from tower A is 89.2°, and the angle of elevation from tower B is 86.5°. If A and B are 1290 km apart, how far is the satellite from A? (Neglect the curvature of the Earth.) See Fig. 9-50.

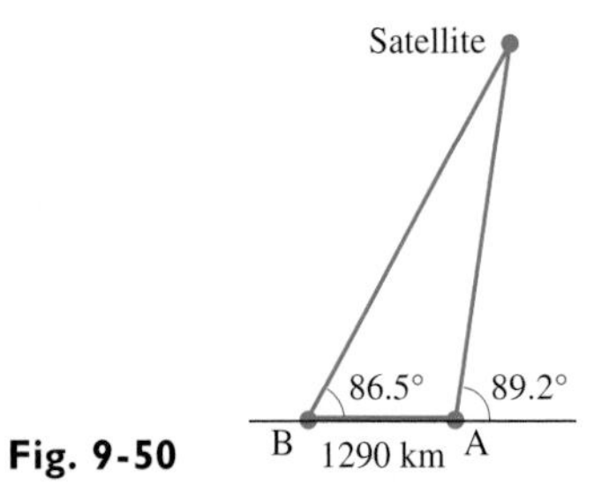

Fig. 9-50

30. Point P on the mechanism shown in Fig. 9-51 is driven back and forth horizontally. If the minimum value of angle θ is 32.0°, what is the distance between extreme positions of P? What is the maximum possible value of angle θ?

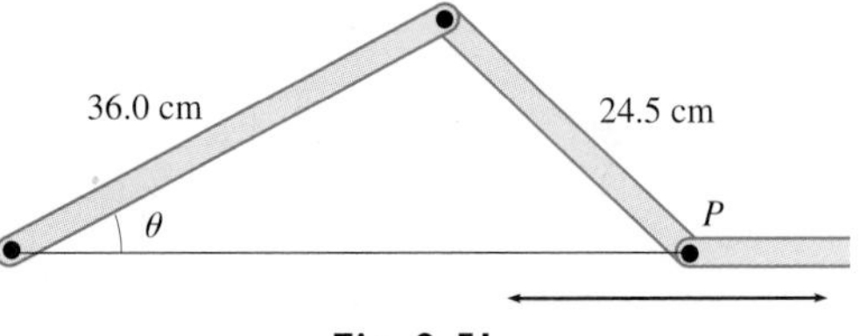

Fig. 9-51

31. A boat owner wishes to cross a river 2.60 km wide and go directly to a point on the opposite side 1.75 km downstream. The boat goes 8.00 km/h in still water, and the stream flows at 3.50 km/h. What should the boat's heading be?

32. A triangular support was measured to have a side of 25.3 cm, a second side of 14.0 cm, and an angle of 36.5° opposite the second side. Find the length of the third side.

9-6 The Law of Cosines

As we noted at the end of the preceding section, the law of sines cannot be used to solve a triangle if the only information given is that of Case 3 (two sides and the included angle) or Case 4 (three sides). Therefore, it is necessary to develop another method of finding at least one more part of the triangle. Therefore, in this section we develop the *law of cosines,* by which we can solve a triangle for Case 3 and Case 4. After obtaining another part of the triangle by the law of cosines, we can then use the law of sines to complete the solution. We do this because the law of sines generally provides a simpler method of solution than does the law of cosines.

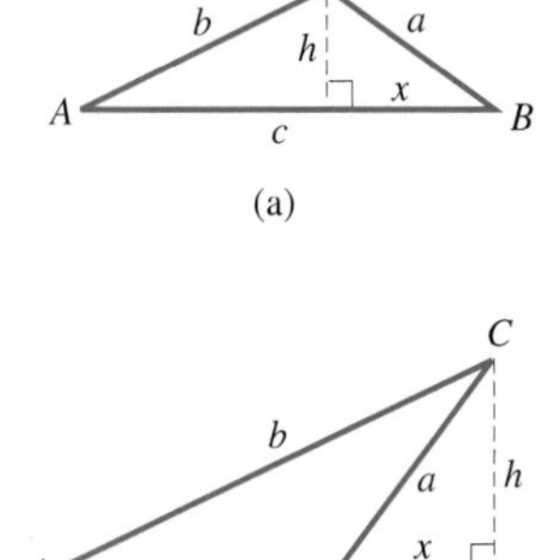

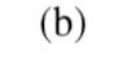
(b)

Fig. 9-52

Consider any oblique triangle, for example, either of the triangles shown in Fig. 9-52. For each of these triangles we see that $h/b = \sin A$, or $h = b \sin A$. Also, by using the Pythagorean theorem, we obtain $a^2 = h^2 + x^2$ for each triangle. Thus (with $(\sin A)^2 = \sin^2 A$), we have

$$a^2 = b^2 \sin^2 A + x^2 \qquad (9\text{-}9)$$

In Fig. 9-52(a), we see that $(c - x)/b = \cos A$, or $c - x = b \cos A$. Solving for x, we have $x = c - b \cos A$. In Fig. 9-52(b), we have $c + x = b \cos A$, and solving for x, we have $x = b \cos A - c$. Substituting these relations into Eq. (9-9), we obtain

$$a^2 = b^2 \sin^2 A + (c - b \cos A)^2$$

and (9-10)

$$a^2 = b^2 \sin^2 A + (b \cos A - c)^2$$

respectively. When expanded, these give

$$a^2 = b^2 \sin^2 A + b^2 \cos^2 A + c^2 - 2bc \cos A$$

and

$$a^2 = b^2(\sin^2 A + \cos^2 A) + c^2 - 2bc \cos A \qquad (9\text{-}11)$$

Recalling the definitions of the trigonometric functions, we know that $\sin\theta = y/r$ and $\cos\theta = x/r$. Thus, $\sin^2\theta + \cos^2\theta = (y^2 + x^2)/r^2$. However, $x^2 + y^2 = r^2$, which means that

$$\sin^2\theta + \cos^2\theta = 1 \qquad (9\text{-}12)$$

This equation is valid for any angle θ, since we have made no assumptions as to any of the properties of θ. Therefore, by substituting Eq. (9-12) into Eq. (9-11), we arrive at the **law of cosines.**

Law of Cosines

$$a^2 = b^2 + c^2 - 2bc \cos A \qquad (9\text{-}13)$$

Using the method above, we may also show that

$$b^2 = a^2 + c^2 - 2ac \cos B$$

and

$$c^2 = a^2 + b^2 - 2ab \cos C$$

Case 3: Two Sides and the Included Angle

If we know two sides and the included angle of a triangle, we see by the forms of the law of cosines that we may directly solve for the side opposite the given angle. Thus, as we noted at the beginning of this section, we may complete the solution by using the law of sines.

EXAMPLE 1 ▸▸ Solve the triangle with $a = 45.0$, $b = 67.0$, and $C = 35.0°$. See Fig. 9-53.

Since angle C is known, we first solve for side c, using the law of cosines in the form

$$c^2 = a^2 + b^2 - 2ab \cos C$$

Substituting, we have

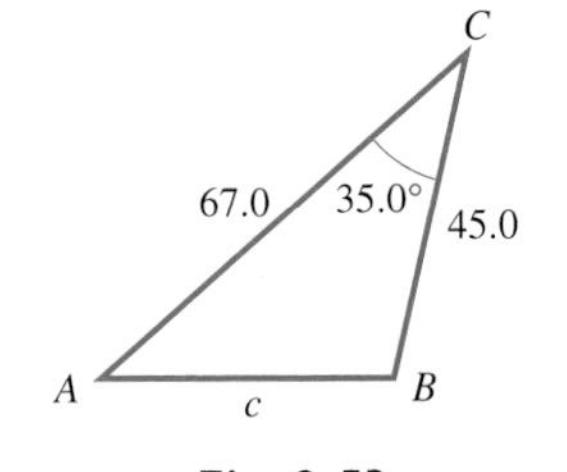

Fig. 9-53

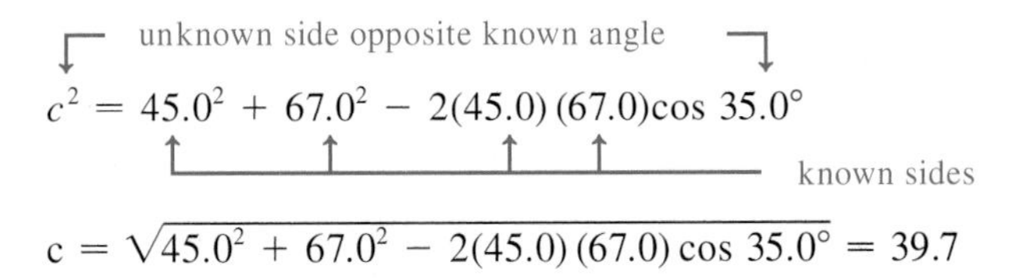

$$c = \sqrt{45.0^2 + 67.0^2 - 2(45.0)(67.0)\cos 35.0°} = 39.7$$

From the law of sines, we now have

$$\frac{45.0}{\sin A} = \frac{67.0}{\sin B} = \frac{39.7}{\sin 35.0°} \quad \begin{matrix} \leftarrow \\ \leftarrow \end{matrix} \text{ sides opposite angles}$$

which leads to

$$\sin A = \frac{45.0 \sin 35.0°}{39.7}, \qquad A = 40.6°$$

In using the calculator, we would store the calculated value of c in memory before rounding it off. We would then use the value from the calculator memory, not 39.7 as shown above. We write 39.7 in the equation only to show how we complete the solution using the law of sines.

We can solve for B from the above relation. A quicker way is to use the fact that the sum of all three angles is 180°. This leads to

$$B = 180° - (35.0° + 40.6°) = 104.4°$$

Therefore, we have found that $c = 39.7$, $A = 40.6°$, and $B = 104.4°$. ◂◂

EXAMPLE 2 ▸▸ Solve the triangle with $a = 0.1762$, $c = 0.5034$, and $B = 129.20°$. See Fig. 9-54.

Again, the given parts are two sides and the included angle, or Case 3. Since B is the known angle, we use the form of the law of cosines that includes angle B. This means that we find b first.

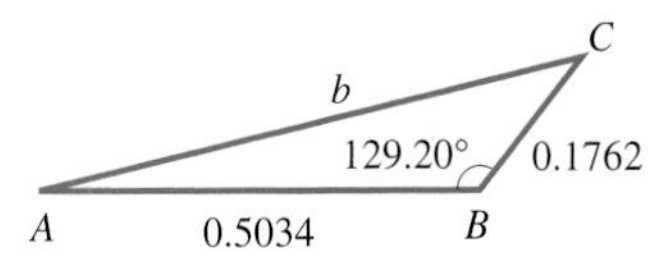

Fig. 9-54

$$\begin{aligned} b^2 &= a^2 + c^2 - 2ac \cos B \\ &= 0.1762^2 + 0.5034^2 - 2(0.1762)(0.5034)\cos 129.20° \\ b &= 0.6297 \end{aligned}$$

From the law of sines, we have

$$\frac{0.1762}{\sin A} = \frac{0.6297}{\sin 129.20°} = \frac{0.5034}{\sin C}$$

Again, in using the calculator we will use the memory value of b. Therefore,

$$\sin A = \frac{0.1762 \sin 129.20°}{0.6297}, \qquad A = 12.52°$$

$$C = 180° - (129.20° + 12.52°) = 38.28°$$

Thus, $b = 0.6297$, $A = 12.52°$, and $C = 38.28°$. ◂◂

Case 4: Three Sides

CAUTION ▶ Given the three sides of a triangle, we may solve for the angle opposite any of the sides using the law of cosines. In solving for an angle, *the best procedure is to find the largest angle first.* This avoids the ambiguous case if we switch to the law of sines and the triangle has an obtuse angle. *The largest angle is always opposite the longest side.* Another procedure is to use the law of cosines to find two angles.

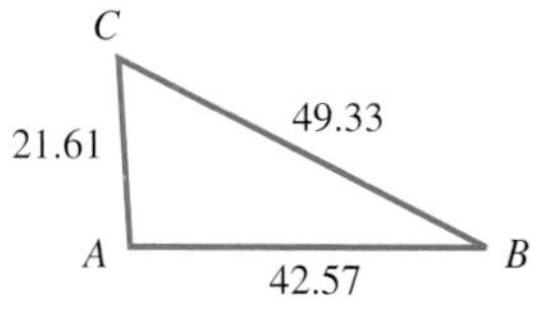

Fig. 9-55

EXAMPLE 3 ▶▶ Solve the triangle given the three sides $a = 49.33$, $b = 21.61$, and $c = 42.57$. See Fig. 9-55.

Since the longest side is $a = 49.33$, we first solve for angle A.

$$a^2 = b^2 + c^2 - 2bc \cos A$$

$$\cos A = \frac{b^2 + c^2 - a^2}{2bc} = \frac{21.61^2 + 42.57^2 - 49.33^2}{2(21.61)(42.57)} = -0.08384$$

$$A = 94.81°$$

NOTE ▶ We do not have to record the value of cos A. However, we note that *the calculator shows an obtuse angle,* which it should since cos A is negative.

From the law of sines, we now have

$$\frac{49.33}{\sin 94.81°} = \frac{21.61}{\sin B} = \frac{42.57}{\sin C}$$

which gives us $B = 25.88°$ and $C = 59.31°$. ◀◀

EXAMPLE 4 ▶▶ Two forces are acting on a bolt. One is a 78.0-N force acting horizontally to the right, and the other is a force of 45.0 N acting upward to the right, 15.0° from the vertical. Find the resultant force **F**. See Fig. 9-56.

Moving the 45.0-N vector to the right, and using the lower triangle with the 105.0° angle, we find the magnitude of **F** as

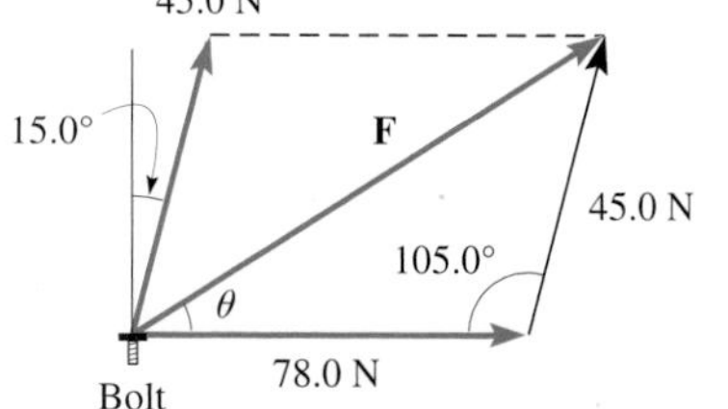

Fig. 9-56

$$F = \sqrt{78.0^2 + 45.0^2 - 2(78.0)(45.0)\cos 105.0°}$$
$$= 99.6 \text{ N}$$

To find θ, we use the law of sines:

$$\frac{45.0}{\sin \theta} = \frac{99.6}{\sin 105.0°}, \qquad \sin \theta = \frac{45.0 \sin 105.0°}{99.6}$$

This gives us $\theta = 25.9°$.

We can also solve this problem using vector components. ◀◀

EXAMPLE 5 ▶▶ A vertical radio antenna is to be built on a hill that makes an angle of 6.0° with the horizontal. Guy wires are to be attached at a point 75.0 m up the antenna and at points 50.0 m from the base of the antenna. What will be the lengths of the guy wires positioned directly up and directly down the hill?

Making an appropriate figure, as in Fig. 9-57, we can set up the equations needed for the solution.

$$L_u^2 = 50.0^2 + 75.0^2 - 2(50.0)(75.0)\cos 84.0°$$
$$L_u = 85.7 \text{ m}$$
$$L_d^2 = 50.0^2 + 75.0^2 - 2(50.0)(75.0)\cos 96.0°$$
$$L_d = 94.4 \text{ m}$$ ◀◀

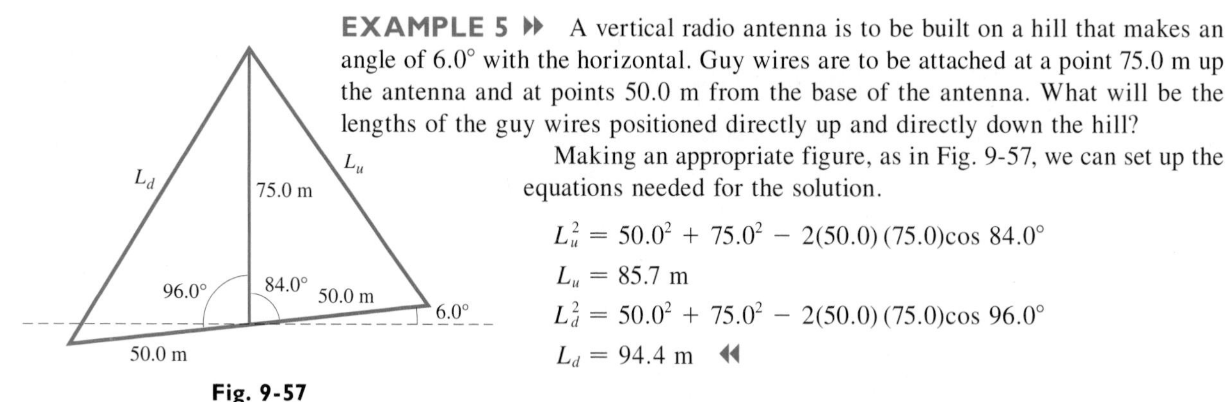

Fig. 9-57

Exercises 9–6

In Exercises 1–20, solve the triangles with the given parts.

1. $a = 6.00, b = 7.56, C = 54.0°$
2. $b = 87.3, c = 34.0, A = 130.0°$
3. $a = 4530, b = 924, C = 98.0°$
4. $a = 0.0845, c = 0.116, B = 85.0°$
5. $a = 39.53, b = 45.22, c = 67.15$
6. $a = 23.31, b = 27.26, c = 29.17$
7. $a = 385.4, b = 467.7, c = 800.9$
8. $a = 0.2433, b = 0.2635, c = 0.1538$
9. $a = 320, b = 847, C = 158.0°$
10. $b = 18.3, c = 27.1, A = 58.7°$
11. $a = 21.4, c = 4.28, B = 86.3°$
12. $a = 11.3, b = 5.10, C = 77.6°$
13. $b = 103.7, c = 159.1, C = 104.67°$
14. $a = 49.32, b = 54.55, B = 114.36°$
15. $a = 0.4937, b = 0.5956, c = 0.6398$
16. $a = 69.72, b = 49.30, c = 56.29$
17. $a = 723, b = 598, c = 158$
18. $a = 1.78, b = 6.04, c = 4.80$
19. $a = 15, A = 15°, B = 140°$
20. $a = 17, b = 24, c = 37$

In Exercises 21–32, use the law of cosines to solve the given problems.

21. A nuclear submarine leaves its base and travels at 23.5 km/h. For 2 hours it travels along a course 32.1° north of west. It then turns an additional 21.5° north of west and travels for another hour. How far from its base is it?

22. The robot arm shown in Fig. 9-58 places packages on a conveyor belt. What is the distance x?

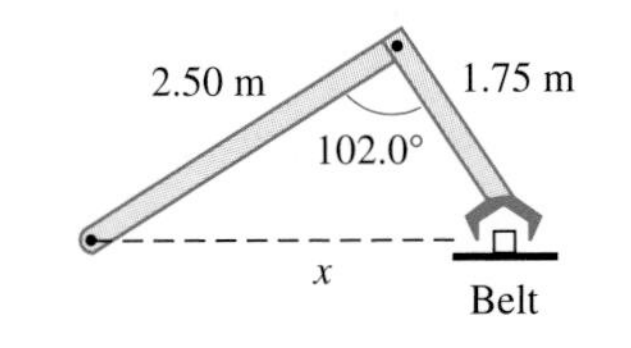

Fig. 9-58

23. In order to get around an obstruction, an oil pipeline is constructed in two straight sections, one 3.756 km long and the other 4.675 km long, with an angle of 168.85° between the sections where they are joined. How much more pipeline was necessary due to the obstruction?

24. In a baseball field the four bases are at the vertices of a square 90.0 ft on a side. The pitching rubber is 60.5 ft from home plate. How far is it from the pitching rubber to first base? See Fig. 9-59. (1 ft = 0.3048 m)

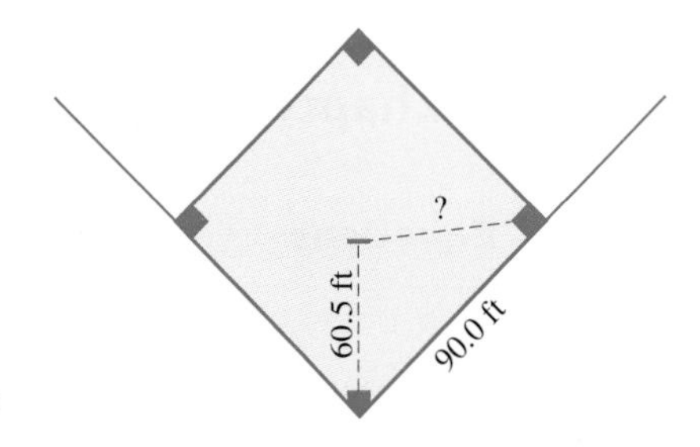

Fig. 9-59

25. A plane leaves an airport and travels 624 km due east. It then turns toward the north and travels another 326 km. It then turns again less than 180° and travels another 846 km directly back to the airport. Through what angles did it turn?

26. The apparent depth of an object submerged in water is less than its actual depth. A coin is actually 5.00 cm from an observer's eye just above the surface, but the distance appears to be only 4.25 cm. The real light ray from the coin makes an angle with the surface that is 8.1° greater than the angle the apparent ray makes. How much deeper is the coin than it appears to be? See Fig. 9-60.

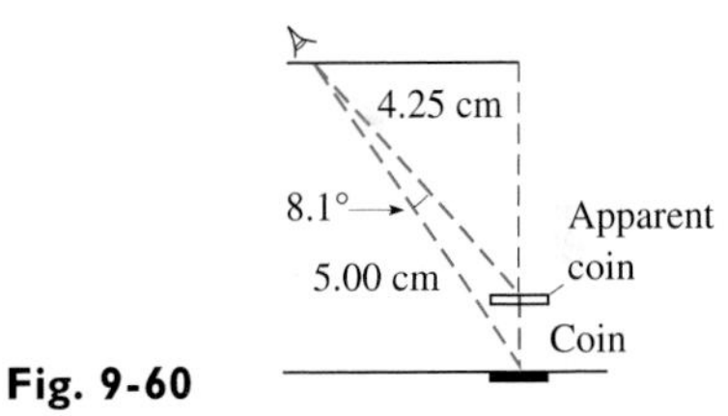

Fig. 9-60

27. A room is in the shape of a regular hexagon (six sides). If each side is 3.20 m, what is the length of the shortest diagonal of the room?

28. Two ropes support a 78.3-N crate from above. The tensions in the ropes are 50.6 N and 37.5 N. What is the angle between the ropes? (See Exercise 22 of Section 9-5.)

29. A ferryboat travels at 11.5 km/h with respect to the water. Because of the river current, it is traveling at 12.7 km/h with respect to the land in the direction of its destination. If the ferryboat's heading is 23.6° from the direction of its destination, what is the velocity of the current?

30. The airline distance from Denver to Dallas is 1060 km. It is 1290 km from Denver to St. Louis and 885 km from Dallas to St. Louis. Find the angle between the routes from Dallas.

31. An air traffic controller sights two planes that are due east from the control tower and headed toward each other. One is 15.8 km from the tower at an angle of elevation of 26.4°, and the other is 32.7 km from the tower at an angle of elevation of 12.4°. How far apart are the planes?

32. A triangular machine part has sides of 5 cm and 8 cm. Explain why the law of sines, or the law of cosines, is used to start the solution of the triangle if the third known part is (a) the third side, (b) the angle opposite the 5-cm side, or (c) the angle between the 5-cm and 8-cm sides.

Chapter Equations, Review Exercises, and Practice Test

Chapter Equations

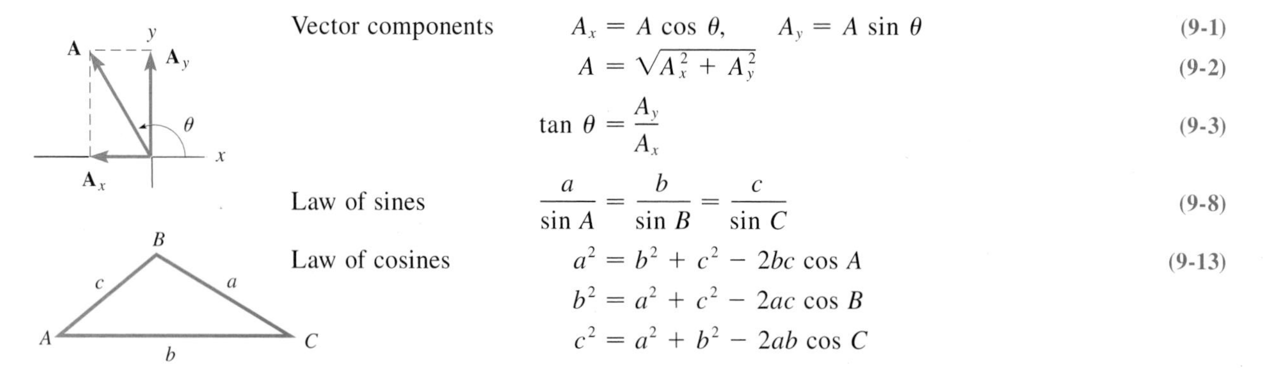

Vector components $\quad A_x = A\cos\theta, \quad A_y = A\sin\theta \quad$ (9-1)

$$A = \sqrt{A_x^2 + A_y^2} \quad (9\text{-}2)$$

$$\tan\theta = \frac{A_y}{A_x} \quad (9\text{-}3)$$

Law of sines $\quad \dfrac{a}{\sin A} = \dfrac{b}{\sin B} = \dfrac{c}{\sin C} \quad$ (9-8)

Law of cosines $\quad a^2 = b^2 + c^2 - 2bc\cos A \quad$ (9-13)

$$b^2 = a^2 + c^2 - 2ac\cos B$$

$$c^2 = a^2 + b^2 - 2ab\cos C$$

Review Exercises

In Exercises 1–4, find the x- and y-components of the given vectors by use of the trigonometric functions.

1. $A = 65.0,\ \theta_A = 28.0°$

2. $A = 8.05,\ \theta_A = 149.0°$

3. $A = 0.9204,\ \theta_A = 215.59°$

4. $A = 657.1,\ \theta_A = 343.74°$

*In Exercises 5–8, vectors **A** and **B** are at right angles. Find the magnitude and direction of the resultant.*

5. $A = 327$, $B = 505$

6. $A = 68$, $B = 29$

7. $A = 4964$, $B = 3298$

8. $A = 26.52$, $B = 89.86$

In Exercises 9–16, add the given vectors by use of the trigonometric functions and the Pythagorean theorem.

9. $A = 780,\ \theta_A = 28.0°$; $B = 346,\ \theta_B = 320.0°$

10. $A = 0.0120,\ \theta_A = 10.5°$; $B = 0.00781,\ \theta_B = 260.0°$

11. $A = 22.51,\ \theta_A = 130.16°$; $B = 7.604,\ \theta_B = 200.09°$

12. $A = 18{,}760,\ \theta_A = 110.43°$; $B = 4835,\ \theta_B = 350.20°$

13. $A = 51.33,\ \theta_A = 12.25°$; $B = 42.61,\ \theta_B = 291.77°$

14. $A = 70.31,\ \theta_A = 122.54°$; $B = 30.29,\ \theta_B = 214.82°$

15. $A = 75.0,\ \theta_A = 15.0°$; $B = 26.5,\ \theta_B = 192.4°$; $C = 54.8,\ \theta_C = 344.7°$

16. $A = 8120,\ \theta_A = 141.9°$; $B = 1540,\ \theta_B = 165.2°$; $C = 3470,\ \theta_C = 296.0°$

In Exercises 17–36, solve the triangles with the given parts.

17. $A = 48.0°,\ B = 68.0°,\ a = 14.5$

18. $A = 132.0°,\ b = 7.50,\ C = 32.0°$

19. $a = 22.8,\ B = 33.5°,\ C = 125.3°$

20. $A = 71.0°,\ B = 48.5°,\ c = 8.42$

21. $A = 17.85°,\ B = 154.16°,\ c = 7863$

22. $a = 1.985,\ b = 4.189,\ c = 3.652$

23. $b = 76.07,\ c = 40.53,\ B = 110.09°$

24. $A = 77.06°,\ a = 12.07,\ c = 5.104$

25. $b = 14.5,\ c = 13.0,\ C = 56.6°$

26. $B = 40.6°,\ b = 7.00,\ c = 18.0$

27. $a = 186,\ B = 130.0°,\ c = 106$

28. $b = 750,\ c = 1100,\ A = 56°$

29. $a = 7.86,\ b = 2.45,\ C = 22.0°$

30. $a = 0.208,\ c = 0.697,\ B = 105.4°$

31. $A = 67.16°,\ B = 96.84°,\ c = 532.9$

32. $A = 43.12°,\ a = 7.893,\ b = 4.113$

33. $a = 17,\ b = 12,\ c = 25$

34. $a = 9064,\ b = 9953,\ c = 1106$

35. $a = 5.30,\ b = 8.75,\ c = 12.5$

36. $a = 47.4,\ b = 40.0,\ c = 45.5$

In Exercises 37–56, solve the given problems.

37. Find the horizontal and vertical components of the force shown in Fig. 9-61.

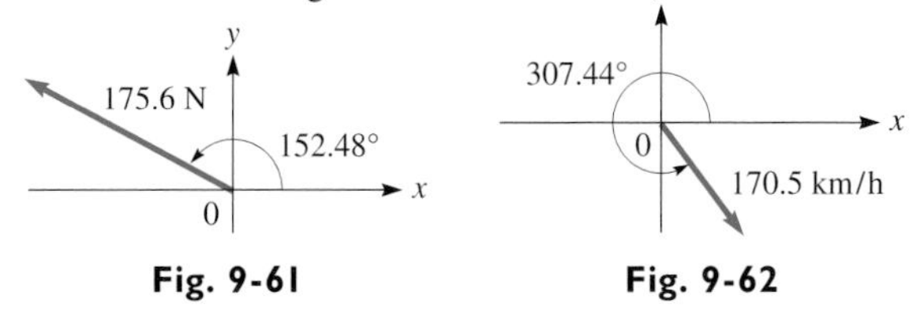

Fig. 9-61 **Fig. 9-62**

38. Find the horizontal and vertical components of the velocity shown in Fig. 9-62.

39. In a ballistics test, a bullet was fired into a block of wood with a velocity of 670 m/s and at an angle of 71.3° with the surface of the block. What was the component of the velocity perpendicular to the surface?

40. A storm cloud is moving at 15 km/h from the northwest. A television tower is 60° south of east of the cloud. What is the component of the cloud's velocity toward the tower?

41. In Fig. 9-63 force **F** represents the total surface tension force around the circumference on the liquid in the capillary tube. The vertical component of **F** holds up the liquid in the tube above the liquid surface outside the tube. What is this vertical component of **F**?

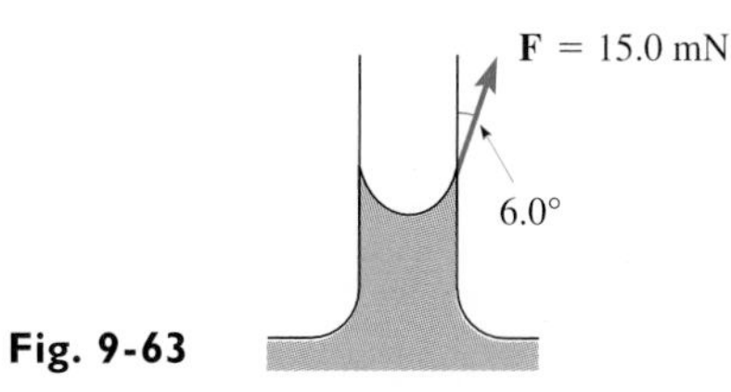

Fig. 9-63

42. During a 3-minute period after taking off, the supersonic jet Concorde traveled at 480 km/h at an angle of 24.0° above the horizontal. What was its gain in altitude during the three minutes?

43. A helium-filled balloon rises vertically at 3.5 m/s as the wind carries it horizontally at 5.0 m/s. What is the resultant velocity of the balloon?

44. A crater on the moon is 150 km in diameter. If the distance to the moon (to each side of the crater) from the earth is 390 000 km, what angle is subtended by the crater at an observer's position on the earth?

45. In Fig. 9-64 a damper mechanism in an air-conditioning system is shown. If $\theta = 27.5°$ when the spring is at its shortest and longest lengths, what are these lengths?

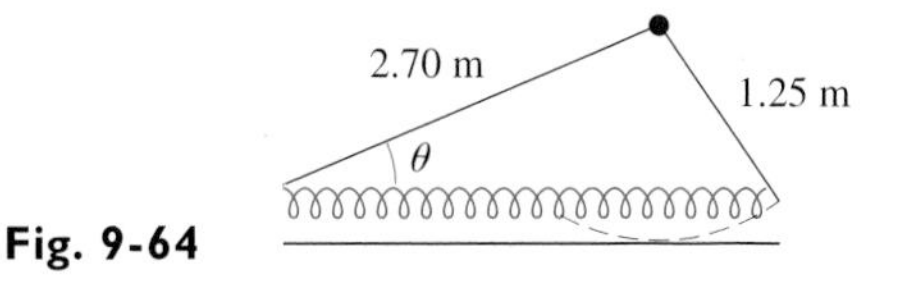

Fig. 9-64

46. A bullet is fired from the ground of a level field at an angle of 39.0° above the horizontal. It travels in a straight line at 670 m/s for 0.20 s when it strikes a target. The sound of the strike is recorded 0.32 s later on the ground. If sound travels at 350 m/s, where is the recording device located?

47. Two satellites are being observed at the same observing station. One is 36 200 km from the station, and the other is 30 100 km away. The angle between their lines of observation is 105.4°. How far apart are the satellites?

48. Find the side x in the truss in Fig. 9-65.

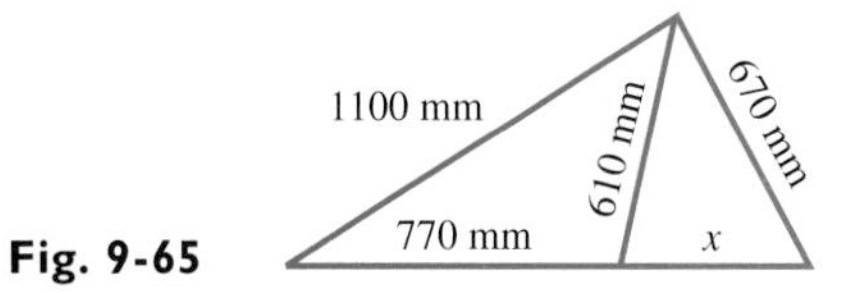

Fig. 9-65

49. The angle of depression of a fire noticed west of a fire tower is 6.2°. The angle of depression of a pond, also west of the tower, is 13.5°. If it is known that the tower is 2.25 km from the pond on a direct line to the pond, how far is the fire from the pond?

50. A surveyor wishes to find the distance between two points between which there is a security-restricted area. The surveyor measures the distance from each of these points to a third point and finds them to be 226.73 m and 185.12 m. If the angle between the lines of sight from the third point to the other points is 126.724°, how far apart are the two points?

51. Atlanta is 467 km and 51.0° south of east from Nashville. The pilot of an airplane due north of Atlanta radios Nashville and finds the plane is on a line 10.5° south of east from Nashville. How far is the plane from Nashville?

52. In going around a storm, a plane flies 125 km south, then 140 km at 30.0° south of west, and finally 225 km at 15.0° north of west. What is the displacement of the plane from its original position?

53. A sailboat is headed due north, and its sail is set perpendicular to the wind, which is from south of west. The component of the force of the wind in the direction of the heading is 480 N, and the component perpendicular to the heading (the *drift* component) is 650 N. What is the force exerted by the wind, and what is the direction of the wind? See Fig. 9-66.

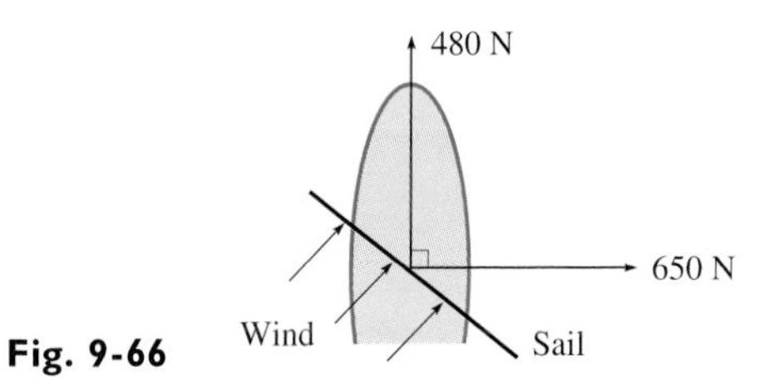

Fig. 9-66

54. Boston is 650 km and 21.0° south of west from Halifax, Nova Scotia. Radio signals locate a ship 10.5° east of south from Halifax and 5.6° north of east from Boston. How far is the ship from each city?

55. One end of a 725-m bridge is sighted from a distance of 1630 m. The angle between the lines of sight of the ends of the bridge is 25.2°. From these data, how far is the observer from the other end of the bridge?

56. A plane is traveling horizontally at 400 m/s. A missile is fired horizontally from it 30.0° from the direction in which the plane is traveling. If the missile leaves the plane at 650 m/s, what is its velocity 10.0 s later if the vertical component is given by $v_V = -9.80\,t$ (in m/s)?

Writing Exercise

57. A laser experiment uses a prism with a triangular base that has sides of 2.00 cm, 3.00 cm, and 4.50 cm. Write two or three paragraphs explaining how to find the angles of the triangle.

Practice Test

In all triangle solutions, sides a, b, c, are opposite angles A, B, C, respectively.

1. By use of a diagram, find the vector sum 2**A** + **B** for the given vectors.

2. For the triangle in which $a = 22.5$, $B = 78.6°$, and $c = 30.9$, find b.

3. A surveyor locates a tree 36.50 m to the northeast of a set position. The tree is 21.38 m north of a utility pole. What is the displacement of the utility pole from the set position?

4. For the triangle in which $A = 18.9°$, $B = 104.2°$, and $a = 426$, find c.

5. Solve the triangle in which $a = 9.84$, $b = 3.29$, and $c = 8.44$.

6. Find the horizontal and vertical components of a vector of magnitude 871 that is directed at a standard-position angle of 284.3°.

7. A ship leaves a port and travels due west. At a certain point it turns 31.5° north of west and travels an additional 42.0 km to a point 63.0 km on a direct line from the port. How far from the port is the point where the ship turned?

8. Find the sum of the vectors for which the magnitudes $A = 449$, $B = 285$, $\theta_A = 74.2°$, and $\theta_B = 208.9°$. Use the trigonometric functions and the Pythagorean theorem.

9. Solve the triangle for which $a = 22.3$, $b = 29.6$, and $A = 36.5°$.

10

Graphs of the Trigonometric Functions

One of the clearest ways of showing the properties of the trigonometric functions is by means of their graphs. In addition, their graphs are very valuable in numerous important areas of application. The applications are found in electronics, communications, acoustics, mechanical vibrations, and in many areas of physics.

These graphs are particularly useful in applications that involve any type of wave motion and periodic values, which repeat on a regular basis. Filtering electronic signals in communications, mixing sounds on a tape in a recording studio, and analyzing ocean waves and tides illustrate some of the applications of this type of periodic motion.

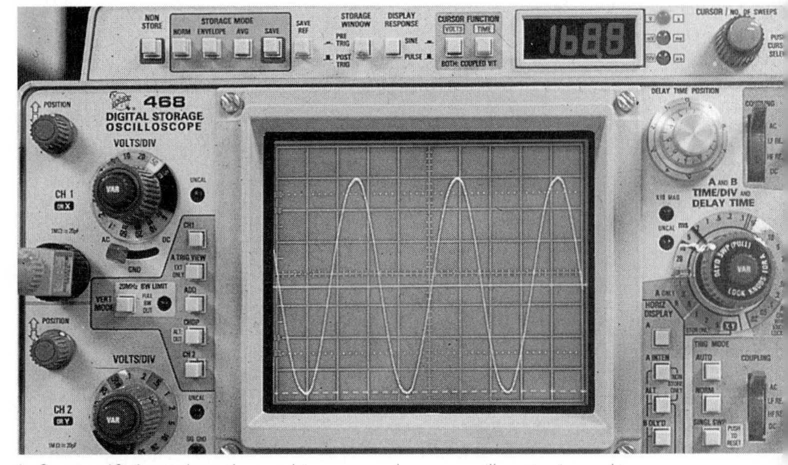

In Section 10-6 we show the resulting curve when an oscilloscope is used to combine and display electric signals.

10-1 Graphs of $y = a \sin x$ and $y = a \cos x$

The graphs of the trigonometric functions are constructed on the rectangular coordinate system. In plotting and sketching the trigonometric functions, *it is normal to express the angle in radians.* By using radians, *x and the trigonometric function of x are expressed as real numbers.*

Therefore, in order that we can plot and sketch the graphs of the trigonometric functions,

it is necessary to be able to readily use angles expressed in radians.

If necessary, review Section 8-3 on radian measure of angles for this purpose.

In this section, the graphs of the sine and cosine functions are shown. We begin by making a table of values of x and y for the function $y = \sin x$, where we are using x and y in the standard way as the *independent variable* and *dependent variable.* We plot the points to obtain the graph in Fig. 10-1.

x	0	$\frac{\pi}{6}$	$\frac{\pi}{3}$	$\frac{\pi}{2}$	$\frac{2\pi}{3}$	$\frac{5\pi}{6}$	π	$\frac{7\pi}{6}$	$\frac{4\pi}{3}$	$\frac{3\pi}{2}$	$\frac{5\pi}{3}$	$\frac{11\pi}{6}$	2π
y	0	0.5	0.87	1	0.87	0.5	0	−0.5	−0.87	−1	−0.87	−0.5	0

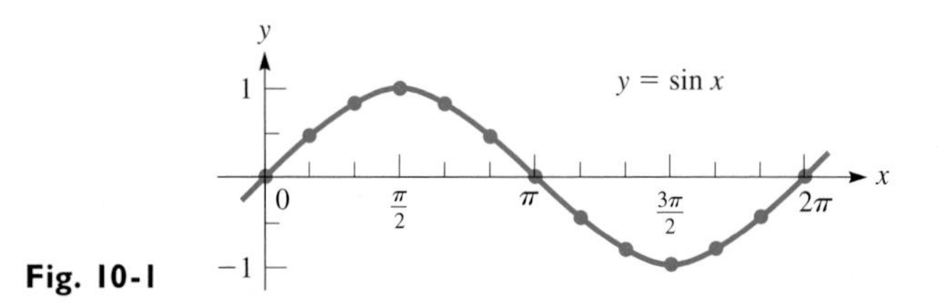

Fig. 10-1

The graph of $y = \cos x$ may be drawn in the same way. The next table gives values for plotting the graph of $y = \cos x$, and the graph is shown in Fig. 10-2.

x	0	$\frac{\pi}{6}$	$\frac{\pi}{3}$	$\frac{\pi}{2}$	$\frac{2\pi}{3}$	$\frac{5\pi}{6}$	π	$\frac{7\pi}{6}$	$\frac{4\pi}{3}$	$\frac{3\pi}{2}$	$\frac{5\pi}{3}$	$\frac{11\pi}{6}$	2π
y	1	0.87	0.5	0	−0.5	−0.87	−1	−0.87	−0.5	0	0.5	0.87	1

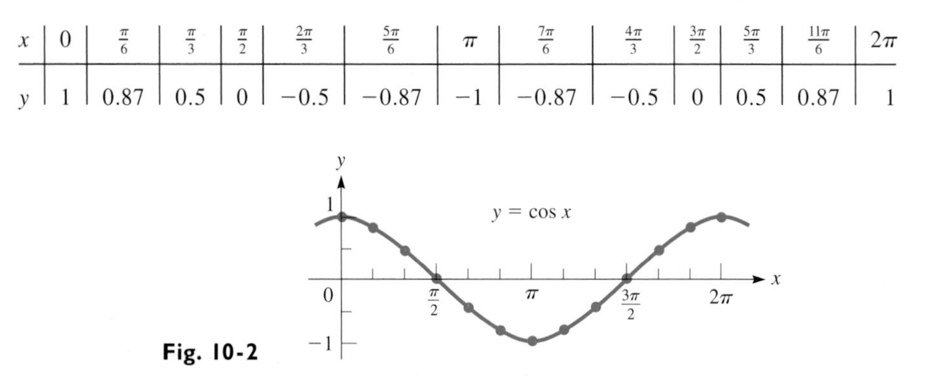

Fig. 10-2

The graphs are continued beyond the values shown in the tables to indicate that *they continue indefinitely in each direction.* To show this more clearly, in Figs. 10-3 and 10-4, we show the graphs of $y = \sin x$ and $y = \cos x$ from $x = -10$ to $x = 10$. (Note that $2\pi \approx 6.3$ for the graphs in Figs. 10-1 and 10-2.)

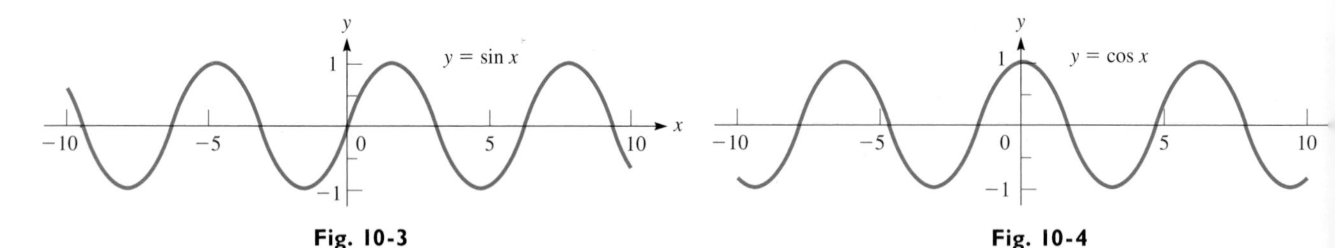

Fig. 10-3

Fig. 10-4

From these tables and graphs, it can be seen that *the graphs of $y = \sin x$ and $y = \cos x$ are of exactly the same shape (called* **sinusoidal***), with the cosine curve displaced $\pi/2$ units to the left of the sine curve.* The shape of these curves should be recognized readily, with special note as to the points at which they cross the axes. This information will be especially valuable in *sketching* similar curves, since the basic sinusoidal shape remains the same. It will not be necessary to plot numerous points every time we wish to sketch such a curve.

Amplitude

To obtain the graph of $y = a \sin x$, we note that all the y-values obtained for the graph of $y = \sin x$ are to be multiplied by the number a. In this case the greatest value of the sine function is $|a|$. *The number $|a|$ is called the* **amplitude** *of the curve and represents the greatest y-value of the curve.* Also, the curve will have no value less than $-|a|$. This is true for $y = a \cos x$ as well as for $y = a \sin x$.

EXAMPLE 1 ▸▸ Plot the graph of $y = 2 \sin x$.

Since $a = 2$, the amplitude of this curve is $|2| = 2$. This means that the maximum value of y is 2 and the minimum value is $y = -2$. The table of values follows, and the curve is shown in Fig. 10-5.

x	0	$\frac{\pi}{6}$	$\frac{\pi}{3}$	$\frac{\pi}{2}$	$\frac{2\pi}{3}$	$\frac{5\pi}{6}$	π
y	0	1	1.73	2	1.73	1	0

x	$\frac{7\pi}{6}$	$\frac{4\pi}{3}$	$\frac{3\pi}{2}$	$\frac{5\pi}{3}$	$\frac{11\pi}{6}$	2π
y	−1	−1.73	−2	−1.73	−1	0

◂◂

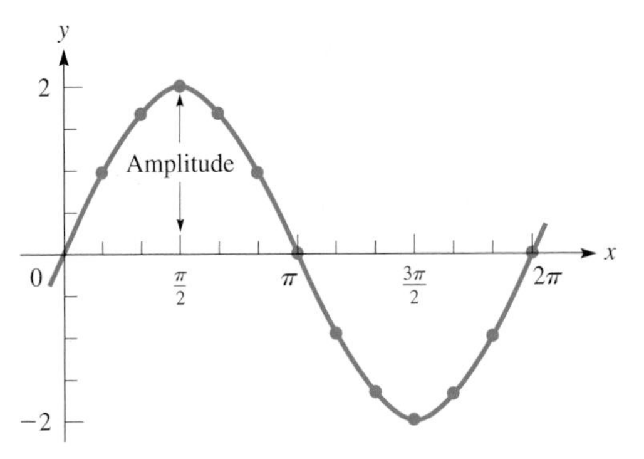

Fig. 10-5

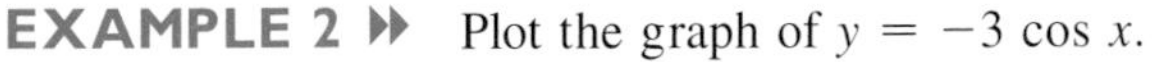

EXAMPLE 2 ▸▸ Plot the graph of $y = -3 \cos x$.

In this case $a = -3$, and this means that the amplitude is $|-3| = 3$. Therefore, the maximum value of y is 3, and the minimum value of y is -3. The table of values follows, and the curve is shown in Fig. 10-6.

x	0	$\frac{\pi}{6}$	$\frac{\pi}{3}$	$\frac{\pi}{2}$	$\frac{2\pi}{3}$	$\frac{5\pi}{6}$	π
y	−3	−2.6	−1.5	0	1.5	2.6	3

x	$\frac{7\pi}{6}$	$\frac{4\pi}{3}$	$\frac{3\pi}{2}$	$\frac{5\pi}{3}$	$\frac{11\pi}{6}$	2π
y	2.6	1.5	0	−1.5	−2.6	−3

◂◂

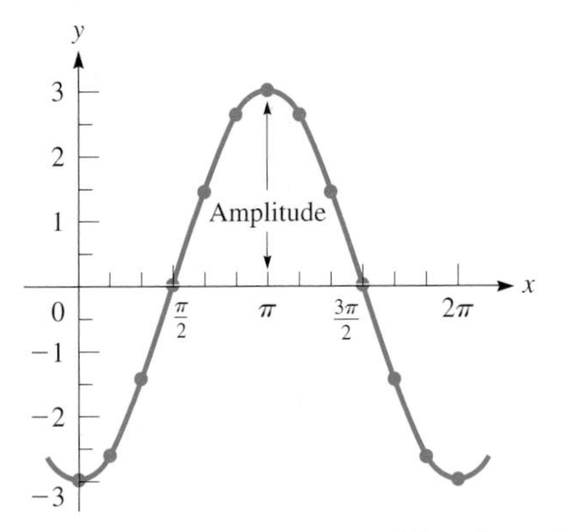

Fig. 10-6

Note from Example 2 that *the effect of the negative sign before the number a is to* ***invert*** *the curve about the x-axis.* The effect of the number a can also be seen readily from these examples.

From the previous examples we see that the function $y = a \sin x$ has zeros for $x = 0, \pi, 2\pi$ and that it has its maximum or minimum values for $x = \pi/2, 3\pi/2$. The function $y = a \cos x$ has its zeros for $x = \pi/2, 3\pi/2$ and its maximum or minimum values for $x = 0, \pi, 2\pi$. This is summarized in Table 10-1. Therefore, by knowing the general shape of the sine curve, where it has its zeros, and what its amplitude is, *we can rapidly* ***sketch*** *curves of the form $y = a \sin x$ and $y = a \cos x$.*

TABLE 10-1

	$x = 0, \pi, 2\pi$	$\frac{\pi}{2}, \frac{3\pi}{2}$
$y = a \sin x$	zeros	max. or min.
$y = a \cos x$	max. or min.	zeros

Since the graphs of $y = a \sin x$ and $y = a \cos x$ can extend indefinitely to the right and to the left, we see that the domain of each is all real numbers. We should note that the key values of $x = 0, \pi/2, \pi, 3\pi/2$, and 2π are those only for x from 0 to 2π. Corresponding values ($x = 5\pi/2, 3\pi$, and their negatives) could also be used. Also from the graphs we can readily see that the range of these functions is $-|a| \leq f(x) \leq |a|$.

EXAMPLE 3 ▸▸ Sketch the graph of $y = 4\cos x$.

First we set up a table of values for the points where the curve has its zeros, maximum points, and minimum points.

x	0	$\frac{\pi}{2}$	π	$\frac{3\pi}{2}$	2π
y	4	0	-4	0	4
	max.		min.		max.

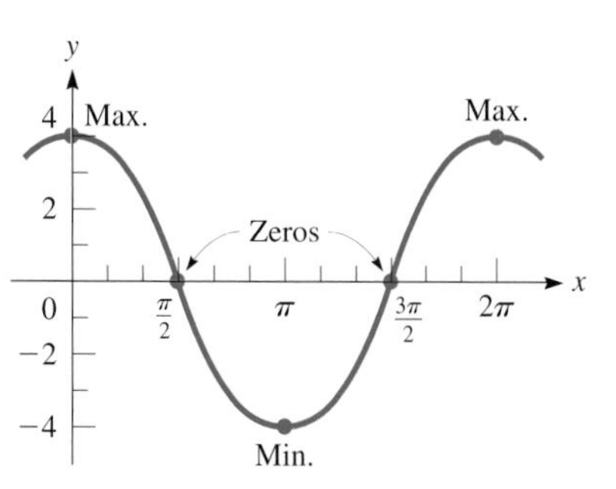

Fig. 10-7

Now we plot these points and join them, knowing the basic sinusoidal shape of the curve. See Fig. 10-7. ◂◂

The graphs of $y = a\sin x$ and $y = a\cos x$ can be displayed easily on a graphing calculator. In the next example we use the graphing calculator, and we will see that the calculator displays the features of the curve we should expect from our previous discussion.

EXAMPLE 4 ▸▸ Display the graph of $y = -2\sin x$ on a graphing calculator.

We can now quickly list the important values of this curve.

x	0	$\frac{\pi}{2}$	π	$\frac{3\pi}{2}$	2π
y	0	-2	0	2	0
		min.		max.	

Fig. 10-8

Knowing these values, we choose values for the RANGE feature as Xmin $= -1$, Xmax $= 7$, Ymin $= -2$, and Ymax $= 2$. *Using radian mode,* the graphing calculator view of this graph is shown in Fig. 10-8. We see the amplitude of 2 and the effect of the negative sign, as expected.

We can show that the curve continues indefinitely to the right and to the left by choosing Xmin values farther to the left and Xmax values farther to the right. ◂◂

Exercises 10–1

In Exercises 1–4, complete the following table for the given functions, and then plot the resulting graphs.

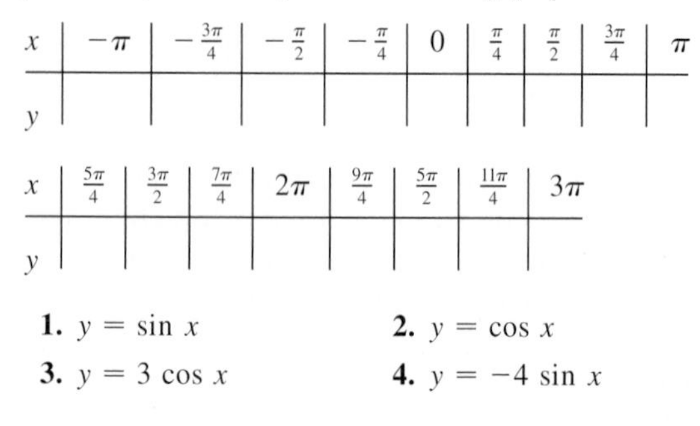

x	$-\pi$	$-\frac{3\pi}{4}$	$-\frac{\pi}{2}$	$-\frac{\pi}{4}$	0	$\frac{\pi}{4}$	$\frac{\pi}{2}$	$\frac{3\pi}{4}$	π
y									

x	$\frac{5\pi}{4}$	$\frac{3\pi}{2}$	$\frac{7\pi}{4}$	2π	$\frac{9\pi}{4}$	$\frac{5\pi}{2}$	$\frac{11\pi}{4}$	3π
y								

1. $y = \sin x$ **2.** $y = \cos x$

3. $y = 3\cos x$ **4.** $y = -4\sin x$

In Exercises 5–20, sketch the graphs of the given functions. Check each using a graphing calculator.

5. $y = 3\sin x$ **6.** $y = 5\sin x$

7. $y = \frac{5}{2}\sin x$ **8.** $y = 0.5\sin x$

9. $y = 2\cos x$ **10.** $y = 3\cos x$

11. $y = 0.8\cos x$ **12.** $y = \frac{3}{2}\cos x$

13. $y = -\sin x$ **14.** $y = -3\sin x$

15. $y = -1.5\sin x$ **16.** $y = -0.2\sin x$

17. $y = -\cos x$ **18.** $y = -8\cos x$

19. $y = -2.5\cos x$ **20.** $y = -0.4\cos x$

Although units of π are often convenient, we must remember that π is only a number. Numbers that are not multiples of π may be used. In Exercises 21–24, plot the indicated graphs by finding the values of y that correspond to values of x of 0, 1, 2, 3, 4, 5, 6, and 7 on a calculator. (Remember, the numbers 0, 1, 2, and so on represent radian measure.)

21. $y = \sin x$ **22.** $y = -3 \sin x$

23. $y = \cos x$ **24.** $y = 2 \cos x$

In Exercises 25–28, the graph of a function of the form $y = a \sin x$ or $y = a \cos x$ is shown. Determine the specific function for each.

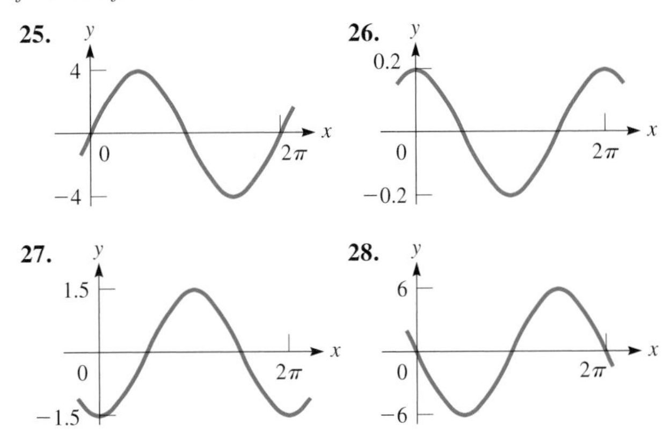

10-2 Graphs of $y = a \sin bx$ and $y = a \cos bx$

Period of a Function

In graphing the function $y = \sin x$, we see that the values of y repeat every 2π units of x. This is because $\sin x = \sin(x + 2\pi) = \sin(x + 4\pi)$, and so forth. For any function F, we say that it has a *period P* if $F(x) = F(x + P)$. For functions that are periodic, such as the sine and the cosine, *the* **period** *is the x-distance between a point and the next corresponding point for which the value of y repeats.*

Let us now plot the curve $y = \sin 2x$. This means that we choose a value for x, multiply this value by two, and find the sine of the result. This leads to the following table of values for this function.

x	0	$\frac{\pi}{8}$	$\frac{\pi}{4}$	$\frac{3\pi}{8}$	$\frac{\pi}{2}$	$\frac{5\pi}{8}$	$\frac{3\pi}{4}$	$\frac{7\pi}{8}$	π	$\frac{9\pi}{8}$	$\frac{5\pi}{4}$
$2x$	0	$\frac{\pi}{4}$	$\frac{\pi}{2}$	$\frac{3\pi}{4}$	π	$\frac{5\pi}{4}$	$\frac{3\pi}{2}$	$\frac{7\pi}{4}$	2π	$\frac{9\pi}{4}$	$\frac{5\pi}{2}$
y	0	0.7	1	0.7	0	−0.7	−1	−0.7	0	0.7	1

Plotting these points, we have the curve shown in Fig. 10-9.

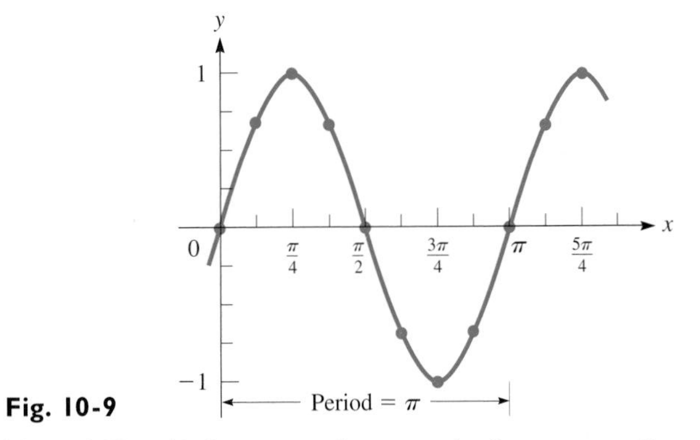

Fig. 10-9

From the table and Fig. 10-9, we see that $y = \sin 2x$ repeats after π units of x. The effect of the 2 is that the period of $y = \sin 2x$ is half the period of the curve of $y = \sin x$. We then conclude that if the period of a function $F(x)$ is P, then the period of $F(bx)$ is P/b. Since each of the functions $\sin x$ and $\cos x$ has a period of 2π, *each of the functions* $\sin bx$ *and* $\cos bx$ *has a period of* $2\pi/b$.

EXAMPLE 1 ▸▸ (a) The period of $\sin 3x$ is $\dfrac{2\pi}{3}$, which means that the curve of the function $y = \sin 3x$ will repeat every $\frac{2\pi}{3}$ (approximately 2.09) units of x.

(b) The period of $\cos 4x$ is $\dfrac{2\pi}{4} = \dfrac{\pi}{2}$.

(c) The period of $\sin \frac{1}{2}x$ is $\dfrac{2\pi}{\frac{1}{2}} = 4\pi$. In this case we see that the period is longer than that of the basic sine curve. ◂◂

EXAMPLE 2 ▸▸ (a) The period of $\sin \pi x$ is $2\pi/\pi = 2$. That is, the curve of the function $\sin \pi x$ repeats every 2 units.

(b) The periods of $\sin 3x$ and $\sin \pi x$ are nearly equal, since π is only slightly greater than 3.

(c) The period of $\cos 3\pi x$ is $\frac{2\pi}{3\pi} = \frac{2}{3}$.

(d) The period of $\sin \frac{\pi}{4}x$ is $2\pi/\frac{\pi}{4} = 8$. ◂◂

Combining the value of the period with the value of the amplitude from Section 10-1, we conclude that *the functions $y = a \sin bx$ and $y = a \cos bx$ each has an amplitude of $|a|$ and a period of $2\pi/b$.* These properties are very useful in sketching these functions.

EXAMPLE 3 ▸▸ Sketch the graph of $y = 3 \sin 4x$ for $0 \le x \le \pi$.

We immediately see that the amplitude is 3 and the period is $2\pi/4 = \pi/2$. Thus, we know that for $x = 0$, then $y = 0$, and for $x = \pi/2$, then $y = 0$. Also, the sine function is zero halfway between $x = 0$ and $x = \pi/2$, which means that for $x = \pi/4$, then $y = 0$. The function reaches its maximum or minimum values halfway between the zeros. This means that for $x = \pi/8$, then $y = 3$, and when $x = 3\pi/8$, then $y = -3$. A table for these important values follows. Note that the values of x that are shown are those for which $4x = 0, \pi/2, \pi, 3\pi/2, 2\pi$, and so on.

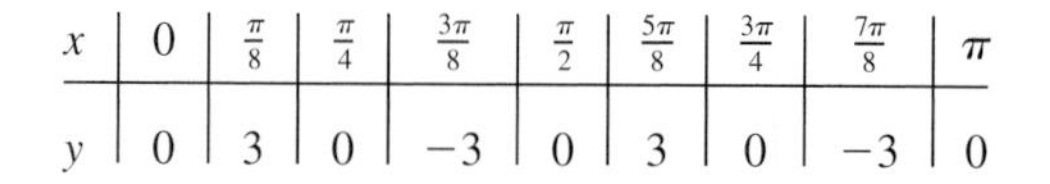

x	0	$\frac{\pi}{8}$	$\frac{\pi}{4}$	$\frac{3\pi}{8}$	$\frac{\pi}{2}$	$\frac{5\pi}{8}$	$\frac{3\pi}{4}$	$\frac{7\pi}{8}$	π
y	0	3	0	−3	0	3	0	−3	0

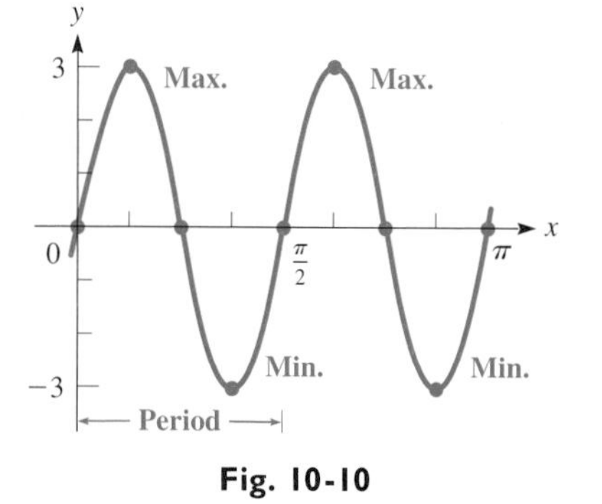

Fig. 10-10

Using the table and the sinusoidal form of the curve, we sketch the function in Fig. 10-10. ◂◂

We see from Example 3 that *an important distance in sketching a sine curve or a cosine curve is one-fourth of the period.* For $y = a \sin bx$, it is one-fourth of the period from the origin to the first value of x where y is at its maximum (or minimum) value. Then we proceed another one-fourth period to a zero, another one-fourth period to the next minimum (or maximum) value, another to the next zero (this is where the period is completed), and so on. Thus,

NOTE ▸ *by finding one-fourth of the period, we can easily find the important values for sketching the curve.*

Similarly, one-fourth of the period is used in sketching $y = a \cos bx$.

> ***Important Values for Sketching $y = a$ sin bx and $y = a$ cos bx***
>
> **1.** *The amplitude:* $|a|$
>
> **2.** *The period:* $2\pi/b$
>
> **3.** *Values of the function for each one-fourth period*

EXAMPLE 4 ▸▸ Sketch the graph of $y = -2 \cos 3x$ for $0 \le x \le 2\pi$.

We note that the amplitude is 2 and the period is $\frac{2\pi}{3}$. This means that one-fourth of the period is $\frac{1}{4} \times \frac{2\pi}{3} = \frac{\pi}{6}$. Since the cosine curve is at a maximum or minimum for $x = 0$, we find that $y = -2$ for $x = 0$ (the negative value is due to the minus sign before the function), which means it is a minimum point. The curve then has a zero at $x = \frac{\pi}{6}$, a maximum value of 2 at $x = 2(\frac{\pi}{6}) = \frac{\pi}{3}$, a zero at $x = 3(\frac{\pi}{6}) = \frac{\pi}{2}$, and its next value of -2 at $x = 4(\frac{\pi}{6}) = \frac{2\pi}{3}$, and so on. Therefore, we have the following table.

x	0	$\frac{\pi}{6}$	$\frac{\pi}{3}$	$\frac{\pi}{2}$	$\frac{2\pi}{3}$	$\frac{5\pi}{6}$	π	$\frac{7\pi}{6}$	$\frac{4\pi}{3}$	$\frac{3\pi}{2}$	$\frac{5\pi}{3}$	$\frac{11\pi}{6}$	2π
y	-2	0	2	0	-2	0	2	0	-2	0	2	0	-2

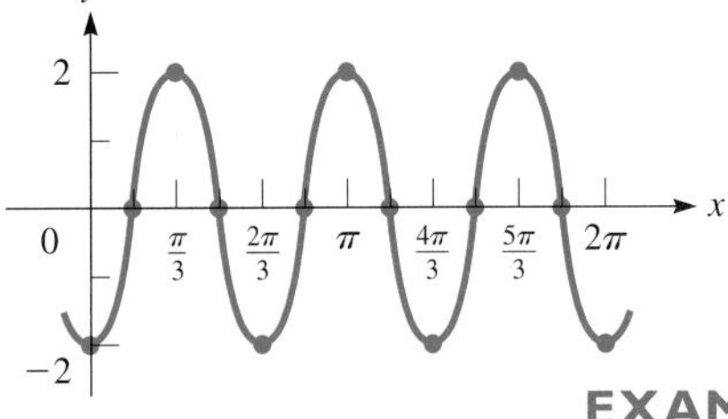

Fig. 10-11

Using this table and the sinusoidal shape of the cosine curve, we sketch the function of Fig. 10-11. ◂◂

EXAMPLE 5 ▸▸ A generator produces a voltage $V = 200 \cos 50\pi t$, where t is the time in seconds (50π has units of rad/s; thus, $50\pi t$ is an angle in radians). Use a graphing calculator to display the graph of V as a function of t for $0 \le t \le 0.06$ s.

Fig. 10-12

The amplitude is 200 V and the period is $\frac{2\pi}{50\pi} = 0.04$ s. Since the period is not in terms of π, it is more convenient to use decimal units for t rather than to use units in terms of π as in the previous graphs. Thus, we have the following table of values:

t (seconds)	0	0.01	0.02	0.03	0.04	0.05	0.06
V (volts)	200	0	-200	0	200	0	-200

For the graphing calculator, we use x for t and y for V. From the amplitude of 200 V and the above table, we choose values for the RANGE feature of Xmin = 0, Xmax = 0.06, Ymin = -200, and Ymax = 200. We do not consider negative values of t, for they have no real meaning in this problem. The view on the graphing calculator is shown in Fig. 10-12. ◂◂

Exercises 10–2

In Exercises 1–20, find the period of each function.

1. $y = 2 \sin 6x$

2. $y = 4 \sin 2x$

3. $y = 3 \cos 8x$

4. $y = \cos 10x$

5. $y = -2 \sin 12x$

6. $y = -\sin 5x$

7. $y = -\cos 16x$

8. $y = -4 \cos 2x$

9. $y = 5 \sin 2\pi x$

10. $y = 2 \sin 3\pi x$

11. $y = 3 \cos 4\pi x$

12. $y = 4 \cos 10\pi x$

13. $y = 3 \sin \frac{1}{3}x$

14. $y = -2 \sin \frac{2}{5}x$

15. $y = -\frac{1}{2} \cos \frac{2}{3}x$

16. $y = \frac{1}{3} \cos \frac{1}{4}x$

17. $y = 0.4 \sin \dfrac{2\pi x}{3}$

18. $y = 1.5 \cos \dfrac{\pi x}{10}$

19. $y = 3.3 \cos \pi^2 x$

20. $y = 2.5 \sin \dfrac{2x}{\pi}$

In Exercises 21–40, sketch the graphs of the given functions. Check each using a graphing calculator. (These are the same functions as in Exercises 1–20.)

21. $y = 2 \sin 6x$ **22.** $y = 4 \sin 2x$

23. $y = 3 \cos 8x$ **24.** $y = \cos 10x$

25. $y = -2 \sin 12x$ **26.** $y = -\sin 5x$

27. $y = -\cos 16x$ **28.** $y = -4 \cos 2x$

29. $y = 5 \sin 2\pi x$ **30.** $y = 2 \sin 3\pi x$

31. $y = 3 \cos 4\pi x$ **32.** $y = 4 \cos 10\pi x$

33. $y = 3 \sin \frac{1}{3}x$ **34.** $y = -2 \sin \frac{2}{5}x$

35. $y = -\frac{1}{2} \cos \frac{2}{3}x$ **36.** $y = \frac{1}{3} \cos \frac{1}{4}x$

37. $y = 0.4 \sin \dfrac{2\pi x}{3}$ **38.** $y = 1.5 \cos \dfrac{\pi x}{10}$

39. $y = 3.3 \cos \pi^2 x$ **40.** $y = 2.5 \sin \dfrac{2x}{\pi}$

In Exercises 41–44, the period is given for a function of the form $y = \sin bx$. Write the function corresponding to the given value of the period.

41. $\dfrac{\pi}{3}$ **42.** $\dfrac{2\pi}{5}$ **43.** 2 **44.** 6

In Exercises 45–48, sketch the indicated graphs.

45. The standard electric voltage in a 60-Hz alternating-current circuit is given by $V = 170 \sin 120\pi t$, where t is the time in seconds. Sketch the graph of V as a function of t for $0 \le t \le 0.05$ s.

46. The end of a tuning fork moves with a displacement given by $y = 1.60 \cos 460\pi t$, where y is in millimetres and t is in seconds. Sketch the graph of y as a function of t for $0 \le t \le 0.02$ s.

47. The velocity of a piston is given by $v = 450 \cos 3600t$, where v is in centimetres per second and t is in seconds. Sketch the graph of v as a function of t for $0 \le t \le 0.006$ s.

48. The displacement y of the end of a robot arm for welding is given by $y = 12.75 \sin 0.419t$, where y is in metres and t is in seconds. Sketch the graph of y as a function of t for $0 \le t \le 15$ s.

In Exercises 49–52, the graph of a function of the form $y = a \sin bx$ or $y = a \cos bx$ is shown. Determine the specific function for each.

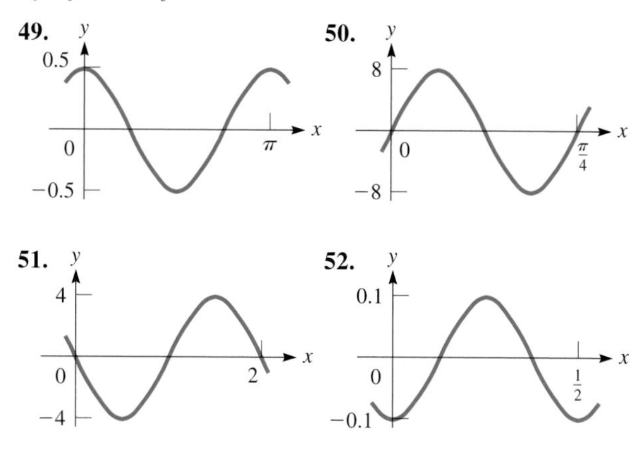

10-3 Graphs of $y = a \sin(bx + c)$ and $y = a \cos(bx + c)$

Another important quantity in graphing the sine and cosine functions is the *phase angle.* In the function $y = a \sin(bx + c)$, *c represents the* **phase angle.** Its meaning is illustrated in the following example.

EXAMPLE 1 ▸▸ Sketch the graph of $y = \sin(2x + \frac{\pi}{4})$.

Note that $c = \pi/4$. Therefore, in order to obtain values for the table, we assume a value for x, multiply it by 2, add $\pi/4$ to this value, and then find the sine of the result. The values that are shown are those for which $2x + \pi/4 = 0, \pi/4, \pi/2, 3\pi/4, \pi$, and so on, which are the important values for $y = \sin 2x$.

x	$-\frac{\pi}{8}$	0	$\frac{\pi}{8}$	$\frac{\pi}{4}$	$\frac{3\pi}{8}$	$\frac{\pi}{2}$	$\frac{5\pi}{8}$	$\frac{3\pi}{4}$	$\frac{7\pi}{8}$	π
y	0	0.7	1	0.7	0	−0.7	−1	−0.7	0	0.7

When we solve $2x + \pi/4 = 0$, we get $x = -\pi/8$, and this gives $y = \sin 0 = 0$. The other values for y are found in the same way. The graph is shown in Fig. 10-13. ◂◂

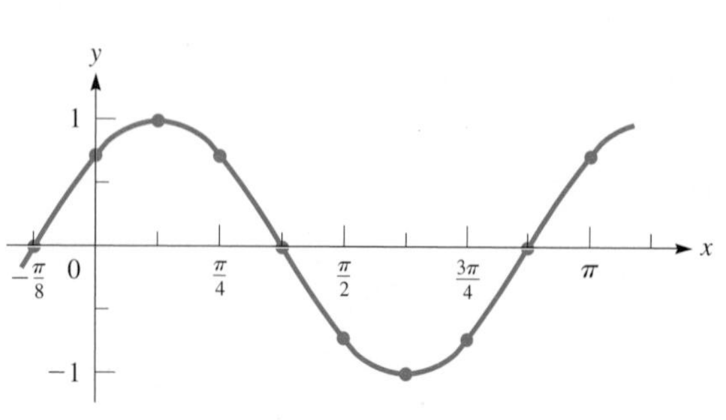

Fig. 10-13

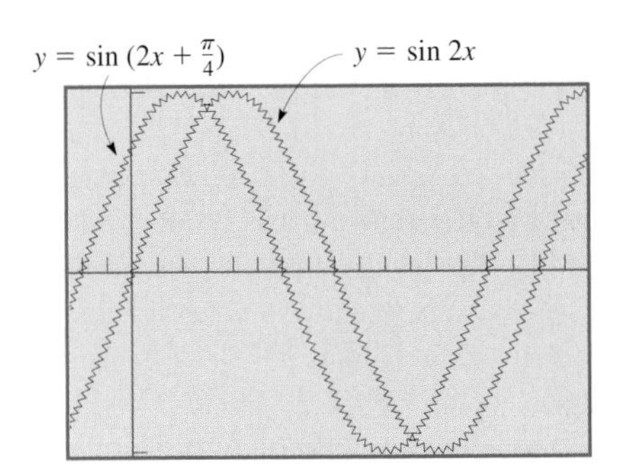

Fig. 10-14

We see from Example 1 that *the graph of* $y = \sin(2x + \frac{\pi}{4})$ *is precisely the same as the graph of* $y = \sin 2x$, *except that it is* ***shifted*** $\pi/8$ *units to the left.* In Fig. 10-14 a graphing calculator view shows the graphs of $y = \sin 2x$ and $y = \sin(2x + \frac{\pi}{4})$. We see that the shapes of the curves are the same and that the graph of $y = \sin(2x + \frac{\pi}{4})$ is about 0.4 unit ($\pi/8 \approx 0.39$) to the left of the graph of $y = \sin 2x$. (The RANGE values we used for x are Xmin $= -0.5$, Xmax $= 3.5$, and Xscl $= 0.2$.)

In general, the effect of c in the equation $y = a\sin(bx + c)$ is to shift the curve of $y = a\sin bx$ to the left if $c > 0$, or shift the curve to the right if $c < 0$. The amount of this shift is given by $-c/b$. Due to its importance in sketching curves, *the quantity* $-c/b$ *is called the* **displacement** (*or* **phase shift**).

We can see the reason that the displacement is $-c/b$ by noting corresponding points on the graphs of $y = \sin bx$ and $y = \sin(bx + c)$. For $y = \sin bx$, when $x = 0$, then $y = 0$. For $y = \sin(bx + c)$, when $x = -c/b$, then $y = 0$. The point $(-c/b, 0)$ on the graph of $y = \sin(bx + c)$ is $-c/b$ units to the left of the point $(0, 0)$ on the graph of $y = \sin x$. In Fig. 10-14, $-c/b = -\pi/8$.

Therefore, we use the displacement combined with the amplitude and period along with the other information from the previous sections to sketch curves of the functions $y = a\sin(bx + c)$ and $y = a\cos(bx + c)$ where $b > 0$.

Important Quantities to Determine for Sketching Graphs of $y = a\sin(bx + c)$ ***and*** $y = a\cos(bx + c)$

$$\text{Amplitude} = |a|$$

$$\text{Period} = \frac{2\pi}{b} \qquad \textbf{(10-1)}$$

$$\text{Displacement} = -\frac{c}{b}$$

By use of these quantities and the one-fourth period distance, the graphs of the sine and cosine functions can be readily sketched. A general illustration of the graph of $y = a\sin(bx + c)$ is shown in Fig. 10-15. Note that

CAUTION ▶

the displacement is **negative** *(to the left) for* $c > 0$ (Fig. 10-15(a))
the displacement is **positive** *(to the right) for* $c < 0$ (Fig. 10-15(b))

Note that we can find the displacement for the graphs of $y = a\sin(bx + c)$ by solving $bx + c = 0$ for x. We see that $x = -c/b$.

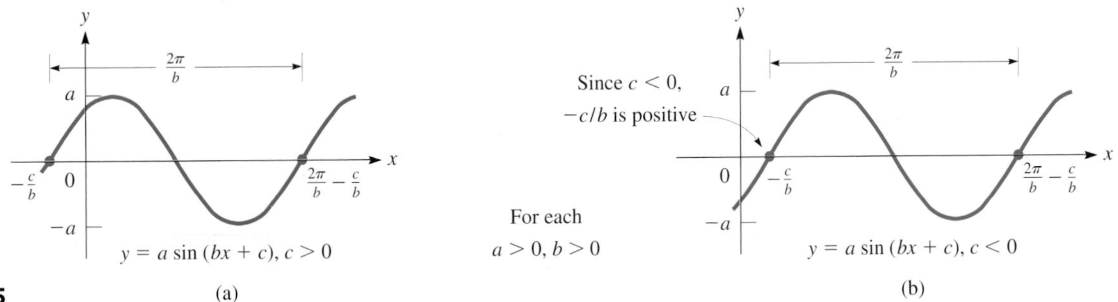

Fig. 10-15

EXAMPLE 2 ▸▸ Sketch the graph of $y = 2\sin(3x - \pi)$.

First we note that $a = 2$, $b = 3$, and $c = -\pi$. Therefore, the amplitude is 2, the period is $2\pi/3$, and the displacement is $-(-\pi/3) = \pi/3$. (We can also get the displacement from $3x - \pi = 0$, $x = \pi/3$.)

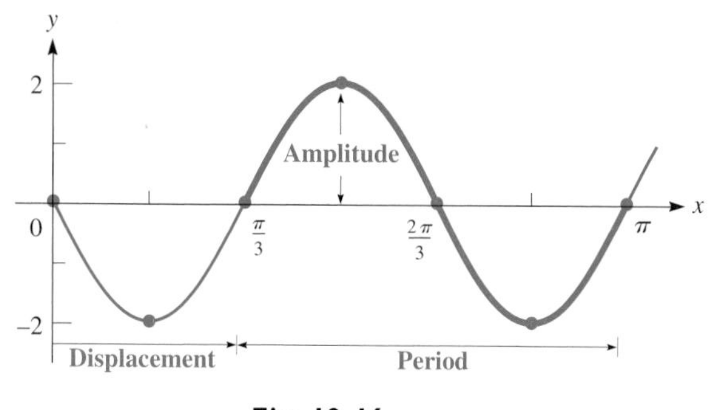

Fig. 10-16

We see that the curve "starts" at $x = \pi/3$ and starts repeating $2\pi/3$ units to the right of this point. Be sure to grasp this point well. *The period tells us the number of units along the x-axis between such corresponding points.* One-fourth of the period is $\frac{1}{4}(\frac{2\pi}{3}) = \frac{\pi}{6}$.

Important values are at $\frac{\pi}{3}$, $\frac{\pi}{3} + \frac{\pi}{6} = \frac{\pi}{2}$, $\frac{\pi}{3} + 2(\frac{\pi}{6}) = \frac{2\pi}{3}$, and so on. We now make the table of important values and sketch the graph shown in Fig. 10-16.

x	0	$\frac{\pi}{6}$	$\frac{\pi}{3}$	$\frac{\pi}{2}$	$\frac{2\pi}{3}$	$\frac{5\pi}{6}$	π
y	0	−2	0	2	0	−2	0

We note that since the period is $2\pi/3$, the curve passes through the origin. ◂◂

EXAMPLE 3 ▸▸ Sketch the graph of the function $y = -\cos(2x + \frac{\pi}{6})$.

First we determine that

(1) the amplitude is 1

(2) the period is $\frac{2\pi}{2} = \pi$

(3) the displacement is $-\frac{\pi}{6} \div 2 = -\frac{\pi}{12}$

We now make a table of important values, noting that the curve starts repeating π units to the right of $-\frac{\pi}{12}$.

x	$-\frac{\pi}{12}$	$\frac{\pi}{6}$	$\frac{5\pi}{12}$	$\frac{2\pi}{3}$	$\frac{11\pi}{12}$
y	−1	0	1	0	−1

From this table we sketch the graph in Fig. 10-17. ◂◂

Fig. 10-17

Each of the heavy portions of the graphs in Figs. 10-16 and 10-17 is called a *cycle* of the curve. *A* **cycle** *is any section of the graph that includes exactly one period.*

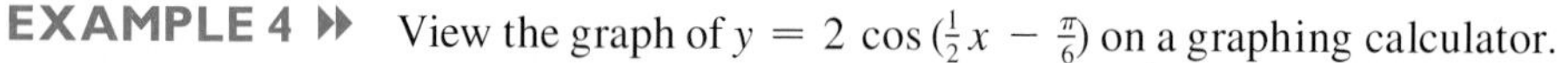

EXAMPLE 4 ▸▸ View the graph of $y = 2\cos(\frac{1}{2}x - \frac{\pi}{6})$ on a graphing calculator.

From the values $a = 2$, $b = 1/2$, and $c = -\pi/6$, we determine that

(1) the amplitude is 2

(2) the period is $2\pi \div \frac{1}{2} = 4\pi$

(3) the displacement is $-(-\frac{\pi}{6}) \div \frac{1}{2} = \frac{\pi}{3}$

We now make a table of important values.

x	$\frac{\pi}{3}$	$\frac{4\pi}{3}$	$\frac{7\pi}{3}$	$\frac{10\pi}{3}$	$\frac{13\pi}{3}$
y	2	0	−2	0	2

Fig. 10-18

This table helps us choose the following values for the RANGE feature.

Xmin = −1 (to start to the left of the y-axis)

Xmax = 14 ($13\pi/3 \approx 13.6$)

Ymin = −2 and Ymax = 2

The graphing calculator view is shown in Fig. 10-18. We see that this shows a little more than one cycle. ◂◂

The following example illustrates the use of the graph of a trigonometric function in an applied problem.

EXAMPLE 5 ▸▸ The cross section of a certain water wave is $y = 0.7 \sin\left(\frac{\pi}{2}x + \frac{\pi}{4}\right)$, where x and y are measured in metres. Display two cycles of y vs. x on a graphing calculator.

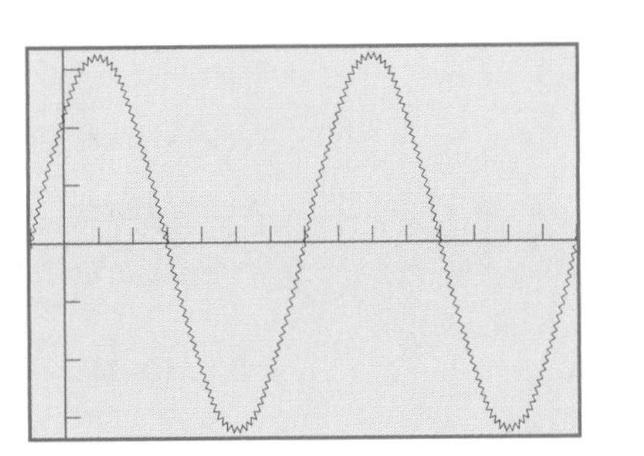

Fig. 10-19

From the values $a = 0.7$ m, $b = \pi/2\ \text{m}^{-1}$ (this means 1/m, or per metre), and $c = \pi/4$, we can find the amplitude, period, and displacement:

(1) amplitude $= 0.7$ m

(2) period $= \dfrac{2\pi}{\frac{\pi}{2}} = 4$ m

(3) displacement $= -\dfrac{\frac{\pi}{4}}{\frac{\pi}{2}} = -0.5$ m

We choose the following values for the RANGE feature:

(1) Xmin $= -0.5$ (the displacement is -0.5 m)

(2) Xmax $= 7.5$ (the period is 4 m, and we want two periods, starting at $x = -0.5$) $(-0.5 + 8 = 7.5)$

(3) Ymin $= -0.7$, Ymax $= 0.7$ (the amplitude is 0.7 m)

The graphing calculator view is shown in Fig. 10-19. The negative values of x have the significance of giving points to the wave to the left of the origin. (When *time* is used, no actual physical meaning is generally given to negative values of t.) ◂◂

Exercises 10–3

In Exercises 1–24, determine the amplitude, period, and displacement for each function. Then sketch the graphs of the functions. Check each using a graphing calculator.

1. $y = \sin\left(x - \dfrac{\pi}{6}\right)$ **2.** $y = 3\sin\left(x + \dfrac{\pi}{4}\right)$

3. $y = \cos\left(x + \dfrac{\pi}{6}\right)$ **4.** $y = 2\cos\left(x - \dfrac{\pi}{8}\right)$

5. $y = 2\sin\left(2x + \dfrac{\pi}{2}\right)$ **6.** $y = -\sin\left(3x - \dfrac{\pi}{2}\right)$

7. $y = -\cos(2x - \pi)$ **8.** $y = 4\cos\left(3x + \dfrac{\pi}{3}\right)$

9. $y = \dfrac{1}{2}\sin\left(\dfrac{1}{2}x - \dfrac{\pi}{4}\right)$ **10.** $y = 2\sin\left(\dfrac{1}{4}x + \dfrac{\pi}{2}\right)$

11. $y = 3\cos\left(\dfrac{1}{3}x + \dfrac{\pi}{3}\right)$ **12.** $y = \dfrac{1}{3}\cos\left(\dfrac{1}{2}x - \dfrac{\pi}{8}\right)$

13. $y = \sin\left(\pi x + \dfrac{\pi}{8}\right)$ **14.** $y = -2\sin(2\pi x - \pi)$

15. $y = \dfrac{3}{4}\cos\left(4\pi x - \dfrac{\pi}{5}\right)$ **16.** $y = 25\cos\left(3\pi x + \dfrac{\pi}{2}\right)$

17. $y = -0.6\sin(2\pi x - 1)$ **18.** $y = 1.8\sin\left(\pi x + \dfrac{1}{3}\right)$

19. $y = 40\cos(3\pi x + 2)$ **20.** $y = 3\cos(6\pi x - 1)$

21. $y = \sin(\pi^2 x - \pi)$ **22.** $y = -\dfrac{1}{2}\sin\left(2x - \dfrac{1}{\pi}\right)$

23. $y = -\dfrac{3}{2}\cos\left(\pi x + \dfrac{\pi^2}{6}\right)$

24. $y = \pi\cos\left(\dfrac{1}{\pi}x + \dfrac{1}{3}\right)$

In Exercises 25 and 26, sketch the indicated curves. In Exercises 27 and 28, use a graphing calculator to view the indicated curves.

25. A wave traveling in a string may be represented by the equation $y = A\sin 2\pi\left(\dfrac{t}{T} - \dfrac{x}{\lambda}\right)$. Here A is the amplitude, t is the time the wave has traveled, x is the distance from the origin, T is the time required for the wave to travel one *wavelength* λ (the Greek letter lambda). Sketch three cycles of the wave for which $A = 2.00$ cm, $T = 0.100$ s, $\lambda = 20.0$ cm, and $x = 5.00$ cm.

26. The electric current i, in microamperes, in a certain circuit is given by $i = 3.8\cos 2\pi(t + 0.20)$, where t is the time in seconds. Sketch three cycles of this function.

27. A certain satellite circles the earth such that its distance y, in kilometres north or south (altitude is not considered) from the equator, is $y = 7200 \cos(0.025t - 0.25)$, where t is the time (in min) after launch. View two cycles of the graph on a graphing calculator.

28. In performing a test on a patient, a medical technician used an ultrasonic signal given by the equation $I = A \sin(\omega t + \theta)$. On a graphing calculator, view two cycles of the graph of I vs. t if $A = 5$ nW/m^2, $\omega = 2 \times 10^5$ rad/s, and $\theta = 0.4$.

In Exercises 29–32, give the specific form of the given equation by evaluating a, b, and c through an inspection of the given curve. Explain how a, b, and c are found.

29. $y = a \sin(bx + c)$ Fig. 10-20

30. $y = a \cos(bx + c)$ Fig. 10-20

31. $y = a \cos(bx + c)$ Fig. 10-21

32. $y = a \sin(bx + c)$ Fig. 10-21

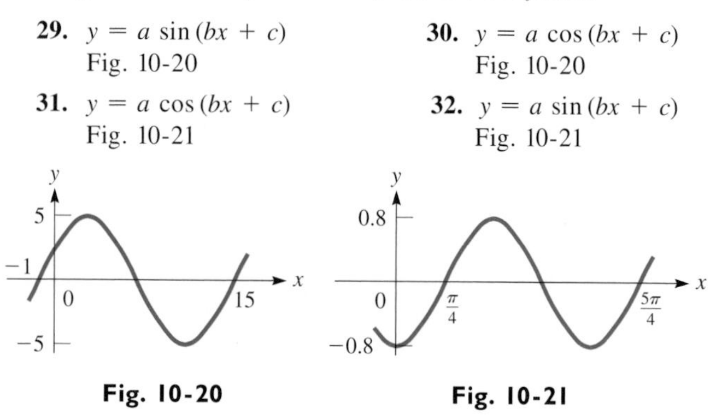

Fig. 10-20 **Fig. 10-21**

10-4 Graphs of y = tan x, y = cot x, y = sec x, y = csc x

In this section we briefly consider the graphs of the other trigonometric functions. We show the basic form of each curve, and then from these we are able to sketch other curves for these functions.

Considering the values of the trigonometric functions we found in Chapter 8, we set up the following table for $y = \tan x$. The graph is shown in Fig. 10-22.

x	0	$\frac{\pi}{6}$	$\frac{\pi}{3}$	$\frac{\pi}{2}$	$\frac{2\pi}{3}$	$\frac{5\pi}{6}$	π
y	0	0.6	1.7	*	−1.7	−0.6	0

x	$\frac{7\pi}{6}$	$\frac{4\pi}{3}$	$\frac{3\pi}{2}$	$\frac{5\pi}{3}$	$\frac{11\pi}{6}$	2π
y	0.6	1.7	*	−1.7	−0.6	0

*Undefined.

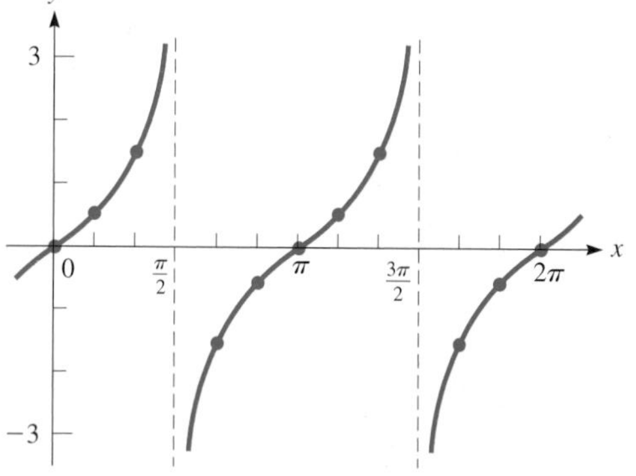

Fig. 10-22

Since the curve is not defined for $x = \pi/2$, $x = 3\pi/2$, and so forth, we use a calculator and find that the value of $\tan x$ becomes very large as x gets closer to $\pi/2$, although *there is no point on the curve for* $x = \pi/2$. For example, if $x = 1.5$, $\tan x = 14.1$ ($\pi/2 \approx 1.57$). We note that *the period of the tangent curve is* π. This differs from the period of the sine and cosine functions.

By knowing the values of $\sin x$, $\cos x$, and $\tan x$, we can find the necessary values of $\csc x$, $\sec x$, and $\cot x$. This is due to the reciprocal relationships among the functions that we showed in Section 4-3. We show these relationships by

$$\csc x = \frac{1}{\sin x} \qquad \sec x = \frac{1}{\cos x} \qquad \cot x = \frac{1}{\tan x} \qquad (10\text{-}2)$$

Thus, to graph $y = \cot x$, $y = \sec x$, and $y = \csc x$, we can obtain the necessary values from the corresponding reciprocal function.

In Figs. 10-23 to 10-26, the graphs of $y = \tan x$, $y = \cot x$, $y = \sec x$, and $y = \csc x$ are shown. The graph of $y = \tan x$ is given again to show it more completely.

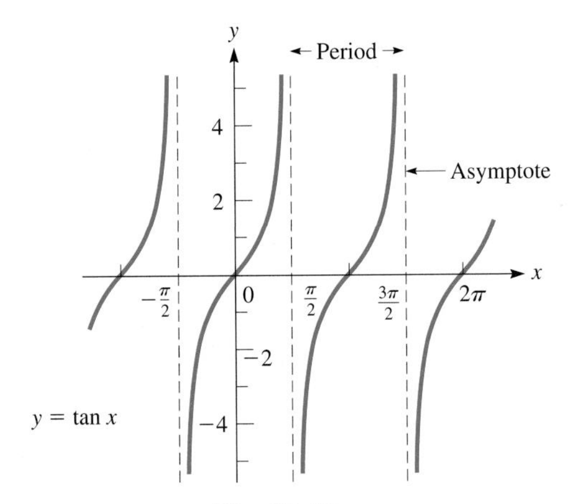

Fig. 10-23

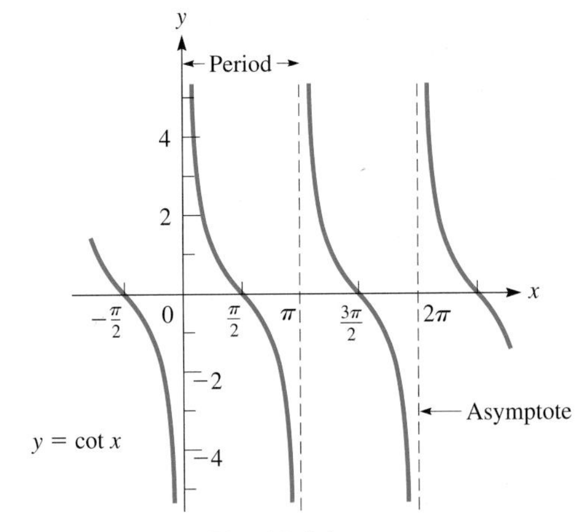

Fig. 10-24

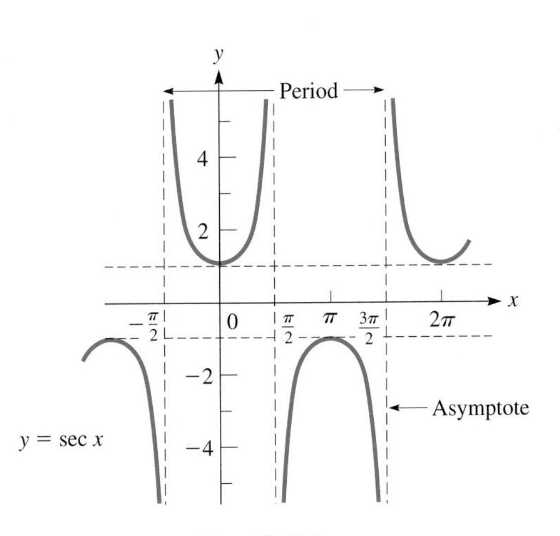

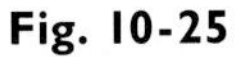
Fig. 10-25

Fig. 10-26

We see from these graphs that the period of $y = \tan x$ and $y = \cot x$ is π and that the period of $y = \sec x$ and $y = \csc x$ is 2π. *The vertical dashed lines in these figures are* **asymptotes** (see Sections 3-4 and 21-6). The curves *approach* these lines, but they never actually touch them.

The functions are not defined for the values of x for which the curve has asymptotes. This means that the domains do not include these values of x. Thus, we see that the domains of $y = \tan x$ and $y = \sec x$ include all real numbers, except the values $x = -\pi/2, \pi/2, 3\pi/2$, and so on. The domain of $y = \cot x$ and $y = \csc x$ include all real numbers except $x = -\pi, 0, \pi, 2\pi$, and so on.

From the graphs we see that the ranges of $y = \tan x$ and $y = \cot x$ are all real numbers, but that the ranges of $y = \sec x$ and $y = \csc x$ do not include the real numbers between -1 and 1.

To sketch functions such as $y = a \sec x$, we may first sketch $y = \sec x$ and then multiply the y-values by a. *Here a is not an amplitude,* since the ranges of these functions are not limited in the same way they are for the sine and cosine functions.

EXAMPLE 1 ▸▸ Sketch the graph of $y = 2 \sec x$.

First we sketch in $y = \sec x$, shown as the light curve in Fig. 10-27. Then we multiply the approximate y-values of the secant function by 2. Although we can do this only approximately, a reasonable graph can be sketched this way. The desired curve is shown in color in Fig. 10-27. ◂◂

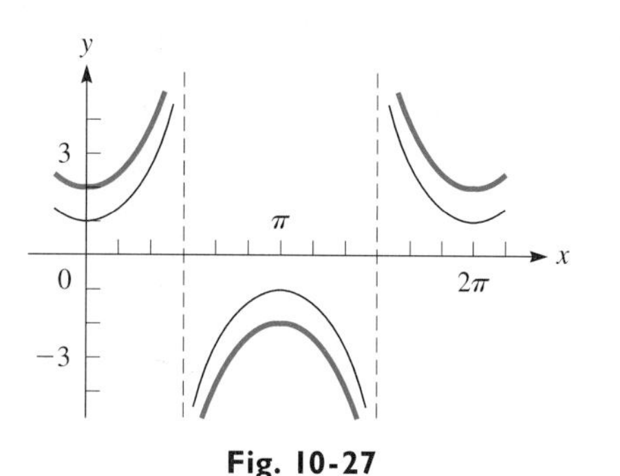

Fig. 10-27

Using a graphing calculator we can get displays of the graphs of these functions more easily and more accurately than by sketching them. By knowing the general shape and the period of the function, the values for the RANGE function can be determined without having to reset them too often.

EXAMPLE 2 ▸▸ View at least two cycles of the graph of $y = 0.5 \cot 2x$ on a graphing calculator.

Since the period of $y = \cot x$ is π, the period of $y = \cot 2x$ is $\pi/2$. Therefore, we set the RANGE values as follows:

Xmin = 0 ($x = 0$ is one asymptote of the curve)

Xmax = 3.2 ($\pi \approx 3.14$; the period is $\pi/2$; two periods is π)

Ymin = −5 (the range of cot x is all x; this should show enough of the curve)

Ymax = 5

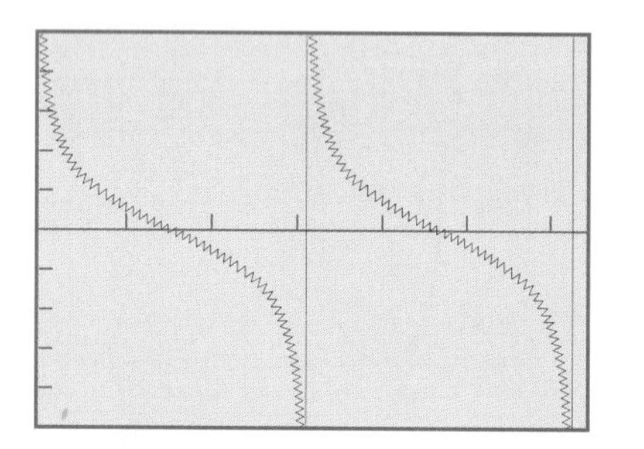

Fig. 10-28

Since there is no cot x key, we must remember to enter the function as $y = 0.5 (\tan 2x)^{-1}$, since $\cot x = (\tan x)^{-1}$. The graphing calculator view is shown in Fig. 10-28.

We can view many more cycles of the curve. Just be sure to get a good view with your choices of Xmin, Xmax, Ymin, and Ymax. ◂◂

EXAMPLE 3 ▸▸ View at least two periods of the graph of $y = 2 \sec (2x - \frac{\pi}{4})$ on a graphing calculator.

Since the period of sec x is 2π, the period of $\sec (2x - \frac{\pi}{4})$ is $2\pi/2 = \pi$. Recalling that $\sec x = (\cos x)^{-1}$, the curve will have the same displacement as $y = \cos (2x - \frac{\pi}{4})$. This displacement is $-\frac{-\pi/4}{2} = \frac{\pi}{8}$. Therefore, we choose these RANGE values:

Xmin = 0 (the displacement is positive)

Xmax = 7 (displacement = $\pi/8$; period = π; $\pi/8 + 2\pi = 17\pi/8 \approx 6.7$)

Ymin = −6 (there is no curve between $y = -2$ and $y = 2$)

Ymax = 6

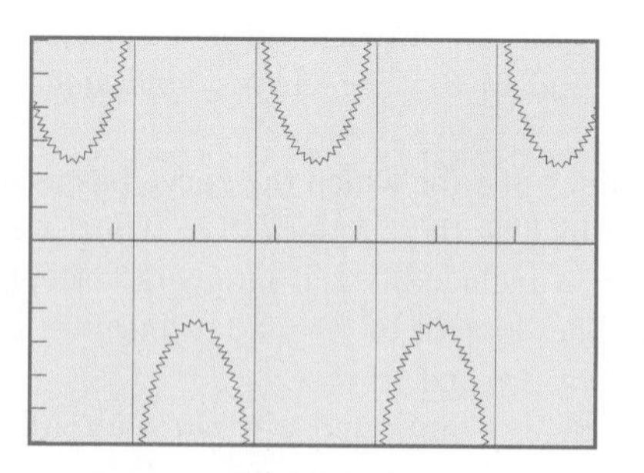

Fig. 10-29

We enter the function as $y = 2(\cos (2x - \pi/4))^{-1}$. The graphing calculator view is shown in Fig. 10-29. ◂◂

Exercises 10–4

In Exercises 1–4, fill in the following table for each function and plot the graph from these points.

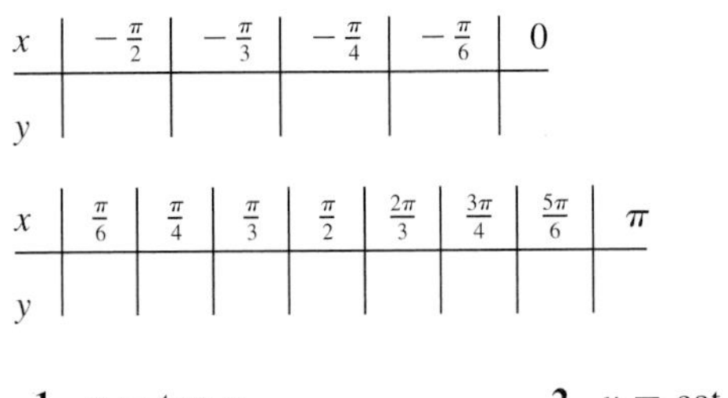

x	$-\frac{\pi}{2}$	$-\frac{\pi}{3}$	$-\frac{\pi}{4}$	$-\frac{\pi}{6}$	0
y					

x	$\frac{\pi}{6}$	$\frac{\pi}{4}$	$\frac{\pi}{3}$	$\frac{\pi}{2}$	$\frac{2\pi}{3}$	$\frac{3\pi}{4}$	$\frac{5\pi}{6}$	π
y								

1. $y = \tan x$

2. $y = \cot x$

3. $y = \sec x$

4. $y = \csc x$

In Exercises 5–12, sketch the graphs of the given functions by use of the basic curve forms (Figs. 10-23, 10-24, 10-25, 10-26). See Example 1.

5. $y = 2 \tan x$

6. $y = 3 \cot x$

7. $y = \frac{1}{2} \sec x$

8. $y = \frac{3}{2} \csc x$

9. $y = -2 \cot x$

10. $y = -\tan x$

11. $y = -3 \csc x$

12. $y = -\frac{1}{2} \sec x$

In Exercises 13–20, view at least two cycles of the graphs of the given functions on a graphing calculator.

13. $y = \tan 2x$

14. $y = 2 \cot 3x$

15. $y = \frac{1}{2} \sec 3x$

16. $y = 4 \csc 2x$

17. $y = 2 \cot\left(2x + \dfrac{\pi}{6}\right)$

18. $y = \tan\left(3x - \dfrac{\pi}{2}\right)$

19. $y = \csc\left(3x - \dfrac{\pi}{3}\right)$

20. $y = 3 \sec\left(2x + \dfrac{\pi}{4}\right)$

In Exercises 21–24, sketch the appropriate graphs. Check each on a graphing calculator.

21. A drafting student draws a circle through the three vertices of a right triangle. The hypotenuse of the triangle is the diameter d of the circle, and from Fig. 10-30, we see that $d = a \sec \theta$. Sketch the graph of d as a function of θ for $a = 3.00$ cm.

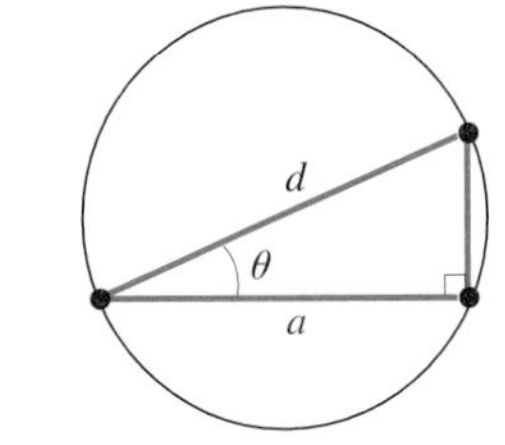

Fig. 10-30

22. At a distance x from the base of a building 200 m high, the angle of elevation θ of the top of the building can be found from the equation $x = 200 \cot \theta$. Sketch x as a function of θ.

23. A mechanism with two springs is shown in Fig. 10-31, where point A is restricted to move horizontally. From the law of sines we see that $b = (a \sin B) \csc A$. Sketch the graph of b as a function of A for $a = 4.00$ cm and $B = \frac{\pi}{4}$.

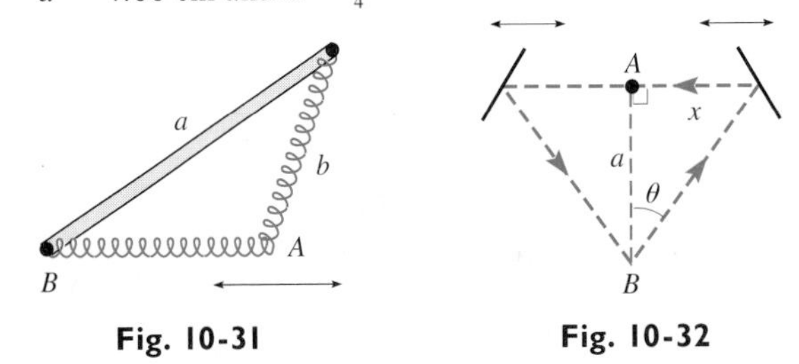

Fig. 10-31 **Fig. 10-32**

24. In a laser experiment, two mirrors move horizontally in equal and opposite distances from point A. The laser path from and to point B is shown in Fig. 10-32. From the figure we see that $x = a \tan \theta$. Sketch the graph of x as a function of θ for $a = 5.00$ cm.

10-5 Applications of the Trigonometric Graphs

In this section we introduce an important physical concept and indicate some of the technical applications.

In Section 8-4 we discussed the velocity of an object moving in a circular path. When this object moves with constant velocity, its *projection* on a diameter moves with what is known as **simple harmonic motion.** For example, consider an object moving around a circle in a plane parallel to rays of light. The movement of the object's shadow on a wall (perpendicular to the light) is simple harmonic motion. The object could be the end of a spoke of a rotating wheel. Another illustration of simple harmonic motion is the displacement of a weight moving up and down on a spring.

EXAMPLE 1 ▸▸ In Fig. 10-33, let us assume that a particle starts at the end of the radius at $(R, 0)$ and moves counterclockwise around the circle with constant angular velocity ω. *The displacement of the projection of the radius on the y-axis is* d and is given by $d = R \sin \theta$. The displacement of this projection is shown for a few different positions of the end of the radius.

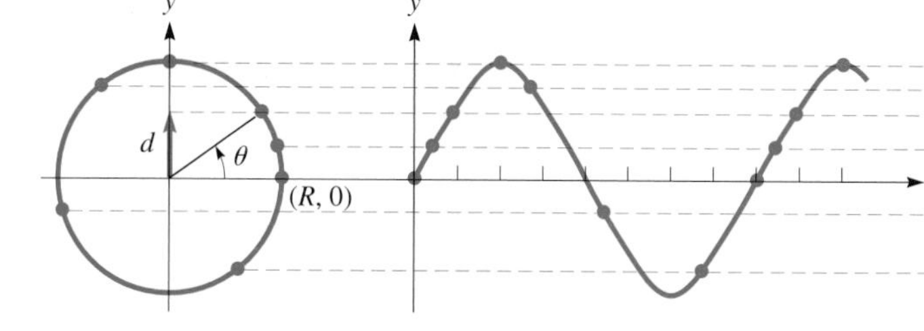

Fig. 10-33

Since $\theta/t = \omega$, or $\theta = \omega t$, we have

$$d = R \sin \omega t \tag{10-3}$$

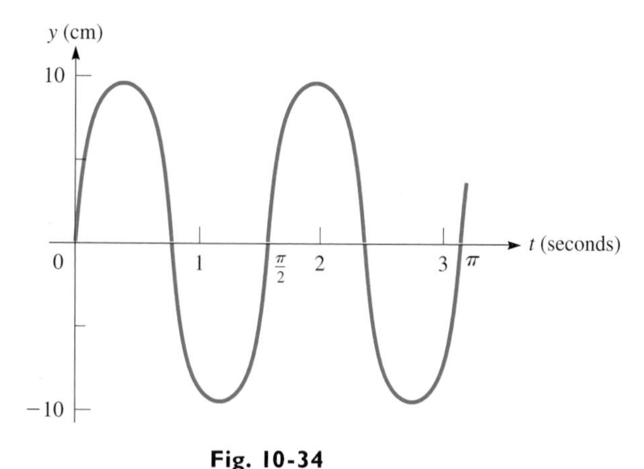

Fig. 10-34

as the equation for the displacement of this projection, with time t as the independent variable.

For the case where $R = 10.0$ cm and $\omega = 4.00$ rad/s, we have

$$d = 10.0 \sin 4.00t$$

By sketching or viewing the graph of this function, we can find the displacement d of the projection for a given time t. The graph is shown in Fig. 10-34. ◂◂

In Example 1, note that *time is the independent variable.* This is motion for which the object (the end of the projection) remains at the same horizontal position $(x = 0)$ and moves only vertically according to a sinusoidal function. In the previous sections, we dealt with functions in which y is a sinusoidal function of the horizontal displacement x. Think of a water wave. At *one point* of the wave, the motion is only vertical and sinusoidal with time. At *one given time,* a picture would indicate a sinusoidal movement from one horizontal position to the next.

EXAMPLE 2 ▸▸ A windmill is used to pump water. The radius of the blade is 2.5 m, and it is moving with constant angular velocity. If the vertical displacement of the end of the blade is timed from the point it is at an angle of 45° ($\pi/4$ rad) from the horizontal [see Fig. 10-35(a)], the displacement d is given by

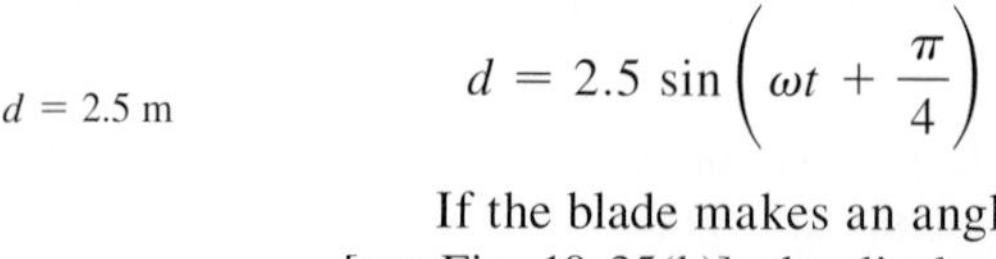

$$d = 2.5 \sin\left(\omega t + \frac{\pi}{4}\right)$$

If the blade makes an angle of 90° ($\pi/2$ rad) when $t = 0$ [see Fig. 10-35(b)], the displacement d is given by

$$d = 2.5 \sin\left(\omega t + \frac{\pi}{2}\right)$$

or $\quad d = 2.5 \cos \omega t$

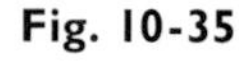

Fig. 10-35

If timing started at the first maximum for the displacement, the resulting curve for the displacement would be that of the cosine function. ◂◂

Other examples of simple harmonic motion are (1) the movement of a pendulum bob through its arc (a very close approximation to simple harmonic motion), (2) the motion of an object "bobbing" in water, and (3) the movement of the end of a vibrating rod (which we hear as sound). Other phenomena that give rise to equations like those for simple harmonic motion are found in the fields of optics, sound, and electricity. Such phenomena have the same mathematical form because they result from vibratory movement or motion in a circle.

EXAMPLE 3 ▸▸ A very important use of the trigonometric curves arises in the study of alternating current, which is caused by the motion of a wire passing through a magnetic field. If the wire is moving in a circular path, with angular velocity ω, the current i in the wire at time t is given by an equation of the form

$$i = I_m \sin(\omega t + \alpha)$$

where I_m is the maximum current attainable and α is the phase angle.

The current may be represented by a sinusoidal wave. Given that $I_m = 6.00$ A, $\omega = 120\pi$ rad/s, and $\alpha = \pi/6$, we have the equation

$$i = 6.00 \sin(120\pi t + \tfrac{\pi}{6})$$

From this equation we see that the amplitude is 6.00 A, the period is $\frac{1}{60}$ s, and the displacement is $-\frac{1}{720}$ s. From these values we draw the graph as shown in Fig. 10-36. Since the current takes on both positive and negative values, we conclude that it moves alternately in one direction and then the other.

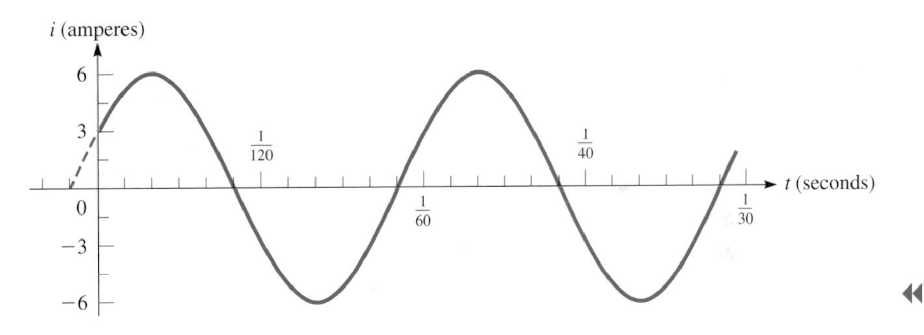

Fig. 10-36 ◂◂

It is a common practice to express the rate of rotation in terms of *the* **frequency** *f, the number of cycles per second,* rather than directly in terms of the angular velocity ω, the number of radians per second. *The unit for frequency is the* **hertz** (Hz), *and* 1 Hz = 1 cycle/s. Since there are 2π rad in one cycle, we have

$$\omega = 2\pi f \quad (10\text{-}4)$$

It is the frequency f which is referred to in electric current, on radio stations, for musical tones, and so on.

EXAMPLE 4 ▸▸ For the electric current in Example 3, $\omega = 120\pi$ rad/s. The corresponding frequency f is

$$f = \frac{120\pi}{2\pi} = 60 \text{ Hz}$$

This means that 120π rad/s corresponds to 60 cycles/s. This is the standard frequency used for alternating current. ◂◂

Exercises 10–5

In the following exercises, a graphing calculator may be used to display the indicated graphs.

In Exercises 1 and 2, sketch two cycles of the curve of the projection of Example 1 as a function of time for the given values.

1. $R = 2.40$ cm, $\omega = 2.00$ rad/s
2. $R = 1.80$ m, $f = 0.250$ Hz

In Exercises 3 and 4, a point on a cam is 8.30 cm *from the center of rotation. The cam is rotating with a constant angular velocity, and the vertical displacement* $d = 8.30$ cm *for* $t = 0$ *s. See Fig. 10-37. Sketch two cycles of d as a function of t for the given values.*

3. $f = 3.20$ Hz
4. $\omega = 3.20$ rad/s

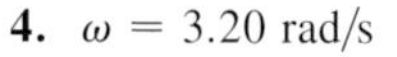

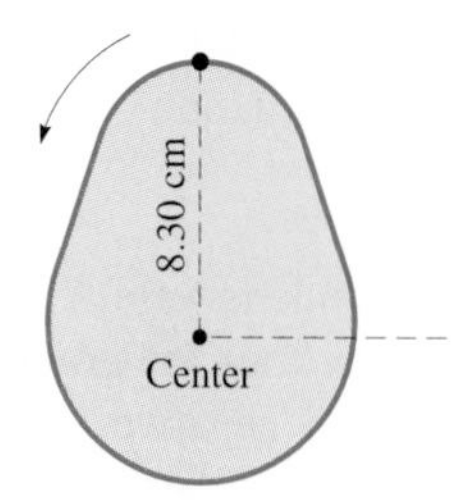

Fig. 10-37

In Exercises 5 and 6, a satellite is orbiting the earth such that its displacement D north of the equator (or south if D is negative) is given by $D = A\sin(\omega t + \alpha)$. *Sketch two cycles of D as a function of t for the given values.*

5. $A = 500$ km, $\omega = 3.60$ rad/h, $\alpha = 0$
6. $A = 850$ km, $f = 1.6 \times 10^{-4}$ Hz, $\alpha = \pi/3$

In Exercises 7 and 8, for an alternating-current circuit in which the voltage e is given by $e = E\cos(\omega t + \alpha)$, *sketch two cycles of the voltage as a function of time for the given values.*

7. $E = 170$ V, $f = 60.0$ Hz, $\alpha = -\pi/3$
8. $E = 80$ V, $\omega = 377$ rad/s, $\alpha = \pi/2$

In Exercises 9 and 10, refer to the wave in the string described in Exercise 25 of Section 10-3. For a point on the string, the displacement y is given by

$$y = A\sin 2\pi\left(\frac{t}{T} - \frac{x}{\lambda}\right)$$

We see that each point on the string moves with simple harmonic motion. Sketch two cycles of y as a function of t for the given values.

9. $A = 3.20$ cm, $T = 0.050$ s, $\lambda = 40.0$ cm, $x = 5.00$ cm
10. $A = 1.45$ cm, $T = 0.250$ s, $\lambda = 24.0$ cm, $x = 20.0$ cm

In Exercises 11 and 12, the air pressure within a plastic container changes above and below the external atmospheric pressure by $p = p_0 \sin 2\pi ft$. *Sketch two cycles of p as a function of t for the given values.*

11. $p_0 = 280$ kPa, $f = 2.30$ Hz
12. $p_0 = 45.0$ kPa, $f = 0.450$ Hz

In Exercises 13–16, sketch the required curves.

13. The displacement of the end of a vibrating rod is given by $y = 1.50\cos 200\pi t$. Sketch two cycles of y (in cm) as a function of t (in seconds).
14. A simple pendulum is started by giving it a velocity from its equilibrium position. The angle θ between the vertical and the pendulum is given by $\theta = \theta_0 \sin\sqrt{g/l}\,t$, where θ_0 is the amplitude, g is the acceleration due to gravity, l is the length of the pendulum, and t is the time of motion. Sketch two cycles of θ as a function of t for the values $\theta_0 = 0.100$ rad, $g = 9.80$ m/s^2, and $l = 2.00$ m.
15. The signal received by a radio is given by $e = 0.014\cos(2\pi ft + \frac{\pi}{4})$, where e is in volts and f is in hertz. Sketch two cycles of e as a function of t for a station that is broadcasting on a frequency of $f = 950$ kHz ("95" on the AM radio dial).
16. The acoustical intensity of a sound wave is given by $I = A\cos(2\pi ft - \alpha)$, where f is the frequency of the sound. Sketch two cycles of I as a function of t if $A = 0.027$ W/cm^2, $f = 240$ Hz, and $\alpha = 0.80$.

10-6 Composite Trigonometric Curves

Many applications involve functions which in themselves are a combination of two or more simpler functions. In this section we discuss methods by which the curve of such a function can be found by combining values from the simpler functions.

EXAMPLE 1 ▸▸ Sketch the graph of $y = 2 + \sin 2x$.

This function is the sum of the simpler functions $y_1 = 2$ and $y_2 = \sin 2x$. We may find the important values for y by adding 2 to each important value of $y_2 = \sin 2x$.

For $y_2 = \sin 2x$, the amplitude is 1 and the period is $\frac{2\pi}{2} = \pi$. Since $y = y_1 + y_2 = 2 + \sin 2x$, we obtain the values in the following table and sketch the graph in Fig. 10-38.

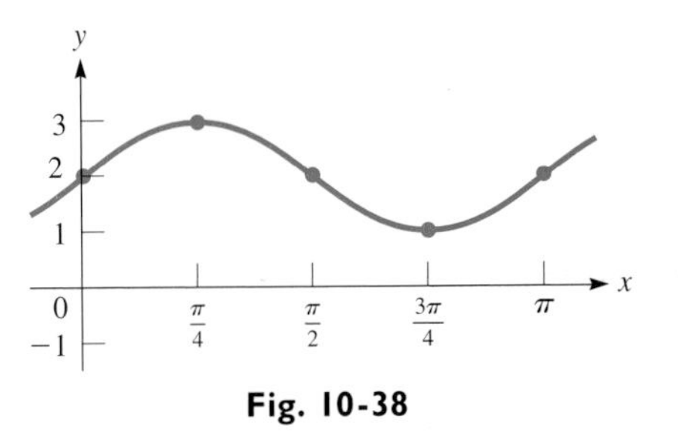

Fig. 10-38

x	0	$\frac{\pi}{4}$	$\frac{\pi}{2}$	$\frac{3\pi}{4}$	π
$\sin 2x$	0	1	0	−1	0
$2 + \sin 2x$	2	3	2	1	2

◂◂

Addition of Ordinates

Another way to sketch the resulting graph is to *first sketch the two simpler curves and then add the y-values graphically. This method is called* **addition of ordinates** and is illustrated in the following example.

EXAMPLE 2 ▸▸ Sketch the graph of $y = 2\cos x + \sin 2x$.

On the same set of coordinate axes we sketch the curves $y = 2\cos x$ and $y = \sin 2x$. These are shown as dashed and solid light curves in Fig. 10-39. For various values of x, we determine the distance above or below the x-axis of each curve and add these distances, noting that those above the axis are positive and those below the axis are negative. We thereby *graphically* ***add*** *the y-values* of these two curves for these values of x to obtain the points on the resulting curve, shown in color in Fig. 10-39.

For example, for the x-value at A we add the two lengths shown (side-by-side for clarity) to get the length for y. At B we see that both lengths are negative, and the value for y is the sum of these two negative values. At C one length is positive and one is negative, and we must subtract the lower length from the upper one to get the length for y.

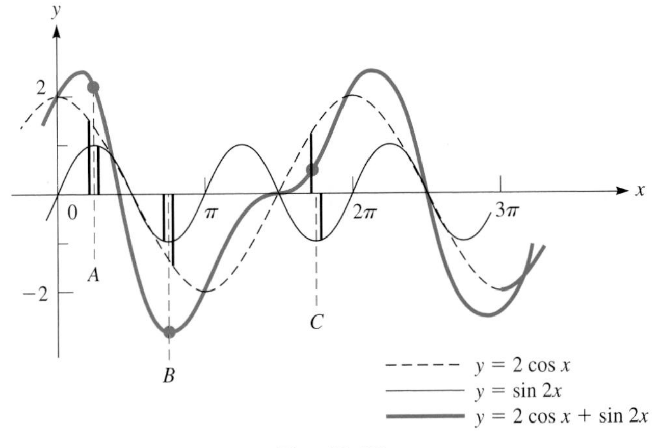

Fig. 10-39

We add (or subtract) these lengths at a sufficient number of x-values to get the proper curve. Some points are easily found. Where one curve crosses the x-axis, its y-value is zero, and therefore the resulting curve has its point on the other curve for this value of x. In this example, $\sin 2x$ is zero at $x = 0, \pi/2, \pi$, and so forth. We see that, for these values of x, the points on the resulting curve lie on the curve of $2\cos x$.

We should also add the values together where each curve is at its maximum or minimum values. In this case, $\sin 2x = 1$ at $\pi/4$, and the two y-values should be added together here to get a point on the resulting curve.

At $x = 5\pi/4$, we must take extra care in adding the values, since

$\sin 2x$ *is* **positive** *and* $2\cos x$ *is* **negative.**

Reasonable care and accuracy are needed to sketch a proper resulting curve.

The graph is shown in Fig. 10-39. ◂◂

We have shown how a fairly complex curve can be sketched graphically. However, it is expected that a graphing calculator (or a computer *grapher*) will be used to view most graphs, particularly ones that are difficult to sketch. The graphing calculator can display such curves much more easily and with much greater accuracy. We can use information about the amplitude, period, and displacement in choosing values for the RANGE feature on the calculator.

EXAMPLE 3 ▸▸ Use a graphing calculator to display the graph of $y = \frac{x}{2} - \cos x$.

Here we note that the curve is a combination of the straight line $y = x/2$ and the trigonometric curve $y = \cos x$. There are several good choices for the RANGE values, depending on how much of the curve is to be viewed. To see a little more than one period of $\cos x$, we can make the following choices:

Xmin $= -1$ (to start to the left of the y-axis)

Xmax $= 7$ (the period of $\cos x$ is $2\pi \approx 6.3$)

Ymin $= -2$ (the line passes through $(0, 0)$; the amplitude of $y = \cos x$ is 1)

Ymax $= 4$ (the slope of the line is $1/2$)

The graphing calculator view of the curve is shown in Fig. 10-40(a). The graph of $y = \frac{x}{2} - \cos x$, $y = \frac{x}{2}$, and $y = -\cos x$ is shown in Fig. 10-40(b).

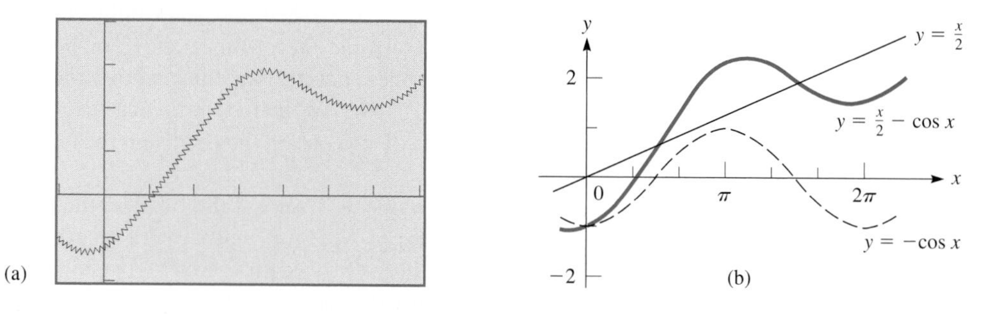

Fig. 10-40 (a)

NOTE ▸ The reason for showing $y = -\cos x$, and not $y = \cos x$, is that if *addition of ordinates* were being used, *it is much easier to add graphic values than to subtract them.* In using the method of addition of ordinates, we could *add* the ordinates of $y = x/2$ and $y = -\cos x$ to get the resulting curve of $y = \frac{x}{2} - \cos x$. ◂◂

EXAMPLE 4 ▸▸ View the graph of $y = \cos \pi x - 2 \sin 2x$ on a graphing calculator.

The combination of $y = \cos \pi x$ and $y = 2 \sin 2x$ leads to the following choices for the RANGE values:

Xmin $= -1$ (to start to the left of the y-axis)

Xmax $= 7$ (the periods are 2 and π; this shows at least two periods of each)

Ymin $= -3$ (the sum of the amplitudes is 3)

Ymax $= 3$

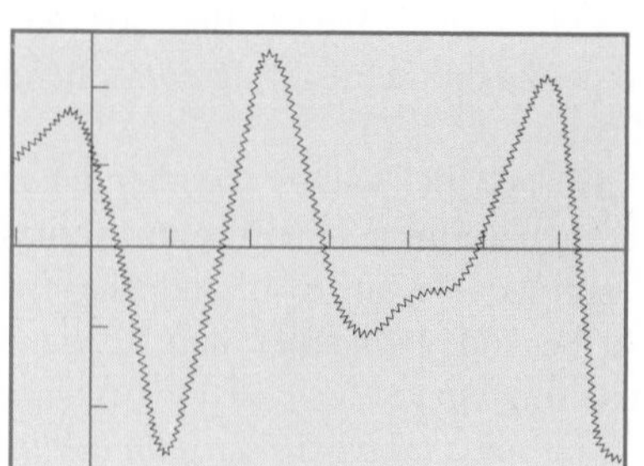

Fig. 10-41

There are many possible choices for Xmin and Xmax to get a good view of the graph on the calculator. However, since the sum of the amplitudes is 3, we know that the curve cannot be below $y = -3$ or above $y = 3$.

The graphing calculator view is shown in Fig. 10-41.

This graph can be constructed by using addition of ordinates, although it is difficult to do very accurately. ◂◂

Lissajous Figures

An important application of trigonometric curves is made when they are added at *right angles*. The methods for doing this are shown in the following examples.

EXAMPLE 5 ▸▸ Plot the graph for which the values of x and y are given by the equations $y = \sin 2\pi t$ and $x = 2 \cos \pi t$. *Equations given in this form, x and y in terms of a third variable, are called* **parametric equations.**

Since both x and y are in terms of t, by assuming values of t we find corresponding values of x and y, and use these values to plot the graph. Since the periods of $\sin 2\pi t$ and $2 \cos \pi t$ are $t = 1$ and $t = 2$, respectively, we will use values of $t = 0, 1/4, 1/2, 3/4, 1$, and so on. These give us convenient values of $0, \pi/4, \pi/2, 3\pi/4, \pi$, and so on to use in the table. We plot the points in Fig. 10-42.

t	0	$\frac{1}{4}$	$\frac{1}{2}$	$\frac{3}{4}$	1	$\frac{5}{4}$	$\frac{3}{2}$	$\frac{7}{4}$	2	$\frac{9}{4}$
x	2	1.4	0	−1.4	−2	−1.4	0	1.4	2	1.4
y	0	1	0	−1	0	1	0	−1	0	1
Point number	1	2	3	4	5	6	7	8	9	10

◂◂

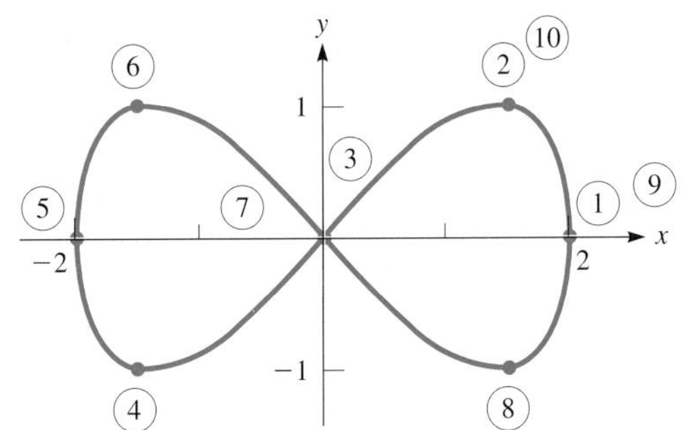

Fig. 10-42

Named for the French physicist Jules Lissajous (1822–1880).

Since x and y are trigonometric functions of a third variable t and since the x- and y-axes are at right angles, values of x and y obtained in this manner result in a combination of two trigonometric curves at right angles. *Figures obtained in this manner are called* **Lissajous figures.** Note that the Lissajous figure in Fig. 10-42 *is not a function* since there are *two* values of y for each value of x (except $x = -2, 0, 2$) in the domain.

In practice, Lissajous figures can be shown by applying different voltages to an ***oscilloscope*** and displaying the electric signals on a screen similar to that on a television set.

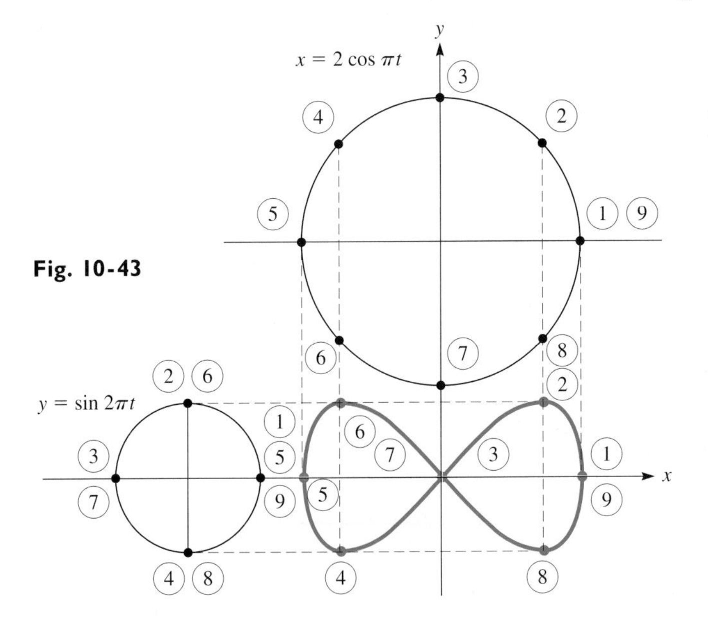

Fig. 10-43

EXAMPLE 6 ▸▸ If we place a circle on the x-axis and another on the y-axis, we may represent the coordinates (x, y) for the curve of Example 5 by the lengths of the projections (see Example 1 of Section 10-5) of a point moving around each circle. A careful study of Fig. 10-43 will clarify this. We note that the radius of the circle giving the x-values is 2 and that the radius of the circle giving the y-values is 1. This is due to the way in which x and y are defined. Also, due to these definitions, the point revolves around the y-circle twice as fast as the corresponding point around the x-circle.

See the chapter introduction.

On an oscilloscope, the curve would result when two electric signals are used. The first would have twice the amplitude and one-half the frequency of the other.

◂◂

Most graphing calculators can be used to display a curve defined by parametric equations. To do this it is necessary to use the MODE feature and make the selection for *parametric equations*. Use the manual for the calculator, as there are some differences in how this is done on the various calculators. In the example that follows, we show the keys and features of the TI-81 graphing calculator.

EXAMPLE 7 ▸▸ Use a graphing calculator to display the graph defined by the parametric equations $x = 2\cos \pi t$ and $y = \sin 2\pi t$. These are the same equations as those used in Examples 5 and 6.

On the TI-81 first press the MODE key, and select Param. Next press the Y = key, and we see XIT = and YIT = . (Note that we could enter three different pairs of parametric equations.) We then enter $2\cos \pi t$ and $\sin 2\pi t$ on the proper lines.

Next press the RANGE key, and make the following choices:

Tmin = 0 (already set, and the usual choice)

Tmax = 2 (the periods are 2 and 1, the longer period is 2)

Tstep = .1047 (already set)

Xmin = −2 (smallest possible value of x)

Xmax = 2 (largest possible value of x) Xscl = 1

Ymin = −1 (smallest possible value of y)

Ymax = 1 (largest possible value of y) Yscl = .5

We press GRAPH and Fig. 10-44 shows the view. ◂◂

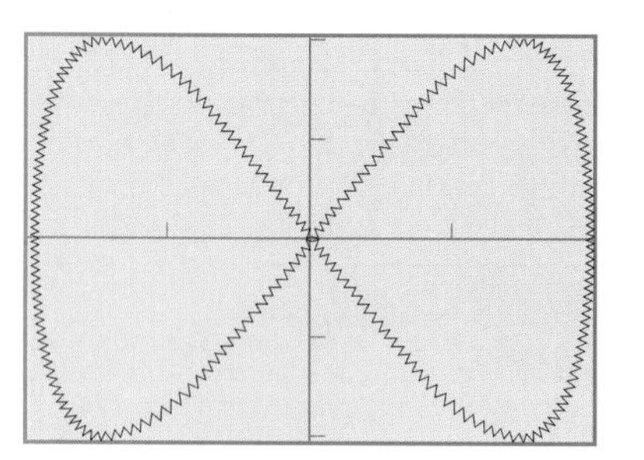

Fig. 10-44

Exercises 10–6

In Exercises 1–8, sketch the curves of the given functions by addition of ordinates.

1. $y = 1 + \sin x$
2. $y = 3 - 2\cos x$
3. $y = \frac{1}{3}x + \sin 2x$
4. $y = x - \sin x$
5. $y = \frac{1}{10}x^2 - \sin \pi x$
6. $y = \frac{1}{4}x^2 + \cos 3x$
7. $y = \sin x + \cos x$
8. $y = \sin x + \sin 2x$

In Exercises 9–20, display the graphs of the given functions on a graphing calculator.

9. $y = x^3 + 10\sin 2x$
10. $y = \dfrac{1}{x^2 + 1} - \cos \pi x$
11. $y = \sin x - \sin 2x$
12. $y = \cos 3x - \sin x$
13. $y = 2\cos 2x + 3\sin x$
14. $y = \frac{1}{2}\sin 4x + \cos 2x$
15. $y = 2\sin x - \cos x$
16. $y = \sin \frac{x}{2} - \sin x$
17. $y = \sin \pi x - \cos 2x$
18. $y = 2\cos 4x - \cos\left(x - \dfrac{\pi}{4}\right)$
19. $y = 2\sin\left(2x - \dfrac{\pi}{6}\right) + \cos\left(2x + \dfrac{\pi}{3}\right)$
20. $y = 3\cos 2\pi x + \sin \frac{\pi}{2}x$

In Exercises 21–24, plot the Lissajous figures.

21. $x = \sin t,\ y = \sin t$
22. $x = 2\cos t,\ y = \cos(t + 4)$
23. $x = \cos \pi t,\ y = \sin \pi t$
24. $x = \cos\left(t + \dfrac{\pi}{4}\right),\ y = \sin 2t$

In Exercises 25–32, use a graphing calculator to display the Lissajous figures.

25. $x = \cos \pi\left(t + \dfrac{1}{6}\right),\quad y = 2\sin \pi t$
26. $x = \sin^2 \pi t,\ y = \cos \pi t$
27. $x = 2\cos 3t,\ y = \cos 2t$
28. $x = 2\sin \pi t,\ y = 3\sin 3\pi t$
29. $x = \sin t,\ y = \sin 5t$
30. $x = 2\cos t,\ y = \sin 5t$
31. $x = 2\cos \pi t,\ y = 3\sin(2\pi t - \frac{\pi}{4})$
32. $x = 2\cos 3\pi t,\ y = \cos 5\pi t$

In Exercises 33–40, sketch the appropriate curves. A graphing calculator may be used.

33. An analysis of temperature records for Montreal indicates that the average daily temperature T (in °C) during the year is given approximately by $T = 6 - 15\cos[\frac{\pi}{6}(x - 0.5)]$, where x is measured in months ($x = 0.5$ is Jan 15, etc.). Sketch the graph of T vs. x for one year.

34. Data showed that the bird population in a certain remote area was given by $P = 8000 + 1500 \sin 3t$, where t is measured in years. Sketch the graph of P as a function of t for the first 5 years for which the data were taken.

35. The vertical displacement y (in dm) of a buoy floating in water is given by $y = 3.0 \cos 0.2t + 1.0 \sin 0.4t$, where t is measured in seconds. Sketch the graph of y as a function of t for the first 40 s.

36. The strain e (dimensionless) on a cable caused by vibration is given by $e = 0.0080 - 0.0020 \sin 30t + 0.0040 \cos 10t$, where t is measured in seconds. Sketch two cycles of e as a function of t.

37. The electric current in a certain circuit is given by $i = 0.32 + 0.50 \sin t - 0.20 \cos 2t$, where i is in milliamperes and t is in milliseconds. Sketch two cycles of i as a function of t.

38. The available solar energy depends on the amount of sunlight, and the available time in a day for sunlight depends on the time of the year. An approximate correction factor, in minutes, to standard time is $C = 10 \sin \frac{1}{29}(n - 80) - 7.5 \cos \frac{1}{58}(n - 80)$, where n is the number of the day of the year. Sketch C as a function of n.

39. Two signals are sent to an oscilloscope and are seen on the oscilloscope as being at right angles. The equations for the displacements of these signals are $x = 4 \cos \pi t$ and $y = 2 \sin 3\pi t$. Sketch the figure that appears on the oscilloscope.

40. In the study of optics, light is said to be elliptically polarized if certain optic vibrations are out of phase. These may be represented by Lissajous figures. Determine the Lissajous figure for two waves of light given by $w_1 = \sin \omega t$, $w_2 = \sin(\omega t + \frac{\pi}{4})$.

Chapter Equations, Review Exercises, and Practice Test

Chapter Equations

For the graphs of $y = a \sin(bx + c)$ and $y = a \cos(bx + c)$

$$\text{Amplitude} = |a|$$

$$\text{Period} = \frac{2\pi}{b} \qquad (10\text{-}1)$$

$$\text{Displacement} = -\frac{c}{b}$$

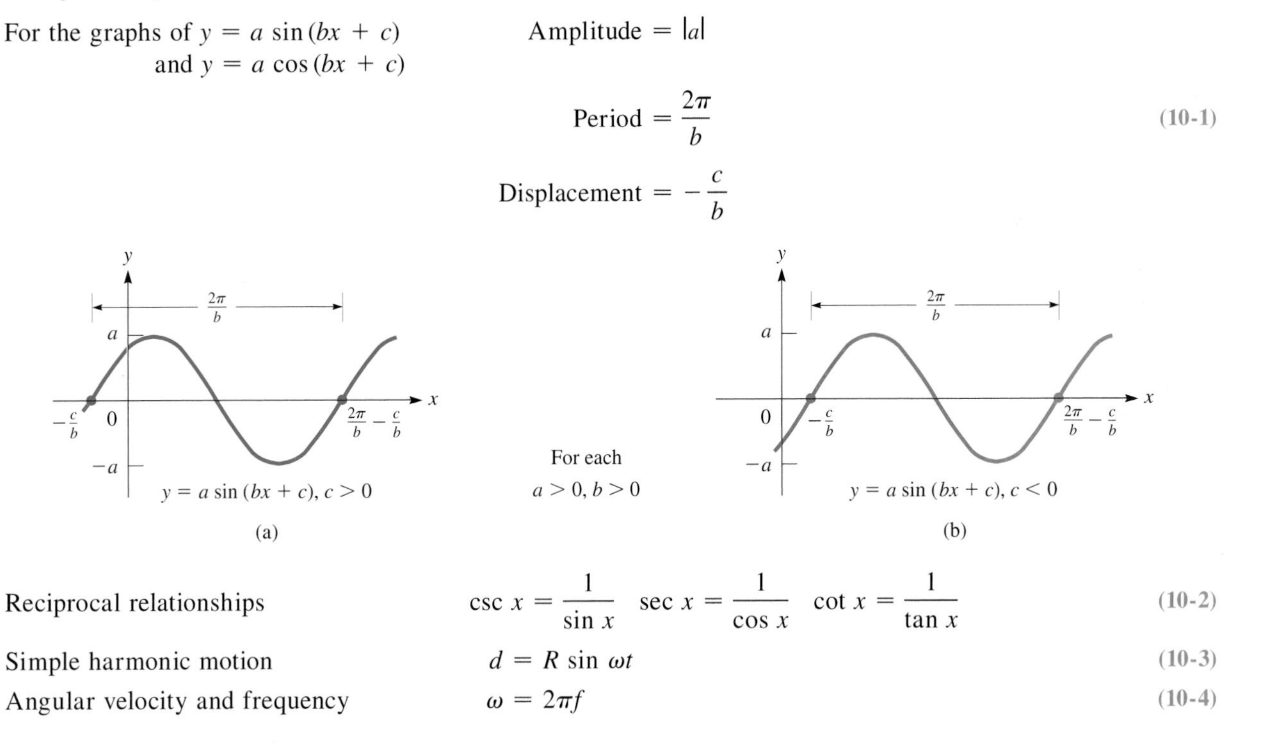

Reciprocal relationships $\quad \csc x = \dfrac{1}{\sin x} \quad \sec x = \dfrac{1}{\cos x} \quad \cot x = \dfrac{1}{\tan x} \qquad (10\text{-}2)$

Simple harmonic motion $\quad d = R \sin \omega t \qquad (10\text{-}3)$

Angular velocity and frequency $\quad \omega = 2\pi f \qquad (10\text{-}4)$

Review Exercises

In Exercises 1–28, sketch the curves of the given trigonometric functions. Check each using a graphing calculator.

1. $y = \frac{2}{3}\sin x$
2. $y = -4\sin x$
3. $y = -2\cos x$
4. $y = 2.3\cos x$
5. $y = 2\sin 3x$
6. $y = 4.5\sin 12x$
7. $y = 2\cos 2x$
8. $y = 24\cos 6x$
9. $y = 3\cos\frac{1}{3}x$
10. $y = 3\sin\frac{1}{2}x$
11. $y = \sin\pi x$
12. $y = 3\sin 4\pi x$
13. $y = 5\cos 2\pi x$
14. $y = -\cos 6\pi x$
15. $y = -0.5\sin\frac{\pi}{6}x$
16. $y = 8\sin\frac{\pi}{4}x$
17. $y = 2\sin\left(3x - \frac{\pi}{2}\right)$
18. $y = 3\sin\left(\frac{x}{2} + \frac{\pi}{2}\right)$
19. $y = -2\cos(4x + \pi)$
20. $y = 0.8\cos\left(\frac{x}{6} - \frac{\pi}{2}\right)$
21. $y = -\sin\left(\pi x + \frac{\pi}{6}\right)$
22. $y = 2\sin(3\pi x - \pi)$
23. $y = 8\cos\left(4\pi x - \frac{\pi}{2}\right)$
24. $y = 3\cos(2\pi x + \pi)$
25. $y = 3\tan x$
26. $y = \frac{1}{4}\sec x$
27. $y = -\frac{1}{3}\csc x$
28. $y = -5\cot x$

In Exercises 29–32, sketch the curves of the given functions by addition of ordinates.

29. $y = 2 + \frac{1}{2}\sin 2x$
30. $y = \frac{1}{2}x - \cos\frac{1}{3}x$
31. $y = \sin 2x + 3\cos x$
32. $y = \sin 3x + 2\cos 2x$

In Exercises 33–36, display the curves of the given functions on a graphing calculator.

33. $y = 2\sin x - \cos 2x$
34. $y = \sin 3x - 2\cos x$
35. $y = \cos\left(x + \frac{\pi}{4}\right) - 2\sin 2x$
36. $y = 2\cos\pi x + \cos(2\pi x - \pi)$

In Exercises 37–40, give the specific form of the indicated equation by evaluating a, b, and c through an inspection of the indicated curve.

37. $y = a\sin(bx + c)$ (Figure 10-45)
38. $y = a\cos(bx + c)$ (Figure 10-45)
39. $y = a\cos(bx + c)$ (Figure 10-46)
40. $y = a\sin(bx + c)$ (Figure 10-46)

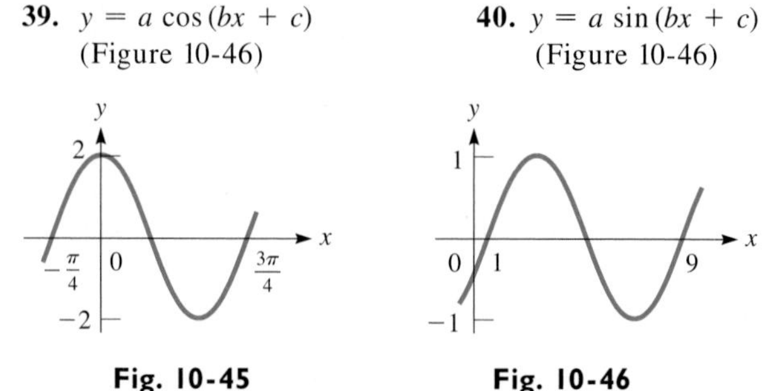

Fig. 10-45 **Fig. 10-46**

In Exercises 41–44, display the Lissajous figures on a graphing calculator.

41. $x = -\cos 2\pi t, \quad y = 2\sin\pi t$
42. $x = \sin\left(t + \frac{\pi}{6}\right), \; y = \sin t$
43. $x = \cos\left(2\pi t + \frac{\pi}{4}\right), \; y = \cos\pi t$
44. $x = \cos\left(t - \frac{\pi}{6}\right), \; y = \cos\left(2t + \frac{\pi}{3}\right)$

In Exercises 45–60, sketch the appropriate curves. A graphing calculator may be used.

45. The range R of a rocket is given by

$$R = \frac{v_0^2 \sin 2\theta}{g}$$

Fig. 10-47

See Fig. 10-47. Sketch R as a function of θ for $v_0 = 1000$ m/s and $g = 9.8$ m/s^2.

46. The blade of a saber saw moves vertically up and down at 18 strokes per second. The vertical displacement y (in cm) is given by $y = 1.2\sin 36\pi t$, where t is in seconds. Sketch at least two cycles of the graph of y vs. t.

47. The velocity v, in centimetres per second, of a piston in a certain engine is given by $v = \omega D\cos\omega t$, where ω is the angular velocity of the crankshaft in radians per second and t is the time in seconds. Sketch the graph of v vs. t if the engine is at 3000 r/min and $D = 3.6$ cm.

48. A light wave for the color yellow can be represented by the equation $y = A\sin 3.4 \times 10^{15}\, t$. With A as a constant, sketch two cycles of y as a function of t, in seconds.

49. The displacement y of the end of a gear tooth is given by $y = A \sin(6t + 0.5)$. Sketch the graph of two cycles of y as a function of t, in seconds, for $A = 5.2$ cm.

50. A circular disk suspended by a thin wire attached to the center at one of its flat faces is twisted through an angle θ. Torsion in the wire tends to turn the disk back in the opposite direction (thus the name *torsion pendulum* is given to this device). The angular displacement as a function of time is given by $\theta = \theta_0 \cos(\omega t + \alpha)$, where θ_0 is the maximum angular displacement, ω is a constant that depends on the properties of the disk and wire, and α is the phase angle. Sketch the graph of θ vs. t if $\theta_0 = 0.100$ rad, $\omega = 2.50$ rad/s, and $\alpha = \frac{\pi}{4}$.

51. The charge q on a certain capacitor as a function of time is given by $q = 0.001(1 - \cos 100t)$. Sketch two cycles of q as a function of t. Charge is measured in coulombs, and time is measured in seconds.

52. The weekly profit p, in thousands of dollars, for a seasonal business is given by $p = 35 + 15 \sin 0.24n$, where n is the number of the week of the year. Sketch p as a function of n for one year.

53. If the upper end of a spring is not fixed and is being moved with a sinusoidal motion, the motion of the bob at the end of the spring is affected. Sketch the curve if the motion of the upper end of a spring is being moved by an external force and the bob moves according to the equation $y = 4 \sin 2t - 2 \cos 2t$.

54. An approximate equation for the loudness L (in decibels) of a fire siren as a function of the time t (in seconds) is $L = 40 - 35 \cos 2t + 60 \sin t$. Sketch this function for $0 \le t \le 10$ s.

55. The path of a roller mechanism used in an assembly line process is given by the equations $x = \theta - \sin\theta$ and $y = 1 - \cos\theta$. Sketch the path for $0 \le \theta \le 2\pi$.

56. Two voltage signals give a resulting curve on an oscilloscope. The equations for these signals are $x = 6 \sin \pi t$ and $y = 4 \cos 4\pi t$. Sketch the graph of the curve displayed by the oscilloscope.

57. The impedance Z, in ohms, and resistance R, in ohms, for an alternating-current circuit are related by $Z = R \sec\theta$, where θ is known as the *phase angle.* Sketch the graph for Z as a function of θ for $-\frac{\pi}{2} < \theta < \frac{\pi}{2}$.

58. For an object sliding down an inclined plane at constant speed, the coefficient of friction μ between the object and the plane is given by $\mu = \tan\theta$, where θ is the angle between the plane and the horizontal. Sketch the graph of μ vs. θ.

59. The length l of a conveyor belt and the height h it rises are related by $l = h \csc\theta$, where θ is the angle of inclination of the belt. Sketch l as a function of θ for $h = 3.2$ m.

60. The instantaneous power in an electric circuit is defined as the product of the instantaneous voltage e and the instantaneous current i. If we have $e = 100 \cos 200t$ and $i = 2 \cos(200t + \frac{\pi}{4})$, plot the graph of the voltage and the graph of the current (in amperes), on the same coordinate system, vs. the time (in seconds). Then sketch the power (in watts) vs. the time by multiplying appropriate values of e and i.

Writing Exercise

61. A wave passing through a string can be described at any instant by the equation $y = a \sin(bx + c)$. Write one or two paragraphs explaining the change in the wave if (a) a is doubled, (b) if b is doubled, and (c) if c is doubled.

Practice Test

In Problems 1–4, sketch the graphs of the given functions.

1. $y = 0.5 \cos \frac{\pi}{2}x$ **2.** $y = 2 + 3 \sin x$

3. $y = 3 \sec x$ **4.** $y = 2 \sin(2x - \frac{\pi}{3})$

5. A wave is traveling in a string. The displacement, as a function of time, from its equilibrium position is given by $y = A \cos(2\pi/T)t$. T is the period (measured in seconds) of the motion. If $A = 0.200$ cm and $T = 0.100$ s, sketch two cycles of the displacement as a function of time.

6. Sketch the graph of $y = 2 \sin x + \cos 2x$ by addition of ordinates.

7. Use a graphing calculator to display the Lissajous figure for which $x = \sin \pi t$ and $y = 2 \cos 2\pi t$.

8. Sketch two cycles of the curve of a projection on the end of a radius on the y-axis. The radius is of length R, and it is rotating counterclockwise about the origin at 2.00 rad/s. It starts at an angle of $\pi/6$ with the positive x-axis.

11

Exponents and Radicals

In finding the rate at which solar radiation changes at a solar energy collector, the following expression is found.

$$\frac{(t^4 + 100)^{1/2} - 2t^3(t + 6)(t^4 + 100)^{-1/2}}{[(t^4 + 100)^{1/2}]^2}$$

In Section 11-2 we show that this expression can be written in a much simpler form.

In Chapter 1 we introduced exponents and radicals. To this point, only a basic understanding of the meaning and elementary operations with them has been necessary. However, in our future work a more detailed understanding of exponents and radicals, and operations with them, will be required. Therefore, in this chapter we shall develop the necessary operations.

We will show the relationship between exponents and radicals when we introduce fractional exponents. In applications, it is common and more convenient to use fractional exponents rather than radicals.

As we develop these algebraic operations, we will show some of their uses in technical areas of application. They are used to help develop various formulas in areas such as electronics, hydrodynamics, optics, and machine design.

11-1 Simplifying Expressions with Integral Exponents

The laws of exponents were given in Section 1-4. For reference, they are:

$$a^m \times a^n = a^{m+n} \tag{11-1}$$

$$\frac{a^m}{a^n} = a^{m-n} \quad \text{or} \quad \frac{a^m}{a^n} = \frac{1}{a^{n-m}}, \qquad a \neq 0 \tag{11-2}$$

$$(a^m)^n = a^{mn} \tag{11-3}$$

$$(ab)^n = a^n b^n, \qquad \left(\frac{a}{b}\right)^n = \frac{a^n}{b^n}, \qquad b \neq 0 \tag{11-4}$$

$$a^0 = 1, \qquad a \neq 0 \tag{11-5}$$

$$a^{-n} = \frac{1}{a^n}, \qquad a \neq 0 \tag{11-6}$$

Although Eqs. (11-1) to (11-4) were originally defined for positive integers as exponents, we showed in Section 1-4 that with the definitions given in Eqs. (11-5) and (11-6), they are valid for all integral exponents. Later in this chapter we shall show how fractions may be used as exponents. Since these equations are very important to the development of the topics in this chapter, they should again be reviewed, and learned thoroughly.

In this section we review the use of exponents while using Eqs. (11-1) to (11-6). Then we show how exponents are used and handled in more involved expressions.

EXAMPLE 1 ▸▸ Applying Eq. (11-1), we have

$$a^5 \times a^{-3} = a^{5+(-3)} = a^{5-3} = a^2$$

Applying Eq. (11-1) and then Eq. (11-6), we have

$$a^3 \times a^{-5} = a^{3-5} = a^{-2} = \frac{1}{a^2}$$

NOTE ▸ We note that *negative exponents are not used in the expression of a final result,* unless specified otherwise. However, they are often used in intermediate steps. ◂◂

EXAMPLE 2 ▸▸ Applying Eq. (11-1), then (11-6), and then (11-4), we have

$$(2^3 \times 2^{-4})^2 = (2^{3-4})^2 = (2^{-1})^2 = \left(\frac{1}{2}\right)^2 = \frac{1}{2^2} = \frac{1}{4}$$

Often, several combinations of the laws can be used to simplify an expression. For example, this expression can be simplified by using Eq. (11-1), then (11-3), and then (11-6), as follows:

$$(2^3 \times 2^{-4})^2 = (2^{3-4})^2 = (2^{-1})^2 = 2^{-2} = \frac{1}{2^2} = \frac{1}{4}$$

The result is in a proper form as either $1/2^2$ or $1/4$. If the exponent is large, then it is common to leave the exponent in the answer. ◂◂

EXAMPLE 3 ▸▸ Applying Eqs. (11-2) and (11-5), we have

$$\frac{a^2b^3c^0}{ab^7} = \frac{a^{2-1}(1)}{b^{7-3}} = \frac{a}{b^4}$$

Applying Eqs. (11-4) and (11-3), we have

$$(x^{-2}y)^3 = (x^{-2})^3(y^3) = x^{-6}y^3 = \frac{y^3}{x^6}$$

Here, the simplification was completed by the use of Eq. (11-6). ◂◂

EXAMPLE 4 ▸▸ $$\left(\frac{4}{a^2}\right)^{-3} = \frac{1}{\left(\frac{4}{a^2}\right)^3} = \frac{1}{\frac{4^3}{a^6}} = \frac{a^6}{4^3} \quad \text{or} \quad \left(\frac{4}{a^2}\right)^{-3} = \frac{4^{-3}}{a^{-6}} = \frac{a^6}{4^3}$$

In the first expression we used Eq. (11-6) first, then Eq. (11-4), and then we inverted the divisor. In the second expression we used Eq. (11-4) and then Eq. (11-6). ◂◂

EXAMPLE 5 ▸▸ $$(x^2y)^2\left(\frac{2}{x}\right)^{-2} = \frac{(x^4y^2)}{\left(\frac{2}{x}\right)^2} = \frac{x^4y^2}{\frac{4}{x^2}} = \frac{x^4y^2}{1} \times \frac{x^2}{4} = \frac{x^6y^2}{4}$$

or

$$(x^2y)^2\left(\frac{2}{x}\right)^{-2} = (x^4y^2)\left(\frac{2^{-2}}{x^{-2}}\right) = (x^4y^2)\left(\frac{x^2}{2^2}\right) = \frac{x^6y^2}{4}$$ ◂◂

The physical units associated with a denominate number can be expressed in terms of negative exponents. This is illustrated in the following example.

EXAMPLE 6 ▸▸ The metric unit for pressure is the *pascal*, where 1 Pa = 1 N/m^2. Using a negative exponent, this can be expressed as

$$1 \text{ Pa} = 1 \text{ N/m}^2 = 1 \text{ N} \cdot \text{m}^{-2}$$

where $1/\text{m}^2 = \text{m}^{-2}$.

The metric unit for energy is the *joule*, where $1 \text{ J} = 1 \text{ kg} \cdot (\text{m} \cdot \text{s}^{-1})^2$, or

$$1 \text{ J} = 1 \text{ kg} \cdot \text{m}^2 \cdot \text{s}^{-2} = 1 \text{ kg} \cdot \text{m}^2/\text{s}^2$$ ◂◂

Care must be taken to apply the laws of exponents properly. Certain common problems are pointed out in the following examples.

EXAMPLE 7 ▸▸ The expression $(-5x)^0$ equals 1, whereas the expression $-5x^0$ equals -5. For $(-5x)^0$ the parentheses show that the expression $-5x$ is raised to the zero power, whereas for $-5x^0$ only x is raised to the zero power and we have

$$-5x^0 = -5(1) = -5$$

CAUTION ▸ Also, $(-5)^0 = 1$, but $-5^0 = -1$. Again, for $(-5)^0$ parentheses show -5 raised to the zero power, whereas for -5^0 only 5 is raised to the zero power.

Similarly, $(-2)^2 = 4$ and $-2^2 = -4$

CAUTION ▸ For the same reasons, $2x^{-1} = \frac{2}{x}$ whereas $(2x)^{-1} = \frac{1}{2x}$ ◂◂

EXAMPLE 8 ▸▸ $$(2a + b^{-1})^{-2} = \frac{1}{(2a + b^{-1})^2} = \frac{1}{\left(2a + \frac{1}{b}\right)^2} = \frac{1}{\left(\frac{2ab + 1}{b}\right)^2}$$

$$= \frac{1}{\frac{(2ab + 1)^2}{b^2}} = \frac{b^2}{(2ab + 1)^2}$$ not necessary to expand the denominator

Another order of operations for simplifying this expression is

$$(2a + b^{-1})^{-2} = \left(2a + \frac{1}{b}\right)^{-2} = \left(\frac{2ab + 1}{b}\right)^{-2}$$

$$= \frac{(2ab + 1)^{-2}}{b^{-2}} = \frac{b^2}{(2ab + 1)^2}$$ positive exponents used in the final result ◂◂

EXAMPLE 9 ▸▸ There is an error which is commonly made in simplifying the type of expression in Example 8. We must be careful to see that

CAUTION ▸ $(2a + b^{-1})^{-2}$ is ***not*** equal to $(2a)^{-2} + (b^{-1})^{-2}$, or $\dfrac{1}{4a^2} + b^2$

Remember: As noted in Section 6-1, when raising a binomial (or any multinomial) to a power, we cannot simply raise each term to the power to obtain the result.

For reference, Eq. (11-4) is $(ab)^n = a^n b^n$.

However, when raising a product of factors to a power, we use Eq. (11-4). Thus,

$$(2ab^{-1})^{-2} = (2a)^{-2}(b^{-1})^{-2} = \frac{b^2}{(2a)^2} = \frac{b^2}{4a^2}$$

We see that we must be careful to distinguish between the power of a sum of terms and the power of a product of factors. ◂◂

From the preceding examples, we see that *when a factor is moved from the denominator to the numerator of a fraction, or conversely, the* **sign** *of the exponent is changed.* We should carefully note the word *factor*; this rule does not apply to moving *terms* in the numerator or the denominator.

EXAMPLE 10 ▸▸

$$3a^{-1} - (2a)^{-2} = \frac{3}{a} - \frac{1}{(2a)^2} = \frac{3}{a} - \frac{1}{4a^2}$$

$$= \frac{12a - 1}{4a^2}$$ ◂◂

EXAMPLE 11 ▸▸

$$3^{-1}\left(\frac{4^{-2}}{3 - 3^{-1}}\right) = \frac{1}{3}\left(\frac{1}{4^2}\right)\left(\frac{1}{3 - \frac{1}{3}}\right) = \frac{1}{3 \times 4^2}\left(\frac{1}{\frac{9 - 1}{3}}\right)$$

$$= \frac{1}{3 \times 4^2}\left(\frac{3}{8}\right) = \frac{1}{128}$$ ◂◂

EXAMPLE 12 ▸▸

terms

$$\frac{1}{x^{-1}}\left(\frac{x^{-1} - y^{-1}}{x^2 - y^2}\right) = \frac{x}{1}\left(\frac{\frac{1}{x} - \frac{1}{y}}{x^2 - y^2}\right) = x\left(\frac{\frac{y - x}{xy}}{x^2 - y^2}\right)$$

$$= \frac{\frac{x(y - x)}{xy}}{(x - y)(x + y)}$$

$$= \frac{x(y - x)}{xy} \times \frac{1}{(x - y)(x + y)}$$

$$= \frac{x(y - x)}{xy(x - y)(x + y)} = \frac{-(x - y)}{y(x - y)(x + y)}$$

$$= -\frac{1}{y(x + y)}$$

CAUTION ▸ Note that in this example ***the x^{-1} and y^{-1} in the numerator could not be moved directly to the denominator with positive exponents*** because they are only terms of the original numerator. ◂◂

EXAMPLE 13 ▸▸ $3(x + 4)^2(x - 3)^{-2} - 2(x - 3)^{-3}(x + 4)^3$

$$= \frac{3(x + 4)^2}{(x - 3)^2} - \frac{2(x + 4)^3}{(x - 3)^3} = \frac{3(x - 3)(x + 4)^2 - 2(x + 4)^3}{(x - 3)^3}$$

$$= \frac{(x + 4)^2[3(x - 3) - 2(x + 4)]}{(x - 3)^3} = \frac{(x + 4)^2(x - 17)}{(x - 3)^3}$$

Expressions such as the one in this example are commonly found in problems in calculus. ◂◂

Exercises 11–1

In Exercises 1–56, express each of the given expressions in simplest form with only positive exponents.

1. x^7x^{-4}

2. y^9y^{-2}

3. a^2a^{-6}

4. ss^{-5}

5. 5×5^{-3}

6. $2 \times 7^4 \times 7^{-2}$

7. $(2^{-1} \times 5)^2$

8. $(3^2 \times 4^{-3})^3$

9. $(2ax^{-1})^2$

10. $(3xy^{-2})^3$

11. $(5an^{-2})^{-1}$

12. $(6s^2t^{-1})^{-2}$

13. $(-4)^0$

14. -4^0

15. $-7x^0$

16. $(-7x)^0$

17. $3x^{-2}$

18. $(3x)^{-2}$

19. $(7ax)^{-3}$

20. $7ax^{-3}$

21. $\left(\frac{2}{n^3}\right)^{-1}$

22. $\left(\frac{3}{x^3}\right)^{-2}$

23. $\left(\frac{a}{b^{-2}}\right)^{-3}$

24. $\left(\frac{2n^{-2}}{m^{-1}}\right)^{-2}$

25. $(a + b)^{-1}$

26. $a^{-1} + b^{-1}$

27. $3x^{-2} + 2y^{-2}$

28. $(3x + 2y)^{-2}$

29. $(2 \times 3^{-2})^2\left(\frac{3}{2}\right)^{-1}$

30. $(3^{-1} \times 7^2)^{-1}\left(\frac{3}{7}\right)^2$

31. $ab^2(3a^{-1})^{-1}$

32. $(3st^{-1})^3\left(\frac{s}{t}\right)$

33. $\left(\frac{3a^2}{4b}\right)^{-3}\left(\frac{4}{a}\right)^{-5}$

34. $(2np^{-2})^{-2}(4^{-1}p^2)^{-1}$

35. $\left(\frac{v^{-1}}{2t}\right)^{-2}\left(\frac{t^2}{v^{-2}}\right)^{-3}$

36. $\left(\frac{a^{-2}}{b^2}\right)^{-3}\left(\frac{a^{-3}}{b^5}\right)^2$

37. $(x^2y^{-1})^2 - x^{-4}$

38. $s^2s^{-4} - (s^2)^{-4}$

39. $2a^{-2} + (2a^{-2})^4$

40. $3(a^{-1}z^2)^{-3} + c^{-2}z^{-1}$

41. $2 \times 3^{-1} + 4 \times 3^{-2}$

42. $5 \times 2^{-2} - 3^{-1} \times 2^3$

43. $(a^{-1} + b^{-1})^{-1}$

44. $(2a - b^{-2})^{-1}$

45. $(n^{-2} - 2n^{-1})^2$

46. $(2^{-3} - 4^{-1})^{-2}$

47. $\frac{3 - 2^{-1}}{3^{-2}}$

48. $\frac{6^{-1}}{4^{-2} + 2}$

49. $\frac{x - y^{-1}}{x^{-1} - y}$

50. $\frac{x^{-2} - y^{-2}}{x^{-1} - y^{-1}}$

51. $\frac{ax^{-2} + a^{-2}x}{a^{-1} + x^{-1}}$

52. $\frac{2x^{-2} - 2y^{-2}}{(xy)^{-3}}$

53. $2t^{-2} + t^{-1}(t + 1)$

54. $3x^{-1} - x^{-3}(y + 2)$

55. $(x - 1)^{-1} + (x + 1)^{-1}$

56. $4(2x - 1)(x + 2)^{-1} - (2x - 1)^2(x + 2)^{-2}$

In Exercises 57–64, perform the indicated operations.

57. Express $4^2 \times 64$ (a) as a power of 4 and (b) as a power of 2.

58. Express 1/81 (a) as a power of 9 and (b) as a power of 3.

59. By use of Eqs. (11-4) and (11-6), show that

$$\left(\frac{a}{b}\right)^{-n} = \left(\frac{b}{a}\right)^n$$

60. Using a calculator, verify the equation in Exercise 59 by evaluating the expression on each side with $a = 3.576$, $b = 8.091$, and $n = 7$.

61. When studying a solar energy system, the units encountered are $\text{kg} \cdot \text{s}^{-1}(\text{m} \cdot \text{s}^{-2})^2$. Simplify these units and include *joules* (see Example 6) and only positive exponents in the final result.

62. An expression encountered in finance is

$$\frac{p(1 + i)^{-1}[(1 + i)^{-n} - 1]}{(1 + i)^{-1} - 1}$$

where n is an integer. Simplify this expression.

63. In analyzing the tuning of an electronic circuit, the expression $[\omega\omega_0^{-1} - \omega_0\omega^{-1}]^2$ is used. Expand and simplify this expression.

64. In optics, the combined focal length F of two lenses is given by $F = [f_1^{-1} + f_2^{-1} + d(f_1f_2)^{-1}]^{-1}$, where f_1 and f_2 are the focal lengths of the lenses and d is the distance between them. Simplify the right side of this equation.

11-2 Fractional Exponents

In Section 11-1 we reviewed the use of integral exponents, including exponents that are negative integers and zero. We now show how rational numbers may be used as exponents. With the appropriate definitions, all the laws of exponents are valid for all rational numbers as exponents.

Equation (11-3) states that $(a^m)^n = a^{mn}$. If we were to let $m = \frac{1}{2}$ and $n = 2$, we would have $(a^{1/2})^2 = a^1$. However, we already have a way of writing a quantity which when squared equals a. This is written as $\sqrt{a}$. To be consistent with previous definitions and to allow the laws of exponents to hold, we define

$$a^{1/n} = \sqrt[n]{a} \qquad (11\text{-}7)$$

In order that Eqs. (11-3) and (11-7) may hold at the same time, we define

$$a^{m/n} = \sqrt[n]{a^m} = (\sqrt[n]{a})^m \qquad (11\text{-}8)$$

It can be shown that these definitions are valid for all the laws of exponents. We must note that Eqs. (11-7) and (11-8) are valid as long as $\sqrt[n]{a}$ does not involve the even root of a negative number. Such numbers are imaginary and are considered in Chapter 12.

For reference, Eq. (11-1) is $a^m \times a^n = a^{m+n}$

Eq. (11-3) is $(a^m)^n = a^{mn}$

EXAMPLE 1 ▸▸ We now verify that Eq. (11-1) holds for the above definitions:

$$a^{1/4}a^{1/4}a^{1/4}a^{1/4} = a^{(1/4)+(1/4)+(1/4)+(1/4)} = a^1$$

Now, $a^{1/4} = \sqrt[4]{a}$ by definition. Also, by definition $\sqrt[4]{a}\,\sqrt[4]{a}\,\sqrt[4]{a}\,\sqrt[4]{a} = a$. Equation (11-1) is thereby verified for $n = 4$ in Eq. (11-7).

Equation (11-3) is verified by the following:

$$(a^{1/4})(a^{1/4})(a^{1/4})(a^{1/4}) = (a^{1/4})^4 = a^1 = (\sqrt[4]{a})^4$$ ◂◂

We may interpret $a^{m/n}$ *in Eq. (11-8) as the mth power of the nth root of a, as well as the nth root of the mth power of a.* This is illustrated in the following example.

EXAMPLE 2 ▸▸

$$8^{2/3} = (\sqrt[3]{8})^2 = (2)^2 = 4 \quad \text{or} \quad 8^{2/3} = \sqrt[3]{8^2} = \sqrt[3]{64} = 4$$ ◂◂

Although both interpretations of Eq. (11-8) are possible, as indicated in Example 2, in evaluating numerical expressions involving fractional exponents without a calculator, it is almost always best to *find the root first, as indicated by the denominator* of the fractional exponent. This allows us to find the root of the smaller number, which is normally easier to find.

EXAMPLE 3 ▸▸ To evaluate $(64)^{5/2}$, we should proceed as follows:

$$(64)^{5/2} = [(64)^{1/2}]^5 = 8^5 = 32\,768$$

If we raised 64 to the fifth power first, we would have

$$(64)^{5/2} = (64^5)^{1/2} = (1\,073\,741\,824)^{1/2}$$

We would now have to evaluate the indicated square root. This demonstrates why it is preferable to find the indicated root first. ◂◂

For reference, basic forms of Eqs. (11-1) to (11-7) are as follows:

$$a^m \times a^n = a^{m+n} \quad \textbf{(11-1)}$$

$$\frac{a^m}{a^n} = a^{m-n} \quad \textbf{(11-2)}$$

$$(a^m)^n = a^{mn} \quad \textbf{(11-3)}$$

$$(ab)^n = a^n b^n \quad \textbf{(11-4)}$$

$$a^0 = 1 \quad \textbf{(11-5)}$$

$$a^{-n} = \frac{1}{a^n} \quad \textbf{(11-6)}$$

$$a^{1/n} = \sqrt[n]{a} \quad \textbf{(11-7)}$$

EXAMPLE 4 ▸▸ (a) $(16)^{3/4} = (16^{1/4})^3 = 2^3 = 8$

(b) $4^{-1/2} = \frac{1}{4^{1/2}} = \frac{1}{2}$ (c) $9^{3/2} = (9^{1/2})^3 = 3^3 = 27$

We note in (b) that Eq. (11-6) must also hold for negative rational exponents. In writing $4^{-1/2}$ as $1/4^{1/2}$, the *sign* of the exponent is changed. ◂◂

Fractional exponents allow us to find roots of numbers on a calculator. By use of the appropriate key (on most calculators $\boxed{x^y}$ or $\boxed{\wedge}$), we may raise any positive number to any power. For roots, we use the equivalent fractional exponent. Powers which are fractions or decimal in form are entered directly.

EXAMPLE 5 ▸▸ The thermodynamic temperature T (in kelvins (K)) is related to the pressure P (in kPa) of a gas by the equation $T = 80.5P^{2/7}$. Find the value of T for $P = 750$ kPa.

Substituting, we have

$$T = 80.5(750)^{2/7}$$

Evaluating this expression on a calculator, we enter the power as (2/7). The result is $T = 534$ K (rounded off). ◂◂

When finding powers of negative numbers, some calculators will show an error. If this is the case, enter the positive value of the number and then enter a negative sign for the result when appropriate. From Section 1-6 we recall that *an even root of a negative number is imaginary, and an odd root of a negative number is negative.* If it is an integral power, the basic laws of signs are used.

EXAMPLE 6 ▸▸ Plot the graph of the function $y = 2x^{1/3}$.

In obtaining points for the graph, we use (1/3) as the power when using the calculator. If your calculator does not evaluate powers of negative numbers, enter positive values for the negative values of x, and then make the results negative. Since $x^{1/3} = \sqrt[3]{x}$, we know that $x^{1/3}$ is negative for negative values of x because we have an odd root of a negative number. We get the following table of values.

x	−3	−2	−1	0	1	2	3
y	−2.9	−2.5	−2.0	0	2.0	2.5	2.9

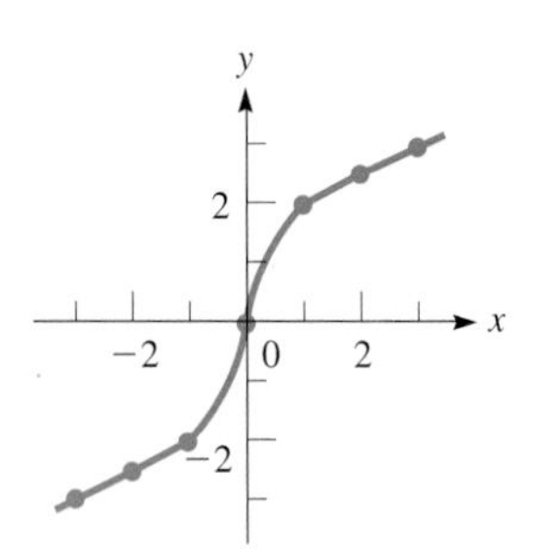

Fig. 11-1

The graph is shown in Fig. 11-1. Of course, this curve can easily be shown on a graphing calculator. ◂◂

Another reason for developing fractional exponents is that they are often easier to use in more complex expressions involving roots. This is true in algebra and in topics from more advanced mathematics. Any expression with radicals can also be expressed with fractional exponents and then simplified. We now show some additional examples with fractional exponents.

EXAMPLE 7 ▸▸ (a) $(8a^2b^4)^{1/3} = [(8^{1/3})(a^2)^{1/3}(b^4)^{1/3}]$ using Eq. (11-4)

$= 2a^{2/3}b^{4/3}$ using Eqs. (11-7) and (11-3)

(b) $a^{3/4}a^{4/5} = a^{3/4+4/5} = a^{31/20}$ using Eq. (11-1) ◂◂

EXAMPLE 8 ▸▸

$$\left(\frac{4^{-3/2}x^{2/3}y^{-7/4}}{2^{3/2}x^{-1/3}y^{3/4}}\right)^{2/3} = \left(\frac{x^{2/3}x^{1/3}}{2^{3/2}4^{3/2}y^{3/4}y^{7/4}}\right)^{2/3} \quad \text{using Eq. (11-6)}$$

$$= \left(\frac{x^{2/3+1/3}}{2^{3/2}4^{3/2}y^{3/4+7/4}}\right)^{2/3} \quad \text{using Eq. (11-1)}$$

$$= \frac{x^{(1)(2/3)}}{2^{(3/2)(2/3)}4^{(3/2)(2/3)}y^{(10/4)(2/3)}} \quad \text{using Eq. (11-4)}$$

$$= \frac{x^{2/3}}{8y^{5/3}}$$ ◂◂

EXAMPLE 9 ▸▸

$$(4x^4)^{-1/2} - 3x^{-3} = \frac{1}{(4x^4)^{1/2}} - \frac{3}{x^3} \quad \text{using Eq. (11-6)}$$

$$= \frac{1}{2x^2} - \frac{3}{x^3} \quad \text{using Eq. (11-7)}$$

$$= \frac{x-6}{2x^3} \quad \text{common denominator}$$ ◂◂

See the chapter introduction.

EXAMPLE 10 ▸▸ The rate R at which solar radiation changes at a solar energy collector during a day is given by the equation

$$R = \frac{(t^4 + 100)^{1/2} - 2t^3(t+6)(t^4+100)^{-1/2}}{[(t^4+100)^{1/2}]^2}$$

Here R is measured in $\text{kW}/(\text{m}^2 \cdot \text{h})$, t is the number of hours from noon, and $-6\text{ h} \le t \le 8\text{ h}$. Express the right side of this equation in simpler form, and find R for $t = 0$ (noon) and for $t = 4$ h (4 P.M.)

Performing the simplification, we have the following steps:

$$R = \frac{(t^4+100)^{1/2} - \dfrac{2t^3(t+6)}{(t^4+100)^{1/2}}}{(t^4+100)} \quad \begin{array}{l}\leftarrow \text{using Eq. (11-6)} \\ \leftarrow \text{using Eq. (11-3)}\end{array}$$

$$= \frac{\dfrac{(t^4+100)^{1/2}(t^4+100)^{1/2} - 2t^3(t+6)}{(t^4+100)^{1/2}}}{(t^4+100)} \quad \leftarrow \text{common denominator}$$

$$= \frac{(t^4+100) - 2t^3(t+6)}{(t^4+100)^{1/2}} \times \frac{1}{t^4+100} \quad \leftarrow \text{invert divisor and multiply}$$

$$= \frac{100 - 12t^3 - t^4}{(t^4+100)^{1/2}(t^4+100)} = \frac{100 - 12t^3 - t^4}{(t^4+100)^{3/2}} \quad \leftarrow \text{using Eq. (11-1)}$$

For $t = 0$: $$R = \frac{100 - 12(0^3) - 0^4}{(0^4+100)^{3/2}} = 0.10\ \text{kW}/(\text{m}^2 \cdot \text{h})$$

For $t = 4$ h: $$R = \frac{100 - 12(4^3) - 4^4}{(4^4+100)^{3/2}} = -0.14\ \text{kW}/(\text{m}^2 \cdot \text{h})$$

We see that the radiation is increasing at noon, and the negative sign tells us that it is decreasing at 4 P.M. ◂◂

Exercises 11–2

In Exercises 1–28, evaluate the given expressions.

1. $(25)^{1/2}$
2. $(49)^{1/2}$
3. $(27)^{1/3}$
4. $(81)^{1/4}$
5. $8^{4/3}$
6. $(125)^{2/3}$
7. $(100)^{25/2}$
8. $(16)^{5/4}$
9. $8^{-1/3}$
10. $16^{-1/4}$
11. $(64)^{-2/3}$
12. $(32)^{-4/5}$
13. $5^{1/2}5^{3/2}$
14. $8^{1/3}4^{-1/2}$
15. $(4^4)^{3/2}$
16. $(3^6)^{2/3}$
17. $\dfrac{121^{-1/2}}{100^{1/2}}$
18. $\dfrac{1000^{1/3}}{400^{-1/2}}$
19. $\dfrac{7^{-1/2}}{6^{-1}7^{1/2}}$
20. $\dfrac{15^{2/3}}{5^2 15^{-1/3}}$
21. $\dfrac{(-27)^{1/3}}{6}$
22. $\dfrac{(-8)^{2/3}}{-2}$
23. $\dfrac{-8}{(-27)^{-1/3}}$
24. $\dfrac{-4}{(-64)^{-2/3}}$
25. $(125)^{-2/3} - (100)^{-3/2}$
26. $32^{0.4} + 25^{-0.5}$
27. $\dfrac{16^{-0.25}}{5} + \dfrac{2^{-0.6}}{2^{0.4}}$
28. $\dfrac{4^{-1}}{(36)^{-1/2}} - \dfrac{5^{-1/2}}{5^{1/2}}$

In Exercises 29–32, evaluate the given expressions by use of a calculator.

29. $(17.98)^{1/4}$
30. $(750.81)^{2/3}$
31. $(4.0187)^{-4/9}$
32. $(0.1863)^{-1/6}$

In Exercises 33–60, simplify the given expressions. Express all answers with positive exponents.

33. $a^{2/3}a^{1/2}$
34. $x^{5/6}x^{-1/3}$
35. $\dfrac{y^{-1/2}}{y^{2/5}}$
36. $\dfrac{2r^{4/5}}{r^{-1}}$
37. $\dfrac{s^{1/4}s^{2/3}}{s^{-1}}$
38. $\dfrac{x^{3/10}}{x^{-1/5}x^2}$
39. $\dfrac{y^{-1}}{y^{1/3}y^{-1/4}}$
40. $\dfrac{a^{-2/5}a^2}{a^{-3/10}}$
41. $(8a^3b^6)^{1/3}$
42. $(8b^{-4}c^2)^{2/3}$
43. $(16a^4b^3)^{-3/4}$
44. $(32x^5y^4)^{-2/5}$
45. $\frac{1}{2}(4x^2+1)^{-1/2}(8x)$
46. $\frac{2}{3}(x^3+1)^{-1/3}(3x^2)$
47. $\left(\dfrac{9t^{-2}}{16}\right)^{3/2}$
48. $\left(\dfrac{a^{5/7}}{a^{2/3}}\right)^{7/4}$
49. $\left(\dfrac{4a^{5/6}b^{-1/5}}{a^{2/3}b^2}\right)^{-1/2}$
50. $\left(\dfrac{a^0b^8c^{-1/8}}{ab^{63/64}}\right)^{32/3}$
51. $\left(\dfrac{6x^{-1/2}y^{2/3}}{18x^{-1}}\right)\left(\dfrac{2y^{1/4}}{x^{1/3}}\right)$
52. $\dfrac{3^{-1}a^{1/2}}{4^{-1/2}b} \div \dfrac{9^{1/2}a^{-1/3}}{2b^{-1/4}}$
53. $(x^{-1}+2x^{-2})^{-1/2}$
54. $(a^{-2}-a^{-4})^{-1/4}$
55. $(a^3)^{-4/3}+a^{-2}$
56. $(4x^6)^{-1/2}-2x^{-1}$
57. $[(a^{1/2}-a^{-1/2})^2+4]^{1/2}$
58. $4x^{1/2}+\frac{1}{2}x^{-1/2}(4x+1)$
59. $x^2(2x-1)^{-1/2}+2x(2x-1)^{1/2}$
60. $(3x-1)^{-2/3}(1-x)-(3x-1)^{1/3}$

In Exercises 61–64, graph the given functions.

61. $f(x) = 3x^{1/2}$
62. $f(x) = 2x^{2/3}$
63. $f(x) = x^{4/5}$
64. $f(x) = 4x^{3/2}$

In Exercises 65–68, perform the indicated operations.

65. A factor used in determining the performance of a solar-energy storage system is $(A/S)^{-1/4}$, where A is the actual storage capacity and S is a standard storage capacity. If this factor is 0.5, explain how to find the ratio A/S.

66. A factor used in measuring the loudness sensed by the human ear is $(I/I_0)^{0.3}$, where I is the intensity of the sound and I_0 is a reference intensity. Evaluate this factor for $I = 3.2 \times 10^{-6}$ W/m² (ordinary conversation) and $I_0 = 10^{-12}$ W/m².

67. The electric current i, in amperes, in a circuit containing a battery of voltage E, a resistance R, and an inductance L is given by $i = \dfrac{E}{R}(1 - e^{-Rt/L})$, where t is the time after the circuit is closed. See Fig. 11-2. Find i for $E = 6.20$ V, $R = 1.20\ \Omega$, $L = 3.24$ H, and $t = 0.00100$ s. (The number e is irrational and can be found from the calculator by using the e^x key with $x = 1$.)

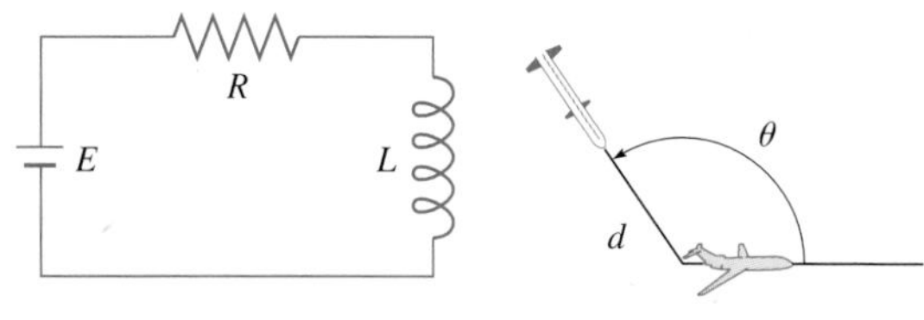

Fig. 11-2 **Fig. 11-3**

68. For a heat-seeking rocket in pursuit of an aircraft, the distance from the rocket to the aircraft is

$$d = \frac{500(\sin\theta)^{1/2}}{(1-\cos\theta)^{3/2}}$$

where θ is shown in Fig. 11-3. Find d, in km, for $\theta = 125.0°$.

11-3 Simplest Radical Form

Radicals were first introduced in Section 1-6 and were used again when we developed the concept of a fractional exponent. As we mentioned in the previous section, it is possible to use fractional exponents for any operation required with radicals. For operations involving multiplication and division, using the fractional exponent form has certain advantages. However, for the addition and subtraction of radicals, there is normally little advantage in changing form.

We shall now define the operations with radicals so that these definitions are consistent with the laws of exponents. This will enable us from now on to use either fractional exponents or radicals, whichever is more convenient for the operation being performed.

$$\sqrt[n]{a^n} = (\sqrt[n]{a})^n = a \qquad (11\text{-}9)$$

$$\sqrt[n]{a}\sqrt[n]{b} = \sqrt[n]{ab} \qquad (11\text{-}10)$$

$$\sqrt[m]{\sqrt[n]{a}} = \sqrt[mn]{a} \qquad (11\text{-}11)$$

$$\frac{\sqrt[n]{a}}{\sqrt[n]{b}} = \sqrt[n]{\frac{a}{b}}, \quad b \neq 0 \qquad (11\text{-}12)$$

The number under the radical is called the **radicand**, *and the number indicating the root being taken is called the* **order** *(or* **index***) of the radical.* To avoid difficulties with imaginary numbers (which are considered in the next chapter), *we shall assume that all letters represent positive numbers.*

NOTE ▶

EXAMPLE 1 ▶▶ Following are illustrations of the use of each of Eqs. (11-9) to (11-12).

(a) $\sqrt[5]{4^5} = (\sqrt[5]{4})^5 = 4$ using Eq. (11-9)

(b) $\sqrt[3]{2}\sqrt[3]{3} = \sqrt[3]{2 \times 3} = \sqrt[3]{6}$ using Eq. (11-10)

(c) $\sqrt[3]{\sqrt{5}} = \sqrt[3\times 2]{5} = \sqrt[6]{5}$ using Eq. (11-11)

(d) $\dfrac{\sqrt{7}}{\sqrt{3}} = \sqrt{\dfrac{7}{3}}$ using Eq. (11-12) ◀◀

EXAMPLE 2 ▶▶ In Example 5 of Section 1-6, we saw that

$\sqrt{16 + 9}$ is ***not*** equal to $\sqrt{16} + \sqrt{9}$

However, using Eq. (11-10),

$$\sqrt{16 \times 9} = \sqrt{16} \times \sqrt{9} = 4 \times 3 = 12$$

NOTE ▶ Therefore, we must *be careful to distinguish between the root of a sum of terms and the root of a product of factors.* This is the same as with powers of sums and powers of products, as shown in Example 9 of Section 11-1. It should be the same, as a root can be interpreted as a fractional exponent. ◀◀

There are certain operations which are performed on radicals in order to put them in their simplest form. The following two examples illustrate one of these operations.

EXAMPLE 3 ⏩ To simplify $\sqrt{75}$, we know that $75 = (25)(3)$ and that $\sqrt{25} = 5$. As in Section 1-6, and now using Eq. (11-10), we write

$$\sqrt{75} = \sqrt{(25)(3)} = \sqrt{25}\sqrt{3} = 5\sqrt{3}$$

(25): perfect square

This illustrates one step which should always be carried out in simplifying radicals: *Always remove all perfect nth-power factors from the radicand of a radical of order n.* ⏪

EXAMPLE 4 ⏩ (a) $\sqrt{72} = \sqrt{(36)(2)} = \sqrt{36}\sqrt{2} = 6\sqrt{2}$

(36): perfect square

(b) $\sqrt{a^3b^2} = \sqrt{(a^2)(a)(b^2)} = \sqrt{a^2}\sqrt{a}\sqrt{b^2} = ab\sqrt{a}$

(a^2), (b^2): perfect squares

(c) cube root ⟶ $\sqrt[3]{40} = \sqrt[3]{(8)(5)} = \sqrt[3]{8}\sqrt[3]{5} = 2\sqrt[3]{5}$

(8): perfect cube

(d) fifth root ⟶ $\sqrt[5]{64x^8y^{12}} = \sqrt[5]{(32)(2)(x^5)(x^3)(y^{10})(y^2)}$

(32), (x^5), (y^{10}): perfect fifth powers

$$= \sqrt[5]{(32)(x^5)(y^{10})}\sqrt[5]{2x^3y^2}$$
$$= 2xy^2\sqrt[5]{2x^3y^2}$$ ⏪

The next two examples illustrate another procedure that is used to simplify certain radicals. This procedure is to *reduce the order of the radical,* when it is possible to do so.

EXAMPLE 5 ⏩ $\sqrt[6]{8} = \sqrt[6]{2^3} = 2^{3/6} = 2^{1/2} = \sqrt{2}$

In this example we started with a sixth root and ended with a square root. Thus, the order of the radical was reduced. Fractional exponents are often helpful when we perform this operation. ⏪

EXAMPLE 6 ⏩ (a) $\sqrt[8]{16} = \sqrt[8]{2^4} = 2^{4/8} = 2^{1/2} = \sqrt{2}$

(b) $\dfrac{\sqrt[4]{9}}{\sqrt{3}} = \dfrac{\sqrt[4]{3^2}}{\sqrt{3}} = \dfrac{3^{2/4}}{3^{1/2}} = 1$

(c) $\dfrac{\sqrt[6]{8}}{\sqrt{7}} = \dfrac{\sqrt[6]{2^3}}{\sqrt{7}} = \dfrac{2^{1/2}}{7^{1/2}} = \sqrt{\dfrac{2}{7}}$

(d) $\sqrt[9]{27x^6y^{12}} = \sqrt[9]{3^3x^6y^9y^3} = 3^{3/9}x^{6/9}y^{9/9}y^{3/9} = 3^{1/3}x^{2/3}yy^{1/3}$

$$= y\sqrt[3]{3x^2y}$$ ⏪

If a radical is to be written in its *simplest form*, the two operations illustrated in the last four examples must be performed. Therefore, we have the following:

> ***Steps to Reduce a Radical to Simplest Form***
>
> **1.** *Remove all perfect nth-power factors from a radical of order n.*
>
> **2.** *If possible, reduce the order of the radical.*

When working with fractions, it has traditionally been the practice to write a fraction with radicals in a form in which the denominator contains no radicals. Such a fraction was not considered to be in simplest form unless this was done. This step of simplification was performed primarily for ease of calculation, but with a calculator it does not matter to any extent that there is a radical in the denominator. However, the procedure of writing a radical in this form, called **rationalizing the denominator,** is at times useful for other purposes. Therefore, the following examples show how the process of rationalizing the denominator is carried out.

EXAMPLE 7 ▸▸ To write $\sqrt{\frac{2}{5}}$ in an equivalent form in which the denominator is not included under the radical sign, we *create a perfect square in the denominator* by multiplying the numerator and the denominator under the radical by 5. This gives us $\sqrt{\frac{10}{25}}$, which may be written as $\frac{1}{5}\sqrt{10}$ or $\frac{\sqrt{10}}{5}$. These steps are written as follows:

$$\sqrt{\frac{2}{5}} = \sqrt{\frac{2 \times 5}{5 \times 5}} = \sqrt{\frac{10}{25}} = \frac{\sqrt{10}}{\sqrt{25}} = \frac{\sqrt{10}}{5}$$

perfect square ◂◂

EXAMPLE 8 ▸▸ (a) $$\frac{5}{\sqrt{18}} = \frac{5}{3\sqrt{2}} = \frac{5(\sqrt{2})}{3\sqrt{2}(\sqrt{2})} = \frac{5\sqrt{2}}{3\sqrt{4}} = \frac{5\sqrt{2}}{6}$$

perfect square

(b) cube root ⟶ $$\sqrt[3]{\frac{2}{3}} = \sqrt[3]{\frac{2 \times 9}{3 \times 9}} = \sqrt[3]{\frac{18}{27}} = \frac{\sqrt[3]{18}}{\sqrt[3]{27}} = \frac{\sqrt[3]{18}}{3}$$

perfect cube

In (a), a perfect square was made by multiplying by $\sqrt{2}$. We can indicate this as we have shown, or by multiplying the numerator by $\sqrt{2}$ and multiplying the 2 under the radical in the denominator by 2, in which case the denominator would be $3\sqrt{2 \times 2}$. In (b), we want a perfect cube, since a cube root is being found. ◂◂

EXAMPLE 9 ▸▸ The period T (in seconds) for one cycle of a simple pendulum is given by $T = 2\pi\sqrt{L/g}$, where L is the length of the pendulum and g is the acceleration due to gravity. Rationalize the denominator on the right side of this equation if $L = 0.915$ m and $g = 9.80$ m/s^2.

Substituting, and then rationalizing, we have

$$T = 2\pi\sqrt{\frac{0.915}{9.80}} = 1.92 \text{ s}$$ ◂◂

EXAMPLE 10 ▸▸ Simplify $\sqrt{\frac{1}{2a^2}+\frac{2}{b^2}}$ and rationalize the denominator.

first combine fractions over lowest common denominator →

$$\sqrt{\frac{1}{2a^2}+\frac{2}{b^2}} = \sqrt{\frac{b^2+4a^2}{2a^2b^2}} = \frac{\sqrt{b^2+4a^2}}{ab\sqrt{2}}$$

← sum of squares—radical cannot be simplified

$$= \frac{\sqrt{b^2+4a^2}\sqrt{2}}{ab\sqrt{2}\times 2} = \frac{\sqrt{2(b^2+4a^2)}}{2ab}$$

◂◂

Exercises 11–3

In Exercises 1–60, write each expression in simplest radical form. Where a radical appears in a denominator, rationalize the denominator.

1. $\sqrt{24}$
2. $\sqrt{150}$
3. $\sqrt{45}$
4. $\sqrt{98}$
5. $\sqrt{x^2y^5}$
6. $\sqrt{s^3t^6}$
7. $\sqrt{pq^2r^7}$
8. $\sqrt{x^2y^4z^3}$
9. $\sqrt{5x^2}$
10. $\sqrt{12ab^2}$
11. $\sqrt{18a^3bc^4}$
12. $\sqrt{54m^5n^3}$
13. $\sqrt[3]{16}$
14. $\sqrt[4]{48}$
15. $\sqrt[5]{96}$
16. $\sqrt[3]{-16}$
17. $\sqrt[3]{8a^2}$
18. $\sqrt[3]{5a^4b^2}$
19. $\sqrt[4]{64r^3s^4t^5}$
20. $\sqrt[5]{16x^5y^3z^{11}}$
21. $\sqrt[5]{8}\sqrt[5]{4}$
22. $\sqrt[7]{4}\sqrt[7]{64}$
23. $\sqrt[3]{ab^4}\sqrt[3]{a^2b}$
24. $\sqrt[6]{3m^5n^8}\sqrt[6]{9mn}$
25. $\sqrt{\frac{3}{2}}$
26. $\sqrt{\frac{6}{5}}$
27. $\sqrt{\frac{a}{b}}$
28. $\sqrt{\frac{a}{b^3}}$
29. $\sqrt[3]{\frac{3}{4}}$
30. $\sqrt[4]{\frac{2}{5}}$
31. $\sqrt[5]{\frac{1}{9}}$
32. $\sqrt[6]{\frac{5}{4}}$
33. $\sqrt[4]{400}$
34. $\sqrt[8]{81}$
35. $\sqrt[6]{64}$
36. $\sqrt[9]{27}$
37. $\sqrt{4\times 10^4}$
38. $\sqrt{4\times 10^5}$
39. $\sqrt{4\times 10^6}$
40. $\sqrt{16\times 10^5}$
41. $\sqrt[4]{4a^2}$
42. $\sqrt[6]{b^2c^4}$
43. $\sqrt[4]{\frac{1}{4}}$
44. $\frac{\sqrt[4]{80}}{\sqrt[4]{5}}$
45. $\sqrt[4]{\sqrt[3]{16}}$
46. $\sqrt[5]{\sqrt[4]{9}}$
47. $\sqrt{\sqrt{\sqrt{2}}}$
48. $\sqrt{b^4\sqrt{a}}$
49. $\sqrt{\frac{1}{2}-\frac{1}{3}}$
50. $\sqrt{\frac{5}{4}-\frac{1}{8}}$
51. $\sqrt{\frac{1}{a^2}+\frac{1}{b}}$
52. $\sqrt{\frac{x}{y}+\frac{y}{x}}$
53. $\sqrt{\frac{x}{2x+1}}$
54. $\sqrt{\frac{x-2}{x+2}}$
55. $\sqrt{a^2+2ab+b^2}$
56. $\sqrt{a^2+b^2}$
57. $\sqrt{4x^2-1}$
58. $\sqrt{9x^2-6x+1}$
59. $\sqrt{x^2+\frac{1}{4}}$
60. $\sqrt{\frac{1}{2}+2r+2r^2}$

In Exercises 61–64, perform the required operation.

61. An approximate equation for the efficiency E, in percent, of an engine is given by $E = 100(1 - 1/\sqrt[5]{R^2})$, where R is the compression ratio. Explain how this equation can be written with fractional exponents, and then find E for $R = 7.35$.

62. When analyzing the velocity of an object that falls through a very great distance, the expression $a\sqrt{2g/a}$ is derived. Show by rationalizing the denominator that this expression takes on a simpler form.

63. A formula for the angular velocity of a disc is $\omega = \sqrt{\frac{576EIg}{WL^3}}$. Rationalize the denominator on the right side of this equation.

64. In analyzing an electronic filter circuit, the expression $\frac{8A}{\pi^2\sqrt{1+(f_0/f)^2}}$ is used. Rationalize the denominator, expressing the answer without the fraction f_0/f.

11-4 Addition and Subtraction of Radicals

When we first introduced the concept of adding algebraic expressions, we found that it was possible to combine similar terms, that is, those which differed only in numerical coefficients. The same is true of adding radicals. *We must have similar radicals in order to perform the addition,* rather than simply to be able to indicate addition. *By* **similar radicals** *we mean radicals that differ only in their numerical coefficients* and which therefore must be of the same order and have the same radicand.

In order to add radicals, we first express each radical in its simplest form, rationalize any denominators, and then combine those which are similar. For those which are not similar, we can only indicate the addition.

EXAMPLE 1 ▸▸ (a) $2\sqrt{7} - 5\sqrt{7} + \sqrt{7} = -2\sqrt{7}$ all similar radicals

This result follows the distributive law, as it should. We can write

$$2\sqrt{7} - 5\sqrt{7} + \sqrt{7} = (2 - 5 + 1)\sqrt{7} = -2\sqrt{7}$$

We can also see that the terms combine just as

$$2x - 5x + x = -2x$$

(b) $\sqrt[5]{6} + 4\sqrt[5]{6} - 2\sqrt[5]{6} = 3\sqrt[5]{6}$ all similar radicals

(c) $\sqrt{5} + 2\sqrt{3} - 5\sqrt{5} = 2\sqrt{3} - 4\sqrt{5}$ answer contains two terms

similar radicals ($\sqrt{5}$ and $5\sqrt{5}$)

We note in (c) that we are only able to indicate the final subtraction since the radicals are not similar. ◂◂

EXAMPLE 2 ▸▸ (a) $\sqrt{2} + \sqrt{8} = \sqrt{2} + \sqrt{4 \times 2} = \sqrt{2} + \sqrt{4}\sqrt{2}$

$$= \sqrt{2} + 2\sqrt{2} = 3\sqrt{2}$$

(b) $\sqrt[3]{24} + \sqrt[3]{81} = \sqrt[3]{8 \times 3} + \sqrt[3]{27 \times 3} = \sqrt[3]{8}\sqrt[3]{3} + \sqrt[3]{27}\sqrt[3]{3}$

$$= 2\sqrt[3]{3} + 3\sqrt[3]{3} = 5\sqrt[3]{3}$$

Notice that $\sqrt{8}$, $\sqrt[3]{24}$, and $\sqrt[3]{81}$ were simplified before performing the addition.

CAUTION ▸ We also note that $\sqrt{2} + \sqrt{8}$ ***is not equal to*** $\sqrt{2 + 8}$. ◂◂

We note in the illustrations of Example 2 that the radicals do not initially appear to be similar. However, after each is simplified we are able to recognize the similar radicals.

EXAMPLE 3 ▸▸ (a) $6\sqrt{7} - \sqrt{28} + 3\sqrt{63} = 6\sqrt{7} - \sqrt{4 \times 7} + 3\sqrt{9 \times 7}$

$$= 6\sqrt{7} - 2\sqrt{7} + 3(3\sqrt{7})$$

$$= 6\sqrt{7} - 2\sqrt{7} + 9\sqrt{7}$$ ← all similar radicals

$$= 13\sqrt{7}$$

(b) $3\sqrt{125} - \sqrt{20} + \sqrt{27} = 3\sqrt{25 \times 5} - \sqrt{4 \times 5} + \sqrt{9 \times 3}$

$$= 3(5\sqrt{5}) - 2\sqrt{5} + 3\sqrt{3}$$

$$= 13\sqrt{5} + 3\sqrt{3}$$ ($3\sqrt{3}$) not similar to others ◂◂

EXAMPLE 4 ▶▶ $\sqrt{24} + \sqrt{\dfrac{3}{2}} = \sqrt{4 \times 6} + \sqrt{\dfrac{3 \times 2}{2 \times 2}} = \sqrt{4}\sqrt{6} + \sqrt{\dfrac{6}{4}}$

$$= 2\sqrt{6} + \frac{\sqrt{6}}{2} = \frac{4\sqrt{6} + \sqrt{6}}{2} = \frac{5}{2}\sqrt{6}$$

One radical was simplified by removing the perfect square factor, and in the other we rationalized the denominator. Note that we would not be able to combine the radicals if we did not rationalize the denominator of the second radical. ◀◀

Our main purpose in this section is to add radicals in radical form. However, a decimal value can be obtained by use of a calculator, and in the following example we use decimal values to verify the result of the addition.

EXAMPLE 5 ▶▶ Perform the addition $\sqrt{32} + 7\sqrt{18} - 2\sqrt{200}$, and use a calculator to verify the result.

$$\begin{aligned}\sqrt{32} + 7\sqrt{18} - 2\sqrt{200} &= \sqrt{16 \times 2} + 7\sqrt{9 \times 2} - 2\sqrt{100 \times 2} \\ &= 4\sqrt{2} + 7(3\sqrt{2}) - 2(10\sqrt{2}) \\ &= 4\sqrt{2} + 21\sqrt{2} - 20\sqrt{2} \\ &= 5\sqrt{2}\end{aligned}$$

Using a calculator we find

$$5\sqrt{2} = 7.071\,067\,812 \quad \text{and} \quad \sqrt{32} + 7\sqrt{18} - 2\sqrt{200} = 7.071\,067\,812$$

which shows that the result has the same value as the original expression. ◀◀

We now show an example of adding radical expressions that contain literal expressions.

EXAMPLE 6 ▶▶

$$\begin{aligned}\sqrt{\frac{2}{3a}} - 2\sqrt{\frac{3}{2a}} &= \sqrt{\frac{2(3a)}{3a(3a)}} - 2\sqrt{\frac{3(2a)}{2a(2a)}} = \sqrt{\frac{6a}{9a^2}} - 2\sqrt{\frac{6a}{4a^2}} \\ &= \frac{1}{3a}\sqrt{6a} - \frac{2}{2a}\sqrt{6a} = \frac{1}{3a}\sqrt{6a} - \frac{1}{a}\sqrt{6a} \\ &= \frac{\sqrt{6a} - 3\sqrt{6a}}{3a} = \frac{-2\sqrt{6a}}{3a} = -\frac{2}{3a}\sqrt{6a}\end{aligned}$$ ◀◀

Exercises 11–4

In Exercises 1–36, express each radical in simplest form, rationalize denominators, and perform the indicated operations.

1. $2\sqrt{3} + 5\sqrt{3}$
2. $8\sqrt{11} - 3\sqrt{11}$
3. $2\sqrt{7} + \sqrt{5} - 3\sqrt{7}$
4. $8\sqrt{6} - 2\sqrt{3} - 5\sqrt{6}$
5. $\sqrt{5} + \sqrt{20}$
6. $\sqrt{7} + \sqrt{63}$
7. $2\sqrt{3} - 3\sqrt{12}$
8. $4\sqrt{2} - \sqrt{50}$
9. $\sqrt{8a} - \sqrt{32a}$
10. $\sqrt{27x} + 2\sqrt{18x}$
11. $2\sqrt{28} + 3\sqrt{175}$
12. $5\sqrt{300} - 7\sqrt{48}$
13. $2\sqrt{20} - \sqrt{125} - \sqrt{45}$
14. $2\sqrt{44} - \sqrt{99} + \sqrt{2}\sqrt{88}$
15. $3\sqrt{75} + 2\sqrt{48} - 2\sqrt{18}$
16. $2\sqrt{28} - \sqrt{108} - 2\sqrt{175}$
17. $\sqrt{60} + \sqrt{\frac{5}{3}}$
18. $\sqrt{84} - \sqrt{\frac{3}{7}}$
19. $\sqrt{\frac{1}{2}} + \sqrt{\frac{25}{2}} - \sqrt{18}$
20. $\sqrt{6} - \sqrt{\frac{2}{3}} - \sqrt{18}$
21. $\sqrt[3]{81} + \sqrt[3]{3000}$
22. $\sqrt[3]{-16} + \sqrt[3]{54}$

23. $\sqrt[4]{32} - \sqrt[8]{4}$

24. $\sqrt[6]{\sqrt{2}} - \sqrt[12]{2^{13}}$

25. $\sqrt{a^3b} - \sqrt{4ab^5}$

26. $\sqrt{2x^2y} + \sqrt{8}\sqrt{y^5}$

27. $\sqrt{6}\sqrt{5}\sqrt{3} - \sqrt{40a^2}$

28. $\sqrt{60b^2n} - b\sqrt{135n}$

29. $\sqrt[3]{24a^2b^4} - \sqrt[3]{3a^5b}$

30. $\sqrt[5]{32a^6b^4} + 3a\sqrt[5]{243ab^9}$

31. $\sqrt{\frac{a}{c^5}} - \sqrt{\frac{c}{a^3}}$

32. $\sqrt{\frac{2x}{3y}} + \sqrt{\frac{27y}{8x}}$

33. $\sqrt[3]{\frac{a}{b}} - \sqrt[3]{\frac{8b^2}{a^2}}$

34. $\sqrt[4]{\frac{c}{b}} - \sqrt[4]{bc}$

35. $\sqrt{\frac{a-b}{a+b}} - \sqrt{\frac{a+b}{a-b}}$

36. $\sqrt{\frac{16}{x} + 8 + x} - \sqrt{1 - \frac{1}{x}}$

In Exercises 37–40, express each radical in simplest form, rationalize denominators, and perform the indicated operations. Then use a calculator to verify the result.

37. $3\sqrt{45} + 3\sqrt{75} - 2\sqrt{500}$

38. $2\sqrt{40} + 3\sqrt{90} - 5\sqrt{250}$

39. $2\sqrt{\frac{2}{3}} + \sqrt{24} - 5\sqrt{\frac{3}{2}}$

40. $\sqrt{\frac{2}{7}} - 2\sqrt{\frac{7}{2}} + 5\sqrt{56}$

In Exercises 41–44, solve the given problems.

41. Find the exact sum of the positive roots of $x^2 - 2x - 2 = 0$ and $x^2 + 2x - 11 = 0$.

42. Find the sum of the two roots of the quadratic equation $ax^2 + bx + c = 0$.

43. A rectangular piece of plywood 1200 mm by 2400 mm has corners cut from it as shown in Fig. 11-4. Find the perimeter of the remaining piece in exact form and in decimal form.

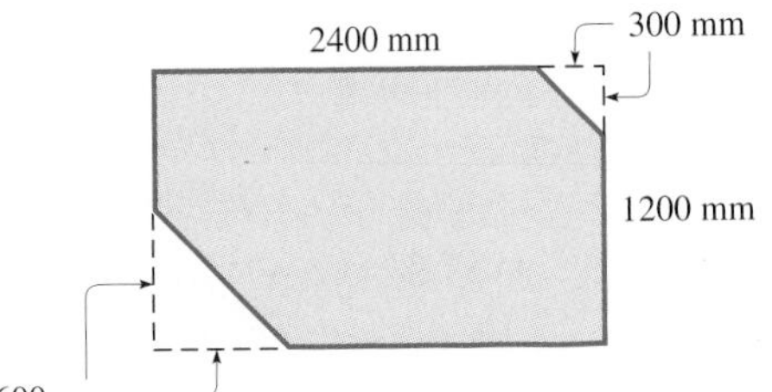

Fig. 11-4

44. In the study of heat transfer, the expression $\sqrt{mk}\left(\sqrt{T_1} - \frac{T_2}{\sqrt{T_1}}\right)$ is found. Combine the terms over a rationalized common denominator.

11-5 Multiplication and Division of Radicals

When multiplying expressions containing radicals, we use Eq. (11-10), along with the normal procedures of algebraic multiplication. *Note that the orders of the radicals being multiplied in Eq. (11-10) are the same.* The following examples illustrate the method.

For reference Eq. (11-10) is $\sqrt[n]{a}\sqrt[n]{b} = \sqrt[n]{ab}$.

EXAMPLE 1 ▸▸ (a) $\sqrt{5}\sqrt{2} = \sqrt{5 \times 2} = \sqrt{10}$

(b) $\sqrt{33}\sqrt{3} = \sqrt{33 \times 3} = \sqrt{99} = \sqrt{9 \times 11} = \sqrt{9}\sqrt{11}$ (perfect square: 9)
$= 3\sqrt{11}$

or $\sqrt{33}\sqrt{3} = \sqrt{33 \times 3} = \sqrt{11 \times 3 \times 3}$
$= 3\sqrt{11}$

Note that we express the resulting radical in simplest form. ◂◂

EXAMPLE 2 ▸▸ (a) $\sqrt[3]{6}\sqrt[3]{4} = \sqrt[3]{6(4)} = \sqrt[3]{24} = \sqrt[3]{8}\sqrt[3]{3}$ (perfect cube: 8)
$= 2\sqrt[3]{3}$

(b) $\sqrt[5]{8a^3b^4}\sqrt[5]{8a^2b^3} = \sqrt[5]{(8a^3b^4)(8a^2b^3)} = \sqrt[5]{64a^5b^7} = \sqrt[5]{32a^5b^5}\sqrt[5]{2b^2}$ (perfect fifth power: $32a^5b^5$)
$= 2ab\sqrt[5]{2b^2}$ ◂◂

EXAMPLE 3 ▸▸ $\sqrt{2}(3\sqrt{5} - 4\sqrt{2}) = 3\sqrt{2}\sqrt{5} - 4\sqrt{2}\sqrt{2} = 3\sqrt{10} - 4\sqrt{4}$
$= 3\sqrt{10} - 4(2) = 3\sqrt{10} - 8$ ◂◂

EXAMPLE 4 ▸▸

$$\begin{aligned}(5\sqrt{7} - 2\sqrt{3})(4\sqrt{7} + 3\sqrt{3}) &= (5\sqrt{7})(4\sqrt{7}) + (5\sqrt{7})(3\sqrt{3}) - (2\sqrt{3})(4\sqrt{7}) - (2\sqrt{3})(3\sqrt{3})\\ &= (5)(4)\sqrt{7}\sqrt{7} + (5)(3)\sqrt{7}\sqrt{3} - (2)(4)\sqrt{3}\sqrt{7} - (2)(3)\sqrt{3}\sqrt{3}\\ &= 20(7) + 15\sqrt{21} - 8\sqrt{21} - 6(3)\\ &= 140 + 7\sqrt{21} - 18\\ &= 122 + 7\sqrt{21}\end{aligned}$$

Using a calculator, we find that the decimal value of the original expression is 154.078 029 9, which is the same as that for $122 + 7\sqrt{21}$. ◂◂

For reference, Eqs. (6-3), (6-4), and (11-10) are
$(x + y)^2 = x^2 + 2xy + y^2$
$(x - y)^2 = x^2 - 2xy + y^2$
$\sqrt[n]{a}\sqrt[n]{b} = \sqrt[n]{ab}$

When raising a single-term radical expression to a power, we use the basic meaning of the power. When raising a binomial to a power, we proceed as with any binomial. We use Eqs. (6-3) and (6-4) along with Eq. (11-10) if the binomial is squared. These are illustrated in the next two examples.

EXAMPLE 5 ▸▸ (a) $(2\sqrt{7})^2 = 2^2(\sqrt{7})^2 = 4(7) = 28$

(b) $(2\sqrt{7})^3 = 2^3(\sqrt{7})^3 = 8(\sqrt{7})^2(\sqrt{7}) = 8(7)\sqrt{7}$
$= 56\sqrt{7}$ ◂◂

EXAMPLE 6 ▸▸ (a) $(3 + \sqrt{5})^2 = 3^2 + 2(3)\sqrt{5} + (\sqrt{5})^2 = 9 + 6\sqrt{5} + 5$
$= 14 + 6\sqrt{5}$

(b) $(\sqrt{a} - \sqrt{b})^2 = (\sqrt{a})^2 - 2\sqrt{a}\sqrt{b} + (\sqrt{b})^2$
$= a + b - 2\sqrt{ab}$ ◂◂

CAUTION ▸ Again, we note that *to multiply radicals and combine them under one radical sign, **it is necessary that the order of the radicals be the same.*** If necessary we can make the order of each radical the same by appropriate operations on each radical separately. Fractional exponents are frequently useful for this purpose.

EXAMPLE 7 ▸▸ (a) $\sqrt[3]{2}\sqrt{5} = 2^{1/3}5^{1/2} = 2^{2/6}5^{3/6} = (2^2 5^3)^{1/6} = \sqrt[6]{500}$

(b)
$$\begin{aligned}\sqrt[3]{4a^2b}\sqrt[4]{8a^3b^2} &= (2^2a^2b)^{1/3}(2^3a^3b^2)^{1/4} = (2^2a^2b)^{4/12}(2^3a^3b^2)^{3/12}\\ &= (2^8a^8b^4)^{1/12}(2^9a^9b^6)^{1/12} = (2^{17}a^{17}b^{10})^{1/12}\\ &= 2a(2^5a^5b^{10})^{1/12}\\ &= 2a\sqrt[12]{32a^5b^{10}}\end{aligned}$$ ◂◂

Division of Radicals

We already have dealt with some cases of division of radicals in the previous sections. When we have had the indicated division of one radical by another, we generally have expressed the answer with no radicals in the denominator by rationalizing the denominator. Although generally no longer important for calculations, the process of rationalizing can make the form of some radical expressions much simpler. Rationalizing *numerators* is also an important step in developing certain expressions in more advanced mathematics. Therefore, if a change is to be made in an expression involving division by a radical, rationalizing the denominator or the numerator is the principal step to be carried out.

When we rationalize the denominator, we change the fraction to an equivalent form in which the denominator is free of radicals. In doing so, multiplication of the numerator and the denominator by the proper quantity is the first important step. We already have dealt with the simpler forms like $a/\sqrt{b}$ in Section 11-3. We now consider fractions in which the denominator is the sum or difference of terms.

If the denominator is the sum (or difference) of two terms, at least one of which is a radical, the fraction can be rationalized by multiplying both the numerator and the denominator by the difference (or sum) of the same two terms, if the radicals are square roots.

EXAMPLE 8 ▸▸ The fraction $\dfrac{1}{\sqrt{3}-\sqrt{2}}$

can be rationalized by multiplying the numerator and the denominator by $\sqrt{3}+\sqrt{2}$. In this way the radicals will be removed from the denominator.

$$\frac{1}{\sqrt{3}-\sqrt{2}} \times \frac{\sqrt{3}+\sqrt{2}}{\sqrt{3}+\sqrt{2}} = \frac{\sqrt{3}+\sqrt{2}}{(\sqrt{3})^2-(\sqrt{2})^2} = \frac{\sqrt{3}+\sqrt{2}}{3-2} = \sqrt{3}+\sqrt{2}$$

change sign

The reason this technique works is that an expression of the form $a^2 - b^2$ is created in the denominator, where a or b (or both) is a radical. We see that the result is a denominator free of radicals. ◂◂

EXAMPLE 9 ▸▸ Rationalize the denominator of $\dfrac{3\sqrt{6}-\sqrt{2}}{\sqrt{6}+\sqrt{2}}$ and simplify the result.

$$\frac{3\sqrt{6}-\sqrt{2}}{\sqrt{6}+\sqrt{2}}\left(\frac{\sqrt{6}-\sqrt{2}}{\sqrt{6}-\sqrt{2}}\right) = \frac{18-4\sqrt{12}+2}{4}$$

change sign

$$= \frac{20-4(2\sqrt{3})}{4} = 5-2\sqrt{3}$$

We note that, after rationalizing the denominator, the result has a much simpler form than the original expression. ◂◂

EXAMPLE 10 ▸▸ In studying the properties of a semiconductor, the expression

$$\frac{\sqrt{1+2V}}{C_1 + C_2\sqrt{1+2V}}$$

is used. Here C_1 and C_2 are constants and V is the voltage across a junction of the semiconductor. Rationalize the denominator of this expression.

Multiplying numerator and denominator by $C_1 - C_2\sqrt{1+2V}$, we have

$$\frac{\sqrt{1+2V}}{C_1 + C_2\sqrt{1+2V}} = \frac{\sqrt{1+2V}(C_1 - C_2\sqrt{1+2V})}{(C_1 + C_2\sqrt{1+2V})(C_1 - C_2\sqrt{1+2V})}$$

$$= \frac{C_1\sqrt{1+2V} - C_2\sqrt{1+2V}\sqrt{1+2V}}{C_1^2 - C_2^2(\sqrt{1+2V})^2}$$

$$= \frac{C_1\sqrt{1+2V} - C_2(1+2V)}{C_1^2 - C_2^2(1+2V)}$$ ◂◂

As we noted earlier, in certain types of algebraic operations it may be necessary to rationalize the numerator of an expression. This procedure is illustrated in the following example.

EXAMPLE 11 ▸▸ Rationalize the numerator of the expression $\dfrac{\sqrt{2x+3}-\sqrt{2x}}{6}$.

We follow the same basic procedure as in rationalizing the denominator of an expression. In this case we multiply the numerator and the denominator of this fraction by $\sqrt{2x+3}+\sqrt{2x}$. This gives us the following solution:

$$\frac{\sqrt{2x+3}-\sqrt{2x}}{6} = \frac{(\sqrt{2x+3}-\sqrt{2x})(\sqrt{2x+3}+\sqrt{2x})}{6(\sqrt{2x+3}+\sqrt{2x})}$$

$$= \frac{(\sqrt{2x+3})^2-(\sqrt{2x})^2}{6(\sqrt{2x+3}+\sqrt{2x})} = \frac{(2x+3)-2x}{6(\sqrt{2x+3}+\sqrt{2x})}$$

$$= \frac{3}{6(\sqrt{2x+3}+\sqrt{2x})} = \frac{1}{2(\sqrt{2x+3}+\sqrt{2x})}$$

Note that the resulting *numerator* does not contain any radicals. ◂◂

Exercises 11–5

In Exercises 1–48, perform the indicated operations, expressing answers in simplest form with rationalized denominators.

1. $\sqrt{3}\sqrt{10}$
2. $\sqrt{2}\sqrt{51}$
3. $\sqrt{6}\sqrt{2}$
4. $\sqrt{7}\sqrt{14}$
5. $\sqrt[3]{4}\sqrt[3]{2}$
6. $\sqrt[5]{4}\sqrt[5]{16}$
7. $(5\sqrt{2})^2$
8. $(3\sqrt{5})^3$
9. $\sqrt{8}\sqrt{\frac{5}{2}}$
10. $\sqrt{\frac{6}{7}}\sqrt{\frac{2}{3}}$
11. $\sqrt{3}(\sqrt{2}-\sqrt{5})$
12. $3\sqrt{5}(\sqrt{15}-2\sqrt{5})$
13. $(2-\sqrt{5})(2+\sqrt{5})$
14. $(2-\sqrt{5})^2$
15. $(3\sqrt{5}-2\sqrt{3})(6\sqrt{5}+7\sqrt{3})$
16. $(3\sqrt{7a}-\sqrt{8})(\sqrt{7a}+\sqrt{2})$
17. $(3\sqrt{11}-\sqrt{x})(2\sqrt{11}+5\sqrt{x})$
18. $(2\sqrt{10}+3\sqrt{15})(\sqrt{10}-7\sqrt{15})$
19. $\sqrt{a}(\sqrt{ab}+\sqrt{c^3})$
20. $\sqrt{3x}(\sqrt{3x}-\sqrt{xy})$
21. $\dfrac{\sqrt{6}-3}{\sqrt{6}}$
22. $\dfrac{5-\sqrt{10}}{\sqrt{10}}$
23. $\dfrac{\sqrt{2a}-b}{\sqrt{a}}$
24. $\dfrac{\sqrt{7x}-\sqrt{14}}{\sqrt{7}}$
25. $(\sqrt{2a}-\sqrt{b})(\sqrt{2a}+3\sqrt{b})$
26. $(2\sqrt{mn}+3\sqrt{n})^2$
27. $\sqrt{2}\sqrt[3]{3}$
28. $\sqrt[5]{16}\sqrt[3]{8}$
29. $\dfrac{1}{\sqrt{7}+\sqrt{3}}$
30. $\dfrac{\sqrt{8}}{2\sqrt{3}-\sqrt{5}}$
31. $\dfrac{3}{2\sqrt{5}-6}$
32. $\dfrac{6\sqrt{5}}{5-2\sqrt{5}}$
33. $\dfrac{\sqrt{2}-1}{\sqrt{7}-3\sqrt{2}}$
34. $\dfrac{2\sqrt{15}-3}{\sqrt{15}+4}$
35. $\dfrac{2\sqrt{3}-5\sqrt{5}}{\sqrt{3}+2\sqrt{5}}$
36. $\dfrac{\sqrt{15}-3\sqrt{5}}{2\sqrt{15}-\sqrt{5}}$
37. $\dfrac{2\sqrt{x}}{\sqrt{x}-\sqrt{y}}$
38. $\dfrac{6\sqrt{a}}{2\sqrt{a}-b}$
39. $\dfrac{8}{3\sqrt{a}-2\sqrt{b}}$
40. $\dfrac{6}{1+2\sqrt{x}}$
41. $(\sqrt[5]{\sqrt{6}-\sqrt{5}})(\sqrt[5]{\sqrt{6}+\sqrt{5}})$
42. $\sqrt[3]{5-\sqrt{17}}\sqrt[3]{5+\sqrt{17}}$
43. $\left(\sqrt{\dfrac{2}{a}}+\sqrt{\dfrac{a}{2}}\right)\left(\sqrt{\dfrac{2}{a}}-2\sqrt{\dfrac{a}{2}}\right)$
44. $(3+\sqrt{6-2a})(2-\sqrt{6-2a})$
45. $\dfrac{\sqrt{2c}+3d}{\sqrt{2c}-d}$
46. $\dfrac{3\sqrt{2x}+2\sqrt{x}}{\sqrt{2x}-\sqrt{x}}$
47. $\dfrac{\sqrt{x+y}}{\sqrt{x-y}-\sqrt{x}}$
48. $\dfrac{\sqrt{1+a}}{a-\sqrt{1-a}}$

In Exercises 49–52, perform the indicated operations, expressing answers in simplest form with rationalized denominators. Then verify the result with a calculator.

49. $(\sqrt{11} + \sqrt{6})(\sqrt{11} - 2\sqrt{6})$

50. $(2\sqrt{5} - \sqrt{7})(3\sqrt{5} + \sqrt{7})$

51. $\dfrac{2\sqrt{6} - \sqrt{5}}{3\sqrt{6} - 4\sqrt{5}}$

52. $\dfrac{\sqrt{7} - 4\sqrt{2}}{5\sqrt{7} - 4\sqrt{2}}$

In Exercises 53–56, combine the terms into a single fraction, but do not rationalize the denominators.

53. $2\sqrt{x} + \dfrac{1}{\sqrt{x}}$

54. $\dfrac{3}{2\sqrt{3x - 4}} - \sqrt{3x - 4}$

55. $\dfrac{x^2}{\sqrt{2x + 1}} + 2x\sqrt{2x + 1}$

56. $4\sqrt{x^2 + 1} - \dfrac{4x}{\sqrt{x^2 + 1}}$

In Exercises 57–60, rationalize the numerator of each fraction.

57. $\dfrac{\sqrt{5} + \sqrt{2}}{3\sqrt{6}}$

58. $\dfrac{\sqrt{19} - 3}{5}$

59. $\dfrac{\sqrt{x + h} - \sqrt{x}}{h}$

60. $\dfrac{\sqrt{3x + 4} + \sqrt{3x}}{8}$

In Exercises 61–68, solve the given problems.

61. By substitution, show that $x = 1 - \sqrt{2}$ is a solution to the equation $x^2 - 2x - 1 = 0$.

62. Determine the product of the two roots of the quadratic equation $ax^2 + bx + c = 0$.

63. For an object oscillating at the end of a spring, and on which there is a force retarding the motion, the equation $m^2 + bm + k^2 = 0$ must be solved. Here, b is a constant related to the retarding force and k is the spring constant. By substitution, show that $m = \frac{1}{2}(\sqrt{b^2 - 4k^2} - b)$ is a solution.

64. A square plastic sheet of side x is stretched by an amount equal to $\sqrt{x}$ in each direction. Find the expression for the percent increase in area of the sheet.

65. The rim velocity of a flywheel is given by the equation $v = \dfrac{\sqrt{gs}}{\sqrt{12w}}$. Rationalize the denominator.

66. An expression used in determining the characteristics of a spur gear is $\dfrac{50}{50 + \sqrt{V}}$. Rationalize the denominator.

67. In finding the time to drain a tank, the expression $\dfrac{h_2 - h_1}{\sqrt{2g}(\sqrt{h_2} - \sqrt{h_1})}$ is used. Rationalize the denominator.

68. In analyzing a tuned electronic amplifier circuit, the expression $\dfrac{2Q}{\sqrt{\sqrt{2} - 1}}$ is used. Rationalize the denominator.

Chapter Equations, Review Exercises, and Practice Test

Chapter Equations

Exponents

$$a^m \times a^n = a^{m+n} \qquad (11\text{-}1)$$

$$\frac{a^m}{a^n} = a^{m-n} \quad \text{or} \quad \frac{a^m}{a^n} = \frac{1}{a^{n-m}}, \qquad a \neq 0 \qquad (11\text{-}2)$$

$$(a^m)^n = a^{mn} \qquad (11\text{-}3)$$

$$(ab)^n = a^n b^n, \qquad \left(\frac{a}{b}\right)^n = \frac{a^n}{b^n}, \qquad b \neq 0 \qquad (11\text{-}4)$$

$$a^0 = 1, \qquad a \neq 0 \qquad (11\text{-}5)$$

$$a^{-n} = \frac{1}{a^n}, \qquad a \neq 0 \qquad (11\text{-}6)$$

Fractional exponents

$$a^{1/n} = \sqrt[n]{a} \qquad (11\text{-}7)$$

$$a^{m/n} = \sqrt[n]{a^m} = (\sqrt[n]{a})^m \qquad (11\text{-}8)$$

Radicals

$$\sqrt[n]{a^n} = (\sqrt[n]{a})^n = a \qquad (11\text{-}9)$$

$$\sqrt[n]{a}\sqrt[n]{b} = \sqrt[n]{ab} \qquad (11\text{-}10)$$

$$\sqrt[m]{\sqrt[n]{a}} = \sqrt[mn]{a} \qquad (11\text{-}11)$$

$$\frac{\sqrt[n]{a}}{\sqrt[n]{b}} = \sqrt[n]{\frac{a}{b}}, \qquad b \neq 0 \qquad (11\text{-}12)$$

Review Exercises

In Exercises 1–28, express each expression in simplest form with only positive exponents.

1. $2a^{-2}b^0$
2. $(2c)^{-1}z^{-2}$
3. $\dfrac{2c^{-1}}{d^{-3}}$
4. $\dfrac{-5x^0}{3y^{-1}}$
5. $3(25)^{3/2}$
6. $32^{2/5}$
7. $400^{-3/2}$
8. $1000^{-2/3}$
9. $\left(\dfrac{3}{t^2}\right)^{-2}$
10. $\left(\dfrac{2x^3}{3}\right)^{-3}$
11. $\dfrac{-8^{2/3}}{49^{-1/2}}$
12. $\dfrac{81^{-0.75}}{6^{-3}}$
13. $(2a^{1/3}b^{5/6})^6$
14. $(ax^{-1/2}y^{1/4})^8$
15. $(-32m^{15}n^{10})^{3/5}$
16. $(27x^{-6}y^9)^{2/3}$
17. $2x^{-2} - y^{-1}$
18. $a^4a^{-1} + (a^4)^{-1}$
19. $\dfrac{2x^{-1}}{2x^{-1} + y^{-1}}$
20. $\dfrac{3a}{(2a)^{-1} - a}$
21. $(a - 3b^{-1})^{-1}$
22. $(2s^{-2} + t)^{-2}$
23. $(x^3 - y^{-3})^{1/3}$
24. $(x^2 + 2xy + y^2)^{-1/2}$
25. $(8a^3)^{2/3}(4a^{-2} + 1)^{1/2}$
26. $\left[\dfrac{(9a)^0(4x^2)^{1/3}(3b^{1/2})}{(2b^0)^2}\right]^{-6}$
27. $2x(x-1)^{-2} - 2(x^2+1)(x-1)^{-3}$
28. $4(1-x^2)^{1/2} - (1-x^2)^{-1/2}$

In Exercises 29–72, perform the indicated operations and express the result in simplest radical form with rationalized denominators.

29. $\sqrt{68}$
30. $\sqrt{96}$
31. $\sqrt{ab^5c^2}$
32. $\sqrt{x^3y^4z^6}$
33. $\sqrt{9a^3b^4}$
34. $\sqrt{8x^5y^2}$
35. $\sqrt{84st^3u^2}$
36. $\sqrt{52x^2y^5}$
37. $\dfrac{5}{\sqrt{2s}}$
38. $\dfrac{3a}{\sqrt{5x}}$
39. $\sqrt{\dfrac{11}{27}}$
40. $\sqrt{\dfrac{7}{8}}$
41. $\sqrt[4]{8m^6n^9}$
42. $\sqrt[3]{9a^7b^{-3}}$
43. $\sqrt[4]{\sqrt[3]{64}}$
44. $\sqrt{a^{-3}\sqrt[5]{b^{12}}}$
45. $\sqrt{200} + \sqrt{32}$
46. $2\sqrt{68x} - \sqrt{153x}$
47. $\sqrt{63} - 2\sqrt{112} - \sqrt{28}$
48. $2\sqrt{20} - \sqrt{80} - 2\sqrt{125}$
49. $a\sqrt{2x^3} + \sqrt{8a^2x^3}$
50. $2\sqrt{m^2n^3} - \sqrt{n^5}$
51. $\sqrt[3]{8a^4} + b\sqrt[3]{a}$
52. $\sqrt[4]{2xy^5} - \sqrt[4]{32xy}$
53. $\sqrt{5}(2\sqrt{5} - \sqrt{11})$
54. $2\sqrt{8}(5\sqrt{2} - \sqrt{6})$
55. $2\sqrt{2}(\sqrt{6} - \sqrt{10})$
56. $3\sqrt{5}(\sqrt{15} + 2\sqrt{35})$
57. $(2 - 3\sqrt{17})(3 + \sqrt{17})$
58. $(5\sqrt{6} - 4)(3\sqrt{6} + 5)$
59. $(2\sqrt{7} - 3\sqrt{a})(3\sqrt{7} + \sqrt{a})$
60. $(3\sqrt{2} - \sqrt{13})(5\sqrt{2} + 3\sqrt{13})$
61. $\dfrac{\sqrt{3x}}{2\sqrt{3x} - \sqrt{y}}$
62. $\dfrac{5\sqrt{a}}{2\sqrt{a} - c}$
63. $\dfrac{\sqrt{2}}{\sqrt{3} - 4\sqrt{2}}$
64. $\dfrac{4}{3 - 2\sqrt{7}}$
65. $\dfrac{\sqrt{7} - \sqrt{5}}{\sqrt{5} + 3\sqrt{7}}$
66. $\dfrac{4 - 2\sqrt{6}}{3 + 2\sqrt{6}}$
67. $\dfrac{2\sqrt{x} - a}{3\sqrt{x} + 5a}$
68. $\dfrac{3\sqrt{y} - \sqrt{z}}{2\sqrt{y} + 5\sqrt{z}}$
69. $\sqrt{4b^2 + 1}$
70. $\sqrt{a^{-2} + \dfrac{1}{b^2}}$
71. $\left(\dfrac{2 - \sqrt{15}}{2}\right)^2 - \left(\dfrac{2 - \sqrt{15}}{2}\right)$
72. $\sqrt{2 + \dfrac{b}{a} + \dfrac{a}{b}} + \sqrt{a^4b^2 + 2a^3b^2 + a^2b^2}$

In Exercises 73–76, perform the indicated operations and express the result in simplest radical form with rationalized denominators. Then verify the result with a calculator.

73. $\sqrt{52} + 4\sqrt{24} - \sqrt{54}$
74. $2\sqrt{5}(6\sqrt{5} - 5\sqrt{6})$
75. $(\sqrt{7} - 2\sqrt{15})(3\sqrt{7} - \sqrt{15})$
76. $\dfrac{2\sqrt{3} - 7\sqrt{14}}{3\sqrt{3} + 2\sqrt{14}}$

In Exercises 77–88, perform the indicated operations.

77. The average annual increase i, in percent, of the cost of living over n years is given by $i = 100[(C_2/C_1)^{1/n} - 1]$, where C_1 is the cost of living index for the first year of the period and C_2 is the cost of living index for the last year of the period. Evaluate i if $C_1 = 107.6$ and $C_2 = 156.2$ are the values for 1985 and 1995, respectively.

78. Kepler's third law of planetary motion may be given as $T = kr^{3/2}$, where T is the time for one revolution of a planet around the sun, r is its mean radius from the sun, and $k = 5.46 \times 10^{-13}$ year/km$^{3/2}$. Find the time for one revolution of Venus about the sun if $r = 1.08 \times 10^8$ km.

79. The speed v of a ship of weight W whose engines produce power P is given by $v = k\sqrt[3]{P/W}$. Express this equation (a) with a fractional exponent and (b) as a radical with the denominator rationalized.

80. In the theory of semiconductors, the expression $km^{3/2}(E - E_1)^{1/2}$ is found. Write this expression in simplified radical form.

81. When studying atomic structure, the expression $\dfrac{v}{n_2^{-2} - n_1^{-2}}$ is used. Express this in simplest form with only positive exponents.

82. In hydrodynamics, the expression $v(4 + 3ar^{-1} - a^3r^{-3})^{-1}$ arises. Express this in simplest form with only positive exponents.

83. A square is decreasing in size on a computer screen. To find how fast the side of the square changes, it is necessary to rationalize the numerator of the expression $\dfrac{\sqrt{A + h} - \sqrt{A}}{h}$. Perform this operation.

84. In thermodynamics, the expression $\left(\dfrac{1}{a^3} + \dfrac{1}{b^3}\right)^{-1/3}$ is used. Combine the fractions and simplify, expressing the answer without negative exponents.

85. In an experiment, a laser beam follows the path shown in Fig. 11-5. Express the length of the path in simplest radical form.

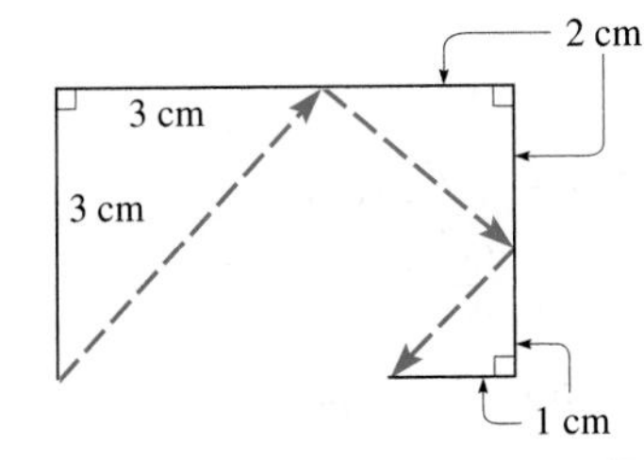

Fig. 11-5

86. The rate of flow V (in m/s) through a storm drain pipe is found from $V = 33.5(0.55)^{2/3}(0.0018)^{1/2}$. Find the value of V.

87. The frequency of a certain electric circuit is given by

$$\frac{1}{2\pi\sqrt{\dfrac{LC_1C_2}{C_1 + C_2}}}$$

Express this in simplest rationalized radical form.

88. A computer analysis of an experiment showed that the fraction f of viruses surviving X-ray dosages was given by $f = \dfrac{20}{d + \sqrt{3d + 400}}$, where d is the dosage. Express this with the denominator rationalized.

Writing Exercise

89. In calculating the forces on a tower by the wind, it is necessary to evaluate $0.18^{0.13}$. Write a paragraph explaining why this form is preferable to the equivalent radical form.

Practice Test

In Problems 1–12, simplify the given expressions. For those with exponents, express each result with only positive exponents. For the radicals, rationalize the denominator where applicable.

1. $2\sqrt{20} - \sqrt{125}$

2. $\dfrac{100^{3/2}}{8^{-2/3}}$

3. $(as^{-1/3}t^{3/4})^{12}$

4. $(2x^{-1} + y^{-2})^{-1}$

5. $(\sqrt{2x} - 3\sqrt{y})^2$

6. $\sqrt[3]{\sqrt[4]{4}}$

7. $\dfrac{3 - 2\sqrt{2}}{2\sqrt{x}}$

8. $\sqrt{27a^4b^3}$

9. $(2x + 3)^{1/2} + (x + 1)(2x + 3)^{-1/2}$

10. $2\sqrt{2}(3\sqrt{10} - \sqrt{6})$

11. $\left(\dfrac{4a^{-1/2}b^{3/4}}{b^{-2}}\right)\left(\dfrac{b^{-1}}{2a}\right)$

12. $\dfrac{2\sqrt{15} + \sqrt{3}}{\sqrt{15} - 2\sqrt{3}}$

13. Express $\dfrac{3^{-1/2}}{2}$ in simplest radical form with a rationalized denominator.

14. In the study of fluid flow in pipes, the expression $0.220N^{-1/6}$ is found. Evaluate this expression for $N = 64 \times 10^6$.

12

Complex Numbers

In Section 12-7 we see that tuning in a radio to any one of a range of stations involves a basic application of complex numbers to electricity.

In Chapter 1, when we were introducing the topic of numbers, imaginary numbers were mentioned. Again, when we considered quadratic equations and their solutions in Chapter 7, we briefly came across this type of number. However, until now we have purposely avoided any extended discussion of imaginary numbers. In this chapter we discuss *complex numbers,* which include both the real numbers and the imaginary numbers.

Despite their names, complex numbers and imaginary numbers have very real and useful applications in many technical areas. One of the most important is in electricity and electronics, where these numbers are used extensively. Other applications are found in mechanical vibrations and in optics.

12-1 Basic Definitions

When we defined radicals we were able to define square roots of positive numbers easily, since any positive or negative number squared equals a positive number. For this reason we can see that it is impossible to square any real number and have the product equal a negative number. We must define a new number system if we wish to include square roots of negative numbers. With the proper definitions, we shall find that these numbers can be used to great advantage in certain applications.

If the radicand in a square root is negative, we can express the indicated root as the product of $\sqrt{-1}$ and the square root of a positive number. *The symbol* $\sqrt{-1}$ *is defined as the* **imaginary unit** *and is denoted by the symbol j.* In keeping with the definition of j, we have

$$j^2 = -1 \tag{12-1}$$

Generally, mathematicians use the symbol i for $\sqrt{-1}$, and therefore most nontechnical textbooks use i. However, one of the major technical applications of complex numbers is in electronics, where i represents electric current. Therefore, we shall use j for $\sqrt{-1}$, which is also the standard symbol in electronics textbooks.

EXAMPLE 1 ▸▸ Express the following square roots in terms of j.

(a) $\sqrt{-9} = \sqrt{(9)(-1)} = \sqrt{9}\sqrt{-1} = 3j$
(b) $\sqrt{-0.25} = \sqrt{0.25}\sqrt{-1} = 0.5j$
(c) $\sqrt{-5} = \sqrt{5}\sqrt{-1} = \sqrt{5}j = j\sqrt{5}$ ← this form is better

When a radical appears in the final result, we write this result in a form with the j before the radical as in the last illustration. This clearly shows that the j is not *under* the radical. ◂◂

EXAMPLE 2 ▸▸ $(\sqrt{-4})^2 = (\sqrt{4}\sqrt{-1})^2 = (j\sqrt{4})^2 = 4j^2 = -4$

The simplification of this expression does not follow Eq. (11-10), which states that $\sqrt{ab} = \sqrt{a}\sqrt{b}$ for square roots. This is the reason it was noted as being valid only if a and b are not negative. Therefore, we see that *Eq. (11-10) does not always hold for negative values of a and b.*

If $(\sqrt{-4})^2$ did follow Eq. (11-10), we would have

$$(\sqrt{-4})^2 = \sqrt{-4}\sqrt{-4} = \sqrt{(-4)(-4)} = \sqrt{16} = 4$$

CAUTION ▸ this step is incorrect if both are negative, as in this case

We obtain 4 and do not obtain the correct result of -4. ◂◂

EXAMPLE 3 ▸▸ To further illustrate the method of handling square roots of negative numbers, consider the difference between $\sqrt{-3}\sqrt{-12}$ and $\sqrt{(-3)(-12)}$. For these expressions we have

$$\begin{aligned}\sqrt{-3}\sqrt{-12} &= (j\sqrt{3})(j\sqrt{12}) = (\sqrt{3}\sqrt{12})j^2 = (\sqrt{36})j^2 \\ &= 6(-1) = -6\end{aligned}$$

and

$$\sqrt{(-3)(-12)} = \sqrt{36} = 6$$

For $\sqrt{-3}\sqrt{-12}$ we have the product of square roots of negative numbers, whereas for $\sqrt{(-3)(-12)}$ we have the product of negative numbers under the radical. We must be careful to note the difference. ◂◂

NOTE ▶ From Examples 2 and 3 we see that *when we are dealing with square roots of negative numbers,* ***each should be expressed in terms of j before proceeding.*** To do this, for any positive real number a we write

$$\sqrt{-a} = j\sqrt{a}, \qquad (a > 0) \tag{12-2}$$

EXAMPLE 4 ▶▶ (a) $\sqrt{-6} = \sqrt{(6)(-1)} = \sqrt{6}\sqrt{-1} = j\sqrt{6}$

this step is correct if only one is negative, as in this case

(b) $-\sqrt{-75} = -\sqrt{(25)(3)(-1)} = -\sqrt{(25)(3)}\sqrt{-1} = -5j\sqrt{3}$

CAUTION ▶ We note that $-\sqrt{-75}$ ***is not equal to*** $\sqrt{75}$. ◀◀

At times we need to raise imaginary numbers to some power. Using the definitions of exponents and of j, we have the following results:

$$\begin{aligned} j &= j, & j^5 &= j^4 j = j \\ j^2 &= -1, & j^6 &= j^4 j^2 = (1)(-1) = -1 \\ j^3 &= j^2 j = -j, & j^7 &= j^4 j^3 = (1)(-j) = -j \\ j^4 &= j^2 j^2 = (-1)(-1) = 1 & j^8 &= j^4 j^4 = (1)(1) = 1 \end{aligned}$$

The powers of j go through the cycle of $j, -1, -j, 1, j, -1, -j, 1$, and so forth. Noting this and the fact that j raised to a power which is a multiple of 4 equals 1 allows us to raise j to any integral power almost on sight.

EXAMPLE 5 ▶▶ (a) $j^{10} = j^8 j^2 = (1)(-1) = -1$
(b) $j^{45} = j^{44} j = (1)(j) = j$
(c) $j^{531} = j^{528} j^3 = (1)(-j) = -j$

exponents 8, 44, 528 are multiples of 4 ◀◀

Rectangular Form of a Complex Number

Using real numbers and the imaginary unit j, we define a new kind of number. *A* **complex number** *is any number that can be written in the form $a + bj$, where a and b are real numbers. If $a = 0$ and $b \neq 0$, we have a number of the form bj, which is a* **pure imaginary number**. *If $b = 0$, then $a + bj$ is a real number. The form $a + bj$ is known as the* **rectangular form** *of a complex number, where a is known as the* **real part** *and b is known as the* **imaginary part**. We see that complex numbers include all real numbers and all pure imaginary numbers.

A comment here about the words *imaginary* and *complex* is in order. The choice of the names of these numbers is historical in nature, and unfortunately it leads to some misconceptions about the numbers. The use of *imaginary* does not imply that the numbers do not exist. Imaginary numbers do in fact exist, as they are defined above. In the same way, the use of *complex* does not imply that the numbers are complicated and therefore difficult to understand. With the proper definitions and operations, we can work with complex numbers, just as with any type of number.

Even negative numbers were not widely accepted by mathematicians until late in the 16th century.

Imaginary numbers were so named because the French mathematician Descartes (1596–1650) referred to them as "imaginaries." Most of the mathematical development of them occurred in the 18th century.

Complex numbers were named by the German mathematician Gauss (1777–1855).

In the following section we will define the basic operations for complex numbers. Before doing this it is necessary to define the equality of complex numbers. Since a complex number is the sum of a real number and an imaginary number, it is not positive or negative in the usual sense, but each real part and each imaginary part are positive or negative. Therefore, *we define two complex numbers to be equal if the real parts are equal and the imaginary parts are equal.* That is,

NOTE ▶ ***two complex numbers, $a + bj$ and $x + yj$, are equal if $a = x$ and $b = y$.***

This is illustrated in the following examples.

EXAMPLE 6 ▶▶ (a) $a + bj = 3 + 4j$ if $a = 3$ and $b = 4$

(b) $x + yj = 5 - 3j$ if $x = 5$ and $y = -3$

(5 labeled "real part"; −3j labeled "imaginary part") ◀◀

EXAMPLE 7 ▶▶ What values of x and y satisfy the equation $4 - 6j - x = j + jy$?

One way to solve this equation is to arrange the terms so that all the known terms are on the right and all the terms containing unknowns x and y are on the left. This leads to

$$-x - jy = -4 + 7j$$

From the definition of equality of complex numbers,

$$-x = -4 \quad \text{and} \quad -y = 7$$
$$x = 4 \quad \text{and} \quad y = -7$$ ◀◀

EXAMPLE 8 ▶▶ What values of x and y satisfy the equation

$$x + 3(xj + y) = 5 - j - jy$$

Rearranging the terms so that the known terms are on the right and the terms containing x and y are on the left, we have

$$x + 3y + 3jx + jy = 5 - j$$

Next, factoring j from the two terms on the left will put the expression on the left into proper form. This leads to

$$(x + 3y) + (3x + y)j = 5 - j$$

Using the definition of equality, we have

$$x + 3y = 5 \quad \text{and} \quad 3x + y = -1$$

We now solve this system of equations. The solution is $x = -1$ and $y = 2$. Actually, the solution can be obtained at any point by writing each side of the equation in the form $a + bj$ and then equating first the real parts and then the imaginary parts. ◀◀

The **conjugate** *of the complex number* $a + bj$ *is the complex number* $a - bj$. We see that the sign of the imaginary part of a complex number is changed to obtain its conjugate.

EXAMPLE 9 ▸▸ (a) $3 - 2j$ is the conjugate of $3 + 2j$. We may also say that $3 + 2j$ is the conjugate of $3 - 2j$. Thus each is the conjugate of the other.

(b) $-2 - 5j$ and $-2 + 5j$ are conjugates.

(c) $6j$ and $-6j$ are conjugates.

(d) 3 is the conjugate of 3 (imaginary part is zero). ◂◂

Exercises 12–1

In Exercises 1–12, express each number in terms of j.

1. $\sqrt{-81}$ **2.** $\sqrt{-121}$

3. $-\sqrt{-4}$ **4.** $-\sqrt{-49}$

5. $\sqrt{-0.36}$ **6.** $-\sqrt{-0.01}$

7. $\sqrt{-8}$ **8.** $\sqrt{-48}$

9. $\sqrt{-\frac{7}{4}}$ **10.** $-\sqrt{-\frac{5}{9}}$

11. $-\sqrt{-\frac{2}{5}}$ **12.** $\sqrt{-\frac{5}{3}}$

In Exercises 13–24, simplify each of the given expressions.

13. (a) $(\sqrt{-7})^2$ (b) $\sqrt{(-7)^2}$

14. (a) $\sqrt{(-15)^2}$ (b) $(\sqrt{-15})^2$

15. (a) $\sqrt{(-2)(-8)}$ (b) $\sqrt{-2}\sqrt{-8}$

16. (a) $\sqrt{-9}\sqrt{-16}$ (b) $\sqrt{(-9)(-16)}$

17. j^7 **18.** j^{49}

19. $-j^{22}$ **20.** j^{408}

21. $j^2 - j^6$ **22.** $2j^5 - j^7$

23. $j^{15} - j^{13}$ **24.** $3j^{48} + j^{200}$

In Exercises 25–36, perform the indicated operations and simplify each complex number to its rectangular form.

25. $2 + \sqrt{-9}$ **26.** $-6 + \sqrt{-64}$

27. $3j - \sqrt{-100}$ **28.** $-\sqrt{64} - \sqrt{-400}$

29. $8 - \sqrt{4} + \sqrt{-4}$ **30.** $5 - j + 2\sqrt{-25}$

31. $2j^2 + 3j$ **32.** $j^3 - 6$

33. $\sqrt{18} - \sqrt{-8}$ **34.** $\sqrt{-27} + \sqrt{12}$

35. $(\sqrt{-2})^2 + j^4$ **36.** $(2\sqrt{2})^2 - (\sqrt{-1})^2$

In Exercises 37–40, find the conjugate of each complex number.

37. (a) $6 - 7j$ (b) $8 + j$

38. (a) $-3 + 2j$ (b) $-9 - j$

39. (a) $2j$ (b) -4

40. (a) 6 (b) $-5j$

In Exercises 41–48, find the values of x and y that satisfy the given equations.

41. $7x - 2yj = 14 + 4j$

42. $2x + 3jy = -6 + 12j$

43. $6j - 7 = 3 - x - yj$

44. $9 - j = xj + 1 - y$

45. $x - y = 1 - xj - yj - j$

46. $2x - 2j = 4 - 2xj - yj$

47. $x + 2 + 7j = yj - 2xj$

48. $2x + 6xj + 3 = yj - y + 7j$

In Exercises 49–56, answer the given questions.

49. Are $8j$ and $-8j$ the solutions to the equation $x^2 + 64 = 0$?

50. Are $2j\sqrt{5}$ and $-2j\sqrt{5}$ the solutions to the equation $x^2 + 20 = 0$?

51. Are $2j$ and $-2j$ solutions to the equation $x^4 + 16 = 0$?

52. Are $3j$ and $-3j$ solutions to the equation $x^3 + 27j = 0$?

53. Evaluate $j + j^2 + j^3 + j^4 + j^5 + j^6 + j^7 + j^8$.

54. What is the smallest possible positive value of n for which $j^{-1} = j^n$?

55. What condition must be satisfied if a complex number and its conjugate are to be equal?

56. What type of number is a complex number if it is equal to the negative of its conjugate?

12-2 Basic Operations with Complex Numbers

The definitions of the operations of addition, subtraction, multiplication, and division for complex numbers are based on the operations for binomials with real coefficients (see Chapters 1 and 6). These operations are performed without regard for the fact that j has a special meaning. However, ***we must be careful to express all complex numbers in terms of j before performing these operations.*** Once this is done, we may proceed as with real numbers. We have the following definitions for these operations on complex numbers.

NOTE ▶

Addition:

$$(a + bj) + (c + dj) = (a + c) + (b + d)j \quad (12\text{-}3)$$

Subtraction:

$$(a + bj) - (c + dj) = (a - c) + (b - d)j \quad (12\text{-}4)$$

Multiplication:

$$(a + bj)(c + dj) = (ac - bd) + (ad + bc)j \quad (12\text{-}5)$$

Division:

$$\frac{a + bj}{c + dj} = \frac{(a + bj)(c - dj)}{(c + dj)(c - dj)} = \frac{(ac + bd) + (bc - ad)j}{c^2 + d^2} \quad (12\text{-}6)$$

(Compare Eq. (12-5) with Eq. (6-6).)

We see from Eqs. (12-3) and (12-4) that *we add and subtract complex numbers by combining the real parts and combining the imaginary parts.* Consider the following examples.

EXAMPLE 1 ▶▶ (a) $(3 - 2j) + (-5 + 7j) = (3 - 5) + (-2 + 7)j$
$= -2 + 5j$

(b) $(7 + 9j) - (6 - 4j) = (7 - 6) + (9 - (-4))j$
$= 1 + 13j$ ◀◀

EXAMPLE 2 ▶▶ $(3\sqrt{-4} - 4) - (6 - 2\sqrt{-25}) - \sqrt{-81}$

$= [3(2j) - 4] - [6 - 2(5j)] - 9j$ write in terms of j

$= [6j - 4] - [6 - 10j] - 9j$

$= 6j - 4 - 6 + 10j - 9j$

$= -10 + 7j$ ◀◀

When complex numbers are multiplied, Eq. (12-5) indicates that *we proceed as in any algebraic multiplication,* properly expressing numbers in terms of j and evaluating the power of j. This is illustrated in the next two examples.

EXAMPLE 3 ▶▶ $(6 - \sqrt{-4})(\sqrt{-9}) = (6 - 2j)(3j)$ write in terms of j

$= 18j - 6j^2 = 18j - 6(-1)$

$= 6 + 18j$ ◀◀

EXAMPLE 4 ▸▸

$$
\begin{aligned}
(-9.4 - 6.2j)(2.5 + 1.5j) &= (-9.4)(2.5) + (-9.4)(1.5j) + (-6.2j)(2.5) + (-6.2j)(1.5j) \\
&= -23.5 - 14.1j - 15.5j - 9.3j^2 \\
&= -23.5 - 29.6j - 9.3(-1) \\
&= -14.2 - 29.6j \quad ◂◂
\end{aligned}
$$

The procedure shown by Eq. (12-6) for dividing by a complex number is the same procedure we used for rationalizing the denominator of a fraction with a radical in the denominator. We use this procedure so that we can express any result in the proper form of a complex number. Therefore, *to divide by a complex number, multiply the numerator and the denominator by the conjugate of the denominator.* This is illustrated in the following examples.

EXAMPLE 5 ▸▸

$$
\begin{aligned}
\frac{7 - 2j}{3 + 4j} &= \frac{(7 - 2j)(3 - 4j)}{(3 + 4j)(3 - 4j)} \quad \text{multiply by conjugate of denominator} \\
&= \frac{21 - 28j - 6j + 8j^2}{9 - 16j^2} = \frac{21 - 34j + 8(-1)}{9 - 16(-1)} \\
&= \frac{13 - 34j}{25}
\end{aligned}
$$

This could be written in the form $a + bj$ as $\frac{13}{25} - \frac{34}{25}j$, but this type of result is generally left as a single fraction. In decimal form, the result would be expressed as $0.52 - 1.36j$. ◂◂

EXAMPLE 6 ▸▸

(a) $\dfrac{6 + j}{2j} = \dfrac{6 + j}{2j}\left(\dfrac{-2j}{-2j}\right) = \dfrac{-12j - 2j^2}{4} = \dfrac{2 - 12j}{4} = \dfrac{1 - 6j}{2}$

(b) $\dfrac{j^3 + 2j}{1 - j^5} = \dfrac{-j + 2j}{1 - j} = \dfrac{j}{1 - j}\left(\dfrac{1 + j}{1 + j}\right) = \dfrac{-1 + j}{2}$ ◂◂

EXAMPLE 7 ▸▸ In an alternating-current circuit, the voltage E is given by $E = IZ$, where I is the current in amperes and Z is the impedance in ohms. Each of these can be represented by complex numbers. Find the complex number representation for I if $E = 4.20 - 3.00j$ volts and $Z = 5.30 + 2.65j$ ohms. (This type of circuit is discussed in more detail in Section 12-7.)

Since $I = E/Z$, we have

$$
\begin{aligned}
I &= \frac{4.20 - 3.00j}{5.30 + 2.65j} = \frac{(4.20 - 3.00j)(5.30 - 2.65j)}{(5.30 + 2.65j)(5.30 - 2.65j)} \quad \text{multiply by conjugate of denominator} \\
&= \frac{22.26 - 11.13j - 15.90j + 7.95j^2}{5.30^2 - 2.65^2j^2} = \frac{22.26 - 7.95 - 27.03j}{5.30^2 + 2.65^2} \\
&= \frac{14.31 - 27.03j}{35.11} = 0.408 - 0.770j \text{ amperes}
\end{aligned}
$$

On a calculator, the result can be found directly from the last expression on the first line, and the intermediate steps need not be written. A good procedure is to evaluate the denominator and store it in memory. Then calculate the two products for the real part (remembering that $j^2 = -1$) and the two products for the imaginary part. Then divide each product by the value of the denominator. ◂◂

Exercises 12–2

In Exercises 1–52, perform the indicated operations, expressing all answers in the form $a + bj$.

1. $(3 - 7j) + (2 - j)$

2. $(-4 - j) + (-7 - 4j)$

3. $(7j - 6) - (3 + j)$

4. $(0.23 + 0.67j) - (0.46 - 0.19j)$

5. $(4 + \sqrt{-16}) + (3 - \sqrt{-81})$

6. $(-1 + 3\sqrt{-4}) + (8 - 4\sqrt{-49})$

7. $(5.4 - 3.4j) - (2.9j + 5.5)$

8. $(5j - 1) - 3j$

9. $j - (j - 7) - 8$

10. $(7 - j) - (4 - 4j) + (6 - j)$

11. $(12j - 21) - (15 - 18j) - 9j$

12. $(0.062j - 0.073) - 0.030j - (0.121 - 0.051j)$

13. $(7 - j)(7j)$

14. $(-2.2j)(1.5j - 4.0)$

15. $\sqrt{-16}(2.8\sqrt{-4.0} + 1.6)$

16. $(\sqrt{-4} - 1)(\sqrt{-9})$

17. $(4 - j)(5 + 2j)$

18. $(3 - 5j)(6 + 7j)$

19. $(20j - 30)(30j + 10)$

20. $(8j - 5)(7 + 4j)$

21. $(\sqrt{-18}\sqrt{-4})(3j)$

22. $\sqrt{-6}\sqrt{-12}\sqrt{3}$

23. $(\sqrt{-5})^5$

24. $(\sqrt{-36})^4$

25. $\sqrt{-108} - \sqrt{-27}$

26. $2\sqrt{-54} + \sqrt{-24}$

27. $3\sqrt{-28} - 2\sqrt{12}$

28. $5\sqrt{24} - 3\sqrt{-45}$

29. $7j^3 - 7\sqrt{-9}$

30. $6j - 5j^2\sqrt{-63}$

31. $j\sqrt{-7} - j^6\sqrt{112} + 3j$

32. $j^2\sqrt{-7} - \sqrt{-28} + 8$

33. $(3 - 7j)^2$

34. $(4j + 5)^2$

35. $(1 - j)^3$

36. $(1 + j)(1 - j)^2$

37. $\dfrac{6j}{2 - 5j}$

38. $\dfrac{4}{3 + 7j}$

39. $\dfrac{0.25}{3.0 - \sqrt{-1.0}}$

40. $\dfrac{\sqrt{-4}}{2 + \sqrt{-9}}$

41. $\dfrac{1 - j}{3j}$

42. $\dfrac{9 - 8j}{-4j}$

43. $\dfrac{j\sqrt{2} - 5}{j\sqrt{2} + 3}$

44. $\dfrac{2 + 3j\sqrt{3}}{5 - j\sqrt{3}}$

45. $\dfrac{\sqrt{-16} - \sqrt{2}}{\sqrt{2} + j}$

46. $\dfrac{1 - \sqrt{-4}}{2 + 9j}$

47. $\dfrac{j^2 - j}{2j - j^8}$

48. $\dfrac{j^5 - j^3}{3 + j}$

49. $\dfrac{(2 + j)(1 - j)}{1 + j}$

50. $\dfrac{7 + 2j}{(4 + 3j)(j - 5)}$

51. $\dfrac{(4 + 5j)(1 - j)}{(8 - 3j)(j + 6)}$

52. $\dfrac{(6j + 5)(2 - 4j)}{(5 - j)(4j + 1)}$

In Exercises 53–60, solve the given problems.

53. Show that $-1 + j$ is a solution to the equation $x^2 + 2x + 2 = 0$.

54. Show that $-1 - j$ is a solution to the equation $x^2 + 2x + 2 = 0$.

55. Show that $1 - j\sqrt{3}$ is a solution to the equation $x^2 + 4 = 2x$.

56. Show that $1 + j\sqrt{3}$ is a solution to the equation $x^2 + 4 = 2x$.

57. Multiply $2 - 3j$ by its conjugate.

58. Multiply $-3 + j$ by its conjugate.

59. Divide $2 - 3j$ by its conjugate.

60. Divide $-3 + j$ by its conjugate.

In Exercises 61–64, solve the given problems. Refer to Example 7.

61. If $I = 0.835 - 0.427j$ amperes and $Z = 250 + 170j$ ohms, find the complex number representation for E.

62. If $E = 5.70 - 3.65j$ volts and $I = 0.360 - 0.525j$ amperes, find the complex number representation for Z.

63. If $E = 85 + 74j$ volts and $Z = 2500 - 1200j$ ohms, find the complex number representation for I.

64. In an alternating-current circuit, two impedances Z_1 and Z_2 have a total impedance Z_T given by $Z_T = \dfrac{Z_1 Z_2}{Z_1 + Z_2}$. Find Z_T for $Z_1 = 2 + 3j$ ohms and $Z_2 = 3 - 4j$ ohms.

In Exercises 65–68, demonstrate the indicated properties.

65. Show that the sum of a complex number and its conjugate is a real number.

66. Explain why the product of a complex number and its conjugate is real and nonnegative.

67. Show that the difference between a complex number and its conjugate is an imaginary number ($b \neq 0$).

68. Show that the reciprocal of the imaginary unit is the negative of the imaginary unit.

12-3 Graphical Representation of Complex Numbers

We showed in Section 1-1 how we could represent real numbers as points on a line. Because complex numbers include all real numbers as well as imaginary numbers, it is necessary to represent them graphically in a different way. Since there are two numbers associated with each complex number (the real part and the imaginary part), we find that we can represent complex numbers by representing the real parts by the x-values of the rectangular coordinate system, and the imaginary parts by the y-values. In this way, *each complex number is represented as a point in the plane,* the point being designated as $a + bj$. *When the rectangular coordinate system is used in this manner, it is called the* **complex plane.** *The horizontal axis is called the* **real axis,** *and the vertical axis is called the* **imaginary axis.**

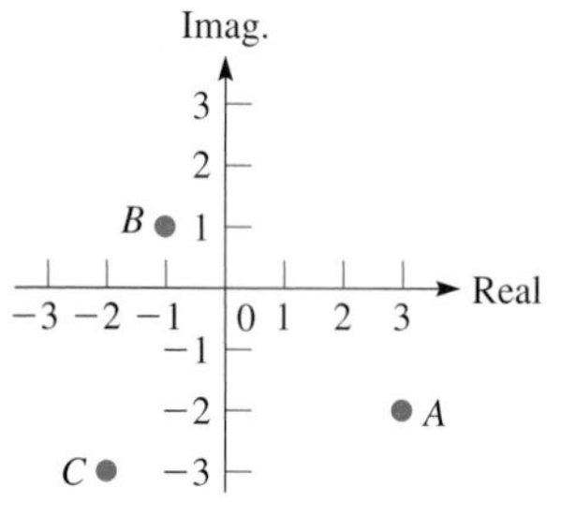

Fig. 12-1

EXAMPLE 1 ▸▸ In Fig. 12-1, point A represents the complex number $3 - 2j$; point B represents $-1 + j$; point C represents $-2 - 3j$. We note that these complex numbers are represented by the points $(3, -2)$, $(-1, 1)$, and $(-2, -3)$, respectively, of the standard rectangular coordinate system.

We must keep in mind that the meaning given to the points representing complex numbers in the complex plane is different from the meaning given to the points in the standard rectangular coordinate system. A point in the complex plane represents a single complex number, whereas a point in the rectangular coordinate system represents a pair of real numbers. ◂◂

Let us represent two complex numbers and their sum in the complex plane. Consider, for example, the two complex numbers $1 + 2j$ and $3 + j$. By algebraic addition the sum is $4 + 3j$. When we draw lines from the origin to these points (see Fig. 12-2), we note that if we think of the complex numbers as being vectors, their sum is the vector sum. Because complex numbers can be used to represent vectors, these numbers are particularly important. *Any complex number can be thought of as representing a vector from the origin to its point in the complex plane.* This leads us to the method used to add complex numbers graphically.

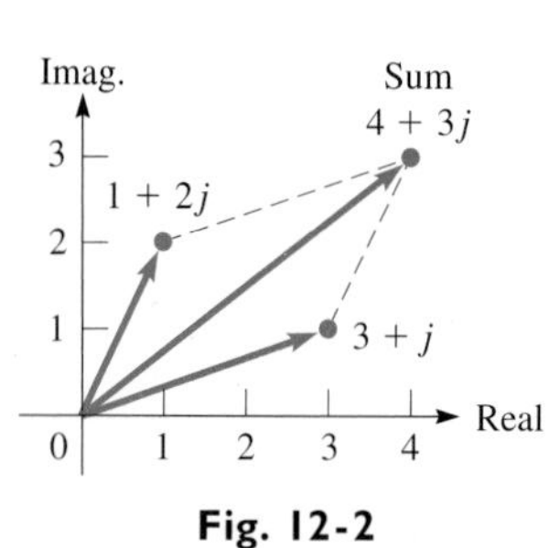

Fig. 12-2

> ***Steps to Add Complex Numbers Graphically***
>
> 1. *Find the point corresponding to one of the numbers, and draw a line from the origin to this point.*
> 2. *Repeat step 1 for the second number.*
> 3. *Complete a parallelogram with the lines drawn as adjacent sides. The resulting fourth vertex is the point representing the sum.*

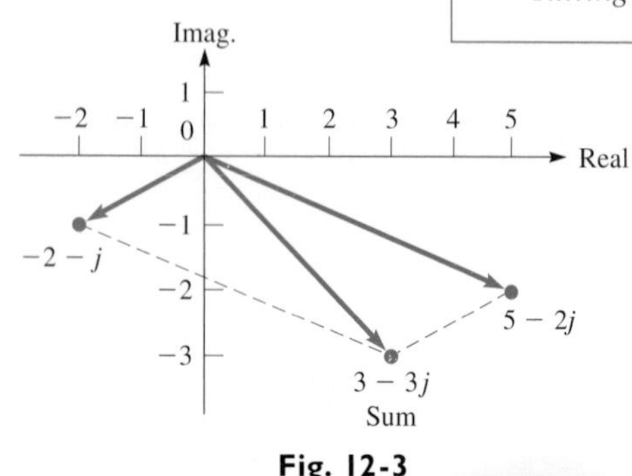

Fig. 12-3

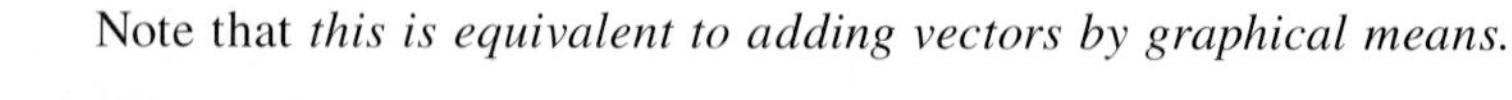

Note that *this is equivalent to adding vectors by graphical means.*

EXAMPLE 2 ▸▸ Add the complex numbers $5 - 2j$ and $-2 - j$ graphically.

The solution is indicated in Fig. 12-3. We can see that the fourth vertex of the parallelogram is at $3 - 3j$. This is, of course, the algebraic sum of these two complex numbers. ◂◂

EXAMPLE 3 ▸▸ Add the complex numbers -3 and $1 + 4j$ graphically.

First we note that $-3 = -3 + 0j$, which means that the point representing -3 is on the negative real axis. In Fig. 12-4, we show the numbers -3 and $1 + 4j$ on the graph and complete the parallelogram. From the graph we see that the sum is $-2 + 4j$. ◂◂

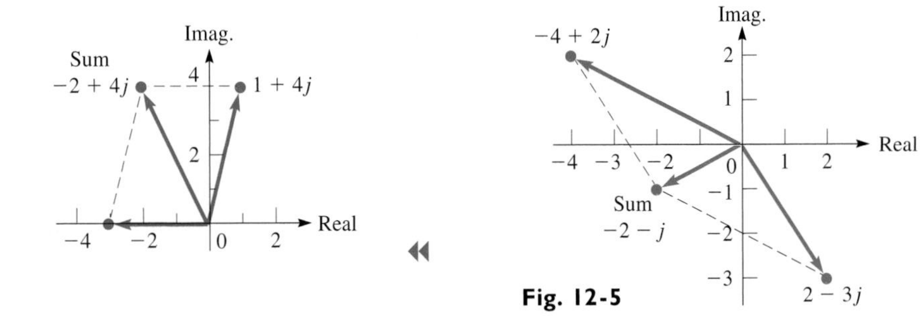

Fig. 12-4

Fig. 12-5

EXAMPLE 4 ▸▸ Subtract $4 - 2j$ from $2 - 3j$ graphically.

Subtracting $4 - 2j$ is equivalent to adding $-4 + 2j$. Therefore, we complete the solution by adding $-4 + 2j$ and $2 - 3j$, as shown in Fig. 12-5. The result is $-2 - j$. ◂◂

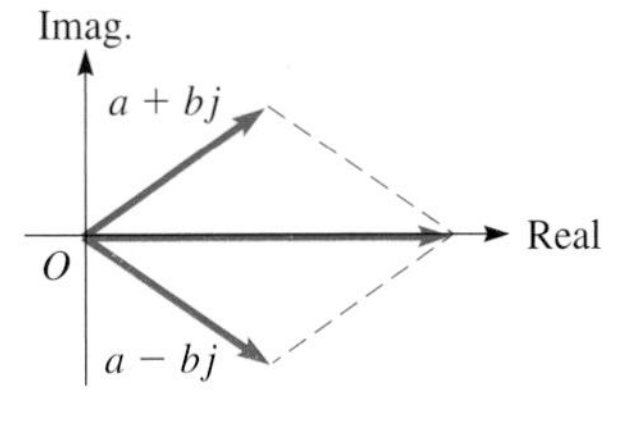

Fig. 12-6

EXAMPLE 5 ▸▸ Show graphically that the sum of a complex number and its conjugate is a real number.

If we choose the complex number $a + bj$, we know that its conjugate is $a - bj$. The imaginary coordinate for the conjugate is as far below the real axis as the imaginary coordinate of $a + bj$ is above it. Therefore, the sum of the imaginary parts must be zero and the sum of the two numbers must therefore lie on the real axis, as shown in Fig. 12-6. Therefore, we have shown that the sum of $a + bj$ and $a - bj$ is real. ◂◂

Exercises 12–3

In Exercises 1–4, locate the given complex numbers in the complex plane.

1. $2 + 6j$ **2.** $-5 + j$

3. $-4 - 3j$ **4.** $3 - 4j$

In Exercises 5–24, perform the indicated operations graphically. Check them algebraically.

5. $2 + (3 + 4j)$ **6.** $2j + (-2 + 3j)$

7. $(5 - j) + (3 + 2j)$ **8.** $(3 - 2j) + (-1 - j)$

9. $5 - (1 - 4j)$ **10.** $(2 - j) - j$

11. $(2 - 4j) + (-2 + j)$ **12.** $(-1 - 2j) + (6 - j)$

13. $(3 - 2j) - (4 - 6j)$ **14.** $(-2 - 4j) - (2 - 5j)$

15. $(1 + 4j) - (3 + j)$ **16.** $(-j - 2) - (-1 - 3j)$

17. $(1.5 - 0.5j) + (3.0 + 2.5j)$

18. $(3.5 + 2.0j) - (-4.0 - 1.5j)$

19. $(3 - 6j) - (-1 + 5j)$ **20.** $(-6 - 3j) + (2 - 7j)$

21. $(2j + 1) - 3j - (j + 1)$ **22.** $(6 - j) - 9 - (2j - 3)$

23. $(j - 6) - j + (j - 7)$ **24.** $j - (1 - j) + (3 + 2j)$

In Exercises 25–28, show the given number, its negative, and its conjugate on the same coordinate system.

25. $3 + 2j$ **26.** $-2 + 4j$

27. $-3 - 5j$ **28.** $5 - j$

In Exercises 29–32, show the number $a + bj$, $3(a + bj)$, and $-3(a + bj)$ on the same coordinate system. The multiplication of a complex number by a real number is called **scalar multiplication** *of the complex number.*

29. $-2 + j$ **30.** $-1 - 3j$

31. $3 - j$ **32.** $2 + j$

12-4 Polar Form of a Complex Number

We have just seen the relationship between complex numbers and vectors. Since we can use one to represent the other, we shall use this fact to write complex numbers in another way. The new form that we develop has certain advantages when the basic operations of multiplication and division are performed on complex numbers. We will discuss these operations later in the chapter.

By drawing a vector from the origin to the point in the complex plane that represents the number $x + yj$, we see the relation between vectors and complex numbers. Further observation indicates that an angle in standard position has been formed. Also, the point $x + yj$ is r units from the origin. In fact, *we can find any point in the complex plane by knowing the angle θ and the value of r.* The necessary equations relating x, y, r, and θ are similar to those we developed for vectors (Eqs. (9-1) to (9-3)). By referring to Fig. 12-7, we can see that

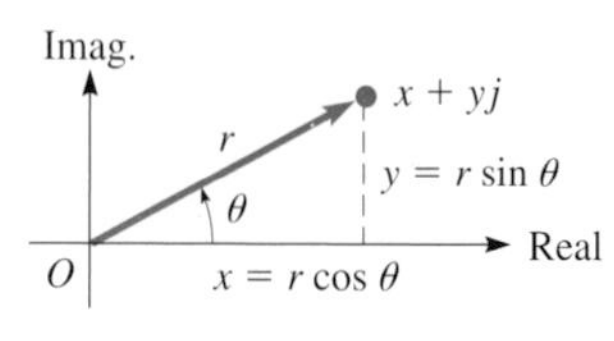

Fig. 12-7

$$x = r\cos\theta \qquad y = r\sin\theta \tag{12-7}$$

$$r^2 = x^2 + y^2 \qquad \tan\theta = \frac{y}{x} \tag{12-8}$$

Substituting Eq. (12-7) into the rectangular form $x + yj$ of a complex number, we have

$$x + yj = r\cos\theta + j(r\sin\theta)$$

or

$$x + yj = r(\cos\theta + j\sin\theta) \tag{12-9}$$

The right side of Eq. (12-9) is called the **polar form** *of a complex number. Sometimes it is referred to as the* **trigonometric form.** An abbreviated form of writing the polar form that is sometimes used is r cis θ. *The length r is called the* **absolute value,** *or the* **modulus,** *and the angle θ is called the* **argument** *of the complex number.* Therefore, Eq. (12-9), along with Eqs. (12-8), defines the polar form of a complex number.

EXAMPLE 1 ▸▸ Represent the complex number $3 + 4j$ graphically, and give its polar form.

From the rectangular form $3 + 4j$, we see that $x = 3$ and $y = 4$. Using Eqs. (12-8), we have

$$r = \sqrt{3^2 + 4^2} = 5, \qquad \tan\theta = \frac{4}{3}, \qquad \theta = 53.1^\circ$$

Thus, the polar form is

$$5(\cos 53.1^\circ + j\sin 53.1^\circ)$$

The graphical representation is shown in Fig. 12-8. ◂◂

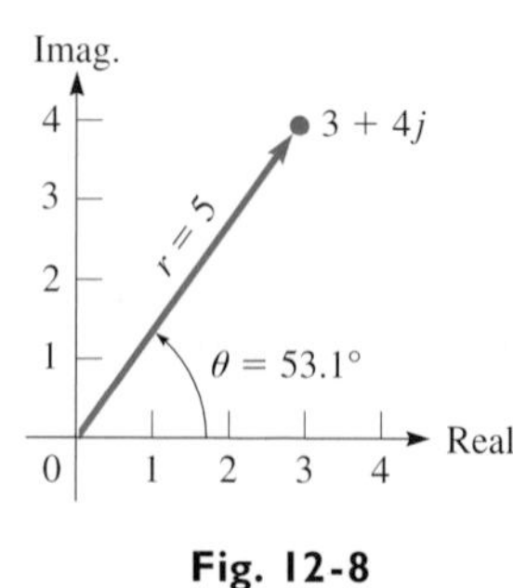

Fig. 12-8

A note on significant digits is in order here. In writing a complex number as $3 + 4j$ in Example 1, no approximate values are intended. However, in expressing the polar form as $5(\cos 53.1° + j \sin 53.1°)$, we rounded off the angle to the nearest $0.1°$, as it is not possible to express the result exactly in degrees. Therefore, in dealing with nonexact numbers, we shall express angles to the nearest $0.1°$. Other results, when approximate, will be expressed to three significant digits, unless a different accuracy is given in the problem. Of course, in applied situations most numbers used are approximate, as they are derived through measurement.

Another convenient and widely used notation for the polar form is $r\underline{/\theta}$. We must remember in using this form that it represents a complex number and is simply a shorthand way of writing $r(\cos \theta + j \sin \theta)$. Therefore,

$$r\underline{/\theta} = r(\cos \theta + j \sin \theta) \tag{12-10}$$

EXAMPLE 2 ▶▶ (a) $3(\cos 40° + j \sin 40°) = 3\underline{/40°}$
(b) $6.26(\cos 217.3° + j \sin 217.3°) = 6.26\underline{/217.3°}$
(c) $5\underline{/120°} = 5(\cos 120° + j \sin 120°)$
(d) $14.5\underline{/306.2°} = 14.5(\cos 306.2° + j \sin 306.2°)$ ◀◀

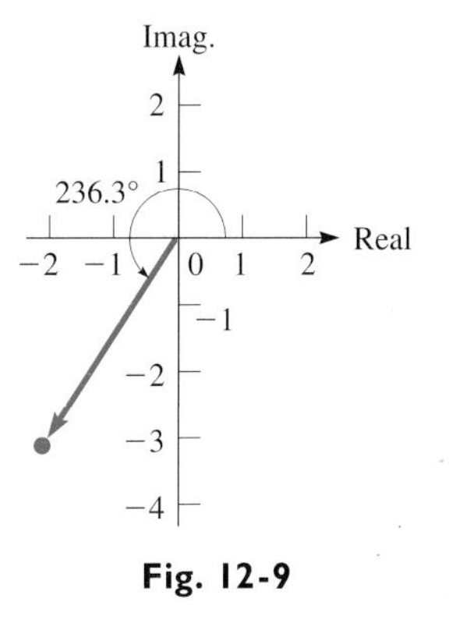

Fig. 12-9

EXAMPLE 3 ▶▶ Represent the complex number $-2.08 - 3.12j$ graphically, and give its polar forms.

The graphical representation is shown in Fig. 12-9. From Eqs. (12-8), we have

$$r = \sqrt{(-2.08)^2 + (-3.12)^2} = 3.75$$

$$\tan \theta_{\text{ref}} = \frac{3.12}{2.08}, \qquad \theta_{\text{ref}} = 56.3°, \qquad \theta = 180° + 56.3° = 236.3°$$

Since both the real and imaginary parts are negative, we know that θ is a third-quadrant angle. Therefore, we found the reference angle before finding θ. This means the polar forms are

$$3.75(\cos 236.3° + j \sin 236.3°) = 3.75\underline{/236.3°}$$ ◀◀

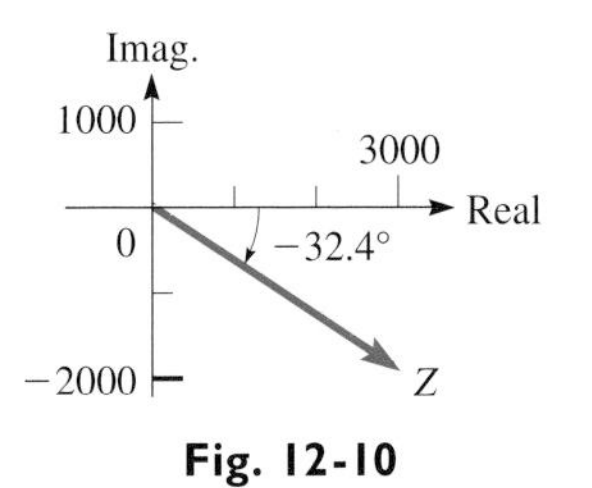

Fig. 12-10

EXAMPLE 4 ▶▶ The impedance Z (in ohms) in an alternating-current circuit is given by $Z = 3560\underline{/-32.4°}$. Express this in rectangular form.

From the polar form we have $r = 3560\ \Omega$ and $\theta = -32.4°$ (it is common to use negative angles in this type of application). This means that we can also write

$$Z = 3560(\cos(-32.4°) + j \sin(-32.4°))$$

See Fig. 12-10. This means that

$$x = 3560 \cos(-32.4°) = 3010$$
$$y = 3560 \sin(-32.4°) = -1910$$

Therefore, the rectangular form is

$$Z = 3010 - 1910j \text{ ohms}$$ ◀◀

Most calculators have keys that will convert directly between rectangular and polar form. See the manual for specific instructions.

EXAMPLE 5 ▸▸ Represent the numbers 5, -5, $7j$, and $-7j$ in polar form.

Since any positive real number lies on the positive real axis in the complex plane, it is expressed in polar form by

$$a = a(\cos 0° + j \sin 0°) = a\underline{/0°}$$

Negative real numbers, being on the negative real axis, are written as

$$a = |a|(\cos 180° + j \sin 180°) = |a|\underline{/180°}$$

Thus,

$$5 = 5(\cos 0° + j \sin 0°) = 5\underline{/0°}$$
$$-5 = 5(\cos 180° + j \sin 180°) = 5\underline{/180°}$$

Positive pure imaginary numbers lie on the positive imaginary axis and are expressed in polar form by

$$bj = b(\cos 90° + j \sin 90°) = b\underline{/90°}$$

Similarly, negative pure imaginary numbers, being on the negative imaginary axis, are written as

$$bj = |b|(\cos 270° + j \sin 270°) = |b|\underline{/270°}$$

Thus,

$$7j = 7(\cos 90° + j \sin 90°) = 7\underline{/90°}$$
$$-7j = 7(\cos 270° + j \sin 270°) = 7\underline{/270°}$$

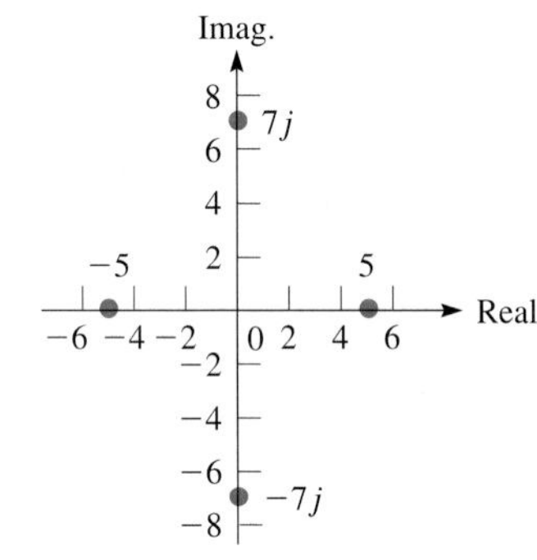

Fig. 12-11

The graphical representations of the *complex numbers* 5, -5, $7j$, and $-7j$ are shown in Fig. 12-11. ◂◂

Exercises 12–4

In Exercises 1–16, represent each complex number graphically, and give the polar form of each.

1. $8 + 6j$
2. $-8 - 15j$
3. $3 - 4j$
4. $-5 + 12j$
5. $-2.00 + 3.00j$
6. $7.00 - 5.00j$
7. $-5.50 - 2.40j$
8. $4.60 - 4.60j$
9. $1 + j\sqrt{3}$
10. $\sqrt{2} - j\sqrt{2}$
11. $3.514 - 7.256j$
12. $6.231 + 9.527j$
13. -3
14. 6
15. $9j$
16. $-2j$

In Exercises 17–36, represent each complex number graphically, and give the rectangular form of each.

17. $5.00(\cos 54.0° + j \sin 54.0°)$
18. $3.00(\cos 232.0° + j \sin 232.0°)$
19. $1.60(\cos 150.0° + j \sin 150.0°)$
20. $2.50(\cos 315.0° + j \sin 315.0°)$
21. $6(\cos 180° + j \sin 180°)$
22. $12(\cos 270° + j \sin 270°)$
23. $8(\cos 360° + j \sin 360°)$
24. $15(\cos 0° + j \sin 0°)$
25. $12.36(\cos 345.56° + j \sin 345.56°)$
26. $220.8(\cos 155.13° + j \sin 155.13°)$
27. $\cos 240.0° + j \sin 240.0°)$
28. $\cos 299.0° + j \sin 299.0°$
29. $4.75\underline{/172.8°}$
30. $1.50\underline{/62.3°}$
31. $0.9326\underline{/229.54°}$
32. $277.8\underline{/-342.63°}$
33. $7.32\underline{/-270°}$
34. $18.3\underline{/180.0°}$
35. $86.42\underline{/94.62°}$
36. $4629\underline{/182.44°}$

In Exercises 37–40, solve the given problems.

37. What are the magnitude and the direction of a force vector that is represented by $25.6 - 34.2j$ newtons?
38. What are the magnitude and the direction of the displacement of a weight at the end of a spring that can be described by $y = 0.395 + 0.148j$ metres?
39. The electric field intensity of a light wave can be described by the complex number $12.4\underline{/78.3°}$ V/m. Write this in rectangular form.
40. The current in a certain microprocessor circuit is represented by the complex number $3.75\underline{/15.0°}$ μA. Write this in rectangular form.

12-5 Exponential Form of a Complex Number

Another important form of a complex number is the *exponential form*. It is commonly used in electronics, engineering, and physics applications. As we will see in the next section, it is also convenient for multiplication and division of complex numbers, as the rectangular form is for addition and subtraction.

The **exponential form** *of a complex number is written as* $re^{j\theta}$, where r and θ have the same meanings as given in the previous section, although θ is expressed in radians. *The number e is a special irrational number and has an approximate value*

$$e = 2.718\,281\,828\,459\,045\,2$$

This number e is very important in mathematics, and we will see it again in the next chapter. For now it is necessary to accept the value for e, although in calculus its meaning is shown along with the reason it has the above value. We can find its value on a calculator by using the e^x key, with $x = 1$.

We now define

$$re^{j\theta} = r(\cos\theta + j\sin\theta) \tag{12-11}$$

By expressing θ in radians, the expression $j\theta$ is an exponent, and it can be shown to obey all the laws of exponents as discussed in Chapter 11. Therefore, we shall always *express θ in radians when using exponential form.*

CAUTION ▶

EXAMPLE 1 ▶▶ Express the number $3 + 4j$ in exponential form.

From Example 1 of Section 12-4, we know that this complex number may be written in polar form as $5(\cos 53.1° + j\sin 53.1°)$. Therefore, we know that $r = 5$. We now express 53.1° in terms of radians as

$$\frac{53.1\pi}{180} = 0.927 \text{ rad}$$

Thus, the exponential form is $5e^{0.927j}$. This means that

$$3 + 4j = 5(\cos 53.1° + j\sin 53.1°) = 5e^{0.927j}$$

degrees to radians

value of r ◀◀

EXAMPLE 2 ▶▶ Express the number $8.50\underline{/136.3°}$ in exponential form.

Since this complex number is in polar form, we note that $r = 8.50$ and that we must express 136.3° in radians. Changing 136.3° to radians, we have

$$\frac{136.3\pi}{180} = 2.38 \text{ rad}$$

Therefore, the required exponential form is $8.50e^{2.38j}$. This means that

$$8.50\underline{/136.3°} = 8.50e^{2.38j}$$

We see that the principal step in changing from polar form to exponential form is to change θ from degrees to radians. ◀◀

EXAMPLE 3 ▸▸ Express the number $3.07 - 7.43j$ in exponential form.

From the rectangular form of the number, we have $x = 3.07$ and $y = -7.43$. Therefore,

$$r = \sqrt{(3.07)^2 + (-7.43)^2} = 8.04$$

$$\tan\theta = \frac{-7.43}{3.07}, \qquad \theta = 292.4°$$

Changing 292.4° to radians, we have 292.4° = 5.10 rad. Therefore, the exponential form is $8.04e^{5.10j}$. This means that

$$3.07 - 7.43j = 8.04e^{5.10j}$$ ◂◂

EXAMPLE 4 ▸▸ Express the complex number $2.00e^{4.80j}$ in polar and rectangular forms.

For reference, Eq. (8-8) is π rad = 180°.

We first express 4.80 rad as 275.0°. From the exponential form we know that $r = 2.00$. Thus, the polar form is

$$2.00(\cos 275.0° + j \sin 275.0°)$$

Next, by use of the distributive law we rewrite the polar form and then evaluate. Thus,

$$\begin{aligned} 2.00e^{4.80j} &= 2.00(\cos 275.0° + j \sin 275.0°) \\ &= 2.00 \cos 275.0° + (2.00 \sin 275.0°)j \\ &= 0.174 - 1.99j \end{aligned}$$ ◂◂

EXAMPLE 5 ▸▸ Express the number $3.408e^{2.457j}$ in polar and rectangular forms.

We first express 2.457 rad as 140.78°. From the exponential form, we know that $r = 3.408$. Thus, the polar form is $3.408\underline{/140.78°}$. Thus,

$$\begin{aligned} 3.408e^{2.457j} &= 3.408(\cos 140.78° + j \sin 140.78°) \\ &= -2.640 + 2.155j \end{aligned}$$ ◂◂

As we noted earlier, an important application of the use of complex numbers is in alternating-current analysis. When an alternating current flows through a given circuit, usually the current and voltage have different phases. That is, they do not reach their peak values at the same time. Therefore, one way of accounting for the magnitude as well as the phase of an electric current or voltage is to write it as a complex number. Here the modulus is the actual magnitude of the current or voltage, and the argument θ is a measure of the phase.

EXAMPLE 6 ▸▸ A current of $2.00 - 4.00j$ amperes flows in a given circuit. Express this current in exponential form, and find the magnitude of the current.

From the rectangular form, we have $x = 2.00$ and $y = -4.00$. Therefore,

$$r = \sqrt{(2.00)^2 + (-4.00)^2} = 4.47$$

Also,

$$\tan\theta = -\tfrac{4.00}{2.00}$$

This means that $\theta = -63.4°$ (it is normal to express the phase in terms of negative angles). Changing 63.4° to radians, we have 63.4° = 1.11 rad. Therefore, the exponential form of the current is $4.47e^{-1.11j}$. The modulus is 4.47, meaning the magnitude of the current is 4.47 A. ◂◂

At this point we shall summarize the three important forms of a complex number. See Fig. 12-12 for the graphical representation.

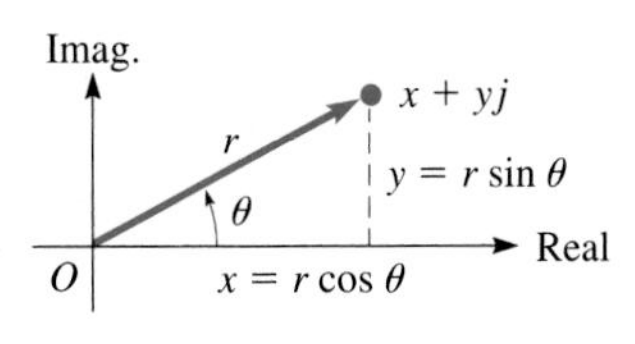

Fig. 12-12

Rectangular:	$x + yj$
Polar:	$r(\cos \theta + j \sin \theta) = r\underline{/\theta}$
Exponential:	$re^{j\theta}$

It follows that

$$x + yj = r(\cos \theta + j \sin \theta) = r\underline{/\theta} = re^{j\theta} \tag{12-12}$$

where

$$r^2 = x^2 + y^2 \qquad \tan \theta = \frac{y}{x} \tag{12-8}$$

In Eq. (12-12) the argument θ is the same for the exponential and polar forms, although it is expressed in radians in exponential form and is usually expressed in degrees in polar form.

Exercises 12–5

In Exercises 1–20, express the given complex numbers in exponential form.

1. $3.00(\cos 60.0^\circ + j \sin 60.0^\circ)$

2. $5.00(\cos 135.0^\circ + j \sin 135.0^\circ)$

3. $4.50(\cos 282.3^\circ + j \sin 282.3^\circ)$

4. $2.10(\cos 228.7^\circ + j \sin 228.7^\circ)$

5. $375.5(\cos 95.46^\circ + j \sin 95.46^\circ)$

6. $16.72[\cos(-7.14^\circ) + j \sin(-7.14^\circ)]$

7. $0.515\underline{/198.3^\circ}$ **8.** $4650\underline{/326.5^\circ}$

9. $4.06\underline{/-61.4^\circ}$ **10.** $0.0192\underline{/76.7^\circ}$

11. $9245\underline{/296.32^\circ}$ **12.** $827.6\underline{/110.09^\circ}$

13. $3 - 4j$ **14.** $-1 - 5j$

15. $-3 + 2j$ **16.** $6 + j$

17. $5.90 + 2.40j$ **18.** $47.3 - 10.9j$

19. $-634.6 - 528.2j$ **20.** $-8573 + 5477j$

In Exercises 21–28, express the given complex numbers in polar and rectangular forms.

21. $3.00e^{0.500j}$ **22.** $2.00e^{1.00j}$

23. $4.64e^{1.85j}$ **24.** $2.50e^{3.84j}$

25. $3.20e^{5.41j}$ **26.** $0.800e^{3.00j}$

27. $0.1724e^{2.391j}$ **28.** $820.7e^{3.492j}$

In Exercises 29–32, perform the indicated operations.

29. The impedance in an antenna circuit is $375 + 110j$ ohms. Write this in exponential form, and find the magnitude of the impedance.

30. The intensity of the signal from a radar microwave signal is $37.0[\cos(-65.3^\circ) + j \sin(-65.3^\circ)]$ V/m. Write this in exponential form.

31. The displacement of a sound wave is $3.50e^{-1.35j}$ μm. Write this in rectangular form.

32. In an electric circuit, the *admittance* is the reciprocal of the impedance. In a transistor circuit, the impedance is $2800 - 1450j$ ohms. Find the exponential form of the admittance.

12-6 Products, Quotients, Powers, and Roots of Complex Numbers

We have previously found products and quotients using the rectangular forms of the given complex numbers. However, these operations can also be performed with complex numbers in polar and exponential forms. These operations not only are convenient but also are useful for purposes of finding powers and roots of complex numbers.

We may find the product of two complex numbers by using the exponential form and the laws of exponents. Multiplying $r_1e^{j\theta_1}$ by $r_2e^{j\theta_2}$, we have

$$[r_1e^{j\theta_1}] \times [r_2e^{j\theta_2}] = r_1r_2e^{j\theta_1+j\theta_2} = r_1r_2^{j(\theta_1+\theta_2)}$$

We use this equation to express the product of two complex numbers in polar form:

$$[r_1e^{j\theta_1}] \times [r_2e^{j\theta_2}] = [r_1(\cos\theta_1 + j\sin\theta_1)] \times [r_2(\cos\theta_2 + j\sin\theta_2)]$$

and

$$r_1r_2e^{j(\theta_1+\theta_2)} = r_1r_2[\cos(\theta_1+\theta_2) + j\sin(\theta_1+\theta_2)]$$

Therefore, the polar expressions are equal, which means that *the product of two complex numbers is*

$$\begin{aligned} r_1(\cos\theta_1 + j\sin\theta_1)r_2(\cos\theta_2 + j\sin\theta_2) \\ = r_1r_2[\cos(\theta_1+\theta_2) + j\sin(\theta_1+\theta_2)] \\ (r_1\underline{/\theta_1})(r_2\underline{/\theta_2}) = r_1r_2\underline{/\theta_1+\theta_2} \end{aligned} \qquad (12\text{-}13)$$

The *magnitudes are multiplied,* and the *angles are added.*

EXAMPLE 1 ▸▸ Multiply the complex numbers $2 + 3j$ and $1 - j$ by using the polar form of each.

For $2 + 3j$: $r_1 = \sqrt{2^2 + 3^2} = 3.61, \quad \tan\theta_1 = \frac{3}{2}, \quad \theta_1 = 56.3°$

For $1 - j$: $r_2 = \sqrt{1^2 + (-1)^2} = 1.41, \quad \tan\theta_2 = \frac{-1}{1}, \quad \theta_2 = 315.0°$

$$(3.61)(\cos 56.3° + j\sin 56.3°)(1.41)(\cos 315.0° + j\sin 315.0°)$$

sum

$$\begin{aligned} &= (3.61)(1.41)[\cos(56.3° + 315.0°) + j\sin(56.3° + 315.0°)] \\ &= 5.09(\cos 371.3° + j\sin 371.3°) \\ &= 5.09(\cos 11.3° + j\sin 11.3°) \end{aligned}$$

Note that in the final result the angle is expressed as 11.3°. The angle is usually expressed between 0° and 360°, unless specified otherwise. As we have seen, in some applications it is common to use negative angles. ◂◂

EXAMPLE 2 ▸▸ When we use the $r\underline{/\theta}$ polar form to multiply the two complex numbers in Example 1, we have

$$r_1 = 3.61 \qquad \theta_1 = 56.3°$$
$$r_2 = 1.41 \qquad \theta_2 = 315.0°$$

$$(3.61\underline{/56.3°})(1.41\underline{/315.0°}) = (3.61)(1.41)\underline{/56.3° + 315.0°}$$
$$= 5.09\underline{/371.3°}$$
$$= 5.09\underline{/11.3°}$$

Again note that the angle in the final result is between 0° and 360°. ◂◂

If we wish to *divide* one complex number in exponential form by another, we arrive at the following result:

$$r_1e^{j\theta_1} \div r_2e^{j\theta_2} = \frac{r_1}{r_2}e^{j(\theta_1-\theta_2)} \tag{12-14}$$

Therefore, *the result of dividing one complex number in polar form by another is given by*

$$\frac{r_1(\cos\theta_1 + j\sin\theta_1)}{r_2(\cos\theta_2 + j\sin\theta_2)} = \frac{r_1}{r_2}[\cos(\theta_1 - \theta_2) + j\sin(\theta_1 - \theta_2)]$$
$$\frac{r_1\underline{/\theta_1}}{r_2\underline{/\theta_2}} = \frac{r_1}{r_2}\underline{/\theta_1 - \theta_2} \tag{12-15}$$

The *magnitudes are divided,* and the *angles are subtracted.*

EXAMPLE 3 ▸▸ Divide the first complex number of Example 1 by the second. Using the polar forms, we have

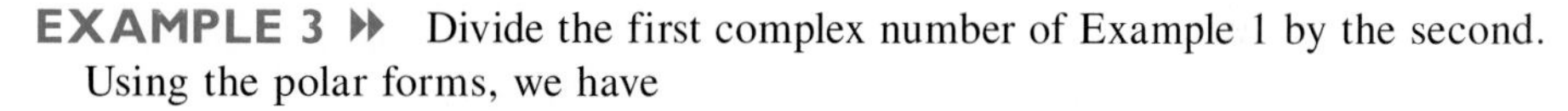

$$\frac{3.61(\cos 56.3° + j\sin 56.3°)}{1.41(\cos 315.0° + j\sin 315.0°)} = \frac{3.61}{1.41}[\cos(56.3° - 315.0°) + j\sin(56.3° - 315.0°)]$$
$$= 2.56[\cos(-258.7°) + j\sin(-258.7°)]$$
$$= 2.56(\cos 101.3° + j\sin 101.3°)$$ ◂◂

EXAMPLE 4 ▸▸ Repeating Example 3 using $r\underline{/\theta}$ polar form, we have

$$\frac{3.61\underline{/56.3°}}{1.41\underline{/315.0°}} = \frac{3.61}{1.41}\underline{/56.3° - 315.0°} = 2.56\underline{/-258.7°}$$
$$= 2.56\underline{/101.3°}$$ ◂◂

CAUTION ▶ We have just seen that multiplying and dividing numbers in polar form can be easily performed. However, if we are to add or subtract numbers in polar form, *we must do the addition or subtraction by using rectangular form.*

EXAMPLE 5 ▸▸ Perform the addition $1.563\underline{/37.56^\circ} + 3.827\underline{/146.23^\circ}$.

In order to do this addition, we must change each number to rectangular form.

$$\begin{aligned}
&1.563\underline{/37.56^\circ} + 3.827\underline{/146.23^\circ}\\
&\quad = 1.563(\cos 37.56^\circ + j\sin 37.56^\circ) + 3.827(\cos 146.23^\circ + j\sin 146.23^\circ)\\
&\quad = 1.2390 + 0.9528j - 3.1813 + 2.1273j\\
&\quad = -1.9423 + 3.0801j
\end{aligned}$$

Now we change this to polar form.

$$r = \sqrt{(-1.9423)^2 + (3.0801)^2} = 3.641$$

$$\tan\theta = \frac{3.0801}{-1.9423}, \qquad \theta = 122.24^\circ$$

Therefore,

$$1.563\underline{/37.56^\circ} + 3.827\underline{/146.23^\circ} = 3.641\underline{/122.24^\circ}$$ ◂◂

DeMoivre's Theorem

For reference, Eq. (11-3) is $(a^m)^n = a^{mn}$.

To raise a complex number to a power, we may use the exponential form of the number in Eq.(11-3). This leads to

$$(re^{j\theta})^n = r^n e^{jn\theta} \tag{12-16}$$

Extending this to polar form, we have

$$\begin{aligned}
[r(\cos\theta + j\sin\theta)]^n &= r^n(\cos n\theta + j\sin n\theta)\\
(r\underline{/\theta})^n &= r^n\underline{/n\theta}
\end{aligned} \tag{12-17}$$

Named for the mathematician Abraham DeMoivre (1667–1754).

Equation (12-17) is known as **DeMoivre's theorem** *and is valid for all real values of n. It is also used for finding roots of complex numbers if n is a fractional exponent.* We note that *the magnitude is raised to the power,* and *the angle is multiplied by the power.*

EXAMPLE 6 ▸▸ Using DeMoivre's theorem, find $(2 + 3j)^3$.

From Example 1 of this section, we know $r = 3.61$ and $\theta = 56.3^\circ$. Therefore,

$$\begin{aligned}
[3.61(\cos 56.3^\circ + j\sin 56.3^\circ)]^3 &= (3.61)^3[\cos(3 \times 56.3^\circ) + j\sin(3 \times 56.3^\circ)]\\
&= 47.0(\cos 168.9^\circ + j\sin 168.9^\circ) = 47.0\underline{/168.9^\circ}
\end{aligned}$$

Expressing θ in radians, we have $\theta = 56.3^\circ = 0.983$ rad. Therefore,

$$\begin{aligned}
(3.61e^{0.983j})^3 &= (3.61)^3 e^{3\times 0.983j} = 47.0e^{2.95j}\\
(2 + 3j)^3 &= 47.0(\cos 168.9^\circ + j\sin 168.9^\circ)\\
&= 47.0\underline{/168.9^\circ} = 47.0e^{2.95j} = -46 + 9j
\end{aligned}$$ ◂◂

EXAMPLE 7 ▸▸ Find the cube root of -1.

Since we know that -1 is a real number, we can find its cube root by means of the definition. That is, $(-1)^3 = -1$. We shall check this by DeMoivre's theorem. Writing -1 in polar form, we have

$$-1 = 1(\cos 180° + j \sin 180°)$$

Applying DeMoivre's theorem, with $n = \frac{1}{3}$, we obtain

$$(-1)^{1/3} = 1^{1/3}(\cos \tfrac{1}{3} 180° + j \sin \tfrac{1}{3} 180°) = \cos 60° + j \sin 60°$$

$$= \frac{1}{2} + j\frac{\sqrt{3}}{2} \quad \text{exact answer}$$

$$= 0.5000 + 0.8660j \quad \text{decimal approximation}$$

Observe that we did not obtain -1 as the answer. If we check the answer, in the form $\frac{1}{2} + j\frac{\sqrt{3}}{2}$, by actually cubing it, we obtain -1! Therefore, it is a correct answer.

We should note that it is possible to take $\frac{1}{3}$ of any angle up to 1080° and still have an angle less than 360°. Since 180° and 540° have the same terminal side, let us try writing -1 as $1(\cos 540° + j \sin 540°)$. Using DeMoivre's theorem, we have

$$(-1)^{1/3} = 1^{1/3}(\cos \tfrac{1}{3} 540° + j \sin \tfrac{1}{3} 540°) = \cos 180° + j \sin 180° = -1$$

We have found the answer we originally anticipated.

Angles of 180° and 900° also have the same terminal side, so we try

$$(-1)^{1/3} = 1^{1/3}(\cos \tfrac{1}{3} 900° + j \sin \tfrac{1}{3} 900°) = \cos 300° + j \sin 300°$$

$$= \frac{1}{2} - j\frac{\sqrt{3}}{2} \quad \text{exact answer}$$

$$= 0.5000 - 0.8660j \quad \text{decimal approximation}$$

Checking this, we find that it is also a correct root. We may try 1260°, but $\frac{1}{3}(1260°) = 420°$, which has the same functional values as 60°, and would give us the answer $0.5000 + 0.8660j$ again.

We have found, therefore, *three cube roots* of -1. They are

$$-1, \frac{1}{2} + j\frac{\sqrt{3}}{2}, \frac{1}{2} - j\frac{\sqrt{3}}{2}$$

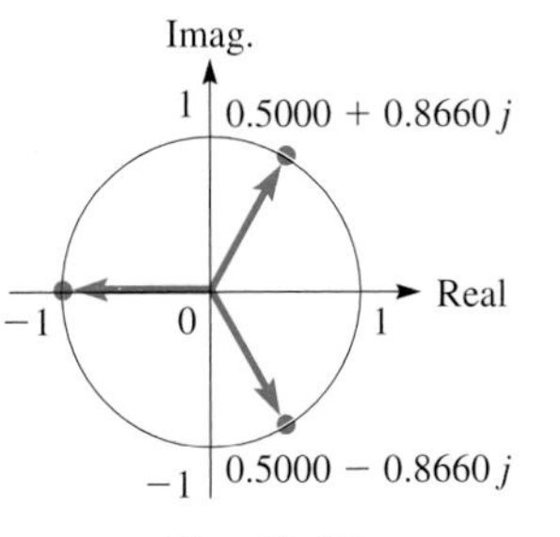

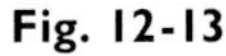
Fig. 12-13

These roots are graphed in Fig. 12-13. Note that they are equally spaced on the circumference of a circle of radius 1. ◂◂

When we generalize on the results of Example 7, it can be proved that *there are n nth roots of a complex number.* Therefore, we have the following method for finding the n roots of a complex number.

> ***Using DeMoivre's Theorem to Find the n nth Roots of a Complex Number***
>
> 1. *Express the number in polar form.*
> 2. *Express the root as a fractional exponent.*
> 3. *Use Eq.* (12-17) *with θ to find one root.*
> 4. *Use Eq.* (12-17) *and add* 360° *to θ, $n - 1$ times, to find the other roots.*

The roots will be on a circle of radius $r^{1/n}$, equally spaced $360°/n$ apart.

EXAMPLE 8 ▸▸ Find the two square roots of $2j$.

We first write $2j$ in polar form as

$$2j = 2(\cos 90° + j \sin 90°)$$

To find square roots, we use the exponent $n = 1/2$. The first square root is

$$(2j)^{1/2} = 2^{1/2}\left(\cos \frac{90°}{2} + j \sin \frac{90°}{2}\right) = \sqrt{2}(\cos 45° + j \sin 45°) = 1 + j$$

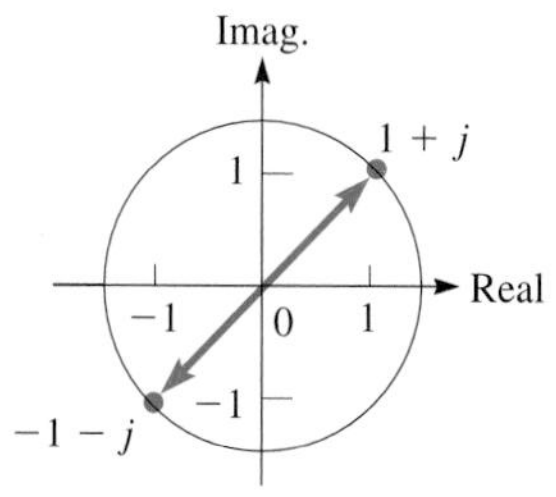

Fig. 12-14

To find the other square root we add 360° to 90°. This gives us

$$(2j)^{1/2} = 2^{1/2}\left(\cos \frac{450°}{2} + j \sin \frac{450°}{2}\right) = \sqrt{2}(\cos 225° + j \sin 225°) = -1 - j$$

Therefore, the two square roots of $2j$ are $1 + j$ and $-1 - j$. We see in Fig. 12-14 that they are on a circle of radius $\sqrt{2}$ and are 180° apart. ◂◂

EXAMPLE 9 ▸▸ Find the six sixth roots of 64.

First we write 64 in polar form as $64 = 64(\cos 0° + j \sin 0°)$. We then note that we use the exponent 1/6 for the sixth root and that $64^{1/6} = \sqrt[6]{64} = 2$.

First root: $$64^{1/6} = 64^{1/6}\left(\cos \frac{0°}{6} + j \sin \frac{0°}{6}\right) = 2(\cos 0° + j \sin 0°) = 2$$

add 360°

Second root: $$64^{1/6} = 64^{1/6}\left(\cos \frac{0° + 360°}{6} + j \sin \frac{0° + 360°}{6}\right)$$
$$= 2(\cos 60° + j \sin 60°) = 1 + j\sqrt{3}$$

add 2 × 360°

Third root: $$64^{1/6} = 64^{1/6}\left(\cos \frac{0° + 720°}{6} + j \sin \frac{0° + 720°}{6}\right)$$
$$= 2(\cos 120° + j \sin 120°) = -1 + j\sqrt{3}$$

add 3 × 360°

Fourth root: $$64^{1/6} = 64^{1/6}\left(\cos \frac{0° + 1080°}{6} + j \sin \frac{0° + 1080°}{6}\right)$$
$$= 2(\cos 180° + j \sin 180°) = -2$$

add 4 × 360°

Fifth root: $$64^{1/6} = 64^{1/6}\left(\cos \frac{0° + 1440°}{6} + j \sin \frac{0° + 1440°}{6}\right)$$
$$= 2(\cos 240° + j \sin 240°) = -1 - j\sqrt{3}$$

add 5 × 360°

Sixth root: $$64^{1/6} = 64^{1/6}\left(\cos \frac{0° + 1800°}{6} + j \sin \frac{0° + 1800°}{6}\right)$$
$$= 2(\cos 300° + j \sin 300°) = 1 - j\sqrt{3}$$

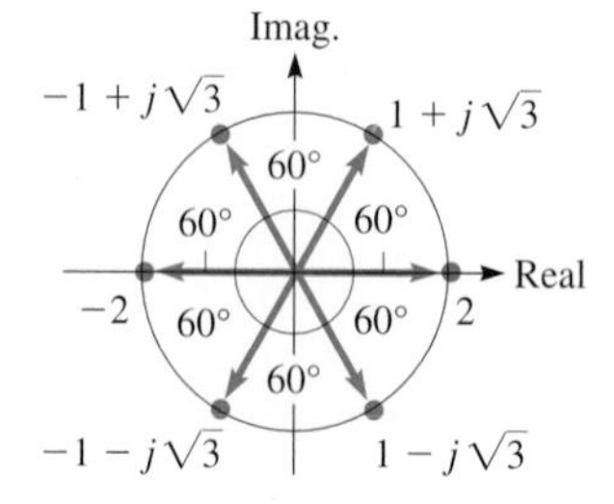

Fig. 12-15

These roots are graphed in Fig. 12-15. Note that they are equally spaced 60° apart on the circumference of a circle of radius 2. ◂◂

At this point we see advantages for the various forms of writing complex numbers. Rectangular form lends itself best to addition and subtraction. Polar form is generally used for multiplying, dividing, raising to powers, and finding roots. Exponential form is used for theoretical purposes (e.g., deriving DeMoivre's theorem).

Exercises 12–6

In Exercises 1–16, perform the indicated operations. Leave the result in polar form.

1. $[4(\cos 60° + j \sin 60°)][2(\cos 20° + j \sin 20°)]$
2. $[3(\cos 120° + j \sin 120°)][5(\cos 45° + j \sin 45°)]$
3. $(0.5\underline{/140°})(6\underline{/110°})$
4. $(0.4\underline{/320°})(5.5\underline{/150°})$
5. $\dfrac{8(\cos 100° + j \sin 100°)}{4(\cos 65° + j \sin 65°)}$
6. $\dfrac{9(\cos 230° + j \sin 230°)}{3(\cos 80° + j \sin 80°)}$
7. $\dfrac{12\underline{/320°}}{5\underline{/210°}}$
8. $\dfrac{2\underline{/90°}}{4\underline{/75°}}$
9. $[2(\cos 35° + j \sin 35°)]^3$
10. $[3(\cos 120° + j \sin 120°)]^4$
11. $(2\underline{/135°})^8$
12. $(1\underline{/142°})^{10}$
13. $\dfrac{(50\underline{/236°})(2\underline{/84°})}{25\underline{/47°}}$
14. $\dfrac{36\underline{/274°}}{(2\underline{/141°})(6\underline{/195°})}$
15. $\dfrac{(4\underline{/24°})(10\underline{/326°})}{(1\underline{/186°})(8\underline{/77°})}$
16. $\dfrac{(25\underline{/194°})(6\underline{/239°})}{(3\underline{/17°})(10\underline{/29°})}$

In Exercises 17–20, perform the indicated operations. Express results in polar form. See Example 5.

17. $2.78\underline{/56.8°} + 1.37\underline{/207.3°}$
18. $15.9\underline{/142.6°} - 18.5\underline{/71.4°}$
19. $7085\underline{/115.62°} - 4667\underline{/296.34°}$
20. $307.5\underline{/326.54°} + 726.3\underline{/96.41°}$

In Exercises 21–32, change each number to polar form and then perform the indicated operations. Express the result in rectangular and polar forms. Check by performing the same operation in rectangular form.

21. $(3 + 4j)(5 - 12j)$
22. $(-2 + 5j)(-1 - j)$
23. $(7 - 3j)(8 + j)$
24. $(1 + 5j)(4 + 2j)$
25. $\dfrac{7}{1 - 3j}$
26. $\dfrac{8j}{7 + 2j}$
27. $\dfrac{3 + 4j}{5 - 12j}$
28. $\dfrac{-2 + 5j}{-1 - j}$
29. $(3 + 4j)^4$
30. $(-1 - j)^8$
31. $(2 + 3j)^5$
32. $(1 - 2j)^6$

In Exercises 33–44, use DeMoivre's theorem to find the indicated roots. Be sure to find all roots.

33. The two square roots of $4(\cos 60° + j \sin 60°)$
34. The three cube roots of $27(\cos 120° + j \sin 120°)$
35. The three cube roots of $3 - 4j$
36. The two square roots of $-5 + 12j$
37. The square roots of $1 + j$
38. The cube roots of $\sqrt{3} + j$
39. The fourth roots of 1
40. The cube roots of 8
41. The cube roots of $-27j$
42. The fourth roots of j
43. The fifth roots of -32
44. The sixth roots of -8

In Exercises 45–48, perform the indicated operations.

45. In Example 7 we showed that one cube root of -1 is $\frac{1}{2} - \frac{1}{2}j\sqrt{3}$. Cube this number in rectangular form, and show that the result is -1.
46. In Example 8 we showed that the square roots of $2j$ are $1 + j$ and $-1 - j$. Square these numbers, and show that the results are both $2j$.
47. The power p, in watts, supplied to an element in an electric circuit is the product of the voltage e and the current i, in amperes. Find the expression for the power supplied if $e = 6.80\underline{/56.3°}$ volts and $i = 7.05\underline{/-15.8°}$ amperes.
48. The displacement d of a weight suspended on a system of two springs is $d = 6.03\underline{/22.5°} + 3.26\underline{/76.0°}$ cm. Perform the addition, and express the answer in polar form.

12-7 An Application to Alternating-Current (ac) Circuits

We shall complete our study of complex numbers by showing their use in one aspect of alternating-current circuit theory. This application will be made to measuring voltage between any two points in a simple ac circuit, similar to the application mentioned in some of the examples and exercises of earlier sections of this chapter. We shall consider a circuit containing a resistance, a capacitance, and an inductance.

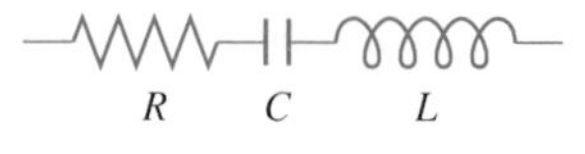

Fig. 12-16

A *resistance* is any part of a circuit that tends to obstruct the flow of electric current through the circuit. It is denoted by R (units in ohms, Ω) and in diagrams by , as shown in Fig. 12-16. A *capacitance* is two nonconnected plates in a circuit; no current actually flows across the gap between them. In an ac circuit, an electric charge is continually going to and from each plate and, therefore, the current in the circuit is not effectively stopped. It is denoted by C (units in farads, F) and in diagrams by (see Fig. 12-16). An *inductance* is basically a coil of wire in which current is induced because the current is continually changing in the circuit. It is denoted by L (units in henrys, H) and in diagrams by (see Fig. 12-16). All of these elements affect the voltage in an alternating-current circuit. We shall state here the relation each has to the voltage and current in the circuit.

In Chapter 10, when we were discussing the graphs of the trigonometric functions, we noted that the current and voltage in an ac circuit could be represented by a sine or cosine curve. Therefore, each reaches peak values periodically. *If they reach their respective peak values at the same time, we say they are* ***in phase.*** *If the voltage reaches its peak before the current, we say that the voltage* ***leads*** *the current. If the voltage reaches its peak after the current, we say that the voltage* ***lags*** *the current.*

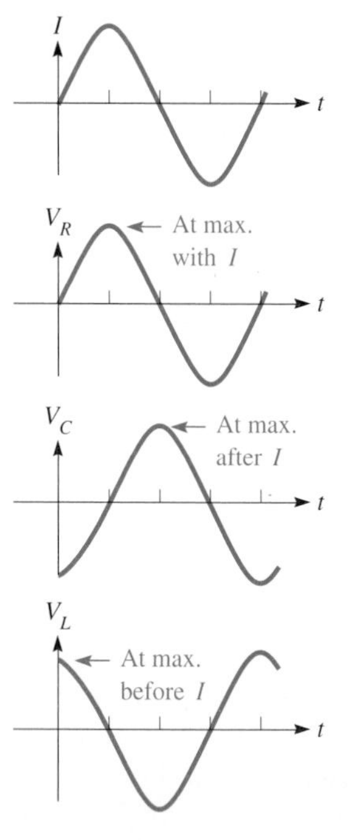

Fig. 12-17

In the study of electricity, it is shown that the voltage across a resistance is in phase with the current. The voltage across a capacitor lags the current by 90°, and the voltage across an inductance leads the current by 90°. This is shown in Fig. 12-17, where, in a given circuit, I represents the current, V_R is the voltage across a resistor, V_C is the voltage across a capacitor, V_L is the voltage across an inductor, and t represents time.

Each element in an ac circuit tends to offer a type of resistance to the flow of current. *The effective resistance of any part of the circuit is called the* **reactance,** and it is denoted by X. The voltage across any part of the circuit whose reactance is X is given by $V = IX$, where I is the current (in amperes) and V is the voltage (in volts). Therefore,

the voltage V_R across a resistor with resistance R,

the voltage V_C across a capacitor with reactance X_C, and

the voltage V_L across an inductor with reactance X_L

are, respectively,

$$V_R = IR \qquad V_C = IX_C \qquad V_L = IX_L \tag{12-18}$$

To determine the voltage across a combination of these elements of a circuit, we must account for the reactance, as well as the phase of the voltage across the individual elements. Since the voltage across a resistor is in phase with the current, we represent V_R along the positive real axis as a real number. Since the voltage across an inductance leads the current by 90°, we represent this voltage as a positive, pure imaginary number. In the same way, by representing the voltage across a capacitor as a negative, pure imaginary number, we show that the voltage *lags* the current by 90°. These representations are meaningful since the positive imaginary axis is +90° from the positive real axis and the negative imaginary axis is −90° from the positive real axis. See Fig. 12-18.

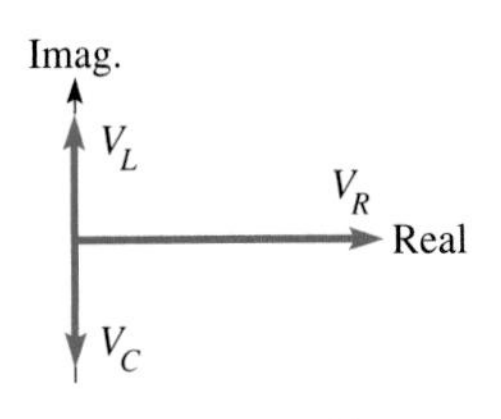

Fig. 12-18

The circuit elements shown in Fig. 12-16 are in *series*, and all circuits we shall consider are series circuits. The total voltage across a series of all three elements is given by $V_R + V_L + V_C$, which we shall represent by V_{RLC}. Therefore,

$$V_{RLC} = IR + IX_L j - IX_C j = I[R + j(X_L - X_C)]$$

This expression is also written as

$$V_{RLC} = IZ \tag{12-19}$$

where the symbol Z is called the **impedance** *of the circuit. It is the total effective resistance to the flow of current by a combination of the elements in the circuit,* taking into account the phase of the voltage in each element. From its definition, we see that Z is a complex number.

$$Z = R + j(X_L - X_C) \tag{12-20}$$

with a magnitude

$$|Z| = \sqrt{R^2 + (X_L - X_C)^2} \tag{12-21}$$

Also, as a complex number, it makes an angle θ with the x-axis, given by

$$\tan\theta = \frac{X_L - X_C}{R} \tag{12-22}$$

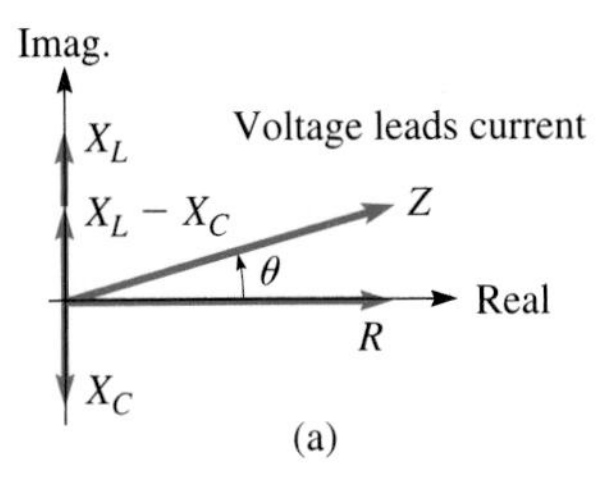

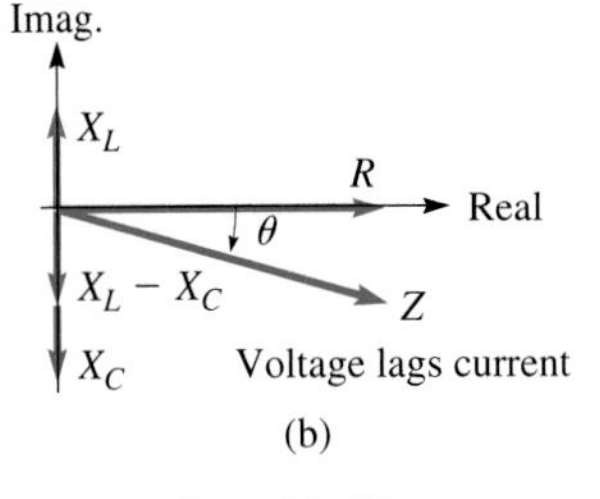

Fig. 12-19

All of these equations are based on phase relations of voltages with respect to the current. Therefore, *the angle θ represents the phase angle between the current and the voltage.* The standard way of expressing θ is to *use a positive angle if the voltage leads the current* and *use a negative angle if the voltage lags the current.*

If the voltage leads the current, then $X_L > X_C$ as shown in Fig. 12-19(a). If the voltage lags the current, then $X_L < X_C$ as shown in Fig. 12-19(b).

In the examples and exercises of this section, the commonly used units and symbols for them are used. For a summary of these units and symbols, including prefixes, see Appendix B.

EXAMPLE 1 ▸▸ In the series circuit shown in Fig. 12-20(a), $R = 12.0\ \Omega$ and $X_L = 5.00\ \Omega$. A current of 2.00 A is in the circuit. Find the voltage across each element, the impedance, the voltage across the combination, and the phase angle between the current and the voltage.

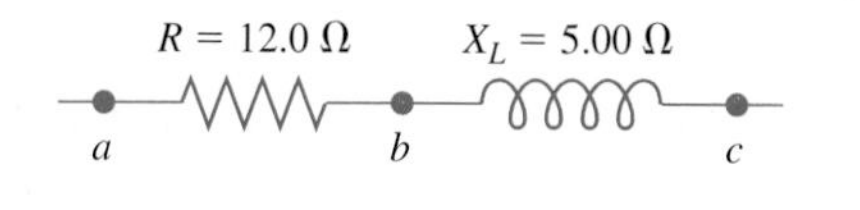

(a)

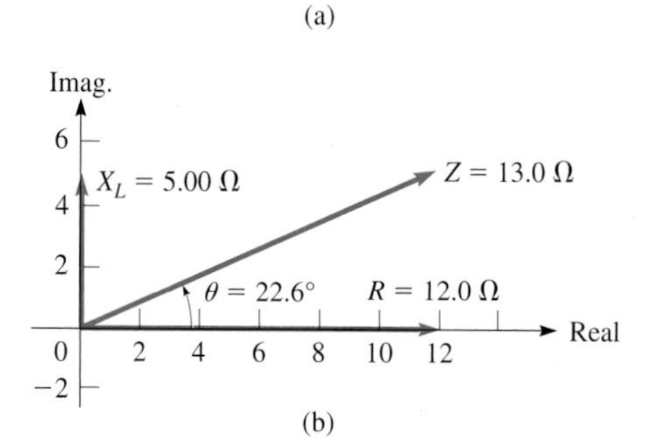

(b)

Fig. 12-20

The voltage across the resistor (between points a and b) is the product of the current and the resistance ($V = IR$). This means $V_R = (2.00)(12.0) = 24.0$ V. The voltage across the inductor (between points b and c) is the product of the current and the reactance, or $V_L = (2.00)(5.00) = 10.0$ V.

To find the voltage across the combination, between points a and c, we must first find the magnitude of the impedance. Note that ***the voltage is not the arithmetic sum of V_R and V_L,*** as we must account for the phase.

By Eq. (12-20), the impedance is (there is no capacitor)

$$Z = 12.0 + 5.00j$$

with magnitude

$$|Z| = \sqrt{R^2 + X_L^2} = \sqrt{(12.0)^2 + (5.00)^2} = 13.0\ \Omega$$

Thus, the magnitude of the voltage across the combination of the resistor and the inductance is

$$|V_{RL}| = (2.00)(13.0) = 26.0\ \text{V}$$

The phase angle between the voltage and the current is found by Eq. (12-22). This gives

$$\tan\theta = \frac{5.00}{12.0}, \qquad \theta = 22.6^\circ$$

The voltage *leads* the current by 22.6°, and this is shown in Fig. 12-20(b). ◂◂

EXAMPLE 2 ▸▸ For a circuit in which $R = 8.00\ \Omega$, $X_L = 7.00\ \Omega$, and $X_C = 13.0\ \Omega$, find the impedance and the phase angle between the current and the voltage.

By the definition of impedance, Eq. (12-20), we have

$$Z = 8.00 + (7.00 - 13.0)j = 8.00 - 6.00j$$

where the magnitude of the impedance is

$$|Z| = \sqrt{(8.00)^2 + (-6.00)^2} = 10.0\ \Omega$$

The phase angle is found by

$$\tan\theta = \frac{-6.00}{8.00}, \qquad \theta = -36.9^\circ$$

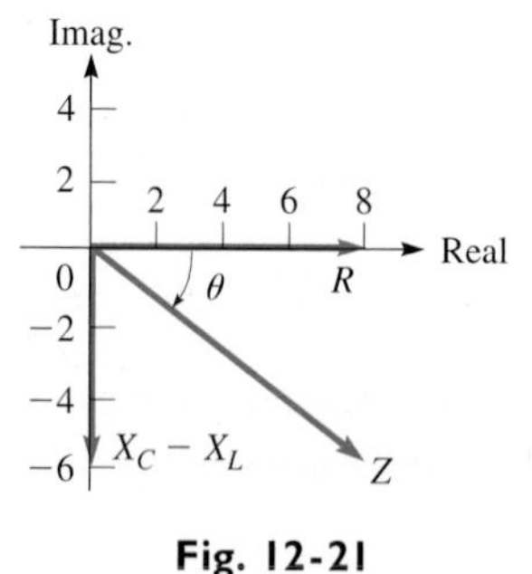

Fig. 12-21

The angle $\theta = -36.9^\circ$ is given directly by the calculator, and it is the angle we want. As we noted after Eq. (12-22), we express θ as a negative angle if the voltage lags the current, as it does in this example. See Fig. 12-21.

From the values above, we write the impedance in polar form as $Z = 10.0\underline{/-36.9^\circ}$ ohms. ◂◂

Note that the resistance is represented in the same way as a vector along the positive x-axis. Actually, resistance is not a vector quantity but is represented in this manner in order to assign an angle as the phase of the current. The important concept in this analysis is that *the phase **difference** between the current and voltage is constant,* and therefore any direction may be chosen arbitrarily for one of them. Once this choice is made, other phase angles are measured with respect to this direction. A common choice, as above, is to make the phase angle of the current zero. If an arbitrary angle is chosen, it is necessary to treat the current, voltage, and impedance as complex numbers.

EXAMPLE 3 ▸▸ In a particular circuit, the current is $2.00 - 3.00j$ amperes and the impedance is $6.00 + 2.00j$ ohms. The voltage across this part of the circuit is

$$\begin{aligned} V &= (2.00 - 3.00j)(6.00 + 2.00j) = 12.0 - 14.0j - 6.00j^2 \\ &= 12.0 - 14.0j + 6.00 \\ &= 18.0 - 14.0j \text{ volts} \end{aligned}$$

The magnitude of the voltage is

$$|V| = \sqrt{(18.0)^2 + (-14.0)^2} = 22.8 \text{ V} \quad ◂◂$$

Since the voltage across a resistor is in phase with the current, this voltage can be represented as having a phase difference of zero with respect to the current. Therefore, the resistance is indicated as an arrow in the positive real direction, denoting the fact that the current and the voltage are in phase. *Such a representation is called a* **phasor**. The arrow denoted by R, as in Fig. 12-20, is actually the phasor representing the voltage across the resistor. Remember, the positive real axis is arbitrarily chosen as the direction of the phase of the current.

To show properly that the voltage across an inductance leads the current by 90°, its reactance (effective resistance) is multiplied by j. We know that there is a positive 90° angle between a positive real number and a positive imaginary number. In the same way, by multiplying the capacitive reactance by $-j$, we show the 90° difference in phase between the voltage and the current in a capacitor, with the current leading. Therefore, jX_L represents the phasor for the voltage across an inductor and $-jX_C$ is the phasor for the voltage across the capacitor. The phasor for the voltage across the combination of the resistance, inductance, and capacitance is Z, where the phase difference between the voltage and the current for the combination is the angle θ.

From this we see that *multiplying a phasor by j means to perform the operation of rotating it through* 90°. For this reason, j is also called the *j-operator.*

EXAMPLE 4 ▸▸ Multiplying a positive real number A by j, we have $A \times j = Aj$, which is a positive imaginary number. In the complex plane, Aj is 90° from A, which means that by multiplying A by j we rotated A by 90°. Similarly, we see that $Aj \times j = Aj^2 = -A$, which is a negative real number, rotated 90° from Aj. Therefore, successive multiplications of A by j give us

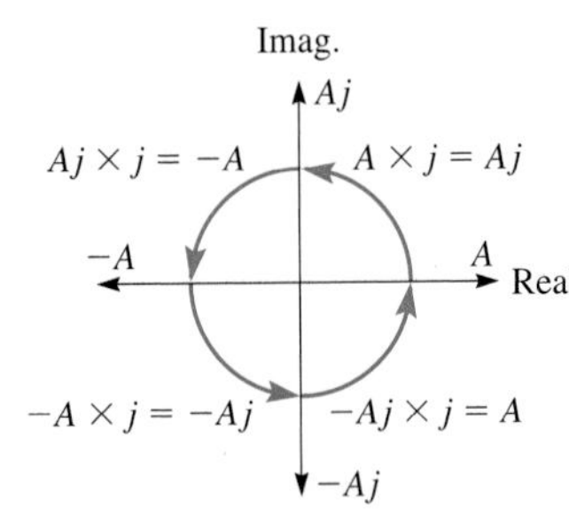

Fig. 12-22

$A \times j = Aj$	positive imaginary number
$Aj \times j = Aj^2 = -A$	negative real number
$-A \times j = -Aj$	negative imaginary number
$-Aj \times j = -Aj^2 = A$	positive real number

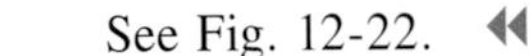

See Fig. 12-22. ◂◂

An alternating current is produced by a coil of wire rotating through a magnetic field. If the angular velocity of the wire is ω, the capacitive and inductive reactances are given by

$$X_C = \frac{1}{\omega C} \quad \text{and} \quad X_L = \omega L \tag{12-23}$$

Therefore, if ω, C, and L are known, the reactance of the circuit can be found.

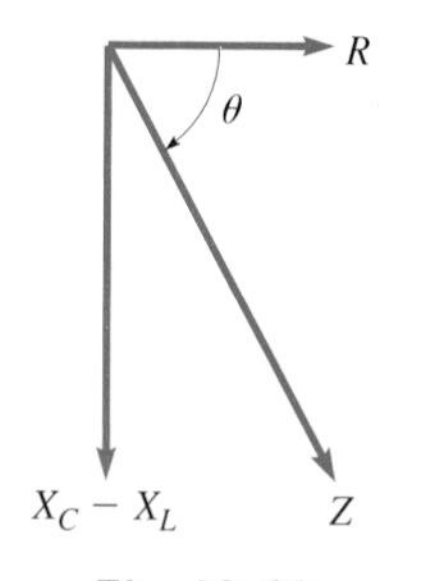

Fig. 12-23

EXAMPLE 5 ▸▸ If $R = 12.0\ \Omega$, $L = 0.300$ H, $C = 250\ \mu$F, and $\omega = 80.0$ rad/s, find the impedance and the phase difference between the current and the voltage.

$$X_C = \frac{1}{(80.0)(250 \times 10^{-6})} = 50.0\ \Omega$$

$$X_L = (0.300)(80.0) = 24.0\ \Omega$$

$$Z = 12.0 + (24.0 - 50.0)j = 12.0 - 26.0j$$

$$|Z| = \sqrt{(12.0)^2 + (-26.0)^2} = 28.6\ \Omega$$

$$\tan\theta = \frac{-26.0}{12.0}, \qquad \theta = -65.2°$$

$$Z = 28.6\underline{/-65.2°}\ \text{ohms}$$

The voltage *lags* the current (see Fig. 12-23). ◂◂

We recall from Section 10-5 that the angular velocity ω is related to the frequency f by the relation $\omega = 2\pi f$. It is very common to use frequency when discussing alternating current.

An important concept in the application of this theory is that of **resonance**. *For resonance, the impedance of any circuit is a minimum, or the total impedance is R.* Thus, $X_L - X_C = 0$. Also, it can be seen that the current and the voltage are in phase under these conditions. Resonance is required for the tuning of radio and television receivers.

See the chapter introduction.

EXAMPLE 6 ▸▸ In the antenna circuit of a radio, the inductance is 4.20 mH and the capacitance is variable. What range of values of capacitance is necessary for the radio to receive the AM band of radio stations, with frequencies from 530 kHz to 1600 kHz?

For proper tuning, the circuit should be in resonance, or $X_L = X_C$. This means that

$$2\pi f L = \frac{1}{2\pi f C} \quad \text{or} \quad C = \frac{1}{(2\pi f)^2 L}$$

From Appendix B, the following prefixes are defined as follows:

Prefix	*Factor*	*Symbol*
pico	10^{-12}	*p*
milli	10^{-3}	*m*
kilo	10^{6}	*k*

For $f_1 = 530$ kHz $= 5.30 \times 10^5$ Hz and $L = 4.20$ mH $= 4.20 \times 10^{-3}$ H,

$$C_1 = \frac{1}{(2\pi)^2(5.30 \times 10^5)^2(4.20 \times 10^{-3})} = 2.15 \times 10^{-11}\ \text{F} = 21.5\ \text{pF}$$

and for $f_2 = 1600$ kHz $= 1.60 \times 10^6$ Hz and $L = 4.20 \times 10^{-3}$ H, we have

$$C_2 = \frac{1}{(2\pi)^2(1.60 \times 10^6)^2(4.20 \times 10^{-3})} = 2.36 \times 10^{-12}\ \text{F} = 2.36\ \text{pF}$$

The capacitance should be capable of varying from 2.36 pF to 21.6 pF. ◂◂

Exercises 12–7

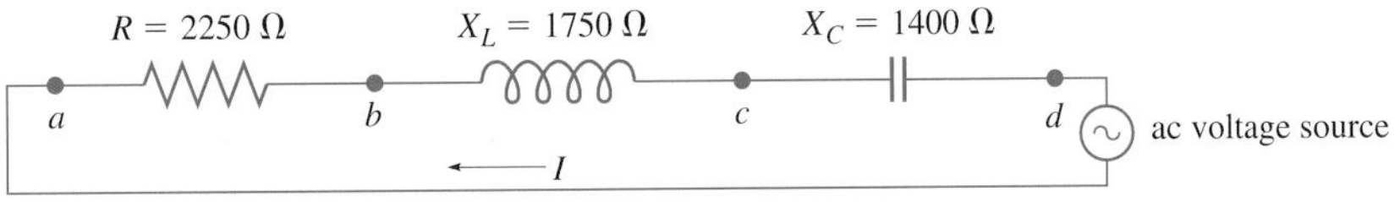

$R = 2250\ \Omega$ $X_L = 1750\ \Omega$ $X_C = 1400\ \Omega$

a b c d ac voltage source

I

Fig. 12-24

In Exercises 1–4, use the circuit shown in Fig. 12-24. The current in the circuit is 5.75 mA. Determine the indicated quantities.

1. The voltage across the resistor (between points a and b).

2. The voltage across the inductor (between points b and c).

3. (a) The magnitude of the impedance across the resistor and the inductor (between points a and c).
(b) The phase angle between the current and the voltage for this combination.
(c) The voltage across this combination.

4. (a) The magnitude of the impedance across the resistor, inductor, and capacitor (between point a and d).
(b) The phase angle between the current and the voltage for this combination.
(c) The voltage across this combination.

In Exercises 5–8, an ac circuit contains the given combinations of circuit elements from among a resistor ($R = 45.0\ \Omega$), a capacitor ($C = 86.2\ \mu$F), and an inductor ($L = 42.9$ mH). If the frequency in the circuit is $f = 60.0$ Hz, find (a) the magnitude of the impedance and (b) the phase angle between the current and the voltage.

5. The circuit has the inductor and the capacitor (an LC circuit).

6. The circuit has the resistor and the capacitor (an RC circuit).

7. The circuit has the resistor and the inductor (an RL circuit).

8. The circuit has the resistor, the inductor, and the capacitor (an RLC circuit).

In Exercises 9–20, solve the given problems.

9. Given that the current in a given circuit is $3.90 - 6.04j$ mA and the impedance is $5.16 + 1.14j$ kΩ, find the magnitude of the voltage.

10. Given that the voltage in a given circuit is $8.375 - 3.140j$ V and the impedance is $2.146 - 1.114j\ \Omega$, find the magnitude of the current.

11. A resistance ($R = 25.3\ \Omega$) and a capacitance ($C = 2.75$ nF) are in an AM radio circuit. If $f = 1200$ kHz, find the impedance across the resistor and the capacitor.

12. A resistance ($R = 64.5\ \Omega$) and an inductance ($L = 1.08$ mH) are in a telephone circuit. If $f = 8.53$ kHz, find the impedance across the resistor and the inductor.

13. The reactance of an inductor is $1200\ \Omega$ for $f = 280$ Hz. What is its inductance?

14. A resistor, an inductor, and a capacitor are connected in series across an ac voltage source. A voltmeter measures 12.0 V, 15.5 V, and 10.5 V, respectively, when placed across each element separately. What is the voltage of the source?

15. An inductance ($L = 25.0$ mH) and a capacitance ($C = 7.18\ \mu$F) are in series. Find the frequency for resonance.

16. A capacitance ($C = 95.2$ nF) and an inductance are in series in the circuit of a receiver for navigation signals. Find the inductance if the frequency for resonance is 50.0 kHz.

17. In Example 6, what should be the capacitance in order to receive a 680-kHz radio signal?

18. A 220-V source with $f = 60.0$ Hz is connected in series to an inductance ($L = 2.05$ H) and a resistance R. Find R if the current is 0.250 A.

19. The power supplied to a series combination of elements in an ac circuit is given by the relation $P = VI \cos \theta$, where P is the power (in watts), V is the effective voltage, I is the effective current, and θ is the phase angle between the current and the voltage. Assuming that the effective voltage across the resistor, capacitor, and inductor combination in Exercise 8 is 225 mV, determine the power supplied to these elements.

20. Explain why the multiplication of a complex number by -1 may be shown as a rotation of the graph of the number about the origin through 180°.

In the development of electromagnetic theory, the existence of radio waves was predicted mathematically in the 1860s by the work of James Clerk Maxwell, a Scottish physicist. They were first produced in the laboratory in 1887 by Heinrich Hertz, a German physicist.

Chapter Equations, Review Exercises, and Practice Test

Chapter Equations for Complex Numbers

Imaginary unit

$$j^2 = -1 \tag{12-1}$$

$$\sqrt{-a} = j\sqrt{a}, \qquad (a > 0) \tag{12-2}$$

Basic operations

$$(a + bj) + (c + dj) = (a + c) + (b + d)j \tag{12-3}$$

$$(a + bj) - (c + dj) = (a - c) + (b - d)j \tag{12-4}$$

$$(a + bj)(c + dj) = (ac - bd) + (ad + bc)j \tag{12-5}$$

$$\frac{a + bj}{c + dj} = \frac{(a + bj)(c - dj)}{(c + dj)(c - dj)} = \frac{(ac + bd) + (bc - ad)j}{c^2 + d^2} \tag{12-6}$$

Complex number forms

Rectangular: $x + yj$

Polar: $r(\cos\theta + j\sin\theta) = r\angle\theta$

Exponential: $re^{j\theta}$

$$x = r\cos\theta \qquad y = r\sin\theta \tag{12-7}$$

$$r^2 = x^2 + y^2 \qquad \tan\theta = \frac{y}{x} \tag{12-8}$$

$$x + yj = r(\cos\theta + j\sin\theta) = r\angle\theta = re^{j\theta} \tag{12-12}$$

Product in polar form

$$r_1(\cos\theta_1 + j\sin\theta_1)r_2(\cos\theta_2 + j\sin\theta_2) = r_1r_2[\cos(\theta_1 + \theta_2) + j\sin(\theta_1 + \theta_2)] \tag{12-13}$$

$$(r_1\angle\theta_1)(r_2\angle\theta_2) = r_1r_2\angle\theta_1 + \theta_2$$

Quotient in polar form

$$\frac{r_1(\cos\theta_1 + j\sin\theta_1)}{r_2(\cos\theta_2 + j\sin\theta_2)} = \frac{r_1}{r_2}[\cos(\theta_1 - \theta_2) + j\sin(\theta_1 - \theta_2)]$$

$$\frac{r_1\angle\theta_1}{r_2\angle\theta_2} = \frac{r_1}{r_2}\angle\theta_1 - \theta_2 \tag{12-15}$$

DeMoivre's theorem

$$[r(\cos\theta + j\sin\theta)]^n = r^n(\cos n\theta + j\sin n\theta)$$

$$(r\angle\theta)^n = r^n\angle n\theta \tag{12-17}$$

Chapter Equations for Alternating-Current Circuits

Voltage, current, reactance

$$V_R = IX_R \qquad V_C = IX_C \qquad V_L = IX_L \tag{12-18}$$

Impedance

$$V_{RLC} = IZ \tag{12-19}$$

$$Z = R + j(X_L - X_C) \tag{12-20}$$

$$|Z| = \sqrt{R^2 + (X_L - X_C)^2} \tag{12-21}$$

Phase angle

$$\tan\theta = \frac{X_L - X_C}{R} \tag{12-22}$$

Capacitive reactance and inductive reactance

$$X_C = \frac{1}{\omega C} \quad \text{and} \quad X_L = \omega L \tag{12-23}$$

Review Exercises

In Exercises 1–16, perform the indicated operations, expressing all answers in simplest rectangular form.

1. $(6 - 2j) + (4 + j)$
2. $(12 + 7j) + (-8 + 6j)$
3. $(18 - 3j) - (12 - 5j)$
4. $(-4 - 2j) - (-6 - \sqrt{-49})$
5. $(2 + j)(4 - j)$
6. $(-5 + 3j)(8 - 4j)$
7. $(2j^5)(6 - 3j)(4 + 3j)$
8. $j(3 - 2j) - (j^3)(5 + j)$

9. $\dfrac{3}{7 - 6j}$ 10. $\dfrac{4j}{2 + 9j}$

11. $\dfrac{6 - \sqrt{-16}}{\sqrt{-4}}$ 12. $\dfrac{3 + \sqrt{-4}}{4 - j}$

13. $\dfrac{5j - (3 - j)}{4 - 2j}$ 14. $\dfrac{2 + (j - 6)}{1 - 2j}$

15. $\dfrac{j(7 - 3j)}{2 + j}$ 16. $\dfrac{(2 - j)(3 + 2j)}{4 - 3j}$

In Exercises 17–20, find the values of x and y for which the equations are valid.

17. $3x - 2j = yj - 2$
18. $2xj - 2y = (y + 3)j - 3$
19. $2x - j + 4 = 6y + 2xj$
20. $3yj + xj = 6 + 3x + y$

In Exercises 21–24, perform the indicated operations graphically. Check them algebraically.

21. $(-1 + 5j) + (4 + 6j)$
22. $(7 - 2j) + (-5 + 4j)$
23. $(9 + 2j) - (5 - 6j)$
24. $(1 + 4j) - (-3 - 3j)$

In Exercises 25–32, give the polar and exponential forms of each of the complex numbers.

25. $1 - j$ 26. $4 + 3j$
27. $-2 - 7j$ 28. $6 - 2j$
29. $1.07 + 4.55j$ 30. $-327 + 158j$
31. 10 32. $-4j$

In Exercises 33–44, give the rectangular form of each of the complex numbers.

33. $2(\cos 225° + j \sin 225°)$
34. $4(\cos 60° + j \sin 60°)$
35. $5.011(\cos 123.82° + j \sin 123.82°)$
36. $2.417(\cos 296.26° + j \sin 296.26°)$
37. $0.62\underline{/-72°}$ 38. $20\underline{/160°}$
39. $27.08\underline{/346.27°}$ 40. $1.689\underline{/194.36°}$
41. $2.00e^{0.25j}$ 42. $e^{3.62j}$
43. $25.37e^{1.906j}$ 44. $44.47e^{6.046j}$

In Exercises 45–60, perform the indicated operations. Leave the result in polar form.

45. $[3(\cos 32° + j \sin 32°)][5(\cos 52° + j \sin 52°)]$
46. $[2.5(\cos 162° + j \sin 162°)][8(\cos 115° + j \sin 115°)]$
47. $(40\underline{/18°})(0.5\underline{/245°})$
48. $(0.1254\underline{/172.38°})(27.17\underline{/204.34°})$
49. $\dfrac{24(\cos 165° + j \sin 165°)}{3(\cos 106° + j \sin 106°)}$
50. $\dfrac{18(\cos 403° + j \sin 403°)}{4(\cos 192° + j \sin 192°)}$
51. $\dfrac{245.6\underline{/326.44°}}{17.19\underline{/192.83°}}$ 52. $\dfrac{100\underline{/206°}}{4\underline{/320°}}$
53. $0.983\underline{/47.2°} + 0.366\underline{/95.1°}$
54. $17.8\underline{/110.4°} - 14.9\underline{/226.3°}$
55. $7644\underline{/294.36°} - 6871\underline{/17.86°}$
56. $4.944\underline{/327.49°} + 8.009\underline{/7.37°}$
57. $[2(\cos 16° + j \sin 16°)]^{10}$
58. $[3(\cos 36° + j \sin 36°)]^6$
59. $(3\underline{/110.5°})^3$ 60. $(5.36\underline{/220.3°})^4$

In Exercises 61–64, change each number to polar form and then perform the indicated operations. Express the final result in rectangular and polar forms. Check by performing the same operation in rectangular form.

61. $(1 - j)^{10}$ 62. $(\sqrt{3} + j)^8(1 + j)^5$

63. $\dfrac{(5 + 5j)^4}{(-1 - j)^6}$ 64. $(\sqrt{3} - j)^{-8}$

In Exercises 65–68, use DeMoivre's theorem to find the indicated roots. Be sure to find all roots.

65. The cube roots of -8.
66. The cube roots of 1.
67. The fourth roots of $-j$.
68. The fifth roots of $32j$.

In Exercises 69–72, determine the rectangular form and the polar form of the complex number for which the graphical representation is shown in the given figure.

69. **70.** **71.** **72.**

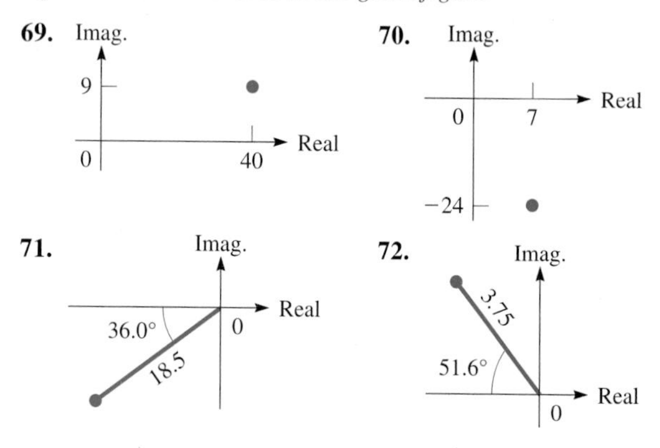

In Exercises 73–84, find the required quantities.

73. A 60-V ac voltage source is connected in series across a resistor, an inductor, and a capacitor. The voltage across the inductor is 60 V, and the voltage across the capacitor is 60 V. What is the voltage across the resistor?

74. In a series ac circuit with a resistor, an inductor, and a capacitor, $R = 6.50\ \Omega$, $X_C = 3.74\ \Omega$, and $Z = 7.50\ \Omega$. Find X_L.

75. In a series ac circuit with a resistor, an inductor, and a capacitor, $R = 6250\ \Omega$, $Z = 6720\ \Omega$, and $X_L = 1320\ \Omega$. Find the phase angle θ.

76. A coil of wire rotates at 120.0 r/s. If the coil generates a current in a circuit containing a resistance of 12.07 Ω, an inductance of 0.1405 H, and an impedance of 22.35 Ω, what must be the value of a capacitor (in farads) in the circuit?

77. What is the frequency f for resonance in a circuit for which $L = 2.65$ H and $C = 18.3\ \mu$F?

78. The displacement of an electromagnetic wave is given by $d = A(\cos \omega t + j \sin \omega t) + B(\cos \omega t - j \sin \omega t)$. Find the expressions for the magnitude and phase angle of the displacement.

79. What are the magnitude and direction of a force that is represented by $590 - 550j$ newtons?

80. What are the magnitude and the direction of a velocity that is represented by $2500 + 1500j$ km/h?

81. In the study of shearing effects in the spinal column, the expression $\dfrac{1}{u + j\omega n}$ is found. Express this in rectangular form.

82. In the theory of light reflection on metals, the expression

$$\frac{\mu(1 - kj) - 1}{\mu(1 - kj) + 1}$$

is encountered. Simplify this expression.

83. Show that $e^{j\pi} = -1$.

84. Show that $(e^{j\pi})^{1/2} = j$.

Writing Exercise

85. A computer programmer is writing a program to determine the n nth roots of a real number. Part of the program is to show the number of real roots and the number of pure imaginary roots. Write one or two paragraphs explaining how these numbers of roots can be determined without actually finding the roots.

Practice Test

1. Add, expressing the result in rectangular form: $(3 - \sqrt{-4}) + (5\sqrt{-9} - 1)$.
2. Multiply, expressing the result in polar form: $(2\underline{/130^\circ})(3\underline{/45^\circ})$.
3. Express $2 - 7j$ in polar form.
4. Express in terms of j: (a) $-\sqrt{-64}$, (b) $-j^{15}$.
5. Add graphically: $(4 - 3j) + (-1 + 4j)$.
6. Simplify, expressing the result in rectangular form: $\dfrac{2 - 4j}{5 + 3j}$.
7. Express $2.56(\cos 125.2^\circ + j \sin 125.2^\circ)$ in exponential form.
8. For an ac circuit in which $R = 3.50\ \Omega$, $X_L = 6.20\ \Omega$, and $X_C = 7.35\ \Omega$, find the impedance and the phase angle between the current and the voltage.
9. Express $3.47 - 2.81j$ in exponential form.
10. Find the values of x and y: $x + 2j - y = yj - 3xj$.
11. What is the capacitance of a radio that has an inductance of 8.75 mH if it is to receive a station with frequency 600 kHz?
12. Find the cube roots of j.

13

Exponential and Logarithmic Functions

In this chapter we introduce two more important types of functions, the *exponential function* and the *logarithmic function.*

Historically, logarithms were developed (in the early 17th century) to help with calculations needed in areas such as navigation and astronomy. With the extensive use of calculators and computers, they are not used directly for such purposes to any extent today. However, these functions are very important in many scientific and technical applications, as well as in areas of advanced mathematics.

To illustrate the importance of exponential functions, many applications may be cited. They are used extensively in electronics, mechanical systems, thermodynamics, and nuclear physics. They are also used in biology in studying population growth and in business to calculate compound interest.

The use of the logarithmic function, which is closely related to the exponential function, is also extensive. The basic units used to measure intensity of sound and those used to measure the intensity of earthquakes are defined in terms of logarithms. In chemistry, the distinction between a base and an acid is defined in terms of logarithms. In electrical transmission lines, power gains and losses are measured in terms of logarithmic units. Many of these applications are illustrated throughout the chapter.

The intensity level of sound is measured on a logarithmic scale. This is shown in Section 13-6.

13-1 The Exponential and Logarithmic Functions

Chapter 11 dealt with exponents in expressions of the form x^n, where we showed that n could be any rational number. Here we shall deal with expressions of the form b^x, where x is any real number. When we look at these expressions, we note the primary difference is that in the second expression *the exponent is variable.* We have not previously dealt with variable exponents. *Thus let us define the* **exponential function** *to be*

$$y = b^x \tag{13-1}$$

In Eq. (13-1), *x is called the* **logarithm** *of the number y to the base b.* In our work with the exponential function we shall restrict all numbers to the real number system. *This leads us to choose the base as a positive number other than* 1. We know that 1 raised to any power will result in 1, which would make y a constant regardless of the value of x. Negative numbers for b would result in imaginary values for y if x were any fractional exponent with an even integer for its denominator.

EXAMPLE 1 ▸▸ $y = 2^x$ is an exponential function, where x is the logarithm of y to the base 2. This means that 2 raised to a given power gives the corresponding value of y.

If $x = 3$, $y = 2^3 = 8$; this means that 3 is the logarithm of 8 to the base 2. If $x = 4$, $y = 2^4 = 16$; this means that 4 is the logarithm of 16 to the base 2.

If $x = \frac{1}{2}$, $y = 2^{1/2} \approx 1.41$. This means that $\frac{1}{2}$ is the logarithm of 1.41 to the base 2. ◂◂

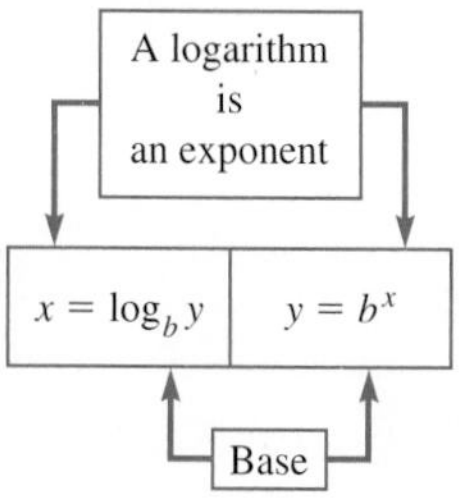

Using the definition of a logarithm, we may express x in terms of y in the form

$$x = \log_b y \tag{13-2}$$

Fig. 13-1

This equation is read in accordance with the definition of x in Eq. (13-1), that is, *x equals the logarithm of y to the base b.* This means that x is the power to which the base b must be raised in order to equal the number y; that is, x is a logarithm, and

CAUTION ▸ ***a logarithm is an exponent.*** Note that Eqs. (13-1) and (13-2) state the same relationship, but in a different way. Equation (13-1) is the **exponential form,** and Eq. (13-2) is the **logarithmic form.** See Fig. 13-1.

EXAMPLE 2 ▸▸ The equation $y = 2^x$ would be written as $x = \log_2 y$ if we put it in logarithmic form. When we choose values of y to find the corresponding values of x from this equation, we ask ourselves "2 raised to what power x gives y?"

This means that if $y = 8$, we know that $2^3 = 8$, and therefore $x = 3$. ◂◂

EXAMPLE 3 ▸▸ (a) $3^2 = 9$ in logarithmic form is $2 = \log_3 9$.
(b) $4^{-1} = 1/4$ in logarithmic form is $-1 = \log_4(1/4)$.

CAUTION ▸ Remember, ***the exponent may be negative.*** The *base* must be positive. ◂◂

EXAMPLE 4 ▸▸ (a) $(64)^{1/3} = 4$ in logarithmic form is $\frac{1}{3} = \log_{64} 4$.
(b) $(32)^{3/5} = 8$ in logarithmic form is $\frac{3}{5} = \log_{32} 8$.
(c) $\log_2 32 = 5$ in exponential form is $32 = 2^5$.
(d) $\log_6 (\frac{1}{36}) = -2$ in exponential form is $\frac{1}{36} = 6^{-2}$. ◂◂

EXAMPLE 5 ▸▸ (a) Find b, given that $-4 = \log_b (\frac{1}{81})$.
Writing this in exponential form, we have $\frac{1}{81} = b^{-4}$. Thus, $\frac{1}{81} = \frac{1}{b^4}$ or $\frac{1}{3^4} = \frac{1}{b^4}$. Therefore, $b = 3$.
(b) Find y, given that $\log_4 y = \frac{1}{2}$.
In exponential form we have $y = 4^{1/2}$, or $y = 2$. ◂◂

We see that exponential form is very useful for determining values written in logarithmic form. For this reason, it is important that you learn to transform readily from one form to the other.

In order to change a function of the form $y = ab^x$ into logarithmic form, we must first write it as $y/a = b^x$. ***The coefficient of b^x must be equal to* 1,** which is the form of Eq. (13-1). In the same way, the coefficient of $\log_b y$ must be 1 in order to change it into exponential form.

CAUTION ▶

EXAMPLE 6 ▸▸ The power supply P (in watts) of a certain satellite is given by $P = 75e^{-0.005t}$, where t is the time (in days) after launch. By writing this equation in logarithmic form, solve for t.

In order to have the equation in the exponential form of Eq. (13-1), we must have only $e^{-0.005t}$ on the right. Therefore, by dividing by 75, we have

$$\frac{P}{75} = e^{-0.005t}$$

Writing this in logarithmic form, we have

$$\log_e \left(\frac{P}{75}\right) = -0.005t$$

or

$$t = \frac{\log_e \left(\frac{P}{75}\right)}{-0.005} = -200 \log_e \left(\frac{P}{75}\right) \qquad \frac{1}{-0.005} = -200$$

We recall from Section 12-5 that e is a special irrational number equal to about 2.718. It is an important number as a base of logarithms. This is discussed in more detail in Section 13-5. ◂◂

When we are working with functions, we must keep in mind that a function is defined by the operation being performed on the independent variable, and not by the letter chosen to represent it. However, for consistency, it is standard practice to let y represent the dependent variable and x represent the independent variable. Therefore, *the* **logarithmic function** *is*

$$y = \log_b x \qquad (13\text{-}3)$$

As with the exponential function, $b > 0$ and $b \neq 1$.

For reference, Eqs. (13-2) and (13-3) are
$x = \log_b y$
$y = \log_b x$

Equations (13-2) and (13-3) do not represent different *functions,* due to the difference in location of the variables, since they represent the *same operation* on the independent variable that appears in each. However, Eq. (13-3) expresses the function with the standard dependent and independent variables.

EXAMPLE 7 ▸▸ For the logarithmic function $y = \log_2 x$, we have the standard independent variable x and the standard dependent variable y.

If $x = 16$, $y = \log_2 16$, which means that $y = 4$, since $2^4 = 16$.
If $x = \frac{1}{16}$, $y = \log_2(\frac{1}{16})$, which means that $y = -4$, since $2^{-4} = \frac{1}{16}$. ◂◂

Exercises 13–1

In Exercises 1–4, evaluate the exponential function $y = 9^x$ for the given values of x.

1. $x = 0.5$
2. $x = 4$
3. $x = -2$
4. $x = -0.5$

In Exercises 5–16, express the given equations in logarithmic form.

5. $3^3 = 27$
6. $5^2 = 25$
7. $4^4 = 256$
8. $8^2 = 64$
9. $4^{-2} = \frac{1}{16}$
10. $3^{-2} = \frac{1}{9}$
11. $2^{-6} = \frac{1}{64}$
12. $(12)^0 = 1$
13. $8^{1/3} = 2$
14. $(81)^{3/4} = 27$
15. $(\frac{1}{4})^2 = \frac{1}{16}$
16. $(\frac{1}{2})^{-2} = 4$

In Exercises 17–28, express the given equations in exponential form.

17. $\log_3 81 = 4$
18. $\log_{11} 121 = 2$
19. $\log_9 9 = 1$
20. $\log_{15} 1 = 0$
21. $\log_{25} 5 = \frac{1}{2}$
22. $\log_8 16 = \frac{4}{3}$
23. $\log_{243} 3 = \frac{1}{5}$
24. $\log_{32}(\frac{1}{8}) = -\frac{3}{5}$
25. $\log_{10} 0.1 = -1$
26. $\log_7(\frac{1}{49}) = -2$
27. $\log_{0.5} 16 = -4$
28. $\log_{1/3} 3 = -1$

In Exercises 29–44, determine the value of the unknown.

29. $\log_4 16 = x$
30. $\log_5 125 = x$
31. $\log_{10} 0.01 = x$
32. $\log_{16}(\frac{1}{4}) = x$
33. $\log_7 y = 3$
34. $\log_8 N = 3$
35. $\log_8 y = -\frac{2}{3}$
36. $\log_7 y = -2$
37. $\log_b 81 = 2$
38. $\log_b 625 = 4$
39. $\log_b 4 = -\frac{1}{3}$
40. $\log_b 4 = \frac{2}{3}$
41. $\log_{10} 10^{0.2} = x$
42. $\log_5 5^{1.3} = x$
43. $\log_3 27^{-1} = x$
44. $\log_b(\frac{1}{4}) = -\frac{1}{2}$

In Exercises 45–48, evaluate the logarithmic function $y = \log_4 x$ for the given values of x.

45. $x = 64$
46. $x = \frac{1}{64}$
47. $x = \frac{1}{2}$
48. $x = 2$

In Exercises 49–56, perform the indicated operations.

49. The value V of a bank account in which A dollars is invested at 10% interest, compounded annually, is given by $V = A(1.1)^t$, where t is the time in years. Solve for t.

50. The intensity I of an earthquake is given by $I = I_0(10)^R$, where I_0 is a minimum intensity for comparison and R is the Richter scale magnitude of the earthquake. Solve for R.

51. The magnitudes (visual brightnesses), m_1 and m_2, of two stars are related to their (actual) brightnesses, b_1 and b_2, by the equation $m_1 - m_2 = 2.5 \log_{10}(b_2/b_1)$. Solve for b_2.

52. The velocity v of a rocket at the point at which its fuel is completely burned is given by $v = u \log_e(w_0/w)$, where u is the exhaust velocity, w_0 is the lift-off weight, and w is the burnout weight. Solve for w.

53. An equation relating the number N of atoms of radium at any time t in terms of the number of atoms at $t = 0$, N_0, is $\log_e(N/N_0) = -kt$, where k is a constant. Solve for N.

54. The charge q on a capacitor is given by $q = q_0(1 - e^{-at})$, where q_0 is the initial charge, a is a constant, and t is the time. Solve for t.

55. A chemical compound decays according to the equation $Q = A(5^{-0.02t})$, where Q is the amount present after time t and A is the initial amount. Solve for t.

56. An equation used in measuring the flow of water in a channel is $C = -a \log_{10}(b/R)$. Solve for R.

13-2 Graphs of $y = b^x$ and $y = \log_b x$

Graphical representation of functions is often valuable when we wish to show their properties. We shall now show representative graphs of the exponential function $y = b^x$ and the logarithmic function $y = \log_b x$.

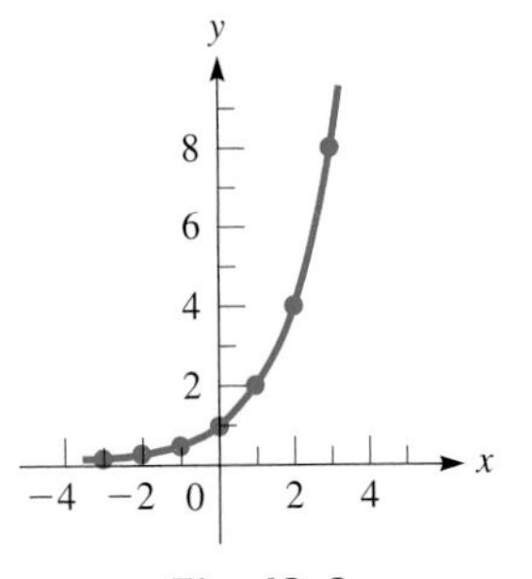

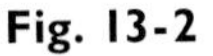
Fig. 13-2

EXAMPLE 1 ▸▸ Plot the graph of $y = 2^x$.

Assuming values for x and then finding the corresponding values for y, we obtain the following table.

x	-3	-2	-1	0	1	2	3
y	$\frac{1}{8}$	$\frac{1}{4}$	$\frac{1}{2}$	1	2	4	8

$2^{-3} = \frac{1}{8}$ $\quad 2^0 = 1 \quad$ $2^3 = 8$

From these values we plot the curve, as shown in Fig. 13-2. We note that the x-axis is an asymptote of the curve. ◂◂

EXAMPLE 2 ▸▸ Plot the graph of $y = \log_3 x$.

We can find the points for this graph more easily if we first put the equation in exponential form: $x = 3^y$. By assuming values for y, we can find the corresponding values for x.

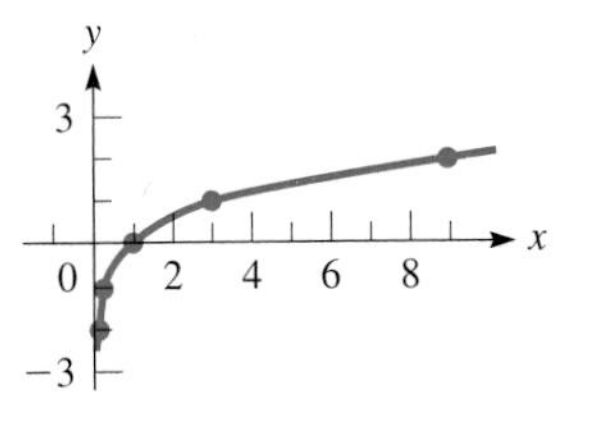

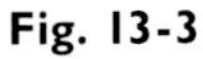
Fig. 13-3

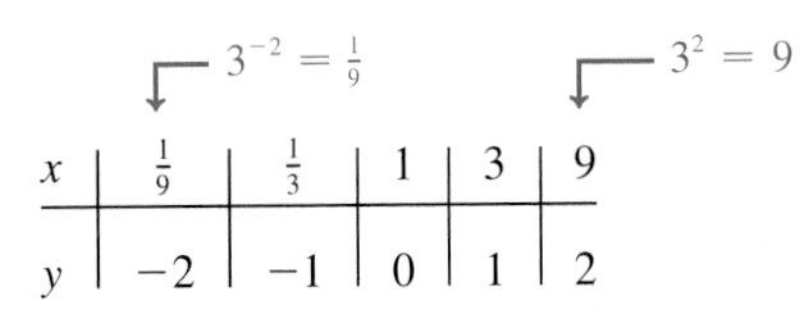

$3^{-2} = \frac{1}{9}$ $\quad 3^2 = 9$

x	$\frac{1}{9}$	$\frac{1}{3}$	1	3	9
y	-2	-1	0	1	2

Using these values, we construct the graph seen in Fig. 13-3. ◂◂

Any exponential or logarithmic curve, where $b > 1$, will be similar in shape to those shown in Examples 1 and 2. From these curves, we can draw certain conclusions:

1. *If* $0 < x < 1$, $\log_b x < 0$; *if* $x = 1$, $\log_b 1 = 0$; *if* $x > 1$, $\log_b x > 0$.
2. *If* $x > 1$, *x increases more rapidly than* $\log_b x$.
3. *For all values of x*, $b^x > 0$.
4. *If* $x > 1$, b^x *increases more rapidly than x.*

From the graphs in Figs. 13-2 and 13-3 and the above analysis, we note that the domain of the exponential function $y = b^x$ is all real numbers and that its range is $y > 0$. For the logarithmic function $y = \log_b x$, the domain is $x > 0$ and the range is all real numbers.

NOTE ▸ We noted above that if $x > 1$, x increases more rapidly than $\log_b x$ and b^x increases more rapidly than x. Actually, as x becomes larger, $\log_b x$ increases very slowly, but b^x increases very rapidly. Using $\log_2 x$ and 2^x and a calculator, we have the following table of values.

x	1	4	16	64
$\log_2 x$	0	2	4	6
2^x	2	16	65 536	1.8×10^{19}

This shows that we must select values of x carefully when graphing these functions.

The graphing calculator also can be used to graph exponential and logarithmic functions. We can use the information about the domain and range, and how the function increases in choosing values for the RANGE feature on the calculator.

Although the bases most important to applications are greater than 1, to understand how the curve of the exponential function differs somewhat if $b < 1$, let us consider the following example.

EXAMPLE 3 ▸▸ Display the graph of $y = (\frac{1}{2})^x$ on a graphing calculator.

We could simply try some values for the RANGE feature to get the graph. However, if we look at the function, we see that

$$\left(\frac{1}{2}\right)^x = \frac{1}{2^x} = 2^{-x}$$

We see that x may be any real number. Also, we note that *as x becomes more negative, y will increase rapidly.* Therefore, we choose the following range values.

Xmin = −3 Xmax = 3

Ymin = 0 (the range is $y > 0$)

Ymax = 8 ($2^3 = 8$)

The exponent x is entered after the [∧] (or [x^y]) key. The view on the graphing calculator is shown in Fig. 13-4. ◂◂

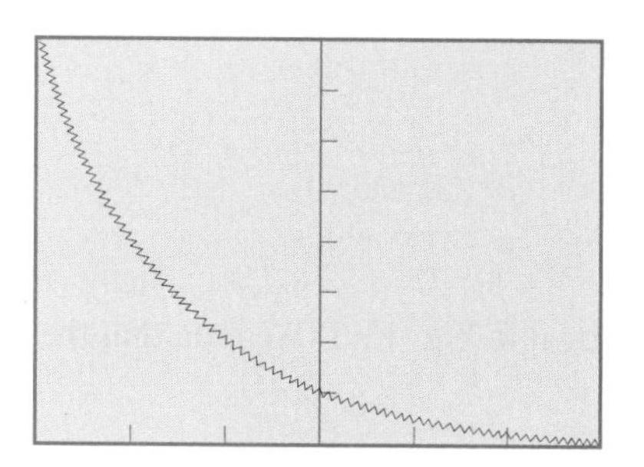

Fig. 13-4

The log keys on the calculator are [log] (for $\log_{10} x$) and [ln] (for $\log_e x$). Therefore, for now we restrict graphing logarithmic functions to bases 10 and e on the graphing calculator. In Section 13-5 we will see how to use the calculator to graph a logarithmic function with any positive real number base.

EXAMPLE 4 ▸▸ Display the graph of $y = 2 \log_{10} x$ on a graphing calculator.

We choose the following values for the RANGE.

Xmin = 0 (the domain is $x > 0$)

Xmax = 20 ($\log_{10} x$ increases very slowly)

Xscl = 5

Ymin = −1 ($\log_{10} 0.1 = -1$)

Ymax = 3 ($\log_{10} 100 = 2$; y does not reach 4 until $x = 100$)

The graphing calculator view is shown in Fig. 13-5. ◂◂

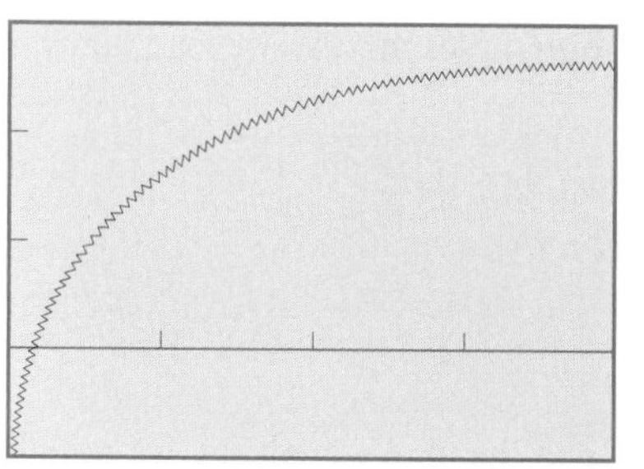

Fig. 13-5

EXAMPLE 5 ▸▸ In an electric circuit in which there is a battery, an inductor, and a resistor, the current i (in amperes) as a function of the time t (in seconds) is $i = 0.8(1 - e^{-4t})$. Display the graph of this function on a graphing calculator.

Here we note that $e^{-4t} = 1$ for $t = 0$, which means that the current $i = 0$ for $t = 0$. Also, e^{-4t} becomes very small in a short time, which means that i cannot be greater than 0.8 A. With these considerations, we choose the following RANGE values, where we use x for t and y for i.

Xmin = 0 Xmax = 1 Xscl = 0.1

Ymin = 0 Ymax = 1 Yscl = 0.1

To enter the function, we use the [e^x] key with exponent $-4x$. The graphing calculator view is shown in Fig. 13-6. ◂◂

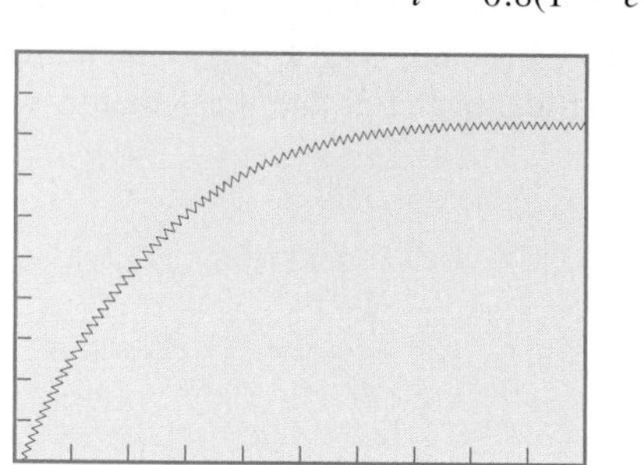

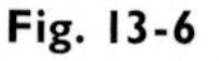

Fig. 13-6

Inverse Functions

For the exponential function $y = b^x$ and the logarithmic function $y = \log_b x$, if we solve for the independent variable in one of the functions by changing the form, then interchange the variables, we obtain the other function. *Such functions are called* **inverse functions.**

This means that the x- and y-coordinates of inverse functions are interchanged. As a result, the graphs of inverse functions are mirror images of each other across the line $y = x$. This is illustrated in the following example.

EXAMPLE 6 ▸▸ The functions $y = 2^x$ and $y = \log_2 x$ are inverse functions. We show this by solving $y = 2^x$ for x, and then interchange x and y.

$y = 2^x$ in logarithmic form is $x = \log_2 y$

Interchanging x and y, we have $y = \log_2 x$, which is the inverse function.

Making a table of values for each function, we have

$y = 2^x$

x	-3	-2	-1	0	1	2	3
y	$\frac{1}{8}$	$\frac{1}{4}$	$\frac{1}{2}$	1	2	4	8

$y = \log_2 x$

x	$\frac{1}{8}$	$\frac{1}{4}$	$\frac{1}{2}$	1	2	4	8
y	-3	-2	-1	0	1	2	3

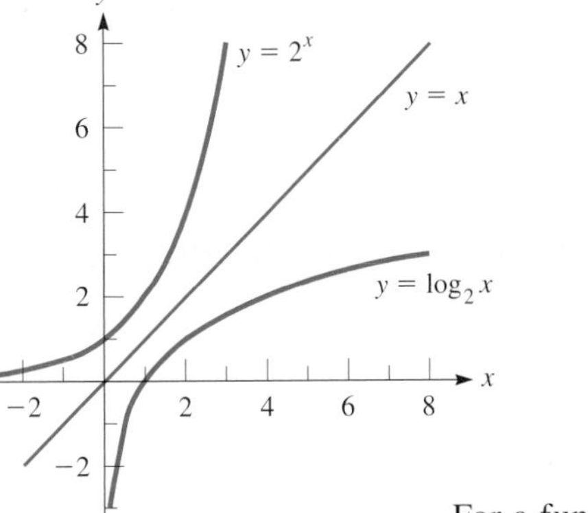

Fig. 13-7

We see that the coordinates are interchanged. The graphs of these two functions and the line $y = x$ are shown in Fig. 13-7. Note that each reflects the other across $y = x$. ◂◂

For a function, there is exactly one value of y in the range for each value of x in the domain. This must also hold for the inverse function. Thus, for a function to have an inverse function, there must be only one x for each y. This is true for $y = b^x$ and $y = \log_b x$, as we have seen earlier in this section.

Exercises 13–2

In Exercises 1–16, plot the graphs of the given functions.

1. $y = 3^x$
2. $y = 4^x$
3. $y = (1.65)^x$
4. $y = (2.72)^x$
5. $y = (\frac{1}{3})^x$
6. $y = (\frac{1}{4})^x$
7. $y = 0.5(3.06)^{-x}$
8. $y = 0.1(10^{-x})$
9. $y = \log_2 x$
10. $y = \log_4 x$
11. $y = \log_{32} x$
12. $y = \log_{0.5} x$
13. $y = 2 \log_3 x$
14. $y = 3 \log_2 x$
15. $y = 0.2 \log_4 x$
16. $y = 5 \log_{10} x$

In Exercises 17–24, display the graphs of the given functions on a graphing calculator.

17. $y = 0.3(2.55)^x$
18. $y = 1.5(4.15)^x$
19. $y = 0.1(0.25)^x$
20. $y = 0.4(0.95)^x$
21. $y = 1.2(6)^{-x}$
22. $y = 0.5e^{-x}$
23. $y = 3 \log_e x$
24. $y = 5 \log_{10} x$

In Exercises 25–32, sketch the indicated graphs. A graphing calculator may be used (except for 29).

25. If an amount of P dollars is invested at an annual interest rate r (expressed as a decimal), the value V of the investment after t years is given by $V = P(1 + r/n)^{nt}$, if interest is compounded n times a year. If \$1000 is invested at an annual interest rate of 6%, compounded semiannually, express V as a function of t and sketch the graph for $0 \le t \le 8$ years.
26. If \$5000 is invested at an annual interest rate of 9%, compounded monthly (see Exercise 25), express V as a function of t and sketch the graph for $0 \le t \le 10$ years.
27. Considering air resistance and other conditions, the velocity (in m/s) of a certain falling object is given by $v = 95(1 - e^{-0.1t})$, where t is the time of fall in seconds. Sketch the graph of this function.

28. The current i, in amperes, in a certain electric circuit is given by $i = 16(1 - e^{-250t})$, where t is the time in seconds. Using appropriate values of t, sketch the graph of this function.

29. The time t, in picoseconds, required for N calculations by a certain computer design is $t = N + \log_2 N$. Sketch the graph of this function.

30. An original amount of 100 mg of radium radioactively decomposes such that N mg remain after t years. The function relating t and N is $t = 2350 (\log_e 100 - \log_e N)$. Sketch the graph of this function.

31. Sketch the graphs of $y = (\frac{1}{3})^x$ and $y = 3^{-x}$, and then compare them.

32. In Exercise 12, the graph of $y = \log_{0.5} x$ is plotted. By inspecting the graph and noting the properties of $\log_{0.5} x$, describe some of the differences of logarithms to a base less than 1 from those to a base greater than 1.

In Exercises 33–36, show that the given functions are inverse functions of each other. Then display the graphs of each function and the line $y = x$ on a graphing calculator, and note that each is the mirror image of the other across $y = x$.

33. $y = 10^{x/2}$ and $y = 2 \log_{10} x$

34. $y = e^x$ and $y = \log_e x$

35. $y = 3x$ and $y = x/3$

36. $y = 2x + 4$ and $y = 0.5x - 2$

13-3 Properties of Logarithms

Since a logarithm is an exponent, it must follow the laws of exponents. Those laws we will find of the greatest importance at this time are now listed here for reference.

$$b^u b^v = b^{u+v} \tag{13-4}$$

$$\frac{b^u}{b^v} = b^{u-v} \tag{13-5}$$

$$(b^u)^n = b^{nu} \tag{13-6}$$

We will use these laws of exponents to derive certain useful properties of logarithms. The following example illustrates the reasoning used in deriving these properties.

EXAMPLE 1 ▸▸ We know that $8 \times 16 = 128$. Writing these numbers as powers of 2, we have

$$8 = 2^3, \qquad 16 = 2^4, \qquad 128 = 2^7 = 2^{3+4}$$

The logarithmic forms can be written as

$$3 = \log_2 8, \qquad 4 = \log_2 16, \qquad 3 + 4 = \log_2 128$$

This means that

$$\log_2 8 + \log_2 16 = \log_2 128$$

where

$$8 \times 16 = 128$$

The *sum of the logarithms* of 8 and 16 equals the logarithm of 128, where the *product* of 8 and 16 equals 128. ◂◂

Following Example 1, if we let $u = \log_b x$ and $v = \log_b y$ and write these equations in exponential form, we have $x = b^u$ and $y = b^v$. Therefore, forming the product of x and y, we obtain

$$xy = b^u b^v = b^{u+v} \quad \text{or} \quad xy = b^{u+v}$$

Writing this last equation in logarithmic form yields

$$u + v = \log_b xy$$

or

Logarithm of a Product

$$\log_b xy = \log_b x + \log_b y \tag{13-7}$$

Equation (13-7) states the property that *the logarithm of the product of two numbers is equal to the sum of the logarithms of the numbers.*

Using the same definitions of u and v to form the quotient of x and y, we then have

$$\frac{x}{y} = \frac{b^u}{b^v} = b^{u-v} \quad \text{or} \quad \frac{x}{y} = b^{u-v}$$

Writing this last equation in logarithmic form, we have

$$u - v = \log_b \left(\frac{x}{y}\right)$$

or

Logarithm of a Quotient

$$\log_b \left(\frac{x}{y}\right) = \log_b x - \log_b y \tag{13-8}$$

Equation (13-8) states the property that *the logarithm of the quotient of two numbers is equal to the logarithm of the numerator minus the logarithm of the denominator.*

If we again let $u = \log_b x$ and write this in exponential form, we have $x = b^u$. To find the nth power of x, we write

$$x^n = (b^u)^n = b^{nu}$$

Expressing this equation in logarithmic form yields

$$nu = \log_b (x^n)$$

or

Logarithm of a Power

$$\log_b (x^n) = n \log_b x \tag{13-9}$$

Equation (13-9) states that *the logarithm of the nth power of a number is equal to n times the logarithm of the number.* The exponent n may be integral or fractional.

In Section 13-1 we showed that the base b of logarithms must be a positive number. Since $x = b^u$ and $y = b^v$, this means that x and y are also positive numbers. Therefore, *the properties of logarithms that have just been derived are valid only for positive values of x and y.*

EXAMPLE 2 ▸▸ (a) Using Eq. (13-7), we may express $\log_4 15$ as a sum of logarithms.

$$\log_4 15 = \log_4(3 \times 5) = \log_4 3 + \log_4 5$$ logarithm of product / sum of logarithms

(b) Using Eq. (13-8), we may express $\log_4(\frac{5}{3})$ as the difference of logarithms.

$$\log_4\left(\frac{5}{3}\right) = \log_4 5 - \log_4 3$$ logarithm of quotient / difference of logarithm

(c) Using Eq. (13-9), we may express $\log_4(t^2)$ as twice $\log_4 t$.

$$\log_4(t^2) = 2\log_4 t$$ logarithm of power / multiple of logarithm

(d) Using Eq. (13-8) and then Eq. (13-7), we have

$$\log_4\left(\frac{xy}{z}\right) = \log_4(xy) - \log_4 z = \log_4 x + \log_4 y - \log_4 z$$ ◂◂

EXAMPLE 3 ▸▸ We may also express a sum or difference of logarithms as the logarithm of a single quantity.

(a) $\log_4 3 + \log_4 x = \log_4(3 \times x) = \log_4 3x$ using Eq. (13-7)

(b) $\log_4 3 - \log_4 x = \log_4\left(\frac{3}{x}\right)$ using Eq. (13-8)

(c) $\log_4 3 + 2\log_4 x = \log_4 3 + \log_4(x^2) = \log_4 3x^2$ using Eqs. (13-7) and (13-9)

(d) $\log_4 3 + 2\log_4 x - \log_4 y = \log_4\left(\frac{3x^2}{y}\right)$ using Eqs. (13-7), (13-8), and (13-9) ◂◂

In Section 13-2 we noted that $\log_b 1 = 0$. Also, since $b = b^1$ in logarithmic form is $\log_b b = 1$, we have $\log_b(b^n) = n\log_b b = n(1) = n$.

Summarizing these properties, we have

$$\log_b 1 = 0, \quad \log_b b = 1 \tag{13-10}$$

$$\log_b(b^n) = n \tag{13-11}$$

These equations may be used to find exact values of certain logarithms.

EXAMPLE 4 ▸▸ (a) We may evaluate $\log_3 9$ using Eq. (13-11).

$$\log_3 9 = \log_3(3^2) = 2$$

We can establish the exact value since the base of logarithms and the number being raised to the power are the same. Of course, this could have been evaluated directly from the definition of a logarithm.

(b) We can evaluate $\log_3(3^{0.4})$ using Eq. (13-11).

$$\log_3(3^{0.4}) = 0.4$$

Although we did not evaluate $3^{0.4}$, we were able to evaluate $\log_3 3^{0.4}$. ◂◂

EXAMPLE 5 ▸▸ (a) $\log_2 6 = \log_2(2 \times 3) = \log_2 2 + \log_2 3 = 1 + \log_2 3$

(b) $\log_5 \frac{1}{5} = \log_5 1 - \log_5 5 = 0 - 1 = -1$

(c) $\log_7\sqrt{7} = \log_7(7^{1/2}) = \frac{1}{2}\log_7 7 = \frac{1}{2}$ ◂◂

EXAMPLE 6 ▸▸ The following illustration shows the evaluation of a logarithm in two different ways.

(a) $\log_5(\frac{1}{25}) = \log_5 1 - \log_5 25 = 0 - \log_5(5^2) = -2$

(b) $\log_5(\frac{1}{25}) = \log_5(5^{-2}) = -2$

Either method is appropriate. ◂◂

EXAMPLE 7 ▸▸ Use the basic properties of logarithms to solve the following equation for y in terms of x: $\log_b y = 2\log_b x + \log_b a$.

Using Eq. (13-9) and then Eq. (13-7), we have

$$\log_b y = \log_b(x^2) + \log_b a = \log_b(ax^2)$$

Since we have the logarithm to the base b of different expressions on each side of the resulting equation, the expressions must be equal. Therefore,

$$y = ax^2$$ ◂◂

EXAMPLE 8 ▸▸ An equation for the current i and the time t in an electric circuit containing a resistance R and a capacitance C is $\log_e i - \log_e I = -t/RC$ (where R and C are both factors of the denominator). Here I is the current for $t = 0$. Solve for i as a function of t.

Using Eq. (13-8), we rewrite the left side of this equation, obtaining

$$\log_e\left(\frac{i}{I}\right) = -\frac{t}{RC}$$

Rewriting this in exponential form, we have

$$\frac{i}{I} = e^{-t/RC} \quad \text{or} \quad i = Ie^{-t/RC}$$ ◂◂

Exercises 13–3

In Exercises 1–12, express each as a sum, difference, or multiple of logarithms. See Example 2.

1. $\log_5 33$
2. $\log_3 14$
3. $\log_7(\frac{5}{3})$
4. $\log_3(\frac{2}{11})$
5. $\log_2(a^3)$
6. $\log_8(n^5)$
7. $\log_6 abc$
8. $\log_2\left(\frac{xy}{z^2}\right)$
9. $\log_5 \sqrt[4]{y}$
10. $\log_4 \sqrt[7]{x}$
11. $\log_2\left(\frac{\sqrt{x}}{a^2}\right)$
12. $\log_3\left(\frac{\sqrt[3]{y}}{8}\right)$

In Exercises 13–20, express each as the logarithm of a single quantity. See Example 3.

13. $\log_b a + \log_b c$
14. $\log_2 3 + \log_2 x$
15. $\log_5 9 - \log_5 3$
16. $\log_8 6 - \log_8 a$
17. $\log_b x^2 - \log_b \sqrt{x}$
18. $\log_4 3^3 + \log_4 9$
19. $2\log_e 2 + 3\log_e n$
20. $\frac{1}{2}\log_b a - 2\log_b 5$

In Exercises 21–28, determine the exact value of each of the given logarithms.

21. $\log_2 \left(\frac{1}{32}\right)$ **22.** $\log_3 \left(\frac{1}{81}\right)$

23. $\log_2 (2^{2.5})$ **24.** $\log_5 (5^{0.1})$

25. $\log_7 \sqrt{7}$ **26.** $\log_6 \sqrt[3]{6}$

27. $\log_3 \sqrt[4]{27}$ **28.** $\log_5 \sqrt[3]{25}$

In Exercises 29–36, express each as a sum, difference, or multiple of logarithms. In each case, part of the logarithm may be determined exactly.

29. $\log_3 18$ **30.** $\log_5 75$

31. $\log_2 \left(\frac{1}{6}\right)$ **32.** $\log_{10} (0.05)$

33. $\log_3 \sqrt{6}$ **34.** $\log_2 \sqrt[3]{24}$

35. $\log_{10} 3000$ **36.** $\log_{10} (40^2)$

In Exercises 37–48, solve for y in terms of x.

37. $\log_b y = \log_b 2 + \log_b x$

38. $\log_b y = \log_b 6 + \log_b x$

39. $\log_4 y = \log_4 x - \log_4 5 + \log_4 3$

40. $\log_3 y = \log_3 7 - 2 \log_3 x$

41. $\log_{10} y = 2 \log_{10} 7 - 3 \log_{10} x$

42. $\log_b y = 3 \log_b \sqrt{x} + 2 \log_b 10$

43. $5 \log_2 y - \log_2 x = 3 \log_2 4 + \log_2 a$

44. $4 \log_2 x - 3 \log_2 y = \log_2 27$

45. $\log_2 x + \log_2 y = 1$

46. $3 \log_4 x + \log_4 y = 1$

47. $2 \log_5 x - \log_5 y = 2$

48. $\log_8 x - 2 \log_8 y = 4$

In Exercises 49–52, display the indicated graphs and perform the indicated operations.

49. Display the graphs of $y = 3 \log_{10} x$ and $y = \log_{10} x^3$ on a graphing calculator, and show that they are the same.

50. Display the graphs of $y = \log_e (e^2 x)$ and $y = 2 + \log_e x$ on a graphing calculator, and explain why they are the same.

51. Under certain conditions, the temperature T, in degrees Celsius, of a cooling object is related to the time t, in minutes, by the equation $\log_e T = \log_e 65.0 - 0.41t$. Solve for T as a function of t.

52. In analyzing the power gain in an electric circuit, the equation

$$N = 10(2 \log_{10} I_1 - 2 \log_{10} I_2 + \log_{10} R_1 - \log_{10} R_2)$$

is used. Express this with a single logarithm on the right side.

13-4 Logarithms to the Base 10

In Section 13-1 we stated that a base of logarithms must be a positive number, not equal to one. In the examples and exercises of the previous sections, we used a number of different bases. There are, however, only two bases that are generally used. They are 10 and e, where e is the irrational number approximately equal to 2.718 that we introduced in Section 12-5 and have used in the previous sections of this chapter.

Base 10 logarithms were developed for calculational purposes and were used a great deal for making calculations until the 1970s, when the modern scientific calculator became widely available. Base 10 logarithms are still used in several scientific measurements, and therefore a need still exists for them. Base e logarithms are used extensively in technical and scientific work; we consider them in detail in the next section.

Logarithms to the base 10 are called **common logarithms.** They may be found directly by use of a calculator, and the [log] key is used for this purpose. This, of course, is the same key we have used with logarithmic functions to the base 10 on the graphing calculator. This calculator key indicates the common notation. *When no base is shown, it is assumed to be the base 10.*

NOTE ▶

EXAMPLE 1 ▸▸ Using a calculator, we find that

$$\log 426 = 2.629$$

↑ no base shown means base is 10

when the result is rounded off. The decimal part of a logarithm is normally expressed to the same accuracy as that of the number of which it is the logarithm, although showing one additional digit in the logarithm is generally acceptable.

Since $10^2 = 100$ and $10^3 = 1000$, and in this case

$$10^{2.629} = 426,$$

we see that the 2.629 power of 10 gives a number between 100 and 1000. ◂◂

EXAMPLE 2 ▸▸ Finding log 0.036 54 on a calculator, we obtain

$$\log 0.03654 = -1.4372$$

We note that the logarithm here is negative. This should be the case when we recall the meaning of a logarithm. Raising 10 to a negative power gives us a number between 0 and 1, and here we have

$$10^{-1.4372} = 0.03654$$ ◂◂

We may also use a calculator to find a number N if we know $\log N$. *In this case we refer to N as the* **antilogarithm** *of log N.* On the calculator we use the [10^x] key, or the sequence [INV] [LOG], depending on the calculator. The [10^x] key shows the basic definition of a logarithm, and the [INV] [LOG] sequence shows that the exponential and logarithmic functions are inverse functions.

EXAMPLE 3 ▸▸ Given $\log N = 1.1854$, a calculator shows that

$$N = 15.32$$

where the result has been rounded off. Since $10^1 = 10$ and $10^2 = 100$, we see that $10^{1.1854}$ is a number between 10 and 100. ◂◂

The following example illustrates an application in which a measurement requires the direct use of the value of a logarithm.

EXAMPLE 4 ▸▸ The power gain G (in decibels) of an electronic device is given by $G = 10 \log (P_0/P_i)$, where P_0 is the output power, in watts, and P_i is the input power. Determine the power gain for an amplifier for which $P_0 = 15.8$ W and $P_i = 0.625$ W.

Substituting the given values, we have

$$\begin{aligned} G &= 10 \log \frac{15.8}{0.625} \\ &= 10 \log 25.28 \\ &= 14.0 \text{ dB} \end{aligned}$$

where the result has been rounded off to three significant digits, the accuracy of the given data. ◂◂

As noted in the chapter introduction, logarithms were developed for calculational purposes. They were first used in the 17th century for making tedious and complicated calculations that arose in astronomy and navigation. These complicated calculations were greatly simplified, since logarithms allowed them to be performed by means of basic additions, subtractions, multiplications, and divisions. Performing calculations in this way provides an opportunity to understand better the meaning and properties of logarithms. Also, certain calculations cannot be done directly on a calculator but can be done by logarithms.

EXAMPLE 5 ▸▸ A certain computer design has 64 different sequences of 10 binary digits so that the total number of possible states is $(2^{10})^{64} = 1024^{64}$. Evaluate 1024^{64} using logarithms.

Since $\log x^n = n \log x$, we know that $\log 1024^{64} = 64 \log 1024$. While most calculators will not directly evaluate 1024^{64}, we can use one to find the value of $64 \log 1024$. Since 1024^{64} is *exact,* we will show ten calculator digits until we round off the result. We therefore evaluate 1024^{64} as follows:

$$\text{Let } N = 1024^{64}$$

$$\log N = \log 1024^{64} = 64 \log 1024 \quad \text{using Eq. (13-9): } \log_b x^n = n \log_b x$$
$$= 64(3.010\,299\,957)$$
$$= 192.659\,197\,2$$

$$N = 10^{192.659\,197\,2} \quad \text{meaning of logarithm}$$
$$= 10^{192} \times 10^{0.659\,197\,2} \quad \text{using Eq. (13-4): } b^u b^v = b^{u+v}$$
$$= (10^{192}) \times (4.5624) \quad \text{antilogarithm of 0.659 197 2 is 4.5624 (rounded off)}$$
$$= 4.5624 \times 10^{192}$$

Note that when we used Eq. (13-4), $10^{0.659\,197\,2}$ is a number between 1 and 10, since $10^0 = 1$ and $10^1 = 10$. This allowed us to write the number immediately in scientific notation.

Although we used a calculator to find 192.659 197 2 and the anitlogarithm of 0.659 197 2, the calculation was done essentially by logarithms. ◂◂

This example shows that calculations using logarithms are based on Eqs. (13-7), (13-8), and (13-9). Multiplication is performed by the addition of logarithms, division is performed by the subtraction of logarithms, and a power is found by a multiple of a logarithm. A root of a number is found by using the fractional exponent form of the power.

Exercises 13–4

In Exercises 1–12, find the common logarithm of each of the given numbers by using a calculator.

1. 567 **2.** 60.5
3. 0.0640 **4.** 0.000 566
5. 9.24×10^6 **6.** 3.19×10^{15}
7. 1.172×10^{-4} **8.** 8.043×10^{-8}
9. $\cos 12.5°$ **10.** $\tan 50.8°$
11. $\sqrt{274}$ **12.** $\sqrt[5]{0.1275}$

In Exercises 13–20, find the antilogarithm of each of the given logarithms by using a calculator.

13. 4.437 **14.** 0.929 **15.** −1.3045 **16.** −6.9788
17. 3.301 12 **18.** 8.824 36 **19.** −2.237 46 **20.** −10.336

In Exercises 21–24, use logarithms to perform the indicated operations.

21. $(5.98)(14.3)$ **22.** $\dfrac{895}{73.4^{86}}$

23. $(\sqrt[10]{7.32})(2470)^{30}$ **24.** $\dfrac{126\,000^{20}}{2.63^{2.5}}$

In Exercises 25–28, find the logarithms of the given numbers.

25. A certain radar signal has a frequency of 1.15×10^9 Hz.

26. The bending moment of a particular concrete column is 4.60×10^6 N · m.

27. One electronvolt of energy equals 1.602×10^{-19} J.

28. In an air sample taken in an urban area, $5/10^6$ of the air was carbon monoxide.

In Exercises 29 and 30, solve the given problems by finding the appropriate logarithms.

29. A stereo amplifier has an input power of 0.750 W and an output power of 25.0 W. What is the power gain? (See Example 4.)

30. Measured on the Richter scale, the magnitude of an earthquake of intensity I is defined as $R = \log(I/I_0)$, where I_0 is a minimum level for comparison. What is the Richter scale reading for an earthquake for which $I = 75\,000 I_0$?

In Exercises 31 and 32, use logarithms to perform the indicated calculations.

31. A certain type of optical switch in a fiber-optic system allows the light signal to continue in either of two fibers. How many possible paths could a light signal follow if it passes through 400 such switches?

32. The peak current I_m, in amperes, in an alternating-current circuit is given by $I_m = \sqrt{\dfrac{2P}{Z \cos\theta}}$, where P is the power developed, Z is the magnitude of the impedance, and θ is the phase angle between the current and the voltage. Evaluate I_m for $P = 5.25$ W, $Z = 320\ \Omega$, and $\theta = 35.4°$.

13-5 Natural Logarithms

As we have noted, another number important as a base of logarithms is the number e. *Logarithms to the base e are called* **natural logarithms.** Since e is an irrational number equal to about 2.718, it may appear to be a very unnatural choice as a base of logarithms. However, in calculus the reason for its choice and the fact that it is a very natural number for a base of logarithms are shown.

Just as $\log x$ refers to logarithms to the base 10, the notation $\ln x$ is used to denote logarithms to the base e. We briefly noted this in Section 13-2 in discussing the graphing calculator. Due to the extensive use of natural logarithms, the notation **$\ln x$** is more convenient than **$\log_e x$**, although they mean the same thing.

Since more than one base is important, at times it is useful to change a logarithm from one base to another. If $u = \log_b x$, then $b^u = x$. Taking logarithms of both sides of this last expression to the base a, we have

$$\log_a b^u = \log_a x$$
$$u \log_a b = \log_a x$$
$$u = \frac{\log_a x}{\log_a b}$$

However, $u = \log_b x$, which means that

$$\log_b x = \frac{\log_a x}{\log_a b} \tag{13-12}$$

Equation (13-12) allows us to change a logarithm in one base to a logarithm in another base. The following examples illustrate the method of performing this operation.

EXAMPLE 1 ▸▸ Change log 20 to a logarithm with base e; that is, find ln 20.

Using Eq. (13-12) with $a = 10$, $b = e$, and $x = 20$, we have

$$\log_e 20 = \frac{\log_{10} 20}{\log_{10} e}$$

$$\text{or}\quad \ln 20 = \frac{\log 20}{\log e} = 2.996 \qquad \text{using a calculator}$$

This means that $e^{2.996} = 20$. ◂◂

EXAMPLE 2 ▸▸ Find $\log_5 560$.

In Eq. (13-12), if we let $a = 10$ and $b = 5$, we have

$$\log_5 x = \frac{\log x}{\log 5}$$

In this example, $x = 560$. Therefore, we have

$$\log_5 560 = \frac{\log 560}{\log 5} = 3.932$$

From the definition of a logarithm, this means that

$$5^{3.932} = 560$$ ◂◂

Since natural logarithms are used extensively, it is often convenient to have Eq. (13-12) written specifically for use with logarithms to the base 10 and natural logarithms. First, using $a = 10$ and $b = e$, we have

$$\ln x = \frac{\log x}{\log e} \tag{13-13}$$

Then, with $a = e$ and $b = 10$, we have

$$\log x = \frac{\ln x}{\ln 10} \tag{13-14}$$

Note that we really found ln 20 in Example 1 by using Eq. (13-13).

Values of natural logarithms can be found directly on a scientific calculator. The LN key is used for this purpose. In order to find the antilogarithm of a natural logarithm we use the e^x key or the key sequence INV LN, depending on the calculator.

EXAMPLE 3 ▸▸ (a) By use of a calculator, we find that

$$\ln 236.5 = 5.4659$$

which means that $e^{5.4659} = 236.5$.

(b) Given that $\ln N = -0.8729$, we determine N by finding $e^{-0.8729}$ on the calculator. This gives us

$$N = 0.4177$$ ◂◂

Using Eq. (13-12), we can display the graph of a logarithmic function with any base on a graphing calculator, as we show in the next example.

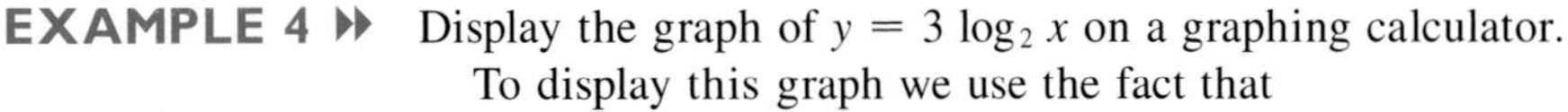

EXAMPLE 4 ▸▸ Display the graph of $y = 3 \log_2 x$ on a graphing calculator.

To display this graph we use the fact that

$$\log_2 x = \frac{\log x}{\log 2} \quad \text{or} \quad \log_2 x = \frac{\ln x}{\ln 2}$$

Therefore, we enter the function

$$y = \frac{3 \log x}{\log 2}$$

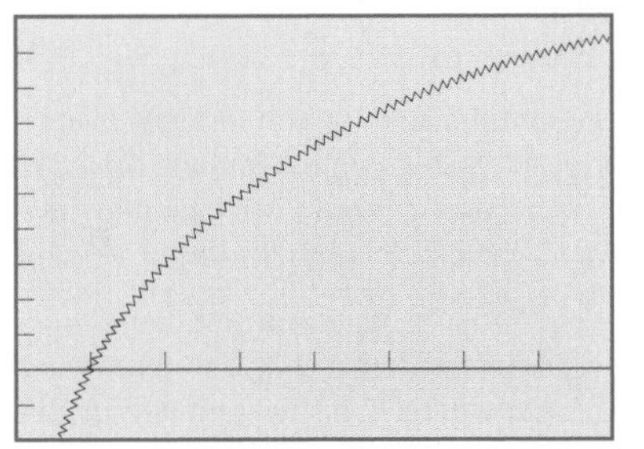

Fig. 13-8

in the calculator. Considering the domain and range of the logarithmic function, we use the following RANGE values:

$$\text{Xmin} = 0,\ \text{Xmax} = 8,\ \text{Ymin} = -2,\ \text{Ymax} = 10$$

The view on the graphing calculator is shown in Fig. 13-8. ◂◂

Applications of natural logarithms are found in many fields of technology. One such application is shown in the next example, and others are found in the exercises.

EXAMPLE 5 ▸▸ Under certain conditions, the electric current i in a circuit containing a resistance and an inductance (see Section 12-7) is given by

$$\ln \frac{i}{I} = -\frac{Rt}{L}$$

where I is the current at $t = 0$, R is the resistance, t is the time, and L is the inductance. Calculate how long (in seconds) it takes i to reach 0.430 A, if $I = 0.750$ A, $R = 7.50\ \Omega$, and $L = 1.25$ H.

Solving for t, we have

$$t = -\frac{L \ln (i/I)}{R} = -\frac{L(\ln i - \ln I)}{R} \qquad \text{either form can be used}$$

Thus, for the given values, we have

$$t = -\frac{1.25(\ln 0.430 - \ln 0.750)}{7.50}$$

$$= 0.0927 \text{ s}$$

Therefore, the current changes from 0.750 A to 0.430 A in 0.0927 s. ◂◂

Exercises 13–5

In Exercises 1–8, use logarithms to the base 10 to find the natural logarithms of the given numbers.

1. 26.0 **2.** 631

3. 1.562 **4.** 45.73

5. 0.5017 **6.** 0.052 94

7. 0.007 326 7 **8.** 0.000 443 48

In Exercises 9–16, use logarithms to the base 10 to find the indicated logarithms.

9. $\log_7 42$ **10.** $\log_2 86$

11. $\log_5 245$ **12.** $\log_3 706$

13. $\log_{12} 122$ **14.** $\log_{20} 86$

15. $\log_{40} 750$ **16.** $\log_{100} 3720$

In Exercises 17–24, find the natural logarithms of the indicated numbers.

17. 51.4 **18.** 293

19. 1.394 **20.** 65.62

21. 0.9917 **22.** 0.002 086

23. 0.012 937 **24.** 0.000 060 808

In Exercises 25–28, use Eq. (13-14) to find the common logarithms of the given numbers.

25. 45.17 **26.** 8765

27. 0.685 28 **28.** 0.001 429 8

In Exercises 29–36, find the natural antilogarithms of the given logarithms.

29. 2.190 **30.** 3.420

31. 0.008 421 0 **32.** 0.632

33. −0.7429 **34.** −2.942 18

35. −23.504 **36.** −0.008 04

In Exercises 37–40, use a graphing calculator to display the indicated graphs.

37. The graph of $y = \log_5 x$.

38. The graph of $y = 2 \log_8 x$.

39. Graphically show that $y = 2^x$ and $y = \log_2 x$ are inverse functions. (See Example 6 of Section 13-2.)

40. Graphically show that $y = 6^{x/2}$ and $y = 2 \log_6 x$ are inverse functions. (See Example 6 of Section 13-2.)

In Exercises 41–48, solve the given problems.

41. Solve for y in terms of x: $\ln y - \ln x = 1.0986$.

42. Solve for y in terms of x: $\ln y + 2 \ln x = 1 + \ln 5$.

43. If interest is compounded continuously (daily compounded interest closely approximates this), with an interest rate i, a bank account will double in t years according to $i = (\ln 2)/t$. Find i if the account is to double in 8.5 years.

44. One approximate formula for world population growth is $T = 50.0 \ln 2$, where T is the number of years for the population to double. According to this formula, how long does it take for the population to double?

45. For the electric circuit of Example 5, find how long it takes the current to reach 0.1 of the initial value of 0.750 A.

46. The intensity I of light decreases from its value I_0 as it passes a distance x through a medium. Given that $x = k(\ln I_0 - \ln I)$, where k is a constant depending on the medium, find x for $I = 0.850I_0$ and $k = 5.00$ cm.

47. The distance x traveled by a motorboat in t seconds after the engine is cut off is given by $x = k^{-1} \ln (kv_0t + 1)$, where v_0 is the velocity of the boat at the time the engine is cut and k is a constant. Find how long it takes a boat to go 150 m if $v_0 = 12.0$ m/s and $k = 6.80 \times 10^{-3}$/m.

48. The electric current i in a circuit containing a 1-H inductor, a 10-Ω resistor, and a 6-V battery is a function of time given by $i = 0.6(1 - e^{-10t})$. Solve for t as a function of i.

13-6 Exponential and Logarithmic Equations

An equation in which the variable occurs in an exponent is called an **exponential equation.** Although some exponential equations may be solved by changing to logarithmic form, they are more generally solved by *taking the logarithm of both sides* and then using the basic properties of logarithms.

EXAMPLE 1 (a) We can solve the exponential equation $2^x = 8$ for x by writing it in logarithmic form. This gives us

$$x = \log_2 8 = 3 \qquad 2^3 = 8$$

This method is good if we can directly evaluate the resulting logarithm.

(b) Since 2^x and 8 are equal, the logarithms of 2^x and 8 are also equal. Therefore, we can also solve $2^x = 8$ in a more general way by taking logarithms (to any proper base) of both sides and equating these logarithms. This gives us

$$\log 2^x = \log 8 \qquad \text{or} \qquad \ln 2^x = \ln 8$$

$$x \log 2 = \log 8 \qquad\qquad x \ln 2 = \ln 8 \qquad \text{using Eq. (13-9)}$$

$$x = \frac{\log 8}{\log 2} = 3 \qquad\qquad x = \frac{\ln 8}{\ln 2} = 3 \qquad \text{using a calculator}$$

For reference, Eq. (13-9) is $\log_b(x^n) = n \log_b x$.

EXAMPLE 2 ▸▸ Solve the equation $3^{x-2} = 5$.

Taking logarithms of both sides and equating them, we have

$$\log 3^{x-2} = \log 5$$

or

$$(x - 2)\log 3 = \log 5 \qquad \text{using Eq. (13-9)}$$

Solving this last equation for x, we have

$$x = 2 + \frac{\log 5}{\log 3} = 3.465$$

This solution means that

$$3^{3.465-2} = 3^{1.465} = 5$$

which can be checked by a calculator. ◂◂

EXAMPLE 3 ▸▸ Solve the equation $2(4^{x-1}) = 17^x$.

By taking logarithms of both sides, we have the following:

For reference, Eqs. (13-7), (13-8), and (13-9) are

$\log_b xy = \log_b x + \log_b y$

$\log_b\left(\frac{x}{y}\right) = \log_b x - \log_b y$

$\log_b(x^n) = n\log_b x$

$$\log 2 + (x - 1)\log 4 = x\log 17 \qquad \text{using Eqs. (13-7) and (13-9)}$$

$$x\log 4 - x\log 17 = \log 4 - \log 2$$

$$x(\log 4 - \log 17) = \log 4 - \log 2$$

$$x = \frac{\log 4 - \log 2}{\log 4 - \log 17} = \frac{\log(4/2)}{\log 4 - \log 17} \qquad \text{using Eq. (13-8)}$$

$$= \frac{\log 2}{\log 4 - \log 17}$$

$$= -0.479$$ ◂◂

EXAMPLE 4 ▸▸ At constant temperature, the atmospheric pressure p (in pascals) at an altitude h (in metres) is given by $p = p_0e^{kh}$, where p_0 is the pressure where $h = 0$ (usually taken as sea level). Given that $p_0 = 101.3$ kPa (atmospheric pressure at sea level) and $p = 68.9$ kPa for $h = 3050$ m, find the value of k.

Since the equation is defined in terms of e, we can solve it most easily by taking natural logarithms of both sides. By doing this we have the following solution.

$$\ln p = \ln(p_0e^{kh}) = \ln p_0 + \ln e^{kh} \qquad \text{using Eq. (13-7)}$$

$$= \ln p_0 + kh\ln e = \ln p_0 + kh \qquad \text{using Eq. (13-9), } \ln e = 1$$

$$\ln p - \ln p_0 = kh$$

$$k = \frac{\ln p - \ln p_0}{h}$$

Substituting the given values, we have

$$k = \frac{\ln(68.9 \times 10^3) - \ln(101.3 \times 10^3)}{3050} = -0.000126/\text{m}$$ ◂◂

Logarithmic Equations

Some of the important measurements in scientific and technical work are defined in terms of logarithms. Using these formulas can lead to solving a **logarithmic equation,** *which is an equation with the logarithm of an expression involving the variable.* In solving logarithmic equations, we use the basic properties of logarithms to help change them into a usable form. There is, however, no general algebraic method for solving such equations, and we shall consider only some special cases.

See the chapter introduction.

EXAMPLE 5 ⏩ It has been found that the human ear responds to sound on a scale which is approximately proportional to the logarithm of the intensity of the sound. Therefore, the loudness of sound (measured in decibels) is defined by the equation $b = 10 \log (I/I_0)$, where I is the intensity of the sound and I_0 is the minimum intensity detectable.

A busy city street has a loudness of 70 dB, and riveting has a loudness of 100 dB. How many times greater is the intensity of the sound of riveting I_r than the sound of the city street I_c?

First, we substitute the decibel readings into the above definition. This gives us

$$70 = 10 \log \left(\frac{I_c}{I_0}\right) \quad \text{and} \quad 100 = 10 \log \left(\frac{I_r}{I_0}\right)$$

To solve these equations for I_c and I_r, we divide each side by 10 and then use the exponential form. Thus, we have

$$7.0 = \log \left(\frac{I_c}{I_0}\right) \quad \text{and} \quad 10 = \log \left(\frac{I_r}{I_0}\right)$$

$$\frac{I_c}{I_0} = 10^{7.0} \qquad \frac{I_r}{I_0} = 10^{10}$$

$$I_c = I_0(10^{7.0}) \qquad I_r = I_0(10^{10})$$

Since we want the number of times I_r is greater than I_c, we divide I_r by I_c. This gives us

$$\frac{I_r}{I_c} = \frac{I_0(10^{10})}{I_0(10^{7.0})} = \frac{10^{10}}{10^{7.0}} = 10^{3.0} \quad \text{or} \quad I_r = 10^{3.0} I_c = 1000 I_c$$

Thus, the sound of riveting is 1000 times as intense as the sound of the city street. This demonstrates that sound intensity levels are considerably greater than loudness levels. (See Exercise 42.) ⏪

For reference, Eqs. (13-7), (13-8), and (13-9) are

$$\log_b xy = \log_b x + \log_b y$$

$$\log_b \left(\frac{x}{y}\right) = \log_b x - \log_b y$$

$$\log_b (x^n) = n \log_b x$$

The following examples illustrate the solution of other logarithmic equations.

EXAMPLE 6 ⏩ Solve the logarithmic equation $\log_2 5 + \log_2 x = 3$.

Using the basic properties of logarithms, we have the following solution.

$$\log_2 5 + \log_2 x = 3$$

$$\log_2 5x = 3 \quad \text{using Eq. (13-7)}$$

$$5x = 2^3 \quad \text{exponential form}$$

$$x = 8/5$$ ⏪

EXAMPLE 7 ▸▸ Solve the logarithmic equation $2 \ln 2 + \ln x = \ln 3$.

Using the properties of logarithms, we have the following solution.

$$2 \ln 2 + \ln x = \ln 3$$

$$\ln 2^2 + \ln x - \ln 3 = 0 \qquad \text{using Eq. (13-9)}$$

$$\ln \frac{4x}{3} = 0 \qquad \text{using Eqs. (13-7) and (13-8)}$$

$$\frac{4x}{3} = e^0 = 1 \qquad \text{exponential form}$$

$$4x = 3$$

$$x = \frac{3}{4} \quad ◂◂$$

EXAMPLE 8 ▸▸ Solve the logarithmic equation $2 \log x - 1 = \log (1 - 2x)$.

$$\log x^2 - \log (1 - 2x) = 1$$

$$\log \frac{x^2}{1 - 2x} = 1 \qquad \text{using Eq. (13-8)}$$

$$\frac{x^2}{1 - 2x} = 10^1 \qquad \text{exponential form}$$

$$x^2 = 10 - 20x$$

$$x^2 + 20x - 10 = 0$$

$$x = \frac{-20 \pm \sqrt{400 + 40}}{2} = -10 \pm \sqrt{110}$$

Since logarithms of negative numbers are not defined, and $-10 - \sqrt{110}$ is negative and cannot be used in the first term of the original equation, we have

$$x = -10 + \sqrt{110} = 0.488 \quad ◂◂$$

Exercises 13–6

In Exercises 1–36, solve the given equations.

1. $2^x = 16$
2. $3^x = \frac{1}{81}$
3. $5^x = 0.3$
4. $6^x = 15$
5. $3^{-x} = 0.525$
6. $15^{-x} = 1.326$
7. $e^{2x} = 3.625$
8. $e^{-x} = 17.54$
9. $6^{x+1} = 10$
10. $5^{x-1} = 2$
11. $4(3^x) = 5$
12. $3(14^x) = 40$
13. $0.8^x = 0.4$
14. $0.6^x = 100$
15. $(15.6)^{x+2} = 23^x$
16. $5^{x+2} = e^{2x}$
17. $3 \log_8 x = -2$
18. $5 \log_{32} x = -3$
19. $2 \ln x = 1$
20. $3 \ln 2x = 2$
21. $\log_2 x + \log_2 7 = \log_2 21$
22. $2 \log_2 3 - \log_2 x = \log_2 45$
23. $2 \log (3 - x) = 1$
24. $3 \log (2x - 1) = 1$
25. $\log 4x + \log x = 2$
26. $\log 12x^2 - \log 3x = 3$
27. $\ln x + \ln 3 = 1$
28. $2 \ln 2 - \ln x = -1$
29. $3 \ln 2 + \ln (x - 1) = \ln 24$
30. $\ln (2x - 1) - 2 \ln 4 = 3 \ln 2$
31. $\frac{1}{2} \log (x + 2) + \log 5 = 1$
32. $\frac{1}{2} \log (x - 1) - \log x = 0$
33. $\log_5 (x - 3) + \log_5 x = \log_5 4$
34. $\log_7 x + \log_7 (2x - 5) = \log_7 3$
35. $\log (2x - 1) + \log (x + 4) = 1$
36. $\log_2 x + \log_2 (x + 2) = 3$

In Exercises 37–48, determine the indicated quantities.

37. In computer design, the number N of bits of memory is often expressed as a power of 2, or $N = 2^x$. Find x if $N = 2.68 \times 10^8$ bits.

38. If half of a radioactive substance decays in a year, the amount N remaining after t years is given by $N = N_0(0.5)^t$, where N_0 is the original amount. Find t if $N = 0.10N_0$.

39. The temperature T, in degrees Celsius, of a certain cooling object is given by $T = T_0 + 70(0.40)^{0.20t}$, where T_0 is the temperature of the surroundings of the object and t is the time in minutes. Determine how long it takes the object to cool to 50°C if $T_0 = 30$°C.

40. The electric current i, in amperes, in a circuit containing a resistance R, an inductor L, and a voltage source E is given by $i = \frac{E}{R}(1 - e^{-Rt/L})$, where t is the time in seconds. Find t if $i = 0.750$ A, $E = 6.00$ V, $R = 4.50\ \Omega$, and $L = 2.50$ H.

41. In chemistry, the pH value of a solution is a measure of its acidity. The pH value is defined by the relation $\text{pH} = -\log(\text{H}^+)$, where H^+ is the hydrogen ion concentration. If the pH of a certain wine is 3.4065, find the hydrogen ion concentration. (If the pH value is less than 7, the solution is acid. If the pH value is above 7, the solution is basic.)

42. Referring to Example 5, show that if the difference in loudness of two sounds is d decibels, the louder sound is $10^{d/10}$ more intense than the quieter sound.

43. Measured on the Richter scale, the magnitude of an earthquake of intensity I is defined as $R = \log(I/I_0)$, where I_0 is a minimum level for comparison. How many times I_0 was the 1906 San Francisco earthquake whose magnitude was 8.3 on the Richter scale?

44. How many times more intense was the 1906 San Francisco earthquake than the 1994 Northridge (Southern California) earthquake, $R = 6.8$? (See Exercise 43.)

45. Pure water is running into a certain brine solution, and the same amount of solution is running out. The number n of kilograms of salt in the solution after t minutes is found by solving the equation $\ln n = -0.04t + \ln 20$. Solve for n as a function of t.

46. In an electric circuit containing a resistor and a capacitor with an initial charge q_0, the charge q on the capacitor at any time t after closing the switch can be found by solving the equation $\ln q = -\frac{t}{RC} + \ln q_0$. Here R is the resistance and C is the capacitance. Solve for q as a function of t.

47. An earth satellite loses 0.1% of its remaining power each week. An equation relating the power P, the initial power P_0, and the time t in weeks is $\ln P = t \ln 0.999 + \ln P_0$. Solve for P as a function of t.

48. An equation relating the distance s through which a falling object moves and its velocity v is $\log s + \log 2g = 2 \log v$, where g is the acceleration due to gravity. Solve for s as a function of v.

To solve more complicated problems, we may use graphical methods. For example, if we wish to solve the equation $2^x + 3^x = 50$, we can set up the function $y = 2^x + 3^x - 50$ and then determine its zeros graphically. Note that the given equation can be written as $2^x + 3^x - 50 = 0$, and therefore the zeros of the function that has been set up will give the desired solution. In Exercises 49–52, solve the given equations in this way by displaying the appropriate graphs on a graphing calculator.

49. $2^x + 3^x = 50$

50. $4^x + x^2 = 25$

51. The curve in which a uniform wire or rope hangs under its own weight is called a *catenary*. An example of a catenary that we see every day is a wire strung between utility poles as shown in Fig. 13-9. For a particular wire, the equation of the catenary it forms is $y = 2(e^{x/4} + e^{-x/4})$, where (x, y) is a point on the curve. Find x for $y = 5.8$ m.

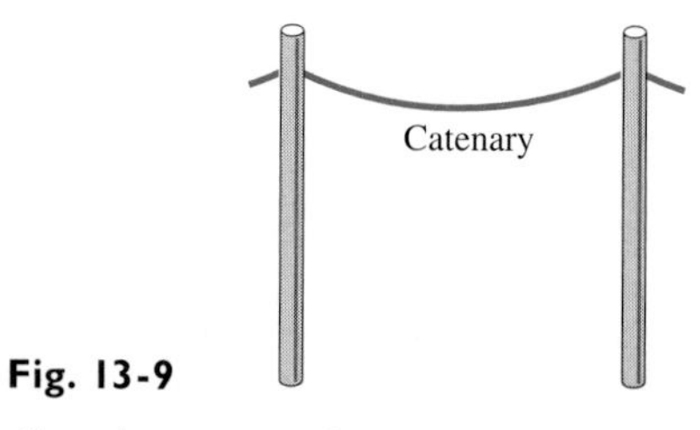

Fig. 13-9

52. In finding the current i in a certain electric circuit, the equation $i = 2 \ln(t + 2) - t$ relates the current and the time t in seconds. Find t for $i = 0.05$ A.

13-7 Graphs on Logarithmic and Semilogarithmic Paper

When constructing the graphs of some functions, one of the variables changes much more rapidly than the other. We saw this in graphing the exponential and logarithmic functions in Section 13-2. The following example illustrates this point.

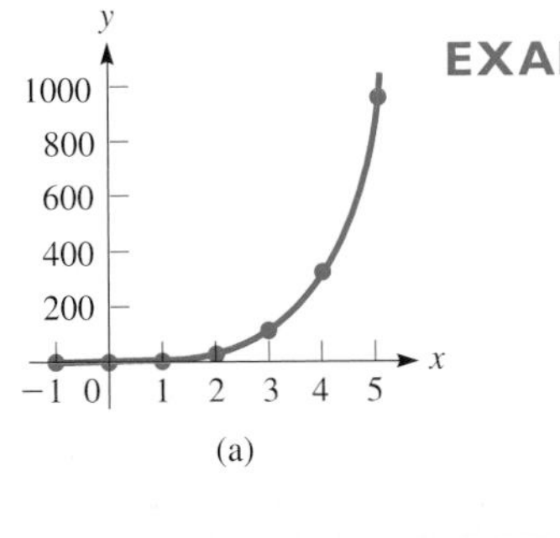

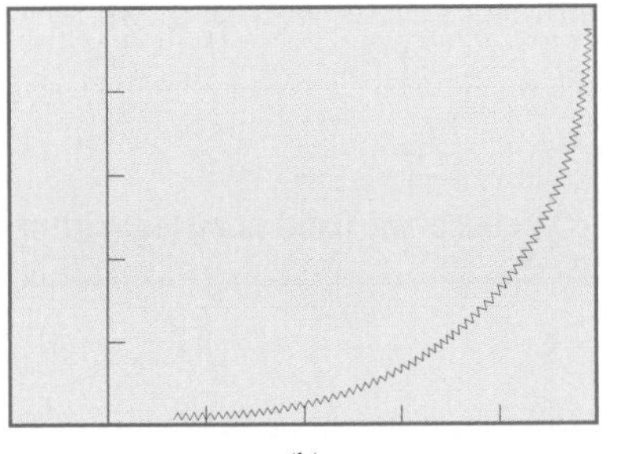

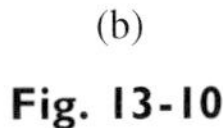
(b)

Fig. 13-10

EXAMPLE 1 ▸▸ Plot the graph of $y = 4(3^x)$.

Constructing the following table of values

x	-1	0	1	2	3	4	5
y	1.3	4	12	36	108	324	972

we then plot these values as shown in Fig. 13-10(a).

We see that as x changes from -1 to 5, y changes much more rapidly, from about 1 to nearly 1000. Also, because of the scale that must be used, we see that it is not possible to show accurately the differences in the y-values on the graph.

Even a graphing calculator cannot show the graph accurately for the values near $x = 0$, if we wish to view all of this part of the curve. In fact, the graphing calculator view shows the curve as being on the axis for these values, as we see in Fig. 13-10(b). In this figure we have used the same RANGE values used in Fig. 13-10(a). ◂◂

It is possible to graph a function with a large change in values, for one or both variables, more accurately than can be done on the standard rectangular coordinate system. This is done by using a scale marked off in distances proportional to the logarithms of the values being represented. *Such a scale is called a* **logarithmic scale.** For example, $\log 1 = 0$, $\log 2 = 0.301$, and $\log 10 = 1$. Thus, on a logarithmic scale the 2 is placed 0.301 unit of distance from the 1 to the 10. Figure 13-11 shows a logarithmic scale with the numbers represented and the distance used for each.

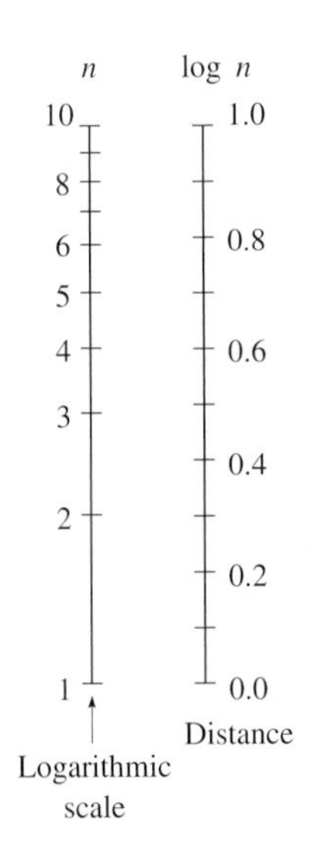

Fig. 13-11

On a logarithmic scale the distances between the integers are not equal, but this scale does allow for a much greater range of values and much greater accuracy for many of the values. There is another advantage to using logarithmic scales. Many equations that would have more complex curves when graphed on the standard rectangular coordinate system will have simpler curves, often straight lines, when graphed using logarithmic scales. In many cases this makes the analysis of the curve much easier.

Zero and negative numbers do not appear on the logarithmic scale. In fact, all numbers used on the logarithmic scale must be positive, since the domain of the logarithmic function includes only positive real numbers. Thus, the logarithmic scale must start at some number greater than zero. This number is a power of 10 and can be very small, say, $10^{-6} = 0.000\,001$, but it is positive.

If we wish to use a large range of values for only one of the variables, we use what is known as **semilogarithmic,** or **semilog,** graph paper. On this graph paper, only one axis (usually the y-axis) uses a logarithmic scale. If we wish to use a large range of values for both variables, we use **logarithmic,** or **log-log,** graph paper. Both axes are marked with logarithmic scales.

The following examples illustrate the use of semilog and log-log graph paper.

EXAMPLE 2 ▸▸ Construct the graph of $y = 4(3^x)$ on semilogarithmic graph paper.

This is the same function as in Example 1, and we repeat the table of values.

x	-1	0	1	2	3	4	5
y	1.3	4	12	36	108	324	972

Again, we see that the range of y-values is large. When we plotted this curve on the rectangular coordinate system in Example 1, we had to use large units along the y-axis. This made the values of 1.3, 4, 12, and 36 appear at practically the same level. However, when we use semilog graph paper, we can label each axis such that all y-values are accurately plotted as well as the x-values.

The logarithmic scale is shown in **cycles,** and we must label the base line of the first cycle as 1 times a power of 10 (0.01, 0.1, 1, 10, 100, and so on) with the following cycle labeled with the next power of 10. The lines between are labeled with 2, 3, 4, and so on, times the proper power of 10. See the vertical scale in Fig. 13-12. We now plot the points in the table on the graph. The resulting graph is a straight line, as we see in Fig. 13-12. Taking logarithms of both sides of the equation, we have

$$\log y = \log [4(3^x)] = \log 4 + \log 3^x \quad \text{using Eq. (13-7)}$$
$$= \log 4 + x \log 3 \quad \text{using Eq. (13-9)}$$

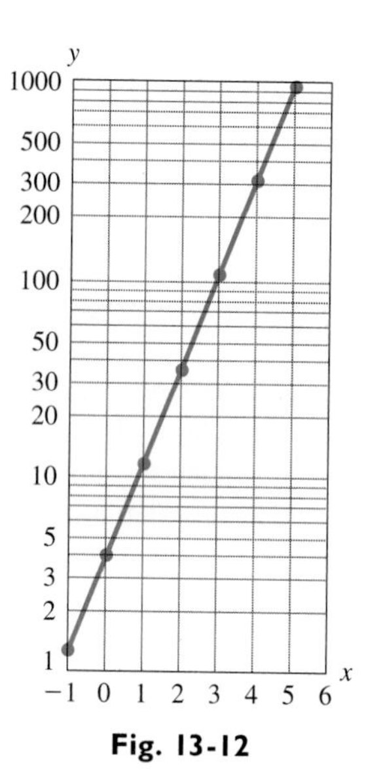

Fig. 13-12

However, since $\log y$ was plotted automatically (because we used semilogarithmic paper), the graph really represents

$$u = \log 4 + x \log 3$$

where $u = \log y$; log 3 and log 4 are constants, and therefore this equation is of the form $u = mx + b$, which is a straight line (see Section 5-2).

The logarithmic scale in Fig. 13-12 has *three cycles,* since all values of three powers of 10 are represented. ◂◂

EXAMPLE 3 ▸▸ Construct the graph of $x^4y^2 = 1$ on logarithmic paper.

First we solve for y and make a table of values. Considering positive values of x and y, we have

$$y = \sqrt{\frac{1}{x^4}} = \frac{1}{x^2}$$

x	0.5	1	2	8	20
y	4	1	0.25	0.0156	0.0025

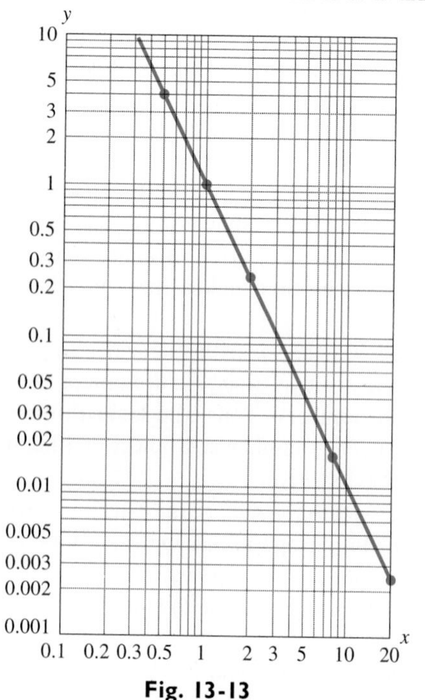

Fig. 13-13

We now plot these values on log-log paper on which both scales are logarithmic, as shown in Fig. 13-13. We again see that we have a straight line. Taking logarithms of both sides of the equation, we have

$$\log (x^4y^2) = \log 1$$
$$\log x^4 + \log y^2 = 0 \quad \text{using Eq. (13-7)}$$
$$4 \log x + 2 \log y = 0 \quad \text{using Eq. (13-9)}$$

If we let $u = \log y$ and $v = \log x$, we then have

$$4v + 2u = 0 \quad \text{or} \quad u = -2v$$

which is the equation of a straight line, as shown in Fig. 13-13. Note, however, that not all graphs on logarithmic paper are straight lines. ◂◂

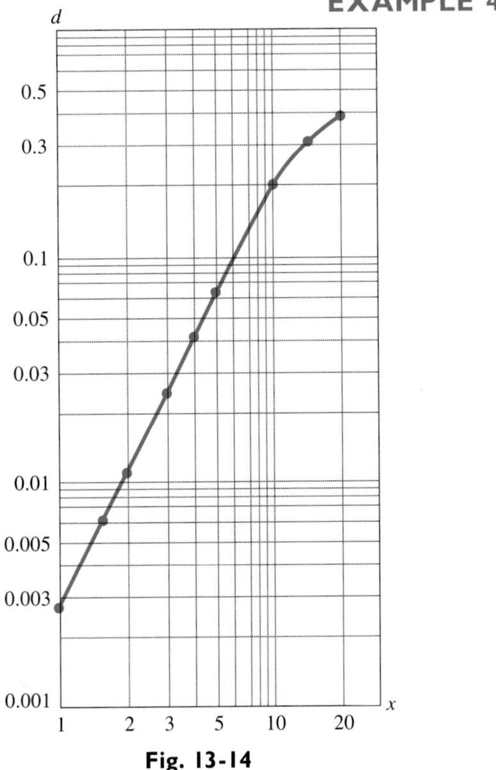

Fig. 13-14

EXAMPLE 4 ▸▸ The deflection (in metres) of a certain cantilever beam as a function of the distance x (in metres) from one end is

$$d = 0.0001(30x^2 - x^3)$$

If the beam is 20.0 m long, plot a graph of d vs. x on log-log paper.

Constructing a table of values we have

x (m)	1.00	1.50	2.00	3.00	4.00
d (m)	0.002 90	0.006 41	0.0112	0.0243	0.0416

x (m)	5.00	10.0	15.0	20.0
d (m)	0.0625	0.200	0.338	0.400

Since the beam is 20.0 m long, there is no meaning to values of x greater than 20.0 m. The graph is shown in Fig. 13-14. ◂◂

Logarithmic and semilogarithmic paper may be useful for plotting data derived from experimentation. Often the data cover too large a range of values to be plotted on ordinary graph paper. The next example illustrates the use of semilogarithmic paper to plot data.

EXAMPLE 5 ▸▸ The vapor pressure of water depends on the temperature. The following table gives the vapor pressure (in kPa) for the corresponding values of temperature (in degrees Celsius).

Temp.	10	20	40	60	80	100	120	140	160
Pressure	1.19	2.33	7.34	19.9	47.3	101	199	361	617

These data are then plotted on semilogarithmic paper, as shown in Fig. 13-15. Intermediate values of temperature and pressure can then be read directly from the graph. ◂◂

Fig. 13-15

Exercises 13–7

In Exercises 1–12, plot the graphs of the given functions on semilogarithmic paper.

1. $y = 2^x$ **2.** $y = 5^x$

3. $y = 2(4^x)$ **4.** $y = 5(10^x)$

5. $y = 3^{-x}$ **6.** $y = 2^{-x}$

7. $y = x^3$ **8.** $y = x^5$

9. $y = 3x^2$ **10.** $y = 2x^4$

11. $y = 2x^3 + 4x$ **12.** $y = 4x^3 + 2x^2$

In Exercises 13–24, plot the graphs of the given functions on log-log paper.

13. $y = 0.01x^4$ **14.** $y = 0.02x^3$

15. $y = \sqrt{x}$ **16.** $y = x^{2/3}$

17. $y = x^2 + 2x$ **18.** $y = x + \sqrt{x}$

19. $xy = 4$ **20.** $xy^2 = 10$

21. $y^2x = 1$ **22.** $x^2y^3 = 1$

23. $x^2y^2 = 25$ **24.** $x^3y = 8$

In Exercises 25–34, plot the indicated graphs.

25. The distance s through which a lead weight falls due to gravity is given by $s = \frac{1}{2}gt^2$, where g is the acceleration due to gravity and t is the time, in seconds, of fall. Given that $g = 9.80\text{ m/s}^2$, plot the graph of s as a function of t for $0 \leq t \leq 10$ s on (a) a regular rectangular coordinate system and (b) a semilogarithmic coordinate system.

26. By pumping, the air pressure within a tank is reduced by 18% each second. Thus, the pressure p, in kilopascals, in the tank is given by $p = 101(0.82)^t$, where t is the time in seconds. Plot the graph of p as a function of t for $0 \leq t \leq 30$ s on (a) a regular rectangular coordinate system and (b) a semilogarithmic coordinate system.

27. Strontium 90 decays according to the equation $N = N_0 e^{-0.028t}$, where N is the amount present after t years and N_0 is the original amount. Plot N as a function of t on semilog paper if $N_0 = 1000$ g.

28. The electric power delivered by a certain battery as a function of the resistance in the circuit is given by $P = \dfrac{100R}{(0.50 + R)^2}$, where P is measured in watts and R is in ohms. Plot P as a function of R on semilog paper, using the logarithmic scale for R and values of R from 0.01 Ω to 10 Ω. Compare the graph with that in Fig. 3-18 of Section 3-4.

29. The acceleration g produced by the gravitational force of the earth on a spacecraft is given by $g = 3.99 \times 10^{14}/r^2$, where r is the distance from the center of the earth to the spacecraft. On log-log paper, graph g (in m/s^2) as a function of r from $r = 6.37 \times 10^6$ m (the earth's surface) to $r = 3.91 \times 10^8$ m (the distance to the moon).

30. In undergoing an adiabatic (no *heat* gained or lost) expansion of a gas, the relation between the pressure p and the volume v is $p^2v^3 = 850$. On log-log paper, graph p (in kPa) as a function of v from $v = 0.10\text{ m}^3$ to $v = 10\text{ m}^3$.

31. The intensity level B, in decibels, and the frequency f, in hertz, for a sound of constant loudness were measured as follows:

f (Hz)	100	200	500	1000	2000	5000	10 000
B (dB)	40	30	22	20	18	24	30

Plot the data for B as a function of f on semilog paper, using the logarithmic scale for f.

32. The atmospheric pressure p at a given altitude h is given in the following table.

h (km)	0	10	20	30	40
p (kPa)	101	25	6.3	2.0	0.53

On semilog paper, plot p as a function of h.

33. One end of a very hot steel bar is sprayed with a stream of cool water. The rate of cooling R as a function of the distance d from the end of the bar is then measured. The following results are obtained:

d (cm)	0.63	1.3	1.9	2.5
R (°C/s)	600	190	100	72

d (cm)	3.8	5.0	7.5	10	15
R (°C/s)	46	29	17	10	6.0

On log-log paper, plot R as a function of d. Such experiments are made to determine the hardenability of steel.

34. The magnetic intensity H (in A/m) and flux density B (in teslas) of annealed iron are given in the following table.

B (T)	0.0042	0.043	0.67	1.01
H (A/m)	10	50	100	150

B (T)	1.18	1.44	1.58	1.72
H (A/m)	200	500	1000	10 000

Plot H as a function of B on logarithmic paper.

In Exercises 35 and 36, plot the indicated semilogarithmic graphs for the following application.

In a particular electric circuit, called a low-pass filter, the input voltage V_i is across a resistor and a capacitor, and the output voltage V_0 is across the capacitor (see Fig. 13-16). The voltage gain G, in decibels, in such a circuit is given by

$$G = 20 \log \frac{1}{\sqrt{1 + (\omega T)^2}}$$

where $\tan \phi = -\omega T$

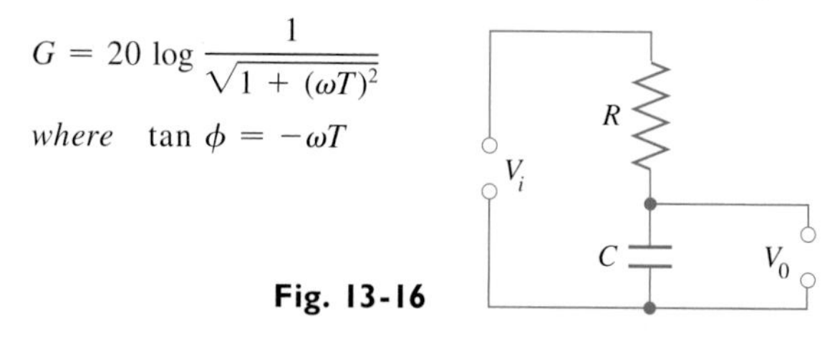

Fig. 13-16

Here ϕ is the phase angle of V_0/V_i. For values of ωT of 0.01, 0.1, 0.3, 1.0, 3.0, 10.0, 30.0, *and* 100, *plot the indicated graphs. These graphs are called a Bode diagram for the circuit.*

35. Calculate values of G for the given values of ωT, and plot a semilogarithmic graph of G vs. ωT.

36. Calculate values of ϕ (as negative angles) for the given values of ωT, and plot a semilogarithmic graph of ϕ vs. ωT.

Chapter Equations, Review Exercises, and Practice Test

Chapter Equations

Exponential function	$y = b^x$	(13-1)
Logarithmic form	$x = \log_b y$	(13-2)
Logarithmic function	$y = \log_b x$	(13-3)
Laws of exponents	$b^u b^v = b^{u+v}$	(13-4)
	$\dfrac{b^u}{b^v} = b^{u-v}$	(13-5)
	$(b^u)^n = b^{nu}$	(13-6)
Properties of logarithms	$\log_b xy = \log_b x + \log_b y$	(13-7)
	$\log_b\left(\dfrac{x}{y}\right) = \log_b x - \log_b y$	(13-8)
	$\log_b(x^n) = n \log_b x$	(13-9)
	$\log_b 1 = 0, \quad \log_b b = 1$	(13-10)
	$\log_b(b^n) = n$	(13-11)
Changing base of logarithms	$\log_b x = \dfrac{\log_a x}{\log_a b}$	(13-12)
	$\ln x = \dfrac{\log x}{\log e}$	(13-13)
	$\log x = \dfrac{\ln x}{\ln 10}$	(13-14)

Review Exercises

In Exercises 1–12, determine the value of x.

1. $\log_{10} x = 4$ **2.** $\log_9 x = 3$

3. $\log_5 x = -1$ **4.** $\log_4 x = -\frac{1}{2}$

5. $\log_2 64 = x$ **6.** $\log_{12} 144 = x$

7. $\log_8 32 = x$ **8.** $\log_9 27 = x$

9. $\log_x 36 = 2$ **10.** $\log_x 243 = 5$

11. $\log_x 10 = \frac{1}{2}$ **12.** $\log_x 8 = \frac{3}{4}$

In Exercises 13–24, express each as a sum, difference, or multiple of logarithms. Wherever possible, evaluate logarithms of the result.

13. $\log_3 2x$ **14.** $\log_5\left(\dfrac{7}{a}\right)$

15. $\log_3(t^2)$ **16.** $\log_6 \sqrt{5}$

17. $\log_2 28$ **18.** $\log_7 98$

19. $\log_3\left(\dfrac{9}{x}\right)$ **20.** $\log_6\left(\dfrac{5}{36}\right)$

21. $\log_4 \sqrt{48}$ **22.** $\log_6 \sqrt{72y}$

23. $\log_{10}(1000x^4)$ **24.** $\log_3(9^2 \times 6^3)$

In Exercises 25–32, solve for y in terms of x.

25. $\log_6 y = \log_6 4 - \log_6 x$

26. $\log_3 y = \frac{1}{2}\log_3 7 + \frac{1}{2}\log_3 x$

27. $\log_2 y + \log_2 x = 3$

28. $6 \log_4 y = 8 \log_4 4 - 3 \log_4 x$

29. $\log_5 x + \log_5 y = \log_5 3 + 1$

30. $\log_7 y = 2 \log_7 5 + \log_7 x + 2$

31. $3(\log_8 y - \log_8 x) = 1$

32. $2(\log_9 y + 2 \log_9 x) = 1$

In Exercises 33–40, graph the given functions. A graphing calculator may be used.

33. $y = 0.5(5^x)$ **34.** $y = 3(2^{-x})$

35. $y = 0.5 \log_4 x$ **36.** $y = 10 \log_{16} x$

37. $y = \log_{3.15} x$ **38.** $y = 0.1 \log_{4.65} x$

39. $y = 1 - e^{-x}$ **40.** $y = 2(1 - e^{-0.2x})$

In Exercises 41–44, use logarithms to the base 10 to find the natural logarithms of the given numbers.

41. 8.86 **42.** 33.0

43. 2.07 **44.** 0.542

In Exercises 45–48, use natural logarithms to find logarithms to the base 10 of the given numbers.

45. 65.89 **46.** 0.0781

47. 0.1197 **48.** 8930

In Exercises 49–56, solve the given equations.

49. $e^{2x} = 5$ **50.** $2(5^x) = 15$

51. $3^{x+2} = 5^x$ **52.** $6^{x+2} = 12^{x-1}$

53. $\log_4 x + \log_4 6 = \log_4 12$

54. $2 \log_3 2 - \log_3(x + 1) = \log_3 5$

55. $\log_8(x + 2) + \log_8 2 = 2$

56. $\log(x + 2) + \log x = 0.4771$

In Exercises 57 and 58, plot the graphs of the given functions on semilogarithmic paper. In Exercises 59 and 60, plot the graphs of the given functions on log-log paper.

57. $y = 6^x$ **58.** $y = 5x^3$

59. $y = \sqrt[3]{x}$ **60.** $xy^4 = 16$

If x is eliminated between Eqs. (13-1) and (13-2), we have

$$y = b^{\log_b y} \tag{13-15}$$

In Exercises 61–64, evaluate the given expressions using Eq. (13-15).

61. $10^{\log 4}$ **62.** $2e^{\ln 7.5}$

63. $3e^{2 \ln 2}$ **64.** $5(10^{2 \log 3})$

In Exercises 65–80, solve the given problems.

65. If A dollars are invested in a certificate of deposit at i percent annual interest, compounded continuously, its value after t years is $V = Ae^{it}$. Solve for t.

66. In studying the plant life in a lake, it was found that the fraction f of surface light intensity at depth d (in cm) is given by $20 \ln f = d \ln 0.9$. Solve for f.

67. In a certain electric circuit, the current i is given by $i = i_0 e^{-5t}$, where i_0 is the current for $t = 0$ and t is the time. Solve for t.

68. A state lottery pays \$500 for a \$1 ticket if a person picks the correct three-digit number determined by the random draw of three numbered balls. The probability of a 50% chance of winning in x drawings with a \$1 ticket is given by $(1 - 0.999^x) = 0.5$. For how many drawings does a person have to buy a ticket to have a 50% chance of winning?

69. An approximate formula for the population (in millions) of a city since 1970 is $P = 2.25e^{0.00486t}$, where t is the number of years since 1970. Sketch the graph of P vs. t for 1970 to 2000.

70. A computer analysis of the luminous efficiency E, in lumens per watt, of a tungsten lamp as a function of its input power P, in watts, is given by $E = 22.0(1 - 0.65e^{-0.008P})$. Sketch the graph of E as a function of P for $0 \leq P \leq 1000$ W.

71. An equation which may be used for the angular velocity ω of the slider mechanism in Fig. 13-17 is $2 \ln \omega = \ln 3g + \ln \sin \theta - \ln l$, where g is the acceleration due to gravity. Solve for $\sin \theta$.

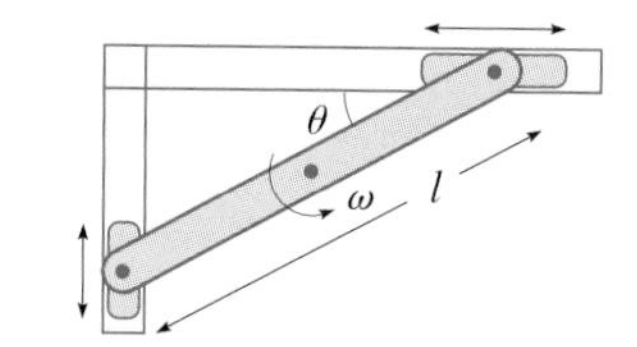

Fig. 13-17

72. Taking into account the weight loss of fuel, the maximum velocity v_m of a rocket is given by

$$v_m = u(\ln m_0 - \ln m_s) - gt_f$$

where m_0 is the initial mass of the rocket and fuel, m_s is the mass of the rocket shell, t_f is the time during which fuel is expended, u is the velocity of the expelled fuel, and g is the acceleration due to gravity. Solve for m_0.

73. An equation used to calculate the capacity C, in bits per second, of a telephone channel is $C = B \log_2(1 + R)$. Solve for R.

74. An equation used in studying the action of a protein molecule is $\ln A = \ln \theta - \ln(1 - \theta)$. Solve for θ.

75. The magnitudes (visual brightnesses), m_1 and m_2, of two stars are related to their (actual) brightnesses, b_1 and b_2, by the equation $m_1 - m_2 = 2.5 \log(b_2/b_1)$. As a result of this definition, magnitudes may be negative, and *magnitudes decrease as brightnesses increase.* The magnitude of the brightest star, Sirius, is -1.4, and the magnitudes of the faintest stars observable with the naked eye are about 6.0. How much brighter is Sirius than these faintest stars?

76. The power gain of an electronic device such as an amplifier is defined as $n = 10 \log(P_0/P_i)$, where n is measured in decibels, P_0 is the power output, and P_i is the power input. If $P_0 = 10.0$ W and $P_i = 0.125$ W, calculate the power gain. (See Example 5 of Section 13-6.)

77. In studying the frictional effects on a flywheel, the revolutions per minute R that it makes as a function of the time t, in minutes, is given by $R = 4520(0.750)^{2.50t}$. Find t for $R = 1950$ r/min.

78. The efficiency e of a gasoline engine as a function of its compression ratio r is given by $e = 1 - r^{1-\gamma}$, where γ is a constant. Find γ for $e = 0.55$ and $r = 7.5$.

79. For a particular solar energy system, the collector area A required to supply a fraction F of the total energy is given by $A = 480\,F^{2.2}$. Plot A, in m^2, as a function of F, from $F = 0.1$ to $F = 0.9$, on semilog paper.

80. The current I and resistance R were measured as follows in a certain microcomputer circuit.

R (Ω)	100	200	500	1000	2000	5000	10 000
I (μA)	81	41	16	8.2	4.0	1.6	0.8

Plot I as a function of R on log-log paper.

Writing Exercise

81. A machine design student noted that the edge of a robotic link was shaped like a logarithmic curve. Using a graphing calculator, the student viewed various logarithmic curves, including $y = \log x^2$ and $y = 2 \log x$, for which the student thought the graphs would be identical, but a difference was observed. Write a paragraph explaining what the difference is and why it occurs.

Practice Test

In Problems 1–4, determine the value of x.

1. $\log_9 x = -\frac{1}{2}$ **2.** $\log_3 x - \log_3 2 = 2$

3. $\log_x 64 = 3$ **4.** $3^{3x+1} = 8$

5. Graph the function $y = 2 \log_4 x$.

6. Graph the function $y = 2(3^x)$ on semilog paper.

7. Express $\log_5\left(\frac{4a^3}{7}\right)$ as a combination of a sum, difference, and multiple of logarithms, including $\log_5 2$.

8. Solve for y in terms of x: $3 \log_7 x - \log_7 y = 2$.

9. An equation used for a certain electric circuit is $\ln i - \ln I = -t/RC$. Solve for i.

10. Evaluate: $\dfrac{2 \ln 0.9523}{\log 6066}$.

11. Evaluate: $\log_5 732$.

12. If A_0 dollars are invested at 8%, compounded continuously for t years, the value A of the investment is given by $A = A_0 e^{0.08t}$. Determine how long it takes for the investment to double in value.

14

Additional Types of Equations and Systems of Equations

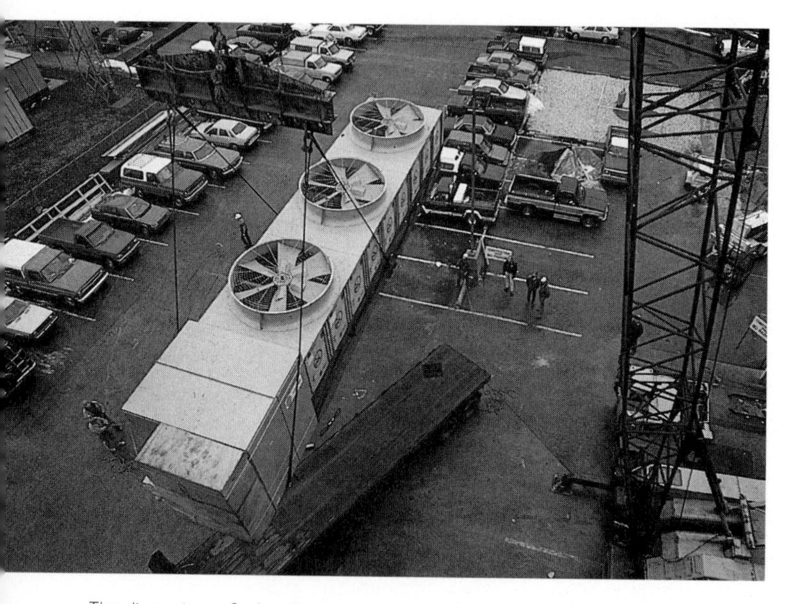

The dimensions of a heating system vent are important to its design. In Section 14-1 we discuss a problem in vent design.

In Chapter 3 we determined how to graph a function, as well as how to solve an equation graphically. Since then we have introduced a number of important types of functions and their graphs. We have also discussed the solutions of linear and quadratic equations as well as systems of linear equations.

In this chapter we discuss graphical and algebraic solutions of systems of equations. In these systems, at least one of the equations is not linear. We also consider solutions of two special types of equations. The first type can be solved by methods developed for quadratic equations, and the second type involves radicals.

Applications of these equations and systems of equations are found in many fields of science and technology. These include physics, electricity, business, and structural design.

14-1 Graphical Solution of Systems of Equations

In this section we first discuss the graphs of a special group of nonlinear equations. Then we will find graphical solutions of systems of equations that involve these equations and other nonlinear equations.

The equations we shall now consider have graphs that are known as the **conic sections.** These curves are the *circle, parabola, ellipse,* and *hyperbola.* We previously introduced the parabola when we discussed the quadratic function in Chapter 7. The following examples illustrate these curves. A more complete discussion of these equations is found in Chapter 21.

EXAMPLE 1 ▸▸ Graph the equation $y = 3x^2 - 6x$.

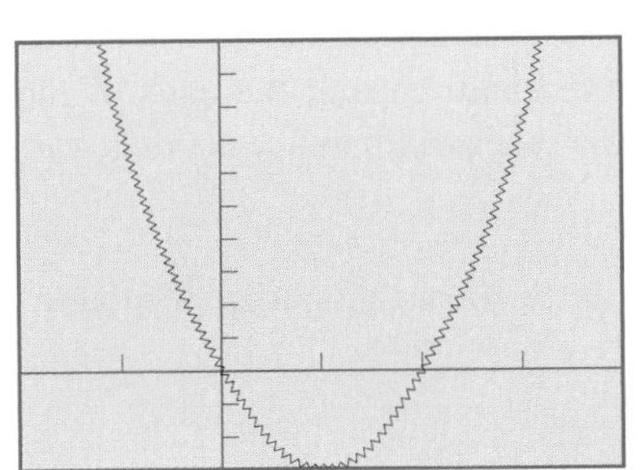

Fig. 14-1

We graphed equations of this form in Section 7-4. Since the general quadratic function is $y = ax^2 + bx + c$, for the function $y = 3x^2 - 6x$, we see that $a = 3$, $b = -6$, and $c = 0$. Therefore, $-b/(2a) = 1$, which tells us that the x-coordinate of the vertex is 1. For $x = 1$, $y = -3$, which means the vertex is $(1, -3)$. Also, we know that it is a minimum point since $a > 0$.

With this information, and the fact that the y-intercept of the curve is $(0, 0)$, we display the graph on a graphing calculator as shown in Fig. 14-1, using the RANGE values

$$\text{Xmin} = -2, \text{Xmax} = 4, \text{Ymin} = -3, \text{Ymax} = 10$$

As we showed in Section 7-4, the curve is a **parabola** and a parabola always results if the equation is of the form of the quadratic function $y = ax^2 + bx + c$. ◂◂

EXAMPLE 2 ▸▸ Plot the graph of the equation $x^2 + y^2 = 25$.

We first solve this equation for y, and we obtain $y = \sqrt{25 - x^2}$, or $y = -\sqrt{25 - x^2}$, which we write as $y = \pm\sqrt{25 - x^2}$. We now assume values for x and find the corresponding values for y.

x	0	±1	±2	±3	±4	±5
y	±5	±4.9	±4.6	±4	±3	0

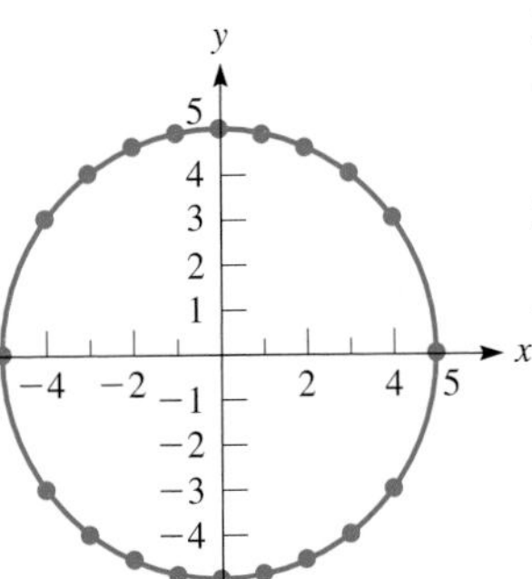

Fig. 14-2

If we try values greater than 5, we have imaginary numbers. These cannot be plotted, for we assume that both x and y are real. (The complex plane is only for *numbers* of the form $a + bj$ and does not represent pairs of numbers representing two variables.) When we give the value $x = \pm 4$ when $y = \pm 3$, this is simply a short way of representing 4 points. These points are $(4, 3)$, $(4, -3)$, $(-4, 3)$, $(-4, -3)$.

In Fig. 14-2 *the resulting curve is a* **circle.** A circle with its center at the origin results from an equation of the form $x^2 + y^2 = r^2$, where r is the radius. ◂◂

From the graph of the circle in Fig. 14-2, we see that *the equation of a circle does not represent a function*. There are *two* values of y for most of the values of x in the domain. We must take this into account when displaying the graph of such an equation on a graphing calculator. This is illustrated in the next example.

EXAMPLE 3 ▸▸ Display the graph of the equation $2x^2 + 5y^2 = 10$ on a graphing calculator.

First solving for y, we get $y = \pm\sqrt{\dfrac{10 - 2x^2}{5}}$.

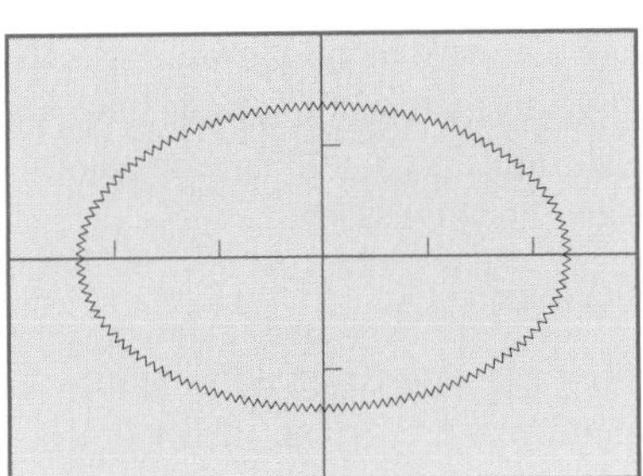

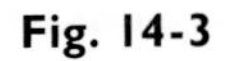

Fig. 14-3

To display this equation, *we must enter both functions* (one with the positive sign and the other with the negative sign).

Trying some RANGE values (or noting that the domain is from $-\sqrt{5}$ to $\sqrt{5}$, and the range is from $-\sqrt{2}$ to $\sqrt{2}$) we use

$$\text{Xmin} = -3, \text{Xmax} = 3, \text{Ymin} = -2, \text{Ymax} = 2$$

and get the view shown in Fig. 14-3.

The curve is an **ellipse.** An ellipse results from an equation of the form $ax^2 + by^2 = c$, where constants a, b, and c must have the same sign and $a \neq b$. ◂◂

NOTE ▶ *Particular care must be used with parentheses when entering the function in the graphing calculator.* In Example 3 the positive function would be entered as $\sqrt{((10 - 2x^2)/5)}$. Also, it is possible that there are small gaps at the ends of the curve. The appearance of such gaps depends on the pixel width and functional values used by the calculator.

EXAMPLE 4 ▶▶ Display the graph of $2x^2 - y^2 = 4$ on a graphing calculator.

Solving for y, we get

$$y = \pm\sqrt{2x^2 - 4}$$

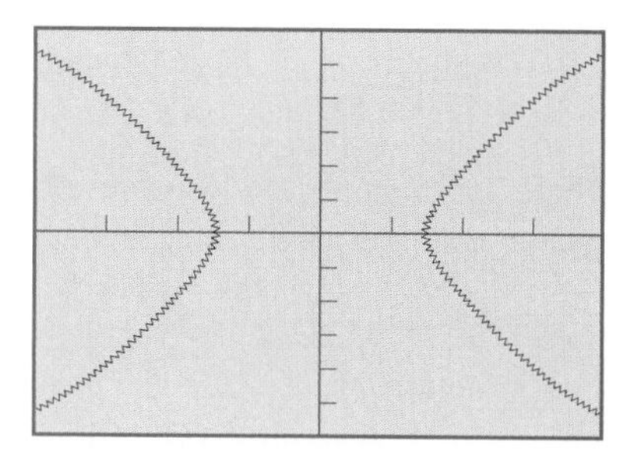

Fig. 14-4

We note that the values $-\sqrt{2} < x < \sqrt{2}$ are not in the domain of either $y = \sqrt{2x^2 - 4}$ or $y = -\sqrt{2x^2 - 4}$, since such values would lead to imaginary values of y. With the RANGE values

$$\text{Xmin} = -4, \text{Xmax} = 4, \text{Ymin} = -6, \text{Ymax} = 6$$

the curve is displayed in Fig. 14-4.

The curve is a **hyperbola,** which results from an equation of the form $ax^2 + by^2 = c$, if a and b have different signs. ◀◀

Solving Systems of Equations

As in solving systems of linear equations, we solve any system by finding the values of x and y that satisfy both equations at the same time. To solve a system graphically, we graph the equations and find the coordinates of all points of intersection. If the curves do not intersect, the system has no real solutions.

See the chapter introduction.

EXAMPLE 5 ▶▶ For proper ventilation, the vent for a hot-air heating system is to have a rectangular cross-sectional area of 2.3 m^2 and is to be made from sheet metal 6.4 m wide. Find the dimensions of this cross-sectional area of the vent.

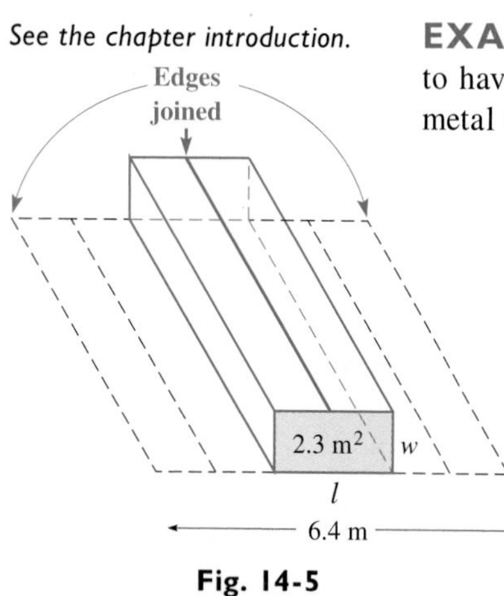

Fig. 14-5

In Fig. 14-5, we have let l = the length and w = the width of the area. Since the area is 2.3 m^2, we have $lw = 2.3$. Also, since the sheet metal is 6.4 m wide, this is the perimeter of the area. This gives us $2l + 2w = 6.4$, or $l + w = 3.2$. This means that the system of equations to be solved is

$$lw = 2.3$$
$$l + w = 3.2$$

Solving each equation for l, we have

$$l = 2.3/w \quad \text{and} \quad l = 3.2 - w$$

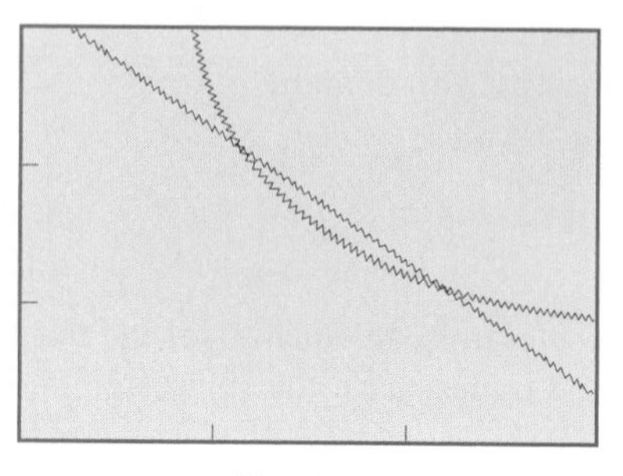

Fig. 14-6

We now display the graphs of these two equations on a graphing calculator, using x for w and y for l. Since negative values of l and w have no meaning to the solution, and since the straight line $l = 3.2 - w$ has intercepts of (3.2, 0) and (0, 3.2), we use the RANGE values

$$\text{Xmin} = 0, \text{Xmax} = 3, \text{Ymin} = 0, \text{Ymax} = 3$$

The graph is displayed in Fig. 14-6. Using the TRACE and ZOOM features, we find that the solutions are approximately (1.1, 2.1) and (2.1, 1.1). Using the length as the longer dimension, we have the solution $l = 2.1$ m and $w = 1.1$ m. We see that this solution checks with the statement of the problem. ◀◀

In Example 5 we graphed $xy = 4.5$. This is also a *hyperbola*, another form of which is $xy = c$. We now show the solutions to two more systems of equations.

EXAMPLE 6 ▸▸ Graphically solve the system of equations

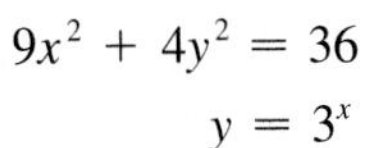

$$9x^2 + 4y^2 = 36$$
$$y = 3^x$$

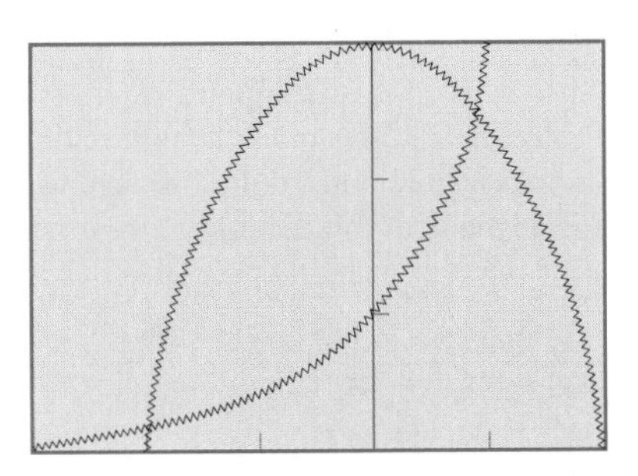

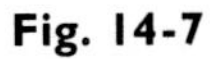

Fig. 14-7

The first equation is of the form represented by an ellipse, as shown in Example 3. The second equation is an exponential function, as discussed in Chapter 13. Solving the first equation for y, we have $y = \pm\frac{1}{2}\sqrt{36 - 9x^2}$. Since this is an ellipse, we note that its domain extends from $x = -2$ to $x = 2$ ($9x^2$ cannot be greater than 36). Also, the exponential curve cannot be negative and increases rapidly. Therefore, using the RANGE values

$$\text{Xmin} = -3,\ \text{Xmax} = 2,\ \text{Ymin} = 0,\ \text{Ymax} = 3$$

we display the graphs as shown in Fig. 14-7.

The points of intersection give the approximate solutions of $x = -2.0$, $y = 0.1$ and $x = 0.9$, $y = 2.7$. ◂◂

EXAMPLE 7 ▸▸ Graphically solve the system of equations

$$x^2 = 2y$$
$$3x - y = 5$$

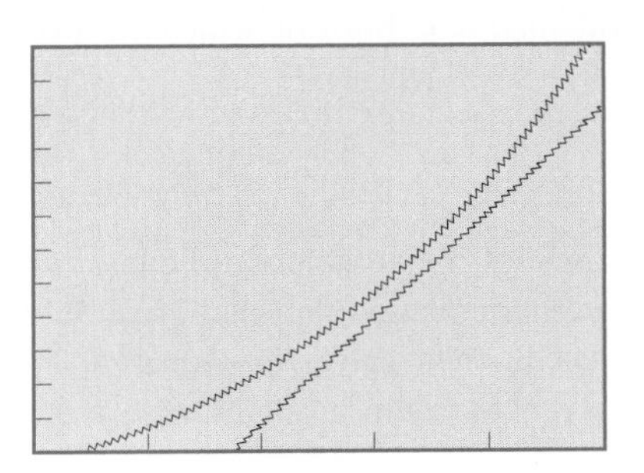

Fig. 14-8

We note that the two curves in this system are a parabola and a straight line. Solving the equation of the parabola for y, we get $y = \frac{1}{2}x^2$. This parabola has its vertex at the origin, and it opens upward since $a > 0$. This means that it is not possible to have a point of intersection for $y < 0$.

The straight line has intercepts of $(0, -5)$ and $(5/3, 0)$. This means that any possible point of intersection must be in the first quadrant. With these considerations, we use the RANGE values

$$\text{Xmin} = 0,\ \text{Xmax} = 5,\ \text{Ymin} = 0,\ \text{Ymax} = 12$$

The graphs are shown in Fig. 14-8. The curves do not intersect, which means that there are no real solutions to the system. ◂◂

Exercises 14–1

In Exercises 1–24, solve the given systems of equations graphically by using a graphing calculator. Find all values to at least the nearest 0.1.

1. $y = 2x$
$x^2 + y^2 = 16$

2. $3x - y = 4$
$y = 6 - 2x^2$

3. $x^2 + 2y^2 = 8$
$x - 2y = 4$

4. $y = 3x - 6$
$xy = 6$

5. $y = x^2 - 2$
$4y = 12x - 17$

6. $x^2 + 4y^2 = 4$
$2y = 12 - x$

7. $y = x^2$
$xy = 4$

8. $y = -2x^2$
$y = x^2 - 6$

9. $y = -x^2 + 4$
$x^2 + y^2 = 9$

10. $y = 2x^2 - 1$
$x^2 + 2y^2 = 16$

11. $x^2 - 4y^2 = 16$
$x^2 + y^2 = 1$

12. $y = 2x^2 - 4x$
$xy = -4$

13. $2x^2 + 3y^2 = 19$
$x^2 + y^2 = 9$

14. $x^2 - y^2 = 4$
$2x^2 + y^2 = 16$

15. $x^2 + y^2 = 1$
$xy = \frac{1}{2}$

16. $x^2 + y^2 = 25$
$x^2 - y^2 = 7$

17. $y = x^2$
$y = \sin x$

18. $y = 4x - x^2$
$y = 2\cos x$

19. $y = e^{-x}$
$x + y = 2$

20. $y = 2^x$
$x^2 + y^2 = 4$

21. $x^2 - y^2 = 1$
$y = \log_2 x$

22. $x^2 + 4y^2 = 16$
$y = 2 \ln x$

23. $y = \ln(x - 1)$
$y = \sin \frac{1}{2}x$

24. $y = \cos x$
$y = \log_3 x$

In Exercises 25–28, set up the indicated systems of equations and solve them graphically.

25. A helicopter is located 5.2 km north of east of a radio tower such that it is three times as far north as it is east of the tower. Find the northern and eastern components of the displacement from the tower.

26. A 4.60-m insulating strip is placed completely around a rectangular solar panel with an area of 1.20 m^2. What are the dimensions of the panel?

27. The power developed in an electric resistor is i^2R, where i is the current. If a first current passes through a 2.0-Ω resistor and a second current passes through a 3.0-Ω resistor, the total power produced is 12 W. If the resistors are reversed, the total power produced is 16 W. Find the currents ($i > 0$).

28. A circular hot tub is located on the square deck of a home. The side of the deck is 7.30 m more than the radius of the hot tub, and there are 72.5 m^2 of deck around the tub. Find the radius of the hot tub and the length of the side of the deck.

14-2 Algebraic Solution of Systems of Equations

Often the graphical method is the easiest way to solve a system of equations. With a graphing calculator it is possible to find the result with good accuracy. However, the graphical method does not usually give the *exact* answer. Using algebraic methods to find exact solutions for some systems of equations is either not possible or quite involved. There are systems, however, for which there are relatively simple algebraic solutions. In this section we consider two useful methods, both of which we discussed before when we were studying systems of linear equations.

Solution by Substitution

The first method is *substitution*. If we can solve one of the equations for one of its variables, we can substitute this solution into the other equation. We then have only one unknown in the resulting equation, and we can then solve this equation by methods discussed in earlier chapters.

EXAMPLE 1 ▸▸ By substitution, solve the system of equations

$$2x - y = 4$$
$$x^2 - y^2 = 4$$

We solve the first equation for y, obtaining $y = 2x - 4$. We now substitute $2x - 4$ for y in the second equation, getting

in second equation, y replaced by $2x - 4$

$$x^2 - (2x - 4)^2 = 4$$

When simplified, this gives a quadratic equation.

$$x^2 - (4x^2 - 16x + 16) = 4$$
$$-3x^2 + 16x - 20 = 0$$
$$x = \frac{-16 \pm \sqrt{256 - 4(-3)(-20)}}{-6} = \frac{-16 \pm \sqrt{16}}{-6} = \frac{-16 \pm 4}{-6} = \frac{10}{3}, 2$$

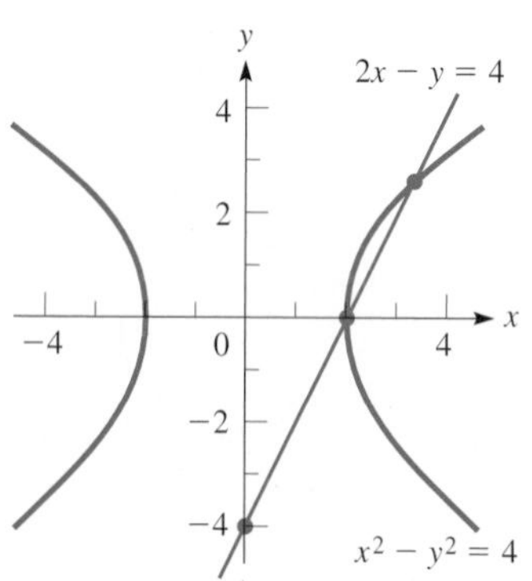

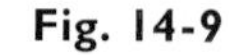
Fig. 14-9

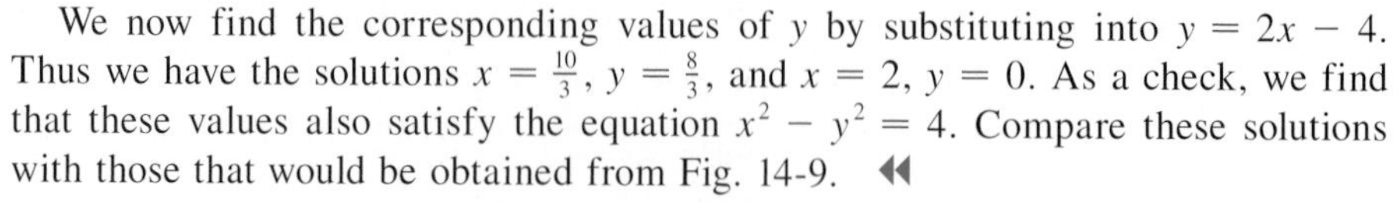
We now find the corresponding values of y by substituting into $y = 2x - 4$. Thus we have the solutions $x = \frac{10}{3}$, $y = \frac{8}{3}$, and $x = 2$, $y = 0$. As a check, we find that these values also satisfy the equation $x^2 - y^2 = 4$. Compare these solutions with those that would be obtained from Fig. 14-9. ◂◂

EXAMPLE 2 ▸▸ By substitution, solve the system of equations

$$xy = -2$$
$$2x + y = 2$$

From the first equation we have $y = -2/x$. Substituting this into the second equation, we have

in second equation, y replaced by $-\frac{2}{x}$

$$2x + \left(-\frac{2}{x}\right) = 2$$
$$2x^2 - 2 = 2x$$
$$x^2 - x - 1 = 0$$
$$x = \frac{1 \pm \sqrt{1 + 4}}{2} = \frac{1 \pm \sqrt{5}}{2}$$

By substituting these values for x into either of the original equations, we find the corresponding values of y, and we have the solutions

$$x = \frac{1 + \sqrt{5}}{2},\ y = 1 - \sqrt{5} \quad \text{and} \quad x = \frac{1 - \sqrt{5}}{2},\ y = 1 + \sqrt{5}$$

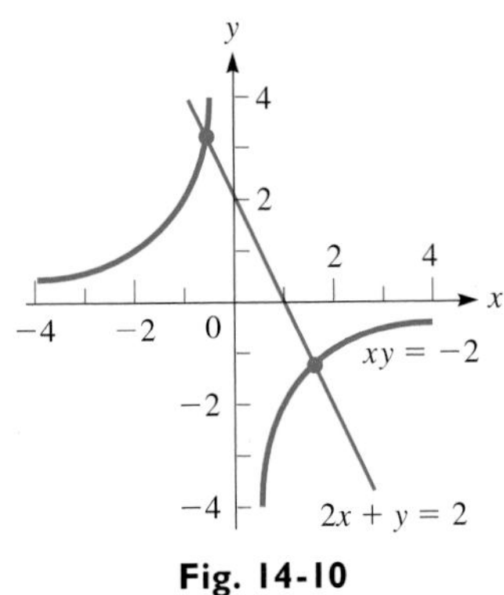

Fig. 14-10

These can be checked by substituting in the original equations. In decimal form they are

$$x \approx 1.618,\ y \approx -1.236 \quad \text{and} \quad x \approx -0.618,\ y \approx 3.236$$

The graphical solutions are shown in Fig. 14-10.

The solutions can also be found by first solving the second equation for y, or either equation for x, and then substituting in the other equation. ◂◂

Solution by Addition or Subtraction

The other algebraic method is that of elimination by *addition or subtraction*. This method is most useful if both equations have only squared terms and constants.

EXAMPLE 3 ▸▸ By addition or subtraction, solve the system of equations

$$2x^2 + y^2 = 9$$
$$x^2 - y^2 = 3$$

We note that if we add the corresponding sides of each equation, y^2 is eliminated. This leads to the solution

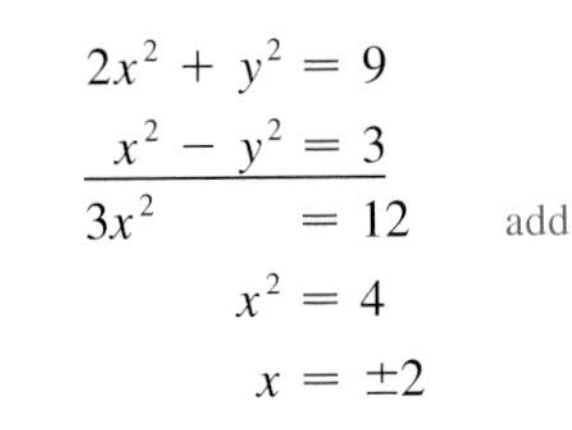

$$\begin{aligned} 2x^2 + y^2 &= 9 \\ x^2 - y^2 &= 3 \\ \hline 3x^2 \qquad &= 12 \quad \text{add} \\ x^2 &= 4 \\ x &= \pm 2 \end{aligned}$$

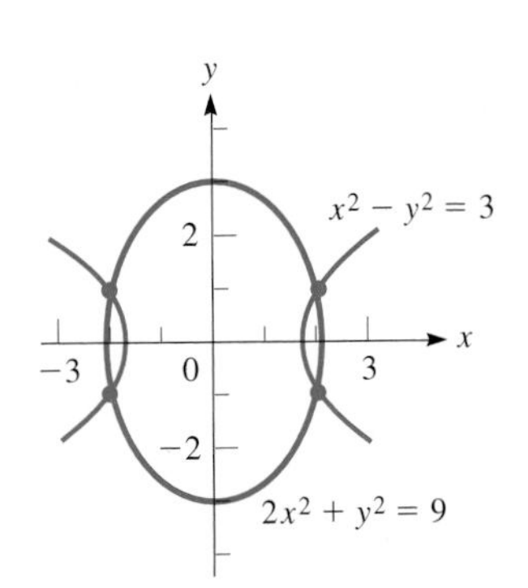

Fig. 14-11

For $x = 2$, we have two corresponding y-values, $y = \pm 1$. Also, for $x = -2$, we have two corresponding y-values, $y = \pm 1$. Thus we have four solutions:

$$x = 2,\ y = 1; \quad x = 2,\ y = -1; \quad x = -2,\ y = 1; \quad \text{and} \quad x = -2,\ y = -1$$

Each of these solutions checks in the original equations. The graphical solutions are shown in Fig. 14-11. ◂◂

EXAMPLE 4 ⏩ By addition or subtraction, solve the system of equations

$$3x^2 - 2y^2 = 5$$
$$x^2 + y^2 = 5$$

If we multiply the second equation by 2 and then add the two resulting equations, we get

$$\begin{aligned} 3x^2 - 2y^2 &= 5 \\ 2x^2 + 2y^2 &= 10 \quad \text{each term of second equation multiplied by 2} \\ \hline 5x^2 \qquad &= 15 \quad \text{add} \\ x^2 &= 3 \qquad x = \pm\sqrt{3} \end{aligned}$$

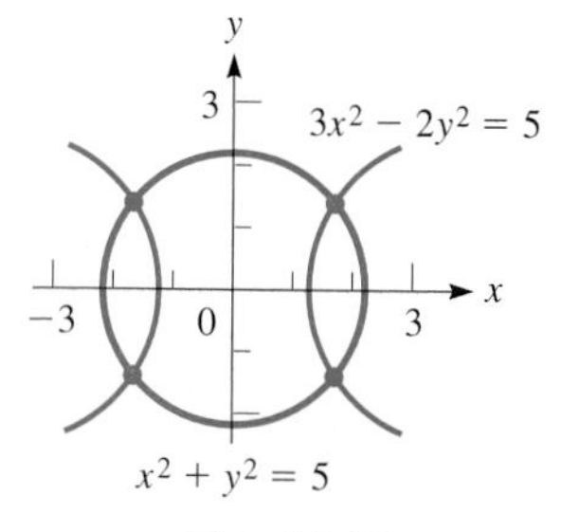

Fig. 14-12

The corresponding values of y for each value of x are $y = \pm\sqrt{2}$. Again we have four solutions:

$$x = \sqrt{3},\ y = \sqrt{2}; \qquad x = \sqrt{3},\ y = -\sqrt{2};$$
$$x = -\sqrt{3},\ y = \sqrt{2}; \quad x = -\sqrt{3},\ y = -\sqrt{2}$$

Each of these solutions checks when substituted in the original equations. The graphical solutions are shown in Fig. 14-12. ⏪

EXAMPLE 5 ⏩ A certain number of machine parts cost \$1000. If they cost \$5 less per part, 10 additional parts could be purchased for the same amount of money. What is the cost of each part?

Since the cost of each part is required, we let $c =$ the cost per part. Also, we let $n =$ the number of parts. From the first statement of the problem, we see that $cn = 1000$. Also, from the second statement, we have $(c - 5)(n + 10) = 1000$. Therefore, we are to solve the system of equations

$$\begin{aligned} cn &= 1000 \\ (c - 5)(n + 10) &= 1000 \end{aligned}$$

Solving the first equation for n, and multiplying out the second equation, we have $n = \dfrac{1000}{c}$ and $cn + 10c - 5n - 50 = 1000$. Now, substituting the expression for n into the second equation, we solve for c.

$$\begin{aligned} c\left(\frac{1000}{c}\right) + 10c - 5\left(\frac{1000}{c}\right) - 50 &= 1000 \\ 1000 + 10c - \frac{5000}{c} - 50 &= 1000 \\ 10c - \frac{5000}{c} - 50 &= 0 \\ c^2 - 5c - 500 &= 0 \\ (c + 20)(c - 25) &= 0 \\ c &= -20, 25 \end{aligned}$$

Since a negative answer has no significance in this particular situation, we see that the solution is $c =$ \$25 per part. Checking with the original statement of the problem, we see that this is correct. ⏪

Exercises 14–2

In Exercises 1–24, solve the given systems of equations algebraically.

1. $y = x + 1$
   $y = x^2 + 1$

2. $y = 2x - 1$
   $y = 2x^2 + 2x - 3$

3. $x + 2y = 3$
   $x^2 + y^2 = 26$

4. $y = x + 1$
   $x^2 + y^2 = 25$

5. $x + y = 1$
   $x^2 - y^2 = 1$

6. $x + y = 2$
   $2x^2 - y^2 = 1$

7. $2x - y = 2$
   $2x^2 + 3y^2 = 4$

8. $6y - x = 6$
   $x^2 + 3y^2 = 36$

9. $xy = 1$
   $x + y = 2$

10. $xy = 2$
    $x + y = 3$

11. $xy = 3$
    $3x - 2y = -7$

12. $xy = -4$
    $2x + y = -2$

13. $y = x^2$
    $y = 3x^2 - 8$

14. $y = x^2 - 1$
    $2x^2 - y^2 = 2$

15. $x^2 - y = -1$
    $x^2 + y^2 = 5$

16. $x^2 + y = 5$
    $x^2 + y^2 = 25$

17. $x^2 - 1 = y$
    $x^2 - 2y^2 = 1$

18. $2y^2 - 4x = 7$
    $y^2 + 2x^2 = 3$

19. $x^2 + y^2 = 25$
    $x^2 - 2y^2 = 7$

20. $3x^2 - y^2 = 4$
    $x^2 + 4y^2 = 10$

21. $y^2 - 2x^2 = 6$
    $5x^2 + 3y^2 = 20$

22. $y^2 - 2x^2 = 17$
    $2y^2 + x^2 = 54$

23. $x^2 + 3y^2 = 37$
    $2x^2 - 9y^2 = 14$

24. $5x^2 - 4y^2 = 15$
    $3y^2 + 4x^2 = 12$

In Exercises 25–32, solve the indicated systems of equations algebraically. In Exercises 27–32, it is necessary to set up the systems of equations properly.

25. A 2-kg block collides with an 8-kg block. Using the physical laws of conservation of energy and conservation of momentum, along with given conditions, the following equations involving the velocities are established.

    $$v_1^2 + 4v_2^2 = 41$$
    $$2v_1 + 8v_2 = 12$$

    Find these velocities (in m/s) if $v_2 > 0$.

26. A rocket is fired from behind a ship and follows the path given by $h = 3x - 0.05x^2$, where h is its altitude, x is the horizontal distance traveled, and distances are in kilometres. A missile fired from the ship follows the path given by $h = 0.8x - 15$. For $h > 0$ and $x > 0$, determine where the paths of the rocket and missile cross.

27. A rectangular computer chip has a surface area of 2.1 cm² and a perimeter of 5.8 cm. Find the length and the width of the chip.

28. The impedance Z in an alternating-current circuit is 2.00 Ω. If the resistance R is numerically equal to the square of the reactance X, find R and X. See Section 12-7.

29. A roof truss is in the shape of a right triangle. If there are 4.60 m of lumber in the truss, and the longest side is 2.20 m long, what are the lengths of the other two sides of the truss?

30. In a certain roller mechanism, the radius of one steel ball is 2.00 cm greater than the radius of a second steel ball. If the difference in their masses is 7100 g, find the radii of the balls. The density of steel is 7.70 g/cm³.

31. A plane travels at 400 km/h relative to the air. It takes the plane 2.5 h longer to travel 2400 km against a certain wind than it does with the wind. Find the velocity of the wind.

32. In a marketing survey, a company found that the total gross income for selling t tables at a price of p dollars each was $35 000. It then increased the price of each table by $100 and found that the total income was only $27 000 because 40 fewer tables were sold. Find p and t.

14-3 Equations in Quadratic Form

Often we encounter equations that can be solved by methods applicable to quadratic equations, even though these equations are not actually quadratic. They do have the property, however, that *with a proper substitution* ***they may be written in the form of a quadratic equation.*** All that is necessary is that the equation have terms including some variable quantity, its square, and perhaps a constant term. The following example illustrates these types of equations.

NOTE ▶

EXAMPLE 1 ▸▸ (a) The equation $x - 2\sqrt{x} - 5 = 0$ is an equation in quadratic form, because if we let $y = \sqrt{x}$, we have $x = (\sqrt{x})^2 = y^2$, and the resulting equation is $y^2 - 2y - 5 = 0$.

(b) $t^{-4} - 5t^{-2} + 3 = 0$

By letting $y = t^{-2}$, we have $y^2 - 5y + 3 = 0$. (y^2 from $(t^{-2})^2$)

(c) $t^3 - 3t^{3/2} - 7 = 0$

By letting $y = t^{3/2}$, we have $y^2 - 3y - 7 = 0$. (y^2 from $(t^{3/2})^2$)

(d) $(x + 1)^4 - (x + 1)^2 - 1 = 0$

By letting $y = (x + 1)^2$, we have $y^2 - y - 1 = 0$. (y^2 from $[(x + 1)^2]^2$)

(e) $x^{10} - 2x^5 + 1 = 0$

By letting $y = x^5$, we have $y^2 - 2y + 1 = 0$. ◂◂ (y^2 from $(x^5)^2$)

The following examples illustrate the method of solving equations in quadratic form.

EXAMPLE 2 ▸▸ Solve the equation $x^4 - 5x^2 + 4 = 0$.

We first let $y = x^2$ and obtain the resulting equivalent equation

$$y^2 - 5y + 4 = 0$$

This may be factored and solved as

$$(y - 4)(y - 1) = 0$$
$$y = 4 \quad \text{or} \quad y = 1$$

Since we want values for x, and since $y = x^2$, we have

$$x^2 = 4 \quad \text{or} \quad x^2 = 1$$
$$x = \pm 2 \quad \text{or} \quad x = \pm 1$$

Substitution in the original equation shows each value to be a solution. ◂◂

EXAMPLE 3 ▸▸ Solve the equation $2x^4 + 7x^2 = 4$.

As in Example 2, we let $y = x^2$ to write the resulting equation in quadratic form.

$$2y^2 + 7y - 4 = 0$$
$$(2y - 1)(y + 4) = 0$$
$$y = \tfrac{1}{2} \quad \text{or} \quad y = -4$$
$$x^2 = \tfrac{1}{2} \quad \text{or} \quad x^2 = -4 \qquad y = x^2$$
$$x = \pm\frac{1}{\sqrt{2}} \quad \text{or} \quad x = \pm 2j$$

Substitution in the original equation shows each value to be a solution. ◂◂

Two of the solutions in Example 3 are complex numbers. We were able to find these solutions directly from the definition of the square root of a negative number. In some cases (see Exercises 23 and 24 of this section), it is necessary to use the method of Section 12-6 to find such complex number solutions.

EXAMPLE 4 ▸▸ Solve the equation $x - \sqrt{x} - 2 = 0$.

By letting $y = \sqrt{x}$, we have

$$y^2 - y - 2 = 0$$
$$(y - 2)(y + 1) = 0$$
$$y = 2 \quad \text{or} \quad y = -1$$

Since $y = \sqrt{x}$, we note that y cannot be negative, and this means $y = -1$ cannot lead to a solution. For $y = 2$, we have $x = 4$. Checking, we find that $x = 4$ satisfies the original equation. Therefore, the only solution is $x = 4$. ◂◂

Extraneous Roots

Example 4 illustrates a very important point. *Whenever an operation involving the unknown is performed on an equation, this operation may introduce roots into a subsequent equation that are not roots of the original equation. Therefore, we must* ***check all answers in the original equation.*** Only operations involving constants—that is, adding, subtracting, multiplying by, or dividing by constants—are certain not to introduce the **extraneous roots.** We first encountered the concept of an extraneous root in Section 6-8, when we discussed equations involving fractions.

CAUTION ▸

EXAMPLE 5 ▸▸ Solve the equation $x^{-2} + 3x^{-1} + 1 = 0$.

By substituting $y = x^{-1}$, we have $y^2 + 3y + 1 = 0$. To solve this equation we may use the quadratic formula:

$$y = \frac{-3 \pm \sqrt{9 - 4}}{2} = \frac{-3 \pm \sqrt{5}}{2}$$

Since $x = 1/y$, we have

$$x = \frac{2}{-3 + \sqrt{5}} \quad \text{or} \quad x = \frac{2}{-3 - \sqrt{5}}$$

These answers in decimal form are

$$x \approx -2.618 \quad \text{or} \quad x \approx -0.382$$

These results check when substituted in the original equation. In checking these decimal answers, it is more accurate to use the calculator values, before rounding them off. This can be done by storing the calculator values in memory. ◂◂

EXAMPLE 6 ▸▸ Solve the equation $(x^2 - x)^2 - 8(x^2 - x) + 12 = 0$.

By substituting $y = x^2 - x$, we have

$$y^2 - 8y + 12 = 0$$
$$(y - 2)(y - 6) = 0$$
$$y = 2 \quad \text{or} \quad y = 6$$

This means that

$$x^2 - x = 2 \quad \text{or} \quad x^2 - x = 6$$

Solving each of these equations, we have

$$x^2 - x - 2 = 0 \qquad x^2 - x - 6 = 0$$
$$(x - 2)(x + 1) = 0 \qquad (x - 3)(x + 2) = 0$$
$$x = 2 \quad \text{or} \quad x = -1 \qquad x = 3 \quad \text{or} \quad x = -2$$

Each of these values checks when substituted in the original equation. ◂◂

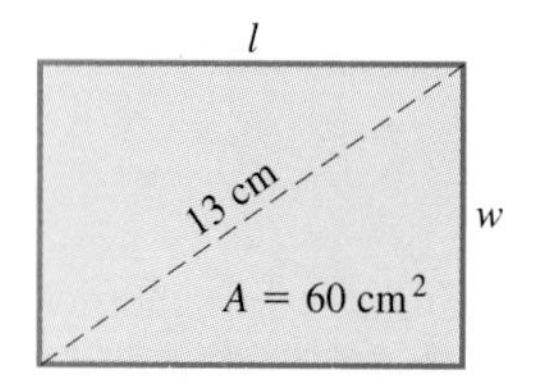

Fig. 14-13

EXAMPLE 7 ▸▸ A rectangular solar cell has an area of 60 cm^2. The diagonal of the cell is 13 cm. Find the length and width of the cell. See Fig. 14-13.

Since the required quantities are the length and width, let l = the length of the cell and w = the width of the cell. Since the area is 60 cm^2, $lw = 60$. Also, using the Pythagorean theorem and the fact that the diagonal is 13 cm, we have $l^2 + w^2 = 169$. Therefore, we are to solve the system of equations

$$lw = 60, \quad l^2 + w^2 = 169$$

Solving the first equation for l, we have $l = 60/w$. Substituting this expression into the second equation, we have

$$\left(\frac{60}{w}\right)^2 + w^2 = 169$$

$$\frac{3600}{w^2} + w^2 = 169$$

$$3600 + w^4 = 169w^2$$

$$w^4 - 169w^2 + 3600 = 0$$

Let $x = w^2$.

$$x^2 - 169x + 3600 = 0$$

$$(x - 144)(x - 25) = 0$$

$$x = 144 \quad \text{or} \quad x = 25$$

Therefore, $w^2 = 144$ or $w^2 = 25$.

Solving for w, we get $w = \pm 12$ or $w = \pm 5$. Only the positive values of w are meaningful in this problem. Therefore, if $w = 5$ cm, then $l = 12$ cm. Normally, we designate the length as the longer dimension, although by letting $w = 12$ we get $l = 5$ cm. Checking with the statement of the problem, we see that these dimensions for the solar cell give an area of 60 cm^2 and a diagonal of 13 cm. ◂◂

Exercises 14–3

In Exercises 1–24, solve the given equations algebraically.

1. $x^4 - 13x^2 + 36 = 0$
2. $x^4 - 20x^2 + 64 = 0$
3. $4x^4 - 5x^2 = 9$
4. $4x^4 + 15x^2 = 4$
5. $x^{-2} - 2x^{-1} - 8 = 0$
6. $10x^{-2} + 3x^{-1} - 1 = 0$
7. $x^{-4} + 2x^{-2} = 24$
8. $x^{-4} + 1 = 2x^{-2}$
9. $2x - 7\sqrt{x} + 5 = 0$
10. $4x + 3\sqrt{x} = 1$
11. $3\sqrt[3]{x} - 5\sqrt[6]{x} + 2 = 0$
12. $\sqrt{x} + 3\sqrt[4]{x} = 28$
13. $x^{2/3} - 2x^{1/3} - 15 = 0$
14. $x^3 + 2x^{3/2} - 80 = 0$
15. $x^{1/2} + x^{1/4} = 20$
16. $4x^{4/3} + 9 = 13x^{2/3}$
17. $(x - 1) - \sqrt{x - 1} - 2 = 0$
18. $(x + 1)^{-2/3} + 5(x + 1)^{-1/3} - 6 = 0$
19. $(x^2 - 2x)^2 - 11(x^2 - 2x) + 24 = 0$
20. $3(x^2 + 3x)^2 - 2(x^2 + 3x) - 5 = 0$
21. $x - 3\sqrt{x - 2} = 6$ (Let $y = \sqrt{x - 2}$.)
22. $(x^2 - 1)^2 + (x^2 - 1)^{-2} = 2$
23. $x^6 + 7x^3 - 8 = 0$
24. $x^6 - 19x^3 = 216$

In Exercises 25–28, solve the given problems algebraically.

25. The equivalent resistance R_T of two resistors R_1 and R_2 in parallel is given by $R_T^{-1} = R_1^{-1} + R_2^{-1}$. If $R_T = 1.00\ \Omega$ and $R_2 = \sqrt{R_1}$, find R_1 and R_2.
26. An equation used in the study of the dispersion of light is $\mu = A + B\lambda^{-2} + C\lambda^{-4}$. Solve for λ.
27. A rectangular TV screen has an area of 2240 cm^2 and a diagonal of 68.6 cm. Find the dimensions of the screen.
28. A metal plate is in the shape of an isosceles triangle. The length of the base equals the square root of one of the equal sides. Determine the lengths of the sides if the perimeter of the plate is 55 cm.

14-4 Equations with Radicals

Equations with radicals in them are normally solved by squaring both sides of the equation if the radical represents a square root or by a similar operation for the other roots. However, when we do this, we often introduce *extraneous roots.* Thus, it is very important that all solutions be checked in the original equation.

EXAMPLE 1 ▸▸ Solve the equation $\sqrt{x-4} = 2$.

By squaring both sides of the equation, we have

$$(\sqrt{x-4})^2 = 2^2$$
$$x - 4 = 4$$
$$x = 8$$

This solution checks when put into the original equation. ◂◂

EXAMPLE 2 ▸▸ Solve the equation $2\sqrt{3x-1} = 3x$.

Squaring both sides of the equation gives us

$$(2\sqrt{3x-1})^2 = (3x)^2 \longleftarrow \text{don't forget to square the 2}$$
$$4(3x-1) = 9x^2$$
$$12x - 4 = 9x^2$$
$$9x^2 - 12x + 4 = 0$$
$$(3x-2)^2 = 0$$
$$x = \frac{2}{3} \quad \text{(double root)}$$

Checking this solution in the original equation, we have

$$2\sqrt{3(\tfrac{2}{3}) - 1} \stackrel{?}{=} 3(\tfrac{2}{3}), \qquad 2\sqrt{2-1} \stackrel{?}{=} 2, \qquad 2 = 2$$

Therefore, the solution $x = \frac{2}{3}$ checks. ◂◂

EXAMPLE 3 ▸▸ Solve the equation $\sqrt[3]{x-8} = 2$.

Cubing both sides of the equation, we have

$$x - 8 = 8$$
$$x = 16$$

Checking this solution in the original equation, we get

$$\sqrt[3]{16-8} \stackrel{?}{=} 2, \quad \text{or} \quad 2 = 2$$

Therefore, the solution checks. ◂◂

In the above examples, the radical is on the left side of the equation and the other terms of the equation are on the right side. However, if one side of the equation contains a radical as well as other terms, we *first isolate the radical.* That is, we rewrite the equation with the radical on one side and collect all other terms on the other side. The examples on the following page illustrate the use of this procedure.

EXAMPLE 4 ▸▸ Solve the equation $\sqrt{x-1} + 3 = x$.

We first isolate the radical by subtracting 3 from each side. This gives us

$$\sqrt{x-1} = x - 3$$

We now square both sides and proceed with the solution.

CAUTION ▶

$$(\sqrt{x-1})^2 = (x-3)^2$$

$$x - 1 = x^2 - 6x + 9 \quad \text{square the expression on each side, not just the terms separately}$$

$$x^2 - 7x + 10 = 0$$

$$(x-5)(x-2) = 0$$

$$x = 5 \quad \text{or} \quad x = 2$$

The solution $x = 5$ checks, but the solution $x = 2$ gives $4 = 2$. Thus the solution is $x = 5$. The value $x = 2$ is an extraneous root. ◂◂

EXAMPLE 5 ▸▸ Solve the equation $\sqrt{x+1} + \sqrt{x-4} = 5$.

This is most easily solved by first isolating one of the radicals by placing the other radical on the right side of the equation. We then square both sides of the resulting equation.

CAUTION ▶

$$\sqrt{x+1} = 5 - \sqrt{x-4} \quad \longleftarrow \text{two terms}$$

$$(\sqrt{x+1})^2 = (5 - \sqrt{x-4})^2$$

$$x + 1 = 25 - 10\sqrt{x-4} + (\sqrt{x-4})^2 \quad \text{be careful!}$$

$$= 25 - 10\sqrt{x-4} + x - 4$$

Now, isolating the radical on one side of the equation and squaring again, we have

$$10\sqrt{x-4} = 20$$

$$\sqrt{x-4} = 2 \quad \text{divide by 10}$$

$$x - 4 = 4 \quad \text{square both sides}$$

$$x = 8$$

This solution checks.

We note again that in squaring $5 - \sqrt{x-4}$, we do not simply square 5 and $\sqrt{x-4}$. This is similar to $(5a-2)^2$ in Example 2 of Section 6-1. ◂◂

EXAMPLE 6 ▸▸ Solve the equation $\sqrt{x} - \sqrt[4]{x} = 2$.

We can solve this most easily by handling it as an equation in quadratic form. By letting $y = \sqrt[4]{x}$, we have

$$y^2 - y - 2 = 0$$

$$(y-2)(y+1) = 0$$

$$y = 2 \quad \text{or} \quad y = -1$$

Since $y = \sqrt[4]{x}$, we know that a negative value of y does not lead to a solution of the original equation. Therefore, we see that $y = -1$ cannot give us a solution. For the other value, $y = 2$, we have

$$\sqrt[4]{x} = 2 \quad \text{or} \quad x = 16$$

This checks, because $\sqrt{16} - \sqrt[4]{16} = 4 - 2 = 2$. Therefore, the only solution of the original equation is $x = 16$. ◂◂

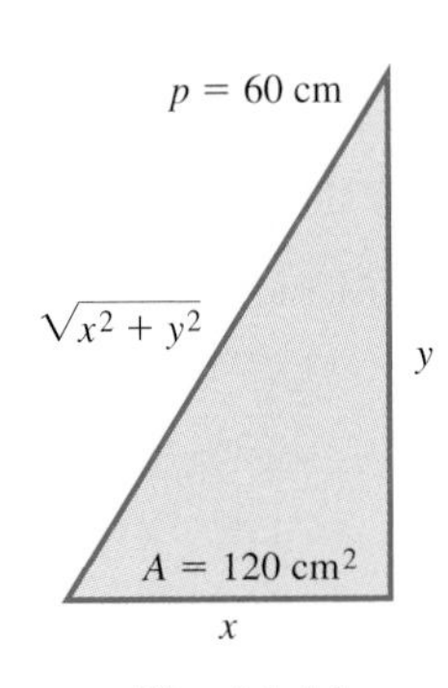

Fig. 14-14

EXAMPLE 7 ▸▸ A piece of sheet metal is cut into the shape of a right triangle. Its perimeter is 60 cm, and its area is 120 cm^2. Find the lengths of the three sides.

If we let the two legs of the triangle be x and y, as shown in Fig. 14-14, from the formulas for the perimeter p and the area A of a triangle, we have

$$p = x + y + \sqrt{x^2 + y^2} \quad \text{and} \quad A = \tfrac{1}{2}xy$$

where the hypotenuse was found by use of the Pythagorean theorem. Using the information given in the statement of the problem, we arrive at the equations

$$x + y + \sqrt{x^2 + y^2} = 60 \quad \text{and} \quad xy = 240$$

Isolating the radical in the first equation and then squaring both sides, we have

$$\begin{aligned} \sqrt{x^2 + y^2} &= 60 - x - y \\ x^2 + y^2 &= 3600 - 120x - 120y + x^2 + 2xy + y^2 \\ 0 &= 3600 - 120x - 120y + 2xy \end{aligned}$$

Solving the second of the original equations for y, we have $y = 240/x$. Substituting, we have

$$\begin{aligned} 0 &= 3600 - 120x - 120\left(\frac{240}{x}\right) + 2x\left(\frac{240}{x}\right) \\ 0 &= 3600x - 120x^2 - 120(240) + 480x && \text{multiply by } x \\ 0 &= 30x - x^2 - 240 + 4x && \text{divide by 120} \\ x^2 - 34x + 240 &= 0 && \text{collect terms on left} \\ (x - 10)(x - 24) &= 0 \\ x = 10 \text{ cm} \quad &\text{or} \quad x = 24 \text{ cm} \end{aligned}$$

If $x = 10$ cm, then $y = 24$ cm, or if $x = 24$ cm, then $y = 10$ cm. Thus the legs are 10 cm and 24 cm, and the hypotenuse is 26 cm. For these sides, $p = 60$ cm and $A = 120$ cm^2, which checks with the statement of the problem. ◂◂

Exercises 14–4

In Exercises 1–32, solve the given equations. In Exercises 17 and 20, explain how the extraneous root is introduced.

1. $\sqrt{x - 8} = 2$
2. $\sqrt{x + 4} = 3$
3. $\sqrt{8 - 2x} = x$
4. $\sqrt{3x + 4} = x$
5. $\sqrt{3x + 2} = 3x$
6. $\sqrt{2x + 6} = 2x$
7. $\sqrt{x - 2} + 3 = x$
8. $\sqrt{5x - 1} + 3 = x$
9. $2\sqrt{3 - x} - x = 5$
10. $x - 3\sqrt{2x + 1} = -5$
11. $\sqrt[3]{y - 5} = 3$
12. $\sqrt[4]{5 - x} = 2$
13. $\sqrt{x} + 12 = x$
14. $\sqrt{x} + 3 = 4x$
15. $5\sqrt{x} + 3 = 2x$
16. $4\sqrt{x} = x + 3$
17. $\sqrt{x + 4} + 8 = x$
18. $\sqrt{x + 15} + 5 = x$
19. $\sqrt{5 + \sqrt{x}} = \sqrt{x} - 1$
20. $\sqrt{13 + \sqrt{x}} = \sqrt{x} + 1$
21. $3\sqrt{1 - 2t} + 1 = 2t$
22. $1 - 2\sqrt{y + 4} = y$
23. $2\sqrt{x + 2} - \sqrt{3x + 4} = 1$
24. $\sqrt{x - 1} + \sqrt{x + 2} = 3$
25. $\sqrt{5x + 1} - 1 = 3\sqrt{x}$
26. $\sqrt{2x + 1} + \sqrt{3x} = 11$
27. $\sqrt{2x - 1} - \sqrt{x + 11} = -1$
28. $\sqrt{5x - 4} - \sqrt{x} = 2$
29. $\sqrt[3]{2x - 1} = \sqrt[3]{x + 5}$
30. $\sqrt[4]{x + 10} = \sqrt{x - 2}$
31. $\sqrt{x - 2} = \sqrt[4]{x - 2} + 12$
32. $\sqrt{3x + \sqrt{3x + 4}} = 4$

In Exercises 33–36, solve for the indicated letter.

33. The resonant frequency f in an electric circuit with an inductance L and a capacitance C is given by $f = \dfrac{1}{2\pi\sqrt{LC}}$. Solve for L.
34. A formula used in calculating the range R for radio communication is $R = \sqrt{2rh + h^2}$. Solve for h.

35. An equation used in analyzing a certain type of concrete beam is $k = \sqrt{2np + (np)^2} - np$. Solve for p.

36. In the study of spur gears in contact, the equation $kC = \sqrt{R_1^2 - R_2^2} + \sqrt{r_1^2 - r_2^2} - A$ is used. Solve for r_1^2.

In Exercises 37–40, set up the proper equations and solve them.

37. A freighter is 5.2 km farther from a Coast Guard station on a straight coast than from the closest point A on the coast. If the station is 8.3 km from A, how far is it from the freighter?

38. The velocity v of an object that falls through a distance h is given by $v = \sqrt{2gh}$, where g is the acceleration due to gravity. Two objects are dropped from heights that differ by 10.0 m such that the sum of their velocities when they strike the ground is 20.0 m/s. Find the heights from which they are dropped if $g = 9.80 \text{ m/s}^2$.

39. An island is 3.0 km offshore from the nearest point P on a straight beach. A person in a motorboat travels straight from the island to a point on the beach x km from P and then travels x km farther along the beach away from P. Find x if the person traveled a total of 8.0 km.

40. The length of the roller belt in Fig. 14-15 is 28.0 m. Find x.

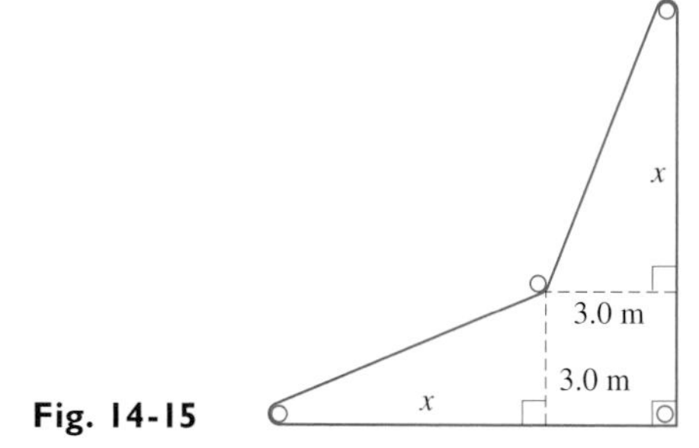

Fig. 14-15

Review Exercises and Practice Test

Review Exercises

In Exercises 1–10, solve the given systems of equations graphically. Use a graphing calculator.

1. $x + 2y = 6$, $y = 4x^2$

2. $x + y = 3$, $x^2 + y^2 = 25$

3. $3x + 2y = 6$, $x^2 + 4y^2 = 4$

4. $x^2 - 2y = 0$, $y = 3x - 5$

5. $y = x^2 + 1$, $2x^2 + y^2 = 4$

6. $\dfrac{x^2}{4} + y^2 = 1$, $x^2 - y^2 = 1$

7. $y = 4 - x^2$, $y = 2x^2$

8. $xy = -2$, $y = 1 - 2x^2$

9. $y = x^2 - 2x$, $y = 1 - e^{-x}$

10. $y = \ln x$, $y = \sin x$

In Exercises 11–20, solve the given systems of equations algebraically.

11. $y = 4x^2$, $y = 8x$

12. $x + y = 2$, $xy = 1$

13. $2y = x^2$, $x^2 + y^2 = 3$

14. $y = x^2$, $2x^2 - y^2 = 1$

15. $4x^2 + y = 3$, $2x + 3y = 1$

16. $2x^2 + y^2 = 3$, $x + 2y = 1$

17. $4x^2 - 7y^2 = 21$, $x^2 + 2y^2 = 99$

18. $3x^2 + 2y^2 = 11$, $2x^2 - y^2 = 30$

19. $4x^2 + 3xy = 4$, $x + 3y = 4$

20. $\dfrac{6}{x} + \dfrac{3}{y} = 4$, $\dfrac{36}{x^2} + \dfrac{36}{y^2} = 13$

In Exercises 21–40, solve the given equations.

21. $x^4 - 20x^2 + 64 = 0$

22. $x^6 - 26x^3 - 27 = 0$

23. $x^{3/2} - 9x^{3/4} + 8 = 0$

24. $x^{1/2} + 3x^{1/4} - 28 = 0$

25. $x^{-2} + 4x^{-1} - 21 = 0$

26. $4x^{-4} + 35x^{-2} = 9$

27. $2x - 3\sqrt{x} - 5 = 0$

28. $x^{-1} + x^{-1/2} = 6$

29. $\left(\dfrac{1}{x + 1}\right)^2 - \dfrac{1}{x + 1} = 2$

30. $(x^2 + 5x)^2 - 5(x^2 + 5x) = 6$

31. $\sqrt{x + 5} = 4$

32. $\sqrt[3]{x - 2} = 3$

33. $\sqrt{5x - 4} = x$

34. $\sqrt{6x + 8} = 3x$

35. $\sqrt{5x + 9} + 1 = x$

36. $2\sqrt{5x - 3} - 1 = 2x$

37. $\sqrt{x + 1} + \sqrt{x} = 2$

38. $\sqrt{3 + x} + \sqrt{3x - 2} = 1$

39. $\sqrt{x + 4} + 2\sqrt{x + 2} = 3$

40. $\sqrt{3x - 2} - \sqrt{x + 7} = 1$

In Exercises 41–46, solve for the indicated quantities.

41. In the study of atomic structure, the equation $L = \dfrac{h}{2\pi}\sqrt{l(l + 1)}$ is used. Solve for $l (l > 0)$.

42. The frequency ω of a certain RLC circuit is given by $\omega = \dfrac{\sqrt{R^2 + 4(L/C)} + R}{2L}$. Solve for C.

43. In the theory dealing with a suspended cable, the equation $y = \sqrt{s^2 - m^2} - m$ is used. Solve for m.

44. The equation $V = e^2cr^{-2} - e^2Zr^{-1}$ is used in spectroscopy. Solve for r.

45. In an experiment, an object is allowed to fall, stops, and then falls for twice the initial time. The total distance the object falls is 392 cm. The equations relating the times t_1 and t_2, in seconds, of fall are $490t_1^2 + 490t_2^2 = 392$ and $t_2 = 2t_1$. Find the times of fall.

46. If two objects collide and the kinetic energy remains constant, the collision is termed perfectly elastic. Under these conditions, if an object of mass m_1 and initial velocity u_1 strikes a second object (initally at rest) of mass m_2, such that the velocities after collision are v_1 and v_2, the following equations are found:

$$m_1u_1 = m_1v_1 + m_2v_2$$
$$\tfrac{1}{2}m_1u_1^2 = \tfrac{1}{2}m_1v_1^2 + \tfrac{1}{2}m_2v_2^2$$

Solve these equations for m_2 in terms of u_1, v_1, and m_1.

In Exercises 47–56, set up the appropriate equations and solve them.

47. A wrench is dropped by a worker at a construction site. Four seconds later the worker hears it hit the ground below. How high is the worker above the ground? (The velocity of sound is 331 m/s, and the distance the wrench falls as a function of time is $s = 4.9t^2$.)

48. A rectangular field is fenced in and divided in half with fencing parallel to the shorter side. The area of the field is 3990 m² and 316 m of fencing are used. What are the dimensions of the field?

49. In a certain electric circuit the impedance Z is twice the square of the reactance X, and the resistance R is 0.800 Ω. Find the impedance and the reactance. See Section 12-7.

50. The perimeter of the machine part shown in Fig. 14-16 is 10.0 cm. Find x.

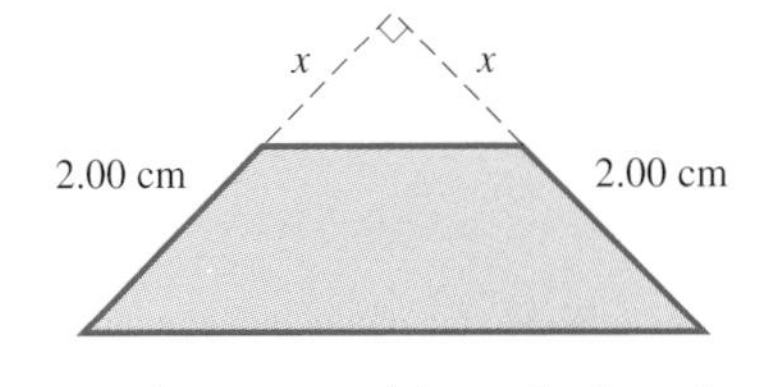

Fig. 14-16

51. The viewing rectangle on a graphing calculator has an area of 1770 mm² and a diagonal of 62 mm. What are the length and width of the rectangle?

52. The circular solar cell and square solar cell shown in Fig. 14-17 have a combined surface area of 40.0 cm². Find the radius of the circular cell and the side of the square cell.

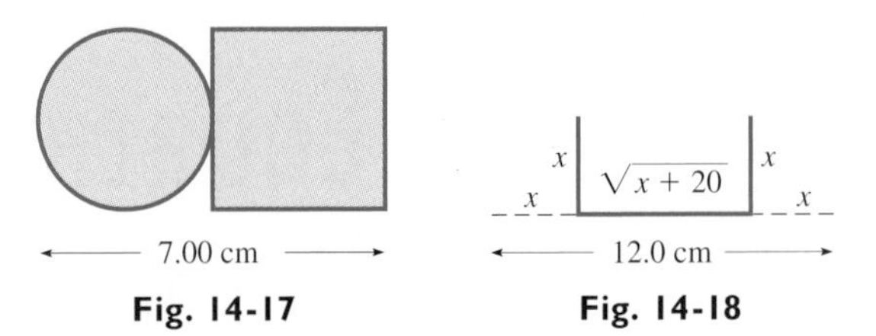

Fig. 14-17 **Fig. 14-18**

53. A trough is made from a piece of sheet metal 12.0 cm wide. The cross section of the trough is shown in Fig. 14-18. Find x.

54. A plastic band 19.0 cm long is bent into the shape of a triangle with sides $\sqrt{x-1}$, $\sqrt{5x-1}$, and 9. Find x.

55. A Coast Guard ship travels from Houston to Mobile, and later it returns to Houston at a speed that is 9.6 km/h faster. If Houston is 820 km from Mobile, and the total travel time is 35 h, find the speed of the ship in each direction.

56. Two trains are approaching the same crossing on tracks that are at right angles to each other. Each is traveling at 60.0 km/h. If one is 6.00 km from the crossing when the other is 3.00 km from it, how much later will they be 4.00 km apart?

Writing Exercise

57. Using a computer, an engineer designs a triangular support structure with sides (in metres) of x, $\sqrt{x-1}$, and 4.00 m. If the perimeter is to be 9.00 m, the equation to be solved is $x + \sqrt{x-1} + 4.00 = 9.00$. This equation can be solved by either of two methods used in this chapter. Write one or two paragraphs identifying the methods and explaining how they are used to solve the equation.

Practice Test

1. Solve for x: $x^{1/2} - 2x^{1/4} = 3$.

2. Solve for x: $3\sqrt{x-2} - \sqrt{x+1} = 1$.

3. Solve for x: $x^4 - 17x^2 + 16 = 0$.

4. Solve for x and y algebraically:

$$x^2 - 2y = 5$$
$$2x + 6y = 1$$

5. The velocity v of an object falling under the influence of gravity in terms of its initial velocity v_0, the acceleration due to gravity g, and the height h fallen is given by $v = \sqrt{v_0^2 + 2gh}$. Solve for h.

6. Solve for x and y graphically: $x^2 - y^2 = 4$, $xy = 2$

7. A rectangular tabletop has a perimeter of 14.0 m and an area of 10.0 m². Find the length and the width of the desktop.

15

Equations of Higher Degree

In Section 15-4 we solve the fourth-degree equation

$$x^4 - 7x^3 + 12x^2 + 4x - 16 = 0$$

in finding the displacement x of a spring system.

In the previous chapters we have discussed methods of solving many types of equations. Except for special cases, however, we have not solved polynomial equations of degree higher than two (a linear equation is a first-degree polynomial equation, and a quadratic equation is a second-degree polynomial equation).

In this chapter we develop certain methods that are useful for solving higher-degree polynomial equations. Since we shall discuss equations involving only polynomials, in this chapter $f(x)$ will be assumed to be a polynomial.

Applications of higher-degree equations arise in a number of technical areas. Included in these are finding the resistance in an electric circuit, determining the dimensions of a structure, and calculating business production costs.

15-1 The Remainder Theorem and the Factor Theorem

As stated above, in this chapter we develop methods of solving higher-degree polynomial equations. In this section we present two theorems that help identify factors and zeros of the polynomial. This, in turn, will help us solve these equations later in the chapter.

Any function of the form

$$f(x) = a_0x^n + a_1x^{n-1} + \cdots + a_n \tag{15-1}$$

where $a_0 \neq 0$ and n is a positive integer or zero is called a **polynomial function.** We will be considering only polynomials in which the coefficients $a_0, a_1, \ldots, a_n$ are real numbers.

If we divide a polynomial by $x - r$, we find a result of the form

$$f(x) = (x - r)q(x) + R \tag{15-2}$$

where $q(x)$ is the quotient and R is the remainder.

EXAMPLE 1 ▸▸ Divide $f(x) = 3x^2 + 5x - 8$ by $x - 2$.

$$\begin{array}{r|l} & 3x + 11 \\ \hline x - 2 & 3x^2 + 5x - 8 \\ & \underline{3x^2 - 6x} \\ & \quad 11x - 8 \\ & \quad \underline{11x - 22} \\ & \qquad\quad 14 \end{array}$$

Thus,

$$3x^2 + 5x - 8 = (x - 2)(3x + 11) + 14$$

where, for this function $f(x)$ with $r = 2$, we identify $q(x)$ and R as

$$q(x) = 3x + 11, \qquad R = 14$$ ◂◂

The Remainder Theorem

If we now set $x = r$ in Eq. (15-2), we have $f(r) = q(r)(r - r) + R$, or

$$f(r) = R \tag{15-3}$$

This leads us to the **remainder theorem,** which states that *if a polynomial $f(x)$ is divided by $x - r$ until a constant remainder R is obtained, then $f(r) = R$.* As we can see, this means the remainder equals the value of the function of x at $x = r$.

EXAMPLE 2 ▸▸ In Example 1, $f(x) = 3x^2 + 5x - 8$, $R = 14$, and $r = 2$

We find that

$$\begin{aligned} f(2) &= 3(2^2) + 5(2) - 8 \\ &= 12 + 10 - 8 \\ &= 14 \end{aligned}$$

Therefore, $f(2) = 14$ verifies that $f(r) = R$ for this example. ◂◂

EXAMPLE 3 ▸▸ By using the remainder theorem, determine the remainder when $3x^3 - x^2 - 20x + 5$ is divided by $x + 4$.

In using the remainder theorem, we determine the remainder when the function is divided by $x - r$ by evaluating the function for $x = r$. To have $x + 4$ in the proper form to identify r, we write it as $x - (-4)$. *This means that* $r = -4$, and we therefore evaluate the function $f(x) = 3x^3 - x^2 - 20x + 5$ for $x = -4$, or find $f(-4)$. Therefore,

$$f(-4) = 3(-4)^3 - (-4)^2 - 20(-4) + 5 = -192 - 16 + 80 + 5$$
$$= -123$$

The remainder is -123 when $3x^3 - x^2 - 20x + 4$ is divided by $x + 4$. ◂◂

In using the remainder theorem and the calculator to evaluate the function, we use the x^y or $\wedge$ key to evaluate the higher powers. As noted in Section 11-2, some calculators show an error if x is negative, as in Example 3. If your calculator does this, use x as a positive value and change the sign if the power is odd.

The Factor Theorem

The remainder theorem leads to another important theorem known as the **factor theorem.** It states that *if* $f(r) = R = 0$, *then* $x - r$ *is a factor of* $f(x)$. We see in Eq. (15-2) that if the remainder $R = 0$, then $f(x) = (x - r)q(x)$, and this shows that $x - r$ is a factor of $f(x)$. Therefore, we have the following meanings for $f(r) = 0$.

> *For the function $f(x)$,*
>
> *if $f(r) = 0$, then $x = r$ is a* ***zero*** *of $f(x)$,*
>
> *$x - r$ is a* ***factor*** *of $f(x)$, and*
>
> *$x = r$ is a* ***root*** *of the equation $f(x) = 0$*

EXAMPLE 4 ▸▸ Is $x + 1$ a factor of $f(x) = x^3 + 2x^2 - 5x - 6$?

Here $r = -1$, and thus

$$f(-1) = -1 + 2 + 5 - 6 = 0$$

Therefore, since $f(-1) = 0$, $x + 1$ is a factor of $f(x)$. ◂◂

EXAMPLE 5 ▸▸ Note that $x + 2$ is not a factor of $f(x)$ in Example 4, since

$$f(-2) = -8 + 8 + 10 - 6 = 4$$

But $x - 2$ is a factor, since $f(2) = 8 + 8 - 10 - 6 = 0$. ◂◂

EXAMPLE 6 ▸▸ Determine if $\frac{2}{3}$ is a zero of $f(x) = 3x^3 + 4x^2 - 16x + 8$.

Evaluating $f(\frac{2}{3})$ we have

$$f\left(\frac{2}{3}\right) = 3\left(\frac{2}{3}\right)^3 + 4\left(\frac{2}{3}\right)^2 - 16\left(\frac{2}{3}\right) + 8$$
$$= \frac{8}{9} + \frac{16}{9} - \frac{32}{3} + 8 = \frac{8 + 16 - 96 + 72}{9} = 0$$

Since $f(\frac{2}{3}) = 0$, $\frac{2}{3}$ is a zero of the function. ◂◂

We now have one way of determining whether or not an expression of the form $x - r$ is a factor of a function $f(x)$. By finding $f(r)$, we can determine whether or not $x - r$ is a factor and whether or not r is a zero of the function.

Exercises 15–1

In Exercises 1–8, find the remainder R by long division and by the remainder theorem.

1. $(x^3 + 2x^2 - x - 2) \div (x - 1)$

2. $(x^3 - 3x^2 - x + 2) \div (x - 2)$

3. $(x^3 + 2x + 3) \div (x + 1)$

4. $(x^4 - 4x^3 - x^2 + x - 100) \div (x + 3)$

5. $(2x^5 - x^2 + 8x + 44) \div (x + 2)$

6. $(x^3 + 4x^2 - 25x - 98) \div (x - 5)$

7. $(3x^4 - 9x^3 - x^2 + 5x - 10) \div (x - 3)$

8. $(2x^4 - 10x^2 + 30x - 60) \div (x + 4)$

In Exercises 9–16, find the remainder using the remainder theorem.

9. $(x^3 + 2x^2 - 3x + 4) \div (x + 1)$

10. $(2x^3 - 4x^2 + x - 1) \div (x + 2)$

11. $(x^4 + x^3 - 2x^2 - 5x + 3) \div (x + 4)$

12. $(4x^4 - x^2 + 5x - 7) \div (x - 3)$

13. $(2x^4 - 7x^3 - x^2 + 8) \div (x - 3)$

14. $(x^4 - 5x^3 + x^2 - 2x + 6) \div (x + 4)$

15. $(x^5 - 3x^3 + 5x^2 - 10x + 6) \div (x - 2)$

16. $(3x^4 - 12x^3 - 60x + 4) \div (x - 5)$

In Exercises 17–24, use the factor theorem to determine whether or not the second expression is a factor of the first expression.

17. $x^2 - 2x - 3, x - 3$

18. $3x^3 + 2x^2 - 3x - 2, x + 2$

19. $4x^3 + x^2 - 16x - 4, x - 2$

20. $3x^3 + 14x^2 + 7x - 4, x + 4$

21. $5x^3 - 3x^2 + 4, x - 2$

22. $x^5 - 2x^4 + 3x^3 - 6x^2 - 4x + 8, x - 2$

23. $x^6 + 1, x + 1$

24. $x^7 - 128, x + 2$

In Exercises 25–28, determine whether or not the given numbers are zeros of the given functions.

25. $f(x) = x^3 - 2x^2 - 9x + 18;\ \ 2$

26. $f(x) = 2x^3 + 3x^2 - 8x - 12;\ \ -\frac{3}{2}$

27. $f(x) = 4x^4 - 4x^3 + 23x^2 + x - 6;\ \ \frac{1}{2}$

28. $f(x) = 2x^4 + 3x^3 - 12x^2 - 7x + 6;\ \ -3$

In Exercises 29–32, answer the given questions.

29. By division, show that $2x - 1$ is a factor of $f(x) = 4x^3 + 8x^2 - x - 2$. May we therefore conclude that $f(1) = 0$? Explain.

30. By division, show that $x^2 + 2$ is a factor of $f(x) = 3x^3 - x^2 + 6x - 2$. May we therefore conclude that $f(-2) = 0$? Explain.

31. For what value of k is $x - 2$ a factor of $f(x) = 2x^3 + kx^2 - x + 14$?

32. For what value of k is $x + 1$ a factor of $f(x) = 3x^4 + 3x^3 + 2x^2 + kx - 4$?

15-2 Synthetic Division

In the sections that follow, we will find that division of a polynomial by the factor $x - r$ is also useful in solving polynomial equations. Therefore, we will now develop a method known as **synthetic division,** which greatly simplifies the process of division. It is an abbreviated form of long division by which we can find the coefficients of the quotient and the remainder. This means that synthetic division allows us to find $f(r)$ by finding the remainder. If the degree of the equation is high, using synthetic division is easier than calculating $f(r)$ directly. The method is developed in the following example.

EXAMPLE 1 ▸▸ Divide $x^4 + 4x^3 - x^2 - 16x - 14$ by $x - 2$.

We shall first perform this division in the usual manner.

$$
\begin{array}{r|l}
 & x^3 + 6x^2 + 11x + 6 \\
\hline
x - 2 & x^4 + 4x^3 - x^2 - 16x - 14 \\
 & \underline{x^4 - 2x^3} \\
 & \quad 6x^3 - x^2 \\
 & \quad \underline{6x^3 - 12x^2} \\
 & \qquad 11x^2 - 16x \\
 & \qquad \underline{11x^2 - 22x} \\
 & \qquad\qquad 6x - 14 \\
 & \qquad\qquad \underline{6x - 12} \\
 & \qquad\qquad\quad -2
\end{array}
$$

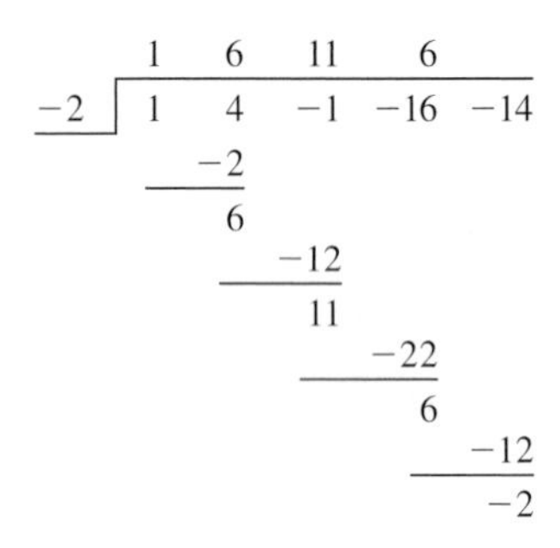

In performing this division we repeated many terms. We should also note that the only quantities of importance in the function being divided are the numerical coefficients. There is no real need to put in the powers of x all of the time. Therefore, at the left we now write the above example without any x's and also eliminate identical terms.

−2	1	4	−1	−16	−14
		−2	−12	−22	−12
		6	11	6	−2

All but the first of the numbers that represent numerical coefficients of the quotients are repeated in the form at the left. Also, all numbers below the dividend may be written in two lines.

1	4	−1	−16	−14	2
	−2	−12	−22	−12	
1	6	11	6	−2	

We now write the first coefficient (in this case, 1) in the bottom line. Also, we change the -2 to 2, which is the actual value of r. Then, in the form at the left, we write the 2 on the right. *In this form the 1, 6, 11, and 6 are the coefficients of the x^3, x^2, x, and constant term of the quotient. The -2 is the remainder.*

1	4	−1	−16	−14	2
	2	12	22	12	
1	6	11	6	−2	

Finally, it is easier to use addition rather than subtraction in the process, so we change the signs of the numbers in the middle row. Remember that originally the bottom line was found by subtraction. Therefore, we have the last form on the left.

In the last form we note the following: The 1 multiplied by the $2(r)$ gives 2, the first number in the middle row. Adding the 4 and 2 (of the second column) gives 6, the second number in the bottom row. This 6 multiplied by 2 (r) is 12, the second number in the middle row. This 12 and -1 give 11. The 11 multiplied by 2 is 22. The 22 added to -16 is 6. This 6 multiplied by 2 gives 12. This 12 added to -14 is -2. In general, *the method is called synthetic division.*

We read the bottom line of the last form as

$1x^3 + 6x^2 + 11x + 6$ with a remainder of -2,

and the last form is the one we shall use in performing synthetic division. ◂◂

Generalizing on Example 1, we have the following steps used in the process of synthetic division.

> ***Procedure for Synthetic Division***
>
> 1. *Write the coefficients of $f(x)$.* Be certain that the powers are in descending order and that zeros are inserted for missing powers.
> 2. *Carry down the left coefficient, then multiply it by r, and place this product under the second coefficient of the top line.*
> 3. *Add the two numbers in the second column and place the result below. Multiply this sum by r and place the product under the third coefficient of the top line.*
> 4. *Continue this process until the bottom row has as many numbers as the top row.*

The last number in the bottom row is the remainder, and the other numbers are the respective coefficients of the quotient. The first term of the quotient is of degree one less than the dividend.

EXAMPLE 2 ▶▶ Divide $x^5 + 2x^4 - 4x^2 + 3x - 4$ by $x + 3$ using synthetic division.

Since the powers of x are in descending order, we write down the coefficients of $f(x)$. In doing so we must be certain to include a zero for the missing x^3 term. Next we note that the divisor is $x + 3$, which means that $r = -3$. The -3 is placed to the right. This gives us a top line of

coefficients ⟶ 1 2 0 −4 3 −4 |−3 ⟵ r

Next we carry the left coefficient, 1, to the bottom line and multiply it by r, -3, placing the product, -3, in the middle line under the second coefficient, 2. We then add the 2 and the -3 and place the result, -1, below. This gives

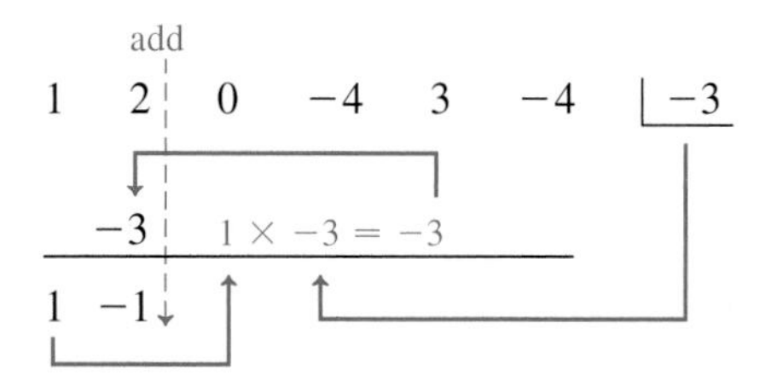

Now we multiply the -1 by r, -3, and place the result, 3, in the middle line under the zero. We now add, and continue the process, obtaining the following result:

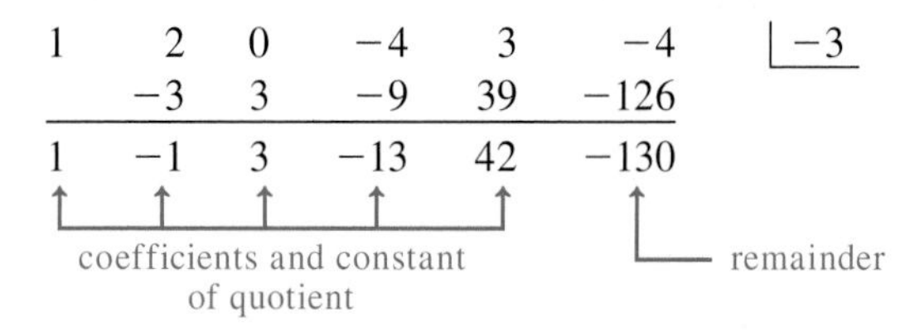

Since the degree of the dividend is 5, the degree of the quotient is 4. This means the quotient is $x^4 - x^3 + 3x^2 - 13x + 42$ and the remainder is -130. In turn, this means that for $f(x) = x^5 + 2x^4 - 4x^2 + 3x - 4$ we have $f(-3) = -130$. ◀◀

EXAMPLE 3 ▸▸ By synthetic division, divide $3x^4 - 5x + 6$ by $x - 4$.

3	0	0	−5	6	⌊4
	12	48	192	748	
3	12	48	187	754	

The quotient is $3x^3 + 12x^2 + 48x + 187$, and the remainder is 754. ◂◂

EXAMPLE 4 ▸▸ By synthetic division, determine whether or not $x - 4$ is a factor of $x^4 + 2x^3 - 15x^2 - 32x - 16$.

1	2	−15	−32	−16	⌊4
	4	24	36	16	
1	6	9	4	0	

Since the remainder is zero, $x - 4$ is a factor. We may also conclude that

$$f(x) = (x - 4)(x^3 + 6x^2 + 9x + 4)$$

since the bottom line gives us the coefficients in the quotient. ◂◂

EXAMPLE 5 ▸▸ By using synthetic division, determine whether $2x - 3$ is a factor of $2x^3 - 3x^2 + 8x - 12$.

CAUTION ▸ We first note that the coefficient of x in the possible factor is not 1. Thus, ***we cannot use $r = 3$, since the factor is not of the form $x - r$.*** However, $2x - 3 = 2(x - \frac{3}{2})$, which means that if $2(x - \frac{3}{2})$ is a factor of the function, $2x - 3$ is a factor. If we use $r = \frac{3}{2}$, and find that the remainder is zero, then $x - \frac{3}{2}$ is a factor.

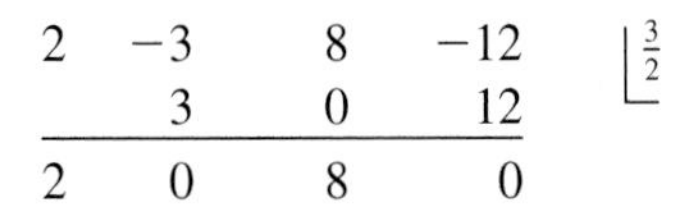

2	−3	8	−12	⌊$\frac{3}{2}$
	3	0	12	
2	0	8	0	

Since the remainder is zero, $x - \frac{3}{2}$ is a factor. Also, the quotient is $2x^2 + 8$, which may be factored into $2(x^2 + 4)$. Thus, 2 is also a factor of the function. This means that $2(x - \frac{3}{2})$ is a factor of the function, and this in turn means that $2x - 3$ is a factor. This tells us that

$$2x^3 - 3x^2 + 8x - 12 = (2x - 3)(x^2 + 4)$$ ◂◂

EXAMPLE 6 ▸▸ By synthetic division, determine whether or not $\frac{1}{3}$ is a zero of the function $3x^3 + 2x^2 - 4x + 1$.

This problem is equivalent to dividing the function by $x - \frac{1}{3}$. If the remainder is zero, $\frac{1}{3}$ is a zero of the function.

3	2	−4	1	⌊$\frac{1}{3}$
	1	1	−1	
3	3	−3	0	

Since the remainder is zero, we conclude that $\frac{1}{3}$ is a zero of the function.

$$\begin{aligned} 3x^3 + 2x^2 - 4x + 1 &= (x - \tfrac{1}{3})(3x^2 + 3x - 3) \\ &= 3(x - \tfrac{1}{3})(x^2 + x - 1) \end{aligned}$$ ◂◂

EXAMPLE 7 ▸▸ Determine whether or not -12.5 is a zero of the function $f(x) = 6x^3 + 61x^2 - 171x + 100$.

If -12.5 is a zero of $f(x)$, then $x - (-12.5)$, or $x + 12.5$, is a factor of $f(x)$, and $f(-12.5) = 0$. We can find the remainder by direct use of the remainder theorem or by synthetic division.

Using synthetic division and a calculator to make the calculations, we have the following setup and calculator sequence.

6	61	−171	100	|−12.5
	−75	175	−50	
6	−14	4	50	

Since the remainder is 50, and not zero, -12.5 is not a zero of $f(x)$. ◂◂

Exercises 15–2

In Exercises 1–20, perform the indicated divisions by synthetic division. Exercises 1–16 are the same as Exercises 1–16 of Section 15-1.

1. $(x^3 + 2x^2 - x - 2) \div (x - 1)$
2. $(x^3 - 3x^2 - x + 2) \div (x - 2)$
3. $(x^3 + 2x + 3) \div (x + 1)$
4. $(x^4 - 4x^3 - x^2 + x - 100) \div (x + 3)$
5. $(2x^5 - x^2 + 8x + 44) \div (x + 2)$
6. $(x^3 + 4x^2 - 25x - 98) \div (x - 5)$
7. $(3x^4 - 9x^3 - x^2 + 5x - 10) \div (x - 3)$
8. $(2x^4 - 10x^2 + 30x - 60) \div (x + 4)$
9. $(x^3 + 2x^2 - 3x + 4) \div (x + 1)$
10. $(2x^3 - 4x^2 + x - 1) \div (x + 2)$
11. $(x^4 + x^3 - 2x^2 - 5x + 3) \div (x + 4)$
12. $(4x^4 - x^2 + 5x - 7) \div (x - 3)$
13. $(2x^4 - 7x^3 - x^2 + 8) \div (x - 3)$
14. $(x^4 - 5x^3 + x^2 - 2x + 6) \div (x + 4)$
15. $(x^5 - 3x^3 + 5x^2 - 10x + 6) \div (x - 2)$
16. $(3x^4 - 12x^3 - 60x + 4) \div (x - 5)$
17. $(x^6 + 2x^2 - 6) \div (x - 2)$
18. $(x^5 + 4x^4 - 8) \div (x + 1)$
19. $(x^7 - 128) \div (x - 2)$
20. $(20x^4 + 11x^3 - 89x^2 + 60x - 77) \div (x + 2.75)$

In Exercises 21–32, use the factor theorem and synthetic division to determine whether or not the second expression is a factor of the first.

21. $x^3 + x^2 - x + 2;\ \ x + 2$
22. $x^3 + 6x^2 + 10x + 6;\ \ x + 3$
23. $x^4 - 6x^2 - 3x - 2;\ \ x - 3$
24. $2x^4 - 5x^3 - 24x^2 + 5;\ \ x - 5$
25. $2x^5 - x^3 + 3x^2 - 4;\ \ x + 1$
26. $x^5 - 3x^4 - x^2 - 6;\ \ x - 3$
27. $4x^3 - 6x^2 + 2x - 2;\ \ x - \frac{1}{2}$
28. $3x^3 - 5x^2 + x + 1;\ \ x + \frac{1}{3}$
29. $2x^4 - x^3 + 2x^2 - 3x + 1;\ \ 2x - 1$
30. $6x^4 + 5x^3 - x^2 + 6x - 2;\ \ 3x - 1$
31. $4x^4 + 2x^3 - 8x^2 + 3x + 12;\ \ 2x + 3$
32. $3x^4 - 2x^3 + x^2 + 15x + 4;\ \ 3x + 4$

In Exercises 33–36, use synthetic division to determine whether or not the given numbers are zeros of the given functions.

33. $x^4 - 5x^3 - 15x^2 + 5x + 14;\ \ 7$
34. $x^4 + 7x^3 + 12x^2 + x + 4;\ \ -4$
35. $85x^3 + 348x^2 - 263x + 120;\ \ -4.8$
36. $2x^3 + 13x^2 + 10x - 4;\ \ \frac{1}{2}$

15-3 The Roots of an Equation

In this section we present certain theorems that are useful in determining the number of roots in the equation $f(x) = 0$ and the nature of some of these roots. In dealing with polynomial equations of higher degree, it is helpful to have as much of this kind of information as is readily obtainable before solving for the roots.

The first of these theorems is so important that it is called the **fundamental theorem of algebra.** It states that

every polynomial equation has at least one (real or complex) root.

The proof of this theorem is of an advanced nature, and therefore we must accept its validity at this time. However, using the fundamental theorem, we can show the validity of other theorems that are useful in solving equations.

Let us now assume that we have a polynomial equation $f(x) = 0$ and that we are looking for its roots. By the fundamental theorem, we know that it has at least one root. Assuming that we can find this root by some means (the factor theorem, for example), we shall call this root r_1. Thus,

$$f(x) = (x - r_1)f_1(x)$$

where $f_1(x)$ is the polynomial quotient found by dividing $f(x)$ by $(x - r_1)$. However, since the fundamental theorem states that any polynomial equation has at least one root, this must apply to $f_1(x) = 0$ as well. Let us assume that $f_1(x) = 0$ has the root r_2. Therefore, this means that $f(x) = (x - r_1)(x - r_2)f_2(x)$. Continuing this process until one of the quotients is a constant a, we have

$$f(x) = a(x - r_1)(x - r_2)\cdots(x - r_n)$$

Note that one linear factor appears each time a root is found and that the degree of the quotient is one less each time. Thus there are n factors, if the degree of $f(x)$ is n.

Therefore, based on the fundamental theorem of algebra, we have the following two related theorems.

1. *A polynomial of the nth degree can be factored into n linear factors.*
2. *A polynomial equation of degree n has exactly n roots.*

These theorems are illustrated in the following example.

EXAMPLE 1 ▶▶ For the equation $f(x) = 2x^4 - 3x^3 - 12x^2 + 7x + 6 = 0$, we are given the factors that we show. In the next section we will see how to find these factors.

For the function $f(x)$, we have

$$\begin{aligned} 2x^4 - 3x^3 - 12x^2 + 7x + 6 &= (x - 3)(2x^3 + 3x^2 - 3x - 2) \\ 2x^3 + 3x^2 - 3x - 2 &= (x + 2)(2x^2 - x - 1) \\ 2x^2 - x - 1 &= (x - 1)(2x + 1) \\ 2x + 1 &= 2(x + \tfrac{1}{2}) \end{aligned}$$

Therefore,

$$2x^4 - 3x^3 - 12x^2 + 7x + 6 = 2(x - 3)(x + 2)(x - 1)(x + \tfrac{1}{2}) = 0$$

The degree of $f(x)$ is 4. There are four linear factors: $(x - 3)$, $(x + 2)$, $(x - 1)$, and $(x + \frac{1}{2})$. There are four roots of the equation: 3, -2, 1, and $-\frac{1}{2}$. Thus, we have verified each of the theorems above for this example. ◀◀

It is not necessary for each root of an equation to be different from the other roots. For example, the equation $(x - 1)^2 = 0$ has two roots, both of which are 1. Such roots are referred to as *multiple* (or *repeated*) *roots.*

When we solve the equation $x^2 + 1 = 0$, the roots are j and $-j$. In fact, for any equation (with real coefficients) that has a root of the form $a + bj$ $(b \neq 0)$ there is also a root of the form $a - bj$. This is so because we can find the solutions of an equation of the form $ax^2 + bx + c = 0$ from the quadratic formula as

$$\frac{-b + \sqrt{b^2 - 4ac}}{2a} \quad \text{and} \quad \frac{-b - \sqrt{b^2 - 4ac}}{2a}$$

and the only difference between these roots is the sign before the radical. Thus we have the following theorem.

If the coefficients of the equation $f(x) = 0$ are real and $a + bj$ $(b \neq 0)$ is a complex root, then its conjugate, $a - bj$, is also a root.

EXAMPLE 2 ▸▸ Consider the equation $f(x) = (x - 1)^3(x^2 + x + 1) = 0$.

The factor $(x - 1)^3$ shows that there is a triple root of 1, and there is a total of five roots, since the highest power term would be x^5 if we were to multiply out the function. To find the other two roots, we use the quadratic formula on the *factor* $(x^2 + x + 1)$. This is permissible, since we are finding the values of x for

$$x^2 + x + 1 = 0$$

For this we have

$$x = \frac{-1 \pm \sqrt{1 - 4}}{2}$$

Thus,

$$x = \frac{-1 + j\sqrt{3}}{2} \quad \text{and} \quad x = \frac{-1 - j\sqrt{3}}{2}$$

Therefore, the roots of $f(x) = 0$ are

$$1, \quad 1, \quad 1, \quad \frac{-1 + j\sqrt{3}}{2}, \quad \frac{-1 - j\sqrt{3}}{2}$$ ◂◂

From Example 2 we can see that *whenever enough roots are known so that the remaining factor is quadratic, it is possible to find the remaining roots from the quadratic formula.* This is true for finding real or complex roots.

3	10	−16	−32	$\lfloor -\frac{4}{3}$
	−4	−8	32	
3	6	−24	0	

EXAMPLE 3 ▸▸ Solve the equation $3x^3 + 10x^2 - 16x - 32 = 0$; $-\frac{4}{3}$ is a root.

Using synthetic division and the given root we have the setup shown at the left. From this we see that

$$3x^3 + 10x^2 - 16x - 32 = (x + \tfrac{4}{3})(3x^2 + 6x - 24)$$

We know that $x + \frac{4}{3}$ is a factor from the given root and that $3x^2 + 6x - 24$ is a factor found from synthetic division. This second factor can be factored as

$$3x^2 + 6x - 24 = 3(x^2 + 2x - 8) = 3(x + 4)(x - 2)$$

Therefore we have

$$3x^3 + 10x^2 - 16x - 32 = 3(x + \tfrac{4}{3})(x + 4)(x - 2)$$

This means the roots are $-\frac{4}{3}$, -4, and 2. ◂◂

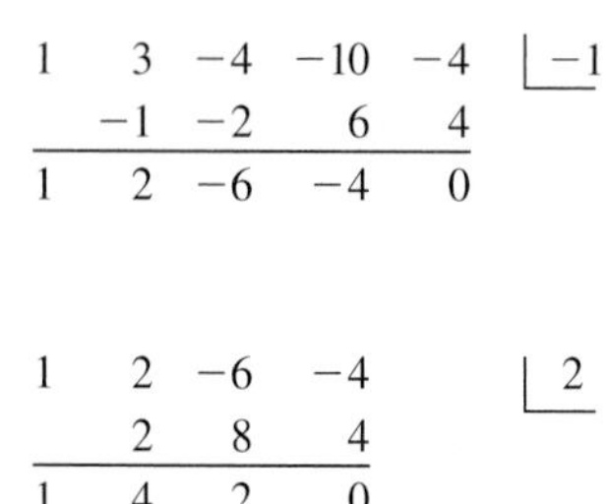

EXAMPLE 4 ▸▸ Solve $x^4 + 3x^3 - 4x^2 - 10x - 4 = 0$; -1 and 2 are roots.

Using synthetic division and the root -1, we have the first setup shown at the left. This tells us that

$$x^4 + 3x^3 - 4x^2 - 10x - 4 = (x + 1)(x^3 + 2x^2 - 6x - 4)$$

We now know that $x - 2$ must be a factor of $x^3 + 2x^2 - 6x - 4$, since it is a factor of the original function. Again, using synthetic division and this time the root 2, we have the second setup at the left. Thus

$$x^4 + 3x^3 - 4x^2 - 10x - 4 = (x + 1)(x - 2)(x^2 + 4x + 2)$$

Since the original equation can now be written as

$$(x + 1)(x - 2)(x^2 + 4x + 2) = 0$$

the remaining two roots are found by solving

$$x^2 + 4x + 2 = 0$$

by the quadratic formula. This gives us

$$x = \frac{-4 \pm \sqrt{16 - 8}}{2} = \frac{-4 \pm 2\sqrt{2}}{2} = -2 \pm \sqrt{2}$$

Therefore, the roots are -1, 2, $-2 + \sqrt{2}$, and $-2 - \sqrt{2}$. ◂◂

3	−26	63	−36	−20	⌊2
	6	−40	46	20	
3	−20	23	10	0	

3	−20	23	10	⌊2
	6	−28	−10	
3	−14	−5	0	

EXAMPLE 5 ▸▸ Solve the equation $3x^4 - 26x^3 + 63x^2 - 36x - 20 = 0$, given that 2 is a double root.

Using synthetic division, we have the first setup at the left. It tells us that

$$3x^4 - 26x^3 + 63x^2 - 36x - 20 = (x - 2)(3x^3 - 20x^2 + 23x + 10)$$

Also, since 2 is a double root, it must be a root of $3x^3 - 20x^2 + 23x + 10 = 0$. Using synthetic division again, we have the second setup at the left. This second quotient $3x^2 - 14x - 5$ factors into $(3x + 1)(x - 5)$. The roots are 2, 2, $-\frac{1}{3}$, and 5.

Since the quotient of the first division is the dividend for the second division, both divisions can be done without rewriting the first quotient as follows:

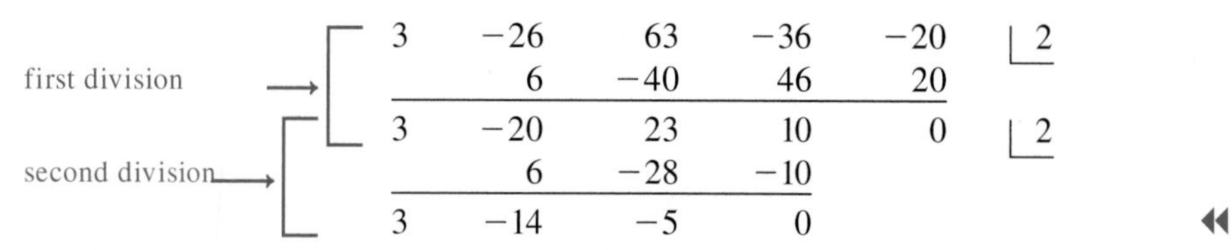

◂◂

EXAMPLE 6 ▸▸ Solve the equation $2x^4 - 5x^3 + 11x^2 - 3x - 5 = 0$, given that $1 + 2j$ is a root.

Since $1 + 2j$ is a root, we know that $1 - 2j$ is also a root. Using synthetic division twice, we can then reduce the remaining factor to a quadratic function.

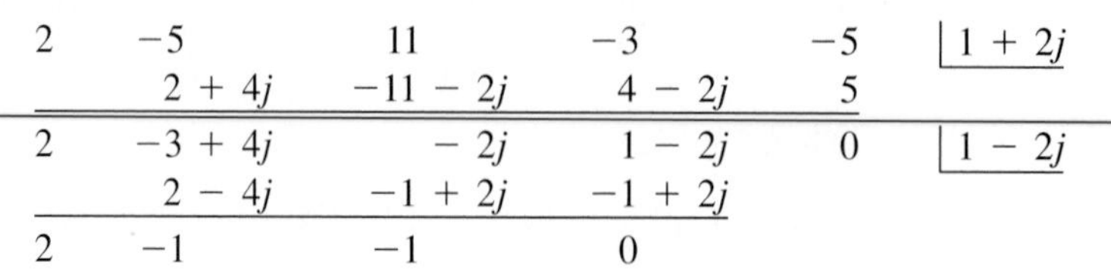

The quadratic factor $2x^2 - x - 1$ factors into $(2x + 1)(x - 1)$. Therefore, the roots of the equation are $1 + 2j$, $1 - 2j$, 1, and $-\frac{1}{2}$. ◂◂

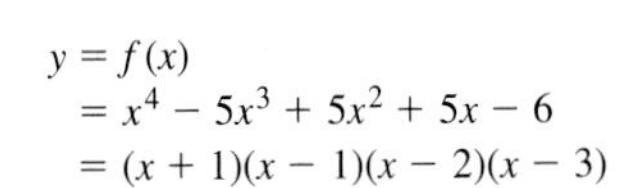

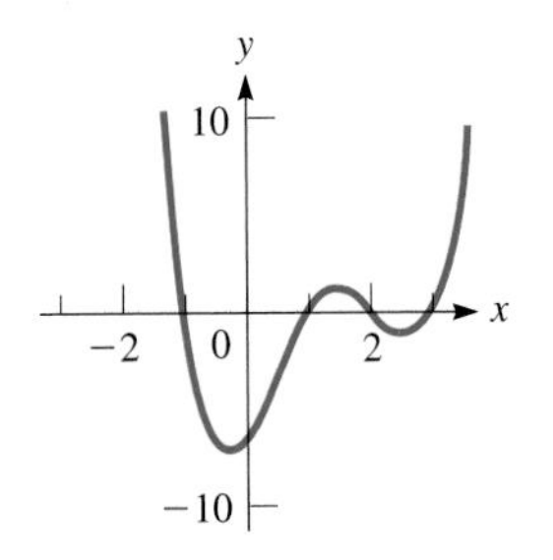

Fig. 15-1

Considering the graphical meaning of the roots of a polynomial equation, we recall that a polynomial equation $f(x) = 0$ of degree n has exactly n roots. If the roots are all real and different, the graph of $f(x)$ crosses the x-axis n times, since $f(x) = 0$ for each value of x where there is a root, and $f(x)$ must change signs as it crosses. See Fig. 15-1, where $n = 4$ and the curve crosses the x-axis four times.

Looking at Fig. 15-1, we see that *the curve must cross the x-axis an odd number of times between two points on the curve and on opposite sides of the x-axis.* Also, *for two points on the curve and on the same side of the x-axis, the curve must cross the x-axis an even number of times between these points, if it crosses at all.*

Considering complex roots and repeated roots, the number of times the curve crosses the x-axis is reduced by 2 for each pair of complex roots (see Fig. 15-2). For a multiple root, the curve crosses the x-axis only once if the multiple is odd, or the curve is tangent to the x-axis if the multiple is even (see Fig. 15-3).

The curve must cross the x-axis at least once if the degree of $f(x)$ is odd, because the range of $f(x)$ includes all real numbers (see Figs. 15-2 and 15-3). The curve may not cross the x-axis at all if the degree of $f(x)$ is even, because the range of $f(x)$ is bounded at a minimum point or a maximum point (as we have seen for the quadratic function) (see Fig. 15-4).

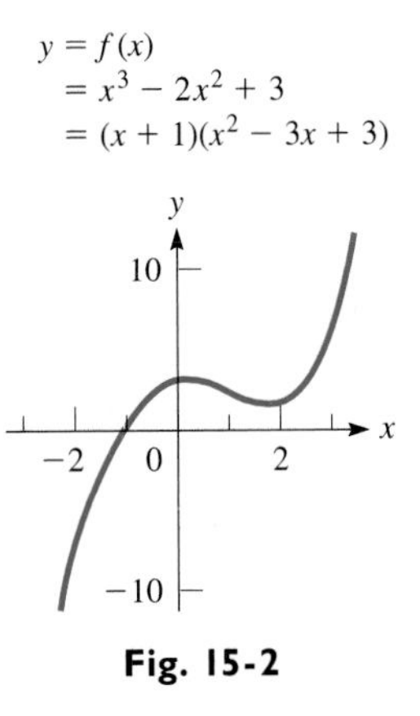

Fig. 15-2

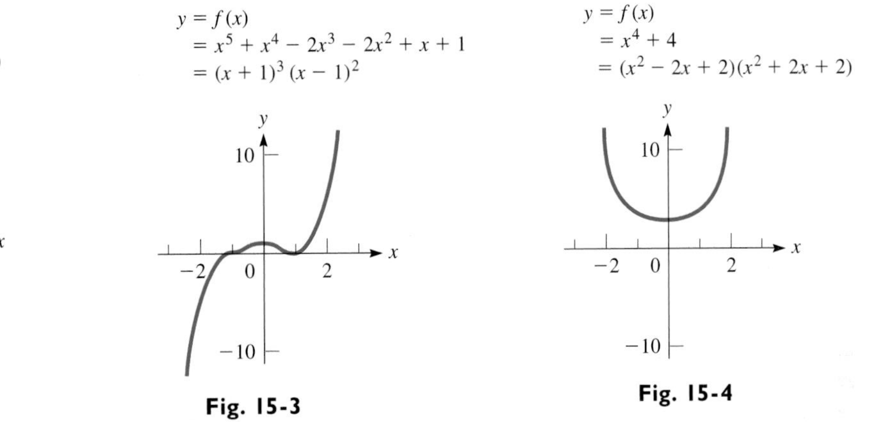

Fig. 15-3

Fig. 15-4

Exercises 15–3

In the following exercises, solve the given equations using synthetic division, given the roots indicated.

1. $x^3 + 2x^2 - x - 2 = 0 \quad (r_1 = 1)$

2. $x^3 + 2x^2 + x + 2 = 0 \quad (r_1 = -2)$

3. $x^3 + x^2 - 8x - 12 = 0 \quad (r_1 = -2)$

4. $x^3 - 1 = 0 \quad (r_1 = 1)$

5. $2x^3 + 11x^2 + 20x + 12 = 0 \quad (r_1 = -\frac{3}{2})$

6. $4x^3 - 20x^2 - x + 5 = 0 \quad (r_1 = \frac{1}{2})$

7. $3x^3 + 2x^2 + 3x + 2 = 0 \quad (r_1 = j)$

8. $x^3 + 5x^2 + 9x + 5 = 0 \quad (r_1 = -2 + j)$

9. $x^4 + x^3 - 2x^2 + 4x - 24 = 0 \quad (r_1 = 2, r_2 = -3)$

10. $x^4 + 2x^3 - 4x^2 - 5x + 6 = 0 \quad (r_1 = 1, r_2 = -2)$

11. $x^4 - 9x^2 + 4x + 12 = 0$ (2 is a double root)

12. $4x^4 + 28x^3 + 61x^2 + 42x + 9 = 0$
(-3 is a double root)

13. $6x^4 + 5x^3 - 15x^2 + 4 = 0 \quad (r_1 = -\frac{1}{2}, r_2 = \frac{2}{3})$

14. $6x^4 - 5x^3 - 14x^2 + 14x - 3 = 0 \quad (r_1 = \frac{1}{3}, r_2 = \frac{3}{2})$

15. $2x^4 - x^3 - 4x^2 + 10x - 4 = 0 \quad (r_1 = 1 + j)$

16. $x^4 - 8x^3 - 72x - 81 = 0 \quad (r_1 = 3j)$

17. $2x^5 + 11x^4 + 16x^3 - 8x^2 - 32x - 16 = 0$
(-2 is a triple root)

18. $x^5 - 3x^4 + 4x^3 - 4x^2 + 3x - 1 = 0$
(1 is a triple root)

19. $2x^5 + x^4 - 15x^3 + 5x^2 + 13x - 6 = 0$
$(r_1 = 1, r_2 = -1, r_3 = \frac{1}{2})$

20. $12x^5 - 7x^4 + 41x^3 - 26x^2 - 28x + 8 = 0$
$(r_1 = 1, r_2 = \frac{1}{4}, r_3 = -\frac{2}{3})$

21. $x^5 - 3x^4 - x + 3 = 0 \quad (r_1 = 3, r_2 = j)$

22. $4x^5 + x^3 - 4x^2 - 1 = 0 \quad (r_1 = 1, r_2 = \frac{1}{2}j)$

23. $x^6 + 2x^5 - 4x^4 - 10x^3 - 41x^2 - 72x - 36 = 0$
(-1 is a double root; $2j$ is a root)

24. $x^6 - x^5 - 2x^3 - 3x^2 - x - 2 = 0$
(j is a double root)

15-4 Rational and Irrational Roots

The product of the factors $(x + 2)(x - 4)(x + 3)$ is $x^3 + x^2 - 14x - 24$. In forming this product, we find that the constant 24 is determined only by the numbers 2, 4, and 3. We see that these numbers represent the roots of the equation if the given function is set equal to zero. In fact, if we find all the integral roots of an equation with integral coefficients and represent the equation in the form

$$f(x) = (x - r_1)(x - r_2)\cdots(x - r_k)f_{k+1}(x) = 0$$

where all the roots indicated are integers, the constant term of $f(x)$ must have factors of $r_1, r_2, \ldots, r_k$. This leads us to the theorem which states that

in a polynomial equation $f(x) = 0$, if the coefficient of the highest power is 1, then any integral roots are factors of the constant term of $f(x)$.

EXAMPLE 1 ▸▸ The equation $x^5 - 4x^4 - 7x^3 + 14x^2 - 44x + 120 = 0$ can be written as

$$(x - 5)(x + 3)(x - 2)(x^2 + 4) = 0$$

We now note that $5(3)(2)(4) = 120$. Thus, the roots 5, -3, and 2 are numerical factors of $|120|$. The theorem states nothing about the signs involved. ◂◂

If the coefficient a_0 of the highest-power term of $f(x)$ is an integer not equal to 1, the polynomial equation $f(x) = 0$ may have rational roots that are not integers. This coefficient a_0 can be factored from every term of $f(x)$. Thus any polynomial equation $f(x) = a_0x^n + a_1x^{n-1} + \cdots + a_n = 0$ with integral coefficients can be written in the form

$$f(x) = a_0\left(x^n + \frac{a_1}{a_0}x^{n-1} + \cdots + \frac{a_n}{a_0}\right) = 0$$

Since a_n and a_0 are integers, a_n/a_0 is a rational number. Using the same reasoning as with integral roots applied to the polynomial within the parentheses, we see that any rational roots are factors of a_n/a_0. This leads to the following theorem:

Any rational root of a polynomial equation (with integral coefficients)

$$f(x) = a_0x^n + a_1x^{n-1} + \cdots + a_n = 0$$

is an integral factor of a_n divided by an integral factor of a_0.

We may show this rational root r_r as

$$r_r = \frac{\text{integral factor of } a_n}{\text{integral factor of } a_0} \qquad (15\text{-}4)$$

EXAMPLE 2 ▸▸ If $f(x) = 4x^3 - 3x^2 - 25x - 6 = 0$, any rational roots, if they exist, must be integral factors of 6 divided by integral factors of 4. The integral factors of 6 are 1, 2, 3, and 6, and the integral factors of 4 are 1, 2, and 4. Forming all possible positive and negative quotients, any rational roots that exist will be found in the following list: $\pm 1, \pm\frac{1}{2}, \pm\frac{1}{4}, \pm 2, \pm 3, \pm\frac{3}{2}, \pm\frac{3}{4}, \pm 6$.

The roots of this equation are -2, 3, and $-\frac{1}{4}$. ◂◂

Named for the French mathematician René Descartes (1596–1650). (See p. 84.)

There are 16 different possible rational roots in Example 2, but we cannot tell which of these are the actual roots. Therefore we now present a rule, known as *Descartes' rule of signs,* which will help us to find these roots.

> **Descartes' Rule of Signs**
>
> 1. *The number of positive roots of a polynomial equation $f(x) = 0$ cannot exceed the number of changes in sign in $f(x)$ in going from one term to the next in $f(x)$.*
> 2. *The number of negative roots cannot exceed the number of sign changes in $f(-x)$.*

We can reason this way: If $f(x)$ has all positive terms, then any positive number substituted in $f(x)$ must give a positive value for $f(x)$. This indicates that the number substituted in the function is not a root. Thus, there must be at least one negative and one positive term in the function for any positive number to be a root. This is not a proof, but it does indicate the type of reasoning used in developing the theorem.

EXAMPLE 3 ▸▸ By Descartes' rule of signs, determine the maximum number of positive and negative roots of $3x^3 - x^2 - x + 4 = 0$.

Here $f(x) = 3x^3 - x^2 - x + 4$. The first term is positive and the second is negative, which indicates a change of sign. The third term is also negative; there is no change of sign from the second to the third term. The fourth term is positive, thus giving us a second change of sign, from the third to the fourth term. Hence there are two changes in sign, which we can show as follows:

$$f(x) = 3x^3 - x^2 - x + 4$$

1 2 ← two sign changes

Since there are *two* changes of sign in $f(x)$, there are *no more than two* positive roots of $f(x) = 0$.

To find the maximum possible number of negative roots, we must find the number of sign changes in $f(-x)$. Thus,

$$f(-x) = 3(-x)^3 - (-x)^2 - (-x) + 4$$
$$= -3x^3 - x^2 + x + 4$$

← one sign change

There is only one change of sign in $f(-x)$; therefore, there is one negative root.

NOTE ▸ *When there is just one change of sign in $f(x)$, there is a positive root, and when there is just one change of sign in $f(-x)$, there is a negative root.* ◂◂

EXAMPLE 4 ▸▸ For the equation $4x^5 - x^4 - 4x^3 + x^2 - 5x - 6 = 0$, we write

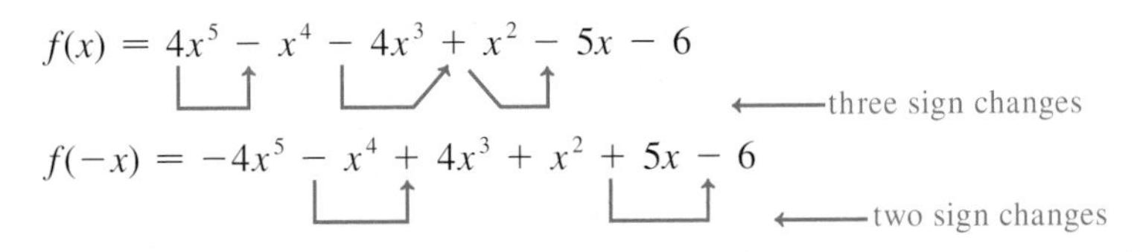

$f(x) = 4x^5 - x^4 - 4x^3 + x^2 - 5x - 6$
three sign changes
$f(-x) = -4x^5 - x^4 + 4x^3 + x^2 + 5x - 6$
two sign changes

Thus, there are no more than three positive and two negative roots. ◂◂

At this point let us summarize the information we can determine about the roots of a polynomial equation $f(x) = 0$ of degree n and with real coefficients.

> ***Roots of a Polynomial Equation of Degree n***
>
> **1.** *There are n roots.*
>
> **2.** *Complex roots appear in conjugate pairs.*
>
> **3.** *Any rational roots must be factors of the constant term divided by factors of the coefficient of the highest-power term.*
>
> **4.** *The maximum number of positive roots is the number of sign changes in $f(x)$, and the maximum number of negative roots is the number of sign changes in $f(-x)$.*
>
> **5.** *Once we find $n - 2$ of the roots, we can find the remaining roots by the quadratic formula.*

Synthetic division is normally used to try possible roots. This is because synthetic division is relatively easy to perform, and when a root is found we have the quotient factor, which is of degree one less than the degree of the dividend. Each root we find makes the ensuing work simpler. The following examples indicate the complete method, as well as two other helpful rules.

EXAMPLE 5 ▶▶ Find the roots of the equation $2x^3 + x^2 + 5x - 3 = 0$.

Since $n = 3$, there are three roots. If we can find one of these roots, we can use the quadratic formula to find the other two. We have

$$f(x) = 2x^3 + x^2 + 5x - 3 \quad \text{and} \quad f(-x) = -2x^3 + x^2 - 5x - 3$$

which shows there is one positive root and no more than two negative roots, which may or may not be rational. The *possible* rational roots are $\pm 1, \pm\frac{1}{2}, \pm\frac{3}{2}, \pm 3$.

2	1	5	−3	1
	2	3	8	
2	3	8	5	

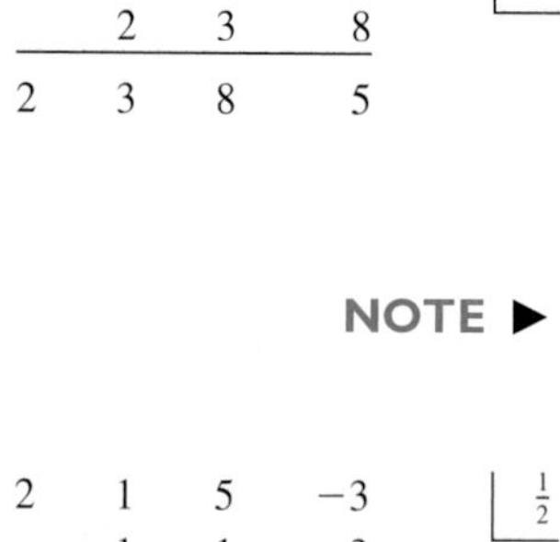

First, trying the root 1 (always a possibility if there are positive roots), we have the synthetic division shown at the left. The remainder of 5 tells us that 1 is not a root, but we have gained some additional information, if we observe closely. If we try any positive number larger than 1, the results in the last row will be larger positive numbers than we now have. The products will be larger, and therefore the sums will also be larger. Thus there is no positive root larger than 1. This leads to the following rule:

NOTE ▶ *When we are trying a positive root, if the bottom row contains all positive numbers, then there are no roots larger than the value tried.* This rule tells us that there is no reason to try $+\frac{3}{2}$ and $+3$ as roots.

2	1	5	−3	$\frac{1}{2}$
	1	1	3	
2	2	6	0	

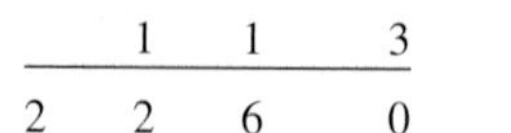

Now let us try $+\frac{1}{2}$, as shown at the left. The zero remainder tells us that $+\frac{1}{2}$ is a root, and the remaining factor is $2x^2 + 2x + 6$, which itself factors to $2(x^2 + x + 3)$. By the quadratic formula we find the remaining roots by solving the equation $x^2 + x + 3 = 0$. This gives us

$$x = \frac{-1 \pm \sqrt{1 - 12}}{2} = \frac{-1 \pm j\sqrt{11}}{2}$$

The three roots are

$$\frac{1}{2}, \quad \frac{-1 + j\sqrt{11}}{2}, \quad \text{and} \quad \frac{-1 - j\sqrt{11}}{2}$$

There are no negative roots, because the nonpositive roots are complex. Proceeding in this way, we did not have to try any negative roots. However, the solutions to all problems may not be so easily determined. ◀◀

See the chapter introduction.

EXAMPLE 6 ▸▸ During a cycle of the movement of the weight on the double spring shown in Fig. 15-5, the force F (in newtons) on the weight by the spring is

$$F = x^4 - 7x^3 + 12x^2 + 4x$$

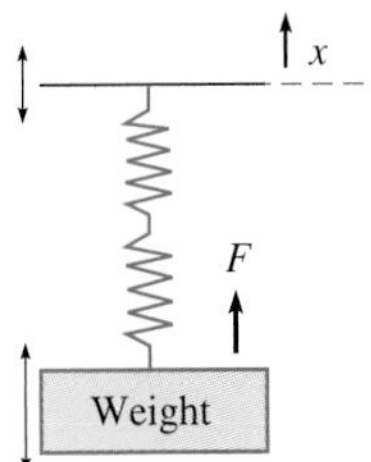

Fig. 15-5

where x is the displacement (in cm) of the top of the double spring. For what values of x is $F = 16$ N?

Substituting 16 for F, we see that we are to solve the equation

$$x^4 - 7x^3 + 12x^2 + 4x - 16 = 0$$

To solve this equation, we write

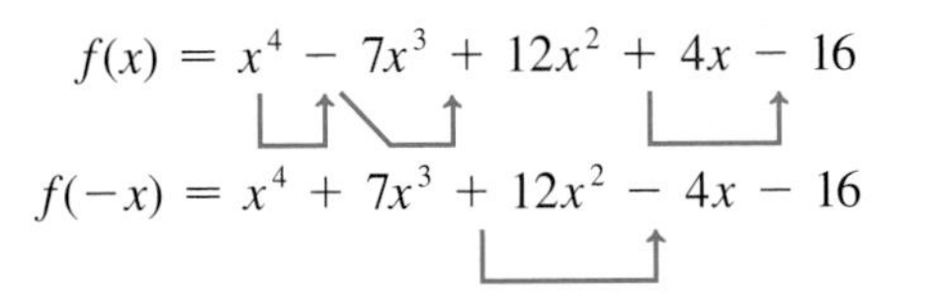

$$f(x) = x^4 - 7x^3 + 12x^2 + 4x - 16$$

$$f(-x) = x^4 + 7x^3 + 12x^2 - 4x - 16$$

We see that there are four roots; there are no more than three positive roots, and there is one negative root. Since the coefficient of x^4 is 1, any possible rational roots must be integers. These possible rational roots are ± 1, ± 2, ± 4, ± 8, and ± 16. Since there is only one negative root, we shall look for this one first. Trying -2, we have

1	−7	+12	+4	−16	$\lfloor -2$
	−2	+18	−60	+112	
1	−9	+30	−56	+96	

NOTE ▸ If we were to try any negative roots less than -2 (remember, -3 is less than -2), we would find that the numbers would still alternate from term to term in the quotient. Thus, we have this rule: *When we are trying a negative root, if the signs alternate in the bottom row, then there are no roots less than the value tried.* In this case, we now know that -4, -8, *and* -16 cannot be roots.

1	−7	+12	+4	−16	$\lfloor -1$
	−1	8	−20	16	
1	−8	20	−16	0	

1	−8	20	−16	$\lfloor 1$
	1	−7	13	
1	−7	13	−3	

1	−8	20	−16	$\lfloor 2$
	2	−12	16	
1	−6	8	0	

Next we try -1, as shown at the left. The remainder of zero tells us that -1 is the negative root.

Now that we have found the one negative root, we look for the positive roots. Trying 1, we have the setup at the left. The remainder of -3 tells us that 1 is not a root. Next we try 2, as shown at the left. We see that 2 is a root.

It is not necessary to find any more roots by trial and error. We may now use the quadratic formula or factoring on the equation $x^2 - 6x + 8 = 0$. The remaining roots are 2 and 4. Thus the roots are -1, 2, 2, and 4. (Note that 2 is a double root.)

These roots now indicate that $F = 16$ N for displacements of -1 cm, 2 cm, and 4 cm. ◂◂

By the methods we have presented, we can look for *all roots* of a polynomial equation. These include any possible *complex roots* and *exact values of the rational and irrational roots,* if they exist. These methods allow us to solve a great many polynomial equations for these roots, but there are numerous other equations for which these methods are not sufficient.

When a polynomial equation has more than two irrational roots, we cannot generally find these roots by the methods we have developed. Approximate values of these roots can be found graphically or by using a scientific calculator.

EXAMPLE 7 ▸▸ Find the roots of the equation $x^3 - 2x^2 - 3x + 2 = 0$ using a graphing calculator.

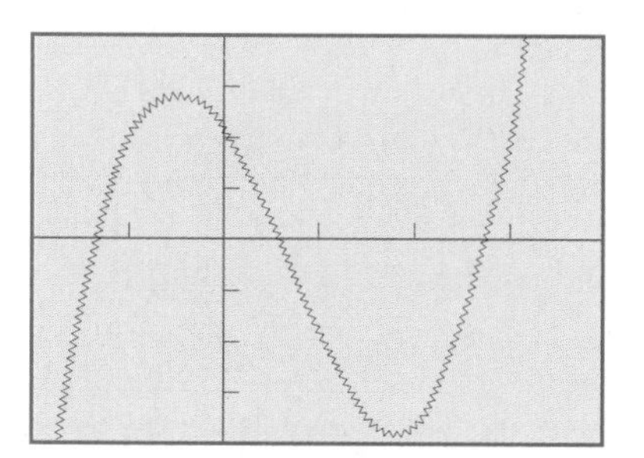

Fig. 15-6

Since the degree of the equation is 3, we know there are three roots, and since the degree is odd there is at least one real root. Using Descartes' rule of signs we have

$$f(x) = x^3 - 2x^2 - 3x + 2 \qquad \text{two changes of sign}$$
$$f(-x) = -x^3 - 2x^2 + 3x + 2 \qquad \text{one change of sign}$$

This means there is one negative root, and no more than two positive roots, if any. Setting

$$y = x^3 - 2x^2 - 3x + 2$$

and using the RANGE values (after two or three trials)

$$\text{Xmin} = -2, \text{Xmax} = 4, \text{Ymin} = -4, \text{Ymax} = 4$$

the view in Fig. 15-6 shows one negative root and two positive roots. Using the TRACE and ZOOM features, these roots are

$$-1.34, 0.53, 2.81 \qquad \text{to the nearest 0.01}$$

Of course, we can use a graphing calculator to help locate and determine the number of real roots of any equation. ◂◂

EXAMPLE 8 ▸▸ A shelf of area 2.50 m^2 is constructed in the shape of a right triangle in which the hypotenuse is 1.00 m longer than one of the legs. Find the lengths of the edges of the shelf.

Let x = the length of the leg and $x + 1.00$ = the length of the hypotenuse. This means the other leg is $\sqrt{(x + 1.00)^2 - x^2} = \sqrt{2.00x + 1.00}$, by using the Pythagorean theorem.

Since the area is one-half the product of the lengths of the legs, we have

$$0.5x\sqrt{2.00x + 1.00} = 2.50 \qquad \text{area} = 2.50 \text{ m}^2$$
$$x\sqrt{2.00x + 1.00} = 5.00 \qquad \text{multiply each side by 2}$$
$$x^2(2.00x + 1.00) = 25.0 \qquad \text{square each side}$$
$$2.00x^3 + 1.00x^2 - 25.0 = 0 \qquad \text{simplify with terms on left}$$

We evaluate $f(x) = 2.00x^3 + 1.00x^2 - 25.0$ for integral values of x by using a scientific calculator and find that $f(2) = -5$ and $f(3) = 38$. Therefore, $f(x) = 0$ between $x = 2$ and $x = 3$. Since $f(2)$ is numerically much smaller than $f(3)$, it is probably nearer $x = 2$. Therefore, we have the following solution.

$$f(2) = -5, \quad f(3) = 38 \qquad \text{different signs mean at least one root between 2 and 3}$$
$$f(2.2) = 1.136 \qquad \text{root between 2.0 and 2.2}$$
$$f(2.16) = -0.179\,008 \qquad \text{root between 2.16 and 2.20}$$
$$f(2.17) = 0.145\,526 \qquad \text{root between 2.16 and 2.17}$$

From these values, $x = 2.17$ m, to three significant digits. Thus the edges are 2.17 m, 2.31 m, and 3.17 m. These values check with the values given in the statement of the problem. ◂◂

Exercises 15–4

In Exercises 1–20, solve the given equations without using a graphing calculator.

1. $x^3 + 2x^2 - x - 2 = 0$

2. $x^3 + x^2 - 5x + 3 = 0$

3. $x^3 + 2x^2 - 5x - 6 = 0$

4. $x^3 + 1 = 0$

5. $2x^3 - 5x^2 - 28x + 15 = 0$

6. $2x^3 - x^2 - 3x - 1 = 0$

7. $3x^3 + 11x^2 + 5x - 3 = 0$

8. $4x^3 - 5x^2 - 23x + 6 = 0$

9. $x^4 - 11x^2 - 12x + 4 = 0$

10. $x^4 + x^3 - 2x^2 - 4x - 8 = 0$

11. $x^4 - 2x^3 - 13x^2 + 14x + 24 = 0$

12. $x^4 - x^3 + 2x^2 - 4x - 8 = 0$

13. $2x^4 - 5x^3 - 3x^2 + 4x + 2 = 0$

14. $2x^4 + 7x^3 + 9x^2 + 5x + 1 = 0$

15. $12x^4 + 44x^3 + 21x^2 - 11x - 6 = 0$

16. $9x^4 - 3x^3 + 34x^2 - 12x - 8 = 0$

17. $x^5 + x^4 - 9x^3 - 5x^2 + 16x + 12 = 0$

18. $x^6 - x^4 - 14x^2 + 24 = 0$

19. $2x^5 - 5x^4 + 6x^3 - 6x^2 + 4x - 1 = 0$

20. $2x^5 + 5x^4 - 4x^3 - 19x^2 - 16x - 4 = 0$

In Exercises 21–24, use a graphing calculator to solve the given equations to the nearest 0.01.

21. $x^3 - 2x^2 - 5x + 4 = 0$

22. $x^4 - x^3 - 2x^2 - x - 3 = 0$

23. $x^4 + 2x^3 - 3x^2 + 2x - 4 = 0$

24. $2x^5 - 3x^4 + 8x^3 - 4x^2 - 4x + 2 = 0$

In Exercises 25–28, use a scientific calculator to find the irrational root (to the nearest 0.01) that lies between the given values. See Example 8.

25. $x^3 - 6x^2 + 10x - 4 = 0$ (0 and 1)

26. $x^4 - x^3 - 3x^2 - x - 4 = 0$ (2 and 3)

27. $3x^3 + 13x^2 + 3x - 4 = 0$ (-1 and 0)

28. $3x^4 - 3x^3 - 11x^2 - x - 4 = 0$ (-2 and -1)

In Exercises 29–40, solve the given problems. Use a graphing calculator in Exercises 37–39.

29. The angular acceleration α (in rad/s^2) of the wheel of a car is given by $\alpha = -0.2t^3 + t^2$, where t is the time in seconds. For what values of t is $\alpha = 2.0$ rad/s^2?

30. In determining one of the dimensions d of the support columns of a building, the equation $3d^3 + 5d^2 - 400d - 18000 = 0$ is found. Determine this dimension (in cm).

31. The deflection y of a beam at a horizontal distance x from one end is given by $y = k(x^4 - 2Lx^3 + L^3x)$, where L is the length of the beam and k is a constant. For what values of x is the deflection zero?

32. The specific gravity s of a sphere of radius r that sinks to a depth h in water is given by $s = \dfrac{3rh^2 - h^3}{4r^3}$. Find the depth to which a spherical buoy of radius 4.0 cm sinks if $s = 0.50$.

33. A company found that the cost C, in dollars, of producing x kilograms of a sealant for a space vehicle is given by $C = 8x^3 - 36x^2 + 90$. Find x for $C = \$36$.

34. The angle θ of a robot arm with the horizontal as a function of time is given by $\theta = 15 + 20t^2 - 4t^3$ for $0 \le t \le 5$ s. Find t for $\theta = 40°$.

35. For electrical resistors connected in parallel, the reciprocal of the combined resistance equals the sum of the reciprocals of the individual resistances. If three resistors are connected in parallel such that the second resistance is 1 Ω more than the first and the third is 4 Ω more than the first, find the resistances for a combined resistance of 1 Ω.

36. The edge of one cubical block of steel is 1.0 cm longer than the edge of another block. Find the length of the edge of each if the sum of their volumes is 91.0 cm^3.

37. A rectangular tray is made from a square piece of sheet metal 10.0 cm on a side by cutting equal squares from each corner, bending up the sides, and then welding them together. How long is the side of the square that must be cut out if the volume of the tray is 70.0 cm^3?

38. A variable electric voltage in a circuit is given by

$$V = 0.1t^4 - 1.0t^3 + 3.5t^2 - 5.0t + 2.3$$

where t is the time in seconds. If the voltage is on for 5.0 s, when is $V = 0$?

39. The pressure difference p (in kPa) at a distance x (in km) from one end of an oil pipeline is given by

$$p = x^5 - 3x^4 - x^2 + 7x$$

If the pipeline is 4 km long, where is $p = 0$?

40. An equation $f(x) = 0$ involves only odd powers of x with positive coefficients. Explain why this equation has no real root except $x = 0$.

Chapter Equations, Review Exercises, and Practice Test

Chapter Equations

Polynomial function $f(x) = a_0x^n + a_1x^{n-1} + \cdots + a_n$ (15-1)

Remainder theorem $f(x) = (x - r)q(x) + R$ (15-2)

$f(r) = R$ (15-3)

Rational roots $r_r = \dfrac{\text{integral factor of } a_n}{\text{integral factor of } a_0}$ (15-4)

Review Exercises

In Exercises 1–4, find the remainder of the indicated division by the remainder theorem.

1. $(2x^3 - 4x^2 - x + 4) \div (x - 1)$

2. $(x^3 - 2x^2 + 9) \div (x + 2)$

3. $(4x^3 + x + 4) \div (x + 3)$

4. $(x^4 - 5x^3 + 8x^2 + 15x - 2) \div (x - 3)$

In Exercises 5–8, use the factor theorem to determine whether or not the second expression is a factor of the first.

5. $x^4 + x^3 + x^2 - 2x - 3;\ \ x + 1$

6. $2x^3 - 2x^2 - 3x - 2;\ \ x - 2$

7. $x^4 + 4x^3 + 5x^2 + 5x - 6;\ \ x + 3$

8. $9x^3 + 6x^2 + 4x + 2;\ \ 3x + 1$

In Exercises 9–16, use synthetic division to perform the indicated divisions.

9. $(x^3 + 3x^2 + 6x + 1) \div (x - 1)$

10. $(3x^3 - 2x^2 + 7) \div (x - 3)$

11. $(2x^3 - 3x^2 - 4x + 3) \div (x + 2)$

12. $(3x^3 - 5x^2 + 7x - 6) \div (x + 4)$

13. $(x^4 - 2x^3 - 3x^2 - 4x - 8) \div (x + 1)$

14. $(x^4 - 6x^3 + x - 8) \div (x - 3)$

15. $(2x^5 - 46x^3 + x^2 - 9) \div (x - 5)$

16. $(x^6 + 63x^3 + 5x^2 - 9x - 8) \div (x + 4)$

In Exercises 17–20, use synthetic division to determine whether or not the given numbers are zeros of the given functions.

17. $x^3 + 8x^2 + 17x - 6;\ \ -3$

18. $2x^3 + x^2 - 4x + 4;\ \ -2$

19. $2x^4 - x^3 + 2x^2 + x - 1;\ \ \frac{1}{2}$

20. $6x^4 - 7x^3 + 2x^2 - 9x - 6;\ \ -\frac{2}{3}$

In Exercises 21–32, find all the roots of the given equations, using synthetic division and the given roots.

21. $x^3 + 8x^2 + 17x + 6 = 0\ \ (r_1 = -3)$

22. $2x^3 + 7x^2 - 6x - 8 = 0\ \ (r_1 = -4)$

23. $3x^4 + 5x^3 + x^2 + x - 10 = 0\ \ (r_1 = 1, r_2 = -2)$

24. $x^4 - x^3 - 5x^2 - x - 6 = 0\ \ (r_1 = 3, r_2 = -2)$

25. $2x^4 + x^3 - 29x^2 - 34x + 24 = 0\ \ (r_1 = -2, r_2 = \frac{1}{2})$

26. $x^4 + x^3 - 11x^2 - 9x + 18 = 0\ \ (r_1 = -3, r_2 = 1)$

27. $4x^4 + 4x^3 + x^2 + 4x - 3 = 0\ \ (r_1 = j)$

28. $x^4 + 2x^3 - 4x - 4 = 0\ \ (r_1 = -1 + j)$

29. $x^5 + 3x^4 - x^3 - 11x^2 - 12x - 4 = 0$ (-1 is a triple root)

30. $24x^5 + 10x^4 + 7x^2 - 6x + 1 = 0\ \ (r_1 = -1, r_2 = \frac{1}{4}, r_3 = \frac{1}{3})$

31. $x^5 + 4x^4 + 5x^3 - x^2 - 4x - 5 = 0\ \ (r_1 = 1, r_2 = -2 + j)$

32. $2x^5 - x^4 + 8x - 4 = 0\ \ (r_1 = \frac{1}{2}, r_2 = 1 + j)$

In Exercises 33–40, solve the given equations.

33. $x^3 + x^2 - 10x + 8 = 0$

34. $x^3 - 8x^2 + 20x - 16 = 0$

35. $2x^3 - x^2 - 8x - 5 = 0$

36. $2x^3 - 3x^2 - 11x + 6 = 0$

37. $6x^3 - x^2 - 12x - 5 = 0$

38. $6x^3 + 19x^2 + 2x - 3 = 0$

39. $2x^4 + x^3 + 3x^2 + 2x - 2 = 0$

40. $2x^4 + 5x^3 - 14x^2 - 23x + 30 = 0$

In Exercises 41–56, determine the required quantities. Where appropriate, set up the required equations.

41. For what value of k is $x + 2$ a factor of $f(x) = 3x^3 + kx^2 - 8x - 8$?

42. For what value of k is $x - 3$ a factor of $f(x) = kx^4 - 15x^2 - 5x - 12$?

43. Where does the graph of the function $f(x) = 6x^4 - 14x^3 + 5x^2 + 5x - 2$ cross the x-axis?

44. Where does the graph of the function $f(x) = 2x^4 - 7x^3 + 11x^2 - 28x + 12$ cross the x-axis?

45. Find the irrational root of the equation $3x^3 - x^2 - 8x - 2 = 0$ that lies between 1 and 2.

46. Find the irrational root of the equation $x^4 + 3x^3 + 6x + 4 = 0$ that lies between -1 and 0.

47. A computer analysis of the number of crimes committed each month in a certain city for the first 10 months of a year showed that $n = x^3 - 9x^2 + 15x + 600$. Here n is the number of monthly crimes and x is the number of the month (as of the last day). In what month were 580 crimes committed?

48. A company determined that the number s (in thousands) of computer chips that it could supply at a price p of less than \$5 is given by $s = 4p^2 - 25$, whereas the demand d (in thousands) for the chips is given by $d = p^3 - 22p + 50$. For what price is the supply equal to the demand?

49. In order to find the diameter d (in cm) of a helical spring subject to given forces, it is necessary to solve the equation $64d^3 - 144d^2 + 108d - 27 = 0$. Solve for d.

50. A cubical tablet for purifying water is wrapped in a sheet of foil 0.500 mm thick. The total volume of tablet and foil is 33.1% greater than the volume of the tablet alone. Find the length of the edge of the tablet.

51. For the mirror shown in Fig. 15-7, the reciprocal of the focal distance f equals the sum of the reciprocals of the object distance p and image distance q. If $q = p + 4$ and $f = \dfrac{p+1}{p}$, find p. Distances are in centimetres.

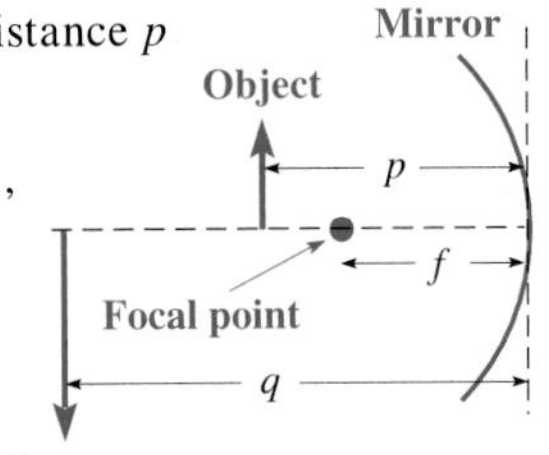

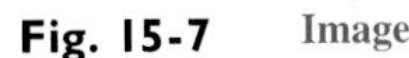

Fig. 15-7

52. Three electric capacitors are connected in series. The capacitance of the second is 1 μF more than that of the first, and the third is 2 μF more than that of the second. The capacitance of the combination is 1.33 μF. The equation used to determine C, the capacitance of the first capacitor, is

$$\frac{1}{C} + \frac{1}{C+1} + \frac{1}{C+3} = \frac{3}{4}$$

Find the values of the capacitances.

53. The height of a cylindrical oil tank is 3.2 m more than the radius. If the volume of the tank is 680 m^3, what are the radius and the height of the tank?

54. A grain storage bin has a square base, each side of which is 5.5 m longer than the height of the bin. If the bin holds 160 m^3 of grain, find its dimensions.

55. A rectangular slab of concrete has a diagonal that is 2.0 m longer than one of the sides. If the area of the slab is 120 m^2, find the lengths of the sides.

56. The radius of one ball bearing is 1.0 mm greater than the radius of a second ball bearing. If the sum of their volumes is 100 mm^3, find the radius of each.

Writing Exercise

57. A computer science student is to write a computer program that will print out the values of n for which $x + r$ is a factor of $x^n + r^n$. Write a paragraph that states which are the values of n and explains how they are found.

Practice Test

1. Is -3 a zero for the function $2x^3 + 3x^2 + 7x - 6$?
2. Find the remaining roots of the equation $x^4 - 2x^3 - 7x^2 + 20x - 12 = 0$; 2 is a double root.
3. Use synthetic division to perform the division $(x^3 - 5x^2 + 4x - 9) \div (x - 3)$.
4. Use the factor theorem and synthetic division to determine whether or not $(2x + 1)$ is a factor of $2x^4 + 15x^3 + 23x^2 - 16$.
5. Use the remainder theorem to find the remainder of the division $(x^3 + 4x^2 + 7x - 9) \div (x + 4)$.
6. Solve: $2x^4 - x^3 + 5x^2 - 4x - 12 = 0$.
7. The ends of a 10-m beam are supported at different levels. The deflection y of the beam is given by $y = kx^2(x^3 + 436x - 4000)$, where x is the horizontal distance from one end and k is a constant. Determine the values of x for which the deflection is zero.
8. A cubical metal block is heated such that its edge increases by 1.0 mm and its volume is doubled. Find the edge of the cube to tenths.

16

Determinants and Matrices

In Section 16-4 we see how to determine material amounts and worker time for the production of machine parts.

In Chapter 5 we first met the concept of a determinant and saw how it is used to solve systems of linear equations. However, at that time we limited our discussion to second- and third-order determinants. In the first two sections of this chapter we show methods of evaluating higher-order determinants. In the remainder of the chapter we develop the related concept of a matrix and its use in solving systems of linear equations.

Methods developed in this chapter are used in solving applied problems in areas such as electric circuits and analysis of forces. Also, these methods are used extensively in business and industry in making appropriate decisions for research, development, and production.

16-1 Determinants; Expansion by Minors

From Section 5-7 we recall that *a third-order* **determinant** *is defined by the equation*

$$\begin{vmatrix} a_1 & b_1 & c_1 \\ a_2 & b_2 & c_2 \\ a_3 & b_3 & c_3 \end{vmatrix} = a_1b_2c_3 + a_3b_1c_2 + a_2b_3c_1 - a_3b_2c_1 - a_1b_3c_2 - a_2b_1c_3 \quad (16\text{-}1)$$

In Eq. (16-1), if we rearrange the terms on the right and factor a_1, $-a_2$, and a_3 from the terms in which they are contained, we have

$$\begin{vmatrix} a_1 & b_1 & c_1 \\ a_2 & b_2 & c_2 \\ a_3 & b_3 & c_3 \end{vmatrix} = a_1(b_2c_3 - b_3c_2) - a_2(b_1c_3 - b_3c_1) + a_3(b_1c_2 - b_2c_1) \quad (16\text{-}2)$$

Recalling the definition of a second-order determinant, we have

$$\begin{vmatrix} a_1 & b_1 & c_1 \\ a_2 & b_2 & c_2 \\ a_3 & b_3 & c_3 \end{vmatrix} = a_1\begin{vmatrix} b_2 & c_2 \\ b_3 & c_3 \end{vmatrix} - a_2\begin{vmatrix} b_1 & c_1 \\ b_3 & c_3 \end{vmatrix} + a_3\begin{vmatrix} b_1 & c_1 \\ b_2 & c_2 \end{vmatrix} \quad (16\text{-}3)$$

In Eq. (16-3) we note that the third-order determinant is expanded with the terms of the expansion as products of the elements of the first column and specific second-order determinants. In each case the elements of the second-order determinant are those elements which are in neither the same row nor the same column as the element from the first column. *These determinants are called* **minors.**

In general, *the minor of a given element of a determinant is the determinant that results by deleting the row and the column in which the element lies.* Consider the following example.

EXAMPLE 1 ▸▸ Consider the determinant $\begin{vmatrix} 1 & 2 & 3 \\ 4 & 5 & 6 \\ 7 & 8 & 9 \end{vmatrix}$

We find the minor of the element 1 by deleting the elements in the first row and first column because the element 1 is located in the first row and in the first column. The minor for the element 6 is formed by deleting the elements in the second row and in the third column, for this is the location of the 6. These minors are shown below.

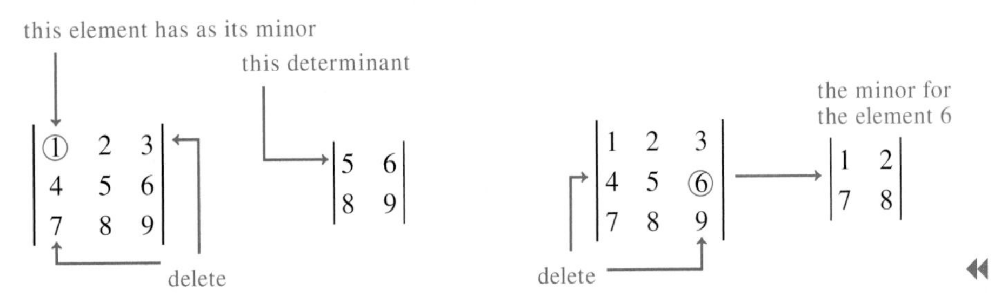

◂◂

We now see that Eq. (16-3) expresses the expansion of a third-order determinant as the sum of the products of the elements of the first column and their minors, with the second term assigned a minus sign. Actually this is only one of several ways of expressing the expansion. However, it does lead to a general theorem regarding the expansion of a determinant of any order. The foregoing provides a basis for this theorem, although it cannot be considered as a proof. The theorem is given on the next page.

Expansion of a Determinant by Minors

The value of a determinant of order n may be found by forming the n products of the elements of any column (or row) and their minors. A product is given a plus sign if the sum of the number of the column and the number of the row in which the element lies is even, and a minus sign if this sum is odd. The algebraic sum of the terms thus obtained is the value of the determinant.

EXAMPLE 2 ▸▸ Evaluate $\begin{vmatrix} 1 & -3 & -2 \\ 4 & -1 & 0 \\ 4 & 3 & -5 \end{vmatrix}$ by expansion by minors.

Since we may expand by any column or row, using the first row, we have:

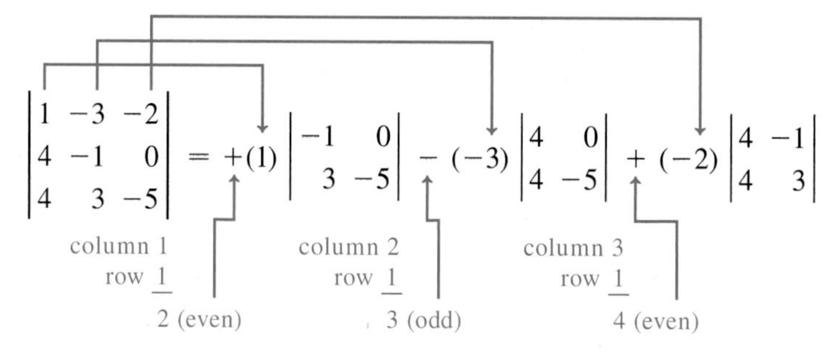

$\begin{vmatrix} 1 & -3 & -2 \\ 4 & -1 & 0 \\ 4 & 3 & -5 \end{vmatrix} = +(1)\begin{vmatrix} -1 & 0 \\ 3 & -5 \end{vmatrix} - (-3)\begin{vmatrix} 4 & 0 \\ 4 & -5 \end{vmatrix} + (-2)\begin{vmatrix} 4 & -1 \\ 4 & 3 \end{vmatrix}$
column 1
row 1
2 (even)
column 2
row 1
3 (odd)
column 3
row 1
4 (even)

For reference, Eq. (5-9) is $\begin{vmatrix} a_1 & b_1 \\ a_2 & b_2 \end{vmatrix} = a_1b_2 - a_2b_1$.

We note that once the first sign has been determined, the others are known since the signs alternate from term to term. Now, using Eq. (5-9), we have

$$\begin{vmatrix} 1 & -3 & -2 \\ 4 & -1 & 0 \\ 4 & 3 & -5 \end{vmatrix} = +(1)(5 - 0) - (-3)(-20 - 0) + (-2)[12 - (-4)]$$

$$= 1(5) + 3(-20) - 2(16) = 5 - 60 - 32 = -87 \quad ◂◂$$

EXAMPLE 3 ▸▸ Evaluate $\begin{vmatrix} 3 & -2 & 0 & 2 \\ 1 & 0 & -1 & 4 \\ -3 & 1 & 2 & -2 \\ 2 & -1 & 0 & -1 \end{vmatrix}$

Expanding by the third column, we have

$$\begin{vmatrix} 3 & -2 & 0 & 2 \\ 1 & 0 & -1 & 4 \\ -3 & 1 & 2 & -2 \\ 2 & -1 & 0 & -1 \end{vmatrix} = +(0)\begin{vmatrix} 1 & 0 & 4 \\ -3 & 1 & -2 \\ 2 & -1 & -1 \end{vmatrix} - (-1)\begin{vmatrix} 3 & -2 & 2 \\ -3 & 1 & -2 \\ 2 & -1 & -1 \end{vmatrix} + (2)\begin{vmatrix} 3 & -2 & 2 \\ 1 & 0 & 4 \\ 2 & -1 & -1 \end{vmatrix} - (0)\begin{vmatrix} 3 & -2 & 2 \\ 1 & 0 & 4 \\ -3 & 1 & -2 \end{vmatrix}$$

$$= \begin{vmatrix} 3 & -2 & 2 \\ -3 & 1 & -2 \\ 2 & -1 & -1 \end{vmatrix} + 2\begin{vmatrix} 3 & -2 & 2 \\ 1 & 0 & 4 \\ 2 & -1 & -1 \end{vmatrix}$$

$$= \left[3\begin{vmatrix} 1 & -2 \\ -1 & -1 \end{vmatrix} - (-3)\begin{vmatrix} -2 & 2 \\ -1 & -1 \end{vmatrix} + 2\begin{vmatrix} -2 & 2 \\ 1 & -2 \end{vmatrix}\right] + 2\left[-(-2)\begin{vmatrix} 1 & 4 \\ 2 & -1 \end{vmatrix} + 0\begin{vmatrix} 3 & 2 \\ 2 & -1 \end{vmatrix} - (-1)\begin{vmatrix} 3 & 2 \\ 1 & 4 \end{vmatrix}\right]$$

expanding first determinant by first column — expanding second determinant by second column

$$= [3(-1 - 2) + 3(2 + 2) + 2(4 - 2)] + 2[2(-1 - 8) + (12 - 2)] = [-9 + 12 + 4] + 2[-18 + 10]$$

$$= +7 + 2(-8) = -9 \quad ◂◂$$

We see in Examples 2 and 3 that minors can be expanded as determinants or that they can be expanded by minors. These examples also illustrate that expansion by minors effectively reduces by one the order of the determinant being evaluated.

Solving Systems of Linear Equations by Determinants

We can use the expansion of determinants by minors to solve systems of linear equations. **Cramer's rule** *for solving systems of linear equations, as stated in Section 5-7, is valid for any system of n equations in n unknowns.*

EXAMPLE 4 ▶▶ Production of a certain computer component is done in three stages, taking a total of 7 h. The second stage is 1 h less than the first, and the third stage is twice as long as the second. How long is each stage of production?

First, we let a = the number of hours of the first production stage, b = the number of hours of the second stage, and c = the number of hours of the third stage.

The total production time of 7 h gives us $a + b + c = 7$. Since the second stage is 1 h less than the first, we then have $b = a - 1$. The fact that the third stage is twice as long as the second gives us $c = 2b$. Stating these equations in standard form for solution, we have

$$\begin{aligned} a + b + c &= 7 \\ a - b \quad\quad &= 1 \\ 2b - c &= 0 \end{aligned}$$

Using Cramer's rule, we have

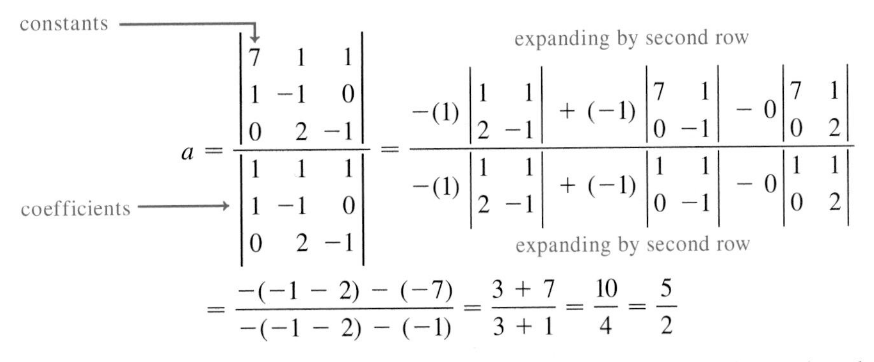

$$a = \frac{\begin{vmatrix} 7 & 1 & 1 \\ 1 & -1 & 0 \\ 0 & 2 & -1 \end{vmatrix}}{\begin{vmatrix} 1 & 1 & 1 \\ 1 & -1 & 0 \\ 0 & 2 & -1 \end{vmatrix}} = \frac{-(1)\begin{vmatrix} 1 & 1 \\ 2 & -1 \end{vmatrix} + (-1)\begin{vmatrix} 7 & 1 \\ 0 & -1 \end{vmatrix} - 0\begin{vmatrix} 7 & 1 \\ 0 & 2 \end{vmatrix}}{-(1)\begin{vmatrix} 1 & 1 \\ 2 & -1 \end{vmatrix} + (-1)\begin{vmatrix} 1 & 1 \\ 0 & -1 \end{vmatrix} - 0\begin{vmatrix} 1 & 1 \\ 0 & 2 \end{vmatrix}}$$

$$= \frac{-(-1-2) - (-7)}{-(-1-2) - (-1)} = \frac{3+7}{3+1} = \frac{10}{4} = \frac{5}{2}$$

Here we expanded each determinant by minors of the second row. Now, using the second equation, we have $\frac{5}{2} - b = 1$, or $b = \frac{3}{2}$. Using the third equation, we have $2(\frac{3}{2}) - c = 0$, or $c = 3$. Thus,

$$a = \frac{5}{2}\text{ h}, \quad b = \frac{3}{2}\text{ h}, \quad \text{and} \quad c = 3\text{ h}$$

Therefore, the first stage takes 2.5 h, the second 1.5 h, and the third 3.0 h. (We are using two significant digits here, although for convenience we used only one in the equations.) These times agree with the given information. ◀◀

NOTE ▶ In Example 4 we found a by determinants and then found the other values by substituting into the equations. Normally, *as soon as we find the values of all but one of the unknowns in any equation, we then find the value of this unknown by substitution.* This is usually easier than evaluating additional determinants.

EXAMPLE 5 Solve the following system of equations.

$$\begin{aligned} x + 2y + z \quad &= 5 \\ 2x \quad + z + 2t &= 1 \\ x - y + 3z + 4t &= -6 \\ 4x - y \quad - 2t &= 0 \end{aligned}$$

constants (first column of the numerator); expanding by fourth row (numerator); expanding by first row (denominator)

$$x = \frac{\begin{vmatrix} 5 & 2 & 1 & 0 \\ 1 & 0 & 1 & 2 \\ -6 & -1 & 3 & 4 \\ 0 & -1 & 0 & -2 \end{vmatrix}}{\begin{vmatrix} 1 & 2 & 1 & 0 \\ 2 & 0 & 1 & 2 \\ 1 & -1 & 3 & 4 \\ 4 & -1 & 0 & -2 \end{vmatrix}} = \frac{-(0)\begin{vmatrix} 2 & 1 & 0 \\ 0 & 1 & 2 \\ -1 & 3 & 4 \end{vmatrix} + (-1)\begin{vmatrix} 5 & 1 & 0 \\ 1 & 1 & 2 \\ -6 & 3 & 4 \end{vmatrix} - (0)\begin{vmatrix} 5 & 2 & 0 \\ 1 & 0 & 2 \\ -6 & -1 & 4 \end{vmatrix} + (-2)\begin{vmatrix} 5 & 2 & 1 \\ 1 & 0 & 1 \\ -6 & -1 & 3 \end{vmatrix}}{(1)\begin{vmatrix} 0 & 1 & 2 \\ -1 & 3 & 4 \\ -1 & 0 & -2 \end{vmatrix} - 2\begin{vmatrix} 2 & 1 & 2 \\ 1 & 3 & 4 \\ 4 & 0 & -2 \end{vmatrix} + (1)\begin{vmatrix} 2 & 0 & 2 \\ 1 & -1 & 4 \\ 4 & -1 & -2 \end{vmatrix} - (0)\begin{vmatrix} 2 & 0 & 1 \\ 1 & -1 & 3 \\ 4 & -1 & 0 \end{vmatrix}}$$

$$= \frac{-(-26) - 2(-14)}{1(0) - 2(-18) + 1(18)} = \frac{26 + 28}{36 + 18} = \frac{54}{54} = 1$$

In solving for x, we evaluated the determinant in the numerator by expanding by the minors of the fourth row, since it contained two zeros. We evaluated the determinant in the denominator by expanding by the minors of the first row. Now we solve for y and again note two zeros in the fourth row of the determinant of the numerator.

expanding by fourth row

$$y = \frac{\begin{vmatrix} 1 & 5 & 1 & 0 \\ 2 & 1 & 1 & 2 \\ 1 & -6 & 3 & 4 \\ 4 & 0 & 0 & -2 \end{vmatrix}}{54} = \frac{-4\begin{vmatrix} 5 & 1 & 0 \\ 1 & 1 & 2 \\ -6 & 3 & 4 \end{vmatrix} + (-2)\begin{vmatrix} 1 & 5 & 1 \\ 2 & 1 & 1 \\ 1 & -6 & 3 \end{vmatrix}}{54}$$

$$= \frac{-4(-26) - 2(-29)}{54} = \frac{104 + 58}{54} = \frac{162}{54} = 3$$

Substituting these values for x and y into the first equation, we can solve for z. This gives $z = -2$. Again, substituting the values for x and y into the fourth equation, we find $t = \frac{1}{2}$. Thus the required solution is $x = 1, y = 3, z = -2, t = \frac{1}{2}$. We can check the solution by substituting these values into either the second or third equation. ◀◀

Exercises 16–1

In Exercises 1–14, evaluate the given determinants by expansion by minors.

1. $\begin{vmatrix} 3 & 0 & 0 \\ -2 & 1 & 4 \\ 4 & -2 & 5 \end{vmatrix}$

2. $\begin{vmatrix} 10 & 0 & -3 \\ -2 & -4 & 1 \\ 3 & 0 & 2 \end{vmatrix}$

3. $\begin{vmatrix} -2 & -4 & 2 \\ 1 & 3 & 0 \\ -4 & 5 & 2 \end{vmatrix}$

4. $\begin{vmatrix} 5 & -1 & 2 \\ 8 & 3 & -4 \\ 0 & 2 & -6 \end{vmatrix}$

5. $\begin{vmatrix} -6 & -1 & 3 \\ 2 & -2 & -3 \\ 10 & 1 & -2 \end{vmatrix}$

6. $\begin{vmatrix} 9 & -3 & 1 \\ -1 & 2 & -1 \\ 2 & -1 & 3 \end{vmatrix}$

7. $\begin{vmatrix} -30 & -25 & 54 \\ 12 & 21 & -14 \\ 37 & -46 & 24 \end{vmatrix}$

8. $\begin{vmatrix} 4.8 & -3.7 & 3.6 \\ -3.8 & 5.7 & 6.5 \\ 2.1 & -1.8 & 2.3 \end{vmatrix}$

9. $\begin{vmatrix} 1 & 0 & 1 & 0 \\ 2 & 4 & -3 & 1 \\ 1 & 1 & 1 & 1 \\ 3 & 5 & 0 & 2 \end{vmatrix}$

10. $\begin{vmatrix} 2 & 0 & 3 & 1 \\ -1 & -1 & 4 & 0 \\ 1 & 2 & 1 & 2 \\ 3 & 3 & -2 & -1 \end{vmatrix}$

11. $\begin{vmatrix} 1 & 2 & -1 & -2 \\ 3 & 1 & 2 & 1 \\ -1 & 3 & -1 & 2 \\ 2 & 1 & 3 & -3 \end{vmatrix}$

12. $\begin{vmatrix} 3 & -1 & 2 & -5 \\ 1 & 4 & 2 & 5 \\ -1 & 1 & 1 & 3 \\ 1 & 2 & -1 & -2 \end{vmatrix}$

13. $\begin{vmatrix} 1 & 2 & 1 & 2 & 1 \\ 1 & 0 & 0 & 1 & 0 \\ 0 & 1 & 1 & 0 & 1 \\ 1 & 1 & 2 & 2 & 1 \\ 0 & 1 & 1 & 0 & 2 \end{vmatrix}$

14. $\begin{vmatrix} 3 & 1 & 1 & 1 & 2 \\ 1 & 1 & 0 & 0 & 1 \\ 1 & 1 & 2 & 2 & 3 \\ 0 & 2 & 1 & 0 & 3 \\ 1 & 1 & 0 & 1 & 0 \end{vmatrix}$

In Exercises 15 and 16, determine the number of terms in the complete expansion of a determinant of the given order. Explain how this number of terms was found.

15. Fourth order

16. Sixth order

In Exercises 17–24, solve the given systems of equations by determinants. Evaluate the determinants by expansion by minors.

17. $2x + y + z = 6$
$x - 2y + 2z = 10$
$3x - y - z = 4$

18. $2x + y = -1$
$4x - 2y - z = 5$
$2x + 3y + 3z = -2$

19. $3x + 6y + 2z = -2$
$x + 3y - 4z = 2$
$2x - 3y - 2z = -2$

20. $x + 3y + z = 4$
$2x - 6y - 3z = 10$
$4x - 9y + 3z = 4$

21. $x + t = 0$
$3x + y + z = -1$
$2y - z + 3t = 1$
$2z - 3t = 1$

22. $2x + y + z = 4$
$2y - 2z - t = 3$
$3y - 3z + 2t = 1$
$6x - y + t = 0$

23. $x + 2y - z = 6$
$y - 2z - 3t = -5$
$3x - 2y + t = 2$
$2x + y + z - t = 0$

24. $2x + 3y + z = 4$
$x - 2y - 3z + 4t = -1$
$3x + y + z - 5t = 3$
$-x + 2y + z + 3t = 2$

In Exercises 25–28, solve the indicated systems by determinants, using methods of this section.

25. In applying Kirchhoff's laws (see Exercise 20 of Section 5-6) to the given electric circuit, the following equations are found. Determine the indicated currents in amperes (see Fig. 16-1).

$$\begin{aligned} I_A + I_B + I_C + I_D &= 0 \\ 2I_A - I_B &= -2 \\ 3I_C - 2I_D &= 0 \\ I_B - 3I_C &= 6 \end{aligned}$$

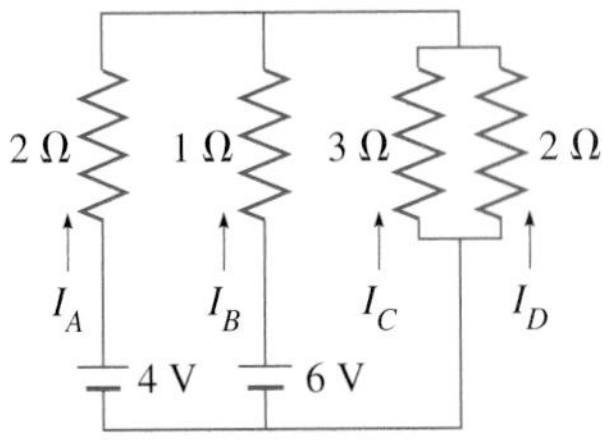

2 Ω 1 Ω 3 Ω 2 Ω
I_A I_B I_C I_D
4 V 6 V

Fig. 16-1

26. In analyzing the motion of four equal particles that are equally spaced along a string, the equation shown below is found. Here, C depends on the string and the mass of each object. Solve for C ($C > 0$).

$$\begin{vmatrix} C & -1 & 0 & 0 \\ -1 & C & -1 & 0 \\ 0 & -1 & C & -1 \\ 0 & 0 & -1 & C \end{vmatrix} = 0$$

27. A firm plans to produce three different appliances, A, B, and C. Total production is to be 1500 appliances each week. A total of 3500 worker-hours are available for production, and 950 worker-hours are available for inspection. Each type A appliance requires 3 worker-hours for production and 1 worker-hour for inspection. Each type B appliance requires 2 worker-hours for production and 20 worker-minutes for inspection, and each type C appliance requires 2 worker-hours for production and 30 worker-minutes for inspection. How many of each are to be produced each week?

28. An alloy is to be made from four other alloys containing copper (Cu), nickel (Ni), zinc (Zn), and iron (Fe). The first is 80% Cu and 20% Ni. The second is 60% Cu, 20% Ni, and 20% Zn. The third is 30% Cu, 60% Ni, and 10% Fe. The fourth is 20% Ni, 40% Zn, and 40% Fe. How much is needed such that the final alloy has 56 g of copper, 28 g of nickel, 10 g of zinc, and 6 g of iron?

16-2 Some Properties of Determinants

Expansion of determinants by minors allows us to evaluate a determinant of any order. However, even a fourth-order determinant usually requires a great deal of calculational work to evaluate. Most graphing calculators and many computer programs can be used to quickly evaluate determinants up to the sixth order, or possibly a higher order. On a graphing calculator, the MATRIX feature is used.

In this section we briefly present some basic properties of determinants with which they can be evaluated, usually with much less work than using minors. These properties illustrate how determinants can be evaluated without a special calculator or computer program. Following are these properties and an illustration of each.

1. *If each element above or each element below the principal diagonal of a determinant is zero, then the product of the elements of the principal diagonal is the value of the determinant.*

EXAMPLE 1 ▸▸ Following property 1, we have

$$\begin{vmatrix} 2 & 1 & 5 & 8 \\ 0 & -5 & 7 & 9 \\ 0 & 0 & 4 & -6 \\ 0 & 0 & 0 & 3 \end{vmatrix} = 2(-5)(4)(3) = -120$$

Since all the elements below the principal diagonal are zero, there is no need to expand the determinant. Note, however, that if the determinant is expanded by the first column, and successive determinants are expanded by their first columns, the same value is found. ◂◂

2. *If **all** corresponding rows and columns of a determinant are interchanged, the value of the determinant is unchanged.*

EXAMPLE 2 ▸▸ Interchanging the first row and first column, the second row and second column, and the third row and third column of the determinant

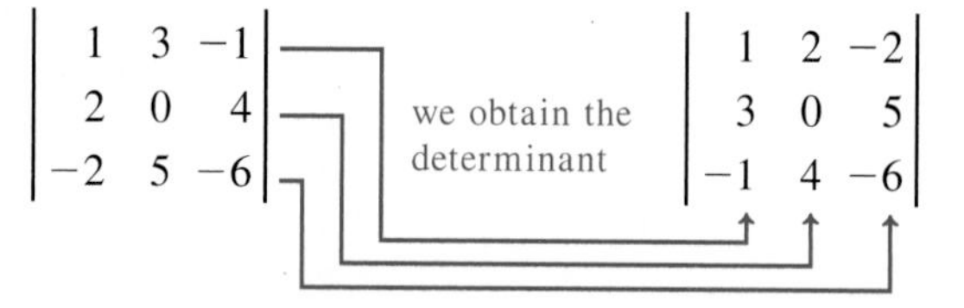

$$\begin{vmatrix} 1 & 3 & -1 \\ 2 & 0 & 4 \\ -2 & 5 & -6 \end{vmatrix} \qquad \begin{vmatrix} 1 & 2 & -2 \\ 3 & 0 & 5 \\ -1 & 4 & -6 \end{vmatrix}$$

By expanding, we can show that the value of each determinant is -18. We obtain very similar expansions with equal values if we expand the first by the first column and the second by the first row. ◂◂

3. *If two columns (or rows) of a determinant are identical, the value of the determinant is zero.*

EXAMPLE 3 ▸▸ Property 3 tells us that

$$\begin{vmatrix} 3 & 5 & 2 \\ -4 & 6 & 9 \\ -4 & 6 & 9 \end{vmatrix} = 0$$

identical (rows 2 and 3)

since the second and third rows are identical. This is verified by expanding by minors of the first row. All of these minors have a value of zero. ◂◂

4. *If two columns (or rows) of a determinant are interchanged, the value of the determinant is changed in sign.*

EXAMPLE 4 ▸▸ The values of the determinants

$$\begin{vmatrix} 3 & 0 & 2 \\ 1 & 1 & 5 \\ 2 & 1 & 3 \end{vmatrix} \quad \text{and} \quad \begin{vmatrix} 2 & 0 & 3 \\ 5 & 1 & 1 \\ 3 & 1 & 2 \end{vmatrix}$$

differ in sign, since the first and third columns are interchanged. By expanding, we find that the value of the first determinant is -8 and the value of the second determinant is 8. ◂◂

5. *If all elements of a column (or row) are multiplied by the same number k, the value of the determinant is multiplied by k.*

EXAMPLE 5 ▸▸ From Property 5 we know that

$$\begin{vmatrix} -1 & 0 & 6 \\ 6 & 3 & -6 \\ 0 & 5 & 3 \end{vmatrix} = 3\begin{vmatrix} -1 & 0 & 6 \\ 2 & 1 & -2 \\ 0 & 5 & 3 \end{vmatrix}$$

since each element of the second row in the left determinant is three times the corresponding element of the second row in the right determinant, and *the other rows are unchanged.*

By expansion, we can show that

$$\begin{vmatrix} -1 & 0 & 6 \\ 6 & 3 & -6 \\ 0 & 5 & 3 \end{vmatrix} = 141 \quad \text{and} \quad \begin{vmatrix} -1 & 0 & 6 \\ 2 & 1 & -2 \\ 0 & 5 & 3 \end{vmatrix} = 47$$

which verifies the original determinant equation, since $141 = 3(47)$. ◂◂

6. *If all the elements of any column (or row) are multiplied by the same number k, and the resulting numbers are added to the corresponding elements of another column (or row), the value of the determinant is unchanged.*

EXAMPLE 6 ▸▸ The value of the following determinant is unchanged if we multiply each element of the first row by 2 and add these numbers to the corresponding elements of the second row. That is,

$4 \times 2 = 8 \qquad -1 \times 2 = -2 \qquad 3 \times 2 = 6$

$$\begin{vmatrix} 4 & -1 & 3 \\ 2 & 2 & 1 \\ 1 & 0 & -3 \end{vmatrix} = \begin{vmatrix} 4 & -1 & 3 \\ 2+8 & 2+(-2) & 1+6 \\ 1 & 0 & -3 \end{vmatrix} = \begin{vmatrix} 4 & -1 & 3 \\ 10 & 0 & 7 \\ 1 & 0 & -3 \end{vmatrix} \quad \text{or} \quad \begin{vmatrix} 4 & -1 & 3 \\ 10 & 0 & 7 \\ 1 & 0 & -3 \end{vmatrix} = \begin{vmatrix} 4 & -1 & 3 \\ 2 & 2 & 1 \\ 1 & 0 & -3 \end{vmatrix}$$

When each determinant is expanded, the value -37 is obtained. The great value in Property 6 is that by its use we can purposely place zeros in the resulting determinant. ◂◂

With the use of these six properties, determinants of higher order can be evaluated much more easily. The technique is to

obtain zeros in a given column (or row) in all positions except one.

We can then expand by this column (or row), thereby reducing the order of the determinant. Property 6 is probably the most valuable for obtaining the zeros.

EXAMPLE 7 ▸▸ Using these properties, we evaluate the following determinant.

$$\begin{vmatrix} 3 & 2 & -1 & 1 \\ -1 & 1 & 2 & 3 \\ 2 & 2 & 1 & 4 \\ 0 & -1 & -2 & 2 \end{vmatrix} = \begin{vmatrix} 0 & 5 & 5 & 10 \\ -1 & 1 & 2 & 3 \\ 2 & 2 & 1 & 4 \\ 0 & -1 & -2 & 2 \end{vmatrix}$$

Each element of the second row is multiplied by 3, and the resulting numbers are added to the corresponding elements of the first row. Here we have used Property 6. In this way a zero has been placed in column 1, row 1.

$$= \begin{vmatrix} 0 & 5 & 5 & 10 \\ -1 & 1 & 2 & 3 \\ 0 & 4 & 5 & 10 \\ 0 & -1 & -2 & 2 \end{vmatrix}$$

Each element of the second row is multiplied by 2, and the resulting numbers are added to the corresponding elements of the third row. Again, we have used Property 6. Also, a zero has been placed in the first column, third row. We now have three zeros in the first column.

$$= -(-1)\begin{vmatrix} 5 & 5 & 10 \\ 4 & 5 & 10 \\ -1 & -2 & 2 \end{vmatrix}$$

Expand the determinant by the first column. We have now reduced the determinant to a third-order determinant.

$$= 5\begin{vmatrix} 1 & 1 & 2 \\ 4 & 5 & 10 \\ -1 & -2 & 2 \end{vmatrix}$$

Factor 5 from each element of the first row. Here we are using Property 5.

$$= 5(2)\begin{vmatrix} 1 & 1 & 1 \\ 4 & 5 & 5 \\ -1 & -2 & 1 \end{vmatrix}$$

Factor 2 from each element of the third column. Again we are using Property 5. Also, by doing this we have reduced the size of the numbers, and the resulting numbers are somewhat easier to work with.

$$= 10\begin{vmatrix} 1 & 1 & 1 \\ 0 & 1 & 1 \\ -1 & -2 & 1 \end{vmatrix}$$

Each element of the first row is multiplied by -4, and the resulting numbers are added to the corresponding elements of the second row. Here we are using Property 6. We have placed a zero in the first column, second row.

$$= 10\begin{vmatrix} 1 & 1 & 1 \\ 0 & 1 & 1 \\ 0 & -1 & 2 \end{vmatrix}$$

Each element of the first row is added to the corresponding element of the third row. Again, we have used Property 6. A zero has been placed in the first column, third row. We now have two zeros in the first column.

$$= 10(1)\begin{vmatrix} 1 & 1 \\ -1 & 2 \end{vmatrix}$$

Expand the determinant by the first column.

$$= 10(2 + 1) = 30$$

Expand the second-order determinant.

A somewhat more systematic method is to place zeros below the principal diagonal and then use Property 1. ◂◂

Exercises 16–2

In Exercises 1–8, evaluate each of the determinants by inspection. Careful observation will allow evaluation by using one or more of the properties of this section.

1. $\begin{vmatrix} 4 & -5 & 8 \\ 0 & 3 & -8 \\ 0 & 0 & -5 \end{vmatrix}$

2. $\begin{vmatrix} 6 & 4 & 0 \\ 0 & -2 & 3 \\ 0 & 0 & -6 \end{vmatrix}$

3. $\begin{vmatrix} -2 & 0 & 0 \\ 15 & 4 & 0 \\ 2 & -7 & 7 \end{vmatrix}$

4. $\begin{vmatrix} 3 & 0 & 0 \\ 0 & 10 & 0 \\ -9 & -1 & -5 \end{vmatrix}$

5. $\begin{vmatrix} -2 & 0 & -1 \\ 5 & 0 & 3 \\ 3 & 0 & -4 \end{vmatrix}$

6. $\begin{vmatrix} -6 & -3 & 1 \\ 1 & 2 & -5 \\ 0 & 0 & 0 \end{vmatrix}$

7. $\begin{vmatrix} 3 & -2 & 4 & 2 \\ 5 & -1 & 2 & -1 \\ 3 & -2 & 4 & 2 \\ 0 & 3 & -6 & 0 \end{vmatrix}$

8. $\begin{vmatrix} -12 & -24 & -24 & 15 \\ 12 & 32 & 32 & -35 \\ -22 & 18 & 18 & 18 \\ 44 & 0 & 0 & -26 \end{vmatrix}$

In Exercises 9–20, evaluate the determinants using the properties of this section. Do not evaluate directly more than one second-order determinant for each.

9. $\begin{vmatrix} 3 & 1 & 0 \\ -2 & 3 & -1 \\ 4 & 2 & 5 \end{vmatrix}$

10. $\begin{vmatrix} 6 & -1 & 3 \\ 0 & 2 & -2 \\ -1 & 4 & 3 \end{vmatrix}$

11. $\begin{vmatrix} 5 & -1 & -2 \\ 3 & -5 & -2 \\ 1 & 4 & 6 \end{vmatrix}$

12. $\begin{vmatrix} -4 & 3 & -2 \\ -2 & 2 & 4 \\ -1 & 5 & -3 \end{vmatrix}$

13. $\begin{vmatrix} 4 & 3 & 6 & 0 \\ 3 & 0 & 0 & 4 \\ 5 & 0 & 1 & 2 \\ 2 & 1 & 1 & 7 \end{vmatrix}$

14. $\begin{vmatrix} -2 & 1 & 3 & 0 \\ 1 & 3 & 0 & 0 \\ 0 & 2 & -3 & -1 \\ 4 & -1 & 2 & 1 \end{vmatrix}$

15. $\begin{vmatrix} 3 & 1 & 2 & -1 \\ 2 & -1 & 3 & -1 \\ 1 & 2 & 1 & 3 \\ 1 & -2 & -3 & 2 \end{vmatrix}$

16. $\begin{vmatrix} 6 & -3 & -6 & 3 \\ -2 & 1 & 2 & -1 \\ 18 & 7 & -1 & 5 \\ 0 & -1 & 10 & 10 \end{vmatrix}$

17. $\begin{vmatrix} 1 & 3 & -3 & 5 \\ 4 & 2 & 1 & 2 \\ 3 & 2 & -2 & 2 \\ 0 & 1 & 2 & -1 \end{vmatrix}$

18. $\begin{vmatrix} -2 & 2 & 1 & 3 \\ 1 & 4 & 3 & 1 \\ 4 & 3 & -2 & -2 \\ 3 & -2 & 1 & 5 \end{vmatrix}$

19. $\begin{vmatrix} 1 & 2 & 0 & 1 & 0 \\ 0 & 2 & 1 & 0 & 1 \\ 1 & 0 & -1 & 1 & -1 \\ -2 & 0 & -1 & 2 & 1 \\ 1 & 0 & 2 & -1 & -2 \end{vmatrix}$

20. $\begin{vmatrix} -1 & 3 & 5 & 0 & -5 \\ 0 & 1 & 7 & 3 & -2 \\ 5 & -2 & -1 & 0 & 3 \\ -3 & 0 & 2 & -1 & 3 \\ 6 & 2 & 1 & -4 & 2 \end{vmatrix}$

In Exercises 21–28, solve the given systems of equations by determinants. Evaluate the determinants by the properties given in this section.

21. $2x - y + z = 5$
$x + 2y + 3z = 10$
$3x + 3y + 2z = 5$

22. $2x + y + z = 5$
$x + 3y - 3z = -13$
$3x + 2y - z = -1$

23. $3x + 2y + z = 1$
$9x + 2z = 5$
$6x - 4y - z = 3$

24. $3x + y + 2z = 4$
$x - y + 4z = 2$
$6x + 3y - 2z = 10$

25. $2x + y + z = 2$
$3y - z + 2t = 4$
$y + 2z + t = 0$
$3x + 2z = 4$

26. $2x + y + z = 0$
$x - y + 2t = 2$
$2y + z + 4t = 2$
$5x + 2z + 2t = 4$

27. $x + y + 2z = 1$
$2x - y + t = -2$
$x - y - z - 2t = 4$
$2x - y + 2z - t = 0$

28. $3x + y + t = 0$
$3z + 2t = 8$
$6x + 2y + 2z + t = 3$
$3x - y - z - t = 0$

In Exercises 29–32, solve the given problems by using determinants.

29. In applying Kirchhoff's laws (see Exercise 20 of Section 5-6) to the circuit shown in Fig. 16-2, the following equations are found. Determine the indicated currents, in amperes.

$$I_A + I_B + I_C + I_D + I_E = 0$$
$$-2I_A + 3I_B = 0$$
$$3I_B - 3I_C = 6$$
$$-3I_C + I_D = 0$$
$$-I_D + 2I_E = 0$$

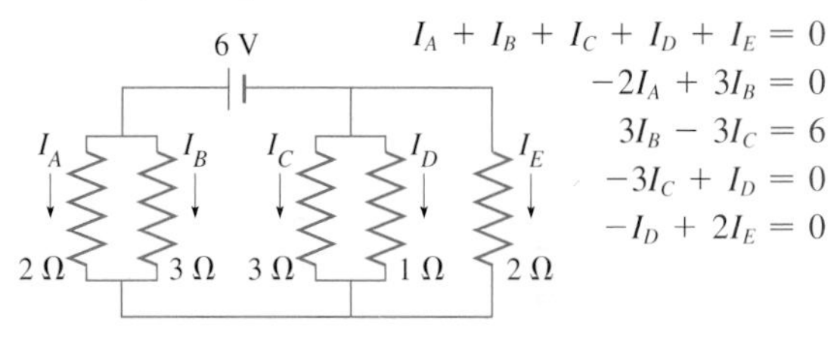

Fig. 16-2

30. In analyzing the forces A, B, C, and D shown on the beam in Fig. 16-3, the following equations are used. Find these forces.

$$A + B = 850$$
$$A + B + 400 = 0.8C + 0.6D$$
$$0.6C = 0.8D$$
$$5A - 5B + 4C - 3D = 0$$

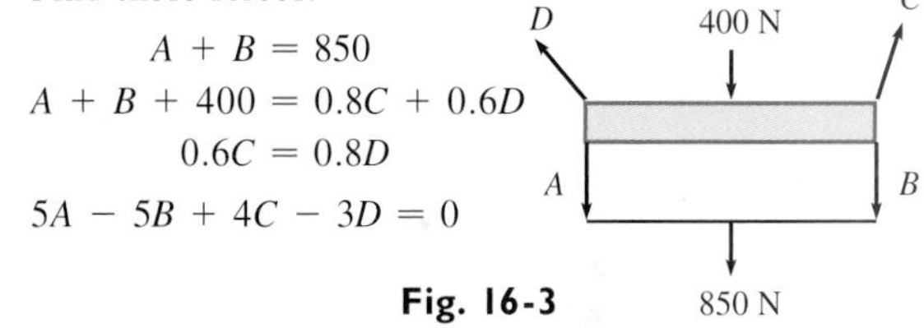

Fig. 16-3

31. In testing for air pollution, a given air sample contained a total of 6.0 parts per million (ppm) of four pollutants, sulfur dioxide (SO_2), nitric oxide (NO), nitrogen dioxide (NO_2), and carbon monoxide (CO). The ppm of CO was 10 times that of SO_2, which in turn equaled those of NO and NO_2. There was a total of 0.8 ppm of SO_2 and NO. How many ppm of each were present in the air sample?

32. A business firm installed a system of 32 computers for \$24 000. Included were four different models costing \$500, \$600, \$1000, and \$1500 each, respectively. There are as many \$600 models as \$1000 models and \$1500 models combined, and twice as many \$500 models as \$1000 models. How many of each model are in the computer system?

16-3 Matrices: Definitions and Basic Operations

Systems of linear equations occur in several areas of important technical and scientific applications. We indicated a few of these in Chapter 5 and in the first two sections of this chapter. The importance of linear systems has led to the development of numerous methods for their solution.

As the use of computers has increased, another mathematical concept that we can use to solve systems of linear equations is now used much more widely than in the past. It is also used in numerous applications other than with systems of equations, in fields such as business, economics, and psychology, as well as the scientific and technical areas. Since it is readily adaptable to use on a computer and on many graphing calculators, its use will continue to increase. At this point, however, we shall only be able to introduce its definitions and basic operations.

A **matrix** *is an ordered rectangular array of numbers.* To distinguish such an array from a determinant, we shall enclose it within parentheses. As with a determinant, *the individual numbers are called* **elements** *of the matrix.*

EXAMPLE 1 ▸▸ Some examples of matrices are shown here.

$$\begin{pmatrix} 2 & 8 \\ 1 & 0 \end{pmatrix} \begin{pmatrix} 2 & -4 & 6 \\ -1 & 0 & 5 \end{pmatrix} \begin{pmatrix} 4 & 6 \\ 0 & -1 \\ -2 & 5 \\ 3 & 0 \end{pmatrix}$$

$$\begin{pmatrix} -1 & 8 & 6 & 7 & 9 \\ 2 & 6 & 0 & 4 & 3 \\ 5 & -1 & 8 & 10 & 2 \end{pmatrix} \quad \begin{pmatrix} -1 & 2 & 0 & 9 \end{pmatrix}$$ ◂◂

As we can see, it is not necessary for the number of columns and number of rows to be the same, although such is the case for a determinant. However, *if the number of rows does equal the number of columns, the matrix is called a* **square matrix.** We shall find that square matrices are of special importance. *If all the elements of a matrix are zero, the matrix is called a* **zero matrix.** It is convenient to designate a given matrix by a capital letter.

CAUTION ▸ We must be careful to distinguish between a matrix and a determinant. ***A matrix is simply any* rectangular array *of numbers, whereas a determinant is a specific value associated with a* square *matrix.***

EXAMPLE 2 ▸▸ Consider the following matrices:

$$A = \begin{pmatrix} 5 & 0 & -1 \\ 1 & 2 & 6 \\ 0 & -4 & -5 \end{pmatrix} \quad B = \begin{pmatrix} 9 \\ 8 \\ 1 \\ 5 \end{pmatrix}, \quad C = \begin{pmatrix} -1 & 6 & 8 & 9 \end{pmatrix}, \quad O = \begin{pmatrix} 0 & 0 \\ 0 & 0 \end{pmatrix}$$

Matrix A is an example of a square matrix, matrix B is an example of a matrix with four rows and one column, matrix C is an example of a matrix with one row and four columns, and matrix O is an example of a zero matrix. ◂◂

To be able to refer to specific elements of a matrix and to give a general representation, a double-subscript notation is usually employed. That is,

$$A = \begin{pmatrix} a_{11} & a_{12} & a_{13} \\ a_{21} & a_{22} & a_{23} \\ a_{31} & a_{32} & a_{33} \end{pmatrix}$$

row ↗↖ column

We see that the first subscript refers to the row in which the element lies, and the second subscript refers to the column in which the element lies.

Two matrices are said to be **equal** *if and only if they are identical.* That is, they must have the same number of columns, the same number of rows, and the elements must respectively be equal.

EXAMPLE 3 ▸▸ (a) $\begin{pmatrix} a_{11} & a_{12} & a_{13} \\ a_{21} & a_{22} & a_{23} \end{pmatrix} = \begin{pmatrix} 1 & -5 & 0 \\ 4 & 6 & -3 \end{pmatrix}$

if and only if $a_{11} = 1$, $a_{12} = -5$, $a_{13} = 0$, $a_{21} = 4$, $a_{22} = 6$, and $a_{23} = -3$.
(b) The matrices

$$\begin{pmatrix} 1 & 2 & 3 \\ -1 & -2 & -5 \end{pmatrix} \text{ and } \begin{pmatrix} 1 & 2 & -5 \\ -1 & -2 & 3 \end{pmatrix}$$

are not equal, since the elements in the third column are reversed.
(c) The matrices

$$\begin{pmatrix} 2 & 3 \\ -1 & 5 \end{pmatrix} \text{ and } \begin{pmatrix} 2 & 3 & 0 \\ -1 & 5 & 0 \end{pmatrix}$$

are not equal, since the number of columns is different. This is true despite the fact that both elements of the third column are zeros. ◂◂

EXAMPLE 4 ▸▸ The forces acting on a bolt are in equilibrium, as shown in Fig. 16-4. Analyzing the horizontal and vertical components as in Section 9-4, we find the following matrix equation. Find forces F_1 and F_2.

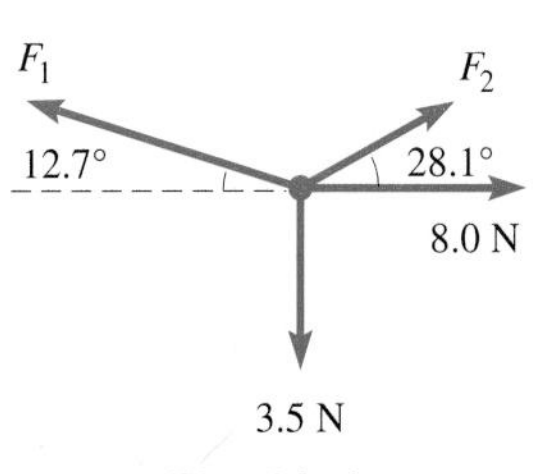

Fig. 16-4

$$\begin{pmatrix} 0.98F_1 - 0.88F_2 \\ 0.22F_1 + 0.47F_2 \end{pmatrix} = \begin{pmatrix} 8.0 \\ 3.5 \end{pmatrix}$$

From the equality of matrices we know that $0.98F_1 - 0.88F_2 = 8.0$ and $0.22F_1 + 0.47F_2 = 3.5$. Therefore, to find the forces F_1 and F_2 we must solve the system of equations

$$0.98F_1 - 0.88F_2 = 8.0$$
$$0.22F_1 + 0.47F_2 = 3.5$$

Using determinants, we have

$$F_1 = \frac{\begin{vmatrix} 8.0 & -0.88 \\ 3.5 & 0.47 \end{vmatrix}}{\begin{vmatrix} 0.98 & -0.88 \\ 0.22 & 0.47 \end{vmatrix}} = \frac{8.0(0.47) - 3.5(-0.88)}{0.98(0.47) - 0.22(-0.88)} = 10.5 \text{ N}$$

Using determinants again, or by substituting this value into either equation, we find that $F_2 = 2.6$ N. These values check when substituted into the original matrix equation. ◂◂

Matrix Addition and Subtraction

If two matrices have the same number of rows and the same number of columns, their **sum** *is defined as the matrix consisting of the sums of the corresponding elements.* If the number of rows or the number of columns of the two matrices is not equal, they cannot be added.

EXAMPLE 5 ▸▸ (a)

$$\begin{pmatrix} 8 & 1 & -5 & 9 \\ 0 & -2 & 3 & 7 \end{pmatrix} + \begin{pmatrix} -3 & 4 & 6 & 0 \\ 6 & -2 & 6 & 5 \end{pmatrix} = \begin{pmatrix} 8+(-3) & 1+4 & -5+6 & 9+0 \\ 0+6 & -2+(-2) & 3+6 & 7+5 \end{pmatrix}$$
$$= \begin{pmatrix} 5 & 5 & 1 & 9 \\ 6 & -4 & 9 & 12 \end{pmatrix}$$

(b) The matrices

$$\begin{pmatrix} 3 & -5 & 8 \\ 2 & 9 & 0 \\ 4 & -2 & 3 \end{pmatrix} \quad \text{and} \quad \begin{pmatrix} 3 & -5 & 8 & 0 \\ 2 & 9 & 0 & 0 \\ 4 & -2 & 3 & 0 \end{pmatrix}$$

cannot be added since the second matrix has one more column than the first matrix. This is true even though the extra column contains only zeros. ◂◂

The product of a number and a matrix (known as **scalar multiplication** *of a matrix) is defined as the matrix whose elements are obtained by multiplying each element of the given matrix by the given number.* That is, kA is the matrix obtained by multiplying the elements of matrix A by k. In this way, $A + A$ and $2A$ will result in the same matrix.

EXAMPLE 6 ▸▸ For the matrix A, where

$$A = \begin{pmatrix} -5 & 7 \\ 3 & 0 \end{pmatrix} \quad \text{we have} \quad 2A = \begin{pmatrix} 2(-5) & 2(7) \\ 2(3) & 2(0) \end{pmatrix} = \begin{pmatrix} -10 & 14 \\ 6 & 0 \end{pmatrix}$$ ◂◂

By combining the definitions for the addition of matrices and for the scalar multiplication of a matrix, we can define the subtraction of matrices. That is, *the* **difference** *of matrices A and B is given by* $A - B = A + (-B)$. Therefore, we change the sign of each element of B, and proceed as in addition.

By the preceding definitions we see that the operations of addition, subtraction, and multiplication of a matrix by a number are like those for real numbers. For these operations, we say that the algebra of matrices is like the algebra of real numbers. We can see that the following laws hold for matrices.

$A + B = B + A$	(commutative law)	**(16-4)**
$A + (B + C) = (A + B) + C$	(associative law)	**(16-5)**
$k(A + B) = kA + kB$		**(16-6)**
$A + O = A$		**(16-7)**

Here we have let O represent the zero matrix. We shall find in the next section that not all laws for matrix operations are like those for real numbers.

Exercises 16–3

In Exercises 1–8, determine the value of the literal numbers in each of the given matrix equalities.

1. $\begin{pmatrix} a & b \\ c & d \end{pmatrix} = \begin{pmatrix} 1 & -3 \\ 4 & 7 \end{pmatrix}$

2. $\begin{pmatrix} x & y & z \\ r & -s & -t \end{pmatrix} = \begin{pmatrix} -2 & 7 & -9 \\ 4 & -4 & 5 \end{pmatrix}$

3. $\begin{pmatrix} x \\ x + y \end{pmatrix} = \begin{pmatrix} 2 \\ 5 \end{pmatrix}$

4. $(a + bj \quad 2c - dj \quad 3e + fj) = (5j \quad a + 6 \quad 3b + c)$ $(j = \sqrt{-1})$

5. $\begin{pmatrix} x & x + y \\ x - z & y + z \\ x + t & y - t \end{pmatrix} = \begin{pmatrix} 2 & 3 \\ 4 & -1 \end{pmatrix}$

6. $\begin{pmatrix} 2x - 3y \\ x + 4y \end{pmatrix} = \begin{pmatrix} 13 \\ 1 \end{pmatrix}$

7. $\begin{pmatrix} x \\ x + 2 \\ 2y - 3 \end{pmatrix} = \begin{pmatrix} 4 \\ y \\ z \end{pmatrix}$

8. $\begin{pmatrix} x & y & z \\ x + y & 2x - y & x + 2 \end{pmatrix} = \begin{pmatrix} 2 & -3 \\ z & t \end{pmatrix}$

In Exercises 9–12, find the indicated sums of matrices.

9. $\begin{pmatrix} 2 & 3 \\ -5 & 4 \end{pmatrix} + \begin{pmatrix} -1 & 7 \\ 5 & -2 \end{pmatrix}$

10. $\begin{pmatrix} 1 & 0 & 9 \\ 3 & -5 & -2 \end{pmatrix} + \begin{pmatrix} 4 & -1 & 7 \\ 2 & 0 & -3 \end{pmatrix}$

11. $\begin{pmatrix} 50 & -82 \\ -34 & 57 \\ -15 & 62 \end{pmatrix} + \begin{pmatrix} -55 & 82 \\ 45 & 14 \\ 26 & -67 \end{pmatrix}$

12. $\begin{pmatrix} 4.7 & 2.1 & -9.6 \\ -6.8 & 4.8 & 7.4 \\ -1.9 & 0.7 & 5.9 \end{pmatrix} + \begin{pmatrix} -4.9 & -9.6 & -2.1 \\ 3.4 & 0.7 & 0.0 \\ 5.6 & 10.1 & -1.6 \end{pmatrix}$

In Exercises 13–20, use the following matrices to find the indicated matrices.

$$A = \begin{pmatrix} -1 & 4 & -7 & 0 \\ 2 & -6 & -1 & 2 \end{pmatrix}, \quad B = \begin{pmatrix} 1 & 5 & -6 & 3 \\ 4 & -1 & 8 & -2 \end{pmatrix},$$

$$C = \begin{pmatrix} 3 & -6 & 9 \\ -4 & 1 & 2 \end{pmatrix}$$

13. $A + B$ 14. $A - B$

15. $A + C$ 16. $B + C$

17. $2A + B$ 18. $2B + A$

19. $A - 2B$ 20. $3A - B$

In Exercises 21–24, use matrices A and B to show that the indicated laws hold for these matrices.

$$A = \begin{pmatrix} -1 & 2 & 3 & 7 \\ 0 & -3 & -1 & 4 \\ 9 & -1 & 0 & -2 \end{pmatrix}, \quad B = \begin{pmatrix} 4 & -1 & -3 & 0 \\ 5 & 0 & -1 & 1 \\ 1 & 11 & 8 & 2 \end{pmatrix}$$

21. $A + B = B + A$ 22. $A + O = A$

23. $-(A - B) = B - A$ 24. $3(A + B) = 3A + 3B$

In Exercises 25 and 26, find the unknown quantities in the given matrix equations.

25. An airplane is flying in a direction 21.0° north of east at 235 km/h but is headed 14.5° north of east. The wind is from the southeast. Find the speed of the wind v_w and the speed of the plane v_p relative to the wind from the given matrix equation. See Fig. 16-5.

$$\begin{pmatrix} v_p \cos 14.5° - v_w \cos 45.0° \\ v_p \sin 14.5° + v_w \sin 45.0° \end{pmatrix} = \begin{pmatrix} 235 \cos 21.0° \\ 235 \sin 21.0° \end{pmatrix}$$

Fig. 16-5

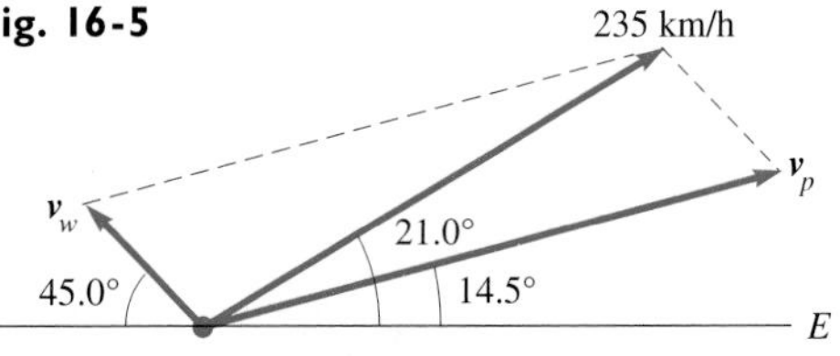

26. Find the electric currents shown in Fig. 16-6 by solving the following matrix equation.

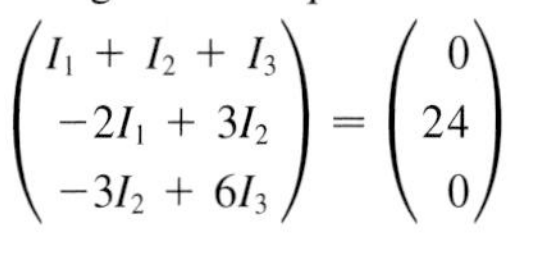

$$\begin{pmatrix} I_1 + I_2 + I_3 \\ -2I_1 + 3I_2 \\ -3I_2 + 6I_3 \end{pmatrix} = \begin{pmatrix} 0 \\ 24 \\ 0 \end{pmatrix}$$

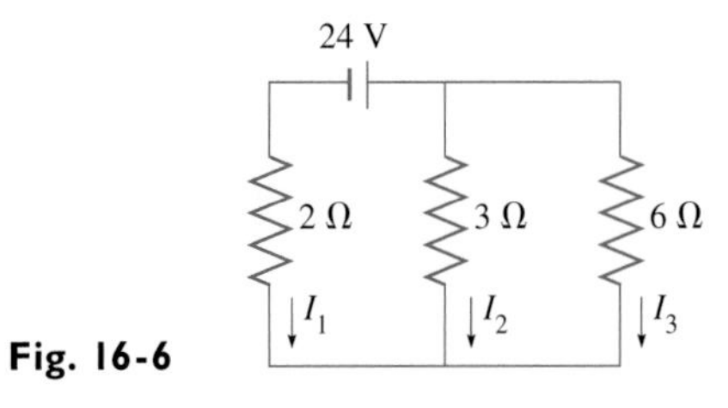

Fig. 16-6

In Exercises 27 and 28, perform the indicated matrix operations.

27. The contractor of a housing development constructs four different types of houses, with either a carport, a one-car garage, or a two-car garage. The following matrix shows the number of houses of each type and the type of garage.

	Type A	Type B	Type C	Type D
Carport	8	6	0	0
1-car garage	5	4	3	0
2-car garage	0	3	5	6

If the contractor builds two additional identical developments, find the matrix showing the total number of each house–garage type built.

28. The inventory of a drug supply company shows that the following numbers of cases of bottles of vitamins C and E are in stock: Vitamin C—25 cases of 100-mg bottles, 10 cases of 250-mg bottles, and 32 cases of 500-mg bottles; vitamin E—30 cases of 100-mg bottles, 18 cases of 250-mg bottles, and 40 cases of 500-mg bottles. This can be represented by matrix A below. After two shipments are sent, each of which can be represented by matrix B below, find the matrix representing the remaining inventory.

$$A = \begin{pmatrix} 25 & 10 & 32 \\ 30 & 18 & 40 \end{pmatrix} \qquad B = \begin{pmatrix} 10 & 5 & 6 \\ 12 & 4 & 8 \end{pmatrix}$$

16-4 Multiplication of Matrices

The definition for the multiplication of matrices does not have an intuitive basis. However, through the solution of a system of linear equations we can, at least in part, show why multiplication is defined as it is. Consider Example 1.

EXAMPLE 1 ▸▸ If we solve the system of equations

$$2x + y = 1$$
$$7x + 3y = 5$$

we get $x = 2$, $y = -3$. Checking this solution in each of the equations, we get

$$2(2) + 1(-3) = 1$$
$$7(2) + 3(-3) = 5$$

Let us represent the coefficients of the equations by the matrix $\begin{pmatrix} 2 & 1 \\ 7 & 3 \end{pmatrix}$ and the solutions by the matrix $\begin{pmatrix} 2 \\ -3 \end{pmatrix}$. If we now indicate the multiplication of these matrices and perform it as shown

$$\begin{pmatrix} 2 & 1 \\ 7 & 3 \end{pmatrix}\begin{pmatrix} 2 \\ -3 \end{pmatrix} = \begin{pmatrix} 2(2) + 1(-3) \\ 7(2) + 3(-3) \end{pmatrix} = \begin{pmatrix} 1 \\ 5 \end{pmatrix}$$

we note that we obtain a matrix which properly represents the right-side values of the equations. (Note the products and sums in the resulting matrix.) ◂◂

Following reasons along the lines indicated in Example 1, we shall now define the **multiplication of matrices.** If the number of columns in a first matrix equals the number of rows in a second matrix, the product of these matrices is formed as follows: *The element in a specified row and a specified column of the product matrix is the sum of the products formed by multiplying each element in the specified row of the first matrix by the corresponding element in the specific column of the second matrix.* The product matrix will have the same number of rows as the first matrix and the same number of columns as the second matrix. Consider the following examples.

EXAMPLE 2 ▸▸ Find the product AB, where

$$A = \begin{pmatrix} 2 & 1 \\ -3 & 0 \\ 1 & 2 \end{pmatrix} \quad \text{and} \quad B = \begin{pmatrix} -1 & 6 & 5 & -2 \\ 3 & 0 & 1 & -4 \end{pmatrix}$$

Since there are two columns in matrix A and two rows in matrix B, the product can be formed. To find the element in the first row and first column of the product, we find the sum of the products of corresponding elements of the first row of A and first column of B. To find the element in the first row and second column of the product, we find the sum of the products of corresponding elements in the first row of A and the second column of B. We continue this process until we have found the three rows (the number of rows in A) and the four columns (the number of columns in B) of the product. The product is formed as follows:

$$\begin{pmatrix} 2 & 1 \\ -3 & 0 \\ 1 & 2 \end{pmatrix}\begin{pmatrix} -1 & 6 & 5 & -2 \\ 3 & 0 & 1 & -4 \end{pmatrix} = \begin{pmatrix} 2(-1) + 1(3) & 2(6) + 1(0) & 2(5) + 1(1) & 2(-2) + 1(-4) \\ -3(-1) + 0(3) & -3(6) + 0(0) & -3(5) + 0(1) & -3(-2) + 0(-4) \\ 1(-1) + 2(3) & 1(6) + 2(0) & 1(5) + 2(1) & 1(-2) + 2(-4) \end{pmatrix}$$

$$= \begin{pmatrix} 1 & 12 & 11 & -8 \\ 3 & -18 & -15 & 6 \\ 5 & 6 & 7 & -10 \end{pmatrix}$$

The specific combination of elements used to form the element in the first row and first column and the element in the third row and second column of the product are outlined in color.

If we attempt to form the product BA, we find that B has four columns and A has three rows. Since the number of columns in B does not equal the number of rows in A, the product BA cannot be formed. In this way we see that $AB \neq BA$, which means that *matrix multiplication is not commutative (except in special cases).* Therefore, we see that matrix multiplication differs from the multiplication of real numbers. ◂◂

EXAMPLE 3 ▸▸ The product of two matrices below may be formed because the first matrix has four columns and the second matrix has four rows. The matrix is formed as shown.

$$\begin{pmatrix} -1 & 9 & 3 & -2 \\ 2 & 0 & -7 & 1 \end{pmatrix}\begin{pmatrix} 6 & -2 \\ 1 & 0 \\ 3 & -5 \\ 3 & 9 \end{pmatrix} = \begin{pmatrix} -1(6) + 9(1) + 3(3) + (-2)(3) & -1(-2) + 9(0) + 3(-5) + (-2)(9) \\ 2(6) + 0(1) + (-7)(3) + 1(3) & 2(-2) + 0(0) + (-7)(-5) + 1(9) \end{pmatrix}$$

$$= \begin{pmatrix} -6 + 9 + 9 - 6 & 2 + 0 - 15 - 18 \\ 12 + 0 - 21 + 3 & -4 + 0 + 35 + 9 \end{pmatrix}$$

$$= \begin{pmatrix} 6 & -31 \\ -6 & 40 \end{pmatrix}$$ ◂◂

Identity Matrix

There are two special matrices of particular importance in the multiplication of matrices. The first of these is the **identity matrix I,** *which is a square matrix with 1's for elements of the principal diagonal with all other elements zero.* (The principal diagonal starts with the element a_{11}.) It has the property that if it is multiplied by another square matrix with the same number of rows and columns, then the second matrix equals the product matrix.

EXAMPLE 4 ▸▸ Show that $AI = IA = A$ for the matrix

$$A = \begin{pmatrix} 2 & -3 \\ 4 & 1 \end{pmatrix}$$

Since A has two rows and two columns, we choose I with two rows and two columns. Therefore, for this case

$$I = \begin{pmatrix} 1 & 0 \\ 0 & 1 \end{pmatrix} \longleftarrow \text{elements of principal diagonal are 1's}$$

Forming the indicated products, we have results as follows:

$$\begin{aligned} AI &= \begin{pmatrix} 2 & -3 \\ 4 & 1 \end{pmatrix}\begin{pmatrix} 1 & 0 \\ 0 & 1 \end{pmatrix} \\ &= \begin{pmatrix} 2(1) + (-3)(0) & 2(0) + (-3)(1) \\ 4(1) + 1(0) & 4(0) + 1(1) \end{pmatrix} = \begin{pmatrix} 2 & -3 \\ 4 & 1 \end{pmatrix} \\ IA &= \begin{pmatrix} 1 & 0 \\ 0 & 1 \end{pmatrix}\begin{pmatrix} 2 & -3 \\ 4 & 1 \end{pmatrix} \\ &= \begin{pmatrix} 1(2) + 0(4) & 1(-3) + 0(1) \\ 0(2) + 1(4) & 0(-3) + 1(1) \end{pmatrix} = \begin{pmatrix} 2 & -3 \\ 4 & 1 \end{pmatrix} \end{aligned}$$

Therefore, we see that $AI = IA = A$. ◂◂

Inverse of a Matrix

For a given square matrix A, its **inverse** A^{-1} is the other important special matrix. *The matrix A and its inverse A^{-1} have the property that*

$$AA^{-1} = A^{-1}A = I \tag{16-8}$$

If the product of two square matrices equals the identity matrix, the matrices are called inverses of each other. Under certain conditions the inverse of a given square matrix may not exist, although for most square matrices the inverse does exist. In the next section we shall develop the procedure for finding the inverse of a square matrix, and the section that follows shows how the inverse is used in the solution of systems of equations. At this point we simply show that the product of certain matrices equals the identity matrix and that therefore these matrices are inverses of each other.

EXAMPLE 5 ▸▸ For the given matrices A and B, show that $AB = BA = I$, and therefore that $B = A^{-1}$.

$$A = \begin{pmatrix} 1 & -3 \\ -2 & 7 \end{pmatrix}, \quad B = \begin{pmatrix} 7 & 3 \\ 2 & 1 \end{pmatrix}$$

Forming the products AB and BA, we have the following:

$$AB = \begin{pmatrix} 1 & -3 \\ -2 & 7 \end{pmatrix}\begin{pmatrix} 7 & 3 \\ 2 & 1 \end{pmatrix} = \begin{pmatrix} 7 - 6 & 3 - 3 \\ -14 + 14 & -6 + 7 \end{pmatrix} = \begin{pmatrix} 1 & 0 \\ 0 & 1 \end{pmatrix}$$

$$BA = \begin{pmatrix} 7 & 3 \\ 2 & 1 \end{pmatrix}\begin{pmatrix} 1 & -3 \\ -2 & 7 \end{pmatrix} = \begin{pmatrix} 7 - 6 & -21 + 21 \\ 2 - 2 & -6 + 7 \end{pmatrix} = \begin{pmatrix} 1 & 0 \\ 0 & 1 \end{pmatrix}$$

Since $AB = I$ and $BA = I$, $B = A^{-1}$ and $A = B^{-1}$. ◂◂

The following example illustrates one kind of application of the multiplication of matrices.

See the chapter introduction.

EXAMPLE 6 ▸▸ A company makes three types of machine parts. In one day it produces 40 of type X, 50 of type Y, and 80 of type Z. Each of type X requires 4 units of material and 1 worker-hour to produce; each of type Y requires 5 units of material and 2 worker-hours to produce; each of type Z requires 3 units of material and 2 worker-hours to produce. By representing the number of each type produced as matrix A, and the material and time requirements as matrix B, we have

$$A = \begin{pmatrix} 40 & 50 & 80 \end{pmatrix} \qquad B = \begin{pmatrix} 4 & 1 \\ 5 & 2 \\ 3 & 2 \end{pmatrix}$$

(In A: columns are type X, type Y, type Z — number of each type produced. In B: columns are units of material and worker-hours; rows are type X, type Y, type Z — material and time required for each.)

The product AB gives the total number of units of material and the total number of worker-hours needed for the day's production in a one-row, two-column matrix.

$$AB = \begin{pmatrix} 40 & 50 & 80 \end{pmatrix}\begin{pmatrix} 4 & 1 \\ 5 & 2 \\ 3 & 2 \end{pmatrix}$$

$$= \begin{pmatrix} 160 + 250 + 240 & 40 + 100 + 160 \end{pmatrix} = \begin{pmatrix} 650 & 300 \end{pmatrix}$$

(650: total units of material; 300: total worker-hours.)

Therefore, 650 units of material and 300 worker-hours are required. ◂◂

We now have seen how multiplication is defined for matrices. We see that *matrix multiplication is not commutative;* that is, $AB \neq BA$ in general. This is a major difference from the multiplication of real numbers. Another difference is that it is possible that $AB = O$, even though neither A nor B is O. There are, however, some similarities. We have seen, for example, that $AI = A$, where we make I and the number 1 equivalent for the two types of multiplication. Also, *the distributive property $A(B + C) = AB + AC$ holds for matrix multiplication.* This points out additional properties of the algebra of matrices.

Exercises 16–4

In Exercises 1–12, perform the indicated multiplications.

1. $(4\ \ -2)\begin{pmatrix} -1 & 0 \\ 2 & 6 \end{pmatrix}$

2. $(-1\ \ 5\ \ -2)\begin{pmatrix} 6 & 3 \\ 2 & -1 \\ 0 & 2 \end{pmatrix}$

3. $\begin{pmatrix} 2 & -3 \\ 5 & -1 \end{pmatrix}\begin{pmatrix} 3 & 0 & -1 \\ 7 & -5 & 8 \end{pmatrix}$

4. $\begin{pmatrix} -7 & 8 \\ 5 & 0 \end{pmatrix}\begin{pmatrix} -9 & 10 \\ 1 & 4 \end{pmatrix}$

5. $\begin{pmatrix} 2 & -3 & 1 \\ 0 & 7 & -3 \end{pmatrix}\begin{pmatrix} 9 \\ -2 \\ 5 \end{pmatrix}$

6. $\begin{pmatrix} 0 & -1 & 2 \\ 4 & 11 & 2 \end{pmatrix}\begin{pmatrix} 3 & -1 \\ 1 & 2 \\ 6 & 1 \end{pmatrix}$

7. $\begin{pmatrix} -8 & 9 \\ 7 & -8 \\ -6 & 4 \end{pmatrix}\begin{pmatrix} -5 & 8 \\ -7 & 5 \end{pmatrix}$

8. $\begin{pmatrix} 12 & -47 \\ 43 & -18 \\ 36 & -22 \end{pmatrix}\begin{pmatrix} 25 \\ 66 \end{pmatrix}$

9. $\begin{pmatrix} -1 & 7 \\ 3 & 5 \\ 10 & -1 \\ -5 & 12 \end{pmatrix}\begin{pmatrix} 2 & 1 \\ 5 & -3 \end{pmatrix}$

10. $(13\ \ 22)\begin{pmatrix} -7 & 10 \\ 9 & 18 \end{pmatrix}$

11. $\begin{pmatrix} -9.2 & 2.3 & 0.5 \\ -3.8 & -2.4 & 9.2 \end{pmatrix}\begin{pmatrix} 6.5 & -5.2 \\ 4.9 & 1.7 \\ -1.8 & 6.9 \end{pmatrix}$

12. $\begin{pmatrix} 1 & 2 & -6 & 6 & 1 \\ -2 & 4 & 0 & 1 & 2 \end{pmatrix}\begin{pmatrix} 1 \\ -1 \\ 0 \\ 5 \\ 2 \end{pmatrix}$

In Exercises 13–16, find, if possible, AB and BA.

13. $A = (1\ \ -3\ \ 8),\quad B = \begin{pmatrix} -1 \\ 5 \\ 7 \end{pmatrix}$

14. $A = \begin{pmatrix} -3 & 2 & 0 \\ 1 & -4 & 5 \end{pmatrix},\quad B = \begin{pmatrix} -2 & 0 \\ 4 & -6 \\ 5 & 1 \end{pmatrix}$

15. $A = \begin{pmatrix} -1 & 2 & 3 \\ 5 & -1 & 0 \end{pmatrix},\quad B = \begin{pmatrix} 1 \\ -5 \\ 2 \end{pmatrix}$

16. $A = \begin{pmatrix} -2 & 1 & 7 \\ 3 & -1 & 0 \\ 0 & 2 & -1 \end{pmatrix},\quad B = (4\ \ -1\ \ 5)$

In Exercises 17–20, show that $AI = IA = A$.

17. $A = \begin{pmatrix} 1 & 8 \\ -2 & 2 \end{pmatrix}$

18. $A = \begin{pmatrix} -3 & 4 \\ 1 & 2 \end{pmatrix}$

19. $A = \begin{pmatrix} 1 & 3 & -5 \\ 2 & 0 & 1 \\ 1 & -2 & 4 \end{pmatrix}$

20. $A = \begin{pmatrix} -1 & 2 & 0 \\ 4 & -3 & 1 \\ 2 & 1 & 3 \end{pmatrix}$

In Exercises 21–24, determine whether or not $B = A^{-1}$.

21. $A = \begin{pmatrix} 5 & -2 \\ -2 & 1 \end{pmatrix},\quad B = \begin{pmatrix} 1 & 2 \\ 2 & 5 \end{pmatrix}$

22. $A = \begin{pmatrix} 3 & -4 \\ 5 & -7 \end{pmatrix},\quad B = \begin{pmatrix} 7 & -4 \\ 5 & -2 \end{pmatrix}$

23. $A = \begin{pmatrix} 1 & -2 & 3 \\ 2 & -5 & 7 \\ -1 & 3 & -5 \end{pmatrix},\quad B = \begin{pmatrix} 4 & -1 & 1 \\ 3 & -2 & -1 \\ 1 & -1 & -1 \end{pmatrix}$

24. $A = \begin{pmatrix} 1 & -1 & 3 \\ 3 & -4 & 8 \\ -2 & 3 & -4 \end{pmatrix},\quad B = \begin{pmatrix} 8 & -5 & -4 \\ 4 & -2 & -1 \\ -1 & 1 & 1 \end{pmatrix}$

In Exercises 25–28, determine by matrix multiplication whether or not A is the proper matrix of solution values.

25. $3x - 2y = -1$, $4x + y = 6$ $\qquad A = \begin{pmatrix} 1 \\ 2 \end{pmatrix}$

26. $4x + y = -5$, $3x + 4y = 6$ $\qquad A = \begin{pmatrix} -2 \\ 3 \end{pmatrix}$

27. $3x + y + 2z = 1$, $x - 3y + 4z = -3$, $2x + 2y + z = 1$ $\qquad A = \begin{pmatrix} -1 \\ 2 \\ 1 \end{pmatrix}$

28. $2x - y + z = 7$, $x - 3y + 2z = 6$, $3x + y - z = 8$ $\qquad A = \begin{pmatrix} 3 \\ -2 \\ -1 \end{pmatrix}$

In Exercises 29–36, perform the indicated matrix multiplications.

29. Using two rows and columns, show that $(-I)^2 = I$.

30. For $J = \begin{pmatrix} j & 0 \\ 0 & j \end{pmatrix}$, where $j = \sqrt{-1}$, show that $J^2 = -I$, $J^3 = -J$, and $J^4 = I$. Note the similarity with j^2, j^3, and j^4.

31. Show that $A^2 - I = (A + I)(A - I)$ for $A = \begin{pmatrix} 2 & 4 \\ 3 & 5 \end{pmatrix}$.

32. In the study of polarized light, the matrix product $\begin{pmatrix} 1 & 0 \\ 0 & -j \end{pmatrix}\begin{pmatrix} 1 & 0 \\ 1 & -j \end{pmatrix}\begin{pmatrix} 1 \\ 1 \end{pmatrix}$ occurs $(j = \sqrt{-1})$. Find this product.

33. In studying the motion of electrons, one of the Pauli spin matrices used is $s_y = \begin{pmatrix} 0 & -j \\ j & 0 \end{pmatrix}$, where $j = \sqrt{-1}$. Show that $s_y^2 = I$.

34. In analyzing the motion of a robotic mechanism, the following matrix multiplication is used. Perform the multiplication and evaluate each element.

$$\begin{pmatrix} \cos 60^\circ & -\sin 60^\circ & 0 \\ \sin 60^\circ & \cos 60^\circ & 0 \\ 0 & 0 & 1 \end{pmatrix}\begin{pmatrix} 2 \\ 4 \\ 0 \end{pmatrix}$$

35. In an *ammeter*, nearly all the electric current flows through a *shunt*, and the remaining known fraction of current is measured by the meter. See Fig. 16-7. From the given matrix equation find voltage v_2 and current i_2 in terms of v_1, i_1, and resistance R, whichever may be applicable.

$$\begin{pmatrix} v_2 \\ i_2 \end{pmatrix} = \begin{pmatrix} 1 & 0 \\ -\frac{1}{R} & 1 \end{pmatrix} \begin{pmatrix} v_1 \\ i_1 \end{pmatrix}$$

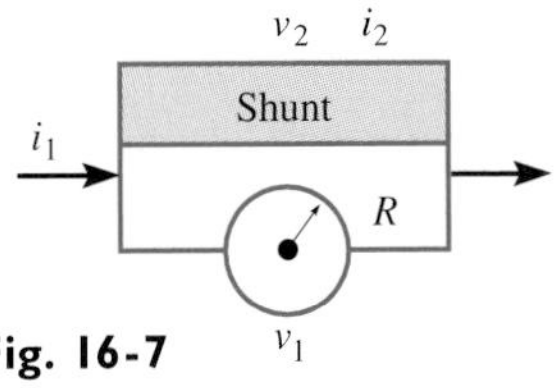

Fig. 16-7

36. In the theory related to the reproduction of color photography, the equations

$$\begin{pmatrix} X \\ Y \\ Z \end{pmatrix} = \begin{pmatrix} 1.0 & 0.1 & 0 \\ 0.5 & 1.0 & 0.1 \\ 0.3 & 0.4 & 1.0 \end{pmatrix} \begin{pmatrix} x \\ y \\ z \end{pmatrix}$$

are found. The X, Y, and Z represent the red, green, and blue densities of the reproductions, respectively, and the x, y, and z represent the red, green, and blue densities, respectively, of the subject. Give the equations relating X, Y, and Z and x, y, and z.

16-5 Finding the Inverse of a Matrix

In this section we show how to find the inverse of a matrix, and in the following section we show how the inverse is used in solving a system of linear equations.

We shall first show two methods of finding the inverse of a two-row, two-column (2×2) matrix. The first method is as follows:

Inverse of a 2×2 Matrix

1. *Interchange the elements on the principal diagonal.*
2. *Change the signs of the off-diagonal elements.*
3. *Divide each resulting element by the determinant of the given matrix.*

This method, which can be used with second-order square matrices *but not with higher-order matrices,* is illustrated in the following example.

EXAMPLE 1 Find the inverse of the matrix

$$A = \begin{pmatrix} 2 & -3 \\ 4 & -7 \end{pmatrix}$$

First we interchange the elements on the principal diagonal and change the signs of the off-diagonal elements. This gives us the matrix

$$\begin{pmatrix} -7 & 3 \\ -4 & 2 \end{pmatrix} \begin{matrix} \text{— signs changed} \\ \text{— elements interchanged} \end{matrix}$$

Now we find the determinant of the original matrix, which means we evaluate

$$\begin{vmatrix} 2 & -3 \\ 4 & -7 \end{vmatrix} = -14 - (-12) = -2$$

We now divide each element of the second matrix by -2. This gives

$$A^{-1} = \frac{1}{-2}\begin{pmatrix} -7 & 3 \\ -4 & 2 \end{pmatrix} = \begin{pmatrix} \frac{-7}{-2} & \frac{3}{-2} \\ \frac{-4}{-2} & \frac{2}{-2} \end{pmatrix} = \begin{pmatrix} \frac{7}{2} & -\frac{3}{2} \\ 2 & -1 \end{pmatrix} \longleftarrow \text{inverse}$$

Checking by multiplication gives

$$AA^{-1} = \begin{pmatrix} 2 & -3 \\ 4 & -7 \end{pmatrix}\begin{pmatrix} \frac{7}{2} & -\frac{3}{2} \\ 2 & -1 \end{pmatrix} = \begin{pmatrix} 7-6 & -3+3 \\ 14-14 & -6+7 \end{pmatrix} = \begin{pmatrix} 1 & 0 \\ 0 & 1 \end{pmatrix} = I$$

Since $AA^{-1} = I$, the matrix A^{-1} is the proper inverse matrix.

Gauss-Jordan Method

Named for the German mathematician Karl Gauss (1777–1855) and the French mathematician Camille Jordan (1838–1922).

The second method, called the *Gauss-Jordan method,* involves *transforming the given matrix into the identity matrix while* ***transforming the identity matrix into the inverse.*** There are three types of steps allowable in making these transformations:

1. *Any two rows may be interchanged.*
2. *Every element in any row may be multiplied by any number other than zero.*
3. *Any row may be replaced by a row whose elements are the sum of a nonzero multiple of itself and a nonzero multiple of another row.*

NOTE ▶ Note that these are **row operations,** not column operations, and are the operations used in solving a system of equations by addition or subtraction.

EXAMPLE 2 ⏩ Find the inverse of the matrix

$$A = \begin{pmatrix} 2 & -3 \\ 4 & -7 \end{pmatrix}$$

First we set up the given matrix with the identity matrix as follows:

$$\left(\begin{array}{cc|cc} 2 & -3 & 1 & 0 \\ 4 & -7 & 0 & 1 \end{array}\right)$$

The vertical line simply shows the separation of the two matrices.

We wish to transform the left matrix into the identity matrix. Therefore, the first requirement is a 1 for element a_{11}. Therefore, we divide all elements of the first row by 2. This gives the following setup:

$$\left(\begin{array}{cc|cc} 1 & -\frac{3}{2} & \frac{1}{2} & 0 \\ 4 & -7 & 0 & 1 \end{array}\right)$$

Next we want to have a zero for element a_{21}. Therefore, we shall subtract 4 times each element of row 1 from the corresponding element in row 2, replacing the elements of row 2. This gives us the following setup:

$$\left(\begin{array}{cc|cc} 1 & -\frac{3}{2} & \frac{1}{2} & 0 \\ 4-4(1) & -7-4(-\frac{3}{2}) & 0-4(\frac{1}{2}) & 1-4(0) \end{array}\right) \quad \text{or} \quad \left(\begin{array}{cc|cc} 1 & -\frac{3}{2} & \frac{1}{2} & 0 \\ 0 & -1 & -2 & 1 \end{array}\right)$$

Next, we want to have 1, not -1, for element a_{22}. Therefore, we multiply each element of row 2 by -1. This gives

$$\left(\begin{array}{cc|cc} 1 & -\frac{3}{2} & \frac{1}{2} & 0 \\ 0 & 1 & 2 & -1 \end{array}\right)$$

Finally, we want zero for element a_{12}. Therefore, we add $\frac{3}{2}$ times each element of row 2 to the corresponding elements of row 1, replacing row 1. This gives

$$\left(\begin{array}{cc|cc} 1+\frac{3}{2}(0) & -\frac{3}{2}+\frac{3}{2}(1) & \frac{1}{2}+\frac{3}{2}(2) & 0+\frac{3}{2}(-1) \\ 0 & 1 & 2 & -1 \end{array}\right) \quad \text{or} \quad \left(\begin{array}{cc|cc} 1 & 0 & \frac{7}{2} & -\frac{3}{2} \\ 0 & 1 & 2 & -1 \end{array}\right)$$

At this point, we have transformed the given matrix into the identity matrix, and the identity matrix into the inverse. Therefore, the matrix to the right of the vertical bar in the last setup is the required inverse. Thus,

$$A^{-1} = \begin{pmatrix} \frac{7}{2} & -\frac{3}{2} \\ 2 & -1 \end{pmatrix}$$

This is the same matrix and inverse as illustrated in Example 1. ⏪

In transforming a matrix into the identity matrix, we work on one column at a time, transforming the columns in order from left to right. It is generally best to make the element on the principal diagonal for the column 1 first and then make all other elements in the column 0. This was done in Example 2, and we now illustrate it with another 2×2 matrix, and then we find the inverse of a 3×3 matrix. The method is applicable for any square matrix.

EXAMPLE 3 ▸▸ Find the inverse of the matrix $\begin{pmatrix} -3 & 6 \\ 4 & 5 \end{pmatrix}$.

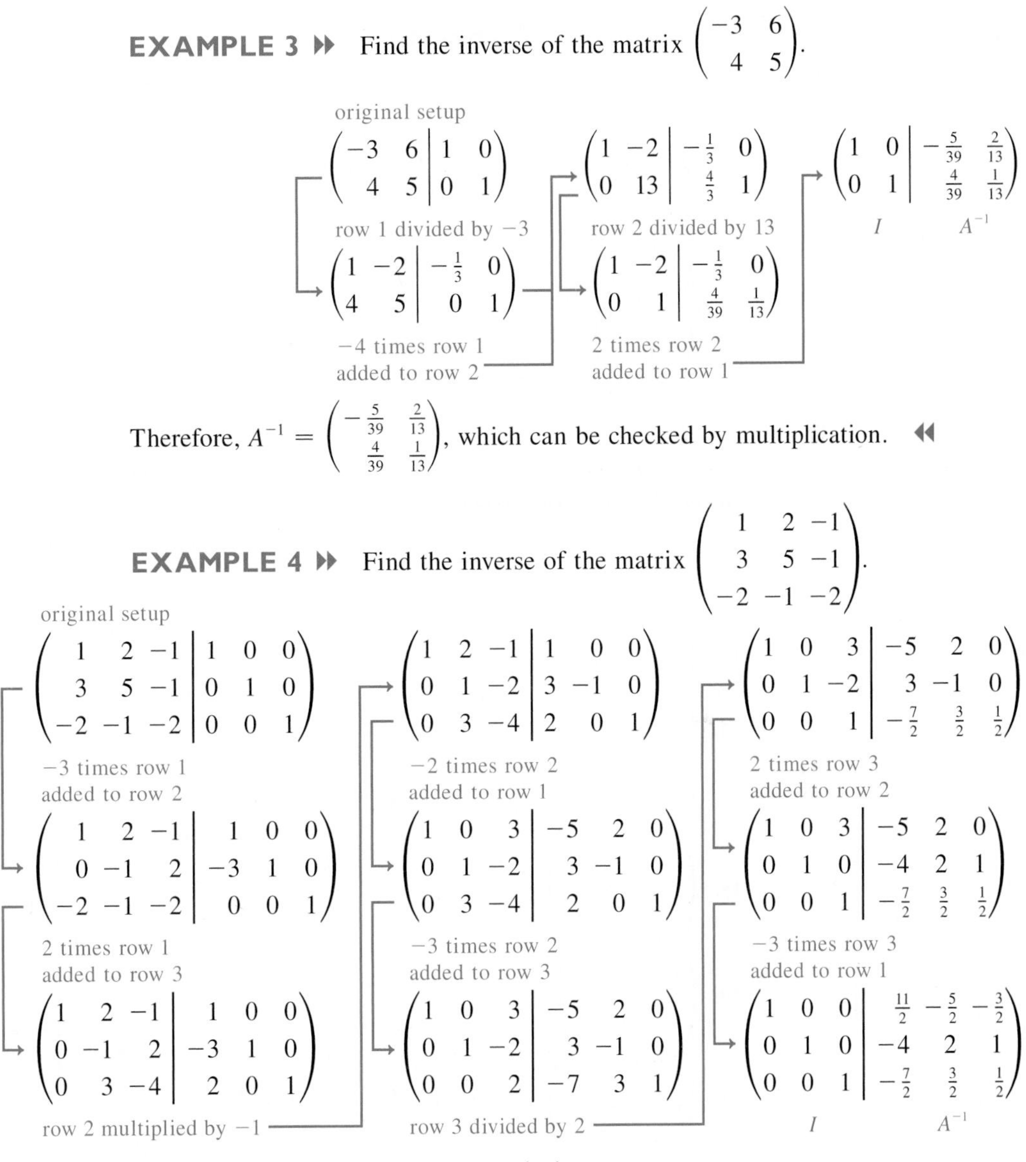

$$\begin{pmatrix} -3 & 6 & | & 1 & 0 \\ 4 & 5 & | & 0 & 1 \end{pmatrix} \rightarrow \begin{pmatrix} 1 & -2 & | & -\frac{1}{3} & 0 \\ 4 & 5 & | & 0 & 1 \end{pmatrix} \rightarrow \begin{pmatrix} 1 & -2 & | & -\frac{1}{3} & 0 \\ 0 & 13 & | & \frac{4}{3} & 1 \end{pmatrix} \rightarrow \begin{pmatrix} 1 & -2 & | & -\frac{1}{3} & 0 \\ 0 & 1 & | & \frac{4}{39} & \frac{1}{13} \end{pmatrix} \rightarrow \begin{pmatrix} 1 & 0 & | & -\frac{5}{39} & \frac{2}{13} \\ 0 & 1 & | & \frac{4}{39} & \frac{1}{13} \end{pmatrix}$$

Therefore, $A^{-1} = \begin{pmatrix} -\frac{5}{39} & \frac{2}{13} \\ \frac{4}{39} & \frac{1}{13} \end{pmatrix}$, which can be checked by multiplication. ◂◂

EXAMPLE 4 ▸▸ Find the inverse of the matrix $\begin{pmatrix} 1 & 2 & -1 \\ 3 & 5 & -1 \\ -2 & -1 & -2 \end{pmatrix}$.

$$\begin{pmatrix} 1 & 2 & -1 & | & 1 & 0 & 0 \\ 3 & 5 & -1 & | & 0 & 1 & 0 \\ -2 & -1 & -2 & | & 0 & 0 & 1 \end{pmatrix} \rightarrow \begin{pmatrix} 1 & 2 & -1 & | & 1 & 0 & 0 \\ 0 & -1 & 2 & | & -3 & 1 & 0 \\ -2 & -1 & -2 & | & 0 & 0 & 1 \end{pmatrix} \rightarrow \begin{pmatrix} 1 & 2 & -1 & | & 1 & 0 & 0 \\ 0 & -1 & 2 & | & -3 & 1 & 0 \\ 0 & 3 & -4 & | & 2 & 0 & 1 \end{pmatrix}$$

$$\rightarrow \begin{pmatrix} 1 & 2 & -1 & | & 1 & 0 & 0 \\ 0 & 1 & -2 & | & 3 & -1 & 0 \\ 0 & 3 & -4 & | & 2 & 0 & 1 \end{pmatrix} \rightarrow \begin{pmatrix} 1 & 0 & 3 & | & -5 & 2 & 0 \\ 0 & 1 & -2 & | & 3 & -1 & 0 \\ 0 & 3 & -4 & | & 2 & 0 & 1 \end{pmatrix} \rightarrow \begin{pmatrix} 1 & 0 & 3 & | & -5 & 2 & 0 \\ 0 & 1 & -2 & | & 3 & -1 & 0 \\ 0 & 0 & 2 & | & -7 & 3 & 1 \end{pmatrix}$$

$$\rightarrow \begin{pmatrix} 1 & 0 & 3 & | & -5 & 2 & 0 \\ 0 & 1 & -2 & | & 3 & -1 & 0 \\ 0 & 0 & 1 & | & -\frac{7}{2} & \frac{3}{2} & \frac{1}{2} \end{pmatrix} \rightarrow \begin{pmatrix} 1 & 0 & 3 & | & -5 & 2 & 0 \\ 0 & 1 & 0 & | & -4 & 2 & 1 \\ 0 & 0 & 1 & | & -\frac{7}{2} & \frac{3}{2} & \frac{1}{2} \end{pmatrix} \rightarrow \begin{pmatrix} 1 & 0 & 0 & | & \frac{11}{2} & -\frac{5}{2} & -\frac{3}{2} \\ 0 & 1 & 0 & | & -4 & 2 & 1 \\ 0 & 0 & 1 & | & -\frac{7}{2} & \frac{3}{2} & \frac{1}{2} \end{pmatrix}$$

Therefore, the required inverse matrix is

$$\begin{pmatrix} \frac{11}{2} & -\frac{5}{2} & -\frac{3}{2} \\ -4 & 2 & 1 \\ -\frac{7}{2} & \frac{3}{2} & \frac{1}{2} \end{pmatrix}$$

which may be checked by multiplication. ◂◂

All of the matrix operations which we have presented can also be done on most graphing calculators and with many computer programs. On a graphing calculator, the MATRIX feature is issued. There are many different calculators with this feature, and the manual for any given model should be used to see how the various operations are performed.

Exercises 16–5

In Exercises 1–8, find the inverse of each of the given matrices by the method of Example 1 of this section.

1. $\begin{pmatrix} 2 & -5 \\ -2 & 4 \end{pmatrix}$

2. $\begin{pmatrix} -6 & 3 \\ 3 & -2 \end{pmatrix}$

3. $\begin{pmatrix} -1 & 5 \\ 4 & 10 \end{pmatrix}$

4. $\begin{pmatrix} 8 & -1 \\ -4 & -5 \end{pmatrix}$

5. $\begin{pmatrix} 0 & -4 \\ 2 & 6 \end{pmatrix}$

6. $\begin{pmatrix} 7 & -2 \\ -6 & 2 \end{pmatrix}$

7. $\begin{pmatrix} -50 & -45 \\ 26 & 80 \end{pmatrix}$

8. $\begin{pmatrix} 7.2 & -3.6 \\ -1.3 & -5.7 \end{pmatrix}$

In Exercises 9–24, find the inverse of each of the given matrices by transforming the identity matrix, as in Examples 2–4.

9. $\begin{pmatrix} 1 & 2 \\ 2 & 3 \end{pmatrix}$

10. $\begin{pmatrix} 1 & 5 \\ -1 & -4 \end{pmatrix}$

11. $\begin{pmatrix} 2 & 4 \\ -1 & -1 \end{pmatrix}$

12. $\begin{pmatrix} -2 & 6 \\ 3 & -4 \end{pmatrix}$

13. $\begin{pmatrix} 2 & 5 \\ -1 & 2 \end{pmatrix}$

14. $\begin{pmatrix} -2 & 3 \\ -3 & 5 \end{pmatrix}$

15. $\begin{pmatrix} 2 & -1 \\ 4 & 6 \end{pmatrix}$

16. $\begin{pmatrix} 1 & -3 \\ 7 & -5 \end{pmatrix}$

17. $\begin{pmatrix} 1 & -3 & -2 \\ -2 & 7 & 3 \\ 1 & -1 & -3 \end{pmatrix}$

18. $\begin{pmatrix} 1 & 2 & -1 \\ 3 & 7 & -5 \\ -1 & -2 & 0 \end{pmatrix}$

19. $\begin{pmatrix} 1 & -1 & -3 \\ 0 & -1 & -2 \\ 2 & 1 & -1 \end{pmatrix}$

20. $\begin{pmatrix} 1 & 4 & 1 \\ -3 & -13 & -1 \\ 0 & -2 & 5 \end{pmatrix}$

21. $\begin{pmatrix} 1 & 3 & 2 \\ -2 & -5 & -1 \\ 2 & 4 & 0 \end{pmatrix}$

22. $\begin{pmatrix} 1 & 3 & 4 \\ -1 & -4 & -2 \\ 4 & 9 & 20 \end{pmatrix}$

23. $\begin{pmatrix} 2 & 4 & 0 \\ 3 & 4 & -2 \\ -1 & 1 & 2 \end{pmatrix}$

24. $\begin{pmatrix} -2 & 6 & 1 \\ 0 & 3 & -3 \\ 4 & -7 & 3 \end{pmatrix}$

In Exercises 25–28, find the inverse of each of the given matrices (same as those for Exercises 21–24) by using a graphing calculator.

25. $\begin{pmatrix} 1 & 3 & 2 \\ -2 & -5 & -1 \\ 2 & 4 & 0 \end{pmatrix}$

26. $\begin{pmatrix} 1 & 3 & 4 \\ -1 & -4 & -2 \\ 4 & 9 & 20 \end{pmatrix}$

27. $\begin{pmatrix} 2 & 4 & 0 \\ 3 & 4 & -2 \\ -1 & 1 & 2 \end{pmatrix}$

28. $\begin{pmatrix} -2 & 6 & 1 \\ 0 & 3 & -3 \\ 4 & -7 & 3 \end{pmatrix}$

In Exercises 29–32, solve the given problems.

29. For the matrix $A = \begin{pmatrix} a & b \\ c & d \end{pmatrix}$, show that

$$\frac{1}{ad - bc}\begin{pmatrix} a & b \\ c & d \end{pmatrix}\begin{pmatrix} d & -b \\ -c & a \end{pmatrix} = \begin{pmatrix} 1 & 0 \\ 0 & 1 \end{pmatrix}.$$

This verifies the method of Example 1.

30. Describe the relationship between the elements of the matrix $\begin{pmatrix} a & 0 & 0 \\ 0 & b & 0 \\ 0 & 0 & c \end{pmatrix}$ and the elements of its inverse.

31. For the *four-terminal network* shown in Fig. 16-8, it can be shown that the voltage matrix V is related to the coefficient matrix A and the current matrix I by $V = A^{-1}I$, where

$$V = \begin{pmatrix} v_1 \\ v_2 \end{pmatrix}, \qquad A = \begin{pmatrix} a_{11} & a_{12} \\ a_{21} & a_{22} \end{pmatrix}, \qquad I = \begin{pmatrix} i_1 \\ i_2 \end{pmatrix}$$

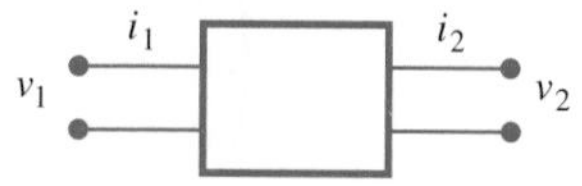

Fig. 16-8

Find the individual equations for v_1 and v_2 that give each in terms of i_1 and i_2.

32. The rotations of a robot arm such as that shown in Fig. 16-9 are often represented by matrices. The values represent trigonometric functions of the angles of rotation. For the following rotation matrix R, find R^{-1}.

$$R = \begin{pmatrix} 0.8 & 0.0 & -0.6 \\ 0.0 & 1.0 & 0.0 \\ 0.6 & 0.0 & 0.8 \end{pmatrix}$$

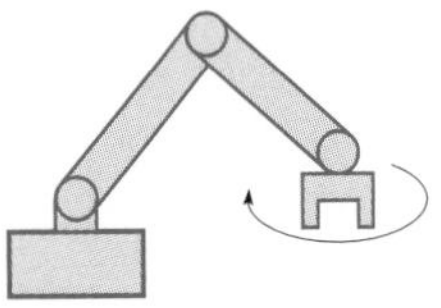

Fig. 16-9

16-6 Matrices and Linear Equations

As we stated earlier, matrices can be used to solve systems of equations. In this section we show one of the methods by which this is done.

Let us consider the system of equations

$$a_1x + b_1y = c_1$$
$$a_2x + b_2y = c_2$$

Recalling the definition of equality of matrices, we can write this system as

$$\begin{pmatrix} a_1x + b_1y \\ a_2x + b_2y \end{pmatrix} = \begin{pmatrix} c_1 \\ c_2 \end{pmatrix} \tag{16-9}$$

If we let

$$A = \begin{pmatrix} a_1 & b_1 \\ a_2 & b_2 \end{pmatrix}, \quad X = \begin{pmatrix} x \\ y \end{pmatrix}, \quad C = \begin{pmatrix} c_1 \\ c_2 \end{pmatrix} \tag{16-10}$$

the left side of Eq. (16-9) can be written as the product of matrices A and X,

$$AX = C \tag{16-11}$$

If we now multiply (on the left) each side of this matrix equation by A^{-1}, we have

$$A^{-1}AX = A^{-1}C$$

Since $A^{-1}A = I$, we have

$$IX = A^{-1}C$$

However, $IX = X$. Therefore,

$$X = A^{-1}C \tag{16-12}$$

NOTE ▶ Equation (16-12) states that *we can solve a system of linear equations by multiplying the one-column matrix of the constants on the right by the inverse of the matrix of the coefficients.* The result is a one-column matrix whose elements are the required values. Note carefully that $X = A^{-1}C$, ***not*** CA^{-1}.

EXAMPLE 1 ▸▸ Use matrices to solve the system of equations

$$2x - y = 7$$
$$5x - 3y = 18$$

We set up the matrix of coefficients and the matrix of constants as

$$A = \begin{pmatrix} 2 & -1 \\ 5 & -3 \end{pmatrix} \quad \text{and} \quad C = \begin{pmatrix} 7 \\ 18 \end{pmatrix}$$

By either of the methods of the previous section, we can determine the inverse of matrix A to be

$$A^{-1} = \begin{pmatrix} 3 & -1 \\ 5 & -2 \end{pmatrix}$$

We now form the matrix product $A^{-1}C$.

$$A^{-1}C = \begin{pmatrix} 3 & -1 \\ 5 & -2 \end{pmatrix}\begin{pmatrix} 7 \\ 18 \end{pmatrix} = \begin{pmatrix} 21 - 18 \\ 35 - 36 \end{pmatrix} = \begin{pmatrix} 3 \\ -1 \end{pmatrix}$$

Since $X = A^{-1}C$, this means that

$$\begin{pmatrix} x \\ y \end{pmatrix} = \begin{pmatrix} 3 \\ -1 \end{pmatrix}$$

Therefore, the required solution is $x = 3$ and $y = -1$, which checks when these values are substituted into the original equations. ◂◂

EXAMPLE 2 ▸▸ For the electric circuit shown in Fig. 16-10, the equations used to find the currents (in amperes) i_1 and i_2 are

$$\begin{aligned} 2.30i_1 + 6.45(i_1 + i_2) &= 15.0 \\ 1.25i_2 + 6.45(i_1 + i_2) &= 12.5 \end{aligned} \quad \text{or} \quad \begin{aligned} 8.75i_1 + 6.45i_2 &= 15.0 \\ 6.45i_1 + 7.70i_2 &= 12.5 \end{aligned}$$

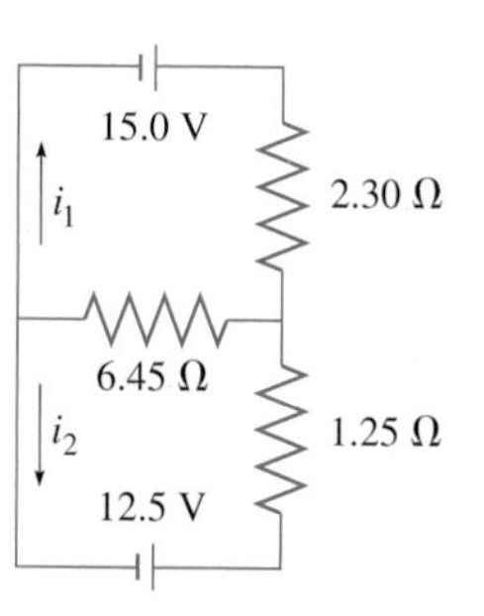

Fig. 16-10

Using matrices to solve this system of equations, we set up the matrix A of coefficients, the matrix C of constants, and the matrix X of currents as

$$A = \begin{pmatrix} 8.75 & 6.45 \\ 6.45 & 7.70 \end{pmatrix}, \quad C = \begin{pmatrix} 15.0 \\ 12.5 \end{pmatrix}, \quad \text{and} \quad X = \begin{pmatrix} i_1 \\ i_2 \end{pmatrix}$$

We now find the inverse of A as

$$A^{-1} = \frac{1}{8.75(7.70) - 6.45(6.45)}\begin{pmatrix} 7.70 & -6.45 \\ -6.45 & 8.75 \end{pmatrix} = \begin{pmatrix} 0.2988 & -0.2503 \\ -0.2503 & 0.3395 \end{pmatrix}$$

Therefore,

$$\begin{aligned} X = A^{-1}C &= \begin{pmatrix} 0.2988 & -0.2503 \\ -0.2503 & 0.3395 \end{pmatrix}\begin{pmatrix} 15.0 \\ 12.5 \end{pmatrix} \\ &= \begin{pmatrix} 0.2988(15.0) - 0.2503(12.5) \\ -0.2503(15.0) + 0.3395(12.5) \end{pmatrix} = \begin{pmatrix} 1.35 \\ 0.49 \end{pmatrix} \end{aligned}$$

Therefore, the required currents are $i_1 = 1.35$ A and $i_2 = 0.49$ A. These values check when substituted into the original equations. ◂◂

EXAMPLE 3 ▸▸ Use matrices to solve the system of equations

$$\begin{aligned} x + 4y - z &= 4 \\ x + 3y + z &= 8 \\ 2x + 6y + z &= 13 \end{aligned}$$

Setting up matrices A, C, and X, we have

$$A = \begin{pmatrix} 1 & 4 & -1 \\ 1 & 3 & 1 \\ 2 & 6 & 1 \end{pmatrix}, \quad C = \begin{pmatrix} 4 \\ 8 \\ 13 \end{pmatrix}, \quad X = \begin{pmatrix} x \\ y \\ z \end{pmatrix}$$

To give another example of finding the inverse of a 3×3 matrix, we shall briefly show the steps for finding A^{-1}.

$$\left(\begin{array}{ccc|ccc} 1 & 4 & -1 & 1 & 0 & 0 \\ 1 & 3 & 1 & 0 & 1 & 0 \\ 2 & 6 & 1 & 0 & 0 & 1 \end{array}\right) \rightarrow \left(\begin{array}{ccc|ccc} 1 & 4 & -1 & 1 & 0 & 0 \\ 0 & -1 & 2 & -1 & 1 & 0 \\ 2 & 6 & 1 & 0 & 0 & 1 \end{array}\right) \rightarrow \left(\begin{array}{ccc|ccc} 1 & 4 & -1 & 1 & 0 & 0 \\ 0 & -1 & 2 & -1 & 1 & 0 \\ 0 & -2 & 3 & -2 & 0 & 1 \end{array}\right)$$

$$\rightarrow \left(\begin{array}{ccc|ccc} 1 & 4 & -1 & 1 & 0 & 0 \\ 0 & 1 & -2 & 1 & -1 & 0 \\ 0 & -2 & 3 & -2 & 0 & 1 \end{array}\right) \rightarrow \left(\begin{array}{ccc|ccc} 1 & 0 & 7 & -3 & 4 & 0 \\ 0 & 1 & -2 & 1 & -1 & 0 \\ 0 & -2 & 3 & -2 & 0 & 1 \end{array}\right) \rightarrow \left(\begin{array}{ccc|ccc} 1 & 0 & 7 & -3 & 4 & 0 \\ 0 & 1 & -2 & 1 & -1 & 0 \\ 0 & 0 & -1 & 0 & -2 & 1 \end{array}\right)$$

$$\rightarrow \left(\begin{array}{ccc|ccc} 1 & 0 & 7 & -3 & 4 & 0 \\ 0 & 1 & -2 & 1 & -1 & 0 \\ 0 & 0 & 1 & 0 & 2 & -1 \end{array}\right) \rightarrow \left(\begin{array}{ccc|ccc} 1 & 0 & 7 & -3 & 4 & 0 \\ 0 & 1 & 0 & 1 & 3 & -2 \\ 0 & 0 & 1 & 0 & 2 & -1 \end{array}\right) \rightarrow \left(\begin{array}{ccc|ccc} 1 & 0 & 0 & -3 & -10 & 7 \\ 0 & 1 & 0 & 1 & 3 & -2 \\ 0 & 0 & 1 & 0 & 2 & -1 \end{array}\right)$$

Thus, $A^{-1} = \begin{pmatrix} -3 & -10 & 7 \\ 1 & 3 & -2 \\ 0 & 2 & -1 \end{pmatrix}$ and

$$X = A^{-1}C = \begin{pmatrix} -3 & -10 & 7 \\ 1 & 3 & -2 \\ 0 & 2 & -1 \end{pmatrix}\begin{pmatrix} 4 \\ 8 \\ 13 \end{pmatrix} = \begin{pmatrix} -12 - 80 + 91 \\ 4 + 24 - 26 \\ 0 + 16 - 13 \end{pmatrix} = \begin{pmatrix} -1 \\ 2 \\ 3 \end{pmatrix}$$

This means that $x = -1$, $y = 2$, and $z = 3$. ◂◂

EXAMPLE 4 ▸▸ Use matrices to solve the system of equations

$$\begin{aligned} x + 2y - z &= -4 \\ 3x + 5y - z &= -5 \\ -2x - y - 2z &= -5 \end{aligned}$$

Setting up matrices A, C, and X, we have

$$A = \begin{pmatrix} 1 & 2 & -1 \\ 3 & 5 & -1 \\ -2 & -1 & -2 \end{pmatrix}, \quad C = \begin{pmatrix} -4 \\ -5 \\ -5 \end{pmatrix}, \quad X = \begin{pmatrix} x \\ y \\ z \end{pmatrix}$$

Finding A^{-1} (see Example 4 of Section 16-5), and solving for X, we have

$$A^{-1} = \begin{pmatrix} \frac{11}{2} & -\frac{5}{2} & -\frac{3}{2} \\ -4 & 2 & 1 \\ -\frac{7}{2} & \frac{3}{2} & \frac{1}{2} \end{pmatrix}$$

$$X = A^{-1}C = \begin{pmatrix} \frac{11}{2} & -\frac{5}{2} & -\frac{3}{2} \\ -4 & 2 & 1 \\ -\frac{7}{2} & \frac{3}{2} & \frac{1}{2} \end{pmatrix}\begin{pmatrix} -4 \\ -5 \\ -5 \end{pmatrix} = \begin{pmatrix} -2 \\ 1 \\ 4 \end{pmatrix}$$

This means that the solution is $x = -2$, $y = 1$, $z = 4$. ◂◂

NOTE ▶ After solving systems of equations in this manner, the reader may feel that the method is much longer and more tedious than previously developed methods. *The principal problem with this method is that a great deal of numerical computation is generally required.* However, methods such as this one are easily programmed for use on a computer, which can do the arithmetic work very rapidly. Therefore, most graphing calculators and many computer programs can perform these operations. Again, on the graphing calculator, the MATRIX feature is used. It is the *method* of solving the system of equations that is of primary importance here.

Exercises 16–6

In Exercises 1–8, solve the given systems of equations by using the inverse of the coefficient matrix. The numbers in parentheses refer to exercises from Section 16-5, where the inverses may be checked.

1. $2x - 5y = -14$ (1)
$-2x + 4y = 11$

2. $-x + 5y = 4$ (3)
$4x + 10y = -4$

3. $2x + 4y = -9$ (11)
$-x - y = 2$

4. $2x + 5y = -6$ (13)
$-x + 2y = -6$

5. $x - 3y - 2z = -8$ (17)
$-2x + 7y + 3z = 19$
$x - y - 3z = -3$

6. $x - y - 3z = -1$ (19)
$-y - 2z = -2$
$2x + y - z = 2$

7. $x + 3y + 2z = 5$ (21)
$-2x - 5y - z = -1$
$2x + 4y = -2$

8. $2x + 4y = -2$ (23)
$3x + 4y - 2z = -6$
$-x + y + 2z = 5$

In Exercises 9–20, solve the given systems of equations by using the inverse of the coefficient matrix.

9. $2x + 7y = 16$
$x + 4y = 9$

10. $4x - 3y = -13$
$-3x + 2y = 9$

11. $2x - 3y = 3$
$4x - 5y = 4$

12. $x + 2y = 3$
$3x + 4y = 11$

13. $5x - 2y = -14$
$3x + 4y = -11$

14. $4x - 3y = -1$
$8x + 3y = 4$

15. $2.5x + 2.8y = -3.0$
$3.5x - 1.6y = 9.6$

16. $12x - 5y = -400$
$31x + 25y = 180$

17. $x + 2y + 2z = -4$
$4x + 9y + 10z = -18$
$-x + 3y + 7z = -7$

18. $x - 4y - 2z = -7$
$-x + 5y + 5z = 18$
$3x - 7y + 10z = 38$

19. $2x + 4y + z = 5$
$-2x - 2y - z = -6$
$-x + 2y + z = 0$

20. $4x + y = 2$
$-2x - y + 3z = -18$
$2x + y - z = 8$

In Exercises 21–24, solve the indicated systems of equations by using the inverse of the coefficient matrix. In Exercises 23 and 24, it is necessary to set up the appropriate equations.

21. Forces A and B hold up a beam that weighs 254 N, as shown in Fig. 16-11. The equations used to find the forces are

$A \sin 47.2° + B \sin 64.4° = 254$
$A \cos 47.2° - B \cos 64.4° = 0$

Find the forces.

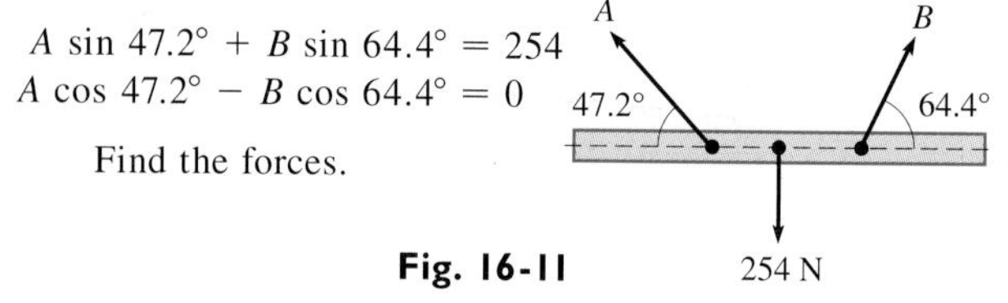

Fig. 16-11

22. In applying Kirchhoff's laws (see Exercise 20 of Section 5-6) to the circuit shown in Fig. 16-12, the following equations are found. Determine the indicated currents, in amperes.

$I_A + I_B + I_C = 0$
$2I_A - 5I_B = 6$
$5I_B - I_C = -3$

Fig. 16-12

23. A research chemist wants to make 10.0 L of gasoline containing 2.0% of a new experimental additive. Gasoline without additive and two mixtures of gasoline with additive, one with 5.0% and the other with 6.0%, are to be used. If four times as much gasoline without additive than the 5.0% mixture is to be used, how much of each is needed?

24. An apartment building has 60 apartments, all of which are rented. The smaller apartments rent for \$900 per month, and the larger ones rent for \$1250 per month. If the total rental income is \$62 400, how many of each type of apartment are there?

Chapter Equations, Review Exercises, and Practice Test

Chapter Equations

Determinants

$$\begin{vmatrix} a_1 & b_1 & c_1 \\ a_2 & b_2 & c_2 \\ a_3 & b_3 & c_3 \end{vmatrix} = a_1b_2c_3 + a_3b_1c_2 + a_2b_3c_1 - a_3b_2c_1 - a_1b_3c_2 - a_2b_1c_3 \quad (16\text{-}1)$$

$$\begin{vmatrix} a_1 & b_1 & c_1 \\ a_2 & b_2 & c_2 \\ a_3 & b_3 & c_3 \end{vmatrix} = a_1(b_2c_3 - b_3c_2) - a_2(b_1c_3 - b_3c_1) + a_3(b_1c_2 - b_2c_1) \quad (16\text{-}2)$$

Minors

$$\begin{vmatrix} a_1 & b_1 & c_1 \\ a_2 & b_2 & c_2 \\ a_3 & b_3 & c_3 \end{vmatrix} = a_1\begin{vmatrix} b_2 & c_2 \\ b_3 & c_3 \end{vmatrix} - a_2\begin{vmatrix} b_1 & c_1 \\ b_3 & c_3 \end{vmatrix} + a_3\begin{vmatrix} b_1 & c_1 \\ b_2 & c_2 \end{vmatrix} \quad (16\text{-}3)$$

Basic laws for matrices

$$A + B = B + A \quad \text{(commutative law)} \quad (16\text{-}4)$$

$$A + (B + C) = (A + B) + C \quad \text{(associative law)} \quad (16\text{-}5)$$

$$k(A + B) = kA + kB \quad (16\text{-}6)$$

$$A + O = A \quad (16\text{-}7)$$

Inverse matrix

$$AA^{-1} = A^{-1}A = I \quad (16\text{-}8)$$

Solving systems of equations by matrices

$$\begin{pmatrix} a_1x + b_1y \\ a_2x + b_2y \end{pmatrix} = \begin{pmatrix} c_1 \\ c_2 \end{pmatrix} \quad (16\text{-}9)$$

$$A = \begin{pmatrix} a_1 & b_1 \\ a_2 & b_2 \end{pmatrix}, \quad X = \begin{pmatrix} x \\ y \end{pmatrix}, \quad C = \begin{pmatrix} c_1 \\ c_2 \end{pmatrix} \quad (16\text{-}10)$$

$$AX = C \quad (16\text{-}11)$$

$$X = A^{-1}C \quad (16\text{-}12)$$

Review Exercises

In Exercises 1–8, evaluate the given determinants by expansion by minors.

1. $\begin{vmatrix} 1 & 2 & -1 \\ 4 & 1 & -3 \\ -3 & -5 & 2 \end{vmatrix}$

2. $\begin{vmatrix} 3 & -1 & 2 \\ 7 & -1 & 4 \\ 2 & 1 & -3 \end{vmatrix}$

3. $\begin{vmatrix} -1 & 3 & -7 \\ 0 & 5 & 4 \\ 4 & -3 & -2 \end{vmatrix}$

4. $\begin{vmatrix} 60 & -54 & -76 \\ -10 & 24 & 40 \\ 25 & -37 & 18 \end{vmatrix}$

5. $\begin{vmatrix} 2 & 6 & 2 & 5 \\ 2 & 0 & 4 & -1 \\ 4 & -3 & 6 & 1 \\ 3 & -1 & 0 & -2 \end{vmatrix}$

6. $\begin{vmatrix} 1 & -2 & 2 & 4 \\ 0 & 1 & 2 & 3 \\ 3 & 2 & 2 & 5 \\ 2 & 1 & -2 & 0 \end{vmatrix}$

7. $\begin{vmatrix} 1 & 3 & -2 & 4 \\ 2 & 0 & 3 & -2 \\ 5 & -1 & 5 & -3 \\ -6 & 4 & -1 & 2 \end{vmatrix}$

8. $\begin{vmatrix} 2 & 3 & -1 & -1 \\ -3 & -2 & 5 & -6 \\ 2 & 1 & -3 & 2 \\ 4 & 0 & -2 & 1 \end{vmatrix}$

In Exercises 9–14, evaluate the determinants of Exercises 1–6 by using the basic properties of determinants.

In Exercises 15–20, evaluate the determinants of Exercises 1–6 by using a graphing calculator. (Use the MATRIX feature.)

In Exercises 21–24, determine the values of the literal numbers.

21. $\begin{pmatrix} 2a \\ a - b \end{pmatrix} = \begin{pmatrix} 8 \\ 5 \end{pmatrix}$

22. $\begin{pmatrix} x - y \\ 2x + 2z \\ 4y + z \end{pmatrix} = \begin{pmatrix} 1 \\ 3 \\ -1 \end{pmatrix}$

23. $\begin{pmatrix} 2x & 3y & 2z \\ x + y & 2y + z & z - x \end{pmatrix} = \begin{pmatrix} 4 & -9 & 5 \\ a & b & c \end{pmatrix}$

24. $\begin{pmatrix} a + bj & b \\ aj & b - aj \end{pmatrix} = \begin{pmatrix} 6j & 2d \\ 2cj & ej^2 \end{pmatrix} \quad (j = \sqrt{-1})$

In Exercises 25–32, use the given matrices and perform the indicated operations.

$$A = \begin{pmatrix} 2 & -3 \\ 4 & 1 \\ -5 & 0 \\ 2 & -3 \end{pmatrix}, \quad B = \begin{pmatrix} -1 & 0 \\ 4 & -6 \\ -3 & -2 \\ 1 & -7 \end{pmatrix}, \quad C = \begin{pmatrix} 5 & -6 \\ 2 & 8 \\ 0 & -2 \end{pmatrix}$$

25. $A + B$ **26.** $2C$

27. $-3B$ **28.** $B - A$

29. $A - C$ **30.** $2C - B$

31. $2A - 3B$ **32.** $2(A - B)$

In Exercises 33–36, perform the indicated matrix multiplications.

33. $\begin{pmatrix} 5 & -1 \\ 3 & 2 \end{pmatrix}\begin{pmatrix} 1 \\ -8 \end{pmatrix}$

34. $\begin{pmatrix} 6 & -4 & 1 & 0 \\ 2 & 0 & -4 & 3 \end{pmatrix}\begin{pmatrix} 7 & -1 & 6 \\ 4 & 0 & 1 \\ 3 & -2 & 5 \\ 9 & 1 & 0 \end{pmatrix}$

35. $\begin{pmatrix} -1 & 7 \\ 2 & 0 \\ 4 & -1 \end{pmatrix}\begin{pmatrix} 1 & -4 & 5 \\ 5 & 1 & 0 \end{pmatrix}$

36. $\begin{pmatrix} 0 & -1 & 6 \\ 8 & 1 & 4 \\ 7 & -2 & -1 \end{pmatrix}\begin{pmatrix} 5 & -1 & 7 & 1 & 5 \\ 0 & 1 & 0 & 4 & 1 \\ 1 & -2 & 3 & 0 & 1 \end{pmatrix}$

In Exercises 37–44, find the inverses of the given matrices.

37. $\begin{pmatrix} 2 & -5 \\ 2 & -4 \end{pmatrix}$ **38.** $\begin{pmatrix} -1 & -6 \\ 2 & 10 \end{pmatrix}$

39. $\begin{pmatrix} 7 & -1 \\ 4 & 8 \end{pmatrix}$ **40.** $\begin{pmatrix} 50 & -12 \\ 42 & -80 \end{pmatrix}$

41. $\begin{pmatrix} 1 & 1 & -2 \\ -1 & -2 & 1 \\ 0 & 3 & 4 \end{pmatrix}$ **42.** $\begin{pmatrix} -1 & -1 & 2 \\ 2 & 3 & 0 \\ 1 & 4 & 1 \end{pmatrix}$

43. $\begin{pmatrix} 2 & -4 & 3 \\ 4 & -6 & 5 \\ -2 & 1 & -1 \end{pmatrix}$ **44.** $\begin{pmatrix} 3 & 1 & -4 \\ -3 & 1 & -2 \\ -6 & 0 & 3 \end{pmatrix}$

In Exercises 45–52, solve the given systems of equations using the inverse of the coefficient matrix.

45. $2x - 3y = -9$
$4x - y = -13$

46. $5x - 7y = 62$
$6x + 5y = -6$

47. $33x + 52y = -450$
$45x - 62y = 1380$

48. $0.24x - 0.26y = -3.1$
$0.40x + 0.34y = -1.3$

49. $2x - 3y + 2z = 7$
$3x + y - 3z = -6$
$x + 4y + z = -13$

50. $2x + 2y - z = 8$
$x + 4y + 2z = 5$
$3x - 2y + z = 17$

51. $x + 2y + 3z = 1$
$3x - 4y - 3z = 2$
$7x - 6y + 6z = 2$

52. $3x + 2y + z = 2$
$2x + 3y - 6z = 3$
$x + 3y + 3z = 1$

In Exercises 53–56, solve the given systems of equations by determinants.

53. $3x - 2y + z = 6$
$2x + 3z = 3$
$4x - y + 5z = 6$

54. $7x + y + 2z = 3$
$4x - 2y + 4z = -2$
$2x + 3y - 6z = 3$

55. $2x - 3y + z - t = -8$
$4x + 3z + 2t = -3$
$2y - 3z - t = 12$
$x - y - z + t = 3$

56. $3x + 2y - 2z - 2t = 0$
$5y + 3z + 4t = 3$
$6y - 3z + 4t = 9$
$6x - y + 2z - 2t = -3$

In Exercises 57 and 58, evaluate the given determinants by minors.

57. $\begin{vmatrix} 1+\sqrt{2} & 2-\sqrt{3} & 0 \\ 3+\sqrt{5} & 7+\sqrt{6} & \sqrt{2} \\ 2+\sqrt{3} & 1-\sqrt{2} & 0 \end{vmatrix}$

58. $\begin{vmatrix} \cos\frac{\pi}{3} & \sin\frac{\pi}{6} & \cos 0 \\ \cos\pi & \sin\pi & \tan\frac{\pi}{4} \\ \sin\frac{\pi}{2} & \cos\frac{\pi}{2} & \sin 0 \end{vmatrix}$

In Exercises 59 and 60, evaluate the given determinants by use of the basic properties of determinants.

59. $\begin{vmatrix} \ln e & \log_3 1 & \frac{1}{2}\log 100 \\ \ln\frac{1}{e} & \log_2 8 & \log 0.1 \\ \ln\sqrt{e} & \log 10 & \log_4 2 \end{vmatrix}$

60. $\begin{vmatrix} j & 1 & 1+j \\ -j & -1+j & -j \\ 2j & 2 & 2+3j \end{vmatrix}$ $(j = \sqrt{-1})$

In Exercises 61 and 62, use the matrix N.

$$N = \begin{pmatrix} 0 & -1 \\ 1 & 0 \end{pmatrix}$$

61. Show that $N^{-1} = -N$. **62.** Show that $N^2 = -I$.

In Exercises 63–66, use matrices A and B.

$$A = \begin{pmatrix} 1 & -2 \\ 0 & 3 \end{pmatrix} \quad B = \begin{pmatrix} -3 & 1 \\ 2 & -1 \end{pmatrix}$$

63. Show that $(A + B)(A - B) \neq A^2 - B^2$.

64. Show that $(A + B)^2 \neq A^2 + 2AB + B^2$.

65. Show that the inverse of $2A$ is $A^{-1}/2$.

66. Show that the inverse of B^2 is $(B^{-1})^2$.

In Exercises 67–70, solve the given systems of equations by any appropriate method of this chapter.

67. Two electric resistors, R_1 and R_2, are tested with currents and voltages such that the following equations are found.

$$2R_1 + 3R_2 = 26$$
$$3R_1 + 2R_2 = 24$$

Find the resistances R_1 and R_2 (in ohms).

68. A company produces two products, each of which is processed in two departments. Considering the worker time available, the numbers x and y of each product produced each week can be found by solving the system of equations

$$4.0x + 2.5y = 1200$$
$$3.2x + 4.0y = 1200$$

Find x and y.

69. A steel beam is supported as shown in Fig. 16-13. Find the force F and the tension T by solving the system of equations

$$0.500F = 0.866T$$
$$0.866F + 0.500T = 350$$

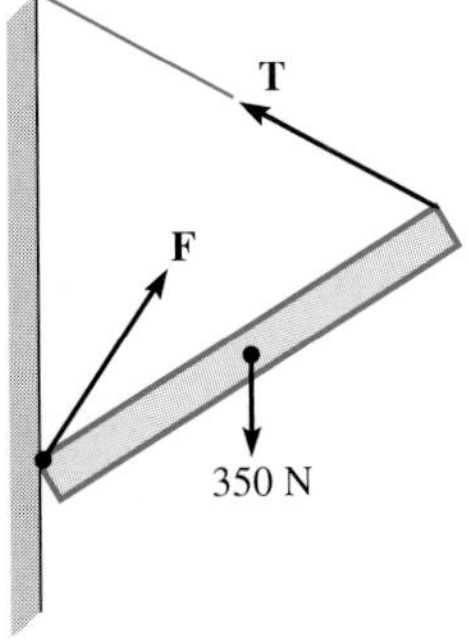

Fig. 16-13

70. To find the electric currents (im amperes) indicated in Fig. 16-14, it is necessary to solve the following equations.

$$I_A + I_B + I_C = 0$$
$$5I_A - 2I_B = -4$$
$$2I_B - I_C = 0$$

Find I_A, I_B, and I_C.

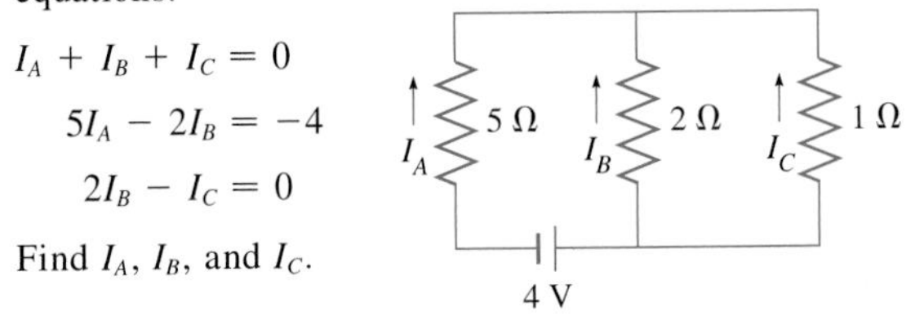

Fig. 16-14

In Exercises 71–76, solve the given problems by using any appropriate method of this chapter.

71. A crime suspect passes an intersection in a car traveling at 180 km/h. The police pass the intersection 3.0 min later in a car traveling at 225 km/h. How long is it before the police overtake the suspect?

72. A contractor needs a backhoe and a generator for two different jobs. Renting the backhoe for 5.0 h and the generator for 6.0 h costs $425 for one job. On the other job, renting the backhoe for 2.0 h and the generator for 8.0 h costs $310. What are the hourly charges for the backhoe and the generator?

73. By mass, three alloys have the following percentages of lead, zinc, and copper.

	Lead	*Zinc*	*Copper*
Alloy A	60%	30%	10%
Alloy B	40%	30%	30%
Alloy C	30%	70%	

How many grams of each of alloys A, B, and C must be mixed to get 100 g of an alloy that is 44% lead, 38% zinc, and 18% copper?

74. Three computer line printers can print a total of 8200 lines/min when printing at the same time. With the first printing for 2 min and the second printing for 3 min, a total of 12 200 lines can be printed. With the first printing for 1 min, the second for 2 min, and the third for 3 min, a total of 17 600 lines can be printed. How many lines per minute can each print?

75. On a 750-km trip that took a total of 5.5 h, a person took a limousine to the airport, then a plane, and finally a car to reach the final destination. The limousine took as long as the final car trip and the time for connections. The limousine averaged 55 km/h, the plane averaged 400 km/h, and the car averaged 40 km/h. The plane traveled four times as far as the limousine and car combined. How long did each part of the trip and the connections take?

76. Refer to Exercise 34 of Section 5-5 (p. 149). By carefully noting the form of the surveyor's formula for a quadrilateral, write the surveyor's formula for a triangle. Then express this formula in the form with one third-order determinant.

In Exercises 77–80, perform the indicated matrix operations.

77. An automobile maker has two assembly plants at which cars with either 4, 6, or 8 cylinders, and with either standard or automatic transmission, are assembled. The annual production at the first plant of cars with the number of cylinders–transmission type (standard, automatic) is as follows:
4—12 000, 15 000; 6—24 000, 8000; 8—4000, 30 000. At the second plant the annual production is 4—15 000, 20 000; 6—12 000, 3000; 8—2000, 22 000. Set up matrices for this production, and by matrix addition find the matrix for the total production by the number of cylinders and type of transmission.

78. Set up a matrix representing the information given in Exercise 73. A given shipment contains 500 g of alloy A, 800 g of alloy B, and 700 g of alloy C. Set up a matrix for this information. By multiplying these matrices, obtain a matrix that gives the total weight of lead, zinc, and copper in the shipment.

79. The matrix equation

$$\left(\begin{pmatrix} R_1 & -R_2 \\ -R_2 & R_1 \end{pmatrix} + R_2\begin{pmatrix} 1 & 0 \\ 0 & 1 \end{pmatrix}\right)\begin{pmatrix} i_1 \\ i_2 \end{pmatrix} = \begin{pmatrix} 6 \\ 0 \end{pmatrix}$$

may be used to represent the system of equations relating the currents and resistances of the circuit in Fig. 16-15. Find this system of equations by performing the indicated matrix operations.

80. In taking inventory, a firm finds that it has in one warehouse 6 pieces of 5-m brass pipe, 8 pieces of 10-m brass pipe, 11 pieces of 15-m brass pipe, 5 pieces of 5-m steel pipe, 10 pieces of 10-m steel pipe, and 15 pieces of 15-m steel pipe. This inventory can be represented by matrix A below. In each of two other warehouses, the inventory of the same items is represented by matrix B below.

$$A = \begin{pmatrix} 6 & 8 & 11 \\ 5 & 10 & 15 \end{pmatrix} \qquad B = \begin{pmatrix} 8 & 3 & 4 \\ 6 & 10 & 5 \end{pmatrix}$$

By matrix addition and scalar multiplication, find the matrix that represents the total number of each item in the three warehouses.

Writing Exercise

81. A hardware company has 60 different retail stores in which 3500 different products are sold. Write a paragraph explaining why matrices provide an efficient method of inventory control for this company.

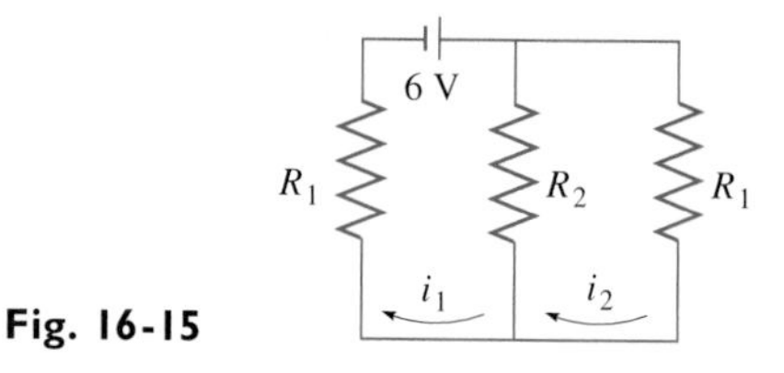

Fig. 16-15

Practice Test

1. For matrices A and B, find $A - 2B$.

$$A = \begin{pmatrix} 3 & -1 & 4 \\ 2 & 0 & -2 \end{pmatrix} \qquad B = \begin{pmatrix} 1 & 4 & 5 \\ -1 & -2 & 3 \end{pmatrix}$$

2. Evaluate using minors:

$$\begin{vmatrix} 4 & 0 & -2 \\ 3 & -3 & 2 \\ -4 & 1 & -1 \end{vmatrix}$$

3. For matrices C and D, find CD and DC.

$$C = \begin{pmatrix} 1 & 0 & 4 \\ 2 & -2 & 1 \\ -1 & 3 & 2 \end{pmatrix} \qquad D = \begin{pmatrix} 2 & -2 \\ 4 & -5 \\ 6 & 1 \end{pmatrix}$$

4. Evaluate using the properties of determinants:

$$\begin{vmatrix} 1 & 0 & 4 & -2 \\ -2 & -1 & 3 & 0 \\ 3 & 2 & -1 & 2 \\ 1 & 1 & -1 & -2 \end{vmatrix}$$

5. For matrix C of Problem 3, find C^{-1}.

6. Solve by using the inverse of the coefficient matrix.

$$2x - 3y = 11$$
$$x + 2y = 2$$

7. Fifty shares of stock A and 30 shares of stock B cost \$2600. Thirty shares of stock A and 40 shares of stock B cost \$2000. What is the price per share of each stock? Solve by setting up the appropriate equations and then using the inverse of the coefficient matrix.

17

Inequalities

Until now we have devoted a great deal of time to the solution of equations. Solving equations is very important in mathematics, but there are also many situations in mathematics and in applications when the solution of an inequality is required.

In mathematics, for example, we find the domain of a function by finding *all values* that give real numbers for the function. Inequalities are also very useful in later topics in mathematics, such as calculus.

In electricity it may be necessary to find the values of a current that are *less than* a given value. In designing a link in a robotic mechanism, it might be necessary to determine the forces that are *greater than* a specific value. Computers often are programmed to switch from one part of a program to another based on a result that is *greater than* (or *less than*) some programmed value.

An important application occurs in business, where inequalities are used to set production levels for maximizing profits or minimizing costs.

In our discussion of inequalities, we will find use of many of the functions and methods developed in earlier chapters.

In Section 17-5 we use inequalities to show how a company can maximize profit in making products such as speaker systems.

17-1 Properties of Inequalities

In Chapter 1 we first introduced the signs of inequality. To this point only a basic understanding of their meanings has been necessary to show certain intervals associated with a variable. In this section we review the meanings and develop certain basic properties of inequalities. We also show the meaning of the solution of an inequality and how it is shown on the number line.

The expression $a < b$ is read as "a is less than b," and the expression $a > b$ is read as "a is greater than b." *These signs define what is known as the* **sense** *(indicated by the direction of the sign) of the inequality.* Two inequalities are said to have the same sense if the signs of inequality point in the same direction. They are said to have the opposite sense if the signs of inequality point in opposite directions. *The two sides of the inequality are called* **members** *of the inequality.*

EXAMPLE 1 ▸▸ The inequalities $x + 3 > 2$ and $x + 1 > 0$ have the same sense, as do the inequalities $3x - 1 < 4$ and $x^2 - 1 < 3$.

The inequalities $x - 4 < 0$ and $x > -4$ have the opposite sense, as do the inequalities $2x + 4 > 1$ and $3x^2 - 7 < 1$. ◂◂

The **solution** *of an inequality consists of all values of the variable that make the inequality a true statement.* Most inequalities with which we shall deal are **conditional inequalities,** *which are true for some, but not all, values of the variable.* Also, *some inequalities are true for all values of the variable, and they are called* **absolute inequalities.** A solution of an inequality consists of only real numbers, as the terms *greater than* or *less than* have not been defined for complex numbers.

EXAMPLE 2 ▸▸ The inequality $x + 1 > 0$ is true for all values of x greater than -1. Therefore, the values of x that satisfy this inequality are written as $x > -1$. This illustrates the difference between the solution of an equation and the solution of an inequality. The solution of an equation normally consists of a few specific numbers, whereas *the solution to an inequality normally consists of an interval of values of the variable.* Any and all values within this interval are termed solutions of the inequality. Since the inequality $x + 1 > 0$ is satisfied only by the values of x in the interval $x > -1$, it is a *conditional inequality.*

NOTE ▶

The inequality $x^2 + 1 > 0$ is true for all real values of x, since x^2 is never negative. It is an *absolute inequality.* ◂◂

There are occasions when it is convenient to combine an inequality with an equality. For such purposes, the symbols $\leq$, meaning *less than or equal to,* and $\geq$, meaning *greater than or equal to,* are used.

EXAMPLE 3 ▸▸ If we wish to state that x is positive, we can write $x > 0$. However, the value zero is not included in the solution. If we wish to state that x is not negative, we write $x \geq 0$. Here zero is part of the solution.

In order to state that x is less than or equal to -5, we write $x \leq -5$. ◂◂

In the sections that follow, we will solve inequalities. It is often useful to show the solution on the number line. The next example shows how this is done.

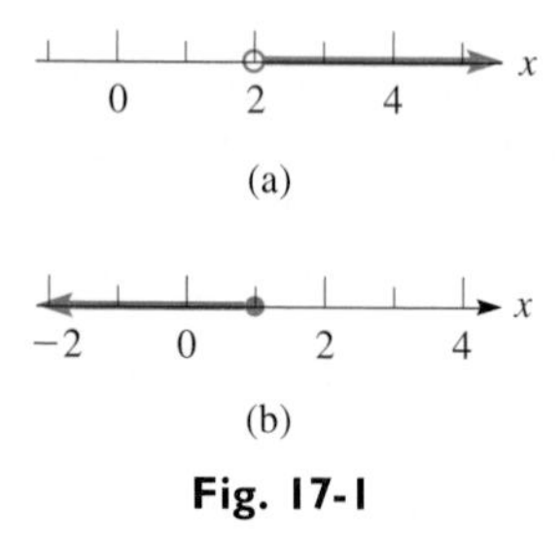

Fig. 17-1

EXAMPLE 4 ▸▸ (a) To graph $x > 2$, we draw a small open circle at 2 on the number line (which is equivalent to the x-axis). Then we draw a solid line to the right of this point with an arrowhead pointing to the right, indicating all values greater than 2. See Fig. 17-1(a). The *open circle* shows that the point is not part of the indicated solution.

(b) To graph $x \leq 1$, we follow the same basic procedure as in part (a), except that we use a solid circle and the arrowhead points to the left. See Fig. 17-1(b). The *solid circle* shows that the point is part of the indicated solution. ◂◂

Properties of Inequalities

We shall now show the basic operations performed on inequalities. These are the same operations as those performed on equations, but in certain cases the results take on a different form. *The following are the* **properties of inequalities.**

1. *The sense of an inequality is not changed when the same number is added to—or subtracted from—both members of the inequality.* Symbolically this may be stated as "if $a > b$, then $a + c > b + c$ and $a - c > b - c$."

EXAMPLE 5 ▸▸ Using Property 1 on the inequality $9 > 6$, we have the following results.

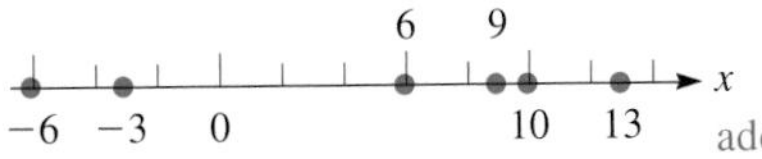

$9 > 6$	$9 > 6$
add 4 to each member	subtract 12 from each member
$9 + 4 > 6 + 4$	$9 - 12 > 6 - 12$
$13 > 10$	$-3 > -6$

Fig. 17-2

In Fig. 17-2 we see that 9 is to the right of 6, 13 is to the right of 10, and -3 is to the right of -6. ◂◂

2. *The sense of an inequality is not changed if both members are multiplied or divided by the same positive number.* Symbolically this is stated as "if $a > b$, then $ac > bc$, and $a/c > b/c$, provided that $c > 0$."

EXAMPLE 6 ▸▸ Using Property 2 on the inequality $8 < 15$, we have the following results.

$8 < 15$	$8 < 15$
multiply both members by 2	divide both members by 2
$2(8) < 2(15)$	$\frac{8}{2} < \frac{15}{2}$
$16 < 30$	$4 < \frac{15}{2}$ ◂◂

CAUTION ▸

3. ***The sense of an inequality is* reversed *if both members are multiplied or divided by the same negative number.*** Symbolically this is stated as "if $a > b$, then $ac < bc$, and $a/c < b/c$, provided that $c < 0$." Be very careful to note that *we obtain different results depending on whether both members are multiplied by a positive number or by a negative number.*

EXAMPLE 7 ▸▸ Using Property 3 on the inequality $4 > -2$, we have the following results.

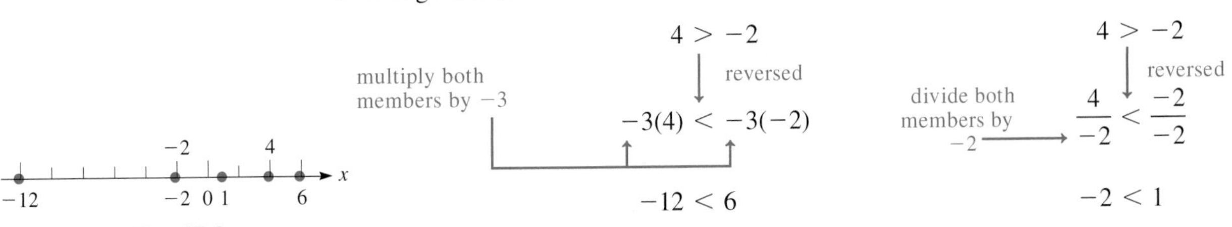

Fig. 17-3

In Fig. 17-3 we see that 4 is to the right of -2, but that -12 *is to the left of* 6 and that -2 *is to the left of 1.* This is consistent with reversing the sense of the inequality when it is multiplied by -3 and when it is divided by -2. ◂◂

4. *If both members of an inequality are positive numbers and n is a positive integer, then the inequality formed by taking the nth power of each member, or the nth root of each member, is in the same sense as the given inequality.* Symbolically this is stated as "if $a > b$, then $a^n > b^n$, and $\sqrt[n]{a} > \sqrt[n]{b}$, provided that $n > 0$, $a > 0$, $b > 0$."

EXAMPLE 8 ⏩ Using Property 4 on the inequality $16 > 9$, we have:

$16 > 9$	$16 > 9$
square both members	take square root of both members
$16^2 > 9^2$	$\sqrt{16} > \sqrt{9}$
$256 > 81$	$4 > 3$ ⏪

Many inequalities have more than two members. In fact, inequalities with three members are common. All the basic properties hold for inequalities with more than two members. Some care must be used, however, in stating these inequalities.

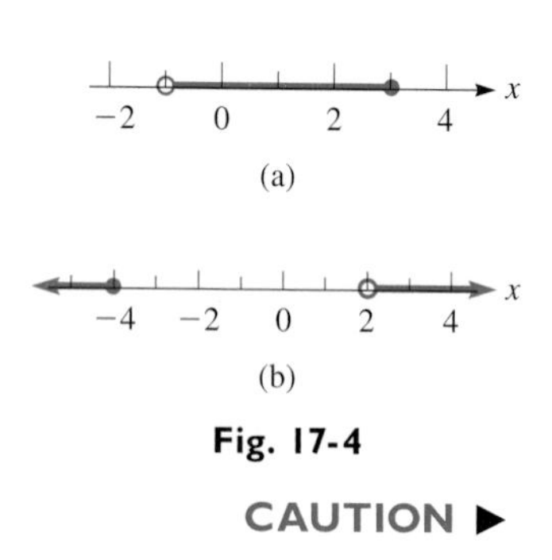

Fig. 17-4

EXAMPLE 9 ⏩ (a) To state that 5 is less than 6, and also greater than 2, we may write $2 < 5 < 6$, or $6 > 5 > 2$. (Generally the *less than* form is preferred.)

(b) To state that a number x may be greater than -1, *and* also less than or equal to 3, we write $-1 < x \leq 3$. (It can also be written as $x > -1$ *and* $x \leq 3$.) This is shown in Fig. 17-4(a). Note the use of the open circle and the solid circle.

(c) By writing $x \leq -4$ *or* $x > 2$, we state that x is less than or equal to -4, *or* greater than 2. **CAUTION ▶** ***It may not be stated as*** $2 < x \leq -4$, for this shows x as being less than -4, and also greater than 2, and *no such numbers exist.* See Fig. 17-4(b). ⏪

NOTE ▶ Notice the use of the words *and* and *or* in Example 9. In stating inequalities, ***and*** is used when the solution consists of values that satisfy ***both*** statements. The word ***or*** is used when the solution consists of values that make ***either*** statement true. (In everyday speech, *or* can sometimes also mean that either one statement is true or another statement is true, but *not* that both statements are true.)

EXAMPLE 10 ⏩ The inequality $x^2 - 3x + 2 > 0$ is satisfied if x is either greater than 2 *or* less than 1. **NOTE ▶** This is written as $x > 2$ *or* $x < 1$, but it is incorrect to state it as $1 > x > 2$. (If we wrote it this way, we would be saying that the same value of x is less than 1 *and* at the same time greater than 2. Of course, as we noted for this type of situation in Example 9, no such number exists.) Any inequality must be valid for all values satisfying it. However, we could say that the inequality is not satisfied for $1 \leq x \leq 2$, which means those values of x greater than *or* equal to 1 *and* less than *or* equal to 2 (between or equal to 1 and 2). ⏪

A

80 cm < l < 90 cm

40 cm < w < 80 cm

Fig. 17-5

EXAMPLE 11 ⏩ The design of a rectangular solar panel shows that the length l is between 80 cm and 90 cm, and the width w between 40 cm and 80 cm. See Fig. 17-5. Find the values of area the panel may have.

Since l is to be less than 90 cm and w less than 80 cm, the area must be less than (90 cm)(80 cm) $= 7200$ cm^2. Also, since l is to be greater than 80 cm and w greater than 40 cm, the area must be greater than (80 cm)(40 cm) $= 3200$ cm^2. Therefore, the area A may be represented as

$$3200 \text{ cm}^2 < A < 7200 \text{ cm}^2$$

This means the area is greater than 3200 cm^2 *and* less than 7200 cm^2. ⏪

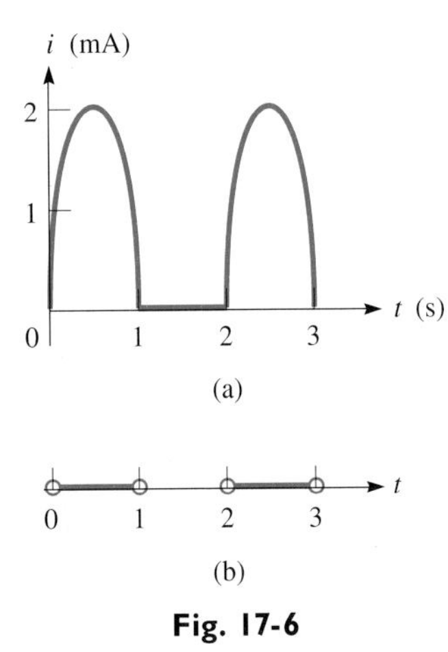

Fig. 17-6

EXAMPLE 12 A semiconductor *diode* has the property that an electric current flows through it in only one direction. If it is an alternating-current circuit, the current in the circuit flows only during the half-cycle when the diode allows it to flow. If a source of current given by $i = 2 \sin \pi t$ (i in mA, t in seconds) is connected in series with a diode, write the inequalities for the current and the time. Assume the source is on for 3.0 s and a positive current passes through the diode.

We are to find the values of t that correspond to $i > 0$. From the properties of the sine function, we know that $2 \sin \pi t$ has a period of $2\pi/\pi = 2.0$ s. Therefore, the current is zero for $t = 0$, 1.0 s, 2.0 s, and 3.0 s.

The source current is positive for $0 < t < 1.0$ s and for $2.0 \text{ s} < t < 3.0$ s.

The source current is negative for $1.0 \text{ s} < t < 2.0$ s.

Therefore, in the circuit

$$i > 0 \quad \text{for} \quad 0 < t < 1.0 \text{ s} \quad \text{and} \quad 2.0 \text{ s} < t < 3.0 \text{ s, and}$$
$$i = 0 \quad \text{for} \quad t = 0,\ 1.0 \text{ s} \leq t \leq 2.0 \text{ s}.$$

A graph of the current in the circuit as a function of time is shown in Fig. 17-6(a). In Fig. 17-6(b) the values of t for which $i > 0$ are shown.

Exercises 17–1

In Exercises 1–8, for the inequality $4 < 9$, state the inequality that results when the given operations are performed on both members.

1. Add 3
2. Subtract 6
3. Multiply by 5
4. Multiply by -2
5. Divide by -1
6. Divide by 2
7. Square both
8. Take square roots

In Exercises 9–20, give the inequalities equivalent to the following statements about the number x.

9. Greater than -2
10. Less than 7
11. Less than or equal to 4
12. Greater than or equal to -6
13. Greater than 1 and less than 7
14. Greater than or equal to -2 and less than 6
15. Less than -9, or greater than or equal to -4
16. Less than or equal to 8, or greater than or equal to 12
17. Less than 1, or greater than 3 and less than or equal to 5
18. Greater than or equal to 0 and less than or equal to 2, or greater than 5
19. Greater than -2 and less than 2, or greater than or equal to 3 and less than 4
20. Less than -4, or greater than or equal to 0 and less than or equal to 1, or greater than or equal to 5

In Exercises 21–24, give verbal statements equivalent to the given inequalities involving the number x.

21. $0 < x \leq 2$
22. $x < 5$ or $x > 7$
23. $x < -1$ or $1 \leq x < 2$
24. $-1 \leq x < 3$ or $5 < x < 7$

In Exercises 25–36, graph the given inequalities on the number line.

25. $x < 3$
26. $x \geq -1$
27. $x \leq 1$ or $x > 3$
28. $x < -3$ or $x \geq 0$
29. $0 \leq x < 5$
30. $-2 < x < 4$
31. $x < -1$ or $1 \leq x < 4$
32. $-1 < x < 2$ or $x > 3$
33. $-3 < x < -1$ or $1 < x \leq 3$
34. $1 < x \leq 2$ or $3 \leq x < 4$
35. $x < -3$ or $x > -3$
36. $x < 1$ or $1 < x \leq 4$

In Exercises 37–44, some applications of inequalities are shown.

37. When Pioneer II left the solar system in 1989, it was 5×10^{12} km from the sun. Assuming it never returns, express its distance d from the sun in the future as an inequality. Graph these values of d.
38. A busy person glances at a digital clock that shows 9:36. Another glance a short time later shows the clock at 9:44. Express the amount of time t (in min) that could have elapsed between glances by use of inequalities. Graph these values of t.
39. An earth satellite put into orbit near the earth's surface will have an elliptic orbit if its velocity v is between 29 000 km/h and 40 000 km/h. State this as an inequality, and graph these values of v.

40. Fossils found in Jurassic rocks indicate that dinosaurs flourished during the Jurassic geological period, 140 MY (million years ago) to 200 MY. Write this as an inequality, with t representing past time. Graph the values of t.

41. In executing a program, a computer must perform a set of calculations. Any one of the calculations takes no more than 2565 steps. Express the number n of steps required for a given calculation by an inequality. (Note that n is a positive *integer*.)

42. A surveyor measures the side of a parcel of land and reports that its length l is $l = 72.37$ m $\pm$ 0.05 m, where the ± 0.05 m gives the possible error in the measurement. Express the length l by an inequality.

43. The electric intensity E within a charged spherical conductor is zero. The intensity on the surface and outside the sphere equals a constant k divided by the square of the distance r from the center of the sphere. State these relations for a sphere of radius a by using inequalities, and graph E as a function of r.

44. If the current from the source in Example 12 is $i = 5 \cos 4\pi t$ and the diode allows only a negative current to flow, write the inequalities and draw the graph for the current in the circuit as a function of time for $0 \leq t \leq 1$ s.

17-2 Solving Linear Inequalities

Using the properties and definitions discussed in Section 17-1, we can now proceed to solve inequalities. In this section we solve linear inequalities in one variable. Similar to linear functions as defined in Chapter 5, *a* **linear inequality** *is one in which each term contains only one variable and the exponent of each variable is 1.* We will consider linear inequalities in two variables in Section 17-5.

The procedure for solving a linear inequality in one variable is like that we used in solving basic equations in Chapter 1. We solve the inequality by isolating the variable, and to do this we perform the same operations on each member of the inequality. The operations are based on the properties given in Section 17-1.

EXAMPLE 1 ▸▸ In each of the following inequalities, by performing the indicated operation we isolate x and thereby solve the inequality.

$x + 2 < 4$	$\frac{x}{2} > 4$	$2x \leq 4$
Subtract 2 from each member.	Multiply each member by 2.	Divide each member by 2.
$x < 2$	$x > 8$	$x \leq 2$

Each solution can be checked by substituting any number in the indicated interval into the original inequality. For example, any value less than 2 will satisfy the first inequality, whereas 2 or any number less than 2 will satisfy the third inequality. ◂◂

EXAMPLE 2 ▸▸ Solve the following inequality: $3 - 2x \geq 15$.

We have the following solution:

$$3 - 2x \geq 15 \quad \text{original inequality}$$

$$-2x \geq 12 \quad \text{subtract 3 from each member}$$

inequality reversed ⟶ ↓

$$x \leq -6 \quad \text{divide each member by } -2$$

CAUTION ▸ Again, carefully note that ***the sign of inequality was reversed when each number was divided by −2.*** We check the solution by substituting −7 in the original inequality, obtaining $17 \geq 15$. ◂◂

EXAMPLE 3 ▸▸ Solve the inequality $2x \le 3 - x$.

The solution proceeds as follows:

$2x \le 3 - x$	original inequality
$3x \le 3$	add x to each member
$x \le 1$	divide each member by 3

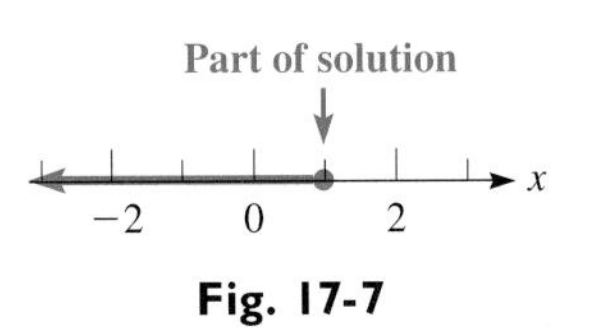

Fig. 17-7

This solution checks and is represented in Fig. 17-7, as we showed in Section 17-1.

This inequality could have been solved by combining x-terms on the right. In doing so, we would obtain $1 \ge x$. Since this might be misread, it is best to combine the variable terms on the left, as we did above. ◂◂

EXAMPLE 4 ▸▸ Solve the inequality $\frac{3}{2}(1 - x) > \frac{1}{4} - x$.

$\frac{3}{2}(1 - x) > \frac{1}{4} - x$	original inequality
$6(1 - x) > 1 - 4x$	multiply each member by 4
$6 - 6x > 1 - 4x$	remove parentheses
$-6x > -5 - 4x$	subtract 6 from each member
$-2x > -5$	add $4x$ to each member
$x < \frac{5}{2}$	divide each member by -2

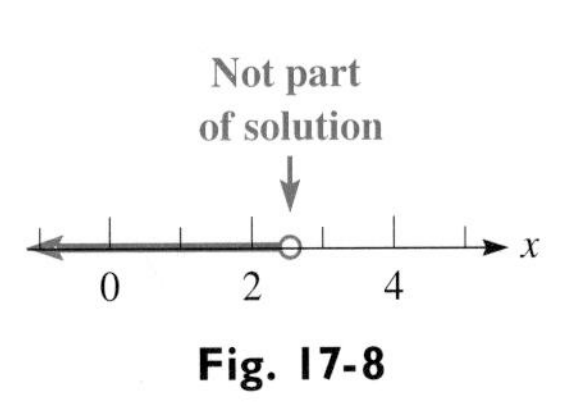

Fig. 17-8

Note that the sense of the inequality was reversed when we divided by -2. This solution is shown in Fig. 17-8. Any value of $x < 5/2$ checks when substituted into the original inequality. ◂◂

The following example illustrates an application that involves the solution of an inequality.

EXAMPLE 5 ▸▸ The velocity v (in m/s) of a missile in terms of the time t (in seconds) is given by $v = 305 - 9.8t$. For how long is the velocity positive? (Since velocity is a vector, this can also be interpreted as asking "how long is the missile moving upward?")

In terms of inequalities, we are asked to find the values of t for which $v > 0$. This means that we must solve the inequality $305 - 9.8t > 0$. The solution is as follows:

$305 - 9.8t > 0$	original inequality
$-9.8t > -305$	subtract 1000 from each member
$t < 31$ s	divide each member by -9.8; round off result

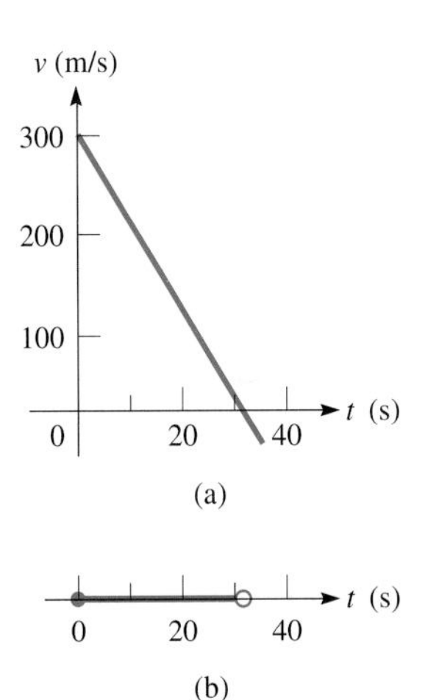

(a)

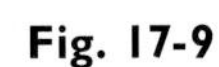

(b)

Fig. 17-9

Negative values of t have no meaning in this problem. Checking $t = 0$, we find that $v = 305$ m/s. Therefore, the complete solution is $0 \le t < 31$ s.

In Fig. 17-9(a) we show the graph of $v = 305 - 9.8t$, and in Fig. 17-9(b) we show the solution $0 \le t < 31$ s on the number line (which is really the t-axis in this case). Note that the values of v are above the t-axis for those values of t that are part of the solution. This shows the relationship of the graph of v as a function of t, and the solution as graphed on the number line (the t-axis). ◂◂

Inequalities with Three Members

EXAMPLE 6 ▸▸ Solve: $-1 < 2x + 3 < 6$.

We have the following solution.

$$-1 < 2x + 3 < 6 \quad \text{original inequality}$$

$$-4 < 2x < 3 \quad \text{subtract 3 from each member}$$

$$-2 < x < \frac{3}{2} \quad \text{divide each member by 2}$$

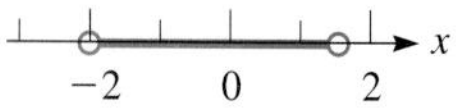

Fig. 17-10

The solution is shown in Fig. 17-10. ◂◂

EXAMPLE 7 ▸▸ Solve the inequality $2x < x - 4 \leq 3x + 8$.

Since we cannot isolate x in the middle member (or in any member), we rewrite the inequality as

$$2x < x - 4 \quad \text{and} \quad x - 4 \leq 3x + 8$$

We then solve each of these inequalities, keeping in mind that the solution must satisfy both of them. Therefore, we have

$$2x < x - 4 \quad \text{and} \quad x - 4 \leq 3x + 8$$

$$-2x \leq 12$$

$$x < -4 \quad \text{and} \quad x \geq -6$$

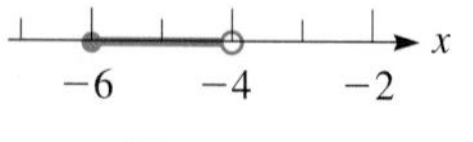

Fig. 17-11

The solution can be written as $-6 \leq x < -4$, and this solution is shown in Fig. 17-11. ◂◂

EXAMPLE 8 ▸▸ In emptying a wastewater tank, one pump can remove no more than 40 L/min. If it operates for 8.0 min and a second pump operates for 5.0 min, what must be the pumping rate of the second pump if 480 L are to be removed?

Let x = the pumping rate of the first pump and y = the pumping rate of the second pump. Since the first operates for 8.0 min and the second for 5.0 min to remove 480 L, we have

first pump, second pump, total ⟵ amounts pumped

$$8.0x + 5.0y = 480$$

Since we know that the first pump can remove no more than 40 L/min, which means that $0 \leq x \leq 40$ L/min, we solve for x, then substitute in this inequality.

$$x = 60 - 0.625y \quad \text{solve for } x$$

$$0 \leq 60 - 0.625y \leq 40 \quad \text{substitute in inequality}$$

$$-60 \leq -0.625y \leq -20 \quad \text{subtract 60 from each member}$$

$$96 \geq y \geq 32 \quad \text{divide each member by } -0.625$$

$$32 \leq y \leq 96 \text{ L/min} \quad \text{use} \leq \text{symbol (optional step)}$$

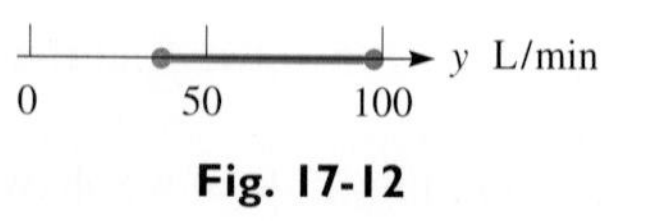

Fig. 17-12

This means that the second pump must be able to pump at least 32 L/min and no more than 96 L/min. See Fig. 17-12.

We note that although this was a three-member inequality and it was combined with equalities, the solution was performed in the same way as with a two-member inequality. ◂◂

Exercises 17–2

In Exercises 1–24, solve the given inequalities. Graph each solution.

1. $x - 3 > -4$

2. $x + 2 \le 6$

3. $\frac{1}{2}x < 3$

4. $4x > -12$

5. $3x - 5 \le -11$

6. $\frac{1}{3}x + 2 \ge 1$

7. $6 - x > 4$

8. $3 - 3x < -1$

9. $4x - 5 \le 2x$

10. $2x \ge 6 - x$

11. $2 - (x + 1) > x + 3$

12. $-2(x + 4) > 1 - 5x$

13. $x + 4 \ge 3(x - 3)$

14. $2x - 7 \le 4 - (x + 2)$

15. $\frac{1}{3} - \frac{x}{2} < x + \frac{3}{2}$

16. $\frac{x}{5} - 2 > \frac{2}{3}(x + 3)$

17. $-1 < 2x + 1 < 3$

18. $2 < 3x + 1 \le 8$

19. $-4 \le 1 - x < -1$

20. $0 \le 3 - 2x \le 6$

21. $2x < x - 1 \le 3x + 5$

22. $x + 1 \le 7 - x < 2x$

23. $2x - 3 < x - 5 < 3x - 3$

24. $x - 1 < 2x + 2 < 3x + 1$

In Exercises 25–36, solve the given problems by setting up and solving appropriate inequalities. Graph each solution.

25. Determine the values of x that are in the domain of the function $f(x) = \sqrt{2x - 10}$.

26. Determine the values of x that are in the domain of the function $f(x) = 1/\sqrt{3 - 0.5x}$.

27. A salesperson is paid a commission of 12% on all monthly sales over \$3600. What must be the sales in a month in order to earn a commission over \$900?

28. The minimum legal speed on a certain highway is 70 km/h, and the maximum legal speed is 110 km/h. What legal distances can a motorist travel in 4 h on this highway without stopping?

29. The value V, in dollars, of each building lot in a development is estimated as $V = 40\,000 + 4000t$, where t is the time in years. For how long is the value of each lot no more than \$64 000?

30. A beam is supported at each end as shown in Fig. 17-13. Analyzing the forces on the beam leads to the equation $F_1 = 13 - 3d$. For what values of d is F_1 more than 6 N?

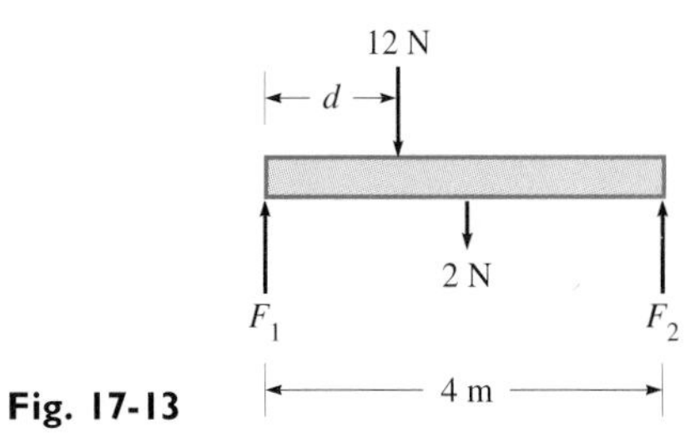

Fig. 17-13

31. The mass m (in grams) of silver plate on a dish is increased by electroplating. The mass of silver on the plate is $m = 125 + 15.0t$, where t is the time (in hours) of electroplating. For what values of t is m between 131 g and 164 g?

32. The amount of heat H (in kJ) required to raise 10 g of ice at 0°C to water at temperature T (in °C) is given by $H = 3.35 + 0.042T$. For what values of T is $H < 5.45$ kJ?

33. During a given rush hour, the numbers of vehicles shown in Fig. 17-14 go in the indicated directions in a one-way street section of a city. By finding the possible values of x and the equation relating x and y, find the possible values of y.

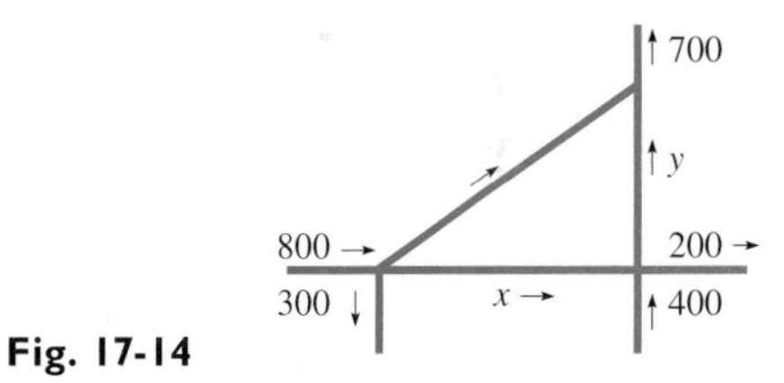

Fig. 17-14

34. One computer line printer can print no more than 2500 lines/min. If it operates for 3.0 min and a second printer operates for 2.0 min, what printing rates must the second printer have in order to print 9000 lines?

35. In obtaining a solution of hydrochloric acid containing 2 L of pure acid, a chemist mixes x litres of a 5% solution with y litres of a 10% solution. If y is no more than 8 L, what are the possible values of x?

36. An oil company plans to install eight storage tanks, each with a capacity of x litres, and five additional tanks, each with a capacity of y litres, such that the total capacity of all tanks is 440 000 L. If capacity y will be at least 40 000 L, what are the values of capacity x?

17-3 Solving Nonlinear Inequalities

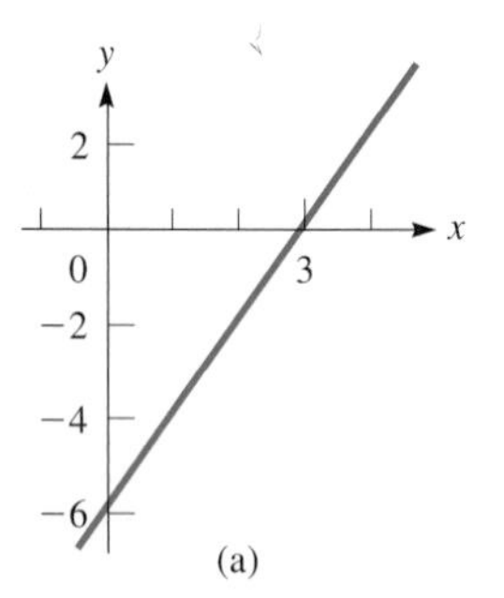

Fig. 17-15

In this section we show how to solve inequalities that involve polynomial and fractional expressions. We will also introduce a graphical method that can be used with nonfactorable and nonalgebraic inequalities. In order to develop the method for polynomial inequalities, we first look again at a linear inequality.

If we graph the linear function $f(x) = ax + b$ ($a \neq 0$), we see that all values of $f(x)$ are positive for all values of x on one side of the point where $f(x) = 0$, and all values of $f(x)$ are negative on the other side of this point. See Fig. 17-15. This leads us to another method of solving a linear inequality. This method is to *express the given inequality with* ***zero*** *on the right side and then determine the* ***sign*** *of the resulting function on either side of the zero of the function.*

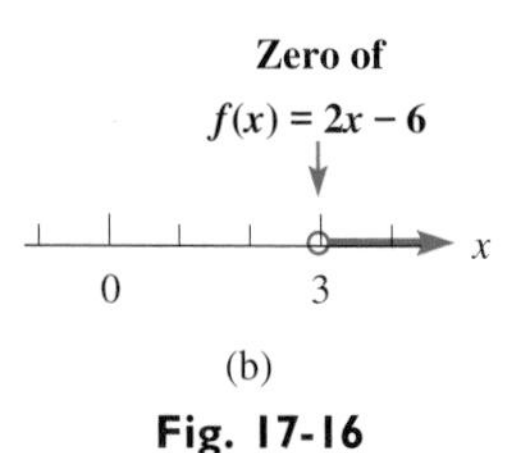

Fig. 17-16

EXAMPLE 1 ▸▸ Solve the inequality $2x - 5 > 1$.

We first find the equivalent inequality with zero on the right. This is done by subtracting 1 from each member. Thus, we have $2x - 6 > 0$. We now set the left member equal to zero. Thus,

$$2x - 6 = 0 \quad \text{for} \quad x = 3$$

which means that 3 is the zero of the function $f(x) = 2x - 6$. We know that the function $f(x)$ has one sign for $x < 3$ and has the opposite sign for $x > 3$. Testing values in these intervals, we find, for example, that

$$f(x) = -2 \quad \text{for} \quad x = 2 \quad \text{and} \quad f(x) = +2 \quad \text{for} \quad x = 4$$

Thus, for $x > 3$, $2x > 6$, and this means the solution to the original inequality is $x > 3$. The solution shown in Fig. 7-16(b) corresponds to the positive values of the function $f(x) = 2x - 6$ shown in Fig. 17-16(a).

We could have solved this inequality by the methods of Section 17-2, but the important idea here is using the *sign* of the function, with *zero* on the right. ◂◂

We can extend this method to solving inequalities with polynomials of higher degree. We first find the equivalent inequality with zero on the right, and then *factor the function on the left into linear factors and any quadratic factors that lead to complex roots.* As all values of x are considered, each linear factor can change sign at the value for which it is zero. The quadratic factors do not change sign.

Critical Values

This method is especially useful in solving inequalities involving fractions. If a linear factor occurs in the numerator, the function is zero at the value of x for which the linear factor is zero. If such a factor appears in the denominator, the function is undefined where the factor is zero. *The values of x for which a function is zero or undefined are called the* **critical values** *of the function.* As all values of x are considered, a function can change sign only at a critical value. Therefore, we have the following method for solving an inequality using its critical values.

Using Critical Values to Solve an Inequality

1. *Determine the equivalent inequality with zero on the right.*
2. *Find all linear factors of the function.*
3. *To find the critical values, set each linear factor equal to zero and solve for x.*
4. *Determine the sign of the function to the left of the leftmost critical value, between critical values, and to the right of the rightmost critical value.*
5. *Those intervals in which the function has the proper sign satisfy the inequality.*

EXAMPLE 2 ▸▸ Solve the inequality $x^2 - 3 > 2x$.

We first find the equivalent inequality with zero on the right. Therefore, we have $x^2 - 2x - 3 > 0$. We then factor the left member and have

$$(x + 1)(x - 3) > 0$$

Setting each of the two factors equal to zero, we find the left critical value is -1 and the right critical value is 3. All values of x to the left of -1 give the same sign for the function. All values between -1 and 3 give the function the same sign. All values of x to the right of 3 give the same sign to the function. Therefore, ***we must determine the sign of $f(x)$ for each of the intervals $x < -1$, $-1 < x < 3$, and $x > 3$.***

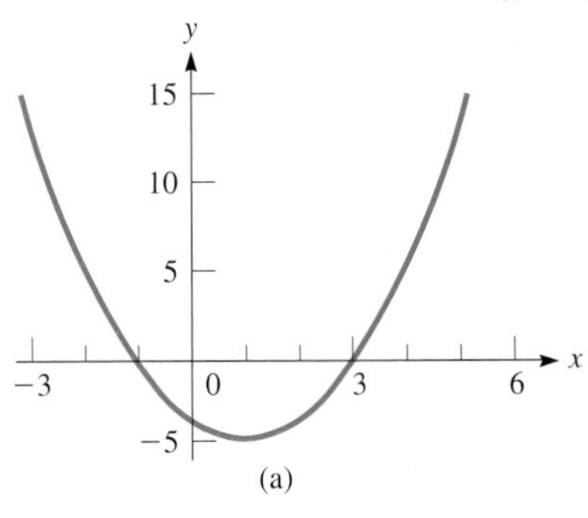

(a)

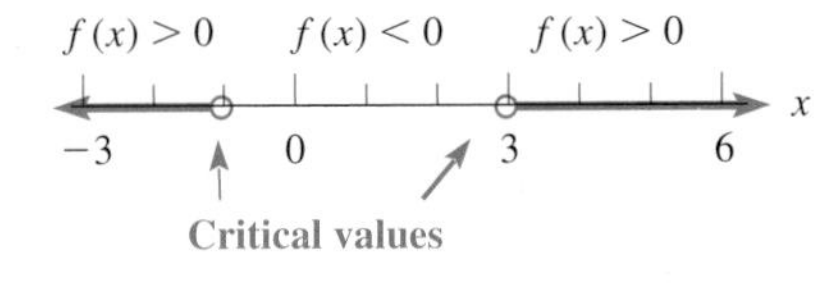

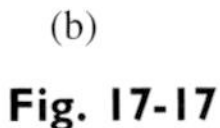

(b)

Fig. 17-17

For $x < -1$, both factors are negative. Since the product of two negative numbers is positive, $(x + 1)(x - 3) > 0$.

For $-1 < x < 3$, the left factor is positive and the right factor is negative. Since the product of a positive number and a negative number is negative, we see that $(x + 1)(x - 3) < 0$.

For $x > 3$, both factors are positive. Therefore, this means that $(x + 1)(x - 3) > 0$.

Summarizing these results, we have

If	$x < -1,$	$(x + 1)(x - 3) > 0$
If	$-1 < x < 3,$	$(x + 1)(x - 3) < 0$
If	$x > 3,$	$(x + 1)(x - 3) > 0$

Therefore, the solution to the inequality is $x < -1$ or $x > 3$.

The solution that is shown in Fig. 17-17(b) corresponds to the positive values of the function $f(x) = x^2 - 2x - 3$ shown in Fig. 17-17(a). ◂◂

EXAMPLE 3 ▸▸ Solve the inequality $x^3 - 4x^2 + x + 6 < 0$.

By methods developed in Chapter 15, we factor the function on the left and obtain $(x + 1)(x - 2)(x - 3) < 0$. The critical values are $-1, 2, 3$. We wish to determine the sign of the left member for the intervals $x < -1$, $-1 < x < 2$, $2 < x < 3$, and $x > 3$. The following table shows each interval, the sign of each factor in each interval, and the resulting sign of the function

$$f(x) = (x + 1)(x - 2)(x - 3)$$

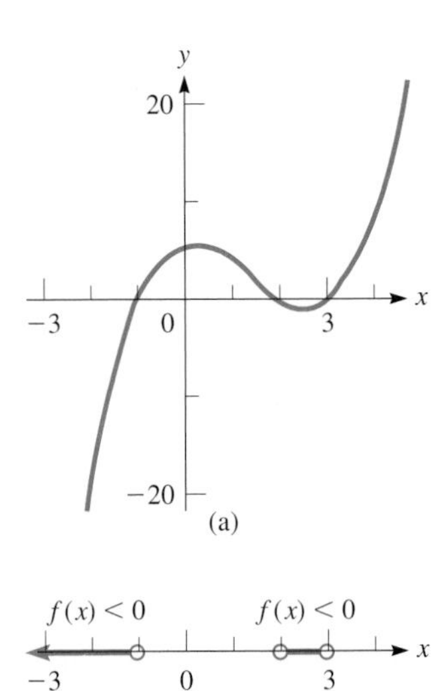

(a)

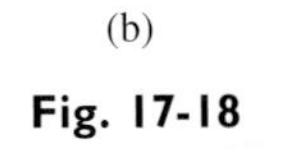

(b)

Fig. 17-18

Interval	$(x + 1)$	$(x - 2)$	$(x - 3)$	*Sign of $f(x)$*
$x < -1$	−	−	−	−
$-1 < x < 2$	+	−	−	+
$2 < x < 3$	+	+	−	−
$x > 3$	+	+	+	+

Since we want $f(x) < 0$, the solution is $x < -1$ or $2 < x < 3$.

The solution shown in Fig. 17-18(b) corresponds to the negative values of the function $f(x)$ shown in Fig. 17-18(a). ◂◂

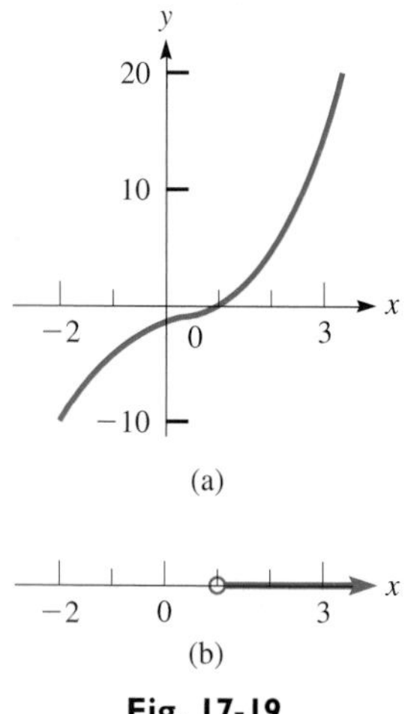

Fig. 17-19

EXAMPLE 4 ▶▶ Solve the inequality $x^3 - x^2 + x - 1 > 0$.

In order to factor this function, we can use the methods of Chapter 15, or we might note that it is factorable by grouping (see Section 6-2). Either method leads to the inequality

$$(x^2 + 1)(x - 1) > 0$$

In this case we have a quadratic factor $x^2 + 1$ that leads to imaginary roots (j and $-j$) and is never negative. This means we have only one linear factor and therefore only one critical value.

Setting $x - 1 = 0$, we get the critical value $x = 1$. Since $x - 1$ is positive for $x > 1$ and negative for $x < 1$, the solution to the inequality is $x > 1$.

The solution is shown in Fig. 17-19(b), and we see that this solution corresponds to the positive values of $f(x) = x^3 - x^2 + x - 1$ shown in Fig. 17-19(a). ◀◀

The following example illustrates an applied situation that involves the solution of an inequality.

EXAMPLE 5 ▶▶ The force F (in newtons) acting on a cam varies according to the time t (in seconds), and it is given by the function $F = 2t^2 - 12t + 20$. For what values of t, $0 \leq t \leq 6$ s, is the force at least 4 N?

For a force of at least 4 N, we know that $F \geq 4$ N, or $2t^2 - 12t + 20 \geq 4$. This means we are to solve the inequality $2t^2 - 12t + 16 \geq 0$, and the solution is as follows:

$$2t^2 - 12t + 16 \geq 0$$
$$t^2 - 6t + 8 \geq 0$$
$$(t - 2)(t - 4) \geq 0$$

The critical values are $t = 2$ and $t = 4$, which lead to the following table.

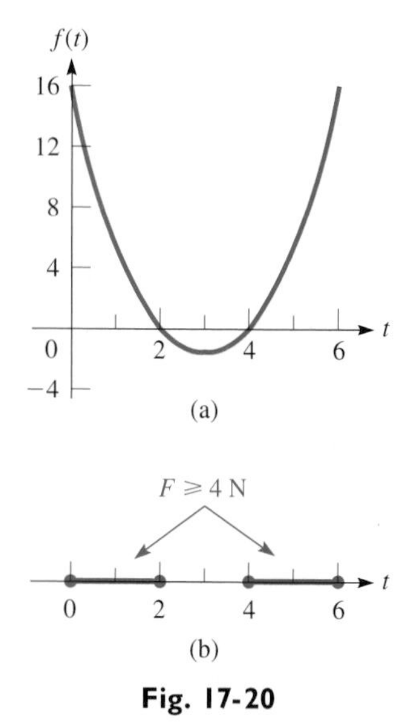

Fig. 17-20

Interval	$(t - 2)(t - 4)$	*Sign of* $(t - 2)(t - 4)$
$0 \leq t < 2$	− −	+
$2 < t < 4$	+ −	−
$4 < t \leq 6$	+ +	+

We see that the values of t which satisfy the *greater than* part of the problem are $0 \leq t < 2$ and $4 < t \leq 6$. Since we know that $(t - 2)(t - 4) = 0$ for $t = 2$ and $t = 4$, the solution is

$$0 \leq t \leq 2 \text{ s}, \qquad 4 \text{ s} \leq t \leq 6 \text{ s}$$

The graph of $f(t) = 2t^2 - 12t + 16$ is shown in Fig. 17-20(a). The solution, shown in Fig. 17-20(b), corresponds to the values of $f(t)$ that are zero or positive or for which $F \geq 4$ N.

We note here that if the cam rotates in 6-s intervals, the force on the cam is periodic, varying from 2 N to 20 N. ◀◀

EXAMPLE 6 ▸▸ Find the values of x for which $\sqrt{\dfrac{x-3}{x+4}}$ represents a real number.

For the expression to represent a real number, the fraction under the radical must be greater than or equal to zero. This means we must solve the inequality

$$\frac{x-3}{x+4} \geq 0$$

The critical values are found from the factors that are in the numerator or in the denominator. Thus, the critical values are -4 and 3. Considering now the *greater than* part of the $\geq$ sign, we set up the following table.

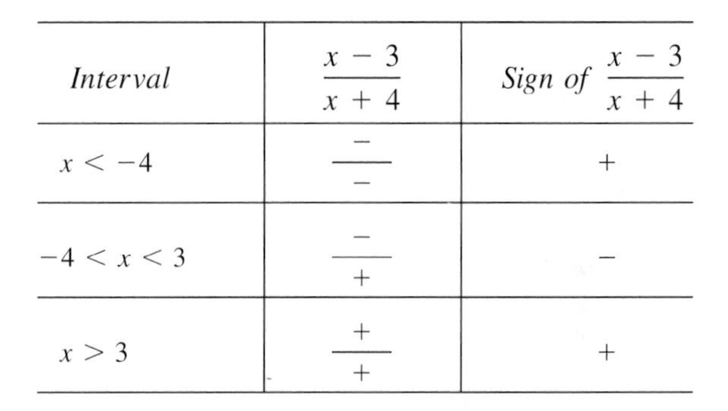

Interval	$\dfrac{x-3}{x+4}$	*Sign of* $\dfrac{x-3}{x+4}$
$x < -4$	$\dfrac{-}{-}$	$+$
$-4 < x < 3$	$\dfrac{-}{+}$	$-$
$x > 3$	$\dfrac{+}{+}$	$+$

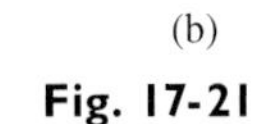

Fig. 17-21

Thus, the values that satisfy the *greater than* part of the problem are those for which $x < -4$ or for which $x > 3$. Now, considering the equality part of the $\geq$ sign, we note that $x = 3$ is valid, for the fraction is zero. However, ***if x = −4, we have division by zero, and thus x may not equal −4.*** Therefore, the inequality is satisfied for $x < -4$ or $x \geq 3$, and these are the values for which the original expression represents a real number. The graph of $f(x) = (x-3)/(x+4)$ is shown in Fig. 17-21(a) (a graphing calculator may be used), and the graph of the solution is shown in Fig. 17-21(b). ◂◂

EXAMPLE 7 ▸▸ Solve the inequality $\dfrac{(x-2)^2(x+3)}{4-x} < 0$.

The critical values are -3, 2, and 4. Thus, we have the following table.

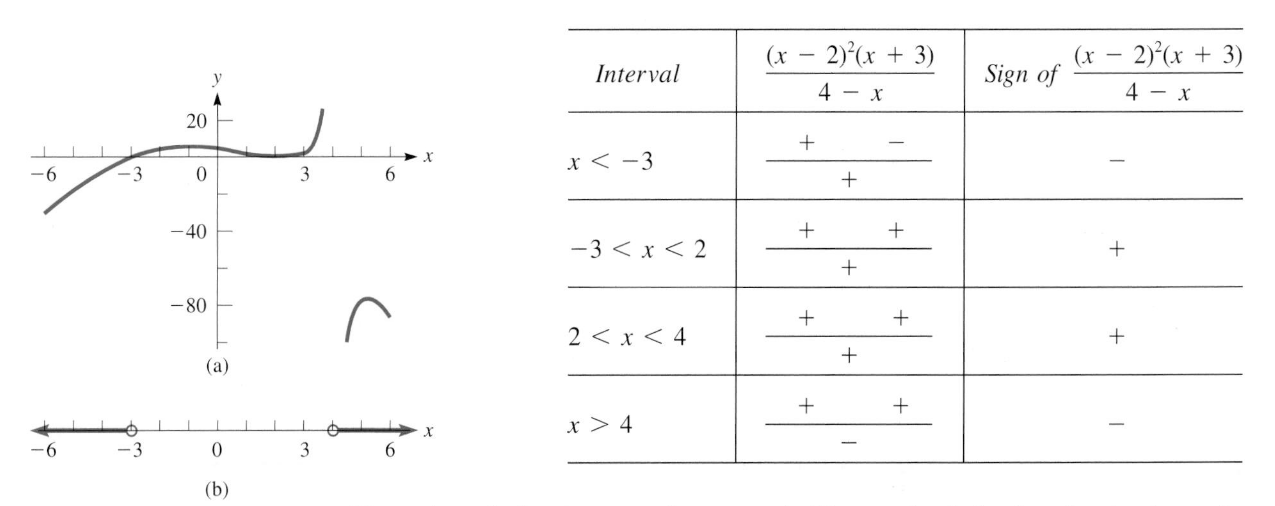

Interval	$\dfrac{(x-2)^2(x+3)}{4-x}$	*Sign of* $\dfrac{(x-2)^2(x+3)}{4-x}$
$x < -3$	$\dfrac{+\quad -}{+}$	$-$
$-3 < x < 2$	$\dfrac{+\quad +}{+}$	$+$
$2 < x < 4$	$\dfrac{+\quad +}{+}$	$+$
$x > 4$	$\dfrac{+\quad +}{-}$	$-$

Fig. 17-22

Thus, the solution is $x < -3$ or $x > 4$. The graph of the function is shown in Fig. 17-22(a), and the graph of the solution is shown in Fig. 17-22(b). ◂◂

Solving Inequalities Graphically

To this point we have considered only those inequalities in which we can factor the function on the left of the sign of inequality, and with zero on the right. Inequalities in which the function is not factorable, including nonalgebraic functions, can be solved approximately by graphing the function. In fact, looking back at several of the examples we have done, we have shown the graph of the function and have seen that the solution of the inequality corresponds to the values of x for which the function is positive or negative (and possibly zero).

The method we will use to solve an inequality graphically is similar to the method we have been using. First we write the equivalent inequality with zero on the right. Then set y equal to the left member, and graph this function. Those values of x corresponding to the proper values of y (either above or below the x-axis) are those which satisfy the inequality. A graphing calculator is very useful for solving inequalities in this manner.

EXAMPLE 8 Use a graphing calculator to solve the inequality $x^3 > x^2 - 3$.

Finding the equivalent inequality with zero on the right, we have $x^3 - x^2 + 3 > 0$. We then let

$$y = x^3 - x^2 + 3$$

and graph this function. Using the RANGE values

$$\text{Xmin} = -3, \text{Xmax} = 3, \text{Ymin} = -5, \text{Ymax} = 5$$

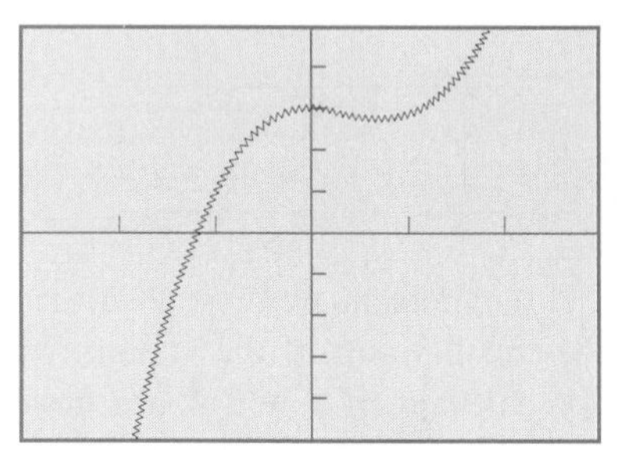

Fig. 17-23

we get the view shown in Fig. 17-23. We can see that the curve crosses the x-axis only once, between $x = -2$ and $x = -1$. Using the TRACE and ZOOM features, we find that this value is about -1.17. Since we want x values that correspond to positive values of y, the solution is $x > -1.17$.

Exercises 17–3

In Exercises 1–28, solve the given inequalities. Graph each solution. It is suggested that you also graph the function on a graphing calculator as a check.

1. $x^2 - 1 < 0$
2. $x^2 + 3x \geq 0$
3. $2x^2 \leq 4x$
4. $x^2 - 4x > 5$
5. $3x^2 + 5x \geq 2$
6. $2x^2 - 12 \leq -5x$
7. $6x^2 + 1 < 5x$
8. $12x^2 + x > 1$
9. $x^2 + 4x \leq -4$
10. $9x^2 + 6x > -1$
11. $x^2 + 4 > 0$
12. $x^4 + 2 < 1$
13. $x^3 + x^2 - 2x > 0$
14. $x^3 - 2x^2 + x \geq 0$
15. $x^3 + 2x^2 - x - 2 \geq 0$
16. $x^4 - 2x^3 - 7x^2 + 8x + 12 < 0$
17. $\dfrac{x - 8}{3 - x} < 0$
18. $\dfrac{x + 5}{x - 1} > 0$
19. $\dfrac{2x - 3}{x + 6} \leq 0$
20. $\dfrac{3x + 1}{x + 3} \geq 0$
21. $\dfrac{2}{x^2 - x - 2} < 0$
22. $\dfrac{-5}{2x^2 + 3x - 2} < 0$
23. $\dfrac{x^2 - 6x - 7}{x + 5} > 0$
24. $\dfrac{4 - x}{3 + 2x - x^2} > 0$
25. $\dfrac{6 - x}{3 - x - 4x^2} \geq 0$
26. $\dfrac{(x - 2)^2(5 - x)}{(4 - x)^3} \leq 0$
27. $\dfrac{x^4(9 - x)(x - 5)(2 - x)}{(4 - x)^5} > 0$
28. $\dfrac{x^3(1 - x)(x - 2)(3 - x)(4 - x)}{(5 - x)^2(x - 6)^3} < 0$

In Exercises 29–32, determine the values of x for which the radicals represent real numbers.

29. $\sqrt{(x - 1)(x + 2)}$
30. $\sqrt{x^2 - 3x}$
31. $\sqrt{-x - x^2}$
32. $\sqrt{\dfrac{x^3 + 6x^2 + 8x}{3 - x}}$

In Exercises 33–40, solve the given inequalities graphically by using a graphing calculator.

33. $x^3 - x > 2$

34. $0.5x^3 < 3 - 2x^2$

35. $x^4 < x^2 - 2x - 1$

36. $3x^4 + x + 1 > 5x^2$

37. $2^x > x + 2$

38. $\log x < 1 - 2x^2$

39. $\sin x < 0.1x^2 - 1$

40. $4 \cos 2x > 2x - 3$

In Exercises 41–48, answer the given questions by solving the appropriate inequalities.

41. The electric power p delivered to part of a circuit is given by $p = 6i - 4i^2$, where i is the current in amperes. For what positive values of i is the power greater than 2 W?

42. The mass m, in Mg, of fuel in a rocket after launch is $m = 2000 - t^2 - 140t$, where t is measured in minutes. During what period of time is the mass of fuel greater than 500 Mg?

43. In programming a computer, the formula $63n = 2^x - 1$ may be used. Here, n is the number of items to be added and x is the number of bits needed to represent the sum. Find x if $n < 100$.

44. The object distance p and image distance q for a camera of focal length 3.00 cm is given by $p = 3.00q/(q - 3.00)$. For what values of q is $p > 12.0$ cm?

45. The length of a microprocessor chip is 2.0 mm more than its width. If its area is less than 35 mm^2, what values are possible for the width if it must be at least 3.0 mm?

46. A laser source is 2.0 cm from the nearest point P on a flat mirror, and the laser beam is directed at a point Q that is on the mirror and is x in. from P. The beam is then reflected to the receiver, which is x in. from Q. What is x if the total length of the beam is greater than 6.5 cm?

47. A plane takes off from Montreal and flies due east at 620 km/h. A second plane takes off from Albany, New York, at the same time and flies due south at 560 km/h. For how long are the planes less than 1000 km apart? Montreal is 310 km due north of Albany.

48. An open box (no top) is formed from a piece of cardboard 8.00 cm square by cutting equal squares from the corners and turning up the resulting sides. Find the edges of the squares that are cut out in order that the volume of the box is greater than 32.0 cm^3.

17-4 Inequalities Involving Absolute Values

Inequalities involving absolute values are often useful in later topics in mathematics such as calculus, and in applications such as the accuracy of measurements. In this section we show the meaning of such inequalities and how they are solved.

By writing $|x| > 1$, we are considering numbers *numerically* larger than 1. This is equivalent to writing $x < -1$ or $x > 1$. In the same way, $|x| < 1$ refers to numbers numerically less than 1, and this can also be written as $-1 < x < 1$. Thus, *an inequality with an absolute value sign can be expressed in terms of inequalities without the absolute value sign.* Following this reasoning, we have the following relationships.

If $|f(x)| > n$, then $f(x) < -n$ or $f(x) > n$. **(17-1)**

If $|f(x)| < n$, then $-n < f(x) < n$. **(17-2)**

EXAMPLE 1 ▸▸ Solve the inequality $|x - 3| < 2$.

Here we want values of x such that $x - 3$ is numerically smaller than 2, or the values of x within 2 units of $x = 3$. These are given by the inequality $1 < x < 5$. Now, using Eq. (17-2), we have

$$-2 < x - 3 < 2$$

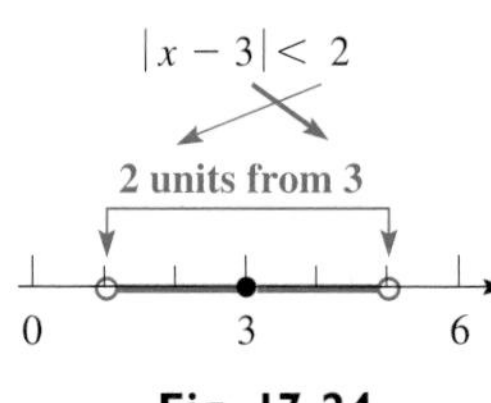

Fig. 17-24

By adding 3 to all three members of this inequality, we have

$$1 < x < 5$$

which is the proper interval. See Fig. 17-24. ◂◂

EXAMPLE 2 ▸▸ Solve the inequality $|2x - 1| > 5$.

By using Eq. (17-1), we have

$$2x - 1 < -5 \quad \text{or} \quad 2x - 1 > 5$$

Completing the solution, we have

$$2x < -4 \quad \text{or} \quad 2x > 6 \qquad \text{add 1 to each member}$$
$$x < -2 \quad \text{or} \quad x > 3 \qquad \text{divide each member by 2}$$

This means that the given inequality is satisfied for $x < -2$ or for $x > 3$. Remember, *we cannot write this as* $3 < x < -2$. The solution is shown in Fig. 17-25.

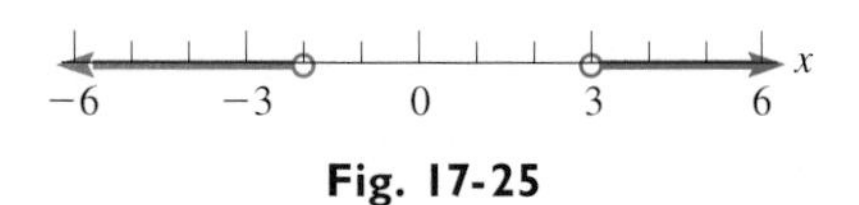

Fig. 17-25

The meaning of the inequality is that the numerical value of $2x - 1$ is greater than 5. We can see that this is true for values of x less than -2 or greater than 3. ◂◂

EXAMPLE 3 ▸▸ Solve the inequality $2\left|\dfrac{2x}{3} + 1\right| \geq 4$.

The solution is as follows:

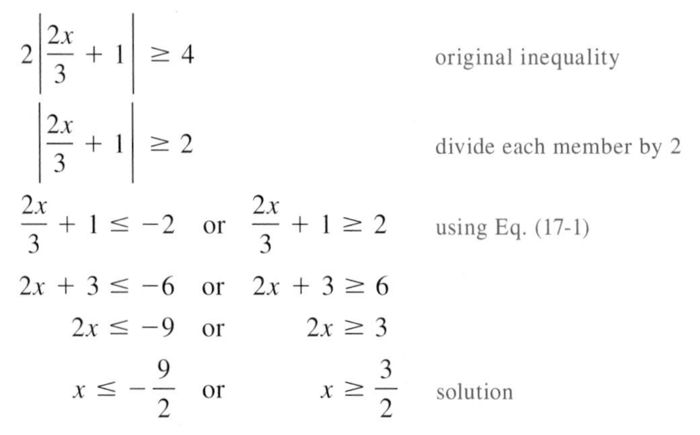

$$2\left|\frac{2x}{3} + 1\right| \geq 4 \qquad \text{original inequality}$$
$$\left|\frac{2x}{3} + 1\right| \geq 2 \qquad \text{divide each member by 2}$$
$$\frac{2x}{3} + 1 \leq -2 \quad \text{or} \quad \frac{2x}{3} + 1 \geq 2 \qquad \text{using Eq. (17-1)}$$
$$2x + 3 \leq -6 \quad \text{or} \quad 2x + 3 \geq 6$$
$$2x \leq -9 \quad \text{or} \quad 2x \geq 3$$
$$x \leq -\frac{9}{2} \quad \text{or} \quad x \geq \frac{3}{2} \qquad \text{solution}$$

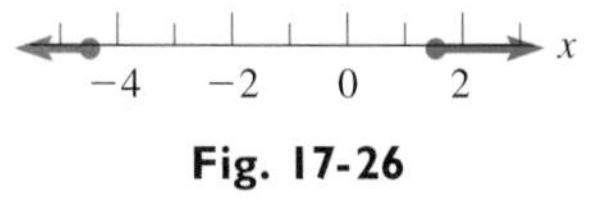

Fig. 17-26

This solution is shown in Fig. 17-26. Note that the sign of equality does not change the method of solution. It simply indicates that $-\frac{9}{2}$ and $\frac{3}{2}$ are included in the solution. ◂◂

EXAMPLE 4 ▸▸ Solve the inequality $|3 - 2x| < 3$.

We have the following solution.

$$|3 - 2x| < 3 \qquad \text{original inequality}$$
$$-3 < 3 - 2x < 3 \qquad \text{using Eq. (17-2)}$$
$$-6 < -2x < 0$$
$$3 > x > 0 \qquad \text{divide by } -2 \text{ and reverse signs of inequality}$$
$$0 < x < 3 \qquad \text{solution}$$

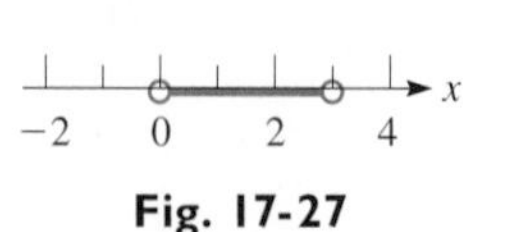

Fig. 17-27

The meaning of the inequality is that the numerical value of $3 - 2x$ is less than 3. This is true for values of x between 0 and 3. The solution is shown in Fig. 17-27. ◂◂

EXAMPLE 5 ▸▸ A technician measures an electric current and reports that it is 0.036 A with a possible error of ± 0.002 A. Write this result for the current i using an inequality with absolute values.

The statement of the problem tells us that the current is no less than 0.034 A and no more than 0.038 A. Another way of stating this is that the numerical difference between the true value of i (unknown exactly) and the measured value, 0.036 A, is less than or equal to 0.002 A. Using an absolute-value inequality, this is written as

$$|i - 0.036| \leq 0.002$$

where values are in amperes.

We can see that this inequality is correct by using (Eq. 17-2).

$$-0.002 \leq i - 0.036 \leq 0.002$$

$$0.034 \leq i \leq 0.038 \quad \text{add 0.036 to each member}$$

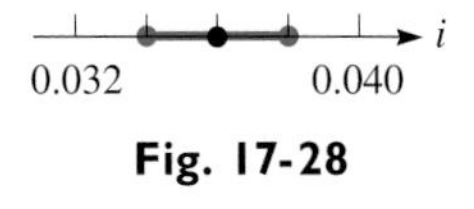

Fig. 17-28

This verifies that i should not be less than 0.034 A or no more than 0.038 A. The solution is shown in Fig. 17-28. ◂◂

Exercises 17–4

In Exercises 1–20, solve the given inequalities. Graph each solution.

1. $|x - 4| < 1$
2. $|x + 1| < 3$
3. $|5x + 4| > 6$
4. $|2x - 1| > 1$
5. $|6x - 5| \leq 4$
6. $|5 - x| \leq 2$
7. $|3 - 4x| > 3$
8. $|3x + 1| \geq 2$
9. $\left|\frac{x + 1}{5}\right| < 3$
10. $\left|\frac{2x - 9}{4}\right| < 1$
11. $|20x + 85| \leq 43$
12. $|2.6x - 9.1| > 10.4$
13. $2|x - 4| > 8$
14. $3|4 - 3x| \leq 10$
15. $4|2 - 5x| \geq 6$
16. $2|7 - 2x| < 12$
17. $\left|\frac{x}{2} + 1\right| < 8$
18. $\left|\frac{4x}{3} - 5\right| \geq 7$
19. $\left|6.5 - \frac{x}{2}\right| \geq 2.3$
20. $\left|27 - \frac{2x}{3}\right| > 17$

In Exercises 21–24, solve the given quadratic inequalities.

21. $|x^2 + x - 4| > 2$
(After using Eq. (17-1), you will have two inequalities. The solution includes the values of x that satisfy *either* of the inequalities.)
22. $|x^2 + 3x - 1| > 3$ (See Exercise 21.)
23. $|x^2 + x - 4| < 2$
[Use Eq. (17-2), then treat the resulting inequality as two inequalities of the form $f(x) > -n$ and $f(x) < n$. The solution includes the values of x that satisfy *both* of the inequalities.]
24. $|x^2 + 3x - 1| < 3$ (See Exercise 23.)

In Exercises 25–28, use inequalities involving absolute values to solve the given problems.

25. Using a vernier micrometer, a technician measured the diameter d of a metal rod and wrote down the result as 0.2537 ± 0.0003 cm. Express the possible values of d using an inequality with absolute values.
26. The production p (in barrels) of an oil refinery for the coming month is estimated at $|p - 2\,000\,000| < 200\,000$. By solving this inequality, describe the anticipated production in a verbal statement.
27. A straight bridge support is 10.0 m below the water level at the shoreline, and it rises 3.0 m for each metre measured horizontally from the shoreline. Set up the vertical distance d from the water level to the support as a function of the horizontal distance x from the shoreline, then find the values of x for which the support is within 6.0 m of water level.
28. A rocket is fired from a plane that is flying horizontally at 3000 m. The height h, in metres, of the rocket above the plane is given by $h = 190t - 4.9t^2$, where t is the time of flight of the rocket in seconds. When is the rocket more than 1300 m above or below the plane?

17-5 Graphical Solution of Inequalities with Two Variables

To this point we have considered inequalities with one variable and certain methods of solving them. We may also graphically solve inequalities involving two variables, such as x and y. In this section we consider the solution of such inequalities, as well as one important type of application.

Let us consider the function $y = f(x)$. We know that the coordinates of points on the graph satisfy the equation $y = f(x)$. However, for points above the graph of the function, we have $y > f(x)$, and for points below the graph of the function we have $y < f(x)$. Consider the following example.

EXAMPLE 1 ▸▸ Consider the linear function $y = 2x - 1$, the graph of which is shown in Fig. 17-29. This equation is satisfied for points on the line. For example, the point (2, 3) is on the line and we have $3 = 2(2) - 1 = 3$. Therefore, for points on the line we have $y = 2x - 1$, or $y - 2x + 1 = 0$.

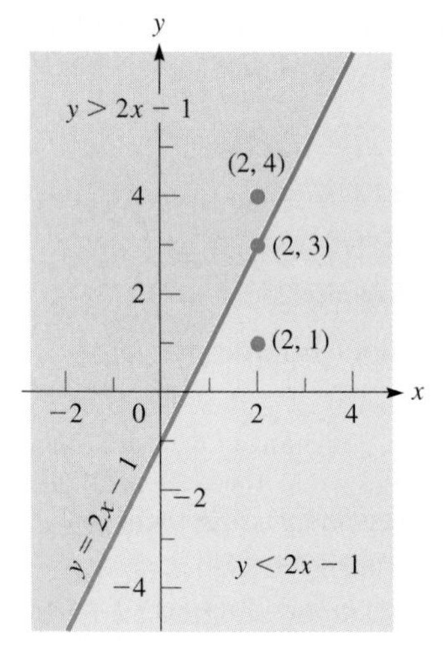

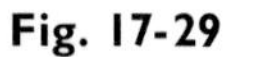
Fig. 17-29

The point (2, 4) is above the line, since we have $4 > 2(2) - 1$, or $4 > 3$. Therefore, for points above the line we have $y > 2x - 1$, or $y - 2x + 1 > 0$. In the same way, for points below the line, $y < 2x - 1$ or $y - 2x + 1 < 0$. We note this is true for the point (2, 1), since $1 < 2(2) - 1$, or $1 < 3$.

The line for which $y = 2x - 1$, and the regions for which $y > 2x - 1$, and for which $y < 2x - 1$ are shown in Fig. 17-29.

Summarizing,

$y > 2x - 1$ for points *above* the line

$y = 2x - 1$ for points *on* the line

$y < 2x - 1$ for points *below* the line ◂◂

The illustration in Example 1 leads us to the graphical method of indicating the points that satisfy an inequality with two variables. First we solve the inequality for y and then determine the graph of the function $y = f(x)$. *If we wish to solve the inequality $y > f(x)$, we indicate the appropriate points by shading in the region above the curve. For the inequality $y < f(x)$, we indicate the appropriate points by shading in the region below the curve.* We note that the complete solution to the inequality consists of all points in an entire region of the plane.

EXAMPLE 2 ▸▸ Draw a sketch of the graph of the inequality $y < x + 3$.

First we graph the function $y = x + 3$, as shown by the dashed line in Fig. 17-30. Since we wish to find all the points that satisfy the inequality $y < x + 3$, we show these points by shading in the region below the line. The line is shown as a *dashed line* to indicate that points on it do not satisfy the inequality. ◂◂

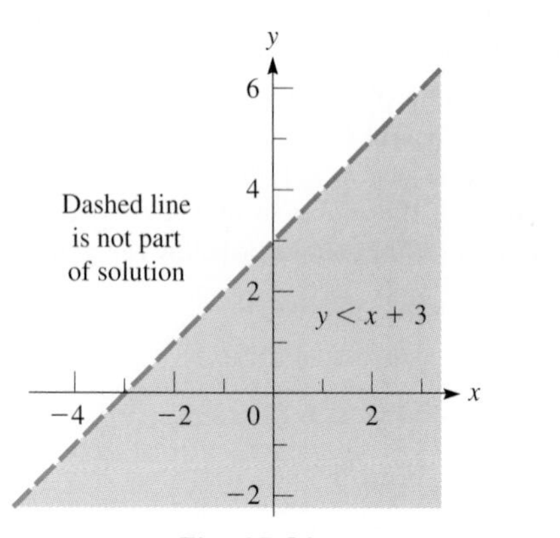

Fig. 17-30

Most graphing calculators can be used to show the solution of an inequality by shading in an area above or below a given curve. The way this is done varies according to the model of the calculator. Therefore, the manual should be used to determine how this is done with any particular model of graphing calculator.

EXAMPLE 3 ▸▸ After a snowstorm, it is estimated that it will take 30 min to plow each kilometre of Route 15 and 45 min to plow each kilometre of Route 80. If no more than 60 plowing-hours are available, what combinations of Route 15 and Route 80 can be plowed?

Let x = kilometres of Route 15 which can be plowed and y = kilometres of Route 80 which can be plowed. The time to plow along each route is the product of the time for each kilometre and the number of kilometres to be plowed. This gives us

time to plow Rt. 15 | time to plow Rt. 80 | max. available time

$$(0.50\ \text{h/km})(x\ \text{km}) + (0.75\ \text{h/km})(y\ \text{km}) \le 60\ \text{h}$$

30 min | 45 min

$$0.50x + 0.75y \le 60$$

$$y \le 80 - 0.67x$$

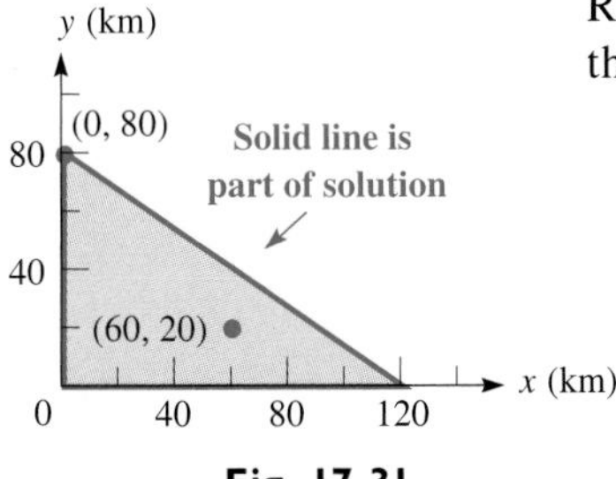

Fig. 17-31

Noting that negative values of x and y do not have meaning, we have the graph in Fig. 17-31, shading in the region below the line since we have $y < 80 - 0.67x$ for that region. Any point in the shaded region, or on the axes or the line around the shaded region, gives a solution. The *solid line* indicates that points on it are part of the solution.

The point (0, 80), for example, is a solution and tells us that 80 km of Route 80 can be plowed if none of Route 15 is plowed. In this case, all 60 h of plowing time are used for Route 80. Another possibility is shown by the point (60, 20), which indicates that 60 km of Route 15 and 20 km of Route 80 can be plowed. In this case, not all of the 60 plowing hours are used. ◂◂

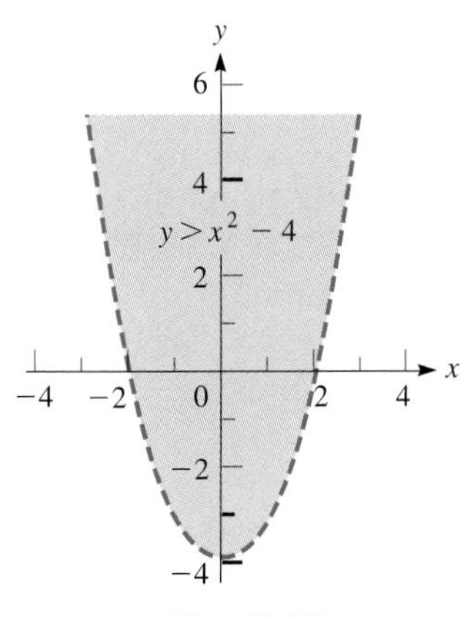

Fig. 17-32

EXAMPLE 4 ▸▸ Draw a sketch of the graph of the inequality $y > x^2 - 4$.

Although the graph of $y = x^2 - 4$ is not a straight line, the method of solution is the same. We graph the function $y = x^2 - 4$ as a dashed curve, since it is not part of the solution, as shown in Fig. 17-32. We then shade in the region above the curve to indicate the points that satisfy the inequality. ◂◂

EXAMPLE 5 ▸▸ Draw a sketch of the region that is defined by the system of inequalities $y \ge -x - 2$ and $y + x^2 < 0$.

Here we want the region common to the graphs of both inequalities. In Fig. 17-33 we first shade in the region above the line $y = -x - 2$, and then shade in the region below the parabola $y = -x^2$. The region defined by this system is the darkly shaded region below the parabola that is above and on the line. (A graphing calculator view is shown in Fig. 17-34.)

The region defined by the system $y \ge -x - 2$ *or* $y + x^2 < 0$ consists of both shaded regions and all points on the line.

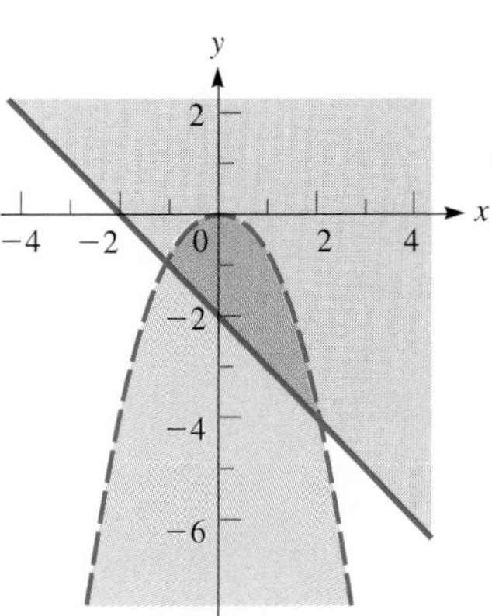

Fig. 17-33

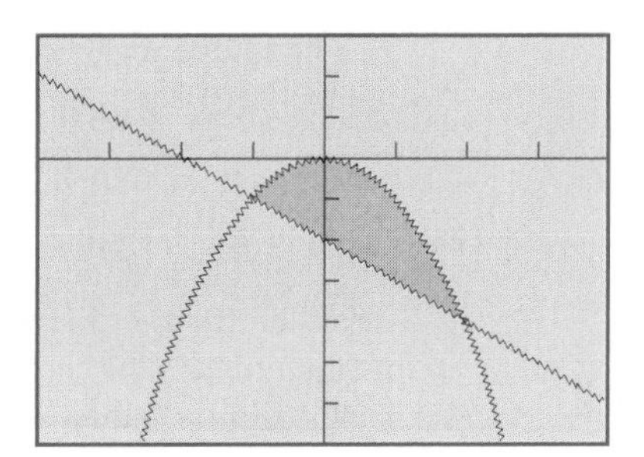
Fig. 17-34

◂◂

Linear Programming

An important area in which graphs of inequalities with two or more variables are used is the branch of mathematics known as **linear programming** (in this context, "programming" has no relation to computer programming). This subject is widely applied in industry, business, economics, and technology. The analysis of many social problems can also be made by the use of linear programming.

Linear programming is used to analyze problems such as those related to maximizing profits, minimizing costs, or the use of materials, with some specified constraints of production. The following is an example of the use of linear programming.

See the chapter introduction.

EXAMPLE 6 ▸▸ A company makes two types of stereo speaker systems, their good-quality system and their highest-quality system. The production of these systems requires assembly of the speaker system itself and the production of the cabinets in which they are installed. The good-quality system requires 3 worker-hours for speaker assembly and 2 worker-hours for cabinet production for each complete system. The highest-quality system requires 4 worker-hours for speaker assembly and 6 worker-hours for cabinet production for each complete system. Available skilled labor allows for a maximum of 480 worker-hours per week for speaker assembly and a maximum of 540 worker-hours per week for cabinet production. It is anticipated that all systems will be sold and that the profit will be \$30 for each good-quality system and \$75 for each highest-quality system. How many of each system should be produced to provide the greatest profit?

First, let x = the number of good-quality systems and y = the number of highest-quality systems made in one week. Thus, the profit p is given by

$$p = 30x + 75y$$

We know that negative numbers are not valid for either x or y, and therefore we have $x \geq 0$ and $y \geq 0$. Also, the number of available worker-hours per week for each part of the production restricts the number of systems that can be made. Both speaker assembly and cabinet production are required for all systems. The number of worker-hours needed to produce the x good-quality systems is $3x$ in the speaker assembly shop. Also, $4y$ worker-hours are required in the speaker assembly shop for the highest-quality systems. Thus,

$$3x + 4y \leq 480$$

since no more than 480 worker-hours are available in the speaker assembly shop. In the cabinet shop, we have

$$2x + 6y \leq 540$$

since no more than 540 worker-hours are available in the cabinet shop.

Therefore, we wish to maximize the profit p under the **constraints**

$x \geq 0, \quad y \geq 0$ — number of systems produced cannot be negative

$3x + 4y \leq 480$ — worker-hours for speaker assembly

$2x + 6y \leq 540$ — worker-hours for cabinet production

In order to do this we sketch the region of points that satisfy this system of inequalities. From the previous examples, we see that the appropriate region is in the first quadrant (since $x \geq 0$ and $y \geq 0$) and under both lines. See Fig. 17-35.

***Any point** in the shaded region defined by the preceding system of inequalities is known as a* **feasible point.** In this case it means that it is possible to produce the number of systems of each type according to the coordinates of the point. For example, the point (50, 25) is in the region, which means that it is possible to produce 50 good-quality systems and 25 highest-quality systems under the given constraints of available skilled labor. However, we wish to find the point that indicates the number of each kind of system which produces the greatest profit.

If we assume values for the profit, the resulting equations are straight lines. Thus, by finding the greatest value of p for which the line passes through a feasible point, we may solve the given problem. If $p = \$3000$, or $p = \$6000$, we have the lines shown. Both are possible with various combinations of speaker systems being produced. However, we note the line for $p = \$6000$ passes through feasible points farther from the origin. It is also clear, since the lines are parallel, that *the greatest profit attainable is given by the line passing through P,* where $3x + 4y = 480$ and $2x + 6y = 540$ intersect. The coordinates of P are (72, 66). Thus, the production should be 72 good-quality systems and 66 highest-quality systems to produce a weekly profit of $p = 30(72) + 75(66) = \$7110$.

For this type of problem, *the solution will be given by one of the vertices of the region.* However, it could be any one of them, which means it is possible that only one type of product should be produced (see Exercise 38 of this section). Therefore, *we can solve the problem by finding the appropriate region and then testing the coordinates of the vertex points.* Here the vertex points are (160, 0), which indicates a profit of \$4800; (0, 90), which indicates a profit of \$6750; and (72, 66), which indicates a profit of \$7110.

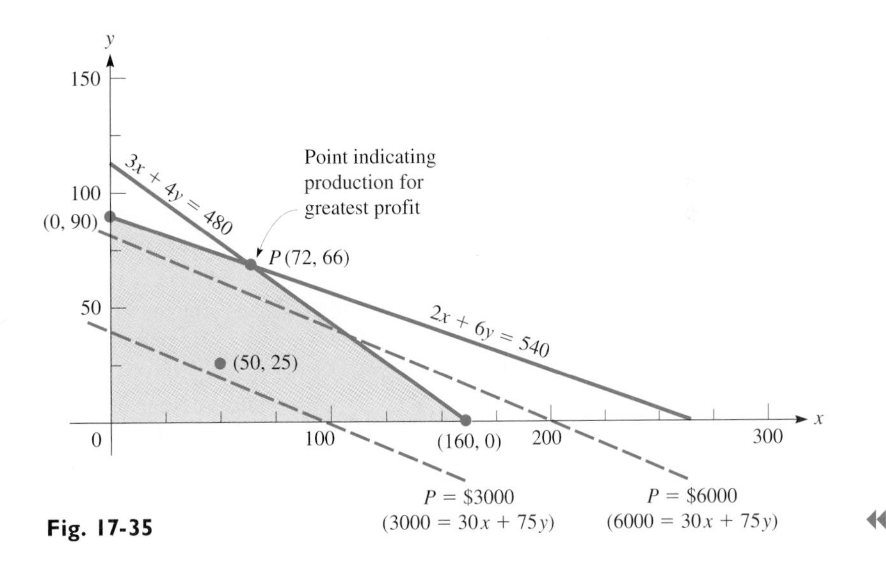

Fig. 17-35

Exercises 17–5

In Exercises 1–24, draw a sketch of the graph of the given inequality.

1. $y > x - 1$
2. $y < 3x - 2$
3. $y \geq 2x + 5$
4. $y \leq 3 - x$
5. $2x + y < 5$
6. $4x - y > 1$
7. $3x + 2y + 6 > 0$
8. $x + 4y - 8 < 0$
9. $y < x^2$
10. $y > -2x^2$
11. $y \geq 1 - x^2$
12. $y \leq 2x^2 - 3$
13. $x^2 + 2x + y < 0$
14. $2x^2 - 4x - y > 0$
15. $y \leq x^3$
16. $y \geq 3x - x^3$
17. $y > x^4 - 8$
18. $y < 32x - x^4$
19. $y \leq \sqrt{2x + 5}$
20. $y > \dfrac{1}{x^2 + 1}$
21. $y < \ln x$
22. $y > \sin 2x$
23. $y \geq 2 \cos \pi x$
24. $y \leq 1 - e^{-x}$

In Exercises 25–32, draw a sketch of the graph of the region in which the points satisfy the given system of inequalities.

25. $y > x$, $y > 1 - x$
26. $y \leq 2x$, $y \geq x - 1$
27. $y \leq 2x^2$, $y > x - 2$
28. $y > x^2$, $y < x + 4$
29. $y > \frac{1}{2}x^2$, $y \leq 4x - x^2$
30. $y > 4 - x$, $y < \sqrt{16 - x^2}$
31. $y \geq 0$, $y \leq \sin x$, $0 \leq x \leq 3\pi$
32. $y > 0$, $y > 1 - x$, $y < e^x$

In Exercises 33–36, set up the necessary inequalities and sketch the graph of the region in which the points satisfy the indicated system of inequalities.

33. A telephone company is installing two types of fiber-optic cable in an area. It is estimated that no more than 300 m of type A cable, and at least 200 m but no more than 400 m of type B cable, are needed. Graph the possible lengths of cable that are needed.

34. A refinery can produce gasoline and diesel fuel, in amounts of any combination, except that equipment restricts maximum total production to 7500 L per day. Graph the different possible production combinations of the two fuels.

35. The elements of an electric circuit dissipate p watts of power. The power p_R dissipated by a resistor in the circuit is given by $p_R = Ri^2$, where R is the resistance and i is the current, in amperes. Graph the possible values of p and i for $p > p_R$ and $R = 0.5\ \Omega$.

36. One pump can remove wastewater at the rate of 250 L/min, and a second pump works at the rate of 150 L/min. Graph the possible values of the time (in min) that each of these pumps operates such that together they pump more than 15 000 L.

In Exercises 37–40, solve the given linear programming problems.

37. A manufacturer makes two types of calculators, a business model and a scientific model. Each model is assembled in two sets of operations, where each operation is in production 8 h each day. The average time required for a business model in the first operation is 3 min, and 6 min is required in the second operation. The scientific model averages 6 min in the first operation and 4 min in the second operation. All calculators can be sold; the profit for a business model is \$8, and the profit for a scientific model is \$10. How many of each model should be made each day in order to maximize profit?

38. Using the information in Example 6, with the one change that profit on each good-quality system is \$60, how many of each system should be made? Explain why this one change in data makes such a change in the solution.

39. Brands A and B of breakfast cereal are both enriched with vitamins P and Q. The necessary information about these cereals is as follows:

	Cereal A	*Cereal B*	*RDA*
Vitamin P	0.1 unit/g	0.2 units/g	10 units
Vitamin Q	0.5 units/g	0.3 units/g	30 units
Cost	0.8¢/g	1.2¢/g	

(RDA is the *Recommended Daily Allowance*.) Find the amount of each cereal that together satisfies the RDA of vitamins P and Q at the lowest cost. (Be careful: We wish to *minimize* cost.)

40. A company makes computer parts A and B in each of two different plants. It costs \$4000 per day to operate the first plant and \$5000 per day to operate the second plant. Each day the first plant produces 100 of part A and 200 of part B, while the second plant produces 250 of A and 100 of B. How many days should each plant operate to produce 2000 of each part and keep operating costs at a minimum?

Chapter Equations, Review Exercises, and Practice Test

Chapter Equations

If $|f(x)| > n$, then $f(x) < -n$ or $f(x) > n$. (17-1)

If $|f(x)| < n$, then $-n < f(x) < n$. (17-2)

Review Exercises

In Exercises 1–32, solve each of the given inequalities algebraically. Graph each solution.

1. $2x - 12 > 0$ **2.** $5 - 3x < 0$

3. $3x + 5 \le 0$ **4.** $\frac{1}{4}x - 2 \ge 3x$

5. $3(x - 7) \ge 5x + 8$ **6.** $2x + 6 < 7(x - 3)$

7. $4 < 2x - 1 < 11$ **8.** $6 \le 4x - 2 < 9$

9. $2x < x + 1 < 4x + 7$

10. $3x < 2x + 1 < x - 5$

11. $5x^2 + 9x < 2$ **12.** $x^2 - 7x \ge 8$

13. $x^2 + 2x > 63$ **14.** $6x^2 - x > 35$

15. $x^3 + 4x^2 - x > 4$ **16.** $2x^3 + 4 \le x^2 + 8x$

17. $\frac{x - 8}{2x + 1} \le 0$ **18.** $\frac{3x + 2}{x - 3} > 0$

19. $\frac{(2x - 1)(3 - x)}{x + 4} > 0$ **20.** $\frac{(3 - x)^2}{2x + 7} \le 0$

21. $x^4 + x^2 \le 0$ **22.** $3x^3 + 7x^2 - 20x < 0$

23. $\frac{1}{x} < 2$ **24.** $\frac{1}{x - 2} < \frac{1}{4}$

25. $|x - 2| > 3$ **26.** $|2x + 1| < 5$

27. $|3x + 2| \le 4$ **28.** $|4 - 3x| \ge 1$

29. $|3 - 5x| > 7$ **30.** $2|2x - 9| < 8$

31. $\left|2 - \frac{x}{2}\right| \le 5$ **32.** $\left|\frac{5x + 1}{5}\right| \ge 4$

In Exercises 33–36, solve the given inequalities graphically by use of a graphing calculator.

33. $x^3 + x + 1 < 0$ **34.** $\frac{1}{x} > 2$

35. $e^{-x} > 0.5$ **36.** $\sin 2x < 0.8 \quad (0 < x < 4)$

In Exercises 37–40, determine the values of x for which the given radicals represent real numbers.

37. $\sqrt{3 - x}$ **38.** $\sqrt{x + 5}$

39. $\sqrt{x^2 + 4x}$ **40.** $\sqrt{\frac{x - 1}{x + 2}}$

In Exercises 41–48, draw a sketch of the given inequality.

41. $y > 4 - x$ **42.** $y < \frac{1}{2}x + 2$

43. $2y - 3x - 4 \le 0$ **44.** $3y - x + 6 \ge 0$

45. $y > x^2 + 1$ **46.** $y \le \frac{1}{x^2 - 4}$

47. $y - x^3 + 1 < 0$ **48.** $2y + 2x^3 + 6x - 3 > 0$

In Exercises 49–52, draw a sketch of the region in which the points satisfy the given systems of inequalities.

49. $y > x + 1$, $y < 4 - x^2$ **50.** $y > 2x - x^2$, $y \ge -2$

51. $y \le \frac{1}{x^2 + 1}$, $y < x - 1$ **52.** $y < \cos\frac{1}{2}x$, $y > \frac{1}{2}e^x$, $-\pi < x < \pi$

In Exercises 53–64, solve the given problems using inequalities. (All data are accurate to at least two significant digits.)

53. The pressure p (in kPa) at a depth d (in metres) in the ocean is given by $p = 101 + 10.1d$. For what values of d is $p > 500$ kPa?

54. After conducting tests, it was determined that the stopping distance x (in metres) of a car traveling 90 km/h was $|x - 95| \le 10$. Express this inequality without absolute values, and determine the interval of stopping distances that were found in the tests.

55. A heating unit with 80% efficiency and a second unit with 90% efficiency deliver 360 MJ of heat to an office complex. If the first unit consumes an amount of fuel that contains no more than 261 MJ, what is the MJ content of the fuel consumed by the second unit?

56. A rectangular parking lot is to have a perimeter of 100 m and an area no greater than 600 m^2. What are the possible dimensions of the lot?

57. The electric power p dissipated in a resistor is given by $p = Ri^2$, where R is the resistance and i is the current in amperes. For a given resistor, $R = 12.0\ \Omega$, and the power varies between 2.50 W and 8.00 W. Find the values of the current.

58. The reciprocal of the total resistance of two resistances in parallel equals the sum of the reciprocals of the resistances. If a 2.0 Ω resistance is in parallel with a resistance R, with a total resistance greater than 0.5 Ω, find R.

59. The efficiency e (in %) of a certain gasoline engine is given by $e = 100(1 - r^{-0.4})$, where r is the *compression ratio* for the engine. For what values of r is $e > 50\%$?

60. A rocket is fired such that its height h is given by $h = 41t - t^2$. For what values of t, in minutes, is the height greater than 400 km?

61. In developing a new product, a company estimates that it will take no more than 1200 min of computer time for research and no more than 1000 min of computer time for development. Graph the possible combinations of the computer times that are needed.

62. A natural-gas supplier has a maximum of 120 worker-hours per week for delivery and for customer service. Graph the possible combinations of times available for these two services.

63. A company produces two types of cameras, the regular model and the deluxe model. For each regular model produced there is a profit of $8, and for each deluxe model the profit is $15. The same amount of materials is used to make each model, but the supply is sufficient only for 450 cameras per day. The deluxe model requires twice the time to produce as the regular model. If only regular models were made, there would be time enough to produce 600 per day. Assuming all models will be sold, how many of each model should be produced if the profit is to be a maximum?

64. A company that manufactures compact disc players gets two different parts, A and B, from two different suppliers. Each package of parts from the first supplier costs $2.00 and contains 6 of each type of part. Each package of parts from the second supplier costs $1.50 and contains 4 of A and 8 of B. How many packages should be bought from each supplier to keep the total cost to a minimum, if production requirements are 600 of A and 900 of B?

Writing Exercise

65. In planning a new city development, an engineer uses a rectangular coordinate system to locate points within the development. A park in the shape of a quadrilateral has corners at $(0, 0)$, $(0, 20)$, $(40, 20)$, and $(20, 40)$ (measurements in metres). Write two or three paragraphs explaining how to describe the park region with inequalities, and find these inequalities.

Practice Test

1. State conditions on x and y in terms of inequalities if the point (x, y) is in the second quadrant.

In Problems 2–6, solve the given inequalities algebraically and graph each solution.

2. $\frac{-x}{2} \geq 3$

3. $3x + 1 < -5$

4. $-1 < 1 - 2x < 5$

5. $\frac{x^2 + x}{x - 2} \leq 0$

6. $|2x + 1| \geq 3$

7. Sketch the region in which the points satisfy the system of inequalities

$$y < x^2$$
$$y \geq x + 1$$

8. Determine the values of x for which $\sqrt{x^2 - x - 6}$ represents a real number.

9. The length of a rectangular lot is 20 m more than its width. If the area is to be at least 4800 m^2, what values may the width be?

10. Type A wire costs $0.10 per metre, and type B wire costs $0.20 per metre. Show the possible combinations of lengths of wire that can be purchased for less than $5.00.

11. The range of the visible spectrum in terms of the wavelength λ of light ranges from about $\lambda = 400$ nm (violet) to about $\lambda = 700$ nm (red). Express these values using an inequality with absolute values.

18

Variation

Experimentation and observation often lead to the discovery of important relationships among quantities. In studying the measured values, it is often possible to see how one quantity changes as the other related quantity also changes.

In this chapter we see how such information can be used to set up general functional relationships among related variables. Then, using known values, we can set up specific functions that relate the variables for specific situations.

We begin this chapter by reviewing the meanings of *ratio* and *proportion*, which were first introduced in Chapter 1. Then we will see how ratio and proportion lead to *variation* and setting up numerous important functional relationships.

Applications of variation are found in all areas of science and technology. It is often used in acoustics, biology, chemistry, computer technology, economics, electronics, environmental technology, hydrodynamics, mechanics, navigation, optics, physics, space technology, thermodynamics, and other fields.

Newton's universal law of gravitation is expressed in the language of variation. See Section 18-2 for a space-age application.

18-1 Ratio and Proportion

In Chapter 1 we introduced the terms *ratio* and *proportion* when we first solved equations. Since then we have seen how they are used in the definitions of the trigonometric functions in Chapter 4 and in a few other specific cases. However, only a basic understanding of their meanings has been necessary to this point. In order to develop the meaning of *variation,* we now review and expand our discussion of these important terms.

From Chapter 1 we recall that *the quotient a/b is called the* **ratio** *of a to b.* Therefore, a fraction is a ratio.

Any measurement made is the ratio of the measured magnitude to an accepted unit of measurement. For example, when we say that an object is 5 metres long, we are saying that the length of that object is five times as long as an accepted unit of length, the metre. Other examples of ratios are density (weight/volume), relative density (density of object/density of water), and pressure (force/area). As these examples illustrate, ratios may compare quantities of the same kind, or they may express a division of magnitudes of different quantities (such a ratio is also called a **rate**).

EXAMPLE 1 ▸▸ The approximate airline distance from Toronto to Los Angeles is 3500 km, and the approximate airline distance from Toronto to Miami is 2000 km. The ratio of these distances is

$$\frac{3500 \text{ km}}{2000 \text{ km}} = \frac{7}{4}$$

Since both units are in kilometres, the resulting ratio is a dimensionless number.

If a jet travels from Toronto to Los Angeles in 5 h, its average speed is

$$\frac{3500 \text{ km}}{5 \text{ h}} = 700 \text{ km/h}$$

In this case we must attach the proper units to the resulting ratio. ◂◂

As we noted in Example 1, we must be careful to attach the proper units to the resulting ratio. Generally, the ratio of measurements of the same kind should be expressed as a dimensionless number. Consider the following example.

EXAMPLE 2 ▸▸ The length of a certain room is 8 m, and the width of the room is 6 m. Therefore, the ratio of the length to the width is $\frac{8}{6}$, or $\frac{4}{3}$.

If the width of the room is expressed as 6000 mm, we have the ratio 8 m/6000 mm = 1 m/750 mm. However, this does not clearly show the ratio. It is better and more meaningful first to change the units of one of the measurements to the units of the other measurement. Changing the length from 8 m to 8000 mm, we express the ratio as $\frac{4}{3}$, as we saw above. From this ratio we can easily see that the length is $\frac{4}{3}$ as long as the width. ◂◂

Dimensionless ratios are often used in definitions in mathematics and in technology. For example, the irrational number π is the dimensionless ratio of the circumference of a circle to its diameter. The specific gravity of a substance is the ratio of its density to the density of water. Other illustrations are found in the exercises for this section.

From Chapter 1 we also recall that *an equation stating that two ratios are equal is called a* **proportion.** By this definition, a proportion is

$$\frac{a}{b} = \frac{c}{d} \tag{18-1}$$

Consider the following example.

EXAMPLE 3 ▸▸ On a certain map, 1 cm represents 10 km. Thus on this map we have a ratio of 1 cm/10 km. To find the distance represented by 3.5 cm, we can set up the proportion

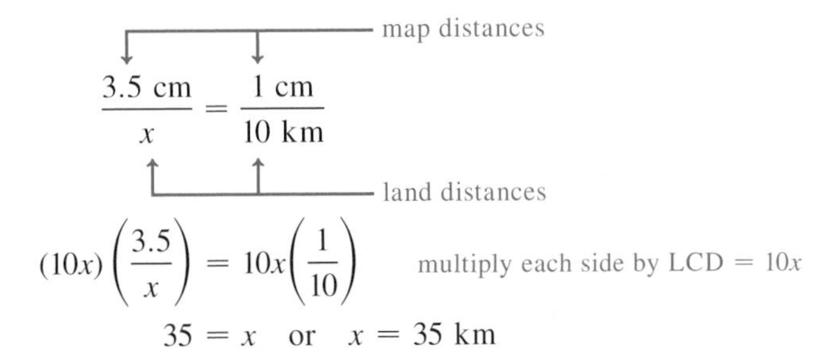

$$(10x)\left(\frac{3.5}{x}\right) = 10x\left(\frac{1}{10}\right) \quad \text{multiply each side by LCD} = 10x$$

$$35 = x \quad \text{or} \quad x = 35 \text{ km}$$

The ratio 1 cm/10 km is the *scale* of the map and has a special meaning, relating map distances in centimetres to land distances in kilometres. In a case like this we should not change either unit to the other, even though they are both units of length. ◂◂

EXAMPLE 4 ▸▸ Given that 1 in. = 2.54 cm (in. is the symbol for the unit of length the *inch*), what is the length in centimetres of the diagonal of a rectangular computer screen that is 10.5 in. long? See Fig. 18-1.

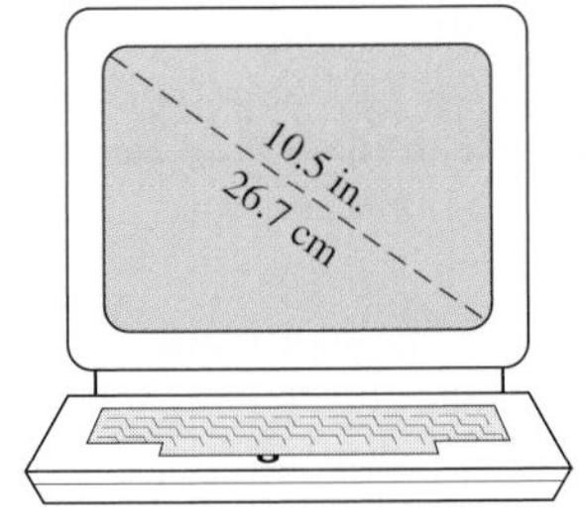

If we equate the ratio of known lengths to the ratio of the given length to the required length, we can find the required length by solving the resulting proportion (which is an equation). This gives us

$$\frac{1 \text{ in.}}{2.54 \text{ cm}} = \frac{10.5 \text{ in.}}{x \text{ cm}}$$

$$x = (10.5)(2.54)$$

$$= 26.7 \text{ cm} \quad \text{rounded off}$$

Therefore, the diagonal of the computer screen is 10.5 in., or 26.7 cm. ◂◂

In Appendix B there are additional illustrations of changing units. Also, a listing of the various units used in the text is included.

EXAMPLE 5 ▸▸ The magnitude of an electric field E is the ratio of the force F on a charge q to the magnitude of q. We can write this as $E = F/q$. If we know the force exerted on a particular charge at some point in the field, we can determine the force that would be exerted on another charge placed at the same point. For example, if we know that a force of 10 nN is exerted on a charge of 4.0 nC, we can then determine the force that would be exerted on a charge of 6.0 nC by the proportion

forces at point

$$\frac{10 \times 10^{-9}}{4.0 \times 10^{-9}} = \frac{F}{6.0 \times 10^{-9}}$$

charges at point

$$F = \frac{(6.0 \times 10^{-9})(10 \times 10^{-9})}{4.0 \times 10^{-9}}$$

$$= 15 \times 10^{-9} = 15 \text{ nN}$$ ◂◂

EXAMPLE 6 ▸▸ A certain alloy is 5 parts tin and 3 parts lead. How many grams of each are there in 40 g of the alloy?

First, we let $x =$ the number of grams of tin in the given amount of the alloy. Next we note that there are 8 total parts of alloy, of which 5 are tin. Thus, 5 is to 8 as x is to 40. This gives the equation

$$\text{parts tin} \longrightarrow \frac{5}{8} = \frac{x}{40} \longleftarrow \text{grams of tin}$$
$$\text{total parts} \longrightarrow 8 \qquad 40 \longleftarrow \text{total grams}$$

$$x = 40\left(\frac{5}{8}\right) = 25 \text{ g}$$

Therefore, there are 25 g of tin and 15 g of lead. The ratio 25 to 15 is the same as 5 to 3. ◂◂

Exercises 18–1

In Exercises 1–8, express the ratios in the simplest form.

1. 18 V to 3 V

2. 27 m to 18 m

3. 96 h to 3 days

4. 120 s to 4 min

5. 48 cm to 3 m

6. 6500 cL to 2.6 L

7. 0.14 kg to 3500 mg

8. 2000 μm to 6 mm

In Exercises 9–20, find the required ratios.

9. The *Mach number* of a moving object is the ratio of its velocity to the velocity of sound (1200 km/h). Find the Mach number of a supersonic jet traveling at 2000 km/h.

10. A virus 3.0×10^{-5} cm long appears to be 1.2 cm long through a microscope. What is the *magnification* (ratio of image length to object length) of the microscope?

11. The *coefficient of friction* for two contacting surfaces is the ratio of the frictional force between them to the perpendicular force that presses them together. If it takes 45 N to overcome friction to move a 110-N crate along the floor, what is the coefficient of friction between the crate and the floor?

12. The *atomic mass* of an atom of carbon is defined to be 12 u. The ratio of the atomic mass of an atom of oxygen to that of an atom of carbon is $\frac{4}{3}$. What is the atomic mass of an atom of oxygen? (The symbol u represents the *unified atomic mass unit,* where $1 \text{ u} = 1.66 \times 10^{-27}$ kg.)

13. An important design feature of an aircraft wing is its *aspect ratio.* It is defined as the ratio of the square of the span of the wing (wingtip to wingtip) to the total area of the wing. If the span of the wing for a certain aircraft is 10.0 m, and the area is 18.0 m^2, find the aspect ratio.

14. For an automobile engine, the ratio of the cylinder volume to compressed volume is the *compression ratio.* If the cylinder volume of 820 cm^3 is compressed to 110 cm^3, find the compression ratio.

15. The *specific gravity* of a substance is the ratio of its density to the density of water. If the density of steel is 7800 kg/m^3 and that of water is 1.0 g/cm^3, what is the specific gravity of steel?

16. The *percent grade* of a road is the ratio of vertical rise to the horizontal change in distance (expressed in percent). If a highway rises 75 m for each 1200 m along the horizontal, what is the percent grade?

17. The *percent error* in a measurement is the ratio of the error in the measurement to the measurement itself, expressed as a percent. When writing a computer program, the memory remaining is determined as 2450 bytes, and then it is correctly found to be 2540 bytes. What is the percent error in the first reading?

18. The electric *current* in a given circuit is the ratio of the voltage to the resistance. What is the current (1 V/1 Ω = 1 A) for a circuit where the voltage is 24.0 V and the resistance is 10.0 Ω?

19. The *mass* of an object is the ratio of its weight to the acceleration g due to gravity. If a space probe weighs 8460 N on earth, where $g = 9.80$ m/s^2, find its mass. (See Appendix B.)

20. *Power* is defined as the ratio of work done to the time required to do the work. If an engine performs 3650 J of work in 15.0 s, find the power developed by the engine. (See Appendix B.)

In Exercises 21–24, find the required quantities from the given proportions.

21. In an electric instrument called a "Wheatstone bridge," electric resistances are related by

$$\frac{R_1}{R_2} = \frac{R_3}{R_4}$$

Find R_2 if $R_1 = 6.00\ \Omega$, $R_3 = 62.5\ \Omega$, and $R_4 = 15.0\ \Omega$. See Fig. 18-2.

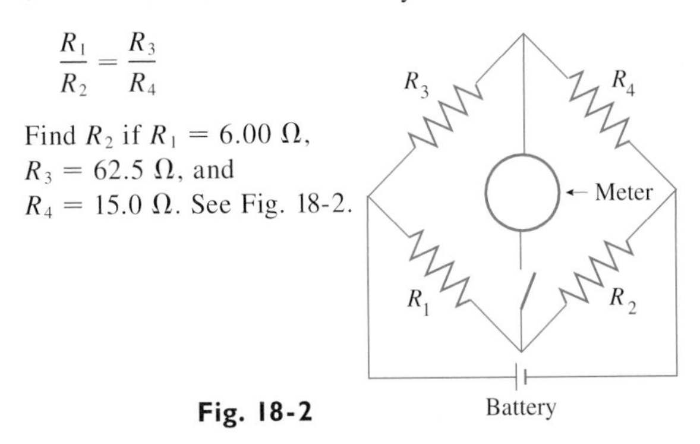

Fig. 18-2

22. For two connected gears, the relation

$$\frac{d_1}{d_2} = \frac{N_1}{N_2}$$

holds, where d is the diameter of the gear and N is the number of teeth. Find N_1 if $d_1 = 2.60$ cm, $d_2 = 11.7$ cm, and $N_2 = 45$. The ratio N_2/N_1 is called the *gear ratio*. See Fig. 18-3.

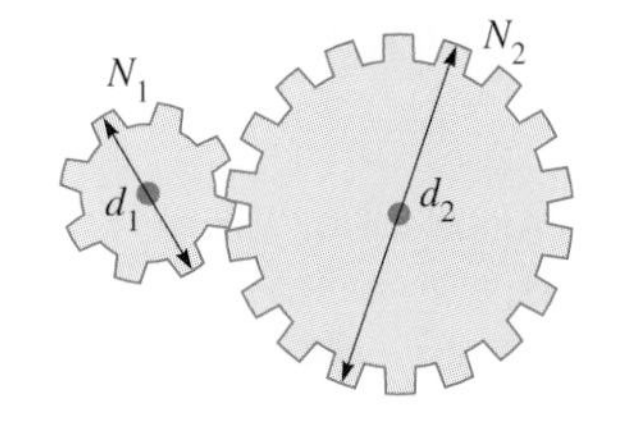

Fig. 18-3

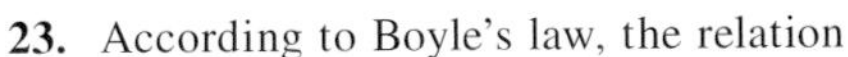

23. According to Boyle's law, the relation

$$\frac{p_1}{p_2} = \frac{V_2}{V_1}$$

holds for pressures p_1 and p_2 and volumes V_1 and V_2 of a gas at constant temperature. Find V_1 if $p_1 = 36.6$ kPa, $p_2 = 84.4$ kPa, and $V_2 = 0.0447\ \text{m}^3$.

24. In a transformer, an electric current in one coil of wire induces a current in a second coil. For a transformer

$$\frac{i_1}{i_2} = \frac{t_2}{t_1}$$

where i is the current and t is the number of windings in each coil, find i_2 for $i_1 = 0.0350$ A, $t_1 = 560$, and $t_2 = 1500$. The ratio t_2/t_1 is called the *turn ratio*.

In Exercises 25–40, answer the given questions by setting up and solving the appropriate proportions.

25. Given that 1 kg = 1000 g, what mass in grams is 20.0 kg?

26. Given that 10 000 m^2 = 1 ha, what area in hectares (ha) is 4500 m^2?

27. Given that 1 W · h = 3.6 kJ, what power in kilojoules is 250 W · h?

28. Given that 0.01 d = 864 s, what time in seconds is 2.75 d?

29. Given that 360° = 2π rad, what angle in degrees is 5.00 rad?

30. Given that $10^4\ \text{cm}^2 = 10^6\ \text{mm}^2$, what area in square centimetres is $2.50 \times 10^5\ \text{mm}^2$?

31. How many metres per second are equivalent to 45.0 km/h?

32. How many kilolitres per hour are equivalent to 540 L/min?

33. A particular type of automobile engine produces 62 500 cm^3 of carbon monoxide in 2.00 min. How much carbon monoxide is produced in 45.0 s?

34. An airplane consumes 140 L of gasoline in flying 680 km. Under similar conditions, how far can it fly on 240 L?

35. By mass, the ratio of chlorine to sodium in table salt is 35.46 to 23.00. How much sodium is contained in 50.00 kg of salt?

36. Ten clicks on an adjustment screw cause an inlet valve opening to change by 0.035 cm. How many clicks are required for a valve adjustment of 0.049 cm?

37. In testing for quality control, it was found that 17 of every 500 computer chips produced by a company in a day were defective. If a total of 595 defective parts were found, what was the total number of chips produced during that day?

38. An electric current of 0.772 mA passes into two wires in which it is divided into currents in the ratio of 2.83 to 1.09. What are the currents in the two wires?

39. One computer line printer can print 2400 lines/min, and a second can print 2800 lines/min. If they print a total of 9100 lines while printing together, how many lines does each print?

40. A couple pays 4.0% of their income in provincial income taxes, and 16.0% in federal income taxes. If the total tax is $8500, how much is the provincial income tax?

18-2 Variation

Scientific laws are often stated in terms of ratios and proportions. For example, Charles' law can be stated as "for a perfect gas under constant pressure, the ratio of any two volumes this gas may occupy equals the ratio of the absolute temperatures." Symbolically this could be stated as $V_1/V_2 = T_1/T_2$. Thus, if the ratio of the volumes and one of the values of the temperature are known, we can easily find the other temperature.

By multiplying both sides of the proportion of Charles' law by V_2/T_1, we can change the form of the proportion to $V_1/T_1 = V_2/T_2$. This statement says that the ratio of the volume to the temperature (for constant pressure) is constant. Thus, if any pair of values of volume and temperature is known, this ratio of V_1/T_1 can be calculated. This ratio of V_1/T_1 can be called a constant k, which means that Charles' law can be written as $V/T = k$. We now have the statement that the ratio of the volume to temperature is always constant; or, as it is normally stated, "The volume is proportional to the temperature." Therefore, we write $V = kT$, the clearest and most informative statement of Charles' law.

Thus, *for any two quantities always in the same proportion, we say that one is* **proportional to** (*or* **varies directly as**) *the second. To show that y is proportional to x (or varies directly as x), we write*

Direct Variation

$$y = kx \tag{18-2}$$

where k is the **constant of proportionality.** This type of relationship is known as **direct variation.**

EXAMPLE 1 ▸▸ The circumference of a circle is proportional to (varies directly as) the radius r. We write this as $c = 2\pi r$. Since we know that $c = 2\pi r$ for a circle, we know in this case that $k = 2\pi$. ◂◂

EXAMPLE 2 ▸▸ The fact that the electric resistance R of a wire varies directly as (is proportional to) its length l is written as $R = kl$. As the length of the wire increases (or decreases), this equation tells us that the resistance increases (or decreases) proportionally. ◂◂

It is very common that, when two quantities are related, the product of the two quantities remains constant. In such a case $yx = k$, or

Inverse Variation

$$y = \frac{k}{x} \tag{18-3}$$

This is read as "y **varies inversely as** *x" or "y* **is inversely proportional to** *x."* This type of relationship is known as **inverse variation.**

EXAMPLE 3 ▸▸ Boyle's law states that "at a given temperature, the pressure p of an ideal gas varies inversely as the volume V." We write this as $p = k/V$. In this case, as the volume of the gas increases, the pressure decreases. ◂◂

In Fig. 18-4(a) the graph of the equation for direct variation $y = kx$ $(x \geq 0)$ is shown. It is a straight line, with slope of k $(k > 0)$ and y-intercept of 0. We see that y increases as x increases. In Fig. 18-4(b) the graph of the equation for inverse variation $y = k/x$ $(k > 0, x > 0)$ is shown. It is a *hyperbola* (a different form of the equation from that of Example 4 of Section 14-1). As x increases, y decreases.

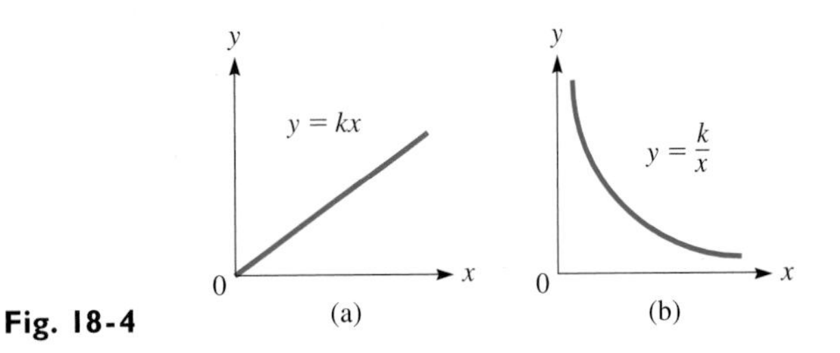

Fig. 18-4

For many relationships, one quantity varies as a specified power of another quantity. The terms *varies directly* and *varies inversely* are used in the following examples with a specified power of the independent variable.

EXAMPLE 4 ▸▸ The statement that the volume V of a sphere varies directly as the cube of its radius r is written as $V = kr^3$. In this case we know that $k = 4\pi/3$. We see that as the radius increases, the volume increases much more rapidly. For example, if $r = 2.00$ cm, $V = 33.5$ cm^3, and if $r = 3.00$ cm, $V = 113$ cm^3. ◂◂

EXAMPLE 5 ▸▸ A company finds that the number n of units of a product which are sold is inversely proportional to the square of the price p of the product. This is written as $n = k/p^2$. As the price of the product is raised, the number of units which are sold decreases much more rapidly. ◂◂

One quantity may vary as the product of two or more other quantities. Such variation is called **joint variation.** *We write*

Joint Variation

$$y = kxz \qquad \textbf{(18-4)}$$

to show that y varies jointly as x and z.

EXAMPLE 6 ▸▸ The cost C of a piece of sheet metal varies jointly as the area A of the piece and the cost c per unit area. This we write as $C = kAc$. Here C increases if the *product* Ac increases. ◂◂

Direct, inverse, and joint variations may be combined. A given relationship may be a combination of two or all three of these types of variation.

Formulated by the great English mathematician and physicist Isaac Newton (1642–1727).

EXAMPLE 7 ▸▸ Newton's *universal law of gravitation* can be stated: "The force F of gravitation between two objects varies jointly as the masses m_1 and m_2 of the objects and inversely as the square of the distance r between their centers." We write this as

$$F = \frac{Gm_1m_2}{r^2}$$

(force varies jointly as masses — numerator; and inversely as the square of the distance — denominator)

where G is the constant of proportionality.

Note the use of the word *and* in this example. It is used to indicate that F varies in more than one way, but *it is* **not** *interpreted as addition.* ◂◂

CAUTION ▸

Calculating the Constant of Proportionality

Once we have used the given statement to set up a general equation in terms of the variables and the constant of proportionality, we may calculate the value of the constant of proportionality if one *complete set of values* of the variables is known. *This value can then be substituted into the general equation to find the specific equation* relating the variables. We can then find the value of any one of the variables for any set of the others.

EXAMPLE 8 ▸▸ If y varies inversely as x, and $x = 15$ when $y = 4$, find the value of y when $x = 12$.

First we write

$$y = \frac{k}{x} \qquad \text{general equation from statement}$$

to show that y varies inversely as x. Next we substitute $x = 15$ and $y = 4$ into the equation. This leads to

$$4 = \frac{k}{15} \quad \text{or} \quad k = 60 \qquad \text{evaluate } k$$

Thus, for this problem the constant of proportionality is 60, and this may be substituted into $y = k/x$, giving

$$y = \frac{60}{x} \qquad \text{specific equation relating } y \text{ and } x$$

as the equation between y and x. Now, for any given value of x, we may find the value of y. For $x = 12$, we have

$$y = \frac{60}{12} = 5 \qquad \text{evaluating } y \text{ for } x = 12$$ ◂◂

EXAMPLE 9 ▸▸ The frequency f of vibration of a wire varies directly as the square root of the tension T of the wire. If $f = 420$ Hz when $T = 1.14$ N, find f when $T = 3.40$ N.

The steps in making this evaluation are outlined below.

$$f = k\sqrt{T} \qquad \text{set up general equation: } f \text{ varies directly as } \sqrt{T}$$

$$420 \text{ Hz} = k\sqrt{1.14 \text{ N}} \qquad \text{substitute given set of values and evaluate } k$$

$$k = 393 \text{ Hz/N}^{1/2}$$

$$f = 393\sqrt{T} \qquad \text{substitute value of } k \text{ to get specific equation}$$

$$f = 393\sqrt{3.40} \qquad \text{evaluate } f \text{ for } T = 3.40 \text{ N}$$

$$= 725 \text{ Hz}$$

We note that k has a set of units associated with it, and this usually will be the case in applied situations. As long as we do not change the units that are used for any of the variables, the units for the final variable that is evaluated will remain the same. ◂◂

EXAMPLE 10 ▸▸ The heat H developed in an electric resistor varies jointly as the time t and the square of the current i in the resistor. If H_0 joules of heat are developed in t_0 seconds with i_0 amperes passing through the resistor, how much heat is developed if both the time and the current are doubled?

$$H = kti^2 \qquad \text{set up general equation}$$

$$H_0\,\text{J} = k(t_0\ \text{s})(i_0\ \text{A})^2 \qquad \text{substitute given values and}$$

$$k = \frac{H_0}{t_0 i_0^2}\text{J}/(\text{s}\cdot\text{A}^2) \qquad \text{evaluate } k$$

$$H = \frac{H_0 t i^2}{t_0 i_0^2} \qquad \text{substitute for } k \text{ to get specific equation}$$

We are asked to determine H when both the time and the current are doubled. This means we are to substitute $t = 2t_0$ and $i = 2i_0$. Making this substitution,

$$H = \frac{H_0(2t_0)(2i_0)^2}{t_0 i_0^2} = \frac{8H_0 t_0 i_0^2}{t_0 i_0^2} = 8H_0$$

The heat developed is eight times that for the original values of i and t. ◂◂

See the chapter introduction.

EXAMPLE 11 ▸▸ In Example 7 we stated Newton's universal law of gravitation. This law was formulated in the late 17th century, but it has numerous modern space-age applications. Use this law to solve the following problem.

A spacecraft is traveling from the earth to the moon, which are 390 000 km apart. The mass of the moon is 0.0123 that of the earth. How far from the earth is the gravitational force of the earth on the spacecraft equal to the gravitational force of the moon on the spacecraft?

From Example 7, we have the gravitational force between two objects as

$$F = \frac{Gm_1m_2}{r^2}$$

where the constant of proportionality G is the same for any two objects. Since we want the force between the earth and the spacecraft to equal the force between the moon and the spacecraft, we have

$$\frac{Gm_sm_e}{r^2} = \frac{Gm_sm_m}{(390\,000 - r)^2}$$

where m_s, m_e, and m_m are the masses of the spacecraft, the earth, and the moon, respectively, r is the distance from the earth to the spacecraft, and $390\,000 - r$ is the distance from the moon to the spacecraft. Since $m_m = 0.0123m_e$, we have

$$\frac{Gm_sm_e}{r^2} = \frac{Gm_s(0.0123m_e)}{(390\,000 - r)^2}$$

$$\frac{1}{r^2} = \frac{0.0123}{(390\,000 - r)^2} \qquad \text{divide each side by } Gm_sm_e$$

$$(390\,000 - r)^2 = 0.0123r^2 \qquad \text{multiply each side by LCD}$$

$$390\,000 - r = 0.111r \qquad \text{take square roots}$$

$$1.111r = 390\,000$$

$$r = 351\,000\ \text{km}$$

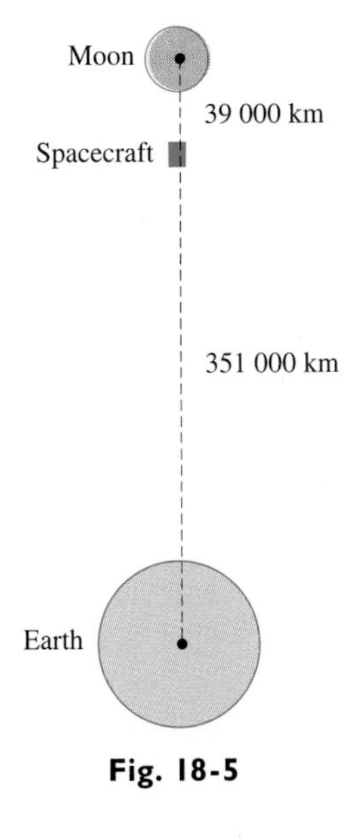

Fig. 18-5

Therefore, the spacecraft is 351 000 km from the earth and 39 000 km from the moon when the gravitational forces are equal. See Fig. 18-5. ◂◂

Exercises 18–2

In Exercises 1–8, set up the general equations from the given statements.

1. y varies directly as z.
2. p varies inversely as q.
3. s varies inversely as the square of t.
4. w is proportional to the cube of L.
5. f is proportional to the square root of x.
6. n is inversely proportional to the $\frac{3}{2}$ power of s.
7. w varies jointly as x and the cube of y.
8. q varies as the square of r and inversely as the fourth power of t.

In Exercises 9–12, express the meaning of the given equation in a verbal statement. (Only k is constant.)

9. $y = kx^4$
10. $s = \dfrac{k}{t^{1.2}}$
11. $n = \dfrac{k\sqrt{t}}{u}$
12. $r = ks^2t$

In Exercises 13–16, give the specific equation relating the variables after evaluating the constant of proportionality for the given set of values.

13. y varies directly as the square root of x, and $y = 2$ when $x = 64$.
14. n is inversely proportional to the square of p, and $n = \frac{1}{27}$ when $p = 3$.
15. p is proportional to q and inversely proportional to the cube of r, and $p = 6$ when $q = 3$ and $r = 2$.
16. v is proportional to t and the square of s, and $v = 80$ when $s = 2$ and $t = 5$.

In Exercises 17–24, find the required value by setting up the general equation and then evaluating.

17. Find y when $x = 10$ if y varies directly as x and $y = 20$ when $x = 8$.
18. Find y when $x = 5$ if y varies directly as the square of x and $y = 6$ when $x = 8$.
19. Find s when $t = 10$ if s is inversely proportional to t and $s = 100$ when $t = 5$.
20. Find p for $q = 0.8$ if p is inversely proportional to the square of q and $p = 18$ when $q = 0.2$.
21. Find y for $x = 6$ and $z = 5$ if y varies directly as x and inversely as z and $y = 60$ when $x = 4$ and $z = 10$.
22. Find r when $n = 16$ if r varies directly as the square root of n and $r = 4$ when $n = 25$.
23. Find f when $p = 2$ and $c = 4$ if f varies jointly as p and the cube of c and $f = 8$ when $p = 4$ and $c = 0.1$.
24. Find v when $r = 2$, $s = 3$, and $t = 4$ if v varies jointly as r and s and inversely as the square of t and $v = 8$ when $r = 2$, $s = 6$, and $t = 6$.

In Exercises 25–48, solve the given applied problems involving variation.

25. The volume V of carbon dioxide (CO_2) that is exhausted from a room in a given time varies directly as the initial volume V_0 that is present. If 75 m^3 of CO_2 are removed in an hour from a room with an initial volume of 160 m^3, how much is removed in an hour if the initial volume is 130 m^3?
26. The amount of heat H required to melt ice is proportional to the mass m of ice that is melted. If it takes 2.93×10^5 J to melt 875 g of ice, how much heat is required to melt 625 g?
27. In electroplating, the mass m of the material deposited varies directly as the time t during which the electric current is on. Set up the equation for this relationship if 2.50 g are deposited in 5.25 h.
28. Hooke's law states that the force needed to stretch a spring is proportional to the amount the spring is stretched. If 10.0 N stretches a certain spring 4.00 cm, how much will the spring be stretched by a force of 6.00 N?
29. The energy output E from an engine is proportional to the input I. The constant of proportionality is the *efficiency* of the engine. Find the efficiency of an engine that has an output of 3600 W for an input of 5400 W.
30. The energy E available daily from a solar collector varies directly as the percent p that the sun shines during the day. If a collector provides 1200 kJ for 75% sunshine, how much does it provide for a day during which there is 35% sunshine?
31. The time t required to empty a wastewater holding tank is inversely proportional to the cross-sectional area A of the drainage pipe. If it takes 2.0 h to empty a tank with a drainage pipe for which $A = 48$ cm^2, how long will it take to empty the tank if $A = 68$ cm^2?
32. The time t required to make a particular trip is inversely proportional to the average speed v. If a jet takes 2.75 h at an average speed of 520 km/h, how long will it take at an average speed of 620 km/h? Explain the meaning of the constant of proportionality.
33. The number N of insects remaining alive t hours after applying an insecticide varies inversely as $t + 1$. If there are initially 8500 insects in the area, how many will remain after 5.00 h?

34. The power P required to propel a ship varies directly as the cube of the speed s of the ship. If 3.88 MW will propel a ship at 19.3 km/h, what power is required to propel it at 24.2 km/h?

35. The lift L of a model airplane wing is directly proportional to the square of its width w. If the lift is 48 N for a wing width of 8.0 cm, what is the lift for a width of 6.0 cm?

36. The f-number lens setting of a camera varies directly as the square root of the time t that the film is exposed. If the f-number is 8 (written as $f/8$) for $t = 0.0200$ s, find the f-number for $t = 0.0098$ s.

37. The force F on the blade of a wind generator varies jointly as the blade area A and the square of the wind velocity v. Find the equation relating F, A, and v if $F = 76.5$ N when $A = 0.372\ \text{m}^2$ and $v = 9.42$ m/s.

38. The escape velocity v a spacecraft needs to leave the gravitational field of a planet varies directly as the square root of the product of the planet's radius R and its acceleration due to gravity g. For Mars and earth, $R_M = 0.533R_e$ and $g_M = 0.400g_e$. Find v_M for Mars if $v_e = 11.2$ km/s.

39. The average speed s of oxygen molecules in the air is directly proportional to the square root of the absolute temperature T. If the speed of the molecules is 460 m/s at 273 K, what is the speed at 300 K?

40. The time t required to test a computer memory unit varies directly as the square of the number n of memory cells in the unit. If a unit with 4800 memory cells can be tested in 15.0 s, how long does it take to test a unit with 8400 memory cells?

41. The electric resistance of a wire varies directly as its length and inversely as its cross-sectional area. Find the relation between resistance, length, and area for a wire that has a resistance of 0.200 Ω for a length of 60.0 m and cross-sectional area of 0.00780 cm^2.

42. The general gas law states that the pressure P of an ideal gas varies directly as the thermodynamic temperature T and inversely as the volume V. If $P = 610$ kPa for $V = 10.0\ \text{cm}^3$ and $T = 290$ K, find V for $P = 400$ kPa and $T = 400$ K.

43. The power in an electric circuit varies jointly as the resistance and the square of the current. Given that the power is 10.0 W when the current is 0.500 A and the resistance is 40.0 Ω, find the power if the current is 2.00 A and the resistance is 20.0 Ω.

44. The difference $m_1 - m_2$ in magnitudes (visual brightnesses) of two stars varies directly as the base 10 logarithm of the ratio b_2/b_1 of their actual brightnesses. For two particular stars, if $b_2 = 100b_1$ for $m_1 = 7$ and $m_2 = 2$, find the equation relating m_1, m_2, b_1, and b_2.

45. The power gain G by a parabolic microwave dish varies directly as the square of the diameter d of the opening and inversely as the square of the wavelength λ of the wave carrier. Find the equation relating G, d, and λ if $G = 5.5 \times 10^4$ for $d = 2.9$ m and $\lambda = 0.030$ m.

46. The intensity I of sound varies directly as the power P of the source and inversely as the square of the distance r from the source. Two sound sources are separated by a distance d, and one has twice the power output of the other. Where should an observer be located on a line between them such that the intensity of each sound is the same?

47. The x-component of the acceleration of an object moving around a circle with constant angular velocity ω varies jointly as $\cos \omega t$ and the square of ω. If the x-component of the acceleration is $-11.4\ \text{cm/s}^2$ when $t = 1.00$ s for $\omega = 0.524$ rad/s, find the x-component of the acceleration when $t = 2.00$ s.

48. The tangent of the proper banking angle θ of the road for a car making a turn is directly proportional to the square of the car's velocity v and inversely proportional to the radius r of the turn. If 7.75° is the proper banking angle for a car traveling at 20.0 m/s around a turn of radius 300 m, what is the proper banking angle for a car traveling at 30.0 m/s around a turn of radius 250 m?

Chapter Equations, Review Exercises, and Practice Test

Chapter Equations

Proportion $\quad \dfrac{a}{b} = \dfrac{c}{d} \quad$ (18-1)

Direct variation $\quad y = kx \quad$ (18-2)

Inverse variation $\quad y = \dfrac{k}{x} \quad$ (18-3)

Joint variation $\quad y = kxz \quad$ (18-4)

Review Exercises

In Exercises 1–8, find the indicated ratios.

1. 4 Mg to 20 kg

2. 300 nm to 6 μm

3. 20 mL to 5 cL

4. 12 ks to 2 h

5. The mechanical advantage of a lever is the ratio of the output force F_0 to the input force F_i. Find the mechanical advantage if $F_0 = 28$ kN and $F_i = 5000$ N.

6. For an automobile, the ratio of the number n_1 of teeth on the ring gear to the number n_2 of teeth on the pinion gear is the *rear axle ratio* of the car. Find this ratio if $n_1 = 64$ and $n_2 = 20$.

7. The pressure p exerted on a surface is the ratio of the force F on the surface to its area A. Find the pressure on a square patch, 2.25 cm on a side, on a tank if the force on the patch is 37.4 N.

8. The resistance R of a resistor is the ratio of the voltage V across the resistor to the current i in the resistor. Find R if $V = 0.632$ V and $i = 2.03$ mA.

In Exercises 9–20, answer the given questions by setting up and solving the appropriate proportions.

9. On a certain map, 1.00 cm represents 16.0 km. What distance on the map represents 52.0 km?

10. Given that 1.000 kg = 1000 g, what is the mass in kilograms of a 14.0-g computer disk?

11. Given that 1.00 kJ = 10^6 mJ, how much heat in kilojoules is produced by a heating element that produces 2660 mJ?

12. Given that 1.00 L = 1000 cm^3, what capacity in litres has a cubical box that is 3.23 cm along an edge?

13. A computer printer can print 3600 characters in 30 s. How many characters can it print in 5.0 min?

14. A solar heater with a collector area of 58.0 m^2 is required to heat 2560 kg of water. Under the same conditions, how much water can be heated by a rectangular solar collector 9.50 m by 8.75 m?

15. The dosage of a certain medicine is 25 mL for each 10 kg of the patient's weight. What is the dosage for a person weighing 56 kg?

16. A woman invests \$50 000 and a man invests \$20 000 in a partnership. If profits are to be shared in the ratio that each invested in the partnership, how much does each receive from \$10 500 in profits?

17. On a certain blueprint, a measurement of 25.0 m is represented by 20.0 mm. What is the actual distance between two points if they are 57.5 mm apart on the blueprint?

18. The chlorine concentration in a water supply is 0.12 part per million. How much chlorine is there in a cylindrical holding tank 4.22 m in radius and 5.82 m high filled from the water supply?

19. One fiber-optic cable carries 60.0% as many messages as another fiber-optic cable. Together they carry 12 000 messages. How many does each carry?

20. Two types of roadbed material, one 50% rock and the other 100% rock, are used in the ratio of 4 to 1 to form a roadbed. If a total of 150 Mg are used, how much rock is in the roadbed?

In Exercises 21–24, give the specific equation relating the variables after evaluating the constant of proportionality for the given set of values.

21. y varies directly as the square of x, and $y = 27$ when $x = 3$.

22. f varies inversely as l, and $f = 5$ when $l = 8$.

23. v is directly proportional to x and inversely proportional to the cube of y, and $v = 10$ when $x = 5$ and $y = 4$.

24. r varies jointly as u, v, and the square of w, and $r = 8$ when $u = 2$, $v = 4$, and $w = 3$.

In Exercises 25–52, solve the given applied problems.

25. For a lever balanced at the fulcrum, the relation

$$\frac{F_1}{F_2} = \frac{L_2}{L_1}$$

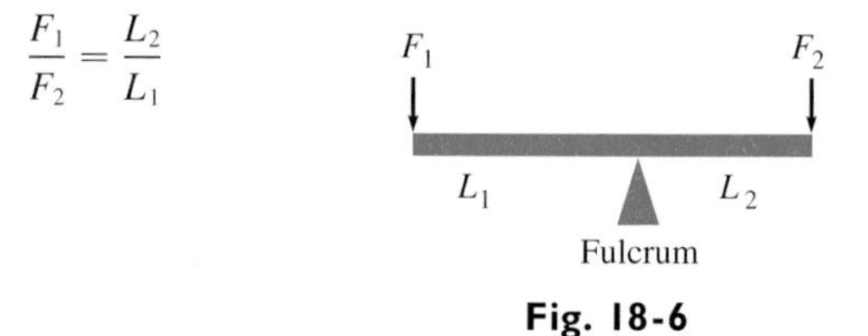

Fig. 18-6

holds, where F_1 and F_2 are forces on opposite sides of the fulcrum at distances L_1 and L_2, respectively. If $F_1 = 4.50$ N, $F_2 = 6.75$ N, and $L_1 = 17.5$ cm, find L_2. See Fig. 18-6.

26. A company finds that the volume V of sales of a certain item and the price P of the item are related by

$$\frac{P_1}{P_2} = \frac{V_2}{V_1}$$

Find V_2 if $P_1 = \$8.00$, $P_2 = \$6.00$, and $V_1 = 3000$ per week.

27. The charge C on a capacitor varies directly as the voltage V across it. If the charge is 6.3 μC with a voltage of 220 V across a capacitor, what is the charge on it with a voltage of 150 V across it?

28. The amount of natural gas burned is proportional to the amount of oxygen consumed. If 24.0 kg of oxygen is consumed in burning 15.0 kg of natural gas, how much air, which is 21.0% oxygen, is consumed to burn 50.0 kg of natural gas?

29. The power of a gas engine is proportional to the area of the piston. If an engine with a piston area of 50.0 cm^2 can develop 22.5 kW, what power is developed by an engine with a piston area of 40.0 cm^2?

30. The volume of hydrogen liberated from hydrochloric acid by zinc is proportional to the amount of zinc used. If 8.28 g of zinc liberates 3.00 L of hydrogen, how much is liberated by 5.36 g of zinc?

31. The distance an object falls under the influence of gravity varies directly as the square of the time of fall. If an object falls 19.6 m in 2.00 s, how far will it fall in 3.00 s?

32. The kinetic energy of a moving object varies jointly as the mass of the object and the square of its velocity. If a 5.00-kg object, traveling at 10.0 m/s, has a kinetic energy of 250 J, find the kinetic energy of an 8.00-kg object traveling at 50.0 m/s.

33. In a particular computer design, N numbers can be sorted in a time proportional to the square of $\log N$. How many times longer does it take to sort 8000 numbers than to sort 2000 numbers?

34. The period of a pendulum is directly proportional to the square root of its length. Given that a pendulum 1.50 m long has a period of 2.46 s, what is the period of a pendulum 2.00 m long?

35. In any given electric circuit containing an inductance and a capacitance, the resonant frequency is inversely proportional to the square root of the capacitance. If the resonant frequency in a circuit is 25.0 Hz and the capacitance is 95.0 μF, what is the resonant frequency of this circuit if the capacitance is 25.0 μF?

36. The rate of emission of radiant energy from the surface of a body is proportional to the fourth power of the thermodynamic temperature. Given that a 25.0-W (the rate of emission) lamp has an operating temperature of 2500 K, what is the operating temperature of a similar 40.0-W lamp?

37. The frequency f of a radio wave is inversely proportional to its wavelength λ. The constant of proportionality is the velocity of the wave, which equals the speed of light. Find this velocity if an FM radio wave has a frequency of 90.9 MHz and a wavelength of 3.29 m.

38. The acceleration of gravity g on a satellite in orbit around the earth varies inversely as the square of its distance from the center of the earth. If $g = 8.7$ m/s^2 for a satellite at an altitude of 400 km above the surface of the earth, find g if it is 1000 km above the surface. The radius of the earth is 6.4×10^6 m.

39. Using *holography* (a method of producing an image without using a lens), an image of concentric circles is formed. The radius r of each circle varies directly as the square root of the wavelength λ of the light used. If $r = 3.56$ cm for $\lambda = 575$ nm, find r if $\lambda = 483$ nm.

40. The velocity of a jet of fluid flowing from an opening in the side of a container is proportional to the square root of the depth of the opening. If the velocity of the jet from an opening at a depth of 1.22 m is 4.88 m/s, what is the velocity of a jet from an opening at a depth of 7.62 m?

41. The heat loss through fiberglass insulation varies directly as the time and inversely as the thickness of the fiberglass. If the loss through 20.0 cm of fiberglass is 1.20 MJ in 30 min, what is the loss through 15.0 cm in 1 h 30 min?

42. Kepler's third law of planetary motion states that the square of the period of any planet is proportional to the cube of the mean radius (about the sun) of that planet, with the constant of proportionality being the same for all planets. Using the fact that the period of the earth is one year and its mean radius is 150×10^6 km, calculate the mean radius for Venus, given that its period is 7.38 months.

43. The range R of a projectile varies jointly as the square of its initial velocity v_0 and the sine of twice the angle θ from the horizontal at which it is fired. See Fig. 18-7. A bullet for which $v_0 = 850$ m/s and $\theta = 22.0°$ has a range of 5.12×10^4 m. Find the range if $v_0 = 750$ m/s and $\theta = 43.2°$.

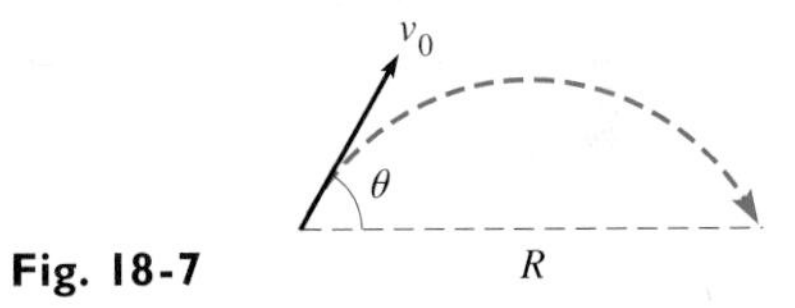

Fig. 18-7

44. The load L that a helical spring can support varies directly as the cube of its wire diameter d and inversely as its coil diameter D. A spring for which $d = 0.120$ cm and $D = 0.953$ cm can support 45.0 N. What is the coil diameter of a similar spring that supports 78.5 N and for which $d = 0.156$ cm?

45. The volume rate of flow of blood through an artery varies directly as the fourth power of the radius of the artery and inversely as the distance along the artery. If an operation is successful in effectively increasing the radius of an artery by 25% and decreasing its length by 2%, by how much is the volume rate of flow increased?

46. The safe uniformly distributed load on a horizontal beam, supported at both ends, varies jointly as the width and the square of the depth and inversely as the distance between supports. Given that one beam has double the dimensions of another, how many times heavier is the safe load it can support than the first can support?

47. A bank statement exactly 30 years old is discovered. It states, "This 10-year-old account is now worth \$185.03 and pays 4% interest compounded annually." An investment with annual compound interest varies directly as $1 + r$ to the power n, where r is the interest rate expressed as a decimal and n is the number of years of compounding. What was the value of the original investment, and what is it worth now?

48. The distance s that an object falls due to gravity varies jointly as the acceleration g due to gravity and the square of the time t of fall. The acceleration due to gravity on the moon is 0.172 of that on earth. If a rock falls for t_0 seconds on earth, how many times farther would the rock fall on the moon in $3t_0$ seconds?

49. The pressure loss p_d in a water pipeline varies directly as the length L and inversely as the radius r of the pipe. If $p_d = 820$ Pa for $L = 750$ m and $r = 0.46$ m, find the pressure loss in 1.2 km of a pipe of half the radius.

50. A quantity important in analyzing the rotation of an object is its *moment of inertia I*. For a ball bearing, the moment of inertia varies directly as its mass m and the square of its radius r. Find the general expression for I if $I = 39.9\ \text{g} \cdot \text{cm}^2$ for $m = 63.8$ g and $r = 1.25$ cm.

51. In the study of polarized light, the intensity I is proportional to the square of the cosine of the angle θ of transmission. If $I = 0.025\ \text{W/m}^2$ for $\theta = 12.0°$, find I for $\theta = 20.0°$.

52. The force F that acts on a pendulum bob is proportional to the mass m of the bob and the sine of the angle θ the pendulum makes with the vertical. If $F = 0.120$ N for $m = 0.350$ kg and $\theta = 2.00°$, find F for $m = 0.750$ kg and $\theta = 3.50°$.

Writing Exercise

53. A fruit packing company plans to reduce the size of its fruit juice can (a right circular cylinder) by 10% and keep the price of each can the same (effectively raising the price). The radius and the height of the new can are to be equally proportional to those of the old can. Write one or two paragraphs explaining how to determine the percent decrease in the radius and the height of the old can that is required to make the new can.

Practice Test

1. Express the ratio of 180 s to 4 min in simplest form.

2. The force F between two parallel wires carrying electric currents is inversely proportional to the distance d between the wires. If a force of 0.750 N exists between wires that are 1.25 cm apart, what is the force between them if they are separated by 1.75 cm?

3. Given that 1 cm $= 10^4\ \mu$m, what length in centimetres is 7.24 μm?

4. The difference p in pressure in a fluid between that at the surface and that at a point below varies jointly as the density d of the fluid and the depth h of the point. The density of water is 1000 kg/m^3, and the density of alcohol is 800 kg/m^3. This difference in pressure at a point 0.200 m below the surface of water is 1.96 kPa. What is the difference in pressure at a point 0.300 m below the surface of alcohol? (All data are accurate to three significant digits.)

5. The perimeter of a rectangular solar panel is 210.0 cm. The ratio of the length to the width is 7 to 3. What are the dimensions of the panel?

6. The crushing load of a pillar varies directly as the fourth power of its radius and inversely as the square of its length. If one pillar has twice the radius and three times the length of a second pillar, what is the ratio of the crushing load of the first pillar to that of the second pillar?

19

Sequences and the Binomial Theorem

In this chapter we study briefly the properties of certain sequences of numbers. A *sequence* is a set of numbers arranged in some specified manner. There are many types of sequences of numbers, and we will develop certain basic sequences, including those used in the expansion of a binomial to a power.

Applications of sequences can be found in many scientific and technical areas. These include physics and chemistry in studying radioactivity, and biology in studying population growth. One of the more important applications comes in business when calculating payments and interest.

Sequences and binomial expansions are also of importance in developing mathematical topics that themselves have wide technical applications.

Sequences are basic to many calculations in business, including compound interest. In Section 19-2 we demonstrate such a calculation.

19-1 Arithmetic Sequences

A *sequence* of numbers may consist of numbers chosen in any way that we may select. This could include a random selection of numbers. However, some sequences are useful in technical applications and in later topics in mathematics, and in this section we consider the first of these, an *arithmetic sequence*. We shall consider only those sequences that consist of real numbers or literal numbers that represent real numbers.

An **arithmetic sequence** (*or* **arithmetic progression**) *is a set of numbers in which each number after the first can be obtained from the preceding one by adding to it a fixed number called the* **common difference.** This definition can be expressed in terms of the *recursion formula*

$$a_n = a_{n-1} + d \qquad (19\text{-}1)$$

where a_n is any term, a_{n-1} is the preceding term, and d is the common difference.

EXAMPLE 1 ▸▸ (a) The sequence 2, 5, 8, 11, 14, . . . , is an arithmetic sequence with a common difference $d = 3$. We can obtain any term by adding 3 to the previous term. We see that the fifth term is $a_5 = a_4 + d$, or $14 = 11 + 3$.

(b) The sequence 7, 2, −3, −8, . . . , is an arithmetic sequence with a common difference of $d = -5$. We can obtain any term after the first by adding −5 to the previous term.

The three dots after the 14 in part (a) and after the −8 in part (b) mean that the sequences continue. ◂◂

If we know the first term of an arithmetic sequence, we can find any other term in the sequence by successively adding the common difference enough times for the desired term to be obtained. This, however, is a very inefficient method, and we can learn more about the sequence if we establish a general way of finding any particular term.

In general, if a_1 is the first term and d is the common difference, the second term is $a_1 + d$, the third term is $a_1 + 2d$, and so forth. If we are looking for the nth term, we note that we need only add d to the first term $n - 1$ times. Thus, *the nth term, a_n, of an arithmetic sequence is given by*

*n*th Term

$$a_n = a_1 + (n - 1)d \qquad (19\text{-}2)$$

Eq. (19-2) can be used to find any given term in any arithmetic sequence. We can refer to a_n as the *last term* of an arithmetic sequence if no terms beyond it are included in the sequence. Such a sequence is called a ***finite*** sequence. If the terms in a sequence continue without end, the sequence is called an ***infinite*** sequence.

EXAMPLE 2 ▸▸ Find the tenth term of the arithmetic sequence 2, 5, 8, ...

By subtracting any given term from the following term, we find that the common difference is $d = 3$. From the terms given, we know that the first term is $a_1 = 2$. From the statement of the problem, the desired term is the tenth, or $n = 10$. Thus, we may find the tenth term, a_{10}, by

$$\begin{aligned} a_{10} &= \underset{\uparrow\, a_1}{2} + (\underset{\uparrow\, n}{10} - 1)\underset{\uparrow\, d}{3} = 2 + (9)(3) \\ &= 29 \end{aligned}$$

Therefore, the tenth term is 29.

The three dots written after the 8 mean that the sequence continues. Since there is no additional information, this would indicate that it is an infinite arithmetic sequence. ◂◂

EXAMPLE 3 ▸▸ Find the common difference between successive terms of the arithmetic sequence for which the first term is 5 and the 32nd term is -119.

We are to find d given that $a_1 = 5$, $a_{32} = -119$, and $n = 32$. Substitution in Eq. (19-2) gives

$$\begin{aligned} -119 &= 5 + (32 - 1)d \\ 31d &= -124 \\ d &= -4 \end{aligned}$$

There is no information as to whether this is a finite or an infinite sequence. The solution is the same in either case. ◂◂

EXAMPLE 4 ▸▸ How many numbers between 10 and 1000 are divisible by 6?

We must first find the smallest and the largest numbers in this range that are divisible by 6. These numbers are 12 and 996. Obviously the common difference between one multiple of 6 and the next is 6. Thus, we can solve this as an arithmetic sequence with $a_1 = 12$, $a_n = 996$, and $d = 6$. Substituting in Eq. (19-2), we have

$$\begin{aligned} 996 &= 12 + (n - 1)6 \\ 6n &= 990 \\ n &= 165 \end{aligned}$$

Thus, 165 numbers between 10 and 1000 are divisible by 6.

All the positive multiples of 6 are included in the infinite arithmetic sequence 6, 12, 18, . . . , whereas those between 10 and 1000 are included in the finite arithmetic sequence 12, 18, 24, . . . , 996. ◂◂

EXAMPLE 5 ▸▸ A package delivery company uses a metal (low-friction) ramp to slide packages from the sorting area to the loading area. If a package is pushed to start it down the ramp at 25 cm/s, and the package accelerates as it slides such that it gains 35 cm/s during each second, after how many seconds is the velocity 305 cm/s? See Fig. 19-1.

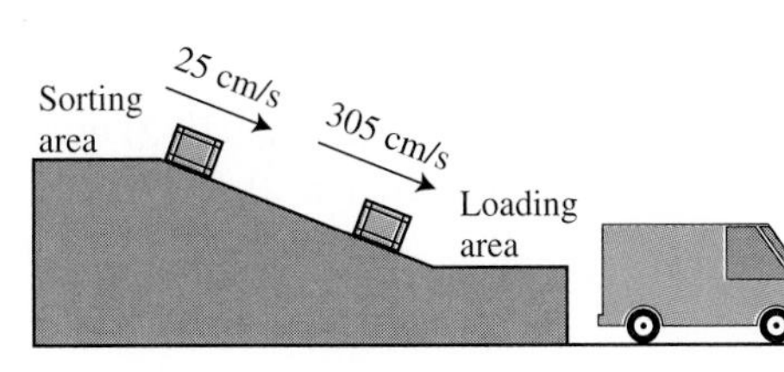

Fig. 19-1

Here we see that the velocity (in cm/s) of the package after each second is

$$60, 95, 130, \ldots, 305, \ldots$$

Therefore, $a_1 = 60$ (the 25 cm/s was at the beginning, that is, after 0 s), $d = 35$, $a_n = 305$, and we are to find n.

$$\begin{aligned} 305 &= 60 + (n - 1)(35) \\ 245 &= 35n - 35 \\ 35n &= 280 \\ n &= 8.0 \end{aligned}$$

This means that the velocity of a package sliding down the ramp is 305 cm/s after 8.0 s. ◂◂

Sum of n Terms

Another important quantity related to an arithmetic sequence is the sum of the first n terms. We can indicate this sum by starting the sum with either the first term or with the last term, as shown by these two equations:

$$S_n = a_1 + (a_1 + d) + (a_1 + 2d) + \cdots + (a_n - d) + a_n$$

or

$$S_n = a_n + (a_n - d) + (a_n - 2d) + \cdots + (a_1 + d) + a_1$$

If we now add the corresponding members of these two equations, we obtain the result

$$2S_n = (a_1 + a_n) + (a_1 + a_n) + (a_1 + a_n) + \cdots + (a_1 + a_n) + (a_1 + a_n)$$

Each term on the right in parentheses has the same expression $(a_1 + a_n)$ and there are n such terms. This tells us that *the sum of the first n terms is given by*

$$S_n = \frac{n}{2}(a_1 + a_n) \qquad \textbf{(19-3)}$$

The use of Eq. (19-3) is illustrated in the following examples.

EXAMPLE 6 ▸▸ Find the sum of the first 1000 positive integers.

The first 1000 integers form a finite arithmetic sequence for which $a_1 = 1$, $a_{1000} = 1000$, $n = 1000$, and $d = 1$. Substituting into Eq. (19-3) (in which we do not use the value of d), we have

$$\begin{aligned} S_{1000} &= \frac{1000}{2}(1 + 1000) = 500(1001) \\ &= 500\,500 \end{aligned}$$ ◂◂

EXAMPLE 7 ▸▸ Find the sum of the first 10 terms of the arithmetic sequence in which the first term is 4 and the common difference is -5.

We are to find S_n, given that $n = 10$, $a_1 = 4$, and $d = -5$. Since Eq. (19-3) uses the value of a_n but not the value of d, we first find a_{10} by using Eq. (19-2). This gives us

$$\begin{aligned} a_{10} &= 4 + (10 - 1)(-5) = 4 - 45 \\ &= -41 \end{aligned}$$

Now we can solve for S_{10} by using Eq. (19-3)

$$\begin{aligned} S_{10} &= \frac{10}{2}(4 - 41) = 5(-37) \\ &= -185 \end{aligned}$$ ◂◂

EXAMPLE 8 ▸▸ For an arithmetic sequence, given that $a_1 = 2$, $d = \frac{3}{2}$, and $S_n = 72$, find n and a_n.

First we substitute the given values in Eqs. (19-2) and (19-3) in order to identify what is known and how we may proceed. Substituting $a_1 = 2$ and $d = \frac{3}{2}$ in Eq. (19-2), we obtain

$$a_n = 2 + (n - 1)\left(\frac{3}{2}\right)$$

Substituting $S_n = 72$ and $a_1 = 2$ in Eq. (19-3), we obtain

$$72 = \frac{n}{2}(2 + a_n)$$

We note that n and a_n appear in both equations, which means that we must solve them simultaneously. Substituting the expression for a_n from the first equation into the second equation, we proceed with the solution.

$$72 = \frac{n}{2}\left[2 + 2 + (n - 1)\left(\frac{3}{2}\right)\right]$$

$$72 = 2n + \frac{3n(n - 1)}{4}$$

$$288 = 8n + 3n^2 - 3n$$

$$3n^2 + 5n - 288 = 0$$

$$n = \frac{-5 \pm \sqrt{25 - 4(3)(-288)}}{6} = \frac{-5 \pm \sqrt{3481}}{6} = \frac{-5 \pm 59}{6}$$

Since n must be a positive integer, we find that $n = \frac{-5 + 59}{6} = 9$. Using this value in the expression for a_n, we find

$$a_9 = 2 + (9 - 1)\left(\frac{3}{2}\right) = 14$$

Therefore, $n = 9$ and $a_9 = 14$. ◂◂

EXAMPLE 9 ▸▸ The voltage across a resistor increases such that during each second the increase is 0.002 mV less than during the previous second. Given that the increase during the first second is 0.350 mV, what is the total voltage increase during the first 10.0 s?

We are asked to find the sum of the voltage increases 0.350 mV, 0.348 mV, 0.346 mV, . . . so as to include 10 increases. This means we want the sum of an arithmetic sequence for which $a_1 = 0.350$, $d = -0.002$, and $n = 10$. Since we need a_n to use Eq. (19-3), we first calculate it using Eq. (19-2).

$$a_{10} = 0.350 + (10 - 1)(-0.002) = 0.332 \text{ mV}$$

Now we use Eq. (19-3) to find the sum with $a_1 = 0.350$, $a_{10} = 0.332$, and $n = 10$.

$$S_{10} = \frac{10}{2}(0.350 + 0.332) = 3.410 \text{ mV}$$

Thus the total voltage increase is 3.410 mV. ◂◂

Exercises 19–1

In Exercises 1–4, write the first five terms of the arithmetic sequence with the given values.

1. $a_1 = 4, d = 2$ **2.** $a_1 = 6, d = -\frac{1}{2}$

3. third term = 5, fifth term = −3

4. second term = −2, fifth term = 7

In Exercises 5–12, find the nth term of the arithmetic sequence with the given values.

5. $1, 4, 7, \ldots n = 8$ **6.** $-6, -4, -2, \ldots n = 10$

7. $18, 13, 8, \ldots n = 17$ **8.** $2, \frac{1}{2}, -1, \ldots n = 25$

9. $a_1 = -7, d = 4, n = 80$

10. $a_1 = \frac{3}{2}, d = \frac{1}{6}, n = 601$

11. $a_1 = b, d = 2b, n = 25$

12. $a_1 = -c, d = 3c, n = 30$

In Exercises 13–16, find the sum of the n terms of the indicated arithmetic sequence.

13. $n = 20, a_1 = 4, a_{20} = 40$

14. $n = 8, a_1 = -12, a_8 = -26$

15. $n = 10, a_1 = -2, d = -\frac{1}{2}$

16. $n = 40, a_1 = 3k, d = \frac{1}{3}k$

In Exercises 17–28, find any of the values of a_1, d, a_n, n, or S_n that are missing for an arithmetic sequence.

17. $a_1 = 5, d = 8, a_n = 45$

18. $a_1 = -2, n = 60, a_n = 28$

19. $a_1 = \frac{5}{3}, n = 20, S_{20} = \frac{40}{3}$

20. $a_1 = 0.1, a_n = -5.9, S_n = -8.7$

21. $d = 3, n = 30, S_{30} = 1875$

22. $d = 9, a_n = 86, S_n = 455$

23. $a_1 = 74, d = -5, a_n = -231$

24. $a_1 = -\frac{9}{7}, n = 19, a_{19} = -\frac{36}{7}$

25. $a_1 = -5k, d = \frac{1}{2}k, S_n = \frac{23}{2}k$

26. $d = -2c, n = 50, S_{50} = 0$

27. $a_1 = -c, a_n = \dfrac{b}{2}, S_n = 2b - 4c$

28. $a_1 = 3b, n = 7, d = \dfrac{b}{3}$

In Exercises 29–48, find the indicated quantities for the appropriate arithmetic sequence.

29. Sixth term = 56, tenth term = 72 (find a_1, d, S_n for $n = 10$).

30. Seventeenth term = −91, second term = −73 (find a_1, d, S_n for $n = 40$).

31. Fourth term = 2, tenth term = 0 (find a_1, d, S_n for $n = 10$).

32. Third term = 1, sixth term = −8 (find a_1, d, S_n for $n = 12$).

33. Find the sum of the first 100 positive integers.

34. Find the sum of the first 100 positive odd integers.

35. Find the sum of the first 200 multiples of 5.

36. Find the number of multiples of 8 between 99 and 999.

37. A beach now has an area of 9500 m^2 but is eroding such that it loses 100 m^2 more of its area each year than during the previous year. If it lost 400 m^2 during the last year, what will be its area 8 years from now?

38. A special antenna is made of 9 parallel straight metal rods. The longest rod is 320 cm long, and each rod is 6 cm shorter than the one before it. How long is the shortest rod? What is the combined length of the 9 rods?

39. There are 12 seats in the first row around a semicircular stage. Each row behind the first has 4 more seats than the row in front of it. How many rows of seats are there if there is a total of 300 seats?

40. A bank loan of $8000 is repaid in annual payments of $1000 plus 10% interest on the unpaid balance. What is the total amount of interest paid?

41. A car depreciates $1800 during the first year after it is purchased. Each year thereafter it depreciates $150 less than the year before. After how many years will it be considered to have no value, and what was the original cost?

42. A clock strikes the number of times of the hour. How many strikes does it make in one day?

43. If a tool dropped from a helicopter falls 4.9 m during the first second, 14.7 m during the second second, 24.5 m during the third second, and so on, how high was the helicopter if the tool takes 10.0 s to reach the ground?

44. In preparing a bid for constructing a new building, a contractor determines that the foundation and basement will cost $605 000 and the first floor will cost $360 000. Each floor above will cost $15 000 more than the one below it. How much will the building cost if it is to be 18 floors high?

45. Derive a formula for S_n in terms of n, a_1, and d.

46. A *harmonic sequence* is a sequence of numbers whose reciprocals form an arithmetic sequence. Is a harmonic sequence also an arithmetic sequence? Explain.

47. Show that the sum of the first n positive integers is $\frac{1}{2}n(n + 1)$.

48. Show that the sum of the first n positive odd integers is n^2.

19-2 Geometric Sequences

A second type of important sequence of numbers is the **geometric sequence** (or **geometric progression**). *In a geometric sequence each number after the first can be obtained from the preceding one by multiplying it by a fixed number, called the* **common ratio.** We can express this definition in terms of the *recursion formula*

$$a_n = ra_{n-1} \tag{19-4}$$

where a_n is any term, a_{n-1} is the preceding term, and r is the common ratio. One important application of geometric sequences is in computing compound interest on savings accounts. Other applications are found in areas such as biology and physics.

EXAMPLE 1 ▸▸ (a) The sequence 2, 4, 8, 16, . . . , is a geometric sequence with a common ratio of 2. Any term after the first can be obtained by multiplying the previous term by 2. We see that the fourth term $a_4 = ra_3$, or $16 = 2(8)$.

(b) The sequence 9, -3, 1, $-1/3$, . . . , is a geometric sequence with a common ratio of $-1/3$. We can obtain any term after the first by multiplying the previous term by $-1/3$. ◂◂

If we know the first term, we can then find any other desired term by multiplying by the common ratio a sufficient number of times. When we do this for a general geometric sequence, we can find the nth term in terms of the first term a_1, the common ratio r, and n. Thus the second term is a_1r, the third term is a_1r^2, and so forth. In general, the expression for the nth term is

*n*th Term

$$a_n = a_1r^{n-1} \tag{19-5}$$

EXAMPLE 2 ▸▸ Find the eighth term of the geometric sequence 8, 4, 2, ...

By dividing any given term by the previous term, we find the common ratio to be $\frac{1}{2}$. From the terms given, we see that $a_1 = 8$. From the statement of the problem, we know that $n = 8$. Thus we substitute into Eq. (19-5) to find a_8.

$$a_8 = \underset{\uparrow a_1}{8}\Big(\underset{\uparrow r}{\frac{1}{2}}\Big)^{\overset{n}{\downarrow}\,8-1} = \frac{8}{2^7} = \frac{1}{16}$$ ◂◂

EXAMPLE 3 ▸▸ Find the tenth term of the geometric sequence for which $a_1 = \frac{8}{625}$ and $r = -\frac{5}{2}$.

Using Eq. (19-5) to find a_{10}, we have

$$a_{10} = \frac{8}{625}\left(-\frac{5}{2}\right)^{10-1} = \frac{8}{625}\left(-\frac{5^9}{2^9}\right) = -\left(\frac{2^3}{5^4}\right)\left(\frac{5^9}{2^9}\right) = -\frac{5^5}{2^6}$$

$$= -\frac{3125}{64}$$ ◂◂

EXAMPLE 4 ▸▸ Find the seventh term of a geometric sequence for which the second term is 3, the fourth term is 9, and $r > 0$.

We can find r if we let $a_1 = 3$, $a_3 = 9$, and $n = 3$. (At this point we are considering a sequence made up of 3, the next number, and 9. These are the second, third, and fourth terms of the original sequence.) Thus

$$9 = 3r^2, \qquad r = \sqrt{3} \qquad (\text{since } r > 0)$$

We can now find a_1 of the original sequence by considering just the first two terms of the sequence, a_1 and $a_2 = 3$.

$$3 = a_1(\sqrt{3})^{2-1}, \qquad a_1 = \sqrt{3}$$

We can now find the seventh term, using $a_1 = \sqrt{3}$, $r = \sqrt{3}$, and $n = 7$.

$$a_7 = \sqrt{3}(\sqrt{3})^{7-1} = \sqrt{3}(3^3) = 27\sqrt{3}$$

We could have shortened this procedure one step by letting the second term be the first term of a new sequence of six terms. If the first term is of no importance in itself, this is acceptable. ◂◂

EXAMPLE 5 ▸▸ In an experiment, 22.0% of a substance changes chemically each 10.0 min. If there is originally 120 g of the substance, how much of it will remain after 45.0 min?

Let P = the portion of the substance remaining after each minute. From the statement of the problem we know that $r = 0.780$, since 78.0% remains after each 10.0-min period. We also know that $a_1 = 120$ g, and we let n represent the number of minutes of elapsed time. This means that $P = 120(0.780)^{n/10.0}$. It is necessary to divide by 10.0 because the ratio is given for a 10.0-min period. In order to find P when $n = 45.0$ min, we write

$$P = 120(0.780)^{45.0/10.0} = 120(0.780)^{4.50}$$
$$= 39.2 \text{ g}$$

This means that 39.2 g remain after 45.0 min. Note that the power 4.50 represents 4.50 ten-minute periods. ◂◂

Sum of n Terms

A general expression for the sum S_n of the first n terms of a geometric sequence may be found by directly forming the sum and multiplying this equation by r.

$$S_n = a_1 + a_1r + a_1r^2 + \cdots + a_1r^{n-1}$$
$$rS_n = a_1r + a_1r^2 + a_1r^3 + \cdots + a_1r^n$$

If we now subtract the second of these equations from the first, we obtain $S_n - rS_n = a_1 - a_1r^n$. All other terms cancel by subtraction. Now, factoring S_n from the terms on the left and a_1 from the terms on the right, we solve for S_n. Thus *the sum S_n of the first n terms of a geometric sequence is*

$$S_n = \frac{a_1(1 - r^n)}{1 - r} \qquad (r \neq 1) \qquad \textbf{(19-6)}$$

EXAMPLE 6 ▸▸ Find the sum of the first seven terms of the geometric sequence in which the first term is 2 and the common ratio is $\frac{1}{2}$.

We are to find S_n given that $a_1 = 2$, $r = \frac{1}{2}$, and $n = 7$. Using Eq. (19-6), we have

$$S_7 = \frac{2(1 - (\frac{1}{2})^7)}{1 - \frac{1}{2}} = \frac{2(1 - \frac{1}{128})}{\frac{1}{2}} = 4\left(\frac{127}{128}\right) = \frac{127}{32}$$ ◂◂

See the chapter introduction.

EXAMPLE 7 ▸▸ If \$100 is invested each year at 5% interest compounded annually, what would be the total amount of the investment after 10 years (before the 11th deposit is made)?

After 1 year the amount invested will have added to it the interest for the year. Therefore, for the last (10th) \$100 invested, its value will become

$$\$100(1 + 0.05) = \$100(1.05) = \$105$$

The next to last \$100 will have interest added twice. After 1 year its value becomes \$100(1.05), and after 2 years it is $\$100(1.05)(1.05) = \$100(1.05)^2$. In the same way, the value of the first \$100 becomes $\$100(1.05)^{10}$, since it will have interest added 10 times. This means that we are to find the sum of the sequence

$$\underset{\text{1 year in account}}{100(1.05)} + \underset{\text{2 years in account}}{100(1.05)^2} + \underset{\text{3 years in account}}{100(1.05)^3} + \cdots + \underset{\text{10 years in account}}{100(1.05)^{10}}$$

or

$$100[1.05 + (1.05)^2 + (1.05)^3 + \cdots + (1.05)^{10}]$$

For the sequence in the brackets we have $a_1 = 1.05$, $r = 1.05$, and $n = 10$. Thus,

$$S_{10} = \frac{1.05[1 - (1.05)^{10}]}{1 - 1.05} = \frac{1.05}{-0.05}(1 - 1.628\,895) = 13.2068$$

The total value of the \$100 investments is $100(13.2068) = \$1320.68$. We see that \$320.68 in interest has been earned. ◂◂

Exercises 19–2

In Exercises 1–4, write down the first five terms of the geometric sequence with the given values.

1. $a_1 = 45, r = \frac{1}{3}$ **2.** $a_1 = 9, r = -\frac{2}{3}$

3. $a_1 = 2, r = 3$ **4.** $a_1 = -3, r = 2$

In Exercises 5–12, find the nth term of the geometric sequence with the given values.

5. $\frac{1}{2}, 1, 2, \ldots \quad (n = 6)$ **6.** $10, 1, 0.1, \ldots \quad (n = 8)$

7. $125, -25, 5, \ldots \quad (n = 7)$

8. $0.1, 0.3, 0.9, \ldots \quad (n = 5)$

9. $a_1 = -27, r = -\frac{1}{3}, n = 10$

10. $a_1 = 48, r = \frac{1}{2}, n = 12$

11. $a_1 = 2, r = 10, n = 7$ **12.** $a_1 = -2, r = 2k, n = 6$

In Exercises 13–16, find the sum of the first n terms of the geometric sequence with the given values.

13. $a_1 = \frac{1}{8}, r = 4, n = 5$ **14.** $a_1 = 162, r = -\frac{1}{3}, n = 6$

15. $a_1 = 192, a_n = 6, n = 6$

16. $a_1 = 9, a_n = -243, n = 4$

In Exercises 17–24, find any of the values of a_1, r, a_n, n, or S_n that are missing.

17. $a_1 = \frac{1}{16}, r = 4, n = 6$

18. $r = 0.2, a_n = 0.000\,32, n = 7$

19. $r = \frac{3}{2}, n = 5, S_5 = 211$

20. $r = -\frac{1}{2}, a_n = \frac{1}{8}, n = 7$

21. $a_n = 27, n = 4, S_4 = 40$

22. $a_1 = 3, n = 7, a_7 = 192$

23. $a_1 = 75, r = \frac{1}{5}, a_n = \frac{3}{625}$

24. $r = -2, n = 6, S_6 = 42$

In Exercises 25–40, find the indicated quantities.

25. Find the tenth term of a geometric sequence if the fourth term is 8 and the seventh term is 16.

26. Find the sum of the first eight terms of the geometric sequence for which the fifth term is 5, the seventh term is 10, and $r > 0$.

27. Each stroke of a pump removes 8.2% of the remaining air from a container. What percent of the air remains after 50 strokes?

28. From 1980 to 1990 a city had a 2.34% average growth rate of population, compounded annually. If the population in 1980 was 314 000, what was the 1990 population?

29. An electric current decreases by 12.5% each 1.00 μs. If the initial current is 3.27 mA, what is the current after 8.20 μs?

30. A copying machine is set to reduce the dimensions of material copied by 10%. A drawing 12.0 cm wide is reduced, and then the copies in turn are reduced. What is the width of the drawing on the sixth reduction?

31. How much is an investment of $250 worth after 8 years if it earns annual interest of 7.2% compounded monthly? (7.2% annual interest compounded monthly means that 0.6% (7.2%/12) interest is added each month.)

32. A chemical spill pollutes a stream. A monitoring device finds 620 ppm (parts per million) of the chemical 1.0 km below the spill, and the readings decrease by 12.5% for each kilometre farther downstream. How far downstream is the reading 100 ppm?

33. During each oscillation, a pendulum swings through 85% of the distance of the previous oscillation. If the pendulum swings through 80.8 cm in the first oscillation, through what total distance does it move in 12 oscillations?

34. A series of deposits, each of value A and made at equal time intervals, earns an interest rate of i for the time interval. The deposits have a total value of $A(1 + i) + A(1 + i)^2 + A(1 + i)^3 + \cdots + A(1 + i)^n$ after n time intervals (just before the next deposit). Find a formula for this sum.

35. A tank with a 22% acid solution has 25% of its contents removed and replaced with water. How many times must this be done in order to have a 7.0% solution remain?

36. The power on a satellite is supplied by a radioactive isotope. On a given satellite the power decreases by 0.2% each day. What percent of the initial power remains after one year?

37. If you decided to save money by putting away 1¢ on a given day, 2¢ one week later, 4¢ a week later, and so on, how much would you have to put away 6 months (26 weeks) after putting away the 1¢?

38. How many direct ancestors (parents, grandparents, and so on) does a person have in the 10 generations that preceded him or her (assuming that no ancestor appears in more than one line of descent)?

39. Derive a formula for S_n in terms of a_1, r, and a_n.

40. Write down several terms of a general geometric sequence. Then take the logarithm of each term. Explain why the resulting sequence is an arithmetic sequence.

19-3 Infinite Geometric Series

In the previous sections we developed formulas for the sum of the first n terms of an arithmetic sequence and of a geometric sequence. *The indicated sum of the terms of a sequence is called a* **series.**

EXAMPLE 1 ▸▸ (a) The infinite arithmetic sequence 2, 5, 8, 11, 14, . . . has the series

$$2 + 5 + 8 + 11 + 14 + \cdots$$

associated with it.

(b) The finite geometric sequence $1, \frac{1}{2}, \frac{1}{4}, \frac{1}{8}$ has the series

$$1 + \frac{1}{2} + \frac{1}{4} + \frac{1}{8}$$

associated with it. ◂◂

The series associated with a finite sequence will sum up to a real number. The series associated with an infinite arithmetic sequence will not sum up to a real number, as the terms being added become larger and larger numerically. The sum is unbounded, as we can see in Example 1(a). The series associated with an infinite geometric sequence may or may not sum up to a real number, as we shall now show.

Let us now consider the sum of the first n terms of the infinite geometric sequence $1, \frac{1}{2}, \frac{1}{4}, \ldots$. This is the sum of the n terms of the associated geometric series

$$1 + \frac{1}{2} + \frac{1}{4} + \cdots + \frac{1}{2^{n-1}}$$

Here $a_1 = 1$ and $r = \frac{1}{2}$, and we find that we get the values of S_n for the given values of n in the following table:

n	2	3	4	5	6	7	8	9	10
S_n	$\frac{3}{2}$	$\frac{7}{4}$	$\frac{15}{8}$	$\frac{31}{16}$	$\frac{63}{32}$	$\frac{127}{64}$	$\frac{255}{128}$	$\frac{511}{256}$	$\frac{1023}{512}$

$1 + \frac{1}{2} + \frac{1}{4} + \frac{1}{8}$ (for $n = 4$)

$1 + \frac{1}{2} + \frac{1}{4} + \frac{1}{8} + \frac{1}{16} + \frac{1}{32} + \frac{1}{64}$ (for $n = 7$)

The series for $n = 4$ and $n = 7$ are shown. We see that as n gets larger, the numerator of each fraction becomes more nearly twice the denominator. In fact, we find that if we continue to compute S_n as n becomes larger, S_n can be found as close to the value 2 as desired, although it will never actually reach the value 2. For example, if $n = 100$, $S_{100} = 2 - 1.6 \times 10^{-30}$, which could be written as

1.999 999 999 999 999 999 999 999 999 998 4

to 32 significant digits. In the formula for the sum of the first n terms of a geometric sequence

$$S_n = a_1 \frac{1 - r^n}{1 - r}$$

the term r^n becomes exceedingly small, and if we consider n as being sufficiently large, we can see that this term is effectively zero. *If this term were exactly zero,* then the sum would be

$$S_n = 1 \frac{1 - 0}{1 - \frac{1}{2}} = 2$$

The only problem is that we cannot find any number large enough for n to make $(\frac{1}{2})^n$ zero. There is, however, an accepted notation for this. This notation is

$$\lim_{n \to \infty} r^n = 0 \qquad (\text{if } |r| < 1)$$

and it is read as "the limit, as n *approaches* infinity, of r to the nth power is zero."

CAUTION ▶ The symbol ∞ is read as **infinity**, *but* ***it must not be thought of as a number.*** It is simply a symbol that stands for a *process* of considering numbers which become large without bound. The number called the **limit** of the sums is simply the number the sums get closer and closer to, as n is considered to approach infinity. This notation and terminology are of particular importance in the calculus.

If we consider values of r such that $|r| < 1$, and let the values of n become unbounded, we find that $\lim_{n\to\infty} r^n = 0$. The formula for *the sum of the terms of an infinite geometric series then becomes*

$$S = \frac{a_1}{1 - r} \qquad (|r| < 1) \tag{19-7}$$

where a_1 is the first term and r is the common ratio. If $|r| \geq 1$, S is unbounded in value.

EXAMPLE 2 ▸▸ Find the sum of the infinite geometric series

$$4 - \frac{1}{2} + \frac{1}{16} - \frac{1}{128} + \cdots$$

Here we see that $a_1 = 4$. We find r by dividing any term by the previous term, and we find that $r = -\frac{1}{8}$. We then find the sum by substituting in Eq. (19-7). This gives us

$$S = \frac{4}{1 - (-\frac{1}{8})} = \frac{4}{1 + \frac{1}{8}}$$
$$= \frac{4}{1} \times \frac{8}{9} = \frac{32}{9}$$ ◂◂

EXAMPLE 3 ▸▸ Find the fraction that has as its decimal form 0.121 212. . . .

This decimal form can be considered as being

$$0.12 + 0.0012 + 0.000\,012 + \cdots$$

which means that we have an infinite geometric series in which $a_1 = 0.12$ and $r = 0.01$. Thus,

$$S = \frac{0.12}{1 - 0.01} = \frac{0.12}{0.99}$$
$$= \frac{4}{33}$$

Therefore, the decimal 0.121 212. . . and the fraction $\frac{4}{33}$ represent the same number. ◂◂

The decimal in Example 3 is called a **repeating decimal**, because *a particular sequence of digits in the decimal form repeats endlessly.* This example verifies the theorem that any repeating decimal represents a rational number. However, not all repeating decimals start repeating immediately. If the numbers never do repeat, the decimal represents an irrational number. For example, there are no repeating decimals that represent π, $\sqrt{2}$, or e. The decimal form of the number e does repeat at one point, but the repetition stops. As we noted in Section 12-5, the decimal form of e to 16 decimal places is 2.7 1828 1828 4590 452. We see that the sequence of digits 1828 repeats only once.

EXAMPLE 4 ▸▸ Find the fraction that has as its decimal form the repeating decimal 0.503 453 453 45 . . . .

We first separate the decimal into the beginning, nonrepeating part, and the infinite repeating decimal, which follows. Thus we have

$$0.503\,453\,453\,45\ldots = 0.50 + 0.003\,453\,453\,45\ldots$$

This means that we are to add $\frac{50}{100}$ to the fraction which represents the sum of the terms of the infinite geometric series $0.003\,45 + 0.000\,003\,45 + \cdots$. For this series, $a_1 = 0.003\,45$ and $r = 0.001$. We find this sum to be

$$S = \frac{0.003\,45}{1 - 0.001} = \frac{0.003\,45}{0.999} = \frac{115}{33\,300} = \frac{23}{6660}$$

Therefore,

$$0.503\,453\,45\ldots = \frac{5}{10} + \frac{23}{6660} = \frac{5(666) + 23}{6660} = \frac{3353}{6660}$$ ◂◂

EXAMPLE 5 ▸▸ Each swing of a certain pendulum bob is 95% as long as the preceding swing. How far does the bob travel in coming to rest if the first swing is 40.0 cm long?

We are to find the sum of the terms of an infinite geometric series for which $a_1 = 40.0$ and $r = 95\% = \frac{19}{20}$. Substituting these values into Eq. (19-7), we obtain

$$S = \frac{40.0}{1 - \frac{19}{20}} = \frac{40.0}{\frac{1}{20}} = (40.0)(20) = 800 \text{ cm}$$

The pendulum bob travels 800 cm in coming to rest. ◂◂

Exercises 19–3

In Exercises 1–12, find the sums of the given infinite geometric series.

1. $4 + 2 + 1 + \frac{1}{2} + \cdots$

2. $6 - 2 + \frac{2}{3} - \frac{2}{9} + \cdots$

3. $5 + 1 + 0.2 + 0.04 + \cdots$

4. $2 + \sqrt{2} + 1 + \cdots$

5. $20 - 1 + 0.05 - \cdots$

6. $9 + 8.1 + 7.29 + \cdots$

7. $1 + \frac{7}{8} + \frac{49}{64} + \cdots$

8. $6 - 4 + \frac{8}{3} - \cdots$

9. $1 + 0.0001 + 0.000\,000\,01 + \cdots$

10. $30 - 9 + 2.7 - \cdots$

11. $(2 + \sqrt{3}) + 1 + (2 - \sqrt{3}) + \cdots$

12. $(1 + \sqrt{2}) - 1 + (\sqrt{2} - 1) - \cdots$

In Exercises 13–24, find the fractions equal to the given decimals.

13. 0.333 33 . . .

14. 0.555 55 . . .

15. 0.404 040 . . .

16. 0.070 707 . . .

17. 0.181 818 . . .

18. 0.336 336 336 . . .

19. 0.273 273 273 . . .

20. 0.822 22 . . .

21. 0.366 666 . . .

22. 0.664 242 42 . . .

23. 0.100 841 841 841 . . .

24. 0.184 561 845 618 456 . . .

In Exercises 25–28, solve the given problems by use of the sum of an infinite geometric series.

25. Liquid is continuously collected in a wastewater holding tank such that during a given hour, only 92.0% as much liquid is collected as in the previous hour. If 28.0 L are collected in the first hour, what must be the minimum capacity of the tank?

26. In coming to rest, a record turntable makes 0.80 as many revolutions in a second as in the previous second. How many revolutions does the turntable make in coming to rest if it makes 0.50 r in the first second after the power is turned off?

27. The amounts of plutonium 237 that decay each day because of radioactivity form a geometric sequence. Given that the amounts which decay during each of the first four days are 5.882 g, 5.782 g, 5.684 g, and 5.587 g, respectively, what total amount will decay?

28. A square has sides of 20 cm each. Another square is inscribed in the original square by joining the midpoints of the sides. Assuming that such inscribed squares can be formed endlessly, find the sum of the areas of all the squares and explain how this sum is found. See Fig. 19-2.

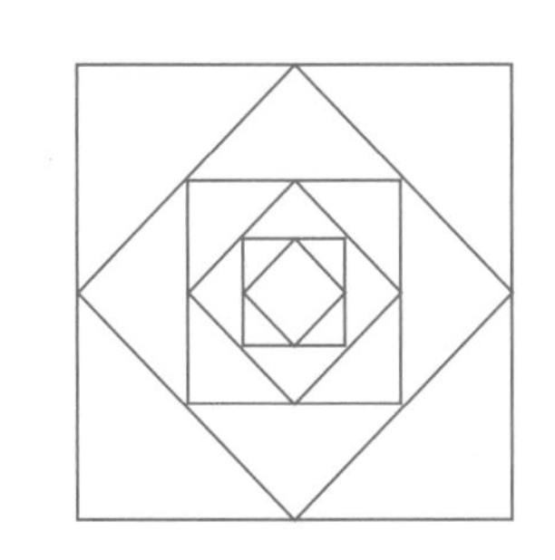

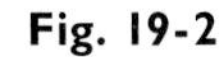

Fig. 19-2

19-4 The Binomial Theorem

If we wish to find the roots of the equation $(x + 2)^5 = 0$, we note that there are five factors of $x + 2$, which in turn tells us that the only distinct root is $x = -2$. However, if we wish to expand the expression $(x + 2)^5$, a number of repeated multiplications would be needed. This would be a relatively tedious operation. In this section we develop *the* **binomial theorem,** *by which it is possible to expand binomials to any given power without direct multiplication.* Such direct expansion can be helpful and labor-saving in developing certain mathematical topics. We may also expand certain expressions where direct multiplication is not actually possible. Also, the binominal theorem is used to develop the necessary expressions for use in certain technical applications.

By direct multiplication, we may obtain the following expansions of the binomial $a + b$.

$$(a + b)^0 = 1$$
$$(a + b)^1 = a + b$$
$$(a + b)^2 = a^2 + 2ab + b^2$$
$$(a + b)^3 = a^3 + 3a^2b + 3ab^2 + b^3$$
$$(a + b)^4 = a^4 + 4a^3b + 6a^2b^2 + 4ab^3 + b^4$$
$$(a + b)^5 = a^5 + 5a^4b + 10a^3b^2 + 10a^2b^3 + 5ab^4 + b^5$$

An inspection indicates certain properties of these expansions, and we shall assume that these properties are valid for the expansion of $(a + b)^n$, where n is any positive integer.

Properties of the binomial $(a + b)^n$

1. *There are $n + 1$ terms.*
2. *The first term is a^n and the final term is b^n.*
3. *Progressing from the first term to the last, the exponent of a decreases by 1 from term to term, the exponent of b increases by 1 from term to term, and the sum of the exponents of a and b in each term is n.*
4. *If the coefficient of any term is multiplied by the exponent of a in that term, and this product is divided by the number of that term, we obtain the coefficient of the next term.*
5. *The coefficients of terms equidistant from the ends are equal.*

EXAMPLE 1 ▸▸ Using the basic properties, develop the expansion for $(a + b)^5$.

Since the exponent of the binomial is 5, we have $n = 5$.

From Property 1, we know that there are six terms.

From Property 2, we know that the first term is a^5 and the final term is b^5.

From Property 3, we know that the factors of a and b in terms 2, 3, 4, and 5 are a^4b, a^3b^2, a^2b^3, and ab^4, respectively.

From Property 4, we obtain the coefficients of terms 2, 3, 4, and 5. In the first term, a^5, the coefficient is 1. Multiplying by 5, the power of a, and dividing by 1, the number of the term, we obtain 5, which is the coefficient of the second term. Thus, the second term is $5a^4b$. Again using Property 4, we obtain the coefficient of the third term. The coefficient of the second term is 5. Multiplying by 4, and dividing by 2, we obtain 10. This means that the third term is $10a^3b^2$.

From Property 5, we know that the coefficient of the fifth term is the same as the second and that the coefficient of the fourth term is the same as the third. These properties are illustrated in the following diagram.

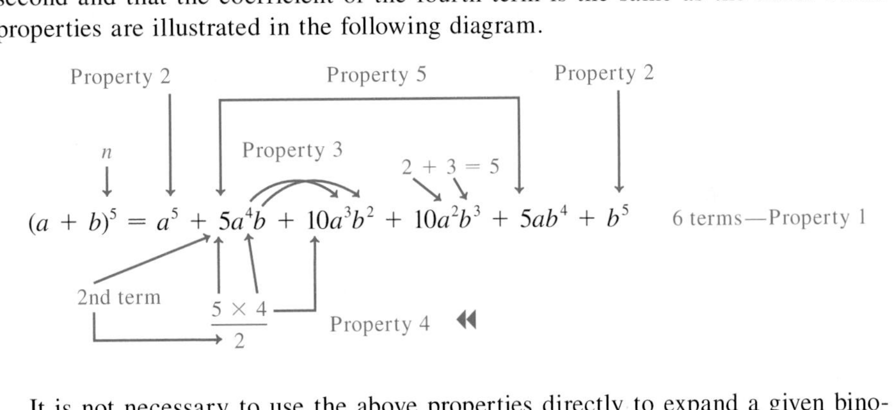

◂◂

It is not necessary to use the above properties directly to expand a given binomial. If they are applied to $(a + b)^n$, a general formula for the expansion of a binomial may be obtained. In developing and stating the general formula, it is convenient to use the **factorial notation $n!$**, where

$$n! = n(n-1)(n-2)\cdots(2)(1) \qquad \textbf{(19-8)}$$

We see that $n!$, read "n factorial," represents the product of the first n positive integers.

EXAMPLE 2 ▸▸ (a) $3! = (3)(2)(1) = 6$

(b) $5! = (5)(4)(3)(2)(1) = 120$

(c) $\dfrac{4!}{2!} = \dfrac{(4)(3)(2)(1)}{(2)(1)} = 12$

We note that $4!/2!$ is *not* $2!$ ◂◂

Binomial Formula

Based on the binomial properties, *the* **binomial theorem** *states that the following* **binomial formula** *is valid for all positive integer values of* n (the binomial theorem is proven through advanced methods).

$$(a + b)^n = a^n + na^{n-1}b + \frac{n(n-1)}{2!}a^{n-2}b^2 + \frac{n(n-1)(n-2)}{3!}a^{n-3}b^3 + \cdots + b^n \qquad \textbf{(19-9)}$$

EXAMPLE 3 ▸▸ Using the binomial formula, expand $(2x + 3)^6$.

In using the binomial formula for $(2x + 3)^6$, we use $2x$ for a, 3 for b, and 6 for n. Thus,

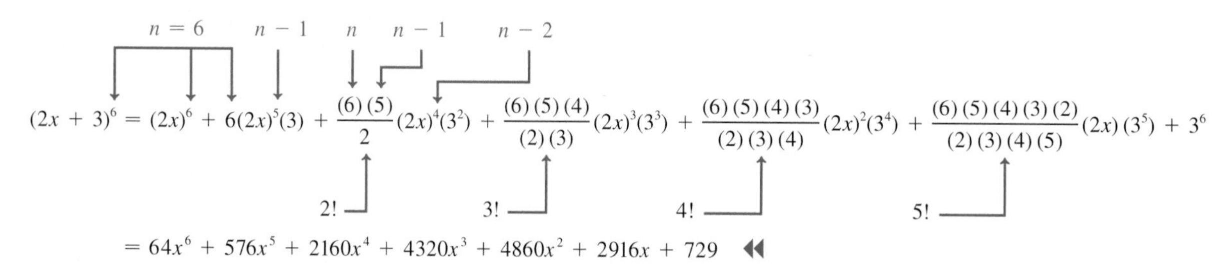

$$(2x + 3)^6 = (2x)^6 + 6(2x)^5(3) + \frac{(6)(5)}{2}(2x)^4(3^2) + \frac{(6)(5)(4)}{(2)(3)}(2x)^3(3^3) + \frac{(6)(5)(4)(3)}{(2)(3)(4)}(2x)^2(3^4) + \frac{(6)(5)(4)(3)(2)}{(2)(3)(4)(5)}(2x)(3^5) + 3^6$$

$$= 64x^6 + 576x^5 + 2160x^4 + 4320x^3 + 4860x^2 + 2916x + 729 \quad ◂◂$$

Pascal's Triangle

Named for the French scientist and mathematician Blaise Pascal (1623–1662).

For the first few integral powers of a binomial $a + b$, the coefficients can be obtained by setting them up in the following pattern, known as **Pascal's triangle**.

$$\begin{array}{lccccccccccccc}
n = 0 & & & & & & & 1 & & & & & & \\
n = 1 & & & & & & 1 & & 1 & & & & & \\
n = 2 & & & & & 1 & & 2 & & 1 & & & & \\
n = 3 & & & & 1 & & 3 & & 3 & & 1 & & & \\
n = 4 & & & 1 & & 4 & & 6 & & 4 & & 1 & & \\
n = 5 & & 1 & & 5 & & 10 & & 10 & & 5 & & 1 & \\
n = 6 & 1 & & 6 & & 15 & & 20 & & 15 & & 6 & & 1
\end{array}$$

See expansions on p. 496

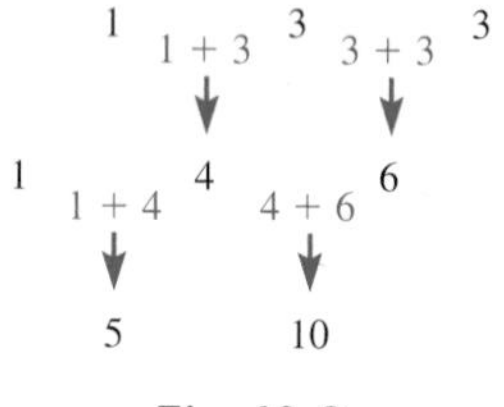

Fig. 19-3

We note that the first and last coefficients shown in each row are 1, and the second and next-to-last coefficients are equal to n. Other coefficients are obtained by adding the two nearest coefficients in the row above, as illustrated in Fig. 19-3 for the indicated section of Pascal's triangle. This pattern may be continued indefinitely, although use of Pascal's triangle is cumbersome for high values of n.

EXAMPLE 4 ▸▸ Using Pascal's triangle, expand $(5s - 2t)^4$.

Here we note that $n = 4$. Thus, the coefficients of the five terms are 1, 4, 6, 4, and 1, respectively. Also, here we use $5s$ for a and $-2t$ for b. We are expanding this expression as $[(5s) + (-2t)]^4$. Therefore,

from Pascal's triangle for $n = 4$

$$(5s - 2t)^4 = (5s)^4 + 4(5s)^3(-2t) + 6(5s)^2(-2t)^2 + 4(5s)(-2t)^3 + (-2t)^4$$
$$= 625s^4 - 1000s^3t + 600s^2t^2 - 160st^3 + 16t^4 \quad ◂◂$$

In certain uses of a binomial expansion it is not necessary to obtain all terms. Only the first few terms are required. The following example illustrates finding the first four terms of an expansion.

EXAMPLE 5 ▸▸ Find the first four terms of the expansion of $(x + 7)^{12}$.

Here we use x for a, 7 for b, and 12 for n. Thus, from the binomial formula we have

$$(x + 7)^{12} = x^{12} + 12x^{11}(7) + \frac{(12)(11)}{2}x^{10}(7^2) + \frac{(12)(11)(10)}{(2)(3)}x^9(7^3) + \cdots$$
$$= x^{12} + 84x^{11} + 3234x^{10} + 75\,460x^9 + \cdots \quad ◂◂$$

If we let $a = 1$ and $b = x$ in the binomial formula, we obtain the **binomial series**

Binomial Series

$$(1 + x)^n = 1 + nx + \frac{n(n-1)}{2!}x^2 + \frac{n(n-1)(n-2)}{3!}x^3 + \cdots \quad \text{(19-10)}$$

which through advanced methods can be shown to be valid for any real number n if $|x| < 1$. When n is either negative or a fraction, we obtain an infinite series. In such a case, we calculate as many terms as may be needed although such a series is not obtainable through direct multiplication. The binomial series may be used to develop important expressions that are used in applications and more advanced mathematics topics.

EXAMPLE 6 ▸▸ In the analysis of forces on beams, the expression $1/(1 + m^2)^{3/2}$ is used. Use the binomial series to find the first four terms of the expansion of this expression.

Using negative exponents, we have

$$1/(1 + m^2)^{3/2} = (1 + m^2)^{-3/2}$$

Now, in using Eq. (19-10), we have $n = -3/2$ and $x = m^2$.

$$(1 + m^2)^{-3/2} = 1 + \left(-\frac{3}{2}\right)(m^2) + \frac{(-\frac{3}{2})(-\frac{3}{2} - 1)}{2!}(m^2)^2 + \frac{(-\frac{3}{2})(-\frac{3}{2} - 1)(-\frac{3}{2} - 2)}{3!}(m^2)^3 + \cdots$$

Therefore,

$$\frac{1}{(1 + m^2)^{3/2}} = 1 - \frac{3}{2}m^2 + \frac{15}{8}m^4 - \frac{35}{16}m^6 + \cdots$$ ◂◂

Exercises 19–4

In Exercises 1–8, expand and simplify the given expressions by use of the binomial formula.

1. $(t + 1)^3$
2. $(x - 2)^3$
3. $(2x - 1)^4$
4. $(x^2 + 3)^4$
5. $(2x + 3)^5$
6. $(xy - z)^5$
7. $(2a - b^2)^6$
8. $\left(\frac{a}{x} + x\right)^6$

In Exercises 9–12, expand and simplify the given expression by use of Pascal's triangle.

9. $(5x - 3)^4$
10. $(b + 4)^5$
11. $(2a + 1)^6$
12. $(x - 3)^7$

In Exercises 13–20, find the first four terms of the indicated expansions.

13. $(x + 2)^{10}$
14. $(x - 3)^8$
15. $(2a - 1)^7$
16. $(3b + 2)^9$
17. $\left(x^2 - \frac{y}{2}\right)^{12}$
18. $\left(2a - \frac{1}{x}\right)^{11}$
19. $\left(b^2 + \frac{1}{2b}\right)^{20}$
20. $\left(2x^2 + \frac{y}{3}\right)^{15}$

In Exercises 21–28, find the first four terms of the indicated expansions by use of the binomial series.

21. $(1 + x)^8$
22. $(1 + x)^{-1/3}$
23. $(1 - x)^{-2}$
24. $(1 - 2x)^9$
25. $\sqrt{1 + x}$
26. $\frac{1}{\sqrt{1 + x}}$
27. $\frac{1}{\sqrt{9 - 9x}}$
28. $\sqrt{4 + x^2}$

In Exercises 29–32, solve the given problems involving factorial notation.

29. Using a calculator (most calculators have an $n!$ feature or key), evaluate (a) $17! + 4!$, (b) $21!$, (c) $17! \times 4!$, (d) $68!$.

30. Using a calculator, evaluate (a) $8! - 7!$, (b) $\frac{8!}{7!}$, (c) $8! \times 7!$, (d) $56!$.

31. Show that $n! = n \times (n - 1)!$ for $n \geq 2$. To use this equation for $n = 1$, explain why it is necessary to define $0! = 1$ (this is a standard definition of 0!).

32. Show that $\dfrac{(n+1)!}{(n-2)!} = n^3 - n$ for $n \geq 2$. See Exercise 31.

In Exercises 33–36, find the indicated terms by use of the following information. The r + 1 term of the expansion of $(a + b)^n$ is given by

$$\frac{n(n-1)(n-2)\cdots(n-r+1)}{r!}a^{n-r}b^r$$

33. The term involving b^5 in $(a + b)^8$

34. The term involving y^6 in $(x + y)^{10}$

35. The fifth term of $(2x - 3b)^{12}$

36. The sixth term of $(a - b)^{14}$

In Exercises 37–40, find the indicated expansions.

37. A company purchases a piece of equipment for A dollars, and the equipment depreciates at a rate of r each year. Its value V after n years is given by $V = A(1 - r)^n$. Expand this expression for $n = 5$.

38. In designing a type of tubing, the expression $(D + 0.1)^4$ is used. Expand this expression.

39. In the theory associated with the magnetic field due to an electric current, the expression $1 - \dfrac{x}{\sqrt{a^2 + x^2}}$ is found. By expanding $(a^2 + x^2)^{-1/2}$ find the first three nonzero terms that could be used to approximate the given expression.

40. In the theory related to the dispersion of light, the expression $1 + \dfrac{A}{1 - \dfrac{\lambda_0^2}{\lambda^2}}$ arises. (a) Let $x = \lambda_0^2/\lambda^2$ and find the first four terms of the expansion of $(1 - x)^{-1}$. (b) Find the same expansion by using long division. (c) Write the original expression in expanded form using the results of (a) and (b).

Chapter Equations, Review Exercises, and Practice Test

Chapter Equations

Arithmetic sequences	Recursion formula	$a_n = a_{n-1} + d$	(19-1)
	nth term	$a_n = a_1 + (n - 1)d$	(19-2)
	Sum of n terms	$S_n = \dfrac{n}{2}(a_1 + a_n)$	(19-3)
Geometric sequences	Recursion formula	$a_n = ra_{n-1}$	(19-4)
	nth term	$a_n = a_1r^{n-1}$	(19-5)
	Sum of n terms	$S_n = \dfrac{a_1(1 - r^n)}{1 - r} \quad (r \neq 1)$	(19-6)
Sum of geometric series		$S = \dfrac{a_1}{1 - r} \quad (\lvert r\rvert < 1)$	(19-7)
Factorial notation		$n! = n(n-1)(n-2)\cdots(2)(1)$	(19-8)
Binomial formula		$(a + b)^n = a^n + na^{n-1}b + \dfrac{n(n-1)}{2!}a^{n-2}b^2 + \dfrac{n(n-1)(n-2)}{3!}a^{n-3}b^3 + \cdots + b^n$	(19-9)
Binomial series		$(1 + x)^n = 1 + nx + \dfrac{n(n-1)}{2!}x^2 + \dfrac{n(n-1)(n-2)}{3!}x^3 + \cdots$	(19-10)

Review Exercises

In Exercises 1–8, find the indicated term of each sequence.

1. $1, 6, 11, \ldots$ (17th) **2.** $1, -3, -7, \ldots$ (21st)

3. $\frac{1}{2}, 0.1, 0.02, \ldots$ (9th)

4. $0.025, 0.01, 0.004, \ldots$ (7th)

5. $8, \frac{7}{2}, -1, \ldots$ (16th)

6. $-1, -\frac{5}{3}, -\frac{7}{3}, \ldots$ (25th)

7. $\frac{3}{4}, \frac{1}{2}, \frac{1}{3}, \ldots$ (7th) **8.** $\frac{2}{3}, 1, \frac{3}{2}, \ldots$ (7th)

In Exercises 9–12, find the sum of each sequence with the indicated values.

9. $a_1 = -4, n = 15, a_{15} = 17$ (arith.)

10. $a_1 = 3, d = -\frac{2}{3}, n = 10$

11. $a_1 = 16, r = -\frac{1}{2}, n = 14$

12. $a_1 = 64, a_n = 729, n = 7$ (geom., $r > 0$)

In Exercises 13–24, find the indicated quantities for the appropriate sequences.

13. $a_1 = 17, d = -2, n = 9, S_9 = ?$

14. $d = \frac{4}{3}, a_1 = -3, a_n = 17, n = ?$

15. $a_1 = 18, r = \frac{1}{2}, n = 6, a_6 = ?$

16. $a_n = \frac{49}{8}, r = -\frac{2}{7}, S_n = \frac{17199}{288}, a_1 = ?$

17. $a_1 = 8, a_n = -2.5, S_n = 22, d = ?$

18. $a_1 = 2, d = 0.2, n = 11, S_{11} = ?$

19. $n = 6, r = -0.25, S_6 = 204.75, a_6 = ?$

20. $a_1 = 10, r = 0.1, S_n = 11.111, n = ?$

21. $a_1 = -1, a_n = 32, n = 12, S_{12} = ?$ (arith.)

22. $a_1 = 1, a_n = 64, S_n = 325, n = ?$ (arith.)

23. $a_1 = 1, n = 7, a_7 = 64, S_7 = ?$

24. $a_1 = \frac{1}{4}, n = 6, a_6 = 8, S_6 = ?$

In Exercises 25–28, find the sums of the given infinite geometric series.

25. $9 + 6 + 4 + \cdots$

26. $80 - 20 + 5 - \cdots$

27. $1 + 1.02^{-1} + 1.02^{-2} + \cdots$

28. $3 - \sqrt{3} + 1 - \cdots$

In Exercises 29–32, find the fractions equal to the given decimals.

29. $0.030\,303\ldots$ **30.** $0.363\,363\ldots$

31. $0.072\,727\,2\ldots$ **32.** $0.253\,993\,993\,99\ldots$

In Exercises 33–36, expand and simplify the given expression. In Exercises 37–40, find the first four terms of the appropriate expansion.

33. $(x - 2)^4$ **34.** $(s + 2t)^4$

35. $(x^2 + 1)^5$ **36.** $(3n - a)^6$

37. $(a + 2b^2)^{10}$ **38.** $\left(\frac{x}{4} - y\right)^{12}$

39. $\left(p^2 - \frac{q}{6}\right)^9$ **40.** $\left(2s^2 - \frac{3}{2t}\right)^{14}$

In Exercises 41–48, find the first four terms of the indicated expansions by use of the binomial series.

41. $(1 + x)^{12}$ **42.** $(1 - x)^{10}$

43. $\sqrt{1 + x^2}$ **44.** $(4 - 4x)^{-1}$

45. $\sqrt{1 - a^2}$ **46.** $\sqrt{1 + b^4}$

47. $(2 - 4x)^{-3}$ **48.** $(1 + 4x)^{-1/4}$

In Exercises 49–72, solve the given problems by use of an appropriate sequence or expansion. All numbers are accurate to at least two significant digits.

49. Find the sum of the first 1000 positive even integers.

50. How many numbers divisible by 4 lie between 23 and 121?

51. Each stroke of a pile driver moves a post 2 cm less than the previous stroke. If the first stroke moves the post 24 cm, which stroke moves the post 4 cm?

52. During each hour, an exhaust fan removes 15.0% of the carbon dioxide present in the air in a room at the beginning of the hour. What percent of the carbon dioxide remains after 10.0 h?

53. Each 1.0 mm of a filter through which light passes reduces the intensity of the light by 12%. How thick should the filter be to reduce the intensity of the light to 20%?

54. A pile of dirt and 10 holes are in a straight line. It is 20 m from the dirt pile to the nearest hole, and the holes are 8 m apart. If a backhoe takes two trips from the dirt pile to fill each hole, how far must it travel in filling all the holes if it starts and ends at the dirt pile?

55. A roof support with equally spaced vertical pieces is shown in Fig. 19-4. Find the total length of the vertical pieces if the shortest one is 254 mm long.

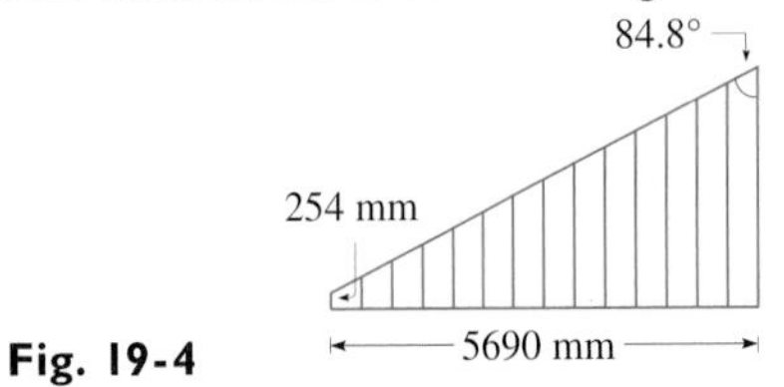

Fig. 19-4

56. During each microsecond the current in an electric circuit decreases by 9.3%. If the initial current is 2.45 mA, how long does it take to reach 0.50 mA?

57. What is the value after three years of a \$2000 investment that earns 6% annual interest compounded quarterly?

58. The level of chemical pollution in a lake is 4.50 ppb (parts per billion). If the level increases by 0.20 ppb in the following month, and by 5.0% less each month thereafter, what will be the maximum level?

59. A piece of paper 0.015 cm thick is cut in half. These two pieces are then placed one on the other and cut in half. If this is repeated such that the paper is cut in half 40 times, how high will the pile be?

60. After the power is turned off, an object on a nearly frictionless surface slows down such that it travels 99.9% as far during one second as during the previous second. If it travels 100 cm during the first second after the power is turned off, how far does it travel while stopping?

61. A person invests \$1000 each year at the beginning of the year. What is the total value of these investments after 20 years if they earn 7.5% annual interest compounded semiannually?

62. In testing a type of insulation, the temperature in a room was made to fall to $\frac{2}{3}$ of the initial temperature after 1.0 h, to $\frac{2}{5}$ of the initial temperature after 2.0 h, to $\frac{2}{7}$ of the initial temperature after 3.0 h, and so on. If the initial temperature was 50.0° C, what was the temperature after 12.0 h? (This is an illustration of a *harmonic sequence.*)

63. Two competing businesses make the same item and sell it initially for \$100. One increases the price by \$8 each year for five years, and the other increases the price by 8% each year for five years. What is the difference in the prices after five years?

64. A well driller charges \$10.00 for drilling the first metre of a well and for each metre thereafter charges 0.20% more than for the preceding metre. How much is charged for drilling a 150-m well?

65. In hydrodynamics, while studying compressible fluid flow, the expression $\left(1 + \frac{a-1}{2}m^2\right)^{a/(a-1)}$ arises. Find the first three terms of the expansion of this expression.

66. In finding the partial pressure P_F of fluorine gas under certain conditions, the equation

$$P_F = \frac{(1 + 2 \times 10^{-10}) - \sqrt{1 + 4 \times 10^{-10}}}{2} \text{atm}$$

is found. By using three terms of the expansion for $\sqrt{1 + x}$, approximate the value of this expression.

67. During one year a beach eroded 1.2 m to a line 48.3 m from the wall of a building. If the erosion is 0.1 m more each year than the previous year, when will the waterline reach the wall?

68. On a highway with a steep incline, a runaway truck ramp is constructed so that a vehicle which has lost its brakes can stop. The ramp is designed to slow a truck in succeeding 20-m distances by 10 km/h, 12 km/h, 14 km/h, ... If the ramp is 160 m long, will it stop a truck moving at 120 km/h when it reaches the ramp?

69. Each application of an insecticide destroys 75% of a certain insect. How many applications are needed to destroy at least 99.9% of the insects?

70. A wire hung between two poles is parabolic in shape. To find the length of wire between two points on the wire, the expression $\sqrt{1 + 0.08x^2}$ is used. Find the first three terms of the binomial expansion of this expression.

71. Do the reciprocals of the terms of a geometric sequence form a geometric sequence?

72. The terms a, $a + 12$, $a + 24$ form an arithmetic sequence, and the terms a, $a + 24$, $a + 12$ form a geometric sequence. Find these sequences.

Writing Exercise

73. Derive a formula for the value V after 1 year of an amount A invested at r% (as a decimal) annual interest, compounded n times during the year. If A = \$1000 and $r = 0.10$ (10%), write two or three paragraphs explaining why the amount of interest increases as n increases and stating your approach to finding the maximum possible amount of interest.

Practice Test

1. Find the sum of the first seven terms of the sequence $6, -2, \frac{2}{3}, \ldots$.

2. For a given sequence $a_1 = 6$, $d = 4$ and $S_n = 126$. Find n.

3. Find the fraction equal to the decimal 0.454 545. . . .

4. Find the first three terms of the expansion of $\sqrt{1 - 4x}$.

5. Expand and simplify the expression $(2x - y)^5$.

6. What is the value after 20 years of an investment of \$2500 if it draws 5% annual interest compounded annually?

7. Find the sum of the first 100 even integers.

8. A ball is dropped from a height of 8.00 m, and on each rebound it rises to $\frac{1}{2}$ of the height it last fell. If it bounces indefinitely, through what total distance will it move?

20

Additional Topics in Trigonometry

The definitions of the trigonometric functions were first introduced in Section 4-2 and were again summarized in Section 8-1. If we take a close look at these definitions, we find that there are many relationships among the various functions. In this chapter we study some of these basic relationships, as well as develop others.

We also develop the concept of the inverse trigonometric functions, which was introduced in Chapter 4.

The trigonometric relationships we develop in this chapter are important for a number of reasons. We already made limited use of some of them in Section 10-4 when we graphed certain trigonometric functions. We also used an important *trigonometric identity* in deriving the law of cosines in Chapter 9.

When we consider equations with trigonometric functions later in the chapter, we will find that the solution of these equations often depends on the proper use of identities.

In the study of calculus, there are certain types of problems that require the use of trigonometric identities for solution (even problems in which trigonometric functions do not appear). Also, they are used to develop expressions and solve equations that are used in a number of technical areas.

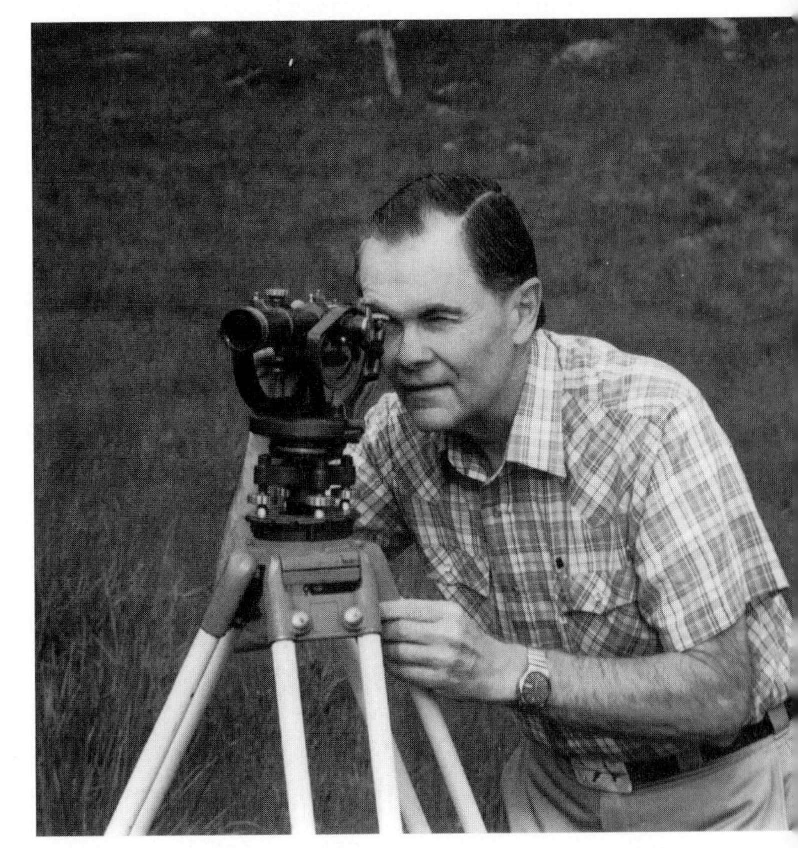

Basic trigonometric relationships can be used to develop useful formulas. In Section 20-3 we derive the formula $A = \frac{1}{4}c^2 \sin 2\theta$, where A is the area of a right triangle, c is the hypotenuse, and θ is either acute angle. It is used in surveying.

20-1 Fundamental Trigonometric Identities

From Chapters 4 and 8, we recall that the definition of the sine of an angle θ is $\sin\theta = y/r$ and that the definition of the cosecant of an angle θ is $\csc\theta = r/y$ (see Fig. 20-1). Since $y/r = 1/(r/y)$, we see that $\sin\theta = 1/\csc\theta$. The definitions hold true for *any* angle, which means this relation between $\sin\theta$ and $\csc\theta$ is true for *any* angle. *This type of relation, which is true for any value of the variable, is called an* **identity.** Of course, values where division by zero would be indicated are excluded.

In this section we develop several important identities among the trigonometric functions. We also show how the basic identities are used to verify other identities.

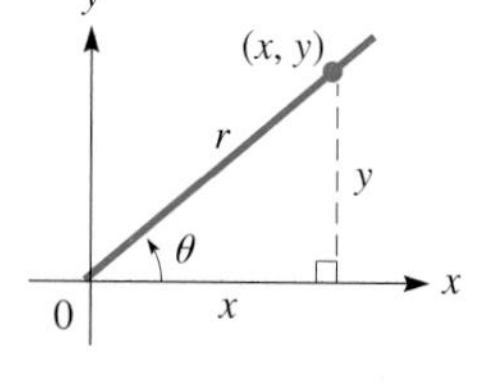

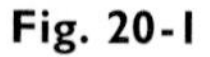
Fig. 20-1

From the definitions, we have

$$\sin\theta\csc\theta = \frac{y}{r}\times\frac{r}{y} = 1 \quad \text{or} \quad \sin\theta = \frac{1}{\csc\theta} \quad \text{or} \quad \csc\theta = \frac{1}{\sin\theta}$$

$$\cos\theta\sec\theta = \frac{x}{r}\times\frac{r}{x} = 1 \quad \text{or} \quad \cos\theta = \frac{1}{\sec\theta} \quad \text{or} \quad \sec\theta = \frac{1}{\cos\theta}$$

$$\tan\theta\cot\theta = \frac{y}{x}\times\frac{x}{y} = 1 \quad \text{or} \quad \tan\theta = \frac{1}{\cot\theta} \quad \text{or} \quad \cot\theta = \frac{1}{\tan\theta}$$

$$\frac{\sin\theta}{\cos\theta} = \frac{y/r}{x/r} = \frac{y}{x} = \tan\theta; \qquad \frac{\cos\theta}{\sin\theta} = \frac{x/r}{y/r} = \frac{x}{y} = \cot\theta$$

Also, from the definitions and the Pythagorean theorem in the form of $x^2 + y^2 = r^2$, we arrive at the following identities.

By dividing the Pythagorean relation through by r^2, we have

$$\left(\frac{x}{r}\right)^2 + \left(\frac{y}{r}\right)^2 = 1 \quad \text{which leads us to} \quad \cos^2\theta + \sin^2\theta = 1$$

By dividing the Pythagorean relation by x^2, we have

$$1 + \left(\frac{y}{x}\right)^2 = \left(\frac{r}{x}\right)^2 \quad \text{which leads us to} \quad 1 + \tan^2\theta = \sec^2\theta$$

By dividing the Pythagorean relation by y^2, we have

$$\left(\frac{x}{y}\right)^2 + 1 = \left(\frac{r}{y}\right)^2 \quad \text{which leads us to} \quad \cot^2\theta + 1 = \csc^2\theta$$

Basic Identities

The term $\cos^2\theta$ is the common way of writing $(\cos\theta)^2$, and it means to square the value of the cosine of the angle. Obviously the same holds true for the other functions.

Summarizing these results, we have the following important identities.

$$\sin\theta = \frac{1}{\csc\theta} \tag{20-1}$$

$$\cos\theta = \frac{1}{\sec\theta} \tag{20-2}$$

$$\tan\theta = \frac{1}{\cot\theta} \tag{20-3}$$

$$\tan\theta = \frac{\sin\theta}{\cos\theta} \tag{20-4}$$

$$\cot\theta = \frac{\cos\theta}{\sin\theta} \tag{20-5}$$

$$\sin^2\theta + \cos^2\theta = 1 \tag{20-6}$$

$$1 + \tan^2\theta = \sec^2\theta \tag{20-7}$$

$$1 + \cot^2\theta = \csc^2\theta \tag{20-8}$$

In using these basic identities, θ may stand for any angle or number or expression representing an angle or a number.

EXAMPLE 1 ▸▸ (a) $\sin(x + 1) = \dfrac{1}{\csc(x + 1)}$ using Eq. (20-1)

(b) $\tan 157^\circ = \dfrac{\sin 157^\circ}{\cos 157^\circ}$ using Eq. (20-4)

(c) $\sin^2\left(\dfrac{\pi}{4}\right) + \cos^2\left(\dfrac{\pi}{4}\right) = 1$ using Eq. (20-6) ◂◂

EXAMPLE 2 ▸▸ We shall check the last two illustrations of Example 1 for the particular values of θ that are used.

(a) Using a calculator, we find that

$$\sin 157^\circ = 0.390\,731\,128\,5 \quad \text{and} \quad \cos 157^\circ = -0.920\,504\,853\,5$$

Considering Eq. (20-4), and illustration (b) in Example 1, by dividing we find that

$$\frac{\sin 157^\circ}{\cos 157^\circ} = \frac{0.390\,731\,128\,5}{-0.920\,504\,853\,5} = -0.424\,474\,816\,2$$

We also find that $\tan 157^\circ = -0.424\,474\,816\,2$, which shows that

$$\tan 157^\circ = \frac{\sin 157^\circ}{\cos 157^\circ}$$

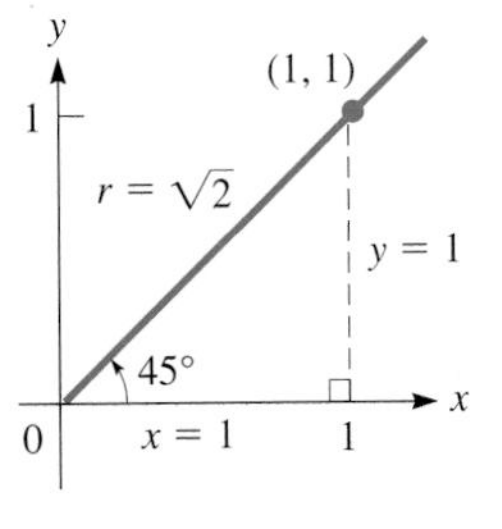

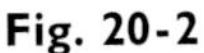
Fig. 20-2

(b) To check illustration (c) of Example 1, we refer to the values found in Example 2 of Section 4-3, or to Fig. 20-2. These tell us that

$$\sin 45^\circ = \frac{1}{\sqrt{2}} = \frac{\sqrt{2}}{2} \quad \text{and} \quad \cos 45^\circ = \frac{\sqrt{2}}{2}$$

Since $\frac{\pi}{4} = 45^\circ$, by adding the squares of sin 45° and cos 45°, we have

$$\sin^2\left(\frac{\pi}{4}\right) + \cos^2\left(\frac{\pi}{4}\right) = \left(\frac{\sqrt{2}}{2}\right)^2 + \left(\frac{\sqrt{2}}{2}\right)^2 = \frac{1}{2} + \frac{1}{2} = 1$$

We see that this checks with Eq. (20-6) for these values. ◂◂

Proving Trigonometric Identities

A great many identities exist among the trigonometric functions. We are going to use the basic identities that have been developed in Eqs. (20-1) through (20-8), along with a few additional ones developed in later sections to prove the validity of still other identities.

CAUTION ▸ ***The ability to prove trigonometric identities depends to a large extent on being very familiar with the basic identities*** so that you can *recognize them in somewhat **different forms.***

If you do not learn these basic identities and learn them well, you will have difficulty in following the examples and doing the exercises. The more readily you recognize these forms, the more easily you will be able to prove such identities.

In proving identities, we should look for combinations that appear in, or are very similar to, those in the basic identities. This is illustrated in the following examples.

EXAMPLE 3 ▸▸ In proving the identity

$$\sin x = \frac{\cos x}{\cot x}$$

we know that $\cot x = \dfrac{\cos x}{\sin x}$. Since $\sin x$ appears on the left, substituting for $\cot x$ on the right will eliminate $\cot x$ and introduce $\sin x$. This should help us proceed in proving the identity. Thus

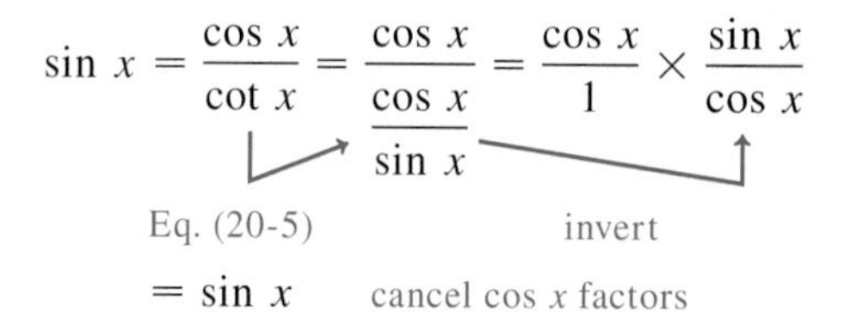

$$\sin x = \frac{\cos x}{\cot x} = \frac{\cos x}{\dfrac{\cos x}{\sin x}} = \frac{\cos x}{1} \times \frac{\sin x}{\cos x}$$

Eq. (20-5) invert

$= \sin x$ cancel $\cos x$ factors

By showing that the right side may be changed exactly to $\sin x$, the expression on the left side, we have proved the identity. ◂◂

Some important points should be made in relation to the proof of the identity of Example 3. We must recognize what basic identities may be useful. The proof of an identity requires the use of basic algebraic operations, and these must be done carefully and correctly. Although in Example 3 we changed the right side to the form on the left, we could have changed the left to the form on the right. From this, and the fact that various substitutions are possible, we see that a variety of procedures can be used to prove any given identity.

EXAMPLE 4 ▸▸ Prove that $\tan\theta \csc\theta = \sec\theta$.

In proving this identity, we know that $\tan\theta = \dfrac{\sin\theta}{\cos\theta}$ and also that $\dfrac{1}{\cos\theta} = \sec\theta$. Thus, by substituting for $\tan\theta$ we introduce $\cos\theta$ in the denominator, which is equivalent to introducing $\sec\theta$ in the numerator. Therefore, changing only the left side, we have

$$\tan\theta \csc\theta = \frac{\sin\theta}{\cos\theta}\csc\theta = \frac{\sin\theta}{\cos\theta}\,\frac{1}{\sin\theta}$$

Eq. (20-4) Eq. (20-1)

$$= \frac{1}{\cos\theta} \quad \text{cancel } \sin\theta \text{ factors}$$

$$= \sec\theta \quad \text{using Eq. (20-2)}$$

Having changed the left side into the form on the right side, we have proven the identity.

Many variations of the preceding steps are possible. Also, we could have changed the right side to obtain the form on the left. For example,

$$\tan\theta \csc\theta = \sec\theta = \frac{1}{\cos\theta} \quad \text{using Eq. (20-2)}$$

$$= \frac{\sin\theta}{\cos\theta \sin\theta} = \frac{\sin\theta}{\cos\theta}\,\frac{1}{\sin\theta} \quad \text{multiply numerator and denominator by } \sin\theta \text{ and rewrite}$$

$$= \tan\theta \csc\theta \quad \text{using Eqs. (20-4) and (20-1)}$$ ◂◂

In proving the identities of Examples 3 and 4 we have shown that the expression on one side of the equal sign can be changed into the expression on the other side. Although making the restriction that we change only one side is not entirely necessary, *we shall restrict the method of proof to changing only one side into the same form as the other side.* In this way we know the form we are to obtain, and by looking ahead we are better able to make the proper changes.

CAUTION ▶ There is no set procedure for working with identities. The most important factors are to (1) ***recognize the proper forms,*** (2) *see what effect a change may have* before performing it, and (3) *perform it correctly.* Normally, *it is easier to change the form of the more complicated side to the same form as the less complicated side.* If the forms of the sides are about the same, a close look often suggests possible steps to use.

EXAMPLE 5 ▶▶ Prove the identity $\dfrac{\cos x \csc x}{\cot^2 x} = \tan x$.

First, we note that the left-hand side has several factors and the right-hand side has only one. Therefore, let us transform the left-hand side. Next, we note that we want $\tan x$ as the final result. We know that $\cot x = 1/\tan x$. Thus

$$\frac{\cos x \csc x}{\cot^2 x} = \frac{\cos x \csc x}{\dfrac{1}{\tan^2 x}} = \cos x \csc x \tan^2 x$$

At this point, we have two factors of $\tan x$ on the left. Since we want only one, let us factor out one. Therefore,

$$\cos x \csc x \tan^2 x = \tan x(\cos x \csc x \tan x)$$

Now, replacing $\tan x$ within the parentheses by $\sin x/\cos x$, we have

$$\tan x(\cos x \csc x \tan x) = \frac{\tan x(\cos x \csc x \sin x)}{\cos x}$$

Now we may cancel $\cos x$. Also, $\csc x \sin x = 1$ from Eq. (19-1). Finally,

$$\begin{aligned}\frac{\tan x\,(\cos x \csc x \sin x)}{\cos x} &= \tan x\left(\frac{\cos x}{\cos x}\right)(\csc x \sin x)\\ &= \tan x\,(1)\,(1) = \tan x\end{aligned}$$

Since we have transformed the left-hand side into $\tan x$, we have proven the identity. Of course, it is not necessary to rewrite expressions as we did in this example. This was done here only to include the explanations. ◀◀

EXAMPLE 6 ▶▶ In finding the radiation rate of an accelerated electric charge, it is necessary to show that $\sin^3\theta = \sin\theta - \sin\theta\cos^2\theta$. Show this by changing the left side.

Since each term on the right has a factor of $\sin\theta$, we see that we can proceed by writing $\sin^3\theta$ as $\sin\theta(\sin^2\theta)$. Then the factor $\sin^2\theta$ and the $\cos^2\theta$ on the right suggest the use of Eq. (20-6). Thus we have

$$\begin{aligned}\sin^3\theta &= \sin\theta(\sin^2\theta) = \sin\theta(1 - \cos^2\theta)\\ &= \sin\theta - \sin\theta\cos^2\theta \quad \text{multiplying}\end{aligned}$$ ◀◀

Since we wanted to substitute for $\sin^2\theta$, we used Eq. (20-6) in the form

$$\sin^2\theta = 1 - \cos^2\theta$$ ◀◀

EXAMPLE 7 ▸▸ Prove the identity $\frac{\sec^2 y}{\cot y} - \tan^3 y = \tan y$.

Here we shall simplify the left side. We can remove $\cot y$ from the denominator, since $\cot y = 1/\tan y$. Also, the presence of $\sec^2 y$ suggests the use of Eq. (20-7). Therefore, we have

$$\frac{\sec^2 y}{\cot y} - \tan^3 y = \frac{\sec^2 y}{\frac{1}{\tan y}} - \tan^3 y = \sec^2 y \tan y - \tan^3 y$$

$$= \tan y(\sec^2 y - \tan^2 y) = \tan y(1)$$

$$= \tan y$$

Here we have used Eq. (20-7) in the form $\sec^2 y - \tan^2 y = 1$. ◂◂

EXAMPLE 8 ▸▸ Prove the identity $\frac{1 - \sin x}{\sin x \cot x} = \frac{\cos x}{1 + \sin x}$.

The combination $1 - \sin x$ also suggests $1 - \sin^2 x$, since multiplying $(1 - \sin x)$ by $(1 + \sin x)$ gives $1 - \sin^2 x$, which can then be replaced by $\cos^2 x$. Thus, changing only the left side, we have

$$\frac{1 - \sin x}{\sin x \cot x} = \frac{(1 - \sin x)(1 + \sin x)}{\sin x \cot x(1 + \sin x)}$$ multiply numerator and denominator by $1 + \sin x$

$$= \frac{1 - \sin^2 x}{\sin x\left(\frac{\cos x}{\sin x}\right)(1 + \sin x)} = \frac{\cos^2 x}{\cos x(1 + \sin x)}$$ Eq. (20-6); cancel $\sin x$

$$= \frac{\cos x}{1 + \sin x}$$ cancel $\cos x$ ◂◂

EXAMPLE 9 ▸▸ Prove the identity $\sec^2 x + \csc^2 x = \sec^2 x \csc^2 x$.

Here we note the presence of $\sec^2 x$ and $\csc^2 x$ on each side. This suggests the possible use of the square relationships. By replacing the $\sec^2 x$ on the right-hand side by $1 + \tan^2 x$, we can create $\csc^2 x$ plus another term. The left-hand side is the $\csc^2 x$ plus another term, so this procedure should help. Thus, changing only the right side,

$$\sec^2 x + \csc^2 x = \sec^2 x \csc^2 x$$

$$= (1 + \tan^2 x)(\csc^2 x)$$ using Eq. (20-7)

$$= \csc^2 x + \tan^2 x \csc^2 x$$ multiplying

$$= \csc^2 x + \left(\frac{\sin^2 x}{\cos^2 x}\right)\left(\frac{1}{\sin^2 x}\right)$$ using Eqs. (20-4) and (20-1)

$$= \csc^2 x + \frac{1}{\cos^2 x}$$ cancel $\sin^2 x$

$$= \csc^2 x + \sec^2 x$$ using Eq. (20-2)

We could have used many other variations of this procedure, and they would have been perfectly valid. ◂◂

EXAMPLE 10 ▸▸ Prove the identity $\dfrac{\csc x}{\tan x + \cot x} = \cos x$.

Here we shall simplify the left-hand side until we have the expression that appears on the right-hand side.

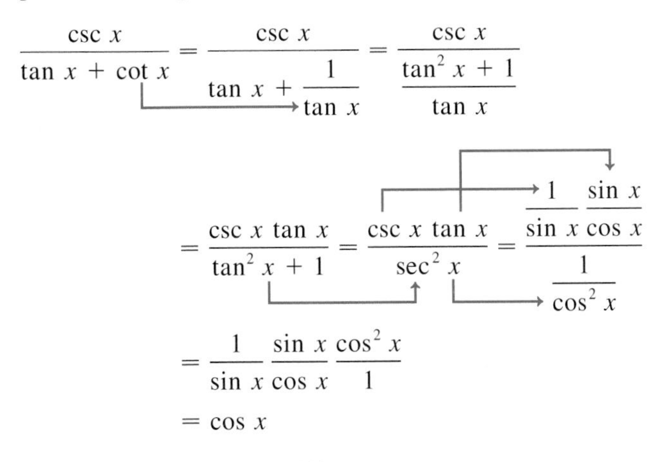

$$\frac{\csc x}{\tan x + \cot x} = \frac{\csc x}{\tan x + \dfrac{1}{\tan x}} = \frac{\csc x}{\dfrac{\tan^2 x + 1}{\tan x}}$$

$$= \frac{\csc x \tan x}{\tan^2 x + 1} = \frac{\csc x \tan x}{\sec^2 x} = \frac{\dfrac{1}{\sin x}\dfrac{\sin x}{\cos x}}{\dfrac{1}{\cos^2 x}}$$

$$= \frac{1}{\sin x}\frac{\sin x}{\cos x}\frac{\cos^2 x}{1}$$

$$= \cos x$$

We have shown that $\dfrac{\csc x}{\tan x + \cot x} = \cos x$, which proves the identity. ◂◂

Exercises 20–1

In Exercises 1–4, check the indicated basic identities for the given angles.

1. Check Eq. (20-3) for $\theta = 56°$.
2. Check Eq. (20-5) for $\theta = 280°$.
3. Check Eq. (20-6) for $\theta = \dfrac{4\pi}{3}$.
4. Check Eq. (20-7) for $\theta = \dfrac{5\pi}{6}$.

In Exercises 5–56, prove the given identities.

5. $\dfrac{\cot\theta}{\cos\theta} = \csc\theta$
6. $\dfrac{\tan y}{\sin y} = \sec y$
7. $\dfrac{\sin x}{\tan x} = \cos x$
8. $\dfrac{\csc\theta}{\sec\theta} = \cot\theta$
9. $\sin y \cot y = \cos y$
10. $\cos x \tan x = \sin x$
11. $\sin x \sec x = \tan x$
12. $\cot\theta \sec\theta = \csc\theta$
13. $\csc^2 x(1 - \cos^2 x) = 1$
14. $\cos^2 x(1 + \tan^2 x) = 1$
15. $\sin x(1 + \cot^2 x) = \csc x$
16. $\sec\theta(1 - \sin^2\theta) = \cos\theta$
17. $\sin x(\csc x - \sin x) = \cos^2 x$
18. $\cos y(\sec y - \cos y) = \sin^2 y$
19. $\tan y(\cot y + \tan y) = \sec^2 y$
20. $\csc x(\csc x - \sin x) = \cot^2 x$
21. $\sin x \tan x + \cos x = \sec x$
22. $\sec x \csc x - \cot x = \tan x$
23. $\cos\theta \cot\theta + \sin\theta = \csc\theta$
24. $\csc x \sec x - \tan x = \cot x$
25. $\sec\theta \tan\theta \csc\theta = \tan^2\theta + 1$
26. $\sin x \cos x \tan x = 1 - \cos^2 x$
27. $\cot\theta \sec^2\theta - \cot\theta = \tan\theta$
28. $\sin y + \sin y \cot^2 y = \csc y$
29. $\tan x + \cot x = \sec x \csc x$
30. $\tan x + \cot x = \tan x \csc^2 x$
31. $\cos^2 x - \sin^2 x = 1 - 2\sin^2 x$
32. $\tan^2 y \sec^2 y - \tan^4 y = \tan^2 y$
33. $\dfrac{\sin x}{1 - \cos x} = \csc x + \cot x$
34. $\dfrac{1 + \cos x}{\sin x} = \dfrac{\sin x}{1 - \cos x}$
35. $\dfrac{\sec x + \csc x}{1 + \tan x} = \csc x$
36. $\dfrac{\cot x + 1}{\cot x} = 1 + \tan x$
37. $\tan^2 x \cos^2 x + \cot^2 x \sin^2 x = 1$
38. $\dfrac{\sin\theta}{\csc\theta} + \dfrac{\cos\theta}{\sec\theta} = 1$

39. $\dfrac{\sec\theta}{\cos\theta} - \dfrac{\tan\theta}{\cot\theta} = 1$ **40.** $\dfrac{\csc\theta}{\sin\theta} - \dfrac{\cot\theta}{\tan\theta} = 1$

41. $\dfrac{1 - 2\cos^2 x}{\sin x \cos x} = \tan x - \cot x$

42. $4\sin x + \tan x = \dfrac{4 + \sec x}{\csc x}$

43. $\cos^3 x \csc^3 x \tan^3 x = \csc^2 x - \cot^2 x$

44. $\dfrac{1 + \tan x}{\sin x} - \sec x = \csc x$

45. $\sec x + \tan x + \cot x = \dfrac{1 + \sin x}{\cos x \sin x}$

46. $\csc\theta(\csc\theta - \cot\theta) = \dfrac{1}{1 + \cos\theta}$

47. $\dfrac{\cos\theta + \sin\theta}{1 + \tan\theta} = \cos\theta$ **48.** $\dfrac{\sec x - \cos x}{\tan x} = \sin x$

49. $(\tan x + \cot x)\sin x \cos x = 1$

50. $2\sin^4 x - 3\sin^2 x + 1 = \cos^2 x(1 - 2\sin^2 x)$

51. $\dfrac{\sin^4 x - \cos^4 x}{1 - \cot^4 x} = \sin^4 x$

52. $\dfrac{1}{2}\sin 5y\left(\dfrac{\sin 5y}{1 - \cos 5y} + \dfrac{1 - \cos 5y}{\sin 5y}\right) = 1$

53. $\sec x(\sec x - \cos x) + \dfrac{\cos x - \sin x}{\cos x} + \tan x = \sec^2 x$

54. $\dfrac{\cot 2y}{\sec 2y - \tan 2y} - \dfrac{\cos 2y}{\sec 2y + \tan 2y} = \sin 2y + \csc 2y$

55. $1 + \sin^2 x + \sin^4 x + \cdots = \sec^2 x$

56. $1 - \tan^2 x + \tan^4 x - \cdots = \cos^2 x \quad (-\frac{\pi}{4} < x < \frac{\pi}{4})$

In Exercises 57–60, solve the given problems involving trigonometric identities.

57. When designing a solar energy collector, it is necessary to account for the latitude and longitude of the location, the angle of the sun, and the angle of the collector. In doing this, the equation

$$\cos\theta = \cos A \cos B \cos C + \sin A \sin B$$

is used. If $\theta = 90°$, show that $\cos C = -\tan A \tan B$.

58. In studying the gravitational force between two objects, the expression $(r - R\cos\theta)^2 + (R\sin\theta)^2$ occurs. Show that this expression can be written as $r^2 - 2rR\cos\theta + R^2$.

59. Show that the length l of the straight brace shown in Fig. 20-3 can be found from the equation

$$l = \frac{a(1 + \tan\theta)}{\sin\theta}$$

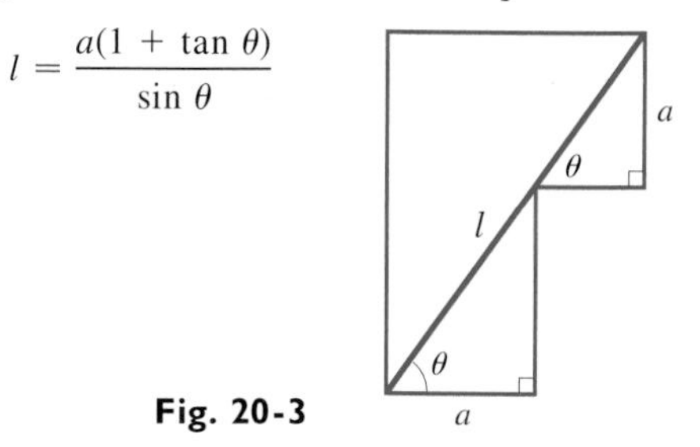

Fig. 20-3

60. In determining the path of least time between two points under certain circumstances, it is necessary to show that

$$\sqrt{\frac{1 + \cos\theta}{1 - \cos\theta}}\sin\theta = 1 + \cos\theta$$

Show this by transforming the left-hand side.

In Exercises 61–64, solve the given problems.

61. Show that $\sin^2 x(1 - \sec^2 x) + \cos^2 x(1 + \sec^4 x)$ has a constant value.

62. Show that $\cot y \csc y \sec y - \csc y \cos y \cot y$ has a constant value.

63. Prove that $\sec^2\theta + \csc^2\theta = \sec^2\theta\csc^2\theta$ by expressing each function in terms of its x, y, and r definition. See Example 9.

64. Prove that $\dfrac{\csc\theta}{\tan\theta + \cot\theta} = \cos\theta$ by expressing each function in terms of its x, y, and r definition. See Example 10.

In Exercises 65–68, use the given substitutions to show that the given equations are valid. In each, $0 < \theta < \frac{\pi}{2}$.

65. If $x = \cos\theta$, show that $\sqrt{1 - x^2} = \sin\theta$.

66. If $x = 3\sin\theta$, show that $\sqrt{9 - x^2} = 3\cos\theta$.

67. If $x = 2\tan\theta$, show that $\sqrt{4 + x^2} = 2\sec\theta$.

68. If $x = 4\sec\theta$, show that $\sqrt{x^2 - 16} = 4\tan\theta$.

20-2 Sine and Cosine of the Sum and Difference of Two Angles

There are other important relations among the trigonometric functions. The most important and useful relations are those that involve twice an angle and half an angle. To obtain these relations, we first derive the expressions for the sine and cosine of the sum and difference of two angles. These expressions will lead directly to the desired relations of double and half angles.

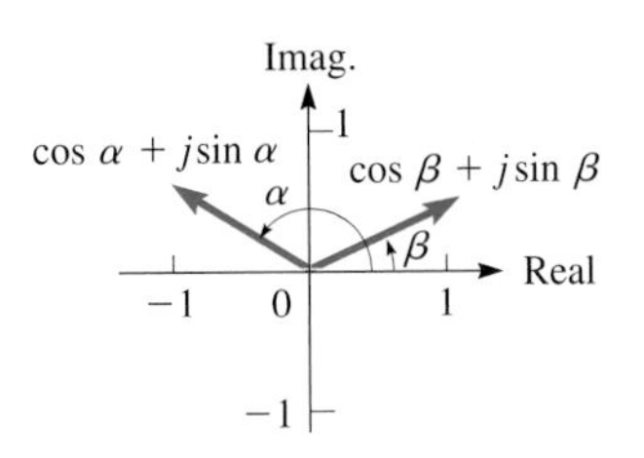

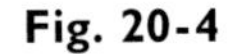
Fig. 20-4

Equation (12-13) gives the polar (or trigonometric) form of the product of two complex numbers. We can use this formula to derive the expression for the sine and cosine of the sum and difference of two angles. These expressions will lead directly to the desired relations of double and half angles in the next section.

Using Eq. (12-13) to find the product of the complex numbers $\cos\alpha + j\sin\alpha$ and $\cos\beta + j\sin\beta$, which are represented in Fig. 20-4, we have

$$(\cos\alpha + j\sin\alpha)(\cos\beta + j\sin\beta) = \cos(\alpha + \beta) + j\sin(\alpha + \beta)$$

Expanding the left side, and then switching sides, we have

$$\cos(\alpha + \beta) + j\sin(\alpha + \beta) = (\cos\alpha\cos\beta - \sin\alpha\sin\beta) + j(\sin\alpha\cos\beta + \cos\alpha\sin\beta)$$

Since two complex numbers are equal if their real parts are equal and their imaginary parts are equal, we have

$$\sin(\alpha + \beta) = \sin\alpha\cos\beta + \cos\alpha\sin\beta \qquad \text{(20-9)}$$

and

$$\cos(\alpha + \beta) = \cos\alpha\cos\beta - \sin\alpha\sin\beta \qquad \text{(20-10)}$$

EXAMPLE 1 ▸▸ Verify that $\sin 90° = 1$, by finding $\sin(60° + 30°)$.

$$\sin 90° = \sin(60° + 30°) = \sin 60°\cos 30° + \cos 60°\sin 30° \qquad \text{using Eq. (20-9)}$$

$$= \frac{\sqrt{3}}{2} \times \frac{\sqrt{3}}{2} + \frac{1}{2} \times \frac{1}{2} = \frac{3}{4} + \frac{1}{4} = 1 \qquad \text{for values, see Section 4-3}$$

We see that this agrees with the result found in Section 8-2.

CAUTION ▸ It should be obvious from this example that **$\sin(\alpha + \beta)$ *is* not *equal to* $\sin\alpha + \sin\beta$,** something that many students assume before they are familiar with the formulas and ideas of this section. If we used such a formula, we would get $\sin 90° = \frac{1}{2}\sqrt{3} + \frac{1}{2} = 1.366$ for the combination $(60° + 30°)$. This is not possible, since the values of the sine never exceed 1 in value. Also, if we used the combination $(45° + 45°)$, we would get 1.414, a different value for the same number, $\sin 90°$. ◂◂

EXAMPLE 2 ▸▸ Given that $\sin\alpha = \frac{5}{13}$ (α in the first quadrant) and $\sin\beta = -\frac{3}{5}$ (for β in the third quadrant), find $\cos(\alpha + \beta)$.

Since $\sin\alpha = \frac{5}{13}$ for α in the first quadrant, from Fig. 20-5 we have $\cos\alpha = \frac{12}{13}$. Also, since $\sin\beta = -\frac{3}{5}$ for β in the third quadrant, from Fig. 20-5 we also have $\cos\beta = -\frac{4}{5}$.

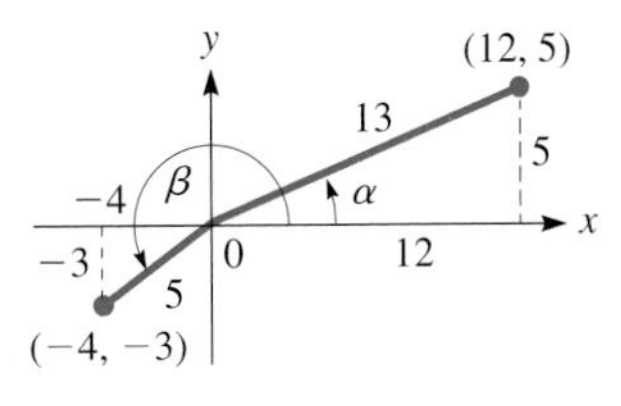

Fig. 20-5

Then, by using Eq. (20-10), we have

$$\cos(\alpha + \beta) = \cos\alpha\cos\beta - \sin\alpha\sin\beta$$

$$= \frac{12}{13}\left(-\frac{4}{5}\right) - \frac{5}{13}\left(-\frac{3}{5}\right)$$

$$= -\frac{48}{65} + \frac{15}{65} = -\frac{33}{65}$$ ◂◂

From Eqs. (20-9) and (20-10) we can easily find expressions for $\sin(\alpha - \beta)$ and $\cos(\alpha - \beta)$. This is done by finding $\sin(\alpha + (-\beta))$ and $\cos(\alpha + (-\beta))$. Thus we have

$$\sin(\alpha - \beta) = \sin(\alpha + (-\beta)) = \sin\alpha\cos(-\beta) + \cos\alpha\sin(-\beta)$$

Since $\cos(-\beta) = \cos\beta$ and $\sin(-\beta) = -\sin\beta$ (see Exercise 53 of Section 8-2), we have

$$\sin(\alpha - \beta) = \sin\alpha\cos\beta - \cos\alpha\sin\beta \qquad (20\text{-}11)$$

In the same manner, we find that

$$\cos(\alpha - \beta) = \cos\alpha\cos\beta + \sin\alpha\sin\beta \qquad (20\text{-}12)$$

EXAMPLE 3 ▸▸ Find $\cos 15°$ from $\cos(45° - 30°)$.

$$\cos 15° = \cos(45° - 30°) = \cos 45°\cos 30° + \sin 45°\sin 30° \qquad \text{using Eq. (20-12)}$$
$$= \frac{\sqrt{2}}{2} \times \frac{\sqrt{3}}{2} + \frac{\sqrt{2}}{2} \times \frac{1}{2} = \frac{\sqrt{6} + \sqrt{2}}{4}$$
$$= 0.9659$$ ◂◂

EXAMPLE 4 ▸▸ Reduce $\sin x\cos(x - y) - \cos x\sin(x - y)$ to a single term.

We could expand $\cos(x - y)$ and $\sin(x - y)$ and simplify. This would be tedious, and it is unnecessary. This expression fits the form of the right side of Eq. (20-11) with $\alpha = x$ and $\beta = x - y$. Thus

$$\sin x\cos(x - y) - \cos x\sin(x - y) = \sin[x - (x - y)]$$
$$= \sin y$$ ◂◂

EXAMPLE 5 ▸▸ Evaluate $\cos 23°\cos 67° - \sin 23°\sin 67°$.

We note that this expression fits the form of the right side of Eq. (20-10), so

$$\cos 23°\cos 67° - \sin 23°\sin 67° = \cos(23° + 67°)$$
$$= \cos 90°$$
$$= 0$$

NOTE ▸ Again, we are able to evaluate this expression by ***recognizing the form*** of the given expression. Evaluation by a calculator will verify the result. ◂◂

By using Eqs. (20-9) and (20-10), we can determine expressions for $\tan(\alpha + \beta)$, $\cot(\alpha + \beta)$, $\sec(\alpha + \beta)$, and $\csc(\alpha + \beta)$. These expressions are used less than those for the sine and cosine, and therefore we shall not derive them here, although the expression for $\tan(\alpha + \beta)$ is found in the exercises at the end of this section. By using Eqs. (20-11) and (20-12), we can find similar expressions for the functions of $(\alpha - \beta)$.

Certain trigonometric identities can be proven by using the formulas derived in this section. The following examples illustrate this use of these formulas.

EXAMPLE 6 ▸▸ Prove that $\sin(180° + x) = -\sin x$.

By using Eq. (20-9) we have

$$\sin(180° + x) = \sin 180° \cos x + \cos 180° \sin x$$

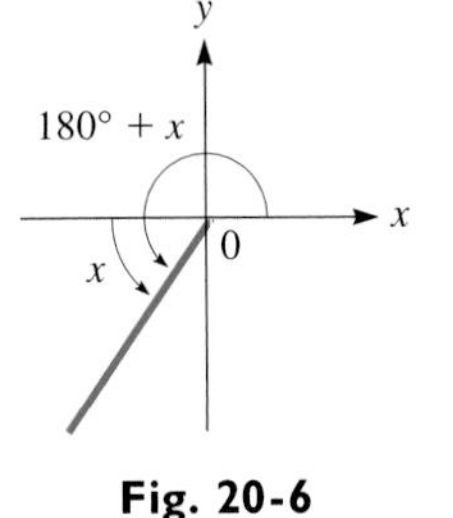

Fig. 20-6

Since $\sin 180° = 0$ and $\cos 180° = -1$, we have

$$\sin 180° \cos x + \cos 180° \sin x = (0) \cos x + (-1) \sin x$$

or

$$\sin(180° + x) = -\sin x$$

Although x may or may not be an acute angle, we see that this agrees with the results in Section 8-2 for the sine of a third-quadrant angle if x is acute. See Fig. 20-6. ◂◂

EXAMPLE 7 ▸▸ Show that $\dfrac{\sin(\alpha - \beta)}{\sin\alpha \sin\beta} = \cot\beta - \cot\alpha$.

$$\frac{\sin(\alpha - \beta)}{\sin\alpha\sin\beta} = \frac{\sin\alpha\cos\beta - \cos\alpha\sin\beta}{\sin\alpha\sin\beta} \qquad \text{using Eq. (20-11)}$$

$$= \frac{\sin\alpha\cos\beta}{\sin\alpha\sin\beta} - \frac{\cos\alpha\sin\beta}{\sin\alpha\sin\beta}$$

$$= \frac{\cos\beta}{\sin\beta} - \frac{\cos\alpha}{\sin\alpha} = \cot\beta - \cot\alpha \qquad \text{using Eq. (20-5)}$$ ◂◂

EXAMPLE 8 ▸▸ Show that $\sin\left(\dfrac{\pi}{4} + x\right)\cos\left(\dfrac{\pi}{4} + x\right) = \dfrac{1}{2}(\cos^2 x - \sin^2 x)$.

The solution is as follows:

$$\sin\left(\frac{\pi}{4} + x\right)\cos\left(\frac{\pi}{4} + x\right) = \left(\sin\frac{\pi}{4}\cos x + \cos\frac{\pi}{4}\sin x\right)\left(\cos\frac{\pi}{4}\cos x - \sin\frac{\pi}{4}\sin x\right) \qquad \text{using Eqs. (20-9) and (20-10)}$$

$$= \sin\frac{\pi}{4}\cos\frac{\pi}{4}\cos^2 x - \sin^2\frac{\pi}{4}\sin x\cos x + \cos^2\frac{\pi}{4}\sin x\cos x - \sin^2 x\sin\frac{\pi}{4}\cos\frac{\pi}{4} \qquad \text{expanding}$$

$$= \frac{\sqrt{2}}{2}\frac{\sqrt{2}}{2}\cos^2 x - \left(\frac{\sqrt{2}}{2}\right)^2\sin x\cos x + \left(\frac{\sqrt{2}}{2}\right)^2\sin x\cos x - \frac{\sqrt{2}}{2}\frac{\sqrt{2}}{2}\sin^2 x \qquad \text{evaluating}$$

$$= \frac{1}{2}\cos^2 x - \frac{1}{2}\sin^2 x = \frac{1}{2}(\cos^2 x - \sin^2 x)$$ ◂◂

EXAMPLE 9 ▸▸ In analyzing the motion of an object at the end of a spring, it is stated that

$$\sin\omega t = \sin(\omega t + \alpha)\cos\alpha - \cos(\omega t + \alpha)\sin\alpha$$

Show that this is correct.

If we let $x = \omega t + \alpha$, we note that the right side of the equation becomes $\sin x\cos\alpha - \cos x\sin\alpha$, which is the form for $\sin(x - \alpha)$. By replacing x with $\omega t + \alpha$, we obtain $\sin(\omega t + \alpha - \alpha)$, which is $\sin\omega t$. These steps (with $x = \omega t + \alpha$) are shown below.

$$\sin\omega t = \sin x\cos\alpha - \cos x\sin\alpha = \sin(x - \alpha)$$

$$= \sin(\omega t + \alpha - \alpha) = \sin\omega t$$

Therefore, the original equation has been shown to be true.

Again, the proper recognition of a basic form leads to the solution. ◂◂

Exercises 20–2

In Exercises 1–4, determine the values of the given functions as indicated.

1. Find $\sin 105°$ by using $105° = 60° + 45°$.
2. Find $\cos 75°$ by using $75° = 30° + 45°$.
3. Find $\cos 15°$ by using $15° = 60° - 45°$.
4. Find $\sin 15°$ by using $15° = 45° - 30°$.

In Exercises 5–8, evaluate the given functions with the following information: $\sin\alpha = \frac{4}{5}$ *(in first quadrant) and* $\cos\beta = -\frac{12}{13}$ *(in second quadrant).*

5. $\sin(\alpha + \beta)$
6. $\cos(\beta - \alpha)$
7. $\cos(\alpha + \beta)$
8. $\sin(\alpha - \beta)$

In Exercises 9–16, reduce each of the given expressions to a single term. Expansion of any term is not necessary; proper recognition of the form of the expression leads to the proper result.

9. $\sin x \cos 2x + \sin 2x \cos x$
10. $\sin 3x \cos x - \sin x \cos 3x$
11. $\cos(x + y)\cos y + \sin(x + y)\sin y$
12. $\cos(2x - y)\cos y - \sin(2x - y)\sin y$
13. $\cos 1 \cos(1 - x) - \sin 1 \sin(1 - x)$
14. $\sin x \cos(x + 1) + \cos x \sin(x + 1)$
15. $\sin 3x \cos(3x - \pi) - \cos 3x \sin(3x - \pi)$
16. $\cos(x + \pi)\cos(x - \pi) + \sin(x + \pi)\sin(x - \pi)$

In Exercises 17–20, evaluate each of the given expressions. Proper recognition of the given form leads to the result. Verify each result by use of a calculator.

17. $\sin 122° \cos 32° - \cos 122° \sin 32°$
18. $\cos 250° \cos 70° + \sin 250° \sin 70°$
19. $\cos 312° \cos 48° - \sin 312° \sin 48°$
20. $\sin 56° \cos 124° + \cos 56° \sin 124°$

In Exercises 21–36, prove the given identities.

21. $\sin(180° - x) = \sin x$
22. $\cos(180° - x) = -\cos x$
23. $\cos(-x) = \cos x$ (*Hint:* $-x = 0 - x$.)
24. $\sin(-x) = -\sin x$
25. $\sin(270° - x) = -\cos x$
26. $\sin(90° + x) = \cos x$
27. $\cos(\frac{\pi}{2} - x) = \sin x$
28. $\cos(\frac{3\pi}{2} + x) = \sin x$
29. $\cos(30° + x) = \dfrac{\sqrt{3}\cos x - \sin x}{2}$
30. $\sin(120° - x) = \dfrac{\sqrt{3}\cos x + \sin x}{2}$
31. $\sin(\frac{\pi}{4} + x) = \dfrac{\sin x + \cos x}{\sqrt{2}}$
32. $\cos(\frac{\pi}{3} + x) = \dfrac{\cos x - \sqrt{3}\sin x}{2}$
33. $\sin(x + y)\sin(x - y) = \sin^2 x - \sin^2 y$
34. $\cos(x + y)\cos(x - y) = \cos^2 x - \sin^2 y$
35. $\cos(\alpha + \beta) + \cos(\alpha - \beta) = 2\cos\alpha\cos\beta$
36. $\cos(x - y) + \sin(x + y) = (\cos x + \sin x)(\cos y + \sin y)$

In Exercises 37–40, additional trigonometric identities are shown. Derive these in the indicated manner. Equations (20-14), (20-15), and (20-16) are known as the product formulas.

37. By dividing Eq. (20-9) by Eq (20-10), show that

$$\tan(\alpha + \beta) = \frac{\tan\alpha + \tan\beta}{1 - \tan\alpha\tan\beta} \quad (20\text{-}13)$$

(*Hint*: Divide numerator and denominator by $\cos\alpha\cos\beta$.)

38. By adding Eqs. (20-9) and (20-11), derive the equation

$$\sin\alpha\cos\beta = \tfrac{1}{2}[\sin(\alpha + \beta) + \sin(\alpha - \beta)] \quad (20\text{-}14)$$

39. By adding Eqs. (20-10) and (20-12), derive the equation

$$\cos\alpha\cos\beta = \tfrac{1}{2}[\cos(\alpha + \beta) + \cos(\alpha - \beta)] \quad (20\text{-}15)$$

40. By subtracting Eq. (20-10) from Eq. (20-12), derive

$$\sin\alpha\sin\beta = \tfrac{1}{2}[\cos(\alpha - \beta) - \cos(\alpha + \beta)] \quad (20\text{-}16)$$

In Exercises 41–44, additional trigonometric identities are shown. Derive them by letting $\alpha + \beta = x$ *and* $\alpha - \beta = y$*, which leads to* $\alpha = \frac{1}{2}(x + y)$ *and* $\beta = \frac{1}{2}(x - y)$*. The resulting equations are known as the factor formulas.*

41. Use Eq. (20-14) and the substitutions above to derive the equation

$$\sin x + \sin y = 2\sin\tfrac{1}{2}(x + y)\cos\tfrac{1}{2}(x - y) \quad (20\text{-}17)$$

42. Use Eqs. (20-9) and (20-11) and the substitutions above to derive the equation

$$\sin x - \sin y = 2\sin\tfrac{1}{2}(x - y)\cos\tfrac{1}{2}(x + y) \quad (20\text{-}18)$$

43. Use Eq. (20-15) and the substitutions above to derive the equation

$$\cos x + \cos y = 2\cos\tfrac{1}{2}(x + y)\cos\tfrac{1}{2}(x - y) \quad (20\text{-}19)$$

44. Use Eq. (20-16) and the substitutions above to derive the equation

$$\cos x - \cos y = -2\sin\tfrac{1}{2}(x + y)\sin\tfrac{1}{2}(x - y) \quad (20\text{-}20)$$

In Exercises 45–48, use the equations of this section to solve the given problems.

45. An alternating electric current i is given by the equation $i = i_0 \sin(\omega t + \alpha)$. Show that this can be written as $i = i_1 \sin \omega t + i_2 \cos \omega t$, where $i_1 = i_0 \cos \alpha$ and $i_2 = i_0 \sin \alpha$.

46. A weight w is held in equilibrium by forces F and T as shown in Fig. 20-7. Equations relating w, F, and T are

$$F \cos \theta = T \sin \alpha$$
$$w + F \sin \theta = T \cos \alpha$$

Show that $w = \dfrac{T \cos(\theta + \alpha)}{\cos \theta}$.

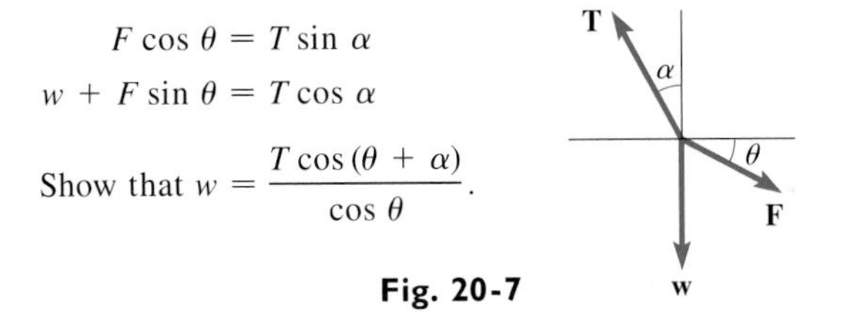

Fig. 20-7

47. For the two bevel gears shown in Fig. 20-8, the equation $\tan \alpha = \dfrac{\sin \beta}{R + \cos \beta}$ is used. Here R is the ratio of gear 1 to gear 2. Show that $R = \dfrac{\sin(\beta - \alpha)}{\sin \alpha}$.

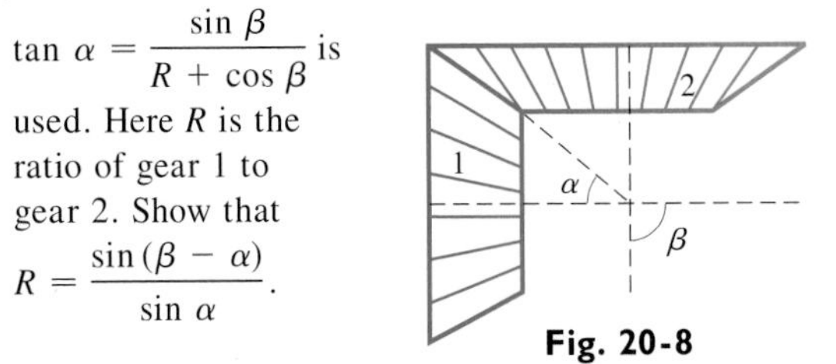

Fig. 20-8

48. In the analysis of the angles of incidence i and reflection r of a light ray subject to certain conditions, the following expression is found: $E_2\left(\dfrac{\tan r}{\tan i} + 1\right) = E_1\left(\dfrac{\tan r}{\tan i} - 1\right)$. Show that $E_2 = E_1 \dfrac{\sin(r - i)}{\sin(r + i)}$.

20-3 Double-Angle Formulas

If we let $\beta = \alpha$ in Eqs. (20-9) and (20-10), we can derive the important double-angle formulas. By making this substitution in Eq. (20-9), we have

$$\sin(\alpha + \alpha) = \sin(2\alpha) = \sin \alpha \cos \alpha + \cos \alpha \sin \alpha = 2 \sin \alpha \cos \alpha$$

Using the same substitution in Eq. (20-10), we have

$$\cos(\alpha + \alpha) = \cos \alpha \cos \alpha - \sin \alpha \sin \alpha = \cos^2 \alpha - \sin^2 \alpha$$

Then using the basic identity (20-6), other forms of this last equation may be derived. Summarizing these forms, we have

$$\sin 2\alpha = 2 \sin \alpha \cos \alpha \quad (20\text{-}21)$$
$$\cos 2\alpha = \cos^2 \alpha - \sin^2 \alpha \quad (20\text{-}22)$$
$$= 2\cos^2 \alpha - 1 \quad (20\text{-}23)$$
$$= 1 - 2\sin^2 \alpha \quad (20\text{-}24)$$

These double-angle formulas are widely used in applications of trigonometry, especially in calculus. They should be recognized quickly in any of the above forms.

EXAMPLE 1 ▸▸ (a) If $\alpha = 30°$, we have

$$\cos 60° = \cos 2(30°) = \cos^2 30° - \sin^2 30° \quad \text{using Eq. (20-22)}$$

(b) If $\alpha = 3x$, we have

$$\sin 6x = \sin 2(3x) = 2 \sin 3x \cos 3x \quad \text{using Eq. (20-21)}$$

(c) If $2\alpha = x$, we may write $\alpha = x/2$, which means that

$$\sin x = \sin 2\left(\frac{x}{2}\right) = 2 \sin \frac{x}{2} \cos \frac{x}{2} \quad \text{using Eq. (20-21)}$$ ◂◂

EXAMPLE 2 ▸▸ Simplify the expression $\cos^2 2x - \sin^2 2x$.

Since this is the difference of the square of the cosine of an angle and the square of the sine of the same angle, it fits the right side of Eq. (20-22). Therefore, letting $\alpha = 2x$, we have

$$\cos^2 2x - \sin^2 2x = \cos 2(2x) = \cos 4x$$ ◂◂

See the chapter introduction.

EXAMPLE 3 ▸▸ To find the area A of a right triangular piece of land, a surveyor may use the formula $A = \frac{1}{4}c^2 \sin 2\theta$, where c is the hypotenuse and θ is *either* of the acute angles. Derive this formula.

In Fig. 20-9 we see that $\sin\theta = a/c$ and $\cos\theta = b/c$, which gives us

$$a = c\sin\theta \quad \text{and} \quad b = c\cos\theta$$

The area is given by $A = \frac{1}{2}ab$, which leads to the solution

$$\begin{aligned} A &= \frac{1}{2}ab = \frac{1}{2}(c\sin\theta)(c\cos\theta) \\ &= \frac{1}{2}c^2\sin\theta\cos\theta = \frac{1}{2}c^2\left(\frac{1}{2}\sin 2\theta\right) \qquad \text{using Eq. (20-21)} \\ &= \frac{1}{4}c^2\sin 2\theta \end{aligned}$$

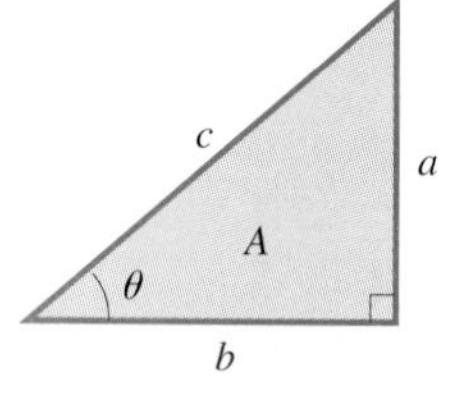

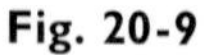
Fig. 20-9

In using Eq. (20-21), we divided both sides by 2 to get $\sin\theta\cos\theta = \frac{1}{2}\sin 2\theta$.

If we had labeled the upper acute angle in Fig. 20-9 as θ, we would then have $a = c\cos\theta$ and $b = c\sin\theta$. Substituting these values in the formula for the area gives the same solution. ◂◂

EXAMPLE 4 ▸▸ (a) Verifying the values of $\sin 90°$ using the functions of 45°, we have

$$\sin 90° = \sin 2(45°) = 2\sin 45°\cos 45° = 2\left(\frac{\sqrt{2}}{2}\right)\left(\frac{\sqrt{2}}{2}\right) = 1 \qquad \text{using Eq. (20-21)}$$

(b) Using Eq. (20-21), $\sin 142° = 2\sin 71°\cos 71°$. Verifying this result by using a calculator, we have

$$\sin 142° = 0.615\,661\,475\,3 \quad \text{and} \quad 2\sin 71°\cos 71° = 0.615\,661\,475\,3$$ ◂◂

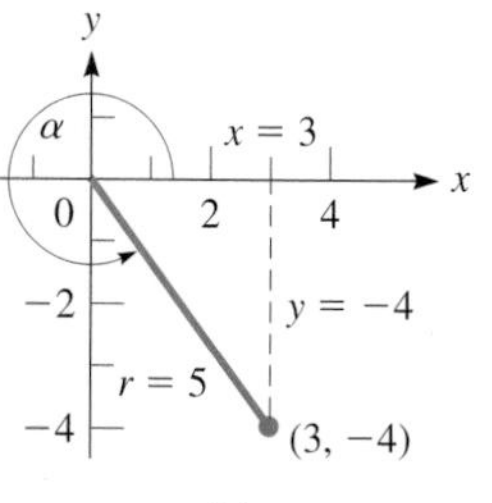

(a)

EXAMPLE 5 ▸▸ Knowing that $\cos\alpha = 3/5$ for an angle in the fourth quadrant, we then determine from Fig. 20-10(a) that

$$\sin\alpha = -4/5$$

Therefore, we have

$$\begin{aligned} \sin 2\alpha &= 2\sin\alpha\cos\alpha \qquad \text{Eq. (20-21)} \\ &= 2\left(-\frac{4}{5}\right)\left(\frac{3}{5}\right) = -\frac{24}{25} \end{aligned}$$

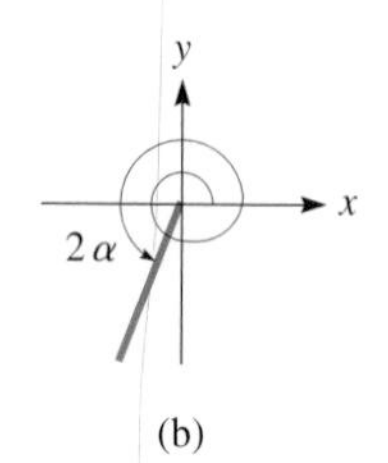

(b)

Fig. 20-10

In Fig. 20-10(b) the angle 2α is shown. It is a third-quadrant angle, which verifies the sign of the result. (Since $\cos\alpha = 3/5$, $\alpha \approx 307°$ and $2\alpha \approx 614°$, which is a third-quadrant angle.) ◂◂

EXAMPLE 6 ▸▸ Prove the identity $\dfrac{2}{1 + \cos 2x} = \sec^2 x$.

$$\frac{2}{1 + \cos 2x} = \frac{2}{1 + (2\cos^2 x - 1)} \qquad \text{using Eq (20-23)}$$

$$= \frac{2}{2\cos^2 x} = \sec^2 x \qquad \text{using Eq. (20-2)}$$ ◂◂

EXAMPLE 7 ▸▸ Show that $\dfrac{\sin 3x}{\sin x} + \dfrac{\cos 3x}{\cos x} = 4\cos 2x$.

Since the left side is the more complex side, we will change it to the form on the right.

$$\frac{\sin 3x}{\sin x} + \frac{\cos 3x}{\cos x} = \frac{\sin 3x \cos x + \cos 3x \sin x}{\sin x \cos x} \qquad \text{combining fractions}$$

$$= \frac{\sin(3x + x)}{\frac{1}{2}\sin 2x} \qquad \begin{matrix}\longleftarrow \text{using Eq. (20-9)} \\ \longleftarrow \text{using Eq. (20-21)}\end{matrix}$$

$$= \frac{2\sin 4x}{\sin 2x} = \frac{2(2\sin 2x\cos 2x)}{\sin 2x} \qquad \text{using Eq. (20-21)}$$

$$= 4\cos 2x$$

Therefore, the expression is shown to be valid. ◂◂

Exercises 20–3

In Exercises 1–4, determine the values of the indicated functions in the given manner.

1. Find sin 60° by using the functions of 30°.
2. Find sin 120° by using the functions of 60°.
3. Find cos 120° by using the functions of 60°.
4. Find cos 60° by using the functions of 30°.

In Exercises 5–8, use a calculator to verify the values found by using the double-angle formulas.

5. Find sin 258° directly and by using functions of 129°.
6. Find sin 84° directly and by using functions of 42°.
7. Find cos 96° directly and by using functions of 48°.
8. Find cos 276° directly and by using functions of 138°.

In Exercises 9–12, evaluate the indicated functions with the given information.

9. Find $\sin 2x$ if $\cos x = \frac{4}{5}$ (in first quadrant).
10. Find $\cos 2x$ if $\sin x = -\frac{12}{13}$ (in third quadrant).
11. Find $\cos 2x$ if $\tan x = -0.5$ (in second quadrant).
12. Find $\sin 4x$ if $\sin x = 0.6$ (in first quadrant).

In Exercises 13–20, simplify the given expressions. Expansion of any given term is not necessary; proper recognition of the form of the expression leads to the proper result.

13. $4\sin 4x \cos 4x$
14. $4\sin^2 x \cos^2 x$
15. $1 - 2\sin^2 4x$
16. $\sin^2 4x - \cos^2 4x$
17. $2\cos^2 \frac{1}{2}x - 1$
18. $2\sin \frac{1}{2}x \cos \frac{1}{2}x$
19. $4\sin^2 2x - 2$
20. $\cos 3x \sin 3x$

In Exercises 21–36, prove the given identities.

21. $\cos^2 \alpha - \sin^2 \alpha = 2\cos^2 \alpha - 1$
22. $\cos^2 \alpha - \sin^2 \alpha = 1 - 2\sin^2 \alpha$
23. $\dfrac{\cos x - \tan x \sin x}{\sec x} = \cos 2x$
24. $2\tan x \cos x = \sec x \sin 2x$
25. $\cos^4 x - \sin^4 x = \cos 2x$
26. $(\sin x + \cos x)^2 = 1 + \sin 2x$
27. $\dfrac{\sin 4\theta}{\sin 2\theta} = 2\cos 2\theta$
28. $2 + \dfrac{\cos 2\theta}{\sin^2 \theta} = \csc^2 \theta$
29. $\dfrac{\sin 2\theta}{1 + \cos 2\theta} = \tan \theta$
30. $\dfrac{2\tan \alpha}{1 + \tan^2 \alpha} = \sin 2\alpha$

31. $\dfrac{1 - \tan^2 x}{\sec^2 x} = \cos 2x$ **32.** $1 - \cos 2\theta = \dfrac{2}{1 + \cot^2 \theta}$

33. $2 \csc 2x \tan x = \sec^2 x$

34. $2 \sin x + \sin 2x = \dfrac{2 \sin^3 x}{1 - \cos x}$

35. $\dfrac{\sin 3x}{\sin x} - \dfrac{\cos 3x}{\cos x} = 2$

36. $\dfrac{\cos 3x}{\sin x} + \dfrac{\sin 3x}{\cos x} = 2 \cot 2x$

In Exercises 37 and 38, prove the given identities. Explain your general approach to the proof.

37. $\sin 3x = 3 \cos^2 x \sin x - \sin^3 x$

38. $\cos 3x = \cos^3 x - 3 \sin^2 x \cos x$

In Exercises 39–44, solve the given problems.

39. In Exercise 37 of Section 20-2, let $\beta = \alpha$, and show that

$$\tan 2\alpha = \frac{2 \tan \alpha}{1 - \tan^2 \alpha} \qquad \text{(20-25)}$$

40. The cross section of a radio wave reflector is defined by $x = \cos 2\theta$, $y = \sin \theta$. Find the relation between x and y by eliminating θ.

41. To find the horizontal range R of a projectile, the equation $R = vt \cos \alpha$ is used, where α is the angle between the line of fire and the horizontal, v is the initial velocity of the projectile, and t is the time of flight. It can be shown that $t = (2v \sin \alpha)/g$, where g is the acceleration due to gravity. Show that $R = (v^2 \sin 2\alpha)/g$.

42. In the theory of reflection of light waves, the expression $\sin(\frac{\pi}{2} - 2\theta)$ arises. Show that this expression can be written as $1 - 2 \sin^2 \theta$.

43. The instantaneous electric power p in an inductor is given by the equation $p = vi \sin \omega t \sin(\omega t - \frac{\pi}{2})$. Show that this equation can be written as $p = -\frac{1}{2} vi \sin 2\omega t$.

44. In the study of the stress at a point in a bar, the equation $s = a \cos^2 \theta + b \sin^2 \theta - 2t \sin \theta \cos \theta$ arises. Show that this equation can be written as $s = \frac{1}{2}(a + b) + \frac{1}{2}(a - b) \cos 2\theta - t \sin 2\theta$.

20-4 Half-Angle Formulas

If we let $\theta = \alpha/2$ in the identity $\cos 2\theta = 1 - 2 \sin^2 \theta$ and then solve for $\sin(\alpha/2)$,

$$\sin \frac{\alpha}{2} = \pm \sqrt{\frac{1 - \cos \alpha}{2}} \qquad \text{(20-26)}$$

Also, with the same substitution in the identity $\cos 2\theta = 2 \cos^2 \theta - 1$, which is then solved for $\cos(\alpha/2)$, we have

$$\cos \frac{\alpha}{2} = \pm \sqrt{\frac{1 + \cos \alpha}{2}} \qquad \text{(20-27)}$$

CAUTION ▶ In each of Eqs. (20-26) and (20-27), ***the sign chosen depends on the quadrant in which $\frac{\alpha}{2}$ lies.***

We can use these half-angle formulas to find values of the functions of angles that are half of those for which the functions are known. The following examples illustrate how these identities are used in evaluations and in identities.

EXAMPLE 1 ▸▸ We can find sin 15° by using the relation

$$\sin 15° = \sqrt{\frac{1 - \cos 30°}{2}} \qquad \text{using Eq. (20-26)}$$

$$= \sqrt{\frac{1 - 0.8660}{2}} = 0.2588$$

Here the plus sign is used, since 15° is in the first quadrant. ◂◂

EXAMPLE 2 ▸▸ We can find cos 165° by use of the relation

$$\cos 165° = -\sqrt{\frac{1+\cos 330°}{2}} \qquad \text{using Eq. (20-27)}$$

$$= -\sqrt{\frac{1+0.8660}{2}} = -0.9659$$

Here the minus sign is used, since 165° is in the second quadrant, and the cosine of a second-quadrant angle is negative. ◂◂

EXAMPLE 3 ▸▸ Simplify $\sqrt{\frac{1-\cos 114°}{2}}$ by expressing the result in terms of one-half the given angle. Then, using a calculator, show that the values are equal.

We note that the given expression fits the form of the right side of Eq. (20-26), which means that

$$\sqrt{\frac{1-\cos 114°}{2}} = \sin \tfrac{1}{2}(114°) = \sin 57°$$

Using a calculator shows that

$$\sqrt{\frac{1-\cos 114°}{2}} = 0.838\,670\,567\,9 \quad \text{and} \quad \sin 57° = 0.838\,670\,567\,9$$

which verifies the equation for these values. ◂◂

EXAMPLE 4 ▸▸ Simplify the expression $\sqrt{\frac{9+9\cos 6x}{2}}$.

$$\sqrt{\frac{9+9\cos 6x}{2}} = \sqrt{\frac{9(1+\cos 6x)}{2}} = 3\sqrt{\frac{1+\cos 6x}{2}}$$

$$= 3\cos \tfrac{1}{2}(6x) \qquad \text{using Eq. (20-27) with } \alpha = 6x$$

$$= 3\cos 3x$$

Noting the original expression, we see that cos 3x cannot be negative. ◂◂

EXAMPLE 5 ▸▸ In the kinetic theory of gases, the expression $\sqrt{(1-\cos\alpha)^2+\sin^2\alpha}$ is found. Show that this expression equals $2\sin\frac{1}{2}\alpha$.

$$\sqrt{(1-\cos\alpha)^2+\sin^2\alpha} = \sqrt{1-2\cos\alpha+\cos^2\alpha+\sin^2\alpha} \qquad \text{expanding}$$

$$= \sqrt{1-2\cos\alpha+1} \qquad \text{using Eq. (20-6)}$$

$$= \sqrt{2-2\cos\alpha}$$

$$= \sqrt{2(1-\cos\alpha)} \qquad \text{factoring}$$

This last expression is very similar to that for $\sin\frac{1}{2}\alpha$, except that no 2 appears in the denominator. Therefore, multiplying the numerator and the denominator under the radical by 2 leads to the solution.

$$\sqrt{2(1-\cos\alpha)} = \sqrt{\frac{4(1-\cos\alpha)}{2}} = 2\sqrt{\frac{1-\cos\alpha}{2}}$$

$$= 2\sin\tfrac{1}{2}\alpha \qquad \text{using Eq. (20-26)}$$

Noting the original expression, we see that $\sin\frac{1}{2}\alpha$ cannot be negative. ◂◂

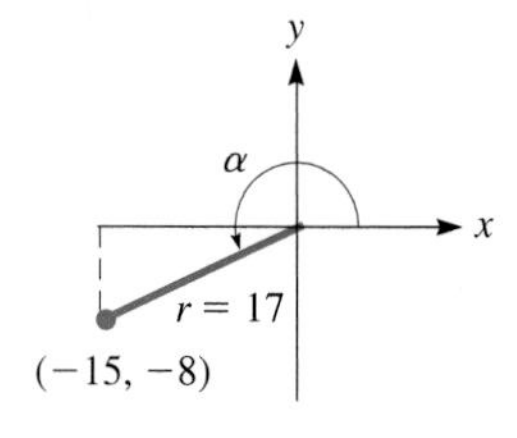

Fig. 20-11

EXAMPLE 6 ▸▸ Given that $\tan\alpha = \frac{8}{15}(180° < \alpha < 270°)$, find $\cos\left(\frac{\alpha}{2}\right)$.

Knowing that $\tan\alpha = \frac{8}{15}$ for a third-quadrant angle, we determine from Fig. 20-11 that $\cos\alpha = -\frac{15}{17}$. This means

$$\cos\frac{\alpha}{2} = -\sqrt{\frac{1+(-15/17)}{2}} = -\sqrt{\frac{2}{34}} \quad \text{using Eq. (20-27)}$$

$$= -\frac{1}{17}\sqrt{17} = -0.2425$$

Since $180° < \alpha < 270°$, we know that $90° < \frac{\alpha}{2} < 135°$, and therefore $\frac{\alpha}{2}$ is in the second quadrant. Since the cosine is negative for second-quadrant angles, we use the negative value of the radical. ◂◂

EXAMPLE 7 ▸▸ Prove the identity $\sec\frac{\alpha}{2} + \csc\frac{\alpha}{2} = \dfrac{2\left(\sin\frac{\alpha}{2} + \cos\frac{\alpha}{2}\right)}{\sin\alpha}$.

By expressing $\sin\alpha$ as $2\sin\frac{\alpha}{2}\cos\frac{\alpha}{2}$, we have

$$\frac{2\left(\sin\frac{\alpha}{2} + \cos\frac{\alpha}{2}\right)}{\sin\alpha} = \frac{2\left(\sin\frac{\alpha}{2} + \cos\frac{\alpha}{2}\right)}{2\sin\frac{\alpha}{2}\cos\frac{\alpha}{2}} \quad \text{see Example 1(c) of Section 20-3}$$

$$= \frac{\sin\frac{\alpha}{2}}{\sin\frac{\alpha}{2}\cos\frac{\alpha}{2}} + \frac{\cos\frac{\alpha}{2}}{\sin\frac{\alpha}{2}\cos\frac{\alpha}{2}}$$

$$= \frac{1}{\cos\frac{\alpha}{2}} + \frac{1}{\sin\frac{\alpha}{2}} = \sec\frac{\alpha}{2} + \csc\frac{\alpha}{2} \quad \text{using Eqs. (20-2) and (20-1)}$$

◂◂

EXAMPLE 8 ▸▸ We can find relations for the other functions of $\frac{\alpha}{2}$ by expressing these functions in terms of $\sin\left(\frac{\alpha}{2}\right)$ and $\cos\left(\frac{\alpha}{2}\right)$. For example,

$$\sec\frac{\alpha}{2} = \frac{1}{\cos\frac{\alpha}{2}} = \pm\frac{1}{\sqrt{\frac{1+\cos\alpha}{2}}} \quad \text{using Eq. (20-27)}$$

$$= \pm\sqrt{\frac{2}{1+\cos\alpha}}$$ ◂◂

EXAMPLE 9 ▸▸ Show that $2\cos^2\frac{x}{2} - \cos x = 1$.

The first step is to substitute for $\cos\frac{x}{2}$, which will result in each term on the left being in terms of x, and no $\frac{x}{2}$ terms will exist. This might allow us to combine terms. So we perform this operation, and we have for the left side

$$2\cos^2\frac{x}{2} - \cos x = 2\left(\frac{1+\cos x}{2}\right) - \cos x \quad \text{using Eq. (20-27) with both sides squared}$$

$$= 1 + \cos x - \cos x = 1$$ ◂◂

Exercises 20–4

In Exercises 1–4, use the half-angle formulas to evaluate the given functions.

1. $\cos 15°$

2. $\sin 22.5°$

3. $\sin 75°$

4. $\cos 112.5°$

In Exercises 5–8, simplify the given expressions by giving the results in terms of one-half the given angle. Then use a calculator to verify the result.

5. $\sqrt{\dfrac{1 - \cos 236°}{2}}$

6. $\sqrt{\dfrac{1 + \cos 98°}{2}}$

7. $\sqrt{1 + \cos 164°}$

8. $\sqrt{2 - 2\cos 328°}$

In Exercises 9–12, use the half-angle formulas to simplify the given expressions.

9. $\sqrt{\dfrac{1 - \cos 6x}{2}}$

10. $\sqrt{\dfrac{4 + 4\cos 8\beta}{2}}$

11. $\sqrt{8 + 8\cos 4x}$

12. $\sqrt{2 - 2\cos 16x}$

In Exercises 13–16, evaluate the indicated functions with the information given.

13. Find the value of $\sin(\frac{\alpha}{2})$ if $\cos\alpha = \frac{12}{13}$ $(0° < \alpha < 90°)$.

14. Find the value of $\cos(\frac{\alpha}{2})$ if $\sin\alpha = -\frac{4}{5}$ $(180° < \alpha < 270°)$.

15. Find the value of $\cos(\frac{\alpha}{2})$ if $\tan\alpha = -0.2917$ $(90° < \alpha < 180°)$.

16. Find the value of $\sin(\frac{\alpha}{2})$ if $\cos\alpha = 0.4706$ $(270° < \alpha < 360°)$.

In Exercises 17–20, derive the required expressions.

17. Derive an expression for $\csc(\frac{\alpha}{2})$ in terms of $\cos\alpha$.

18. Derive an expression for $\sec(\frac{\alpha}{2})$ in terms of $\sec\alpha$.

19. Derive an expression for $\tan(\frac{\alpha}{2})$ in terms of $\sin\alpha$ and $\cos\alpha$.

20. Derive an expression for $\cot(\frac{\alpha}{2})$ in terms of $\sin\alpha$ and $\cos\alpha$.

In Exercises 21–28, prove the given identities.

21. $\sin\dfrac{\alpha}{2} = \dfrac{1 - \cos\alpha}{2\sin\dfrac{\alpha}{2}}$

22. $2\cos\dfrac{x}{2} = (1 + \cos x)\sec\dfrac{x}{2}$

23. $2\sin^2\dfrac{x}{2} + \cos x = 1$

24. $2\cos^2\dfrac{\theta}{2}\sec\theta = \sec\theta + 1$

25. $\cos\dfrac{\theta}{2} = \dfrac{\sin\theta}{2\sin\dfrac{\theta}{2}}$

26. $\cos^2\dfrac{x}{2}\left[1 + \left(\dfrac{\sin x}{1 + \cos x}\right)^2\right] = 1$

27. $2\sin^2\dfrac{\alpha}{2} - \cos^2\dfrac{\alpha}{2} = \dfrac{1 - 3\cos\alpha}{2}$

28. $\tan\dfrac{\alpha}{2} = \dfrac{\sin\alpha}{1 + \cos\alpha}$

In Exercises 29–32, use the half-angle formulas to solve the given problems.

29. In electronics, in order to find the *root-mean-square current* in a circuit, it is necessary to express $\sin^2 \omega t$ in terms of $\cos 2\omega t$. Show how this is done.

30. In studying interference patterns of radio signals, the expression $2E^2 - 2E^2\cos(\pi - \theta)$ arises. Show that this can be written as $4E^2\cos^2(\theta/2)$.

31. The index of refraction n, the angle of a prism A, and the minimum angle of deflection ϕ are related by

$$n = \frac{\sin\frac{1}{2}(A + \phi)}{\sin\frac{1}{2}A}$$

See Fig. 20-12. Show that an equivalent expression is

$$n = \sqrt{\frac{1 - \cos A\cos\phi + \sin A\sin\phi}{1 - \cos A}}$$

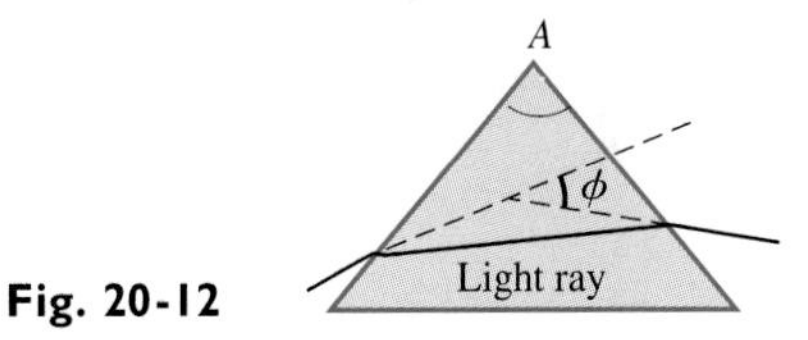

Fig. 20-12

32. For the structure shown in Fig. 20-13, show that $x = 2l\sin^2\frac{1}{2}\theta$.

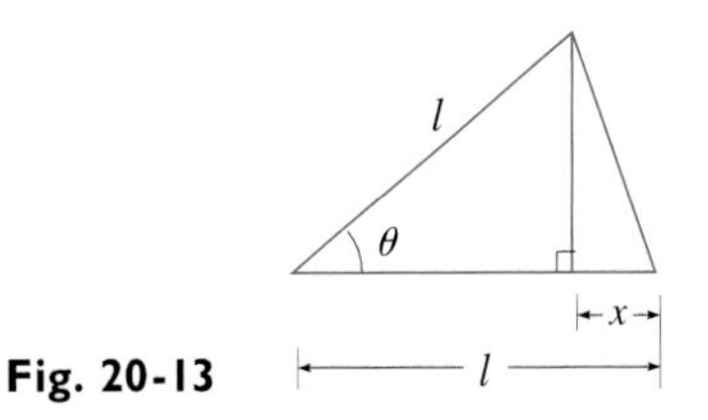

Fig. 20-13

20-5 Solving Trigonometric Equations

One of the most important uses of the trigonometric identities is in the solution of equations involving trigonometric functions. The solution of this type of equation consists of the angles that satisfy the equation. When solving for the angle, we generally first solve for a value of a function of the angle and then find the angle from this value of the function.

When equations are written in terms of more than one function, the identities provide a way of changing many of them to equations or factors involving only one function of the same angle. Thus, *the solution is found by using algebraic methods and trigonometric identities and values.* From Chapter 8 we recall that we must be careful regarding the sign of the value of a trigonometric function in finding the angle. Figure 20-14 shows again the quadrants in which the functions are positive. Functions not listed are negative.

NOTE ▶

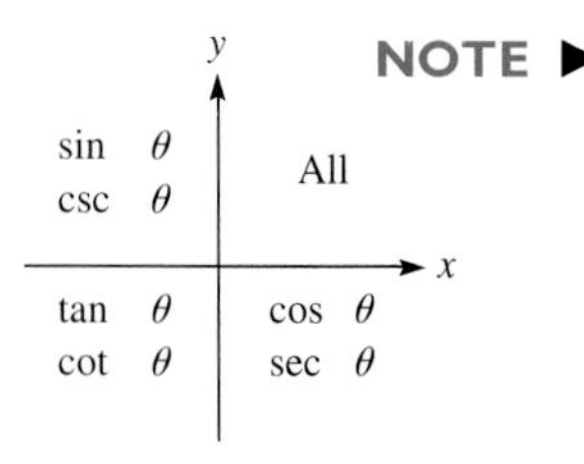

Positive functions

Fig. 20-14

EXAMPLE 1 ▸▸ Solve the equation $2\cos\theta - 1 = 0$ for all values of θ such that $0 \le \theta < 2\pi$.

Solving the equation for $\cos\theta$, we obtain $\cos\theta = \frac{1}{2}$. The problem asks for all values of θ from 0 to 2π that satisfy the equation. We know that the cosines of angles in the first and fourth quadrants are positive. Also, we know that $\cos\frac{\pi}{3} = \frac{1}{2}$, which means that $\frac{\pi}{3}$ is the reference angle. Therefore, the solution proceeds as follows:

$$2\cos\theta - 1 = 0$$

$$2\cos\theta = 1 \qquad \text{solve for } \cos\theta$$

$$\cos\theta = \frac{1}{2}$$

$$\theta = \frac{\pi}{3}, \frac{5\pi}{3} \qquad \theta \text{ in quadrants I and IV}$$ ◂◂

EXAMPLE 2 ▸▸ Solve the equation $2\cos^2 x - \sin x - 1 = 0$ $(0 \le x < 2\pi)$.

By use of the identity $\sin^2 x + \cos^2 x = 1$, this equation may be put in terms of $\sin x$ only. Thus, we have

$$2(1 - \sin^2 x) - \sin x - 1 = 0 \qquad \text{use identity}$$

$$-2\sin^2 x - \sin x + 1 = 0 \qquad \text{solve for } \sin x$$

$$2\sin^2 x + \sin x - 1 = 0$$

or

$$(2\sin x - 1)(\sin x + 1) = 0$$

Just as in solving algebraic equations, we can set each factor equal to zero to find valid solutions. Thus, $\sin x = \frac{1}{2}$ or $\sin x = -1$. For the domain from 0 to 2π, the value $\sin x = \frac{1}{2}$ gives values of x as $\frac{\pi}{6}$ and $\frac{5\pi}{6}$, and $\sin x = -1$ gives the value $x = \frac{3\pi}{2}$. Thus, the complete solution is

$$x = \frac{\pi}{6}, \frac{5\pi}{6}, \frac{3\pi}{2}$$

These values check when substituted in the original equation. ◂◂

Graphical Solutions

We can also graphically solve equations involving trigonometric functions. As with algebraic equations, graphical solutions are approximate, whereas with analytical solutions we can often get exact solutions. In finding graphical solutions we will follow the same procedure we introduced in Section 3-5. That is, we *collect all terms on the left of the equal sign, with zero on the right.* We then *graph the function on the left and determine the zeros of this function by finding the values of x where the curve crosses the x-axis.* We also used this method in Section 7-4 in solving quadratic equations, and in Section 15-4 in solving higher-degree equations.

EXAMPLE 3 ▸▸ Graphically solve the equation $2\cos^2 x - \sin x - 1 = 0$ such that $0 \leq x < 2\pi$ by using a graphing calculator. (This is the same equation as in Example 2.)

Since all terms are on the left, with zero on the right, we set $y = 2\cos^2 x - \sin x - 1$. We then use a graphing calculator with the RANGE values

Xmin = 0, Xmax = 6.5, Ymin = −3, Ymax = 3

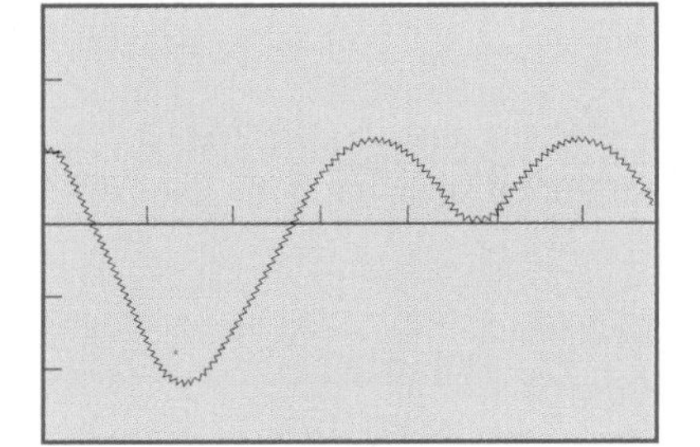

Fig. 20-15

(We chose Xmax = 6.5 since $2\pi \approx 6.3$ and angles are in radians.) The view is shown in Fig. 20-15. Using the TRACE and ZOOM features, we find that $y = 0$ for

$$x = 0.52, 2.62, 4.71$$

These values agree with those found in Example 2. We note that for $x = 4.71$ the curve *touches* the x-axis but does not cross it. It is *tangent* to the x-axis. ◂◂

In the examples that follow, we show the analytical solution to each equation and a graph that could be used to determine the graphical solution. The graph also can be used as an approximate check on the analytical solution.

EXAMPLE 4 ▸▸ Solve the equation $\sec^2 x + 2\tan x - 6 = 0$ $(0 \leq x < 2\pi)$.

By use of the identity $1 + \tan^2 x = \sec^2 x$ we may express this equation in terms of $\tan x$ only. Therefore,

$$1 + \tan^2 x + 2\tan x - 6 = 0 \quad \text{use identity}$$
$$\tan^2 x + 2\tan x - 5 = 0$$

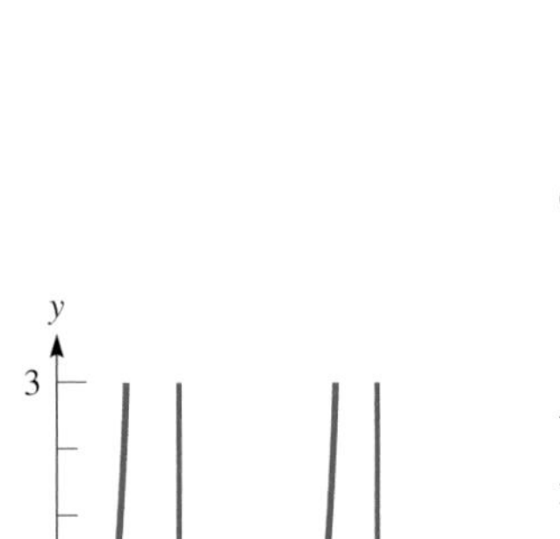

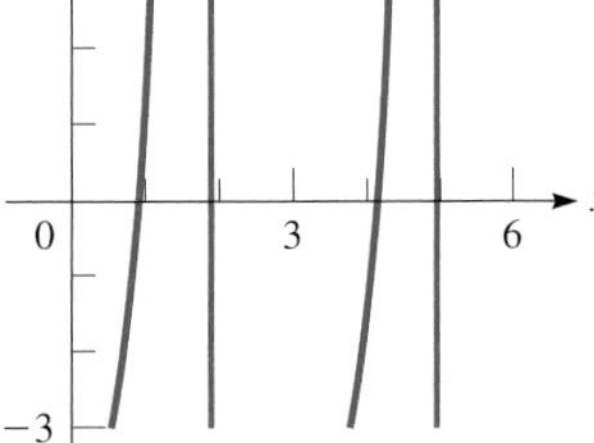

Fig. 20-16

We note that this expression is not factorable. Therefore, using the quadratic formula, we solve for $\tan x$.

$$\tan x = \frac{-2 \pm \sqrt{4 + 20}}{2} = -1 \pm 2.4495$$

Therefore, $\tan x = 1.4495$ and $\tan x = -3.4495$. In radians, we find that $\tan x = 1.4495$ for $x = 0.9669$. Since $\tan x$ is also positive in the third quadrant, we have $x = 4.108$ as well. In the same way, using $\tan x = -3.4495$, we obtain $x = 1.853$ and $x = 4.995$. Therefore, the correct solutions are

$$x = 0.9669, 1.853, 4.108, 4.995$$

These values check in the original equation. Also, they agree with the values of x for which the graph of $y = \sec^2 x + 2\tan x - 6$ crosses the x-axis in Fig. 20-16 (which can be viewed on a graphing calculator.). ◂◂

EXAMPLE 5 ▸▸ The vertical displacement y of an object at the end of a spring, which itself is being moved up and down, is given by $y = 3.50 \sin t + 1.20 \sin 2t$. Find the first two values of t (in seconds) for which $y = 0$.

Using the double-angle formula for $\sin 2t$ leads to the solution.

$$3.50 \sin t + 1.20 \sin 2t = 0 \quad \text{setting } y = 0$$
$$3.50 \sin t + 2.40 \sin t \cos t = 0 \quad \text{using identities}$$
$$\sin t(3.50 + 2.40 \cos t) = 0 \quad \text{factoring}$$
$$\sin t = 0 \quad \text{or} \quad \cos t = -1.46$$
$$t = 0.00, 3.14, \ldots$$

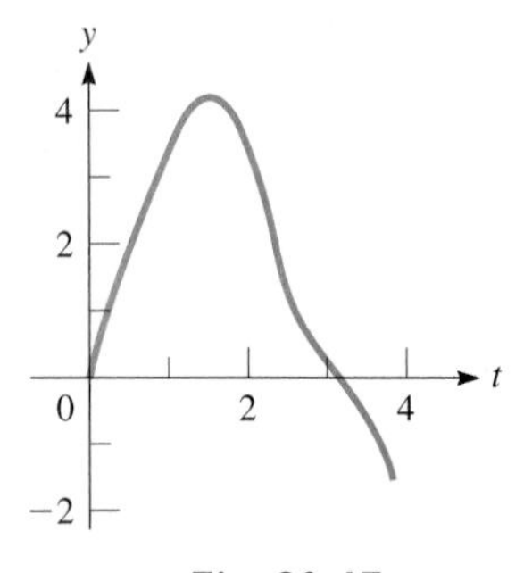

Fig. 20-17

Since $\cos t$ cannot be numerically larger than 1, there are no values of t for which $\cos t = -1.46$. Thus, the required times are $t = 0.00$ s, 3.14 s.

We can see that these values agree with the values of t for which the graph of $y = 3.50 \sin t + 1.20 \sin 2t$ crosses the t-axis in Fig. 20-17. (In using a graphing calculator, use x for t.) ◂◂

EXAMPLE 6 ▸▸ Solve the equation $\cos \frac{x}{2} = 1 + \cos x\,(0 \le x < 2\pi)$.

By using the half-angle formula for $\cos(x/2)$ and then squaring both sides of the resulting equation, this equation can be solved.

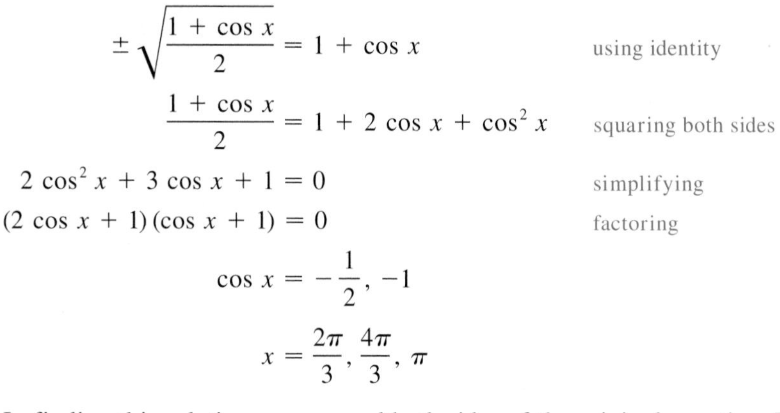

$$\pm\sqrt{\frac{1 + \cos x}{2}} = 1 + \cos x \quad \text{using identity}$$
$$\frac{1 + \cos x}{2} = 1 + 2\cos x + \cos^2 x \quad \text{squaring both sides}$$
$$2\cos^2 x + 3\cos x + 1 = 0 \quad \text{simplifying}$$
$$(2\cos x + 1)(\cos x + 1) = 0 \quad \text{factoring}$$
$$\cos x = -\frac{1}{2}, -1$$
$$x = \frac{2\pi}{3}, \frac{4\pi}{3}, \pi$$

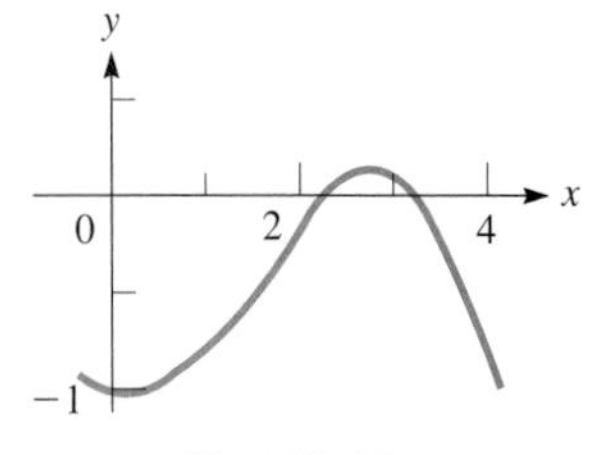

Fig. 20-18

In finding this solution, we squared both sides of the original equation. In doing this we may have introduced extraneous solutions (see Section 14-3). Thus we must check each solution in the original equation to see if it is valid. Hence

$$\cos\frac{\pi}{3} \stackrel{?}{=} 1 + \cos\frac{2\pi}{3} \quad \text{or} \quad \frac{1}{2} \stackrel{?}{=} 1 + \left(-\frac{1}{2}\right) \quad \text{or} \quad \frac{1}{2} = \frac{1}{2}$$
$$\cos\frac{2\pi}{3} \stackrel{?}{=} 1 + \cos\frac{4\pi}{3} \quad \text{or} \quad -\frac{1}{2} \stackrel{?}{=} 1 + \left(-\frac{1}{2}\right) \quad \text{or} \quad -\frac{1}{2} \ne \frac{1}{2}$$
$$\cos\frac{\pi}{2} \stackrel{?}{=} 1 + \cos\pi \quad \text{or} \quad 0 \stackrel{?}{=} 1 - 1 \quad \text{or} \quad 0 = 0$$

Thus the apparent solution $x = \frac{4\pi}{3}$ is not a solution of the original equation. The correct solutions are $x = \frac{2\pi}{3}$ and $x = \pi$.

We can see that these values agree with the values of x for which the graph of $y = \cos(x/2) - 1 - \cos x$ crosses the x-axis in Fig. 20-18. In using a graphing calculator, be sure to enter the function as we have shown it here. ◂◂

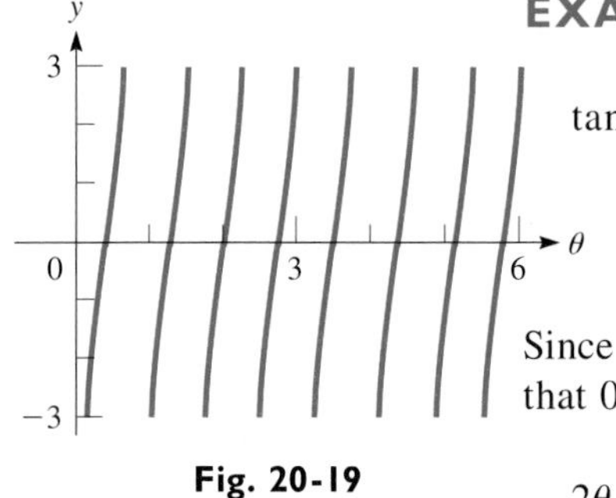

Fig. 20-19

EXAMPLE 7 ▸▸ Solve the equation $\tan 2\theta - \cot 2\theta = 0$ $(0 \le \theta < 2\pi)$.

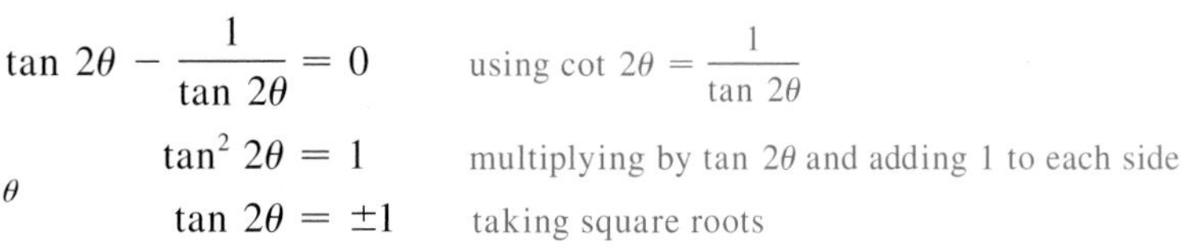

$$\tan 2\theta - \frac{1}{\tan 2\theta} = 0 \qquad \text{using } \cot 2\theta = \frac{1}{\tan 2\theta}$$

$$\tan^2 2\theta = 1 \qquad \text{multiplying by } \tan 2\theta \text{ and adding 1 to each side}$$

$$\tan 2\theta = \pm 1 \qquad \text{taking square roots}$$

Since we require values of θ such that $0 \le \theta < 2\pi$, we must have values of 2θ such that $0 \le 2\theta < 4\pi$. Therefore,

$$2\theta = \frac{\pi}{4}, \frac{3\pi}{4}, \frac{5\pi}{4}, \frac{7\pi}{4}, \frac{9\pi}{4}, \frac{11\pi}{4}, \frac{13\pi}{4}, \frac{15\pi}{4}$$

This means that the solutions are

$$\theta = \frac{\pi}{8}, \frac{3\pi}{8}, \frac{5\pi}{8}, \frac{7\pi}{8}, \frac{9\pi}{8}, \frac{11\pi}{8}, \frac{13\pi}{8}, \frac{15\pi}{8}$$

These values satisfy the original equation. Since we multiplied through by $\tan 2\theta$ in the solution, any value of θ that leads to $\tan 2\theta = 0$ would not be valid, since this would indicate division by zero in the original equation.

We see that these solutions agree with the values of θ for which the graph of $y = \tan 2\theta - \cot 2\theta$ crosses the θ axis in Fig. 20-19. (In using a graphing calculator, use x for θ.) ◂◂

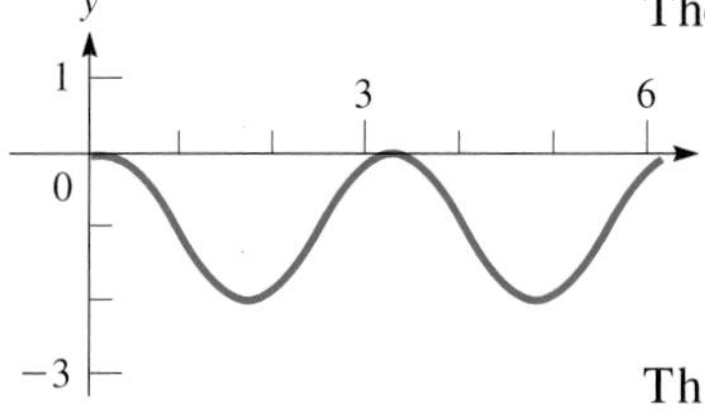

Fig. 20-20

EXAMPLE 8 ▸▸ Solve the equation $\cos 3x \cos x + \sin 3x \sin x = 1$ $(0 \le x < 2\pi)$.

The left side of this equation is of the general form $\cos(A - x)$, where $A = 3x$. Therefore,

$$\cos 3x \cos x + \sin 3x \sin x = \cos(3x - x) = \cos 2x$$

The original equation becomes

$$\cos 2x = 1$$

This equation is satisfied if $2x = 0$ or $2x = 2\pi$. The solutions are $x = 0$ and $x = \pi$. Only through recognition of the proper trigonometric form can we readily solve this equation.

We see that these solutions agree with the two values of θ for which the graph of $y = \cos 3x \cos x + \sin 3x \sin x - 1$ touches the x-axis in Fig. 20-20. ◂◂

EXAMPLE 9 ▸▸ Solve the equation $\sin 2x + 3 = x^2$.

Although we can substitute $2 \sin x \cos x$ for $\sin 2x$, we cannot express x^2 in terms of a trigonometric function. However, we can find an approximate graphical solution to this equation.

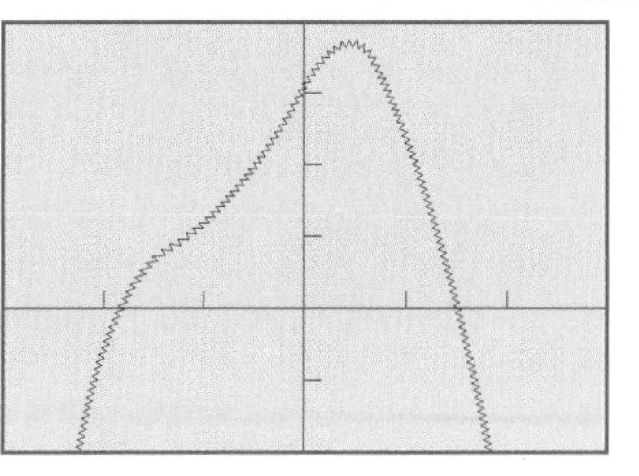

Fig. 20-21

We subtract x^2 from each side and then let

$$y = \sin 2x + 3 - x^2$$

Using a graphing calculator with the RANGE values

$$\text{Xmin} = -3,\ \text{Xmax} = 3,\ \text{Ymin} = -2,\ \text{Ymax} = 4$$

we have the view shown in Fig. 20-21. Using the TRACE and ZOOM features, we find that the solutions are

$$x = -1.90,\ 1.67$$

These are the only values for which $y = 0$. ◂◂

Exercises 20–5

In Exercises 1–16, solve the given trigonometric equations analytically (using identities when necessary for exact values when possible) for values of x for $0 \leq x < 2\pi$.

1. $\sin x - 1 = 0$ **2.** $2 \cos x + 1 = 0$

3. $2(2 + \cos x) = 3 + \cos x$

4. $4 \tan x + 2 = 3(1 + \tan x)$

5. $4 \cos^2 x - 1 = 0$ **6.** $3 \tan^2 x - 1 = 0$

7. $2 \sin^2 x - \sin x = 0$ **8.** $\sin 4x - \sin 2x = 0$

9. $\sin 2x \sin x + \cos x = 0$

10. $\sin x - \sin \frac{x}{2} = 0$

11. $2 \cos^2 x - 2 \cos 2x - 1 = 0$

12. $\tan^2 x - 5 \tan x + 6 = 0$

13. $4 \tan x - \sec^2 x = 0$ **14.** $\sin x \sin \frac{1}{2}x = 1 - \cos x$

15. $\sin 2x \cos x - \cos 2x \sin x = 0$

16. $\cos 3x \cos x - \sin 3x \sin x = 0$

In Exercises 17–32, solve the given trigonometric equations analytically and by use of a graphing calculator. Compare results. $(0 \leq x < 2\pi)$

17. $\tan x + 1 = 0$ **18.** $2 \sin x + 1 = 0$

19. $3 - 4 \cos x = 7 - (2 - \cos x)$

20. $7 \sin x - 2 = 3(2 - \sin x)$

21. $4 \sin^2 x - 3 = 0$

22. $\sin^2 x - 1 = 0$

23. $\sin 4x - \cos 2x = 0$ **24.** $3 \cos x - 4 \cos^2 x = 0$

25. $2 \sin x - \tan x = 0$

26. $\cos 2x + \sin^2 x = 0$

27. $\sin^2 x - 2 \sin x - 1 = 0$

28. $2 \cos^2 2x + 1 = 3 \cos 2x$

29. $\tan x + 3 \cot x = 4$

30. $\tan^2 x - 2 \sec^2 x + 4 = 0$

31. $\sin 2x + \cos 2x = 0$ **32.** $2 \sin 4x + \csc 4x = 3$

In Exercises 33–36, solve the indicated equations analytically.

33. To find the angle θ subtended by a certain object on a camera film, it is necessary to solve the equation

$$\frac{p^2 \tan \theta}{0.0063 + p \tan \theta} = 1.6$$

where p is the distance from the camera to the object. Find θ if $p = 4.8$ m.

34. In finding the maximum illuminance from a point source of light, it is necessary to solve the equation $\cos \theta \sin 2\theta - \sin^3 \theta = 0$. Find θ if $0 < \theta < 90°$.

35. The vertical displacement y of the end of a robot arm is given by $y = 2.30 \cos 0.1t - 1.35 \sin 0.2t$. Find the first four values of t, in seconds, for which $y = 0$.

36. A building 41.0 m high is constructed on a vertical 40.0-m cliff at the edge of a river. How far out on the river from the base of the cliff will the building and the cliff subtend equal angles?

In Exercises 37–44, solve the given equations graphically.

37. $3 \sin x - x = 0$ **38.** $4 \cos x + 3x = 0$

39. $2 \sin 2x = x^2 + 1$ **40.** $\sqrt{x} - \sin 3x = 1$

41. $2 \ln x = 1 - \cos 2x$ **42.** $e^x = 1 + \sin x$

43. In finding the frequencies of vibration of a vibrating wire, the equation $x \tan x = 2.00$ occurs. Find x if $0 < x < \frac{\pi}{2}$.

44. An equation used in astronomy is $\theta - e \sin \theta = M$. Solve for θ for $e = 0.25$ and $M = 0.75$.

20-6 The Inverse Trigonometric Functions

When we studied the exponential and logarithmic functions, we often found it useful to change an expression from exponential to logarithmic form or from logarithmic to exponential form. Each of these forms has its advantages for particular purposes. We also showed that the exponential function $y = b^x$ can be written in logarithmic form with x as a function of y, or $x = \log_b y$. We then represented both functions with y in terms of x, saying that the letters used for the dependent and independent variables did not matter when we wished to express a functional relationship. Since it is standard practice to use y as the dependent variable and x as the independent variable, we wrote the logarithmic function as $y = \log_b x$.

These two functions, the exponential function $y = b^x$ and the logarithmic function $y = \log_b x$, are called **inverse functions.** *This means that if we solve for the independent variable in terms of the dependent variable in one function, we will arrive at the functional relationship expressed by the other.* It also means that, for every value of x, there is only one corresponding value of y.

Just as we are able to solve $y = b^x$ for the exponent by writing it in logarithmic form, at times it is necessary to solve for the independent variable (the angle) in trigonometric functions. Therefore, we define the **inverse sine function**

$$y = \sin^{-1} x \qquad \left(-\frac{\pi}{2} \leq y \leq \frac{\pi}{2}\right) \tag{20-28}$$

where y *is the angle whose sine is* x. This means that x is the value of the sine of the angle y, or $x = \sin y$. (It is necessary to show the range as $-\pi/2 \leq y \leq \pi/2$, as we will see shortly.)

CAUTION ▶ *In Eq. (20-28), the* -1 *is* **not** *an exponent.*

The -1 *in* $\sin^{-1} x$ *is the notation showing the inverse function.* We first introduced this notation in Chapter 4 when we were finding the angle with a known value of one of the functions.

The notations Arcsin x, arcsin x, $\text{Sin}^{-1} x$ are also used to designate the inverse sine. Some calculators use the INV key and then the SIN key to find values of the inverse sine. However, since most calculators use the notation $\sin^{-1}$ as a second function on the SIN key, we will continue to use $\sin^{-1} x$ for the inverse sine.

Similar definitions are used for the other inverse trigonometric functions. They also have meanings similar to that of Eq. (20-28).

EXAMPLE 1 ▸▸ (a) $y = \cos^{-1} x$ is read as "y is the angle whose cosine is x." In this case, $x = \cos y$.

(b) $y = \tan^{-1} 2x$ is read as "y is the angle whose tangent is $2x$." In this case, $2x = \tan y$.

(c) $y = \csc^{-1}(1 - x)$ is read as "y is the angle whose cosecant is $1 - x$." In this case, $1 - x = \csc y$, or $x = 1 - \csc y$. ◂◂

We have seen that $y = \sin^{-1} x$ means that $x = \sin y$. From our previous work with the trigonometric functions, we know that there is an unlimited number of possible values of y for a given value of x in $x = \sin y$. Consider the following example.

EXAMPLE 2 ▸▸ (a) For $x = \sin y$, we know that

$$\sin\frac{\pi}{6} = \frac{1}{2} \quad \text{and} \quad \sin\frac{5\pi}{6} = \frac{1}{2}$$

In fact, $x = \frac{1}{2}$ also for values of y of $-\frac{7\pi}{6}, \frac{13\pi}{6}, \frac{17\pi}{6}$, and so on.

(b) For $x = \cos y$, we know that

$$\cos 0 = 1 \quad \text{and} \quad \cos 2\pi = 1$$

In fact, the $\cos y = 1$ for y equal to any even multiple of π. ◂◂

From Chapter 3, we know that *to have a properly defined* ***function,*** *there must be only one value of the dependent variable for a given value of the independent variable.* (A *relation,* on the other hand, may have more than one such value.) Therefore, as in Eq. (20-28), in order to have only one value of y for each value of x in the domain of the inverse trigonometric functions, it is not possible to include all values of y in the range. For this reason, *the range of each of the* **inverse trigonometric functions** *is defined as follows:*

$$-\frac{\pi}{2} \leq \sin^{-1} x \leq \frac{\pi}{2} \qquad 0 \leq \cos^{-1} x \leq \pi \qquad -\frac{\pi}{2} < \tan^{-1} x < \frac{\pi}{2} \qquad 0 < \cot^{-1} x < \pi$$

$$0 \leq \sec^{-1} x \leq \pi \quad \left(\sec^{-1} x \neq \frac{\pi}{2}\right) \qquad -\frac{\pi}{2} \leq \csc^{-1} x \leq \frac{\pi}{2} \quad (\csc^{-1} x \neq 0) \tag{20-29}$$

For any of the inverse trigonometric functions $y = f(x)$, we must use a value of y in the range as defined in Eqs. (20-29) that corresponds to a given value of x in the domain. We will discuss the domains and the reasons for these definitions, along with the graphs of the inverse trigonometric functions, following the next two examples.

EXAMPLE 3 ▸▸ (a) $\sin^{-1}\left(\frac{1}{2}\right) = \frac{\pi}{6}$ first-quadrant angle

This is the only value of the function that lies within the defined range. The value $\frac{5\pi}{6}$ is not correct, even though $\sin(\frac{5\pi}{6}) = \frac{1}{2}$, since $\frac{5\pi}{6}$ lies outside the defined range.

(b) $\cos^{-1}\left(-\frac{1}{2}\right) = \frac{2\pi}{3}$ second-quadrant angle

Other values such as $\frac{4\pi}{3}$ and $-\frac{2\pi}{3}$ are not correct, since they are not within the defined range for the function $\cos^{-1} x$. ◂◂

EXAMPLE 4 ▸▸ $\tan^{-1}(-1) = -\frac{\pi}{4}$ fourth-quadrant angle

This is the only value within the defined range for the function $\tan^{-1} x$. We must remember that ***when x is negative for*** $\sin^{-1} x$ ***and*** $\tan^{-1} x$***, the value of y is a fourth-quadrant angle, expressed as a*** **negative angle.** This is a direct result of the definition. (The single exception is $\sin^{-1}(-1) = -\pi/2$, which is a quadrantal angle and is not *in* the fourth quadrant.) ◂◂

CAUTION ▸

In choosing these values to be the ranges of the inverse trigonometric functions, we first note that the *domain* of $y = \sin^{-1} x$ and $y = \cos^{-1} x$ are each $-1 \leq x \leq 1$, since the sine and cosine functions take on only these values. Therefore, for each value in this domain we use only one value of y in the range of the function. Although the domain of $y = \tan^{-1} x$ is all real numbers, we still use only one value of y in the range.

The ranges of the inverse trigonometric functions are chosen so that if x is positive, the resulting value is an angle in the first quadrant. However, care must be taken in choosing the range for negative values of x.

We cannot choose second-quadrant angles for $\sin^{-1} x$. Since the sine of a second-quadrant angle is also positive, this would lead to ambiguity. The sine is negative for fourth-quadrant angles, and to have a continuous range of values we express the fourth-quadrant angles in the form of negative angles. This range is also chosen for $\tan^{-1} x$, for similar reasons. However, the range for $\cos^{-1} x$ cannot be chosen in this way, since the cosine of a fourth-quadrant angle is also positive. Thus, to keep a continuous range of values for $\cos^{-1} x$, the second-quadrant angles are chosen for negative values of x.

As for the values for the other functions, we chose values such that if x is positive, the result is also an angle in the first quadrant. As for negative values of x, it rarely makes any difference, since either positive values of x arise, or we can use one of the other functions. Our definitions, however, are those which are generally used.

The graphs of the inverse trigonometric functions can be used to show the domains and ranges. We can obtain the graph of the inverse sine function by first sketching the sine curve $x = \sin y$ *along the y-axis.* We then mark the specific part of this curve for which $-\frac{\pi}{2} \le y \le \frac{\pi}{2}$ as the graph of the inverse sine function. The graphs of the other inverse trigonometric functions are found in the same manner. In Figs. 20-22, 20-23, and 20-24, the graphs of $x = \sin y$, $x = \cos y$, and $x = \tan y$, respectively, are shown. The heavier, colored portions indicate the graphs of the respective inverse trigonometric functions.

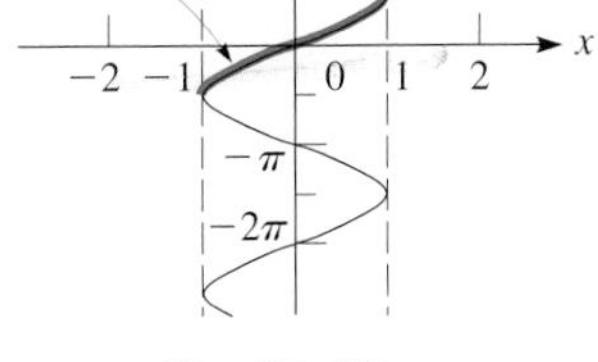

Fig. 20-22

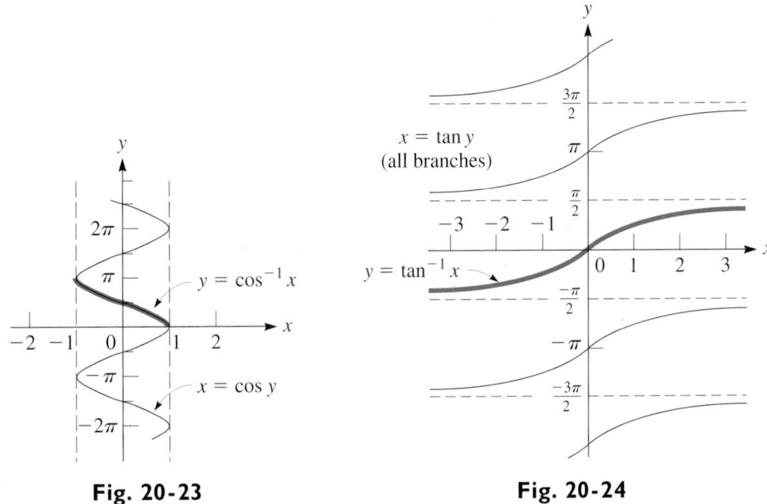

Fig. 20-23

Fig. 20-24

The following examples further illustrate the values and meanings of the inverse trigonometric functions.

EXAMPLE 5 ▸▸ (a) $\sin^{-1}(-\sqrt{3}/2) = -\pi/3$ (b) $\cos^{-1}(-1) = \pi$
(c) $\tan^{-1} 0 = 0$ (d) $\tan^{-1}(\sqrt{3}) = \pi/3$

Using a calculator in radian mode, we find the following values:

(e) $\sin^{-1}0.6294 = 0.6808$ (f) $\sin^{-1}(-0.1568) = -0.1574$
(g) $\cos^{-1}(-0.8026) = 2.5024$ (h) $\tan^{-1}(-1.9268) = -1.0921$

We note that the calculator gives values that are in the defined range for each function. ◂◂

EXAMPLE 6 ▸▸ Given that $y = \pi - \sec^{-1} 2x$, solve for x.

We first find the expression for $\sec^{-1} 2x$ and then use the meaning of the inverse secant. The solution follows.

$$\begin{aligned} y &= \pi - \sec^{-1} 2x \\ \sec^{-1} 2x &= \pi - y && \text{solve for } \sec^{-1} 2x \\ 2x &= \sec(\pi - y) && \text{use meaning of inverse secant} \\ x &= -\frac{1}{2}\sec y && \sec(\pi - y) = -\sec y \end{aligned}$$

As $\sec 2x$ and $2 \sec x$ are different functions, $\sec^{-1} 2x$ and $2 \sec^{-1} x$ are also different functions. Since the values of $\sec^{-1} 2x$ are restricted, so are the resulting values of y. ◂◂

EXAMPLE 7 ▸▸ The instantaneous power p in an electric inductor is given by the equation $p = vi \sin \omega t \cos \omega t$. Solve for t.

Noting the product $\sin \omega t \cos \omega t$ suggests using $\sin 2\alpha = 2 \sin \alpha \cos \alpha$. Then, using the meaning of the inverse sine, we can complete the solution.

$$\begin{aligned} p &= vi \sin \omega t \cos \omega t \\ &= \frac{1}{2} vi \sin 2\omega t && \text{using double-angle formula} \\ \sin 2\omega t &= \frac{2p}{vi} \\ 2\omega t &= \sin^{-1}\left(\frac{2p}{vi}\right) && \text{using meaning of inverse sine} \\ t &= \frac{1}{2\omega} \sin^{-1}\left(\frac{2p}{vi}\right) \end{aligned}$$ ◂◂

If we know the value of one of the inverse functions, we can find the trigonometric functions of the angle. If general relations are desired, a representative triangle is very useful. The following examples illustrate these methods.

EXAMPLE 8 ▸▸ Find $\cos(\sin^{-1} 0.5)$.

Knowing that the values of inverse trigonometric functions are *angles*, we see that $\sin^{-1} 0.5$ is a first-quadrant angle. Thus we find $\sin^{-1} 0.5 = \pi/6$. The problem is now to find $\cos(\pi/6)$. This is, of course, $\sqrt{3}/2$, or 0.8660. Thus,

$$\cos(\sin^{-1} 0.5) = \cos(\pi/6) = 0.8660.$$ ◂◂

EXAMPLE 9 ▸▸ (a) $\sin(\cot^{-1} 1) = \sin(\pi/4)$ first-quadrant angle

$$= \frac{\sqrt{2}}{2} = 0.7071$$

(b) $\tan[\cos^{-1}(-1)] = \tan \pi$ quadrantal angle

$$= 0$$

(c) $\cos[\sin^{-1}(-0.2395)] = 0.9709$ using a calculator ◂◂

EXAMPLE 10 ▸▸ Find $\sin(\tan^{-1} x)$.

We know that $\tan^{-1} x$ is another way of stating "the angle whose tangent is x." Thus, let us draw a right triangle (as in Fig. 20-25) and label one of the acute angles as θ, the side opposite θ as x, and the side adjacent to θ as 1. In this way we see that, by definition, $\tan\theta = \frac{x}{1}$, or $\theta = \tan^{-1} x$, which means θ is the desired angle. By the Pythagorean theorem, the hypotenuse of this triangle is $\sqrt{x^2 + 1}$. Now we find that the $\sin\theta$, which is the same as $\sin(\tan^{-1} x)$, is $x/\sqrt{x^2 + 1}$, from the definition of the sine. Thus,

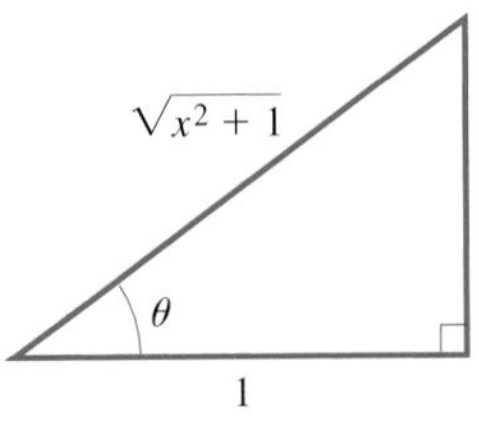

$$\sin(\tan^{-1} x) = \frac{x}{\sqrt{x^2 + 1}}$$ ◂◂

Fig. 20-25

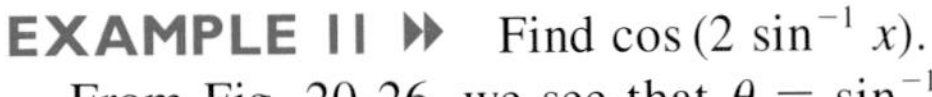

EXAMPLE 11 ▸▸ Find $\cos(2\sin^{-1} x)$.

From Fig. 20-26, we see that $\theta = \sin^{-1} x$. From the double-angle formulas, we have

$$\cos 2\theta = 1 - 2\sin^2\theta$$

Thus, since $\sin\theta = x$, we have

$$\cos(2\sin^{-1} x) = 1 - 2x^2$$ ◂◂

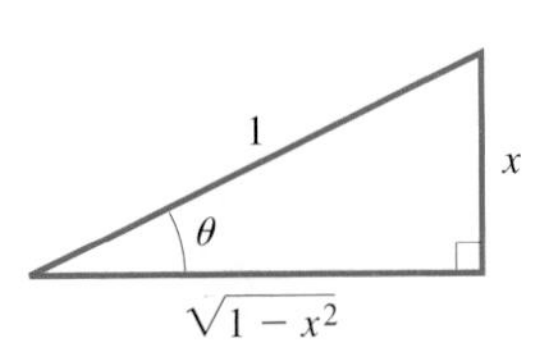

Fig. 20-26

EXAMPLE 12 ▸▸ A triangular brace of sides a, b, and c supports a shelf, as shown in Fig. 20-27. Find the expression for the angle between sides b and c.

The law of cosines leads to the solution, which follows.

$$a^2 = b^2 + c^2 - 2bc\cos A \quad \text{law of cosines}$$

$$2bc\cos A = b^2 + c^2 - a^2 \quad \text{solving for } \cos A$$

$$\cos A = \frac{b^2 + c^2 - a^2}{2bc}$$

$$A = \cos^{-1}\left(\frac{b^2 + c^2 - a^2}{2bc}\right) \quad \text{using meaning of inverse cosine}$$ ◂◂

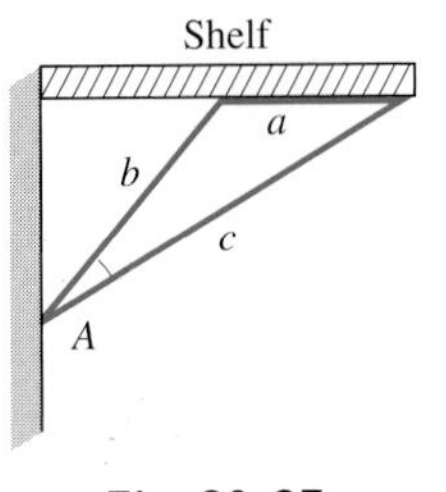

Fig. 20-27

Exercises 20-6

In Exercises 1–8, write down the meaning of each of the given equations. See Example 1.

1. $y = \tan^{-1} x$ **2.** $y = \sec^{-1} x$

3. $y = \cot^{-1} 3x$ **4.** $y = \csc^{-1} 4x$

5. $y = 2\sin^{-1} x$ **6.** $y = 3\tan^{-1} x$

7. $y = 5\cos^{-1}(2x - 1)$ **8.** $y = 4\sin^{-1}(3x + 2)$

In Exercises 9–32, evaluate the given expressions.

9. $\cos^{-1} 0.5$ **10.** $\sin^{-1} 1$

11. $\sin^{-1} 0$ **12.** $\cos^{-1} 0$

13. $\tan^{-1}(-\sqrt{3})$ **14.** $\sin^{-1}(-0.5)$

15. $\sec^{-1} 2$ **16.** $\cot^{-1}\sqrt{3}$

17. $\tan^{-1}(\sqrt{3}/3)$ **18.** $\tan^{-1} 1$

19. $\sin^{-1}(-\sqrt{2}/2)$ **20.** $\cos^{-1}(-\sqrt{3}/2)$

21. $\csc^{-1}\sqrt{2}$ **22.** $\cot^{-1} 1$

23. $\sin^{-1}(-\sqrt{3}/2)$ **24.** $\cot^{-1}(-\sqrt{3})$

25. $\tan(\sin^{-1} 0)$ **26.** $\csc(\tan^{-1} 1)$

27. $\sin(\tan^{-1}\sqrt{3})$ **28.** $\tan[\sin^{-1}(\sqrt{2}/2)]$

29. $\cos[\tan^{-1}(-1)]$ **30.** $\sec[\cos^{-1}(-0.5)]$

31. $\cos(2\sin^{-1} 1)$ **32.** $\sin(2\tan^{-1} 2)$

In Exercises 33–44, use a calculator to evaluate the given expressions.

33. $\tan^{-1}(-3.7321)$ **34.** $\cos^{-1}(-0.6561)$

35. $\sin^{-1}(-0.8326)$ **36.** $\tan^{-1} 0.2846$

37. $\cos^{-1} 0.1291$ **38.** $\sin^{-1} 0.2119$

39. $\tan^{-1} 8.2614$ **40.** $\sin^{-1}(-0.8881)$

41. $\tan[\cos^{-1}(-0.6281)]$ **42.** $\cos[\tan^{-1}(-1.2256)]$

43. $\sin[\tan^{-1}(-0.2297)]$ **44.** $\tan[\sin^{-1}(-0.3019)]$

In Exercises 45–52, solve the given equations for x.

45. $y = \sin 3x$ **46.** $y = \cos(x - \pi)$

47. $y = \tan^{-1}(x/4)$ **48.** $y = 2\sin^{-1}(x/6)$

49. $y = 1 + 3\sec 3x$ **50.** $4y = 5 - 2\csc 8x$

51. $1 - y = \cos^{-1}(1 - x)$ **52.** $2y = \cot^{-1} 3x - 5$

In Exercises 53–60, find an algebraic expression for each of the expressions given.

53. $\tan(\sin^{-1} x)$ **54.** $\sin(\cos^{-1} x)$

55. $\cos(\sec^{-1} x)$ **56.** $\cot(\cot^{-1} x)$

57. $\sec(\csc^{-1} 3x)$ **58.** $\tan(\sin^{-1} 2x)$

59. $\sin(2\sin^{-1} x)$ **60.** $\cos(2\tan^{-1} x)$

In Exercises 61–64, solve the given problems with the use of the inverse trigonometric functions.

61. In the analysis of ocean tides, the equation $y = A\cos 2(\omega t + \phi)$ is used. Solve for t.

62. For an object of weight w on an inclined plane that is at an angle θ to the horizontal, the equation relating w and θ is $\mu w \cos\theta = w \sin\theta$, where μ is the coefficient of friction between the surfaces in contact. Solve for θ.

63. The electric current in a certain circuit is given by $i = I_m[\sin(\omega t + \alpha)\cos\phi + \cos(\omega t + \alpha)\sin\phi]$. Solve for t.

64. The time t as a function of the displacement d of a piston is given by the equation $t = \dfrac{1}{2\pi f}\cos^{-1}\dfrac{d}{A}$. Solve for d.

In Exercises 65 and 66, prove that the given expressions are equal. Use the relation for $\sin(\alpha + \beta)$ and show that the sine of the sum of the angles on the left equals the sine of the angle on the right.

65. $\sin^{-1}\dfrac{3}{5} + \sin^{-1}\dfrac{5}{13} = \sin^{-1}\dfrac{56}{65}$

66. $\tan^{-1}\dfrac{1}{3} + \tan^{-1}\dfrac{1}{2} = \dfrac{\pi}{4}$

In Exercises 67 and 68, verify the given expressions.

67. $\sin^{-1} 0.5 + \cos^{-1} 0.5 = \pi/2$

68. $\tan^{-1}\sqrt{3} + \cot^{-1}\sqrt{3} = \pi/2$

In Exercises 69 and 70, solve for the angle A for the triangles in the given figures in terms of the given sides and angles. Explain your method.

69. Fig. 20-28 **70.** Fig. 20-29

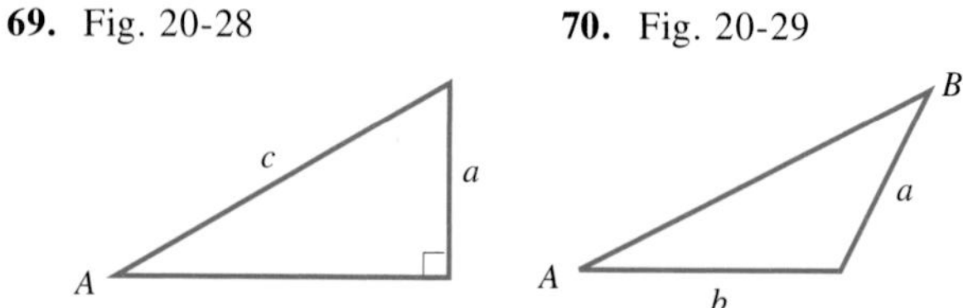

In Exercises 71 and 72, derive the given expressions.

71. The height of the Statue of Liberty is 46.0 m. See Fig. 20-30. From the deck of a boat at a horizontal distance d from the statue, the angles of elevation of the top of the statue and the top of its pedestal are α and β, respectively. Show that

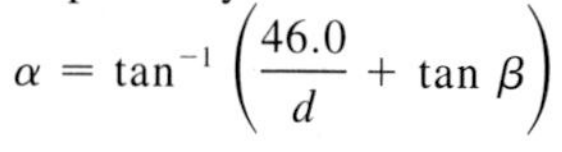

$$\alpha = \tan^{-1}\left(\frac{46.0}{d} + \tan\beta\right)$$

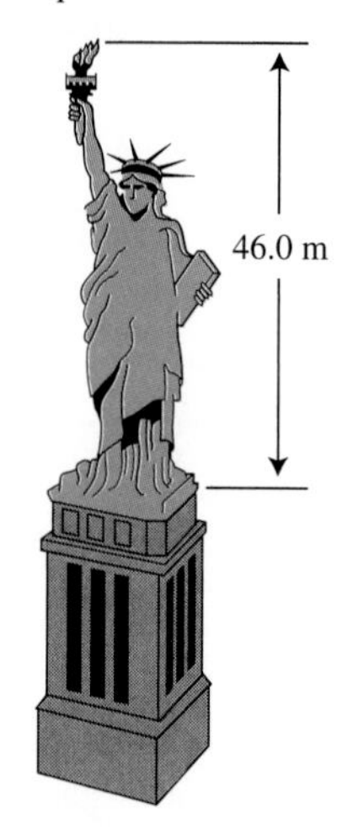

Fig. 20-30

72. Show that the length L of the pulley belt shown in Fig. 20-31 is $L = 24 + 11\pi + 10\sin^{-1}\dfrac{5}{13}$.

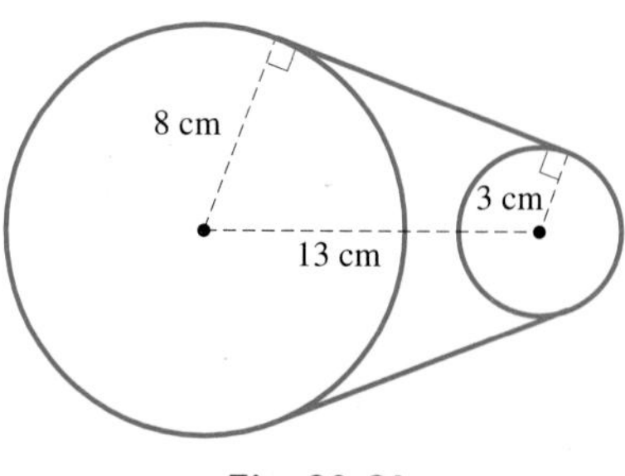

Fig. 20-31

Chapter Equations, Review Exercises, and Practice Test

Chapter Equations

Basic trigonometric identities

$$\sin\theta = \frac{1}{\csc\theta} \qquad (20\text{-}1)$$

$$\cos\theta = \frac{1}{\sec\theta} \qquad (20\text{-}2)$$

$$\tan\theta = \frac{1}{\cot\theta} \qquad (20\text{-}3)$$

$$\tan\theta = \frac{\sin\theta}{\cos\theta} \qquad (20\text{-}4)$$

$$\cot\theta = \frac{\cos\theta}{\sin\theta} \qquad (20\text{-}5)$$

$$\sin^2\theta + \cos^2\theta = 1 \qquad (20\text{-}6)$$

$$1 + \tan^2\theta = \sec^2\theta \qquad (20\text{-}7)$$

$$1 + \cot^2\theta = \csc^2\theta \qquad (20\text{-}8)$$

Sum and difference identities

$$\sin(\alpha + \beta) = \sin\alpha\cos\beta + \cos\alpha\sin\beta \quad (20\text{-}9)$$

$$\cos(\alpha + \beta) = \cos\alpha\cos\beta - \sin\alpha\sin\beta \quad (20\text{-}10)$$

$$\sin(\alpha - \beta) = \sin\alpha\cos\beta - \cos\alpha\sin\beta \quad (20\text{-}11)$$

$$\cos(\alpha - \beta) = \cos\alpha\cos\beta + \sin\alpha\sin\beta \quad (20\text{-}12)$$

Double-angle formulas

$$\sin 2\alpha = 2\sin\alpha\cos\alpha \quad (20\text{-}21)$$

$$\cos 2\alpha = \cos^2\alpha - \sin^2\alpha \quad (20\text{-}22)$$

$$= 2\cos^2\alpha - 1 \quad (20\text{-}23)$$

$$= 1 - 2\sin^2\alpha \quad (20\text{-}24)$$

Half-angle formulas

$$\sin\frac{\alpha}{2} = \pm\sqrt{\frac{1 - \cos\alpha}{2}} \quad (20\text{-}26)$$

$$\cos\frac{\alpha}{2} = \pm\sqrt{\frac{1 + \cos\alpha}{2}} \quad (20\text{-}27)$$

Inverse trigonometric functions

$$y = \sin^{-1} x \quad \left(-\frac{\pi}{2} \le y \le \frac{\pi}{2}\right) \quad (20\text{-}28)$$

$$-\frac{\pi}{2} \le \sin^{-1} x \le \frac{\pi}{2} \qquad 0 \le \cos^{-1} x \le \pi \qquad -\frac{\pi}{2} < \tan^{-1} x < \frac{\pi}{2} \qquad 0 < \cot^{-1} x < \pi$$

$$0 \le \sec^{-1} x \le \pi \quad \left(\sec^{-1} x \ne \frac{\pi}{2}\right) \qquad -\frac{\pi}{2} \le \csc^{-1} x \le \frac{\pi}{2} \quad (\csc^{-1} x \ne 0) \quad (20\text{-}29)$$

Review Exercises

In Exercises 1–8, determine the values of the indicated functions in the given manner.

1. Find sin 120° by using 120° = 90° + 30°.
2. Find cos 30° by using 30° = 90° − 60°.
3. Find sin 135° by using 135° = 180° − 45°.
4. Find cos 225° by using 225° = 180° + 45°.
5. Find cos 180° by using 180° = 2(90°).
6. Find sin 180° by using 180° = 2(90°).
7. Find sin 45° by using 45° = $\frac{1}{2}$(90°).
8. Find cos 45° by using 45° = $\frac{1}{2}$(90°).

In Exercises 9–16, simplify the given expressions by using one of the basic formulas of the chapter. Then use a calculator to verify the result by finding the value of the original expression and the value of the simplified expression.

9. $\sin 14^\circ \cos 38^\circ + \cos 14^\circ \sin 38^\circ$
10. $\cos^2 148^\circ - \sin^2 148^\circ$
11. $2 \sin 46^\circ \cos 46^\circ$
12. $\cos 73^\circ \cos 142^\circ + \sin 73^\circ \sin 142^\circ$
13. $\sqrt{\frac{1 + \cos 12^\circ}{2}}$
14. $\cos 3^\circ \cos 215^\circ - \sin 3^\circ \sin 215^\circ$
15. $1 - 2\sin^2 82^\circ$
16. $\sqrt{\frac{1 - \cos 166^\circ}{2}}$

In Exercises 17–24, simplify each of the given expressions. Expansion of any term is not necessary; recognition of the form of the expression leads to the proper result.

17. $\sin 2x \cos 3x + \cos 2x \sin 3x$
18. $\cos 7x \cos 3x + \sin 7x \sin 3x$
19. $8 \sin 6x \cos 6x$
20. $10 \sin 5x \cos 5x$
21. $2 - 4\sin^2 6x$
22. $\cos^2 2x - \sin^2 2x$
23. $\sqrt{2 + 2\cos 2x}$
24. $\sqrt{32 - 32\cos 4x}$

In Exercises 25–32, evaluate the given expressions.

25. $\sin^{-1}(-1)$
26. $\sec^{-1}\sqrt{2}$
27. $\cos^{-1} 0.9659$
28. $\tan^{-1}(-0.6249)$

29. $\tan[\sin^{-1}(-0.5)]$ **30.** $\cos[\tan^{-1}(-\sqrt{3})]$

31. $\sin^{-1}(\tan \pi)$ **32.** $\cos^{-1}[\tan(-\pi/4)]$

In Exercises 33–60, prove the given identities.

33. $\dfrac{\sec y}{\csc y} = \tan y$ **34.** $\cos\theta \csc\theta = \cot\theta$

35. $\sin x(\csc x - \sin x) = \cos^2 x$

36. $\cos y(\sec y - \cos y) = \sin^2 y$

37. $\dfrac{1}{\sin\theta} - \sin\theta = \cot\theta\cos\theta$

38. $\sin\theta\sec\theta\csc\theta\cos\theta = 1$

39. $\cos\theta\cot\theta + \sin\theta = \csc\theta$

40. $\dfrac{\sin x\cot x + \cos x}{\cot x} = 2\sin x$

41. $\dfrac{\sec^4 x - 1}{\tan^2 x} = 2 + \tan^2 x$

42. $\cos^2 y - \sin^2 y = \dfrac{1 - \tan^2 y}{1 + \tan^2 y}$

43. $2\csc 2x\cot x = 1 + \cot^2 x$

44. $\sin x\cot^2 x = \csc x - \sin x$

45. $\dfrac{1 - \sin^2\theta}{1 - \cos^2\theta} = \cot^2\theta$ **46.** $\dfrac{1 + \cos 2\theta}{\cos^2\theta} = 2$

47. $\dfrac{\cos 2\theta}{\cos^2\theta} = 1 - \tan^2\theta$ **48.** $\dfrac{\sin 2\theta\sec\theta}{2} = \sin\theta$

49. $\sin\dfrac{\theta}{2}\cos\dfrac{\theta}{2} = \dfrac{\sin\theta}{2}$ **50.** $\sin\dfrac{x}{2} = \dfrac{\sec x - 1}{2\sec x\sin\left(\dfrac{x}{2}\right)}$

51. $\sec x + \tan x = \dfrac{\cos x}{1 - \sin x}$

52. $\dfrac{\cos\theta - \sin\theta}{\cos\theta + \sin\theta} = \dfrac{\cot\theta - 1}{\cot\theta + 1}$

53. $\cos(x - y)\cos y - \sin(x - y)\sin y = \cos x$

54. $\sin 3y\cos 2y - \cos 3y\sin 2y = \sin y$

55. $\sin 4x(\cos^2 2x - \sin^2 2x) = \dfrac{\sin 8x}{2}$

56. $\csc 2x + \cot 2x = \cot x$

57. $\dfrac{\sin x}{\csc x - \cot x} = 1 + \cos x$

58. $\cos x - \sin\dfrac{x}{2} = \left(1 - 2\sin\dfrac{x}{2}\right)\left(1 + \sin\dfrac{x}{2}\right)$

59. $\dfrac{\sin(x + y) + \sin(x - y)}{\cos(x + y) + \cos(x - y)} = \tan x$

60. $\sec\dfrac{x}{2} + \csc\dfrac{x}{2} = \dfrac{2\left(\sin\dfrac{x}{2} + \cos\dfrac{x}{2}\right)}{\sin x}$

In Exercises 61–64, solve for x.

61. $y = 2\cos 2x$ **62.** $y - 2 = 2\tan\left(x - \dfrac{\pi}{2}\right)$

63. $y = \dfrac{\pi}{4} - 3\sin^{-1} 5x$ **64.** $2y = \sec^{-1} 4x - 2$

In Exercises 65–76, solve the given equations for x such that $0 \le x < 2\pi$.

65. $3(\tan x - 2) = 1 + \tan x$

66. $5\sin x = 3 - (\sin x + 2)$

67. $2(1 - 2\sin^2 x) = 1$

68. $\sec^2 x = 2\tan x$

69. $\cos^2 2x - 1 = 0$

70. $2\sin 2x + 1 = 0$

71. $4\cos^2 x - 3 = 0$

72. $\cos 2x = \sin x$

73. $\sin^2 x - \cos^2 x + 1 = 0$

74. $\cos 3x\cos x + \sin 3x\sin x = 0$

75. $\sin^2\left(\dfrac{x}{2}\right) - \cos x + 1 = 0$

76. $\sin x + \cos x = 1$

In Exercises 77–80, solve the given equations graphically.

77. $x + \ln x - 3\cos^2 x = 2$

78. $e^{\sin x} - 2 = x\cos^2 x$

79. $2\tan^{-1} x + x^2 = 3$

80. $3\sin^{-1} x = 6\sin x + 1$

In Exercises 81–84, find an algebraic expression for each of the given expressions.

81. $\tan(\cot^{-1} x)$ **82.** $\cos(\csc^{-1} x)$

83. $\sin(2\cos^{-1} x)$ **84.** $\cos(\pi - \tan^{-1} x)$

In Exercises 85–88, use the given substitutions to show that the equations are valid for $0 \le \theta < \pi/2$.

85. If $x = 2\cos\theta$, show that $\sqrt{4 - x^2} = 2\sin\theta$.

86. If $x = 2\sec\theta$, show that $\sqrt{x^2 - 4} = 2\tan\theta$.

87. If $x = \tan\theta$, show that $\dfrac{x}{\sqrt{1 + x^2}} = \sin\theta$.

88. If $x = \cos\theta$, show that $\dfrac{\sqrt{1 - x^2}}{x} = \tan\theta$.

In Exercises 89–104, use the formulas and methods of this chapter to solve the given problems.

89. Show that $\sin 2x\csc x\sec x$ has a constant value.

90. Show that $(\cos 2\alpha + \sin^2\alpha)\sec^2\alpha$ has a constant value.

91. Show that $y = A\sin 2t + B\cos 2t$ may be written as $y = C\sin(2t + \alpha)$ where $C = \sqrt{A^2 + B^2}$ and $\tan\alpha = B/A$. (*Hint:* Let $A/C = \cos\alpha$ and $B/C = \sin\alpha$.)

92. In a right triangle with sides a and b, and hypotenuse c, show that $\sin(A/2) = \sqrt{(c - b)/2c}$, where angle A is opposite a.

93. Forces **A** and **B** act on a bolt such that **A** makes an angle θ with the x-axis and **B** makes an angle θ with the y-axis as shown in Fig. 20-32. The resultant **R** has components $R_x = A\cos\theta - B\sin\theta$ and $R_y = A\sin\theta + B\cos\theta$. Using these components, show that $R = \sqrt{A^2 + B^2}$.

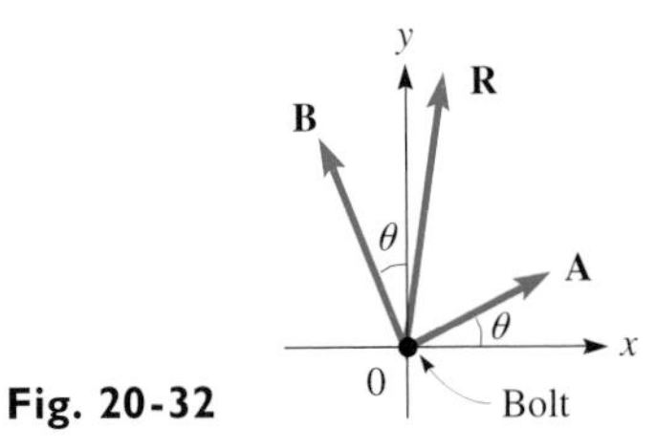

Fig. 20-32

94. In surveying, when determining an azimuth (a measure used for reference purposes), it might be necessary to simplify the expression

$$\frac{1}{2\cos\alpha\cos\beta} - \tan\alpha\tan\beta$$

Perform this operation by expressing it in the simplest possible form when $\alpha = \beta$.

95. Some comets follow a parabolic path that can be described by the equation $r = (k/2)\csc^2(\theta/2)$, where r is the distance to the sun and k is a constant. Show that this equation can be written as $r = k/(1 - \cos\theta)$.

96. In studying the interference of light waves, the identity $\frac{\sin\frac{3}{2}x}{\sin\frac{1}{2}x}\sin x = \sin x + \sin 2x$ is used. Prove this identity. (*Hint:* $\sin\frac{3}{2}x = \sin(x + \frac{1}{2}x)$.)

97. In the study of chemical spectroscopy, the equation $\omega t = \sin^{-1}\frac{\theta - \alpha}{R}$ arises. Solve for θ.

98. The power p in a certain electric circuit is given by $p = 2.5[\cos\alpha\sin(\omega t + \phi) - \sin\alpha\cos(\omega t + \phi)]$. Solve for t.

99. In finding the pressure exerted by soil on a retaining wall, the expression $1 - \cos 2\phi - \sin 2\phi\tan\alpha$ is found. Show that this expression can also be written as $2\sin\phi(\sin\phi - \cos\phi\tan\alpha)$.

100. In analyzing the motion of an automobile universal joint, the equation $\sec^2 A - \sin^2 B\tan^2 A = \sec^2 C$ is used. Show that this equation is true if $\tan A\cos B = \tan C$.

101. Determining the angle between two sections of a robot arm requires the solution of the equation $1.20\cos\theta + 0.135\cos 2\theta = 0$. Find the required angle θ if $0 < \theta < 180°$.

102. The angle of elevation of the top of the Washington Monument from a point on level ground 76.2 m from the center of its base is twice the angle of elevation of the top from a point 186 m farther away. Find the height of the monument.

103. If a plane surface inclined at angle θ moves horizontally, the angle for which the lifting force of the air is a maximum is found by solving the equation $2\sin\theta\cos^2\theta - \sin^3\theta = 0$, where $0 < \theta < 90°$. Solve for θ.

104. The electric current as a function of the time for a particular circuit is given by $i = 8.00\,e^{-20t}(1.73\cos 10.0t - \sin 10.0t)$. Find the time, in seconds, when the current is first zero.

Writing Exercise

105. Analyzing the forces on a bridge support, an engineer finds the equation $\tan\theta = 0.4250$. Write a paragraph explaining the difference in solving this equation and evaluating the expression $\tan^{-1} 0.4250$.

Practice Test

1. Prove that $\sec\theta - \frac{\tan\theta}{\csc\theta} = \cos\theta$.
2. Solve for x $(0 \le x < 2\pi)$ analytically, using trigonometric relations where necessary: $\sin 2x + \sin x = 0$
3. Find an algebraic expression for $\cos(\sin^{-1} x)$.
4. The angular displacement θ of a certain pendulum in terms of the time t is given by $\theta = e^{-0.1t}(\cos 2t + 3\sin 2t)$. What is the smallest value of t for which the displacement is zero?
5. Prove that $\frac{\tan\alpha + \tan\beta}{\tan\alpha - \tan\beta} = \frac{\sin(\alpha + \beta)}{\sin(\alpha - \beta)}$.
6. Prove that $\cot^2 x - \cos^2 x = \cot^2 x\cos^2 x$.
7. Find $\cos\frac{1}{2}x$ if $\sin x = -\frac{3}{5}$ and $270° < x < 360°$.
8. The intensity of a certain type of polarized light is given by $I = I_0\sin 2\theta\cos 2\theta$. Solve for θ.
9. Solve graphically: $x - 2\cos x = 5$.

21

Plane Analytic Geometry

In Section 21-4 we show an important application of analytic geometry in the design of automobile headlights.

We first introduced the graph of a function in Chapter 3. Since then we have seen that the graph of a specific type of function fits a particular form. For example, the graph of a linear function is a straight line, the graph of the quadratic function is a parabola, and the graph of the sine function is a periodic wave form.

The underlying principle of analytic geometry is the relationship of an algebraic equation and the geometric properties of the curve that represents the equation. In this chapter we develop equations for a number of important geometric curves, and show how to find the properties of these curves through an analysis of the equation. Most important among these curves are the *conic sections*, which we briefly introduced in Chapter 14.

The concepts developed in analytic geometry are very useful in the study of calculus. Also, there are a great many technical and scientific applications, which include projectile motion, planetary orbits, and fluid motion. Other important applications range from the design of gears, airplane wings, and automobile headlights to the construction of bridges and nuclear cooling towers.

21-1 Basic Definitions

In this section we develop certain basic concepts that will be needed for future use in establishing the proper relationships between an equation and a curve.

The Distance Formula

The first of these concepts involves the distance between any two points in the coordinate plane. If these points lie on a line parallel to the x-axis, the **directed distance** from the first point $A(x_1, y)$ to the second point $B(x_2, y)$ is denoted as $\overline{AB}$ and is defined as $x_2 - x_1$. We see that *$\overline{AB}$ is positive if B is to the right of A, and it is negative if B is to the left of A.* Similarly, the directed distance between two points $C(x, y_1)$ and $D(x, y_2)$ on a line parallel to the y-axis is $y_2 - y_1$. We see that *$\overline{CD}$ is positive if D is above C, and it is negative if D is below C.*

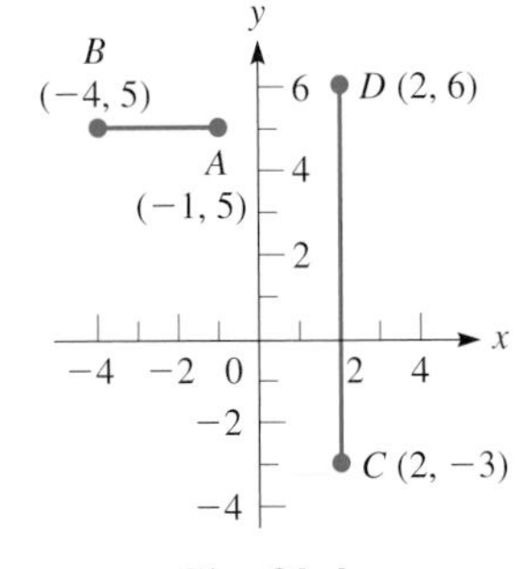

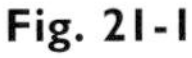
Fig. 21-1

EXAMPLE 1 ▸▸ The line segment joining $A(-1, 5)$ and $B(-4, 5)$ in Fig. 21-1 is parallel to the x-axis. Therefore, the directed distance

$$\overline{AB} = -4 - (-1) = -3$$

and the directed distance

$$\overline{BA} = -1 - (-4) = 3$$

Also in Fig. 21-1, the line segment joining $C(2, -3)$ and $D(2, 6)$ is parallel to the y-axis. The directed distance

$$\overline{CD} = 6 - (-3) = 9$$

and the directed distance

$$\overline{DC} = -3 - 6 = -9$$ ◂◂

We now wish to find the length of a line segment joining any two points in the plane. If these points are on a line that is not parallel to either axis (see Fig. 21-2), we use the Pythagorean theorem to find the distance between them. By making a right triangle with the line segment joining the points as the hypotenuse, and line segments parallel to the axes as legs, we have *the* **distance formula**, *which gives the distance between any two points in the plane.* This formula is

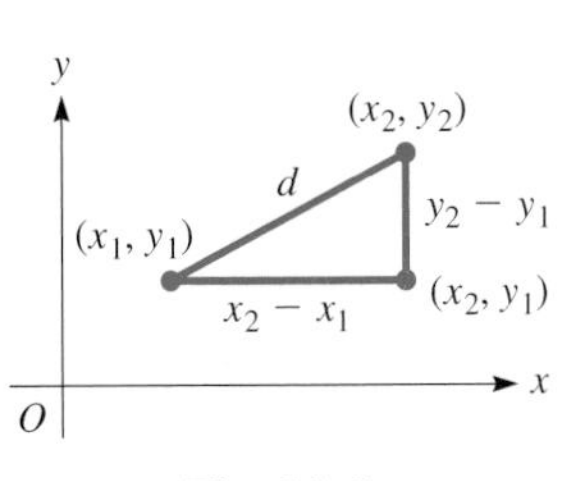

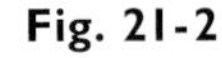
Fig. 21-2

$$d = \sqrt{(x_2 - x_1)^2 + (y_2 - y_1)^2} \tag{21-1}$$

Here we choose the positive square root since we are concerned only with the magnitude of the length of the line segment.

EXAMPLE 2 ▸▸ The distance between $(3, -1)$ and $(-2, -5)$ is given by

$$\begin{aligned} d &= \sqrt{[(-2) - 3]^2 + [(-5) - (-1)]^2} \\ &= \sqrt{(-5)^2 + (-4)^2} = \sqrt{25 + 16} \\ &= \sqrt{41} = 6.403 \end{aligned}$$

See Fig. 21-3.

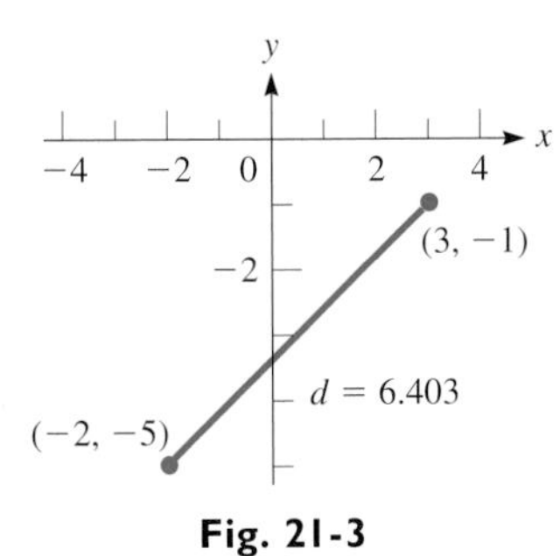

Fig. 21-3

It makes no difference which point is chosen as (x_1, y_1) and which is chosen as (x_2, y_2), since the differences in the x-coordinates and the y-coordinates are squared. We obtain the same value for the distance when we calculate it as

$$\begin{aligned} d &= \sqrt{[3 - (-2)]^2 + [(-1) - (-5)]^2} \\ &= \sqrt{5^2 + 4^2} = \sqrt{41} = 6.403 \end{aligned}$$ ◂◂

The Slope of a Line

Another important quantity for a line is its *slope*, which we defined in Chapter 5. Here we give it a somewhat more general definition and develop its meaning in more detail, as well as review the basic meaning.

The **slope** *gives a measure of the direction of a line, and is defined as the vertical directed distance from one point to another on the same straight line, divided by the horizontal directed distance from the first point to the second.* Thus, the slope, m, is given by

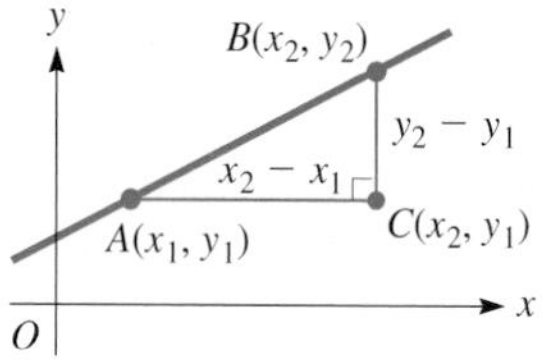

$$m = \frac{y_2 - y_1}{x_2 - x_1} \qquad \textbf{(21-2)}$$

Fig. 21-4

See Fig. 21-4. When the line is horizontal, $y_2 = y_1$ and $m = 0$. When the line is vertical, $x_2 = x_1$ and the slope is undefined.

EXAMPLE 3 ▸▸ The slope of a line joining $(3, -5)$ and $(-2, -6)$ is

$$m = \frac{-6 - (-5)}{-2 - 3} = \frac{-6 + 5}{-5} = \frac{1}{5}$$

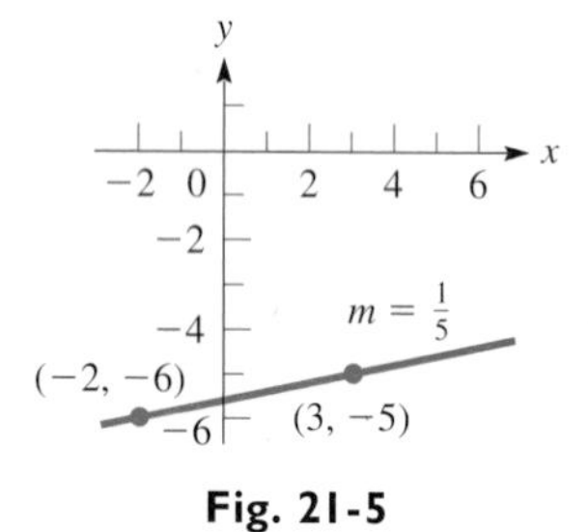

Fig. 21-5

See Fig. 21-5. Again we may interpret either of the points as (x_1, y_1) and the other as (x_2, y_2). We can also obtain the slope of this same line from

$$m = \frac{-5 - (-6)}{3 - (-2)} = \frac{1}{5} \quad ◂◂$$

The larger the numerical value of the slope of a line, the more nearly vertical is the line. Also, *a line rising to the right has a positive slope, and a line falling to the right has a negative slope.*

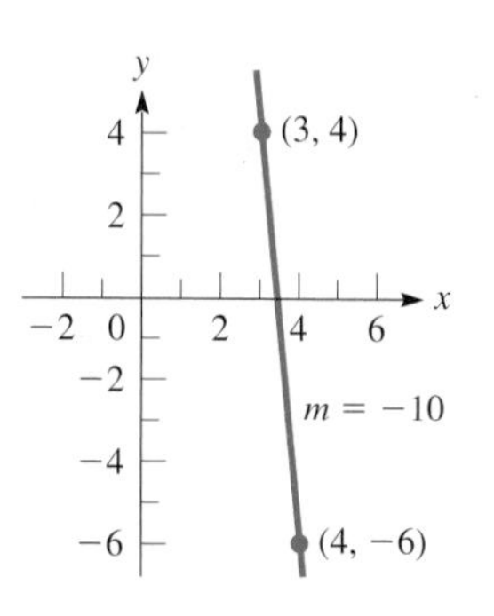

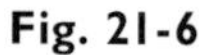
Fig. 21-6

EXAMPLE 4 ▸▸ (a) The line in Example 3 has a positive slope, which is numerically small. From Fig. 21-5 it can be seen that the line rises slightly to the right.

(b) The line joining $(3, 4)$ and $(4, -6)$ has a slope of

$$m = \frac{4 - (-6)}{3 - 4} = -10 \qquad \text{or} \qquad m = \frac{-6 - 4}{4 - 3} = -10$$

This line falls sharply to the right, as shown in Fig. 21-6. ◂◂

If a given line is extended indefinitely in either direction, it must cross the x-axis at some point unless it is parallel to the x-axis. *The angle measured from the x-axis in a positive direction to the line is called the* **inclination** *of the line* (see Fig. 21-7). The inclination of a line parallel to the x-axis is defined to be zero. *An alternate definition of slope, in terms of the inclination α, is*

$$m = \tan\alpha \quad (0° \leq \alpha < 180°) \qquad \textbf{(21-3)}$$

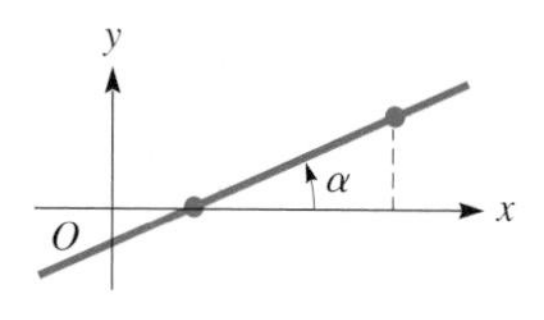

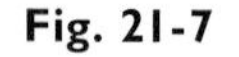
Fig. 21-7

Since the slope can be defined in terms of any two points on the line, we can choose the x-intercept and any other point. Therefore, from the definition of the tangent of an angle, we see that Eq. (21-3) is in agreement with Eq. (21-2).

EXAMPLE 5 ▸▸ (a) The slope of a line with an inclination of 45° is

$$m = \tan 45° = 1.000$$

(b) The slope of a line having an inclination of 120° is

$$m = \tan 120° = -\sqrt{3} = -1.732$$

See Fig. 21-8.

We see that if the inclination is an acute angle, the slope is positive and the line rises to the right. If the inclination is obtuse, the slope is negative and the line falls to the right. ◂◂

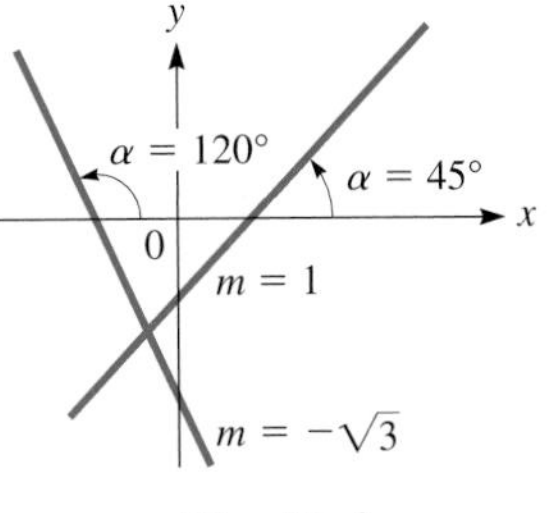

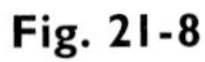
Fig. 21-8

Any two parallel lines crossing the x-axis have the same inclination. Therefore, as shown in Fig. 21-9, the *slopes of parallel lines are equal*. This can be stated as

$$m_1 = m_2 \qquad \text{(for } \parallel \text{ lines)} \qquad (21\text{-}4)$$

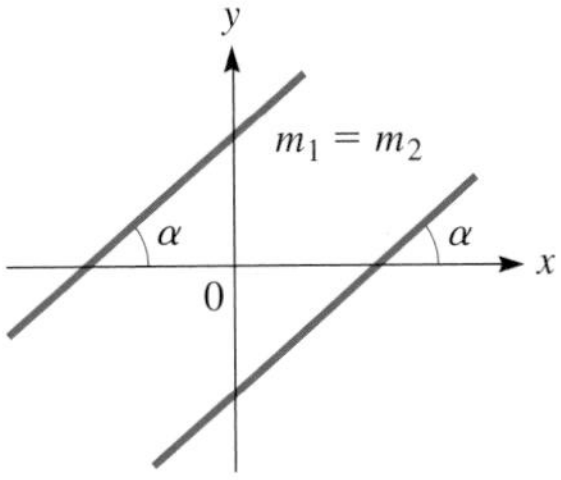

Fig. 21-9

If two lines are perpendicular, this means that there must be 90° between their inclinations (Fig. 21-10). The relation between their inclinations is

$$\alpha_2 = \alpha_1 + 90°$$

which can be written as

$$90° - \alpha_2 = -\alpha_1$$

If neither line is vertical (the slope of a vertical line is undefined), and we take the tangent in this last relation, we have

$$\tan(90° - \alpha_2) = \tan(-\alpha_1)$$

or

$$\cot \alpha_2 = -\tan \alpha_1$$

since a function of the complement of an angle equals the cofunction of that angle (see Section 4-4) and since $\tan(-\alpha) = -\tan \alpha$ (see Exercise 53 of Section 8-2). But $\cot \alpha = 1/\tan \alpha$, which means $1/\tan \alpha_2 = -\tan \alpha_1$. Using the inclination definition of slope, we have as *the relation between slopes of perpendicular lines,*

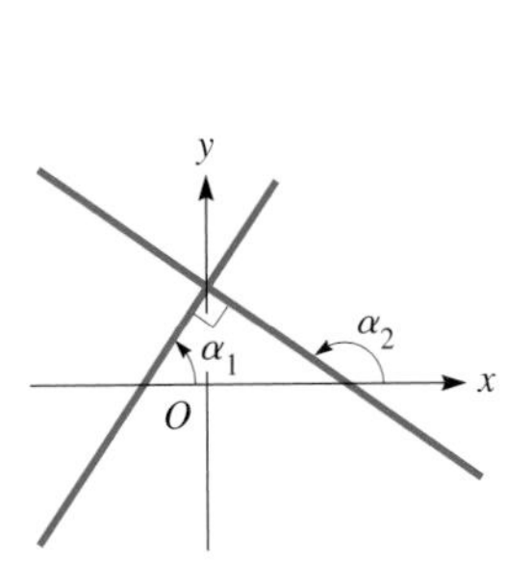

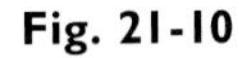
Fig. 21-10

$$m_2 = -\frac{1}{m_1} \quad \text{or} \quad m_1 m_2 = -1 \qquad \text{(for } \perp \text{ lines)} \qquad (21\text{-}5)$$

EXAMPLE 6 ▸▸ The line through (3, −5) and (2, −7) has a slope of

$$m_1 = \frac{-5 + 7}{3 - 2} = 2$$

The line through (4, −6) and (2, −5) has a slope of

$$m_2 = \frac{-6 - (-5)}{4 - 2} = -\frac{1}{2}$$

Since the slopes of the two lines are negative reciprocals, we know that the lines are perpendicular. See Fig. 21-11. ◂◂

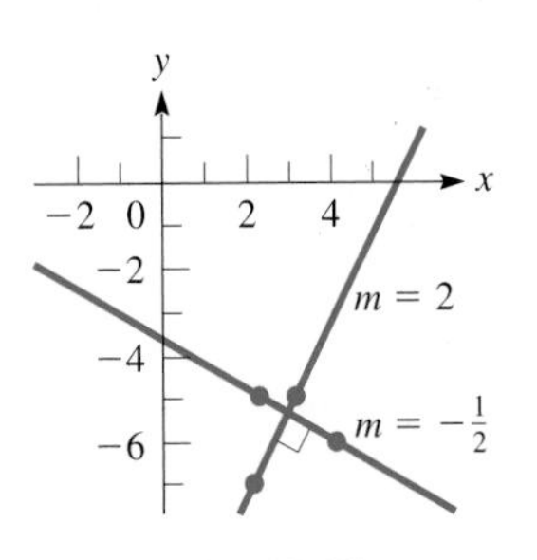

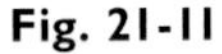
Fig. 21-11

Using the formulas for distance and slope, we can show certain basic geometric relationships. The following examples illustrate the use of the formulas and thereby show the use of algebra in solving problems that are basically geometric. This illustrates the methods of analytic geometry.

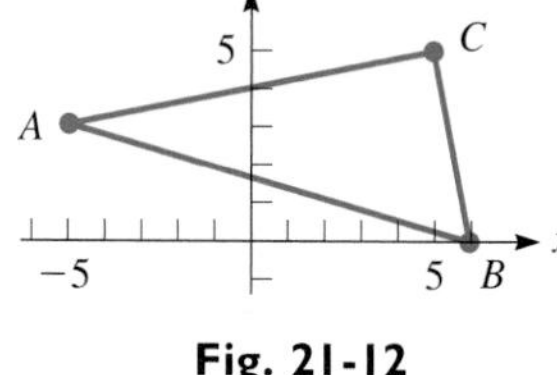

Fig. 21-12

EXAMPLE 7 ▸▸ Show that the line segments joining $A(-5,3)$, $B(6,0)$, and $C(5,5)$ form a right triangle. See Fig. 21-12.

If these points are vertices of a right triangle, the slopes of two of the sides must be negative reciprocals. This would show perpendicularity. Thus, we find the slopes of the three lines to be

$$m_{AB} = \frac{3-0}{-5-6} = -\frac{3}{11}, \qquad m_{AC} = \frac{3-5}{-5-5} = \frac{1}{5}, \qquad m_{BC} = \frac{0-5}{6-5} = -5$$

We see that the slopes of AC and BC are negative reciprocals, which means that $AC \perp BC$. From this we can conclude that the triangle is a right triangle. ◂◂

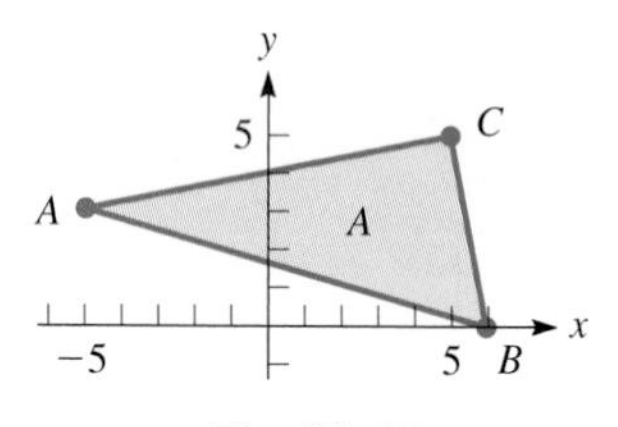

Fig. 21-13

EXAMPLE 8 ▸▸ Find the area of the triangle in Example 7. See Fig. 21-13.

Since the right angle is at C, the legs of the triangle are AC and BC. The area is one-half the product of the lengths of the legs of a right triangle. The lengths of the legs are

$$d_{AC} = \sqrt{(-5-5)^2 + (3-5)^2} = \sqrt{104} = 2\sqrt{26}$$
$$d_{BC} = \sqrt{(6-5)^2 + (0-5)^2} = \sqrt{26}$$

Therefore, the area is

$$A = \frac{1}{2}(2\sqrt{26})(\sqrt{26}) = 26$$ ◂◂

Exercises 21–1

In Exercises 1–10, find the distance between the given pairs of points.

1. $(3,8)$ and $(-1,-2)$
2. $(-1,3)$ and $(-8,-4)$
3. $(4,-5)$ and $(4,-8)$
4. $(-3,7)$ and $(2,10)$
5. $(-12,20)$ and $(32,-13)$
6. $(23,-9)$ and $(-25,11)$
7. $(-4,-3)$ and $(3,-3)$
8. $(-2,5)$ and $(-2,-2)$
9. $(1.22,-3.45)$ and $(-1.07,-5.16)$
10. $(-5.6,2.3)$ and $(8.2,-7.5)$

In Exercises 11–20, find the slopes of the lines through the points in Exercises 1–10.

In Exercises 21–24, find the slopes of the lines with the given inclinations.

21. 30° **22.** 62.5° **23.** 132.7° **24.** 135°

In Exercises 25–28, find the inclinations of the lines with the given slopes.

25. 0.364 **26.** 0.824 **27.** −6.691 **28.** −1.428

In Exercises 29–32, determine whether the lines through the two pairs of points are parallel or perpendicular.

29. $(6,-1)$ and $(4,3)$; $(-5,2)$ and $(-7,6)$
30. $(-3,9)$ and $(4,4)$; $(9,-1)$ and $(4,-8)$
31. $(-1,-4)$ and $(2,3)$; $(-5,2)$ and $(-19,8)$
32. $(-1,-2)$ and $(3,6)$; $(2,-6)$ and $(5,0)$

In Exercises 33–36, determine the value of k.

33. The distance between $(-1,3)$ and $(11,k)$ is 13.
34. The distance between $(k,0)$ and $(0,2k)$ is 10.
35. Points $(6,-1)$, $(3,k)$, and $(-3,-7)$ are on the same line.
36. The points in Exercise 35 are the vertices of a right triangle, with the right angle at $(3,k)$.

In Exercises 37–40, show that the given points are vertices of the given geometric figures.

37. $(2,3)$, $(4,9)$, and $(-2,7)$ are vertices of an isosceles triangle.

38. $(-1, 3)$, $(3, 5)$, and $(5, 1)$ are the vertices of a right triangle.

39. $(-5, -4)$, $(7, 1)$, $(10, 5)$ and $(-2, 0)$ are the vertices of a parallelogram.

40. $(-5, 6)$, $(0, 8)$, $(-3, 1)$, and $(2, 3)$ are the vertices of a square.

In Exercises 41–44, find the indicated areas and perimeters.

41. Find the area of the triangle in Exercise 38.

42. Find the area of the square in Exercise 40.

43. Find the perimeter of the triangle in Exercise 37.

44. Find the perimeter of the parallelogram in Exercise 39.

In Exercises 45–48, use the following definition to find the midpoints between the given points on a straight line.

The *midpoint* between points (x_1, y_1) and (x_2, y_2) on a straight line is the point

$$\left(\frac{x_1 + x_2}{2}, \frac{y_1 + y_2}{2}\right)$$

45. $(-4, 9)$ and $(6, 1)$

46. $(-1, 6)$ and $(-13, -8)$

47. $(-12.4, 25.7)$ and $(6.8, -17.3)$

48. $(2.6, 5.3)$ and $(-4.2, -2.7)$

21-2 The Straight Line

In Chapter 5 we derived the *slope-intercept form* of the equation of the straight line. Here we extend the development to include other forms of the equation of a straight line. Also, other methods of finding and applying these equations are shown. For completeness, we review some of the material in Chapter 5.

Using the definition of slope, we can derive the general type of equation that represents a straight line. This is another basic method of analytic geometry. That is, equations of a particular form can be shown to represent a particular type of curve. When we recognize the form of the equation, we know the kind of curve it represents. As we have seen, this is of great assistance in sketching the graph.

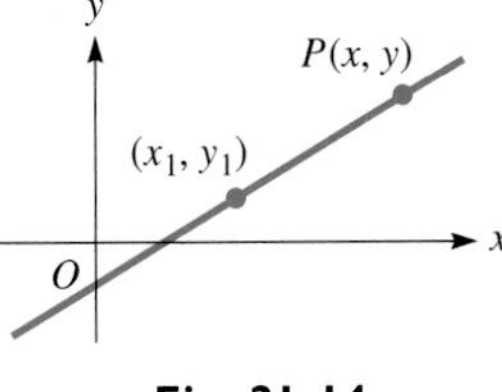

Fig. 21-14

A straight line can be defined as a *curve* with a constant slope. This means that the value for the slope is the same for any two different points on the line that might be chosen. Thus, considering point (x_1, y_1) on a line to be fixed (Fig. 21-14), and another point $P(x, y)$ that *represents* any other point on the line, we have

$$m = \frac{y - y_1}{x - x_1}$$

which can be written as

$$y - y_1 = m(x - x_1) \qquad (21\text{-}6)$$

Equation (21-6) is the **point-slope form** *of the equation of a straight line.* It is useful when we know the slope of a line and some point through which the line passes.

EXAMPLE 1 ▸▸ Find the equation of the line that passes through $(-4, 1)$ with a slope of -2. See Fig. 21-15.

Substituting in Eq. (21-6), we find that

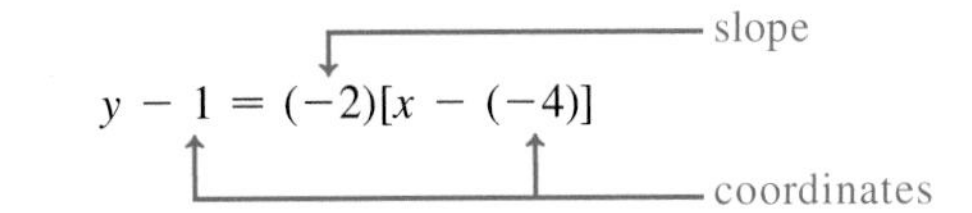

which can be simplified to

$$y + 2x + 7 = 0$$ ◂◂

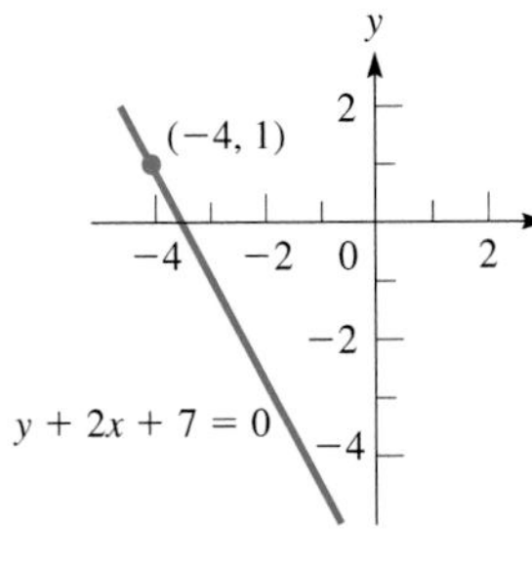

Fig. 21-15

EXAMPLE 2 ▸▸ Find the equation of the line through $(2, -1)$ and $(6, 2)$.

We first find the slope of the line through these points:

$$m = \frac{2+1}{6-2} = \frac{3}{4}$$

Then, by using either of the two known points and Eq. (21-6), we can find the equation of the line:

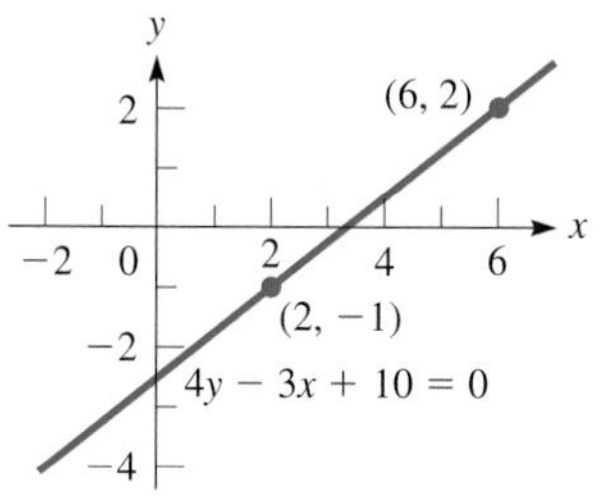

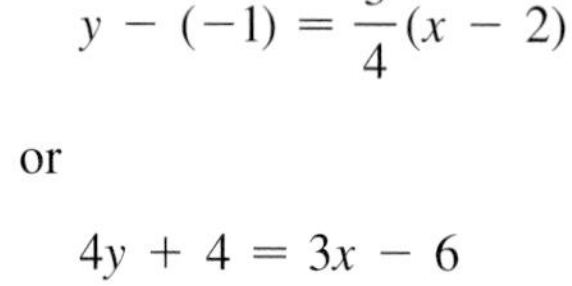

Fig. 21-16

$$y - (-1) = \frac{3}{4}(x - 2)$$

or

$$4y + 4 = 3x - 6$$

or

$$4y - 3x + 10 = 0$$

This line is shown in Fig. 21-16. ◂◂

Equation (21-6) can be used for any line except for one parallel to the y-axis. Such a line has an undefined slope. However, it does have the property that all points have the same x-coordinate, regardless of the y-coordinate. *We represent a line parallel to the y-axis (see Fig. 21-17) as*

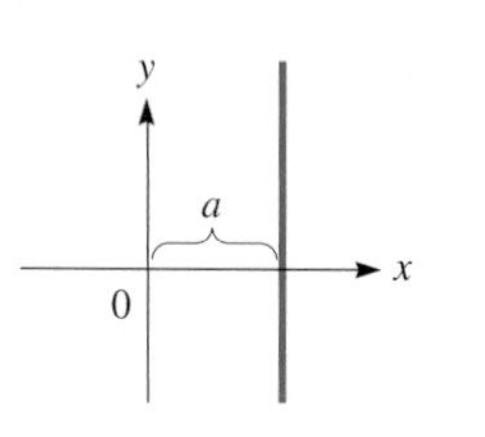

Fig. 21-17

$$x = a \tag{21-7}$$

A line parallel to the x-axis has a slope of zero. From Eq. (21-6), we can find its equation to be $y = y_1$. To keep the same form as Eq. (21-7), we normally write this as

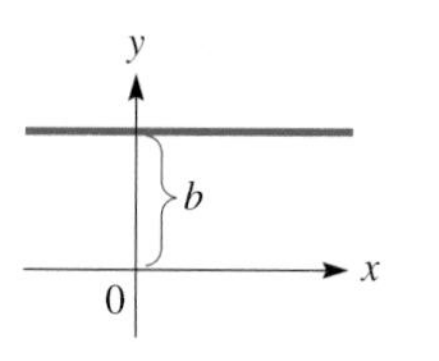

Fig. 21-18

$$y = b \tag{21-8}$$

See Fig. 21-18.

EXAMPLE 3 ▸▸ (a) The line $x = 2$ is a line parallel to the y-axis and 2 units to the right of it. This line is shown in Fig. 21-19.

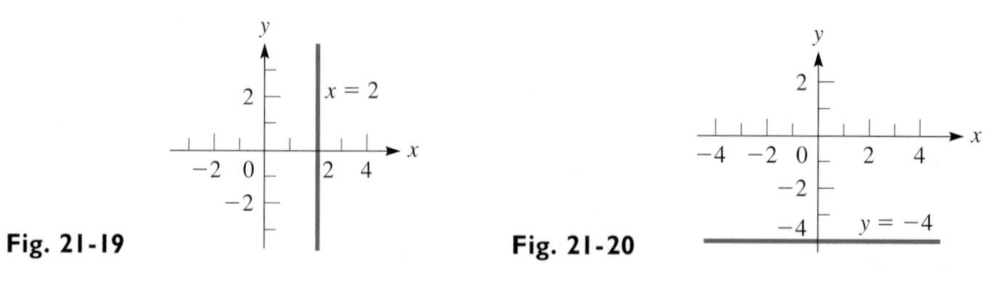

Fig. 21-19

Fig. 21-20

(b) The line $y = -4$ is a line parallel to the x-axis and 4 units below it. This line is shown in Fig. 21-20. ◂◂

If we choose the special point $(0, b)$, which is the y-intercept of the line, as the point to use in Eq. (21-6), we have $y - b = m(x - 0)$, or

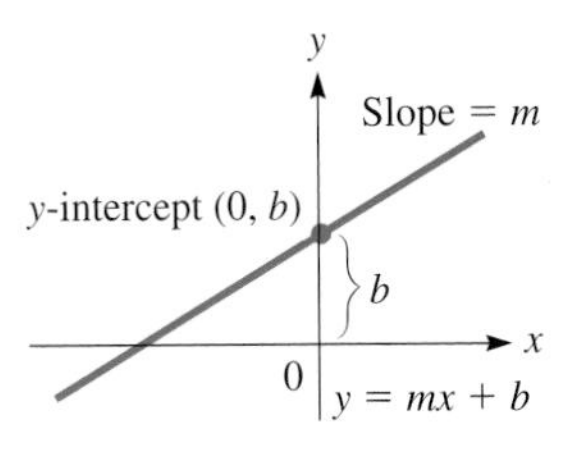

Fig. 21-21

$$y = mx + b \tag{21-9}$$

Equation (21-9) is the **slope-intercept form** *of the equation of a straight line,* and we first derived it in Chapter 5. Its primary usefulness lies in the fact that once we find the equation of a line and then write it in slope-intercept form, we know that the slope of the line is the coefficient of the x-term and that it crosses the y-axis at the coordinate indicated by the constant term. See Fig. 21-21.

EXAMPLE 4 ▸▸ Find the slope and the y-intercept of the straight line whose equation is $2y + 4x - 5 = 0$.

We write this equation in slope-intercept form:

$$2y = -4x + 5$$

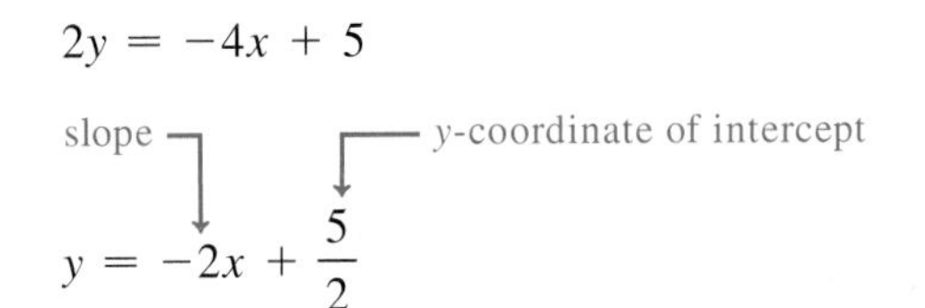

$$y = -2x + \frac{5}{2}$$

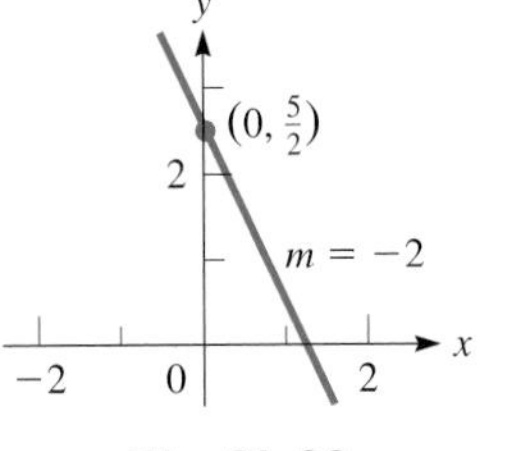

Fig. 21-22

Since the coefficient of x in this form is -2, the slope is -2. The constant on the right is $5/2$, which means that the y-intercept is $(0, 5/2)$. See Fig. 21-22. ◂◂

EXAMPLE 5 ▸▸ The pressure p_0 at the surface of a body of water (due to the atmosphere) is 101 kPa. The pressure p at a depth of 10.0 m is 199 kPa. In general, the pressure difference $p - p_0$ varies directly as the depth h. Sketch a graph of p as a function of h.

The solution is as follows:

$p - p_0 = kh$	direct variation
$199 - 101 = k(10.0)$	substitute given values
$k = 9.80$ kPa/m	
$p - 101 = 9.80h$	substitute in first equation
$p = 9.80h + 101$	

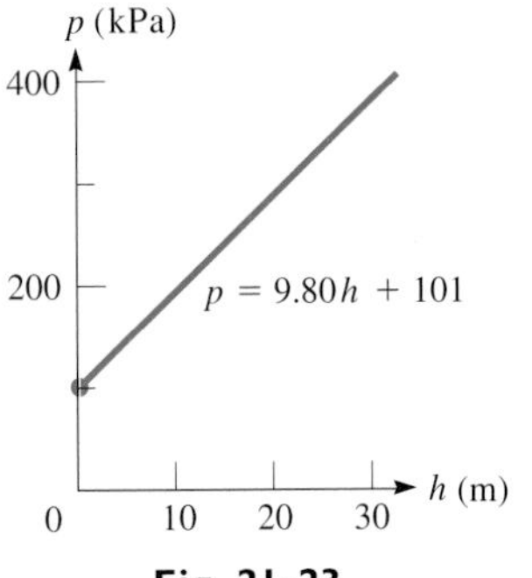

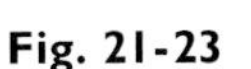
Fig. 21-23

We see that this is the equation of a straight line. The slope is 9.80, and the p-intercept is $(0, 101)$. Negative values do not have any physical meaning. The graph is shown in Fig. 21-23. ◂◂

From Eqs. (21-6) and (21-9), and from the examples of this section, we see that the equation of the straight line has certain characteristics: We have a term in y, a term in x, and a constant term if we simplify as much as possible. *This form is represented by the equation*

$$Ax + By + C = 0 \tag{21-10}$$

which is known as the **general form** *of the equation of the straight line.* We saw this form before in Chapter 5. Now we have shown why it represents a straight line.

EXAMPLE 6 ▸▸ Find the general form of the equation of the line parallel to the line $3x + 2y - 6 = 0$ and which passes through the point $(-1, 2)$.

Since the line whose equation we want is parallel to the line $3x + 2y - 6 = 0$, it has the same slope. Thus, writing $3x + 2y - 6 = 0$ in slope-intercept form,

$$\begin{aligned} 3x + 2y - 6 &= 0 \\ 2y &= -3x + 6 \\ y &= -\frac{3}{2}x + 3 \end{aligned}$$

Since the slope of $3x + 2y - 6 = 0$ is $-3/2$, the slope of the required line is also $-3/2$. Using $m = -3/2$, the point $(-1, 2)$, and the point-slope form, we have

$$\begin{aligned} y - 2 &= -\frac{3}{2}(x + 1) \\ 2y - 4 &= -3(x + 1) \\ 2y - 4 &= -3x - 3 \\ 3x + 2y - 1 &= 0 \end{aligned}$$

Fig. 21-24

This is the general form of the equation. Both lines are shown in Fig. 21-24. ◂◂

In many physical situations, a linear relationship exists between variables. A few examples of this are (1) the distance traveled by an object and the elapsed time, when the velocity is constant, (2) the amount a spring stretches and the force applied, (3) the change in electric resistance and the change in temperature, (4) the force applied to an object and the resulting acceleration, and (5) the pressure at a certain point within a liquid and the depth of the point.

EXAMPLE 7 ▸▸ Under the condition of constant acceleration, the velocity of an object varies linearly with the time. If after 1.0 s a car has a velocity of 40 m/s, and 3.0 s later it has a velocity of 55 m/s, find the equation relating the velocity v and the time t, and graph the equation. From the graph, find the initial velocity (when the timing was started, or when $t = 0$) and the velocity after 6.0 s.

With v as dependent variable and t as independent variable, the slope is

$$m = \frac{v_2 - v_1}{t_2 - t_1}$$

Using the information given in the statement of the problem, we have

$$m = \frac{55 - 40}{4.0 - 1.0} = 5.0$$

Then, using the point-slope form of the equation of a straight line, we have

$$\begin{aligned} v - 40 &= 5.0(t - 1.0) \\ v &= 5.0t + 35 \end{aligned}$$

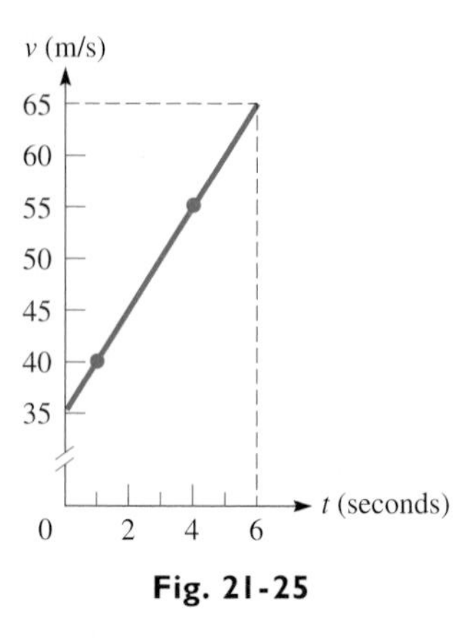

Fig. 21-25

This is the required equation. The graph is shown in Fig. 21-25. For purposes of graphing the line, the values given are sufficient. Of course, there is no need to include negative values of t, since these have no physical meaning. From the graph we see that the line crosses the v-axis at 35. This means that the initial velocity is 35 m/s. Also, when $t = 6.0$ s we see that $v = 65$ m/s. ◂◂

Exercises 21–2

In Exercises 1–20, find the equation of each of the lines with the given properties. Sketch the graph of each line.

1. Passes through $(-3, 8)$ with a slope of 4

2. Passes through $(-2, -1)$ with a slope of -2

3. Passes through $(2, -5)$ and $(4, 2)$

4. Passes through $(-3, 5)$ and $(-2, 3)$

5. Passes through $(1, 3)$ and has an inclination of $45°$

6. Has a y-intercept $(0, -2)$ and an inclination of $120°$

7. Passes through $(5.3, -2.7)$ and is parallel to the x-axis

8. Passes through $(-4, -2)$; perpendicular to x-axis

9. Is parallel to the y-axis and is 3 units to the left of it

10. Is parallel to the x-axis and is 4.1 units below it

11. Has an x-intercept $(4, 0)$ and a y-intercept of $(0, -6)$

12. Has an x-intercept of $(-3, 0)$ and a slope of 2

13. Is perpendicular to a line with a slope of 3 and passes through $(1, -2)$

14. Is perpendicular to a line with a slope of -4 and has a y-intercept of $(0, 3)$

15. Is parallel to a line with a slope of $-\frac{1}{2}$ and has an x-intercept of $(4, 0)$

16. Is parallel to a line through $(-1, 2)$ and $(3, 1)$ and passes through $(1, 2)$

17. Is parallel to a line through $(7, -1)$ and $(4, 3)$ and has a y-intercept of $(0, -2)$

18. Is perpendicular to the line joining $(4, 2)$ and $(3, -5)$ and passes through $(4, 2)$

19. Is perpendicular to the line $6.0x - 2.4y - 3.9 = 0$ and passes through $(7.5, -4.7)$

20. Is parallel to the line $2y - 6x - 5 = 0$ and passes through $(-4, -5)$

In Exercises 21–24, draw the lines with the given equations.

21. $4x - y = 8$

22. $2x - 3y - 6 = 0$

23. $3x + 5y - 10 = 0$

24. $4y = 6x - 9$

In Exercises 25–28, reduce the equations to slope-intercept form and find the slope and the y-intercept.

25. $3x - 2y - 1 = 0$

26. $4x + 2y - 5 = 0$

27. $11.2x - 3.2y + 1.6 = 0$

28. $11.5x + 4.60y - 5.98 = 0$

In Exercises 29–32, determine the value of k.

29. What is the value of k if the lines $4x - ky = 6$ and $6x + 3y + 2 = 0$ are parallel?

30. What is the value of k in Exercise 29 if the given lines are perpendicular?

31. What is the value of k if the lines $3x - y = 9$ and $kx + 3y = 5$ are perpendicular? Explain how this value is found.

32. What is the value of k in Exercise 31 if the given lines are parallel? Explain how this value is found.

In Exercises 33 and 34, show that the given lines are parallel. In Exercises 35 and 36, show that the given lines are perpendicular.

33. $3x - 2y + 5 = 0$ and $4y = 6x - 1$

34. $3y - 2x = 4$ and $6x - 9y = 5$

35. $6x - 3y - 2 = 0$ and $x + 2y - 4 = 0$

36. $4.5x - 1.8y = 1.7$ and $2.4x + 6.0y = 0.3$

In Exercises 37–40, find the equations of the given lines.

37. The line that has an x-intercept of $(4, 0)$ and a y-intercept the same as the line $2y - 3x - 4 = 0$.

38. The line that is perpendicular to the line $8x + 2y - 3 = 0$ and has the same x-intercept as this line.

39. The line with a slope of -3 that also passes through the intersection of the lines $5x - y = 6$ and $x + y = 12$.

40. The line that passes through the point of intersection of $2x + y - 3 = 0$ and $x - y - 3 = 0$ and through the point $(4, -3)$.

In Exercises 41–52, some applications involving straight lines are shown.

41. The velocity v of a box sliding down a long ramp is given by $v = v_0 + at$, where v_0 is the initial velocity, a is the acceleration, and t is the time. If $v_0 = 3.35$ m/s and $v = 9.87$ m/s when $t = 4.50$ s, find v as a function of t. Sketch the graph.

42. The voltage V across part of an electric circuit is given by $V = E - iR$, where E is a battery voltage, i is the current, and R is the resistance. If $E = 6.00$ V and $V = 4.35$ V for $i = 9.17$ mA, find V as a function of i. Sketch the graph (i and V may be negative).

43. The length l of a pulley belt is 8 cm longer than twice the circumference c of one of the pulley wheels. Express l as a function of c.

44. An acid solution is made from x litres of a 20% solution and y litres of a 30% solution. If the final solution contains 20 L of acid, find the equation relating x and y.

45. One computer printer can print x characters per second, and a second printer can print y characters per second. If the first prints for 50 s and the second for 60 s and they print a total of 12 200 characters, find the equation relating x and y.

46. A wall is 15 cm thick. At the outside, the temperature is 3°C and at the inside, it is 23°C. If the temperature changes at a constant rate through the wall, write an equation of the temperature T in the wall as a function of the distance x from the outside to the inside of the wall.

47. One heating unit uses x litres of fuel at 72% efficiency, and another heating unit uses y litres at 90% efficiency. If 135 MJ of heat is delivered by these units together, express y as a function of x.

48. The length of a rectangular solar cell is 10 cm more than the width. Express the perimeter p of the cell as a function of the width w.

49. A light beam is reflected off the edge of an optic fiber at an angle of 0.0032°. The diameter of the fiber is 48 μm. Find the equation of the reflected beam with the x-axis (at the center of the fiber) and the y-axis as shown in Fig. 21-26.

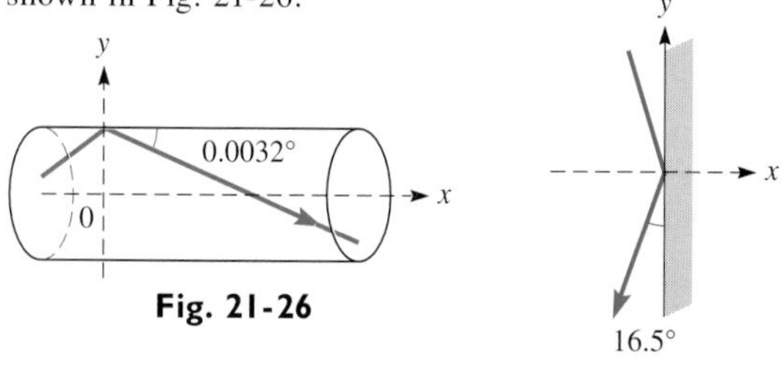

Fig. 21-26

Fig. 21-27

50. A police report stated that a bullet caromed downward off a wall at an angle of 16.5° with the wall, as shown in Fig. 21-27. What is the equation of the path of the bullet after impact?

51. A survey of the traffic on a particular highway showed that the number of cars passing a particular point each minute varied linearly from 6:30 A.M. to 8:30 A.M. on workday mornings. The study showed that an average of 45 cars passed the point in one minute at 7 A.M. and that 115 cars passed in one minute at 8 A.M. If n is the number of cars passing the point in one minute and t is the number of minutes after 6:30 A.M., find the equation relating n and t, and graph the equation. From the graph, determine n at 6:30 A.M. and at 8:30 A.M.

52. In a research project on cancer, a tumor was determined to weigh 30 mg when first discovered. While being treated, it grew smaller by 2 mg each month. Find the equation relating the weight w of the tumor as a function of the time t in months. Graph the equation.

In Exercises 53–56, treat the given nonlinear functions as linear functions in order to sketch their graphs. As an example, $y = 2 + 3x^2$ can be sketched as a straight line by graphing y as a function of x^2. A table of values for this graph is shown along with the corresponding graph in Fig. 21-28.

x	0	1	2	3	4	5
x^2	0	1	4	9	16	25
y	2	5	14	29	50	77

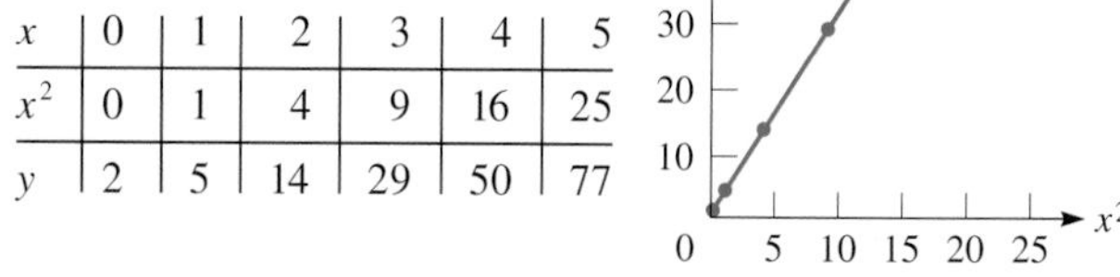

Fig. 21-28

53. The number n of memory cells of a certain computer that can be tested in t seconds is given by $n = 1200\sqrt{t}$. Sketch n as a function of $\sqrt{t}$.

54. The force F, in newtons, applied to a lever to balance a certain weight on the opposite side of the fulcrum is given by $F = 40/d$, where d is the distance, in metres, of the force from the fulcrum. Sketch F as a function of $1/d$.

55. A spacecraft is launched such that its altitude h, in kilometres, is given by $h = 300 + 2t^{3/2}$ for $0 \le t < 100$ s. Sketch this as a linear function.

56. The current i, in amperes, in a certain electric circuit is given by $i = 6(1 - e^{-t})$. Sketch this as a linear function.

In Exercises 57–60, show that the given nonlinear functions are linear when plotted on semilogarithmic or logarithmic paper. In Section 13-7, we noted that graphs plotted on this paper often become straight lines.

57. A function of the form $y = ax^n$ is straight when plotted on logarithmic paper, since $\log y = \log a + n \log x$ is in the form of a straight line. The variables are $\log y$ and $\log x$; the slope can be found from $(\log y - \log a)/\log x = n$, and the intercept is a. (To get the slope from the graph, it is necessary to measure vertical and horizontal distances between two points. The log y-intercept is found where $\log x = 0$, and this occurs when $x = 1$.) Plot $y = 3x^4$ on logarithmic paper to verify this analysis.

58. A function of the form $y = a(b^x)$ is a straight line on semilogarithmic paper, since $\log y = \log a + x \log b$ is in the form of a straight line. The variables are $\log y$ and x, the slope is $\log b$, and the intercept is a. (To get the slope from the graph, we calculate $(\log y - \log a)/x$ for some set of values x and y. The intercept is read directly off the graph where $x = 0$.) Plot $y = 3(2^x)$ on semilogarithmic paper to verify this analysis.

59. If experimental data are plotted on logarithmic paper and the points lie on a straight line, it is possible to determine the function (see Exercise 57). The following data come from an experiment to determine the functional relationship between the pressure p and the volume V of a gas undergoing an adiabatic (no heat loss) change. From the graph on logarithmic paper, determine p as a function of V.

V (m^3)	0.100	0.500	2.00	5.00	10.0
p (kPa)	20.1	2.11	0.303	0.0840	0.0318

60. If experimental data are plotted on semilogarithmic paper, and the points lie on a straight line, it is possible to determine the function (see Exercise 58). The following data come from an experiment designed to determine the relationship between the voltage across an inductor and the time, after the switch is opened. Determine v as a function of t.

v (volts)	40	15	5.6	2.2	0.8
t (milliseconds)	0.0	20	40	60	80

21-3 The Circle

We have found that we can obtain a general equation which represents a straight line by considering a fixed point on the line and then a general point $P(x, y)$ which can represent any other point on the same line. Mathematically we can state this as "the line is the **locus** of a point $P(x, y)$ that *moves* from a fixed point with constant slope along the line." That is, the point $P(x, y)$ can be considered as a variable point that moves along the line.

In this way we can define a number of important curves. *A* **circle** *is defined as the locus of a point $P(x, y)$ which moves so that it is always equidistant from a fixed point. We call this fixed distance the* **radius**, *and we call the fixed point the* **center** *of the circle.* Thus, using this definition, calling the fixed point (h, k) and the radius r, we have

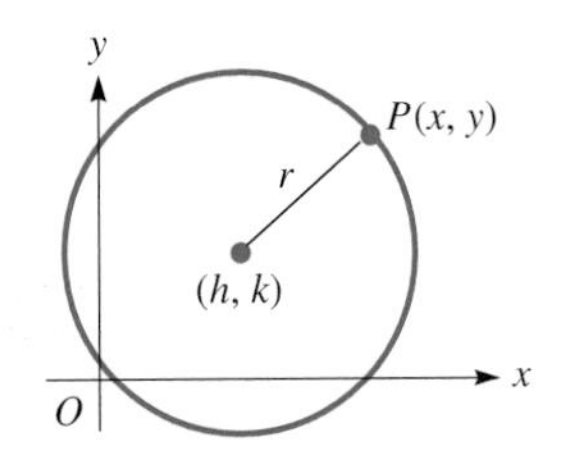

Fig. 21-29

$$\sqrt{(x - h)^2 + (y - k)^2} = r$$

or, by squaring both sides, we have

$$(x - h)^2 + (y - k)^2 = r^2 \qquad (21\text{-}11)$$

Equation (21-11) is called the **standard equation** *of a circle with center at (h, k) and radius r.* See Fig. 21-29.

EXAMPLE 1 ▸▸ The equation $(x - 1)^2 + (y + 2)^2 = 16$ represents a circle with center at $(1, -2)$ and a radius of 4. We determine these values by considering the equation of the circle to be in the form of Eq. (21-11) as

form requires − signs

$$(x - 1)^2 + [y - (-2)]^2 = 4^2$$

coordinates of center   radius

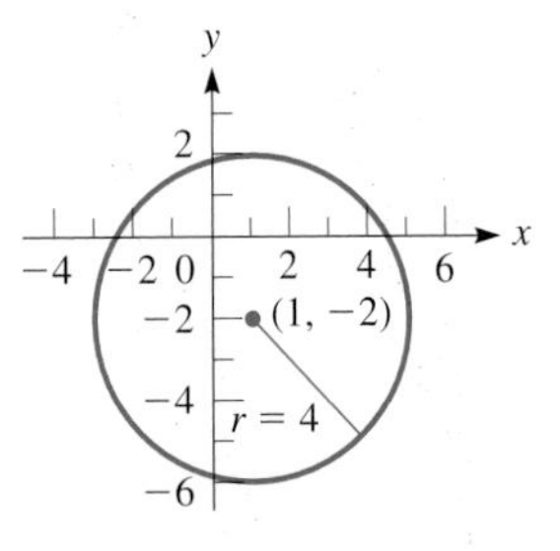

Fig. 21-30

Note carefully the way in which we found the y-coordinate of the center. *We must have a minus sign before each of the coordinates.* Here, to get the y-coordinate, we had to write $+2$ as $-(-2)$. This circle is shown in Fig. 21-30. ◂◂

EXAMPLE 2 ▸▸ Find the equation of the circle with center at (2, 1) and which passes through (4, 8).

In Eq. (21-11) we can determine the equation if we can find h, k, and r for this circle. From the given information, $h = 2$ and $k = 1$. To find r, we use the fact that *all points on the circle must satisfy the equation of the circle.* The point (4, 8) must satisfy Eq. (21-11), with $h = 2$ and $k = 1$. Thus, $(4 - 2)^2 + (8 - 1)^2 = r^2$. From this relation we find $r^2 = 53$. The equation of the circle is

$$(x - 2)^2 + (y - 1)^2 = 53$$

This circle is shown in Fig. 21-31. ◂◂

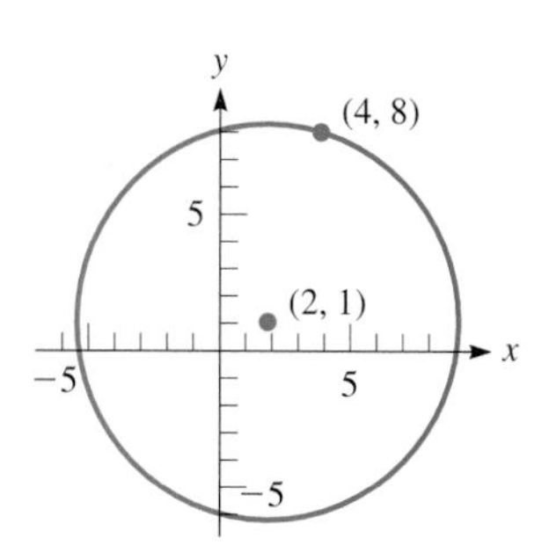

Fig. 21-31

If the center of the circle is at the origin, which means that the coordinates of the center are (0, 0), the equation of the circle (see Fig. 21-32) becomes

$$x^2 + y^2 = r^2 \qquad (21\text{-}12)$$

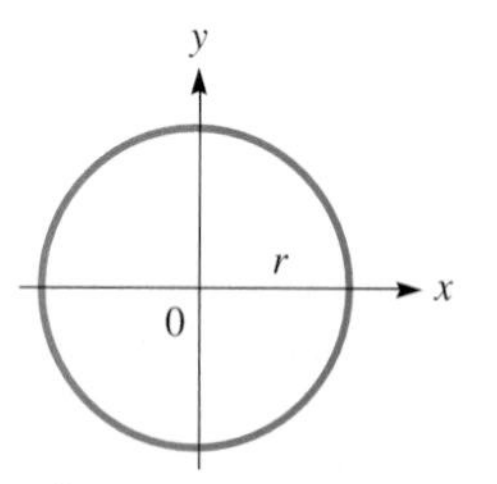

Fig. 21-32

The following example illustrates an application using this type of circle and one with its center not at the origin.

EXAMPLE 3 ▸▸ A student is drawing a friction drive in which two circular disks are in contact with each other. They are represented by circles in the drawing. The first has a radius of 10.0 cm, and the second has a radius of 12.0 cm. What is the equation of each circle if the origin is at the center of the first circle and the positive x-axis passes through the center of the second circle? See Fig. 21-33.

Since the center of the smaller circle is at the origin, we can use Eq. (21-12). Given that the radius is 10.0 cm, we have as its equation

$$x^2 + y^2 = 100$$

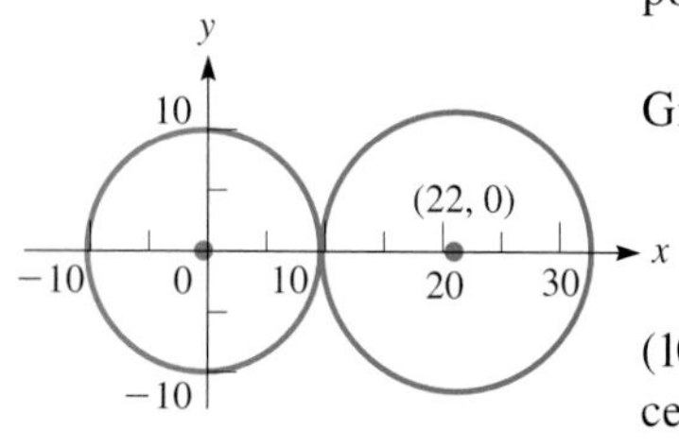

Fig. 21-33

The fact that the two disks are in contact tells us that they meet at the point (10.0, 0). Knowing that the radius of the larger circle is 12.0 cm tells us that its center is at (22.0, 0). Thus, using Eq. (21-11) with $h = 22.0$, $k = 0$, and $r = 12.0$,

$$(x - 22.0)^2 + (y - 0)^2 = 12.0^2$$

or

$$(x - 22.0)^2 + y^2 = 144$$

as the equation of the larger circle. ◂◂

Symmetry

A circle with its center at the origin exhibits an important property of the graphs of many equations. *It is* **symmetrical** *to the x-axis and also to the y-axis.* Symmetry to the x-axis can be thought of as meaning that the lower half of the curve is a reflection of the upper half, and conversely. It can be shown that *if $-y$ can replace y in an equation without changing the equation, the graph of the equation is* ***symmetrical to the x-axis***. Symmetry to the y-axis is similar. *If $-x$ can replace x in the equation without changing the equation, the graph is symmetrical to the y-axis.*

This type of circle is also symmetrical to the origin as well as being symmetrical to both axes. The meaning of symmetry to the origin is that the origin is the midpoint of any two points (x, y) and $(-x, -y)$ that are on the curve. Thus, *if $-x$ can replace x, and $-y$ can replace y at the same time, without changing the equation, the graph of the equation is symmetrical to the origin.*

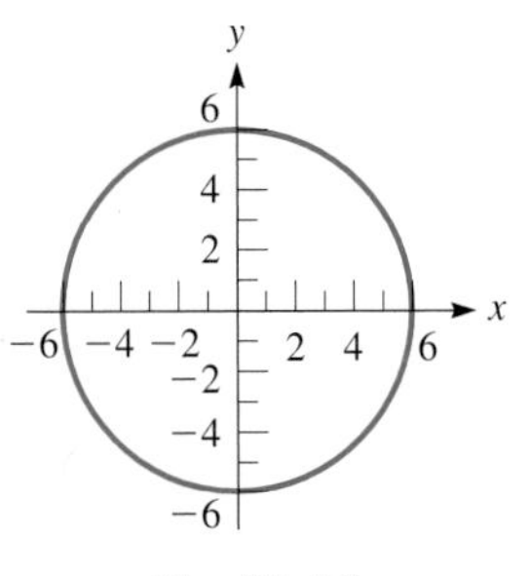

Fig. 21-34

EXAMPLE 4 ▸▸ The equation of the circle with its center at the origin and with a radius of 6 is $x^2 + y^2 = 36$.

The symmetry of this circle can be shown analytically by the substitutions mentioned above. Replacing x by $-x$, we obtain $(-x)^2 + y^2 = 36$. Since $(-x)^2 = x^2$, this equation can be rewritten as $x^2 + y^2 = 36$. Since this substitution did not change the equation, the graph is symmetrical to the y-axis.

Replacing y by $-y$, we obtain $x^2 + (-y)^2 = 36$, which is the same as $x^2 + y^2 = 36$. This means that the curve is symmetrical to the x-axis.

Replacing x by $-x$, and simultaneously replacing y by $-y$, we obtain $(-x)^2 + (-y)^2 = 36$, which is the same as $x^2 + y^2 = 36$. This means that the curve is symmetrical to the origin. This circle is shown in Fig. 21-34. ◂◂

If we multiply out each of the terms in Eq. (21-11), we may combine the resulting terms to obtain

$$x^2 - 2hx + h^2 + y^2 - 2ky + k^2 = r^2$$

$$\mathbf{x^2 + y^2 - 2hx - 2ky + (h^2 + k^2 - r^2) = 0} \qquad (21\text{-}13)$$

Since each of h, k, and r is constant for any given circle, the coefficients of x and y and the term within parentheses in Eq. (21-13) are constants. Equation (21-13) can then be written as

$$x^2 + y^2 + Dx + Ey + F = 0 \qquad (21\text{-}14)$$

Equation (21-14) is called the **general equation** *of the circle.* It tells us that any equation which can be written in that form will represent a circle.

EXAMPLE 5 ▸▸ Find the center and radius of the circle

$$x^2 + y^2 - 6x + 8y - 24 = 0$$

We can find this information if we write the given equation in standard form. To do so, ***we must complete the square in the x-terms and also in the y-terms***. This is done by first writing the equation in the form

CAUTION ▸

$$(x^2 - 6x \qquad) + (y^2 + 8y \qquad) = 24$$

To complete the square of the x-terms, we take half of -6, which is -3, square it, and add the result, 9, to each side of the equation. In the same way, we complete the square of the y-terms by adding 16 to each side of the equation, which gives

$$\left(\frac{-6}{2}\right)^2 \qquad \left(\frac{8}{2}\right)^2 \qquad \text{add to both sides}$$

$$(x^2 - 6x + 9) + (y^2 + 8y + 16) = 24 + 9 + 16$$

$$(x - 3)^2 + (y + 4)^2 = 49$$

$$(x - 3)^2 + (y - (-4))^2 = 7^2$$

coordinates of center — radius

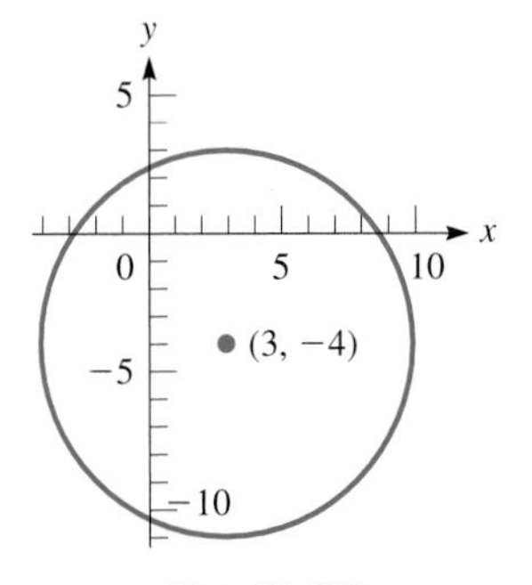

Fig. 21-35

Thus, the center is $(3, -4)$, and the radius is 7 (see Fig. 21-35). ◂◂

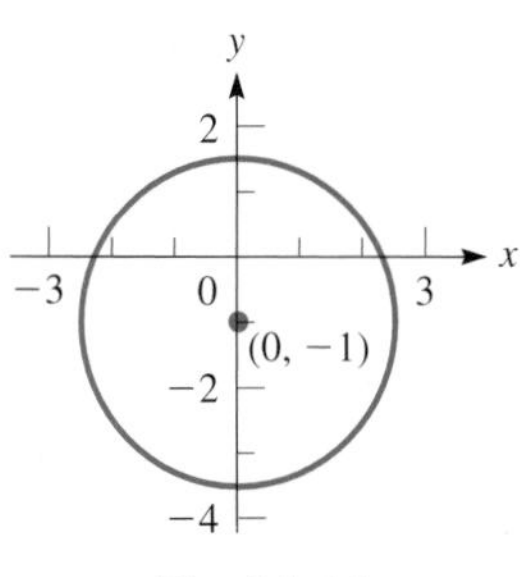

Fig. 21-36

EXAMPLE 6 ▸▸ The equation $3x^2 + 3y^2 + 6y - 20 = 0$ can be seen to represent a circle by writing it in general form. This is done by dividing through by 3. In this way we have $x^2 + y^2 + 2y - 20/3 = 0$. To find the center and radius, we write the equation in standard form and complete the necessary squares. This leads to

$$x^2 + (y^2 + 2y + 1) = 20/3 + 1$$
$$x^2 + (y + 1)^2 = 23/3$$

Therefore, the center is $(0, -1)$, and the radius is $\sqrt{23/3} \approx 2.77$. See Fig. 21-36.

We can also see that this circle is symmetrical to the y-axis, but it is not symmetrical to the x-axis or to the origin. If we replace x by $-x$, the equation does not change, but if we replace y by $-y$, the term $6y$ in the original equation becomes negative, and the equation *does* change. ◂◂

NOTE ▸ In Section 14-1 we noted that the equation of a circle does not represent a *function* since there are two values of y for most values of x in the domain. In fact, *it might be necessary to use the quadratic formula to find the two functions to enter into a graphing calculator* in order to view the curve. This is illustrated in the following example.

EXAMPLE 7 ▸▸ Display the graph of the circle $3x^2 + 3y^2 + 6y - 20 = 0$ on a graphing calculator. (This is the same circle as in Example 6.)

To fit the form of a quadratic equation in y, we write

$$3y^2 + 6y + (3x^2 - 20) = 0$$

Now, using the quadratic formula to solve for y, we let

$$a = 3,\ b = 6, \text{ and } c = 3x^2 - 20$$

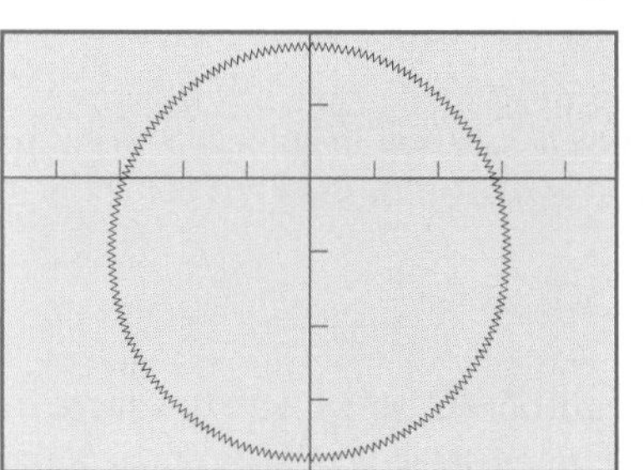

Fig. 21-37

Therefore, we get the *two* functions

$$y = \frac{-6 + \sqrt{276 - 36x^2}}{6} \quad \text{and} \quad y = \frac{-6 - \sqrt{276 - 36x^2}}{6}$$

which are entered into the calculator to get the view shown in Fig. 21-37. We have used the RANGE values

$$\text{Xmin} = -4.5,\ \text{Xmax} = 4.5,\ \text{Ymin} = -4,\ \text{Ymax} = 2$$

for which the length along the x-axis is about 1.5 times that along the y-axis, to have less distortion in the circle. There may be gaps at the left and right ends of the circle. ◂◂

Exercises 21–3

In Exercises 1–4, determine the center and the radius of each circle.

1. $(x - 2)^2 + (y - 1)^2 = 25$

2. $(x - 3)^2 + (y + 4)^2 = 49$

3. $(x + 1)^2 + y^2 = 4$ **4.** $x^2 + (y - 6)^2 = 64$

In Exercises 5–20, find the equation of each of the circles from the given information.

5. Center at $(0, 0)$, radius 3 **6.** Center at $(0, 0)$, radius 1

7. Center at $(2, 2)$, radius 4 **8.** Center at $(0, 2)$, radius 2

9. Center at $(-2, 5)$, radius $\sqrt{5}$

10. Center at $(-3, -5)$, radius $2\sqrt{3}$

11. Center at $(12, -15)$, radius 18
12. Center at $(\frac{3}{2}, -2)$, radius $\frac{5}{2}$
13. Center at $(2, 1)$, passes through $(4, -1)$
14. Center at $(-1, 4)$, passes through $(-2, 3)$
15. Center at $(-3, 5)$, tangent to the x-axis
16. Center at $(2, -4)$, tangent to the y-axis
17. Tangent to both axes and the lines $y = 4$ and $x = 4$
18. Tangent to both axes, radius 4, in the second quadrant
19. Center on the line $5x = 2y$, radius 5, tangent to the x-axis
20. The points $(3, 8)$ and $(-3, 0)$ are the ends of a diameter

In Exercises 21–32, determine the center and radius of each circle. Sketch each circle.

21. $x^2 + (y - 3)^2 = 4$
22. $(x - 2)^2 + (y + 3)^2 = 49$
23. $4(x + 1)^2 + 4(y - 5)^2 = 81$
24. $2(x + 4)^2 + 2(y + 3)^2 = 25$
25. $x^2 + y^2 - 25 = 0$
26. $x^2 + y^2 - 12 = 0$
27. $x^2 + y^2 - 2x - 8 = 0$
28. $x^2 + y^2 - 4x - 6y - 12 = 0$
29. $x^2 + y^2 + 4.20x - 2.60y - 3.51 = 0$
30. $x^2 + y^2 + 22x + 14y + 1 = 0$
31. $2x^2 + 2y^2 - 4x - 8y - 1 = 0$
32. $3x^2 + 3y^2 - 12x + 4 = 0$

In Exercises 33–36, determine whether the circles with the given equations are symmetrical to either axis or to the origin.

33. $x^2 + y^2 = 100$
34. $x^2 + y^2 - 4x - 5 = 0$
35. $x^2 + y^2 + 8y - 9 = 0$
36. $x^2 + y^2 - 2x + 4y - 3 = 0$

In Exercises 37–48, solve the given problems.

37. Determine whether the circle $x^2 - 6x + y^2 - 7 = 0$ crosses the x-axis.
38. Find the points of intersection of the circle $x^2 + y^2 - x - 3y = 0$ and the line $y = x - 1$.
39. Find the locus of a point $P(x, y)$ which moves so that its distance from $(2, 4)$ is twice its distance from $(0, 0)$. Describe the locus.
40. Find the equation of the locus of a point $P(x, y)$ which moves so that the line joining it and $(2, 0)$ is always perpendicular to the line joining it and $(-2, 0)$. Describe the locus.
41. Use a graphing calculator to view the circle $x^2 + y^2 + 5y - 4 = 0$.
42. Use a graphing calculator to view the circle $2x^2 + 2y^2 + 2y - x - 1 = 0$.
43. A pendulum swings through an arc of the circle $2x^2 + 2y^2 - 6.80y - 1.90 = 0$. How long (in metres) is the pendulum, and from what point is it swinging?
44. The design of a machine part shows it as a circle represented by the equation $x^2 + y^2 = 42.5$ (measured in centimetres), with a circular hole represented by $x^2 + y^2 + 3.06y - 1.24 = 0$ cut out. What is the least distance from the edge of the hole to the edge of the machine part?
45. A wire is rotating in a circular path through a magnetic field to induce an electric current in the wire. The wire is rotating at 60.0 Hz with a constant velocity of 37.7 m/s. Taking the origin at the center of rotation, find the equation of the path of the wire.
46. A communications satellite remains stationary at an altitude of 36 200 km over a point on the earth's equator. It therefore rotates once each day about the earth's center. Its velocity is constant, but the horizontal and vertical components, v_H and v_V, of the velocity constantly change. Show that the equation relating v_H and v_V, measured in km/h, is that of a circle. The radius of the earth is 6370 km.
47. In analyzing the stress on a beam, *Mohr's circle* is often used. To form it, normal stress is plotted as the x-coordinate and shear stress is plotted as the y-coordinate. The center of the circle is midway between the minimum and maximum values of normal stress on the x-axis. Find the equation of Mohr's circle if the minimum normal stress is 100×10^{-6} and the maximum normal stress is 900×10^{-6} (stress is unitless). Sketch the graph.
48. A Norman window has the form of a rectangle surmounted by a semicircle. An architect designs a Norman window on a coordinate system as shown in Fig. 21-38. If the circumference of the circular part of the window is on the circle $x^2 + y^2 - 3.00y + 1.25 = 0$, find the area of the window. Measurements are in metres.

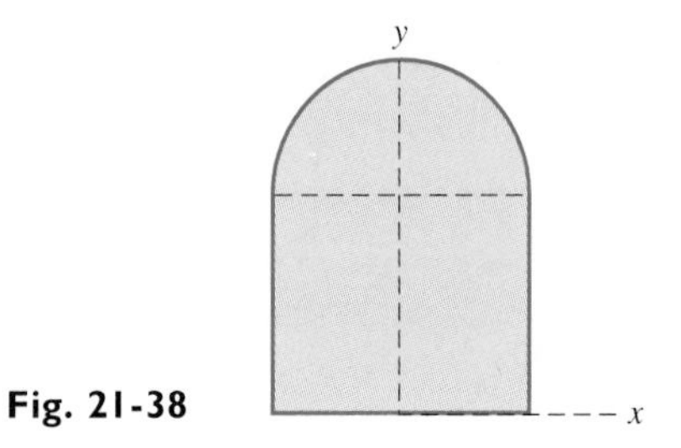

Fig. 21-38

21-4 The Parabola

Another important curve is the parabola. We came across this curve several times in earlier chapters. In Chapter 7 we showed that the graph of a quadratic function is a parabola. In this section we define the parabola more generally and thereby find the general form of its equation.

A **parabola** *is defined as the locus of a point* $P(x, y)$ *which moves so that it is always equidistant from a given line and a given point. The given line is called the* **directrix,** *and the given point is called the* **focus**. The line through the focus that is perpendicular to the directrix is called the **axis** of the parabola. The point midway between the directrix and focus is the **vertex** of the parabola. Using this definition, we shall find the equation of the parabola for which the focus is the point $(p, 0)$ and the directrix is the line $x = -p$. By choosing the focus and directrix in this manner, we can find a general representation of the equation of a parabola with its vertex at the origin.

According to the definition of the parabola, the distance from a point $P(x, y)$ on the parabola to the focus $(p, 0)$ must equal the distance from $P(x, y)$ to the directrix $x = -p$. The distance from P to the focus can be found by using the distance formula. The distance from P to the directrix is the perpendicular distance, and this can be found as the distance between two points on a line parallel to the x-axis. These distances are indicated in Fig. 21-39.

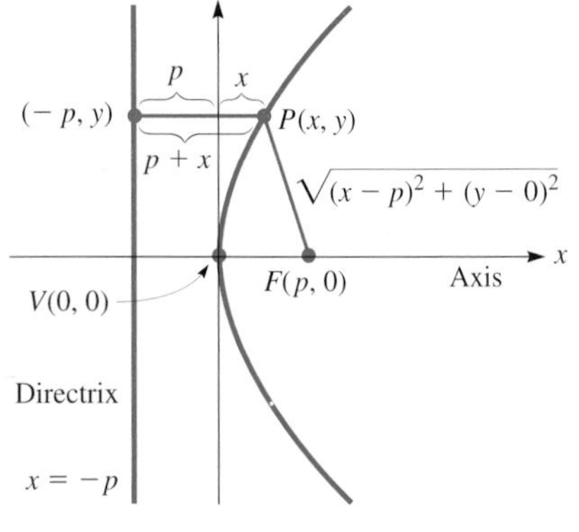

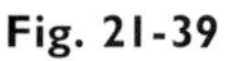
Fig. 21-39

Thus, we have

$$\sqrt{(x - p)^2 + (y - 0)^2} = x + p$$

Squaring both sides of this equation, we have

$$(x - p)^2 + y^2 = (x + p)^2$$

or

$$x^2 - 2px + p^2 + y^2 = x^2 + 2px + p^2$$

Simplifying, we obtain

$$y^2 = 4px \tag{21-15}$$

Equation (21-15) is called the **standard form** *of the equation of a parabola with its axis along the x-axis and the vertex at the origin.* Its symmetry to the x-axis can be proven since $(-y)^2 = 4px$ is the same as $y^2 = 4px$.

EXAMPLE 1 ▸▸ Find the coordinates of the focus and the equation of the directrix, and sketch the graph of the parabola $y^2 = 12x$.

Since the equation of this parabola fits the form of Eq. (21-15), we know that the vertex is at the origin. The coefficient of 12 tells us that

$$4p = 12, \qquad p = 3$$

Since $p = 3$, the focus is the point $(3, 0)$ and the directrix is the line $x = -3$, as shown in Fig. 21-40. ◂◂

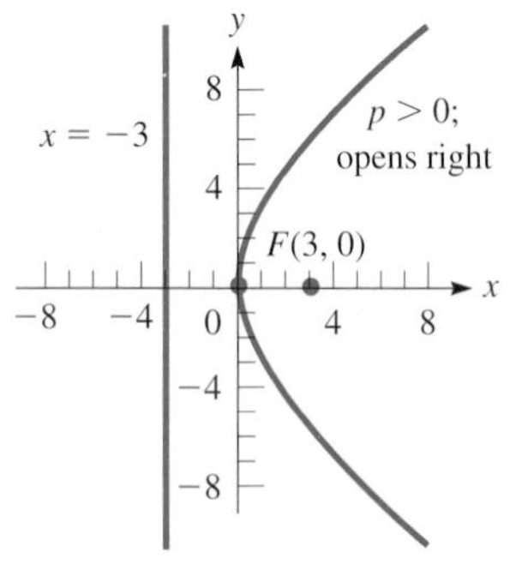

Fig. 21-40

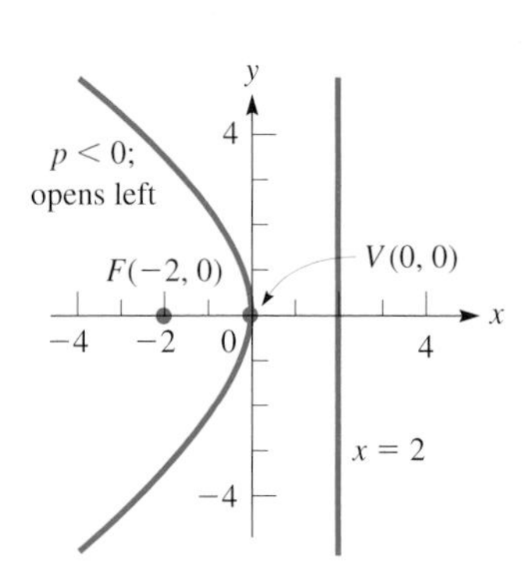

Fig. 21-41

EXAMPLE 2 ▸▸ If the focus is to the left of the origin, with the directrix an equal distance to the right, the coefficient of the x-term is negative. This tells us that the parabola opens to the left, rather than to the right, as is the case when the focus is to the right of the origin. For example, the parabola $y^2 = -8x$ has its vertex at the origin, its focus at $(-2, 0)$, and the line $x = 2$ as its directrix. We determine this from the equation as follows:

$$y^2 = -8x, \qquad 4p = -8 \qquad p = -2$$

Focus $(p, 0)$ is $(-2, 0)$; directrix $x = -p$ is $x = -(-2) = 2$. The parabola opens to the left, as shown in Fig. 21-41. ◂◂

If we chose the focus as the point $(0, p)$ and the directrix as the line $y = -p$ (see Fig. 21-42), we would find that the resulting equation is

$$x^2 = 4py \tag{21-16}$$

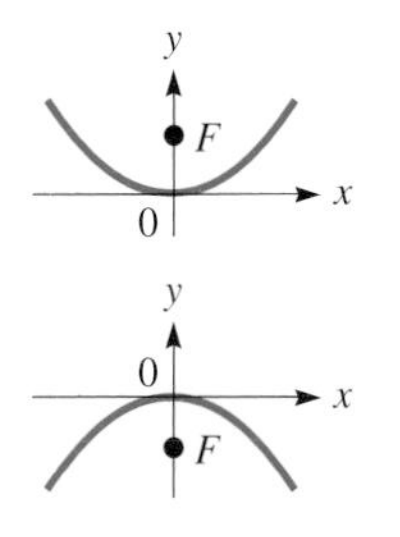

Fig. 21-42

This is the standard form of the equation of a parabola with the y-axis as its axis and the vertex at the origin. Its symmetry to the y-axis can be proved, since $(-x)^2 = 4py$ is the same as $x^2 = 4py$. We note that ***the difference between this equation and Eq. (21-15) is that x is squared and y appears to the first power in Eq. (21-16), rather than the reverse, as in Eq. (21-15).***

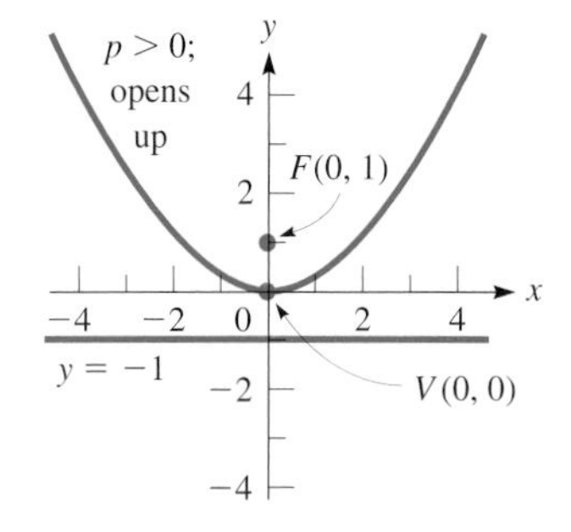

Fig. 21-43

EXAMPLE 3 ▸▸ The parabola $x^2 = 4y$ fits the form of Eq. (21-16). Therefore, its axis is along the y-axis and its vertex is at the origin. From the equation, we find the value of p, which in turn tells us the location of the vertex and the directrix. Therefore, we have

$$x^2 = 4y, \qquad 4p = 4, \qquad p = 1$$

Focus $(0, p)$ is $(0, 1)$; directrix $y = -p$ is $y = -1$. The parabola is shown in Fig. 21-43, and we see in this case that it opens upward. ◂◂

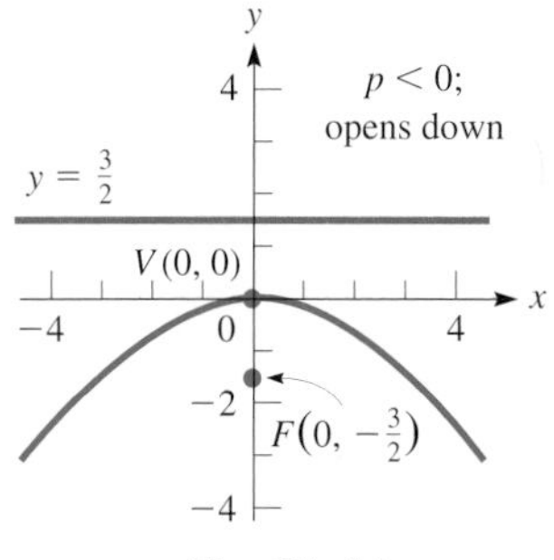

Fig. 21-44

EXAMPLE 4 ▸▸ The parabola $x^2 = -6y$ fits the form of Eq. (21-16) with $4p = -6$. Therefore, its axis is along the y-axis and its vertex is at the origin. Since $4p = -6$, we have

$$p = -\frac{3}{2}, \quad \text{focus}\left(0, -\frac{3}{2}\right), \quad \text{directrix } y = \frac{3}{2}$$

The parabola opens downward, as shown in Fig. 21-44. ◂◂

See the chapter introduction.

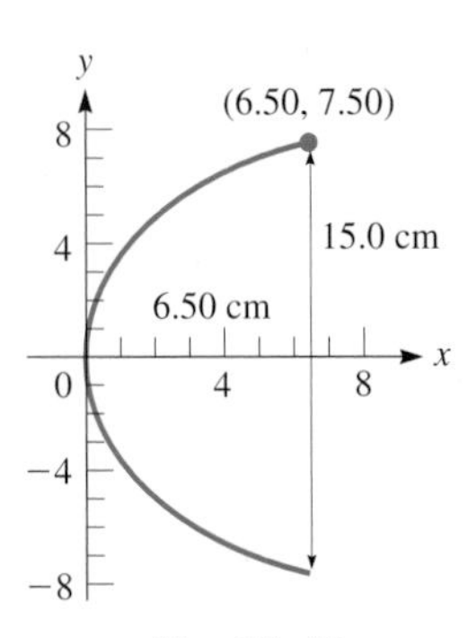

Fig. 21-45

EXAMPLE 5 ▸▸ In calculus it can be shown that a light ray coming from the focus of a parabola will be reflected off the parabolic surface parallel to the axis of the parabola. This property of a parabola involving light reflection has many very useful applications. One of these is in the design of automobile headlights.

An automobile headlight reflector is designed such that the cross sections of the reflecting surface are equal parabolas. It has an opening of 15.0 cm and is 6.50 cm deep, as shown in Fig. 21-45. Determine where the filament of the bulb should be located so that reflected light rays are parallel in order to create a light beam.

By finding the equation of the parabola we can find the location of the focus, which is the desired location of the filament. Knowing that the parabolic opening is 15.0 cm wide and 6.50 cm deep tells us that the point (6.50, 7.50) is on the parabola.

Since we have placed the parabola with its vertex at the origin and its axis along the x-axis, its general form is given by Eq. (21-15), or $y^2 = 4px$. We can find the value of p by using the fact that (6.50, 7.50) is a point on the parabola and the fact that the coordinates must satisfy the equation. This means that

$$7.50^2 = 4p(6.50) \quad \text{or} \quad p = 2.16$$

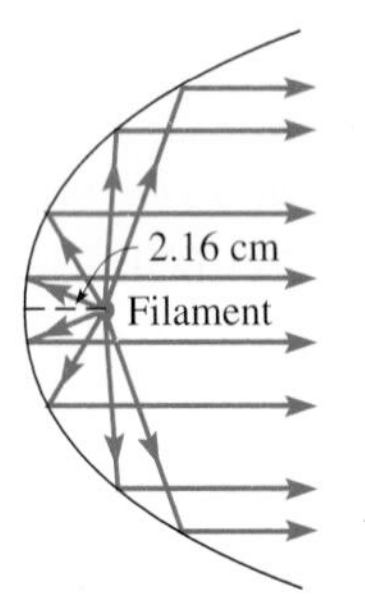

Fig. 21-46

Therefore, the equation of the parabola is $y^2 = 8.64x$. The filament should be located at the focus, which means it should be on the axis of the parabola, 2.16 cm from the vertex. In this way the reflected rays are parallel to the axis, as shown in Fig. 21-46. ◂◂

Equations (21-15) and (21-16) give us the general form of the equation of a parabola with its vertex at the origin and its focus on one of the coordinate axes. The next example shows the use of the definition to find the equation of a parabola that has its vertex at a point other than at the origin.

EXAMPLE 6 ▸▸ Using the definition of the parabola, find the equation of the parabola with its focus at (2, 3) and its directrix the line $y = -1$. See Fig. 21-47.

Choosing a general point $P(x, y)$ on the parabola, and equating the distances from this point to (2, 3) and to the line $y = -1$, we have

$$\sqrt{(x-2)^2 + (y-3)^2} = y + 1$$

distance P to F = distance P to $y = -1$

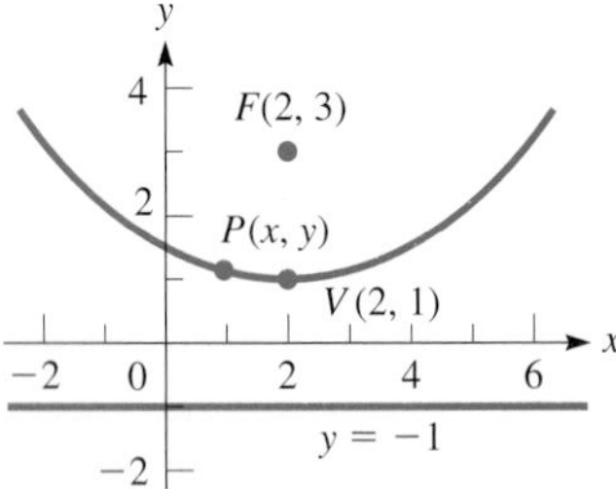

Fig. 21-47

Squaring both sides of this equation and simplifying, we have

$$(x-2)^2 + (y-3)^2 = (y+1)^2$$
$$x^2 - 4x + 4 + y^2 - 6y + 9 = y^2 + 2y + 1$$

or

$$8y = 12 - 4x + x^2$$

We note that this type of equation has appeared frequently in earlier chapters. The x-term and the constant (12 in this case) are characteristic of a parabola that does not have its vertex at the origin if the directrix is parallel to the x-axis.

We can readily view this parabola on a graphing calculator. However, if the axis of the parabola is parallel to the x-axis (the equation would contain a y^2-term and a y-term, and no x^2-term), we would have to use the quadratic formula to find the two functions to use, as we did in Example 7 of Section 21-3. ◂◂

We can conclude that *the equation of a parabola is characterized by the presence of the square of either (but not both) x or y, and a first power term in the other.* We will consider further the equation of the parabola in Sections 21-7 and 21-8.

The parabola has numerous technical applications. The reflection property illustrated in Example 5 has other important applications, such as the design of a radar antenna. The path of a projectile is parabolic. The cables of a suspension bridge are parabolic. These and other applications are illustrated in the exercises.

Exercises 21–4

In Exercises 1–12, determine the coordinates of the focus and the equation of the directrix of the given parabolas. Sketch each curve.

1. $y^2 = 4x$ **2.** $y^2 = 16x$ **3.** $y^2 = -4x$

4. $y^2 = -16x$ **5.** $x^2 = 8y$ **6.** $x^2 = 10y$

7. $x^2 = -4y$ **8.** $x^2 = -12y$ **9.** $y^2 = 2x$

10. $x^2 = 14y$ **11.** $y = 0.48x^2$ **12.** $x = 7.6y^2$

In Exercises 13–20, find the equations of the parabolas satisfying the given conditions.

13. Focus $(3, 0)$, directrix $x = -3$

14. Focus $(-2, 0)$, directrix $x = 2$

15. Focus $(0, 4)$, vertex $(0, 0)$

16. Focus $(-3, 0)$, vertex $(0, 0)$

17. Vertex $(0, 0)$, directrix $y = -1$

18. Vertex $(0, 0)$, directrix $y = 2.3$

19. Vertex $(0, 0)$, axis along the y-axis, passes through $(-1, 8)$

20. Vertex $(0, 0)$, axis along the x-axis, passes through $(2, -1)$

In Exercises 21–40, solve the given problems.

21. Find the equation of the parabola with focus $(6, 1)$ and directrix $x = 0$ by use of the definition. Sketch the curve.

22. Find the equation of the parabola with focus $(1, 1)$ and directrix $y = 5$ by use of the definition. Sketch the curve.

23. Use a graphing calculator to view the parabola $y^2 + 2x + 8y + 13 = 0$.

24. Use a graphing calculator to view the parabola $y^2 - 2x - 6y + 19 = 0$.

25. The equation of a parabola with vertex (h, k) and axis parallel to the x-axis is $(y - k)^2 = 4p(x - h)$. (This is shown in Section 21-7.) Sketch the parabola for which (h, k) is $(2, -3)$ and $p = 2$.

26. The equation of a parabola with vertex (h, k) and axis parallel to the y-axis is $(x - h)^2 = 4p(y - k)$. (This is shown in Section 21-7.) Sketch the parabola for which (h, k) is $(-1, 2)$ and $p = -3$.

27. The chord of a parabola that passes through the focus and is parallel to the directrix is called the *latus rectum* of the parabola. Find the length of the latus rectum of the parabola $y^2 = 4px$.

28. Find the equation of the circle that has the focus and the vertex of the parabola $x^2 = 8y$ as the ends of a diameter.

29. The Golden Gate Bridge at San Francisco Bay is a suspension bridge, and its supporting cables are parabolic. See Fig. 21-48. With the origin at the low point of the cable, what equation represents the cable if the towers are 1280 m apart and the maximum sag is 90 m?

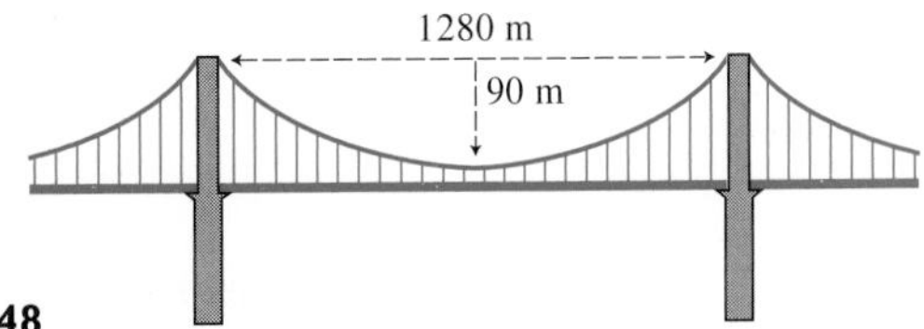

Fig. 21-48

30. The entrance to a building is a parabolic arch 5.6 m high at the center and 7.4 m wide at the base. What equation represents the arch if the vertex is at the top of the arch?

31. The rate of development of heat H (measured in watts) in a resistor of resistance R (measured in ohms) of an electric circuit is given by $H = Ri^2$, where i is the current (measured in amperes) in the resistor. Sketch the graph of H versus i, if $R = 6.0\ \Omega$.

32. What is the length of the horizontal bar across the parabolically shaped window shown in Fig. 21-49?

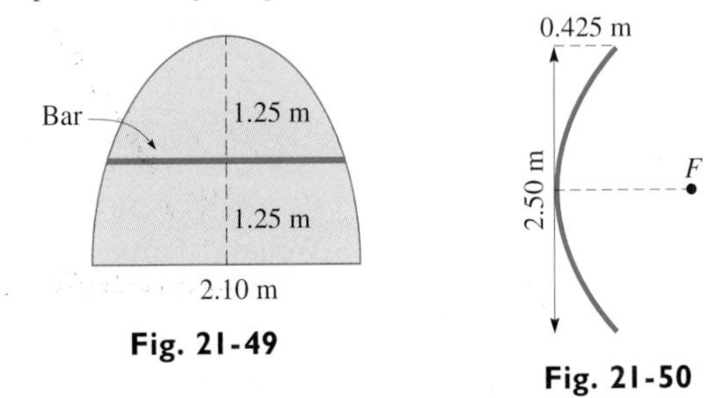

Fig. 21-49

Fig. 21-50

33. A linear solar reflector has a parabolic cross section, which is shown in Fig. 21-50. From the given dimensions, what is the focal length (vertex to focus) of the reflector?

34. A rocket is fired horizontally from a plane. Its horizontal distance x and vertical distance y from the point at which it was fired are given by $x = v_0 t$ and $y = \frac{1}{2}gt^2$, where v_0 is the initial velocity of the rocket, t is the time, and g is the acceleration due to gravity. Express y as a function of x, and show that it is the equation of a parabola.

35. A wave entering parallel to the axis of a radio wave antenna with a parabolic cross section is reflected through the focus. What is the equation of the parabola for the antenna with the reflected wave shown in Fig. 21-51?

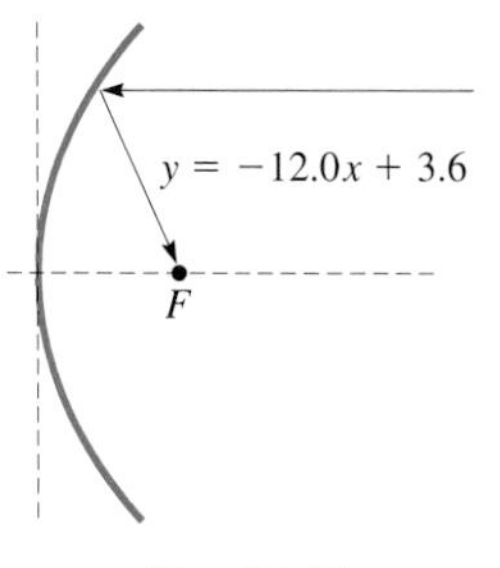

Fig. 21-51

36. A wire is fastened 12.0 m up on each of two telephone poles that are 60.0 m apart. Halfway between the poles the wire is 10.0 m above the ground. Assuming the wire is parabolic, find the height of the wire 15.0 m from either pole.

37. The total annual fraction f of energy supplied by solar energy to a home is given by $f = 0.065\sqrt{A}$, where A is the area of the solar collector. Sketch the graph of f as a function of A ($0 < A \leq 200$ m^2).

38. The velocity v of a jet of water flowing from an opening in the side of a certain container is given by $v = 4.4\sqrt{h}$, where h is the depth of the opening. Sketch a graph of v (in m/s) versus h (in metres).

39. A small island is 4 km from a straight shoreline. A ship channel is equidistant between the island and the shoreline. Write an equation for the channel.

40. Under certain circumstances, the maximum power P in an electric circuit varies as the square of the voltage of the source E_0 and inversely as the internal resistance R_i of the source. If 10 W is the maximum power for a source of 2.0 V and internal resistance of 0.10 Ω, sketch the graph of P versus E_0 if R_i remains constant.

21-5 The Ellipse

The next important curve is the ellipse. *An* **ellipse** *is defined as the locus of a point $P(x, y)$ which moves so that the sum of its distances from two fixed points is constant. These fixed points are the* **foci** *of the ellipse.* Letting this sum of distances be $2a$, and the foci be the points $(-c, 0)$ and $(c, 0)$, we have

$$\sqrt{(x-c)^2 + y^2} + \sqrt{(x+c)^2 + y^2} = 2a$$

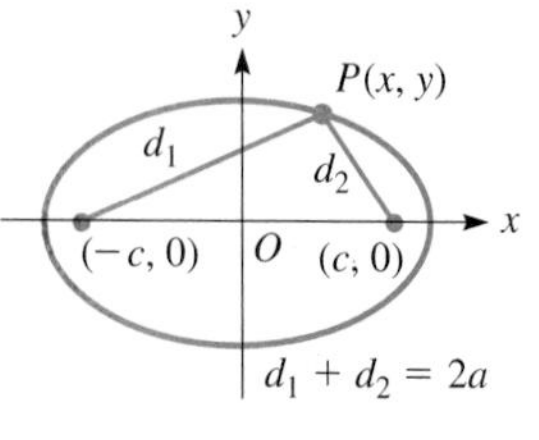

Fig. 21-52

See Fig. 21-52. The ellipse has its center at the origin such that c is the length of the line segment from the center to a focus. We shall also see that a has a special meaning. Now, from Section 14-4, we see that we should move one radical to the right and then square each side. This leads to the following steps.

$$\begin{aligned}
\sqrt{(x+c)^2 + y^2} &= 2a - \sqrt{(x-c)^2 + y^2} \\
(x+c)^2 + y^2 &= 4a^2 - 4a\sqrt{(x-c)^2 + y^2} + (\sqrt{(x-c)^2 + y^2})^2 \\
x^2 + 2cx + c^2 + y^2 &= 4a^2 - 4a\sqrt{(x-c)^2 + y^2} + x^2 - 2cx + c^2 + y^2 \\
4a\sqrt{(x-c)^2 + y^2} &= 4a^2 - 4cx \\
a\sqrt{(x-c)^2 + y^2} &= a^2 - cx \\
a^2(x^2 - 2cx + c^2 + y^2) &= a^4 - 2a^2cx + c^2x^2 \\
(a^2 - c^2)x^2 + a^2y^2 &= a^2(a^2 - c^2)
\end{aligned}$$

We now define $a^2 - c^2 = b^2$ (the reason will be shown presently). Therefore,

$$b^2x^2 + a^2y^2 = a^2b^2$$

Dividing through by a^2b^2, we have

$$\frac{x^2}{a^2} + \frac{y^2}{b^2} = 1 \qquad (21\text{-}17)$$

The x-intercepts are $(-a, 0)$ and $(a, 0)$. This means that $2a$ (the sum of distances used in the derivation) is also the distance between the x-intercepts. *The points* $(a, 0)$ *and* $(-a, 0)$ *are the* **vertices** *of the ellipse, and the line between them is the* **major axis** [see Fig. 21-53(a)]. Thus, *a is the length of the* **semimajor axis.**

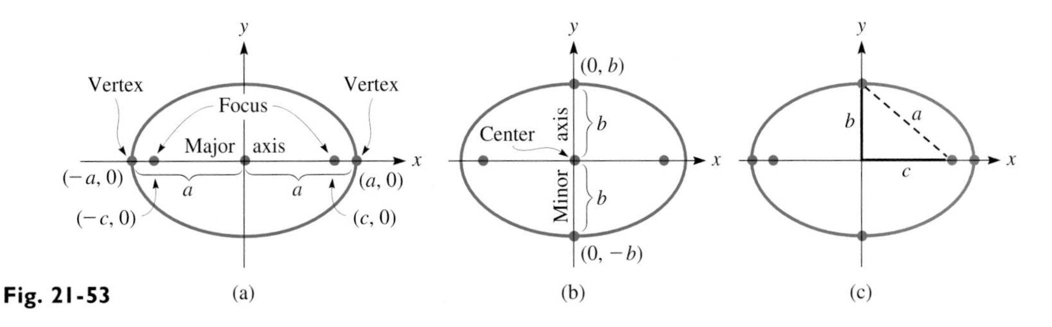

Fig. 21-53

We can now state that *Eq. (21-17) is called the* **standard equation** *of the ellipse with its major axis along the x-axis and its center at the origin.*

The y-intercepts of this ellipse are $(0, -b)$ and $(0, b)$. *The line joining these intercepts is called the* **minor axis** *of the ellipse* [Fig. 21-53(b)], *which means b is the length of the* **semiminor** axis. The intercept $(0, b)$ is equidistant from $(-c, 0)$ and $(c, 0)$. Since the sum of the distances from these points to $(0, b)$ is $2a$, the distance from $(c, 0)$ to $(0, b)$ must be a. Thus, we have a right triangle with line segments of lengths a, b, and c, with a as hypotenuse [Fig. 21-53(c)]. Therefore,

$$a^2 = b^2 + c^2 \tag{21-18}$$

is the relation between distances a, b, and c. This also shows why b was defined as it was in the derivation of Eq. (21-17).

If we choose points on the y-axis as the foci, *the standard equation of the ellipse, with its center at the origin and its major axis along the y-axis, is*

$$\frac{y^2}{a^2} + \frac{x^2}{b^2} = 1 \tag{21-19}$$

Fig. 21-54

In this case the vertices are $(0, a)$ and $(0, -a)$, the foci are $(0, c)$ and $(0, -c)$, and the ends of the minor axis are $(b, 0)$ and $(-b, 0)$. See Fig. 21-54.

The ellipses represented by Eqs. (21-17) and (21-19) are both symmetrical to both axes and to the origin.

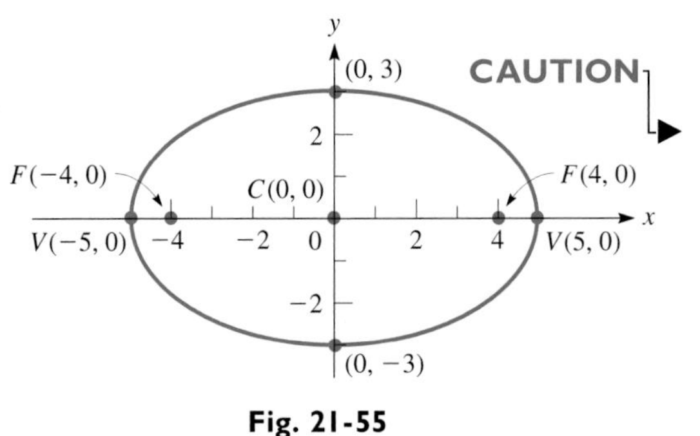

Fig. 21-55

EXAMPLE 1 ▸▸ The ellipse $\frac{x^2}{25} + \frac{y^2}{9} = 1$ seems to fit the form of either Eq. (21-17) or Eq. (21-19). Since $a^2 = b^2 + c^2$, we know that ***a is always larger than b.*** Since the square of the larger number appears under x^2, we know the equation is in the form of Eq. (21-17). Therefore, $a^2 = 25$ and $b^2 = 9$, or $a = 5$ and $b = 3$. This means that the vertices are $(5, 0)$ and $(-5, 0)$ and the minor axis extends from $(0, -3)$ to $(0, 3)$. See Fig. 21-55.

CAUTION

We find c from the relation $c^2 = a^2 - b^2$. This means that $c^2 = 16$, and the foci are $(4, 0)$ and $(-4, 0)$. ◂◂

EXAMPLE 2 ▸▸ The ellipse

$$\frac{x^2}{4} + \frac{y^2}{9} = 1$$

(b^2 under x^2, a^2 under y^2)

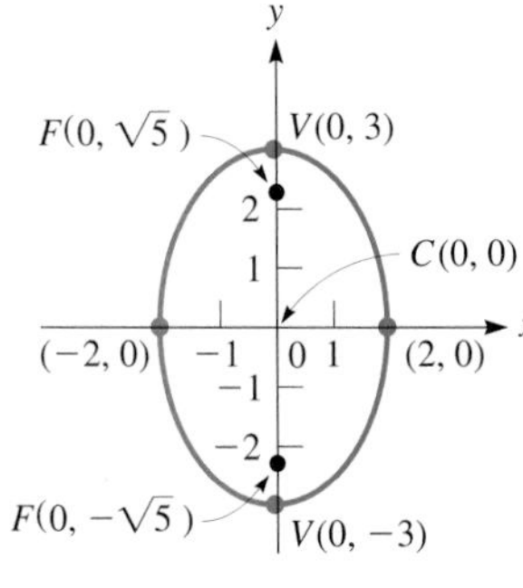

Fig. 21-56

has vertices $(0, 3)$ and $(0, -3)$. The minor axis extends from $(-2, 0)$ to $(2, 0)$. The equation fits the form of Eq. (21-19) since the larger number appears under y^2. Therefore, $a^2 = 9$, $b^2 = 4$, and $c^2 = 5$. The foci are $(0, \sqrt{5})$ and $(0, -\sqrt{5})$. This ellipse is shown in Fig. 21-56. ◂◂

EXAMPLE 3 ▸▸ Find the coordinates of the vertices, the ends of the minor axis, and the foci of the ellipse $4x^2 + 16y^2 = 64$.

This equation must be put in standard form first, which we do by dividing through by 64. When this is done, we obtain

$$\frac{x^2}{16} + \frac{y^2}{4} = 1$$

(form requires + and 1)

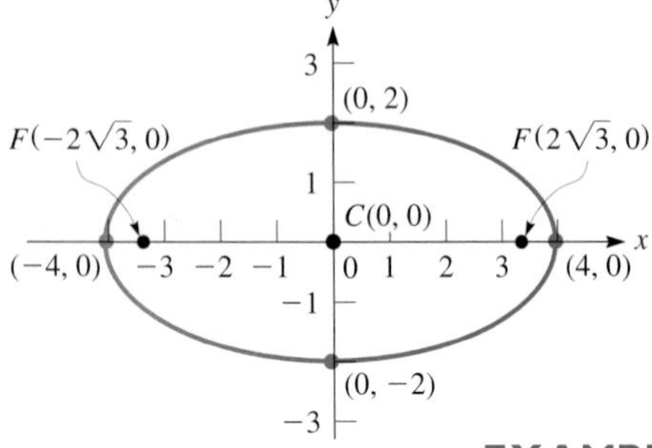

Fig. 21-57

We see that $a^2 = 16$ and $b^2 = 4$, which tells us that $a = 4$ and $b = 2$. Then, $c = \sqrt{16 - 4} = \sqrt{12} = 2\sqrt{3}$. Since a^2 appears under x^2, the vertices are $(4, 0)$ and $(-4, 0)$. The ends of the minor axis are $(0, 2)$ and $(0, -2)$, and the foci are $(2\sqrt{3}, 0)$ and $(-2\sqrt{3}, 0)$. See Fig. 21-57. ◂◂

EXAMPLE 4 ▸▸ A satellite to study the earth's atmosphere has a minimum altitude of 1000 km and a maximum altitude of 3200 km. If the path of the satellite about the earth is an ellipse with the center of the earth at one focus, what is the equation of its path? Assume the radius of the earth is 6400 km.

We set up the coordinate system such that the center of the ellipse is at the origin and the center of the earth is at the right focus, as shown in Fig. 21-58. We know that the distance between vertices is

$$2a = 3200 + 6400 + 6400 + 1000 = 17\,000 \text{ km}$$
$$a = 8500 \text{ km}$$

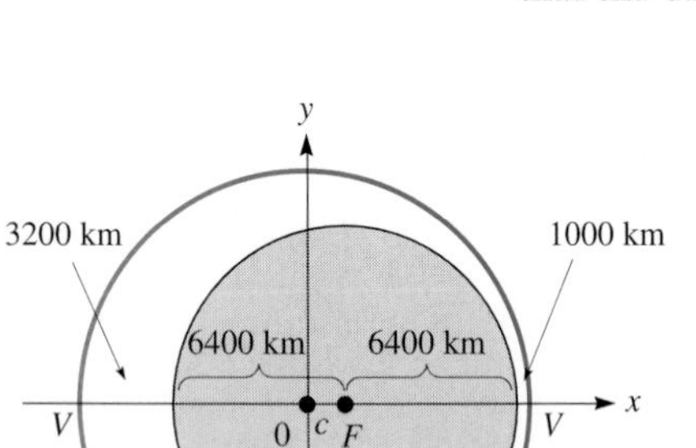

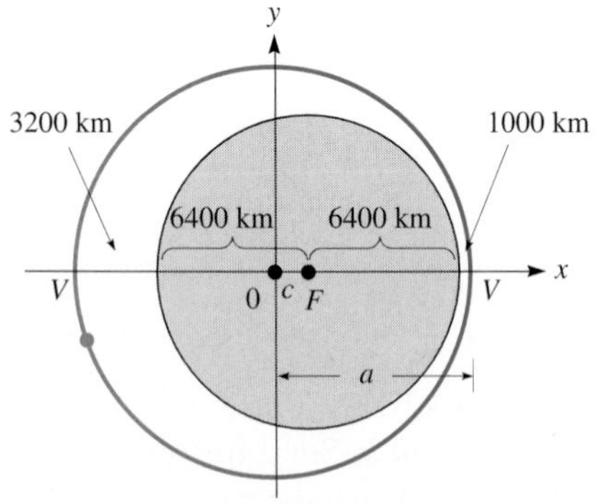

Fig. 21-58

From the right focus to the right vertex is 7400 km. This tells us

$$c = a - 7400 = 8500 - 7400 = 1100 \text{ km}$$

We can now calculate b^2 as

$$b^2 = a^2 - c^2 = 8500^2 - 1100^2 = 7.10 \times 10^7 \text{ km}^2$$

Since $a^2 = 8500^2 = 7.23 \times 10^7 \text{ km}^2$, the equation is

$$\frac{x^2}{7.23 \times 10^7} + \frac{y^2}{7.10 \times 10^7} = 1$$

or

$$7.10x^2 + 7.23y^2 = 5.13 \times 10^8$$ ◂◂

EXAMPLE 5 ▸▸ Find the equation of the ellipse with its center at the origin and an end of its minor axis at $(2, 0)$ and which passes through $(-1, \sqrt{6})$.

Since the center is at the origin and an end of the minor axis is at $(2, 0)$, we know that the ellipse is of the form of Eq. (21-19) and that $b = 2$. Thus we have

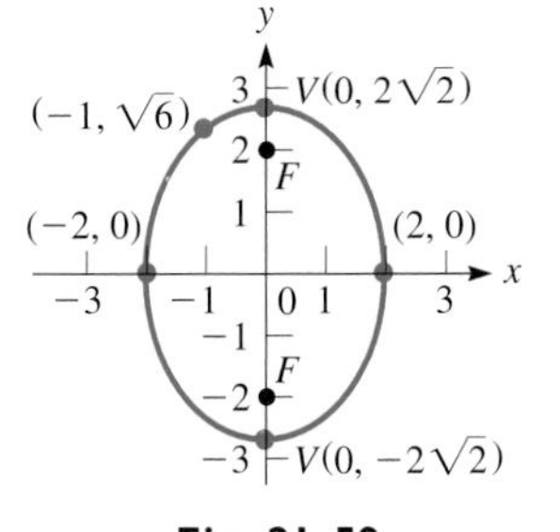

Fig. 21-59

$$\frac{y^2}{a^2} + \frac{x^2}{2^2} = 1$$

In order to find a^2, we use the fact that the ellipse passes through $(-1, \sqrt{6})$. This means that these coordinates satisfy the equation of the ellipse. This gives

$$\frac{(\sqrt{6})^2}{a^2} + \frac{(-1)^2}{4} = 1, \qquad \frac{6}{a^2} = \frac{3}{4}, \qquad a^2 = 8$$

Therefore, the equation of the ellipse, shown in Fig. 21-59, is

$$\frac{y^2}{8} + \frac{x^2}{4} = 1 \quad ◂◂$$

The following example illustrates the use of the definition of the ellipse to find the equation of an ellipse with its center at a point other than the origin.

EXAMPLE 6 ▸▸ Using the definition, find the equation of the ellipse with foci at $(1, 3)$ and $(9, 3)$, with major axis of 10.

Recalling that the sum of distances in the definition equals the length of the major axis, we now use the same method as in the derivation of Eq. (21-17).

$$\sqrt{(x-1)^2 + (y-3)^2} + \sqrt{(x-9)^2 + (y-3)^2} = 10 \qquad \text{use definition of ellipse}$$

$$\sqrt{(x-1)^2 + (y-3)^2} = 10 - \sqrt{(x-9)^2 + (y-3)^2} \qquad \text{isolate a radical}$$

$$x^2 - 2x + 1 + y^2 - 6y + 9 = 100 - 20\sqrt{(x-9)^2 + (y-3)^2} + x^2 - 18x + 81 + y^2 - 6y + 9 \qquad \text{square both sides and simplify}$$

$$20\sqrt{(x-9)^2 + (y-3)^2} = 180 - 16x \qquad \text{isolate radical}$$

$$5\sqrt{(x-9)^2 + (y-3)^2} = 45 - 4x \qquad \text{divide by 4}$$

$$25(x^2 - 18x + 81 + y^2 - 6y + 9) = 2025 - 360x + 16x^2 \qquad \text{square both sides}$$

$$9x^2 - 90x + 25y^2 - 150y + 225 = 0 \qquad \text{simplify}$$

The additional x- and y-terms are characteristic of the equation of an ellipse whose center is not at the origin (see Fig. 21-60).

To view this ellipse on a graphing calculator as shown in Fig. 21-61, we solve for y to get the two functions needed. The solutions are $y = \dfrac{15 \pm 3\sqrt{10x - x^2}}{5}$.

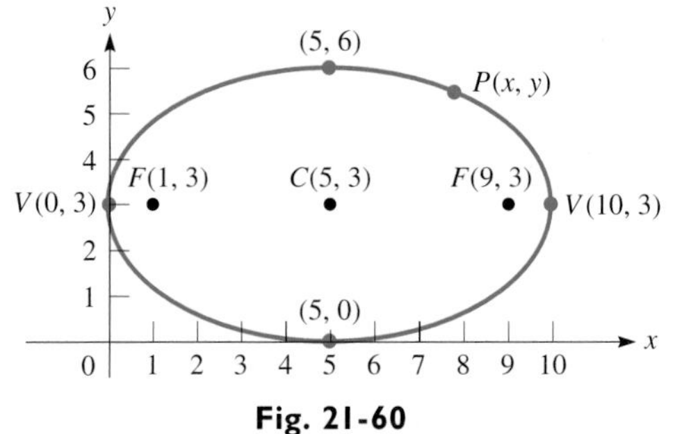

Fig. 21-60

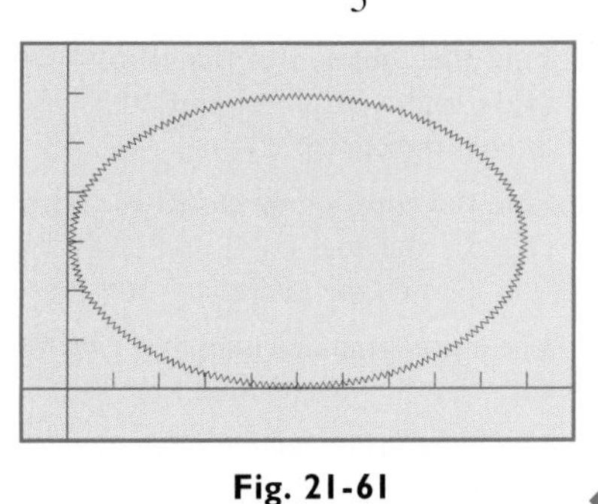

Fig. 21-61

◂◂

We can conclude that *the equation of an ellipse is characterized by the presence of both an x^2-term and a y^2-term, having different coefficients (in value but not in sign).* The difference between the equation of an ellipse and that of a circle is that the coefficients of the squared terms in the equation of the circle are the same, whereas those of the ellipse differ. We will consider the equation of the ellipse further in Sections 21-7 and 21-8.

The ellipse has many applications. The orbits of the planets about the sun are elliptical. Gears, cams, and springs are often elliptical in shape. Arches are often constructed in the form of a semiellipse. These and other applications are illustrated in the exercises.

Exercises 21–5

In Exercises 1–12, find the coordinates of the vertices and foci of the given ellipses. Sketch each curve.

1. $\frac{x^2}{4} + \frac{y^2}{1} = 1$

2. $\frac{x^2}{100} + \frac{y^2}{64} = 1$

3. $\frac{x^2}{25} + \frac{y^2}{36} = 1$

4. $\frac{x^2}{49} + \frac{y^2}{81} = 1$

5. $4x^2 + 9y^2 = 36$

6. $x^2 + 36y^2 = 144$

7. $49x^2 + 4y^2 = 196$

8. $25x^2 + y^2 = 25$

9. $8x^2 + y^2 = 16$

10. $2x^2 + 3y^2 = 6$

11. $4x^2 + 25y^2 = 0.25$

12. $9x^2 + 4y^2 = 0.09$

In Exercises 13–20, find the equations of the ellipses satisfying the given conditions. The center of each is at the origin.

13. Vertex $(15, 0)$, focus $(9, 0)$

14. Minor axis 8, vertex $(0, -5)$

15. Focus $(0, 2)$, major axis 6

16. Semiminor axis 2, focus $(3, 0)$

17. Vertex $(8, 0)$, passes through $(2, 3)$

18. Focus $(0, 2)$, passes through $(-1, \sqrt{3})$

19. Passes through $(2, 2)$ and $(1, 4)$

20. Passes through $(-2, 2)$ and $(1, \sqrt{6})$

In Exercises 21–40, solve the given problems.

21. Find the equation of the ellipse with foci $(-2, 1)$ and $(4, 1)$, and a major axis of 10 by use of the definition. Sketch the curve.

22. Find the equation of the ellipse with vertices $(1, 5)$ and $(1, -1)$, and foci $(1, 4)$ and $(1, 0)$ by use of the definition. Sketch the curve.

23. Use a graphing calculator to view the ellipse $4x^2 + 3y^2 + 16x - 18y + 31 = 0$.

24. Use a graphing calculator to view the ellipse $4x^2 + 8y^2 + 4x - 24y + 1 = 0$.

25. The equation of an ellipse with center (h, k) and major axis parallel to the x-axis is $\frac{(x-h)^2}{a^2} + \frac{(y-k)^2}{b^2} = 1$. (This is shown in Section 21-7.) Sketch the ellipse that has a major axis of 6, a minor axis of 4, and for which (h, k) is $(2, -1)$.

26. The equation of an ellipse with center (h, k) and major axis parallel to the y-axis is $\frac{(y-k)^2}{a^2} + \frac{(x-h)^2}{b^2} = 1$. (This is shown in Section 21-7.) Sketch the ellipse that has a major axis of 8, a minor axis of 6, and for which (h, k) is $(1, 3)$.

27. For what values of k does the ellipse $x^2 + ky^2 = 1$ have its vertices on the y-axis? Explain how these values are found.

28. For what value of k does the ellipse $x^2 + k^2y^2 = 25$ have a focus at $(3, 0)$? Explain how this value is found.

29. Show that the ellipse $2x^2 + 3y^2 - 8x - 4 = 0$ is symmetrical to the x-axis.

30. Show that the ellipse $5x^2 + y^2 - 3y - 7 = 0$ is symmetrical to the y-axis.

31. The *eccentricity* e of an ellipse is defined as $e = c/a$. A cam in the shape of an ellipse can be described by the equation $x^2 + 9y^2 = 81$. Find the eccentricity of this elliptical cam.

32. The planet Pluto moves about the sun in an elliptical orbit, with the sun at one focus. The closest that Pluto approaches the sun is 4.8 Tm, and the farthest it gets from the sun is 7.4 Tm. Find the eccentricity of Pluto's orbit. (See Exercise 31.)

33. A draftsman draws a series of triangles with a base from $(-3, 0)$ to $(3, 0)$ and a perimeter of 14 cm (all measurements in centimetres). Find the equation of the curve on which all of the third vertices of the triangles are located.

34. The electric power P dissipated in a resistance R is given by $P = Ri^2$, where i is the current in the resistor. Find the equation for the total power of 64 W dissipated in two resistors, with resistances 2.0 Ω and 8.0 Ω, respectively, and with currents i_1 and i_2, respectively. Sketch the graph, assuming that negative values of current are meaningful.

35. An ellipse has a focal property such that a light ray or sound wave emanating from one focus will be reflected through the other focus. Many buildings, such as Statuary Hall in the U.S. Capitol and the Taj Mahal, are built with elliptical ceilings with the property that a sound from one focus is easily heard at the other focus. If a building has a ceiling whose cross sections are part of an ellipse that can be described by the equation $36x^2 + 225y^2 = 8100$ (measurements in metres), how far apart must two persons stand in order to whisper to each other using this focal property?

36. An airplane wing is designed such that a certain cross section is an ellipse 2.80 m wide and 0.40 m thick. Find an equation that can be used to describe the perimeter of this cross section.

37. A road passes through a tunnel with a semielliptical cross section 19.6 m wide and 5.5 m high at the center. What is the height of the tallest vehicle that can pass through the tunnel at a point 6.7 m from the center? See Fig. 21-62.

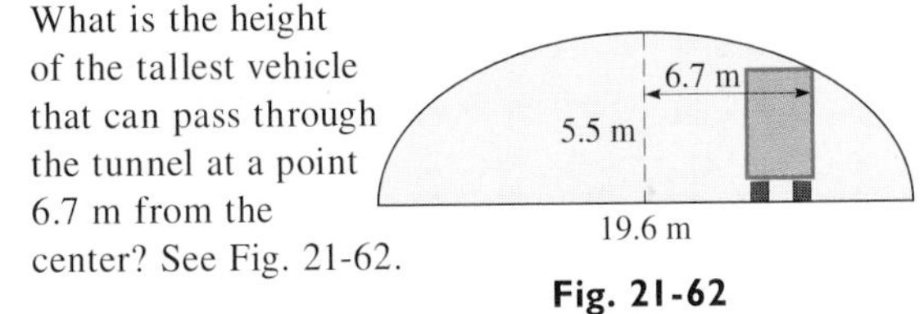

Fig. 21-62

38. An architect designs a window in the shape of an ellipse 1.50 m wide and 1.10 m high. Find the perimeter of the window from the formula $p = \pi(a + b)$. This formula gives a good *approximation* for the perimeter when a and b are nearly equal.

39. The ends of a horizontal tank 6.00 m long are ellipses, which can be described by the equation $x^2 + 6y^2 = 6$, where x and y are measured in metres. The area of an ellipse is $A = \pi ab$. Find the volume of the tank.

40. A laser beam 6.80 mm in diameter is incident on a plane surface at an angle of 62.0°, as shown in Fig. 21-63. What is the elliptical area that the laser covers on the surface? (See Exercise 39.)

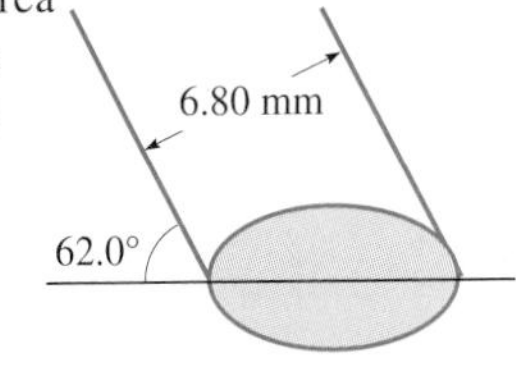

Fig. 21-63

21-6 The Hyperbola

The final curve we shall discuss in detail is the hyperbola. *A* **hyperbola** *is defined as the locus of a point P(x, y) which moves so that the difference of the distances from two fixed points is a constant. These fixed points are the* **foci** *of the hyperbola.* We choose the foci of the hyperbola as the points $(-c, 0)$ and $(c, 0)$ (see Fig. 21-64) and the constant difference to be $2a$. As with the ellipse, these choices make c the length of the line segment from the center to a focus, and a (as we will see) the length of the line segment from the center to a vertex. Therefore,

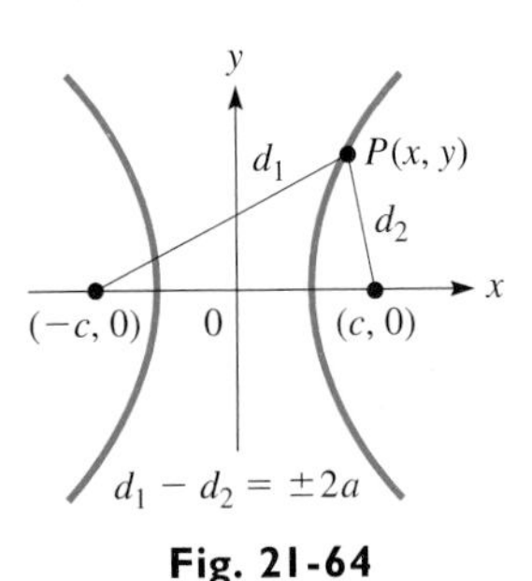

Fig. 21-64

$$\sqrt{(x + c)^2 + y^2} - \sqrt{(x - c)^2 + y^2} = 2a$$

Following the same procedure as in the preceding section, we find the equation of the hyperbola to be

$$\frac{x^2}{a^2} - \frac{y^2}{b^2} = 1 \qquad \textbf{(21-20)}$$

CAUTION ▶ When we derive this equation, ***we have a definition of the relation between, a, b, and c which is different from that for the ellipse.*** This relation is

$$c^2 = a^2 + b^2 \qquad \textbf{(21-21)}$$

For reference, Eq. (21-20) is $\frac{x^2}{a^2} - \frac{y^2}{b^2} = 1$

In Eq. (21-20), if we let $y = 0$, we find that the x-intercepts are $(-a, 0)$ and $(a, 0)$, just as they are for the ellipse. *These are the* **vertices** *of the hyperbola.* For $x = 0$, we find that we have imaginary solutions for y, which means there are no points on the curve that correspond to a value of $x = 0$.

To find the meaning of b, we solve Eq. (21-20) for y in a special form:

$$\frac{y^2}{b^2} = \frac{x^2}{a^2} - 1$$
$$= \frac{x^2}{a^2} - \frac{a^2x^2}{a^2x^2}$$
$$= \frac{x^2}{a^2}\left(1 - \frac{a^2}{x^2}\right)$$

Multiplying through by b^2 and then taking the square root of each side, we have

$$y^2 = \frac{b^2x^2}{a^2}\left(1 - \frac{a^2}{x^2}\right)$$
$$y = \pm\frac{bx}{a}\sqrt{1 - \frac{a^2}{x^2}} \qquad (21\text{-}22)$$

We note that, if large values of x are assumed in Eq. (21-22), the quantity under the radical becomes approximately 1. In fact, the larger x becomes, the nearer 1 this expression becomes, since the x^2 in the denominator of a^2/x^2 makes this term nearly zero. Thus, for large values of x, Eq. (21-22) is approximately

$$y = \pm\frac{bx}{a} \qquad (21\text{-}23)$$

NOTE ▶ Equation (21-23) is seen to represent two straight lines, each of which passes through the origin. One has a slope of b/a, and the other has a slope of $-b/a$. ***These lines are called the* asymptotes *of the hyperbola.*** *An* **asymptote** *is a line that the curve approaches as one of the variables approaches some particular value.* The graph of the tangent function also has asymptotes, as we saw in Fig. 10-23. We can designate this limiting procedure with notation introduced in Chapter 19 by saying that

$$y \to \frac{bx}{a} \quad \text{as} \quad x \to \pm\infty$$

Since straight lines are easily sketched, the easiest way to sketch a hyperbola is to draw its asymptotes and then to draw the hyperbola out from each vertex so that it comes closer and closer to each of these asymptotes as x becomes numerically larger. To draw in the asymptotes, the usual procedure is to first draw a small rectangle $2a$ by $2b$, with the origin at the center, as shown in Fig. 21-65. Then straight lines are drawn through opposite vertices of the rectangle. These straight lines are the asymptotes of the hyperbola. Therefore, we see that the significance of the value of b lies in the slope of the asymptotes of the hyperbola.

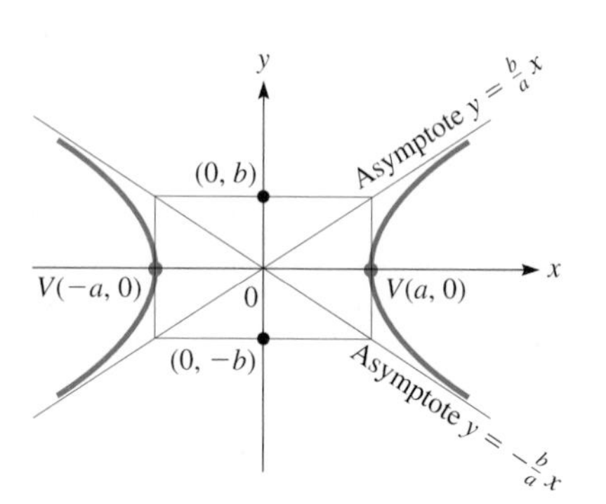

Fig. 20-65

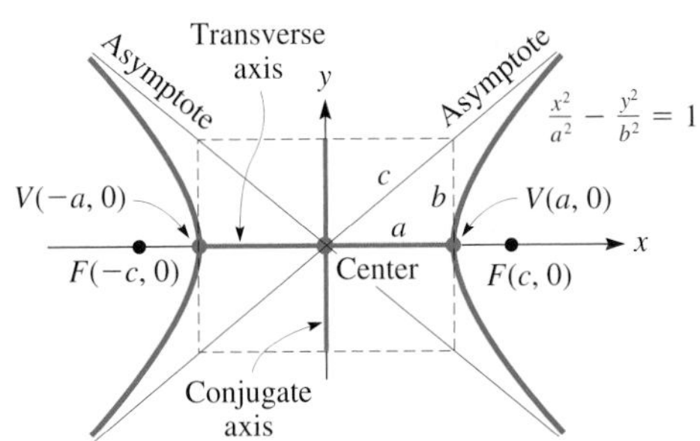

Fig. 21-66

Equation (21-20) is called the **standard equation** *of the hyperbola with its center at the origin. It has a* **transverse axis** *of length 2a along the x-axis and a* **conjugate axis** *of length 2b along the y-axis.* This means that a represents the length of the semitransverse axis, and b represents the length of the semiconjugate axis. See Fig. 21-66. From the definition of c, it is the length of the line segment from the center to a focus. Also, c is the length of the semidiagonal as shown in Fig. 21-66. This shows us the geometric meaning of the relation among a, b, and c given in Eq. (21-21).

If the tranverse axis is along the y-axis and the conjugate axis is along the x-axis, the equation of a hyperbola with its center at the origin (see Fig. 21-67) is

$$\frac{y^2}{a^2} - \frac{x^2}{b^2} = 1 \qquad (21\text{-}24)$$

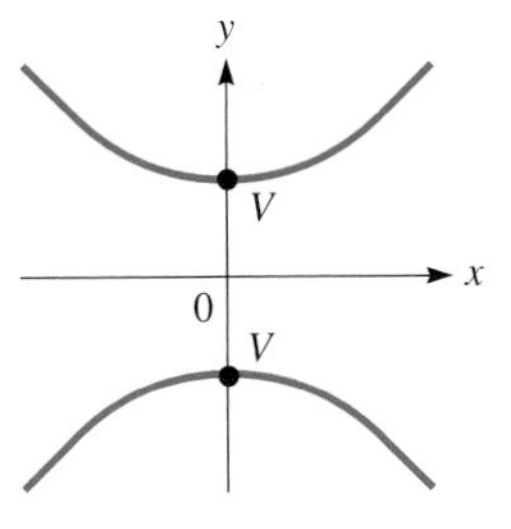

Fig. 21-67

The hyperbolas represented by Eqs. (21-20) and (21-24) are both symmetrical to both axes and to the origin.

EXAMPLE 1 ▸▸ The hyperbola $\frac{x^2}{16} - \frac{y^2}{9} = 1$ (with $a^2 = 16$, $b^2 = 9$ indicated)

fits the form of Eq. (21-20). We know that it fits Eq. (21-20) and not Eq. (21-24) since the x^2-term is the positive term with 1 on the right. From the equation we see that $a^2 = 16$ and $b^2 = 9$, or $a = 4$ and $b = 3$. In turn this means the vertices are $(4, 0)$ and $(-4, 0)$ and the conjugate axis extends from $(0, -3)$ to $(0, 3)$.

Since $c^2 = a^2 + b^2$, we find that $c^2 = 25$, or $c = 5$. The foci are $(-5, 0)$ and $(5, 0)$.

Drawing the rectangle and the asymptotes in Fig. 21-68, we then sketch in the hyperbola from each vertex toward each asymptote. ◂◂

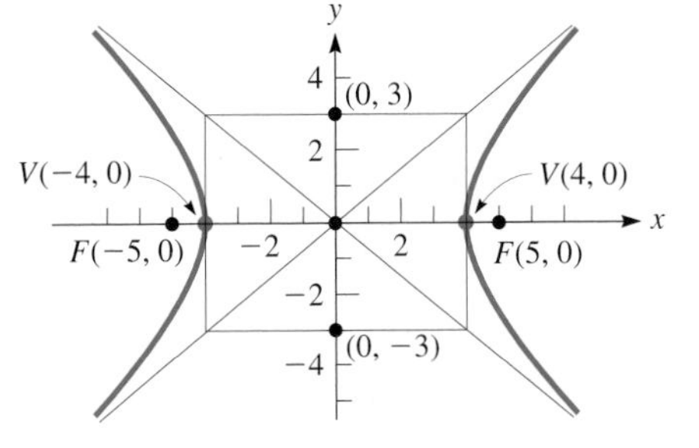

Fig. 21-68

EXAMPLE 2 ▸▸ The hyperbola $\frac{y^2}{4} - \frac{x^2}{16} = 1$ (with $a^2 = 4$, $b^2 = 16$ indicated)

has vertices at $(0, -2)$ and $(0, 2)$. Its conjugate axis extends from $(-4, 0)$ to $(4, 0)$. The foci are $(0, -2\sqrt{5})$ and $(0, 2\sqrt{5})$. We find this directly from the equation since the y^2-term is the positive term with 1 on the right. This means the equation fits the form of Eq. (21-24) with $a^2 = 4$ and $b^2 = 16$. Also, $c^2 = 20$, which means that $c = \sqrt{20} = 2\sqrt{5}$.

Since $2a$ extends along the y-axis, we see that the equations of the asymptotes are $y = \pm(a/b)x$. This is not a contradiction of Eq. (21-23) but an extension of it for a hyperbola with its transverse axis along the y-axis. The ratio a/b gives the slope of the asymptote. The hyperbola is shown in Fig. 21-69. ◂◂

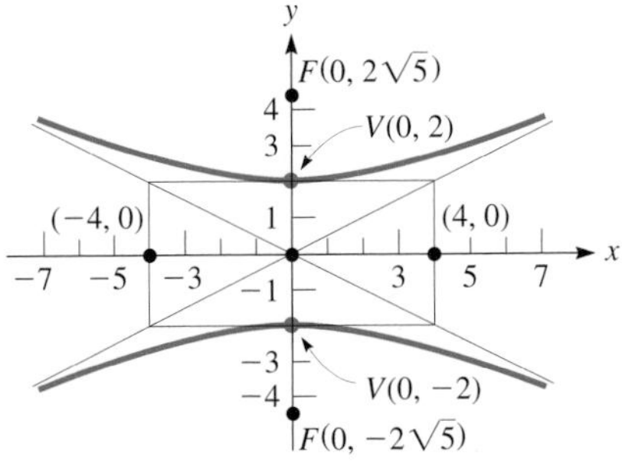

Fig. 21-69

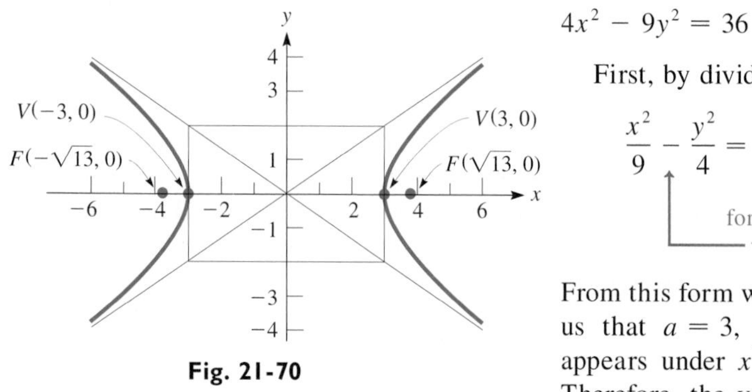

Fig. 21-70

EXAMPLE 3 ▸▸ Determine the coordinates of the vertices of the hyperbola

$$4x^2 - 9y^2 = 36$$

First, by dividing through by 36, we have

$$\frac{x^2}{9} - \frac{y^2}{4} = 1$$

From this form we see that $a^2 = 9$ and $b^2 = 4$. In turn this tells us that $a = 3$, $b = 2$, and $c = \sqrt{9 + 4} = \sqrt{13}$. Since a^2 appears under x^2, the equation fits the form of Eq. (21-20). Therefore, the vertices are $(-3, 0)$ and $(3, 0)$ and the foci are $(-\sqrt{13}, 0)$ and $(\sqrt{13}, 0)$. The hyperbola is shown in Fig. 21-70. ◂◂

EXAMPLE 4 ▸▸ In physics it is shown that where the velocity of a fluid is greatest, the pressure is the least. In designing an experiment to study this effect in the flow of water, a pipe is constructed such that its lengthwise cross section is hyperbolic. The pipe is 1.0 m long, 0.2 m in diameter at the narrowest point in the middle, and 0.4 m in diameter at each end. What is the equation that represents the cross section of the pipe as shown in Fig. 21-71?

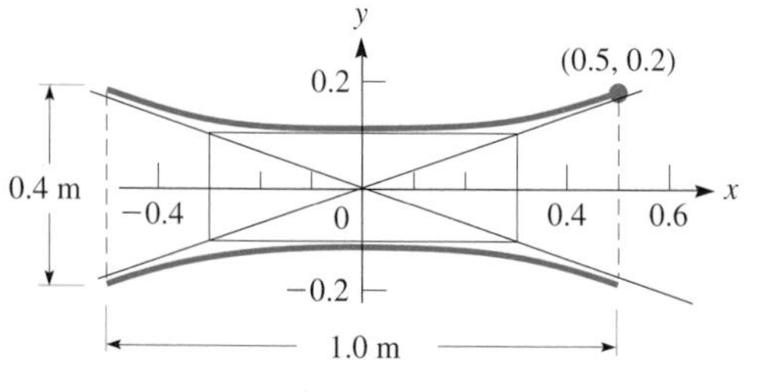

Fig. 21-71

As shown, the hyperbola has its transverse axis along the y-axis and its center at the origin. This means the general equation is given by Eq. (21-24). Since the radius at the middle of the pipe is 0.1 m, we know that $a = 0.1$ m. Also, since it is 1.0 m long and the radius at the end is 0.2 m, we know the point $(0.5, 0.2)$ is on the hyperbola. This point must satisfy the equation.

$$\frac{y^2}{a^2} - \frac{x^2}{b^2} = 1 \qquad \text{Eq. (21-24)}$$

point (0.5, 0.2) satisfies equation

$$a = 0.1 \rightarrow \frac{0.2^2}{0.1^2} - \frac{0.5^2}{b^2} = 1$$

$$4 - \frac{0.25}{b^2} = 1, \qquad 3b^2 = 0.25, \qquad b^2 = 0.083$$

$$\frac{y^2}{0.1^2} - \frac{x^2}{0.083} = 1 \qquad \text{substituting } a = 0.1,\ b^2 = 0.083 \text{ in Eq. (21-24)}$$

$$100y^2 - 12x^2 = 1 \qquad \text{equation of cross section}$$ ◂◂

Equations (21-20) and (21-24) give us the standard forms of the equation of the hyperbola with its center at the origin and its foci on one of the coordinate axes. There is another important equation form that represents a hyperbola, and it is

$$xy = c \qquad \textbf{(21-25)}$$

The asymptotes of this hyperbola are the coordinate axes, and the foci are on the line $y = x$ if c is positive or on the line $y = -x$ if c is negative.

The hyperbola represented by Eq. (21-25) is symmetrical to the origin, for if $-x$ replaces x, and $-y$ replaces y at the same time, we obtain $(-x)(-y) = c$, or $xy = c$. The equation is unchanged. However, if $-x$ replaces x, or if $-y$ replaces y, but not both, the sign on the left is changed. This means it is not symmetrical to either axis. Here c represents a constant and is not related to the focus.

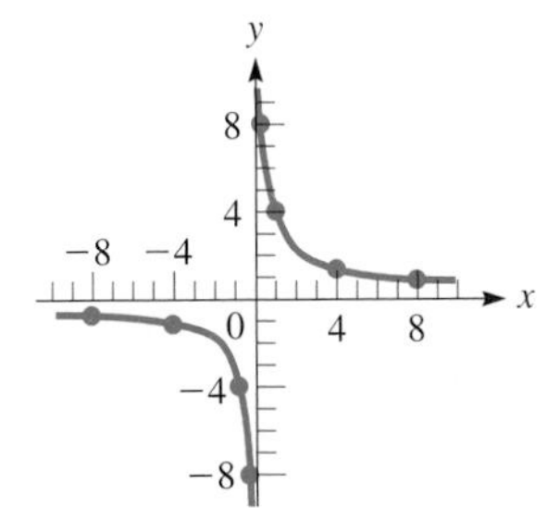

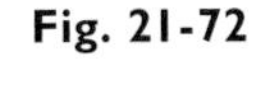

Fig. 21-72

EXAMPLE 5 ▸▸ Plot the graph of the equation $xy = 4$.

We find the values in the table below and then plot the appropriate points. Here it is permissible to use a limited number of points, since we know the equation represents a hyperbola. Therefore, using $y = 4/x$, we obtain the values

x	-8	-4	-1	$-\frac{1}{2}$	$\frac{1}{2}$	1	4	8
y	$-\frac{1}{2}$	-1	-4	-8	8	4	1	$\frac{1}{2}$

Note that neither x nor y may equal zero. The hyperbola is shown in Fig. 21-72.

If the constant on the right is negative (for example, if $xy = -4$), then the two branches of the hyperbola are in the second and fourth quadrants. ◂◂

EXAMPLE 6 ▸▸ One statement of Boyle's law is that the product of the pressure p and the volume V, at constant temperature, remains constant for a perfect gas. If $p = 300$ kPa when $V = 8.0$ L for a certain gas, graph p as a function of V.

From the statement of Boyle's law, we know that $pV = c$, where c is a constant. From the given values we have

$$(300 \text{ kPa})(8.0 \text{ L}) = c \qquad \text{or} \qquad c = 2400 \text{ kPa} \cdot \text{L}$$

Thus we are to sketch the graph of $pV = 2400$. Solving for p as $p = 2400/V$, and since only positive values have meaning, we find the following values.

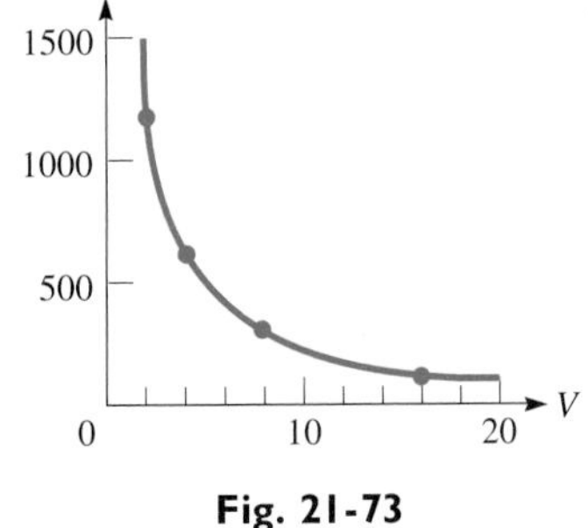

Fig. 21-73

V (L)	2.0	4.0	8.0	16.0
p (kPa)	1200	600	300	150

See Fig. 21-73. ◂◂

We can conclude that *the equation of a hyperbola is characterized by the presence of both an x^2-term and a y^2-term, having different signs, or by the presence of an xy-term with no squared terms.* We will consider the equation of the hyperbola further in Sections 21-7 and 21-8.

The hyperbola has some very useful applications. The LORAN radio navigation system is based on the use of hyperbolic paths. Some reflecting telescopes use hyperbolic mirrors. The paths of comets that never return to pass by the sun are hyperbolic. Some applications are illustrated in the exercises.

Exercises 21–6

In Exercises 1–12, find the coordinates of the vertices and the foci of the given hyperbolas. Sketch each curve.

1. $\dfrac{x^2}{25} - \dfrac{y^2}{144} = 1$

2. $\dfrac{x^2}{16} - \dfrac{y^2}{4} = 1$

3. $\dfrac{y^2}{9} - \dfrac{x^2}{1} = 1$

4. $\dfrac{y^2}{2} - \dfrac{x^2}{2} = 1$

5. $4x^2 - y^2 = 4$

6. $x^2 - 9y^2 = 81$

7. $2y^2 - 5x^2 = 10$

8. $3y^2 - 2x^2 = 6$

9. $4x^2 - y^2 + 4 = 0$

10. $9x^2 - y^2 - 9 = 0$

11. $4x^2 - y^2 = 0.64$

12. $9y^2 - x^2 = 0.36$

In Exercises 13–20, find the equations of the hyperbolas satisfying the given conditions. The center of each is at the origin.

13. Vertex $(3, 0)$, focus $(5, 0)$

14. Vertex $(0, 1)$, focus $(0, \sqrt{3})$

15. Conjugate axis $= 12$, vertex $(0, 10)$

16. Focus $(8, 0)$, transverse axis $= 4$

17. Passes through $(2, 3)$, focus $(2, 0)$

18. Passes through $(8, \sqrt{3})$, vertex $(4, 0)$

19. Passes through $(5, 4)$ and $(3, \frac{4}{5}\sqrt{5})$

20. Passes through $(1, 2)$ and $(2, 2\sqrt{2})$

In Exercises 21–36, solve the given problems.

21. Sketch the graph of the hyperbola $xy = 2$.

22. Sketch the graph of the hyperbola $xy = -4$.

23. Find the equation of the hyperbola with foci $(1, 2)$ and $(11, 2)$, and a transverse axis of 8, by use of the definition. Sketch the curve.

24. Find the equation of the hyperbola with vertices $(-2, 4)$ and $(-2, -2)$, and a conjugate axis of 4, by use of the definition. Sketch the curve.

25. Use a graphing calculator to view the hyperbola $x^2 - 4y^2 + 4x + 32y - 64 = 0$.

26. Use a graphing calculator to view the hyperbola $5y^2 - 4x^2 + 8x + 40y + 56 = 0$.

27. The equation of a hyperbola with center (h, k) and transverse axis parallel to the x-axis is $\frac{(x-h)^2}{a^2} - \frac{(y-k)^2}{b^2} = 1$. (This is shown in Section 21-7.) Sketch the hyperbola that has a transverse axis of 4, a conjugate axis of 6, and for which (h, k) is $(-3, 2)$.

28. The equation of a hyperbola with center (h, k) and transverse axis parallel to the y-axis is $\frac{(y-k)^2}{a^2} - \frac{(x-h)^2}{b^2} = 1$. (This is shown in Section 21-7.) Sketch the hyperbola that has a transverse axis of 2, a conjugate axis of 8, and for which (h, k) is $(5, 0)$.

29. Two concentric hyperbolas are called conjugate hyperbolas if the transverse and conjugate axes of one are respectively the conjugate and transverse axes of the other. What is the equation of the hyperbola conjugate to the hyperbola in Exercise 14?

30. As with an ellipse, the *eccentricity* e of a hyperbola is defined as $e = c/a$. Find the eccentricity of the hyperbola $2x^2 - 3y^2 = 24$.

31. A plane flying at a constant altitude of 2000 m is observed from the control tower of an airport. Show that the equation relating the horizontal distance x and direct line distance l from the tower to the plane is that of a hyperbola. Sketch the graph of l as a function of x. See Fig. 21-74.

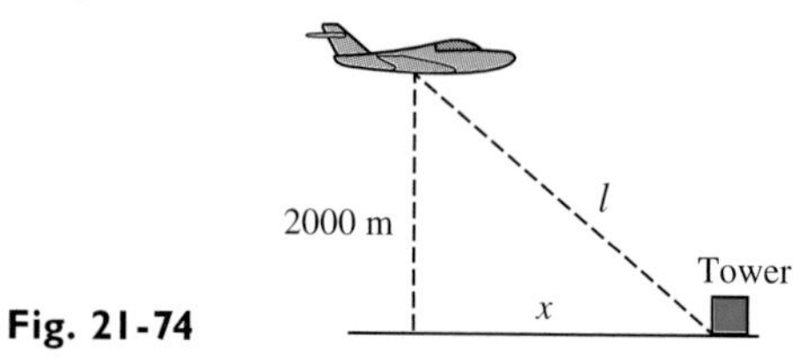

Fig. 21-74

32. Two holes of radius r are drilled from a circular area of radius R such that 24 cm^2 of material remains. Show that the equation relating R and r is that of a hyperbola.

33. Ohm's law in electricity states that the product of the current i and the resistance R equals the voltage V across the resistance. If a battery of 6.00 V is placed across a variable resistor R, find the equation relating i and R and sketch the graph of i as a function of R.

34. A ray of light directed at one focus of a hyperbolic mirror is reflected toward the other focus. Find the equation for the hyperbolic mirror shown in Fig. 21-75.

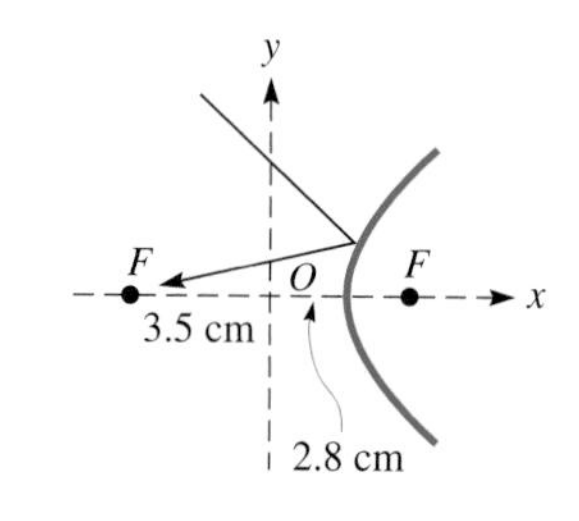

Fig. 21-75

35. A radio signal is sent simultaneously from stations A and B 600 km apart on the Carolina coast. A ship receives the signal from A 1.20 ms before it receives the signal from B. Given that radio signals travel at 300 km/ms, draw a graph showing the possible locations of the ship. This problem illustrates the basis of LORAN.

36. For monochromatic (single-color) light coming from two point sources, curves of maximum intensity occur where the difference in the distances from the sources is an integral number of wavelengths. If a thin translucent film is placed in the plane of the sources, find the equation of the curves of maximum intensity in the film where the difference in paths is two wavelengths and the sources are separated by four wavelengths. Assume the sources are on the x-axis with the origin midway between, and use units of one wavelength for both x and y.

21-7 Translation of Axes

The equations we have considered for the parabola, the ellipse, and the hyperbola are those for which the center of the ellipse or hyperbola, or vertex of the parabola, is at the origin. In this section we consider, without specific use of the definition, the equations of these curves for the cases in which the axis of the curve is parallel to one of the coordinate axes. This is done by **translation of axes.**

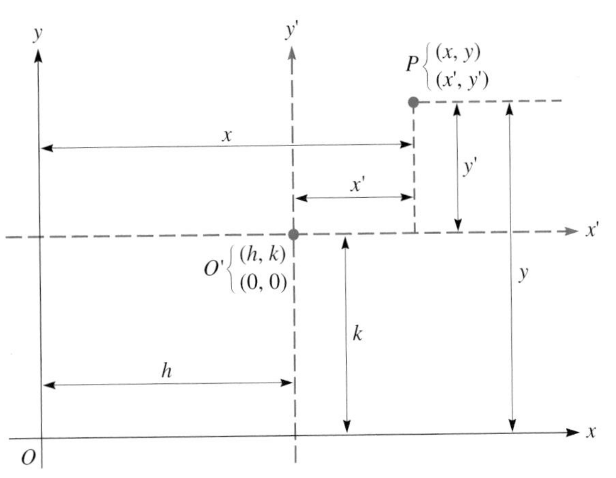

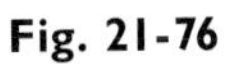
Fig. 21-76

In Fig. 21-76, we choose a point (h, k) in the xy-coordinate plane as the origin of another coordinate system, the $x'y'$-coordinate system. The x'-axis is parallel to the x-axis, and the y'-axis is parallel to the y-axis. Every point in the plane now has two sets of coordinates, (x, y) and (x', y'). We see that

$$x = x' + h \quad \text{and} \quad y = y' + k \tag{21-26}$$

Equations (21-26) can also be written in the form

$$x' = x - h \quad \text{and} \quad y' = y - k \tag{21-27}$$

EXAMPLE 1 ▸▸ Find the equation of the parabola with vertex $(2, 4)$ and focus $(4, 4)$.

If we let the origin of the $x'y'$-coordinate system be the point $(2, 4)$, then the point $(4, 4)$ is the point $(2, 0)$ in the $x'y'$-system. This means $p = 2$ and $4p = 8$. See Fig. 21-77. In the $x'y'$-system, the equation is

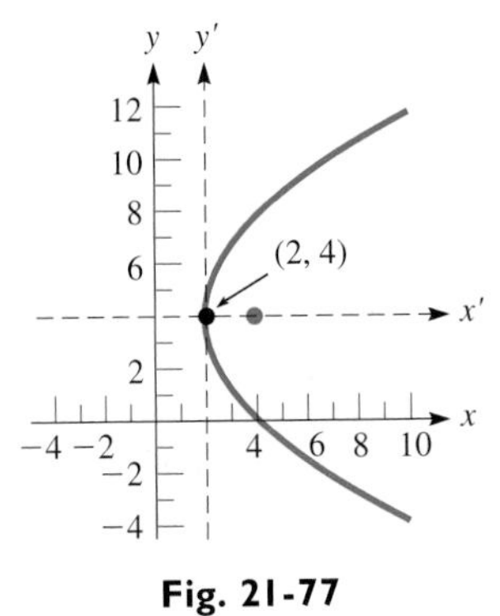

Fig. 21-77

$$(y')^2 = 8(x')$$

Using Eqs. (21-27), we have

$$(y - 4)^2 = 8(x - 2)$$

coordinates of vertex (2, 4)

as the equation of the parabola in the xy-coordinate system. ◂◂

Following the method of Example 1, by writing the equation of the curve in the $x'y'$-system and then using Eqs. (21-27), we have the following more general forms of the equations of the parabola, ellipse, and hyperbola.

Parabola, vertex (h, k):	$(y - k)^2 = 4p(x - h)$	(axis parallel to x-axis)	**(21-28)**
	$(x - h)^2 = 4p(y - k)$	(axis parallel to y-axis)	**(21-29)**
Ellipse, center (h, k):	$\dfrac{(x - h)^2}{a^2} + \dfrac{(y - k)^2}{b^2} = 1$	(major axis parallel to x-axis)	**(21-30)**
	$\dfrac{(y - k)^2}{a^2} + \dfrac{(x - h)^2}{b^2} = 1$	(major axis parallel to y-axis)	**(21-31)**
Hyperbola, center (h, k):	$\dfrac{(x - h)^2}{a^2} - \dfrac{(y - k)^2}{b^2} = 1$	(transverse axis parallel to x-axis)	**(21-32)**
	$\dfrac{(y - k)^2}{a^2} - \dfrac{(x - h)^2}{b^2} = 1$	(transverse axis parallel to y-axis)	**(21-33)**

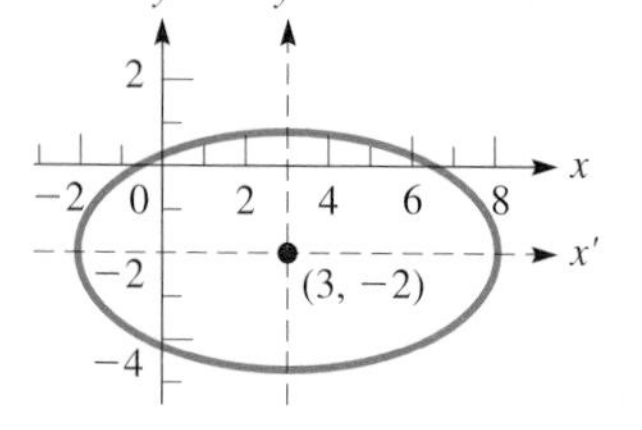

Fig. 21-78

EXAMPLE 2 ▸▸ Describe the curve of the equation

$$\frac{(x-3)^2}{25}+\frac{(y+2)^2}{9}=1$$

We see that this equation fits the form of Eq. (21-30) with $h = 3$ and $k = -2$. It is the equation of an ellipse with its center at $(3, -2)$ and its major axis parallel to the x-axis. The semimajor axis is $a = 5$, and the semiminor axis is $b = 3$. The ellipse is shown in Fig. 21-78. ◂◂

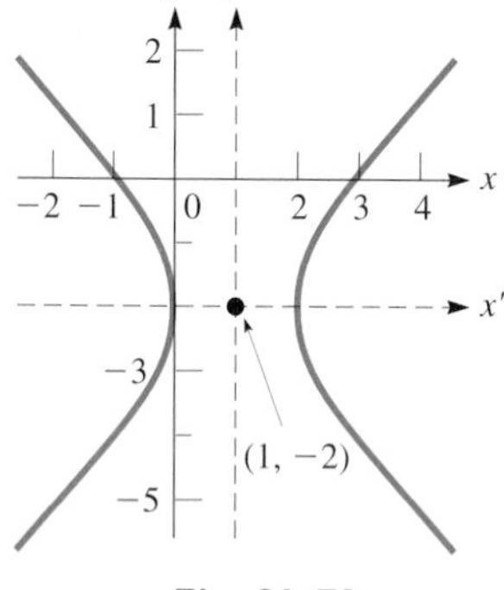

Fig. 21-79

EXAMPLE 3 ▸▸ Find the center of the hyperbola $2x^2 - y^2 - 4x - 4y - 4 = 0$.

To analyze this curve, we first complete the square in the x-terms and in the y-terms. This will allow us to recognize properly the choice of h and k.

$$2x^2 - 4x - y^2 - 4y = 4$$

$$2(x^2 - 2x \quad) - (y^2 + 4y \quad) = 4$$

$$2(x^2 - 2x + 1) - (y^2 + 4y + 4) = 4 + 2 - 4$$

CAUTION ▸ We note here that when we added 1 to complete the square of the x-terms within the parentheses, ***we were actually adding 2 to the left side.*** Thus, we added 2 to the right side. Similarly, when we added 4 to the y-terms within the parentheses, ***we were actually subtracting 4 from the left side.*** Continuing, we have

$$2(x-1)^2 - (y+2)^2 = 2$$

coordinates of center $(1, -2)$

$$\frac{(x-1)^2}{1}-\frac{(y+2)^2}{2}=1$$

Therefore, the center of the hyperbola is $(1, -2)$. See Fig. 21-79. ◂◂

EXAMPLE 4 ▸▸ Cylindrical glass beakers are to be made with a height of 3 cm. Express the surface area in terms of the radius of the base, and sketch the curve.

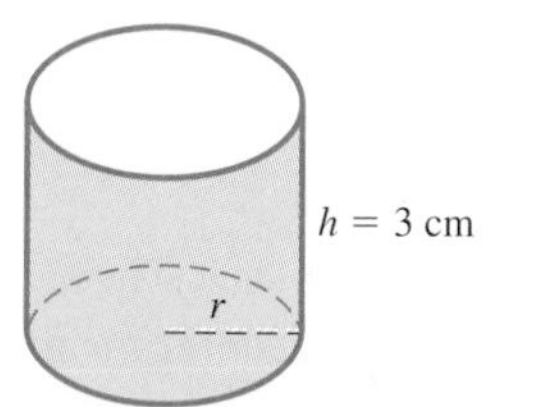

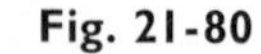
Fig. 21-80

The total surface area S of a beaker is the sum of the area of the base and the lateral surface area of the side. In general, S in terms of the radius r of the base and height h of the side is $S = \pi r^2 + 2\pi rh$. Since $h = 3$ cm, we have

$$S = \pi r^2 + 6\pi r$$

which is the desired relationship. See Fig. 21-80.

To get the equation relating S and r, we complete the square of the r terms.

$$S = \pi(r^2 + 6r)$$

$$S + 9\pi = \pi(r^2 + 6r + 9) \quad \text{complete the square}$$

$$S + 9\pi = \pi(r+3)^2$$

vertex $(-3, -9\pi)$

$$(r+3)^2 = \frac{1}{\pi}(S + 9\pi)$$

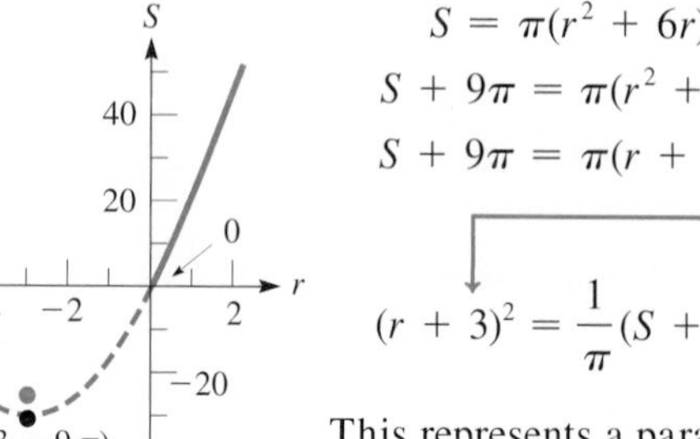

Fig. 21-81

This represents a parabola with vertex $(-3, -9\pi)$. Since $4p = 1/\pi$, $p = 1/(4\pi)$, the focus is $(-3, \frac{1}{4\pi} - 9\pi)$ as shown in Fig. 21-81. The part of the graph for negative r is dashed since only positive values have meaning. ◂◂

Exercises 21–7

In Exercises 1–8, describe the curve represented by each equation. Identify the type of curve and its center (or vertex if it is a parabola). Sketch each curve.

1. $(y-2)^2 = 4(x+1)$
2. $\dfrac{(x+4)^2}{4} + \dfrac{(y-1)^2}{1} = 1$
3. $\dfrac{(x-1)^2}{4} - \dfrac{(y-2)^2}{9} = 1$
4. $(y+5)^2 = -8(x-2)$
5. $\dfrac{(x+1)^2}{1} + \dfrac{y^2}{9} = 1$
6. $\dfrac{(y-4)^2}{16} - \dfrac{(x+2)^2}{4} = 1$
7. $(x+3)^2 = -12(y-1)$
8. $\dfrac{x^2}{16} + \dfrac{(y+1)^2}{1} = 1$

In Exercises 9–20, find the equation of each of the curves described by the given information.

9. Parabola: vertex $(-1, 3)$, focus $(3, 3)$
10. Parabola: vertex $(2, -1)$, directrix $y = 3$
11. Parabola: vertex $(-3, 2)$, focus $(-3, 3)$
12. Parabola: focus $(2, 4)$, directrix $x = 6$
13. Ellipse: center $(-2, 2)$, focus $(-5, 2)$, vertex $(-7, 2)$
14. Ellipse: center $(0, 3)$, focus $(12, 3)$, major axis 26 units
15. Ellipse: vertices $(-2, -3)$ and $(-2, 5)$, end of minor axis $(0, 1)$
16. Ellipse: foci $(1, -2)$ and $(1, 10)$, minor axis 5 units
17. Hyperbola: vertex $(-1, 1)$, focus $(-1, 4)$, center $(-1, 2)$
18. Hyperbola: foci $(2, 1)$ and $(8, 1)$, conjugate axis 6 units
19. Hyperbola: vertices $(2, 1)$ and $(-4, 1)$, focus $(-6, 1)$
20. Hyperbola: center $(1, -4)$, focus $(1, 1)$, transverse axis 8 units

In Exercises 21–28, determine the center (or vertex if the curve is a parabola) of the given curve. Sketch each curve.

21. $x^2 + 2x - 4y - 3 = 0$
22. $y^2 - 2x - 2y - 9 = 0$
23. $4x^2 + 9y^2 + 24x = 0$
24. $2x^2 + 9y^2 + 8x - 72y + 134 = 0$
25. $9x^2 - y^2 + 8y - 7 = 0$
26. $5x^2 - 4y^2 + 20x + 8y = 4$
27. $2x^2 - 4x = 9y - 2$
28. $0.04x^2 + 0.16y^2 = 0.01y$

In Exercises 29–36, solve the given problems.

29. Find the equation of the hyperbola with asymptotes $x - y = -1$ and $x + y = -3$ and vertex $(3, -1)$.
30. The circle $x^2 + y^2 + 4x - 5 = 0$ passes through the foci and the ends of the minor axis of an ellipse that has its major axis along the x-axis. Find the equation of the ellipse.
31. A first parabola has its vertex at the focus of a second parabola and its focus at the vertex of the second parabola. If the equation of the second parabola is $y^2 = 4x$, find the equation of the first parabola.
32. Identify the curve represented by the equation $4y^2 - x^2 - 6x - 2y - 14 = 0$ and then view its graph on a graphing calculator.
33. The stream of water from a fire hose follows a parabolic curve. If the stream from a hose nozzle fastened at the ground reaches a maximum height of 18 m at a horizontal distance of 28 m from the nozzle, find the equation that describes the stream. Take the origin at the location of the nozzle. Sketch the graph of the stream.
34. For a constant capacitive reactance and a constant resistance, sketch the graph of the impedance and inductive reactance (as abscissas) for an alternating-current circuit. See Section 12-7.
35. Two wheels in a friction drive assembly are equal ellipses, as shown in Fig. 21-82. The wheels are always in contact, with the center of the left wheel fixed in position and with the right wheel able to move horizontally. Find the equation that can be used to describe the circumference of each wheel in the position shown.

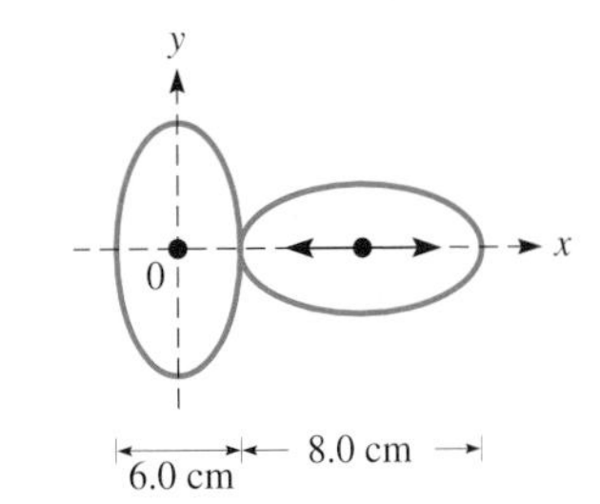

Fig. 21-82

36. An agricultural test station plans to divide a large tract of land into rectangular sections such that the perimeter of each section is 480 m. Express the area A of each section in terms of its width w. Identify the type of curve the equation represents, and sketch the graph of A as a function of w. For what value of w is A the greatest?

21-8 The Second-Degree Equation

The equations of the circle, parabola, ellipse, and hyperbola are all special cases of the same general equation. In this section we discuss this equation and how to identify the particular form it takes when it represents a specific type of curve.

Each of these curves can be represented by a **second-degree equation** *of the form*

$$Ax^2 + Bxy + Cy^2 + Dx + Ey + F = 0 \qquad (21\text{-}34)$$

The coefficients of the second-degree equation terms determine the type of curve that results. Recalling the discussions of the general forms of the equations of the circle, parabola, ellipse, and hyperbola from the previous sections of this chapter, Eq. (21-34) represents the indicated curve for given conditions of A, B, and C, as follows:

1. If $A = C$, $B = 0$, a circle.
2. If $A \neq C$ (but they have the same sign), $B = 0$, an ellipse.
3. If A and C have different signs, $B = 0$, a hyperbola.
4. If $A = 0$, $C = 0$, $B \neq 0$, a hyperbola.
5. If either $A = 0$ or $C = 0$ (but not both), $B = 0$, a parabola.
   (Special cases, such as a single point or no real locus, can also result.)

Another conclusion about Eq. (21-34) is that, if either $D \neq 0$ or $E \neq 0$ (or both), the center of the curve (or the vertex of a parabola) is not at the origin. If $B \neq 0$, the axis of the curve has been rotated. We have considered only one such case (the hyperbola $xy = c$) in this chapter. Rotation of axes is covered as one of the supplementary topics following the final chapter.

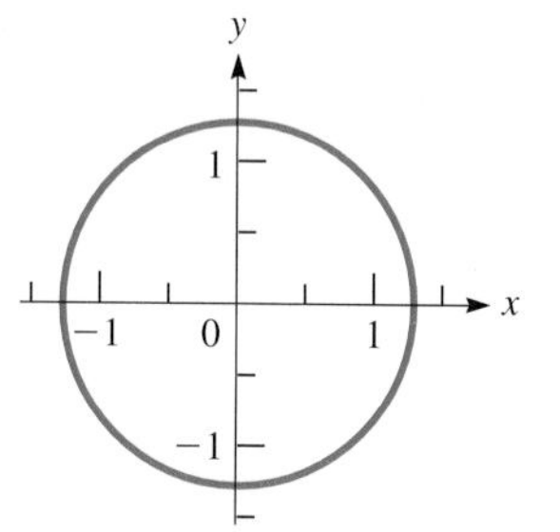

Fig. 21-83

EXAMPLE 1 ▸▸ The equation $2x^2 = 3 - 2y^2$ represents a circle. This can be seen by putting the equation in the form of Eq. (21-34). This form is

$$2x^2 + 2y^2 - 3 = 0$$

$A = 2$ (coefficient of x^2), $C = 2$ (coefficient of y^2)

We see that $A = C$. Also, since there is no xy-term, we know that $B = 0$. This means that the equation represents a circle. If we write it as $x^2 + y^2 = \frac{3}{2}$, we see that it fits the form of Eq. (21-12). The circle is shown in Fig. 21-83. ◂◂

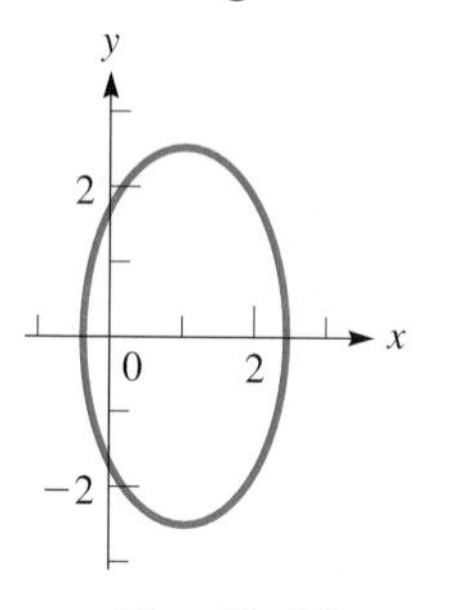

Fig. 21-84

EXAMPLE 2 ▸▸ The equation $3x^2 = 6x - y^2 + 3$ represents an ellipse. Before we analyze the equation, we should put it in the form of Eq. (21-34). For this equation, this form is

$$3x^2 + y^2 - 6x - 3 = 0$$

$A = 3$ (coefficient of x^2), $C = 1$ (coefficient of y^2)

Here we see that $B = 0$, A and C have the same sign, and $A \neq C$. Therefore, it is an ellipse. The $-6x$ term indicates that the center of the ellipse is not at the origin. The ellipse is shown in Fig. 21-84. ◂◂

EXAMPLE 3 ▸▸ Identify the curve represented by $2x^2 + 12x = y^2 - 14$. Determine the appropriate quantities for the curve, and sketch the graph.

Writing this equation in the form of Eq. (21-34), we have

$$2x^2 - y^2 + 12x + 14 = 0$$

$A = 2$ (coefficient of x^2), $C = -1$ (coefficient of y^2)

We identify this equation as representing a hyperbola, since A and C have different signs, and $B = 0$. We now write it in the standard form of a hyperbola.

$$2x^2 + 12x - y^2 = -14$$

$$2(x^2 + 6x \quad\quad) - y^2 = -14 \qquad \text{complete the square}$$

$$2(x^2 + 6x + 9) - y^2 = -14 + 18$$

$$2(x + 3)^2 - y^2 = 4$$

center is $(-3, 0)$

$$\frac{(x+3)^2}{2} - \frac{y^2}{4} = 1, \qquad \frac{x'^2}{2} - \frac{y'^2}{4} = 1$$

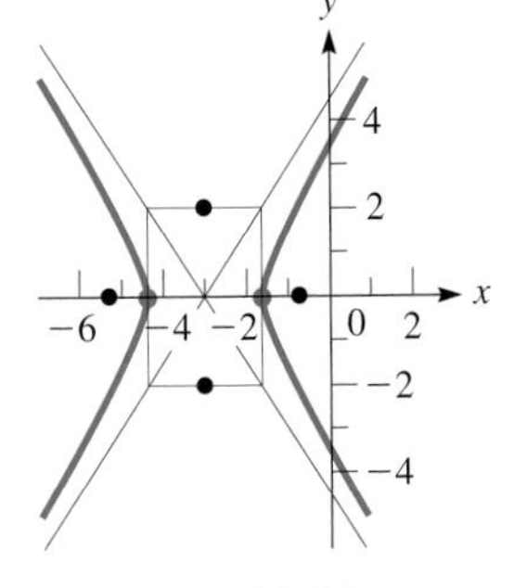

Fig. 21-85

Thus, we see that the center (h, k) of the hyperbola is the point $(-3, 0)$. Also, $a = \sqrt{2}$ and $b = 2$. This means that the vertices are $(-3 + \sqrt{2}, 0)$ and $(-3 - \sqrt{2}, 0)$, and the conjugate axis extends from $(-3, 2)$ to $(-3, -2)$. Also, $c^2 = 2 + 4 = 6$, which means that $c = \sqrt{6}$. The foci are $(-3 + \sqrt{6}, 0)$ and $(-3 - \sqrt{6}, 0)$. The graph is shown in Fig. 21-85. ◂◂

EXAMPLE 4 ▸▸ Identify the curve represented by $4y^2 - 23 = 4(4x + 3y)$, and find the appropriate important quantities. Then view it on a graphing calculator.

Writing the equation in the form of Eq. (21-34), we have

$$4y^2 - 16x - 12y - 23 = 0$$

Therefore, we recognize the equation as representing a parabola, since $A = 0$ and $B = 0$. Now, writing the equation in the standard form of a parabola, we have

$$4y^2 - 12y = 16x + 23$$

$$4(y^2 - 3y \quad\quad) = 16x + 23 \qquad \text{complete the square}$$

$$4\left(y^2 - 3y + \frac{9}{4}\right) = 16x + 23 + 9$$

$$4\left(y - \frac{3}{2}\right)^2 = 16(x + 2)$$

vertex $\left(-2, \frac{3}{2}\right)$

$$\left(y - \frac{3}{2}\right)^2 = 4(x + 2) \qquad \text{or} \qquad y'^2 = 4x'$$

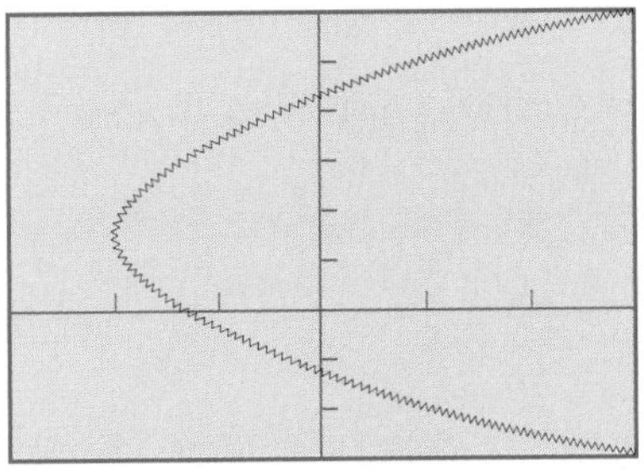

Fig. 21-86

We now note that the vertex is $(-2, 3/2)$ and that $p = 1$. This means that the focus is $(-1, 3/2)$ and the directrix is $x = -3$.

To view this equation on a graphing calculator, we first solve the equation for y and get $y = (3 \pm 4\sqrt{x + 2})/2$. Entering these two functions in the calculator we get the view shown in Fig. 21-86. ◂◂

In Chapter 14, when these curves were first introduced, they were referred to as **conic sections.** If a plane is passed through a cone, the intersection of the plane and the cone results in one of these curves; the curve formed depends on the angle of the plane with respect to the axis of the cone. This is shown in Fig. 21-87.

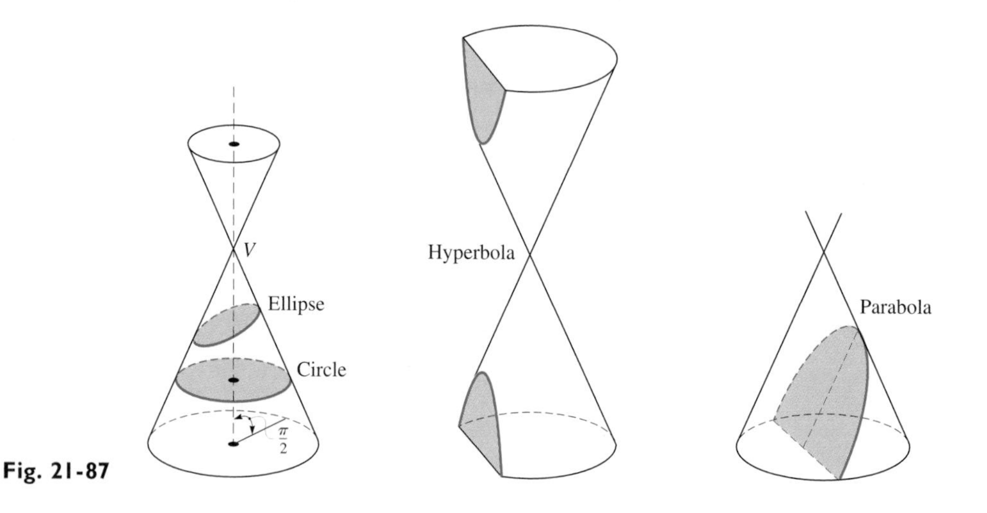

Fig. 21-87

Exercises 21–8

In Exercises 1–20, identify each of the equations as representing either a circle, a parabola, an ellipse, or a hyperbola.

1. $x^2 + 2y^2 - 2 = 0$ **2.** $x^2 - y = 0$

3. $2x^2 - y^2 - 1 = 0$ **4.** $3x^2 + 3y^2 - 1 = 0$

5. $2x^2 + 2y^2 - 3y - 1 = 0$

6. $x^2 - 2y^2 - 3x - 1 = 0$

7. $2.2x^2 - x - y = 1.6$

8. $2x^2 + 4y^2 - y - 2x = 4$

9. $x^2 = y^2 - 1$ **10.** $32x^2 = 21y - 47y^2$

11. $3.6x^2 = 1.1y - 3.6y^2$ **12.** $y = 3 - 6x^2$

13. $y(3 - 2y) = 2(x^2 - y^2)$ **14.** $x(13 - 5x) = 5y^2$

15. $2xy + x - 3y = 6$ **16.** $(y + 1)^2 = x^2 + y^2 - 1$

17. $2x(x - y) = y(3 - y - 2x)$

18. $2x^2 = x(x - 1) + 4y^2$

19. $x(y + 3x) = x^2 + xy - y^2 + 1$

20. $4x(x - 1) = 2x^2 - 2y^2 + 3$

In Exercises 21–28, identify the curve represented by each of the given equations. Determine the appropriate important quantities for the curve, and sketch the graph.

21. $x^2 = 8(y - x - 2)$ **22.** $x^2 = 6x - 4y^2 - 1$

23. $y^2 = 2(x^2 - 2x - 2y)$ **24.** $4x^2 + 4 = 9 - 8x - 4y^2$

25. $y^2 + 42 = 2x(10 - x)$

26. $x^2 - 4y = y^2 + 4(1 - x)$

27. $4(y^2 - 4x - 2) = 5(4y - 5)$

28. $2(2x^2 - y) = 8 - y^2$

In Exercises 29–32, view the curve for each equation on a graphing calculator. In Exercises 29 and 30, identify the type of curve before viewing it. In Exercises 31 and 32, the axis of each curve has been rotated.

29. $x^2 + 2y^2 - 4x + 12y + 14 = 0$

30. $4y^2 - x^2 + 40y - 4x + 60 = 0$

31. $x^2 + 6xy + 9y^2 - 2x + 14y - 10 = 0$

32. $x^2 - xy + y^2 - 6 = 0$

In Exercises 33–36, use the given values to determine the type of curve represented.

33. For the equation $x^2 + ky^2 = a^2$, what type of curve is represented if (a) $k = 1$, (b) $k < 0$, and (c) if $k > 0$ $(k \neq 1)$?

34. For the equation $\frac{x^2}{4 - C} - \frac{y^2}{C} = 1$, what type of curve is represented if (a) $C < 0$, (b) $0 < C < 4$? (For $C > 4$, see Exercise 36.)

35. In Eq. (21-34), if $A = C \neq 0$ and $B = D = E = F = 0$, describe the locus of the equation.

36. For the equation in Exercise 34, describe the locus of the equation if $C > 4$.

In Exercises 37–40, determine the type of curve from the given information.

37. The diagonal brace in a rectangular metal frame is 3.0 cm longer than the length of one of the sides. Determine the type of equation relating the lengths of the sides of the frame.

38. One circular solar cell has a radius that is 2.0 cm less than the radius r of a second circular solar cell. Determine the type of curve represented by the equation relating the total area A of both cells and r.

39. A flashlight emits a cone of light onto the floor. What type of curve is the perimeter of the lighted area on the floor, if the floor cuts completely through the cone of light?

40. The supersonic jet airliner Concorde creates a conical shock wave behind it. What type of curve is outlined on the surface of a lake by the shock wave if the Concorde is flying horizontally?

21-9 Polar Coordinates

Thus far we have graphed all curves in one coordinate system. This system, the rectangular coordinate system, is probably the most useful and widely applicable system. However, for certain types of curves, other coordinate systems prove to be better adapted. These coordinate systems are widely used, especially when certain applications of higher mathematics are involved. We shall discuss one of these systems here.

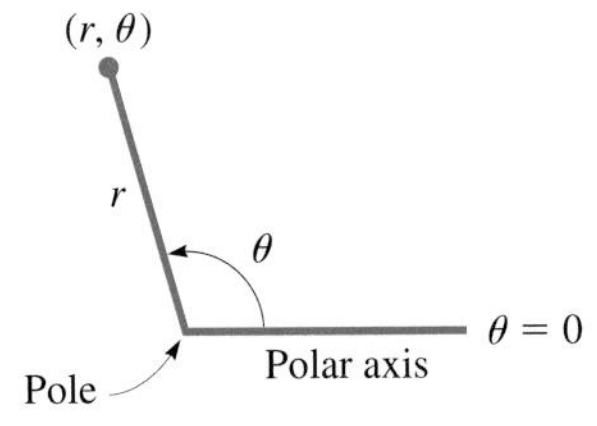

Fig. 21-88

Instead of designating a point by its x- and y-coordinates, we can specify its location by its radius vector and the angle the radius vector makes with the x-axis. Thus, the r and θ that are used in the definitions of the trigonometric functions can also be used as the coordinates of points in the plane. The important aspect of choosing coordinates is that, for each set of values, there must be only one point which corresponds to this set. We can see that this condition is satisfied by the use of r and θ as coordinates. *In* **polar coordinates,** *the origin is called the* **pole,** *and the half-line for which the angle is zero (equivalent to the positive x-axis) is called the* **polar axis.** The coordinates of a point are designated as (r, θ). We shall use radians when measuring the value of θ. See Fig. 21-88.

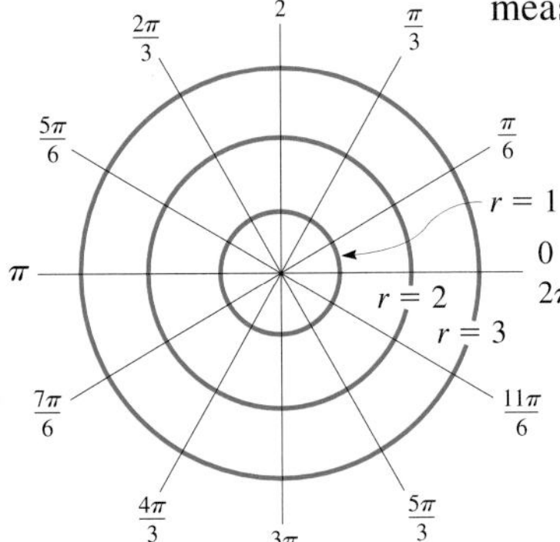

When using polar coordinates, we generally label the lines for some of the values of θ; namely, those for $\theta = 0$ (the polar axis), $\theta = \pi/2$ (equivalent to the positive y-axis), $\theta = \pi$ (equivalent to the negative x-axis), $\theta = 3\pi/2$ (equivalent to the negative y-axis), and possibly others. In Fig. 21-89, these lines and those for multiples of $\pi/6$ are shown. Also, the circles for $r = 1$, $r = 2$, and $r = 3$ are shown in this figure.

Fig. 21-89

EXAMPLE 1 ▸▸ (a) If $r = 2$ and $\theta = \pi/6$, we have the point as shown in Fig. 21-90. The coordinates (r, θ) of this point are written as $(2, \pi/6)$ when polar coordinates are used. This point corresponds to $(\sqrt{3}, 1)$ in rectangular coordinates.

(b) In Fig. 21-90, the polar coordinate point $(1, 3\pi/4)$ is also shown. It is equivalent to the point $(-\sqrt{2}/2, \sqrt{2}/2)$ in rectangular coordinates.

(c) In Fig. 21-90, the polar coordinate point $(2, 5)$ is also shown. It is equivalent approximately to the point $(0.6, -1.9)$ in rectangular coordinates. Remember, the 5 is an angle in radian measure. ◂◂

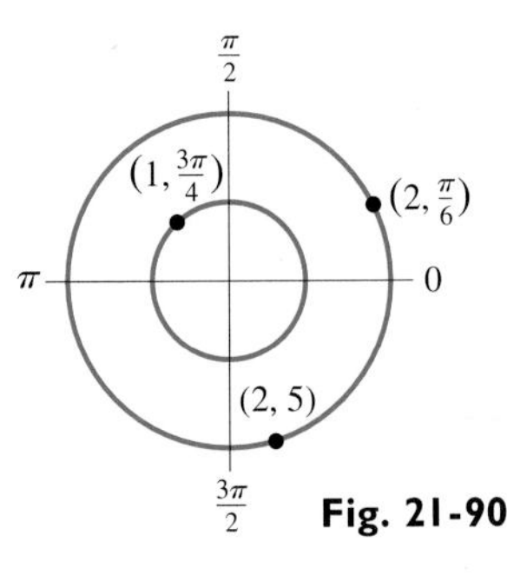

Fig. 21-90

One difference between rectangular coordinates and polar coordinates is that, for each point in the plane, there are limitless possibilities for the polar coordinates of that point. For example, the point $(2, \frac{\pi}{6})$ can also be represented by $(2, \frac{13\pi}{6})$ since the angles $\frac{\pi}{6}$ and $\frac{13\pi}{6}$ are coterminal. We also remove one restriction on r that we imposed in the definition of the trigonometric functions. That is, r is allowed to take on positive and negative values. If r is considered negative, then ***the point is found on the opposite side of the pole*** from that on which it is positive.

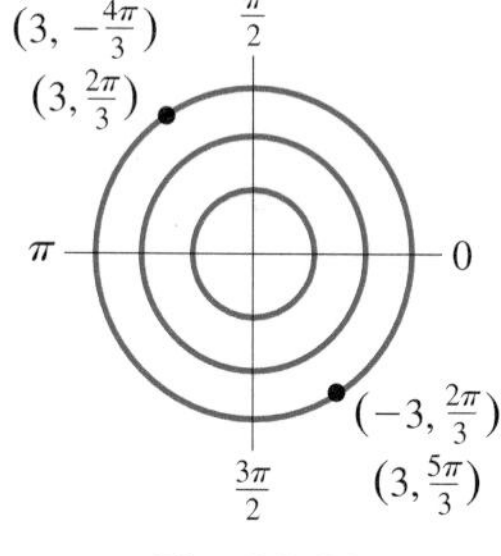

Fig. 21-91

EXAMPLE 2 ▸▸ The coordinates $(3, 2\pi/3)$ and $(3, -4\pi/3)$ represent the same point. However, the point $(-3, 2\pi/3)$ is on the opposite side of the pole, three units from the pole. Another possible set of coordinates for the point $(-3, 2\pi/3)$ is $(3, 5\pi/3)$. See Fig. 21-91. ◂◂

When plotting a point in polar coordinates, it is generally easier to *first locate the terminal side of θ and then measure r along this terminal side.* This is illustrated in the following example.

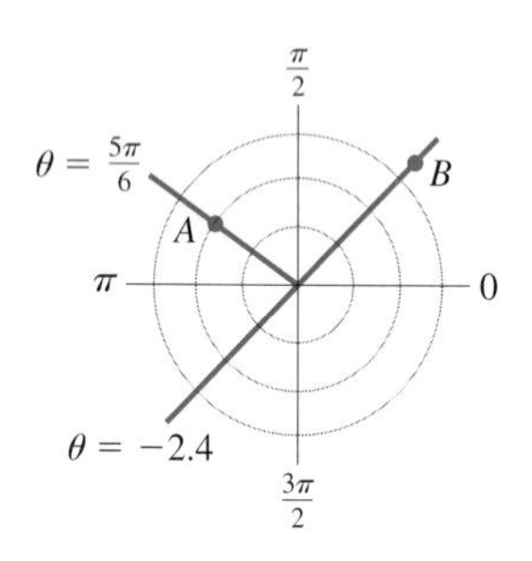

Fig. 21-92

EXAMPLE 3 ▸▸ Plot the points $A(2, 5\pi/6)$ and $B(-3.2, -2.4)$ in the polar coordinate system.

To locate A we determine the terminal side of $\theta = 5\pi/6$ and then determine $r = 2$. See Fig. 21-92.

To locate B we find the terminal side of $\theta = -2.4$, measuring clockwise from the polar axis (and recalling that $\pi = 3.14 = 180°$). Then we locate $r = -3.2$ on the opposite side of the pole. See Fig. 21-92.

We will find that points with negative values of r occur frequently when plotting curves in polar coordinates. ◂◂

Polar and Rectangular Coordinates

The relationships between the polar coordinates of a point and the rectangular coordinates of the same point come from the definitions of the trigonometric functions. Those most commonly used are (see Fig. 21-93):

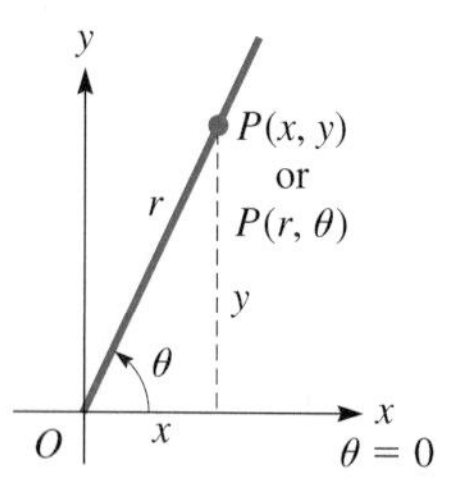

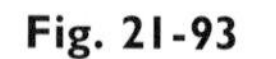
Fig. 21-93

$$x = r\cos\theta, \qquad y = r\sin\theta \tag{21-35}$$

$$\tan\theta = \frac{y}{x}, \qquad r = \sqrt{x^2 + y^2} \tag{21-36}$$

The following examples show the use of Eqs. (21-35) and (21-36) in changing coordinates in one system to coordinates in the other system. Also, these equations are used to transform equations from one system to the other.

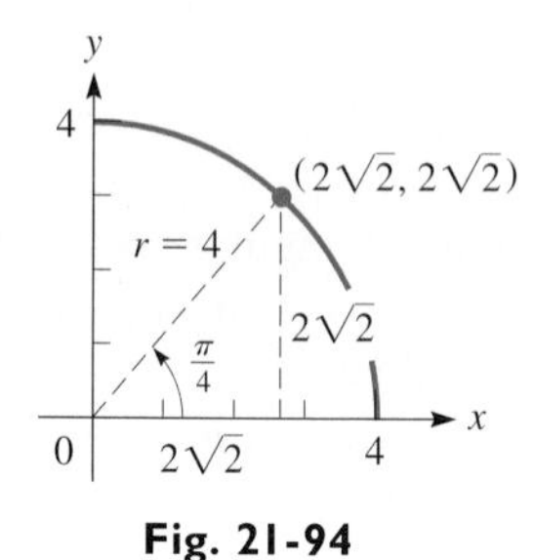

Fig. 21-94

EXAMPLE 4 ▸▸ Using Eqs. (21-35), we can transform the polar coordinates of $(4, \pi/4)$ into the rectangular coordinates $(2\sqrt{2}, 2\sqrt{2})$, since

$$x = 4\cos\frac{\pi}{4} = 4\left(\frac{\sqrt{2}}{2}\right) = 2\sqrt{2} \qquad \text{and} \qquad y = 4\sin\frac{\pi}{4} = 4\left(\frac{\sqrt{2}}{2}\right) = 2\sqrt{2}$$

See Fig. 21-94. ◂◂

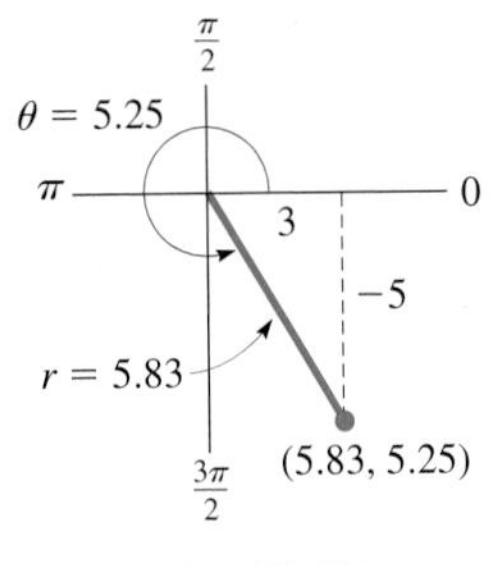

Fig. 21-95

EXAMPLE 5 ▸▸ Using Eqs. (21-36), we can transform the rectangular coordinates $(3, -5)$ into polar coordinates.

$$\tan\theta = -\frac{5}{3}, \qquad \theta = 5.25 \qquad (\text{or } -1.03)$$

$$r = \sqrt{3^2 + (-5)^2} = 5.83$$

We know that θ is a fourth-quadrant angle since x is positive and y is negative. Therefore, the point $(3, -5)$ in rectangular coordinates can be expressed as the point $(5.83, 5.25)$ in polar coordinates (see Fig. 21-95). Other polar coordinates for the point are also possible. ◂◂

EXAMPLE 6 ▸▸ If an electrically charged particle enters a magnetic field at right angles to the field, the particle follows a circular path. This fact is used in the design of nuclear particle accelerators.

A proton (positively charged) enters a magnetic field such that its path may be described by the rectangular equation $x^2 + y^2 = 2x$, where measurements are in metres. Find the polar equation of this circle.

We change this equation expressed in the rectangular coordinates x and y into an equation expressed in the polar coordinates r and θ by using the relations $r^2 = x^2 + y^2$ and $x = r\cos\theta$ as follows:

$$x^2 + y^2 = 2x \qquad \text{rectangular equation}$$

$$r^2 = 2r\cos\theta \qquad \text{substitute}$$

$$r = 2\cos\theta \qquad \text{divide by } r$$

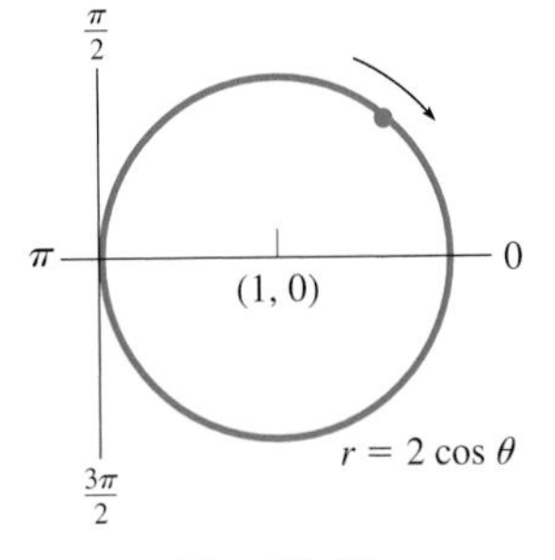

Fig. 21-96

This is the polar equation of the circle, which is shown in Fig. 21-96. ◂◂

EXAMPLE 7 ▸▸ Find the rectangular equation of the *rose* $r = 4\sin 2\theta$.

Using the trigonometric identity $\sin 2\theta = 2\sin\theta\cos\theta$ and Eqs. (21-35) and (21-36) leads to the solution.

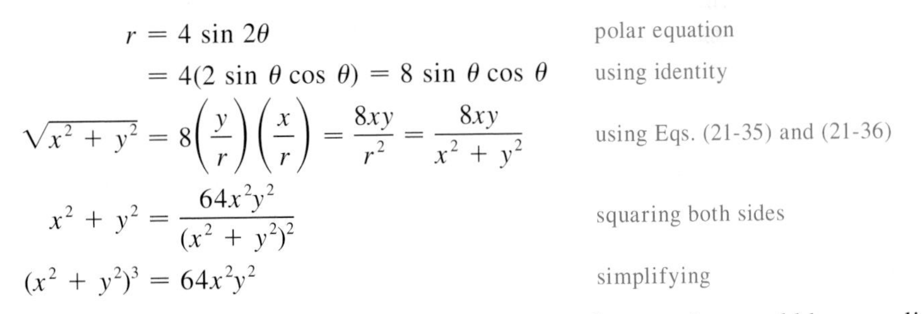

$$r = 4\sin 2\theta \qquad \text{polar equation}$$

$$= 4(2\sin\theta\cos\theta) = 8\sin\theta\cos\theta \qquad \text{using identity}$$

$$\sqrt{x^2 + y^2} = 8\left(\frac{y}{r}\right)\left(\frac{x}{r}\right) = \frac{8xy}{r^2} = \frac{8xy}{x^2 + y^2} \qquad \text{using Eqs. (21-35) and (21-36)}$$

$$x^2 + y^2 = \frac{64x^2y^2}{(x^2 + y^2)^2} \qquad \text{squaring both sides}$$

$$(x^2 + y^2)^3 = 64x^2y^2 \qquad \text{simplifying}$$

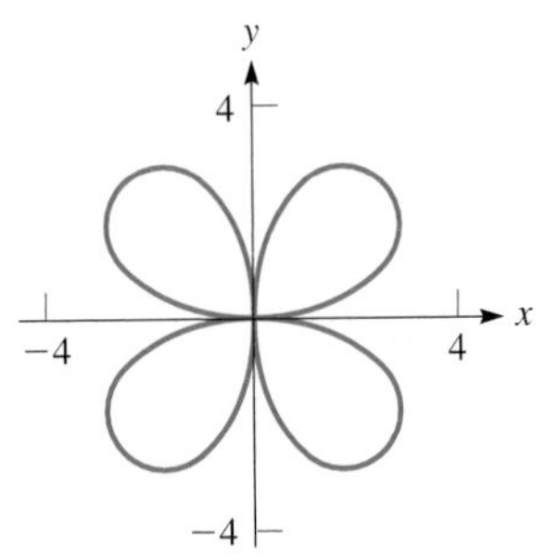

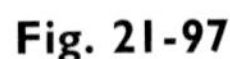
Fig. 21-97

Plotting the graph of this equation from the rectangular equation would be complicated. However, as we will see in the next section, plotting this graph in polar coordinates is quite simple. The curve is shown in Fig. 21-97. ◂◂

Exercises 21–9

In Exercises 1–12, plot the given polar coordinate points on polar coordinate paper.

1. $\left(3, \frac{\pi}{6}\right)$ **2.** $(2, \pi)$ **3.** $\left(\frac{5}{2}, -\frac{2\pi}{5}\right)$ **4.** $\left(5, -\frac{\pi}{3}\right)$ **5.** $\left(-2, \frac{7\pi}{6}\right)$ **6.** $\left(-5, \frac{\pi}{4}\right)$

7. $\left(-3, -\frac{5\pi}{4}\right)$ **8.** $\left(-4, -\frac{5\pi}{3}\right)$ **9.** $\left(0.5, -\frac{8\pi}{3}\right)$

10. $(2.2, -6\pi)$ **11.** $(2, 2)$ **12.** $(-1, -1)$

In Exercises 13–16, find a set of polar coordinates for each of the points given in rectangular coordinates.

13. $(\sqrt{3}, 1)$ **14.** $(-1, -1)$

15. $\left(-\frac{\sqrt{3}}{2}, -\frac{1}{2}\right)$ **16.** $(-5, 4)$

In Exercises 17–20, find the rectangular coordinates corresponding to the points for which the polar coordinates are given.

17. $\left(8, \frac{4\pi}{3}\right)$ **18.** $(-4, -\pi)$

19. $(3.0, -0.40)$ **20.** $(-1.0, 1.0)$

In Exercises 21–28, find the polar equation of each of the given rectangular equations.

21. $x = 3$ **22.** $y = 2$

23. $x^2 + y^2 = 0.81$ **24.** $x^2 + y^2 = 4y$

25. $y^2 = 4x$ **26.** $x^2 - y^2 = 0.01$

27. $x^2 + 4y^2 = 4$ **28.** $y = x^2$

In Exercises 29–36, find the rectangular equation of each of the given polar equations.

29. $r = \sin\theta$ **30.** $r = 4\cos\theta$

31. $r\cos\theta = 4$ **32.** $r\sin\theta = -2$

33. $r = 2(1 + \cos\theta)$ **34.** $r = 1 - \sin\theta$

35. $r^2 = \sin 2\theta$ **36.** $r^2 = 16\cos 2\theta$

In Exercises 37–40, find the required equations.

37. Under certain conditions, the x- and y-components of a magnetic field B are given by the equations

$$B_x = \frac{-ky}{x^2 + y^2} \quad \text{and} \quad B_y = \frac{kx}{x^2 + y^2}$$

Write these equations in terms of polar coordinates.

38. In designing a domed roof for a building, an architect uses the equation $x^2 + \frac{y^2}{k^2} = 1$, where k is a constant. Write this equation in polar form.

39. The shape of a cam can be described by the polar equation $r = 3 - \sin\theta$. Find the rectangular equation for the shape of the cam.

40. The polar equation of the path of a weather satellite of the earth is

$$r = \frac{7600}{1 + 0.14\cos\theta}$$

where r is measured in kilometres. Find the rectangular equation of the path of this satellite. The path is an ellipse, with the earth at one of the foci.

21-10 Curves in Polar Coordinates

The basic method for finding a curve in polar coordinates is the same as in rectangular coordinates. We assume values of θ and then find the corresponding values of r. These points are plotted and joined, thus forming the curve that represents the function. However, there are certain basic curves that can be sketched directly from the equation.

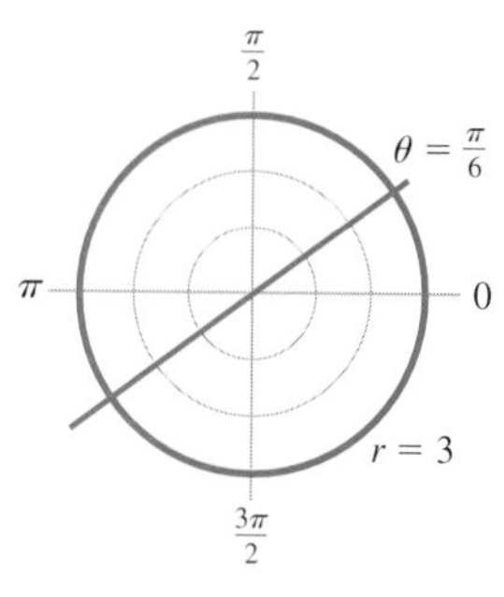

Fig. 21-98

EXAMPLE 1 ▸▸ (a) The graph of the polar equation $r = 3$ is a circle of radius 3, with center at the pole. This is the case, since $r = 3$ for all values of θ. It is not necessary to find specific points for this circle, which is shown in Fig. 21-98.

(b) The graph of $\theta = \pi/6$ is a straight line through the pole. It represents all points for which $\theta = \pi/6$ for all values of r. This line is shown in Fig. 21-98. ◂◂

EXAMPLE 2 ▸▸ Plot the graph of $r = 1 + \cos\theta$.

We find the following values of r corresponding to the chosen values of θ.

θ	0	$\frac{\pi}{4}$	$\frac{\pi}{2}$	$\frac{3\pi}{4}$	π	$\frac{5\pi}{4}$	$\frac{3\pi}{2}$	$\frac{7\pi}{4}$	2π
r	2	1.7	1	0.3	0	0.3	1	1.7	2
Point number	1	2	3	4	5	6	7	8	9

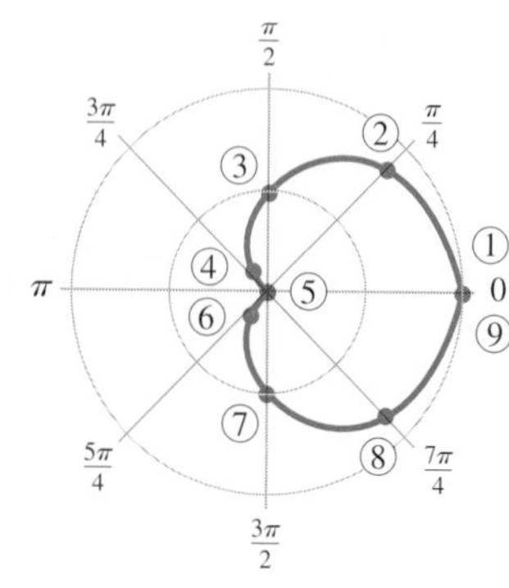

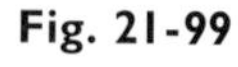

Fig. 21-99

We now see that the points start repeating, and it is unnecessary to find additional points. The curve is called a **cardioid** and is shown in Fig. 21-99. ◂◂

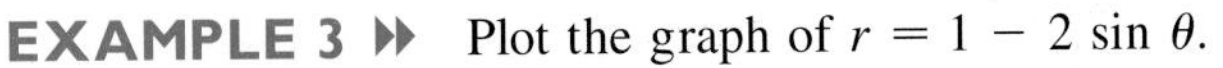

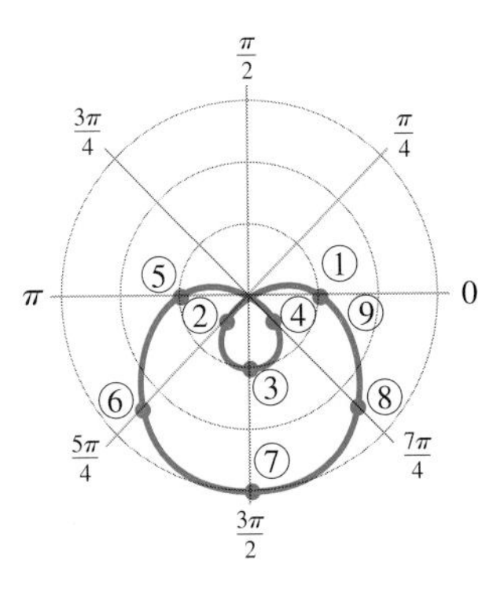

Fig. 21-100

EXAMPLE 3 ▸▸ Plot the graph of $r = 1 - 2\sin\theta$.

Choosing values of θ, and then finding the corresponding values of r, we find the following table of values.

θ	0	$\frac{\pi}{4}$	$\frac{\pi}{2}$	$\frac{3\pi}{4}$	π	$\frac{5\pi}{4}$	$\frac{3\pi}{2}$	$\frac{7\pi}{4}$	2π
r	1	−0.4	−1	−0.4	1	2.4	3	2.4	1
Point number	1	2	3	4	5	6	7	8	9

Particular care should be taken in plotting the points for which r is negative. This curve is known as a **limaçon** and is shown in Fig. 21-100. ◂◂

EXAMPLE 4 ▸▸ A cam is shaped such that the edge of the upper "half" is represented by the equation $r = 2.0 + \cos\theta$, and the lower "half" by the equation $r = \dfrac{3.0}{2.0 - \cos\theta}$, where measurements are in inches. Plot the curve that represents the shape of the cam.

We get the points for the edge of the cam by using values of θ from 0 to π for the upper "half" and from π to 2π for the lower half. The table of values follows.

$r = 2.0 + \cos\theta$

θ	0	$\frac{\pi}{4}$	$\frac{\pi}{2}$	$\frac{3\pi}{4}$	π
r	3.0	2.7	2.0	1.3	1.0
Point number	1	2	3	4	5

$r = \dfrac{3.0}{2.0 - \cos\theta}$

θ	π	$\frac{5\pi}{4}$	$\frac{3\pi}{2}$	$\frac{7\pi}{4}$	2π
r	1.0	1.1	1.5	2.3	3.0
Point number	6	7	8	9	10

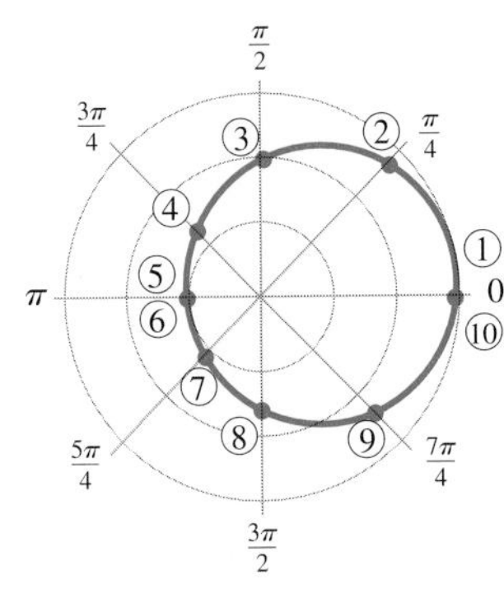

Fig. 21-101

The upper "half" is part of a limaçon, and the lower "half" is a semiellipse. The cam is shown in Fig. 21-101. ◂◂

EXAMPLE 5 ▸▸ Plot the graph of $r = 2\cos 2\theta$.

In finding values of r we must be careful first to multiply the values of θ by 2 before finding the cosine of the angle. Also, for this reason, we take values of θ as multiples of $\pi/12$, so as to get enough useful points. The table of values follows.

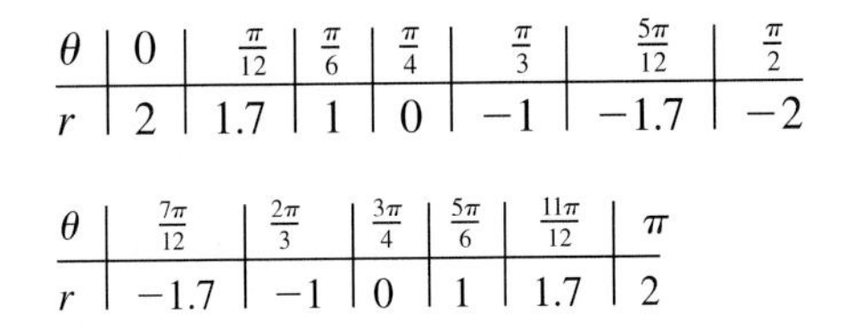

θ	0	$\frac{\pi}{12}$	$\frac{\pi}{6}$	$\frac{\pi}{4}$	$\frac{\pi}{3}$	$\frac{5\pi}{12}$	$\frac{\pi}{2}$
r	2	1.7	1	0	−1	−1.7	−2

θ	$\frac{7\pi}{12}$	$\frac{2\pi}{3}$	$\frac{3\pi}{4}$	$\frac{5\pi}{6}$	$\frac{11\pi}{12}$	π
r	−1.7	−1	0	1	1.7	2

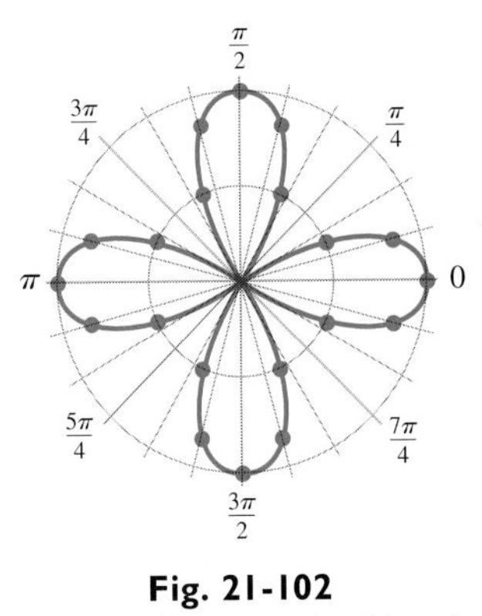

Fig. 21-102

For values of θ starting with π, the values of θ repeat. We have a four-leaf **rose,** as shown in Fig. 21-102. ◂◂

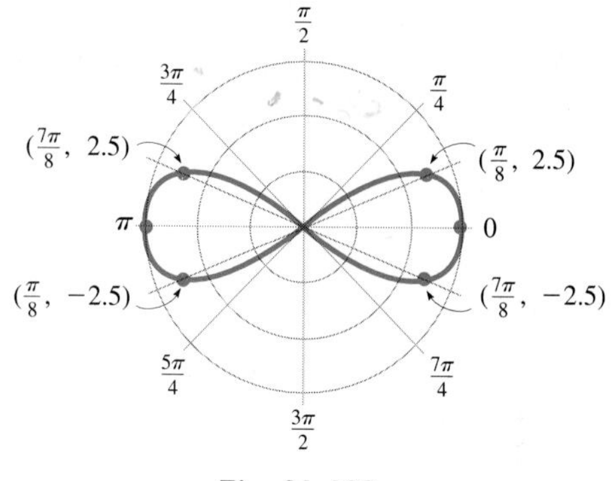

Fig. 21-103

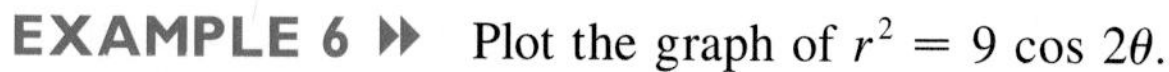

EXAMPLE 6 ▸▸ Plot the graph of $r^2 = 9 \cos 2\theta$.

Choosing the indicated values of θ, we get the values of r shown in the following table of values.

θ	0	$\frac{\pi}{8}$	$\frac{\pi}{4}$	...	$\frac{3\pi}{4}$	$\frac{7\pi}{8}$	π
r	± 3	± 2.5	0		0	± 2.5	± 3

There are no values of r corresponding to values of θ in the range $\pi/4 < \theta < 3\pi/4$, since twice these angles are in the second and third quadrants, and the cosine is negative for such angles. The value of r^2 cannot be negative. Also, the values of r repeat for $\theta > \pi$. The figure is called a **lemniscate** and is shown in Fig. 21-103. ◂◂

EXAMPLE 7 ▸▸ View the graph of $r = 1 - 2 \cos \theta$ on a graphing calculator.

A polar equation can be viewed on most graphing calculators. Using the MODE feature, the polar graph option or parametric graph option is used, depending on the calculator. (Review the manual for the calculator.) With the polar graph option, the function is entered directly.

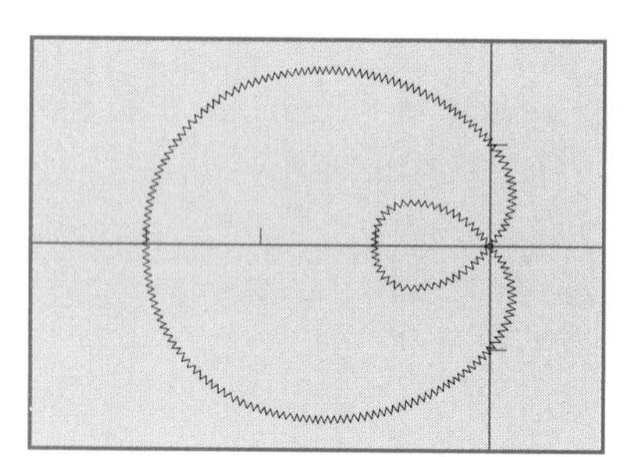

Fig. 21-104

With the parametric graph option, to graph $r = f(\theta)$, we note that $x = r \cos \theta$, and $y = r \sin \theta$, which tells us that

$$x = f(\theta) \cos \theta$$
$$y = f(\theta) \sin \theta$$

Thus, for $r = 1 - 2 \cos \theta$, by using

$$x = (1 - 2 \cos \theta)\cos \theta$$
$$y = (1 - 2 \cos \theta)\sin \theta$$

the graph can be displayed, as shown in Fig. 21-104. ◂◂

Exercises 21-10

In Exercises 1–28, plot the curves of the given polar equations in polar coordinates.

1. $r = 4$ **2.** $r = 2$ **3.** $\theta = 3\pi/4$

4. $\theta = -1.5$ **5.** $r = 4 \sec \theta$ **6.** $r = 4 \csc \theta$

7. $r = 2 \sin \theta$ **8.** $r = 3 \cos \theta$

9. $r = 1 - \cos \theta$ (cardioid)

10. $r = \sin \theta - 1$ (cardioid)

11. $r = 2 - \cos \theta$ (limaçon)

12. $r = 2 + 3 \sin \theta$ (limaçon)

13. $r = 4 \sin 2\theta$ (rose) **14.** $r = 2 \sin 3\theta$ (rose)

15. $r^2 = 4 \sin 2\theta$ (lemniscate)

16. $r^2 = 2 \sin \theta$

17. $r = 2^{\theta}$ (spiral) **18.** $r = 1.5^{-\theta}$ (spiral)

19. $r = -4 \sin 4\theta$ (rose) **20.** $r = -\cos 3\theta$ (rose)

21. $r = \dfrac{1}{2 - \cos \theta}$ (ellipse)

22. $r = \dfrac{1}{1 - \cos \theta}$ (parabola)

23. $r = \dfrac{6}{1 - 2 \cos \theta}$ (hyperbola)

24. $r = \dfrac{6}{3 - 2 \sin \theta}$ (ellipse)

25. $r = 4 \cos \frac{1}{2}\theta$ **26.** $r = 2 + \cos 3\theta$

27. $r = 3 - \sin 3\theta$ **28.** $r = 4 \tan \theta$

In Exercises 29–32, view the curves of the given polar equations on a graphing calculator.

29. $r = \theta \quad (0 \le \theta \le 10)$ **30.** $r = \theta \quad (-20 \le \theta \le 20)$

31. $r = 4 + 2 \sin 2\theta$ **32.** $r = 1 - 2 \cos 3\theta$

In Exercises 33–36, sketch the indicated graphs.

33. An architect designs a patio which is shaped such that it can be described as the area within the polar curve $r = 4.0 - \sin \theta$, where measurements are in metres. Sketch the perimeter of the patio.

34. A missile is fired at an airplane and is always directed toward the airplane. The missile is traveling at twice the speed of the airplane. An equation that describes the distance r between the missile and the airplane is $r = \dfrac{70 \sin \theta}{(1 - \cos \theta)^2}$, where θ is the angle between their directions at any time, and the plane is assumed to be at the pole at all times. See Fig. 21-105. This is a *relative pursuit curve*. Sketch the graph of this equation for $\pi/4 \le \theta \le \pi$.

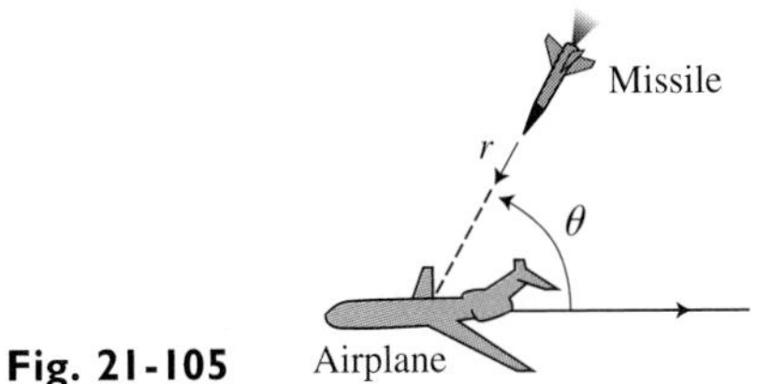

Fig. 21-105

35. In studying the photoelectric effect, an equation used for the rate R at which photoelectrons are ejected at various angles θ is $R = \dfrac{\sin^2 \theta}{(1 - 0.5 \cos \theta)^2}$. Sketch the graph.

36. Sketch the graph of the rectangular equation $4(x^6 + 3x^4y^2 + 3x^2y^4 + y^6 - x^4 - 2x^2y^2 - y^4) + y^2 = 0$ (*Hint:* The equation can be written as $4(x^2 + y^2)^3 - 4(x^2 + y^2)^2 + y^2 = 0$. Transform to polar coordinates, and then sketch the curve.)

Chapter Equations, Review Exercises, and Practice Test

Chapter Equations

Distance formula	Fig. 21-2	$d = \sqrt{(x_2 - x_1)^2 + (y_2 - y_1)^2}$	(21-1)
Slope	Fig. 21-4	$m = \dfrac{y_2 - y_1}{x_2 - x_1}$	(21-2)
	Fig. 21-7	$m = \tan \alpha \quad (0° \le \alpha < 180°)$	(21-3)
	Fig. 21-9	$m_1 = m_2 \quad (\text{for} \parallel \text{lines})$	(21-4)
	Fig. 21-10	$m_2 = -\dfrac{1}{m_1} \quad \text{or} \quad m_1m_2 = -1 \quad (\text{for} \perp \text{lines})$	(21-5)
Straight line	Fig. 21-14	$y - y_1 = m(x - x_1)$	(21-6)
	Fig. 21-17	$x = a$	(21-7)
	Fig. 21-18	$y = b$	(21-8)
	Fig. 21-21	$y = mx + b$	(21-9)
		$Ax + By + C = 0$	(21-10)
Circle	Fig. 21-29	$(x - h)^2 + (y - k)^2 = r^2$	(21-11)
	Fig. 21-32	$x^2 + y^2 = r^2$	(21-12)
		$x^2 + y^2 + Dx + Ey + F = 0$	(21-14)
Parabola	Fig. 21-39	$y^2 = 4px$	(21-15)
	Fig. 21-42	$x^2 = 4py$	(21-16)
Ellipse	Fig. 21-53	$\dfrac{x^2}{a^2} + \dfrac{y^2}{b^2} = 1$	(21-17)
	Fig. 21-53	$a^2 = b^2 + c^2$	(21-18)
	Fig. 21-54	$\dfrac{y^2}{a^2} + \dfrac{x^2}{b^2} = 1$	(21-19)

Hyperbola	Fig. 21-66	$\frac{x^2}{a^2} - \frac{y^2}{b^2} = 1$	(21-20)
	Fig. 21-66	$c^2 = a^2 + b^2$	(21-21)
	Fig. 21-65	$y = \pm\frac{bx}{a}$ (asymptotes)	(21-23)
	Fig. 21-67	$\frac{y^2}{a^2} - \frac{x^2}{b^2} = 1$	(21-24)
	Fig. 21-72	$xy = c$	(21-25)
Translation of axes	Fig. 21-76	$x = x' + h$ and $y = y' + k$	(21-26)
		$x' = x - h$ and $y' = y - k$	(21-27)
Parabola, vertex (h, k):		$(y - k)^2 = 4p(x - h)$ (axis parallel to x-axis)	(21-28)
		$(x - h)^2 = 4p(y - k)$ (axis parallel to y-axis)	(21-29)
Ellipse, center (h, k):		$\frac{(x-h)^2}{a^2} + \frac{(y-k)^2}{b^2} = 1$ (major axis parallel to x-axis)	(21-30)
		$\frac{(y-k)^2}{a^2} + \frac{(x-h)^2}{b^2} = 1$ (major axis parallel to y-axis)	(21-31)
Hyperbola, center (h, k):		$\frac{(x-h)^2}{a^2} - \frac{(y-k)^2}{b^2} = 1$ (transverse axis parallel to x-axis)	(21-32)
		$\frac{(y-k)^2}{a^2} - \frac{(x-h)^2}{b^2} = 1$ (transverse axis parallel to y-axis)	(21-33)
Second-degree equation		$Ax^2 + Bxy + Cy^2 + Dx + Ey + F = 0$	(21-34)
Polar coordinates	Fig. 21-93	$x = r\cos\theta$, $y = r\sin\theta$	(21-35)
		$\tan\theta = \frac{y}{x}$, $r = \sqrt{x^2 + y^2}$	(21-36)

Review Exercises

In Exercises 1–12, find the equation of the indicated curve subject to the given conditions. Sketch each curve.

1. Straight line: passes through $(1, -7)$ with a slope of 4

2. Straight line: passes through $(-1, 5)$ and $(-2, -3)$

3. Straight line: perpendicular to $3x - 2y + 8 = 0$ and has a y-intercept of $(0, -1)$

4. Straight line: parallel to $2x - 5y + 1 = 0$ and has an x-intercept of $(2, 0)$

5. Circle: center at $(1, -2)$, passes through $(4, -3)$

6. Circle: tangent to the line $x = 3$, center at $(5, 1)$

7. Parabola: focus $(3, 0)$, vertex $(0, 0)$

8. Parabola: directrix $y = -5$, vertex $(0, 0)$

9. Ellipse: vertex $(10, 0)$, focus $(8, 0)$, center $(0, 0)$

10. Ellipse: center $(0, 0)$, passes through $(0, 3)$ and $(2, 1)$

11. Hyperbola: $V(0, 13)$, $C(0, 0)$, conj. axis of 24

12. Hyperbola: $F(0, 10)$, $F(0, -10)$, $V(0, 8)$

In Exercises 13–24, find the indicated quantities for each of the given equations. Sketch each curve.

13. $x^2 + y^2 + 6x - 7 = 0$, center and radius

14. $x^2 + y^2 - 4x + 2y - 20 = 0$, center and radius

15. $x^2 = -20y$, focus and directrix

16. $y^2 = 24x$, focus and directrix

17. $16x^2 + y^2 = 16$, vertices and foci

18. $2y^2 - 9x^2 = 18$, vertices and foci

19. $2x^2 - 5y^2 = 8$, vertices and foci

20. $2x^2 + 25y^2 = 50$, vertices and foci

21. $x^2 - 8x - 4y - 16 = 0$, vertex and focus

22. $y^2 - 4x + 4y + 24 = 0$, vertex and directrix

23. $4x^2 + y^2 - 16x + 2y + 13 = 0$, center

24. $x^2 - 2y^2 + 4x + 4y + 6 = 0$, center

In Exercises 25–32, plot the given curves in polar coordinates.

25. $r = 4(1 + \sin\theta)$ **26.** $r = 1 - 3\cos\theta$

27. $r = 4\cos 3\theta$ **28.** $r = -3\sin\theta$

29. $r = \cot\theta$ **30.** $r = \dfrac{1}{2(\sin\theta - 1)}$

31. $r = 2\sin\left(\dfrac{\theta}{2}\right)$ **32.** $r = 1 - \cos 2\theta$

In Exercises 33–36, find the polar equation of each of the given rectangular equations.

33. $y = 2x$ **34.** $2xy = 1$

35. $x^2 - y^2 = 16$ **36.** $x^2 + y^2 = 7 - 6y$

In Exercises 37–40, find the rectangular equation of each of the given polar equations.

37. $r = 2\sin 2\theta$ **38.** $r^2 = \sin\theta$

39. $r = \dfrac{4}{2 - \cos\theta}$ **40.** $r = \dfrac{2}{1 - \sin\theta}$

In Exercises 41–44, determine the number of real solutions of the given systems of equations by sketching the curves. (See Section 14-1.)

41. $x^2 + y^2 = 9$
$4x^2 + y^2 = 16$

42. $y = e^x$
$x^2 - y^2 = 1$

43. $x^2 + y^2 - 4y - 5 = 0$
$y^2 - 4x^2 - 4 = 0$

44. $x^2 - 4y^2 + 2x - 3 = 0$
$y^2 - 4x - 4 = 0$

In Exercises 45–48, view the curves of the given equations on a graphing calculator.

45. $x^2 - 4y^2 + 4x + 24y - 48 = 0$

46. $x^2 + 2xy + y^2 - 3x + 8y = 0$

47. $r = 3\cos(3\theta/2)$

48. $r = 5 - 2\sin 4\theta$

In Exercises 49–80, solve the given problems.

49. Find the points of intersection of the ellipses $25x^2 + 4y^2 = 100$ and $4x^2 + 9y^2 = 36$.

50. Find the points of intersection of the hyperbola $y^2 - x^2 = 1$ and the ellipse $x^2 + 25y^2 = 25$.

51. In two ways show that the line segments joining $(-3, 11)$, $(2, -1)$, and $(14, 4)$ form a right triangle.

52. Show that the altitudes of the triangle with vertices $(2, -4)$, $(3, -1)$, and $(-2, 5)$ meet at a single point.

53. By means of the definition of a parabola, find the equation of the parabola with focus at $(3, 1)$ and directrix the line $y = -3$. Find the same equation by the method of translation of axes.

54. For what value of k does $x^2 - ky^2 = 1$ represent an ellipse with vertices on the y-axis?

55. The total resistance R_T of two resistances in series in an electric circuit is the sum of the resistances. If a variable resistor R is in series with a 2.5-Ω resistor, express R_T as a function of R and sketch the graph.

56. The acceleration of an object is defined as the change in velocity v divided by the corresponding change in time t. Find the equation relating the velocity v and time t for an object for which the acceleration is 6.0 m/s^2 and $v = 5.0$ m/s when $t = 0$ s.

57. One computer line printer prints 2500 lines/min for x min, and a second printer prints 1500 lines/min for y min. If they print a total of 37 500 lines together, express y as a function of x and sketch the graph.

58. An airplane touches down when landing at 150 km/h. Its velocity v while coming to a stop is given by $v = 150 - 20\,000t$, where t is the time in hours. Sketch the graph of v vs. t.

59. It takes 2.010 kJ of heat to raise the temperature of 1.000 kg of steam by 1.000°C. In a steam generator, a total of y kJ is used to raise the temperature of 50.00 kg of steam from 100°C to T°C. Express y as a function of T and sketch the graph.

60. An organism is spread over an area of 4 m^2. It then spreads such that it covers an area of 6 m^2 after one day. If the equation relating area A covered and time t is linear, determine the equation.

61. A solar panel reflector is circular and has a circumference of 4.50 m. Taking the origin of a coordinate system as the center of the reflector, what is the equation of the circumference?

62. A friction disk drive wheel is tangent to another wheel, as shown in Fig. 21-106. The radii of the wheels are 2 cm and 6 cm. What is the equation of the circumference of each wheel if the coordinate system is located at the center of the drive wheel as shown?

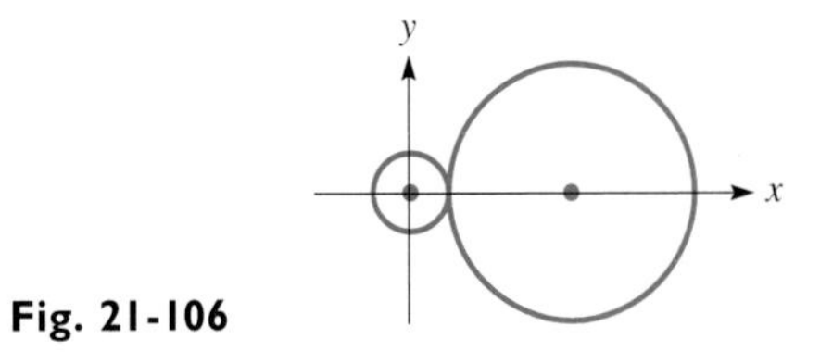

Fig. 21-106

63. The top horizontal cross section of a dam is parabolic. The open area within this cross section is 80 m across and 50 m from front to back. Find the equation of the edge of the open area with the vertex at the origin of the coordinate system and the axis along the x-axis.

64. The *quality factor* Q of a series resonant electric circuit with resistance R, inductance L, and capacitance C is given by $Q = \frac{1}{R}\sqrt{\frac{L}{C}}$. Sketch the graph of Q and L for a circuit in which $R = 1000\ \Omega$ and $C = 4.00\ \mu F$.

65. A rectangular parking lot is to have a perimeter of 600 m. Express the area A in terms of the width w and sketch the graph.

66. At very low temperatures, certain metals have an electric resistance of zero. This phenomenon is called *superconductivity*. A magnetic field also affects the superconductivity. A certain level of magnetic field H_T, the threshold field, is related to the thermodynamic temperature T by $H_T/H_0 = 1 - (T/T_0)^2$, where H_0 and T_0 are specifically defined values of magnetic field and temperature. Sketch H_T/H_0 vs. T/T_0.

67. The electric power P, in watts, supplied by a battery is given by $P = 12.0i - 0.500i^2$, where i is the current. Sketch the graph of P and i.

68. The speed v (in cm/s) of liquid flowing in a pipe of radius r (in cm) is $v = 2(r^2 - d^2)$, where d is the distance from the center of the pipe. Sketch the graph of v vs. d for $r = 8.0$ cm.

69. A study indicated that the fraction f of cells destroyed by various dosages d of X rays is given by the graph in Fig. 21-107. Assuming that the curve is a quarter-ellipse, find the equation relating f and d for $0 \leq f \leq 1$ and $0 < d \leq 10$ units.

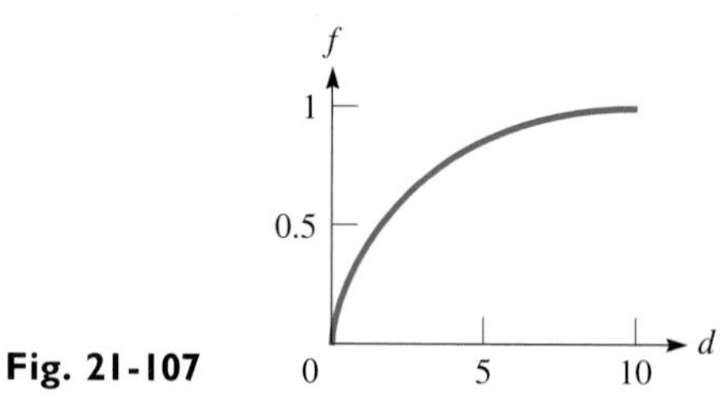

Fig. 21-107

70. The Colosseum in Rome is in the shape of an ellipse 188 m long and 156 m wide. Find the area of the Colosseum. ($A = \pi ab$ for an ellipse.)

71. A machine-part designer wishes to make a model for an elliptical cam by placing two pins in a design board, putting a loop of string over the pins, and marking off the outline by keeping the string taut. (Note that the definition of the ellipse is being used.) If the cam is to measure 10 cm by 6 cm, how long should the loop of string be and how far apart should the pins be?

72. Soon after reaching the vicinity of the moon, *Apollo 11* (the first spacecraft to land a man on the moon) went into an elliptical lunar orbit. The closest the craft was to the moon in this orbit was 110 km, and the farthest it was from the moon was 310 km. What was the equation of the path if the center of the moon was at one of the foci of the ellipse? Assume that the major axis is along the x-axis and that the center of the ellipse is at the origin. The radius of the moon is 1740 km.

73. The vertical cross section of the cooling tower of a nuclear power plant is hyperbolic, as shown in Fig. 21-108. Find the radius r of the smallest circular horizontal cross section.

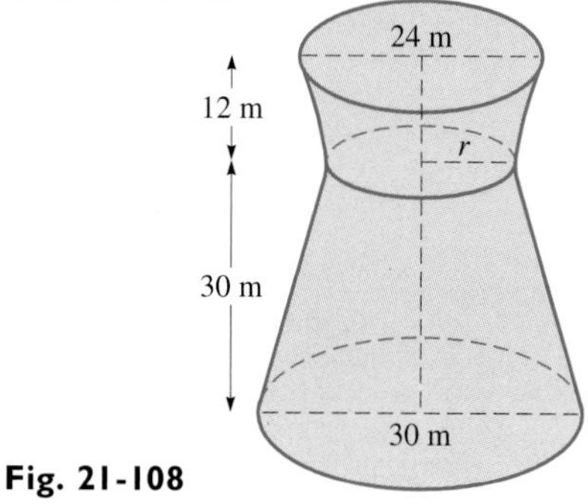

Fig. 21-108

74. Tremors from an earthquake are recorded at the California Institute of Technology (Pasadena, California) 36 s before they are recorded at Stanford University (Palo Alto, California). If the seismographs are 510 km apart, and the shock waves from the tremors travel at 5.0 km/s, what is the curve on which lies the point where the earthquake occurred?

75. An automobile grille is in the shape of an ellipse 90.0 cm wide and 24.0 cm high. Find the width of the grille 10.0 cm from the top.

76. A 20-m rope passes over a pulley 4 m above the ground, and a crate on the ground is attached at one end. A man holds the other end at a level of 1 m above the ground. If the man walks away from the pulley, express the height of the crate above the ground in terms of the distance the man is from directly below the object. Sketch the graph of distance and height. See Fig. 21-109. (Neglect the thickness of the crate.)

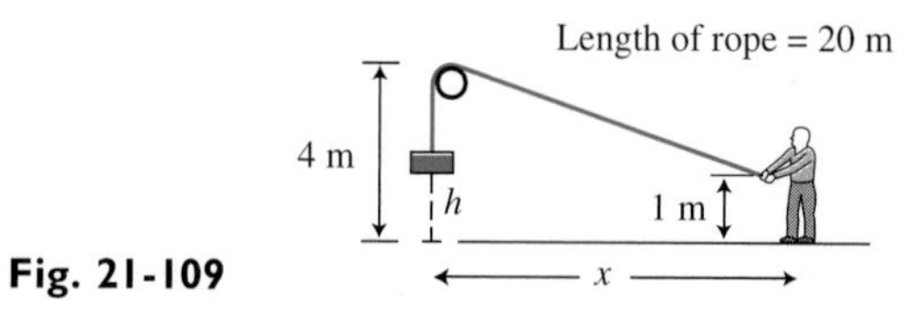

Fig. 21-109

77. A satellite at a height proper to make one revolution per day around the earth will have for an excellent approximation of its projection on the earth of its path the curve $r^2 = R^2 \cos 2(\theta + \frac{\pi}{2})$, where R is the radius of the earth. Sketch the path of the projection.

78. The vertical cross sections of two pipes as drawn on a drawing board are shown in Fig. 21-110. Find the polar equation of each.

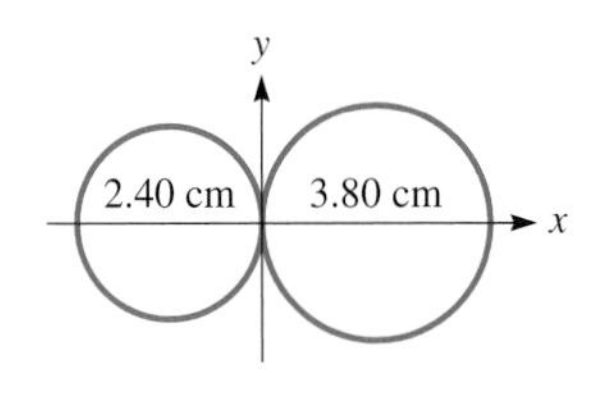

Fig. 21-110

79. An electronic instrument located at point P records the sound of a rifle shot and the impact of the bullet striking the target at the same instant. Show that P lies on a branch of a hyperbola.

80. The sound produced by a jet engine was measured at a distance of 100 m in all directions. The loudness of the sound (in decibels) was found to be $d = 115 + 10 \cos \theta$, where the 0° line for the angle θ is directed in front of the engine. Sketch the graph of d vs. θ in polar coordinates (use d as r).

Writing Exercise

81. Under a force that varies inversely as the square of the distance from an attracting object (such as the sun exerts on the earth), it can be shown that the equation of the path an object follows is given in general by

$$\frac{1}{r} = a + b \cos \theta$$

where a and b are constants for a particular path. First transform this equation into rectangular coordinates. Then write one or two paragraphs explaining why this equation represents one of the conic sections, depending on the values of a and b. It is through this kind of analysis that we know the paths of the planets and comets are conic sections.

Practice Test

1. Identify the type of curve represented by the equation $2(x^2 + x) = 1 - y^2$.
2. Sketch the graph of the straight line $4x - 2y + 5 = 0$ by finding its slope and y-intercept.
3. Find the polar equation of the curve whose rectangular equation is $x^2 = 2x - y^2$.
4. Find the vertex and the focus of the parabola $x^2 = -12y$. Sketch the graph.
5. Find the equation of the circle with center at $(-1, 2)$ and which passes through $(2, 3)$.
6. Find the equation of the straight line that passes through $(-4, 1)$ and $(2, -2)$.
7. Where is the focus of a parabolic reflector that is 12.0 cm across and 4.00 cm deep?
8. A hallway 6.0 m wide has a ceiling whose cross section is a semiellipse. The ceiling is 3.0 m high at the walls and 4.0 m high at the center. Find the height of the ceiling 1.0 m from each wall.
9. Plot the polar curve $r = 3 + \cos \theta$.
10. Find the center and vertices of the conic section $4y^2 - x^2 - 4x - 8y - 4 = 0$. Show completely the sketch of the curve.

22

Introduction to Statistics and Empirical Curve Fitting

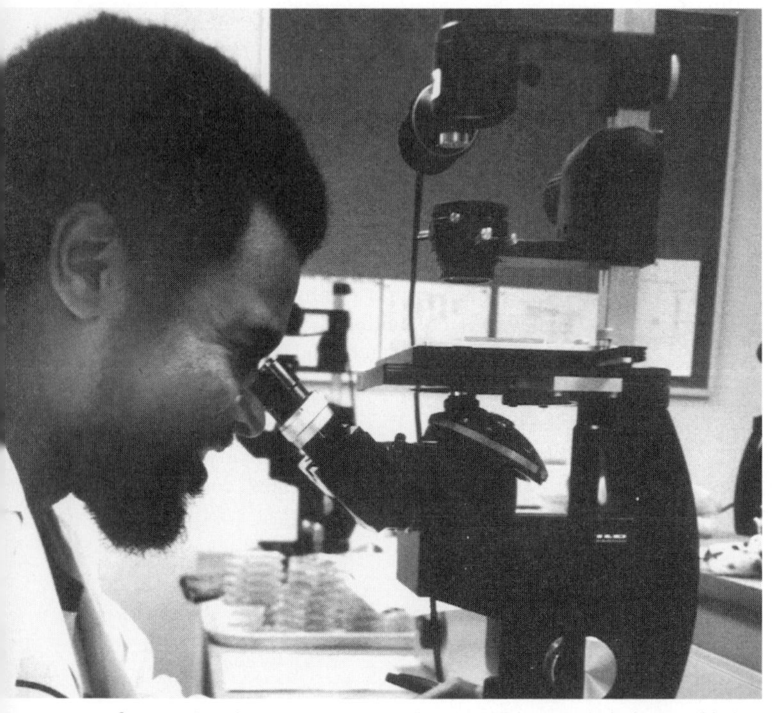

Statistical analysis is used extensively in medical research. In Section 22-4, a use of curve fitting in this field is shown.

In finding the volume of a cylindrical support column, an engineer uses the formula $V = \pi r^2 h$, which is true for all right circular cylinders. However, to find the safe load a cable can support, it is not possible simply to write down a known equation. The safe load depends on the diameter of the cable, the quality of the material of which it is made, and many other possible differences of this cable from other cables.

Thus, to determine the load the cable can support, the engineer may test cables of particular specifications. This testing will show that most cables can support approximately the same load, but that there is some variation and occasionally perhaps a great variation for some reason or other.

The engineer, in testing a number of cables and thereby selecting a certain type of cable, is making use of the basic methods of statistics. That is, the engineer (1) collects the data, (2) then analyzes the data, and (3) interprets the data. In the first three sections of this chapter, we discuss some of the basic methods of tabulating and analyzing this type of statistical information. Applications of statistics occur in nearly all fields of study, including science and technology.

The sections that follow show how to start with a set of points and find an equation that best "fits" data obtained through experimentation and observation. Such an equation can show a basic relationship between the variables of the experiment and can make it possible to analyze the experiment better. Curve fitting can be particularly useful in research and development leading to new and improved products.

This chapter gives an introduction to some of the basic concepts of statistics and empirical curve fitting. In a more complete coverage, a number of other useful methods in statistics with important applications are developed.

22-1 Frequency Distributions

In statistics we deal with various sets of numbers. These could be measurements of length, test scores, weights of objects, or numerous other possibilities. We could deal with the entire set of numbers, but this is often too large to be practical. In such a case we deal with a selected (assumed to be representative) sample.

EXAMPLE 1 ▸▸ A company produces a precision machine part designed to be 3.8750 cm long. If 10 000 are produced daily, for purposes of quality control it could be impractical to test each one to determine if it meets specifications. Therefore, it might be decided to test every tenth, or hundredth, part.

If only ten such parts were produced daily, it might be possible to test each one for specifications. ◂◂

One way of organizing data in order to develop some understanding of it is to *tabulate the number of occurrences for each particular value within the set. This is called a* **frequency distribution.**

EXAMPLE 2 ▸▸ A test station measured the loudness of the sound of jet aircraft taking off from a certain airport. The decibel readings of the first twenty jets were as follows: 110, 95, 100, 115, 105, 110, 120, 110, 115, 105, 90, 95, 105, 110, 100, 115, 105, 120, 95, 110.

We can see that it is difficult to determine any pattern to the readings. Therefore, we set up the following table to show the frequency distribution.

Decibel reading	90	95	100	105	110	115	120
Frequency	1	3	2	4	5	3	2

We note that the distribution and pattern of readings is clearer in this form where they are listed from lowest to highest, and the number of each is shown. ◂◂

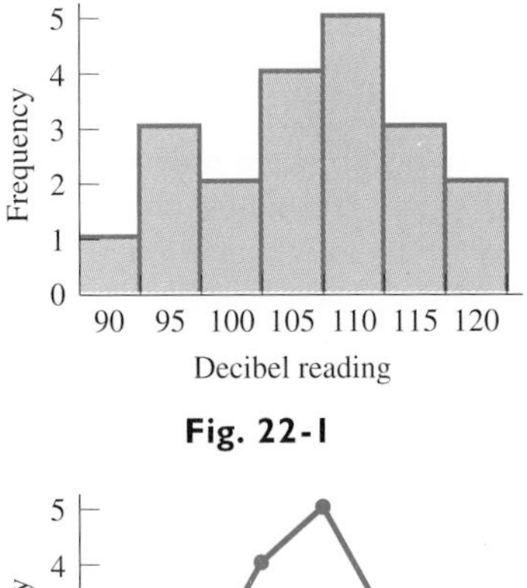

Fig. 22-1

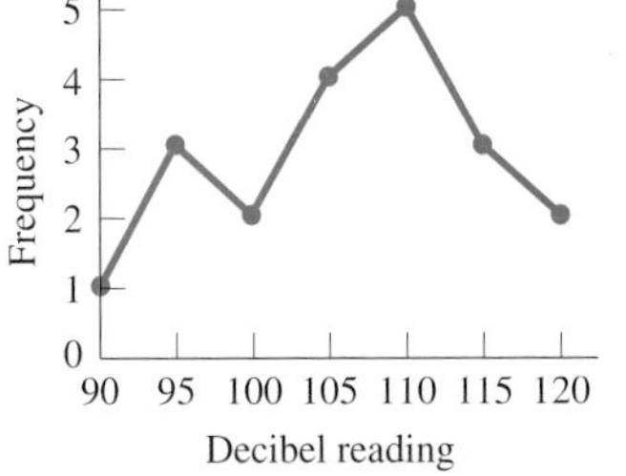

Fig. 22-2

Just as graphs are useful in representing algebraic functions, so also are they a very convenient method of representing frequency distributions. There are several useful methods of graphing such distributions, *the most important of which are the* **histogram** *and the* **frequency polygon.** The following examples illustrate these graphical representations.

EXAMPLE 3 ▸▸ *A histogram represents a particular set of data by use of rectangles.* The width of each rectangle is the same and is labeled for one of the readings. The height of the rectangle represents the number of values of a given reading. In Fig. 22-1, a histogram representing the data of the frequency of distribution of Example 2 is shown. ◂◂

EXAMPLE 4 ▸▸ *The frequency polygon is used to represent a set of data by plotting as abscissas (x-values) the values of the readings, and as ordinates (y-values) the number of occurrences of each reading (the frequency).* The resulting points are joined by straight-line segments. Figure 22-2 shows a frequency polygon of the data of Example 2. ◂◂

Often the number of measurements is so large that it is not practical to tabulate them for each value. In other cases such a table does not give a clear idea of the distribution. In these cases we may designate certain intervals and tabulate the number of values within each interval. Frequency distributions, histograms, and frequency polygons can be used to represent data tabulated in this way.

EXAMPLE 5 ▸▸ Eighty students were enrolled in a mathematics course. After the final exam, a numerical average (based on 100) was found for each student. Following is a list of numerical grades, with the number of students receiving each.

22—1, 37—1, 40—2, 44—1, 47—1, 53—1, 55—3, 56—1, 60—3,
61—1, 63—4, 65—2, 66—1, 67—5, 68—2, 70—2, 71—4, 72—3,
74—5, 75—4, 77—7, 78—4, 79—1, 81—2, 82—4, 84—1, 85—1,
86—3, 87—2, 88—1, 90—2, 92—1, 93—2, 95—1, 97—1

We can observe from this listing that it is difficult to see just how the grades were distributed. Therefore the instructors grouped the grades into intervals, which included five possible grades in each interval. This led to the following table:

Interval	20–24	25–29	30–34	35–39	40–44	45–49	50–54	55–59
Number in interval	1	0	0	1	3	1	1	4
Interval	**60–64**	**65–69**	**70–74**	**75–79**	**80–84**	**85–89**	**90–94**	**95–99**
Number in interval	8	10	14	16	7	7	5	2

We see that the distribution of grades becomes much clearer in this table. Finally, since the school graded students with the letters A, B, C, D, and F, the grades were grouped in intervals below 60, from 60 to 69, from 70 to 79, from 80 to 89, and from 90 to 100, and assigned the appropriate letter grade as follows:

Grade	A	B	C	D	F
Number receiving this grade	7	14	30	18	11

We can see that if the number of intervals were reduced much further, it would be difficult to draw any reasonable conclusions about the distribution of grades.

Histograms showing the numerical grades and letter grades are shown in Figs. 22-3 and 22-4. In Fig. 22-3, each interval is represented by the middle value. Figure 22-5 shows a frequency polygon with each interval represented by the middle value.

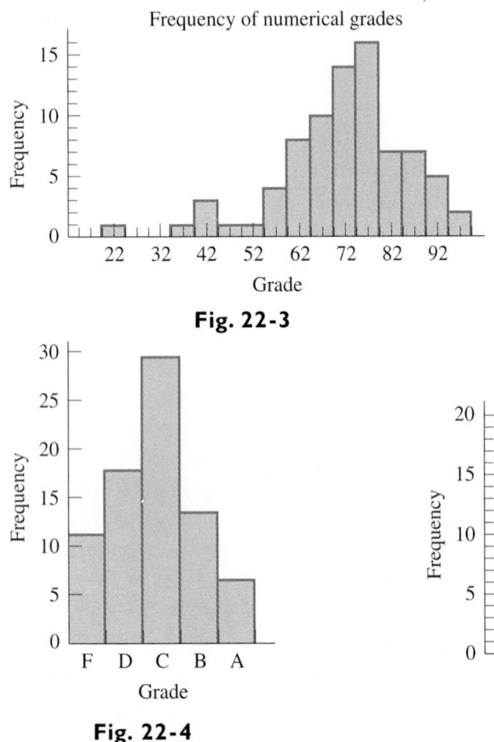

Fig. 22-3

Fig. 22-4

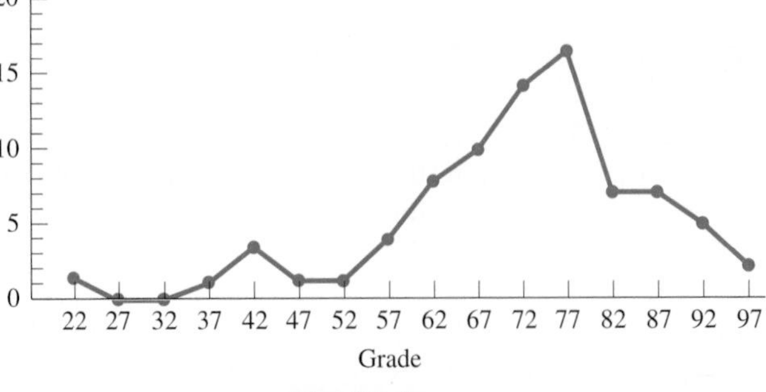

Fig. 22-5 ◂◂

We see that care must be taken in grouping data. If the number of intervals is too small, important features of the data may not be apparent. If the number of intervals is too large, it may be difficult to see the distribution and patterns.

Most graphing calculators may be used to display histograms and frequency polygons. Consult the manual to see how this is done on any particular model.

Exercises 22–1

In Exercises 1–16, use the following sets of numbers.

A: 3, 6, 4, 2, 5, 4, 7, 6, 3, 4, 6, 4, 5, 7, 3

B: 25, 26, 23, 24, 25, 28, 26, 27, 23, 28, 25

C: 0.48, 0.53, 0.49, 0.45, 0.55, 0.49, 0.47, 0.55, 0.48, 0.57, 0.51, 0.46

D: 105, 108, 103, 108, 106, 104, 109, 104, 110, 108, 108, 104, 113, 106, 107, 106, 107, 109, 105, 111, 109, 108

In Exercises 1–4, set up a frequency distribution table, indicating the frequency of each number given in the indicated set.

1. Set A **2.** Set B

3. Set C **4.** Set D

In Exercises 5–8, set up a frequency distribution table, indicating the frequency of numbers for the given intervals of the given sets.

5. Intervals 2–3, 4–5, and 6–7 for set A

6. Intervals 22–24, 25–27, and 28–30 for set B

7. Intervals 0.43–0.45, 0.46–0.48, 0.49–0.51, 0.52–0.54, 0.55–0.57 for set C

8. Intervals 101–105, 106–110, and 111–115 for set D

In Exercises 9–12, draw histograms for the data in the given exercise.

9. Exercise 1 **10.** Exercise 4

11. Exercise 7 **12.** Exercise 8

In Exercises 13–16, draw frequency polygons for the data in the given exercise.

13. Exercise 1 **14.** Exercise 4

15. Exercise 7 **16.** Exercise 8

In Exercises 17–32, find the indicated quantities.

17. In testing a computer system, the number of instructions it could perform in 1 ns was measured at different points in a program. The numbers of instructions recorded were as follows:

19, 21, 22, 25, 22, 20, 18, 21, 20, 19, 22, 21, 19, 23, 21

Form a frequency distribution table indicating the frequency of each number recorded in this data.

18. For the data of Exercise 17, draw a histogram.

19. For the data of Exercise 17, draw a frequency polygon.

20. For the data of Exercise 17, form a frequency distribution table for intervals 17–19, 20–22, and 23–25. Then draw a histogram to represent these data.

21. A strobe light is designed to flash every 2.25 s at a certain setting. Sample bulbs were tested with the results in the following table:

Number of bulbs	2	7	18	41	56
Time between flashes (seconds)	2.21	2.22	2.23	2.24	2.25

Number of bulbs	32	8	3	3
Time between flashes (seconds)	2.26	2.27	2.28	2.29

Draw a histogram for these data.

22. Draw a frequency polygon for the data of Exercise 21.

23. In testing a braking system, the distance required to stop a car from 110 km/h was measured in 120 trials. The results are in the following table:

Stopping distance (m)	47–49	50–52	53–55	56–58
Times car stopped	2	15	32	36

Stopping distance (m)	59–61	62–64	65–67
Times car stopped	24	10	1

Draw a frequency polygon for these data.

24. Draw a histogram for the data of Exercise 23.

25. The dosage, in millisieverts (mSv), given by a particular X-ray machine was measured 20 times, with the following readings:

0.425, 0.436, 0.396, 0.421, 0.444, 0.383, 0.437, 0.427, 0.433, 0.434, 0.415, 0.390, 0.441, 0.451, 0.418, 0.426, 0.429, 0.409, 0.436, 0.423

Form a histogram for the intervals 0.380–0.389, 0.390-0.399, and so on.

26. Form a histogram for the data given in Exercise 25 for the intervals 0.371–0.385, 0.386–0.400, and so on.

27. The life of a certain type of battery was measured for a sample of batteries with the following results (in number of hours):

34, 30, 32, 35, 31, 28, 29, 30, 32, 25, 31, 30,
28, 36, 33, 34, 30, 33, 31, 34, 29, 30, 32

Draw a frequency polygon for these data.

28. Form a histogram for the data given in Exercise 27 for the intervals 25–27, 28–30, 31–33, and 34–36.

29. The diameters of a sample of fiber-optic cables were measured for a sample of cables with the following results:

Diam. (mm)	0.0055	0.0056	0.0057	0.0058	0.0059	0.0060
No. cables	4	15	32	36	59	64

Diam. (mm)	0.0061	0.0062	0.0063	0.0064	0.0065	0.0066
No. cables	22	18	10	12	4	4

Draw a histogram for these data.

30. Draw a histogram for the data of Exercise 29 for the intervals 0.0055–0.0057, 0.0058–0.0060, 0.0061–0.0063, and 0.0064–0.0066.

31. Toss four coins 50 times, and tabulate the number of heads that appear for each toss. Draw a frequency polygon showing the number of tosses for which 0, 1, 2, 3, or 4 heads appeared. Describe the distribution (is it about what should be expected?).

32. Most calculators can generate random numbers (between 0 and 1). Using the random number key on a calculator, display 50 random numbers and record the first digit. Draw a histogram showing the number of times for which each first digit (0, 1, 2, . . . , 9) appeared. Describe the distribution (is it about what should be expected?).

22-2 Measures of Central Tendency

Tables and graphical representations give a general description of data. However, it is often profitable and convenient to find representative values for the location of the center of the distribution, and other numbers to give a measure of the deviation from this central value. In this way we can obtain an arithmetical description of the data. We shall now discuss the values commonly used to measure the location of the center of the distribution. These are referred to as *measures of central tendency.*

Median

The first of these measures of central tendency is the **median.** *The median is the middle number, that number for which there are as many above it as below it in the distribution.* If there is no middle number, the median is that number halfway between the two numbers nearest the middle of the distribution.

EXAMPLE 1 ▸▸ Given the numbers 5, 2, 6, 4, 7, 4, 7, 2, 8, 9, 4, 11, 9, 1, 3, we first arrange them in numerical order. This arrangement is

middle number ↓

1, 2, 2, 3, 4, 4, 4, 5, 6, 7, 7, 8, 9, 9, 11

Since there are 15 numbers, the middle number is the eighth. Since the eighth number is 5, the median is 5.

If the number 11 is not included in this set of numbers, and there are only 14 numbers in all, the median is that number halfway between the seventh and eighth numbers. Since the seventh is 4 and the eighth is 5, the median is 4.5. ◂◂

EXAMPLE 2 ▸▸ In the distribution of grades given in Example 5 of Section 22-1, the median is 74. There are 80 grades in all, and if they are listed in numerical order we find that the 39th through the 43rd grade is 74. This means that the 40th and 41st grades were both 74. The number halfway between the 40th and 41st grades would be the median. The fact that both are 74 means that the median is also 74. ◂◂

Arithmetic Mean

Another very widely applied measure of central tendency is the **arithmetic mean.** *The mean is calculated by finding the sum of all the values and then dividing by the number of values.* (The arithmetic mean is the number most people call the "average." However, in statistics the word *average* has the more general meaning of a measure of central tendency.)

EXAMPLE 3 ▸▸ The arithmetic mean of the numbers given in Example 1 is determined by finding the sum of all the numbers and dividing by 15. Therefore, by letting $\bar{x}$ (read as "x bar") represent the mean, we have

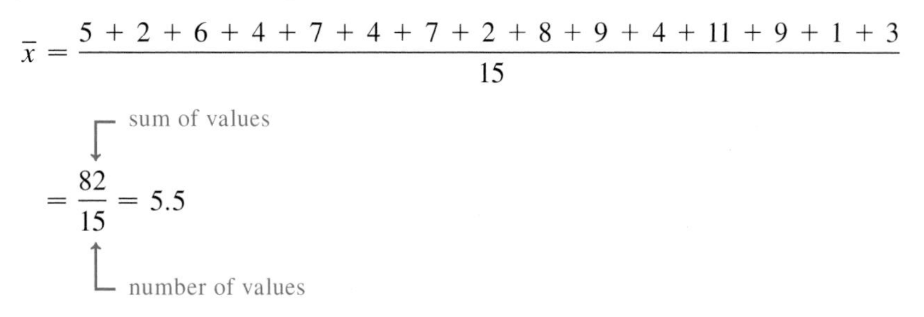

$$\bar{x} = \frac{5 + 2 + 6 + 4 + 7 + 4 + 7 + 2 + 8 + 9 + 4 + 11 + 9 + 1 + 3}{15}$$

$$= \frac{82}{15} = 5.5$$

Thus the mean is 5.5. ◂◂

If we wish to find the arithmetic mean of a large number of values, and if some of them appear more than once, the calculation can be somewhat simplified. The mean can be calculated by multiplying each value by its *frequency* (the number of times it occurs), adding these results, and then dividing by the total number of values considered (the sum of the frequencies). Letting $\bar{x}$ represent the mean of the values $x_1, x_2, \ldots, x_n$, which occur with frequencies $f_1, f_2, \ldots, f_n$, respectively, we have

$$\bar{x} = \frac{x_1 f_1 + x_2 f_2 + \cdots + x_n f_n}{f_1 + f_2 + \cdots + f_n} \qquad (22\text{-}1)$$

EXAMPLE 4 ▸▸ Using Eq. (22-1) to find the arithmetic mean of the numbers of Example 1, we first set up a table of values and their respective frequencies, as follows:

Value	1	2	3	4	5	6	7	8	9	11
Frequency	1	2	1	3	1	1	2	1	2	1

We now calculate the arithmetic mean $\bar{x}$ by using Eq. (22-1):

multiply each value by its frequency and add results

$$\bar{x} = \frac{1(1) + 2(2) + 3(1) + 4(3) + 5(1) + 6(1) + 7(2) + 8(1) + 9(2) + 11(1)}{1 + 2 + 1 + 3 + 1 + 1 + 2 + 1 + 2 + 1}$$

sum of frequencies

$$= \frac{82}{15} = 5.5$$

We see that this agrees with the result of Example 3. ◂◂

Summations such as those in Eq. (22-1) occur frequently in statistics and other branches of mathematics. In order to simplify writing these sums, the symbol Σ is used to indicate the process of summation. (Σ is the Greek capital letter sigma.) Σx means the sum of the x's.

EXAMPLE 5 ▸▸ We can show the sum of the numbers $x_1, x_2, x_3, \ldots, x_n$ as

$$\Sigma x = x_1 + x_2 + x_3 + \cdots + x_n$$

If these numbers are 3, 7, 2, 6, 8, 4, and 9, we have

$$\Sigma x = 3 + 7 + 2 + 6 + 8 + 4 + 9 = 39$$ ◂◂

Using the summation symbol Σ, we can write Eq. (22-1) for the arithmetic mean as

$$\bar{x} = \frac{x_1 f_1 + x_2 f_2 + x_3 f_3 + \cdots + x_n f_n}{f_1 + f_2 + f_3 + \cdots + f_n} = \frac{\sum xf}{\sum f} \tag{22-1}$$

The summation notation Σx is an abbreviated form of the more general notation $\sum_{i=1}^{n} x_i$. This more general form can be used to indicate the sum of the first n numbers of a sequence or to indicate the sum of a certain set within the sequence. For example, for a set of at least 5 numbers, $\sum_{i=3}^{5} x_i$ indicates the sum of the third through the fifth of these numbers (in Example 5, $\sum_{i=3}^{5} x_i = 16$). We will use the abbreviated form Σx to indicate the sum of all the numbers being considered.

EXAMPLE 6 ▸▸ We find the arithmetic mean of the grades in Example 5 of Section 22-1 by

$$\bar{x} = \frac{\sum xf}{\sum f} = \frac{22(1) + (37)(1) + (40)(2) + \cdots + (67)(5) + \cdots + (97)(1)}{80}$$

$$= \frac{5733}{80} = 71.7$$ ◂◂

Mode

Another measure of central tendency is *the* **mode,** *which is the value that appears most frequently.* If two or more values appear with the same greatest frequency, each is a mode. If no value is repeated, there is no mode.

EXAMPLE 7 ▸▸ (a) The mode of the numbers in Example 1 is 4, since it appears three times and no other value appears more than twice.

(b) The modes of the numbers

1, 2, 2, 4, 5, 5, 6, 7

are 2 and 5, since each appears twice and no other number is repeated.

(c) There is no mode for the values

1, 2, 5, 6, 7, 9

since none of the values is repeated. ◂◂

EXAMPLE 8 ▸▸ To find the frictional force between two specially designed surfaces, the force to move a block with one surface along an inclined plane with the other surface is measured 10 times. The results, with forces in newtons, are

2.2, 2.4, 2.1, 2.2, 2.5, 2.2, 2.4, 2.7, 2.1, 2.5

Find the mean, median, and mode of these forces.

To find the mean we sum the values of the forces and divide this total by 10. This gives

$$\overline{F} = \frac{\sum F}{10} = \frac{2.2 + 2.4 + 2.1 + 2.2 + 2.5 + 2.2 + 2.4 + 2.7 + 2.1 + 2.5}{10}$$
$$= \frac{23.3}{10} = 2.3 \text{ N}$$

The median is found by arranging the values in order and finding the middle value. The values in order are

2.1, 2.1, 2.2, 2.2, 2.2, 2.4, 2.4, 2.5, 2.5, 2.7

Since there are 10 values, we see that the fifth value is 2.2 and the sixth is 2.4. The value midway between these is 2.3, which is the median. Therefore, the median force is 2.3 N.

The mode is 2.2 N, since this value appears three times, which is more than any other value. ◂◂

Exercises 22–2

In Exercises 1–12, use the following sets of numbers. They are the same as those used in Exercises 22–1.

A: 3, 6, 4, 2, 5, 4, 7, 6, 3, 4, 6, 4, 5, 7, 3

B: 25, 26, 23, 24, 25, 28, 26, 27, 23, 28, 25

C: 0.48, 0.53, 0.49, 0.45, 0.55, 0.49, 0.47, 0.55, 0.48, 0.57, 0.51, 0.46

D: 105, 108, 103, 108, 106, 104, 109, 104, 110, 108, 108, 104, 113, 106, 107, 106, 107, 109, 105, 111, 109, 108

In Exercises 1–4, determine the median of the numbers of the given set.

1. Set A **2.** Set B

3. Set C **4.** Set D

In Exercises 5–8, determine the arithmetic mean of the numbers of the given set.

5. Set A **6.** Set B

7. Set C **8.** Set D

In Exercises 9–12, determine the mode of the numbers of the given set.

9. Set A **10.** Set B

11. Set C **12.** Set D

In Exercises 13–28, the required sets of numbers are those in Section 22-1. Find the indicated measures of central tendency.

13. Median of computer instructions in Exercise 17

14. Mean of computer instructions in Exercise 17

15. Mode of computer instructions in Exercise 17

16. Median of times in Exercise 21

17. Mean of times in Exercise 21

18. Mode of times in Exercise 21

19. Median of stopping distances in Exercise 23. (Use middle value for each interval.)

20. Mean of stopping distances in Exercise 23. (Use middle value for each interval.)

21. Mean of X-ray dosages in Exercise 25

22. Median of X-ray dosages in Exercise 25

23. Mode of X-ray dosages in Exercise 25

24. Mean of battery lives in Exercise 27

25. Median of battery lives in Exercise 27

26. Mode of battery lives in Exercise 27

27. Mean of cable diameters in Exercise 29

28. Median of cable diameters in Exercise 29

In Exercises 29–40, find the indicated measures of central tendency.

29. The weekly salaries (in dollars) for the workers in a small factory are as follows:

250, 350, 275, 225, 175, 300, 200,
350, 275, 400, 300, 225, 250, 300

Find the median and the mode of the salaries.

30. Find the mean salary for the salaries in Exercise 29.

31. In a particular month the electrical usage, rounded to the nearest 400 MJ, of 1000 homes in a certain city was summarized as follows:

No. homes	22	80	106	185	380	122	90	15
Usage (MJ)	2000	2400	2800	3200	3600	4000	4400	4800

Find the mean of the electrical usage.

32. Find the median and mode of electrical usage in Exercise 31.

33. A test of air pollution in a city gave the following readings of the concentration of sulfur dioxide (in parts per million) for 18 consecutive days:

0.14, 0.18, 0.27, 0.19, 0.15, 0.22, 0.20, 0.18, 0.15,
0.17, 0.24, 0.23, 0.22, 0.18, 0.32, 0.26, 0.17, 0.23

Find the median and mode of these readings.

34. Find the mean of the readings in Exercise 33.

35. The *midrange,* another measure of central tendency, is found by finding the sum of the lowest and the highest values and dividing this sum by 2. Find the midrange of the salaries in Exercise 29.

36. Find the midrange of the sulfur dioxide readings in Exercise 33. (See Exercise 35.)

37. Add $100 to each of the salaries in Exercise 29. Then find the median, mean, and mode of the resulting salaries. State any conclusion that might be drawn from the results.

38. Multiply each of the salaries in Exercise 29 by 2. Then find the median, mean, and mode of the resulting salaries. State any conclusion that might be drawn from the results.

39. Change the final salary in Exercise 29 to $4000, with all other salaries being the same. Then find the mean of these salaries. State any conclusion that might be drawn from the result. (The $4000 here is called an *outlier,* which is an extreme value.)

40. Find the median and mode of the salaries indicated in Exercise 39. State any conclusion that might be drawn from the results.

22-3 Standard Deviation

In the preceding section we discussed measures of central tendency of sets of data. However, regardless of the measure that may be used, it does not tell us whether the data are grouped closely together or spread over a large range of values. Therefore, we also need some measure of the deviation, or dispersion, of the values from the median or mean. If the dispersion is small and the numbers are grouped closely together, the measure of central tendency is more reliable and descriptive of the data than the case in which the spread is greater.

There are several measures of dispersion, and we discuss in this section one that is widely used: the *standard deviation.*

The **standard deviation** *of a set of n numbers is given by the equation*

$$s = \sqrt{\frac{\sum (x - \overline{x})^2}{n}} \qquad (22\text{-}2)$$

The definition of s shows that the following steps are used in computing its value.

1. Find the arithmetic mean $\overline{x}$ of the set of numbers.
2. Subtract the mean from each of the numbers of the set.
3. Square these differences.
4. Find the arithmetic mean of these squares.
5. Find the square root of this last arithmetic mean.

The standard deviation s is a positive number. It is a *deviation from the mean,* regardless of whether or not the individual numbers are greater than or less than the mean. Numbers close together will have a small standard deviation, whereas numbers farther apart have a larger standard deviation. Therefore, *the standard deviation becomes larger as the spread of data increases.*

In many textbooks, standard deviation is defined such that in Step 4, the sum of the squares found in Step 3 is divided by $n - 1$ rather than n, where n is the number of values. Either definition is correct, and the choice of n or $n - 1$ is determined by considerations beyond the scope of this chapter.

Following the steps listed after Eq. (22-2), we use Eq. (22-2) for the calculation of standard deviation in the following examples.

EXAMPLE 1 ▸▸ Find the standard deviation of the following numbers: 1, 5, 4, 2, 6, 2, 1, 1, 5, 3.

A table of the necessary values is shown below, and steps 1–5 are indicated.

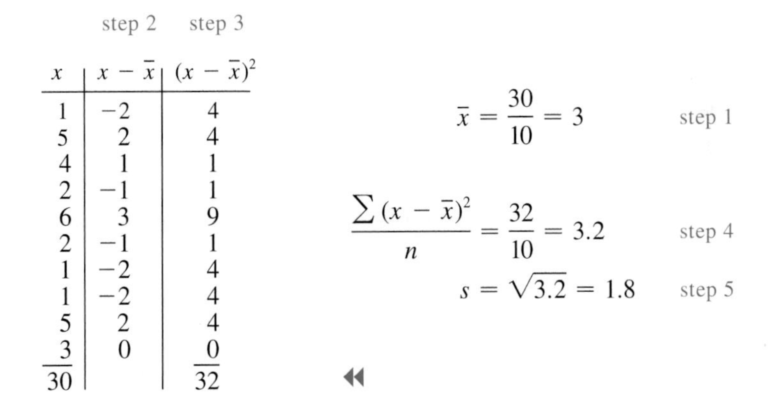

x	$x - \bar{x}$ (step 2)	$(x - \bar{x})^2$ (step 3)
1	−2	4
5	2	4
4	1	1
2	−1	1
6	3	9
2	−1	1
1	−2	4
1	−2	4
5	2	4
3	0	0
30		32

$$\bar{x} = \frac{30}{10} = 3 \quad \text{step 1}$$

$$\frac{\sum (x - \bar{x})^2}{n} = \frac{32}{10} = 3.2 \quad \text{step 4}$$

$$s = \sqrt{3.2} = 1.8 \quad \text{step 5}$$

◂◂

EXAMPLE 2 ▸▸ Find the standard deviation of the numbers in Example 1 of Section 22-2.

Since several of the numbers appear more than once, it is helpful to use the frequency of each number in the table, as follows:

x	f	fx	$x - \bar{x}$ (step 2)	$(x - \bar{x})^2$ (step 3)	$f(x - \bar{x})^2$
1	1	1	−4.5	20.25	20.25
2	2	4	−3.5	12.25	24.50
3	1	3	−2.5	6.25	6.25
4	3	12	−1.5	2.25	6.75
5	1	5	−0.5	0.25	0.25
6	1	6	0.5	0.25	0.25
7	2	14	1.5	2.25	4.50
8	1	8	2.5	6.25	6.25
9	2	18	3.5	12.25	24.50
11	1	11	5.5	30.25	30.25
	15	82			123.75

step 1

$$\bar{x} = \frac{82}{15} = 5.5$$

$$\frac{\sum f(x - \bar{x})^2}{n} = \frac{123.75}{15} = 8.25 \quad \text{step 4}$$

$$s = \sqrt{8.25} = 2.9 \quad \text{step 5}$$

◂◂

It is possible to reduce the computational work required to find the standard deviation. We now show how this may be done.

Two sets of numbers, x_i and y_i, where each set has n values, have a sum $(x_i + y_i)$. Finding the mean of this sum, we have

$$\overline{x + y} = \frac{\sum (x + y)}{n} = \frac{(x_1 + y_1) + (x_2 + y_2) + \cdots + (x_n + y_n)}{n}$$

$$= \frac{x_1 + x_2 + \cdots + x_n + y_1 + y_2 + \cdots + y_n}{n} = \frac{\Sigma x}{n} + \frac{\Sigma y}{n} = \bar{x} + \bar{y}$$

or

$$\overline{x + y} = \bar{x} + \bar{y} \qquad (22\text{-}3)$$

If all numbers have a constant factor, it can be factored before the mean is found.

$$\overline{kx} = k\bar{x} \qquad (22\text{-}4)$$

Expanding the expression under the radical in Eq. (22-2) and using Eq. (22-4) leads to

$$\frac{\sum (x - \bar{x})^2}{n} = \overline{(x - \bar{x})^2} = \overline{x^2 - 2x\bar{x} + \bar{x}^2} = \overline{x^2} - \overline{2x\bar{x}} + \overline{\bar{x}^2}$$

The 2 and $\bar{x}$ are constants for any given problem. Thus, using Eq. (22-4), we have

$$\overline{x^2} - \overline{2x\bar{x}} + \overline{\bar{x}^2} = \overline{x^2} - 2\bar{x}(\bar{x}) + \bar{x}^2(1) = \overline{x^2} - \bar{x}^2$$

Substituting this result into Eq. (22-2), we have

$$s = \sqrt{\overline{x^2} - \bar{x}^2} \qquad (22\text{-}5)$$

Equation (22-5) uses the following steps in calculating the standard deviation s.

1. Calculate the mean of the squares $(\overline{x^2})$.
2. Calculate the square of the mean $(\bar{x}^2)$.
3. Subtract result 2 from result 1 $(\overline{x^2} - \bar{x}^2)$.
4. Take the square root $(\sqrt{\overline{x^2} - \bar{x}^2})$.

Using Eq. (22-5) eliminates the step of subtracting $\bar{x}$ from x, a step required by the definition in Eq. (22-2). Obviously, a calculator is useful for these calculations.

EXAMPLE 3 ▸▸ Using Eq. (22-5), find s for the numbers in Example 1.

x	x^2
1	1
5	25
4	16
2	4
6	36
2	4
1	1
1	1
5	25
3	9
30	122

$\overline{x^2} = \frac{122}{10} = 12.2$ step 1

$\bar{x} = \frac{30}{10} = 3, \quad \bar{x}^2 = 9$ step 2

$\overline{x^2} - \bar{x}^2 = 12.2 - 9 = 3.2$ step 3

$s = \sqrt{3.2} = 1.8$ step 4

◂◂

EXAMPLE 4 ▸▸ An ammeter measures the electric current flowing in a circuit. In an ammeter, two resistances are connected in parallel. Most of the current passing through the meter goes through a very low resistance called the *shunt*. To determine the accuracy of the resistance of the shunts being made for ammeters, a manufacturer tested a sample of 100 shunts. The measured resistances are shown in the following table. Calculate the standard deviation of the resistances of the shunts.

R (ohms)	f	fR	fR^2
0.200	1	0.200	0.0400
0.210	3	0.630	0.1323
0.220	5	1.100	0.2420
0.230	10	2.300	0.5290
0.240	17	4.080	0.9792
0.250	40	10.000	2.5000
0.260	13	3.380	0.8788
0.270	6	1.620	0.4374
0.280	3	0.840	0.2352
0.290	2	0.580	0.1682
	100	24.730	6.1421

$$\overline{R^2} = \frac{\sum fR^2}{\sum f} = \frac{6.1421}{100} = 0.061421 \qquad \text{step 1}$$

$$\overline{R} = \frac{\sum fR}{\sum f} = \frac{24.73}{100} = 0.2473 \qquad \text{step 2}$$

$$\overline{R}^2 = 0.061\,157\,29$$

$$s = \sqrt{\overline{R^2} - \overline{R}^2} = 0.016 \qquad \text{steps 3, 4}$$

The arithmetic mean of the resistances is 0.247 Ω, with a standard deviation of 0.016 Ω. ◂◂

EXAMPLE 5 ▸▸ Find the standard deviation of the grades in Example 5 of Section 22-1. Use the frequency distribution grouped in intervals 20–24, 25–29, and so forth, and then assume that each value in the interval is equal to the representative value (middle value) of the interval. (This method is not exact, but when a problem involves a large number of values, the method provides a good approximation and eliminates a great deal of arithmetic work.)

Interval	x	f	fx	fx^2
20–24	22	1	22	484
25–29	27	0	0	0
30–34	32	0	0	0
35–39	37	1	37	1 369
40–44	42	3	126	5 292
45–49	47	1	47	2 209
50–54	52	1	52	2 704
55–59	57	4	228	12 996
60–64	62	8	496	30 752
65–69	67	10	670	44 890
70–74	72	14	1 008	72 576
75–79	77	16	1 232	94 864
80–84	82	7	574	47 068
85–89	87	7	609	52 983
90–94	92	5	460	42 320
95–99	97	2	194	18 818
		80	5 755	429 325

$$\overline{x^2} = \frac{\sum fx^2}{80} = \frac{429\,325}{80} = 5366.5625 \qquad \text{step 1}$$

$$\overline{x} = \frac{\sum fx}{\sum f} = \frac{5755}{80} = 71.9375 \qquad \text{step 2}$$

$$\overline{x}^2 = 5175.0039$$

$$s = \sqrt{\overline{x^2} - \overline{x}^2} = 13.8 \qquad \text{steps 3, 4}$$

◂◂

Many calculators are programmed to give the arithmetic mean and standard deviation when data are entered. Also, these statistical measures are readily programmed to be determined by a computer.

Standard Normal Distribution

We have seen that the standard deviation is a measure of the dispersion of a set of data. Generally, if we make a great many measurements of a given type, we would expect to find that most of them are close to the arithmetic mean. If this is the case, the distribution probably follows, at least to a reasonable extent, the **standard normal distribution curve**. See Fig. 22-6. This curve gives a theoretical distribution about the arithmetic mean (with $\overline{x} = 0$ and $s = 1$), assuming that the number of values in the distribution becomes infinite. Its equation is

$$y = \frac{1}{\sqrt{2\pi}} e^{-x^2/2} \qquad (22\text{-}6)$$

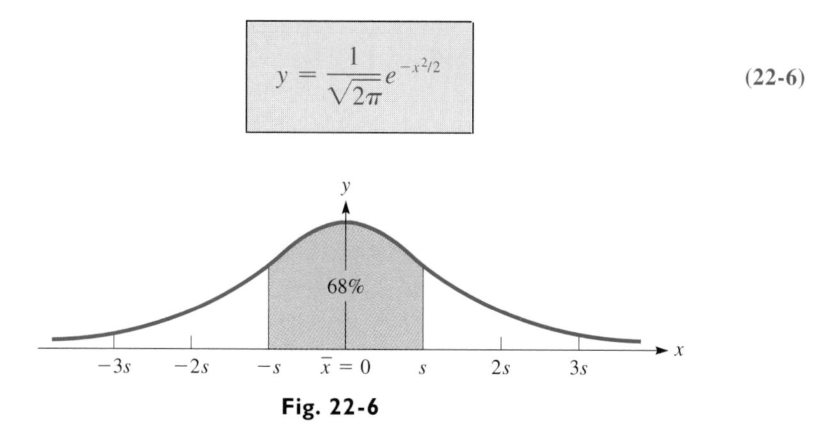

Fig. 22-6

An important feature of the standard deviation is that *in a normal distribution, about* 68% *of the values are found within the interval* $\overline{x} - s$ *to* $\overline{x} + s$. Also, approximately 95.4% of the numbers are within the interval $\overline{x} - 2s$ to $\overline{x} + 2s$, and 99.7% of the values are in the interval $\overline{x} - 3s$ to $\overline{x} + 3s$.

EXAMPLE 6 ▸▸ (a) In Example 2, 9 of the 15 values, or 60%, are between 2.6 and 8.4. Here, $\overline{x} = 5.5$ and $s = 2.9$, which tells us that $\overline{x} - s = 2.6$ and $\overline{x} + s = 8.4$. Therefore, we see that the percent of values in this interval is slightly less than in a normal distribution.

(b) In Example 5, using the values $\overline{x} = 72$ and $s = 14$, we have $\overline{x} - s = 58$ and $\overline{x} \pm s = 86$. We then note that 59 of the 80 values are in the interval from 58 to 86, which means that 74% of the values are in this interval. ◂◂

EXAMPLE 7 ▸▸ Referring to Example 4, we see that the arithmetic mean of the resistances is 0.247 Ω, with a standard deviation of 0.016 Ω.

The standard deviation can be used in making quality control decisions about the manufacture of the shunts. For example, for the accuracy requirements of the ammeter being manufactured, it may be required that the shunts have a resistance of 0.247 Ω ± 0.012 Ω (about a 5% tolerance). This means that each should have a resistance of at least 0.235 Ω and no more than 0.259 Ω.

Since $\overline{x} - s = 0.231$ Ω and $\overline{x} + s = 0.263$ Ω includes about 68% of the shunts, we see that less than 68% have the required resistances. Therefore, an improvement in the manufacture of these shunts would be needed. Most manufacturers would require nearly 100% to be within the required values. ◂◂

We have introduced a few of the measures used in statistics. In using these measures, we have considered sets of numbers for which these measures give a reasonable description of the center and distribution of the numbers. However, we must be careful in using and interpreting such measures.

EXAMPLE 8 (a) The numbers 1, 2, 3, 4, 5 have a mean of 3 and a median of 3. The standard deviation is 1.4. These values describe fairly well the center and distribution of the numbers in the set.

(b) The numbers 1, 2, 3, 4, 100 have a mean of 22, a median of 3, and a standard deviation of 39. These values do not describe this set of numbers well. Most of the values in the interval $\bar{x} - s$ to $\bar{x} + s$ (-17 to 61) are not even in the set, and the one outside the interval is 39 units from it. In a case like this, the 100 should be double-checked to see if its value is in error.

Example 8 illustrates that the statistical measures can be misleading if the numbers in a set are unevenly distributed. Misleading statistics can also come from the source of the data. Consider the probable results of a survey to find the percent of persons in favor of raising income taxes for the wealthy if the survey is taken at the entrance to a welfare office or if it is taken at the entrance to a stock brokerage firm. There are many other considerations in the proper use and interpretation of statistical measures.

Exercises 22–3

In Exercises 1–20, use the following sets of numbers. They are the same as those used in Exercises 22-1 and 22-2.

A: 3, 6, 4, 2, 5, 4, 7, 6, 3, 4, 6, 4, 5, 7, 3

B: 25, 26, 23, 24, 25, 28, 26, 27, 23, 28, 25

C: 0.48, 0.53, 0.49, 0.45, 0.55, 0.49, 0.47, 0.55, 0.48, 0.57, 0.51, 0.46

D: 105, 108, 103, 108, 106, 104, 109, 104, 110, 108, 108, 104, 113, 106, 107, 106, 107, 109, 105, 111, 109, 108

In Exercises 1–4, use Eq. (22-2) to find the standard deviation s for the indicated sets of numbers.

1. Set A **2.** Set B

3. Set C **4.** Set D

In Exercises 5–8, use Eq. (22-5) to find the standard deviation s for the indicated sets of numbers.

5. Set A **6.** Set B

7. Set C **8.** Set D

In Exercises 9–16, find the standard deviation s for the indicated sets of numbers from Exercises 22-1 and 22-2.

9. The computer instructions in Exercise 17 of Section 22-1

10. The X-ray dosages in Exercise 25 of Section 22-1

11. The battery lives in Exercise 27 of Section 22-1

12. The salaries of Exercise 29 of Section 22-2

13. The strobe light times in Exercise 21 of Section 22-1

14. The stopping distances in Exercise 23 of Section 22-1

15. The fiber-optic cable diameters in Exercise 29 of Section 22-1

16. The electric power usages in Exercise 31 of Section 22-2

In Exercises 17–24, find the percent of values in the interval $\bar{x} - s$ to $\bar{x} + s$ for the indicated sets of numbers.

17. Set A **18.** Set B

19. Set C **20.** Set D

21. The computer instructions in Exercise 17 of Section 22-1

22. The X-ray dosages in Exercise 25 of Section 22-1

23. The strobe light times in Exercise 21 of Section 22-1

24. The stopping distances in Exercise 23 of Section 22-1

In Exercises 25–28, refer to Exercises 37 and 39 of Section 22-2.

25. Find s for the salaries in Exercise 37.

26. Find the interval $\bar{x} - s$ to $\bar{x} + s$ for the salaries of Exercise 37. What conclusion can you draw?

27. Find s for the salaries in Exercise 39.

28. Find the interval $\bar{x} - s$ to $\bar{x} + s$ for the salaries of Exercise 39. What conclusion can you draw?

22-4 Fitting a Straight Line to a Set of Points

We have considered statistical methods for dealing with one variable. We have discussed methods of tabulating, graphing, and measuring the central tendency and the deviations from this value for one variable. We now discuss how to obtain a relationship between two variables for which a set of points is known.

In this section we show a method of "fitting" a straight line to a given set of points. In the following section, we will discuss fitting suitable nonlinear curves to given sets of points. Some of the reasons for doing this are (1) to express a concise relationship between the variables, (2) to use the equation to predict certain fundamental results, (3) to determine the reliability of certain sets of data, and (4) to use the data for testing theoretical concepts.

We shall assume for the examples and exercises of the remainder of this chapter that there is some relationship between the variables. Often when we are analyzing statistics for variables between which we think a relationship might exist, the points are so scattered as to give no reasonable idea regarding the possible functional relationship. We shall assume here that such combinations of variables have been discarded in the analysis.

EXAMPLE 1 ▸▸ All of the students enrolled in the mathematics course referred to in Example 5 of Section 22-1 took an entrance test in mathematics. To study the reliability of this test, an instructor tabulated the test scores of 10 students (selected at random), along with their course averages, and made a graph of the figures. See the table below and Fig. 22-7.

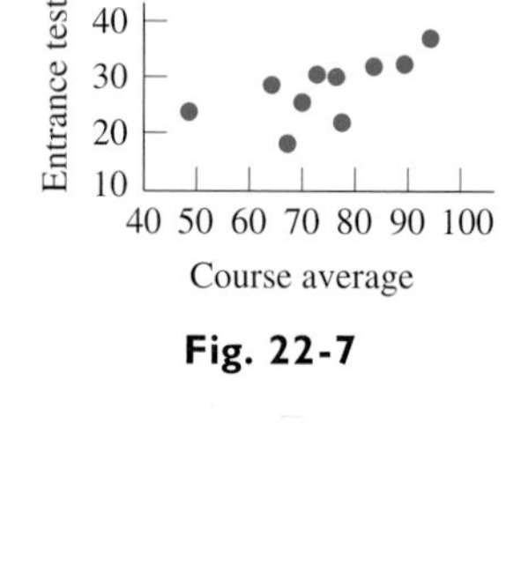

Fig. 22-7

Student	*Entrance test score, based on 40*	*Course average, based on 100*
A	29	63
B	33	88
C	22	77
D	17	67
E	26	70
F	37	93
G	30	72
H	32	81
I	23	47
J	30	74

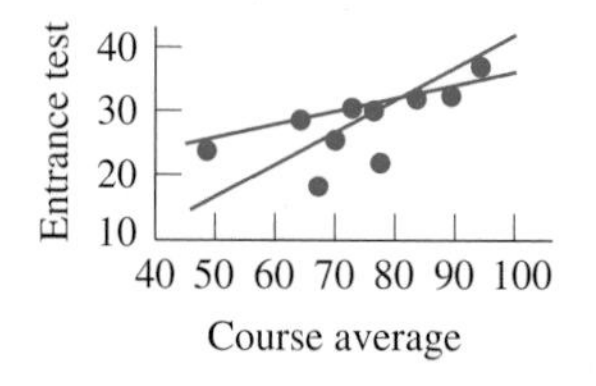

Fig. 22-8

We now ask whether or not there is a functional relationship between the test scores and the course grades. Certainly no clear-cut relationship exists, but in general we see that the higher the test score, the higher the course grade. This leads to the possibility that there might be some straight line, from which none of the points would vary too significantly. If such a line could be found, then it could be the basis of predictions as to the possible success a student might have in the course, on the basis of his or her grade on the entrance test. Assuming that such a straight line exists, the problem is to find the equation of this line. Figure 22-8 shows two such possible lines. ◂◂

There are a number of different methods of determining the straight line that best fits the given data points. We shall employ the method that is most widely used: the **method of least squares**. *The basic principle of this method is that the sum of the squares of the deviations of all data points from the best line (in accordance with this method) has the least value possible. By* **deviation** *we mean the difference between the y-value of the line and the y-value for the point (of original data) for a particular value of x.*

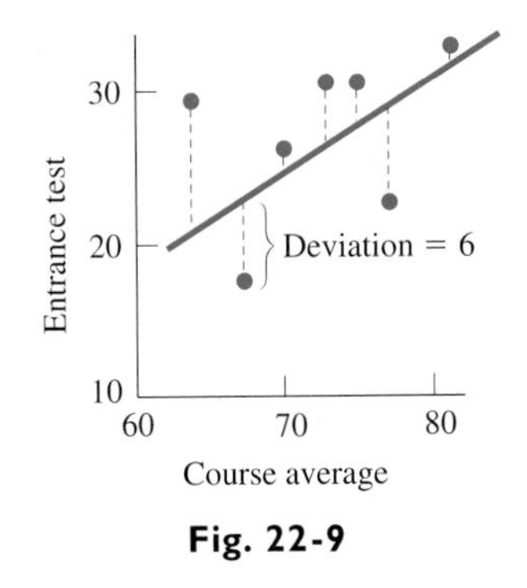

Fig. 22-9

EXAMPLE 2 In Fig. 22-9 the deviations of some of the points of Example 1 are shown. The point (67, 17) (student D of Example 1) has a deviation of 6 from the indicated line in the figure. Thus we square the value of this deviation to obtain 36. In order to find the equation of the straight line that best fits the given points, the method of least squares requires that the sum of all such squares be a minimum.

Therefore, in applying this method of least squares, it is necessary to use the equation of a straight line and the coordinates of the points of the data. The deviations of all of these data points are determined, and these values are then squared. It is then necessary to determine the constants for the slope m and the y-intercept b in the equation of a straight line $y = mx + b$ for which the sum of the squares is a minimum. To do this requires certain methods of advanced mathematics.

Using the methods that are required from advanced mathematics, we show that *the equation of the* **least squares line**

$$y = mx + b \tag{22-7}$$

can be found by calculating the values of the slope m and the y-intercept b by using the formulas

$$m = \frac{n\sum xy - \left(\sum x\right)\left(\sum y\right)}{n\sum x^2 - \left(\sum x\right)^2} \tag{22-8}$$

and

$$b = \frac{\left(\sum x^2\right)\left(\sum y\right) - \left(\sum xy\right)\left(\sum x\right)}{n\sum x^2 - \left(\sum x\right)^2} \tag{22-9}$$

CAUTION ▶ In Eqs. (22-8) and (22-9), ***the x's and y's are those of the points in the given data,*** and n is the number of points of data. We can reduce the calculational work in finding the values of m and b by noting that the denominators in Eqs. (22-8) and (22-9) are the same. Therefore, in using a calculator, the value of this denominator can be stored in memory.

The following three examples illustrate the use of Eqs. (22-7), (22-8), and (22-9) in finding the equation of the least squares line for the given data points.

EXAMPLE 3 ▸▸ Find the equation of the least-squares line for the points indicated in the following table. Graph the line and data points on the same graph.

x	1	2	3	4	5
y	3	6	6	8	12

We see from Eqs. (22-8) and (22-9) that we need the sums of x, y, xy, and x^2 in order to find m and b. Thus we set up a table for these values, along with the necessary calculations, as follows:

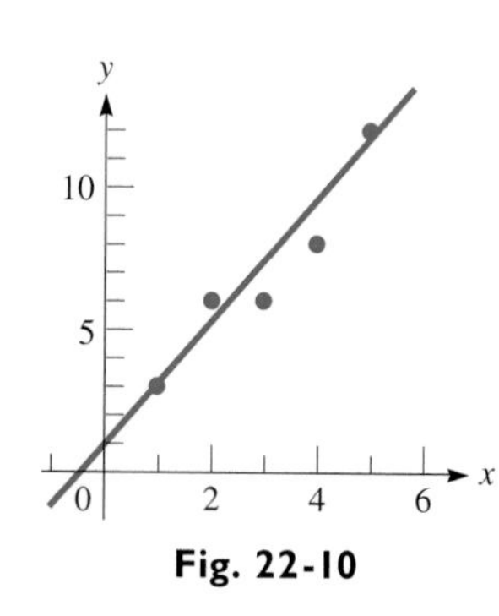

Fig. 22-10

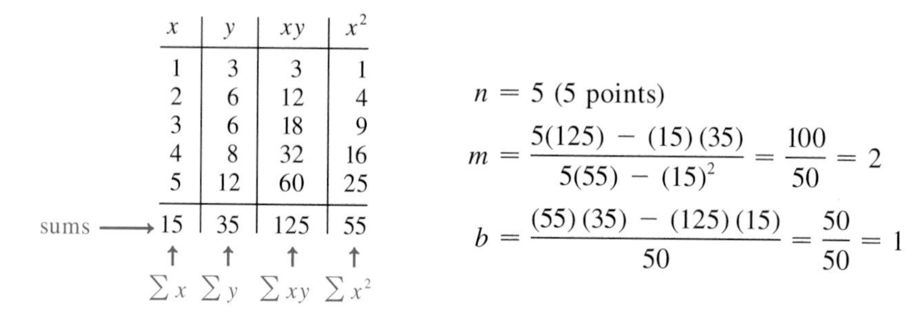

	x	y	xy	x^2
	1	3	3	1
	2	6	12	4
	3	6	18	9
	4	8	32	16
	5	12	60	25
sums ⟶	15	35	125	55
	↑ $\sum x$	↑ $\sum y$	↑ $\sum xy$	↑ $\sum x^2$

$n = 5$ (5 points)

$$m = \frac{5(125) - (15)(35)}{5(55) - (15)^2} = \frac{100}{50} = 2$$

$$b = \frac{(55)(35) - (125)(15)}{50} = \frac{50}{50} = 1$$

This means that the equation of the least-squares line is $y = 2x + 1$. This line and the data points are shown in Fig. 22-10. ◂◂

EXAMPLE 4 ▸▸ Find the least-squares line for the data of Example 1.

Here the y-values will be the entrance-test scores, and the x-values the course averages.

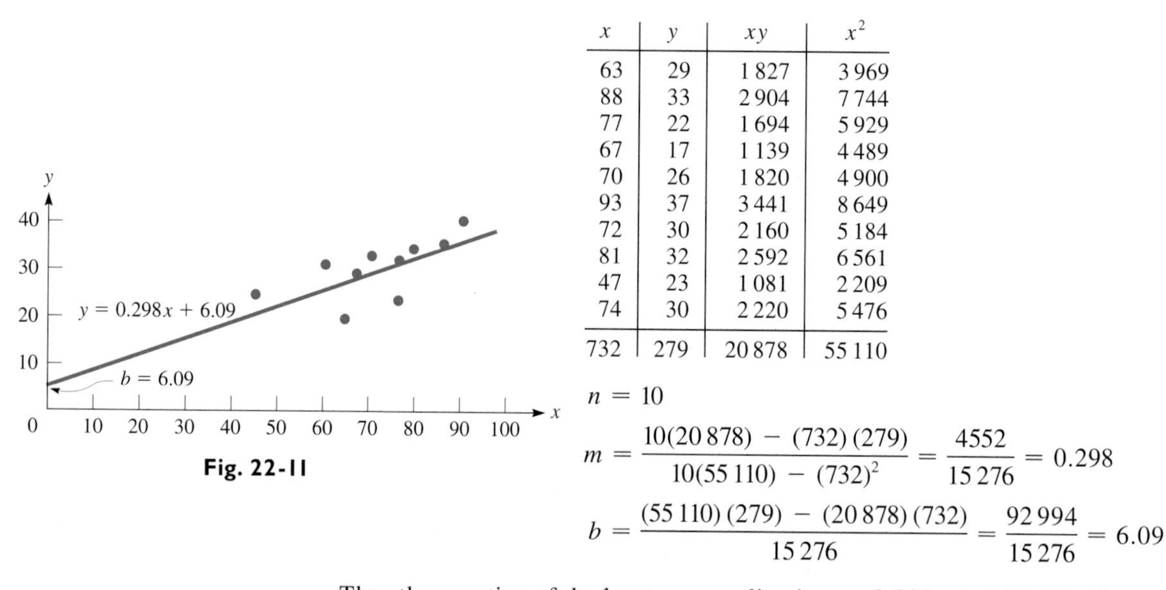

x	y	xy	x^2
63	29	1 827	3 969
88	33	2 904	7 744
77	22	1 694	5 929
67	17	1 139	4 489
70	26	1 820	4 900
93	37	3 441	8 649
72	30	2 160	5 184
81	32	2 592	6 561
47	23	1 081	2 209
74	30	2 220	5 476
732	279	20 878	55 110

Fig. 22-11

$n = 10$

$$m = \frac{10(20\,878) - (732)(279)}{10(55\,110) - (732)^2} = \frac{4552}{15\,276} = 0.298$$

$$b = \frac{(55\,110)(279) - (20\,878)(732)}{15\,276} = \frac{92\,994}{15\,276} = 6.09$$

Thus the equation of the least-squares line is $y = 0.298x + 6.09$. The line and data points are shown in Fig. 22-11. This line best fits the data, although the fit is obviously approximate. It can be used to predict the approximate course average that a student might be expected to attain, based on the entrance test. ◂◂

See the chapter introduction.

EXAMPLE 5 ▸▸ In a research project to determine the amount of a drug that remains in the bloodstream after a given dosage, the amounts y (in mg of drug/dL of blood) were recorded after t hours.

t (h)	1.0	2.0	4.0	8.0	10.0	12.0
y (mg/dL)	7.6	7.2	6.1	3.8	2.9	2.0

Find the least-squares line for these data, expressing y as a function of t. Sketch the graph of the line and data points.

The table and calculations are shown below.

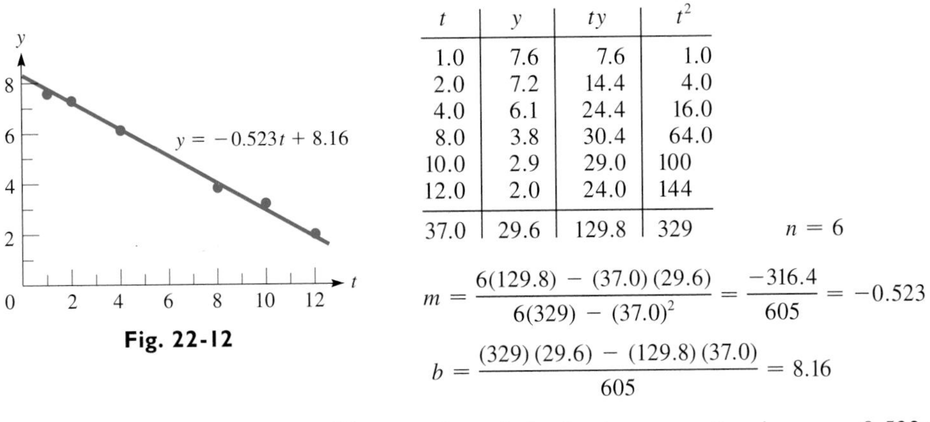

Fig. 22-12

t	y	ty	t^2
1.0	7.6	7.6	1.0
2.0	7.2	14.4	4.0
4.0	6.1	24.4	16.0
8.0	3.8	30.4	64.0
10.0	2.9	29.0	100
12.0	2.0	24.0	144
37.0	29.6	129.8	329

$n = 6$

$$m = \frac{6(129.8) - (37.0)(29.6)}{6(329) - (37.0)^2} = \frac{-316.4}{605} = -0.523$$

$$b = \frac{(329)(29.6) - (129.8)(37.0)}{605} = 8.16$$

The equation of the least-squares line is $y = -0.523t + 8.16$. The line and data points are shown in Fig. 22-12. This line is useful in determining the effectiveness of the drug. It can also be used to determine when additional medication may be administered. ◂◂

Exercises 22–4

In Exercises 1–12, find the equation of the least-squares line for the given data. Graph the line and data points on the same graph.

1. The points in the following table:

x	4	6	8	10	12
y	1	4	5	8	9

2. The points in the following table:

x	1	2	3	4	5	6	7
y	10	17	28	37	49	56	72

3. The points in the following table:

x	20	26	30	38	48	60
y	160	145	135	120	100	90

4. The points in the following table:

x	1	3	6	5	8	10	4	7	3	8
y	15	12	10	8	9	2	11	9	11	7

5. In an electrical experiment, the following data were found for the values of current and voltage for a particular element of the circuit. Find the voltage as a function of the current.

Current (mA)	15.0	10.8	9.30	3.55	4.60
Voltage (volts)	3.00	4.10	5.60	8.00	10.50

6. A particular muscle was tested for its speed of shortening as a function of the force applied to it. The results appear below. Find the speed as a function of the force.

Force (N)	60.0	44.2	37.3	24.2	19.5
Speed (m/s)	1.25	1.67	1.96	2.56	3.05

7. The altitude h of a rocket was measured at several positions at a horizontal distance x from the launch site, shown in the table. Find the least-squares line for h as a function of x.

x (metres)	0	500	1000	1500	2000	2500
h (metres)	0	1130	2250	3360	4500	5600

8. In testing an air-conditioning system, the temperature T in a building was measured during the afternoon hours with the results shown in the table. Find the least-squares line for T as a function of the time t from noon.

t (hours)	0.0	1.0	2.0	3.0	4.0	5.0
T (°C)	20.5	20.6	20.9	21.3	21.7	22.0

9. The pressure p was measured along an oil pipeline at different distances from a reference point, with results as shown. Find the least-squares line for p as a function of x.

x (m)	0	50	100	150	200
p (kPa)	4370	4240	4070	3970	3840

10. The heat loss L per hour through various thicknesses of a particular type of insulation was measured as shown in the table. Find the least-squares line for L as a function of t.

t (cm)	3.0	4.0	5.0	6.0	7.0
L (MJ)	5.90	4.80	3.90	3.10	2.45

11. In an experiment on the photoelectric effect, the frequency of the light being used was measured as a function of the stopping potential (the voltage just sufficient to stop the photoelectric current) with the results given below. Find the least-squares line for V as a function of f. The frequency for $V = 0$ is known as the *threshold frequency*. From the graph, determine the threshold frequency.

f (PHz)	0.550	0.605	0.660	0.735	0.805	0.880
V (volts)	0.350	0.600	0.850	1.10	1.45	1.80

12. If gas is cooled under conditions of constant volume, it is noted that the pressure falls nearly proportionally as the temperature. If this were to happen until there was no pressure, the theoretical temperature for this case is referred to as *absolute zero*. In an elementary experiment, the following data were found for pressure and temperature for a gas under constant volume.

T (°C)	0.0	20	40	60	80	100
P (kPa)	133	143	153	162	172	183

Find the least-squares line for P as a function of T, and, from the graph, determine the value of absolute zero found in this experiment.

The linear coefficient of correlation, a measure of the relatedness of two variables, is defined by $r = m(s_x/s_y)$, where s_x and s_y are the standard deviations of the x-values and y-values, respectively. Due to its definition, the values of r lie in the range $-1 \le r \le 1$. If r is near 1, the correlation is considered good. For values of r between $-\frac{1}{2}$ and $+\frac{1}{2}$, the correlation is poor. If r is near -1, the variables are said to be negatively correlated; that is, one increases while the other decreases. In Exercises 13–16, compute r for the given sets of data.

13. Exercise 1

14. Exercise 2

15. Exercise 4

16. Example 1

22-5 Fitting Nonlinear Curves to Data

If the experimental points do not appear to be on a straight line, but we recognize them as being approximately on some other type of curve, the method of least squares can be extended to use on these other curves. For example, if the points are apparently on a parabola, we could use the function $y = a + bx^2$. To use the above method, we shall extend the least-squares line to

$$y = m[f(x)] + b \tag{22-10}$$

Here, $f(x)$ must be calculated first, and then the problem can be treated as a least-squares line to find the values of m and b. Some of the functions $f(x)$ that may be considered for use are x^2, $1/x$, and 10^x.

EXAMPLE 1 ▸▸ Find the least-squares curve $y = mx^2 + b$ for the following points.

x	0	1	2	3	4	5
y	1	5	12	24	53	76

In using Eq. (22-10), $f(x) = x^2$. Our first step is to calculate values of x^2, and then we use x^2 as we used x in finding the equation of the least-squares line.

x	$f(x) = x^2$	y	x^2y	$(x^2)^2$
0	0	1	0	0
1	1	5	5	1
2	4	12	48	16
3	9	24	216	81
4	16	53	848	256
5	25	76	1900	625
	55	171	3017	979

$$n = 6$$

$$m = \frac{6(3017) - (55)(171)}{6(979) - (55)^2} = \frac{8697}{2849} = 3.05$$

$$b = \frac{(979)(171) - (3017)(55)}{2849} = 0.52$$

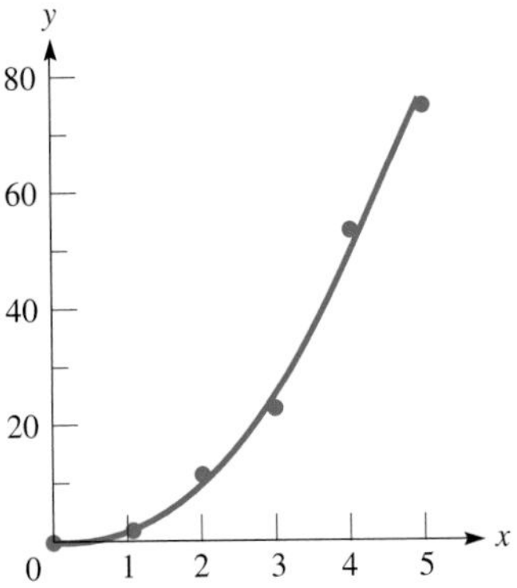

Fig. 22-13

Therefore, the required equation is $y = 3.05x^2 + 0.52$. The graph of this equation and the data points are shown in Fig. 22-13. ◂◂

EXAMPLE 2 ▸▸ In a physics experiment, the pressure p and volume V of a gas were measured at constant temperature. When the points were plotted, they were seen to approximate the hyperbola $y = c/x$. Find the least-squares approximation to the hyperbola $y = m(1/x) + b$ for the given data. See Fig. 22-14.

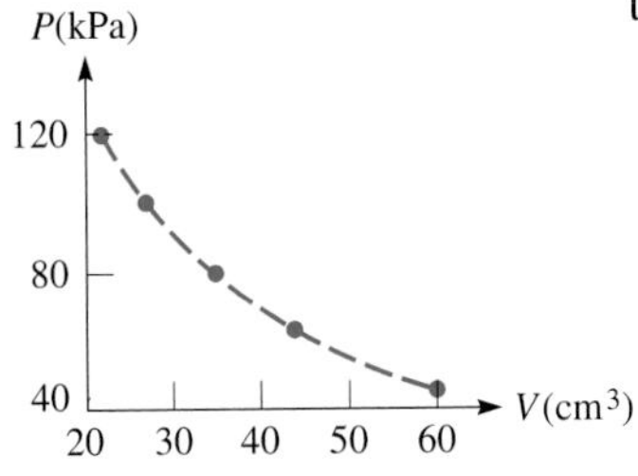

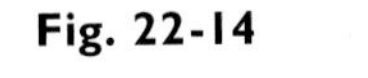
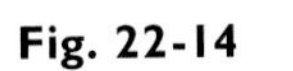
Fig. 22-14

P (kPa)	V (cm^3)
120.0	21.0
99.2	25.0
81.3	31.8
60.6	41.1
42.7	60.1

$x\,(= V)$	$f(x) = \frac{1}{x}$	$y\,(= P)$	$(\frac{1}{x})y$	$(\frac{1}{x})^2$
21.0	0.047 619 0	120.0	5.714 285 7	0.002 267 6
25.0	0.040 000 0	99.2	3.968 000 0	0.001 600 0
31.8	0.031 446 5	81.3	2.556 603 8	0.000 988 9
41.1	0.024 330 9	60.6	1.474 452 6	0.000 592 0
60.1	0.016 638 9	42.7	0.710 482 5	0.000 276 9
	0.160 035 3	403.8	14.423 824 6	0.005 725 4

(*Calculator note:* The final digits for the values shown may vary depending on the calculator and how the values are used. Here, all individual values are shown with eight digits (rounded off), although more digits were used. The value of $1/x$ was found from the value of x, with the eight digits shown. However, the values of $(1/x)y$ and $(1/x)^2$ were found from the value of $1/x$ using the extra digits. The sums were found using the rounded-off values shown. However, since the data contains only three digits, any variation in the final digits for $1/x$, $(1/x)y$, or $(1/x)^2$ will not matter.)

$$m = \frac{5(14.423\,824\,6) - (0.160\,035\,3)(403.8)}{5(0.005\,725\,4) - (0.160\,035\,3)^2} = \frac{7.496\,868\,9}{0.003\,015\,7} = 2490$$

$$b = \frac{(0.005\,725\,4)(403.8) - (14.423\,824\,6)(0.160\,035\,3)}{0.003\,015\,7} = 1.2$$

The equation of the hyperbola $y = m(1/x) + b$ is

$$y = \frac{2490}{x} + 1.2$$

This hyperbola and the data points are shown in Fig. 22-15. ◂◂

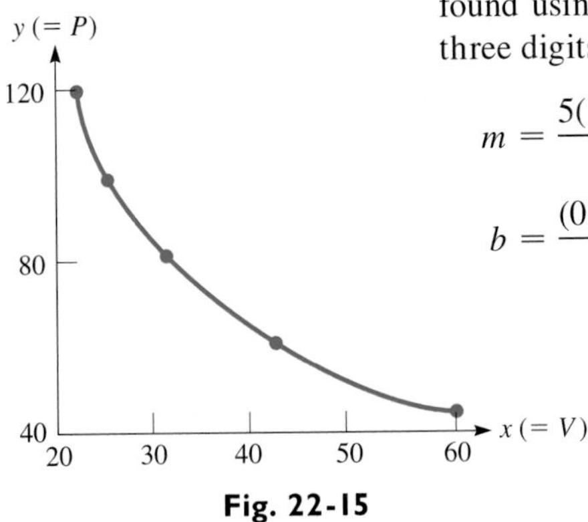

Fig. 22-15

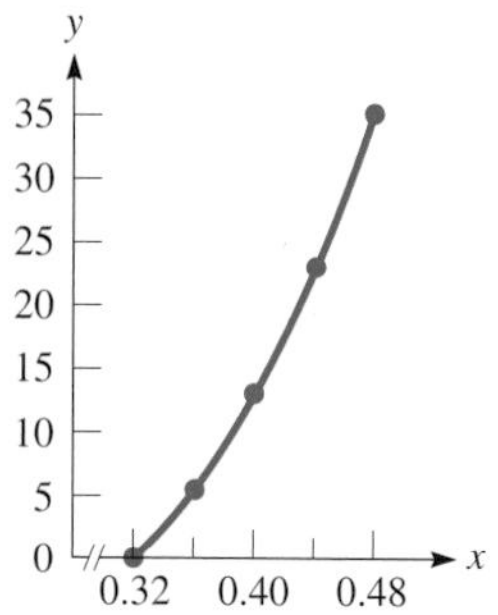

EXAMPLE 3 ▶▶ It has been found experimentally that the tensile strength of brass (a copper–zinc alloy) increases (within certain limits) with the percent of zinc. The following table shows the values that have been found. See Fig. 22-16.

Tensile strength (GPa)	0.32	0.36	0.40	0.44	0.48
Percent of zinc	0	5	13	22	34

Fig. 22-16

Fit a curve of the form $y = m(10^x) + b$ to the data. Let x = tensile strength ($\times 10^5$) and y = percent of zinc.

x	$f(x) = 10^x$	y	$(10^x)y$	$(10^x)^2$
0.32	2.089 296 1	0	0.000 000	4.365 158 3
0.36	2.290 867 7	5	11.454 338	5.248 074 6
0.40	2.511 886 4	13	32.654 524	6.309 573 4
0.44	2.754 228 7	22	60.593 031	7.585 775 8
0.48	3.019 951 7	34	102.678 36	9.120 108 4
	12.666 230 6	74	207.380 25	32.628 690 5

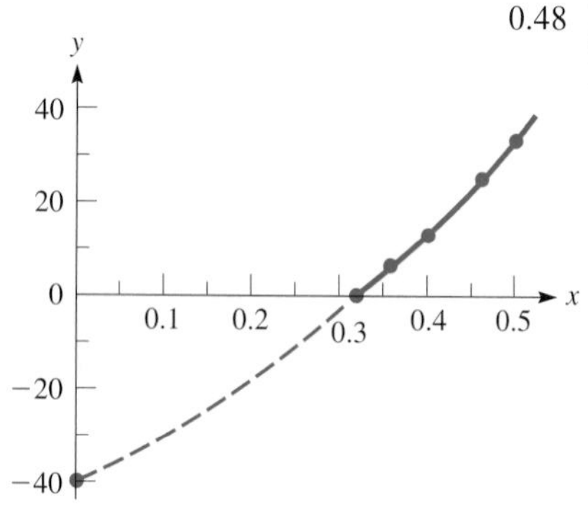

(See the note on calculator use in Example 2.)

$$m = \frac{5(207.380\,25) - (12.666\,230\,6)(74)}{5(32.628\,690\,5) - (12.666\,230\,6)^2}$$

$$= \frac{99.600\,186}{2.710\,054\,9} = 36.8$$

$$b = \frac{(32.628\,690\,5)(74) - (207.380\,25)(12.666\,230\,6)}{2.710\,054\,9} = -78.3$$

Fig. 22-17

The equation of the curve is $y = 36.8(10^x) - 78.3$. It must be remembered that for practical purposes, y must be positive. The graph of the equation is shown in Fig. 22-17, with the solid portion denoting the meaningful part of the curve. The points of the data are also shown. ◀◀

Most graphing calculators are programmed to give the equation of the least-squares line, and to graph it, after the data are entered. They can fit points to other specific types of curves, such as $y = ax^b$, as well. Also, they show the coefficient of correlation (see Exercises 13–16 of Section 22-4) to determine how well the equation fits the points. Review the manual to use this feature for any particular calculator.

Exercises 22–5

In the following exercises, find the indicated least-squares curve. Sketch the curve and plot the data points on the same graph.

1. For the points in the following table, find the least-squares curve $y = mx^2 + b$.

x	2	4	6	8	10
y	12	38	72	135	200

2. For the points in the following table, find the least-squares curve $y = m\sqrt{x} + b$.

x	0	4	8	12	16
y	1	9	11	14	15

3. For the points in the following table, find the least-squares curve $y = m(1/x) + b$.

x	1.10	2.45	4.04	5.86	6.90	8.54
y	9.85	4.50	2.90	1.75	1.48	1.30

4. For the points in the following table, find the least-squares curve $y = m(10^x) + b$.

x	0.00	0.200	0.500	0.950	1.325
y	6.00	6.60	8.20	14.0	26.0

5. The following data were found for the distance y that an object rolled down an inclined plane in time t. Determine the least-squares curve $y = mt^2 + b$.

t (seconds)	1.0	2.0	3.0	4.0	5.0
y (cm)	6.0	23	55	98	148

6. The increase in length of a certain metallic rod was measured in relation to particular increases in temperature. If y represents the increase in length for the corresponding increase in temperature x, find the least-squares curve $y = mx^2 + b$ for these data.

x (°C)	50.0	100	150	200	250
y (cm)	1.00	4.40	9.40	16.4	24.0

7. The pressure p at which Freon, a refrigerant, vaporizes for temperature T is given in the following table. Find the least-squares curve $p = me^{0.01T} + b$ for these data.

T (°C)	0	10	20	30	40
p (kPa)	480	600	830	1040	1400

8. A fraction f of annual hot-water loads at a certain facility are heated by solar energy. The fractions f for certain values of the collector area A are given in the following table. Find the least-squares curve $f = m\sqrt{A} + b$ for these data.

A (m²)	0	12	27	56	90
f	0.0	0.2	0.4	0.6	0.8

9. The makers of a special blend of coffee found that the demand for the coffee depended on the price charged. The price P per kilogram and the monthly sales S are shown in the following table. Find the least-squares curve $P = m(\frac{1}{S}) + b$ for these data.

S (thousands)	240	305	420	480	560
P (dollars)	11.20	8.80	6.40	5.60	4.80

10. The resonant frequency of an electric circuit containing a 4-μF capacitor was measured as a function of an inductance in the circuit. The following data were found. Find the least-squares curve $f = m(1/\sqrt{L}) + b$.

L (henrys)	1.0	2.0	4.0	6.0	9.0
f (hertz)	490	360	250	200	170

11. The displacement y of an object at the end of a spring at given times t is shown in the following table. Find the least-squares curve $y = me^{-t} + b$.

t (seconds)	0.0	0.5	1.0	1.5	2.0	3.0
y (cm)	6.1	3.8	2.3	1.3	0.7	0.3

12. The average daily temperatures T (in °C) for each month in Minneapolis are given in the following table.

t	J	F	M	A	M	J	J	A	S	O	N	D
T (°C)	−12	−8	−2	8	14	20	23	22	16	10	1	−7

Find the least-squares curve $T = m\cos[\frac{\pi}{6}(t - 0.5)] + b$ for these data. Assume the average temperature is for the 15th of each month. Then values of t (in months) are 0.5, 1.5, . . . , 11.5. (The fit is fairly good.)

Chapter Equations, Review Exercises, and Practice Test

Chapter Equations

Arithmetic mean
$$\bar{x} = \frac{x_1f_1 + x_2f_2 + \cdots + x_nf_n}{f_1 + f_2 + \cdots + f_n} = \frac{\sum xf}{\sum f} \qquad \text{(22-1)}$$

Standard deviation
$$s = \sqrt{\frac{\sum (x - \bar{x})^2}{n}} \qquad \text{(22-2)} \qquad s = \sqrt{\overline{x^2} - \bar{x}^2} \qquad \text{(22-5)}$$

Normal distribution
$$y = \frac{1}{\sqrt{2\pi}}e^{-x^2/2} \qquad \text{(22-6)}$$

Least-squares line
$$y = mx + b \qquad \text{(22-7)}$$
$$m = \frac{n\sum xy - \left(\sum x\right)\left(\sum y\right)}{n\sum x^2 - \left(\sum x\right)^2} \qquad \text{(22-8)} \qquad b = \frac{\left(\sum x^2\right)\left(\sum y\right) - \left(\sum xy\right)\left(\sum x\right)}{n\sum x^2 - \left(\sum x\right)^2} \qquad \text{(22-9)}$$

Nonlinear curves
$$y = m[f(x)] + b \qquad \text{(22-10)}$$

Review Exercises

In Exercises 1–4, use the following set of numbers:

2.3, 2.6, 4.2, 3.6, 3.5, 4.1, 4.8, 2.5, 3.0, 4.1, 3.8

1. Determine the median of the set of numbers.
2. Determine the arithmetic mean of the set of numbers.
3. Determine the standard deviation of the set of numbers.
4. Construct a frequency table with intervals 2.0–2.9, 3.0–3.9, and 4.0–4.9.

In Exercises 5–12, use the following set of numbers:

109, 103, 113, 110, 113, 106, 114, 101, 108, 101,
112, 106, 115, 102, 107, 110, 102, 115, 106, 105

5. Construct a frequency distribution table with intervals 101–103, 104–106, and so on.
6. Determine the median.
7. Determine the mode.
8. Determine the mean.
9. Determine the standard deviation.
10. Draw a frequency polygon for the data in Exercise 5.
11. Draw a histogram for the data in Exercise 5.
12. Construct a frequency distribution table with intervals of 101–105, 106–110, and 111–115. Then draw a histogram for these data.

In Exercises 13–18, use the following data: An important property of oil is its coefficient of viscosity, which gives a measure of how well it flows. In order to determine the viscosity of a certain motor oil, a refinery took samples from 12 different storage tanks and tested them at 50°C. *The results (in pascal-seconds) were* 0.24, 0.28, 0.29, 0.26, 0.27, 0.26, 0.25, 0.27, 0.28, 0.26, 0.26, 0.25.

13. Determine the arithmetic mean.
14. Determine the median.
15. Determine the standard deviation.
16. Make a histogram.
17. Make a frequency polygon.
18. Determine the mode.

In Exercises 19–24, use the following data: A group of wind generators was tested for power output when the wind speed was 30 km/h. *The following table gives the number of generators producing the powers shown.*

Power (watts)	650	660	670	680	690
No. generators	3	2	7	12	27

Power (watts)	700	710	720	730
No. generators	34	15	16	5

19. Determine the arithmetic mean.
20. Determine the median.
21. Determine the mode.
22. Make a histogram.
23. Determine the standard deviation.
24. Make a frequency polygon.

In Exercises 25–28, use the following data: A Geiger counter records the presence of high-energy nuclear particles. Even though no apparent radioactive source is present, a certain number of particles will be recorded. These are primarily cosmic rays, which are caused by very high energy particles from outer space. In an experiment to measure the amount of cosmic radiation, the number of counts were recorded during 200 5-s *intervals. The following table gives the number of counts and the number of 5-s intervals having this number of counts. Draw a frequency curve for these data.*

Counts	0	1	2	3	4	5	6	7	8	9	10
Intervals	3	10	25	45	29	39	26	11	7	2	3

25. Determine the median.
26. Determine the arithmetic mean.
27. Make a histogram.
28. Make a frequency polygon.

In Exercises 29–32, use the following data: Police radar on a city street recorded the speeds (to the nearest 5 km/h) *of 110 cars in a* 55 km/h *zone with the following results:*

Speed (km/h)	30	35	40	45	50	55	60	65	70	75
No. cars	3	4	4	5	8	22	48	10	4	2

29. Find the arithmetic mean.
30. Find the median.
31. Find the standard deviation.
32. Make a histogram.

In Exercises 33–40, find the indicated least-squares curve.

33. In a certain experiment, the resistance of a certain resistor was measured as a function of the temperature. The data found were as follows:

T (°C)	0.0	20.0	40.0	60.0	80.0	100
R (ohms)	25.0	26.8	28.9	31.2	32.8	34.7

Find the least-squares line for these data, expressing R as a function of T. Sketch the line and data points on the same graph.

34. An air-pollution monitoring station took samples of air each hour during the later morning hours and tested each sample for the number n of parts per million (ppm) of carbon monoxide. The results are shown in the table, where t is the number of hours after 6 A.M. Find the least-squares line for n as a function of t.

t (hours)	0.0	1.0	2.0	3.0	4.0	5.0	6.0
n (ppm)	8.0	8.2	8.8	9.5	9.7	10.0	10.7

35. The *Mach number* of a moving object is the ratio of its speed to the speed of sound (1200 km/h). The following table shows the speed s of a jet aircraft, in terms of Mach numbers, and the time t after it starts to accelerate. Find the least-squares line of s as a function of t.

t (minutes)	0.00	0.60	1.20	1.80	2.40	3.00
s (Mach number)	0.88	0.97	1.03	1.11	1.19	1.25

36. In an experiment to determine the relation between the load on a spring and the length of the spring, the following data were found. Find the least-squares line which expresses the length as a function of the load.

Load (kg)	0.0	1.0	2.0	3.0	4.0	5.0
Length (cm)	10.0	11.2	12.3	13.4	14.6	15.9

37. The distance s of a missile above the ground at time t after being released from a plane is given by the following table. Find the least-squares curve of the form $s = mt^2 + b$ for these data.

t (seconds)	0.0	3.0	6.0	9.0	12.0	15.0	18.0
s (metres)	3000	2960	2820	2600	2290	1900	1410

38. In an elementary experiment that measured the wavelength of sound as a function of the frequency, the following results were obtained.

Frequency (Hz)	240	320	400	480	560
Wavelength (cm)	140	107	81.0	70.0	60.0

Find the least-squares curve of the form $y = m(\frac{1}{x}) + b$ for these data, expressing wavelength as y and frequency as x.

39. After being heated, the temperature T of an insulated liquid is measured at times t as follows:

t (hours)	0	2	4	6	8	10
T (°C)	100	85	72	63	54	48

Plot these points and choose an appropriate function $f(x)$ for $y = m[f(x)] + b$. Then find the equation of the least-squares curve and graph it on the same graph with the points.

40. The vertical distance y of the cable of a suspension bridge above the surface of the bridge is measured at a horizontal distance x along the bridge from its center. See Fig. 22-18. The results are as follows:

x (metres)	0	100	200	300	400	500
y (metres)	15	17	23	33	47	65

Plot these points and choose an appropriate function $f(x)$ for $y = m[f(x)] + b$. Then find the equation of the least-squares curve and graph it on the same graph with the points.

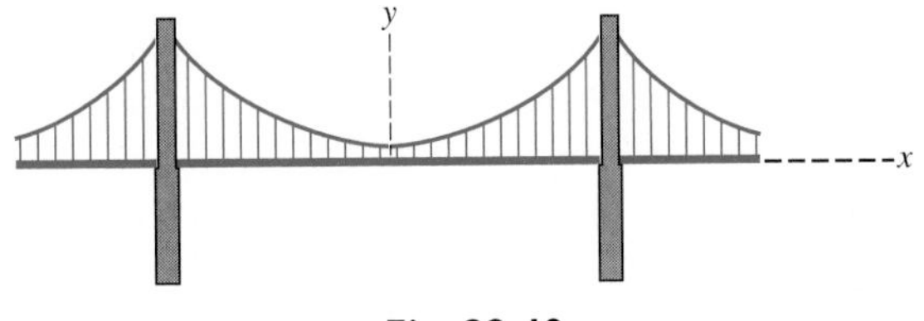

Fig. 22-18

In Exercises 41–44, solve the given problems.

41. The nth root of the product of n positive numbers is the *geometric mean* of the numbers. Find the geometric mean of the carbon monoxide readings in Exercise 34.

42. One use of the geometric mean (see Exercise 41) is to find an average ratio. By finding the geometric mean, find the average Mach number for the jet in Exercise 35.

43. Show that Eqs. (22-8) and (22-9) can be written as

$$m = \frac{\overline{xy} - \bar{x}\,\bar{y}}{s_x^2} \quad \text{and} \quad b = \frac{\overline{x^2}\,\bar{y} - \overline{xy}\,\bar{x}}{s_x^2}$$

where s_x is the standard deviation of the x-values.

44. By using the expressions for m and b from Exercise 43, show that the point $(\bar{x}, \bar{y})$ satisfies Eq. (22-7).

Writing Exercise

45. A study is to be made of the effectiveness of a treatment for glaucoma (a severe eye disorder) depending on the age of the person treated. Write two or three paragraphs explaining what data could be found and how it can be analyzed and then used to predict the effects of the treatment on future patients.

Practice Test

In Problems 1–3, use the following set of numbers.

5, 6, 1, 4, 9, 5, 7, 3, 8, 10, 5, 8, 4, 9, 6

1. Find the median.
2. Find the mode.
3. Draw a histogram for the intervals 0–2, 3–5, and so on.

In Problems 4–6, use the following data: Two machine parts are considered satisfactorily assembled if their total thickness (to the nearest 0.01 cm*) is between or equal to* 0.92 cm *and* 0.94 cm. *One hundred sample assemblies are tested, and the thicknesses, to the nearest* 0.01 cm, *are given in the following table.*

Total thickness	0.90	0.91	0.92	0.93	0.94	0.95	0.96
Number	3	9	31	38	12	5	2

4. Find the mean.
5. Find the standard deviation.
6. Draw a frequency polygon.
7. Find the equation of the least-squares line for the points indicated in the following table. Graph the line and data points on the same graph.

x	1	3	5	7	9
y	5	11	17	20	27

8. The velocity (in m/s) of an object moving down an inclined plane was measured as a function of the distance (in metres) it moves, with the following results:

Distance (m)	1.00	3.00	5.00	7.00	9.00
Velocity (m/s)	1.10	1.90	2.50	2.90	3.30

Find the equation of the least-squares curve of the form $y = m\sqrt{x} + b$, which expresses the velocity as a function of the distance.

23

The Derivative

The problems that can be solved by the methods of algebra and trigonometry are numerous. There are, however, a great many problems that arise in the various fields of technology which require for their solution methods beyond those available from algebra and trigonometry. These traditional topics remain of definite importance, but it is necessary to develop additional methods of analyzing and solving problems.

In this chapter we start developing the methods of **differential calculus.** This subject deals with the important problem involving the *rate of change* of one quantity with respect to another. Examples of rates of change are velocity (the rate of change of distance with respect to time), the rate of change of the length of a metal rod with respect to temperature, the rate of change of light intensity with respect to the distance from the source, the rate of change of electric current with respect to time, and the rate of change of production costs with respect to the number of units a business produces.

Determining how fast an object is moving is important in physics and in many areas of technology. In Section 23-4 we develop the method of finding the instantaneous velocity of a moving object.

Another principal type of problem that calculus allows us to solve is that of finding a function when its rate of change is known. This is **integral calculus.** One of its principal applications comes from electricity, where current is the time rate of change of electric charge. Integral calculus also leads to the solution of a great many apparently unrelated problems, including the determination of plane areas, volumes, and the physical concepts of work and pressure. The study of integral calculus starts in Chapter 25.

Isaac Newton (1642–1727), an English mathematician and physicist, and Gottfried Wilhelm Leibniz (1646–1716), a German mathematician and philosopher, are credited with the creation of calculus. In order to solve problems in fields such as geometry and astronomy, each independently developed the basic methods of calculus.

23-1 Limits

Before dealing with the rate of change of a function, we first take up the concept of a *limit.* We encountered a limit with infinite geometric series and with the asymptotes of a hyperbola. It is necessary to develop this concept further.

Continuity

To help develop the concept of a limit, we consider briefly the **continuity** of a function. *For a function to be* **continuous at a point,** *the function must exist at the point, and any small change in x produces only a small change in f(x).* In fact, the change in $f(x)$ can be made as small as we wish by restricting the change in x sufficiently, if the function is continuous. Also, *a function is said to be* **continuous over an interval** *if it is continuous at each point in the interval.*

EXAMPLE 1 ▸▸ The function $f(x) = 3x^2$ is continuous for all values of x. That is, $f(x)$ is defined for all values of x, and a small change in x for any given value of x produces only a small change in $f(x)$. If we choose $x = 2$ and then let x change by 0.1, 0.01, and so on, we obtain the values in the following table.

x	2	2.1	2.01	2.001
$f(x)$	12	13.23	12.1203	12.012 003
Change in x		0.1	0.01	0.001
Change in $f(x)$		1.23	0.1203	0.012 003

We can see that the change in $f(x)$ is made smaller by the smaller changes in x. This shows that $f(x)$ is continuous at $x = 2$. Since this type of result would be obtained for any other x we may choose, we see that $f(x)$ is continuous for all values, and therefore it is continuous over the interval of all values of x. ◂◂

EXAMPLE 2 ▸▸ The function $f(x) = \dfrac{1}{x - 2}$ is not continuous at $x = 2$. When we substitute 2 for x, we have division by zero. This means the function is not defined. The condition that the function must exist is not satisfied. ◂◂

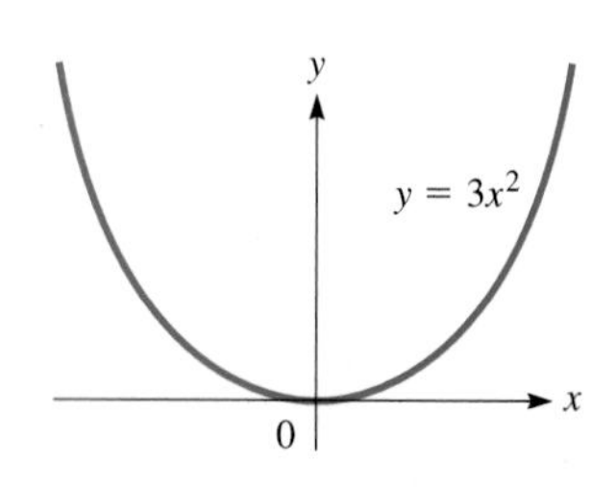

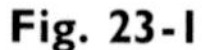
Fig. 23-1

From a graphical point of view, a function that is continuous over an interval has no "breaks" in its graph over that interval. The function is continuous over the interval if we can draw its graph without lifting the marker from the paper. If the function is not continous (*discontinuous*), a break occurs because the function is not defined, or the definition of the function leads to an instantaneous "jump" in its values.

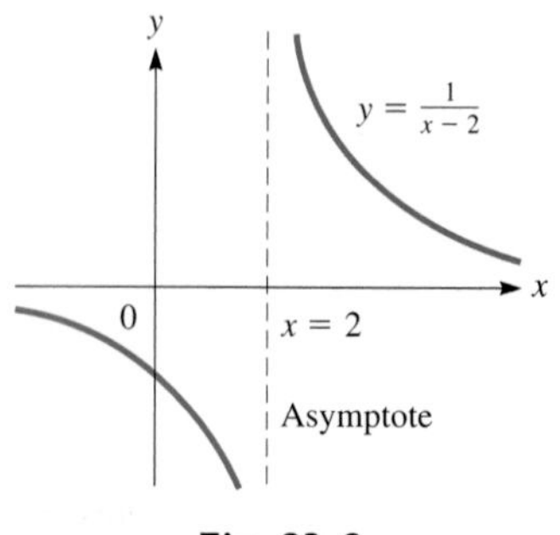

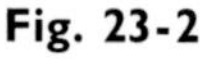
Fig. 23-2

EXAMPLE 3 ▸▸ (a) The graph of the function $f(x) = 3x^2$, which we determined to be continuous for all values of x in Example 1, is shown in Fig. 23-1. We see that there are no breaks in the curve.

(b) The graph of $f(x) = \dfrac{1}{x - 2}$, which we determined not to be continuous at $x = 2$ in Example 2, is shown in Fig. 23-2. We see that there is a break in the curve for $x = 2$, and this shows the fact that $f(x)$ does not exist *at* $x = 2$. The curve is a hyperbola with an asymptote $x = 2$. ◂◂

EXAMPLE 4 ▸▸ (a) The function represented in Fig. 23-3 is continuous in the interval for which $x > 1$. The graph does not exist for values $x \leq 1$. The open circle point at $x = 1$ indicates that the point is not part of the graph.

(b) The function represented by the graph in Fig. 23-4 is not continuous at $x = 1$. The function is defined (by the solid circle point) for $x = 1$. However, a small change from $x = 1$ may result in a change of at least 1.5 in $f(x)$, regardless of how small a change in x is made. The small change condition is not satisfied.

(c) The function represented by the graph in Fig. 23-5 is not continuous for $x = -2$. The open circle shows that the point is not part of the graph, and therefore $f(x)$ is not defined for $x = -2$. ◂◂

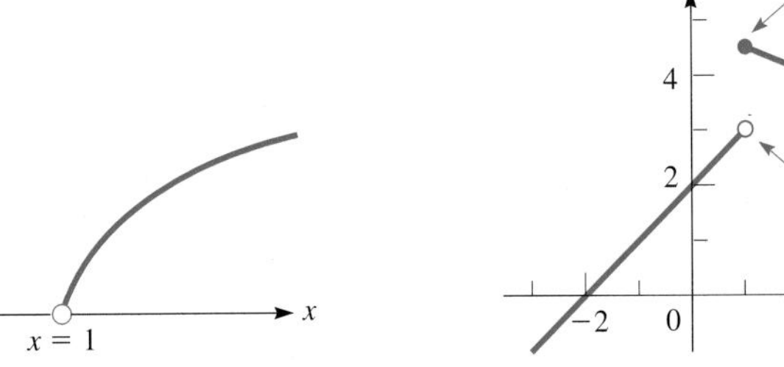

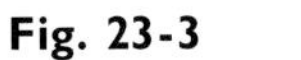
Fig. 23-3

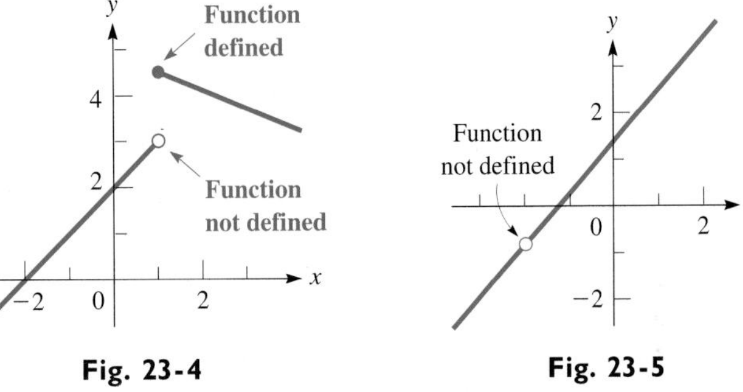

Fig. 23-4

Fig. 23-5

EXAMPLE 5 ▸▸ (a) We can define the function in Fig. 23-4 as

$$f(x) = \begin{cases} x + 2 & \text{for } x < 1 \\ -\frac{1}{2}x + 5 & \text{for } x \geq 1 \end{cases}$$

where we note that the equation differs for different parts of the domain.

(b) The graph of the function

$$g(x) = \begin{cases} 2x - 1 & \text{for } x \leq 2 \\ -x + 5 & \text{for } x > 2 \end{cases}$$

is shown in Fig. 23-6. We see that it is a continuous function even though the equation for $x \leq 2$ is different from that for $x > 2$. ◂◂

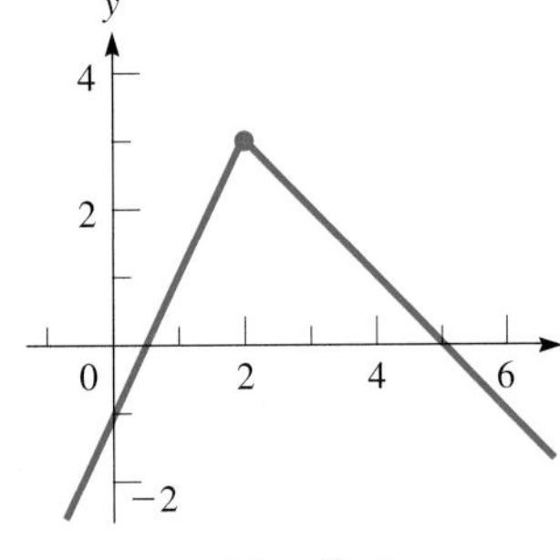

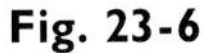
Fig. 23-6

In our earlier discussions of infinite geometric series and the asymptotes of a hyperbola, we used the symbol $\rightarrow$, which means "approaches." *When we say that $x \rightarrow 2$, we mean that x may take on any value as close to 2 as desired, but it also* **CAUTION ▸** *distinctly means that* ***x cannot be set equal to 2.***

EXAMPLE 6 ▸▸ Consider the behavior of $f(x) = 2x + 1$ as $x \rightarrow 2$.

Since we are not to use $x = 2$, we use a calculator to set up tables in order to determine values of $f(x)$, as x gets close to 2.

x	1.000	1.500	1.900	1.990	1.999
$f(x)$	3.000	4.000	4.800	4.980	4.998

x	3.000	2.500	2.100	2.010	2.001
$f(x)$	7.000	6.000	5.200	5.020	5.002

values approach 5

We can see that $f(x)$ approaches 5, as x approaches 2, from above 2 and from below 2. ◂◂

Limit of a Function

In Example 6, since $f(x) \to 5$ as $x \to 2$, the number 5 is called the limit of $f(x)$ as $x \to 2$. This leads to the meaning of the limit of a function. In general, *the* **limit of a function $f(x)$** *is that value which the function approaches as x approaches the given value a.* This is written as

$$\lim_{x \to a} f(x) = L \tag{23-1}$$

where L is the value of the limit of the function. Remember, in approaching a, x may come as arbitrarily close as desired to a, but it may not equal a.

An important conclusion can be drawn from the limit in Example 6. The function $f(x)$ is a continuous function, and $f(2)$ equals the value of the limit as $x \to 2$. In general, it is true that

if $f(x)$ is continuous at $x = a$, then the limit as $x \to a$ equals $f(a)$.

In fact, looking back at our definition of continuity, we see that this is what the definition means. That is, a function $f(x)$ is continuous at $x = a$ if *all three* of the following conditions are satisfied:

1. $f(a)$ exists, **2.** $\lim_{x\to a} f(x)$ exists, and **3.** $\lim_{x\to a} f(x) = f(a)$.

Although we can evaluate the limit for a continuous function as $x \to a$ by evaluating $f(a)$, it is possible that a function is not continuous at $x = a$ and that the limit exists and can be determined. Thus, we must be able to determine the value of a limit without finding $f(a)$. The following example illustrates the evaluation of such a limit.

EXAMPLE 7 ▸▸ Find $\lim_{x\to 2} \dfrac{2x^2 - 3x - 2}{x - 2}$.

We note immediately that the function is not continuous at $x = 2$, for division by zero is indicated. Thus, we cannot evaluate the limit by substituting $x = 2$ into the function. By setting up tables, we determine the value that $f(x)$ approaches, as x approaches 2.

x	1.000	1.500	1.900	1.990	1.999
$f(x)$	3.000	4.000	4.800	4.980	4.998

x	3.000	2.500	2.100	2.010	2.001
$f(x)$	7.000	6.000	5.200	5.020	5.002

values approach 5

We see that the values obtained are identical to those in Example 6. Since $f(x) \to 5$ as $x \to 2$, we have

$$\lim_{x\to 2} \frac{2x^2 - 3x - 2}{x - 2} = 5$$

Therefore, we see that the limit exists as $x \to 2$, although the function does not exist at $x = 2$. ◂◂

The reason that the functions in Examples 6 and 7 have the same limit is shown in the following example.

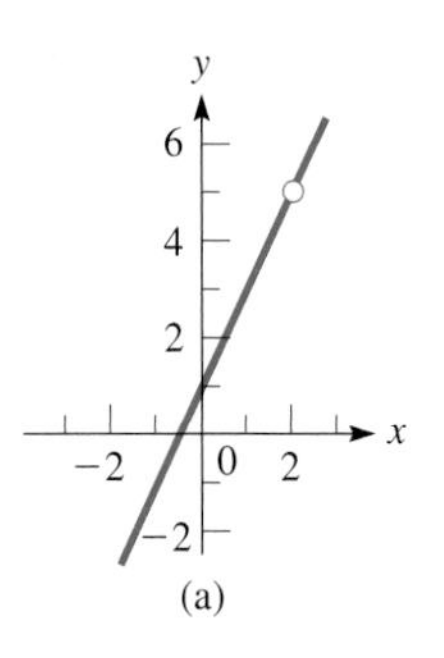

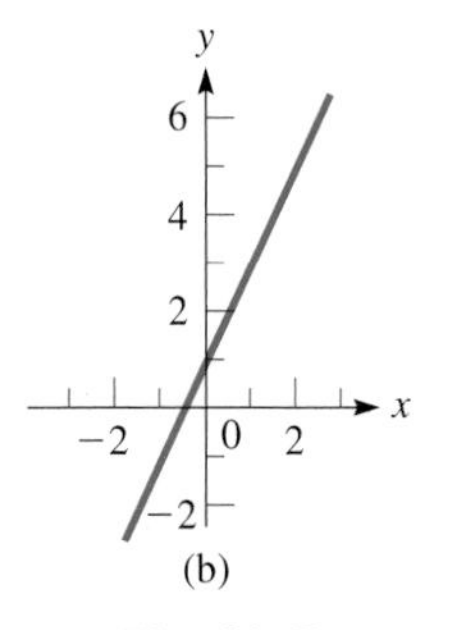

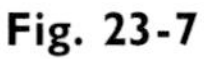
Fig. 23-7

EXAMPLE 8 ▸▸ The function $\dfrac{2x^2 - 3x - 2}{x - 2}$ in Example 7 is the same as the function $2x + 1$ in Example 6, except when $x = 2$. By factoring the numerator of the function of Example 7, we have

$$\frac{2x^2 - 3x - 2}{x - 2} = \frac{(2x + 1)(x - 2)}{x - 2} = 2x + 1$$

The cancellation here is valid, as long as x does not equal 2, for we have division by zero at $x = 2$. Also, in finding the limit as $x \to 2$, we do not use the value $x = 2$. Therefore,

$$\lim_{x \to 2} \frac{2x^2 - 3x - 2}{x - 2} = \lim_{x \to 2}(2x + 1) = 5$$

The limits of the two functions are equal, since, again, in finding the limit, we do not let $x = 2$. The graphs of the two functions are shown in Fig. 23-7(a) and (b). We can see from the graphs that the limits are the same, although one of the functions is not continuous.

If $f(x) = 5$ for $x = 2$ is added to the definition of the function in Example 7, it is then the same as $2x + 1$, and its graph is that in Fig. 23-7(b). ◂◂

The limit of the function in Example 7 was determined by calculating values near $x = 2$ and by means of an algebraic change in the function. This illustrates that limits may be found through the meaning and definition, and through other procedures when the function is not continuous. The following example illustrates a function for which the limit does not exist as x approaches the indicated value.

EXAMPLE 9 ▸▸ In trying to find

$$\lim_{x \to 2} \frac{1}{x - 2}$$

we note that $f(x)$ is not defined for $x = 2$, since we would have division by zero. Therefore, we set up the following table to see how $f(x)$ behaves as $x \to 2$.

x	3	2.5	2.1	2.01	2.001	
$f(x)$	1	2	10	100	1000	$f(x) \to +\infty$

x	1	1.5	1.9	1.99	1.999	
$f(x)$	−1	−2	−10	−100	−1000	$f(x) \to -\infty$

Fig. 23-2

We see that $f(x)$ gets larger as $x \to 2$ from above 2, and $f(x)$ gets smaller (large negative values) as $x \to 2$ from below 2. This may be written as $f(x) \to +\infty$ as $x \to 2^+$ and $f(x) \to -\infty$ as $x \to 2^-$, but we must remember that ∞ is not a real number. Therefore, the limit as $x \to 2$ does not exist. The graph of this function is shown in Fig. 23-2, which is shown again for reference. ◂◂

The following examples further illustrate the evaluation of limits.

EXAMPLE 10 ▸▸ Find $\lim_{x\to 4}(x^2 - 7)$.

Since the function $x^2 - 7$ is continuous at $x = 4$, we may evaluate this limit by substitution. For $f(x) = x^2 - 7$, we have $f(4) = 9$. This means that

$$\lim_{x\to 4}(x^2 - 7) = 9$$ ◂◂

EXAMPLE 11 ▸▸ Find

$$\lim_{x\to 2}\left(\frac{x^2 - 4}{x - 2}\right)$$

Since

$$\frac{x^2 - 4}{x - 2} = \frac{(x - 2)(x + 2)}{x - 2} = x + 2$$

is valid as long as $x \neq 2$, we find that

$$\lim_{x\to 2}\left(\frac{x^2 - 4}{x - 2}\right) = \lim_{x\to 2}(x + 2) = 4$$

Again, we do not have to concern ourselves with the fact that the cancellation is not valid for $x = 2$. In finding the limit we do not consider the value of $f(x)$ at $x = 2$. ◂◂

EXAMPLE 12 ▸▸ Find $\lim_{x\to 0}(x\sqrt{x - 3})$.

We see that $x\sqrt{x - 3} = 0$ if $x = 0$, but this function does not have real values for values of x less than 3 other than $x = 0$. *Since x cannot approach 0, $f(x)$ does not approach 0 and the limit does not exist.* The point of this example is that even if $f(a)$ exists, we cannot evaluate the limit by finding $f(a)$, unless $f(x)$ is continuous at $x = a$. Here, $f(a)$ exists but the limit does not exist. ◂◂

Returning briefly to the discussion of continuity, we again see the need for all three conditions of continuity given on p. 612. If $f(a)$ does not exist, the function is discontinuous at $x = a$, and if $f(a)$ does not equal $\lim_{x\to a} f(x)$, a small change in x will not result in a small change in $f(x)$. Also, Example 12 shows another reason to carefully consider the domain of the function in finding the limit.

Limits as x Approaches Infinity

Limits as x approaches infinity are also of importance. However, when dealing with these limits, we must remember that

CAUTION ▸ ***∞ does not represent a real number and that algebraic operations may not be performed on it.***

Therefore, when we write $x \to \infty$, we know we are to consider values of x that are becoming large without bound. We first encountered this concept in Chapter 19 when we discussed infinite geometric series. The following examples illustrate the evaluation of this type of limit.

EXAMPLE 13 ▸▸ The efficiency E of an engine is given by $E = 1 - Q_2/Q_1$, where Q_1 is the heat taken in, and Q_2 is the heat ejected by the engine. ($Q_1 - Q_2$ is the work done by the engine.) If, in an engine cycle, $Q_2 = 500$ kJ, determine E as Q_1 becomes large without bound.

We are to find

$$\lim_{Q_1 \to \infty} \left(1 - \frac{500}{Q_1}\right)$$

As Q_1 becomes larger and larger, $500/Q_1$ becomes smaller and smaller and approaches zero. This means $f(Q_1) \to 1$ as $Q_1 \to \infty$. Thus,

$$\lim_{Q_1 \to \infty} \left(1 - \frac{500}{Q_1}\right) = 1$$

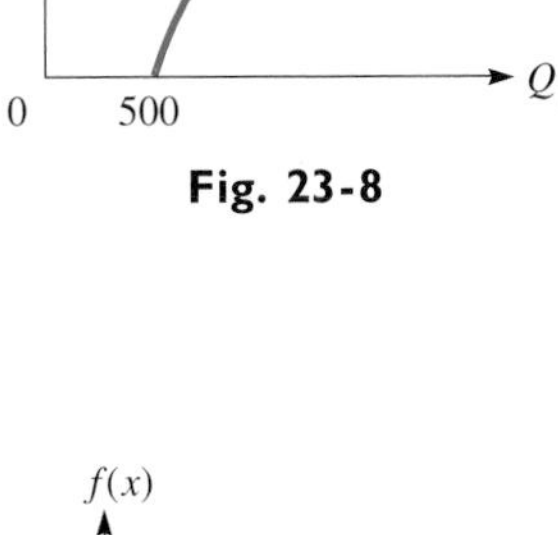

Fig. 23-8

We can verify our reasoning and the value of the limit by making a table of values for Q_1 and E as Q_1 becomes large.

Q_1	500	5000	50 000	500 000	
E	0	0.9	0.99	0.999	values approach 1

Again we see that $E \to 1$ as $Q_1 \to \infty$. See Fig. 23-8.

This is primarily a theoretical consideration, as there are obvious practical limitations as to how much heat can be supplied to an engine. An engine for which $E = 1$ would operate at 100% efficiency. ◂◂

EXAMPLE 14 ▸▸ Find $\lim_{x \to \infty} \dfrac{x^2 + 1}{2x^2 + 3}$.

We note that as $x \to \infty$, both the numerator and the denominator become large without bound. Therefore, we use a calculator to make a table to see how $f(x)$ behaves as x becomes very large.

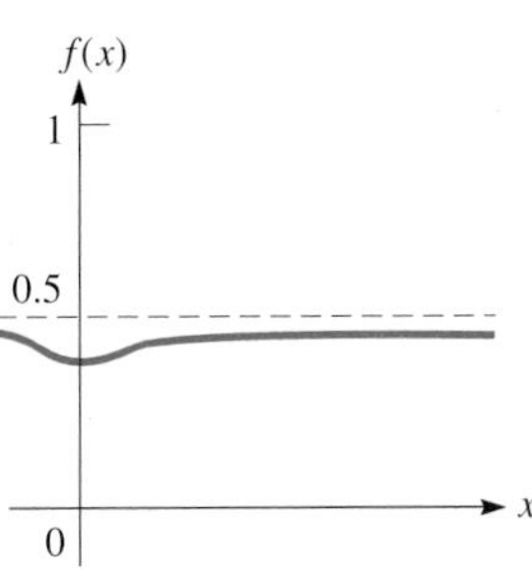

Fig. 23-9

x	1	10	100	1000	
$f(x)$	0.4	0.497 536 9	0.499 975 0	0.499 999 8	values approach 0.5

From this table we see that $f(x) \to 0.5$ as $x \to \infty$. See Fig. 23-9.

This limit can also be found through algebraic operations and an examination of the resulting algebraic form. If we divide both the numerator and the denominator of the function by x^2, ***which is the largest power of x that appears in either the numerator or the denominator,*** we have

NOTE ▸

$$\frac{x^2 + 1}{2x^2 + 3} = \frac{1 + \dfrac{1}{x^2}}{2 + \dfrac{3}{x^2}}$$

terms → 0 as $x \to \infty$

Here we see that $1/x^2$ and $3/x^2$ both approach zero as $x \to \infty$. This means that the numerator approaches 1 and the denominator approaches 2. Thus,

$$\lim_{x \to \infty} \frac{x^2 + 1}{2x^2 + 3} = \lim_{x \to \infty} \frac{1 + \dfrac{1}{x^2}}{2 + \dfrac{3}{x^2}} = \frac{1}{2}$$

which agrees with our previous result. ◂◂

The definitions and development of continuity and of a limit presented in this section are not mathematically rigorous. However, the development is consistent with a more rigorous development, and the concept of a limit is the principal concern.

Exercises 23–1

In Exercises 1–6, determine the values of x for which the function is continuous. If the function is not continuous, determine the reason.

1. $f(x) = 3x - 2$

2. $f(x) = 9 - x^2$

3. $f(x) = \dfrac{1}{x + 3}$

4. $f(x) = \dfrac{2}{x^2 - x}$

5. $f(x) = \dfrac{1}{\sqrt{x}}$

6. $f(x) = \dfrac{\sqrt{x + 2}}{x}$

In Exercises 7–12, determine the values of x for which the function, as represented by the graph in Fig. 23-10, is continuous. If the function is not continuous, determine the reason.

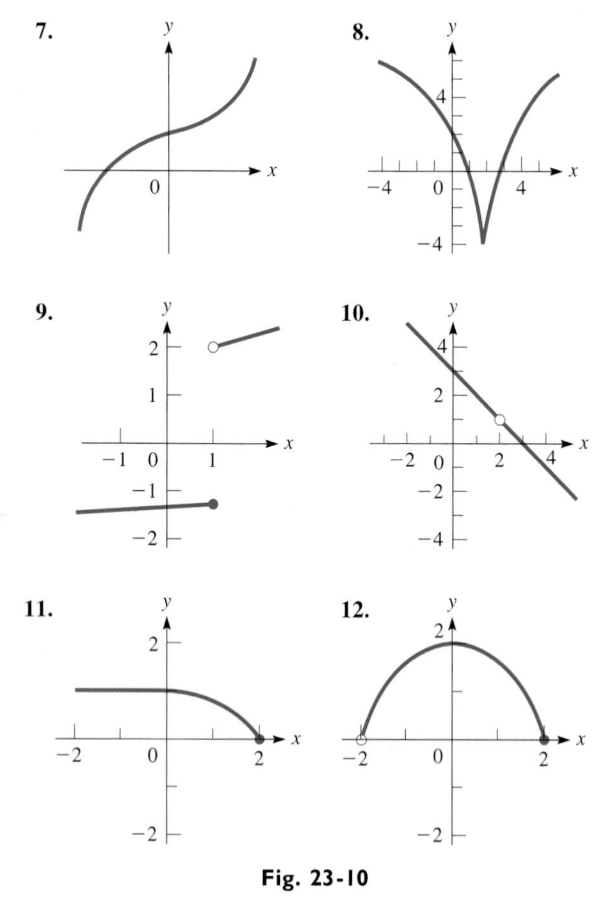

Fig. 23-10

In Exercises 13–16, graph the given functions and determine the values of x for which the functions are continuous. If the function is not continuous, determine the reason.

13. $f(x) = \begin{cases} x^2 & \text{for } x < 2 \\ 2 & \text{for } x \geq 2 \end{cases}$

14. $f(x) = \begin{cases} \dfrac{x^3 - x^2}{x - 1} & \text{for } x \neq 1 \\ 1 & \text{for } x = 1 \end{cases}$

15. $f(x) = \begin{cases} \dfrac{2x^2 - 18}{x - 3} & \text{for } x < 3 \text{ or } x > 3 \\ 12 & \text{for } x = 3 \end{cases}$

16. $f(x) = \begin{cases} \dfrac{x + 2}{x^2 - 4} & \text{for } x < -2 \\ \dfrac{x}{8} & \text{for } x > -2 \end{cases}$

In Exercises 17–24, evaluate the given function for the values of x shown in the table. Do not change the form of the function. Then, by observing the values obtained, find the indicated limit.

17. Evaluate $f(x) = 3x - 2$.
Find $\lim_{x \to 3} (3x - 2)$.

x	2.900	2.990	2.999	3.001	3.010	3.100
$f(x)$						

18. Evaluate $f(x) = x^2 - 7$.
Find $\lim_{x \to 4} (x^2 - 7)$.

x	3.900	3.990	3.999	4.001	4.010	4.100
$f(x)$						

19. Evaluate $f(x) = \dfrac{x^3 - x}{x - 1}$.

Find $\lim_{x \to 1} \dfrac{x^3 - x}{x - 1}$.

x	0.900	0.990	0.999	1.001	1.010	1.100
$f(x)$						

20. Evaluate $f(x) = \frac{x^3 + 2x^2 - 2x + 3}{x + 3}$.

Find $\lim_{x \to -3} \frac{x^3 + 2x^2 - 2x + 3}{x + 3}$.

x	−3.100	−3.010	−3.001	−2.999	−2.990	−2.900
$f(x)$						

21. Evaluate $f(x) = \frac{2 - \sqrt{x + 2}}{x - 2}$. Find $\lim_{x \to 2} \frac{2 - \sqrt{x + 2}}{x - 2}$.

x	1.900	1.990	1.999	2.001	2.010	2.100
$f(x)$						

22. Evaluate $f(x) = \frac{2 - \sqrt{x}}{4 - x}$. Find $\lim_{x \to 4} \frac{2 - \sqrt{x}}{4 - x}$.

x	3.900	3.990	3.999	4.001	4.010	4.100
$f(x)$						

23. Evaluate $f(x) = \frac{2x + 1}{5x - 3}$.

Find $\lim_{x \to \infty} \frac{2x + 1}{5x - 3}$.

x	10	100	1000
$f(x)$			

24. Evaluate $f(x) = \frac{1 - x^2}{8x^2 + 5}$.

Find $\lim_{x \to \infty} \frac{1 - x^2}{8x^2 + 5}$.

x	10	100	1000
$f(x)$			

In Exercises 25–44, evaluate the indicated limits by direct evaluation as in Examples 10–14. Change the form of the function where necessary.

25. $\lim_{x \to 3} (3x - 2)$

26. $\lim_{x \to 4} (x^2 - 7)$

27. $\lim_{x \to 2} \frac{x^2 - 1}{x + 1}$

28. $\lim_{x \to 5} \left(\frac{3}{x^2 + 2} \right)$

29. $\lim_{x \to 0} \frac{x^2 + x}{x}$

30. $\lim_{x \to 2} \frac{x^2 - 2x}{x - 2}$

31. $\lim_{x \to -1} \frac{x^2 - 1}{x + 1}$

32. $\lim_{x \to 3} \frac{x^2 - 2x - 3}{3 - x}$

33. $\lim_{x \to 1} \frac{x^3 - x}{x - 1}$

34. $\lim_{x \to 1/3} \frac{3x - 1}{3x^2 + 5x - 2}$

35. $\lim_{x \to 1} \frac{(2x - 1)^2 - 1}{2x - 2}$

36. $\lim_{x \to 0} \frac{(2 + x)^2 - 4}{x}$

37. $\lim_{x \to -1} \sqrt{x}(x + 1)$

38. $\lim_{x \to 1} (x - 1)\sqrt{x^2 - 4}$

39. $\lim_{x \to \infty} \frac{2/x}{1 - 2x}$

40. $\lim_{x \to \infty} \frac{6}{1 + \frac{2}{x^2}}$

41. $\lim_{x \to \infty} \frac{3x^2 + 5}{x^2 - 2}$

42. $\lim_{x \to \infty} \frac{x - 1}{7x + 4}$

43. $\lim_{x \to \infty} \frac{2x - 6}{x^2 - 9}$

44. $\lim_{x \to \infty} \frac{1 - x^2}{8x^2 + 5}$

In Exercises 45 and 46, evaluate the function at 0.1, 0.01, and 0.001 from both sides of the value it approaches. In Exercises 47 and 48, evaluate the function for values of x of 10, 100, and 1000. From these values, determine the limit. Then, by using an appropriate change of algebraic form, evaluate the limit directly and compare values.

45. $\lim_{x \to 0} \frac{x^2 - 3x}{x}$

46. $\lim_{x \to 3} \frac{2x^2 - 6x}{x - 3}$

47. $\lim_{x \to \infty} \frac{2x^2 + x}{x^2 - 3}$

48. $\lim_{x \to \infty} \frac{x^2 + 6x + 5}{2x^2 + 1}$

In Exercises 49–56, solve the given problems involving limits.

49. Velocity can be found by dividing the displacement s of an object by the elapsed time t in moving through the displacement. In a certain experiment, the following values were measured for the displacements and elapsed times for the motion of an object. Determine the limiting value of the velocity.

s (cm)	0.480 000	0.280 000	0.029 800	0.002 998 0	0.000 299 98
t (s)	0.200 000	0.100 000	0.010 000	0.001 000 0	0.000 100 00

50. A rectangular solar panel is to be designed to have an area of 520 cm^2. Express the perimeter p as a function of the width w. Find $\lim_{w \to 20} p$ by evaluating the function for the following values of w (in cm): 19.0, 19.9, 19.99, 20.01, 20.1, and 21.

51. A certain object, after being heated, cools at such a rate that its temperature T (in °C) decreases 10% each minute. If the object is originally heated to 100°C, find $\lim_{t \to 10} T$ and $\lim_{t \to \infty} T$, where t is the time (in min).

52. A 5-Ω resistor and a variable resistor of resistance R are placed in parallel. The expression for the resulting resistance R_T is given by $R_T = \frac{5R}{5 + R}$. Determine the limiting value of R_T as $R \to \infty$.

53. Using a calculator, find $\lim_{x \to \infty} (1 + x)^{1/x}$. Do you recognize the limiting value?

54. Using a calculator in radian mode, find $\lim_{x \to 0} \frac{\sin x}{x}$.

55. Using a calculator, find $\lim_{x \to \infty} \frac{\sqrt{4x^2 + 100}}{x}$.

56. Explain why $\lim_{x \to 0^+} 2^{1/x} \neq \lim_{x \to 0^-} 2^{1/x}$ where $\lim_{x \to 0^+}$ means to find the limit as x approaches zero from the right only and $\lim_{x \to 0^-}$ means to find the limit as x approaches zero from the left only.

23-2 The Slope of a Tangent to a Curve

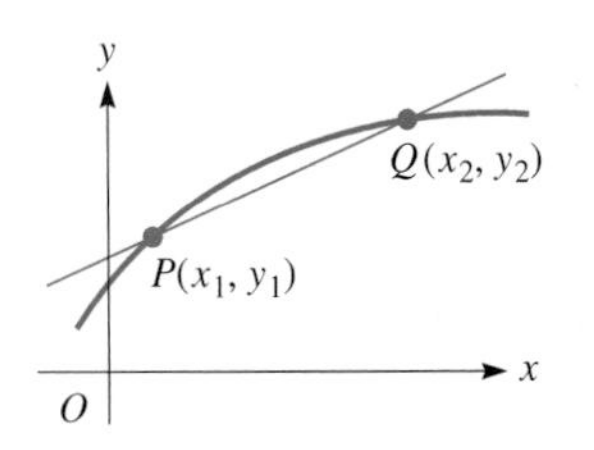

Fig. 23-11

Having developed the basic operations with functions and the concept of a limit, we now turn our attention to a graphical interpretation of the rate of change of a function. This interpretation, basic to an understanding of the calculus, deals with the slope of a line tangent to the curve of a function.

Consider the points $P(x_1, y_1)$ and $Q(x_2, y_2)$ in Fig. 23-11. From Chapter 21 we know that the slope of the line through these points is given by

$$m = \frac{y_2 - y_1}{x_2 - x_1}$$

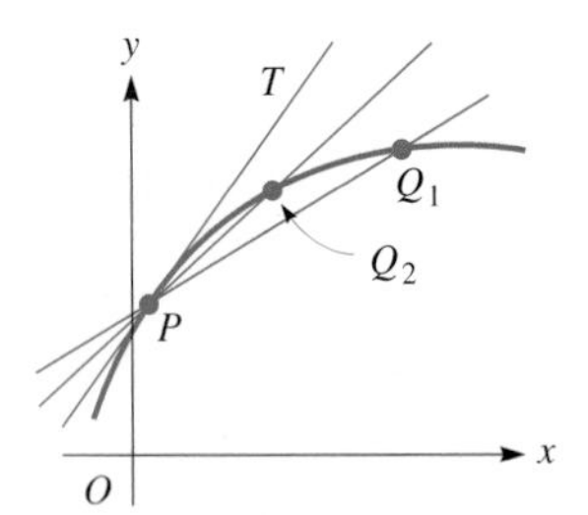

Fig. 23-12

This, however, represents the slope of the line through P and Q and no other line. If we now allow Q to be a point closer to P, the slope of PQ will more closely approximate the slope of a line drawn tangent to the curve at P (see Fig. 23-12). In fact, the closer Q is to P, the better this approximation becomes. It is not possible to allow Q to coincide with P, for then it would not be possible to define the slope of PQ in terms of two points. *The slope of the tangent line, often referred to as the slope of the curve, is the limiting value of the slope of PQ as Q approaches P.*

EXAMPLE 1 ▶▶ Find the slope of a line tangent to the curve $y = x^2 + 3x$ at the point $P(2, 10)$ by finding the limit of slopes of lines PQ as Q approaches P.

We shall let point Q have the x-values of 3.0, 2.5, 2.1, 2.01, and 2.001. Then, using a calculator, we tabulate the necessary values. Since P is the point (2, 10), $x_1 = 2$ and $y_1 = 10$. Thus, using the values of x_2, we tabulate the values of y_2, $y_2 - 10$, $x_2 - 2$ and thereby the values of the slope m.

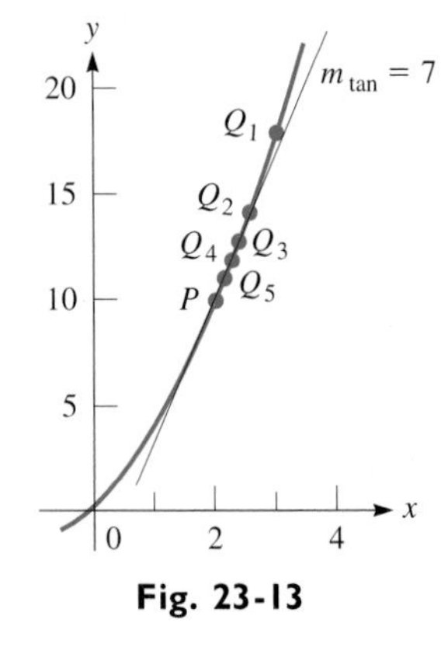

Fig. 23-13

Point	Q_1	Q_2	Q_3	Q_4	Q_5	P
x_2	3.0	2.5	2.1	2.01	2.001	2
y_2	18.0	13.75	10.71	10.0701	10.007 001	10
$y_2 - 10$	8.0	3.75	0.71	0.0701	0.007 001	
$x_2 - 2$	1.0	0.5	0.1	0.01	0.001	
$m = \frac{y_2 - 10}{x_2 - 2}$	8.0	7.5	7.1	7.01	7.001	

We see that the slope of PQ approaches the value of 7 as Q approaches P. Therefore, the slope of the tangent line at (2, 10) is 7. See Fig. 23-13. ◀◀

With the proper notation, it is possible to express the coordinates of Q in terms of the coordinates of P. If we define the quantities Δx ("delta" x) and Δy ("delta" y) by the equations

$$\Delta x = x_2 - x_1 \quad (23\text{-}2)$$

$$\Delta y = y_2 - y_1 \quad (23\text{-}3)$$

the coordinates of $Q(x_2, y_2)$ become $(x_1 + \Delta x, y_1 + \Delta y)$. As we see, the quantities Δx and Δy represent the difference in the coordinates of P and Q. **CAUTION ▶** ***The quantity Δx is not to be thought of as "Δ times x,"*** for the symbol Δ used here has no meaning by itself. *The name* **increment** *is given to the difference of the coordinates of two points, and therefore Δx and Δy are the increments in x and y, respectively.*

Using Eqs. (23-2) and (23-3), along with the definition of slope, we can express the slope of PQ as

$$m_{PQ} = \frac{(y_1 + \Delta y) - y_1}{(x_1 + \Delta x) - x_1} = \frac{\Delta y}{\Delta x} \qquad (23\text{-}4)$$

By previous discussion, as Q approaches P, the slope of the tangent line is more *nearly* approximated by $\Delta y/\Delta x$.

EXAMPLE 2 ▸▸ Find the slope of a line tangent to the curve $y = x^2 + 3x$ at the point (2, 10) by the increment method indicated in Eq. (23-4). (This is the same slope as calculated in Example 1.)

As in Example 1, point P has the coordinates (2, 10). Thus the coordinates of any other point Q can be expressed as $(2 + \Delta x,\ 10 + \Delta y)$. See Fig. 23-14. The slope of PQ then becomes

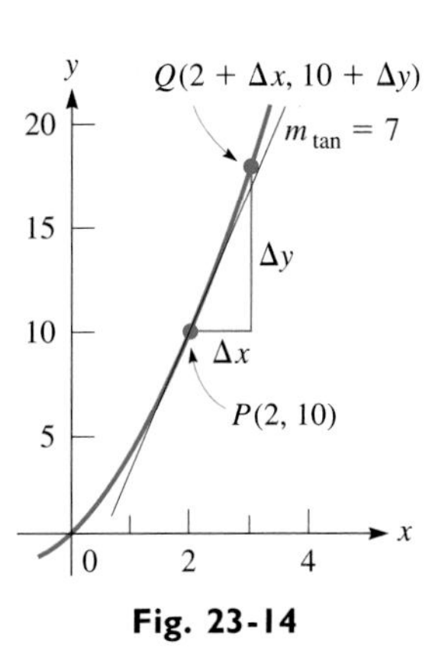

Fig. 23-14

$$m_{PQ} = \frac{(10 + \Delta y) - 10}{(2 + \Delta x) - 2} = \frac{\Delta y}{\Delta x}$$

This expression itself does not enable us to find the slope, since values of Δx and Δy are not known. If, however, we can express Δy in terms of Δx, we might derive more information. Both P and Q are on the curve of the function, which means that the coordinates of each must satisfy the function. Using the coordinates of Q, we have

$$\begin{aligned}(10 + \Delta y) &= (2 + \Delta x)^2 + 3(2 + \Delta x)\\ &= 4 + 4\Delta x + (\Delta x)^2 + 6 + 3\Delta x\end{aligned}$$

Subtracting 10 from each side, we have

$$\Delta y = 7\Delta x + (\Delta x)^2$$

As Q approaches P, Δx becomes smaller and smaller. Calculating Δy and the ratio $\Delta y/\Delta x$ for increasingly small Δx, we have the following values.

Δx	0.1	0.01	0.001
Δy	0.71	0.0701	0.007 001
$\frac{\Delta y}{\Delta x}$	7.1	7.01	7.001

$\Delta y = 7\Delta x + (\Delta x)^2$

We note that the values of $\Delta y/\Delta x$ are the same as those for the slope for points Q_3, Q_4, and Q_5 in Example 1. We also see that as $\Delta x \to 0$, $\Delta y/\Delta x \to 7$.

Now, substituting the expression for Δy into the expression for m_{PQ}, we have

$$m_{PQ} = \frac{7\Delta x + (\Delta x)^2}{\Delta x} = 7 + \Delta x$$

From this expression we can see directly that as $\Delta x \to 0$, $m_{PQ} \to 7$. Using this fact, we can see that the slope of the tangent line is

$$m_{\tan} = \lim_{\Delta x \to 0} m_{PQ} = 7$$

We see that this result agrees with that found in Example 1. ◂◂

EXAMPLE 3 ▸▸ Find the slope of a line tangent to the curve $y = 4x - x^2$ at the point (x_1, y_1).

The points P and Q are $P(x_1, y_1)$ and $Q(x_1 + \Delta x, y_1 + \Delta y)$. Using the coordinates of Q in the function, we obtain

$$\begin{aligned}(y_1 + \Delta y) &= 4(x_1 + \Delta x) - (x_1 + \Delta x)^2 \\ &= 4x_1 + 4\,\Delta x - x_1^2 - 2x_1\,\Delta x - (\Delta x)^2\end{aligned}$$

Using the coordinates of P, we obtain

$$y_1 = 4x_1 - x_1^2$$

Subtracting the second from the first to solve for Δy, we obtain

$$\begin{aligned}(y_1 + \Delta y) - y_1 &= [4x_1 + 4\,\Delta x - x_1^2 - 2x_1\,\Delta x - (\Delta x)^2] - (4x_1 - x_1^2) \\ \Delta y &= 4\,\Delta x - 2x_1\,\Delta x - (\Delta x)^2\end{aligned}$$

Dividing through by Δx to obtain an expression for $\Delta y/\Delta x$, we obtain

$$\frac{\Delta y}{\Delta x} = 4 - 2x_1 - \Delta x$$

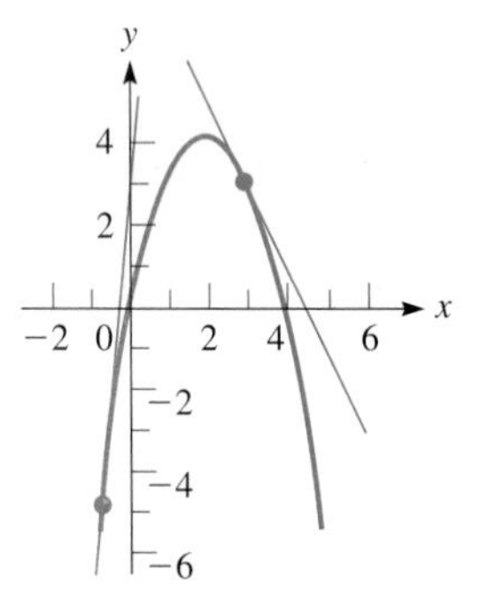

Fig. 23-15

In this last equation, the desired expression is on the left, but all we can determine from $\Delta y/\Delta x$ itself is that the ratio will become one very small number divided by another very small number as $\Delta x \to 0$. The right side, however, approaches $4 - 2x_1$ as $\Delta x \to 0$. This indicates that the slope of a tangent at the point (x_1, y_1) is given by

$$m_{\tan} = 4 - 2x_1$$

This method has an advantage over that used in Example 2. We now have a general expression for the slope of a tangent line for any value x_1. If $x_1 = -1$, $m_{\tan} = 6$, and if $x_1 = 3$, $m_{\tan} = -2$. The tangent lines for these values are shown in Fig. 23-15. ◂◂

EXAMPLE 4 ▸▸ Find the expression for the slope of a line tangent to the curve $y = x^3 + 2$ at the general point (x_1, y_1), and use this expression to find the slope when $x = 1/2$.

Using the coordinates of the points $P(x_1, y_1)$ and $Q(x_1 + \Delta x,\ y_1 + \Delta y)$ in the function, we have the following steps:

$$\begin{aligned}y_1 + \Delta y &= (x_1 + \Delta x)^3 + 2 && \text{substitute coordinates of } Q \\ &= x_1^3 + 3x_1^2\,\Delta x + 3x_1(\Delta x)^2 + (\Delta x)^3 + 2 \\ y_1 &= x_1^3 + 2 && \text{substitute coordinates of } P \\ \Delta y &= 3x_1^2\,\Delta x + 3x_1(\Delta x)^2 + (\Delta x)^3 && \text{subtract } y_1 \text{ from } y_1 + \Delta y \\ \frac{\Delta y}{\Delta x} &= 3x_1^2 + 3x_1\,\Delta x + (\Delta x)^2 && \text{divide by } \Delta x\end{aligned}$$

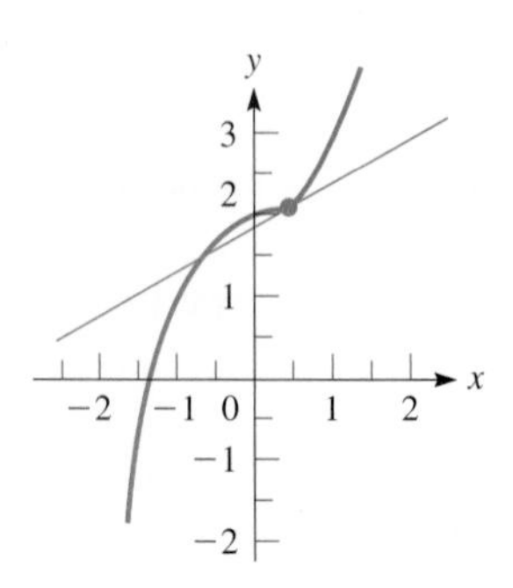

Fig. 23-16

As $\Delta x \to 0$, the right side approaches the value $3x_1^2$. This means that

$$m_{\tan} = 3x_1^2$$

When $x_1 = \frac{1}{2}$, we find that the slope of the tangent is $3(\frac{1}{4}) = \frac{3}{4}$. The curve and this tangent line are indicated in Fig. 23-16. ◂◂

In interpreting this analysis as the rate of change of a function, we see that if $\Delta x = 1$ and the corresponding value of Δy is found, it can be said that y changes by the amount Δy as x changes one unit. If x changed by some lesser amount, we could still calculate a ratio for the amount of change in y for the given change in x. Therefore, as long as x changes at all, there will be a corresponding change in y. In this way, *the ratio* $\Delta y/\Delta x$ *is the* **average rate of change** *of* y*, which is a function of* x*, with respect to* x*. As* $\Delta x \to 0$ *the limit of the ratio* $\Delta y/\Delta x$ *gives us the* **instantaneous rate of change** *of* y *with respect to* x.

EXAMPLE 5 ▸▸ In Example 1 let us consider points $P(2, 10)$ and $Q_2(2.5, 13.75)$. From P to Q_2, x changes by 0.5 unit and y changes by 3.75 units. This means that *the average change in* y *for a* 1*-unit change in* x is $3.75/0.5 = 7.5$ units. However, this is not the rate at which y is changing with respect to x *at* most points within this interval.

At point P, the slope of the tangent line of 7 tells us that y is changing 7 units for a one-unit change in x. However, ***this is an instantaneous rate of change* at *point P*** and tells us the rate at which y is changing with respect to x ***at P and only at P.***

NOTE ▶

Exercises 23–2

In Exercises 1–4, use the method of Example 1 to calculate the slope of a line tangent to the curve of each of the given functions. Let Q_1*,* Q_2*,* Q_3*, and* Q_4 *have the indicated x-values. Sketch the curve and tangent lines.*

1. $y = x^2$; P is $(2, 4)$; let Q have x-values of 1.5, 1.9, 1.99, 1.999.

2. $y = 1 - \frac{1}{2}x^2$; P is $(2, -1)$; let Q have x-values of 1.5, 1.9, 1.99, 1.999.

3. $y = x^2 + 2x$; P is $(-3, 3)$; let Q have x-values of -2.5, -2.9, -2.99, -2.999.

4. $y = x^3 + 1$; P is $(-1, 0)$; let Q have x-values of -0.5, -0.9, -0.99, -0.999.

In Exercises 5–8, use the method of Example 2 (divide the expression for Δy *by* Δx *and use the simplified expression for* m_{PQ}*) to calculate the slope of a line tangent to the curve of each of the given functions for the given points P. (These are the same functions and points as for Exercises 1–4.)*

5. $y = x^2$; P is $(2, 4)$.

6. $y = 1 - \frac{1}{2}x^2$; P is $(2, -1)$.

7. $y = x^2 + 2x$; P is $(-3, 3)$.

8. $y = x^3 + 1$; P is $(-1, 0)$.

In Exercises 9–24, use the method of Example 3 to find a general expression for the slope of a tangent line to each of the indicated curves. Then find the slopes for the given values of x. Sketch the curves and tangent lines.

9. $y = x^2$; $x = 2$, $x = -1$

10. $y = 1 - \frac{1}{2}x^2$; $x = 2$, $x = -2$

11. $y = x^2 + 2x$; $x = -3$, $x = 1$

12. $y = 4 - 3x^2$; $x = 0$, $x = 2$

13. $y = x^2 + 4x + 5$; $x = -3$, $x = 2$

14. $y = 2x^2 - 4x$; $x = 1$, $x = 1.5$

15. $y = 6x - x^2$; $x = -2$, $x = 3$

16. $y = 2x - 3x^2$; $x = 0$, $x = 0.5$

17. $y = x^3 - 2x$; $x = -1$, $x = 0$, $x = 1$

18. $y = 3x - x^3$; $x = -2$, $x = 0$, $x = 2$

19. $y = x^4$; $x = 0$, $x = 0.5$, $x = 1$

20. $y = 1 - x^4$; $x = 0$, $x = 1$, $x = 2$

21. $y = x^5$; $x = 0$, $x = 0.5$, $x = 1$

22. $y = x^6$; $x = 0$, $x = 0.5$, $x = 1$

23. $y = \dfrac{1}{x}$; $x = 0.5$, $x = 1$, $x = 2$

24. $y = \dfrac{1}{x + 1}$; $x = -0.5$, $x = 0$, $x = 1$

In Exercises 25–28, find the average rate of change of y with respect to x from P to Q. Then compare this with the instantaneous rate of change of y with respect to x at P by finding $m_{\tan}$ *at P.*

25. $y = x^2 + 2$, $P(2, 6)$, $Q(2.1, 6.41)$

26. $y = 1 - 2x^2$, $P(1, -1)$, $Q(1.1, -1.42)$

27. $y = 9 - x^3$, $P(2, 1)$, $Q(2.1, -0.261)$

28. $y = x^3 - 6x$, $P(3, 9)$, $Q(3.1, 11.191)$

23-3 The Derivative

We are now ready to establish one of the fundamental definitions of calculus. Generalizing on the method of Example 2 of the preceding section, if the point $P(x, y)$ is held constant while the point $Q(x + \Delta x, y + \Delta y)$ approaches it, then both Δx and Δy approach zero. Both points P and Q lie on the curve of the function $f(x)$, which means the coordinates of each point must satisfy the equation. For P this means $y = f(x)$, and for Q this means $y + \Delta y = f(x + \Delta x)$. Subtracting the first expression from the second, we have

$$(y + \Delta y) - y = f(x + \Delta x) - f(x)$$

$$\Delta y = f(x + \Delta x) - f(x) \tag{23-5}$$

In the previous section, we saw that the slope of a line tangent to $f(x)$ at $P(x, y)$ was found as the limiting value of the ratio $\Delta y/\Delta x$ as Δx approaches zero. Formally, *this limiting value of the ratio of $\Delta y/\Delta x$ is known as the* **derivative** *of the function.* Therefore, the derivative of a function $f(x)$ is defined as

$$\lim_{\Delta x \to 0} \frac{\Delta y}{\Delta x} = \lim_{\Delta x \to 0} \frac{f(x + \Delta x) - f(x)}{\Delta x} \tag{23-6}$$

The process of finding the derivative of a function from its definition is called the *delta-process. The general process of finding a derivative is called* **differentiation.**

EXAMPLE 1 ▸▸ Find the derivative of $y = 2x^2 + 3x$ by the delta-process.

To find Δy, we must first derive the quantity $f(x + \Delta x) - f(x)$. Thus, we first find $y + \Delta y = f(x + \Delta x)$ by replacing y by $y + \Delta y$ and x by $x + \Delta x$.

$$y + \Delta y = 2(x + \Delta x)^2 + 3(x + \Delta x)$$

Then we subtract the original function:

$$y + \Delta y - y = 2(x + \Delta x)^2 + 3(x + \Delta x) - (2x^2 + 3x)$$

Simplifying, we find that the result is

$$\Delta y = 4x\,\Delta x + 2(\Delta x)^2 + 3\,\Delta x$$

Next, dividing through by Δx, we obtain

$$\frac{\Delta y}{\Delta x} = 4x + 2\,\Delta x + 3$$

As Δx approaches zero, the $2\,\Delta x$-term on the right approaches zero. Therefore,

$$\lim_{\Delta x \to 0} \frac{\Delta y}{\Delta x} = 4x + 3$$

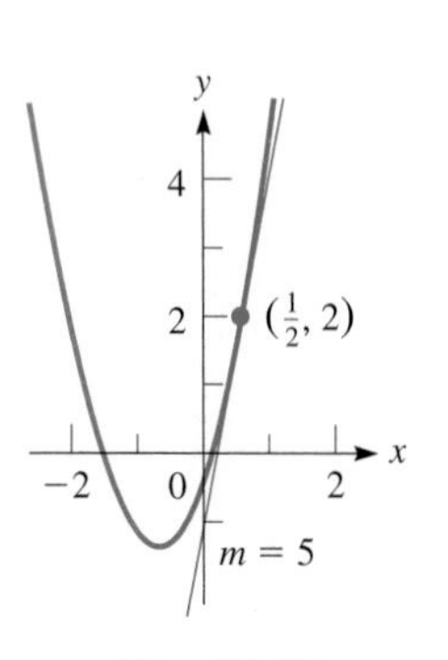

Fig. 23-17

We see that the derivative of the function $2x^2 + 3x$ is the function $4x + 3$. From the definition of the derivative and from the previous section, this means we can find the slope of a tangent line for any point on the curve of $y = 2x^2 + 3x$ by substituting the x-coordinate into the expression $4x + 3$. For example, the slope of a tangent line is 5 if $x = 1/2$ (at the point $(1/2, 2)$). See Fig. 23-17. ◂◂

EXAMPLE 2 ▸▸ Find the derivative of the function $y = 6x - 2x^3$ by using the delta-process.

By first replacing y by $y + \Delta y$ and x by $x + \Delta x$, we then have the following steps.

$$\begin{aligned}
y + \Delta y &= 6(x + \Delta x) - 2(x + \Delta x)^3 && y + \Delta y = f(x + \Delta x) \\
&= 6x + 6(\Delta x) - 2[x^3 + 3x^2(\Delta x) + 3x(\Delta x)^2 + (\Delta x)^3] \\
&= 6x + 6(\Delta x) - 2x^3 - 6x^2(\Delta x) - 6x(\Delta x)^2 - 2(\Delta x)^3 \\
y + \Delta y - y &= [6x + 6(\Delta x) - 2x^3 - 6x^2(\Delta x) - 6x(\Delta x)^2 - 2(\Delta x)^3] - (6x - 2x^3) && \text{subtract } y = 6x - 2x^3 \\
\Delta y &= 6(\Delta x) - 6x^2(\Delta x) - 6x(\Delta x)^2 - 2(\Delta x)^3 \\
\frac{\Delta y}{\Delta x} &= \frac{6(\Delta x) - 6x^2(\Delta x) - 6x(\Delta x)^2 - 2(\Delta x)^3}{\Delta x} && \text{divide by } \Delta x \\
&= 6 - 6x^2 - 6x(\Delta x) - 2(\Delta x)^2 \\
\lim_{\Delta x \to 0} \frac{\Delta y}{\Delta x} &= \lim_{\Delta x \to 0} [6 - 6x^2 - 6x(\Delta x) - 2(\Delta x)^2] && \text{find limit as } \Delta x \to 0 \\
&= 6 - 6x^2
\end{aligned}$$

Therefore, the derivative of $y = 6x - 2x^3$ is $6 - 6x^2$. ◂◂

The derivative of a function is itself a function, and it is possible that it may not be defined for all values of x. *If the value x_0 is in the domain of the derivative, then the function is said to be* ***differentiable*** *at x_0*. The examples that follow illustrate functions which are not differentiable for all values of x.

EXAMPLE 3 ▸▸ Find the derivative of $y = \dfrac{1}{x}$ by the delta-process.

CAUTION ▸

$$\begin{aligned}
y + \Delta y &= \frac{1}{x + \Delta x} && y + \Delta y = f(x + \Delta x) \\
y + \Delta y - y &= \frac{1}{x + \Delta x} - \frac{1}{x} && \text{subtract } y = \frac{1}{x} \\
\Delta y &= \frac{x - (x + \Delta x)}{x(x + \Delta x)} = \frac{-\Delta x}{x(x + \Delta x)} && \text{combine fractions} \\
\frac{\Delta y}{\Delta x} &= \frac{-1}{x(x + \Delta x)} && \text{divide by } \Delta x \\
\lim_{\Delta x \to 0} \frac{\Delta y}{\Delta x} &= \lim_{\Delta x \to 0} \frac{-1}{x(x + \Delta x)} && \text{find limit as } \Delta x \to 0 \\
&= \frac{-1}{x^2}
\end{aligned}$$

We note that neither the function nor the derivative is defined for $x = 0$. This means the function is not differentiable at $x = 0$. ◂◂

EXAMPLE 4 ▸▸ Find the derivative of $y = x^2 + \frac{1}{x+1}$ by the delta-process.

$$y + \Delta y = (x + \Delta x)^2 + \frac{1}{x + \Delta x + 1} \qquad y + \Delta y = f(x + \Delta x)$$

$$y + \Delta y - y = x^2 + 2x\Delta x + (\Delta x)^2 + \frac{1}{x + \Delta x + 1} - x^2 - \frac{1}{x+1} \qquad \text{subtract } y = x^2 + \frac{1}{x+1}$$

$$\Delta y = 2x\Delta x + (\Delta x)^2 + \frac{1}{x + \Delta x + 1} - \frac{1}{x+1}$$

algebra handled most easily if fractions combined separately from other terms

$$= 2x\Delta x + (\Delta x)^2 + \frac{(x+1) - (x + \Delta x + 1)}{(x + \Delta x + 1)(x+1)}$$

$$= 2x\Delta x + (\Delta x)^2 - \frac{\Delta x}{(x + \Delta x + 1)(x+1)}$$

$$\frac{\Delta y}{\Delta x} = 2x + \Delta x - \frac{1}{(x + \Delta x + 1)(x+1)} \qquad \text{divide by } \Delta x$$

$$\lim_{\Delta x \to 0} \frac{\Delta y}{\Delta x} = 2x - \frac{1}{(x+1)^2} \qquad \text{find limit as } \Delta x \to 0$$

We note that this function is not differentiable at $x = -1$. ◂◂

The notation of the definition of the derivative in terms of the limit is somewhat awkward for general use. Therefore, a number of other notations are commonly used. They include

$$y', \qquad D_x y, \qquad f'(x), \qquad \frac{dy}{dx}$$

The following example illustrates the use of some of these basic notations for the derivative.

EXAMPLE 5 ▸▸ In Example 2, $y = 6x - 2x^3$, and we found that the derivative is $6 - 6x^2$. Therefore, we may write

$$y' = 6 - 6x^2 \qquad \text{or} \qquad \frac{dy}{dx} = 6 - 6x^2$$

If we had written $f(x) = 6x - 2x^3$, we would then write $f'(x) = 6 - 6x^2$.

If we wish to find the value of the derivative at some point, such as $(-2, 4)$, we write

$$\frac{dy}{dx} = 6 - 6x^2$$

$$\left.\frac{dy}{dx}\right|_{x=-2} = 6 - 6(-2)^2 = 6 - 24$$

$$= -18$$

We note that only the x-coordinate of the point was needed in the evaluation of the derivative. ◂◂

EXAMPLE 6 ▸▸ Find dy/dx of the function $y = \sqrt{x}$ by the delta-process.

We first square both sides of the equation, thus obtaining $y^2 = x$. (This is valid only for $y \geq 0$, since the original function $y = \sqrt{x}$ is not defined for $y < 0$.) Replacing y by $y + \Delta y$ and x by $x + \Delta x$, we have

$$(y + \Delta y)^2 = x + \Delta x$$

We complete the solution as follows:

$$y^2 + 2y\Delta y + (\Delta y)^2 - y^2 = x + \Delta x - x \qquad \text{expand left side and subtract } y^2 = x$$

$$2y\Delta y + (\Delta y)^2 = \Delta x$$

$$2y\frac{\Delta y}{\Delta x} + \Delta y\frac{\Delta y}{\Delta x} = 1 \qquad \text{divide by } \Delta x$$

$$\frac{\Delta y}{\Delta x} = \frac{1}{2y + \Delta y} \qquad \text{solve for } \frac{\Delta y}{\Delta x}$$

$$\lim_{\Delta x \to 0} \frac{\Delta y}{\Delta x} = \lim_{\Delta x \to 0} \frac{1}{2y + \Delta y} = \frac{1}{2y} \qquad \text{find limit as } \Delta x \to 0$$

The Δy is omitted, since $\Delta y \to 0$ as $\Delta x \to 0$. Now, substituting in the original function $y = \sqrt{x}$, we obtain the final result:

$$\frac{dy}{dx} = \frac{1}{2\sqrt{x}}$$

The domain of the function is $x \geq 0$. However, since x appears in the denominator of the derivative, the domain of the derivative is $x > 0$. This means that the function is differentiable for $x > 0$. ◂◂

One might ask why, when we are finding a derivative, we take a limit as Δx approaches zero and do not simply let Δx equal zero. If we did this, the ratio $\Delta y/\Delta x$ would be exactly 0/0, which would then require division by zero. As we know, this is an undefined operation in mathematics, and therefore **Δx *cannot* equal *zero.*** However, it can equal any value as near zero as necessary. This idea is basic in the meaning of the word *limit.*

CAUTION ▶

Exercises 23–3

In Exercises 1–24, find the derivative of each of the functions by using the delta-process.

1. $y = 3x - 1$
2. $y = 6x + 3$
3. $y = 1 - 2x$
4. $y = 2 - 5x$
5. $y = x^2 - 1$
6. $y = 4 - x^2$
7. $y = 5x^2$
8. $y = -6x^2$
9. $y = x^2 - 7x$
10. $y = x^2 + 4x$

10. $y = 8x - 2x^2$

12. $y = 3x - \frac{1}{2}x^2$
13. $y = x^3 + 4x - 6$
14. $y = 2x - 4x^3$
15. $y = \dfrac{1}{x + 2}$
16. $y = \dfrac{1}{x + 1}$
17. $y = x + \dfrac{1}{x}$
18. $y = \dfrac{x}{x - 1}$
19. $y = \dfrac{2}{x^2}$
20. $y = \dfrac{2}{x^2 + 4}$
21. $y = x^4 + x^3 + x^2 + x$
22. $y = \frac{1}{3}x^3 + \frac{1}{2}x^2 + x$
23. $y = x^4 - \dfrac{2}{x}$
24. $y = \dfrac{1}{x} + \dfrac{1}{x^2}$

In Exercises 25–28, find the derivative of each function by using the delta-process. Then evaluate the derivative at the given point.

25. $y = 3x^2 - 2x, (-1, 5)$ **26.** $y = 9x - x^3, (2, 10)$

27. $y = \dfrac{6}{x + 3}, (3, 1)$ **28.** $y = x^2 - \dfrac{2}{x}, (-2, 5)$

In Exercises 29–32, find the derivative of each function by using the delta-process. Then determine the values for which the function is differentiable.

29. $y = 1 + \dfrac{2}{x}$ **30.** $y = \dfrac{2}{x - 4}$

31. $y = \dfrac{2}{x^2 - 1}$ **32.** $y = \dfrac{2}{x^2 + 1}$

In Exercises 33–36, solve the given problems.

33. Find dy/dx for $y = \sqrt{x + 1}$ by the method of Example 6.

34. Find dy/dx for $y = \sqrt{x^2 + 3}$ by the method of Example 6.

35. Find dy/dx for $y = \sqrt{x}$ by using the delta-process and the function directly. Do not square both sides. (*Hint:* In the expression for Δy, multiply and divide by $\sqrt{x} + \sqrt{x + \Delta x}$. This is rationalizing the numerator.)

36. Find dy/dx for $y = \sqrt{x - 2}$ by the method outlined in Exercise 35. For what values of x is the function differentiable? Explain.

23-4 The Derivative as an Instantaneous Rate of Change

In Section 23-2 we saw that the slope of a line tangent to a curve at point P was the limiting value of the slope of the line through points P and Q as Q approaches P. In Section 23-3 we defined the limit of the ratio $\Delta y/\Delta x$ as $\Delta x \to 0$ as the derivative. Therefore, *the first meaning we have given to the derivative is the slope of a line tangent to a curve,* as we noted in Example 1 of Section 23-3. The following example further illustrates this meaning of the derivative.

EXAMPLE 1 ▸▸ Find the slope of a line tangent to the curve of $y = 4x - x^2$ at the point $(1, 3)$.

We first find the derivative and then evaluate it at the given point.

$$y + \Delta y = 4(x + \Delta x) - (x + \Delta x)^2$$

$$y + \Delta y - y = 4x + 4\Delta x - x^2 - 2x\Delta x - (\Delta x)^2 - (4x - x^2)$$

$$\Delta y = 4\Delta x - 2x\Delta x - (\Delta x)^2$$

$$\frac{\Delta y}{\Delta x} = 4 - 2x - \Delta x$$

$$\lim_{\Delta x \to 0} \frac{\Delta y}{\Delta x} = 4 - 2x \quad \text{derivative}$$

$$\left.\frac{dy}{dx}\right|_{(1,3)} = 4 - 2(1) = 2 \quad \text{evaluate derivative}$$

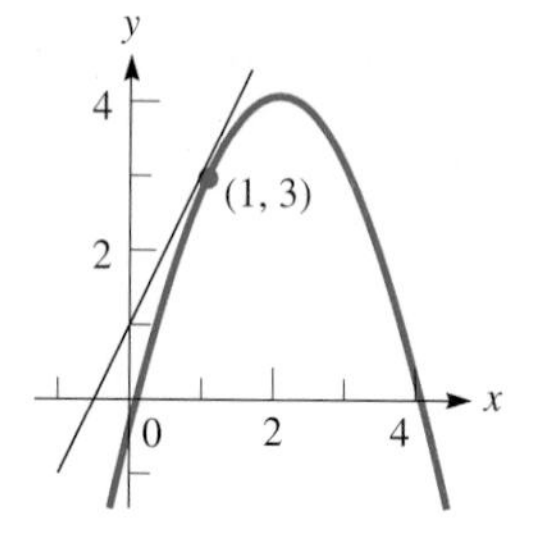

Fig. 23-18

The slope of the tangent line at $(1, 3)$ is 2. Note that only $x = 1$ was needed for the evaluation. The curve and tangent line are shown in Fig. 23-18. ◂◂

At the end of Section 23-2, we discussed the idea that $\Delta y/\Delta x$ indicates the rate of change of y with respect to x. In defining the derivative as the limit of the ratio of $\Delta y/\Delta x$ as $\Delta x \to 0$, it is a measure of the rate of change of y with respect to x at point P. However, P may represent any point, which means that the value of the derivative changes from one point on a curve to another point.

We therefore interpret the derivative as the ***instantaneous rate of change*** *of y with respect to x.*

EXAMPLE 2 ▸▸ In Examples 1 and 2 of Section 23-2, y is changing at the rate of 7 units for every 1-unit change in x, *when x is* 2, *and only when x is* 2. Then, in Example 3 of Section 23-2, y is increasing 6 units for every increase of 1 unit of x *when* $x = -1$. When $x = 3$, y is decreasing 2 units for a 1-unit increase in x. ◂◂

This gives us a more general meaning of the derivative. If a functional relationship exists between any two variables, then one can be taken to be varying with respect to the other and the derivative gives us the instantaneous rate of change. There are many applications of this principle, one of which is the velocity of an object. We consider here the case of *rectilinear motion,* that is, motion along a straight line.

As we have seen, the velocity of an object is found by dividing the change in displacement by the time required for this change. This, however, gives a value only for the **average velocity** for the specified time interval. If the time interval considered becomes smaller and smaller, then the average velocity that is calculated more nearly approximates the **instantaneous velocity** at some particular time. In the limit, the value of the average velocity gives the value of the instantaneous velocity. Using the symbols as defined for the derivative, *the instantaneous velocity of an object moving in rectilinear motion at a particular time t is given by*

Instantaneous Velocity

$$v = \lim_{\Delta t \to 0} \frac{\Delta s}{\Delta t} \quad (23\text{-}7)$$

where s is the displacement.

In Eq. (23-7) the derivative gives us the instantaneous rate of change of s with respect to t. Therefore, the units of this derivative would be in units of displacement divided by units of time. In general, *units of the derivative of* $y = f(x)$ *are in units of y divided by units of x.*

See the chapter introduction.

EXAMPLE 3 ▸▸ Find the instantaneous velocity, when $t = 4$ s (exactly), of a falling object for which the displacement s (in cm), which is the distance fallen, is given by $s = 490t^2$ by calculating values of $\Delta s/\Delta t$ and finding the limit as Δt approaches zero.

Here we shall let t take on values of 3.5, 3.9, 3.99, and 3.999 s. When $t = 4$ s, $s = 7840$ cm. Therefore, we calculate Δt by subtracting values of t from 4, and Δs is calculated by subtracting values from 7840. The values of velocity are then calculated by using $v = \Delta s/\Delta t$.

	t (s)	3.5	3.9	3.99	3.999
	s (cm)	6002.5	7452.9	7800.849	7836.0805
$\mathbf{7840 - s}$	Δs (cm)	1837.5	387.1	39.15	3.9195
$\mathbf{4 - t}$	Δt (s)	0.5	0.1	0.01	0.001
$\mathbf{\frac{\Delta s}{\Delta t}}$	v (cm/s)	3675.0	3871.0	3915.0	3919.5

We can see that the value of v is approaching 3920 cm/s, which is therefore the instantaneous velocity when $t = 4$ s. ◂◂

See the chapter introduction.

EXAMPLE 4 ▸▸ Find the expression for the instantaneous velocity of the object of Example 3, for which $s = 490t^2$, where s is the displacement (in cm) and t is the time (in seconds). Determine the instantaneous velocity for $t = 2$ s and $t = 4$ s.

The required expression is the derivative of s with respect to t.

$$s + \Delta s = 490(t + \Delta t)^2 = 490t^2 + 980t\Delta t + 490(\Delta t)^2$$

$$\Delta s = 980t\Delta t + 490(\Delta t)^2$$

$$\frac{\Delta s}{\Delta t} = 980t + 490\Delta t$$

$$v = \lim_{\Delta t \to 0} \frac{\Delta s}{\Delta t} = \frac{ds}{dt} = 980t \qquad \text{expression for instantaneous velocity}$$

$$\left.\frac{ds}{dt}\right|_{t=2} = 980(2) = 1960 \text{ cm/s} \quad \text{and} \quad \left.\frac{ds}{dt}\right|_{t=4} = 980(4) = 3920 \text{ cm/s}$$

We see that the second result agrees with that found in Example 3. ◂◂

By finding $\lim_{\Delta x \to 0} \Delta y/\Delta x$, we can find the instantaneous rate of change of y with respect to x. The expression $\lim_{\Delta t \to 0} \Delta s/\Delta t$ gives the velocity, or instantaneous rate of change of displacement with respect to time. Generalizing, we can say that

the derivative can be interpreted as the instantaneous rate of change of the dependent variable with respect to the independent variable.

This is true for a differentiable function, no matter what the variables represent.

EXAMPLE 5 ▸▸ A spherical balloon is being inflated. Find the expression for the instantaneous rate of change of the volume with respect to the radius. Evaluate this rate of change for a radius of 2.00 m.

$$V = \frac{4}{3}\pi r^3 \qquad \text{volume of sphere}$$

$$V + \Delta V = \frac{4\pi}{3}(r + \Delta r)^3 \qquad \text{find derivative}$$

$$= \frac{4\pi}{3}[r^3 + 3r^2\Delta r + 3r(\Delta r)^2 + (\Delta r)^3]$$

$$\Delta V = \frac{4\pi}{3}[3r^2\Delta r + 3r(\Delta r)^2 + (\Delta r)^3]$$

$$\frac{\Delta V}{\Delta r} = \frac{4\pi}{3}[3r^2 + 3r\Delta r + (\Delta r)^2]$$

$$\lim_{\Delta r \to 0} \frac{\Delta V}{\Delta r} = 4\pi r^2 \qquad \text{expression for instantaneous rate of change}$$

$$\left.\frac{dV}{dr}\right|_{r=2.00 \text{ m}} = 4\pi(2.00)^2 = 16.0\pi = 50.3 \text{ m}^2 \qquad \text{instantaneous rate of change when } r = 2.00 \text{ m}$$

The instantaneous rate of change of the volume with respect to the radius (dV/dr) for $r = 2.00$ m is 50.3 m^3/m (this way of showing the units is more meaningful).

As r increases, dV/dr also increases. This should be expected as the volume of a sphere varies directly as the cube of the radius. ◂◂

EXAMPLE 6 ▸▸ The power P produced by an electric current i in a resistor varies directly as the square of the current. Given that 1.2 W of power are produced by a current of 0.50 A in a certain resistor, find an expression for the instantaneous rate of change of power with respect to current. Evaluate this rate of change for $i = 2.5$ A.

We must first find the functional relationship between power and current, by solving the indicated problem in variation:

$$P = ki^2, \qquad 1.2 = k(0.50)^2, \qquad k = 4.8 \text{ W/A}^2, \qquad P = 4.8i^2$$

Now, knowing the function, we may determine the expression for the instantaneous rate of change of P with respect to i by using the delta-process:

$$P + \Delta P = 4.8(i + \Delta i)^2 = 4.8[i^2 + 2i(\Delta i) + (\Delta i)^2]$$

$$\Delta P = 4.8[2i(\Delta i) + (\Delta i)^2]$$

$$\frac{\Delta P}{\Delta i} = 4.8(2i + \Delta i)$$

$$\lim_{\Delta i \to 0} \frac{\Delta P}{\Delta i} = 9.6i \qquad \text{expression for instantaneous rate of change}$$

$$\left.\frac{dP}{di}\right|_{i=2.5\text{ A}} = 9.6(2.5) = 24 \text{ W/A} \qquad \text{instantaneous rate of change when } i = 2.5 \text{ A}$$

This tells us that when $i = 2.5$ A, the rate of change of power with respect to current is 24 W/A. Also, we see that the larger the current is, the greater is the increase in power. This should be expected, since the power varies directly as the square of the current. ◂◂

Exercises 23–4

In Exercises 1–4, find the slope of a line tangent to the curve of the given equation at the given point. Sketch the curve and the tangent line.

1. $y = x^2 - 1;\ (2, 3)$

2. $y = 2x - x^2;\ (-1, -3)$

3. $y = \dfrac{4}{x + 1};\ (-3, -2)$

4. $y = 3 - \dfrac{16}{x^2};\ (2, -1)$

In Exercises 5–8, calculate the instantaneous velocity for the indicated value of the time (in seconds) of an object for which the displacement (in metres) is given by the indicated function. Use the method of Example 3, and calculate values of $\Delta s/\Delta t$ for the given values of t and find the limit as Δt approaches zero.

5. $s = 4t + 10$; when $t = 3$; use values of t of 2.0, 2.5, 2.9, 2.99, 2.999

6. $s = 6 - 3t$; when $t = 4$; use values of t of 3.0, 3.5, 3.9, 3.99, 3.999

7. $s = 3t^2 - 4t$; when $t = 2$; use values of t of 1.0, 1.5, 1.9, 1.99, 1.999

8. $s = 40t - 4.9t^2$; when $t = 0.5$; use values of t of 0.4, 0.45, 0.49, 0.499, 0.4999

In Exercises 9–12, use the delta-process to find an expression for the instantaneous velocity of an object moving with rectilinear motion according to the given functions (the same as those for Exercises 5–8) relating s (in metres) and t (in seconds). Then calculate the instantaneous velocity for the given value of t.

9. $s = 4t + 10;\ t = 3$

10. $s = 6 - 3t;\ t = 4$

11. $s = 3t^2 - 4t;\ t = 2$

12. $s = 40t - 4.9t^2;\ t = 0.5$

In Exercises 13–16, use the delta-process to find an expression for the instantaneous velocity of an object moving with rectilinear motion according to the given functions relating s and t.

13. $s = 3t - \dfrac{2}{t}$

14. $s = \dfrac{2t}{t + 2}$

15. $s = 3t^2 - 2t^3$

16. $s = s_0 + v_0 t - \dfrac{1}{2}at^2$ (s_0, v_0, and a are constants.)

In Exercises 17–20, use the delta-process to find an expression for the instantaneous acceleration a of an object moving with rectilinear motion according to the given functions. The instantaneous acceleration of an object is defined as the instantaneous rate of change of velocity with respect to time.

17. $v = 6t^2 - 4t + 2$

18. $v = \sqrt{2t + 1}$

19. $s = t^3 + 2t$ (Find v, then find a.)

20. $s = s_0 + v_0 t - \frac{1}{2}at^2$ (s_0, v_0, and a are constants.) (Find v, then find a.)

In Exercises 21–32, find the indicated instantaneous rates of change.

21. The electric current i at a point in an electric circuit is the instantaneous rate of change of the electric charge q which passes the point, with respect to the time t. Find i in a circuit for which $q = 30 - 2t$.

22. A load L (in newtons) is distributed along a beam 10 m long such that $L = 5x - 0.5x^2$, where x is the distance from one end of the beam. Find the expression for the instantaneous rate of change of L with respect to x.

23. A rectangular metal plate contracts while cooling. Find the expression for the instantaneous rate of change of the area A of the plate with respect to its width w, if the length of the plate is constantly three times as long as the width.

24. A circular oil spill is increasing in size. Find the instantaneous rate of change of the area A of the spill with respect to its radius r for $r = 240$ m.

25. The total power P, in watts, transmitted by an AM radio station is given by $P = 500 + 250m^2$, where m is the modulation index. Find the instantaneous rate of change of P with respect to m for $m = 0.92$.

26. The number N, in thousands, of bacteria present in a certain culture is given by $N = t^3 - 20t^2 + 100t + 20$, where t is the time in hours. Find the instantaneous rate of change of N with respect to t for $t = 6.0$ h.

27. The total solar radiation H, in watts per square metre, on a particular surface during an average clear day is given by $H = \frac{5000}{t^2 + 10}$, where t $(-6 \leq t \leq 6)$ is the number of hours from noon (6 A.M. is equivalent to $t = -6$ h). Find the instantaneous rate of change of H with respect to t at 3 P.M.

28. The value, in thousands of dollars, of a certain car is given by the function $V = \frac{48}{t + 3}$, where t is measured in years. Find a general expression for the instantaneous rate of change of V with respect to t, then evaluate this expression when $t = 3$ years.

29. Oil in a certain machine is stored in a conical reservoir, for which the radius and height are both 4 cm. Find the instantaneous rate of change of the volume V of oil in the reservoir with respect to the depth d of the oil. See Fig. 23-19.

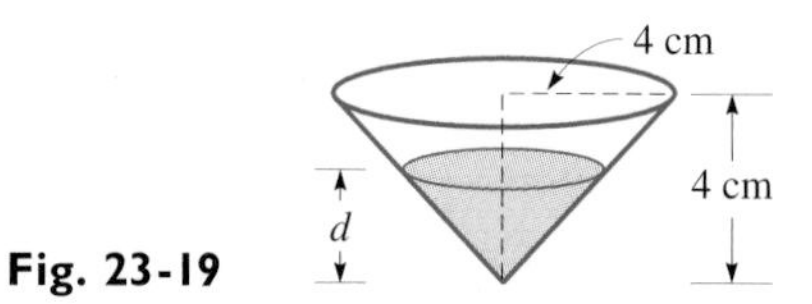

Fig. 23-19

30. The time t required to test a computer memory unit is directly proportional to the square of the number n of memory cells in the unit. For a particular type of unit, $n = 6400$ for $t = 25.0$ s. Find the instantaneous rate of change of t with respect to n for this type of unit for $n = 8000$.

31. A *holograph* (an image formed without using a lens) of concentric circles is formed. The radius r of each circle varies directly as the square root of the wavelength λ of the light used. If $r = 3.72$ cm for $\lambda = 592$ nm, find the expression for the instantaneous rate of change of r with respect to λ.

32. The force F between two electric charges varies inversely as the square of the distance r between them. For two charged particles, $F = 0.12$ N for $r = 0.060$ m. Find the instantaneous rate of change of F with respect to r for $r = 0.120$ m.

23-5 Derivatives of Polynomials

The task of finding the derivative of a function can be considerably shortened from that involved in the direct use of the delta-process. We can use the delta-process to derive certain basic formulas for finding derivatives of particular types of functions. These formulas will then be used to find the derivatives. In this section we derive the formulas for finding the derivatives of polynomial functions of the form $f(x) = a_0x^n + a_1x^{n-1} + \cdots + a_n$.

First we find the derivative of a constant. By letting $y = c$ and applying the delta-process to this function, we obtain the desired result:

$$y = c, \qquad y + \Delta y = c, \qquad \Delta y = 0, \qquad \frac{\Delta y}{\Delta x} = 0, \qquad \lim_{\Delta x \to 0} \frac{\Delta y}{\Delta x} = 0$$

From this we conclude that *the derivative of a constant is zero.* This result holds for all constants. Therefore, if $y = c$, $dy/dx = 0$, or

Derivative of a Constant

$$\frac{dc}{dx} = 0 \tag{23-8}$$

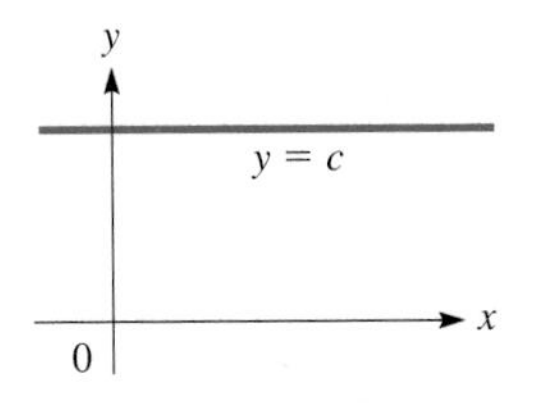

Fig. 23-20

Graphically this means that for any function of the type $y = c$ the slope is always zero. We know that $y = c$ represents a straight line parallel to the x-axis. From the definition of slope, we know that any line parallel to the x-axis has a slope of zero. See Fig. 23-20. We see that the two results are consistent.

Next we find the derivative of any integral power of x. If $y = x^n$, where n is an integer, by using the binomial theorem we have

$$\begin{aligned} (y + \Delta y) &= (x + \Delta x)^n \\ &= x^n + nx^{n-1}\Delta x + \frac{n(n-1)}{2}x^{n-2}(\Delta x)^2 + \cdots + (\Delta x)^n \end{aligned}$$

$$\Delta y = nx^{n-1}\Delta x + \frac{n(n-1)}{2}x^{n-2}(\Delta x)^2 + \cdots + (\Delta x)^n$$

$$\frac{\Delta y}{\Delta x} = nx^{n-1} + \underbrace{\frac{n(n-1)}{2}x^{n-2}\Delta x + \cdots + (\Delta x)^{n-1}}_{\text{each term} \to 0 \text{ as } \Delta x \to 0}$$

$$\lim_{\Delta x \to 0} \frac{\Delta y}{\Delta x} = nx^{n-1}$$

Thus, *the derivative of the nth power of x is*

Derivative of Power of x

$$\frac{dx^n}{dx} = nx^{n-1} \tag{23-9}$$

EXAMPLE 1 ▸▸ Find the derivative of the function $y = -5$.

Since -5 is a constant, applying Eq. (23-8), we have

$$\frac{dy}{dx} = \frac{d(-5)}{dx} = 0$$ ◂◂

EXAMPLE 2 ▸▸ Find the derivative of $y = x^3$.

Using Eq. (23-9), we have

$$\frac{dy}{dx} = \frac{d(x^3)}{dx} = 3x^{3-1} = 3x^2$$

This result is consistent with those found previously in this chapter. ◂◂

EXAMPLE 3 ▸▸ Find the derivative of the function $y = x$.

In using Eq. (23-9), we have $n = 1$ since $x = x^1$. This means

$$\frac{dy}{dx} = \frac{d(x)}{dx} = (1)x^{1-1} = (1)(x^0)$$

Since $x^0 = 1$, we have

$$\frac{dy}{dx} = 1$$

Thus, the derivative of $y = x$ is 1, which means that the slope of the line $y = x$ is always 1. This is consistent with our previous discussion of the slope of a straight line. ◂◂

EXAMPLE 4 ▸▸ Find the derivative of $y = x^{10}$.

We find that

$$\frac{dy}{dx} = \frac{d(x^{10})}{dx} = 10x^{10-1}$$
$$= 10x^9$$ ◂◂

Next we shall find the derivative of a constant times a function of x. If $y = cu$, where $u = f(x)$, we have the following result:

$$y + \Delta y = c(u + \Delta u)$$

(As x increases by Δx, u increases by Δu, since u is a function of x.) Then

$$\Delta y = c\,\Delta u, \qquad \frac{\Delta y}{\Delta x} = c\frac{\Delta u}{\Delta x}, \qquad \lim_{\Delta x \to 0} \frac{\Delta y}{\Delta x} = c \lim_{\Delta x \to 0} \frac{\Delta u}{\Delta x}$$

Therefore, *the derivative of the product of a constant and a differentiable function of x is the product of the constant and the derivative of the function of x.* This is written as

Derivative of a Constant Times a Function

$$\frac{d(cu)}{dx} = c\frac{du}{dx} \tag{23-10}$$

EXAMPLE 5 ▸▸ Find the derivative of $y = 3x^2$.

In this case, $c = 3$ and $u = x^2$. Thus, $du/dx = 2x$. Therefore,

$$\frac{dy}{dx} = \frac{d(3x^2)}{dx} = 3\frac{d(x^2)}{dx} = 3(2x)$$
$$= 6x$$ ◂◂

Occasionally the derivative of a constant times a function of x is confused with the derivative of a constant that stands alone. It is necessary to clearly distinguish between a constant that multiplies a function and an isolated constant.

Finally, if the types of functions for which we have found derivatives are added, the result is a polynomial function with more than one term. The derivative of such a function is found by letting $y = u + v$, where u and v are functions of x. Applying the delta-process, since u and v are functions of x, each has an increment corresponding to an increment in x. Thus, we have the following result:

$$y + \Delta y = (u + \Delta u) + (v + \Delta v)$$
$$\Delta y = \Delta u + \Delta v$$
$$\frac{\Delta y}{\Delta x} = \frac{\Delta u}{\Delta x} + \frac{\Delta v}{\Delta x}$$
$$\lim_{\Delta x \to 0} \frac{\Delta y}{\Delta x} = \lim_{\Delta x \to 0} \frac{\Delta u}{\Delta x} + \lim_{\Delta x \to 0} \frac{\Delta v}{\Delta x}$$

This tells us that *the derivative of the sum of differentiable functions of x is the sum of the derivatives of the functions.* This is written as

Derivative of a Sum

$$\frac{d(u + v)}{dx} = \frac{du}{dx} + \frac{dv}{dx} \qquad (23\text{-}11)$$

EXAMPLE 6 ▶▶ Find the derivative of $y = 4x^2 + 5$.

Here $u = 4x^2$ and $v = 5$. Thus, $du/dx = 8x$ and $dv/dx = 0$. Hence we have

$$\frac{dy}{dx} = \frac{d(4x^2)}{dx} + \frac{d(5)}{dx}$$
$$= 8x + 0$$
$$= 8x \quad ◀◀$$

EXAMPLE 7 ▶▶ Evaluate the derivative of $y = 2x^4 - 6x^2 - 8x - 9$ at $(-2, 15)$.

First, finding the derivative, we have

$$\frac{dy}{dx} = \frac{d(2x^4)}{dx} - \frac{d(6x^2)}{dx} - \frac{d(8x)}{dx} - \frac{d(9)}{dx}$$
$$= 8x^3 - 12x - 8$$

Since the derivative is a function only of x, we now evaluate it for $x = -2$.

$$\left.\frac{dy}{dx}\right|_{x=-2} = 8(-2)^3 - 12(-2) - 8 = -48 \quad ◀◀$$

EXAMPLE 8 ▶▶ Find the slope of a line tangent to the curve of $y = 4x^7 - x^4$ at the point $(1, 3)$. See Fig. 23-21.

We must find, and then evaluate, the derivative for the value $x = 1$.

$$\frac{dy}{dx} = 28x^6 - 4x^3 \qquad \text{find derivative}$$
$$\left.\frac{dy}{dx}\right|_{x=1} = 28(1) - 4(1) \qquad \text{evaluate derivative}$$
$$= 24$$

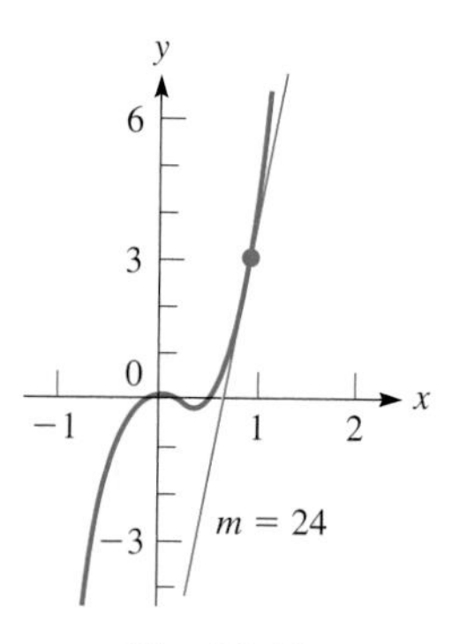

Fig. 23-21

Thus, the slope of the tangent line is 24. Again, we note that ***the substitution $x = 1$ must be made after the differentiation has been performed.*** ◀◀

CAUTION ▶

EXAMPLE 9 ▸▸ For each 4.0-s cycle, the displacement s (in cm) of a piston is given by the equation $s = t^3 - 6t^2 + 8t$, where t is the time. Find the instantaneous velocity of the piston for $t = 2.6$ s.

$$s = t^3 - 6t^2 + 8t$$

$$\frac{ds}{dt} = 3t^2 - 12t + 8 \qquad \text{find derivative}$$

$$\left.\frac{ds}{dt}\right|_{t=2.6} = 3(2.6)^2 - 12(2.6) + 8 \qquad \text{evaluate derivative}$$

$$= -2.9 \text{ cm/s}$$

This tells us that the piston is moving at -2.9 cm/s (it is moving in a negative direction) when $t = 2.6$ s. ◂◂

Exercises 23–5

In Exercises 1–16, find the derivative of each of the given functions.

1. $y = x^5$
2. $y = x^{12}$
3. $y = -4x^9$
4. $y = -7x^6$
5. $y = x^4 - 6$
6. $y = 3x^5 - 1$
7. $y = x^2 + 2x$
8. $y = x^3 - 2x^2$
9. $y = 5x^3 - x - 1$
10. $y = 6x^2 - 6x + 5$
11. $y = x^8 - 4x^7 - x$
12. $y = 4x^4 - 2x + 9$
13. $y = -6x^7 + 5x^3 + 2^3$
14. $y = 13x^4 - 6x^3 - x - 1$
15. $y = \frac{1}{3}x^3 + \frac{1}{2}x^2$
16. $y = -\frac{1}{4}x^8 + \frac{1}{2}x^4 - 3^2$

In Exercises 17–20, evaluate the derivative of each of the given functions at the given point.

17. $y = 6x^2 - 8x + 1 \quad (2, 9)$
18. $y = x^3 - 5x^2 + 4 \quad (-1, -2)$
19. $y = 2x^3 + 9x - 7 \quad (-2, -41)$
20. $y = x^4 - 9x^2 - 5x \quad (3, -15)$

In Exercises 21–24, find the slope of a line tangent to the curve of each of the given functions for the given values of x. View the function on a graphing calculator to see if the slope is reasonable for the given value of x.

21. $y = 2x^6 - 4x^2 \quad (x = -1)$
22. $y = 3x^3 - 9x \quad (x = 1)$
23. $y = 35x - 2x^4 \quad (x = 2)$
24. $y = x^4 - \frac{1}{2}x^2 + 2 \quad (x = -2)$

In Exercises 25–28, determine an expression for the instantaneous velocity of objects moving with rectilinear motion according to the functions given, if s represents displacement in terms of time t.

25. $s = 6t^5 - 5t + 2$
26. $s = 20 + 60t - 4.9t^2$
27. $s = 2 - 6t - 2t^3$
28. $s = s_0 + v_0t + \frac{1}{2}at^2$

In Exercises 29–32, s represents the displacement and t represents the time for objects moving with rectilinear motion according to the given functions. Find the instantaneous velocity for the given times.

29. $s = 2t^3 - 4t^2, t = 4$
30. $s = 120 + 80t - 16t^2, t = 2.5$
31. $s = 0.5t^4 - 1.5t^2 + 2.5, t = 3$
32. $s = 8t^2 - 10t + 6, t = 5$

In Exercises 33–48, solve the given problems by finding the appropriate derivative.

33. For what value(s) of x is the tangent to the curve of $y = 3x^2 - 6x$ parallel to the x-axis? (That is, where is the slope zero?)
34. Find the value of a if the tangent to the curve of $y = ax^2 + 2x$ has a slope of -4 for $x = 2$.
35. For what point(s) on the curve of $y = 3x^2 - 4x$ is the slope of a tangent line equal to 8?
36. Explain why the curve $y = 5x^3 + 4x - 3$ does not have a tangent line with a slope less than 4.
37. For what value(s) of x is the slope of a line tangent to the curve of $y = 4x^2 + 3x$ equal to the slope of a line tangent to the curve of $y = 5 - 2x^2$?
38. For what value(s) of t is the instantaneous velocity of an object moving according to $s = 5t - 2t^2$ equal to the instantaneous velocity of an object moving according to $s = 3t^2 + 4$?
39. A cylindrical metal container is heated and then allowed to cool. If the radius always equals the height, find an expression for the instantaneous rate of change of the volume with respect to the radius.

40. As an ice cube melts uniformly, find the expression for the instantaneous rate of change of the surface area A of the cube with respect to the edge e.

41. The electric power P, in watts, as a function of the current i, in amperes, in a certain circuit is given by $P = 16i^2 + 60i$. Find the instantaneous rate of change of P with respect to i for $i = 0.75$ A.

42. The torque T on the arm of a robotic control mechanism varies directly as the cube of the diameter d of the arm. If $T = 850$ N · cm for $d = 0.925$ cm, find the expression for the instantaneous rate of change of T with respect to d.

43. The electric polarization P of a light wave for high values of the electric field E is given by $P = a(c_1E + c_2E^2 + c_3E^3)$, where a, c_1, c_2, and c_3 are constants. Find the expression for the instantaneous rate of change of P with respect to E.

44. The ends of a 10-m beam are supported at different levels. The deflection y of the beam is given by $y = kx^2(x^3 + 450x - 3500)$, where x is the horizontal distance from one end and k is a constant. Determine the expression for the instantaneous rate of change of deflection with respect to x.

45. The force F, in newtons, exerted by a cam on a lever is given by $F = x^4 - 12x^3 + 46x^2 - 60x + 25$, where x $(1 \leq x \leq 5)$ is the distance (in cm) from the center of rotation of the cam to the edge of the cam in contact with the lever (see Fig. 23-22). Find the instantaneous rate of change of F with respect to x when $x = 4.0$ cm.

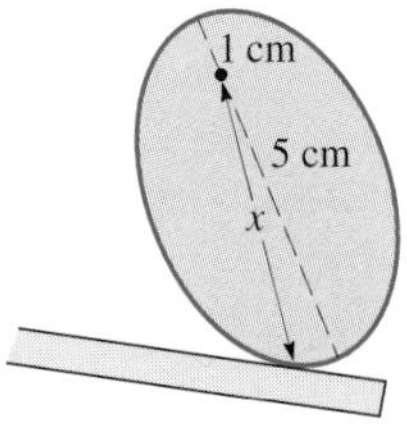

1 cm
5 cm
x

Fig. 23-22

46. A company determines that the cost C, in dollars, of producing x machine parts per day is given by $C = x^3 - 300x + 100$. Find the value of the instantaneous rate of change of C with respect to x when $x = 12$ parts.

47. Two ball bearings wear down such that the radius r of one is constantly 1.20 mm less than the radius of the other. Find the instantaneous rate of change of the total volume V_T of the two ball bearings with respect to r for $r = 3.30$ mm.

48. An open-top container is to be made from a rectangular piece of cardboard 6.00 cm by 8.00 cm. Equal squares of side x are to be cut from each corner, then the sides are to be bent up and taped together. Find the instantaneous rate of change of the volume V of the container with respect to x for $x = 1.75$ cm.

23-6 Derivatives of Products and Quotients of Functions

The formulas developed in the previous section are valid for polynomial functions. However, many functions are not polynomial in form. Some functions can best be expressed as the product of two or more simpler functions, others are the quotient of two simpler functions, and some are expressed as powers of a function. In this section we develop the formula for the derivative of a product of functions and the formula for the derivative of the quotient of two functions.

EXAMPLE 1 ▸▸ The functions $f(x) = x^2 + 2$ and $g(x) = 3 - 2x$ can be combined to form new functions of the types mentioned above. For example, the function

$$p(x) = f(x)g(x) = (x^2 + 2)(3 - 2x)$$

is an example of a function expressed as the product of two simpler functions.

The function

$$q(x) = \frac{g(x)}{f(x)} = \frac{3 - 2x}{x^2 + 2}$$

is an example of a rational function that is the quotient of two other functions.

The function

$$F(x) = [g(x)]^3 = (3 - 2x)^3$$

is an example of a power of a function. ◂◂

If u and v both represent differentiable functions of x, the derivative of the product of u and v is found by applying the delta-process. This leads to the following result:

$$y = uv$$
$$y + \Delta y = (u + \Delta u)(v + \Delta v) = uv + u\Delta v + v\Delta u + \Delta u\Delta v$$

(Since u and v are functions of x, each has an increment corresponding to an increment in x.) Then we have

$$\Delta y = u\Delta v + v\Delta u + \Delta u\Delta v$$
$$\frac{\Delta y}{\Delta x} = u\frac{\Delta v}{\Delta x} + v\frac{\Delta u}{\Delta x} + \Delta u\frac{\Delta v}{\Delta x}$$
$$\lim_{\Delta x \to 0}\frac{\Delta y}{\Delta x} = u\lim_{\Delta x \to 0}\frac{\Delta v}{\Delta x} + v\lim_{\Delta x \to 0}\frac{\Delta u}{\Delta x} + \lim_{\Delta x \to 0}\left(\Delta u\frac{\Delta v}{\Delta x}\right)$$

(The functions u and v are not affected by Δx approaching zero, but Δu and Δv both approach zero as Δx approaches 0.) Thus,

$$\frac{dy}{dx} = u\frac{dv}{dx} + v\frac{du}{dx} + 0\frac{dv}{dx}$$

We conclude that *the derivative of the product of two differentiable functions equals the first function times the derivative of the second function plus the second function times the derivative of the first function.* This is written as

Derivative of a Product

$$\frac{d(uv)}{dx} = u\frac{dv}{dx} + v\frac{du}{dx} \tag{23-12}$$

EXAMPLE 2 ▸▸ Find the derivative of the product function in Example 1.

$$y = (x^2 + 2)(3 - 2x), \qquad u = x^2 + 2, \qquad v = 3 - 2x$$

$$\frac{d(uv)}{dx} = \quad u \quad \frac{dv}{dx} \quad + \quad v \quad \frac{du}{dx}$$

$$\frac{dy}{dx} = (x^2 + 2)(-2) + (3 - 2x)(2x) = -2x^2 - 4 + 6x - 4x^2$$
$$= -6x^2 + 6x - 4 \quad ◂◂$$

EXAMPLE 3 ▸▸ Find the derivative of the function $y = (3 - x - 2x^2)(x^4 - x)$.
In this problem, $u = 3 - x - 2x^2$ and $v = x^4 - x$. Hence

$$\frac{dy}{dx} = (3 - x - 2x^2)(4x^3 - 1) + (x^4 - x)(-1 - 4x)$$
$$= 12x^3 - 3 - 4x^4 + x - 8x^5 + 2x^2 - x^4 - 4x^5 + x + 4x^2$$
$$= -12x^5 - 5x^4 + 12x^3 + 6x^2 + 2x - 3 \quad ◂◂$$

In both of these examples, we could have multiplied the functions first and then taken the derivative as a polynomial. However, we shall soon meet functions for which this latter method would not be applicable.

We shall now find the derivative of the quotient of two differentiable functions by applying the delta-process to the function $y = u/v$ as follows:

$$y + \Delta y = \frac{u + \Delta u}{v + \Delta v}$$

$$\Delta y = \frac{u + \Delta u}{v + \Delta v} - \frac{u}{v} = \frac{vu + v\Delta u - uv - u\Delta v}{v(v + \Delta v)}$$

$$\frac{\Delta y}{\Delta x} = \frac{v(\Delta u/\Delta x) - u(\Delta v/\Delta x)}{v(v + \Delta v)}$$

$$\lim_{\Delta x \to 0} \frac{\Delta y}{\Delta x} = \frac{v \lim\limits_{\Delta x \to 0} (\Delta u/\Delta x) - u \lim\limits_{\Delta x \to 0} (\Delta v/\Delta x)}{\lim\limits_{\Delta x \to 0} v(v + \Delta v)}$$

$$\frac{dy}{dx} = \frac{v(du/dx) - u(dv/dx)}{v^2}$$

Therefore, *the derivative of the quotient of two differentiable functions equals the denominator times the derivative of the numerator minus the numerator times the derivative of the denominator, all divided by the square of the denominator.*

Derivative of a Quotient

$$\frac{d\dfrac{u}{v}}{dx} = \frac{v\dfrac{du}{dx} - u\dfrac{dv}{dx}}{v^2} \qquad (23\text{-}13)$$

EXAMPLE 4 ▸▸ Find the derivative of the quotient indicated in Example 1.

$$y = \frac{3 - 2x}{x^2 + 2}, \qquad u = 3 - 2x, \qquad v = x^2 + 2$$

$$\frac{dy}{dx} = \frac{\overbrace{(x^2 + 2)}^{v}\overbrace{(-2)}^{du/dx} - \overbrace{(3 - 2x)}^{u}\overbrace{(2x)}^{dv/dx}}{\underbrace{(x^2 + 2)^2}_{v^2}} = \frac{-2x^2 - 4 - 6x + 4x^2}{(x^2 + 2)^2}$$

$$= \frac{2(x^2 - 3x - 2)}{(x^2 + 2)^2}$$ ◂◂

EXAMPLE 5 ▸▸ Evaluate the derivative of $y = \dfrac{3x^2 + x}{1 - 4x}$ at $(2, -2)$.

$$\frac{dy}{dx} = \frac{(1 - 4x)(6x + 1) - (3x^2 + x)(-4)}{(1 - 4x)^2}$$

$$= \frac{6x + 1 - 24x^2 - 4x + 12x^2 + 4x}{(1 - 4x)^2}$$

$$= \frac{-12x^2 + 6x + 1}{(1 - 4x)^2}$$

$$\left.\frac{dy}{dx}\right|_{x=2} = \frac{-12(2^2) + 6(2) + 1}{[1 - 4(2)]^2} = \frac{-48 + 12 + 1}{49} = \frac{-35}{49}$$

$$= -\frac{5}{7}$$ ◂◂

EXAMPLE 6 ▸▸ The stress S on a hollow tube is given by

$$S = \frac{16DT}{\pi(D^4 - d^4)}$$

where T is the tension, D is the outer diameter, and d is the inner diameter of the tube. Find the expression for the instantaneous rate of change of S with respect to D, with the other values being constant.

We are to find the derivative of S with respect to D, and it is found as follows:

$$\frac{dS}{dD} = \frac{\pi(D^4 - d^4)(16T) - 16DT(\pi)(4D^3)}{\pi^2(D^4 - d^4)^2} = \frac{16\pi T(D^4 - d^4 - 4D^4)}{\pi^2(D^4 - d^4)^2}$$

$$= \frac{-16T(3D^4 + d^4)}{\pi(D^4 - d^4)^2}$$ ◂◂

Exercises 23–6

In Exercises 1–8, find the derivative of each function by using Eq. (23-12). Do not find the product before finding dy/dx.

1. $y = x^2(3x + 2)$
2. $y = 3x(x^3 + 1)$
3. $y = 6x(3x^2 - 5x)$
4. $y = 2x^3(3x^4 + x)$
5. $y = (x + 2)(2x - 5)$
6. $y = (3x + 1)(x^2 + 1)$
7. $y = (x^4 - 3x^2 + 3)(1 - 2x^3)$
8. $y = (x^3 - 6x)(2 - 4x^3)$

In Exercises 9–12, find the derivative of each function by using Eq. (23-12). Then multiply out each function and find the derivative by treating it as a polynomial. Compare results.

9. $y = (2x - 7)(5 - 2x)$
10. $y = (x^2 + 2)(2x^2 - 1)$
11. $y = (x^3 - 1)(2x^2 - x - 1)$
12. $y = (3x^2 - 4x + 1)(5 - 6x^2)$

In Exercises 13–24, find the derivative of each function by using Eq. (23-13).

13. $y = \frac{x}{2x + 3}$
14. $y = \frac{2x}{x + 1}$
15. $y = \frac{1}{x^2 + 1}$
16. $y = \frac{x + 2}{2x + 3}$
17. $y = \frac{x^2}{3 - 2x}$
18. $y = \frac{2}{3x^2 - 5x}$
19. $y = \frac{2x - 1}{3x^2 + 2}$
20. $y = \frac{2x^3}{4 - x}$
21. $y = \frac{x + 8}{x^2 + x + 2}$
22. $y = \frac{3x}{4x^5 - 3x - 4}$
23. $y = \frac{2x^2 - x - 1}{x^3 + 2x^2}$
24. $y = \frac{3x^3 - x}{2x^2 - 5x + 4}$

In Exercises 25–32, evaluate the derivatives of the given functions for the given values of x. In Exercises 25–28, find the derivative by using Eq. (23-12).

25. $y = (3x - 1)(4 - 7x)$, $x = 3$
26. $y = (3x^2 - 5)(2x^2 - 1)$, $x = -1$
27. $y = (2x^2 - x + 1)(4 - 2x - x^2)$, $x = -3$
28. $y = (4x^4 + 0.5x^2 + 1)(3x - 2x^2)$, $x = 0.5$
29. $y = \frac{3x - 5}{2x + 3}$, $x = -2$
30. $y = \frac{2x^2 - 5x}{3x + 2}$, $x = 2$
31. $y = \frac{x^3 - 2x + 8}{2x - x^4}$, $x = -1$
32. $y = \frac{2x^3 - x^2 - 2}{4x + 3}$, $x = 0.5$

In Exercises 33–48, solve the given problems by finding the appropriate derivatives.

33. Find the derivative of $y = \frac{x^2(1 - 2x)}{3x - 7}$ in each of the following two ways. (1) Do not multiply out the numerator before finding the derivative. (2) Multiply out the numerator before finding the derivative. Compare the results.

34. Find the derivative of $y = 4x^2 - \dfrac{1}{x-1}$ in each of the following two ways. (1) Do not combine the terms over a common denominator before finding the derivative. (2) Combine the terms over a common denominator before finding the derivative. Compare the results.

35. Find the slope of a tangent line to the curve of $y = (4x + 1)(x^4 - 1)$ at the point $(-1, 0)$. Do not multiply the factors together before taking the derivative. View the function on a graphing calculator to see if the calculated value of the slope is reasonable for $x = -1$.

36. Find the slope of a line tangent to the curve of the function $y = (3x + 4)(1 - 4x)$ at $(2, -70)$. Do not multiply the factors together before taking the derivative. View the function on a graphing calculator to see if the calculated value of the slope is reasonable for $x = 2$.

37. For what value(s) of x is the slope of a tangent to the curve of $y = \dfrac{x}{x^2 + 1}$ equal to zero? View the function on a graphing calculator to verify the values found.

38. Determine the sign of the derivative of the function $y = \dfrac{2x - 1}{1 - x^2}$ for the following values of x: -2, -1, 0, 1, 2. Is the slope of a tangent line to this curve ever negative? View the function on a graphing calculator to verify your conclusion.

39. During each cycle, the displacement s of the end of a robot arm is given by $s = (t^2 - 8t)(2t^2 + t + 1)$, where t is the time. Find the expression for the instantaneous velocity of the end of the robot arm.

40. The sales S of a product as a function of the time t, in weeks, is given by $S = 5000 - \dfrac{2000}{t}$ $(t \geq 1)$. Find the rate of change of S with respect to t for $t = 10$ weeks. Use the quotient rule.

41. The voltage V at the junction of a 3-Ω resistance and a variable resistance R in a circuit is given by $V = \dfrac{6R + 25}{R + 3}$. Find the rate of change of V with respect to R for $R = 7\ \Omega$.

42. During a chemical change the number n of grams of a compound being formed is given by $n = \dfrac{6t^2}{2t^2 + 3}$, where t is measured in seconds. How many grams per second are being formed after 3.0 s?

43. A computer, using data from a refrigeration plant, estimated that in the event of a power failure, the temperature T (in °C) in the freezers would be given by $T = \dfrac{2t}{0.05t + 1} - 20$, where t is the number of hours after the power failure. Find the time rate of change of temperature after 6.0 h.

44. The voltage across a resistor in an electric circuit is the product of the resistance and the current. If the current I (in amperes) varies with time (in seconds) according to the relation $I = 5.00 + 0.01t^2$, and the resistance varies with time according to the relation $R = 15.00 - 0.10t$, find the time rate of change of the voltage when $t = 5.00$ s.

45. The frictional radius r_f of a disc clutch is given by $r_f = \dfrac{2(R^2 + Rr + r^2)}{3(R + r)}$, where R and r are the outer radius and inner radius of the clutch, respectively. Find the derivative of r_f with respect to R with r constant.

46. In thermodynamics, an equation relating the thermodynamic temperature T, pressure p, and volume V of a gas is $T = \left(p + \dfrac{a}{V^2}\right)\left(\dfrac{V - b}{R}\right)$, where a, b, and R are constants. Find the derivative of T with respect to V, assuming p is constant.

47. The electric power produced by a certain source is given by $P = \dfrac{E^2 r}{R^2 + 2Rr + r^2}$, where E is the voltage of the source, R is the resistance of the source, and r is the resistance in the circuit. Find the derivative of P with respect to r, assuming that the other quantities remain constant.

48. In the theory of lasers, the power P radiated is given by $P = \dfrac{kf^2}{\omega^2 - 2\omega f + f^2 + a^2}$, where f is the field frequency and a, k, and ω are constants. Find the derivative of P with respect to f.

23-7 The Derivative of a Power of a Function

In Example 1 of Section 23-6, we illustrated $y = (3 - 2x)^3$ as the third power of a function of x, where $3 - 2x$ is the function. If we let $u = 3 - 2x$, we can write

$$y = u^3 \quad \text{where } u = 3 - 2x$$

Writing it this way, y is a function of u and u is a function of x. This means that y *is a function of a function of x, referred to as a* **composite function.** However, y is still a function of x, since u is a function of x.

Since we will often need to find the derivative of a power of a function, we now develop the necessary formula. For $y = f(u)$, where $u = g(x)$, we can express the derivative dy/dx in terms of dy/du and du/dx. If Δx is the increment in x, then Δy and Δu are the corresponding increments in y and u, respectively. We may write

$$\frac{\Delta y}{\Delta x} = \frac{\Delta y}{\Delta u}\frac{\Delta u}{\Delta x}$$

When Δx approaches zero, Δu and Δy both approach zero, for u and y are functions of x. Thus,

$$\lim_{\Delta x \to 0}\frac{\Delta y}{\Delta x} = \left(\lim_{\Delta u \to 0}\frac{\Delta y}{\Delta u}\right)\left(\lim_{\Delta x \to 0}\frac{\Delta u}{\Delta x}\right)$$

Chain Rule

$$\frac{dy}{dx} = \frac{dy}{du}\frac{du}{dx} \tag{23-14}$$

(Here we have assumed $\Delta u \neq 0$, although it can be shown that this condition is not necessary.) *Equation (23-14) is known as the* **chain rule** *for derivatives.*

Using Eq. (23-14) for $y = u^n$, where u is a differentiable function of x, we have

$$\frac{dy}{dx} = \frac{d(u^n)}{du}\frac{du}{dx}$$

Derivative of a Power of a Function of x

$$\frac{du^n}{dx} = nu^{n-1}\left(\frac{du}{dx}\right) \tag{23-15}$$

By using Eq. (22-15) we may find the derivative of a power of a differentiable function of x.

EXAMPLE 1 ▸▸ Find the derivative of $y = (3 - 2x)^3$.

For this function $n = 3$ and $u = 3 - 2x$. Therefore, $du/dx = -2$. This means

$$\frac{du^n}{dx} = n \quad u \quad {}^{n-1} \quad \left(\frac{du}{dx}\right)$$

$$\frac{dy}{dx} = 3(3 - 2x)^2(-2)$$

$$= -6(3 - 2x)^2$$

CAUTION ▸ A common type of error in finding this type of derivative is to omit the du/dx factor; in this case it is the -2. ***The derivative is incomplete, and therefore incorrect, without this factor.*** ◂◂

EXAMPLE 2 ▸▸ Find the derivative of $y = (1 - 3x^2)^4$.

In this example, $n = 4$ and $u = 1 - 3x^2$. Hence

$$\frac{dy}{dx} = 4(1 - 3x^2)^3(-6x) = -24x(1 - 3x^2)^3$$

CAUTION ▸ ***(We must not forget the −6x.)*** ◂◂

EXAMPLE 3 ▸▸ Find the derivative of $y = \dfrac{(3x - 1)^3}{1 - x}$.

NOTE ▸ Here we must ***use the quotient rule in combination with the power rule.*** We find the derivative of the numerator by using the power rule.

derivative of numerator

$$\frac{dy}{dx} = \frac{(1 - x)3(3x - 1)^2(3) - (3x - 1)^3(-1)}{(1 - x)^2}$$

$$= \frac{(3x - 1)^2(9 - 9x + 3x - 1)}{(1 - x)^2} = \frac{(3x - 1)^2(8 - 6x)}{(1 - x)^2}$$

$$= \frac{2(3x - 1)^2(4 - 3x)}{(1 - x)^2}$$ ◂◂

It is normally better to have the derivative in a factored, simplified form, since this is the only form from which useful information may readily be obtained. In this form we can determine where the derivative is undefined (denominator equal to zero) or where the slope is zero (numerator equal to zero); we can also make other required analyses of the derivative. Thus, *all derivatives should be in simplest algebraic form.*

So far, we have derived the formulas for the derivatives of powers of x and for differentiable functions of x for integral powers. We shall now establish that these formulas are also valid for any rational number used as an exponent. If $y = u^{p/q}$, and if each side is raised to the qth power, we have $y^q = u^p$. Applying the power rule to each side of this equation, we have

$$qy^{q-1}\left(\frac{dy}{dx}\right) = pu^{p-1}\left(\frac{du}{dx}\right)$$

Solving for dy/dx, we have

$$\frac{dy}{dx} = \frac{pu^{p-1}(du/dx)}{qy^{q-1}} = \frac{p}{q}\,\frac{u^{p-1}}{(u^{p/q})^{q-1}}\,\frac{du}{dx} = \frac{p}{q}\,\frac{u^{p-1}}{u^{p-p/q}}\,\frac{du}{dx}$$

$$= \frac{p}{q}u^{p-1-p+(p/q)}\frac{du}{dx}$$

Thus

$$\frac{du^{p/q}}{dx} = \frac{p}{q}u^{(p/q)-1}\frac{du}{dx} \qquad (23\text{-}16)$$

We see that in finding the derivative we multiply the function by the rational exponent and subtract 1 from it to find the exponent of the function in the derivative. *This is the same rule as derived for integral exponents in Eq. (23-15).*

For reference, Eq. (23-9) is $\frac{dx^n}{dx} = nx^{n-1}$

In deriving Eqs. (23-15) and (23-16) we used Eq. (23-9). In deriving Eq. (23-9) we used the binomial theorem, in which the exponent n can be any positive or negative integer. Together, these tell us that

the power rule for derivatives, Eq. (23-15), can be extended to include all rational exponents, positive or negative.

This, of course, includes all integral exponents, positive and negative. Also, we note that Eq. (23-9) is equivalent to Eq. (23-15) with $u = x$ (since $du/dx = 1$).

EXAMPLE 4 ▸▸ We can now find the derivative of $y = \sqrt{x^2 + 1}$.

By using Eq. (23-16) or Eq. (23-15), and writing the square root as the fractional exponent 1/2, we can derive the result.

$$y = (x^2 + 1)^{1/2}$$

$$\frac{dy}{dx} = \frac{1}{2}(x^2 + 1)^{-1/2}(2x)$$

$$= \frac{x}{(x^2 + 1)^{1/2}}$$

To avoid introducing apparently significant factors into the numerator, we do not usually rationalize such fractions. ◂◂

Having shown that we may use fractional exponents to find derivatives of roots of functions of x, we may also use them to find derivatives of roots of x itself. Consider the following example.

EXAMPLE 5 ▸▸ Find the derivative of $y = 6\sqrt[3]{x}$.

We can write this function as $y = 6x^{1/3}$. In finding the derivative we may use Eq. (23-9) with $n = \frac{1}{3}$. This gives us

$$y = 6x^{1/3}$$

$$\frac{dy}{dx} = 6\left(\frac{1}{3}\right)x^{-2/3} = \frac{2}{x^{2/3}} \qquad \left(\tfrac{1}{3} - 1\right)$$

We could also use Eq. (23-15) with $u = x$ and $n = \frac{1}{3}$. This gives us

$$\frac{du}{dx} = \frac{dx}{dx} = 1$$

$$\frac{dy}{dx} = 6\left(\frac{1}{3}\right)x^{-2/3}(1) = \frac{2}{x^{2/3}}$$

This shows us why Eq. (23-9) is equivalent to Eq. (23-15) with $u = x$.

Note that the domain of the function is all real numbers, but the function is not differentiable for $x = 0$. ◂◂

We now show the use of Eqs. (23-9) and (23-15) with n being a negative exponent.

EXAMPLE 6 ▸▸ The electric resistance R of a wire varies inversely as the square of its radius r. For a given wire, $R = 4.66\ \Omega$ for $r = 0.105$ mm. Find the derivative of R with respect to r for this wire.

Since R varies inversely as the square of r, we have $R = k/r^2$. Then, using the fact that $R = 4.66\ \Omega$ for $r = 0.150$ mm, we have

$$4.66 = \frac{k}{(0.150)^2}, \qquad k = 0.105\ \Omega \cdot \text{mm}^2$$

which means that

$$R = \frac{0.105}{r^2}$$

We could find the derivative by the quotient rule. However, by using negative exponents the derivative is easily found.

$$R = \frac{0.105}{r^2} = 0.105r^{-2}$$

CAUTION ▸

$$\frac{dR}{dr} = 0.105(-2)r^{-3} \longleftarrow -2 - 1 = -3$$

$$= -\frac{0.210}{r^3}$$

Here we used Eq. (23-9) directly. ◂◂

EXAMPLE 7 ▸▸ Find the derivative of $y = \dfrac{1}{(1 - 4x)^5}$.

The derivative is found as follows:

$$y = \frac{1}{(1 - 4x)^5} = (1 - 4x)^{-5} \qquad \text{use negative exponent}$$

$$\frac{dy}{dx} = (-5)(1 - 4x)^{-6}(-4) \qquad \text{use Eq. (23-15)}$$

$$= \frac{20}{(1 - 4x)^6} \qquad \text{express result with positive exponent}$$

CAUTION ▸ Remember: ***Subtracting* 1 *from* −5 *gives* −6.** ◂◂

We now see the value of fractional exponents in calculus. They are useful in many algebraic operations, but they are almost essential in calculus. Without fractional exponents, it would be necessary to develop additional formulas to find the derivatives of radical expressions. In order to find the derivative of an algebraic function, we need only those formulas we have already developed. Often it is necessary to combine these formulas, as we saw in Example 3. Actually, most derivatives are combinations. The problem in finding the derivative is ***recognizing the form of the function*** with which you are dealing. When you have recognized the form, completing the problem is only a matter of mechanics and algebra. You should now see the importance of being able to handle algebraic operations with ease.

NOTE ▸

EXAMPLE 8 ▸▸ Evaluate the derivative of

$$y = \frac{x}{\sqrt{1 - 4x}}$$

for $x = -2$.

Here we have a quotient, and in order to find the derivative of this quotient, we must also use the power rule (and a derivative of a polynomial form). With sufficient practice in taking derivatives, we can recognize the rule to use almost automatically. Thus, we find the derivative:

$$\frac{dy}{dx} = \frac{(1 - 4x)^{1/2}(1) - x(\frac{1}{2})(1 - 4x)^{-1/2}(-4)}{1 - 4x}$$

$$= \frac{(1 - 4x)^{1/2} + \dfrac{2x}{(1 - 4x)^{1/2}}}{1 - 4x} = \frac{\dfrac{(1 - 4x)^{1/2}(1 - 4x)^{1/2} + 2x}{(1 - 4x)^{1/2}}}{1 - 4x}$$

$$= \frac{(1 - 4x) + 2x}{(1 - 4x)^{1/2}(1 - 4x)} = \frac{1 - 2x}{(1 - 4x)^{3/2}}$$

Now, evaluating the derivative for $x = -2$, we have

$$\left.\frac{dy}{dx}\right|_{x=-2} = \frac{1 - 2(-2)}{[1 - 4(-2)]^{3/2}} = \frac{1 + 4}{(1 + 8)^{3/2}} = \frac{5}{9^{3/2}}$$

$$= \frac{5}{27}$$ ◂◂

Many models of graphing calculators can be used to evaluate a derivative for a given value of x. See the manual as to how this is done on any particular model.

Exercises 23–7

In Exercises 1–24, find the derivative of each of the given functions.

1. $y = \sqrt{x}$

2. $y = \sqrt[4]{x}$

3. $y = \frac{1}{x^2}$

4. $y = \frac{2}{x^4}$

5. $y = \frac{3}{\sqrt[3]{x}}$

6. $y = \frac{1}{\sqrt{x}}$

7. $y = x\sqrt{x} - \frac{1}{x}$

8. $y = 2x^{-3} - x^{-2}$

9. $y = (x^2 + 1)^5$

10. $y = (1 - 2x)^4$

11. $y = 2(7 - 4x^3)^8$

12. $y = 3(8x^2 - 1)^6$

13. $y = (2x^3 - 3)^{1/3}$

14. $y = (1 - 6x)^{3/2}$

15. $y = \frac{1}{(1 - x^2)^4}$

16. $y = \frac{4}{\sqrt{1 - 3x}}$

17. $y = 4(2x^4 - 5)^{3/4}$

18. $y = 5(3x^7 - 4)^{2/3}$

19. $y = \sqrt[4]{1 - 8x^2}$

20. $y = \sqrt[3]{4x^6 + 2}$

21. $y = x\sqrt{8x + 5}$

22. $y = x^2(1 - 3x)^5$

23. $y = \frac{\sqrt{4x + 3}}{8x + 1}$

24. $y = \frac{x\sqrt{x - 1}}{1 - 2x}$

In Exercises 25–28, evaluate the derivatives of the given functions at the given values of x.

25. $y = \sqrt{3x + 4}$, $x = 7$

26. $y = (4 - x^2)^{-1}$, $x = -1$

27. $y = \frac{\sqrt{x}}{1 - x}$, $x = 4$

28. $y = x^2\sqrt[3]{3x + 2}$, $x = 2$

In Exercises 29–44, solve the given problems by finding the appropriate derivatives.

29. Find the derivative of $y = 1/x^3$ as (a) a quotient and (b) a negative power of x, and show that the results are the same.

30. Find the derivative of $y = \dfrac{2}{4x + 3}$ as (a) a quotient and (b) a negative power of $4x + 3$, and show that the results are the same.

31. Find any values of x for which the derivative of $y = \dfrac{x^2}{\sqrt{x^2 + 1}}$ is zero. View the curve of the function on a graphing calculator to verify the values found.

32. Find any values of x for which the derivative of $y = \dfrac{x}{\sqrt{4x - 1}}$ is zero. View the curve of the function on a graphing calculator to verify the values found.

33. Find the slope of a line tangent to the parabola $y^2 = 4x$ at the point $(1, 2)$. View the graph of the parabola on a graphing calculator to see if the calculated value of the slope is reasonable for $x = 1$.

34. Find the slope of a line tangent to the circle $x^2 + y^2 = 25$ at the point $(4, 3)$. View the graph of the circle on a graphing calculator to see if the calculated value of the slope is reasonable for $x = 4$.

35. The displacement s (in cm) of a linkage joint of a robot is given by $s = (8t - t^2)^{2/3}$, where t is in seconds. Find the velocity of the joint for $t = 6.25$ s.

36. An analysis of a company's records shows that the profit p, in dollars, of producing x units of a product is $p = 3000\sqrt{x^3 - 12x^2 + 36x} - 2000$. Find dp/dx for $x = 4$ units.

37. When the volume of a gas changes very rapidly, an approximate relation is that the pressure varies inversely as the $\frac{3}{2}$ power of the volume. If P is 300 kPa when $V = 100\ \text{cm}^3$, find the derivative of P with respect to V. Evaluate this derivative for $V = 100\ \text{cm}^3$.

38. The power gain G of a certain antenna is inversely proportional to the square of the wavelength λ, in metres, of the carrier wave. If $G = 5.0 \times 10^4$ for $\lambda = 0.11$ m, find the derivative of G with respect to λ for $\lambda = 0.11$ m.

39. The total solar radiation H (in W/m^2) on a certain surface during an average clear day is given by $H = \dfrac{4000}{\sqrt{t^6 + 100}}$, where t $(-6 < t < 6)$ is the number of hours from noon. Find the rate at which H is changing with time at 4 P.M.

40. In determining the time for a laser beam to go from S to P (see Fig. 23-23), which are in different mediums, it is necessary to find the derivative of the time

$$t = \frac{\sqrt{a^2 + x^2}}{v_1} + \frac{\sqrt{b^2 + (c - x)^2}}{v_2}$$

with respect to x, where a, b, c, v_1, and v_2 are constants. Here v_1 and v_2 are the velocities of the laser in each medium. Find this derivative.

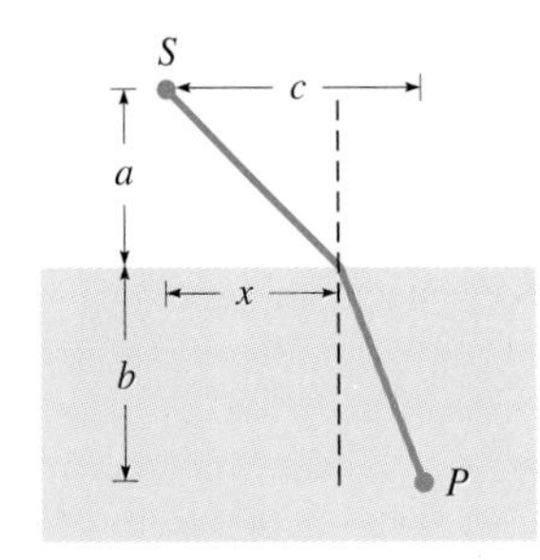

Fig. 23-23

41. The radio waveguide wavelength λ_r is related to its free-space wavelength λ by

$$\lambda_r = \frac{2a\lambda}{\sqrt{4a^2 - \lambda^2}}$$

where a is a constant. Find $d\lambda_r/d\lambda$.

42. The current in a circuit containing a resistance R and an inductance L is found from the expression

$$I = \frac{V}{\sqrt{R^2 + (\omega L)^2}}$$

Find the expression for the rate of change of current with respect to L, assuming that the other quantities remain constant.

43. The length l of a rectangular microprocessor chip is 2 mm longer than its width w. Find the derivative of the diagonal D with respect to w.

44. The trapezoidal structure shown in Fig. 23-24 has an internal support of length l. Find the derivative of l with respect to x.

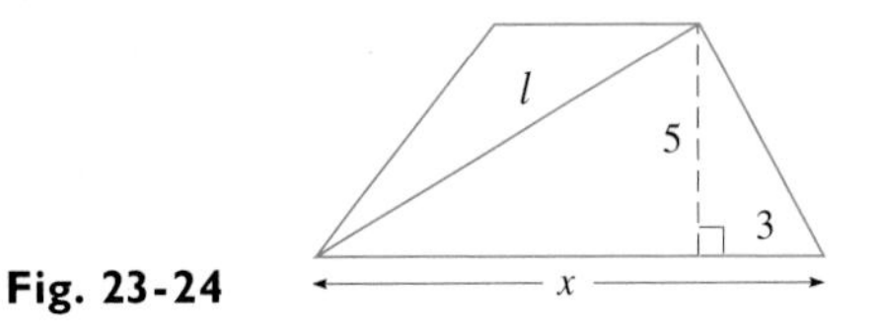

Fig. 23-24

23-8 Differentiation of Implicit Functions

To this point the functions we have differentiated have been of the form $y = f(x)$. There are, however, occasions when we need to find the derivative of a function determined by an equation that does not express the dependent variable explicitly in terms of the independent variable.

An equation in which y is not expressed explicitly in terms of x may determine one or more functions. *Any such function, where y is defined implicitly as a function of x, is called an* **implicit function.** Some equations defining implicit functions may be solved to determine the explicit functions, and for others it is not possible to solve for the explicit functions. Also, not all such equations define y as a function of x for real values of x.

EXAMPLE 1 ▸▸ (a) The equation $3x + 4y = 5$ is an equation that defines a function, although it is not in explicit form. In solving for y as $y = -\frac{3}{4}x + \frac{5}{4}$, we have the explicit form of the function.

(b) The equation $y^2 + x = 3$ is an equation that defines two functions, although we do not have the explicit forms. When we solve for y, we obtain the explicit functions $y = \sqrt{3 - x}$ and $y = -\sqrt{3 - x}$.

(c) The equation $y^5 + xy^2 + 3x^2 = 5$ defines y as a function of x, although we cannot actually solve for the explicit algebraic form of the function.

(d) The equation $x^2 + y^2 + 4 = 0$ is not satisfied by any pair of real values of x and y. ◂◂

Even when it is possible to determine the explicit form of a function given in implicit form, it is not always desirable to do so. In some cases the implicit form is more convenient than the explicit form.

The derivative of an implicit function may be found directly without having to solve for the explicit function. Thus, *to find dy/dx when y is defined as an implicit function of x, we differentiate each term of the equation with respect to x, regarding y as a differentiable function of x.* We then solve for dy/dx, which will usually be in terms of x and y.

EXAMPLE 2 ▸▸ Find dy/dx if $y^2 + 2x^2 = 5$.

Here we find the derivative of each term and then solve for dy/dx. Thus

$$\frac{d(y^2)}{dx} + \frac{d(2x^2)}{dx} = \frac{d(5)}{dx}$$

$$2y^{2-1}\frac{dy}{dx} + 2\left(2x^{2-1}\frac{dx}{dx}\right) = 0$$

$$2y\frac{dy}{dx} + 4x = 0$$

$$\frac{dy}{dx} = -\frac{2x}{y}$$

For reference, Eq. (23-15) is $\frac{du^n}{dx} = nu^{n-1}\frac{du}{dx}$.

CAUTION ▸ ***The factor dy/dx arises from the derivative of the first term as a result of using the derivative of a power of a function of x (Eq. 23-15). The factor dy/dx corresponds to the du/dx of the formula.*** In the second term, no factor of dy/dx appears, since there are no y factors in the term. ◂◂

EXAMPLE 3 ▶▶ Find dy/dx if $3y^4 + xy^2 + 2x^3 - 6 = 0$.

In finding the derivative, we note that the second term is a product, and we must use the product rule for derivatives on it. Thus, we have

$$\frac{d(3y^4)}{dx} + \frac{d(xy^2)}{dx} + \frac{d(2x^3)}{dx} - \frac{d(6)}{dx} = \frac{d(0)}{dx}$$

using product rule

$$12y^3\frac{dy}{dx} + \left[x\left(2y\frac{dy}{dx}\right) + y^2(1)\right] + 6x^2 - 0 = 0$$

$$12y^3\frac{dy}{dx} + 2xy\frac{dy}{dx} + y^2 + 6x^2 = 0 \qquad \text{solve for } \frac{dy}{dx}$$

$$(12y^3 + 2xy)\frac{dy}{dx} = -y^2 - 6x^2$$

$$\frac{dy}{dx} = \frac{-y^2 - 6x^2}{12y^3 + 2xy}$$ ◀◀

EXAMPLE 4 ▶▶ Find dy/dx if $2x^3y + (y^2 + x)^3 = x^4$.

In this case we use the product rule on the first term and the power rule on the second term.

$$\frac{d(2x^3y)}{dx} + \frac{d(y^2 + x)^3}{dx} = \frac{d(x^4)}{dx}$$

product / power

$$2x^3\left(\frac{dy}{dx}\right) + y(6x^2) + 3(y^2 + x)^2\left(2y\frac{dy}{dx} + 1\right) = 4x^3$$

$$2x^3\frac{dy}{dx} + 6x^2y + 3(y^2 + x)^2\left(2y\frac{dy}{dx}\right) + 3(y^2 + x)^2 = 4x^3$$

$$[2x^3 + 6y(y^2 + x)^2]\frac{dy}{dx} = 4x^3 - 6x^2y - 3(y^2 + x)^2$$

$$\frac{dy}{dx} = \frac{4x^3 - 6x^2y - 3(y^2 + x)^2}{2x^3 + 6y(y^2 + x)^2}$$ ◀◀

EXAMPLE 5 ▶▶ Find the slope of a line tangent to the curve of $2y^3 + xy + 1 = 0$ at the point $(-3, 1)$.

Here we must find dy/dx and evaluate it for $x = -3$ and $y = 1$.

$$\frac{d(2y^3)}{dx} + \frac{d(xy)}{dx} + \frac{d(1)}{dx} = \frac{d(0)}{dx}$$

$$6y^2\frac{dy}{dx} + x\frac{dy}{dx} + y + 0 = 0$$

$$\frac{dy}{dx} = \frac{-y}{6y^2 + x}$$

$$\left.\frac{dy}{dx}\right|_{(-3,1)} = \frac{-1}{6(1^2) - 3} = \frac{-1}{6 - 3} = -\frac{1}{3}$$

Thus, the slope is $-\frac{1}{3}$. ◀◀

Exercises 23–8

In Exercises 1–20, find dy/dx by differentiating implicitly. When applicable, express the result in terms of x and y.

1. $3x + 2y = 5$
2. $6x - 3y = 4$
3. $4y - 3x^2 = x$
4. $x^5 - 5y = 6 - x$
5. $x^2 - y^2 - 9 = 0$
6. $x^2 + 2y^2 - 11 = 0$
7. $y^5 = x^2 - 1$
8. $y^4 = 3x^3 - x$
9. $y^2 + y = x^2 - 4$
10. $2y^3 - y = 7 - x^4$
11. $y + 3xy - 4 = 0$
12. $8y - xy - 7 = 0$
13. $xy^3 + 3y + x^2 = 9$
14. $y^2x - \dfrac{y}{x+1} + 3x = 4$
15. $\dfrac{3x^2}{y^2+1} + y = 3x + 1$
16. $2x - x^3y^2 = y - x^2 - 1$
17. $(2y - x)^4 + x^2 = y + 3$
18. $(y^2 + 2)^3 = x^4y + 11$
19. $2(x^2 + 1)^3 + (y^2 + 1)^2 = 17$
20. $(2x + 1)(1 - 3y) + y^2 = 13$

In Exercises 21–24, evaluate the derivatives of the given functions at the given points.

21. $3x^3y^2 - 2y^3 = -4 \quad (1, 2)$
22. $2y + 5 - x^2 - y^3 = 0 \quad (2, -1)$
23. $5y^4 + 7 = x^4 - 3y \quad (3, -2)$
24. $(xy - y^2)^3 = 5y^2 + 22 \quad (4, 1)$

In Exercises 25–32, solve the given problems by using implicit differentiation.

25. Find the slope of a line tangent to the curve of $xy + y^2 + 2 = 0$ at the point $(-3, 1)$. View the curve of the function on a graphing calculator to see if the calculated slope is reasonable at this point.
26. Oil moves through a pipeline such that the distance s it moves and the time t are related by $s^3 - t^2 = 7t$. Find the velocity of the oil for $s = 4.01$ m and $t = 5.25$ s.
27. The shelf support shown in Fig. 23-25 is 0.75 m long. Find the expression for dy/dx in terms of x and y.

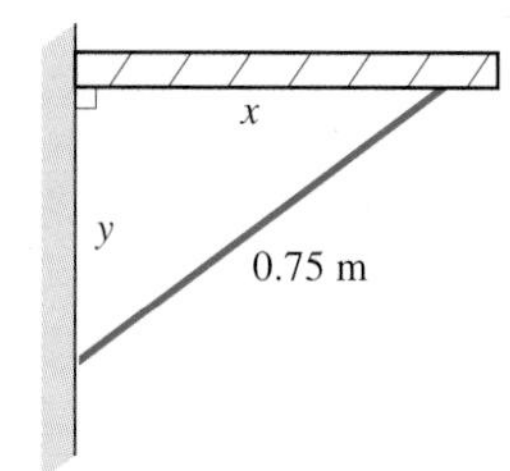

Fig. 23-25

28. An open (no top) right circular cylindrical container of radius r and height h has a total surface area of 940 cm^2. Find dr/dh in terms of r and h.
29. Two resistors, with resistances r and $r + 2$, are connected in parallel. Their combined resistance R is related to r by the equation $r^2 = 2rR + 2R - 2r$. Find dR/dr.
30. The polar moment of inertia I of a rectangular slab of concrete is given by $I = \frac{1}{12}(b^3h + bh^3)$, where b and h are the base and the height, respectively, of the slab. If I is constant, find the expression for db/dh.
31. A formula relating the length L and radius of gyration r of a steel column is $24C^3Sr^3 = 40C^3r^3 + 9LC^2r^2 - 3L^3$, where C and S are constants. Find dL/dr.
32. A computer is programmed to draw the graph of $(x^2 + y^2)^3 = 64x^2y^2$ (see Example 7 of Section 21-9). Find the slope of a line tangent to this curve at (2.00, 0.56) and at (2.00, 3.07). The graph is shown in Fig. 23-26.

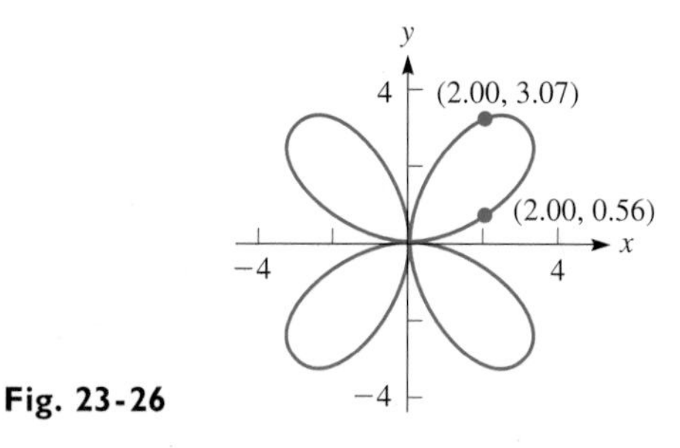

Fig. 23-26

23-9 Higher Derivatives

In the previous sections of this chapter we developed certain important formulas for finding the derivative of a function. The derivative itself is a function, and we may therefore take its derivative. In technology and in mathematics there are many applications that require the use of the derivative of a derivative. Therefore, in this section we develop the concept and necessary notation, as well as show some of the applications.

The Second Derivative

The derivative of a function is called the **first derivative** *of the function. The derivative of the first derivative is called the* **second derivative.** Since the second derivative is a function, we may find its derivative, which is called the third derivative. We may continue to find the fourth derivative, fifth derivative, and so on (provided each derivative is defined). *The second derivative, third derivative, and so on, are known as* **higher derivatives.**

The notations used for higher derivatives follow closely those used for the first derivative. As shown in Section 23-3, the notations for the first derivative are y', $D_x y$, $f'(x)$, and dy/dx. The notations for the second derivative are y'', $D_x^2 y$, $f''(x)$, and d^2y/dx^2. Similar notations are used for other higher derivatives.

EXAMPLE 1 ▸▸ Find the higher derivatives of $y = 5x^3 - 2x$.

We find the first derivative as

$$\frac{dy}{dx} = 15x^2 - 2 \quad \text{or} \quad y' = 15x^2 - 2$$

Next, we obtain the second derivative by finding the derivative of the first derivative:

$$\frac{d^2y}{dx^2} = 30x \quad \text{or} \quad y'' = 30x$$

Continuing to find the successive derivatives, we have

$$\frac{d^3y}{dx^3} = 30 \quad \text{or} \quad y''' = 30$$

$$\frac{d^4y}{dx^4} = 0 \quad \text{or} \quad y^{(4)} = 0$$

Since the third derivative is a constant, the fourth derivative and all successive derivatives will be zero. This can be shown as $d^ny/dx^n = 0$ for $n \geq 4$. ◂◂

EXAMPLE 2 ▸▸ Find the higher derivatives of $f(x) = x(x^2 - 1)^2$.

For reference, Eq. (23-12) is $\frac{d(uv)}{dx} = u\frac{dv}{dx} + v\frac{du}{dx}$.

Using the product rule, Eq. (23-12), to find the first derivative, we have

$$\begin{aligned} f'(x) &= x(2)(x^2 - 1)(2x) + (x^2 - 1)^2(1) \\ &= (x^2 - 1)(4x^2 + x^2 - 1) = (x^2 - 1)(5x^2 - 1) \\ &= 5x^4 - 6x^2 + 1 \end{aligned}$$

Continuing to find the higher derivatives, we have

$$\begin{aligned} f''(x) &= 20x^3 - 12x \\ f'''(x) &= 60x^2 - 12 \\ f^{(4)}(x) &= 120x \\ f^{(5)}(x) &= 120 \\ f^{(n)}(x) &= 0 \quad \text{for } n \geq 6 \end{aligned}$$

All derivatives after the fifth derivative are equal to zero. ◂◂

EXAMPLE 3 ▸▸ Evaluate the second derivative of $y = \frac{2}{1 - x}$ for $x = -2$.

We write the function as $y = 2(1 - x)^{-1}$ and then find the derivatives.

$$y = 2(1 - x)^{-1}$$

$$\frac{dy}{dx} = 2(-1)(1 - x)^{-2}(-1) = 2(1 - x)^{-2}$$

$$\frac{d^2y}{dx^2} = 2(-2)(1 - x)^{-3}(-1) = 4(1 - x)^{-3} = \frac{4}{(1 - x)^3}$$

Evaluating the second derivative for $x = -2$, we have

$$\left.\frac{d^2y}{dx^2}\right|_{x=-2} = \frac{4}{(1 + 2)^3} = \frac{4}{27}$$

The function is not differentiable for $x = 1$. Also, if we continue to find higher derivatives, the expressions will not become zero, as in Examples 1 and 2. ◂◂

EXAMPLE 4 ▸▸ Find y'' for the implicit function defined by $2x^2 + 3y^2 = 6$.

Differentiating with respect to x, we have

$$2(2x) + 3(2yy') = 0$$

$$4x + 6yy' = 0 \quad \text{or} \quad 2x + 3yy' = 0 \tag{1}$$

CAUTION ▸ Before differentiating again we see that **$3yy'$ *is a product,*** and we note that the derivative of y' is y''. Thus, differentiating again, we have

differentiation of $3yy'$

$$2 + 3yy'' + 3y'(y') = 0$$

$$2 + 3yy'' + 3(y')^2 = 0 \tag{2}$$

Now, solving Eq. (1) for y' and substituting this into Eq. (2), we have

$$y' = -\frac{2x}{3y}$$

$$2 + 3yy'' + 3\left(-\frac{2x}{3y}\right)^2 = 0$$

$$2 + 3yy'' + \frac{4x^2}{3y^2} = 0$$

$$6y^2 + 9y^3y'' + 4x^2 = 0$$

$$y'' = \frac{-4x^2 - 6y^2}{9y^3} = \frac{-2(2x^2 + 3y^2)}{9y^3}$$

Since $2x^2 + 3y^2 = 6$, we have

$$y'' = \frac{-2(6)}{9y^3} = -\frac{4}{3y^3}$$ ◂◂

As mentioned earlier, higher derivatives are useful in certain applications. This is particularly true of the second derivative. The first and second derivatives are used in the next chapter for several types of applications, and higher derivatives are used when we discuss infinite series in Chapter 29. An important technical application of the second derivative is shown in the example that follows.

In Section 23-4 we briefly discussed the instantaneous velocity of an object, and in the exercises we mentioned acceleration. From that discussion we recall that the instantaneous velocity is the time rate of change of the displacement, and *the* **instantaneous acceleration** *is the time rate of change of the instantaneous velocity.* Therefore, we see that the *acceleration is found from the second derivative of the displacement with respect to time.* Consider the following example.

EXAMPLE 5 ▸▸ For the first 12 s after launch, the height s (in metres) of a certain rocket is given by $s = 10\sqrt{t^4 + 25} - 50$. Find the vertical acceleration of the rocket when $t = 10.0$ s.

Since the velocity is found from the first derivative and the acceleration is found from the second derivative, we must find the second derivative and then evaluate it for $t = 10.0$ s.

$$s = 10\sqrt{t^4 + 25} - 50$$

$$v = \frac{ds}{dt} = 10\left(\frac{1}{2}\right)(t^4 + 25)^{-1/2}(4t^3) = \frac{20t^3}{(t^4 + 25)^{1/2}}$$

$$a = \frac{dv}{dt} = \frac{d^2s}{dt^2} = \frac{(t^4 + 25)^{1/2}(60t^2) - 20t^3(\frac{1}{2})(t^4 + 25)^{-1/2}(4t^3)}{t^4 + 25}$$

$$= \frac{(t^4 + 25)(60t^2) - 40t^6}{(t^4 + 25)^{3/2}} = \frac{20t^6 + 1500t^2}{(t^4 + 25)^{3/2}}$$

$$= \frac{20t^2(t^4 + 75)}{(t^4 + 25)^{3/2}}$$

Finding the value of the acceleration when $t = 10.0$ s, we have

$$a\Big|_{t=10.0} = \frac{20(10.0^2)(10.0^4 + 75)}{(10.0^4 + 25)^{3/2}} = 20.1 \text{ m/s}^2$$ ◂◂

Exercises 23–9

In Exercises 1–8, find all the higher derivatives of the given functions.

1. $y = x^3 + x^2$
2. $f(x) = 3x - x^4$
3. $f(x) = x^3 - 6x^4$
4. $y = 2x^5 + 5x^2$
5. $y = (1 - 2x)^4$
6. $f(x) = (3x + 2)^3$
7. $f(x) = x(2x + 1)^3$
8. $y = x(x - 1)^3$

In Exercises 9–28, find the second derivative of each of the given functions.

9. $y = 2x^7 - x^6 - 3x$
10. $y = 6x - 2x^5$
11. $y = 2x + \sqrt{x}$
12. $y = x^2 - \dfrac{1}{\sqrt{x}}$
13. $f(x) = \sqrt[4]{8x - 3}$
14. $f(x) = \sqrt[3]{6x + 5}$
15. $f(x) = \dfrac{4}{\sqrt{1 - 2x}}$
16. $f(x) = \dfrac{5}{\sqrt{3 - 4x}}$
17. $y = 2(2 - 5x)^4$
18. $y = (4x + 1)^6$
19. $y = (3x^2 - 1)^5$
20. $y = 3(2x^3 + 3)^4$
21. $f(x) = \dfrac{2x}{1 - x}$
22. $f(x) = \dfrac{1 - x}{1 + x}$
23. $y = \dfrac{x^2}{x + 1}$
24. $y = \dfrac{x}{\sqrt{1 - x^2}}$
25. $x^2 - y^2 = 9$
26. $xy + y^2 = 4$
27. $x^2 - xy = 1 - y^2$
28. $xy = y^2 + 1$

In Exercises 29–34, evaluate the second derivative of the given function for the given value of x.

29. $f(x) = \sqrt{x^2 + 9}, x = 4$
30. $f(x) = x - \dfrac{2}{x^3}, x = -1$
31. $y = 3x^{2/3} - \dfrac{2}{x}, x = -8$
32. $y = 3(1 + 2x)^4, x = \dfrac{1}{2}$
33. $y = x(1 - x)^5, x = 2$
34. $y = \dfrac{x}{2 - 3x}, x = -\dfrac{1}{3}$

In Exercises 35–40, solve the given problems by finding the appropriate derivatives.

35. What is the instantaneous rate of change of the first derivative of y with respect to x for $y = (1 - 2x)^4$ for $x = 1$?

36. What is the instantaneous rate of change of the first derivative of y with respect to x for $2xy + y = 1$ for $x = 0.5$?

37. A bullet is fired vertically upward. Its distance s, in metres, above the ground is given by $s = 670t - 4.9t^2$, where t is measured in seconds. Find the acceleration of the bullet.

38. In testing the brakes on a new model automobile, it was found that the distance s, in metres, which it traveled under specified conditions after the brakes were applied was given by $s = 19.2t - 0.40t^3$. What were the velocity and acceleration of the automobile for $t = 4.00$ s?

39. The voltage V, in volts, induced in an inductor in an electric circuit is given by $V = L(d^2q/dt^2)$, where L is the inductance in henrys (H). Find the expression of the voltage induced in a 1.60-H inductor if $q = \sqrt{2t + 1} - 1$.

40. How fast is the rate of change of solar radiation changing on the surface in Exercise 27 of Section 23-4 at 3 P.M.?

Chapter Equations, Review Exercises, and Practice Test

Chapter Equations

Limit of function
$$\lim_{x \to a} f(x) = L \tag{23-1}$$

Increments
$$\Delta x = x_2 - x_1 \tag{23-2}$$
$$\Delta y = y_2 - y_1 \tag{23-3}$$

Slope
$$m_{PQ} = \frac{(y_1 + \Delta y) - y_1}{(x_1 + \Delta x) - x_1} = \frac{\Delta y}{\Delta x} \tag{23-4}$$
$$\Delta y = f(x + \Delta x) - f(x) \tag{23-5}$$

Definition of derivative
$$\lim_{\Delta x \to 0} \frac{\Delta y}{\Delta x} = \lim_{\Delta x \to 0} \frac{f(x + \Delta x) - f(x)}{\Delta x} \tag{23-6}$$

Instantaneous velocity
$$v = \lim_{\Delta t \to 0} \frac{\Delta s}{\Delta t} \tag{23-7}$$

Derivatives of polynomials
$$\frac{dc}{dx} = 0 \tag{23-8}$$
$$\frac{dx^n}{dx} = nx^{n-1} \tag{23-9}$$
$$\frac{d(cu)}{dx} = c\frac{du}{dx} \tag{23-10}$$
$$\frac{d(u + v)}{dx} = \frac{du}{dx} + \frac{dv}{dx} \tag{23-11}$$

Derivative of product
$$\frac{d(uv)}{dx} = u\frac{dv}{dx} + v\frac{du}{dx} \tag{23-12}$$

Derivative of quotient
$$\frac{d\frac{u}{v}}{dx} = \frac{v\frac{du}{dx} - u\frac{dv}{dx}}{v^2} \tag{23-13}$$

Chain rule
$$\frac{dy}{dx} = \frac{dy}{du}\frac{du}{dx} \tag{23-14}$$

Derivative of power
$$\frac{du^n}{dx} = nu^{n-1}\left(\frac{du}{dx}\right) \quad (23\text{-}15)$$

$$\frac{du^{p/q}}{dx} = \frac{p}{q}u^{(p/q)-1}\frac{du}{dx} \quad (23\text{-}16)$$

Review Exercises

In Exercises 1–12, evaluate the given limits.

1. $\lim_{x\to 4}(8 - 3x)$
2. $\lim_{x\to 3}(2x^2 - 10)$
3. $\lim_{x\to -3}\frac{2x+5}{x-1}$
4. $\lim_{x\to 1}\frac{x^2-1}{x+1}$
5. $\lim_{x\to 2}\frac{4x-8}{x^2-4}$
6. $\lim_{x\to 5}\frac{x^2-25}{3x-15}$
7. $\lim_{x\to 2}\frac{x^2+3x-10}{x^2-x-2}$
8. $\lim_{x\to 0}\frac{(x-3)^2-9}{x}$
9. $\lim_{x\to\infty}\frac{2+\frac{1}{x+4}}{3-\frac{1}{x^2}}$
10. $\lim_{x\to\infty}\left(7-\frac{1}{x+1}\right)$
11. $\lim_{x\to\infty}\frac{x-2x^3}{1+x^3}$
12. $\lim_{x\to\infty}\frac{2x+5}{3x^3-2x}$

In Exercises 13–20, use the delta-process to find the derivative of each of the given functions.

13. $y = 7 + 5x$
14. $y = 6x - 2$
15. $y = 6 - 2x^2$
16. $y = 2x^2 - x^3$
17. $y = \frac{2}{x^2}$
18. $y = \frac{1}{1-4x}$
19. $y = \sqrt{x+5}$
20. $y = \frac{1}{\sqrt{x}}$

In Exercises 21–36, find the derivative of each of the given functions.

21. $y = 2x^7 - 3x^2 + 5$
22. $y = 8x^7 - 2^5 - x$
23. $y = 4\sqrt{x} - \frac{3}{x} + \sqrt{3}$
24. $y = \frac{3}{x^2} - 8\sqrt[4]{x}$
25. $y = \frac{x}{1-x}$
26. $y = \frac{2x-1}{x^2+1}$
27. $y = (2-3x)^4$
28. $y = (2x^2-3)^6$
29. $y = \frac{3}{(5-2x^2)^{3/4}}$
30. $y = \frac{7}{(3x-1)^3}$
31. $y = x^2\sqrt{1-6x}$
32. $y = (x-1)^3(x^2-2)^2$
33. $y = \frac{\sqrt{4x+3}}{2x}$
34. $y = \frac{x}{\sqrt{x^2+1}}$
35. $(2x-3y)^3 = x^2 - y$
36. $x^2y^2 = x^2 + y^2$

In Exercises 37–40, evaluate the derivatives of the given functions for the given values of x.

37. $y = \frac{4}{x} + 2\sqrt[3]{x},\ x = 8$
38. $y = (3x-5)^4,\ x = -2$
39. $y = 2x\sqrt{4x+1},\ x = 6$
40. $y = \frac{\sqrt{2x^2+1}}{3x},\ x = 2$

In Exercises 41–44, find the second derivative of each of the given functions.

41. $y = 3x^4 - \frac{1}{x}$
42. $y = \sqrt{1-8x}$
43. $y = \frac{1-3x}{1+4x}$
44. $y = 2x(6x+5)^4$

In Exercises 45–72, solve the given problems.

45. View the graph of $y = \frac{2(x^2-4)}{x-2}$ on a graphing calculator with Xmin $= -1$ (or 0), Xmax $= 4$, Ymin $= 0$, Ymax $= 10$ such that y can be evaluated for $x = 2$ exactly. Determine the value of y for $x = 2$. Comment on the accuracy of the view and the value found.
46. A continuous function $f(x)$ is positive at $x = 0$ and negative for $x = 1$. How many solutions does $f(x) = 0$ have between $x = 0$ and $x = 1$?
47. The number n of bacteria in a certain culture is found to be $n = 8000 + 4000/(t+1)^2$, where t is the time (in hours) after the culture is started. What are (1) the initial number of bacteria and (2) the number of bacteria as $t \to \infty$?

48. Two lenses of focal lengths f_1 and f_2, separated by a distance d, are used in the study of lasers. The combined focal length f of this lens combination is $f = \dfrac{f_1 f_2}{f_1 + f_2 - d}$. If f_2 and d remain constant, find the limiting value of f as f_1 continues to increase in value.

49. Find the slope of a line tangent to the curve of $y = 7x^4 - x^3$ at $(-1, 8)$. View the graph of the curve on a graphing calculator to see if the calculated value of the slope is reasonable for $x = -1$.

50. Find the slope of a line tangent to the curve of $y = \sqrt[3]{3 - 8x}$ at $(-3, 3)$. View the graph of the curve on a graphing calculator to see if the calculated value of the slope is reasonable for $x = -3$.

51. Find the value of c if the curve of $y = x^2 + c$ is tangent to the line $y = 2x$.

52. The displacement s (in cm) of a piston during each 8-second cycle is given by $s = 8t - t^2$, where t is the time in seconds. For what value(s) of t is the velocity of the piston 4 cm/s?

53. The reliability R of a computer system measures the probability that the system will be operating properly after t hours. For one system, $R = 1 - kt + \dfrac{k^2t^2}{2} - \dfrac{k^3t^3}{6}$, where k is a constant. Find the expression for the instantaneous rate of change of R with respect to t.

54. The distance s (in metres) traveled by a subway train after the brakes are applied is given by $s = 20t - 2t^2$. How far does it travel, after the brakes are applied, in coming to a stop?

55. The electric field E at a distance r from a point charge is $E = k/r^2$, where k is a constant. Find an expression for the instantaneous rate of change of the electric field with respect to r.

56. The velocity of an object moving with constant acceleration can be found from the equation $v = \sqrt{v_0^2 + 2as}$, where v_0 is the initial velocity, a is the acceleration, and s is the distance traveled. Find dv/ds.

57. The voltage induced in an inductor L is given by $E = L\dfrac{dI}{dt}$, where I is the current in the circuit and t is the time. Find the voltage induced in a 0.4-H inductor if the current I (in amperes) is related to the time (in seconds) by $I = t(0.01t + 1)^3$.

58. In studying the energy used by a mechanical robotic device, the equation $v = \dfrac{z}{\alpha(1 - z^2) - \beta}$ is used. If α and β are constants, find dv/dz.

59. The frictional radius r_f of a collar used in a braking system is given by $r_f = \dfrac{2(R^3 - r^3)}{3(R^2 - r^2)}$ where R is the outer radius and r is the inner radius. Find dr_f/dR if r is constant.

60. Water is being drained from a pond such that the volume of water in the pond (in m^3) after t hours is given by $V = 5000(60 - t)^2$. Find the rate at which the pond is being drained after 4.00 h.

61. The frequency f of a certain electronic oscillator is given by $f = \dfrac{1}{2\pi\sqrt{C(L + 2)}}$, where C is a capacitance and L is an inductance. If C is constant, find df/dL.

62. The volume V of fluid produced in the retina of the eye in reaction to exposure to light of intensity I is given by $V = \dfrac{aI^2}{b - I}$, where a and b are constants. Find dV/dI.

63. Under certain conditions, the efficiency of an internal combustion engine is given by

$$\text{eff(in percent)} = 100\left(1 - \frac{1}{(V_1/V_2)^{0.4}}\right)$$

where V_1 and V_2 are the maximum and minimum volumes of air in a cylinder, respectively. Assuming that V_2 is kept constant, find the expression for the rate of change of efficiency with respect to V_1.

64. The temperature T (in °C) in a freezer as a function of the time t (in hours) is given by $T = \dfrac{10(1 - t)}{0.5t + 1}$. Find dT/dt.

65. The deflection y, in metres, of a 5.00-m beam is given by $y = 0.0001(x^5 - 25x^2)$, where x is the distance, in metres, from one end. Find the value of d^2y/dx^2 (the rate at which the slope of the beam changes) where $x = 3.00$ m.

66. The number n of grams of a compound formed during a certain chemical reaction is given by $n = \dfrac{2t}{t + 1}$, where t is the time in minutes. Evaluate d^2n/dt^2 (the rate of increase of the amount of the compound being formed) when $t = 4.00$ min.

67. The area of a rectangular patio is to be 75 m^2. Express the perimeter p of the patio as a function of its width w and find dp/dw.

68. A water tank is being designed in the shape of a right circular cylinder with a volume of 100 m^3. Find the expression for the instantaneous rate of change of the total surface area A of the tank with respect to the radius r of the base.

69. An arch over a walkway can be described by the first-quadrant part of the parabola $y = 4 - x^2$. In order to determine the size and shape of rectangular objects that can pass under the arch, express the area A of a rectangle inscribed under the parabola in terms of x. Find dA/dx.

70. An airplane flies over an observer with a velocity of 400 km/h and at an altitude of 500 m. If the plane flies horizontally in a straight line, find the rate at which the distance from the observer to the plane is changing 0.600 min after the plane passes over the observer. See Fig. 23-27.

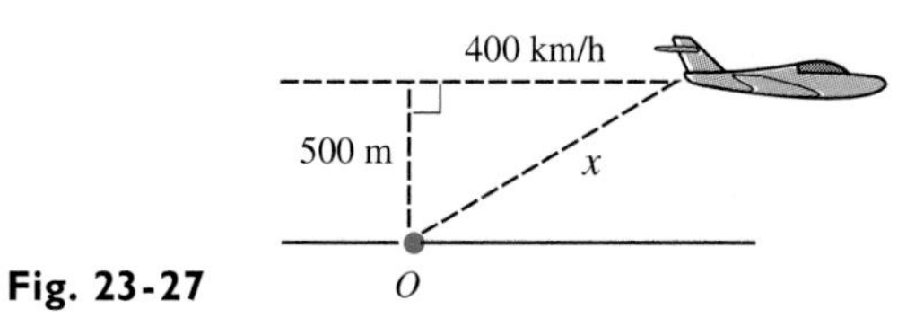

Fig. 23-27

71. A computer analysis showed that a specialized piece of machinery has a value, in dollars, given by $V = 1\,500\,000/(2t + 10)$, where t is the number of years after purchase. Calculate the value of dV/dt and d^2V/dt^2 for $t = 5$ years. What is the meaning of these values?

72. The *radius of curvature* of $y = f(x)$ at the point (x, y) on a curve is given by

$$R = \frac{[1 + (y')^2]^{3/2}}{|y''|}$$

A certain roadway follows the parabola $y = 1.2x - x^2$ for $0 < x < 1.2$, where x is measured in kilometres. Find R for $x = 0.2$ km and $x = 0.6$ km. See Fig. 23-28.

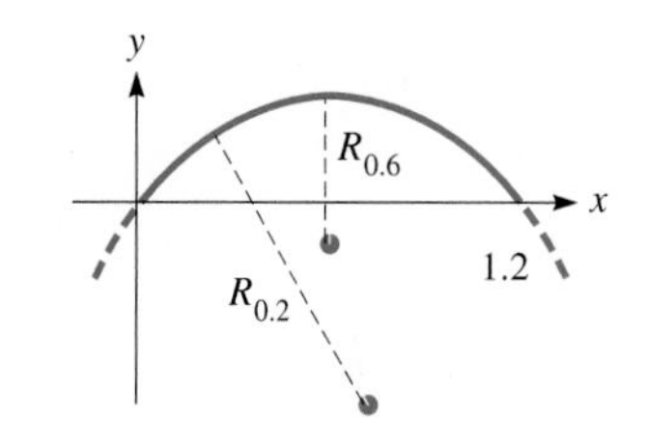

Fig. 23-28

Writing Exercises

73. An engineer designing military rockets uses computer simulation to find the path of a rocket as $y = f(x)$ and the path of an aircraft to be $y = g(x)$. Write two or three paragraphs explaining how the engineer can determine the angle at which the path of the rocket crosses the path of the aircraft.

Practice Test

1. Find $\lim_{x \to 1} \dfrac{x^2 - x}{x^2 - 1}$.

2. Find $\lim_{x \to \infty} \dfrac{1 - 4x^2}{x + 2x^2}$.

3. Find the slope of a line tangent to the curve of $y = 3x^2 - \dfrac{4}{x^2}$ at $(2, 11)$.

4. The displacement s (in cm) of a pumping machine piston in each cycle is given by $s = t\sqrt{10 - 2t}$. Find its velocity for $t = 4.00$ s.

5. Find dy/dx: $(1 + y^2)^3 - x^2y = 7x$.

6. Under certain conditions, due to the presence of a charge q, the electric potential V along a line is given by

$$V = \frac{kq}{\sqrt{x^2 + b^2}}$$

where k is a constant and b is the minimum distance from the charge to the line. Find the expression for the rate of change of V with respect to x.

7. Find the second derivative of $y = \dfrac{2}{3x + 2}$.

8. By using the delta-process, find the derivative of $y = 5x - 2x^2$ with respect to x.

24

Applications of the Derivative

In Section 24-7 we see how to use the derivative to find the dimensions of a beam of greatest strength.

In Chapter 23 we developed the meaning of the derivative of a function and then went on to find several formulas by which we can differentiate functions. We also established the concept of the derivative as an instantaneous rate of change.

Numerous applications of the derivative were indicated in the examples and exercises of Chapter 23. It was not necessary to develop any of these applications in detail, since finding the derivative was the primary concern. There are, however, certain types of problems in many areas of technology in which the derivative plays a key role in the solution.

These applications include the analysis of the motion of objects and finding the maximum values or minimum values of functions. Such values are useful, for example, in finding maximum possible income from production and the least amount of material needed in making a product. In this chapter we consider some of these types of applications of the derivative.

24-1 Tangents and Normals

The first application of the derivative we consider involves finding the equation of a line that is *tangent* to a given curve and the equation of a line that is *normal* (perpendicular) to a given curve.

Tangent Line

To find the equation of a line tangent to a curve at a given point, we first find the derivative of the function. The derivative is then evaluated at the point, and this gives us the slope of a line tangent to the curve at the point. Then, by using the point-slope form of the equation of a straight line, we find the equation of the tangent line. The following examples illustrate the method.

EXAMPLE 1 ▸▸ Find the equation of the line tangent to the parabola $y = x^2 - 1$ at the point $(-2, 3)$.

The derivative of this function is

$$\frac{dy}{dx} = 2x$$

The value of this derivative for the value $x = -2$ (the y-value of 3 is not used since the derivative does not directly contain y) is

$$\frac{dy}{dx} = -4$$

which means that the slope of the tangent line at $(-2, 3)$ is -4. Thus, by using the point-slope form of the equation of the straight line we obtain the desired equation. Thus we have

$$y - 3 = -4(x + 2)$$
$$y - 3 = -4x - 8$$

or

$$y = -4x - 5$$

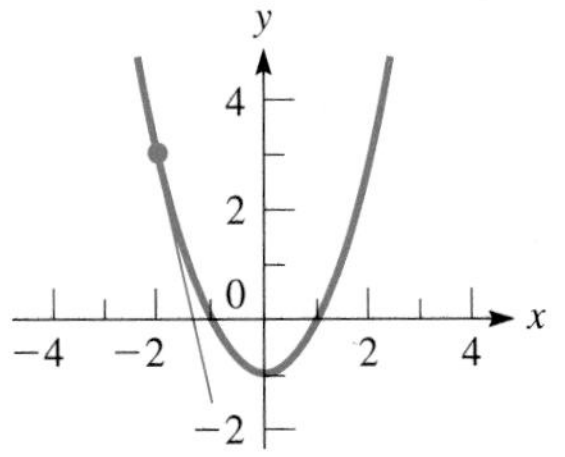

The parabola and the tangent line $y = -4x - 5$ are shown in Fig. 24-1. ◂◂

Fig. 24-1

EXAMPLE 2 ▸▸ Find the equation of the line tangent to the ellipse $4x^2 + 9y^2 = 40$ at the point $(1, 2)$.

The easiest method of finding the derivative of this equation is to treat the equation as an implicit function. In this way we have the following solution:

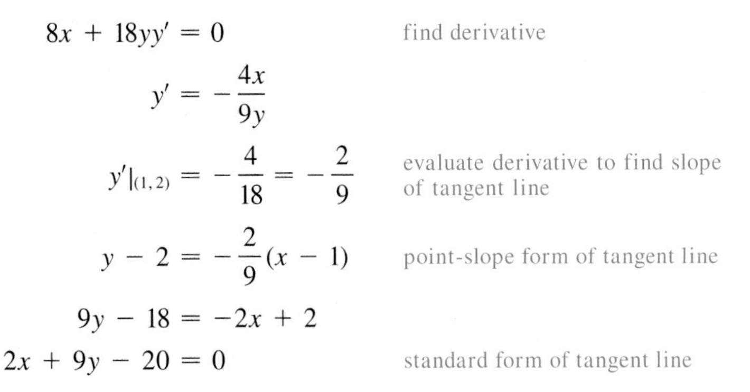

$$8x + 18yy' = 0 \quad \text{find derivative}$$
$$y' = -\frac{4x}{9y}$$
$$y'|_{(1,2)} = -\frac{4}{18} = -\frac{2}{9} \quad \text{evaluate derivative to find slope of tangent line}$$
$$y - 2 = -\frac{2}{9}(x - 1) \quad \text{point-slope form of tangent line}$$
$$9y - 18 = -2x + 2$$
$$2x + 9y - 20 = 0 \quad \text{standard form of tangent line}$$

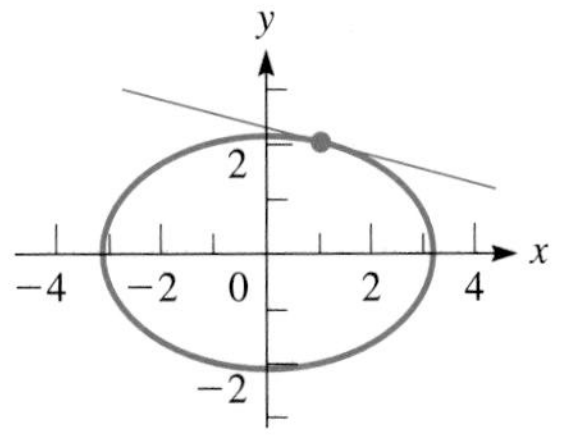

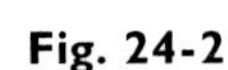

Fig. 24-2

The ellipse and the tangent line $2x + 9y - 20 = 0$ are shown in Fig. 24-2.

The derivative could also have been found by solving for y. In this way we would have to differentiate $y = \frac{2}{3}\sqrt{10 - x^2}$. ◂◂

Normal Line

NOTE ▶

If we wish to obtain the equation of a line normal (perpendicular to a tangent) to a curve, we recall that the slopes of perpendicular lines are negative reciprocals. Thus the derivative is found and evaluated at the specified point. Since this gives the slope of a tangent line, ***we take the negative reciprocal of this number to find the slope of the normal line.*** Then, by using the point-slope form of the equation of a straight line we find the equation of the normal. The following examples illustrate the method.

EXAMPLE 3 ▶▶ Find the equation of the line normal to the hyperbola $y = 2/x$ at the point $(2, 1)$.

The derivative of this function is $dy/dx = -2/x^2$, which evaluated at $x = 2$ gives $dy/dx = -1/2$. Therefore, the slope of a line normal to the curve at the point $(2, 1)$ is 2. The equation of the normal line is then

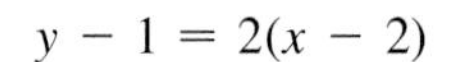

$$y - 1 = 2(x - 2)$$

or

$$y = 2x - 3$$

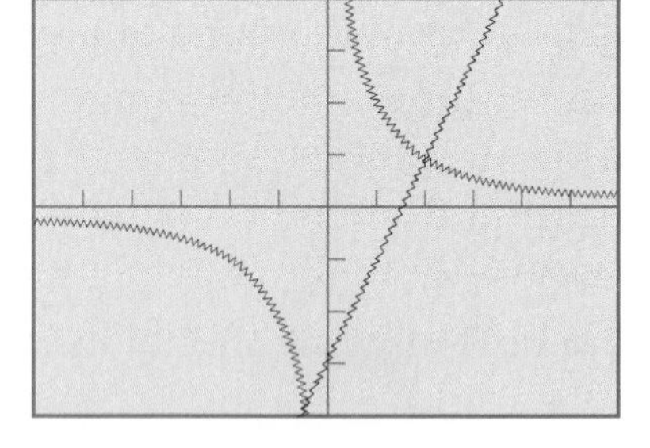

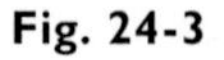

Fig. 24-3

The hyperbola and the normal line are shown in the graphing-calculator view shown in Fig. 24-3. The RANGE values used are

Xmin = −6, Xmax = 6, Ymin = −4, Ymax = 4 ◀◀

EXAMPLE 4 ▶▶ Find the y-intercept of the line that is normal to the curve of $y = 2x - \frac{1}{3}x^3$, where $x = 3$.

First we find the equation of the normal line. We find the y-intercept by writing the equation in slope-intercept form. The solution proceeds as follows:

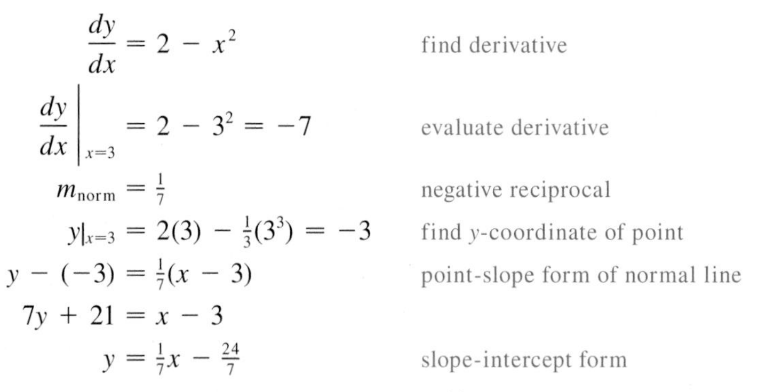

$$\frac{dy}{dx} = 2 - x^2 \qquad \text{find derivative}$$

$$\left.\frac{dy}{dx}\right|_{x=3} = 2 - 3^2 = -7 \qquad \text{evaluate derivative}$$

$$m_{\text{norm}} = \frac{1}{7} \qquad \text{negative reciprocal}$$

$$y|_{x=3} = 2(3) - \frac{1}{3}(3^3) = -3 \qquad \text{find } y\text{-coordinate of point}$$

$$y - (-3) = \frac{1}{7}(x - 3) \qquad \text{point-slope form of normal line}$$

$$7y + 21 = x - 3$$

$$y = \frac{1}{7}x - \frac{24}{7} \qquad \text{slope-intercept form}$$

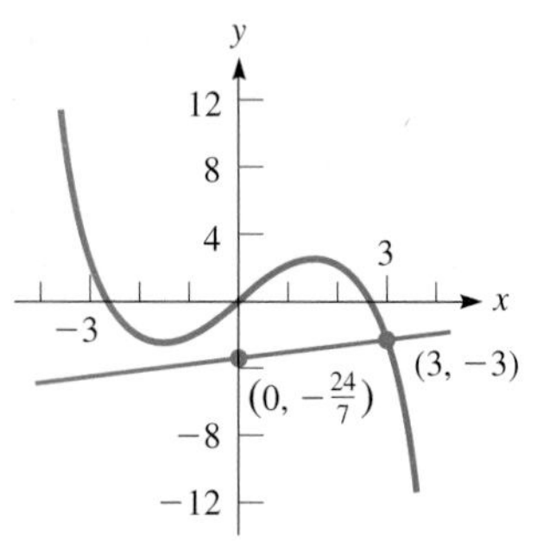

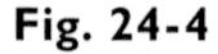

Fig. 24-4

This tells us that the y-intercept is $(0, -\frac{24}{7})$. The curve, normal line, and intercept are shown in Fig. 24-4. ◀◀

Many of the applications of tangents and normals are geometric. However, there are certain applications in technology, and one of these is shown in the following example. Others are shown in the exercises.

EXAMPLE 5 ▸▸ In Fig. 24-5 the cross section of a parabolic solar reflector is shown, along with an incident ray of light and the reflected ray. The angle of incidence i is equal to the angle of reflection r where both angles are measured with respect to the normal to the surface. If the incident ray strikes at the point where the slope of the normal is -1, and the equation of the parabola is $4y = x^2$, what is the equation of the normal line?

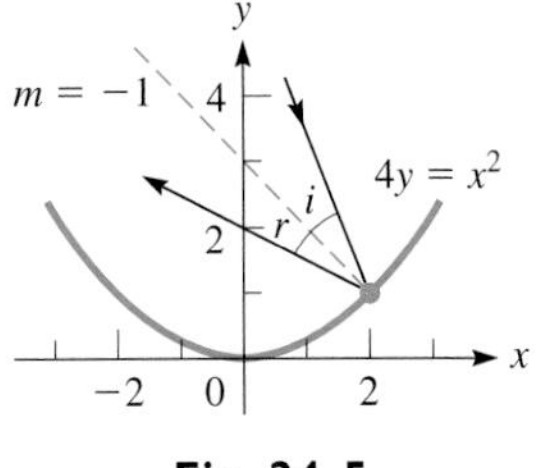

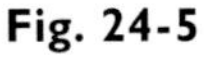
Fig. 24-5

If the slope of the normal line is -1, then the slope of a tangent line is $-(\frac{1}{-1}) = 1$. Therefore, we know that the value of the derivative at the point of reflection is 1. This allows us to find the coordinates of the point.

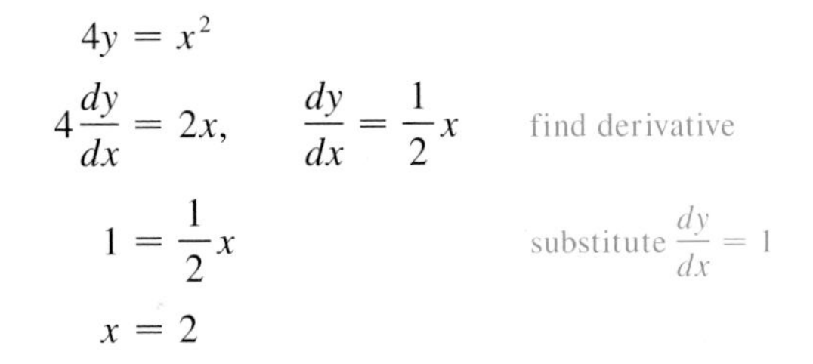

$$4y = x^2$$

$$4\frac{dy}{dx} = 2x, \quad \frac{dy}{dx} = \frac{1}{2}x \qquad \text{find derivative}$$

$$1 = \frac{1}{2}x \qquad \text{substitute } \frac{dy}{dx} = 1$$

$$x = 2$$

This means that the x-coordinate of the point of reflection is 2. We can find the y-coordinate by substituting $x = 2$ into the equation of the parabola. Thus the point is $(2, 1)$. Since the slope is -1, the equation is

$$y - 1 = (-1)(x - 2)$$
$$y = -x + 3$$

We might note that if the incident ray is vertical, for which $i = 45°$, the reflected ray passes through $(0, 1)$, which is the focus of the parabola. This shows the important reflection property of a parabola that *any incident ray parallel to the axis of a parabola passes through the focus.* We first noted this property in our discussion of the parabola in Example 5 of Section 21-4. ◂◂

Exercises 24–1

In Exercises 1–4, find the equations of the lines tangent to the indicated curves at the given points. In Exercises 1 and 4, sketch the curve and tangent line. In Exercises 2 and 3, use a graphing calculator to view the curve and tangent line.

1. $y = x^2 + 2$ at $(2, 6)$

2. $y = \frac{1}{3}x^3 - 5x$ at $(3, -6)$

3. $y = \dfrac{1}{x^2 + 1}$ at $(1, \frac{1}{2})$

4. $x^2 + y^2 = 25$ at $(3, 4)$

In Exercises 5–8, find the equations of the lines normal to the indicated curves at the given points. In Exercises 5 and 8, sketch the curve and normal line. In Exercises 6 and 7, use a graphing calculator to view the curve and normal line.

5. $y = 6x - 2x^2$ at $(2, 4)$

6. $y = 8 - x^3$ at $(-1, 9)$

7. $y = \dfrac{1}{(x^2 + 1)^2}$ at $(1, \frac{1}{4})$

8. $x^2 - y^2 = 8$ at $(3, 1)$

In Exercises 9–12, find the equations of the tangent lines and the normal lines to the indicated curves. In Exercises 9 and 10, use a graphing calculator to view the curve and the lines. In Exercises 11 and 12, sketch the curve and lines.

9. $y = \dfrac{1}{\sqrt{x^2 + 1}}$, where $x = \sqrt{3}$

10. $y = \dfrac{4}{(5 - 2x)^2}$, where $x = 2$

11. The parabola with vertex at $(0, 3)$ and focus at $(0, 0)$, where $x = -1$

12. The ellipse with focus at $(4, 0)$, vertex at $(5, 0)$, and center at $(0, 0)$, where $x = 2$

In Exercises 13–16, find the equations of the lines tangent or normal to the given curves and with the given slopes. View the curves and lines on a graphing calculator.

13. $y = x^2 - 2x$, tangent line with slope 2

14. $y = \sqrt{2x - 9}$, tangent line with slope 1

15. $y = (2x - 1)^3$, normal line with slope $-\frac{1}{24}$, $x > 0$

16. $y = \frac{1}{2}x^4 + 1$, normal line with slope 4

In Exercises 17–24, solve the given problems by finding the equation of the appropriate tangent or normal line.

17. Find the x-intercept of the line tangent to the parabola $y = 4x^2 - 8x$ at $(-1, 12)$.

18. Find the y-intercept of the line normal to the curve $y = x^{3/4}$, where $x = 16$.

19. A certain suspension cable with supports on the same level is closely approximated as being parabolic in shape. If the supports are 80 m apart and the sag at the center is 10 m, what is the equation of the line along which the tension acts (tangentially) at the right support? (Choose the origin of the coordinate system at the lowest point of the cable.)

20. A laser source is 2.00 cm from a spherical surface of radius 3.00 cm, and the laser beam is tangent to the spherical surface. By placing the center of the sphere at the origin and the source on the positive x-axis, find the equation of the line along which the beam shown in Fig. 24-6 is directed.

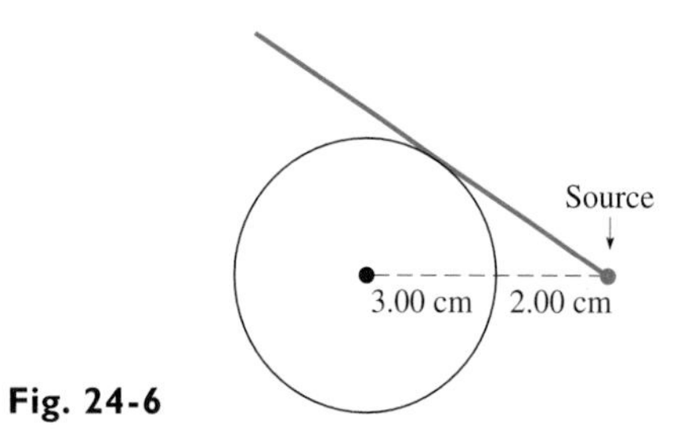

Fig. 24-6

21. In an electric field, the lines of force are perpendicular to the curves of equal electric potential. In a certain electric field, a curve of equal potential is $y = \sqrt{2x^2 + 8}$. If the line along which the force acts on an electron has an inclination of 135°, find its equation.

22. A radio wave reflects from a reflecting surface in the same way as a light wave (see Example 5). A certain horizontal radio wave reflects off a parabolic reflector such that the reflected wave is 43.60° below the horizontal, as shown in Fig. 24-7. If the equation of the parabola is $y^2 = 8x$, what is the equation of the normal line through the point of reflection?

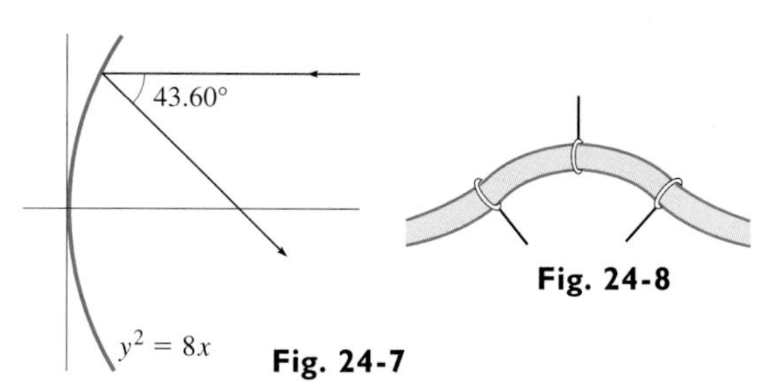

Fig. 24-7

Fig. 24-8

23. In designing a flexible tubing system, the supports for the tubing must be perpendicular to the tubing. If a section of the tubing follows the curve $y = \dfrac{4}{x^2 + 1}$ (units in dm, with $-2 < x < 2$), along which lines must the supports be directed if they are located at $x = -1$, $x = 0$, and $x = 1$? See Fig. 24-8.

24. On a particular drawing, a pulley wheel can be described by the equation $x^2 + y^2 = 100$ (units in cm). The pulley belt is directed along the lines $y = -10$ and $4y - 3x - 50 = 0$ when first and last making contact with the wheel. What are the first and last points on the wheel where the belt makes contact?

24-2 Newton's Method for Solving Equations

As we know, finding the roots of an equation $f(x) = 0$ is very important in mathematics and in many types of applications, and we have developed methods of solving many types of equations in the previous chapters. However, for a great many algebraic and nonalgebraic equations there is no method for finding the roots exactly.

We have shown how equations can be solved graphically, and by using a graphing calculator the roots can be found with great accuracy. In this section we show a method, known as **Newton's method,** which uses the derivative to locate approximately, but very accurately, the real roots of many kinds of equations. It can be used with polynomial equations of any degree and with other algebraic and nonalgebraic equations.

Newton's method is an example of an **iterative method.** In using this type of method, we start with a reasonable guess for a root of the equation. By using the method we obtain a new value, which is a better approximation. This, in turn, gives a still better approximation. Continuing in this way, using a calculator we can obtain an approximate answer with the required accuracy. Iterative methods in general are easily programmable for use on a computer.

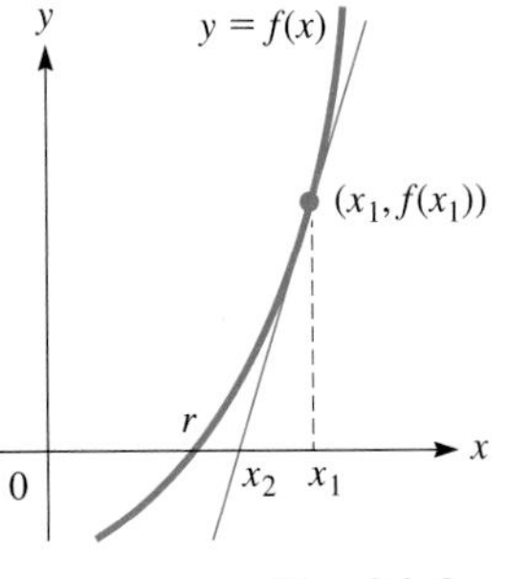

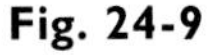

Fig. 24-9

Let us consider a section of the curve of $y = f(x)$ that (a) crosses the x-axis, (b) always has either a positive slope or a negative slope, and (c) has a slope that either becomes greater or becomes less as x increases. See Fig. 24-9. The curve in the figure crosses the x-axis at $x = r$, which means that $x = r$ is a root of the equation $f(x) = 0$. If x_1 is sufficiently close to r, a line tangent to the curve at $[x_1, f(x_1)]$ will cross the x-axis at a point $(x_2, 0)$, which is closer to r than is x_1.

We know that the slope of the tangent line is the value of the derivative at x_1, or $m_{\tan} = f'(x_1)$. Therefore, the equation of the tangent line is

$$y - f(x_1) = f'(x_1)(x - x_1)$$

For the point $(x_2, 0)$ on this line, we have

$$-f(x_1) = f'(x_1)(x_2 - x_1)$$

Solving for x_2, we have the formula

$$x_2 = x_1 - \frac{f(x_1)}{f'(x_1)} \tag{24-1}$$

Here, x_2 is a second approximation to the root. We can then replace x_1 in Eq. (24-1) by x_2 and find a closer approximation, x_3. This process can be repeated as many times as needed to find the root to the required accuracy. This method lends itself well to the use of a calculator or a computer for finding the root.

EXAMPLE 1 ▸▸ Find the root of $x^2 - 3x + 1 = 0$ between $x = 0$ and $x = 1$.

Here, $f(x) = x^2 - 3x + 1$. Therefore, $f(0) = 1$ and $f(1) = -1$, which indicates that the root may be near the middle of the interval. Since x_1 must be within the interval, we choose $x_1 = 0.5$.

The derivative is

$$f'(x) = 2x - 3$$

Therefore, $f(0.5) = -0.25$ and $f'(0.5) = -2$, which gives us

$$x_2 = 0.5 - \frac{-0.25}{-2} = 0.375$$

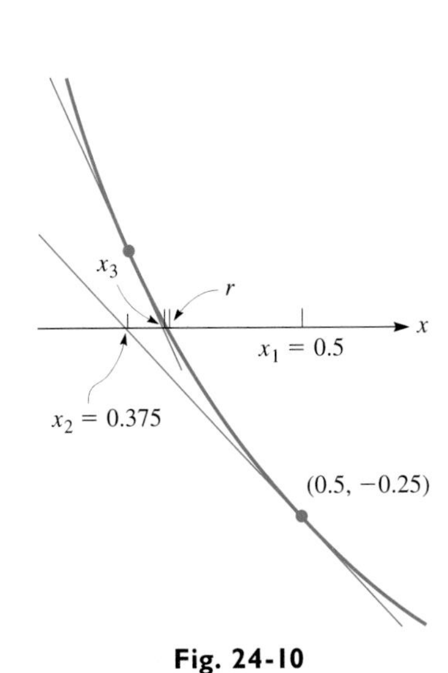

Fig. 24-10

This is a second approximation, which is closer to the actual value of the root. See Fig. 24-10. We can get an even better approximation, x_3, by using the method again with $x_2 = 0.375$, $f(0.375) = 0.015\,625$, and $f'(0.375) = -2.25$. This gives us

$$x_3 = 0.375 - \frac{0.015\,625}{-2.25} = 0.381\,944\,4$$

Since this is a quadratic equation, we can check this result by using the quadratic formula. Using this formula we find the root is $x = 0.381\,966\,0$. Our result using Newton's method is good to three decimal places. Additional accuracy may be obtained by using the method again as many times as needed. ◂◂

EXAMPLE 2 ▸▸ Find the root of $x^4 - 5x^3 + 6x^2 - 5x + 5 = 0$ between 1 and 2 to five decimal places.

Here

$$f(x) = x^4 - 5x^3 + 6x^2 - 5x + 5$$

and

$$f'(x) = 4x^3 - 15x^2 + 12x - 5$$

Since $f(1) = 2$ and $f(2) = -5$, it appears that the root is possibly closer to 1 than to 2. Therefore, we let $x_1 = 1.3$. Setting up a table (we will use 7 decimal places in the table, although many calculators display at least 10 digits), we have these values:

n	x_n	$f(x_n)$	$f'(x_n)$	$x_n - \dfrac{f(x_n)}{f'(x_n)}$
1	1.3	0.5111	−5.962	1.385 726 3
2	1.385 726 3	−0.024 512 8	−6.531 151 3	1.381 973 1
3	1.381 973 1	−0.000 046 1	−6.506 624 2	1.381 966 0

Since $x_4 = x_3 = 1.381\,97$ to five decimal places, this is the value of the required root. (Actually, by finding x_5 we see that x_4 is accurate to the value shown.)

If your calculator can store more than one value in memory, the results shown above can be more easily obtained by storing at least x_n, $f(x_n)$, and $f'(x_n)$ in memory. ◂◂

EXAMPLE 3 ▸▸ Solve the equation $x^2 - 1 = \sqrt{4x - 1}$.

We can see approximately where the root is by sketching the graphs of $y_1 = x^2 - 1$ and $y_2 = \sqrt{4x - 1}$ or by viewing the graphs on a graphing calculator, as shown in Fig. 24-11. From this view we see that they intersect between $x = 1$ and $x = 2$. Therefore, we choose $x_1 = 1.5$. With

$$f(x) = x^2 - 1 - \sqrt{4x - 1}$$

$$f'(x) = 2x - \frac{2}{\sqrt{4x - 1}}$$

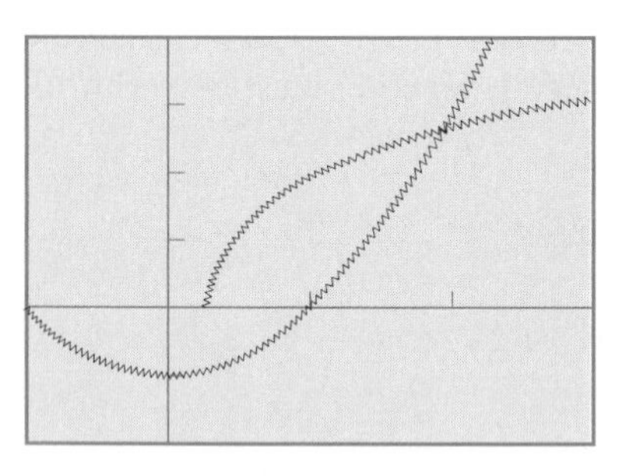

Fig. 24-11

we now find the values in the following table:

n	x_n	$f(x_n)$	$f'(x_n)$	$x_n - \dfrac{f(x_n)}{f'(x_n)}$
1	1.5	−0.986 067 98	2.105 572 8	1.968 313 4
2	1.968 313 4	0.252 568 59	3.173 759 8	1.888 733 2
3	1.888 733 2	0.007 052 69	2.996 295 7	1.886 379 4
4	1.886 379 4	0.000 006 20	2.991 026 5	1.886 377 3

Since $x_5 = x_4 = 1.886\,38$ to five decimal places, this is the required solution. (Here, rounded-off values of x_n are shown, although additional digits were carried and used.) This value can be verified on the graphing calculator by using the TRACE and ZOOM features. ◂◂

Exercises 24–2

In Exercises 1–4, find the indicated roots of the given quadratic equations by finding x_3 from Newton's method. Compare this root with that obtained by using the quadratic formula.

1. $x^2 - 2x - 5 = 0$ (between 3 and 4)
2. $2x^2 - x - 2 = 0$ (between 1 and 2)
3. $3x^2 - 5x - 1 = 0$ (between -1 and 0)
4. $x^2 + 4x + 2 = 0$ (between -4 and -3)

In Exercises 5–16, find the indicated roots of the given equations to at least four decimal places by using Newton's method. Compare with the value of the root found using a graphing calculator.

5. $x^3 - 6x^2 + 10x - 4 = 0$ (between 0 and 1)
6. $x^3 - 3x^2 - 2x + 3 = 0$ (between 0 and 1)
7. $x^3 + 5x^2 + x - 1 = 0$ (the positive root)
8. $2x^3 + 2x^2 - 11x + 3 = 0$ (the larger positive root)
9. $x^4 - x^3 - 3x^2 - x - 4 = 0$ (between 2 and 3)
10. $2x^4 - 2x^3 - 5x^2 - x - 3 = 0$ (between 2 and 3)
11. $x^4 - 2x^3 - 8x - 16 = 0$ (the negative root)
12. $3x^4 - 3x^3 - 11x^2 - x - 4 = 0$ (the negative root)
13. $2x^2 = \sqrt{2x + 1}$ (the positive real solution)
14. $x^3 = \sqrt{x + 1}$ (the real solution)
15. $x = \dfrac{1}{\sqrt{x + 2}}$ (the real solution)
16. $x^{3/2} = \dfrac{1}{2x + 1}$ (the real solution)

In Exercises 17–24, determine the required values by using Newton's method.

17. Find all the real roots of $x^3 - 2x^2 - 5x + 4 = 0$.
18. Find all the real roots of $x^3 - 2x^2 - 2x - 7 = 0$.
19. Explain how to find $\sqrt[3]{4}$ by using Newton's method.
20. Find the point of intersection of the curves $y = 1 - x^3$ and $y = \sqrt{2x - 1}$.
21. A dome in the shape of a spherical segment is to be placed over the top of a sports stadium. If the radius r of the dome is to be 60.0 m and the volume V within the dome is 180 000 m^3, find the height h of the dome. See Fig. 24-12.
$(V = \frac{1}{6}\pi h(h^2 + 3r^2).)$

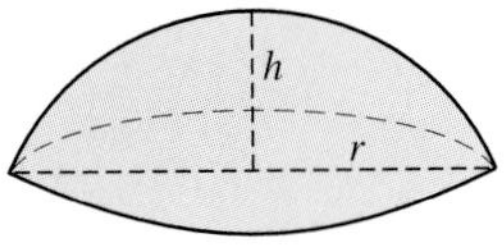

Fig. 24-12

22. The capacitances (in μF) of three capacitors in series are C, $C + 1.00$, and $C + 2.00$. If their combined capacitance is 1.00 μF, their individual values can be found by solving the equation
$$\frac{1}{C} + \frac{1}{C + 1.00} + \frac{1}{C + 2.00} = 1.00$$
Find these capacitances.
23. An oil storage tank has the shape of a right circular cylinder with a hemisphere at each end. See Fig. 24-13. If the volume of the tank is 50.0 m^3 and the length l is 4.00 m, find the radius r.

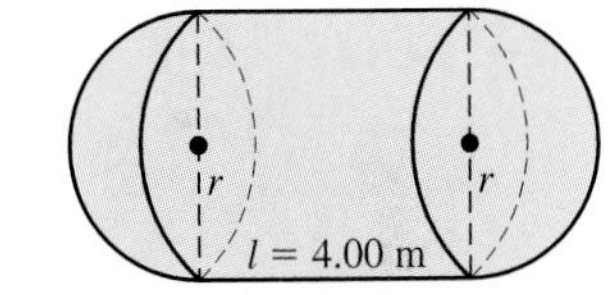

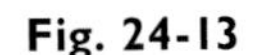
Fig. 24-13

24. A rectangular block of plastic with edges 2.00 cm, 2.00 cm, and 4.00 cm is heated until its volume doubles. By how much does each edge increase?

24-3 Curvilinear Motion

A great many phenomena in technology involve the time rate of change of certain quantities. We encountered one of the most fundamental and most important of these when we discussed velocity, the time rate of change of displacement, in Section 23-4. In this section we further develop the concept of velocity and certain other concepts necessary for the discussion. Other time-rate-of-change problems are discussed in the next section.

When velocity was introduced in Section 23-4, the discussion was limited to rectilinear motion, or motion along a straight line. A more general discussion of velocity is necessary when we discuss the motion of an object in a plane. There are many important applications of motion in a plane, a principal one being the motion of a projectile.

An important concept in developing this topic is that of a vector. The necessary fundamentals related to vectors are taken up in Chapter 9. Although vectors can be used to represent many physical quantities, we shall restrict our attention to their use in describing the velocity and acceleration of an object moving in a plane along a specified path. Such motion is called **curvilinear motion.**

In describing an object undergoing curvilinear motion, it is common to express the x- and y-coordinates of its position separately as functions of time. Equations given in this form—that is, *x and y both given in terms of a third variable (in this case t)—are said to be in* **parametric form,** which we encountered in Section 10-6. *The third variable, t, is called the* **parameter.**

To find the velocity of an object whose coordinates are given in parametric form, we find its x-component of velocity v_x by determining dx/dt and its y-component of velocity v_y by determining dy/dt. These are then evaluated, and the resultant velocity is found from $v = \sqrt{v_x^2 + v_y^2}$. The direction in which the object is moving is found from $\tan\theta = v_y/v_x$.

EXAMPLE 1 ▸▸ If the horizontal distance x that an object has moved is given by $x = 3t^2$, and the vertical distance y is given by $y = 1 - t^2$, find the resultant velocity when $t = 2$.

To find the resultant velocity, we must find v and θ, by first finding v_x and v_y. After the derivatives are found, they are evaluated for $t = 2$. Therefore,

$$v_x = \frac{dx}{dt} = 6t \qquad v_x|_{t=2} = 12 \qquad \text{find velocity components}$$

$$v_y = \frac{dy}{dt} = -2t \qquad v_y|_{t=2} = -4$$

$$v = \sqrt{12^2 + (-4)^2} = 12.6 \qquad \text{magnitude of velocity}$$

$$\tan\theta = \frac{-4}{12} \qquad \theta = -18.4^\circ \qquad \text{direction of motion}$$

The path and velocity vectors are shown in Fig. 24-14. ◂◂

Fig. 24-14

EXAMPLE 2 ▸▸ Find the velocity and direction of motion when $t = 2$ of an object moving such that its x- and y-coordinates of position are given by $x = 1 + 2t$ and $y = t^2 - 3t$.

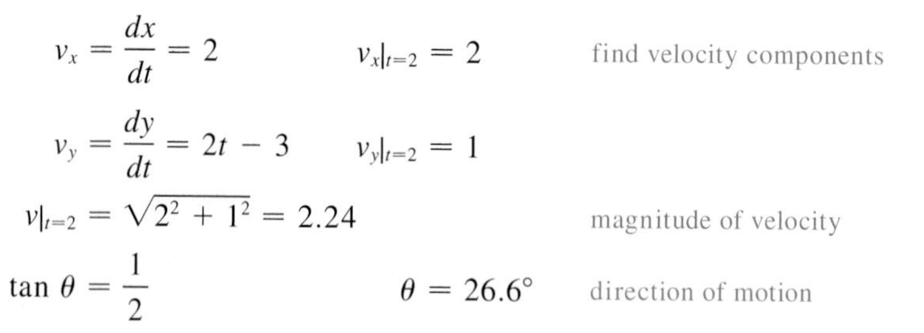

$$v_x = \frac{dx}{dt} = 2 \qquad v_x|_{t=2} = 2 \qquad \text{find velocity components}$$

$$v_y = \frac{dy}{dt} = 2t - 3 \qquad v_y|_{t=2} = 1$$

$$v|_{t=2} = \sqrt{2^2 + 1^2} = 2.24 \qquad \text{magnitude of velocity}$$

$$\tan\theta = \frac{1}{2} \qquad \theta = 26.6^\circ \qquad \text{direction of motion}$$

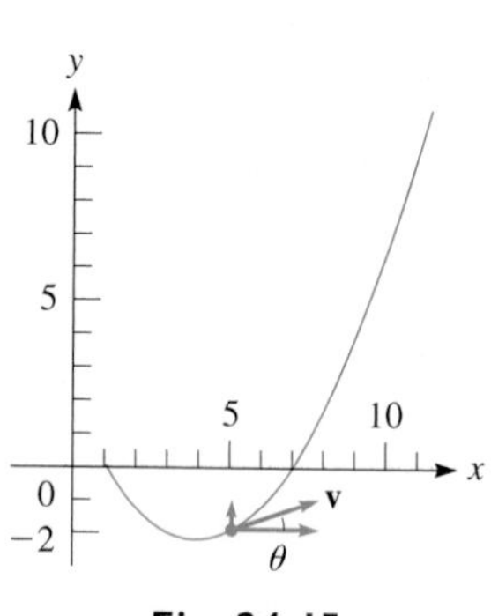

Fig. 24-15

These quantities are shown in Fig. 24-15. ◂◂

CAUTION ▸ In these examples we note that ***we first find the necessary derivatives and then we evaluate them.*** This procedure should always be followed. When a derivative is to be found, it is incorrect to take the derivative of the expression that is the evaluated function.

Acceleration *is the time rate of change of velocity.* Therefore, if the velocity, or its components, is known as a function of time, the acceleration of an object can be found by taking the derivative of the velocity with respect to time. If the displacement is known, the acceleration is found by finding the second derivative with respect to time. Finding the acceleration of an object is illustrated in the following example.

EXAMPLE 3 ▶▶ Find the magnitude and direction of the acceleration when $t = 2$ for an object which is moving such that its x- and y-coordinates of position are given by $x = t^3$ and $y = 1 - t^2$.

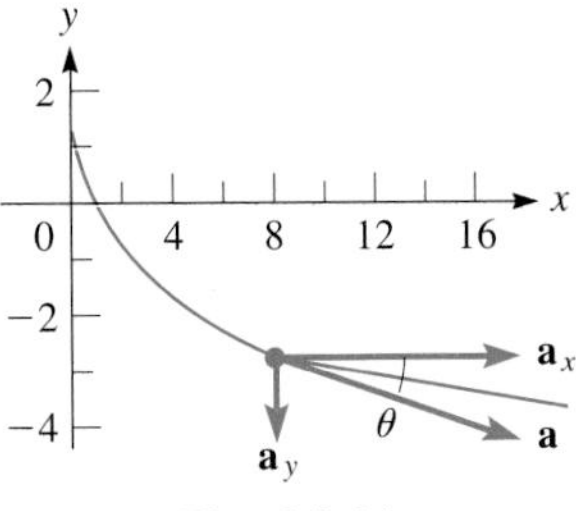

Fig. 24-16

$$v_x = \frac{dx}{dt} = 3t^2 \qquad a_x = \frac{dv_x}{dt} = \frac{d^2x}{dt^2} = 6t \qquad a_x|_{t=2} = 12$$

take second derivatives to find acceleration components

$$v_y = \frac{dy}{dt} = -2t \qquad a_y = \frac{dv_y}{dt} = \frac{d^2y}{dt^2} = -2 \qquad a_y|_{t=2} = -2$$

$$a|_{t=2} = \sqrt{12^2 + (-2)^2} = 12.2 \qquad \text{magnitude of acceleration}$$

$$\tan\theta = \frac{a_y}{a_x} = -\frac{2}{12} \qquad \theta = -9.5° \qquad \text{direction of acceleration}$$

CAUTION ▶ The quadrant in which θ lies is determined from the fact that ***a_y* is negative and *a_x* is positive.** Thus, θ must be a fourth-quadrant angle (see Fig. 24-16). We see from this example that the magnitude and direction of acceleration are found from its components just as with velocity. ◀◀

We now summarize the equations used to find the velocity and acceleration of an object for which the displacement is a function of time. They indicate how to find the components, as well as the magnitude and direction, of each.

$$v_x = \frac{dx}{dt} \qquad v_y = \frac{dy}{dt} \qquad \text{velocity components} \tag{24-2}$$

$$a_x = \frac{dv_x}{dt} = \frac{d^2x}{dt^2} \qquad a_y = \frac{dv_y}{dt} = \frac{d^2y}{dt^2} \qquad \text{acceleration components} \tag{24-3}$$

$$v = \sqrt{v_x^2 + v_y^2} \qquad a = \sqrt{a_x^2 + a_y^2} \qquad \text{magnitude} \tag{24-4}$$

$$\tan\theta_v = \frac{v_y}{v_x} \qquad \tan\theta_a = \frac{a_y}{a_x} \qquad \text{direction} \tag{24-5}$$

CAUTION ▶ If the curvilinear path an object follows is given with y as a function of x, ***the velocity (and acceleration) is found by taking derivatives of each term of the equation with respect to time.*** It is assumed that both x and y are functions of time, although these functions are not stated. When finding derivatives we must be careful in using the power rule, Eq. (23-15), so that the factor du/dx is not neglected. In the following examples, we illustrate the use of Eqs. (24-2) to (24-5) in applied situations for which we know the equation of the path of the motion. Again, we must be careful to find the direction of the vector as well as its magnitude in order to have a complete solution.

For reference, Eq. (23-15) is $\frac{du^n}{dx} = nu^{n-1}\frac{du}{dx}$.

EXAMPLE 4 ▸▸ In a physics experiment, a small sphere is constrained to move along a parabolic path described by $y = \frac{1}{3}x^2$. If the horizontal velocity v_x is constant at 6.00 cm/s, find the velocity at the point (2.00, 1.33). See Fig. 24-17.

Since both y and x change with time, both can be considered to be functions of time. Therefore, we can take derivatives of $y = \frac{1}{3}x^2$ with respect to time.

CAUTION ▸

$$\frac{dy}{dt} = \frac{1}{3}\left(2x\frac{dx}{dt}\right) \longleftarrow \frac{dx^2}{dt} = 2x\frac{dx}{dt}$$

$$v_y = \frac{2}{3}xv_x \qquad \text{using Eqs. (24-2)}$$

$$v_y = \frac{2}{3}(2.00)(6.00) = 8.00 \text{ cm/s} \qquad \text{substituting}$$

$$v = \sqrt{6.00^2 + 8.00^2} = 10.0 \text{ cm/s} \qquad \text{magnitude [Eqs. (24-4)]}$$

$$\tan\theta = \frac{8.00}{6.00}, \qquad \theta = 53.1° \qquad \text{direction [Eqs. (24-5)]}$$ ◂◂

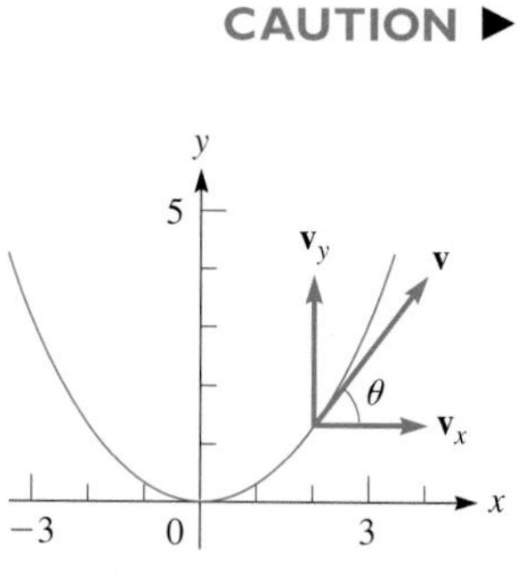

Fig. 24-17

EXAMPLE 5 ▸▸ A helicopter is flying at 18.0 m/s and at an altitude of 120 m when a rescue marker is released from it. The marker maintains a horizontal velocity and follows a path given by $y = 120 - 0.0151x^2$, as shown in Fig. 24-18. Find the magnitude and direction of the velocity and of the acceleration of the marker 3.00 s after release. This is a typical problem in projectile motion.

From the given information we know that $v_x = dx/dt = 18.0$ m/s. Taking derivatives with respect to time leads to this solution:

$$y = 120 - 0.0151x^2$$

$$\frac{dy}{dt} = -0.0302x\frac{dx}{dt} \qquad \text{taking derivatives}$$

$$v_y = -0.0302xv_x \qquad \text{using Eqs. (24-2)}$$

$$x = (3.00)(18.0) = 54.0 \text{ m} \qquad \text{evaluating at } t = 3.00 \text{ s}$$

$$v_y = -0.0302(54.0)(18.0) = -29.35 \text{ m/s}$$

$$v = \sqrt{18.0^2 + (-29.35)^2} = 34.4 \text{ m/s} \qquad \text{magnitude}$$

$$\tan\theta = \frac{-29.35}{18.0}, \qquad \theta = -58.5° \qquad \text{direction}$$

Fig. 24-18

The velocity is 34.4 m/s and is directed at an angle of 58.5° below the horizontal.

To find the acceleration, we return to the equation $v_y = -0.0302xv_x$. Since v_x is constant, we can substitute 18.0 for v_x to get

$$v_y = -0.5436x$$

Again taking derivatives with respect to time, we have

$$\frac{dv_y}{dt} = -0.5436\frac{dx}{dt}$$

$$a_y = -0.5436v_x \qquad \text{using Eqs. (24-3) and (24-2)}$$

$$a_y = -0.5436(18.0) = -9.78 \text{ m/s}^2 \qquad \text{evaluating}$$

We know that v_x is constant, which means that $a_x = 0$. Therefore, the acceleration is 9.78 m/s^2 and is directed vertically downward. ◂◂

Exercises 24–3

In Exercises 1–4, given that the x- and y-coordinates of a moving particle are given by the indicated parametric equations, find the magnitude and direction of the velocity for the specific value of t. Sketch the curves and show the appropriate components of the velocity.

1. $x = 3t,\ y = 1 - t,\ t = 4$

2. $x = \dfrac{5t}{2t + 1},\ y = 0.1(t^2 + t),\ t = 2$

3. $x = t(2t + 1)^2,\ y = \dfrac{6}{\sqrt{4t + 3}},\ t = 0.5$

4. $x = \sqrt{1 + 2t},\ y = t - t^2,\ t = 4$

In Exercises 5–8, use the parametric equations and values of t of Exercises 1–4 to find the magnitude and direction of the acceleration in each case.

In Exercises 9–24, find the indicated velocities and accelerations.

9. The water from a valve at the bottom of a water tank follows a path described by $y = 4.0 - 0.20x^2$, where units are in metres. If the velocity v_x is constant at 5.0 m/s, find the resultant velocity at the point (4.0, 0.80).

10. A roller mechanism follows a path described by $y = \sqrt{4x + 1}$, where units are in metres. If $v_x = 2x$, find the resultant velocity (in m/s) at the point (2.0, 3.0).

11. A float is used to test the flow pattern of a stream. It follows a path described by $x = 0.2t^2$, $y = -0.1t^3$, where units of x and y are in metres and t is in minutes. Find the acceleration of the float after 2.0 min.

12. A car on a test track goes into a turn described by $x = 0.2t^3$, $y = 20t - 2t^2$, where x and y are measured in metres and t is in seconds. Find the acceleration of the car at $t = 3.0$ s.

13. A golf ball moves according to the equations $x = 32t$ and $y = 42t - 4.9t^2$, where distances are in metres and time is in seconds. Find the resultant velocity and acceleration of the golf ball for $t = 6.0$ s.

14. A package of relief supplies is dropped and moves according to the parametric equations $x = 45t$ and $y = -4.9t^2$, where distances are in metres and time is in seconds. Find the magnitude and direction of the velocity and of the acceleration when $t = 3.0$ s.

15. A spacecraft moves such that it follows the equation $x = 10(\sqrt{1 + t^4} - 1)$, $y = 40t^{3/2}$ for the first hundred seconds after launch. Here, x and y are measured in metres and t is measured in seconds. Find the magnitude and direction of the velocity of the spacecraft 10.0 s and 100 s after launch.

16. An electron moves in an electric field according to the parametric equations

$$x = \frac{20}{\sqrt{1 + t^2}} \quad \text{and} \quad y = \frac{20t}{\sqrt{1 + t^2}}$$

where distances are in metres and time is in seconds. Find the magnitude and direction of the velocity when $t = 1.0$ s.

17. Find the resultant acceleration of the spacecraft in Exercise 15 for the specified times.

18. Find the resultant acceleration of the electron in Exercise 16 for $t = 10$ s.

19. A rocket follows a path given by $y = x - \frac{1}{90}x^3$ (distances in kilometres). If the horizontal velocity is given by $v_x = x$, find the magnitude and direction of the velocity when the rocket hits the ground (assume level terrain) if time is in minutes.

20. A shipping route around an island is described by $y = 3x^2 - 0.2x^3$. A ship on this route is moving such that $v_x = 1.2$ km/h, where $x = 3.5$ km. Find the velocity of the ship at this point.

21. A personal computer's hard disk is 88.9 mm in diameter and rotates at 3600 r/min. Set up the equation for the circumference of the disk, with its center at the origin. Using this equation, find the components v_x and v_y of a point on the circumference for $x = 30.5$ mm, $y > 0$, and $v_x > 0$.

22. A robot arm joint moves in an elliptical path. The horizontal major axis of the ellipse is 8.0 cm long and the minor axis is 4.0 cm long. For $-2 < x < 2$ and $y > 0$, the joint moves such that $v_x = 2.5$ cm/s. On this part of its path, find its velocity if $x = -1.5$ cm. Assume the center of the path is at the origin.

23. An airplane ascends such that its gain h in altitude is proportional to the square root of the change x in horizontal distance traveled. If $h = 280$ m for $x = 400$ m and v_x is constant at 350 m/s, find the velocity at this point.

24. A meteor traveling toward the earth has a velocity inversely proportional to the square root of the distance from the earth's center. State how its acceleration is related to its distance from the center of the earth.

24-4 Related Rates

Any two variables that vary with respect to time and between which a relation is known to exist can have the time rate of change of one expressed in terms of the time rate of change of the other. We do this by taking the derivative with respect to time of the expression that relates the variables, as we did in Examples 4 and 5 of Section 24-3. Since the rates of change are related, this type of problem is referred to as a **related-rate** problem. The following examples illustrate the basic method of solution.

EXAMPLE 1 The voltage of a certain thermocouple as a function of the temperature is given by $E = 2.800T + 0.006T^2$. If the temperature is increasing at the rate of 1.00 °C/min, how fast is the voltage increasing when $T = 100$°C?

Since we are asked to find the time rate of change of voltage, we first take derivatives with respect to time. This gives us

$$\frac{dE}{dt} = 2.800\frac{dT}{dt} + 0.012T\frac{dT}{dt} \longleftarrow \frac{d}{dt}(0.006T^2) = 0.006\left(2T\frac{dT}{dt}\right)$$

CAUTION ▶ ***again being careful to include the factor dT/dt.*** From the given information we know that $dT/dt = 1.00$°C/min and that we wish to know dE/dt when $T = 100$°C. Thus

$$\left.\frac{dE}{dt}\right|_{T=100} = 2.800(1.00) + 0.012(100)(1.00) = 4.00 \text{ V/min}$$

The derivative must be taken before values are substituted. In this problem we are finding the time rate of change of the voltage for a specified value of T. For other values of T, dE/dt would have different values.

EXAMPLE 2 The distance q that an image is from a certain lens in terms of p, the distance of the object from the lens, is given by

$$q = \frac{10p}{p - 10}$$

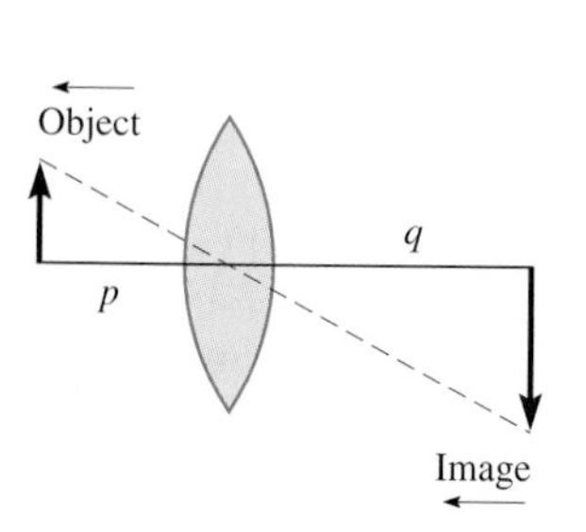

Fig. 24-19

If the object distance is increasing at the rate of 0.200 cm/s, how fast is the image distance changing when $p = 15.0$ cm? See Fig. 24-19.

Taking derivatives with respect to time, we have

$$\frac{dq}{dt} = \frac{(p - 10)\left(10\frac{dp}{dt}\right) - 10p\left(\frac{dp}{dt}\right)}{(p - 10)^2} = \frac{-100\frac{dp}{dt}}{(p - 10)^2}$$

(don't forget the $\frac{dp}{dt}$)

Now, substituting $p = 15.0$ and $dp/dt = 0.200$, we have

$$\left.\frac{dq}{dt}\right|_{p=15} = \frac{-100(0.200)}{(15.0 - 10)^2} = -0.800 \text{ cm/s}$$

Thus the image distance is decreasing (the significance of the minus sign) at the rate of 0.800 cm/s when $p = 15.0$ cm.

In many related-rate problems the function is not given but must be set up according to the statement of the problem. The following examples illustrate this type of problem.

EXAMPLE 3 ▸▸ A spherical balloon is being blown up such that its volume increases at the constant rate of 2.00 m^3/min. Find the rate at which the radius is increasing when it is 3.00 m. See Fig. 24-20.

We are asked to find the relation between the rate of change of the volume of a sphere with respect to time and the corresponding rate of change of the radius with respect to time. Therefore, we are to ***take derivatives of the expression for the volume of a sphere with respect to time.***

CAUTION ▶

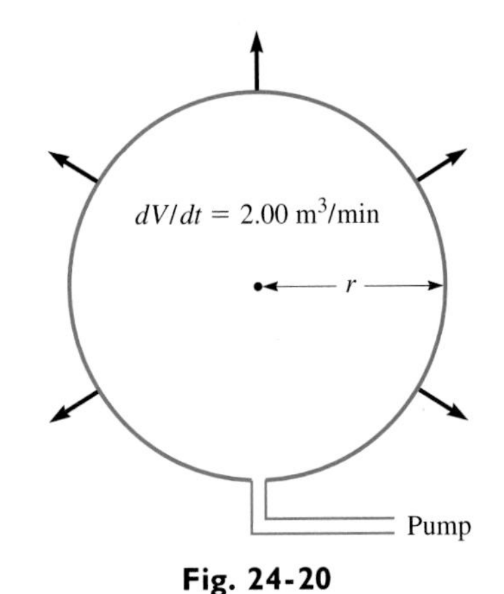

Fig. 24-20

$$V = \frac{4}{3}\pi r^3 \qquad \text{volume of sphere}$$

$$\frac{dV}{dt} = 4\pi r^2\left(\frac{dr}{dt}\right) \qquad \text{take derivatives with respect to time}$$

$$2.00 = 4\pi(3.00)^2\left(\frac{dr}{dt}\right) \qquad \text{substitute } \frac{dV}{dt} = 2.00 \text{ m}^3/\text{min and } r = 3.00 \text{ m}$$

$$\left.\frac{dr}{dt}\right|_{r=3} = \frac{1}{18.0\pi} \qquad \text{solve for } \frac{dr}{dt}$$

$$= 0.0177 \text{ m/min}$$ ◂◂

EXAMPLE 4 ▸▸ The force F of gravity of the earth on a spacecraft varies inversely as the square of the distance r of the spacecraft from the center of the earth. A particular spacecraft weighs 4500 N on the launchpad ($F = 4500$ N for $r = 6370$ km). Find the rate at which F changes later as the spacecraft moves away from the earth at the rate of 12 km/s where $r = 8500$ km.

First setting up the equation, we have the following solution.

$$F = \frac{k}{r^2} \qquad \text{inverse variation}$$

$$4500 = \frac{k}{6370^2} \qquad \text{substitute } F = 4500 \text{ N}, r = 6370 \text{ km}$$

$$k = 1.83 \times 10^{11} \text{ N}\cdot\text{km}^2 \qquad \text{solve for } k$$

$$F = \frac{1.83 \times 10^{11}}{r^2} \qquad \text{substitute for } k \text{ in equation}$$

$$\frac{dF}{dt} = (1.83 \times 10^{11})(-2)(r^{-3})\frac{dr}{dt} \qquad \text{take derivatives with respect to time}$$

$$= \frac{-3.66 \times 10^{11}}{r^3}\frac{dr}{dt}$$

$$\left.\frac{dF}{dt}\right|_{r=8500 \text{ km}} = \frac{-3.66 \times 10^{11}}{8500^3}(12) \qquad \text{evaluate derivative for } r = 8500 \text{ km}, dr/dt = 12 \text{ km/s}$$

$$= -7.2 \text{ N/s}$$

Therefore, the gravitational force is decreasing at the rate of 7.2 N/s. ◂◂

EXAMPLE 5 ▸▸ Two cruise ships leave Vancouver, British Columbia, at noon. Ship A travels west at 12.0 km/h (before turning toward Alaska), and ship B travels south at 16.0 km/h (toward Seattle). How fast are they separating at 2 P.M.?

In Fig. 24-21 we let $x =$ the distance traveled by A and $y =$ the distance traveled by B. We can find the distance between them, z, from the Pythagorean theorem. Therefore, we are to find dz/dt for $t = 2.00$ h. Even though there are three variables, each is a function of time. This means we can find dz/dt by taking derivatives of each term with respect to time. This gives us

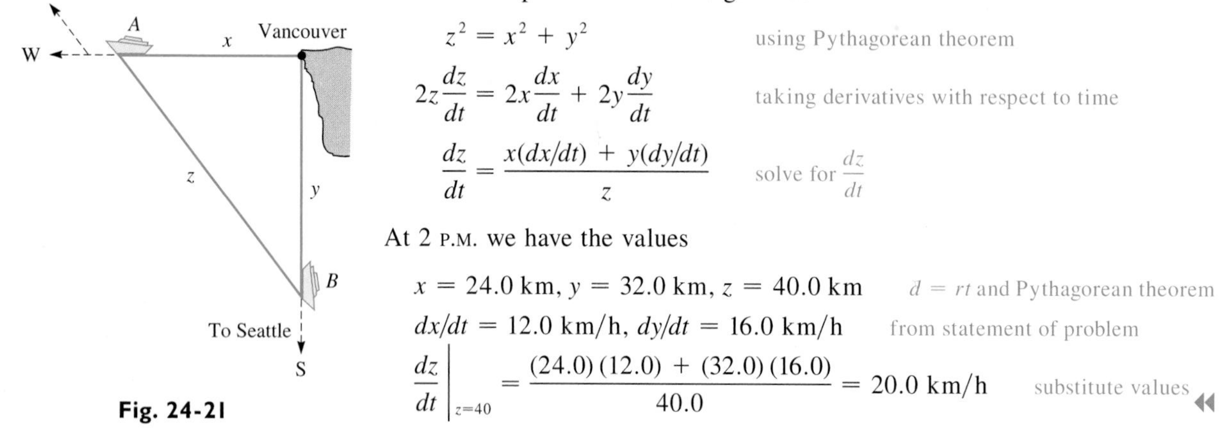

Fig. 24-21

$$z^2 = x^2 + y^2 \qquad \text{using Pythagorean theorem}$$

$$2z\frac{dz}{dt} = 2x\frac{dx}{dt} + 2y\frac{dy}{dt} \qquad \text{taking derivatives with respect to time}$$

$$\frac{dz}{dt} = \frac{x(dx/dt) + y(dy/dt)}{z} \qquad \text{solve for } \frac{dz}{dt}$$

At 2 P.M. we have the values

$x = 24.0$ km, $y = 32.0$ km, $z = 40.0$ km $d = rt$ and Pythagorean theorem

$dx/dt = 12.0$ km/h, $dy/dt = 16.0$ km/h   from statement of problem

$$\left.\frac{dz}{dt}\right|_{z=40} = \frac{(24.0)(12.0) + (32.0)(16.0)}{40.0} = 20.0 \text{ km/h} \qquad \text{substitute values}$$ ◂◂

From these examples, we see that we have the following method of solving a related-rates problem.

Steps for Solving Related-Rates Problems

1. *Identify the variables and rates* in the problem.
2. If possible, *make a sketch* showing the variables.
3. *Determine the equation* relating the variables.
4. *Differentiate with respect to time.*
5. *Solve for the required rate.*

Exercises 24–4

Solve the following problems in related rates.

1. The electric resistance of a certain resistor as a function of temperature is given by $R = 4.000 + 0.003T^2$, where R is measured in ohms and T in degrees Celsius. If the temperature is increasing at 0.100°C/s, find how fast the resistance changes when $T = 150$°C.

2. The kinetic energy K of an object is given by $K = \frac{1}{2}mv^2$, where m is the mass of the object and v is its velocity. If a 250-kg wrecking ball accelerates at 5.00 m/s², how fast is the kinetic energy (in joules) changing when $v = 30.0$ m/s?

3. A firm found that its profit was p dollars for the production of x tonnes per week of a product according to the function $p = 30\sqrt{10x - x^2} - 50$. Find the rate of change of profit if $dx/dt = 0.200$ t/week² when $x = 4.00$ t/week.

4. A variable resistor R and an 8-Ω resistor in parallel have a combined resistance R_T given by $R_T = \dfrac{8R}{8 + R}$. If R is changing at 0.30 Ω/min, find the rate at which R_T is changing when $R = 6.0$ Ω.

5. The radius r of a ring of a certain holograph (an image produced without using a lens) is given by $r = \sqrt{0.4\lambda}$, where λ is the wavelength of the light being used. If λ is changing at the rate of 0.10×10^{-7} m/s when $\lambda = 6.0 \times 10^{-7}$ m, find the rate at which r is changing.

6. An earth satellite moves in a path that can be described by $\dfrac{x^2}{72.5} + \dfrac{y^2}{71.5} = 1$, where x and y are in thousands of kilometres. If $dx/dt = 12\,900$ km/h for $x = 3200$ km and $y > 0$, find dy/dt.

7. The magnetic field B due to a magnet of length l at a distance r is given by $B = \dfrac{k}{[r^2 + (l/2)^2]^{3/2}}$, where k is a constant for a given magnet. Find the expression for the time rate of change of B in terms of the time rate of change of r.

8. An approximate relationship between the pressure p and volume v of the vapor in a diesel engine cylinder is $pv^{1.4} = k$, where k is a constant. At a certain instant, $p = 4200$ kPa, $v = 75$ cm^3, and the volume is increasing at the rate of 850 cm^3/s. What is the time rate of change of the pressure at this instant?

9. Fatty deposits have decreased the circular cross-sectional opening of a person's artery. A test drug reduces these deposits such that the radius of the opening increases at the rate of 0.020 mm/month. Find the rate at which the area of the opening increases when $r = 1.2$ mm.

10. A computer program increases the side of a square image on the screen at the rate of 0.25 cm/s. Find the rate at which the area of the image increases when the edge is 6.50 cm.

11. A metal cube dissolves in acid such that an edge of the cube decreases by 0.50 mm/min. How fast is the volume of the cube changing when the edge is 8.20 mm?

12. A light in a garage is 2.90 m above the floor and 3.65 m behind the door. If the garage door descends vertically at 0.45 m/s, how fast is the door's shadow moving toward the garage when the door is 0.60 m above the floor?

13. One statement of Boyle's law is that the pressure of a gas varies inversely as the volume for constant temperature. If a certain gas occupies 650 cm^3 when the pressure is 230 kPa and the volume is increasing at the rate of 20.0 cm^3/min, how fast is the pressure changing when the volume is 810 cm^3?

14. The tuning frequency f of an electronic tuner is inversely proportional to the square root of the capacitance C in the circuit. If $f = 920$ kHz for $C = 3.5$ pF, find how fast f is changing at this frequency if $dC/dt = 0.3$ pF/s.

15. A spherical metal object is ejected from an earth satellite and reenters the atmosphere. It heats up (until it burns) so that the radius increases at the rate of 5.00 mm/s. What is the rate of change of volume when the radius is 225 mm?

16. The acceleration due to the gravity g on a spacecraft is inversely proportional to its distance from the center of the earth. At the surface of the earth, $g = 9.80$ m/s^2. Given that the radius of the earth is 6370 km, how fast is g changing on a spacecraft approaching the earth at 1400 m/s at a distance of 41 000 km from the surface?

17. A tank in the shape of an inverted cone has a height of 3.60 m and a radius at the top of 1.15 m. Water is flowing into the tank at the rate of 0.50 m^3/min. How fast is the level rising when it is 1.80 m deep?

18. A ladder is slipping down along a vertical wall. If the ladder is 4.00 m long, and the top of it is slipping at the constant rate of 3.00 m/s, how fast is the bottom of the ladder moving along the ground when the bottom is 2.00 m from the wall?

19. A rope attached to a boat is being pulled in at a rate of 2.50 m/s. If the water is 5.00 m below the level at which the rope is being drawn in, how fast is the boat approaching the wharf when 13.0 m of rope are yet to be pulled in? See Fig. 24-22.

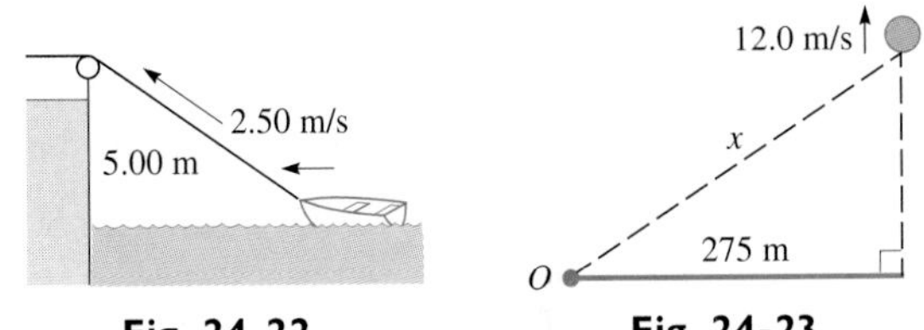

Fig. 24-22

Fig. 24-23

20. A weather balloon leaves the ground 275 m from an observer and rises vertically at 12.0 m/s. How fast is the line of sight from the observer to the balloon increasing when the balloon is 450 m high? See Fig. 24-23.

21. A supersonic jet leaves an airfield traveling due east at 1600 km/h. A second jet leaves the same airfield at the same time and travels at 1800 km/h along a line north of east such that it remains due north of the first jet. After a half-hour, how fast are the jets separating?

22. A car passes over a bridge at 15.0 m/s at the same time a boat passes under the bridge at a point 10.5 m directly below the car. If the boat is moving perpendicularly to the bridge at 4.0 m/s, how fast are the car and the boat separating 5.0 s later?

23. A man 1.80 m tall approaches a street light 4.50 m above the ground at the rate of 1.50 m/s. How fast is the end of the man's shadow moving when he is 3.00 m from the base of the light? See Fig. 24-24.

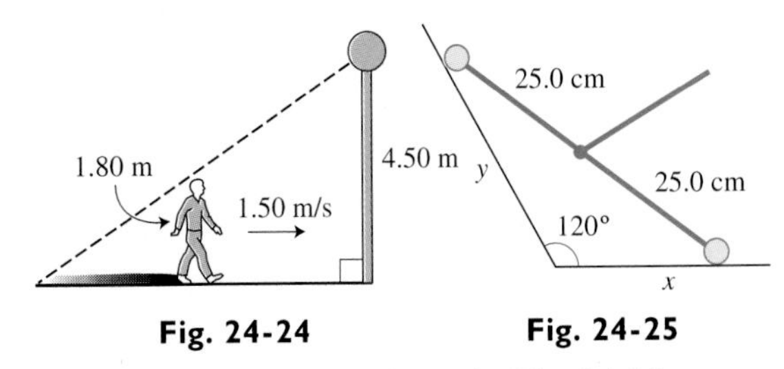

Fig. 24-24

Fig. 24-25

24. A roller mechanism, as shown in Fig. 24-25, moves such that the right roller is always in contact with the bottom surface and the left roller is always in contact with the left surface. If the right roller is moving to the right at 1.50 cm/s when $x = 10.0$ cm, how fast is the left roller moving?

24-5 Using Derivatives in Curve Sketching

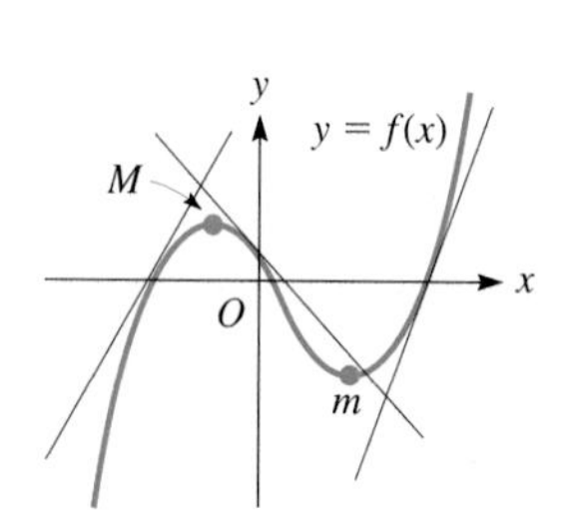

Fig. 24-26

Derivatives can be used effectively in sketching curves. An analysis of the first two derivatives can provide useful information as to the graph of a function. This information, possibly along with two or three key points on the curve, is often sufficient to obtain a good, although approximate, graph of the function. Graphs where this information must be supplemented with other analyses are the subject of the next section.

Considering the function $f(x)$ as shown in Fig. 24-26, we see that as x increases (from left to right) the y-values also increase until the point M is reached. From M to m the values of y decrease. To the right of m the values of y again increase. We also note that any tangent line to the left of M or to the right of m will have a positive slope. Any tangent line between M and m will have a negative slope. Since the derivative of a function determines the slope of a tangent line, we can conclude that, *as x increases, y increases if the derivative is positive and decreases if the derivative is negative.* This can be stated as

Function Increasing

$$f(x) \text{ increases if } f'(x) > 0$$

and

Function Decreasing

$$f(x) \text{ decreases if } f'(x) < 0$$

CAUTION ▶ ***It is always assumed that x is increasing.*** *Also, we assume in our present analysis that $f(x)$ and its derivatives are continuous over the indicated interval.*

EXAMPLE 1 ▶▶ Find those values of x for which the function $f(x) = x^3 - 3x^2$ is increasing and those values for which it is decreasing.

We solve this problem by finding those values of x for which the derivative is positive and those values for which it is negative. The derivative is

$$f'(x) = 3x^2 - 6x = 3x(x - 2)$$

This now becomes a problem of solving an inequality. To find the values of x for which $f(x)$ is increasing, we must solve the inequality

$$3x(x - 2) > 0$$

We now recall that the solution of an inequality consists of *all* values of x which may satisfy it. Normally, this consists of certain intervals of values of x. *These intervals are found by first setting the left side of the inequality equal to zero, thus obtaining the* **critical values** *of x.* The function on the left will have the same *sign* for all values of x less than the leftmost critical value. The sign of the function will also be the same within any given interval between critical values and to the right of the rightmost critical value. Those intervals that give the proper sign will satisfy the inequality. In this case the critical values are $x = 0$ and $x = 2$. Therefore, we have the following analysis:

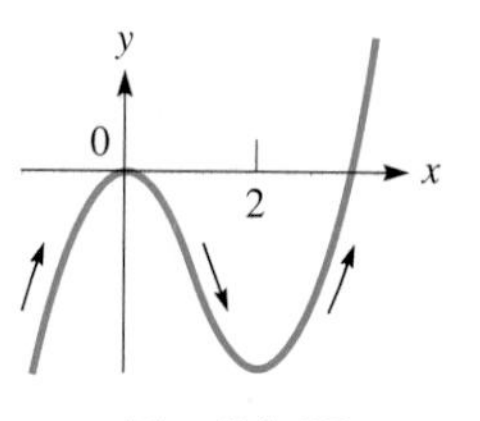

Fig. 24-27

If $x < 0, 3x(x - 2) > 0$ or $f'(x) > 0$. $f(x)$ increasing
If $0 < x < 2, 3x(x - 2) < 0$ or $f'(x) < 0$. $f(x)$ decreasing
If $x > 2, 3x(x - 2) > 0$ or $f'(x) > 0$. $f(x)$ increasing

Therefore, the solution of the above inequality is $x < 0$ or $x > 2$, which means that for these values $f(x)$ is increasing. We can also see that for $0 < x < 2, f'(x) < 0$, which means that $f(x)$ is decreasing for these values of x. The solution is now complete. The graph of $f(x) = x^3 - 3x^2$ is shown in Fig. 24-27. ◀◀

Maximum Points
Minimum Points

The points M and m in Fig. 24-26 are called a **relative maximum point** and a **relative minimum point,** respectively. *This means that M has a greater y-value than any other point near it and that m has a smaller y-value than any point near it.* This does not necessarily mean that M has the greatest y-value of any point on the curve or that m has the least y-value of any point on the curve. However, the points M and m are the greatest or least values of y for that part of the curve (that is why we use the word "relative"). Examination of Fig. 24-26 verifies this point. *The characteristic of both M and m is that the derivative is zero at each point.* (We see that this is so since a tangent line would have a slope of zero at each.) *This is how relative maximum and relative minimum points are located. The derivative is found and then set equal to zero. The solutions of the resulting equation give the x-coordinates of the maximum and minimum points.*

First Derivative Test for
Maxima and Minima

It remains now to determine whether a given value of x, for which the derivative is zero, is the coordinate of a maximum or a minimum point (or neither, which is also possible). From the discussion of increasing and decreasing values for y, we see that *the derivative changes sign from plus to minus when passing through a relative maximum point and from minus to plus when passing through a relative minimum point.* Thus we find maximum and minimum points by determining those values of x for which the derivative is zero and by properly analyzing the sign change of the derivative. If the sign of the derivative does not change, it is neither a maximum nor a minimum point. This is known as the **first-derivative test for maxima and minima.**

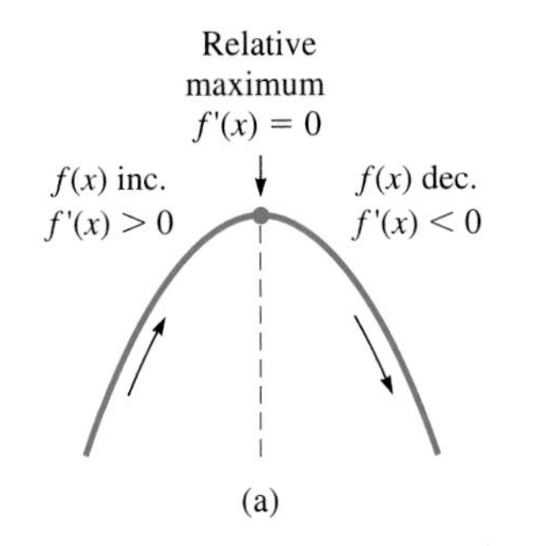

Fig. 24-28

In Fig. 24-28 a diagram for the first derivative test is shown. The test for a relative maximum is shown in Fig. 24-28(a), and that for a relative minimum is shown in Fig. 24-28(b). For the curves shown in Fig. 24-28, $f(x)$ and $f'(x)$ are continuous throughout the interval shown. (Although $f(x)$ must be continuous, $f'(x)$ may be discontinuous at the maximum point or the minimum point, and the sign changes of the first-derivative test remain valid.)

EXAMPLE 2 ▸▸ Find any maximum points and minimum points on the graph of the function

$$y = 3x^5 - 5x^3$$

Finding the derivative and setting it equal to zero, we have

$$y' = 15x^4 - 15x^2 = 15x^2(x^2 - 1)$$

Therefore,

$$15x^2(x - 1)(x + 1) = 0 \quad \text{for } x = 0, \quad x = 1, \quad \text{and} \quad x = -1$$

Thus, the sign of the derivative is the same for all points to the left of $x = -1$. For these values, $y' > 0$ (thus y is increasing). For values of x between -1 and 0, $y' < 0$. For values of x between 0 and 1, $y' < 0$. For values of x greater than 1, $y' > 0$. Thus the curve has a maximum at $(-1, 2)$ and a minimum at $(1, -2)$. The point $(0, 0)$ is neither a maximum nor a minimum, since the sign of the derivative did not change for this value of x. The graph of $y = 3x^5 - 5x^3$ is shown in Fig. 24-29. ◂◂

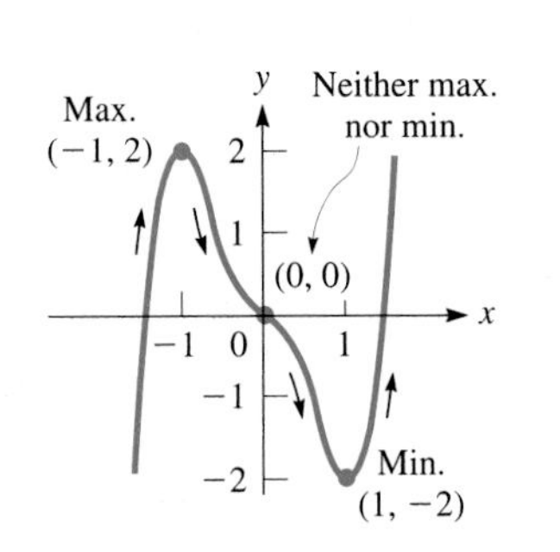

Fig. 24-29

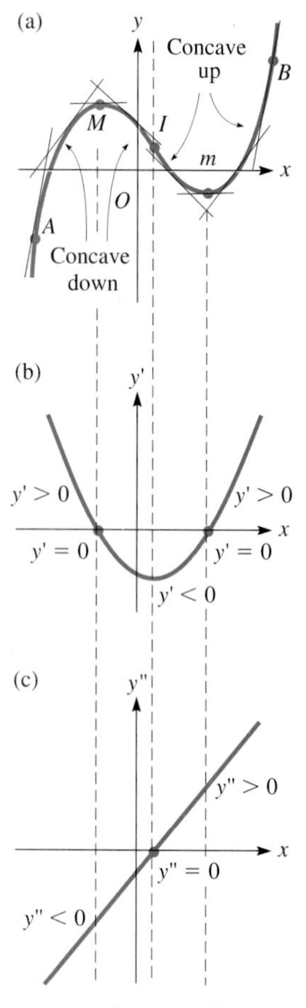

Fig. 24-30

We now look again at the slope of a tangent drawn to a curve. In Fig. 24-30(a), consider the *change* in the values of the slope of a tangent at a point as the point moves from A to B. At A the slope is positive, and as the point moves toward M, the slope remains positive but becomes smaller until it becomes zero at M. To the right of M the slope is negative and becomes more negative until it reaches I. Therefore, *from A to I the slope continually decreases.* To the right of I the slope remains negative but increases until it becomes zero again at m. To the right of m the slope becomes positive and increases to point B. Therefore, *from I to B, the slope continually increases.* We say that *the curve is* **concave down** *from A to I and* **concave up** *from I to B.*

The curve in Fig. 24-30(b) is that of the derivative, and it therefore indicates the values of the slope of $f(x)$. If the slope changes, we are dealing with the rate of change of slope, or the rate of change of the derivative. This function is the second derivative. The curve in Fig. 24-30(c) is that of the second derivative. We see that *where the second derivative of a function is* ***negative,*** *the slope is decreasing, or the curve is* ***concave down*** *(opens down). Where the second derivative is* ***positive,*** *the slope is increasing, or the curve is* ***concave up*** *(opens up).* This may be summarized as follows:

If $f''(x) > 0$, the curve is concave up.

If $f''(x) < 0$, the curve is concave down.

We can also now use this information in the determination of maximum and minimum points. By the nature of the definition of maximum and minimum points and of concavity, it is apparent that *a curve is concave down at a maximum point and concave up at a minimum point.* We can see these properties when we make a close analysis of the curve in Fig. 24-30. Therefore, at $x = a$,

if $f'(a) = 0$ and $f''(a) < 0$,

then $f(x)$ has a relative maximum at $x = a$, or

if $f'(a) = 0$ and $f''(a) > 0$,

then $f(x)$ has a relative minimum at $x = a$.

Second Derivative Test for Maxima and Minima

These statements comprise what is known as the **second-derivative test for maxima and minima.** This test is often easier to use than the first-derivative test. However, it can happen that $y'' = 0$ at a maximum or minimum point, and in such cases it is necessary that we use the first-derivative test.

In using the second-derivative test we should note that $f''(x)$ is ***negative*** at a ***maximum*** point and ***positive*** at a ***minimum*** point. This is contrary to a natural inclination to think of "maximum" and "positive" together or "minimum" and "negative" together.

Points of Inflection

The points at which the curve changes from concave up to concave down, or from concave down to concave up, are known as **points of inflection.** Thus point I in Fig. 24-30 is a point of inflection. Inflection points are found by determining those values of x for which the second derivative changes sign. This is analogous to finding maximum and minimum points by the first-derivative test. In Fig. 24-31, various types of points of inflection are illustrated.

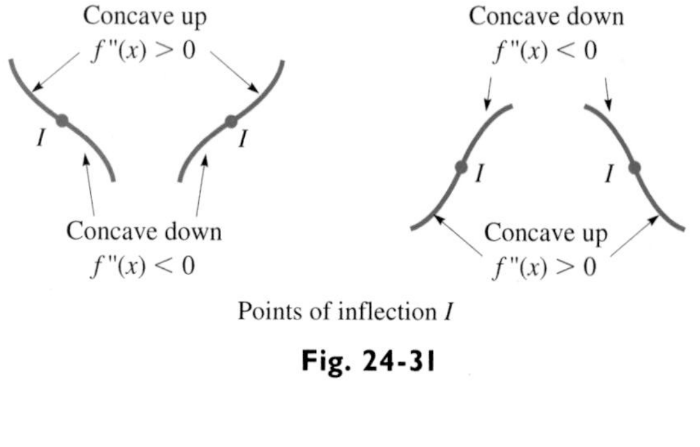

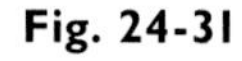
Points of inflection I

Fig. 24-31

EXAMPLE 3 ▸▸ Determine the concavity and find any points of inflection of the function $y = x^3 - 3x$.

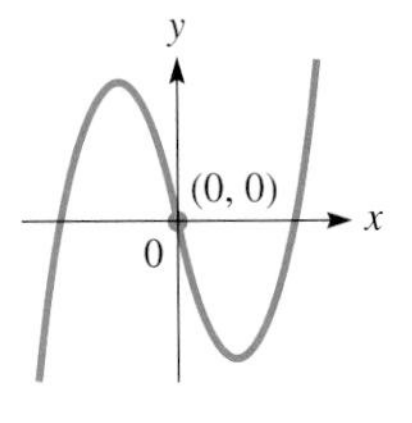

Fig. 24-32

This requires an inspection and analysis of the second derivative. Therefore, we find the first two derivatives.

$$y' = 3x^2 - 3$$
$$y'' = 6x$$

The second derivative is positive where the function is concave up, and this occurs if $x > 0$. The curve is concave down for $x < 0$, since y'' is negative. Thus, $(0, 0)$ is a point of inflection, since the concavity changes there. The graph of $y = x^3 - 3x$ is shown in Fig. 24-32. ◂◂

At this point we summarize the information found from the derivatives of a function $f(x)$. See Fig. 24-33.

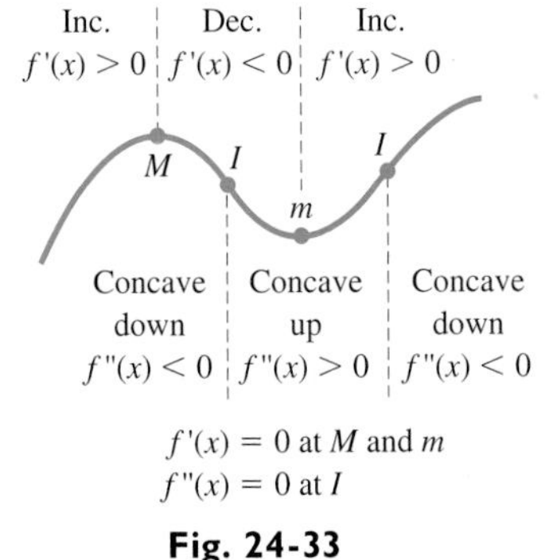

Fig. 24-33

$f'(x) > 0$ where $f(x)$ increases; $f'(x) < 0$ where $f(x)$ decreases.

$f''(x) > 0$ where the graph of $f(x)$ is concave up; $f''(x) < 0$ where the graph of $f(x)$ is concave down.

If $f'(x) = 0$ at $x = a$, there is a maximum point if $f'(x)$ changes from + to − or if $f''(a) < 0$.

If $f'(x) = 0$ at $x = a$, there is a minimum point if $f'(x)$ changes from − to + or if $f''(a) > 0$.

If $f''(x) = 0$ at $x = a$, there is a point of inflection if $f''(x)$ changes from + to − or from − to +.

The following examples illustrate how the above information is put together to obtain the graph of a function.

EXAMPLE 4 ▸▸ Sketch the graph of $y = 6x - x^2$.

Finding the first two derivatives, we have

$$y' = 6 - 2x = 2(3 - x)$$
$$y'' = -2$$

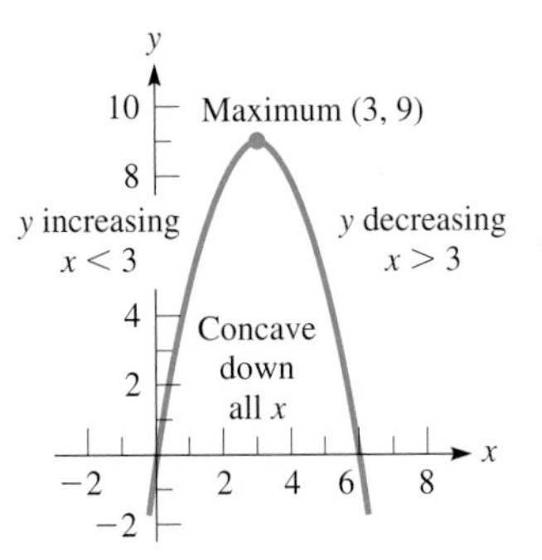

Fig. 24-34

We now note that $y' = 0$ for $x = 3$. For $x < 3$ we see that $y' > 0$, which means that y is increasing over this interval. Also, for $x > 3$, we note that $y' < 0$, which means that y is decreasing over this interval.

Since y' changes from positive on the left of $x = 3$ to negative on the right of $x = 3$, the curve has a maximum point where $x = 3$. Since $y = 9$ for $x = 3$, this maximum point is $(3, 9)$.

Since $y'' = -2$, this means that its value remains constant for all values of x. Therefore, there are no points of inflection and the curve is concave down for all values of x. This also shows that the point $(3, 9)$ is a maximum point.

Summarizing, we know that y is increasing for $x < 3$, y is decreasing for $x > 3$, there is a maximum point at $(3, 9)$, and the curve is always concave down. Using this information, we sketch the curve shown in Fig. 24-34.

From the equation, we know this curve is a parabola. We could also find the maximum point from the material of Section 7-4 or Section 21-7. However, using derivatives we can find this kind of important information about the graphs of a great many types of functions. ◂◂

By finding the information about a graph from the derivative, we are doing more than just finding the shape of the curve. We are also identifying key points where the curve changes in some important respect. This usually cannot be done with great accuracy from the view of the graph on a graphing calculator. At a maximum or minimum point, the view usually shows a short set of pixels parallel to the x-axis. Therefore, the exact location of the point is not obvious. Also, on a calculator it is difficult to see just where the curve changes concavity, as the graph often appears to be straight at a point of inflection. Of course, calculators or computer programs that can find derivatives can locate these points.

EXAMPLE 5 ▸▸ Sketch the graph of $y = 2x^3 + 3x^2 - 12x$.

Finding the first two derivatives, we have

$$y' = 6x^2 + 6x - 12 = 6(x + 2)(x - 1)$$
$$y'' = 12x + 6 = 6(2x + 1)$$

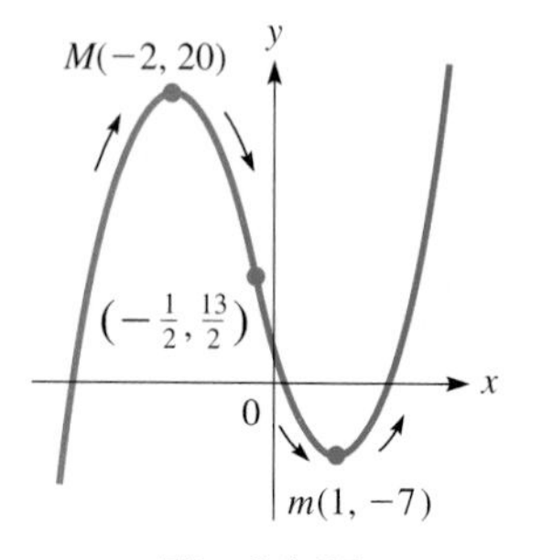

Fig. 24-35

We note that $y' = 0$ when $x = -2$ and $x = 1$. Using these values in the second derivative, we find that y'' is negative (-18) for $x = -2$ and y'' is positive $(+18)$ when $x = 1$. When $x = -2$, $y = 20$; and when $x = 1$, $y = -7$. Therefore, $(-2, 20)$ is a maximum point and $(1, -7)$ is a minimum point.

Next we see that $y' > 0$ if $x < -2$ or $x > 1$. Also, $y' < 0$ for the interval $-2 < x < 1$. Therefore, y is increasing if $x < -2$ or $x > 1$, and y is decreasing if $-2 < x < 1$.

Now we note that $y'' = 0$ when $x = -\frac{1}{2}$, $y'' < 0$ when $x < -\frac{1}{2}$, and $y'' > 0$ when $x > -\frac{1}{2}$. When $x = -\frac{1}{2}$, $y = \frac{13}{2}$. Therefore, there is a point of inflection at $(-\frac{1}{2}, \frac{13}{2})$, the curve is concave down if $x < -\frac{1}{2}$, and the curve is concave up if $x > -\frac{1}{2}$.

Finally, by locating the points $(-2, 20)$, $(-\frac{1}{2}, \frac{13}{2})$, and $(1, -7)$, we draw the curve *up* to $(-2, 20)$ and then *down* to $(-\frac{1}{2}, \frac{13}{2})$, with the curve *concave down.*

CAUTION ▸ ***Continuing* down, *but* concave up,** we draw the curve to $(1, -7)$, at which point we start *up* and continue up. We now know the key points and the shape of the curve. See Fig. 24-35. For more precision, additional points may be used. ◂◂

EXAMPLE 6 ▸▸ Sketch the graph of $y = x^5 - 5x^4$.

The first two derivatives are

$$y' = 5x^4 - 20x^3 = 5x^3(x - 4)$$
$$y'' = 20x^3 - 60x^2 = 20x^2(x - 3)$$

We now see that $y' = 0$ when $x = 0$ and $x = 4$. For $x = 0$, $y'' = 0$ also, which

CAUTION ▸ means ***we cannot use the second-derivative test*** for maximum and minimum points for $x = 0$ in this case. For $x = 4$, $y'' > 0$ $(+320)$, which means that $(4, -256)$ is a minimum point.

Next we note that

$$y' > 0 \quad \text{for } x < 0 \quad \text{or} \quad x > 4 \qquad y' < 0 \quad \text{for } 0 < x < 4$$

Thus, by the first-derivative test, there is a maximum point at $(0, 0)$. Also, y is increasing for $x < 0$ or $x > 4$ and decreasing for $0 < x < 4$.

The second derivative indicates that there is a point of inflection at $(3, -162)$. It also indicates that the curve is concave down for $x < 3$ $(x \neq 0)$ and concave up for $x > 3$. There is no point of inflection at $(0, 0)$ since the second derivative does not change sign at $x = 0$.

From this information, we sketch the curve in Fig. 24-36. ◂◂

Fig. 24-36

Exercises 24–5

In Exercises 1–4, find those values of x for which the given functions are increasing and those values of x for which they are decreasing.

1. $y = x^2 + 2x$
2. $y = 4 - x^2$
3. $y = 12x - x^3$
4. $y = x^4 - 6x^2$

In Exercises 5–8, find any maximum or minimum points of the given functions. (These are the same functions as in Exercises 1–4.)

5. $y = x^2 + 2x$
6. $y = 4 - x^2$
7. $y = 12x - x^3$
8. $y = x^4 - 6x^2$

In Exercises 9–12, find the values of x for which the given function is concave up, the values of x for which it is concave down, and any points of inflection. (These are the same functions as in Exercises 1–4.)

9. $y = x^2 + 2x$
10. $y = 4 - x^2$
11. $y = 12x - x^3$
12. $y = x^4 - 6x^2$

In Exercises 13–16, use the information from Exercises 1–12 to sketch the graphs of the given functions. Use a graphing calculator to check the graph.

13. $y = x^2 + 2x$
14. $y = 4 - x^2$
15. $y = 12x - x^3$
16. $y = x^4 - 6x^2$

In Exercises 17–28, sketch the graphs of the given functions by determining the appropriate information and points from the first and second derivatives. Use a graphing calculator to check the graph.

17. $y = 12x - 2x^2$
18. $y = 3x^2 - 1$
19. $y = 2x^3 + 6x^2$
20. $y = x^3 - 9x^2 + 15x + 1$
21. $y = x^3 + 3x^2 + 3x + 2$
22. $y = x^3 - 12x + 12$
23. $y = 4x^3 - 24x^2 + 36x$
24. $y = -2x^3 - x^2 + 8x - 5$
25. $y = 4x^3 - 3x^4$
26. $y = x^5 - 20x^2$
27. $y = x^5 - 5x$
28. $y = x^4 + 8x + 2$

In Exercises 29 and 30, view the graphs of y, y′, and y″ together on a graphing calculator. State how the graphs of y′ and y″ are related to the graph of y.

29. $y = x^3 - 12x$
30. $y = 24x - 9x^2 - 2x^3$

In Exercises 31–40, sketch the indicated curves by the methods of this section. You may check the graphs by using a graphing calculator.

31. A batter hits a baseball that follows a path given by $y = x - 0.025x^2$, where distances are in metres. Sketch the graph of the path of the baseball.
32. The angle θ, in degrees, of a robot arm with the horizontal as a function of the time t, in seconds, is given by $\theta = 10 + 12t^2 - 2t^3$. Sketch the graph for $0 \le t \le 6$ s.
33. An electric circuit is designed such that the resistance R, in ohms, is a function of the current i, in milliamperes, according to $R = 75 - 18i^2 + 8i^3 - i^4$. Sketch the graph if $R \ge 0$ and i can be positive or negative.
34. The deflection y of a beam at a horizontal distance x from one end is given by $y = k(x^3 - 60x^2)$. Sketch the curve representing the beam if it is 10 m long and $k = 1/5000$.
35. A rectangular box is made from a piece of cardboard 8 cm by 12 cm by cutting equal squares from each corner and bending up the sides. Express the volume of the box as a function of the side of the square that is cut out, then sketch the curve of the resulting equation.
36. A rectangular planter with a square end is to be made from 8.0 m² of redwood. Express the volume of soil the planter can hold as a function of the side of the square of the end. Sketch the curve of the resulting function.
37. Sketch a continuous curve having the following characteristics:

    $f(1) = 0, \quad f'(x) > 0 \text{ for all } x, \quad f''(x) < 0 \text{ for all } x$

38. Sketch a continuous curve having the following characteristics:

    $f(0) = 1, \quad f'(x) < 0 \text{ for all } x$
    $f''(x) < 0 \text{ for } x < 0, \quad f''(x) > 0 \text{ for } x > 0$

39. Sketch a continuous curve having the following characteristics:

    $f(-1) = 0, \quad f(2) = 2; \quad f'(x) < 0 \text{ for } x < -1;$
    $f'(x) > 0 \text{ for } x > -1; \quad f''(x) < 0 \text{ for } 0 < x < 2;$
    $f''(x) > 0 \text{ for } x < 0 \text{ or } x > 2$

40. Sketch a continuous curve having the following characteristics:

    $f(0) = -1; f'(x) > 0 \text{ for } x < 0 \text{ or } x > 2;$
    $f'(x) < 0 \text{ for } 0 < x < 2;$
    $f''(x) < 0 \text{ for } x < 1; \quad f''(x) > 0 \text{ for } x > 1$

24-6 More on Curve Sketching

At this point we combine the information from the derivative with information obtainable from the function itself to sketch the graph. We determine intercepts, symmetry, the behavior of the curve as x becomes large, the vertical asymptotes, and the domain and range of the function. Also, continuity is important in sketching certain functions. We will find that some of these features are of more value than others in graphing any particular curve.

EXAMPLE 1 ▸▸ Sketch the graph of $y = \dfrac{2}{x^2 + 1}$.

Intercepts: If $x = 0$, $y = 2$, which means $(0, 2)$ is an intercept. If $y = 0$, there is no corresponding value of x, since $2/(x^2 + 1)$ is a fraction greater than zero for all x. This also indicates that all points on the curve are above the x-axis.

Symmetry: For a review of symmetry, see Section 21-3.

The curve is symmetric to the y-axis since $y = \dfrac{2}{(-x)^2 + 1}$ is the same as $y = \dfrac{2}{x^2 + 1}$.

The curve is not symmetric to the x-axis since $-y = \dfrac{2}{x^2 + 1}$ is not the same as $y = \dfrac{2}{x^2 + 1}$.

The curve is not symmetric to the origin since $-y = \dfrac{2}{(-x)^2 + 1}$ is not the same as $y = \dfrac{2}{x^2 + 1}$.

The value in knowing the symmetry is that we should find those portions of the curve on either side of the y-axis reflections of the other. It is possible to use this fact directly or to use it as a check.

Behavior as x becomes large: We note that as $x \to \infty$, $y \to 0$ since $2/(x^2 + 1)$ is always a fraction that is greater than zero but which becomes smaller as x becomes larger. Therefore, we see that $y = 0$ is an asymptote. From either the symmetry or the function, we also see that $y \to 0$ as $x \to -\infty$.

Vertical asymptotes: From the discussion of the hyperbola we recall that an asymptote is a line that a curve approaches. We have already noted that $y = 0$ is an asymptote for this curve. This asymptote, the x-axis, is a horizontal line. *Vertical asymptotes, if any exist, are found by determining those values of x for which the denominator of any term is zero.* Such a value of x makes y undefined. Since $x^2 + 1$ cannot be zero, this curve has no vertical asymptotes. The next example illustrates a curve that has a vertical asymptote.

Domain and range: Since the denominator $x^2 + 1$ cannot be zero, x can take on any value. This means the domain of the function is all values of x. Also, we have noted that $2/(x^2 + 1)$ is a fraction greater than zero. Since $x^2 + 1$ is 1 or greater, y is 2 or less. This tells us that the range of the function is $0 < y \leq 2$.

Derivatives: Since $y = \dfrac{2}{x^2 + 1} = 2(x^2 + 1)^{-1}$

$$y' = -2(x^2 + 1)^{-2}(2x) = \frac{-4x}{(x^2 + 1)^2}$$

Since $(x^2 + 1)^2$ is positive for all values of x, the sign of y' is determined by the numerator. Thus we note that $y' = 0$ for $x = 0$ and that $y' > 0$ for $x < 0$ and $y' < 0$ for $x > 0$. The curve, therefore, is increasing for $x < 0$, is decreasing for $x > 0$, and has a maximum point at $(0, 2)$.

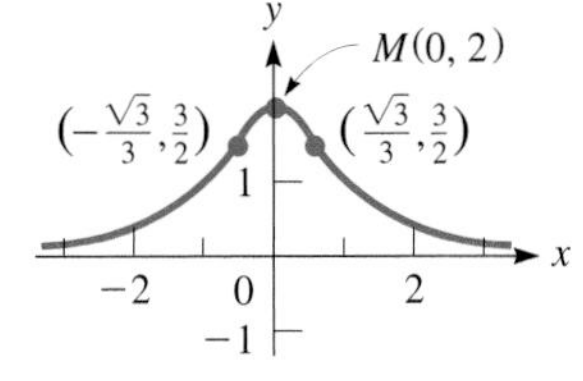

Fig. 24-37

Now finding the second derivative, we have

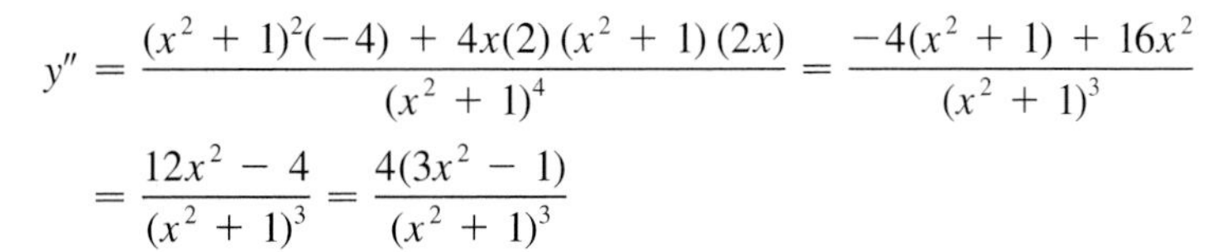

$$y'' = \frac{(x^2+1)^2(-4) + 4x(2)(x^2+1)(2x)}{(x^2+1)^4} = \frac{-4(x^2+1) + 16x^2}{(x^2+1)^3}$$

$$= \frac{12x^2 - 4}{(x^2+1)^3} = \frac{4(3x^2-1)}{(x^2+1)^3}$$

We note that y'' is negative for $x = 0$, which confirms that $(0, 2)$ is a maximum point. Also, points of inflection are found for the values of x satisfying $3x^2 - 1 = 0$. Thus, $(-\frac{1}{3}\sqrt{3}, \frac{3}{2})$ and $(\frac{1}{3}\sqrt{3}, \frac{3}{2})$ are points of inflection. The curve is concave up if $x < -\frac{1}{3}\sqrt{3}$ or $x > \frac{1}{3}\sqrt{3}$, and the curve is concave down if $-\frac{1}{3}\sqrt{3} < x < \frac{1}{3}\sqrt{3}$.

Putting this information together, we sketch the curve shown in Fig. 24-37. Note that this curve could have been sketched primarily by use of the fact that $y \to 0$ as $x \to +\infty$ and as $x \to -\infty$ and the fact that a maximum point exists at $(0, 2)$. However, the other parts of the analysis, such as symmetry and concavity, serve as checks and make the curve more accurate. ◀◀

EXAMPLE 2 ▶▶ Sketch the graph of $y = x + \dfrac{4}{x}$.

Intercepts: If we set $x = 0$, y is undefined. This means that the curve is not *continuous at* $x = 0$ and there are no y-intercepts. If we set $y = 0$, $x + 4/x = (x^2 + 4)/x$ cannot be zero since $x^2 + 4$ cannot be zero. Therefore, there are no intercepts. This may seem to be of little value, but we must realize *this curve does not cross either axis.* This will be of value when we sketch the curve in Fig. 24-38.

Symmetry: In testing for symmetry, we find that the curve is not symmetric to either axis. However, this curve does possess symmetry to the origin. This is determined by the fact that when $-x$ replaces x and at the same time $-y$ replaces y, the equation does not change.

Behavior as x becomes large: As $x \to +\infty$, and as $x \to -\infty$, $y \to x$ since $4/x \to 0$. Thus, $y = x$ is an asymptote of the curve.

Vertical asymptotes: As we noted in Example 1, vertical asymptotes exist for values of x for which y is undefined. In this equation, $x = 0$ makes the second term on the right undefined and therefore y is undefined. In fact, as $x \to 0$ from the positive side, $y \to +\infty$, and as $x \to 0$ from the negative side, $y \to -\infty$. This is derived from the sign of $4/x$ in each case.

Domain and range: Since x cannot be zero, the domain of the function is all x except zero. As for the range, the analysis from the derivatives will show it to be $y \le -4$, $y \ge 4$.

Derivatives: Finding the first derivative, we have

$$y' = 1 - \frac{4}{x^2} = \frac{x^2 - 4}{x^2}$$

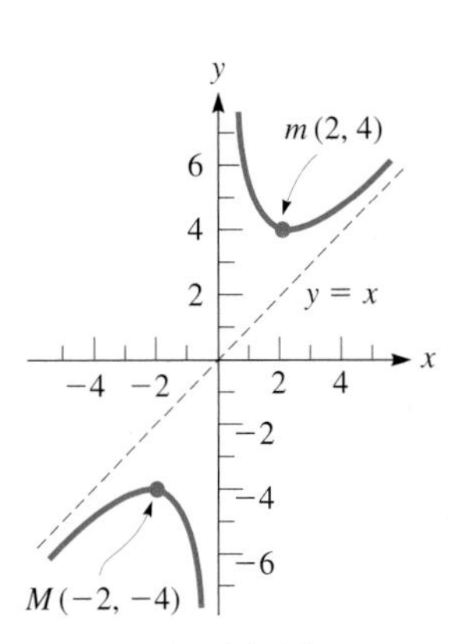

Fig. 24-38

The x^2 in the denominator indicates that the sign of the first derivative is the same as its numerator. The numerator is zero if $x = -2$ or $x = 2$. If $x < -2$ or $x > 2$, then $y' > 0$; and if $-2 < x < 2$, $x \neq 0$, $y' < 0$. Thus, y is increasing if $x < -2$ or $x > 2$, and also y is decreasing if $-2 < x < 2$, except at $x = 0$ (y is undefined). Also, $(-2, -4)$ is a maximum point and $(2, 4)$ is a minimum point. The second derivative is $y'' = 8/x^3$. This cannot be zero, but it is negative if $x < 0$ and positive if $x > 0$. Thus, the curve is concave down if $x < 0$ and concave up if $x > 0$. Using this information, we have the curve shown in Fig. 24-38. ◀◀

EXAMPLE 3 ▸▸ Sketch the graph of $y = \dfrac{1}{\sqrt{1 - x^2}}$.

Intercepts: If $x = 0$, $y = 1$. If $y = 0$, $1/\sqrt{1 - x^2}$ would have to be zero, but it cannot since it is a fraction with 1 as the numerator for all values of x. Thus, $(0, 1)$ is an intercept.

Symmetry: The curve is symmetric to the y-axis.

Behavior as x becomes large: The values of x cannot be considered beyond 1 or -1, for any value of $x < -1$ or $x > 1$ gives imaginary values for y. Thus the curve does not exist for values of $x < -1$ or $x > 1$.

Vertical asymptotes: If $x = 1$ or $x = -1$, y is undefined. In each case, as $x \to 1$ and as $x \to -1$, $y \to +\infty$.

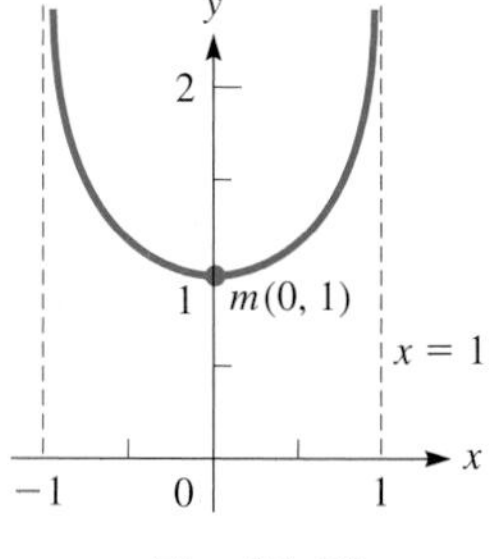

Fig. 24-39

Domain and range: From the analysis of x becoming large and of the vertical asymptotes, we see that the domain is $-1 < x < 1$. Also, since $\sqrt{1 - x^2}$ is 1 or less, $1/\sqrt{1 - x^2}$ is 1 or more, which means the range is $y \geq 1$.

Derivatives:

$$y' = -\frac{1}{2}(1 - x^2)^{-3/2}(-2x) = \frac{x}{(1 - x^2)^{3/2}}$$

We see that $y' = 0$ if $x = 0$. If $-1 < x < 0$, $y' < 0$, and also if $0 < x < 1$, $y' > 0$. Thus the curve is decreasing if $-1 < x < 0$ and increasing if $0 < x < 1$. There is a minimum point at $(0, 1)$.

$$y'' = \frac{(1 - x^2)^{3/2} - x(\frac{3}{2})(1 - x^2)^{1/2}(-2x)}{(1 - x^2)^3} = \frac{(1 - x^2) + 3x^2}{(1 - x^2)^{5/2}}$$

$$= \frac{2x^2 + 1}{(1 - x^2)^{5/2}}$$

The second derivative cannot be zero since $2x^2 + 1$ is positive for all values of x. The second derivative is also positive for all permissible values of x, which means the curve is concave up for these values.

Using this information, we sketch the graph in Fig. 24-39. ◂◂

EXAMPLE 4 ▸▸ Sketch the graph of $y = \dfrac{x}{x^2 - 4}$.

Intercepts: If $x = 0$, $y = 0$, and if $y = 0$, $x = 0$. The only intercept is $(0, 0)$.

Symmetry: The curve is not symmetric to either axis. However, since $-y = -x/[(-x)^2 - 4]$ is the same as $y = x/(x^2 - 4)$, it is symmetric to the origin.

Behavior as x becomes large: As $x \to +\infty$ and as $x \to -\infty$, $y \to 0$. This means that $y = 0$ is an asymptote.

Vertical asymptotes: If $x = -2$ or $x = 2$, y is undefined. As $x \to -2$, $y \to -\infty$ if $x < -2$ since $x^2 - 4$ is positive, and $y \to +\infty$ if $x > -2$ since $x^2 - 4$ is negative. As $x \to 2$, $y \to -\infty$ if $x < 2$, and $y \to +\infty$ if $x > 2$.

Domain and range: The domain is all real values of x except -2 and 2. As for the range, if $x < -2$, $y < 0$ (the numerator is negative and the denominator is positive). If $x > 2$, $y > 0$ (both numerator and denominator are positive). Since $(0, 0)$ is an intercept, we see that the range is all values of y.

Derivatives:

$$y' = \frac{(x^2 - 4)(1) - x(2x)}{(x^2 - 4)^2} = -\frac{x^2 + 4}{(x^2 - 4)^2}$$

Since $y' < 0$ for all values of x except -2 and 2, the curve is decreasing for all values in the domain.

$$y'' = -\frac{(x^2 - 4)^2(2x) - (x^2 + 4)(2)(x^2 - 4)(2x)}{(x^2 - 4)^4}$$

$$= -\frac{2x(x^2 - 4) - 4x(x^2 + 4)}{(x^2 - 4)^3} = \frac{2x^3 + 24x}{(x^2 - 4)^3} = \frac{2x(x^2 + 12)}{(x^2 - 4)^3}$$

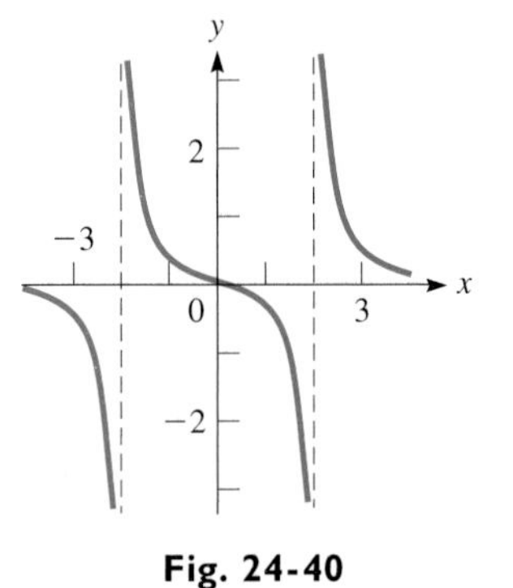

Fig. 24-40

The sign of y'' depends on x and $(x^2 - 4)^3$. If $x < -2$, $y'' < 0$. If $-2 < x < 0$, $y'' > 0$. If $0 < x < 2$, $y'' < 0$. If $x > 2$, $y'' > 0$. This means the curve is concave down for $x < -2$ or $0 < x < 2$ and is concave up for $-2 < x < 0$ or $x > 2$.

The curve is sketched in Fig. 24-40. ◀◀

Exercises 24–6

In the following exercises, use the method of the examples of this section to sketch the indicated curves. Use a graphing calculator to check the graph.

1. $y = \frac{4}{x^2}$

2. $y = \frac{2}{x^3}$

3. $y = \frac{2}{x + 1}$

4. $y = \frac{x}{x - 2}$

5. $y = x^2 + \frac{2}{x}$

6. $y = x + \frac{4}{x^2}$

7. $y = x - \frac{1}{x}$

8. $y = 3x + \frac{1}{x^3}$

9. $y = \frac{x^2}{x + 1}$

10. $y = \frac{9x}{x^2 + 9}$

11. $y = \frac{1}{x^2 - 1}$

12. $y = \frac{8}{4 - x^2}$

13. $y = \frac{4}{x} - \frac{4}{x^2}$

14. $y = 4x + \frac{1}{\sqrt{x}}$

15. $y = x\sqrt{1 - x^2}$

16. $y = \frac{x - 1}{x^2 - 2x}$

17. $y = \frac{9x}{9 - x^2}$

18. $y = \frac{x^2 - 4}{x^2 + 4}$

19. The combined capacitance C_T, in microfarads, of a 6-μF capacitance and a variable capacitance C in series is given by $C_T = \frac{6C}{6 + C}$. Sketch the graph.

20. Sketch a graph of P vs. V from van der Waals' equation (see Exercise 46 of Section 23–6), assuming the following values: $R = T = a = 1$ and $b = 0$. For many gases the value of a is much greater than that of b. Even though the values of $R = T = 1$ are not realistic, the *shape* of the curve will be correct for the assumed value of $b = 0$.

21. The reliability R of a computer model is found to be $R = \frac{200}{\sqrt{t^2 + 40\,000}}$, where t is the time of operation in hours. ($R = 1$ is perfect reliability, and $R = 0.5$ means there is a 50% chance of a malfunction.) Sketch the graph.

22. The electric power P, in watts, produced by a source is given by $P = \frac{36R}{R^2 + 2R + 1}$, where R is the resistance in the circuit. Sketch the graph.

23. A cylindrical oil drum is to be made such that it will contain 20 kL. Sketch the area of sheet metal required for construction as a function of the radius of the drum.

24. A fence is to be constructed to enclose a rectangular area of 20 000 m^2. A previously constructed wall is to be used for one side. Sketch the length of fence to be built as a function of the side of the fence parallel to the wall. See Fig. 24-41.

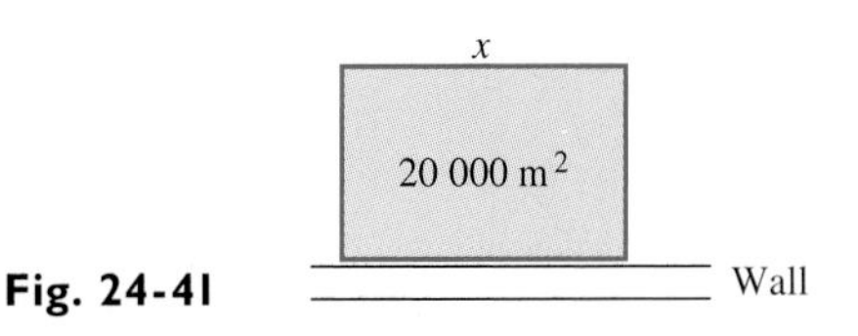

Fig. 24-41

24-7 Applied Maximum and Minimum Problems

Problems from various applied situations frequently occur that require finding a maximum or minimum value of some function. If the function is known, the methods we have already discussed can be used directly. This is discussed in the following example.

EXAMPLE 1 ▸▸ An automobile manufacturer, in testing a new engine on one of its new models, found that the efficiency e of the engine as a function of the speed s of the car was given by $e = 0.768s - 0.000\,04s^3$. Here, e is measured in percent and s is measured in km/h. What is the maximum efficiency of the engine?

In order to find a maximum value, we find the derivative of e with respect to s.

$$\frac{de}{ds} = 0.768 - 0.000\,12s^2$$

We then set the derivative equal to zero in order to find the value of s for which a maximum may occur.

$$\begin{aligned} 0.768 - 0.000\,12s^2 &= 0 \\ 0.000\,12s^2 &= 0.768 \\ s^2 &= 6400 \\ s &= 80.0 \text{ km/h} \end{aligned}$$

We know that s must be positive to have meaning in this problem. Therefore, the apparent solution of $s = -80$ is discarded. The second derivative is

$$\frac{d^2e}{ds^2} = -0.000\,24s$$

which is negative for any positive value of s. Therefore, we have a maximum for $s = 80.0$. Substituting $s = 80.0$ in the function for e, we obtain

$$e = 0.768(80.0) - 0.000\,04(80.0^3) = 61.44 - 20.48 = 40.96$$

The maximum efficiency is about 41.0%, which occurs for $s = 80.0$ km/h. ◂◂

In many problems for which a maximum or minimum value is to be found, the function is not given. To solve such a problem, we use these steps:

Steps in Solving Applied Maximum and Minimum Problems

1. *Determine the quantity Q to be maximized or minimized.*
2. If possible, *draw a figure illustrating the problem.*
3. *Write an equation for Q in terms of another variable of the problem.*
4. *Take the derivative of the function in step 3.*
5. *Set the derivative equal to zero, and solve the resulting equation.*
6. *Check as to whether the value found in step 5 makes Q a maximum or a minimum.* This might be clear from the statement of the problem, or it might require one of the derivative tests.
7. *Be sure the stated answer is the one the problem required.* Some problems require the maximum or minimum value, and others require values of other variables that give the maximum or minimum value.

CAUTION ▶ ***The principal difficulty that arises in these problems is finding the proper function.*** We must carefully read the problem to find the information needed to set up the function. For some problems it is more convenient to set up more than one function and find a derivative with respect to the same variable. Any unwanted variables or derivatives can be eliminated by substitution. The following examples illustrate several types of stated problems involving maximum and minimum values.

EXAMPLE 2 ▶▶ Find the number that exceeds its square by the greatest amount.

The quantity to be maximized is the difference D between a number x and its square x^2. Therefore, the required function is

$$D = x - x^2$$

Since we want D to be a maximum, we find dD/dx, which is

$$\frac{dD}{dx} = 1 - 2x$$

Setting the derivative equal to zero and solving for x, we have

$$0 = 1 - 2x, \qquad x = \tfrac{1}{2}$$

The second derivative gives $d^2D/dx^2 = -2$, which tells us that the second derivative is always negative. This means that whenever the first derivative is zero, it represents a maximum. In many problems it is not necessary to test for maximum or minimum, since the nature of the problem will indicate which must be the case. For example, in this problem we know that numbers greater than 1 do not exceed their squares at all. The same is true for all negative numbers. Thus the answer must be between 0 and 1; in this case it is $x = 1/2$. ◀◀

EXAMPLE 3 ▶▶ A rectangular corral is to be enclosed with 1600 m of fencing. Find the maximum possible area of the corral.

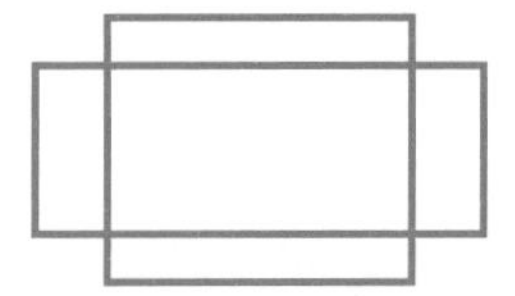

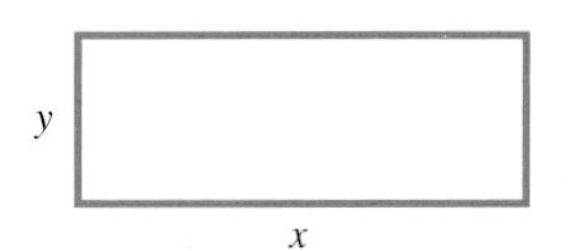

Fig. 24-42

There are limitless possibilities for rectangles of a perimeter of 1600 m and differing areas. See Fig. 24-42. For example, if the sides are 700 m and 100 m, the area is 70 000 m^2, or if the sides are 600 m and 200 m, the area is 120 000 m^2. Therefore, we set up a function for the area of a rectangle in terms of its sides x and y.

$$A = xy$$

Another important fact is that the perimeter of the corral is 1600 m. Therefore, $2x + 2y = 1600$. Solving for y, we have $y = 800 - x$. By using this expression for y, we can express the area in terms of x only. This gives us

$$A = x(800 - x) = 800x - x^2$$

We complete the solution as follows:

$$\frac{dA}{dx} = 800 - 2x \qquad \text{take derivative}$$

$$800 - 2x = 0 \qquad \text{set derivative equal to zero}$$

$$x = 400 \text{ m}$$

By checking values of the derivative near 400, or by finding the second derivative, we can show that we have a maximum for $x = 400$. This means that $x = 400$ m and $y = 400$ m give the maximum area of 160 000 m^2 for the corral. ◀◀

See the chapter introduction.

EXAMPLE 4 ▸▸ The strength S of a beam with a rectangular cross section is directly proportional to the product of its width w and the square of its depth d. Find the dimensions of the strongest beam that can be cut from a log with a circular cross section which is 16.0 cm in diameter. See Fig. 24-43.

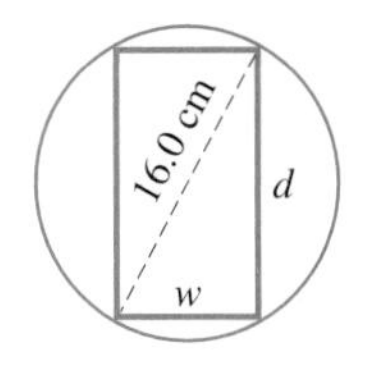

Fig. 24-43

The solution proceeds as follows:

$$
\begin{aligned}
S &= kwd^2 && \text{direct variation} \\
d^2 &= 256 - w^2 && \text{Pythagorean theorem} \\
S &= kw(256 - w^2) && \text{substituting} \\
&= k(256w - w^3) && S = f(w) \\
\frac{dS}{dw} &= k(256 - 3w^2) && \text{take derivative} \\
0 &= k(256 - 3w^2) && \text{set derivative equal to zero} \\
3w^2 &= 256 && \text{solve for } w \\
w &= \frac{16}{\sqrt{3}} = 9.24 \text{ in.} && \\
d &= \sqrt{256 - \frac{256}{3}} && \text{solve for } d \\
&= 13.1 \text{ cm} &&
\end{aligned}
$$

This means that the strongest beam is about 9.24 cm wide and 13.1 cm deep. Since $d^2S/dw^2 = -6kw$ and is negative for $w > 0$ (which are the only possible values with meaning in this problem), these dimensions give the maximum strength for the beam. ◂◂

EXAMPLE 5 ▸▸ Find the point on the parabola $y = x^2$ that is nearest to the point $(6, 3)$.

In this example we must set up a function for this distance between a general point (x, y) on the parabola and the point $(6, 3)$. This relation is

$$D = \sqrt{(x - 6)^2 + (y - 3)^2}$$

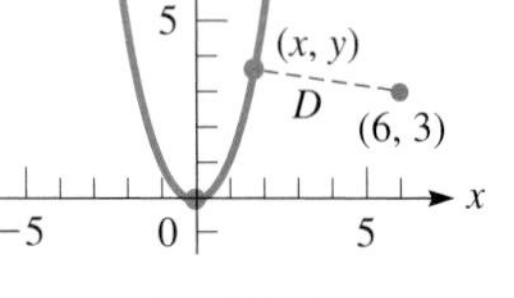

Fig. 24-44

However, to make it easier to take derivatives, we shall square both sides of this expression. If a function is a minimum, then so is its square. We shall also use the fact that the point (x, y) is on $y = x^2$ by replacing y by x^2. Thus we have

$$
\begin{aligned}
D^2 &= (x - 6)^2 + (x^2 - 3)^2 = x^2 - 12x + 36 + x^4 - 6x^2 + 9 \\
&= x^4 - 5x^2 - 12x + 45 \\
\frac{dD^2}{dx} &= 4x^3 - 10x - 12 && \text{take derivative} \\
0 &= 2x^3 - 5x - 6 && \text{set derivative equal to zero}
\end{aligned}
$$

Using synthetic division or some similar method, we find that the solution to this equation is $x = 2$. Thus the required point on the parabola is $(2, 4)$. (See Fig. 24-44.) We can show that we have a minimum by analyzing the first derivative, by analyzing the second derivative, or by noting that points at much greater distances exist (therefore, it cannot be a maximum). ◂◂

EXAMPLE 6 ▸▸ A company determines that it can sell 1000 units of a product per month if the price is \$5 for each unit. It also estimates that for each 1¢ reduction in unit price, 10 more units can be sold. Under these conditions, what is the maximum possible income and what price per unit gives this income?

If we let x = the number of units over 1000 sold, the total number of units sold is $1000 + x$. The price for each unit is \$5 less 1¢ (\$0.01) for each block of 10 units over 1000 that are sold. Thus the price for each unit is

$$5 - 0.01\left(\frac{x}{10}\right) \quad \text{or} \quad 5 - 0.001x \text{ dollars}$$

The income I is the number of units sold times the price of each unit. Therefore, we have

$$I = (1000 + x)(5 - 0.001x)$$

Multiplying and finding the first derivative, we have

$$I = 5000 + 4x - 0.001x^2$$

$$\frac{dI}{dx} = 4 - 0.002x \quad \text{take derivative}$$

$$0 = 4 - 0.002x \quad \text{set derivative equal to zero}$$

$$x = 2000$$

We note that if $x < 2000$, the derivative is positive, and if $x > 2000$, the derivative is negative. Therefore, if $x = 2000$, I is at a maximum. This means that the maximum income is derived if 2000 units over 1000 are sold, or 3000 units in all. This in turn means that the maximum income is \$9000 and the price per unit is \$3. These values are found by substituting $x = 2000$ into the expression for I and for the price. ◂◂

EXAMPLE 7 ▸▸ Find the most economical dimensions for a cylindrical cup that holds a specified volume.

When we analyze the wording of the problem carefully, we see that we are to minimize the surface area of a right circular cylinder that has a bottom, but no top. Also the volume is to be considered as constant. Thus we set up expressions for the surface area and volume:

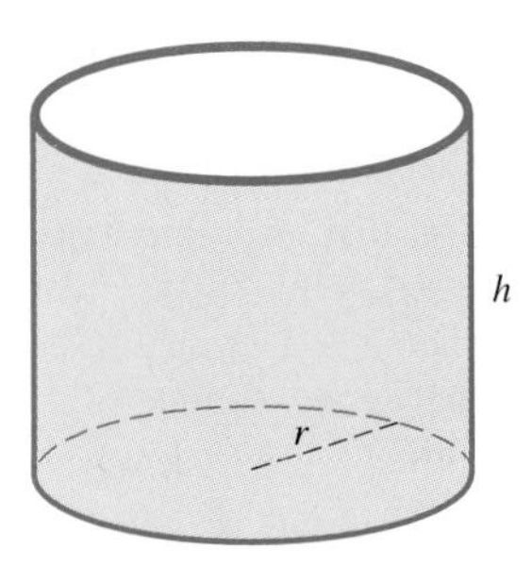

Fig. 24-45

$$A = \pi r^2 + 2\pi rh, \qquad V = \pi r^2 h$$

Since the volume is not specified numerically, it is easier to take the derivative of each of these formulas separately. The derivatives may be taken with respect to either r or h. Let us choose r and write

$$\frac{dA}{dr} = 2\pi r + 2\pi r\frac{dh}{dr} + 2\pi h, \qquad \frac{dV}{dr} = 0 = 2\pi rh + \pi r^2\left(\frac{dh}{dr}\right)$$

Setting dA/dr equal to zero, and at the same time eliminating the derivative dh/dr by substitution, we have

$$0 = 2\pi r + 2\pi r\left(-\frac{2h}{r}\right) + 2\pi h \quad \text{or} \quad 0 = r - 2h + h$$

This last equation tells us that $h = r$, which is the relation between dimensions for the most economical cup (see Fig. 24-45). ◂◂

EXAMPLE 8 ▸▸ The illuminance of a light source at any point equals the strength of the source divided by the square of the distance from the source. Two sources, of strengths 8 units and 1 unit, respectively, are 100 m apart. Determine at what point between them the illuminance is the least, assuming that the illuminance at any point is the sum of the illuminances of the two sources.

Let I = the sum of the illuminances and

x = the distance from the source of strength 8.

Then we find that

$$I = \frac{8}{x^2} + \frac{1}{(100 - x)^2}$$

is the function between the illuminance and the distance from the source of strength 8. We must now take a derivative of I with respect to x, set it equal to zero, and solve for x to find the point at which the illuminance is a minimum:

$$\frac{dI}{dx} = -\frac{16}{x^3} + \frac{2}{(100 - x)^3} = \frac{-16(100 - x)^3 + 2x^3}{x^3(100 - x)^3}$$

This function will be zero if the numerator is zero. Therefore, we have

$$2x^3 - 16(100 - x)^3 = 0 \quad \text{or} \quad x^3 = 8(100 - x)^3$$

Taking cube roots of each side, we have

$$x = 2(100 - x) \quad \text{or} \quad x = 66.7 \text{ m}$$

The point where the illuminance is a minimum is 66.7 m from the 8-unit source of illuminance. ◂◂

Exercises 24–7

In the following exercises, solve the given maximum and minimum problems.

1. The height s (in metres) of a flare shot vertically upward from the ground is given by $s = 34.3t - 4.9t^2$, where t is measured in seconds. What is the greatest height to which the flare goes?
2. A small oil refinery estimates that its daily profit P, in dollars, from refining x barrels of oil is $P = 8x - 0.02x^2$. How many barrels should be refined for maximum daily profit, and what is the maximum profit?
3. The power output P of a battery of voltage E and internal resistance R is $P = EI - RI^2$, where I is the current. Find the current for which the power is a maximum.
4. The velocity v, in milligrams per second, at which a certain chemical reaction takes place is related to the amount x, in milligrams, of the chemical produced by $v = k(100x - x^2)$, where k is a constant. For what value of x is v a maximum?
5. A company projects that its total savings S, in dollars, by converting to a solar heating system with a solar collector area A (in m^2) will be $S = 360A - 0.10A^3$. Find the area that should give the maximum savings, and find the amount of the maximum savings.
6. The altitude h, in metres, of a jet that goes into a dive and then again turns upward is given by $h = 16t^3 - 240t^2 + 10\,000$, where t is the time, in seconds, of the dive and turn. What is the altitude of the jet when it turns up out of the dive?
7. The impedance in an electric circuit is given by $Z = \sqrt{R^2 + (X_L - X_C)^2}$. If $R = 2500\ \Omega$ and $X_L = 1500\ \Omega$, what value of X_C makes the impedance a minimum?
8. The electric potential V on the line $3x + 2y = 6$ is given by $V = 3x^2 + 2y^2$. At what point on this line is the potential a minimum?
9. The ratio of one positive number to the reciprocal of another is 64. Find the numbers if their sum is a minimum.

10. The sum of two negative numbers is -40. Find the numbers such that the sum of their reciprocals is a maximum.

11. A rectangular microprocessor chip is designed to have an area of 25 mm^2. What must be its dimensions if its perimeter is to be a minimum?

12. A rectangular storage area is to be constructed along the side of a tall building. A security fence is required along the remaining three sides of the area. What is the maximum area that can be enclosed with 800 m of fencing?

13. Ship A is traveling due east at 18.0 km/h as it passes a point 40.0 km due south of ship B, which is traveling due south at 16.0 km/h. How much later are the ships nearest each other?

14. An architect is designing a rectangular building in which the front wall costs twice as much per linear metre as the other three walls. The building is to cover 1350 m^2. What dimensions must it have such that the cost of the walls is a minimum?

15. A computer is programmed to display a slowly changing right triangle with its hypotenuse always equal to 12.0 cm. What are the legs of the triangle when it has its maximum area?

16. Canadian postal regulations require that the sum of the three dimensions of a rectangular package not exceed 3 m. What are the dimensions of the largest rectangular box with square ends that can be mailed?

17. A culvert designed with a semicircular cross section of diameter 2.40 m is redesigned to have an isosceles trapezoidal cross section by inscribing the trapezoid in the semicircle. See Fig. 24-46. What is the length of the bottom base b of the trapezoid if its area is to be maximum?

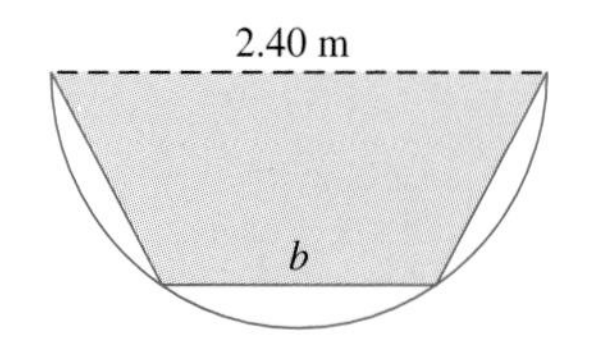

Fig. 24-46

18. A lap pool (a pool for swimming laps) is designed to be seven times as long as it is wide. If the area of the sides and bottom is 90.0 m^2, what are the dimensions of the pool if the volume of water it can hold is a maximum?

19. What is the maximum slope of the curve $y = 6x^2 - x^3$?

20. What is the minimum slope of the curve $y = x^5 - 10x^2$?

21. The deflection y of a beam of length L at a horizontal distance x from one end is given by $y = k(2x^4 - 5Lx^3 + 3L^2x^2)$, where k is a constant. For what value of x does the maximum deflection occur?

22. The potential energy E of an electric charge q due to another charge q_1 at a distance r_1 is proportional to q_1 and inversely proportional to r_1. If charge q is placed directly between two charges of 2.00 nC and 1.00 nC that are separated by 10.0 mm, find the point at which the total potential energy (the sum due to the other two charges) of q is a minimum.

23. An open box is to be made from a square piece of cardboard whose sides are 20.0 cm long by cutting equal squares from the corners and bending up the sides. Determine the side of the square which is to be cut out so that the volume of the box may be a maximum. See Fig. 24-47.

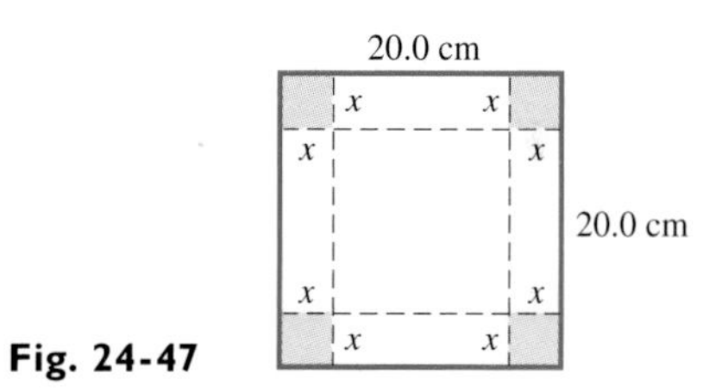

Fig. 24-47

24. A cone-shaped paper cup is to hold 100 cm^3 of water. Find the height and radius of the cup that can be made from the least amount of paper.

25. A race track 400 m long is to be built around an area that is a rectangle with a semicircle at each end. Find the open side of the rectangle if the area of the rectangle is to be a maximum. See Fig. 24-48.

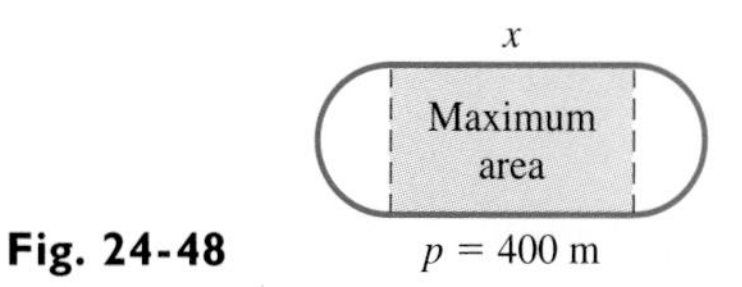

Fig. 24-48

26. A company finds that there is a net profit of $10 for each of the first 1000 units produced each week. For each unit over 1000 produced, there is 2 cents less profit per unit. How many units should be produced each week to net the greatest profit?

27. A beam of rectangular cross section is to be cut from a log 1.00 m in diameter. The stiffness varies directly as the width and the cube of the depth. What dimensions will give the beam maximum stiffness? See Fig. 24-49.

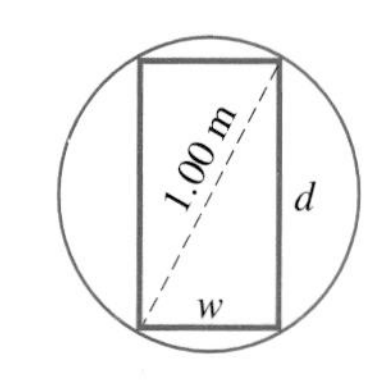

Fig. 24-49

28. An alpha particle moves through a magnetic field along the parabolic path $y = x^2 - 4$. Determine the closest that the particle comes to the origin.

29. An oil pipeline is to be built from a refinery to a tanker loading area. The loading area is 10.0 km downstream from the refinery and on the opposite side of a river 2.5 km wide. The pipeline is to run along the river and then cross to the loading area. If the pipeline costs \$50 000 per kilometre alongside the river and \$80 000 per kilometre across the river, find the point P (see Fig. 24-50) at which the pipeline should be turned to cross the river if construction costs are to be a minimum.

30. A light ray follows a path of least time. If a ray starts at point A (see Fig. 24-51) and is reflected off a plane mirror to point B, show that the angle of incidence α equals the angle of reflection β. (*Hint:* Set up the expression in terms of x, which will lead to $\sin \alpha = \sin \beta$.)

31. A rectangular building covering 7000 m^2 is to be built on a rectangular lot as shown in Fig. 24-52. If the building is to be 10.0 m from the lot boundary on each side and 20.0 m from the boundary in front and back, find the dimensions of the building if the area of the lot is a minimum.

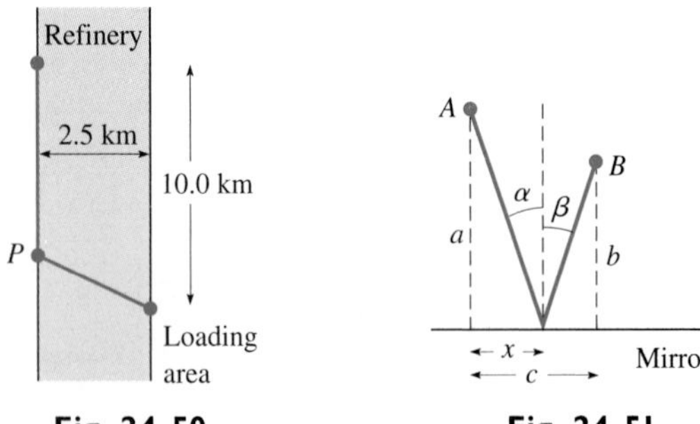

Fig. 24-50

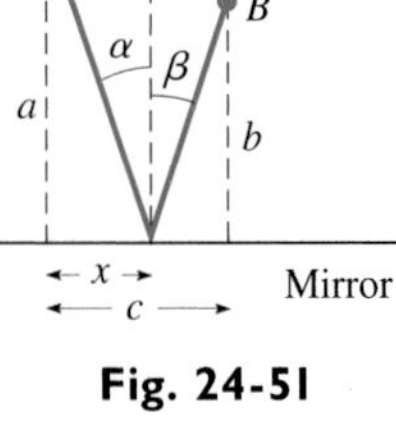

Fig. 24-51

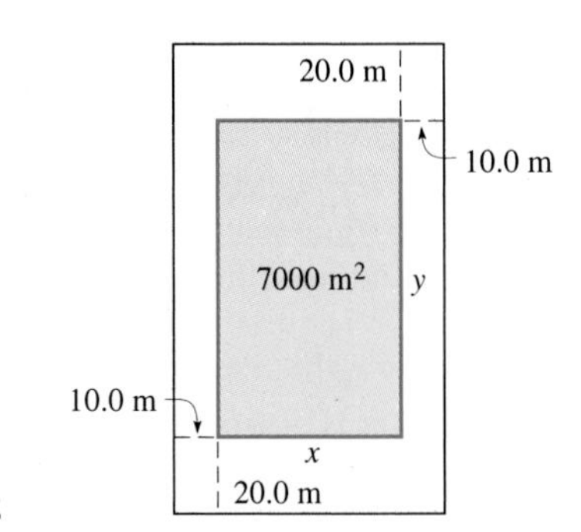

Fig. 24-52

32. A cylindrical wastewater holding tank has a capacity of 200 kL. What are its radius and height if its total surface area (top, bottom, and lateral area) is a minimum?

Chapter Equations, Review Exercises, and Practice Test

Chapter Equations

Newton's method
$$x_2 = x_1 - \frac{f(x_1)}{f'(x_1)} \quad (24\text{-}1)$$

Curvilinear motion
$$v_x = \frac{dx}{dt} \qquad v_y = \frac{dy}{dt} \quad (24\text{-}2)$$

$$a_x = \frac{dv_x}{dt} = \frac{d^2x}{dt^2} \qquad a_y = \frac{dv_y}{dt} = \frac{d^2y}{dt^2} \quad (24\text{-}3)$$

$$v = \sqrt{v_x^2 + v_y^2} \qquad a = \sqrt{a_x^2 + a_y^2} \quad (24\text{-}4)$$

$$\tan \theta_v = \frac{v_y}{v_x} \qquad \tan \theta_a = \frac{a_y}{a_x} \quad (24\text{-}5)$$

Curve sketching and maximum and minimum values

$f'(x) > 0$ where $f(x)$ increases; $f'(x) < 0$ where $f(x)$ decreases.
$f''(x) > 0$ where the graph of $f(x)$ is concave up; $f''(x) < 0$ where the graph of $f(x)$ is concave down.
If $f'(x) = 0$ at $x = a$, there is a maximum point if $f'(x)$ changes from + to − or if $f''(a) < 0$.
If $f'(x) = 0$ at $x = a$, there is a minimum point if $f'(x)$ changes from − to + or if $f''(a) > 0$.
If $f''(x) = 0$ at $x = a$, there is a point of inflection if $f''(x)$ changes from + to − or from − to +.

Review Exercises

In Exercises 1–6, find the equations of the tangent and normal lines. Use a graphing calculator to view the curve and the line.

1. Find the equation of the line tangent to the parabola $y = 3x - x^2$ at the point $(-1, -4)$.

2. Find the equation of the line tangent to the curve $y = x^2 - \frac{6}{x}$ at the point $(2, 1)$.

3. Find the equation of the line normal to $y = \frac{x}{4x - 1}$ at the point $(1, \frac{1}{3})$.

4. Find the equation of the line normal to $y = \frac{1}{\sqrt{x - 2}}$ at the point $(6, \frac{1}{2})$.

5. Find the equation of the line tangent to the curve $y = \sqrt{x^2 + 3}$ and which has a slope of $\frac{1}{2}$.

6. Find the equation of the line normal to the curve $y = \frac{1}{2x + 1}$ and which has a slope of $\frac{1}{2}$ if $x \geq 0$.

In Exercises 7–12, find the indicated velocities and accelerations.

7. Given that the x- and y-coordinates of a moving particle are given as a function of time by the parametric equations $x = \sqrt{t} + t$, $y = \frac{1}{12}t^3$, find the magnitude and direction of the velocity when $t = 4$.

8. If the x- and y-coordinates of a moving object as functions of time are given by $x = 0.1t^2 + 1$, $y = \sqrt{4t + 1}$, find the magnitude and direction of the velocity when $t = 6$.

9. An object moves along the curve $y = 0.5x^2 + x$ such that $v_x = 0.5\sqrt{x}$. Find v_y at $(2, 4)$.

10. A particle moves along the curve of $y = \frac{1}{x + 2}$ with a constant velocity in the x-direction of 4 cm/s. Find v_y at $(2, \frac{1}{4})$.

11. Find the magnitude and direction of the acceleration for the particle in Exercise 7.

12. Find the magnitude and direction of the acceleration of the particle in Exercise 10.

In Exercises 13–16, find the indicated roots of the given equations to at least four decimal places by use of Newton's method.

13. $x^3 - 3x^2 - x + 2 = 0$ (between 0 and 1)

14. $x^3 + 3x^2 - 6x - 2 = 0$ (between 1 and 2)

15. $3x^3 - x^2 - 8x - 2 = 0$ (between 1 and 2)

16. $x^4 + 3x^3 + 6x + 4 = 0$ (between -1 and 0)

In Exercises 17–24, sketch the graphs of the given functions by information obtained from the function as well as information obtained from the derivatives. Use a graphing calculator to check the graph.

17. $y = 4x^2 + 16x$

18. $y = 2x^2 + x - 1$

19. $y = 27x - x^3$

20. $y = x^3 + 2x^2 + x + 1$

21. $y = x^4 - 32x$

22. $y = x^4 + 4x^3 - 16x$

23. $y = \frac{x}{x + 1}$

24. $y = x^3 + \frac{3}{x}$

In Exercises 25–52, solve the given problems.

25. The parabolas $y = x^2 + 2$ and $y = 4x - x^2$ are tangent to each other. Find the equation of the line tangent to them at the point of tangency.

26. Find the equation of the line tangent to $y = x^4 - 8x$ and parallel to $y + 4x + 3 = 0$.

27. The deflection y of a beam at a horizontal distance x from one end is given by $y = k(x^4 - 30x^3 + 1000x)$, where k is a constant. Observing the equation and using Newton's method, find the values of x where the deflection is zero, if the beam is 10.000 m long.

28. The edges of a rectangular water tank are 3.00 m, 5.00 m, and 8.00 m. By Newton's method, determine by how much each edge should be increased equally to double the volume of the tank.

29. A parachutist descends (after the parachute opens) in a path that can be described by $x = 8t$ and $y = -0.15t^2$, where distances are in metres and time is in seconds. Find the parachutist's velocity upon landing if landing occurs when $t = 12$ s.

30. An electron is moving along the path $y = 1/x$ at the constant velocity of 100 m/s. Find the velocity in the x-direction when the electron is at the point $(2, \frac{1}{2})$.

31. In Fig. 24-53, the tension T supports the 40.0-N weight. The relation between the tension T and the deflection d is $d = \frac{1000}{\sqrt{T^2 - 400}}$. If the tension is increasing at 2.00 N/s when $T = 28.0$ N, how fast is the deflection, in centimetres, changing?

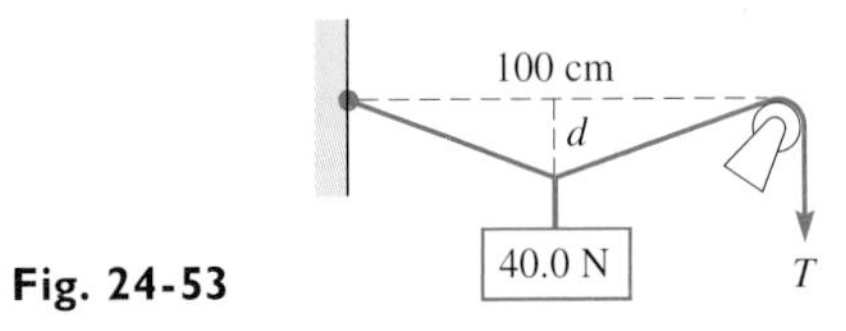

Fig. 24-53

32. The impedance Z, in ohms, in a particular electric circuit is given by $Z = \sqrt{48 + R^2}$, where R is the resistance. If R is increasing at the rate of 0.45 Ω/min for $R = 6.5$ Ω, find the rate at which Z is changing.

33. A calculator manufacturer determines that the monthly profit p, in dollars, from the sale of x calculators is given by $p = -0.25x^2 + 80x - 4000$. What is the maximum monthly profit that can be made?

34. The altitude h, in feet, of a rocket as a function of time after launching is given by $h = 1600t - 16t^2$. What is the maximum altitude the rocket attains?

35. Sketch a continuous curve having these characteristics:

$f(0) = 2 \quad f'(x) < 0 \text{ for } x < 0 \quad f''(x) > 0 \text{ for all } x$
$f'(x) > 0 \text{ for } x > 0$

36. Sketch a continuous curve having these characteristics:

$f(0) = 1 \quad f'(0) = 0 \quad f''(x) < 0 \text{ for } x < 0$
$f'(x) > 0 \text{ for } |x| > 0 \quad f''(x) > 0 \text{ for } x > 0$

37. The angle between two equal edges of a triangular plastic sheet is 120°. The plastic sheet is being stretched such that each of these edges is increasing at the rate of 0.750 cm/min. How fast is the third edge changing?

38. The current I through a circuit with a resistance R and a battery whose voltage is E and whose internal resistance is r is given by $I = E/(R + r)$. If R changes at the rate of 0.250 Ω/min, how fast is the current changing when $R = 6.25$ Ω, if $E = 3.10$ V and $r = 0.230$ Ω?

39. The radius of a circular oil spill is increasing at the rate of 15 m/min. How fast is the area of the spill changing when the radius is 400 m?

40. A baseball diamond is a square 90.0 ft on a side. See Fig. 24-54. As a player runs from first base toward second base at 18.0 ft/s, at what rate is the player's distance from home plate increasing when the player is 40.0 ft from first base? (1 ft = 0.3048 m)

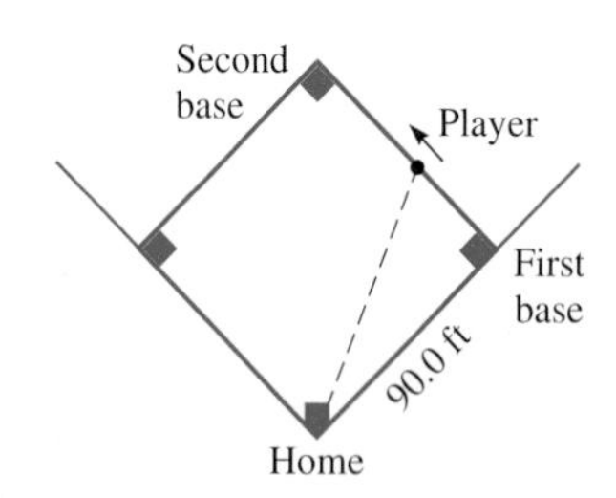

Fig. 24-54

41. A special insulation strip is to be sealed completely around three edges of a rectangular solar panel. If 200 cm of the strip are used, what is the maximum area of the panel?

42. A swimming pool with a rectangular surface of 130 m² is to have a cement border area that is 4.00 m wide at each end and 2.75 m wide at the sides. Find the surface dimensions of the pool if the total area covered is to be a minimum.

43. A study showed that the percent y of persons surviving burns to x percent of the body is given by
$y = \dfrac{300}{0.0005x^2 + 2} - 50$. Sketch the graph.

44. A company estimates that the sales S, in dollars, of a new product will be $S = 5000t/(t + 4)^2$, where t is the time, in months, after it is put into production. Sketch the graph of S vs. t.

45. An airplane flying horizontally at 2400 m is moving toward a radar installation at 1110 km/h. If the plane is directly over a point on the ground 8.00 km from the radar installation, what is its actual speed? See Fig. 24-55.

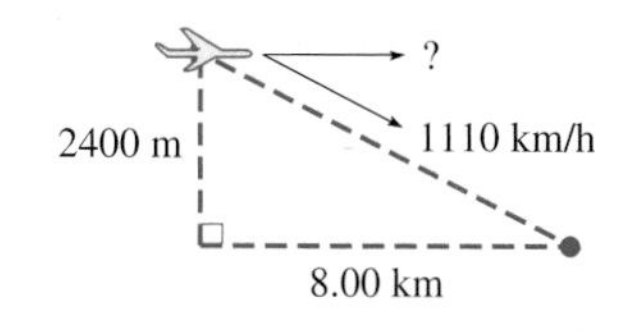

Fig. 24-55

46. The base of a conical machine part is being milled such that the height is decreasing at the rate of 0.05 cm/min. If the part originally had a radius of 1.0 cm and a height of 3.0 cm, how fast is the volume changing when the height is 2.8 cm?

47. The reciprocal of the total resistance R_T of electric resistances in parallel equals the sum of the reciprocals of the individual resistances. If the sum of two resistances is 12 Ω, find their values if their total resistance in parallel is a maximum.

48. A cable is to be from point A to point B on a wall and then to point C. See Fig. 24-56. Where is B located if the total length of cable is a minimum?

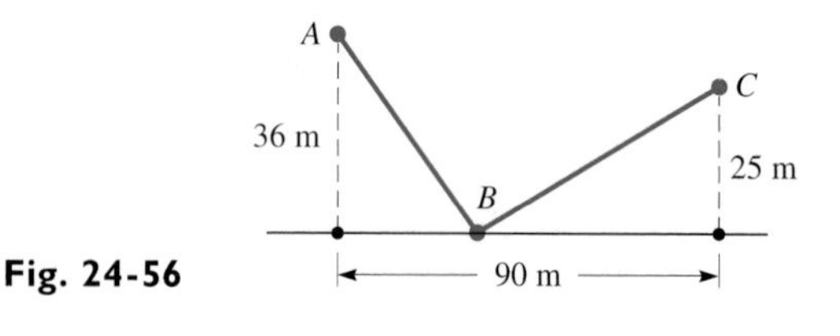

Fig. 24-56

49. A box with a square base and an open top is to be made of 27 dm^2 of cardboard. What is the maximum volume that can be contained within the box?

50. A machine part is to be in the shape of a circular sector of radius r and central angle θ. Find r and θ if the area is one unit and the perimeter is a minimum. See Fig. 24-57.

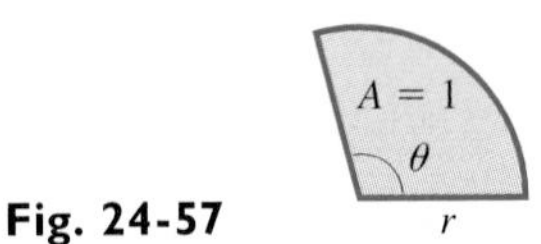

Fig. 24-57

51. An open drawer for small tools is to be made from a rectangular piece of heavy sheet metal 36.0 cm by 30.0 cm by cutting out equal squares from two corners and bending up the three sides, as shown in Fig. 24-58. Determine the side of the square that should be cut out so that the volume of the drawer is a maximum.

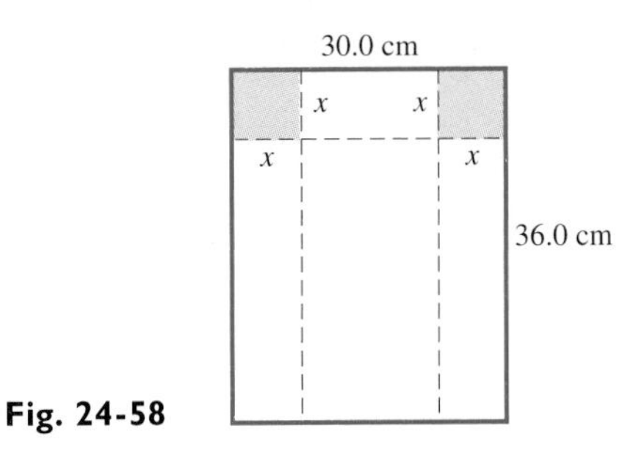

Fig. 24-58

52. A Norman window has the form of a rectangle surmounted by a semicircle. Find the dimensions (radius of circular part and height of rectangular part) of the window that will admit the most light if the perimeter of the window is 4.00 m. See Fig. 24-59.

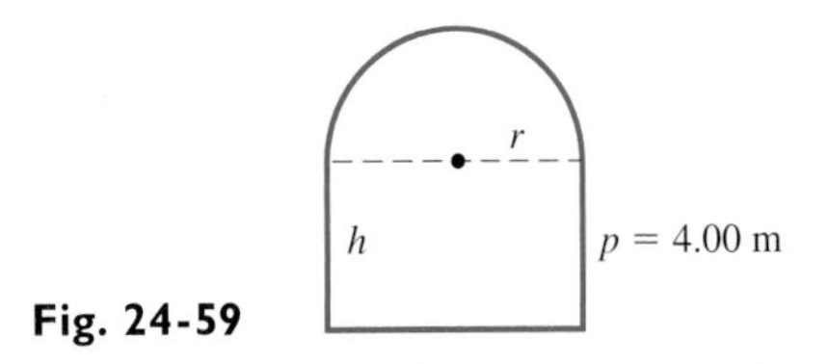

Fig. 24-59

Writing Exercise

53. A container manufacturer makes various sizes of closed cylindrical plastic containers for shipping liquid products. For each size, write two or three paragraphs to explain how to determine the ratio of the height to radius of the container such that the least amount of plastic is used. Include the reason why it is not necessary to specify the volume of the container in finding this ratio.

Practice Test

1. Find the equation of the line tangent to the curve $y = x^4 - 3x^2$ at the point $(1, -2)$.

2. If the x- and y-coordinates of a moving object as functions of time are given by the parametric equations $x = 3t^2$, $y = 2t^3 - t^2$, find the magnitude and direction of the acceleration when $t = 2$.

3. The electric power (in watts) produced by a certain source is given by $P = \dfrac{144r}{(r + 0.6)^2}$, where r is the resistance in the circuit. For what value of r is the power a maximum?

4. Find the root of the equation $x^2 - \sqrt{4x + 1} = 0$ between 1 and 2 to four decimal places by use of Newton's method. Use $x_1 = 1.5$ and find x_3.

5. Sketch the graph of $y = x^3 + 6x^2$ by finding the values of x for which the function is increasing, decreasing, concave up, and concave down, and by finding any maximum points, minimum points, and points of inflection.

6. Sketch the graph of $y = \dfrac{4}{x^2} - x$ by finding the same information as required in Problem 5, as well as intercepts, symmetry, behavior as x becomes large, vertical asymptotes, and the domain and range.

7. Trash is being compacted into a cubical volume. The edge of the cube is decreasing at the rate of 0.10 m/s. When an edge of the cube is 1.25 m, how fast is the volume changing?

8. A rectangular field is to be fenced and then divided in half by a fence parallel to two opposite sides. If a total of 6000 m of fencing is used, what is the maximum area that can be fenced?

25

Integration

In Section 25-5, integration is used to find the rate of flow of water over an obstacle.

In the study of physical and technical applications, we frequently find information related to the rate of change of a variable. With such information we have to reverse the process of differentiation in order to determine the functional relationship between the variables. This leads us to consider the problem of finding the function when we know its derivative. This process is known as *integration.*

As we study integration in this chapter and the next, we shall see that it has many applications in science and technology. It can be applied to finding areas and volumes, as well as to the physical concepts of work, pressure, and center of mass. Although these applications appear to be quite different, we will show that they have a very similar mathematical formulation. A few of these applications are shown in this chapter, although most of them are developed in the following chapter.

25-1 Differentials

Although the primary concern of this chapter is the development of the inverse process of differentiation, it is necessary to first introduce certain concepts and notation in this section before proceeding to the inverse process. The material presented here is necessary to properly relate differentiation and the methods we shall develop.

We define the **differential** *of a function* $y = f(x)$ *as*

$$dy = f'(x)\,dx \tag{25-1}$$

In Eq. (25-1), the quantity *dy is the differential of y, and dx is the differential of x. The differential dx is defined as equal to* Δx*, the increment in x.* We define it purposely in this way so that $f'(x) = dy/dx$. In this way we can interpret the derivative as the ratio of the differential of y to the differential of x. That is, now the derivative can be considered as a fraction. Although we had previously used the notation dy/dx for the derivative, we had not interpreted it as a fraction.

EXAMPLE 1 ▸▸ Find the differential of $y = 3x^5 - x$.

In this example, $f(x) = 3x^5 - x$, which in turn means that $f'(x) = 15x^4 - 1$. Thus

$$dy = (15x^4 - 1)\,dx$$

($15x^4 - 1$ is $f'(x)$; dx is the differential of x) ◂◂

EXAMPLE 2 ▸▸ Find the differential of $y = (2x^3 - 1)^4$.

$$\begin{aligned} dy &= 4(2x^3 - 1)^3(6x^2)\,dx && \text{using derivative power rule} \\ &= 24x^2(2x^3 - 1)^3\,dx \end{aligned}$$ ◂◂

EXAMPLE 3 ▸▸ Find the differential of $y = \dfrac{4x}{x^2 + 4}$.

$$\begin{aligned} dy &= \frac{(x^2 + 4)(4) - (4x)(2x)}{(x^2 + 4)^2}\,dx && \text{using derivative quotient rule} \\ &= \frac{4x^2 + 16 - 8x^2}{(x^2 + 4)^2}\,dx = \frac{-4x^2 + 16}{(x^2 + 4)^2}\,dx \\ &= \frac{-4(x^2 - 4)}{(x^2 + 4)^2}\,dx \end{aligned}$$

(don't forget the dx) ◂◂

We shall find that the definition of the differential will clarify the connection between the various interpretations of the inverse process of differentiation. Although the principal purpose in introducing it here is to be able to use the notation, there are useful direct applications of the differential.

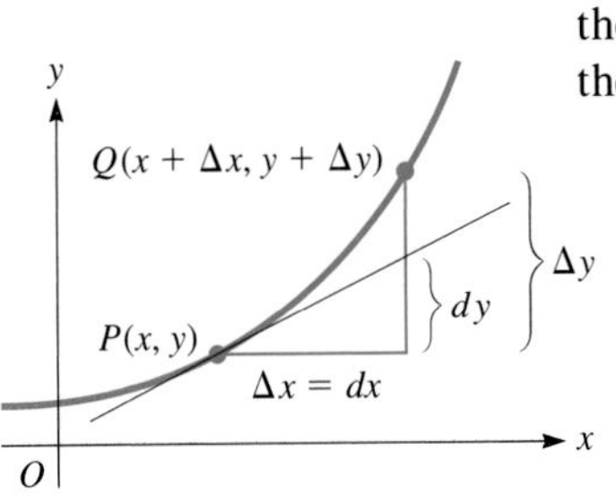

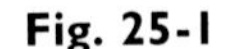
Fig. 25-1

The applications of the differential are based on the fact that the differential of y, dy, closely approximates the increment in y, Δy, if the differential of x, dx, is small. To understand this statement, let us look at Fig. 25-1. Recalling the meaning of Δx and Δy, we see that points $P(x, y)$ and $Q(x + \Delta x, y + \Delta y)$ lie on the curve of $f(x)$. However, $f'(x) = dy/dx$ at P, which means that if we draw a tangent line at P, its slope may be indicated by dy/dx. By choosing $\Delta x = dx$, we can see the difference between Δy and dy. We see that as dx becomes smaller, Δy more nearly equals dy.

For given changes in x, it is necessary to use the delta-process to find the exact change, Δy, in y. However, *for small values of* Δx, dy *can be used to approximate* Δy *closely.* Generally dy is much more easily determined than is Δy.

EXAMPLE 4 ▸▸ Calculate Δy and dy for $y = x^3 - 2x$ for $x = 3$ and $\Delta x = 0.1$.

By the delta-process, we find

$$y + \Delta y = (x + \Delta x)^3 - 2(x + \Delta x)$$
$$\Delta y = 3x^2 \Delta x + 3x(\Delta x)^2 + (\Delta x)^3 - 2\Delta x$$

Using the given values, we find

$$\Delta y = 3(9)(0.1) + 3(3)(0.01) + (0.001) - 2(0.1) = 2.591$$

The differential of y is

$$dy = (3x^2 - 2)\,dx$$

Since $dx = \Delta x$, we have

$$dy = [3(9) - 2](0.1) = 2.5$$

Thus, $\Delta y = 2.591$ and $dy = 2.5$. In this case, dy is very nearly equal to Δy. ◂◂

EXAMPLE 5 ▸▸ If finding Δy by the delta-process is complicated or lengthy, we can use a calculator to find the difference between Δy and dy. Therefore, for the function

$$y = f(x) = \sqrt{8x + 3}$$

the differential is

$$dy = \frac{1}{2}(8x + 3)^{-1/2}(8)\,dx = \frac{4\,dx}{\sqrt{8x + 3}}$$

For $x = 2$ and $\Delta x = 0.003$,

$$f(x) = \sqrt{8(2) + 3} = \sqrt{19} = 4.358\,898\,9$$

and

$$f(x + \Delta x) = \sqrt{8(2.003) + 3} = \sqrt{19.024} = 4.361\,651\,1$$

This means that

$$\Delta y = f(x + \Delta x) - f(x)$$
$$= 4.361\,651\,1 - 4.358\,898\,9 = 0.002\,752\,2$$

Now, calculating the value of dy,

$$dy = \frac{4(0.003)}{\sqrt{8(2) + 3}} = 0.002\,753\,0$$

Again, in this example the values of dy and Δy are very nearly equal. ◂◂

Estimating Errors in Measurement

The fact that dy can be used to approximate Δy is useful in finding the error in a result from a measurement, if the data are in error, or the equivalent problem of finding the change in the result if a change is made in the data. Even though such changes can be found by using a calculator, the differential can be used to set up a general expression for the change of a particular function.

EXAMPLE 6 ▸▸ The edge of a cube of gold was measured to be 3.850 cm. From this value the volume was found. Later it was discovered that the value of the edge was 0.020 cm too small. By approximately how much was the volume in error?

The volume V of a cube, in terms of an edge, e, is $V = e^3$. Since we wish to find the change in V for a given change in e, we want the value of dV for $e = 3.850$ cm and $de = 0.020$ cm.

First, finding the general expression for dV, we have

$$dV = 3e^2\,de$$

Now, evaluating this expression for the given values, we have

$$dV = 3(3.850)^2(0.020) = 0.89 \text{ cm}^3$$

In this case the volume was in error by about 0.89 cm^3. As long as de is small compared with e, we can calculate an error or change in the volume of a cube by calculating the value of $3e^2\,de$. ◂◂

Often when considering the error of a given value or result, *the actual numerical value of the error, the* **absolute error,** is not as important as its size in relation to the size of the quantity itself. *The ratio of the absolute error to the size of the quantity itself is known as the* **relative error.** The relative error is commonly expressed as a percent.

EXAMPLE 7 ▸▸ Referring to Example 6, we see that the absolute error in the edge was 0.020 cm. The relative error in the edge was

$$\frac{de}{e} = \frac{0.020}{3.85} = 0.0052 = 0.52\%$$

The absolute error in the volume was 0.89 cm^3, and the original value of the volume was $3.850^3 = 57.07$ cm^3. This means the relative error in the volume was

$$\frac{dV}{V} = \frac{0.89}{57.07} = 0.016 = 1.6\%$$ ◂◂

Exercises 25–1

In Exercises 1–12, find the differential of each of the given functions.

1. $y = x^5 + x$

2. $y = 3x^2 + 6$

3. $y = \frac{2}{x^5} + 3$

4. $y = 2\sqrt{x} - \frac{1}{x}$

5. $y = (x^2 - 1)^4$

6. $y = 5(4 + 3x)^{1/3}$

7. $y = \frac{2}{3x^2 + 1}$

8. $y = \frac{1}{\sqrt{1 - x^3}}$

9. $y = x^2(1 - x)^3$

10. $y = 6x\sqrt{1 - 4x}$

11. $y = \frac{x}{5x + 2}$

12. $y = \frac{3x + 1}{\sqrt{2x - 1}}$

In Exercises 13–16, find the values of Δy and dy for the given values of x and dx.

13. $y = 7x^2 + 4x,\ x = 4,\ \Delta x = 0.2$

14. $y = 2x^2 - 3x + 1,\ x = 5,\ \Delta x = 0.15$

15. $y = 2x^3 - 4x,\ x = 2.5,\ \Delta x = 0.05$

16. $y = x - x^4,\ x = 3.2,\ \Delta x = 0.08$

In Exercises 17–20, determine the value of dy for the given values of x and Δx. Compare with values of $f(x + \Delta x) - f(x)$ found by using a calculator.

17. $y = (1 - 3x)^5,\ x = 1,\ \Delta x = 0.01$

18. $y = (x^2 + 2x)^3,\ x = 7,\ \Delta x = 0.02$

19. $y = x\sqrt{1 + 4x},\ x = 12,\ \Delta x = 0.06$

20. $y = \dfrac{x}{\sqrt{6x - 1}},\ x = 3.5,\ \Delta x = 0.025$

In Exercises 21–32, solve the given problems by finding the appropriate differential.

21. The side of a square microprocessor chip is measured as 0.950 cm, and later it is measured as 0.952 cm. What is the difference in the calculations of the area due to the difference in the measurements of the side? See Fig. 25-2.

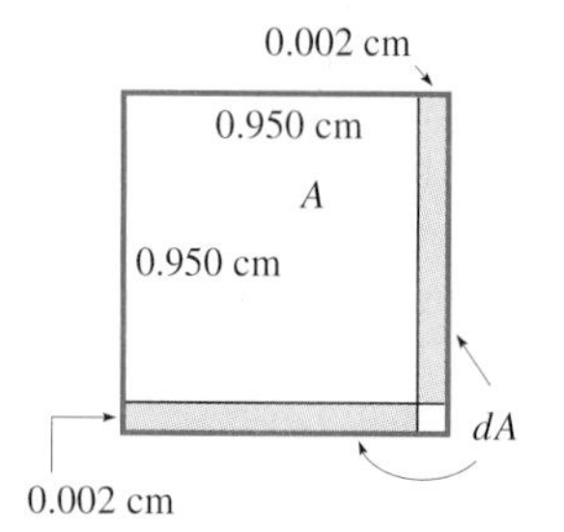

Fig. 25-2

22. A circular solar cell is made with a radius measured to be 12.00 ± 0.05 cm. (This means that the radius is 12.00 cm with a possible error of no more than 0.05 cm.) What is the maximum possible relative error in the calculation of the area of the cell?

23. The capacitance C, in microfarads, in an element of an electronic tuner is $C = \dfrac{3.6}{\sqrt{1 + 2V}}$, where V is the voltage. What is the approximate change in C if V changes from 0.220 V to 0.224 V?

24. The velocity of an object rolling down a certain inclined plane is given by $v = \sqrt{40 + 4.9h}$, where h is the distance traveled along the plane by the object. What is the increase in velocity (in m/s) of an object in moving from 20.0 m to 20.5 m along the plane? What is the relative change in the velocity?

25. An insulated cylindrical hot water heater has an inner diameter of 48.3 cm, an outer diameter of 55.4 cm, an inner height of 132.1 cm, and an outer height of 139.2 cm. What is the volume of the material of which the heater is made?

26. A precisely measured 10.0-Ω resistor (any error in its value will be considered negligible) is put in parallel with a variable resistor of resistance R. The combined resistance of the two resistors is $R_T = \dfrac{10R}{10 + R}$. What is the relative error of the combined resistance if R is measured at 40.0 Ω, with a possible error of 1.5 Ω?

27. The radius r of a holograph is directly proportional to the square root of the wavelength λ of the light used. Show that $dr/r = \frac{1}{2}d\lambda/\lambda$.

28. The gravitational force F of the earth on an object is inversely proportional to the square of the distance r of the object from the center of the earth. Show that $dF/F = -2dr/r$.

29. Show that an error of 2% in the measurement of the side of a square results in an error of approximately 4% in the calculation of the area.

30. Show that the relative error in the calculation of the volume of a sphere is approximately three times the relative error in the measurement of the radius.

31. Using differentials, evaluate 2.03^4. (*Hint:* Let $y = x^4$.)

32. Explain how to evaluate $\sqrt{9.02}$ using differentials.

25-2 Antiderivatives

Since many kinds of problems in many areas, including science and technology, can be solved by reversing the process of finding a derivative or a differential, we shall introduce the basic technique of this procedure in this section. *This reverse process is known as* **antidifferentiation.** In the next section we shall formalize the process, but it is only the basic idea that is the topic of this section. Many of the applications are found in Chapter 26. The following example illustrates the method.

EXAMPLE 1 ▸▸ Find a function for which the derivative is $8x^3$. That is, find an antiderivative of $8x^3$.

We know that, when we set out to find the derivative of a polynomial, we reduce the power of x by 1. Also we multiply the coefficient by the power of x. Thus the power of the function must have been 4, since the power in the derivative is 3. A factor of 4 of the derivative must also be divided out in the process of finding the function. If we write the derivative as $2(4x^3)$, we recognize $4x^3$ as the derivative of x^4. Therefore, the desired function is $2x^4$, which means that an antiderivative of $8x^3$ is $2x^4$. We can verify our result by finding the derivative. ◂◂

EXAMPLE 2 ▸▸ Find an antiderivative of $x^2 + 2x$.

As for the x^2, we know that the power of x required in an antiderivative is 3. Also, to make the coefficient correct, we must multiply by $\frac{1}{3}$. The $2x$ should be recognized as the derivative of x^2. Therefore, we have as an antiderivative $\frac{1}{3}x^3 + x^2$. ◂◂

In Examples 1 and 2, we note that we could add any constant to the antiderivative given as the result and still have a correct antiderivative. This is due to the fact that the derivative of a constant is zero. This is considered further in the following section. For the examples and exercises in this section, we will not include any such constants in the results.

We note that when we find an antiderivative of a given function, we obtain another function. Thus *we can define an* **antiderivative** *of the function $f(x)$ to be a function $F(x)$ such that $F'(x) = f(x)$.*

EXAMPLE 3 ▸▸ Find an antiderivative of the function $f(x) = \sqrt{x} - \dfrac{2}{x^3}$.

Since we wish to find an antiderivative of $f(x)$, we know that $f(x)$ is the derivative of the required function.

Considering the term $\sqrt{x}$, we first write it as $x^{1/2}$. To have x to the $\frac{1}{2}$ power in the derivative, we must have x to the $\frac{3}{2}$ power in the antiderivative. Knowing that the derivative of $x^{3/2}$ is $\frac{3}{2}x^{1/2}$, we write $x^{1/2}$ as $\frac{2}{3}(\frac{3}{2}x^{1/2})$. Thus the first term of the antiderivative is $\frac{2}{3}x^{3/2}$.

As for the term $-2/x^3$, we write it as $-2x^{-3}$. This we recognize as the derivative of x^{-2}, or $1/x^2$.

This means that an antiderivative of the function

function ↓ $\qquad f(x) = \sqrt{x} - \dfrac{2}{x^3}$ ⟵ derivative

is the function

antiderivative ⟶ $F(x) = \dfrac{2}{3}x^{3/2} + \dfrac{1}{x^2}$ $\qquad$ function ↑ ◂◂

A great many functions of which we must find an antiderivative are not polynomials or simple powers of x. It is these functions that may cause more difficulty in the general process of antidifferentiation. Pay special attention to the following examples, for they illustrate a type of problem that you will find to be very important.

EXAMPLE 4 ▸▸ Find an antiderivative of the function $f(x) = 3(x^3 - 1)^2(3x^2)$.

Noting that we have a power of $x^3 - 1$ in the derivative, it is reasonable that the antiderivative may include a power of $x^3 - 1$. Since, in the derivative, $x^3 - 1$ is raised to the power 2, the antiderivative would then have $x^3 - 1$ raised to the power 3. Noting that the derivative of $(x^3 - 1)^3$ is $3(x^3 - 1)^2(3x^2)$, the desired antiderivative is

$$F(x) = (x^3 - 1)^3$$

CAUTION ▸ We note that ***the factor of*** $3x^2$ ***does not appear in the antiderivative,*** for it was included from the process of finding a derivative. Therefore, it must be present for $(x^3 - 1)^3$ to be the proper antiderivative, but it must also be excluded in the process of antidifferentiation. ◂◂

EXAMPLE 5 ▸▸ Find an antiderivative of the function $f(x) = (2x + 1)^{1/2}$.

Here we note a power of $2x + 1$ in the derivative, which infers that the antiderivative has a power of $2x + 1$. Since in finding a derivative 1 is subtracted from the power of $2x + 1$, we should add 1 in finding the antiderivative. Thus we should have $(2x + 1)^{3/2}$ as part of the antiderivative. Finding a derivative of $(2x + 1)^{3/2}$, we obtain $\frac{3}{2}(2x + 1)^{1/2}(2) = 3(2x + 1)^{1/2}$. This differs from the given derivative by the factor of 3. Thus, if we write $(2x + 1)^{1/2} = \frac{1}{3}[3(2x + 1)^{1/2}]$, we have the required antiderivative as

$$F(x) = \tfrac{1}{3}(2x + 1)^{3/2}$$

Checking, the derivative of $\frac{1}{3}(2x + 1)^{3/2}$ is $\frac{1}{3}(\frac{3}{2})(2x + 1)^{1/2}(2) = (2x + 1)^{1/2}$. ◂◂

Exercises 25–2

In the following exercises, find antiderivatives of the given functions.

1. $f(x) = 3x^2$

2. $f(x) = 5x^4$

3. $f(x) = 12x^5$

4. $f(x) = 30x^9$

5. $f(x) = 6x^3 + 1$

6. $f(x) = 12x^5 + 2x$

7. $f(x) = 2x^2 - x$

8. $f(x) = x^2 - 5$

9. $f(x) = \frac{5}{2}x^{3/2}$

10. $f(x) = \frac{4}{3}x^{1/3}$

11. $f(x) = 2\sqrt{x} + 3$

12. $f(x) = 3\sqrt[3]{x} - 4$

13. $f(x) = -\frac{1}{x^2}$

14. $f(x) = -\frac{7}{x^6}$

15. $f(x) = \frac{6}{x^4}$

16. $f(x) = \frac{8}{x^5}$

17. $f(x) = 2x^4 + 1$

18. $f(x) = 3x^3 - 5x^2 - 3$

19. $f(x) = 6x + \frac{1}{x^4}$

20. $f(x) = \frac{1}{2\sqrt{x}} + 4$

21. $f(x) = x^2 + 2 + x^{-2}$

22. $f(x) = x\sqrt{x} - x^{-3}$

23. $f(x) = 6(2x + 1)^5(2)$

24. $f(x) = 3(x^2 + 1)^2(2x)$

25. $f(x) = 4(x^2 - 1)^3(2x)$

26. $f(x) = 5(2x^4 + 1)^4(8x^3)$

27. $f(x) = x^3(2x^4 + 1)^4$

28. $f(x) = x(1 - x^2)^7$

29. $f(x) = \frac{3}{2}(6x + 1)^{1/2}(6)$

30. $f(x) = \frac{5}{4}(1 - x)^{1/4}(-1)$

31. $f(x) = (3x + 1)^{1/3}$

32. $f(x) = (4x + 3)^{1/2}$

25-3 The Indefinite Integral

In the previous section, in developing the basic technique of finding an antiderivative, we noted that the results given are not unique. That is, we could have added any constant to the answers and the result would still have been correct. Again, this is the case since the derivative of a constant is zero.

EXAMPLE 1 ▸▸ The derivatives of x^3, $x^3 + 4$, $x^3 - 7$, and $x^3 + 4\pi$ are all $3x^2$. This means that any of the functions listed, as well as others, would be a proper answer to the problem of finding an antiderivative of $3x^2$. ◂◂

From Section 25-1, we know that the differential of a function $F(x)$ can be written as $d[F(x)] = F'(x)\,dx$. Therefore, since finding a differential of a function is closely related to finding the derivative, so is the antiderivative closely related to the process of finding the function for which the differential is known.

The notation used for finding the general form of the antiderivative, the **indefinite integral,** *is written in terms of the differential.* Thus *the indefinite integral of a function $f(x)$, for which $dF(x)/dx = f(x)$, or $dF(x) = f(x)\,dx$, is defined as*

$$\int f(x)\,dx = F(x) + C \qquad (25\text{-}2)$$

Here $f(x)$ is called the **integrand,** *$F(x) + C$ is the indefinite integral, and C is an arbitrary constant, called the* **constant of integration.** It represents any of the constants that may be attached to an antiderivative to have a proper result. We must have additional information beyond a knowledge of the differential to assign a specific value to C. *The symbol $\int$ is the* **integral sign,** *and it indicates that the inverse of the differential is to be found. Determining the indefinite integral is called* **integration,** which we can see is essentially the same as finding an antiderivative.

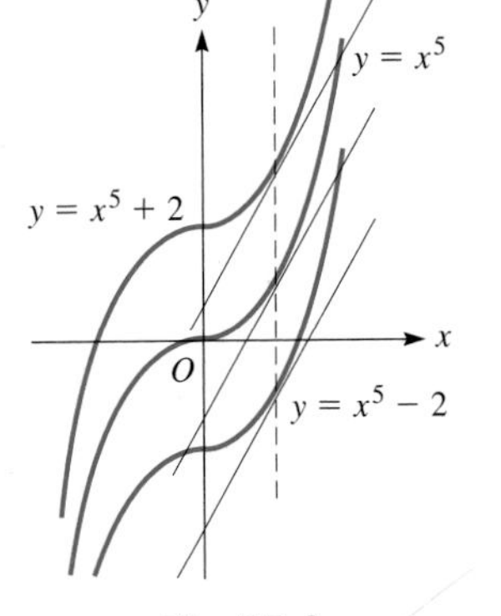

Fig. 25-3

EXAMPLE 2 ▸▸ In performing the integration

$$\underbrace{\int 5x^4\,dx}_{\text{integrand: } 5x^4} = \underbrace{x^5 + C}_{\text{indefinite integral}}$$

(constant of integration: C)

we might think that the inclusion of this constant C would affect the derivative of the function x^5. However, the only effect of the C is to raise or lower the curve. The slope of $x^5 + 2$, $x^5 - 2$, or any function of the form $x^5 + C$ is the same for any given value of x. As Fig. 25-3 shows, tangents drawn to the curves are all parallel for the same value of x. ◂◂

At this point we shall derive some basic formulas for integration. Since

$$\frac{d(cu)}{dx} = c\frac{du}{dx}$$

where u is a function of x and c is a constant, we can write

$$\int c\,du = c\int du = cu + C \qquad (25\text{-}3)$$

Also, since the derivative of a sum of functions equals the sum of the derivatives, we write

$$\int (du + dv) = u + v + C \qquad (25\text{-}4)$$

To find the derivative of a power of a function, we multiply by the power and subtract 1 from it. *To find the integral, we reverse this by adding 1 to the power of $f(x)$ in $f(x)\,dx$ and dividing by the new power. The power formula for integration is*

$$\int u^n\,du = \frac{u^{n+1}}{n+1} + C \qquad (n \neq -1) \tag{25-5}$$

We must be able to recognize the proper form and the component parts to use these formulas. Unless you do this, and have a good knowledge of differentiation, you will have trouble using Eq. (25-5). Most of the difficulty, if it exists, arises from an improper identification of du.

EXAMPLE 3 ▸▸ Integrate: $\int 6x\,dx$.

We must identify u, n, du, and any multiplying constants. Noting that 6 is a multiplying constant, we identify x as u, which means dx must be du and $n = 1$.

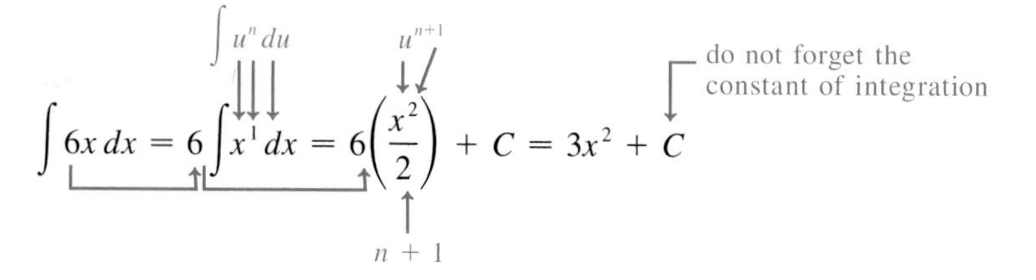

We see that our result checks since the differential of $3x^2 + C$ is $6x\,dx$. ◂◂

EXAMPLE 4 ▸▸ Integrate: $\int (5x^3 - 6x^2 + 1)\,dx$.

Here we must use a combination of Eqs. (25-3), (25-4), and (25-5). Therefore,

$$\int (5x^3 - 6x^2 + 1)\,dx = \int 5x^3\,dx + \int (-6x^2)\,dx + \int dx$$

$$= 5\int x^3\,dx - 6\int x^2\,dx + \int dx$$

In the first integral, $u = x$, $n = 3$, and $du = dx$. In the second, $u = x$, $n = 2$, and $du = dx$. The third uses Eq. (25-3) directly, with $c = 1$ and $du = dx$. This means

$$5\int x^3\,dx - 6\int x^2\,dx + \int dx = 5\left(\frac{x^4}{4}\right) - 6\left(\frac{x^3}{3}\right) + x + C$$

$$= \frac{5}{4}x^4 - 2x^3 + x + C \quad ◂◂$$

EXAMPLE 5 ▸▸ Integrate: $\int \left(\sqrt{x} - \frac{1}{x^3}\right) dx$.

In order to use Eq. (25-5) we must first write $\sqrt{x} = x^{1/2}$ and $1/x^3 = x^{-3}$.

$$\int \left(\sqrt{x} - \frac{1}{x^3}\right) dx = \int x^{1/2}\,dx - \int x^{-3}\,dx = \frac{1}{\frac{3}{2}}x^{3/2} - \frac{1}{-2}x^{-2} + C$$

$$= \frac{2}{3}x^{3/2} + \frac{1}{2}x^{-2} + C = \frac{2}{3}x^{3/2} + \frac{1}{2x^2} + C \quad ◂◂$$

EXAMPLE 6 ▸▸ Integrate: $\int (x^2 + 1)^3(2x\,dx)$.

We first note that $n = 3$, for this is the power involved in the function being integrated. If $n = 3$, then $x^2 + 1$ must be u. If $u = x^2 + 1$, then $du = 2x\,dx$. Thus the integral is in proper form for integration *as it stands*. Using the power formula,

$$\int (x^2 + 1)^3(2x\,dx) = \frac{(x^2 + 1)^4}{4} + C$$

A good knowledge of differential forms is very important in this problem and makes recognition of this type possible. *It cannot be overemphasized that the entire* CAUTION ▶ ***quantity (2x dx) must be equated to du** if we are to integrate properly.* Normally u and n are recognized first, and then the proper form of du is derived from u.

Showing the use of u directly, we can write the integration as

$$\int (x^2 + 1)^3(2x\,dx) = \int u^3\,du = \frac{1}{4}u^4 + C = \frac{(x^2 + 1)^4}{4} + C$$ ◂◂

EXAMPLE 7 ▸▸ Integrate: $\int x^2\sqrt{x^3 + 2}\,dx$.

We first note that $n = \frac{1}{2}$ and u is then $x^3 + 2$. Since $u = x^3 + 2$, $du = 3x^2\,dx$. To integrate properly, we group the quantity $3x^2\,dx$ as du, with other factors isolated from this quantity. Since there is no 3 under the integral sign, we introduce one. In order not to change the numerical value, we also introduce $\frac{1}{3}$, normally before the CAUTION ▶ integral sign. In this way we take advantage of the fact that ***a constant (and only a constant) factor may be moved across the integral sign.***

$$\int x^2\sqrt{x^3 + 2}\,dx = \frac{1}{3}\int 3x^2\sqrt{x^3 + 2}\,dx = \frac{1}{3}\int \sqrt{x^3 + 2}\,(3x^2\,dx)$$

Here we indicate the proper grouping to have the proper form of Eq. (25-5).

$$\int x^2\sqrt{x^3 + 2}\,dx = \frac{1}{3}\int \sqrt{x^3 + 2}\,(3x^2\,dx) = \frac{1}{3}\left(\frac{2}{3}\right)(x^3 + 2)^{3/2} + C$$

$$= \frac{2}{9}(x^3 + 2)^{3/2} + C$$

The $1/\frac{3}{2}$ was written as $\frac{2}{3}$, since this form is generally more convenient when we are working with fractions.

With $u = x^3 + 2$, and using u directly in the integration, we can write

$$\int x^2\sqrt{x^3 + 2}\,dx = \int (x^3 + 2)^{1/2}(x^2\,dx)$$

$$= \int u^{1/2}\left(\frac{1}{3}du\right) = \frac{1}{3}\int u^{1/2}\,du \qquad \text{integrating in terms of } u$$

$$= \frac{1}{3}\left(\frac{2}{3}\right)u^{3/2} + C = \frac{2}{9}u^{3/2} + C$$

$$= \frac{2}{9}(x^3 + 2)^{3/2} + C \qquad \text{substituting } x^3 + 2 = u$$ ◂◂

Evaluating the Constant of Integration

We mentioned earlier that, in addition to knowing the differential, we need more information to find the constant of integration. Such information usually consists of a set of values that the function is known to satisfy. That is, a point through which the curve of the function passes would provide the necessary information. This is illustrated in the following examples.

EXAMPLE 8 ▸▸ Find y in terms of x, given that $dy/dx = 3x - 1$ and the curve passes through (1, 4).

We write the equation as

$$dy = (3x - 1)\,dx$$

and then indicate and perform the integration:

$$\int dy = \int (3x - 1)\,dx$$

$$y = \frac{3}{2}x^2 - x + C$$

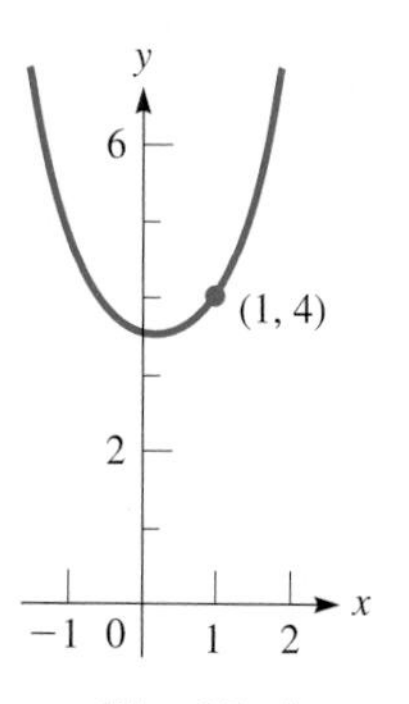

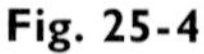
Fig. 25-4

We know that the required curve passes through (1, 4), which means that the coordinates of this point must satisfy the equation. Thus

$$4 = \frac{3}{2} - 1 + C, \quad \text{or} \quad C = \frac{7}{2}$$

This means that the solution is

$$y = \frac{3}{2}x^2 - x + \frac{7}{2} \quad \text{or} \quad 2y = 3x^2 - 2x + 7$$

The graph of this parabola is shown in Fig. 25-4. ◂◂

EXAMPLE 9 ▸▸ The time rate of change of the displacement (velocity) of a robot arm is $ds/dt = t\sqrt{9 - t^2}$. Find the expression for the displacement as a function of time if $s = 0$ cm when $t = 0$ s.

We start the solution by writing

$$ds = t\sqrt{9 - t^2}\,dt$$

$$\int ds = \int t(9 - t^2)^{1/2}\,dt$$

To integrate the expression on the right, we recognize that $n = \frac{1}{2}$ and $u = 9 - t^2$, and therefore $du = -2t\,dt$. This means we need a -2 with the t and dt to form the proper du, which in turn means we place a $-\frac{1}{2}$ before the integral sign.

$$\int ds = \int t(9 - t^2)^{1/2}\,dt = -\frac{1}{2}\int (9 - t^2)^{1/2}(-2t\,dt)$$

$$s = -\frac{1}{2}\left(\frac{2}{3}\right)(9 - t^2)^{3/2} + C$$

$$= -\frac{1}{3}(9 - t^2)^{3/2} + C$$

Since $s = 0$ cm when $t = 0$ s, we have

$$0 = -\frac{1}{3}(9)^{3/2} + C, \qquad C = 9$$

This means that the expression for the displacement is

$$s = -\frac{1}{3}(9 - t^2)^{3/2} + 9$$ ◂◂

We have discussed the integration of certain basic functions. Many other methods are used to integrate other functions, and some of these are discussed in Chapter 28. Also, there are many functions that cannot be integrated.

Exercises 25–3

In Exercises 1–32, integrate each of the given expressions.

1. $\int 2x\,dx$
2. $\int 5x^4\,dx$
3. $\int x^7\,dx$
4. $\int x^5\,dx$
5. $\int 2x^{3/2}\,dx$
6. $\int 6\sqrt[3]{x}\,dx$
7. $\int x^{-4}\,dx$
8. $\int \frac{4}{\sqrt{x}}\,dx$
9. $\int (x^2 - x^5)\,dx$
10. $\int (1 - 3x)\,dx$
11. $\int (9x^2 + x + 3)\,dx$
12. $\int (4x^2 - 2x + 5)\,dx$
13. $\int \left(\frac{1}{x^3} + \frac{1}{2}\right) dx$
14. $\int \left(4x - \frac{2}{x^3}\right) dx$
15. $\int \sqrt{x}(x^2 - x)\,dx$
16. $\int (x\sqrt{x} - 5x^2)\,dx$
17. $\int (2x^{-2/3} + 3^{-2})\,dx$
18. $\int (x^{1/3} + x^{1/5} + x^{-1/7})\,dx$
19. $\int (1 + 2x)^2\,dx$
20. $\int (x^2 + 4x + 4)^{1/3}\,dx$
21. $\int (x^2 - 1)^5(2x\,dx)$
22. $\int (x^3 - 2)^6(3x^2\,dx)$
23. $\int (x^4 + 3)^4(4x^3\,dx)$
24. $\int (1 - 2x)^{1/3}(-2\,dx)$
25. $\int (x^5 + 4)^7 x^4\,dx$
26. $\int 6x^2(1 - x^3)^{4/3}\,dx$
27. $\int \sqrt{8x + 1}\,dx$
28. $\int \sqrt[3]{4 - 3x}\,dx$
29. $\int \frac{x\,dx}{\sqrt{6x^2 + 1}}$
30. $\int \frac{2x^2\,dx}{\sqrt{2x^3 + 1}}$
31. $\int \frac{x - 1}{\sqrt{x^2 - 2x}}\,dx$
32. $\int (x^2 - x)\left(x^3 - \frac{3}{2}x^2\right)^8 dx$

In Exercises 33–36, find y in terms of x.

33. $\frac{dy}{dx} = 6x^2$, curve passes through $(0, 2)$
34. $\frac{dy}{dx} = 8x + 1$, curve passes through $(-1, 4)$
35. $\frac{dy}{dx} = x^2(1 - x^3)^5$, curve passes through $(1, 5)$
36. $\frac{dy}{dx} = 2x^3(x^4 - 6)^4$, curve passes through $(2, 10)$

In Exercises 37–44, find the required equations.

37. Find the equation of the curve whose slope is $-x\sqrt{1 - 4x^2}$ and which passes through $(0, 7)$.
38. Find the equation of the curve whose slope is $\sqrt{6x - 3}$ and which passes through $(2, -1)$.
39. The time rate of change of electric current in a circuit is given by $di/dt = 4t - 0.6t^2$. Find the expression for the current as a function of time if $i = 2$ A when $t = 0$ s.
40. The time rate of change of velocity (acceleration) of a roller mechanism is $dv/dt = 3\sqrt{t} - 2$. Find the expression for the velocity as a function of time if $v = 2$ cm/s when $t = 4$ s.
41. At a given site, the rate of change of the annual fraction f of energy supplied by solar energy with respect to the solar collector area A (in m^2) is $\frac{df}{dA} = \frac{0.005}{\sqrt{0.01A + 1}}$. Find f as a function of A if $f = 0$ for $A = 0\ m^2$.
42. The rate of change of the frequency f of an electronic oscillator with respect to the inductance L is $df/dL = 80(4 + L)^{-3/2}$. Find f as a function of L if $f = 80$ Hz for $L = 0$ H.
43. Find the equation of the curve for which the second derivative is 6. The curve passes through $(1, 2)$ with a slope of 8.
44. The second derivative of a function is $12x^2$. Explain how to find the function if its curve passes through the points $(1, 6)$ and $(2, 21)$. Find the function.

25-4 The Area Under a Curve

Another basic problem which can be solved by integration is that of finding the area under a curve. In geometry, there are methods and formulas for finding the area of regular figures. By means of the calculus we will find it possible to find the area between curves for which we know the equations. First let us look at an example that illustrates the basic idea behind the method.

EXAMPLE 1 ▸▸ Approximate the area in the first quadrant to the left of the line $x = 4$ and under the parabola $y = x^2 + 1$. First make this approximation by inscribing two rectangles of equal width under the parabola and finding the sum of the areas of these rectangles. Then improve the approximation by repeating the process with eight rectangles.

The area to be approximated is shown in Fig. 25-5(a). The area with two rectangles inscribed under the curve is shown in Fig. 25-5(b). The first approximation, admittedly small, of the area can be found by adding the areas of the two rectangles. Both rectangles have a width of 2. The left rectangle is 1 unit high, and the right rectangle is 5 units high. Thus the area of the two rectangles is

$$A = 2(1 + 5) = 12$$

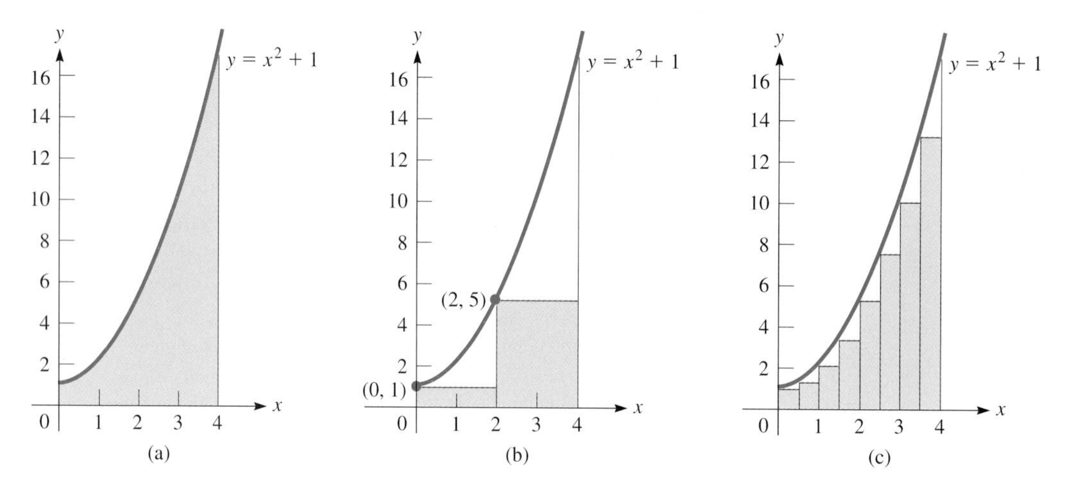

Fig. 25-5

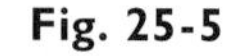

TABLE 25-1

Number of rectangles n	Total area of rectangles
8	21.5
100	25.0144
1 000	25.301 344
10 000	25.330 134

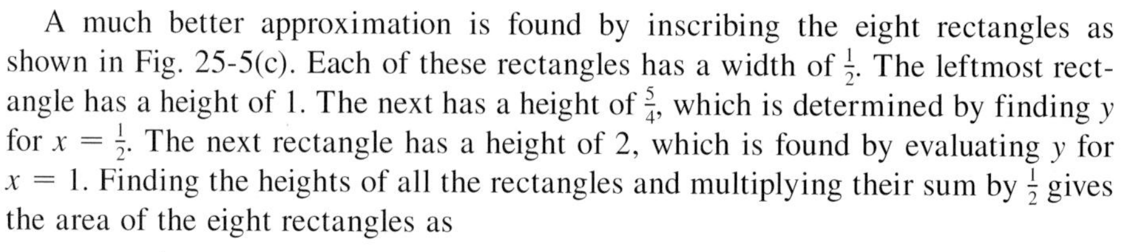

A much better approximation is found by inscribing the eight rectangles as shown in Fig. 25-5(c). Each of these rectangles has a width of $\frac{1}{2}$. The leftmost rectangle has a height of 1. The next has a height of $\frac{5}{4}$, which is determined by finding y for $x = \frac{1}{2}$. The next rectangle has a height of 2, which is found by evaluating y for $x = 1$. Finding the heights of all the rectangles and multiplying their sum by $\frac{1}{2}$ gives the area of the eight rectangles as

$$A = \frac{1}{2}\left(1 + \frac{5}{4} + 2 + \frac{13}{4} + 5 + \frac{29}{4} + 10 + \frac{53}{4}\right) = \frac{43}{2} = 21.5$$

An even better approximation could be obtained by inscribing more rectangles under the curve. The greater the number of rectangles, the more nearly the sum of their areas equals the area under the curve. See Table 25-1. By a method involving integration developed later in this section, we determine the *exact* area to be $\frac{76}{3} = 25\frac{1}{3}$. ◂◂

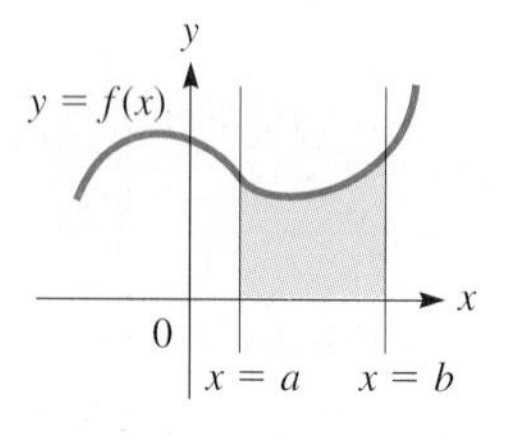

Fig. 25-6

We now develop the basic method used to find the area under a curve, which is the area bounded by the curve, the x-axis, and the lines $x = a$ and $x = b$. See Fig. 25-6. We assume here that $f(x)$ is never negative in the interval $a < x < b$. In Chapter 26 we will extend the method such that $f(x)$ may be negative.

In finding the area under a curve we consider the sum of the areas of inscribed rectangles, as the number of rectangles is assumed to increase without bound. The reason for this last condition is that, as we saw in Example 1, as the number of rectangles increases, the approximation of the area is better.

EXAMPLE 2 ▸▸ Find the area under the straight line $y = 2x$, above the x-axis, and to the left of the line $x = 4$.

Since this figure is a right triangle, the area can easily be found. However, the *method* we shall use here is the important concept. We first subdivide the interval from $x = 0$ to $x = 4$ into n inscribed rectangles of Δx in width. The extremities of the intervals are labeled $a, x_1, x_2, \ldots, b\ (= x_n)$, as shown in Fig. 25-7, where

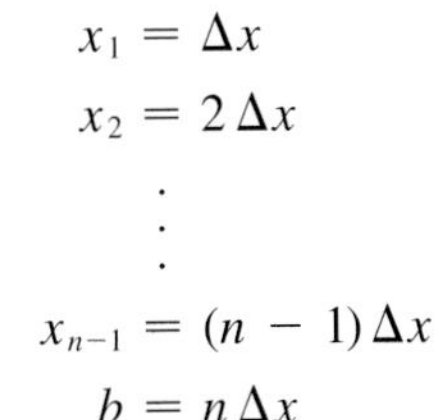

$$\begin{aligned} x_1 &= \Delta x \\ x_2 &= 2\,\Delta x \\ &\ \vdots \\ x_{n-1} &= (n - 1)\,\Delta x \\ b &= n\,\Delta x \end{aligned}$$

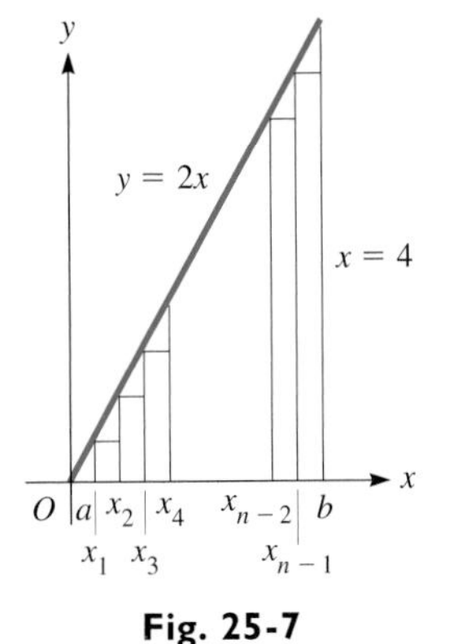

Fig. 25-7

The area of each of these n rectangles is as follows:

First $f(a)\,\Delta x$, where $f(a) = f(0) = 2(0) = 0$ is the height.
Second $f(x_1)\,\Delta x$, where $f(x_1) = 2(\Delta x) = 2\,\Delta x$ is the height.
Third $f(x_2)\,\Delta x$, where $f(x_2) = 2(2\,\Delta x) = 4\,\Delta x$ is the height.
Fourth $f(x_3)\,\Delta x$, where $f(x_3) = 2(3\,\Delta x) = 6\,\Delta x$ is the height.

$\vdots$

Last $f(x_{n-1})\,\Delta x$, where $f[(n - 1)\,\Delta x] = 2(n - 1)\,\Delta x$ is the height.

These areas are summed up as follows:

NOTE ▸

$$\begin{aligned} A_n &= \boxed{f(a)\,\Delta x + f(x_1)\,\Delta x + f(x_2)\,\Delta x + \cdots + f(x_{n-1})\,\Delta x} \\ &= 0 + 2\,\Delta x(\Delta x) + 4\,\Delta x(\Delta x) + \cdots + 2[n - 1)\,\Delta x]\,\Delta x \\ &= 2(\Delta x)^2[1 + 2 + 3 + \cdots + (n - 1)] \end{aligned}$$

Now $b = n\,\Delta x$, or $4 = n\,\Delta x$, or $\Delta x = 4/n$. Thus

$$A_n = 2\left(\frac{4}{n}\right)^2[1 + 2 + 3 + \cdots + (n - 1)]$$

The sum of the arithmetic sequence $1 + 2 + 3 + \cdots + n - 1$ is

$$s = \frac{n - 1}{2}(1 + n - 1) = \frac{n(n - 1)}{2} = \frac{n^2 - n}{2}$$

Now the expression for the sum of the areas can be written as

$$A_n = \frac{32}{n^2}\left(\frac{n^2 - n}{2}\right) = 16\left(1 - \frac{1}{n}\right)$$

This expression is an approximation of the actual area under consideration. The larger n becomes, the better the approximation. If we let $n \to \infty$ (which is equivalent to letting $\Delta x \to 0$), the limit of this sum will equal the area in question.

$$A = \lim_{n \to \infty} 16\left(1 - \frac{1}{n}\right) = 16 \qquad 1/n \to 0 \text{ as } n \to \infty$$

NOTE ▸ (This checks with the geometric result.) *The area under the curve is the limit of the sum of the areas of the inscribed rectangles, as the number of rectangles approaches infinity.* ◂◂

The method indicated in Example 2 illustrates the interpretation of finding an area as a summation process, although it should not be considered as a proof. However, we shall find that integration proves to be a much more useful method for finding an area. Let us now see how integration can be used directly.

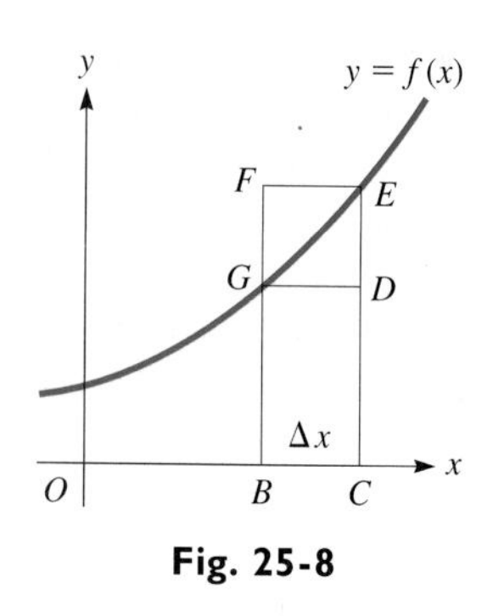

Fig. 25-8

Let ΔA represent the area $BCEG$ under the curve, as indicated in Fig. 25-8. We see that the following inequality is true for the indicated areas:

$$A_{BCDG} < \Delta A < A_{BCEF}$$

If the point G is now designated as (x, y) and E as $(x + \Delta x, y + \Delta y)$, we have $y\Delta x < \Delta A < (y + \Delta y)\Delta x$. Dividing through by Δx, we have

$$y < \frac{\Delta A}{\Delta x} < y + \Delta y$$

Now we take the limit as $\Delta x \to 0$ (Δy then also approaches 0). This results in

$$\frac{dA}{dx} = y \qquad \textbf{(25-6)}$$

This is true since the left member of the inequality is y and the right member approaches y. Also remember that

$$\lim_{\Delta x \to 0} \frac{\Delta A}{\Delta x} = \frac{dA}{dx}$$

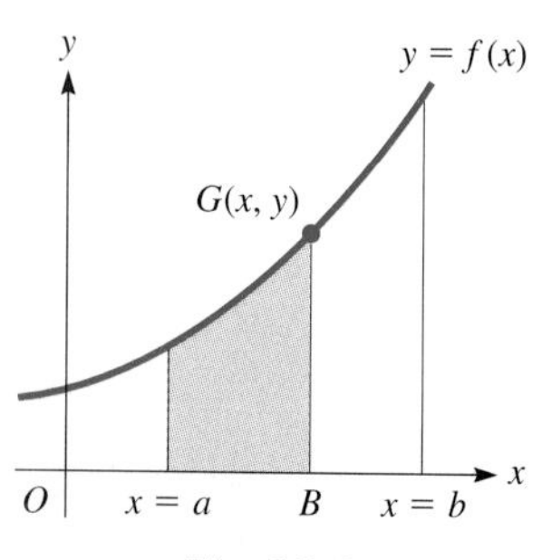

Fig. 25-9

We shall now use Eq. (25-6) to show the method of finding the complete area under a curve. We now let $x = a$ be the left boundary of the desired area and $x = b$ be the right boundary (Fig. 25-9). The area under the curve to the right of $x = a$ and bounded on the right by the line GB is now designated as A_{ax}. From Eq. (25-6) we have

$$dA_{ax} = [y\,dx]_a^x \quad \text{or} \quad A_{ax} = \left[\int y\,dx\right]_a^x$$

where $[\ ]_a^x$ is the notation used to indicate the boundaries of the area. Thus

$$A_{ax} = \left[\int f(x)\,dx\right]_a^x = [F(x) + C]_a^x \qquad \textbf{(25-7)}$$

But we know that if $x = a$, then $A_{aa} = 0$. Thus, $0 = F(a) + C$, or $C = -F(a)$. Therefore,

$$A_{ax} = \left[\int f(x)\,dx\right]_a^x = F(x) - F(a) \qquad \textbf{(25-8)}$$

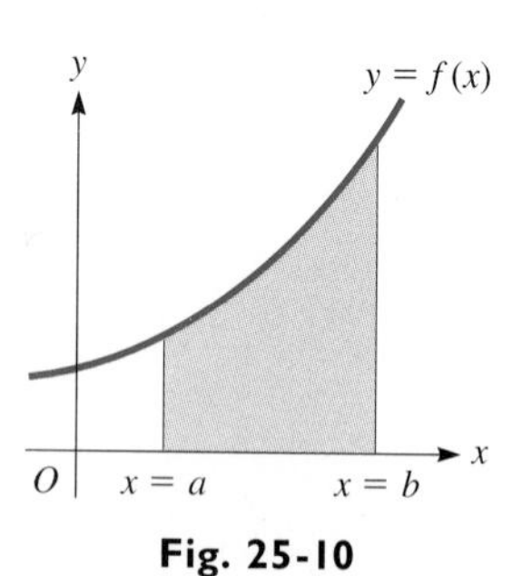

Fig. 25-10

Now, to find the area under the curve that reaches from a to b, we write

$$A_{ab} = F(b) - F(a) \qquad \textbf{(25-9)}$$

Thus the area under the curve that reaches from a to b is given by

$$A_{ab} = \left[\int f(x)\,dx\right]_a^b = F(b) - F(a) \qquad \textbf{(25-10)}$$

NOTE ▶ This shows that *the area under the curve may be found by integrating the function $f(x)$ to find the function $F(x)$, which is then evaluated at each boundary value. The area is the difference between these values of $F(x)$.* See Fig. 25-10.

Integration as Summation

In Example 2, we found an area under a curve by finding the limit of the sum of the areas of the inscribed rectangles as the number of rectangles approaches infinity. Equation (25-10) expresses the area under a curve in terms of integration. We can now see that we have obtained an area by summation and also expressed it in terms of integration. Therefore, we conclude that

summations can be evaluated by integration.

Also, we have seen the connection between the problem of finding the slope of a tangent to a curve (differentiation) and the problem of finding an area under a curve (integration). We would not normally suspect that these two problems would have solutions that lead to reverse processes. We have also seen that the definition of integration has much more application than originally anticipated.

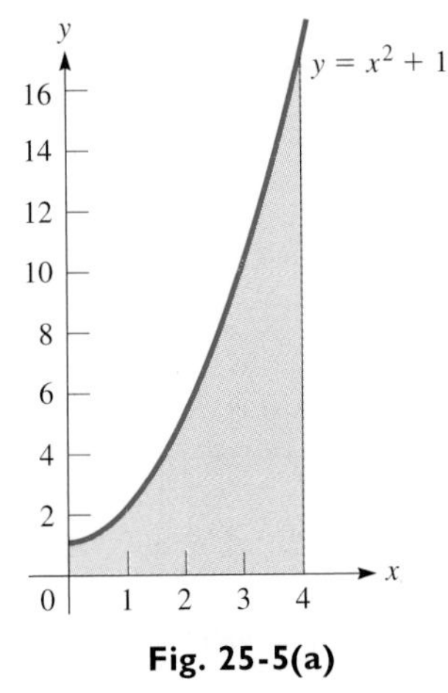

Fig. 25-5(a)

EXAMPLE 3 ▸▸ Find the area under the curve $y = x^2 + 1$ between the y-axis and the line $x = 4$. This is the area of Example 1 shown in Fig. 25-5(a), which is shown again here for reference.

Using Eq. (25-10), we note that $f(x) = x^2 + 1$. This means that

$$\int (x^2 + 1)\,dx = \frac{1}{3}x^3 + x + C$$

Therefore, with $F(x) = \frac{1}{3}x^3 + x$, the area is given by

$$A_{0,4} = F(4) - F(0) \qquad \text{using Eq. (25-10)}$$

$$= \left[\frac{1}{3}(4^3) + 4\right] - \left[\frac{1}{3}(0^3) + 0\right] \qquad \text{evaluating } F(x) \text{ at } x = 4 \text{ and } x = 0$$

$$= \frac{1}{3}(64) + 4 = \frac{76}{3}$$

We note that this is about 4 square units greater than the value obtained by the approximation in Example 1. This result means that the exact area is $25\frac{1}{3}$, as stated at the end of Example 1. ◂◂

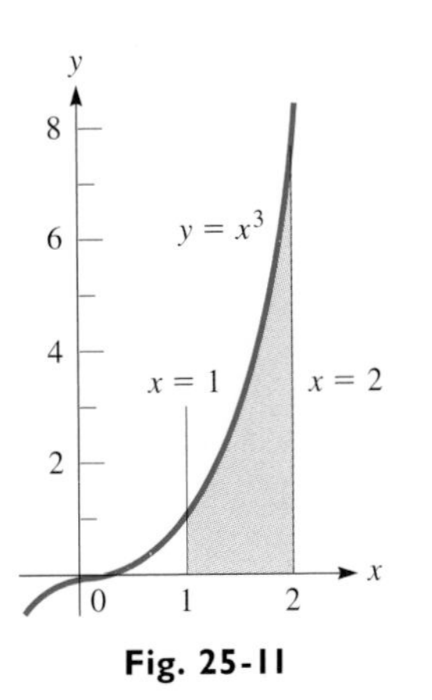

Fig. 25-11

EXAMPLE 4 ▸▸ Find the area under the curve $y = x^3$ that is between the lines $x = 1$ and $x = 2$.

In Eq. (25-10), $f(x) = x^3$. Therefore,

$$\int x^3\,dx = \underbrace{\frac{1}{4}x^4}_{F(x)} + C$$

$$A_{1,2} = F(2) - F(1) = \left[\frac{1}{4}(2^4)\right] - \left[\frac{1}{4}(1^4)\right] \qquad \text{using Eq. (25-10) and evaluating}$$

$$= 4 - \frac{1}{4} = \frac{15}{4}$$

Therefore, we see that the required area is 15/4. Again, as in Example 3, this is an exact area, not an approximation. The curve and the area are shown in Fig. 25-11. ◂◂

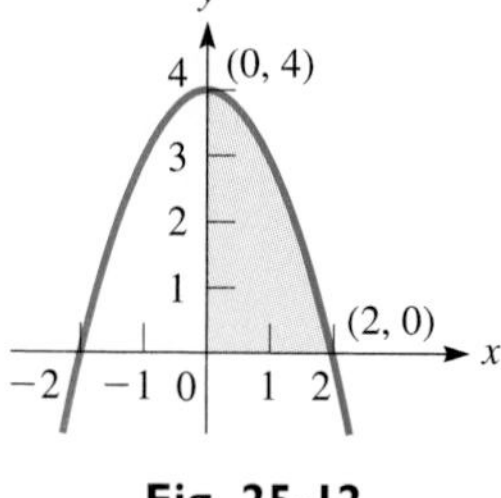

Fig. 25-12

EXAMPLE 5 ▸▸ Find the area under the curve of $y = 4 - x^2$ that lies in the first quadrant.

By solving the equation $4 - x^2 = 0$, we determine that the area to be found extends from $x = 0$ to $x = 2$ (see Fig. 25-12). Thus

$$\int (4 - x^2)\,dx = \underbrace{4x - \frac{x^3}{3}}_{F(x)} + C$$

$$A_{0,2} = \left(8 - \frac{8}{3}\right) - (0 - 0) = \frac{16}{3} \qquad \text{using Eq. (25-10) and evaluating}$$

Therefore, the exact area is 16/3. ◂◂

We can see from these examples that we do not have to include the constant of integration when we are finding areas. Any constant that might be added to $F(x)$ would cancel out when $F(a)$ is subtracted from $F(b)$.

Exercises 25–4

In Exercises 1–10, find the approximate area under the curves of the given equations by dividing the indicated intervals into n subintervals, and add up the areas of the inscribed rectangles. There are two values of n for each exercise, and therefore two approximations for each area. The height of each rectangle may be found by evaluating the function for the proper value of x. See Example 1.

1. $y = 3x$, between $x = 0$ and $x = 3$, for (a) $n = 3$ $(\Delta x = 1)$, (b) $n = 10$ $(\Delta x = 0.3)$

2. $y = 2x + 1$, between $x = 0$ and $x = 2$, for (a) $n = 4$ $(\Delta x = 0.5)$, (b) $n = 10$ $(\Delta x = 0.2)$

3. $y = x^2$, between $x = 0$ and $x = 2$, for (a) $n = 5$ $(\Delta x = 0.4)$, (b) $n = 10$ $(\Delta x = 0.2)$

4. $y = x^2 + 2$, between $x = 0$ and $x = 3$, for (a) $n = 3$ $(\Delta x = 1)$, (b) $n = 10$ $(\Delta x = 0.3)$

5. $y = 4x - x^2$, between $x = 1$ and $x = 4$, for (a) $n = 6$, (b) $n = 10$

6. $y = 1 - x^2$, between $x = 0.5$ and $x = 1$, for (a) $n = 5$, (b) $n = 10$

7. $y = \frac{1}{x^2}$, between $x = 1$ and $x = 5$, for (a) $n = 4$, (b) $n = 8$

8. $y = \sqrt{x}$, between $x = 1$ and $x = 4$, for (a) $n = 3$, (b) $n = 12$

9. $y = \frac{1}{\sqrt{x + 1}}$, between $x = 3$ and $x = 8$, for (a) $n = 5$, (b) $n = 10$

10. $y = 2x\sqrt{x^2 + 1}$, between $x = 0$ and $x = 6$, for (a) $n = 6$, (b) $n = 12$

In Exercises 11–20, find the exact area under the given curves between the indicated values of x. The functions are the same as those for which approximate areas were found in Exercises 1–10.

11. $y = 3x$, between $x = 0$ and $x = 3$

12. $y = 2x + 1$, between $x = 0$ and $x = 2$

13. $y = x^2$, between $x = 0$ and $x = 2$

14. $y = x^2 + 2$, between $x = 0$ and $x = 3$

15. $y = 4x - x^2$, between $x = 1$ and $x = 4$

16. $y = 1 - x^2$, between $x = 0.5$ and $x = 1$

17. $y = \frac{1}{x^2}$, between $x = 1$ and $x = 5$

18. $y = \sqrt{x}$, between $x = 1$ and $x = 4$

19. $y = \frac{1}{\sqrt{x + 1}}$, between $x = 3$ and $x = 8$

20. $y = 2x\sqrt{x^2 + 1}$, between $x = 0$ and $x = 6$

25-5 The Definite Integral

Using reasoning similar to that in the preceding section, *we define the* **definite integral** *of a function $f(x)$ as*

$$\int_a^b f(x)\,dx = F(b) - F(a) \tag{25-11}$$

where $F'(x) = f(x)$. ***We call this a* definite integral *because the final result of integrating and evaluating is a* number.** (The *indefinite* integral had an arbitrary constant in the result.) *The numbers a and b are called the* **lower limit** *and the* **upper limit,** *respectively. We can see that the value of a definite integral is found by evaluating the function (found by integration) at the upper limit and subtracting the value of this function at the lower limit.*

Limits of Integration

We know, from the analysis in the preceding section, that *this definite integral can be interpreted as a summation process,* where the size of the subdivision approaches a limit of zero. This fact explains the choice of the $\int$ symbol for integration: It is an elongated S, representing the sum. It is this interpretation of integration that we shall apply to many kinds of problems.

EXAMPLE 1 ▸▸ Evaluate the integral $\int_0^2 x^4\,dx$.

$$\int_0^2 x^4\,dx = \frac{x^5}{5}\bigg|_0^2 = \frac{2^5}{5} - 0 = \frac{32}{5}$$

upper limit (2); lower limit (0); $f(x)$ (the integrand x^4); $F(x)$ (the result $\frac{x^5}{5}$)

Note that a vertical line—with the limits written at the top and the bottom—is the way the value is indicated after integration, but before evaluation. ◂◂

EXAMPLE 2 ▸▸ Evaluate $\int_1^3 (x^{-2} - 1)\,dx$.

$$\int_1^3 (x^{-2} - 1)\,dx = -\frac{1}{x} - x\bigg|_1^3 = \left(-\frac{1}{3} - 3\right) - (-1 - 1)$$

upper limit: $\left(-\frac{1}{3} - 3\right)$; subtract: $-$; lower limit: $(-1 - 1)$

$$= -\frac{10}{3} + 2 = -\frac{4}{3}$$ ◂◂

EXAMPLE 3 ▸▸ Evaluate $\int_0^1 5x(x^2 + 1)^5\,dx$.

For purposes of integration, $n = 5$, $u = x^2 + 1$, and $du = 2x\,dx$. Hence

$$\int_0^1 5x(x^2 + 1)^5\,dx = \frac{5}{2}\int_0^1 (x^2 + 1)^5(2x\,dx)$$

$$= \frac{5}{2}\left(\frac{1}{6}\right)(x^2 + 1)^6\bigg|_0^1 \quad \text{integrate}$$

$$= \frac{5}{12}(2^6 - 1^6) = \frac{5(63)}{12} = \frac{105}{4} \quad \text{evaluate}$$ ◂◂

EXAMPLE 4 ▸▸ Evaluate $\int_{-1}^{3} x(1-3x^2)^{1/3}\,dx$.

For purposes of integration, $n = \frac{1}{3}$, $u = 1 - 3x^2$, and $du = -6x\,dx$. Thus

$$\begin{aligned}
\int_{-1}^{3} x(1-3x^2)^{1/3}\,dx &= -\frac{1}{6}\int_{-1}^{3}(1-3x^2)^{1/3}(-6x\,dx) \\
&= -\frac{1}{6}\left(\frac{3}{4}\right)(1-3x^2)^{4/3}\Big|_{-1}^{3} && \text{integrate} \\
&= -\frac{1}{8}(-26)^{4/3} + \frac{1}{8}(-2)^{4/3} && \text{evaluate} \\
&= \frac{1}{8}(2\sqrt[3]{2} - 26\sqrt[3]{26}) = \frac{1}{4}(\sqrt[3]{2} - 13\sqrt[3]{26})
\end{aligned}$$

The result shown is the exact value. The approximate value to three decimal places is -9.313. ◂◂

EXAMPLE 5 ▸▸ Evaluate $\int_{0.1}^{2.7} \frac{dx}{\sqrt{4x+1}}$.

In order to integrate we have $n = -\frac{1}{2}$, $u = 4x + 1$, and $du = 4\,dx$. Therefore,

$$\begin{aligned}
\int_{0.1}^{2.7} \frac{dx}{\sqrt{4x+1}} &= \int_{0.1}^{2.7}(4x+1)^{-1/2}\,dx = \frac{1}{4}\int_{0.1}^{2.7}(4x+1)^{-1/2}(4\,dx) \\
&= \frac{1}{4}\left(\frac{1}{\frac{1}{2}}\right)(4x+1)^{1/2}\Big|_{0.1}^{2.7} = \frac{1}{2}(4x+1)^{1/2}\Big|_{0.1}^{2.7} && \text{integrate} \\
&= \frac{1}{2}(\sqrt{11.8} - \sqrt{1.4}) = 1.126 && \text{evaluate}
\end{aligned}$$

◂◂

EXAMPLE 6 ▸▸ Evaluate $\int_{0}^{4} \frac{x+1}{(x^2+2x+2)^3}\,dx$

For purposes of integration,

$$n = -3, \qquad u = x^2 + 2x + 2, \quad \text{and} \quad du = (2x+2)\,dx$$

Therefore,

$$\begin{aligned}
\int_{0}^{4}(x^2+2x+2)^{-3}(x+1)\,dx &= \frac{1}{2}\int_{0}^{4}(x^2+2x+2)^{-3}[2(x+1)\,dx] \\
&= \frac{1}{2}\left(\frac{1}{-2}\right)(x^2+2x+2)^{-2}\Big|_{0}^{4} && \text{integrate} \\
&= -\frac{1}{4}(16+8+2)^{-2} + \frac{1}{4}(0+0+2)^{-2} && \text{evaluate} \\
&= \frac{1}{4}\left(-\frac{1}{26^2} + \frac{1}{2^2}\right) = \frac{1}{4}\left(\frac{1}{4} - \frac{1}{676}\right) \\
&= \frac{1}{4}\left(\frac{168}{676}\right) = \frac{21}{338}
\end{aligned}$$

◂◂

The definition of the definite integral is valid regardless of the source of $f(x)$. That is, we may apply the definite integral whenever we want to sum a function in a manner similar to that which we use to find an area.

The following example illustrates an application of the definite integral. In Chapter 26 we shall see that the definite integral has applications in many areas of science and technology.

See the chapter introduction.

EXAMPLE 7 ▸▸ The rate of flow Q (in m^3/s) of water over a certain dam is found by evaluating the definite integral in the equation $Q = \int_0^{1.25} 240\sqrt{1.50 - y}\,dy$. See Fig. 25-13. Find Q.

The solution is as follows:

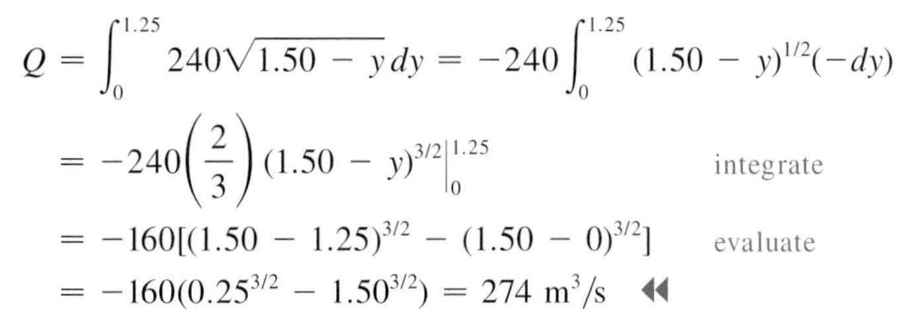

$$Q = \int_0^{1.25} 240\sqrt{1.50 - y}\,dy = -240\int_0^{1.25} (1.50 - y)^{1/2}(-dy)$$

$$= -240\left(\frac{2}{3}\right)(1.50 - y)^{3/2}\Big|_0^{1.25} \qquad \text{integrate}$$

$$= -160[(1.50 - 1.25)^{3/2} - (1.50 - 0)^{3/2}] \qquad \text{evaluate}$$

$$= -160(0.25^{3/2} - 1.50^{3/2}) = 274 \text{ m}^3/\text{s}$$ ◂◂

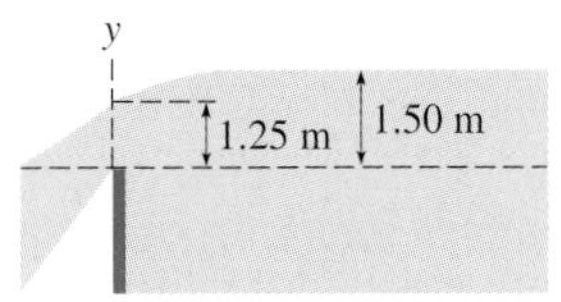

Fig. 25-13

Exercises 25–5

In Exercises 1–32, evaluate the given definite integrals.

1. $\int_0^1 2x\,dx$

2. $\int_0^2 3x^2\,dx$

3. $\int_1^4 x^{5/2}\,dx$

4. $\int_4^9 (x^{3/2} - 1)\,dx$

5. $\int_3^6 \left(\frac{1}{\sqrt{x}} + 2\right) dx$

6. $\int_{1.2}^{1.6} \left(5 + \frac{6}{x^4}\right) dx$

7. $\int_{-1.6}^{0.7} (1 - x)^{1/3}\,dx$

8. $\int_1^5 \sqrt{2x - 1}\,dx$

9. $\int_0^3 (x^4 - x^3 + x^2)\,dx$

10. $\int_1^2 (3x^5 - 2x^3)\,dx$

11. $\int_{0.5}^{2.2} (\sqrt[3]{x} - 2)\,dx$

12. $\int_{2.7}^{5.3} \left(\frac{1}{x\sqrt{x}} + 4\right) dx$

13. $\int_0^4 (1 - \sqrt{x})^2\,dx$

14. $\int_1^4 \frac{x^2 + 1}{\sqrt{x}}\,dx$

15. $\int_{-2}^{-1} 2x(4 - x^2)^3\,dx$

16. $\int_0^1 x(3x^2 - 1)^3\,dx$

17. $\int_0^4 \frac{x\,dx}{\sqrt{x^2 + 9}}$

18. $\int_{0.2}^{0.7} x^2(x^3 + 2)^{3/2}\,dx$

19. $\int_{2.75}^{3.25} \frac{dx}{\sqrt[3]{6x + 1}}$

20. $\int_2^6 \frac{2dx}{\sqrt{4x + 1}}$

21. $\int_1^3 \frac{2x\,dx}{(2x^2 + 1)^3}$

22. $\int_{12.6}^{17.2} \frac{3\,dx}{(6x - 1)^2}$

23. $\int_3^7 \sqrt{16x^2 - 8x + 1}\,dx$

24. $\int_{-5}^1 \sqrt{6 - 2x}\,dx$

25. $\int_0^2 2x(9 - 2x^2)^2\,dx$

26. $\int_{-1}^0 x^3(1 - 2x^4)^3\,dx$

27. $\int_0^1 (x^2 + 3)(x^3 + 9x + 6)^2\,dx$

28. $\int_2^3 \frac{x^2 + 1}{(x^3 + 3x)^2}\,dx$

29. $\int_{-1}^2 \frac{8x - 2}{(2x^2 - x + 1)^3}\,dx$

30. $\int_{-3}^{-2} (3x^2 - 2)\sqrt[3]{2x^3 - 4x + 1}\,dx$

31. $\int_3^{11} (x + 3 + \sqrt[4]{2x - 6})\,dx$

32. $\int_{-2}^0 (\sqrt{2x + 4} - \sqrt[3]{3x + 8})\,dx$

In Exercises 33–36, solve the given problems.

33. The work W (in N · m) in winding up an 80-m cable is $W = \int_0^{80} (1000 - 5x)\,dx$. Evaluate W.

34. The total volume V of liquid flowing through a certain pipe of radius R is $V = k(R^2 \int_0^R r\,dr - \int_0^R r^3\,dr)$, where k is a constant. Evaluate V and explain why R, but not r, can be to the left of the integral sign.

35. The surface area A (in m^2) of a certain parabolic radio wave reflector is $A = 4\pi \int_0^2 \sqrt{3x + 9}\,dx$. Evaluate A.

36. The total force (in newtons) on the circular end of a water tank is $F = 19\,600 \int_0^5 y\sqrt{25 - y^2}\,dy$. Evaluate F.

25-6 Numerical Integration: The Trapezoidal Rule

For data and functions that cannot be directly integrated by available methods, it is possible to develop numerical methods of integration. These numerical methods are of greater importance today since they are readily adaptable for use on a calculator or computer. There are a great many such numerical techniques for approximating the value of an integral. In this section we develop one of these, the trapezoidal rule. In the following section, another numerical method is discussed.

We know from Sections 25-4 and 25-5 that we can interpret a definite integral as the area under a curve. We shall therefore show how to approximate the value of the integral by approximating the appropriate area by a set of inscribed trapezoids. The basic idea here is very similar to that used when rectangles were inscribed under a curve. However, the use of trapezoids reduces the error and provides a better approximation.

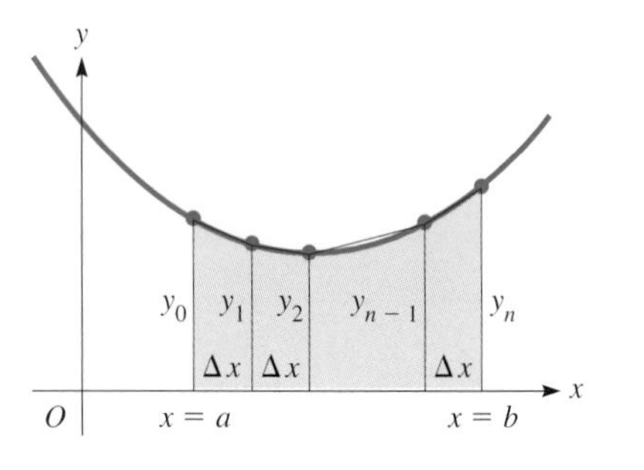

Fig. 25-14

The area to be found is subdivided into n intervals of equal width. Perpendicular lines are then dropped from the curve (or points, if only a given set of numbers is available). If the points on the curve are joined by straight-line segments, the area of successive parts under the curve is approximated by finding the area of each of the trapezoids formed. However, if these points are not too far apart, the approximation will be very good (see Fig. 25-14). From geometry we recall that the area of a trapezoid equals one-half the product of the sum of the bases times the altitude. For these trapezoids the bases are the y-coordinates and the altitudes are Δx. Therefore, when we indicate the sum of these trapezoidal areas, we have

$$A_T = \frac{1}{2}(y_0 + y_1)\Delta x + \frac{1}{2}(y_1 + y_2)\Delta x + \frac{1}{2}(y_2 + y_3)\Delta x + \cdots + \frac{1}{2}(y_{n-2} + y_{n-1})\Delta x + \frac{1}{2}(y_{n-1} + y_n)\Delta x$$

We note, when this addition is performed, that the result is

$$A_T = \left(\frac{1}{2}y_0 + y_1 + y_2 + \cdots + y_{n-1} + \frac{1}{2}y_n\right)\Delta x \tag{25-12}$$

The y-values to be used either are derived from the function $y = f(x)$ or are the y-coordinates of a set of data.

Since A_T approximates the area under the curve, it also approximates the value of the definite integral, or

$$\int_a^b f(x)\,dx \approx \frac{\Delta x}{2}(y_0 + 2y_1 + 2y_2 + \cdots + 2y_{n-1} + y_n) \tag{25-13}$$

Equation (25-13) is known as the **trapezoidal rule.** With $\Delta x = h$, this is the same rule as Eq. (2-12), which we used to measure irregular areas (see p. 65). We can now use it to find the approximate value of an integral.

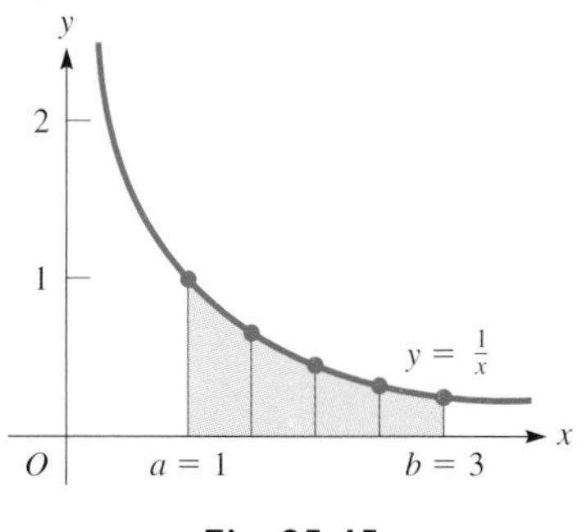

Fig. 25-15

EXAMPLE 1 ▸▸ Approximate the value of $\int_1^3 \frac{1}{x}\,dx$ by the trapezoidal rule. Let $n = 4$.

We are to approximate the area under $y = 1/x$ from $x = 1$ to $x = 3$ by dividing the area into four trapezoids. This area is found by applying Eq. (25-12), which is the approximate value of the integral, as shown in Eq. (25-13). Figure 25-15 shows the graph. In this example, $f(x) = 1/x$, and

$$\Delta x = \frac{3-1}{4} = \frac{1}{2}, \qquad y_0 = f(a) = f(1) = 1$$

$$y_1 = f\left(\frac{3}{2}\right) = \frac{2}{3}, \qquad y_2 = f(2) = \frac{1}{2}$$

$$y_3 = f\left(\frac{5}{2}\right) = \frac{2}{5}, \qquad y_n = y_4 = f(b) = f(3) = \frac{1}{3}$$

$$A_T = \frac{1/2}{2}\left[1 + 2\left(\frac{2}{3}\right) + 2\left(\frac{1}{2}\right) + 2\left(\frac{2}{5}\right) + \frac{1}{3}\right]$$

$$= \frac{1}{4}\left(\frac{15 + 20 + 15 + 12 + 5}{15}\right) = \frac{1}{4}\left(\frac{67}{15}\right) = \frac{67}{60}$$

TABLE 25-2

Number of trapezoids n	*Total area of trapezoids*
4	1.116 666 7
100	1.098 641 9
1 000	1.098 612 6
10 000	1.098 612 3

Therefore,

$$\int_1^3 \frac{1}{x}\,dx \approx \frac{67}{60}$$

We cannot perform this integration directly by methods developed up to this point. As we increase the number of trapezoids, the value becomes more accurate. See Table 25-2. The actual value to seven decimal places is 1.098 612 3. ◂◂

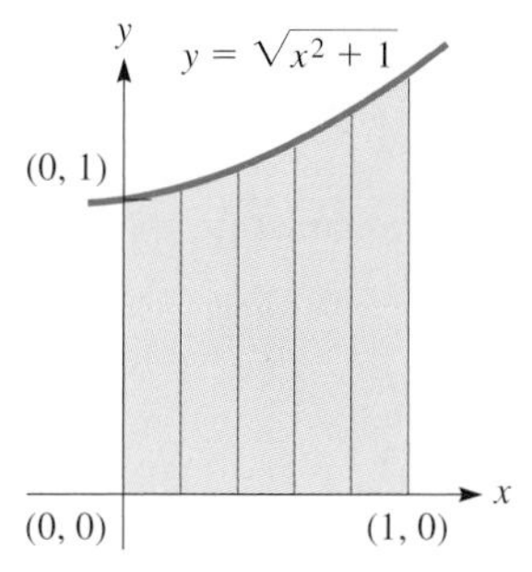

Fig. 25-16

EXAMPLE 2 ▸▸ Approximate the value of $\int_0^1 \sqrt{x^2 + 1}\,dx$ by the trapezoidal rule. Let $n = 5$.

Figure 25-16 shows the graph. In this example,

$$\Delta x = \frac{1-0}{5} = 0.2$$

$$y_0 = f(0) = 1 \qquad y_1 = f(0.2) = \sqrt{1.04} = 1.019\,803\,9$$

$$y_2 = f(0.4) = \sqrt{1.16} = 1.077\,033\,0 \qquad y_3 = f(0.6) = \sqrt{1.36} = 1.166\,190\,4$$

$$y_4 = f(0.8) = \sqrt{1.64} = 1.280\,624\,8 \qquad y_5 = f(1) = \sqrt{2.00} = 1.414\,213\,6$$

Hence we have

$$A_T = \frac{0.2}{2}[1 + 2(1.019\,803\,9) + 2(1.077\,033\,0) + 2(1.166\,190\,4)$$

$$+ 2(1.280\,624\,8) + 1.414\,213\,6]$$

$$= 1.150 \qquad \text{(rounded off)}$$

This means that

$$\int_0^1 \sqrt{x^2 + 1}\,dx \approx 1.150$$

(The actual value is 1.148 to three decimal places.) The above values of y_0, y_1, and so on do not have to be tabulated when using a calculator. The entire calculation can be done directly on a calculator. ◂◂

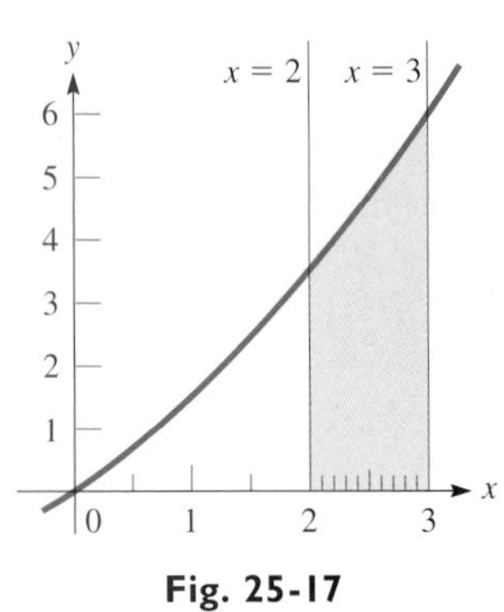

Fig. 25-17

EXAMPLE 3 ▸▸ Approximate the value of the integral $\int_2^3 x\sqrt{x+1}\,dx$ by using the trapezoidal rule. Use $n = 10$.

In Fig. 25-17 the graph of the function and the area used in the trapezoidal rule are shown. From the given values we have $\Delta x = \frac{3-2}{10} = 0.1$. Therefore,

$$y_0 = f(2) = 2\sqrt{3} = 3.464\,101\,6 \qquad y_1 = f(2.1) = 2.1\sqrt{3.1} = 3.697\,431\,5$$
$$y_2 = f(2.2) = 2.2\sqrt{3.2} = 3.935\,479\,6 \qquad y_3 = f(2.3) = 2.3\sqrt{3.3} = 4.178\,157\,5$$
$$y_4 = f(2.4) = 2.4\sqrt{3.4} = 4.425\,381\,3 \qquad y_5 = f(2.5) = 2.5\sqrt{3.5} = 4.677\,071\,7$$
$$y_6 = f(2.6) = 2.6\sqrt{3.6} = 4.933\,153\,2 \qquad y_7 = f(2.7) = 2.7\sqrt{3.7} = 5.193\,553\,7$$
$$y_8 = f(2.8) = 2.8\sqrt{3.8} = 5.458\,204\,8 \qquad y_9 = f(2.9) = 2.9\sqrt{3.9} = 5.727\,041\,1$$
$$y_{10} = f(3) = 3\sqrt{4} = 6.000\,000\,0$$

$$A_T = \frac{0.1}{2}[3.464\,101\,6 + 2(3.697\,431\,5) + \cdots + 2(5.727\,041\,1) + 6.000\,000\,0]$$
$$= 4.6958$$

Therefore, $\int_2^3 x\sqrt{x+1}\,dx \approx 4.6958$. (The actual value, to five significant digits, is 4.6954.) We note again that the entire calculation can be done directly on a calculator. ◂◂

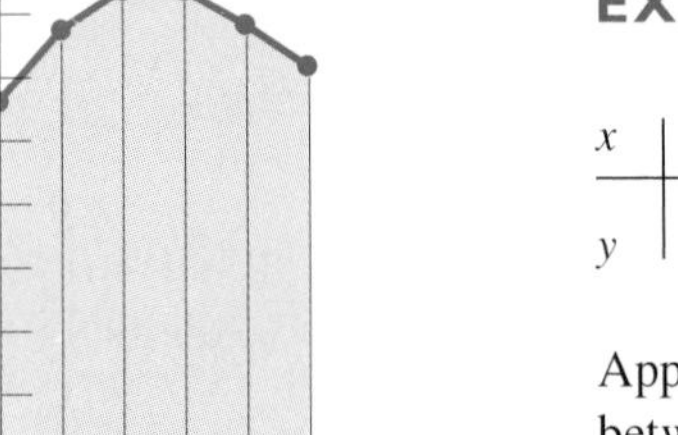

Fig. 25-18

EXAMPLE 4 ▸▸ The following points were found empirically.

x	0	1	2	3	4	5
y	5.68	6.75	7.32	7.35	6.88	6.24

Approximate the value of the integral of the function defined by these points between $x = 0$ and $x = 5$ by the trapezoidal rule.

In order to find A_T, we use the values of y_0, y_1, and so on, directly from the table. We also note that $\Delta x = 1$. The graph is shown in Fig. 25-18. Therefore, we have

$$A_T = \frac{1}{2}[5.68 + 2(6.75) + 2(7.32) + 2(7.35) + 2(6.88) + 6.24] = 34.26$$

Although we do not know the algebraic form of the function, we can state that

$$\int_0^5 f(x)\,dx \approx 34.26$$ ◂◂

Exercises 25–6

In Exercises 1–4, (a) approximate the value of each of the given integrals by use of the trapezoidal rule, using the given value of n, and (b) check by direct integration.

1. $\int_0^2 2x^2\,dx,\ n = 4$

2. $\int_0^1 (1 - x^2)\,dx,\ n = 3$

3. $\int_1^4 (1 + \sqrt{x})\,dx,\ n = 6$

4. $\int_3^8 \sqrt{1 + x}\,dx,\ n = 5$

In Exercises 5–12, approximate the value of each of the given integrals by use of the trapezoidal rule, using the given value of n.

5. $\int_2^3 \frac{1}{2x}\,dx,\ n = 2$

6. $\int_2^6 \frac{dx}{x+3},\ n = 4$

7. $\int_0^5 \sqrt{25 - x^2}\,dx,\ n = 5$

8. $\int_0^2 \sqrt{x^3 + 1}\,dx,\ n = 4$

9. $\int_1^5 \frac{1}{x^2 + x}\,dx,\ n = 10$

10. $\int_2^4 \frac{1}{x^2 + 1}\,dx,\ n = 10$

11. $\int_0^4 2^x\,dx,\ n = 12$

12. $\int_0^{1.5} 10^x\,dx,\ n = 15$

In Exercises 13 and 14, approximate the values of the integrals defined by the given sets of points.

13. $\int_2^{14} y\,dx$

x	2	4	6	8	10	12	14
y	0.67	2.34	4.56	3.67	3.56	4.78	6.87

14. $\int_{1.4}^{3.2} y\,dx$

x	1.4	1.7	2.0	2.3	2.6	2.9	3.2
y	0.18	7.87	18.23	23.53	24.62	20.93	20.76

In Exercises 15 and 16, solve the given problems by using the trapezoidal rule.

15. A force F that a distributed electric charge has on a point charge is $F = k\int_0^2 \frac{dx}{(4 + x^2)^{3/2}}$, where x is the distance along the distributed charge and k is a constant. With $n = 8$, evaluate F in terms of k.

16. The length L of telephone wire needed (considering the sag) between two poles exactly 100 m apart is $L = 2\int_0^{100} \sqrt{1.6 \times 10^{-7}x^2 + 1}\,dx$. With $n = 10$, evaluate L (to six significant digits).

25-7 Simpson's Rule

The numerical method of integration developed in this section is also readily programmable for use on a computer or easily usable with the necessary calculations done on a calculator. It is obtained by interpreting the definite integral as the area under a curve, as we did in developing the trapezoidal rule, and by approximating the curve by a set of parabolic arcs. The use of parabolic arcs, rather than chords as with the trapezoidal rule, usually gives a better approximation.

Since we will be using parabolic arcs, we first derive a formula for the area that is under a parabolic arc. The curve shown in Fig. 25-19 represents the parabola $y = ax^2 + bx + c$. The points shown on this curve are $(-h, y_0)$, $(0, y_1)$, and (h, y_2). The area under the parabola is given by

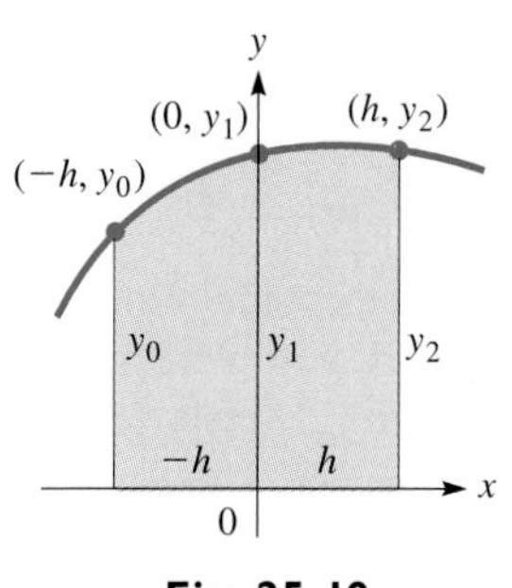

Fig. 25-19

$$A = \int_{-h}^{h} y\,dx = \int_{-h}^{h} (ax^2 + bx + c)\,dx = \frac{ax^3}{3} + \frac{bx^2}{2} + cx\Big|_{-h}^{h}$$

$$= \frac{2}{3}ah^3 + 2ch$$

or

$$A = \frac{h}{3}(2ah^2 + 6c) \tag{25-14}$$

The coordinates of the three points also satisfy the equation $y = ax^2 + bx + c$. This means that

$$y_0 = ah^2 - bh + c$$
$$y_1 = c$$
$$y_2 = ah^2 + bh + c$$

By finding the sum of $y_0 + 4y_1 + y_2$, we have

$$y_0 + 4y_1 + y_2 = 2ah^2 + 6c \tag{25-15}$$

Substituting Eq. (25-15) into Eq. (25-14), we have

$$A = \frac{h}{3}(y_0 + 4y_1 + y_2) \tag{25-16}$$

We note that the area depends only on the distance h and the three y-coordinates.

Now let us consider the area under the curve in Fig. 25-20. If a parabolic arc is passed through the points (x_0, y_0), (x_1, y_1), and (x_2, y_2), we may use Eq. (25-16) to approximate the area under the curve between x_0 and x_2. We also note that the distance h used in finding Eq. (25-16) is the difference in the x-coordinates, or $h = \Delta x$. Therefore, the area under the curve between x_0 and x_2 is

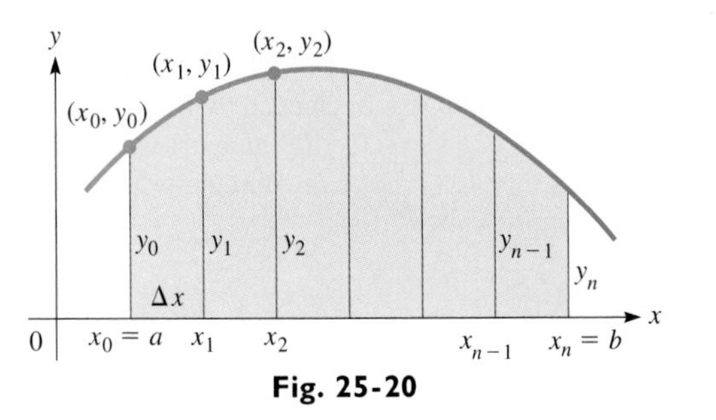

Fig. 25-20

$$A_1 = \frac{\Delta x}{3}(y_0 + 4y_1 + y_2)$$

Similarly, if a parabolic arc is passed through the three points starting with (x_2, y_2), the area between x_2 and x_4 is

$$A_2 = \frac{\Delta x}{3}(y_2 + 4y_3 + y_4)$$

The sum of these areas is

$$A_1 + A_2 = \frac{\Delta x}{3}(y_0 + 4y_1 + 2y_2 + 4y_3 + y_4) \tag{25-17}$$

We can continue this procedure until the approximate value of the entire area has been found. We must note, however, that ***the number of intervals n of width* Δx *must be even.*** Therefore, generalizing on Eq. (25-17) and recalling again that the value of the definite integral is the area under the curve, we have

CAUTION ▶

$$\int_a^b f(x)\,dx \approx \frac{\Delta x}{3}(y_0 + 4y_1 + 2y_2 + 4y_3 + 2y_4 + \cdots + 4y_{n-1} + y_n) \tag{25-18}$$

Equation (25-18) is known as **Simpson's rule.** As with the trapezoidal rule, we used Simpson's rule in Chapter 2 to measure irregular areas (see p. 67). We now see how it is derived. Simpson's rule is also used in many calculator models for the evaluation of definite integrals.

EXAMPLE 1 ▸▸ Approximate the value of the integral $\int_0^1 \frac{dx}{x+1}$ by Simpson's rule. Let $n = 2$.

In Fig. 25-21 the graph of the function and the area used are shown. We are to approximate the integral by using Eq. (25-18). We therefore note that $f(x) = 1/(x + 1)$. Also, $x_0 = a = 0$, $x_1 = 0.5$, and $x_2 = b = 1$. This is due to the fact that $n = 2$ and $\Delta x = 0.5$ since the total interval is 1 unit (from $x = 0$ to $x = 1$). Therefore,

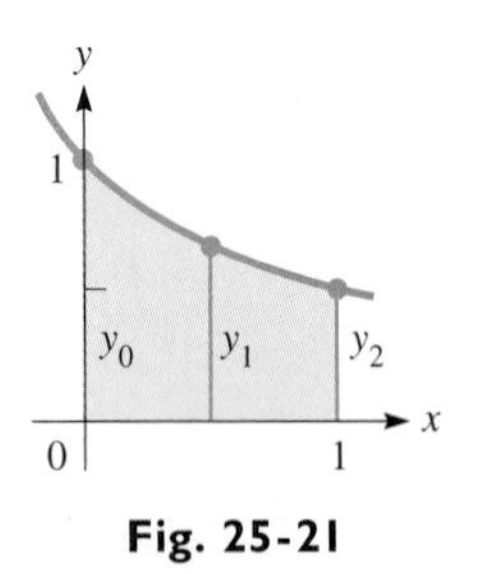

Fig. 25-21

$$y_0 = \frac{1}{0+1} = 1.0000, \qquad y_1 = \frac{1}{0.5+1} = 0.6667, \qquad y_2 = \frac{1}{1+1} = 0.5000$$

Substituting, we have

$$\int_0^1 \frac{dx}{x+1} = \frac{0.5}{3}[1.0000 + 4(0.6667) + 0.5000]$$
$$= 0.694$$

To three decimal places, the actual value of the integral is 0.693. We will consider the method of integrating this function in a later chapter. ◂◂

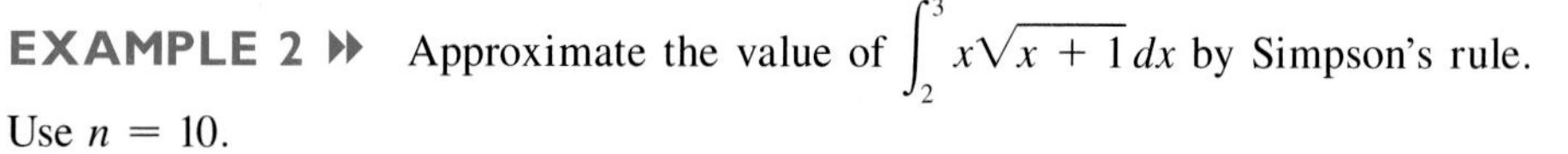

EXAMPLE 2 ▸▸ Approximate the value of $\int_2^3 x\sqrt{x+1}\,dx$ by Simpson's rule. Use $n = 10$.

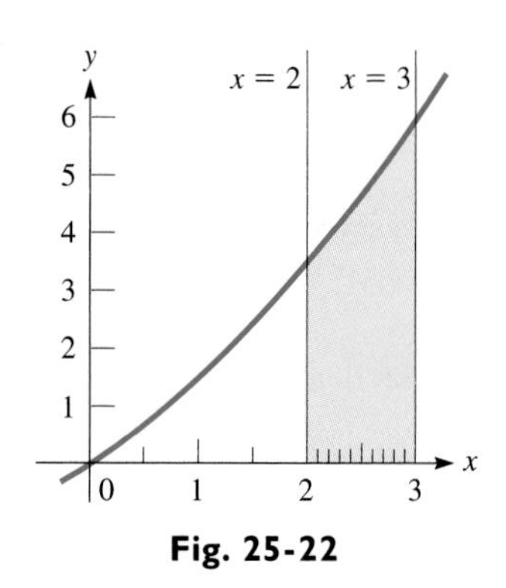

Fig. 25-22

Since the necessary values for this function are shown in Example 3 of Section 25-6, we shall simply tabulate them here. ($\Delta x = 0.1$.) See Fig. 25-22.

$$y_0 = 3.464\,101\,6 \quad y_1 = 3.697\,431\,5 \quad y_2 = 3.935\,479\,6 \quad y_3 = 4.178\,157\,5$$
$$y_4 = 4.425\,381\,3 \quad y_5 = 4.677\,071\,7 \quad y_6 = 4.933\,153\,2 \quad y_7 = 5.193\,553\,7$$
$$y_8 = 5.458\,204\,8 \quad y_9 = 5.727\,041\,1 \quad y_{10} = 6.000\,000\,0$$

Therefore, we evaluate the integral as follows:

$$\int_2^3 x\sqrt{x+1}\,dx = \frac{0.1}{3}[3.464\,101\,6 + 4(3.697\,431\,5) + 2(3.935\,479\,6)$$
$$+ 4(4.178\,157\,5) + 2(4.425\,381\,3) + 4(4.677\,071\,7)$$
$$+ 2(4.933\,153\,2) + 4(5.193\,553\,7) + 2(5.458\,204\,8)$$
$$+ 4(5.727\,041\,1) + 6.000\,000\,0]$$
$$= \frac{0.1}{3}(140.861\,56) = 4.695\,385\,4$$

This result agrees with the actual value to the eight significant digits shown. The value we obtained with the trapezoidal rule was 4.6958. ◂◂

EXAMPLE 3 ▸▸ The rear stabilizer of a certain aircraft is shown in Fig. 25-23. The area A (in m²) of one side of the stabilizer is $A = \int_0^3 (3x^2 - x^3)^{0.6}\,dx$. Find this area using Simpson's rule with $n = 6$.

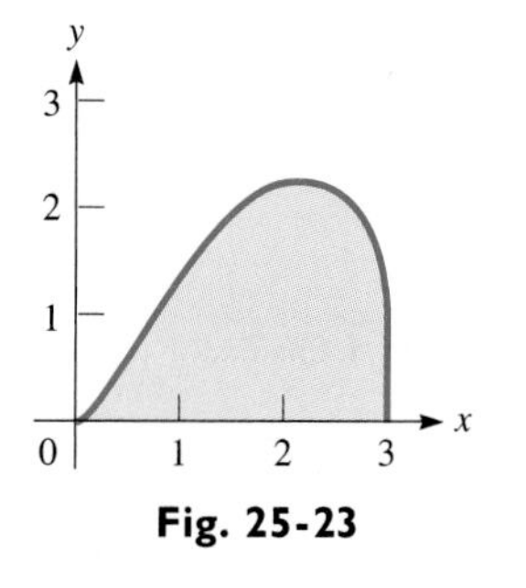

Fig. 25-23

Here we note that $f(x) = (3x^2 - x^3)^{0.6}$, $a = 0$, $b = 3$, and $\Delta x = \frac{3-0}{6} = 0.5$. Therefore,

$$y_0 = f(0) = [3(0)^2 - 0^3]^{0.6} = 0$$
$$y_1 = f(0.5) = [3(0.5)^2 - 0.5^3]^{0.6} = 0.754\,272\,0$$
$$y_2 = f(1) = [3(1)^2 - 1^3]^{0.6} = 1.515\,716\,6$$
$$y_3 = f(1.5) = [3(1.5)^2 - 1.5^3]^{0.6} = 2.074\,742\,8$$
$$y_4 = f(2) = [3(2)^2 - 2^3]^{0.6} = 2.297\,396\,7$$
$$y_5 = f(2.5) = [3(2.5)^2 - 2.5^3]^{0.6} = 1.981\,116\,5$$
$$y_6 = f(3) = [3(3)^2 - 3^3]^{0.6} = 0$$
$$A = \int_0^3 (3x^2 - x^3)^{0.6}\,dx = \frac{0.5}{3}[0 + 4(0.754\,272\,0) + 2(1.515\,716\,6)$$
$$+ 4(2.074\,742\,8) + 2(2.297\,396\,7) + 4(1.981\,116\,5) + 0]$$
$$= 4.477\,792\,0 \text{ m}^2$$

Thus the area of one side of the stabilizer is 4.478 m² (rounded off). ◂◂

Exercises 25–7

In Exercises 1–4, (a) approximate the value of each of the given integrals by use of Simpson's rule, using the given value of n, and (b) check by direct integration.

1. $\int_0^2 (1 + x^3)\,dx,\ n = 2$

2. $\int_0^8 x^{1/3}\,dx,\ n = 2$

3. $\int_1^4 (2x + \sqrt{x})\,dx,\ n = 6$

4. $\int_0^2 x\sqrt{x^2+1}\,dx,\ n = 4$

In Exercises 5–12, approximate the value of each of the given integrals by use of Simpson's rule, using the given values of n. Exercises 5–10 are the same as Exercises 5–10 (except Exercise 7) of Section 25-6.

5. $\int_2^3 \frac{1}{2x}\,dx,\ n = 2$

6. $\int_2^6 \frac{dx}{x+3},\ n = 4$

7. $\int_0^5 \sqrt{25 - x^2}\,dx,\ n = 4$

8. $\int_0^2 \sqrt{x^3 + 1}\,dx,\ n = 4$

9. $\int_1^5 \frac{1}{x^2 + x}\,dx,\ n = 10$

10. $\int_2^4 \frac{1}{x^2 + 1}\,dx,\ n = 10$

11. $\int_{-4}^5 (2x^4 + 1)^{0.1}\,dx,\ n = 6$

12. $\int_0^{2.4} \frac{dx}{(4 + \sqrt{x})^{3/2}},\ n = 8$

In Exercises 13 and 14, approximate the values of the integrals defined by the given sets of points. These are the same as Exercises 13 and 14 of Section 25-6.

13. $\int_2^{14} y\,dx$

x	2	4	6	8	10	12	14
y	0.67	2.34	4.56	3.67	3.56	4.78	6.87

14. $\int_{1.4}^{3.2} y\,dx$

x	1.4	1.7	2.0	2.3	2.6	2.9	3.2
y	0.18	7.87	18.23	23.53	24.62	20.93	20.76

In Exercises 15 and 16, solve the given problems using Simpson's rule.

15. The distance $\overline{x}$ (in cm) from one end of a barrel plug (with vertical circular cross sections) to its center of mass, as shown in Fig. 25-24, is $\overline{x} = 0.9129 \int_0^3 x\sqrt{0.3 - 0.1x}\,dx$. Find $\overline{x}$ with $n = 12$.

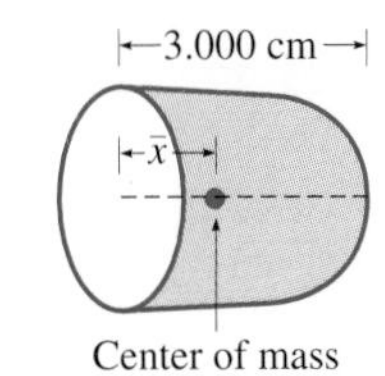

Fig. 25-24

16. The average value of the electric current i_{av}, in amperes, in a circuit for the first 4 s is $i_{av} = \frac{1}{4}\int_0^4 (4t - t^2)^{0.2}\,dt$. Find i_{av} with $n = 10$.

Chapter Equations, Review Exercises, and Practice Test

Chapter Equations

Differential $\quad dy = f'(x)\,dx \quad$ (25-1)

Indefinite integral $\quad \int f(x)\,dx = F(x) + C \quad$ (25-2)

Integrals $\quad \int c\,du = c\int du = cu + C \quad$ (25-3)

$\int (du + dv) = u + v + C \quad$ (25-4)

Power formula $\quad \int u^n\,du = \frac{u^{n+1}}{n+1} + C \quad (n \neq -1) \quad$ (25-5)

Area under a curve $\quad A_{ab} = \left[\int f(x)\,dx\right]_a^b = F(b) - F(a) \quad$ (25-10)

Definite integral $\quad \int_a^b f(x)\,dx = F(b) - F(a) \quad$ (25-11)

Trapezoidal rule $\quad \int_a^b f(x)\,dx \approx \frac{\Delta x}{2}(y_0 + 2y_1 + 2y_2 + \cdots + 2y_{n-1} + y_n) \quad$ (25-13)

Simpson's rule $\quad \int_a^b f(x)\,dx \approx \frac{\Delta x}{3}(y_0 + 4y_1 + 2y_2 + 4y_3 + 2y_4 + \cdots + 4y_{n-1} + y_n) \quad$ (25-18)

Review Exercises

In Exercises 1–24, evaluate the given integrals.

1. $\int (4x^3 - x)\,dx$ **2.** $\int (5 + 3x^2)\,dx$

3. $\int 2(x - x^{3/2})\,dx$ **4.** $\int x(x - 3x^4)\,dx$

5. $\int_1^4 \left(\sqrt{x} + \frac{1}{\sqrt{x}}\right) dx$ **6.** $\int_1^2 \left(x + \frac{1}{x^2}\right) dx$

7. $\int_0^2 x(4 - x)\,dx$ **8.** $\int_0^1 x(x^3 + 1)\,dx$

9. $\int \left(3 + \frac{2}{x^3}\right) dx$ **10.** $\int \left(3\sqrt{x} + \frac{1}{2\sqrt{x}} - \frac{1}{4}\right) dx$

11. $\int_{-2}^{5} \frac{dx}{\sqrt[3]{x^2 + 6x + 9}}$ **12.** $\int_{0.35}^{0.85} x(\sqrt{1 - x^2} + 1)\,dx$

13. $\int \frac{dx}{(2 - 5x)^2}$ **14.** $\int \frac{1}{x^2}\sqrt{1 + \frac{1}{x}}\,dx$

15. $\int 3(7 - 2x)^{3/4}\,dx$ **16.** $\int (3x + 1)^{1/3}\,dx$

17. $\int_0^2 \frac{3x\,dx}{\sqrt[3]{1 + 2x^2}}$ **18.** $\int_1^6 \frac{2\,dx}{(3x - 2)^{3/4}}$

19. $\int x^2(1 - 2x^3)^4\,dx$ **20.** $\int 3x^3(1 - 5x^4)^{1/3}\,dx$

21. $\int \frac{(2 - 3x^2)\,dx}{(2x - x^3)^2}$ **22.** $\int \frac{x^2 - 3}{\sqrt{6 + 9x - x^3}}\,dx$

23. $\int_1^3 (x^2 + x + 2)(2x^3 + 3x^2 + 12x)\,dx$

24. $\int_0^2 (4x + 18x^2)(x^2 + 3x^3)^2\,dx$

In Exercises 25–32, find the differential of each of the given functions.

25. $y = 4x^3 + \frac{1}{x}$ **26.** $y = 2\sqrt{x} - \frac{3}{\sqrt{x}}$

27. $y = \frac{1}{(x^2 - 1)^3}$ **28.** $y = \frac{1}{(2x - 1)^2}$

29. $y = x\sqrt[3]{1 - 3x}$ **30.** $y = \frac{3 + x}{4 - x^2}$

31. $y = \frac{2x^2}{2 - x}$ **32.** $y = 2x^2(1 - 4x)^4$

In Exercises 33 and 34, evaluate $\Delta y - dy$ for the given functions and values.

33. $y = x^3,\ x = 2,\ \Delta x = 0.1$

34. $y = 6x^2 - x,\ x = 3,\ \Delta x = 0.2$

In Exercises 35 and 36, find the required equations.

35. Find the equation of the curve that passes through $(-1, 3)$ for which the slope is given by $3 - x^2$.

36. Find the equation of the curve that passes through $(1, -2)$ for which the slope is $x(x^2 + 1)^2$.

In Exercises 37 and 38, perform the indicated integrations as directed.

37. Perform the integration $\int (1 - 2x)\,dx$ (a) term by term, labeling the constant of integration as C_1, and then (b) by letting $u = 1 - 2x$, using the general power rule, and labeling the constant of integration as C_2. Compare C_1 and C_2.

38. Following methods (a) and (b) in Exercise 37, perform the integration $\int (3x + 2)\,dx$. In (b) let $u = 3x + 2$. Compare the constants of integration C_1 and C_2.

In Exercises 39 and 40, use Eq. (25-10) to find the indicated areas.

39. The area under $y = 6x - 1$ between $x = 1$ and $x = 3$

40. The first-quadrant area under $y = 8x - x^4$

In Exercises 41 and 42, solve the given problems by using the trapezoidal rule.

41. Approximate $\int_1^3 \frac{dx}{2x - 1}$ with $n = 4$.

42. Approximate the value of the integral defined by the following set of points:

x	6.0	9.0	12	15	18	21
y	2.0	1.2	0.2	1.0	6.0	12

In Exercises 43 and 44, solve the given problems by using Simpson's rule.

43. Approximate $\int_1^3 \frac{dx}{2x - 1}$ with $n = 4$ (see Exercise 41).

44. Approximate the value of the integral of the function defined by the following points:

x	1.0	1.4	1.8	2.2	2.6	3.0	3.4
y	1.45	1.89	2.66	3.50	3.22	3.04	2.44

In Exercises 45 and 46, use the function $y = \dfrac{x}{x^2 + 2}$ *and approximate the area under the curve in the first quadrant to the left of the line* $x = 5$ *by the indicated method.*

45. Inscribe five rectangles, and find the sum of the areas of the rectangles.

46. Use the trapezoidal rule with $n = 5$.

In Exercises 47 and 48, use the function $y = x\sqrt{x^3 + 1}$ *and approximate the area under the curve from* $x = 1$ *to* $x = 3$ *by the indicated method. See Fig. 25-25.*

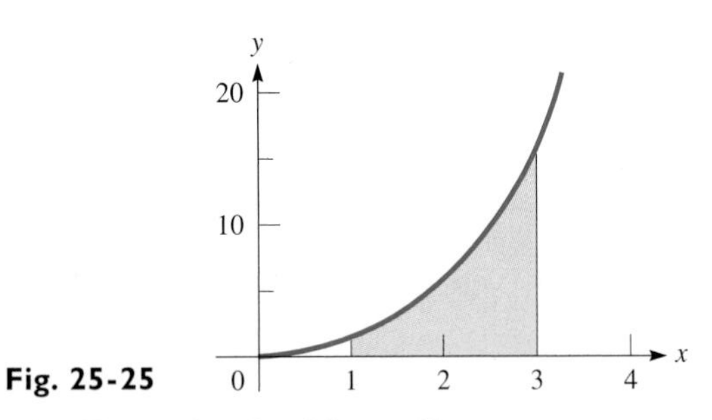

Fig. 25-25

47. Use Simpson's rule with $n = 10$.

48. Inscribe ten rectangles, and find the sum of the areas of the rectangles.

In Exercises 49–52, use the function $y = x\sqrt[3]{9 - x^2}$ *and approximate the area under the curve between* $x = 0$ *and* $x = 3$ *by the indicated method. See Fig. 25-26.*

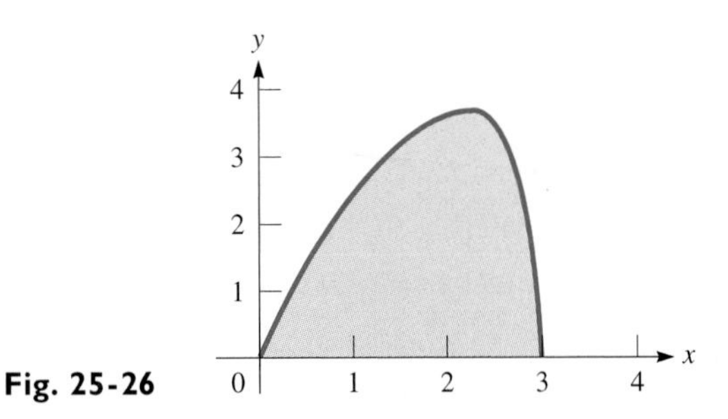

Fig. 25-26

49. Inscribe six rectangles, and find the sum of the areas of the rectangles.

50. Use the trapezoidal rule with $n = 6$.

51. Use Simpson's rule with $n = 6$.

52. By integration (for the exact area).

In Exercises 53–56, solve the given problems by finding the appropriate differentials.

53. A weather balloon 3.500 m in radius becomes covered with a uniform layer of ice 1.2 cm thick. What is the volume of the ice?

54. The total power P, in watts, transmitted by an AM radio transmitter is $P = 460 + 230m^2$, where m is the modulation index. What is the change in power if m changes from 0.86 to 0.89?

55. The impedance Z of an electric circuit as a function of the resistance R and the reactance X is given by $Z = \sqrt{R^2 + X^2}$. Derive an expression of the relative error in impedance for an error in R and a given value of X.

56. Show that the relative error of the *nth* root of a given measurement equals approximately $1/n$ of the relative error of the measurement.

In Exercises 57–60, solve the given problems by integration.

57. The deflection y of a certain beam at a distance x from one end is $dy/dx = k(2L^3 - 12Lx + 2x^4)$, where k is a constant and L is the length of the beam. Find y as a function of x if $y = 0$ for $x = 0$.

58. The total electric charge Q on a charged sphere is $Q = k\int\left(r^2 - \dfrac{r^3}{R}\right)dr$, where k is a constant, r is the distance from the center of the sphere, and R is the radius of the sphere. Find Q as a function of r if $Q = Q_0$ for $r = R$.

59. Part of the deck of a boat is the parabolic area shown in Fig. 25-27. The area A (in m^2) is $A = 2\int_0^5 \sqrt{5 - y}\,dy$. Evaluate A.

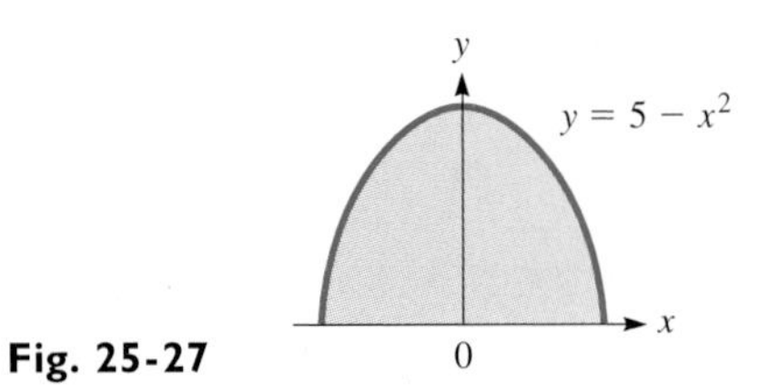

Fig. 25-27

60. The distance s (in cm) through which a cam follower moves in 4 s is $s = \int_0^4 t\sqrt{4 + 9t^2}\,dt$. Evaluate s.

Writing Exercise

61. A computer science student is writing a program to find a good approximation for the value of π by using the formula $A = \pi r^2$ for a circle. The value of π is to be found by approximating the area of a circle with a given radius. Write two or three paragraphs explaining how the value of π can be approximated in this way. Include any equations and values that may be used, but do not actually make the calculations.

Practice Test

1. For $y = 3x^2 - x$, evaluate (a) Δy, (b) dy, and (c) $\Delta y - dy$ for $x = 3$ and $\Delta x = 0.1$.

2. Integrate: $\int x\sqrt{1 - 2x^2}\,dx$.

3. Evaluate: $\int_2^5 (6 - x)^4\,dx$.

4. Approximate the area under $y = \frac{1}{x + 2}$ between $x = 1$ and $x = 4$ (above the x-axis) by inscribing six rectangles and finding the sum of their areas.

5. Evaluate $\int_1^4 \frac{dx}{x + 2}$ by using the trapezoidal rule with $n = 6$.

6. Evaluate the definite integral of Problem 5 using Simpson's rule with $n = 6$.

7. The inside radius of a pipe is 4.50 cm. By using differentials, find the approximate volume of metal in the pipe if it is 20.0 cm long and the metal is 0.30 cm thick.

8. The total electric current (in amperes) to pass a point in the circuit between $t = 1$ s and $t = 3$ s is

   $i = \int_1^3 \left(t^2 + \frac{1}{t^2}\right) dt$. Evaluate i.

26

Applications of Integration

In Section 26-6, integration is used to find the force water exerts on the floodgate of a dam.

Integration is useful in many areas of science, engineering, and technology. It has numerous important applications in electricity, mechanics, hydrostatics, architecture, machine design, and business as well as in geometry and other areas of physics.

Some of these applications were indicated in the previous chapter. In this chapter we develop the necessary concepts for setting up integrals for some of these types of applications.

In the first section of this chapter we present some important applications of the indefinite integral. In the remaining sections we present applications of the definite integral in geometry and certain technical areas.

26-1 Applications of the Indefinite Integral

In the examples of this section we show two basic applications of the indefinite integral. Some other applications are shown in the exercises.

Velocity and Displacement

The first application deals with the motion of an object. The concepts of velocity as a first derivative and acceleration as a second derivative were introduced in Chapter 24. Here we apply integration to the problem of finding the displacement and velocity as a function of time, when we know the relationship between acceleration and time, and certain values of displacement and velocity. As we saw in Section 25-3, these values are needed for finding the values of the constants of integration that are introduced.

Recalling now that the acceleration a of an object is given by $a = dv/dt$, we can find the expression for the velocity v in terms of a, the time t, and the constant of integration. We write

$$dv = a\,dt$$

or

$$v = \int a\,dt \tag{26-1}$$

If the acceleration is constant, we have

$$v = at + C_1 \tag{26-2}$$

Of course, Eq. (26-1) can be used in general to find the velocity as a function of time so long as we know the acceleration as a function of time. However, since the case of constant acceleration is often encountered, Eq. (26-2) is often encountered. If the velocity is known for a specified time, the constant C_1 may be evaluated.

EXAMPLE 1 ▸▸ Find the expression for the velocity if $a = 12t$, given that $v = 8$ when $t = 1$.

Using Eq. (26-1), we have

$$v = \int (12t)\,dt = 6t^2 + C_1$$

Substituting the known values, we obtain

$$8 = 6 + C_1 \quad \text{or} \quad C_1 = 2$$

Thus, $v = 6t^2 + 2$. ◂◂

EXAMPLE 2 ▸▸ For an object falling under the influence of gravity, the acceleration due to gravity is essentially constant. Its value is -9.8 m/s^2. (The negative sign is chosen so that *all quantities directed up are positive* and ***all quantities directed down are negative.***) Find the expression for the velocity of an object under the influence of gravity if $v = v_0$ when $t = 0$.

NOTE ▸

We write

$$\begin{aligned}
v &= \int (-9.8)\,dt && \text{substitute } a = -9.8 \text{ into Eq. (26-1)} \\
&= -9.8t + C_1 && \text{integrate} \\
v_0 &= -9.8(0) + C_1 && \text{substitute given values} \\
C_1 &= v_0 && \text{solve for } C_1 \\
v &= v_0 - 9.8t && \text{substitute}
\end{aligned}$$

The velocity v_0 is called the *initial velocity*. If the object is given an initial upward velocity of 40 m/s, $v_0 = 40$ m/s. If the object is dropped, $v_0 = 0$. If the object is given an initial downward velocity of 40 m/s, $v_0 = -40$ m/s. ◂◂

Once we have the expression for velocity, we can then integrate to find the expression for displacement s in terms of the time. Since $v = ds/dt$, we can write $ds = v\,dt$, or

$$s = \int v\,dt \tag{26-3}$$

Consider the following examples.

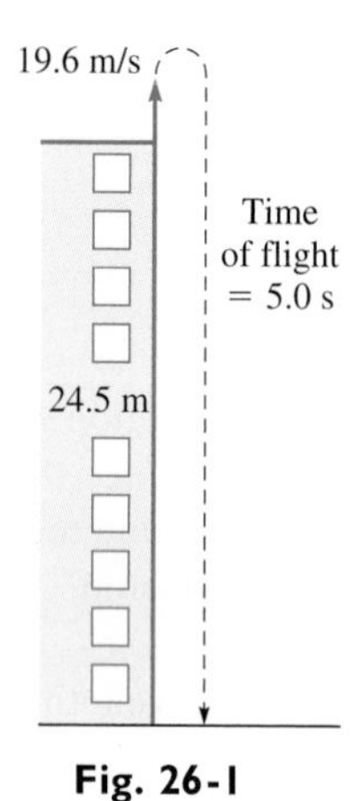

Fig. 26-1

EXAMPLE 3 ▸▸ A ball is thrown vertically from the top of a building 24.5 m high and hits the ground 5.0 s later. What initial velocity was the ball given?

Measuring vertical distances from the ground, we know that $s = 24.5$ m when $t = 0$ and that $v = v_0 - 9.8t$. Thus

$$s = \int (v_0 - 9.8t)\,dt = v_0 t - 4.9t^2 + C \quad \text{integrate}$$

$$24.5 = v_0(0) - 4.9(0) + C, \qquad C = 24.5 \quad \text{evaluate } C$$

$$s = v_0 t - 4.9t^2 + 24.5$$

We also know that $s = 0$ when $t = 5.0$ s. Thus

$$0 = v_0(5.0) - 4.9(5.0)^2 + 24.5 \quad \text{substitute given values}$$

$$5.0v_0 = 98.0$$

$$v_0 = 19.6 \text{ m/s}$$

This means that the initial velocity was 19.6 m/s upward. See Fig. 26-1. ◂◂

EXAMPLE 4 ▸▸ During the initial stage of launching a spacecraft vertically, the acceleration a (in m/s^2) of the spacecraft is $a = 6t^2$. Find the height s of the spacecraft after 6.0 s if $s = 12$ m for $t = 0.0$ s and $v = 16$ m/s for $t = 2.0$ s.

First we use Eq. (26-1) to get an expression for the velocity:

$$v = \int 6t^2\,dt = 2t^3 + C_1 \quad \text{integrate}$$

$$16 = 2(2.0)^3 + C_1 \qquad C_1 = 0 \quad \text{evaluate } C_1$$

$$v = 2t^3$$

We now use Eq. (26-3) to get an expression for the displacement:

$$s = \int 2t^3\,dt = \tfrac{1}{2}t^4 + C_2 \quad \text{integrate}$$

$$12 = \tfrac{1}{2}(0.0)^4 + C_2, \qquad C_2 = 12 \quad \text{evaluate } C_2$$

$$s = \tfrac{1}{2}t^4 + 12$$

Now, finding s for $t = 6.0$ s, we have

$$s = \tfrac{1}{2}(6.0)^4 + 12 = 660 \text{ m} \quad ◂◂$$

Voltage Across a Capacitor

The second basic application of the indefinite integral we shall discuss comes from the field of electricity. By definition, *the current i in an electric circuit equals the time rate of change of the charge q (in coulombs) that passes a given point in the circuit, or*

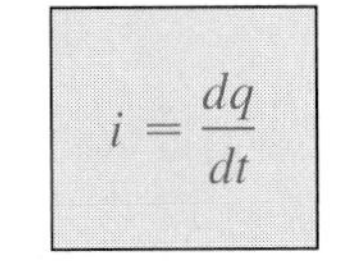

$$i = \frac{dq}{dt} \tag{26-4}$$

Rewriting this expression in differential notation as $dq = i\,dt$ and integrating both sides of the equation, we have

$$q = \int i\,dt \tag{26-5}$$

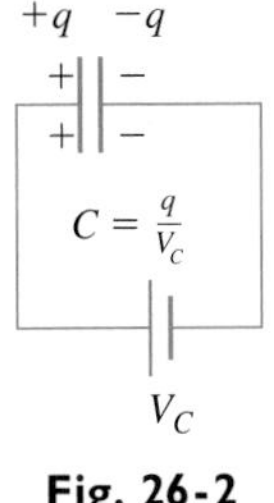

Fig. 26-2

Now, the voltage V_C across a capacitor with capacitance C (see Fig. 26-2) is given by $V_C = q/C$. By combining equations, the voltage V_C is given by

$$V_C = \frac{1}{C}\int i\,dt \tag{26-6}$$

Here, V_C is measured in volts, C in farads, i in amperes, and t in seconds.

EXAMPLE 5 ▸▸ The current in a certain electric circuit as a function of time is given by $i = 6t^2 + 4$. Find an expression for the amount of charge q that passes a point in the circuit as a function of time. Assuming that $q = 0$ when $t = 0$, determine the total charge that passes the point in 2 s.

Since $q = \int i\,dt$, we have

$$q = \int (6t^2 + 4)\,dt \quad \text{substitute into Eq. (26-5)}$$

$$= 2t^3 + 4t + C \quad \text{integrate}$$

This last expression is the desired expression giving charge as a function of time. We note that when $t = 0$, then $q = C$, which means that the constant of integration represents the initial charge, or the charge that passed a given point before we started timing. Using q_0 to represent this charge, we have

$$q = 2t^3 + 4t + q_0$$

Now, returning to the second part of the problem, we see that $q_0 = 0$. Therefore, evaluating q for $t = 2$ s, we have

$$q = 2(8) + 4(2) = 24 \text{ C}$$

(Here, the symbol C represents coulombs and is not the C for capacitance of Eq. (26-6) or the constant of integration.) This is the charge that passes any specified point in the circuit in 2 s. ◂◂

EXAMPLE 6 ▸▸ The voltage across a 5.0-μF capacitor is zero. What is the voltage after 20 ms if a current of 75 mA charges the capacitor?

Since the current is 75 mA, we know that $i = 0.075\text{ A} = 7.5 \times 10^{-2}\text{ A}$. We see

CAUTION ▸ that ***we must use the proper power of 10 that corresponds to each prefix.*** Since $5.0\ \mu\text{F} = 5.0 \times 10^{-6}\text{ F}$, we have

$$V_C = \frac{1}{5.0 \times 10^{-6}} \int 7.5 \times 10^{-2}\, dt \qquad \text{substituting into Eq. (26-6)}$$

$$= (1.5 \times 10^4) \int dt$$

$$= (1.5 \times 10^4)t + C_1 \qquad \text{integrate}$$

From the given information we know that $V_C = 0$ when $t = 0$. Thus

$$0 = (1.5 \times 10^4)(0) + C_1 \quad \text{or} \quad C_1 = 0 \qquad \text{evaluate } C_1$$

This means that

$$V_C = (1.5 \times 10^4)t$$

Evaluating this expression for $t = 20 \times 10^{-3}$ s, we have

$$V_C = (1.5 \times 10^4)(20 \times 10^{-3})$$
$$= 30 \times 10 = 300\text{ V} \quad ◂◂$$

EXAMPLE 7 ▸▸ A certain capacitor is measured to have a voltage of 100 V across it. At this instant a current as a function of time given by $i = 0.06\sqrt{t}$ is sent through the circuit. After 0.25 s, the voltage is measured to be 140 V. What is the capacitance of the capacitor?

Substituting $i = 0.06\sqrt{t}$, we find that

$$V_C = \frac{1}{C} \int (0.06\sqrt{t}\, dt) = \frac{0.06}{C} \int t^{1/2}\, dt \qquad \text{using Eq. (26-6)}$$

$$= \frac{0.04}{C} t^{3/2} + C_1 \qquad \text{integrate}$$

From the given information we know that $V_C = 100$ V when $t = 0$. Thus

$$100 = \frac{0.04}{C}(0) + C_1 \quad \text{or} \quad C_1 = 100\ V \qquad \text{evaluate } C_1$$

This means that

$$V_C = \frac{0.04}{C} t^{3/2} + 100$$

We also know that $V_C = 140$ V when $t = 0.25$ s. Therefore,

$$140 = \frac{0.04}{C}(0.25)^{3/2} + 100$$

$$40 = \frac{0.04}{C}(0.125)$$

or

$$C = 1.25 \times 10^{-4}\text{ F} = 125\ \mu\text{F} \quad ◂◂$$

Exercises 26–1

1. What is the velocity (in m/s) of a wrench 2.5 s after it is dropped from a building platform?

2. A hoop is started upward along an inclined plane at 5.0 m/s. If the acceleration of the hoop is 2.0 m/s^2 downward along the plane, find the velocity of the hoop after 6.0 s.

3. A conveyor belt 8.00 m long moves at 0.25 m/s. If a package is placed at one end, find its displacement from the other end as a function of time.

4. During each cycle, the velocity v (in mm/s) of a piston is $v = 6t - 6t^2$, where t is the time in seconds. Find the displacement s of the piston after 0.75 s if the initial displacement is zero.

5. While in the barrel of a tennis ball machine, the acceleration a (in m/s^2) of a ball is $a = 30\sqrt{1 - 4t}$, where t is the time in seconds. If $v = 0$ for $t = 0$, find the velocity of the ball as it leaves the barrel at $t = 0.25$ s.

6. A proton moves in an electric field such that its acceleration (in cm/s^2) is $a = -20(1 + 2t)^{-2}$, where t is the time in seconds. Find the velocity as a function of time if $v = 30$ cm/s when $t = 0$ s.

7. A rocket is fired vertically upward. When it reaches an altitude of 16 500 m, the engines cut off and the rocket is moving upward at 450 m/s. What will be its altitude 3.00 s later ($a = -9.80$ m/s^2)?

8. A flare is ejected vertically upward from the ground at 15 m/s. Find the height of the flare after 2.5 s.

9. What must be the nozzle velocity of the water from a fire hose if it is to reach a point 30 m directly above the nozzle?

10. An arrow is shot upward with a vertical velocity of 40.0 m/s from the edge of a cliff. If it hits the ground below after 9.0 s, how high is the cliff?

11. In coming to a stop, the acceleration of a car is $-4t$. If it is traveling at 32.0 m/s when the brakes are applied, how far does it travel while stopping?

12. A hoist mechanism raises a crate with an acceleration (in m/s^2) $a = \sqrt{1 + 0.2t}$, where t is the time in seconds. Find the displacement of the crate as a function of time if $v = 0$ m/s and $s = 2$ m for $t = 0$ s.

13. The electric current in a microprocessor circuit is 0.230 μA. How many coulombs pass a given point in the circuit in 1.50 ms?

14. The electric current (in mA) in a computer circuit as a function of time is $i = 0.3 - 0.2t$. What total charge passes a point in the circuit in 0.050 s?

15. In an amplifier circuit the current (in amperes) changes with time (in seconds) according to $i = 0.06t\sqrt{1 + t^2}$. If 0.015 C of charge have passed a point in the circuit at $t = 0$, find the total charge to have passed the point at $t = 0.25$ s.

16. The current i, in microamperes, in a certain microprocessor circuit is given by $i = 8 - t$, where t is the time in microseconds and $0 \le t \le 20$ μs. If $q_0 = 0$, for what value of t, greater than zero, is $q = 0$? What interpretation can be given to this result?

17. The voltage across a 2.5-μF capacitor in a copying machine is zero. What is the voltage after 12 ms if a current of 25 mA charges the capacitor?

18. The voltage across an 8.50-nF capacitor in an FM receiver circuit is zero. Find the voltage after 2.00 μs if a current (in mA) $i = 0.042t$ charges the capacitor.

19. The voltage across a 3.75-μF capacitor in a television circuit is 4.50 mV. Find the voltage after 0.565 ms if a current (in μA) $i = \sqrt[3]{1 + 6t}$ further charges the capacitor.

20. A current $i = t/\sqrt{t^2 + 1}$ (in amperes) is sent through an electric dryer circuit containing a previously uncharged 2.0-μF capacitor. How long does it take for the capacitor voltage to reach 120 V?

21. The angular velocity ω is the time rate of change of the angular displacement θ of a rotating object. See Fig. 26-3. In testing the shaft of an engine, its angular velocity is $\omega = 16t + 0.50t^2$, where t is the time of rotation. Find the angular displacement through which the shaft goes in 10.0 s.

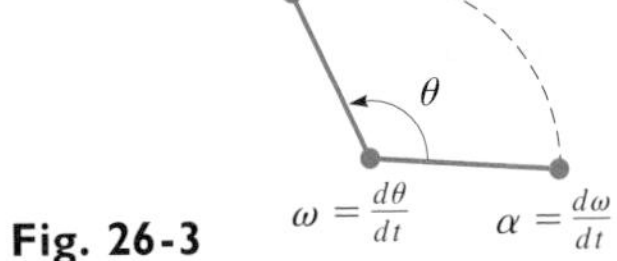

Fig. 26-3

22. The angular acceleration α is the time rate of change of angular velocity ω of a rotating object. See Fig. 26-3. When starting up, the angular acceleration of a helicopter blade is $\alpha = \sqrt{8t + 1}$. Find the expression for θ if $\omega = 0$ and $\theta = 0$ for $t = 0$.

23. An inductor in an electric circuit is essentially a coil of wire in which the voltage is affected by a changing current. By definition, the voltage caused by the changing current is given by $V_L = L(di/dt)$, where L is the inductance and is measured in henrys. If $V_L = 12.0 - 0.2t$ for a 3.0-H inductor, find the current in the circuit after 20 s if the initial current was zero.

24. If the inner and outer walls of a container are at different temperatures, the rate of change of temperature with respect to the distance from one wall is a function of the distance from the wall. Symbolically this is stated as $dT/dx = f(x)$, where T is the temperature. If x is measured from the outer wall, at 20°C, and $f(x) = 72x^2$, find the temperature at the inner wall if the container walls are 0.5 cm thick.

25. Surrounding an electrically charged particle is an electric field. The rate of change of electric potential with respect to the distance from the particle creating the field equals the negative of the value of the electric field. That is, $dV/dx = -E$, where E is the electric field. If $E = k/x^2$, where k is a constant, find the electric potential at a distance x_1 from the particle, if $V \to 0$ as $x \to \infty$.

26. The rate of change of the vertical deflection y with respect to the horizontal distance x from one end of a beam is a function of x. For a particular beam, this function is $k(x^5 + 1350x^3 - 7000x^2)$, where k is a constant. Find y as a function of x.

27. Fresh water is flowing into a brine solution, with an equal volume of mixed solution flowing out. The amount of salt in the solution decreases, but more slowly as time increases. Under certain conditions the time rate of change of mass of salt (in grams per minute) is given by $-1/\sqrt{t+1}$. Find the mass of salt as a function of time if 1000 g were originally present. Under these conditions, how long would it take for all the salt to be removed?

28. A holograph of a circle is formed. The rate of change of the radius r of the circle with respect to the wavelength λ of the light used is inversely proportional to the square root of λ. If $dr/d\lambda = 3.55 \times 10^4$ and $r = 4.08$ cm for $\lambda = 574$ nm, find r as a function of λ.

26-2 Areas by Integration

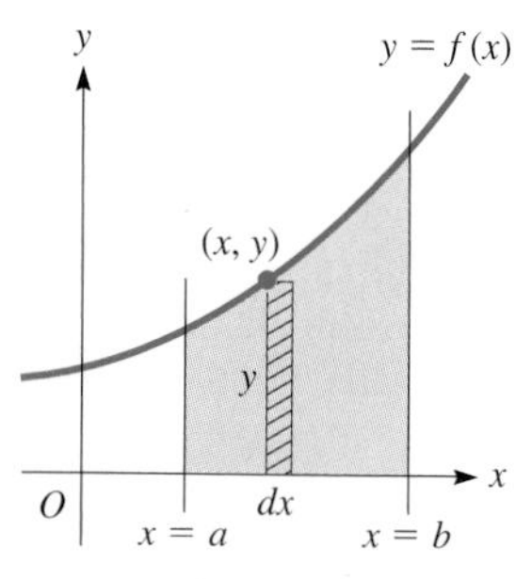

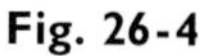

Fig. 26-4

In Section 25-4 we introduced the method of finding the area under a curve by integration. We also showed that the area can be found by a summation process on the rectangles inscribed under the curve, which means that integration can be interpreted as a summation process. *The applications of the definite integral use this summation interpretation of the integral.* We shall now develop a general procedure for finding the area for which the bounding curves are known by summing the areas of inscribed rectangles and using integration for the summation.

The first step is to make a sketch of the area. Next a representative **element of area** dA (a typical rectangle) is drawn. In Fig. 26-4 the width of the element is dx. The length of the element is determined by the y-coordinate (of the vertex of the element) of the point on the curve. Thus the length is y. The area of this element is $y\,dx$, which in turn means that $dA = y\,dx$, or

$$A = \int_a^b y\,dx = \int_a^b f(x)\,dx \tag{26-7}$$

This equation states that the elements are to be summed (this is the meaning of the integral sign) from a (the left boundary) to b (the right boundary).

Fig. 26-5

EXAMPLE 1 ▸▸ Find the area bounded by $y = 2x^2$, $y = 0$, $x = 1$, and $x = 2$.

This area is shown in Fig. 26-5. The rectangle shown is the representative element. Its area is $y\,dx$. The elements are to be summed from $x = 1$ to $x = 2$.

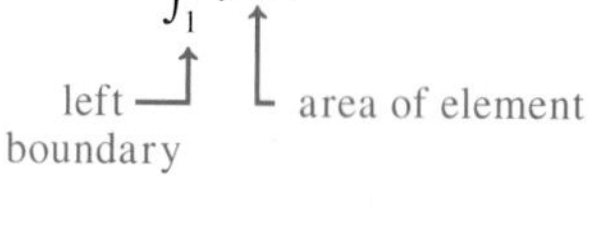

$$A = \int_1^2 y\,dx = \int_1^2 2x^2\,dx \qquad \text{substitute } 2x^2 \text{ for } y$$

$$= \frac{2}{3}x^3\Big|_1^2 = \frac{2}{3}(8) - \frac{2}{3}(1) \qquad \text{integrate and evaluate}$$

$$= \frac{14}{3} \quad ◂◂$$

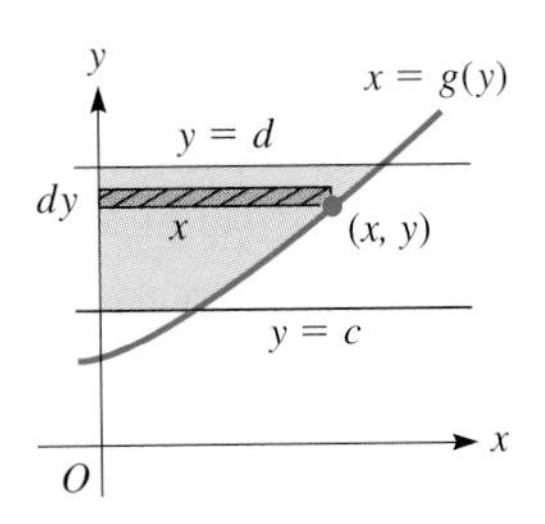

Fig. 26-6

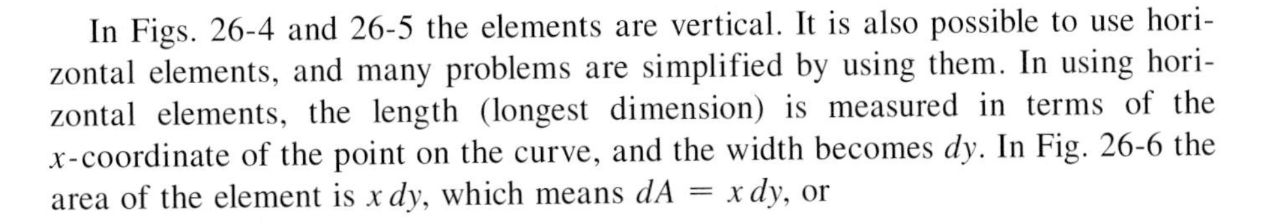
In Figs. 26-4 and 26-5 the elements are vertical. It is also possible to use horizontal elements, and many problems are simplified by using them. In using horizontal elements, the length (longest dimension) is measured in terms of the x-coordinate of the point on the curve, and the width becomes dy. In Fig. 26-6 the area of the element is $x\,dy$, which means $dA = x\,dy$, or

$$A = \int_c^d x\,dy = \int_c^d g(y)\,dy \qquad (26\text{-}8)$$

In using Eq. (26-8) the elements are summed from c (the lower boundary) to d (the upper boundary). In the following example, the area is found by use of both vertical and horizontal elements of area.

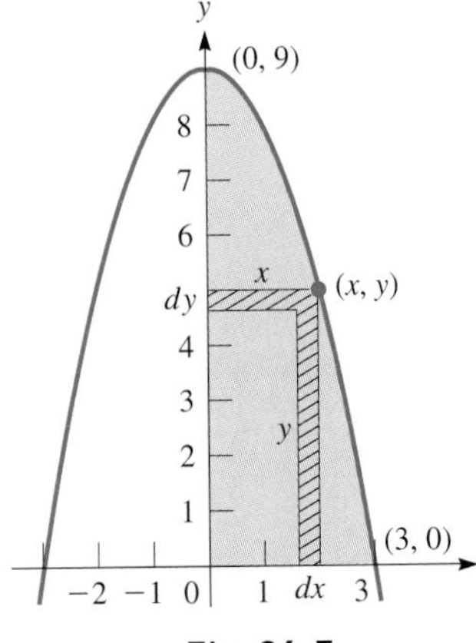

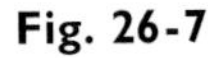
Fig. 26-7

EXAMPLE 2 ▸▸ Find the area in the first quadrant bounded by $y = 9 - x^2$.

The area to be found is shown in Fig. 26-7. First using the vertical element of length y and width dx, we have

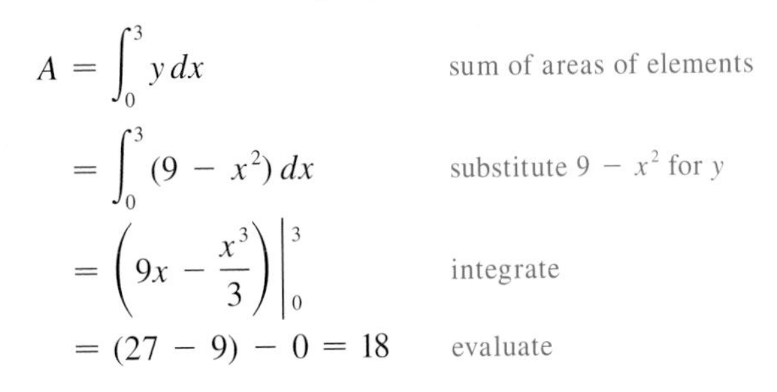

$$\begin{aligned} A &= \int_0^3 y\,dx && \text{sum of areas of elements} \\ &= \int_0^3 (9 - x^2)\,dx && \text{substitute } 9 - x^2 \text{ for } y \\ &= \left(9x - \frac{x^3}{3}\right)\Bigg|_0^3 && \text{integrate} \\ &= (27 - 9) - 0 = 18 && \text{evaluate} \end{aligned}$$

Now, using the horizontal element of length x and width dy, we have

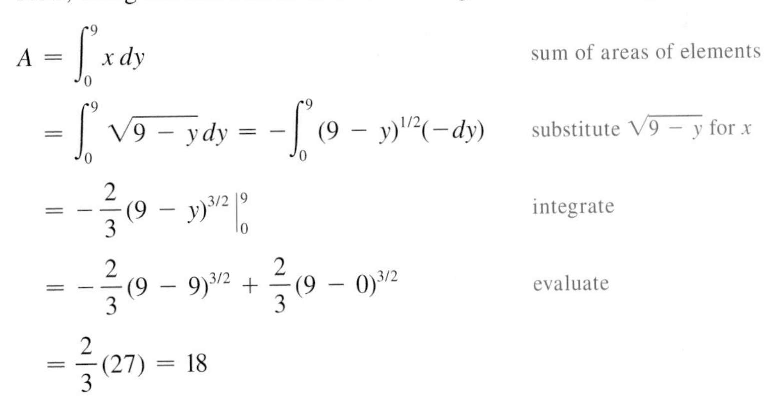

$$\begin{aligned} A &= \int_0^9 x\,dy && \text{sum of areas of elements} \\ &= \int_0^9 \sqrt{9 - y}\,dy = -\int_0^9 (9 - y)^{1/2}(-dy) && \text{substitute } \sqrt{9 - y} \text{ for } x \\ &= -\frac{2}{3}(9 - y)^{3/2}\Bigg|_0^9 && \text{integrate} \\ &= -\frac{2}{3}(9 - 9)^{3/2} + \frac{2}{3}(9 - 0)^{3/2} && \text{evaluate} \\ &= \frac{2}{3}(27) = 18 \end{aligned}$$

Note that the limits for the vertical elements were 0 and 3, while those for the horizontal elements were 0 and 9. These limits are determined by the direction in which the elements are summed. As we have noted, ***vertical elements are summed from left to right, and horizontal elements are summed from bottom to top.*** Doing it in this way means that the summation will be done in a positive direction. ◂◂

CAUTION ▸

The choice of vertical or horizontal elements is determined by (1) which one leads to the simplest solution or (2) the form of the resulting integral. In some problems it makes little difference which is chosen. However, our present methods of integration do not include many types of integrals.

Area Between Two Curves

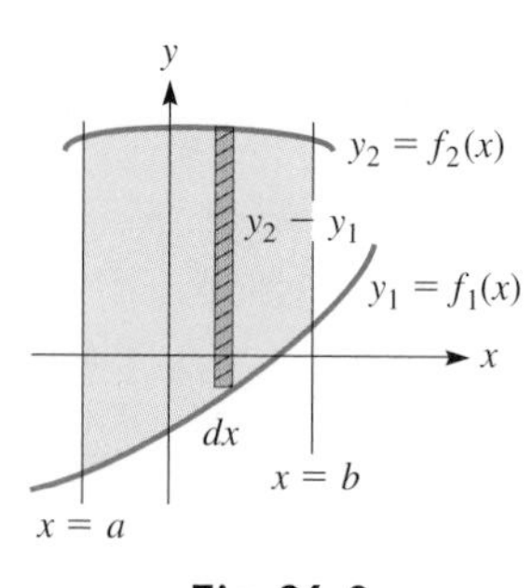

Fig. 26-8

It is also possible to find the area between two curves when one of the curves is not an axis. In such a case, the length of the element of area becomes the difference in the y- or x-coordinates, depending on whether a vertical element or a horizontal element is used.

In Fig. 26-8, by using vertical elements, the element of area is bounded on the bottom by $y_1 = f_1(x)$ and on the top by $y_2 = f_2(x)$. The length of the element is $y_2 - y_1$, and its width is dx. Thus the area is

$$A = \int_a^b (y_2 - y_1)\,dx \tag{26-9}$$

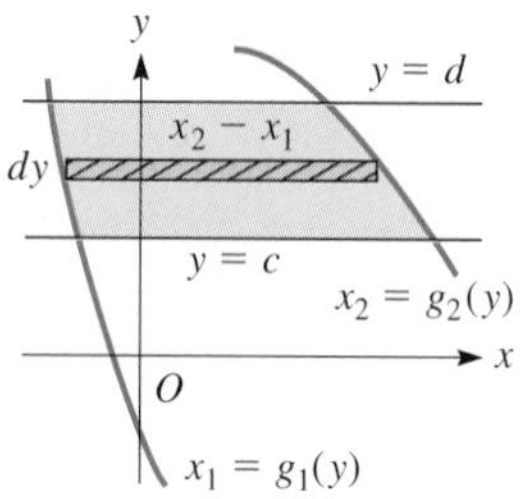

Fig. 26-9

In Fig. 26-9, by using horizontal elements, the element of area is bounded on the left by $x_1 = g_1(y)$ and on the right by $x_2 = g_2(y)$. The length of the element is $x_2 - x_1$, and its width is dy. Thus the area is

$$A = \int_c^d (x_2 - x_1)\,dy \tag{26-10}$$

The following examples show the use of Eqs. (26-9) and (26-10) to find the indicated areas.

EXAMPLE 3 ▸▸ Find the area bounded by $y = x^2$ and $y = x + 2$.

$$x^2 = x + 2$$
$$x^2 - x - 2 = 0$$
$$(x + 1)(x - 2) = 0$$
$$x = -1, 2$$

First, by sketching each curve, we see that the area to be found is that shown in Fig. 26-10. The points of intersection of these curves are found by solving the equations simultaneously. The solution for the x-values is shown at the left. We then find the y-coordinates by substituting into either equation.

Here we choose vertical elements, since they are all bounded at the top by the line $y = x + 2$ and at the bottom by the parabola $y = x^2$. If we were to choose horizontal elements, the bounding curves are different above $(-1, 1)$ from below this point. Choosing horizontal elements would then require two separate integrals for solution. Therefore, using vertical elements, we have

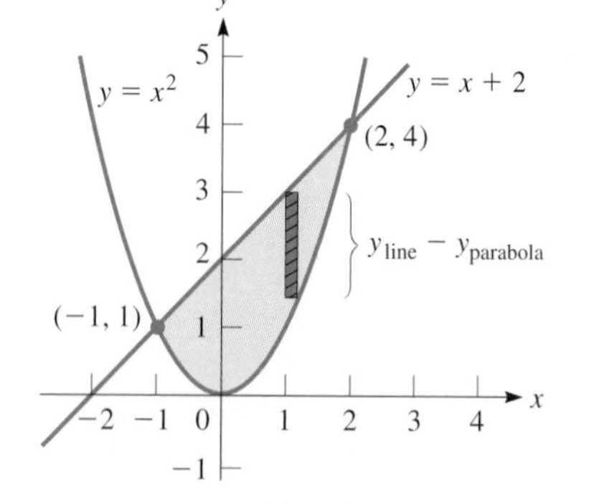

Fig. 26-10

$$A = \int_{-1}^{2} (y_{\text{line}} - y_{\text{parabola}})\,dx \quad \text{using Eq. (26-9)}$$
$$= \int_{-1}^{2} (x + 2 - x^2)\,dx = \left(\frac{x^2}{2} + 2x - \frac{x^3}{3}\right)\Bigg|_{-1}^{2}$$
$$= \left(2 + 4 - \frac{8}{3}\right) - \left(\frac{1}{2} - 2 + \frac{1}{3}\right)$$
$$= \frac{10}{3} + \frac{7}{6} = \frac{27}{6}$$
$$= \frac{9}{2}$$ ◂◂

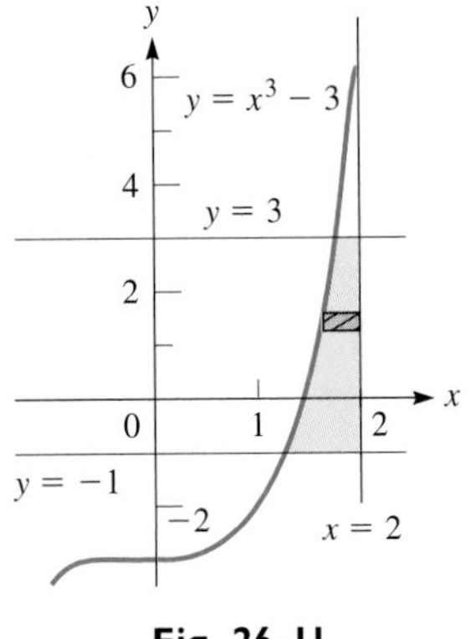

Fig. 26-11

EXAMPLE 4 ▶▶ Find the area bounded by the curve $y = x^3 - 3$ and the lines $x = 2$, $y = -1$, and $y = 3$.

Sketching the curve and lines, we show the area in Fig. 26-11. Horizontal elements are better, since they avoid having to evaluate the area in two parts. Therefore, we have

$$A = \int_{-1}^{3} (x_{\text{line}} - x_{\text{cubic}})\,dy = \int_{-1}^{3} (2 - \sqrt[3]{y+3})\,dy \qquad \text{using Eq. (26-10)}$$

$$= 2y - \frac{3}{4}(y+3)^{4/3}\Big|_{-1}^{3} = \left[6 - \frac{3}{4}(6^{4/3})\right] - \left[-2 - \frac{3}{4}(2^{4/3})\right]$$

$$= 8 - \frac{9}{2}\sqrt[3]{6} + \frac{3}{2}\sqrt[3]{2} = 1.713$$

As we see, the choice of horizontal elements leads to limits of -1 and 3. If we had chosen vertical elements, the limits would have been $\sqrt[3]{2}$ and $\sqrt[3]{6}$ for the area to the left of $(\sqrt[3]{6}, 3)$, and $\sqrt[3]{6}$ and 2 to the right of this point. ◀◀

CAUTION ▶ It is important to set up the element of area so that its length is positive. If the difference is taken incorrectly, the result will show a negative area. ***Getting positive lengths can be ensured for vertical elements if we subtract y of the lower curve from y of the upper curve. For horizontal elements we should subtract x of the left curve from x of the right curve.*** This important point is illustrated in the following example.

EXAMPLE 5 ▶▶ Find the area bounded by $y = x^3 - 3x - 2$ and the x-axis.

Sketching the graph, we find that $y = x^3 - 3x - 2$ has a maximum point at $(-1, 0)$, a minimum point at $(1, -4)$, and an intercept at $(2, 0)$. The graph is shown in Fig. 26-12, and we see that the area is *below* the x-axis. In using vertical elements, we see that the top is the x-axis ($y = 0$) and the bottom is the curve of $y = x^3 - 3x - 2$. Therefore, we have

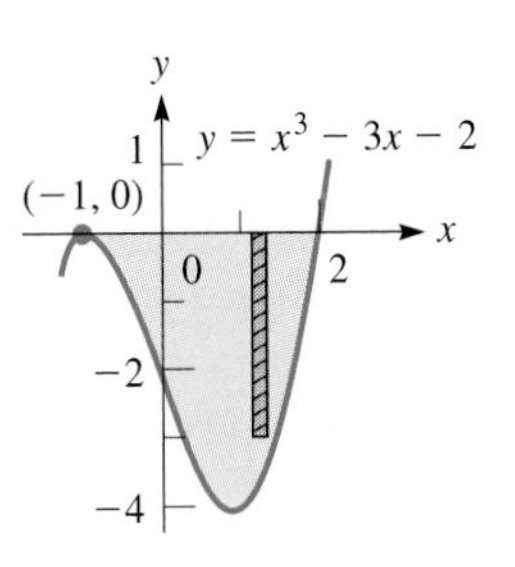

Fig. 26-12

$$A = \int_{-1}^{2} [0 - (x^3 - 3x - 2)]\,dx = \int_{-1}^{2} (-x^3 + 3x + 2)\,dx$$

$$= -\frac{1}{4}x^4 + \frac{3}{2}x^2 + 2x\Big|_{-1}^{2}$$

$$= \left[-\frac{1}{4}(2^4) + \frac{3}{2}(2^2) + 2(2)\right] - \left[-\frac{1}{4}(-1)^4 + \frac{3}{2}(-1)^2 + 2(-1)\right]$$

$$= \frac{27}{4} = 6.75$$

If we had simply set up the area as $A = \int_{-1}^{2} (x^3 - 3x - 2)\,dx$, we would have found $A = -6.75$. The negative sign shows that the area is below the x-axis. Again, we avoid any complications with negative areas by making the length of the element positive.

Also note that since

$$0 - (x^3 - 3x - 2) = -(x^3 - 3x - 2)$$

an area bounded on top by the x-axis can be found by setting up the area as being "under" the curve and using the negative of the function. ◀◀

We must be particularly careful if the bounding curves of an area cross. In such a case, for part of the area one curve is above the area, and for a different part of the area this curve is below the area. When this happens, the area can be found by using two integrals. The following example illustrates the necessity of using this procedure.

EXAMPLE 6 ▸▸ Find the area between $y = x^3 - x$ and the x-axis.

We note from Fig. 26-13 that the area to the left of the origin is above the axis and the area to the right is below. If we find the area from

$$A = \int_{-1}^{1} (x^3 - x)\,dx = \left.\frac{x^4}{4} - \frac{x^2}{2}\right|_{-1}^{1}$$
$$= \left(\frac{1}{4} - \frac{1}{2}\right) - \left(\frac{1}{4} - \frac{1}{2}\right) = 0$$

Fig. 26-13

we see that the apparent area is zero. From the figure we know this is not correct. Noting that the y-values (of the area) are negative to the right of the origin, we set up the integrals

$$A = \int_{-1}^{0} (x^3 - x)\,dx + \int_{0}^{1} [0 - (x^3 - x)]\,dx$$
$$= \left.\left(\frac{x^4}{4} - \frac{x^2}{2}\right)\right|_{-1}^{0} - \left.\left(\frac{x^4}{4} - \frac{x^2}{2}\right)\right|_{0}^{1}$$
$$= 0 - \left(\frac{1}{4} - \frac{1}{2}\right) - \left(\frac{1}{4} - \frac{1}{2}\right) + 0 = \frac{1}{2}$$ ◂◂

The area under a curve can be applied to various kinds of functions. This is illustrated in the following example and in some of the exercises that follow.

EXAMPLE 7 ▸▸ Measurements of solar radiation on a particular surface indicated that the rate r, in joules per hour, at which solar energy is received during the day is given by the equation $r = 3600(12t^2 - t^3)$, where t is the time in hours. Since r is a rate, we may write $r = dE/dt$, where E is the energy, in joules, received at the surface. This means that $dE = 3600(12t^2 - t^3)\,dt$, and we can find the total energy by evaluating the definite integral

$$E = 3600\int_{0}^{12} (12t^2 - t^3)\,dt$$

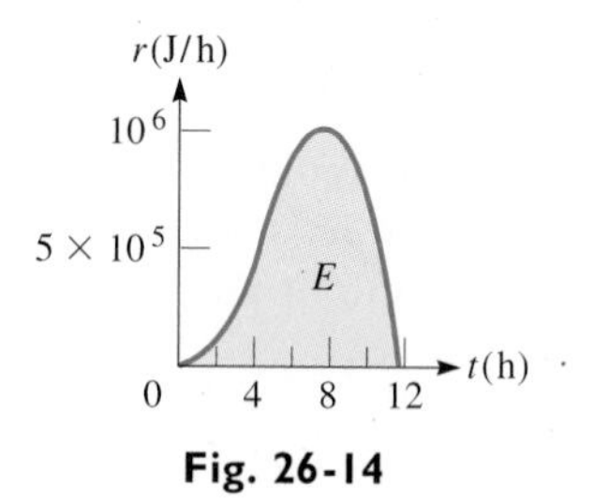

Fig. 26-14

This integral can be interpreted as being the area under $f(t) = 3600(12t^2 - t^3)$ from $t = 0$ to $t = 12$, as shown in Fig. 26-14 (the limits of integration are found from the t-intercepts of the curve). Evaluating this integral, we have

$$E = 3600\int_{0}^{12} (12t^2 - t^3)\,dt = 3600\left.\left(4t^3 - \frac{1}{4}t^4\right)\right|_{0}^{12}$$
$$= 3600\left[4(12^3) - \frac{1}{4}(12^4) - 0\right]$$
$$= 6.22 \times 10^6 \text{ J}$$

Therefore, 6.22 MJ of energy were received in 12 h. ◂◂

Exercises 26–2

In Exercises 1–28, find the areas bounded by the indicated curves.

1. $y = 4x, y = 0, x = 1$
2. $y = 4 - 2x, y = 0, x = 0$
3. $y = x^2, y = 0, x = 2$
4. $y = 3x^2, y = 0, x = 3$
5. $y = 6 - 4x, x = 0, y = 0, y = 3$
6. $y = 8x, x = 0, y = 4$
7. $y = x^2 + 2, x = 0, y = 4 \quad (x > 0)$
8. $y = x^3, x = 0, y = 3$
9. $y = x^2 - 4, y = 0, x = 4$
10. $y = x^2 - 2x, y = 0$
11. $y = x^{-2}, y = 0, x = 2, x = 3$
12. $y = 16 - x^2, y = 0, x = 1, x = 2$
13. $y = \sqrt{x}, x = 0, y = 1, y = 3$
14. $y = 2\sqrt{x + 1}, x = 0, y = 4$
15. $y = 2/\sqrt{x}, x = 0, y = 1, y = 4$
16. $x = y^2 - y, x = 0$
17. $y = 4 - 2x, x = 0, y = 0, y = 3$
18. $y = x, y = 2 - x, x = 0$
19. $y = x^2, y = 2 - x, x = 0 \quad (x \geq 0)$
20. $y = x^2, y = 2 - x, y = 1$
21. $y = x^4, y = 8x$
22. $y = x^4 - 8x^2 + 16, y = 16 - x^4$
23. $y = \sqrt{x - 1}, y = 3 - x, y = 0$
24. $y = x^2 + 5x, y = 3 - x^2$
25. $y = x^3, y = x^2 + 4, x = -1$
26. $y = 4x, y = x^3$
27. $y = x^5, x = -1, x = 2, y = 0$
28. $y = x^2 + 2x - 8, y = x + 4$

In Exercises 29–36, some applications of areas are shown.

29. Certain physical quantities are often represented as an area under a curve. By definition, power is defined as the time rate of change of performing work. Thus, $p = dw/dt$, or $dw = p\,dt$. Therefore, if $p = 12t - 4t^2$, find the work (in joules) performed in 3 s by finding the area under the curve of p vs. t. See Fig. 26-15.

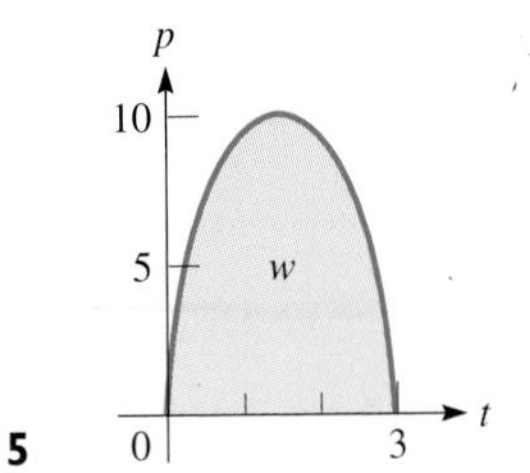

Fig. 26-15

30. The total electric charge Q, in coulombs, to pass a point in the circuit from time t_1 to t_2 is $Q = \int_{t_1}^{t_2} i\,dt$, where i is the current in amperes. Find Q if $t_1 = 1$ s, $t_2 = 4$ s, and $i = 0.0032t\sqrt{t^2 + 1}$.
31. Since the displacement s, velocity v, and time t of a moving object are related by $s = \int v\,dt$, it is possible to represent the change in displacement as an area. A rocket is launched such that its vertical velocity v (in km/s) as a function of time in seconds is $v = 1 - 0.01\sqrt{2t + 1}$. Find the change in vertical displacement from $t = 10$ s to $t = 100$ s.
32. The total cost of production can be interpreted as an area. If the cost per unit C', in dollars per unit, of producing x units is given by $\dfrac{100}{(0.01x + 1)^2}$, find the total cost C of producing 100 units by finding the area under the curve of C' versus x.
33. A cam is designed such that one face of it is described as being the area between the curves $y = x^3 - 2x^2 - x + 2$ and $y = x^2 - 1$, with units in centimetres. Show that this description does not uniquely describe the face of the cam. Find the area of the face of the cam, if a complete description requires that $x \leq 1$.
34. Using CAD (computer-assisted design), an architect programs a computer to sketch the shape of a swimming pool designed between the curves $y = \dfrac{800x}{(x^2 + 10)^2}$, $y = 0.5x^2 - 4x$, and $x = 8$, with dimensions in metres. Find the area of the surface of the pool.
35. A window is designed to be the area between a parabolic section and a straight base, as shown in Fig. 26-16. What is the area of the window?

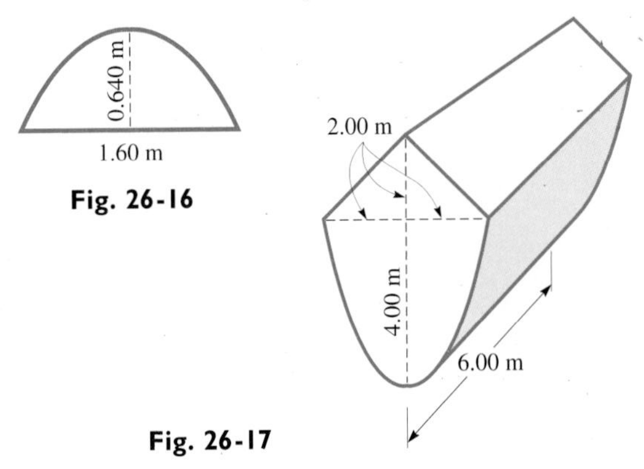

Fig. 26-16

Fig. 26-17

36. The vertical ends of a fuel storage tank have a parabolic bottom section and a triangular top section, as shown in Fig. 26-17. What volume does the tank hold?

26-3 Volumes by Integration

Consider an area and its representative element, as shown in Fig. 26-18(a), to be revolved about the x-axis. When an area is revolved in this manner, it is said to generate a volume, which is also shown in the figure. We shall now show methods of finding volumes that are generated in this way.

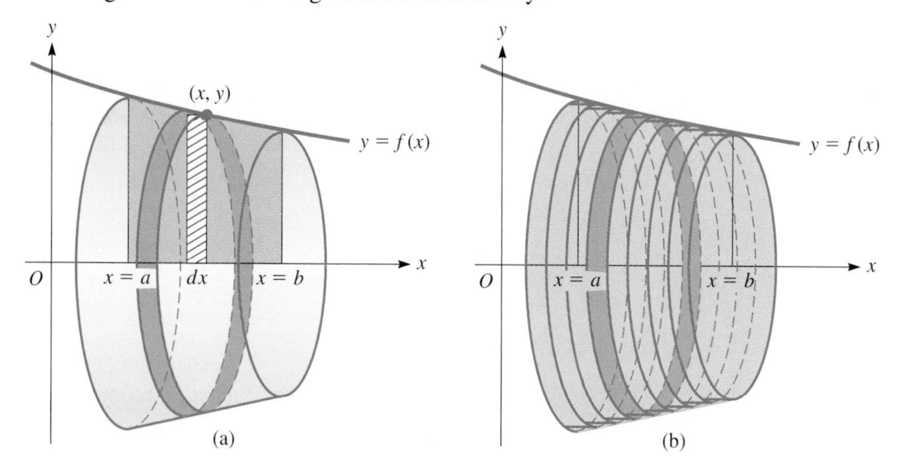

Fig. 26-18

As the area revolves about the x-axis, so does its representative element. The element generates a solid for which the volume is known, namely, a thin cylindrical **disk.** We know that the volume of a right circular cylinder is π times the square of the radius times the height (in this case the thickness) of the cylinder. We must now determine the radius and thickness of the disk. Since the element is revolved about the x-axis, the y-coordinate of the point on the curve that touches the element must represent the radius. Also, the thickness is dx (the disk is on its side). This disk, which is the representative **element of volume,** has a volume of $dV = \pi y^2\,dx$. Summing these elements of volume from left to right, as illustrated in Fig. 25-18(b), we have for the total volume

$$V = \pi\int_a^b y^2\,dx = \pi\int_a^b [f(x)]^2\,dx \qquad (26\text{-}11)$$

The element of volume is a ***disk,*** *and by use of Eq. (26-11) we can find the volume generated by an area bounded by the x-axis, which is revolved about the x-axis.*

EXAMPLE 1 ▸▸ Find the volume generated by revolving the area bounded by $y = x^2$, $x = 2$, and $y = 0$ about the x-axis. See Fig. 26-19.

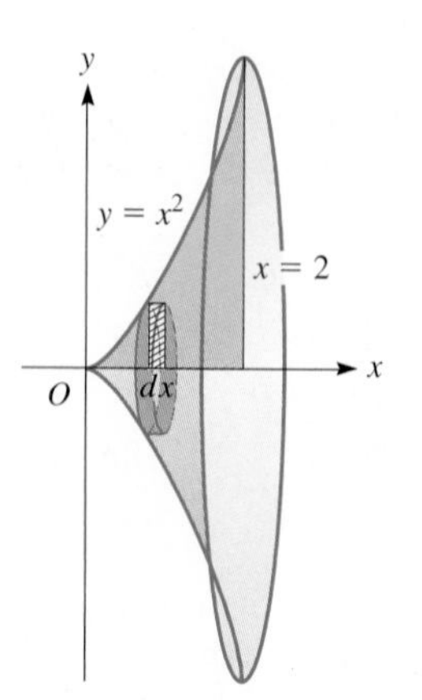

Fig. 26-19

From the figure we see that the radius of the disk is y and its thickness is dx. The elements are summed from left ($x = 0$) to right ($x = 2$).

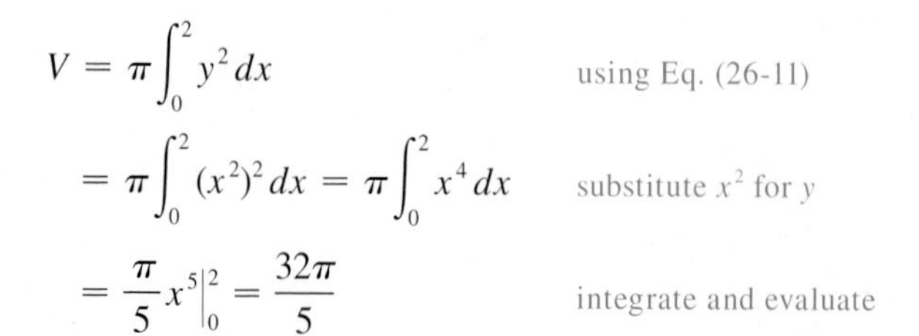

$$V = \pi\int_0^2 y^2\,dx \qquad \text{using Eq. (26-11)}$$

$$= \pi\int_0^2 (x^2)^2\,dx = \pi\int_0^2 x^4\,dx \qquad \text{substitute } x^2 \text{ for } y$$

$$= \frac{\pi}{5}x^5\Big|_0^2 = \frac{32\pi}{5} \qquad \text{integrate and evaluate}$$

Since π is used in Eq. (26-11), it is common to leave results in terms of π. In applied problems, a decimal result would normally be given. ◂◂

If an area bounded by the y-axis is revolved about the y-axis, the volume generated is given by

$$V = \pi \int_c^d x^2\,dy \qquad (26\text{-}12)$$

In this case the radius of the element of volume, a ***disk,*** is the x-coordinate of the point on the curve, and the height of the disk is dy, as shown in Fig. 26-20. One should always be careful to identify the radius and the height properly.

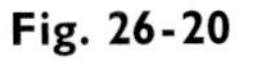

Fig. 26-20

EXAMPLE 2 ▸▸ Find the volume generated by revolving the area bounded by $y = 2x$, $y = 6$, and $x = 0$ about the y-axis.

Figure 26-21 shows the volume to be found. Note that the radius of the disk is x and its thickness is dy.

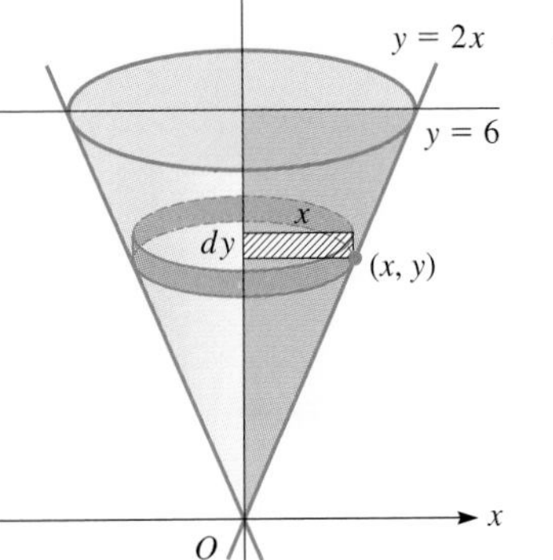

Fig. 26-21

$$V = \pi \int_0^6 x^2\,dy \qquad \text{using Eq. (26-12)}$$

$$= \pi \int_0^6 \left(\frac{y}{2}\right)^2 dy = \frac{\pi}{4}\int_0^6 y^2\,dy \qquad \text{substitute } \frac{y}{2} \text{ for } x$$

$$= \frac{\pi}{12} y^3 \Big|_0^6 = 18\pi \qquad \text{integrate and evaluate}$$

Since this volume is a right circular cone, it is possible to check the result:

$$V = \frac{1}{3}\pi r^2 h = \frac{1}{3}\pi(3^2)(6) = 18\pi$$ ◂◂

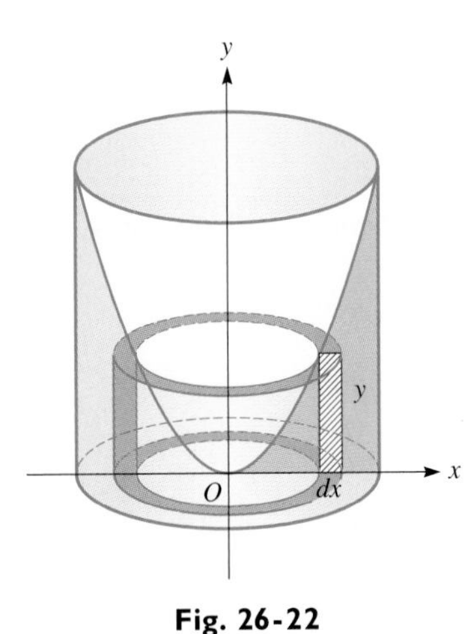

Fig. 26-22

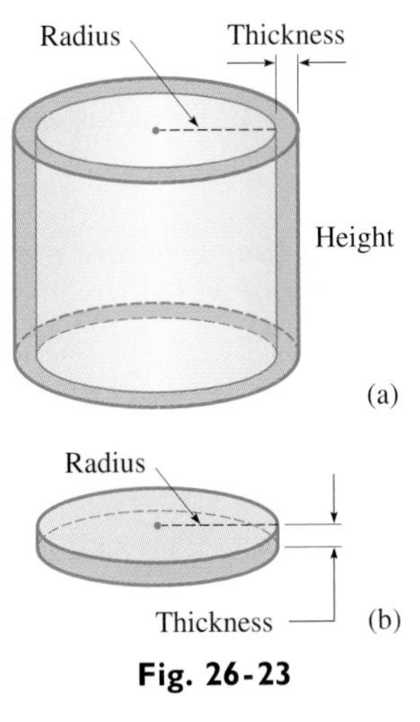

Fig. 26-23

If the area in Fig. 26-19 is revolved about the y-axis, the element of area $y\,dx$ generates a different element of volume from that when it is revolved about the x-axis. In Fig. 26-22, this element of volume is a **cylindrical shell.** *The total volume is made up of an infinite number of concentric shells.* When the volumes of these shells are summed, we have the total volume generated. Thus we must now find the approximate volume dV of the representative shell. By finding the circumference of the base and multiplying this by the height, we obtain an expression for the surface area of the shell. Then, by multiplying this by the thickness of the shell, we find its volume. The volume of the representative ***shell*** shown in Fig. 26-23(a) is

Shell

$$dV = 2\pi(\text{radius}) \times (\text{height}) \times (\text{thickness}) \qquad (26\text{-}13)$$

Similarly, the volume of a ***disk*** is [see Fig. 26-23(b)]

Disk

$$dV = \pi(\text{radius})^2 \times (\text{thickness}) \qquad (26\text{-}14)$$

It is generally better to remember the formulas for the elements of volume in the general forms given in Eqs. (26-13) and (26-14), and not in the specific forms such as Eqs. (26-11) and (26-12) (both of these use *disks*). If we remember the formulas in this way, we can readily apply these methods to finding any such volume of a solid of revolution.

EXAMPLE 3 ▸▸ Use the method of cylindrical shells to find the volume generated by revolving the area bounded by $y = 4 - x^2$, $x = 0$, and $y = 0$ about the y-axis.

From Fig. 26-24, we identify the radius, the height, and the thickness of the shell.

radius $= x$, height $= y$, thickness $= dx$

CAUTION ▸ ***The fact that the elements of area that generate the shells go from $x = 0$ to $x = 2$ determines the limits of integration as 0 and 2.*** Therefore,

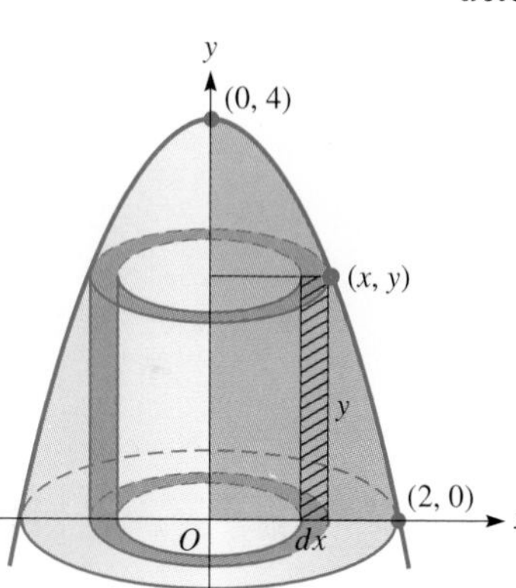

Fig. 26-24

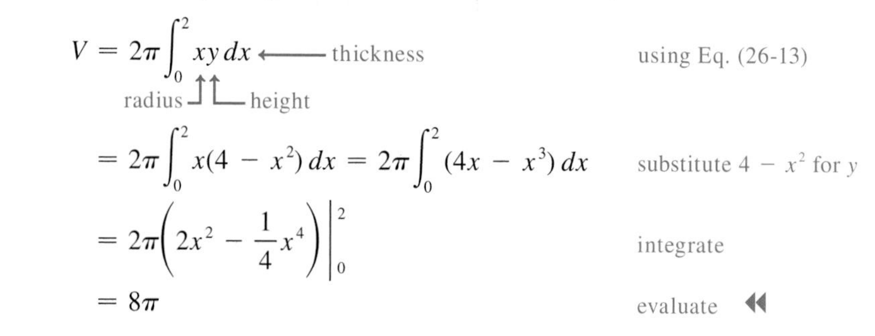

$$V = 2\pi\int_0^2 xy\,dx \quad \text{(radius } x\text{, height } y\text{, thickness } dx) \qquad \text{using Eq. (26-13)}$$

$$= 2\pi\int_0^2 x(4 - x^2)\,dx = 2\pi\int_0^2 (4x - x^3)\,dx \qquad \text{substitute } 4 - x^2 \text{ for } y$$

$$= 2\pi\left(2x^2 - \frac{1}{4}x^4\right)\Bigg|_0^2 \qquad \text{integrate}$$

$$= 8\pi \qquad \text{evaluate} \quad ◂◂$$

We can find the volume shown in Example 3 by using disks, as we show in the following example.

EXAMPLE 4 ▸▸ Use the method of disks to find the volume indicated in Example 3.

From Fig. 26-25, we identify the radius and the thickness of the disk.

radius $= x$, thickness $= dy$

CAUTION ▸ ***Since the elements of area that generate the disks go from $y = 0$ to $y = 4$, the limits of integration are 0 and 4.*** *Thus*

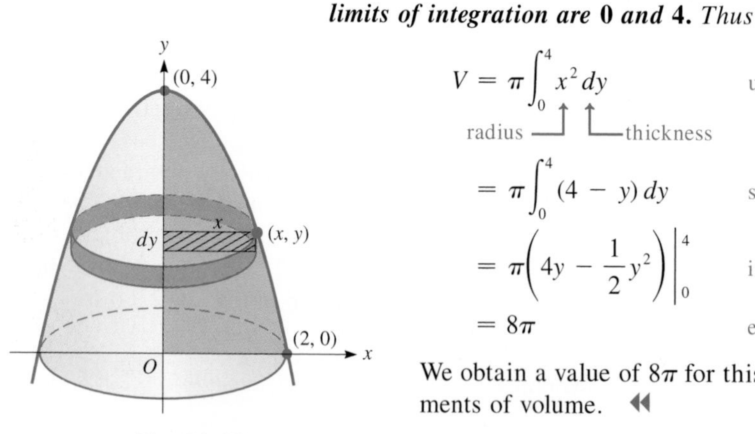

$$V = \pi\int_0^4 x^2\,dy \quad \text{(radius } x\text{, thickness } dy) \qquad \text{using Eq. (26-14)}$$

$$= \pi\int_0^4 (4 - y)\,dy \qquad \text{substitute } \sqrt{4 - y} \text{ for } x$$

$$= \pi\left(4y - \frac{1}{2}y^2\right)\Bigg|_0^4 \qquad \text{integrate}$$

$$= 8\pi \qquad \text{evaluate}$$

Fig. 26-25

We obtain a value of 8π for this volume using either shells or disks as elements of volume. ◂◂

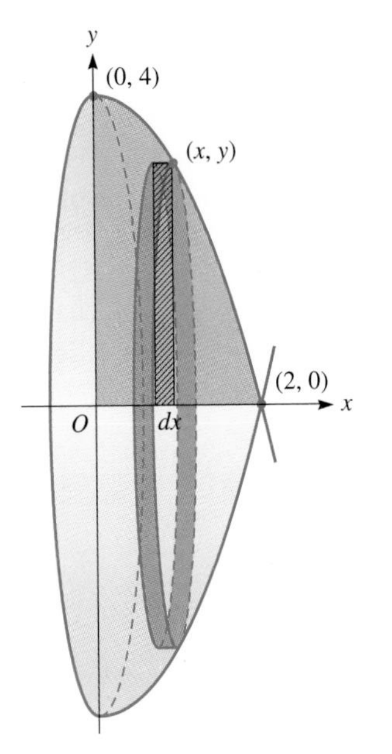

Fig. 26-26

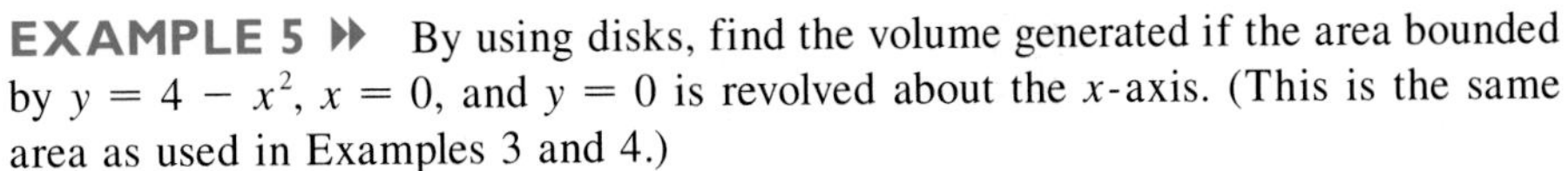

EXAMPLE 5 ▸▸ By using disks, find the volume generated if the area bounded by $y = 4 - x^2$, $x = 0$, and $y = 0$ is revolved about the x-axis. (This is the same area as used in Examples 3 and 4.)

For the disk in Fig. 26-26, we have

$$\text{radius} = y, \qquad \text{thickness} = dx$$

and the limits of integration are $x = 0$ and $x = 2$. This gives us

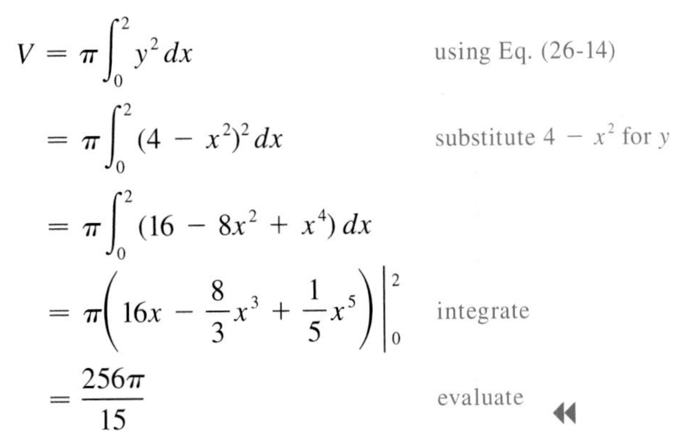

$$\begin{aligned}
V &= \pi \int_0^2 y^2\,dx && \text{using Eq. (26-14)} \\
&= \pi \int_0^2 (4 - x^2)^2\,dx && \text{substitute } 4 - x^2 \text{ for } y \\
&= \pi \int_0^2 (16 - 8x^2 + x^4)\,dx \\
&= \pi\left(16x - \frac{8}{3}x^3 + \frac{1}{5}x^5\right)\Bigg|_0^2 && \text{integrate} \\
&= \frac{256\pi}{15} && \text{evaluate}
\end{aligned}$$

◂◂

We will now show how to set up the integral to find the volume shown in Example 5 by using cylindrical shells. As it turns out, we are not able at this point to integrate the expression that arises, but we are still able to set up the proper integral.

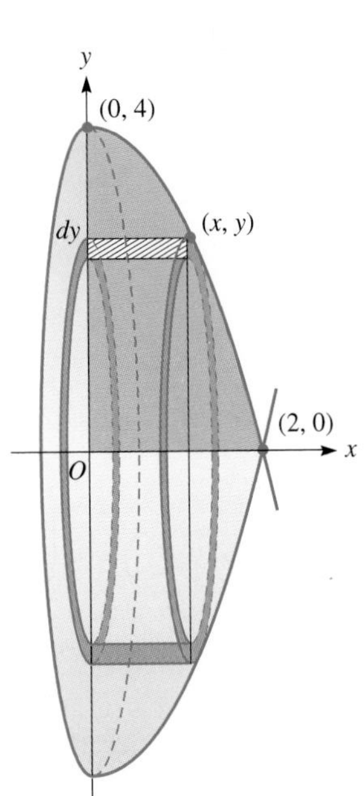

Fig. 26-27

EXAMPLE 6 ▸▸ Use the method of cylindrical shells to find the volume indicated in Example 5.

From Fig. 26-27, we see for the shell we have

$$\text{radius} = y, \qquad \text{height} = x, \qquad \text{thickness} = dy$$

Since the elements go from $y = 0$ to $y = 4$, the limits of integration are 0 and 4. Hence

$$\begin{aligned}
V &= 2\pi \int_0^4 xy\,dy && \text{using Eq. (26-13)} \\
&= 2\pi \int_0^4 \sqrt{4 - y}\,(y\,dy) && \text{substitute } \sqrt{4 - y} \text{ for } x \\
&= \frac{256\pi}{15}
\end{aligned}$$

The method of performing the integration $\int \sqrt{4 - y}\,(y\,dy)$ has not yet been discussed. We present the answer here for the reader's information to show that the volume found in this example is the same as that found in Example 5. ◂◂

In the next example we show how to find the volume generated if an area is revolved about a line other than one of the axes. We will see that a proper choice of the radius, height, and thickness for Eq. (26-13) leads to the result.

EXAMPLE 7 Find the volume generated if the area in Example 3 is revolved about the line $x = 2$.

Shells are convenient, since the volume of a shell can be expressed as a single integral. We can find the radius, height, and thickness of the shell from Fig. 26-28.

CAUTION ▶ We carefully note that ***the radius is not x but 2 − x,*** since the area is revolved about $x = 2$. We see that

$$\text{radius} = 2 - x, \qquad \text{height} = y, \qquad \text{thickness} = dx$$

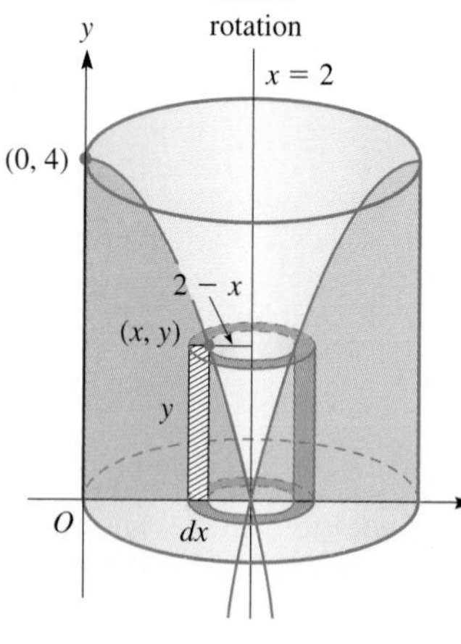

Fig. 26-28

Since the elements that generate the shells go from $x = 0$ to $x = 2$, the limits of integration are 0 and 2. This means we have

$$V = 2\pi \int_0^2 (2 - x)y\,dx \quad \text{(radius: } 2-x\text{; height: } y\text{; thickness: } dx\text{)} \qquad \text{using Eq. (25-13)}$$

$$= 2\pi \int_0^2 (2 - x)(4 - x^2)\,dx \qquad \text{substitute } 4 - x^2 \text{ for } y$$

$$= 2\pi \int_0^2 (8 - 2x^2 - 4x + x^3)\,dx$$

$$= 2\pi\left(8x - \frac{2}{3}x^3 - 2x^2 + \frac{1}{4}x^4\right)\Bigg|_0^2 \qquad \text{integrate}$$

$$= \frac{40\pi}{3} \qquad \text{evaluate}$$

(If the area had been revolved about the line $x = 3$, the only difference in the integral would have been that $r = 3 - x$. Everything else, including the limits, would have remained the same.)

Exercises 26–3

In Exercises 1–12, find the volume generated by the areas bounded by the given curves if they are revolved about the x-axis. Use the indicated method in each case.

1. $y = 1 - x, x = 0, y = 0$ (disks)
2. $y = x, y = 0, x = 2$ (disks)
3. Area of Exercise 1 (shells)
4. $y = \sqrt{x}, x = 0, y = 2$ (shells)
5. $y = 3\sqrt{x}, y = 0, x = 4$ (disks)
6. $y = 2x - x^2, y = 0$ (disks)
7. $y = x^3, y = 8, x = 0$ (shells)
8. $y = x^2, y = x$ (shells)
9. $y = x^2 + 1, x = 0, x = 3, y = 0$ (disks)
10. $y = 6 - x - x^2, x = 0, y = 0$ (quadrant I) (disks)
11. $x = 4y - y^2 - 3, x = 0$ (shells)
12. $y = x^4, x = 0, y = 1, y = 2$ (shells)

In Exercises 13–24, find the volume generated by the areas bounded by the given curves if they are revolved about the y-axis. Use the indicated method in each case.

13. Area of Exercise 1 (disks)
14. $y = x^{1/3}, x = 0, y = 2$ (disks)
15. Area of Exercise 1 (shells)
16. $y = \sqrt{x - 1}, y = 0, x = 3$ (shells)
17. $y = 2\sqrt{x}, x = 0, y = 2$ (disks)
18. $y^2 = x, y = 4, x = 0$ (disks)
19. $x^2 - 4y^2 = 4, x = 3$ (shells)
20. $y = 3x^2 - x^3, y = 0$ (shells)
21. $x = 6y - y^2, x = 0$ (disks)
22. $x^2 + 4y^2 = 4$ (quadrant I) (disks)
23. $y = \sqrt{4 - x^2}$ (quadrant I) (shells)
24. $y = 8 - x^3, x = 0, y = 0$ (shells)

In Exercises 25–32, find the indicated volumes by integration.

25. Find the volume generated if the area of Exercise 6 is revolved about the line $x = 2$.

26. Find the volume generated if the area bounded by $y = \sqrt{x}$ and $y = x/2$ is revolved about the line $y = 4$.

27. Derive the formula for the volume of a right circular cone of radius r and height h by revolving the area bounded by $y = (r/h)x$, $y = 0$, and $x = h$ about the x-axis.

28. Explain how to derive the formula for the volume of a sphere by using the disk method.

29. The capillary tube shown in Fig. 26-29 has circular horizontal cross sections of inner radius 1.1 mm. What is the volume of the liquid in the tube above the level of liquid outside the tube if the top of the liquid in the center vertical cross section is described by the equation $y = x^4 + 1.5$, as shown?

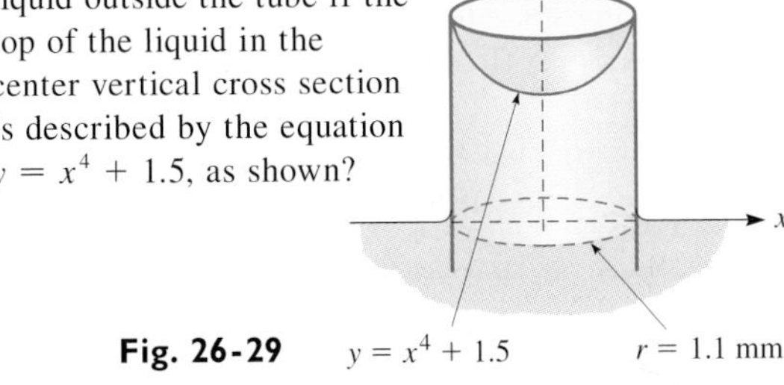

Fig. 26-29 $y = x^4 + 1.5$ $r = 1.1$ mm

30. A commercial dirigible used for outdoor advertising has a helium-filled balloon in the shape of an ellipse revolved about its major axis. If the balloon is 41.3 m long and 12.0 m in diameter, what volume of helium is required to fill it? See Fig. 26-30.

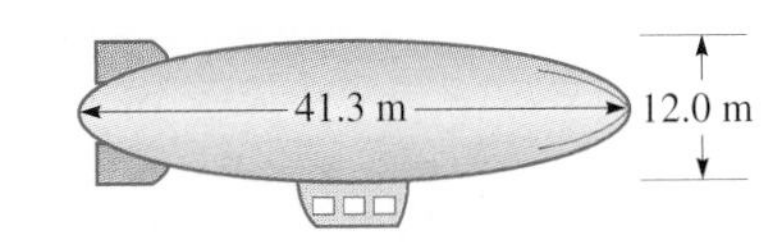

Fig. 26-30

31. A hole 2.00 cm in diameter is drilled through the center of a spherical lead weight 6.00 cm in diameter. How much lead is removed?

32. All horizontal cross sections of a keg 1.2 m tall are circular, and the sides of the keg are parabolic. The diameter at the top and the bottom is 0.80 m, and the diameter in the middle is 1.0 m. Find the volume that the keg holds.

26-4 Centroids

In the study of mechanics, a very important property of an object is its center of mass. In this section we explain the meaning of center of mass and then show how integration is used to determine the center of mass for areas and solids of rotation.

If a mass m is at a distance d from a specified point O, the **moment** *of the mass about O is defined as md.* If several masses $m_1, m_2, \ldots, m_n$ are at distances $d_1, d_2, \ldots, d_n$, respectively, from point O, their moment (as a group) about O is defined as $m_1d_1 + m_2d_2 + \cdots + m_nd_n$. If all the masses could be concentrated at one point $\overline{d}$ units from O, the moment would be $(m_1 + m_2 + \cdots + m_n)\overline{d}$, and this is what is meant by $m_1d_1 + m_2d_2 + \cdots + m_nd_n$. Therefore, we may write

$$m_1d_1 + m_2d_2 + \cdots + m_nd_n = (m_1 + m_2 + \cdots + m_n)\overline{d} \qquad \text{(26-15)}$$

In Eq. (26-15), $\overline{d}$ is the distance from O to the **center of mass.** The moment of a mass is a measure of its tendency to rotate about a point. A weight far from the point of balance of a long rod is more likely to make the rod turn than if the same weight were placed near the point of balance. It is easier to open a door if you push near the doorknob than if you push near the hinges. This is the type of physical property that the moment of mass measures.

EXAMPLE 1 ▸▸ One of the simplest and most basic illustrations of moments and center of mass is seen in balancing a long rod with masses of different sizes, one on either side of the balance point.

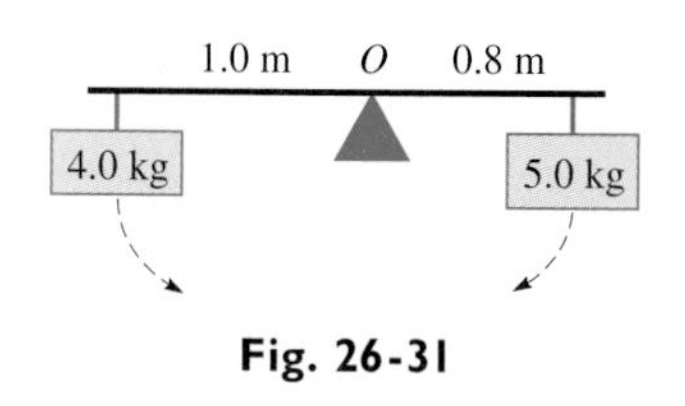

Fig. 26-31

In Fig. 26-31, a mass of 5.0 kg is hung from the rod 0.8 m to the right of point O. We see that this 5.0-kg mass tends to turn the rod clockwise. A mass placed on the opposite side of O will tend to turn the rod counterclockwise. Neglecting the mass of the rod, in order to balance the rod at O, the moments must be equal in magnitude but opposite in sign. Therefore, a 4.0-kg mass would have to be placed 1.0 m to the left.

Thus, with $d_1 = 0.8$ m, $d_2 = -1.0$ m, we see that

$$(5.0 + 4.0)\overline{d} = 5.0(0.8) + 4.0(-1.0) = 4.0 - 4.0$$
$$\overline{d} = 0.0 \text{ m}$$

CAUTION ▶ The center of mass of the combination of the 5.0-kg mass and the 4.0-kg mass is at O. Also, note that ***we must use* directed distances *in finding moments.*** ◂◂

EXAMPLE 2 ▸▸ On the x-axis a mass of 3 units is placed at $(2, 0)$, another of 6 units at $(5, 0)$, and a third of 7 units at $(6, 0)$. Find the center of mass of these three masses.

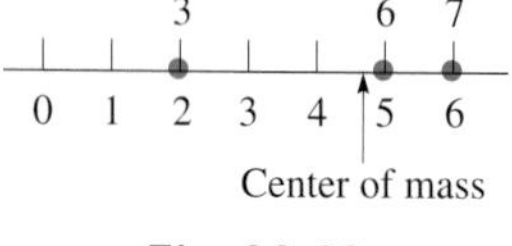

Fig. 26-32

Taking the reference point as the origin, we find $d_1 = 2$, $d_2 = 5$, and $d_3 = 6$. Thus, $m_1d_1 + m_2d_2 + m_3d_3 = (m_1 + m_2 + m_3)\overline{d}$ becomes

$$3(2) + 6(5) + 7(6) = (3 + 6 + 7)\overline{d} \quad \text{or} \quad \overline{d} = 4.88$$

This means that the center of mass of the three objects is at $(4.88, 0)$. Therefore, a mass of 16 units placed at this point has the same moment as the three masses as a unit (see Fig. 26-32). ◂◂

EXAMPLE 3 ▸▸ Find the center of mass of the area shown in Fig. 26-33.

We first note that the center of mass is not *on* either axis. This can be seen from the fact that the major portion of the area is in the first quadrant. *We shall therefore measure the moments with respect to each axis to find the point that is the center of mass. This point is also called the* **centroid** *of the area.*

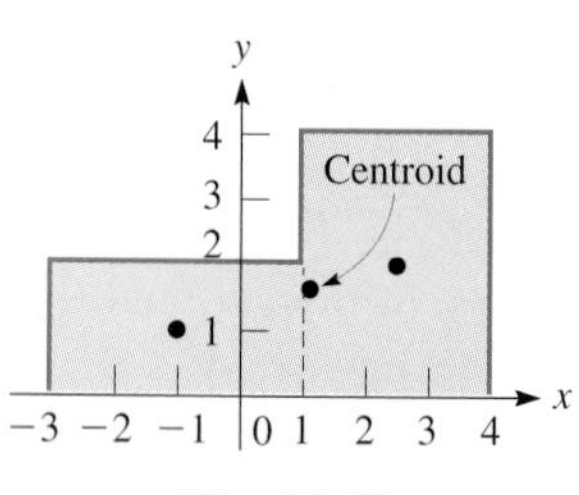

Fig. 26-33

The easiest method of finding the centroid is to divide the area into rectangles, as indicated by the dashed line in Fig. 26-33, and assume that we may consider the mass of each rectangle to be concentrated at its center. In this way the left rectangle has its center at $(-1, 1)$ and the right rectangle has its center at $(\frac{5}{2}, 2)$. The mass of each rectangle, assumed uniform, is proportional to the area. The area of the left rectangle is 8 units and that of the right rectangle is 12 units. Thus, taking moments with respect to the y-axis, we have

$$8(-1) + 12\left(\frac{5}{2}\right) = (8 + 12)\overline{x}$$

where $\overline{x}$ is the x-coordinate of the centroid. Solving for $\overline{x}$, we have $\overline{x} = \frac{11}{10}$.

Now, taking moments with respect to the x-axis, we have

$$8(1) + 12(2) = (8 + 12)\overline{y}$$

where $\overline{y}$ is the y-coordinate of the centroid. Thus, $\overline{y} = \frac{8}{5}$. This means that the coordinates of the centroid, the center of mass, are $(\frac{11}{10}, \frac{8}{5})$. This may be interpreted as meaning that an area of this shape would balance on a single support under this point. As an approximate check, we note from the figure that this point appears to be a reasonable balance point for the area. ◂◂

Centroid of an Area by Integration

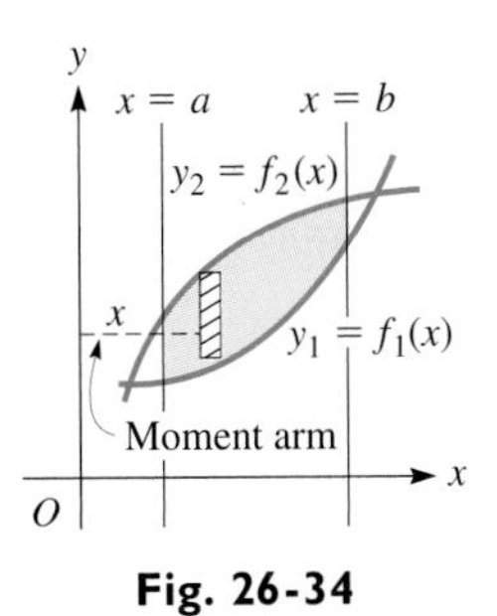

Fig. 26-34

If an area is bounded by the curves of the functions $y_1 = f_1(x)$, $f_2(x)$, $x = a$, and $x = b$, as shown in Fig. 26-34, the moment of the element of area about the y-axis is given by $(k\,dA)x$, where k is the mass per unit area. In this expression, $k\,dA$ is the mass of the element, and x is its distance (moment arm) from the y-axis. The element dA may be written as $(y_2 - y_1)\,dx$, which means that the moment may be written as $kx(y_2 - y_1)\,dx$. If we then sum up the moments of all the elements and express this as an integral (which, of course, means sum), we have $k\int_a^b x(y_2 - y_1)\,dx$. If we consider all the mass of the area to be concentrated at one point $\overline{x}$ units from the y-axis, the moment would be $(kA)\overline{x}$, where kA is the mass of the entire area, and $\overline{x}$ is the distance the center of mass is from the y-axis. By the previous discussion these two expressions should be equal. This means $k\int_a^b x(y_2 - y_1)\,dx = kA\overline{x}$. Since k appears on each side of the equation, we divide it out (we are assuming that the mass per unit area is constant). The area A is found by the integral $\int_a^b (y_2 - y_1)\,dx$. Therefore, the x-coordinate of the centroid is given by

$$\overline{x} = \frac{\displaystyle\int_a^b x(y_2 - y_1)\,dx}{\displaystyle\int_a^b (y_2 - y_1)\,dx} \qquad (26\text{-}16)$$

Equation (26-16) gives us the x-coordinate of the centroid of an area if vertical elements are used. Note that

CAUTION ▶ ***the two integrals in Eq. (26-16) must be evaluated separately.***

We cannot cancel out the apparent common factor $y_2 - y_1$, and we cannot combine quantities and perform only one integration. The two integrals must be evaluated separately first. Then any possible cancellations of factors common to the numerator and the denominator may be made.

Following the same reasoning that we used in developing Eq. (26-16), if an area is bounded by the functions $x_1 = g_1(y)$, $x_2 = g_2(y)$, $y = c$, and $y = d$, as shown in Fig. 26-35, the y-coordinate of the centroid of the area is given by the equation

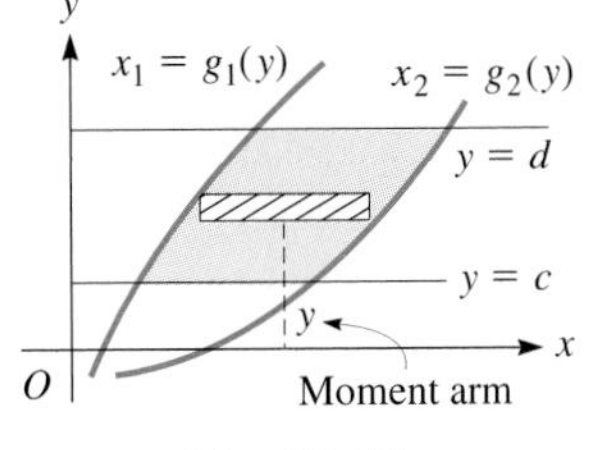

Fig. 26-35

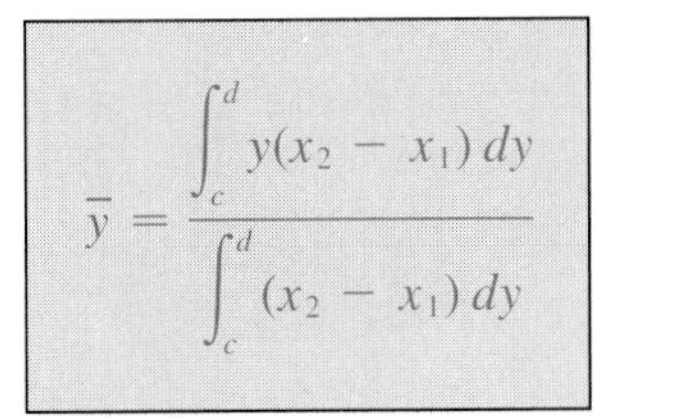

$$\overline{y} = \frac{\displaystyle\int_c^d y(x_2 - x_1)\,dy}{\displaystyle\int_c^d (x_2 - x_1)\,dy} \qquad (26\text{-}17)$$

In this equation, horizontal elements are used.

In applying Eqs. (26-16) and (26-17), we should keep in mind that each denominator of the right-hand sides gives the area and that once we have found this area, we may use it for both $\overline{x}$ and $\overline{y}$. In this way, we can avoid having to set up and perform one of the indicated integrations. Also, in finding the coordinates of the centroid, we should look for and utilize any symmetry a curve may have.

EXAMPLE 4 ▸▸ Find the coordinates of the centroid of the area bounded by the parabola $y = x^2$ and the line $y = 4$.

We sketch a graph indicating the area and an element of area (see Fig. 26-36). The curve is a parabola whose axis is the y-axis. Since the area is symmetrical to the y-axis, the centroid must be on this axis. This means that the x-coordinate of the centroid is zero, or $\bar{x} = 0$. To find the y-coordinate of the centroid, we have

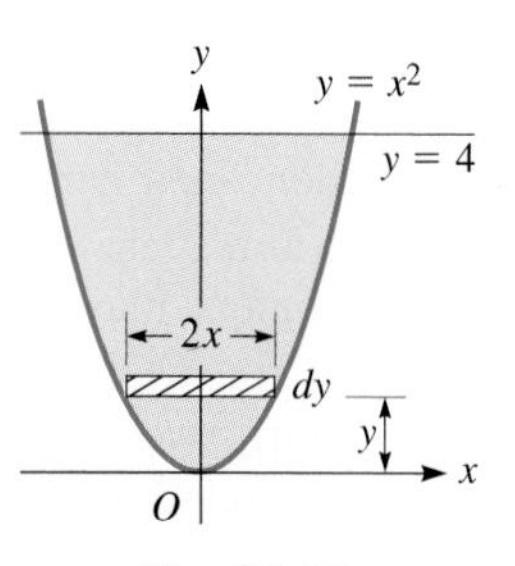

Fig. 26-36

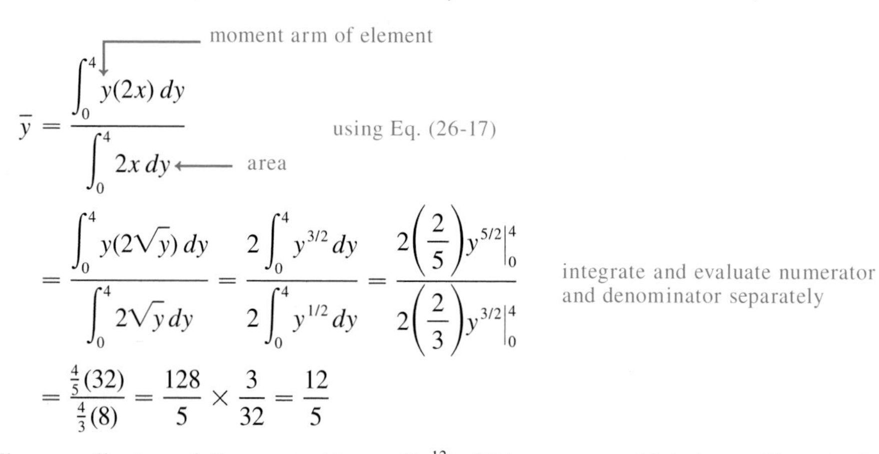

$$\bar{y} = \frac{\int_0^4 y(2x)\,dy}{\int_0^4 2x\,dy} \quad \text{using Eq. (26-17)}$$

(moment arm of element: y; area: $2x\,dy$)

$$= \frac{\int_0^4 y(2\sqrt{y})\,dy}{\int_0^4 2\sqrt{y}\,dy} = \frac{2\int_0^4 y^{3/2}\,dy}{2\int_0^4 y^{1/2}\,dy} = \frac{2\left(\frac{2}{5}\right)y^{5/2}\Big|_0^4}{2\left(\frac{2}{3}\right)y^{3/2}\Big|_0^4} \quad \text{integrate and evaluate numerator and denominator separately}$$

$$= \frac{\frac{4}{5}(32)}{\frac{4}{3}(8)} = \frac{128}{5} \times \frac{3}{32} = \frac{12}{5}$$

The coordinates of the centroid are $(0, \frac{12}{5})$. This area would balance if a single pointed support were to be put under this point. ◂◂

EXAMPLE 5 ▸▸ Find the coordinates of the centroid of an isosceles right triangle with side a.

We must first set up this area in the xy-plane. One choice is to place the triangle with one vertex at the origin and the right angle on the x-axis (see Fig. 26-37). Since each side is a, the hypotenuse passes through the point (a, a). The equation of the hypotenuse is $y = x$. The x-coordinate of the centroid is found by using Eq. (26-16):

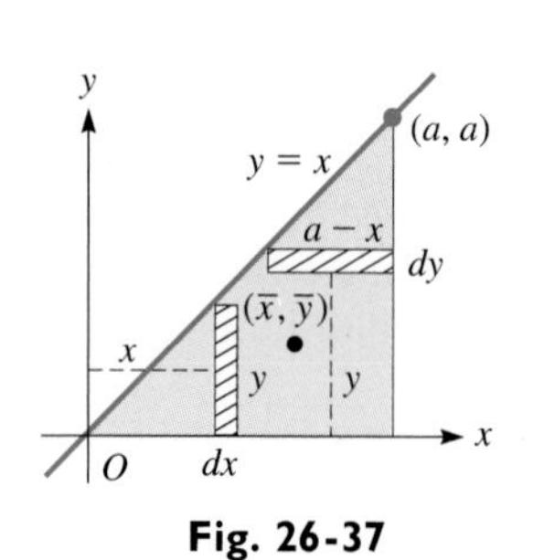

Fig. 26-37

$$\bar{x} = \frac{\int_0^a xy\,dx}{\int_0^a y\,dx} = \frac{\int_0^a x(x)\,dx}{\int_0^a x\,dx} = \frac{\int_0^a x^2\,dx}{\frac{1}{2}x^2\Big|_0^a} = \frac{\frac{1}{3}x^3\Big|_0^a}{\frac{a^2}{2}} = \frac{\frac{a^3}{3}}{\frac{a^2}{2}} = \frac{2a}{3}$$

The y-coordinate of the centroid is found by using Eq. (26-17):

$$\bar{y} = \frac{\int_0^a y(a - x)\,dy}{\frac{a^2}{2}} = \frac{\int_0^a y(a - y)\,dy}{\frac{a^2}{2}} = \frac{\int_0^a (ay - y^2)\,dy}{\frac{a^2}{2}}$$

$$= \frac{\left.\frac{ay^2}{2} - \frac{y^3}{3}\right|_0^a}{\frac{a^2}{2}} = \frac{\frac{a^3}{6}}{\frac{a^2}{2}} = \frac{a}{3}$$

Thus the coordinates of the centroid are $(\frac{2}{3}a, \frac{1}{3}a)$. The results indicate that the center of mass is $\frac{1}{3}a$ units from each of the equal sides. ◂◂

Centroid of a Solid of Revolution

Another figure for which we wish to find the centroid is a solid of revolution. If the density of the solid is constant, the centroid is on the axis of revolution. The problem that remains is to find just where on the axis the centroid is located.

If a given area bounded by the x-axis, as shown in Fig. 26-38, is revolved about the x-axis, a vertical element of area generates a disk element of volume. The center of mass of the disk is at its center, and therefore, for purposes of finding moments, we may consider its mass concentrated there. The moment about the y-axis of a typical element is $x(k)(\pi y^2\,dx)$, where x is the moment arm, k is the density, and $\pi y^2\,dx$ is the volume. The sum of the moments of the elements can be expressed as an integral; it equals the volume times the density times the x-coordinate of the centroid of the volume. Since the density k and the factor of π would appear on each side of the equation, they need not be written. Therefore,

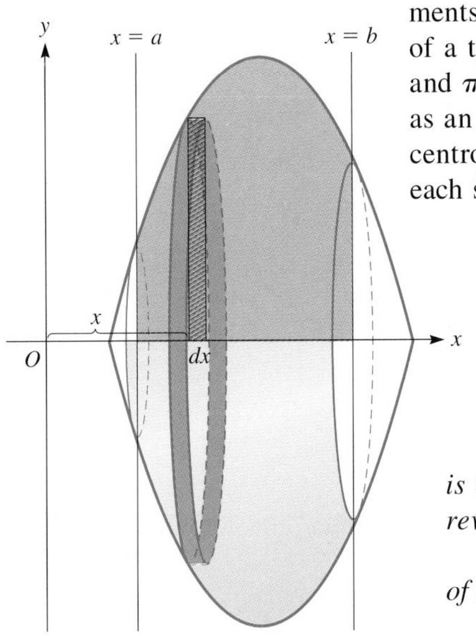

Fig. 26-38

$$\overline{x} = \frac{\int_a^b xy^2\,dx}{\int_a^b y^2\,dx} \qquad (26\text{-}18)$$

is the equation giving the x-coordinate of the centroid of a volume of a solid of revolution about the x-axis.

In the same manner, we may find *the y-coordinate of the centroid of a volume of a solid of revolution about the y-axis. It is*

$$\overline{y} = \frac{\int_c^d yx^2\,dy}{\int_c^d x^2\,dy} \qquad (26\text{-}19)$$

EXAMPLE 6 ▸▸ Find the coordinates of the centroid of the volume generated by revolving the first-quadrant area under the curve of $y = 4 - x^2$ about the x-axis.

Since the curve (see Fig. 26-39) is rotated about the x-axis, the centroid is on the x-axis, which means that $\overline{y} = 0$. We find the x-coordinate as follows:

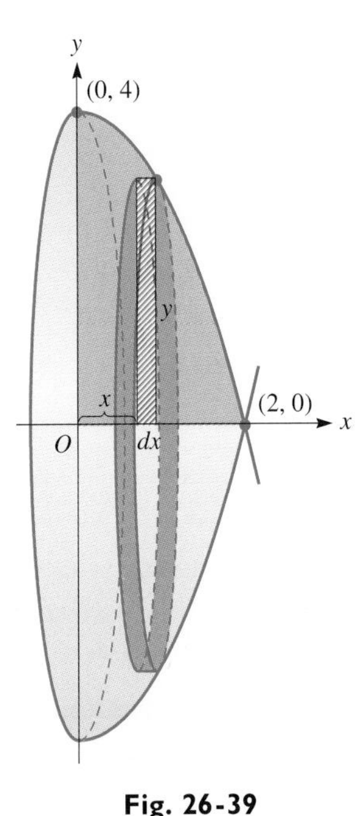

Fig. 26-39

$$\overline{x} = \frac{\int_0^2 xy^2\,dx}{\int_0^2 y^2\,dx} \qquad \text{(moment arm: } x\text{)} \qquad \text{using Eq. (26-18)}$$

$$= \frac{\int_0^2 x(4 - x^2)^2\,dx}{\int_0^2 (4 - x^2)^2\,dx} = \frac{\int_0^2 (16x - 8x^3 + x^5)\,dx}{\int_0^2 (16 - 8x^2 + x^4)\,dx}$$

$$= \frac{8x^2 - 2x^4 + \frac{1}{6}x^6\Big|_0^2}{16x - \frac{8}{3}x^3 + \frac{1}{5}x^5\Big|_0^2} = \frac{32 - 32 + \frac{64}{6}}{32 - \frac{64}{3} + \frac{32}{5}} = \frac{5}{8}$$

The coordinates of the centroid are $(\frac{5}{8}, 0)$. ◂◂

EXAMPLE 7 ▸▸ Find the coordinates of the centroid of the volume generated by revolving the first-quadrant area under the curve of $y = 4 - x^2$ about the y-axis. (This is the same area as in Example 6.)

Since the curve is rotated about the y-axis, $\overline{x} = 0$ (see Fig. 26-40). The y-coordinate is

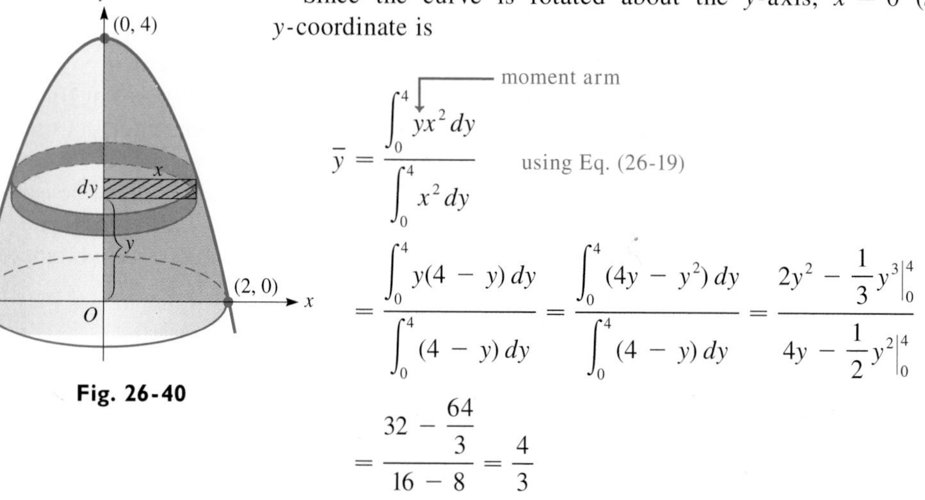

Fig. 26-40

$$\overline{y} = \frac{\int_0^4 yx^2\,dy}{\int_0^4 x^2\,dy} \quad \text{using Eq. (26-19)}$$

(moment arm: y)

$$= \frac{\int_0^4 y(4 - y)\,dy}{\int_0^4 (4 - y)\,dy} = \frac{\int_0^4 (4y - y^2)\,dy}{\int_0^4 (4 - y)\,dy} = \frac{2y^2 - \frac{1}{3}y^3\Big|_0^4}{4y - \frac{1}{2}y^2\Big|_0^4}$$

$$= \frac{32 - \frac{64}{3}}{16 - 8} = \frac{4}{3}$$

The coordinates of the centroid are $(0, \frac{4}{3})$. ◂◂

EXAMPLE 8 ▸▸ Find the centroid of a solid right circular cone of radius a and altitude h.

To generate a right circular cone, we may revolve a right triangle about one of its legs (Fig. 26-41). Placing the leg of length h along the x-axis, we rotate the right triangle whose hypotenuse is given by $y = (a/h)x$ about the x-axis. The x-coordinate of the centroid is

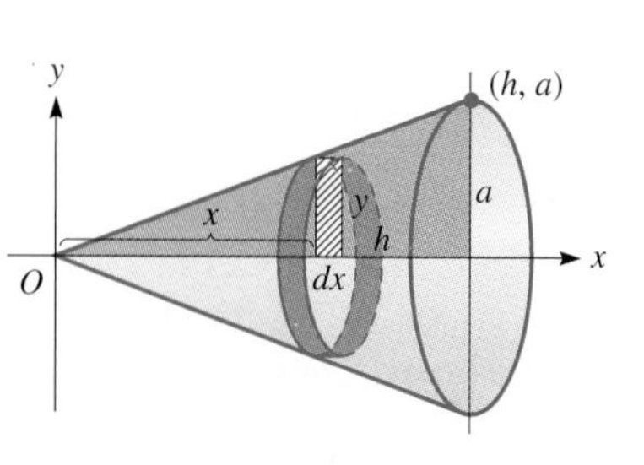

Fig. 26-41

$$\overline{x} = \frac{\int_0^h xy^2\,dx}{\int_0^h y^2\,dx} \quad \text{using Eq. (26-18)}$$

(moment arm: x)

$$= \frac{\int_0^h x\left[\left(\frac{a}{h}\right)x\right]^2 dx}{\int_0^h \left[\left(\frac{a}{h}\right)x\right]^2 dx} = \frac{\left(\frac{a^2}{h^2}\right)\left(\frac{1}{4}x^4\right)\Big|_0^h}{\left(\frac{a^2}{h^2}\right)\left(\frac{1}{3}x^3\right)\Big|_0^h} = \frac{3}{4}h$$

Therefore, the centroid is located along the altitude $\frac{3}{4}$ of the way from the vertex, or $\frac{1}{4}$ of the way from the base. ◂◂

Exercises 26–4

In Exercises 1–4, find the center of mass of the particles of the given masses located at the given points.

1. 5 units at $(1, 0)$, 10 units at $(4, 0)$, 3 units at $(5, 0)$
2. 2 units at $(2, 0)$, 9 units at $(3, 0)$, 3 units at $(8, 0)$, 1 unit at $(12, 0)$
3. 4 units at $(-3, 0)$, 2 units at $(0, 0)$, 1 unit at $(2, 0)$, 8 units at $(3, 0)$
4. 2 units at $(-4, 0)$, 1 unit at $(-3, 0)$, 5 units at $(1, 0)$, 4 units at $(4, 0)$

In Exercises 5–8, find the coordinates of the centroid of the area shown.

5. Fig. 26-42(a) **6.** Fig. 26-42(b)

7. Fig. 26-42(c) **8.** Fig. 26-42(d)

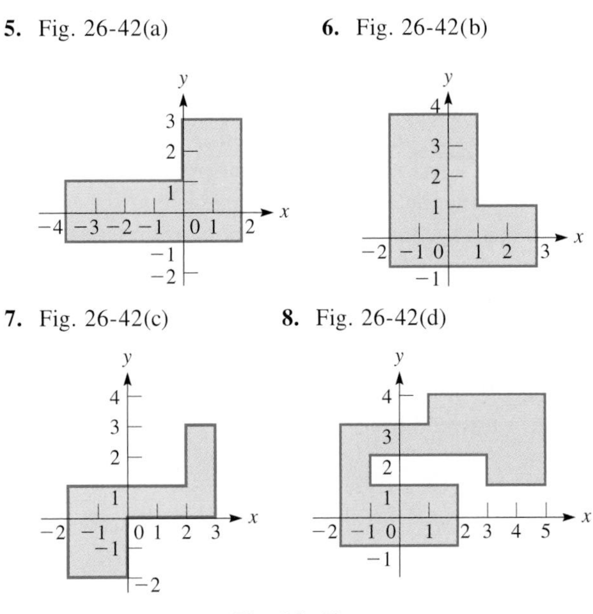

Fig. 26-42

In Exercises 9–28, find the coordinates of the centroids of the given figures.

9. Find the coordinates of the centroid of the area bounded by $y = x^2$ and $y = 2$.

10. Find the coordinates of the centroid of a semicircular area.

11. Find the coordinates of the centroid of the area bounded by $y = 4 - x$ and the axes.

12. Find the coordinates of the centroid of the area bounded by $y = x^3$, $x = 2$, and the x-axis.

13. Find the coordinates of the centroid of the area bounded by $y = x^2$ and $y = x^3$.

14. Find the coordinates of the centroid of the area bounded by $y^2 = x$, $y = 2$, $x = 0$.

15. Find the coordinates of the centroid of the area bounded by $y = 6x - x^2$ and $y = 5$.

16. Find the coordinates of the centroid of the area bounded by $y = x^{2/3}$, $x = 8$, and $y = 0$.

17. Find the coordinates of the centroid of the volume generated by revolving the area bounded by $y = x^3$, $y = 0$, and $x = 1$ about the x-axis.

18. Find the coordinates of the centroid of the volume generated by revolving the area bounded by $y = 2 - 2x$, $x = 0$, and $y = 0$ about the y-axis.

19. Find the coordinates of the centroid of the volume generated by revolving the area in the first quadrant bounded by $y^2 = 4x$, $y = 0$, and $x = 1$ about the y-axis.

20. Find the coordinates of the centroid of the volume generated by revolving the area bounded by $y = x^2$, $x = 2$, and the x-axis about the x-axis.

21. Find the coordinates of the centroid of the volume generated by revolving the area bounded by $y^2 = 4x$ and $x = 1$ about the x-axis.

22. Find the coordinates of the centroid of the volume generated by revolving the area bounded by $x^2 - y^2 = 9$, $y = 4$, and the x-axis about the y-axis.

23. Explain how to find the centroid of a right triangle with legs a and b. Find the location of the centroid.

24. Find the location of the centroid of a hemisphere of radius a.

25. A lens with semielliptical vertical cross sections and circular horizontal cross sections is shown in Fig. 26-43. For proper installation in an optical device, its centroid must be known. Locate its centroid.

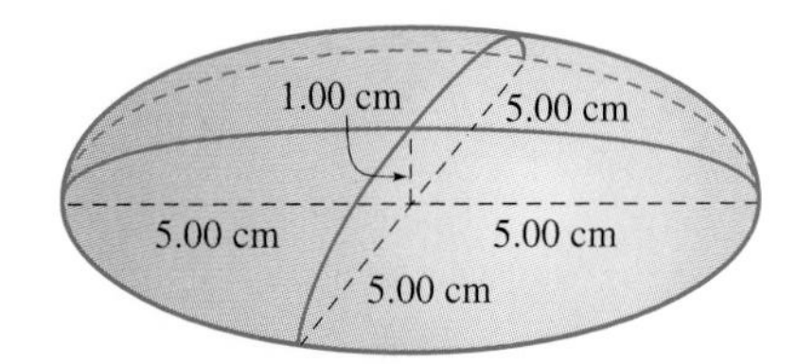

Fig. 26-43

26. A sanding machine disc can be described as the volume generated by rotating the area bounded $y^2 = 4/x$, $y = 1$, $y = 2$, and the y-axis about the y-axis, where measurements are in centimetres. Locate the centroid of the disc.

27. A highway marking pylon has the shape of a frustum of a cone. Find its centroid if the radii of its bases are 5.00 cm and 20.0 cm and the height between bases is 60.0 cm.

28. A floodgate is in the shape of an isosceles trapezoid. Find the location of the centroid of the floodgate if the upper base is 20 m, the lower base is 12 m, and the height between bases is 6.0 m.

26-5 Moments of Inertia

In the discussion of rotational motion in physics, an important quantity is the **moment of inertia** of an object. The moment of inertia of an object rotating about an axis is analogous to the mass of a moving object. In each case, *the moment of inertia or mass is the measure of the tendency of the object to resist a change in motion.*

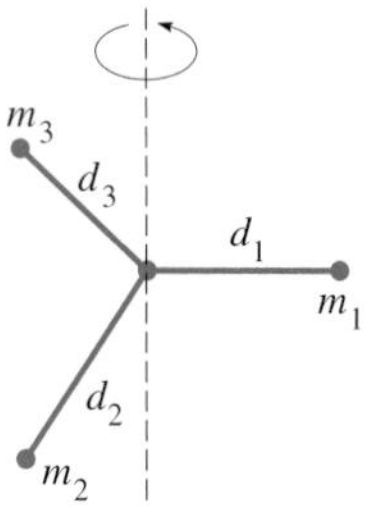

Suppose that a particle of mass m is rotating about some point: We define its moment of inertia as md^2, where d is the distance from the particle to the point. If a group of particles of masses $m_1, m_2, \ldots, m_n$ are rotating about an axis, as shown in Fig. 26-44, the moment of inertia I with respect to the axis of the group is

Axis of rotation

Fig. 26-44

$$I = m_1d_1^2 + m_2d_2^2 + \cdots + m_nd_n^2$$

where the d's are the respective distances of the particles from the axis. If all the masses were at the same distance R from the axis of rotation, so that the total moment of inertia were the same, we would have

$$m_1d_1^2 + m_2d_2^2 + \cdots + m_nd_n^2 = (m_1 + m_2 + \cdots + m_n)R^2 \qquad \text{(26-20)}$$

Radius of Gyration

where R is called the **radius of gyration.**

EXAMPLE 1 ▸▸ Find the moment of inertia and the radius of gyration of three masses, one of 3 units at $(-2, 0)$, the second of 5 units at $(1, 0)$, and the third of 4 units at $(4, 0)$ with respect to the origin. See Fig. 26-45.

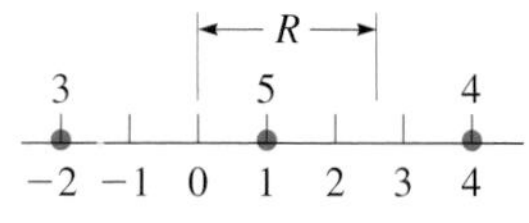

Fig. 26-45

The moment of inertia of the group is

$$I = 3(-2)^2 + 5(1)^2 + 4(4)^2 = 81$$

The radius of gyration is found from $I = (m_1 + m_2 + m_3)R^2$. Thus

$$81 = (3 + 5 + 4)R^2 \qquad R^2 = \frac{81}{12} \quad \text{or} \quad R = 2.60$$

Therefore, a mass of 12 units placed at $(2.60, 0)$ (or $(-2.60, 0)$) has the same rotational inertia about the origin as the three objects as a unit. ◂◂

Moment of Inertia of an Area

If an area is bounded by the curves of the functions $y_1 = f_1(x)$, $y_2 = f_2(x)$ and the lines $x = a$ and $x = b$, as shown in Fig. 26-46, the moment of inertia of this area with respect to the y-axis, I_y, is given by the sum of the moments of inertia of the individual elements. The mass of each element is $k(y_2 - y_1)\,dx$, where k is the mass per unit area and $(y_2 - y_1)\,dx$ is the area of the element. The distance of the element from the y-axis is x. Representing this sum as an integral, we have

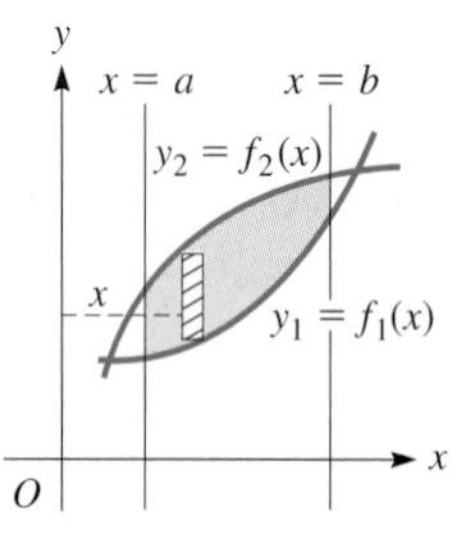

Fig. 26-46

$$I_y = k\int_a^b x^2(y_2 - y_1)\,dx \qquad \text{(26-21)}$$

To find the radius of gyration of the area with respect to the y-axis, R_y, we would first find the moment of inertia, divide this by the mass of the area, and take the square root of this result.

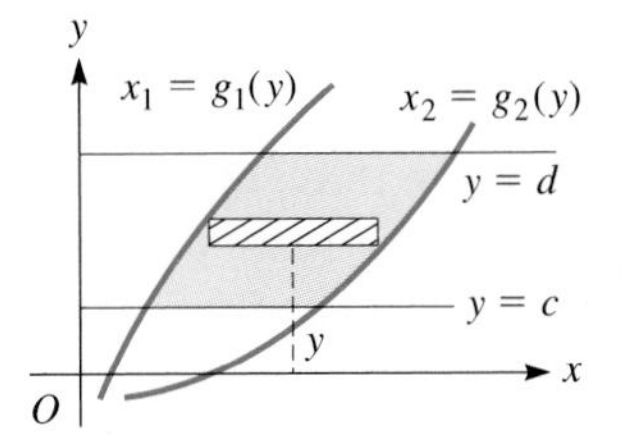

Fig. 26-47

In the same manner, the moment of inertia of an area, with respect to the x-axis, bounded by $x_1 = g_1(y)$ and $x_2 = g_2(y)$ is given by

$$I_x = k\int_c^d y^2(x_2 - x_1)\,dy \tag{26-22}$$

We find the radius of gyration of the area with respect to the x-axis, R_x, in the same manner as we find it with respect to the y-axis (see Fig. 26-47).

EXAMPLE 2 ▸▸ Find the moment of inertia and the radius of gyration of the area bounded by $y = 4x^2$, $x = 1$, and the x-axis with respect to the y-axis.

We find the moment of inertia of this area (see Fig. 26-48) as follows:

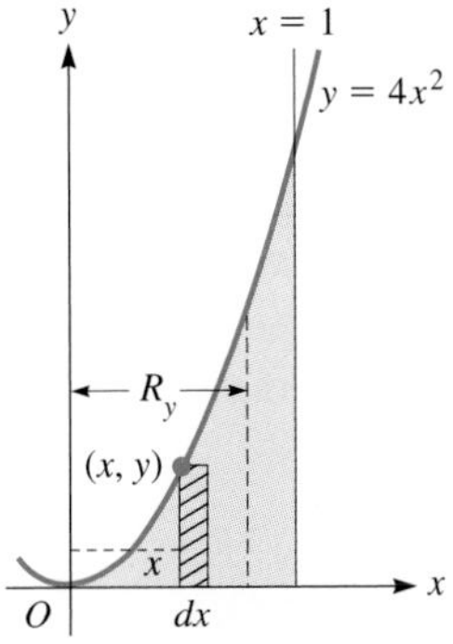

Fig. 26-48

$$I_y = k\int_0^1 x^2 y\,dx \qquad \text{using Eq. (26-21)}$$

(distance from element to axis: x^2)

$$= k\int_0^1 x^2(4x^2)\,dx = 4k\int_0^1 x^4\,dx$$

$$= 4k\left(\frac{1}{5}x^5\right)\Bigg|_0^1 = \frac{4k}{5}$$

To find the radius of gyration, we first determine the mass of the area:

$$m = k\int_0^1 y\,dx = k\int_0^1 (4x^2)\,dx \qquad m = kA$$

$$= 4k\left(\frac{1}{3}x^3\right)\Bigg|_0^1 = \frac{4k}{3}$$

$$R_y^2 = \frac{I_y}{m} = \frac{4k}{5} \times \frac{3}{4k} = \frac{3}{5} \qquad R_y^2 = I_y/m$$

$$R_y = \sqrt{\frac{3}{5}} = \frac{\sqrt{15}}{5} \quad ◂◂$$

EXAMPLE 3 ▸▸ Find the moment of inertia of a right triangle with sides a and b with respect to side b. Assume that $k = 1$.

Placing the triangle as shown in Fig. 26-49, we see that the equation of the hypotenuse is $y = (a/b)x$. The moment of inertia is

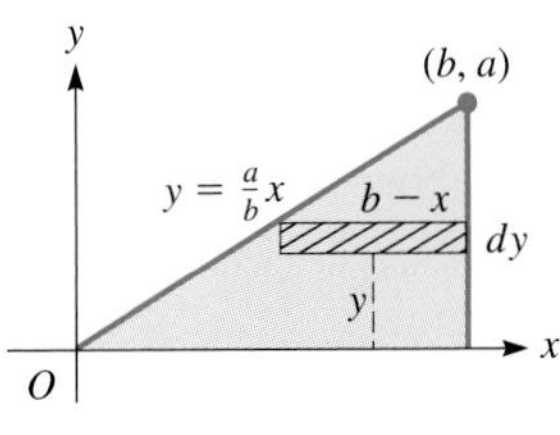

Fig. 26-49

$$I_x = \int_0^a y^2(b - x)\,dy \qquad \text{using Eq. (26-22)}$$

(distance from element to axis: y^2)

$$= \int_0^a y^2\left(b - \frac{b}{a}y\right)dy = b\int_0^a \left(y^2 - \frac{1}{a}y^3\right)dy$$

$$= b\left(\frac{1}{3}y^3 - \frac{1}{4a}y^4\right)\Bigg|_0^a$$

$$= b\left(\frac{a^3}{3} - \frac{a^3}{4}\right) = \frac{ba^3}{12} \quad ◂◂$$

Moment of Inertia of a Volume

CAUTION ▶ In applications, among the most important moments of inertia are those of solids of revolution. Since ***all parts of an element of mass should be at the same distance from the axis,*** the most convenient element of volume to use is the cylindrical shell. In Fig. 26-50, if the area bounded by the curves $y = f(x)$, $x = a$, $x = b$, and the x-axis is revolved about the y-axis, the moment of inertia of the element of volume is $k(2\pi xy\,dx)(x^2)$, where k is the density, $2\pi xy\,dx$ is the volume of the element, and x^2 is the square of its distance from the y-axis. Expressing the sum of the moments of the elements as an integral, *the moment of inertia of the volume with respect to the y-axis, I_y, is*

$$I_y = 2\pi k \int_a^b yx^3\,dx \qquad \textbf{(26-23)}$$

The radius of gyration of the volume with respect to the y-axis, R_y, is found by determining (1) the moment of inertia, (2) the mass of the volume, and (3) the square root of the quotient of the moment of inertia divided by the mass.

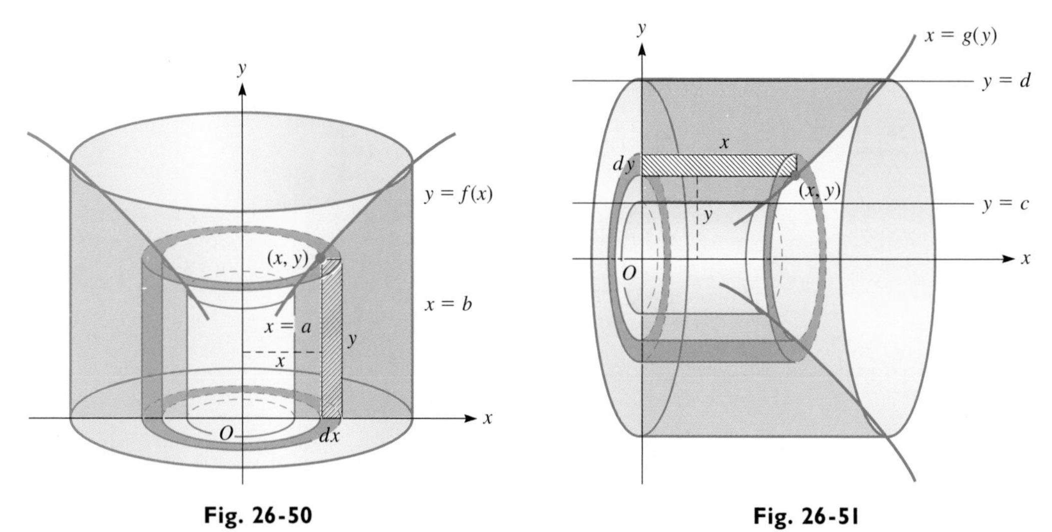

Fig. 26-50 **Fig. 26-51**

The moment of inertia of the volume (see Fig. 26-51) generated by rotating the area bounded by $x = g(y)$, $y = c$, $y = d$, and the y-axis about the x-axis, I_x, is given by

$$I_x = 2\pi k \int_c^d xy^3\,dy \qquad \textbf{(26-24)}$$

The radius of gyration of the volume with respect to the x-axis, R_x, is found in the same manner as R_y.

EXAMPLE 4 ▸▸ Find the moment of inertia and the radius of gyration with respect to the x-axis of the solid generated by revolving the area bounded by the curves of $y^3 = x$, $y = 2$, and the y-axis about the x-axis. See Fig. 26-52.

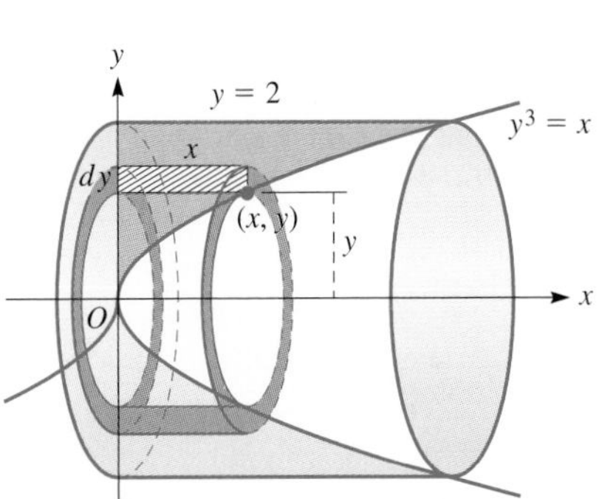

Fig. 26-52

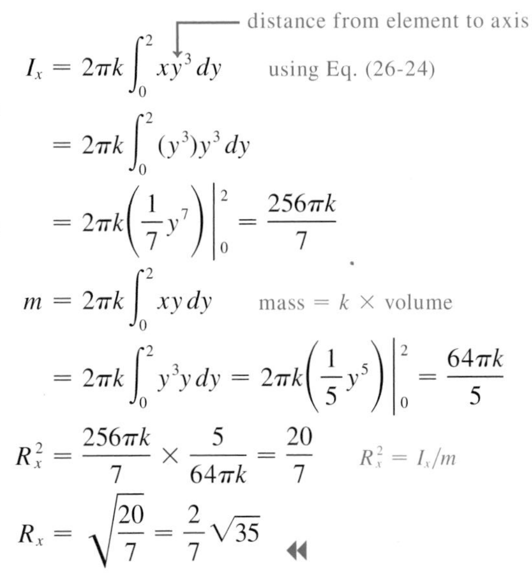

distance from element to axis

$$I_x = 2\pi k \int_0^2 xy^3\,dy \quad \text{using Eq. (26-24)}$$

$$= 2\pi k \int_0^2 (y^3)y^3\,dy$$

$$= 2\pi k \left(\frac{1}{7}y^7\right)\Bigg|_0^2 = \frac{256\pi k}{7}$$

$$m = 2\pi k \int_0^2 xy\,dy \quad \text{mass} = k \times \text{volume}$$

$$= 2\pi k \int_0^2 y^3 y\,dy = 2\pi k\left(\frac{1}{5}y^5\right)\Bigg|_0^2 = \frac{64\pi k}{5}$$

$$R_x^2 = \frac{256\pi k}{7} \times \frac{5}{64\pi k} = \frac{20}{7} \quad R_x^2 = I_x/m$$

$$R_x = \sqrt{\frac{20}{7}} = \frac{2}{7}\sqrt{35}$$ ◂◂

EXAMPLE 5 ▸▸ As noted at the beginning of this section, the moment of inertia is important when studying the rotational motion of an object. For this reason, the moments of inertia of various objects are calculated, and the formulas tabulated. Such formulas are usually expressed in terms of the mass of the object.

Among the objects for which the moment of inertia is important is a solid disk. Find the moment of inertia of a disk with respect to its axis and in terms of its mass.

To generate a disk (see Fig. 26-53), we rotate the area bounded by the axes, $x = r$, and $y = b$, about the y-axis. We then have

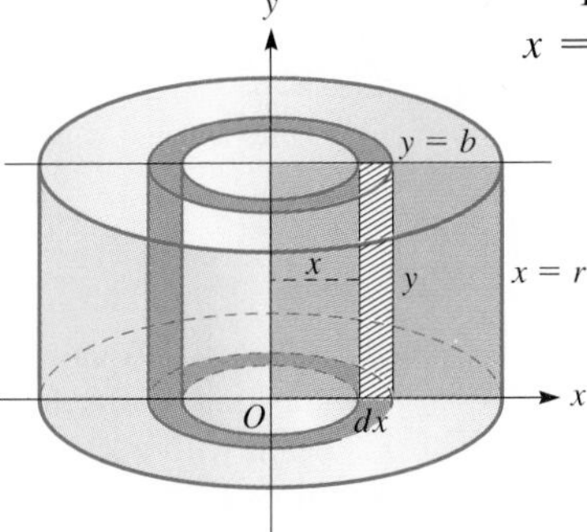

Fig. 26-53

distance from element to axis

$$I_y = 2\pi k \int_0^r x^3 y\,dx \quad \text{using Eq. (26-23)}$$

$$= 2\pi k \int_0^r x^3(b)\,dx = 2\pi kb \int_0^r x^3\,dx$$

$$= 2\pi kb\left(\frac{1}{4}x^4\right)\Bigg|_0^r = \frac{\pi kbr^4}{2}$$

The mass of the disk is $k(\pi r^2)b$. Rewriting the expression for I_y, we have

$$I_y = \frac{(\pi kbr^2)r^2}{2} = \frac{mr^2}{2}$$ ◂◂

Due to the limited methods of integration available at this point, we cannot integrate the expressions for the moments of inertia of circular areas or of a sphere. These will be introduced in Section 28-8 in the exercises, by which point the proper method of integration will have been developed.

Exercises 26–5

In Exercises 1–4, find the moment of inertia and radius of gyration with respect to the origin of the given masses at the given points.

1. 5 units at $(2, 0)$ and 3 units at $(6, 0)$
2. 3 units at $(-1, 0)$, 6 units at $(3, 0)$, 2 units at $(4, 0)$
3. 4 units at $(-4, 0)$, 10 units at $(0, 0)$, 6 units at $(5, 0)$
4. 6 units at $(-3, 0)$, 5 units at $(-2, 0)$, 9 units at $(1, 0)$, 2 units at $(8, 0)$

In Exercises 5–24, find the indicated moment of inertia or radius of gyration.

5. Find the moment of inertia of the area bounded by $y^2 = x, x = 4$, and the x-axis with respect to the x-axis.
6. Find the moment of inertia of the area bounded by $y = 2x$, $x = 1$, $x = 2$, and the x-axis with respect to the y-axis.
7. Find the radius of gyration of the area bounded by $y = x^3, x = 2$, and the x-axis with respect to the y-axis.
8. Find the radius of gyration of the first-quadrant area bounded by $y^2 = 1 - x$ with respect to the x-axis.
9. Find the moment of inertia of a right triangle with sides a and b with respect to side a in terms of the mass of the area.
10. Find the moment of inertia of a rectangular area of sides a and b with respect to side a. Express the result in terms of the mass.
11. Find the radius of gyration of the area bounded by $y = x^2, x = 2$, and the x-axis with respect to the x-axis.
12. Find the radius of gyration of the area bounded by $y^2 = x^3$, $y = 8$, and the y-axis with respect to the y-axis.
13. Find the radius of gyration of the area of Exercise 12 with respect to the x-axis.
14. Find the radius of gyration of the first-quadrant area bounded by $x = 1$, $y = 2 - x$, and the y-axis with respect to the y-axis.
15. Find the moment of inertia with respect to its axis of the solid generated by revolving the area bounded by $y^2 = x$, $y = 2$, and the y-axis about the x-axis.
16. Find the radius of gyration with respect to its axis of the volume generated by revolving the first-quadrant area under the curve $y = 4 - x^2$ about the y-axis.
17. Find the radius of gyration with respect to its axis of the volume generated by revolving the area bounded by $y = 2x - x^2$ and the x-axis about the y-axis.
18. Find the radius of gyration with respect to its axis of the volume generated by revolving the area bounded by $y = 2x$ and $y = x^2$ about the y-axis.
19. Find the moment of inertia in terms of its mass of a right circular cone of radius r and height h with respect to its axis. See Fig. 26-54.

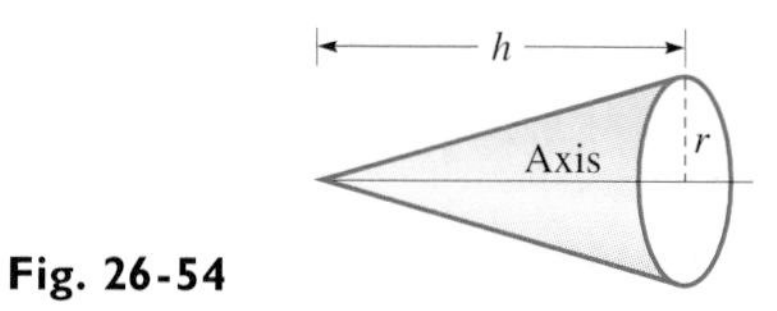

Fig. 26-54

20. Find the moment of inertia in terms of its mass of a circular hoop of radius r and of negligible thickness with respect to its center.
21. A rotating drill head is in the shape of a right circular cone. Find the moment of inertia of the drill head with respect to its axis if its radius is 0.600 cm, its height is 0.800 cm, and its mass is 3.00 g. (See Exercise 19.)
22. Find the moment of inertia (in $kg \cdot m^2$) of a rectangular door 2 m high and 1 m wide with respect to its hinges if $k = 3\ kg/m^2$. (See Exercise 10.)
23. Find the moment of inertia of a flywheel with respect to its axis if its inner radius is 4.0 cm, its outer radius is 6.0 cm, and its mass is 1.2 kg. See Fig. 26-55.

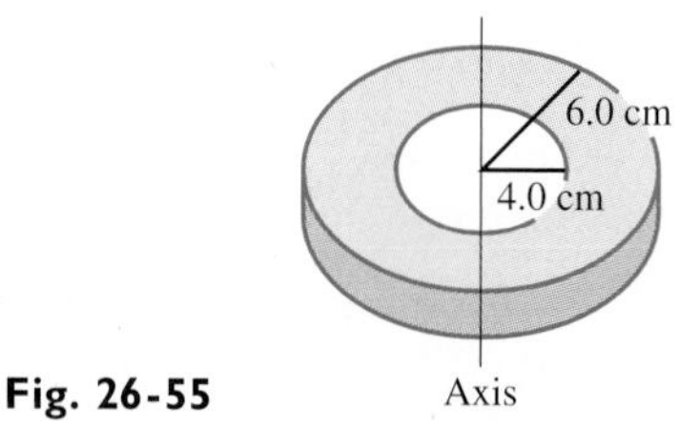

Fig. 26-55

24. A cantilever beam is supported only at its left end, as shown in Fig. 26-56. Explain how to find the formula for the moment of inertia of this beam with respect to a vertical axis through its left end if its length is L and its mass is m. (Consider the mass to be distributed evenly along the beam. This is not an area or volume type of problem.) Find the formula for the moment of inertia.

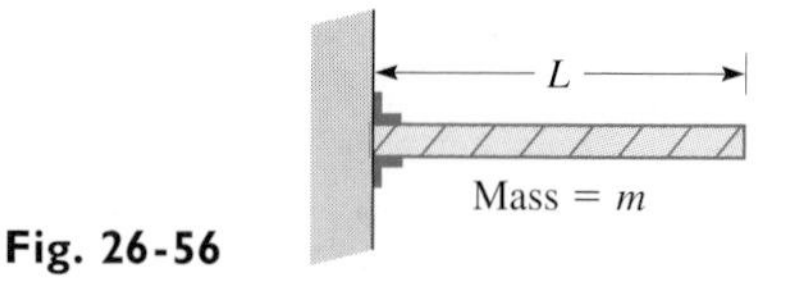

Fig. 26-56

26-6 Other Applications

To illustrate the great variety of applications of integration, we now show three more types of applied problems in the examples of this section. Certain other applications are shown in the exercises.

Work by a Variable Force

In physics, **work** *is defined as the product of a constant force times the distance through which it acts.* When we consider the work done in stretching a spring, the first thing we recognize is that the more the spring is stretched, the greater is the force necessary to stretch it. Thus the force varies. However, if we are stretching the spring a distance Δx, where we are considering the limit as $\Delta x \to 0$, the force can be considered as approaching a constant over Δx. Adding the products of force$_1$ times Δx_1, force$_2$ times Δx_2, and so forth, we see that the total work is the sum of these products. Thus the work can be expressed as a definite integral in the form

$$W = \int_a^b f(x)\,dx \tag{26-25}$$

where $f(x)$ is the force as a function of the distance the spring is stretched. The

CAUTION ▶ ***limits a and b refer to the initial and final distances the spring is stretched* from its normal length.**

One problem remains: We must find the function $f(x)$. From physics we learn that the force required to stretch a spring is proportional to the amount it is stretched (Hooke's law). If a spring is stretched x units from its normal length, then $f(x) = kx$. From conditions stated for a particular spring, the value of k may be determined. Thus, $W = \int_a^b kx\,dx$ is the formula for finding the total work done in stretching a spring.

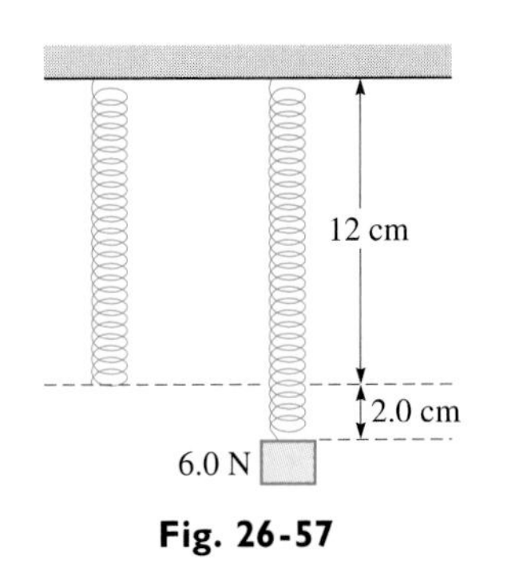

Fig. 26-57

EXAMPLE 1 ▶▶ A spring of natural length 12 cm requires a force of 6.0 N to stretch it 2.0 cm. See Fig. 26-57. Find the work done in stretching it 6.0 cm.

From Hooke's law, we find the constant k for the spring as

$$f(x) = kx, \qquad 6.0 = k(2.0), \qquad k = 3.0 \text{ N/cm}$$

Since the spring is to be stretched 6.0 cm, $a = 0$ (it starts unstretched) and $b = 6.0$ (it is 6.0 cm longer than its normal length). Therefore, the work done in stretching it is

$$W = \int_0^{6.0} 3.0x\,dx = 1.5x^2\Big|_0^{6.0} \quad \text{using Eq. (26-25)}$$
$$= 54 \text{ N}\cdot\text{cm} = 540 \text{ mJ} \quad ◀◀$$

Problems involving work by a variable force arise in many fields of technology. An illustration from electricity is found in the motion of an electric charge through an electric field created by another electric charge.

Electric charges are of two types, designated as positive and negative. A basic law is that charges of the same sign repel each other and charges of opposite signs attract each other. *The force between charges is proportional to the product of their charges, and inversely proportional to the square of the distance between them.*

The force $f(x)$ between electric charges is therefore given by

$$f(x) = \frac{kq_1q_2}{x^2} \tag{26-26}$$

when q_1 and q_2 are the charges (in coulombs), x is the distance (in metres), the force is in newtons, and $k = 9.0 \times 10^9\ \text{N} \cdot \text{m}^2/\text{C}^2$. For other systems of units, the numerical value of k is different. We can find the work done when electric charges move toward each other or when they separate by use of Eq. (26-26) in Eq. (26-25), which can be used in general for the work done by a variable force $f(x)$ acting through the distance from $x = a$ to $x = b$.

EXAMPLE 2 ▸▸ Find the work done when two α-particles, $q = 0.32$ aC each, move until they are 10 nm apart, if they were originally separated by 1.0 m.

From the given information, we have for each α-particle

$$q = 0.32\ \text{aC} = 0.32 \times 10^{-18}\ \text{C} = 3.2 \times 10^{-19}\ \text{C}$$

Since the particles start 1.0 m apart and are moved to 10 nm apart, $a = 1.0$ m and $b = 10 \times 10^{-9}\ \text{m} = 10^{-8}$ m. The work done is

$f(x)$ from Eq. (26-26)

$$W = \int_{1.0}^{10^{-8}} \frac{9.0 \times 10^9(3.2 \times 10^{-19})^2}{x^2}\,dx \quad \text{using Eq. (26-25)}$$

$$= 9.2 \times 10^{-28} \int_{1.0}^{10^{-8}} \frac{dx}{x^2} = 9.2 \times 10^{-28}\left(-\frac{1}{x}\right)\Bigg|_{1.0}^{10^{-8}}$$

$$= -9.2 \times 10^{-28}(10^8 - 1) = -9.2 \times 10^{-20}\ \text{J}$$

Since $10^8 \gg 1$, where $\gg$ means "much greater than," the 1 may be neglected in the calculation. The meaning of the minus sign in the result is that work must be done *on* the system to move the particles together. If free to move, they would tend to separate. ◂◂

The following example illustrates another type of problem in which work by a variable force is involved.

EXAMPLE 3 ▸▸ Find the work done in winding up 60.0 m of a 100-m cable that weighs 4.00 N/m. See Fig. 26-58.

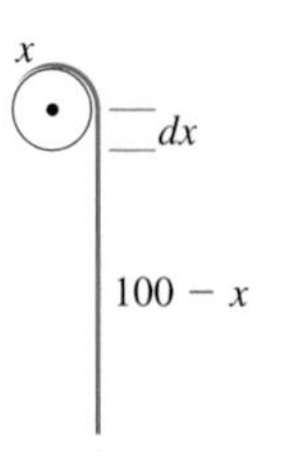

Fig. 26-58

First we let x denote the length of cable that has been wound up at any time. Then the force required to raise the remaining cable equals the weight of the cable that has not yet been wound up. This weight is the product of the unwound cable length, $100 - x$, and its weight per unit length, 4.00 N/m, or

$$f(x) = 4.00(100 - x)$$

Since 60.0 m of cable are to be wound up, $a = 0$ (none is initially wound up) and $b = 60.0$ m. The work done is

$$W = \int_0^{60.0} 4.00(100 - x)\,dx \quad \text{using Eq. (26-25)}$$

$$= \int_0^{60.0} (400 - 4.00x)\,dx = 400x - 2.00x^2\Big|_0^{60.0} = 16\,800\ \text{N} \cdot \text{m}$$ ◂◂

Force Due to Liquid Pressure

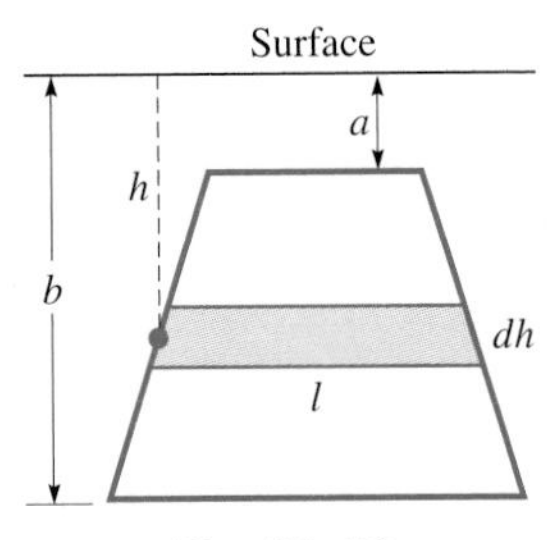

Fig. 26-59

The second application of integration in this section deals with the force due to liquid pressure. The force F on an area A at a depth h in a liquid of density w is $F = whA$. Let us assume that the plate shown in Fig. 26-59 is submerged vertically in water. Using integration to sum the forces on the elements of area, *the total force on the plate is given by*

$$F = w\int_a^b lh\,dh \tag{26-27}$$

Here l is the length of the element of area, h is the depth of the element of area, w is the weight per unit volume of the liquid, a is the depth of the top, and b is the depth of the bottom of the area on which the force is exerted.

See the chapter introduction.

EXAMPLE 4 ▸▸ A vertical floodgate of a dam is 3.00 m wide and 2.00 m high. Find the force on the floodgate if its upper edge is 1.00 m below the surface of the water. See Fig. 26-60.

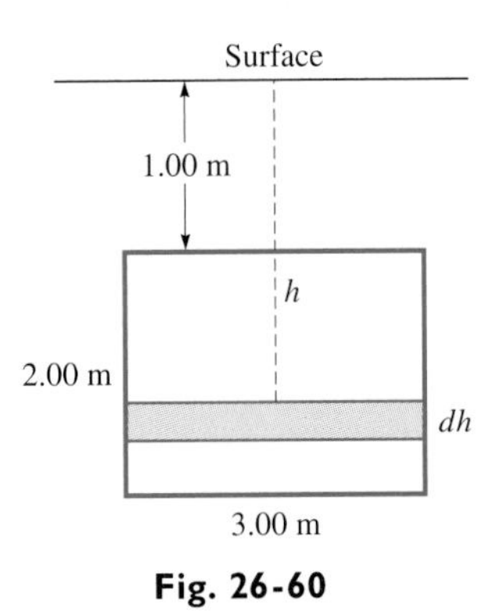

Fig. 26-60

Each element of area of the floodgate has a length of 3.00 m, which means that $l = 3.00$ m. Since the top of the gate is 1.00 m below the surface, $a = 1.00$ m, and since the gate is 2.00 m high, $b = 3.00$ m. Using the weight density as $w = 9800$ N/m^3, we have the force on the gate as

$$F = 9800\int_{1.00}^{3.00} 3.00h\,dh \qquad \text{using Eq. (26-27)}$$

$$= 29\,400\int_{1.00}^{3.00} h\,dh$$

$$= 14\,700h^2\Big|_{1.00}^{3.00} = 14\,700(9.00 - 1.00)$$

$$= 118\,000 \text{ N} = 118 \text{ kN} \quad ◂◂$$

EXAMPLE 5 ▸▸ The vertical end of a tank of water is in the shape of a right triangle as shown in Fig. 26-61. What is the force on the end of the tank?

NOTE ▶ In setting up the figure, it is convenient to use coordinate axes. It is also convenient to have the y-axis directed downward, since we integrate from the top to the bottom of the area. The equation of the line OA is $y = \frac{1}{2}x$. Thus we see that the length of an element of area of the end of the tank is $4.0 - x$, the depth of the element of area is y, the top of the tank is $y = 0$, and the bottom is $y = 2.0$ m. Therefore, the force on the end of the tank is ($w = 9800$ N/m^3)

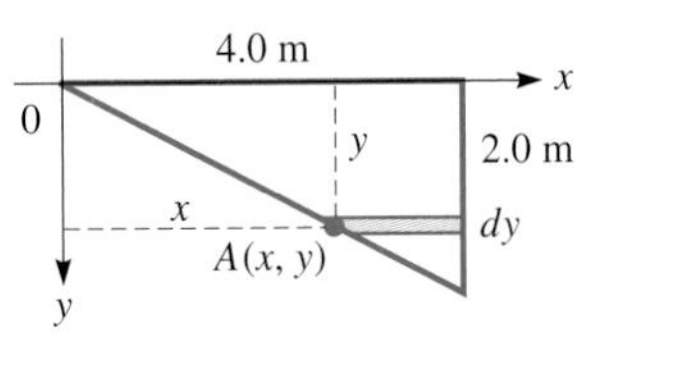

Fig. 26-61

$$F = 9800\int_0^{2.0} \underset{\text{length}}{(4.0 - x)}\,\underset{\text{depth}}{(y)}\,(dy) \qquad \text{using Eq. (26-27)}$$

$$= 9800\int_0^{2.0} (4.0 - 2y)(y\,dy)$$

$$= 19\,600\int_0^{2.0} (2.0y - y^2)\,dy = 19\,600\left(1.0y^2 - \frac{1}{3}y^3\right)\Big|_0^{2.0}$$

$$= 26\,100 \text{ N} \quad ◂◂$$

Average Value of a Function

The third application of integration shown in this section is that of the *average value* of a function. In general, *an average is found by summing up the quantities to be averaged and then dividing by the total number of them.* Generalizing on this and using integration for the summation, *the* **average value** *of a function y with respect to x from x = a to x = b is given by*

$$y_{av} = \frac{\int_a^b y\,dx}{b - a} \tag{26-28}$$

The following examples illustrate the applications of the average value of a function.

EXAMPLE 6 ▸▸ The velocity of an object falling under the influence of gravity as a function of time is given by $v = 9.80t$, where t is the time in seconds. What is the average velocity (in m/s) of the object with respect to time for the first 3.00 s?

In this case, we want the average value of the function v from $t = 0$ to $t = 3.00$ s. This gives us

$$v_{av} = \frac{\int_0^{3.00} v\,dt}{3.00 - 0} \quad \text{using Eq. (26-28)}$$

$$= \frac{\int_0^{3.00} 9.80t\,dt}{3.00} = \frac{4.90t^2}{3.00}\Bigg|_0^{3.00}$$

$$= 14.7 \text{ m/s}$$

This result can be interpreted as meaning that an average velocity of 14.7 m/s for 3.0 s would result in the same distance, 44.1 m, being traveled by the object as that with the variable velocity. Since $s = \int v\,dt$, the numerator represents the distance traveled. ◂◂

EXAMPLE 7 ▸▸ The power P developed in a certain resistor as a function of the current i is $P = 6i^2$. What is the average power (in watts) with respect to the current as the current changes from 2.0 A to 5.0 A?

In this case, we are to find the average value of the function P from $i = 2.0$ A to $i = 5.0$ A. This average value of P is

$$P_{av} = \frac{\int_{2.0}^{5.0} P\,di}{5.0 - 2.0} \quad \text{using Eq. (26-28)}$$

$$= \frac{6\int_{2.0}^{5.0} i^2\,di}{3.0} = \frac{2i^3}{3.0}\Bigg|_{2.0}^{5.0} = \frac{2(125 - 8.0)}{3.0} = 78 \text{ W}$$ ◂◂

In general, it might be noted that the average value of y with respect to x is that value of y which, when multiplied by the length of the interval for x, gives the same area as that under the curve of y as a function of x.

Exercises 26–6

1. The spring of a spring balance is 8.0 cm long when there is no weight on the balance, and it is 9.5 cm long with 6.0 N hung from the balance. How much work is done in stretching it from 8.0 cm to a length of 10.0 cm?

2. A force of 1200 N compresses a spring from its natural length of 18 cm to a length of 16 cm. How much work is done in compressing it from 16 cm to 14 cm?

3. A force F of 25 N on the spring in the lever-spring mechanism shown in Fig. 26-62 stretches the spring by 16 mm. How much work is done by the 25-N force in stretching the spring?

4. An electron has a 1.6×10^{-19} C negative charge. How much work is done in separating two electrons from 1.0 pm to 4.0 pm?

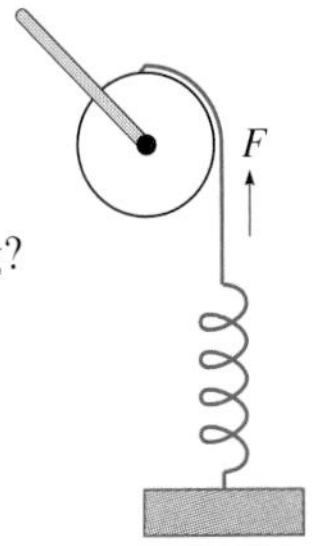

Fig. 26-62

5. How much work is done in separating an electron (see Exercise 4) and an oxygen nucleus, which has a positive charge of 1.3×10^{-18} C, from a distance of 2.0 μm to a distance of 1.0 m?

6. The gravitational force (in N) of attraction between two objects is given by $F = k/x^2$. If two objects are 10 m apart, find the work required to separate them until they are 100 m apart. Express the result in terms of k.

7. Find the work done in winding up 20 m of a 25-m rope on which the force of gravity is 6.00 N/m.

8. Find the work done in winding up a 50.0-m cable that weighs 40.0 N.

9. At lift-off, a rocket weighs 32.5 tonnes, including the weight of its fuel. It is fired vertically, and, during the first stage of ascent, the fuel is consumed at the rate of 1.25 tonnes per 1000 m of ascent. How much work is done in lifting the rocket to an altitude of 12 000 m?

10. While descending, a 550-N weather balloon enters a zone of freezing rain in which ice forms on it at the rate of 7.50 N per 100 m of descent. Find the work done on the balloon during the first 1000 m of descent through the freezing rain.

11. Find the work done in pumping the water out of the top of a cylindrical tank 2.0 m in radius and 4.0 m high, given that water weighs 9.8 kN/m^3. (*Hint:* If horizontal slices dx ft thick are used, each element weighs $9800(4.0\pi\, dx)$ N, and each element must be raised $4.0 - x$ metres, if x is the distance from the base to the element. In this way the force, the weight of the slice, and the distance are determined. Thus the products of force and distance are summed by integration.)

12. A hemispherical tank of radius 3.0 m is full of water. Find the work done in pumping the water to the top of the tank. (See Exercise 11. This problem is similar, except that the weight of each element is $9800\pi(\text{radius})^2(\text{thickness})$, where the radius of each element is different. If we let x be the radius of an element and y be the distance the element must be raised, we have $9800\pi x^2\, dy$, with $x^2 + y^2 = 9.0$.)

13. One end of a spa is a vertical rectangular wall 4.0 m wide. What is the force exerted on this wall by the water if it is 0.80 m deep?

14. Find the force on one side of a cubical container 6.0 cm on an edge if the container is filled with mercury. The weight density of mercury is 133 kN/m^3.

15. A rectangular sea aquarium observation window is 3.00 m wide and 2.00 m high. What is the force on this window if the upper edge is 1.50 m below the surface of the water? The weight density of seawater is 10.1 kN/m^3.

16. A horizontal tank has vertical circular ends, each with a radius of 4.00 m. It is filled to a depth of 4.00 m with oil of weight density 9.40 kN/m^3. Find the force on one end of the tank.

17. A right triangular plate of base 2.0 m and height 1.0 m is submerged vertically, as shown in Fig. 26-63. Find the force on one side of the plate.

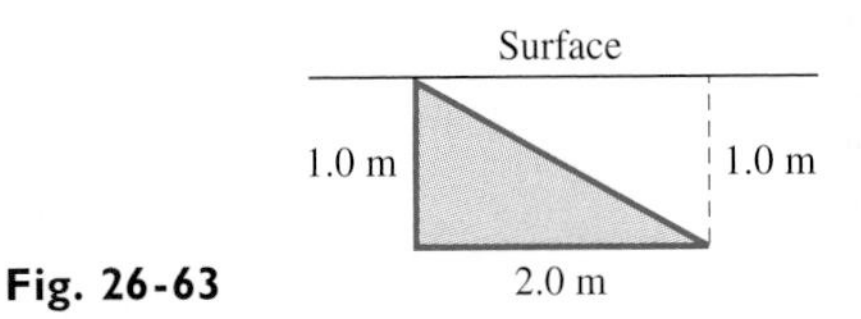

Fig. 26-63

18. Find the force on the plate in Exercise 17 if the top vertex is 3.0 m below the surface.

19. A small dam is in the shape of the area bounded by $y = x^2$ and $y = 20$. Find the force on the area below $y = 4$ if the surface of the water is at the top of the dam. Assume all distances are in metres.

20. A horizontal cylindrical tank has vertical elliptical ends with the minor axis horizontal. The major axis is 2.0 m and the minor axis is 1.3 m. Find the force on one end if the tank is half-filled with fuel oil of weight density 7.8 kN/m^3.

21. The electric current as a function of time for a certain circuit is given by $i = 4t - t^2$. Find the average value of the current (in amperes), with respect to time, for the first 4.0 s.

22. The temperature T (in °C) recorded in a city during a given day approximately followed the curve of $T = 0.001\,00t^4 - 0.280t^2 + 25.0$, where t is the number of hours from noon ($-12 \text{ h} \le t \le 12 \text{ h}$). What was the average temperature during the day?

23. The efficiency e (in percent) of an automobile engine is $e = 0.768s - 0.000\,04s^3$, where s is the speed (in km/h) of the car. Find the average efficiency with respect to the speed for $s = 30.0$ km/h to $s = 90.0$ km/h. (See Example 1 of Section 24-7.)

24. Find the average value of the volume of a sphere with respect to the radius. Explain the meaning of the result.

25. The length of arc s of a curve from $x = a$ to $x = b$ is

$$s = \int_a^b \sqrt{1 + \left(\frac{dy}{dx}\right)^2}\, dx$$

The cable of a bridge can be described by the equation $y = 0.04x^{3/2}$ from $x = 0$ to $x = 100$ m. Find the length of the cable. See Fig. 26-64.

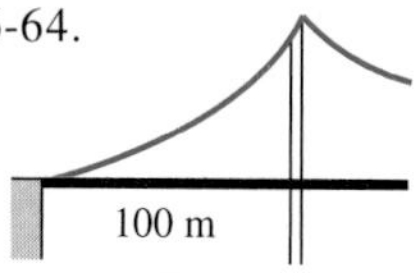

100 m

Fig. 26-64

26. A rocket takes off in a path described by the equation $y = \frac{2}{3}(x^2 - 1)^{3/2}$. Find the distance traveled by the rocket for $x = 1.0$ km to $x = 3.0$ km. (See Exercise 25.)

27. The area of a surface of revolution generated by revolving an arc from $x = a$ to $x = b$ is given by $S = 2\pi \int_a^b y\sqrt{1 + (dy/dx)^2}\, dx$. Find the formula for the lateral surface area of a right circular cone of radius r and height h.

28. The grinding surface of a grinding machine can be described as the surface generated by rotating the curve $y = 0.2x^3$ from $x = 0$ to $x = 2.0$ cm about the x axis. Find the grinding surface area. (See Exercise 27.)

Chapter Equations, Review Exercises, and Practice Test

Chapter Equations

Velocity
$$v = \int a\, dt \tag{26-1}$$

$$v = at + C_1 \tag{26-2}$$

Displacement
$$s = \int v\, dt \tag{26-3}$$

Electric current
$$i = \frac{dq}{dt} \tag{26-4}$$

Electric charge
$$q = \int i\, dt \tag{26-5}$$

Voltage across capacitor
$$V_C = \frac{1}{C}\int i\, dt \tag{26-6}$$

Area
$$A = \int_a^b y\, dx = \int_a^b f(x)\, dx \tag{26-7}$$

$$A = \int_c^d x\, dy = \int_c^d g(y)\, dy \tag{26-8}$$

$$A = \int_a^b (y_2 - y_1)\, dx \tag{26-9}$$

$$A = \int_c^d (x_2 - x_1)\, dy \tag{26-10}$$

Volume
$$V = \pi\int_a^b y^2\, dx = \pi\int_a^b [f(x)]^2\, dx \tag{26-11}$$

$$V = \pi\int_c^d x^2\, dy \tag{26-12}$$

Shell
$$dV = 2\pi(\text{radius}) \times (\text{height}) \times (\text{thickness}) \tag{26-13}$$

Disk
$$dV = \pi(\text{radius})^2 \times (\text{thickness}) \tag{26-14}$$

Center of mass $$m_1d_1 + m_2d_2 + \cdots + m_nd_n = (m_1 + m_2 + \cdots + m_n)\overline{d} \qquad (26\text{-}15)$$

Centroid of area $$\overline{x} = \frac{\int_a^b x(y_2 - y_1)\,dx}{\int_a^b (y_2 - y_1)\,dx} \qquad (26\text{-}16)$$

$$\overline{y} = \frac{\int_c^d y(x_2 - x_1)\,dy}{\int_c^d (x_2 - x_1)\,dy} \qquad (26\text{-}17)$$

Centroid of volume $$\overline{x} = \frac{\int_a^b xy^2\,dx}{\int_a^b y^2\,dx} \qquad (26\text{-}18)$$

$$\overline{y} = \frac{\int_c^d yx^2\,dy}{\int_c^d x^2\,dy} \qquad (26\text{-}19)$$

Radius of gyration $$m_1d_1^2 + m_2d_2^2 + \cdots + m_nd_n^2 = (m_1 + m_2 + \cdots + m_n)R^2 \qquad (26\text{-}20)$$

Moment of inertia of area $$I_y = k\int_a^b x^2(y_2 - y_1)\,dx \qquad (26\text{-}21)$$

$$I_x = k\int_c^d y^2(x_2 - x_1)\,dy \qquad (26\text{-}22)$$

Moment of inertia of volume $$I_y = 2\pi k\int_a^b yx^3\,dx \qquad (26\text{-}23)$$

$$I_x = 2\pi k\int_c^d xy^3\,dy \qquad (26\text{-}24)$$

Work $$W = \int_a^b f(x)\,dx \qquad (26\text{-}25)$$

Force between electric charges $$f(x) = \frac{kq_1q_2}{x^2} \qquad (26\text{-}26)$$

Force due to liquid pressure $$F = w\int_a^b lh\,dh \qquad (26\text{-}27)$$

Average value $$y_{av} = \frac{\int_a^b y\,dx}{b - a} \qquad (26\text{-}28)$$

Review Exercises

1. How long after it is dropped does a baseball reach a speed of 42 m/s (the speed of a good fastball)?

2. If the velocity v (in m/s) of a subway train after the brakes are applied can be expressed as $v = \sqrt{400 - 20t}$, where t is the time in seconds, how far does it travel in coming to a stop?

3. A weather balloon is rising at the rate of 10.0 m/s when a small metal part drops off. If the balloon is 60.0 m high at this instant, when will the part hit the ground?

4. A float is dropped into a river at a point where it is flowing at 1.5 m/s. How far does the float travel in 30 s if it accelerates downstream at 0.010 m/s^2?

5. The electric current i, in amperes, in a circuit as a function of the time t, in seconds, is $i = 0.25(2\sqrt{t} - t)$. Find the total charge to pass a point in the circuit in 2.0 s.

6. The current in a certain electric circuit is given by $i = \sqrt{1 + 4t}$, where i is in amperes and t is in seconds. Find the charge that passes a given point during the first two seconds.

7. The voltage across a 5.5-nF capacitor in an FM radio receiver is zero. What is the voltage after 25 μs if a current of 12 mA charges the capacitor?

8. The initial voltage across a capacitor is zero, and $V_C = 2.50$ V after 8.00 ms. If a current $i = t/\sqrt{t^2 + 1}$, where i is in amperes and t is in seconds, charges the capacitor, find the capacitance C of the capacitor.

9. The distribution of weight on a cable is not uniform. If the slope of the cable at any point is given by $dy/dx = 20 + \frac{1}{40}x^2$, and if the origin of the coordinate system is at the lowest point, find the equation that gives the curve described by the cable.

10. The time rate of change of the reliability R, in percent, of a computer system is $dR/dt = -2.5(0.05t + 1)^{-1.5}$, where t is in hours. If $R = 100$ for $t = 0$, find R for $t = 100$ h.

11. Find the area between $y = \sqrt{1 - x}$ and the coordinate axes.

12. Find the area bounded by $y = 3x^2 - x^3$ and the x-axis.

13. Find the area bounded by $y^2 = 2x$ and $y = x - 4$.

14. Find the area bounded by $y = 1/(2x + 1)^2$, $y = 0$, $x = 1$, and $x = 2$.

15. Find the area between $y = x^2$ and $y = x^3 - 2x^2$.

16. Find the area between $y = x^2 + 2$ and $y = 3x^2$.

17. Find the volume generated by revolving the area bounded by $y = 3 + x^2$ and the line $y = 4$ about the x-axis.

18. Find the volume generated by revolving the area bounded by $y = 8x - x^4$ and the x-axis about the x-axis.

19. Find the volume generated by revolving the area bounded by $y = x^3 - 4x^2$ and the x-axis about the y-axis.

20. Find the volume generated by revolving the area bounded by $y = x$ and $y = 3x - x^2$ about the y-axis.

21. Find the volume generated by revolving an ellipse about its major axis.

22. A hole of radius 1 cm is bored along the diameter of a sphere of radius 4 cm. Find the volume of the material that is removed from the sphere.

23. Find the centroid of the area bounded by $y^2 = x^3$ and $y = 2x$.

24. Find the centroid of the area bounded by $y = 2x - 4$, $x = 1$, and $y = 0$.

25. Find the centroid of the volume generated by revolving the area bounded by $y = \sqrt{x}$, $x = 1$, $x = 4$, and $y = 0$ about the x-axis.

26. Find the centroid of the volume generated by revolving the area bounded by $yx^4 = 1$, $y = 1$, and $y = 4$ about the y-axis.

27. Find the moment of inertia of the area bounded by $y = 3x - x^2$ and $y = x$ with respect to the y-axis.

28. Find the radius of gyration of the first-quadrant area bounded by $y = 8 - x^3$ with respect to the y-axis.

29. Find the moment of inertia with respect to its axis of the solid generated by revolving the area bounded by $y = x^{1/2}$, $y = 0$, and $x = 8$ about the x-axis.

30. Find the radius of gyration with respect to its axis of the solid generated by revolving the area bounded by $xy = 1$, $x = 1$, $x = 3$, and $y = \frac{1}{3}$ about the x-axis.

31. A pail and its contents weigh 80 N. The pail is attached to the end of a 30-m rope that weighs 20 N and is hanging vertically. How much work is done in winding up the rope with the pail attached?

32. The gravitational force, in newtons, of the earth on a satellite (the weight of the satellite) is given by $F = 10^{11}/x^2$, where x is the vertical distance, in kilometres, from the center of the earth to the satellite. How much work is done in moving the satellite from the earth's surface to an altitude of 3000 km? The radius of the earth is 6370 km.

33. The rear stabilizer of a certain aircraft can be described as the area under the curve $y = 3x^2 - x^3$, as shown in Fig. 26-65. Find the x-coordinate, in metres, of the centroid of the stabilizer.

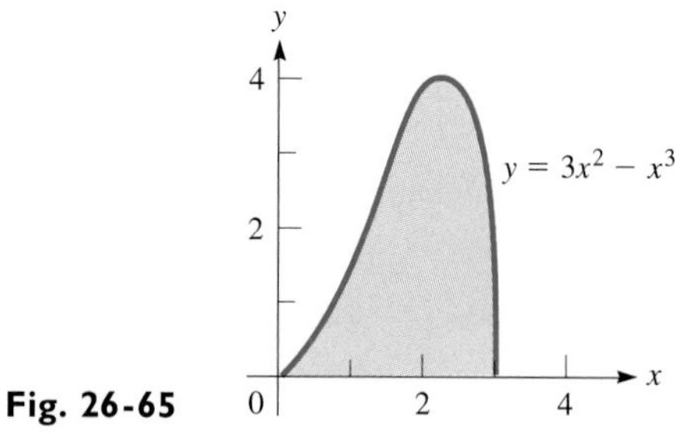

Fig. 26-65

34. The water in a spherical tank 20.0 m in radius is 15.0 m deep at the deepest point. How much water is in the tank?

35. The nose cone of a rocket has the shape of a semiellipse revolved about its major axis, as shown in Fig. 26-66. What is the volume of the nose cone?

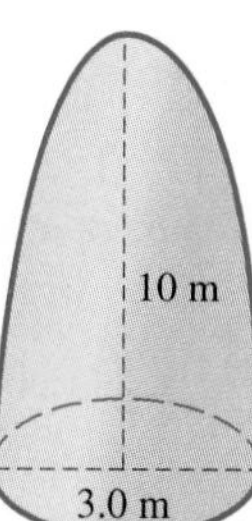

Fig. 26-66

36. The deck area of a boat is a parabolic section as shown in Fig. 26-67. What is the area of the deck?

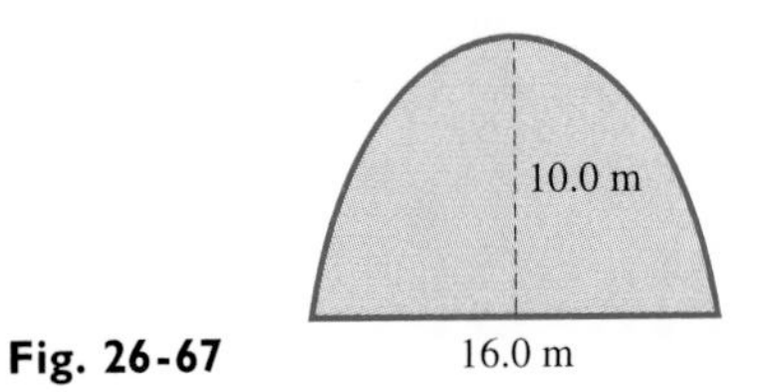

Fig. 26-67

37. A cylindrical chemical waste holding tank 4.50 m in radius has a depth of 3.25 m. Find the total force on the circular side of the tank when it is filled with liquid with a weight density of 10.6 kN/m^3.

38. A section of a dam is in the shape of a right triangle. The base of the triangle is 6.00 m and is in the surface of the water. If the triangular section goes to a depth of 4.00 m, find the force on it. See Fig. 26-68.

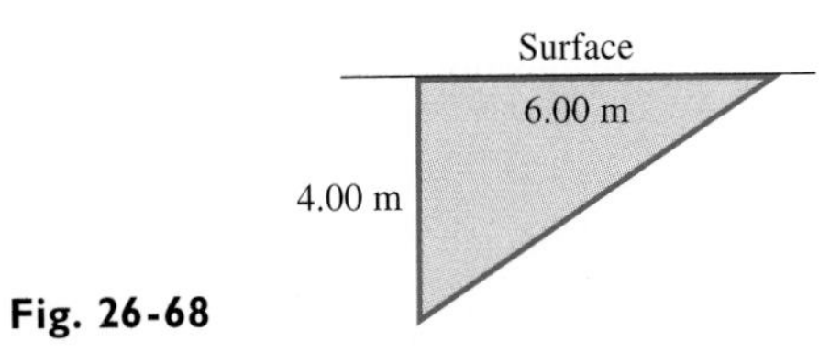

Fig. 26-68

39. The electric resistance of a wire is inversely proportional to the square of its radius. If a certain wire has a resistance of 0.30 Ω when its radius is 2.0 mm, find the average value of the resistance with respect to the radius if the radius changes from 2.0 mm to 2.1 mm.

40. A horizontal straight section of pipe is supported at its center by a vertical wire as shown in Fig. 26-69. Find the formula for the moment of inertia of the pipe with respect to an axis along the wire if the pipe is of length L and mass m.

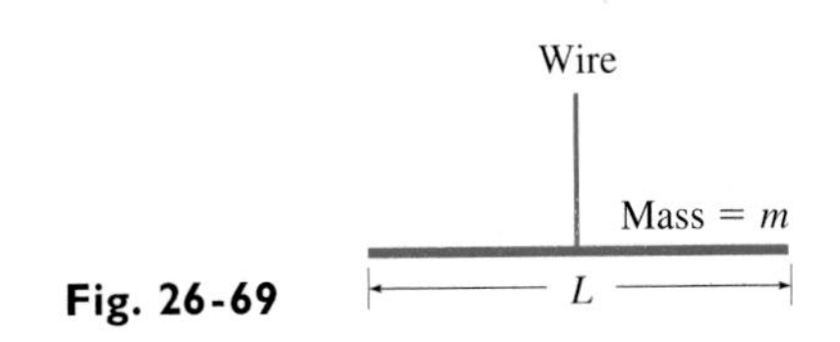

Fig. 26-69

Writing Exercise

41. A tub for holding liquids has a circular top of radius a. All cross sections of the tub that are perpendicular to a fixed diameter of the top are squares. Write one or two paragraphs explaining how to derive the formula that gives the volume of the tub. What is the formula?

Practice Test

In Problems 1–3, use the area bounded by $y = \frac{1}{4}x^2$, $y = 0$, *and* $x = 2$.

1. Find the area.
2. Find the coordinates of the centroid of the area.
3. Find the volume if the given area is revolved about the x-axis.

In Problems 4 and 5, use the first-quadrant area bounded by $y = x^2$, $x = 0$, *and* $y = 9$.

4. Find the volume if the given area is revolved about the x-axis.
5. Find the moment of inertia of the area with respect to the y-axis.
6. The velocity v of an object as a function of the time t is $v = 60 - 4t$. Find the expression for the displacement s if $s = 10$ for $t = 0$.
7. The natural length of a spring is 8.0 cm. A force of 12 N stretches it to a length of 10.0 cm. How much work is done in stretching it from a length of 10.0 cm to a length of 14.0 cm?
8. A vertical rectangular floodgate is 6.00 m wide and 2.00 m high. Find the force on the gate if its upper edge is 1.00 m below the surface of the water ($w = 9.80\ kN/m^3$).

27

Differentiation of Transcendental Functions

In Sections 27-4 and 27-7 we use derivatives of transcendental functions in analyzing the motion of a rocket.

In our development of differentiation and integration in the previous chapters, we have used only algebraic functions. We have not yet used the trigonometric, inverse trigonometric, exponential, or logarithmic functions, for the derivative of each of these is a special form. In this chapter we find formulas for the derivatives of these functions, which are the most important of the **transcendental** (nonalgebraic) **functions.** In the next chapter we shall take up integration involving these functions.

As we have seen in the earlier chapters, there are many technical and scientific applications of these functions. These applications are found in alternating-current theory, harmonic motion, rocket motion, monetary interest calculations, population growth, and many other types of problems.

27-1 Derivatives of the Sine and Cosine Functions

If we find the derivative of the sine function, we can use it to find the derivatives of the other trigonometric and inverse trigonometric functions. Therefore, we now find the derivative of the sine function by using the delta-process.

Let $y = \sin u$, where u is a function of x and is expressed in radians. If u changes by Δu, y then changes by Δy. Thus,

$$y + \Delta y = \sin(u + \Delta u)$$

$$\Delta y = \sin(u + \Delta u) - \sin u$$

$$\frac{\Delta y}{\Delta u} = \frac{\sin(u + \Delta u) - \sin u}{\Delta u}$$

For reference, Eq. (20-18) is $\sin x - \sin y = 2 \sin \frac{1}{2}(x - y) \cos \frac{1}{2}(x + y)$.

Referring now to Eq. (20-18), we have

$$\frac{\Delta y}{\Delta u} = \frac{2 \sin \frac{1}{2}(u + \Delta u - u) \cos \frac{1}{2}(u + \Delta u + u)}{\Delta u}$$

$$= \frac{\sin(\Delta u/2) \cos[u + (\Delta u/2)]}{\Delta u/2}$$

Looking ahead to the next step of letting $\Delta u \to 0$, we see that the numerator and denominator both approach zero. This situation is precisely the same as that in which we were finding the derivatives of the algebraic functions. To find the limit, we must find

$$\lim_{\Delta u \to 0} \frac{\sin(\Delta u/2)}{\Delta u/2}$$

since these are the factors that cause the numerator and the denominator to approach zero.

In finding this limit, we let $\theta = \Delta u/2$ for convenience of notation. This means that we are to determine $\lim_{\theta \to 0} \frac{\sin \theta}{\theta}$. Of course, it would be convenient to know before proceeding if this limit does actually exist. Therefore, by using a calculator, we can develop a table of values of $\frac{\sin \theta}{\theta}$ as θ becomes very small.

θ (radians)	0.5	0.1	0.05	0.01	0.001
$\frac{\sin \theta}{\theta}$	0.958 851 1	0.998 334 2	0.999 583 4	0.999 983 3	0.999 999 8

We see from this table that the limit of $\frac{\sin \theta}{\theta}$, as $\theta \to 0$, appears to be 1.

In order to prove that $\lim_{\theta \to 0} \frac{\sin \theta}{\theta} = 1$, we use a geometric approach. Considering Fig. 27-1, we see that the following inequality is true:

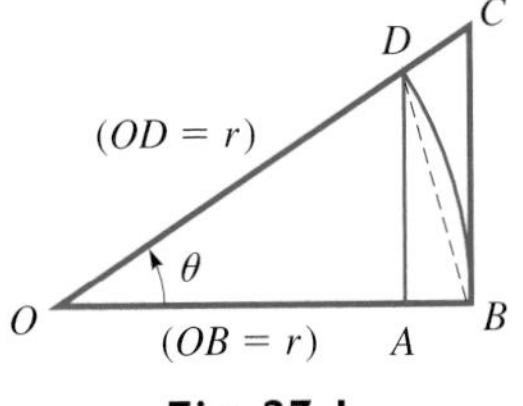

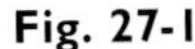
Fig. 27-1

Area triangle OBD < area sector OBD < area triangle OBC

$$\frac{1}{2}r(r \sin \theta) < \frac{1}{2}r^2\theta < \frac{1}{2}r(r \tan \theta) \quad \text{or} \quad \sin \theta < \theta < \tan \theta$$

Remembering that we want to find the limit of $(\sin \theta)/\theta$, we next divide through by $\sin \theta$ and then take reciprocals:

$$1 < \frac{\theta}{\sin \theta} < \frac{1}{\cos \theta} \quad \text{or} \quad 1 > \frac{\sin \theta}{\theta} > \cos \theta$$

When we consider the limit as $\theta \to 0$, we see that the left member remains 1 and the right member approaches 1. Thus, $(\sin \theta)/\theta$ must approach 1. This means

$$\lim_{\theta \to 0} \frac{\sin \theta}{\theta} = \lim_{\Delta u \to 0} \frac{\sin(\Delta u/2)}{\Delta u/2} = 1 \qquad \textbf{(27-1)}$$

Using the result in Eq. (27-1) in the expression for $\Delta y/\Delta u$, we have

$$\lim_{\Delta u \to 0} \frac{\Delta y}{\Delta u} = \lim_{\Delta u \to 0} \left[\cos\left(u + \frac{\Delta u}{2} \right) \frac{\sin(\Delta u/2)}{\Delta u/2} \right] = \cos u$$

or

$$\frac{dy}{du} = \cos u \tag{27-2}$$

However, we want the derivative of y with respect to x. This requires the use of the chain rule, Eq. (23-14), which we repeat here for reference.

$$\boxed{\frac{dy}{dx} = \frac{dy}{du}\frac{du}{dx}} \tag{27-3}$$

Combining Eqs. (27-2) and (27-3), we have ($y = \sin u$)

$$\boxed{\frac{d(\sin u)}{dx} = \cos u \frac{du}{dx}} \tag{27-4}$$

EXAMPLE 1 ▸▸ Find the derivative of $y = \sin 2x$.

In this example, $u = 2x$. Thus

$$\frac{dy}{dx} = \frac{d(\sin 2x)}{dx} = \cos 2x \frac{d(2x)}{dx} = (\cos 2x)\overset{\frac{du}{dx}\downarrow}{(2)} \qquad \text{using Eq. (27-4)}$$

$$= 2 \cos 2x \quad ◂◂$$

EXAMPLE 2 ▸▸ Find the derivative of $y = 2 \sin(x^2)$.

In this example, $u = x^2$, which means that $du/dx = 2x$. Hence

$$\frac{dy}{dx} = 2[\cos(x^2)](2x) \qquad \text{using Eq. (27-4)}$$

$$= 4x \cos(x^2)$$

CAUTION ▸ It is important here, just as it is ***in finding the derivatives of powers of all functions, to remember to include the factor du/dx.*** ◂◂

EXAMPLE 3 ▸▸ Find the derivative of $y = \sin^2 x$.

This example is a combination of the use of the power rule, Eq. (23-15), and the derivative of the sine function Eq. (27-4). Since $\sin^2 x$ means $(\sin x)^2$, in using the power rule we have $u = \sin x$. Thus

For reference, Eq. (23-15) is $\frac{du^n}{dx} = nu^{n-1}\left(\frac{du}{dx}\right)$.

$$\frac{dy}{dx} = 2(\sin x)\frac{d \sin x}{dx} \qquad \text{using Eq. (23-15)}$$

$$= 2 \sin x \cos x \qquad \text{using Eq. (27-4)}$$

$$= \sin 2x \qquad \text{using identity (Eq. (20-21))} \quad ◂◂$$

EXAMPLE 4 ▸▸ Find the derivative of $y = 2\sin^3(2x^4)$.

In the general power rule, $u = \sin(2x^4)$. In the derivative of the sine function, $u = 2x^4$. Thus we have

$$\frac{dy}{dx} = 2(3)\sin^2(2x^4)\frac{d(\sin 2x^4)}{dx} \qquad \text{using Eq. (23-15)}$$

$$= 6\sin^2(2x^4)\cos(2x^4)\frac{d(2x^4)}{dx} \qquad \text{using Eq. (27-4)}$$

$$= 6\sin^2(2x^4)\cos(2x^4)(8x^3)$$

$$= 48x^3\sin^2(2x^4)\cos(2x^4) \quad ◂◂$$

In order to find the derivative of the cosine function, we write it in the form $\cos u = \sin(\frac{\pi}{2} - u)$. Thus, if $y = \sin(\frac{\pi}{2} - u)$, we have

$$\frac{dy}{dx} = \cos\left(\frac{\pi}{2} - u\right)\frac{d(\frac{\pi}{2} - u)}{dx} = \cos\left(\frac{\pi}{2} - u\right)\left(-\frac{du}{dx}\right)$$

$$= -\cos\left(\frac{\pi}{2} - u\right)\frac{du}{dx}$$

Since $\cos(\frac{\pi}{2} - u) = \sin u$, we have

$$\frac{d(\cos u)}{dx} = -\sin u\frac{du}{dx} \qquad (27\text{-}5)$$

EXAMPLE 5 ▸▸ The electric power p developed in a resistor of an amplifier circuit is $p = 25\cos^2 120\pi t$, where t is the time. Find the expression for the time rate of change of power.

From Chapter 23, we know that we are to find the derivative dp/dt. Therefore,

$$p = 25\cos^2 120\pi t$$

$$\frac{dp}{dt} = 25(2\cos 120\pi t)\frac{d\cos 120\pi t}{dt} \qquad \text{using Eq. (23-15)}$$

$$= 50\cos 120\pi t(-\sin 120\pi t)\frac{d(120\pi t)}{dt} \qquad \text{using Eq. (27-5)}$$

$$= (-50\cos 120\pi t\sin 120\pi t)(120\pi)$$

$$= -6000\pi\cos 120\pi t\sin 120\pi t$$

$$= -3000\pi\sin 240\pi t \qquad \text{using Eq. (20-21)} \quad ◂◂$$

For reference, Eq. (20-21) is $\sin 2\alpha = 2\sin\alpha\cos\alpha$.

EXAMPLE 6 ▸▸ Find the derivative of $y = \sqrt{1 + \cos 2x}$.

$$y = (1 + \cos 2x)^{1/2}$$

$$\frac{dy}{dx} = \frac{1}{2}(1 + \cos 2x)^{-1/2}\frac{d(1 + \cos 2x)}{dx} \qquad \text{using Eq. (23-15)}$$

$$= \frac{1}{2}(1 + \cos 2x)^{-1/2}(-\sin 2x)(2) \qquad \text{using Eq. (27-5)}$$

$$= -\frac{\sin 2x}{\sqrt{1 + \cos 2x}} \quad ◂◂$$

EXAMPLE 7 ▸▸ Find the differential of $y = \sin 2x \cos x^2$.

From Section 25-1, we recall that the differential of a function $y = f(x)$ is $dy = f'(x)\,dx$. Thus, using the derivative product rule and the derivatives of the sine and cosine functions, we arrive at the following result:

$$y = \sin 2x \cos x^2 \qquad y = (\sin 2x)(\cos x^2)$$
$$dy = [\sin 2x(-\sin x^2)(2x) + \cos x^2(\cos 2x)(2)]\,dx$$
$$= (-2x \sin 2x \sin x^2 + 2 \cos 2x \cos x^2)\,dx$$ ◂◂

EXAMPLE 8 ▸▸ Find the slope of a line tangent to the curve of $y = 5 \sin 3x$, where $x = 0.2$.

Here we are to find the derivative of $y = 5 \sin 3x$ and then evaluate the derivative for $x = 0.2$. Therefore, we have the following:

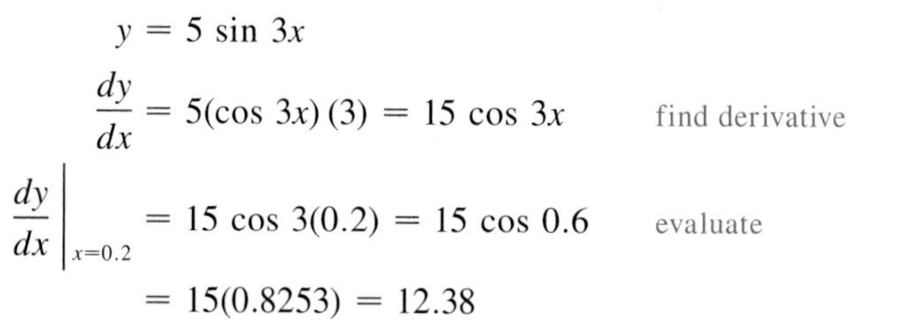

$$y = 5 \sin 3x$$
$$\frac{dy}{dx} = 5(\cos 3x)(3) = 15 \cos 3x \qquad \text{find derivative}$$
$$\left.\frac{dy}{dx}\right|_{x=0.2} = 15 \cos 3(0.2) = 15 \cos 0.6 \qquad \text{evaluate}$$
$$= 15(0.8253) = 12.38$$

Fig. 27-2

CAUTION ▶ In evaluating the slope we must remember that $x = 0.2$ means the ***values are in radians.*** Therefore, the slope is 12.38. See Fig. 27-2. ◂◂

We also note here that many graphing calculators may be used to evaluate derivatives. The manual of the calculator should be consulted to determine how this is done.

Exercises 27–1

In Exercises 1–32, find the derivatives of the given functions.

1. $y = \sin(x + 2)$
2. $y = 3 \sin 4x$
3. $y = 2 \sin(2x^3 - 1)$
4. $y = 5 \sin(3 - x)$
5. $y = 6 \cos \frac{1}{2}x$
6. $y = \cos(1 - x)$
7. $y = 2 \cos(3x - 1)$
8. $y = 4 \cos(6x^2 + 5)$
9. $y = \sin^2 4x$
10. $y = 3 \sin^3(2x^4 + 1)$
11. $y = 3 \cos^3(5x + 2)$
12. $y = 4 \cos^2 \sqrt{x}$
13. $y = x \sin 3x$
14. $y = x^2 \sin 2x$
15. $y = 3x^3 \cos 5x$
16. $y = 5x \cos 2x^3$
17. $y = \sin x^2 \cos 2x$
18. $y = 6 \sin x \cos 4x$
19. $y = \sqrt{1 + \sin 4x}$
20. $y = (x - \cos^2 x)^4$
21. $y = \dfrac{\sin 3x}{x}$
22. $y = \dfrac{2x + 3}{\sin 4x}$
23. $y = \dfrac{2 \cos x^2}{3x - 1}$
24. $y = \dfrac{5x}{\cos x}$
25. $y = 2 \sin^2 3x \cos 2x$
26. $y = \cos^3 4x \sin^2 2x$
27. $y = \dfrac{\cos^2 3x}{1 + 2 \sin^2 2x}$
28. $y = \dfrac{\sin 5x}{3 - \cos^2 3x}$
29. $y = \sin^3 x - \cos 2x$
30. $y = x \sin x + \cos x$
31. $y = x - \dfrac{1}{3} \sin^3 4x$
32. $y = 2x \sin x + 2 \cos x - x^2 \cos x$

In Exercises 33–52, solve the given problems.

33. Using a graphing calculator, (a) display the graph of $y = (\sin x)/x$, to verify that $(\sin \theta)/\theta \to 1$ as $\theta \to 0$, and (b) verify the values for $(\sin \theta)/\theta$ in the table on p. 761.
34. Evaluate $\lim_{\theta \to 0} (\tan \theta)/\theta$. (Use the fact that $\lim_{\theta \to 0} (\sin \theta)/\theta = 1$.)

35. On a calculator, find the values of (a) cos 1.0000 and (b) (sin 1.0001 − sin 1.0000)/0.0001. Compare the values and give the meaning of each in relation to the derivative of the sine function, where $x = 1$.

36. On a calculator, find the values of (a) −sin 1.0000 and (b) (cos 1.0001 − cos 1.0000)/0.0001. Compare the values and give the meaning of each in relation to the derivative of the cosine function, where $x = 1$.

37. On the graph of $y = \sin x$ in Fig. 27-3, draw tangent lines at the indicated points and determine the slopes of these tangent lines. Then plot the values of these slopes for the same values of x and join the points with a smooth curve. Compare the resulting curve with $y = \cos x$. (Note the meaning of the derivative as the slope of a tangent line.)

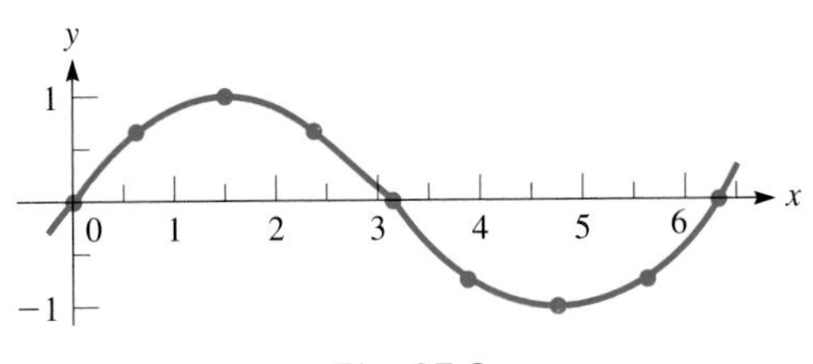

Fig. 27-3

38. Repeat the instructions given in Exercise 37 for the graph of $y = \cos x$ in Fig. 27-4. Compare the resulting curve with $y = \sin x$. (Be careful in this comparison, and remember the difference between $y = \sin x$ and the derivative of $y = \cos x$. As in Exercise 37, note the meaning of the derivative as the slope of a tangent line.)

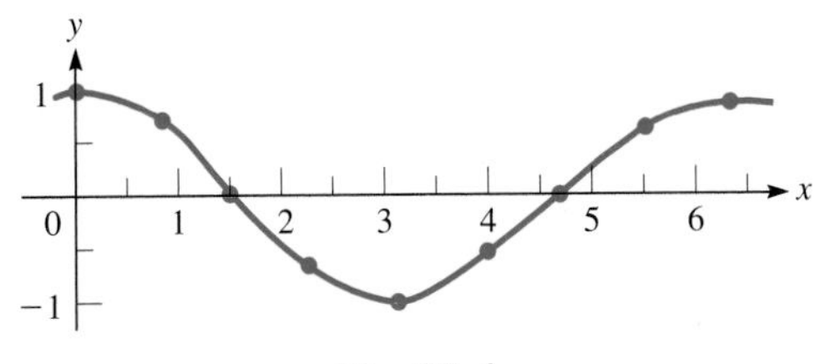

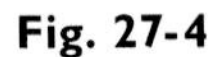

Fig. 27-4

39. Find the derivative of the implicit function $\sin(xy) + \cos 2y = x^2$.

40. Find the derivative of the implicit function $x \cos 2y + \sin x \cos y = 1$.

41. Show that $\dfrac{d^4 \sin x}{dx^4} = \sin x$.

42. If $y = \cos 2x$, show that $\dfrac{d^2y}{dx^2} = -4y$.

43. Find the derivative of each member of the identity $\cos 2x = 2\cos^2 x - 1$ and thereby obtain another trigonometric identity.

44. Find the derivative of each member of the identity $2\sin^2 x = 1 - \cos 2x$ and thereby obtain another trigonometric identity.

45. Evaluate the differential of $y = 3 \sin 2x$ for $x = \frac{\pi}{8}$ and $dx = 0.02$.

46. Evaluate the differential of $y = \sin(\cos x)$ for $x = 0.50$ and $dx = 0.04$.

47. Find the slope of a line tangent to the curve of $y = x \cos 2x$ where $x = 1.20$. Verify the result by using the derivative-evaluating feature of a graphing calculator.

48. Find the slope of a line tangent to the curve of $y = \dfrac{2 \sin 3x}{x}$ where $x = 0.15$. Verify the result by using the derivative-evaluating feature of a graphing calculator.

49. The blade of a saber saw moves vertically up and down, and its displacement y (in cm) is given by $y = 1.85 \sin 36\pi t$, where t is the time in seconds. Find the velocity of the blade for $t = 0.0250$ s.

50. The current i (in amperes) in an amplifier circuit, as a function of time t (in seconds) is given by $i = 0.10 \cos(120\pi t + \pi/6)$. Find the expression for the voltage across a 2.0-mH inductor in the circuit. (See Exercise 23 of Section 26-1.)

51. The distance r (in km) from an aircraft to a rocket directed at the aircraft is $r = \dfrac{100}{1 - \cos\theta}$, where θ is the angle between their directions. Find $dr/d\theta$ for $\theta = 120°$.

52. The number N of reflections of a light ray passing through an optic fiber of length L and diameter d is $N = \dfrac{L \sin\theta}{d\sqrt{n^2 - \sin^2\theta}}$. Here n is the index of refraction of the fiber and θ is the angle between the light ray and the fiber's axis. Find $dN/d\theta$.

27-2 Derivatives of the Other Trigonometric Functions

We can find the derivatives of the other trigonometric functions by expressing the functions in terms of the sine and cosine. After we perform the differentiation, we use trigonometric relations to put the derivative in a convenient form.

We obtain the derivative of tan u by expressing tan u as sin u/cos u. Therefore, letting $y = \sin u/\cos u$, by employing the quotient rule we have

$$\frac{dy}{dx} = \frac{\cos u[\cos u(du/dx)] - \sin u[-\sin u(du/dx)]}{\cos^2 u}$$

$$= \frac{\cos^2 u + \sin^2 u}{\cos^2 u}\frac{du}{dx} = \frac{1}{\cos^2 u}\frac{du}{dx} = \sec^2 u\frac{du}{dx}$$

$$\frac{d(\tan u)}{dx} = \sec^2 u\frac{du}{dx} \tag{27-6}$$

We find the derivative of cot u by letting $y = \cos u/\sin u$ and again using the quotient rule:

$$\frac{dy}{dx} = \frac{\sin u[-\sin u(du/dx)] - \cos u[\cos u(du/dx)]}{\sin^2 u}$$

$$= \frac{-\sin^2 u - \cos^2 u}{\sin^2 u}\frac{du}{dx}$$

$$\frac{d(\cot u)}{dx} = -\csc^2 u\frac{du}{dx} \tag{27-7}$$

To obtain the derivative of sec u, we let $y = 1/\cos u$. Then

$$\frac{dy}{dx} = -(\cos u)^{-2}\left[(-\sin u)\left(\frac{du}{dx}\right)\right] = \frac{1}{\cos u}\frac{\sin u}{\cos u}\frac{du}{dx}$$

$$\frac{d(\sec u)}{dx} = \sec u \tan u\frac{du}{dx} \tag{27-8}$$

We obtain the derivative of csc u by letting $y = 1/\sin u$. And so

$$\frac{dy}{dx} = -(\sin u)^{-2}\left(\cos u\frac{du}{dx}\right) = -\frac{1}{\sin u}\frac{\cos u}{\sin u}\frac{du}{dx}$$

$$\frac{d(\csc u)}{dx} = -\csc u \cot u\frac{du}{dx} \tag{27-9}$$

Note the convenient forms of these derivatives that are obtained by using basic trigonometric identities.

EXAMPLE 1 ▸▸ Find the derivative of $y = 2 \tan 8x$.

The derivative is

$$\frac{dy}{dx} = 2(\sec^2 8x)\underbrace{(8)}_{du/dx} \qquad \text{using Eq. (27-6)}$$

$$= 16 \sec^2 8x \quad ◂◂$$

EXAMPLE 2 ▸▸ Find the derivative of $y = 3 \sec^2 4x$.

Using the power rule and Eq. (27-8), we have

$$\frac{dy}{dx} = 3(2)(\sec 4x)\frac{d(\sec 4x)}{dx} \qquad \text{using } \frac{du^n}{dx} = nu^{n-1}\frac{du}{dx}$$

$$= 6(\sec 4x)(\sec 4x \tan 4x)(4) \qquad \text{using } \frac{d \sec u}{dx} = \sec u \tan u \frac{du}{dx}$$

$$= 24 \sec^2 4x \tan 4x \quad ◂◂$$

EXAMPLE 3 ▸▸ Find the derivative of $y = x \csc^3 2x$.

Using the power rule, the product rule, and Eq. (27-9), we have

$$\frac{dy}{dx} = x(3 \csc^2 2x)(-\csc 2x \cot 2x)(2) + (\csc^3 2x)(1)$$

$$= \csc^3 2x(-6x \cot 2x + 1) \quad ◂◂$$

EXAMPLE 4 ▸▸ Find the derivative of $y = (\tan 2x + \sec 2x)^3$.

Using the power rule and Eqs. (27-6) and (27-8), we have

$$\frac{dy}{dx} = 3(\tan 2x + \sec 2x)^2[\sec^2 2x(2) + \sec 2x \tan 2x(2)]$$

$$= 3(\tan 2x + \sec 2x)^2(2 \sec 2x)(\sec 2x + \tan 2x)$$

$$= 6 \sec 2x(\tan 2x + \sec 2x)^3 \quad ◂◂$$

EXAMPLE 5 ▸▸ Find the differential of $y = \sin 2x \tan x^2$.

Here we are to find the derivative of the given function and multiply by dx. Therefore, using the product rule along with Eqs. (27-4) and (27-6), we have

$$dy = [(\sin 2x)(\sec^2 x^2)(2x) + (\tan x^2)(\cos 2x)(2)]\, dx$$

$$= (2x \sin 2x \sec^2 x^2 + 2 \cos 2x \tan x^2)\, dx \quad ◂◂$$

EXAMPLE 6 ▸▸ Find dy/dx if $\cot 2x - 3 \csc xy = y^2$.

In finding the derivative of this implicit function, we must be careful not to forget the factor dy/dx when it occurs. The derivative is found as follows:

$$\cot 2x - 3 \csc xy = y^2$$

$$(-\csc^2 2x)(2) - 3(-\csc xy \cot xy)\left(x\frac{dy}{dx} + y\right) = 2y\frac{dy}{dx}$$

$$3x \csc xy \cot xy\frac{dy}{dx} - 2y\frac{dy}{dx} = 2 \csc^2 2x - 3y \csc xy \cot xy$$

$$\frac{dy}{dx} = \frac{2 \csc^2 2x - 3y \csc xy \cot xy}{3x \csc xy \cot xy - 2y} \quad ◂◂$$

EXAMPLE 7 ▸▸ Evaluate the derivative of $y = \dfrac{2x}{1 - \cot 3x}$, where $x = 0.25$.

Finding the derivative, we have

$$\frac{dy}{dx} = \frac{(1 - \cot 3x)(2) - 2x(\csc^2 3x)(3)}{(1 - \cot 3x)^2}$$
$$= \frac{2 - 2\cot 3x - 6x\csc^2 3x}{(1 - \cot 3x)^2}$$

Now, substituting $x = 0.25$, we have

$$\left.\frac{dy}{dx}\right|_{x=0.25} = \frac{2 - 2\cot 0.75 - 6(0.25)\csc^2 0.75}{(1 - \cot 0.75)^2}$$
$$= -626.1$$

In using the calculator we recall that we must have it in radian mode. Also, to evaluate cot 0.75 and csc 0.75, we must use reciprocals of tan 0.75 and sin 0.75, respectively. ◂◂

Exercises 27–2

In Exercises 1–32, find the derivatives of the given functions.

1. $y = \tan 5x$
2. $y = 3\tan(3x + 2)$
3. $y = \cot(1 - x)^2$
4. $y = 3\cot 6x$
5. $y = 3\sec 2x$
6. $y = \sec\sqrt{1 - x}$
7. $y = -3\csc\sqrt{2x + 3}$
8. $y = \csc(1 - 2x)$
9. $y = 5\tan^2 3x$
10. $y = 2\tan^2(x^2)$
11. $y = 2\cot^4 \frac{1}{2}x$
12. $y = \cot^2(1 - x^2)$
13. $y = \sqrt{\sec 4x}$
14. $y = \sec^3 x$
15. $y = 3\csc^4 7x$
16. $y = \csc^2(2x^2)$
17. $y = x^2 \tan x$
18. $y = 3x\sec 4x$
19. $y = 4\cos x \csc x^2$
20. $y = \frac{1}{2}\sin 2x \sec x$
21. $y = \dfrac{\csc x}{x}$
22. $y = \dfrac{\cot 4x}{2x}$
23. $y = \dfrac{2\cos 4x}{1 + \cot 3x}$
24. $y = \dfrac{\tan^2 3x}{2 + \sin x^2}$
25. $y = \frac{1}{3}\tan^3 x - \tan x$
26. $y = \csc 2x - 2\cot 2x$
27. $y = \tan 2x - \sec 2x$
28. $y = x\tan x + \sec^2 2x$
29. $y = \sqrt{2x + \tan 4x}$
30. $y = (1 - \csc^2 3x)^3$
31. $x\sec y - 2y = \sin 2x$
32. $3\cot(x + y) = \cos y^2$

In Exercises 33–36, find the differentials of the given functions.

33. $y = 4\tan^2 3x$
34. $y = 5\sec^3 2x$
35. $y = \tan 4x \sec 4x$
36. $y = 2x\cot 3x$

In Exercises 37–48, solve the given problems.

37. On a calculator, find the values of (a) $\sec^2 1.0000$ and (b) $(\tan 1.0001 - \tan 1.0000)/0.0001$. Compare the values and give the meaning of each in relation to the derivative of $\tan x$, where $x = 1$.
38. On a calculator, find the values of (a) $\sec 1.0000 \tan 1.0000$ and (b) $(\sec 1.0001 - \sec 1.0000)/0.0001$. Compare the values and give the meaning of each in relation to the derivative of $\sec x$, where $x = 1$.
39. Find the derivative of each member of the identity $1 + \tan^2 x = \sec^2 x$, and show that the results are equal.
40. Find the derivative of each member of the identity $1 + \cot^2 x = \csc^2 x$, and show that the results are equal.
41. Find the slope of a line tangent to the curve of $y = 2\cot 3x$ where $x = \pi/12$. Verify the result by using the derivative-evaluating feature of a graphing calculator.
42. Find the slope of a line normal to the curve of $y = \csc\sqrt{2x + 1}$ where $x = 0.45$. Verify the result by using the derivative-evaluating feature of a graphing calculator.

21. A person observes an object dropped from the top of a building 40.0 m away. If the building is 60.0 m high, how fast is the angle of elevation of the object changing after 1.0 s? (The distance the object drops is given by $s = 4.9t^2$.) See Fig. 27-18.

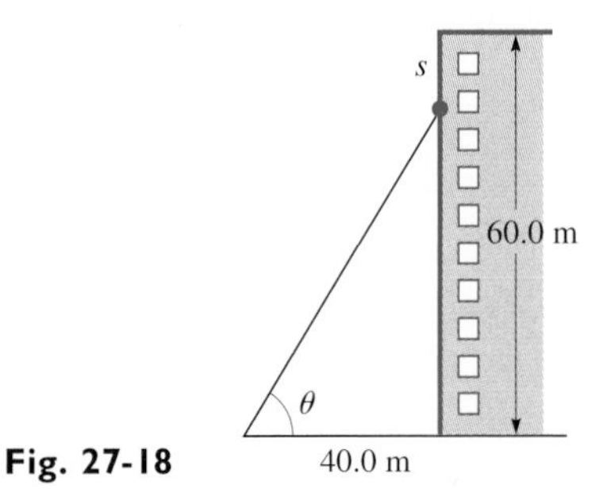

Fig. 27-18

22. A car passes directly under a police helicopter 150 m above a straight and level highway. After the car has traveled another 20.0 m, the angle of depression of the car from the helicopter is decreasing at 0.215 rad/s. What is the speed of the car?

23. A searchlight is 225 m from a straight wall. As the beam moves along the wall, the angle between the beam and the perpendicular to the wall is increasing at 1.5°/s. How fast is the length of the beam increasing when it is 315 m long? See Fig. 27-19.

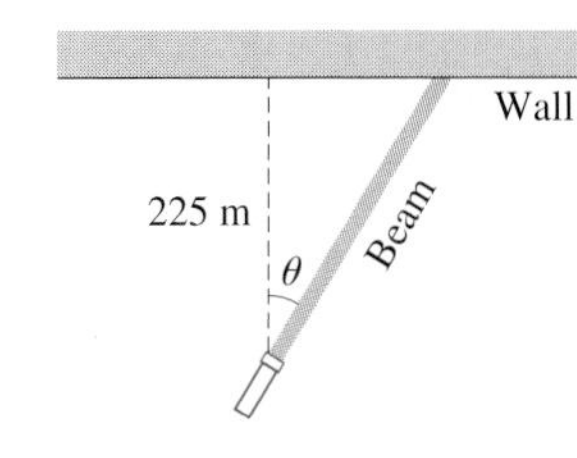

Fig. 27-19

24. In a modern hotel, where the elevators are directly observable from the lobby area (and a person can see from the elevators), a person in the lobby observes one of the elevators rising at the rate of 4.00 m/s. If the person was 16.0 m from the elevator when it left the lobby, how fast is the angle of elevation of the line of sight to the elevator increasing 10.0 s later?

25. If a block is placed on a plane inclined with the horizontal at an angle θ such that the block just moves down the plane, the coefficient of friction μ is given by $\mu = \tan\theta$. Use differentials to find the change in μ if θ changes from 20° to 21°.

26. The electric power p, in watts, developed in a resistor in an FM receiver circuit is $p = 0.0307\cos^2 120\pi t$, where t is the time in seconds. Use differentials to find the approximate change in p between $t = 10.0$ ms and $t = 12.2$ ms.

27. A surveyor measures two sides and the included angle of a triangular parcel of land to be 82.04 m, 75.37 m, and 38.38°. What error is caused in the calculation of the third side by an error of 0.15° in the angle?

28. To connect the four vertices of a square with the minimum amount of electric wire requires using the wiring pattern shown in Fig. 27-20. Find θ for the total length of wire ($L = 4x + y$) to be a minimum.

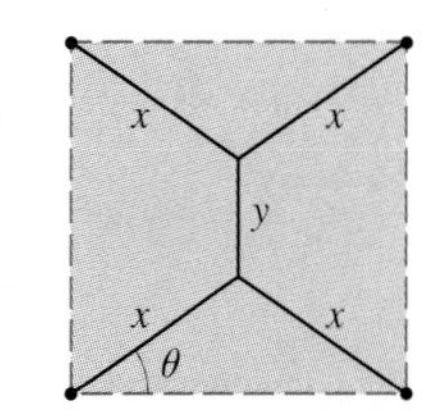

Fig. 27-20

29. The strength S of a rectangular beam is directly proportional to the product of its width w and the square of its depth d. Use trigonometric functions to find the dimensions of the strongest beam that can be cut from a circular log 16.0 cm in diameter. (See Example 4 on p. 684).

30. An architect is designing a window in the shape of an isosceles triangle with a perimeter of 150 cm. What is the vertex angle of the window of greatest area?

31. A wall is 1.8 m high and 1.2 m from a building. What is the length of the shortest pole that can touch the building and the ground beyond the wall? (*Hint:* From Fig. 27-21 it can be shown that

$y = 1.8\csc\theta + 1.2\sec\theta$.)

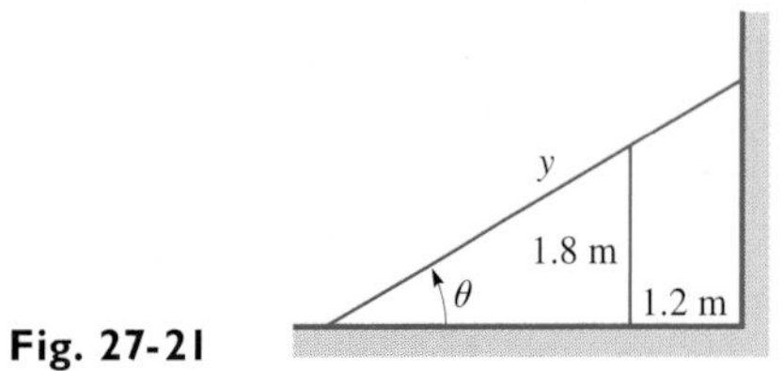

Fig. 27-21

32. The television screen at a sports arena is vertical and 2.4 m high. The lower edge is 8.5 m above an observer's eye level. If the best view of the screen is obtained when the angle subtended by the screen at eye level is a maximum, how far from directly below the screen must the observer be? See Fig. 27-22.

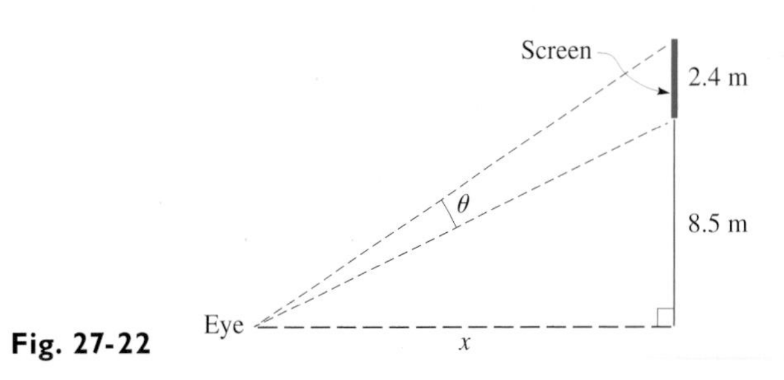

Fig. 27-22

27-5 Derivative of the Logarithmic Function

The next function for which we shall find the derivative is the logarithmic function. Again, we shall use the delta-process.

If we let $y = \log_b u$, where u is a function of x, we have

$$y + \Delta y = \log_b(u + \Delta u)$$

$$\Delta y = \log_b(u + \Delta u) - \log_b u = \log_b\left(\frac{u + \Delta u}{u}\right) = \log_b\left(1 + \frac{\Delta u}{u}\right)$$

$$\frac{\Delta y}{\Delta u} = \frac{\log_b(1 + \Delta u/u)}{\Delta u} = \frac{1}{u}\frac{u}{\Delta u}\log_b\left(1 + \frac{\Delta u}{u}\right)$$

$$= \frac{1}{u}\log_b\left(1 + \frac{\Delta u}{u}\right)^{u/\Delta u}$$

(We multiply and divide by u for purposes of evaluating the limit, as we shall now show.)

Before we can evaluate $\lim_{\Delta u \to 0} \Delta y/\Delta u$, we must determine

$$\lim_{\Delta u \to 0}\left(1 + \frac{\Delta u}{u}\right)^{u/\Delta u}$$

We can see that the exponent becomes unbounded, but the number being raised to this exponent approaches 1. Therefore, we shall investigate this limiting value.

To find an approximate value, let us graph the function $y = (1 + x)^{1/x}$ (for purposes of graphing we let $\Delta u/u = x$). We construct a table of values and then graph the function (see Fig. 27-23).

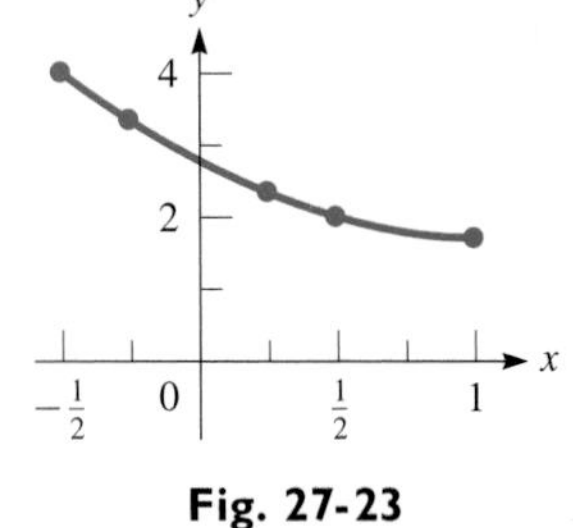

Fig. 27-23

x	-0.5	-0.25	$+0.25$	$+0.50$	$+1.00$
y	4.00	3.16	2.44	2.25	2.00

Only these values are shown, since we are interested in the y-value corresponding to $x = 0$. We see from the graph that this value is approximately 2.7. Choosing very small values of x, we may obtain these values:

x	0.1	0.01	0.001	0.0001
y	2.5937	2.7048	2.7169	2.718 15

NOTE ▶ By methods developed in Chapter 29, it can be shown that this value is about 2.718 281 8. ***The limiting value is the irrational number e.*** This is the same number used in the exponential form of a complex number in Chapter 12 and as the base of natural logarithms in Chapter 13.

Returning to the derivative of the logarithmic function, we have

$$\lim_{\Delta u \to 0}\frac{\Delta y}{\Delta u} = \lim_{\Delta u \to 0}\left[\frac{1}{u}\log_b\left(1 + \frac{\Delta u}{u}\right)^{u/\Delta u}\right] = \frac{1}{u}\log_b e$$

Therefore,

$$\frac{dy}{du} = \frac{1}{u}\log_b e$$

For reference, Eq. (27-3) is $\frac{dy}{dx} = \frac{dy}{du}\frac{du}{dx}$.

Combining this equation with Eq. (27-3), we have

$$\frac{d(\log_b u)}{dx} = \frac{1}{u}\log_b e\frac{du}{dx} \tag{27-13}$$

At this point we see that if we choose e as the basis of a system of logarithms, the above formula becomes

$$\frac{d(\ln u)}{dx} = \frac{1}{u}\frac{du}{dx} \tag{27-14}$$

The choice of e as the base b makes $\log_e e = 1$; thus, this factor does not appear in Eq. (27-14). We now see why the number e is chosen as the base for a system of logarithms, the natural logarithms. The notation $\ln u$ is the same as that used in Chapter 13 for natural logarithms.

EXAMPLE 1 ▸▸ Find the derivative of $y = \log 4x$.

Using Eq. (27-13), we have

$$\frac{dy}{dx} = \frac{1}{4x}(\log e)(4)$$

(u = $4x$; $\frac{du}{dx}$ = 4)

$$= \frac{1}{x}\log e \qquad (\log e = 0.4343)$$ ◂◂

EXAMPLE 2 ▸▸ Find the derivative of $y = \ln 3x^4$.

Using Eq. (27-14), we have (with $u = 3x^4$)

$$\frac{dy}{dx} = \frac{1}{3x^4}(12x^3)$$

($\frac{du}{dx}$ = $12x^3$)

$$= \frac{4}{x}$$ ◂◂

EXAMPLE 3 ▸▸ Find the derivative of $y = \ln \tan 4x$.

Using Eq. (27-14), along with the derivative of the tangent, we have

$$\frac{dy}{dx} = \frac{1}{\tan 4x}(\sec^2 4x)(4)$$

($\frac{d \tan 4x}{dx}$ = $(\sec^2 4x)(4)$)

$$= \frac{\cos 4x}{\sin 4x}\frac{4}{\cos^2 4x} \qquad \text{using trigonometric relations}$$

$$= \frac{1}{\sin 4x}\frac{4}{\cos 4x}$$

$$= 4 \csc 4x \sec 4x$$ ◂◂

NOTE ▶ Frequently we can find the derivative of a logarithmic function more simply if we ***use the properties of logarithms to simplify the expression before the derivative is found.*** The following examples illustrate this.

For reference, Eqs. (13-7), (13-8), and (13-9) are:

$\log_b xy = \log_b x + \log_b y$

$\log_b\left(\frac{x}{y}\right) = \log_b x - \log_b y$

$\log_b x^n = n \log_b x$

EXAMPLE 4 ▶▶ Find the derivative of $y = \ln\frac{x-1}{x+1}$.

In this example it is easier to write y in the form

$$y = \ln(x-1) - \ln(x+1)$$

by using the properties of logarithms (Eq. (13-8)). Hence,

$$\frac{dy}{dx} = \frac{1}{x-1} - \frac{1}{x+1} = \frac{x+1-x+1}{(x-1)(x+1)}$$

$$= \frac{2}{x^2-1} \quad ◀◀$$

EXAMPLE 5 ▶▶ (a) Find the derivative of $y = \ln(1-2x)^3$.

First, using Eq. (13-9), we rewrite the equation as $y = 3\ln(1-2x)$. Then we have

$$\frac{dy}{dx} = 3\left(\frac{1}{1-2x}\right)(-2) = \frac{-6}{1-2x}$$

(b) Find the derivative of $y = \ln^3(1-2x)$.

First we note that

$$y = \ln^3(1-2x) = [\ln(1-2x)]^3$$

where $\ln^3(1-2x)$ is usually the preferred notation.

Next, we must be careful to distinguish this function from that in part (a). For $y = \ln^3(1-2x)$, it is the logarithm of $1-2x$ that is being cubed, whereas for $y = \ln(1-2x)^3$, it is $1-2x$ that is being cubed.

Now, finding the derivative of $y = \ln^3(1-2x)$, we have

$$\frac{dy}{dx} = 3[\ln^2(1-2x)]\left(\frac{1}{1-2x}\right)(-2)$$

$$= -\frac{6\ln^2(1-2x)}{1-2x} \qquad \frac{d\ln(1-2x)}{dx} \quad ◀◀$$

EXAMPLE 6 ▶▶ Evaluate the derivative of $y = \ln[(\sin 2x)(\sqrt{x^2+1})]$ for $x = 0.375$.

First, using the properties of logarithms, we rewrite the function as

$$y = \ln\sin 2x + \frac{1}{2}\ln(x^2+1)$$

Now we have

$$\frac{dy}{dx} = \frac{1}{\sin 2x}(\cos 2x)(2) + \frac{1}{2}\left(\frac{1}{x^2+1}\right)(2x) \qquad \text{take the derivative}$$

$$= 2\cot 2x + \frac{x}{x^2+1}$$

$$\left.\frac{dy}{dx}\right|_{x=0.375} = 2\cot 0.750 + \frac{0.375}{0.375^2+1} = 2.48 \qquad \text{evaluate} \quad ◀◀$$

Exercises 27–5

In Exercises 1–32, find the derivatives of the given functions.

1. $y = \log x^2$
2. $y = \log_2 6x$
3. $y = 2 \log_5 (3x + 1)$
4. $y = 3 \log_7 (x^2 + 1)$
5. $y = \ln (1 - 3x)$
6. $y = 2 \ln (3x^2 - 1)$
7. $y = 2 \ln \tan 2x$
8. $y = \ln \sin^2 x$
9. $y = \ln\sqrt{x}$
10. $y = 5 \ln\sqrt{4x - 3}$
11. $y = \ln (x^2 + 2x)^3$
12. $y = \ln (2x^3 - x)^2$
13. $y = x \ln x^2$
14. $y = x^2 \ln 2x$
15. $y = \dfrac{3x}{\ln (2x + 1)}$
16. $y = \dfrac{8 \ln x}{x}$
17. $y = \ln (\ln x)$
18. $y = \ln \cos x^2$
19. $y = \ln \dfrac{2x}{1 + x}$
20. $y = \ln (x\sqrt{x + 1})$
21. $y = \sin \ln x$
22. $y = \tan^{-1} \ln 2x$
23. $y = 3 \ln^2 2x$
24. $y = x \ln^3 x$
25. $y = \ln (x \tan x)$
26. $y = \ln (x + \sqrt{x^2 - 1})$
27. $y = \ln \dfrac{x^2}{x + 2}$
28. $y = \sqrt{x + \ln 3x}$
29. $y = \sqrt{x^2 + 1} - \ln \dfrac{1 + \sqrt{x^2 + 1}}{x}$
30. $3 \ln xy + \sin y = x^2$
31. $y = x - \ln^2 (x + y)$
32. $y = \ln (x + \ln x)$

In Exercises 33–48, solve the given problems.

33. On a calculator find the value of $(\ln 2.0001 - \ln 2.0000)/0.0001$ and compare it with 0.5. Give the meanings of the value found and 0.5 in relation to the derivative of $\ln x$, where $x = 2$.
34. On a calculator find the value of $(\ln 0.5001 - \ln 0.5000)/0.0001$ and compare it with 2. Give the meanings of the value found and 2 in relation to the derivative of $\ln x$, where $x = 0.5$.
35. Using a graphing calculator, (a) display the graph $y = (1 + x)^{1/x}$ to verify that $(1 + x)^{1/x} \to 2.718$ as $x \to 0$, and (b) verify the values for $(1 + x)^{1/x}$ in the tables on p. 778.
36. Find the second derivative of the function $y = x^2 \ln x$.
37. Evaluate the derivative of $y = \sin^{-1} 2x + \sqrt{1 - 4x^2}$, where $x = 0.250$.
38. Evaluate the derivative of $y = \ln \sqrt{\dfrac{2x + 1}{3x + 1}}$, where $x = 2.75$.
39. Find the differential of the function $y = \ln \cos^2 x - 2 \ln \tan x$.
40. Find the differential of the function $y = 6 \log_x 2$.
41. Find the slope of a line tangent to the curve of $y = \tan^{-1} 2x + \ln(4x^2 + 1)$ where $x = 0.625$. Verify the result by using the derivative-evaluating feature of a graphing calculator.
42. Find the slope of a line tangent to the curve of $y = x \ln 2x$ at $x = 2$. Verify the result by using the derivative-evaluating feature of a graphing calculator.
43. Find the derivative of $y = x^x$ by first taking logarithms of each side of the equation. Explain why Eq. (23-15) cannot be used to find the derivative of this function.
44. Find the derivative of $y = (\sin x)^x$ by first taking logarithms of each side of the equation. Explain why Eq. (23-15) cannot be used to find the derivative of this function.
45. If the loudness b (in decibels) of a sound of intensity I is given by $b = 10 \log (I/I_0)$ where I_0 is a constant, find the expression for db/dt in terms of dI/dt.
46. The time t for a particular computer system to process n bits of data is directly proportional to $N \ln N$. Find the expression for dt/dN.
47. When air friction is considered, the time t (in seconds) it takes a certain falling object to attain a velocity v (in m/s) is given by $t = 5 \ln \dfrac{5}{5 - 0.1v}$. Find dt/dv for $v = 10.0$ m/s.
48. The electric potential V at a distance x from an electric charge distributed along a wire of length $2a$ is $V = k \ln \dfrac{\sqrt{a^2 + x^2} + a}{\sqrt{a^2 + x^2} - a}$, where k is a constant. Find the expression for the electric field E, which is defined as $E = -dV/dx$.

27-6 Derivative of the Exponential Function

To obtain the derivative of the exponential function, we let $y = b^u$ and then take natural logarithms of both sides:

$$\ln y = \ln b^u = u \ln b$$

$$\frac{1}{y}\frac{dy}{dx} = \ln b \frac{du}{dx}$$

$$\frac{dy}{dx} = y \ln b \frac{du}{dx}$$

Thus

$$\frac{d(b^u)}{dx} = b^u \ln b\left(\frac{du}{dx}\right) \tag{27-15}$$

If we let $b = e$, Eq. (27-15) becomes

$$\frac{d(e^u)}{dx} = e^u\left(\frac{du}{dx}\right) \tag{27-16}$$

The simplicity of Eq. (27-16) compared with Eq. (27-15) again shows the advantage of choosing e as the basis of natural logarithms. It is for this reason that e appears so often in applications of calculus.

EXAMPLE 1 ▶▶ Find the derivative of $y = e^x$.

Using Eq. (27-16), we have

$$\frac{dy}{dx} = e^x\overset{\frac{du}{dx}}{\overset{\downarrow}{(1)}} = e^x$$

We see that the derivative of the function e^x equals itself. This exponential function is widely used in applications of calculus. ◀◀

For reference, Eq. (23-15) is $\frac{du^n}{dx} = nu^{n-1}\left(\frac{du}{dx}\right)$.

We should note carefully that Eq. (23-15) is used with a variable raised to a constant exponent, whereas with Eqs. (27-15) and (27-16) we are finding the derivative of a constant raised to a variable exponent.

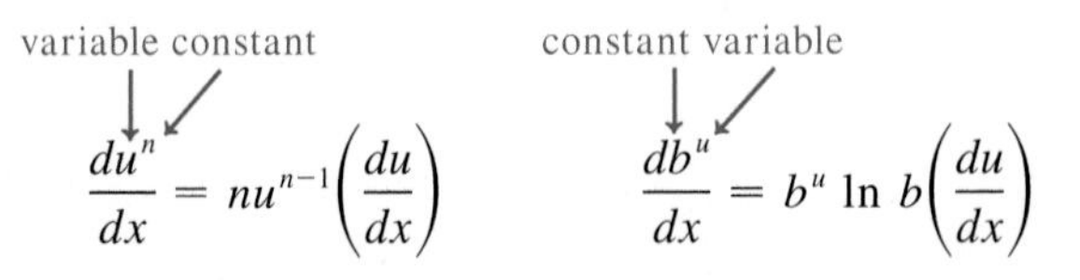

$$\frac{d\overset{\substack{\text{variable}\\\downarrow}}{u}\overset{\substack{\text{constant}\\\swarrow}}{{}^{n}}}{dx} = nu^{n-1}\left(\frac{du}{dx}\right) \qquad \frac{d\overset{\substack{\text{constant}\\\downarrow}}{b}\overset{\substack{\text{variable}\\\swarrow}}{{}^{u}}}{dx} = b^u \ln b\left(\frac{du}{dx}\right)$$

In the following example we must note carefully this difference in the type of function that leads us to use either Eq. (23-15) or Eq. (27-15) in order to find the derivative.

EXAMPLE 2 ▸▸ Find the derivatives of $y = (4x)^2$ and $y = 2^{4x}$.

Using Eq. (23-15), we have

$$y = (4x)^2$$

$$\frac{dy}{dx} = 2(4x)^1(4) \quad \frac{du}{dx}$$

$$= 32x$$

Using Eq. (27-15), we have

$$y = 2^{4x}$$

$$\frac{dy}{dx} = 2^{4x}(\ln 2)(4) \quad \frac{du}{dx}$$

$$= (4 \ln 2)(2^{4x})$$ ◂◂

We continue now with additional examples of the use of Eq. (27-16).

EXAMPLE 3 ▸▸ Find the derivative of $y = \ln \cos e^{2x}$.

Using Eq. (27-16), along with the derivatives of the logarithmic and cosine functions, we have

$$\frac{dy}{dx} = \frac{1}{\cos e^{2x}} \frac{d \cos e^{2x}}{dx} \qquad \text{using } \frac{d \ln u}{dx} = \frac{1}{u}\frac{du}{dx}$$

$$= \frac{1}{\cos e^{2x}}(-\sin e^{2x})\frac{de^{2x}}{dx} \qquad \text{using } \frac{d \cos u}{dx} = -\sin u \frac{du}{dx}$$

$$= -\frac{\sin e^{2x}}{\cos e^{2x}}(e^{2x})(2) \qquad \text{using } \frac{de^u}{dx} = e^u \frac{du}{dx}$$

$$= -2e^{2x} \tan e^{2x} \qquad \text{using } \frac{\sin \theta}{\cos \theta} = \tan \theta$$ ◂◂

EXAMPLE 4 ▸▸ Find the derivative of $y = xe^{\tan x}$.

Here we use Eq. (27-16) with the derivatives of a product and the tangent.

$$\frac{dy}{dx} = xe^{\tan x}(\sec^2 x) + e^{\tan x}(1)$$

$$= e^{\tan x}(x \sec^2 x + 1)$$ ◂◂

EXAMPLE 5 ▸▸ Find the derivative of $y = (e^{1/x})^2$.

In this example we use Eqs. (23-15) and (27-16).

$$\frac{dy}{dx} = 2(e^{1/x})(e^{1/x})\left(-\frac{1}{x^2}\right)$$

using Eqs. (27-16) and (23-15) to find $\frac{du}{dx}$ of Eq. (23-15)

$$= \frac{-2(e^{1/x})^2}{x^2} = \frac{-2e^{2/x}}{x^2}$$

This problem could have also been solved by first writing the function as $y = e^{2/x}$, which is an equivalent form determined by the laws of exponents. When we use this form, the derivative becomes

using Eq. (23-15) to find $\frac{du}{dx}$ of Eq. (27-16)

$$\frac{dy}{dx} = e^{2/x}\left(-\frac{2}{x^2}\right) = \frac{-2e^{2/x}}{x^2}$$

This change in form of the function simplifies the steps necessary for finding the derivative. ◂◂

EXAMPLE 6 ▸▸ Find the derivative of $y = (3e^{4x} + 4x^2 \ln x)^3$.

Using the general power rule (Eq. (23-15)) for derivatives, the derivative of the exponential function (Eq. (27-16)), the derivative of a product (Eq. (23-12)), and the derivative of a logarithm (Eq. (27-14)), we have

$$\frac{dy}{dx} = 3(3e^{4x} + 4x^2 \ln x)^2\left[12e^{4x} + 4x^2\left(\frac{1}{x}\right) + (\ln x)(8x)\right]$$
$$= 3(3e^{4x} + 4x^2 \ln x)^2(12e^{4x} + 4x + 8x \ln x)$$ ◂◂

EXAMPLE 7 ▸▸ Find the slope of a line tangent to the curve of $y = \frac{3e^{2x}}{x^2 + 1}$ where $x = 1.275$.

Here we are to find the derivative and then evaluate it for $x = 1.275$. The solution is as follows:

$$\frac{dy}{dx} = \frac{(x^2 + 1)(3e^{2x})(2) - 3e^{2x}(2x)}{(x^2 + 1)^2} \quad \text{take the derivative}$$
$$= \frac{6e^{2x}(x^2 - x + 1)}{(x^2 + 1)^2}$$
$$\left.\frac{dy}{dx}\right|_{x=1.275} = \frac{6e^{2(1.275)}(1.275^2 - 1.275 + 1)}{(1.275^2 + 1)^2} \quad \text{evaluate}$$
$$= 15.05$$

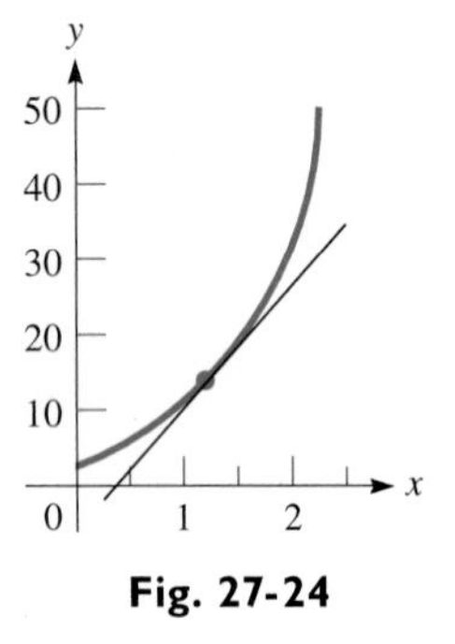

Fig. 27-24

The graph of the function and the tangent line is shown in Fig. 27-24. ◂◂

Exercises 27–6

In Exercises 1–32, find the derivatives of the given functions.

1. $y = 3^{2x}$
2. $y = 3^{1-x}$
3. $y = 4^{6x}$
4. $y = 10^{x^2}$
5. $y = e^{6x}$
6. $y = 3e^{x^2}$
7. $y = e^{\sqrt{x}}$
8. $y = e^{2x^4}$
9. $y = xe^{-x}$
10. $y = 5x^2e^{2x}$
11. $y = xe^{\sin x}$
12. $y = 4e^x \sin \frac{1}{2}x$
13. $y = \frac{3e^{2x}}{x + 1}$
14. $y = \frac{e^x}{x}$
15. $y = e^{-3x} \sin 4x$
16. $y = (\cos 2x)(e^{x^2-1})$
17. $y = \frac{2e^{3x}}{4x + 3}$
18. $y = \frac{7 \ln 2x}{e^{2x} + 2}$
19. $y = \ln(e^{x^2} + 4)$
20. $y = (3e^{2x} + x)^3$
21. $y = (2e^{2x})^3 \sin x^2$
22. $y = (e^{3/x} \cos x)^2$
23. $y = (\ln 2x + e^{2x})^2$
24. $y = (2e^{x^2} + x^2)^3$
25. $y = xe^{xy} + \sin y$
26. $y = 4e^{-2/x} \ln y + 1$
27. $y = 3e^{2x} \ln x$
28. $y = e^{x^2} \ln \cos x$
29. $y = \ln \sin 2e^{6x}$
30. $y = 6 \tan e^{x+1}$
31. $y = 2 \sin^{-1} e^{2x}$
32. $y = \tan^{-1} e^{3x}$

In Exercises 33–48, solve the given problems.

33. On a calculator, find the values of (a) e and (b) $(e^{1.0001} - e^{1.0000})/0.0001$. Compare the values and give the meaning of each in relation to the derivative of e^x where $x = 1$.
34. On a calculator, find the values of (a) e^2 and (b) $(e^{2.0001} - e^{2.0000})/0.0001$. Compare the values and give the meaning of each in relation to the derivative of e^x where $x = 2$.
35. Find the slope of a line tangent to the curve of $y = e^{-2x} \cos 2x$ for $x = 0.625$. Verify the result by using the derivative-evaluating feature of a graphing calculator.

36. Find the slope of a line tangent to the curve of $y = \dfrac{e^{-x}}{1 + \ln 4x}$ for $x = 1.842$. Verify the result by using the derivative-evaluating feature of a graphing calculator.

37. Find the differential of the function $y = 4e^{x/2}(x + \ln x)$.

38. Find the differential of the function $y = 2e^{4x}/(x + 2)$.

39. Use a graphing calculator to display the graph of $y = e^x$. By roughly estimating slopes of tangent lines, note that it is reasonable that these values are equal to the y-coordinates of the points at which these estimates are made. (*Remember:* For $y = e^x$, $dy/dx = e^x$ also.)

40. Use a graphing calculator to display the graphs of $y = e^{-x}$ and $y = -e^{-x}$. By roughly estimating slopes of tangent lines of $y = e^{-x}$, note that $y = -e^{-x}$ gives reasonable values for the derivative of $y = e^{-x}$.

41. Show that $y = xe^{-x}$ satisfies the equation $(dy/dx) + y = e^{-x}$.

42. Show that $y = e^{-x} \sin x$ satisfies the equation

$$\frac{d^2y}{dx^2} + 2\frac{dy}{dx} + 2y = 0$$

43. For $y = \dfrac{e^{2x} - 1}{e^{2x} + 1}$ show that $\dfrac{dy}{dx} = 1 - y^2$.

44. If $e^x + e^y = e^{x+y}$, show that $dy/dx = -e^{y-x}$.

45. The electric current i in a certain circuit is given by $i = 2te^{-0.5t}$, where t is the time. Find the expression for the instantaneous time rate of change of i.

46. The Beer-Lambert law of light absorption may be expressed as $I = I_0 e^{-\alpha x}$, where I/I_0 is that fraction of the incident light beam which is transmitted, α is a constant, and x is the distance the light travels through the medium. Find the expression for the instantaneous rate of change of I with respect to x.

47. The reliability R $(0 \le R \le 1)$ of a certain computer system is $R = e^{-0.002t}$, where t is the time of operation in hours. Find dR/dt for $t = 100$ h.

48. An equation used in analyzing biological cells is $n = N(1 - e^{-at})$, where a and N are constants and t is the time. Express dn/dt as a function of n.

In Exercises 49–52, use the following information.

The **hyperbolic sine** *of u is defined as*

$$\sinh u = \frac{1}{2}(e^u - e^{-u})$$

Figure 27-25 shows the graph of $y = \sinh x$.

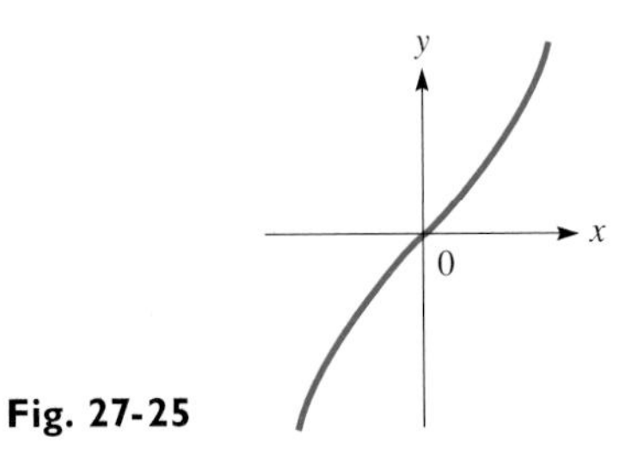

Fig. 27-25

The **hyperbolic cosine** *of u is defined as*

$$\cosh u = \frac{1}{2}(e^u + e^{-u})$$

Figure 27-26 shows the graph of $y = \cosh x$.

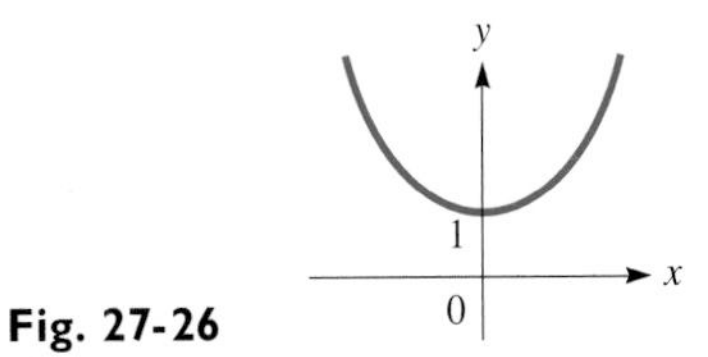

Fig. 27-26

These functions are called hyperbolic *functions since, if $x = \cosh u$ and $y = \sinh u$, x and y satisfy the equation of the hyperbola $x^2 - y^2 = 1$.*

49. Verify the fact that the expressions for the hyperbolic sine and hyperbolic cosine satisfy the equation of the hyperbola.

50. Show that $\sinh u$ and $\cosh u$ satisfy the identity $\cosh^2 u - \sinh^2 u = 1$.

51. Show that

$$\frac{d}{dx}\sinh u = \cosh u\frac{du}{dx} \quad \text{and}$$

$$\frac{d}{dx}\cosh u = \sinh u\frac{du}{dx}$$

where u is a function of x.

52. Show that

$$\frac{d^2 \sinh x}{dx^2} = \sinh x \quad \text{and} \quad \frac{d^2 \cosh x}{dx^2} = \cosh x$$

27-7 Applications

The following examples show applications of the logarithmic and exponential functions to curve tracing, Newton's method, and time-rate-of-change problems. Certain other applications are indicated in the exercises.

EXAMPLE 1 ▸▸ Sketch the graph of the function $y = x \ln x$.

First we note that x cannot be zero since $\ln x$ is not defined at $x = 0$. Since $\ln 1 = 0$, we have an intercept at $(1, 0)$. There is no symmetry to the axes or origin, and there are no vertical asymptotes. Also, because $\ln x$ is defined only for $x > 0$, the domain is $x > 0$.

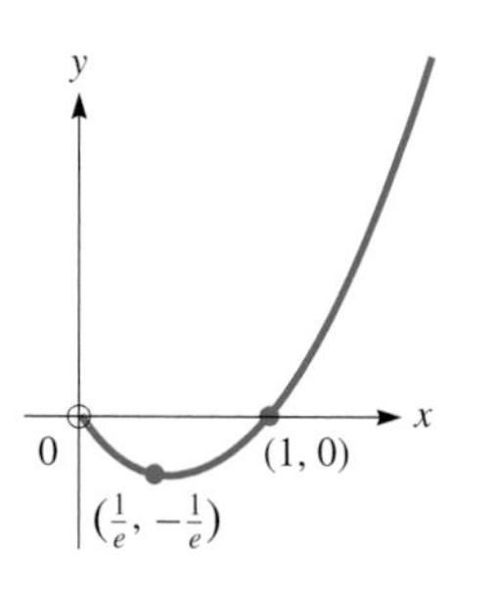

Fig. 27-27

Finding the first two derivatives, we have

$$\frac{dy}{dx} = x\left(\frac{1}{x}\right) + \ln x = 1 + \ln x, \qquad \frac{d^2y}{dx^2} = \frac{1}{x}$$

The first derivative is zero if $\ln x = -1$, or $x = e^{-1}$. The second derivative is positive for this value of x. Thus, there is a minimum point at $(1/e, -1/e)$. Since the domain is $x > 0$, the second derivative indicates that the curve is always concave up. In turn, we now see that the range of the function is $y \geq -1/e$. The graph is shown in Fig. 27-27.

Although the curve approaches the origin as x approaches zero, the origin is not included on the graph of the function. ◂◂

EXAMPLE 2 ▸▸ Sketch the graph of the function $y = e^{-x} \cos x \, (0 \leq x \leq 2\pi)$.

This curve has intercepts for all values for which $\cos x$ is zero. Those values in the domain $0 \leq x \leq 2\pi$ for which $\cos x = 0$ are $x = \frac{\pi}{2}$ and $x = \frac{3\pi}{2}$. The factor e^{-x} is always positive, and $e^{-x} = 1$ for $x = 0$, which means $(0, 1)$ is also an intercept. There is no symmetry to the axes or the origin, and there are no vertical asymptotes.

Next, finding the first derivative, we have

$$\frac{dy}{dx} = -e^{-x} \sin x - e^{-x} \cos x = -e^{-x}(\sin x + \cos x)$$

Setting the derivative equal to zero, since e^{-x} is always positive, we have

$$\sin x + \cos x = 0, \qquad \tan x = -1, \qquad x = \frac{3\pi}{4}, \frac{7\pi}{4}$$

Now, finding the second derivative, we have

$$\frac{d^2y}{dx^2} = -e^{-x}(\cos x - \sin x) - e^{-x}(-1)(\sin x + \cos x) = 2e^{-x}\sin x$$

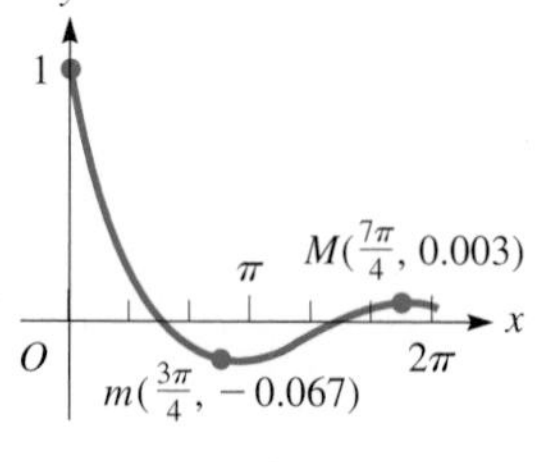

Fig. 27-28

The sign of the second derivative depends only on $\sin x$. Thus, $d^2y/dx^2 > 0$ for $x = \frac{3\pi}{4}$ and $d^2y/dx^2 < 0$ for $x = \frac{7\pi}{4}$. Hence, $(\frac{3\pi}{4}, -0.067)$ is a minimum and $(\frac{7\pi}{4}, 0.003)$ is a maximum. Also, from the second derivative, points of inflection occur for $x = 0$, π, and 2π since $\sin x = 0$ for these values. The graph is shown in Fig. 27-28. ◂◂

EXAMPLE 3 ▸▸ Find the root of the equation $e^{2x} - 4 \cos x = 0$ which lies between 0 and 1 by using Newton's method.

Here

$$f(x) = e^{2x} - 4 \cos x$$
$$f'(x) = 2e^{2x} + 4 \sin x$$

This means that $f(0) = -3$ and $f(1) = 5.2$. Therefore, we choose $x_1 = 0.5$. Using Eq. (24-1), which is

$$x_2 = x_1 - \frac{f(x_1)}{f'(x_1)}$$

we have these values:

$$f(x_1) = e^{2(0.5)} - 4 \cos 0.5 = -0.792\,048\,4$$
$$f'(x_1) = 2e^{2(0.5)} + 4 \sin 0.5 = 7.354\,265\,8$$
$$x_2 = 0.5 - \frac{-0.792\,048\,4}{7.354\,265\,8} = 0.607\,699\,2$$

Using the method again, we find $x_3 = 0.597\,975\,1$, which is correct to three decimal places. ◂◂

EXAMPLE 4 ▸▸ A good model for population growth is that the population P at time t is given by $P = P_0e^{kt}$, where P_0 is the *initial* population ($t = 0$, when timing starts for the population being considered) and k is a constant. Show that the instantaneous time rate of change of population is directly proportional to the population present at time t.

To find the time rate of change, we find the derivative dP/dt:

$$\frac{dP}{dt} = (P_0e^{kt})(k) = kP_0e^{kt}$$
$$= kP \qquad \text{since } P = P_0e^{kt}$$

Thus we see that population growth increases as the population increases. ◂◂

See the chapter introduction.

EXAMPLE 5 ▸▸ A rocket is moving such that the only force on it is due to gravity and its mass is decreasing at a constant rate r. If it moves vertically, its velocity v as a function of the time t is given by

$$v = v_0 - gt - k \ln\left(1 - \frac{rt}{m_0}\right)$$

where v_0 is the initial velocity, g is the acceleration due to gravity, t is the time, m_0 is the initial mass, and k is a constant. Determine the expression for the acceleration.

Since the acceleration is the time rate of change of the velocity, we must find dv/dt. Therefore,

$$\frac{dv}{dt} = -g - k\frac{1}{1 - \frac{rt}{m_0}}\left(\frac{-r}{m_0}\right) = -g + \frac{km_0}{m_0 - rt}\left(\frac{r}{m_0}\right)$$
$$= \frac{kr}{m_0 - rt} - g$$ ◂◂

Exercises 27–7

In Exercises 1–12, sketch the graphs of the given functions. Check each by displaying the graph on a graphing calculator.

1. $y = \ln \cos x$ **2.** $y = \dfrac{\ln x}{x}$

3. $y = xe^{-x}$ **4.** $y = \dfrac{e^x}{x}$

5. $y = \ln \dfrac{1}{x^2 + 1}$ **6.** $y = \ln \dfrac{1}{x}$

7. $y = e^{-x^2}$ **8.** $y = x - e^x$

9. $y = \ln x - x$ **10.** $y = e^{-x} \sin x$

11. $y = \frac{1}{2}(e^x - e^{-x})$ (See Exercise 49 of Section 27-6.)

12. $y = \frac{1}{2}(e^x + e^{-x})$ (See Exercise 49 of Section 27-6.)

In Exercises 13–32, solve the given problems by finding the appropriate derivative.

13. Find the equation of the line tangent to the curve of $y = x^2 \ln x$ at the point $(1, 0)$.

14. Find the equation of the line tangent to the curve of $y = \tan^{-1} 2x$, where $x = 1$.

15. Find the equation of the line normal to the curve of $y = 2 \sin \frac{1}{2}x$, where $x = \frac{3}{2}\pi$.

16. Find the equation of the line normal to the curve of $y = e^{2x}/x$ at $x = 1$.

17. By Newton's method, solve the equation $x^2 - 2 + \ln x = 0$. Check the solution by displaying the graph on a graphing calculator and then using the TRACE and ZOOM features.

18. By Newton's method, solve the equation $e^{-2x} - \tan^{-1} x = 0$. Check the solution by displaying the graph on a graphing calculator and then using the TRACE and ZOOM features.

19. The power supply P (in watts) in a satellite is given by $P = 100e^{-0.005t}$, where t is measured in days. Find the time rate of change of power after 100 days.

20. The number N of atoms of radium at any time t is given in terms of the number at $t = 0$, N_0, by $N = N_0 e^{-kt}$. Show that the time rate of change of N is proportional to N.

21. The vapor pressure p and thermodynamic temperature T of a gas are related by the equation $\ln p = \dfrac{a}{T} + b \ln T + c$, where a, b, and c are constants. Find the expression for dp/dT.

22. The charge on a capacitor in a circuit containing the capacitor of capacitance C, a resistance R, and a source of voltage E is given by $q = CE(1 - e^{-t/RC})$. Show that this equation satisfies the equation $R\dfrac{dq}{dt} + \dfrac{q}{C} = E$.

23. Assuming that force is proportional to acceleration, show that a particle moving along the x-axis, so that its displacement $x = ae^{kt} + be^{-kt}$, has a force acting on it which is proportional to its displacement.

24. The radius of curvature at a point on a curve is given by

$$R = \frac{[1 + (dy/dx)^2]^{3/2}}{d^2y/dx^2}$$

A roller mechanism moves along the path defined by $y = \ln \sec x$, $-1.5 \text{ dm} \le x \le 1.5 \text{ dm}$. Find the radius of curvature of this path for $x = 0.85$ dm.

25. Sketch the graph of $y = \ln \sec x$, marking that part which is the path of the roller mechanism of Exercise 24.

26. In an electronic device, the maximum current density i_m as a function of the temperature T is given by $i_m = AT^2 e^{k/T}$, where A and k are constants. If the temperature is changing with time, find the expression for the time rate of change of i_m.

27. In the development of the theory dealing with the friction between a pulley wheel and the pulley belt, the ratio of the tensions in the belt on either side of the wheel is given by $R_T = e^{k \csc(\theta/2)}$, where k is a constant and θ is the angle of the opening of the pulley wheel. Find the expression for a small change in the ratio of tensions for a small change in the angle θ.

28. The reliability R $(0 \le R \le 1)$ of a certain computer system for t hours of operation is found from the equation $R = 3e^{-0.004t} - 2e^{-0.006t}$. Use Newton's method to find how long the system operates to have a reliability of 0.8 (80% probability that there will be no system failure).

29. An object on the end of a spring is moving so that its displacement (in cm) from the equilibrium position is given by $y = e^{-0.5t}(0.4 \cos 6t - 0.2 \sin 6t)$. Find the expression for the velocity of the object. What is the velocity when $t = 0.26$ s? The motion described by this equation is called *damped harmonic motion*.

30. A package of weather instruments is propelled into the air to an altitude of about 7 km. A parachute then opens, and the package returns to the surface. The altitude y of the package as a function of the time t, in minutes, is given by $y = \dfrac{10t}{e^{0.4t} + 1}$. Find the vertical velocity of the package for $t = 8.0$ min.

31. The speed s of signaling by use of a certain communications cable is directly proportional to $x^2 \ln \frac{1}{x}$, where x is the ratio of the radius of the core of the cable to the thickness of the surrounding insulation. For what value of x is s a maximum?

32. A computer is programmed to inscribe a series of rectangles in the first quadrant under the curve of $y = e^{-x}$. What is the area of the largest rectangle that can be inscribed? See Fig. 27-29.

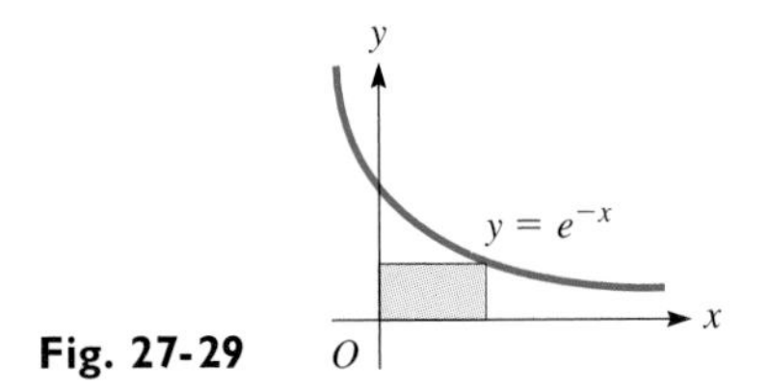

Fig. 27-29

Chapter Equations, Review Exercises, and Practice Test

Chapter Equations

Limit of $\dfrac{\sin\theta}{\theta}$ as $\theta \to 0$

$$\lim_{\theta\to 0}\frac{\sin\theta}{\theta} = \lim_{\Delta u\to 0}\frac{\sin(\Delta u/2)}{\Delta u/2} = 1 \qquad (27\text{-}1)$$

Chain rule

$$\frac{dy}{dx} = \frac{dy}{du}\frac{du}{dx} \qquad (27\text{-}3)$$

Derivatives

$$\frac{d(\sin u)}{dx} = \cos u\,\frac{du}{dx} \qquad (27\text{-}4)$$

$$\frac{d(\cos u)}{dx} = -\sin u\,\frac{du}{dx} \qquad (27\text{-}5)$$

$$\frac{d(\tan u)}{dx} = \sec^2 u\,\frac{du}{dx} \qquad (27\text{-}6)$$

$$\frac{d(\cot u)}{dx} = -\csc^2 u\,\frac{du}{dx} \qquad (27\text{-}7)$$

$$\frac{d(\sec u)}{dx} = \sec u\tan u\frac{du}{dx} \qquad (27\text{-}8)$$

$$\frac{d(\csc u)}{dx} = -\csc u\cot u\frac{du}{dx} \qquad (27\text{-}9)$$

$$\frac{d(\sin^{-1} u)}{dx} = \frac{1}{\sqrt{1-u^2}}\frac{du}{dx} \qquad (27\text{-}10)$$

$$\frac{d(\cos^{-1} u)}{dx} = -\frac{1}{\sqrt{1-u^2}}\frac{du}{dx} \qquad (27\text{-}11)$$

$$\frac{d(\tan^{-1} u)}{dx} = \frac{1}{1+u^2}\frac{du}{dx} \qquad (27\text{-}12)$$

Derivatives

$$\frac{d(\log_b u)}{dx} = \frac{1}{u}\log_b e\frac{du}{dx} \tag{27-13}$$

$$\frac{d(\ln u)}{dx} = \frac{1}{u}\frac{du}{dx} \tag{27-14}$$

$$\frac{d(b^u)}{dx} = b^u \ln b\frac{du}{dx} \tag{27-15}$$

$$\frac{d(e^u)}{dx} = e^u\frac{du}{dx} \tag{27-16}$$

Review Exercises

In Exercises 1–40, find the derivative of each of the given functions.

1. $y = 3\cos(4x - 1)$
2. $y = 4\sec(1 - x^3)$
3. $y = \tan\sqrt{3 - x}$
4. $y = 5\sin(1 - 6x)$
5. $y = \csc^2(3x + 2)$
6. $y = \cot^2 5x$
7. $y = 3\cos^4 x^2$
8. $y = 2\sin^3\sqrt{x}$
9. $y = (e^{x-3})^2$
10. $y = 0.5e^{\sin 2x}$
11. $y = 3\ln(x^2 + 1)$
12. $y = \ln(3 + \sin x^2)$
13. $y = 3\tan^{-1}\left(\frac{x}{3}\right)$
14. $y = 4\cos^{-1}(2x + 3)$
15. $y = \ln\sin^{-1} 4x$
16. $y = \sin(\tan^{-1} x)$
17. $y = \sqrt{\csc 4x + \cot 4x}$
18. $y = \cos^2(\tan x)$
19. $y = 7\ln(x - e^{-x})^2$
20. $y = \ln\sqrt{\sin 2x}$
21. $y = \frac{\cos^2 x}{e^{3x} + 1}$
22. $y = \sqrt{\frac{1 + \cos 2x}{2}}$
23. $y = \frac{x^2}{\tan^{-1} 2x}$
24. $y = \frac{\sin^{-1} x}{4x}$
25. $y = \ln(\csc x^2)$
26. $y = 2e^{\sqrt{1-x}}$
27. $y = \ln^2(3 + \sin x)$
28. $y = \ln(3 + \sin x)^2$
29. $y = e^{-2x}\sec x$
30. $y = 5e^{3x}\ln x$
31. $y = \sqrt{\sin 2x + e^{4x}}$
32. $x + y\ln 2x = y^2$
33. $\tan^{-1}\frac{y}{x} = x^2e^y$
34. $x^2\ln y = y + x$
35. $y = x^2(e^{\cos^2 x})^2$
36. $y = (\ln 4x - \tan 4x)^3$
37. $\ln xy + ye^{-x} = 1$
38. $y = x(\sin^{-1} x)^2 + 2\sqrt{1 - x^2}\sin^{-1} x - 2x$
39. $y = x\cos^{-1} x - \sqrt{1 - x^2}$
40. $y = \ln(4x^2 + 1) + \tan^{-1} 2x$

In Exercises 41–44, sketch the graphs of the given functions. Check each by displaying the graph on a graphing calculator.

41. $y = x - \cos x$
42. $y = 4\sin x + \cos 2x$
43. $y = x(\ln x)^2$
44. $y = \ln(1 + x)$

In Exercises 45–48, find the equations of the indicated tangent or normal lines.

45. Find the equation of the line tangent to the curve of $y = 4\cos^2(x^2)$ at $x = 1$.
46. Find the equation of the line tangent to the curve of $y = \ln\cos x$ at $x = \frac{\pi}{6}$.
47. Find the equation of the line normal to the curve of $y = e^{x^2}$ at $x = \frac{1}{2}$.
48. Find the equation of the line normal to the curve of $y = \tan^{-1} x$ at $x = 1$.

In Exercises 49–80, solve the given problems.

49. Find the derivative of each member of the identity $\sin^2 x + \cos^2 x = 1$, and show that the results are equal.
50. Find the derivative of each member of the identity

$$\sin(x + 1) = \sin x\cos 1 + \cos x\sin 1$$

and show that the results are equal.

51. By Newton's method, solve the equation $e^x - x^2 = 0$. Check the solution by displaying the graph on a graphing calculator and then using the TRACE and ZOOM features.

52. By Newton's method, solve the equation $x^2 = \tan^{-1} x$. Check the solution by displaying the graph on a graphing calculator and then using the TRACE and ZOOM features.

53. The vertical displacement y (in cm) of an object at the end of a spring is given by $y = 3.5 \sin(0.75\pi t + 0.50)$, where t is the time in seconds. Find the velocity of the object for $t = 1.50$ s.

54. An earth-orbiting satellite is launched such that its altitude y (in km) is given by $y = 240(1 - e^{-0.05t})$, where t is the time in minutes. Find the vertical velocity of the satellite for $t = 10.0$ min.

55. Power can be defined as the time rate of doing work. If work is being done in an electric circuit according to $W = 10 \cos 2t$, find P as a function of t.

56. The value V of a bank account in which \$1000 is deposited and then earns 6% annual interest, compounded continuously (daily compounding approximates this, and some banks actually use continuous compounding), is $V = 1000e^{0.06t}$ after t years. How fast is the account growing after exactly 2 years?

57. In determining how to divide files on the hard disk of a computer, we can use the equation $n = xN \log_x N$. Sketch the graph of n vs. x for $1 < x \leq 10$ if $N = 8$.

58. Under certain conditions, the potential V (in volts) due to a magnet is given by $V = -k \ln\left(1 + \frac{L}{x}\right)$, where L is the length of the magnet and x is the distance from the point where the potential is measured. Find the expression for dV/dx.

59. In the theory of making images by holography, an expression used for the light-intensity distribution is $I = kE_0^2 \cos^2 \frac{1}{2}\theta$, where k and E_0 are constants and θ is the phase angle between two light waves. Find the expression for $dI/d\theta$.

60. If we neglect air resistance, the range R of a bullet fired at an angle θ with the horizontal is $R = \frac{v_0^2}{g} \sin 2\theta$, where v_0 is the initial velocity and g is the acceleration due to gravity. Find θ for the maximum range. See Fig. 27-30.

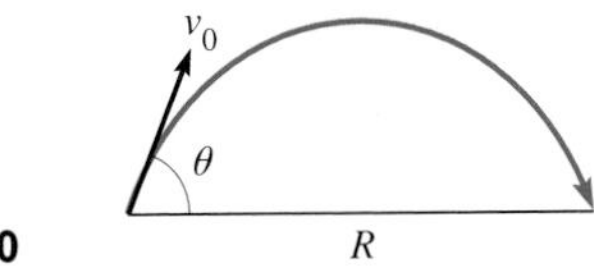

Fig. 27-30

61. In the design of a cone-type clutch, an equation that relates the cone angle θ and the applied force F is $\theta = \sin^{-1} \frac{Ff}{R}$, where R is the frictional resistance and f is the coefficient of friction. For constant R and f, find $d\theta/dF$.

62. If inflation makes the dollar worth 5% less each year, then the value of \$100 in t years will be $V = 100(0.95)^t$. What is the approximate change in the value during the fourth year?

63. An object attached to a cord of length l, as shown in Fig. 27-31, moves in a circular path. The angular velocity ω is given by $\omega = \sqrt{g/(l \cos \theta)}$. By use of differentials, find the approximate change in ω if θ changes from 32.50° to 32.75°, given that $g = 9.800$ m/s² and $l = 0.6375$ m.

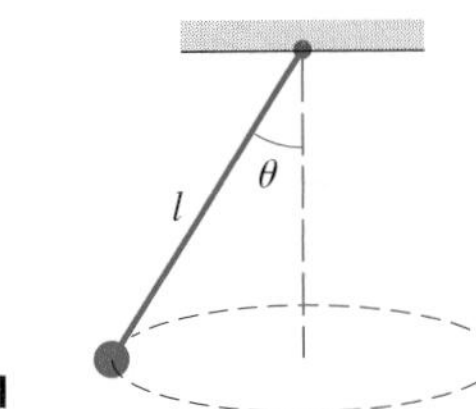

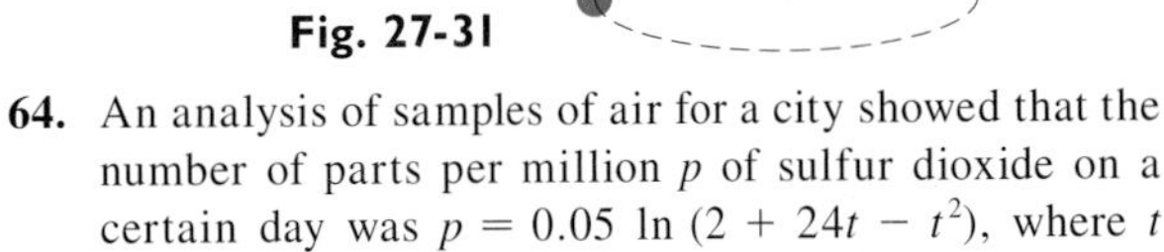

Fig. 27-31

64. An analysis of samples of air for a city showed that the number of parts per million p of sulfur dioxide on a certain day was $p = 0.05 \ln(2 + 24t - t^2)$, where t is the hour of the day. Using differentials, find the approximate change in the amount of sulfur dioxide between 10 A.M. and noon.

65. According to Newton's law of cooling (Isaac Newton, again), the rate at which a body cools is proportional to the difference in temperature between it and the surrounding medium. By use of this law, the temperature T (in °C) of an engine coolant as a function of the time t (in min) is $T = 30 + 60(0.5)^{0.2t}$. The coolant was initially at 90°C, and the air temperature was 30°C. Find dT/dt for $t = 5.00$ min.

66. The charge q on a certain capacitor in an amplifier circuit as a function of time t is given by $q = e^{-0.1t}(0.2 \sin 120\pi t + 0.8 \cos 120\pi t)$. The current i in the circuit is the instantaneous time rate of change of the charge. Find the expression for i as a function of t.

67. An object is dropped from a weather balloon. The distance (in metres) it falls, assuming a resisting force of the air on the object, is given by $y = 98.0(t + 4e^{-0.1t} - 4)$. Find the velocity after 10.0 s.

68. A football is thrown horizontally (very little arc) at 18 m/s parallel to the sideline. A TV camera is 31 m from the path of the football. Find $d\theta/dt$, the rate at which the camera must turn to follow the ball when $\theta = 15°$. See Fig. 27-32.

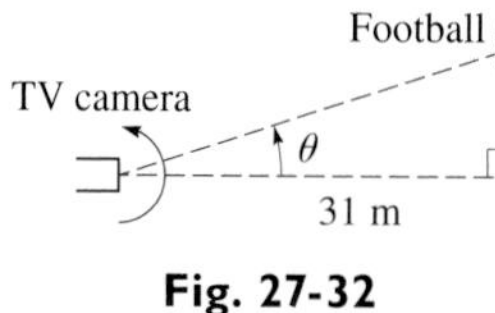

Fig. 27-32

69. An architect designs an arch of height y (in m) over a walkway by the curve of the equation $y = 3e^{-0.5x^2}$. What are the dimensions of the largest rectangular passage area under the arch?

70. A force P at an angle θ above the horizontal drags a 50-N box across a level floor. The coefficient of friction between the floor and the box is constant and equals 0.20. The magnitude of the force P is given by $P = \dfrac{(0.20)(50)}{0.20 \sin\theta + \cos\theta}$. Find θ such that P is a minimum.

71. A jet is flying at 220 m/s directly away from the control tower of an airport. If the jet is at a constant altitude of 1700 m, how fast is the angle of elevation of the jet from the control tower changing when it is 13.0°?

72. The current i in an electric circuit with a resistance R and an inductance L is $i = i_0 e^{-Rt/L}$, where i_0 is the initial current. Show that the time rate of change of the current is directly proportional to the current.

73. When a wheel rolls along a straight line, a point P on the circumference traces a curve called a *cycloid*. See Fig. 27-33. The equations of a cycloid are $x = r(\theta - \sin\theta)$ and $y = r(1 - \cos\theta)$. Find the velocity of the point on the rim of a wheel for which $r = 5.500$ cm and $d\theta/dt = 0.12$ rad/s for $\theta = 35.0°$. (An inverted cycloid is the path of least time of descent (the *brachistochrone*) of an object acted on only by gravity.)

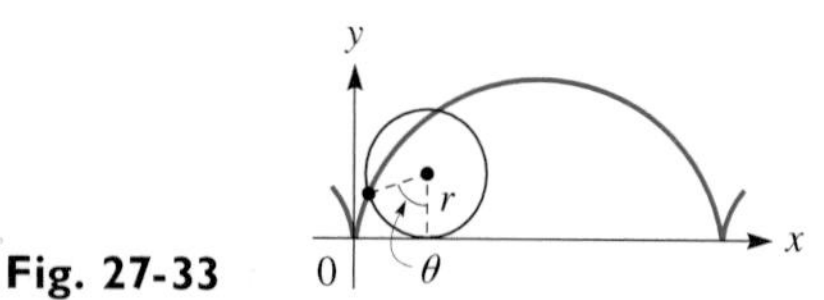

Fig. 27-33

74. In the study of atomic spectra, it is necessary to solve the equation $x = 5(1 - e^{-x})$ for x. Use Newton's method to find the solution.

75. The illuminance from a point source of light varies directly as the cosine of the angle of incidence (measured from the perpendicular) and inversely as the square of the distance r from the source. How high above the center of a circle of radius 10.0 cm should a light be placed so that the illuminance at the circumference will be a maximum? See Fig. 27-34.

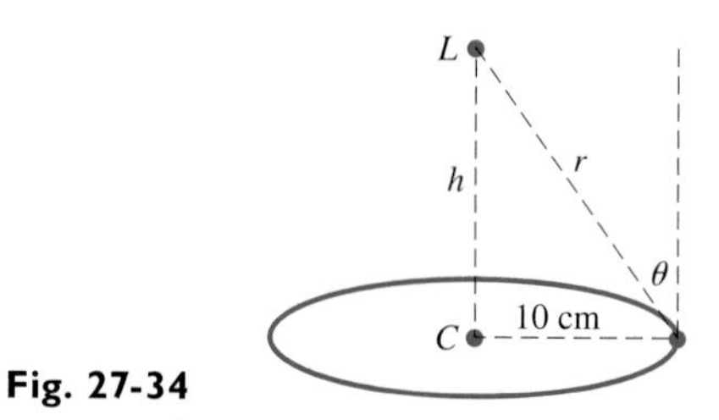

Fig. 27-34

76. A Y-shaped metal bracket is to be made such that its height is 10.0 cm and its width across the top is 6.00 cm. What shape will require the least amount of material? See Fig. 27-35.

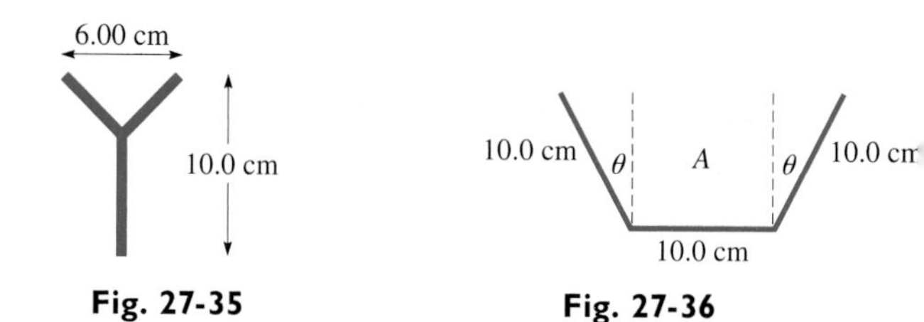

Fig. 27-35

Fig. 27-36

77. A gutter is to be made from a sheet of metal 30.0 cm wide by turning up strips of width 10.0 cm along each side to make equal angles θ with the vertical. Sketch a graph of the cross-sectional area A as a function of θ. See Fig. 27-36.

78. Show that the equation of the hyperbolic cosine function

$$y = \frac{H}{w}\cosh\frac{wx}{H} \quad (w \text{ and } H \text{ are constants})$$

satisfies the equation

$$\frac{d^2y}{dx^2} = \frac{w}{H}\sqrt{1 + \left(\frac{dy}{dx}\right)^2}$$

(see Exercise 49 on p. 785). A *catenary* (see Exercise 51 on p. 370) is the curve of a uniform cable hanging under its own weight and is in the shape of a hyperbolic cosine curve. Also, this shape (inverted) was chosen for the St. Louis Gateway Arch (shown in Fig. 27-37) and makes the arch self-supporting.

Fig. 27-37

79. A company determines that the gross income I it receives by selling x items per week is $I = 100xe^{-x/10}$. How many items should be sold each week to make I a maximum?

80. A conical filter is made from a circular piece of wire mesh of radius 24.0 cm by cutting out a sector with central angle θ and then taping the cut edges of the remaining piece together (see Fig. 27-38). What is the maximum possible volume the resulting filter can hold?

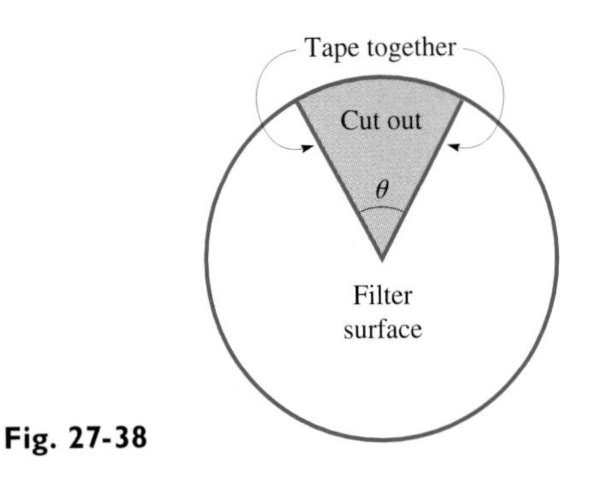

Fig. 27-38

Writing Exercise

81. To find the area of the largest rectangular microprocessor chip with a perimeter of 40 mm, it is possible to use either an algebraic function or a trigonometric function. Write two or three paragraphs to explain how each type of function can be used to find the required area.

Practice Test

In Problems 1–3, find the derivative of each of the given functions.

1. $y = \tan^3 2x + \tan^{-1} 2x$ **2.** $y = 2(3 + \cot 4x)^3$

3. $y \sec 2x = \sin^{-1} 3y$

4. Find the differential of the function $y = \dfrac{\cos^2(3x + 1)}{x}$.

5. Find the slope of a tangent to the curve of $y = \ln\dfrac{2x - 1}{1 + x^2}$ for $x = 2$.

6. Find the expression for the time rate of change of electric current that is given by the equation $i = 8e^{-t}\sin 10t$, where t is the time.

7. Sketch the graph of the function $y = xe^x$.

8. A balloon leaves the ground 250 m from an observer and rises at the rate of 5.0 m/s. How fast is the angle of elevation of the balloon increasing after 8.0 s?

28

Methods of Integration

In Section 28-5 we show an application of integration that is important in the design of electric appliances.

Having developed the derivatives of the basic transcendental functions, we can now expand considerably the functions that we are able to integrate. In addition to the transcendental functions, we will find that it is now possible to integrate many algebraic functions that we were unable to integrate previously.

In this chapter we expand the use of the general power formula for integration for use with integrands that have transcendental functions. We then develop additional standard forms for algebraic and transcendental integrands. Also, certain basic methods of integration, including the use of tables, are shown. In using all the standard forms and methods, *recognition of the integral form* will be of great importance.

As we have seen in Chapters 25 and 26, there are numerous applications of integration in geometry, science, and technology. Additional examples of these applications are found in the examples and exercises of this chapter.

28-1 The General Power Formula

The first integration formula we discuss is the general power formula. We first used this with basic algebraic functions in Chapter 25.

For reference, we repeat the general power formula for integration, Eq. (25-5):

$$\int u^n\,du = \frac{u^{n+1}}{n+1} + C \qquad (n \neq -1) \tag{28-1}$$

CAUTION ▶ The general power formula will now be applied to transcendental integrands as well as algebraic integrands. When applying this formula, ***we must properly recognize the quantities u, n, and du.*** This requires familiarity with the differential forms presented in Chapters 23 and 27. The following examples illustrate the use of the general power formula.

EXAMPLE 1 ▶▶ Integrate: $\int \sin^3 x \cos x\, dx$.

Since $d(\sin x) = \cos x\, dx$, we note that this integral fits the form of Eq. (28-1) for $u = \sin x$. Thus, with $u = \sin x$, we have $du = \cos x\, dx$, which means that this integral is of the form $\int u^3\, du$. Therefore, the integration can now be completed.

$$\int \sin^3 x \cos x\, dx = \int \sin^3 x(\underbrace{\cos x\, dx}_{du})$$

$$= \frac{1}{4}\sin^4 x + C$$

← do not forget the constant of integration

CAUTION ▶ We note here that ***the factor* cos *x is a necessary part of the du*** in order to have the proper form of integration ***and therefore does not appear in the final result.***

We also see that our result checks by finding the derivative of $\frac{1}{4}\sin^4 x + C$. This derivative is

$$\frac{d}{dx}\left(\frac{1}{4}\sin^4 x + C\right) = \frac{1}{4}(4)\sin^3 x \cos x = \sin^3 x \cos x$$ ◀◀

EXAMPLE 2 ▶▶ Integrate: $\int 2\sqrt{1 + \tan x}\sec^2 x\, dx$.

Here we note that $d(\tan x) = \sec^2 x\, dx$, which means that the integral fits the form of Eq. (28-1) with

$$u = 1 + \tan x, \qquad du = \sec^2 x\, dx, \qquad n = \tfrac{1}{2}$$

The integral is of the form $\int u^{1/2}\, du$. Thus

$$\int 2\sqrt{1 + \tan x}(\sec^2 x\, dx) = 2\int \underbrace{(1 + \tan x)}_{u}{}^{1/2}(\underbrace{\sec^2 x\, dx}_{du})$$

$$= 2\left(\frac{2}{3}\right)(1 + \tan x)^{3/2} + C = \frac{4}{3}(1 + \tan x)^{3/2} + C$$

◀◀

EXAMPLE 3 ▶▶ Integrate: $\int \ln x\left(\frac{dx}{x}\right)$.

By noting that $d(\ln x) = \frac{dx}{x}$, we have

$$u = \ln x, \qquad du = \frac{dx}{x}, \qquad n = 1$$

This means that the integral is of the form $\int u\, du$. Thus

$$\int \underbrace{\ln x}_{u}\underbrace{\left(\frac{dx}{x}\right)}_{du} = \frac{1}{2}(\ln x)^2 + C = \frac{1}{2}\ln^2 x + C$$ ◀◀

EXAMPLE 4 ▸▸ Find the value of $\int_0^{0.5} \frac{\sin^{-1} x}{\sqrt{1-x^2}}\, dx$.

For purposes of integrating, we see that

$$u = \sin^{-1} x, \qquad du = \frac{dx}{\sqrt{1-x^2}}, \qquad n = 1$$

Therefore,

$$\int_0^{0.5} \frac{\sin^{-1} x}{\sqrt{1-x^2}}\, dx = \int_0^{0.5} \sin^{-1} x\left(\frac{dx}{\sqrt{1-x^2}}\right) \qquad \int u\, du$$

$$= \left.\frac{(\sin^{-1} x)^2}{2}\right|_0^{0.5} \qquad \text{integrate}$$

$$= \frac{(\frac{\pi}{6})^2}{2} - 0 = \frac{\pi^2}{72} \qquad \text{evaluate}$$ ◂◂

EXAMPLE 5 ▸▸ Find the first-quadrant area bounded by $y = \frac{e^{2x}}{\sqrt{e^{2x}+1}}$ and $x = 1.5$.

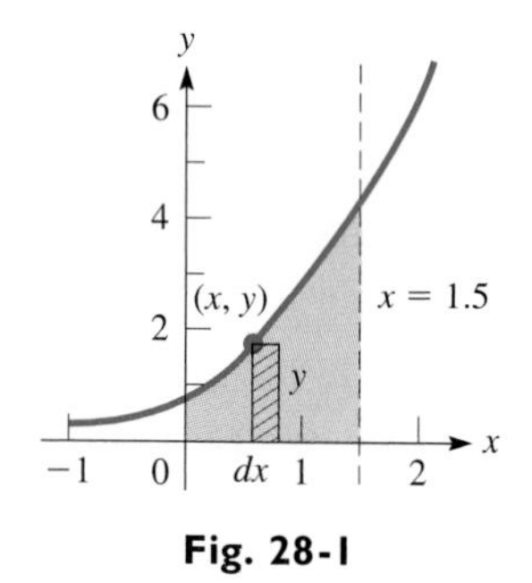

Fig. 28-1

The area to be found is shown in Fig. 28-1. The area of the representative element is $y\,dx$. Therefore, the area is found by evaluating the integral

$$\int_0^{1.5} \frac{e^{2x}\, dx}{\sqrt{e^{2x}+1}}$$

For the purpose of integration, $n = -\frac{1}{2}$, $u = e^{2x} + 1$, and $du = 2e^{2x}\, dx$. Therefore,

$$\int_0^{1.5} (e^{2x}+1)^{-1/2} e^{2x}\, dx = \frac{1}{2}\int_0^{1.5} (e^{2x}+1)^{-1/2}(2e^{2x}\, dx)$$

$$= \frac{1}{2}(2)\left.(e^{2x}+1)^{1/2}\right|_0^{1.5} = \left.(e^{2x}+1)^{1/2}\right|_0^{1.5}$$

$$= \sqrt{e^3+1} - \sqrt{2} = 3.178$$

This means that the area is 3.178 square units. ◂◂

Exercises 28–1

In Exercises 1–24, integrate each of the given functions.

1. $\int \sin^4 x \cos x\, dx$
2. $\int \cos^5 x(-\sin x\, dx)$
3. $\int \sqrt{\cos x}\, \sin x\, dx$
4. $\int 8 \sin^{1/3} x \cos x\, dx$
5. $\int 4 \tan^2 x \sec^2 x\, dx$
6. $\int \sec^3 x(\sec x \tan x)\, dx$
7. $\int_0^{\pi/8} \cos 2x \sin 2x\, dx$
8. $\int_{\pi/6}^{\pi/4} 3\sqrt{\cot x}\, \csc^2 x\, dx$
9. $\int (\sin^{-1} x)^3\left(\frac{dx}{\sqrt{1-x^2}}\right)$
10. $\int \frac{(\cos^{-1} 2x)^4\, dx}{\sqrt{1-4x^2}}$
11. $\int \frac{5 \tan^{-1} 5x}{1+25x^2}\, dx$
12. $\int \frac{\sin^{-1} 4x\, dx}{\sqrt{1-16x^2}}$
13. $\int [\ln(x+1)]^2 \frac{dx}{x+1}$
14. $\int (3 + \ln 2x)^3 \frac{dx}{x}$
15. $\int_0^{1/2} \frac{\ln(2x+3)}{2x+3}\, dx$
16. $\int_1^{e} \frac{(1 - 2\ln x)\, dx}{x}$
17. $\int (4 + e^x)^3 e^x\, dx$
18. $\int 2\sqrt{1 - e^{-x}}\,(e^{-x}\, dx)$
19. $\int (2e^{2x} - 1)^{1/3} e^{2x}\, dx$
20. $\int \frac{(1 + 3e^{-2x})^4\, dx}{e^{2x}}$
21. $\int (1 + \sec^2 x)^4(\sec^2 x \tan x\, dx)$
22. $\int (e^x + e^{-x})^{1/4}(e^x - e^{-x})\, dx$
23. $\int_0^{\pi/6} \frac{\tan x}{\cos^2 x}\, dx$
24. $\int_{\pi/3}^{\pi/2} \frac{\sin\theta\, d\theta}{\sqrt{1+\cos\theta}}$

In Exercises 25–32, solve the given problems by integration.

25. Find the area under the curve $y = \frac{1 + \tan^{-1} 2x}{1 + 4x^2}$ from $x = 0$ to $x = 2$.

26. Find the first-quadrant area bounded by $y = \frac{\ln(4x + 1)}{4x + 1}$ and $x = 2$.

27. The general expression for the slope of a given curve is $(\ln x)^2/x$. If the curve passes through $(1, 2)$, find its equation.

28. Find the equation of the curve for which $dy/dx = (1 + \tan 2x)^2 \sec^2 2x$ if the curve passes through $(2, 1)$.

29. In the development of the expression for the total pressure P on a wall due to molecules with mass m and velocity v striking the wall, the following equation is found: $P = mnv^2 \int_0^{\pi/2} \sin\theta \cos^2\theta \, d\theta$. The symbol n represents the number of molecules per unit volume, and θ represents the angle between a perpendicular to the wall and the direction of the molecule. Find the final expression for P.

30. The solar energy E passing through a hemispherical surface per unit time, per unit area, is given by $E = 2\pi I \int_0^{\pi/2} \cos\theta \sin\theta \, d\theta$, where I is the solar intensity and θ is the angle at which it is directed (from the perpendicular). Evaluate this integral.

31. After an electric power interruption, the current i in a circuit is given by $i = 3(1 - e^{-t})^2(e^{-t})$, where t is the time. Find the expression for the total electric charge q to pass a point in the circuit if $q = 0$ for $t = 0$.

32. A space vehicle is launched vertically from the ground such that its velocity v (in km/s) is given by $v = [\ln^2(t^3 + 1)]\frac{t^2}{t^3 + 1}$, where t is the time in seconds. Find the altitude of the vehicle after 10.0 s.

28-2 The Basic Logarithmic Form

The general power formula for integration, Eq. (28-1), is valid for all values of n except $n = -1$. If n were set equal to -1, this would cause the result to be undefined. When we obtained the derivative of the logarithmic function, we found

$$\frac{d(\ln u)}{dx} = \frac{1}{u}\frac{du}{dx}$$

which means the differential of the logarithmic form is $d(\ln u) = du/u$. Reversing the process, we then determine that $\int du/u = \ln u + C$. In other words, when the exponent of the expression being integrated is -1, the expression is a logarithmic form.

Logarithms are defined only for positive numbers. Thus, $\int du/u = \ln u + C$ is valid if $u > 0$. If $u < 0$, then $-u > 0$. In this case, $d(-u) = -du$, or $\int(-du)/(-u) = \ln(-u) + C$. However, $\int du/u = \int(-du)/(-u)$. These results can be combined into a single form using the absolute value of u. Therefore,

$$\int \frac{du}{u} = \ln|u| + C \tag{28-2}$$

EXAMPLE 1 ▸▸ Integrate: $\int \frac{dx}{x + 1}$.

Since $d(x + 1) = dx$, this integral fits the form of Eq. (28-2) with $u = x + 1$ and $du = dx$. Therefore, we have

$$\int \frac{dx}{x + 1} = \ln|x + 1| + C$$

(with dx labeled du and $x + 1$ labeled u) ◂◂

EXAMPLE 2 ▸▸ Newton's law of cooling states that the rate at which an object cools is directly proportional to the difference in its temperature T and the temperature of the surrounding medium. By use of this law, the time t (in min) a certain object takes to cool from 80°C to 20°C in air is found to be

$$t = -9.8\int_{80}^{50} \frac{dT}{T - 20}$$

Find the value of t.

We see that the integral fits Eq. (28-2) with $u = T - 20$ and $du = dT$. Thus

$$\begin{aligned} t &= -9.8\int_{80}^{50} \frac{dT}{T - 20} \quad \begin{matrix}\leftarrow du \\ \leftarrow u\end{matrix} \\ &= -9.8 \ln|T - 20|\Big|_{80}^{50} && \text{integrate} \\ &= -9.8(\ln 30 - \ln 60) && \text{evaluate} \\ &= -9.8 \ln\frac{30}{60} = -9.8 \ln (0.50) && \ln x - \ln y = \ln\frac{x}{y} \\ &= 6.8 \text{ min} \end{aligned}$$

◂◂

EXAMPLE 3 ▸▸ Integrate: $\int \frac{\cos x}{\sin x}\,dx$.

We note that $d(\sin x) = \cos x\,dx$. This means that this integral fits the form of Eq. (28-2) with $u = \sin x$ and $du = \cos x\,dx$. Thus

$$\begin{aligned} \int \frac{\cos x}{\sin x}\,dx &= \int \frac{\cos x\,dx}{\sin x} \quad \begin{matrix}\leftarrow du \\ \leftarrow u\end{matrix} \\ &= \ln|\sin x| + C \end{aligned}$$

◂◂

EXAMPLE 4 ▸▸ Integrate: $\int \frac{x\,dx}{4 - x^2}$.

This integral fits the form of Eq. (28-2) with $u = 4 - x^2$ and $du = -2x\,dx$. This means that we must introduce a factor of -2 into the numerator and a factor of $-\frac{1}{2}$ before the integral. Therefore,

$$\begin{aligned} \int \frac{x\,dx}{4 - x^2} &= -\frac{1}{2}\int \frac{-2x\,dx}{4 - x^2} \quad \begin{matrix}\leftarrow du \\ \leftarrow u\end{matrix} \\ &= -\frac{1}{2}\ln|4 - x^2| + C \end{aligned}$$

CAUTION ▸ We should note that if the quantity $4 - x^2$ were raised to any power other than that in the example, we would have to employ the general power formula for integration. For example,

$$\int \frac{x\,dx}{(4 - x^2)^2} = -\frac{1}{2}\int \frac{-2x\,dx}{(4 - x^2)^2} \quad \begin{matrix}\leftarrow du \\ \leftarrow u^2\end{matrix}$$

$$= -\frac{1}{2}\frac{(4 - x^2)^{-1}}{-1} + C = \frac{1}{2(4 - x^2)} + C$$

◂◂

EXAMPLE 5 ▸▸ Integrate: $\int \frac{e^{4x}\,dx}{1 + 3e^{4x}}$.

Since $d(e^{4x})/dx = 4e^{4x}\,dx$, we see that we can use Eq. (28-2) with $u = 1 + 3e^{4x}$ and $du = 12e^{4x}\,dx$. Therefore, we write

$$\int \frac{e^{4x}\,dx}{1 + 3e^{4x}} = \frac{1}{12}\int \frac{12e^{4x}\,dx}{1 + 3e^{4x}} \quad \text{introduce factors of 12}$$

$$= \frac{1}{12}\ln|1 + 3e^{4x}| + C \quad \text{integrate}$$

$$= \frac{1}{12}\ln(1 + 3e^{4x}) + C \quad 1 + 3e^{4x} > 0 \text{ for all } x$$ ◂◂

EXAMPLE 6 ▸▸ Evaluate: $\int_0^{\pi/8} \frac{\sec^2 2x}{1 + \tan 2x}\,dx$.

Since $d(\tan 2x)/dx = 2\sec^2 2x\,dx$, we see that we can use Eq. (28-2) with $u = 1 + \tan 2x$ and $du = 2\sec^2 2x\,dx$. Therefore, we have

$$\int_0^{\pi/8} \frac{\sec^2 2x}{1 + \tan 2x}\,dx = \frac{1}{2}\int_0^{\pi/8} \frac{2\sec^2 2x\,dx}{1 + \tan 2x} \quad \text{introduce factors of 2}$$

$$= \frac{1}{2}\ln|1 + \tan 2x|\Big|_0^{\pi/8} \quad \text{integrate}$$

$$= \frac{1}{2}\ln|2| - \frac{1}{2}\ln|1| \quad \text{evaluate}$$

$$= \frac{1}{2}\ln 2 - \frac{1}{2}\ln 1 = \frac{1}{2}\ln 2 - 0$$

$$= \frac{1}{2}\ln 2$$ ◂◂

EXAMPLE 7 ▸▸ Find the volume generated by revolving the area bounded by the curve of $y = \frac{3}{\sqrt{4x + 3}}$, $x = 2.5$, and the axes about the x-axis.

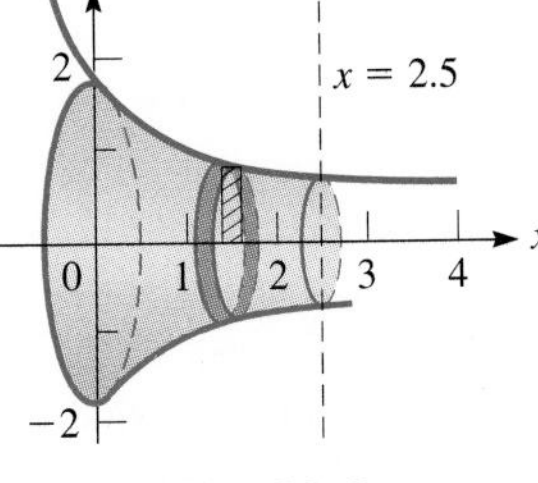

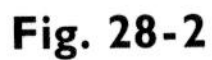
Fig. 28-2

Figure 28-2 shows the volume to be found. As shown, the element of volume is a disk. The volume is found as follows:

$$V = \pi\int_0^{2.5} y^2\,dx = \pi\int_0^{2.5}\left(\frac{3}{\sqrt{4x + 3}}\right)^2 dx$$

$$= \pi\int_0^{2.5} \frac{9\,dx}{4x + 3} = \frac{9\pi}{4}\int_0^{2.5} \frac{4\,dx}{4x + 3}$$

$$= \frac{9\pi}{4}\ln(4x + 3)\Big|_0^{2.5} = \frac{9\pi}{4}(\ln 13 - \ln 3)$$

$$= \frac{9\pi}{4}\ln\frac{13}{3} = 10.36$$

Therefore, the volume is about 10.36 cubic units. ◂◂

Exercises 28–2

In Exercises 1–28, integrate each of the given functions.

1. $\int \frac{dx}{1 + 4x}$

2. $\int \frac{dx}{1 - 4x}$

3. $\int \frac{2x\,dx}{4 - 3x^2}$

4. $\int \frac{x^2\,dx}{1 - x^3}$

5. $\int_0^2 \frac{dx}{8 - 3x}$

6. $\int_{-1}^3 \frac{2x^3\,dx}{x^4 + 1}$

7. $\int \frac{\csc^2 2x}{\cot 2x}\,dx$

8. $\int \frac{7 \sin x}{\cos x}\,dx$

9. $\int_0^{\pi/2} \frac{\cos x\,dx}{1 + \sin x}$

10. $\int_0^{\pi/4} \frac{\sec^2 x\,dx}{4 + \tan x}$

11. $\int \frac{e^{-x}}{1 - e^{-x}}\,dx$

12. $\int \frac{5e^{3x}}{1 - e^{3x}}\,dx$

13. $\int \frac{1 + e^x}{x + e^x}\,dx$

14. $\int \frac{e^x - e^{-x}}{e^x + e^{-x}}\,dx$

15. $\int \frac{\sec x \tan x\,dx}{1 + 4 \sec x}$

16. $\int \frac{\sin 2x}{1 - \cos^2 x}\,dx$

17. $\int_1^3 \frac{1 + x}{4x + 2x^2}\,dx$

18. $\int_1^2 \frac{4x + 6x^2}{x^2 + x^3}\,dx$

19. $\int \frac{dx}{x \ln x}$

20. $\int \frac{dx}{x(1 + 2 \ln x)}$

21. $\int \frac{2 + \sec^2 x}{2x + \tan x}\,dx$

22. $\int \frac{x + \cos 2x}{x^2 + \sin 2x}\,dx$

23. $\int \frac{2\,dx}{\sqrt{1 - 2x}}$

24. $\int \frac{4x\,dx}{(1 + x^2)^2}$

25. $\int \frac{x + 2}{x^2}\,dx$

26. $\int \frac{2 - 3x^2}{x^3}\,dx$

27. $\int_0^{\pi/12} \frac{\sec^2 3x}{4 + \tan 3x}\,dx$

28. $\int_1^2 \frac{x^2 + 1}{x^3 + 3x}\,dx$

In Exercises 29–40, solve the given problems by integration.

29. Find the area bounded by $y(x + 1) = 1$, $x = 0$, $y = 0$, and $x = 2$. See Fig. 28-3.

30. Find the area bounded by $xy = 9$, $x = 1$, $x = 2$, and $y = 0$.

31. Find the volume generated by revolving the area bounded by $y = 1/(x^2 + 1)$, $x = 0$, $x = 1$, $y = 0$ about the y-axis. (Use shells.)

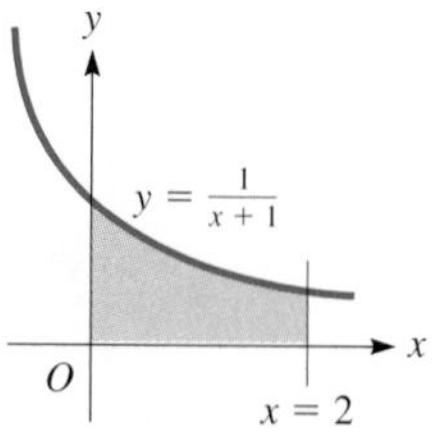

Fig. 28-3

32. Find the volume of the solid generated by revolving the area bounded by $y = 2/\sqrt{3x + 1}$, $x = 0$, $x = 3.5$, and $y = 0$ about the x-axis.

33. The general expression for the slope of a curve is $\sin x/(3 + \cos x)$. If the curve passes through $(\frac{\pi}{3}, 2)$, find its equation.

34. Find the average value of the function $xy = 4$ from $x = 1$ to $x = 2$.

35. Under ideal conditions, the natural law of population growth is that population increases at a rate proportional to the population P present at any time t. This leads to the equation $t = \frac{1}{k}\int \frac{dP}{P}$. Assuming ideal conditions for the United States, if $P = 227$ million in 1980 ($t = 0$) and $P = 249$ million in 1990 ($t = 10$ years), find the population projected in 2010 ($t = 30$ years).

36. In determining the temperature that is absolute zero (0 K, or about $-273°C$), we use the equation $\ln T = -\int \frac{dr}{r - 1}$. Here T is the thermodynamic temperature and r is the ratio between certain specific vapor pressures. If $T = 273.16$ K for $r = 1.3361$, find T as a function of r (if $r > 1$ for all T).

37. The time t and electric current i for a circuit with a voltage E, a resistance R, and an inductance L is given by $t = L\int \frac{di}{E - iR}$. If $t = 0$ for $i = 0$, integrate and express i as a function of t.

38. Conditions are often such that a force proportional to the velocity tends to retard the motion of an object. Under such conditions, the acceleration of a certain object moving down an inclined plane is given by $20 - v$. This leads to the equation $t = \int \frac{dv}{20 - v}$. If the object starts from rest, find the expression for the velocity as a function of time.

39. An architect designs a wall panel that can be described as the first-quadrant area bounded by $y = \frac{50}{x^2 + 20}$ and $x = 3.00$. If the area of the panel is 6.61 m^2, find the x-coordinate (in m) of the centroid of the panel.

40. The electric power p developed in a certain resistor is given by $p = 3\int \frac{\sin \pi t}{2 + \cos \pi t}\,dt$, where t is the time. Express p as a function of t.

28-3 The Exponential Form

In deriving the derivative for the exponential function we obtained the result $de^u/dx = e^u(du/dx)$. This means that the differential of the exponential form is $d(e^u) = e^u\,du$. Reversing this form to find the proper form of the integral for the exponential function, we have

$$\int e^u\,du = e^u + C \tag{28-3}$$

EXAMPLE 1 ▸▸ Integrate: $\int xe^{x^2}\,dx$.

Since $d(x^2) = 2x\,dx$, we can write this integral in the form of Eq. (28-3) with $u = x^2$ and $du = 2x\,dx$. Thus

$$\begin{aligned}\int xe^{x^2}\,dx &= \frac{1}{2}\int e^{\overbrace{x^2}^{u}}(\underbrace{2x\,dx}_{du})\\ &= \frac{1}{2}e^{x^2} + C \quad ◂◂\end{aligned}$$

EXAMPLE 2 ▸▸ For an electric circuit containing a direct voltage source E, a resistance R, and an inductance L, the current i and time t are related by $ie^{Rt/L} = \dfrac{E}{L}\displaystyle\int e^{Rt/L}\,dt$. See Fig. 28-4. If $i = 0$ for $t = 0$, perform the integration and then solve for i as a function of t.

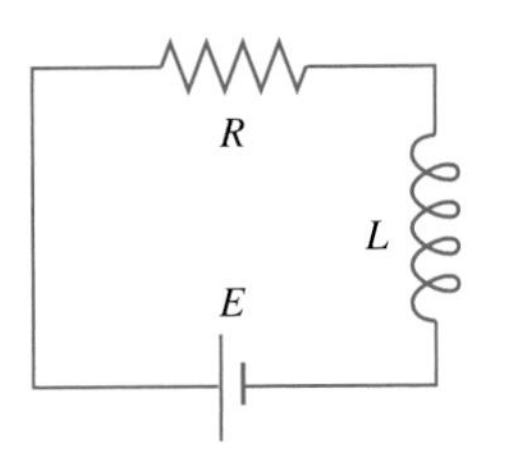

Fig. 28-4

For this integral we see that $u = \dfrac{Rt}{L}$, which means that $du = \dfrac{R\,dt}{L}$. The solution is then as follows:

$$\begin{aligned}
ie^{Rt/L} &= \frac{E}{L}\int e^{Rt/L}\,dt = \frac{E}{L}\left(\frac{L}{R}\right)\int e^{Rt/L}\left(\frac{R\,dt}{L}\right) && \text{introduce factor } \frac{R}{L}\\
&= \frac{E}{R}e^{Rt/L} + C && \text{integrate}\\
0(e^0) &= \frac{E}{R}e^0 + C, \qquad C = -\frac{E}{R} && i = 0 \text{ for } t = 0\text{; evaluate } C\\
ie^{Rt/L} &= \frac{E}{R}e^{Rt/L} - \frac{E}{R} && \text{substitute for } C\\
i &= \frac{E}{R} - \frac{E}{R}e^{-Rt/L} = \frac{E}{R}(1 - e^{-Rt/L}) && \text{solve for } i \quad ◂◂
\end{aligned}$$

EXAMPLE 3 ▸▸ Integrate: $\displaystyle\int \frac{dx}{e^{3x}}$.

This integral can be put in proper form by writing it as $\int e^{-3x}\,dx$. In this form $u = -3x$, $du = -3\,dx$. Thus

$$\int \frac{dx}{e^{3x}} = \int e^{-3x}\,dx = -\frac{1}{3}\int e^{-3x}(-3\,dx) = -\frac{1}{3}e^{-3x} + C \quad ◂◂$$

EXAMPLE 4 ▸▸ Integrate: $\int \frac{4e^{3x} - 3e^{x}}{e^{x+1}}\,dx$.

This can be put in the proper form for integration, and then integrated, as follows:

For reference, Eq. (11-2) is $\frac{a^m}{a^n} = a^{m-n}$.

$$\int \frac{4e^{3x} - 3e^{x}}{e^{x+1}}\,dx = \int \frac{4e^{3x}}{e^{x+1}}\,dx - \int \frac{3e^{x}}{e^{x+1}}\,dx$$

$$= 4\int e^{3x-(x+1)}\,dx - 3\int e^{x-(x+1)}\,dx \qquad \text{using Eq. (11-2)}$$

$$= 4\int e^{2x-1}\,dx - 3\int e^{-1}\,dx$$

$$= \frac{4}{2}\int e^{2x-1}(2\,dx) - \frac{3}{e}\int dx$$

$$= 2e^{2x-1} - \frac{3}{e}x + C$$ ◂◂

EXAMPLE 5 ▸▸ Evaluate: $\int_0^{\pi/2} (\sin 2x)(e^{\cos 2x})\,dx$.

With $u = \cos 2x$, $du = -2\sin 2x\,dx$, we have

$$\int_0^{\pi/2} (\sin 2x)(e^{\cos 2x})\,dx = -\frac{1}{2}\int_0^{\pi/2} (e^{\cos 2x})(-2\sin 2x\,dx)$$

$$= -\frac{1}{2}e^{\cos 2x}\Big|_0^{\pi/2} \qquad \text{integrate}$$

$$= -\frac{1}{2}\left(\frac{1}{e} - e\right) = 1.175 \qquad \text{evaluate}$$ ◂◂

EXAMPLE 6 ▸▸ Find the equation of the curve for which $\frac{dy}{dx} = \frac{e^{\sqrt{x+1}}}{\sqrt{x+1}}$ if the curve passes through (0, 1).

The solution of this problem requires that we integrate the given function and then evaluate the constant of integration. Hence

$$dy = \frac{e^{\sqrt{x+1}}}{\sqrt{x+1}}\,dx, \qquad \int dy = \int \frac{e^{\sqrt{x+1}}}{\sqrt{x+1}}\,dx$$

For purposes of integrating the right-hand side,

$$u = \sqrt{x+1} \quad \text{and} \quad du = \frac{1}{2\sqrt{x+1}}\,dx$$

$$y = 2\int e^{\sqrt{x+1}}\left(\frac{1}{2\sqrt{x+1}}\,dx\right)$$

$$= 2e^{\sqrt{x+1}} + C$$

Letting $x = 0$ and $y = 1$, we have $1 = 2e + C$, or $C = 1 - 2e$. This means that the equation is

$$y = 2e^{\sqrt{x+1}} + 1 - 2e$$

The graph of this function is shown in Fig. 28-5. ◂◂

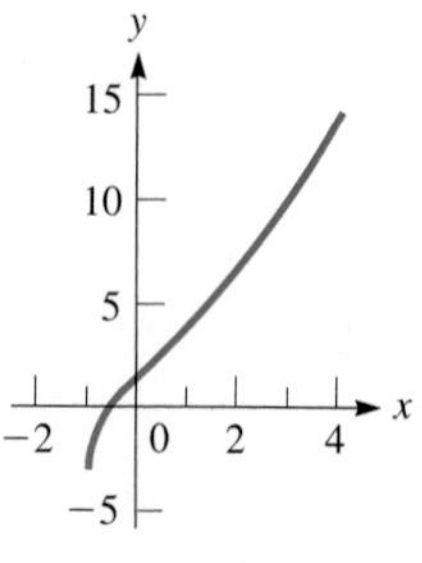

Fig. 28-5

Exercises 28–3

In Exercises 1–24, integrate each of the given functions.

1. $\int e^{7x}(7\,dx)$
2. $\int e^{x^4}(4x^3\,dx)$
3. $\int e^{2x+5}\,dx$
4. $\int 2e^{-4x}\,dx$
5. $\int_0^2 e^{x/2}\,dx$
6. $\int_1^2 3e^{4x}\,dx$
7. $\int 6x^2e^{x^3}\,dx$
8. $\int xe^{-x^2}\,dx$
9. $\int_1^4 \frac{e^{\sqrt{x}}}{\sqrt{x}}\,dx$
10. $\int_0^1 4x^3e^{2x^4}\,dx$
11. $\int (\sec x \tan x)e^{2\sec x}\,dx$
12. $\int (\sec^2 x)e^{\tan x}\,dx$
13. $\int \frac{(3 - e^x)\,dx}{e^{2x}}$
14. $\int (e^x - e^{-x})^2\,dx$
15. $\int_1^3 3e^{2x}(e^{-2x} - 1)\,dx$
16. $\int_0^{0.5} \frac{3e^{3x+1}}{e^x}\,dx$
17. $\int \frac{2\,dx}{\sqrt{x}\,e^{\sqrt{x}}}$
18. $\int \frac{4\,dx}{\sec x\; e^{\sin x}}$
19. $\int \frac{e^{\tan^{-1} x}}{x^2 + 1}\,dx$
20. $\int \frac{e^{\sin^{-1} 2x}\,dx}{\sqrt{1 - 4x^2}}$
21. $\int \frac{e^{\cos 3x}\,dx}{\csc 3x}$
22. $\int \frac{e^{2/x}\,dx}{x^2}$
23. $\int_0^{\pi} (\sin 2x)e^{\cos^2 x}\,dx$
24. $\int_{-1}^{1} \frac{dx}{e^{2-3x}}$

In Exercises 25–36, solve the given problems by integration.

25. Find the area bounded by $y = 3e^x$, $x = 0$, $y = 0$, and $x = 2$.

26. Find the area bounded by $x = a$, $x = b$, $y = 0$, and $y = e^x$. Explain the meaning of the result.

27. Find the volume generated by revolving the area bounded by $y = e^{x^2}$, $x = 1$, $y = 0$, and $x = 2$ about the y-axis. See Fig. 28-6.

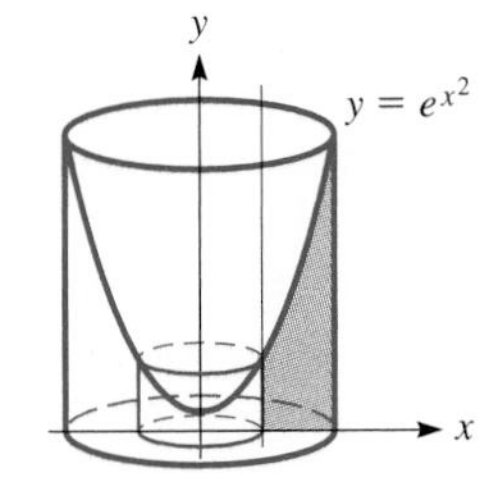

Fig. 28-6

28. Find the equation of the curve for which $dy/dx = \sqrt{e^{x+3}}$ if the curve passes through $(1, 0)$.

29. Find the average value of the function $y = e^{2x}$ from $x = 0$ to $x = 4$.

30. Find the moment of inertia with respect to the y-axis of the first-quadrant area bounded by $y = e^{x^3}$, $x = 1$, and the axes.

31. Using Eq. (27-15), show that $\int b^u\,du = \frac{b^u}{\ln b} + C$ $(b > 0, b \neq 1)$.

32. Find the first-quadrant area bounded by $y = 2^x$ and $x = 3$. See Exercise 31.

33. For an electric circuit containing a voltage source E, a resistance R, and a capacitance C, an equation relating the charge q on the capacitor and the time t is $qe^{t/RC} = \frac{E}{R}\int e^{t/RC}\,dt$. See Fig. 28-7. If $q = 0$ for $t = 0$, perform the integration and then solve for q as a function of t.

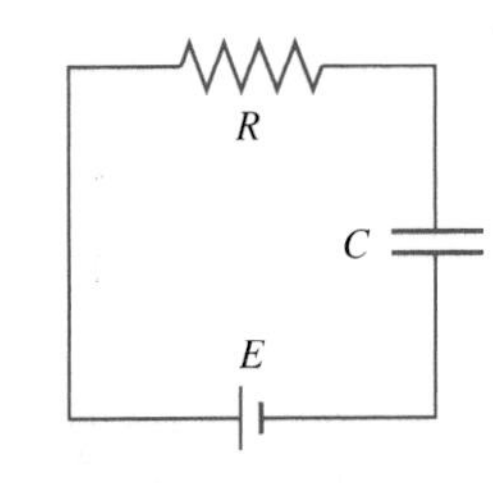

Fig. 28-7

34. In the theory dealing with energy propagation of lasers, the equation $E = a\int_0^{I_0} e^{-Tx}\,dx$ is used. Here a, I_0, and T are constants. Evaluate this integral.

35. An object at the end of a spring is immersed in liquid. Its velocity (in cm/s) is then described by the equation $v = 2e^{-2t} + 3e^{-5t}$, where t is the time in seconds. Such motion is called *overdamped*. Find the displacement s as a function of t if $s = -1.6$ cm for $t = 0$.

36. The force F (in newtons) exerted by a robot programmed to staple carton sections together is given by $F = 6\int e^{\sin \pi t}\cos \pi t\,dt$, where t is the time in seconds. Find F as a function of t if $F = 0$ for $t = 1.5$ s.

28-4 Basic Trigonometric Forms

In this section we discuss the integrals of the six trigonometric functions and the trigonometric integrals that arise directly from reversing the formulas for differentiation. Other trigonometric forms will be discussed later.

By directly reversing the differentiation formulas, the following integral formulas are obtained:

$$\int \sin u\,du = -\cos u + C \tag{28-4}$$

$$\int \cos u\,du = \sin u + C \tag{28-5}$$

$$\int \sec^2 u\,du = \tan u + C \tag{28-6}$$

$$\int \csc^2 u\,du = -\cot u + C \tag{28-7}$$

$$\int \sec u \tan u\,du = \sec u + C \tag{28-8}$$

$$\int \csc u \cot u\,du = -\csc u + C \tag{28-9}$$

EXAMPLE 1 ▸▸ Integrate: $\int x \sec^2 x^2\,dx$.

With $u = x^2$, $du = 2x\,dx$, we have

$$\int x \sec^2 x^2\,dx = \frac{1}{2}\int (\sec^2 \underbrace{x^2}_{u})(\overbrace{2x\,dx}^{du})$$

$$= \frac{1}{2}\tan x^2 + C \qquad \text{using Eq. (28-6)}$$ ◂◂

EXAMPLE 2 ▸▸ Integrate: $\displaystyle\int \frac{\tan 2x}{\cos 2x}\,dx$.

By using the basic identity $\sec\theta = 1/\cos\theta$, we can transform this integral to form $\int \sec 2x \tan 2x\,dx$. In this form $u = 2x$, $du = 2\,dx$. Therefore,

$$\int \frac{\tan 2x}{\cos 2x}\,dx = \int \sec 2x \tan 2x\,dx = \frac{1}{2}\int \sec \underbrace{2x}_{u} \tan \underbrace{2x}_{u}(\overbrace{2\,dx}^{du})$$

$$= \frac{1}{2}\sec 2x + C \qquad \text{using Eq. (28-8)}$$ ◂◂

EXAMPLE 3 ▸▸ The vertical velocity v (in cm/s) of the end of a vibrating rod is given by $v = 80 \cos 20\pi t$, where t is the time in seconds. Find the vertical displacement y (in cm) as a function of t if $y = 0$ for $t = 0$.

Since $v = dy/dt$, we have the following solution.

$$\frac{dy}{dt} = 80 \cos 20\pi t$$

$$\int dy = \int 80 \cos 20\pi t\,dt = \frac{80}{20\pi}\int (\cos 20\pi t)(20\pi\,dt) \qquad \text{set up integration}$$

$$y = \frac{4}{\pi}\sin 20\pi t + C \qquad \text{using Eq. (28-5)}$$

$$0 = 1.27 \sin 0 + C, \quad C = 0 \qquad \text{evaluate } C$$

$$y = 1.27 \sin 20\pi t \qquad \text{solution}$$ ◂◂

To find the integrals for the other trigonometric functions, we must change them to a form for which the integral can be determined by methods previously discussed. We can accomplish this by using the basic trigonometric relations.

The formula for $\int \tan u\, du$ is found by expressing the integral in the form $\int (\sin u/\cos u)\, du$. We recognize this as being a logarithmic form, where the u of the logarithmic form is $\cos u$ in this integral. The differential of $\cos u$ is $-\sin u\, du$. Therefore, we have

$$\int \tan u\, du = \int \frac{\sin u}{\cos u}\, du = -\int \frac{-\sin u\, du}{\cos u} = -\ln|\cos u| + C$$

The formula for $\int \cot u\, du$ is found by writing it in the form $\int (\cos u/\sin u)\, du$. In this manner we obtain the result

$$\int \cot u\, du = \int \frac{\cos u}{\sin u}\, du = \int \frac{\cos u\, du}{\sin u} = \ln|\sin u| + C$$

The formula for $\int \sec u\, du$ is found by writing it in the form

$$\int \frac{\sec u(\sec u + \tan u)}{\sec u + \tan u}\, du$$

We see that this form is also a logarithmic form, since

$$d(\sec u + \tan u) = (\sec u \tan u + \sec^2 u)\, du$$

The right side of this equation is the expression appearing in the numerator of the integral. Thus

$$\int \sec u\, du = \int \frac{\sec u(\sec u + \tan u)\, du}{\sec u + \tan u} = \int \frac{\sec u \tan u + \sec^2 u}{\sec u + \tan u}\, du$$
$$= \ln|\sec u + \tan u| + C$$

To obtain the formula for $\int \csc u\, du$, we write it in the form

$$\int \frac{\csc u(\csc u - \cot u)\, du}{\csc u - \cot u}$$

Thus we have

$$\int \csc u\, du = \int \frac{\csc u(\csc u - \cot u)}{\csc u - \cot u}\, du$$
$$= \int \frac{(-\csc u \cot u + \csc^2 u)\, du}{\csc u - \cot u}$$
$$= \ln|\csc u - \cot u| + C$$

Summarizing these results, we have the following integrals:

$$\int \tan u\, du = -\ln|\cos u| + C \tag{28-10}$$

$$\int \cot u\, du = \ln|\sin u| + C \tag{28-11}$$

$$\int \sec u\, du = \ln|\sec u + \tan u| + C \tag{28-12}$$

$$\int \csc u\, du = \ln|\csc u - \cot u| + C \tag{28-13}$$

EXAMPLE 4 ▶▶ Integrate: $\int \tan 4x\,dx$.

Noting that $u = 4x$, $du = 4\,dx$, we have

$$\int \tan 4x\,dx = \frac{1}{4}\int \tan 4x(4\,dx) \qquad \text{introducing factors of 4}$$

$$= -\frac{1}{4}\ln|\cos 4x| + C \qquad \text{using Eq. (28-10)}$$ ◀◀

EXAMPLE 5 ▶▶ Integrate: $\int \frac{\sec e^{-x}\,dx}{e^x}$.

In this integral, $u = e^{-x}$, $du = -e^{-x}\,dx$. Therefore,

$$\int \frac{\sec e^{-x}\,dx}{e^x} = -\int (\sec e^{-x})(-e^{-x}\,dx) \qquad \text{introducing} - \text{sign}$$

$$= -\ln|\sec e^{-x} + \tan e^{-x}| + C \qquad \text{using Eq. (28-12)}$$ ◀◀

EXAMPLE 6 ▶▶ Evaluate: $\int_{\pi/6}^{\pi/4} \frac{1 + \cos x}{\sin x}\,dx$.

The solution is as follows:

$$\int_{\pi/6}^{\pi/4} \frac{1 + \cos x}{\sin x}\,dx = \int_{\pi/6}^{\pi/4} \csc x\,dx + \int_{\pi/6}^{\pi/4} \cot x\,dx \qquad \text{using Eqs. (20-1) and (20-5)}$$

$$= \ln|\csc x - \cot x|\Big|_{\pi/6}^{\pi/4} + \ln|\sin x|\Big|_{\pi/6}^{\pi/4} \qquad \text{integrating}$$

$$= \ln|\sqrt{2} - 1| - \ln|2 - \sqrt{3}| + \ln\left|\frac{1}{2}\sqrt{2}\right| - \ln\left|\frac{1}{2}\right| \qquad \text{evaluating}$$

$$= \ln\frac{(\frac{1}{2}\sqrt{2})(\sqrt{2} - 1)}{(\frac{1}{2})(2 - \sqrt{3})} = \ln\frac{2 - \sqrt{2}}{2 - \sqrt{3}}$$

$$= 0.782$$ ◀◀

Exercises 28–4

In Exercises 1–24, integrate each of the given functions.

1. $\int \cos 2x\,dx$
2. $\int 4\sin(2 - x)\,dx$
3. $\int \sec^2 3x\,dx$
4. $\int \csc 2x \cot 2x\,dx$
5. $\int \sec\frac{1}{2}x \tan\frac{1}{2}x\,dx$
6. $\int e^x \csc^2(e^x)\,dx$
7. $\int_{0.5}^{1} x^2 \cot x^3\,dx$
8. $\int_0^1 6\tan\frac{1}{2}x\,dx$
9. $\int 3x \sec x^2\,dx$
10. $\int 2\csc 3x\,dx$
11. $\int \frac{\sin(1/x)}{x^2}\,dx$
12. $\int \frac{3\,dx}{\sin 4x}$
13. $\int_0^{\pi/6} \frac{dx}{\cos^2 2x}$
14. $\int_0^1 2e^x \cos e^x\,dx$
15. $\int \frac{\sec 5x}{\cot 5x}\,dx$
16. $\int \frac{\sin 2x}{\cos^2 x}\,dx$
17. $\int \sqrt{\tan^2 2x + 1}\,dx$
18. $\int 5(1 + \cot x)^2\,dx$
19. $\int \frac{1 + \sin 2x}{\tan 2x}\,dx$
20. $\int \frac{1 - \cot^2 x}{\cos^2 x}\,dx$
21. $\int \frac{1 - \sin x}{1 + \cos x}\,dx$
22. $\int \frac{1 + \sec^2 x}{x + \tan x}\,dx$
23. $\int_0^{\pi/9} \sin 3x(\csc 3x + \sec 3x)\,dx$
24. $\int_{\pi/4}^{\pi/3} (1 + \sec x)^2\,dx$

In Exercises 25–32, solve the given problems by integration.

25. Find the area bounded by $y = 2\tan x$, $x = \frac{\pi}{4}$, and $y = 0$.
26. Find the area under the curve $y = \sin x$ from $x = 0$ to $x = \pi$.

27. Find the volume generated by revolving the area bounded by $y = \sec x$, $x = 0$, $x = \frac{\pi}{3}$, and $y = 0$ about the x-axis.

28. Find the volume generated by revolving the area bounded by $y = \cos x^2$, $x = 0$, $y = 0$, and $x = 1$ about the y-axis.

29. The angular velocity ω (in rad/s) of a pendulum is $\omega = -0.25 \sin 2.5t$. Find the angular displacement θ as a function of t if $\theta = 0.10$ for $t = 0$.

30. If the current in a certain electric circuit is $i = 110 \cos 377t$, find the expression for the voltage across a 500-μF capacitor as a function of time. The initial voltage is zero. Show that the voltage across the capacitor is 90° out of phase with the current.

31. A fin on a wind-direction indicator has a shape that can be described as the area bounded by $y = \tan x^2$, $y = 0$, and $x = 1$. Find the x-coordinate (in m) of the centroid of the fin if its area is 0.3984 m^2.

32. A force is given as a function of the distance from the origin as $F = \dfrac{2 + \tan x}{\cos x}$. Express the work done by this force as a function of x if $W = 0$ for $x = 0$.

28-5 Other Trigonometric Forms

The basic trigonometric relations developed in Chapter 20 provide the means by which many other integrals involving trigonometric functions may be integrated. By use of the square relations, Eqs. (20-6), (20-7), and (20-8), and the equations for the cosine of the double angle, Eqs. (20-22), (20-23), and (20-24), it is possible to transform integrals involving powers of the trigonometric functions into integrable form. We repeat these equations here for reference:

$$\cos^2 x + \sin^2 x = 1 \quad (28\text{-}14)$$

$$1 + \tan^2 x = \sec^2 x \quad (28\text{-}15)$$

$$1 + \cot^2 x = \csc^2 x \quad (28\text{-}16)$$

$$2\cos^2 x = 1 + \cos 2x \quad (28\text{-}17)$$

$$2\sin^2 x = 1 - \cos 2x \quad (28\text{-}18)$$

CAUTION ▶ *To integrate a product of powers of the sine and cosine, we use Eq. (28-14)* ***if at least one of the powers is odd.*** The method is based on transforming the integral so that it is made up of powers of either the sine or cosine and the first power of the other. In this way this first power becomes a factor of *du*.

EXAMPLE 1 ▸▸ Integrate: $\int \sin^3 x \cos^2 x\,dx$.

Since $\sin^3 x = \sin^2 x \sin x = (1 - \cos^2 x)\sin x$, it is possible to write this integral with powers of $\cos x$ along with $\sin x\,dx$. In this way we can have $-\sin x\,dx$ as the necessary du of the integral. Therefore,

$$\int \sin^3 x \cos^2 x\,dx = \int (1 - \cos^2 x)(\sin x)(\cos^2 x)\,dx \quad \text{using Eq. (28-14)}$$

$$= \int (\cos^2 x - \cos^4 x)(\sin x\,dx)$$

$$= \int \cos^2 x(\sin x\,dx) - \int \cos^4 x(\sin x\,dx)$$

du

$$= -\int \cos^2 x(-\sin x\,dx) + \int \cos^4 x(-\sin x\,dx)$$

$$= -\frac{1}{3}\cos^3 x + \frac{1}{5}\cos^5 x + C \quad ◂◂$$

EXAMPLE 2 ▸▸ Integrate: $\int \cos^5 2x\,dx$.

Since $\cos^5 2x = \cos^4 2x \cos 2x = (1 - \sin^2 2x)^2 \cos 2x$, it is possible to write this integral with powers of $\sin 2x$ along with $\cos 2x\,dx$. Thus, with the introduction of a factor of 2, $(\cos 2x)(2\,dx)$ is the necessary du of the integral. Thus

$$\int \cos^5 2x\,dx = \int (1 - \sin^2 2x)^2 \cos 2x\,dx \qquad \text{using Eq. (28-14)}$$

$$= \int (1 - 2\sin^2 2x + \sin^4 2x)\cos 2x\,dx$$

$$= \int \cos 2x\,dx - \int 2\sin^2 2x \cos 2x\,dx + \int \sin^4 2x \cos 2x\,dx$$

$$= \frac{1}{2}\int \underbrace{\cos 2x(2\,dx)}_{du} - \int \sin^2 2x\underbrace{(2\cos 2x\,dx)}_{du} + \frac{1}{2}\int \sin^4 2x\underbrace{(2\cos 2x\,dx)}_{du}$$

$$= \frac{1}{2}\sin 2x - \frac{1}{3}\sin^3 2x + \frac{1}{10}\sin^5 2x + C$$ ◂◂

CAUTION ▸ *In products of powers of the sine and cosine,* ***if the powers to be integrated are even, we use Eqs. (28-17) and (28-18) to transform the integral.*** Those most commonly met are $\int \cos^2 u\,du$ and $\int \sin^2 u\,du$. Consider the following examples.

EXAMPLE 3 ▸▸ Integrate: $\int \sin^2 2x\,dx$.

Using Eq. (28-18) in the form $\sin^2 2x = \frac{1}{2}(1 - \cos 4x)$, this integral can be transformed into a form that can be integrated. (Here we note ***the x of Eq. (28-18) is treated as 2x*** for this integral.) Therefore, we write

CAUTION ▸

$$\int \sin^2 2x\,dx = \int \left[\frac{1}{2}(1 - \cos 4x)\right] dx \qquad \text{using Eq. (28-18)}$$

$$= \frac{1}{2}\int dx - \frac{1}{8}\int \cos 4x(4\,dx)$$

$$= \frac{x}{2} - \frac{1}{8}\sin 4x + C$$ ◂◂

To integrate even powers of the secant, powers of the tangent, or products of the secant and tangent, we use Eq. (28-15) to transform the integral. In transforming, the forms we look for are ***powers of the tangent with*** $\mathbf{sec^2\,x}$, which becomes part of du, or ***powers of the secant along with*** $\mathbf{sec\,x\,tan\,x}$, which becomes part of du in this case. Similar transformations are made when we integrate powers of the cotangent and cosecant, with the use of Eq. (28-16).

NOTE ▸

EXAMPLE 4 ▸▸ Integrate: $\int \sec^3 x \tan x\,dx$.

By writing $\sec^3 x \tan x$ as $\sec^2 x(\sec x \tan x)$, we can use the $\sec x \tan x\,dx$ as the du of the integral. Thus

$$\int \sec^3 x \tan x\,dx = \int (\sec^2 x)\overbrace{(\sec x \tan x\,dx)}^{du}$$

$$= \frac{1}{3}\sec^3 x + C$$ ◂◂

EXAMPLE 5 ▸▸ Integrate: $\int \tan^5 x\,dx$.

Since $\tan^5 x = \tan^3 x \tan^2 x = \tan^3 x(\sec^2 x - 1)$, we can write this integral with powers of $\tan x$ along with $\sec^2 x\,dx$. Thus, $\sec^2 x\,dx$ becomes the necessary du of the integral. It is necessary to replace $\tan^2 x$ with $\sec^2 x - 1$ twice during the integration. Therefore,

$$\int \tan^5 x\,dx = \int \tan^3 x(\sec^2 x - 1)\,dx \qquad \text{using Eq. (28-15)}$$

$$= \int \tan^3 x(\sec^2 x\,dx) - \int \tan^3 x\,dx$$

$$= \frac{1}{4}\tan^4 x - \int \tan x(\sec^2 x - 1)\,dx \qquad \text{using Eq. (28-15) again}$$

$$= \frac{1}{4}\tan^4 x - \int \tan x(\sec^2 x\,dx) + \int \tan x\,dx$$

$$= \frac{1}{4}\tan^4 x - \frac{1}{2}\tan^2 x - \ln|\cos x| + C$$ ◂◂

EXAMPLE 6 ▸▸ Integrate: $\int \csc^4 2x\,dx$.

By writing $\csc^4 2x = \csc^2 2x \csc^2 2x = \csc^2 2x(1 + \cot^2 2x)$, we can write this integral with powers of $\cot 2x$ along with $\csc^2 2x\,dx$, which becomes part of the necessary du of the integral. Thus

$$\int \csc^4 2x\,dx = \int \csc^2 2x(1 + \cot^2 2x)\,dx \qquad \text{using Eq. (28-16)}$$

$$= \frac{1}{2}\int \csc^2 2x(2\,dx) - \frac{1}{2}\int \cot^2 2x(\underbrace{-2\csc^2 2x\,dx}_{du})$$

$$= -\frac{1}{2}\cot 2x - \frac{1}{6}\cot^3 2x + C$$ ◂◂

EXAMPLE 7 ▸▸ Integrate: $\displaystyle\int_0^{\pi/4} \frac{\tan^3 x}{\sec^3 x}\,dx$.

This integral requires the use of several trigonometric relationships to obtain integrable forms.

$$\int_0^{\pi/4} \frac{\tan^3 x}{\sec^3 x}\,dx = \int_0^{\pi/4} \frac{(\sec^2 x - 1)\tan x}{\sec^3 x}\,dx \qquad \text{using Eq. (28-15)}$$

$$= \int_0^{\pi/4} \frac{\tan x}{\sec x}\,dx - \int_0^{\pi/4} \frac{\tan x\,dx}{\sec^3 x} \qquad \frac{\tan x}{\sec x} = \frac{\sin x}{\cos x \sec x} = \sin x$$

$$= \int_0^{\pi/4} \sin x\,dx - \int_0^{\pi/4} \cos^2 x \sin x\,dx \qquad \frac{\tan x}{\sec^3 x} = \frac{\tan x}{\sec x \sec^2 x} = \sin x \cos^2 x$$

$$= -\cos x + \frac{1}{3}\cos^3 x\Big|_0^{\pi/4} \qquad \text{integrate}$$

$$= -\frac{\sqrt{2}}{2} + \frac{1}{3}\left(\frac{\sqrt{2}}{2}\right)^3 - \left(-1 + \frac{1}{3}\right) \qquad \text{evaluate}$$

$$= \frac{8 - 5\sqrt{2}}{12} = 0.0774$$ ◂◂

See the chapter introduction.

EXAMPLE 8 The *root-mean-square value of a function* with respect to x is defined by

$$y_{\text{rms}} = \sqrt{\frac{1}{T}\int_0^T y^2\,dx} \tag{28-19}$$

Usually the value of T that is of importance is the period of the function. Find the root-mean-square value of an electric current $i = 3\cos \pi t$ for one period.

The period is $\frac{2\pi}{\pi} = 2$ s. Therefore, we must find the square root of the integral

$$\frac{1}{2}\int_0^2 (3\cos \pi t)^2\,dt = \frac{9}{2}\int_0^2 \cos^2 \pi t\,dt$$

Evaluating this integral, we have

$$\frac{9}{2}\int_0^2 \cos^2 \pi t\,dt = \frac{9}{4}\int_0^2 (1 + \cos 2\pi t)\,dt \qquad \cos^2 \pi t = \tfrac{1}{2}(1 + \cos 2\pi t)$$

$$= \frac{9}{4}t\Big|_0^2 + \frac{9}{8\pi}\int_0^2 \cos 2\pi t(2\pi\,dt)$$

$$= \frac{9}{2} + \frac{9}{8\pi}\sin 2\pi t\Big|_0^2 = \frac{9}{2}$$

Thus the root-mean-square current is

$$i_{\text{rms}} = \sqrt{\frac{9}{2}} = \frac{3\sqrt{2}}{2} = 2.12 \text{ A}$$

This value of the current, often referred to as the *effective current,* is the value of direct current that would produce the same quantity of heat energy in the same time. It is important in the design of electric appliances.

Exercises 28–5

In Exercises 1–28, integrate each of the given functions.

1. $\int \sin^2 x \cos x\,dx$
2. $\int \sin x \cos^5 x\,dx$
3. $\int \sin^3 2x\,dx$
4. $\int 3\cos^3 x\,dx$
5. $\int 2\sin^2 x \cos^3 x\,dx$
6. $\int \sin^3 x \cos^6 x\,dx$
7. $\int_0^{\pi/4} 5\sin^5 x\,dx$
8. $\int_{\pi/3}^{\pi/2} \sqrt{\cos x}\sin^3 x\,dx$
9. $\int \sin^2 x\,dx$
10. $\int \cos^2 2x\,dx$
11. $\int \cos^2 3x\,dx$
12. $\int_0^1 \sin^2 4x\,dx$
13. $\int \tan^3 x\,dx$
14. $\int 6\cot^3 x\,dx$
15. $\int_0^{\pi/4} \tan x \sec^4 x\,dx$
16. $\int \cot 4x \csc^4 4x\,dx$
17. $\int \tan^4 2x\,dx$
18. $\int 4\cot^4 x\,dx$
19. $\int \tan^3 3x \sec^3 3x\,dx$
20. $\int \sqrt{\tan x}\sec^4 x\,dx$
21. $\int (\sin x + \cos x)^2\,dx$
22. $\int (\tan 2x + \cot 2x)^2\,dx$
23. $\int \frac{1 - \cot x}{\sin^4 x}\,dx$
24. $\int \frac{1 + \sin x}{\cos^4 x}\,dx$
25. $\int_{\pi/6}^{\pi/4} \cot^5 x\,dx$
26. $\int_{\pi/6}^{\pi/3} \frac{2\,dx}{1 + \sin x}$
27. $\int \sec^6 x\,dx$
28. $\int \tan^7 x\,dx$

In Exercises 29–40, solve the given problems by integration.

29. Find the volume generated by revolving the area bounded by $y = \sin x$ and $y = 0$, from $x = 0$ to $x = \pi$, about the x-axis.
30. Find the volume generated by revolving the area bounded by $y = \tan^3 (x^2)$, $y = 0$, and $x = \frac{\pi}{4}$ about the y-axis.

31. Find the area bounded by $y = \sin x$, $y = \cos x$, and $x = 0$ in the first quadrant.

32. Find the length of the curve $y = \ln \cos x$ from $x = 0$ to $x = \frac{\pi}{3}$. (See Exercise 25 of Section 26-6.)

33. Show that $\int \sin x \cos x\, dx$ can be integrated in two ways. Explain the difference in the answers.

34. Show that $\int \sec^2 x \tan x\, dx$ can be integrated in two ways. Explain the difference in the answers.

35. In the study of the rate of radiation by an accelerated charge, the following integral must be evaluated: $\int_0^{\pi} \sin^3 \theta\, d\theta$. Find the value of the integral.

36. In finding the volume of a special O-ring for a space vehicle, we must evaluate the integral $\int \frac{\sin^2 \theta}{\cos^2 \theta}\, d\theta$. Perform this integration.

37. Find the root-mean-square value of the voltage for one period of standard house current, which is given by $V = 170 \sin 120\pi t$.

38. For a current $i = i_0 \sin \omega t$, show that the root-mean-square value of the current for one period is $i_0/\sqrt{2}$.

39. In the analysis of the intensity of light from a certain source, the equation $I = A \int_{-a/2}^{a/2} \cos^2 [b\pi(c - x)]\, dx$ is used. Here, A, a, b, and c are constants. Evaluate this integral. (The simplification of the result is lengthy.)

40. In the study of the lifting force L due to a stream of fluid passing around a cylinder, the equation $L = k \int_0^{2\pi} (a \sin \theta + b \sin^2 \theta - b \sin^3 \theta)\, d\theta$ is used. Here, k, a, and b are constants and θ is the angle from the direction of flow. Evaluate the integral.

28-6 Inverse Trigonometric Forms

For reference, Eq. (27-10) is $\frac{d(\sin^{-1} u)}{dx} = \frac{1}{\sqrt{1 - u^2}} \frac{du}{dx}$.

Using Eq. (27-10), we can find the differential of $\sin^{-1} (u/a)$, where a is constant:

$$d\left(\sin^{-1} \frac{u}{a}\right) = \frac{1}{\sqrt{1 - (u/a)^2}} \frac{du}{a} = \frac{a}{\sqrt{a^2 - u^2}} \frac{du}{a} = \frac{du}{\sqrt{a^2 - u^2}}$$

Reversing this differentiation formula, we have the important integration formula

$$\int \frac{du}{\sqrt{a^2 - u^2}} = \sin^{-1} \frac{u}{a} + C \qquad (28\text{-}20)$$

By finding the differential of $\tan^{-1} (u/a)$, we have

$$d\left(\tan^{-1} \frac{u}{a}\right) = \frac{1}{1 + (u/a)^2} \frac{du}{a} = \frac{a^2}{a^2 + u^2} \frac{du}{a} = \frac{a\, du}{a^2 + u^2}$$

Now, reversing this differential, we have

$$\int \frac{du}{a^2 + u^2} = \frac{1}{a} \tan^{-1} \frac{u}{a} + C \qquad (28\text{-}21)$$

This shows one of the principal uses of the inverse trigonometric functions: They provide a solution to the integration of important algebraic functions.

EXAMPLE 1 ▸▸ Integrate: $\int \frac{dx}{\sqrt{9 - x^2}}$.

This integral fits the form of Eq. (28-20) with $u = x$, $du = dx$, and $a = 3$. Thus

$$\int \frac{dx}{\sqrt{9 - x^2}} = \int \frac{dx}{\sqrt{3^2 - x^2}}$$

$$= \sin^{-1} \frac{x}{3} + C$$ ◂◂

EXAMPLE 2 ▸▸ The volume flow rate Q (in m^3/s) of a constantly flowing liquid is given by $Q = 24\int_0^2 \frac{dx}{6 + x^2}$, where x is the distance from the center of flow. Find the value of Q.

For the integral, we see that it fits Eq. (28-21) with $u = x$, $du = dx$, and $a = \sqrt{6}$.

$$Q = 24\int_0^2 \frac{dx}{6 + x^2} = 24\int_0^2 \frac{dx}{(\sqrt{6})^2 + x^2}$$

$$= \frac{24}{\sqrt{6}} \tan^{-1} \frac{x}{\sqrt{6}}\Big|_0^2 = \frac{24}{\sqrt{6}}\left(\tan^{-1}\frac{2}{\sqrt{6}} - \tan^{-1} 0\right)$$

$$= 9.80 \tan^{-1} 0.8165$$

$$= 6.71 \text{ m}^3/\text{s}$$ ◂◂

EXAMPLE 3 ▸▸ Integrate: $\int \frac{dx}{\sqrt{25 - 4x^2}}$.

This integral fits the form of Eq. (28-20) with $u = 2x$, $du = 2\,dx$, and $a = 5$. Thus, in order to have the proper du, we must include a factor of 2 in the numerator, and therefore we also place a $\frac{1}{2}$ before the integral. This leads to

$$\int \frac{dx}{\sqrt{25 - 4x^2}} = \frac{1}{2}\int \frac{2\,dx \leftarrow du}{\sqrt{5^2 - \underbrace{(2x)^2}_{u}}}$$

$$= \frac{1}{2} \sin^{-1} \frac{2x}{5} + C$$ ◂◂

EXAMPLE 4 ▸▸ Integrate: $\int_{-1}^{3} \frac{dx}{x^2 + 6x + 13}$.

At first glance it does not appear that this integral fits any of the forms presented up to this point. However, by writing the denominator in the form

CAUTION ▸ $(x^2 + 6x + 9) + 4 = (x + 3)^2 + 2^2$, ***we recognize that*** $u = x + 3$, $du = dx$, and $a = 2$. Thus,

$$\int_{-1}^{3} \frac{dx}{x^2 + 6x + 13} = \int_{-1}^{3} \frac{dx \leftarrow du}{\underbrace{(x + 3)^2}_{u} + 2^2}$$

$$= \frac{1}{2} \tan^{-1} \frac{x + 3}{2}\Big|_{-1}^{3} \qquad \text{integrate}$$

$$= \frac{1}{2}(\tan^{-1} 3 - \tan^{-1} 1) \qquad \text{evaluate}$$

$$= 0.2318$$

Now we can see the use of completing the square when we are transforming integrals into proper form. ◂◂

EXAMPLE 5 ▸▸ Integrate: $\int \frac{2x + 5}{x^2 + 9} dx$.

By writing this integral as the sum of two integrals, we may integrate each of these separately:

$$\int \frac{2x + 5}{x^2 + 9} dx = \int \frac{2x\,dx}{x^2 + 9} + \int \frac{5\,dx}{x^2 + 9}$$

The first integral is a logarithmic form, and the second is an inverse tangent form. For the first, $u = x^2 + 9$, $du = 2x\,dx$. For the second, $u = x$, $du = dx$, $a = 3$.

$$\int \frac{2x\,dx}{x^2 + 9} + 5\int \frac{dx}{x^2 + 9} = \ln|x^2 + 9| + \frac{5}{3}\tan^{-1}\frac{x}{3} + C$$ ◂◂

CAUTION ▸ The inverse trigonometric integral forms show very well the importance of ***proper recognition of the form of the integral.*** It is important that these forms are not confused with those of the general power rule or the logarithmic form.

EXAMPLE 6 ▸▸ The integral $\int \frac{dx}{\sqrt{1 - x^2}}$ is of the inverse sine form with $u = x$, $du = dx$, and $a = 1$. Thus

$$\int \frac{dx}{\sqrt{1 - x^2}} = \sin^{-1} x + C$$

The integral $\int \frac{x\,dx}{\sqrt{1 - x^2}}$ is not of the inverse sine form due to the factor of x in the numerator. It is integrated by use of the general power rule, with $u = 1 - x^2$, $du = -2x\,dx$, and $n = -\frac{1}{2}$. Thus

$$\int \frac{x\,dx}{\sqrt{1 - x^2}} = -\sqrt{1 - x^2} + C$$

The integral $\int \frac{x\,dx}{1 - x^2}$ is of the basic logarithmic form with $u = 1 - x^2$ and $du = -2x\,dx$. If $1 - x^2$ is raised to any power other than 1 in the denominator, we would use the general power rule. To be of the inverse sine form, we would have the square root of $1 - x^2$ and no factor of x, as in the first illustration. Thus

$$\int \frac{x\,dx}{1 - x^2} = -\frac{1}{2}\ln|1 - x^2| + C$$ ◂◂

EXAMPLE 7 ▸▸ The following integrals are of the form indicated.

$\int \frac{dx}{1 + x^2}$	Inverse tangent form	$u = x$, $du = dx$
$\int \frac{x\,dx}{1 + x^2}$	Logarithmic form	$u = 1 + x^2$, $du = 2x\,dx$
$\int \frac{x\,dx}{\sqrt{1 + x^2}}$	General power form	$u = 1 + x^2$, $du = 2x\,dx$
$\int \frac{dx}{1 + x}$	Logarithmic form	$u = 1 + x$, $du = dx$ ◂◂

There are a number of integrals whose forms appear to be similar to those in Examples 6 and 7, but which do not fit the forms we have discussed. They include

$$\int \frac{dx}{\sqrt{x^2 - 1}}, \quad \int \frac{dx}{\sqrt{1 + x^2}}, \quad \int \frac{dx}{1 - x^2}, \quad \text{and} \quad \int \frac{dx}{x\sqrt{1 + x^2}}$$

We will develop methods to integrate some of these forms, and all of them can be integrated by the tables discussed in Section 28-9.

Exercises 28–6

In Exercises 1–24, integrate each of the given functions.

1. $\int \frac{dx}{\sqrt{4 - x^2}}$
2. $\int \frac{dx}{\sqrt{49 - x^2}}$
3. $\int \frac{dx}{64 + x^2}$
4. $\int \frac{4\,dx}{4 + x^2}$
5. $\int \frac{dx}{\sqrt{1 - 16x^2}}$
6. $\int_0^1 \frac{2\,dx}{\sqrt{9 - 4x^2}}$
7. $\int_0^2 \frac{3\,dx}{1 + 9x^2}$
8. $\int_1^3 \frac{dx}{49 + 4x^2}$
9. $\int_0^{0.4} \frac{2\,dx}{\sqrt{4 - 5x^2}}$
10. $\int \frac{dx}{2\sqrt{x}\sqrt{1 - x}}$
11. $\int \frac{8x\,dx}{9x^2 + 16}$
12. $\int \frac{3\,dx}{25 + 16x^2}$
13. $\int_1^2 \frac{dx}{5x^2 + 7}$
14. $\int_0^1 \frac{4x\,dx}{1 + x^4}$
15. $\int \frac{e^x\,dx}{\sqrt{1 - e^{2x}}}$
16. $\int \frac{\sec^2 x\,dx}{\sqrt{1 - \tan^2 x}}$
17. $\int \frac{dx}{x^2 + 2x + 2}$
18. $\int \frac{2\,dx}{x^2 + 8x + 17}$
19. $\int \frac{4\,dx}{\sqrt{-4x - x^2}}$
20. $\int \frac{dx}{\sqrt{2x - x^2}}$
21. $\int_{\pi/6}^{\pi/2} \frac{\cos 2x}{1 + \sin^2 2x}\,dx$
22. $\int_{-4}^0 \frac{dx}{x^2 + 4x + 5}$
23. $\int \frac{2 - x}{\sqrt{4 - x^2}}\,dx$
24. $\int \frac{3 - 2x}{1 + 4x^2}\,dx$

In Exercises 25–28, identify the form of each integral as being inverse sine, inverse tangent, logarithmic, or general power, as in Examples 6 and 7. Do not integrate. In each part (a), explain how the choice was made.

25. (a) $\int \frac{2\,dx}{4 + 9x^2}$ (b) $\int \frac{2\,dx}{4 + 9x}$ (c) $\int \frac{2x\,dx}{\sqrt{4 + 9x^2}}$
26. (a) $\int \frac{2x\,dx}{4 - 9x^2}$ (b) $\int \frac{2\,dx}{\sqrt{4 - 9x}}$ (c) $\int \frac{2x\,dx}{4 + 9x^2}$
27. (a) $\int \frac{2x\,dx}{\sqrt{4 - 9x^2}}$ (b) $\int \frac{2\,dx}{\sqrt{4 - 9x^2}}$ (c) $\int \frac{2\,dx}{4 - 9x}$
28. (a) $\int \frac{2\,dx}{9x^2 + 4}$ (b) $\int \frac{2x\,dx}{\sqrt{9x^2 - 4}}$ (c) $\int \frac{2x\,dx}{9x^2 - 4}$

In Exercises 29–36, solve the given problems by integration.

29. Find the area bounded by $y(1 + x^2) = 1$, $x = 0$, $y = 0$, and $x = 2$.
30. Find the area bounded by $y\sqrt{4 - x^2} = 1$, $x = 0$, $y = 0$, and $x = 1$. See Fig. 28-8.

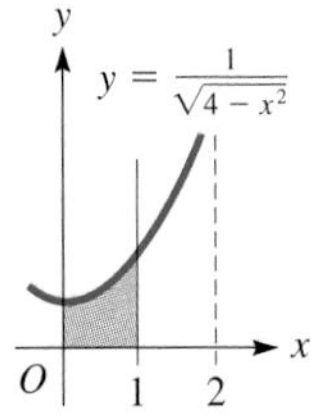

Fig. 28-8

31. To find the electric field E from an electric charge distributed uniformly over the entire xy-plane at a distance d from the plane, it is necessary to evaluate the integral $kd \int \frac{dx}{d^2 + x^2}$. Here x is the distance from the origin to the element of charge. Perform the indicated integration.
32. An oil storage tank can be described as the volume generated by revolving the area bounded by $y = 24/\sqrt{16 + x^2}$, $x = 0$, $y = 0$, and $x = 3$ about the x-axis. Find the volume (in m^3) of the tank.
33. In dealing with the theory for simple harmonic motion, it is necessary to solve the equation
$$\frac{dx}{\sqrt{A^2 - x^2}} = \sqrt{\frac{k}{m}}\,dt \quad \text{(with } k, m, \text{ and } A \text{ as constants)}$$
Determine the solution to this equation if $x = x_0$ when $t = 0$.
34. During each cycle, the velocity v (in m/s) of a robotic welding device is given by $v = 2t - \frac{12}{2 + t^2}$, where t is the time in seconds. Find the expression for the displacement s (in metres) as a function of t if $s = 0$ for $t = 0$.

35. Find the moment of inertia with respect to the y-axis for the area bounded by $y = 1/(1 + x^6)$, the x-axis, $x = 1$, and $x = 2$.

36. Find the length of arc along the curve $y = \sqrt{1 - x^2}$ between $x = 0$ and $x = 1$. (See Exercise 25 of Section 26-6.)

28-7 Integration by Parts

There are many methods of transforming integrals into forms that can be integrated by one of the basic formulas. In the preceding section we saw that completing the square and trigonometric identities can be used for this purpose. In this section and the following one, we develop two general methods. The method of integration by parts is discussed in this section.

Since the derivative of a product of functions is found by use of the formula

$$\frac{d(uv)}{dx} = u\frac{dv}{dx} + v\frac{du}{dx}$$

the differential of a product of functions is given by $d(uv) = u\,dv + v\,du$. Integrating both sides of this equation, we have $uv = \int u\,dv + \int v\,du$. Solving for $\int u\,dv$, we obtain

$$\int u\,dv = uv - \int v\,du \qquad (28\text{-}22)$$

Integration by use of Eq. (28-22) is called **integration by parts.**

EXAMPLE 1 ▸▸ Integrate: $\int x \sin x\,dx$.

This integral does not fit any of the previous forms we have discussed, since neither x nor $\sin x$ can be made a factor of a proper du. However, by choosing $u = x$ and $dv = \sin x\,dx$, integration by parts may be used. Thus

$$u = x, \qquad dv = \sin x\,dx$$

By finding the differential of u and integrating dv, we find du and v. This gives us

$$du = dx, \qquad v = -\cos x + C_1$$

Now, substituting in Eq. (28-22), we have

$$\int \underset{\downarrow}{u}\ \underset{\downarrow}{dv} = \underset{\downarrow}{u}\ \underset{\downarrow}{v} - \int \underset{\downarrow}{v}\ \underset{\downarrow}{du}$$

$$\begin{aligned}\int (x)(\sin x\,dx) &= (x)(-\cos x + C_1) - \int (-\cos x + C_1)(dx)\\ &= -x\cos x + C_1x + \int \cos x\,dx - \int C_1\,dx\\ &= -x\cos x + C_1x + \sin x - C_1x + C\\ &= -x\cos x + \sin x + C\end{aligned}$$

Other choices of u and dv may be made, but they are not useful. For example, if we choose $u = \sin x$ and $dv = x\,dx$, then $du = \cos x\,dx$ and $v = \frac{1}{2}x^2 + C_2$. This makes $\int v\,du = \int (\frac{1}{2}x^2 + C_2)(\cos x\,dx)$, which is more complex than the integrand of the original problem.

We also note that the constant C_1 that was introduced when we integrated dv does not appear in the final result. This constant will always cancel out, and therefore *we will not show any constant of integration when finding v.* ◂◂

NOTE ▸

As in Example 1, there is often more than one choice as to the part of the integrand that is selected to be u and the part that is selected to be dv. There are no set rules that may be stated for the best choice of u and dv, but two guidelines may be stated.

CAUTION ▶

> ***Guidelines for Choosing u and dv***
>
> **1.** *The quantity u is normally chosen such that du/dx is of simpler form than u.*
>
> **2.** *The differential dv is normally chosen such that $\int dv$ is easily integrated.*

Working examples, and thereby gaining experience in methods of integration, is the best way to determine when this method should be used and how to use it.

EXAMPLE 2 ▶▶ Integrate: $\int x\sqrt{1-x}\,dx$.

We see that this form does not fit the general power rule, for $x\,dx$ is not a factor of the differential of $1 - x$. By choosing $u = x$ and $dv = \sqrt{1-x}\,dx$ we have $du/dx = 1$, and v can readily be determined. Thus

$$u = x, \qquad dv = \sqrt{1-x}\,dx = (1-x)^{1/2}\,dx$$

$$du = dx, \qquad v = -\frac{2}{3}(1-x)^{3/2}$$

Substituting in Eq. (28-22), we have

$$\int \overset{u}{\downarrow}\hspace{-0.5em}x[\overset{dv}{(1-x)^{1/2}\,dx}] = \overset{u}{x}\left[\overset{v}{-\frac{2}{3}(1-x)^{3/2}}\right] - \int \left[\overset{v}{-\frac{2}{3}(1-x)^{3/2}}\right]\overset{du}{dx}$$

At this point we see that we can complete the integration. Thus

$$\begin{aligned}
\int x(1-x)^{1/2}\,dx &= -\frac{2x}{3}(1-x)^{3/2} + \frac{2}{3}\int (1-x)^{3/2}\,dx \\
&= -\frac{2x}{3}(1-x)^{3/2} + \frac{2}{3}\left(-\frac{2}{5}\right)(1-x)^{5/2} + C \\
&= -\frac{2}{3}(1-x)^{3/2}\left[x + \frac{2}{5}(1-x)\right] + C \\
&= -\frac{2}{15}(1-x)^{3/2}(2+3x) + C
\end{aligned}$$ ◀◀

EXAMPLE 3 ▶▶ Integrate: $\int \sqrt{x}\ln x\,dx$.

For this integral we have

$$u = \ln x, \qquad dv = x^{1/2}\,dx$$

$$du = \frac{1}{x}\,dx, \qquad v = \frac{2}{3}x^{3/2}$$

This leads to

$$\begin{aligned}
\int \sqrt{x}\ln x\,dx &= \frac{2}{3}x^{3/2}\ln x - \frac{2}{3}\int x^{1/2}\,dx \\
&= \frac{2}{3}x^{3/2}\ln x - \frac{4}{9}x^{3/2} + C
\end{aligned}$$ ◀◀

EXAMPLE 4 ▸▸ Integrate: $\int \sin^{-1} x\,dx$.

We write

$$u = \sin^{-1} x, \qquad dv = dx$$

$$du = \frac{dx}{\sqrt{1 - x^2}}, \qquad v = x$$

$$\int \sin^{-1} x\,dx = x \sin^{-1} x - \int \frac{x\,dx}{\sqrt{1 - x^2}}$$

$$= x \sin^{-1} x + \frac{1}{2}\int \frac{-2x\,dx}{\sqrt{1 - x^2}}$$

$$= x \sin^{-1} x + \sqrt{1 - x^2} + C$$ ◂◂

EXAMPLE 5 ▸▸ Integrate: $\int_0^1 xe^{-x}\,dx$.

We write

$$u = x, \qquad dv = e^{-x}\,dx$$

$$du = dx, \qquad v = -e^{-x}$$

$$\int_0^1 xe^{-x}\,dx = -xe^{-x}\Big|_0^1 + \int_0^1 e^{-x}\,dx = -xe^{-x} - e^{-x}\Big|_0^1$$

$$= -e^{-1} - e^{-1} + 1$$

$$= 1 - \frac{2}{e}$$ ◂◂

EXAMPLE 6 ▸▸ Integrate: $\int e^x \sin x\,dx$.

Let $u = \sin x$, $dv = e^x\,dx$, $du = \cos x\,dx$, $v = e^x$.

$$\int e^x \sin x\,dx = e^x \sin x - \int e^x \cos x\,dx$$

At first glance it appears that we have made no progress in applying the method of integration by parts. We note, however, that when we integrated $\int e^x \sin x\,dx$, part of the result was a term of $\int e^x \cos x\,dx$. This implies that ***if*** $\int e^x \cos x\,dx$ ***were integrated, a term of*** $\int e^x \sin x\,dx$ ***might result.*** Thus the method of integration by parts is now applied to the integral $\int e^x \cos x\,dx$:

NOTE ▸

$$u = \cos x, \qquad dv = e^x\,dx, \qquad du = -\sin x\,dx, \qquad v = e^x$$

And so $\int e^x \cos x\,dx = e^x \cos x + \int e^x \sin x\,dx$. Substituting this expression into the expression for $\int e^x \sin x\,dx$, we obtain

$$\int e^x \sin x\,dx = e^x \sin x - \left(e^x \cos x + \int e^x \sin x\,dx\right)$$

$$= e^x \sin x - e^x \cos x - \int e^x \sin x\,dx$$

$$2\int e^x \sin x\,dx = e^x(\sin x - \cos x) + 2C$$

$$\int e^x \sin x\,dx = \frac{e^x}{2}(\sin x - \cos x) + C$$

Thus, by combining integrals of like form, we obtain the desired result. ◂◂

Exercises 28–7

In Exercises 1–16, integrate each of the given functions.

1. $\int x \cos x\, dx$
2. $\int x \sin 2x\, dx$
3. $\int xe^{2x}\, dx$
4. $\int 3xe^{x}\, dx$
5. $\int x \sec^2 x\, dx$
6. $\int_0^{\pi/4} x \sec x \tan x\, dx$
7. $\int 2 \tan^{-1} x\, dx$
8. $\int \ln x\, dx$
9. $\int \frac{4x\, dx}{\sqrt{1-x}}$
10. $\int x\sqrt{x+1}\, dx$
11. $\int x \ln x\, dx$
12. $\int x^2 \ln 4x\, dx$
13. $\int x^2 \sin 2x\, dx$
14. $\int x^2 e^{2x}\, dx$
15. $\int_0^{\pi/2} e^x \cos x\, dx$
16. $\int e^{-x} \sin 2x\, dx$

In Exercises 17–28, solve the given problems by integration.

17. Find the area bounded by $y = xe^{-x}$, $y = 0$, and $x = 2$. See Fig. 28-9.

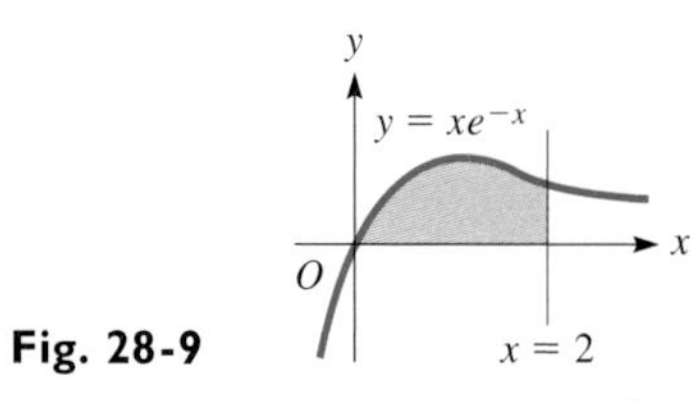

Fig. 28-9

18. Find the area bounded by $y = 2(\ln x)/x^2$, $y = 0$, and $x = 3$.
19. Find the volume generated by revolving the area bounded by $y = \tan^2 x$, $y = 0$, and $x = 0.5$ about the y-axis.
20. Find the volume generated by revolving the area bounded by $y = \sin x$ and $y = 0$ (from $x = 0$ to $x = \pi$) about the y-axis.
21. Find the x-coordinate of the centroid of the area bounded by $y = \cos x$ and $y = 0$ for $0 \le x \le \frac{\pi}{2}$.
22. Find the moment of inertia with respect to its axis of the volume generated by revolving the area bounded by $y = e^x$, $x = 1$, and the coordinate axes about the y-axis.
23. Find the root-mean-square value of the function $y = \sqrt{\sin^{-1} x}$ between $x = 0$ and $x = 1$. (See Example 8 of Section 28-5.)
24. The general expression for the slope of a curve is $dy/dx = x^3\sqrt{1+x^2}$. Find the equation of the curve if it passes through the origin.
25. Computer simulation shows that the velocity v (in m/s) of a test car is $v = t^3/\sqrt{t^2+1}$ from $t = 0$ to $t = 8.0$ s. Find the expression for the distance traveled by the car in t seconds.
26. The nose cone of a rocket has the shape of the volume that is generated by revolving the area bounded by $y = \ln x$, $y = 0$, and $x = 9.5$ about the x-axis, where dimensions are in metres. Find the volume of the nose cone.
27. The current in a given circuit is given by $i = e^{-2t} \cos t$. Find an expression for the amount of charge that passes a given point in the circuit as a function of the time, if $q_0 = 0$.
28. In finding the average length $\bar{x}$ (in nm) of a certain type of large molecule, we use the equation $\bar{x} = \lim_{b\to\infty} [0.1 \int_0^b x^3 e^{-x^2/8}\, dx]$. Evaluate the integral and then use a calculator to show that $\bar{x} \to 3.2$ nm as $b \to \infty$.

28-8 Integration by Trigonometric Substitution

For reference, Eqs. (28-14), (28-15), and (28-16) are
$\cos^2 x + \sin^2 x = 1$
$1 + \tan^2 x = \sec^2 x$
$1 + \cot^2 x = \csc^2 x$

We have seen that trigonometric relations provide a means of transforming trigonometric integrals into forms that can be integrated. As we will show in this section, they can also be useful for integrating certain types of algebraic integrals. Substitutions based on Eqs. (28-14), (28-15), and (28-16) prove to be particularly useful for integrating expressions that involve radicals. The following examples illustrate the method.

EXAMPLE 1 ⏩ Integrate: $\int \frac{dx}{x^2\sqrt{1-x^2}}$.

If we let $x = \sin\theta$, the radical becomes $\sqrt{1-\sin^2\theta} = \cos\theta$. Therefore, by making this substitution, the integral can be transformed into a trigonometric integral. We must be careful to ***replace all factors of the integral by proper expressions in terms of θ.*** With $x = \sin\theta$, by finding the differential of x, we have $dx = \cos\theta\,d\theta$.

CAUTION ▶

Now, by substituting $x = \sin\theta$ and $dx = \cos\theta\,d\theta$ into the integral, we have

$$\int \frac{dx}{x^2\sqrt{1-x^2}} = \int \frac{\cos\theta\,d\theta}{\sin^2\theta\sqrt{1-\sin^2\theta}} \qquad \text{substituting}$$

$$= \int \frac{\cos\theta\,d\theta}{\sin^2\theta\cos\theta} = \int \csc^2\theta\,d\theta \qquad \text{using trigonometric relations}$$

This last integral can be integrated by using Eq. (28-7). This leads to

$$\int \csc^2\theta\,d\theta = -\cot\theta + C$$

We have now performed the integration, but the answer is expressed in terms of a variable different from the original integral. We must express the result in terms of x. Making a right triangle with an angle θ such that $\sin\theta = x/1$ (see Fig. 28-10), we may express any of the trigonometric functions in terms of x. (This is the method used with inverse trigonometric functions.) Thus

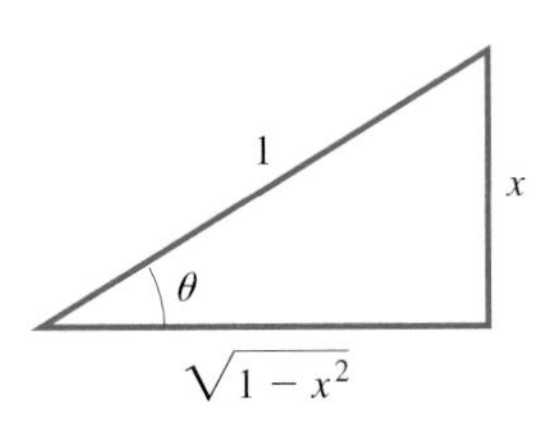

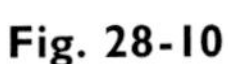
Fig. 28-10

$$\cot\theta = \frac{\sqrt{1-x^2}}{x}$$

Therefore, the result of the integration becomes

$$\int \frac{dx}{x^2\sqrt{1-x^2}} = -\cot\theta + C = -\frac{\sqrt{1-x^2}}{x} + C$$ ⏪

EXAMPLE 2 ⏩ Integrate: $\int \frac{dx}{\sqrt{x^2+4}}$.

If we let $x = 2\tan\theta$, the radical in this integral becomes

$$\sqrt{x^2+4} = \sqrt{4\tan^2\theta + 4} = 2\sqrt{\tan^2\theta + 1} = 2\sqrt{\sec^2\theta} = 2\sec\theta$$

Therefore, with $x = 2\tan\theta$ and $dx = 2\sec^2\theta\,d\theta$, we have

$$\int \frac{dx}{\sqrt{x^2+4}} = \int \frac{2\sec^2\theta\,d\theta}{\sqrt{4\tan^2\theta+4}} = \int \frac{2\sec^2\theta\,d\theta}{2\sec\theta} \qquad \text{substituting}$$

$$= \int \sec\theta\,d\theta = \ln|\sec\theta + \tan\theta| + C \qquad \text{using Eq. (28-12)}$$

$$= \ln\left|\frac{\sqrt{x^2+4}}{2} + \frac{x}{2}\right| + C = \ln\left|\frac{\sqrt{x^2+4}+x}{2}\right| + C \qquad \text{see Fig. 28-11}$$

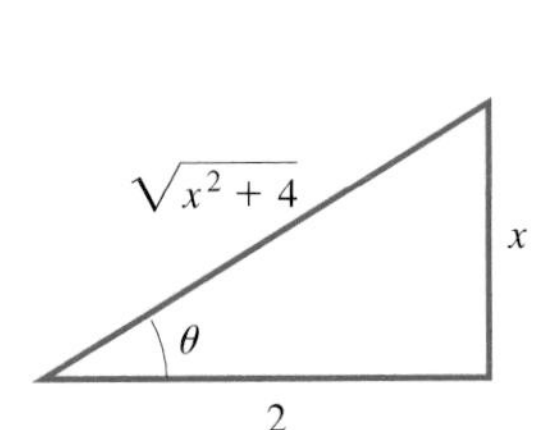

Fig. 28-11

This answer is acceptable, but by using the properties of logarithms, we have

$$\ln\left|\frac{\sqrt{x^2+4}+x}{2}\right| + C = \ln\left|\sqrt{x^2+4}+x\right| + (C - \ln 2)$$

$$= \ln\left|\sqrt{x^2+4}+x\right| + C'$$

Combining constants, as we did in writing $C' = C - \ln 2$, is a common practice in integration problems. ⏪

EXAMPLE 3 ▸▸ Integrate: $\int \frac{2\,dx}{x\sqrt{x^2 - 9}}$.

If we let $x = 3\sec\theta$, the radical in this integral becomes

$$\sqrt{x^2 - 9} = \sqrt{9\sec^2\theta - 9} = 3\sqrt{\sec^2\theta - 1} = 3\sqrt{\tan^2\theta} = 3\tan\theta$$

Therefore, with $x = 3\sec\theta$ and $dx = 3\sec\theta\tan\theta\,d\theta$, we have

$$\int \frac{2\,dx}{x\sqrt{x^2 - 9}} = 2\int \frac{3\sec\theta\tan\theta\,d\theta}{3\sec\theta\sqrt{9\sec^2\theta - 9}} = 2\int \frac{\tan\theta\,d\theta}{3\tan\theta}$$
$$= \frac{2}{3}\int d\theta = \frac{2}{3}\theta + C = \frac{2}{3}\sec^{-1}\frac{x}{3} + C$$

It is not necessary to refer to a triangle to express the result in terms of x. The solution is found by solving $x = 3\sec\theta$ for θ, as indicated. ◂◂

These examples show that by making the proper substitution, we can integrate algebraic functions by using the equivalent trigonometric forms. In summary, for the indicated radical form, the following trigonometric substitutions are used:

Basic Trigonometric Substitutions

For	$\sqrt{a^2 - x^2}$	use	$x = a\sin\theta$
For	$\sqrt{a^2 + x^2}$	use	$x = a\tan\theta$
For	$\sqrt{x^2 - a^2}$	use	$x = a\sec\theta$

(28-23)

EXAMPLE 4 ▸▸ Evaluate: $\int_1^4 \frac{\sqrt{9 + x^2}}{x}\,dx$.

Since we have the form $\sqrt{a^2 + x^2}$, where $a = 3$, we make the substitution $x = 3\tan\theta$. This means that $dx = 3\sec^2\theta\,d\theta$. Thus

$$\int \frac{\sqrt{9 + x^2}}{x}\,dx = \int \frac{\sqrt{9 + 9\tan^2\theta}}{3\tan\theta}(3\sec^2\theta\,d\theta) \quad \text{substituting}$$
$$= 3\int \frac{\sec\theta}{\tan\theta}(1 + \tan^2\theta)\,d\theta = 3\left(\int \frac{\sec\theta}{\tan\theta}\,d\theta + \int \tan\theta\sec\theta\,d\theta\right) \quad \text{using trig. relations}$$
$$= 3\left(\int \csc\theta\,d\theta + \int \tan\theta\sec\theta\,d\theta\right)$$
$$= 3[\ln|\csc\theta - \cot\theta| + \sec\theta] + C \quad \text{using Eqs. (28-13) and (28-8)}$$
$$= 3\left[\ln\left|\frac{\sqrt{x^2 + 9}}{x} - \frac{3}{x}\right| + \frac{\sqrt{x^2 + 9}}{3}\right] + C \quad \text{see Fig. 28-12}$$

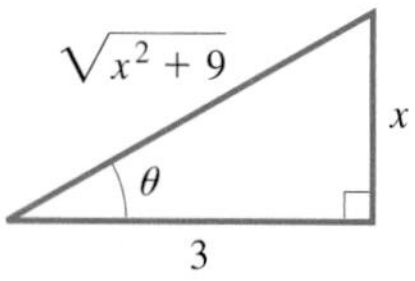

$\sqrt{x^2 + 9}$, x, θ, 3

Fig. 28-12

Limits have not been included, due to the changes in variables. The actual evaluation may now be completed:

$$\int_1^4 \frac{\sqrt{9 + x^2}}{x}\,dx = 3\left[\ln\left|\frac{\sqrt{x^2 + 9} - 3}{x}\right| + \frac{\sqrt{x^2 + 9}}{3}\right]_1^4$$
$$= 3\left[\left(\ln\frac{1}{2} - \ln\frac{\sqrt{10} - 3}{1}\right) + \left(\frac{5}{3} - \frac{\sqrt{10}}{3}\right)\right]$$
$$= 3\left[\ln\frac{1}{2(\sqrt{10} - 3)} + \frac{5 - \sqrt{10}}{3}\right] = 5.21 \quad ◂◂$$

Exercises 28–8

In Exercises 1–16, integrate each of the given functions.

1. $\int \frac{\sqrt{1-x^2}}{x^2}\,dx$
2. $\int \frac{dx}{(x^2+9)^{3/2}}$
3. $\int \frac{2\,dx}{\sqrt{x^2-4}}$
4. $\int \frac{\sqrt{x^2-25}}{x}\,dx$
5. $\int \frac{dx}{x^2\sqrt{x^2+9}}$
6. $\int \frac{3\,dx}{x\sqrt{4-x^2}}$
7. $\int \frac{4\,dx}{(4-x^2)^{3/2}}$
8. $\int \frac{2x^3\,dx}{\sqrt{9+x^2}}$
9. $\int_0^{0.5} \frac{x^3\,dx}{\sqrt{1-x^2}}$
10. $\int_4^5 \frac{\sqrt{x^2-16}}{x^2}\,dx$
11. $\int \frac{5\,dx}{\sqrt{x^2+2x+2}}$
12. $\int \frac{dx}{\sqrt{x^2+2x}}$
13. $\int \frac{dx}{x\sqrt{4x^2-9}}$
14. $\int \sqrt{16-x^2}\,dx$
15. $\int \frac{2\,dx}{\sqrt{e^{2x}-1}}$
16. $\int \frac{\sec^2 x\,dx}{(4-\tan^2 x)^{3/2}}$

In Exercises 17–24, solve the given problems by integration.

17. Find the area of a circle of radius 1.
18. Find the area bounded by $y = \frac{1}{x^2\sqrt{x^2-1}}$, $x = \sqrt{2}$, $x = \sqrt{5}$, and $y = 0$. See Fig. 28-13.

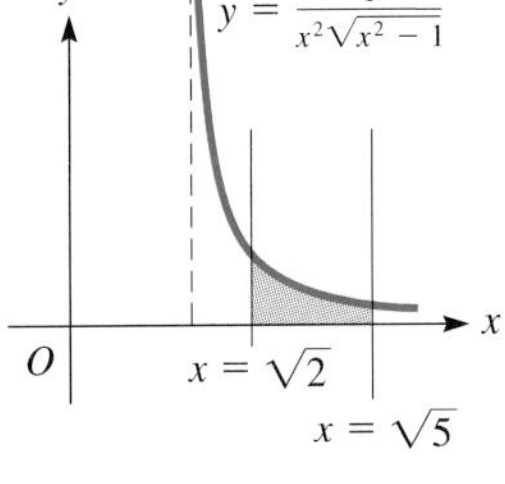

Fig. 28-13

19. Find the moment of inertia with respect to the y-axis of the first-quadrant area under the circle $x^2 + y^2 = a^2$ in terms of its mass.
20. Find the moment of inertia of a sphere of radius a with respect to its axis in terms of its mass.
21. Find the volume generated by revolving the area bounded by $y = \frac{\sqrt{x^2-16}}{x^2}$, $y = 0$, and $x = 5$ about the y-axis.
22. Find the length of arc along the curve of $y = \ln x$ from $x = 1$ to $x = 3$. (See Exercise 25 of Section 26-6.)
23. If an electric charge Q is distributed along a straight wire of length $2a$, the electric potential V at a point P, which is at a distance b from the center of the wire, is $V = kQ\int_{-a}^{a} \frac{dx}{\sqrt{b^2+x^2}}$. Here k is a constant and x is the distance along the wire. Evaluate the integral.
24. An electric insulating ring for a machine part can be described as the volume generated by revolving the area bounded by $y = x^2\sqrt{x^2-4}$, $y = 0$, and $x = 2.5$ about the y-axis. Find the volume (in cm^3) of material in the ring.

28-9 Integration by Use of Tables

In this chapter we have introduced certain basic integrals and have also brought in some methods of reducing other integrals to these basic forms. Often this transformation and integration requires a number of steps to be performed, and therefore integrals are tabulated for reference. The integrals found in tables have been derived by using the methods introduced thus far, as well as many other methods that can be used. Therefore, an understanding of the basic forms and some of the basic methods is very useful in finding integrals from tables. Such an understanding forms a basis for proper recognition of the forms that are used in the tables, as well as the types of results that may be expected. Therefore, ***the use of the tables depends on proper recognition of the form and the variables and constants of the integral.*** The following examples illustrate the use of the table of integrals found in Appendix D. More extensive tables are available in other sources.

CAUTION ▶

EXAMPLE 1 ▸▸ Integrate: $\int \frac{x\,dx}{\sqrt{2+3x}}$.

We first note that this integral fits the form of Formula 6 of Appendix D, with $u = x$, $a = 2$, and $b = 3$. Therefore,

$$\int \frac{x\,dx}{\sqrt{2+3x}} = -\frac{2(4-3x)\sqrt{2+3x}}{27} + C$$ ◂◂

EXAMPLE 2 ▸▸ Integrate: $\int \frac{\sqrt{4-9x^2}}{x}\,dx$.

This fits the form of Formula 18, with proper identification of constants; $u = 3x$, $du = 3\,dx$, $a = 2$. Hence

$$\int \frac{\sqrt{4-9x^2}}{x}\,dx = \int \frac{\sqrt{4-9x^2}}{3x}\,3\,dx$$

$$= \sqrt{4-9x^2} - 2\ln\left(\frac{2+\sqrt{4-9x^2}}{3x}\right) + C$$ ◂◂

EXAMPLE 3 ▸▸ Integrate: $\int 5\sec^3 2x\,dx$.

This fits the form of Formula 37; $n = 3$, $u = 2x$, $du = 2\,dx$. And so

For reference, Eq. (28-12) is

$$\int \sec u\,du = \ln|\sec u + \tan u| + C.$$

$$\int 5\sec^3 2x\,dx = 5\left(\frac{1}{2}\right)\int \sec^3 2x(2\,dx)$$

$$= \frac{5}{2}\,\frac{\sec 2x\tan 2x}{2} + \frac{5}{2}\left(\frac{1}{2}\right)\int \sec 2x(2\,dx)$$

To complete this integral, we must use the basic form of Eq. (28-12). Thus we complete it by

$$\int 5\sec^3 2x\,dx = \frac{5\sec 2x\tan 2x}{4} + \frac{5}{4}\ln|\sec 2x + \tan 2x| + C$$ ◂◂

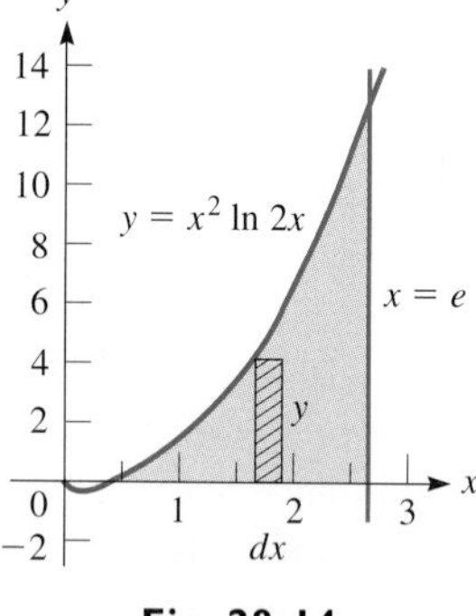

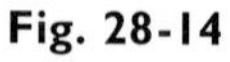
Fig. 28-14

EXAMPLE 4 ▸▸ Find the area bounded by $y = x^2\ln 2x$, $y = 0$, and $x = e$.

From Fig. 28-14 we see that the area is

$$A = \int_{0.5}^{e} x^2\ln 2x\,dx$$

This integral fits the form of Formula 46 if $u = 2x$. Thus we have

$$A = \frac{1}{8}\int_{0.5}^{e} (2x)^2\ln 2x(2\,dx) = \frac{1}{8}(2x)^3\left[\frac{\ln 2x}{3} - \frac{1}{9}\right]_{0.5}^{e}$$

$$= e^3\left(\frac{\ln 2e}{3} - \frac{1}{9}\right) - \frac{1}{8}\left(\frac{\ln 1}{3} - \frac{1}{9}\right) = e^3\left(\frac{3\ln 2e - 1}{9}\right) + \frac{1}{72}$$

$$= 9.118$$ ◂◂

NOTE ▸ The proper identification of u and du is the key step in the use of tables. Therefore, for the integrals in the following example the proper u and du, along with the appropriate formula from the table, are identified, but the integrations are not performed.

EXAMPLE 5 (a) $\int x\sqrt{1-x^4}\,dx \qquad u = x^2, \qquad du = 2x\,dx,$ Formula 15

(b) $\int \frac{(4x^6-9)^{3/2}}{x}\,dx \qquad u = 2x^3, \qquad du = 6x^2\,dx,$ Formula 22
Introduce a factor of x^2 into numerator and denominator

(c) $\int x^3 \sin x^2\,dx \qquad u = x^2, \qquad du = 2x\,dx,$ Formula 47

Following is a brief summary of the approach to integration we have used to obtain the exact result. Also, definite integrals may be approximated by methods such as the trapezoidal rule or Simpson's rule.

Basic Approach to Integrating a Function

1. *Write the integral such that it fits an integral form. Either a basic form as developed in this chapter or a form from a table of integrals may be used.*
2. *Use a method of transforming the integral such that an integral form may be used. Appropriate methods are covered in this chapter or in other sources.*

Exercises 28–9

In the following exercises, integrate each function by using the table in Appendix D.

1. $\int \frac{3x\,dx}{2+5x}$
2. $\int \frac{4x\,dx}{(1+x)^2}$
3. $\int_2^7 4x\sqrt{2+x}\,dx$
4. $\int \frac{dx}{x^2-4}$
5. $\int \sqrt{4-x^2}\,dx$
6. $\int_0^{\pi/3} \sin^3 x\,dx$
7. $\int \sin 2x \sin 3x\,dx$
8. $\int 6\sin^{-1} 3x\,dx$
9. $\int \frac{\sqrt{4x^2-9}}{x}\,dx$
10. $\int \frac{(9x^2+16)^{3/2}}{x}\,dx$
11. $\int \cos^5 4x\,dx$
12. $\int \tan^2 x\,dx$
13. $\int \tan^{-1} x^2(x\,dx)$
14. $\int 5xe^{4x}\,dx$
15. $\int_1^2 (4-x^2)^{3/2}\,dx$
16. $\int \frac{3\,dx}{9-16x^2}$
17. $\int \frac{dx}{x\sqrt{4x^2+1}}$
18. $\int \frac{\sqrt{4+x^2}}{x}\,dx$
19. $\int \frac{8\,dx}{x\sqrt{1-4x^2}}$
20. $\int \frac{dx}{x(1+4x)^2}$
21. $\int \sin x \cos 5x\,dx$
22. $\int_0^2 x^2e^{3x}\,dx$
23. $\int x^5 \cos x^3\,dx$
24. $\int 2\sin^3 x\cos^2 x\,dx$
25. $\int \frac{2x\,dx}{(1-x^4)^{3/2}}$
26. $\int \frac{dx}{x(1-4x)}$
27. $\int_1^3 \frac{\sqrt{3+5x^2}\,dx}{x}$
28. $\int_0^1 \frac{\sqrt{9-4x^2}}{x}\,dx$
29. $\int x^3 \ln x^2\,dx$
30. $\int \frac{x\,dx}{x^2\sqrt{x^4-9}}$
31. $\int \frac{9x^2\,dx}{(x^6-1)^{3/2}}$
32. $\int x^7\sqrt{x^4+4}\,dx$
33. Find the length of arc of the curve $y = x^2$ from $x = 0$ to $x = 1$. (See Exercise 25 of Section 26-6.)
34. Find the moment of inertia with respect to its axis of the volume generated by revolving the area bounded by $y = 3\ln x$, $x = e$, and the x-axis about the y-axis.
35. Find the area bounded by $y = \tan^{-1} 2x$, $x = 2$, and the x-axis.
36. Find the area bounded by $y = (x^2+4)^{-3/2}$, $x = 3$, and the axes.
37. Find the force (in newtons) on the area bounded by $x = 1/\sqrt{1+y}$, $y = 0$, $y = 3$, and the y-axis, if the surface of the water is at the upper edge of the area.
38. If 6.00 g of a chemical are placed in water, the time t (in min) it takes to dissolve half of the chemical is given by $t = 560\int_3^6 \frac{dx}{x(x+4)}$, where x is the amount of undissolved chemical at any time. Evaluate t.
39. The force F, in newtons, required to move a particular lever mechanism is a function of the displacement s, in feet, according to $F = 4s\sqrt{4s+3}$. Find the work done if the displacement changes from zero to 2.50 m.
40. If an electric charge Q is distributed along a wire of length $2a$, the force F exerted on an electric charge q placed at point P is $F = kqQ\int \frac{b\,dx}{(b^2+x^2)^{3/2}}$. Integrate to find F as a function of x.

Chapter Equations, Review Exercises, and Practice Test

Chapter Equations

Integrals

$$\int u^n\,du = \frac{u^{n+1}}{n+1} + C \qquad (n \neq -1) \tag{28-1}$$

$$\int \frac{du}{u} = \ln|u| + C \tag{28-2}$$

$$\int e^u\,du = e^u + C \tag{28-3}$$

$$\int \sin u\,du = -\cos u + C \tag{28-4}$$

$$\int \cos u\,du = \sin u + C \tag{28-5}$$

$$\int \sec^2 u\,du = \tan u + C \tag{28-6}$$

$$\int \csc^2 u\,du = -\cot u + C \tag{28-7}$$

$$\int \sec u \tan u\,du = \sec u + C \tag{28-8}$$

$$\int \csc u \cot u\,du = -\csc u + C \tag{28-9}$$

$$\int \tan u\,du = -\ln|\cos u| + C \tag{28-10}$$

$$\int \cot u\,du = \ln|\sin u| + C \tag{28-11}$$

$$\int \sec u\,du = \ln|\sec u + \tan u| + C \tag{28-12}$$

$$\int \csc u\,du = \ln|\csc u - \cot u| + C \tag{28-13}$$

Trigonometric relations

$$\cos^2 x + \sin^2 x = 1 \tag{28-14}$$

$$1 + \tan^2 x = \sec^2 x \tag{28-15}$$

$$1 + \cot^2 x = \csc^2 x \tag{28-16}$$

$$2\cos^2 x = 1 + \cos 2x \tag{28-17}$$

$$2\sin^2 x = 1 - \cos 2x \tag{28-18}$$

Root-mean-square value

$$y_{\text{rms}} = \sqrt{\frac{1}{T}\int_0^T y^2\,dx} \tag{28-19}$$

Integrals

$$\int \frac{du}{\sqrt{a^2 - u^2}} = \sin^{-1}\frac{u}{a} + C \tag{28-20}$$

$$\int \frac{du}{a^2 + u^2} = \frac{1}{a}\tan^{-1}\frac{u}{a} + C \tag{28-21}$$

$$\int u\,dv = uv - \int v\,du \tag{28-22}$$

Trigonometric substitutions

$$\begin{aligned} &\text{For } \sqrt{a^2 - x^2} \quad \text{use} \quad x = a\sin\theta \\ &\text{For } \sqrt{a^2 + x^2} \quad \text{use} \quad x = a\tan\theta \\ &\text{For } \sqrt{x^2 - a^2} \quad \text{use} \quad x = a\sec\theta \end{aligned} \tag{28-23}$$

Review Exercises

In Exercises 1–36, integrate the given functions without using a table of integrals.

1. $\int e^{-2x}\,dx$
2. $\int e^{\cos 2x}\sin x\cos x\,dx$
3. $\int \frac{dx}{x(\ln 2x)^2}$
4. $\int x^{1/3}\sqrt{x^{4/3}+1}\,dx$
5. $\int \frac{4\cos x\,dx}{1+\sin x}$
6. $\int \frac{\sec^2 x\,dx}{2+\tan x}$
7. $\int \frac{2\,dx}{25+49x^2}$
8. $\int \frac{dx}{\sqrt{1-4x^2}}$
9. $\int_0^{\pi/2}\cos^3 2x\,dx$
10. $\int_0^{\pi/8}\sec^3 2x\tan 2x\,dx$
11. $\int_0^2 \frac{x\,dx}{4+x^2}$
12. $\int_1^e \frac{\ln x^2\,dx}{x}$
13. $\int \sec^4 3x\tan 3x\,dx$
14. $\int \frac{\sin^3 x\,dx}{\sqrt{\cos x}}$
15. $\int \frac{e^x\,dx}{1+e^{2x}}$
16. $\int 3\sec 4x\,dx$
17. $\int \sec^4 3x\,dx$
18. $\int \frac{\sin^2 2x\,dx}{1+\cos 2x}$
19. $\int_0^{0.5}\frac{2e^{2x}-3e^x}{e^{2x}}\,dx$
20. $\int \frac{4-e^{\sqrt{x}}}{\sqrt{x}e^{\sqrt{x}}}\,dx$
21. $\int \frac{3x\,dx}{4+x^4}$
22. $\int_1^3 \frac{2\,dx}{\sqrt{x}(1+x)}$
23. $\int \frac{4\,dx}{\sqrt{4x^2-9}}$
24. $\int \frac{x^2\,dx}{\sqrt{9-x^2}}$
25. $\int \frac{e^{2x}\,dx}{\sqrt{e^{2x}+1}}$
26. $\int \frac{(4+\ln 2x)^3\,dx}{x}$
27. $\int 3\sin^2 3x\,dx$
28. $\int \sin^4 x\,dx$
29. $\int x\csc^2 2x\,dx$
30. $\int x\tan^{-1}x\,dx$
31. $\int e^{2x}\cos e^{2x}\,dx$
32. $\int \frac{3\,dx}{x^2+6x+10}$
33. $\int_1^e \frac{(\ln x)^2\,dx}{x}$
34. $\int_1^3 \frac{2\,dx}{x^2-2x+5}$
35. $\int \frac{x^2-1}{x+2}\,dx$
36. $\int \frac{\log_x 2\,dx}{x\ln x}$

In Exercises 37–64, solve the given problems by integration.

37. Show that $\int e^x(e^x+1)^2\,dx$ can be integrated in two ways. Explain the difference in the answers.

38. Show that $\int \frac{1}{x}(1+\ln x)\,dx$ can be integrated in two ways. Explain the difference in the answers.

39. Evaluate $\int \sin^2 x\,dx$ (a) by using Eq. (28-18) and (b) by using Eq. (28-22). Show that the results are equivalent.

40. Find the equation of the curve for which $dy/dx = e^x(2-e^x)^2$, if the curve passes through (0, 4).

41. Find the equation of the curve for which $dy/dx = \sec^4 x$, if the curve passes through the origin.

42. Find the equation of the curve for which $\frac{dy}{dx} = \frac{\sqrt{4+x^2}}{x^4}$, if the curves passes through (2, 1).

43. Find the area bounded by $y = 4e^{2x}$, $x = 1.5$, and the axes.

44. Find the area bounded by $y = x/(1+x)^2$, the x-axis, and the line $x = 4$.

45. Find the area inside the circle $x^2 + y^2 = 25$ and to the right of the line $x = 3$.

46. Find the area bounded by $y = x\sqrt{x+4}$, $y = 0$, and $x = 5$.

47. Find the volume generated by revolving the area bounded by $y = xe^x$, $y = 0$, and $x = 2$ about the y-axis.

48. Find the volume generated by revolving about the y-axis the area bounded by $y = x + \sqrt{x+1}$, $x = 3$, and the axes.

49. Find the volume of the solid generated by revolving the area bounded by $y = e^x\sin x$ and the x-axis between $x = 0$ and $x = \pi$ about the x-axis.

50. Find the centroid of the area bounded by $y = \ln x$, $x = 2$, and the x-axis.

51. Find the length of arc along the curve of $y = \ln\sin x$ from $x = \frac{1}{3}\pi$ to $x = \frac{2}{3}\pi$. See Exercise 25 of Section 26-6.

52. Find the area of the surface generated by revolving the curve of $y = \sqrt{4-x^2}$ from $x = -2$ to $x = 2$ about the x-axis. See Exercise 27 of Section 26-6.

53. The change in the thermodynamic entity of entropy ΔS may be expressed as $\Delta S = \int (c_v/T)\,dT$, where c_v is the heat capacity at constant volume and T is the temperature. For increased accuracy, c_v is often given by the equation $c_v = a + bT + cT^2$, where a, b, and c are constants. Express ΔS as a function of temperature.

54. A second-order chemical reaction leads to the equation $dt = \frac{k_1\,dx}{a-x} + \frac{k_2\,dx}{b-x}$, where k_1 and k_2 are constants, a and b are initial concentrations, t is the time, and x is the decrease in concentration. Solve for t as a function of x.

55. When we consider the resisting force of the air, the velocity v (in m/s) of a falling brick in terms of the time t (in seconds) is given by $dv/(9.8 - 0.1v) = dt$. If $v = 0$ when $t = 0$, find v as a function of t.

56. The power delivered to an electric circuit is given by $P = ei$, where e and i are the instantaneous voltage and the instantaneous current in the circuit, respectively. The mean power, averaged over a period $2\pi/\omega$, is given by $P_{av} = \dfrac{\omega}{2\pi}\displaystyle\int_0^{2\pi/\omega} ei\,dt$. If $e = 20\cos 2t$ and $i = 3\sin 2t$, find the average power over a period of $\pi/4$.

57. Find the root-mean-square value of the electric current i if $i = 2\sin t$.

58. In atomic theory, when finding the number n of atoms per unit volume of a substance, we use the equation $n = A\int_0^{\pi} e^{a\cos\theta}\sin\theta\,d\theta$. Perform the indicated integration.

59. In the study of the effects of an electric field on molecular orientation, the integral $\int_0^{\pi}(1 + k\cos\theta)\cos\theta\sin\theta\,d\theta$ is used. Evaluate this integral.

60. In finding the lift of the air flowing around an airplane wing, we use the integral $\int_{-\pi/2}^{\pi/2}\theta^2\cos\theta\,d\theta$. Evaluate this integral.

61. Find the volume of the piece of tubing in an oil distribution line shown in Fig. 28-15. All cross sections are circular.

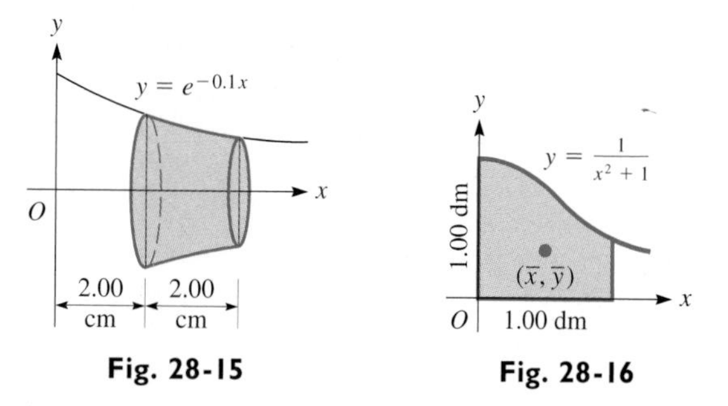

Fig. 28-15

Fig. 28-16

62. A metal plate has the shape shown in Fig. 28-16. Find the x-coordinate of the centroid of the plate.

63. The nose cone of a space vehicle is to be covered with a heat shield. The cone is designed such that a cross section x metres from the tip and perpendicular to its axis is a circle of radius $1.5x^{2/3}$ metres. Find the surface area of the heat shield if the nose cone is 4.00 m long. See Fig. 28-17. (See Exercise 27 of Section 26-6.)

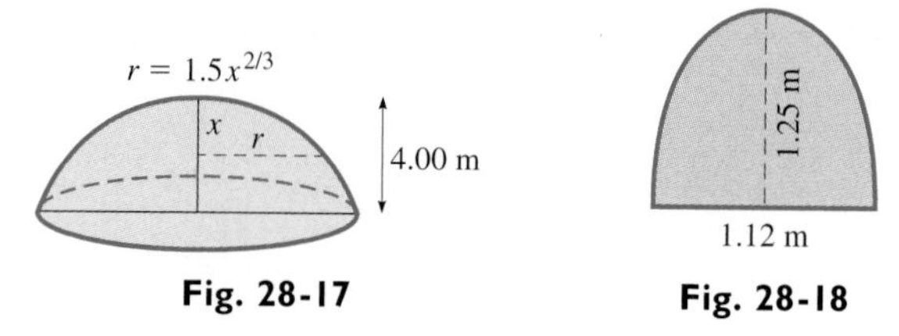

Fig. 28-17

Fig. 28-18

64. A window has the shape of a semiellipse, as shown in Fig. 28-18. What is the area of the window?

Writing Exercise

65. The side of a cutting blade designed using CAD (computer-assisted design) can be described as the area bounded by $y = 4/(1 + e^x)$, $x = 2.8$, and the axes. Write two or three paragraphs explaining how this area may be found by *algebraically* changing the form of the appropriate integral in either one of two ways. (The evaluation of the integral requires only the use of basic forms of this chapter.)

Practice Test

In Problems 1–6, evaluate the given integrals.

1. $\displaystyle\int(\sec x - \sec^3 x\tan x)\,dx$

2. $\displaystyle\int\sin^3 x\,dx$

3. $\displaystyle\int\tan^3 2x\,dx$

4. $\displaystyle\int\cos^2 4x\,dx$

5. $\displaystyle\int\frac{dx}{x^2\sqrt{4 - x^2}}$

6. $\displaystyle\int xe^{-2x}\,dx$

7. The electric current in a certain circuit is given by $i = \displaystyle\int\frac{6t + 1}{4t^2 + 9}\,dt$, where t is the time. Integrate and find the resulting function if $i = 0$ for $t = 0$.

8. Find the first-quadrant area bounded by $y = \dfrac{1}{\sqrt{16 - x^2}}$ and $x = 3$.

29

Expansion of Functions in Series

The values of the trigonometric functions can be determined exactly only for a few particular angles. The value of e can be approximated only in decimal form. The question arises as to how these values may be found, particularly if a specified degree of accuracy is necessary. In this chapter we show how a given function may be expressed in terms of a polynomial. Once this polynomial that approximates the function has been found, we shall be able to evaluate the function to any desired degree of accuracy. A number of applications and other uses of this polynomial are shown as well.

In Section 29-6 we show how certain types of functions can be represented by a series of sine and cosine terms. Such series are very useful in the study of functions and applications that are periodic in nature, such as mechanical vibrations and the currents and voltages in an ac electric circuit.

We begin the chapter by reviewing the meanings and properties of sequences and series.

Many types of electronic devices are used to control the current in a circuit. In Section 29-6 we see how series are used to analyze one such device.

29-1 Infinite Series

In Chapter 19 we discussed arithmetic and geometric sequences. Also, we introduced the concept of an infinite geometric series. In this section we further develop these topics for use in the sections that follow.

As well as arithmetic sequences and geometric sequences, there are many other ways of generating sequences of numbers. The squares of the integers 1, 4, 9, 16, 25 . . . form a sequence. Also, the successive approximations $x_1, x_2, x_3, \ldots$ found by using Newton's method in solving a particular equation form a sequence.

In general, *a* **sequence** (*or* **infinite sequence**) *is an infinite succession of numbers. Each of the numbers is a* **term** *of the sequence.* Each term of the sequence is associated with a positive integer, although at times it is convenient to associate the first term with zero (or some specified positive integer). We shall use a_n to designate the term of the sequence corresponding to the integer n.

EXAMPLE 1 ▶▶ Find the first three terms of the sequence for which $a_n = 2n + 1$, $n = 1, 2, 3, \ldots$.

Substituting the values of n, we obtain the values

$$a_1 = 2(1) + 1 = 3,\ a_2 = 2(2) + 1 = 5,\ a_3 = 2(3) + 1 = 7, \ldots$$

Therefore, we have the sequence

$$3, 5, 7, \ldots$$

If we are given $a_n = 2n + 1$ for $n = 0, 1, 2, \ldots$, the sequence is

$$1, 3, 5, \ldots$$ ◀◀

As we stated in Chapter 19, *the indicated sum of the terms of a sequence is called an* **infinite series.** Thus, for the sequence

$$a_1, a_2, a_3, \ldots, a_n, \ldots$$

the associated infinite series is

$$a_1 + a_2 + a_3 + \cdots + a_n + \cdots$$

Using the summation sign Σ (see Section 22-2) to indicate the sum, we have

Infinite Series

$$\sum_{n=1}^{\infty} a_n = a_1 + a_2 + a_3 + \cdots + a_n + \cdots \qquad (29\text{-}1)$$

We must realize that an infinite series, as shown in Eq. (29-1), does not have a sum in the ordinary sense of the word, for it is not possible actually to carry out the addition of infinitely many terms. Therefore, we define the sum for an infinite series in terms of a limit.

For the infinite series of Eq. (29-1), we let S_n represent the sum of the first n terms. Therefore,

$$S_1 = a_1$$
$$S_2 = a_1 + a_2$$
$$S_3 = a_1 + a_2 + a_3$$
$$S_n = a_1 + a_2 + a_3 + \cdots + a_n$$

Partial Sum

The numbers $S_1, S_2, S_3, \ldots, S_n, \ldots$ form a sequence. *Each term of this sequence is called a* **partial sum.** *We say that the infinite series, Eq. (29-1), is* **convergent** *and has the sum S given by*

$$S = \lim_{n\to\infty} S_n = \lim_{n\to\infty} \sum_{i=1}^{n} a_i \qquad (29\text{-}2)$$

if this limit exists. If the limit does not exist, the series is **divergent.**

EXAMPLE 2 ▸▸ For the infinite series

$$\sum_{n=0}^{\infty} \frac{1}{5^n} = \frac{1}{5^0} + \frac{1}{5^1} + \frac{1}{5^2} + \cdots + \frac{1}{5^n} + \cdots$$

the first six partial sums are

$$S_0 = 1 \qquad \text{first term}$$

$$S_1 = 1 + \frac{1}{5} = 1.2 \qquad \text{sum of first two terms}$$

$$S_2 = 1 + \frac{1}{5} + \frac{1}{25} = 1.24 \qquad \text{sum of first three terms}$$

$$S_3 = 1 + \frac{1}{5} + \frac{1}{25} + \frac{1}{125} = 1.248$$

$$S_4 = 1 + \frac{1}{5} + \frac{1}{25} + \frac{1}{125} + \frac{1}{625} = 1.2496$$

$$S_5 = 1 + \frac{1}{5} + \frac{1}{25} + \frac{1}{125} + \frac{1}{625} + \frac{1}{3125} = 1.249\,92$$

These values are easily found by using a calculator. Here it appears that the sequence of partial sums approaches the value 1.25. We therefore conclude that this infinite series converges and that its sum is approximately 1.25. (In Example 4 of this section, we will show that this infinite series does in fact have a sum of 1.25.) ◂◂

EXAMPLE 3 ▸▸ (a) The infinite series

$$\sum_{n=1}^{\infty} 5^n = 5 + 5^2 + 5^3 + \cdots + 5^n + \cdots$$

is a divergent series. The first four partial sums are

$$S_1 = 5, \qquad S_2 = 30, \qquad S_3 = 155, \quad \text{and} \quad S_4 = 780$$

Obviously they are increasing without bound.

(b) The infinite series

$$\sum_{n=0}^{\infty} (-1)^n = 1 + (-1) + 1 + (-1) + \cdots + (-1)^n + \cdots$$

has as its first five partial sums

$$S_0 = 1, \qquad S_1 = 0, \qquad S_2 = 1, \qquad S_3 = 0, \quad \text{and} \quad S_4 = 1$$

The values of these partial sums do not approach a limiting value, and therefore the series diverges. ◂◂

Since convergent series are those that have a value associated with them, they are the ones that are of primary use to us. However, generally it is not easy to determine whether a given series is convergent, and many types of tests have been developed for this purpose. These tests for convergence may be found in most textbooks that include the more advanced topics in calculus.

One important series for which we are able to determine the convergence, and its sum if convergent, is the geometric series. For this series the nth partial sum is

$$S_n = a_1 + a_1 r + a_1 r^2 + \cdots + a_1 r^{n-1}$$

where r is the fixed number by which we multiply a given term to get the next term. In Chapter 19 we determined that if $|r| < 1$, the sum S of the infinite geometric series is

$$S = \lim_{n \to \infty} S_n = \frac{a_1}{1 - r} \tag{29-3}$$

If $r = 1$, we see that the series is $a_1 + a_1 + a_1 + \cdots + a_1 + \cdots$ and is therefore divergent. If $r = -1$, the series is $a_1 - a_1 + a_1 - a_1 + \cdots$ and is also divergent. If $|r| > 1$, $\lim_{n \to \infty} r^n$ is unbounded. Therefore, ***the geometric series is convergent only if*** $|r| < 1$ *and has the value given by Eq. (29-3).*

CAUTION ▶

EXAMPLE 4 ▶▶ Show that the infinite series

$$\sum_{n=0}^{\infty} \frac{1}{5^n} = \frac{1}{5^0} + \frac{1}{5^1} + \frac{1}{5^2} + \cdots + \frac{1}{5^n} + \cdots$$

is convergent and find its sum. This is the same series as in Example 2.

We see that this is a geometric series with $r = \frac{1}{5}$. Since $|r| < 1$, the series is convergent. We find the sum to be

$$S = \frac{1}{1 - \frac{1}{5}} = \frac{1}{\frac{4}{5}} = \frac{5}{4} = 1.25 \qquad \text{using Eq. (29-3)}$$

This value agrees with the conclusion in Example 2. ◀◀

Exercises 29–1

In Exercises 1–4, give the first four terms of the sequences for which a_n is given.

1. $a_n = n^2, \quad n = 1, 2, 3, \ldots$

2. $a_n = \dfrac{2}{3^n}, \quad n = 1, 2, 3, \ldots$

3. $a_n = \dfrac{1}{n + 2}, \quad n = 0, 1, 2, \ldots$

4. $a_n = \dfrac{n^2 + 1}{2n + 1}, \quad n = 0, 1, 2, \ldots$

In Exercises 5–8, give (a) the first four terms of the sequence for which a_n is given and (b) the first four terms of the infinite series associated with the sequence.

5. $a_n = \left(-\dfrac{2}{5}\right)^n, \quad n = 1, 2, 3, \ldots$

6. $a_n = \dfrac{1}{n} + \dfrac{1}{n + 1}, \quad n = 1, 2, 3, \ldots$

7. $a_n = 1 + (-1)^n, \quad n = 0, 1, 2, \ldots$

8. $a_n = \dfrac{1}{n(n + 1)}, \quad n = 2, 3, 4, \ldots$

In Exercises 9–12, find the nth term of the given infinite series for which $n = 1, 2, 3, \ldots$.

9. $\dfrac{1}{2} + \dfrac{1}{3} + \dfrac{1}{4} + \dfrac{1}{5} + \cdots$

10. $\dfrac{1}{2} + \dfrac{1}{4} + \dfrac{1}{8} + \dfrac{1}{16} + \cdots$

11. $\dfrac{1}{2 \times 3} + \dfrac{1}{3 \times 4} + \dfrac{1}{4 \times 5} + \dfrac{1}{5 \times 6} + \cdots$

12. $-\dfrac{2}{3} + \dfrac{4}{9} - \dfrac{8}{27} + \dfrac{16}{81} - \cdots$

In Exercises 13–20, find the first five partial sums of the given series and determine whether the series appears to be convergent or divergent. If it is convergent, find its approximate sum.

13. $1 + \frac{1}{8} + \frac{1}{27} + \frac{1}{64} + \frac{1}{125} + \cdots$

14. $1 + 2 + 5 + 10 + 17 + \cdots$

15. $1 + \frac{1}{2} + \frac{2}{3} + \frac{3}{4} + \frac{4}{5} + \cdots$

16. $\frac{1}{3} - \frac{1}{9} + \frac{1}{27} - \frac{1}{81} + \frac{1}{243} - \cdots$

17. $\sum_{n=0}^{\infty} (-n)$

18. $\sum_{n=1}^{\infty} \frac{2}{n(n+1)}$

19. $\sum_{n=1}^{\infty} \frac{2n+1}{n^2(n+1)^2}$

20. $\sum_{n=1}^{\infty} \frac{n}{2n+1}$

In Exercises 21–28, test each of the given geometric series for convergence or divergence. Find the sum of each convergent series.

21. $1 + 2 + 4 + \cdots + 2^n + \cdots$

22. $1 + \frac{1}{2} + \frac{1}{4} + \cdots + \frac{1}{2^n} + \cdots$

23. $1 - \frac{1}{3} + \frac{1}{9} - \cdots + \left(-\frac{1}{3}\right)^n + \cdots$

24. $1 - \frac{3}{2} + \frac{9}{4} - \cdots + \left(-\frac{3}{2}\right)^n + \cdots$

25. $10 + 9 + 8.1 + 7.29 + 6.561 + \cdots$

26. $4 + 1 + \frac{1}{4} + \frac{1}{16} + \frac{1}{64} + \cdots$

27. $512 - 64 + 8 - 1 + \frac{1}{8} - \cdots$

28. $16 + 12 + 9 + \frac{27}{4} + \frac{81}{16} + \cdots$

In Exercises 29–36, solve the given problems as indicated.

29. Using a calculator, take successive square roots of 2 and find at least 20 approximate values for the terms of the sequence $2^{1/2}, 2^{1/4}, 2^{1/8}, \ldots$. From the values obtained, (a) what do you observe about the value of $\lim_{n\to\infty} 2^{1/2^n}$? (b) Determine whether the infinite series for this sequence converges or diverges.

30. Using a calculator, take successive square roots of 0.01. Then take successive square roots of 100. From these sequences of square roots, state any general conclusions that might be drawn.

31. Referring to Chapter 19, we see that the sum of the first n terms of a geometric sequence is

$$S_n = a_1(1 - r^n)/(1 - r) \quad (r \neq 1) \qquad \text{Eq. (19-6)}$$

where a_1 is the first term and r is the common ratio. We can visualize the corresponding infinite series by graphing the function
$f(x) = a_1(1 - r^x)/(1 - r) \quad (r \neq 1)$
The graph represents the sequence of partial sums for values where $x = n$, since $f(n) = S_n$.

Use a graphing calculator to visualize the first five partial sums of the series

$$\frac{1}{2} + \frac{1}{4} + \frac{1}{8} + \cdots$$

What value does the infinite series approach? (Remember, only points for which x is an integer have real meaning.)

32. Following Exercise 31, use a graphing calculator to show that the sum of the infinite series of Example 4 is 1.25. (Be careful: Because of the definition of the series, $x = 1$ corresponds to $n = 0$.)

33. The value V (in dollars) of a certain investment after n years can be expressed as
$V = 100(1.05 + 1.05^2 + 1.05^3 + \cdots + 1.05^n)$. (a) By finding partial sums, determine whether this series converges or diverges. (b) Following Exercise 31, use a graphing calculator to visualize the first ten partial sums. (See Example 7 of Section 19-2.)

34. If an electric discharge is passed through hydrogen gas, a spectrum of isolated parallel lines, called the Balmer series, is formed. See Fig. 29-1. The wavelengths λ (in nm) of the light for these lines is given by the formula

$$\frac{1}{\lambda} = 1.097 \times 10^{-2}\left(\frac{1}{2^2} - \frac{1}{n^2}\right), \qquad n = 3, 4, 5, \ldots$$

Find the wavelengths of the first three lines and the shortest wavelength of all the lines of the series.

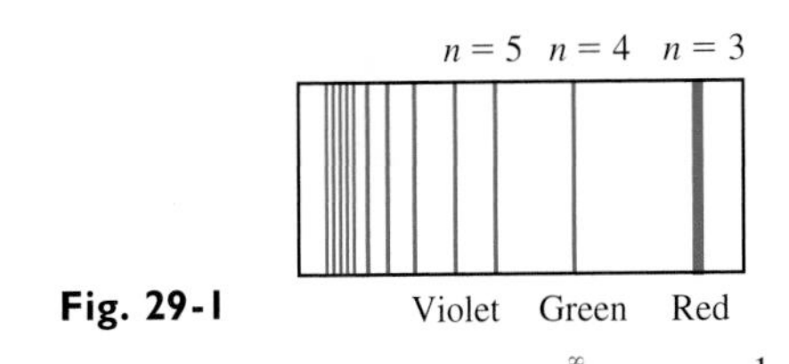

Fig. 29-1

35. Use geometric series to show that $\sum_{n=0}^{\infty} x^n = \frac{1}{1-x}$ for $|x| < 1$.

36. Use geometric series to show that $\sum_{n=0}^{\infty} (-1)^n x^n = \frac{1}{1+x}$ for $|x| < 1$.

29-2 Maclaurin Series

In this section we develop a very important basic polynomial form of a function. Before developing the method using calculus, we shall review how this can be done for some functions algebraically.

$$
\begin{array}{r}
1 + \frac{x}{2} \\
2 - x \overline{)\,2\qquad} \\
2 - x \\
\hline
x \\
x - \frac{x^2}{2} \\
\hline
\frac{x^2}{2}
\end{array}
$$

EXAMPLE 1 ▸▸ By using long division (as started at the left), we have

$$\frac{2}{2-x} = 1 + \frac{1}{2}x + \frac{1}{4}x^2 + \cdots + \left(\frac{1}{2}x\right)^{n-1} + \cdots \tag{1}$$

where n is the number of the term of the expression on the right. Since x represents a number, the right-hand side of Eq. (1) becomes a geometric series.

From Eq. (29-3) we know that the sum of a geometric series with first term a_1 and common ratio r is

$$S = \frac{a_1}{1-r}$$

where $|r| < 1$ and the series converges.

If $x = 1$, the right-hand side of Eq. (1) is

$$1 + \frac{1}{2} + \frac{1}{4} + \cdots + \left(\frac{1}{2}\right)^{n-1} + \cdots$$

For this series, $r = \frac{1}{2}$ and $a_1 = 1$, which means that the series converges and $S = 2$. If $x = 3$, the right-hand side of Eq. (1) is

$$1 + \frac{3}{2} + \frac{9}{4} + \cdots + \left(\frac{3}{2}\right)^{n-1} + \cdots$$

which diverges since $r > 1$. Referring to the left side of Eq. (1), we see that it also equals 2 when $x = 1$. Thus we see that the two sides agree for $x = 1$, but that the series diverges for $x = 3$. In fact, as long as $|x| < 2$, the series will converge to the value of the function on the left. From this we conclude that the series on the right properly represents the function on the left, as long as $|x| < 2$. ◂◂

From Example 1 we see that an algebraic function may be properly represented by a function of the form

Power Series

$$f(x) = a_0 + a_1x + a_2x^2 + \cdots + a_nx^n + \cdots \tag{29-4}$$

Equation (29-4) is known as a **power-series expansion** *of the function f(x).* The problem now arises as to whether or not functions in general may be represented in this form. If such a representation were possible, it would provide a means of evaluating the transcendental functions for the purpose of making tables of values. Also, since a power-series expansion is in the form of a polynomial, it makes algebraic operations much simpler due to the properties of polynomials. A further study of calculus shows many other uses of power series.

In Example 1 we saw that the function could be represented by a power series as long as $|x| < 2$. That is, if we substitute any value of x in this interval into the series and also into the function, the series will converge to the value of the function. *This interval of values for which the series converges is called the* **interval of convergence.**

EXAMPLE 2 ▸▸ In Example 1 the interval of convergence for the series

$$1 + \frac{1}{2}x + \frac{1}{4}x^2 + \cdots + \left(\frac{1}{2}x\right)^{n-1} + \cdots$$

is $|x| < 2$. We saw that the series converges for $x = 1$, with $S = 2$, and that the value of the function is 2 for $x = 1$. This verifies that $x = 1$ is in the interval of convergence.

Also, we saw that the series diverges for $x = 3$, which verifies that $x = 3$ is not in the interval of convergence. ◂◂

At this point we shall assume that unless otherwise noted the functions with which we shall be dealing may be properly represented by a power-series expansion (it takes more advanced methods to prove that this is generally possible), for appropriate intervals of convergence. We shall find that the methods of calculus are very useful in developing the method of general representation. Thus, writing a general power series, along with the first few derivatives, we have

$$f(x) = a_0 + a_1x + a_2x^2 + a_3x^3 + a_4x^4 + a_5x^5 + \cdots + a_nx^n + \cdots$$
$$f'(x) = a_1 + 2a_2x + 3a_3x^2 + 4a_4x^3 + 5a_5x^4 + \cdots + na_nx^{n-1} + \cdots$$
$$f''(x) = 2a_2 + 2(3)a_3x + 3(4)a_4x^2 + 4(5)a_5x^3 + \cdots + (n-1)na_nx^{n-2} + \cdots$$
$$f'''(x) = 2(3)a_3 + 2(3)(4)a_4x + 3(4)(5)a_5x^2 + \cdots + (n-2)(n-1)na_nx^{n-3} + \cdots$$
$$f^{\text{iv}}(x) = 2(3)(4)a_4 + 2(3)(4)(5)a_5x + \cdots + (n-3)(n-2)(n-1)na_nx^{n-4} + \cdots$$

NOTE ▸ Regardless of the values of the constants a_n for any power series, ***if $x = 0$, the left and right sides must be equal,*** and all the terms on the right are zero except the first. Thus, setting $x = 0$ in each of the above equations, we have

$$f(0) = a_0 \qquad f'(0) = a_1 \qquad f''(0) = 2a_2$$
$$f'''(0) = 2(3)a_3 \qquad f^{\text{iv}}(0) = 2(3)(4)a_4$$

Solving each of these for the constants a_n, we have

$$a_0 = f(0) \qquad a_1 = f'(0) \qquad a_2 = \frac{f''(0)}{2!} \qquad a_3 = \frac{f'''(0)}{3!} \qquad a_4 = \frac{f^{\text{iv}}(0)}{4!}$$

Substituting these into the expression for $f(x)$, we have

Maclaurin Series

$$f(x) = f(0) + f'(0)x + \frac{f''(0)x^2}{2!} + \frac{f'''(0)x^3}{3!} + \cdots + \frac{f^{(n)}(0)x^n}{n!} + \cdots \qquad (29\text{-}5)$$

Equation (29-5) is known as the **Maclaurin series expansion** *of a function.* For a function to be represented by a Maclaurin expansion, the function and all of its derivatives must exist at $x = 0$. Also, we note that the factorial notation introduced in Section 19-4 is used in writing the Maclaurin series expansion.

As we mentioned earlier, one of the uses we will make of series expansions is that of determining the values of functions for particular values of x. If x is sufficiently small, successive terms become smaller and smaller and the series will converge rapidly. This is considered in the sections that follow.

EXAMPLE 3 ▸▸ Find the first four terms of the Maclaurin series expansion of $f(x) = \frac{2}{2 - x}$.

This is written as

$$f(x) = \frac{2}{2-x} \qquad f(0) = 1$$

$$f'(x) = \frac{2}{(2-x)^2} \qquad f'(0) = \frac{1}{2}$$

$$f''(x) = \frac{4}{(2-x)^3} \qquad f''(0) = \frac{1}{2}$$

$$f'''(x) = \frac{12}{(2-x)^4} \qquad f'''(0) = \frac{3}{4}$$

find derivatives and evaluate each at $x = 0$

$$f(x) = 1 + \frac{1}{2}x + \frac{1}{2}\left(\frac{x^2}{2!}\right) + \frac{3}{4}\left(\frac{x^3}{3!}\right) + \cdots$$

using Eq. (29-5)

or

$$\frac{2}{2-x} = 1 + \frac{1}{2}x + \frac{1}{4}x^2 + \frac{1}{8}x^3 + \cdots$$

We see that this result agrees with that obtained by direct division. ◂◂

EXAMPLE 4 ▸▸ Find the first four terms of the Maclaurin series expansion of $f(x) = e^{-x}$.

We write

$$f(x) = e^{-x} \qquad f(0) = 1$$

$$f'(x) = -e^{-x} \qquad f'(0) = -1$$

$$f''(x) = e^{-x} \qquad f''(0) = 1$$

$$f'''(x) = -e^{-x} \qquad f'''(0) = -1$$

find derivitives and evaluate each at $x = 0$

$$f(x) = 1 + (-1)x + 1\left(\frac{x^2}{2!}\right) + (-1)\left(\frac{x^3}{3!}\right) + \cdots$$

using Eq. (29-5)

or

$$e^{-x} = 1 - x + \frac{x^2}{2!} - \frac{x^3}{3!} + \cdots \quad ◂◂$$

EXAMPLE 5 ▸▸ Find the first three nonzero terms of the Maclaurin series expansion of $f(x) = \sin 2x$.

We have

$$f(x) = \sin 2x \qquad f(0) = 0 \qquad f'''(x) = -8\cos 2x \qquad f'''(0) = -8$$

$$f'(x) = 2\cos 2x \qquad f'(0) = 2 \qquad f^{\text{iv}}(x) = 16\sin 2x \qquad f^{\text{iv}}(0) = 0$$

$$f''(x) = -4\sin 2x \qquad f''(0) = 0 \qquad f^{\text{v}}(x) = 32\cos 2x \qquad f^{\text{v}}(0) = 32$$

$$f(x) = 0 + 2x + 0 + (-8)\frac{x^3}{3!} + 0 + 32\frac{x^5}{5!} + \cdots$$

$$\sin 2x = 2x - \frac{4}{3}x^3 + \frac{4}{15}x^5 - \cdots$$

This series is called an **alternating series,** *since every other term is negative.* ◂◂

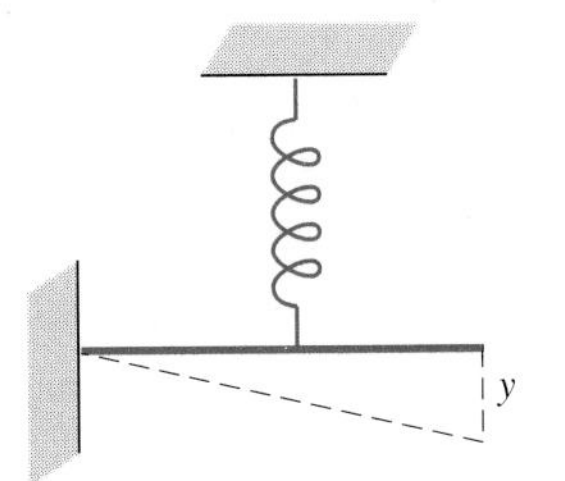

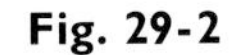

Fig. 29-2

EXAMPLE 6 ▸▸ A lever is attached to a spring as shown in Fig. 29-2. Frictional forces in the spring are just sufficient so that the lever does not oscillate after being depressed. Such motion is called *critically damped.* The displacement y of the lever as a function of the time t for one such case is $y = (1 + t)e^{-t}$. In order to study the motion for small values of t, a polynomial form of $y = f(t)$ is to be used. Find the first four nonzero terms of the expansion.

$$\begin{aligned} f(t) &= (1 + t)e^{-t} & f(0) &= 1 \\ f'(t) &= (1 + t)e^{-t}(-1) + e^{-t} = -te^{-t} & f'(0) &= 0 \\ f''(t) &= te^{-t} - e^{-t} & f''(0) &= -1 \\ f'''(t) &= -te^{-t} + e^{-t} + e^{-t} = 2e^{-t} - te^{-t} & f'''(0) &= 2 \\ f^{\text{iv}}(t) &= -2e^{-t} + te^{-t} - e^{-t} = te^{-t} - 3e^{-t} & f^{\text{iv}}(0) &= -3 \end{aligned}$$

$$f(t) = 1 + 0 + (-1)\frac{t^2}{2!} + 2\frac{t^3}{3!} + (-3)\frac{t^4}{4!} + \cdots$$

or

$$(1 + t)e^{-t} = 1 - \frac{t^2}{2} + \frac{t^3}{3} - \frac{t^4}{8} + \cdots$$ ◂◂

Exercises 29–2

In Exercises 1–16, find the first three nonzero terms of the Maclaurin expansion of the given functions.

1. $f(x) = e^x$

2. $f(x) = \sin x$

3. $f(x) = \cos x$

4. $f(x) = \ln(1 + x)$

5. $f(x) = \sqrt{1 + x}$

6. $f(x) = \dfrac{1}{(1 - x)^{1/3}}$

7. $f(x) = e^{-2x}$

8. $f(x) = \dfrac{1}{2}(e^x + e^{-x})$

9. $f(x) = \cos 4x$

10. $f(x) = e^x \sin x$

11. $f(x) = \dfrac{1}{1 - x}$

12. $f(x) = \dfrac{1}{(1 + x)^2}$

13. $f(x) = \ln(1 - 2x)$

14. $f(x) = (1 + x)^{3/2}$

15. $f(x) = \cos \frac{1}{2}x$

16. $f(x) = \ln(1 + 4x)$

In Exercises 17–24, find the first two nonzero terms of the Maclaurin expansion of the given functions.

17. $f(x) = \tan^{-1} x$

18. $f(x) = \cos x^2$

19. $f(x) = \tan x$

20. $f(x) = \sec x$

21. $f(x) = \ln \cos x$

22. $f(x) = xe^{\sin x}$

23. $f(x) = \sin^2 x$

24. $f(x) = e^{-x^2}$

In Exercises 25–32, solve the given problems.

25. Is it possible to find a Maclaurin expansion for (a) $f(x) = \csc x$ or (b) $f(x) = \ln x$? Explain.

26. Is it possible to find a Maclaurin expansion for (a) $f(x) = \sqrt{x}$ or (b) $f(x) = \sqrt{1 + x}$? Explain.

27. Find the first three nonzero terms of the Maclaurin expansion for (a) $f(x) = e^x$ and (b) $f(x) = e^{x^2}$. Compare these expansions.

28. By finding the Maclaurin expansion of $f(x) = (1 + x)^n$, derive the first four terms of the binomial series, which is Eq. (19-10). Its interval of convergence is $|x| < 1$ for all values of n.

29. If $f(x) = x^3$, show that this function is obtained when a Maclaurin expansion is found.

30. If $f(x) = x^4 + 2x^2$, show that this function is obtained when a Maclaurin expansion is found.

31. The reliability R $(0 \le R \le 1)$ of a certain computer system is $R = e^{-0.001t}$, where t is the time of operation in minutes. Express $R = f(t)$ in polynomial form by using the first three terms of the Maclaurin expansion.

32. In the analysis of the optical paths of light from a narrow slit S to a point P as shown in Fig. 29-3, the law of cosines is used to obtain the equation

$$c^2 = a^2 + (a + b)^2 - 2a(a + b)\cos\frac{s}{a}$$

where s is part of the circular arc $\widehat{AB}$. By using two nonzero terms of the Maclaurin expansion of $\cos \frac{s}{a}$, simplify the right side of the equation. (In finding the expansion, let $x = \frac{s}{a}$ and then substitute back into the expansion.)

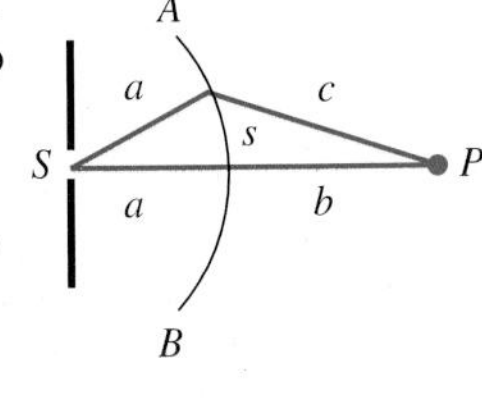

Fig. 29-3

29-3 Certain Operations with Series

The series found in the first four exercises and Exercise 28 (the binomial series) of Section 29-2 are of particular importance. They are used to evaluate exponential functions, trigonometric functions, logarithms, powers, and roots, as well as develop other series. For reference, we give them here with their intervals of convergence.

$$e^x = 1 + x + \frac{x^2}{2!} + \frac{x^3}{3!} + \cdots \quad \text{(all } x\text{)} \tag{29-6}$$

$$\sin x = x - \frac{x^3}{3!} + \frac{x^5}{5!} - \cdots \quad \text{(all } x\text{)} \tag{29-7}$$

$$\cos x = 1 - \frac{x^2}{2!} + \frac{x^4}{4!} - \cdots \quad \text{(all } x\text{)} \tag{29-8}$$

$$\ln(1 + x) = x - \frac{x^2}{2} + \frac{x^3}{3} - \frac{x^4}{4} + \cdots \quad (|x| < 1) \tag{29-9}$$

$$(1 + x)^n = 1 + nx + \frac{n(n-1)}{2!}x^2 + \cdots \quad (|x| < 1) \tag{29-10}$$

In the next section we shall see how to use these series in finding values of functions. In this section we see how new series are developed by using the above basic series, and we also show other uses of series.

When we discussed functions in Chapter 3, we mentioned functions such as $f(2x)$ and $f(-x)$. *By using functional notation and the preceding series, we can find the series expansions of many other series without using direct expansion.* This can often save time in finding a desired series.

EXAMPLE 1 ▸▸ Find the Maclaurin expansion of e^{2x}.

From Eq. (29-6), we know the expansion of e^x. Hence

$$f(x) = 1 + x + \frac{x^2}{2!} + \frac{x^3}{3!} + \cdots$$

Since $e^{2x} = f(2x)$, we have

$$f(2x) = 1 + (2x) + \frac{(2x)^2}{2!} + \frac{(2x)^3}{3!} + \cdots \qquad \text{in } f(x)\text{, replace } x \text{ by } 2x$$

$$e^{2x} = 1 + 2x + 2x^2 + \frac{4x^3}{3} + \cdots$$ ◂◂

EXAMPLE 2 ▸▸ Find the Maclaurin expansion of $\sin x^2$.

From Eq. (29-7), we know the expansion of $\sin x$. Therefore,

$$f(x) = x - \frac{x^3}{3!} + \frac{x^5}{5!} - \cdots$$

$$f(x^2) = (x^2) - \frac{(x^2)^3}{3!} + \frac{(x^2)^5}{5!} - \cdots \qquad \text{in } f(x)\text{, replace } x \text{ by } x^2$$

$$\sin x^2 = x^2 - \frac{x^6}{3!} + \frac{x^{10}}{5!} - \cdots$$

Direct expansion of this series is quite lengthy. ◂◂

The basic algebraic operations may be applied to series in the same manner they are applied to polynomials. That is, we may add, subtract, multiply, or divide series in order to obtain other series. The interval of convergence for the resulting series is that which is common to those of the series being used. The multiplication of series is illustrated in the following example.

EXAMPLE 3 ▸▸ Multiply the series for e^x and $\cos x$ in order to obtain the series expansion for $e^x \cos x$.

Using the series expansion for e^x and $\cos x$ as shown in Eqs. (29-6) and (29-8), we have the following indicated multiplication:

$$e^x \cos x = \left(1 + x + \frac{x^2}{2!} + \frac{x^3}{3!} + \frac{x^4}{4!} + \cdots\right)\left(1 - \frac{x^2}{2!} + \frac{x^4}{4!} - \cdots\right)$$

By multiplying the series on the right, we have the following result, considering through the x^4 terms in the product.

$$1\left(1 - \frac{x^2}{2!} + \frac{x^4}{4!}\right) \quad x\left(1 - \frac{x^2}{2!}\right) \quad \frac{x^2}{2!}\left(1 - \frac{x^2}{2!}\right) \quad \left(\frac{x^3}{3!} + \frac{x^4}{4!}\right)(1)$$

$$e^x \cos x = 1 - \frac{x^2}{2} + \frac{x^4}{24} + x - \frac{x^3}{2} + \frac{x^2}{2} - \frac{x^4}{4} + \frac{x^3}{6} + \frac{x^4}{24} + \cdots$$

$$= 1 + x - \frac{1}{3}x^3 - \frac{1}{6}x^4 + \cdots \quad ◂◂$$

It is also possible to use the operations of differentiation and integration to obtain series expansions, although the proof of this is found in more advanced texts. Consider the following example.

EXAMPLE 4 ▸▸ Show that by differentiating the expansion for $\ln(1 + x)$ term by term, the result is the same as the expansion for $\frac{1}{1+x}$.

The series for $\ln(1 + x)$ is shown in Eq. (29-9) as

$$\ln(1 + x) = x - \frac{x^2}{2} + \frac{x^3}{3} - \frac{x^4}{4} + \cdots$$

Differentiating, we have

$$\frac{1}{1 + x} = 1 - \frac{2x}{2} + \frac{3x^2}{3} - \frac{4x^3}{4} + \cdots$$

$$= 1 - x + x^2 - x^3 + \cdots$$

Using the binomial expansion for $\frac{1}{1+x} = (1 + x)^{-1}$, we have

$$(1 + x)^{-1} = 1 + (-1)x + \frac{(-1)(-2)}{2!}x^2 + \frac{(-1)(-2)(-3)}{3!}x^3 + \cdots$$

using Eq. (29-10) with $n = -1$

$$= 1 - x + x^2 - x^3 + \cdots$$

We see that the results are the same. ◂◂

For reference, Eq. (12-11) is $re^{j\theta} = r(\cos\theta + j\sin\theta)$.

We can use algebraic operations on series to verify that the definition of the exponential form of a complex number, as shown in Eq. (12-11), is consistent with other definitions. The only assumption required here is that the Maclaurin expansions for e^x, $\sin x$, and $\cos x$ are also valid for complex numbers. This is shown in advanced calculus. Thus

$$e^{j\theta} = 1 + j\theta + \frac{(j\theta)^2}{2!} + \frac{(j\theta)^3}{3!} + \cdots = 1 + j\theta - \frac{\theta^2}{2!} - j\frac{\theta^3}{3!} + \cdots \tag{29-11}$$

$$j\sin\theta = j\theta - j\frac{\theta^3}{3!} + \cdots \tag{29-12}$$

$$\cos\theta = 1 - \frac{\theta^2}{2!} + \cdots \tag{29-13}$$

When we add the terms of Eq. (29-12) to those of Eq. (29-13), the result is the series given in Eq. (29-11). Thus

$$e^{j\theta} = \cos\theta + j\sin\theta \tag{29-14}$$

A comparison of Eqs. (12-11) and (29-14) indicates the reason for the choice of the definition of the exponential form of a complex number.

An additional use of power series is now shown. Many integrals that occur in practice cannot be integrated by methods given in the preceding chapters. However, power series can be very useful in giving excellent approximations to some definite integrals.

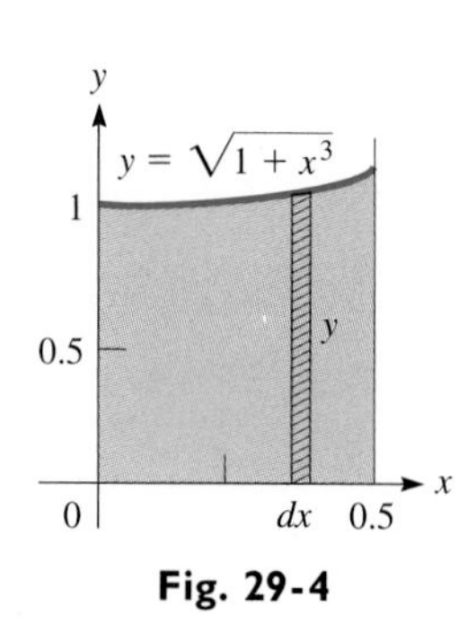

Fig. 29-4

EXAMPLE 5 ▸▸ Find the first-quadrant area bounded by $y = \sqrt{1 + x^3}$ and $x = 0.5$.

From Fig. 29-4, we see that the area is

$$A = \int_0^{0.5} \sqrt{1 + x^3}\,dx$$

This integral does not fit any form we have used. However, its value can be closely approximated by using the binomial expansion for $\sqrt{1 + x^3}$ and then integrating.

Using the binomial expansion to find the first three terms of the expansion for $\sqrt{1 + x^3}$, we have

$$\sqrt{1 + x^3} = (1 + x^3)^{0.5} = 1 + 0.5x^3 + \frac{0.5(-0.5)}{2}(x^3)^2 + \cdots$$

$$= 1 + 0.5x^3 - 0.125x^6 + \cdots$$

Substituting in the integral, we have

$$A = \int_0^{0.5} (1 + 0.5x^3 - 0.125x^6 + \cdots)\,dx$$

$$= x + \frac{0.5}{4}x^4 - \frac{0.125}{7}x^7 + \cdots \Big|_0^{0.5}$$

$$= 0.5 + 0.0078125 - 0.0001395 + \cdots = 0.507673 + \cdots$$

We can see that each of the terms omitted was very small. The result shown is correct to four decimal places, or $A = 0.5077$. Additional accuracy can be obtained by using more terms of the expansion. ◂◂

EXAMPLE 6 ▸▸ Evaluate: $\int_0^{0.1} e^{-x^2}\,dx$.

We write

$$e^{-x^2} = 1 + (-x^2) + \frac{(-x^2)^2}{2!} + \cdots \qquad \text{using Eq. (29-6)}$$

Thus

$$\begin{aligned}\int_0^{0.1} e^{-x^2}\,dx &= \int_0^{0.1}\left(1 - x^2 + \frac{x^4}{2} - \cdots\right)dx && \text{substitute}\\ &= \left(x - \frac{x^3}{3} + \frac{x^5}{10} - \cdots\right)\Bigg|_0^{0.1} && \text{integrate}\\ &= 0.1 - \frac{0.001}{3} + \frac{0.000\,01}{10} = 0.099\,667\,7 && \text{evaluate}\end{aligned}$$

This answer is correct to the indicated accuracy. ◂◂

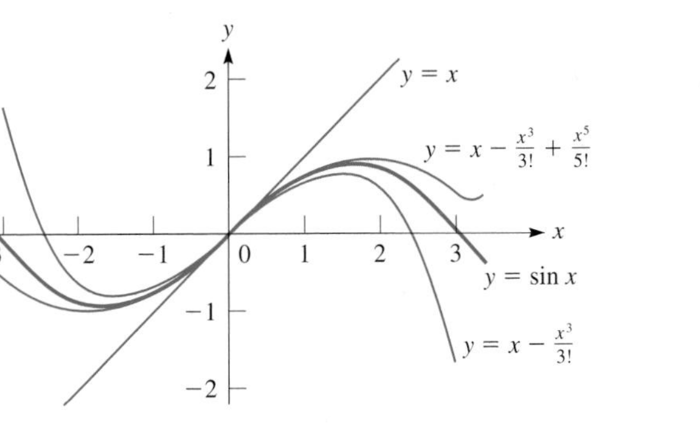

Fig. 29-5

The question of accuracy now arises. The integrals just evaluated indicate that the more terms used, the greater the accuracy of the result. To graphically show the accuracy involved, Fig. 29-5 depicts the graphs of $y = \sin x$ and the graphs of

$$y = x, \qquad y = x - \frac{x^3}{3!}, \quad \text{and} \quad y = x - \frac{x^3}{3!} + \frac{x^5}{5!}$$

which are the first three approximations of $y = \sin x$. We can see that each term added gives a better fit to the curve of $y = \sin x$. Also, this gives a graphical representation of the meaning of a series expansion.

We have just shown that the more terms included, the more accurate the result. For small values of x, a Maclaurin series gives good accuracy with a very few terms. In this case the series *converges* rapidly, as we mentioned earlier. For this reason a Maclaurin series is of particular use for small values of x. For larger values of x, usually a function is expanded in a Taylor series (see Section 29-5). Of course, if we omit any term in a series, there is some error in the calculation.

Exercises 29–3

In Exercises 1–8, find the first four nonzero terms of the Maclaurin expansions of the given functions by using Eqs. (29-6) to (29-10).

1. $f(x) = e^{3x}$ **2.** $f(x) = e^{-2x}$

3. $f(x) = \sin \frac{1}{2}x$ **4.** $f(x) = \sin x^4$

5. $f(x) = \cos 4x$ **6.** $f(x) = \sqrt{1 - x^4}$

7. $f(x) = \ln(1 + x^2)$ **8.** $f(x) = \ln(1 - x)$

In Exercises 9–12, evaluate the given integrals by using three terms of the appropriate series.

9. $\int_0^1 \sin x^2\,dx$ **10.** $\int_0^{0.4} \sqrt[4]{1 - 2x^2}\,dx$

11. $\int_0^{0.2} \cos\sqrt{x}\,dx$ **12.** $\int_{0.1}^{0.2} \frac{\cos x - 1}{x}\,dx$

In Exercises 13–20, find the indicated series by the given operation.

13. Find the first four nonzero terms of the expansion of the function $f(x) = \frac{1}{2}(e^x + e^{-x})$ by adding the terms of the appropriate series. The result is the series for $\cosh x$. (See Exercise 49 of Section 27-6.)

14. Find the first four nonzero terms of the expansion of the function $f(x) = \frac{1}{2}(e^x - e^{-x})$ by subtracting the terms of the appropriate series. The result is the series for $\sinh x$. (See Exercise 49 of Section 27-6.)

15. Find the first three terms of the expansion for $e^x \sin x$ by multiplying the proper expansions together, term by term.

16. Find the first three nonzero terms of the expansion for $f(x) = \tan x$ by dividing the series for $\sin x$ by that for $\cos x$.

17. Show that by differentiating term by term the expansion for $\sin x$, the result is the expansion for $\cos x$.

18. Show that by differentiating term by term the expansion for e^x, the result is also the expansion for e^x.

19. Show that by integrating term by term the expansion for $\cos x$, the result is the expansion for $\sin x$.

20. Show that by integrating term by term the expansion for $-1/(1-x)$ (see Exercise 11 of Section 29-2), the result is the expansion for $\ln(1-x)$.

In Exercises 21–28, solve the given problems.

21. Evaluate $\int_0^1 e^x dx$ directly and compare the result obtained by using four terms of the series for e^x and then integrating.

22. Evaluate $\lim_{x\to 0} \dfrac{\sin x}{x}$ by using the series expansion for $\sin x$. Compare the result with Eq. (27-1).

23. Find the approximate value of the area bounded by $y = x^2 e^x$, $x = 0.2$, and the x-axis by using three terms of the appropriate Maclaurin series.

24. Find the approximate area bounded by $y = e^{-x^2}$, $x = -1$, $x = 1$, and $y = 0$ by using three terms of the appropriate series. See Fig. 29-6 and Fig. 22-6.

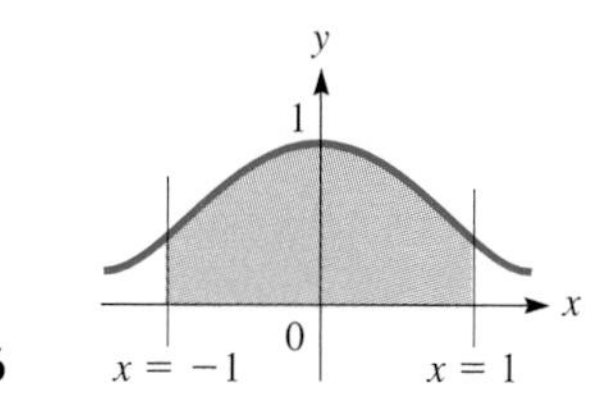

Fig. 29-6

25. The *Fresnel integral* $\int_0^x \cos t^2\,dt$ is used in the analysis of beam displacements (and in optics). Evaluate this integral for $x = 0.2$ by using two terms of the appropriate series.

26. Find the volume generated by revolving the area bounded by $y = e^{-x}$, $y = 0$, $x = 0$, and $x = 0.1$ about the y-axis by using three terms of the appropriate series.

27. The vertical displacement y of a mass at the end of a spring is given by $y = \sin 3t - \cos 2t$, where t is the time. By subtraction of series, find the first four nonzero terms of the series for y.

28. The charge q on a capacitor in a certain electric circuit is given by $q = ce^{-at}\sin 6at$, where t is the time. By multiplication of series, find the first four nonzero terms of the expansion for q.

In Exercises 29–32, use a graphing calculator to display (a) the given function and (b) the first three series approximations of the function in the same display. Each display will be similar to that in Fig. 29-5 for the function $y = \sin x$ and its first three approximations. Be careful in choosing appropriate values for the RANGE *feature.*

29. $y = e^x$

30. $y = \cos x$

31. $y = \ln(1 + x)$ $(|x| < 1)$

32. $y = \sqrt{1 + x}$ $(|x| < 1)$

29-4 Computations by Use of Series Expansions

As we mentioned at the beginning of the previous section, power-series expansions can be used to compute numerical values of exponential functions, trigonometric functions, logarithms, powers, and roots. By including a sufficient number of terms in the expansion, we can calculate these values to any degree of accuracy that may be required.

It is through such calculations that tables of values can be made, and decimal approximations of numbers such as e and π can be found. Also, many of the values found on a calculator or a computer are calculated by using series expansions that have been programmed into the chip which is in the calculator or computer.

EXAMPLE 1 ▸▸ Calculate the value of $e^{0.1}$.

In order to evaluate $e^{0.1}$, we substitute 0.1 for x in the expansion for e^x. The more terms that are used, the more accurate a value we can obtain. The limit of the partial sums would be the actual value. However, since $e^{0.1}$ is irrational, we cannot express the exact value in decimal form.

Therefore, the value is found as follows:

$$e^x = 1 + x + \frac{x^2}{2!} + \cdots \qquad \text{Eq. (29-6)}$$

$$e^{0.1} = 1 + 0.1 + \frac{(0.1)^2}{2} + \cdots \qquad \text{substitute 0.1 for } x$$

$$= 1.105 \qquad \text{using 3 terms}$$

Using a calculator, we find that $e^{0.1} = 1.105\,170\,918$, which shows that our answer is valid to the accuracy shown. ◂◂

EXAMPLE 2 ▸▸ Calculate the value of sin 2°.

CAUTION ▶

In finding trigonometric values, we must be careful to ***express the angle in radians.*** Thus the value of sin 2° is found as follows:

$$\sin x = x - \frac{x^3}{3!} + \cdots \qquad \text{Eq. (29-7)}$$

$$\sin 2^\circ = \left(\frac{\pi}{90}\right) - \frac{(\pi/90)^3}{6} + \cdots \qquad 2^\circ = \frac{\pi}{90}\text{ rad}$$

$$= 0.034\,899\,496\,3 \qquad \text{using 2 terms}$$

A calculator gives the value 0.034 899 496 7. Here we note that the second term is much smaller than the first. In fact, a good approximation of 0.0349 can be found by using just one term. We now see that $\sin\theta \approx \theta$ for small values of θ, as we noted in Section 8-4. ◂◂

EXAMPLE 3 ▸▸ Calculate the value of cos 0.5429.

Since the angle is expressed in radians, we have

$$\cos 0.5429 = 1 - \frac{0.5429^2}{2} + \frac{0.5429^4}{4!} - \cdots \qquad \text{using Eq. (29-8)}$$

$$= 0.856\,249\,5 \qquad \text{using 3 terms}$$

A calculator shows that cos 0.5429 = 0.856 214 082 4. Since the angle is not small, additional terms are needed to obtain this accuracy. With one more term, the value 0.856 213 9 is obtained. ◂◂

EXAMPLE 4 ▸▸ Calculate the value of ln 1.2.

$$\ln(1 + x) = x - \frac{x^2}{2} + \frac{x^3}{3} - \cdots \qquad \text{Eq. (29-9)}$$

$$\ln 1.2 = \ln(1 + 0.2)$$

$$= 0.2 - \frac{(0.2)^2}{2} + \frac{(0.2)^3}{3} - \cdots = 0.1827$$

To four significant digits, ln (1.2) = 0.1823. One more term is required to obtain this accuracy. ◂◂

We now illustrate the use of series in error calculations. We also discussed this as an application of differentials. A series solution allows as close a value of the calculated error as needed, whereas only one term can be found using differentials.

EXAMPLE 5 ▸▸ The velocity v of an object that has fallen h metres is $v = 4.43\sqrt{h}$. Find the approximate error in calculating the velocity of an object that has fallen 100.0 m, with a possible error of 2.0 m.

NOTE ▶ If we ***let*** $v = 4.43\sqrt{100.0 + x}$***, where x is the error in h,*** we may express v as a Maclaurin expansion in x:

$$f(x) = 4.43(100.0 + x)^{1/2} \qquad f(0) = 44.3$$
$$f'(x) = 2.22(100.0 + x)^{-1/2} \qquad f'(0) = 0.222$$
$$f''(x) = -1.11(100.0 + x)^{-3/2} \qquad f''(0) = -0.00100$$

Therefore,

$$v = 4.43\sqrt{100.0 + x} = 44.3 + 0.222x - 0.00056x^2 + \cdots$$

Since the calculated value of v for $x = 0$ is 44.3, the error e in the value of v is

$$e = 0.222x - 0.00056x^2 + \cdots$$

Calculating, the error for $x = 2.0$ is

$$e = 0.222(2.0) - 0.00056(4.0) = 0.444 - 0.002 = 0.442 \text{ m/s}$$

The value 0.444 is that which is found using differentials. The additional terms are corrections to this term. The additional term in this case shows that the first term is a good approximation to the error. Although this problem can be done numerically, a series solution allows us to find the error for any value of x. ◂◂

EXAMPLE 6 ▸▸ Assuming that the earth is a perfect sphere 6400 km in radius, find how far the end of a tangent line 2 km long is from the surface.

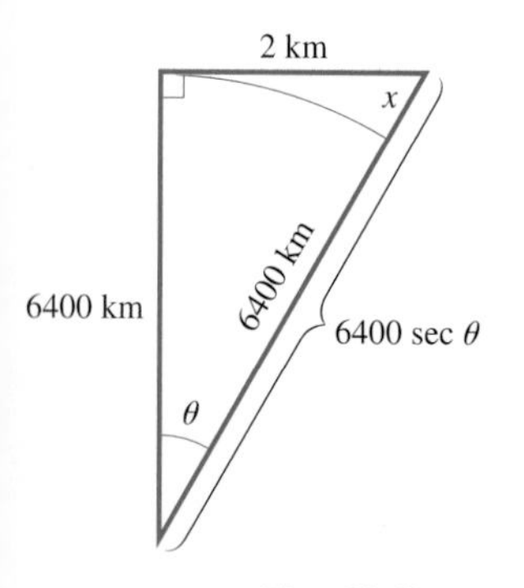

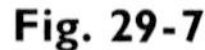
Fig. 29-7

From Fig. 29-7, we see that

$$x = 6400 \sec\theta - 6400$$

Finding the series for $\sec\theta$, we have

$$f(\theta) = \sec\theta \qquad f(0) = 1$$
$$f'(\theta) = \sec\theta\tan\theta \qquad f'(0) = 0$$
$$f''(\theta) = \sec^3\theta + \sec\theta\tan^2\theta \qquad f''(0) = 1$$

Thus the first two nonzero terms are $\sec\theta = 1 + (\theta^2/2)$. Therefore,

$$x = 6400(\sec\theta - 1)$$
$$= 6400\left(1 + \frac{\theta^2}{2} - 1\right) = 3200\,\theta^2$$

The first two terms of the expansion for $\tan\theta$ are $\theta + \theta^3/3$, which means that $\tan\theta \approx \theta$, since θ is small (see Section 8-4). From Fig. 29-7, $\tan\theta = 2/6400$, and therefore $\theta = 1/3200$. Therefore, we have

$$x = 3200\left(\frac{1}{3200}\right)^2 = \frac{1}{3200} = 0.0003 \text{ km}$$

This means that the end of a 2-km line drawn tangent to the earth would be only about 30 cm from the surface! ◂◂

Exercises 29–4

In Exercises 1–16, calculate the value of each of the given functions. Use the indicated number of terms of the appropriate series. Compare with the value found directly on a calculator.

1. $e^{0.2}$ (3)

2. $e^{-0.5}$ (3)

3. $\sin 0.1$ (2)

4. $\cos 0.05$ (2)

5. e (7)

6. $1/\sqrt{e}$ (5)

7. $\cos 3°$ (2)

8. $\sin 4°$ (2)

9. $\ln(1.4)$ (4)

10. $\ln(0.95)$ (4)

11. $\sin 0.3625$ (3)

12. $\cos 0.4072$ (3)

13. $\ln 0.9861$ (3)

14. $\ln 1.0534$ (3)

15. $e^{-0.3165}$ (4)

16. $e^{0.2651}$ (4)

In Exercises 17–20, calculate the value of each of the given functions. In Exercises 17 and 18, use the expansion for $\sqrt{1+x}$, and in Exercises 19 and 20 use the expansion for $\sqrt[3]{1+x}$. Use three terms of the appropriate series.

17. $\sqrt{1.1076}$

18. $\sqrt{0.7915}$

19. $\sqrt[3]{0.9628}$

20. $\sqrt[3]{1.1392}$

In Exercises 21–24, calculate the maximum error of the values indicated. If a series is alternating (every other term is negative), the maximum possible error in the calculated value is the value of the first term omitted.

21. The value found in Exercise 3

22. The value found in Exercise 2

23. The value found in Exercise 7

24. The value found in Exercise 9

In Exercises 25–32, solve the given problems by using series expansions.

25. We can evaluate π by use of $\frac{1}{4}\pi = \tan^{-1}\frac{1}{2} + \tan^{-1}\frac{1}{3}$ (see Exercise 66 of Section 20-6), along with the series expansion for $\tan^{-1} x$. The first three terms are $\tan^{-1} x = x - \frac{1}{3}x^3 + \frac{1}{5}x^5$. Using these terms, expand $\tan^{-1}\frac{1}{2}$ and $\tan^{-1}\frac{1}{3}$, and approximate the value of π.

26. Use the fact that $\frac{1}{4}\pi = \tan^{-1}\frac{1}{7} + 2\tan^{-1}\frac{1}{3}$ to approximate the value of π. (See Exercise 25.)

27. The time t, in years, for an investment to increase by 10% when the interest rate is 6% is given by $t = \dfrac{\ln 1.1}{0.06}$. Evaluate this expression by using the first four terms of the appropriate series.

28. The period T of a pendulum of length L is given by

$$T = 2\pi\sqrt{\frac{L}{g}}\left(1 + \frac{1}{4}\sin^2\frac{\theta}{2} + \frac{9}{64}\sin^4\frac{\theta}{2} + \cdots\right)$$

where g is the acceleration due to gravity and θ is the maximum angular displacement. If $L = 1.000$ m and $g = 9.800\ \text{m/s}^2$, calculate T for $\theta = 10.0°$ (a) if only one term (the 1) of the series is used and (b) if two terms of the indicated series are used. In the second term, substitute one term of the series for $\sin^2(\theta/2)$.

29. The electric current in a circuit containing a resistance R, an inductance L, and a battery whose voltage is E is given by $i = \dfrac{E}{R}(1 - e^{-Rt/L})$, where t is the time. Approximate this expression by using the first three terms of the appropriate exponential series. Under what conditions will this approximation be valid?

30. The image distance q from a certain lens as a function of the object distance p is given by $q = 20p/(p - 20)$. Find the first three nonzero terms of the expansion of the right side. From this expression, calculate q for $p = 2$ cm and compare with the value found by substituting 2 in the original expression.

31. At what height above the shoreline of Lake Ontario must an observer be in order to see a point 15 km distant on the surface of the lake? (The radius of the earth is 6400 km.)

32. The efficiency (in percent) of an internal combustion engine in terms of its compression ratio c is given by $E = 100(1 - c^{-0.4})$. Determine the possible approximate error in the efficiency for a compression ratio measured to be 6.00 with a possible error of 0.50. (*Hint:* Set up a series for $(6 + x)^{-0.4}$.)

29-5 Taylor Series

To obtain accurate values of a function for values of x that are not close to zero, it is usually necessary to use many terms of a Maclaurin expansion. However, we can use another type of series, called a **Taylor series,** *which is a more general expansion than a Maclaurin expansion.* Also, functions for which a Maclaurin series may not be found may have a Taylor series.

The basic assumption in formulating a Taylor expansion is that a function may be expanded in a polynomial of the form

$$f(x) = c_0 + c_1(x - a) + c_2(x - a)^2 + \cdots \quad (29\text{-}15)$$

Following the same line of reasoning as in deriving the Maclaurin expansion, we may find the constants $c_0, c_1, c_2, \ldots$. That is, derivatives of Eq. (29-15) are taken, and the function and its derivatives are evaluated at $x = a$. This leads to

$$f(x) = f(a) + f'(a)(x - a) + \frac{f''(a)(x - a)^2}{2!} + \cdots \quad (29\text{-}16)$$

Equation (29-16) is the **Taylor series expansion** *of a function.* It converges rapidly for values of x that are close to a, and this is illustrated in Examples 3 and 4.

EXAMPLE 1 ▸▸ Expand $f(x) = e^x$ in a Taylor series with $a = 1$.

$$\begin{aligned} f(x) &= e^x & f(1) &= e & &\text{find derivatives and evaluate each at } x = 1 \\ f'(x) &= e^x & f'(1) &= e \\ f''(x) &= e^x & f''(1) &= e \\ f'''(x) &= e^x & f'''(1) &= e \end{aligned}$$

$$f(x) = e + e(x - 1) + e\frac{(x - 1)^2}{2!} + e\frac{(x - 1)^3}{3!} + \cdots \quad \text{using Eq. (29-16)}$$

$$e^x = e\left[1 + (x - 1) + \frac{(x - 1)^2}{2} + \frac{(x - 1)^3}{6} + \cdots\right]$$

This series can be used in evaluating e^x for values of x near 1. ◂◂

EXAMPLE 2 ▸▸ Expand $f(x) = \sqrt{x}$ in powers of $(x - 4)$.

Another way of stating this is to find the Taylor series for $f(x) = \sqrt{x}$, with $a = 4$. Thus

$$\begin{aligned} f(x) &= x^{1/2} & f(4) &= 2 & &\text{find derivatives and evaluate each at } x = 4 \\ f'(x) &= \frac{1}{2x^{1/2}} & f'(4) &= \frac{1}{4} \\ f''(x) &= -\frac{1}{4x^{3/2}} & f''(4) &= -\frac{1}{32} \\ f'''(x) &= \frac{3}{8x^{5/2}} & f'''(4) &= \frac{3}{256} \end{aligned}$$

$$f(x) = 2 + \frac{1}{4}(x - 4) - \frac{1}{32}\frac{(x - 4)^2}{2!} + \frac{3}{256}\frac{(x - 4)^3}{3!} - \cdots \quad \text{using Eq. (29-16)}$$

$$\sqrt{x} = 2 + \frac{(x - 4)}{4} - \frac{(x - 4)^2}{64} + \frac{(x - 4)^3}{512} - \cdots$$

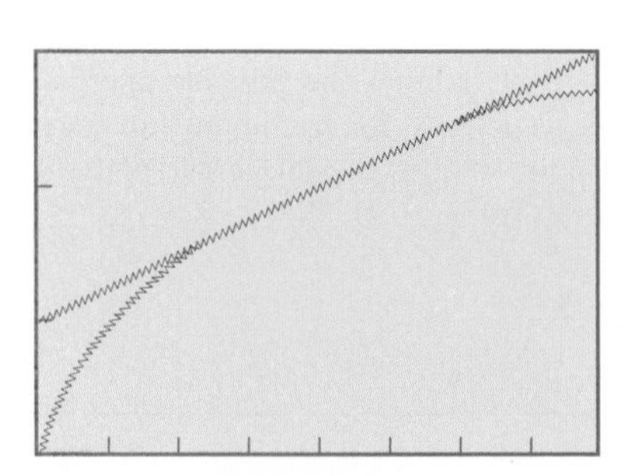

Fig. 29-8

This series would be used to evaluate square roots of numbers near 4.

In Fig. 29-8 we show a graphing calculator view of $y = \sqrt{x}$ and $y = 1 + x/4$, which are the first two terms of the Taylor series. Each passes through (4, 2), and they have nearly equal values of y for values of x near 4. ◂◂

In the last section we evaluated functions by using Maclaurin series. In the following examples we use Taylor series to evaluate functions.

EXAMPLE 3 ▸▸ By using Taylor series, evaluate $\sqrt{4.5}$.

Using the four terms of the series found in Example 2, we have

$$\sqrt{4.5} = 2 + \frac{(4.5-4)}{4} - \frac{(4.5-4)^2}{64} + \frac{(4.5-4)^3}{512} \qquad \text{substitute 4.5 for } x$$

$$= 2 + \frac{(0.5)}{4} - \frac{(0.5)^2}{64} + \frac{(0.5)^3}{512}$$

$$= 2.121\,337\,891$$

The value found directly on a calculator is 2.121 320 344. Therefore, the value found by these terms of the series expansion is correct to four decimal places. ◂◂

In Example 3 we saw that successive terms become small rapidly. If a value of x is chosen such that $x - a$ is larger, the successive terms may not become small rapidly and many terms may be required. Therefore, ***we should choose the value of a as conveniently close to the x-values that will be used.*** Also, we should note that a Maclaurin expansion for $\sqrt{x}$ cannot be used since the derivatives of $\sqrt{x}$ are not defined for $x = 0$.

CAUTION ▶

EXAMPLE 4 ▸▸ Calculate the approximate value of sin 29° by using three terms of the appropriate Taylor expansion.

Since the value of sin 30° is known to be $\frac{1}{2}$, if we let $a = \frac{\pi}{6}$ (remember, we must use values expressed in radians), when we evaluate the expansion for $x = 29°$ (when expressed in radians) the quantity $(x - a)$ is $-\frac{\pi}{180}$ (equivalent to $-1°$). This means that its numerical values are small and become smaller when it is raised to higher powers. Therefore,

$$f(x) = \sin x \qquad f\left(\frac{\pi}{6}\right) = \frac{1}{2} \qquad \text{find derivatives and evaluate each at } x = \frac{\pi}{6}$$

$$f'(x) = \cos x \qquad f'\left(\frac{\pi}{6}\right) = \frac{\sqrt{3}}{2}$$

$$f''(x) = -\sin x \qquad f''\left(\frac{\pi}{6}\right) = -\frac{1}{2}$$

$$f(x) = \frac{1}{2} + \frac{\sqrt{3}}{2}\left(x - \frac{\pi}{6}\right) - \frac{1}{4}\left(x - \frac{\pi}{6}\right)^2 - \cdots \qquad \text{using Eq. (29-16)}$$

$$\sin x = \frac{1}{2} + \frac{\sqrt{3}}{2}\left(x - \frac{\pi}{6}\right) - \frac{1}{4}\left(x - \frac{\pi}{6}\right)^2 - \cdots \qquad f(x) = \sin x$$

$$\sin 29° = \sin\left(\frac{\pi}{6} - \frac{\pi}{180}\right) \qquad 29° = 30° - 1° = \frac{\pi}{6} - \frac{\pi}{180}$$

$$= \frac{1}{2} + \frac{\sqrt{3}}{2}\left(\frac{\pi}{6} - \frac{\pi}{180} - \frac{\pi}{6}\right) - \frac{1}{4}\left(\frac{\pi}{6} - \frac{\pi}{180} - \frac{\pi}{6}\right)^2 - \cdots \qquad \text{substitute } \frac{\pi}{6} - \frac{\pi}{180} \text{ for } x$$

$$= \frac{1}{2} + \frac{\sqrt{3}}{2}\left(-\frac{\pi}{180}\right) - \frac{1}{4}\left(-\frac{\pi}{180}\right)^2 - \cdots$$

$$= 0.484\,808\,850\,9$$

The value found directly on a calculator is 0.484 809 620 2. ◂◂

Exercises 29–5

In Exercises 1–8, evaluate the given functions by using the series developed in the examples of this section.

1. $e^{1.2}$ (Use $e = 2.7183$.) **2.** $e^{0.7}$

3. $\sqrt{4.2}$ **4.** $\sqrt{3.5}$

5. $\sin 31°$ **6.** $\sin 28°$

7. $\sin 29.53°$ **8.** $\sqrt{3.8527}$

In Exercises 9–16, find the first three nonzero terms of the Taylor expansion for the given function and given value of a.

9. e^{-x} $(a = 2)$ **10.** $\cos x$ $(a = \frac{\pi}{4})$

11. $\sin x$ $(a = \frac{\pi}{3})$ **12.** $\ln x$ $(a = 3)$

13. $\sqrt[3]{x}$ $(a = 8)$ **14.** $\frac{1}{x}$ $(a = 2)$

15. $\tan x$ $(a = \frac{\pi}{4})$ **16.** $\ln \sin x$ $(a = \frac{\pi}{2})$

In Exercises 17–24, evaluate the given functions by using three terms of the appropriate Taylor series.

17. $e^{-2.2}$ (Use $e^{-2} = 0.1353$.)

18. $\ln (3.1)$ (Use $\ln 3 = 1.0986$.)

19. $\sqrt{9.3}$ **20.** $\sqrt{15}$

21. $\sqrt[3]{8.3}$ **22.** $\tan 46°$

23. $\sin 61°$ **24.** $\cos 42°$

In Exercises 25–28, solve the given problems.

25. By completing the steps indicated before Eq. (29-16) in the text, complete the derivation of Eq. (29-16).

26. Calculate $e^{0.9}$ by using four terms of the Maclaurin expansion for e^x. Also calculate $e^{0.9}$ by using the first three terms of the Taylor expansion in Example 1, using $e = 2.7183$. Compare the accuracy of the values obtained with that found directly on a calculator.

27. Calculate $\sin 31°$ by using three terms of the Maclaurin expansion for $\sin x$. Also calculate $\sin 31°$ by using three terms of the Taylor expansion in Example 4. Compare the accuracy of the values obtained with that found directly on a calculator.

28. In the analysis of the electric potential of an electric charge distributed along a straight wire of length L, the expression $\ln \frac{x + L}{x}$ is used. Find three terms of the Taylor expansion of this expression in powers of $(x - L)$.

In Exercises 29–32, use a graphing calculator to display (a) the function in the indicated exercise of this set and (b) the first two terms of the Taylor series found for that exercise in the same display. Describe how closely the graph in part (b) fits the graph in part (a). Use the given values of x for Xmin and Xmax.

29. Exercise 11 ($\sin x$), $x = 0$ to $x = 2$

30. Exercise 13 ($\sqrt[3]{x}$), $x = 0$ to $x = 16$

31. Exercise 14 ($1/x$), $x = 0$ to $x = 4$

32. Exercise 15 ($\tan x$), $x = 0$ to $x = 1.5$

29-6 Fourier Series

Many problems encountered in the various fields of science and technology involve functions that are periodic. *A periodic function is one for which $F(x + P) = F(x)$, where P is the period.* We noted that the trigonometric functions are periodic when we discussed their graphs in Chapter 10. Illustrations of applied problems that involve periodic functions are alternating-current voltages and mechanical oscillations.

Therefore, in this section we use a series made of terms of sines and cosines. This allows us to represent complicated periodic functions in terms of the simpler sines and cosines. It also provides us a good approximation over a greater interval than Maclaurin and Taylor series, which give good approximations with a few terms only near a specific value. Illustrations of applications of this type of series are given in Example 3 and in the exercises.

We shall assume that a function $f(x)$ may be represented by the series of sines and cosines as indicated:

$$f(x) = a_0 + a_1 \cos x + a_2 \cos 2x + \cdots + a_n \cos nx + \cdots$$
$$+ b_1 \sin x + b_2 \sin 2x + \cdots + b_n \sin nx + \cdots \qquad (29\text{-}17)$$

Since all the sines and cosines indicated in this expansion have a period of 2π (the period of any given term may be less than 2π, but all do repeat every 2π units—for example, $\sin 2x$ has a period of π, but it also repeats every 2π), the series expansion indicated in Eq. (29-17) will also have a period of 2π. *This series is called a* **Fourier series.**

The principal problem to be solved is that of finding the coefficients a_n and b_n. Derivatives proved to be useful in finding the coefficients for a Maclaurin expansion. We use the properties of certain integrals to find the coefficients of a Fourier series. To utilize these properties, we multiply all terms of Eq. (29-17) by $\cos mx$ and then evaluate from $-\pi$ to π (in this way we take advantage of the period 2π). Thus we have

$$\int_{-\pi}^{\pi} f(x) \cos mx\, dx = \int_{-\pi}^{\pi} (a_0 + a_1 \cos x + a_2 \cos 2x + \cdots)(\cos mx)\, dx$$
$$+ \int_{-\pi}^{\pi} (b_1 \sin x + b_2 \sin 2x + \cdots)(\cos mx)\, dx \qquad (29\text{-}18)$$

Using the methods of integration of Chapter 28, we now find the values of the coefficients a_n and b_n. For the coefficients a_n we find that the values differ depending on whether or not $n = m$. Therefore, first considering the case for which $n \neq m$, we have

$$\int_{-\pi}^{\pi} a_0 \cos mx\, dx = \frac{a_0}{m} \sin mx \Big|_{-\pi}^{\pi}$$
$$= \frac{a_0}{m}(0 - 0) = 0 \qquad (29\text{-}19)$$

$$\int_{-\pi}^{\pi} a_n \cos nx \cos mx\, dx$$
$$= a_n \left(\frac{\sin (n - m)x}{2(n - m)} + \frac{\sin (n + m)x}{2(n + m)} \right) \Big|_{-\pi}^{\pi} = 0 \qquad (n \neq m) \qquad (29\text{-}20)$$

These values are all equal to zero since the sine of any multiple of π is zero.

Now, considering the case for which $n = m$, we have

$$\int_{-\pi}^{\pi} a_n \cos nx \cos nx\, dx = \int_{-\pi}^{\pi} a_n \cos^2 nx\, dx$$
$$= \left(\frac{a_n x}{2} + \frac{a_n}{2n} \sin nx \cos nx \right) \Big|_{-\pi}^{\pi}$$
$$= \frac{a_n x}{2} \Big|_{-\pi}^{\pi} = \pi a_n \qquad (29\text{-}21)$$

On the next page, we continue by finding the values of the coefficient b_n in Eq. (29-18).

Now, finding the values of the coefficient b_n, we have

$$\int_{-\pi}^{\pi} b_n \sin nx \cos mx\, dx = b_n\left(-\frac{\cos(n-m)x}{2(n-m)} - \frac{\cos(n+m)x}{2(n+m)}\right)\Bigg|_{-\pi}^{\pi}$$

$$= b_n\left(-\frac{\cos(n-m)\pi}{2(n-m)} - \frac{\cos(n+m)\pi}{2(n+m)} + \frac{\cos(n-m)(-\pi)}{2(n-m)} + \frac{\cos(n+m)(-\pi)}{2(n+m)}\right)$$

$$= 0 \text{ [since } \cos\theta = \cos(-\theta)] \qquad (n \neq m) \qquad (29\text{-}22)$$

$$\int_{-\pi}^{\pi} b_n \sin nx \cos nx\, dx = \frac{b_n}{2n}\sin^2 nx\Big|_{-\pi}^{\pi} = 0 \qquad (29\text{-}23)$$

These integrals are seen to be zero, except for the one specific case of $\int_{-\pi}^{\pi} a_n \cos nx \cos mx\, dx$ when $n = m$, for which the result is indicated in Eq. (29-21). Using these results in Eq. (29-18), we have

$$\int_{-\pi}^{\pi} f(x)\cos nx\, dx = a_n\int_{-\pi}^{\pi}\cos^2 nx\, dx = \pi a_n$$

$$a_n = \frac{1}{\pi}\int_{-\pi}^{\pi} f(x)\cos nx\, dx \qquad (29\text{-}24)$$

This equation allows us to find the coefficients a_n, except a_0. We find the term a_0 by direct integration of Eq. (29-17) from $-\pi$ to π. When we perform this integration, all the sine and cosine terms integrate to zero, thereby giving the result

$$\int_{-\pi}^{\pi} f(x)\, dx = \int_{-\pi}^{\pi} a_0\, dx = a_0 x\Big|_{-\pi}^{\pi} = 2\pi a_0$$

$$a_0 = \frac{1}{2\pi}\int_{-\pi}^{\pi} f(x)\, dx \qquad (29\text{-}25)$$

By multiplying all terms of Eq. (29-17) by $\sin mx$ and then integrating from $-\pi$ to π, we find the coefficients b_n. We obtain the result

$$b_n = \frac{1}{\pi}\int_{-\pi}^{\pi} f(x)\sin nx\, dx \qquad (29\text{-}26)$$

We can restate our equations for the Fourier series of a function $f(x)$:

$$f(x) = a_0 + a_1\cos x + a_2\cos 2x + \cdots + a_n\cos nx + \cdots + b_1\sin x + b_2\sin 2x + \cdots + b_n\sin nx + \cdots \qquad (29\text{-}17)$$

where the coefficients are found by

$$a_0 = \frac{1}{2\pi}\int_{-\pi}^{\pi} f(x)\, dx \qquad (29\text{-}25)$$

$$a_n = \frac{1}{\pi}\int_{-\pi}^{\pi} f(x)\cos nx\, dx \qquad (29\text{-}24)$$

$$b_n = \frac{1}{\pi}\int_{-\pi}^{\pi} f(x)\sin nx\, dx \qquad (29\text{-}26)$$

EXAMPLE 1 ▸▸ Find the Fourier series for the square wave function

$$f(x) = \begin{cases} 0 & \text{for } -\pi \leq x < 0 \\ 1 & \text{for } \quad 0 \leq x < \pi \end{cases}$$

(Many of the functions we shall expand in Fourier series are discontinuous (not continuous) like this one. See Section 23-1 for a discussion of continuity.)

CAUTION ▸ Since $f(x)$ is defined differently for the intervals of x indicated, ***it requires two integrals for each coefficient:***

$$a_0 = \frac{1}{2\pi}\int_{-\pi}^{0} 0\,dx + \frac{1}{2\pi}\int_{0}^{\pi} (1)\,dx = \frac{1}{2} \qquad \text{using Eq. (29-25)}$$

$$a_n = \frac{1}{\pi}\int_{-\pi}^{0} 0\,dx + \frac{1}{\pi}\int_{0}^{\pi} (1)\cos nx\,dx = \frac{1}{n\pi}(\sin nx)\Big|_0^{\pi} = 0 \qquad \text{using Eq. (29-24)}$$

for all values of n, since $\sin n\pi = 0$;

$$b_1 = \frac{1}{\pi}\int_{-\pi}^{0} 0\,dx + \frac{1}{\pi}\int_{0}^{\pi} \sin x\,dx = -\frac{1}{\pi}(\cos x)\Big|_0^{\pi} \qquad \text{using Eq. (29-26) with } n = 1$$

$$= -\frac{1}{\pi}(-1 - 1) = \frac{2}{\pi}$$

$$b_2 = \frac{1}{\pi}\int_{-\pi}^{0} 0\,dx + \frac{1}{\pi}\int_{0}^{\pi} \sin 2x\,dx = -\frac{1}{2\pi}(\cos 2x)\Big|_0^{\pi} \qquad \text{using Eq. (29-26) with } n = 2$$

$$= -\frac{1}{2\pi}(1 - 1) = 0$$

$$b_3 = \frac{1}{\pi}\int_{-\pi}^{0} 0\,dx + \frac{1}{\pi}\int_{0}^{\pi} \sin 3x\,dx = -\frac{1}{3\pi}(\cos 3x)\Big|_0^{\pi} \qquad \text{using Eq. (29-26) with } n = 3$$

$$= -\frac{1}{3\pi}(-1 - 1) = \frac{2}{3\pi}$$

In general, if n is even, $b_n = 0$, and if n is odd, then $b_n = 2/n\pi$. Therefore,

$$f(x) = \frac{1}{2} + \frac{2}{\pi}\sin x + \frac{2}{3\pi}\sin 3x + \cdots$$

A graph of the function as defined, and the curve found by using the first three terms of the Fourier series, are shown in Fig. 29-9.

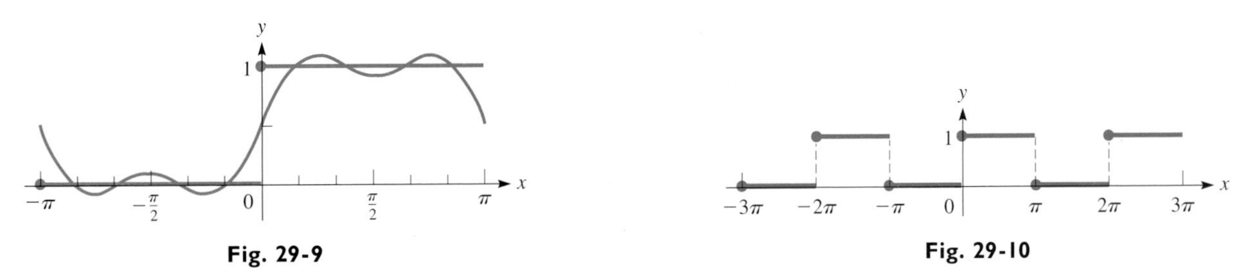

Fig. 29-9

Fig. 29-10

Since functions found by Fourier series have a period of 2π, they can represent functions with this period. If the function $f(x)$ were defined to be periodic with period 2π, with the same definitions as originally indicated, we would graph the function as shown in Fig. 29-10. The Fourier series representation would follow it as in Fig. 29-9. If more terms were used, the fit would be closer. ◂◂

EXAMPLE 2 ▸▸ Find the Fourier series for the function

$$f(x) = \begin{cases} 1 & \text{for } -\pi \le x < 0 \\ x & \text{for } \ \ 0 \le x < \pi \end{cases}$$

For the periodic function, let $f(x + 2\pi) = f(x)$ for all x.

A graph of three periods of this function is shown in Fig. 29-11.

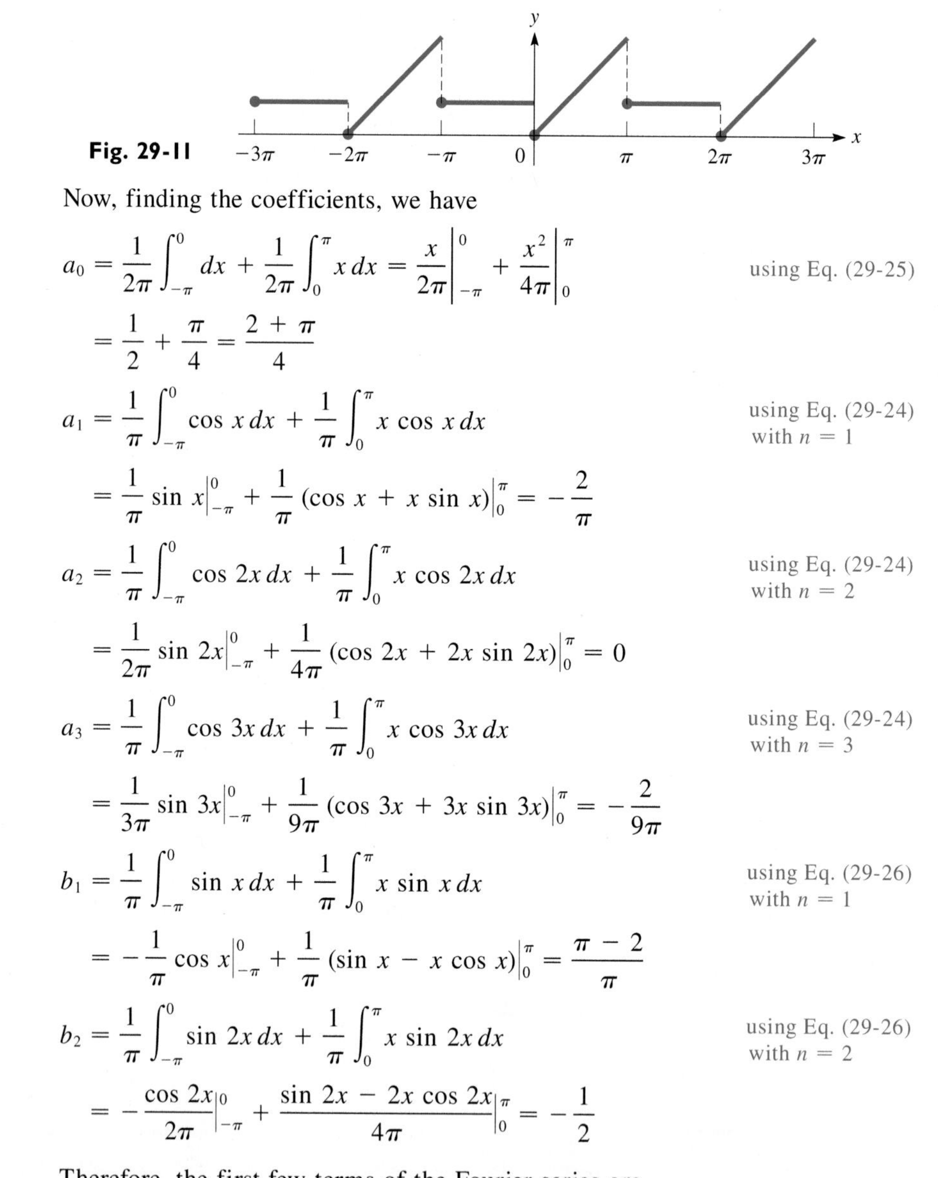

Fig. 29-11

Now, finding the coefficients, we have

$$a_0 = \frac{1}{2\pi}\int_{-\pi}^{0} dx + \frac{1}{2\pi}\int_{0}^{\pi} x\,dx = \frac{x}{2\pi}\Bigg|_{-\pi}^{0} + \frac{x^2}{4\pi}\Bigg|_{0}^{\pi} \qquad \text{using Eq. (29-25)}$$

$$= \frac{1}{2} + \frac{\pi}{4} = \frac{2+\pi}{4}$$

$$a_1 = \frac{1}{\pi}\int_{-\pi}^{0} \cos x\,dx + \frac{1}{\pi}\int_{0}^{\pi} x\cos x\,dx \qquad \text{using Eq. (29-24) with } n = 1$$

$$= \frac{1}{\pi}\sin x\Big|_{-\pi}^{0} + \frac{1}{\pi}(\cos x + x\sin x)\Big|_{0}^{\pi} = -\frac{2}{\pi}$$

$$a_2 = \frac{1}{\pi}\int_{-\pi}^{0} \cos 2x\,dx + \frac{1}{\pi}\int_{0}^{\pi} x\cos 2x\,dx \qquad \text{using Eq. (29-24) with } n = 2$$

$$= \frac{1}{2\pi}\sin 2x\Big|_{-\pi}^{0} + \frac{1}{4\pi}(\cos 2x + 2x\sin 2x)\Big|_{0}^{\pi} = 0$$

$$a_3 = \frac{1}{\pi}\int_{-\pi}^{0} \cos 3x\,dx + \frac{1}{\pi}\int_{0}^{\pi} x\cos 3x\,dx \qquad \text{using Eq. (29-24) with } n = 3$$

$$= \frac{1}{3\pi}\sin 3x\Big|_{-\pi}^{0} + \frac{1}{9\pi}(\cos 3x + 3x\sin 3x)\Big|_{0}^{\pi} = -\frac{2}{9\pi}$$

$$b_1 = \frac{1}{\pi}\int_{-\pi}^{0} \sin x\,dx + \frac{1}{\pi}\int_{0}^{\pi} x\sin x\,dx \qquad \text{using Eq. (29-26) with } n = 1$$

$$= -\frac{1}{\pi}\cos x\Big|_{-\pi}^{0} + \frac{1}{\pi}(\sin x - x\cos x)\Big|_{0}^{\pi} = \frac{\pi - 2}{\pi}$$

$$b_2 = \frac{1}{\pi}\int_{-\pi}^{0} \sin 2x\,dx + \frac{1}{\pi}\int_{0}^{\pi} x\sin 2x\,dx \qquad \text{using Eq. (29-26) with } n = 2$$

$$= -\frac{\cos 2x}{2\pi}\Bigg|_{-\pi}^{0} + \frac{\sin 2x - 2x\cos 2x}{4\pi}\Bigg|_{0}^{\pi} = -\frac{1}{2}$$

Therefore, the first few terms of the Fourier series are

$$f(x) = \frac{2+\pi}{4} - \frac{2}{\pi}\cos x - \frac{2}{9\pi}\cos 3x - \cdots + \left(\frac{\pi - 2}{\pi}\right)\sin x - \frac{1}{2}\sin 2x + \cdots$$

◂◂

See the chapter introduction.

EXAMPLE 3 ▸▸ Certain electronic devices allow an electric current to pass through in only one direction. When an alternating current is applied to the circuit, the current exists for only half the cycle. Figure 29-12 is a representation of such a current as a function of time. This type of electronic device is called a *half-wave rectifier.* Derive the Fourier series for a rectified wave for which half is defined by $f(t) = \sin t\ (0 \le t \le \pi)$ and for which the other half is defined by $f(t) = 0$.

In finding the Fourier coefficients, we first find a_0 as

$$a_0 = \frac{1}{2\pi}\int_0^\pi \sin t\, dt = \frac{1}{2\pi}(-\cos t)\Big|_0^\pi = \frac{1}{2\pi}(1 + 1) = \frac{1}{\pi}$$

In the previous example we evaluated each of the coefficients individually. Here we show how to set up a general expression for a_n and another for b_n. Once we have determined these, we can substitute values of n in the formula to obtain the individual coefficients:

$$a_n = \frac{1}{\pi}\int_0^\pi \sin t \cos nt\, dt = -\frac{1}{2\pi}\left[\frac{\cos(1-n)t}{1-n} + \frac{\cos(1+n)t}{1+n}\right]_0^\pi$$

$$= -\frac{1}{2\pi}\left[\frac{\cos(1-n)\pi}{1-n} + \frac{\cos(1+n)\pi}{1+n} - \frac{1}{1-n} - \frac{1}{1+n}\right]$$

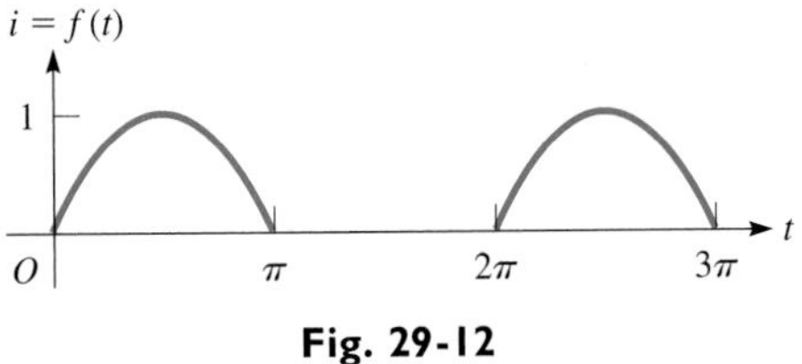

Fig. 29-12

See Formula 40 in the table of integrals in Appendix D. It is valid for all values of n except $n = 1$. Now we write

$$a_1 = \frac{1}{\pi}\int_0^\pi \sin t \cos t\, dt = \frac{1}{2\pi}\sin^2 t\Big|_0^\pi = 0$$

$$a_2 = -\frac{1}{2\pi}\left(\frac{-1}{-1} + \frac{-1}{3} - \frac{1}{-1} - \frac{1}{3}\right) = -\frac{2}{3\pi}$$

$$a_3 = -\frac{1}{2\pi}\left(\frac{1}{-2} + \frac{1}{4} - \frac{1}{-2} - \frac{1}{4}\right) = 0$$

$$a_4 = -\frac{1}{2\pi}\left(\frac{-1}{-3} + \frac{-1}{5} - \frac{1}{-3} - \frac{1}{5}\right) = -\frac{2}{15\pi}$$

$$b_n = \frac{1}{\pi}\int_0^\pi \sin t \sin nt\, dt = \frac{1}{2\pi}\left[\frac{\sin(1-n)t}{1-n} - \frac{\sin(1+n)t}{1+n}\right]_0^\pi$$

$$= \frac{1}{2\pi}\left[\frac{\sin(1-n)\pi}{1-n} - \frac{\sin(1+n)\pi}{1+n}\right]$$

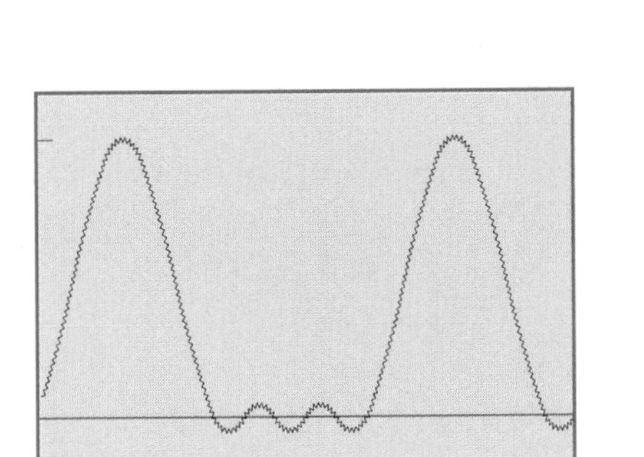

Fig. 29-13

See Formula 39 in Appendix D. It is valid for all values of n except $n = 1$.

Therefore, we have

$$b_1 = \frac{1}{\pi}\int_0^\pi \sin t \sin t\, dt = \frac{1}{\pi}\int_0^\pi \sin^2 t\, dt = \frac{1}{2\pi}(t - \sin t \cos t)\Big|_0^\pi = \frac{1}{2}$$

We see that $b_n = 0$ if $n > 1$, since each is evaluated in terms of the sine of a multiple of π.

Therefore, the Fourier series for the rectified wave is

$$f(t) = \frac{1}{\pi} + \frac{1}{2}\sin t - \frac{2}{\pi}\left(\frac{1}{3}\cos 2t + \frac{1}{15}\cos 4t + \cdots\right)$$

The graph of these terms of the Fourier series is shown in Fig. 29-13. ◂◂

The standard form of a Fourier series we have considered to this point is defined over the interval from $x = -\pi$ to $x = \pi$. At times it is preferable to have a series that is defined over a different interval.

Noting that

$$\sin \frac{n\pi}{L}(x + 2L) = \sin n\left(\frac{\pi x}{L} + 2\pi\right) = \sin \frac{n\pi x}{L}$$

we see that $\sin(n\pi x/L)$ has a period of $2L$. Thus, by using $\sin(n\pi x/L)$ and $\cos(n\pi x/L)$ and the same method of derivation, the following equations are found for the coefficients for the Fourier series for the interval from $x = -L$ to $x = L$.

$$a_0 = \frac{1}{2L}\int_{-L}^{L} f(x)\,dx \tag{29-27}$$

$$a_n = \frac{1}{L}\int_{-L}^{L} f(x)\cos\frac{n\pi x}{L}\,dx \tag{29-28}$$

$$b_n = \frac{1}{L}\int_{-L}^{L} f(x)\sin\frac{n\pi x}{L}\,dx \tag{29-29}$$

EXAMPLE 4 ▸▸ Find the Fourier series for the function

$$f(x) = \begin{cases} 0 & \text{for } -4 \le x < 0 \\ 2 & \text{for } \quad 0 \le x < 4 \end{cases}$$

and for which the period is 8. See Fig. 29-14.

Here $L = 4$. The coefficients are as follows:

$$a_0 = \frac{1}{2(4)}\int_{-4}^{0} 0\,dx + \frac{1}{2(4)}\int_{0}^{4} 2\,dx = \frac{1}{8}(2x)\Big|_0^4 = 1$$

using Eq. (29-27)

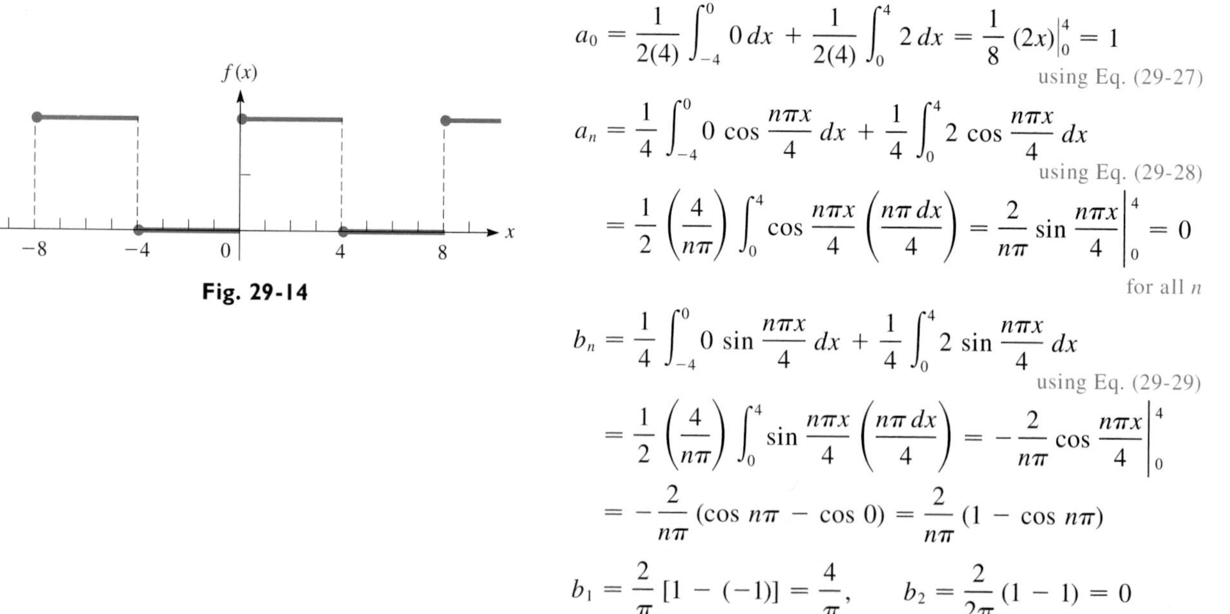

Fig. 29-14

$$a_n = \frac{1}{4}\int_{-4}^{0} 0\cos\frac{n\pi x}{4}\,dx + \frac{1}{4}\int_{0}^{4} 2\cos\frac{n\pi x}{4}\,dx$$

using Eq. (29-28)

$$= \frac{1}{2}\left(\frac{4}{n\pi}\right)\int_0^4 \cos\frac{n\pi x}{4}\left(\frac{n\pi\,dx}{4}\right) = \frac{2}{n\pi}\sin\frac{n\pi x}{4}\Big|_0^4 = 0$$

for all n

$$b_n = \frac{1}{4}\int_{-4}^{0} 0\sin\frac{n\pi x}{4}\,dx + \frac{1}{4}\int_{0}^{4} 2\sin\frac{n\pi x}{4}\,dx$$

using Eq. (29-29)

$$= \frac{1}{2}\left(\frac{4}{n\pi}\right)\int_0^4 \sin\frac{n\pi x}{4}\left(\frac{n\pi\,dx}{4}\right) = -\frac{2}{n\pi}\cos\frac{n\pi x}{4}\Big|_0^4$$

$$= -\frac{2}{n\pi}(\cos n\pi - \cos 0) = \frac{2}{n\pi}(1 - \cos n\pi)$$

$$b_1 = \frac{2}{\pi}[1 - (-1)] = \frac{4}{\pi}, \qquad b_2 = \frac{2}{2\pi}(1 - 1) = 0$$

$$b_3 = \frac{2}{3\pi}[1 - (-1)] = \frac{4}{3\pi}, \qquad b_4 = \frac{2}{4\pi}(1 - 1) = 0$$

Therefore, the Fourier series is

$$f(x) = 1 + \frac{4}{\pi}\sin\frac{\pi x}{4} + \frac{4}{3\pi}\sin\frac{3\pi x}{4} + \cdots$$ ◂◂

All the types of periodic functions included in this section (as well as many others) may actually be seen on an oscilloscope when the proper signal is sent into it. In this way the oscilloscope may be used to analyze the periodic nature of such phenomena as sound waves and electric currents.

Exercises 29–6

In Exercises 1–12, find a few terms of the Fourier series for the given functions, and sketch at least three periods of the function. In Exercises 1–10, $f(x+2\pi)=f(x)$.

1. $f(x)=\begin{cases}1 & -\pi\le x<0\\ 0 & 0\le x<\pi\end{cases}$

2. $f(x)=\begin{cases}-1 & -\pi\le x<0\\ 1 & 0\le x<\pi\end{cases}$

3. $f(x)=\begin{cases}1 & -\pi\le x<0\\ 2 & 0\le x<\pi\end{cases}$

4. $f(x)=\begin{cases}0 & -\pi\le x<0,\ \frac{\pi}{2}<x<\pi\\ 1 & 0\le x\le\frac{\pi}{2}\end{cases}$

5. $f(x)=\begin{cases}0 & -\pi\le x<0\\ x & 0\le x<\pi\end{cases}$

6. $f(x)=x,\ -\pi\le x<\pi$

7. $f(x)=\begin{cases}-1 & -\pi\le x<0\\ 0 & 0\le x<\frac{\pi}{2}\\ 1 & \frac{\pi}{2}\le x<\pi\end{cases}$

8. $f(x)=x^2,\ -\pi\le x<\pi$

9. $f(x)=\begin{cases}-x & -\pi\le x<0\\ x & 0\le x<\pi\end{cases}$

10. $f(x)=\begin{cases}0 & -\pi\le x<0\\ x^2 & 0\le x<\pi\end{cases}$

11. $f(x)=\begin{cases}5 & -3\le x<0\\ 0 & 0\le x<3,\quad \text{period}=6\end{cases}$

12. $f(x)=\begin{cases}-x & -4\le x<0\\ x & 0\le x<4,\quad \text{period}=8\end{cases}$

In Exercises 13–16, use a graphing calculator to display the terms of the Fourier series given in the indicated example or answer for the indicated exercise. Compare with the sketch of the function. For each calculator display use Xmin = −8 and Xmax = 8.

13. Example 1
14. Example 4
15. Exercise 5
16. Exercise 9

In Exercises 17–20, solve the given problems. In Exercises 17 and 18, $f(x+2\pi)=f(x)$.

17. Find the Fourier expansion of the electronic device known as a *full-wave rectifier.* This is found by using as the function for the current $f(t)=-\sin t$ for $-\pi\le t\le 0$ and $f(t)=\sin t$ for $0<t\le\pi$. See Fig. 29-15. The portion of the curve to the left of the $f(t)$-axis is dashed because from a physical point of view we can give no significance to this part of the wave, although mathematically we can derive the proper form of the expansion by using it.

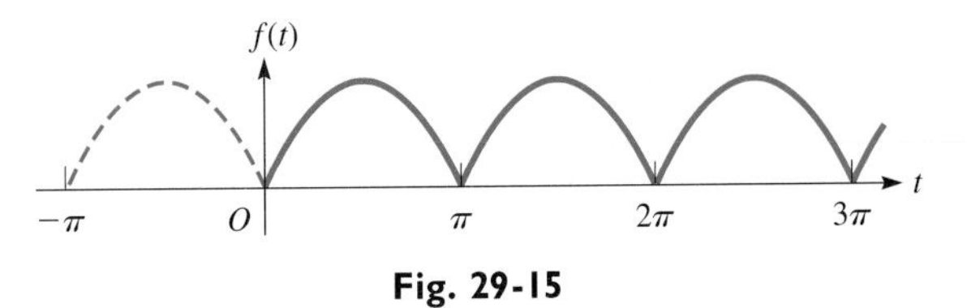

Fig. 29-15

18. The loudness L, in decibels, of a certain siren as a function of time, in seconds, can be described by the function $L=0$ for $-\pi\le t<0$, $L=120t$ for $0\le t<\frac{\pi}{2}$, and $L=120(\pi-t)$ for $\frac{\pi}{2}\le t<\pi$, with a period of 2π seconds (where only positive values of t have physical significance). Find the Fourier expansion for the loudness of the siren.

19. Each pulse of a pulsating force F of a pressing machine is 8 N. The force lasts for 1 s, followed by a 3-s pause. Thus it can be represented by $F=0$ for $-2\le t<0$ and $1\le t\le 2$, and $F=8$ for $0\le t<1$, with a period of 4 s (only positive values of t have physical significance). Find the Fourier expansion for the force.

20. A function for which $f(x)=f(-x)$ is called an **even function,** and a function for which $f(x)=-f(-x)$ is called an **odd function.** From our discussion of symmetry in Section 21-3, we see that an even function is symmetrical to the y-axis and that an odd function is symmetrical to the origin. It can be proven that the Fourier expansion of an even function contains no sine terms and that of an odd function contains no cosine terms. Show that the function in Exercise 9 is an even function and the function in Exercise 6 is an odd function. Note and compare the Fourier expansions of these functions.

Chapter Equations, Review Exercises, and Practice Test

Chapter Equations

Infinite series
$$\sum_{n=1}^{\infty} a_n = a_1 + a_2 + a_3 + \cdots + a_n + \cdots \tag{29-1}$$

Sum of series
$$S = \lim_{n\to\infty} S_n = \lim_{n\to\infty} \sum_{i=1}^{n} a_i \tag{29-2}$$

Sum of geometric series
$$S = \lim_{n\to\infty} S_n = \frac{a_1}{1-r} \tag{29-3}$$

Power series
$$f(x) = a_0 + a_1 x + a_2 x^2 + \cdots + a_n x^n + \cdots \tag{29-4}$$

Maclaurin series
$$f(x) = f(0) + f'(0)x + \frac{f''(0)x^2}{2!} + \frac{f'''(0)x^3}{3!} + \cdots + \frac{f^{(n)}(0)x^n}{n!} + \cdots \tag{29-5}$$

Special series
$$e^x = 1 + x + \frac{x^2}{2!} + \frac{x^3}{3!} + \cdots \qquad (\text{all } x) \tag{29-6}$$

$$\sin x = x - \frac{x^3}{3!} + \frac{x^5}{5!} - \cdots \qquad (\text{all } x) \tag{29-7}$$

$$\cos x = 1 - \frac{x^2}{2!} + \frac{x^4}{4!} - \cdots \qquad (\text{all } x) \tag{29-8}$$

$$\ln(1+x) = x - \frac{x^2}{2} + \frac{x^3}{3} - \frac{x^4}{4} + \cdots \qquad (|x| < 1) \tag{29-9}$$

$$(1+x)^n = 1 + nx + \frac{n(n-1)}{2!}x^2 + \cdots \qquad (|x| < 1) \tag{29-10}$$

Complex number
$$e^{j\theta} = \cos\theta + j\sin\theta \tag{29-14}$$

Taylor series
$$f(x) = f(a) + f'(a)(x-a) + \frac{f''(a)(x-a)^2}{2!} + \cdots \tag{29-16}$$

Fourier series
$$f(x) = a_0 + a_1\cos x + a_2\cos 2x + \cdots + a_n\cos nx + \cdots + b_1\sin x + b_2\sin 2x + \cdots + b_n\sin nx + \cdots \tag{29-17}$$

Period $= 2\pi$
$$a_0 = \frac{1}{2\pi}\int_{-\pi}^{\pi} f(x)\,dx \tag{29-25}$$

$$a_n = \frac{1}{\pi}\int_{-\pi}^{\pi} f(x)\cos nx\,dx \tag{29-24}$$

$$b_n = \frac{1}{\pi}\int_{-\pi}^{\pi} f(x)\sin nx\,dx \tag{29-26}$$

Period $= 2L$
$$a_0 = \frac{1}{2L}\int_{-L}^{L} f(x)\,dx \tag{29-27}$$

$$a_n = \frac{1}{L}\int_{-L}^{L} f(x)\cos\frac{n\pi x}{L}\,dx \tag{29-28}$$

$$b_n = \frac{1}{L}\int_{-L}^{L} f(x)\sin\frac{n\pi x}{L}\,dx \tag{29-29}$$

Review Exercises

In Exercises 1–8, find the first three nonzero terms of the Maclaurin expansion of the given functions.

1. $f(x) = \dfrac{1}{1 + e^x}$ **2.** $f(x) = e^{\cos x}$

3. $f(x) = \sin 2x^2$ **4.** $f(x) = \dfrac{1}{(1 - x)^2}$

5. $f(x) = (x + 1)^{1/3}$ **6.** $f(x) = \dfrac{2}{2 - x^2}$

7. $f(x) = \sin^{-1} x$ **8.** $f(x) = \dfrac{1}{1 - \sin x}$

In Exercises 9–20, calculate the value of each of the given functions. Use three terms of the appropriate series.

9. $e^{-0.2}$ **10.** $\ln(1.10)$ **11.** $\sqrt[3]{1.3}$

12. $\sin 3.5°$ **13.** $\sqrt{1.07}$ **14.** $e^{0.4173}$

15. $\ln 0.8172$ **16.** $\cos 0.1376$ **17.** $\tan 43.62°$

18. $\sqrt[4]{260}$ **19.** $\sqrt{148}$ **20.** $\cos 47°$

In Exercises 21 and 22, evaluate the given integrals by using three terms of the appropriate series.

21. $\displaystyle\int_{0.1}^{0.2} \frac{\cos x}{\sqrt{x}}\,dx$ **22.** $\displaystyle\int_0^{0.1} \sqrt[3]{1 + x^2}\,dx$

In Exercises 23 and 24, find the first three terms of the Taylor expansion for the given function and value of a.

23. $\cos x \quad (a = \pi/3)$ **24.** $\ln \cos x \quad (a = \pi/4)$

In Exercises 25–28, find a few terms of the Fourier series for the given functions with the given periods. Sketch three periods of the function. Use a graphing calculator to view the Fourier series and compare with the sketch.

25. $f(x) = \begin{cases} 0 & -\pi \le x < -\dfrac{\pi}{2} \text{ and } \dfrac{\pi}{2} < x < \pi, \\ 1 & -\dfrac{\pi}{2} \le x \le \dfrac{\pi}{2} \qquad \text{period} = 2\pi \end{cases}$

26. $f(x) = \begin{cases} -x & -\pi \le x < 0, \quad \text{period} = 2\pi \\ 0 & 0 \le x < \pi \end{cases}$

27. $f(x) = x, \quad -2 \le x < 2, \quad \text{period} = 4$

28. $f(x) = \begin{cases} -2 & -3 \le x < 0, \quad \text{period} = 6 \\ 2 & 0 \le x < 3 \end{cases}$

In Exercises 29–52, solve the given problems.

29. Find the sum of the series $64 + 48 + 36 + 27 + \cdots$.

30. Find the first five partial sums of the series $\displaystyle\sum_{n=1}^{\infty} \frac{n}{3n + 1}$ and determine whether it appears to be convergent or divergent.

31. If h is small, show that $\sin(x + h) - \sin(x - h) = 2h \cos x$.

32. Find the first three nonzero terms of the Maclaurin expansion of the function $\sin x + x \cos x$ by differentiating the expansion term by term for $x \sin x$.

33. Using the properties of logarithms and Eq. (29-9), find four terms of the Maclaurin expansion of $y = \ln(1 + x)^4$.

34. By multiplication of series, show that the first two terms of the Maclaurin series for $2 \sin x \cos x$ are the same as those of the series for $\sin 2x$.

35. Find the first four nonzero terms of the expansion for $\cos^2 x$ by using the identity $\cos^2 x = \frac{1}{2}(1 + \cos 2x)$ and the series for $\cos x$.

36. Evaluate the integral $\int_0^1 x \sin x\,dx$ (a) by methods of Chapter 28 and (b) by using three terms of the series for $\sin x$. Compare results.

37. Find the first three terms of the Maclaurin expansion for $\sec x$ by finding the reciprocal of the series for $\cos x$.

38. By simplifying the sum of the squares of the Maclaurin series for $\sin x$ and $\cos x$ through the x^4 terms, verify (to this extent) that $\sin^2 x + \cos^2 x = 1$.

39. From the Maclaurin series for $f(x) = \dfrac{1}{1 - x}$ (see Exercise 11 of Section 29-2), find the series for $\dfrac{1}{1 + x}$.

40. Show that the Maclaurin expansions for $\cos x$ and $\cos(-x)$ are the same.

41. Find the approximate area between the curve $y = \dfrac{x - \sin x}{x^2}$ and the x-axis between $x = 0.1$ and $x = 0.2$.

42. Find the approximate value of the moment of inertia with respect to its axis of the volume generated by revolving the smaller area bounded by $y = \sin x$, $x = 0.3$, and the x-axis about the y-axis. Use two terms of the appropriate series.

43. Find the first three terms of the Maclaurin series for $\tan^{-1} x$ by integrating the series for $1/(1 + x^2)$ term by term.

44. The displacement y (in metres) of a water wave as a function of t (in seconds) is $y = 0.5 \sin 0.5t$. Find the first three terms of the Maclaurin series for the displacement.

45. The number N of radioactive nuclei in a radioactive sample is $N = N_0 e^{-\lambda t}$. Here t is the time, N_0 is the number at $t = 0$, and λ is the *decay constant.* By using four terms of the appropriate series, express the right side of this equation as a polynomial.

46. The length of Lake Erie is a great circle arc of 390 km. If the lake is assumed to be flat, use series to find the error in calculating the distance from the center of the lake to the center of the earth. The radius of the earth is 6400 km.

47. The electric potential V at a distance x along a certain surface is given by $V = \ln \dfrac{1+x}{1-x}$. Find the first four terms of the Maclaurin series for V.

48. If a mass M is hung from a spring of mass m, the ratio of masses is $m/M = k\omega \tan k\omega$, where k is a constant and ω is a measure of the frequency of vibration. By using two terms of the appropriate series, express m/M as a polynomial in terms of ω.

49. In the study of electromagnetic radiation, the expression $\dfrac{N_0}{1 - e^{-k/T}}$ is used. Here T is the thermodynamic temperature and N_0 and k are constants. Show that this expression can be written as $N_0(1 + e^{-k/T} + e^{-2k/T} + \cdots)$. (*Hint:* Let $x = e^{-k/T}$.)

50. In the analysis of reflection from a spherical mirror, it is necessary to express the x-coordinate on the surface shown in Fig. 29-16 in terms of the y-coordinate and the radius R. Using the equation of the semicircle shown, solve for x (note that $x \leq R$). Then express the result as a series. (Note that the first approximation gives a parabolic surface.)

51. A certain electric current is pulsating so that the current as a function of time is given by $f(t) = 0$ if $-\pi \leq t < 0$ and $\pi/2 < t < \pi$. If $0 < t < \pi/2$, $f(t) = \sin t$. Find the Fourier expansion for this pulsating current, and sketch three periods.

52. The force F applied to a spring system as a function of the time t is given by $F = x/\pi$ if $0 \leq t \leq \pi$ and $F = 0$ if $\pi < t < 2\pi$. If the period of the force is 2π, find the first few terms of the Fourier series that represents the force.

Writing Exercise

53. A computer science class is assigned to write a program to make a table of values of the sine, cosine, and tangent of an angle in degrees to the nearest 0.1°. Write a few paragraphs explaining how these values may be found using the known values for 0°, 30°, and 45° (see p. 112), trigonometric relations in Chapter 20, and series from this chapter, without using many terms of any series.

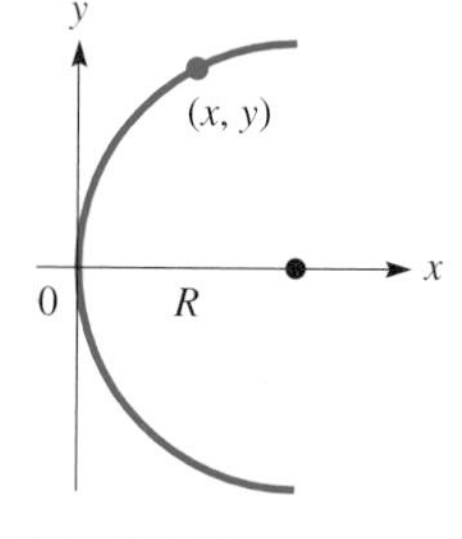

Fig. 29-16

Practice Test

1. By direct expansion, find the first four nonzero terms of the Maclaurin expansion for $f(x) = (1 + e^x)^2$.
2. Find the first three nonzero terms of the Taylor expansion for $f(x) = \cos x$, with $a = \pi/3$.
3. Evaluate ln 0.96 by using four terms of the expansion for $\ln(1 + x)$.
4. Find the first three nonzero terms of the expansion for $f(x) = \dfrac{1}{\sqrt{1-2x}}$ by using the binomial series.
5. Evaluate $\int_0^1 x \cos x \, dx$ by using three terms of the appropriate series.
6. An electric current is pulsating such that the current is a function of time with a period of 2π. If $f(t) = 2$ for $0 \leq t < \pi$ and $f(t) = 0$ for the other half-cycle, find the first three nonzero terms of the Fourier series for this current.

30

Differential Equations

A great many problems that arise in science, technology, and engineering involve rates of change. It is for this reason that equations which contain derivatives, called *differential equations,* are of considerable importance in nearly all areas of application. Among the areas that use the methods of differential equations for solution are those involving velocities, chemical reactions, interest calculations, changes in pressure and temperature, population growth, forces on beams and structures, electric currents and voltages, nuclear energy, and many others.

This chapter gives an introduction to differential equations and in it we develop a few methods of solving some basic types of differential equations. Actually, we already solved a few simple differential equations in earlier chapters when we started a solution with the expression for the slope of a tangent line or the velocity of an object.

In Section 30–5 we show how differential equations are used in the study of radioactivity, which is important in the development of nuclear energy.

30-1 Solutions of Differential Equations

A **differential equation** *is an equation that contains derivatives or differentials.* In this section we introduce the basic meaning of the solution of a differential equation. In the sections that follow, we consider certain methods of finding such solutions and show some applications.

The types of differential equations we shall consider are those that contain first and second derivatives. *An equation that contains only first derivatives is called a* **first-order** *differential equation. An equation that contains second derivatives, and possibly first derivatives, is called a* **second-order** *differential equation. In general, the* **order** *of a differential equation is that of the highest derivative in the equation, and the* **degree** *is the highest power of that derivative.*

EXAMPLE 1 ▸▸ (a) The equation $dy/dx + x = y$ is a first-order differential equation since it contains only a first derivative.

(b) The equations

$$\overset{\downarrow}{\frac{d^2y}{dx^2}} + y = 3x^2 \quad \overset{\text{order}}{\text{and}} \quad \overset{\downarrow}{\frac{d^2y}{dx^2}} + 2\frac{dy}{dx} = x$$

are second-order equations since each contains a second derivative and no higher derivatives. The dy/dx in the second equation does not affect the order. ◂◂

EXAMPLE 2 ▸▸ The equation

$$\frac{d^2y}{dx^2} + \left(\frac{dy}{dx}\right)^4 - y = 6$$

is a differential equation of the second order and the first degree. That is, the highest derivative that appears is the second, and it is raised to the first power. Since the second derivative appears, the fourth power of the first derivative does not affect the degree. ◂◂

In our discussion of differential equations, we shall restrict our attention to equations of the first degree.

General Solution

Particular Solution

A **solution** *of a differential equation is a relation between the variables that satisfies the differential equation.* That is, when this relation is substituted into the differential equation, an algebraic identity results. *A solution containing a number of independent arbitrary constants equal to the order of the differential equation is called the* **general solution** *of the equation. When specific values are given to at least one of these constants, the solution is called a* **particular solution**.

EXAMPLE 3 ▸▸ Any coefficients that are not specified numerically after like terms have been combined are independent arbitrary contants. In the expression $c_1x + c_2 + c_3x$, there are only two arbitrary constants since the x-terms may be combined; $c_2 + c_4x$ is an equivalent expression with $c_4 = c_1 + c_3$. ◂◂

EXAMPLE 4 ▸▸ $y = c_1e^{-x} + c_2e^{2x}$ is the general solution of the differential equation

$$\frac{d^2y}{dx^2} - \frac{dy}{dx} = 2y$$

The order of this differential equation is 2, and there are two independent arbitrary constants in the solution. The equation $y = 4e^{-x}$ is a particular solution. It can be derived from the general solution by letting $c_1 = 4$ and $c_2 = 0$. Each of these solutions can be shown to satisfy the differential equation by taking two derivatives and substituting. ◂◂

To solve a differential equation, we have to find some method of transforming the equation so that each term may be integrated. Some of these methods will be considered after this section. The purpose here is to show that a given equation is a solution of a differential equation *by taking the required derivatives and to show that an identity results after substitution.*

NOTE ▶

EXAMPLE 5 ▶▶ Show that $y = c_1 \sin x + c_2 \cos x$ is the general solution of the differential equation $y'' + y = 0$.

The function and its first two derivatives are

$$y = c_1 \sin x + c_2 \cos x$$
$$y' = c_1 \cos x - c_2 \sin x$$
$$y'' = -c_1 \sin x - c_2 \cos x$$

Substituting these into the differential equation, we have

$$\underbrace{(-c_1 \sin x - c_2 \cos x)}_{y''} + \underbrace{(c_1 \sin x + c_2 \cos x)}_{y} = 0 \quad \text{or} \quad 0 = 0$$

We know that this must be the general solution, since there are two independent arbitrary constants and the order of the differential equation is 2. ◀◀

EXAMPLE 6 ▶▶ Show that $y = cx + x^2$ is a solution of the differential equation $xy' - y = x^2$.

Taking one derivative of the function and substituting into the differential equation, we have

$$y = cx + x^2$$
$$xy' - y = x^2$$
$$y' = c + 2x$$
$$x(c + 2x) - (cx + x^2) = x^2 \quad \text{or} \quad x^2 = x^2$$ ◀◀

Exercises 30–1

In Exercises 1–4, determine whether the given equation is the general solution or a particular solution of the given differential equation.

1. $\frac{dy}{dx} + 2xy = 0, \quad y = e^{-x^2}$

2. $y' \ln x - \frac{y}{x} = 0, \quad y = c \ln x$

3. $y'' + 3y' - 4y = 3e^x, \quad y = c_1e^x + c_2e^{-4x} + \frac{3}{5}xe^x$

4. $\frac{d^2y}{dx^2} + 4y = 0, \quad y = c_1 \sin 2x + 3 \cos 2x$

In Exercises 5–28, show that the given equation is a solution of the given differential equation.

5. $\frac{dy}{dx} = 1, \quad y = x + 3$

6. $\frac{dy}{dx} = 2x, \quad y = x^2 + 1$

7. $\frac{dy}{dx} - y = 1, \quad y = e^x - 1$

8. $\frac{dy}{dx} - 3 = 2x, \quad y = x^2 + 3x$

9. $xy' = 2y, \quad y = cx^2$

10. $y' = 2xy^2, \quad y = -\frac{1}{x^2 + c}$

11. $y' + 2y = 2x, \quad y = ce^{-2x} + x - \frac{1}{2}$

12. $y' - 3x^2 = 1, \quad y = x^3 + x + c$

13. $\frac{d^2y}{dx^2} + 4y = 0, \quad y = 3 \cos 2x$

14. $y'' + 9y = 4 \cos x, \quad 2y = \cos x$

15. $y'' - 4y' + 4y = e^{2x}, \quad y = e^{2x}\left(c_1 + c_2x + \frac{x^2}{2}\right)$

16. $\frac{d^2y}{dx^2} = 2\frac{dy}{dx}$, $y = c_1 + c_2e^{2x}$

17. $x^2y' + y^2 = 0$, $xy = cx + cy$

18. $xy' - 3y = x^2$, $y = cx^3 - x^2$

19. $x\frac{d^2y}{dx^2} + \frac{dy}{dx} = 0$, $y = c_1 \ln x + c_2$

20. $y'' + 4y = 10e^x$, $y = c_1 \sin 2x + c_2 \cos 2x + 2e^x$

21. $y' + y = 2 \cos x$, $y = \sin x + \cos x - e^{-x}$

22. $(x + y) - xy' = 0$, $y = x \ln x - cx$

23. $y'' + y' = 6 \sin 2x$, $y = e^{-x} - \frac{3}{5}\cos 2x - \frac{6}{5}\sin 2x$

24. $xy'' + y' = 16x^3$, $y = x^4 + c_1 + c_2 \ln x$

25. $\cos x \frac{dy}{dx} + \sin x = 1 - y$, $y = \frac{x + c}{\sec x + \tan x}$

26. $2xyy' + x^2 = y^2$, $x^2 + y^2 = cx$

27. $(y')^2 + xy' = y$, $y = cx + c^2$

28. $x^4(y')^2 - xy' = y$, $y = c^2 + \frac{c}{x}$

30-2 Separation of Variables

We shall now solve differential equations of the first order and first degree. Of the many methods for solving such equations, a few are presented in this and the next two sections. The first of these is *the method of* **separation of variables**.

A differential equation of the first order and first degree contains the first derivative to the first power. That is, it may be written as $dy/dx = f(x, y)$. This type of equation is more commonly expressed in its differential form,

$$M(x, y)\,dx + N(x, y)\,dy = 0 \tag{30-1}$$

where $M(x, y)$ and $N(x, y)$ may represent constants, functions of either x or y, or functions of x and y.

To solve an equation of the form of Eq. (30-1), we must integrate. However, if $M(x, y)$ contains y, the first term cannot be integrated. Also, if $N(x, y)$ contains x, the second term cannot be integrated. If it is possible to rewrite Eq. (30-1) as

$$A(x)\,dx + B(y)\,dy = 0 \tag{30-2}$$

NOTE ▶ where $A(x)$ does not contain y and $B(y)$ does not contain x, then *we may find the solution by integrating each term and adding the constant of integration*. (In rewriting Eq. (30-1), if division is used, the solution is not valid for values that make the divisor zero.) Many differential equations can be solved in this way.

EXAMPLE 1 ▶▶ Solve the differential equation $dx - 4xy^3dy = 0$.

We can write this equation as

(1) $dx + (-4xy^3)\,dy = 0$

which means that $M(x, y) = 1$ and $N(x, y) = -4xy^3$.

CAUTION ▶ We must remove the x from the coefficient of dy ***without introducing y into the coefficient of dx***. We do this by dividing each term by x, which gives us

$dx/x - 4y^3dy = 0$

It is now possible to integrate each term. Performing this integration, we have

$\ln|x| - y^4 = c$

The constant of integration c becomes the arbitrary constant of the solution. ◀◀

In Example 1 we showed the integration of dx/x as $\ln|x|$, which follows our discussion in Section 28-2. We know $\ln|x| = \ln x$ if $x > 0$ and $\ln|x| = \ln(-x)$ if $x < 0$. Since we know the values being used when we find a particular solution, *we generally will not use the absolute value notation when integrating logarithmic forms.* We would show the integration of dx/x as $\ln x$, with the understanding that we know $x > 0$. When using negative values of x, we would express is as $\ln(-x)$.

NOTE ▶

EXAMPLE 2 ▶▶ Solve the differential equation $xy\,dx + (x^2 + 1)dy = 0$.

In order to integrate each term, it is necessary to divide each term by $y(x^2 + 1)$. When this is done, we have

$$\frac{x\,dx}{x^2 + 1} + \frac{dy}{y} = 0$$

Integrating, we obtain the solution

$$\frac{1}{2}\ln(x^2 + 1) + \ln y = c$$

It is possible to make use of the properties of logarithms to make the form of this solution neater. If we write the constant of integration as $\ln c_1$, rather than c, we have $\frac{1}{2}\ln(x^2 + 1) + \ln y = \ln c_1$. Multiplying through by 2 and using the property of logarithms given by Eq. (13-9), we have $\ln(x^2 + 1) + \ln y^2 = \ln c_1^2$. Next, using the property of logarithms given by Eq. (13-7), we then have $\ln(x^2 + 1)y^2 = \ln c_1^2$, which means

For reference, Eq. (13-9) is $\log_b x^n = n \log_b x$ and Eq. (13-7) is $\log_b xy = \log_b x + \log_b y$.

$$(x^2 + 1)y^2 = c_1^2$$

This form of the solution is more compact and generally would be preferred. However, *any expression that represents a constant may be chosen as the constant of integration* and leads to a correct solution. In checking answers, we must remember that a different choice of constant will lead to a different form of the solution. Thus two different-appearing answers may both be correct. *Often there is more than one reasonable choice of a constant, and different forms of the solution may be expected.* ◀◀

CAUTION ▶

EXAMPLE 3 ▶▶ Solve the differential equation $\dfrac{dy}{dx} = \dfrac{y}{x^2 + 4}$.

The solution proceeds as follows:

$$\frac{dy}{y} = \frac{dx}{x^2 + 4} \qquad \text{separate variables by multiplying by } dx \text{ and dividing by } y$$

$$\ln y = \frac{1}{2}\tan^{-1}\frac{x}{2} + \frac{c}{2} \qquad \text{integrate}$$

$$2\ln y = \tan^{-1}\frac{x}{2} + c$$

$$\ln y^2 = \tan^{-1}\frac{x}{2} + c$$

Note the different forms of the result using $c/2$ as the constant of integration. These forms would differ somewhat had we chosen c as the constant.

The choice of $\ln c$ as the constant of integration (on the left) is also reasonable. It would lead to the result $2\ln cy = \tan^{-1}(x/2)$. ◀◀

In order to separate the variables of a differential equation that contains exponential functions, it may be necessary to use the properties of exponents. Also, when trigonometric functions are involved, the basic trigonometric identities may be needed.

EXAMPLE 4 ▸▸ Solve the differential equation $2e^{3x}\sin y\,dx + e^x\csc y\,dy = 0$.

In the dx-term we want only a function of x, which means that we must divide by $\sin y$. Also, the dy-term indicates that we must divide by e^x. Thus,

$$\frac{2e^{3x}\sin y\,dx}{e^x\sin y} + \frac{e^x\csc y\,dy}{e^x\sin y} = 0 \quad \text{divide by } e^x \sin y$$

$$2e^{2x}\,dx + \csc^2 y\,dy = 0 \quad \text{variables separated}$$

$$e^{2x}(2\,dx) + \csc^2 y\,dy = 0 \quad \text{form for integrating}$$

$$e^{2x} - \cot y = c \quad \text{integrate}$$ ◂◂

Finding Particular Solutions

In order to find a particular solution of a differential equation, we must have information that allows us to evaluate the constant of integration. The following examples show how particular solutions of differential equations are found. Also, we will show graphically the difference between the general solution and the particular solution.

EXAMPLE 5 ▸▸ Solve the differential equation $(x^2+1)^2\,dy + 4x\,dx = 0$, subject to the condition that $x = 1$ when $y = 3$.

Separating variables, we have

$$dy + \frac{4x\,dx}{(x^2+1)^2} = 0 \quad \text{dividing by } (x^2+1)^2$$

$$y - \frac{2}{x^2+1} = c \quad \text{integrating}$$

$$y = \frac{2}{x^2+1} + c \quad \text{general solution}$$

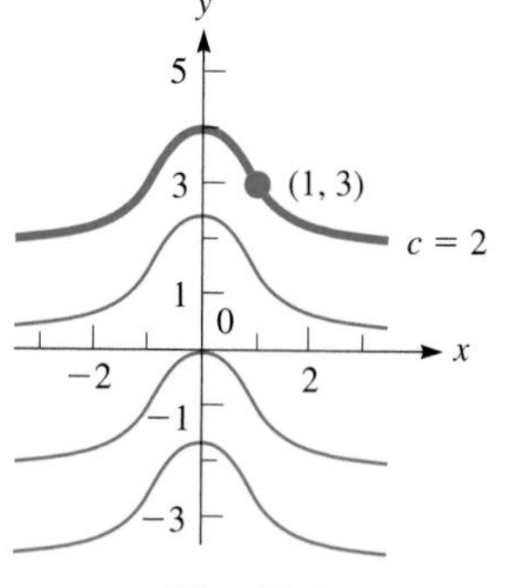

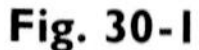

Fig. 30-1

Since a specific set of values is given, we can evaluate the constant of integration and thereby get a particular solution. Using the values $x = 1$ and $y = 3$, we have

$$3 = \frac{2}{1+1} + c, \qquad c = 2 \quad \text{evaluate } c$$

which gives us

$$y = \frac{2}{x^2+1} + 2 \quad \text{particular solution}$$

The general solution defines a *family* of curves, one member of the family for each value of c that may be considered. A few of these curves are shown in Fig. 30-1. When c is specified as in the particular solution, we have the specific (darker) curve shown in Fig. 30-1. ◂◂

EXAMPLE 6 ▶▶ Find the particular solution in Example 2 if the function is subject to the condition that $x = 0$ when $y = e$.

Using the solution $\frac{1}{2}\ln(x^2 + 1) + \ln y = c$, we have

$$\frac{1}{2}\ln(0 + 1) + \ln e = c, \qquad \frac{1}{2}\ln 1 + 1 = c \quad \text{or} \quad c = 1$$

The particular solution is then

$$\frac{1}{2}\ln(x^2 + 1) + \ln y = 1 \qquad \text{substitute } c = 1$$

$$\ln(x^2 + 1) + 2\ln y = 2$$

$$\ln y^2(x^2 + 1) = 2 \qquad \text{using properties of logarithms}$$

$$y^2(x^2 + 1) = e^2 \qquad \text{exponential form}$$

Using the general solution $(x^2 + 1)y^2 = c_1^2$, we have

$$(0 + 1)e^2 = c_1^2, \qquad c_1^2 = e^2$$

$$y^2(x^2 + 1) = e^2$$

NOTE ▶ which is precisely the same solution as above. We see, therefore, that *the choice of the form of the constant does not affect the final result and that the constant is truly arbitrary.* ◀◀

Exercises 30–2

In Exercises 1–28, solve the given differential equations.

1. $2x\,dx + dy = 0$
2. $y^2\,dy + x^3\,dx = 0$
3. $y^2\,dx + dy = 0$
4. $y\,dx + x\,dy = 0$
5. $\dfrac{dV}{dP} = -\dfrac{V}{P^2}$
6. $\dfrac{2\,dy}{dx} = \dfrac{y(x + 1)}{x}$
7. $x^2 + (x^3 + 5)y' = 0$
8. $xyy' + \sqrt{1 + y^2} = 0$
9. $dy + \ln xy\,dx = (4x + \ln y)\,dx$
10. $r\sqrt{1 - \theta^2}\dfrac{dr}{d\theta} = \theta + 4$
11. $e^{x^2}\,dy = x\sqrt{1 - y}\,dx$
12. $\sqrt{1 + 4x^2}\,dy = y^3x\,dx$
13. $e^{x+y}\,dx + dy = 0$
14. $e^{2x}\,dy + e^x\,dx = 0$
15. $y' - y = 4$
16. $y' + 4x = 3$
17. $x\dfrac{dy}{dx} = y^2 + y^2\ln x$
18. $(yx^2 + y)\dfrac{dy}{dx} = \tan^{-1} x$
19. $y\tan x\,dx + \cos^2 x\,dy = 0$
20. $\sin x\sec y\,dx = dy$
21. $yx^2\,dx = y\,dx - x^2\,dy$
22. $e^{\sin x}\,dx + \sec x\,dy = 0$
23. $y\sqrt{1 - x^2}\,dy + 2\,dx = 0$
24. $(x^3 + x^2)\,dx + (x + 1)y\,dy = 0$
25. $2\ln t\,dt + t\,di = 0$
26. $2y(x^3 + 1)\,dy + 3x^2(y^2 - 1)\,dx = 0$
27. $y^2e^x + (e^x + 1)\dfrac{dy}{dx} = 0$
28. $y + 1 + \sec x(\sin x + 1)\dfrac{dy}{dx} = 0$

In Exercises 29–36, find the particular solution of the given differential equation for the indicated values.

29. $\dfrac{dy}{dx} + yx^2 = 0$; $x = 0$ when $y = 1$
30. $\dfrac{dy}{dx} + 2y = 6$; $x = 0$ when $y = 1$
31. $(xy^2 + x)\dfrac{dy}{dx} = \ln x$; $x = 1$ when $y = 0$
32. $\dfrac{ds}{dt} = \sec s$; $t = 0$ when $s = 0$
33. $y' = (1 - y)\cos x$; $x = \pi/6$ when $y = 0$
34. $x\,dy = y\ln y\,dx$; $x = 2$ when $y = e$
35. $y^2e^x\,dx + e^{-x}\,dy = y^2\,dx$; $x = 0$ when $y = 2$
36. $2y\cos y\,dy - \sin y\,dy = y\sin y\,dx$; $x = 0$ when $y = \pi/2$

30-3 Integrable Combinations

Many differential equations cannot be solved by the method of separation of variables. Many other methods have been developed for solving such equations. One of these methods is based on the fact that *certain combinations of basic differentials can be integrated together as a unit.* The following differentials suggest some of these combinations that may occur:

$$d(xy) = x\,dy + y\,dx \tag{30-3}$$

$$d(x^2 + y^2) = 2(x\,dx + y\,dy) \tag{30-4}$$

$$d\left(\frac{y}{x}\right) = \frac{x\,dy - y\,dx}{x^2} \tag{30-5}$$

$$d\left(\frac{x}{y}\right) = \frac{y\,dx - x\,dy}{y^2} \tag{30-6}$$

Equation (30-3) suggests that if the combination $x\,dy + y\,dx$ occurs in a differential equation, we should look for a function of xy as a solution. Equation (30-4) suggests that if the combination $x\,dx + y\,dy$ occurs, we should look for a function of $x^2 + y^2$. Equations (30-5) and (30-6) suggest that if either of the combinations $x\,dy - y\,dx$ or $y\,dx - x\,dy$ occurs, we should look for a function of y/x or x/y.

EXAMPLE 1 ▸▸ Solve the differential equation $x\,dy + y\,dx + xy\,dy = 0$.

By dividing through by xy, we have

$$\frac{x\,dy + y\,dx}{xy} + dy = 0$$

The left term is the differential of xy divided by xy. Thus it integrates to $\ln xy$.

$$\frac{d(xy)}{xy} + dy = 0$$

for which the solution is

$$\ln xy + y = c$$ ◂◂

EXAMPLE 2 ▸▸ Solve the differential equation $y\,dx - x\,dy + x\,dx = 0$.

The combination of $y\,dx - x\,dy$ suggests that this equation might make use of either Eq. (30-5) or (30-6). This would require dividing through by x^2 or y^2. If we divide by y^2, the last term cannot be integrated, but ***division by x^2 still allows integration of the last term.*** Performing this division, we obtain

CAUTION ▸

$$\frac{y\,dx - x\,dy}{x^2} + \frac{dx}{x} = 0$$

This left combination is the negative of Eq. (30-5). Thus we have

$$-d\left(\frac{y}{x}\right) + \frac{dx}{x} = 0$$

for which the solution is $-\dfrac{y}{x} + \ln x = c$. ◂◂

EXAMPLE 3 ▸▸ Solve the differential equation $(x^2 + y^2 + x)\,dx + y\,dy = 0$.

Regrouping the terms of this equation, we have

$$(x^2 + y^2)\,dx + (x\,dx + y\,dy) = 0$$

By dividing through by $x^2 + y^2$, we have

$$dx + \frac{x\,dx + y\,dy}{x^2 + y^2} = 0$$

The right term now can be put in the form of du/u (with $u = x^2 + y^2$) by multiplying each of the terms of the numerator by 2. This leads to

$$dx + \left(\frac{1}{2}\right)\frac{\overbrace{2x\,dx + 2y\,dy}^{d(x^2+y^2)}}{x^2 + y^2} = 0$$

$$x + \frac{1}{2}\ln(x^2 + y^2) = \frac{c}{2} \quad \text{or} \quad 2x + \ln(x^2 + y^2) = c$$ ◂◂

EXAMPLE 4 ▸▸ Find the particular solution of the differential equation

$$(x^3 + xy^2 + 2y)\,dx + (y^3 + x^2y + 2x)\,dy = 0$$

which satisfies the condition that $x = 1$ when $y = 0$.

Regrouping the terms of the equation, we have

$$x(x^2 + y^2)\,dx + y(x^2 + y^2)\,dy + 2(y\,dx + x\,dy) = 0$$

Factoring $x^2 + y^2$ from each of the first two terms gives

$$(x^2 + y^2)(x\,dx + y\,dy) + 2(y\,dx + x\,dy) = 0$$

$$\frac{1}{2}(x^2 + y^2)\overbrace{(2x\,dx + 2y\,dy)}^{d(x^2+y^2)} + 2\overbrace{(y\,dx + x\,dy)}^{d(xy)} = 0$$

$$\frac{1}{2}\left(\frac{1}{2}\right)(x^2 + y^2)^2 + 2xy + \frac{c}{4} = 0 \quad \text{integrating}$$

$$(x^2 + y^2)^2 + 8xy + c = 0$$

Using the given condition gives $(1 + 0)^2 + 0 + c = 0$, or $c = -1$. The particular solution is then

$$(x^2 + y^2)^2 + 8xy = 1$$ ◂◂

NOTE ▸ The use of integrable combinations depends on proper recognition of the forms. It may take two or three arrangements to find the combination that leads to the solution. Of course, many equations cannot be arranged so as to give integrable combinations in all terms.

Exercises 30–3

In Exercises 1–16, solve the given differential equations.

1. $x\,dy + y\,dx + x\,dx = 0$

2. $(2y + x)\,dy + y\,dx = 0$

3. $y\,dx - x\,dy + x^3\,dx = 2\,dx$

4. $x\,dy - y\,dx + y^2\,dx = 0$

5. $A^3\,dr + A^2r\,dA + r\,dA - A\,dr = 0$

6. $\sec(xy)\,dx + (x\,dy + y\,dx) = 0$

7. $x^3y^4(x\,dy + y\,dx) = 3\,dy$

8. $x\,dy + y\,dx + 4xy^3\,dy = 0$

9. $\sqrt{x^2 + y^2}\,dx - 2y\,dy = 2x\,dx$

10. $R\,dR + (R^2 + T^2 + T)\,dT = 0$

11. $\tan(x^2 + y^2)\,dy + x\,dx + y\,dy = 0$

12. $(x^2 + y^3)^2\,dy + 2x\,dx + 3y^2\,dy = 0$

13. $y\,dy - x\,dx + (y^2 - x^2)\,dx = 0$

14. $e^{x+y}(dx + dy) + 4x\,dx = 0$

15. $10x\,dy + 5y\,dx + 3y\,dy = 0$

16. $x^2\,dy + 3xy\,dx + 2\,dx = 0$

In Exercises 17–20, find the particular solutions to the given differential equations that satisfy the given conditions.

17. $2(x\,dy + y\,dx) + 3x^2\,dx = 0$; $x = 1$ when $y = 2$

18. $t\,dt + s\,ds = 2(t^2 + s^2)\,dt$; $t = 1$ when $s = 0$

19. $y\,dx - x\,dy = y^3\,dx + y^2x\,dy$; $x = 2$ when $y = 4$

20. $e^{x/y}(x\,dy - y\,dx) = y^4\,dy$; $x = 0$ when $y = 2$

30-4 The Linear Differential Equation of the First Order

There is one type of differential equation of the first order and first degree for which an integrable combination can always be found. *It is the* **linear differential equation** *of the first order and is of the form*

$$dy + Py\,dx = Q\,dx \tag{30-7}$$

where P and Q are functions of x only. This type of equation occurs widely in applications.

If each side of Eq. (30-7) is multiplied by $e^{\int P\,dx}$ it becomes integrable, since the left side becomes of the form du with $u = ye^{\int P\,dx}$ and the right side is a function of x only. This is shown by finding the differential of $ye^{\int P\,dx}$. Thus

$$d(ye^{\int P\,dx}) = e^{\int P\,dx}(dy + Py\,dx)$$

In finding the differential of $\int P\,dx$ we use the fact that, by definition, these are reverse processes. Thus, $d(\int P\,dx) = P\,dx$. Therefore, if each side is multiplied by $e^{\int P\,dx}$, the left side may be immediately integrated to $ye^{\int P\,dx}$ and the right-side integration may be indicated. The solution becomes

$$ye^{\int P\,dx} = \int Qe^{\int P\,dx}\,dx + c \tag{30-8}$$

EXAMPLE 1 ▸▸ Solve the differential equation $dy + \left(\frac{2}{x}\right)y\,dx = 4x\,dx$.

This equation fits the form of Eq. (30-7) with $P = 2/x$ and $Q = 4x$. The first expression to find is $e^{\int P\,dx}$. In this case this is

$$e^{\int(2/x)dx} = e^{2\ln x} = e^{\ln x^2} = x^2 \qquad \text{see text comments following example}$$

The left side integrates to yx^2, while the right side becomes $\int 4x(x^2)\,dx$. Thus

$$ye^{\int P\,dx} = \int Qe^{\int P\,dx}\,dx + c$$

$$y(x^2) = \int (4x)(x^2)\,dx + c \qquad \text{using Eq. (30-8)}$$

$$yx^2 = \int 4x^3\,dx + c = x^4 + c \qquad \text{integrating}$$

$$y = x^2 + cx^{-2}$$ ◂◂

As in Example 1, in finding the factor $e^{\int P\,dx}$ we often obtain an expression of the form $e^{\ln u}$. Using the properties of logarithms, we now show that $e^{\ln u} = u$.

NOTE ▶

$$\text{Let } y = e^{\ln u}$$
$$\ln y = \ln e^{\ln u} = \ln u(\ln e) = \ln u$$
$$y = u, \quad \text{or} \quad e^{\ln u} = u$$

NOTE ▶ Also, in finding $e^{\int P\,dx}$, the constant of integration in the exponent $\int P\,dx$ can always be taken as zero, as we did in Example 1. To show why this is so, let $P = 2/x$ as in Example 1.

$$e^{\int (2/x)\,dx} = e^{\ln x^2 + c} = (e^{\ln x^2})(e^c) = x^2 e^c$$

The solution to the differential equation, as given in Eq. (30-8), is then

$$y(x^2)(e^c) = \int 4x(x^2)(e^c)\,dx + c_1 e^c$$

Regardless of the value of c, the factor e^c can be divided out. Therefore, it is convenient to let $c = 0$ and have $e^c = 1$.

EXAMPLE 2 ▶▶ Solve the differential equation $x\,dy - 3y\,dx = x^3\,dx$.

Putting this equation in the form of Eq. (30-7) by dividing through by x gives $dy - (3/x)y\,dx = x^2\,dx$. Here $P = -3/x$, $Q = x^2$, and the factor $e^{\int P\,dx}$ becomes

$$e^{\int(-3/x)\,dx} = e^{-3\ln x} = e^{\ln x^{-3}} = x^{-3}$$

Therefore,

$$\underbrace{ye^{\int P\,dx}} = \int Q\underbrace{e^{\int P\,dx}}\,dx + c$$
$$yx^{-3} = \int x^2(x^{-3})\,dx + c \quad \text{using Eq. (30-8)}$$
$$= \int x^{-1}\,dx + c = \ln x + c$$
$$y = x^3(\ln x + c) \quad ◀◀$$

EXAMPLE 3 ▶▶ Solve the differential equation $dy + y\,dx = x\,dx$.

Here, $P = 1$, $Q = x$, and $e^{\int P\,dx} = e^{\int (1)dx} = e^x$

Therefore,

$$ye^x = \int xe^x\,dx + c = e^x(x - 1) + c \quad \text{using Eq. (30-8) and integrating by parts or tables}$$
$$y = x - 1 + ce^{-x} \quad ◀◀$$

EXAMPLE 4 ▶▶ Solve the differential equation $x^2\,dy + 2xy\,dx = \sin x\,dx$.

This equation is first written in the form of Eq. (30-7). This gives us

$$dy + \left(\frac{2}{x}\right)y\,dx = \frac{1}{x^2}\sin x\,dx$$

This shows us that $P = 2/x$ and $e^{\int P\,dx} = e^{\int (2/x)dx} = x^2$. Therefore,

$$yx^2 = \int \sin x\,dx + c = -\cos x + c \quad \text{using Eq. (30-8)}$$
$$yx^2 + \cos x = c \quad ◀◀$$

EXAMPLE 5 ▸▸ Solve the differential equation $\cos x \dfrac{dy}{dx} = 1 - y \sin x$.

Writing this in the form of Eq. (30-7), we have

$$dy + y \tan x\, dx = \sec x\, dx \qquad \text{dividing by } \cos x$$

Thus, with $P = \tan x$, we have

$$e^{\int P\,dx} = e^{\int \tan x\,dx} = e^{-\ln \cos x} = \sec x \qquad \text{see the first NOTE after Example 1}$$

The solution is

$$y \sec x = \int \sec^2 x\, dx = \tan x + c \qquad \text{using Eq. (30-8)}$$

$$y = \sin x + c \cos x$$ ◂◂

EXAMPLE 6 ▸▸ For the differential equation $dy = (1 - 2y)x\,dx$, find the particular solution such that $x = 0$ when $y = 2$.

The solution proceeds as follows:

$$dy + 2xy\,dx = x\,dx \qquad \text{form of Eq. (30-7)}$$

$$e^{\int P\,dx} = e^{\int 2x\,dx} = e^{x^2} \qquad \text{find } e^{\int P\,dx}$$

$$ye^{x^2} = \int xe^{x^2}\,dx \qquad \text{using Eq. (30-8)}$$

$$= \frac{1}{2}e^{x^2} + c \qquad \text{general solution}$$

$$(2)(e^0) = \frac{1}{2}(e^0) + c \qquad 2 = \frac{1}{2} + c \qquad c = \frac{3}{2} \qquad x = 0,\ y = 2;\ \text{evaluate } c$$

$$ye^{x^2} = \frac{1}{2}e^{x^2} + \frac{3}{2} \qquad \text{substitute } c = \frac{3}{2}$$

$$y = \frac{1}{2}(1 + 3e^{-x^2}) \qquad \text{particular solution}$$ ◂◂

Exercises 30–4

In Exercises 1–24, solve the given differential equations.

1. $dy + y\,dx = e^{-x}\,dx$
2. $dy + 3y\,dx = e^{-3x}\,dx$
3. $dy + 2y\,dx = e^{-4x}\,dx$
4. $di + i\,dt = e^{-t}\cos t\,dt$
5. $\dfrac{dy}{dx} - 2y = 4$
6. $2\dfrac{dy}{dx} = 5 - 6y$
7. $x\,dy - y\,dx = 3x\,dx$
8. $x\,dy + 3y\,dx = dx$
9. $2x\,dy + y\,dx = 8x^3\,dx$
10. $3x\,dy - y\,dx = 9x\,dx$
11. $dr + r\cot\theta\,d\theta = d\theta$
12. $y' = x^2y + 3x^2$
13. $\sin x\dfrac{dy}{dx} = 1 - y\cos x$
14. $\dfrac{dy}{dx} - \dfrac{y}{x} = \ln x$
15. $y' + y = 3$
16. $y' + 2y = \sin x$
17. $ds = (te^{4t} + 4s)dt$
18. $y' - 2y = 2e^{2x}$
19. $y' = x^3(1 - 4y)$
20. $y' + y\tan x = -\sin x$
21. $x\dfrac{dy}{dx} = y + (x^2 - 1)^2$
22. $2x(dy - dx) + y\,dx = 0$
23. $x\,dy + (1 - 3x)y\,dx = 3x^2e^{3x}\,dx$
24. $(1 + x^2)\,dy + xy\,dx = x\,dx$

In Exercises 25 and 26, solve the given differential equations. Explain how each can be solved using either of two different methods.

25. $y' = 2(1 - y)$
26. $x\,dy = (2x - y)dx$

In Exercises 27–32, find the indicated particular solutions of the given differential equations.

27. $\dfrac{dy}{dx} + 2y = e^{-x}$; $x = 0$ when $y = 1$
28. $dq - 4q\,du = 2\,du$; $q = 2$ when $u = 0$
29. $y' + 2y\cot x = 4\cos x$; $x = \pi/2$ when $y = 1/3$
30. $y'\sqrt{x} + \frac{1}{2}y = e^{\sqrt{x}}$; $x = 1$ when $y = 3$
31. $(\sin x)y' + y = \tan x$; $x = \pi/4$ when $y = 0$
32. $f(x)\,dy + 2yf'(x)\,dx = f(x)f'(x)\,dx$; $f(x) = -1$ when $y = 3$

30-5 Elementary Applications

The differential equations of the first order and first degree we have discussed thus far have numerous applications in geometry and the various fields of technology. In this section we illustrate some of these applications.

EXAMPLE 1 The slope of a given curve is given by the expression $6xy$. Find the equation of the curve if it passes through the point (2, 1).

Since the slope is $6xy$, the differential equation for the curve is

$$\frac{dy}{dx} = 6xy$$

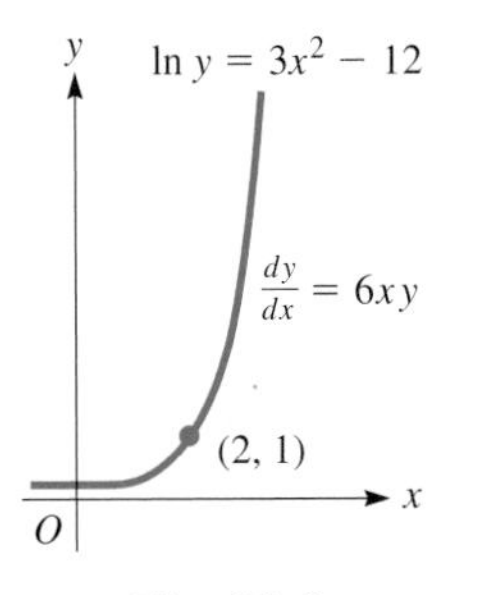

Fig. 30-2

We now want to find the particular solution of this equation for which $x = 2$ when $y = 1$. The solution follows:

$$\frac{dy}{y} = 6x\,dx \qquad \text{separate variables}$$

$$\ln y = 3x^2 + c \qquad \text{general solution}$$

$$\ln 1 = 3(2^2) + c \qquad \text{evaluate } c$$

$$0 = 12 + c, \qquad c = -12$$

$$\ln y = 3x^2 - 12 \qquad \text{particular solution}$$

The graph of this solution is shown in Fig. 30-2.

EXAMPLE 2 *A curve that intersects all members of a family of curves at right angles is called an* **orthogonal trajectory** *of the family.* Find the equations of the orthogonal trajectories of the parabolas $x^2 = cy$. As before, each value of c gives us a particular member of the family.

CAUTION ▶ The derivative of the given equation is $dy/dx = 2x/c$. This equation contains the constant c, which depends on the point (x, y) on the parabola. ***Eliminating this constant between the equations of the parabolas and the derivative***, we have

$$c = \frac{x^2}{y}; \quad \frac{dy}{dx} = \frac{2x}{c} = \frac{2x}{x^2/y} \quad \text{or} \quad \frac{dy}{dx} = \frac{2y}{x}$$

This equation gives a general expression for the slope of any of the members of the family. For a curve to be perpendicular, its slope must equal the negative reciprocal of this expression, or the slope of the orthogonal trajectories must be

$$\left.\frac{dy}{dx}\right|_{\text{OT}} = -\frac{x}{2y} \longleftarrow \text{this equation must not contain the constant } c$$

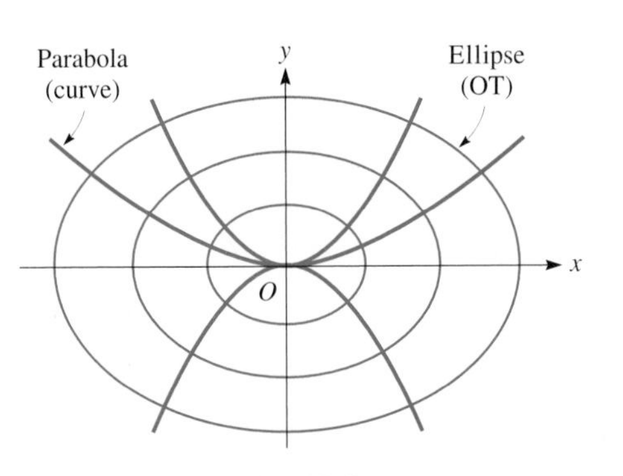

Fig. 30-3

Solving this differential equation gives the family of orthogonal trajectories.

$$2y\,dy = -x\,dx$$

$$y^2 = -\frac{x^2}{2} + \frac{c}{2}$$

$$2y^2 + x^2 = c \qquad \text{orthogonal trajectories}$$

Thus the orthogonal trajectories are ellipses. Note in Fig. 30-3 that each parabola intersects each ellipse at right angles.

See the chapter introduction.

EXAMPLE 3 ▸▸ Radioactive elements decay at rates that are proportional to the amount of the element present. Uranium 231, which is found among the fission products of nuclear reactors, changes form such that one-half of an original amount decays into other elements in 4.2 days. Determine the equation relating the amount present with the time, and determine the fraction that remains after a week (7.0 days).

Let N_0 be the original amount and N be the amount present at any time t (in days). Since we know information related to the rate of decay, we express this rate as a derivative. Therefore, since the rate of change of N is proportional to N, we have the equation

$$\frac{dN}{dt} = kN$$

Solving this differential equation, we have

$$\begin{aligned}
\frac{dN}{N} &= k\,dt && \text{separate variables} \\
\ln N &= kt + \ln c && \text{general solution} \\
\ln N_0 &= k(0) + \ln c && N = N_0 \text{ for } t = 0 \\
c &= N_0 && \text{solve for } c \\
\ln N &= kt + \ln N_0 && \text{substitute } N_0 \text{ for } c \\
\ln N - \ln N_0 &= kt && \text{use properties of logarithms} \\
\ln \frac{N}{N_0} &= kt \\
N &= N_0 e^{kt} && \text{exponential form}
\end{aligned}$$

Now, using the condition that half this isotope decays in 4.2 days, we have $N = N_0/2$ when $t = 4.2$ days. This gives

$$\frac{N_0}{2} = N_0 e^{4.2k} \quad \text{or} \quad \frac{1}{2} = (e^k)^{4.2}$$

CAUTION ▸

$$e^k = 0.5^{1/4.2} = 0.5^{0.24}$$

Therefore, the equation relating N and t is

$$N = N_0(0.5)^{0.24t}$$

In order to determine the fraction remaining after 7.0 days, we evaluate this equation for $t = 7.0$. This gives

$$\begin{aligned}
N &= N_0(0.5)^{0.24(7.0)} = N_0(0.5)^{1.68} \\
&= 0.31N_0
\end{aligned}$$

This tells us that about 31% of the original amount of uranium 231 remains unchanged after a week. ◂◂

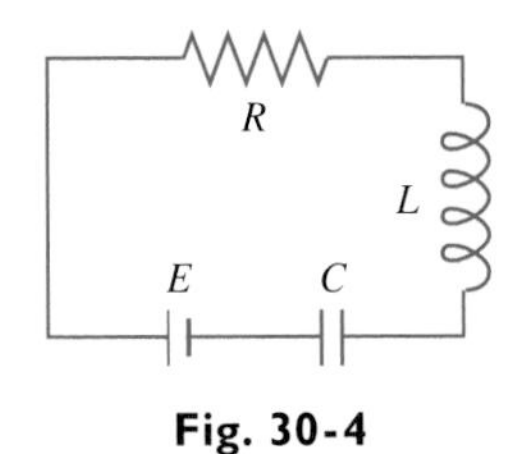

Fig. 30-4

EXAMPLE 4 ▸▸ The general equation relating the current i, voltage E, inductance L, capacitance C, and resistance R of a simple electric circuit (see Fig. 30-4) is

$$L\frac{di}{dt} + Ri + \frac{q}{C} = E \tag{30-9}$$

where q is the charge on the capacitor. Find the general expression for the current in a circuit containing an inductance, a resistance, and a voltage source if $i = 0$ when $t = 0$.

The differential equation for this circuit is

$$L\frac{di}{dt} + Ri = E$$

Using the method of the linear differential equation of the first order, we have the equation

$$di + \frac{R}{L}i\,dt = \frac{E}{L}dt$$

The factor $e^{\int P\,dt}$ is $e^{\int (R/L)dt} = e^{(R/L)t}$. This gives

$$ie^{(R/L)t} = \frac{E}{L}\int e^{(R/L)t}\,dt = \frac{E}{R}e^{(R/L)t} + c$$

Letting the current be zero for $t = 0$, we have $c = -E/R$. The result is

$$ie^{(R/L)t} = \frac{E}{R}e^{(R/L)t} - \frac{E}{R}$$

$$i = \frac{E}{R}(1 - e^{-(R/L)t})$$

(We can see that $i \to E/R$ as $t \to \infty$. In practice the exponential term becomes negligible very quickly.) ◂◂

EXAMPLE 5 ▸▸ Fifty litres of brine originally containing 3.00 kg of salt are in a tank into which 2.00 L of water run each minute with the same amount of mixture running out each minute. How much salt is in the tank after 10.0 min?

CAUTION ▸

Let x = the number of kilograms of salt in the tank after t minutes. Each litre of brine contains $x/50$ kg of salt, and in time dt, *2 dt litres of mixture leave the tank* ***with (x/50) (2 dt) kg*** *of salt.* The amount of salt that is leaving may also be written as $-dx$ (the minus sign is included to show that x is decreasing). Thus

$$-dx = \frac{2x\,dt}{50}, \quad \text{or} \quad \frac{dx}{x} = -\frac{dt}{25}$$

This leads to $\ln x = -(t/25) + \ln c$. Using the fact that $x = 3.00$ kg when $t = 0$, we find that $\ln 3.00 = \ln c$, or $c = 3.00$. Therefore,

$$x = 3.00e^{-t/25}$$

is the general expression for the amount of salt in the tank at time t. Therefore, when $t = 10.0$ min, we have

$$x = 3.00e^{-10/25} = 3.00e^{-0.4} = 3.00(0.670) = 2.01 \text{ kg}$$

There are 2.01 kg of salt in the tank after 10.0 min. (Although the data were given with three significant digits, we did not use all significant digits in writing the equations that were used.) ◂◂

EXAMPLE 6 ▸▸ An object moving through (or across) a resisting medium often experiences a retarding force that is approximately proportional to the velocity as well as the force that causes the motion. An example of this is a ball that falls due to the force of gravity and air resistance produces a retarding force. Applying Newton's laws of motion (from physics) to this ball leads to the equation

$$m\frac{dv}{dt} = F - kv \qquad \textbf{(30-10)}$$

where m is the mass of the object, v is the velocity of the object, t is the time, F is the force causing the motion, and k $(k > 0)$ is a constant. The quantity kv is the retarding force.

We assume that these conditions hold for a certain falling object whose mass is 5.00 kg and which experiences a force (its own weight) of 49.0 N. The object starts from rest, and the air causes a retarding force that is numerically equal to 0.200 times the velocity.

Substituting in Eq. (30-10), we have

$$5\frac{dv}{dt} = 49 - 0.2v$$

Solving this differential equation, we have

$$\frac{5\,dv}{49 - 0.2v} = dt \qquad \text{separate variables}$$

$$-25\ln(49 - 0.2v) = t - 25\ln c \qquad \text{integrate}$$

$$\ln(49 - 0.2v) = -\frac{t}{25} + \ln c \qquad \text{solve for } v$$

$$49 - 0.2v = ce^{-t/25}$$

$$v = 5(49 - ce^{-t/25}) \qquad \text{general solution}$$

Since the object started from rest, $v = 0$ when $t = 0$. Thus

$$0 = 5(49 - c), \quad \text{or} \quad c = 49 \qquad \text{evaluate } c$$

$$v = 245(1 - e^{-t/25}) \qquad \text{particular solution}$$

Evaluating v for $t = 5.00$ s, we have

$$v = 245(1 - e^{-0.200}) = 245(1 - 0.8187) = 44.4 \text{ m/s}$$

Therefore, after 5.00 s the velocity of the object is 44.4 m/s. Without the air resistance, the velocity of the object would be about 49.0 m/s after 5.00 s. (The data were given to three significant digits, but all significant digits were not used in writing the equations.) ◂◂

Exercises 30–5

In Exercises 1–4, find the equation of the curve for the given slope and point through which it passes. Use a graphing calculator to display the curve.

1. Slope given by $2x/y$; passes through $(2, 3)$
2. Slope given by $-y/(x + y)$; passes through $(-1, 3)$
3. Slope given by $y + x$; passes through $(0, 1)$
4. Slope given by $-2y + e^{-x}$; passes through $(0, 2)$

In Exercises 5–8, find the equation of the orthogonal trajectories of the curves for the given equations. Use a graphing calculator to display at least two members of the family of curves and at least two of the orthogonal trajectories.

5. The exponential curves $y = ce^x$

6. The cubic curves $y = cx^3$

7. The curves $y = c(\sec x + \tan x)$

8. The family of circles, all with centers at the origin

In Exercises 9–40, solve the given problems by solving the appropriate differential equation.

9. The isotope neon 23 decays such that half of an original amount disintegrates in 40.0 s. Find the relation between the amount present and the time, then find the percent remaining after 60.0 s.

10. Radium 226 decays such that 10% of an original amount disintegrates in 246 years. Find the half-life (the time for one-half of it to disintegrate) of radium 226.

11. Carbon 14 has a half-life of about 5730 years. The analysis of some wood at the site of the ancient city of Troy showed that the concentration of carbon 14 was 67.8% of the concentration which new wood would have. How old is this wood at the site of Troy? (This method, called *carbon dating*, is used to determine the dates of prehistoric events.)

12. A radioactive element leaks from a nuclear power plant at a constant rate r, and it decays at a rate proportional to the amount present. Find the relation between the amount N present in the environment and the time t, if $N = 0$ for $t = 0$.

13. The rate of change of the radial stress S on the walls of a pipe with respect to the distance r from the axis of the pipe is given by $r\dfrac{dS}{dr} = 2(a - S)$, where a is a constant. Solve for S as a function of r.

14. The velocity v of a meteor approaching the earth is given by $v\dfrac{dv}{dr} = -\dfrac{GM}{r^2}$, where r is its distance from the center of the earth, M is the mass of the earth, and G is a universal gravitational constant. If $v = 0$ for $r = r_0$, solve for v as a function of r.

15. Assume that the rate at which highway construction increases is directly proportional to the total mileage M of all highways already completed at time t (in years). Solve for M as a function of t if $M = 5250$ km for a certain county when $t = 0$ and $M = 5460$ km for $t = 2.00$ years.

16. The marginal profit function gives the change in the total profit P of a business due to a change in the business, such as adding new machinery. A company determines that the marginal profit dP/dx is $e^{-x^2} - 2Px$, where x is the amount invested in new machinery. Determine the total profit (in thousands of dollars) as a function of x, if $P = 0$ for $x = 0$.

17. According to Newton's law of cooling, the rate at which a body cools is proportional to the difference in temperature between it and the surrounding medium. How long will it take a cup of hot water, initially at 90°C, to cool to 40°C if the room temperature is 25°C, if it cools to 60°C in 5.0 min?

18. An object whose temperature is 100°C is placed in a medium whose temperature is 20°C. The temperature of the object falls to 50°C in 10 min. Express the temperature, T, of the object as a function of time. (See Exercise 17.)

19. If interest in a bank account is compounded continuously, the amount grows at a rate that is proportional to the amount present. An account earning daily interest very closely approximates this situation. Determine the amount in an account after one year if \$1000 is placed in an account that pays 4% interest per year, compounded continuously.

20. If interest is compounded continuously (see Exercise 19), how long will it take a bank account to double in value if the rate of interest is 5% per year?

21. If the current in an RL circuit with a voltage source E is zero when $t = 0$ (see Example 4), show that $\lim_{t\to\infty} i = E/R$. See Fig. 30-5.

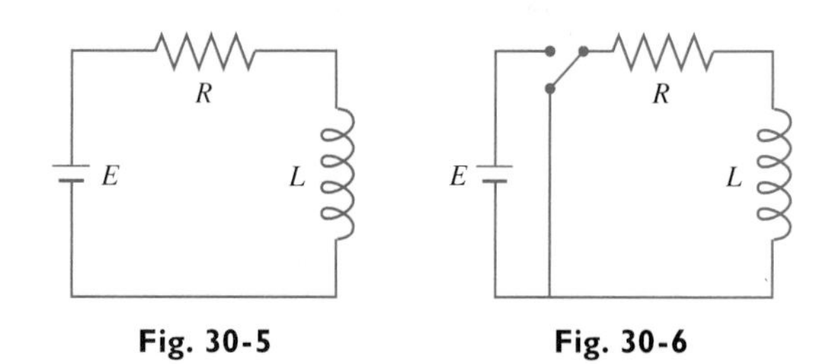

Fig. 30-5 **Fig. 30-6**

22. If a circuit contains only an inductance and a resistance, with $L = 2.0$ H and $R = 30\ \Omega$, find the current i as a function of the time t if $i = 0.020$ A when $t = 0$. See Fig. 30-6.

23. An amplifier circuit contains a resistance R, an inductance L, and a voltage source $E \sin \omega t$. Express the current in the circuit as a function of the time t if the initial current is zero.

24. A radio transmitter circuit contains a resistance of 2.0 Ω, a variable inductor of $100 - t$ henrys, and a voltage source of 4.0 V. Find the current i in the circuit as a function of the time t for $0 \le t \le 100$ s if the initial current is zero.

25. If a circuit contains only a resistance and a capacitance, find the expression relating the charge on the capacitor in terms of the time if $i = dq/dt$ and $q = q_0$ when $t = 0$. See Fig. 30-7.

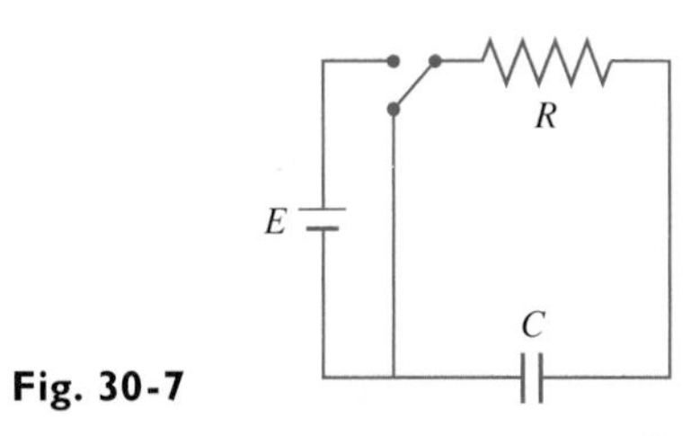

Fig. 30-7

26. A circuit contains a 4.0-μF capacitor, a 450-Ω resistor, and a 20.0-mV voltage source. If the charge on the capacitor is 20.0 nC when $t = 0$, find the charge after 0.010 s. ($i = dq/dt$.)

27. One hundred litres of brine originally containing 4.0 kg of salt are in a tank into which 5.0 L of water run each minute. The same amount of mixture from the tank leaves each minute. How much salt is in the tank after 20 min?

28. Repeat Exercise 27 with the change that the water entering the tank contains 0.10 kg of salt per litre.

29. An object falling under the influence of gravity has a variable acceleration given by $9.8 - v$, where v represents the velocity. If the object starts from rest, find an expression for the velocity in terms of the time. Also find the limiting value of the velocity (find $\lim_{t \to \infty} v$).

30. In a ballistics test, a bullet is fired into a sandbag. The acceleration of the bullet is $-15\sqrt{v}$, where v is the velocity in m/s. When will the bullet stop if it enters the sandbag at 300 m/s?

31. A boat with a mass of 150 kg is being towed at 8.0 km/h. The tow rope is then cut, and a motor that exerts a force of 80 N on the boat is started. If the water exerts a retarding force that numerically equals twice the velocity, what is the velocity of the boat 3.0 min later?

32. A parachutist is falling at the rate of 60.0 m/s when her parachute opens. If the air resists the fall with a force equal to $5v^2$, find the velocity as a function of time. The woman and her equipment have a combined mass of 100 kg (weight is 980 N).

33. For each cycle, a roller mechanism follows a path described by $y = 2x - x^2$, $y \ge 0$, such that $dx/dt = 6t - 3t^2$. Find x and y (in cm) in terms of t (in seconds) if x and y are zero for $t = 0$.

34. In studying the flow of water in a stream, it is found that an object follows the hyperbolic path $y(x + 1) = 10$ such that $(t + 1)dx = (x - 2)dt$. Find x and y (in m) in terms of t (in seconds) if $x = 4$ m and $y = 2$ m for $t = 0$.

35. The rate of change of air pressure with respect to height is approximately proportional to the pressure. If the pressure is 100 kPa when $h = 0$, and $p = 80$ kPa when $h = 2000$ m, find the expression relating pressure and height.

36. Water flows from a vertical cylindrical storage tank through a hole of area A at the bottom of the tank. The rate of flow is $2.6\,A\sqrt{h}$, where h is the distance from the surface of the water to the hole. If h changes from 9.0 m to 8.0 m in 16 min, how long will it take the tank to empty?

37. Assume that the rate of depreciation of an object is proportional to its value at any time t. If a car costs \$16 500 new, and its value 3 years later is \$9850, what is its value 11 years after it was purchased?

38. Assume that sugar dissolves at a rate proportional to the undissolved amount. If there are initially 525 g of sugar and 225 g remain after 4.00 min, how long does it take to dissolve 375 g?

39. Fresh air is being circulated into a room whose volume is 120 m³. Under specified conditions the number of cubic metres x of carbon dioxide present at any time t is found by solving the differential equation $\frac{dx}{dt} = 1 - 5.0\,x$. Find x as a function of t if $x = 0.35$ m³ when $t = 0$ min.

40. The lines of equal potential in a field of force are all at right angles to the lines of force. In an electric field of force caused by charged particles, the lines of force are given by $x^2 + y^2 = cx$. Find the equation of the lines of equal potential. Use a graphing calculator to view a few members of the lines of force and those of equal potential.

30-6 Second-Order Homogeneous Equations

Higher-Order Differential Equations

Another important type of differential equation is the linear differential equation of the second order with constant coefficients. Before restricting our attention to this specific type, we shall briefly describe the general higher-order equation and the notation we shall use with this type of equation.

The general **linear differential equation of the *n*th order** *is of the form*

$$a_0\frac{d^ny}{dx^n} + a_1\frac{d^{n-1}y}{dx^{n-1}} + \cdots + a_{n-1}\frac{dy}{dx} + a_ny = b \tag{30-11}$$

where the a's and b are either functions of x or constants.

For convenience of notation, the nth derivative with respect to x will be denoted by D^n. Here D is referred to as an **operator**, since it denotes the *operation* of differentiation with respect to x. Using this notation, the general linear differential equation becomes

$$a_0D^ny + a_1D^{n-1}y + \cdots + a_{n-1}Dy + a_ny = b \tag{30-12}$$

If $b = 0$ the general linear equation is called **homogeneous**, *and if $b \neq 0$ it is called* **nonhomogeneous**. Both types of equations have important applications.

EXAMPLE 1 ▸▸ Using the operator form of Eq. (30-12), the differential equation

$$\frac{d^3y}{dx^3} - 3\frac{d^2y}{dx^2} + 4\frac{dy}{dx} - 2y = e^x \sec x$$

is written as

$$D^3y - 3D^2y + 4Dy - 2y = e^x \sec x$$

This equation is nonhomogeneous since $b = e^x \sec x$. ◂◂

Although the a's may be functions of x, we shall restrict our attention to the case where they are all constants. This section and the following three sections are devoted to second-order homogeneous and nonhomogeneous equations, although the methods may be applied to equations of higher order.

Second-Order Homogeneous Equations with Constant Coefficients

Using the operator notation, *a second-order, linear, homogeneous differential equation with constant coefficients is one of the form*

$$a_0D^2y + a_1Dy + a_2y = 0 \tag{30-13}$$

where the a's are constants. The following example indicates the kind of solution we should expect for this type of equation.

EXAMPLE 2 ▸▸ Solve the differential equation $D^2y - Dy - 2y = 0$.

First we put this equation in the form $(D^2 - D - 2)y = 0$. This is another way of saying that we are to take the second derivative of y, subtract the first derivative, and finally subtract twice the function. This expression may now be factored as $(D - 2)(D + 1)y = 0$. (We shall not develop the algebra of the operator D. However, most such algebraic operations can be shown to be valid.) This formula tells us to find the first derivative of the function and add this to the function. Then twice this result is to be subtracted from the derivative of this result. If we let $z = (D + 1)y$, which is valid since $(D + 1)y$ is a function of x, we have $(D - 2)z = 0$. This equation is easily solved by separation of variables. Thus

$$\frac{dz}{dx} - 2z = 0, \qquad \frac{dz}{z} - 2\,dx = 0, \qquad \ln z - 2x = \ln c_1$$

$$\ln\frac{z}{c_1} = 2x \quad \text{or} \quad z = c_1e^{2x}$$

Replacing z by $(D + 1)y$, we have

$$(D + 1)y = c_1e^{2x}$$

This is a linear equation of the first order. Then

$$dy + y\,dx = c_1e^{2x}\,dx$$

The factor $e^{\int P\,dx}$ is $e^{\int dx} = e^x$. And so

$$ye^x = \int c_1e^{3x}\,dx = \frac{c_1}{3}e^{3x} + c_2 \qquad \text{using Eq. (30-8)}$$

or

$$y = c_1'e^{2x} + c_2e^{-x}$$

where $c_1' = \frac{1}{3}c_1$. This example indicates that solutions of the form e^{mx} result for this equation. ◂◂

Based on the result of Example 2, assume that an equation of the form of Eq. (30-13) has a particular solution ce^{mx}. Substituting this into Eq. (30-13) gives

$$a_0cm^2e^{mx} + a_1cme^{mx} + a_2ce^{mx} = 0$$

Since the exponential function $e^{mx} > 0$ for all real x, this equation will be satisfied if m is a root of the equation

Auxiliary Equation

$$a_0m^2 + a_1m + a_2 = 0 \qquad \textbf{(30-14)}$$

Equation (30-14) is called the **auxiliary equation** *of Eq. (30-13). Note that it may be formed directly by inspection from Eq. (30-13).*

There are two roots of the auxiliary Eq. (30-14), and there are two arbitrary constants in the solution of Eq. (30-13). These factors lead us to *the general solution of Eq. (30-13), which is*

General Solution

$$y = c_1e^{m_1x} + c_2e^{m_2x} \qquad \textbf{(30-15)}$$

where m_1 and m_2 are the solutions of Eq. (30-14). We see that this is in agreement with the results of Example 2.

EXAMPLE 3 ▸▸ Solve the differential equation $D^2y - 5Dy + 6y = 0$.

From this operator form of the differential equation, we write the auxiliary equation

$$m^2 - 5m + 6 = 0$$

Solving the auxiliary equation, we have

$$(m - 3)(m - 2) = 0$$

$$m_1 = 3, \qquad m_2 = 2$$

Now, using Eq. (30-15), we write the solution of the differential equation as

$$y = c_1e^{3x} + c_2e^{2x}$$

Obviously it makes no difference which constant is written with each exponential function. ◂◂

EXAMPLE 4 ▸▸ Solve the differential equation $y'' = 6y'$.

We first rewrite this equation using the D notation for derivatives. Also, we want to write it in the proper form of a homogeneous equation. This gives us

$$D^2y - 6Dy = 0$$

Proceeding with the solution, we have

$$m^2 - 6m = 0 \qquad \text{auxiliary equation}$$

$$m(m - 6) = 0 \qquad \text{solve for } m$$

$$m_1 = 0, \qquad m_2 = 6$$

$$y = c_1e^{0x} + c_2e^{6x} \qquad \text{using Eq. (30-15)}$$

Since $e^{0x} = 1$, we have

$$y = c_1 + c_2e^{6x} \qquad \text{general solution}$$ ◂◂

EXAMPLE 5 ▸▸ Solve the differential equation $2\dfrac{d^2y}{dx^2} + \dfrac{dy}{dx} - 7y = 0$.

First, writing this equation with the D notation for derivatives, we have

$$2D^2y + Dy - 7y = 0$$

This means that the auxiliary equation is

$$2m^2 + m - 7 = 0$$

Since this equation is not factorable, we solve it by using the quadratic formula. This gives us

$$m = \frac{-1 \pm \sqrt{1 + 56}}{4} = \frac{-1 \pm \sqrt{57}}{4}$$

Therefore, using Eq. (30-15), we have

$$y = c_1e^{\frac{-1+\sqrt{57}}{4}x} + c_2e^{\frac{-1-\sqrt{57}}{4}x}$$

A somewhat better form can be obtained by observing that $e^{-x/4}$ is a factor of each term. Thus

$$y = e^{-x/4}(c_1e^{\sqrt{57}x/4} + c_2e^{-\sqrt{57}x/4})$$ ◂◂

EXAMPLE 6 ▸▸ Solve the differential equation $D^2y - 2Dy - 15y = 0$, and find the particular solution which satisfies the conditions that $Dy = 2$, $y = -1$, when $x = 0$. (It is necessary to give two conditions since there are two constants to evaluate.)

We have

$$m^2 - 2m - 15 = 0, \qquad (m - 5)(m + 3) = 0$$

$$m_1 = 5, \qquad m_2 = -3$$

$$y = c_1e^{5x} + c_2e^{-3x}$$

This equation is the general solution. In order to evaluate the constants c_1 and c_2,

CAUTION ▸ ***we use the given conditions to find two simultaneous equations in c_1 and c_2.*** These are then solved to determine the particular solution. Thus

$$y' = 5c_1e^{5x} - 3c_2e^{-3x}$$

Using the given conditions in the general solution and its derivative, we have

$$c_1 + c_2 = -1 \qquad y = -1 \text{ when } x = 0$$

$$5c_1 - 3c_2 = 2 \qquad Dy = 2 \text{ when } x = 0$$

The solution to this system of equations is $c_1 = -\frac{1}{8}$ and $c_2 = -\frac{7}{8}$. The particular solution becomes

$$y = -\frac{1}{8}e^{5x} - \frac{7}{8}e^{-3x} \quad \text{or} \quad 8y + e^{5x} + 7e^{-3x} = 0$$ ◂◂

To solve the auxiliary equation for differential equations of higher order, we use methods developed in Chapter 15. With these roots, we form the solutions in the same way as we do those for second-order equations.

Exercises 30–6

In Exercises 1–20, solve the given differential equations.

1. $\frac{d^2y}{dx^2} - \frac{dy}{dx} - 6y = 0$
2. $\frac{d^2y}{dx^2} + \frac{dy}{dx} = 0$
3. $3\frac{d^2y}{dx^2} + 4\frac{dy}{dx} + y = 0$
4. $\frac{d^2y}{dx^2} - 2\frac{dy}{dx} - 8y = 0$
5. $D^2y - 3Dy = 0$
6. $D^2y + 7Dy + 6y = 0$
7. $3D^2y + 12y = 20Dy$
8. $4D^2y + 12Dy = 7y$
9. $3y'' + 8y' - 3y = 0$
10. $8y'' + 6y' - 9y = 0$
11. $3y'' + 2y' - y = 0$
12. $2y'' - 7y' + 6y = 0$
13. $2\frac{d^2y}{dx^2} - 4\frac{dy}{dx} + y = 0$
14. $\frac{d^2y}{dx^2} + \frac{dy}{dx} - 5y = 0$
15. $4D^2y - 3Dy - 2y = 0$
16. $2D^2y - 3Dy - y = 0$
17. $y'' = 3y' + y$
18. $5y'' - y' = 3y$
19. $y'' + y' = 8y$
20. $8y'' = y' + y$

In Exercises 21–24, find the particular solutions of the given differential equations that satisfy the given conditions.

21. $D^2y - 4Dy - 21y = 0$; $Dy = 0$ and $y = 2$ when $x = 0$
22. $4D^2y - Dy = 0$; $Dy = 2$ and $y = 4$ when $x = 0$
23. $D^2y - Dy - 12y = 0$; $y = 0$ when $x = 0$ and $y = 1$ when $x = 1$
24. $2D^2y + 5Dy = 0$; $y = 0$ when $x = 0$ and $y = 2$ when $x = 1$

In Exercises 25–28, solve the given differential equations. The auxiliary equations will be third degree, and the solutions will have three arbitrary constants.

25. $y''' - 2y'' - 3y' = 0$
26. $2y''' + 3y'' - 2y' = 0$
27. $y''' - 6y'' + 11y' - 6y = 0$
28. $y''' - 2y'' - y' + 2y = 0$

30-7 Auxiliary Equation with Repeated or Complex Roots

The way to solve a second-order homogeneous linear differential equation was shown in the previous section. However, we purposely avoided repeated and complex roots of the auxiliary equation. In this section we develop the solutions for such equations. The following example indicates the type of solution that results from the case of repeated roots.

EXAMPLE 1 ▸▸ Solve the differential equation $D^2y - 4Dy + 4y = 0$.

Using the method of Example 2 of the previous section, we have the following steps:

$$(D^2 - 4D + 4)y = 0, \qquad (D - 2)(D - 2)y = 0, \qquad (D - 2)z = 0$$

where $z = (D - 2)y$. The solution to $(D - 2)z = 0$ is found by separation of variables. And so

$$\frac{dz}{dx} - 2z = 0, \qquad \frac{dz}{z} - 2\,dx = 0$$

$$\ln z - 2x = \ln c_1 \quad \text{or} \quad z = c_1e^{2x}$$

Substituting back, we have $(D - 2)y = c_1e^{2x}$, which is a linear equation of the first order. Then

$$dy - 2y\,dx = c_1e^{2x}\,dx, \qquad e^{\int -2\,dx} = e^{-2x}$$

This leads to

$$ye^{-2x} = c_1\int dx = c_1x + c_2 \quad \text{or} \quad y = c_1xe^{2x} + c_2e^{2x}$$

This example indicates the type of solution that results when the auxiliary equation has repeated roots. If the method of the previous section were to be used, the solution of the above example would be $y = c_1e^{2x} + c_2e^{2x}$. This would not be the general solution, since both terms are similar, which means that there is only one independent constant. The constants can be combined to give a solution of the form $y = ce^{2x}$, where $c = c_1 + c_2$. This solution would contain only one constant for a second-order equation. ◂◂

For reference, Eq. (30-13) is $a_0D^2y + a_1Dy + a_2y = 0$ and Eq. (30-14) is $a_0m^2 + a_1m + a_2 = 0$.

Based on the above example, *the solution to Eq. (30-13) when the auxiliary Eq. (30-14) has repeated roots is*

Solution with Repeated Roots

$$y = e^{mx}(c_1 + c_2x) \tag{30-16}$$

where m is the double root. (In Example 1, this double root is 2.)

EXAMPLE 2 ▸▸ Solve the differential equation $(D + 2)^2y = 0$.

The auxiliary equation is $(m + 2)^2 = 0$, for which the solutions are $m = -2, -2$. Since we have repeated roots, the solution of the differential equation is

$$y = e^{-2x}(c_1 + c_2x) \qquad \text{using Eq. (30-16)}$$ ◂◂

EXAMPLE 3 ▸▸ Solve the differential equation $\frac{d^2y}{dx^2} - 10\frac{dy}{dx} + 25y = 0$.

The solution is as follows:

$$D^2y - 10Dy + 25y = 0 \qquad \text{using operator } D \text{ notation}$$
$$m^2 - 10m + 25 = 0 \qquad \text{auxiliary equation}$$
$$(m - 5)^2 = 0 \qquad \text{solve for } m$$
$$m = 5, 5 \qquad \text{double root}$$
$$y = e^{5x}(c_1 + c_2x) \qquad \text{using Eq. (30-16)}$$ ◂◂

When the auxiliary equation has complex roots, it can be solved by the method of the previous section and the solution can be put in a more useful form. For complex roots of the auxiliary equation $m = \alpha \pm j\beta$, the solution is of the form

$$y = c_1e^{(\alpha+j\beta)x} + c_2e^{(\alpha-j\beta)x} = e^{\alpha x}(c_1e^{j\beta x} + c_2e^{-j\beta x})$$

Using the exponential form of a complex number, Eq. (12-11), we have

$$\begin{aligned} y &= e^{\alpha x}[c_1 \cos \beta x + jc_1 \sin \beta x + c_2 \cos (-\beta x) + jc_2 \sin (-\beta x)] \\ &= e^{\alpha x}(c_1 \cos \beta x + c_2 \cos \beta x + jc_1 \sin \beta x - jc_2 \sin \beta x) \\ &= e^{\alpha x}(c_3 \cos \beta x + c_4 \sin \beta x) \end{aligned}$$

where $c_3 = c_1 + c_2$ and $c_4 = jc_1 - jc_2$.

Therefore, *if the auxiliary equation has complex roots of the form* $\alpha \pm j\beta$,

Solution with Complex Roots

$$y = e^{\alpha x}(c_1 \sin \beta x + c_2 \cos \beta x) \tag{30-17}$$

is the solution to Eq. (30-13). The c_1 and c_2 here are not the same as those above. They are simply the two arbitrary constants of the solution.

EXAMPLE 4 ▸▸ Solve the differential equation $D^2y - Dy + y = 0$.

We have the following solution:

$$m^2 - m + 1 = 0 \qquad \text{auxiliary equation}$$
$$m = \frac{1 \pm j\sqrt{3}}{2} \qquad \text{complex roots}$$
$$\alpha = \frac{1}{2}, \quad \beta = \frac{\sqrt{3}}{2} \qquad \text{identify } \alpha \text{ and } \beta$$
$$y = e^{x/2}\left(c_1 \sin \frac{\sqrt{3}}{2}x + c_2 \cos \frac{\sqrt{3}}{2}x\right) \qquad \text{using Eq. (30-17)}$$ ◂◂

EXAMPLE 5 ▸▸ Solve the differential equation $D^2y + 4y = 0$.

In this case we have

$$m^2 + 4 = 0 \qquad \text{auxiliary equation}$$
$$m_1 = 2j, \quad m_2 = -2j \qquad \text{complex roots}$$
$$\alpha = 0, \quad \beta = 2 \qquad \text{identify } \alpha \text{ and } \beta$$
$$y = e^{0x}(c_1 \sin 2x + c_2 \cos 2x) \qquad \text{using Eq. (30-17)}$$
$$= c_1 \sin 2x + c_2 \cos 2x \qquad e^0 = 1$$ ◂◂

EXAMPLE 6 ▸▸ Solve the differential equation $y'' - 2y' + 12y = 0$, if $y' = 2$ and $y = 1$ when $x = 0$.

$$D^2y - 2Dy + 12y = 0 \qquad \text{use operator } D \text{ notation}$$

$$m^2 - 2m + 12 = 0 \qquad \text{auxiliary equation}$$

$$m = \frac{2 \pm \sqrt{4 - 48}}{2} = \frac{2 \pm 2j\sqrt{11}}{2} \qquad \text{solve for } m$$

$$= 1 \pm j\sqrt{11} \qquad \text{complex roots: } \alpha = 1, \beta = \sqrt{11}$$

$$y = e^x(c_1 \cos \sqrt{11}x + c_2 \sin \sqrt{11}x) \qquad \text{general solution}$$

Using the condition that $y = 1$ when $x = 0$, we have

$$1 = e^0(c_1 \cos 0 + c_2 \sin 0) \quad \text{or} \quad c_1 = 1$$

Since the other condition involves the value of y', we find the derivative and then evaluate c_2.

$$y' = e^x(c_1 \cos \sqrt{11}x + c_2 \sin \sqrt{11}x - \sqrt{11}c_1 \sin \sqrt{11}x + \sqrt{11}c_2 \cos \sqrt{11}x)$$

$$2 = e^0(\cos 0 + c_2 \sin 0 - \sqrt{11} \sin 0 + \sqrt{11}c_2 \cos 0) \qquad y' = 2 \text{ when } x = 0$$

$$2 = 1 + \sqrt{11}c_2, \quad c_2 = \frac{1}{11}\sqrt{11} \qquad \text{solve for } c_2$$

$$y = e^x\left(\cos \sqrt{11}x + \frac{1}{11}\sqrt{11} \sin \sqrt{11}x\right) \qquad \text{particular solution}$$ ◂◂

Now that we have determined the various types of possible solutions of second-order homogeneous linear differential equations, we can see that it is possible to find the differential equation if the solution is known. Consider the following example.

EXAMPLE 7 ▸▸ (a) If the solution of a differential equation is $y = c_1e^x + c_2e^{2x}$, we then know that the auxiliary equation is $(m - 1)(m - 2) = 0$, since it gives the solutions $m_1 = 1$ and $m_2 = 2$. Therefore, the simplest form of the differential equation is $(D^2 - 3D + 2)y = 0$.

(b) For the solution $y = c_1e^{2x} + c_2xe^{2x}$, the auxiliary equation is $(m - 2)^2 = 0$, for repeated roots are indicated by the second terms of the solution. The differential equation would be $(D^2 - 4D + 4)y = 0$.

(c) For the solution $y = c_1 \sin 2x + c_2 \cos 2x$, the auxiliary equation would be $m^2 + 4 = 0$, for imaginary roots are indicated by the terms $\sin 2x$ and $\cos 2x$. The differential equation would be $(D^2 + 4)y = 0$. ◂◂

Exercises 30–7

In Exercises 1–24, solve the given differential equations.

1. $\frac{d^2y}{dx^2} - 2\frac{dy}{dx} + y = 0$
2. $\frac{d^2y}{dx^2} - 6\frac{dy}{dx} + 9y = 0$
3. $D^2y + 12Dy + 36y = 0$
4. $16D^2y + 8Dy + y = 0$
5. $\frac{d^2y}{dx^2} + 9y = 0$
6. $\frac{d^2y}{dx^2} + y = 0$
7. $D^2y + Dy + 2y = 0$
8. $D^2y - 2Dy + 4y = 0$
9. $D^2y = 0$
10. $4D^2y = 12Dy - 9y$
11. $4D^2y + y = 0$
12. $9D^2y + 4y = 0$
13. $16y'' - 24y' + 9y = 0$
14. $9y'' - 24y' + 16y = 0$
15. $25y'' + 2y = 0$
16. $y'' - 4y' + 5y = 0$
17. $2D^2y + 5y = 4Dy$
18. $D^2y + 4Dy + 6y = 0$
19. $25y'' + 16y = 40y'$
20. $9y'' + 0.6y' + 0.01y = 0$
21. $2D^2y - 3Dy - y = 0$
22. $D^2y - 5Dy - 4y = 0$
23. $3D^2y + 12Dy = 2y$
24. $36D^2y = 25y$

In Exercises 25–28, find the particular solutions of the given differential equations that satisfy the given conditions.

25. $y'' + 2y' + 10y = 0$; $y = 0$ when $x = 0$ and $y = e^{-1}$ when $x = \pi/6$

26. $9D^2y + 16y = 0$; $Dy = 0$ and $y = 2$ when $x = \pi/2$

27. $D^2y - 8Dy + 16y = 0$; $Dy = 2$ and $y = 4$ when $x = 0$

28. $4y'' + 20y' + 25y = 0$; $y = 0$ when $x = 0$ and $y = e$ when $x = -\frac{2}{5}$

In Exercises 29–32, find the simplest form of the second-order homogeneous linear differential equation that has the given solution. In Exercises 30 and 31, explain how the equation is found.

29. $y = c_1e^{3x} + c_2e^{-3x}$

30. $y = c_1e^{3x} + c_2xe^{3x}$

31. $y = c_1 \cos 3x + c_2 \sin 3x$

32. $y = c_1e^{2x} \cos x + c_2e^{2x} \sin x$

30-8 Solutions of Nonhomogeneous Equations

We now consider the solution of a nonhomogeneous linear equation of the form

$$a_0D^2y + a_1Dy + a_2y = b \tag{30-18}$$

where b is a function of x or is a constant. When the solution is substituted into the left side, we must obtain b. Solutions found from the methods of Sections 30-6 and 30-7 give zero when substituted into the left side, but they do contain the arbitrary constants necessary in the solution. If we could find a particular solution that when substituted into the left side produced b, it could be added to the solution containing the arbitrary constants. Therefore, *the solution is of the form*

$$y = y_c + y_p \tag{30-19}$$

where y_c, called the **complementary solution,** *is obtained by solving the corresponding homogeneous equation and where y_p is the particular solution necessary to produce the expression b of Eq. (30-18).*

EXAMPLE 1 ▶▶ The differential equation $D^2y - Dy - 6y = e^x$ has the solution

$$y = c_1e^{3x} + c_2e^{-2x} - \tfrac{1}{6}e^x,$$

where the complementary solution y_c and particular solution y_p are

$$y_c = c_1e^{3x} + c_2e^{-2x} \qquad y_p = -\tfrac{1}{6}e^x$$

The complementary solution y_c is obtained by solving the corresponding homogeneous equation $D^2y - Dy - 6y = 0$, and we shall discuss below the method of finding y_p. We can see that y_p alone will satisfy the differential equation, but since it has no arbitrary constants, it cannot be the general solution. ◀◀

By inspecting the form of b on the right side of the equation, we can find the form which the particular solution must have. Since a combination of the particular solution and its derivatives must form the function b,

CAUTION ▶ ***y_p is an expression that contains all possible forms of b and its derivatives.***

The method that is used to find the exact form of y_p is called the **method of undetermined coefficients.**

EXAMPLE 2 ▸▸ (a) If the function b is $4x$, we choose the particular solution y_p to be of the form $y_p = A + Bx$. The Bx-term is included to account for the $4x$. Since the derivative of Bx is a constant, the A-term is included to account for any first derivative of the Bx-term that may be present. Since the derivative of A is zero, no other terms are needed to account for higher-derivative terms of the Bx-term.

(b) If the function b is e^{2x}, we choose the form of the particular solution to be $y_p = Ce^{2x}$. *Since all derivatives of Ce^{2x} are a constant times e^{2x}, no other forms appear in the derivatives, and no other forms are needed in y_p.*

NOTE ▸

(c) If the function b is $4x + e^{2x}$, we choose the form of the particular solution to be $y_p = A + Bx + Ce^{2x}$. ◂◂

EXAMPLE 3 ▸▸ (a) If b is of the form $x^2 + e^{-x}$, we choose the particular solution to be of the form $y_p = A + Bx + Cx^2 + Ee^{-x}$.

(b) If b is of the form $xe^{-2x} - 5$, we choose the form of the particular solution to be $y_p = Ae^{-2x} + Bxe^{-2x} + C$.

(c) If b is of the form $x \sin x$, we choose the form of the particular solution to be $y_p = A \sin x + B \cos x + Cx \sin x + Ex \cos x$. All of these types of terms occur in the derivatives of $x \sin x$. ◂◂

EXAMPLE 4 ▸▸ (a) If b is of the form $e^x + xe^x$, we would then choose y_p to be of the form $y_p = Ae^x + Bxe^x$. These terms occur for xe^x and its derivatives. *Since the form of the e^x-term of b is already included in Ae^x, we do not include another e^x-term in y_p.*

CAUTION ▸

(b) In the same way, if b is of the form $2x + 4x^2$, we choose the form of y_p to be $y_p = A + Bx + Cx^2$. *These are the only forms that occur in either $2x$ or $4x^2$ and their derivatives.* ◂◂

NOTE ▸

Once we have determined the form of y_p, we have to find the numerical values of the coefficients $A, B, \ldots$. The *method of undetermined coefficients* is to

NOTE ▸

substitute the chosen form of y_p into the differential equation and equate the coefficients of like terms.

EXAMPLE 5 ▸▸ Solve the differential equation $D^2y - Dy - 6y = e^x$.

In this case the solution of the auxiliary equation $m^2 - m - 6 = 0$ gives us the roots $m_1 = 3$ and $m_2 = -2$. Thus

$$y_c = c_1e^{3x} + c_2e^{-2x}$$

The proper form of y_p is $y_p = Ae^x$. This means that $Dy_p = Ae^x$ and $D^2y_p = Ae^x$. Substituting y_p and its derivatives into the differential equation, we have

$$Ae^x - Ae^x - 6Ae^x = e^x$$

To produce equality, the coefficients of e^x must be the same on each side of the equation. Thus

$$-6A = 1, \text{ or } A = -1/6$$

Therefore, $y_p = -\frac{1}{6}e^x$. This gives the complete solution $y = y_c + y_p$,

$$y = c_1e^{3x} + c_2e^{-2x} - \frac{1}{6}e^x \qquad \text{see Example 1}$$

This solution checks when substituted into the original differential equation. ◂◂

EXAMPLE 6 ▸▸ Solve the differential equation $D^2y + 4y = x - 4e^{-x}$.

In this case we have $m^2 + 4 = 0$, which gives us $m_1 = 2j$ and $m_2 = -2j$. Therefore, $y_c = c_1 \sin 2x + c_2 \cos 2x$.

The proper form of the particular solution is $y_p = A + Bx + Ce^{-x}$. Finding two derivatives and then substituting into the differential equation gives

$$y_p = A + Bx + Ce^{-x}, \quad Dy_p = B - Ce^{-x}, \quad D^2y_p = Ce^{-x}$$

$$D^2y + 4y = x - 4e^{-x} \quad \text{differential equation}$$

$$(Ce^{-x}) + 4(A + Bx + Ce^{-x}) = x - 4e^{-x} \quad \text{substituting}$$

$$Ce^{-x} + 4A + 4Bx + 4Ce^{-x} = x - 4e^{-x}$$

$$(4A) + (4B)x + (5C)e^{-x} = 0 + (1)x + (-4)e^{-x} \quad \text{note coefficients}$$

CAUTION ▸ ***Equating the constants, and the coefficients*** of x and e^{-x} on either side gives

$$4A = 0, \quad 4B = 1, \quad 5C = -4$$
$$A = 0, \quad B = 1/4, \quad C = -4/5$$

This means that the particular solution is

$$y_p = \frac{1}{4}x - \frac{4}{5}e^{-x}$$

In turn this tells us that the complete solution is

$$y = c_1 \sin 2x + c_2 \cos 2x + \frac{1}{4}x - \frac{4}{5}e^{-x}$$

Substitution into the original differential equation verifies this solution. ◂◂

EXAMPLE 7 ▸▸ Solve the differential equation $D^2y - 3Dy + 2y = 2 \sin x$.

$$m^2 - 3m + 2 = 0, \quad (m - 1)(m - 2) = 0, \quad m_1 = 1, \quad m_2 = 2 \quad \text{auxiliary equation}$$

$$y_c = c_1e^x + c_2e^{2x} \quad \text{complementary solution}$$

We now find the particular solution:

$$y_p = A \sin x + B \cos x \quad \text{particular solution form}$$

$$Dy_p = A \cos x - B \sin x \quad \text{find two derivatives}$$

$$D^2y_p = -A \sin x - B \cos x$$

$$(-A \sin x - B \cos x) - 3(A \cos x - B \sin x) + 2(A \sin x + B \cos x) = 2 \sin x \quad \text{substitute into differential equation}$$

$$(A + 3B) \sin x + (B - 3A) \cos x = 2 \sin x$$

$$A + 3B = 2, \quad -3A + B = 0 \quad \text{equate coefficients}$$

The solution of this system is $A = \frac{1}{5}$ and $B = \frac{3}{5}$. Thus

$$y_p = \frac{1}{5} \sin x + \frac{3}{5} \cos x \quad \text{particular solution}$$

$$y = c_1e^x + c_2e^{2x} + \frac{1}{5} \sin x + \frac{3}{5} \cos x \quad \text{complete general solution}$$

Check this solution in the differential equation. ◂◂

EXAMPLE 8 ▸▸ Find the particular solution of $y'' + 16y = 2e^{-x}$ if $Dy = -2$ and $y = 1$ when $x = 0$.

In this case we must not only find y_c and y_p, we must also evaluate the constants of y_c from the given conditions. The solution is as follows:

$$D^2y + 16y = 2e^{-x} \qquad \text{operator } D \text{ form}$$

$$m^2 + 16 = 0, \qquad m = \pm 4j \qquad \text{auxiliary equation}$$

$$y_c = c_1 \sin 4x + c_2 \cos 4x \qquad \text{complementary solution}$$

$$y_p = Ae^{-x} \qquad \text{particular solution form}$$

$$Dy_p = -Ae^{-x}, \quad D^2y_p = Ae^{-x}$$

$$Ae^{-x} + 16Ae^{-x} = 2e^{-x} \qquad \text{substituting}$$

$$17Ae^{-x} = 2e^{-x}, \quad A = \frac{2}{17} \qquad \text{equate coefficients}$$

$$y_p = \frac{2}{17}e^{-x}$$

$$y = c_1 \sin 4x + c_2 \cos 4x + \frac{2}{17}e^{-x} \qquad \text{complete general solution}$$

We now evaluate c_1 and c_2 from the given conditions:

$$Dy = 4c_1 \cos 4x - 4c_2 \sin 4x - \frac{2}{17}e^{-x}$$

$$1 = c_1(0) + c_2(1) + \frac{2}{17}(1), \qquad c_2 = \frac{15}{17} \qquad y = 1 \text{ when } x = 0$$

$$-2 = 4c_1(1) - 4c_2(0) - \frac{2}{17}(1), \qquad c_1 = -\frac{8}{17} \qquad Dy = -2 \text{ when } x = 0$$

$$y = -\frac{8}{17}\sin 4x + \frac{15}{17}\cos 4x + \frac{2}{17}e^{-x} \qquad \text{required particular solution}$$

This solution checks when substituted into the differential equation. ◂◂

If any terms of the form required for y_p are of the same form as any of the terms of y_c, it is necessary to multiply such terms of y_p by a power of x. We have not included any equations of this type to this point, although we will be able to get solutions by methods presented in Section 30-11.

Exercises 30–8

In Exercises 1–12, solve the given differential equations. The form of y_p is given.

1. $D^2y - Dy - 2y = 4$ (Let $y_p = A$.)
2. $D^2y - Dy - 6y = 4x$ (Let $y_p = A + Bx$.)
3. $4D^2y + y = x^2$ (Let $y_p = A + Bx + Cx^2$.)
4. $D^2y - y = 2 + x^2$ (Let $y_p = A + Bx + Cx^2$.)
5. $D^2y + 4Dy + 3y = 2 + e^x$ (Let $y_p = A + Be^x$.)
6. $2D^2y + Dy + y = e^{2x}$ (Let $y_p = Ae^{2x}$.)
7. $y'' - 3y' = 2e^x + xe^x$ (Let $y_p = Ae^x + Bxe^x$.)
8. $y'' + y' - 2y = 8 + 4x + 2xe^{2x}$ (Let $y_p = A + Bx + Ce^{2x} + Exe^{2x}$.)
9. $9D^2y - y = \sin x$ (Let $y_p = A \sin x + B \cos x$.)
10. $D^2y + 4y = \sin x + 4$ (Let $y_p = A + B \sin x + C \cos x$.)
11. $\dfrac{d^2y}{dx^2} + 9y = 9x + 5 \cos 2x + 10 \sin 2x$ (Let $y_p = A + Bx + C \sin 2x + E \cos 2x$.)
12. $\dfrac{d^2y}{dx^2} - 2\dfrac{dy}{dx} + y = 2x + x^2 + \sin 3x$ (Let $y_p = A + Bx + Cx^2 + E \sin 3x + F \cos 3x$.)

In Exercises 13–24, solve the given differential equations.

13. $\dfrac{d^2y}{dx^2} - \dfrac{dy}{dx} - 30y = 10$

14. $2\dfrac{d^2y}{dx^2} + 11\dfrac{dy}{dx} - 6y = 8x$

15. $3\dfrac{d^2y}{dx^2} + 13\dfrac{dy}{dx} - 10y = 14e^{3x}$

16. $\dfrac{d^2y}{dx^2} + 4y = 2 \sin 3x$

17. $D^2y - 4y = \sin x + 2 \cos x$

18. $2D^2y + 5Dy - 3y = e^x + 4e^{2x}$

19. $D^2y + y = 4 + \sin 2x$

20. $D^2y - Dy + y = x + \sin x$

21. $D^2y + 5Dy + 4y = xe^x + 4$

22. $3D^2y + Dy - 2y = 4 + 2x + e^x$

23. $y'' + 6y' + 9y = e^{2x} - e^{-2x}$

24. $y'' + 8y' - y = x^2 + 4e^{-2x}$

In Exercises 25–28, find the particular solution of each differential equation for the given conditions.

25. $D^2y - Dy - 6y = 5 - e^x$; $Dy = 4$ and $y = 2$ when $x = 0$

26. $3y'' - 10y' + 3y = xe^{-2x}$; $Dy = 0$ and $y = -1$ when $x = 0$

27. $y'' + y = x + \sin 2x$; $Dy = 1$ and $y = 0$ when $x = \pi$

28. $D^2y - 2Dy + y = xe^{2x} - e^{2x}$; $Dy = 4$ and $y = -2$ when $x = 0$

30-9 Applications of Second-Order Equations

Linear differential equations of the second order have many important applications. We shall restrict our attention to two of these. In this section we apply second-order differential equations to solve problems in simple harmonic motion and simple electric circuits.

EXAMPLE 1 ▸▸ Simple harmonic motion may be defined as motion in a straight line for which the acceleration is proportional to the displacement and in the opposite direction. Examples of this type of motion are a weight on a spring, a simple pendulum, and an object bobbing in water. If x represents the displacement, d^2x/dt^2 is the acceleration.

Using the definition of simple harmonic motion, we have

$$\frac{d^2x}{dt^2} = -k^2x$$

(We chose k^2 for convenience of notation in the solution.) We write this equation in the form

$$D^2x + k^2x = 0$$

where $D = d/dt$. The roots of the auxiliary equation are kj and $-kj$, and the solution is

$$x = c_1 \sin kt + c_2 \cos kt$$

This solution indicates an oscillating motion, which is known to be the case. If, for example, $k = 4$ and we know that $x = 2$ and $Dx = 0$ (which means the velocity is zero) for $t = 0$, we have

$$Dx = 4c_1 \cos 4t - 4c_2 \sin 4t$$

$$2 = c_1(0) + c_2(1) \qquad x = 2 \text{ for } t = 0$$

$$0 = 4c_1(1) - 4c_2(0) \qquad Dx = 0 \text{ for } t = 0$$

which gives $c_1 = 0$ and $c_2 = 2$. Therefore,

$$x = 2 \cos 4t$$

is the equation relating the displacement and time; Dx is the velocity and D^2x is the acceleration. See Fig. 30-8. ◂◂

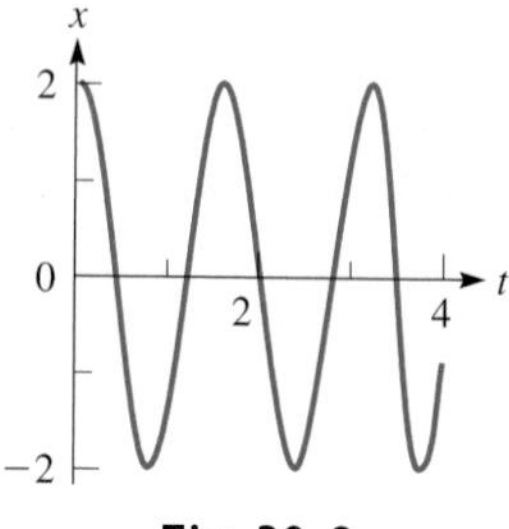

Fig. 30-8

EXAMPLE 2 ▸▸ In practice, an object moving with simple harmonic motion will in time cease to move due to unavoidable frictional forces. A "freely" oscillating object has a retarding force that is approximately proportional to the velocity. The differential equation for this case is $D^2x = -k^2x - bDx$. This results from applying (from physics) Newton's second law of motion, which states that the net force acting on an object equals its mass times its acceleration. The term D^2x represents the acceleration, the term $-k^2x$ is a measure of the restoring force (of the spring, for example), and the term $-bDx$ represents the retarding (damping) force. This equation can be written as

$$D^2x + bDx + k^2x = 0$$

The auxiliary equation is $m^2 + bm + k^2 = 0$, for which the roots are

$$m = \frac{-b \pm \sqrt{b^2 - 4k^2}}{2}$$

If $k = 3$ and $b = 4$, $m = -2 \pm j\sqrt{5}$, which means the solution is

$$\mathbf{x = e^{-2t}(c_1 \sin \sqrt{5}t + c_2 \cos \sqrt{5}t)} \tag{1}$$

Here, $4k^2 > b^2$, and this case is called **underdamped.** In this case the object oscillates as the amplitude becomes smaller.

If $k = 2$ and $b = 5$, $m = -1, -4$, which means the solution is

$$\mathbf{x = c_1e^{-t} + c_2e^{-4t}} \tag{2}$$

Here $4k^2 < b^2$, and the case is called **overdamped.** It will be noted that the motion is not oscillatory, since no sine or cosine terms appear. In this case the object returns slowly to equilibrium without oscillating.

If $k = 2$ and $b = 4$, $m = -2, -2$, which means the solution is

$$\mathbf{x = e^{-2t}(c_1 + c_2t)} \tag{3}$$

Here $4k^2 = b^2$, and the case is called **critically damped.** Again the motion is not oscillatory. In this case there is just enough damping to prevent any oscillations. The object returns to equilibrium in the minimum time.

See Fig. 30-9, in which Eqs. (1), (2), and (3) are represented in general. Of course, the actual values depend on c_1 and c_2, which in turn depend on the conditions imposed on the motion.

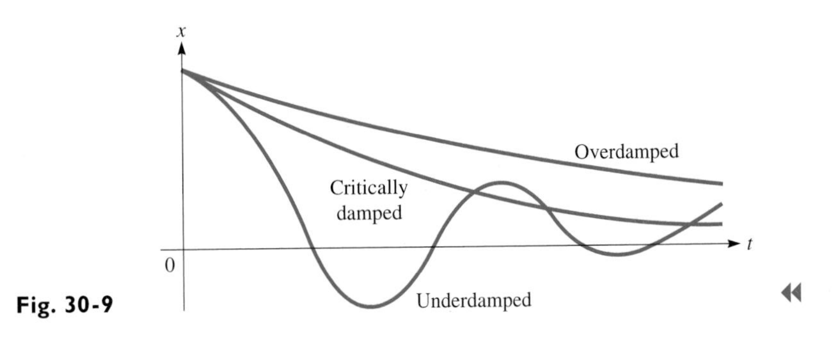

Fig. 30-9 ◂◂

EXAMPLE 3 ▸▸ In testing the characteristics of a particular type of spring, it is found that a weight of 4.90 N stretches the spring 0.490 m when the weight and spring are placed in a fluid that resists the motion with a force equal to twice the velocity. If the weight is brought to rest and then given a velocity of 12.0 m/s, find the equation of motion. See Fig. 30-10.

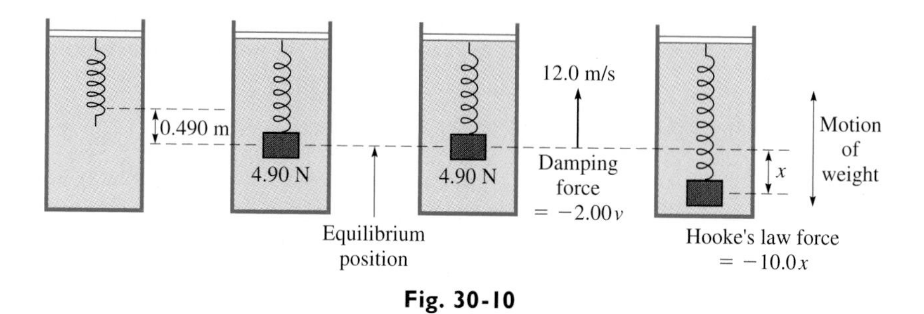

Fig. 30-10

In order to find the equation of motion, we use Newton's second law of motion (see Example 2). The weight (one force) at the end of the spring is offset by the equilibrium position force exerted by the spring, in accordance with Hooke's law (see Section 26-6). Therefore, the net force acting on the weight is the sum of the Hooke's law force due to the displacement from the equilibrium position and the resisting force. Using Newton's second law, we have

mass × acceleration = resisting force + Hooke's law force

$$mD^2x = -2.00\,Dx - kx$$

The mass of an object is its weight divided by the acceleration due to gravity. The weight is 4.90 N, and the acceleration due to gravity is 9.80 m/s^2. Thus the mass m is

$$m = \frac{4.90 \text{ N}}{9.80 \text{ m/s}^2} = 0.500 \text{ kg}$$

where the kilogram is the unit of mass.

The constant k for the Hooke's law force is found from the fact that the spring stretches 0.490 m for a force of 4.90 N. Thus, using Hooke's law,

$$4.90 = k(0.490), \qquad k = 10.0 \text{ N/m}$$

This means that the differential equation to be solved is

$$0.500\,D^2x + 2.00\,Dx + 10.0x = 0$$

or

$$1.00\,D^2x + 4.00\,Dx + 20.0x = 0$$

Solving this equation, we have

$$1.00m^2 + 4.00m + 20.0 = 0 \qquad \text{auxiliary equation}$$

$$m = \frac{-4.00 \pm \sqrt{16.0 - 4(20.0)(1.00)}}{2.00}$$

$$= -2.00 \pm 4.00j \qquad \text{complex roots}$$

$$x = e^{-2.00t}(c_1 \cos 4.00t + c_2 \sin 4.00t) \qquad \text{general solution}$$

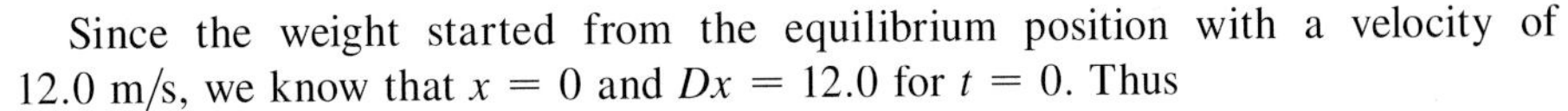

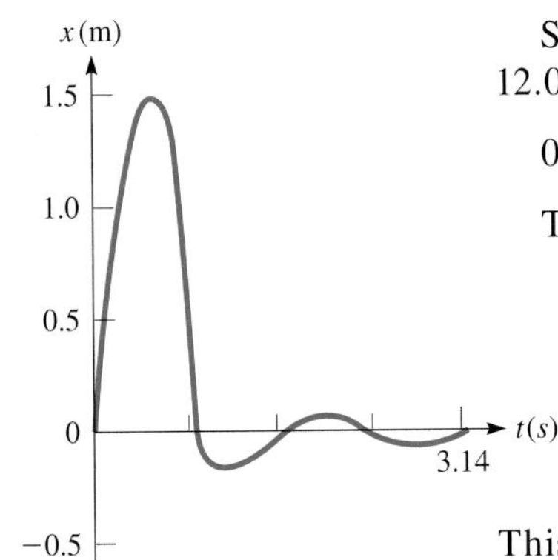

Fig. 30-11

Since the weight started from the equilibrium position with a velocity of 12.0 m/s, we know that $x = 0$ and $Dx = 12.0$ for $t = 0$. Thus

$$0 = e^0(c_1 + 0c_2), \quad \text{or} \quad c_1 = 0 \qquad x = 0 \text{ for } t = 0$$

Thus, since $c_1 = 0$, we have

$$\begin{aligned}
x &= c_2e^{-2.00t} \sin 4.00t \\
Dx &= c_2e^{-2.00t}(\cos 4.00t)(4.00) + c_2 \sin 4.00t(e^{-2.00t})(-2.00) \\
12.0 &= c_2e^0(1)(4.00) + c_2(0)(e^0)(-2.00) \qquad Dx = 12.0 \text{ for } t = 0 \\
c_2 &= 3.00
\end{aligned}$$

This means that the equation of motion is

$$x = 3.00e^{-2.00t} \sin 4.00t$$

This motion is underdamped; the graph is shown in Fig. 30-11. ◀◀

It is possible to have an additional force acting on a weight such as the one in Example 3. For example, a vibratory force may be applied to the support of the spring. In such a case, called ***forced vibrations,*** this additional external force is added to the other net force. This means that the added force $F(t)$ becomes a nonzero function on the right side of the differential equation, and we must then solve a nonhomogeneous equation.

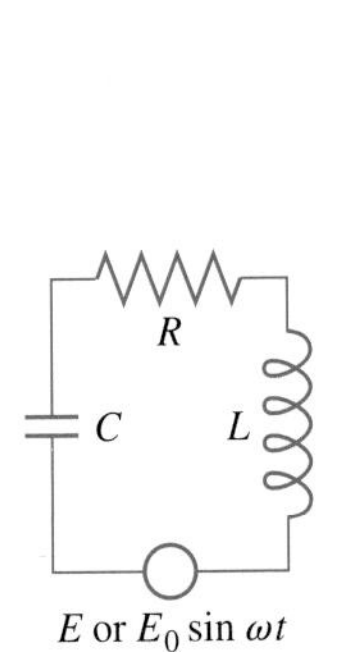

Fig. 30-12

EXAMPLE 4 ▶▶ The impressed voltage in an electric circuit equals the sum of the voltages across the components of the circuit. For a circuit with a resistance R, an inductance L, a capacitance C, and a voltage source E (see Fig. 30-12), we have

$$L\frac{d^2q}{dt^2} + R\frac{dq}{dt} + \frac{q}{C} = E \tag{30-20}$$

By definition q represents the electric charge, $dq/dt = i$ is the current, and d^2q/dt^2 is the time rate of change of current. This equation may be written as

$$LD^2q + RDq + q/C = E$$

The auxiliary equation is $Lm^2 + Rm + 1/C = 0$. The roots are

$$\begin{aligned}
m &= \frac{-R \pm \sqrt{R^2 - 4L/C}}{2L} \\
&= -\frac{R}{2L} \pm \sqrt{\frac{R^2}{4L^2} - \frac{1}{LC}}
\end{aligned}$$

If we let $a = R/2L$ and $\omega = \sqrt{1/LC - R^2/4L^2}$, we have (assuming complex roots, which corresponds to realistic values of R, L, and C)

$$q_c = e^{-at}(c_1 \sin \omega t + c_2 \cos \omega t)$$

This indicates an oscillating charge, or an alternating current. However, the exponential term usually is such that the current dies out rapidly unless there is a source of voltage in the circuit. ◀◀

If there is no source of voltage in the circuit of Example 4, we have a homogeneous differential equation to solve. If we have a constant voltage source, the particular solution is of the form $q_p = A$. If there is an alternating voltage source, the particular solution is of the form $q_p = A \sin \omega_1 t + B \cos \omega_1 t$, where ω_1 is the angular velocity of the source. After a very short time, the exponential factor in the complementary solution makes it negligible. For this reason it is referred to as the **transient** term, and *the particular solution is the* **steady-state** *solution*. Therefore, to find the steady-state solution, we need find only the particular solution.

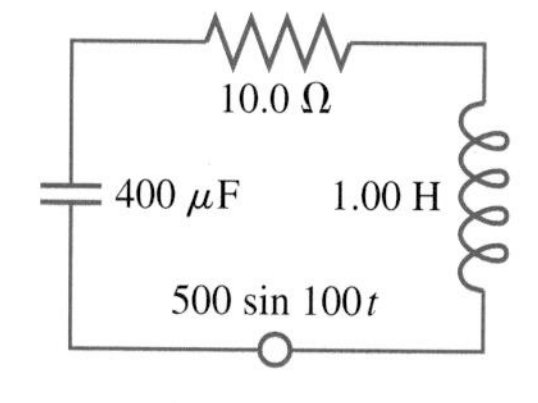

Fig. 30-13

EXAMPLE 5 ▸▸ Find the steady-state solution for the current in a circuit containing the following elements: $C = 400\ \mu\text{F}$, $L = 1.00$ H, $R = 10.0\ \Omega$, and a voltage source of $500 \sin 100t$. See Fig. 30-13.

This means the differential equation to be solved is

$$\frac{d^2q}{dt^2} + 10\frac{dq}{dt} + \frac{10^4}{4}q = 500 \sin 100t$$

Since we wish to find the steady-state solution, we must find q_p, from which we may find i_p by finding a derivative. The solution now follows:

$$q_p = A \sin 100t + B \cos 100t \qquad \text{particular solution form}$$

$$\frac{dq_p}{dt} = 100A \cos 100t - 100B \sin 100t$$

$$\frac{d^2q_p}{dt^2} = -10^4A \sin 100t - 10^4B \cos 100t$$

$$-10^4A \sin 100t - 10^4B \cos 100t + 10^3A \cos 100t - 10^3B \sin 100t \qquad \text{substitute into differential equation}$$

$$+ \frac{10^4}{4}A \sin 100t + \frac{10^4}{4}B \cos 100t = 500 \sin 100t$$

$$(-0.75 \times 10^4A - 10^3B) \sin 100t + (-0.75 \times 10^4B + 10^3A) \cos 100t = 500 \sin 100t$$

$$-7.5 \times 10^3A - 10^3B = 500 \qquad \text{equate coefficients of } \sin 100t$$

$$10^3A - 7.5 \times 10^3B = 0 \qquad \text{equate coefficients of } \cos 100t$$

Solving these equations, we obtain

$$B = -8.73 \times 10^{-3} \quad \text{and} \quad A = -65.5 \times 10^{-3}$$

Therefore,

$$q_p = -65.5 \times 10^{-3} \sin 100t - 8.73 \times 10^{-3} \cos 100t$$

$$i_p = \frac{dq_p}{dt} = -6.55 \cos 100t + 0.87 \sin 100t$$

which is the required solution. (We assumed three significant digits for the data but did not use all of them in most equations of the solution.) ◂◂

It should be noted that the complementary solutions of the mechanical and electric cases are of identical form. There is also an equivalent mechanical case to that of an impressed sinusoidal voltage source in the electric case. This arises in the case of forced vibrations, when an external force affecting the vibrations is applied to the system. Thus we may have transient and steady-state solutions to mechanical and other nonelectric situations.

Exercises 30–9

1. When the angular displacement θ of a pendulum is small (less than about 6°), the pendulum moves with simple harmonic motion closely approximated by $D^2\theta + \frac{g}{l}\theta = 0$. Here $D = d/dt$, g is the acceleration due to gravity, and l is the length of the pendulum. Find θ as a function of time if $g = 9.8\ \mathrm{m/s^2}$, $l = 1.0$ m, $\theta = 0.1$, and $D\theta = 0$ when $t = 0$. Sketch the curve.

2. Find θ as a function of time for the pendulum in Exercise 1 if it moves according to the equation $D^2\theta + 0.2\,D\theta + \frac{g}{l}\theta = 0$. Sketch the curve.

3. A car suspension is depressed from its equilibrium position such that its equation of motion is $D^2y + bDy + 25y = 0$, where $D = d/dt$ and y is the displacement. What must be the value of b if the motion is critically damped?

4. A block of wood floating in oil is depressed from its equilibrium position such that its equation of motion is $D^2y + 8Dy + 3y = 0$, where y is the displacement (in inches) and $D = d/dt$. Find its displacement after 12 s if $y = 6.0$ cm and $Dy = 0$ when $t = 0$.

5. A 4.00-N weight stretches a certain spring 5.00 cm. With this weight attached, the spring is pulled 10.0 cm longer than its equilibrium length and released. Find the equation of the resulting motion, assuming no damping.

6. Find the solution for the spring of Exercise 5 if a damping force numerically equal to the velocity is present.

7. Find the solution for the spring of Exercise 5 if no damping is present, but an external force of 4 sin $2t$ is acting on the spring.

8. Find the solution for the spring of Exercise 5 if the damping force of Exercise 6 and the impressed force of Exercise 7 are both acting.

9. A mass of 0.820 kg stretches a given spring by 0.250 m. The mass is pulled down 0.150 m below the equilibrium position and released. Find the equation of motion of the mass if there is no damping.

10. Find the equation of motion for the mass of Exercise 9 if a damping force numerically equal to three times the velocity is also present.

11. Find the equation relating the charge and the time in an electric circuit with the following elements: $L = 0.200$ H, $R = 8.00\ \Omega$, $C = 1.00\ \mu$F, $E = 0$. In this circuit, $q = 0$ and $i = 0.500$ A when $t = 0$.

12. For a given circuit, $L = 2$ mH, $R = 0$, $C = 50$ nF, and $E = 0$. Find the equation relating the charge and the time if $q = 10^5$ C and $i = 0$ when $t = 0$.

13. For a given circuit, $L = 0.100$ H, $R = 0$, $C = 100\ \mu$F, and $E = 100$ V. Find the equation relating the charge and the time if $q = 0$ and $i = 0$ when $t = 0$.

14. Find the relation between the current and the time for the circuit of Exercise 13.

15. For a radio tuning circuit, $L = 0.500$ H, $R = 10.0\ \Omega$, $C = 200\ \mu$F, and $E = 120 \sin 120\pi t$. Find the equation relating charge and time.

16. Find the steady-state current for the circuit of Exercise 15.

17. In a given circuit $L = 8.00$ mH, $R = 0$, $C = 0.500\ \mu$F, and $E = 20.0e^{-200t}$ mV. Find the relation between the current and the time if $q = 0$ and $i = 0$ for $t = 0$.

18. Find the current as a function of time for a circuit in which $L = 0.400$ H, $R = 60.0\ \Omega$, $C = 0.200\ \mu$F, and $E = 0.800e^{-100t}$ V, if $q = 0$ and $i = 5.00$ mA for $t = 0$.

19. Find the steady-state current for a circuit with $L = 1.00$ H, $R = 5.00\ \Omega$, $C = 150\ \mu$F, and $E = 120 \sin 100t$ V.

20. Find the steady-state solution for the current in an electric circuit containing the following elements: $C = 20.0\ \mu$F, $L = 2.00$ H, $R = 20.0\ \Omega$, and $E = 200 \sin 10t$.

30-10 Laplace Transforms

Named for the French mathematician and astronomer Pierre Laplace (1749–1827).

The final method of solving differential equations that we shall discuss is by means of **Laplace transforms.** As we shall see, *Laplace transforms provide an algebraic method of obtaining a* ***particular*** *solution of a differential equation from stated initial conditions.* Since this is often what is desired in practice, Laplace transforms are often preferred for the solution of differential equations in engineering and electronics. In this section we discuss the meaning of the transform and its operations. These methods will be applied to solving differential equations in the following section. The treatment in this text is intended only as an introduction to the topic of Laplace transforms.

The Laplace transform of a function $f(t)$ is defined as the function $F(s)$ by the equation

$$F(s) = \int_0^{\infty} e^{-st} f(t)\, dt \tag{30-21}$$

By writing the transform as $F(s)$, we show that the result of integrating and evaluating is a function of s. To denote that we are dealing with "the Laplace transform of the function $f(t)$," the notation $L(f)$ is used. Thus

$$F(s) = L(f) = \int_0^{\infty} e^{-st} f(t)\, dt \tag{30-22}$$

We shall see that both notations are quite useful.

A note regarding the form of the integral in Eqs. (30-21) and (30-22) is in order at this point. Since the upper limit is ∞, which means that it is unbounded, this integral is one type of what is known as an **improper integral.** In evaluating this integral at the upper limit, it is necessary to find the limit of the resulting function as the upper limit approaches infinity. This may be denoted by

$$\lim_{c \to \infty} \int_0^{c} e^{-st} f(t)\, dt$$

where we substitute c for t in the resulting function and determine the limit as $c \to \infty$ to determine the result for the upper limit. This also means that the Laplace transform, $F(s)$, is defined only for those values of s for which the limit is defined.

EXAMPLE 1 ▸▸ Find the Laplace transform of the function $f(t) = t$, $t > 0$.

By the definition of the Laplace transform

$$L(f) = L(t) = \int_0^{\infty} e^{-st} t\, dt$$

This may be integrated by parts or by Formula (44) in Appendix D. Using the formula, we have

$$\begin{aligned} L(t) &= \int_0^{\infty} te^{-st}\, dt = \lim_{c \to \infty} \int_0^{c} te^{-st}\, dt = \lim_{c \to \infty} \frac{e^{-st}(-st - 1)}{s^2}\Bigg|_0^c \\ &= \lim_{c \to \infty} \left[\frac{e^{-sc}(-sc - 1)}{s^2}\right] + \frac{1}{s^2} \end{aligned}$$

Now, for $s > 0$, as $c \to \infty$, $e^{-sc} \to 0$ and $sc \to \infty$. However, although we cannot prove it here, $e^{-sc} \to 0$ much faster than $sc \to \infty$. We can see that this is reasonable, for $ce^{-c} = 4.5 \times 10^{-4}$ for $c = 10$ and $ce^{-c} = 3.7 \times 10^{-42}$ for $c = 100$. Thus the value at the upper limit approaches zero, which means the limit is zero. This means that

$$L(t) = \frac{1}{s^2}$$

As we noted, this transform is defined for $s > 0$. ◂◂

EXAMPLE 2 ▸▸ Find the Laplace transform of the function $f(t) = \cos at$.

By definition,

$$L(f) = L(\cos at) = \int_0^\infty e^{-st} \cos at\, dt$$

Using Formula (50) in Appendix D, we have

$$\begin{aligned} L(\cos at) &= \int_0^\infty e^{-st} \cos at\, dt = \lim_{c\to\infty} \int_0^c e^{-st} \cos at\, dt \\ &= \lim_{c\to\infty} \left. \frac{e^{-st}(-s \cos at + a \sin at)}{s^2 + a^2} \right|_0^c \\ &= \lim_{c\to\infty} \frac{e^{-sc}(-s \cos ac + a \sin ac)}{s^2 + a^2} - \left(-\frac{s}{s^2 + a^2}\right) \\ &= 0 + \frac{s}{s^2 + a^2} = \frac{s}{s^2 + a^2} \quad (s > 0) \end{aligned}$$

Therefore, the Laplace transform of the function $\cos at$ is

$$L(\cos at) = \frac{s}{s^2 + a^2}$$

In both examples the resulting transform was an algebraic function of s. ◂◂

We now present a short table of Laplace transforms. They are sufficient for our work in this chapter. More complete tables are available in many references.

TABLE OF LAPLACE TRANSFORMS

	$f(t) = L^{-1}(F)$	$L(f) = F(s)$		$f(t) = L^{-1}(F)$	$L(f) = F(s)$
1.	1	$\frac{1}{s}$	11.	te^{-at}	$\frac{1}{(s+a)^2}$
2.	$\frac{t^{n-1}}{(n-1)!}$	$\frac{1}{s^n} (n = 1, 2, 3, \ldots)$	12.	$t^{n-1}e^{-at}$	$\frac{(n-1)!}{(s+a)^n}$
3.	e^{-at}	$\frac{1}{s+a}$	13.	$e^{-at}(1 - at)$	$\frac{s}{(s+a)^2}$
4.	$1 - e^{-at}$	$\frac{a}{s(s+a)}$	14.	$[(b-a)t + 1]e^{-at}$	$\frac{s+b}{(s+a)^2}$
5.	$\cos at$	$\frac{s}{s^2+a^2}$	15.	$\sin at - at \cos at$	$\frac{2a^3}{(s^2+a^2)^2}$
6.	$\sin at$	$\frac{a}{s^2+a^2}$	16.	$t \sin at$	$\frac{2as}{(s^2+a^2)^2}$
7.	$1 - \cos at$	$\frac{a^2}{s(s^2+a^2)}$	17.	$\sin at + at \cos at$	$\frac{2as^2}{(s^2+a^2)^2}$
8.	$at - \sin at$	$\frac{a^3}{s^2(s^2+a^2)}$	18.	$t \cos at$	$\frac{s^2-a^2}{(s^2+a^2)^2}$
9.	$e^{-at} - e^{-bt}$	$\frac{b-a}{(s+a)(s+b)}$	19.	$e^{-at} \sin bt$	$\frac{b}{(s+a)^2+b^2}$
10.	$ae^{-at} - be^{-bt}$	$\frac{s(a-b)}{(s+a)(s+b)}$	20.	$e^{-at} \cos bt$	$\frac{s+a}{(s+a)^2+b^2}$

An important property of transforms is the **linearity property,**

$$L[af(t) + bg(t)] = aL(f) + bL(g) \tag{30-23}$$

We state this property here since it determines that the transform of a sum of functions is the sum of the transforms. This is of definite importance when dealing with a sum of functions. This property is a direct result of the definition of the Laplace transform.

Another Laplace transform important to the solution of a differential equation is the transform of the derivative of a function. Let us first find the Laplace transform of the first derivative of a function.

By definition,

$$L(f') = \int_0^{\infty} e^{-st} f'(t)\, dt$$

To integrate by parts, let $u = e^{-st}$ and $dv = f'(t)\,dt$, so $du = -se^{-st}\,dt$ and $v = f(t)$ (the integral of the derivative of a function is the function). Therefore,

$$\begin{aligned} L(f') &= e^{-st}f(t)\Big|_0^{\infty} + s\int_0^{\infty} e^{-st} f(t)\, dt \\ &= 0 - f(0) + sL(f) \end{aligned}$$

It is noted that the integral in the second term on the right is the Laplace transform of $f(t)$ by definition. Therefore, *the Laplace transform of the first derivative of a function is*

$$L(f') = sL(f) - f(0) \tag{30-24}$$

Applying the same analysis, we may find *the Laplace transform of the second derivative of a function. It is*

$$L(f'') = s^2L(f) - sf(0) - f'(0) \tag{30-25}$$

Here it is necessary to integrate by parts twice to derive the result. The transforms of higher derivatives are found in a similar manner.

Equations (30-24) and (30-25) allow us to express the transform of each derivative in terms of s and the transform itself. This is illustrated in the following example.

EXAMPLE 3 ▸▸ Given that $f(0) = 0$ and $f'(0) = 1$, express the transform of $f''(t) - 2f'(t)$ in terms of s and the transform of $f(t)$.

By using the linearity property and the transforms of the derivatives, we have

$$\begin{aligned} L[f''(t) - 2f'(t)] &= L(f'') - 2L(f') && \text{using Eq. (30-23)} \\ &= [s^2L(f) - sf(0) - f'(0)] - 2[sL(f) - f(0)] && \text{using Eqs. (30-25) and (30-24)} \\ &= [s^2L(f) - s(0) - 1] - 2[sL(f) - 0] && \text{substitute given values} \\ &= (s^2 - 2s)L(f) - 1 \end{aligned}$$ ◂◂

If the Laplace transform of a function is known, it is then possible to find the function by finding the **inverse transform,**

$$L^{-1}(F) = f(t) \tag{30-26}$$

where L^{-1} denotes the inverse transform.

EXAMPLE 4 ▸▸ If $F(s) = \dfrac{s}{s^2 + a^2}$, from Transform (5) of the table we see that

$$L^{-1}(F) = L^{-1}\left(\frac{s}{s^2 + a^2}\right) = \cos at$$

or

$$f(t) = \cos at \quad ◂◂$$

EXAMPLE 5 ▸▸ If $(s^2 - 2s)L(f) - 1 = 0$, then

$$L(f) = \frac{1}{s^2 - 2s} \quad \text{or} \quad F(s) = \frac{1}{s(s - 2)}$$

Therefore, we have

$$\begin{aligned} f(t) = L^{-1}(F) &= L^{-1}\left[\frac{1}{s(s-2)}\right] && \text{inverse transform} \\ &= -\frac{1}{2}L^{-1}\left[\frac{-2}{s(s-2)}\right] && \text{fit form of Transform (4)} \\ &= -\frac{1}{2}(1 - e^{2t}) && \text{use Transform (4)} \quad ◂◂ \end{aligned}$$

EXAMPLE 6 ▸▸ If $F(s) = \dfrac{s + 5}{s^2 + 6s + 10}$, then

$$L^{-1}(F) = L^{-1}\left[\frac{s + 5}{s^2 + 6s + 10}\right]$$

It appears that this function does not fit any of the forms given. However,

$$s^2 + 6s + 10 = (s^2 + 6s + 9) + 1 = (s + 3)^2 + 1$$

By writing $F(s)$ as

$$F(s) = \frac{(s + 3) + 2}{(s + 3)^2 + 1} = \frac{s + 3}{(s + 3)^2 + 1} + \frac{2}{(s + 3)^2 + 1}$$

we can find the inverse of each term. Therefore,

$$L^{-1}(F) = e^{-3t}\cos t + 2e^{-3t}\sin t \qquad \text{using Transforms (20) and (19)}$$

or

$$f(t) = e^{-3t}(\cos t + 2 \sin t) \quad ◂◂$$

Exercises 30–10

In Exercises 1–4, verify the indicated transforms given in the table.

1. Transform 1
2. Transform 3
3. Transform 6
4. Transform 11

In Exercises 5–12, find the transforms of the given functions by use of the table.

5. $f(t) = e^{3t}$

6. $f(t) = 1 - \cos 2t$

7. $f(t) = t^3e^{-2t}$

8. $f(t) = 2e^{-3t} \sin 4t$

9. $f(t) = \cos 2t - \sin 2t$

10. $f(t) = 2t \sin 3t + e^{-3t} \cos t$

11. $f(t) = 3 + 2t \cos 3t$

12. $f(t) = t^3 - 3te^{-t}$

In Exercises 13–16, express the transforms of the given expressions in terms of s and L(f).

13. $y'' + y', f(0) = 0, f'(0) = 0$

14. $y'' - 3y', f(0) = 2, f'(0) = -1$

15. $2y'' - y' + y, f(0) = 1, f'(0) = 0$

16. $y'' - 3y' + 2y, f(0) = -1, f'(0) = 2$

In Exercises 17–24, find the inverse transforms of the given functions of s.

17. $F(s) = \dfrac{2}{s^3}$

18. $F(s) = \dfrac{3}{s^2 + 4}$

19. $F(s) = \dfrac{1}{2s + 6}$

20. $F(s) = \dfrac{3}{s^4 + 4s^2}$

21. $F(s) = \dfrac{1}{s^3 + 3s^2 + 3s + 1}$

22. $F(s) = \dfrac{s^2 - 1}{s^4 + 2s^2 + 1}$

23. $F(s) = \dfrac{s + 2}{(s^2 + 9)^2}$

24. $F(s) = \dfrac{s + 3}{s^2 + 4s + 13}$

30-11 Solving Differential Equations by Laplace Transforms

We will now show how certain differential equations can be solved by using Laplace transforms. *It must be remembered that these solutions are the* ***particular*** *solutions of the equations subject to the given conditions.* The necessary operations were developed in the preceding section. The following examples illustrate the method.

EXAMPLE 1 ▸▸ Solve the differential equation $2y' - y = 0$ if $y(0) = 1$. (Note that we are using y to denote the function.)

Taking transforms of each term in the equation, we have

$$L(2y') - L(y) = L(0)$$
$$2L(y') - L(y) = 0$$

$L(0) = 0$ by direct use of the definition of the transform. Now, using Eq. (30-24), $L(y') = sL(y) - y(0)$, we have

$$2[sL(y) - 1] - L(y) = 0 \quad y(0) = 1$$

Solving for $L(y)$, we obtain

$$2sL(y) - L(y) = 2$$
$$L(y) = \frac{2}{2s - 1} = \frac{1}{s - \frac{1}{2}}$$

Finding the inverse transform, we have

$$y = e^{t/2} \quad \text{using Transform (3)}$$

The reader should check this solution with that obtained by methods developed earlier. Also, it should be noted that the solution was essentially an algebraic one.

NOTE ▸ This points out the power and usefulness of Laplace transforms. ***We are able to translate a differential equation into an algebraic form,*** which can in turn be translated into the solution of the differential equation. Thus we are able to solve a differential equation by using algebra and specific algebraic forms. ◂◂

EXAMPLE 2 ▸▸ Solve the differential equation $y'' + 2y' + 2y = 0$, if $y(0) = 0$ and $y'(0) = 1$.

Using the same steps as outlined in Example 1, we have

$$L(y'') + 2L(y') + 2L(y) = 0 \qquad \text{take transforms}$$

$$[s^2L(y) - sy(0) - y'(0)] + 2[sL(y) - y(0)] + 2L(y) = 0 \qquad \text{using Eqs. (30-25) and (30-24)}$$

$$[s^2L(y) - s(0) - 1] + 2[sL(y) - 0] + 2L(y) = 0 \qquad \text{substitute given values}$$

$$s^2L(y) - 1 + 2sL(y) + 2L(y) = 0$$

$$(s^2 + 2s + 2)L(y) = 1 \qquad \text{solve for } L(y)$$

$$L(y) = \frac{1}{s^2 + 2s + 2} = \frac{1}{(s + 1)^2 + 1} \qquad \text{take inverse transform}$$

$$y = e^{-t} \sin t \qquad \text{using Transform (19)}$$ ◂◂

EXAMPLE 3 ▸▸ Solve the differential equation $y'' + y = \cos t$ if $y(0) = 1$ and $y'(0) = 2$.

Using transforms, we have the following solution.

$$L(y'') + L(y) = L(\cos t) \qquad \text{take transforms}$$

$$[s^2L(y) - s(1) - 2] + L(y) = \frac{s}{s^2 + 1} \qquad \text{using Eq. (30-25) and Transform (5)}$$

$$(s^2 + 1)L(y) = \frac{s}{s^2 + 1} + s + 2$$

$$L(y) = \frac{s}{(s^2 + 1)^2} + \frac{s}{s^2 + 1} + \frac{2}{s^2 + 1}$$

$$y = \frac{t}{2} \sin t + \cos t + 2 \sin t \qquad \text{using Transforms (16), (5), (6)}$$ ◂◂

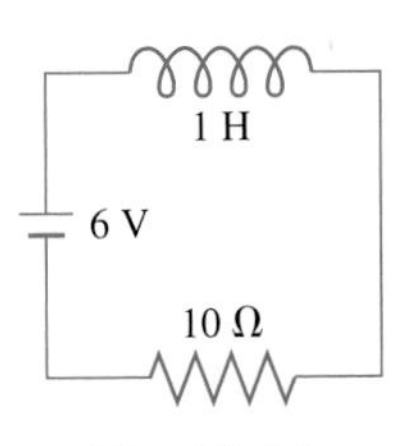

Fig. 30-14

EXAMPLE 4 ▸▸ An electric circuit contains a 1-H inductor, a 10-Ω resistor, and a 6-V battery. See Fig. 30-14. If the initial current is zero, find the current i of the circuit as a function of time t.

The differential equation for this circuit is

$$\frac{di}{dt} + 10i = 6 \qquad \text{using Eq. (30-20)}$$

Following the procedures outlined in the previous examples, the solution is found:

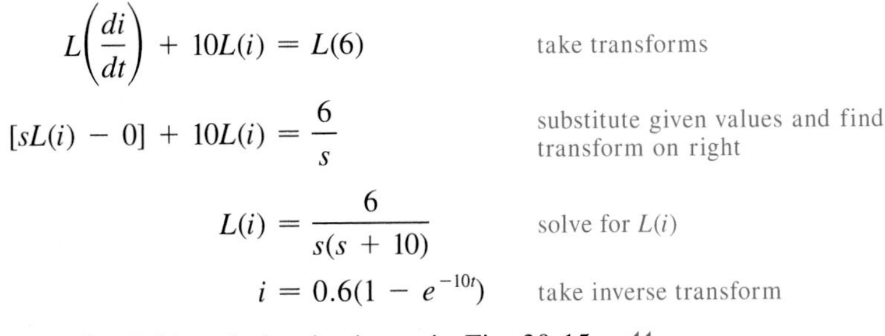

$$L\left(\frac{di}{dt}\right) + 10L(i) = L(6) \qquad \text{take transforms}$$

$$[sL(i) - 0] + 10L(i) = \frac{6}{s} \qquad \text{substitute given values and find transform on right}$$

$$L(i) = \frac{6}{s(s + 10)} \qquad \text{solve for } L(i)$$

$$i = 0.6(1 - e^{-10t}) \qquad \text{take inverse transform}$$

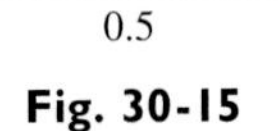
Fig. 30-15

The graph of this solution is shown in Fig. 30-15. ◂◂

EXAMPLE 5 ▸▸ A spring is stretched 0.31 m by a weight of 4.9 N (mass of 0.50 kg). The medium resists the motion of the object with a force of $4v$, where v is the velocity of motion. The differential equation describing the displacement y is

$$\frac{1}{2}\frac{d^2y}{dt^2} + 4\frac{dy}{dt} + 16y = 0$$

Find y as a function of time t, if $y(0) = 1$ and $dy/dt = 0$ for $t = 0$.

Clearing fractions and denoting derivatives by y'' and y', we have the following differential equation and solution.

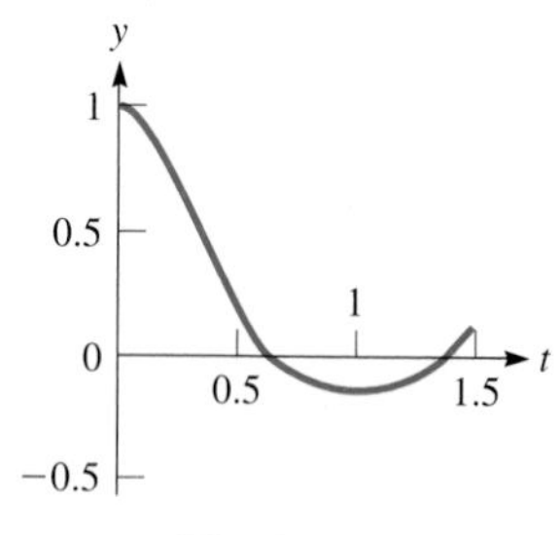

Fig. 30-16

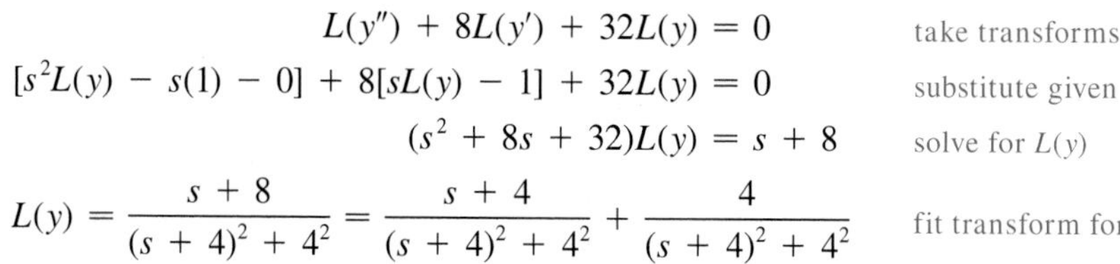

$$y'' + 8y' + 32y = 0$$

$$L(y'') + 8L(y') + 32L(y) = 0 \quad \text{take transforms}$$

$$[s^2L(y) - s(1) - 0] + 8[sL(y) - 1] + 32L(y) = 0 \quad \text{substitute given values}$$

$$(s^2 + 8s + 32)L(y) = s + 8 \quad \text{solve for } L(y)$$

$$L(y) = \frac{s+8}{(s+4)^2 + 4^2} = \frac{s+4}{(s+4)^2 + 4^2} + \frac{4}{(s+4)^2 + 4^2} \quad \text{fit transform forms}$$

$$y = e^{-4t}\cos 4t + e^{-4t}\sin 4t$$

$$= e^{-4t}(\cos 4t + \sin 4t)$$

The graph of this solution is shown in Fig. 30-16. ◂◂

Exercises 30–11

In the following exercises, solve the given differential equations by using Laplace transforms, where the function is subject to the given conditions.

1. $y' + y = 0;\ y(0) = 1$
2. $y' - 2y = 0,\ y(0) = 2$
3. $2y' - 3y = 0,\ y(0) = -1$
4. $y' + 2y = 1,\ y(0) = 0$
5. $y' + 3y = e^{-3t},\ y(0) = 1$
6. $y' + 2y = te^{-2t},\ y(0) = 0$
7. $y'' + 4y = 0,\ y(0) = 0,\ y'(0) = 1$
8. $y'' - 4y = 0,\ y(0) = 2,\ y'(0) = 0$
9. $y'' + 2y' = 0,\ y(0) = 0,\ y'(0) = 2$
10. $y'' + 2y' + y = 0,\ y(0) = 0,\ y'(0) = -2$
11. $y'' - 4y' + 5y = 0,\ y(0) = 1,\ y'(0) = 2$
12. $4y'' + 4y' + y = 0,\ y(0) = 1,\ y'(0) = 0$
13. $y'' + y = 1,\ y(0) = 1,\ y'(0) = 1$
14. $y'' + 4y = 2t,\ y(0) = 0,\ y'(0) = 0$
15. $y'' + 2y' + y = e^{-t},\ y(0) = 1,\ y'(0) = 2$
16. $y'' + 4y = \sin 2t,\ y(0) = 0,\ y'(0) = 0$
17. A constant force of 6 N moves a 2-kg mass through a medium that resists the motion with a force equal to v, where v is the velocity. The differential equation relating the velocity and the time is $2\frac{dv}{dt} = 6 - v$. Find v as a function of time if the object starts from rest.
18. A pendulum moves with simple harmonic motion according to the differential equation $D^2\theta + 20\theta = 0$, where θ is the angular displacement and $D = d/dt$. Find θ as a function of t if $\theta = 0$ and $D\theta = 0.40$ rad/s when $t = 0$.
19. A 50-Ω resistor, a 4.0-μF capacitor, and a 40-V battery are connected in series. Find the charge as a function of the time if the initial charge on the capacitor is zero.
20. A 2-H inductor, an 80-Ω resistor, and an 8-V battery are connected in series. Find the current in the circuit as a function of time if the initial current is zero.
21. A 10-H inductor, a 40-μF capacitor, and a voltage supply whose voltage is given by $100 \sin 50t$ are connected in series in an electric circuit. Find the current as a function of time if the initial charge on the capacitor and the initial current are zero.
22. A 20-mH inductor, a 40-Ω resistor, a 50-μF capacitor, and a voltage source of $100e^{-1000t}$ are connected in series in a circuit. Find the charge on the capacitor as a function of time if $q = 0$ and $i = 0$ when $t = 0$.
23. The weight on a spring undergoes forced vibrations according to the equation $D^2y + 9y = 18 \sin 3t$. Find its displacement y as a function of the time t, if $y = 0$ and $Dy = 0$ when $t = 0$.
24. A spring is such that it is stretched 1 m by a 2-kg (20-N) weight. The spring is stretched 0.5 m below the equilibrium position with the weight on it and is released. If the spring is in a medium that resists the motion with a force equal to $12v$, find the displacement y as a function of time.

Chapter Equations, Review Exercises, and Practice Test

Chapter Equations

Separation of variables	$M(x, y)\,dx + N(x, y)\,dy = 0$	(30-1)
	$A(x)\,dx + B(y)\,dy = 0$	(30-2)
Integrable combinations	$d(xy) = x\,dy + y\,dx$	(30-3)
	$d(x^2 + y^2) = 2(x\,dx + y\,dy)$	(30-4)
	$d\left(\dfrac{y}{x}\right) = \dfrac{x\,dy - y\,dx}{x^2}$	(30-5)
	$d\left(\dfrac{x}{y}\right) = \dfrac{y\,dx - x\,dy}{y^2}$	(30-6)
Linear differential equation of first order	$dy + Py\,dx = Q\,dx$	(30-7)
	$ye^{\int P\,dx} = \int Qe^{\int P\,dx}\,dx + c$	(30-8)
Electric circuit	$L\dfrac{di}{dt} + Ri + \dfrac{q}{C} = E$	(30-9)
Motion in resisting medium	$m\dfrac{dv}{dt} = F - kv$	(30-10)
General linear differential equation	$a_0\dfrac{d^ny}{dx^n} + a_1\dfrac{d^{n-1}y}{dx^{n-1}} + \cdots + a_{n-1}\dfrac{dy}{dx} + a_ny = b$	(30-11)
	$a_0D^ny + a_1D^{n-1}y + \cdots + a_{n-1}Dy + a_ny = b$	(30-12)
Homogeneous linear differential equation	$a_0D^2y + a_1Dy + a_2y = 0$	(30-13)
Auxiliary equation	$a_0m^2 + a_1m + a_2 = 0$	(30-14)
Distinct roots	$y = c_1e^{m_1x} + c_2e^{m_2x}$	(30-15)
Repeated roots	$y = e^{mx}(c_1 + c_2x)$	(30-16)
Complex roots	$y = e^{\alpha x}(c_1 \sin \beta x + c_2 \cos \beta x)$	(30-17)
Nonhomogeneous linear differential equation	$a_0D^2y + a_1Dy + a_2y = b$	(30-18)
	$y = y_c + y_p$	(30-19)
Electric circuit	$L\dfrac{d^2q}{dt^2} + R\dfrac{dq}{dt} + \dfrac{q}{C} = E$	(30-20)
Laplace transforms	$F(s) = \int_0^\infty e^{-st}f(t)\,dt$	(30-21)
	$F(s) = L(f) = \int_0^\infty e^{-st}f(t)\,dt$	(30-22)
	$L[af(t) + bg(t)] = aL(f) + bL(g)$	(30-23)
	$L(f') = sL(f) - f(0)$	(30-24)
	$L(f'') = s^2L(f) - sf(0) - f'(0)$	(30-25)
Inverse transform	$L^{-1}(F) = f(t)$	(30-26)

Review Exercises

In Exercises 1–28, find the general solution of the given differential equations.

1. $4xy^3\,dx + (x^2 + 1)\,dy = 0$
2. $\frac{dy}{dx} = e^{x-y}$
3. $\frac{dy}{dx} + 2y = e^{-2x}$
4. $x\,dy + y\,dx = y\,dy$
5. $2D^2y + Dy = 0$
6. $2D^2y - 5Dy + 2y = 0$
7. $y'' + 2y' + y = 0$
8. $y'' + 2y' + 2y = 0$
9. $(x + y)\,dx + (x + y^3)\,dy = 0$
10. $y \ln x\,dx = x\,dy$
11. $x\frac{dy}{dx} - 3y = x^2$
12. $dy - 2y\,dx = (x - 2)e^x\,dx$
13. $dy = 2y\,dx + y^2\,dx$
14. $x^2y\,dy = (1 + x)\csc y\,dx$
15. $D^2y + 2Dy + 6y = 0$
16. $4D^2y - 4Dy + y = 0$
17. $y' + 4y = 2$
18. $2xy\,dx = (2y - \ln y)\,dy$
19. $\sin x\frac{dy}{dx} + y\cos x + x = 0$
20. $y\,dy = (x^2 + y^2 - x)\,dx$
21. $2\frac{d^2y}{dx^2} + \frac{dy}{dx} - 3y = 6$
22. $\frac{d^2y}{dx^2} + 6\frac{dy}{dx} + 9y = 3x$
23. $y'' + y' - y = 2e^x$
24. $4D^2y + 9y = xe^x$
25. $9D^2y - 18Dy + 8y = 16 + 4x$
26. $y'' + y = 4\cos 2x$
27. $D^2y + 9y = \sin x$
28. $y'' + y' = e^x + \cos 2x$

In Exercises 29–36, find the indicated particular solution of the given differential equations.

29. $3y' = 2y\cot x;\ \ x = \frac{\pi}{2}$ when $y = 2$
30. $x\,dy - y\,dx = y^3\,dy;\ \ x = 1$ when $y = 3$
31. $y' = 4x - 2y;\ \ x = 0$ when $y = -2$
32. $xy^2\,dx + e^x\,dy = 0;\ \ x = 0$ when $y = 2$
33. $\frac{d^2y}{dx^2} + \frac{dy}{dx} + 4y = 0;\ \ Dy = \sqrt{15},\ y = 0$ when $x = 0$
34. $5y'' + 7y' - 6y = 0;\ \ y' = 10,\ y = 2$ when $x = 0$
35. $(D^2 + 4D + 4)y = 4\cos x;\ \ Dy = 1,\ y = 0$ when $x = 0$
36. $y'' - 2y' + y = e^{2x} + x;\ \ Dy = 0,\ y = 2$ when $x = 0$

In Exercises 37–44, solve the given differential equations by using Laplace transforms where the function is subject to the given conditions.

37. $4y' - y = 0,\ y(0) = 1$
38. $2y' - y = 4,\ y(0) = 1$
39. $y' - 3y = e^t,\ y(0) = 0$
40. $y' + 2y = e^{-2t},\ y(0) = 2$
41. $y'' + y = 0,\ y(0) = 0,\ y'(0) = -4$
42. $y'' + 4y' + 5y = 0,\ y(0) = 1,\ y'(0) = 1$
43. $y'' + 9y = 3t,\ y(0) = 0,\ y'(0) = -1$
44. $y'' + 4y' + 4y = 2e^{-2t},\ y(0) = 0,\ y'(0) = 1$

In Exercises 45–72, solve the given problems.

45. The time rate of change of volume of an evaporating substance is proportional to the surface area. Express the radius of an evaporating sphere of ice as a function of time. Let $r = r_0$ when $t = 0$. (*Hint:* Express both V and A in terms of the radius r.)
46. An insulated tank is filled with a solution containing radioactive cobalt. Due to the radioactivity, energy is released and the temperature T (in °C) of the solution rises with the time t (in hours). The following differential equation expresses the relation between temperature and time for a specific case:

$$56\,600 = 262(T - 70) + 20\,200\frac{dT}{dt}$$

If the initial temperature is 70°C, what is the temperature 24 h later?
47. An object with a mass of 1.00 kg slides down a long inclined plane. The effective force of gravity is 4.00 N, and the motion is retarded by a force numerically equal to the velocity. If the object starts from rest, what is its velocity (in m/s) 4.00 s later?
48. A 760-N object falls from rest under the influence of gravity. Find the equation for the velocity at any time t if the air resists the motion with a force numerically equal to twice the velocity.
49. Initially there are 500 mg of the radioactive hydrogen isotope tritium. After 4.02 years, 400 mg remain. What is the half-life of tritium?
50. When a gas undergoes an adiabatic change (no gain or loss of heat), the rate of change of pressure with respect to volume is directly proportional to the pressure and inversely proportional to the volume. Express the pressure in terms of the volume.

51. Under ideal conditions, the natural law of population change is that the population increases at a rate proportional to the population at any time. Under these conditions, project the population of the world in the year 2000 if it reached 4.0 billion in 1976 and 5.0 billion in 1987.

52. A spherical balloon is being blown up such that its volume V increases at a rate directly proportional to its surface area. Show that this leads to the differential equation $dV/dt = kV^{2/3}$, and solve for V as a function of t.

53. Find the orthogonal trajectories of the family of curves $y = cx^5$.

54. Find the equation of the curves such that their normals at all points are in the direction with the lines connecting the points and the origin.

55. If a circuit contains a resistance R, a capacitance C, and a source of voltage E, express the charge q on the capacitor as a function of time.

56. A 2-H inductor, a 40-Ω resistor, and a 20-V battery are connected in series. Find the current in the circuit as a function of time if the initial current is zero.

57. A certain spring stretches 0.50 m by a 40-N weight. With this weight suspended on it the spring is stretched 0.50 m beyond the equilibrium position and released. Find the equation of the resulting motion if the medium in which the weight is suspended retards the motion with a force equal to 16 times the velocity. Classify the motion as underdamped, critically damped, or overdamped.

58. The end of a vibrating rod moves according to the equation $D^2y + 0.2Dy + 4000y = 0$, where y is the displacement and $D = d/dt$. Find y as a function of t if $y = 3.00$ cm and $Dy = -0.300$ cm/s when $t = 0$.

59. A 0.5-H inductor, a 6-Ω resistor, and a 20-mF capacitor are connected in series with a generator for which $E = 24 \sin 10t$. Find the charge on the capacitor as a function of time if the initial charge and current are zero.

60. A 5.00-mH inductor and a 10.0-μF capacitor are connected in series with a voltage source of $0.200e^{-200t}$ V. Find the charge on the capacitor as a function of time if $q = 0$ and $i = 4.00$ mA when $t = 0$.

61. Find the equation for the current as a function of time if a resistance of 20 Ω, an inductor of 4 H, a capacitor of 100 μF, and a battery of 100 V are in series. The initial charge on the capacitor is 10 mC, and the initial current is zero.

62. If a circuit contains an inductor of inductance L, a capacitor of capacitance C, and a sinusoidal source of voltage $E_0 \sin \omega t$, express the charge q on the capacitor as a function of the time. Assume $q = 0$, $i = 0$ when $t = 0$.

63. The differential equation relating current and time for a certain electric circuit is $2\dfrac{di}{dt} + i = 12$. Solve this equation by use of Laplace transforms given that the initial current is zero. Evaluate the current for $t = 0.300$ s.

64. A 6.0-H inductor and a 30-Ω resistor are connected in series. Find the current as a function of time if the initial current is 15 A. Use Laplace transforms.

65. A 0.25-H inductor, a 4.0-Ω resistor, and a 100-μF capacitor are connected in series. If the initial charge on the capacitor is 400μC and the initial current is zero, find the charge on the capacitor as a function of time. Use Laplace transforms.

66. An inductor of 0.5 H, a resistor of 6 Ω, and a capacitor of 200 μF are connected in series. If the initial charge on the capacitor is 10 mC and the initial current is zero, find the charge on the capacitor as a function of time after the switch is closed. Use Laplace transforms.

67. A mass of 0.25 kg stretches a spring for which $k = 16$ N/m. An external force of $\cos 8t$ is applied to the spring. Express the displacement y of the object as a function of time if the initial displacement and velocity are zero. Use Laplace transforms.

68. A spring is stretched 1.00 m by a mass of 5.00 kg (assume the weight to be 50.0 N). Find y, the displacement of the object, as a function of time if $y(0) = 1$ and $dy/dt = 0$ when $t = 0$. Use Laplace transforms.

69. The approximate differential equation relating the displacement y of a beam at a horizontal distance x from one end is $EI\dfrac{d^2y}{dx^2} = M$, where E is the modulus of elasticity, I is the moment of inertia of the cross section of the beam perpendicular to its axis, and M is the bending moment at the cross section. If $M = 2000x - 40x^2$ for a particular beam of length L for which $y = 0$ when $x = 0$ and when $x = L$, express y in terms of x. Consider E and I as constants.

70. Air containing 20% oxygen passes into a 5.00-L container initially filled with 100% oxygen. A uniform mixture of the air and oxygen then passes from the container at the same rate. What volume of oxygen is in the container after 5.00 L of air have passed into it?

71. When a circular disk of mass m and radius r is suspended by a wire at the center on one of its flat faces and the disk is twisted through an angle θ, torsion in the wire tends to turn the disk back in the opposite direction. The differential equation for this case is $\frac{1}{2}mr^2\frac{d^2\theta}{dt^2} = -k\theta$, where k is a constant. Determine the equation of motion if $\theta = \theta_0$ and $d\theta/dt = \omega_0$ when $t = 0$. See Fig. 30-17.

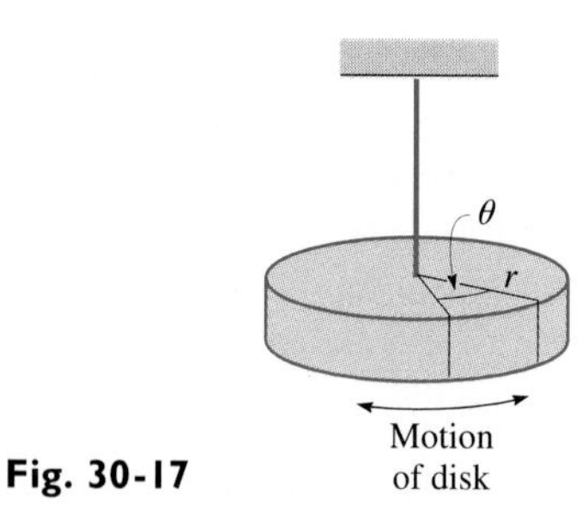

Fig. 30-17

72. The gravitational acceleration of an object is inversely proportional to the square of its distance r from the center of the earth. Use the chain rule, Eq. (23-14), to show that the acceleration is $\frac{dv}{dt} = v\frac{dv}{dr}$, where $v = \frac{dr}{dt}$ is the velocity of the object. Then solve for v as a function r if $dv/dt = -g$ and $v = v_0$ for $r = R$, where R is the radius of the earth. Finally, show that a spacecraft must have a velocity of at least $v_0 = \sqrt{2gR}$ (= 11 km/s) in order to escape from the earth's gravitation. (Note the expression for v^2 as $r \to \infty$.)

Writing Exercise

73. An electric circuit contains an inductor L, a resistor R, and a battery of voltage E. The initial current in the circuit is zero. Write three or four paragraphs explaining how the differential equation for the current in the circuit is solved using (a) separation of variables, (b) the linear differential equation of the first order, and (c) Laplace transforms.

Practice Test

In Problems 1–5, find the general solution of each of the given differential equations.

1. $x\frac{dy}{dx} + 2y = 4$
2. $y'' + 2y' + 5y = 0$
3. $x\,dx + y\,dy = x^2\,dx + y^2\,dx$
4. $2D^2y - Dy = 2\cos x$
5. $\frac{d^2y}{dx^2} - 4\frac{dy}{dx} + 4y = 3x$
6. Find the particular solution of the differential equation $(xy + y)\frac{dy}{dx} = 2$, if $y = 2$ when $x = 0$.
7. If interest in a bank account is compounded continuously, the amount grows at a rate that is proportional to the amount present. Derive the equation for the amount A in an account in which the initial amount is A_0 and the interest rate is r as a function of the time t after A_0 is deposited.
8. Using Laplace transforms, solve the differential equation $y'' + 9y = 9$, if $y(0) = 0$ and $y'(0) = 1$.
9. Find the equation for the current as a function of the time (in seconds) in a circuit containing a 2-H inductance, an 8-Ω resistance, and a 6-V battery in series, if $i = 0$ for $t = 0$.
10. A mass of 0.50 kg stretches a spring for which $k = 32$ N/m. With this weight attached, the spring is pulled 0.3 m longer than its equilibrium length and released. Find the equation of the resulting motion, assuming no damping. (The acceleration due to gravity is 9.8 m/s^2.)

Supplementary Topics

S-1 Gaussian Elimination

In Chapters 5 and 16 we developed several methods of solving systems of simultaneous linear equations. These included algebraic methods, determinants, and the use of the inverse of a matrix.

We now show a general method that can be used to solve a system of linear equations. The procedure is similar to that used in finding the inverse of a matrix in Section 16-5 and is known as **Gaussian elimination**. It is commonly used in computer programs, particularly when there are several equations in the system.

Gaussian elimination is based on the use of the following two operations on the equations of a system.

1. *Both sides of an equation may be multiplied by a constant.*
2. *A multiple of one equation may be added to another equation.*

Using three linear equations in three unknowns as an example, by using the above operations we can change the system

$$\begin{aligned} a_1x + b_1y + c_1z &= d_1 \\ a_2x + b_2y + c_2z &= d_2 \\ a_3x + b_3y + c_3z &= d_3 \end{aligned} \qquad \text{(S-1)}$$

into the equivalent system

$$\begin{aligned} x + b_4y + c_4z &= d_4 \\ y + c_5z &= d_5 \\ z &= d_6 \end{aligned} \qquad \text{(S-2)}$$

The solution is now completed by substituting the known value of z into the second equation, and then substituting the known values of y and z into the first equation.

EXAMPLE 1 Solve the given system of equations by Gaussian elimination.

Given system of equations (equations and solution at left)

$$\begin{aligned} 2x + y &= 4 \\ 3x - 2y &= 3 \end{aligned}$$

We want the coefficient of x in the first equation to be 1. Therefore, divide the first equation by 2, the coefficient of x.

$$\begin{aligned} x + \tfrac{1}{2}y &= 2 \\ 3x - 2y &= 3 \end{aligned}$$

We next eliminate x in the second equation by subtracting 3 times the first equation from the second equation.

$$\begin{aligned} x + \tfrac{1}{2}y &= 2 \\ -\tfrac{7}{2}y &= -3 \end{aligned}$$

Now we solve the second equation for y by dividing by $-\frac{7}{2}$.

$$\begin{aligned} x + \tfrac{1}{2}y &= 2 \\ y &= \tfrac{6}{7} \end{aligned}$$

To find the value of x, we substitute the value of $y = \frac{6}{7}$ into the first equation.

$$\begin{aligned} x + \tfrac{1}{2}(\tfrac{6}{7}) &= 2 \\ x &= \tfrac{11}{7} \end{aligned}$$

The solution is $x = \frac{11}{7}$, $y = \frac{6}{7}$, which checks when substituted into the original equations.

EXAMPLE 2 ▸▸ Solve the given system of equations by Gaussian elimination.

$$\begin{aligned} x + 3y - 2z &= -5 \\ 2x - y + 4z &= 7 \\ -3x + 2y - 3z &= -1 \end{aligned}$$

Given system of equations (equations and solution at left)

$$\begin{aligned} x + 3y - 2z &= -5 \\ -7y + 8z &= 17 \\ 11y - 9z &= -16 \end{aligned}$$

Since the coefficient of x in the first equation is 1, we proceed to the next step. Subtract 2 times first equation from second equation and add three times first equation to third equation to eliminate x from the second and third equations.

$$\begin{aligned} x + 3y - 2z &= -5 \\ y - \tfrac{8}{7}z &= -\tfrac{17}{7} \\ 11y - 9z &= -16 \end{aligned}$$

To get the coefficient of y equal to 1 in the second equation, we divide it by -7.

$$\begin{aligned} x + 3y - 2z &= -5 \\ y - \tfrac{8}{7}z &= -\tfrac{17}{7} \\ \tfrac{25}{7}z &= \tfrac{75}{7} \end{aligned}$$

To eliminate y from the third equation, we subtract 11 times the second equation from the third equation.

$$\begin{aligned} x + 3y - 2z &= -5 \\ y - \tfrac{8}{7}z &= -\tfrac{17}{7} \\ z &= 3 \end{aligned}$$

We solve for z by multiplying the third equation by $\frac{7}{25}$ (or dividing by $\frac{25}{7}$).

$$\begin{aligned} y - \tfrac{8}{7}(3) &= -\tfrac{17}{7} \\ y &= 1 \end{aligned}$$

The value of y is found by substituting $z = 3$ into the second equation.

$$\begin{aligned} x + 3(1) - 2(3) &= -5 \\ x &= -2 \end{aligned}$$

The value of x is found by substituting $y = 1$ and $z = 3$ back into the first equation.

The solution is $x = -2$, $y = 1$, $z = 3$, which checks when substituted into the original equations. ◂◂

EXAMPLE 3 ▸▸ Solve the given system of equations by Gaussian elimination.

$$\begin{aligned} 4y + z &= 2 \\ 2x + 6y - 2z &= 3 \\ 4x + 8y - 5z &= 4 \end{aligned}$$

Given system of equations (equations and solution at left)

$$\begin{aligned} x + 3y - z &= \tfrac{3}{2} \\ 4y + z &= 2 \\ 4x + 8y - 5z &= 4 \end{aligned}$$

Since the first equation does not contain x, which means that $a_1 = 0$, we cannot divide by a_1. Therefore, we interchange the first and second equations, then divide the new first equation by 2.

$$\begin{aligned} x + 3y - z &= \tfrac{3}{2} \\ 4y + z &= 2 \\ -4y - z &= -2 \end{aligned}$$

We eliminate x from the third equation by subtracting 4 times the first equation from the third equation.

$$\begin{aligned} x + 3y - z &= \tfrac{3}{2} \\ y + \tfrac{1}{4}z &= \tfrac{1}{2} \\ -4y - z &= -2 \end{aligned}$$

Make the coefficient of y in the second equation equal to 1 by dividing the second equation by 4.

$$\begin{aligned} x + 3y - z &= \tfrac{3}{2} \\ y + \tfrac{1}{4}z &= \tfrac{1}{2} \\ 0 &= 0 \end{aligned}$$

Eliminate y from the third equation by adding 4 times the second equation to the third equation.

Since the third equation, $0 = 0$, is correct, we can continue. Although there is no specific value for z, it is possible to express both x and y in terms of z.

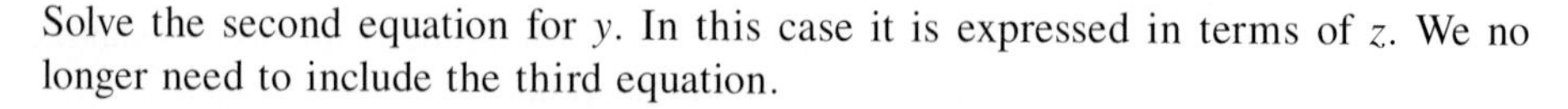

$$\begin{aligned} x + 3y - z &= \tfrac{3}{2} \\ y &= \tfrac{1}{2} - \tfrac{1}{4}z \end{aligned}$$

Solve the second equation for y. In this case it is expressed in terms of z. We no longer need to include the third equation.

$$\begin{aligned} x + 3\left(\tfrac{1}{2} - \tfrac{1}{4}z\right) - z &= \tfrac{3}{2} \\ x &= \tfrac{7}{4}z \end{aligned}$$

Substitute the solution for y in the first equation, and solve for x in terms of z.

We have expressed the solution as $x = \frac{7}{4}z$ and $y = \frac{1}{2} - \frac{1}{4}z$. Since both x and y are expressed in terms of z, and there is no specific value of z, the value of z can be chosen arbitrarily. This means ***there is an unlimited number of solutions.*** For example, if $z = 4$, $x = 7$ and $y = -\frac{1}{2}$. If $z = -2$, $x = -\frac{7}{2}$ and $y = 1$. ◂◂

CAUTION ▸

When we solved systems of linear equations in Chapters 5 and 16, we found that not all systems have unique solutions, as in Examples 1 and 2. In Example 3 we illustrated the use of Gaussian elimination on a system of equations for which the solution is not unique.

In Example 3 one of the equations became $0 = 0$, and there was an unlimited number of solutions. If any of the equations of a system becomes $0 = a$, $a \neq 0$, then the system is inconsistent and there is no solution.

If a system of equations has more unknowns than equations, or if it can be written in this way, as in Example 3, it usually has an unlimited number of solutions. It is possible, however, that such a system is inconsistent.

If a system of equations has more equations than unknowns, it is inconsistent unless enough equations become $0 = 0$ such that at least one solution is found. The following example illustrates two systems of equations in which there are more equations than unknowns.

EXAMPLE 4 ▸▸ Solve the following systems of equations by Gaussian elimination.

First system	Second system
$x + 2y = 5$	$x + 2y = 5$
$3x - y = 1$	$3x - y = 1$
$4x + y = 6$	$4x + y = 2$
$x + 2y = 5$	$x + 2y = 5$
$-7y = -14$	$-7y = -14$
$-7y = -14$	$-7y = -18$
$x + 2y = 5$	$x + 2y = 5$
$y = 2$	$y = 2$
$-7y = -14$	$-7y = -18$
$x + 2y = 5$	$x + 2y = 5$
$y = 2$	$y = 2$
$0 = 0$	$0 = -4$
$x = 1$	

The solutions are shown at the left. We note that each system has three equations and two unknowns.

In the solution of the first system, the third equation becomes $0 = 0$ and only two equations are needed to find the solution $x = 1$, $y = 2$.

In the solution of the second system, the third equation becomes $0 = -4$, which means the system is inconsistent and there is no solution.

The solutions are shown graphically in Figs. S1-1 and S1-2. In Fig. S1-1 each of the three lines passes through the point (1, 2), whereas in Fig. S1-2 there is no point common to the three lines. ◂◂

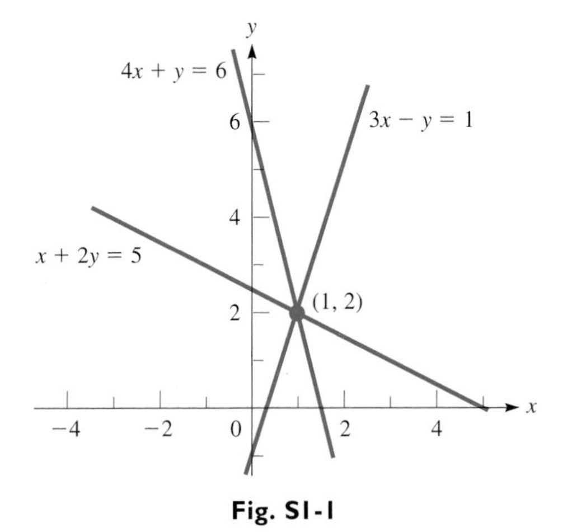

Fig. S1-1

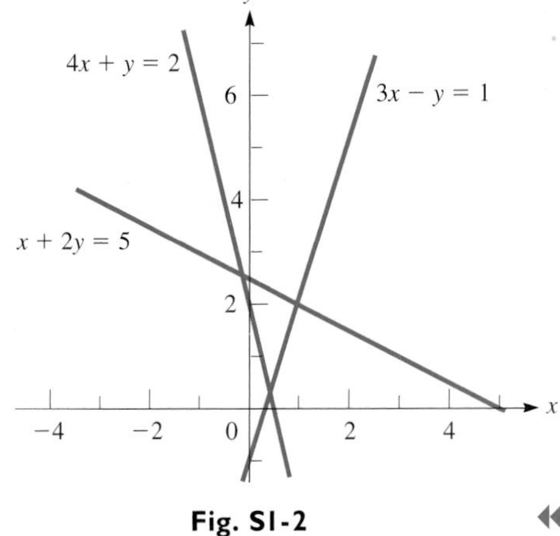

Fig. S1-2 ◂◂

Exercises S–1

In Exercises 1–20, solve the given systems of equations by Gaussian elimination. If there is an unlimited number of solutions, find two of them.

1. $\begin{aligned} x + 2y &= 4 \\ 3x - y &= 5 \end{aligned}$

2. $\begin{aligned} 2x + y &= 1 \\ 5x + 2y &= 1 \end{aligned}$

3. $\begin{aligned} 5x - 3y &= 2 \\ -2x + 4y &= 3 \end{aligned}$

4. $\begin{aligned} -3x + 2y &= 4 \\ 4x + y &= -5 \end{aligned}$

5. $\begin{aligned} 2x + y - z &= 0 \\ 4x + y + z &= 2 \\ -2x - 2y + 3z &= 0 \end{aligned}$

6. $\begin{aligned} 2y + 2z &= -1 \\ 3x - 4y + 3z &= 1 \\ 4x + 2y + 5z &= 4 \end{aligned}$

7. $\begin{aligned} x + 3y + 3z &= -3 \\ 2x + 2y + z &= -5 \\ -2x - y + 4z &= 6 \end{aligned}$

8. $\begin{aligned} 3x - y + 2z &= 3 \\ 4x - 2y + z &= 3 \\ 6x + 6y + 3z &= 4 \end{aligned}$

9. $\begin{aligned} w + 2x - y + 3z &= 12 \\ 2w - 2y - z &= 3 \\ 3x - y - z &= -1 \\ -w + 2x + y + 2z &= 3 \end{aligned}$

10. $\begin{aligned} 2x - 3y + 2z + 2t &= 3 \\ 4x + 2y - 3z &= -4 \\ 2x - y + 3z + 2t &= 3 \\ 6x + 3y - 2z - t &= 2 \end{aligned}$

11. $\begin{aligned} x - 4y + z &= 2 \\ 3x - y + 4z &= -4 \end{aligned}$

12. $\begin{aligned} 4x + z &= 6 \\ 2x - y - 2z &= -2 \end{aligned}$

13. $\begin{aligned} 2x - y + z &= 5 \\ 3x + 2y - 2z &= 4 \\ 5x + 8y - 8z &= 5 \end{aligned}$

14. $\begin{aligned} 3x + 2y - z &= 3 \\ 2x - y - 3z &= 2 \\ -x + 4y + 5z &= -1 \end{aligned}$

15. $\begin{aligned} 2x - 4y &= 7 \\ 3x + 5y &= -6 \\ 9x - 7y &= 15 \end{aligned}$

16. $\begin{aligned} 4x - y &= 5 \\ 2x + 2y &= 3 \\ 6x - 4y &= 7 \\ 2x + y &= 4 \end{aligned}$

17. $\begin{aligned} 3x + 5y &= -2 \\ 24x - 18y &= 13 \\ 15x - 33y &= 19 \\ 6x + 68y &= -33 \end{aligned}$

18. $\begin{aligned} x + 3y - z &= 1 \\ 3x - y + 4z &= 4 \\ -2x + 2y + 3z &= 17 \\ 3x + 7y + 5z &= 23 \end{aligned}$

19. $\begin{aligned} x - 2y - 2z &= 3 \\ 2x + y + 3z &= 4 \\ -2x - y - z &= 5 \\ 3x + 3y - 2z &= 2 \end{aligned}$

20. $\begin{aligned} 2x - y - 2z - t &= 4 \\ 4x + 2y + 3z + 2t &= 3 \\ -2x - y + 4z &= -2 \end{aligned}$

In Exercises 21–24, set up systems of equations and solve by Gaussian elimination.

21. One personal computer can perform x calculations per second, and a second personal computer can perform y calculations per second. If each operates for two seconds, 25.0 million calculations are performed. If the first operates for four seconds and the second for three seconds, 43.2 million calculations are performed. Find x and y.

22. The voltage across an electric resistor equals the current times the resistance. If a current of 3.00 A passes through each of two resistors, the sum of the voltages is 10.5 V. If 2.00 A passes through the first resistor and 4.00 A passes through the second resistor, the sum of the voltages is 13.0 V. Find the resistances (in Ω).

23. Three machines together produce 650 parts each hour. Twice the production of the second machine is 10 parts/h more than the sum of the other two machines. If the first operates for three hours and the others operate for two hours, 1550 parts are produced. Find the production rate of each machine.

24. A total of $12,000 is invested, part at 6.5%, part at 6.0%, and part at 5.5%, yielding a total annual interest of $726. The income from the 6.5% part yields $128 less than that for the other two parts combined. How much is invested at each rate?

S-2 Rotation of Axes

In Chapter 21 we discussed the circle, parabola, ellipse, and hyperbola and how these curves are represented by the second-degree equation

$$Ax^2 + Bxy + Cy^2 + Dx + Ey + F = 0 \qquad \text{(S2-1)}$$

Our discussion included the properties of the curves and their equations with center (vertex of a parabola) at the origin. However, except for the special case of the hyperbola $xy = c$, we did not cover what happens when the axes are rotated about the origin.

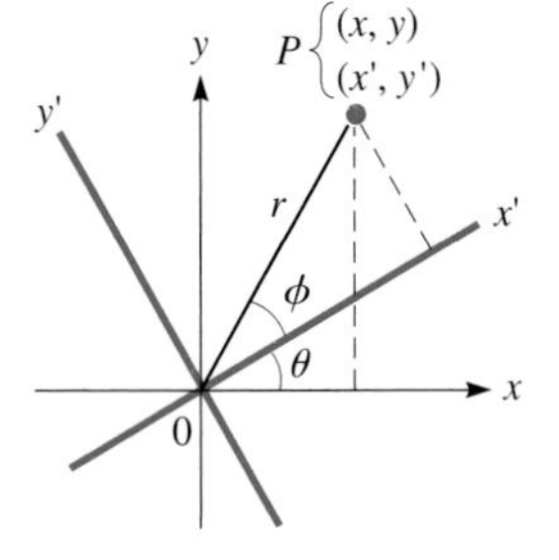

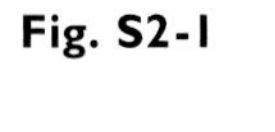

Fig. S2-1

If a set of axes is rotated about the origin through an angle θ, as shown in Fig. S2-1, we say that there has been a **rotation of axes.** In this case each point P in the plane has two sets of coordinates, (x, y) in the original system and (x', y') in the rotated system.

If we now let r equal the distance from the origin O to point P, and let ϕ be the angle between the x'-axis and the line OP, we have

$$x' = r\cos\phi, \qquad y' = r\sin\phi \tag{S2-2}$$

$$x = r\cos(\theta + \phi), \qquad y = r\sin(\theta + \phi) \tag{S2-3}$$

Using the cosine and sine of the sum of two angles, we can write Eqs. (S2-3) as

$$\begin{aligned} x &= r\cos\phi\cos\theta - r\sin\phi\sin\theta \\ y &= r\cos\phi\sin\theta + r\sin\phi\cos\theta \end{aligned} \tag{S2-4}$$

Now, using Eqs. (S2-2), we have

$$\begin{aligned} x &= x'\cos\theta - y'\sin\theta \\ y &= x'\sin\theta + y'\cos\theta \end{aligned} \tag{S2-5}$$

In our derivation, we have used the special case when θ is acute and P is in the first quadrant of both sets of axes. When simplifying equations of curves using Eqs. (S2-5), we find that a rotation through a positive acute angle θ is sufficient. It can be shown, however, that Eqs. (S2-5) hold for any θ and position of P.

EXAMPLE 1 ▸▸ Transform $x^2 - y^2 + 8 = 0$ by rotating the axes through 45°.

When $\theta = 45°$, the rotation equations (S2-5) become

$$x = x'\cos 45° - y'\sin 45° = \frac{x'}{\sqrt{2}} - \frac{y'}{\sqrt{2}}$$

$$y = x'\sin 45° + y'\cos 45° = \frac{x'}{\sqrt{2}} + \frac{y'}{\sqrt{2}}$$

Substituting into the equation $x^2 - y^2 + 8 = 0$ gives

$$\left(\frac{x'}{\sqrt{2}} - \frac{y'}{\sqrt{2}}\right)^2 - \left(\frac{x'}{\sqrt{2}} + \frac{y'}{\sqrt{2}}\right)^2 + 8 = 0$$

$$\frac{1}{2}x'^2 - x'y' + \frac{1}{2}y'^2 - \frac{1}{2}x'^2 - x'y' - \frac{1}{2}y'^2 + 8 = 0$$

$$x'y' = 4$$

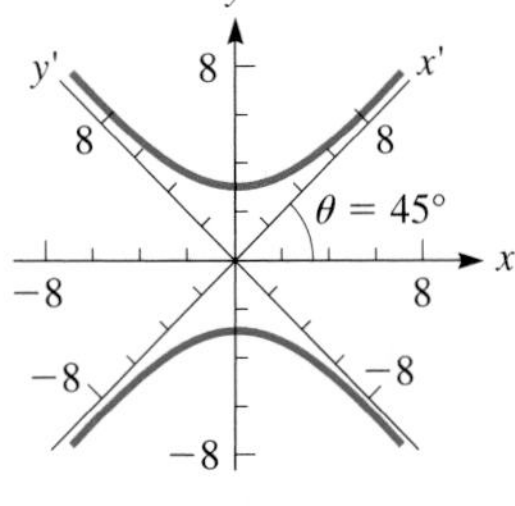

Fig. S2-2

The graph and both sets of axes are shown in Fig. S2-2. The original equation represents a hyperbola. We have the $xy = c$ form with rotation through 45°. ◂◂

When we showed the type of curve represented by the second-degree equation Eq. (S2-1) in Section 21-8, the standard forms of the parabola, ellipse, and hyperbola required that $B = 0$. This means that there is no xy-term in the equation. In Section 21-8, we considered only one case ($xy = c$) for which $B \neq 0$.

If we can remove the xy-term from a second-degree equation, the analysis of the graph is simplified. By a proper rotation of axes we find that Eq. (S2-1) can be transformed into an equation that has no $x'y'$-term.

By substituting Eqs. (S2-5) into Eq. (S2-1) and then simplifying, we have

$$(A\cos^2\theta + B\sin\theta\cos\theta + C\sin^2\theta)x'^2 + [B\cos 2\theta - (A - C)\sin 2\theta]x'y'$$
$$+ (A\sin^2\theta - B\sin\theta\cos\theta + C\cos^2\theta)y'^2 + (D\cos\theta + E\sin\theta)x' + (E\cos\theta - D\sin\theta)y' + F = 0$$

If there is to be no $x'y'$-term, its coefficient must be zero. This means that $B\cos 2\theta - (A - C)\sin 2\theta = 0$, or

Angle of Rotation

$$\tan 2\theta = \frac{B}{A - C} \qquad (A \neq C) \tag{S2-6}$$

Eq. (S2-6) gives the angle of rotation except when $A = C$. In this case the coefficient of the $x'y'$-term is $B\cos 2\theta$, which is zero if $2\theta = 90°$. Thus

$$\theta = 45° \qquad (A = C) \tag{S2-7}$$

Consider the following example.

EXAMPLE 2 ▸▸ By rotation of axes, transform $8x^2 + 4xy + 5y^2 = 9$ into a form without an xy-term. Identify and sketch the curve.

For reference: Eq. (20-2) is

$$\cos\theta = \frac{1}{\sec\theta}$$

Eq. (20-7) is

$$1 + \tan^2\theta = \sec^2\theta$$

Eq. (20-26) is

$$\sin\frac{\alpha}{2} = \pm\sqrt{\frac{1 - \cos\alpha}{2}}$$

Eq. (20-27) is

$$\cos\frac{\alpha}{2} = \pm\sqrt{\frac{1 + \cos\alpha}{2}}$$

Here $A = 8$, $B = 4$, and $C = 5$. Therefore, using Eq. (S2-6), we have

$$\tan 2\theta = \frac{4}{8 - 5} = \frac{4}{3}$$

Since $\tan 2\theta$ is positive, we may take 2θ as an acute angle, which means θ is also acute. For the transformation we need $\sin\theta$ and $\cos\theta$. We find these values by first finding the value of $\cos 2\theta$ and then using the half-angle formulas.

$$\cos 2\theta = \frac{1}{\sec 2\theta} = \frac{1}{\sqrt{1 + \tan^2 2\theta}} = \frac{1}{\sqrt{1 + (\frac{4}{3})^2}} = \frac{3}{5} \qquad \text{using Eqs. (20-2) and (20-7)}$$

Now, using the half-angle formulas, Eqs. (20-26) and (20-27), we have

$$\sin\theta = \sqrt{\frac{1 - \cos 2\theta}{2}} = \sqrt{\frac{1 - \frac{3}{5}}{2}} = \frac{1}{\sqrt{5}}, \quad \cos\theta = \sqrt{\frac{1 + \cos 2\theta}{2}} = \sqrt{\frac{1 + \frac{3}{5}}{2}} = \frac{2}{\sqrt{5}}$$

Here, θ is about 26.6°. Now substituting these values into Eqs. (S2-5), we have

$$x = x'\left(\frac{2}{\sqrt{5}}\right) - y'\left(\frac{1}{\sqrt{5}}\right) = \frac{2x' - y'}{\sqrt{5}}, \quad y = x'\left(\frac{1}{\sqrt{5}}\right) + y'\left(\frac{2}{\sqrt{5}}\right) = \frac{x' + 2y'}{\sqrt{5}}$$

Now, substituting into the equation $8x^2 + 4xy + 5y^2 = 9$ gives

$$8\left(\frac{2x' - y'}{\sqrt{5}}\right)^2 + 4\left(\frac{2x' - y'}{\sqrt{5}}\right)\left(\frac{x' + 2y'}{\sqrt{5}}\right) + 5\left(\frac{x' + 2y'}{\sqrt{5}}\right)^2 = 9$$

$$8(4x'^2 - 4x'y' + y'^2) + 4(2x'^2 + 3x'y' - 2y'^2) + 5(x'^2 + 4x'y' + 4y'^2) = 45$$

$$45x'^2 + 20y'^2 = 45$$

$$\frac{x'^2}{1} + \frac{y'^2}{\frac{9}{4}} = 1$$

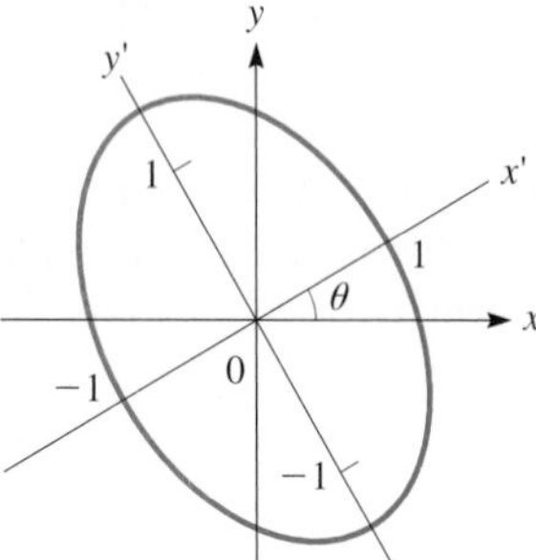

Fig. S2-3

This is an ellipse with semimajor axis of 3/2 and semiminor axis of 1. See Fig. S2-3. ◂◂

In Example 2, $\tan 2\theta$ was positive, and we made 2θ and θ positive. If, when using Eq. (S2-6), $\tan 2\theta$ is negative, we then make 2θ obtuse ($90° < 2\theta < 180°$). In this case $\cos 2\theta$ will be negative, but θ will be acute ($45° < \theta < 90°$).

In Section 21-7 we showed the use of translation of axes in writing an equation in standard form if $B = 0$. In this section we have seen how rotation of axes is used to eliminate the xy-term. It is possible that both a translation of axes and a rotation of axes are needed to write an equation in standard form.

In Section 21-8 we identified a conic section by inspecting the values of A and C when $B = 0$. If $B \neq 0$, these curves are identified as follows:

1. If $B^2 - 4AC = 0$, a parabola.

2. If $B^2 - 4AC < 0$, an ellipse.

3. If $B^2 - 4AC > 0$, a hyperbola.

Special cases such as a point, parallel or intersecting lines, or no curve may result.

EXAMPLE 3 ▸▸ For the equation $16x^2 - 24xy + 9y^2 + 20x - 140y - 300 = 0$, identify the curve and simplify it to standard form. Sketch the graph and display it on a graphing calculator.

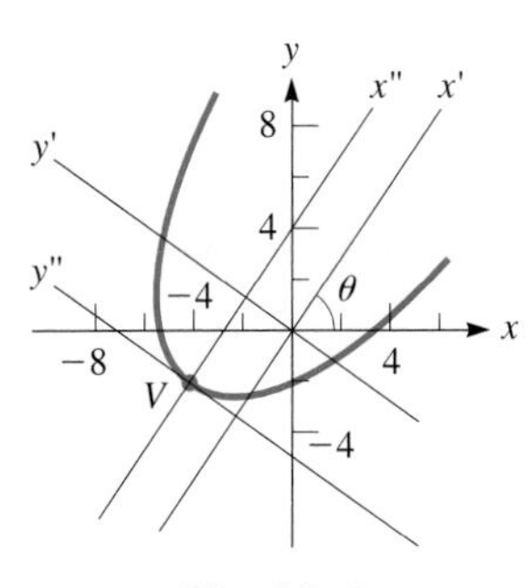

Fig. S2-4

With $A = 16$, $B = -24$, and $C = 9$, using Eq. (S2-6), we have

$$\tan 2\theta = \frac{-24}{16 - 9} = -\frac{24}{7}$$

In this case $\tan 2\theta$ is negative, and we take 2θ as obtuse. We then find that $\cos 2\theta = -7/25$. In turn we find that $\sin\theta = 4/5$ and $\cos\theta = 3/5$. Here θ is about $53.1°$. Using these values in Eqs. (S2-5), we find that

$$x = \frac{3x' - 4y'}{5}, \qquad y = \frac{4x' + 3y'}{5}$$

Substituting these into the original equation and simplifying, we get

$$y'^2 - 4x' - 4y' - 12 = 0$$

This equation represents a parabola with its axis parallel to the x'-axis. The vertex is found by completing the square:

$$(y' - 2)^2 = 4(x' + 4)$$

The vertex is the point $(-4, 2)$ in the $x'y'$-rotated system. Therefore,

$$y''^2 = 4x''$$

is the equation in the $x''y''$-rotated and then translated system. The graph and the coordinate systems are shown in Fig. S2-4.

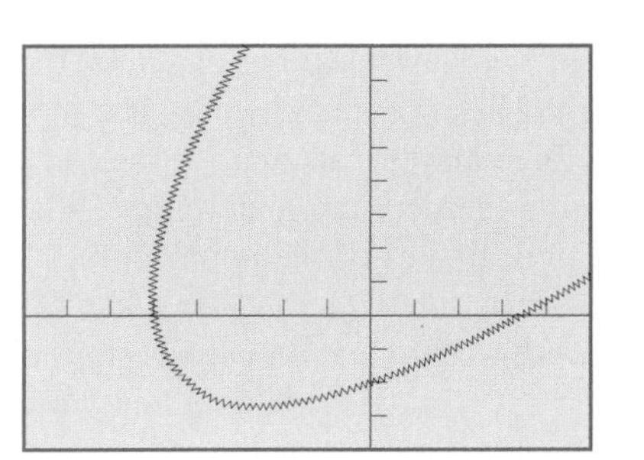

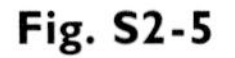
Fig. S2-5

To display the curve on a graphing calculator, we solve for y by using the quadratic formula. Writing the equation as

$$9y^2 + (-24x - 140)y + (16x^2 + 20x - 300) = 0$$

we see that in using the quadratic formula, $a = 9$, $b = -24x - 140$, and $c = 16x^2 + 20x - 300$. Now, solving for y, we have

$$y = \frac{24x + 140 \pm \sqrt{(-24x - 140)^2 - 4(9)(16x^2 + 20x - 300)}}{18}$$

CAUTION ▶

We now ***enter both functions indicated by the ± sign*** and get the display shown in Fig. S2-5. ◂◂

Exercises S–2

In Exercises 1–4, transform the given equations by rotating the axes through the given angle. Identify and sketch each curve.

1. $x^2 - y^2 = 25$, $\theta = 45°$
2. $x^2 + y^2 = 16$, $\theta = 60°$
3. $8x^2 - 4xy + 5y^2 = 36$, $\theta = \tan^{-1} 2$
4. $2x^2 + 24xy - 5y^2 = 8$, $\theta = \tan^{-1} \frac{3}{4}$

In Exercises 5–10, transform each equation to a form without an xy-term by a rotation of axes. Identify and sketch each curve. Then display each curve on a graphing calculator.

5. $x^2 + 2xy + y^2 - 2x + 2y = 0$
6. $5x^2 - 6xy + 5y^2 = 32$
7. $3x^2 + 4xy = 4$
8. $9x^2 - 24xy + 16y^2 - 320x - 240y = 0$
9. $11x^2 - 6xy + 19y^2 = 20$
10. $x^2 + 4xy - 2y^2 = 6$

In Exercises 11 and 12, transform each equation to a form without an xy-term by a rotation of axes. Then transform the equation to a standard form by a translation of axes. Identify and sketch each curve. Then display each curve on a graphing calculator.

11. $16x^2 - 24xy + 9y^2 - 60x - 80y + 400 = 0$
12. $73x^2 - 72xy + 52y^2 + 100x - 200y + 100 = 0$

S-3 Functions of Two Variables

Many situations in science, engineering, and technology involve functions with more than one independent variable. Although some of these applications involve three or more independent variables, we shall concern ourselves primarily with functions of two variables.

In this section we establish the meaning of a function of two variables, and in the section which follows we discuss the graph of this type of function.

Many familiar formulas express one variable in terms of two or more other variables. The following example illustrates one from geometry.

EXAMPLE 1 The total surface area of a right circular cylinder is a function of the radius and the height of the cylinder. That is, the area will change if either or both of these change. The formula for the total surface area is

$$A = 2\pi r^2 + 2\pi rh$$

We say that A is a function of r and h. See Fig. S3-1.

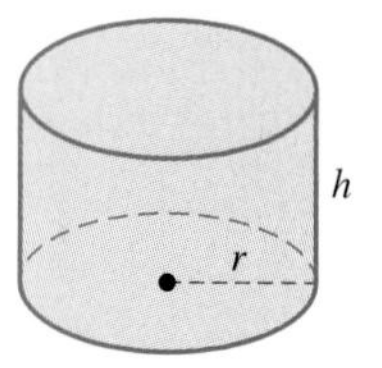

Fig. S3-1

There are numerous applications of functions of two variables. For example, the voltage in a simple electric circuit depends on the resistance and current. The moment of a force depends on the magnitude of the force and the distance of the force from the axis. The pressure of a gas depends on its temperature and volume. The temperature at a point on a plate depends on the coordinates of the point.

We define a function of two variables as follows: *If z is uniquely determined for given values of x and y, then z is a function of x and y.* The notation used is similar to that used for one independent variable. It is $z = f(x, y)$, where both x and y are independent variables. Therefore, it follows that $f(a, b)$ *means "the value of the function when $x = a$ and $y = b$."*

EXAMPLE 2 If $f(x, y) = 3x^2 + 2xy - y^3$, find $f(-1, 2)$. Substituting, we have

$$f(-1, 2) = 3(-1)^2 + 2(-1)(2) - (2)^3$$
$$= 3 - 4 - 8 = -9$$

EXAMPLE 3 ▸▸ If $f(x, y) = 2xy^2 - y$, find $f(x, 2x) - f(x, x^2)$.

We note that in each evaluation the x factor remains as x, but that we are to substitute $2x$ for y and subtract the function for which x^2 is substituted for y.

$$\begin{aligned} f(x, 2x) - f(x, x^2) &= [2x(2x)^2 - (2x)] - [2x(x^2)^2 - x^2] \\ &= [8x^3 - 2x] - [2x^5 - x^2] \\ &= 8x^3 - 2x - 2x^5 + x^2 \\ &= -2x^5 + 8x^3 + x^2 - 2x \end{aligned}$$

This type of difference of functions is important in Section S-6. ◂◂

Restricting values of the function to real numbers means that certain restrictions may be placed on the independent variables. *Values of either x or y or both that lead to division by zero or to imaginary values for z are not permissible.*

EXAMPLE 4 ▸▸ If $f(x, y) = \dfrac{3xy}{(x - y)(x + 3)}$

all values of x and y are permissible except those for which $x = y$ and $x = -3$. Each of these would indicate division by zero.

If $f(x, y) = \sqrt{4 - x^2 - y^2}$, neither x nor y may be greater than 2 in absolute value, and the sum of their squares may not exceed 4. Otherwise, imaginary values of the function would result. ◂◂

One of the primary difficulties students have with functions is setting them up from stated conditions. Although many examples of this have been encountered in our previous work, an example of setting up a function of two variables should prove helpful. Although no general rules can be given for this procedure, a careful analysis of the statement should lead to the desired function.

EXAMPLE 5 ▸▸ An open rectangular metal box is to be made to contain 2 m^3. The cost of sheet metal is c cents per square metre. Express the cost of the sheet metal needed to construct the box as a function of the length and width of the box.

The cost in question depends on the surface area of the sides of the box. An "open" box is one that has no top. Thus the surface area of the box is

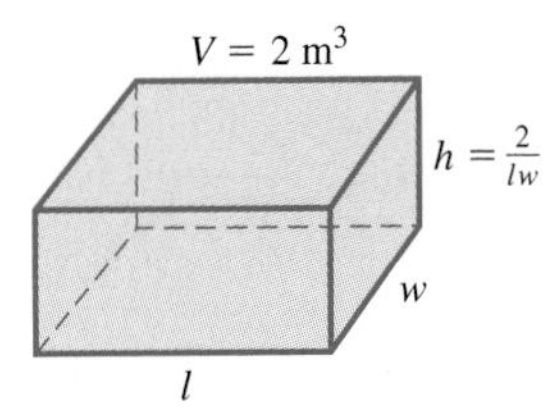

Fig. S3-2

$$S = lw + 2lh + 2hw$$

where l is the length, w the width, and h the height of the box (see Fig. S3-2). However, this expression contains three independent variables. Using the condition that the volume of the box is 2 m^3, we have $lwh = 2$. Since we wish to have only l and w, we solve this expression for h, and find that $h = 2/lw$. Substituting for h in the expression for the surface area, we have

$$\begin{aligned} S &= lw + 2l\left(\frac{2}{lw}\right) + 2\left(\frac{2}{lw}\right)w \\ &= lw + \frac{4}{w} + \frac{4}{l} \end{aligned}$$

Since the cost C is given by $C = cS$, we have the expression for the cost as

$$C = c\left(lw + \frac{4}{w} + \frac{4}{l}\right)$$ ◂◂

Exercises S–3

In Exercises 1–4, determine the indicated functions.

1. Express the volume of a right circular cone as a function of the radius of the base and the height.

2. A cylindrical can is to be made to contain a volume V. Express the total surface area of the can as a function of V and the radius of the can.

3. The angle between two forces $\mathbf{F}_1$ and $\mathbf{F}_2$ is 30°. Express the magnitude of the resultant $\mathbf{R}$ in terms of $\mathbf{F}_1$ and $\mathbf{F}_2$. See Fig. S3-3.

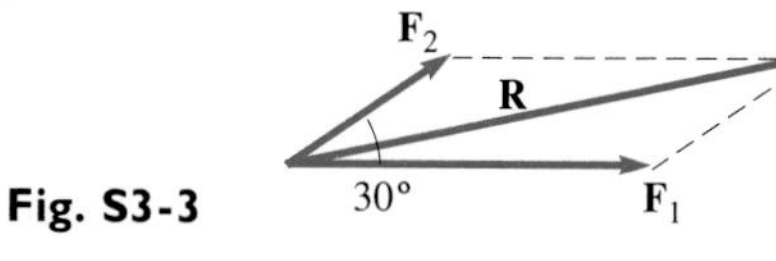

Fig. S3-3

4. A computer leasing firm charges a monthly fee F based on the length of time a corporation has used the service plus \$100 for every hour the computer is used during the month. Express the total monthly charge T as a function of F and the number of hours h the computer is used.

In Exercises 5–12, evaluate the given functions.

5. $f(x, y) = 2x - 6y$; find $f(0, -4)$

6. $g(r, s) = r - 2rs - r^2s$; find $g(-2, 1)$

7. $Y(y, t) = \dfrac{2 - 3y}{t - 1} + 2y^2t$; find $Y(y, 2)$

8. $X(x, t) = -6xt + xt^2 - t^3$; find $X(x, -t)$

9. $H(p, q) = p - \dfrac{p - 2q^2 - 5q}{p + q}$; find $H(p, q + k)$

10. $f(x, y) = x^2 - 2xy - 4x$; find $f(x + h, y + k) - f(x, y)$

11. $f(x, y) = xy + x^2 - y^2$; find $f(x, x) - f(x, 0)$

12. $g(y, z) = 3y^3 - y^2z + 5z^2$; find $g(3z^2, z) - g(z, z)$

In Exercises 13 and 14, determine which values of x and y, if any, are not permissible.

13. $f(x, y) = \dfrac{\sqrt{y}}{2x}$

14. $f(x, y) = \sqrt{x^2 + y^2 - x^2y - y^3}$

In Exercises 15–20, solve the given problems.

15. For a certain electric circuit, the current i, in amperes, in terms of the voltage E and resistance R is given by $i = \dfrac{E}{R + 0.25}$. Find the current for $E = 1.50$ V and $R = 1.20\ \Omega$ and for $E = 1.60$ V and $R = 1.05\ \Omega$.

16. The reciprocal of the image distance q from a lens as a function of the object distance p and the focal length f of the lens is given by

$$\frac{1}{q} = \frac{1}{f} - \frac{1}{p}$$

Find the image distance of an object 20 cm from a lens whose focal length is 5 cm.

17. A rectangular solar cell panel has a perimeter p and a width w. Express the area A of the panel in terms of p and w, and evaluate the area for $p = 250$ cm and $w = 55$ cm. See Fig. S3-4.

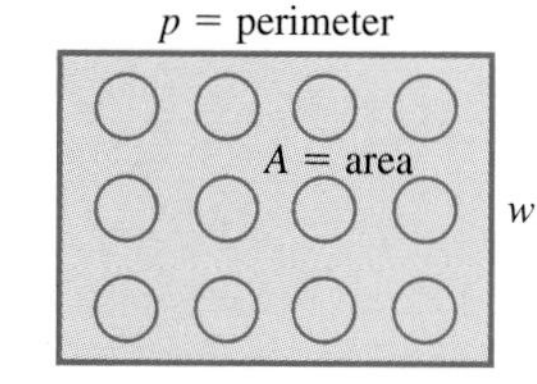

Fig. S3-4

18. A gasoline storage tank is in the shape of a right circular cylinder with a hemisphere at each end, as shown in Fig. S3-5. Express the volume V of the tank in terms of r and h, and then evaluate the volume for $r = 1.25$ m and $h = 4.17$ m.

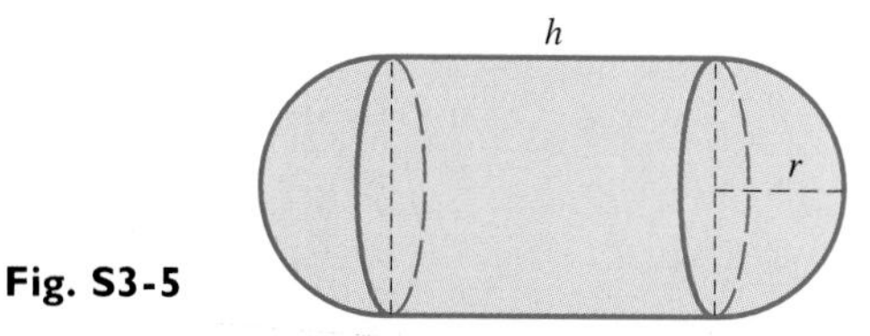

Fig. S3-5

19. The crushing load L of a pillar varies as the fourth power of its radius and inversely as the square of its length. Express L as a function of r and l for a pillar 6.0 m tall and 1.0 m in diameter that is crushed by a load of 20 Mg.

20. The resonant frequency of an electric circuit containing an inductance L and capacitance C is inversely proportional to the square root of the product of the inductance and the capacitance. If the resonant frequency of a circuit containing a 4-H inductor and a 64-μF capacitor is 10 Hz, express f as a function of L and C.

S-4 Curves and Surfaces in Three Dimensions

We will now undertake a brief description of the graphical representation of a function of two variables. We shall show first a method of representation in the rectangular coordinate system in two dimensions. The following example illustrates the method.

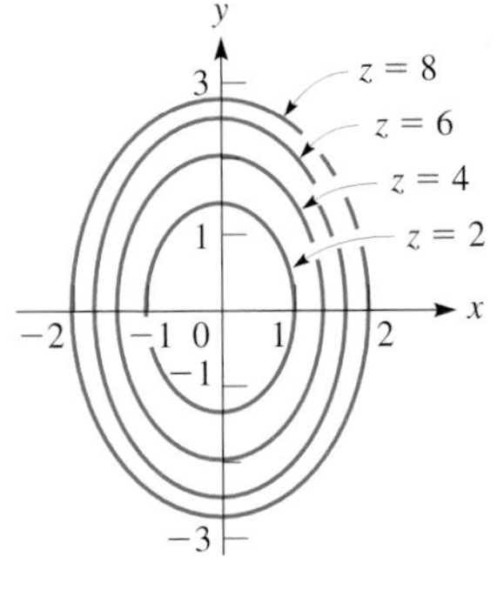

Fig. S4-1

EXAMPLE 1 ▸▸ In order to represent $z = 2x^2 + y^2$, we will assume various values of z and sketch the resulting equation in the xy-plane. For example, if $z = 2$ we have

$$2x^2 + y^2 = 2$$

We recognize this as an ellipse with its major axis along the y-axis and vertices at $(0, \sqrt{2})$ and $(0, -\sqrt{2})$. The ends of the minor axis are at $(1, 0)$ and $(-1, 0)$. However, the ellipse $2x^2 + y^2 = 2$ represents the function $z = 2x^2 + y^2$ only for the value of $z = 2$. If $z = 4$, we have $2x^2 + y^2 = 4$, which is another ellipse. In fact, for all positive values of z, an ellipse is the resulting curve. Negative values of z are not possible, since neither x^2 nor y^2 may be negative. Figure S4-1 shows the ellipses that are obtained by using the indicated values of z. ◂◂

The method of representation illustrated in Example 1 is useful if only a few specific values of z are to be used, or at least if the various curves do not intersect in such a way that they cannot be distinguished. If a general representation of z as a function of x and y is desired, it is necessary to use three coordinate axes, one each for x, y, and z. The most widely applicable system of this kind is to place a third coordinate axis at right angles to each of the x- and y-axes. In this way we employ three dimensions for the representation.

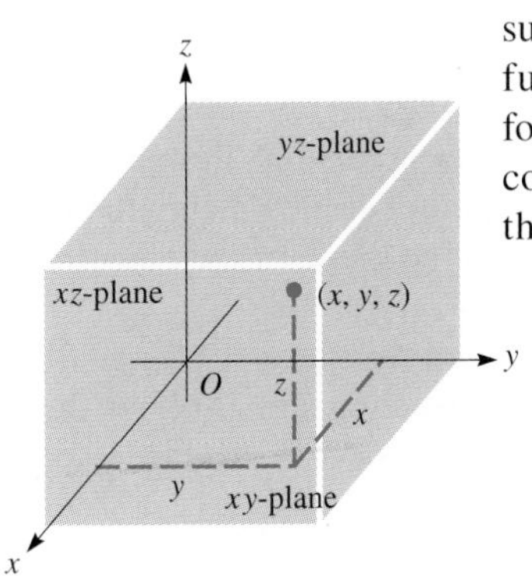

Fig. S4-2

The three mutually perpendicular coordinate axes, the x-axis, the y-axis, and the z-axis are the basis of the **rectangular coordinate system in three dimensions.** Together they form three mutually perpendicular planes in space, the xy-plane, the yz-plane, and the xz-plane. To every point in space of the coordinate system is associated the set of numbers (x, y, z). The point at which the axes meet is the *origin*. The positive directions of the axes are indicated in Fig. S4-2. *That part of space in which all values of the coordinates are positive is called the first* **octant.** Numbers are not assigned to the other octants.

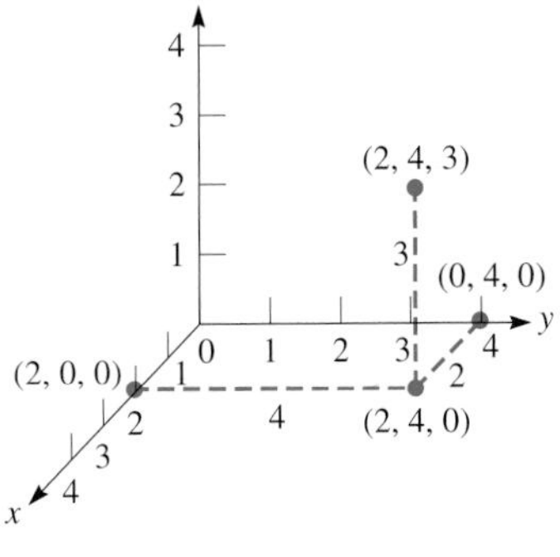

Fig. S4-3

EXAMPLE 2 ▸▸ Represent the point (2, 4, 3) in rectangular coordinates.

We first note that a certain distortion is necessary to represent values of x reasonably, since the x-axis "comes out" of the plane of the page. Units $\sqrt{2}/2$ ($= 0.7$) as long as those used on the other axes give a good representation. With this in mind, we draw a line four units long from the point $(2, 0, 0)$ on the x-axis in the xy-plane. This locates the point $(2, 4, 0)$. From this point a line 3 units long is drawn vertically upward. This locates the desired point, $(2, 4, 3)$. The point may be located by starting from $(0, 4, 0)$, proceeding two units *parallel* to the x-axis to $(2, 4, 0)$, and then proceeding vertically three units to $(2, 4, 3)$. It may also be located by starting from $(0, 0, 3)$ (see Fig. S4-3). ◂◂

We now show certain basic techniques by which three-dimensional figures may be drawn. We start by showing the general equation of a plane.

In Chapter 21 we showed that the graph of the equation $Ax + By + C = 0$ in two dimensions is a straight line. By the following example, we will verify that *the graph of the equation*

Equation of a Plane

$$Ax + By + Cz + D = 0 \tag{S4-1}$$

is a **plane** *in three dimensions.*

EXAMPLE 3 ▸▸ Show that the graph of $2x + 3y + z - 6 = 0$ in three dimensions is a plane.

NOTE ▶ *If we let any of the three variables take on a specific value, we obtain a linear equation in the other two variables.* For example, the point $(\frac{1}{2}, 1, 2)$ satisfies the equation and therefore lies on the graph of the equation. For $x = \frac{1}{2}$, we have

$$3y + z - 5 = 0$$

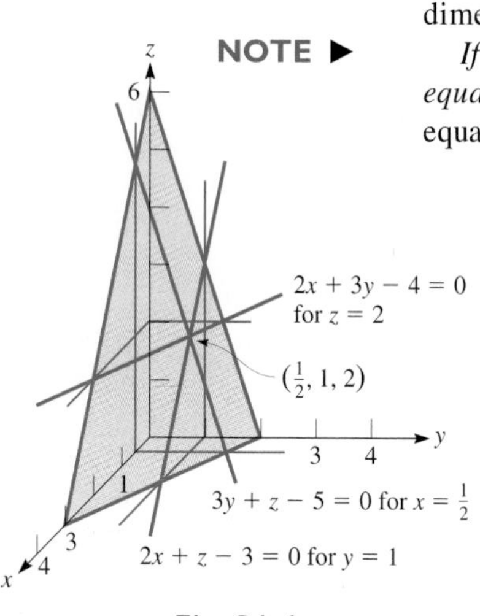

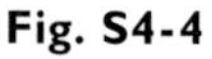
Fig. S4-4

which is the equation of a straight line. This means that all pairs of values of y and z which satisfy this equation, along with $x = \frac{1}{2}$, satisfy the given equation. Thus, for $x = \frac{1}{2}$, the straight line $3y + z - 5 = 0$ lies on the graph of the equation.

For $z = 2$, we have

$$2x + 3y - 4 = 0$$

which is also a straight line. By similar reasoning this line lies on the graph of the equation. Since two lines through a point define a plane, these lines through $(\frac{1}{2}, 1, 2)$ define a plane. This plane is the graph of the equation (see Fig. S4-4).

For any point on the graph, there is a straight line on the graph that is parallel to one of the coordinate planes. Therefore, there are intersecting straight lines through the point. Thus, the graph is a plane. A similar analysis can be made for any equation of the same form. ◂◂

Since we know that the graph of an equation of the form of Eq. (S4-1) is a plane, its graph can be found by determining its three intercepts, and the plane can then be represented by drawing in the lines between these intercepts. If the plane passes through the origin, by letting two of the variables in turn be zero, two straight lines that define the plane are found.

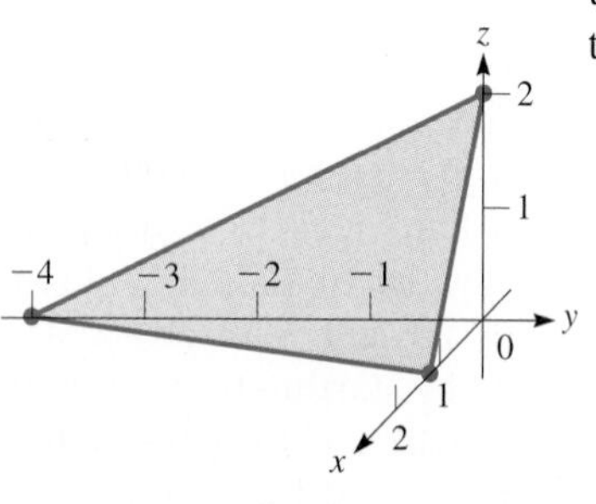

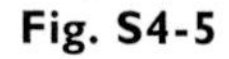
Fig. S4-5

EXAMPLE 4 ▸▸ Sketch the graph of $3x - y + 2z - 4 = 0$.

The intercepts of the graph of an equation are those points where it crosses the respective axes. Thus, by letting two of the variables at a time equal zero, we obtain the intercepts. For the given equation the intercepts are $(\frac{4}{3}, 0, 0)$, $(0, -4, 0)$, and $(0, 0, 2)$. These points are located (see Fig. S4-5), and lines are drawn between them to represent the plane. ◂◂

The graph of an equation in three variables, which is essentially equivalent to a function with two independent variables, is a **surface** *in space.* This is seen for the plane and will be verified for other equations in the examples that follow.

The intersection of two surfaces is a **curve** *in space.* This has been seen in Examples 3 and 4, since the intersections of the given planes and the coordinate planes are lines (which in the general sense are curves). *We define the* **traces** *of a surface to be the curves of intersection of the surface and the coordinate planes.* The traces of a plane are those lines drawn between the intercepts to represent the plane. Many surfaces may be sketched by finding their traces and intercepts.

Traces

EXAMPLE 5 ▸▸ Find the intercepts and traces, and sketch the graph of the equation $z = 4 - x^2 - y^2$.

The intercepts of the graph of the equation are $(2, 0, 0)$, $(-2, 0, 0)$, $(0, 2, 0)$, $(0, -2, 0)$, and $(0, 0, 4)$.

Since the traces of a surface lie within the coordinate planes, for each trace one of the variables is zero. Thus, by ***letting each variable in turn be zero, we find the trace of the surface in the plane of the other two variables.*** Therefore, the traces of this surface are

in the yz-plane: $z = 4 - y^2$ (a parabola)
in the xz-plane: $z = 4 - x^2$ (a parabola)
in the xy-plane: $x^2 + y^2 = 4$ (a circle)

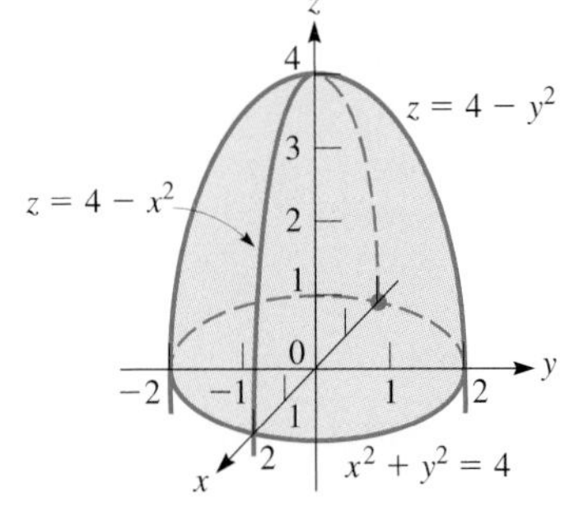

Fig. S4-6

Using the intercepts and sketching the traces, we obtain the surface represented by the equation as shown in Fig. S4-6. This figure is called a **circular paraboloid.** ◂◂

There are numerous techniques for analyzing the equation of a surface in order to obtain its graph. Another which we shall discuss here, which is closely associated with a trace, is that of a **section**. *By assuming a specific value of one of the variables, we obtain an equation in two variables, the graph of which lies in a plane parallel to the coordinate plane of the two variables.* The following example illustrates sketching a surface by use of intercepts, traces, and sections.

Section

EXAMPLE 6 ▸▸ Sketch the graph of $4x^2 + y^2 - z^2 = 4$.

The intercepts are $(1, 0, 0)$, $(-1, 0, 0)$, $(0, 2, 0)$, and $(0, -2, 0)$. We note that there are no intercepts on the z-axis, for this would necessitate $z^2 = -4$.

The traces are

in the yz-plane: $y^2 - z^2 = 4$ (a hyperbola)
in the xz-plane: $4x^2 - z^2 = 4$ (a hyperbola)
in the xy-plane: $4x^2 + y^2 = 4$ (an ellipse)

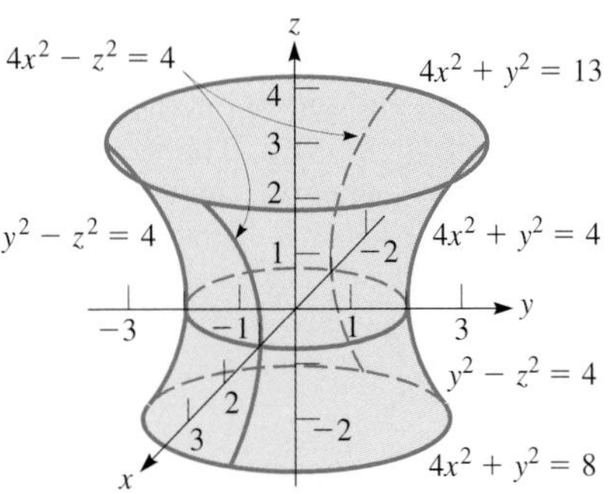

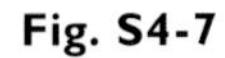
Fig. S4-7

The surface is reasonably defined by these curves, but by assuming suitable values of z we may indicate its shape better. For example, if $z = 3$, we have $4x^2 + y^2 = 13$, which is an ellipse. In using it we must remember that it is valid for $z = 3$ and therefore should be drawn three units above the xy-plane. If $z = -2$, we have $4x^2 + y^2 = 8$, which is also an ellipse. Thus we have the following sections:

for $z = 3$: $4x^2 + y^2 = 13$ (an ellipse)
for $z = -2$: $4x^2 + y^2 = 8$ (an ellipse)

Other sections could be found, but these are sufficient to obtain a good sketch of the graph (see Fig. S4-7). The figure is called an **elliptic hyperboloid.** ◂◂

Having developed the rectangular coordinate system in three dimensions, we can compare the graph of a function using two dimensions and three dimensions. The next example shows the surface for the function of Example 1.

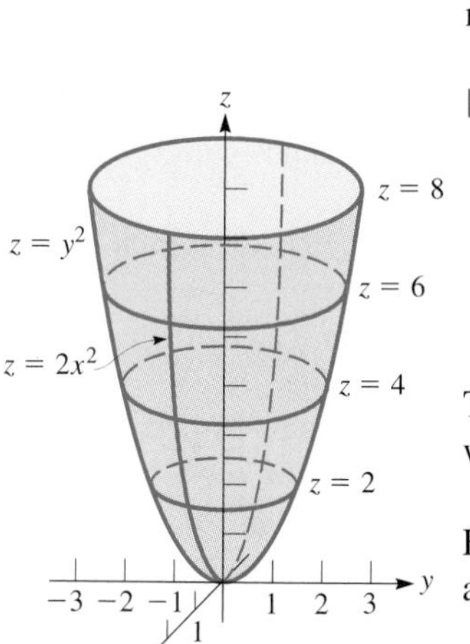

Fig. S4-8

EXAMPLE 7 ▸▸ Sketch the graph of $z = 2x^2 + y^2$.

The only intercept is $(0, 0, 0)$. The traces are

in the yz-plane: $z = y^2$ (a parabola)

in the xz-plane: $z = 2x^2$ (a parabola)

in the xy-plane: the origin

The trace in the xy-plane is only the point at the origin, since $2x^2 + y^2 = 0$ may be written as $y^2 = -2x^2$, which is true only for $x = 0$ and $y = 0$.

To get a better graph we should use some positive values for z. As we noted in Example 1, negative values of z cannot be used. Since we used $z = 2$, $z = 4$, $z = 6$, and $z = 8$ in Example 1, we shall use these values here. Therefore,

for $z = 2$: $2x^2 + y^2 = 2$, for $z = 4$: $2x^2 + y^2 = 4$

for $z = 6$: $2x^2 + y^2 = 6$, for $z = 8$: $2x^2 + y^2 = 8$

Each of these sections is an ellipse. The surface, called an **elliptic paraboloid,** is shown in Fig. S4-8. Compare with Fig. S4-1. ◂◂

Example 7 illustrates how *topographic maps* may be drawn. These maps represent three-dimensional terrain in two dimensions. For example, if Fig. S4-8 represents an excavation in the surface of the earth, then Fig. S4-1 represents the curves of constant elevation, or *contours*, with equally spaced elevations measured from the bottom of the excavation.

An equation with only two variables may represent a surface in space. Since only two variables are included in the equation, the surface is independent of the other variable. Another interpretation is that all sections, for all values of the variable not included, are the same. That is, ***all sections parallel to the coordinate plane of the included variables are the same as the trace in that plane.***

NOTES ▸

EXAMPLE 8 ▸▸ Sketch the graph of $x + y = 2$ in the rectangular coordinate system in three dimensions and also in two dimensions.

Since z does not appear in the equation, we can consider the equation to be $x + y + 0z = 2$. Therefore, we see that *for any value of* z the section is the straight line $x + y = 2$. Thus the graph is a plane as shown in Fig. S4-9(a). The graph as a straight line in two dimensions is shown in Fig. S4-9(b). ◂◂

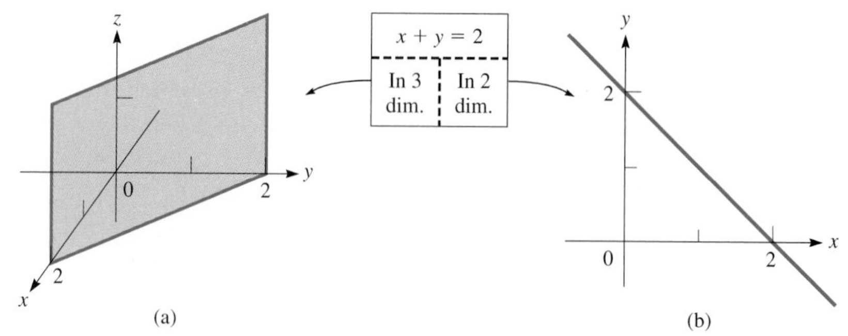

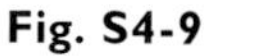
Fig. S4-9

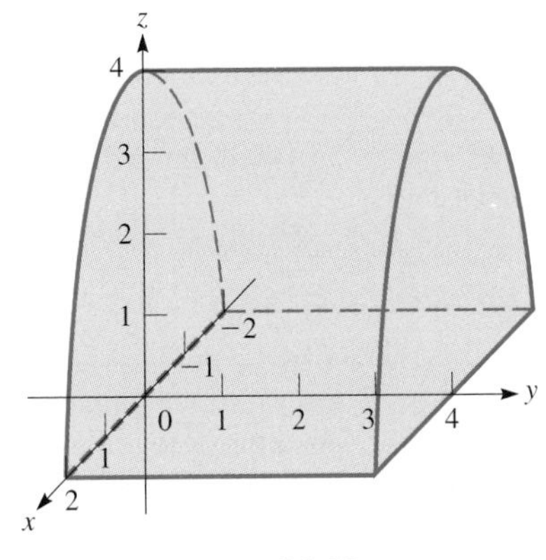

Fig. S4-10

EXAMPLE 9 ▸▸ The graph of the equation $z = 4 - x^2$ in three dimensions is a surface whose sections, for all values of y, are given by the parabola $z = 4 - x^2$. The surface is shown in Fig. S4-10. ◂◂

The surface in Example 9 is known as a **cylindrical surface.** In general, a cylindrical surface is one that can be generated by a line moving parallel to a fixed line while passing through a plane curve.

It must be realized that most of the figures shown extend beyond the ranges indicated by the traces and sections. However, these traces and sections are convenient for representing and visualizing the surfaces.

A computer can be programmed to draw surfaces by drawing sections that are perpendicular to both the x- and y-axes or by sections that are perpendicular to only one axis. Such computer-drawn surfaces are of great value for all types of surfaces, especially those of a complex nature. Figure S4-11 shows the graph of $z = \sin(x^2 + y^2)$ drawn by a computer in these two ways.

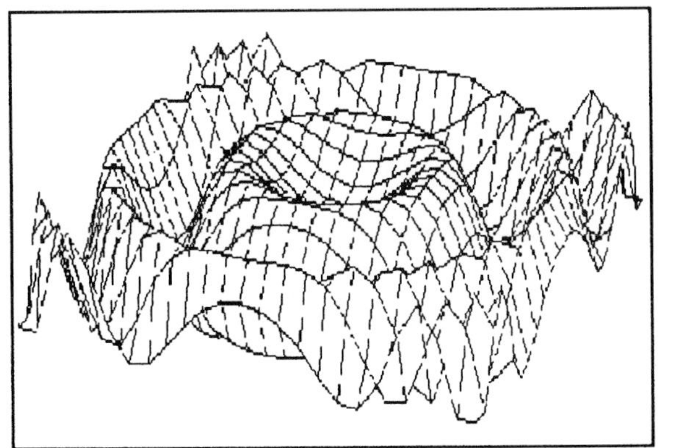

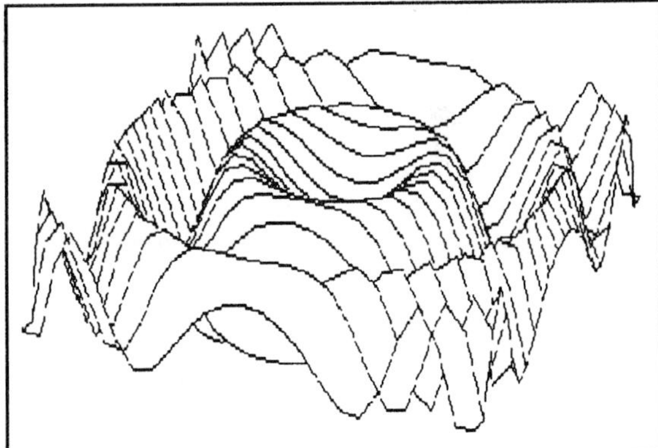

Fig. S4-11

Exercises S–4

In Exercises 1 and 2, use the method of Example 1 and draw the graphs of the given equations for the given values of z.

1. $z = x^2 + y^2$, $z = 1$, $z = 4$, $z = 9$

2. $z = y - x^2$, $z = 0$, $z = 2$, $z = 4$

In Exercises 3–14, sketch the graphs of the given equations in the rectangular coordinate system in three dimensions.

3. $x + y + 2z - 4 = 0$

4. $4x - 2y + z - 8 = 0$

5. $z = y - 2x - 2$

6. $x + 2y = 4$

7. $x^2 + y^2 + z^2 = 4$

8. $z = x^2 + y^2$

9. $z = 4 - 4x^2 - y^2$

10. $x^2 + y^2 - 4z^2 = 4$

11. $z = 2x^2 + y^2 + 2$

12. $x^2 - y^2 - z^2 = 9$

13. $x^2 + y^2 = 16$

14. $y^2 + 9z^2 = 9$

In Exercises 15–20, sketch the indicated curves and surfaces.

15. Curves that represent a constant temperature are called *isotherms*. The temperature at a point (x, y) of a flat plate is t degrees Celsius, where $t = 4x - y^2$. In two dimensions draw the isotherms for $t = -4$, 0, and 8.

16. At a point (x, y) in the xy-plane the electric potential V is given by $V = y^2 - x^2$, where V is measured in volts. Draw the lines of equal potential for $V = -9$, 0, and 9.

17. An electric charge is so distributed that the electric potential at all points on an imaginary surface is the same. Such a surface is called an *equipotential surface*. Sketch the graph of the equipotential surface whose equation is $2x^2 + 2y^2 + 3z^2 = 6$.

18. The surface of a small hill can be roughly approximated by the equation $z(2x^2 + y^2 + 100) = 1500$, where the units are metres. Draw the surface of the hill and the contours for $z = 3$ m, $z = 6$ m, $z = 9$ m, $z = 12$ m, and $z = 15$ m.

19. Sketch the line in space defined by the intersection of the planes $x + 2y + 3z - 6 = 0$ and $2x + y + z - 4 = 0$.

20. Sketch the graph of $x^2 + y^2 - 2y = 0$ in three dimensions and in two dimensions.

S-5 Partial Derivatives

In Chapter 23, when we showed that the derivative is the instantaneous rate of change of one variable with respect to another, only one independent variable was involved. To extend the derivative to functions of two (or more) variables, we find the derivative of the function with respect to one of the independent variables, while the other is held constant.

If $z = f(x, y)$, and y is held constant, z becomes a function of x alone. The derivative of this function with respect to x is termed the **partial derivative** *of z with respect to x.* Similarly, *if x is held constant, the derivative of the function with respect to y is the* **partial derivative** *of z with respect to y.*

For the function $z = f(x, y)$, the notations used for the partial derivative of z with respect to x include

$$\frac{\partial z}{\partial x}, \qquad \frac{\partial f}{\partial x}, \qquad f_x, \qquad \frac{\partial}{\partial x} f(x, y), \qquad f_x(x, y)$$

Similarly, $\partial z/\partial y$ denotes the partial derivative of z with respect to y. In speaking, this is often shortened to "the partial of z with respect to y."

EXAMPLE 1 ▸▸ If $z = 4x^2 + xy - y^2$, find $\partial z/\partial x$ and $\partial z/\partial y$.

To find the partial derivative of z with respect to x, we treat y as a constant.

treat as constant

$$z = 4x^2 + xy - y^2$$
$$\frac{\partial z}{\partial x} = 8x + y$$

To find the partial derivative of z with respect to y, we treat x as a constant.

treat as constant

$$z = 4x^2 + xy - y^2$$
$$\frac{\partial z}{\partial y} = x - 2y$$ ◂◂

EXAMPLE 2 ▸▸ If $z = \dfrac{x \ln y}{x^2 + 1}$, find $\partial z/\partial x$ and $\partial z/\partial y$.

$$\frac{\partial z}{\partial x} = \frac{(x^2 + 1)(\ln y) - (x \ln y)(2x)}{(x^2 + 1)^2} = \frac{(1 - x^2) \ln y}{(1 + x^2)^2}$$
$$\frac{\partial z}{\partial y} = \left(\frac{x}{x^2 + 1}\right)\left(\frac{1}{y}\right) = \frac{x}{y(x^2 + 1)}$$

We note that in finding $\partial z/\partial x$ it is necessary to use the quotient rule, since x appears in both numerator and denominator. However, when finding $\partial z/\partial y$ the only derivative needed is that of $\ln y$. ◂◂

EXAMPLE 3 ▸▸ For the function $f(x, y) = x^2y\sqrt{2 + xy^2}$, find $f_y(2, 1)$.

The notation $f_y(2, 1)$ means the partial derivative of f with respect to y, evaluated for $x = 2$ and $y = 1$. Thus, first finding $f_y(x, y)$, we have

$$f(x, y) = x^2y(2 + xy^2)^{1/2}$$

$$f_y(x, y) = x^2y\left(\frac{1}{2}\right)(2 + xy^2)^{-1/2}(2xy) + (2 + xy^2)^{1/2}(x^2)$$

$$= \frac{x^3y^2}{(2 + xy^2)^{1/2}} + x^2(2 + xy^2)^{1/2}$$

$$= \frac{x^3y^2 + x^2(2 + xy^2)}{(2 + xy^2)^{1/2}} = \frac{2x^2 + 2x^3y^2}{(2 + xy^2)^{1/2}}$$

$$f_y(2, 1) = \frac{2(4) + 2(8)(1)}{(2 + 2)^{1/2}} = 12$$ ◂◂

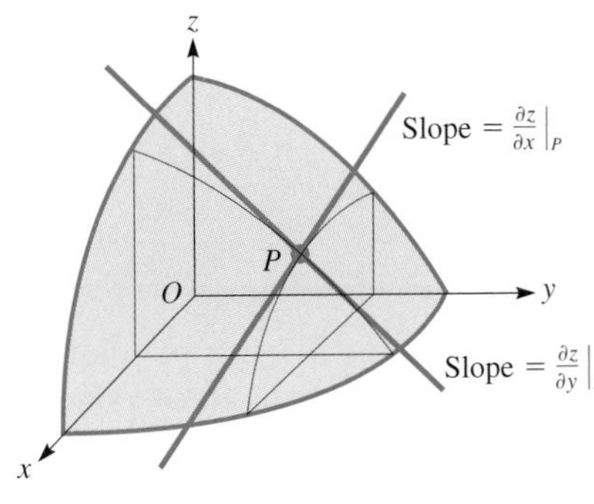

Fig. S5-1

To determine the geometric interpretation of a partial derivative, assume that $z = f(x, y)$ is the surface shown in Fig. S5-1. Choosing a point P on the surface, we then draw a plane through P parallel to the xz-plane. On this plane through P, the value of y is constant. The intersection of this plane and the surface is the curve as indicated. *The partial derivative of z with respect to x represents the slope of a line tangent to this curve.* When the values of the coordinates of point P are substituted into the expression for this partial derivative, it gives the slope of the tangent line at that point. In the same way, the partial derivative of z with respect to y, evaluated at P, gives the slope of the line tangent to the curve that is found from the intersection of the surface and the plane parallel to the yz-plane through P.

EXAMPLE 4 ▸▸ Find the slope of a line tangent to the surface $2z = x^2 + 2y^2$ and parallel to the xz-plane at the point (2, 1, 3). Also find the slope of a line tangent to this surface and parallel to the yz-plane at the same point.

Finding the partial derivative with respect to x, we have

$$2\frac{\partial z}{\partial x} = 2x \quad \text{or} \quad \frac{\partial z}{\partial x} = x$$

This derivative, evaluated at the point (2, 1, 3), will give us the slope of the line tangent that is also parallel to the xz-plane. Therefore, the first required slope is

$$\left.\frac{\partial z}{\partial x}\right|_{(2,1,3)} = 2$$

The partial derivative of z with respect to y, evaluated at (2, 1, 3), will give us the second required slope. Thus

$$\frac{\partial z}{\partial y} = 2y, \quad \left.\frac{\partial z}{\partial y}\right|_{(2,1,3)} = 2$$

Therefore, both slopes are 2. See Fig. S5-2. ◂◂

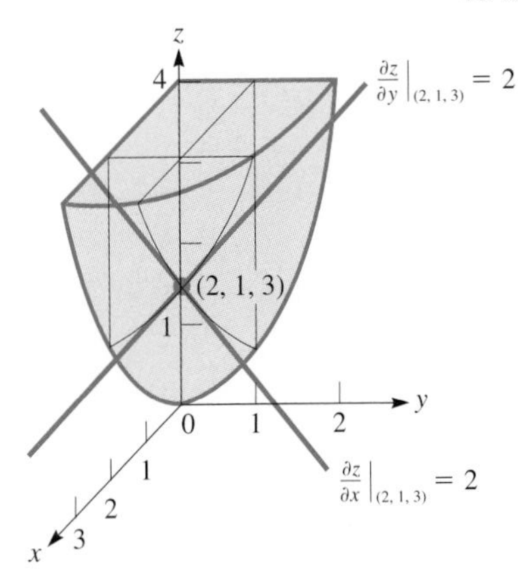

Fig. S5-2

The general interpretation of the partial derivative follows that of a derivative of a function with one independent variable. *The partial derivative $f_x(x_0, y_0)$ is the instantaneous rate of change of the function $f(x, y)$ with respect to x, with y held constant at the value of y_0.* This holds regardless of what the variables represent.

Applications of partial derivatives are found in many fields of technology. We show here an application from electricity, and others are found in the exercises.

EXAMPLE 5 ▸▸ An electric circuit in a microwave transmitter has parallel resistances r and R. The current through r can be found from

$$i = \frac{IR}{r + R}$$

where I is the total current for the two branches. Assuming that I is constant at 85.4 mA, find $\partial i/\partial r$ and evaluate it for $R = 0.150\ \Omega$ and $r = 0.032\ \Omega$.

Substituting for I and finding the partial derivative, we have the following:

$$i = \frac{85.4R}{r + R} = 85.4R(r + R)^{-1}$$

$$\frac{\partial i}{\partial r} = (-1)(85.4R)(r + R)^{-2}(1) = \frac{-85.4R}{(r + R)^2}$$

$$\left.\frac{\partial i}{\partial r}\right|_{\substack{r=0.032\\R=0.150}} = \frac{-85.4(0.150)}{(0.032 + 0.150)^2} = -387\ \text{mA}/\Omega$$

This result tells us that the current is decreasing at the rate of 387 mA per ohm of change in the smaller resistor at the instant when $r = 0.032\ \Omega$, for a fixed value of $R = 0.150\ \Omega$. ◂◂

Since the partial derivatives $\partial f/\partial x$ and $\partial f/\partial y$ are functions of x and y, we can take partial derivatives of each of them. This gives rise to **partial derivatives of higher order,** in a manner similar to the higher derivatives of a function of one independent variable. *The possible* **second-order partial derivatives** *of a function $f(x, y)$ are*

$$\frac{\partial^2 f}{\partial x^2} = \frac{\partial}{\partial x}\left(\frac{\partial f}{\partial x}\right) \qquad \frac{\partial^2 f}{\partial y^2} = \frac{\partial}{\partial y}\left(\frac{\partial f}{\partial y}\right)$$

$$\frac{\partial^2 f}{\partial x\,\partial y} = \frac{\partial}{\partial x}\left(\frac{\partial f}{\partial y}\right) \qquad \frac{\partial^2 f}{\partial y\,\partial x} = \frac{\partial}{\partial y}\left(\frac{\partial f}{\partial x}\right)$$

EXAMPLE 6 ▸▸ Find the second-order partial derivatives of $z = x^3y^2 - 3xy^3$.

First we find $\partial z/\partial x$ and $\partial z/\partial y$:

$$\frac{\partial z}{\partial x} = 3x^2y^2 - 3y^3 \qquad \frac{\partial z}{\partial y} = 2x^3y - 9xy^2$$

Therefore, we have the following second-order partial derivatives:

$$\frac{\partial^2 z}{\partial x^2} = \frac{\partial}{\partial x}\left(\frac{\partial z}{\partial x}\right) = 6xy^2 \qquad \frac{\partial^2 z}{\partial y^2} = \frac{\partial}{\partial y}\left(\frac{\partial z}{\partial y}\right) = 2x^3 - 18xy$$

$$\frac{\partial^2 z}{\partial x\,\partial y} = \frac{\partial}{\partial x}\left(\frac{\partial z}{\partial y}\right) = 6x^2y - 9y^2 \qquad \frac{\partial^2 z}{\partial y\,\partial x} = \frac{\partial}{\partial y}\left(\frac{\partial z}{\partial x}\right) = 6x^2y - 9y^2$$ ◂◂

In Example 6 we note that

$$\frac{\partial^2 z}{\partial x\,\partial y} = \frac{\partial^2 z}{\partial y\,\partial x} \tag{S5-1}$$

In general, this is true if the function and partial derivatives are continuous.

EXAMPLE 7 ▶▶ For $f(x, y) = \tan^{-1}\dfrac{y}{x^2}$, show that $\dfrac{\partial^2 f}{\partial x\,\partial y} = \dfrac{\partial^2 f}{\partial y\,\partial x}$.

Finding $\partial f/\partial x$ and $\partial f/\partial y$, we have

$$\frac{\partial f}{\partial x} = \frac{1}{1 + \left(\dfrac{y}{x^2}\right)^2}\left(\frac{-2y}{x^3}\right) = \frac{x^4}{x^4 + y^2}\left(-\frac{2y}{x^3}\right) = \frac{-2xy}{x^4 + y^2}.$$

$$\frac{\partial f}{\partial y} = \frac{1}{1 + \left(\dfrac{y}{x^2}\right)^2}\left(\frac{1}{x^2}\right) = \frac{x^4}{x^4 + y^2}\left(\frac{1}{x^2}\right) = \frac{x^2}{x^4 + y^2}$$

Now, finding $\partial^2 f/\partial x\,\partial y$ and $\partial^2 f/\partial y\,\partial x$, we have

$$\frac{\partial^2 f}{\partial x\,\partial y} = \frac{(x^4 + y^2)(2x) - x^2(4x^3)}{(x^4 + y^2)^2} = \frac{-2x^5 + 2xy^2}{(x^4 + y^2)^2}$$

$$\frac{\partial^2 f}{\partial y\,\partial x} = \frac{(x^4 + y^2)(-2x) - (-2xy)(2y)}{(x^4 + y^2)^2} = \frac{-2x^5 + 2xy^2}{(x^4 + y^2)^2}$$

We see that they are equal. ◀◀

Exercises S–5

In Exercises 1–12, find the partial derivative of the dependent variable or function with respect to each of the independent variables.

1. $z = 5x + 4x^2y$
2. $z = \dfrac{x^2}{y} - 2xy$
3. $f(x, y) = xe^{2y}$
4. $f(x, y) = \dfrac{2 + \cos x}{1 - \sec 3y}$
5. $\phi = r\sqrt{1 + 2rs}$
6. $z = (x^2 + xy^3)^4$
7. $z = \sin xy$
8. $y = \ln(r^2 + s)$
9. $f(x, y) = \dfrac{2\sin^3 2x}{1 - 3y}$
10. $z = \dfrac{\sin^{-1} xy}{3 + x^2}$
11. $z = \sin x + \cos xy - \cos y$
12. $f(x, y) = e^x \cos xy + e^{-2x}\tan y$

In Exercises 13 and 14, evaluate the indicated partial derivatives at the given points.

13. $z = 3xy - x^2 + y,\quad \left.\dfrac{\partial z}{\partial x}\right|_{(1,-2,-9)}$
14. $z = e^y \ln xy,\quad \left.\dfrac{\partial z}{\partial y}\right|_{(e,1,e)}$

In Exercises 15 and 16, find all the second partial derivatives.

15. $z = 2xy^3 - 3x^2y$
16. $f(x, y) = \dfrac{2 + \cos y}{1 + x^2}$

In Exercises 17–24, solve the given problems.

17. Find the slope of a line tangent to the surface $z = 9 - x^2 - y^2$ and parallel to the yz-plane that passes through (1, 2, 4). Repeat the instructions for the line through (2, 2, 1). Draw an appropriate figure.
18. In quality testing, a rectangular sheet of vinyl is stretched. Set up the length of the diagonal d of the sheet as a function of its sides x and y. Find the rate of change of d with respect to x for $x = 6.50$ m if y remains constant at 4.75 m.
19. Two resistors R_1 and R_2, placed in parallel, have a combined resistance R_T given by $\dfrac{1}{R_T} = \dfrac{1}{R_1} + \dfrac{1}{R_2}$. Find $\partial R_T/\partial R_1$.

20. A metallic machine part contracts while cooling. It is in the shape of a hemisphere attached to a cylinder, as shown in Fig. S5-3. Find the rate of change of volume with respect to r when $r = 2.65$ cm and $h = 4.20$ cm.

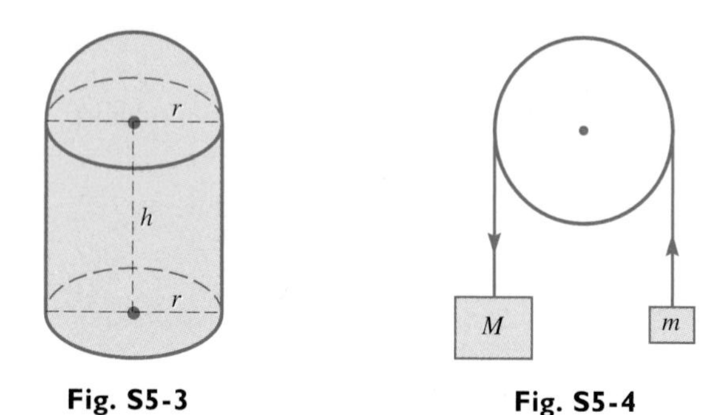

Fig. S5-3

Fig. S5-4

21. Two masses M and m are attached as shown in Fig. S5-4. If $M > m$, the downward acceleration a of mass M is given by $a = \dfrac{M - m}{M + m}g$, where g is the acceleration due to gravity. Show that $M\dfrac{\partial a}{\partial M} + m\dfrac{\partial a}{\partial m} = 0$.

22. If an observer and a source of sound are moving toward or away from each other, the observed frequency of sound is different from that emitted. This is known as the *Doppler effect*. The equation relating the frequency f_0 the observer hears and the frequency f_s emitted by the source (a constant) is $f_0 = f_s\left(\dfrac{v + v_0}{v - v_s}\right)$, where v is the velocity of sound in air (a constant), v_0 is the velocity of the observer, and v_s is the velocity of the source. Show that $f_s\dfrac{\partial f_0}{\partial v_s} = f_0\dfrac{\partial f_0}{\partial v_0}$.

23. The *mutual conductance*, measured in reciprocal ohms, of a certain electronic device is defined as $g_m = \partial i_b/\partial e_c$. Under certain circumstances, the current i_b, measured in microamperes, is given by $i_b = 50(e_b + 5e_c)^{1.5}$. Find g_m when $e_b = 200$ V and $e_c = -20$ V.

24. The displacement y at any point in a taut, flexible string depends on the distance x from one end of the string and the time t. Show that $y(x, t) = 2 \sin 2x \cos 4t$ satisfies the *wave equation* $\dfrac{\partial^2 y}{\partial t^2} = a^2\dfrac{\partial^2 y}{\partial x^2}$ with $a = 2$.

S-6 Double Integrals

We now turn our attention to integration in the case of a function of two variables. The analysis has similarities to that of partial differentiation, in that an operation is performed while holding one of the independent variables constant.

If $z = f(x, y)$ and we wish to integrate with respect to x and y, we first consider either x or y constant and integrate with respect to the other. After this integral is evaluated, we then integrate with respect to the variable first held constant. We shall now define this type of integral and then give an appropriate geometric interpretation.

If $z = f(x, y)$ the **double integral** *of the function over x and y is defined as*

$$\int_a^b \left[\int_{g(x)}^{G(x)} f(x, y)\, dy\right] dx$$

NOTE ▶ *It will be noted that the limits on the inner integral are functions of x and those on the outer integral are explicit values of x.* In performing the integration, ***x* is held constant while the inner integral is found and evaluated.** This results in a function of x only. This function is then integrated and evaluated.

It is customary not to include the brackets in stating a double integral. Therefore, we write

$$\int_a^b \left[\int_{g(x)}^{G(x)} f(x, y)\, dy\right] dx = \int_a^b \int_{g(x)}^{G(x)} f(x, y)\, dy\, dx \qquad \text{(S6-1)}$$

EXAMPLE 1 ▸▸ Evaluate: $\int_0^1 \int_{x^2}^x xy\,dy\,dx$.

First we integrate the inner integral with y as the variable and x as a constant.

treat as constant

$$\int_{x^2}^{x} xy\,dy = \left(x\frac{y^2}{2}\right)\Bigg|_{x^2}^{x} = x\left(\frac{x^2}{2} - \frac{x^4}{2}\right) = \frac{1}{2}(x^3 - x^5)$$

This means

$$\int_0^1 \int_{x^2}^x xy\,dy\,dx = \int_0^1 \frac{1}{2}(x^3 - x^5)\,dx = \frac{1}{2}\left(\frac{x^4}{4} - \frac{x^6}{6}\right)\Bigg|_0^1$$

$$= \frac{1}{2}\left(\frac{1}{4} - \frac{1}{6}\right) - \frac{1}{2}(0) = \frac{1}{24}$$ ◂◂

EXAMPLE 2 ▸▸ Evaluate: $\int_0^{\pi/2} \int_0^{\sin y} e^{2x} \cos y\,dx\,dy$.

CAUTION ▸ Since the inner differential is dx *(the inner limits must then be functions of y, which may be constant)*, we first integrate with x as the variable and y as a constant. The second integration is with y as the variable.

$$\int_0^{\pi/2} \int_0^{\sin y} e^{2x} \cos y\,dx\,dy = \int_0^{\pi/2} \left[\frac{1}{2}e^{2x}\cos y\right]_0^{\sin y} dy$$

$$= \frac{1}{2}\int_0^{\pi/2} (e^{2\sin y}\cos y - \cos y)\,dy$$

$$= \frac{1}{2}\left[\frac{1}{2}e^{2\sin y} - \sin y\right]_0^{\pi/2} = \frac{1}{2}\left(\frac{1}{2}e^2 - 1\right) - \frac{1}{2}\left(\frac{1}{2} - 0\right)$$

$$= \frac{1}{4}e^2 - \frac{1}{2} - \frac{1}{4} = \frac{1}{4}(e^2 - 3) = 1.097$$ ◂◂

For the geometric interpretation of a double integral, consider the surface shown in Fig. S6-1(a). An **element of volume** (dimensions of dx, dy, and z) extends from the xy-plane to the surface. With x a constant, sum (integrate) these elements of volume from the left boundary, $y = g(x)$, to the right boundary, $y = G(x)$. Now the volume of the vertical slice is a function of x, as shown in Fig. S6-1(b). By summing (integrating) the volumes of these slices from $x = a$ ($x = 0$ in the figure) to $x = b$, we have the complete volume as shown in Fig. S6-1(c).

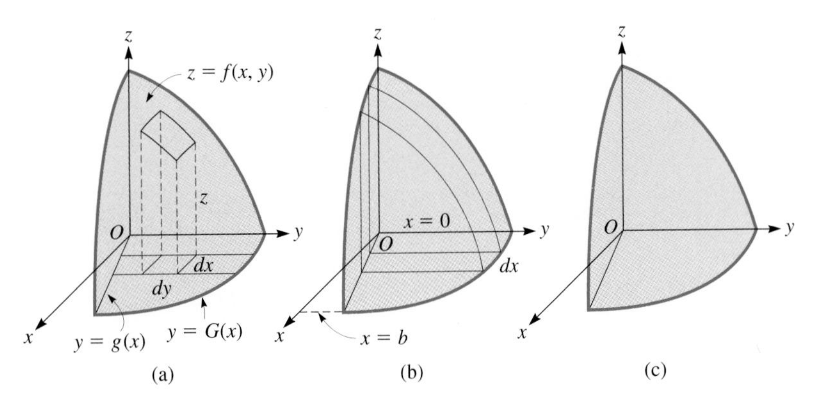

Fig. S6-1

Thus, *we may interpret a double integral as the* **volume under a surface,** in the same way as the integral was interpreted as the area of a plane figure. We may find the volume of a more general figure, as this volume is not necessarily a volume of revolution, as discussed in Section 26-3.

EXAMPLE 3 ▸▸ Find the volume that is in the first octant and under the plane $x + 2y + 4z - 8 = 0$. See Fig. S6-2.

This figure is a tetrahedron, for which $V = \frac{1}{3}Bh$. Assuming the base is in the xy-plane, $B = \frac{1}{2}(4)(8) = 16$, and $h = 2$. Therefore, $V = \frac{1}{3}(16)(2) = \frac{32}{3}$ cubic units. We shall use this value to check the one we find by double integration.

To find $z = f(x, y)$, we solve the given equation for z. Thus

$$z = \frac{8 - x - 2y}{4}$$

CAUTION ▸ Next, we must find the limits on y and x. Choosing to integrate over y first, we see that y goes from $y = 0$ to $y = (8 - x)/2$. ***This last limit is the trace of the surface in the xy-plane.*** Next we note that x goes from $x = 0$ to $x = 8$. Therefore, we set up and evaluate the integral:

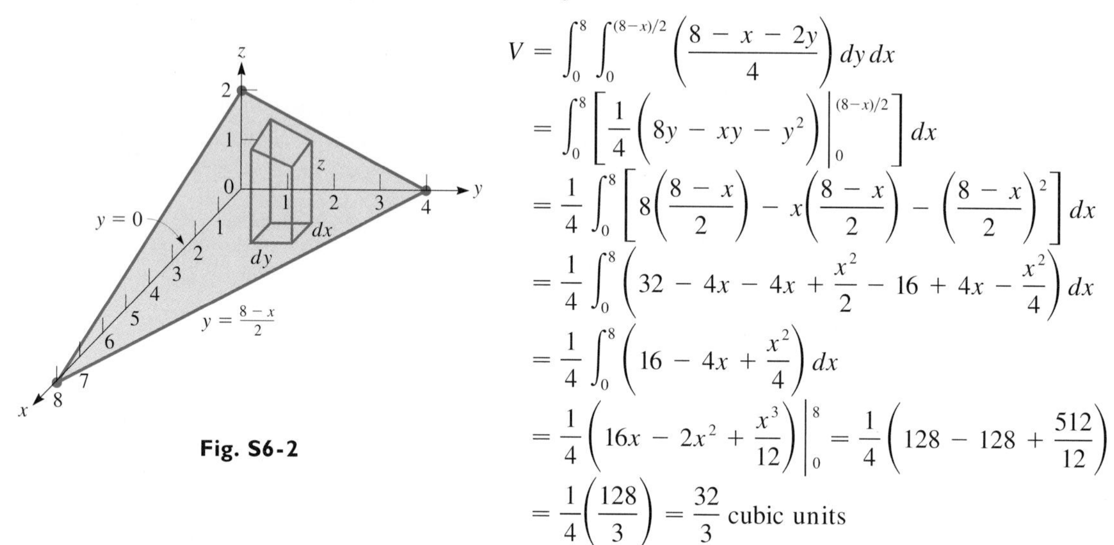

$$\begin{aligned}
V &= \int_0^8 \int_0^{(8-x)/2} \left(\frac{8 - x - 2y}{4}\right) dy\,dx \\
&= \int_0^8 \left[\frac{1}{4}\left(8y - xy - y^2\right)\Bigg|_0^{(8-x)/2}\right] dx \\
&= \frac{1}{4}\int_0^8 \left[8\left(\frac{8 - x}{2}\right) - x\left(\frac{8 - x}{2}\right) - \left(\frac{8 - x}{2}\right)^2\right] dx \\
&= \frac{1}{4}\int_0^8 \left(32 - 4x - 4x + \frac{x^2}{2} - 16 + 4x - \frac{x^2}{4}\right) dx \\
&= \frac{1}{4}\int_0^8 \left(16 - 4x + \frac{x^2}{4}\right) dx \\
&= \frac{1}{4}\left(16x - 2x^2 + \frac{x^3}{12}\right)\Bigg|_0^8 = \frac{1}{4}\left(128 - 128 + \frac{512}{12}\right) \\
&= \frac{1}{4}\left(\frac{128}{3}\right) = \frac{32}{3} \text{ cubic units}
\end{aligned}$$

Fig. S6-2

We see that the values obtained by the two different methods agree. ◂◂

EXAMPLE 4 ▸▸ Find the volume above the xy-plane, below the surface $z = xy$, and enclosed by the cylinder $y = x^2$ and the plane $y = x$.

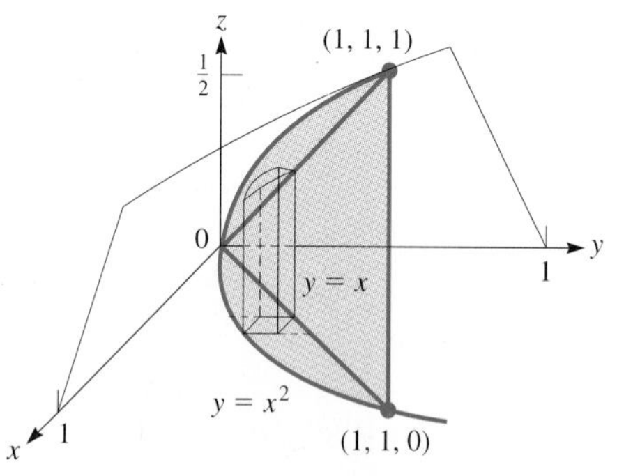

Fig. S6-3

Constructing the figure, shown in Fig. S6-3, we now note that $z = xy$ is the desired function of x and y. Integrating over y first, the limits on y are $y = x^2$ to $y = x$. The corresponding limits on x are $x = 0$ to $x = 1$. Therefore, the double integral to be evaluated is

$$V = \int_0^1 \int_{x^2}^{x} xy\,dy\,dx$$

This integral has already been evaluated in Example 1 of this section, and we can now see the geometric interpretation of that integral. Using the result from Example 1, we see that the required volume is 1/24 cubic unit. ◂◂

EXAMPLE 5 ▸▸ Find the volume that is in the first octant under the surface $z = 4 - x^2 - y^2$ and that is between the cylinder $x^2 = 3y$ and the plane $y = 1$. See Fig. S6-4.

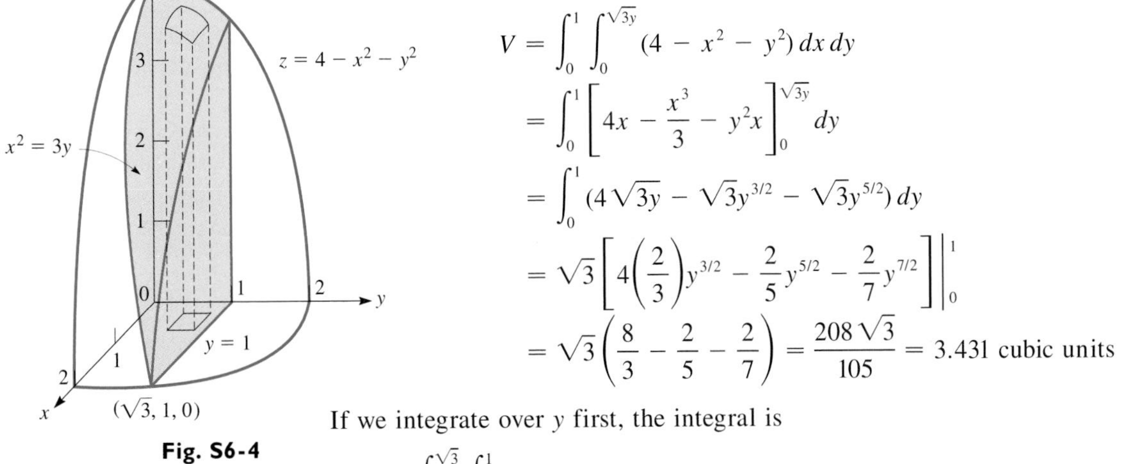

Fig. S6-4

Setting up the integration such that we integrate over x first, we have

$$V = \int_0^1 \int_0^{\sqrt{3y}} (4 - x^2 - y^2)\,dx\,dy$$

$$= \int_0^1 \left[4x - \frac{x^3}{3} - y^2x\right]_0^{\sqrt{3y}} dy$$

$$= \int_0^1 (4\sqrt{3y} - \sqrt{3}y^{3/2} - \sqrt{3}y^{5/2})\,dy$$

$$= \sqrt{3}\left[4\left(\frac{2}{3}\right)y^{3/2} - \frac{2}{5}y^{5/2} - \frac{2}{7}y^{7/2}\right]\Bigg|_0^1$$

$$= \sqrt{3}\left(\frac{8}{3} - \frac{2}{5} - \frac{2}{7}\right) = \frac{208\sqrt{3}}{105} = 3.431 \text{ cubic units}$$

If we integrate over y first, the integral is

$$V = \int_0^{\sqrt{3}} \int_{x^2/3}^1 (4 - x^2 - y^2)\,dy\,dx$$

Integrating and evaluating this integral, we arrive at the same result. ◂◂

Exercises S–6

In Exercises 1–8, evaluate the given double integrals.

1. $\int_2^4 \int_0^1 xy^2\,dx\,dy$
2. $\int_0^4 \int_1^{\sqrt{y}} (x - y)dx\,dy$
3. $\int_0^1 \int_0^{\sqrt{1-x^2}} y\,dy\,dx$
4. $\int_4^9 \int_0^x \sqrt{x - y}\,dy\,dx$
5. $\int_1^e \int_1^y \frac{1}{x}\,dx\,dy$
6. $\int_0^{\pi/6} \int_0^1 y \sin x\,dy\,dx$
7. $\int_0^{\ln 3} \int_0^x e^{2x+3y}\,dy\,dx$
8. $\int_0^{1/2} \int_y^{y^2} \frac{dx\,dy}{\sqrt{y^2 - x^2}}$

In Exercises 9–14, find the indicated volumes by double integration.

9. The first-octant volume under the plane $x + y + z - 4 = 0$
10. The volume above the xy-plane and under the surface $z = 4 - x^2 - y^2$
11. The first-octant volume bounded by the xy-plane, the planes $x = y$, $y = 2$, and $z = 2 + x^2 + y^2$
12. The first-octant volume under the plane $z = x + y$ and inside the cylinder $x^2 + y^2 = 9$
13. A wedge is to be made in the shape shown in Fig. S6-5 (all vertical cross sections are equal right triangles). By double integration, find the volume of the wedge.

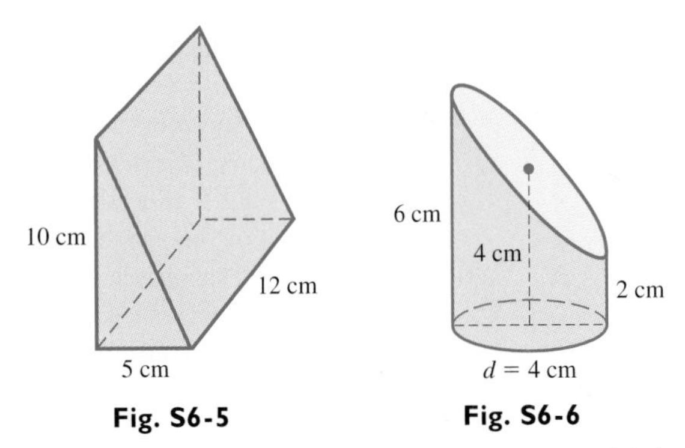

Fig. S6-5　　Fig. S6-6

14. A circular piece of pipe is cut as shown in Fig. S6-6. Find the volume within the pipe. (*Hint*: Place the z-axis along the axis of the pipe, with the base of the pipe in the xy-plane.)

In Exercises 15 and 16, draw the appropriate figure.

15. Draw an appropriate figure indicating a volume that is found from the integral

$$\int_0^{1/2} \int_{x^2}^1 (4 - x - 2y)\,dy\,dx$$

16. Repeat Exercise 15 for the integral

$$\int_1^2 \int_0^{2-y} \sqrt{1 + x^2 + y^2}\,dx\,dy$$

S-7 Integration by Partial Fractions: Nonrepeated Linear Factors

In Chapter 28 we developed basic integral forms and methods of integration. There are many additional methods by which integrals can be put in a form for integration. In this section and the following section we develop one more useful method of integration.

In algebra we combine fractions into a single fraction by means of addition. However, if we wish to integrate an expression that contains a rational fraction, in which both the numerator and the denominator are polynomials, it is often advantageous to reverse the operation of addition and express the rational fraction as the sum of simpler fractions.

EXAMPLE 1 ▸▸ In attempting to integrate $\int \frac{7-x}{x^2+x-2}\,dx$, we find that it does not fit any of the standard forms in Chapter 28. However, we can show that

$$\frac{7-x}{x^2+x-2} = \frac{2}{x-1} - \frac{3}{x+2}$$

This means that

$$\begin{aligned}\int \frac{7-x}{x^2+x-2}\,dx &= \int \frac{2\,dx}{x-1} - \int \frac{3\,dx}{x+2}\\ &= 2\ln|x-1| - 3\ln|x+2| + C\end{aligned}$$ ◂◂

In Example 1 we saw that the integral is readily determined once the rational fraction $(7-x)/(x^2+x-2)$ is replaced by the simpler fractions. In this section and the next we describe how certain rational fractions can be expressed in terms of simpler fractions and thereby be integrated. *This technique is called the* **method of partial fractions.**

In order to express the rational fraction $f(x)/g(x)$ in terms of simpler partial fractions, *the degree of the numerator $f(x)$ must be less than that of the denominator $g(x)$.* If this is not the case, we divide numerator by denominator until the remainder is of the proper form. Then the denominator $g(x)$ is factored into a product of linear and quadratic factors. The method of determining the partial fractions depends on the factors that are obtained. In advanced algebra the form of the partial fractions is shown (but we shall not show the proof here).

There are four cases for the types of factors of the denominator. They are (1) *nonrepeated linear factors,* (2) *repeated linear factors,* (3) *nonrepeated quadratic factors,* and (4) *repeated quadratic factors.* In this section we consider the case of nonrepeated linear factors, and the other cases are discussed in the next section.

For the case of nonrepeated linear factors we use the fact that *corresponding to each linear factor $ax+b$, occurring once in the denominator, there will be a partial fraction of the form*

$$\frac{A}{ax+b}$$

where A is a constant to be determined. The following examples illustrate the method.

EXAMPLE 2 ▸▸ Integrate: $\int \frac{7 - x}{x^2 + x - 2} dx$. (This is the same integral as in Example 1. Here we will see how the partial fractions are found.)

First we note that the degree of the numerator is 1 (the highest power is x) and that of the denominator is 2 (the highest power is x^2). Since the degree of the denominator is higher, we may proceed to factoring it. Thus

$$\frac{7 - x}{x^2 + x - 2} = \frac{7 - x}{(x - 1)(x + 2)}$$

There are two linear factors, $(x - 1)$ and $(x + 2)$, in the denominator and they are different. This means that there are two partial fractions. Therefore, we write

$$\frac{7 - x}{(x - 1)(x + 2)} = \frac{A}{x - 1} + \frac{B}{x + 2} \tag{1}$$

We are to determine constants A and B so that Eq. (1) is an identity. In finding A and B we clear Eq. (1) of fractions by multiplying both sides by $(x - 1)(x + 2)$.

$$7 - x = A(x + 2) + B(x - 1) \tag{2}$$

Equation (2) is also an identity, which means that there are two ways of determining the values of A and B.

NOTE ▸ *Solution by substitution:* Since Eq. (2) is an identity, ***it is true for any value of x.*** Thus, in turn we pick $x = -2$ and $x = 1$, for each of these values makes a factor on the right equal to zero, and the values of B and A are easily found. Therefore,

$$\text{for } x = -2\text{:} \quad 7 - (-2) = A(-2 + 2) + B(-2 - 1)$$
$$9 = -3B, \qquad B = -3$$
$$\text{for } x = 1\text{:} \quad 7 - 1 = A(1 + 2) + B(1 - 1)$$
$$6 = 3A, \qquad A = 2$$

NOTE ▸ *Solution by equating coefficients:* Since Eq. (2) is an identity, another way of finding the constants A and B is to ***equate coefficients of like powers of x from each side.*** Thus, writing Eq. (2) as

$$7 - x = (2A - B) + (A + B)x$$

we have

$$2A - B = 7 \quad \text{(equating constants: } x^0 \text{ terms)}$$
$$A + B = -1 \quad \text{(equating coefficients of } x\text{)}$$

$$\begin{aligned} 2A - B &= 7 \\ A + B &= -1 \\ \hline 3A &= 6 \\ A &= 2 \\ 2(2) - B &= 7 \\ -B &= 3 \\ B &= -3 \end{aligned}$$

Now, using the values $A = 2$ and $B = -3$ (as found at the left), we have

$$\frac{7 - x}{(x - 1)(x + 2)} = \frac{2}{x - 1} - \frac{3}{x + 2}$$

Therefore, the integral is found as in Example 1.

$$\int \frac{7 - x}{x^2 + x - 2} dx = \int \frac{7 - x}{(x - 1)(x + 2)} dx = \int \frac{2\,dx}{x - 1} - \int \frac{3\,dx}{x + 2}$$
$$= 2 \ln |x - 1| - 3 \ln |x + 2| + C$$

Using the properties of logarithms, we may write this as

$$\int \frac{7 - x}{x^2 + x - 2} dx = \ln \left| \frac{(x - 1)^2}{(x + 2)^3} \right| + C$$ ◂◂

EXAMPLE 3 ▸▸ Integrate: $\int \frac{6x^2 - 14x - 11}{(x + 1)(x - 2)(2x + 1)} dx$.

The denominator is factored and is of degree 3 (when multiplied out, the highest power of x is x^3). This means we have three nonrepeated linear factors.

$$\frac{6x^2 - 14x - 11}{(x + 1)(x - 2)(2x + 1)} = \frac{A}{x + 1} + \frac{B}{x - 2} + \frac{C}{2x + 1}$$

Multiplying through by $(x + 1)(x - 2)(2x + 1)$, we have

$$6x^2 - 14x - 11 = A(x - 2)(2x + 1) + B(x + 1)(2x + 1) + C(x + 1)(x - 2)$$

We now substitute the values of 2, $-\frac{1}{2}$, and -1 for x. Again, these are chosen because they make factors of the coefficients of A, B, or C equal to zero, although any values may be chosen. Therefore,

for $x = 2$: $\quad 6(4) - 14(2) - 11 = A(0)(5) + B(3)(5) + C(3)(0), \quad B = -1$

for $x = -\frac{1}{2}$: $\quad 6\left(\frac{1}{4}\right) - 14\left(-\frac{1}{2}\right) - 11 = A\left(-\frac{5}{2}\right)(0) + B\left(\frac{1}{2}\right)(0) + C\left(\frac{1}{2}\right)\left(-\frac{5}{2}\right), \quad C = 2$

for $x = -1$: $\quad 6(1) - 14(-1) - 11 = A(-3)(-1) + B(0)(-1) + C(0)(-3), \quad A = 3$

Therefore,

$$\int \frac{6x^2 - 14x - 11}{(x + 1)(x - 2)(2x + 1)} dx = \int \frac{3\,dx}{x + 1} - \int \frac{dx}{x - 2} + \int \frac{2\,dx}{2x + 1}$$

$$= 3 \ln|x + 1| - \ln|x - 2| + \ln|2x + 1| + C_1 = \ln\left|\frac{(2x + 1)(x + 1)^3}{x - 2}\right| + C_1$$

Here we have let the constant of integration be C_1 since we used C as the numerator of the third partial fraction. ◂◂

EXAMPLE 4 ▸▸ Integrate: $\int \frac{2x^4 - x^3 - 9x^2 + x - 12}{x^3 - x^2 - 6x} dx$.

Since the numerator is of a higher degree than the denominator, we must first divide the numerator by the denominator. This gives

$$\frac{2x^4 - x^3 - 9x^2 + x - 12}{x^3 - x^2 - 6x} = 2x + 1 + \frac{4x^2 + 7x - 12}{x^3 - x^2 - 6x}$$

We must now express this rational fraction in terms of its partial fractions.

$$\frac{4x^2 + 7x - 12}{x^3 - x^2 - 6x} = \frac{4x^2 + 7x - 12}{x(x + 2)(x - 3)} = \frac{A}{x} + \frac{B}{x + 2} + \frac{C}{x - 3}$$

Clearing fractions, we have

$$4x^2 + 7x - 12 = A(x + 2)(x - 3) + Bx(x - 3) + Cx(x + 2)$$

Now, using values of x of -2, 3, and 0 for substitution, we obtain the values of $B = -1$, $C = 3$, and $A = 2$, respectively. Therefore,

$$\int \frac{2x^4 - x^3 - 9x^2 + x - 12}{x^3 - x^2 - 6x} dx = \int \left(2x + 1 + \frac{2}{x} - \frac{1}{x + 2} + \frac{3}{x - 3}\right) dx$$

$$= x^2 + x + 2 \ln|x| - \ln|x + 2| + 3 \ln|x - 3| + C_1$$

$$= x^2 + x + \ln\left|\frac{x^2(x - 3)^3}{x + 2}\right| + C_1$$ ◂◂

In Example 2 we showed two ways of finding the values of A and B. The method of substitution is generally easier to use with linear factors, and we used it in Examples 3 and 4. However, as we will see in the next section, the method of equating coefficients can be very useful for other cases with partial fractions.

Exercises S–7

In Exercises 1–12, integrate the given functions.

1. $\displaystyle\int \frac{x+3}{(x+1)(x+2)}\,dx$
2. $\displaystyle\int \frac{x+2}{x(x+1)}\,dx$
3. $\displaystyle\int \frac{dx}{x^2-4}$
4. $\displaystyle\int \frac{x-9}{2x^2-3x+1}\,dx$
5. $\displaystyle\int \frac{x^2+3}{x^2+3x}\,dx$
6. $\displaystyle\int \frac{x^3}{x^2+3x+2}\,dx$
7. $\displaystyle\int_0^1 \frac{2x+4}{3x^2+5x+2}\,dx$
8. $\displaystyle\int_1^3 \frac{x-1}{4x^2+x}\,dx$
9. $\displaystyle\int \frac{4x^2-10}{x(x+1)(x-5)}\,dx$
10. $\displaystyle\int \frac{6x^2-2x-1}{4x^3-x}\,dx$
11. $\displaystyle\int_2^3 \frac{dx}{x^3-x}$
12. $\displaystyle\int \frac{dx}{(x^2-4)(x^2-9)}$

In Exercises 13–16, solve the given problems by integration.

13. The current i, in amperes, as a function of the time, in seconds, in a certain electric circuit is given by $i = (4t+3)/(2t^2+3t+1)$. Find the total charge that passes a given point in the circuit during the first second.

14. The force F, in newtons, applied by a stamping machine in making a certain computer part is given by $F = 4x/(x^2+3x+2)$, where x is the distance (in cm) through which the force acts. Find the work done by the force from $x = 0$ to $x = 0.500$ cm.

15. Find the first-quadrant area bounded by $y = 1/(x^3+3x^2+2x)$, $x = 1$, and $x = 3$. See Fig. S7-1.

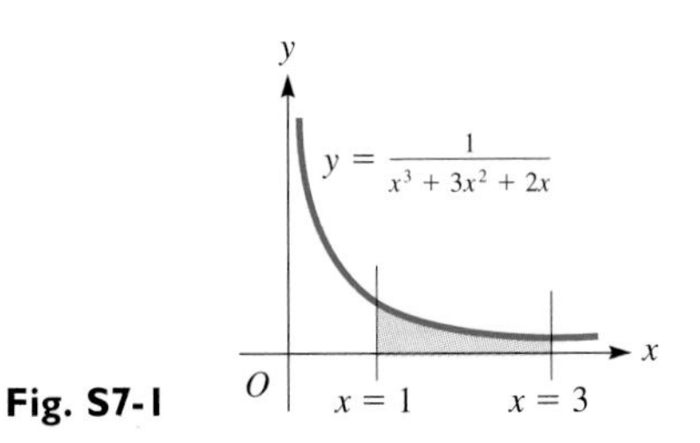

Fig. S7-1

16. Find the volume generated if the area of Exercise 15 is revolved about the y-axis.

S-8 Integration by Partial Fractions: Other Cases

In the previous section we introduced the method of partial fractions and considered the case of nonrepeated linear factors. In this section we develop the use of partial fractions for the cases of repeated linear factors and nonrepeated quadratic factors. We also briefly discuss the case of repeated quadratic factors.

Repeated Linear Factors

For the case of repeated linear factors we use the fact that *corresponding to each linear factor $ax + b$ that occurs n times in the denominator there will be n partial fractions*

$$\frac{A_1}{ax+b} + \frac{A_2}{(ax+b)^2} + \cdots + \frac{A_n}{(ax+b)^n}$$

where $A_1, A_2, \ldots, A_n$ are constants to be determined.

EXAMPLE 1 ⏩ Integrate: $\int \frac{dx}{x(x+3)^2}$.

Here we see that the denominator has a factor of x and two factors of $x + 3$. For the factor of x we use a partial fraction as in the previous section, for it is a nonrepeated factor. For the factor $x + 3$ we need two partial fractions, one with a denominator of $x + 3$ and the other with a denominator of $(x + 3)^2$. Thus we write

NOTE ▶

$$\frac{1}{x(x+3)^2} = \frac{A}{x} + \frac{B}{x+3} + \frac{C}{(x+3)^2}$$

Multiplying each side by $x(x + 3)^2$, we have

$$1 = A(x+3)^2 + Bx(x+3) + Cx \qquad (1)$$

Using the values of x of -3 and 0, we have

$$\text{for } x = -3{:} \quad 1 = A(0^2) + B(-3)(0) + (-3)C, \qquad C = -\frac{1}{3}$$

$$\text{for } x = 0{:} \quad 1 = A(3^2) + B(0)(3) + C(0), \qquad A = \frac{1}{9}$$

Since there are no other values that make a factor in Eq. (1) equal to zero, we must either choose some other value of x or equate coefficients of some power of x in Eq. (1). We will let $x = 1$.

Since Eq. (1) is an identity, we may choose any value of x. With $x = 1$, we have

$$1 = A(4^2) + B(1)(4) + C(1)$$

$$1 = 16A + 4B + C$$

Using the known values of A and C, we have

$$1 = 16\left(\frac{1}{9}\right) + 4B - \frac{1}{3}, \qquad B = -\frac{1}{9}$$

This means that

$$\frac{1}{x(x+3)^2} = \frac{\frac{1}{9}}{x} + \frac{-\frac{1}{9}}{x+3} + \frac{-\frac{1}{3}}{(x+3)^2}$$

or

$$\begin{aligned}\int \frac{dx}{x(x+3)^2} &= \frac{1}{9}\int \frac{dx}{x} - \frac{1}{9}\int \frac{dx}{x+3} - \frac{1}{3}\int \frac{dx}{(x+3)^2} \\ &= \frac{1}{9}\ln|x| - \frac{1}{9}\ln|x+3| - \frac{1}{3}\left(\frac{1}{-1}\right)(x+3)^{-1} + C_1 \\ &= \frac{1}{9}\ln\left|\frac{x}{x+3}\right| + \frac{1}{3(x+3)} + C_1\end{aligned}$$

We have used the properties of logarithms in order to write the final form of the answer. ⏪

EXAMPLE 2 ▸▸ Integrate: $\int \frac{3x^3 + 15x^2 + 21x + 15}{(x-1)(x+2)^3} dx$.

First we set up the partial fractions as

$$\frac{3x^3 + 15x^2 + 21x + 15}{(x-1)(x+2)^3} = \frac{A}{x-1} + \frac{B}{x+2} + \frac{C}{(x+2)^2} + \frac{D}{(x+2)^3}$$

Next we clear fractions.

$$3x^3 + 15x^2 + 21x + 15 = A(x+2)^3 + B(x-1)(x+2)^2 + C(x-1)(x+2) + D(x-1) \qquad (1)$$

For $x = 1$: $\quad 3 + 15 + 21 + 15 = 27A, \quad 54 = 27A, \quad A = 2$

For $x = -2$: $\quad 3(-8) + 15(4) + 21(-2) + 15 = -3D, \quad 9 = -3D, \quad D = -3$

To find B and C we equate coefficients of powers of x. Therefore, we write Eq. (1) as

$$3x^3 + 15x^2 + 21x + 15 = (A + B)x^3 + (6A + 3B + C)x^2 + (12A + C + D)x + (8A - 4B - 2C - D)$$

Coefficients of x^3: $\quad 3 = A + B, \quad 3 = 2 + B, \quad B = 1$

Coefficients of x^2: $\quad 15 = 6A + 3B + C, \quad 15 = 12 + 3 + C, \quad C = 0$

$$\frac{3x^3 + 15x^2 + 21x + 15}{(x-1)(x+2)^3} = \frac{2}{x-1} + \frac{1}{x+2} + \frac{0}{(x+2)^2} + \frac{-3}{(x+2)^3}$$

$$\int \frac{3x^3 + 15x^2 + 21x + 15}{(x-1)(x+2)^3} dx = 2\int \frac{dx}{x-1} + \int \frac{dx}{x+2} - 3\int \frac{dx}{(x+2)^3}$$

$$= 2\ln|x-1| + \ln|x+2| - 3\left(\frac{1}{-2}\right)(x+2)^{-2} + C_1$$

$$= \ln|(x-1)^2(x+2)| + \frac{3}{2(x+2)^2} + C_1 \quad ◂◂$$

If there is one repeated factor in the denominator and it is the only factor present in the denominator, a substitution is easier and more convenient than using partial fractions. This is illustrated in the following example.

EXAMPLE 3 ▸▸ Integrate: $\int \frac{x\,dx}{(x-2)^3}$.

This could be integrated by first setting up the appropriate partial fractions. However, the solution is more easily found by using the substitution $u = x - 2$. Using this, we have

$$u = x - 2, \quad x = u + 2, \quad dx = du$$

$$\int \frac{x\,dx}{(x-2)^3} = \int \frac{(u+2)(du)}{u^3} = \int \frac{du}{u^2} + 2\int \frac{du}{u^3}$$

$$= \int u^{-2}du + 2\int u^{-3}du = \frac{1}{-u} + \frac{2}{-2}u^{-2} + C$$

$$= -\frac{1}{u} - \frac{1}{u^2} + C = -\frac{u+1}{u^2} + C$$

$$= -\frac{x-2+1}{(x-2)^2} + C = \frac{1-x}{(x-2)^2} + C \quad ◂◂$$

Quadratic Factors

For the case of nonrepeated quadratic factors, we use the fact that *corresponding to each irreducible (cannot be further factored) quadratic factor* $ax^2 + bx + c$ *that occurs once in the denominator there is a partial fraction of the form*

$$\frac{Ax + B}{ax^2 + bx + c}$$

where A and B are constants to be determined. This case is illustrated in the following examples.

EXAMPLE 4 ▸▸ Integrate: $\displaystyle\int \frac{4x + 4}{x^3 + 4x}\,dx.$

In setting up the partial fractions, we note that the denominator factors as $x^3 + 4x = x(x^2 + 4)$. Here the factor $x^2 + 4$ cannot be further factored. This means we have

NOTE ▶

$$\frac{4x + 4}{x^3 + 4x} = \frac{4x + 4}{x(x^2 + 4)} = \boxed{\frac{A}{x} + \frac{Bx + C}{x^2 + 4}}$$

Clearing fractions, we have

$$\begin{aligned} 4x + 4 &= A(x^2 + 4) + Bx^2 + Cx \\ &= (A + B)x^2 + Cx + 4A \end{aligned}$$

Equating coefficients of powers of x gives us

for x^2: $\quad 0 = A + B$

for x: $\quad 4 = C$

for constants: $\quad 4 = 4A, \quad$ or $\quad A = 1$

Therefore, we easily find that $B = -1$ from the first equation. This means that

$$\frac{4x + 4}{x^3 + 4x} = \frac{1}{x} + \frac{-x + 4}{x^2 + 4}$$

and

$$\begin{aligned} \int \frac{4x + 4}{x^3 + 4x}\,dx &= \int \frac{1}{x}\,dx + \int \frac{-x + 4}{x^2 + 4}\,dx \\ &= \int \frac{1}{x}\,dx - \int \frac{x\,dx}{x^2 + 4} + \int \frac{4\,dx}{x^2 + 4} \\ &= \ln|x| - \frac{1}{2}\ln|x^2 + 4| + 2\tan^{-1}\frac{x}{2} + C_1 \end{aligned}$$

We could use the properties of logarithms to combine the first two terms of the result. ◂◂

EXAMPLE 5 ▸▸ Integrate: $\displaystyle\int \frac{x^3 + 3x^2 + 2x + 4}{x^2(x^2 + 2x + 2)}\,dx$.

In the denominator we have a repeated linear factor, x^2, and a quadratic factor. Therefore,

$$\frac{x^3 + 3x^2 + 2x + 4}{x^2(x^2 + 2x + 2)} = \frac{A}{x} + \frac{B}{x^2} + \frac{Cx + D}{x^2 + 2x + 2}$$

$$\begin{aligned} x^3 + 3x^2 + 2x + 4 &= Ax(x^2 + 2x + 2) + B(x^2 + 2x + 2) + Cx^3 + Dx^2 \\ &= (A + C)x^3 + (2A + B + D)x^2 + (2A + 2B)x + 2B \end{aligned}$$

Equating coefficients, we find that

$$A = -1,\ B = 2,\ C = 2, \text{ and } D = 3$$

Therefore,

$$\frac{x^3 + 3x^2 + 2x + 4}{x^2(x^2 + 2x + 2)} = -\frac{1}{x} + \frac{2}{x^2} + \frac{2x + 3}{x^2 + 2x + 2}$$

$$\begin{aligned} \int \frac{x^3 + 3x^2 + 2x + 4}{x^2(x^2 + 2x + 2)}\,dx &= -\int \frac{dx}{x} + 2\int \frac{dx}{x^2} + \int \frac{2x + 3}{x^2 + 2x + 2}\,dx \\ &= -\ln|x| - 2\left(\frac{1}{x}\right) + \int \frac{2x + 2 + 1}{x^2 + 2x + 2}\,dx \\ &= -\ln|x| - \frac{2}{x} + \int \frac{2x + 2}{x^2 + 2x + 2}\,dx + \int \frac{dx}{(x^2 + 2x + 1) + 1} \\ &= -\ln|x| - \frac{2}{x} + \ln|x^2 + 2x + 2| + \tan^{-1}(x + 1) + C_1 \end{aligned}$$

CAUTION ▸ ***Note the manner in which the integral with the quadratic denominator was handled for the purpose of integration.*** First, the numerator, $2x + 3$, was written in the form $(2x + 2) + 1$ so that we could fit the logarithmic form with the $2x + 2$. Then we completed the square in the denominator of the final integral so that it then fit an inverse tangent form. ◂◂

Repeated Quadratic Factors

Finally, considering the case of repeated quadratic factors, we use the fact that *corresponding to each irreducible quadratic factor $ax^2 + bx + c$ that occurs n times in the denominator there will be n partial fractions*

$$\frac{A_1x + B_1}{ax^2 + bx + c} + \frac{A_2x + B_2}{(ax^2 + bx + c)^2} + \cdots + \frac{A_nx + B_n}{(ax^2 + bx + c)^n}$$

where $A_1, A_2, \ldots, A_n, B_1, B_2, \ldots, B_n$ are constants to be determined. The procedures that lead to the solution are the same as those for the other cases. Exercises 11 and 12 in the following set are solved by using these partial fractions for repeated quadratic factors.

Exercises S–8

In Exercises 1–12, integrate each of the given functions.

1. $\int \frac{x-8}{x^3-4x^2+4x}\,dx$

2. $\int \frac{dx}{x^3-x^2}$

3. $\int \frac{2\,dx}{x^2(x^2-1)}$

4. $\int_1^3 \frac{3x^3+8x^2+10x+2}{x(x+1)^3}\,dx$

5. $\int_1^2 \frac{2x\,dx}{(x-3)^3}$

6. $\int \frac{4\,dx}{(x+1)^2(x-1)^2}$

7. $\int_0^2 \frac{x^2+x+5}{(x+1)(x^2+4)}\,dx$

8. $\int \frac{x^2+x-1}{(x^2+1)(x-2)}\,dx$

9. $\int \frac{5x^2+8x+16}{x^2(x^2+4x+8)}\,dx$

10. $\int \frac{10x^3+40x^2+22x+7}{(4x^2+1)(x^2+6x+10)}\,dx$

11. $\int \frac{2x^3}{(x^2+1)^2}\,dx$

12. $\int \frac{-x^3+x^2+x+3}{(x+1)(x^2+1)^2}\,dx$

In Exercises 13–16, solve the given problems by integration.

13. Find the area bounded by $y=(x-3)/(x^3+x^2)$, $y=0$, and $x=1$.

14. Find the volume generated by revolving the first-quadrant area bounded by $y=4/(x^4+6x^2+5)$ and $x=2$ about the y-axis.

15. Under certain conditions the velocity v, in metres per second, of an object moving along a straight line as a function of time, in seconds, is given by $v=\frac{t^2+14t+27}{(2t+1)(t+5)^2}$. Find the distance traveled by the object during the first 2.00 s.

16. By a computer analysis the electric current i, in amperes, in a certain integrated circuit is given by $i=\frac{0.001(7t^2+16t+48)}{(t+4)(t^2+16)}$, where t is the time in seconds. Find the total charge that passes a point in the circuit in the first 0.250 s.

A

Study Aids

A-1 Introduction

The primary objective of this text is to give you an understanding of mathematics so that you can use it effectively as a tool in your technology. Without an understanding of the basic methods, knowledge is usually short-lived. However, if you do understand, you will find your work much more enjoyable and rewarding. This is true in any course you may take, be it in mathematics or in any other field.

Mathematics is an indispensable tool in almost all scientific fields of study. You will find it used to a greater and greater degree as you work in your chosen field. Generally, in the introductory portions of allied courses, it is enough to have a grasp of elementary concepts in algebra and geometry. However, as you progress in your field, the need for more mathematics will be apparent. This text is designed to develop these necessary tools so that they will be available to you in your technical courses. *You cannot derive the full benefit from your mathematics course unless you devote the necessary amount of time to develop a sound understanding of the subject.*

Your mathematics background probably includes some geometry and algebra. Therefore, some of the topics covered in this book may seem familiar to you, especially in the earlier chapters. However, it is likely that your background in some of these areas is not complete, either because you have not studied mathematics for a while or because you did not fully understand the topics when you first encountered them. *If a topic is familiar, take the opportunity to clarify any points on which you are not certain,* and do not reason that there is no sense in studying it again. In almost every topic you probably will find certain points that can use further study. *If the topic is new to you, use your time effectively to develop an understanding of the methods involved,* and do not simply memorize problems of a certain type.

There is only one good way to develop the understanding and working knowledge necessary in any course, and that is to **work with it**. Many students consider mathematics difficult. They say that it is their lack of mathematical ability and the complexity of the material that makes it difficult. Some of the topics in mathematics, especially in the more advanced areas, do require a certain aptitude for full comprehension. However, *a large proportion of poor grades in elementary mathematics courses results from the fact that* ***the student is not willing to put in the necessary time to develop a full understanding.*** The student takes a quick glance through the material, tries a few exercises, is largely unsuccessful, and then decides that the material is "impossible." A detailed reading of the text, following the illustrative examples carefully, and then solving the exercises would lead to more success and therefore make the work much more enjoyable and rewarding. No matter what text is used, what methods are used in the course, or what other variables may be introduced, *if you do not put in an adequate amount of time studying, you will not derive the proper results.* More detailed suggestions for study are included in the following section.

A-2 Suggestions for Study

When you are studying the material presented in this text, the following suggestions may help you to derive full benefit from the time you devote to it.

1. Before attempting to do the exercises, read through the material preceding them.
2. Follow the illustrative examples carefully, being certain that you know how to proceed from step to step. You should then have a good idea of the methods involved.
3. Work through the exercises, spending a reasonable amount of time on each problem. If you cannot solve a certain problem in a reasonable amount of time, leave it and go to the next. Return to this problem later. If you find many problems difficult, you should reread the explanatory material and the examples to determine what point or points you have not understood.
4. When you have completed the exercises, or at least most of them, glance back through the explanatory material to be sure you understand the methods and principles.
5. If you have gone through the first four steps and certain points still elude you, ask to have these points clarified in class. Do not be afraid to ask questions; only be sure that you have made a sincere effort on your own before you ask them.

Some study habits that are useful not only here but in all of your other subjects are the following:

1. Put in the time required to develop the material fully, being certain that you are making effective use of your time. A good place to study helps immeasurably.
2. Learn the *methods and principles* being presented. Memorize as little as possible, for although certain basic facts are more expediently learned by memorization, these should be kept to a minimum.
3. Keep up with the material in all of your courses. Do not let yourself get so behind in your studies that it becomes difficult to make up the time. Usually the time is never really made up. Studying only before tests is a poor way of learning and is usually rather ineffective.
4. When you are taking examinations, always read each question carefully before attempting the solution. Solve those you find easiest first, and do not spend too much time on any one problem. Also, use all the time available for the examination. If you finish early, use the remainder of the time to check your work.

If you consider these suggestions carefully, and follow good study habits, you should enjoy a successful learning experience in this course as well as in other courses.

A-3 Problem Analysis

Drill-type problems require a working knowledge of the methods presented. However, they do not require, in general, much analysis before being put in proper form for solution. Stated problems, on the other hand, do require proper interpretation before they can be put in a form for solution. The remainder of this section is devoted to some suggestions for solving stated problems.

We have to put stated problems in symbolic form before we attempt to solve them. It is this step that most students find difficult. Because such problems require the student to do more than merely go through a certain routine, they demand more analysis and thus appear more "difficult." There are several reasons for the student's difficulty, some of them being (1) unsuccessful previous attempts at solving such problems, leading the student to believe that all stated problems are "impossible"; (2) failure to read the problem carefully; (3) a poorly organized approach to the solution; and (4) improper and incomplete interpretation of the statement given. The first two of these can be overcome only with the proper attitude and care.

There are over 120 completely worked examples of stated problems (as well as numerous other examples that indicate a similar analysis) throughout this text, illustrating proper interpretations and approaches to these problems. Therefore, we shall not include specific examples here. However, we shall set forth the method of analysis of any stated problem. Such an analysis generally follows these steps:

Procedure for Solving Stated Problems

1. *Read the problem carefully.*
2. *Carefully identify the known and unknown quantities.*
3. *Draw a figure when appropriate* (which is quite often the case).
4. *Write, in symbols, the relations given in the statements.* A careful analysis of the statements is necessary for this step.
5. *Solve for the desired quantities.*

(These steps are similar to those outlined on pg. 42, when stated problems are first covered in our study of algebra.)

If you follow this step-by-step method and write out the solution neatly, you should find that stated problems lend themselves to solution more readily than you had previously found.

B

Units of Measurement; the Metric System

B-1 The Metric System (SI)

The solution of most technical problems involves the use of the basic arithmetic operations on numbers. However, many of these numbers represent some type of measurement or calculation. *Such numbers are called* **denominate numbers,** *and associated with these denominate numbers are* **units of measurement**. In order for the number or calculation to be meaningful, we must know which units are being used. For example, if we measure the length of an object to be 15, we must know whether it is being measured in metres, centimetres, or some other specified unit of length.

Two basic systems of units, the **metric system** and the **United States Customary** system, are in use today. (The U.S. Customary system is also known as the *British system,* but the metric system is now used in Great Britain.) Nearly every country in the world now uses the metric system. In the United States both systems are used, and international trade has led many major U.S. industrial firms to convert their products to the metric system. Also due to world trade, to further promote conversion to the metric system, Congress has passed legislation which states that the metric system is the preferred system, and which requires Federal agencies to use the metric system in their business-related activities.

The metric system was first adopted in France in the late eighteenth century. Since then several systems using metric units were developed, each adding some specific units, and some used British units in certain cases. Therefore, a wide variety of metric units was being used.

In order to simplify the metric system, a long series of international discussions were held, and in 1960 a modernized metric system was established. It is the **International System of Units,** called SI, and it is the system that has now replaced all previous systems of units. In this text, the metric system and SI are synonymous, and *all units are SI units or are acceptable for use with SI.*

In the SI system there are seven **base units** that are used to measure fundamental quantities. These base units form the foundation of the system. The base units are: (1) metre (length), (2) kilogram (mass), (3) second (time), (4) ampere (electric current), (5) kelvin (thermodynamic temperature), (6) mole (amount of substance), and (7) candela (luminous intensity).

The SI system also has **supplementary units** (used for measuring plane and solid angles), **derived units** (which are used for numerous quantities), as well as other units that are acceptable for use. The only supplementary unit used in this text is the *radian* (see Section 8-3).

Many derived units have special names, and many others do not. For example, the volt is defined as a metre^2-kilogram/second^3-ampere, which is in terms of (a unit of length)^2 (a unit of mass)/(a unit of time)^3 (a unit of electric current). The unit for acceleration has no special name, and is left in terms of the base units: metre per second squared.

Among the units permitted for use with the SI system are the units of time, the *minute, hour, day,* and *year.* Also permitted are the *degree, minute,* and *second* measurements for an angle, the *litre* for capacity, and the *degree Celsius* for temperature.

In designating units of measurement it is convenient to use designated **symbols,** rather than the complete name of the unit each time. In Table B-2 a list of quantities, the units of measurement, and the appropriate symbols are listed.

We note that the kilogram is the base unit of *mass.* The distinction between mass and *force* is very significant in physics, and this distinction causes difficulty when discussing the *weight* of an object. *Weight, which is the force with which an object is attracted to the earth, is different from mass, which is a measure of the amount of material in an object.* Although they are different quantities, mass and weight are very closely related. Near the surface of the earth the weight of an object is directly proportional to its mass. However, at great distance from the earth the weight of an object will be essentially zero, whereas its mass does not change.

The kilogram is commonly used for weight, although it is mass that is actually being denoted. The metric unit of force, the *newton,* is also used for weight. Thus, it is preferable to *specify the mass of an object in kilograms* and *the force of gravity of an object in newtons,* and thereby avoid the use of the term *weight* unless its meaning is completely clear.

As for temperature, the *kelvin* is the base unit for thermodynamic temperature, but *degrees Celsius* (formerly *Centigrade*) is the unit that is used for most general purposes. For temperature intervals, 1°C = 1 K, although on a scale, 0°C = 273.15 K.

The base units and derived units are often not of a convenient size for many types of measurements. Therefore, other units of more convenient sizes are used. To denote these other units, the SI system employs certain prefixes to denote different orders of magnitude. These prefixes, along with their meanings and symbols, are shown in Table B-1.

TABLE B-1 Metric Prefixes

Prefix	*Factor*	*Symbol*	*Prefix*	*Factor*	*Symbol*
exa	10^{18}	E	deci	10^{-1}	d
peta	10^{15}	P	centi	10^{-2}	c
tera	10^{12}	T	milli	10^{-3}	m
giga	10^{9}	G	micro	10^{-6}	μ
mega	10^{6}	M	nano	10^{-9}	n
kilo	10^{3}	k	pico	10^{-12}	p
hecto	10^{2}	h	femto	10^{-15}	f
deca	10^{1}	da	atto	10^{-18}	a

TABLE B-2 Quantities and Their Associated SI Units

Quantity	*Quantity Symbol*	*Metric (SI) Unit*		
		Name	*Symbol*	*In Terms of Other SI Units*
Length	s	**metre**	m	
Mass	m	**kilogram**	kg	
Force	F	newton	N	$\mathrm{m \cdot kg/s^2}$
Time	t	**second**	s	
Area	A		$\mathrm{m^2}$	
Volume	V		$\mathrm{m^3}$	
Capacity	V	litre	L	$(1\ \mathrm{L} = 1\ \mathrm{dm^3})$
Velocity	v		m/s	
Acceleration	a		$\mathrm{m/s^2}$	
Density	d, ρ		$\mathrm{kg/m^3}$	
Pressure	p	pascal	Pa	$\mathrm{N/m^2}$
Energy, work	E, W	joule	J	$\mathrm{N \cdot m}$
Power	P	watt	W	J/s
Period	T		s	
Frequency	f	hertz	Hz	1/s
Angle	θ	radian	rad	
Electric current	I, i	**ampere**	A	
Electric charge	q	coulomb	C	$\mathrm{A \cdot s}$
Electric potential	V, E	volt	V	$\mathrm{J/(A \cdot s)}$
Capacitance	C	farad	F	$\mathrm{s/\Omega}$
Inductance	L	henry	H	$\mathrm{\Omega \cdot s}$
Resistance	R	ohm	Ω	V/A
Thermodynamic temperature	T	**kelvin**	K	
Temperature	T	degree Celsius	°C	(1°C = 1 K)
Quantity of heat	Q	joule	J	
Amount of substance	n	**mole**	mol	
Luminous intensity	I	**candela**	cd	

Special Notes:

1. The SI base units are shown in boldface type.
2. The unit symbols shown above are those that are used in the text. Many of them were adopted with the adoption of the SI system. This means, for example, that we use s rather than sec for seconds, and A rather than amp for amperes. Also, other units such as volt are not spelled out, a common practice in the past.
3. The litre and degree Celsius are not actually SI units. However, they are recognized for use with the SI system due to their practical importance. Also, the symbol for litre has several variations. Presently L is recognized for use in the United States and Canada, l is recognized by the International Committee of Weights and Measures, and ℓ is also recognized for use in several countries.
4. Other units of time, along with their symbols, which are recognized for use with the SI system and are used in this text, are minute, min; hour, h; day, d.
5. Many additional specialized units are used with the SI system. However, most of those that appear in this text are shown in the table. A few of the specialized units are noted when used in the text. One which is frequently used is that for revolution, r.
6. There are a number of units that were used with the metric system prior to the development of the SI system. However, many of these are not to be used with the SI system. Among those that were commonly used are the dyne, erg, and calorie.

EXAMPLE 1 ▸▸ Some commonly used units that use prefixes in Table B-2, along with their meanings, are shown below.

Unit	*Symbol*	*Meaning*	*Unit*	*Symbol*	*Meaning*
megohm	MΩ	10^6 ohms	milligram	mg	10^{-3} gram
kilometre	km	10^3 metres	microfarad	μF	10^{-6} farad
centimetre	cm	10^{-2} metre	nanosecond	ns	10^{-9} second

(Mega is shortened to meg when used with "ohm.") ◂◂

It is the relationship among units within the metric system that makes it convenient to use. *Units of different magnitudes differ by a power of ten, and therefore only a shift of the decimal point is often the change necessary when changing units.* No specific arrangement exists within the U.S. system.

There are a number of matters of style that are employed with denominate numbers and their units when using the SI system. Following are some of the basic rules of style, although many others may be found in standard reference sources: (1) Symbols remain unaltered if the unit is in the plural. For example, either volt or volts is designated by the symbol V. (2) When designating square or cubic units, we use exponents in the designation. For example, we use m^2 rather than sq m. (3) The multiplication of unit symbols is designated by a dot of multiplication. For example, m · N means metre newton (whereas mN means millinewton). (4) A solidus (/) or negative exponent is used to designate the division of one unit symbol by another. For example, a metre per second would be designated by m/s or $m \cdot s^{-1}$. Generally we have used the solidus in this text. (5) In long numbers, the digits are separated into groups of three, to the left and right of the decimal point. A space is not necessary if four digits appear. For example, we write 56 000 rather than 56,000 (that is, use only a space, and not a comma). Also, we write 0.000 000 7 rather than 0.0000007. We may write either 0.1234 or 0123 4, although we should be consistent. In this text, no space is used with four-digit numbers.

B-2 Changing Units

When using denominate numbers, it may be necessary to change from one set of units to another. *A change within a given system is called a* **reduction,** and *a change from one system to another is called a* **conversion**. Some calculators are programmed to do conversions.

NOTE ▸ To change a given number of one set of units into another set of units, *we perform algebraic operations with units in the same manner as we do with any algebraic symbol*. Consider the following example.

EXAMPLE 1 ▸▸ If we had a number representing metres per second to be multiplied by another number representing seconds per minute, as far as the units are concerned, we have

$$\frac{\text{m}}{\text{s}} \times \frac{\text{s}}{\text{min}} = \frac{\text{m} \times \cancel{\text{s}}}{\cancel{\text{s}} \times \text{min}} = \frac{\text{m}}{\text{min}}$$

This means that the final result would be in metres per minute. ◂◂

In changing a number of one set of units to another set of units, we use reduction and conversion factors and the principle illustrated in Example 1. The convenient way to use the values in the tables is in the form of fractions. Since the given values are equal to each other, their quotient is 1. For example, since 100 cm = 1 m,

NOTE ▸

$$\frac{100\text{ cm}}{1\text{ m}} = 1 \quad \text{or} \quad \frac{1\text{ m}}{100\text{ cm}} = 1$$

since each represents the division of a certain length by itself. Multiplying a quantity by 1 does not change its value. The following examples illustrate changing units.

EXAMPLE 2 ▸▸ Change 20 kg to milligrams.

$$\begin{aligned} 20\text{ kg} &= 20\,\cancel{\text{kg}}\left(\frac{10^3\,\cancel{\text{g}}}{1\,\cancel{\text{kg}}}\right)\left(\frac{10^3\text{ mg}}{1\,\cancel{\text{g}}}\right) \\ &= 20 \times 10^6\text{ mg} \\ &= 2.0 \times 10^7\text{ mg} \end{aligned}$$

We note that this result is found essentially by moving the decimal point three places when changing from kilograms to grams, and another three places when changing from grams to milligrams. ◂◂

EXAMPLE 3 ▸▸ Change 30 m/h to centimetres per second.

$$\begin{aligned} 30\frac{\text{m}}{\text{h}} &= \left(30\frac{\cancel{\text{m}}}{\cancel{\text{h}}}\right)\left(\frac{100\text{ cm}}{1\,\cancel{\text{m}}}\right)\left(\frac{1\,\cancel{\text{h}}}{60\,\cancel{\text{min}}}\right)\left(\frac{1\,\cancel{\text{min}}}{60\text{ s}}\right) \\ &= \frac{(30)(100)\text{ cm}}{(60)(60)\text{ s}} = 0.833\frac{\text{cm}}{\text{s}} \end{aligned}$$

The only units remaining after the division are those required. ◂◂

EXAMPLE 4 ▸▸ Change 575 g/cm^3 to kilograms per cubic metre.

$$\begin{aligned} 575\frac{\text{g}}{\text{cm}^3} &= \left(575\frac{\text{g}}{\text{cm}^3}\right)\left(\frac{100\text{ cm}}{1\text{ m}}\right)^3\left(\frac{1\text{ kg}}{1000\text{ g}}\right) \\ &= \left(575\frac{\cancel{\text{g}}}{\cancel{\text{cm}^3}}\right)\left(\frac{10^6\,\cancel{\text{cm}^3}}{1\text{ m}^3}\right)\left(\frac{1\text{ kg}}{10^3\,\cancel{\text{g}}}\right) \\ &= 575 \times 10^3\frac{\text{kg}}{\text{m}^3} = 5.75 \times 10^5\frac{\text{kg}}{\text{m}^3} \end{aligned}$$ ◂◂

EXAMPLE 5 ▸▸ Change 62.8 kPa to newtons per square centimetre.

$$62.8 \text{ kPa} = \left(62.8 \times 10^3 \frac{\text{N}}{\text{m}^2}\right)\left(\frac{1 \text{ m}}{10^2 \text{ cm}}\right)^2$$
$$= \left(62.8 \times 10^3 \frac{\text{N}}{\cancel{\text{m}^2}}\right)\left(\frac{1 \cancel{\text{m}^2}}{10^4 \text{ cm}^2}\right)$$
$$= \frac{62.8 \times 10^3}{10^4} \frac{\text{N}}{\text{cm}^2} = 6.28 \frac{\text{N}}{\text{cm}^2}$$ ◂◂

Exercises for Appendix B

In Exercises 1–4, give the symbol and the meaning for the given unit.

1. megahertz **2.** kilowatt
3. millimetre **4.** picosecond

In Exercises 5–8, give the name and the meaning for the units whose symbols are given.

5. kV **6.** GΩ
7. mA **8.** pF

In Exercises 9–40, make the indicated changes of units.

9. Change 1 km to centimetres.

10. Change 1 kg to milligrams.

11. Change 20 s to megaseconds.

12. Change 800 Pa to kilopascals.

13. Change 810.15 K to degrees Celsius.

14. Change 31.50°C to kelvins.

15. Change 250 mm^2 to square metres.

16. Change 1.75 m^2 to square centimetres.

17. Change 80.0 m^3 to cubic decimetres.

18. Change 360 μm^3 to cubic centimetres.

19. Change 50 mL to litres.

20. Change 0.125 L to millilitres.

21. Change 45.0 m/s to centimetres per second.

22. Change 200 mm/s to metres per second.

23. Change 38.2 cm/s to metres per hour.

24. Change 1.32 km/h to centimetres per minute.

25. Change 9.80 m/s^2 to centimetres per minute squared.

26. Change 5.10 g/cm^3 to kilograms per cubic millimetre.

27. Change 5.25 mV to watts per ampere.

28. Change 15.0 μF to millicoulombs per volt.

29. A weather satellite orbiting the earth has a mass of 2200 kg. How many megagrams is this?

30. At sea level, atmospheric pressure is about 101 300 Pa. How many kilopascals is this?

31. A car's gasoline tank holds 56 L. What is this capacity in cubic centimetres?

32. A hockey puck has a mass of about 0.160 kg. What is its mass in milligrams?

33. The density of water is 1000 kg/m^3. Change this to grams per litre.

34. Water flows from a kitchen faucet at the rate of 8500 mL/min. What is this rate in litres per second?

35. The speed of sound is about 332 m/s. Change this speed to kilometres per hour.

36. The acceleration due to gravity is about 980 cm/s^2. Convert this to metres per squared hour.

37. Fifteen grams of a medication are to be dissolved in 0.060 L of water. Express this concentration in milligrams per decilitre.

38. The earth's surface receives energy from the sun at the rate of 1.35 kW/m^2. Change this to joules per second square centimetre.

39. The average density of the earth is about 5.52 g/cm^3. Change this to kilograms per cubic metre.

40. The moon travels about 2 400 000 km in about 28 d in one rotation about the earth. Express its velocity in metres per second.

C

Scientific and Graphing Calculators

C-1 Introduction

Since the development of the first electronic calculators in the early 1970s and graphing calculators in the late 1980s, there has been a wide variety of different types of calculators. This appendix includes a brief reference for the use of scientific calculators. Some of the many uses of graphing calculators are also noted, but it is necessary to refer to the calculator manual as to how to use the particular model for these features.

In Section 1-3 calculators were briefly introduced, and throughout the text there are notes related to their use. In Section 3-5 there is an introduction to the use of graphing calculators for their specific use in graphing.

There are many types and models of scientific and graphing calculators. Some models now available can do some symbolic operations. However, the discussions in this appendix are based on the operations that can be performed by using the basic keys of a scientific calculator and some of the basic features of a graphing calculator. These discussions are general enough to apply to most such calculators, and we note some of the variations and special features that may be found. All calculators come with a manual that can be used to learn the operations of the calculator. This manual should be used for any special features a given calculator may have.

In this appendix we discuss the scientific calculator keys and operations that are used for data entry, arithmetic operations, special functions (squares, square roots, reciprocals, powers, and roots), trigonometric and inverse trigonometric functions, exponential and logarithmic functions, and calculator memory. Examples are included.

Most models of calculators use many of the keys for more than one purpose, and labeling of some keys varies from one model to another. We note some of the variations that are used.

When a number is entered, or result calculated, it shows on the display at the top of the calculator. In this appendix we use a ten-digit display, as well as a possible display of the exponent of ten for scientific notation. In the discussions in the book, eight-digit and ten-digit displays are used. If the result contains more digits than the display can show, the result shown will be rounded off.

In using a calculator, we must keep in mind that certain operations do not have defined results. If such an operation is attempted, the calculator will show an error display. Operations that can result in an error display include division by zero, square root of a negative number, logarithm of a negative number, and the inverse trigonometric function of a value outside of the defined interval. Also, when using a graphing calculator, an error display will result if an improper sequence of keys is used.

As we noted in Section 1-3, the calculator logic determines the order in which entries must be made in order to obtain the desired result. We also noted that *in this text we assume that the calculator uses algebraic logic*. However, although algebraic logic is used, some keys are used in a different order on a scientific calculator from the order used on a graphing calculator. *The user must become acquainted with the logic of the particular calculator being used.*

C-2 Scientific Calculator Keys and Operations

Following is a listing of certain basic keys that will be found on a scientific calculator and which can be of use in performing calculations for the exercises in this text. Other keys and other calculational capabilities are also found on many scientific calculators. Along with each key is a description and an illustration of its basic use. In the examples shown at the right, to perform the indicated operation, use the given sequence of keys and entries. The entry of a number is shown as one operation. The final display is also shown.

Many types of problems require a number of calculator steps and keys for their solutions. The use of the calculator in solving these problems is discussed in the appropriate sections throughout the text. Although the calculator is very useful in solving such problems, it is still necessary to learn the required mathematics in order to set up and understand them.

The use of many of the keys may be beyond the scope of the reader until the appropriate text material has been covered.

Keys	*Examples*
[+/−] **Change Sign Key** This key is used to change the sign of the number on the display or to change the sign of the exponent when scientific notation is used. (May be designated as [CHS].)	To enter: −375.14 Sequence: [3] [7] [5] [.] [1] [4] [+/−] Display: −375.14
[π] **Pi Key** This key is used to enter π to the number of digits of the display. On some calculators a key will have a secondary purpose of providing the value of π. See the second function key.	To enter: π Sequence: [π] 3.141592654
[EE] **Enter Exponent Key** This key is used to enter an exponent when scientific notation is used. After the key is pressed, the exponent is entered. For a negative exponent, the [+/−] key is pressed after the exponent is entered. (May be designated as [EEX] or [EXP].)	To enter: 2.936×10^{8} Sequence: 2.936 [EE] 8 2.936 08 To enter: -2.936×10^{-8} Sequence: 2.936 [+/−] [EE] 8 [+/−] −2.936 −08
[=] **Equals Key** This key is used to complete a calculation to give the required result.	See the following examples for illustrations of the use of this key.

Keys	*Examples*
[+] **Add Key** This key is used to add the next entry to the previous entry or result.	Evaluate: 37.56 + 241.9 Sequence: 37.56 [+] 241.9 [=] [279.46]
[−] **Subtract Key** This key is used to subtract the next entry from the previous entry or result.	Evaluate: 37.56 − 241.9 Sequence: 37.56 [−] 241.9 [=] [−204.34]
[×] **Multiply Key** This key is used to multiply the previous entry or result by the next entry.	Evaluate: 8.75 × 30.92 Sequence: 8.75 [×] 30.92 [=] [270.55]
[÷] **Divide Key** This key is used to divide the previous entry or result by the next entry.	Evaluate: 8.75 ÷ 30.92 Sequence: 8.75 [÷] 30.92 [=] [0.282988357]
[(] [)] **Parentheses Keys** These keys are used to group specified numbers in a calculation.	Evaluate: 3.586(30.72 + 47.92) Sequence: 3.586 [×] [(] 30.72 [+] 47.92 [)] [=] [282.00304]
[CE] **Clear Entry Key** This key is used to clear the last entry. Its use will not affect any other part of a calculation. On some calculators one press of the [C/CE] or [CL] key is used for this purpose.	Evaluate 37.56 + 241.9, with an improper entry of 242.9 Sequence: 37.56 [+] 242.9 [CE] 241.9 [=] [279.46]
[C] **Clear Key** This key is used to clear the display and information being calculated (not including memory) so that a new calculation may be started. For calculators with a [C/CE] or [C] key, and no [CE] key, a second press on these keys is used for this purpose.	To clear previous calculation: Sequence: [C] [0.]
[2ND] **Second Function Key** This key is used on calculators on which many of the keys serve dual purposes. It is pressed before the second key functions are activated. The [INV] key may also serve as a second function key.	Evaluate $(37.4)^2$ on a calculator where x^2 is a second use of a key Sequence: 37.4 [2ND] [x^2] [1398.76]
[x^2] **Square Key** This key is used to square the number on the display.	Evaluate: $(37.4)^2$ Sequence: 37.4 [x^2] [1398.76]
[$\sqrt{x}$] **Square Root Key** This key is used to find the square root of the number on the display.	Evaluate: $\sqrt{37.4}$ Sequence: 37.4 [$\sqrt{x}$] [6.115553941]

Keys	Examples
[1/x] **Reciprocal Key** This key is used to find the reciprocal of the number on the display.	Evaluate: $\frac{1}{37.4}$ Sequence: 37.4 [1/x] [0.026737968]
[x^y] **x to the y Power Key** This key is used to raise x, the first entry, to the y power, the second entry. (This key may be labeled as y^x).	Evaluate: $(3.73)^{1.5}$ Sequence: 3.73 [x^y] 1.5 [=] [7.203826553]
[DRG] **Degree-Radian-Grad Key** This key is used to designate a displayed angle as being measured in degrees, radians, or grads. (100 grads = 90°.)	This key should show the appropriate angle measurement before the calculation is started. (On some calculators this key is also used to convert from one angle measure to another.)
[DMS] **Degrees-Minutes-Seconds Key** This key is used to convert an angle in degrees, minutes, and seconds to an angle in decimal degrees. By use of the [INV] key the conversion is done in reverse.	Convert 52°8′17″ to decimal degrees. Sequence: 52.0817 [DMS] [52.13805556] Convert 52.138 056° to degrees, minutes, and seconds. Sequence: 52.138 056 [INV] [DMS] [52.0817] ← 52°8′17″
[sin] **Sine Key** This key is used to find the sine of the angle on the display.	Evaluate: sin 37.4° Sequence: [ⒹRG] 37.4 [sin] [0.60737584]
[cos] **Cosine Key** This key is used to find the cosine of the angle on the display.	Evaluate: cos 2.475 (rad) Sequence: [DⓇG] 2.475 [cos] [−0.785933026]
[tan] **Tangent Key** This key is used to find the tangent of the angle on the display.	Evaluate: tan(−24.9°) Sequence: [ⒹRG] 24.9 [+/−] [tan] [−0.464184545]
Cotangent, Secant, Cosecant These functions are found through their reciprocal relation with the tangent, cosine, and sine functions, respectively.	Evalute: cot 2.841 (rad) Sequence: [DⓇG] 2.841 [tan] [1/x] [−3.225954921]
[INV] **Inverse Function Key** On some models this key is used to find the inverse of the primary function on a key. On other models it is the second function key and many of the paired functions are not inverses. For inverse trigonometric functions, most models have [$\sin^{-1}$], [$\cos^{-1}$], and [$\tan^{-1}$] keys that are used after the [INV] key.	
[$\sin^{-1}$] **Inverse Sine Key** This key is used to find the angle for which the value of the sine function is the number on display.	Evaluate: $\sin^{-1}$ 0.1758 (the angle whose sine is 0.1758) in degrees. Sequence: [ⒹRG] 0.1758 [$\sin^{-1}$] [10.12521652]

Keys	Examples
[cos⁻¹] **Inverse Cosine Key** This key is used to find the angle for which the value of the cosine function is the number on display.	Evaluate: $\cos^{-1}(-0.7828)$ in degrees. Sequence: [ⒹRG] .7828 [+/−] [cos⁻¹] [141.5176605]
[tan⁻¹] **Inverse Tangent Key** This key is used to find the angle for which the value of the tangent function is the number on display.	Evaluate: $\tan^{-1}(-1.4062)$ in radians. Sequence: [DⓇG] 1.4062 [+/−] [tan⁻¹] [−0.952635308]
[LOG] **Common Logarithm Key** This key is used to find the logarithm to the base 10 of the number on the display.	Evaluate: log 37.45 Sequence: 37.45 [LOG] [1.573451822]
[LN] **Natural Logarithm Key** This key is used to find the logarithm to the base e of the number on the display.	Evaluate: ln 0.8421 Sequence: 0.8421 [LN] [−0.171856507]
[10^x] **10 to the x Power Key** This key is used to find antilogarithms (base 10) of the number on the display. (The [x^y] key can be used for this purpose, if the calculator does not have this key.)	Evaluate: Antilog 0.7265 (or $10^{0.7265}$) Sequence: 0.7265 [10^x] [5.327212243]
[e^x] ***e* to the *x* Power Key** This key is used to raise e to the power on the display.	Evaluate: $e^{-4.05}$ Sequence: 4.05 [+/−] [e^x] [0.017422375]
[!] **Factorial Key** This key is used to calculate the value of n!	Evaluate: 40! Sequence: 40 [!] [8.159152832 47]
[STO] **Store in Memory Key** This key is used to store the displayed number in the memory.	Store in memory: 56.02 Sequence: 56.02 [STO]
[RCL] **Recall from Memory Key** This key is used to recall the number in the memory to the display. (May be designated as [MR].)	Recall from memory: 56.02 Sequence: [RCL] [56.02]
[M] **Other Memory Keys** Some calculators use an [M] key to store a number in the memory. It also may add the entry to the number in the memory. On such calculators, [CM] (Clear Memory) is used to clear the memory. There are also keys for other operations on the number in the memory.	

C-3 The Graphing Calculator

Since their introduction in the late 1980s, many types and models of graphing calculators have been developed. All models can display graphs of functions and perform all the calculational operations of a scientific calculator. Most models also have many additional computing and graphing capabilities. In this section we briefly note some of the differences between scientific calculators and graphing calculators, and those features of a graphing calculator to which reference is made in the text. *To determine how the features of a particular model are used, refer to the manual for that model.*

One major advantage a graphing calculator has in performing calculations is that the entries can be seen in the *viewing window*. This allows a check of entered data for possible errors.

In performing calculations, most models of graphing calculators use *algebraic logic*. However, for many of the operations the order in which entries are made differs from the order on a scientific calculator. *In evaluating most functions on a graphing calculator the function is entered first, and then the functional value is entered.* The display in the viewing window appears just as it would be written. On a scientific calculator, the reverse order is used. This is true when using the following basic keys:

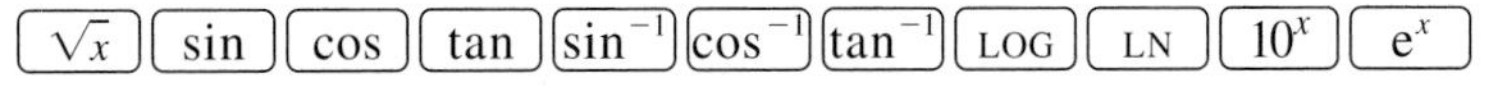

For the [x^2] and [x^{-1}] keys, both types of calculators use the same order.

EXAMPLE 1 ▸▸ The following orders of entries are used.

To evaluate	Order of entries on a scientific calculator	Order of entries on a graphing calculator
56.3^2	56.3 [x^2]	56.3 [x^2]
sin 39.2°	39.2 [sin]	[sin] 39.2
$\cos^{-1} 0.7519$	.7519 [$\cos^{-1}$]	[$\cos^{-1}$] .7519
log 389	389 [LOG]	[LOG] 389 ◂◂

Another difference in making entries on a graphing calculator is that it is not necessary to use the multiplication key [×] if entries are adjacent (except adjacent numbers). Again, the display of entries in the viewing window appears as it would be written.

EXAMPLE 2 ▸▸ (a) To evaluate 3.5 tan 18°, the following orders of entries are used to obtain the final display as shown.

scientific calculator: 3.5 [×] 18 [tan] [=] [1.137218937]
graphing calculator: 3.5 [tan] 18 [ENTER]
1.137218937

(b) To evaluate 6(2 + 3 ln 7), the following orders of entries are used.

scientific calculator: 6 [×] [( ] 2 [+] 3 [×] 7 [LN] [ )] [=] [47.02638268]
graphing calculator: 6 [( ] 2 [+] 3 [LN] 7 [ )] [ENTER]
47.02638268

On some scientific calculators, the [×] key does not have to be used before parentheses. ◂◂

In using a graphing calculator, negative numbers are entered by using the (−) key, and the negative sign is entered *before* the number. On a scientific calculator, the +/− key is used *after* the number is entered. There is additional discussion of this on p. 11 in Section 1-3.

Graphing calculators are *programmable*. That is, *programs* like those used on a computer can be stored in the calculator memory to automatically perform specific operations or calculations. For example, it is possible to enter a program that will automatically calculate and display the circumference and the area of a circle if the value of the radius is entered. The manual should be consulted in order to see how this feature is used.

Many of the keys on a graphing calculator are used for two or three purposes. The additional uses are activated by using the ALPHA and 2ND (or SHIFT) keys. The ALPHA key is used for entering alphabetical characters when writing a program. It is also used on some calculators to enter the variable x when using the graphing feature.

Of course, one of the most important uses of a graphing calculator is that it can display the graph of a function in the display window. This is introduced in Section 3-5 on p. 92 and is used in numerous examples throughout the book.

Following is a list of four basic features on a graphing calculator that are used for graphing. Here we indicate the general use of the feature, but again, consult the manual to see how each feature is used on a particular model.

Feature	*Use*
GRAPH	On some models it is used to display the graph after the function is entered. On other models it puts the calculator into graphing mode.
RANGE	Used to set the boundaries of the viewed portion of the graph.
TRACE	Used to move the cursor from pixel to pixel along a displayed graph.
ZOOM	Used to adjust the viewing window.

The following list shows the indicated uses of these graphing features on the given reference pages. Use of these features also appears on many other pages.

Uses Shown in Text	*Page*
Graphing a function	92
Solving an equation graphically	94
Finding the range of a function	95
Graphing a straight line	134
Solving a system of linear equations	136
Graphing a quadratic function	211, 379
Graphing a sine or cosine function	274, 277
Graphing other trigonometric functions	284
Graphing an addition of ordinates type of function	290
Graphing parametric equations	292
Graphing exponential and logarithmic functions	354
Graphing a circle, parabola, ellipse, or hyperbola (two functions)	379, 380, 550, 559, 571, 909
Solving a system of equations	380
Finding the roots of an equation	410
Solving an inequality	458
Solving a trigonometric equation	523
Displaying a curve and its normal line	658

The MODE feature determines how numbers are calculated and displayed and how graphs are displayed. Some calculators use a specific key sequence to set a particular mode. Most calculators display a *menu* that shows particular settings and allows these to be changed. The mode can be set, for example, (1) to display angles measured in degrees, radians, or grads (see pp. 113 and 229); (2) for the number of decimal places displayed in numbers; (3) to display the graph of a function or of parametric equations (see p. 292); (4) to display a graph in rectangular or polar coordinates (see p. 578). All graphing calculators have many other possible mode settings for specific operations and displays.

Most graphing calculators have a MATRIX feature that is used to perform all the basic matrix operations (see p. 436). These include the evaluation of determinants (see p. 159) and solving systems of linear equations (see p. 440). Again, a *menu* is used to select the operation to be performed.

Graphing calculators can calculate various statistical measures such as the arithmetic mean and standard deviation (see p. 596). They also can be used to display statistical graphs such as a histogram (see p. 587). Another important use is finding an equation that fits a given set of points (see p. 604). All of these topics are covered in Chapter 22. The many possible statistical operations are accessed from a menu of the MODE feature on some calculators and of the STAT feature on other calculators.

There are many other possible uses of a graphing calculator. Among those noted in the text is solving a system of inequalities by shading in the appropriate area (see p. 462). This is accessed from a menu of the MODE feature on some calculators and from the DRAW feature on other calculators.

Many graphing calculators can evaluate the derivative of a function at a point (see pg. 644). Also, they can be used to find the maximum or minimum point of a function over a particular interval (see p. 676), and they can be used to evaluate definite integrals (see p. 716).

As we have noted, the graphing calculator has a great many useful calculating and graphing capabilities. Most calculators can perform many operations in addition to those mentioned.

The exercises that follow provide an opportunity for practice in performing calculations on a scientific calculator or a graphing calculator. No exercises included here are intended specifically for a graphing calculator. Such exercises are found in the appropriate sections throughout the text.

Exercises for Appendix C

In Exercises 1–36, perform all calculations on a scientific calculator or a graphing calculator.

1. $47.08 + 8.94$

2. $654.1 + 407.7$

3. $4724 - 561.9$

4. $0.9365 - 8.077$

5. 0.0396×471

6. 26.31×0.9393

7. $76.7 \div 194$

8. $52{,}060 \div 75.09$

9. 3.76^2

10. 0.986^2

11. $\sqrt{0.2757}$

12. $\sqrt{60.36}$

13. $\dfrac{1}{0.0749}$

14. $\dfrac{1}{607.9}$

15. $(19.66)^{2.3}$

16. $(8.455)^{1.75}$

17. $\sin 47.3°$

18. $\sin 1.15$

19. $\cos 3.85$

20. $\cos 119.1°$

21. $\tan 306.8°$

22. $\tan 0.537$

23. $\sec 6.11$

24. $\csc 242.0°$

25. $\sin^{-1} 0.6607$ (in degrees)

26. $\cos^{-1} (-0.8311)$ (in radians)

27. $\tan^{-1} (-2.441)$ (in radians)

28. $\sin^{-1} 0.0737$ (in degrees)

29. $\log 3.857$

30. $\log 0.9012$

31. $\ln 808$

32. $\ln 70.5$

33. $10^{0.545}$

34. $10^{-0.0915}$

35. $e^{-5.17}$

36. $e^{1.672}$

In Exercises 37–92, perform all calculations on a scientific calculator or a graphing calculator. In these exercises, some of the combined operations encountered in certain types of problems are given. For specific types of applications, solve problems from the appropriate sections of the text.

37. $(4.38 + 9.07) \div 6.55$

38. $(382 + 964) \div 844$

39. $4.38 + (9.07 \div 6.55)$

40. $382 + (964 \div 844)$

41. $\dfrac{5.73 \times 10^{11}}{20.61 - 7.88}$

42. $\dfrac{7.09 \times 10^{23}}{284 + 839}$

43. $50.38\pi^2$

44. $\dfrac{5\pi}{14.6}$

45. $\sqrt{1.65^2 + 6.44^2}$

46. $\sqrt{0.735^2 + 0.409^2}$

47. $3(3.5)^4 - 4(3.5)^2$

48. $\dfrac{3(-1.86)}{(-1.86)^2 + 1}$

49. $29.4 \cos 72.5°$

50. $\dfrac{477}{\sin 58.7°}$

51. $\dfrac{4 + \sqrt{(-4)^2 - 4(3)(-9)}}{2(3)}$

52. $\dfrac{-5 - \sqrt{5^2 - 4(4)(-7)}}{2(4)}$

53. $\dfrac{0.176(180)}{\pi}$

54. $\dfrac{209.6\pi}{180}$

55. $\dfrac{1}{2}\left(\dfrac{51.4\pi}{180}\right)(7.06)^2$

56. $\dfrac{1}{2}\left(\dfrac{148.2\pi}{180}\right)(49.13)^2$

57. $\sin^{-1} \dfrac{27.3 \sin 36.5°}{46.8}$

58. $\dfrac{0.684 \sin 76.1°}{\sin 39.5°}$

59. $\sqrt{3924^2 + 1762^2 - 2(3924)(1762) \cos 106.2°}$

60. $\cos^{-1} \dfrac{8.09^2 + 4.91^2 - 9.81^2}{2(8.09)(4.91)}$

61. $\sqrt{5.81 \times 10^8} + \sqrt[3]{7.06 \times 10^{11}}$

62. $(6.074 \times 10^{-7})^{2/5} - (1.447 \times 10^{-5})^{4/9}$

63. $\dfrac{3}{2\sqrt{7} - \sqrt{6}}$

64. $\dfrac{7\sqrt{5}}{4\sqrt{5} - \sqrt{11}}$

65. $\tan^{-1} \dfrac{7.37}{5.06}$

66. $\tan^{-1} \dfrac{46.3}{-25.5}$

67. $2 + \dfrac{\log 12}{\log 7}$

68. $\dfrac{10^{0.4115}}{\pi}$

69. $\dfrac{26}{2}(-1.450 + 2.075)$

70. $\dfrac{4.55(1 - 1.08^{15})}{1 - 1.08}$

71. $\sin^2\left(\dfrac{\pi}{7}\right) + \cos^2\left(\dfrac{\pi}{7}\right)$

72. $\sec^2\left(\dfrac{2}{9}\pi\right) - \tan^2\left(\dfrac{2}{9}\pi\right)$

73. $\sin 31.6° \cos 58.4° + \sin 58.4° \cos 31.6°$

74. $\cos^2 296.7° - \sin^2 296.7°$

75. $\sqrt{(1.54 - 5.06)^2 + (-4.36 - 8.05)^2}$

76. $\sqrt{(7.03 - 2.94)^2 + (3.51 - 6.44)^2}$

77. $\dfrac{(4.001)^2 - 16}{4.001 - 4}$

78. $\dfrac{(2.001)^2 + 3(2.001) - 10}{2.001 - 2}$

79. $\dfrac{4\pi}{3}(8.01^3 - 8.00^3)$

80. $4\pi(76.3^2 - 76.0^2)$

81. $0.01\left(\dfrac{1}{2}\sqrt{2} + \sqrt{2.01} + \sqrt{2.02} + \dfrac{1}{2}\sqrt{2.03}\right)$

82. $0.2\left[\dfrac{1}{2}(3.5)^2 + 3.7^2 + 3.9^2 + \dfrac{1}{2}(4.1)^2\right]$

83. $\dfrac{e^{0.45} - e^{-0.45}}{e^{0.45} + e^{-0.45}}$

84. $\ln \sin 2e^{-0.055}$

85. $\ln \dfrac{2 - \sqrt{2}}{2 - \sqrt{3}}$

86. $\sqrt{\dfrac{9}{2} + \dfrac{9 \sin 0.2\pi}{8\pi}}$

87. $2 + \dfrac{0.3}{4} - \dfrac{(0.3)^2}{64} + \dfrac{(0.3)^2}{512}$

88. $\dfrac{1}{2} + \dfrac{\pi\sqrt{3}}{360} - \dfrac{1}{4}\left(\dfrac{\pi}{180}\right)^2$

89. $160(1 - e^{-1.50})$

90. $e^{-3.60}(\cos 1.20 + 2 \sin 1.20)$

91. $\dfrac{(10)(9)(8)(7)(6)}{5!}$

92. $\dfrac{20! - 15!}{20! + 15!}$

D

A Table of Integrals

The basic forms of Chapter 28 are not included. The constant of integration is omitted.

Forms containing $a + bu$ and $\sqrt{a + bu}$

1. $$\int \frac{u\,du}{a + bu} = \frac{1}{b^2}[(a + bu) - a \ln (a + bu)]$$

2. $$\int \frac{du}{u(a + bu)} = -\frac{1}{a} \ln \frac{a + bu}{u}$$

3. $$\int \frac{u\,du}{(a + bu)^2} = \frac{1}{b^2}\left(\frac{a}{a + bu} + \ln (a + bu)\right)$$

4. $$\int \frac{du}{u(a + bu)^2} = \frac{1}{a(a + bu)} - \frac{1}{a^2} \ln \frac{a + bu}{u}$$

5. $$\int u\sqrt{a + bu}\,du = -\frac{2(2a - 3bu)(a + bu)^{3/2}}{15b^2}$$

6. $$\int \frac{u\,du}{\sqrt{a + bu}} = -\frac{2(2a - bu)\sqrt{a + bu}}{3b^2}$$

7. $$\int \frac{du}{u\sqrt{a + bu}} = \frac{1}{\sqrt{a}} \ln \left(\frac{\sqrt{a + bu} - \sqrt{a}}{\sqrt{a + bu} + \sqrt{a}}\right), \qquad a > 0$$

8. $$\int \frac{\sqrt{a + bu}}{u}\,du = 2\sqrt{a + bu} + a \int \frac{du}{u\sqrt{a + bu}}$$

Forms containing $\sqrt{u^2 \pm a^2}$ and $\sqrt{a^2 - u^2}$

9. $$\int \frac{du}{u^2 - a^2} = \frac{1}{2a} \ln \frac{u - a}{u + a}$$

10. $$\int \frac{du}{\sqrt{u^2 \pm a^2}} = \ln (u + \sqrt{u^2 \pm a^2})$$

11. $$\int \frac{du}{u\sqrt{u^2 + a^2}} = -\frac{1}{a} \ln \left(\frac{a + \sqrt{u^2 + a^2}}{u}\right)$$

12. $$\int \frac{du}{u\sqrt{u^2 - a^2}} = \frac{1}{a} \sec^{-1} \frac{u}{a}$$

13. $$\int \frac{du}{u\sqrt{a^2 - u^2}} = -\frac{1}{a} \ln \left(\frac{a + \sqrt{a^2 - u^2}}{u}\right)$$

14. $$\int \sqrt{u^2 \pm a^2}\,du = \frac{u}{2}\sqrt{u^2 \pm a^2} \pm \frac{a^2}{2} \ln (u + \sqrt{u^2 \pm a^2})$$

15. $$\int \sqrt{a^2 - u^2}\,du = \frac{u}{2}\sqrt{a^2 - u^2} + \frac{a^2}{2} \sin^{-1} \frac{u}{a}$$

16. $\displaystyle\int \frac{\sqrt{u^2 + a^2}}{u}\,du = \sqrt{u^2 + a^2} - a \ln\left(\frac{a + \sqrt{u^2 + a^2}}{u}\right)$

17. $\displaystyle\int \frac{\sqrt{u^2 - a^2}}{u}\,du = \sqrt{u^2 - a^2} - a \sec^{-1}\frac{u}{a}$

18. $\displaystyle\int \frac{\sqrt{a^2 - u^2}}{u}\,du = \sqrt{a^2 - u^2} - a \ln\left(\frac{a + \sqrt{a^2 - u^2}}{u}\right)$

19. $\displaystyle\int (u^2 \pm a^2)^{3/2}\,du = \frac{u}{4}(u^2 \pm a^2)^{3/2} \pm \frac{3a^2u}{8}\sqrt{u^2 \pm a^2} + \frac{3a^4}{8}\ln\left(u + \sqrt{u^2 \pm a^2}\right)$

20. $\displaystyle\int (a^2 - u^2)^{3/2}\,du = \frac{u}{4}(a^2 - u^2)^{3/2} + \frac{3a^2u}{8}\sqrt{a^2 - u^2} + \frac{3a^4}{8}\sin^{-1}\frac{u}{a}$

21. $\displaystyle\int \frac{(u^2 + a^2)^{3/2}}{u}\,du = \frac{1}{3}(u^2 + a^2)^{3/2} + a^2\sqrt{u^2 + a^2} - a^3 \ln\left(\frac{a + \sqrt{u^2 + a^2}}{u}\right)$

22. $\displaystyle\int \frac{(u^2 - a^2)^{3/2}}{u}\,du = \frac{1}{3}(u^2 - a^2)^{3/2} - a^2\sqrt{u^2 - a^2} + a^3 \sec^{-1}\frac{u}{a}$

23. $\displaystyle\int \frac{(a^2 - u^2)^{3/2}}{u}\,du = \frac{1}{3}(a^2 - u^2)^{3/2} - a^2\sqrt{a^2 - u^2} + a^3 \ln\left(\frac{a + \sqrt{a^2 - u^2}}{u}\right)$

24. $\displaystyle\int \frac{du}{(u^2 \pm a^2)^{3/2}} = \pm\frac{u}{a^2\sqrt{u^2 \pm a^2}}$

25. $\displaystyle\int \frac{du}{(a^2 - u^2)^{3/2}} = \frac{u}{a^2\sqrt{a^2 - u^2}}$

26. $\displaystyle\int \frac{du}{u(u^2 + a^2)^{3/2}} = \frac{1}{a^2\sqrt{u^2 + a^2}} - \frac{1}{a^3}\ln\left(\frac{a + \sqrt{u^2 + a^2}}{u}\right)$

27. $\displaystyle\int \frac{du}{u(u^2 - a^2)^{3/2}} = -\frac{1}{a^2\sqrt{u^2 - a^2}} - \frac{1}{a^3}\sec^{-1}\frac{u}{a}$

28. $\displaystyle\int \frac{du}{u(a^2 - u^2)^{3/2}} = \frac{1}{a^2\sqrt{a^2 - u^2}} - \frac{1}{a^3}\ln\left(\frac{a + \sqrt{a^2 - u^2}}{u}\right)$

Trigonometric forms

29. $\displaystyle\int \sin^2 u\,du = \frac{u}{2} - \frac{1}{2}\sin u \cos u$

30. $\displaystyle\int \sin^3 u\,du = -\cos u + \frac{1}{3}\cos^3 u$

31. $\displaystyle\int \sin^n u\,du = -\frac{1}{n}\sin^{n-1}u \cos u + \frac{n-1}{n}\int \sin^{n-2}u\,du$

32. $\displaystyle\int \cos^2 u\,du = \frac{u}{2} + \frac{1}{2}\sin u \cos u$

33. $\displaystyle\int \cos^3 u\,du = \sin u - \frac{1}{3}\sin^3 u$

34. $\displaystyle\int \cos^n u\,du = \frac{1}{n}\cos^{n-1}u \sin u + \frac{n-1}{n}\int \cos^{n-2}u\,du$

35. $\displaystyle\int \tan^n u\,du = \frac{\tan^{n-1}u}{n-1} - \int \tan^{n-2}u\,du$

36. $\int \cot^n u\, du = -\dfrac{\cot^{n-1} u}{n-1} - \int \cot^{n-2} u\, du$

37. $\int \sec^n u\, du = \dfrac{\sec^{n-2} u \tan u}{n-1} + \dfrac{n-2}{n-1}\int \sec^{n-2} u\, du$

38. $\int \csc^n u\, du = -\dfrac{\csc^{n-2} u \cot u}{n-1} + \dfrac{n-2}{n-1}\int \csc^{n-2} u\, du$

39. $\int \sin au \sin bu\, du = \dfrac{\sin(a-b)u}{2(a-b)} - \dfrac{\sin(a+b)u}{2(a+b)}$

40. $\int \sin au \cos bu\, du = -\dfrac{\cos(a-b)u}{2(a-b)} - \dfrac{\cos(a+b)u}{2(a+b)}$

41. $\int \cos au \cos bu\, du = \dfrac{\sin(a-b)u}{2(a-b)} + \dfrac{\sin(a+b)u}{2(a+b)}$

42. $\int \sin^m u \cos^n u\, du = \dfrac{\sin^{m+1} u \cos^{n-1} u}{m+n} + \dfrac{n-1}{m+n}\int \sin^m u \cos^{n-2} u\, du$

43. $\int \sin^m u \cos^n u\, du = -\dfrac{\sin^{m-1} u \cos^{n+1} u}{m+n} + \dfrac{m-1}{m+n}\int \sin^{m-2} u \cos^n u\, du$

Other forms

44. $\int ue^{au} du = \dfrac{e^{au}(au-1)}{a^2}$

45. $\int u^2 e^{au} du = \dfrac{e^{au}}{a^3}(a^2u^2 - 2au + 2)$

46. $\int u^n \ln u\, du = u^{n+1}\left(\dfrac{\ln u}{n+1} - \dfrac{1}{(n+1)^2}\right)$

47. $\int u \sin u\, du = \sin u - u \cos u$

48. $\int u \cos u\, du = \cos u + u \sin u$

49. $\int e^{au} \sin bu\, du = \dfrac{e^{au}(a \sin bu - b \cos bu)}{a^2 + b^2}$

50. $\int e^{au} \cos bu\, du = \dfrac{e^{au}(a \cos bu + b \sin bu)}{a^2 + b^2}$

51. $\int \sin^{-1} u\, du = u \sin^{-1} u + \sqrt{1-u^2}$

52. $\int \tan^{-1} u\, du = u \tan^{-1} u - \dfrac{1}{2}\ln(1+u^2)$

Answers to Odd-Numbered Exercises

Although a graphing calculator does not show labels on axes, labels will be shown in many answers showing graphing calculator displays in order to indicate appropriate values to be used for the RANGE feature.

Exercises 1-1, page 5

1. integer, rational, real; irrational, real

3. imaginary; irrational, real

5. $3, \frac{7}{2}$ **7.** $\frac{6}{7}, \sqrt{3}$ **9.** $6 < 8$ **11.** $\pi > -1$

13. $-4 < -3$ **15.** $-\frac{1}{3} > -\frac{1}{2}$ **17.** $\frac{1}{3}, -\frac{1}{2}$

19. $-\frac{\pi}{5}, \frac{1}{x}$ **21.**

$-\frac{1}{2}$ 2.5
−3 −2 −1 0 1 2 3

23.

$-\frac{\sqrt{2}}{2}$ 2π
−3 −2 −1 0 1 2 3 4 5 6 7

25. $-18, -|-3|, -1, \sqrt{5}, \pi, |-8|, 9$

27. (a) positive integer (b) negative integer (c) positive rational number less than 1

29. (a) yes (b) no **31.** (a) to right of origin (b) to left of -4

33. between 0 and 1

35. L, t are variables; a is constant

37. $N = 1000an$ **39.** yes

Exercises 1-2, page 10

1. 4 **3.** 6 **5.** −3 **7.** −24

9. 35 **11.** 20 **13.** 40 **15.** −1

17. 9 **19.** undefined **21.** 20 **23.** −5

25. −9 **27.** 24 **29.** −6 **31.** 3

33. commutative law of multiplication

35. distributive law

37. associative law of addition

39. associative law of multiplication

41. d **43.** b **45.** positive

47. $\frac{4}{2} = 2; \frac{2}{4} = \frac{1}{2}; 2 \neq \frac{1}{2}$

49. 100 m + 200 m = 200 m + 100 m, commutative law of addition

51. 8($2000 + $1000), distributive law

Exercises 1-3, page 15

1. 8 is exact; 55 is approx. **3.** 1 and 19.3 are approx.

5. 3, 4 **7.** 3, 4 **9.** 3, 3 **11.** 1, 6

13. (a) 3.764 (b) 3.764 **15.** (a) 0.01 (b) 30.8

17. (a) same (b) 78.0 **19.** (a) 0.004 (b) same

21. (a) 4.94 (b) 4.9 **23.** (a) 50 900 (b) 51 000

25. (a) 9550 (b) 9500 **27.** (a) 0.950 (b) 0.95

29. 51.2 **31.** 62.1 **33.** 0.148 **35.** 0.0114

37. 28.7 **39.** −0.0022 **41.** 0.1356 **43.** 6.086

45. 15.8788 **47.** 204.2 **49.** 128.25 m, 128.35 m

51. Too many sig. digits; time has only 2 sig. digits

53. $\pi = 3.141\,592\,65\ldots$ **55.** error

57. (a) 0.242 424 242 4 (b) 3.141 592 654 **59.** 196 m

61. 262 144 bytes **63.** 59.14

Exercises 1-4, page 21

1. x^7 **3.** $2b^6$ **5.** m^2 **7.** $\frac{1}{n^4}$ **9.** a^8

11. t^{20} **13.** $8n^3$ **15.** a^2x^8 **17.** $\frac{8}{b^3}$

19. $\frac{x^8}{16}$ **21.** 1 **23.** −3 **25.** $\frac{1}{6}$ **27.** s^2

29. $-t^{14}$ **31.** $64x^{12}$ **33.** 1 **35.** $-b^2$

37. $\frac{1}{8}$ **39.** 1 **41.** $\frac{a}{x^2}$ **43.** $\frac{x^3}{64a^3}$ **45.** $64g^2s^6$

47. $\frac{5a}{n}$ **49.** −53 **51.** 253 **53.** −0.421

55. 9990 **57.** $\frac{r}{6}$ **59.** 69 W

Exercises 1-5, page 23

1. 45 000 **3.** 0.002 01 **5.** 3.23 **7.** 18.6

9. 4×10^4 **11.** 8.7×10^{-3} **13.** 6×10^0

15. 6.3×10^{-2} **17.** 5.6×10^{13} **19.** 2.2×10^8

21. 4.85×10^{10} **23.** 1.59×10^7 **25.** 9.965×10^{-3}

27. 2.9874×10^{-3} **29.** 6.5×10^6 kW

31. 0.000 000 000 001 6 W **33.** 192 000 000 Hz

35. 3×10^{-6} W

37. 0.000 000 000 000 000 000 000 000 000 000 000 000 000 000 24

39. 3.6×10^4 km **41.** 2.57×10^{14} cm^2

43. 3.433 Ω

Exercises 1-6, page 26

1. 5 **3.** −11 **5.** −7 **7.** 20 **9.** 5
11. −6 **13.** 5 **15.** 31 **17.** $3\sqrt{2}$ **19.** $2\sqrt{3}$
21. $4\sqrt{21}$ **23.** 7 **25.** 10 **27.** $3\sqrt{10}$
29. 9.24 **31.** 0.6877 **33.** (a) 60 (b) 84
35. (a) 0.0388 (b) 0.0246 **37.** 2.66 s
39. 1450 m/s **41.** 48.3 cm
43. no, not true if $a < 0$

Exercises 1-7, page 30

1. $8x$ **3.** $y + 4x$ **5.** $a + c - 2$
7. $-a^2b - a^2b^2$ **9.** $4s + 4$ **11.** $5x - y - 4$
13. $5a - 5$ **15.** $-5a + 2$ **17.** $-2t + 5u$
19. $7r + 8s$ **21.** $-50 + 19c$ **23.** $-9 + 3n$
25. $18 - 2t^2$ **27.** $6a$ **29.** $2a\sqrt{xy} + 1$
31. $4c - 6$ **33.** $8p - 5q$ **35.** $-4x^2 + 22$
37. $4a - 3$ **39.** $-6t + 13$ **41.** $2D + d$
43. $-b + 4c - 3a$

Exercises 1-8, page 32

1. a^3x **3.** $-a^2c^3x^3$ **5.** $-8a^3x^5$ **7.** $2a^8x^3$
9. $a^2x + a^2y$ **11.** $-3s^3 + 15st$ **13.** $5m^3n + 15m^2n$
15. $-3x^2 - 3xy + 6x$ **17.** $a^2b^2c^5 - ab^3c^5 - a^2b^3c^4$
19. $acx^4 + acx^3y^3$ **21.** $x^2 + 2x - 15$
23. $2x^2 + 9x - 5$ **25.** $6a^2 - 7ab + 2b^2$
27. $6s^2 + 11st - 35t^2$ **29.** $2x^3 + 5x^2 - 2x - 5$
31. $x^3 + 2x^2 - 8x$ **33.** $x^3 - 2x^2 - x + 2$
35. $x^5 - x^4 - 6x^3 + 4x^2 + 8x$ **37.** $2a^2 - 16a - 18$
39. $2x^3 + 6x^2 - 8x$ **41.** $4x^2 - 20x + 25$
43. $x^2 + 6ax + 9a^2$ **45.** $x^2y^2z^2 - 4xyz + 4$
47. $2x^2 + 32x + 128$ **49.** $-x^3 + 2x^2 + 5x - 6$
51. $6x^4 + 21x^3 + 12x^2 - 12x$
53. $n^2 + 200n + 10\,000$ **55.** $R^2 - r^2$

Exercises 1-9, page 34

1. $-4x^2y$ **3.** $\frac{4t^4}{r^2}$ **5.** $4x^2$ **7.** $-6a$
9. $a^2 + 4y$ **11.** $t - 2rt^2$ **13.** $q + 2p - 4q^3$
15. $\frac{2L}{R} - R$ **17.** $\frac{1}{3a} - \frac{2b}{3a} + 1$ **19.** $x^2 + a$
21. $2x + 1$ **23.** $x - 1$ **25.** $4x^2 - x - 1, R = -3$
27. $x + 5$ **29.** $x^2 + x - 6$
31. $2x^2 + 4x + 2, R = 4x + 4$ **33.** $x^2 - 2x + 4$
35. $x - y$ **37.** $A + \frac{\mu^2E^2}{2A} - \frac{\mu^4E^4}{8A^3}$
39. $3T^2 - 2T - 4$

Exercises 1-10, page 38

1. 9 **3.** −1 **5.** 10 **7.** −5 **9.** −3
11. 1 **13.** $-\frac{7}{2}$ **15.** 8 **17.** $\frac{10}{3}$
19. $-\frac{13}{3}$ **21.** 2 **23.** 0 **25.** 9.5 **27.** −1.5
29. 5.7 **31.** 0.85 **33.** 1.3 km/h **35.** 750 L
37. 60 mg **39.** true for all x

Exercises 1-11, page 41

1. $\frac{b}{a}$ **3.** $\frac{4m - 1}{4}$ **5.** $\frac{c + 6}{a}$ **7.** $2a + 8$
9. $\theta - kA$ **11.** $\frac{E}{I}$ **13.** $\frac{P}{2\pi f}$ **15.** $\frac{p - p_a}{dg}$
17. $\frac{APV}{R}$ **19.** $\frac{s + 16t^2}{t}$ **21.** $\frac{C_0^2 - C_1^2}{2C_1^2}$
23. $\frac{a + PV^2}{V}$ **25.** $\frac{Q_1 + PQ_1}{P}$ **27.** $\frac{N + N_2 - N_2T}{T}$
29. $\frac{L - \pi r_2 - 2x_1 - x_2}{\pi}$ **31.** $\frac{gJP + V_1^2}{V_1}$
33. 10.6 m **35.** 204 K

Exercises 1-12, page 44

1. \$138, \$252
3. 1.9 million the first year, 2.6 million the second year
5. 20 hectares at \$20 000/hectare, 50 hectares at \$10 000/hectare
7. 20 girders **9.** −2.3 μA, −4.6 μA, 6.9 μA
11. 6.9 km, 9.5 km **13.** \$11 000
15. 387 s, first car **17.** 84.2 km/h, 92.2 km/h
19. 900 m **21.** 146 km from A **23.** 180 g

Review Exercises for Chapter 1, page 46

1. −10 **3.** −20 **5.** 2 **7.** −25
9. −4 **11.** 5 **13.** $4r^2t^4$ **15.** $-\frac{6m^2}{nt^2}$
17. $\frac{8t^3}{s^2}$ **19.** $3\sqrt{5}$ **21.** (a) 3 (b) 8800
23. (a) 4 (b) 9.0 **25.** 18.0 **27.** 1.3×10^{-4}
29. $-a - 2ab$ **31.** $5xy + 3$ **33.** $2x^2 + 9x - 5$
35. $x^2 + 16x + 64$ **37.** $hk - 3h^2k^4$ **39.** $7a - 6b$
41. $13xy - 10z$ **43.** $2x^3 - x^2 - 7x - 3$
45. $-3x^2y + 24xy^2 - 48y^3$ **47.** $-9p^2 + 3pq + 18p^2q$
49. $\frac{6q}{p} - 2 + \frac{3q^4}{p^3}$ **51.** $2x - 5$ **53.** $x^2 - 2x + 3$
55. $4x^3 - 2x^2 + 6x$, R = −1 **57.** $15r - 3s - 3t$

59. $y^2 + 5y - 1, R = 4$ **61.** $-\frac{9}{2}$ **63.** $\frac{21}{10}$

65. $-\frac{7}{3}$ **67.** 3 **69.** $-\frac{19}{5}$ **71.** 1

73. 4×10^4 km/h **75.** 1 020 000 000 Hz

77. 1.2×10^{-6} cm^2 **79.** 0.18 kg/m^3

81. $\frac{5a - 2}{3}$ **83.** $\frac{4 - 2n}{3}$ **85.** $\frac{R}{n^2}$ **87.** $\frac{I - P}{Pr}$

89. $\frac{dV - m}{dV}$ **91.** $\frac{2C - mN_2}{m}$ **93.** $J - E + ES$

95. $\frac{d - 3kbx^2 + kx^3}{3kx^2}$ **97.** 110 m **99.** 0.0188 Ω

101. $2rV - aV - bV$

103. $4t + 4h - 2t^2 - 4th - 2h^2$

105. 2.1 megabytes, 8.4 megabytes

107. 1900 Ω, 3100 Ω **109.** 37 N **111.** 1.4 h

113. 400 L, 600 L **115.** 27 m^2

Exercises 2-1, page 51

1. $\angle EBD, \angle DBC$ **3.** 25° **5.** 140° **7.** 145°

9. 62° **11.** 28° **13.** 46° **15.** 136°

17. 4.53 cm **19.** 3.40 cm **21.** 133° **23.** 233 m

Exercises 2-2, page 57

1. 56° **3.** 48° **5.** 9.9 mm **7.** 64.5 cm

9. 8.4 m^2 **11.** 32 300 cm^2 **13.** 4.41 cm^2

15. 0.390 m^2 **17.** 26.6 mm **19.** 52.2 cm

21. 67° **23.** 227.2 cm

25. $\angle K = \angle N = 90°$; $\angle LMK = \angle OMN$; $\angle KLM = \angle NOM$; $\Delta MKL \sim \Delta MNO$

27. 8 **29.** 71° **31.** 148 m^2 **33.** 930 m

35. 38 m

Exercises 2-3, page 61

1. 2.6 m **3.** 167.8 mm **5.** 12.8 m **7.** 214.4 dm

9. 7.3 mm^2 **11.** 0.174 km^2 **13.** 9.3 m^2

15. 2000 dm^2 **17.** $p = 4a + 2b$ **19.** $A = bh + a^2$

21. 12 cm **23.** 6.30 m^2 **25.** 344 m

27. 75 m by 95 m

Exercises 2-4, page 64

1. 17.3 cm **3.** 72.6 mm **5.** 0.0285 km^2

7. 4.26 m^2 **9.** 25° **11.** 25° **13.** 120°

15. 40° **17.** 0.393 rad **19.** 2.185 rad

21. $p = \frac{1}{2}\pi r + 2r$ **23.** $A = \frac{1}{4}\pi r^2$ **25.** 40 000 km

27. 35.7 cm **29.** 52.1 m^2

31. horizontally and opposite to original direction

Exercises 2-5, page 67

1. 84 m^2 **3.** 0.45 m^2 **5.** 9.8 km^2 **7.** 550 cm^2

9. 11 800 m^2 **11.** 8100 m^2

13. 2.73 cm^2 The trapezoids are inside the boundary and do not include some of the area.

15. 2.98 cm^2 The ends of the areas are curved so that they can get closer to the boundary.

Exercises 2-6, page 70

1. 366 dm^3 **3.** 399 m^2 **5.** 2.83 m^3

7. 20 500 cm^2 **9.** 1100 dm^3 **11.** 3.358 m^2

13. 0.15 cm^3 **15.** 0.825 cm^2 **17.** 604 cm^2

19. 1.4×10^{-3} km^3 $= 1.4 \times 10^6$ m^3 **21.** 2.6×10^6 m^3

23. 66 600 m^3

Review Exercises for Chapter 2, page 73

1. 32° **3.** 32° **5.** 41 **7.** 42 **9.** 7.36

11. 21.1 **13.** 25.5 mm **15.** 3.06 m^2

17. 309 mm **19.** 2580 cm^2 **21.** 6190 cm^3

23. 160 000 m^3 **25.** 162 m^2 **27.** 1230 mm^2

29. 25° **31.** 65° **33.** 53° **35.** 2.4

37. $p = \pi a + b + \sqrt{4a^2 + b^2}$

39. $A = ab + \frac{1}{2}\pi a^2$ **41.** 7.9 m **43.** 30 m

45. 215 m^2 **47.** 5.91 km **49.** 42 000 km

51. 2.7×10^6 mm^2 **53.** 1.0×10^6 m^2 **55.** 873 L

Exercises 3-1, page 78

1. (a) $A = \pi r^2$ (b) $A = \frac{1}{4}\pi d^2$ **3.** $V = \frac{1}{6}\pi d^3$

5. $A = 5l$ **7.** $A = s^2, s = \sqrt{A}$ **9.** 3, −1

11. 11, 3.8 **13.** −18, 70 **15.** $\frac{5}{2}, -\frac{1}{2}$

17. $\frac{1}{4}a + \frac{1}{2}a^2, 0$ **19.** $3s^2 + s + 6, 12s^2 - 2s + 6$

21. 62.9, 260 **23.** −0.2998

25. Square the value of the independent variable and add 2.

27. Cube the value of the independent variable and subtract this from 6 times the value of the independent variable.

29. $y = x^2, f(x) = x^2$

31. $P = 40(p - 24), f(p) = 40(p - 24)$

33. 10.4 m **35.** 13.2 m, $0.4v + 0.032v^2$, 40.8 m, 40.8 m

Exercises 3-2, page 82

1. domain: all real numbers; range: all real numbers

3. domain: all real numbers except 0; range: all real numbers except 0

5. domain: all real numbers except 0; range: all real numbers $f(s) > 0$

7. domain: all real numbers $h \geq 0$; range: all real numbers $H(h) \geq 1$

9. all real numbers $y > 2$

11. all real numbers except 2 and -4

13. 2, not defined

15. $2, \frac{3}{4}$ **17.** $d = 80 + 55t$ **19.** $w = 5500 - 2t$

21. $m = 0.5h - 390$ **23.** $C = 5l + 250$

25. $y = 3000 - 0.25x$ **27.** $A = \frac{1}{16}p^2 + \frac{(60 - p)^2}{4\pi}$

29. $A = \pi(6 - x)^2$, domain $0 \leq x \leq 6$, range: $0 \leq A \leq 36\pi$

31. $s = \frac{300}{t}$, domain: $t > 0$, range: $s > 0$ (upper limits depend on truck)

33. Domain is all values of $C > 0$, with some upper limit depending on the circuit.

35. $m = \begin{cases} 0.5h - 390 \text{ for } h > 1000 \\ 110 \text{ for } 0 \leq h \leq 1000 \end{cases}$

Exercises 3-3, page 86

1. (2, 1), (−1, 2), (−2, −3) **3.**

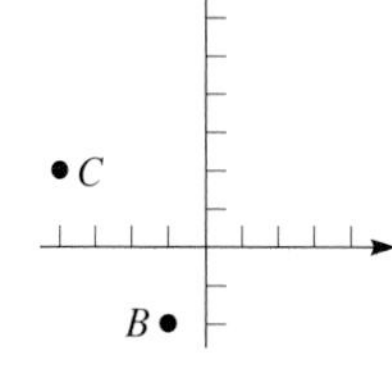

5. isosceles triangle **7.** rectangle **9.** (5, 4)

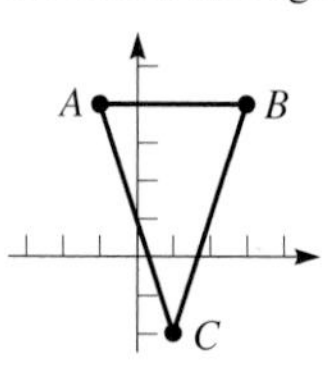

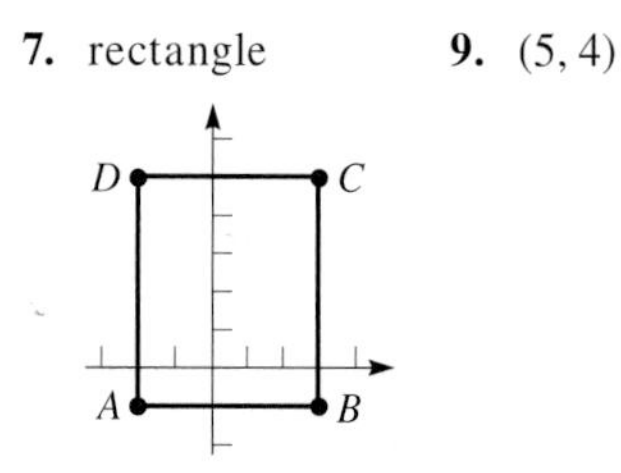

11. (3, −2)

13. on a line parallel to the y-axis, one unit to the right

15. on a line parallel to the x-axis, 3 units above

17. on a line bisecting the first and third quadrants

19. 0 **21.** to the right of the y-axis **23.** to the left of a line that is parallel to the y-axis, 1 unit to its left

25. on the negative y-axis **27.** first, third

Exercises 3-4, page 91

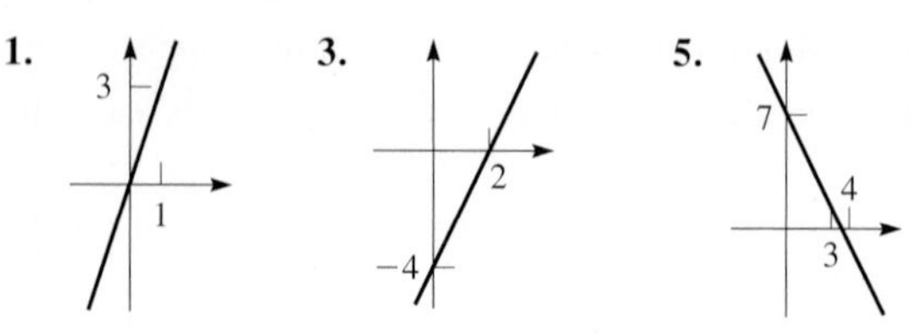

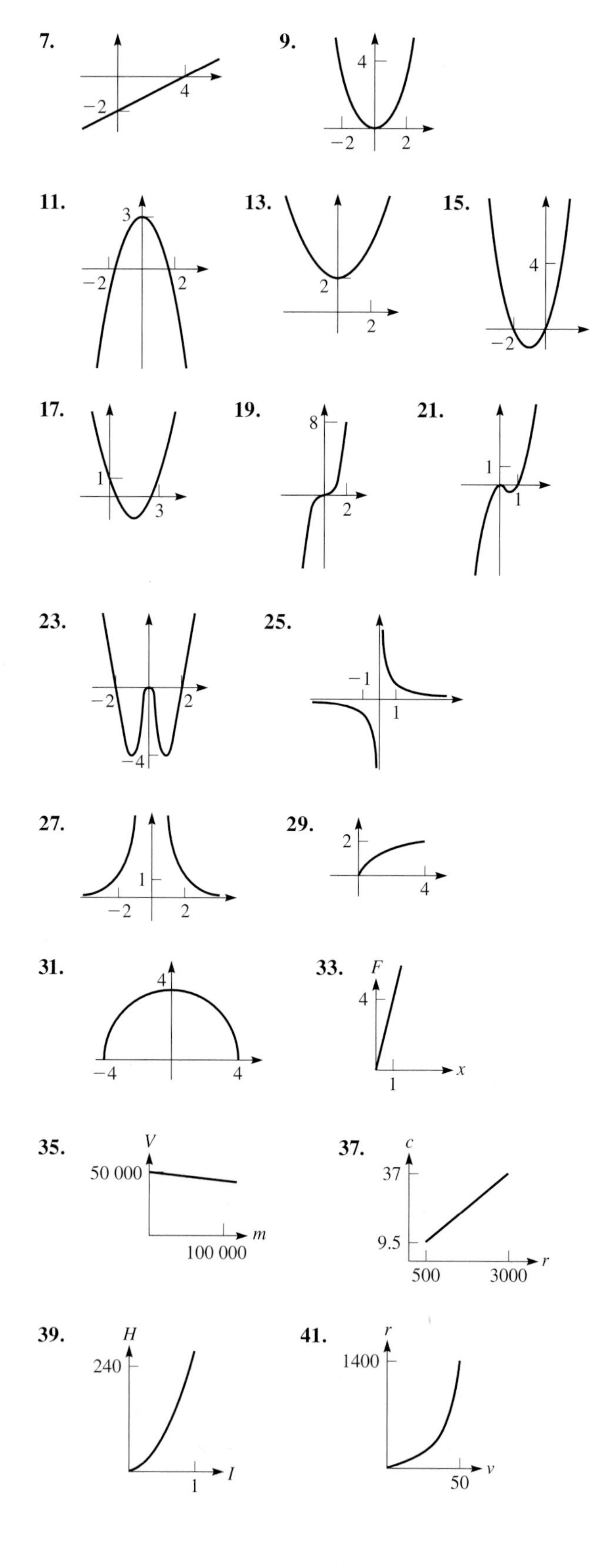

43. **45.** $A = 30w - w^2$

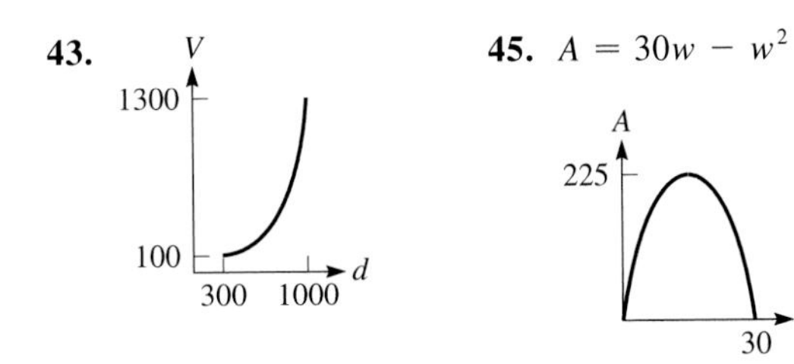

47.

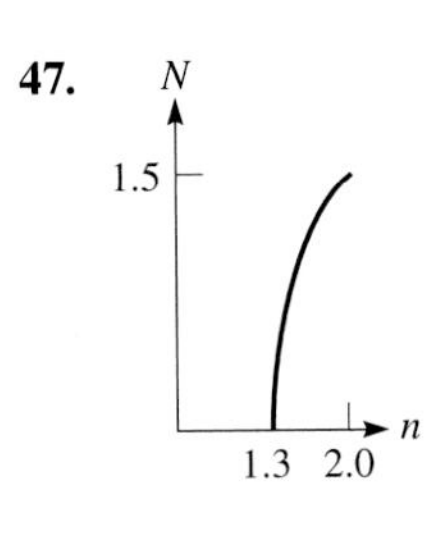

49.

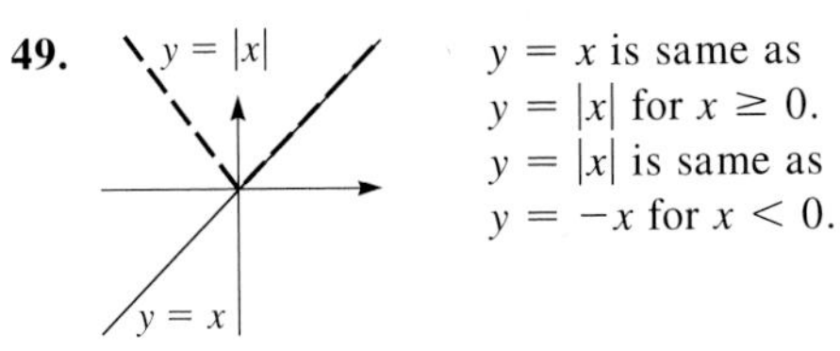

$y = x$ is same as $y = |x|$ for $x \geq 0$. $y = |x|$ is same as $y = -x$ for $x < 0$.

51.

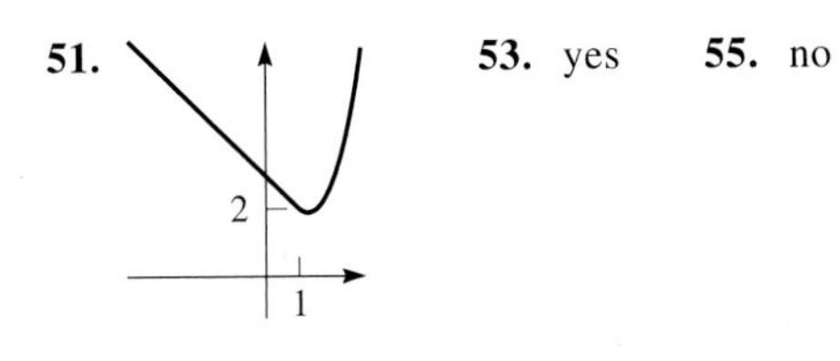

53. yes **55.** no

Exercises 3-5, page 96

1. 0.7

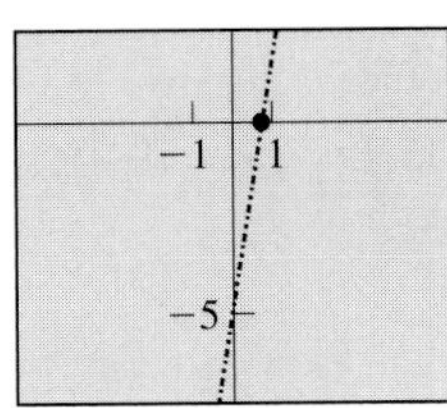

3. −6.4, 6.4

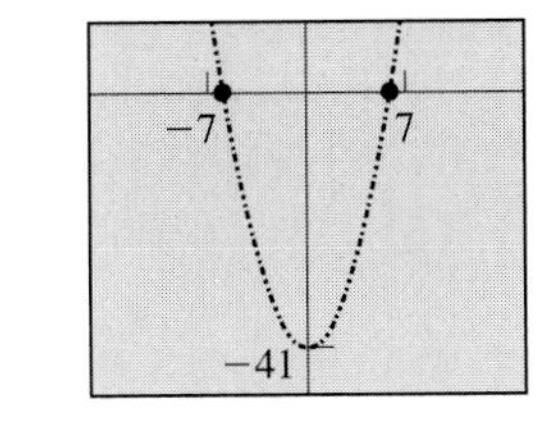

5. −2.8, 1.8

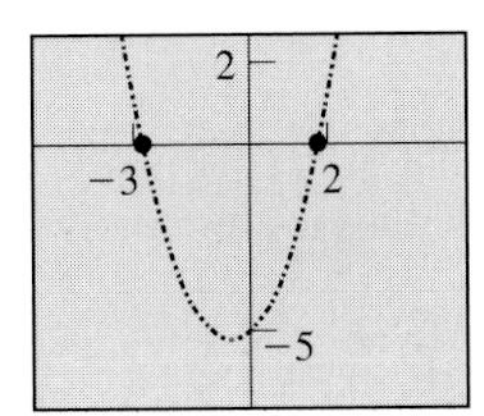

7. −1.6, 2.1

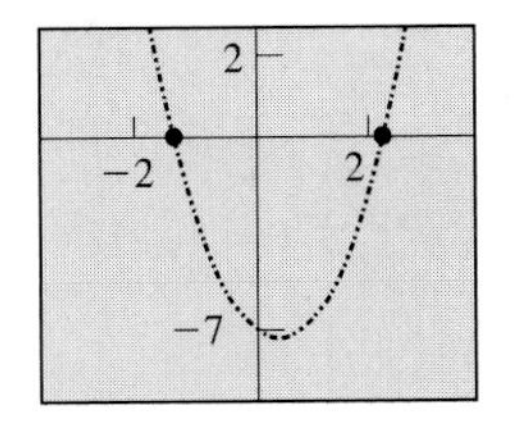

9. −2.0, 0.0, 2.0

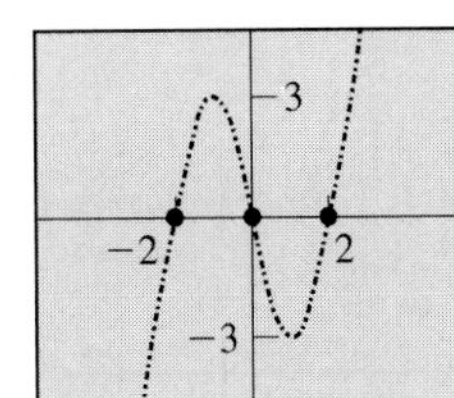

11. 0.0, 1.3

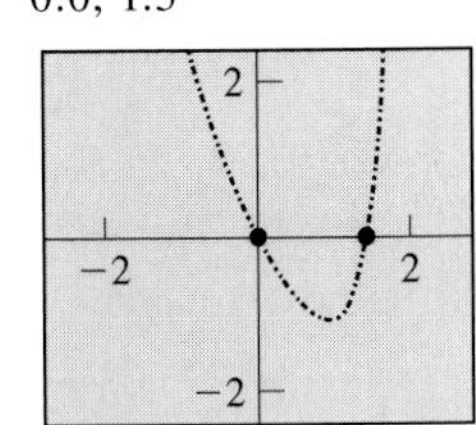

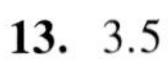

13. 3.5

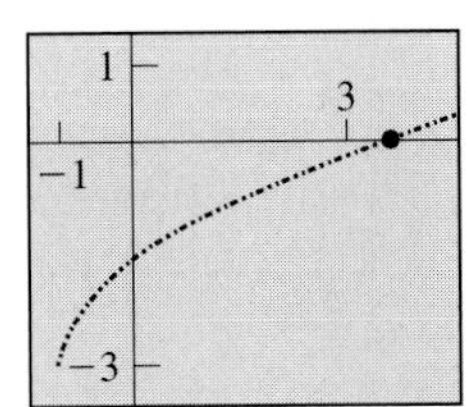

15. no values

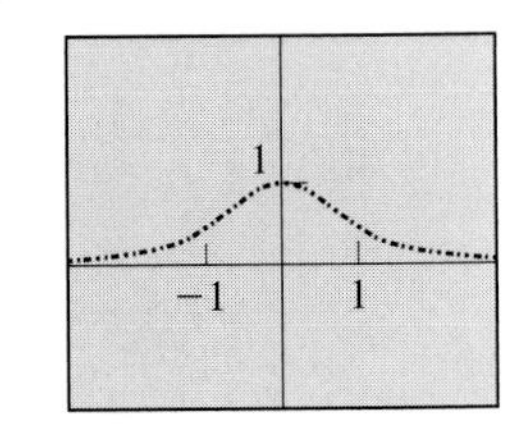

17. all real numbers $y > 0$ or $y \leq -1$

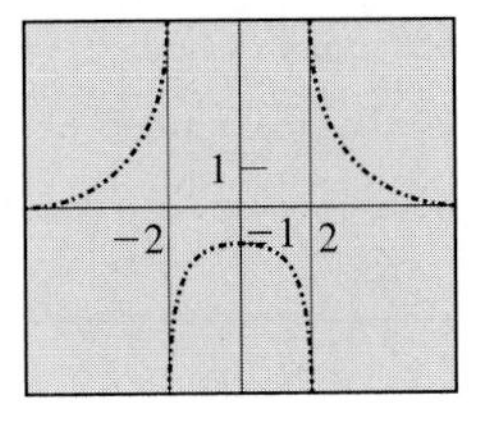

19. all real numbers $y \geq 0$ or $y \leq -4$

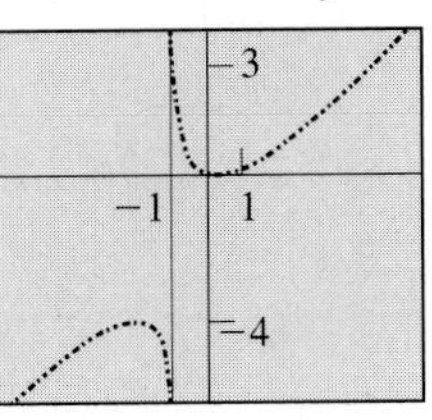

21. all real numbers $Y(y) > 3.46$ (approx.)

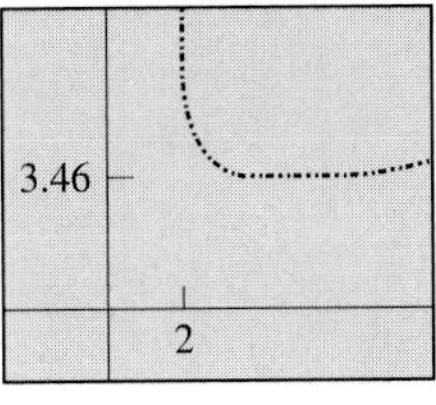

23. all real numbers **25.** 6.0 V **27.** 0.30 m

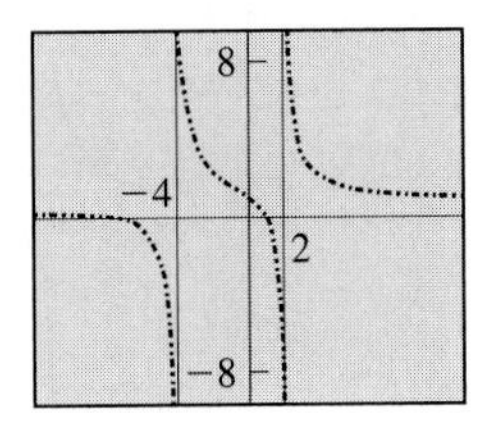

29. 18 cm, 30 cm **31.** 67 m **33.** −0.414, 2.414

35. 0.25 cm/s

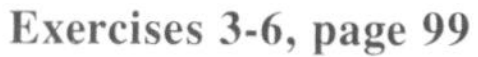

Exercises 3-6, page 99

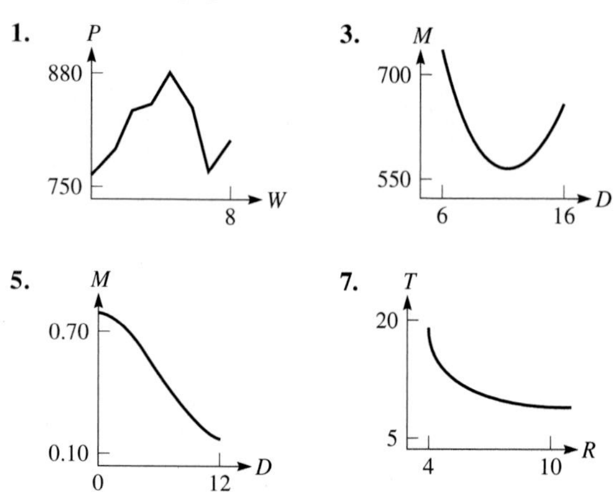

9. (a) 132.1°C (b) 0.7 min **11.** (a) 7.8 V (b) 43°C

13. 0.30 H **15.** 7.2% **17.** (a) 170 cm (b) 2.4 m^3/s

19. 1.3 m^3/s **21.** 0.34

23. 76 m^2 **25.** 130.3°C

27. 3.7 m^3/s

33. 0.4

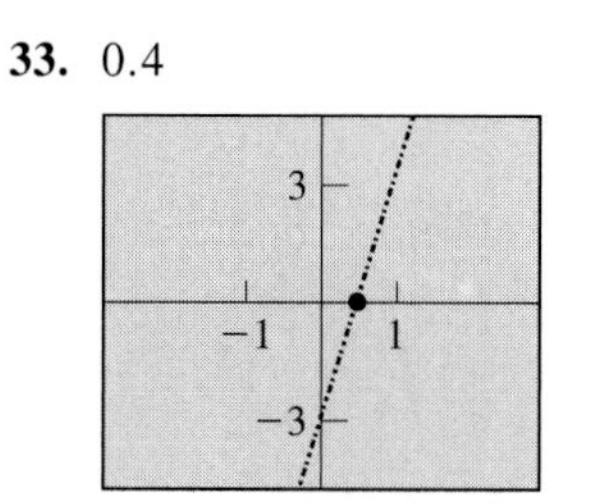

35. 0.2, 5.8

37. 1.4

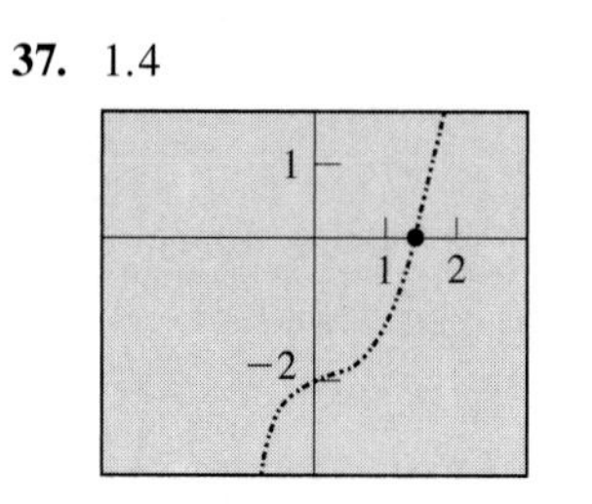

39. −0.7, 0.7

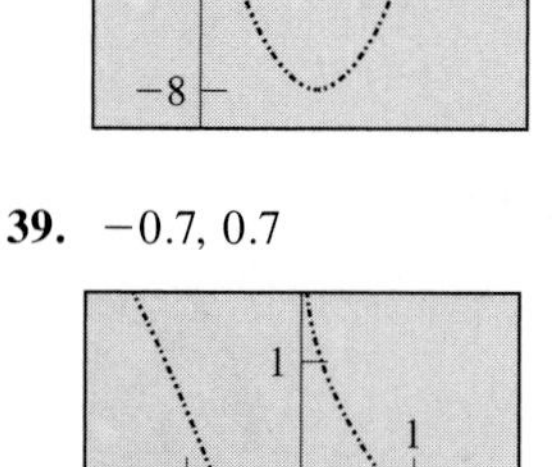

41. all real numbers $y \geq -6.25$

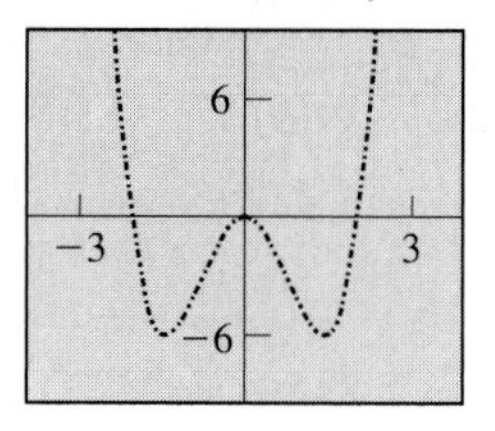

43. all real numbers $y \leq -2.83$ or $y \geq 2.83$

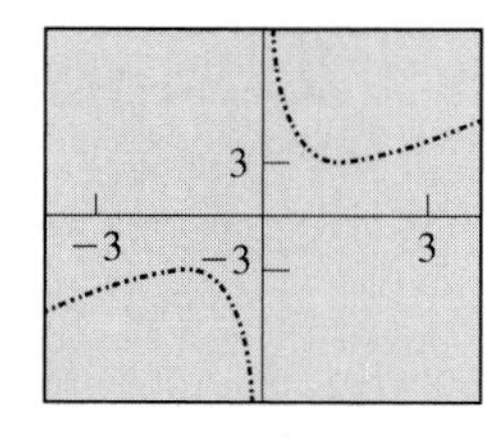

Review Exercises for Chapter 3, page 100

1. $A = 4\pi t^2$ **3.** $y = -\dfrac{10}{9}x + \dfrac{250}{9}$

5. 16, −47 **7.** −5, −1.08 **9.** 3, $\sqrt{1-4h}$

11. $3h^2 + 6hx - 2h$ **13.** −3.67, 16.7

15. 0.165 72, −0.215 66

17. domain: all real numbers; range: all real numbers $f(x) \geq 1$

19. domain: all real numbers $t > -4$; range: all real numbers $g(t) > 0$

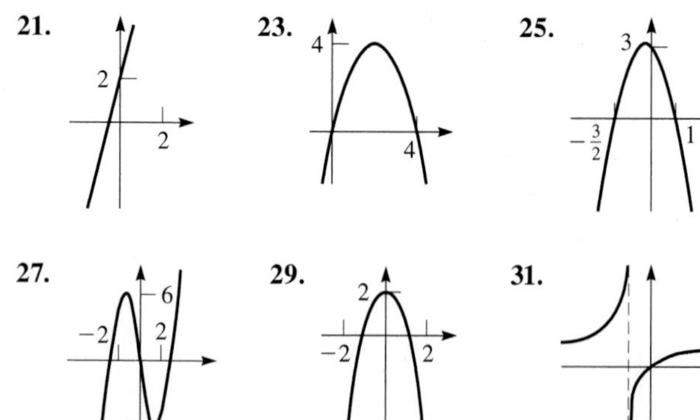

45. either a or b is positive, the other is negative

47. $(1, \sqrt{3})$ or $(1, -\sqrt{3})$ **49.** 13.4 **51.** 72.0°

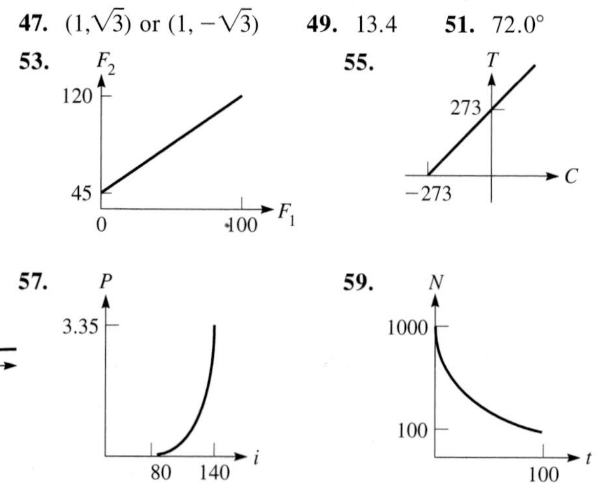

61. **63.** $f(10) = 204$

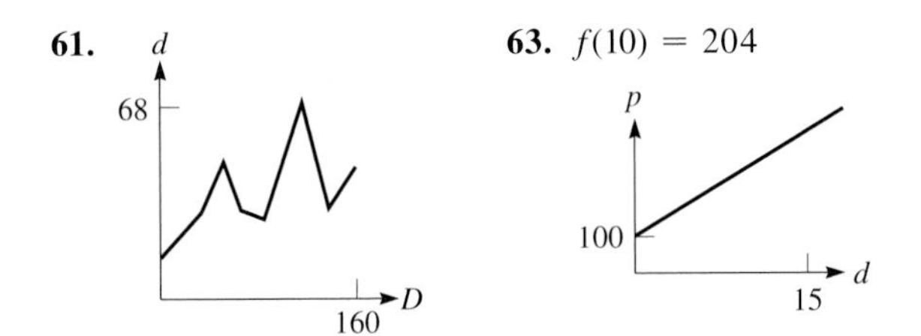

65. 5.1 N **67.** 33°C **69.** 0.39 m **71.** 6.5 h

Exercises 4-1, page 107

1. **3.**

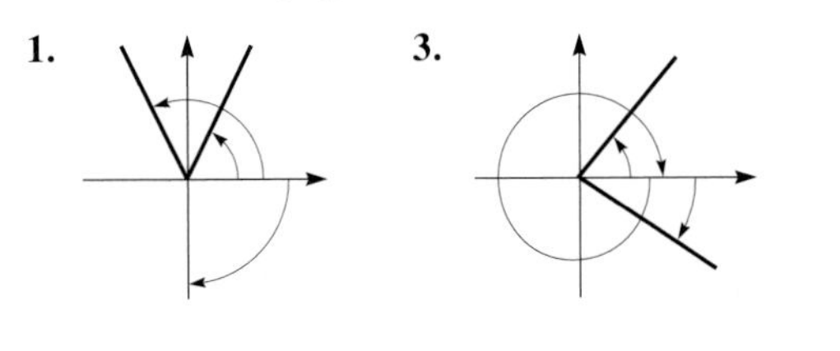

5. 405°, −315° **7.** 210°, −510°

9. 430° 30′, −289° 30′ **11.** 638.1°, −81.9°

13. 15.18° **15.** 82.91° **17.** 15.20° **19.** 86.05°

21. 301.27° **23.** −96.13° **25.** 47° 30′

27. 19° 45′ **29.** −5° 37′

31. 24° 55′ **33.** (4, 2) θ **35.** θ (−3, −5)

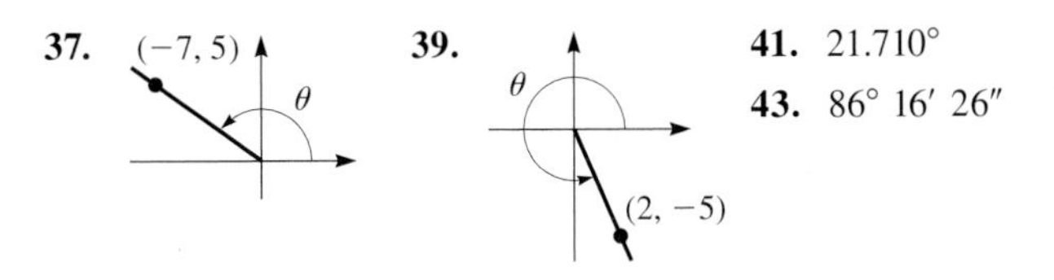

41. 21.710°

43. 86° 16′ 26″

Exercises 4-2, page 111

1. $\sin\theta = \frac{4}{5}$, $\cos\theta = \frac{3}{5}$, $\tan\theta = \frac{4}{3}$, $\cot\theta = \frac{3}{4}$, $\sec\theta = \frac{5}{3}$, $\csc\theta = \frac{5}{4}$

3. $\sin\theta = \frac{8}{17}$, $\cos\theta = \frac{15}{17}$, $\tan\theta = \frac{8}{15}$, $\cot\theta = \frac{15}{8}$, $\sec\theta = \frac{17}{15}$, $\csc\theta = \frac{17}{8}$

5. $\sin\theta = \frac{40}{41}$, $\cos\theta = \frac{9}{41}$, $\tan\theta = \frac{40}{9}$, $\cot\theta = \frac{9}{40}$, $\sec\theta = \frac{41}{9}$, $\csc\theta = \frac{41}{40}$

7. $\sin\theta = \frac{\sqrt{15}}{4}$, $\cos\theta = \frac{1}{4}$, $\tan\theta = \sqrt{15}$, $\cot\theta = \frac{1}{\sqrt{15}}$, $\sec\theta = 4$, $\csc\theta = \frac{4}{\sqrt{15}}$

9. $\sin\theta = \frac{1}{\sqrt{2}}$, $\cos\theta = \frac{1}{\sqrt{2}}$, $\tan\theta = 1$, $\cot\theta = 1$, $\sec\theta = \sqrt{2}$, $\csc\theta = \sqrt{2}$

11. $\sin\theta = \frac{2}{\sqrt{29}}$, $\cos\theta = \frac{5}{\sqrt{29}}$, $\tan\theta = \frac{2}{5}$, $\cot\theta = \frac{5}{2}$, $\sec\theta = \frac{\sqrt{29}}{5}$, $\csc\theta = \frac{\sqrt{29}}{2}$

13. $\sin\theta = 0.846$, $\cos\theta = 0.534$, $\tan\theta = 1.58$, $\cot\theta = 0.631$, $\sec\theta = 1.87$, $\csc\theta = 1.18$

15. $\sin\theta = 0.1521$, $\cos\theta = 0.9884$, $\tan\theta = 0.1539$, $\cot\theta = 6.498$, $\sec\theta = 1.012$, $\csc\theta = 6.575$

17. $\frac{5}{13}, \frac{12}{5}$ **19.** $\frac{1}{\sqrt{2}}, \sqrt{2}$ **21.** 0.882, 1.33

23. 0.246, 3.94 **25.** $\sin\theta = \frac{4}{5}$, $\tan\theta = \frac{4}{3}$

27. $\tan\theta = \frac{1}{2}$, $\sec\theta = \frac{\sqrt{5}}{2}$ **29.** $\sec\theta$

31. $\frac{x}{y} \times \frac{y}{r} = \frac{x}{r} = \cos\theta$

Exercises 4-3, page 115

1. $\sin 40° = 0.64$, $\cos 40° = 0.77$, $\tan 40° = 0.84$, $\cot 40° = 1.19$, $\sec 40° = 1.31$, $\csc 40° = 1.56$

3. $\sin 15° = 0.26$, $\cos 15° = 0.97$, $\tan 15° = 0.27$, $\cot 15° = 3.73$, $\sec 15° = 1.04$, $\csc 15° = 3.86$

5. 0.381 **7.** 1.58 **9.** 0.9626 **11.** 0.99

13. 0.4085 **15.** 1.57 **17.** 1.32 **19.** 0.070 63

21. 70.97° **23.** 65.70° **25.** 11.7° **27.** 49.453°

29. 53.44° **31.** 81.79° **33.** 74.1° **35.** 17.85°

37. 0.8885 **39.** 0.936 14 **41.** 87 dB **43.** 70.6°

Exercises 4-4, page 119

1. **3.**

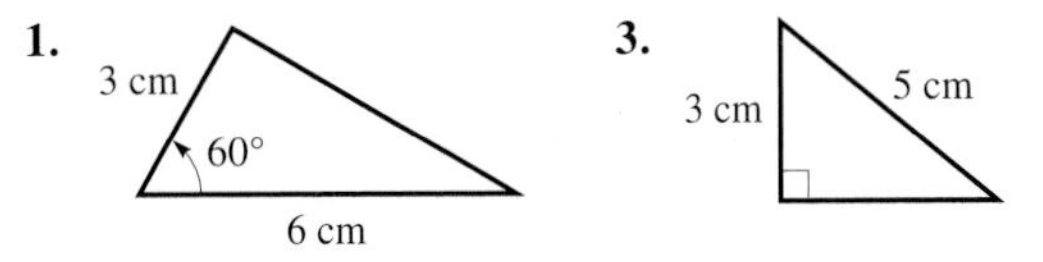

5. $B = 12.2°$, $b = 1450$, $c = 6850$

7. $A = 25.8°$, $B = 64.2°$, $b = 311$

9. $A = 57.9°$, $a = 20.2$, $b = 12.6$

11. $A = 21°$, $a = 32$, $B = 69°$

13. $a = 30.21$, $B = 57.90°$, $b = 48.16$

15. $A = 52.15°$, $B = 37.85°$, $c = 71.85$

17. $A = 52.5°$, $b = 0.661$, $c = 1.09$

19. $A = 15.82°$, $a = 0.5239$, $c = 1.922$

21. $A = 65.886°$, $B = 24.114°$, $c = 648.46$

23. $a = 3.3621$, $B = 77.025°$, $c = 14.974$

25. 4.45 **27.** 40.24°

29. $a = c \sin A$, $b = c \cos A$, $B = 90° - A$

31. $A = 90° - B$, $b = a \tan B$, $c = \dfrac{a}{\cos B}$

Exercises 4-5, page 122

1. 97 m **3.** 44.0 m **5.** 0.4° **7.** 850.1 cm
9. 7610 mm **11.** 26.6°, 63.4°, 90.0° **13.** 3.4°
15. 51.3° **17.** 8.1° **19.** 3.07 cm **21.** 651 m
23. 30.2° **25.** 47.3 m **27.** 642 m

Review Exercises for Chapter 4, page 124

1. 377.0°, −343.0° **3.** 142.5°, −577.5° **5.** 31.9°
7. 38.1° **9.** 17° 30′ **11.** 49° 42′

13. $\sin\theta = \dfrac{7}{25}$, $\cos\theta = \dfrac{24}{25}$, $\tan\theta = \dfrac{7}{24}$, $\cot\theta = \dfrac{24}{7}$, $\sec\theta = \dfrac{25}{24}$, $\csc\theta = \dfrac{25}{7}$

15. $\sin\theta = \dfrac{1}{\sqrt{2}}$, $\cos\theta = \dfrac{1}{\sqrt{2}}$, $\tan\theta = 1$, $\cot\theta = 1$, $\sec\theta = \sqrt{2}$, $\csc\theta = \sqrt{2}$

17. 0.923, 2.40 **19.** 0.447, 1.12 **21.** 0.952
23. 1.853 **25.** 1.05 **27.** 8.074 **29.** 18.2°
31. 57.57° **33.** 12.25° **35.** 66.8°
37. $a = 1.83$, $B = 73.0°$, $c = 6.27$
39. $A = 51.5°$, $B = 38.5°$, $c = 104$
41. $B = 52.5°$, $b = 15.6$, $c = 19.7$
43. $A = 31.61°$, $a = 4.006$, $B = 58.39°$
45. $a = 0.6292$, $B = 40.33°$, $b = 0.5341$
47. $A = 48.813°$, $B = 41.187°$, $b = 10.196$
49. 44.4 N **51.** 679.2 m^2 **53.** 12.0°
55. 4.92 km **57.** 56% **59.** 20 m
61. 4430 mm **63.** 0.57 km **65.** 10.2 cm

67. middle: $\dfrac{580}{\tan 1.1°} = 30\,200$ m; end: $\dfrac{1160}{\tan 2.2°} = 30\,200$ m

69. 3800 m **71.** 1.83 km **73.** 464 m
75. 733 mm

Exercises 5-1, page 130

1. yes; no **3.** yes; yes **5.** −1; −16

7. $\dfrac{1}{4}$, −0.6 **9.** yes **11.** no **13.** no

15. yes **17.** yes **19.** no

Exercises 5-2, page 134

1. 4 **3.** −5 **5.** $\dfrac{2}{7}$ **7.** $\dfrac{1}{2}$

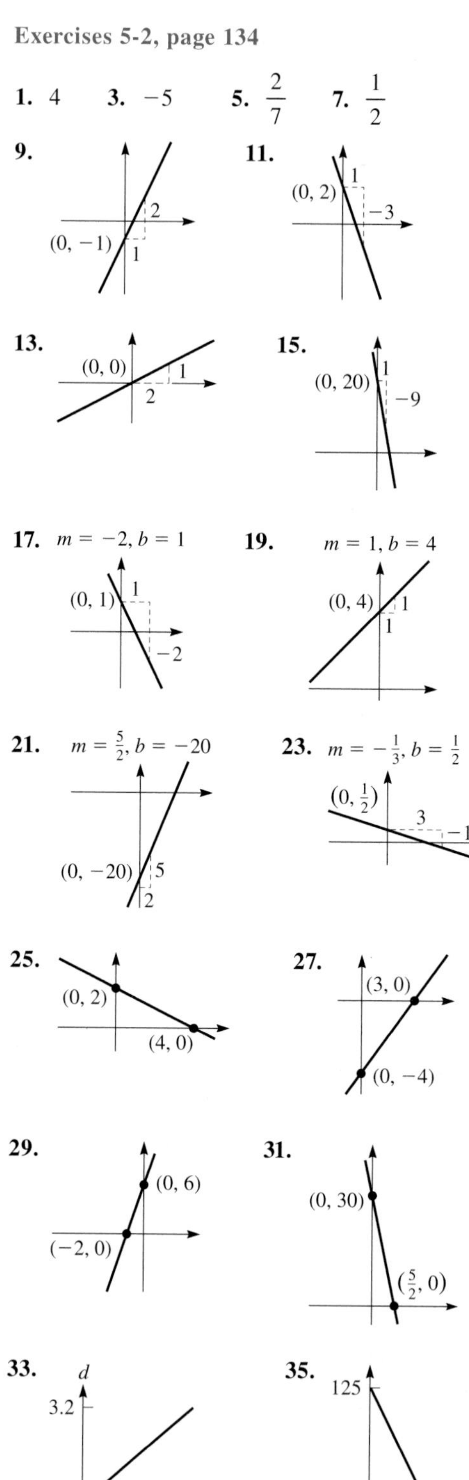

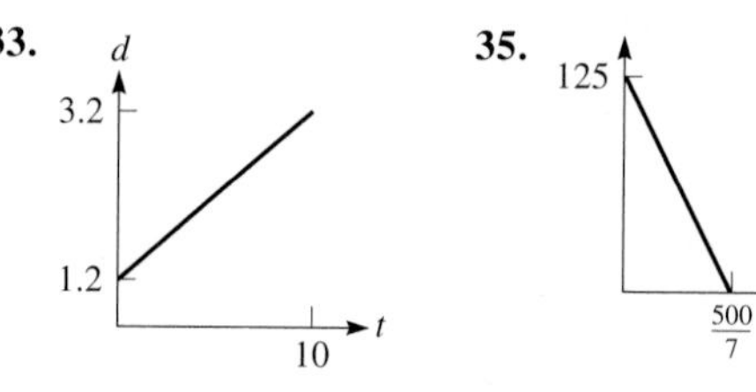

Exercises 5-3, page 138

1. $x = 3.0, y = 1.0$ **3.** $x = 3.0, y = 0.0$
5. $x = 2.2, y = -0.3$ **7.** $x = -0.9, y = -2.3$
9. $s = 1.1, t = -1.7$ **11.** $x = 0.0, y = 3.0$
13. $x = -14.0, y = -5.0$ **15.** $r_1 = 4.0, r_2 = 7.5$
17. $x = -3.6, y = -1.4$ **19.** $x = 1.1, y = 0.8$
21. dependent **23.** $t = -1.9, v = -3.2$
25. $x = 1.5, y = 4.5$ **27.** inconsistent
29. 50 N, 47 N **31.** 2.6 m, 3.4 m

Exercises 5-4, page 143

1. $x = 1, y = -2$ **3.** $V = 7, p = 3$
5. $x = -1, y = -4$ **7.** $x = \frac{1}{2}, y = 2$
9. $x = -\frac{1}{3}, y = 4$ **11.** $p = \frac{9}{22}, n = -\frac{16}{11}$
13. $x = 3, y = 1$ **15.** $x = -1, y = -2$
17. $x = 1, y = 2$ **19.** inconsistent
21. $x = -\frac{14}{5}, y = -\frac{16}{5}$ **23.** $x = 2.38, y = 0.45$
25. $x = \frac{1}{2}, y = -4$ **27.** $x = -\frac{2}{3}, y = 0$
29. dependent **31.** $C = -1, V = -2$
33. $V_1 = 9.0$ V, $V_2 = 6.0$ V
35. $x = 6250$ L, $y = 3750$ L **37.** $t_1 = 32$ s, $t_2 = 20$ s
39. 4.0×10^6 calc/s, 2.5×10^6 calc/s
41. 10 m, 8.0 m, 4.0 m
43. incorrect conclusion or error in sales figures; system of equations is inconsistent

Exercises 5-5, page 149

1. -10 **3.** 29 **5.** 32 **7.** 93 **9.** 0.9
11. 96 **13.** $x = 3, y = 1$ **15.** $x = -1, y = -2$
17. $x = 1, y = 2$ **19.** inconsistent
21. $x = -\frac{14}{5}, y = -\frac{16}{5}$ **23.** $i_1 = 2.38, i_2 = 0.45$
25. $x = 0.32, y = -4.5$ **27.** $R = 4.0, t = -3.2$
29. $x = -11.2, y = -9.26$ **31.** $x = -1.0, y = -2.0$
33. $F_1 = 15$ N, $F_2 = 6.0$ N
35. $x = 73.6$ L, $y = 70.4$ L **37.** 2.5 h, 2.1 h
39. 4200, 1400 **41.** 0.8 L, 1.2 L **43.** $V = 4.5i - 3.2$

Exercises 5-6, page 153

1. $x = 2, y = -1, z = 1$ **3.** $x = 4, y = -3, z = 3$
5. $x = \frac{1}{2}, y = \frac{2}{3}, z = \frac{1}{6}$
7. $x = \frac{2}{3}, y = -\frac{1}{3}, z = 1$
9. $x = \frac{4}{15}, y = -\frac{3}{5}, z = \frac{1}{3}$
11. $p = -2, q = \frac{2}{3}, r = \frac{1}{3}$
13. $x = \frac{3}{4}, y = 1, z = -\frac{1}{2}$
15. $r = 0, s = 0, t = 0, u = -1$
17. $P = 800$ h, $M = 125$ h, $I = 225$ h
19. $F_1 = 9.43$ N, $F_2 = 8.33$ N, $F_3 = 1.67$ N
21. 2200 nuts, 1800 bolts, 3600 washers
23. 70 kg, 100 kg, 30 kg
25. unlimited: $x = -10, y = -6, z = 0$
27. no solution

Exercises 5-7, page 159

1. 122 **3.** 651 **5.** -439 **7.** 202 **9.** 128
11. 0.128 **13.** $x = -1, y = 2, z = 0$
15. $x = 2, y = -1, z = 1$ **17.** $x = 4, y = -3, z = 3$
19. $x = \frac{1}{2}, y = \frac{2}{3}, z = \frac{1}{6}$
21. $x = \frac{2}{3}, y = -\frac{1}{3}, z = 1$
23. $x = \frac{4}{15}, y = -\frac{3}{5}, z = \frac{1}{3}$
25. $p = -2, q = \frac{2}{3}, r = \frac{1}{3}$
27. $x = \frac{3}{4}, y = 1, z = -\frac{1}{2}$
29. $A = 125$ N, $B = 60$ N, $F = 75$ N
31. 30.9 km/h, 45.9 km/h, 551 km/h

Review Exercises for Chapter 5, page 160

1. -17 **3.** -1485 **5.** -4 **7.** $\frac{2}{7}$
9. $m = -2, b = 4$ **11.** $m = 4, b = -\frac{5}{2}$

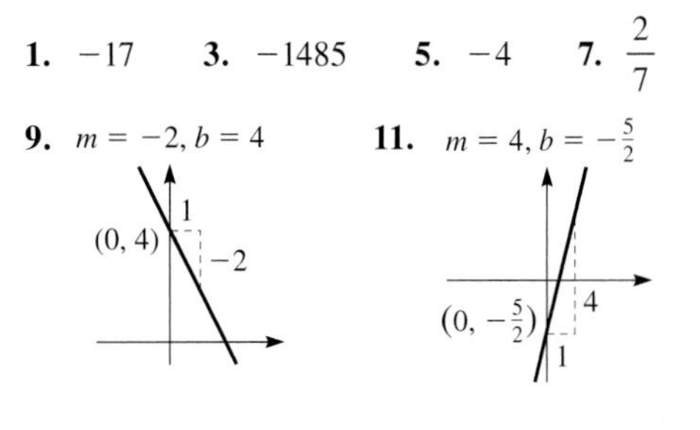

13. $x = 2.0, y = 0.0$ **15.** $x = 2.2, y = 2.7$
17. $x = 1.5, y = -1.9$ **19.** $x = 1.5, y = 0.4$

21. $x = 1, y = 2$ **23.** $x = \frac{1}{2}, y = -2$

25. $i = -\frac{1}{3}, v = \frac{7}{4}$ **27.** $x = \frac{11}{19}, y = -\frac{26}{19}$

29. $x = -\frac{6}{19}, y = \frac{36}{19}$ **31.** $x = 1.10, y = 0.54$

33. $x = 1, y = 2$ **35.** $x = \frac{1}{2}, y = -2$

37. $i = -\frac{1}{3}, v = \frac{7}{4}$ **39.** $x = \frac{11}{19}, y = -\frac{26}{19}$

41. $x = -\frac{6}{19}, y = \frac{36}{19}$ **43.** $x = 1.10, y = 0.54$

45. -115 **47.** 230.08 **49.** $x = 2, y = -1, z = 1$

51. $x = \frac{2}{3}, y = -\frac{1}{2}, z = 0$ **53.** $r = 3, s = -1, t = \frac{3}{2}$

55. $x = -0.17, y = 0.16, z = 2.4$

57. $x = 2, y = -1, z = 1$

59. $x = \frac{2}{3}, y = -\frac{1}{2}, z = 0$ **61.** $r = 3, s = -1, t = \frac{3}{2}$

63. $x = -0.17, y = 0.16, z = 2.4$

65. $x = \frac{8}{3}, y = -8$ **67.** $x = 1, y = 3$ **69.** -6

71. $-\frac{4}{3}$ **73.** $F_1 = 21\,000$ N, $F_2 = 2400$ N, $F_3 = 18\,000$ N

75. $a = 440$ m·°C, $b = 9.6$ °C

77. 22 800 km/h, 1400 km/h

79. $R_1 = 0.50\ \Omega$, $R_2 = 1.5\ \Omega$

81. $L = 10$ N, $w = 40$ N

83. 425 TVs, 475 VCRs, 850 CDs

Exercisess 6-1, page 167

1. $40x - 40y$ **3.** $2x^3 - 8x^2$ **5.** $y^2 - 36$

7. $9v^2 - 4$ **9.** $16x^2 - 25y^2$ **11.** $144 - 25a^2b^2$

13. $25f^2 + 40f + 16$ **15.** $4x^2 + 28x + 49$

17. $x^2 - 2x + 1$ **19.** $16a^2 + 56axy + 49x^2y^2$

21. $16x^2 - 16xy + 4y^2$ **23.** $36s^2 - 12st + t^2$

25. $x^2 + 6x + 5$ **27.** $c^2 + 9c + 18$

29. $6x^2 + 13x - 5$ **31.** $20x^2 - 21x - 5$

33. $20v^2 + 13v - 15$ **35.** $6x^2 - 13xy - 63y^2$

37. $2x^2 - 8$ **39.** $8a^3 - 2a$

41. $6ax^2 + 24abx + 24ab^2$

43. $20n^4 + 100n^3 + 125n^2$ **45.** $16a^3 - 48a^2 + 36a$

47. $x^2 + y^2 + 2xy + 2x + 2y + 1$

49. $x^2 + y^2 + 2xy - 6x - 6y + 9$

51. $125 - 75t + 15t^2 - t^3$

53. $8x^3 + 60x^2t + 150xt^2 + 125t^3$

55. $x^2 + 2xy + y^2 - 1$ **57.** $x^3 + 8$

59. $64 - 27x^3$ **61.** $P_0P_1c + P_1G$

63. $4p^2 + 8pDA + 4D^2A^2$ **65.** $\frac{1}{2}\pi R^2 - \frac{1}{2}\pi r^2$

67. $\frac{L}{6}x^3 - \frac{L}{2}ax^2 + \frac{L}{2}a^2x - \frac{L}{6}a^3$ **69.** $4x^2 - 9$

71. $(-2 - 3)^3 = (-2)^3 - 3(-2)^2(3) + 3(-2)(3^2) - 3^3 = -125$

Exercises 6-2, page 171

1. $6(x + y)$ **3.** $5(a - 1)$ **5.** $3x(x - 3)$

7. $7b(by - 4)$ **9.** $6n(2n + 1)$ **11.** $2(x + 2y - 4z)$

13. $3ab(b - 2 + 4b^2)$ **15.** $4pq(3q - 2 - 7q^2)$

17. $2(a^2 - b^2 + 2c^2 - 3d^2)$ **19.** $(x + 2)(x - 2)$

21. $(10 + y)(10 - y)$ **23.** $(6a + 1)(6a - 1)$

25. $(9s + 5t)(9s - 5t)$ **27.** $(12n + 13p^2)(12n - 13p^2)$

29. $(x + y + 3)(x + y - 3)$ **31.** $2(x + 2)(x - 2)$

33. $3(x + 3z)(x - 3z)$ **35.** $2(a - 1)(a - 5)$

37. $(x^2 + 4)(x + 2)(x - 2)$

39. $(x^4 + 1)(x^2 + 1)(x + 1)(x - 1)$

41. $\frac{3 + b}{2 - b}$ **43.** $\frac{3}{2(t - 1)}$ **45.** $(3 + b)(x - y)$

47. $(a - b)(a + x)$ **49.** $(x + 2)(x - 2)(x + 3)$

51. $(x - y)(x + y + 1)$ **53.** $Rv(1 + v + v^2)$

55. $a(D_1 + D_2)(D_1 - D_2)$ **57.** $Pb(L + b)(L - b)$

59. $\frac{ER}{A(T_0 - T_1)}$

Exercises 6-3, page 176

1. $(x + 1)(x + 4)$ **3.** $(s - 7)(s + 6)$

5. $(t + 8)(t - 3)$ **7.** $(x + 1)^2$ **9.** $(x - 2y)^2$

11. $(3x + 1)(x - 2)$ **13.** $(3y + 1)(y - 3)$

15. $(2s + 11)(s + 1)$ **17.** $(3f - 1)(f - 5)$

19. $(2t - 3)(t + 5)$ **21.** $(3t - 4u)(t - u)$

23. $(4x - 7)(x + 1)$ **25.** $(9x - 2y)(x + y)$

27. $(2m + 5)^2$ **29.** $(2x - 3)^2$ **31.** $(3t - 4)(3t - 1)$

33. $(8b - 1)(b + 4)$ **35.** $(4p - q)(p - 6q)$

37. $(12x - y)(x + 4y)$ **39.** $2(x - 1)(x - 6)$

41. $2(2x - 1)(x + 4)$ **43.** $ax(x + 6a)(x - 2a)$

45. $(a + b + 2)(a + b - 2)$

47. $(5a + 5x + y)(5a - 5x - y)$ **49.** $4(s + 1)(s + 3)$

51. $100(2n + 3)(n - 12)$ **53.** $wx^2(x - 2L)(x - 3L)$

55. $Ad(3u - v)(u - v)$

Exercises 6-4, page 178

1. $(x + 1)(x^2 - x + 1)$ **3.** $(2 - t)(4 + 2t + t^2)$

5. $(3x - 2a)(9x^2 + 6ax + 4a^2)$

7. $2(x+2)(x^2-2x+4)$ **9.** $6a(a+1)(a^2-a+1)$

11. $6x^3y(1-y)(1+y+y^2)$

13. $x^3y^3(x+y)(x^2-xy+y^2)$

15. $3a^2(a^2+1)(a+1)(a-1)$

17. $(a+b-4)(a^2+2ab+b^2+4a+4b+16)$

19. $(4+x^2)(16-4x^2+x^4)$

21. $2(x+5)(x^2-5x+25)$

23. $D(D-d)(D^2+Dd+d^2)$

Exercises 6-5, page 181

1. $\frac{14}{21}$ **3.** $\frac{2ax^2}{2xy}$ **5.** $\frac{2x-4}{x^2+x-6}$

7. $\frac{ax^2-ay^2}{x^2-xy-2y^2}$ **9.** $\frac{7}{11}$ **11.** $\frac{2xy}{4y^2}$

13. $\frac{2}{x+1}$ **15.** $\frac{x-5}{2x-1}$ **17.** $\frac{1}{4}$ **19.** $\frac{3x}{4}$

21. $\frac{1}{5a}$ **23.** $\frac{3a-2b}{2a-b}$

25. $\frac{4x^2+1}{(2x+1)(2x-1)}$ (cannot be reduced) **27.** $3x$

29. $\frac{1}{2y^2}$ **31.** $\frac{x-4}{x+4}$ **33.** $\frac{2x-1}{x+8}$ **35.** $\frac{5x+4}{x(x+3)}$

37. $(x^2+4)(x-2)$ **39.** $\frac{x^2y^2(y+x)}{y-x}$ **41.** $\frac{x+3}{x-3}$

43. $-\frac{1}{2}$ **45.** $-\frac{2x-1}{x}$ **47.** $\frac{(x+5)(x-3)}{(5-x)(x+3)}$

49. $\frac{x^2-xy+y^2}{2}$ **51.** $\frac{2x}{9x^2-3x+1}$

53. (a) $\frac{x^2(x+2)}{x^2+4}$ (b) $\frac{x^2}{x^2-4}$ Numerator and denominator have no common factor. In each, x^2 is not a factor of the denominator.

55. (a) $\frac{(x-2)(x+1)}{x(x-1)}$ (b) $\frac{x-2}{x}$ Numerator and denominator have no common factor. In each, x is not a factor of the numerator.

57. $u+v$ **59.** $\frac{E^2(R-r)}{(R+r)^3}$

Exercises 6-6, page 184

1. $\frac{3}{28}$ **3.** $6xy$ **5.** $\frac{7}{18}$ **7.** $\frac{xy^2}{bz^2}$ **9.** $4t$

11. $3(u+v)(u-v)$ **13.** $\frac{10}{3(a+4)}$ **15.** $\frac{x-3}{x(x+3)}$

17. $\frac{3x}{5a}$ **19.** $\frac{(x+1)(x-1)(x-4)}{4(x+2)}$ **21.** $\frac{x^2}{a+x}$

23. $\frac{15}{4}$ **25.** $\frac{3}{4x+3}$ **27.** $\frac{7x-1}{3x+5}$ **29.** $\frac{7x^4}{3a^4}$

31. $\frac{4t(2t-1)(t+5)}{(2t+1)^2}$ **33.** $\frac{x+y}{2}$

35. $(x+y)(3p+7q)$ **37.** $\frac{na^2}{v(1-a)}$ **39.** $\frac{\pi}{2}$

Exercises 6-7, page 189

1. $\frac{9}{5}$ **3.** $\frac{8}{x}$ **5.** $\frac{5}{4}$ **7.** $\frac{3+7ax}{4x}$ **9.** $\frac{ax-b}{x^2}$

11. $\frac{30+ax^2}{25x^3}$ **13.** $\frac{14-a^2}{10a}$

15. $\frac{-x^2+4x+xy+y-2}{xy}$ **17.** $\frac{7}{2(2x-1)}$

19. $\frac{5-3x}{2x(x+1)}$ **21.** $\frac{-3}{4(s-3)}$ **23.** $\frac{7x+6}{3(x+3)(x-3)}$

25. $\frac{2x-5}{(x-4)^2}$ **27.** $\frac{x+27}{(x-5)(x+5)(x-6)}$

29. $\frac{9x^2+x-2}{(3x-1)(x-4)}$ **31.** $\frac{13t^2+27t}{(t-3)(t+2)(t+3)^2}$

33. $\frac{-2x^3+x^2-x}{(x+1)(x^2-x+1)}$ **35.** $\frac{1}{x-1}$ **37.** $\frac{x-y}{y}$

39. $-\frac{(x+1)(x-1)(2x+1)}{x^2(x+2)}$ **41.** $-\frac{(3x+4)(x-1)}{2x}$

43. $\frac{2s}{r-s}$ **45.** $\frac{h}{(x+1)(x+h+1)}$

47. $\frac{-2hx-h^2}{x^2(x+h)^2}$ **49.** $\frac{y^2-rx+r^2}{r^2}$ **51.** $\frac{2a-1}{a^2}$

53. $\frac{a^2+2a-1}{a+1}$ **55.** $\frac{a+b}{\frac{1}{a}+\frac{1}{b}}=ab$

57. $\frac{3(H-H_0)}{4\pi H}$ **59.** $\frac{n(2n+1)}{2(n+2)(n-1)}$

61. $\frac{R^4-2r^2R^2+r^4}{R^4}$ **63.** $\frac{sL+R}{s^2LC+sRC+1}$

Exercises 6-8, page 194

1. 4 **3.** -3 **5.** $\frac{7}{2}$ **7.** $\frac{16}{21}$ **9.** -9

11. $-\frac{2}{13}$ **13.** $\frac{5}{3}$ **15.** -2 **17.** $\frac{3}{4}$ **19.** $\frac{37}{6}$

21. -5 **23.** $\frac{63}{8}$ **25.** no solution **27.** $\frac{2}{3}$

29. $\frac{3b}{1-2b}$ **31.** $\frac{(2b-1)(b+6)}{2(b-1)}$ **33.** $\frac{n_1V-nV}{n_1}$

35. $\frac{V_rA-V_0}{V_0A}$ **37.** $\frac{jX}{1-g_mZ}$

39. $\frac{PV^3-bPV^2+aV-ab}{RV^2}$ **41.** $\frac{kA_1A_2R-A_1L_2}{A_2}$

43. $\dfrac{fnR_2 - fR_2}{R_2 + f - fn}$ **45.** 2.4 h **47.** 3.2 min

49. 80 km/h **51.** 2.5 L

Review Exercises for Chapter 6, page 195

1. $12ax + 15a^2$ **3.** $4a^2 - 49b^2$

5. $4a^2 + 4a + 1$ **7.** $b^2 + 3b - 28$

9. $2x^2 - 13x - 45$ **11.** $16c^2 + 6cd - d^2$

13. $3(s + 3t)$ **15.** $a^2(x^2 + 1)$

17. $(x + 12)(x - 12)$

19. $(4x + 8 + t^2)(4x + 8 - t^2)$ **21.** $(3t - 1)^2$

23. $(5t + 1)^2$ **25.** $(x + 8)(x - 7)$

27. $(t - 9)(t + 4)$ **29.** $(2x - 9)(x + 4)$

31. $(2x + 5)(2x - 7)$ **33.** $(5b - 1)(2b + 5)$

35. $4(x + 4y)(x - 4y)$

37. $2(5 - 2y^2)(25 + 10y^2 + 4y^4)$

39. $(2x + 3)(4x^2 - 6x + 9)$ **41.** $(a - 3)(b^2 + 1)$

43. $(x + 5)(n - x + 5)$ **45.** $\dfrac{16x^2}{3a^2}$ **47.** $\dfrac{3x + 1}{2x - 1}$

49. $\dfrac{16}{5x(x - y)}$ **51.** $\dfrac{6}{5 - x}$ **53.** $\dfrac{x + 2}{2x(7x - 1)}$

55. $\dfrac{1}{x - 1}$ **57.** $\dfrac{16x - 15}{36x^2}$ **59.** $\dfrac{5y + 6}{2xy}$

61. $\dfrac{-2(2a + 3)}{a(a + 2)}$ **63.** $\dfrac{2x^2 - x + 1}{x(x + 3)(x - 1)}$

65. $\dfrac{12x^2 - 7x - 4}{2(x - 1)(x + 1)(4x - 1)}$ **67.** $\dfrac{x^3 + 6x^2 - 2x + 2}{x(x - 1)(x + 3)}$

69. 2 **71.** $\dfrac{7}{2c + 4}$ **73.** $-\dfrac{(a - 1)^2}{2a}$ **75.** 6

77. $\dfrac{1}{4}[(x + y)^2 - (x - y)^2]$

$= \dfrac{1}{4}(x^2 + 2xy + y^2 - x^2 + 2xy - y^2) = \dfrac{1}{4}(4xy)$

$= xy$

79. $2zS^2 + 2zS$ **81.** $\pi l(r_1 + r_2)(r_1 - r_2)$

83. $16(2 + t)(8 - t)$ **85.** $(t + 1)(1 - t)$

87. $8n^6 + 36n^5 + 66n^4 + 63n^3 + 33n^2 + 9n + 1$

89. $10aT - 10at + aT^2 - 2aTt + at^2$

91. $4(3x^2 + 12x + 16)$ **93.** $\dfrac{12\pi^2wv^2D}{gn^2t}$

95. $\dfrac{60t + 9t^2 + 2t^3}{6}$ **97.** $\dfrac{120 - 60d^2 + 5d^4 - d^6}{120}$

99. $\dfrac{2\pi^2CN + 2\pi^2Cn + N^2 - 2nN + n^2}{4\pi^2C}$

101. $\dfrac{4r^3 - 3ar^2 - a^3}{4r^3}$ **103.** $\dfrac{c^2u^2 - 2gc^2x}{1 - c^2u^2 + 2gc^2x}$

105. $\dfrac{q_2D - fd}{D + d}$ **107.** $\dfrac{RHw}{w - RH}$

109. $\dfrac{-mb^2s^2 - kL^2}{b^2s}$ **111.** $\dfrac{HT_1}{H - RT_1X}$ **113.** 3.4 h

115. 15 s **117.** 11.3 **119.** 18 Ω

Exercises 7-1, page 203

1. $a = 1, b = -8, c = 5$ **3.** $a = 1, b = -2, c = -4$

5. not quadratic **7.** $a = 1, b = -1, c = 0$

9. $2, -2$ **11.** $\frac{3}{2}, -\frac{3}{2}$ **13.** $-1, 9$ **15.** $3, 4$

17. $0, -2$ **19.** $\frac{1}{3}, -\frac{1}{3}$ **21.** $\frac{1}{3}, 4$ **23.** $-4, -4$

25. $\frac{2}{3}, \frac{3}{2}$ **27.** $\frac{1}{2}, -\frac{3}{2}$ **29.** $2, -1$ **31.** $2b, -2b$

33. $0, \frac{5}{2}$ **35.** $\frac{5}{2}, -\frac{9}{2}$ **37.** $0, -2$

39. $b - a, -b - a$ **41.** 0, L **43.** 2 A.M., 10 A.M.

45. $\frac{3}{2}, 4$ **47.** $-2, \frac{1}{2}$ **49.** 3 N/cm, 6 N/cm

51. 800 km/h

Exercises 7-2, page 206

1. $-5, 5$ **3.** $-\sqrt{7}, \sqrt{7}$ **5.** $-3, 7$ **7.** $-3 \pm \sqrt{7}$

9. $2, -4$ **11.** $-2, -1$ **13.** $2 \pm \sqrt{2}$ **15.** $-5, 3$

17. $-3, \frac{1}{2}$ **19.** $\frac{1}{6}(3 \pm \sqrt{33})$ **21.** $\frac{1}{4}(1 \pm \sqrt{17})$

23. $-b \pm \sqrt{b^2 - c}$

Exercises 7-3, page 210

1. $2, -4$ **3.** $-2, -1$ **5.** $2 \pm \sqrt{2}$ **7.** $-5, 3$

9. $-3, \frac{1}{2}$ **11.** $\frac{1}{6}(3 \pm \sqrt{33})$ **13.** $\frac{1}{4}(1 \pm \sqrt{17})$

15. $-\frac{8}{5}, \frac{5}{6}$ **17.** $\frac{1}{2}(-5 \pm \sqrt{-5})$ **19.** $\frac{1}{6}(1 \pm \sqrt{109})$

21. $\frac{3}{2}, -\frac{3}{2}$ **23.** $\frac{3}{4}, -\frac{5}{8}$ **25.** $-0.54, 0.74$

27. $-0.26, 2.43$ **29.** $-c \pm \sqrt{c^2 + 1}$

31. $\dfrac{b + 1 \pm \sqrt{-3b^2 + 2b + 4b^2a + 1}}{2b^2}$ **33.** 4.376 cm

35. $\dfrac{-R \pm \sqrt{R^2 - 4L/C}}{2L}$ **37.** imaginary, unequal

39. real, irrational, unequal **41.** 11.0 m by 23.8 m

43. 1.1 m

Exercises 7-4, page 214

1. **3.**

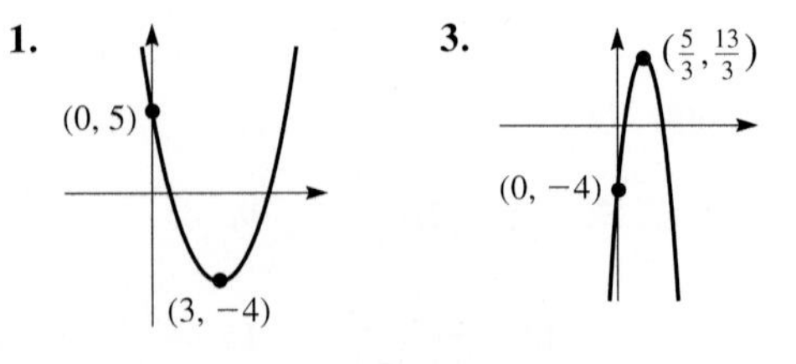

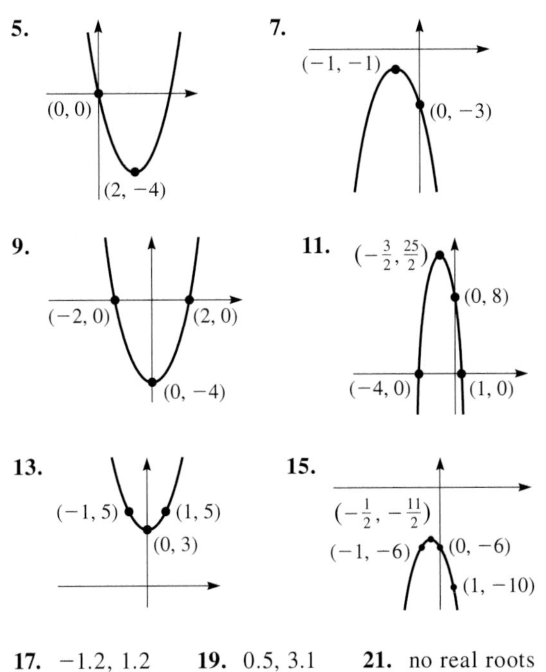

17. −1.2, 1.2 **19.** 0.5, 3.1 **21.** no real roots

23. −2.4, 1.2

25.

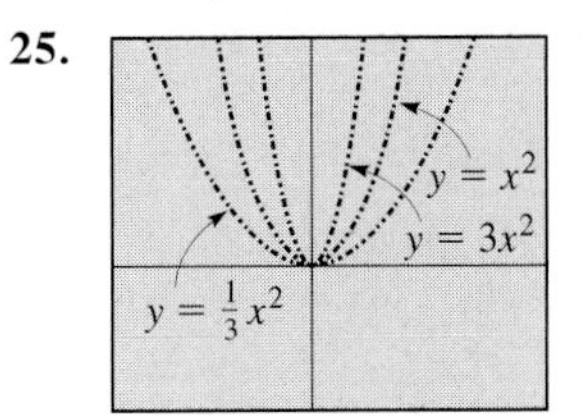

The parabola $y = 3x^2$ rises more quickly, and the parabola $y = \frac{1}{3}x^2$ rises more slowly.

27.

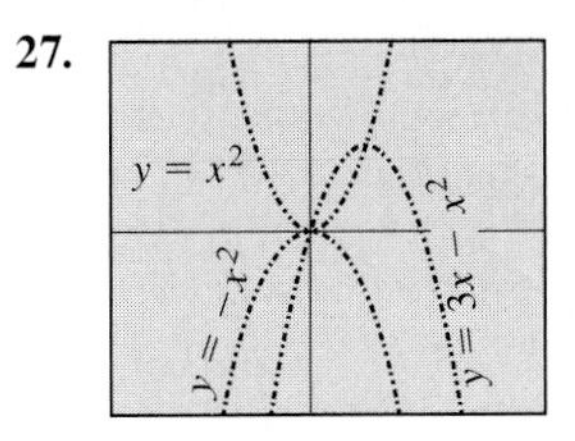

The parabola $y = -x^2$ is inverted, and the parabola $y = 3x - x^2$ is inverted with a different vertex.

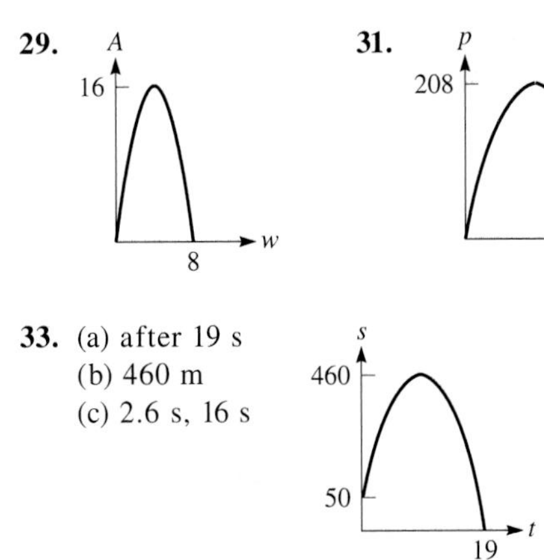

33. (a) after 19 s
(b) 460 m
(c) 2.6 s, 16 s

35. 62 m by 32 m, or 22 m by 93 m

Review Exercises for Chapter 7, page 215

1. −4, 1 **3.** 2, 8 **5.** $\frac{1}{3}, -4$ **7.** $\frac{1}{2}, \frac{5}{3}$ **9.** $0, \frac{25}{6}$

11. $-\frac{3}{2}, \frac{7}{2}$ **13.** −10, 11 **15.** $-1 \pm \sqrt{6}$

17. $-4, \frac{9}{2}$ **19.** $\frac{1}{8}(3 \pm \sqrt{41})$

21. $\frac{1}{42}(-23 \pm \sqrt{-4091})$ **23.** $\frac{1}{6}(-2 \pm \sqrt{58})$

25. $-2 \pm 2\sqrt{2}$ **27.** $\frac{1}{3}(-4 \pm \sqrt{10})$ **29.** $-1, \frac{5}{4}$

31. $\frac{1}{4}(-3 \pm \sqrt{-47})$ **33.** $\dfrac{-1 \pm \sqrt{-1}}{a}$

35. $\dfrac{-3 \pm \sqrt{9 + 4a^3}}{2a}$ **37.** −5, 6 **39.** $\frac{1}{4}(1 \pm \sqrt{33})$

41. $3 \pm \sqrt{7}$ **43.** 0 (3 is not a solution)

45. (0, −1), $(\frac{1}{4}, -\frac{9}{8})$ **47.** (0, 0), $(\frac{1}{6}, \frac{1}{12})$

49. −1.7, 1.2 **51.** no real roots **53.** 1.2 cm, 4.0 cm

55. 0.6 s, 2.2 s **57.** 8000

59. $\dfrac{-\pi h \pm \sqrt{\pi^2 h^2 + 2\pi A}}{2\pi}$ **61.** p: 0.135; t: 6

63. 61 m, 81 m

65. 9.9 cm

67. 40.7 cm, 55.2 cm

69. 25

71. p: 0.268, 0.017; h: 24 — 6 h, 18 h

Exercises 8-1, page 220

1. +, −, − **3.** +, +, − **5.** +, +, +

7. +, −, +

9. $\sin\theta = \dfrac{1}{\sqrt{5}}$, $\cos\theta = \dfrac{2}{\sqrt{5}}$, $\tan\theta = \dfrac{1}{2}$, $\cot\theta = 2$,
$\sec\theta = \dfrac{1}{2}\sqrt{5}$, $\csc\theta = \sqrt{5}$

11. $\sin\theta = -\dfrac{3}{\sqrt{13}}$, $\cos\theta = -\dfrac{2}{\sqrt{13}}$, $\tan\theta = \dfrac{3}{2}$,
$\cot\theta = \dfrac{2}{3}$, $\sec\theta = -\dfrac{1}{2}\sqrt{13}$, $\csc\theta = -\dfrac{1}{3}\sqrt{13}$

13. $\sin\theta = \dfrac{12}{13}$, $\cos\theta = -\dfrac{5}{13}$, $\tan\theta = -\dfrac{12}{5}$,
$\cot\theta = -\dfrac{5}{12}$, $\sec\theta = -\dfrac{13}{5}$, $\csc\theta = \dfrac{13}{12}$

15. $\sin\theta = -\frac{2}{\sqrt{29}}$, $\cos\theta = \frac{5}{\sqrt{29}}$, $\tan\theta = -\frac{2}{5}$, $\cot\theta = -\frac{5}{2}$, $\sec\theta = \frac{1}{5}\sqrt{29}$, $\csc\theta = -\frac{1}{2}\sqrt{29}$

17. II **19.** II **21.** IV **23.** III

Exercises 8-2, page 226

1. sin 20°; −cos 40° **3.** −tan 75°; −csc 58°
5. −sin 57°; −cot 6° **7.** cos 40°; −tan 40°
9. −sin 15° = −0.26 **11.** −cos 73.7° = −0.281
13. tan 39.15° = 0.8141 **15.** sec 31.67° = 1.175
17. −0.523 **19.** −0.7620 **21.** −0.34
23. −3.910 **25.** 237.99°, 302.01°
27. 66.40°, 293.60° **29.** 15.8°, 195.8°
31. 102.0°, 282.0° **33.** 119.5° **35.** 263°
37. 306.21° **39.** 299.24° **41.** −0.7003
43. −0.777 **45.** < **47.** = **49.** 0.0183 A
51. 12.6 cm
53. The y-coordinate has the opposite sign for $\sin(-\theta)$ than it has for $\sin\theta$. $\cos(-\theta) = \frac{x}{r}$, $\tan(-\theta) = \frac{-y}{x}$, $\cot(-\theta) = \frac{x}{-y}$, $\sec(-\theta) = \frac{r}{x}$, $\csc(-\theta) = \frac{r}{-y}$
55. (a) 5.7 (b) −1.4

Exercises 8-3, page 231

1. $\frac{\pi}{12}, \frac{5\pi}{6}$ **3.** $\frac{5\pi}{12}, \frac{11\pi}{6}$ **5.** $\frac{7\pi}{6}, \frac{3\pi}{2}$ **7.** $\frac{8\pi}{9}, \frac{13\pi}{9}$
9. 72°, 270° **11.** 10°, 315° **13.** 170°, 300°
15. 15°, 27° **17.** 0.401 **19.** 4.40 **21.** 5.821
23. 3.115 **25.** 43.0° **27.** 195.2° **29.** 140°
31. 940.8° **33.** 0.7071 **35.** 3.732
37. −0.8660 **39.** −8.327 **41.** 0.9056
43. −0.89 **45.** −0.48 **47.** −0.15
49. 0.3141, 2.827 **51.** 2.932, 6.074
53. 0.8309, 5.452 **55.** 2.442, 3.841
57. 0.030 N · m **59.** 2900 m

Exercises 8-4, page 235

1. 3.46 cm **3.** 1010 mm **5.** 0.9449 km
7. 43 cm^2 **9.** 0.0119 m^2 **11.** 2.04 = 117.0°
13. 350 m **15.** 5570 cm^2 **17.** 0.382 = 21.9°
19. 627 m^2 **21.** 0.52 rad/s **23.** 34.73 m^2
25. 0.704 m **27.** 22.6 m^2 **29.** 369 m^3
31. 0.4 km **33.** 8.1 r/min **35.** 150 m/min
37. 9.41 m **39.** 35.9 m/min **41.** 0.433 rad/s
43. 3000 m/min **45.** 250 rad **47.** 14.9 m^3
49. 4.848×10^{-6} (all three values) **51.** 1.15×10^8 km

Review Exercises for Chapter 8, page 238

1. $\sin\theta = \frac{4}{5}$, $\cos\theta = \frac{3}{5}$, $\tan\theta = \frac{4}{3}$, $\cot\theta = \frac{3}{4}$, $\sec\theta = \frac{5}{3}$, $\csc\theta = \frac{5}{4}$

3. $\sin\theta = -\frac{2}{\sqrt{53}}$, $\cos\theta = \frac{7}{\sqrt{53}}$, $\tan\theta = -\frac{2}{7}$, $\cot\theta = -\frac{7}{2}$, $\sec\theta = \frac{\sqrt{53}}{7}$, $\csc\theta = -\frac{\sqrt{53}}{2}$

5. −cos 48°, tan 14° **7.** −sin 71°, sec 15°
9. $\frac{2\pi}{9}, \frac{17\pi}{20}$ **11.** $\frac{4\pi}{15}, \frac{9\pi}{8}$ **13.** 252°, 130°
15. 12°, 330° **17.** 32.1° **19.** 206.7° **21.** 1.78
23. 0.3534 **25.** 4.5736 **27.** 2.377 **29.** −0.415
31. −0.47 **33.** −1.080 **35.** −0.4264
37. −1.64 **39.** 4.140 **41.** −0.5878
43. −0.8660 **45.** 0.5569 **47.** 1.197
49. 10.30°, 190.30° **51.** 118.23°, 241.77°
53. 0.5759, 5.707 **55.** 4.187, 5.238 **57.** 223.76°
59. 246.78° **61.** 0.0562 W **63.** 0.800 = 45.8°
65. 18.98 cm **67.** 4710 cm/s **69.** 2.70 m^2
71. 1.81×10^6 cm/s **73.** 51.27 m
75. 3.58×10^5 km

Exercises 9-1, page 244

1. (a) vector: magnitude and direction are specified; (b) scalar: only magnitude is specified
3. (a) vector: magnitude and direction are specified; (b) scalar: only magnitude is specified

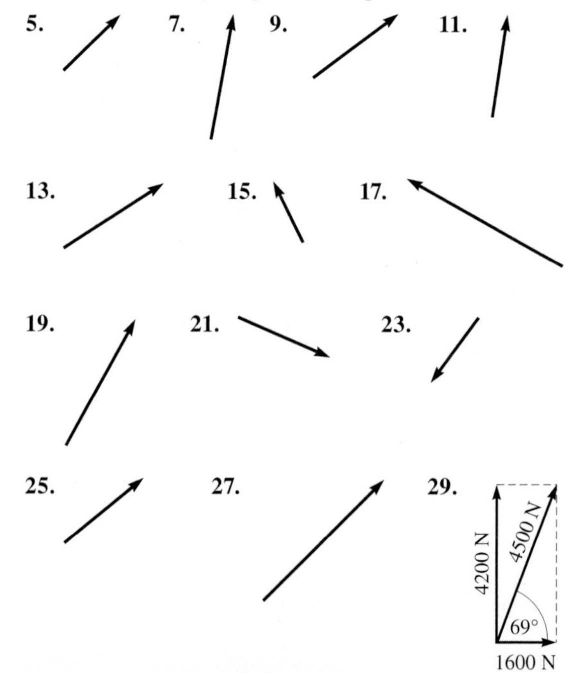

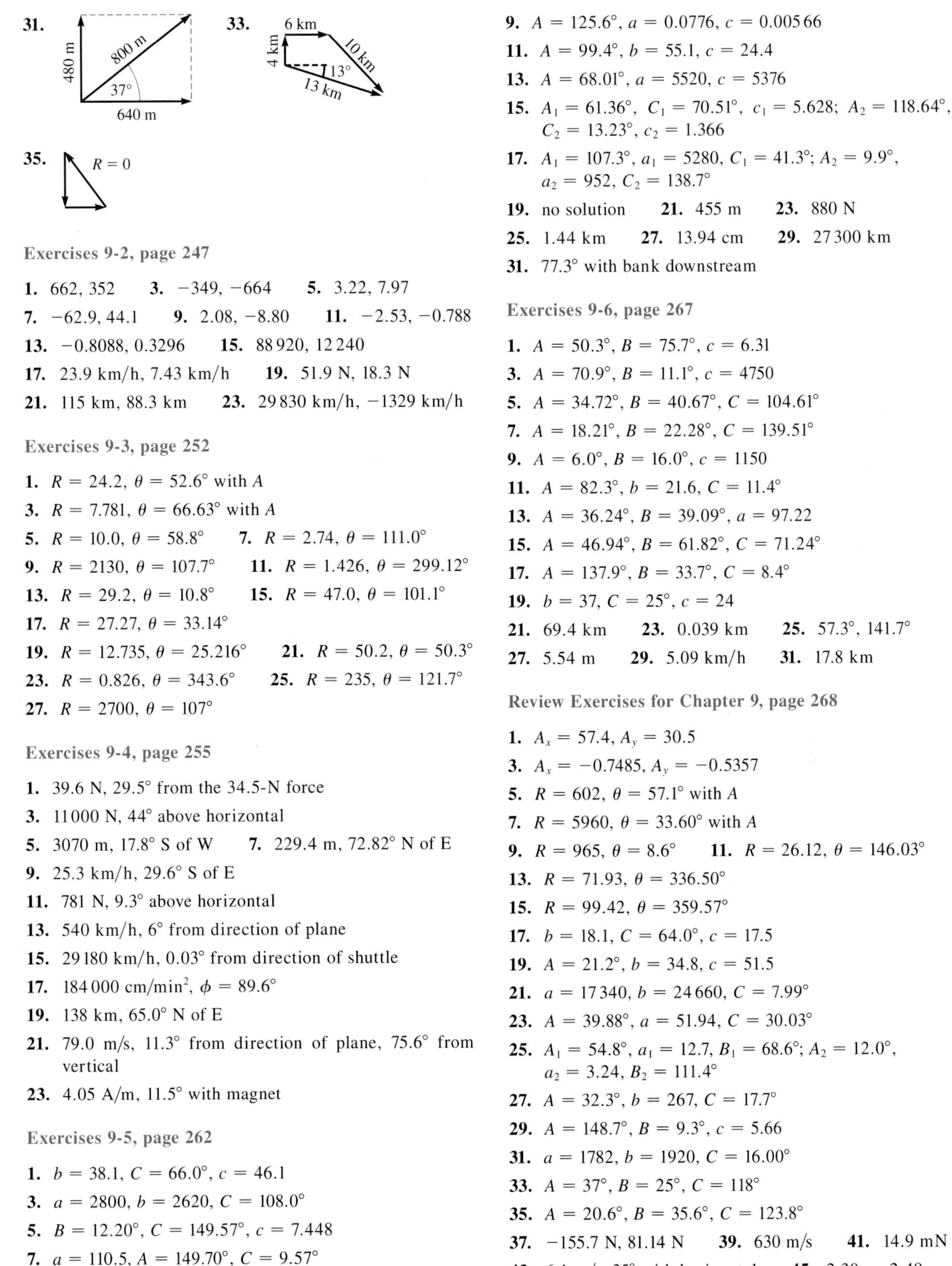

31.

33.

35.

Exercises 9-2, page 247

1. 662, 352 **3.** −349, −664 **5.** 3.22, 7.97

7. −62.9, 44.1 **9.** 2.08, −8.80 **11.** −2.53, −0.788

13. −0.8088, 0.3296 **15.** 88 920, 12 240

17. 23.9 km/h, 7.43 km/h **19.** 51.9 N, 18.3 N

21. 115 km, 88.3 km **23.** 29 830 km/h, −1329 km/h

Exercises 9-3, page 252

1. $R = 24.2$, $\theta = 52.6°$ with A

3. $R = 7.781$, $\theta = 66.63°$ with A

5. $R = 10.0$, $\theta = 58.8°$ **7.** $R = 2.74$, $\theta = 111.0°$

9. $R = 2130$, $\theta = 107.7°$ **11.** $R = 1.426$, $\theta = 299.12°$

13. $R = 29.2$, $\theta = 10.8°$ **15.** $R = 47.0$, $\theta = 101.1°$

17. $R = 27.27$, $\theta = 33.14°$

19. $R = 12.735$, $\theta = 25.216°$ **21.** $R = 50.2$, $\theta = 50.3°$

23. $R = 0.826$, $\theta = 343.6°$ **25.** $R = 235$, $\theta = 121.7°$

27. $R = 2700$, $\theta = 107°$

Exercises 9-4, page 255

1. 39.6 N, 29.5° from the 34.5-N force

3. 11 000 N, 44° above horizontal

5. 3070 m, 17.8° S of W **7.** 229.4 m, 72.82° N of E

9. 25.3 km/h, 29.6° S of E

11. 781 N, 9.3° above horizontal

13. 540 km/h, 6° from direction of plane

15. 29 180 km/h, 0.03° from direction of shuttle

17. 184 000 cm/min^2, $\phi = 89.6°$

19. 138 km, 65.0° N of E

21. 79.0 m/s, 11.3° from direction of plane, 75.6° from vertical

23. 4.05 A/m, 11.5° with magnet

Exercises 9-5, page 262

1. $b = 38.1$, $C = 66.0°$, $c = 46.1$

3. $a = 2800$, $b = 2620$, $C = 108.0°$

5. $B = 12.20°$, $C = 149.57°$, $c = 7.448$

7. $a = 110.5$, $A = 149.70°$, $C = 9.57°$

9. $A = 125.6°$, $a = 0.0776$, $c = 0.00566$

11. $A = 99.4°$, $b = 55.1$, $c = 24.4$

13. $A = 68.01°$, $a = 5520$, $c = 5376$

15. $A_1 = 61.36°$, $C_1 = 70.51°$, $c_1 = 5.628$; $A_2 = 118.64°$, $C_2 = 13.23°$, $c_2 = 1.366$

17. $A_1 = 107.3°$, $a_1 = 5280$, $C_1 = 41.3°$; $A_2 = 9.9°$, $a_2 = 952$, $C_2 = 138.7°$

19. no solution **21.** 455 m **23.** 880 N

25. 1.44 km **27.** 13.94 cm **29.** 27 300 km

31. 77.3° with bank downstream

Exercises 9-6, page 267

1. $A = 50.3°$, $B = 75.7°$, $c = 6.31$

3. $A = 70.9°$, $B = 11.1°$, $c = 4750$

5. $A = 34.72°$, $B = 40.67°$, $C = 104.61°$

7. $A = 18.21°$, $B = 22.28°$, $C = 139.51°$

9. $A = 6.0°$, $B = 16.0°$, $c = 1150$

11. $A = 82.3°$, $b = 21.6$, $C = 11.4°$

13. $A = 36.24°$, $B = 39.09°$, $a = 97.22$

15. $A = 46.94°$, $B = 61.82°$, $C = 71.24°$

17. $A = 137.9°$, $B = 33.7°$, $C = 8.4°$

19. $b = 37$, $C = 25°$, $c = 24$

21. 69.4 km **23.** 0.039 km **25.** 57.3°, 141.7°

27. 5.54 m **29.** 5.09 km/h **31.** 17.8 km

Review Exercises for Chapter 9, page 268

1. $A_x = 57.4$, $A_y = 30.5$

3. $A_x = -0.7485$, $A_y = -0.5357$

5. $R = 602$, $\theta = 57.1°$ with A

7. $R = 5960$, $\theta = 33.60°$ with A

9. $R = 965$, $\theta = 8.6°$ **11.** $R = 26.12$, $\theta = 146.03°$

13. $R = 71.93$, $\theta = 336.50°$

15. $R = 99.42$, $\theta = 359.57°$

17. $b = 18.1$, $C = 64.0°$, $c = 17.5$

19. $A = 21.2°$, $b = 34.8$, $c = 51.5$

21. $a = 17340$, $b = 24660$, $C = 7.99°$

23. $A = 39.88°$, $a = 51.94$, $C = 30.03°$

25. $A_1 = 54.8°$, $a_1 = 12.7$, $B_1 = 68.6°$; $A_2 = 12.0°$, $a_2 = 3.24$, $B_2 = 111.4°$

27. $A = 32.3°$, $b = 267$, $C = 17.7°$

29. $A = 148.7°$, $B = 9.3°$, $c = 5.66$

31. $a = 1782$, $b = 1920$, $C = 16.00°$

33. $A = 37°$, $B = 25°$, $C = 118°$

35. $A = 20.6°$, $B = 35.6°$, $C = 123.8°$

37. −155.7 N, 81.14 N **39.** 630 m/s **41.** 14.9 mN

43. 6.1 m/s, 35° with horizontal **45.** 2.30 m, 2.49 m

47. 52 700 km **49.** 2.65 km **51.** 299 km

53. 810 N, 36° N of E **55.** 1280 m or 1680 m (ambiguous)

Exercises 10-1, page 274

1. 0, −0.7, −1, −0.7, 0, 0.7, 1, 0.7, 0, −0.7, −1, −0.7, 0, 0.7, 1, 0.7, 0

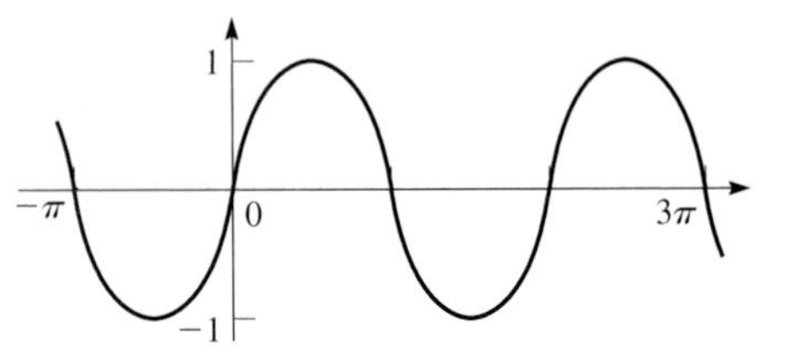

3. −3, −2.1, 0, 2.1, 3, 2.1, 0, −2.1, −3, −2.1, 0, 2.1, 3, 2.1, 0, −2.1, −3

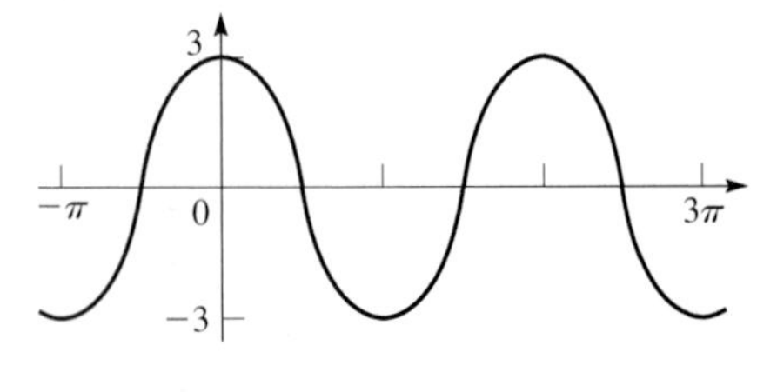

5.

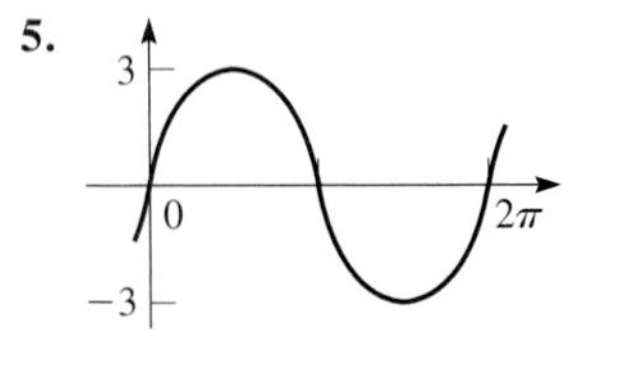

7.

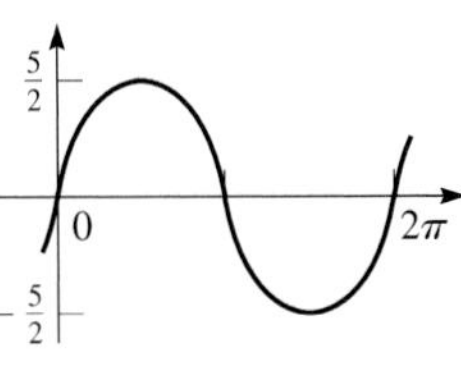

9.

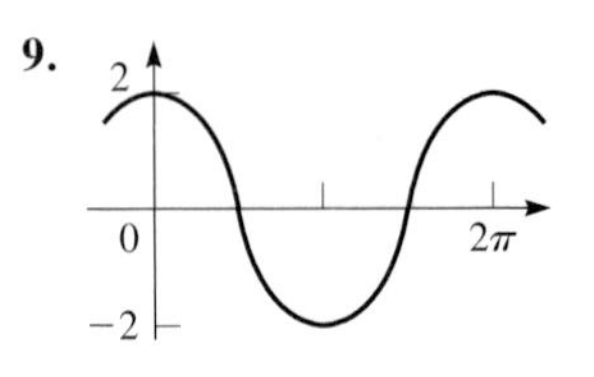

11.

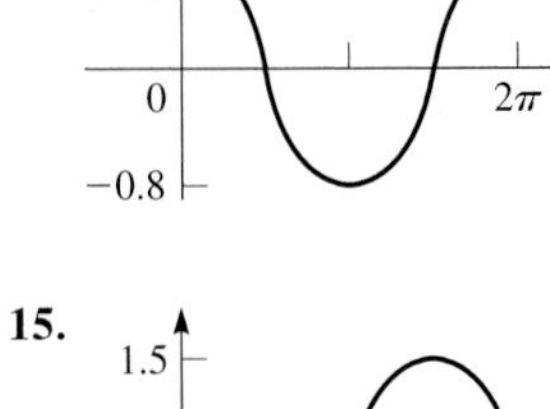

13.

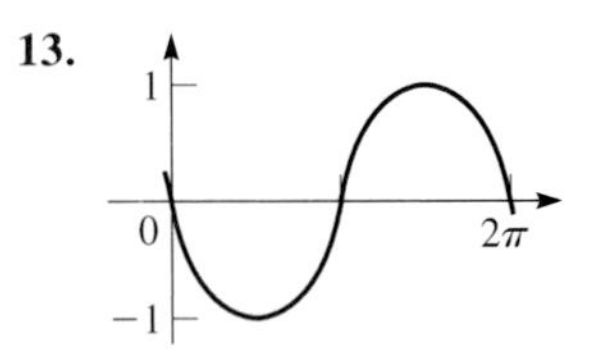

15.

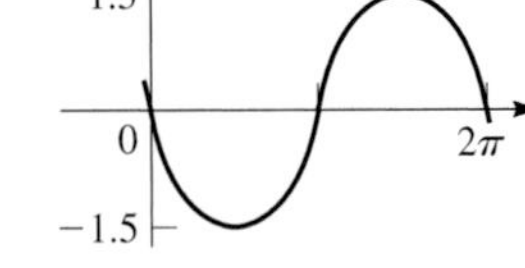

17.

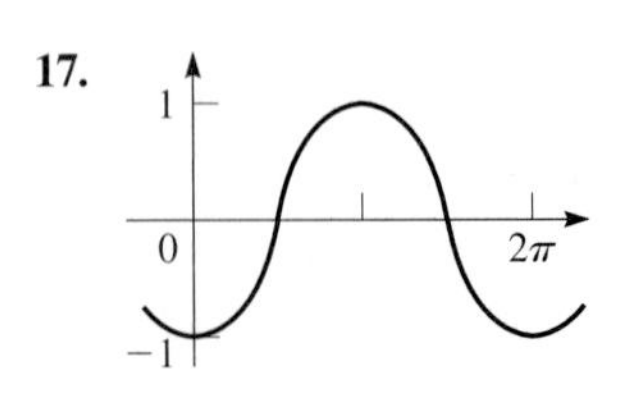

19.

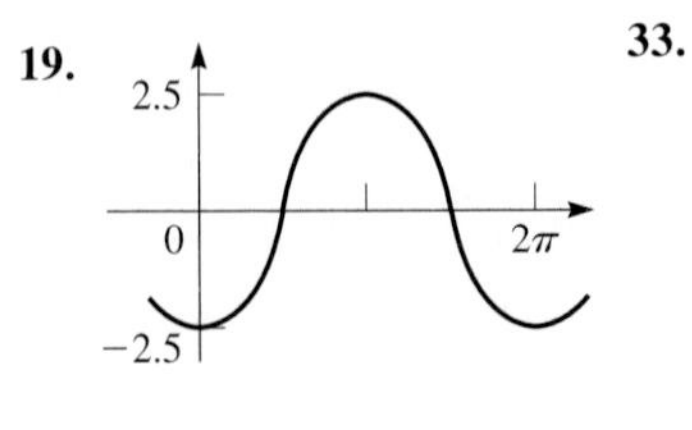

21. 0, 0.84, 0.91, 0.14, −0.76, −0.96, −0.28, 0.66

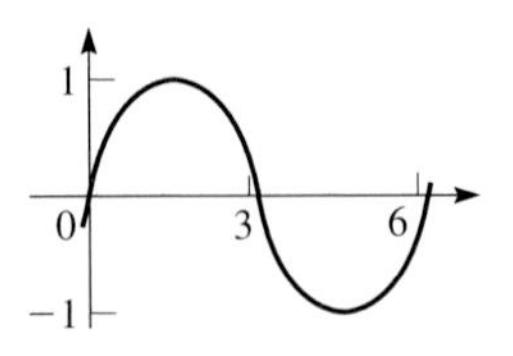

23. 1, 0.54, −0.42, −0.99, −0.65, 0.28, 0.96, 0.75

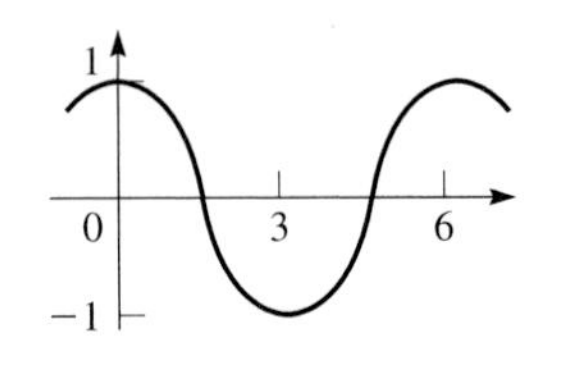

25. $y = 4 \sin x$ **27.** $y = -1.5 \cos x$

Exercises 10-2, page 277

1. $\frac{\pi}{3}$ **3.** $\frac{\pi}{4}$ **5.** $\frac{\pi}{6}$ **7.** $\frac{\pi}{8}$ **9.** 1

11. $\frac{1}{2}$ **13.** 6π **15.** 3π **17.** 3 **19.** $\frac{2}{\pi}$

21. **23.** **25.** **27.**

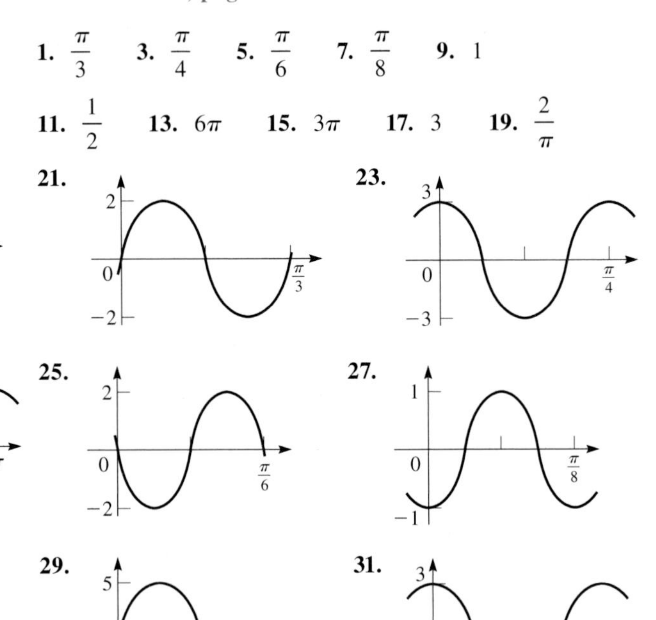

29.

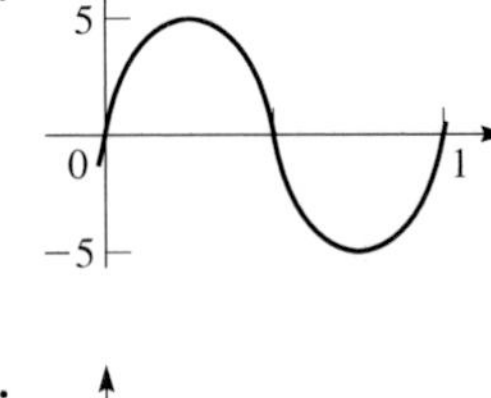

31.

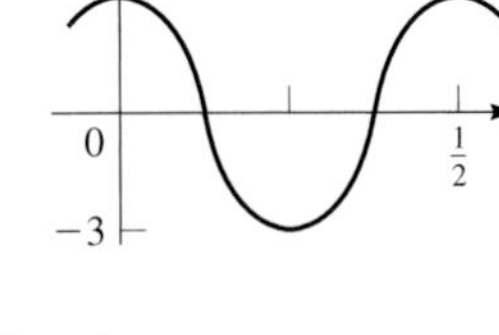

33.

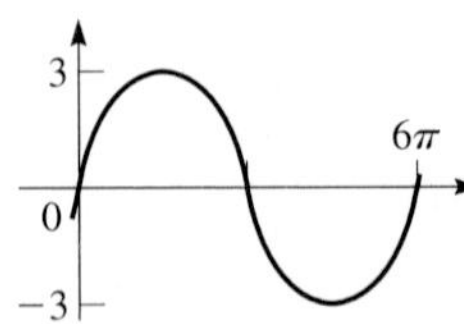

35.

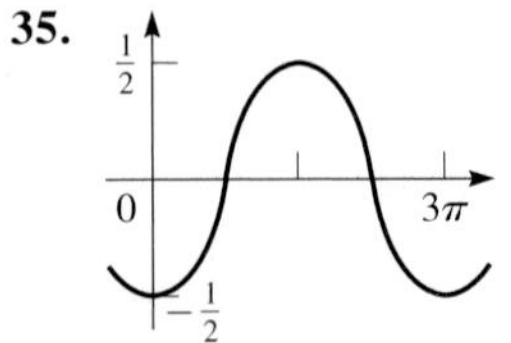

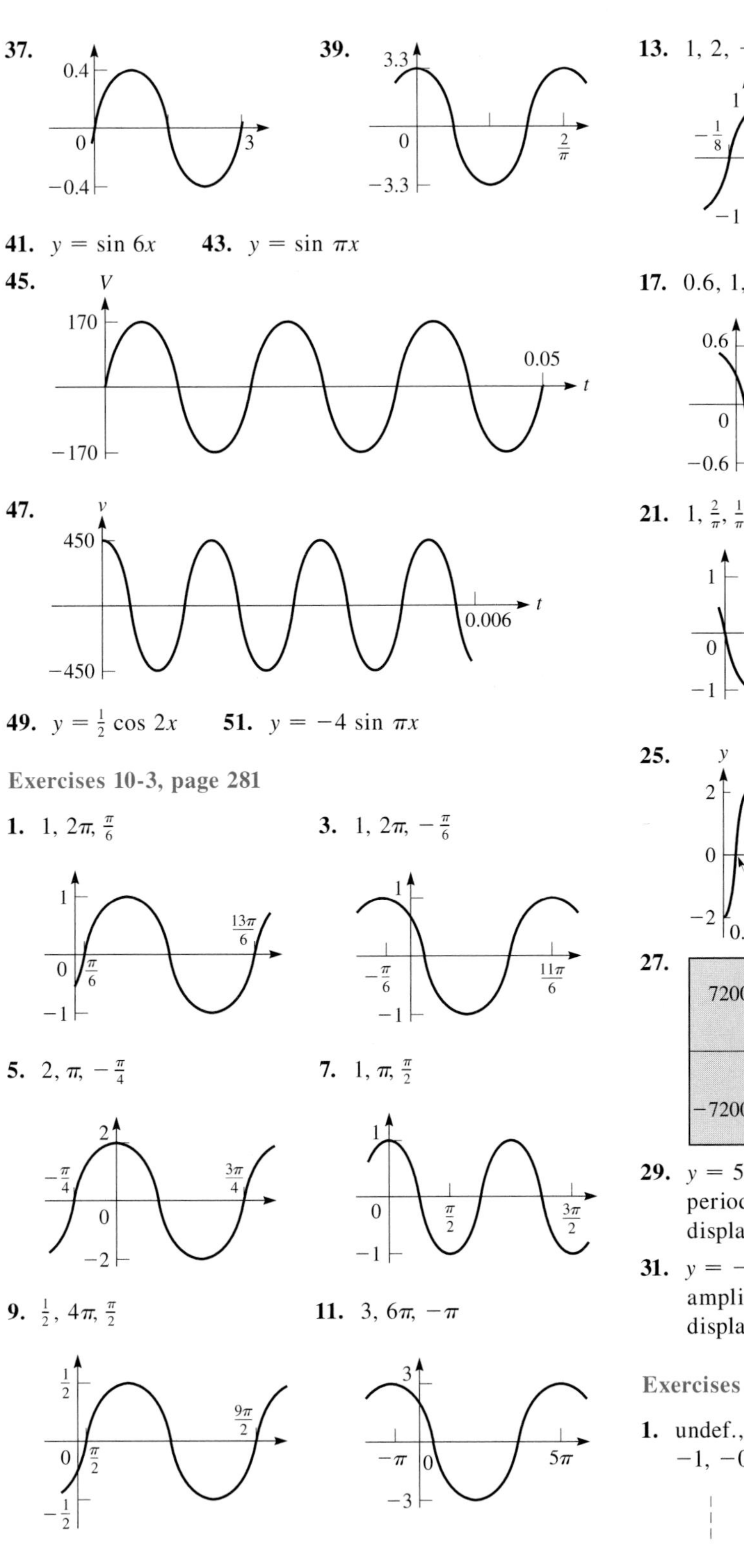

37.

39.

41. $y = \sin 6x$ **43.** $y = \sin \pi x$

45.

47.

49. $y = \frac{1}{2} \cos 2x$ **51.** $y = -4 \sin \pi x$

Exercises 10-3, page 281

1. $1, 2\pi, \frac{\pi}{6}$

3. $1, 2\pi, -\frac{\pi}{6}$

5. $2, \pi, -\frac{\pi}{4}$

7. $1, \pi, \frac{\pi}{2}$

9. $\frac{1}{2}, 4\pi, \frac{\pi}{2}$

11. $3, 6\pi, -\pi$

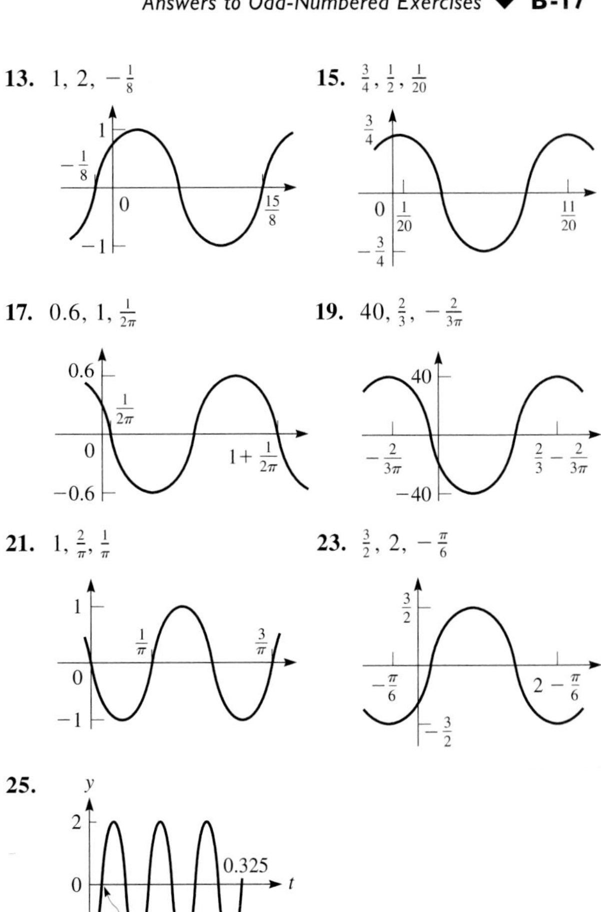

13. $1, 2, -\frac{1}{8}$

15. $\frac{3}{4}, \frac{1}{2}, \frac{1}{20}$

17. $0.6, 1, \frac{1}{2\pi}$

19. $40, \frac{2}{3}, -\frac{2}{3\pi}$

21. $1, \frac{2}{\pi}, \frac{1}{\pi}$

23. $\frac{3}{2}, 2, -\frac{\pi}{6}$

25.

27.

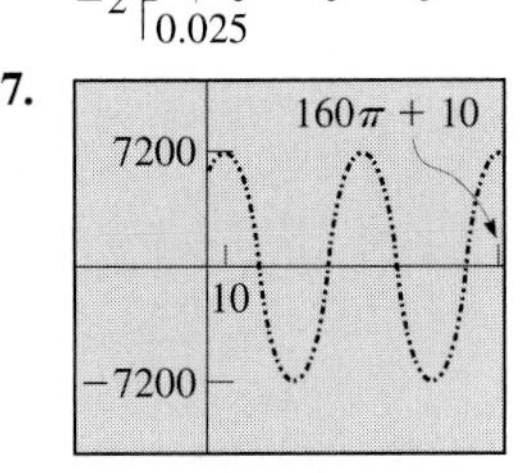

29. $y = 5 \sin(\frac{\pi}{8}x + \frac{\pi}{8})$. As shown: amplitude = 5; period $= \frac{2\pi}{b} = 16$, $b = \frac{\pi}{8}$; displacement $= -\frac{c}{b} = -1$, $c = \frac{\pi}{8}$

31. $y = -0.8 \cos 2x$. As shown: amplitude $= |-0.8| = 0.8$; period $= \frac{2\pi}{b} = \pi$, $b = 2$; displacement $= -\frac{c}{b} = 0$, $c = 0$

Exercises 10-4, page 285

1. undef., −1.7, −1, −0.58, 0, 0.58, 1, 1.7, undef., −1.7, −1, −0.58, 0

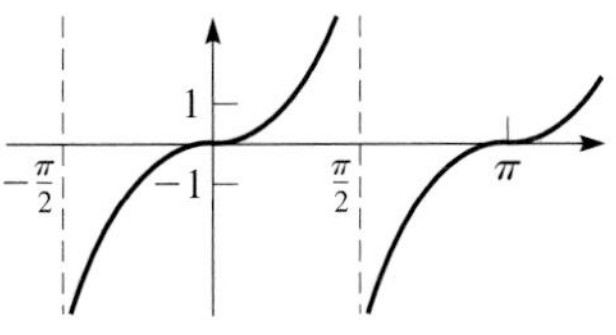

3. undef., 2, 1.4, 1.2, 1, 1.2, 1.4, 2, undef., −2, −1.4, −1.2, −1

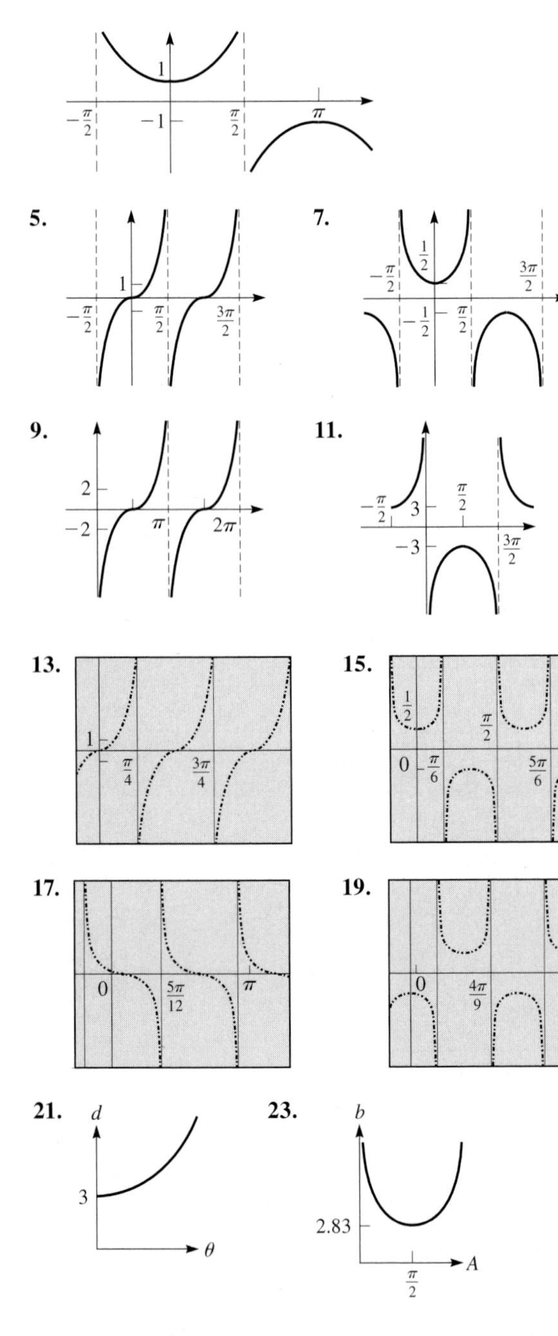

Exercises 10-5, page 288

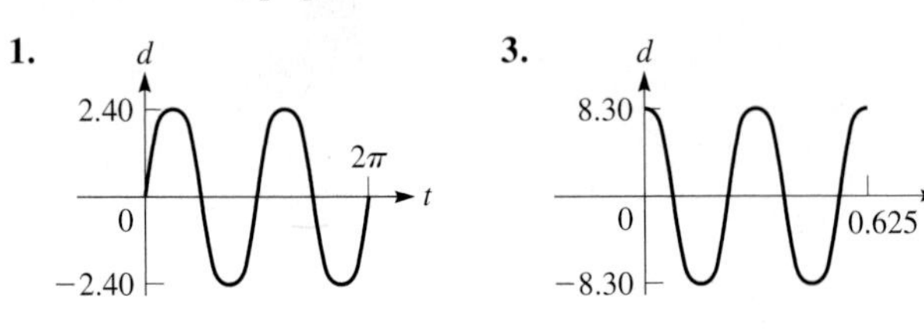

5. 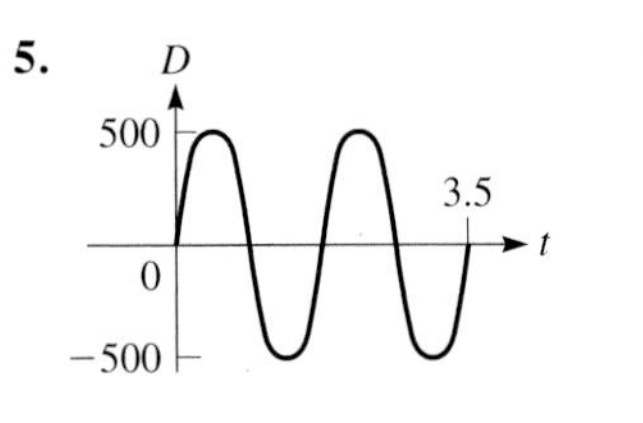

7. 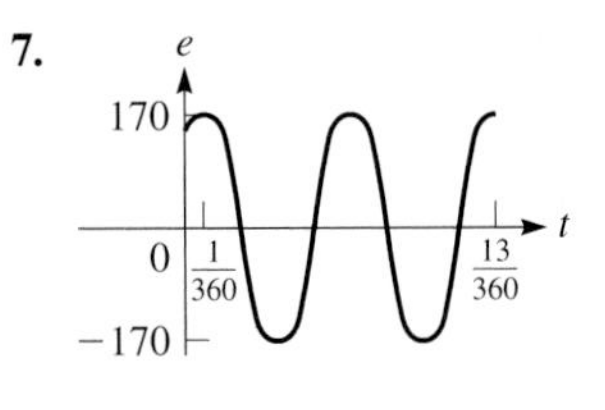

9. 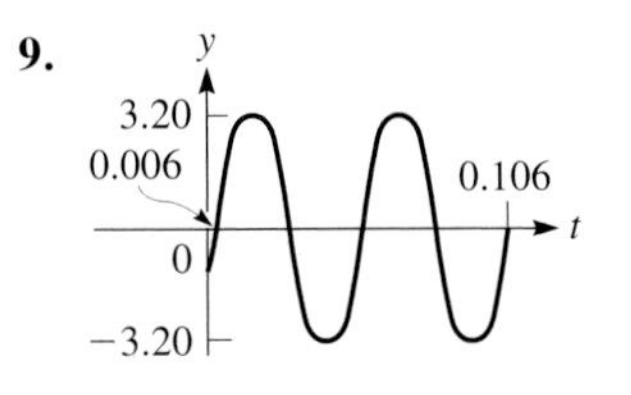

11. 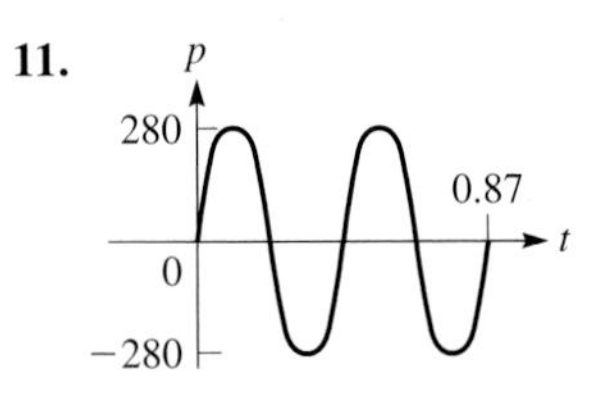

13. 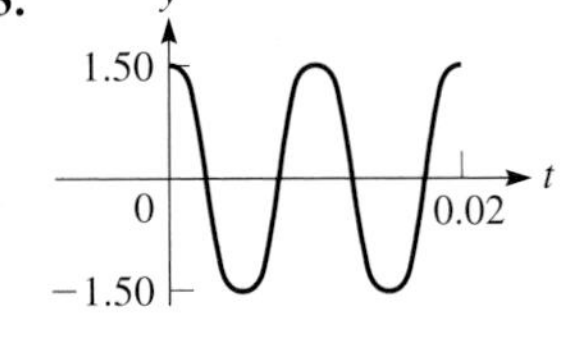

15. 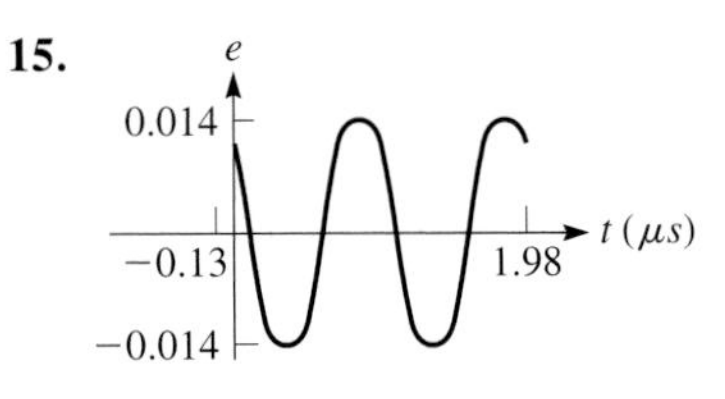

Exercises 10-6, page 292

1. 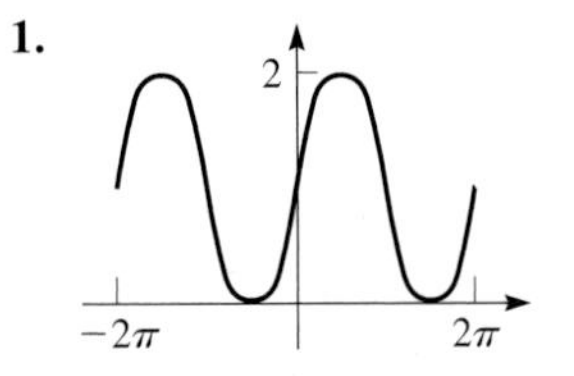

3. 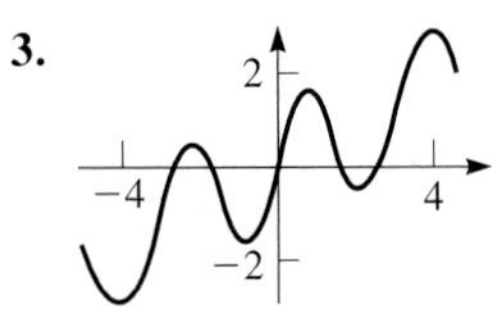

5. 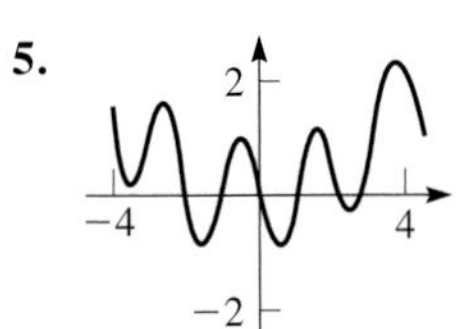

7. 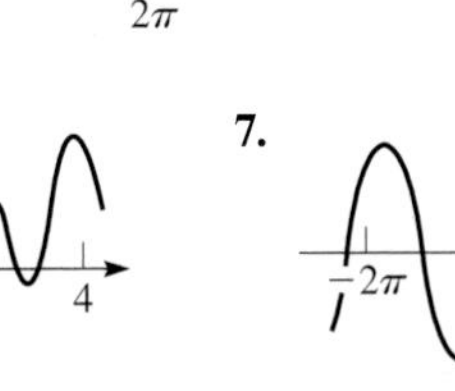

9. 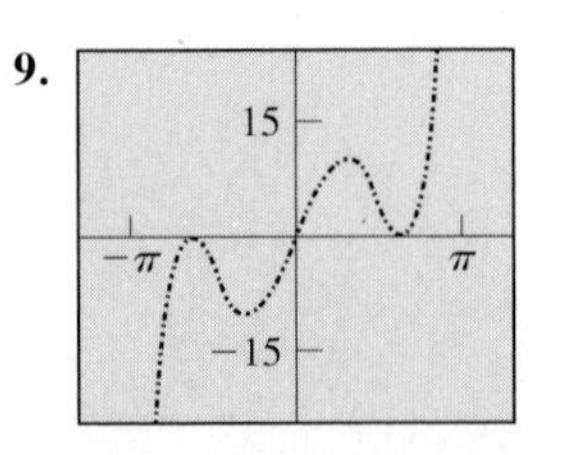

11. 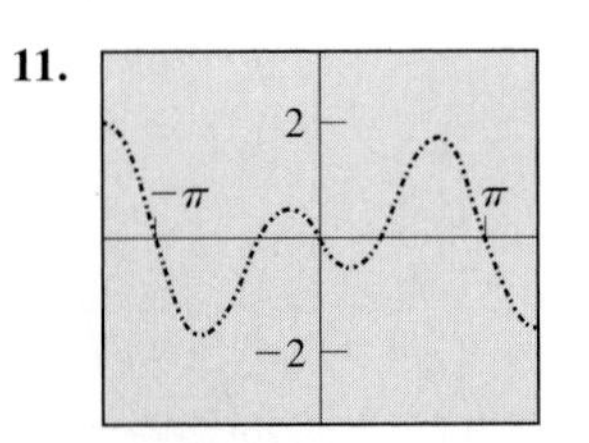

13. 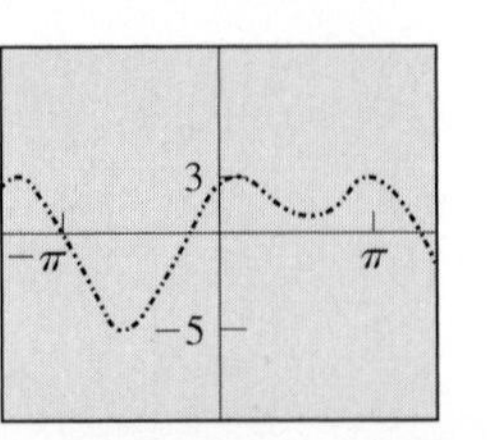

15. 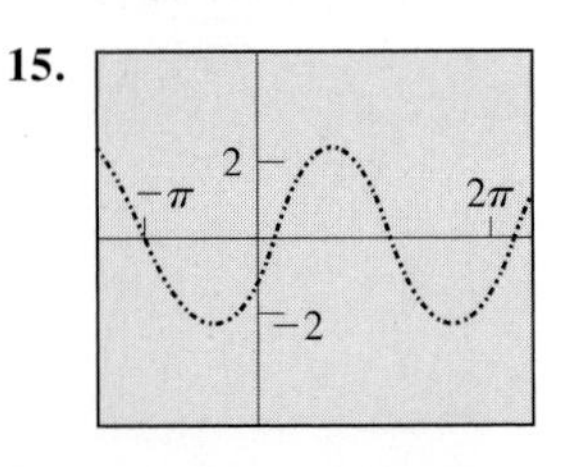

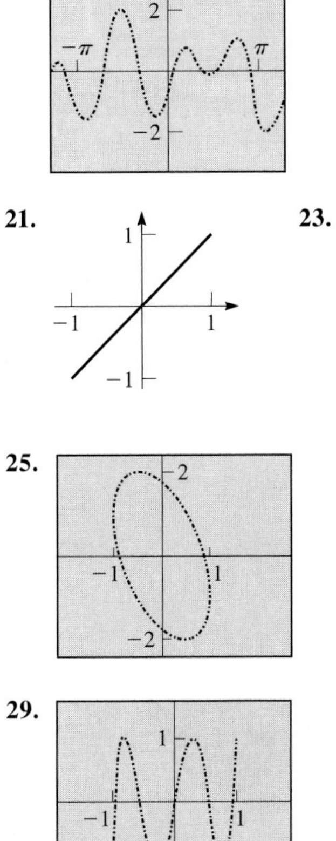

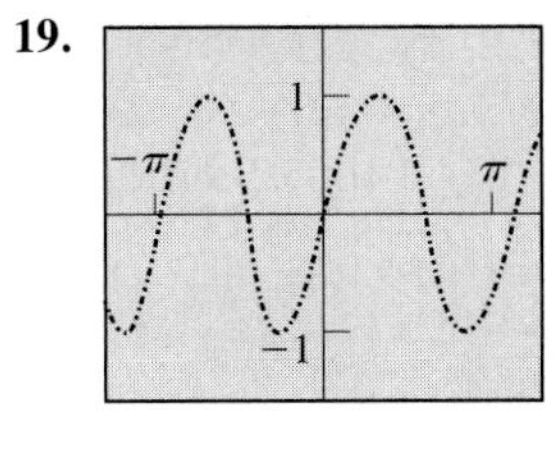

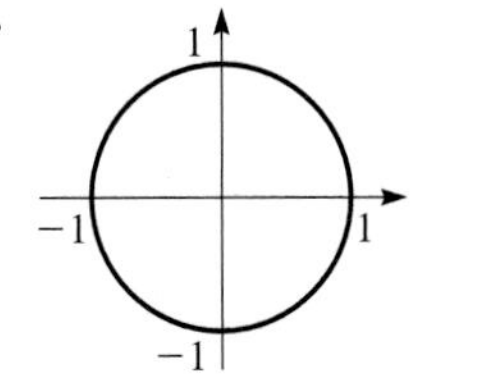

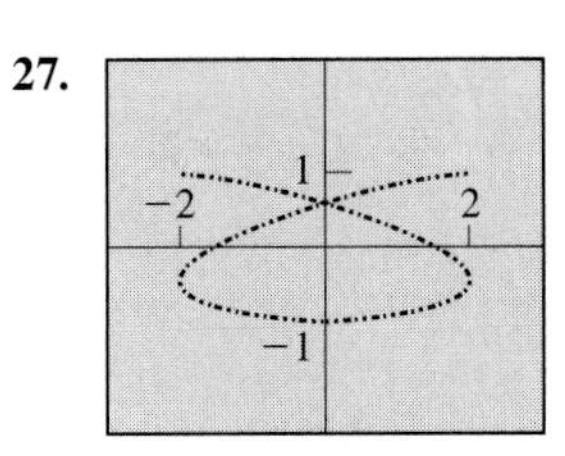

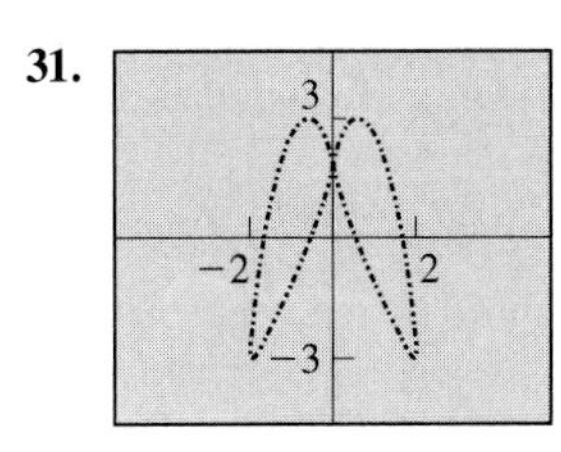

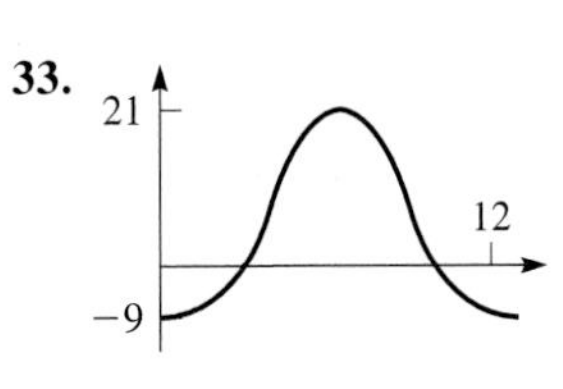

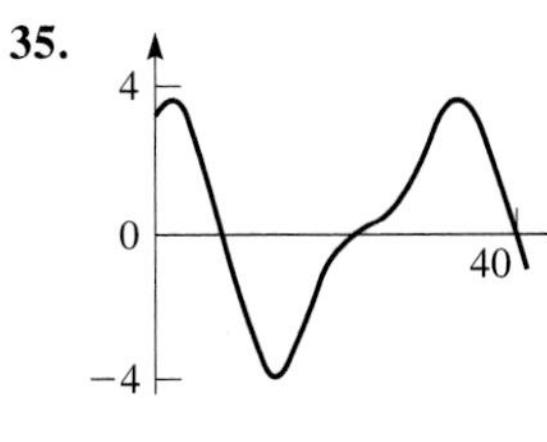

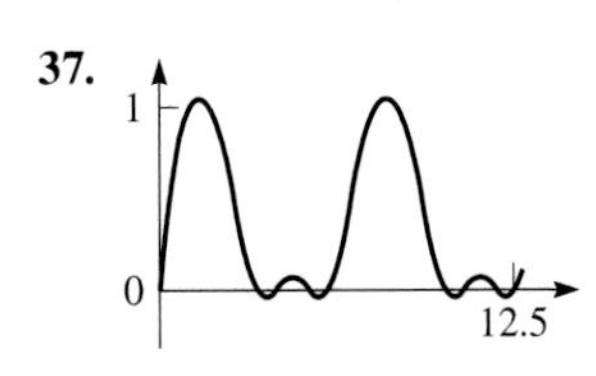

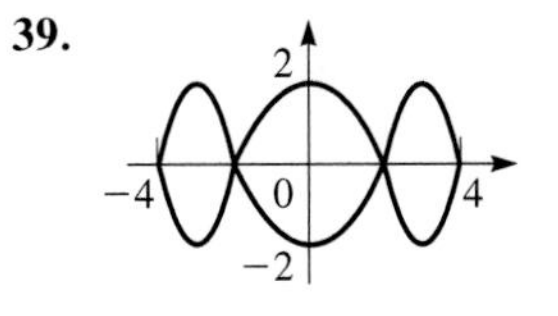

Review Exercises for Chapter 10, page 294

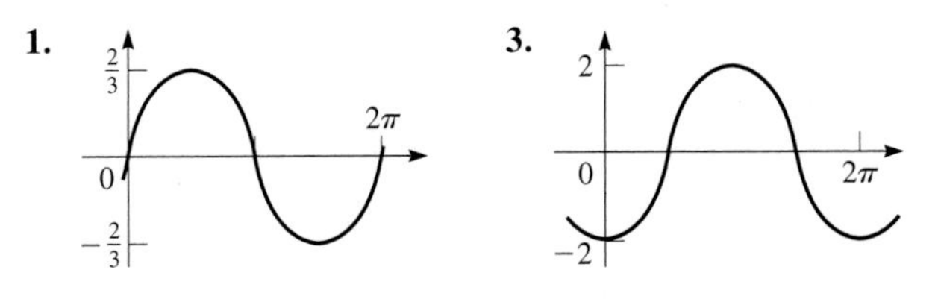

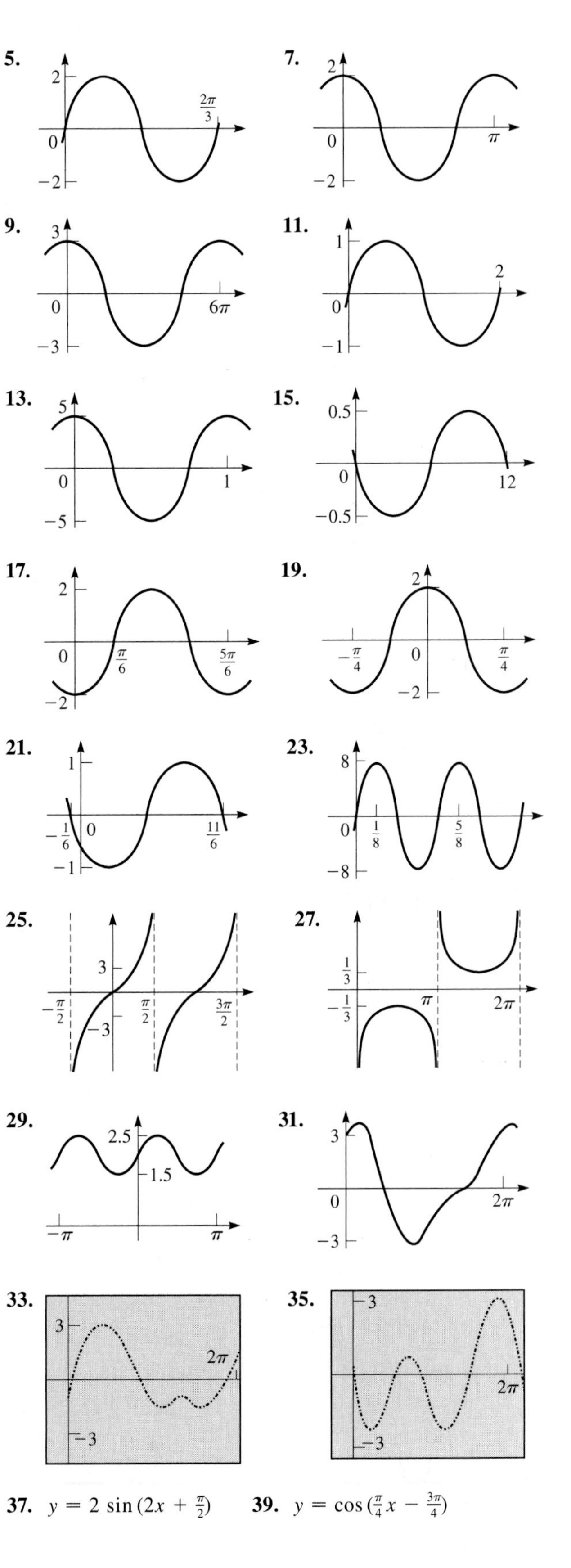

37. $y = 2\sin\left(2x + \frac{\pi}{2}\right)$ **39.** $y = \cos\left(\frac{\pi}{4}x - \frac{3\pi}{4}\right)$

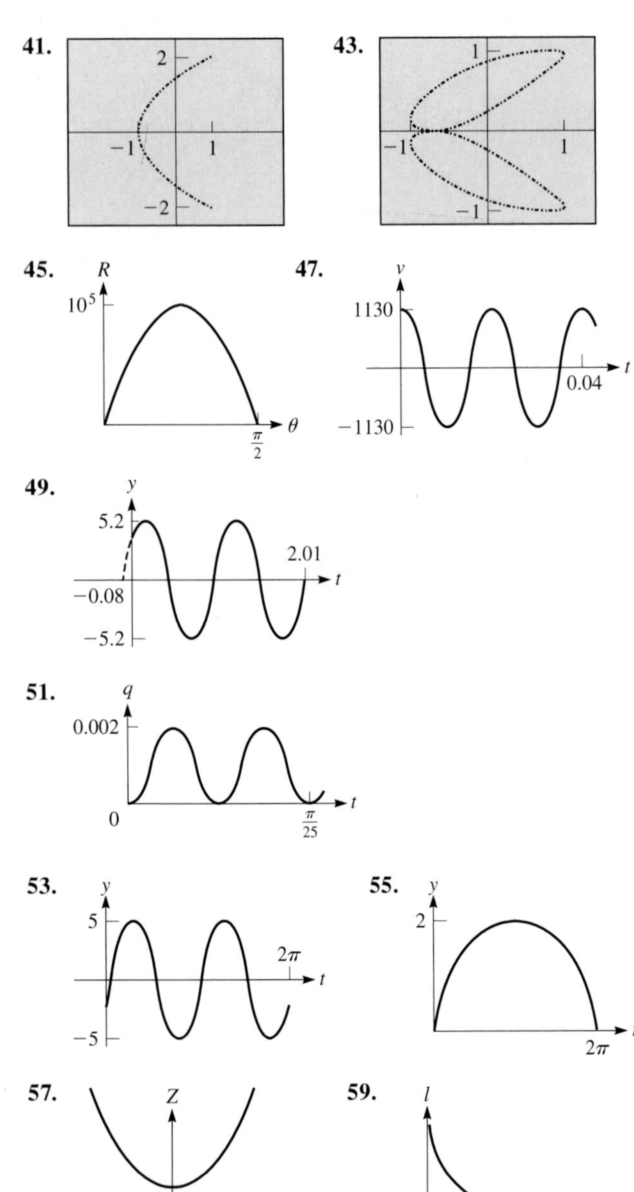

Exercises 11-1, page 300

1. x^3 **3.** $\frac{1}{a^4}$ **5.** $\frac{1}{25}$ **7.** $\frac{25}{4}$ **9.** $\frac{4a^2}{x^2}$

11. $\frac{n^2}{5a}$ **13.** 1 **15.** −7 **17.** $\frac{3}{x^2}$

19. $\frac{1}{7^3a^3x^3}$ **21.** $\frac{n^3}{2}$ **23.** $\frac{1}{a^3b^6}$ **25.** $\frac{1}{a+b}$

27. $\frac{2x^2+3y^2}{x^2y^2}$ **29.** $\frac{2^3}{3^5}$ **31.** $\frac{a^2b^2}{3}$ **33.** $\frac{b^3}{432a}$

35. $\frac{4}{t^4v^4}$ **37.** $\frac{x^8-y^2}{x^4y^2}$ **39.** $\frac{2a^6+16}{a^8}$ **41.** $\frac{10}{9}$

43. $\frac{ab}{a+b}$ **45.** $\frac{4n^2-4n+1}{n^4}$ **47.** $\frac{45}{2}$ **49.** $-\frac{x}{y}$

51. $\frac{a^2-ax+x^2}{ax}$ **53.** $\frac{t^2+t+2}{t^2}$

55. $\frac{2x}{(x+1)(x-1)}$ **57.** (a) 4^5 (b) 2^{10}

59. $\left(\frac{a}{b}\right)^{-n} = \frac{1}{\left(\frac{a}{b}\right)^n} = \frac{1}{\frac{a^n}{b^n}} = \frac{b^n}{a^n} = \left(\frac{b}{a}\right)^n$

61. J/s^3 **63.** $\frac{\omega^4-2\omega^2\omega_0^2+\omega_0^4}{\omega^2\omega_0^2}$

Exercises 11-2, page 304

1. 5 **3.** 3 **5.** 16 **7.** 10^{25} **9.** $\frac{1}{2}$ **11.** $\frac{1}{16}$

13. 25 **15.** 4096 **17.** $\frac{1}{110}$ **19.** $\frac{6}{7}$ **21.** $-\frac{1}{2}$

23. 24 **25.** $\frac{39}{1000}$ **27.** $\frac{3}{5}$ **29.** 2.059

31. 0.538 91 **33.** $a^{7/6}$ **35.** $\frac{1}{y^{9/10}}$ **37.** $s^{23/12}$

39. $\frac{1}{y^{13/12}}$ **41.** $2ab^2$ **43.** $\frac{1}{8a^3b^{9/4}}$

45. $\frac{4x}{(4x^2+1)^{1/2}}$ **47.** $\frac{27}{64t^3}$ **49.** $\frac{b^{11/10}}{2a^{1/12}}$

51. $\frac{2}{3}x^{1/6}y^{11/12}$ **53.** $\frac{x}{(x+2)^{1/2}}$ **55.** $\frac{a^2+1}{a^4}$

57. $\frac{a+1}{a^{1/2}}$ **59.** $\frac{5x^2-2x}{(2x-1)^{1/2}}$

61. 6, 4 **63.** 3, −4, 4

65. If $(A/S)^{-1/4} = 0.5 = 1/2$, then $(A/S)^{1/4} = 2$. Raise each to the fourth power and get $A/S = 16$. **67.** 1.91 mA

Exercises 11-3, page 308

1. $2\sqrt{6}$ **3.** $3\sqrt{5}$ **5.** $xy^2\sqrt{y}$ **7.** $qr^3\sqrt{pr}$

9. $x\sqrt{5}$ **11.** $3ac^2\sqrt{2ab}$ **13.** $2\sqrt[3]{2}$ **15.** $2\sqrt[5]{3}$

17. $2\sqrt[3]{a^2}$ **19.** $2st\sqrt[4]{4r^3t}$ **21.** 2 **23.** $ab\sqrt[3]{b^2}$

25. $\frac{1}{2}\sqrt{6}$ **27.** $\frac{\sqrt{ab}}{b}$ **29.** $\frac{1}{2}\sqrt[3]{6}$ **31.** $\frac{1}{3}\sqrt[5]{27}$

33. $2\sqrt{5}$ **35.** 2 **37.** 200 **39.** 2000

41. $\sqrt{2a}$ 43. $\frac{1}{2}\sqrt{2}$ 45. $\sqrt[3]{2}$ 47. $\sqrt[8]{2}$

49. $\frac{1}{6}\sqrt{6}$ 51. $\frac{\sqrt{b(a^2+b)}}{ab}$ 53. $\frac{\sqrt{2x^2+x}}{2x+1}$

55. $a+b$ 57. $\sqrt{4x^2-1}$ 59. $\frac{1}{2}\sqrt{4x^2+1}$

61. Write $1/\sqrt[5]{R^2}$ as $R^{-2/5}$. $E = 100(1 - R^{-2/5}) = 55.0$.

63. $\frac{24\sqrt{EIgWL}}{WL^2}$

Exercises 11-4, page 310

1. $7\sqrt{3}$ 3. $\sqrt{5}-\sqrt{7}$ 5. $3\sqrt{5}$ 7. $-4\sqrt{3}$

9. $-2\sqrt{2a}$ 11. $19\sqrt{7}$ 13. $-4\sqrt{5}$

15. $23\sqrt{3}-6\sqrt{2}$ 17. $\frac{7}{3}\sqrt{15}$ 19. 0

21. $13\sqrt[3]{3}$ 23. $\sqrt[4]{2}$ 25. $(a-2b^2)\sqrt{ab}$

27. $(3-2a)\sqrt{10}$ 29. $(2b-a)\sqrt[3]{3a^2b}$

31. $\frac{(a^2-c^3)\sqrt{ac}}{a^2c^3}$ 33. $\frac{(a-2b)\sqrt[3]{ab^2}}{ab}$

35. $\frac{2b\sqrt{a^2-b^2}}{b^2-a^2}$ 37. $15\sqrt{3}-11\sqrt{5} = 1.384\,014\,4$

39. $\frac{1}{6}\sqrt{6} = 0.408\,248\,3$ 41. $3\sqrt{3}$

43. $5400 + 900\sqrt{2} = 6670$ mm

Exercises 11-5, page 314

1. $\sqrt{30}$ 3. $2\sqrt{3}$ 5. 2 7. 50 9. $2\sqrt{5}$

11. $\sqrt{6}-\sqrt{15}$ 13. -1 15. $48+9\sqrt{15}$

17. $66+13\sqrt{11x}-5x$ 19. $a\sqrt{b}+c\sqrt{ac}$

21. $\frac{2-\sqrt{6}}{2}$ 23. $\frac{a\sqrt{2}-b\sqrt{a}}{a}$

25. $2a-3b+2\sqrt{2ab}$ 27. $\sqrt[6]{72}$

29. $\frac{1}{4}(\sqrt{7}-\sqrt{3})$ 31. $-\frac{3}{8}(\sqrt{5}+3)$

33. $\frac{1}{11}(\sqrt{7}+3\sqrt{2}-6-\sqrt{14})$

35. $\frac{1}{17}(-56+9\sqrt{15})$ 37. $\frac{2x+2\sqrt{xy}}{x-y}$

39. $\frac{8(3\sqrt{a}+2\sqrt{b})}{9a-4b}$ 41. 1 43. $\frac{2-a-a^2}{a}$

45. $\frac{2c+4d\sqrt{2c}+3d^2}{2c-d^2}$

47. $-\frac{\sqrt{x^2-y^2}+\sqrt{x^2+xy}}{y}$

49. $-1-\sqrt{66} = -9.124\,038\,4$

51. $-\frac{16+5\sqrt{30}}{26} = -1.668\,697\,2$

53. $\frac{2x+1}{\sqrt{x}}$ 55. $\frac{5x^2+2x}{\sqrt{2x+1}}$ 57. $\frac{1}{\sqrt{30}-2\sqrt{3}}$

59. $\frac{1}{\sqrt{x+h}+\sqrt{x}}$

61. $(1-\sqrt{2})^2 - 2(1-\sqrt{2}) - 1 = 1 - 2\sqrt{2} + 2 - 2 + 2\sqrt{2} - 1 = 0$

63. $\left[\frac{1}{2}(\sqrt{b^2-4k^2}-b)\right]^2 + b\left[\frac{1}{2}(\sqrt{b^2-4k^2}-b)\right] + k^2 = \frac{1}{4}(b^2-4k^2) - \frac{b}{2}\sqrt{b^2-4k^2} + \frac{1}{4}b^2 + \frac{b}{2}\sqrt{b^2-4k^2} - \frac{1}{2}b^2 + k^2 = 0$

65. $\frac{\sqrt{3gsw}}{6w}$ 67. $\frac{\sqrt{2g}(\sqrt{h_2}+\sqrt{h_1})}{2g}$

Review Exercises for Chapter 11, page 316

1. $\frac{2}{a^2}$ 3. $\frac{2d^3}{c}$ 5. 375 7. $\frac{1}{8000}$ 9. $\frac{t^4}{9}$

11. -28 13. $64a^2b^5$ 15. $-8m^9n^6$ 17. $\frac{2y-x^2}{x^2y}$

19. $\frac{2y}{x+2y}$ 21. $\frac{b}{ab-3}$ 23. $\frac{(x^3y^3-1)^{1/3}}{y}$

25. $4a(a^2+4)^{1/2}$ 27. $\frac{-2(x+1)}{(x-1)^3}$ 29. $2\sqrt{17}$

31. $b^2c\sqrt{ab}$ 33. $3ab^2\sqrt{a}$ 35. $2tu\sqrt{21st}$

37. $\frac{5\sqrt{2s}}{2s}$ 39. $\frac{1}{9}\sqrt{33}$ 41. $mn^2\sqrt[4]{8m^2n}$

43. $\sqrt{2}$ 45. $14\sqrt{2}$ 47. $-7\sqrt{7}$ 49. $3ax\sqrt{2x}$

51. $(2a+b)\sqrt[4]{a}$ 53. $10-\sqrt{55}$ 55. $4\sqrt{3}-4\sqrt{5}$

57. $-45-7\sqrt{17}$ 59. $42-7\sqrt{7a}-3a$

61. $\frac{6x+\sqrt{3xy}}{12x-y}$ 63. $-\frac{8+\sqrt{6}}{29}$ 65. $\frac{13-2\sqrt{35}}{29}$

67. $\frac{6x-13a\sqrt{x}+5a^2}{9x-25a^2}$ 69. $\sqrt{4b^2+1}$

71. $\frac{15-2\sqrt{15}}{4}$ 73. $2\sqrt{13}+5\sqrt{6} = 19.458\,551$

75. $51-7\sqrt{105} = -20.728\,655$ 77. 3.797%

79. (a) $v = k(P/W)^{1/3}$ (b) $v = \frac{k\sqrt[3]{PW^2}}{W}$

81. $\frac{n_1^2n_2^2v}{n_1^2-n_2^2}$ 83. $\frac{1}{\sqrt{A+h}+\sqrt{A}}$

85. $6\sqrt{2}$ cm 87. $\frac{\sqrt{LC_1C_2(C_1+C_2)}}{2\pi LC_1C_2}$

Exercises 12-1, page 322

1. $9j$ **3.** $-2j$ **5.** $0.6j$ **7.** $2j\sqrt{2}$ **9.** $\frac{1}{2}j\sqrt{7}$

11. $-j\sqrt{\frac{2}{5}} = -\frac{1}{5}j\sqrt{10}$ **13.** (a) -7 (b) 7

15. (a) 4 (b) -4 **17.** $-j$ **19.** 1 **21.** 0

23. $-2j$ **25.** $2 + 3j$ **27.** $-7j$ **29.** $6 + 2j$

31. $-2 + 3j$ **33.** $3\sqrt{2} - 2j\sqrt{2}$ **35.** -1

37. (a) $6 + 7j$ (b) $8 - j$ **39.** (a) $-2j$ (b) -4

41. $x = 2, y = -2$ **43.** $x = 10, y = -6$

45. $x = 0, y = -1$ **47.** $x = -2, y = 3$

49. yes **51.** no **53.** 0 **55.** It is a real number.

Exercises 12-2, page 325

1. $5 - 8j$ **3.** $-9 + 6j$ **5.** $7 - 5j$

7. $-0.1 - 6.3j$ **9.** -1 **11.** $-36 + 21j$

13. $7 + 49j$ **15.** $-22.4 + 6.4j$ **17.** $22 + 3j$

19. $-900 - 700j$ **21.** $-18j\sqrt{2}$ **23.** $25j\sqrt{5}$

25. $3j\sqrt{3}$ **27.** $-4\sqrt{3} + 6j\sqrt{7}$ **29.** $-28j$

31. $3\sqrt{7} + 3j$ **33.** $-40 - 42j$ **35.** $-2 - 2j$

37. $\frac{1}{29}(-30 + 12j)$ **39.** $0.075 + 0.025j$

41. $-\frac{1}{3}(1 + j)$ **43.** $\frac{1}{11}(-13 + 8j\sqrt{2})$

45. $\frac{1}{3}(2 + 5j\sqrt{2})$ **47.** $\frac{1}{5}(-1 + 3j)$

49. $1 - 2j$ **51.** $\frac{1}{2701}(449 + 141j)$

53. $(-1 + j)^2 + 2(-1 + j) + 2 = 1 - 2j + j^2 - 2 + 2j + 2 = 0$

55. $(1 - j\sqrt{3})^2 + 4 = 1 - 2j\sqrt{3} + 3j^2 + 4 = 2 - 2j\sqrt{3} = 2(1 - j\sqrt{3})$

57. 13 **59.** $-\frac{1}{13}(5 + 12j)$ **61.** $281 + 35.2j$ volts

63. $0.016 + 0.037j$ amperes

65. $(a + bj) + (a - bj) = 2a$

67. $(a + bj) - (a - bj) = 2bj$

Exercises 12-3, page 327

1. **3.** **5.** $5 + 4j$

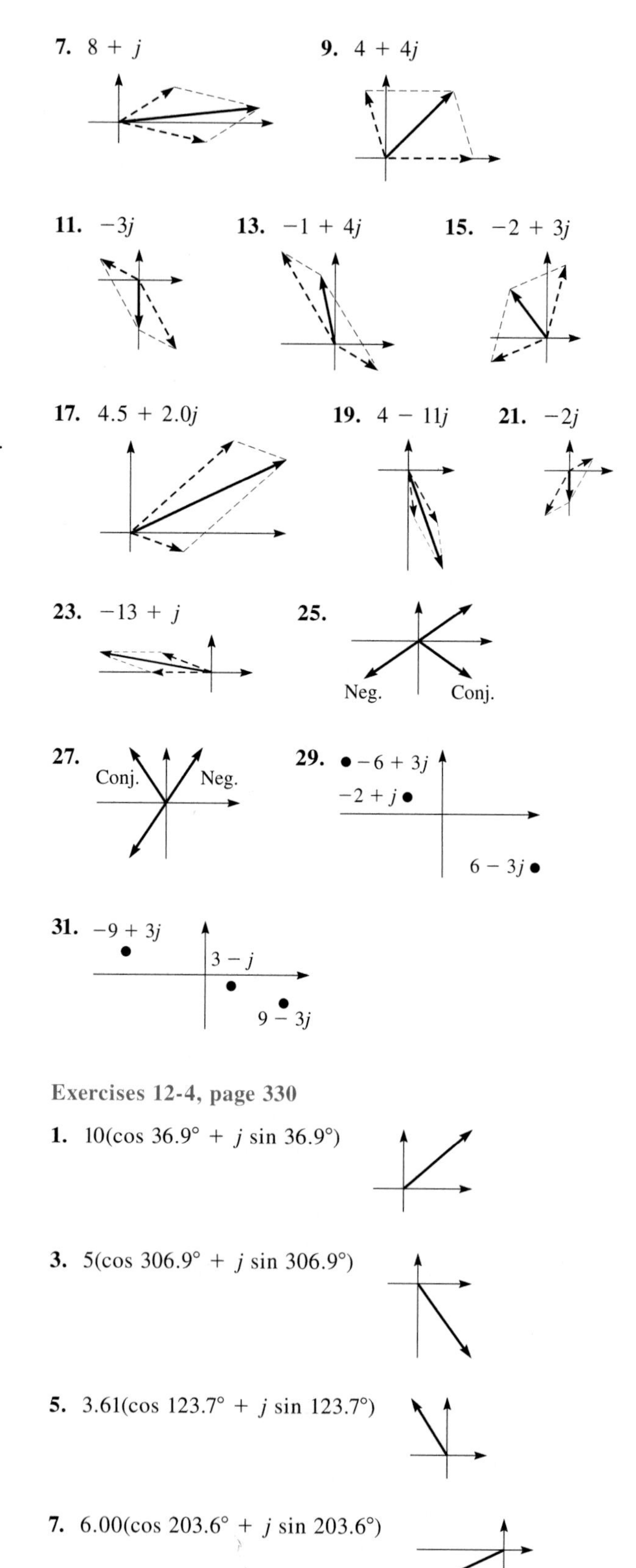

7. $8 + j$ **9.** $4 + 4j$

11. $-3j$ **13.** $-1 + 4j$ **15.** $-2 + 3j$

17. $4.5 + 2.0j$ **19.** $4 - 11j$ **21.** $-2j$

23. $-13 + j$ **25.**

27.

29.

31.

Exercises 12-4, page 330

1. $10(\cos 36.9° + j \sin 36.9°)$

3. $5(\cos 306.9° + j \sin 306.9°)$

5. $3.61(\cos 123.7° + j \sin 123.7°)$

7. $6.00(\cos 203.6° + j \sin 203.6°)$

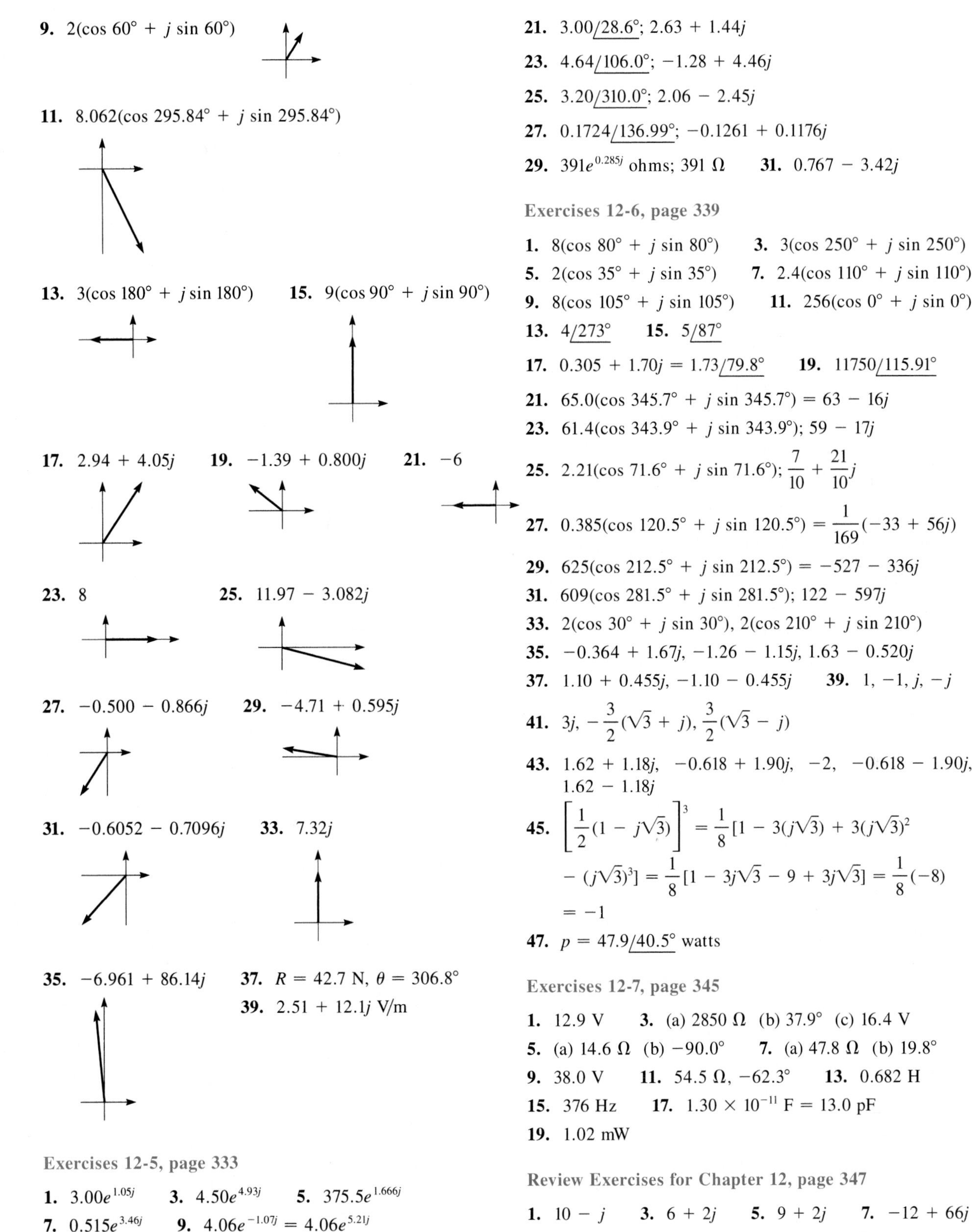

9. $2(\cos 60° + j \sin 60°)$

11. $8.062(\cos 295.84° + j \sin 295.84°)$

13. $3(\cos 180° + j \sin 180°)$ **15.** $9(\cos 90° + j \sin 90°)$

17. $2.94 + 4.05j$ **19.** $-1.39 + 0.800j$ **21.** -6

23. 8 **25.** $11.97 - 3.082j$

27. $-0.500 - 0.866j$ **29.** $-4.71 + 0.595j$

31. $-0.6052 - 0.7096j$ **33.** $7.32j$

35. $-6.961 + 86.14j$ **37.** $R = 42.7$ N, $\theta = 306.8°$
39. $2.51 + 12.1j$ V/m

Exercises 12-5, page 333

1. $3.00e^{1.05j}$ **3.** $4.50e^{4.93j}$ **5.** $375.5e^{1.666j}$
7. $0.515e^{3.46j}$ **9.** $4.06e^{-1.07j} = 4.06e^{5.21j}$
11. $9245e^{5.172j}$ **13.** $5.00e^{5.36j}$ **15.** $3.61e^{2.55j}$
17. $6.37e^{0.386j}$ **19.** $825.7e^{3.836j}$
21. $3.00\underline{/28.6°}$; $2.63 + 1.44j$
23. $4.64\underline{/106.0°}$; $-1.28 + 4.46j$
25. $3.20\underline{/310.0°}$; $2.06 - 2.45j$
27. $0.1724\underline{/136.99°}$; $-0.1261 + 0.1176j$
29. $391e^{0.285j}$ ohms; 391 Ω **31.** $0.767 - 3.42j$

Exercises 12-6, page 339

1. $8(\cos 80° + j \sin 80°)$ **3.** $3(\cos 250° + j \sin 250°)$
5. $2(\cos 35° + j \sin 35°)$ **7.** $2.4(\cos 110° + j \sin 110°)$
9. $8(\cos 105° + j \sin 105°)$ **11.** $256(\cos 0° + j \sin 0°)$
13. $4\underline{/273°}$ **15.** $5\underline{/87°}$
17. $0.305 + 1.70j = 1.73\underline{/79.8°}$ **19.** $11750\underline{/115.91°}$
21. $65.0(\cos 345.7° + j \sin 345.7°) = 63 - 16j$
23. $61.4(\cos 343.9° + j \sin 343.9°)$; $59 - 17j$
25. $2.21(\cos 71.6° + j \sin 71.6°)$; $\frac{7}{10} + \frac{21}{10}j$
27. $0.385(\cos 120.5° + j \sin 120.5°) = \frac{1}{169}(-33 + 56j)$
29. $625(\cos 212.5° + j \sin 212.5°) = -527 - 336j$
31. $609(\cos 281.5° + j \sin 281.5°)$; $122 - 597j$
33. $2(\cos 30° + j \sin 30°)$, $2(\cos 210° + j \sin 210°)$
35. $-0.364 + 1.67j$, $-1.26 - 1.15j$, $1.63 - 0.520j$
37. $1.10 + 0.455j$, $-1.10 - 0.455j$ **39.** $1, -1, j, -j$
41. $3j,\ -\frac{3}{2}(\sqrt{3} + j),\ \frac{3}{2}(\sqrt{3} - j)$
43. $1.62 + 1.18j,\ -0.618 + 1.90j,\ -2,\ -0.618 - 1.90j,\ 1.62 - 1.18j$
45. $\left[\frac{1}{2}(1 - j\sqrt{3})\right]^3 = \frac{1}{8}[1 - 3(j\sqrt{3}) + 3(j\sqrt{3})^2 - (j\sqrt{3})^3] = \frac{1}{8}[1 - 3j\sqrt{3} - 9 + 3j\sqrt{3}] = \frac{1}{8}(-8) = -1$
47. $p = 47.9\underline{/40.5°}$ watts

Exercises 12-7, page 345

1. 12.9 V **3.** (a) 2850 Ω (b) 37.9° (c) 16.4 V
5. (a) 14.6 Ω (b) −90.0° **7.** (a) 47.8 Ω (b) 19.8°
9. 38.0 V **11.** 54.5 Ω, −62.3° **13.** 0.682 H
15. 376 Hz **17.** 1.30×10^{-11} F = 13.0 pF
19. 1.02 mW

Review Exercises for Chapter 12, page 347

1. $10 - j$ **3.** $6 + 2j$ **5.** $9 + 2j$ **7.** $-12 + 66j$
9. $\frac{1}{85}(21 + 18j)$ **11.** $-2 - 3j$ **13.** $\frac{1}{10}(-12 + 9j)$

15. $\frac{1}{5}(13 + 11j)$ **17.** $x = -\frac{2}{3}, y = -2$

19. $x = -\frac{1}{2}, y = \frac{1}{2}$

21. $3 + 11j$ **23.** $4 + 8j$

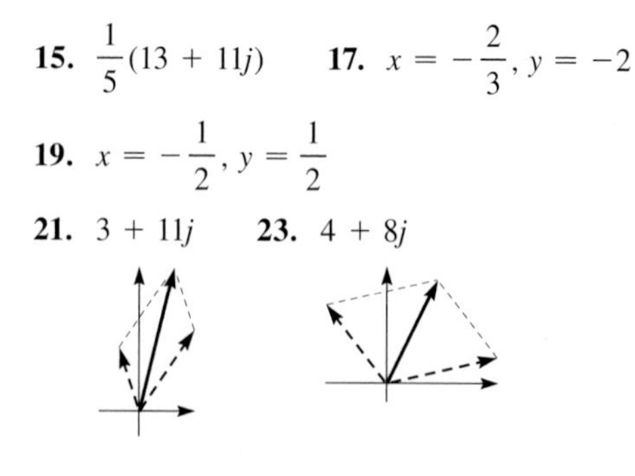

25. $1.41(\cos 315° + j \sin 315°) = 1.41e^{5.50j}$

27. $7.28(\cos 254.1° + j \sin 254.1°) = 7.28e^{4.43j}$

29. $4.67(\cos 76.8° + j \sin 76.8°)$; $4.67e^{1.34j}$

31. $10(\cos 0° + j \sin 0°)$; $10e^{0j}$ **33.** $-1.41 - 1.41j$

35. $-2.789 + 4.163j$ **37.** $0.19 - 0.59j$

39. $26.31 - 6.427j$ **41.** $1.94 + 0.495j$

43. $-8.346 + 23.96j$ **45.** $15(\cos 84° + j \sin 84°)$

47. $20\underline{/263°}$ **49.** $8(\cos 59° + j \sin 59°)$

51. $14.29\underline{/133.61°}$ **53.** $1.26\underline{/59.7°}$

55. $9682\underline{/249.52°}$ **57.** $1024(\cos 160° + j \sin 160°)$

59. $27\underline{/331.5°}$ **61.** $32(\cos 270° + j \sin 270°) = -32j$

63. $\frac{625}{2}(\cos 270° + j \sin 270°) = -\frac{625}{2}j$

65. $1.00 + 1.73j, -2, 1.00 - 1.73j$

67. $\cos 67.5° + j \sin 67.5°, \cos 157.5° + j \sin 157.5°, \cos 247.5° + j \sin 247.5°, \cos 337.5° + j \sin 337.5°$

69. $40 + 9j, 41(\cos 12.7° + j \sin 12.7°)$

71. $-15.0 - 10.9j, 18.5(\cos 216.0° + j \sin 216.0°)$

73. 60 V **75.** $-21.6°$ **77.** 22.9 Hz

79. 807 N, 317.0° **81.** $\frac{u - j\omega n}{u^2 + \omega^2 n^2}$

83. $e^{j\pi} = \cos \pi + j \sin \pi = -1$

Exercises 13-1, page 352

1. 3 **3.** $\frac{1}{81}$ **5.** $\log_3 27 = 3$ **7.** $\log_4 256 = 4$

9. $\log_4 \left(\frac{1}{16}\right) = -2$ **11.** $\log_2 \left(\frac{1}{64}\right) = -6$

13. $\log_8 2 = \frac{1}{3}$ **15.** $\log_{1/4} \left(\frac{1}{16}\right) = 2$ **17.** $81 = 3^4$

19. $9 = 9^1$ **21.** $5 = 25^{1/2}$ **23.** $3 = 243^{1/5}$

25. $0.1 = 10^{-1}$ **27.** $16 = (0.5)^{-4}$ **29.** 2

31. -2 **33.** 343 **35.** $\frac{1}{4}$ **37.** 9 **39.** $\frac{1}{64}$

41. 0.2 **43.** -3 **45.** 3 **47.** $-\frac{1}{2}$

49. $t = \log_{1.1}(V/A)$ **51.** $b_2 = b_1 10^{0.4(m_1 - m_2)}$

53. $N = N_0 e^{-kt}$ **55.** $t = -50(5^{Q/A})$

Exercises 13-2, page 355

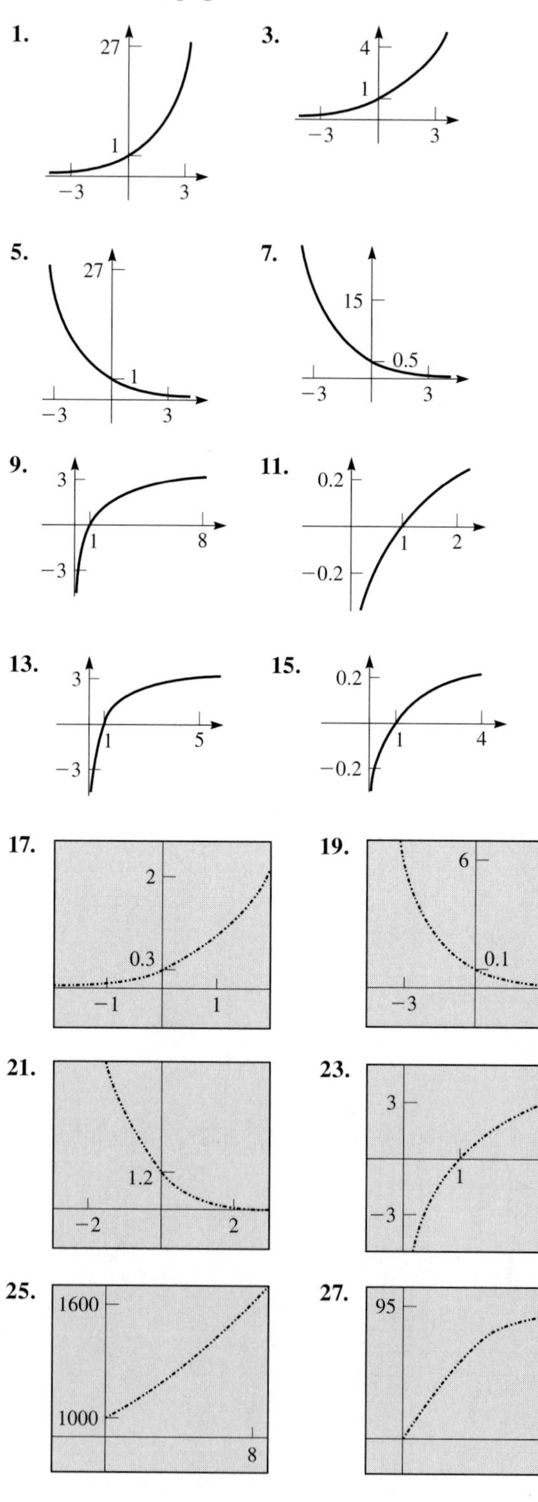

29. **31.** They are the same.

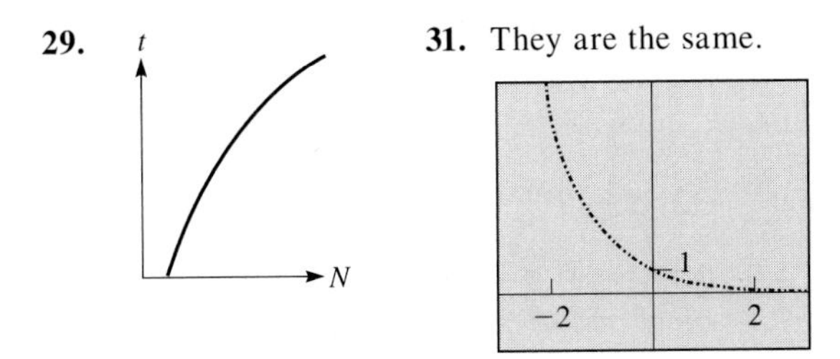

33. $x/2 = \log_{10} y$; $x = 2 \log_{10} y$; $y = 2 \log_{10} x$

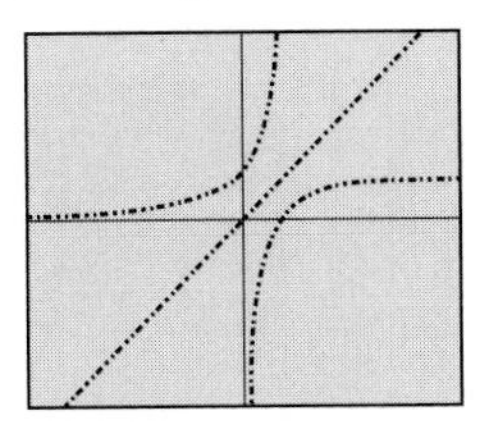

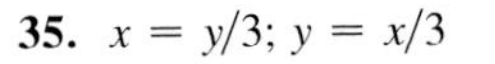

35. $x = y/3$; $y = x/3$

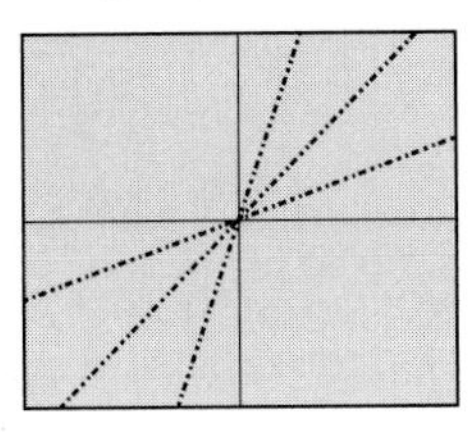

Exercises 13-3, page 359

1. $\log_5 3 + \log_5 11$ **3.** $\log_7 5 - \log_7 3$ **5.** $3 \log_2 a$

7. $\log_6 a + \log_6 b + \log_6 c$ **9.** $\frac{1}{4}\log_5 y$

11. $\frac{1}{2}\log_2 x - 2 \log_2 a$ **13.** $\log_b ac$ **15.** $\log_5 3$

17. $\log_b x^{3/2}$ **19.** $\log_e 4n^3$ **21.** -5

23. 2.5 **25.** $\dfrac{1}{2}$ **27.** $\dfrac{3}{4}$ **29.** $2 + \log_3 2$

31. $-1 - \log_2 3$ **33.** $\frac{1}{2}(1 + \log_3 2)$

35. $3 + \log_{10} 3$ **37.** $y = 2x$ **39.** $y = \dfrac{3x}{5}$

41. $y = \dfrac{49}{x^3}$ **43.** $y = 2(2ax)^{1/5}$ **45.** $y = \dfrac{2}{x}$

47. $y = \dfrac{x^2}{25}$ **49.**

51. $T = 65e^{-0.41t}$

Exercises 13-4, page 362

1. 2.754 **3.** −1.194 **5.** 6.966 **7.** −3.9311

9. −0.0104 **11.** 1.219 **13.** 27 400

15. 0.049 60 **17.** 2000.4 **19.** 0.005 788 2

21. 85.5 **23.** 7.37×10^{101} **25.** 9.0607

27. −18.7953 **29.** 15.2 dB **31.** $2^{400} = 2.58 \times 10^{120}$

Exercises 13-5, page 365

1. 3.258 **3.** 0.4460 **5.** −0.6898 **7.** −4.916 23

9. 1.92 **11.** 3.418 **13.** 1.933 **15.** 1.795

17. 3.940 **19.** 0.3322 **21.** −0.008 335

23. −4.347 66 **25.** 1.6549 **27.** −0.164 13

29. 8.94 **31.** 1.0085 **33.** 0.4757

35. 6.20×10^{-11}

37. **39.**

41. $y = 3x$ **43.** 8.155% **45.** 0.384 s **47.** 21.7 s

Exercises 13-6, page 369

1. 4 **3.** −0.748 **5.** 0.587 **7.** 0.6439

9. 0.285 **11.** 0.203 **13.** 4.11 **15.** 14.2

17. $\dfrac{1}{4}$ **19.** 1.649 **21.** 3 **23.** −0.162 **25.** 5

27. 0.906 **29.** 4 **31.** 2 **33.** 4 **35.** 1.42

37. 28.0 **39.** 6.8 min **41.** 3.922×10^{-4}

43. $10^{8.3} = 2.0 \times 10^8$ **45.** $n = 20e^{-0.04t}$

47. $P = P_0(0.999)^t$ **49.**

51. ±3.7 m

Exercises 13-7, page 373

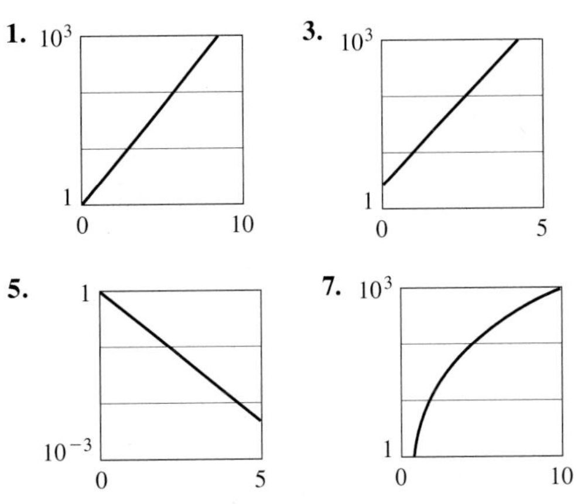

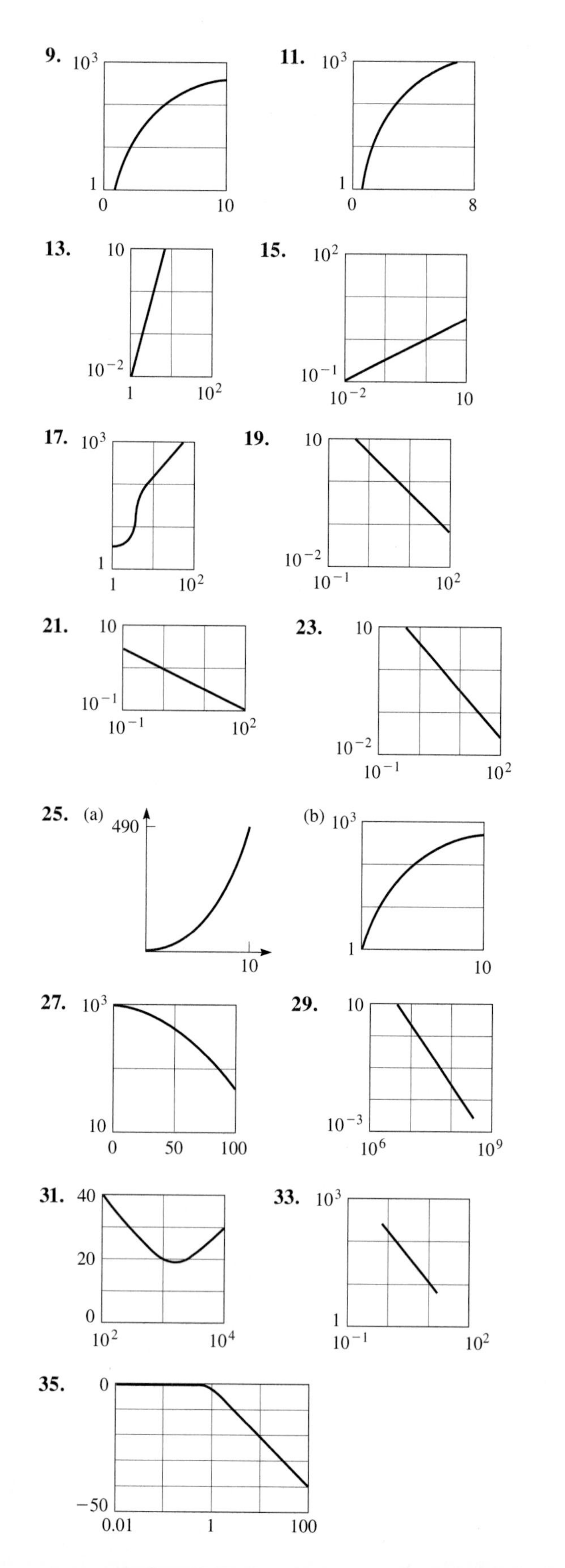

Review Exercises for Chapter 13, page 375

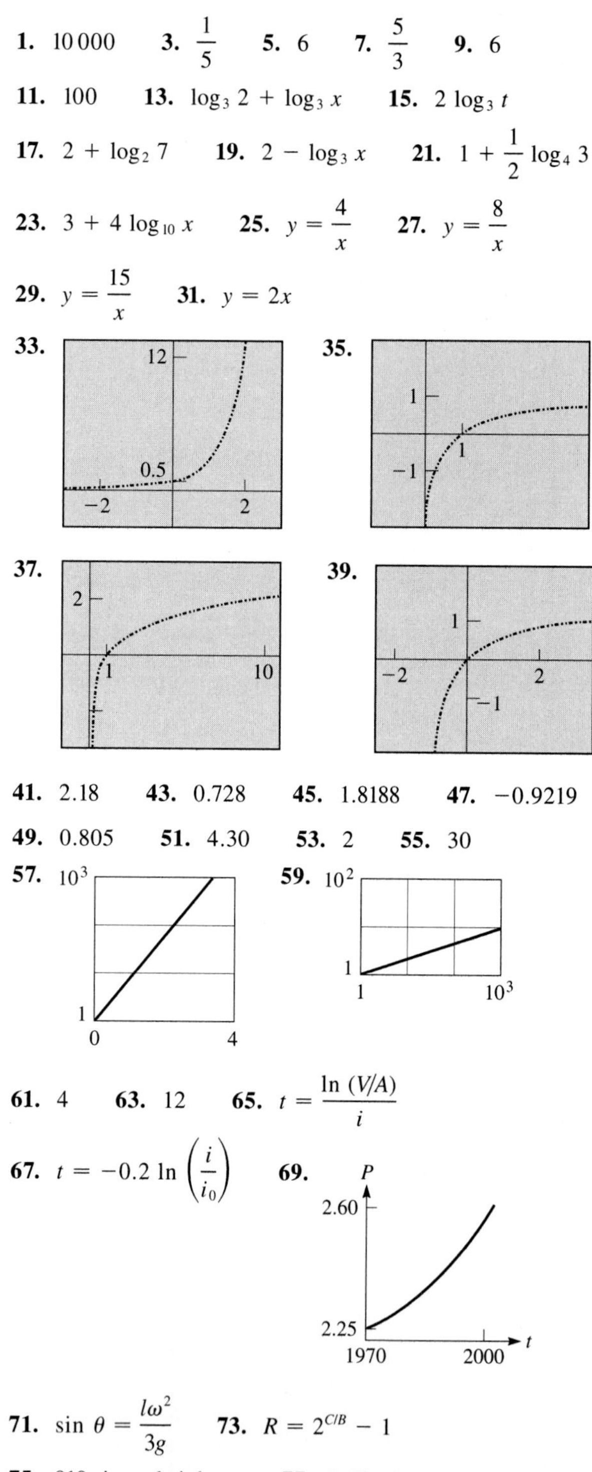

1. 10 000 **3.** $\frac{1}{5}$ **5.** 6 **7.** $\frac{5}{3}$ **9.** 6

11. 100 **13.** $\log_3 2 + \log_3 x$ **15.** $2 \log_3 t$

17. $2 + \log_2 7$ **19.** $2 - \log_3 x$ **21.** $1 + \frac{1}{2} \log_4 3$

23. $3 + 4 \log_{10} x$ **25.** $y = \frac{4}{x}$ **27.** $y = \frac{8}{x}$

29. $y = \frac{15}{x}$ **31.** $y = 2x$

41. 2.18 **43.** 0.728 **45.** 1.8188 **47.** −0.9219

49. 0.805 **51.** 4.30 **53.** 2 **55.** 30

61. 4 **63.** 12 **65.** $t = \frac{\ln (V/A)}{i}$

67. $t = -0.2 \ln \left(\frac{i}{i_0}\right)$ **69.**

71. $\sin \theta = \frac{l\omega^2}{3g}$ **73.** $R = 2^{C/B} - 1$

75. 910 times brighter **77.** 1.17 min

79. 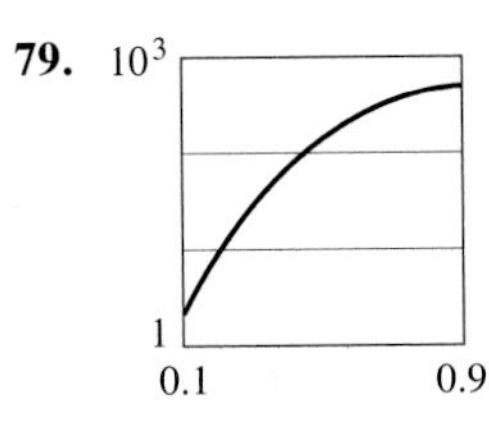

Exercises 14-1, page 381

1. $x = 1.8, y = 3.6; x = -1.8, y = -3.6$

3. $x = 0.0, y = -2.0; x = 2.7, y = -0.7$

5. $x = 1.5, y = 0.2$ **7.** $x = 1.6, y = 2.5$

9. $x = 1.1, y = 2.8; x = -1.1, y = 2.8;$
$x = 2.4, y = -1.8; x = -2.4, y = -1.8$

11. no solution

13. $x = -2.8, y = -1.0; x = 2.8, y = 1.0;$
$x = 2.8, y = -1.0; x = -2.8, y = 1.0$

15. $x = 0.7, y = 0.7; x = -0.7, y = -0.7$

17. $x = 0.0, y = 0.0; x = 0.9, y = 0.8$

19. $x = -1.1, y = 3.0; x = 1.8, y = 0.2$

21. $x = 1.0, y = 0.0$ **23.** $x = 3.6, y = 1.0$

25. 4.9 km N, 1.6 km E **27.** 2.2 A, 0.9 A

Exercises 14-2, page 385

1. $x = 0, y = 1; x = 1, y = 2$

3. $x = -\frac{19}{5}, y = \frac{17}{5}; x = 5, y = -1$ **5.** $x = 1, y = 0$

7. $x = \frac{2}{7}(3 + \sqrt{2}), y = \frac{2}{7}(-1 + 2\sqrt{2});$
$x = \frac{2}{7}(3 - \sqrt{2}), y = \frac{2}{7}(-1 - 2\sqrt{2})$

9. $x = 1, y = 1$ **11.** $x = \frac{2}{3}, y = \frac{9}{2}; x = -3, y = -1$

13. $x = -2, y = 4; x = 2, y = 4$

15. $x = 1, y = 2; x = -1, y = 2$

17. $x = 1, y = 0; x = -1, y = 0; x = \frac{1}{2}\sqrt{6}, y = \frac{1}{2};$
$x = -\frac{1}{2}\sqrt{6}, y = \frac{1}{2}$

19. $x = \sqrt{19}, y = \sqrt{6}; x = \sqrt{19}, y = -\sqrt{6};$
$x = -\sqrt{19}, y = \sqrt{6}; x = -\sqrt{19}, y = -\sqrt{6}$

21. $x = \frac{1}{11}\sqrt{22}, y = \frac{1}{11}\sqrt{770}; x = \frac{1}{11}\sqrt{22},$
$y = -\frac{1}{11}\sqrt{770}; x = -\frac{1}{11}\sqrt{22}, y = \frac{1}{11}\sqrt{770};$
$x = -\frac{1}{11}\sqrt{22}, y = -\frac{1}{11}\sqrt{770}$

23. $x = -5, y = -2; x = -5, y = 2; x = 5, y = -2;$
$x = 5, y = 2$

25. $v_1 = -4.0$ m/s, $v_2 = 2.5$ m/s **27.** 1.5 cm, 1.4 cm

29. 2.19 m, 0.21 m **31.** 80 km/h

Exercises 14-3, page 388

1. $-3, -2, 2, 3$ **3.** $\frac{3}{2}, -\frac{3}{2}, j, -j$ **5.** $-\frac{1}{2}, \frac{1}{4}$

7. $-\frac{1}{2}, \frac{1}{2}, \frac{1}{6}j\sqrt{6}, -\frac{1}{6}j\sqrt{6}$ **9.** $1, \frac{25}{4}$ **11.** $\frac{64}{729}, 1$

13. $-27, 125$ **15.** 256 **17.** 5

19. $-2, -1, 3, 4$ **21.** 18

23. $1, -2, 1 + j\sqrt{3}, 1 - j\sqrt{3}, \frac{1}{2}(-1 + j\sqrt{3}),$
$-\frac{1}{2}(1 + j\sqrt{3})$

25. $R_1 = 2.62\ \Omega, R_2 = 1.62\ \Omega$ **27.** 40.4 cm, 55.4 cm

Exercises 14-4, page 391

1. 12 **3.** 2 **5.** $\frac{2}{3}$ **7.** $\frac{1}{2}(7 + \sqrt{5})$

9. -1 **11.** 32 **13.** 16 **15.** 9

17. 12 (Extraneous root introduced in squaring both sides of $\sqrt{x + 4} = x - 8$.)

19. 16 **21.** $\frac{1}{2}$ **23.** $7, -1$ **25.** 0 **27.** 5

29. 6 **31.** 258 **33.** $L = \frac{1}{4\pi^2 f^2 C}$

35. $\frac{k^2}{2n(1 - k)}$ **37.** 9.2 km **39.** 3.4 km

Review Exercises for Chapter 14, page 392

1. $x = -0.9, y = 3.5; x = 0.8, y = 2.6$

3. $x = 2.0, y = 0.0; x = 1.6, y = 0.6$

5. $x = 0.8, y = 1.6; x = -0.8, y = 1.6$

7. $x = -1.2, y = 2.7; x = 1.2, y = 2.7$

9. $x = 0.0, y = 0.0; x = 2.4, y = 0.9$

11. $x = 0, y = 0; x = 2, y = 16$

13. $x = \sqrt{2}, y = 1; x = -\sqrt{2}, y = 1$

15. $x = \frac{1}{12}(1 + \sqrt{97}), y = \frac{1}{18}(5 - \sqrt{97});$
$x = \frac{1}{12}(1 - \sqrt{97}), y = \frac{1}{18}(5 + \sqrt{97})$

17. $x = 7, y = 5; x = 7, y = -5; x = -7, y = 5;$
$x = -7, y = -5$

19. $x = -2, y = 2; x = \frac{2}{3}, y = \frac{10}{9}$

21. $-4, -2, 2, 4$ **23.** 1, 16 **25.** $\frac{1}{3}, -\frac{1}{7}$

27. $\frac{25}{4}$ **29.** $-\frac{1}{2}, -2$ **31.** 11 **33.** 1, 4

35. 8 **37.** $\frac{9}{16}$ **39.** $\frac{1}{3}(11 - 4\sqrt{15})$

41. $l = \frac{1}{2}(-1 + \sqrt{1 + 16\pi^2L^2/h^2}$

43. $m = \frac{1}{2}(-y \pm \sqrt{2s^2 - y^2})$ **45.** 0.40 s, 0.80 s

47. 70 m **49.** $Z = 1.09\ \Omega$, $X = 0.738\ \Omega$

51. 34 mm, 52 mm **53.** 3.57 cm

55. 42.5 km/h, 52.1 km/h

Exercises 15-1, page 397

1. 0 **3.** 0 **5.** −40 **7.** −4 **9.** 8

11. 183 **13.** −28 **15.** 14 **17.** yes **19.** yes

21. no **23.** no **25.** yes **27.** yes

29. $(4x^3 + 8x^2 - x - 2) \div (2x - 1) = 2x^2 + 5x + 2$; No, because the coefficient of x in $2x - 1$ is 2, not 1.

31. −7

Exercises 15-2, page 401

1. $x^2 + 3x + 2, R = 0$ **3.** $x^2 - x + 3, R = 0$

5. $2x^4 - 4x^3 + 8x^2 - 17x + 42, R = -40$

7. $3x^3 - x + 2, R = -4$ **9.** $x^2 + x - 4, R = 8$

11. $x^3 - 3x^2 + 10x - 45, R = 183$

13. $2x^3 - x^2 - 4x - 12, R = -28$

15. $x^4 + 2x^3 + x^2 + 7x + 4, R = 14$

17. $x^5 + 2x^4 + 4x^3 + 8x^2 + 18x + 36, R = 66$

19. $x^6 + 2x^5 + 4x^4 + 8x^3 + 16x^2 + 32x + 64, R = 0$

21. yes **23.** no **25.** no **27.** no **29.** yes

31. no **33.** yes **35.** yes

Exercises 15-3, page 405

(Note: Unknown roots listed)

1. −2, −1 **3.** −2, 3 **5.** −2, −2 **7.** $-j, -\frac{2}{3}$

9. $2j, -2j$ **11.** −1, −3 **13.** −2, 1

15. $1 - j, \frac{1}{2}, -2$ **17.** $\frac{1}{4}(1 + \sqrt{17}), \frac{1}{4}(1 - \sqrt{17})$

19. 2, −3 **21.** $-j, -1, 1$ **23.** $-2j, 3, -3$

Exercises 15-4, page 411

1. 1, −1, −2 **3.** 2, −1, −3 **5.** $\frac{1}{2}, 5, -3$

7. $\frac{1}{3}, -3, -1$ **9.** $-2, -2, 2 \pm \sqrt{3}$

11. 2, 4, −1, −3 **13.** $1, -\frac{1}{2}, 1 \pm \sqrt{3}$

15. $\frac{1}{2}, -\frac{2}{3}, -3, -\frac{1}{2}$ **17.** 2, 2, −1, −1, −3

19. $\frac{1}{2}, 1, 1, j, -j$ **21.** −1.86, 0.68, 3.18

23. −3.24, 1.24 **25.** 0.59 **27.** −0.77

29. 1.8 s, 4.5 s **31.** 0, L **33.** 1.5 kg, 4.1 kg

35. 2 Ω, 3 Ω, 6 Ω **37.** 1.23 cm or 2.14 cm

39. 0.0 km, 1.6 km, 2.8 km

Review Exercises for Chapter 15, page 412

1. 1 **3.** −107 **5.** yes **7.** no

9. $x^2 + 4x + 10, R = 11$

11. $2x^2 - 7x + 10, R = -17$

13. $x^3 - 3x^2 - 4, R = -4$

15. $2x^4 + 10x^3 + 4x^2 + 21x + 105, R = 516$

17. no **19.** yes **21.** (unlisted roots) $\frac{1}{2}(-5 \pm \sqrt{17})$

23. (unlisted roots) $\frac{1}{3}(-1 \pm j\sqrt{14})$

25. (unlisted roots) 4, −3

27. (unlisted roots) $-j, \frac{1}{2}, -\frac{3}{2}$

29. (unlisted roots) 2, −2

31. (unlisted roots) $-2 - j, \frac{1}{2}(-1 \pm j\sqrt{3})$

33. 1, 2, −4 **35.** $-1, -1, \frac{5}{2}$ **37.** $\frac{5}{3}, -\frac{1}{2}, -1$

39. $\frac{1}{2}, -1, j\sqrt{2}, -j\sqrt{2}$ **41.** 4

43. 1, 0.4, 1.5, −0.6 (last three are irrational)

45. 1.91 **47.** April **49.** 0.75 cm **51.** 2 cm

53. 5.1 m, 8.3 m **55.** 8.0 m, 15 m

Exercises 16-1, page 418

1. 39 **3.** 30 **5.** 50 **7.** −47416

9. −6 **11.** 118 **13.** −2

15. 24; Since a complete expansion of a third-order determinant has 6 terms, expanding by a row or column of a fourth-order determinant gives $4 \times 6 = 24$ terms.

17. $x = 2, y = -1, z = 3$

19. $x = -1, y = \frac{1}{3}, z = -\frac{1}{2}$

21. $x = -1, y = 0, z = 2, t = 1$

23. $x = 1, y = 2, z = -1, t = 3$

25. $\frac{2}{7}$ A, $\frac{18}{7}$ A, $-\frac{8}{7}$ A, $-\frac{12}{7}$ A **27.** 500, 300, 700

Exercises 16-2, page 422

1. -60 **3.** -56 **5.** 0 **7.** 0 **9.** 57
11. -124 **13.** -13 **15.** -118 **17.** -72
19. 0 **21.** $x = 0, y = -1, z = 4$
23. $x = \frac{1}{3}, y = -\frac{1}{2}, z = 1$
25. $x = 2, y = -1, z = -1, t = 3$
27. $x = 1, y = 2, z = -1, t = -2$
29. $\frac{33}{16}$ A, $\frac{11}{8}$ A, $-\frac{5}{8}$ A, $-\frac{15}{8}$ A, $-\frac{15}{16}$ A
31. ppm of SO_2: 0.5, NO: 0.3, NO_2: 0.2, CO: 5.0

Exercises 16-3, page 427

1. $a = 1, b = -3, c = 4, d = 7$ **3.** $x = 2, y = 3$
5. Elements cannot be equated; different number of rows
7. $x = 4, y = 6, z = 9$
9. $\begin{pmatrix} 1 & 10 \\ 0 & 2 \end{pmatrix}$ **11.** $\begin{pmatrix} -5 & 0 \\ 11 & 71 \\ 11 & -5 \end{pmatrix}$
13. $\begin{pmatrix} 0 & 9 & -13 & 3 \\ 6 & -7 & 7 & 0 \end{pmatrix}$ **15.** cannot be added
17. $\begin{pmatrix} -1 & 13 & -20 & 3 \\ 8 & -13 & 6 & 2 \end{pmatrix}$ **19.** $\begin{pmatrix} -3 & -6 & 5 & -6 \\ -6 & -4 & -17 & 6 \end{pmatrix}$
21. $A + B = B + A = \begin{pmatrix} 3 & 1 & 0 & 7 \\ 5 & -3 & -2 & 5 \\ 10 & 10 & 8 & 0 \end{pmatrix}$
23. $-(A - B) = B - A = \begin{pmatrix} 5 & -3 & -6 & -7 \\ 5 & 3 & 0 & -3 \\ -8 & 12 & 8 & 4 \end{pmatrix}$
25. $v_w = 31.0$ km/h, $v_p = 249$ km/h
27. $\begin{pmatrix} 24 & 18 & 0 & 0 \\ 15 & 12 & 9 & 0 \\ 0 & 9 & 15 & 18 \end{pmatrix}$

Exercises 16-4, page 432

1. $(-8 \ -12)$ **3.** $\begin{pmatrix} -15 & 15 & -26 \\ 8 & 5 & -13 \end{pmatrix}$
5. $\begin{pmatrix} 29 \\ -29 \end{pmatrix}$ **7.** $\begin{pmatrix} -23 & -19 \\ 21 & 16 \\ 2 & -28 \end{pmatrix}$ **9.** $\begin{pmatrix} 33 & -22 \\ 31 & -12 \\ 15 & 13 \\ 50 & -41 \end{pmatrix}$
11. $\begin{pmatrix} -49.43 & 55.2 \\ -53.02 & 79.16 \end{pmatrix}$
13. $AB = (40)$, $BA = \begin{pmatrix} -1 & 3 & -8 \\ 5 & -15 & 40 \\ 7 & -21 & 56 \end{pmatrix}$
15. $AB = \begin{pmatrix} -5 \\ 10 \end{pmatrix}$, BA not defined
17. $AI = IA = A$ **19.** $AI = IA = A$
21. $B = A^{-1}$ **23.** $B = A^{-1}$ **25.** yes
27. no **29.** $\begin{pmatrix} -1 & 0 \\ 0 & -1 \end{pmatrix}\begin{pmatrix} -1 & 0 \\ 0 & -1 \end{pmatrix} = \begin{pmatrix} 1 & 0 \\ 0 & 1 \end{pmatrix}$
31. $A^2 - I = (A + I)(A - I) = \begin{pmatrix} 15 & 28 \\ 21 & 36 \end{pmatrix}$
33. $\begin{pmatrix} 0 & -j \\ j & 0 \end{pmatrix}\begin{pmatrix} 0 & -j \\ j & 0 \end{pmatrix} = \begin{pmatrix} 1 & 0 \\ 0 & 1 \end{pmatrix}$
35. $v_2 = v_1$, $i_2 = -v_1/R + i_1$

Exercises 16-5, page 436

1. $\begin{pmatrix} -2 & -\frac{5}{2} \\ -1 & -1 \end{pmatrix}$ **3.** $\begin{pmatrix} -\frac{1}{3} & \frac{1}{6} \\ \frac{2}{15} & \frac{1}{30} \end{pmatrix}$ **5.** $\begin{pmatrix} \frac{3}{4} & \frac{1}{2} \\ -\frac{1}{4} & 0 \end{pmatrix}$
7. $\begin{pmatrix} -\frac{8}{283} & -\frac{9}{566} \\ \frac{13}{1415} & \frac{5}{283} \end{pmatrix}$ **9.** $\begin{pmatrix} -3 & 2 \\ 2 & -1 \end{pmatrix}$
11. $\begin{pmatrix} -\frac{1}{2} & -2 \\ \frac{1}{2} & 1 \end{pmatrix}$ **13.** $\begin{pmatrix} \frac{2}{9} & -\frac{5}{9} \\ \frac{1}{9} & \frac{2}{9} \end{pmatrix}$ **15.** $\begin{pmatrix} \frac{3}{8} & \frac{1}{16} \\ -\frac{1}{4} & \frac{1}{8} \end{pmatrix}$
17. $\begin{pmatrix} -18 & -7 & 5 \\ -3 & -1 & 1 \\ -5 & -2 & 1 \end{pmatrix}$ **19.** $\begin{pmatrix} 3 & -4 & -1 \\ -4 & 5 & 2 \\ 2 & -3 & -1 \end{pmatrix}$
21. $\begin{pmatrix} 2 & 4 & \frac{7}{2} \\ -1 & -2 & -\frac{3}{2} \\ 1 & 1 & \frac{1}{2} \end{pmatrix}$ **23.** $\begin{pmatrix} \frac{5}{2} & -2 & -2 \\ -1 & 1 & 1 \\ \frac{7}{4} & -\frac{3}{2} & -1 \end{pmatrix}$
25. $\begin{pmatrix} 2 & 4 & 3.5 \\ -1 & -2 & -1.5 \\ 1 & 1 & 0.5 \end{pmatrix}$ **27.** $\begin{pmatrix} 2.5 & -2 & -2 \\ -1 & 1 & 1 \\ 1.75 & -1.5 & -1 \end{pmatrix}$
29. $\frac{1}{ad - bc}\begin{pmatrix} ad - bc & -ba + ab \\ cd - dc & -bc + ad \end{pmatrix} = \begin{pmatrix} 1 & 0 \\ 0 & 1 \end{pmatrix}$
31. $v_1 = a_{22}i_1 - a_{12}i_2$
$v_2 = -a_{21}i_1 + a_{11}i_2$

Exercises 16-6, page 440

1. $x = \frac{1}{2}, y = 3$ **3.** $x = \frac{1}{2}, y = -\frac{5}{2}$
5. $x = -4, y = 2, z = -1$
7. $x = -1, y = 0, z = 3$ **9.** $x = 1, y = 2$

11. $x = -\frac{3}{2}, y = -2$ **13.** $x = -3, y = -\frac{1}{2}$

15. $x = 1.6, y = -2.5$ **17.** $x = 2, y = -4, z = 1$

19. $x = 2, y = -\frac{1}{2}, z = 3$

21. $A = 118$ N, $B = 186$ N **23.** 6.4 L, 1.6 L, 2.0 L

Review Exercises for Chapter 16, page 441

1. 6 **3.** 186 **5.** −438 **7.** 44 **9.** 6

11. 186 **13.** −438 **15.** 6 **17.** 186

19. −438 **21.** $a = 4, b = -1$

23. $x = 2, y = -3, z = \frac{5}{2}, a = -1, b = -\frac{7}{2}, c = \frac{1}{2}$

25. $\begin{pmatrix} 1 & -3 \\ 8 & -5 \\ -8 & -2 \\ 3 & -10 \end{pmatrix}$ **27.** $\begin{pmatrix} 3 & 0 \\ -12 & 18 \\ 9 & 6 \\ -3 & 21 \end{pmatrix}$

29. cannot be subtracted **31.** $\begin{pmatrix} 7 & -6 \\ -4 & 20 \\ -1 & 6 \\ 1 & 15 \end{pmatrix}$

33. $\begin{pmatrix} 13 \\ -13 \end{pmatrix}$ **35.** $\begin{pmatrix} 34 & 11 & -5 \\ 2 & -8 & 10 \\ -1 & -17 & 20 \end{pmatrix}$ **37.** $\begin{pmatrix} -2 & \frac{5}{2} \\ -1 & 1 \end{pmatrix}$

39. $\begin{pmatrix} \frac{2}{15} & \frac{1}{60} \\ -\frac{1}{15} & \frac{7}{60} \end{pmatrix}$ **41.** $\begin{pmatrix} 11 & 10 & 3 \\ -4 & -4 & -1 \\ 3 & 3 & 1 \end{pmatrix}$

43. $\begin{pmatrix} \frac{1}{2} & -\frac{1}{2} & -1 \\ -3 & 2 & 1 \\ -4 & 3 & 2 \end{pmatrix}$ **45.** $x = -3, y = 1$

47. $x = 10, y = -15$ **49.** $x = -1, y = -3, z = 0$

51. $x = 1, y = \frac{1}{2}, z = -\frac{1}{3}$ **53.** $x = 3, y = 1, z = -1$

55. $x = 1, y = 2, z = -3, t = 1$ **57.** $2\sqrt{2}$ **59.** 0

61. $N^{-1} = -N = \begin{pmatrix} 0 & 1 \\ -1 & 0 \end{pmatrix}$

63. $(A + B)(A - B) = \begin{pmatrix} -6 & 2 \\ 4 & 2 \end{pmatrix}, A^2 - B^2 = \begin{pmatrix} -10 & -4 \\ 8 & 6 \end{pmatrix}$

65. $\begin{pmatrix} \frac{1}{2} & \frac{1}{3} \\ 0 & \frac{1}{6} \end{pmatrix} = \frac{1}{2}\begin{pmatrix} 1 & \frac{2}{3} \\ 0 & \frac{1}{3} \end{pmatrix}$

67. $R_1 = 4\ \Omega, R_2 = 6\ \Omega$ **69.** $F = 303$ N, $T = 175$ N

71. 0.20 h after police pass intersection

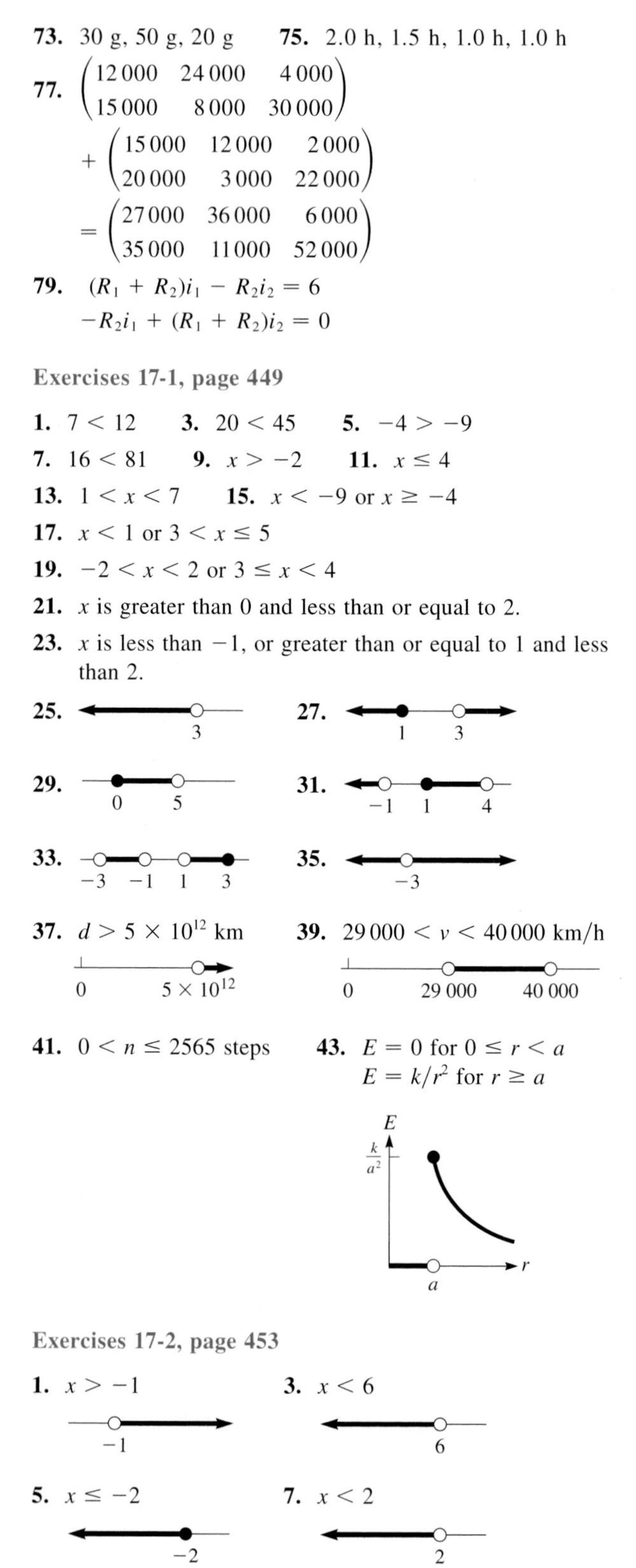

73. 30 g, 50 g, 20 g **75.** 2.0 h, 1.5 h, 1.0 h, 1.0 h

77. $\begin{pmatrix} 12\,000 & 24\,000 & 4\,000 \\ 15\,000 & 8\,000 & 30\,000 \end{pmatrix} + \begin{pmatrix} 15\,000 & 12\,000 & 2\,000 \\ 20\,000 & 3\,000 & 22\,000 \end{pmatrix} = \begin{pmatrix} 27\,000 & 36\,000 & 6\,000 \\ 35\,000 & 11\,000 & 52\,000 \end{pmatrix}$

79. $(R_1 + R_2)i_1 - R_2 i_2 = 6$
$-R_2 i_1 + (R_1 + R_2)i_2 = 0$

Exercises 17-1, page 449

1. $7 < 12$ **3.** $20 < 45$ **5.** $-4 > -9$

7. $16 < 81$ **9.** $x > -2$ **11.** $x \leq 4$

13. $1 < x < 7$ **15.** $x < -9$ or $x \geq -4$

17. $x < 1$ or $3 < x \leq 5$

19. $-2 < x < 2$ or $3 \leq x < 4$

21. x is greater than 0 and less than or equal to 2.

23. x is less than −1, or greater than or equal to 1 and less than 2.

25. 3 **27.** 1 3

29. 0 5 **31.** −1 1 4

33. −3 −1 1 3 **35.** −3

37. $d > 5 \times 10^{12}$ km (0, 5×10^{12}) **39.** $29\,000 < v < 40\,000$ km/h (0, 29 000, 40 000)

41. $0 < n \leq 2565$ steps **43.** $E = 0$ for $0 \leq r < a$
$E = k/r^2$ for $r \geq a$

E, $\frac{k}{a^2}$, r, a

Exercises 17-2, page 453

1. $x > -1$ (−1) **3.** $x < 6$ (6)

5. $x \leq -2$ (−2) **7.** $x < 2$ (2)

9. $x \le \frac{5}{2}$ **11.** $x < -1$

13. $x \le \frac{13}{2}$ **15.** $x > -\frac{7}{9}$

17. $-1 < x < 1$ **19.** $2 < x \le 5$

21. $-3 \le x < -1$ **23.** no values

25. $x \ge 5$ **27.** $s > \$11\,100$

29. $0 \le t \le 6$ years **31.** $0.4 < t < 2.6$ h

33. $0 \le x \le 500$
$200 \le y \le 700$

35. $24 \le x \le 40$ L

Exercises 17-3, page 458

1. $-1 < x < 1$ **3.** $0 \le x \le 2$

5. $x \le -2, x \ge \frac{1}{3}$ **7.** $\frac{1}{3} < x < \frac{1}{2}$

9. $x = -2$ **11.** all x

13. $-2 < x < 0, x > 1$ **15.** $-2 \le x \le -1, x \ge 1$

17. $x < 3, x > 8$ **19.** $-6 < x \le \frac{3}{2}$

21. $-1 < x < 2$ **23.** $-5 < x < -1, x > 7$

25. $-1 < x < \frac{3}{4}, x \ge 6$ **27.** $2 < x < 4, 5 < x < 9$

29. $x \le -2, x \ge 1$ **31.** $-1 \le x \le 0$

33. $x > 1.52$ **35.** $-1.39 < x < -0.43$

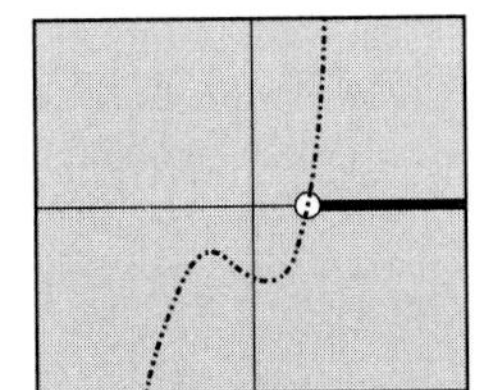

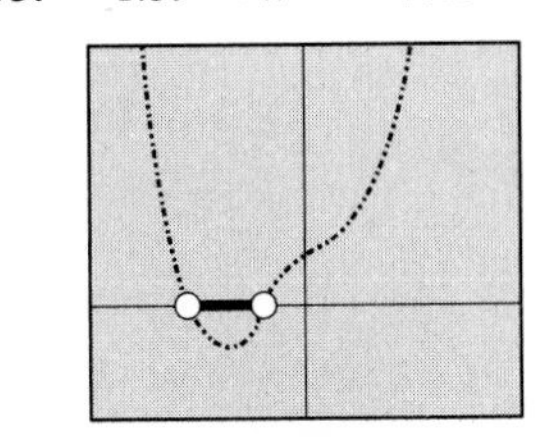

37. $x < -1.69, x > 2.00$

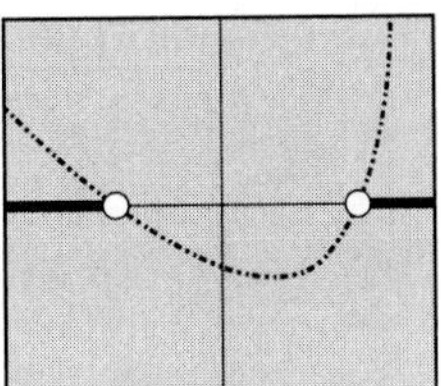

39. $x < -4.43, -3.11 < x < -1.08,$
$x > 3.15$

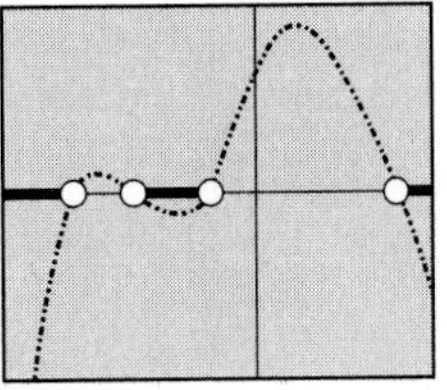

41. $0.5 < i < 1$ A **43.** $x \le 12$

45. $3.0 \le w < 5.0$ mm **47.** $0 \le t < 0.92$ h

Exercises 17-4, page 461

1. $3 < x < 5$ **3.** $x < -2, x > \frac{2}{5}$

5. $\frac{1}{6} \le x \le \frac{3}{2}$ **7.** $x < 0, x > \frac{3}{2}$

9. $-16 < x < 14$ **11.** $-6.4 \le x \le -2.1$

13. $x < 0, x > 8$ **15.** $x \le \frac{1}{10}, x \ge \frac{7}{10}$

17. $-18 < x < 14$

19. $x \leq 8.4,\ x \geq 17.6$

21. $x < -3,\ -2 < x < 1,\ x > 2$

23. $-3 < x < -2,\ 1 < x < 2$

25. $|d - 0.2537| \leq 0.0003$ cm

27. $d = 3.0x - 10.0$; $|d| < 6.0$ m for $1.3 < x < 5.3$ m

Exercises 17-5, page 466

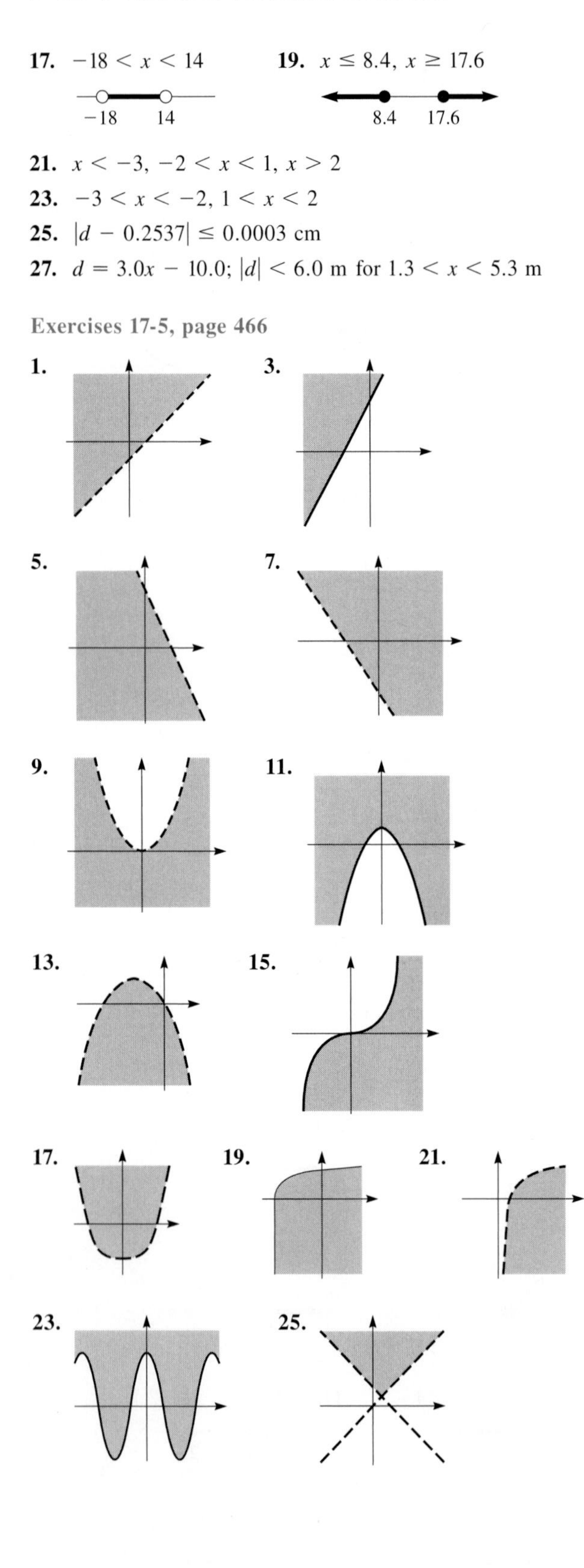

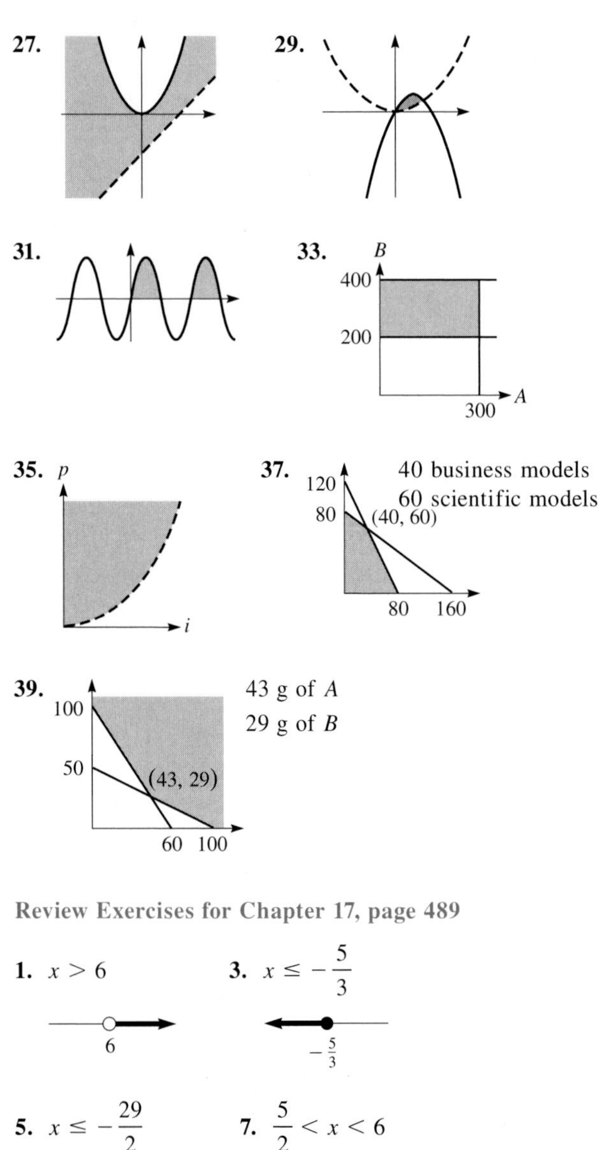

37. 40 business models
60 scientific models

39. 43 g of A
29 g of B

Review Exercises for Chapter 17, page 489

1. $x > 6$

3. $x \leq -\frac{5}{3}$

5. $x \leq -\frac{29}{2}$

7. $\frac{5}{2} < x < 6$

9. $-2 < x < 1$

11. $-2 < x < \frac{1}{5}$

13. $x < -9,\ x > 7$

15. $-4 < x < -1,\ x > 1$

17. $-\frac{1}{2} < x \leq 8$

19. $x < -4,\ \frac{1}{2} < x < 3$

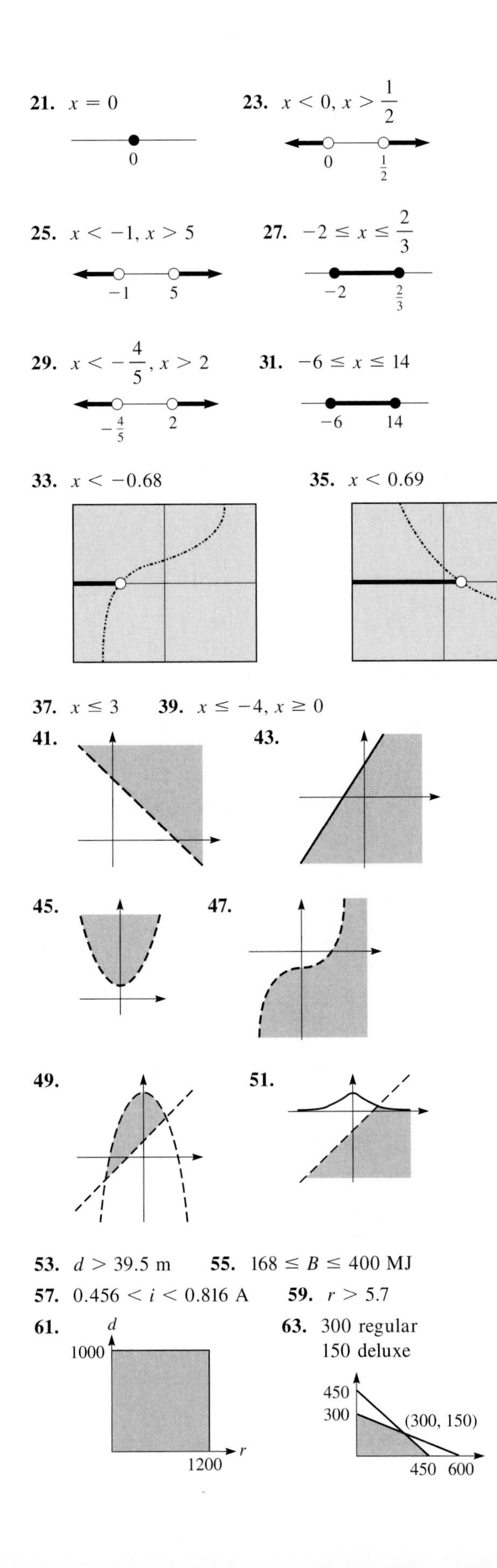

21. $x = 0$ **23.** $x < 0, x > \frac{1}{2}$

25. $x < -1, x > 5$ **27.** $-2 \le x \le \frac{2}{3}$

29. $x < -\frac{4}{5}, x > 2$ **31.** $-6 \le x \le 14$

33. $x < -0.68$ **35.** $x < 0.69$

37. $x \le 3$ **39.** $x \le -4, x \ge 0$

41. **43.**

45. **47.**

49. **51.**

53. $d > 39.5$ m **55.** $168 \le B \le 400$ MJ

57. $0.456 < i < 0.816$ A **59.** $r > 5.7$

61. **63.** 300 regular 150 deluxe

Exercises 18-1, page 472

1. 6 **3.** $\frac{4}{3}$ **5.** $\frac{4}{25}$ **7.** 40 **9.** 1.7 **11.** 0.41

13. 5.56 **15.** 7.8 **17.** 3.5% **19.** 863 kg

21. 1.44 Ω **23.** 0.103 m^3 **25.** 20 000 g

27. 900 kJ **29.** 286° **31.** 12.5 m/s

33. 23 400 cm^3 **35.** 19.67 kg **37.** 17 500 chips

39. 4200 lines, 4900 lines

Exercises 18-2, page 478

1. $y = kz$ **3.** $s = \frac{k}{t^2}$ **5.** $f = k\sqrt{x}$ **7.** $w = kxy^3$

9. y varies directly as the fourth power of x.

11. n varies as the square root of t and inversely as u.

13. $y = \frac{1}{4}\sqrt{x}$ **15.** $p = \frac{16q}{r^3}$ **17.** 25 **19.** 50

21. 180 **23.** 2.56×10^5 **25.** 61 m^3

27. $m = 0.476t$ **29.** 0.67 **31.** 1.4 h **33.** 1400

35. 27 N **37.** $F = 2.32Av^2$ **39.** 480 m/s

41. $R = \frac{2.60 \times 10^{-5} l}{A}$ **43.** 80.0 W **45.** $G = \frac{5.9d^2}{\lambda^2}$

47. -6.57 cm/s^2

Review Exercises for Chapter 18, page 480

1. 200 **3.** $\frac{2}{5}$ **5.** 5.6 **7.** 7.39 N/cm^2

9. 3.25 cm **11.** 2.66×10^{-3} kJ

13. 36 000 characters **15.** 310 mL **17.** 71.9 m

19. 4500, 7500 **21.** $y = 3x^2$ **23.** $v = \frac{128x}{y^3}$

25. 11.7 cm **27.** 4.3 μC **29.** 18.0 kW **31.** 44.1 m

33. 1.4 **35.** 48.7 Hz **37.** 2.99×10^8 m/s

39. 3.26 cm **41.** 4.8 MJ **43.** 5.73×10^4 m

45. 150% **47.** \$125.00, \$600.13 **49.** 2600 Pa

51. 0.023 W/m^2

Exercises 19-1, page 488

1. 4, 6, 8, 10, 12 **3.** 13, 9, 5, 1, −3 **5.** 22

7. −62 **9.** 309 **11.** 49b **13.** 440

15. $-\frac{85}{2}$ **17.** $n = 6, S_6 = 150$

19. $d = -\frac{2}{19}, a_{20} = -\frac{1}{3}$ **21.** $a_1 = 19, a_{30} = 106$

23. $n = 62, S_{62} = -4867$ **25.** $n = 23, a_{23} = 6k$

27. $n = 8, d = \frac{1}{14}(b + 2c)$

29. $a_1 = 36, d = 4, S_{10} = 540$

31. $a_1 = 3, d = -\frac{1}{3}, S_{10} = 15$ **33.** 5050

35. 100 500 **37.** 2700 m^2 **39.** 10 rows

41. 12 years, \$11 700 **43.** 490 m

45. $S_n = \frac{1}{2}n[2a_1 + (n-1)d]$ **47.** $a_1 = 1, a_n = n$

Exercises 19-2, page 491

1. $45, 15, 5, \frac{5}{3}, \frac{5}{9}$ **3.** 2, 6, 18, 54, 162 **5.** 16

7. $\frac{1}{125}$ **9.** $\frac{1}{729}$ **11.** 2×10^6 **13.** $\frac{341}{8}$

15. 378 **17.** $a_6 = 64, S_6 = \frac{1365}{16}$

19. $a_1 = 16, a_5 = 81$ **21.** $a_1 = 1, r = 3$

23. $n = 7, S_n = \frac{58\,593}{625}$ **25.** 32 **27.** 1.4%

29. 1.09 mA **31.** \$443.96 **33.** 462 cm **35.** 4

37. \$671 088.64 **39.** $S_n = \frac{a_1 - ra_n}{1 - r}$

Exercises 19-3, page 495

1. 8 **3.** $\frac{25}{4}$ **5.** $\frac{400}{21}$ **7.** 8 **9.** $\frac{10\,000}{9999}$

11. $\frac{1}{2}(5 + 3\sqrt{3})$ **13.** $\frac{1}{3}$ **15.** $\frac{40}{99}$ **17.** $\frac{2}{11}$

19. $\frac{91}{333}$ **21.** $\frac{11}{30}$ **23.** $\frac{100\,741}{999\,000}$ **25.** 350 L

27. 346 g

Exercises 19-4, page 499

1. $t^3 + 3t^2 + 3t + 1$ **3.** $16x^4 - 32x^3 + 24x^2 - 8x + 1$

5. $32x^5 + 240x^4 + 720x^3 + 1080x^2 + 810x + 243$

7. $64a^6 - 192a^5b^2 + 240a^4b^4 - 160a^3b^6 + 60a^2b^8 - 12ab^{10} + b^{12}$

9. $625x^4 - 1500x^3 + 1350x^2 - 540x + 81$

11. $64a^6 + 192a^5 + 240a^4 + 160a^3 + 60a^2 + 12a + 1$

13. $x^{10} + 20x^9 + 180x^8 + 960x^7 + \cdots$

15. $128a^7 - 448a^6 + 672a^5 - 560a^4 + \cdots$

17. $x^{24} - 6x^{22}y + \frac{33}{2}x^{20}y^2 - \frac{55}{2}x^{18}y^3 + \cdots$

19. $b^{40} + 10b^{37} + \frac{95}{2}b^{34} + \frac{285}{2}b^{31} + \cdots$

21. $1 + 8x + 28x^2 + 56x^3 + \cdots$

23. $1 + 2x + 3x^2 + 4x^3 + \cdots$

25. $1 + \frac{1}{2}x - \frac{1}{8}x^2 + \frac{1}{16}x^3 - \cdots$

27. $\frac{1}{3}\left[1 + \frac{1}{2}x + \frac{3}{8}x^2 + \frac{5}{16}x^3 + \cdots\right]$

29. (a) 3.557×10^{14} (b) 5.109×10^{19} (c) 8.536×10^{15} (d) 2.480×10^{96}

31. $n! = n(n-1)(n-2)(\cdots)(2)(1) = n \times (n-1)!$; for $n = 1$, $1! = 1 \times 0!$ Since $1! = 1$, $0!$ must $= 1$.

33. $56a^3b^5$ **35.** $10\,264\,320x^8b^4$

37. $V = A(1 - 5r + 10r^2 - 10r^3 + 5r^4 - r^5)$

39. $1 - \frac{x}{a} + \frac{x^3}{2a^3} - \cdots$

Review Exercises for Chapter 19, page 501

1. 81 **3.** 1.28×10^{-6} **5.** $-\frac{119}{2}$ **7.** $\frac{16}{243}$

9. $\frac{195}{2}$ **11.** $\frac{16\,383}{1536}$ **13.** 81 **15.** $\frac{9}{16}$

17. −1.5 **19.** −0.25 **21.** 186

23. $\frac{455}{2}$ (as), 127 (gs), or 43 (gs) **25.** 27 **27.** 51

29. $\frac{1}{33}$ **31.** $\frac{8}{110}$ **33.** $x^4 - 8x^3 + 24x^2 - 32x + 16$

35. $x^{10} + 5x^8 + 10x^6 + 10x^4 + 5x^2 + 1$

37. $a^{10} + 20a^9b^2 + 180a^8b^4 + 960a^7b^6 + \cdots$

39. $p^{18} - \frac{3}{2}p^{16}q + p^{14}q^2 - \frac{7}{18}p^{12}q^3 + \cdots$

41. $1 + 12x + 66x^2 + 220x^3 + \cdots$

43. $1 + \frac{1}{2}x^2 - \frac{1}{8}x^4 + \frac{1}{16}x^6 - \cdots$

45. $1 - \frac{1}{2}a^2 - \frac{1}{8}a^4 - \frac{1}{16}a^6 - \cdots$

47. $\frac{1}{8} + \frac{3}{4}x + 3x^2 + 10x^3 + \cdots$ **49.** 1 001 000

51. 11th **53.** 12.6 mm **55.** 7694 mm

57. \$2391.24 **59.** 1.65×10^{10} cm = 165 000 km

61. \$47 340.80 **63.** \$6.93

65. $1 + \frac{1}{2}am^2 + \frac{1}{8}am^4$ **67.** 21 years

69. 5 applications **71.** yes

Exercises 20-1, page 509

(Note: "Answers" to trigonometric identities are intermediate steps of suggested reductions of the left member.)

1. $1.483 = \frac{1}{0.6745}$

3. $\left(-\frac{1}{2}\sqrt{3}\right)^2 + \left(-\frac{1}{2}\right)^2 = \frac{3}{4} + \frac{1}{4} = 1$

5. $\frac{\cos\theta}{\sin\theta}\left(\frac{1}{\cos\theta}\right) = \frac{1}{\sin\theta}$

7. $\frac{\sin x}{\frac{\sin x}{\cos x}} = \frac{\sin x}{1}\left(\frac{\cos x}{\sin x}\right)$ **9.** $\sin y\left(\frac{\cos y}{\sin y}\right)$

11. $\sin x\left(\frac{1}{\cos x}\right)$ **13.** $\csc^2 x\,(\sin^2 x)$

15. $\sin x\,(\csc^2 x) = (\sin x)(\csc x)(\csc x)$
$= \sin x\left(\frac{1}{\sin x}\right)\csc x$

17. $\sin x \csc x - \sin^2 x = 1 - \sin^2 x$

19. $\tan y \cot y + \tan^2 y = 1 + \tan^2 y$

21. $\sin x\left(\frac{\sin x}{\cos x}\right) + \cos x = \frac{\sin^2 x + \cos^2 x}{\cos x} = \frac{1}{\cos x}$

23. $\cos\theta\left(\frac{\cos\theta}{\sin\theta}\right) + \sin\theta = \frac{\cos^2\theta + \sin^2\theta}{\sin\theta} = \frac{1}{\sin\theta}$

25. $\sec\theta\left(\frac{\sin\theta}{\cos\theta}\right)\csc\theta = \sec\theta\left(\frac{1}{\cos\theta}\right)(\sin\theta\csc\theta)$
$= \sec\theta\,(\sec\theta)\,(1)$

27. $\cot\theta\,(\sec^2\theta - 1) = \cot\theta\tan^2\theta = (\cot\theta\tan\theta)\tan\theta$

29. $\frac{\sin x}{\cos x} + \frac{\cos x}{\sin x} = \frac{\sin^2 x + \cos^2 x}{\cos x \sin x} = \frac{1}{\cos x \sin x}$

31. $(1 - \sin^2 x) - \sin^2 x$

33. $\frac{\sin x\,(1 + \cos x)}{1 - \cos^2 x} = \frac{1 + \cos x}{\sin x}$

35. $\frac{(1/\cos x) + (1/\sin x)}{1 + (\sin x/\cos x)} = \frac{(\sin x + \cos x)/(\cos x \sin x)}{(\cos x + \sin x)/\cos x}$
$= \frac{\cos x}{\cos x \sin x}$

37. $\frac{\sin^2 x}{\cos^2 x}\cos^2 x + \frac{\cos^2 x}{\sin^2 x}\sin^2 x = \sin^2 x + \cos^2 x$

39. $\frac{\sec\theta}{\frac{1}{\sec\theta}} - \frac{\tan\theta}{\frac{1}{\tan\theta}} = \sec^2\theta - \tan^2\theta$

41. $\frac{\sin^2 x + \cos^2 x - 2\cos^2 x}{\sin x \cos x} = \frac{\sin^2 x - \cos^2 x}{\sin x \cos x}$
$= \frac{\sin x}{\cos x} - \frac{\cos x}{\sin x}$

43. $\cos^3 x\left(\frac{1}{\sin^3 x}\right)\left(\frac{\sin^3 x}{\cos^3 x}\right) = 1$

45. $\frac{1}{\cos x} + \frac{\sin x}{\cos x} + \frac{\cos x}{\sin x} = \frac{\sin x + \cos^2 x + \sin^2 x}{\sin x \cos x}$

47. $\frac{\cos\theta + \sin\theta}{1 + \frac{\sin\theta}{\cos\theta}} = \frac{\cos\theta + \sin\theta}{\frac{\cos\theta + \sin\theta}{\cos\theta}}$

49. $\left(\frac{\sin x}{\cos x} + \frac{\cos x}{\sin x}\right)\sin x \cos x$
$= \frac{\sin^2 x \cos x}{\cos x} + \frac{\sin x \cos^2 x}{\sin x}$

51. $\frac{(\sin^2 x - \cos^2 x)(\sin^2 x + \cos^2 x)}{(1 - \cot^2 x)(1 + \cot^2 x)}$
$= \frac{\sin^2 x - \cos^2 x}{[1 - (\cos^2 x/\sin^2 x)]\csc^2 x}$

53. $\sec^2 x - 1 + 1 - \tan x + \tan x$

55. infinite series: $\frac{1}{1 - \sin^2 x} = \frac{1}{\cos^2 x}$

57. $0 = \cos A \cos B \cos C + \sin A \sin B$,
$\cos C = -\frac{\sin A \sin B}{\cos A \cos B}$

59. $l = a\csc\theta + a\sec\theta = a\left(\frac{1}{\sin\theta} + \frac{\tan\theta}{\sin\theta}\right)$

61. $\sin^2 x - \sin^2 x \sec^2 x + \cos^2 x + \cos^2 x \sec^4 x$
$= \sin^2 x - \tan^2 x + \cos^2 x + \sec^2 x = 2$

63. $\left(\frac{r}{x}\right)^2 + \left(\frac{r}{y}\right)^2 = \frac{r^2(x^2 + y^2)}{x^2 y^2}$

65. $\sqrt{1 - \cos^2\theta} = \sqrt{\sin^2\theta}$

67. $\sqrt{4 + 4\tan^2\theta} = 2\sqrt{1 + \tan^2\theta}$

Exercises 20-2, page 514

1. $\sin 105° = \sin 60° \cos 45° + \cos 60° \sin 45°$
$= \frac{\sqrt{3}}{2}\frac{\sqrt{2}}{2} + \frac{1}{2}\frac{\sqrt{2}}{2} = 0.9659$

3. $\cos 15° = \cos(60° - 45°)$
$= \cos 60° \cos 45° + \sin 60° \sin 45°$
$= \left(\frac{1}{2}\right)\left(\frac{1}{2}\sqrt{2}\right) + \left(\frac{1}{2}\sqrt{3}\right)\left(\frac{1}{2}\sqrt{2}\right)$
$= \frac{1}{4}\sqrt{2} + \frac{1}{4}\sqrt{6} = \frac{1}{4}(\sqrt{2} + \sqrt{6}) = 0.9659$

5. $-\frac{33}{65}$ **7.** $-\frac{56}{65}$ **9.** $\sin 3x$ **11.** $\cos x$

13. $\cos(2 - x)$ **15.** 0 **17.** 1 **19.** 1

21. $\sin(180° - x) = \sin 180° \cos x - \cos 180° \sin x$
$= (0)\cos x - (-1)\sin x$

23. $\cos(0° - x) = \cos 0° \cos x + \sin 0° \sin x$
$= (1)\cos x + (0)\sin x$

25. $\sin(270° - x) = \sin 270° \cos x - \cos 270° \sin x$
$= (-1)\cos x - 0\,(\sin x)$

27. $\cos\left(\frac{1}{2}\pi - x\right) = \cos\frac{1}{2}\pi\cos x + \sin\frac{1}{2}\pi\sin x$
$= 0\,(\cos x) + 1\,(\sin x)$

29. $\cos(30° + x) = \cos 30° \cos x - \sin 30° \sin x$
$= \frac{1}{2}\sqrt{3}\cos x - \frac{1}{2}\sin x = \frac{1}{2}(\sqrt{3}\cos x - \sin x)$

31. $\sin\left(\frac{\pi}{4} + x\right) = \sin\frac{\pi}{4}\cos x + \cos\frac{\pi}{4}\sin x$
$= \frac{1}{2}\sqrt{2}\cos x + \frac{1}{2}\sqrt{2}\sin x$

33. $(\sin x \cos y + \cos x \sin y)(\sin x \cos y - \cos x \sin y)$
$= \sin^2 x \cos^2 y - \cos^2 x \sin^2 y$
$= \sin^2 x(1 - \sin^2 y) - (1 - \sin^2 x)\sin^2 y$

35. $(\cos\alpha\cos\beta - \sin\alpha\sin\beta)$
$+ (\cos\alpha\cos\beta + \sin\alpha\sin\beta)$

37, 39, 41, 43. Use the indicated method.

45. $i_0\sin(\omega t + \alpha) = i_0(\sin\omega t\cos\alpha + \cos\omega t\sin\alpha)$

47. $\tan\alpha(R + \cos\beta) = \sin\beta,\ R = \dfrac{\sin\beta - \tan\alpha\cos\beta}{\tan\alpha}$
$= \dfrac{\sin\beta\cos\alpha - \cos\beta\sin\alpha}{\cos\alpha\tan\alpha}$

Exercises 20-3, page 517

1. $\sin 60° = \sin 2(30°) = 2\sin 30°\cos 30°$
$= 2\left(\frac{1}{2}\right)\left(\frac{1}{2}\sqrt{3}\right) = \frac{1}{2}\sqrt{3}$

3. $\cos 120° = \cos 2(60°) = \cos^2 60° - \sin^2 60°$
$= \left(\frac{1}{2}\right)^2 - \left(\frac{1}{2}\sqrt{3}\right)^2 = -\frac{1}{2}$

5. $\sin 258° = 2\sin 129°\cos 129° = -0.9781476$

7. $\cos 96° = \cos^2 48° - \sin^2 48° = -0.1045285$

9. $\frac{24}{25}$ **11.** 0.6 **13.** $2\sin 8x$ **15.** $\cos 8x$

17. $\cos x$ **19.** $-2\cos 4x$

21. $\cos^2\alpha - (1 - \cos^2\alpha)$

23. $\dfrac{\cos x - (\sin x/\cos x)\sin x}{1/\cos x} = \cos^2 x - \sin^2 x$

25. $(\cos^2 x - \sin^2 x)(\cos^2 x + \sin^2 x)$
$= (\cos^2 x - \sin^2 x)(1)$

27. $\dfrac{2\sin 2\theta\cos 2\theta}{\sin 2\theta}$ **29.** $\dfrac{2\sin\theta\cos\theta}{1 + 2\cos^2\theta - 1} = \dfrac{\sin\theta}{\cos\theta}$

31. $\dfrac{1}{\sec^2 x} - \dfrac{\tan^2 x}{\sec^2 x} = \cos^2 x - \dfrac{\sin^2 x}{\cos^2 x\sec^2 x}$

33. $\dfrac{2\tan x}{\sin 2x} = \dfrac{2(\sin x/\cos x)}{2\sin x\cos x} = \dfrac{1}{\cos^2 x}$

35. $\dfrac{\sin 3x\cos x - \cos 3x\sin x}{\sin x\cos x} = \dfrac{\sin 2x}{\frac{1}{2}\sin 2x}$

37. $\sin(2x + x) = \sin 2x\cos x + \cos 2x\sin x$
$= (2\sin x\cos x)(\cos x) + (\cos^2 x - \sin^2 x)\sin x$

39. Use the indicated method.

41. $R = v\left(\dfrac{2v\sin\alpha}{g}\right)\cos\alpha = \dfrac{v^2(2\sin\alpha\cos\alpha)}{g}$

43. $vi\sin\omega t\sin\left(\omega t - \frac{\pi}{2}\right)$
$= vi\sin\omega t\left(\sin\omega t\cos\frac{\pi}{2} - \cos\omega t\sin\frac{\pi}{2}\right)$
$= vi\sin\omega t[-(\cos\omega t)(1)] = -\frac{1}{2}vi(2\sin\omega t\cos\omega t)$

Exercises 20-4, page 521

1. $\cos 15° = \cos\frac{1}{2}(30°) = \sqrt{\dfrac{1 + \cos 30°}{2}} = \sqrt{\dfrac{1.8660}{2}}$
$= 0.9659$

3. $\sin 75° = \sin\frac{1}{2}(150°) = \sqrt{\dfrac{1 - \cos 150°}{2}}$
$= \sqrt{\dfrac{1.8660}{2}} = 0.9659$

5. $\sin 118° = 0.8829476$

7. $\sqrt{2\left(\dfrac{1 + \cos 164°}{2}\right)} = \sqrt{2}\cos 82° = 0.1968205$

9. $\sin 3x$ **11.** $4\cos 2x$ **13.** $\frac{1}{26}\sqrt{26}$

15. 0.1414 **17.** $\pm\sqrt{\dfrac{2}{1 - \cos\alpha}}$

19. $\tan\frac{1}{2}\alpha = \dfrac{1 - \cos\alpha}{\sin\alpha} = \dfrac{\sin\alpha}{1 + \cos\alpha}$

21. $\dfrac{1 - \cos\alpha}{2\sin\frac{1}{2}\alpha} = \dfrac{1 - \cos\alpha}{2\sqrt{\frac{1}{2}(1 - \cos\alpha)}} = \sqrt{\dfrac{1 - \cos\alpha}{2}}$

23. $2\left(\dfrac{1 - \cos x}{2}\right) + \cos x$

25. $\sqrt{\dfrac{(1 + \cos\theta)(1 - \cos\theta)}{2(1 - \cos\theta)}} = \dfrac{\sin\theta}{\sqrt{4\left(\dfrac{1 - \cos\theta}{2}\right)}}$

27. $2\left(\dfrac{1 - \cos\alpha}{2}\right) - \left(\dfrac{1 + \cos\alpha}{2}\right)$

29. $\sin\omega t = \pm\sqrt{\dfrac{1 - \cos 2\omega t}{2}}$,

$\sin^2\omega t = \frac{1}{2}(1 - \cos 2\omega t)$

31. $\dfrac{\sqrt{\dfrac{1 - \cos(A + \phi)}{2}}}{\sqrt{\dfrac{1 - \cos A}{2}}} = \sqrt{\dfrac{1 - \cos(A + \phi)}{1 - \cos A}}$

Exercises 20-5, page 526

1. $\frac{\pi}{2}$ **3.** π **5.** $\frac{\pi}{3}, \frac{2\pi}{3}, \frac{4\pi}{3}, \frac{5\pi}{3}$

7. $0, \frac{\pi}{6}, \frac{5\pi}{6}, \pi$ **9.** $\frac{\pi}{2}, \frac{3\pi}{2}$

11. $\frac{\pi}{4}, \frac{3\pi}{4}, \frac{5\pi}{4}, \frac{7\pi}{4}$ **13.** 0.2618, 1.309, 3.403, 4.451

15. $0, \pi$ **17.** $\frac{3\pi}{4} = 2.36, \frac{7\pi}{4} = 5.50$

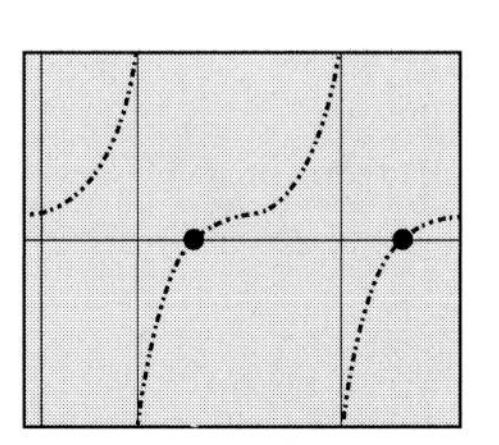

19. 1.9823, 4.3009

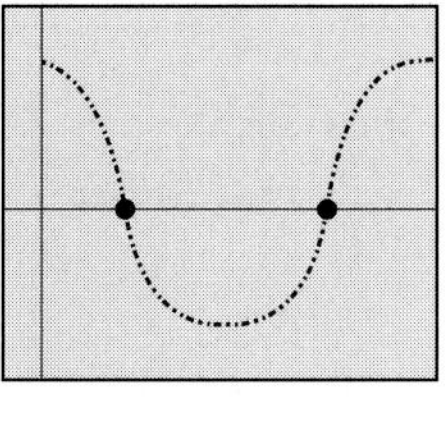

21. $\frac{\pi}{3} = 1.05, \frac{2\pi}{3} = 2.09, \frac{4\pi}{3} = 4.19, \frac{5\pi}{3} = 5.24$

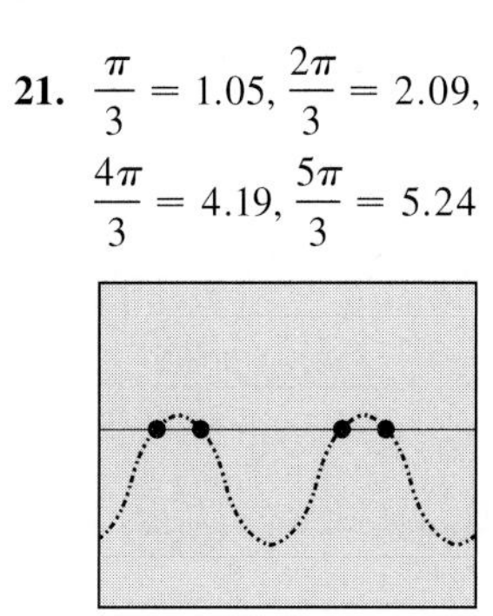

23. $\frac{\pi}{12} = 0.26, \frac{\pi}{4} = 0.79, \frac{5\pi}{12} = 1.31, \frac{3\pi}{4} = 2.36, \frac{13\pi}{12} = 3.40, \frac{5\pi}{4} = 3.93, \frac{17\pi}{12} = 4.45, \frac{7\pi}{4} = 5.50$

25. $0 = 0.00, \frac{\pi}{3} = 1.05, \pi = 3.14, \frac{5\pi}{3} = 5.24$

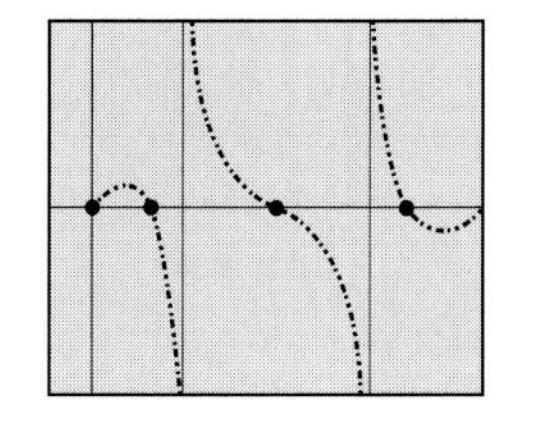

27. 3.569, 5.856

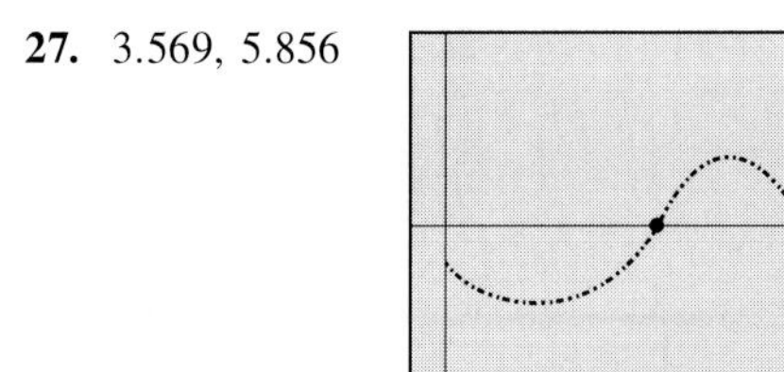

29. 0.7854, 1.249, 3.927, 4.391

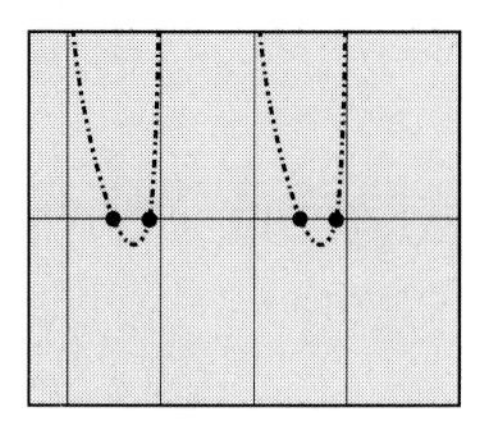

31. $\frac{3\pi}{8} = 1.18, \frac{7\pi}{8} = 2.75, \frac{11\pi}{8} = 4.32, \frac{15\pi}{8} = 5.89$

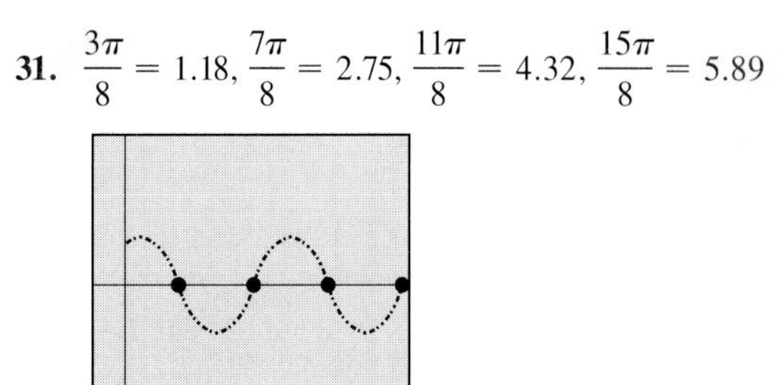

33. 6.56×10^{-4} **35.** 10.2 s, 15.7 s, 21.2 s, 47.1 s

37. −2.28, 0.00, 2.28 **39.** 0.29, 0.95

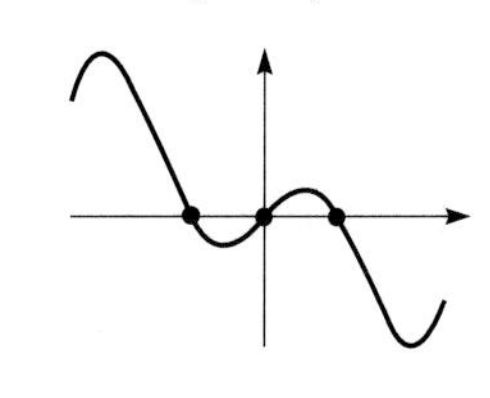

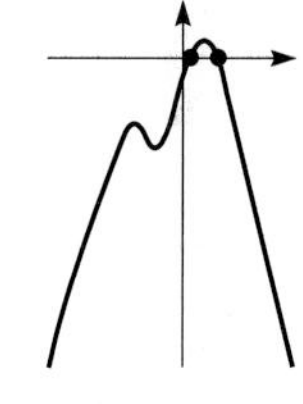

41. 2.10 **43.** 1.08

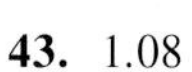

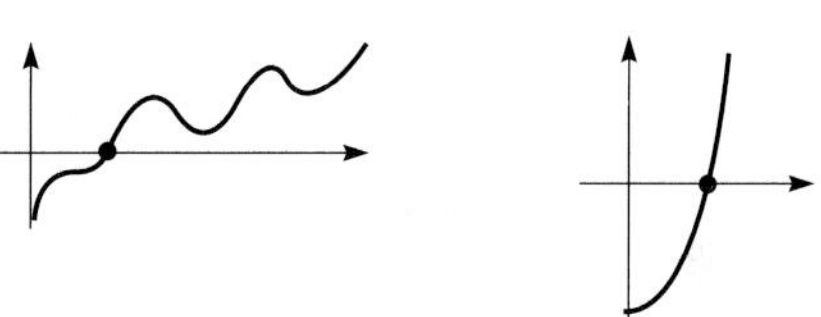

Exercises 20-6, page 531

1. y is the angle whose tangent is x.

3. y is the angle whose cotangent is $3x$.

5. y is twice the angle whose sine is x.

7. y is 5 times the angle whose cosine is $2x - 1$.

9. $\frac{\pi}{3}$ **11.** 0 **13.** $-\frac{\pi}{3}$ **15.** $\frac{\pi}{3}$ **17.** $\frac{\pi}{6}$

19. $-\frac{\pi}{4}$ **21.** $\frac{\pi}{4}$ **23.** $-\frac{\pi}{3}$ **25.** 0

27. $\frac{1}{2}\sqrt{3}$ **29.** $\frac{1}{2}\sqrt{2}$ **31.** −1 **33.** −1.3090

35. −0.9838 **37.** 1.4413 **39.** 1.4503

41. −1.2389 **43.** −0.2239 **45.** $x = \frac{1}{3}\sin^{-1} y$

47. $x = 4 \tan y$ **49.** $x = \frac{1}{3}\sec^{-1}\left(\frac{y-1}{3}\right)$

51. $x = 1 - \cos(1 - y)$ **53.** $\frac{x}{\sqrt{1 - x^2}}$ **55.** $\frac{1}{x}$

57. $\frac{3x}{\sqrt{9x^2 - 1}}$ **59.** $2x\sqrt{1 - x^2}$

61. $t = \frac{1}{2\omega} \cos^{-1} \frac{y}{A} - \frac{\phi}{\omega}$

63. $t = \frac{1}{\omega}\left(\sin^{-1}\frac{i}{I_m} - \alpha - \phi\right)$

65. $\sin\left(\sin^{-1}\frac{3}{5} + \sin^{-1}\frac{5}{13}\right) = \frac{3}{5} \cdot \frac{12}{13} + \frac{4}{5} \cdot \frac{5}{13} = \frac{56}{65}$

67. $\frac{\pi}{6} + \frac{\pi}{3} = \frac{\pi}{2}$ **69.** $\sin^{-1}\left(\frac{a}{c}\right)$

71. Let y = height to top of pedestal; $\tan \alpha = \frac{46.0 + y}{d}$, $\tan \beta = \frac{y}{d}$; $\tan \alpha = \frac{46.0 + d \tan \beta}{d}$

Review Exercises for Chapter 20, page 533

1. $\sin(90° + 30°) = \sin 90° \cos 30° + \cos 90° \sin 30°$
$= (1)\left(\frac{1}{2}\sqrt{3}\right) + (0)\left(\frac{1}{2}\right) = \frac{1}{2}\sqrt{3}$

3. $\sin(180° - 45°) = \sin 180° \cos 45° - \cos 180° \sin 45°$
$= 0\left(\frac{1}{2}\sqrt{2}\right) - (-1)\left(\frac{1}{2}\sqrt{2}\right) = \frac{1}{2}\sqrt{2}$

5. $\cos 2(90°) = \cos^2 90° - \sin^2 90° = 0 - 1 = -1$

7. $\sin \frac{1}{2}(90°) = \sqrt{\frac{1}{2}(1 - \cos 90°)} = \sqrt{\frac{1}{2}(1 - 0)}$
$= \frac{1}{2}\sqrt{2}$

9. $\sin 52° = 0.788\,010\,8$ **11.** $\sin 92° = 0.999\,390\,8$

13. $\cos 6° = 0.994\,521\,9$ **15.** $\cos 164° = -0.961\,261\,7$

17. $\sin 5x$ **19.** $4 \sin 12x$ **21.** $2 \cos 12x$

23. $2 \cos x$ **25.** $-\frac{\pi}{2}$ **27.** 0.2619

29. $-\frac{1}{3}\sqrt{3}$ **31.** 0

33. $\frac{\frac{1}{\cos y}}{\frac{1}{\sin y}} = \frac{1}{\cos y} \cdot \frac{\sin y}{1}$

35. $\sin x \csc x - \sin^2 x = 1 - \sin^2 x$

37. $\frac{1 - \sin^2 \theta}{\sin \theta} = \frac{\cos^2 \theta}{\sin \theta}$

39. $\cos \theta\left(\frac{\cos \theta}{\sin \theta}\right) + \sin \theta = \frac{\cos^2 \theta + \sin^2 \theta}{\sin \theta}$

41. $\frac{(\sec^2 x - 1)(\sec^2 x + 1)}{\tan^2 x} = \sec^2 x + 1$

43. $2\left(\frac{1}{\sin 2x}\right)\left(\frac{\cos x}{\sin x}\right) = 2\left(\frac{1}{2 \sin x \cos x}\right)\left(\frac{\cos x}{\sin x}\right)$
$= \frac{1}{\sin^2 x}$

45. $\frac{\cos^2 \theta}{\sin^2 \theta}$ **47.** $\frac{\cos^2 \theta - \sin^2 \theta}{\cos^2 \theta} = 1 - \frac{\sin^2 \theta}{\cos^2 \theta}$

49. $\frac{1}{2}\left(2 \sin \frac{\theta}{2} \cos \frac{\theta}{2}\right)$

51. $\frac{1}{\cos x} + \frac{\sin x}{\cos x} = \frac{(1 + \sin x)(1 - \sin x)}{\cos x(1 - \sin x)}$
$= \frac{1 - \sin^2 x}{\cos x(1 - \sin x)}$

53. $\cos[(x - y) + y]$ **55.** $\sin 4x(\cos 4x)$

57. $\frac{\sin x}{\frac{1}{\sin x} - \frac{\cos x}{\sin x}} = \frac{\sin^2 x}{1 - \cos x} = \frac{1 - \cos^2 x}{1 - \cos x}$

59. $\frac{\sin x \cos y + \cos x \sin y + \sin x \cos y - \cos x \sin y}{\cos x \cos y - \sin x \sin y + \cos x \cos y + \sin x \sin y}$
$= \frac{2 \sin x \cos y}{2 \cos x \cos y}$

61. $x = \frac{1}{2} \cos^{-1} \frac{1}{2} y$ **63.** $x = \frac{1}{5} \sin \frac{1}{3}\left(\frac{1}{4}\pi - y\right)$

65. 1.2925, 4.4341 **67.** $\frac{\pi}{6}, \frac{5\pi}{6}, \frac{7\pi}{6}, \frac{11\pi}{6}$

69. $0, \frac{\pi}{2}, \pi, \frac{3\pi}{2}$ **71.** $\frac{\pi}{6}, \frac{5\pi}{6}, \frac{7\pi}{6}, \frac{11\pi}{6}$ **73.** $0, \pi$

75. 0

77. 1.56, 2.16, 3.46 **79.** −2.31, 1.14

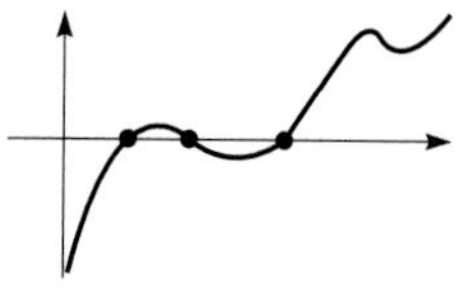

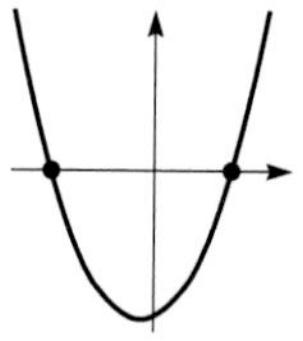

81. $\frac{1}{x}$ **83.** $2x\sqrt{1 - x^2}$ **85.** $2\sqrt{1 - \cos^2\theta}$

87. $\frac{\tan \theta}{\sqrt{1 + \tan^2\theta}} = \frac{\tan \theta}{\sec \theta}$

89. $2 \sin x \cos x \frac{1}{\sin x} \cdot \frac{1}{\cos x} = 2$

91. $C\left(\frac{A}{C} \sin 2t + \frac{B}{C} \cos 2t\right)$
$= C(\cos \alpha \sin 2t + \sin \alpha \cos 2t)$

93. $R = \sqrt{(A \cos \theta - B \sin \theta)^2 + (A \sin \theta + B \cos \theta)^2}$
$= \sqrt{A^2(\cos^2\theta + \sin^2\theta) + B^2(\sin^2\theta + \cos^2\theta)}$

95. $\frac{k}{2} \cdot \frac{1}{\sin^2\left(\frac{\theta}{2}\right)} = \frac{k}{2} \cdot \frac{1}{\left(\frac{1 - \cos \theta}{2}\right)}$

97. $\theta = \alpha + R \sin \omega t$

99. $1 - (1 - 2\sin^2 \phi) - (2 \sin \phi \cos \phi) \tan \alpha$
$= 2 \sin^2 \phi - 2 \sin \phi \cos \phi \tan \alpha$

101. 83.7° **103.** 54.7°

Exercises 21-1, page 540

1. $2\sqrt{29}$ **3.** 3 **5.** 55 **7.** 7 **9.** 2.86

11. $\frac{5}{2}$ **13.** undefined **15.** $-\frac{3}{4}$ **17.** 0

19. 0.747 **21.** $\frac{1}{3}\sqrt{3}$ **23.** −1.084 **25.** 20.0°

27. 98.50° **29.** parallel **31.** perpendicular

33. 8, −2 **35.** −3 **37.** two sides equal $2\sqrt{10}$

39. $m_1 = \frac{5}{12}, m_2 = \frac{4}{3}$ **41.** 10

43. $4\sqrt{10} + 4\sqrt{2} = 18.3$ **45.** (1, 5)

47. (−2.8, 4.2)

Exercises 21-2, page 545

1. $4x - y + 20 = 0$ **3.** $7x - 2y - 24 = 0$

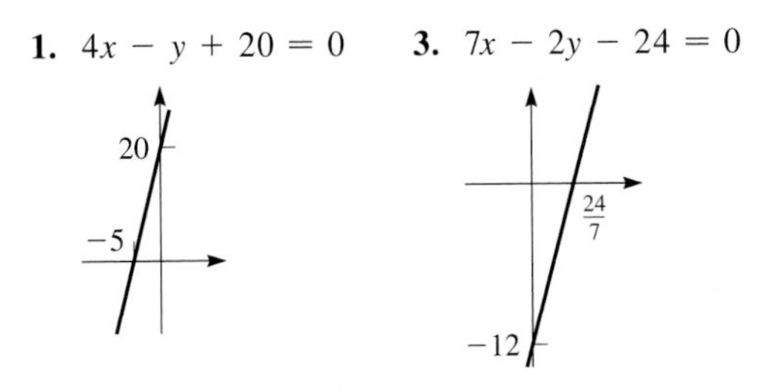

5. $x - y + 2 = 0$ **7.** $y = -2.7$

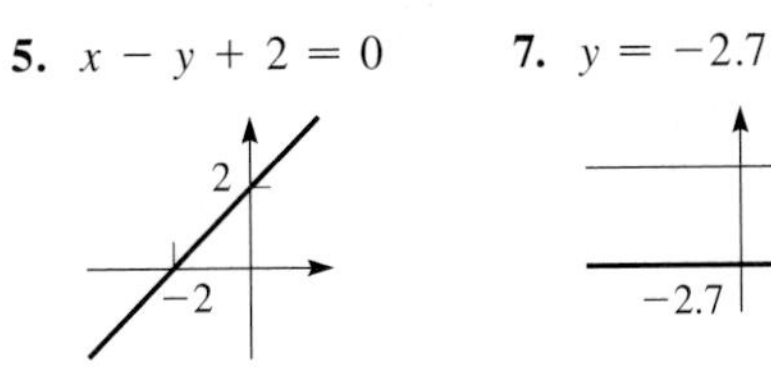

−2.7

9. $x = -3$ **11.** $3x - 2y - 12 = 0$

−3

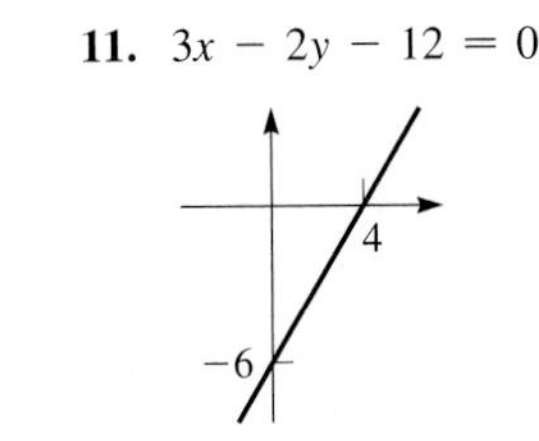

13. $x + 3y + 5 = 0$ **15.** $x + 2y - 4 = 0$

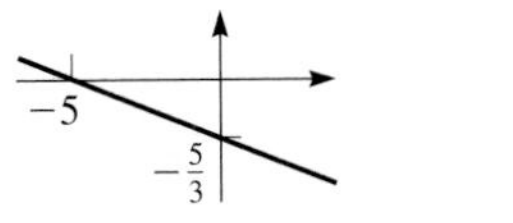

2
4

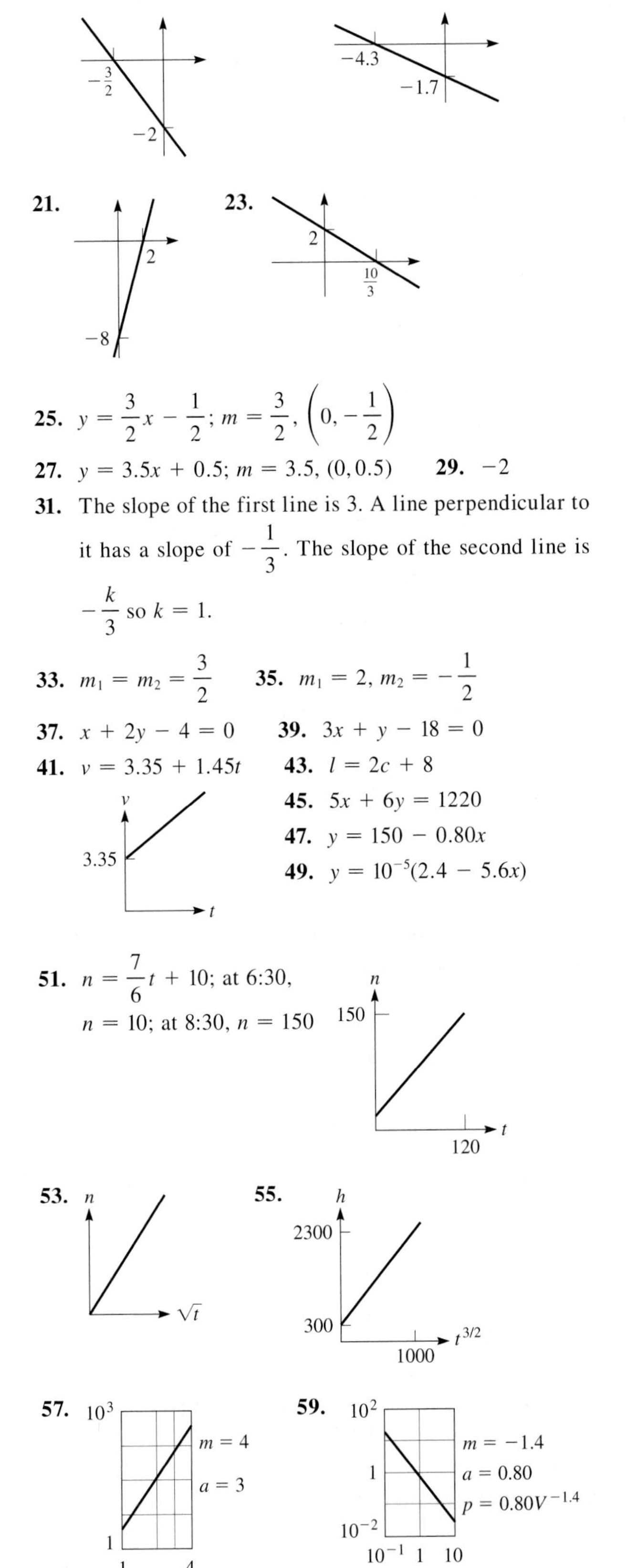

17. $4x + 3y + 6 = 0$ **19.** $0.4x + y + 1.7 = 0$

21. **23.**

25. $y = \frac{3}{2}x - \frac{1}{2}$; $m = \frac{3}{2}$, $\left(0, -\frac{1}{2}\right)$

27. $y = 3.5x + 0.5$; $m = 3.5$, $(0, 0.5)$ **29.** −2

31. The slope of the first line is 3. A line perpendicular to it has a slope of $-\frac{1}{3}$. The slope of the second line is $-\frac{k}{3}$ so $k = 1$.

33. $m_1 = m_2 = \frac{3}{2}$ **35.** $m_1 = 2, m_2 = -\frac{1}{2}$

37. $x + 2y - 4 = 0$ **39.** $3x + y - 18 = 0$

41. $v = 3.35 + 1.45t$ **43.** $l = 2c + 8$

45. $5x + 6y = 1220$

47. $y = 150 - 0.80x$

49. $y = 10^{-5}(2.4 - 5.6x)$

51. $n = \frac{7}{6}t + 10$; at 6:30, $n = 10$; at 8:30, $n = 150$

53. **55.**

57. **59.**

Exercises 21-3, page 550

1. $(2, 1)$, $r = 5$ **3.** $(-1, 0)$, $r = 2$ **5.** $x^2 + y^2 = 9$

7. $(x - 2)^2 + (y - 2)^2 = 16$, or $x^2 + y^2 - 4x - 4y - 8 = 0$

9. $(x + 2)^2 + (y - 5)^2 = 5$, or $x^2 + y^2 + 4x - 10y + 24 = 0$

11. $(x - 12)^2 + (y + 15)^2 = 324$, or $x^2 + y^2 - 24x + 30y + 45 = 0$

13. $(x - 2)^2 + (y - 1)^2 = 8$, or $x^2 + y^2 - 4x - 2y - 3 = 0$

15. $(x + 3)^2 + (y - 5)^2 = 25$, or $x^2 + y^2 + 6x - 10y + 9 = 0$

17. $(x - 2)^2 + (y - 2)^2 = 4$, or $x^2 + y^2 - 4x - 4y + 4 = 0$

19. $(x - 2)^2 + (y - 5)^2 = 25$, or $x^2 + y^2 - 4x - 10y + 4 = 0$; and $(x + 2)^2 + (y + 5)^2 = 25$, or $x^2 + y^2 + 4x + 10y + 4 = 0$

21. $(0, 3)$, $r = 2$ **23.** $(-1, 5)$, $r = \frac{9}{2}$

C C

25. $(0, 0)$, $r = 5$ **27.** $(1, 0)$, $r = 3$

C C

29. $(-2.1, 1.3)$, $r = 3.1$ **31.** $(1, 2)$, $r = \frac{1}{2}\sqrt{22}$

C C

33. symmetrical to both axes and origin

35. symmetrical to y-axis **37.** $(7, 0)$, $(-1, 0)$

39. $3x^2 + 3y^2 + 4x + 8y - 20 = 0$, circle

41.

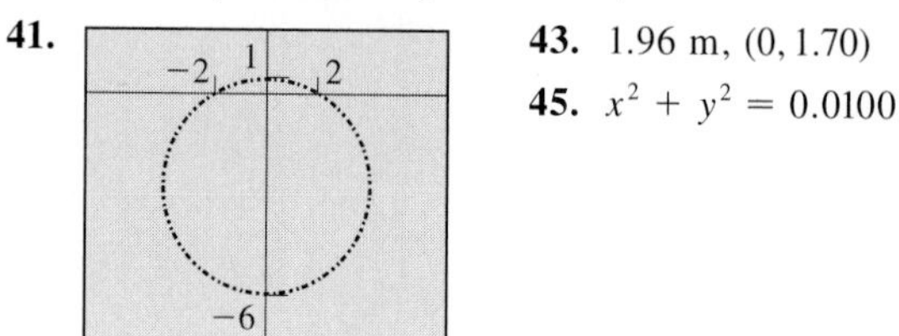

43. 1.96 m, $(0, 1.70)$

45. $x^2 + y^2 = 0.0100$

47. $(x - 500 \times 10^{-6})^2 + y^2 = 0.16 \times 10^{-6}$

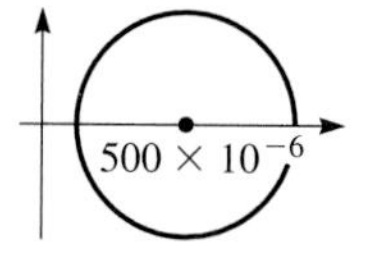

Exercises 21-4, page 555

1. $F(1, 0)$, $x = -1$ **3.** $F(-1, 0)$, $x = 1$

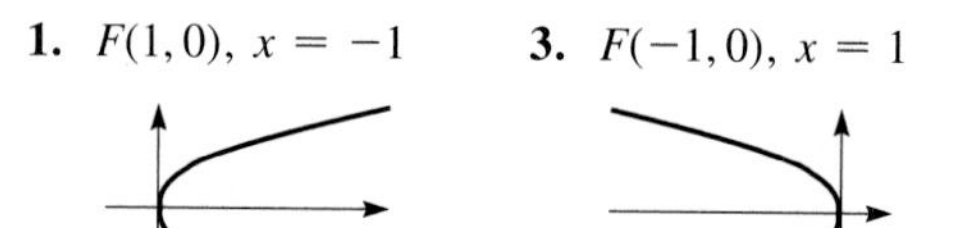

5. $F(0, 2)$, $y = -2$ **7.** $F(0, -1)$, $y = 1$

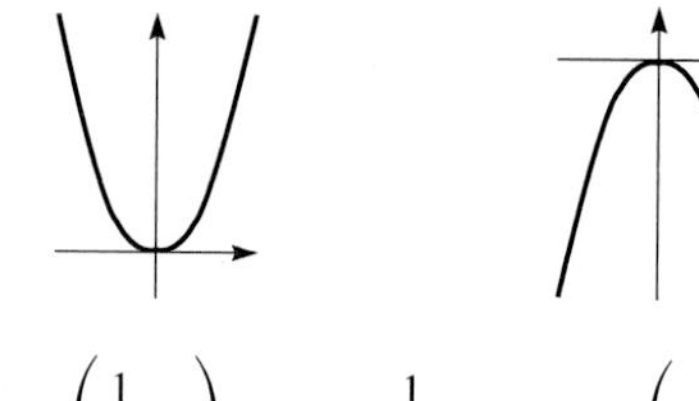

9. $F\left(\frac{1}{2}, 0\right)$, $x = -\frac{1}{2}$ **11.** $F\left(0, \frac{25}{48}\right)$, $y = -\frac{25}{48}$

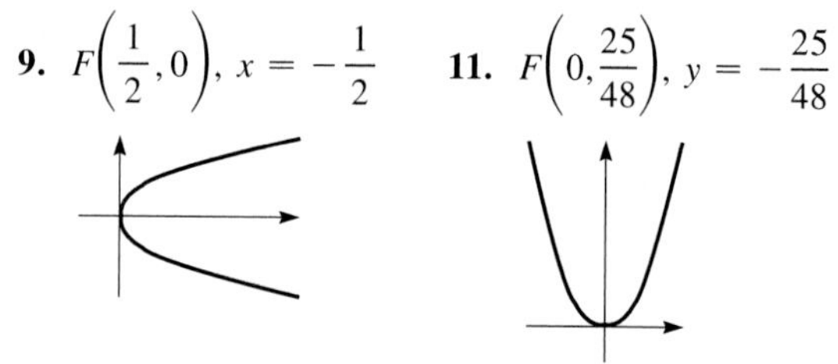

13. $y^2 = 12x$ **15.** $x^2 = 16y$ **17.** $x^2 = 4y$

19. $x^2 = \frac{1}{8}y$ **21.** $y^2 - 2y - 12x + 37 = 0$

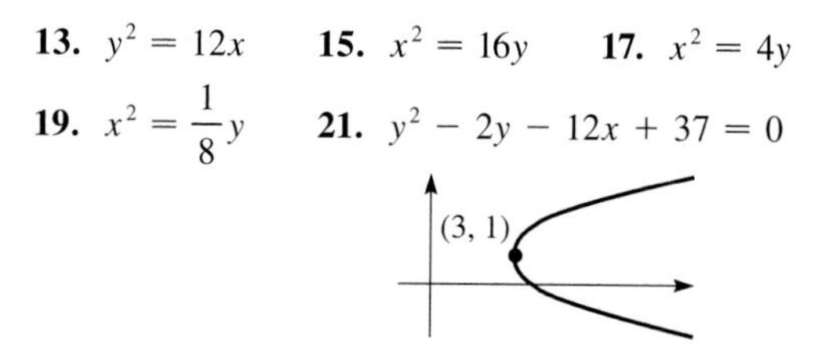

23.

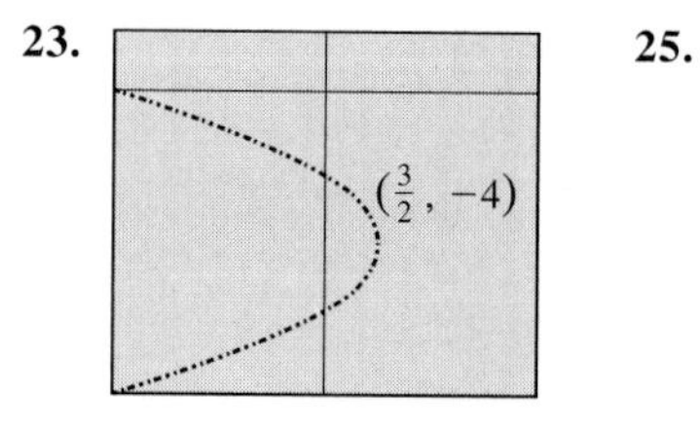

25.

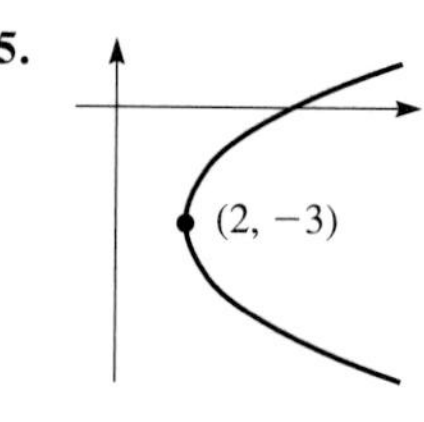

27. $4p$ **29.** $x^2 = 4550y$ **31.**

H

i

33. 0.919 m **35.** $y^2 = 1.2x$ **37.**

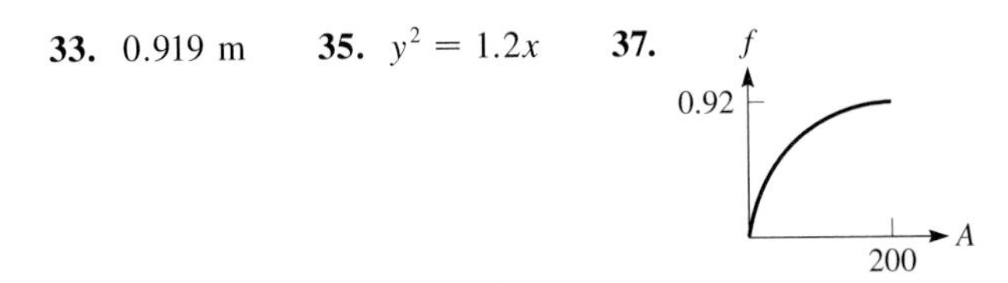

39. $y^2 = 8x$ or $x^2 = 8y$ with vertex midway between island and shore

Exercises 21-5, page 560

1. $V(2, 0)$, $V(-2, 0)$, $F(\sqrt{3}, 0)$, $F(-\sqrt{3}, 0)$

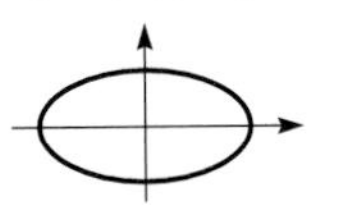

3. $V(0, 6)$, $V(0, -6)$, $F(0, \sqrt{11})$, $F(0, -\sqrt{11})$

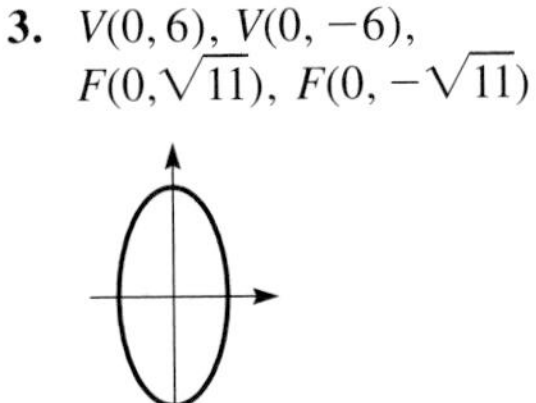

5. $V(3, 0)$, $V(-3, 0)$, $F(\sqrt{5}, 0)$, $F(-\sqrt{5}, 0)$

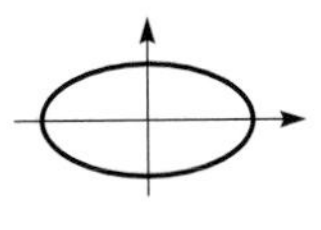

7. $V(0, 7)$, $V(0, -7)$, $F(0, \sqrt{45})$, $F(0, -\sqrt{45})$

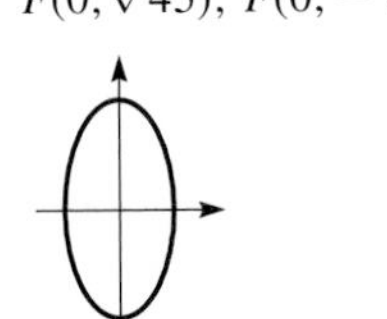

9. $V(0, 4)$, $V(0, -4)$, $F(0, \sqrt{14})$, $F(0, -\sqrt{14})$

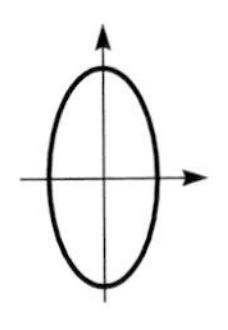

11. $V(0.25, 0)$, $V(-0.25, 0)$, $F(0.23, 0)$, $F(-0.23, 0)$

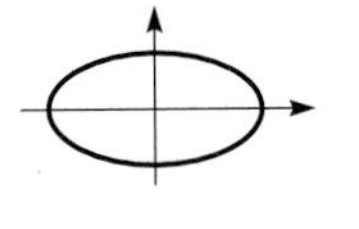

13. $\frac{x^2}{225} + \frac{y^2}{144} = 1$, or $144x^2 + 225y^2 = 32\,400$

15. $\frac{y^2}{9} + \frac{x^2}{5} = 1$, or $9x^2 + 5y^2 = 45$

17. $\frac{x^2}{64} + \frac{15y^2}{144} = 1$, or $3x^2 + 20y^2 = 192$

19. $\frac{x^2}{5} + \frac{y^2}{20} = 1$, or $4x^2 + y^2 = 20$

21. $16x^2 + 25y^2 - 32x - 50y - 359 = 0$

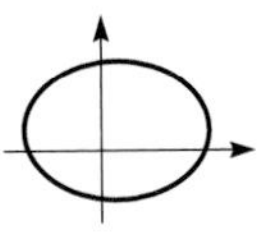

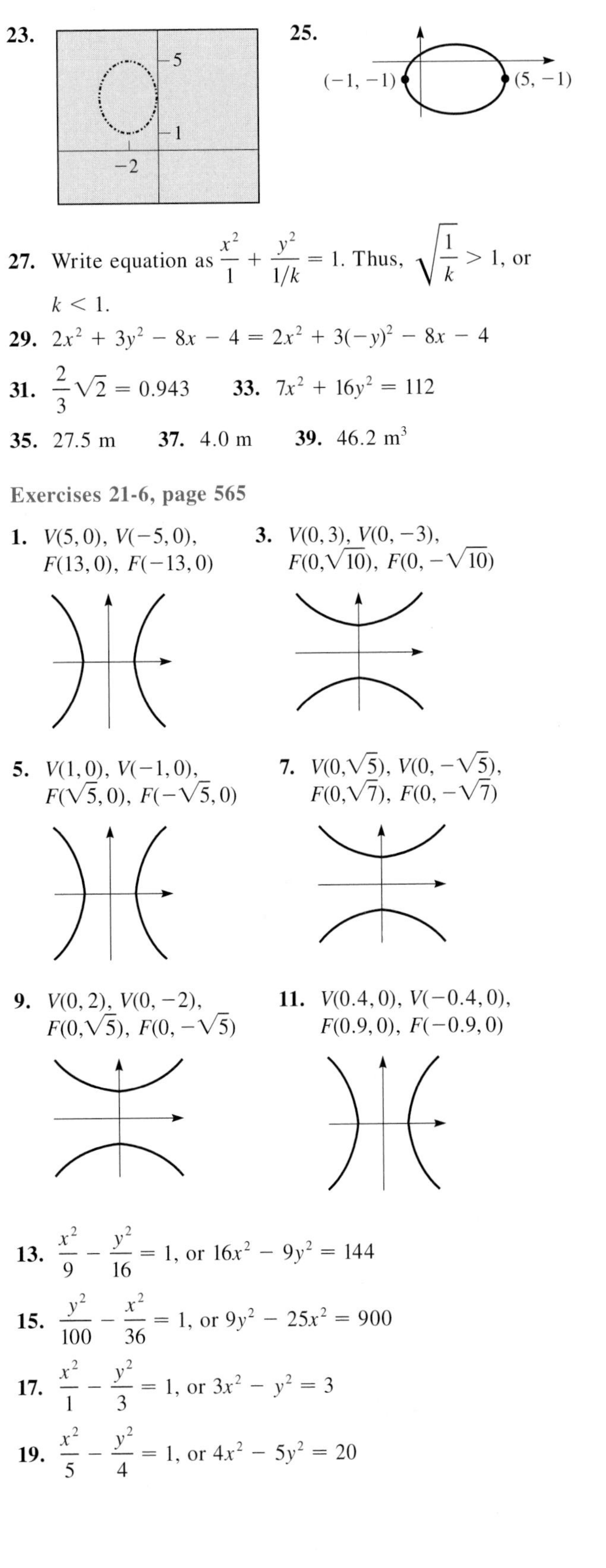

23.

25.

27. Write equation as $\frac{x^2}{1} + \frac{y^2}{1/k} = 1$. Thus, $\sqrt{\frac{1}{k}} > 1$, or $k < 1$.

29. $2x^2 + 3y^2 - 8x - 4 = 2x^2 + 3(-y)^2 - 8x - 4$

31. $\frac{2}{3}\sqrt{2} = 0.943$ **33.** $7x^2 + 16y^2 = 112$

35. 27.5 m **37.** 4.0 m **39.** 46.2 m^3

Exercises 21-6, page 565

1. $V(5, 0)$, $V(-5, 0)$, $F(13, 0)$, $F(-13, 0)$

3. $V(0, 3)$, $V(0, -3)$, $F(0, \sqrt{10})$, $F(0, -\sqrt{10})$

5. $V(1, 0)$, $V(-1, 0)$, $F(\sqrt{5}, 0)$, $F(-\sqrt{5}, 0)$

7. $V(0, \sqrt{5})$, $V(0, -\sqrt{5})$, $F(0, \sqrt{7})$, $F(0, -\sqrt{7})$

9. $V(0, 2)$, $V(0, -2)$, $F(0, \sqrt{5})$, $F(0, -\sqrt{5})$

11. $V(0.4, 0)$, $V(-0.4, 0)$, $F(0.9, 0)$, $F(-0.9, 0)$

13. $\frac{x^2}{9} - \frac{y^2}{16} = 1$, or $16x^2 - 9y^2 = 144$

15. $\frac{y^2}{100} - \frac{x^2}{36} = 1$, or $9y^2 - 25x^2 = 900$

17. $\frac{x^2}{1} - \frac{y^2}{3} = 1$, or $3x^2 - y^2 = 3$

19. $\frac{x^2}{5} - \frac{y^2}{4} = 1$, or $4x^2 - 5y^2 = 20$

21.

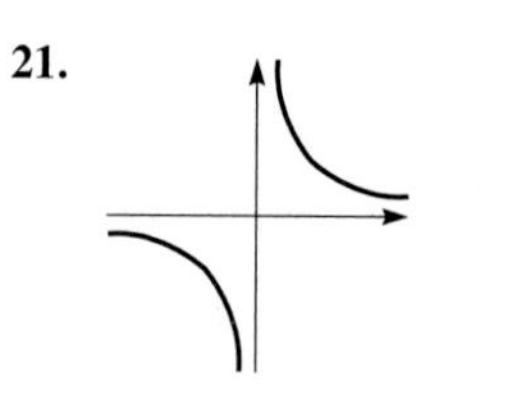

23. $9x^2 - 16y^2 - 108x + 64y + 116 = 0$

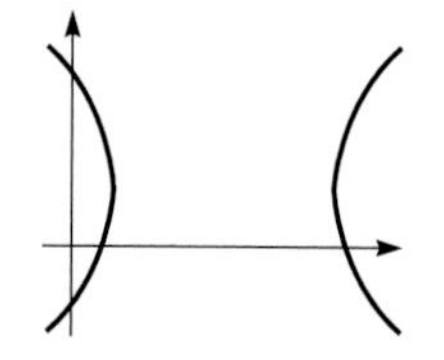

25. **27.**

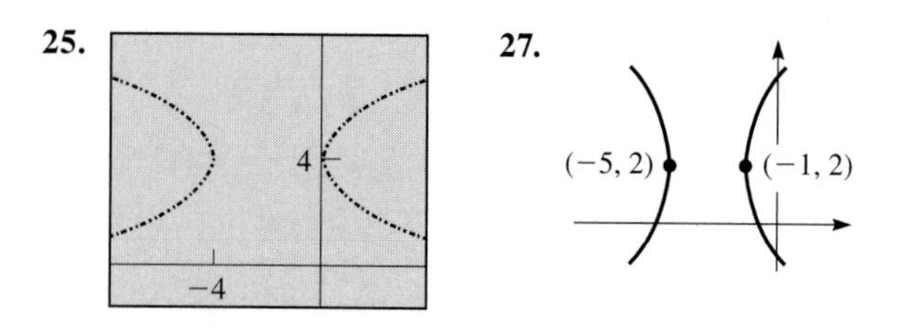

29. $x^2 - 2y^2 = 2$ **31.** $l^2 - x^2 = 2000^2$

33. $i = 6.00/R$ **35.**

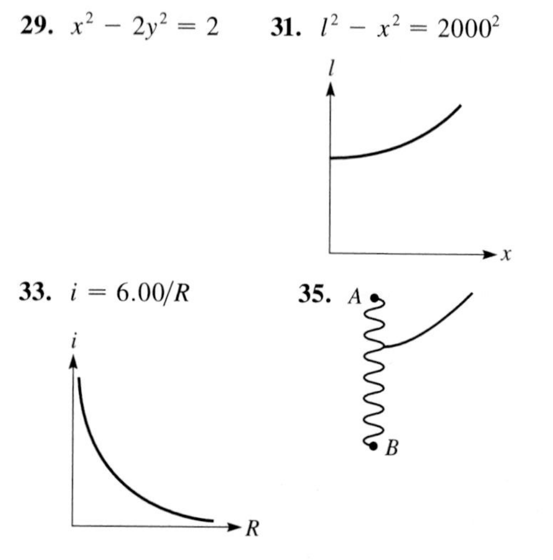

Exercises 21-7, page 569

1. parabola, $(-1, 2)$ **3.** hyperbola, $(1, 2)$

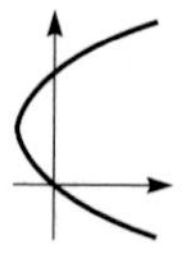

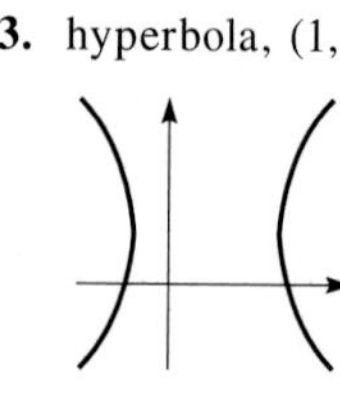

5. ellipse, $(-1, 0)$

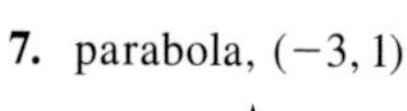

9. $(y - 3)^2 = 16(x + 1)$, or $y^2 - 6y - 16x - 7 = 0$

11. $(x + 3)^2 = 4(y - 2)$, or $x^2 + 6x - 4y + 17 = 0$

13. $\frac{(x + 2)^2}{25} + \frac{(y - 2)^2}{16} = 1$, or
$16x^2 + 25y^2 + 64x - 100y - 236 = 0$

15. $\frac{(y - 1)^2}{16} + \frac{(x + 2)^2}{4} = 1$, or
$4x^2 + y^2 + 16x - 2y + 1 = 0$

17. $\frac{(y - 2)^2}{1} - \frac{(x + 1)^2}{3} = 1$, or
$x^2 - 3y^2 + 2x + 12y - 8 = 0$

19. $\frac{(x + 1)^2}{9} - \frac{(y - 1)^2}{16} = 1$, or
$16x^2 - 9y^2 + 32x + 18y - 137 = 0$

21. parabola, $(-1, -1)$ **23.** ellipse, $(-3, 0)$

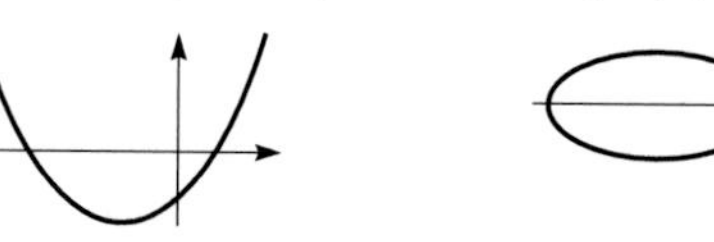

25. hyperbola, $(0, 4)$ **27.** parabola, $(1, 0)$

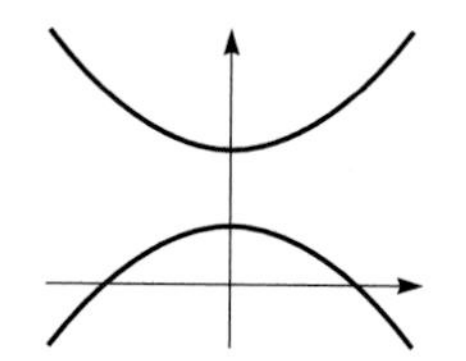

29. $x^2 - y^2 + 4x - 2y - 22 = 0$

31. $y^2 + 4x - 4 = 0$

33. $(x - 28)^2 = -\frac{28^2}{18}(y - 18)$

35. $\frac{x^2}{9.0} + \frac{y^2}{16} = 1, \frac{(x - 7.0)^2}{16} + \frac{y^2}{9.0} = 1$

Exercises 21-8, page 572

1. ellipse **3.** hyperbola **5.** circle **7.** parabola
9. hyperbola **11.** circle **13.** parabola
15. hyperbola **17.** ellipse **19.** ellipse
21. parabola; $V(-4, 0)$; $F(-4, 2)$

23. hyperbola; $C(1, -2)$; $V(1, -2 \pm \sqrt{2})$

25. ellipse; $C(5, 0)$; $V(5, \pm 2\sqrt{2})$

27. parabola; $V\left(-\frac{1}{2}, \frac{5}{2}\right)$; $F = \left(\frac{1}{2}, \frac{5}{2}\right)$

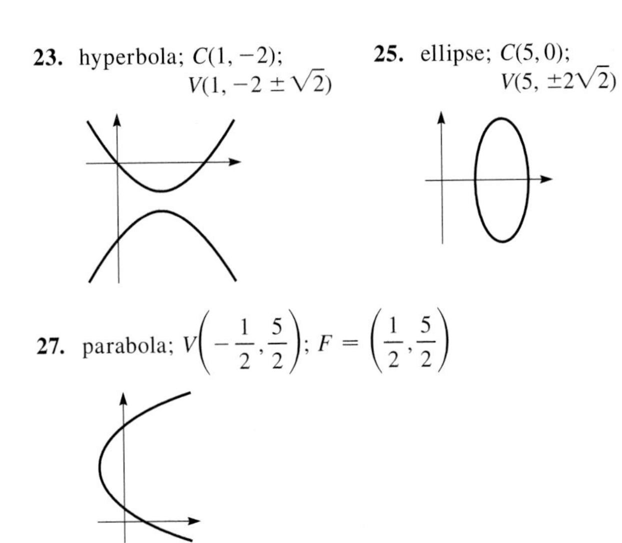

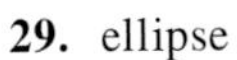

29. ellipse

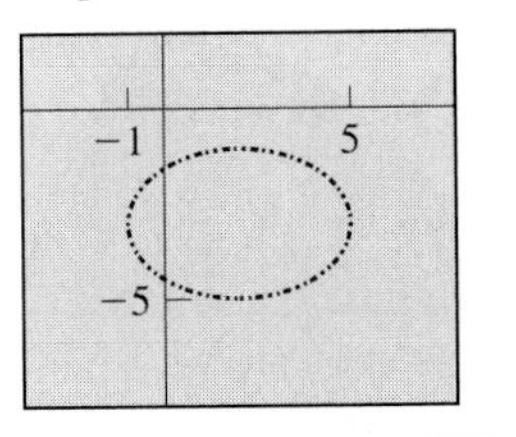

31. $y = \dfrac{-3x - 7 \pm \sqrt{60x + 139}}{9}$

Enter both functions (from quadratic formula).

33. (a) circle (b) hyperbola (c) ellipse

35. a point at the origin **37.** parabola

39. Circle if light beam is perpendicular to floor; otherwise, an ellipse.

Exercises 21-9, page 575

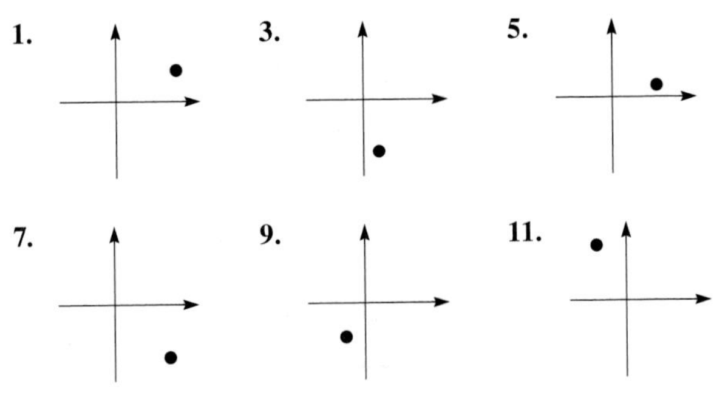

13. $\left(2, \frac{\pi}{6}\right)$ **15.** $\left(1, \frac{7\pi}{6}\right)$ **17.** $(-4, -4\sqrt{3})$

19. $(2.76, -1.17)$ **21.** $r = 3 \sec \theta$ **23.** $r = 0.9$

25. $r = 4 \cot \theta \csc \theta$ **27.** $r^2 = \dfrac{4}{1 + 3 \sin^2 \theta}$

29. $x^2 + y^2 - y = 0$ **31.** $x = 4$

33. $x^4 + y^4 - 4x^3 + 2x^2y^2 - 4xy^2 - 4y^2 = 0$

35. $(x^2 + y^2)^2 = 2xy$

37. $B_x = -\dfrac{k \sin \theta}{r}$, $B_y = \dfrac{k \cos \theta}{r}$

39. $x^4 + y^4 + 2x^2y^2 + 2x^2y + 2y^3 - 9x^2 - 8y^2 = 0$

Exercises 21-10, page 578

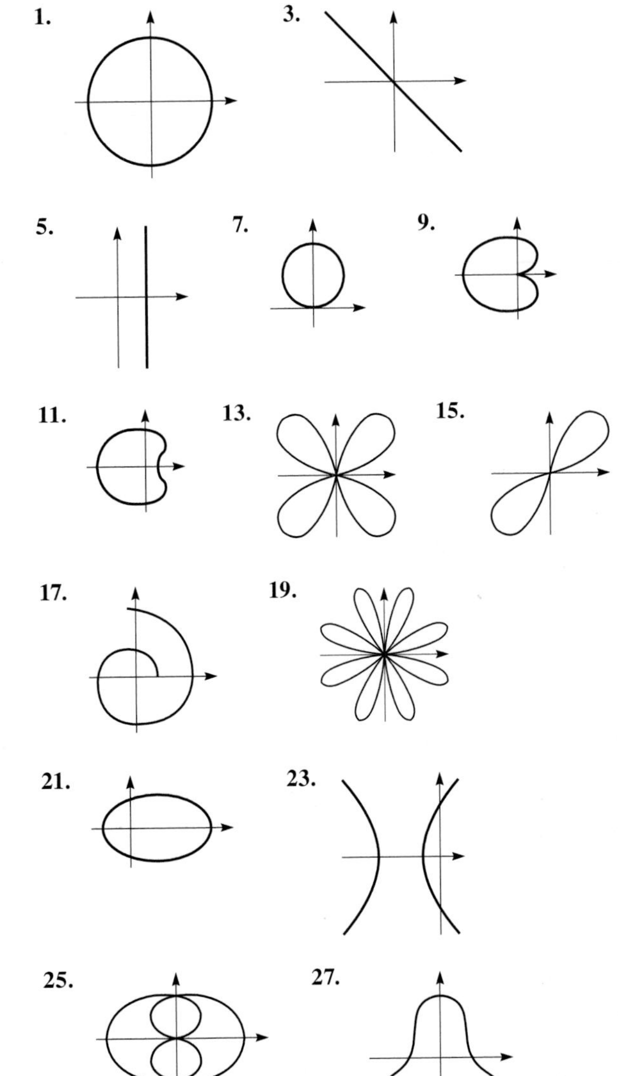

29. **31.**

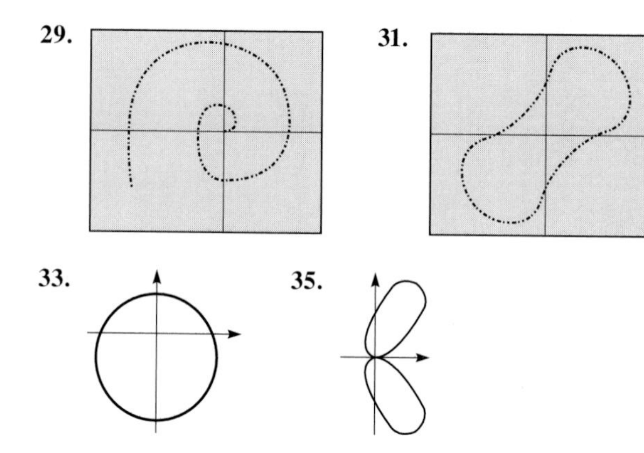

33. **35.**

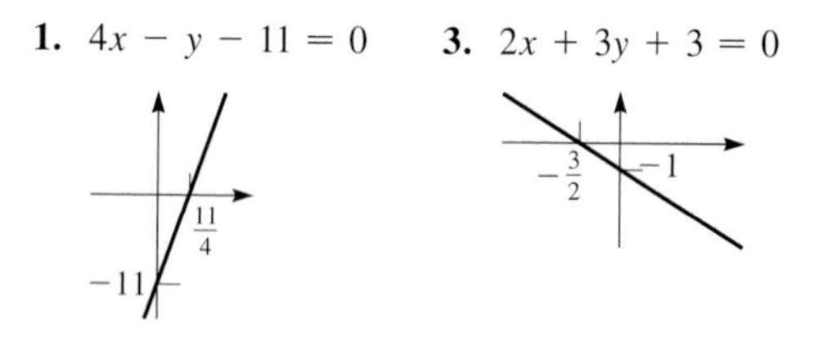

Review Exercises for Chapter 21, page 580

1. $4x - y - 11 = 0$ **3.** $2x + 3y + 3 = 0$

5. $x^2 + y^2 - 2x + 4y - 5 = 0$ **7.** $y^2 = 12x$

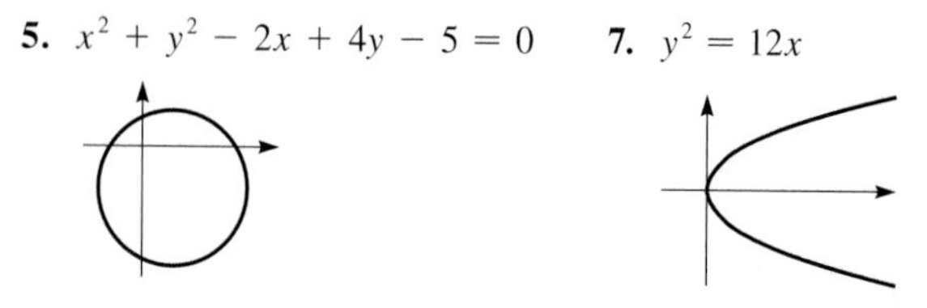

9. $9x^2 + 25y^2 = 900$ **11.** $144y^2 - 169x^2 = 24{,}336$

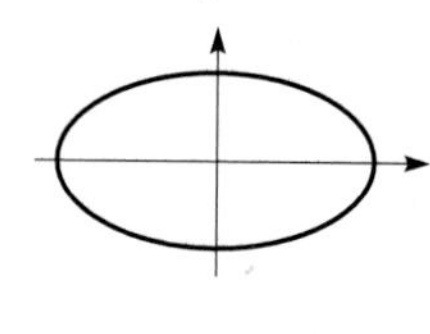

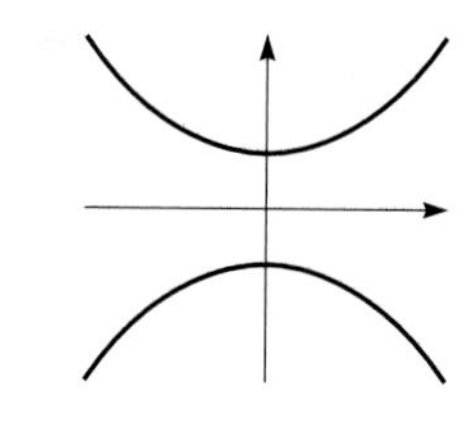

13. $(-3, 0)$, $r = 4$ **15.** $(0, -5)$, $y = 5$

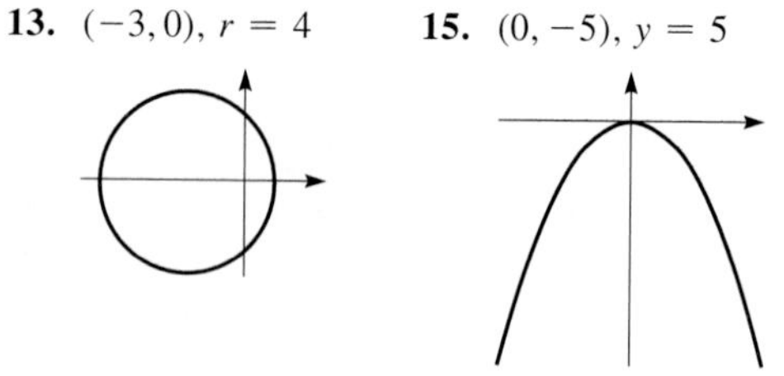

17. $V(0, 4)$, $V(0, -4)$, $F(0, \sqrt{15})$, $F(0, -\sqrt{15})$

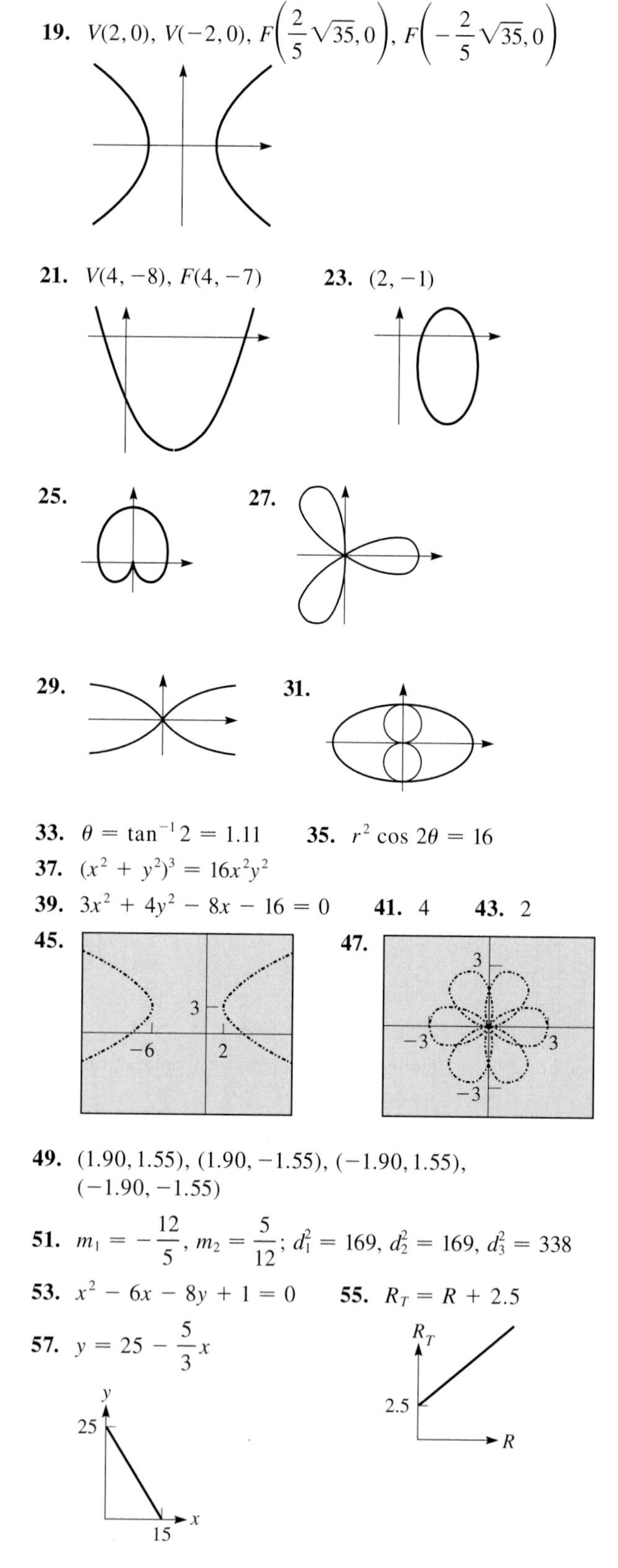

19. $V(2, 0)$, $V(-2, 0)$, $F\left(\frac{2}{5}\sqrt{35}, 0\right)$, $F\left(-\frac{2}{5}\sqrt{35}, 0\right)$

21. $V(4, -8)$, $F(4, -7)$ **23.** $(2, -1)$

25. **27.**

29. **31.**

33. $\theta = \tan^{-1} 2 = 1.11$ **35.** $r^2 \cos 2\theta = 16$

37. $(x^2 + y^2)^3 = 16x^2y^2$

39. $3x^2 + 4y^2 - 8x - 16 = 0$ **41.** 4 **43.** 2

45. **47.**

49. $(1.90, 1.55)$, $(1.90, -1.55)$, $(-1.90, 1.55)$, $(-1.90, -1.55)$

51. $m_1 = -\frac{12}{5}$, $m_2 = \frac{5}{12}$; $d_1^2 = 169$, $d_2^2 = 169$, $d_3^2 = 338$

53. $x^2 - 6x - 8y + 1 = 0$ **55.** $R_T = R + 2.5$

57. $y = 25 - \frac{5}{3}x$

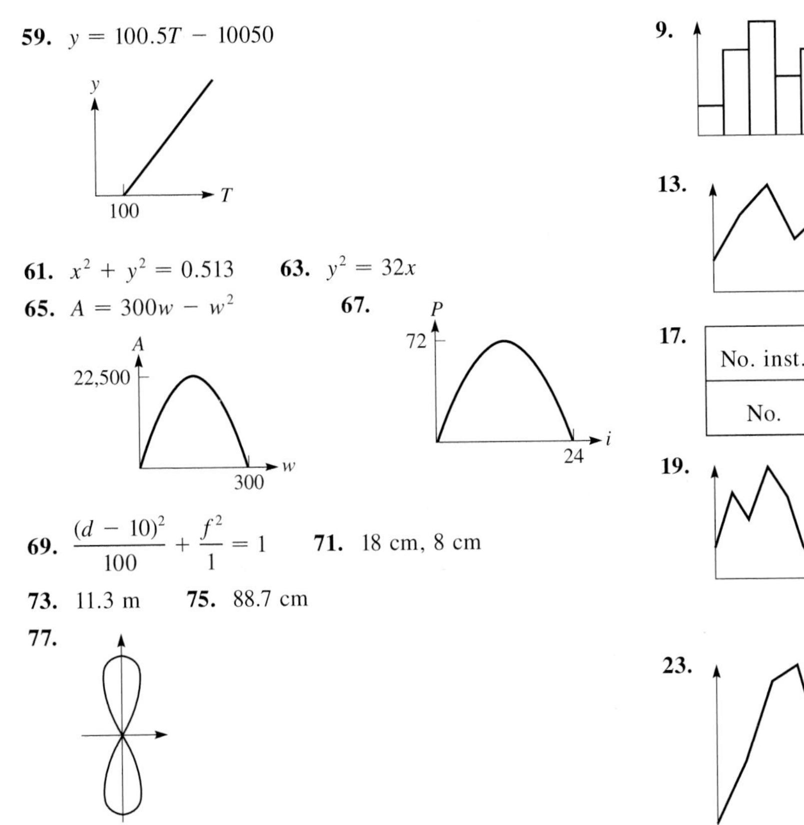

59. $y = 100.5T - 10050$

61. $x^2 + y^2 = 0.513$ **63.** $y^2 = 32x$

65. $A = 300w - w^2$ **67.**

69. $\frac{(d-10)^2}{100} + \frac{f^2}{1} = 1$ **71.** 18 cm, 8 cm

73. 11.3 m **75.** 88.7 cm

77.

79. Dist. from rifle to P − dist. from target to P = constant (related to dist. from rifle to target).

Exercises 22-1, page 587

1.

No.	2	3	4	5	6	7
Freq.	1	3	4	2	3	2

3.

No.	0.45	0.46	0.47	0.48	0.49	0.50	0.51	0.52	0.53	0.54	0.55	0.56	0.57
Freq.	1	1	1	2	2	0	1	0	1	0	2	0	1

5.

Int.	2–3	4–5	6–7
Freq.	4	6	5

7.

Int.	0.43–0.45	0.46–0.48	0.49–0.51	0.52–0.54	0.55–0.57
Freq.	1	4	3	1	3

9.

11.

13.

15.

17.

No. inst.	18	19	20	21	22	23	24	25
No.	1	3	2	4	3	1	0	1

19.

21.

23.

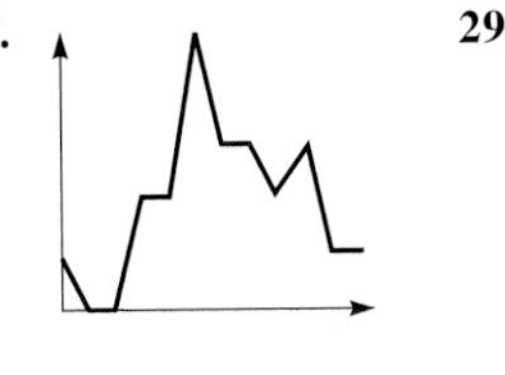

25.

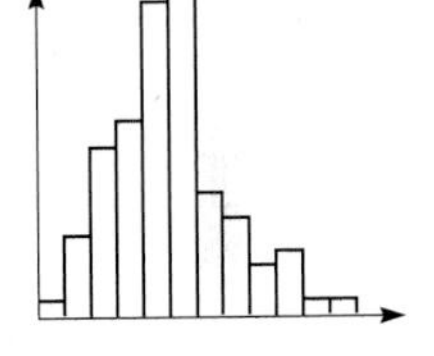

27.

29.

31. The greatest frequency should be at 2 and the least at 0 and 4.

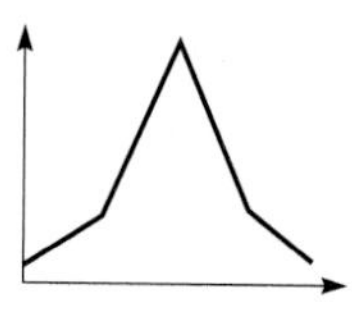

(Graphs will generally have approximately the shape shown.)

Exercises 22-2, page 591

1. 4 **3.** 0.49 **5.** 4.6 **7.** 0.503 **9.** 4
11. 0.48, 0.49, 0.55 **13.** 21 **15.** 21
17. 2.248 s **19.** 57 m **21.** 0.4237 mSv
23. 0.436 mSv **25.** 31 h **27.** 0.005 95 mm
29. \$275, \$300 **31.** 3450 MJ **33.** 0.195, 0.18
35. \$287.50
37. \$375, \$377, \$400; if each value is increased by the same amount, the median, mean, and mode are also increased by this amount.
39. \$541; an outlier can make the mean a poor measure of the center of the distribution.

Exercises 22-3, page 597

1. 1.50 **3.** 0.037 **5.** 1.50 **7.** 0.037 **9.** 1.7
11. 2.5 h **13.** 0.014 s **15.** 0.000 22 mm
17. 60% **19.** 58% **21.** 60% **23.** 76%
25. 60 **27.** 960

Exercises 22-4, page 601

1. $y = 1.0x - 2.6$ **3.** $y = -1.77x + 191$

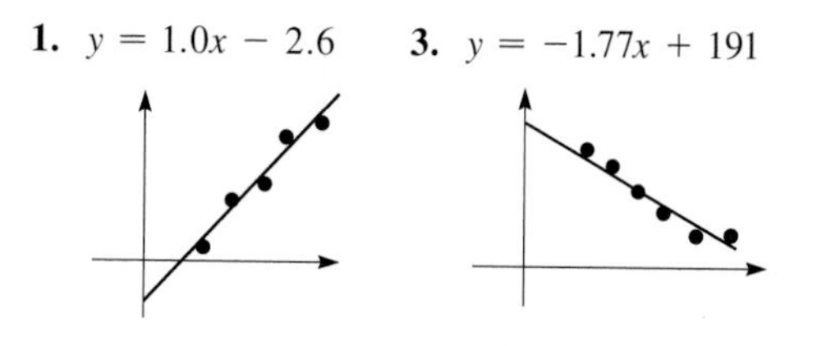

5. $V = -0.590i + 11.3$ **7.** $h = 2.24x + 5.2$

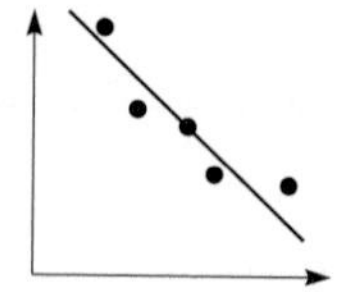

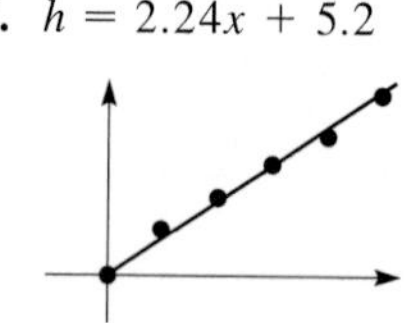

9. $p = -2.66x + 4364$ **11.** $V = 4.32 \times 10^{-15} f - 2.03$
$f_0 = 0.470$ PHz

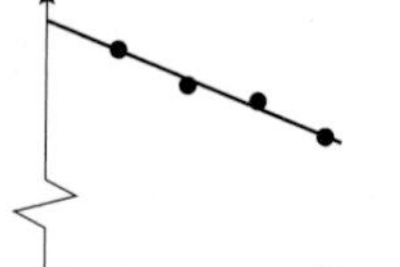

13. 0.985 **15.** −0.901

Exercises 22-5, page 604

1. $y = 1.97x^2 + 4.8$ **3.** $y = 10.9/x$

5. $y = 5.97t^2 + 0.38$ **7.** $p = 2030e^{0.01T} - 1640$

9. $p = \dfrac{2686}{S}$ **11.** $y = 6.20e^{-t} - 0.05$

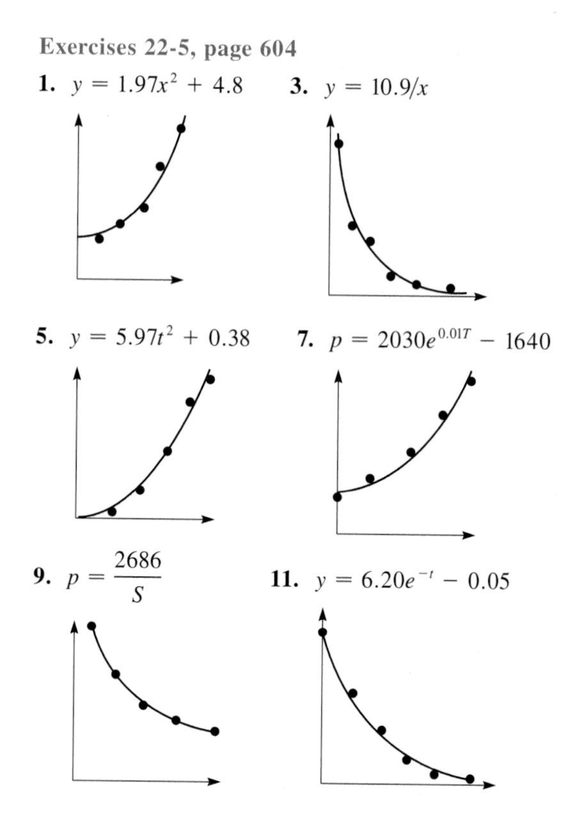

Review Exercises for Chapter 22, page 606

1. 3.6 **3.** 0.77

5.

Int.	101–103	104–106	107–109	110–112	113–115
Freq.	5	4	3	3	5

7. 106 **9.** 4.6 **11.**
13. 0.264 Pa · s
15. 0.014 Pa · s

17.

19. 697 W **21.** 700 W
23. 17 W **25.** 4

27.

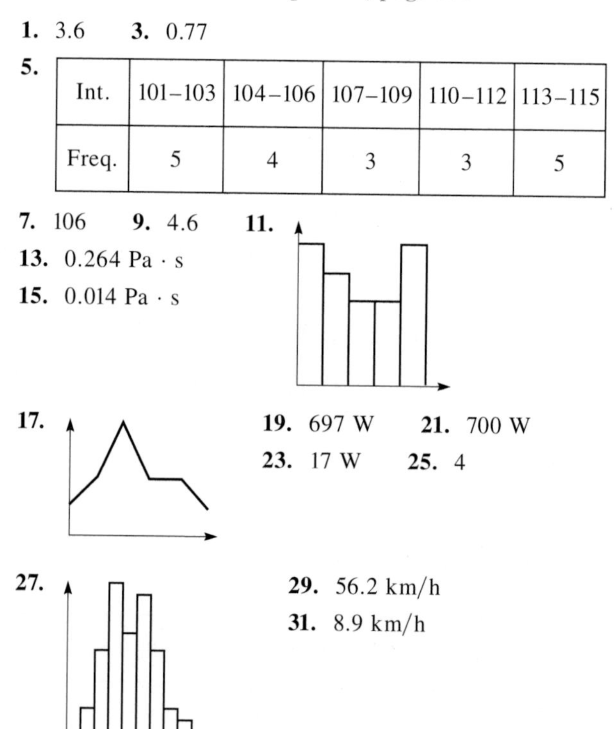

29. 56.2 km/h
31. 8.9 km/h

33. $R = 0.0983T + 25.0$

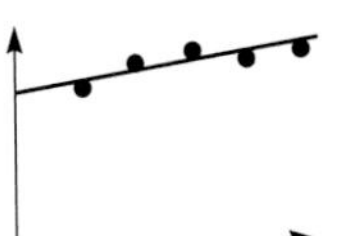

35. $s = 0.123t + 0.887$

37. $s = -4.90t^2 + 3000$

39. $y = -5.2x + 96.2$ (Curve has a good fit to a straight line, but fits $T = 82.6e^{-0.1t} + 17.3$ better.)

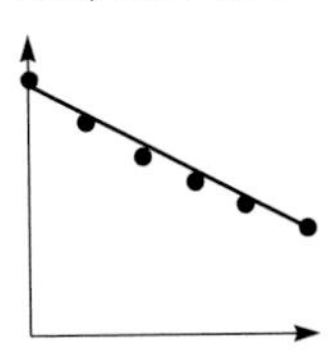

41. 9.2 ppm

43. Divide numerator and denominator by n^2.

Exercises 23-1, page 616

1. cont. all x **3.** not cont. $x = -3$, div. by zero

5. cont. $x > 0$ **7.** cont. all x

9. not cont. $x = 1$, small change **11.** cont. $x < 2$

13. not cont. $x = 2$, small change **15.** cont. all x

17.

x	2.900	2.990	2.999	3.001	3.010	3.100
$f(x)$	6.700	6.970	6.997	7.003	7.030	7.300

$\lim_{x \to 3} f(x) = 7$

19.

x	0.900	0.990	0.999	1.001	1.010	1.100
$f(x)$	1.7100	1.9701	1.9970	2.0030	2.0301	2.3100

$\lim_{x \to 1} f(x) = 2$

21.

x	1.900	1.990	1.999	2.001	2.010	2.100
$f(x)$	−0.2516	−0.2502	−0.250 02	−0.249 98	−0.2498	−0.2485

$\lim_{x \to 2} f(x) = -0.25$

23.

x	10	100	1000
$f(x)$	0.4468	0.4044	0.4004

$\lim_{x \to \infty} f(x) = 0.4$

25. 7 **27.** 1 **29.** 1 **31.** −2 **33.** 2 **35.** 2

37. Does not exist **39.** 0 **41.** 3 **43.** 0

45.

x	−0.1	−0.01	−0.001	0.001	0.01	0.1
$f(x)$	−3.1	−3.01	−3.001	−2.999	−2.99	−2.9

$\lim_{x \to 0} f(x) = -3$

47.

x	10	100	1000
$f(x)$	2.1649	2.0106	2.0010

$\lim_{x \to \infty} f(x) = 2$

49. 3 cm/s **51.** 34.9° C, 0° C **53.** e **55.** 2

Exercises 23-2, page 621

1. (slopes) 3.5, 3.9, 3.99 3.999; $m = 4$

3. (slopes) −3.5, −3.9, −3.99, −3.999; $m = -4$

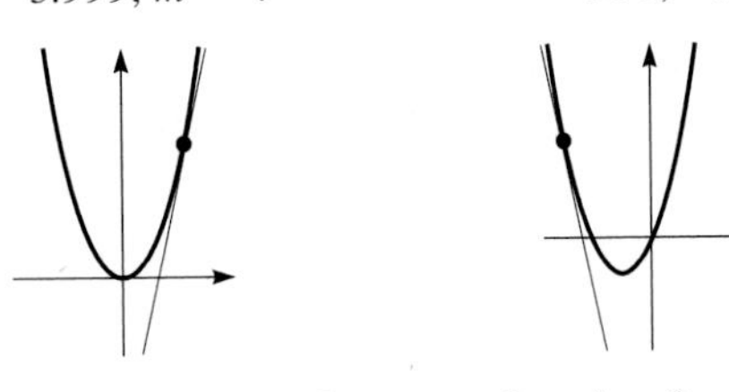

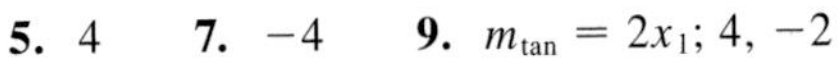

5. 4 **7.** −4 **9.** $m_{tan} = 2x_1$; 4, −2

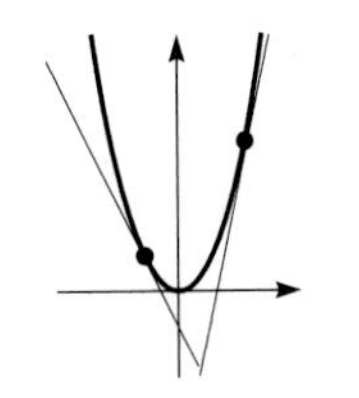

11. $m_{\tan} = 2x_1 + 2;\ -4, 4$ **13.** $m_{\tan} = 2x_1 + 4;\ -2, 8$

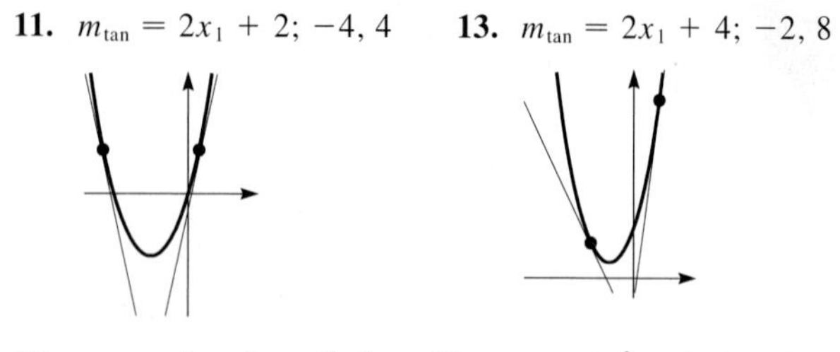

15. $m_{\tan} = 6 - 2x_1;\ 10, 0$ **17.** $m_{\tan} = 3x_1^2 - 2;\ 1, -2, 1$

19. $m_{\tan} = 4x_1^3;\ 0, 0.5, 4$ **21.** $m_{\tan} = 5x_1^4;\ 0, 0.31, 5$

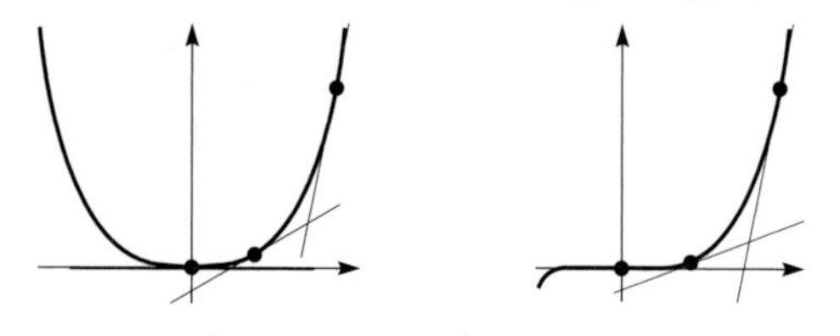

23. $m_{\tan} = -\dfrac{1}{x_1^2};\ -4, -1, -\dfrac{1}{4}$

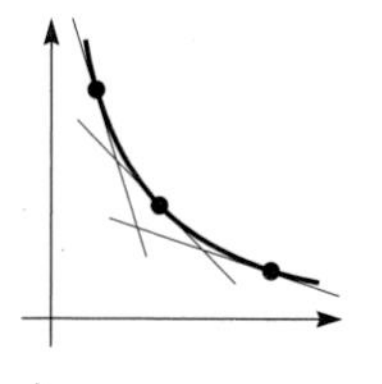

25. $\dfrac{\Delta y}{\Delta x} = 4.1,\ m_{\tan} = 4$

27. $\dfrac{\Delta y}{\Delta x} = -12.61,\ m_{\tan} = -12$

Exercises 23-3, page 625

1. 3 **3.** -2 **5.** $2x$ **7.** $10x$ **9.** $2x - 7$

11. $8 - 4x$ **13.** $3x^2 + 4$ **15.** $-\dfrac{1}{(x + 2)^2}$

17. $1 - \dfrac{1}{x^2}$ **19.** $-\dfrac{4}{x^3}$ **21.** $4x^3 + 3x^2 + 2x + 1$

23. $4x^3 + \dfrac{2}{x^2}$ **25.** $6x - 2;\ -8$

27. $-\dfrac{6}{(x + 3)^2};\ -\dfrac{1}{6}$ **29.** $\dfrac{-2}{x^2}$, all real numbers except 0

31. $\dfrac{-4x}{(x^2 - 1)^2}$, all real numbers except -1 and 1

33. $\dfrac{1}{2\sqrt{x + 1}}$ **35.** $\dfrac{1}{2\sqrt{x}}$

Exercises 23-4, page 629

1. $m = 4$ **3.** $m = -1$

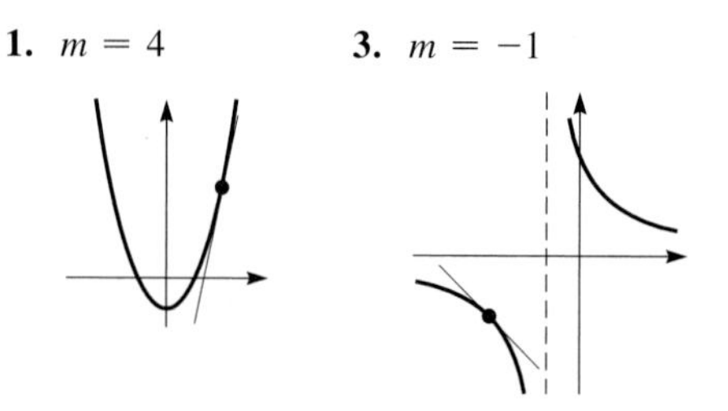

5. 4.00, 4.00, 4.00, 4.00, 4.00; $\lim_{t\to 3} v = 4$ m/s

7. 5, 6.5, 7.7, 7.97, 7.997; $\lim_{t\to 2} v = 8$ m/s **9.** 4; 4 m/s

11. $6t - 4$; 8 m/s **13.** $3 + \dfrac{2}{t^2}$ **15.** $6t - 6t^2$

17. $12t - 4$ **19.** $6t$ **21.** -2 **23.** $6w$

25. 460 W **27.** -83.1 W/(m^2·h) **29.** πd^2

31. $24.2/\sqrt{\lambda}$

Exercises 23-5, page 634

1. $5x^4$ **3.** $-36x^8$ **5.** $4x^3$ **7.** $2x + 2$

9. $15x^2 - 1$ **11.** $8x^7 - 28x^6 - 1$

13. $-42x^6 + 15x^2$ **15.** $x^2 + x$ **17.** 16 **19.** 33

21. -4 **23.** -29

25. $30t^4 - 5$ **27.** $-6 - 6t^2$ **29.** 64 **31.** 45

33. 1 **35.** (2, 4) **37.** $-\frac{1}{4}$ **39.** $3\pi r^2$

41. 84 W/A **43.** $a(c_1 + 2c_2E + 3c_3E^2)$

45. -12 N/cm **47.** 391 mm^2

Exercises 23-6, page 638

1. $x^2(3) + (3x + 2)(2x) = 9x^2 + 4x$

3. $6x(6x - 5) + (3x^2 - 5x)(6) = 54x^2 - 60x$

5. $(x + 2)(2) + (2x - 5)(1) = 4x - 1$

7. $(x^4 - 3x^2 + 3)(-6x^2) + (1 - 2x^3)(4x^3 - 6x)$
$= -14x^6 + 30x^4 + 4x^3 - 18x^2 - 6x$

9. $(2x - 7)(-2) + (5 - 2x)(2) = -8x + 24$

11. $(x^3 - 1)(4x - 1) + (2x^2 - x - 1)(3x^2)$
$= 10x^4 - 4x^3 - 3x^2 - 4x + 1$

13. $\dfrac{3}{(2x + 3)^2}$ **15.** $\dfrac{-2x}{(x^2 + 1)^2}$ **17.** $\dfrac{6x - 2x^2}{(3 - 2x)^2}$

19. $\dfrac{-6x^2 + 6x + 4}{(3x^2 + 2)^2}$ **21.** $\dfrac{-x^2 - 16x - 6}{(x^2 + x + 2)^2}$

23. $\dfrac{-2x^4 + 2x^3 + 5x^2 + 4x}{(x^3 + 2x^2)^2}$ **25.** -107 **27.** 75

29. 19 **31.** $-\frac{19}{3}$

33. (1) $\dfrac{-12x^3 + 45x^2 - 14x}{(3x - 7)^2}$ (2) $\dfrac{-12x^3 + 45x^2 - 14x}{(3x - 7)^2}$

35. 12 **37.** 1, -1

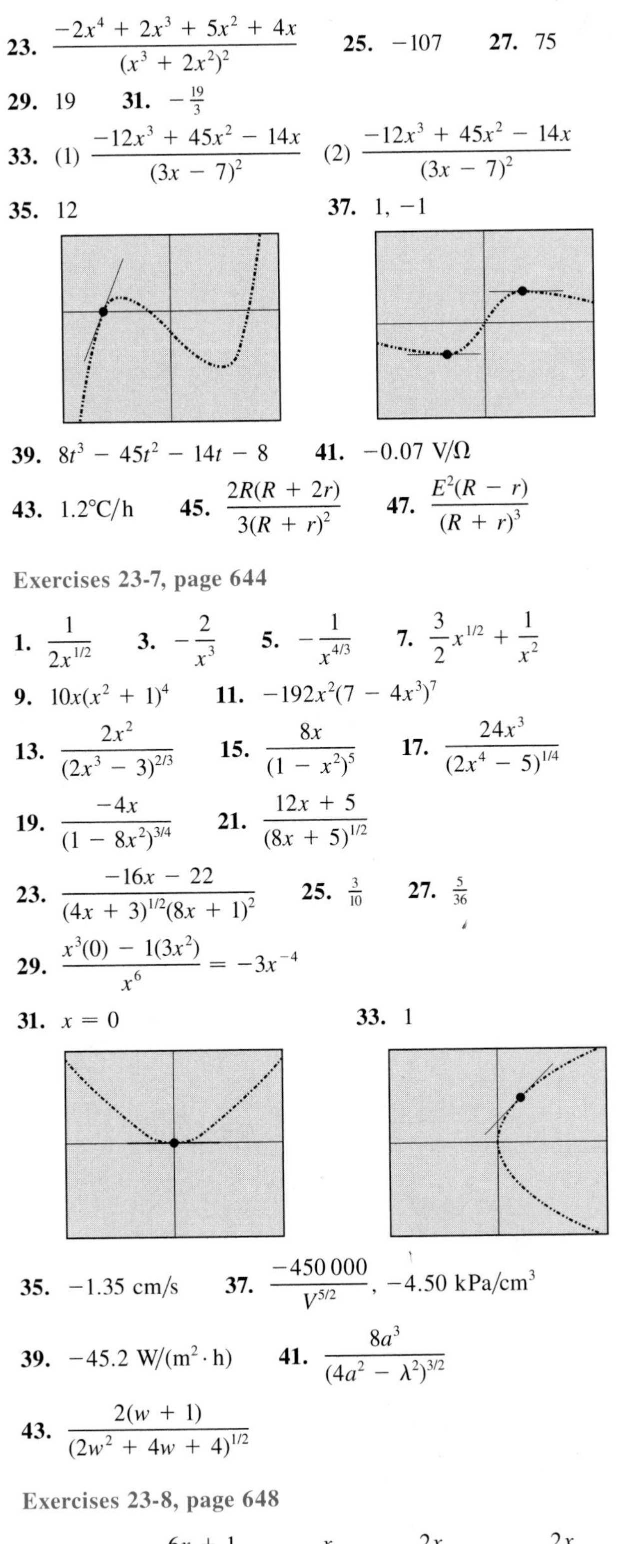

39. $8t^3 - 45t^2 - 14t - 8$ **41.** -0.07 V/Ω

43. 1.2°C/h **45.** $\dfrac{2R(R + 2r)}{3(R + r)^2}$ **47.** $\dfrac{E^2(R - r)}{(R + r)^3}$

Exercises 23-7, page 644

1. $\dfrac{1}{2x^{1/2}}$ **3.** $-\dfrac{2}{x^3}$ **5.** $-\dfrac{1}{x^{4/3}}$ **7.** $\dfrac{3}{2}x^{1/2} + \dfrac{1}{x^2}$

9. $10x(x^2 + 1)^4$ **11.** $-192x^2(7 - 4x^3)^7$

13. $\dfrac{2x^2}{(2x^3 - 3)^{2/3}}$ **15.** $\dfrac{8x}{(1 - x^2)^5}$ **17.** $\dfrac{24x^3}{(2x^4 - 5)^{1/4}}$

19. $\dfrac{-4x}{(1 - 8x^2)^{3/4}}$ **21.** $\dfrac{12x + 5}{(8x + 5)^{1/2}}$

23. $\dfrac{-16x - 22}{(4x + 3)^{1/2}(8x + 1)^2}$ **25.** $\frac{3}{10}$ **27.** $\frac{5}{36}$

29. $\dfrac{x^3(0) - 1(3x^2)}{x^6} = -3x^{-4}$

31. $x = 0$ **33.** 1

35. -1.35 cm/s **37.** $\dfrac{-450\,000}{V^{5/2}}$, -4.50 kPa/cm^3

39. -45.2 W/(m$^2\cdot$h) **41.** $\dfrac{8a^3}{(4a^2 - \lambda^2)^{3/2}}$

43. $\dfrac{2(w + 1)}{(2w^2 + 4w + 4)^{1/2}}$

Exercises 23-8, page 648

1. $-\frac{3}{2}$ **3.** $\dfrac{6x + 1}{4}$ **5.** $\dfrac{x}{y}$ **7.** $\dfrac{2x}{5y^4}$ **9.** $\dfrac{2x}{2y + 1}$

11. $\dfrac{-3y}{3x + 1}$ **13.** $\dfrac{-2x - y^3}{3xy^2 + 3}$

15. $\dfrac{3(y^2 + 1)(y^2 - 2x + 1)}{(y^2 + 1)^2 - 6x^2y}$ **17.** $\dfrac{4(2y - x)^3 - 2x}{8(2y - x)^3 - 1}$

19. $\dfrac{-3x(x^2 + 1)^2}{y(y^2 + 1)}$ **21.** 3 **23.** $-\frac{108}{157}$

25. 1 **27.** $-\dfrac{x}{y}$ **29.** $\dfrac{r - R + 1}{r + 1}$

31. $\dfrac{2C^2r(12CSr - 20Cr - 3L)}{3(C^2r^2 - L^2)}$

Exercises 23-9, page 651

1. $y' = 3x^2 + 2x$, $y'' = 6x + 2$, $y''' = 6$, $y^{(n)} = 0\ (n \geq 4)$

3. $f'(x) = 3x^2 - 24x^3$, $f''(x) = 6x - 72x^2$,
$f'''(x) = 6 - 144x$, $f^{(4)}(x) = -144$, $f^{(n)}(x) = 0\ (n \geq 5)$

5. $y' = -8(1 - 2x)^3$, $y'' = 48(1 - 2x)^2$,
$y''' = -192(1 - 2x)$, $y^{(4)} = 384$, $y^{(n)} = 0\ (n \geq 5)$

7. $f'(x) = (8x + 1)(2x + 1)^2$, $f''(x) = 12(2x + 1)(4x + 1)$,
$f'''(x) = 24(8x + 3)$, $f^{(4)}(x) = 192$, $f^{(n)}(x) = 0\ (n \geq 5)$

9. $84x^5 - 30x^4$ **11.** $-\dfrac{1}{4x^{3/2}}$ **13.** $-\dfrac{12}{(8x - 3)^{7/4}}$

15. $\dfrac{12}{(1 - 2x)^{5/2}}$ **17.** $600(2 - 5x)^2$

19. $30(27x^2 - 1)(3x^2 - 1)^3$ **21.** $\dfrac{4}{(1 - x)^3}$

23. $\dfrac{2}{(x + 1)^3}$ **25.** $-\dfrac{9}{y^3}$ **27.** $-\dfrac{6(x^2 - xy + y^2)}{(2y - x)^3}$

29. $\frac{9}{125}$ **31.** $-\frac{13}{384}$ **33.** -50 **35.** 48

37. -9.8 m/s^2 **39.** $-\dfrac{1.60}{(2t + 1)^{3/2}}$

Review Exercises for Chapter 23, page 653

1. -4 **3.** $\frac{1}{4}$ **5.** 1 **7.** $\frac{7}{3}$ **9.** $\frac{2}{3}$ **11.** -2

13. 5 **15.** $-4x$ **17.** $-\dfrac{4}{x^3}$ **19.** $\dfrac{1}{2\sqrt{x + 5}}$

21. $14x^6 - 6x$ **23.** $\dfrac{2}{x^{1/2}} + \dfrac{3}{x^2}$ **25.** $\dfrac{1}{(1 - x)^2}$

27. $-12(2 - 3x)^3$ **29.** $\dfrac{9x}{(5 - 2x^2)^{7/4}}$

31. $\dfrac{-15x^2 + 2x}{(1 - 6x)^{1/2}}$ **33.** $\dfrac{-2x - 3}{2x^2(4x + 3)^{1/2}}$

35. $\dfrac{2x - 6(2x - 3y)^2}{1 - 9(2x - 3y)^2}$ **37.** $\frac{5}{48}$ **39.** $\frac{74}{5}$

41. $36x^2 - \frac{2}{x^3}$ **43.** $\frac{56}{(1 + 4x)^3}$

45. It appears to be 8, but using TRACE there is no value shown for $x = 2$.

Point (2, 8) is missing

47. (a) 12 000 (b) 8000 **49.** -31

51. 1 **53.** $-k + k^2t - \frac{1}{2}k^3t^2$ **55.** $-\frac{2k}{r^3}$

57. $0.4(0.01t + 1)^2(0.04t + 1)$

59. $\frac{2R(R + 2r)}{3(R + r)^2}$ **61.** $-\frac{1}{4\pi\sqrt{C}(L + 2)^{3/2}}$

63. $\frac{40V_2^{0.4}}{V_1^{1.4}}$ **65.** 0.049/m

67. $p = 2w + \frac{150}{w}, \frac{dp}{dw} = 2 - \frac{150}{w^2}$

69. $A = 4x - x^3, \frac{dA}{dx} = 4 - 3x^2$

71. At $t = 5$ years, $dV/dt = -\$7500$/year (rate of appreciation is decreasing); $d^2V/dt^2 = \$1500/\text{year}^2$ (rate at which appreciation changes is increasing) (Machinery is depreciating, but depreciation is lessening.)

Exercises 24-1, page 659

1. $4x - y - 2 = 0$ **3.** $2y + x - 2 = 0$

5. $x - 2y + 6 = 0$ **7.** $8x - 4y - 7 = 0$

9. $\sqrt{3}x + 8y - 7 = 0$
$16x - 2\sqrt{3}y - 15\sqrt{3} = 0$

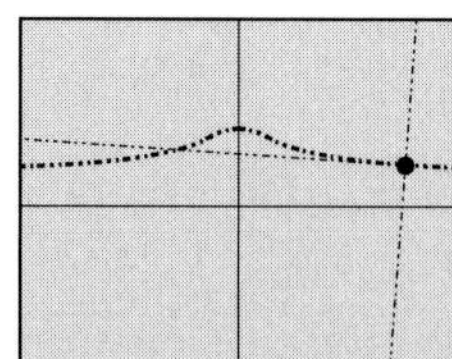

11. $2x - 12y + 37 = 0$
$72x + 12y + 37 = 0$

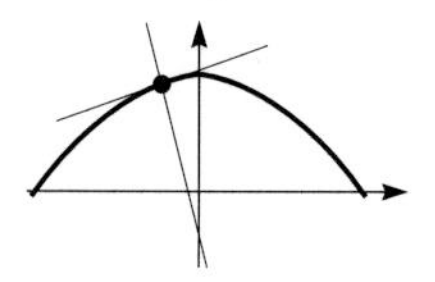

13. $y = 2x - 4$

15. $y - 8 = -\frac{1}{24}(x - \frac{3}{2})$, or $2x + 48y - 387 = 0$

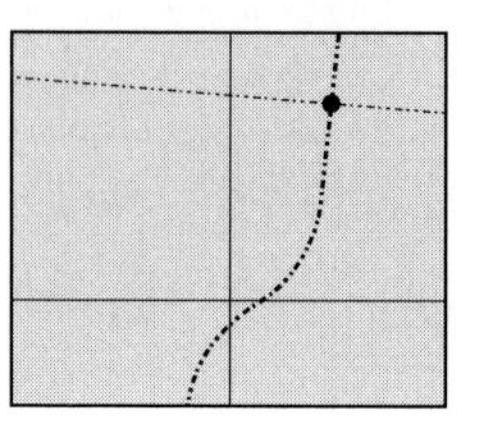

17. $(-\frac{1}{4}, 0)$ **19.** $x - 2y - 20 = 0$

21. $x + y - 6 = 0$

23. $x + 2y - 3 = 0, x = 0, x - 2y + 3 = 0$

Exercises 24-2, page 663

1. 3.449 489 7 **3.** $-0.180\,460\,4$ **5.** 0.585 786 4

7. 0.348 894 2 **9.** 2.561 552 8 **11.** $-1.236\,068\,0$

13. 0.917 543 3 **15.** 0.618 034 0

17. $-1.855\,772\,5$, 0.678 362 8, 3.177 409 7

19. Find the real root of $x^3 - 4 = 0$; 1.587 401 1

21. 29.4 m **23.** 1.61 m

Exercises 24-3, page 667

1. 3.16, 341.6° **3.** 8.07, 352.4°

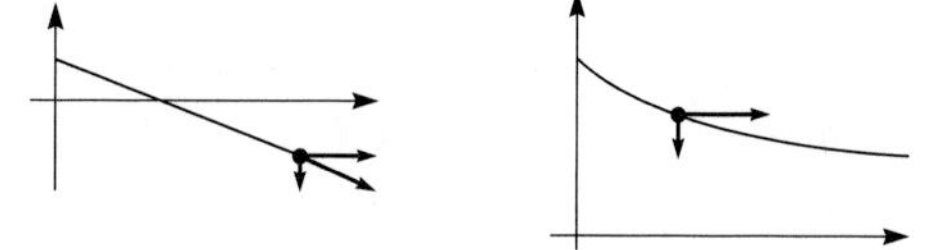

5. $a = 0$ **7.** 20.0, 3.7° **9.** 9.4 m/s, 302°

11. 1.3 m/min², 288° **13.** 36 m/s, 332°; 9.8 m/s², 270°

15. 276 m/s, 43.5°; 2090 m/s, 16.7°

17. 22.1 m/s², 25.4°; 20.2 m/s², 8.5°

19. 21.2 km/min, 296.6°

21. $x^2 + y^2 = 44.45^2$; $v_x = 731$ m/min, $v_y = -690$ m/min

23. 370 m/s, 19°

Exercises 24-4, page 670

1. 0.0900 Ω/s **3.** \$1.22/week **5.** 4.1×10^{-6} m/s

7. $\frac{dB}{dt} = \frac{-3kr(dr/dt)}{[r^2 + (l/2)^2]^{5/2}}$ **9.** 0.15 mm²/month

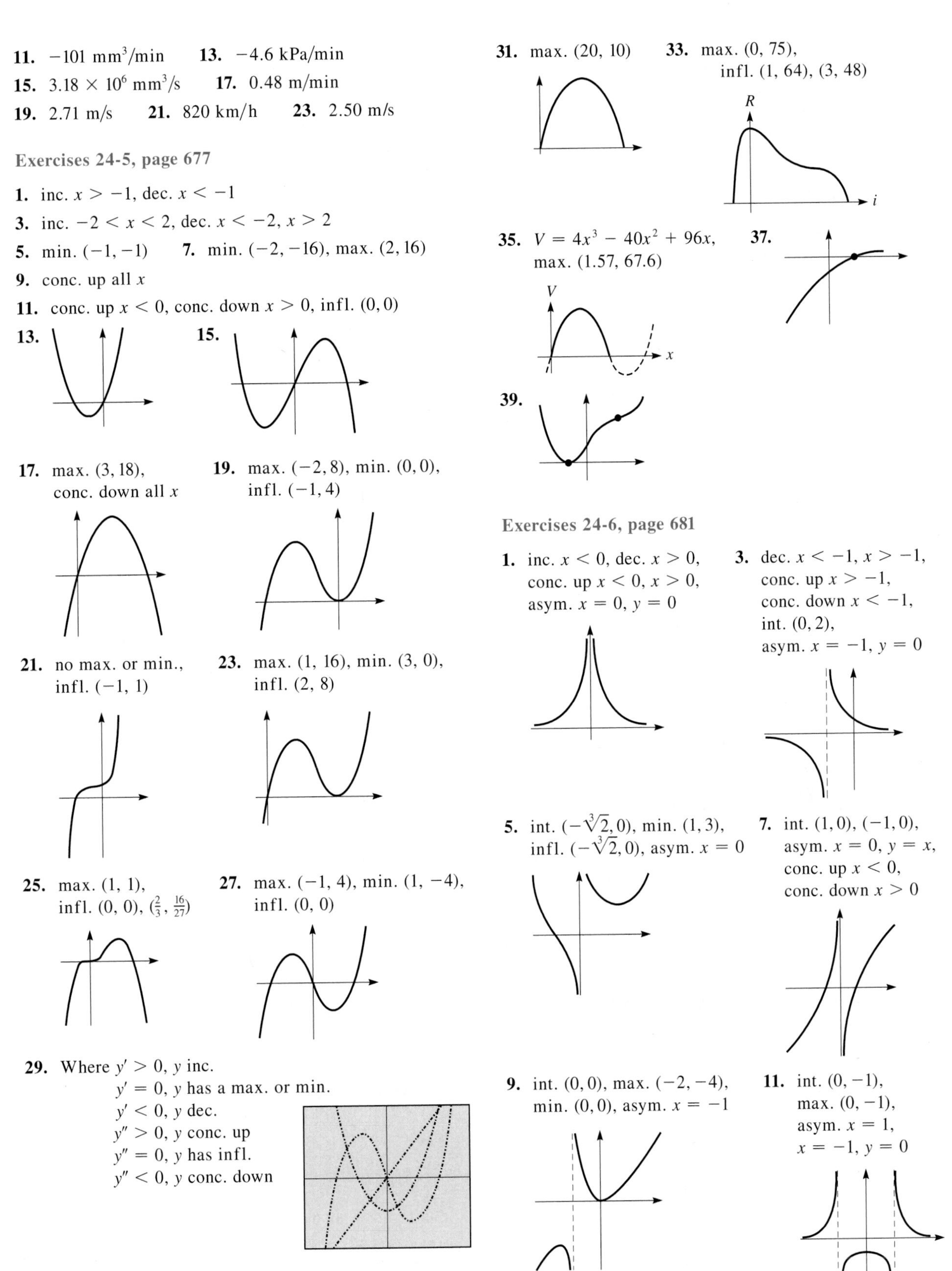

11. $-101\ \text{mm}^3/\text{min}$ **13.** -4.6 kPa/min
15. $3.18 \times 10^6\ \text{mm}^3/\text{s}$ **17.** 0.48 m/min
19. 2.71 m/s **21.** 820 km/h **23.** 2.50 m/s

Exercises 24-5, page 677

1. inc. $x > -1$, dec. $x < -1$
3. inc. $-2 < x < 2$, dec. $x < -2$, $x > 2$
5. min. $(-1, -1)$ **7.** min. $(-2, -16)$, max. $(2, 16)$
9. conc. up all x
11. conc. up $x < 0$, conc. down $x > 0$, infl. $(0, 0)$
13.
15.
17. max. $(3, 18)$, conc. down all x
19. max. $(-2, 8)$, min. $(0, 0)$, infl. $(-1, 4)$
21. no max. or min., infl. $(-1, 1)$
23. max. $(1, 16)$, min. $(3, 0)$, infl. $(2, 8)$
25. max. $(1, 1)$, infl. $(0, 0)$, $(\frac{2}{3}, \frac{16}{27})$
27. max. $(-1, 4)$, min. $(1, -4)$, infl. $(0, 0)$
29. Where $y' > 0$, y inc.
$y' = 0$, y has a max. or min.
$y' < 0$, y dec.
$y'' > 0$, y conc. up
$y'' = 0$, y has infl.
$y'' < 0$, y conc. down
31. max. $(20, 10)$
33. max. $(0, 75)$, infl. $(1, 64)$, $(3, 48)$
35. $V = 4x^3 - 40x^2 + 96x$, max. $(1.57, 67.6)$
37.
39.

Exercises 24-6, page 681

1. inc. $x < 0$, dec. $x > 0$, conc. up $x < 0$, $x > 0$, asym. $x = 0$, $y = 0$
3. dec. $x < -1$, $x > -1$, conc. up $x > -1$, conc. down $x < -1$, int. $(0, 2)$, asym. $x = -1$, $y = 0$
5. int. $(-\sqrt[3]{2}, 0)$, min. $(1, 3)$, infl. $(-\sqrt[3]{2}, 0)$, asym. $x = 0$
7. int. $(1, 0)$, $(-1, 0)$, asym. $x = 0$, $y = x$, conc. up $x < 0$, conc. down $x > 0$
9. int. $(0, 0)$, max. $(-2, -4)$, min. $(0, 0)$, asym. $x = -1$
11. int. $(0, -1)$, max. $(0, -1)$, asym. $x = 1$, $x = -1$, $y = 0$

13. int. (1, 0), max. (2, 1), infl. $(3, \frac{8}{9})$, asym. $x = 0$, $y = 0$

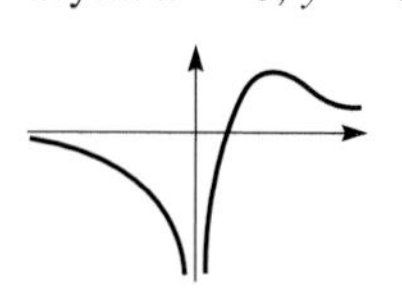

15. int. (0, 0), (1, 0), (−1, 0), max. $(\frac{1}{2}\sqrt{2}, \frac{1}{2})$, min. $(-\frac{1}{2}\sqrt{2}, -\frac{1}{2})$

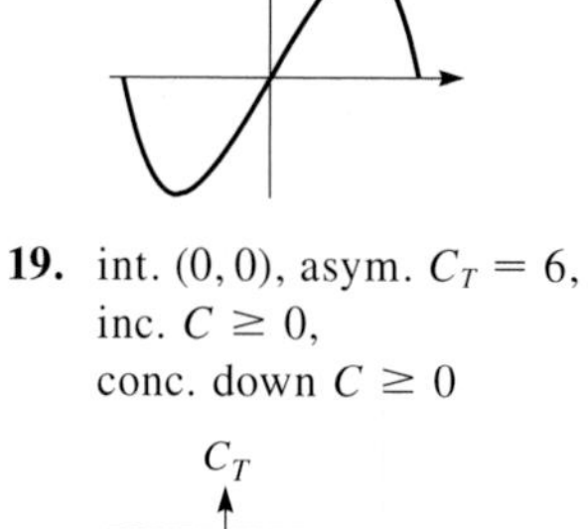

17. int. (0, 0), infl. (0, 0) asym. $x = -3$, $x = 3$, $y = 0$

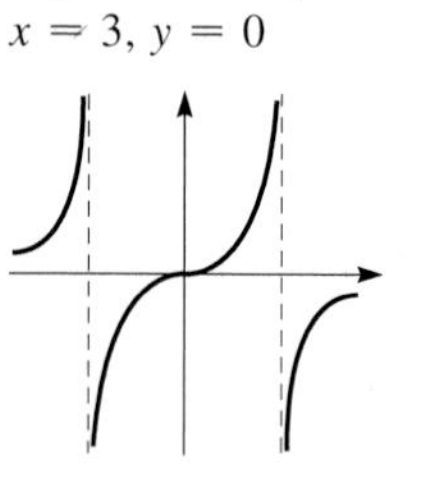

19. int. (0, 0), asym. $C_T = 6$, inc. $C \geq 0$, conc. down $C \geq 0$

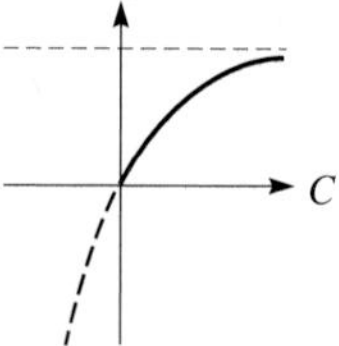

21. int. (0, 1), max. (0, 1) infl. (141, 0.82), asym. $R = 0$

23. $A = 2\pi r^2 + \dfrac{40}{r}$, min. (1.47, 40.8), asym. $r = 0$

Exercises 24-7, page 686

1. 60 m **3.** $\dfrac{E}{2R}$ **5.** 35 m^2, $8300 **7.** 1500 Ω

9. 8, 8 **11.** 5 mm, 5 mm **13.** 1.1 h

15. 8.49 cm, 8.49 cm **17.** 1.20 m **19.** 12

21. $0.58L$ **23.** 3.3 cm **25.** 100 m

27. $w = 0.50$ m, $d = 0.87$ m **29.** 8.0 km from refinery

31. 59.2 m, 118 m

Review Exercises for Chapter 24, page 689

1. $5x - y + 1 = 0$

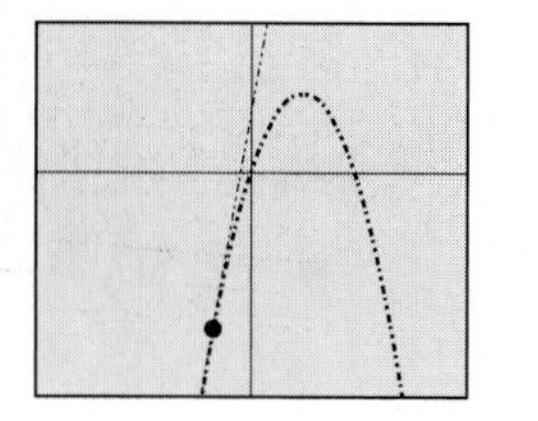

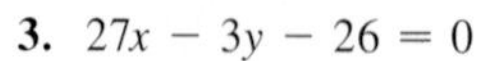

3. $27x - 3y - 26 = 0$

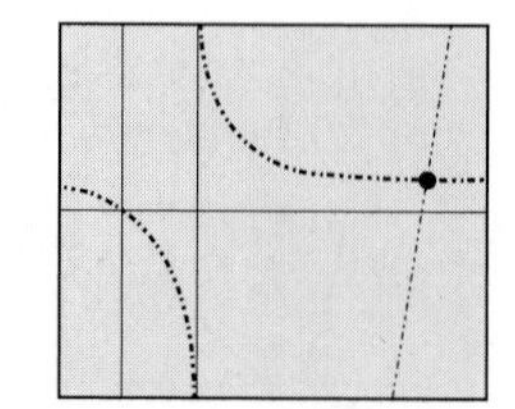

5. $x - 2y + 3 = 0$

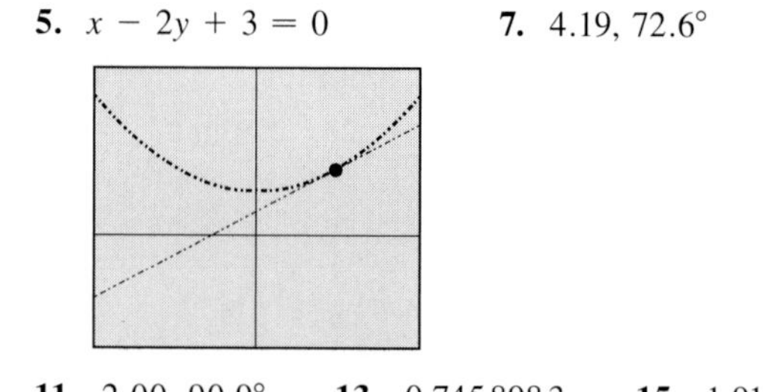

7. 4.19, 72.6° **9.** 2.12

11. 2.00, 90.9° **13.** 0.745 898 3 **15.** 1.911 164 3

17. min. (−2, −16), conc. up all x

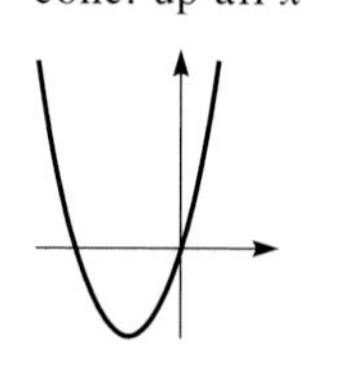

19. int. (0, 0), $(\pm 3\sqrt{3}, 0)$, max. (3, 54), min. (−3, −54), infl. (0, 0)

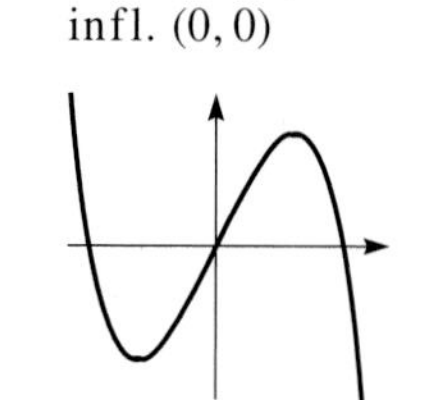

21. min. (2, −48), conc. up $x < 0$, $x > 0$

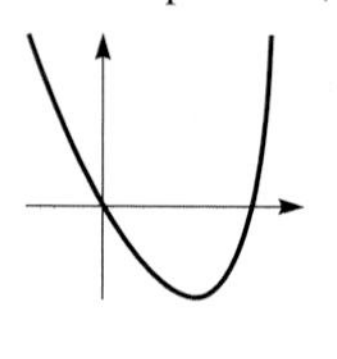

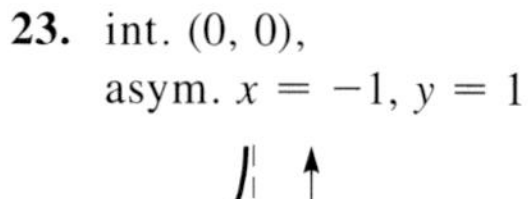

23. int. (0, 0), asym. $x = -1$, $y = 1$

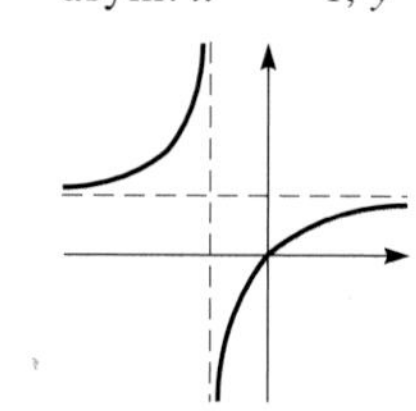

25. $2x - y + 1 = 0$ **27.** 0.0 m, 6.527 m

29. 8.8 m/s, 336° **31.** −7.44 cm/s **33.** $2400

35.

37. 1.30 cm/min

39. 38 000 m^2/min **41.** 5000 cm^2

43. max. (0, 100); infl. (37, 63); int. (0, 100), (89, 0)

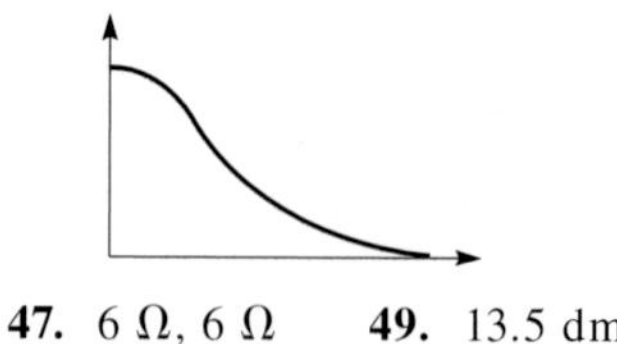

45. 1160 km/h

47. 6 Ω, 6 Ω **49.** 13.5 dm^3 **51.** 6.6 cm

Exercises 25-1, page 695

1. $(5x^4 + 1)\,dx$ **3.** $\dfrac{-10\,dx}{x^6}$ **5.** $8x(x^2 - 1)^3\,dx$

7. $\dfrac{-12x\,dx}{(3x^2 + 1)^2}$ **9.** $x(1 - x)^2(-5x + 2)\,dx$

11. $\frac{2\,dx}{(5x+2)^2}$ **13.** 12.28, 12 **15.** 1.712 75, 1.675
17. −2.4, −2.473 088 1 **19.** 0.6257, 0.626 490 3
21. 0.0038 cm^2 **23.** −8.3 nF **25.** $84\,200\text{ cm}^3$
27. $\frac{dr}{r} = \frac{kd\lambda/(2\lambda^{1/2})}{k\lambda^{1/2}}$ **29.** $\frac{dA}{A} = \frac{2ds}{s}$ **31.** 16.96

Exercises 25-2, page 698

1. x^3 **3.** $2x^6$ **5.** $\frac{3}{2}x^4 + x$ **7.** $\frac{2}{3}x^3 - \frac{1}{2}x^2$
9. $x^{5/2}$ **11.** $\frac{4}{3}x^{3/2} + 3x$ **13.** $\frac{1}{x}$ **15.** $-\frac{2}{x^3}$
17. $\frac{2}{5}x^5 + x$ **19.** $3x^2 - \frac{1}{3x^3}$ **21.** $\frac{1}{3}x^3 + 2x - \frac{1}{x}$
23. $(2x+1)^6$ **25.** $(x^2-1)^4$ **27.** $\frac{1}{40}(2x^4+1)^5$
29. $(6x+1)^{3/2}$ **31.** $\frac{1}{4}(3x+1)^{4/3}$

Exercises 25-3, page 703

1. $x^2 + C$ **3.** $\frac{1}{8}x^8 + C$ **5.** $\frac{4}{5}x^{5/2} + C$
7. $-\frac{1}{3x^3} + C$ **9.** $\frac{1}{3}x^3 - \frac{1}{6}x^6 + C$
11. $3x^3 + \frac{1}{2}x^2 + 3x + C$ **13.** $-\frac{1}{2x^2} + \frac{1}{2}x + C$
15. $\frac{2}{7}x^{7/2} - \frac{2}{5}x^{5/2} + C$ **17.** $6x^{1/3} + \frac{1}{9}x + C$
19. $\frac{1}{6}(1+2x)^3 + C$ **21.** $\frac{1}{6}(x^2-1)^6 + C$
23. $\frac{1}{5}(x^4+3)^5 + C$ **25.** $\frac{1}{40}(x^5+4)^8 + C$
27. $\frac{1}{12}(8x+1)^{3/2} + C$ **29.** $\frac{1}{6}\sqrt{6x^2+1} + C$
31. $\sqrt{x^2-2x} + C$ **33.** $y = 2x^3 + 2$
35. $y = 5 - \frac{1}{18}(1-x^3)^6$ **37.** $12y = 83 + (1-4x^2)^{3/2}$
39. $i = 2t^2 - 0.2t^3 + 2$ **41.** $f = \sqrt{0.01A+1} - 1$
43. $y = 3x^2 + 2x - 3$

Exercises 25-4, page 708

1. 9, 12.15 **3.** 1.92, 2.28 **5.** 7.625, 8.208
7. 0.464, 0.5995 **9.** 1.92, 1.96 **11.** 13.5
13. $\frac{8}{3}$ **15.** 9 **17.** 0.8 **19.** 2

Exercises 25-5, page 711

1. 1 **3.** $\frac{254}{7}$ **5.** $6 + 2\sqrt{6} - 2\sqrt{3}$ **7.** 2.53
9. $\frac{747}{20}$ **11.** $\frac{33}{20}\sqrt[3]{\frac{11}{5}} - \frac{3}{8}\sqrt[3]{\frac{1}{2}} - \frac{17}{5} = -1.552$ **13.** $\frac{4}{3}$
15. $-\frac{81}{4}$ **17.** 2 **19.** $\frac{1}{4}(20.5^{2/3} - 17.5^{2/3}) = 0.1875$
21. $\frac{88}{3249} = 0.0271$ **23.** 76 **25.** $\frac{364}{3}$ **27.** $\frac{3880}{9}$
29. $\frac{33}{784} = 0.0421$ **31.** $\frac{464}{5}$ **33.** $64\,000\text{ N}\cdot\text{m}$
35. 86.8 m^2

Exercises 25-6, page 714

1. $\frac{11}{2} = 5.50$, $\frac{16}{3} = 5.33$ **3.** 7.661, $\frac{23}{3} = 7.667$
5. 0.2042 **7.** 18.98 **9.** 0.5205 **11.** 21.74
13. 45.36 **15.** $0.177k$

Exercises 25-7, page 717

1. (a) 6 (b) 6 **3.** (a) 19.67 (b) 19.67 **5.** 0.2028
7. 19.27 **9.** 0.5114 **11.** 13.147 **13.** 44.63
15. 1.200 cm

Review Exercises for Chapter 25, page 719

1. $x^4 - \frac{1}{2}x^2 + C$ **3.** $x^2 - \frac{4}{5}x^{5/2} + C$ **5.** $\frac{20}{3}$
7. $\frac{16}{3}$ **9.** $3x - \frac{1}{x^2} + C$ **11.** 3
13. $\frac{1}{5(2-5x)} + C$ **15.** $-\frac{6}{7}(7-2x)^{7/4} + C$
17. $\frac{9}{8}(3\sqrt[3]{3} - 1)$ **19.** $-\frac{1}{30}(1-2x^3)^5 + C$
21. $-\frac{1}{2x-x^3} + C$ **23.** $\frac{3350}{3}$ **25.** $\left(12x^2 - \frac{1}{x^2}\right)dx$
27. $\frac{-6x\,dx}{(x^2-1)^4}$ **29.** $\frac{(1-4x)\,dx}{(1-3x)^{2/3}}$ **31.** $\frac{(8x-2x^2)\,dx}{(2-x)^2}$
33. 0.061 **35.** $y = 3x - \frac{1}{3}x^3 + \frac{17}{3}$
37. (a) $x - x^2 + C_1$
(b) $-\frac{1}{4}(1-2x)^2 + C_2 = x - x^2 + C_2 - \frac{1}{4}$;
$C_1 = C_2 - \frac{1}{4}$
39. 22 **41.** 0.842 **43.** 0.811 **45.** 1.01
47. 13.77 **49.** 4.64 **51.** 6.72 **53.** 1.85 m^3
55. $\frac{R\,dR}{R^2 + X^2}$ **57.** $y = k(2L^3x - 6Lx^2 + \frac{2}{5}x^5)$
59. 14.9 m^2

Exercises 26-1, page 727

1. 24.5 m/s **3.** $s = 8.00 - 0.25t$ **5.** 5.0 m/s
7. 17 800 m **9.** 24 m/s **11.** 85.3 m
13. 0.345 nC **15.** 0.017 C **17.** 120 V
19. 4.65 mV **21.** 970 rad **23.** 66.7 A **25.** $\frac{k}{x_1}$
27. $m = 1002 - 2\sqrt{t+1}$, 2.51×10^5 min

Exercises 26-2, page 733

1. 2 **3.** $\frac{8}{3}$ **5.** $\frac{27}{8}$ **7.** $\frac{4}{3}\sqrt{2}$ **9.** $\frac{32}{3}$ **11.** $\frac{1}{6}$
13. $\frac{26}{3}$ **15.** 3 **17.** $\frac{15}{4}$ **19.** $\frac{7}{6}$ **21.** $\frac{48}{5}$ **23.** $\frac{7}{6}$
25. $\frac{45}{4}$ **27.** $\frac{65}{6}$ **29.** 18.0 J **31.** 80.8 km
33. 4 cm^2 **35.** 0.683 m^2

Exercises 26-3, page 738

1. $\frac{1}{3}\pi$ **3.** $\frac{1}{3}\pi$ **5.** 72π **7.** $\frac{768}{7}\pi$ **9.** $\frac{348}{5}\pi$

11. $\frac{16}{3}\pi$ **13.** $\frac{1}{3}\pi$ **15.** $\frac{1}{3}\pi$ **17.** $\frac{2}{5}\pi$

19. $\frac{10\pi}{3}\sqrt{5}$ **21.** $\frac{1296}{5}\pi$ **23.** $\frac{16}{3}\pi$ **25.** $\frac{8}{3}\pi$

27. $\frac{1}{3}\pi r^2 h$ **29.** $7.56\ \text{mm}^3$ **31.** $18.3\ \text{cm}^3$

Exercises 26-4, page 744

1. $(\frac{10}{3}, 0)$ **3.** $(\frac{14}{15}, 0)$ **5.** $(-\frac{1}{2}, \frac{1}{2})$

7. $(\frac{7}{22}, \frac{5}{22})$ **9.** $(0, \frac{6}{5})$ **11.** $(\frac{4}{3}, \frac{4}{3})$ **13.** $(\frac{3}{5}, \frac{12}{35})$

15. $(3, \frac{33}{5})$ **17.** $(\frac{7}{8}, 0)$ **19.** $(0, \frac{5}{6})$ **21.** $(\frac{2}{3}, 0)$

23. $(\frac{2}{3}b, \frac{1}{3}a)$. Place triangle with a vertex at origin, side b and right angle on x-axis. Equation of hypotenuse is $y = ax/b$. Use Eqs. (26-16) and (26-17).

25. 0.375 cm above center of base

27. 19.3 cm from larger base

Exercises 26-5, page 750

1. 128, 4 **3.** 214, 3.27 **5.** $\frac{64}{15}k$ **7.** $\frac{2}{3}\sqrt{6}$

9. $\frac{1}{6}mb^2$ **11.** $\frac{4}{7}\sqrt{7}$ **13.** $\frac{8}{11}\sqrt{55}$ **15.** $\frac{64}{3}\pi k$

17. $\frac{2}{5}\sqrt{10}$ **19.** $\frac{3}{10}mr^2$ **21.** $0.324\ \text{g}\cdot\text{cm}^2$

23. $31.2\ \text{kg}\cdot\text{cm}^2$

Exercises 26-6, page 755

1. 8.0 N · cm **3.** 200 N · mm

5. 9.4×10^{-22} N · m **7.** 1800 N · m

9. 3.00×10^5 m · tonne **11.** 9.85×10^5 N · m

13. 12.5 kN **15.** 152 kN **17.** 6500 N

19. 1.84 MN **21.** 2.7 A **23.** 35.3% **25.** 109 m

27. $A = \pi r\sqrt{r^2 + h^2}$

Review Exercises for Chapter 26, page 757

1. 4.3 s **3.** 4.7 s **5.** 0.44 C **7.** 55 V

9. $y = 20x + \frac{1}{120}x^3$ **11.** $\frac{2}{3}$ **13.** 18 **15.** $\frac{27}{4}$

17. $\frac{48}{5}\pi$ **19.** $\frac{512}{5}\pi$ **21.** $\frac{4}{3}\pi ab^2$ **23.** $(\frac{40}{21}, \frac{10}{3})$

25. $(\frac{14}{5}, 0)$ **27.** $\frac{8}{5}k$ **29.** $\frac{256}{3}\pi k$ **31.** 2700 N · m

33. 1.8 m **35.** $47\ \text{m}^3$ **37.** 1580 kN

39. 0.29 Ω

Exercises 27-1, page 764

1. $\cos(x + 2)$ **3.** $12x^2\cos(2x^3 - 1)$ **5.** $-3\sin\frac{1}{2}x$

7. $-6\sin(3x - 1)$ **9.** $8\sin 4x\cos 4x = 4\sin 8x$

11. $-45\cos^2(5x + 2)\sin(5x + 2)$

13. $\sin 3x + 3x\cos 3x$ **15.** $9x^2\cos 5x - 15x^3\sin 5x$

17. $2x\cos x^2\cos 2x - 2\sin x^2\sin 2x$

19. $\dfrac{2\cos 4x}{\sqrt{1 + \sin 4x}}$ **21.** $\dfrac{3x\cos 3x - \sin 3x}{x^2}$

23. $\dfrac{4x(1 - 3x)\sin x^2 - 6\cos x^2}{(3x - 1)^2}$

25. $4\sin 3x(3\cos 3x\cos 2x - \sin 3x\sin 2x)$

27. $\dfrac{-2\cos 3x[3\sin 3x(1 + 2\sin^2 2x) + 4\cos 3x\sin 2x\cos 2x]}{(1 + 2\sin^2 2x)^2}$

29. $3\sin^2 x\cos x + 2\sin 2x$

31. $1 - 4\sin^2 4x\cos 4x$ **33.** (a)

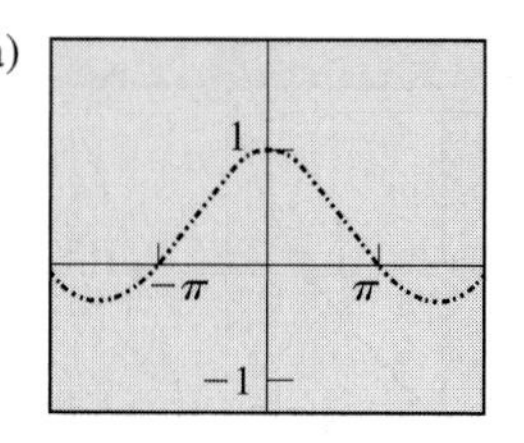

(b) See the table.

35. (a) 0.540 302 3, value of derivative
(b) 0.540 260 2, slope of secant line

37. Resulting curve is $y = \cos x$.

39. $\dfrac{2x - y\cos xy}{x\cos xy - 2\sin 2y}$

41. $\dfrac{d\sin x}{dx} = \cos x,\ \dfrac{d^2\sin x}{dx^2} = -\sin x,$
$\dfrac{d^3\sin x}{dx^3} = -\cos x,\ \dfrac{d^4\sin x}{dx^4} = \sin x$

43. $\sin 2x = 2\sin x\cos x$ **45.** 0.085 **47.** −2.36

49. −199 cm/s **51.** −38.5 km

Exercises 27-2, page 768

1. $5\sec^2 5x$ **3.** $2(1 - x)\csc^2(1 - x)^2$

5. $6\sec 2x\tan 2x$

7. $\dfrac{3}{\sqrt{2x + 3}}\csc\sqrt{2x + 3}\cot\sqrt{2x + 3}$

9. $30\tan 3x\sec^2 3x$ **11.** $-4\cot^3\frac{1}{2}x\csc^2\frac{1}{2}x$

13. $2\tan 4x\sqrt{\sec 4x}$ **15.** $-84\csc^4 7x\cot 7x$

17. $x^2\sec^2 x + 2x\tan x$

19. $-4\csc x^2(2x\cos x\cot x^2 + \sin x)$

21. $-\dfrac{\csc x(x\cot x + 1)}{x^2}$

23. $\dfrac{2(-4\sin 4x - 4\sin 4x\cot 3x + 3\cos 4x\csc^2 3x)}{(1 + \cot 3x)^2}$

25. $\sec^2 x(\tan^2 x - 1)$ **27.** $2\sec 2x(\sec 2x - \tan 2x)$

29. $\dfrac{1 + 2\sec^2 4x}{\sqrt{2x + \tan 4x}}$ **31.** $\dfrac{2\cos 2x - \sec y}{x\sec y\tan y - 2}$

33. $24\tan 3x\sec^2 3x\,dx$

35. $4\sec 4x(\tan^2 4x + \sec^2 4x)\,dx$

37. (a) 3.425 518 8, value of derivative
(b) 3.426 052 4, slope of secant line

39. $2 \tan x \sec^2 x = 2 \sec x(\sec x \tan x)$ **41.** -12

43. $2 \sec^2 x - \sec x \tan x = \dfrac{2}{\cos^2 x} - \dfrac{\sin x}{\cos^2 x}$

45. -8.4 cm/s **47.** 140 m/s

Exercises 27-3, page 772

1. $\dfrac{2x}{\sqrt{1-x^4}}$ **3.** $\dfrac{18x^2}{\sqrt{1-9x^6}}$ **5.** $-\dfrac{1}{\sqrt{4-x^2}}$

7. $\dfrac{1}{\sqrt{(x-1)(2-x)}}$ **9.** $\dfrac{1}{2\sqrt{x}(1+x)}$ **11.** $-\dfrac{6}{x^2+1}$

13. $\dfrac{5x}{\sqrt{1-x^2}} + 5 \sin^{-1} x$ **15.** $\dfrac{4x}{1+4x^2} + 2 \tan^{-1} 2x$

17. $\dfrac{3\sqrt{1-4x^2} \sin^{-1} 2x - 6x + 2}{\sqrt{1-4x^2}(\sin^{-1} 2x)^2}$

19. $\dfrac{2(\cos^{-1} 2x + \sin^{-1} 2x)}{\sqrt{1-4x^2}(\cos^{-1} 2x)^2}$ **21.** $\dfrac{-24(\cos^{-1} 4x)^2}{\sqrt{1-16x^2}}$

23. $\dfrac{4 \sin^{-1}(4x+1)}{\sqrt{-4x^2-2x}}$ **25.** $\dfrac{18(\tan^{-1} 2x)^2}{1+4x^2}$

27. $\dfrac{-2(2x+1)^2}{(1+4x^2)^2}$ **29.** $\dfrac{18(4-\cos^{-1} 2x)^2}{\sqrt{1-4x^2}}$

31. $-\dfrac{x^2y^2 + 2y + 1}{2x}$

33. (a) 1.154 700 5, value of derivative
(b) 1.154 739 0, slope of secant line

35. $\dfrac{3(\sin^{-1} x)^2\, dx}{\sqrt{1-x^2}}$ **37.** 0.41

39.

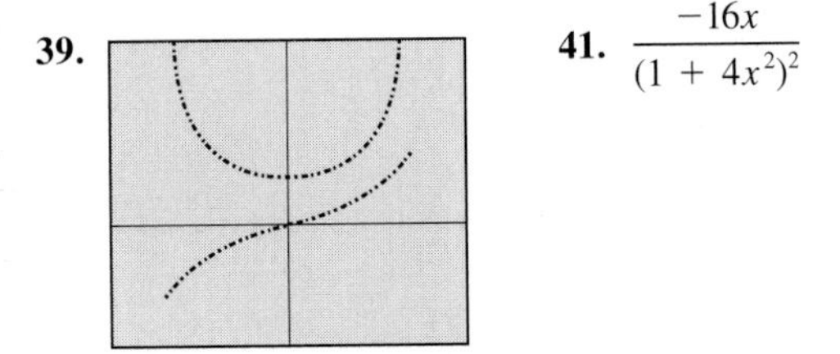

41. $\dfrac{-16x}{(1+4x^2)^2}$

43. Let $y = \sec^{-1} u$; solve for u; take derivatives; substitute.

45. $\dfrac{E - A}{\omega m\sqrt{m^2E^2 - (A-E)^2}}$

47. $\theta = \tan^{-1} \dfrac{h}{x}; \dfrac{d\theta}{dx} = \dfrac{-h}{x^2 + h^2}$

Exercises 27-4, page 776

1. $d \sin x/dx = \cos x$ and $d \cos x/dx = -\sin x$, and $\sin x = \cos x$ at points of intersection.

3. $\dfrac{1}{x^2+1}$ is always positive.

5.

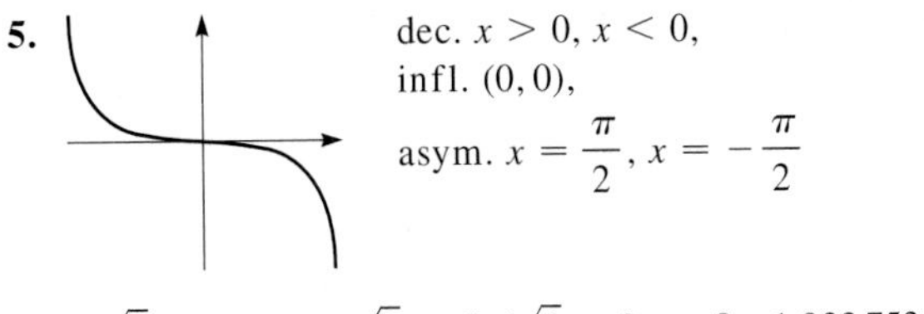

dec. $x > 0, x < 0$,
infl. $(0, 0)$,
asym. $x = \dfrac{\pi}{2}, x = -\dfrac{\pi}{2}$

7. $8\sqrt{2}x + 8y + 4\sqrt{2} - 5\pi\sqrt{2} = 0$ **9.** 1.933 753 8

11. 10 **13.** 0.58 m/s, -1.7 m/s^2 **15.** -0.072 N/s

17. 2510 mm/s, 270° **19.** 94 800 mm/s^2, 0°

21. -0.085 rad/s **23.** 8.08 m/s **25.** 0.020

27. 0.19 m **29.** $w = 9.24$ cm, $d = 13.1$ cm

31. 4.2 m

Exercises 27-5, page 781

1. $\dfrac{2 \log e}{x}$ **3.** $\dfrac{6 \log_5 e}{3x+1}$ **5.** $\dfrac{-3}{1-3x}$

7. $\dfrac{4 \sec^2 2x}{\tan 2x} = 4 \sec 2x \csc 2x$ **9.** $\dfrac{1}{2x}$

11. $\dfrac{6(x+1)}{x^2+2x}$ **13.** $2(1 + \ln x)$

15. $\dfrac{3(2x+1)\ln(2x+1) - 6x}{(2x+1)[\ln(2x+1)]^2}$ **17.** $\dfrac{1}{x \ln x}$

19. $\dfrac{1}{x^2+x}$ **21.** $\dfrac{\cos \ln x}{x}$ **23.** $\dfrac{6 \ln 2x}{x}$

25. $\dfrac{x \sec^2 x + \tan x}{x \tan x}$ **27.** $\dfrac{x+4}{x(x+2)}$

29. $\dfrac{\sqrt{x^2+1}}{x}$ **31.** $\dfrac{x + y - 2\ln(x+y)}{x + y + 2\ln(x+y)}$

33. 0.5 is value of derivative; 0.499 987 5 is slope of secant line.

35. (a)

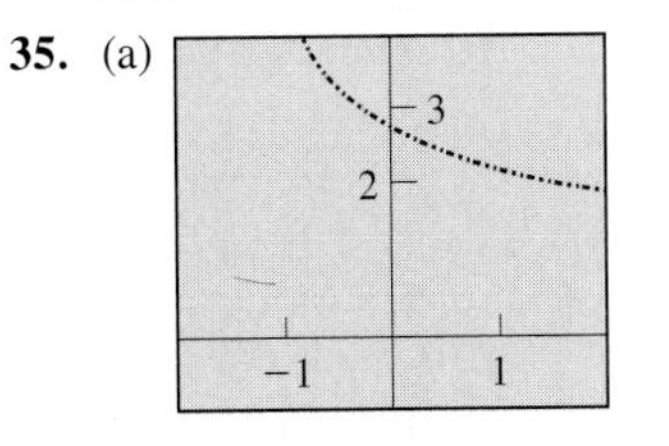

(b) See the table.

37. 1.15 **39.** $-2(\tan x + \sec x \csc x)\, dx$ **41.** 2.73

43. $x^x(\ln x + 1)$. In Eq. (23-15) the exponent is constant. For x^x, both the base and the exponent are variables.

45. $\dfrac{10 \log e}{I} \dfrac{dI}{dt}$ **47.** 0.125 s^2/m

Exercises 27-6, page 784

1. $(2 \ln 3)3^{2x}$ **3.** $(6 \ln 4)4^{6x}$ **5.** $6e^{6x}$ **7.** $\dfrac{e^{\sqrt{x}}}{2\sqrt{x}}$

9. $e^{-x}(1-x)$ **11.** $e^{\sin x}(x \cos x + 1)$

13. $\dfrac{3e^{2x}(2x+1)}{(x+1)^2}$ **15.** $e^{-3x}(4 \cos 4x - 3 \sin 4x)$

17. $\dfrac{2e^{3x}(12x + 5)}{(4x + 3)^2}$ **19.** $\dfrac{2xe^{x^2}}{e^{x^2} + 4}$

21. $16e^{6x}(x \cos x^2 + 3 \sin x^2)$

23. $2(\ln 2x + e^{2x})\left(\dfrac{1}{x} + 2e^{2x}\right)$ **25.** $\dfrac{e^{xy}(xy + 1)}{1 - x^2e^{xy} - \cos y}$

27. $\dfrac{3e^{2x}}{x} + 6e^{2x} \ln x$ **29.** $12e^{6x} \cot 2e^{6x}$

31. $\dfrac{4e^{2x}}{\sqrt{1 - e^{4x}}}$

33. (a) 2.718 281 8, value of derivative
(b) 2.718 417 8, slope of secant line

35. -0.724 **37.** $2e^{x/2}\left(\dfrac{2 + 2x + x^2 + x \ln x}{x}\right) dx$

39.

41. $(-xe^{-x} + e^{-x}) + (xe^{-x}) = e^{-x}$

43. $\dfrac{2e^{2x}(e^{2x} + 1) - 2e^{2x}(e^{2x} - 1)}{(e^{2x} + 1)^2} = \dfrac{4e^{2x}}{(e^{2x} + 1)^2}$
$= \dfrac{(e^{2x} + 1)^2 - (e^{2x} - 1)^2}{(e^{2x} + 1)^2}$

45. $(2 - t)e^{-0.5t}$ **47.** -0.00164/h

49. Substitute and simplify.

51. $\dfrac{d}{dx}\left[\dfrac{1}{2}(e^u - e^{-u})\right] = \dfrac{1}{2}(e^u + e^{-u})\dfrac{du}{dx}$;
$\dfrac{d}{dx}\left[\dfrac{1}{2}(e^u + e^{-u})\right] = \dfrac{1}{2}(e^u - e^{-u})\dfrac{du}{dx}$

Exercises 27-7, page 788

1. int. (0, 0), max. (0, 0), not defined for $\cos x < 0$, asym. $x = -\frac{1}{2}\pi, \frac{1}{2}\pi, \ldots$

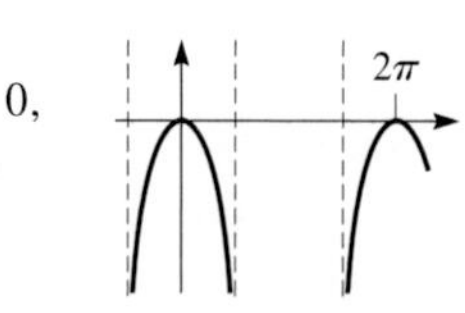

3. int. (0, 0), max. $\left(1, \dfrac{1}{e}\right)$, infl. $\left(2, \dfrac{2}{e^2}\right)$, asym. $y = 0$

5. int. (0, 0), max. (0, 0), infl. $(-1, -\ln 2)$, $(1, -\ln 2)$

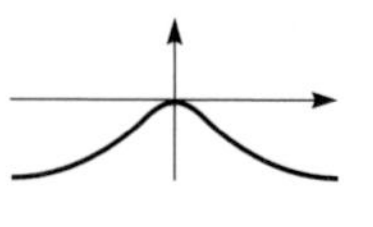

7. int. (0, 1), max. (0, 1), infl. $\left(\dfrac{1}{2}\sqrt{2}, \dfrac{1}{e}\sqrt{e}\right)$, $\left(-\dfrac{1}{2}\sqrt{2}, \dfrac{1}{e}\sqrt{e}\right)$, asym. $y = 0$

9. max. $(1, -1)$, asym. $x = 0$ **11.** int. (0, 0), infl. (0, 0), inc. all x

13. $y = x - 1$ **15.** $2\sqrt{2}x - 2y + 2\sqrt{2} - 3\pi\sqrt{2} = 0$

17. 1.314 096 8 **19.** -0.303 W/day

21. $\dfrac{p(-a + bT)}{T^2}$ **23.** $a = k^2x$

25. int. (0, 0), min. $(0, 0), (2\pi, 0), \ldots,$ asym. $x = -\dfrac{\pi}{2}, \dfrac{\pi}{2}, \dfrac{3\pi}{2}, \ldots$

2π

27. $dR_T = -\dfrac{1}{2}ke^{k \csc(\theta/2)} \csc \dfrac{\theta}{2} \cot \dfrac{\theta}{2} d\theta$

29. $v = -e^{-0.5t}(1.4 \cos 6t + 2.3 \sin 6t)$, -2.03 cm/s

31. $1/\sqrt{e} = 0.607$

Review Exercises for Chapter 27, page 790

1. $-12 \sin(4x - 1)$ **3.** $-\dfrac{\sec^2\sqrt{3 - x}}{2\sqrt{3 - x}}$

5. $-6 \csc^2(3x + 2) \cot(3x + 2)$

7. $-24x \cos^3 x^2 \sin x^2$ **9.** $2e^{2(x-3)}$ **11.** $\dfrac{6x}{x^2 + 1}$

13. $\dfrac{9}{9 + x^2}$ **15.** $\dfrac{4}{(\sin^{-1} x)(\sqrt{1 - 16x^2})}$

17. $(-2 \csc 4x)\sqrt{\csc 4x + \cot 4x}$ **19.** $\dfrac{14(1 + e^{-x})}{x - e^{-x}}$

21. $\dfrac{-\cos x(2e^{3x} \sin x + 3e^{3x} \cos x + 2 \sin x)}{(e^{3x} + 1)^2}$

23. $\dfrac{2x(1 + 4x^2)(\tan^{-1} 2x) - 2x^2}{(1 + 4x^2)(\tan^{-1} 2x)^2}$ **25.** $-2x \cot x^2$

27. $\dfrac{2 \cos x \ln(3 + \sin x)}{3 + \sin x}$ **29.** $e^{-2x} \sec x(\tan x - 2)$

31. $\dfrac{\cos 2x + 2e^{4x}}{\sqrt{\sin 2x + e^{4x}}}$ **33.** $\dfrac{2x^3e^y + 2xy^2e^y + y}{x - x^4e^y - x^2y^2e^y}$

35. $2x(e^{2\cos^2 x})(1 - 2x \sin x \cos x)$ **37.** $\dfrac{y(xye^{-x} - 1)}{x(ye^{-x} + 1)}$

39. $\cos^{-1} x$ **41.** infl. $(\frac{1}{2}\pi, \frac{1}{2}\pi)$, $(\frac{3}{2}\pi, \frac{3}{2}\pi)$

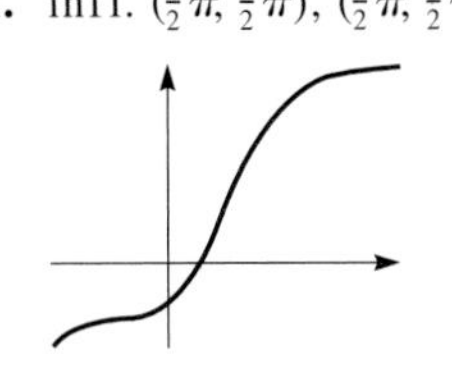

43. max. $(e^{-2}, 4e^{-2})$, min. $(1, 0)$, infl. (e^{-1}, e^{-1}) **45.** $7.27x + y - 8.44 = 0$

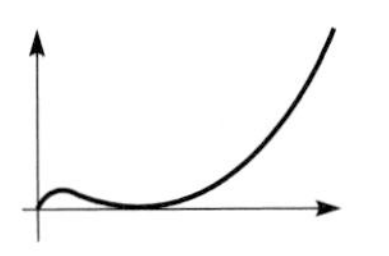

47. $2x + 2.57y - 4.30 = 0$

49. $2 \sin x \cos x - 2 \cos x \sin x = 0$

51. -0.7034674 **53.** -5.17 cm/s **55.** $-20 \sin 2t$

57.

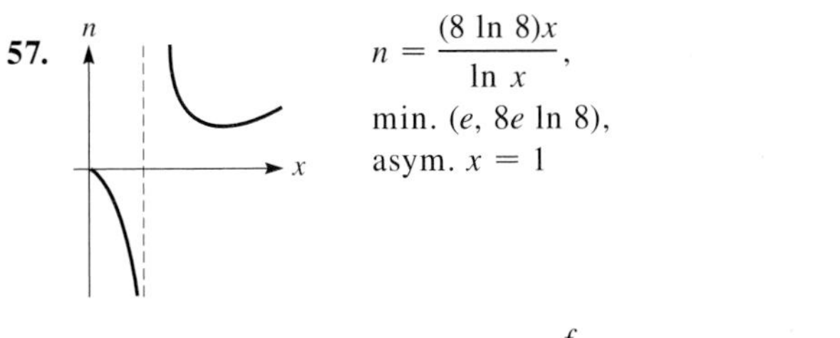

$n = \dfrac{(8 \ln 8)x}{\ln x}$, min. $(e, 8e \ln 8)$, asym. $x = 1$

59. $-kE_0^2 \cos \frac{1}{2}\theta \sin \frac{1}{2}\theta$ **61.** $\dfrac{f}{\sqrt{R^2 - F^2 f^2}}$

63. 0.005 934 rad/s **65.** -4.16°C/min **67.** 83.6 m/s

69. 2.00 m wide, 1.82 m high **71.** -0.0065 rad/s

73. 0.40 cm/s, 72.5° **75.** 7.07 cm

77. $A = 100 \cos \theta(1 + \sin \theta)$ **79.** 10

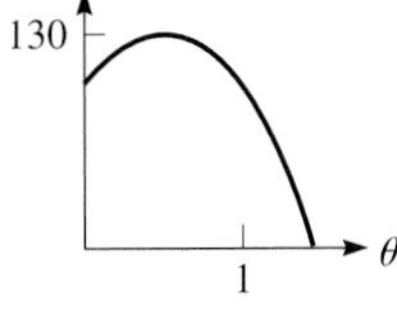

Exercises 28-1, page 796

1. $\frac{1}{5}\sin^5 x + C$ **3.** $-\frac{2}{3}(\cos x)^{3/2} + C$

5. $\frac{4}{3}\tan^3 x + C$ **7.** $\frac{1}{8}$ **9.** $\frac{1}{4}(\sin^{-1} x)^4 + C$

11. $\frac{1}{2}(\tan^{-1} 5x)^2 + C$ **13.** $\frac{1}{3}[\ln(x + 1)]^3 + C$

15. 0.179 **17.** $\frac{1}{4}(4 + e^x)^4 + C$

19. $\frac{3}{16}(2e^{2x} - 1)^{4/3} + C$ **21.** $\frac{1}{10}(1 + \sec^2 x)^5 + C$

23. $\frac{1}{6}$ **25.** 1.102 **27.** $y = \frac{1}{3}(\ln x)^3 + 2$

29. $\frac{1}{3}mnv^2$ **31.** $q = (1 - e^{-t})^3$

Exercises 28-2, page 800

1. $\frac{1}{4}\ln|1 + 4x| + C$ **3.** $-\frac{1}{3}\ln|4 - 3x^2| + C$

5. $\frac{1}{3}\ln 4 = 0.462$ **7.** $-\frac{1}{2}\ln|\cot 2x| + C$

9. $\ln 2 = 0.693$ **11.** $\ln|1 - e^{-x}| + C$

13. $\ln|x + e^x| + C$ **15.** $\frac{1}{4}\ln|1 + 4\sec x| + C$

17. $\frac{1}{4}\ln 5 = 0.402$ **19.** $\ln|\ln x| + C$

21. $\ln|2x + \tan x| + C$ **23.** $-2\sqrt{1 - 2x} + C$

25. $\ln|x| - \frac{2}{x} + C$ **27.** $\frac{1}{3}\ln(\frac{5}{4}) = 0.0744$ **29.** 1.10

31. $\pi \ln 2 = 2.18$ **33.** $y = \ln \dfrac{3.5}{3 + \cos x} + 2$

35. 300 million **37.** $i = \dfrac{E}{R}(1 - e^{-Rt/L})$ **39.** 1.41 m

Exercises 28-3, page 803

1. $e^{7x} + C$ **3.** $\frac{1}{2}e^{2x+5} + C$ **5.** $2(e - 1) = 3.44$

7. $2e^{x^3} + C$ **9.** $2(e^2 - e) = 9.34$ **11.** $\frac{1}{2}e^{2\sec x} + C$

13. $\dfrac{2e^x - 3}{2e^{2x}} + C$ **15.** $6 - \dfrac{3(e^6 - e^2)}{2} = -588.06$

17. $-\dfrac{4}{e^{\sqrt{x}}} + C$ **19.** $e^{\tan^{-1} x} + C$ **21.** $-\frac{1}{3}e^{\cos 3x} + C$

23. 0 **25.** $3e^2 - 3 = 19.2$ **27.** $\pi(e^4 - e) = 163$

29. $\frac{1}{8}(e^8 - 1) = 372$ **31.** $\ln b \int b^u du = b^u + C_1$

33. $q = EC(1 - e^{-t/RC})$ **35.** $s = -e^{-2t} - 0.6e^{-5t}$

Exercises 28-4, page 806

1. $\frac{1}{2}\sin 2x + C$ **3.** $\frac{1}{3}\tan 3x + C$ **5.** $2\sec\frac{1}{2}x + C$

7. 0.6365 **9.** $\frac{3}{2}\ln|\sec x^2 + \tan x^2| + C$

11. $\cos\left(\dfrac{1}{x}\right) + C$ **13.** $\frac{1}{2}\sqrt{3}$ **15.** $\frac{1}{5}\sec 5x + C$

17. $\frac{1}{2}\ln|\sec 2x + \tan 2x| + C$

19. $\frac{1}{2}(\ln|\sin 2x| + \sin 2x) + C$

21. $\csc x - \cot x - \ln|\csc x - \cot x| + \ln|\sin x| + C$

23. $\frac{1}{9}\pi + \frac{1}{3}\ln 2 = 0.580$ **25.** 0.693

27. $\pi\sqrt{3} = 5.44$ **29.** $\theta = 0.10 \cos 2.5t$

31. 0.7726 m

Exercises 28-5, page 810

1. $\frac{1}{3}\sin^3 x + C$ **3.** $-\frac{1}{2}\cos 2x + \frac{1}{6}\cos^3 2x + C$

5. $\frac{2}{3}\sin^3 x - \frac{2}{5}\sin^5 x + C$

7. $\frac{1}{24}(64 - 43\sqrt{2}) = 0.1329$

9. $\frac{1}{2}x - \frac{1}{4}\sin 2x + C$ **11.** $\frac{1}{2}x + \frac{1}{12}\sin 6x + C$

13. $\frac{1}{2}\tan^2 x + \ln|\cos x| + C$ **15.** $\frac{3}{4}$

17. $\frac{1}{6}\tan^3 2x - \frac{1}{2}\tan 2x + x + C$

19. $\frac{1}{15}\sec^5 3x - \frac{1}{9}\sec^3 3x + C$ **21.** $x - \frac{1}{2}\cos 2x + C$

23. $\frac{1}{4}\cot^4 x - \frac{1}{3}\cot^3 x + \frac{1}{2}\cot^2 x - \cot x + C$

25. $1 + \frac{1}{2}\ln 2 = 1.347$

27. $\frac{1}{5}\tan^5 x + \frac{2}{3}\tan^3 x + \tan x + C$

29. $\frac{1}{2}\pi^2 = 4.935$ **31.** $\sqrt{2} - 1 = 0.414$

33. $\int \sin x \cos x\, dx = \frac{1}{2}\sin^2 x + C_1 = -\frac{1}{2}\cos^2 x + C_2$; $C_2 = C_1 + \frac{1}{2}$

35. $\frac{4}{3}$ **37.** 120 V **39.** $\frac{aA}{2} + \frac{A}{2b\pi}\sin ab\pi \cos 2bc\pi$

Exercises 28-6, page 814

1. $\sin^{-1}\frac{1}{2}x + C$ **3.** $\frac{1}{8}\tan^{-1}\frac{1}{8}x + C$

5. $\frac{1}{4}\sin^{-1} 4x + C$ **7.** $\tan^{-1} 6 = 1.41$

9. $\frac{2}{5}\sqrt{5}\sin^{-1}\frac{1}{5}\sqrt{5} = 0.415$ **11.** $\frac{4}{9}\ln|9x^2 + 16| + C$

13. $\frac{1}{35}\sqrt{35}(\tan^{-1}\frac{2}{7}\sqrt{35} - \tan^{-1}\frac{1}{7}\sqrt{35}) = 0.057$

15. $\sin^{-1} e^x + C$ **17.** $\tan^{-1}(x + 1) + C$

19. $4\sin^{-1}\frac{1}{2}(x + 2) + C$ **21.** -0.357

23. $2\sin^{-1}(\frac{1}{2}x) + \sqrt{4 - x^2} + C$

25. (a) inverse tangent, $\int \frac{du}{a^2 + u^2}$ where $u = 3x$, $du = 3\,dx$, $a = 2$; numerator cannot fit du of denominator. Positive $9x^2$ leads to inverse tangent form. (b) logarithmic, $\int \frac{du}{u}$ where $u = 4 + 9x$, $du = 9\,dx$ (c) general power, $\int u^{-1/2}\,du$ where $u = 4 + 9x^2$, $du = 18x\,dx$

27. (a) general power, $\int u^{-1/2}\,du$ where $u = 4 - 9x^2$, $du = -18x\,dx$; numerator can fit du of denominator. Square root becomes $-1/2$ power. Does not fit inverse sine form. (b) inverse sine, $\int \frac{du}{\sqrt{a^2 - u^2}}$ where $u = 3x$, $du = 3dx$, $a = 2$ (c) logarithmic, $\int \frac{du}{u}$ where $u = 4 - 9x$, $du = -9dx$

29. $\tan^{-1} 2 = 1.11$ **31.** $k\tan^{-1}\frac{x}{d} + C$

33. $\sin^{-1}\frac{x}{A} = \sqrt{\frac{k}{m}}\,t + \sin^{-1}\frac{x_0}{A}$ **35.** $0.22k$

Exercises 28-7, page 818

1. $\cos x + x\sin x + C$ **3.** $\frac{1}{2}xe^{2x} - \frac{1}{4}e^{2x} + C$

5. $x\tan x + \ln|\cos x| + C$

7. $2x\tan^{-1}x - 2\ln\sqrt{1 + x^2} + C$

9. $-8x\sqrt{1 - x} - \frac{16}{3}(1 - x)^{3/2} + C$

11. $\frac{1}{2}x^2\ln x - \frac{1}{4}x^2 + C$

13. $\frac{1}{2}x\sin 2x - \frac{1}{4}(2x^2 - 1)\cos 2x + C$

15. $\frac{1}{2}(e^{\pi/2} - 1) = 1.91$ **17.** $1 - \frac{3}{e^2} = 0.594$

19. 0.1104 **21.** $\frac{1}{2}\pi - 1 = 0.571$ **23.** 0.756

25. $s = \frac{1}{3}[(t^2 - 2)\sqrt{t^2 + 1} + 2]$

27. $q = \frac{1}{5}[e^{-2t}(\sin t - 2\cos t) + 2]$

Exercises 28-8, page 821

1. $-\frac{\sqrt{1 - x^2}}{x} - \sin^{-1} x + C$

3. $2\ln|x + \sqrt{x^2 - 4}| + C$ **5.** $-\frac{\sqrt{x^2 + 9}}{9x} + C$

7. $\frac{x}{\sqrt{4 - x^2}} + C$ **9.** $\frac{16 - 9\sqrt{3}}{24} = 0.017$

11. $5\ln|\sqrt{x^2 + 2x + 2} + x + 1| + C$

13. $\frac{1}{3}\sec^{-1}\frac{2}{3}x + C$ **15.** $2\sec^{-1} e^x + C$ **17.** π

19. $\frac{1}{4}ma^2$ **21.** 2.68 **23.** $kQ\ln\frac{\sqrt{a^2 + b^2} + a}{\sqrt{a^2 + b^2} - a}$

Exercises 28-9, page 823

1. $\frac{3}{25}[2 + 5x - 2\ln|2 + 5x|] + C$ **3.** $\frac{3544}{15} = 236.3$

5. $\frac{1}{2}x\sqrt{4 - x^2} + 2\sin^{-1}\frac{1}{2}x + C$

7. $\frac{1}{2}\sin x - \frac{1}{10}\sin 5x + C$

9. $\sqrt{4x^2 - 9} - 3\sec^{-1}\left(\frac{2x}{3}\right) + C$

11. $\frac{1}{20}\cos^4 4x\sin 4x + \frac{1}{5}\sin 4x - \frac{1}{15}\sin^3 4x + C$

13. $\frac{1}{2}x^2\tan^{-1} x^2 - \frac{1}{4}\ln(1 + x^4) + C$

15. $\frac{1}{4}(8\pi - 9\sqrt{3}) = 2.386$

17. $-\ln\left(\frac{1 + \sqrt{4x^2 + 1}}{2x}\right) + C$

19. $-8\ln\left(\frac{1 + \sqrt{1 - 4x^2}}{2x}\right) + C$

21. $\frac{1}{8}\cos 4x - \frac{1}{12}\cos 6x + C$

23. $\frac{1}{3}(\cos x^3 + x^3\sin x^3) + C$ **25.** $\frac{x^2}{\sqrt{1 - x^4}} + C$

27. 4.892 **29.** $\frac{1}{4}x^4(\ln x^2 - \frac{1}{2}) + C$

31. $-\frac{3x^3}{\sqrt{x^6 - 1}} + C$

33. $\frac{1}{4}[2\sqrt{5} + \ln(2 + \sqrt{5})] = 1.479$

35. $\frac{1}{4}(8\tan^{-1} 4 - \ln 17) = 1.943$ **37.** 32.7 kN

39. 38.5 N · m

Review Exercises for Chapter 28, page 825

1. $-\frac{1}{2}e^{-2x} + C$ **3.** $-\frac{1}{\ln 2x} + C$

5. $4\ln(1 + \sin x) + C$ **7.** $\frac{2}{35}\tan^{-1}\frac{7}{5}x + C$ **9.** 0

11. $\frac{1}{2}\ln 2 = 0.3466$ **13.** $\frac{1}{12}\sec^4 3x + C$

15. $\tan^{-1} e^x + C$ **17.** $\frac{1}{9}\tan^3 3x + \frac{1}{3}\tan 3x + C$

19. $\frac{3}{\sqrt{e}} - 2 = -0.1804$ **21.** $\frac{3}{4}\tan^{-1}\frac{x^2}{2} + C$

23. $2\ln|2x + \sqrt{4x^2 - 9}| + C$ **25.** $\sqrt{e^{2x} + 1} + C$

27. $\frac{1}{2}(3x - \sin 3x\cos 3x) + C$

29. $-\frac{1}{2}x \cot 2x + \frac{1}{4}\ln|\sin 2x| + C$ **31.** $\frac{1}{2}\sin e^{2x} + C$

33. $\frac{1}{3}$ **35.** $\frac{1}{2}x^2 - 2x + 3\ln|x + 2| + C$

37. $\frac{1}{3}(e^x + 1)^3 + C_1 = \frac{1}{3}e^{3x} + e^{2x} + e^x + C_2;\ C_2 = C_1 + \frac{1}{3}$

39. (a) $\dfrac{1}{2}\displaystyle\int(1 - \cos 2x)\,dx = \dfrac{x}{2} - \dfrac{1}{4}\sin 2x + C_1$

(b) $\displaystyle\int \sin x(\sin x\,dx) = -\sin x \cos x + \int \cos^2 x\,dx$

$= -\sin x \cos x + \displaystyle\int (1 - \sin^2 x)\,dx$

41. $y = \frac{1}{3}\tan^3 x + \tan x$ **43.** $2(e^3 - 1) = 38.17$

45. 11.18 **47.** $4\pi(e^2 - 1) = 80.29$

49. $\frac{1}{8}\pi(e^{2\pi} - 1) = 209.9$ **51.** ln 3

53. $\Delta S = a\ln T + bT + \frac{1}{2}cT^2 + C$

55. $v = 98(1 - e^{-0.1t})$ **57.** $\sqrt{2}$ **59.** $\frac{2}{3}k$

61. 3.47 cm^3 **63.** 73.0 m^2

Exercises 29-1, page 830

1. 1, 4, 9, 16 **3.** $\frac{1}{2}, \frac{1}{3}, \frac{1}{4}, \frac{1}{5}$

5. (a) $-\frac{2}{5}, \frac{4}{25}, -\frac{8}{125}, \frac{16}{625}$ (b) $-\frac{2}{5} + \frac{4}{25} - \frac{8}{125} + \frac{16}{625} - \cdots$

7. (a) 2, 0, 2, 0 (b) $2 + 0 + 2 + 0 + \cdots$

9. $a_n = \dfrac{1}{n + 1}$ **11.** $a_n = \dfrac{1}{(n + 1)(n + 2)}$

13. 1, 1.125, 1.162 037 0, 1.177 662 0, 1.185 662 0; convergent; 1.2

15. 1, 1.5, 2.166 666 7, 2.916 666 7, 3.716 666 7; divergent

17. 0, −1, −3, −6, −10; divergent

19. 0.75, 0.888 888 9, 0.937 500 0, 0.960 000 0, 0.972 222 2; convergent; 1 **21.** divergent

23. convergent, $S = \frac{3}{4}$ **25.** convergent, $S = 100$

27. convergent, $S = \frac{4096}{9}$ **29.** (a) 1 (b) diverges

31. 1 **33.** (a) diverges (b) $y = 2100(1.05^x - 1)$

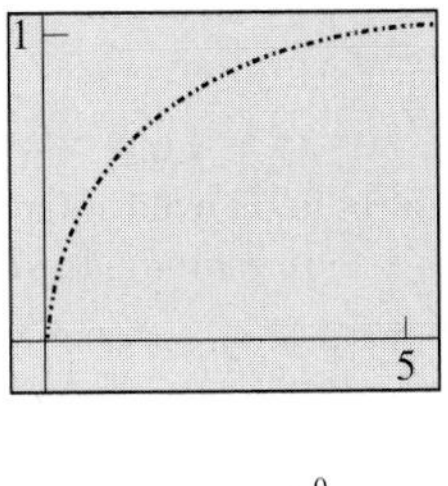

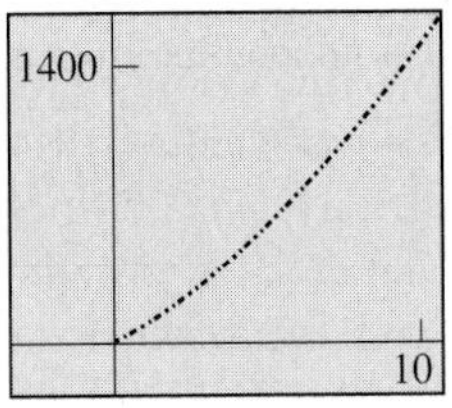

35. $r = x;\ S = \dfrac{x^0}{1 - x} = \dfrac{1}{1 - x}$

Exercises 29-2, page 835

1. $1 + x + \frac{1}{2}x^2 + \cdots$ **3.** $1 - \frac{1}{2}x^2 + \frac{1}{24}x^4 - \cdots$

5. $1 + \frac{1}{2}x - \frac{1}{8}x^2 + \cdots$ **7.** $1 - 2x + 2x^2 + \cdots$

9. $1 - 8x^2 + \frac{32}{3}x^4 - \cdots$ **11.** $1 + x + x^2 + \cdots$

13. $-2x - 2x^2 - \frac{8}{3}x^3 - \cdots$ **15.** $1 - \dfrac{x^2}{8} + \dfrac{x^4}{384} - \cdots$

17. $x - \frac{1}{3}x^3 + \cdots$ **19.** $x + \frac{1}{3}x^3 + \cdots$

21. $-\frac{1}{2}x^2 - \frac{1}{12}x^4 - \cdots$ **23.** $x^2 - \frac{1}{3}x^4 + \cdots$

25. No, functions are not defined at $x = 0$.

27. $e^x = 1 + x + \dfrac{x^2}{2} + \cdots,\ e^{x^2} = 1 + x^2 + \dfrac{x^4}{2} + \cdots$

29. $f^n(0) = 0$ except $f'''(0) = 6$

31. $R = e^{-0.001t} = 1 - 0.001t + (5 \times 10^{-7})t^2 - \cdots$

Exercises 29-3, page 839

1. $1 + 3x + \frac{9}{2}x^2 + \frac{9}{2}x^3 + \cdots$

3. $\dfrac{x}{2} - \dfrac{x^3}{2^3 3!} + \dfrac{x^5}{2^5 5!} - \dfrac{x^7}{2^7 7!} + \cdots$

5. $1 - 8x^2 + \frac{32}{3}x^4 - \frac{256}{45}x^6 + \cdots$

7. $x^2 - \frac{1}{2}x^4 + \frac{1}{3}x^6 - \frac{1}{4}x^8 + \cdots$ **9.** 0.3103

11. 0.1901 **13.** $1 + \frac{1}{2}x^2 + \frac{1}{24}x^4 + \frac{1}{720}x^6 + \cdots$

15. $x + x^2 + \frac{1}{3}x^3 + \cdots$

17. $\dfrac{d}{dx}\left(x - \dfrac{1}{6}x^3 + \dfrac{1}{120}x^5 - \cdots\right) = 1 - \dfrac{1}{2}x^2 + \dfrac{1}{24}x^4 - \cdots$

19. $\displaystyle\int \cos x\,dx = x - \dfrac{x^3}{3!} + \cdots$

21. $\displaystyle\int_0^1 e^x\,dx = 1.718\,281\,8,$

$\displaystyle\int_0^1 (1 + x + \tfrac{1}{2}x^2 + \tfrac{1}{6}x^3)\,dx = 1.708\,333\,3$

23. 0.003 099 **25.** 0.199 968

27. $y = -1 + 3t + 2t^2 - \frac{9}{2}t^3 - \cdots$

29. **31.**

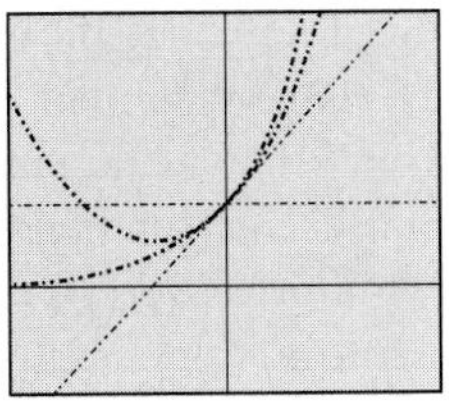

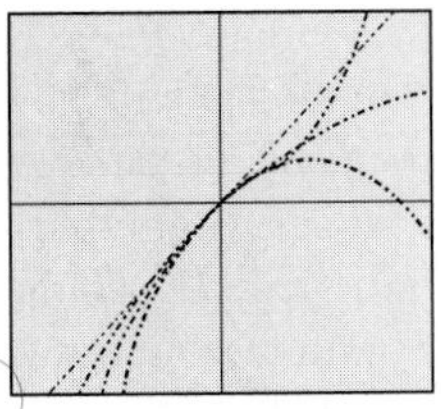

Exercises 29-4, page 843

1. 1.22, 1.221 402 8 **3.** 0.099 833 3, 0.099 833 4

5. 2.718 055 6, 2.718 281 8 **7.** 0.998 629 2, 0.998 629 5

9. 0.334 933 3, 0.336 472 2 **11.** 0.354 613 0, 0.354 612 9

13. −0.013 997 5, −0.013 997 5

15. 0.728 302 0, 0.728 695 0 **17.** 1.052 352 8

19. 0.987 446 2 **21.** 8.3×10^{-8} **23.** 3.1×10^{-7}

25. 3.146 **27.** 1.59 years

29. $i = \dfrac{E}{L}\left(t - \dfrac{Rt^2}{2L}\right)$; small values of t **31.** 18 m

Exercises 29-5, page 846

1. 3.32 **3.** 2.049 **5.** 0.5150 **7.** 0.492 88

9. $e^{-2}\left[1 - (x - 2) + \dfrac{(x - 2)^2}{2!} - \cdots\right]$

11. $\frac{1}{2}\left[\sqrt{3} + \left(x - \frac{1}{3}\pi\right) - \frac{\sqrt{3}}{2!}\left(x - \frac{1}{3}\pi\right)^2 - \cdots\right]$

13. $2 + \frac{1}{12}(x - 8) - \frac{1}{288}(x - 8)^2 + \cdots$

15. $1 + 2(x - \frac{1}{4}\pi) + 2(x - \frac{1}{4}\pi)^2 + \cdots$ **17.** 0.111

19. 3.0496 **21.** 2.0247 **23.** 0.874 62

25. Use the indicated method.

27. 0.515 040 8, 0.515 038 8, 0.515 038 1

29. Graph of part (b) fits well near $x = \pi/3$.

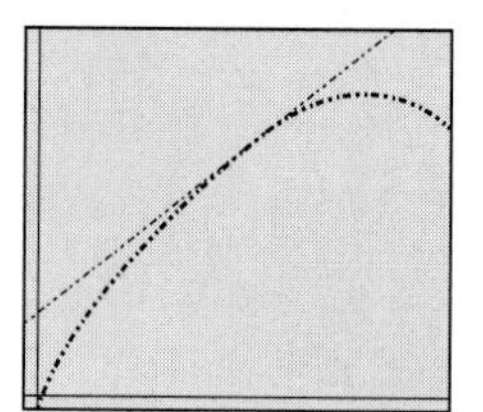

31. Graph of part (b) fits well near $x = 2$.

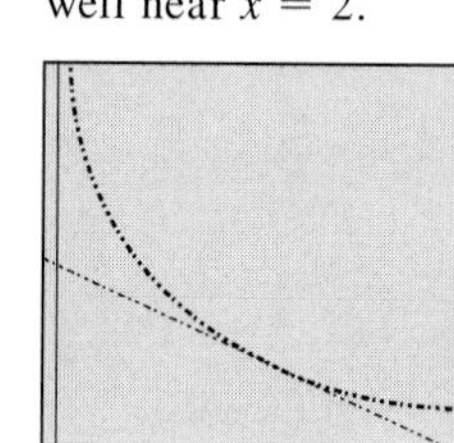

Exercises 29-6, page 853

1. $f(x) = \frac{1}{2} - \frac{2}{\pi}\sin x - \frac{2}{3\pi}\sin 3x - \cdots$

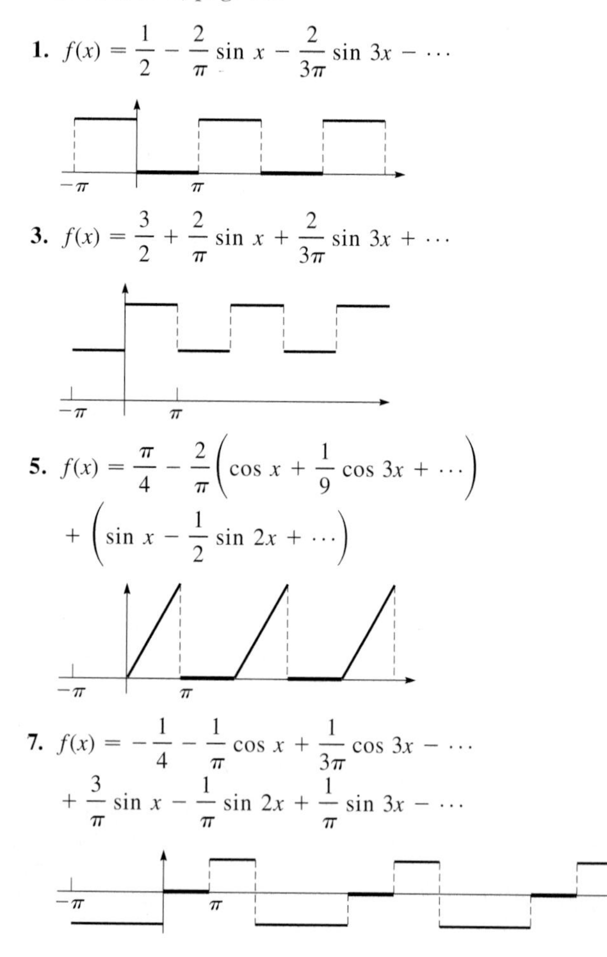

3. $f(x) = \frac{3}{2} + \frac{2}{\pi}\sin x + \frac{2}{3\pi}\sin 3x + \cdots$

5. $f(x) = \frac{\pi}{4} - \frac{2}{\pi}\left(\cos x + \frac{1}{9}\cos 3x + \cdots\right) + \left(\sin x - \frac{1}{2}\sin 2x + \cdots\right)$

7. $f(x) = -\frac{1}{4} - \frac{1}{\pi}\cos x + \frac{1}{3\pi}\cos 3x - \cdots + \frac{3}{\pi}\sin x - \frac{1}{\pi}\sin 2x + \frac{1}{\pi}\sin 3x - \cdots$

9. $f(x) = \frac{\pi}{2} - \frac{4}{\pi}\cos x - \frac{4}{9\pi}\cos 3x - \cdots$

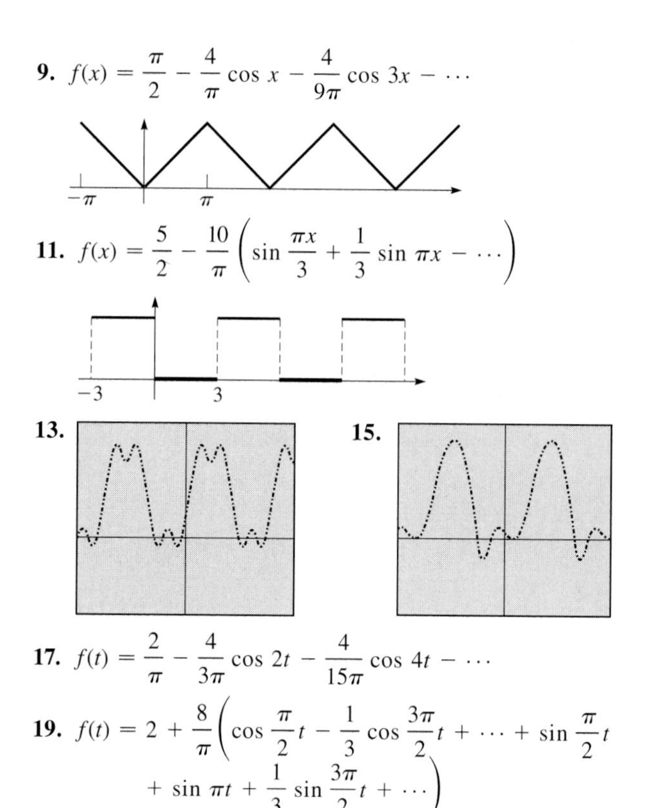

11. $f(x) = \frac{5}{2} - \frac{10}{\pi}\left(\sin\frac{\pi x}{3} + \frac{1}{3}\sin \pi x - \cdots\right)$

13.

15.

17. $f(t) = \frac{2}{\pi} - \frac{4}{3\pi}\cos 2t - \frac{4}{15\pi}\cos 4t - \cdots$

19. $f(t) = 2 + \frac{8}{\pi}\left(\cos\frac{\pi}{2}t - \frac{1}{3}\cos\frac{3\pi}{2}t + \cdots + \sin\frac{\pi}{2}t + \sin \pi t + \frac{1}{3}\sin\frac{3\pi}{2}t + \cdots\right)$

Review Exercises for Chapter 29, page 855

1. $\frac{1}{2} - \frac{1}{4}x + \frac{1}{48}x^3 - \cdots$ **3.** $2x^2 - \frac{4}{3}x^6 + \frac{4}{15}x^{10} - \cdots$

5. $1 + \frac{1}{3}x - \frac{1}{9}x^2 + \cdots$ **7.** $x + \frac{1}{6}x^3 + \frac{3}{40}x^5 + \cdots$

9. 0.82 **11.** 1.09 **13.** 1.0344 **15.** −0.2015

17. 0.952 99 **19.** 12.1655 **21.** 0.259

23. $\frac{1}{2} - \frac{1}{2}\sqrt{3}(x - \frac{1}{3}\pi) - \frac{1}{4}(x - \frac{1}{3}\pi)^2 + \cdots$

25. $f(x) = \frac{1}{2} + \frac{2}{\pi}(\cos x - \frac{1}{3}\cos 3x + \cdots)$

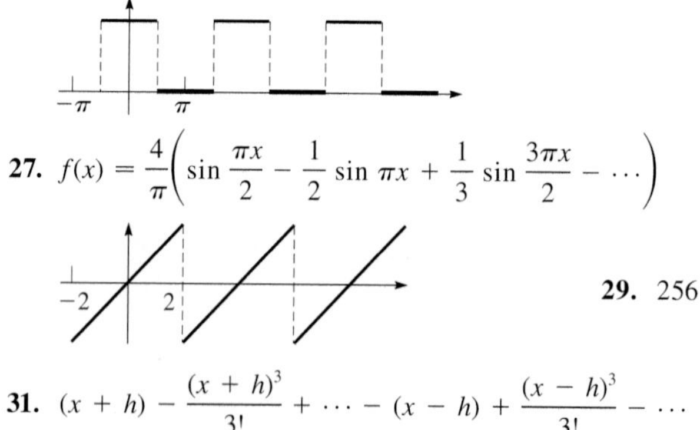

27. $f(x) = \frac{4}{\pi}\left(\sin\frac{\pi x}{2} - \frac{1}{2}\sin \pi x + \frac{1}{3}\sin\frac{3\pi x}{2} - \cdots\right)$

29. 256

31. $(x + h) - \frac{(x + h)^3}{3!} + \cdots - (x - h) + \frac{(x - h)^3}{3!} - \cdots = 2h - \frac{2hx^2}{2!} + \cdots = 2h\left(1 - \frac{x^2}{2!} + \cdots\right)$

33. $4x - 2x^2 + \frac{4}{3}x^3 - x^4 + \cdots$

35. $1 - x^2 + \frac{1}{3}x^4 - \frac{2}{45}x^6 + \cdots$

37. $1 + \frac{x^2}{2} + \frac{5x^4}{24} + \cdots$ **39.** $1 - x + x^2 - \cdots$

41. 0.002 496 88 **43.** $x - \frac{x^3}{3} + \frac{x^5}{5} - \cdots$

45. $N_0\left(1 - \lambda t + \frac{\lambda^2 t^2}{2} - \frac{\lambda^3 t^3}{6} + \cdots\right)$

47. $2x + \frac{2}{3}x^3 + \frac{2}{5}x^5 + \frac{2}{7}x^7 + \cdots$

49. $N_0[1 + e^{-k/T} + (e^{-k/T})^2 + \cdots]$
$= N_0(1 + e^{-k/T} + e^{-2k/T} + \cdots)$

51. $f(t) = \frac{1}{2\pi} + \frac{1}{\pi}\left(\frac{1}{2}\cos t - \frac{1}{3}\cos 2t + \cdots\right)$
$+ \frac{1}{4}\sin t + \frac{2}{3\pi}\sin 2t + \cdots$

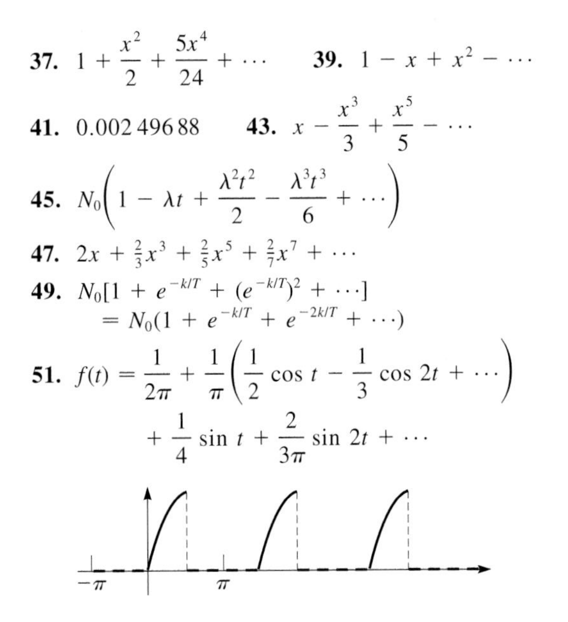

Exercises 30-1, page 859

1. particular solution **3.** general solution

(The following "answers" are the unsimplified expressions obtained by substituting functions and derivatives.)

5. $1 = 1$ **7.** $e^x - (e^x - 1) = 1$ **9.** $x(2cx) = 2(cx^2)$

11. $(-2ce^{-2x} + 1) + 2(ce^{-2x} + x - \frac{1}{2}) = 2x$

13. $(-12\cos 2x) + 4(3\cos 2x) = 0$

15. $[e^{2x}(1 + 4c_1 + 4c_2 + 4x + 4c_2x + 2x^2)]$
$- 4[e^{2x}(2c_1 + c_2 + x + 2c_2x + x^2)]$
$+ 4[e^{2x}(c_1 + c_2x + x^2/2)] = e^{2x}$

17. $x^2\left[-\frac{c^2}{(x-c)^2}\right] + \left[\frac{cx}{(x-c)}\right]^2 = 0$

19. $x\left(-\frac{c_1}{x^2}\right) + \frac{c_1}{x} = 0$

21. $(\cos x - \sin x + e^{-x}) + (\sin x + \cos x - e^{-x})$
$= 2\cos x$

23. $(e^{-x} + \frac{12}{5}\cos 2x + \frac{24}{5}\sin 2x) +$
$(-e^{-x} + \frac{6}{5}\sin 2x - \frac{12}{5}\cos 2x) = 6\sin 2x$

25. $\cos x\left[\frac{(\sec x + \tan x) - (x + c)(\sec x \tan x + \sec^2 x)}{(\sec x + \tan x)^2}\right]$
$+ \sin x = 1 - \frac{x + c}{\sec x + \tan x}$

27. $c^2 + cx = cx + c^2$

Exercises 30-2, page 863

1. $y = c - x^2$ **3.** $x - \frac{1}{y} = c$ **5.** $\ln V = \frac{1}{P} + c$

7. $\ln(x^3 + 5) + 3y = c$ **9.** $y = 2x^2 + x - x\ln x + c$

11. $4\sqrt{1 - y} = e^{-x^2} + c$ **13.** $e^x - e^{-y} = c$

15. $\ln(y + 4) = x + c$ **17.** $y(1 + \ln x)^2 + cy + 2 = 0$

19. $\tan^2 x + 2\ln y = c$

21. $x^2 + 1 + x\ln y + cx = 0$ **23.** $y^2 + 4\sin^{-1} x = c$

25. $i = c - (\ln t)^2$ **27.** $\ln(e^x + 1) - \frac{1}{y} = c$

29. $3\ln y + x^3 = 0$ **31.** $\frac{1}{3}y^3 + y = \frac{1}{2}\ln^2 x$

33. $2\ln(1 - y) = 1 - 2\sin x$ **35.** $e^{2x} - \frac{2}{y} = 2(e^x - 1)$

Exercises 30-3, page 865

1. $2xy + x^2 = c$ **3.** $x^3 - 2y = cx - 4$

5. $A^2r - r = cA$ **7.** $(xy)^4 = 12\ln y + c$

9. $2\sqrt{x^2 + y^2} = x + c$

11. $y = c - \frac{1}{2}\ln\sin(x^2 + y^2)$

13. $\ln(y^2 - x^2) + 2x = c$ **15.** $5xy^2 + y^3 = c$

17. $2xy + x^3 = 5$ **19.** $2x = 2xy^2 - 15y$

Exercises 30-4, page 868

1. $y = e^{-x}(x + c)$ **3.** $y = -\frac{1}{2}e^{-4x} + ce^{-2x}$

5. $y = -2 + ce^{2x}$ **7.** $y = x(3\ln x + c)$

9. $y = \frac{8}{7}x^3 + \frac{c}{\sqrt{x}}$ **11.** $r = -\cot\theta + c\csc\theta$

13. $y = (x + c)\csc x$ **15.** $y = 3 + ce^{-x}$

17. $2s = e^{4t}(t^2 + c)$ **19.** $y = \frac{1}{4} + ce^{-x^4}$

21. $3y = x^4 - 6x^2 - 3 + cx$ **23.** $xy = (x^3 + c)e^{3x}$

25. Can solve by separation of variables: $\frac{dy}{1 - y} = 2dx$.
Can also solve as linear differential equation of first order: $dy + 2y\,dx = 2\,dx$; $y = 1 + ce^{-2x}$

27. $y = e^{-x}$ **29.** $y = \frac{4}{3}\sin x - \csc^2 x$

31. $y(\csc x - \cot x) = \ln\frac{(\sqrt{2} - 1)(\csc 2x - \cot 2x)}{\csc x - \cot x}$

Exercises 30-5, page 872

1. $y^2 = 2x^2 + 1$

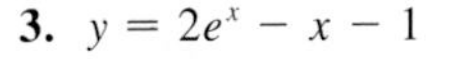

3. $y = 2e^x - x - 1$

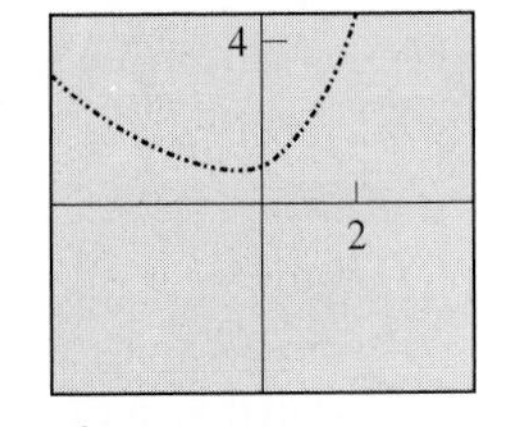

5. $y^2 = c - 2x$ **7.** $y^2 = c - 2\sin x$

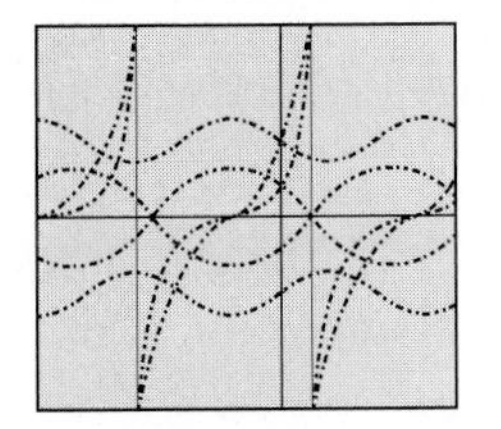

9. $N = N_0(0.5)^{t/40}$, 35.4% **11.** 3210 years

13. $S = a + \dfrac{c}{r^2}$ **15.** $5250e^{0.0196t}$ **17.** 12 min

19. \$1040.81 **21.** $\lim_{t\to\infty} \dfrac{E}{R}(1 - e^{-Rt/L}) = \dfrac{E}{R}$

23. $i = \dfrac{E}{R^2 + \omega^2 L^2}(R \sin \omega t - \omega L \cos \omega t + \omega L e^{-Rt/L})$

25. $q = q_0 e^{-t/RC}$ **27.** 1.47 kg

29. $v = 9.8(1 - e^{-t})$, 9.8 **31.** 4.0 m/s

33. $x = 3t^2 - t^3$, $y = 6t^2 - 2t^3 - 9t^4 + 6t^5 - t^6$

35. $p = 100(0.8)^{h/2000}$ **37.** \$2490

39. $x = 0.20 + 0.15e^{-2.0t}$

Exercises 30-6, page 878

1. $y = c_1 e^{3x} + c_2 e^{-2x}$ **3.** $y = c_1 e^{-x} + c_2 e^{-x/3}$

5. $y = c_1 + c_2 e^{3x}$ **7.** $y = c_1 e^{6x} + c_2 e^{2x/3}$

9. $y = c_1 e^{x/3} + c_2 e^{-3x}$ **11.** $y = c_1 e^{x/3} + c_2 e^{-x}$

13. $y = e^x(c_1 e^{x\sqrt{2}/2} + c_2 e^{-x\sqrt{2}/2})$

15. $y = e^{3x/8}(c_1 e^{x\sqrt{41}/8} + c_2 e^{-x\sqrt{41}/8})$

17. $y = e^{3x/2}(c_1 e^{x\sqrt{13}/2} + c_2 e^{-x\sqrt{13}/2})$

19. $y = e^{-x/2}(c_1 e^{x\sqrt{33}/2} + c_2 e^{-x\sqrt{33}/2})$

21. $y = \frac{1}{5}(3e^{7x} + 7e^{-3x})$ **23.** $y = \dfrac{e^3}{e^7 - 1}(e^{4x} - e^{-3x})$

25. $y = c_1 + c_2 e^{-x} + c_3 e^{3x}$ **27.** $y = c_1 e^x + c_2 e^{2x} + c_3 e^{3x}$

Exercises 30-7, page 881

1. $y = (c_1 + c_2 x)e^x$ **3.** $y = (c_1 + c_2 x)e^{-6x}$

5. $y = c_1 \sin 3x + c_2 \cos 3x$

7. $y = e^{-x/2}(c_1 \sin \frac{1}{2}\sqrt{7}x + c_2 \cos \frac{1}{2}\sqrt{7}x)$

9. $y = c_1 + c_2 x$ **11.** $y = c_1 \sin \frac{1}{2}x + c_2 \cos \frac{1}{2}x$

13. $y = (c_1 + c_2 x)e^{3x/4}$

15. $y = c_1 \sin \frac{1}{5}\sqrt{2}x + c_2 \cos \frac{1}{5}\sqrt{2}x$

17. $y = e^x(c_1 \cos \frac{1}{2}\sqrt{6}x + c_2 \sin \frac{1}{2}\sqrt{6}x)$

19. $y = (c_1 + c_2 x)e^{4x/5}$

21. $y = e^{3x/4}(c_1 e^{x\sqrt{17}/4} + c_2 e^{-x\sqrt{17}/4})$

23. $y = c_1 e^{x(-6+\sqrt{42})/3} + c_2 e^{x(-6-\sqrt{42})/3}$

25. $y = e^{(\pi/6 - 1 - x)} \sin 3x$ **27.** $y = (4 - 14x)e^{4x}$

29. $(D^2 - 9)y = 0$

31. $(D^2 + 9)y = 0$. The sum of cos $3x$ and sin $3x$ with no exponential factor indicates imaginary roots with $\alpha = 0$ and $\beta = 3$.

Exercises 30-8, page 885

1. $y = c_1 e^{2x} + c_2 e^{-x} - 2$

3. $y = c_1 \sin \frac{1}{2}x + c_2 \cos \frac{1}{2}x + x^2 - 8$

5. $y = c_1 e^{-x} + c_2 e^{-3x} + \frac{1}{8}e^x + \frac{2}{3}$

7. $y = c_1 + c_2 e^{3x} - \frac{3}{4}e^x - \frac{1}{2}xe^x$

9. $y = c_1 e^{x/3} + c_2 e^{-x/3} - \frac{1}{10} \sin x$

11. $y = c_1 \sin 3x + c_2 \cos 3x + x + 2 \sin 2x + \cos 2x$

13. $y = c_1 e^{-5x} + c_2 e^{6x} - \frac{1}{3}$

15. $y = c_1 e^{2x/3} + c_2 e^{-5x} + \frac{1}{4}e^{3x}$

17. $y = c_1 e^{2x} + c_2 e^{-2x} - \frac{1}{5} \sin x - \frac{2}{5} \cos x$

19. $y = c_1 \sin x + c_2 \cos x - \frac{1}{3} \sin 2x + 4$

21. $y = c_1 e^{-x} + c_2 e^{-4x} - \frac{7}{100}e^x + \frac{1}{10}xe^x + 1$

23. $y = (c_1 + c_2 x)e^{-3x} + \frac{1}{25}e^{2x} - e^{-2x}$

25. $y = \frac{1}{6}(11e^{3x} + 5e^{-2x} + e^x - 5)$

27. $y = -\frac{2}{3} \sin x + \pi \cos x + x - \frac{1}{3} \sin 2x$

Exercises 30-9, page 891

1. $\theta = 0.1 \cos 3.1t$ **3.** 10

5. $y = 0.100 \cos 14.0t$

7. $y = 0.100 \cos 14.0t + 0.050 \sin 2.00t - 0.007 \sin 14.0t$

9. $x = 0.150 \cos 6.26t$

11. $q = 2.23 \times 10^{-4} e^{-20t} \sin 2240t$

13. $q = 0.01(1 - \cos 316t)$

15. $q = e^{-10t}(c_1 \sin 99.5t + c_2 \cos 99.5t) - 1.81 \times 10^{-3} \sin 120\pi t - 1.03 \times 10^{-4} \cos 120\pi t$

17. $i = 10^{-6}(2.00 \cos (1.58 \times 10^4 t) + 158 \sin (1.58 \times 10^4 t) - 2.00e^{-200t})$

19. $i_p = 0.528 \sin 100t - 3.52 \cos 100t$

Exercises 30-10, page 895

1. $F(s) = \int_0^\infty e^{-st}\,dt = -\dfrac{1}{s}e^{-st}\Big|_0^\infty = \dfrac{1}{s}$

3. $F(s) = \int_0^\infty e^{-st} \sin at\,dt = \dfrac{e^{-st}(-s \sin at - a \cos at)}{s^2 + a^2}\Big|_0^\infty = \dfrac{a}{s^2 + a^2}$

5. $\dfrac{1}{s - 3}$ **7.** $\dfrac{6}{(s + 2)^4}$ **9.** $\dfrac{s - 2}{s^2 + 4}$

11. $\dfrac{3}{s} + \dfrac{2(s^2 - 9)}{(s^2 + 9)^2}$ **13.** $s^2 L(f) + sL(f)$

15. $(2s^2 - s + 1)L(f) - 2s + 1$ **17.** t^2 **19.** $\frac{1}{2}e^{-3t}$

21. $\frac{1}{2}t^2 e^{-t}$ **23.** $\frac{1}{54}(9t \sin 3t + 2 \sin 3t - 6t \cos 3t)$

Exercises 30-11, page 898

1. $y = e^{-t}$ **3.** $y = -e^{3t/2}$ **5.** $y = (1 + t)e^{-3t}$

7. $y = \frac{1}{2} \sin 2t$ **9.** $y = 1 - e^{-2t}$ **11.** $y = e^{2t} \cos t$

13. $y = 1 + \sin t$ **15.** $y = e^{-t}(\frac{1}{2}t^2 + 3t + 1)$

17. $v = 6(1 - e^{-t/2})$ **19.** $q = 1.6 \times 10^{-4}(1 - e^{-5000t})$

21. $i = 5t \sin 50t$ **23.** $y = \sin 3t - 3t \cos 3t$

Review Exercises for Chapter 30, page 900

1. $2 \ln(x^2 + 1) - \dfrac{1}{2y^2} = c$ **3.** $ye^{2x} = x + c$

5. $y = c_1 + c_2e^{-x/2}$ **7.** $y = (c_1 + c_2x)e^{-x}$

9. $2x^2 + 4xy + y^4 = c$ **11.** $y = cx^3 - x^2$

13. $y = c(y + 2)e^{2x}$

15. $y = e^{-x}(c_1 \sin\sqrt{5}x + c_2 \cos\sqrt{5}x)$

17. $y = \frac{1}{2}(1 + ce^{-4x})$ **19.** $y = \frac{1}{2}(c - x^2) \csc x$

21. $y = c_1e^x + c_2e^{-3x/2} - 2$

23. $y = e^{-x/2}(c_1e^{x\sqrt{5}/2} + c_2e^{-x\sqrt{5}/2}) + 2e^x$

25. $y = c_1e^{2x/3} + c_2e^{4x/3} + \frac{1}{2}x + \frac{25}{8}$

27. $y = c_1 \sin 3x + c_2 \cos 3x + \frac{1}{8} \sin x$

29. $y^3 = 8 \sin^2 x$ **31.** $y = 2x - 1 - e^{-2x}$

33. $y = 2e^{-x/2} \sin(\frac{1}{2}\sqrt{15}x)$

35. $y = \frac{1}{25}[16 \sin x + 12 \cos x - 3e^{-2x}(4 + 5x)]$

37. $y = e^{t/4}$ **39.** $y = \frac{1}{2}(e^{3t} - e^t)$ **41.** $y = -4 \sin t$

43. $y = \frac{1}{3}t - \frac{4}{9} \sin 3t$ **45.** $r = r_0 + kt$

47. 3.93 m/s **49.** 12.5 years **51.** 6.5 billion

53. $5y^2 + x^2 = c$ **55.** $q = c_1e^{-t/RC} + EC$

57. $y = 0.25e^{-2t}(2 \cos 4t + \sin 4t)$, underdamped

59. $q = e^{-6t}(0.4 \cos 8t + 0.3 \sin 8t) - 0.4 \cos 10t$

61. $i = 0$ **63.** $i = 12(1 - e^{-t/2})$; $i(0.3) = 1.67$ A

65. $q = 10^{-4}e^{-8t}(4.0 \cos 200t + 0.16 \sin 200t)$

67. $y = 0.25t \sin 8t$

69. $y = \dfrac{10}{3EI}[100x^3 - x^4 + xL^2(L - 100)]$

71. $\theta = \theta_0 \cos \omega t + \dfrac{\omega_0}{\omega} \sin \omega t$; $\omega = \sqrt{\dfrac{2k}{mr^2}}$

Exercises S-1, page 906

1. $x = 2, y = 1$ **3.** $x = \dfrac{17}{14}, y = \dfrac{19}{14}$

5. $x = -1, y = 4, z = 2$

7. $x = -2, y = -\dfrac{2}{3}, z = \dfrac{1}{3}$

9. $w = 1, x = 0, y = -2, z = 3$

11. unlimited: $x = -3, y = -1, z = 1$; $x = 12, y = 0, z = -10$

13. inconsistent **15.** $x = \dfrac{1}{2}, y = -\dfrac{3}{2}$

17. $x = \dfrac{1}{6}, y = -\dfrac{1}{2}$ **19.** inconsistent

21. 5.7 calc/s, 6.8 calc/s

23. 250 parts/h, 220 parts/h, 180 parts/h

Exercises S-2, page 910

1. hyperbola; $2x'y' + 25 = 0$

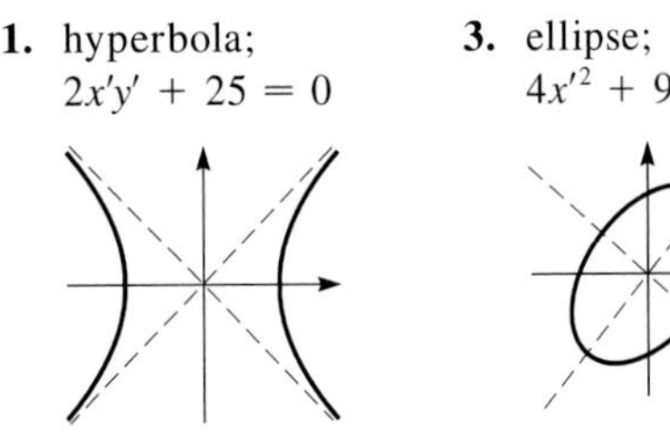

3. ellipse; $4x'^2 + 9y'^2 = 36$

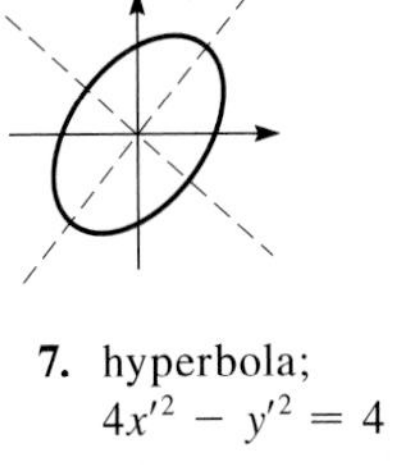

5. parabola; $x'^2 + \sqrt{2}y' = 0$

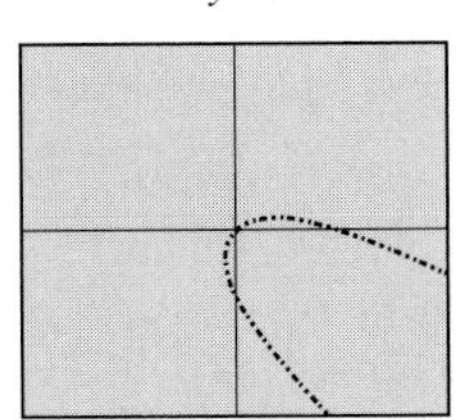

7. hyperbola; $4x'^2 - y'^2 = 4$

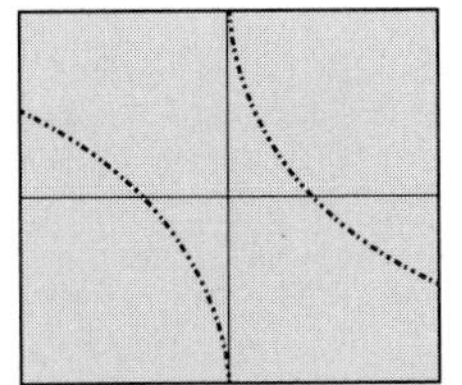

9. ellipse; $x'^2 + 2y'^2 = 2$

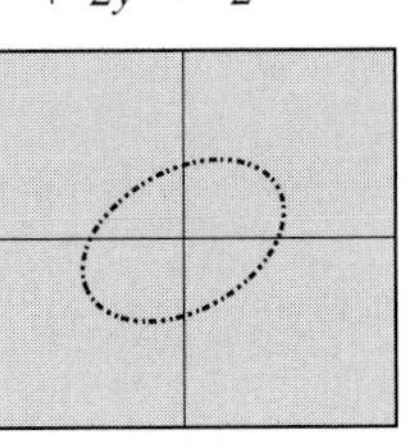

11. parabola; $y'^2 - 4x' + 16 = 0$
$y''^2 = 4x''$

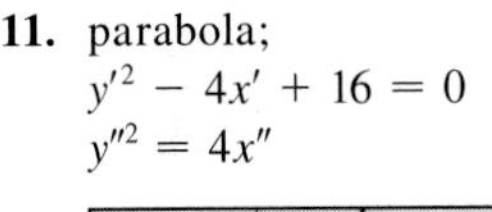

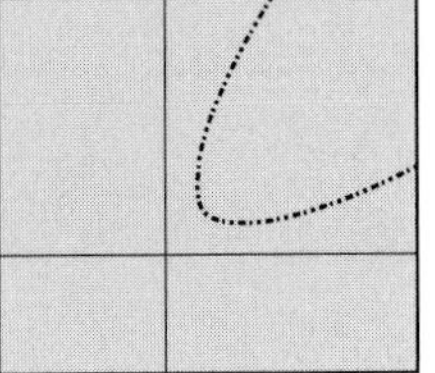

Exercises S-3, page 912

1. $V = \frac{1}{3}\pi r^2h$ **3.** $R = \sqrt{F_1^2 + F_2^2 + 1.732F_1F_2}$

5. 24 **7.** $2 - 3y + 4y^2$

9. $\dfrac{p^2 + pq + kp - p + 2q^2 + 4kq + 2k^2 + 5q + 5k}{p + q + k}$

11. 0 **13.** $x \neq 0, y \geq 0$ **15.** 1.03 A, 1.23 A

17. $A = \dfrac{pw - 2w^2}{2}$, 3850 cm² **19.** $L = \dfrac{(1.15 \times 10^4)r^4}{l^2}$

Exercises S-4, page 917

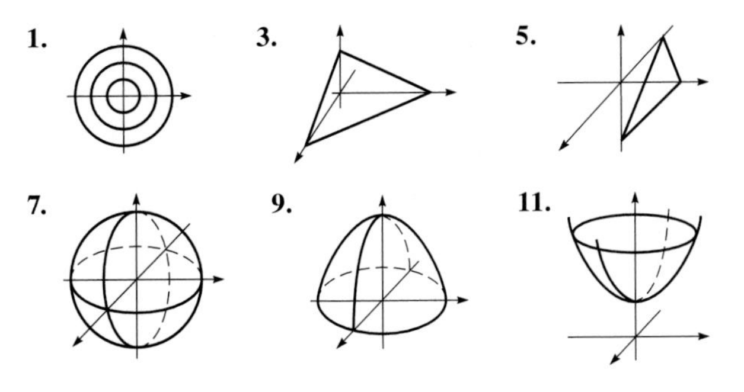

13. 15. 17.

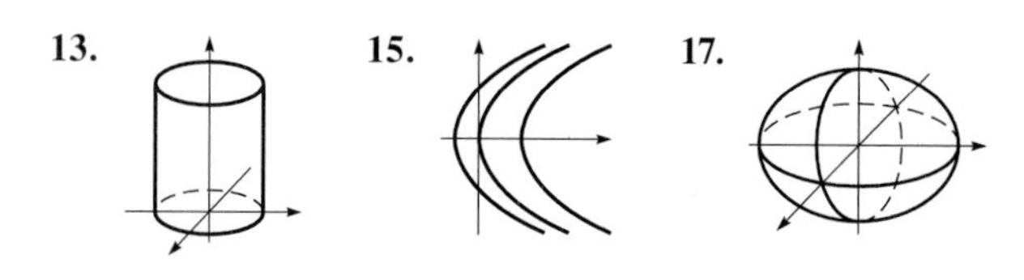

19.

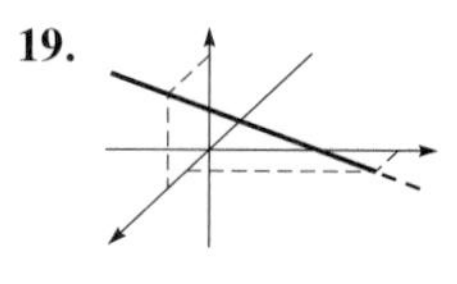

Exercises S-5, page 921

1. $\frac{\partial z}{\partial x} = 5 + 8xy, \frac{\partial z}{\partial y} = 4x^2$ **3.** $\frac{\partial f}{\partial x} = e^{2y}, \frac{\partial f}{\partial y} = 2xe^{2y}$

5. $\frac{\partial \phi}{\partial r} = \frac{1 + 3rs}{\sqrt{1 + 2rs}}, \frac{\partial \phi}{\partial s} = \frac{r^2}{\sqrt{1 + 2rs}}$

7. $\frac{\partial z}{\partial x} = y \cos xy, \frac{\partial z}{\partial y} = x \cos xy$

9. $\frac{\partial f}{\partial x} = \frac{12 \sin^2 2x \cos 2x}{1 - 3y}, \frac{\partial f}{\partial y} = \frac{6 \sin^3 2x}{(1 - 3y)^2}$

11. $\frac{\partial z}{\partial x} = \cos x - y \sin xy, \frac{\partial z}{\partial y} = -x \sin xy + \sin y$

13. -8 **15.** $\frac{\partial^2 z}{\partial x^2} = -6y, \frac{\partial^2 z}{\partial y^2} = 12xy, \frac{\partial^2 z}{\partial x \partial y} = 6y^2 - 6x$

17. $-4, -4$ **19.** $\left(\frac{R_2}{R_1 + R_2}\right)^2$

21. $M\left(\frac{2mg}{(M + m)^2}\right) + m\left(\frac{-2Mg}{(M + m)^2}\right) = 0$

23. 3.75×10^{-3} 1/Ω

Exercises S-6, page 925

1. $\frac{28}{3}$ **3.** $\frac{1}{3}$ **5.** 1 **7.** $\frac{74}{5}$ **9.** $\frac{32}{3}$ **11.** $\frac{28}{3}$

13. 300 cm^3 **15.**

Exercises S-7, page 929

1. $\ln\left|\frac{(x + 1)^2}{x + 2}\right| + C$ **3.** $\frac{1}{4} \ln\left|\frac{x - 2}{x + 2}\right| + C$

5. $x + \ln\left|\frac{x}{(x + 3)^4}\right| + C$ **7.** 1.057

9. $\ln\left|\frac{x^2(x - 5)^3}{x + 1}\right| + C$ **11.** 0.084 95 **13.** 1.79 C

15. $\frac{1}{2} \ln \frac{5}{4} = 0.1116$

Exercises S-8, page 934

1. $\frac{3}{x - 2} + 2 \ln\left|\frac{x - 2}{x}\right| + C$ **3.** $\frac{2}{x} + \ln\left|\frac{x - 1}{x + 1}\right| + C$

5. $-\frac{5}{4}$ **7.** $\frac{1}{8}\pi + \ln 3 = 1.491$

9. $-\frac{2}{x} + \frac{3}{2} \tan^{-1} \frac{x + 2}{2} + C$

11. $\ln(x^2 + 1) + \frac{1}{x^2 + 1} + C$

13. $2 + 4 \ln \frac{2}{3} = 0.3781$ **15.** 0.9190 m

Exercises for Appendix B, page A-9

1. MHz, 1 MHz = 10^6 Hz **3.** mm, 1 mm = 10^{-3} m

5. kilovolt, 1 kV = 10^3 V

7. milliampere, 1 mA = 10^{-3} A **9.** 10^5 cm

11. 2.0×10^{-5} Ms **13.** 537.00°C

15. 2.50×10^{-4} m^2 **17.** 8.00×10^4 dm^3

19. 0.050 L **21.** 4.50×10^3 cm/s

23. 1.38×10^3 m/h **25.** 3.53×10^6 cm/min^2

27. 5.25×10^{-3} W/A **29.** 2.2 Mg

31. 5.6×10^4 cm^3 **33.** 1000 g/L

35. 1.20×10^3 km/h **37.** 25 000 mg/dL

39. 5.52×10^3 kg/m^3

Exercises for Appendix C, page A-17

(Most answers have been rounded off to four significant digits.)

1. 56.02 **3.** 4162.1 **5.** 18.65 **7.** 0.3954

9. 14.14 **11.** 0.5251 **13.** 13.35 **15.** 944.6

17. 0.7349 **19.** −0.7594 **21.** −1.337

23. 1.015 **25.** 41.35° **27.** −1.182 **29.** 0.5862

31. 6.695 **33.** 3.508 **35.** 0.005 685 **37.** 2.053

39. 5.765 **41.** 4.501×10^{10} **43.** 497.2

45. 6.648 **47.** 401.2 **49.** 8.841 **51.** 2.523

53. 10.08 **55.** 22.36 **57.** 20.3° **59.** 4729

61. 3.301×10^4 **63.** 1.056 **65.** 55.5°

67. 3.277 **69.** 8.125 **71.** 1.000 **73.** 1.000

75. 12.90 **77.** 8.001 **79.** 8.053 **81.** 0.042 59

83. 0.4219 **85.** 0.7822 **87.** 2.073 646 5

89. 124.3 **91.** 252

Solutions to Practice Test Problems

Chapter 1

1. $\sqrt{9+16}=\sqrt{25}=5$

2. $\dfrac{(7)(-3)(-2)}{(-6)(0)}$ is undefined (division by zero).

3. $\dfrac{3.372\times10^{-3}}{7.526\times10^{12}}=4.480\times10^{-16}$

(Scientific calculator sequence:
3.372 [EE] 3 [+/−] [÷] 7.526 [EE] 12 [=]
[4.480467712 − 16])
(Graphing calculator sequence:
3.372 [EE] [(−)] 3 [÷] 7.526 [EE] 12 [ENTER] (or [EXE])
[4.480467712E −16])

4. $\dfrac{(+6)(-2)-3(-1)}{5-2}=\dfrac{-12-(-3)}{3}=\dfrac{-12+3}{3}$
$=\dfrac{-9}{3}=-3$

5. $\dfrac{346.4-23.5}{287.7}-\dfrac{0.944^3}{(3.46)(0.109)}=-1.108$

(Calculator sequence:
[(] 346.4 [−] 23.5 [)] [÷] 287.7 [−] .944 [x^y] 3
[÷] [(] 3.46 [×] .109 [)] [=] [−1.10820746])

6. $(2a^0b^{-2}c^3)^{-3}=2^{-3}a^{0(-3)}b^{(-2)(-3)}c^{3(-3)}=2^{-3}a^0b^6c^{-9}$
$=\dfrac{b^6}{8c^9}$

7. $(2x+3)^2=(2x+3)(2x+3)$
$=2x(2x)+2x(3)+3(2x)+3(3)$
$=4x^2+6x+6x+9$
$=4x^2+12x+9$

8. $3m^2(am-2m^3)=3m^2(am)+3m^2(-2m^3)$
$=3am^3-6m^5$

9. $\dfrac{8a^3x^2-4a^2x^4}{-2ax^2}=\dfrac{8a^3x^2}{-2ax^2}-\dfrac{4a^2x^4}{-2ax^2}=-4a^2-(-2ax^2)$
$=-4a^2+2ax^2$

10.
$$\begin{array}{r|l} & 3x-5 \quad \text{(quotient)} \\ 2x-1 & 6x^2-13x+7 \\ & 6x^2-3x \\ \hline & -10x+7 \\ & -10x+5 \\ \hline & 2 \quad \text{(remainder)}\end{array}$$

11. $(2x-3)(x+7)$
$=2x(x)+2x(7)+(-3)(x)+(-3)(7)$
$=2x^2+14x-3x-21=2x^2+11x-21$

12. $3x-[4x-(3-2x)]=3x-[4x-3+2x]$
$=3x-[6x-3]$
$=3x-6x+3=-3x+3$

13. $5y-2(y-4)=7$
$5y-2y+8=7$
$3y+8-8=7-8$
$3y=-1$
$y=-1/3$

14. $3(x-3)=x-(2-3d)$
$3x-9=x-2+3d$
$3x-9-x+9=x-2+3d-x+9$
$2x=3d+7$
$x=\dfrac{3d+7}{2}$

15. $0.000\,003\,6=3.6\times10^{-6}$ (6 places to right)

16.

$-\pi$	-3	0.3	$\sqrt{2}$	$\lvert-4\rvert$	(order)
-3.14	-3	0.3	1.41	4	(value)

17. $3(5+8)=3(5)+3(8)$ illustrates distributive law.

18. (a) 5, (b) 3.0 (zero is significant)

19. Evaluation:
$1000(1+0.05/2)^{2(3)}=1000(1.025)^6=\$1159.69$

20. $8(100-x)^2+x^2=8(100-x)(100-x)+x^2$
$=8(10\,000-200x+x^2)+x^2$
$=80\,000-1600x+8x^2+x^2$
$=80\,000-1600x+9x^2$

21. $L=L_0[1+\alpha(t_2-t_1)]$
$=L_0[1+\alpha t_2-\alpha t_1]$
$=L_0+\alpha L_0t_2-\alpha L_0t_1$
$L-L_0+\alpha L_0t_1=\alpha L_0t_2$
$t_2=\dfrac{L-L_0+\alpha L_0t_1}{\alpha L_0}$

22. Let n = number of newtons of second alloy.
$0.3(20)+0.8n=0.6(n+20)$
$6+0.8n=0.6n+12$
$0.2n=6$
$n=30$ N

Chapter 2

1. $\angle1+\angle3+90^\circ=180^\circ$ (sum of angles of a triangle)
$\angle3=52^\circ$ (vertical angles)
$\angle1=180^\circ-90^\circ-52^\circ=38^\circ$

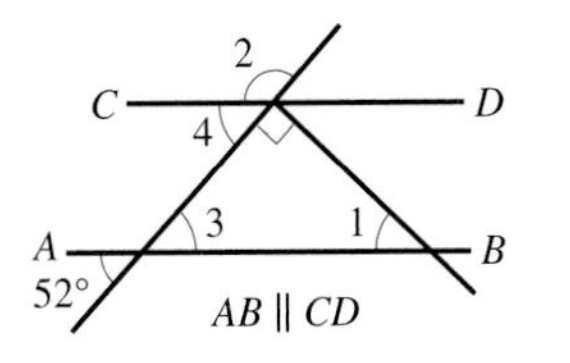

2. $\angle2+\angle4=180^\circ$ (straight angle)
$\angle4=52^\circ$ (corresponding angles)
$\angle2+52^\circ=180^\circ$
$\angle2=128^\circ$

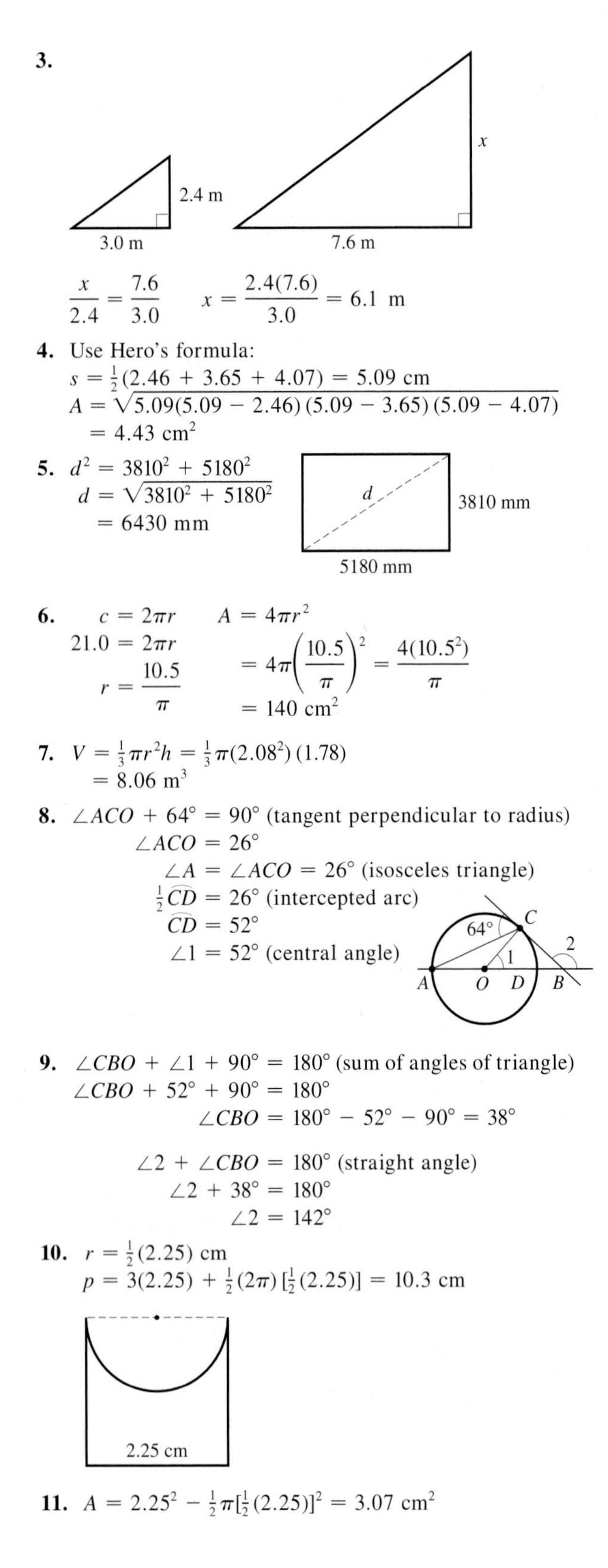

3.

$\frac{x}{2.4} = \frac{7.6}{3.0}$ $\quad x = \frac{2.4(7.6)}{3.0} = 6.1\text{ m}$

4. Use Hero's formula:
$s = \frac{1}{2}(2.46 + 3.65 + 4.07) = 5.09\text{ cm}$
$A = \sqrt{5.09(5.09 - 2.46)(5.09 - 3.65)(5.09 - 4.07)}$
$= 4.43\text{ cm}^2$

5. $d^2 = 3810^2 + 5180^2$
$d = \sqrt{3810^2 + 5180^2}$
$= 6430\text{ mm}$

6. $c = 2\pi r \qquad A = 4\pi r^2$
$21.0 = 2\pi r \qquad = 4\pi\left(\frac{10.5}{\pi}\right)^2 = \frac{4(10.5^2)}{\pi}$
$r = \frac{10.5}{\pi} \qquad = 140\text{ cm}^2$

7. $V = \frac{1}{3}\pi r^2 h = \frac{1}{3}\pi(2.08^2)(1.78)$
$= 8.06\text{ m}^3$

8. $\angle ACO + 64° = 90°$ (tangent perpendicular to radius)
$\angle ACO = 26°$
$\angle A = \angle ACO = 26°$ (isosceles triangle)
$\frac{1}{2}\overset{\frown}{CD} = 26°$ (intercepted arc)
$\overset{\frown}{CD} = 52°$
$\angle 1 = 52°$ (central angle)

9. $\angle CBO + \angle 1 + 90° = 180°$ (sum of angles of triangle)
$\angle CBO + 52° + 90° = 180°$
$\angle CBO = 180° - 52° - 90° = 38°$

$\angle 2 + \angle CBO = 180°$ (straight angle)
$\angle 2 + 38° = 180°$
$\angle 2 = 142°$

10. $r = \frac{1}{2}(2.25)\text{ cm}$
$p = 3(2.25) + \frac{1}{2}(2\pi)[\frac{1}{2}(2.25)] = 10.3\text{ cm}$

11. $A = 2.25^2 - \frac{1}{2}\pi[\frac{1}{2}(2.25)]^2 = 3.07\text{ cm}^2$

12. $A = \frac{1}{2}(50)[0 + 2(90) + 2(145) + 2(260) + 2(205) + 2(110) + 20]$
$= 41\,000\text{ m}^2$

Chapter 3

1. $f(x) = 2x - x^2 + \frac{8}{x}$

$f(-4) = 2(-4) - (-4)^2 + \frac{8}{-4}$
$= -8 - 16 - 2$
$= -26$

$f(2.385) = 2(2.385) - (2.385)^2 + \frac{8}{2.385} = 2.436$

(Calculator sequence: 2 [×] 2.385 [−] 2.385 [x^2] [+] 8 [÷] 2.385 [=] [2.436072694])

2. $w = 2000 - 10t$

3. $f(x) = 4 - 2x$

$y = 4 - 2x$
$y = 4 - 2(-1) = 6$
$y = 4 - 2(0) = 4$
$y = 4 - 2(1) = 2$
$y = 4 - 2(2) = 0$
$y = 4 - 2(3) = -2$
$y = 4 - 2(4) = -4$

x	y
−1	6
0	4
1	2
2	0
3	−2
4	−4

4. $y = 2x^2 - 3x - 3$

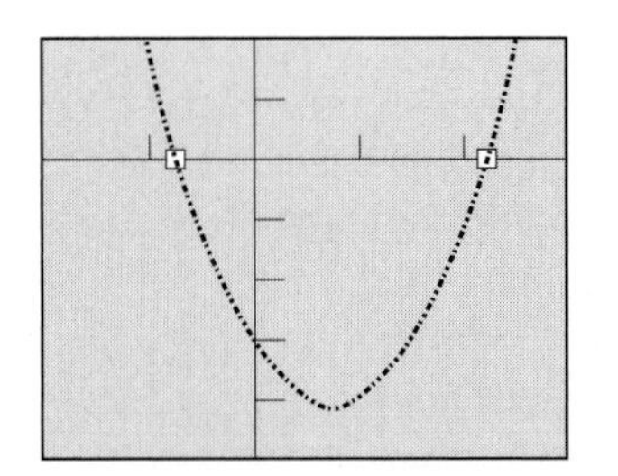

$x = -0.7$ and $x = 2.2$

5. $y = \sqrt{4 + 2x}$

$y = \sqrt{4 + 2(-2)} = 0$
$y = \sqrt{4 + 2(-1)} = 1.4$
$y = \sqrt{4 + 2(0)} = 2$
$y = \sqrt{4 + 2(1)} = 2.4$
$y = \sqrt{4 + 2(2)} = 2.8$
$y = \sqrt{4 + 2(4)} = 3.5$

x	y
−2	0
−1	1.4
0	2
1	2.4
2	2.8
4	3.5

6. On negative x-axis

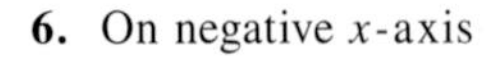

7. $f(x) = \sqrt{6 - x}$
Domain: $x \leq 6$; x cannot be greater than 6 to have real values of $f(x)$.
Range: $f(x) \geq 0$; $\sqrt{6 - x}$ is the principal square root of $6 - x$ and cannot be negative.

8. Let r = radius of circular part.

square semicircle
$$A = (2r)(2r) + \tfrac{1}{2}(\pi r^2)$$
$$= 4r^2 + \tfrac{1}{2}\pi r^2$$

9.

Voltage V	10.0	20.0	30.0	40.0	50.0	60.0
Current i	145	188	220	255	285	315

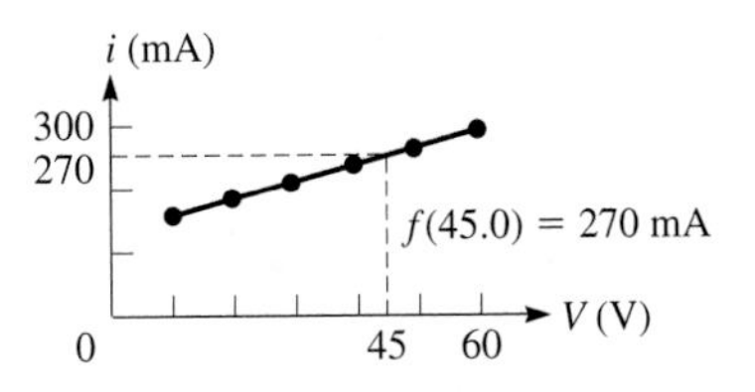

10. $V = 20.0$ V for $i = 188$ mA.
$V = 30.0$ V for $i = 220$ mA.
$$\frac{x}{10} = \frac{12}{32}, \qquad x = 3.8 \quad \text{(rounded off)}$$
$V = 20.0 + 3.8 = 23.8$ V
for $i = 200$ mA

Chapter 4

1. $39' = (\tfrac{39}{60})^\circ = 0.65^\circ$
$37^\circ 39' = 37.65^\circ$

2. $\cos\theta = 0.3726$ (Scientific calculator sequence:
.3726 [cos^{-1}] [68.12394435])
$\theta = 68.12^\circ$ (Graphing calculator sequence:
[cos^{-1}] .3726 [ENTER] [68.12394435])

3. Let x = distance from course to east.
$$\frac{x}{22.62} = \sin 4.05^\circ$$
$$x = 22.62 \sin 4.05^\circ$$
$$= 1.598 \text{ km}$$

4. $\sin\theta = \dfrac{2}{3}$
$$x = \sqrt{3^2 - 2^2} = \sqrt{5}$$
$$\tan\theta = \frac{2}{\sqrt{5}}$$

5. $\tan\theta = 1.294$ (Scientific calculator sequence:
1.294 [tan^{-1}] [sin] [x^{-1}] [1.263810119])
$\csc\theta = 1.264$ (Graphing calculator sequence:
[(] [sin] [tan^{-1}] 1.294 [)] [x^{-1}] [ENTER]
[1.263810119])

6. $B = 90^\circ - 37.4^\circ = 52.6^\circ$
$$\frac{a}{52.8} = \tan 37.4^\circ$$
$$a = 52.8 \tan 37.4^\circ$$
$$= 40.4$$
$$\frac{52.8}{c} = \cos 37.4^\circ$$
$$c = \frac{52.8}{\cos 37.4^\circ}$$
$$= 66.5$$

7. $2.49^2 + b^2 = 3.88^2$
$b = \sqrt{3.88^2 - 2.49^2} = 2.98$
$\sin A = \dfrac{2.49}{3.88}$ (Scientific calculator sequence:
2.49 [÷] 3.88 [=] [sin^{-1}] [39.92262926])
(Graphing calculator sequence:
[sin^{-1}] [(] 2.49 [÷] 3.88 [)] [ENTER]
[39.92262926])
$A = 39.9^\circ$, $\quad B = 90^\circ - 39.9^\circ = 50.1^\circ$

8. $\lambda = d \sin\theta$
$= 30.05 \sin 1.167^\circ$
$= 0.6120\ \mu\text{m}$

9. $r = \sqrt{5^2 + 2^2} = \sqrt{29}$
$$\sin\theta = \frac{2}{\sqrt{29}} = 0.3714, \qquad \csc\theta = \frac{\sqrt{29}}{2} = 2.693$$
$$\cos\theta = \frac{5}{\sqrt{29}} = 0.9285, \qquad \sec\theta = \frac{\sqrt{29}}{5} = 1.077$$
$$\tan\theta = \frac{2}{5} = 0.4000, \qquad \cot\theta = \frac{5}{2} = 2.500$$

10. Distance between points is $x - y$.

$$\frac{18.525}{x} = \tan 13.500°, \quad \frac{18.525}{y} = \tan 21.375°$$

$$x - y = \frac{18.525}{\tan 13.500°} - \frac{18.525}{\tan 21.375°} = 29.831 \text{ m}$$

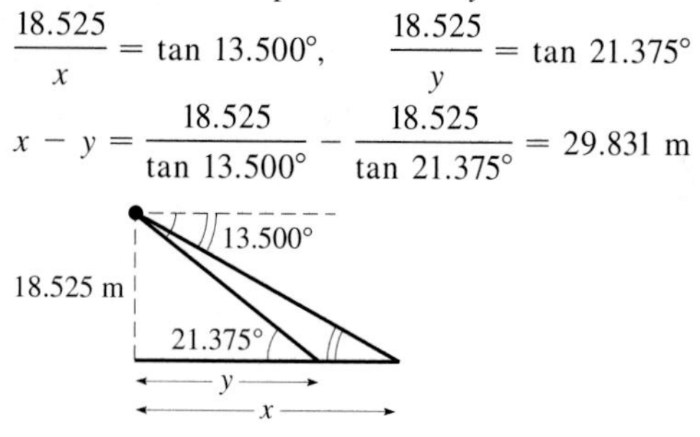

Chapter 5

1. Points $(2, -5)$ and $(-1, 4)$

$$m = \frac{4 - (-5)}{-1 - 2} = \frac{9}{-3} = -3$$

2.
$$x + 2y = 5$$
$$4y = 3 - 2x$$
$$x = 5 - 2y$$

$$4y = 3 - 2(5 - 2y)$$
$$4y = 3 - 10 + 4y$$
$$0 = -7$$

Inconsistent

3.
$$3x - 2y = 4$$
$$2x + 5y = -1$$

$$x = \frac{\begin{vmatrix} 4 & -2 \\ -1 & 5 \end{vmatrix}}{\begin{vmatrix} 3 & -2 \\ 2 & 5 \end{vmatrix}} = \frac{20 - 2}{15 - (-4)} = \frac{18}{19}$$

$$y = \frac{\begin{vmatrix} 3 & 4 \\ 2 & -1 \end{vmatrix}}{19} = \frac{-3 - 8}{19} = -\frac{11}{19}$$

4.
$$2x + y = 4$$
$$y = -2x + 4$$
$$m = -2, \quad b = 4$$

5.
$$2l + 2w = 24 \text{ (perimeter)}$$
$$l = w + 6.0$$

$$l + w = 12$$
$$l - w = 6.0$$

$$2l = 18$$
$$l = 9 \text{ km}$$
$$9 = w + 6$$
$$w = 3 \text{ km}$$

6.
$2x - 3y = 6$ | $4x + y = 4$

$x = 0$: $y = -2$ | $x = 0$: $y = 4$

$y = 0$: $x = 3$ | $y = 0$: $x = 1$

Int: $(0, -2)$, $(3, 0)$ | Int: $(0, 4)$, $(1, 0)$

$x = 1.3, \quad y = -1.1$

7. Let x = vol. of first alloy,
y = vol. of second alloy.
z = vol. of third alloy.

$$x + y + z = 100 \quad \text{total vol.}$$
$$0.6x + 0.5y + 0.3z = 40 \quad \text{copper}$$
$$0.3x + 0.3y = 15 \quad \text{zinc}$$

$$x + y + z = 100$$
$$6x + 5y + 3z = 400$$
$$3x + 3y = 150$$

$$3x + 2y = 100$$
$$3x + 3y = 150$$

$$y = 50 \text{ cm}^3$$
$$x = 0 \text{ cm}^3$$
$$z = 50 \text{ cm}^3$$

8.
$$3x + 2y - z = 4$$
$$2x - y + 3z = -2$$
$$x + 4z = 5$$

$$y = \frac{\begin{vmatrix} 3 & 4 & -1 \\ 2 & -2 & 3 \\ 1 & 5 & 4 \end{vmatrix}}{\begin{vmatrix} 3 & 2 & -1 \\ 2 & -1 & 3 \\ 1 & 0 & 4 \end{vmatrix}}$$

$$= \frac{-24 + 12 - 10 - 2 - 45 - 32}{-12 + 6 + 0 - 1 - 0 - 16}$$

$$= \frac{-101}{-23} = \frac{101}{23}$$

Chapter 6

1.
$$2x(2x - 3)^2 = 2x[(2x)^2 - 2(2x)(3) + 3^2]$$
$$= 2x(4x^2 - 12x + 9)$$
$$= 8x^3 - 24x^2 + 18x$$

2.
$$\frac{1}{R} = \frac{1}{R_1 + r} + \frac{1}{R_2}$$

$$\frac{RR_2(R_1 + r)}{R} = \frac{RR_2(R_1 + r)}{R_1 + r} + \frac{RR_2(R_1 + r)}{R_2}$$

$$R_2(R_1 + r) = RR_2 + R(R_1 + r)$$
$$R_1 R_2 + rR_2 = RR_2 + RR_1 + rR$$
$$R_1 R_2 - RR_1 = RR_2 + rR - rR_2$$
$$R_1(R_2 - R) = RR_2 + rR - rR_2$$
$$R_1 = \frac{RR_2 + rR - rR_2}{R_2 - R}$$

3.
$$\frac{2x^2 + 5x - 3}{2x^2 + 12x + 18} = \frac{(2x - 1)(x + 3)}{2(x + 3)^2} = \frac{2x - 1}{2(x + 3)}$$

4. $4x^2 - 16y^2 = 4(x^2 - 4y^2) = 4(x + 2y)(x - 2y)$

5. $$\frac{3}{4x^2} - \frac{2}{x^2 - x} - \frac{x}{2x - 2} = \frac{3}{4x^2} - \frac{2}{x(x-1)} - \frac{x}{2(x-1)}$$
$$= \frac{3(x-1) - 2(4x) - x(2x^2)}{4x^2(x-1)}$$
$$= \frac{3x - 3 - 8x - 2x^3}{4x^2(x-1)}$$
$$= \frac{-2x^3 - 5x - 3}{4x^2(x-1)}$$

6. $$\frac{x^2 + x}{2 - x} \div \frac{x^2}{x^2 - 4x + 4} = \frac{x^2 + x}{2 - x} \times \frac{x^2 - 4x + 4}{x^2}$$
$$= \frac{x(x+1)}{2 - x} \times \frac{(x-2)^2}{x^2}$$
$$= -\frac{x(x+1)(x-2)^2}{(x-2)(x^2)}$$
$$= -\frac{(x+1)(x-2)}{x}$$

7. $$\frac{1 - \frac{3}{2x + 2}}{\frac{x}{5} - \frac{1}{2}} = \frac{\frac{2(x+1) - 3}{2(x+1)}}{\frac{2x - 5}{10}}$$
$$= \frac{2x + 2 - 3}{2(x+1)} \times \frac{10}{2x - 5}$$
$$= \frac{5(2x - 1)}{(x+1)(2x-5)}$$

8. Let t = time working together.
$$\frac{t}{12} + \frac{t}{16} = 1$$
LCD of 12 and 16 is 48.
$$\frac{48t}{12} + \frac{48t}{16} = 48$$
$$4t + 3t = 48$$
$$t = \frac{48}{7} = 6.9 \text{ days}$$

Chapter 7

1. $x^2 - 3x - 5 = 0$
Not factorable
$a = 1, \quad b = -3, \quad c = -5$
$$x = \frac{-(-3) \pm \sqrt{(-3)^2 - 4(1)(-5)}}{2(1)}$$
$$= \frac{3 \pm \sqrt{29}}{2}$$

2. $2x^2 = 9x - 4$
$2x^2 - 9x + 4 = 0$
$(2x - 1)(x - 4) = 0$
$2x - 1 = 0, \quad x - 4 = 0$
$x = \frac{1}{2}$ or $x = 4$

3. $$\frac{3}{x} - \frac{2}{x + 2} = 1$$
$$\frac{3x(x+2)}{x} - \frac{2x(x+2)}{x+2} = x(x+2)$$
$$3(x+2) - 2x = x^2 + 2x$$
$$3x + 6 - 2x = x^2 + 2x$$
$$0 = x^2 + x - 6$$
$$(x+3)(x-2) = 0$$
$$x = -3, 2$$

4. $2x^2 - x = 6 - 2x(3 - x)$
$2x^2 - x = 6 - 6x + 2x^2$
$5x = 6$ (not quadratic)
$x = \frac{6}{5}$

5. $y = 2x^2 + 8x + 5$
$$\frac{-b}{2a} = \frac{-8}{2(2)} = -2$$
$y = 2(-2)^2 + 8(-2) + 5 = -3$
Min. pt. $(a > 0)$ is $(-2, -3)$,
$c = 5$, y-intercept is $(0, 5)$.

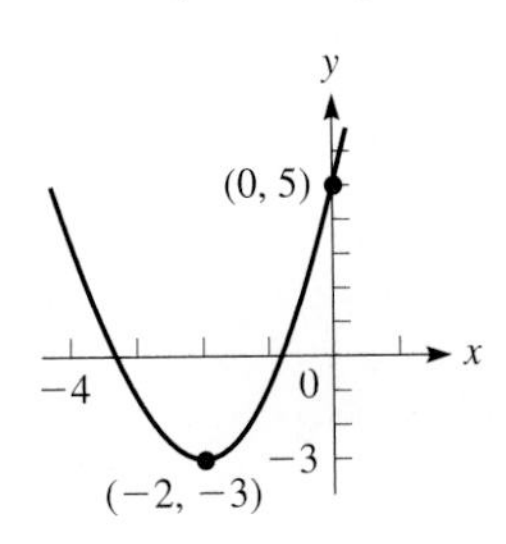

6. $P = EI - RI^2$
$RI^2 - EI + P = 0$
$$I = \frac{-(-E) \pm \sqrt{(-E)^2 - 4RP}}{2R}$$
$$= \frac{E \pm \sqrt{E^2 - 4RP}}{2R}$$

7. $x^2 - 6x - 9 = 0$
$x^2 - 6x = 9$
$x^2 - 6x + 9 = 9 + 9$
$(x - 3)^2 = 18$
$x - 3 = \pm\sqrt{18}$
$x = 3 \pm 3\sqrt{2}$

8. Let w = width of window,
h = height of window.
$2w + 2h = 8.4$ (perimeter)
$w + h = 4.2, \quad h = 4.2 - w$
$w(4.2 - w) = 3.8$ (area)
$-w^2 + 4.2w = 3.8$
$w^2 - 4.2w + 3.8 = 0$
$$w = \frac{-(-4.2) \pm \sqrt{(-4.2)^2 - 4(3.8)}}{2}$$
$= 2.88, 1.32$
$w = 1.3$ m, $h = 2.9$ m or
$w = 2.9$ m, $h = 1.3$ m

Chapter 8

1. $150° = \left(\frac{\pi}{180}\right)(150) = \frac{5\pi}{6}$

2. $\sin 205° = -\sin(205° - 180°) = -\sin 25°$

3. $x = -9, \quad y = 12$
$r = \sqrt{(-9)^2 + 12^2} = 15$
$\sin\theta = \frac{12}{15} = \frac{4}{5}$
$\sec\theta = \frac{15}{-9} = -\frac{5}{3}$

4. $r = 1.40$ m
$\omega = 2200$ r/min $= (2200$ r/min$)(2\pi$ rad/r$)$
$= 4400\pi$ rad/min
$v = \omega r = (4400\pi)(1.40) = 19\,000$ m/min

5. $3.572 = 3.572\left(\frac{180^\circ}{\pi}\right) = 204.7^\circ$

6. $\tan\theta = 0.2396$
$\theta_{\text{ref}} = 13.47^\circ$
$\theta = 13.47^\circ$ or
$\theta = 180^\circ + 13.47^\circ = 193.47^\circ$

7. $\cos\theta = -0.8244$, $\csc\theta < 0$;
$\cos\theta$ negative, $\csc\theta$ negative,
θ in third quadrant
$\theta_{\text{ref}} = 0.6017$, $\theta = 3.7432$

8. $s = r\theta$, $16.0 = 4.25\theta$
$\theta = \frac{16.0}{4.25} = 3.76$ rad
$A = \frac{1}{2}\theta r^2 = \frac{1}{2}(3.76)(4.25)^2$
$= 34.0\text{ m}^2$

Chapter 9

1.

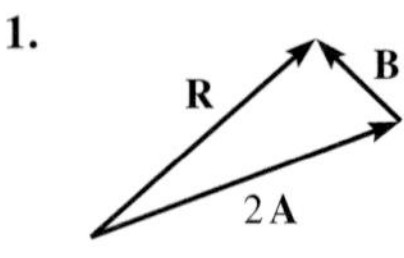

2. $b^2 = a^2 + c^2 - 2ac\cos B$
$b = \sqrt{22.5^2 + 30.9^2 - 2(22.5)(30.9)\cos 78.6^\circ} = 34.4$

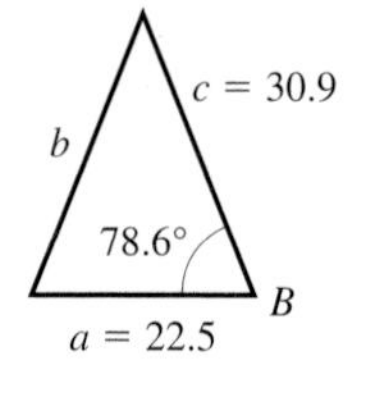

3. $x^2 = 36.50^2 + 21.38^2 - 2(36.50)(21.38)\cos 45.00^\circ$,
$x = 26.19$ m, $\quad \dfrac{21.38}{\sin\alpha} = \dfrac{26.19}{\sin 45.00^\circ}$,
$\sin\alpha = \dfrac{21.38\sin 45.00^\circ}{26.19}$, $\quad \alpha = 35.26^\circ$
$\theta = 45.00^\circ - 35.26^\circ = 9.74^\circ$
Displacement is 26.19 m,
9.74° N of E.

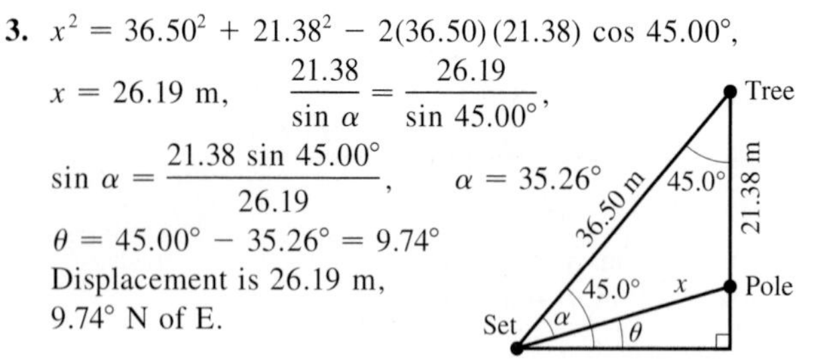

4. $C = 180^\circ - (18.9^\circ + 104.2^\circ) = 56.9^\circ$
$\dfrac{c}{\sin C} = \dfrac{a}{\sin A} \qquad c = \dfrac{426\sin 56.9^\circ}{\sin 18.9^\circ} = 1100$

5. Since a is longest side, find A first.
$a^2 = b^2 + c^2 - 2bc\cos A$
$\cos A = \dfrac{b^2 + c^2 - a^2}{2bc} = \dfrac{3.29^2 + 8.44^2 - 9.84^2}{2(3.29)(8.44)}$
$A = 105.4^\circ$
$\dfrac{b}{\sin B} = \dfrac{a}{\sin A}$
$\sin B = \dfrac{b\sin A}{a} = \dfrac{3.29\sin 105.4^\circ}{9.84}$
$B = 18.8^\circ$
$C = 180^\circ - (105.4^\circ + 18.8^\circ) = 55.8^\circ$

6. $A_x = 871\cos 284.3^\circ = 215$
$A_y = 871\sin 284.3^\circ = -844$

7. $\dfrac{63.0}{\sin 148.5^\circ} = \dfrac{42.0}{\sin A} \qquad \dfrac{x}{\sin 11.1^\circ} = \dfrac{63.0}{\sin 148.5^\circ}$
$\sin A = \dfrac{42.0\sin 148.5^\circ}{63.0} \qquad x = \dfrac{63.0\sin 11.1^\circ}{\sin 148.5^\circ}$
$A = 20.4^\circ \qquad = 23.2$ km
$C = 180^\circ - (148.5^\circ + 20.4^\circ) = 11.1^\circ$

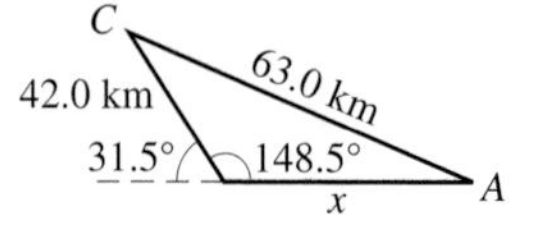

8. $A_x = 449\cos 74.2^\circ$, $\quad B_x = 285\cos 208.9^\circ$
$A_y = 449\sin 74.2^\circ$, $\quad B_y = 285\sin 208.9^\circ$
$R_x = A_x + B_x = 449\cos 74.2^\circ + 285\cos 208.9^\circ$
$= -127.3$
$R_y = A_y + B_y = 449\sin 74.2^\circ + 285\sin 208.9^\circ$
$= 294.3$
$R = \sqrt{(-127.3)^2 + 294.3^2} = 321$
$\tan\theta_{\text{ref}} = \dfrac{294.3}{127.3}$, $\quad \theta_{\text{ref}} = 66.6^\circ$, $\quad \theta = 113.4^\circ$
θ is in second quadrant, since R_x is negative and R_y is positive.

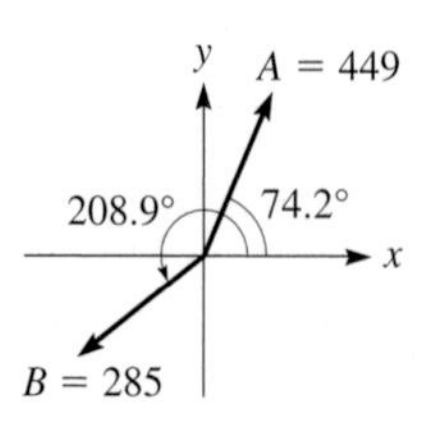

9. $29.6 \sin 36.5° = 17.6$

$17.6 < 22.3 < 29.6$ means two solutions.

$$\frac{29.6}{\sin B} = \frac{22.3}{\sin 36.5°}, \quad \sin B = \frac{29.6 \sin 36.5°}{22.3}$$

$B_1 = 52.1°$, $C_1 = 180° - 36.5° - 52.1° = 91.4°$

$B_2 = 180° - 52.1° = 127.9°$,

$C_2 = 180° - 36.5° - 127.9° = 15.6°$

$$\frac{c_1}{\sin 91.4°} = \frac{22.3}{\sin 36.5°}, \quad c_1 = \frac{22.3 \sin 91.4°}{\sin 36.5°} = 37.5$$

$$\frac{c_2}{\sin 15.6°} = \frac{22.3}{\sin 36.5°}, \quad c_2 = \frac{22.3 \sin 15.6°}{\sin 36.5°} = 10.1$$

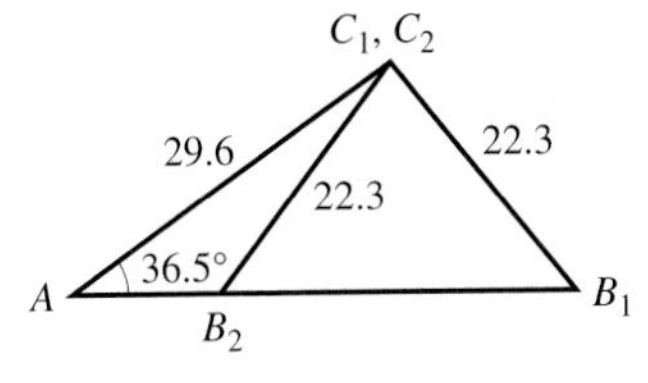

Chapter 10

1. $y = 0.5 \cos \frac{\pi}{2}x$

Amp. = 0.5, disp. = 0,

$$\text{per.} = \frac{2\pi}{\pi/2} = 4.$$

x	0	1	2	3	4
y	0.5	0	−0.5	0	0.5

2. $y = 2 + 3 \sin x$

For $y_1 = 3 \sin x$,

amp. = 3, per. = 2π,

disp. = 0.

x	0	$\frac{\pi}{2}$	π	$\frac{3\pi}{2}$	2π
$y_1 = 3 \sin x$	0	3	0	−3	0
$y = 2 + 3 \sin x$	2	5	2	−1	2

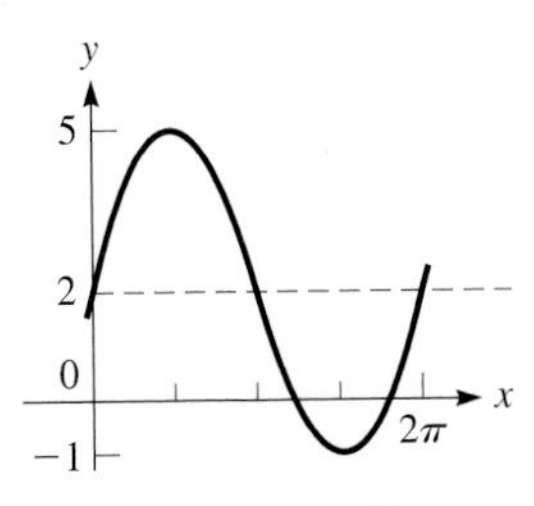

3. $y = 3 \sec x$

$$\sec x = \frac{1}{\cos x}$$

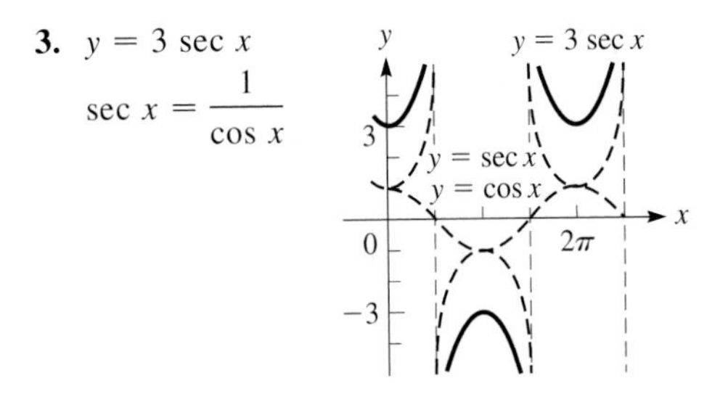

4. $y = 2 \sin (2x - \frac{\pi}{3})$

Amp. = 2, per. = $\frac{2\pi}{2} = \pi$

$$\text{disp.} = -\frac{-\pi/3}{2} = \frac{\pi}{6}.$$

x	$\frac{\pi}{6}$	$\frac{5\pi}{12}$	$\frac{2\pi}{3}$	$\frac{11\pi}{12}$	$\frac{7\pi}{6}$
y	0	2	0	−2	0

$\frac{\pi}{6} + \frac{\pi}{4} = \frac{5\pi}{12}$, $\frac{\pi}{6} + \frac{\pi}{2} = \frac{2\pi}{3}$, $\frac{\pi}{6} + \frac{3\pi}{4} = \frac{11\pi}{12}$

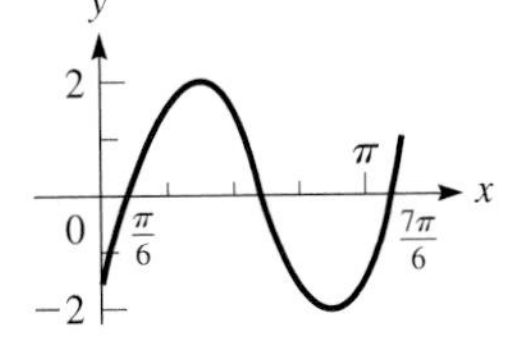

5. $y = A \cos \frac{2\pi}{T} t$

$A = 0.200$ cm, $T = 0.100$ s,

$y = 0.200 \cos 20\pi t$,

amp. = 0.200 cm, per. = 0.100 s.

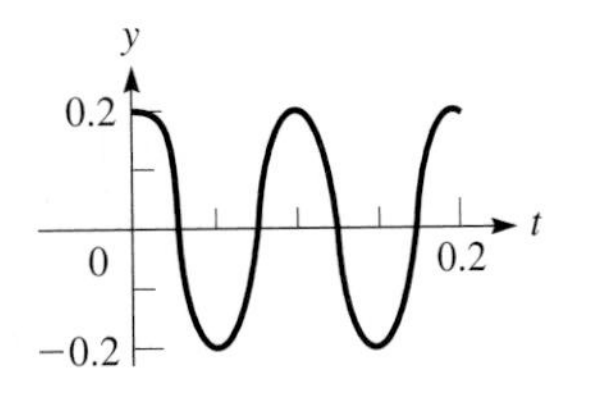

6. $y = 2 \sin x + \cos 2x$

For $y_1 = 2 \sin x$,

amp. = 2, per. = 2π, disp. = 0.

For $y_2 = \cos 2x$,

amp. = 1, per. = $\frac{2\pi}{2} = \pi$, disp. = 0.

7. $x = \sin \pi t,\ y = 2\cos 2\pi t$

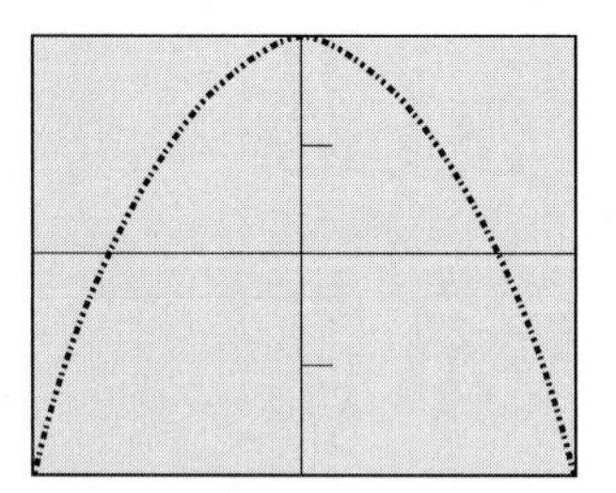

8. $d = R\sin(\omega t + \frac{\pi}{6})$
$\omega = 2.00$ rad/s
$d = R\sin(2.00t + \frac{\pi}{6})$
Amp. $= R$, per. $= \dfrac{2\pi}{2.00} = \pi$ s $= 3.14$ s,
disp. $= \dfrac{-\pi/6}{2.00} = -\dfrac{\pi}{12}$ s $= -0.26$ s.

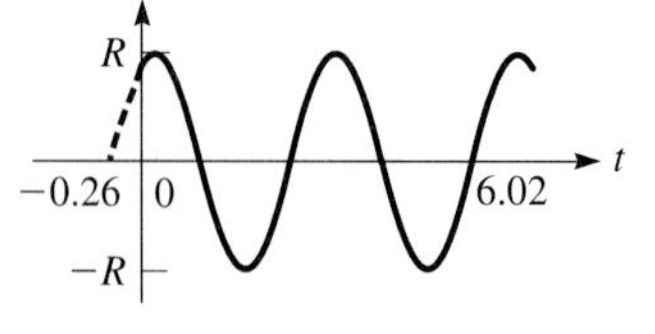

Chapter 11

1. $2\sqrt{20} - \sqrt{125} = 2\sqrt{4\times 5} - \sqrt{25\times 5} = 2(2\sqrt{5}) - 5\sqrt{5} = 4\sqrt{5} - 5\sqrt{5} = -\sqrt{5}$

2. $\dfrac{100^{3/2}}{8^{-2/3}} = (100^{3/2})(8^{2/3}) = [(100^{1/2})^3][(8^{1/3})^2] = (10^3)(2^2) = 4000$

3. $(as^{-1/3}t^{3/4})^{12} = a^{12}s^{(-1/3)(12)}t^{(3/4)(12)} = a^{12}s^{-4}t^9 = \dfrac{a^{12}t^9}{s^4}$

4. $(2x^{-1} + y^{-2})^{-1} = \dfrac{1}{2x^{-1} + y^{-2}} = \dfrac{1}{\frac{2}{x} + \frac{1}{y^2}} = \dfrac{1}{\frac{2y^2 + x}{xy^2}} = \dfrac{xy^2}{2y^2 + x}$

5. $(\sqrt{2x} - 3\sqrt{y})^2 = (\sqrt{2x})^2 - 2\sqrt{2x}(3\sqrt{y}) + (3\sqrt{y})^2 = 2x - 6\sqrt{2xy} + 9y$

6. $\sqrt[3]{\sqrt[4]{4}} = \sqrt[12]{4} = \sqrt[12]{2^2} = 2^{2/12} = 2^{1/6} = \sqrt[6]{2}$

7. $\dfrac{3 - 2\sqrt{2}}{2\sqrt{x}} = \dfrac{3 - 2\sqrt{2}}{2\sqrt{x}} \times \dfrac{\sqrt{x}}{\sqrt{x}} = \dfrac{3\sqrt{x} - 2\sqrt{2x}}{2x}$

8. $\sqrt{27a^4b^3} = \sqrt{9\times 3\times (a^2)^2(b^2)(b)} = 3a^2b\sqrt{3b}$

9. $(2x + 3)^{1/2} + (x + 1)(2x + 3)^{-1/2}$
$= (2x + 3)^{1/2} + \dfrac{x + 1}{(2x + 3)^{1/2}}$
$= \dfrac{(2x + 3)^{1/2}(2x + 3)^{1/2} + x + 1}{(2x + 3)^{1/2}} = \dfrac{2x + 3 + x + 1}{(2x + 3)^{1/2}}$
$= \dfrac{3x + 4}{(2x + 3)^{1/2}}$

10. $2\sqrt{2}(3\sqrt{10} - \sqrt{6}) = 2\sqrt{2}(3\sqrt{10}) - 2\sqrt{2}(\sqrt{6}) = 6\sqrt{20} - 2\sqrt{12} = 6(2\sqrt{5}) - 2(2\sqrt{3}) = 12\sqrt{5} - 4\sqrt{3}$

11. $\left(\dfrac{4a^{-1/2}b^{3/4}}{b^{-2}}\right)\left(\dfrac{b^{-1}}{2a}\right) = \dfrac{4a^{-1/2}b^{3/4-1}}{2ab^{-2}} = \dfrac{2b^{2-1/4}}{a^{1+1/2}} = \dfrac{2b^{7/4}}{a^{3/2}}$

12. $\dfrac{2\sqrt{15} + \sqrt{3}}{\sqrt{15} - 2\sqrt{3}} = \dfrac{2\sqrt{15} + \sqrt{3}}{\sqrt{15} - 2\sqrt{3}} \times \dfrac{\sqrt{15} + 2\sqrt{3}}{\sqrt{15} + 2\sqrt{3}}$
$= \dfrac{2(15) + 4\sqrt{45} + \sqrt{45} + 2(3)}{15 - 4(3)}$
$= \dfrac{36 + 15\sqrt{5}}{3}$
$= 12 + 5\sqrt{5}$

13. $\dfrac{3^{-1/2}}{2} = \dfrac{1}{2\times 3^{1/2}} = \dfrac{1}{2\sqrt{3}} = \dfrac{\sqrt{3}}{2\sqrt{3}\sqrt{3}} = \dfrac{\sqrt{3}}{6}$

14. $0.220N^{-1/6},\quad N = 64\times 10^6$
$0.220(64\times 10^6)^{-1/6} = \dfrac{0.220}{(64\times 10^6)^{1/6}} = \dfrac{0.220}{2\times 10} = 0.011$

Chapter 12

1. $(3 - \sqrt{-4}) + (5\sqrt{-9} - 1) = (3 - 2j) + [5(3j) - 1] = 3 - 2j + 15j - 1 = 2 + 13j$

2. $(2\angle 130^\circ)(3\angle 45^\circ) = (2)(3)\angle 130^\circ + 45^\circ = 6\angle 175^\circ$

3. $2 - 7j$: $r = \sqrt{2^2 + (-7)^2} = \sqrt{53} = 7.28$
$\tan\theta = \dfrac{-7}{2} = -3.500,\qquad \theta_{\text{ref}} = 74.1^\circ,\qquad \theta = 285.9^\circ$
$2 - 7j = 7.28(\cos 285.9^\circ + j\sin 285.9^\circ) = 7.28\angle 285.9^\circ$

4. (a) $-\sqrt{-64} = -(8j) = -8j$,
(b) $-j^{15} = -j^{12}j^3 = (-1)(-j) = j$

5. $(4 - 3j) + (-1 + 4j) = 3 + j$

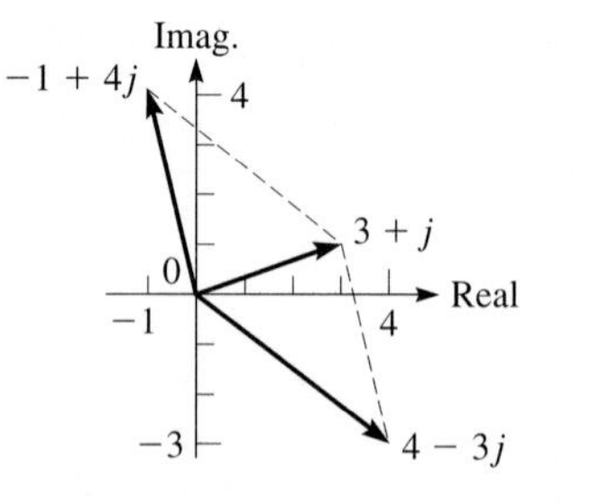

6. $\dfrac{2 - 4j}{5 + 3j} = \dfrac{(2 - 4j)(5 - 3j)}{(5 + 3j)(5 - 3j)} = \dfrac{10 - 26j + 12j^2}{25 - 9j^2} = \dfrac{10 - 26j - 12}{25 - 9(-1)} = \dfrac{-2 - 26j}{34} = -\dfrac{1 + 13j}{17}$

7. $2.56(\cos 125.2° + j \sin 125.2°) = 2.56e^{2.185j}$

$$125.2° = \frac{125.2\pi}{180} = 2.185 \text{ rad}$$

8. $R = 3.50\ \Omega,\ X_L = 6.20\ \Omega,\ X_C = 7.35\ \Omega$

$$|Z| = \sqrt{R^2 + (X_L - X_C)^2}$$
$$= \sqrt{3.50^2 + (6.20 - 7.35)^2} = 3.68\ \Omega$$
$$\tan\theta = \frac{X_L - X_C}{R} = \frac{6.20 - 7.35}{3.50} = \frac{-1.15}{3.50}$$
$$\theta = -18.2°$$

9. $3.47 - 2.81j = 4.47e^{5.60j}$

$$R = \sqrt{3.47^2 + (-2.81)^2} = 4.47$$
$$\tan\theta = \frac{-2.81}{3.47}, \quad \theta_{ref} = 39.0°$$
$$\theta = 321.0° = 5.60 \text{ rad}$$

10. $x + 2j - y = yj - 3xj$

$$x - y + 3xj - yj = -2j$$
$$(x - y) + (3x - y)j = 0 - 2j$$
$$x - y = 0$$
$$3x - y = -2$$
$$2x = -2$$
$$x = -1, \quad y = -1$$

11. $L = 8.75 \text{ mH} = 8.75 \times 10^{-3}$ H

$f = 600 \text{ kHz} = 6.00 \times 10^5$ Hz

$$2\pi fL = \frac{1}{2\pi fC}$$
$$C = \frac{1}{(2\pi f)^2 L} = \frac{1}{(2\pi)^2(6.00 \times 10^5)^2(8.75 \times 10^{-3})}$$
$$= 8.04 \times 10^{-12} = 8.04 \text{ pF}$$

12. $j = 1(\cos 90° + j \sin 90°)$

$$j^{1/3} = 1^{1/3}\left(\cos\frac{90°}{3} + j\sin\frac{90°}{3}\right)$$
$$= \cos 30° + j \sin 30° = 0.8660 + 0.5000j$$
$$= 1^{1/3}\left(\cos\frac{90° + 360°}{3} + j\sin\frac{90° + 360°}{3}\right)$$
$$= \cos 150° + j \sin 150° = -0.8660 + 0.5000j$$
$$= 1^{1/3}\left(\cos\frac{90° + 720°}{3} + j\sin\frac{90° + 720°}{3}\right)$$
$$= \cos 270° + j \sin 270° = -j$$

Cube roots of j: $0.8660 + 0.5000j$

$-0.8660 + 0.5000j$

$-j$

Chapter 13

1. $\log_9 x = -\frac{1}{2}$

$$x = 9^{-1/2}$$
$$= \frac{1}{9^{1/2}} = \frac{1}{3}$$

2. $\log_3 x - \log_3 2 = 2$

$$\log_3 \frac{x}{2} = 2$$
$$\frac{x}{2} = 3^2$$
$$x = 2(3^2) = 18$$

3. $\log_x 64 = 3$

$$64 = x^3$$
$$4^3 = x^3$$
$$x = 4$$

4. $3^{3x+1} = 8$

$$(3x + 1)\log 3 = \log 8$$
$$3x + 1 = \frac{\log 8}{\log 3}$$
$$x = \frac{1}{3}\left(\frac{\log 8}{\log 3} - 1\right) = 0.298$$

5. $y = 2 \log_4 x$

x	$\frac{1}{4}$	1	4	16
y	-2	0	2	4

$\log_4 \frac{1}{4} = -1,\ \log_4 16 = 2$

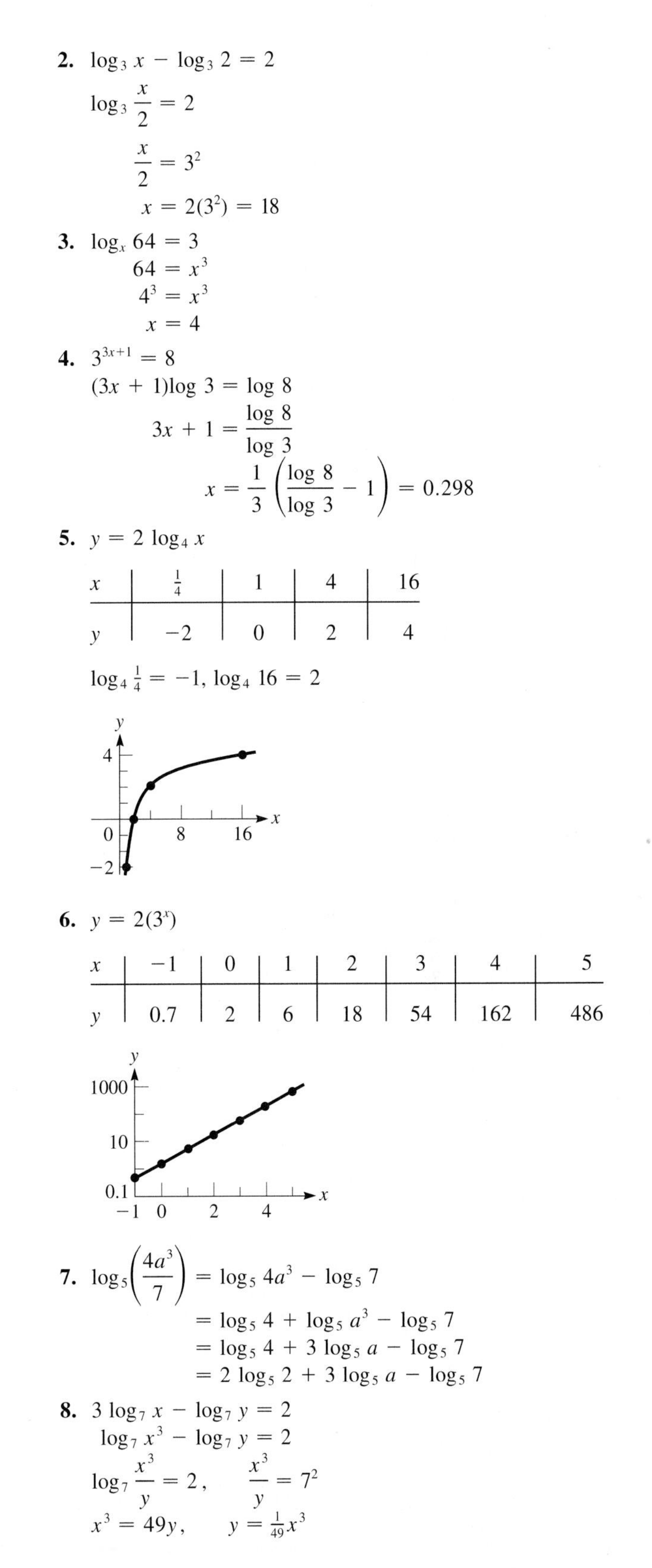

6. $y = 2(3^x)$

x	-1	0	1	2	3	4	5
y	0.7	2	6	18	54	162	486

7. $\log_5\left(\frac{4a^3}{7}\right) = \log_5 4a^3 - \log_5 7$

$$= \log_5 4 + \log_5 a^3 - \log_5 7$$
$$= \log_5 4 + 3\log_5 a - \log_5 7$$
$$= 2\log_5 2 + 3\log_5 a - \log_5 7$$

8. $3\log_7 x - \log_7 y = 2$

$$\log_7 x^3 - \log_7 y = 2$$
$$\log_7 \frac{x^3}{y} = 2, \quad \frac{x^3}{y} = 7^2$$
$$x^3 = 49y, \quad y = \tfrac{1}{49}x^3$$

9. $\ln i - \ln I = -t/RC$

$$\ln \frac{i}{I} = -t/RC$$

$$\frac{i}{I} = e^{-t/RC} \qquad i = Ie^{-t/RC}$$

10. $\dfrac{2 \ln 0.9523}{\log 6066} = -0.025\,84$

11. $\log_b x = \dfrac{\log_a x}{\log_a b}$

$$\log_5 732 = \frac{\log 732}{\log 5} = 4.098$$

12. $A = A_0 e^{0.08t}, \qquad A = 2A_0$

$$2A_0 = A_0 e^{0.08t}, \qquad 2 = e^{0.08t}$$

$$\ln 2 = \ln e^{0.08t} = 0.08t \qquad t = \frac{\ln 2}{0.08} = 8.66 \text{ years}$$

Chapter 14

1. $x^{1/2} - 2x^{1/4} = 3$

Let $y = x^{1/4}$

$y^2 - 2y - 3 = 0$

$(y - 3)(y + 1) = 0$

$y = 3, -1$

$x^{1/4} \neq -1$

$x^{1/4} = 3, \qquad x = 81$

Check: $81^{1/2} - 2(81^{1/4}) = 3$

$9 - 6 = 3$

Solution: $x = 81$

2. $3\sqrt{x-2} - \sqrt{x+1} = 1$

$$3\sqrt{x-2} = 1 + \sqrt{x+1}$$

$$9(x-2) = 1 + 2\sqrt{x+1} + (x+1)$$

$$8x - 20 = 2\sqrt{x+1}$$

$$4x - 10 = \sqrt{x+1}$$

$16x^2 - 80x + 100 = x + 1$

$16x^2 - 81x + 99 = 0$

$$x = \frac{81 \pm \sqrt{81^2 - 4(16)(99)}}{32}$$

$$= \frac{81 \pm 15}{32} = 3, \frac{33}{16}$$

Check: $x = 3$: $3\sqrt{3-2} - \sqrt{3+1} \stackrel{?}{=} 1$

$3 - 2 = 1$

$x = \frac{33}{16}$: $3\sqrt{\frac{33}{16} - 2} - \sqrt{\frac{33}{16} + 1} \stackrel{?}{=} 1$

$\frac{3}{4} - \frac{7}{4} \neq 1$

Solution: $x = 3$

3. $x^4 - 17x^2 + 16 = 0$

Let $y = x^2$.

$y^2 - 17y + 16 = 0$

$(y - 1)(y - 16) = 0$

$y = 1, 16$

$x^2 = 1, 16$

$x = -1, 1, -4, 4$

All values check.

4. $x^2 - 2y = 5$

$2x + 6y = 1$

$y = \frac{1}{2}(x^2 - 5)$

$2x + 6(\frac{1}{2})(x^2 - 5) = 1$

$3x^2 + 2x - 16 = 0$

$(3x + 8)(x - 2) = 0$

$x = -\frac{8}{3}, 2$

$x = -\frac{8}{3}$: $y = \frac{1}{2}(\frac{64}{9} - 5) = \frac{19}{18}$

$x = 2$: $y = \frac{1}{2}(4 - 5) = -\frac{1}{2}$

$x = -\frac{8}{3}, y = \frac{19}{18}$

or $x = 2, y = -\frac{1}{2}$

5. $v = \sqrt{v_0^2 + 2gh}$

$v^2 = v_0^2 + 2gh$

$2gh = v^2 - v_0^2$

$$h = \frac{v^2 - v_0^2}{2g}$$

6. $x^2 - y^2 = 4, \qquad xy = 2$

$y = \pm\sqrt{x^2 - 4}, \qquad y = \dfrac{2}{x}$

x	y
± 2	0
± 3	± 2.2
± 4	± 3.5

x	y
-4	$-\frac{1}{2}$
-2	-1
-1	-2
0	—
1	2
2	1
4	$\frac{1}{2}$

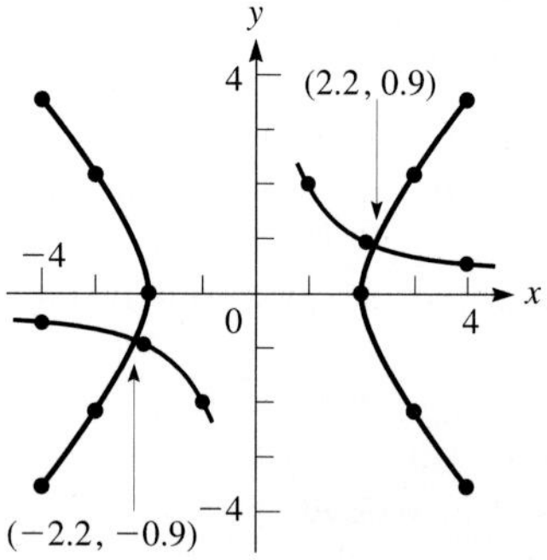

$x = 2.2, \quad y = 0.9; \qquad x = -2.2, \quad y = -0.9$

7. Let l = length, w = width.

$2l + 2w = 14$

$lw = 10$

$l + w = 7$

$l = 7 - w$

$(7 - w)w = 10$

$7w - w^2 = 10$

$w^2 - 7w + 10 = 0$

$(w - 5)(w - 2) = 0$

$w = 5, 2$

$l = 5.00$ m, $w = 2.00$ m

(or $l = 2.00$ m, $w = 5.00$ m)

Chapter 15

1. $f(x) = 2x^3 + 3x^2 + 7x - 6$
$f(-3) = 2(-3)^3 + 3(-3)^2 + 7(-3) - 6$
$= -54 + 27 - 21 - 6 = -54$
$f(-3) \neq 0, \quad -3$ is not a zero.

2. $x^4 - 2x^3 - 7x^2 + 20x - 12 = 0$

$$\begin{array}{rrrrr|l} 1 & -2 & -7 & 20 & -12 & \underline{2} \\ & 2 & 0 & -14 & 12 & \\ \hline 1 & 0 & -7 & 6 & & \underline{2} \\ & 2 & 4 & -6 & & \\ \hline 1 & 2 & -3 & & & \end{array}$$

$x^2 + 2x - 3 = (x + 3)(x - 1)$
Other roots: $x = -3, 1$.

3. $(x^3 - 5x^2 + 4x - 9) \div (x - 3)$

$$\begin{array}{rrrr|l} 1 & -5 & 4 & -9 & \underline{3} \\ & 3 & -6 & -6 & \\ \hline 1 & -2 & -2 & -15 & \end{array}$$

Quotient: $x^2 - 2x - 2$.
Remainder $= -15$.

4. $f(x) = 2x^4 + 15x^3 + 23x^2 - 16$
$2x + 1 = 2(x + \frac{1}{2})$

$$\begin{array}{rrrrr|l} 2 & 15 & 23 & 0 & -16 & \underline{-\frac{1}{2}} \\ & -1 & -7 & -8 & 4 & \\ \hline 2 & 14 & 16 & -8 & -12 & \end{array}$$

Remainder is not zero;
$2x + 1$ is not a factor.

5. $(x^3 + 4x^2 + 7x - 9) \div (x + 4)$
$f(x) = x^3 + 4x^2 + 7x - 9$
$f(-4) = (-4)^3 + 4(-4)^2 + 7(-4) - 9$
$= -64 + 64 - 28 - 9 = -37$
Remainder $= -37$.

6. $2x^4 - x^3 + 5x^2 - 4x - 12 = 0$
$f(x) = 2x^4 - x^3 + 5x^2 - 4x - 12$
$f(-x) = 2x^4 + x^3 + 5x^2 + 4x - 12$
$n = 4$; 4 roots;
no more than 3 positive roots;
one negative root;
rational roots: factors of 12 divided by factors of 2;
possible rational roots: $\pm 1, \pm 2, \pm 3, \pm 4, \pm 6, \pm 12, \pm\frac{1}{2}, \pm\frac{3}{2}$

$$\begin{array}{rrrrr|l} 2 & -1 & 5 & -4 & -12 & \underline{2} \\ & 4 & 6 & 22 & 36 & \\ \hline 2 & 3 & 11 & 18 & 24 & \end{array}$$

2 is too large.

$$\begin{array}{rrrrr|l} 2 & -1 & 5 & -4 & -12 & \underline{\frac{3}{2}} \\ & 3 & 3 & 12 & 12 & \\ \hline 2 & 2 & 8 & 8 & 0 & \underline{-1} \\ & -2 & 0 & -8 & & \\ \hline 2 & 0 & 8 & 0 & & \end{array}$$

$2x^2 + 8 = 0, \qquad x^2 + 4 = 0, \qquad x = \pm 2j$
roots: $\frac{3}{2}, -1, 2j, -2j$

7. $y = kx^2(x^3 + 436x - 4000)$
$y = 0,\ kx^2(x^3 + 436x - 4000) = 0$
$x = 0,\ x^3 + 436x - 4000 = 0$

$$\begin{array}{rrrr|l} 1 & 0 & 436 & -4000 & \underline{8} \\ & 8 & 64 & 4000 & \\ \hline 1 & 8 & 500 & 0 & \end{array}$$

$y = 0$ for $x = 0$ m, $x = 8$ m

8. Let $x =$ length of edge.
$V = x^3, \qquad 2V = (x + 1)^3$
$2x^3 = x^3 + 3x^2 + 3x + 1$
$x^3 - 3x^2 - 3x - 1 = 0$
$y = x^3 - 3x^2 - 3x - 1$

$x = 3.8$ mm

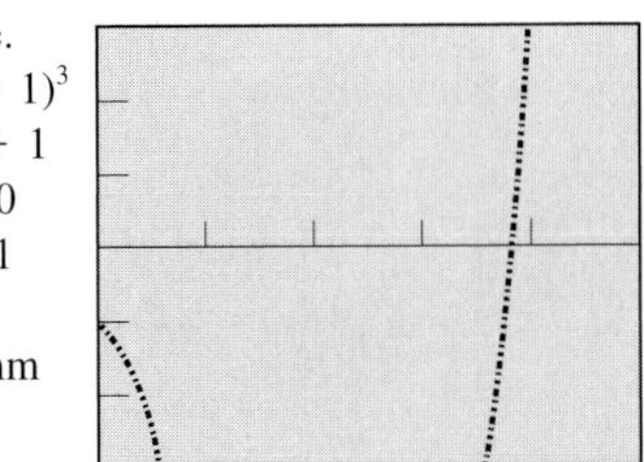

Chapter 16

1. $A = \begin{pmatrix} 3 & -1 & 4 \\ 2 & 0 & -2 \end{pmatrix}, \qquad B = \begin{pmatrix} 1 & 4 & 5 \\ -1 & -2 & 3 \end{pmatrix},$

$2B = \begin{pmatrix} 2 & 8 & 10 \\ -2 & -4 & 6 \end{pmatrix}$

$A - 2B = \begin{pmatrix} 3 - 2 & -1 - 8 & 4 - 10 \\ 2 + 2 & 0 + 4 & -2 - 6 \end{pmatrix}$
$= \begin{pmatrix} 1 & -9 & -6 \\ 4 & 4 & -8 \end{pmatrix}$

2. $\begin{vmatrix} 4 & 0 & -2 \\ 3 & -3 & 2 \\ -4 & 1 & -1 \end{vmatrix} = 4\begin{vmatrix} -3 & 2 \\ 1 & -1 \end{vmatrix} - 0\begin{vmatrix} 3 & 2 \\ -4 & -1 \end{vmatrix} + (-2)\begin{vmatrix} 3 & -3 \\ -4 & 1 \end{vmatrix}$ using first row

$= 4(3 - 2) - 0 - 2(3 - 12)$
$= 4 - 2(-9) = 4 + 18 = 22$

$= -0\begin{vmatrix} 3 & 2 \\ -4 & -1 \end{vmatrix} + (-3)\begin{vmatrix} 4 & -2 \\ -4 & -1 \end{vmatrix} - (1)\begin{vmatrix} 4 & -2 \\ 3 & 2 \end{vmatrix}$
using second column
$= 0 - 3(-4 - 8) - (8 + 6) = -3(-12) - 14 = 22$

3. $CD = \begin{pmatrix} 1 & 0 & 4 \\ 2 & -2 & 1 \\ -1 & 3 & 2 \end{pmatrix}\begin{pmatrix} 2 & -2 \\ 4 & -5 \\ 6 & 1 \end{pmatrix}$

$= \begin{pmatrix} 2 + 0 + 24 & -2 + 0 + 4 \\ 4 - 8 + 6 & -4 + 10 + 1 \\ -2 + 12 + 12 & 2 - 15 + 2 \end{pmatrix}$

$= \begin{pmatrix} 26 & 2 \\ 2 & 7 \\ 22 & -11 \end{pmatrix}$

$DC = \begin{pmatrix} 2 & -2 \\ 4 & -5 \\ 6 & 1 \end{pmatrix}\begin{pmatrix} 1 & 0 & 4 \\ 2 & -2 & 1 \\ -1 & 3 & 2 \end{pmatrix}$ not defined, since D has 2 columns and C has 3 rows

4. $\begin{vmatrix} 1 & 0 & 4 & -2 \\ -2 & -1 & 3 & 0 \\ 3 & 2 & -1 & 2 \\ 1 & 1 & -1 & -2 \end{vmatrix} = \begin{vmatrix} 1 & 0 & 0 & 0 \\ -2 & -1 & 11 & -4 \\ 3 & 2 & -13 & 8 \\ 1 & 1 & -5 & 0 \end{vmatrix}$

$= 1\begin{vmatrix} -1 & 11 & -4 \\ 2 & -13 & 8 \\ 1 & -5 & 0 \end{vmatrix} = \begin{vmatrix} -1 & 11 & -4 \\ 0 & 9 & 0 \\ 0 & 6 & -4 \end{vmatrix}$

$= -1\begin{vmatrix} 9 & 0 \\ 6 & -4 \end{vmatrix} = -(-36 - 0) = 36$

5. $\left(\begin{array}{rrr|rrr} 1 & 0 & 4 & 1 & 0 & 0 \\ 2 & -2 & 1 & 0 & 1 & 0 \\ -1 & 3 & 2 & 0 & 0 & 1 \end{array}\right) \to \left(\begin{array}{rrr|rrr} 1 & 0 & 4 & 1 & 0 & 0 \\ 0 & -2 & -7 & -2 & 1 & 0 \\ 0 & 3 & 6 & 1 & 0 & 1 \end{array}\right) \to$

$\left(\begin{array}{rrr|rrr} 1 & 0 & 4 & 1 & 0 & 0 \\ 0 & 1 & \frac{7}{2} & 1 & -\frac{1}{2} & 0 \\ 0 & 3 & 6 & 1 & 0 & 1 \end{array}\right) \to \left(\begin{array}{rrr|rrr} 1 & 0 & 4 & 1 & 0 & 0 \\ 0 & 1 & \frac{7}{2} & 1 & -\frac{1}{2} & 0 \\ 0 & 0 & -\frac{9}{2} & -2 & \frac{3}{2} & 1 \end{array}\right) \to$

$\left(\begin{array}{rrr|rrr} 1 & 0 & 4 & 1 & 0 & 0 \\ 0 & 1 & \frac{7}{2} & 1 & -\frac{1}{2} & 0 \\ 0 & 0 & 1 & \frac{4}{9} & -\frac{1}{3} & -\frac{2}{9} \end{array}\right) \to \left(\begin{array}{rrr|rrr} 1 & 0 & 0 & -\frac{7}{9} & \frac{4}{3} & \frac{8}{9} \\ 0 & 1 & 0 & -\frac{5}{9} & \frac{2}{3} & \frac{7}{9} \\ 0 & 0 & 1 & \frac{4}{9} & -\frac{1}{3} & -\frac{2}{9} \end{array}\right)$

$C^{-1} = \begin{pmatrix} -\frac{7}{9} & \frac{4}{3} & \frac{8}{9} \\ -\frac{5}{9} & \frac{2}{3} & \frac{7}{9} \\ \frac{4}{9} & -\frac{1}{3} & -\frac{2}{9} \end{pmatrix}$

6. $2x - 3y = 11$
$x + 2y = 2$

$A = \begin{pmatrix} 2 & -3 \\ 1 & 2 \end{pmatrix} \qquad A^{-1} = \begin{pmatrix} \frac{2}{7} & \frac{3}{7} \\ -\frac{1}{7} & \frac{2}{7} \end{pmatrix} \qquad C = \begin{pmatrix} 11 \\ 2 \end{pmatrix}$

$A^{-1}C = \begin{pmatrix} \frac{2}{7} & \frac{3}{7} \\ -\frac{1}{7} & \frac{2}{7} \end{pmatrix}\begin{pmatrix} 11 \\ 2 \end{pmatrix} = \begin{pmatrix} \frac{22}{7} + \frac{6}{7} \\ -\frac{11}{7} + \frac{4}{7} \end{pmatrix} = \begin{pmatrix} 4 \\ -1 \end{pmatrix}$

$x = 4, \quad y = -1$

7. Let A = number of shares of stock A,
B = number of shares of stock B.

$50A + 30B = 2600$
$30A + 40B = 2000$

$5A + 3B = 260$
$3A + 4B = 200$

Let C = coefficient matrix.

$C = \begin{pmatrix} 5 & 3 \\ 3 & 4 \end{pmatrix}, \qquad \begin{vmatrix} 5 & 3 \\ 3 & 4 \end{vmatrix} = 20 - 9 = 11$

$C^{-1} = \frac{1}{11}\begin{pmatrix} 4 & -3 \\ -3 & 5 \end{pmatrix} = \begin{pmatrix} \frac{4}{11} & -\frac{3}{11} \\ -\frac{3}{11} & \frac{5}{11} \end{pmatrix}$

$\begin{pmatrix} A \\ B \end{pmatrix} = \begin{pmatrix} \frac{4}{11} & -\frac{3}{11} \\ -\frac{3}{11} & \frac{5}{11} \end{pmatrix}\begin{pmatrix} 260 \\ 200 \end{pmatrix} = \begin{pmatrix} \frac{4}{11}(260) - \frac{3}{11}(200) \\ -\frac{3}{11}(260) + \frac{5}{11}(200) \end{pmatrix}$

$= \begin{pmatrix} 40 \\ 20 \end{pmatrix}$

$A = 40$ shares, $\quad B = 20$ shares

Chapter 17

1. $x < 0, y > 0$

2. $\frac{-x}{2} \geq 3$
$-x \geq 6$
$x \leq -6$

−6

3. $3x + 1 < -5$
$3x < -6$
$x < -2$

−2

4. $-1 < 1 - 2x < 5$
$-2 < -2x < 4$
$-2 < x < 1$

−2 1

5. $\frac{x^2 + x}{x - 2} \leq 0, \frac{x(x + 1)}{x - 2} \leq 0$

Interval	$\frac{x(x+1)}{x-2}$	Sign
$x < -1$	$\frac{- \; -}{-}$	$-$
$-1 < x < 0$	$\frac{- \; +}{-}$	$+$
$0 < x < 2$	$\frac{+ \; +}{-}$	$-$
$x > 2$	$\frac{+ \; +}{+}$	$+$

Solution: $x \leq -1$ or
$0 \leq x < 2$
(x cannot equal 2)

−1 0 2

6. $|2x + 1| \geq 3$
$2x + 1 \geq 3, \qquad 2x + 1 \leq -3$
$2x \geq 2 \qquad 2x \leq -4$
$x \geq 1$ or $\quad x \leq -2$

−2 1

7.

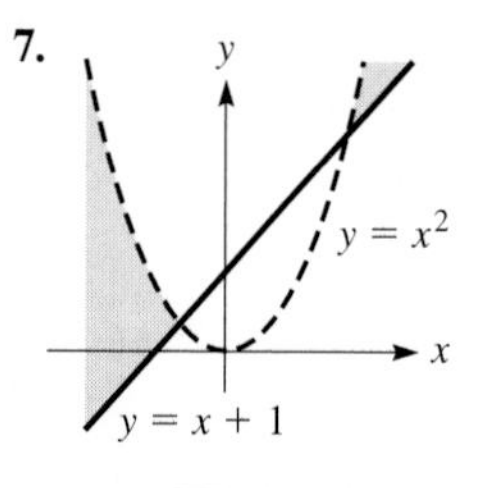

8. If $\sqrt{x^2 - x - 6}$ is real,
then $x^2 - x - 6 \geq 0$.
$(x - 3)(x + 2) \geq 0$
$x \leq -2$ or $x \geq 3$

9. Let w = width, l = length.

$$l = w + 20 \qquad w^2 + 20w - 4800 \geq 0$$
$$wl \geq 4800 \qquad (w + 80)(w - 60) \geq 0$$
$$w(w + 20) \geq 4800 \qquad w \geq 60 \text{ m}$$

10. Let A = length of type A wire,
B = length of type B wire.

$$0.10A + 0.20B < 5.00$$
$$A + 2B < 50$$

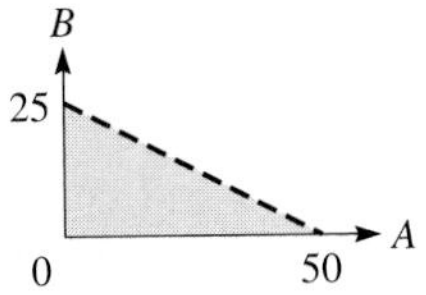

11. $|\lambda - 550 \text{ nm}| < 150 \text{ nm}$ (within 150 nm of $\lambda = 550$ nm)

Chapter 18

1. $\dfrac{180 \text{ s}}{4 \text{ min}} = \dfrac{180 \text{ s}}{240 \text{ s}} = \dfrac{3}{4}$

2. $F = \dfrac{k}{d}$; $\quad 0.750 = \dfrac{k}{1.25}$, $\quad k = 0.938 \text{ N}\cdot\text{cm}$

$F = \dfrac{0.938}{d}$; $\quad F = \dfrac{0.938}{1.75} = 0.536 \text{ N}$

3. $\dfrac{1 \text{ cm}}{10^4\ \mu\text{m}} = \dfrac{x}{7.24\ \mu\text{m}}$

$x = 7.24 \times 10^{-4}$ cm

4. $p = kdh$

Using values for water,

$1.96 = k(1000)(0.200)$

$k = 0.009\,80 \text{ kPa}\cdot\text{m}^2/\text{kg}$

For alcohol,

$p = 0.009\,80(800)(0.300)$

$= 2.35$ kPa

5.

$$2l + 2w = 210.0$$
$$l = 105.0 - w$$
$$\frac{l}{w} = \frac{7}{3}$$
$$\frac{105.0 - w}{w} = \frac{7}{3}$$
$$315.0 - 3w = 7w$$
$$10w = 315.0$$
$$w = 31.5 \text{ cm}$$
$$l = 105.0 - 31.5 = 73.5 \text{ cm}$$

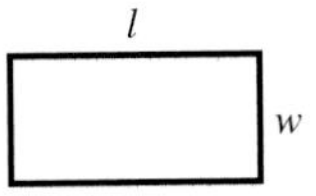

6. Let L_1 = crushing load of first pillar,
L_2 = crushing load of second pillar.

$$L_2 = \frac{kr_2^4}{l_2^2}, \qquad L_1 = \frac{k(2r_2)^4}{(3l_2)^2};$$

$$\frac{L_1}{L_2} = \frac{\dfrac{k(2r_2)^4}{(3l_2)^2}}{\dfrac{kr_2^4}{l_2^2}} = \frac{16kr_2^4}{9l_2^2} \times \frac{l_2^2}{kr_2^4} = \frac{16}{9}$$

Chapter 19

1. $6, -2, \frac{2}{3}, \ldots$;
geometric sequence

$a_1 = 6, \qquad r = \dfrac{-2}{6} = -\dfrac{1}{3}$

$$S_7 = \frac{6[1 - (-\frac{1}{3})^7]}{1 - (-\frac{1}{3})}$$
$$= \frac{6\left(1 + \dfrac{1}{3^7}\right)}{\frac{4}{3}} = \frac{1094}{243}$$

2. $a_1 = 6$, $d = 4$, $s_n = 126$;
arithmetic sequence

$$126 = \tfrac{n}{2}(6 + a_n)$$
$$a_n = 6 + (n - 1)4 = 2 + 4n$$
$$126 = \tfrac{n}{2}[6 + (2 + 4n)]$$
$$252 = n(8 + 4n) = 4n^2 + 8n$$
$$n^2 + 2n - 63 = 0$$
$$(n + 9)(n - 7) = 0$$
$$n = 7$$

3. $0.454\,545\ldots = 0.45 + 0.0045 + 0.000\,045 + \cdots$

$a = 0.45, \qquad r = 0.01$

$$S = \frac{0.45}{1 - 0.01} = \frac{0.45}{0.99} = \frac{5}{11}$$

4. $\sqrt{1 - 4x} = (1 - 4x)^{1/2}$

$$= 1 + (\tfrac{1}{2})(-4x) + \frac{\frac{1}{2}(\frac{1}{2} - 1)}{2}(-4x)^2 + \cdots$$
$$= 1 - 2x - 2x^2 + \cdots$$

5. $(2x - y)^5 = (2x)^5 + 5(2x)^4(-y) + \dfrac{5(4)}{2}(2x)^3(-y)^2$

$$+ \frac{5(4)(3)}{2(3)}(2x)^2(-y)^3$$
$$+ \frac{5(4)(3)(2)}{2(3)(4)}(2x)(-y)^4 + (-y)^5$$
$$= 32x^5 - 80x^4y + 80x^3y^2 - 40x^2y^3 + 10xy^4 - y^5$$

6. $5\% = 0.05$

Value after 1 year is

$2500 + 2500(0.05) = 2500(1.05)$

$V_{20} = 2500(1.05)^{20}$

$= \$6633.24$

7. $2 + 4 + \cdots + 200$

$a_1 = 2, \qquad a_{100} = 200, \qquad n = 100$

$S_{100} = \frac{100}{2}(2 + 200)$

$= 10{,}100$

8. Ball falls 8.00 m, rises 4.00 m, falls 4.00 m, etc.

Distance $= 8.00 + (4.00 + 4.00)$
$+ (2.00 + 2.00) + \cdots$
$= 8.00 + 8.00 + 4.00 + 2.00 + \cdots$

$$= 8.00 + \frac{8.00}{1 - 0.5} = 8.00 + 16.0 = 24.0 \text{ m}$$

Chapter 20

1. $\sec\theta - \dfrac{\tan\theta}{\csc\theta} = \dfrac{1}{\cos\theta} - \dfrac{\dfrac{\sin\theta}{\cos\theta}}{\dfrac{1}{\sin\theta}} = \dfrac{1}{\cos\theta} - \dfrac{\sin^2\theta}{\cos\theta}$

$= \dfrac{1-\sin^2\theta}{\cos\theta} = \dfrac{\cos^2\theta}{\cos\theta} = \cos\theta;$

$\cos\theta = \cos\theta$

2. $\sin 2x + \sin x = 0$

$2\sin x\cos x + \sin x = 0$

$\sin x(2\cos x + 1) = 0$

$\sin x = 0, \quad \cos x = -\dfrac{1}{2}$

$x = 0, \pi, \frac{2\pi}{3}, \frac{4\pi}{3}$

3. $\theta = \sin^{-1} x$

$\cos\theta = \cos(\sin^{-1} x)$

$= \sqrt{1-x^2}$

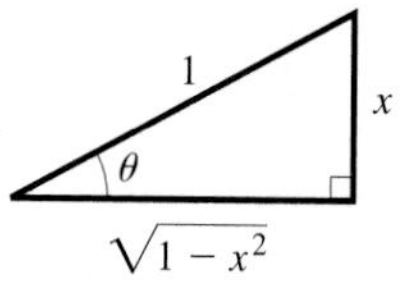

1, x, θ, $\sqrt{1-x^2}$

4. $\theta = e^{-0.1t}(\cos 2t + 3\sin 2t)$

$e^{-0.1t}(\cos 2t + 3\sin 2t) = 0$

$\cos 2t + 3\sin 2t = 0$

$3\sin 2t = -\cos 2t$

$3\tan 2t = -1, \quad \tan 2t = -\dfrac{1}{3}$

$2t = 2.82, \quad t = 1.41$

5. $\dfrac{\tan\alpha + \tan\beta}{\tan\alpha - \tan\beta} = \dfrac{\dfrac{\sin\alpha}{\cos\alpha} + \dfrac{\sin\beta}{\cos\beta}}{\dfrac{\sin\alpha}{\cos\alpha} - \dfrac{\sin\beta}{\cos\beta}}$

$= \dfrac{\sin\alpha\cos\beta + \sin\beta\cos\alpha}{\sin\alpha\cos\beta - \sin\beta\cos\alpha}$

$= \dfrac{\sin(\alpha+\beta)}{\sin(\alpha-\beta)}$

6. $\cot^2 x - \cos^2 x = \dfrac{\cos^2 x}{\sin^2 x} - \cos^2 x =$

$\cos^2 x(\csc^2 x - 1) = \cos^2 x\cot^2 x;$

$\cos^2 x\cot^2 x = \cos^2 x\cot^2 x$

7. $\sin x = -\frac{3}{5}$

$\cos x = \frac{4}{5}$

$\cos\dfrac{x}{2} = -\sqrt{\dfrac{1+(\frac{4}{5})}{2}} = -\sqrt{\dfrac{9}{10}} = -0.9487$

since $270° < x < 360°$, $135° < \frac{x}{2} < 180°$;

$\frac{x}{2}$ is in second quadrant, where $\cos\frac{x}{2}$ is negative

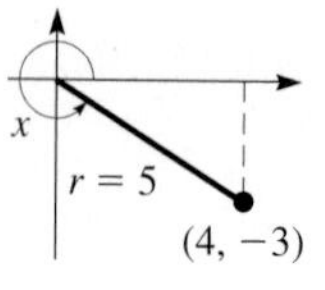

x, $r = 5$, $(4, -3)$

8. $I = I_0 \sin 2\theta\cos 2\theta$

$\dfrac{2I}{I_0} = 2\sin 2\theta\cos 2\theta = \sin 4\theta$

$4\theta = \sin^{-1}\dfrac{2I}{I_0}$

$\theta = \dfrac{1}{4}\sin^{-1}\dfrac{2I}{I_0}$

9. $y = x - 2\cos x - 5$

$x = 3.02, 4.42, 6.77$

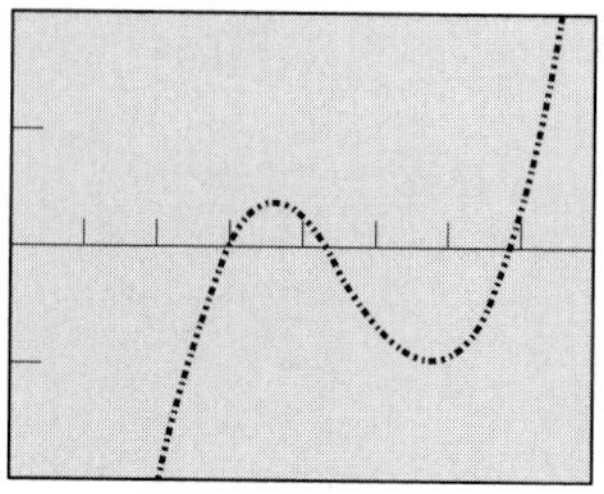

Chapter 21

1. $2(x^2 + x) = 1 - y^2$

$2x^2 + y^2 + 2x - 1 = 0$

$A \neq C$ (same sign)

$B = 0$: ellipse

2. $4x - 2y + 5 = 0$

$y = 2x + \frac{5}{2}$

$m = 2, \quad b = \frac{5}{2}$

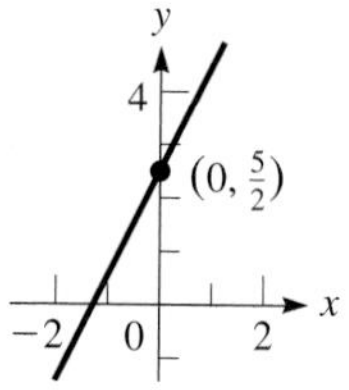

y, x, 4, $(0, \frac{5}{2})$, −2, 0, 2

3. $x^2 = 2x - y^2$

$x^2 + y^2 = 2x$

$r^2 = 2r\cos\theta$

$r = 2\cos\theta$

4. $x^2 = -12y$

$4p = -12, \quad p = -3$

$V(0, 0), \quad F(0, -3)$

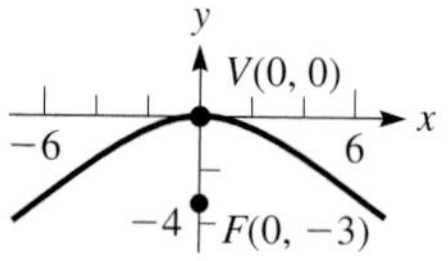

y, $V(0, 0)$, x, −6, 6, −4, $F(0, -3)$

5. Center $(-1, 2)$; $h = -1$, $k = 2$

$r = \sqrt{(2+1)^2 + (3-2)^2}$

$= \sqrt{10}$

$(x+1)^2 + (y-2)^2 = 10$

or

$x^2 + y^2 + 2x - 4y - 5 = 0$

6. $m = \dfrac{-2-1}{2+4} = -\dfrac{1}{2}$

$y - 1 = -\frac{1}{2}(x+4)$

$2y - 2 = -x - 4$

$x + 2y + 2 = 0$

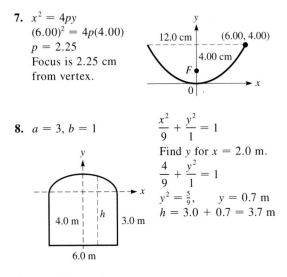

7. $x^2 = 4py$

$(6.00)^2 = 4p(4.00)$

$p = 2.25$

Focus is 2.25 cm from vertex.

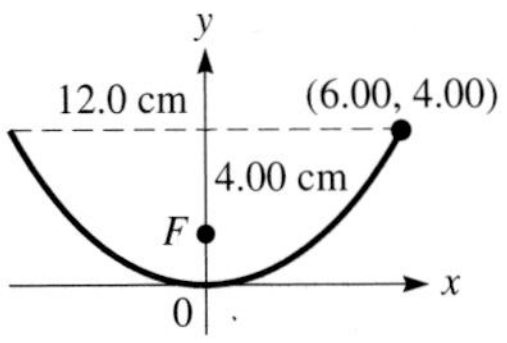

8. $a = 3,\ b = 1$

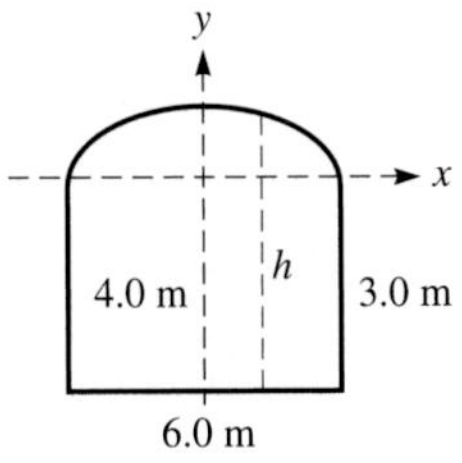

$$\frac{x^2}{9} + \frac{y^2}{1} = 1$$

Find y for $x = 2.0$ m.

$$\frac{4}{9} + \frac{y^2}{1} = 1$$

$y^2 = \frac{5}{9}, \qquad y = 0.7$ m

$h = 3.0 + 0.7 = 3.7$ m

9. $r = 3 + \cos\theta$

r	0	$\frac{\pi}{4}$	$\frac{\pi}{2}$	$\frac{3\pi}{4}$	π	$\frac{5\pi}{4}$	$\frac{3\pi}{2}$	$\frac{7\pi}{4}$	2π
θ	4.0	3.7	3.0	2.3	2.0	2.3	3.0	3.7	4.0

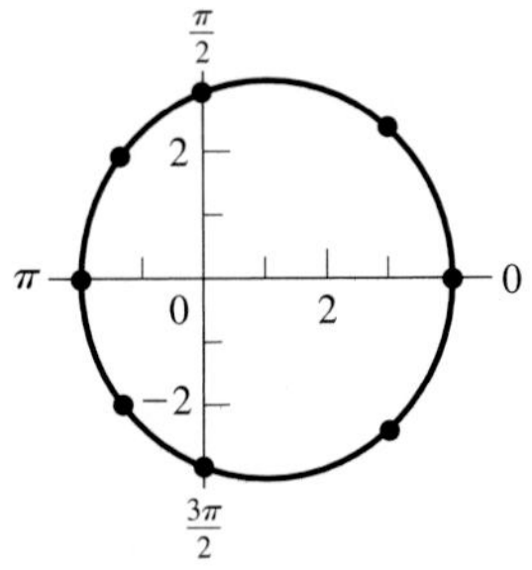

10. $4y^2 - x^2 - 4x - 8y - 4 = 0$

$4(y^2 - 2y \qquad) - (x^2 + 4x \qquad) = 4$

$4(y^2 - 2y + 1) - (x^2 + 4x + 4) = 4 + 4 - 4$

$$\frac{(y-1)^2}{1^2} - \frac{(x+2)^2}{2^2} = 1$$

$C(-2, 1), \qquad V(-2, 0), \qquad V(-2, 2)$

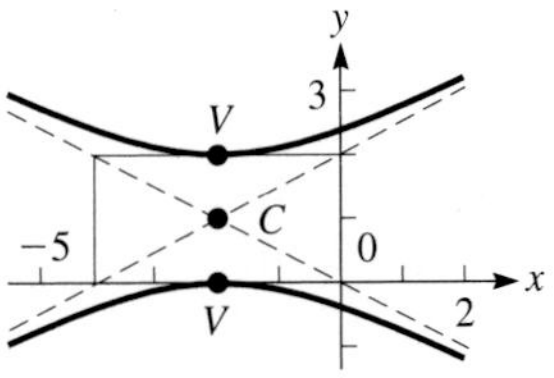

Chapter 22

1.

Number	1	2	3	4	5	6	7	8	9	10
Frequency	1	0	1	2	3	2	1	2	2	1

$\sum f = 15$

Median is eighth number; median = 6.

2. 5 appears 3 times, and no other number appears more than twice; mode = 5.

3.

Number	0–2	3–5	6–8	9–11
Frequency	1	6	5	3

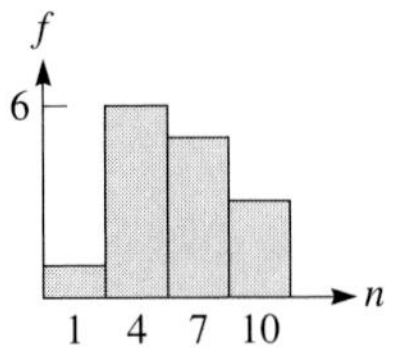

4. Let t = thickness.

$$\bar{t} = \frac{3(0.90) + 9(0.91) + 31(0.92) + 38(0.93) + 12(0.94) + 5(0.95) + 2(0.96)}{3 + 9 + 31 + 38 + 12 + 5 + 2}$$

$$= \frac{92.7}{100} = 0.927 \text{ cm}$$

5. $\overline{t^2} = \dfrac{\Sigma ft^2}{\Sigma f} = \dfrac{85.9464}{100} = 0.859\,464$

$s_t = \sqrt{\overline{t^2} - \bar{t}^2} = \sqrt{0.859\,464 - 0.927^2}$

$= 0.012$ cm

6.

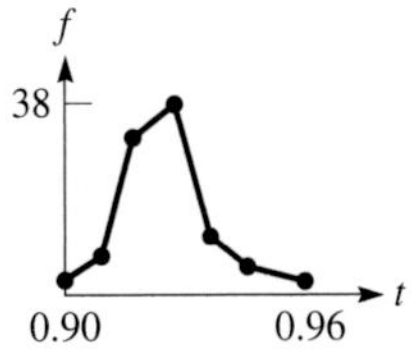

7.

x	y	xy	x^2
1	5	5	1
3	11	33	9
5	17	85	25
7	20	140	49
9	27	243	81
25	80	506	165

$n = 5$

$$m = \frac{5(506) - (25)(80)}{5(165) - 25^2}$$

$= 2.65$

$$b = \frac{165(80) - (506)(25)}{5(165) - 25^2}$$

$= 2.75$

$y = 2.65x + 2.75$

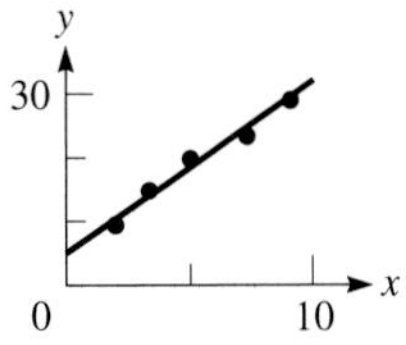

8.

x	$\sqrt{x}$	y	$\sqrt{x}y$	$(\sqrt{x})^2 = x$
1.00	1.000	1.10	1.1000	1.000
3.00	1.732	1.90	3.2908	3.000
5.00	2.236	2.50	5.5900	5.000
7.00	2.646	2.90	7.6734	7.000
9.00	3.000	3.30	9.9000	9.000
	10.614	11.70	27.5542	25.000

$n = 5$

$$m = \frac{5(27.5542) - (10.614)(11.70)}{5(25.000) - (10.614)^2} = 1.10$$

$$b = \frac{(25.000)(11.70) - (27.5542)(10.614)}{5(25.000) - (10.614)^2} = 0.00$$

(rounded off)

$y = 1.10\sqrt{x}$

Chapter 23

1. $\lim\limits_{x\to 1}\dfrac{x^2-x}{x^2-1}=\lim\limits_{x\to 1}\dfrac{x(x-1)}{(x+1)(x-1)}=\lim\limits_{x\to 1}\dfrac{x}{x+1}=\dfrac{1}{2}$

2. $\lim\limits_{x\to\infty}\dfrac{1-4x^2}{x+2x^2}=\lim\limits_{x\to\infty}\dfrac{\dfrac{1}{x^2}-4}{\dfrac{1}{x}+2}=-2$

3. $y=3x^2-\dfrac{4}{x^2}$

$$\frac{dy}{dx}=6x+\frac{8}{x^3}$$

$$\left.\frac{dy}{dx}\right|_{x=2}=6(2)+\frac{8}{2^3}$$

$$=13$$

$$m_{\tan}=13$$

4. $s=t\sqrt{10-2t}$

$$v=\frac{ds}{dt}=t(\tfrac{1}{2})(10-2t)^{-1/2}(-2)+(10-2t)^{1/2}(1)$$

$$=\frac{-t}{(10-2t)^{1/2}}+(10-2t)^{1/2}=\frac{10-3t}{(10-2t)^{1/2}}$$

$$v\Big|_{t=4.00}=\frac{10-3(4.00)}{[10-2(4.00)]^{1/2}}=\frac{10-12.00}{2.00^{1/2}}=-1.41\text{ cm/s}$$

5. $(1+y^2)^3-x^2y=7x$

$$3(1+y^2)^2(2yy')-x^2y'-y(2x)=7$$

$$6y(1+y^2)^2y'-x^2y'=7+2xy$$

$$y'=\frac{7+2xy}{6y(1+y^2)^2-x^2}$$

6. $V=\dfrac{kq}{\sqrt{x^2+b^2}}=kq(x^2+b^2)^{-1/2}$

$$\frac{dV}{dx}=kq\left(-\frac{1}{2}\right)(x^2+b^2)^{-3/2}(2x)=\frac{-kqx}{(x^2+b^2)^{3/2}}$$

7. $y=\dfrac{2}{3x+2}=2(3x+2)^{-1}$

$$\frac{dy}{dx}=2(-1)(3x+2)^{-2}(3)=-6(3x+2)^{-2}$$

$$\frac{d^2y}{dx^2}=-6(-2)(3x+2)^{-3}(3)=\frac{36}{(3x+2)^3}$$

8. $y=5x-2x^2$

$$y+\Delta y=5(x+\Delta x)-2(x+\Delta x)^2$$

$$\Delta y=5\,\Delta x-4x\,\Delta x-2\,\Delta x^2$$

$$\frac{\Delta y}{\Delta x}=5-4x-2\,\Delta x$$

$$\lim_{\Delta x\to 0}\frac{\Delta y}{\Delta x}=5-4x$$

Chapter 24

1. $y=x^4-3x^2$

$$\frac{dy}{dx}=4x^3-6x$$

$$\left.\frac{dy}{dx}\right|_{x=1}=4(1^3)-6(1)$$

$$=-2$$

$$y-(-2)=-2(x-1)$$

$$y=-2x$$

2. $x=3t^2$ $\qquad y=2t^3-t^2$

$$v_x=\frac{dx}{dt}=6t \qquad v_y=\frac{dy}{dt}=6t^2-2t$$

$$a_x=\frac{dv_x}{dt}=\frac{d^2x}{dt^2}=6 \qquad a_y=\frac{dv_y}{dt}=\frac{d^2y}{dt^2}=12t-2$$

$$a_x\Big|_{t=2}=6 \qquad a_y\Big|_{t=2}=12(2)-2=22$$

$$a\Big|_{t=2}=\sqrt{6^2+22^2}=22.8 \qquad \tan\theta=\tfrac{22}{6},\qquad \theta=74.7^\circ$$

3. $P=\dfrac{144r}{(r+0.6)^2}$

$$\frac{dP}{dr}=\frac{144[(r+0.6)^2(1)-r(2)(r+0.6)(1)]}{(r+0.6)^4}$$

$$=\frac{144[(r+0.6)-2r]}{(r+0.6)^3}=\frac{144(0.6-r)}{(r+0.6)^3}$$

$$\frac{dP}{dr}=0;\qquad 0.6-r=0,\qquad r=0.6\ \Omega$$

$$\left(r<0.6,\ \frac{dP}{dr}>0;\ r>0.6,\ \frac{dP}{dr}<0\right)$$

4. $x^2-\sqrt{4x+1}=0;\qquad f(x)=x^2-\sqrt{4x+1}$

$$f'(x)=2x-\tfrac{1}{2}(4x+1)^{-1/2}(4)=2x-\frac{2}{(4x+1)^{1/2}}$$

n	x_n	$f(x_n)$	$f'(x_n)$	$x_n-\dfrac{f(x_n)}{f'(x_n)}$
1	1.5	−0.395 751 3	2.244 071 1	1.676 354 2
2	1.676 354 2	0.034 300 1	2.632 211 8	1.663 323 3

$x_3=1.6633$

5. $y=x^3+6x^2$

$$y'=3x^2+12x=3x(x+4)$$

$$y''=6x+12=6(x+2)$$

$x<-4$	y inc.	Max. $(-4,32)$
$-4<x<0$	y dec.	Min. $(0,0)$
$x>0$	y inc.	Infl. $(-2,16)$
$x<-2$	y conc. down	
$x>-2$	y conc. up	

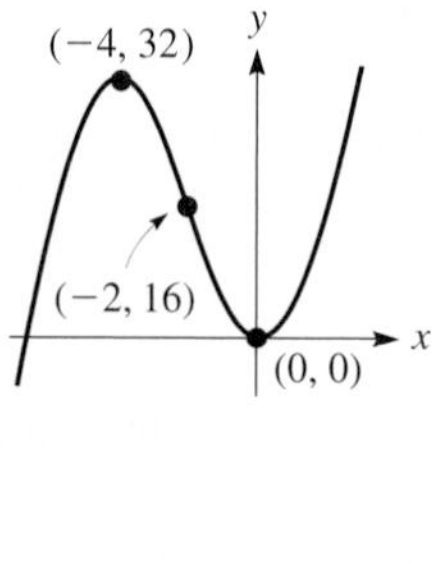

(−4, 32)
(−2, 16)
(0, 0)

6. $y = \dfrac{4}{x^2} - x$

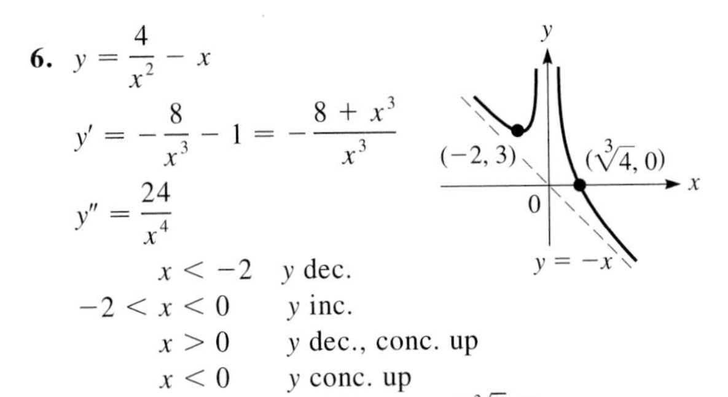

$y' = -\dfrac{8}{x^3} - 1 = -\dfrac{8 + x^3}{x^3}$

$y'' = \dfrac{24}{x^4}$

$x < -2$ y dec.
$-2 < x < 0$ y inc.
$x > 0$ y dec., conc. up
$x < 0$ y conc. up

Min. $(-2, 3)$, no infl., int. $(\sqrt[3]{4}, 0)$, sym. none; as $x \to \pm\infty$, $y \to -x$, asym. $y = -x$, $x = 0$. Domain: all real x except 0; range: all real y.

7. Let V = volume of cube,
e = edge of cube.

$V = e^3; \quad \dfrac{dV}{dt} = 3e^2\dfrac{de}{dt}$

$$\left.\frac{dV}{dt}\right|_{e=1.25} = 3(1.25)^2(-0.10) = -0.47 \text{ m}^3/\text{s}$$

8.

$3x + 2y = 6000$

$y = \dfrac{6000 - 3x}{2}$

$A = xy = x\left(\dfrac{6000 - 3x}{2}\right) = 3000x - \frac{3}{2}x^2$

$\dfrac{dA}{dx} = 3000 - 3x$

$3000 - 3x = 0$, $x = 1000$ m
$y = 1500$ m
$A_{max} = (1000)(1500) = 1.5 \times 10^6 \text{ m}^2$
$\left(x < 1000, \dfrac{dA}{dx} > 0; x > 1000, \dfrac{dA}{dx} < 0\right)$

Chapter 25

1. $y = 3x^2 - x$
$y + \Delta y = 3(x + \Delta x)^2 - (x + \Delta x)$
$\Delta y = 6x\Delta x + 3\Delta x^2 - \Delta x$
$dy = (6x - 1)\,dx$
For $x = 3$, $\Delta x = 0.1$.
$\Delta y = 6(3)(0.1) + 3(0.1)^2 - (0.1) = 1.73$
$dy = [6(3) - 1](0.1) = 1.7$
$\Delta y - dy = 0.03$

2. $\displaystyle\int x\sqrt{1 - 2x^2}\,dx = \int x(1 - 2x^2)^{1/2}\,dx$

$u = 1 - 2x^2, \quad du = -4x\,dx, \quad n = \frac{1}{2}, \quad n + 1 = \frac{3}{2}$

$$\int x(1 - 2x^2)^{1/2}\,dx = -\frac{1}{4}\int (1 - 2x^2)^{1/2}(-4x\,dx) = -\frac{1}{4}\left(\frac{2}{3}\right)(1 - 2x^2)^{3/2} + C = -\frac{1}{6}(1 - 2x^2)^{3/2} + C$$

3. $$\int_2^5 (6 - x)^4\,dx = -\int_2^5 (6 - x)^4(-dx) = -\frac{1}{5}(6 - x)^5\Big|_2^5 = -\frac{1}{5}(1^5 - 4^5) = \frac{1023}{5}$$

4. $y = \dfrac{1}{x + 2}, \quad n = 6, \quad \Delta x = \dfrac{4 - 1}{6} = \dfrac{1}{2}$

$A = \frac{1}{2}\left(\frac{2}{7} + \frac{1}{4} + \frac{2}{9} + \frac{1}{5} + \frac{2}{11} + \frac{1}{6}\right) = 0.6532$

x	1	$\frac{3}{2}$	2	$\frac{5}{2}$	3	$\frac{7}{2}$	4
y	$\frac{1}{3}$	$\frac{2}{7}$	$\frac{1}{4}$	$\frac{2}{9}$	$\frac{1}{5}$	$\frac{2}{11}$	$\frac{1}{6}$

5. (See values for Problem 4.)

$$\int_1^4 \frac{dx}{x + 2} = \frac{1}{4}\left[\frac{1}{3} + 2\left(\frac{2}{7}\right) + 2\left(\frac{1}{4}\right) + 2\left(\frac{2}{9}\right) + 2\left(\frac{1}{5}\right) + 2\left(\frac{2}{11}\right) + \frac{1}{6}\right] = 0.6949$$

6. (See values for Problem 4.)

$$\int_1^4 \frac{dx}{x + 2} = \frac{1}{6}\left[\frac{1}{3} + 4\left(\frac{2}{7}\right) + 2\left(\frac{1}{4}\right) + 4\left(\frac{2}{9}\right) + 2\left(\frac{1}{5}\right) + 4\left(\frac{2}{11}\right) + \frac{1}{6}\right] = 0.6932$$

7. $A = \pi r^2$
$dA = 2\pi r\,dr$
For $r = 4.50$ cm, $dr = 0.30$ cm.
$dA = 2\pi(4.50)(0.30) = 8.48 \text{ cm}^2$
$V = 20.0(8.48) = 170 \text{ cm}^3$

8. $$i = \int_1^3 \left(t^2 + \frac{1}{t^2}\right)dt = \frac{1}{3}t^3 - \frac{1}{t}\Big|_1^3 = \frac{1}{3}(27) - \frac{1}{3} - \left(\frac{1}{3} - 1\right) = 9.3 \text{ A}$$

Chapter 26

1. $$A = \int_0^2 \frac{1}{4}x^2\,dx = \frac{1}{12}x^3\Big|_0^2 = \frac{2}{3}$$

2. $\bar{x} = \dfrac{\int_0^2 x(\frac{1}{4}x^2)\,dx}{\frac{2}{3}} = \dfrac{\frac{1}{4}\int_0^2 x^3\,dx}{\frac{2}{3}} = \dfrac{\frac{1}{16}x^4\big|_0^2}{\frac{2}{3}} = \dfrac{1}{\frac{2}{3}} = \dfrac{3}{2}$

$\bar{y} = \dfrac{\int_0^1 y(2 - 2\sqrt{y})\,dy}{\frac{2}{3}} = \dfrac{2\int_0^1 (y - y^{3/2})\,dy}{\frac{2}{3}}$

$= \dfrac{2(\frac{1}{2}y^2 - \frac{2}{5}y^{5/2})\big|_0^1}{\frac{2}{3}}$

$= \dfrac{1 - \frac{4}{5}}{\frac{2}{3}} = \dfrac{1}{5} \times \dfrac{3}{2} = \dfrac{3}{10}$

3. $V = \pi\int_0^2 \left(\dfrac{1}{4}x^2\right)^2 dx = \dfrac{\pi}{16}\int_0^2 x^4\,dx = \dfrac{\pi}{80}x^5\Big|_0^2$

$= \dfrac{32\pi}{80} = \dfrac{2\pi}{5}$

4. $V = \pi\int_0^3 9^2\,dx - \pi\int_0^3 (x^2)^2\,dx = 81\pi x\Big|_0^3 - \dfrac{\pi}{5}x^5\Big|_0^3$

$= 243\pi - \dfrac{243\pi}{5} = \dfrac{972\pi}{5}$

or $V = 2\pi\int_0^9 xy\,dy = 2\pi\int_0^9 y^{1/2}y\,dy = \dfrac{4\pi}{5}y^{5/2}\Big|_0^9$

$= \dfrac{4\pi}{5}(3^5 - 0) = \dfrac{972\pi}{5}$

5. $I_y = k\int_0^3 x^2(9 - x^2)\,dx = k\int_0^3 (9x^2 - x^4)\,dx$

$= k\left(3x^3 - \dfrac{1}{5}x^5\right)\Big|_0^3 = k\left(81 - \dfrac{243}{5}\right) = \dfrac{162k}{5}$

6. $s = \int (60 - 4t)\,dt = 60t - 2t^2 + C$

$s = 10$ for $t = 0$, $10 = 60(0) - 2(0^2) + C$, $C = 10$

$s = 60t - 2t^2 + 10$

7. $F = kx$, $12 = k(2.0)$, $k = 6.0$ N/cm

$W = \int_{2.0}^{6.0} 6.0x\,dx = 3.0x^2\Big|_{2.0}^{6.0} = 3.0(36.0 - 4.0)$

$= 96$ N · cm

8. $F = 9.80\int_{1.00}^{3.00} 6.00h\,dh = (9.80)(6.00)\frac{1}{2}h^2\Big|_{1.00}^{3.00}$

$= (9.80)(3.00)(9.00 - 1.00) = 235$ kN

Chapter 27

1. $y = \tan^3 2x + \tan^{-1} 2x$

$\dfrac{dy}{dx} = 3(\tan^2 2x)(\sec^2 2x)(2) + \dfrac{2}{1 + (2x)^2}$

$= 6\tan^2 2x \sec^2 2x + \dfrac{2}{1 + 4x^2}$

2. $y = 2(3 + \cot 4x)^3$

$\dfrac{dy}{dx} = 2(3)(3 + \cot 4x)^2(-\csc^2 4x)(4)$

$= -24(3 + \cot 4x)^2 \csc^2 4x$

3. $y \sec 2x = \sin^{-1} 3y$

$y(\sec 2x \tan 2x)(2) + (\sec 2x)(y') = \dfrac{3y'}{\sqrt{1 - (3y)^2}}$

$\sqrt{1 - 9y^2}\sec 2x(2y\tan 2x + y') = 3y'$

$y' = \dfrac{2y\sqrt{1 - 9y^2}\sec 2x \tan 2x}{3 - \sqrt{1 - 9y^2}\sec 2x}$

4. $y = \dfrac{\cos^2(3x + 1)}{x}$

$dy = \dfrac{x\{2\cos(3x + 1)[-\sin(3x + 1)(3)]\} - \cos^2(3x + 1)(1)}{x^2}\,dx$

$= \dfrac{-6x\cos(2x + 1)\sin(3x + 1) - \cos^2(3x + 1)}{x^2}\,dx$

5. $y = \ln\dfrac{2x - 1}{1 + x^2}$

$= \ln(2x - 1) - \ln(1 + x^2)$

$\dfrac{dy}{dx} = \dfrac{2}{2x - 1} - \dfrac{2x}{1 + x^2}$

$m_{\tan} = \dfrac{dy}{dx}\Big|_{x=2} = \dfrac{2}{4 - 1} - \dfrac{4}{1 + 4} = \dfrac{2}{3} - \dfrac{4}{5} = -\dfrac{2}{15}$

6. $i = 8e^{-t}\sin 10t$

$\dfrac{di}{dt} = 8[e^{-t}\cos 10t(10) + \sin 10t(-e^{-t})]$

$= 8e^{-t}(10\cos 10t - \sin 10t)$

7. $y = xe^x$

$y' = xe^x + e^x = e^x(x + 1)$

$y'' = xe^x + e^x + e^x = e^x(x + 2)$

$x < -1$ y dec.

$x > -1$ y inc.

$x < -2$ y conc. down

$x > -2$ y conc. up

Int. $(0, 0)$, min. $(-1, -e^{-1})$,

infl. $(-2, -2e^{-2})$, asym. $y = 0$.

8. For $t = 8.0$ s, $x = 40$ m.

$\theta = \tan^{-1}\dfrac{x}{250}$

$\dfrac{d\theta}{dt} = \dfrac{1}{1 + \dfrac{x^2}{250^2}}\,\dfrac{dx/dt}{250} = \dfrac{250^2}{250^2 + x^2}\,\dfrac{dx/dt}{250}$

$\dfrac{d\theta}{dt}\Big|_{t=8.0} = \dfrac{250}{250^2 + 40^2}(5.0) = 0.020$ rad/s

Chapter 28

1. $\int (\sec x - \sec^3 x \tan x)\, dx$

$$= \int \sec x\, dx - \int \sec^2 x(\sec x \tan x)\, dx$$

$$= \ln|\sec x + \tan x| - \tfrac{1}{3}\sec^3 x + C$$

2. $\int \sin^3 x\, dx = \int \sin^2 x \sin x\, dx$

$$= \int (1 - \cos^2 x)\sin x\, dx$$

$$= \int \sin x\, dx - \int \cos^2 x \sin x\, dx$$

$$= -\cos x + \tfrac{1}{3}\cos^3 x + C$$

3. $\int \tan^3 2x\, dx = \int \tan 2x(\tan^2 2x)\, dx$

$$= \int \tan 2x(\sec^2 2x - 1)\, dx$$

$$= \tfrac{1}{2}\int \tan 2x \sec^2 2x(2\, dx)$$

$$- \tfrac{1}{2}\int \tan 2x(2\, dx)$$

$$= \tfrac{1}{4}\tan^2 2x + \tfrac{1}{2}\ln|\cos 2x| + C$$

4. $\int \cos^2 4x\, dx = \tfrac{1}{2}\int (1 + \cos 8x)\, dx$

$$= \tfrac{1}{2}\int dx + \tfrac{1}{16}\int \cos 8x(8\, dx)$$

$$= \tfrac{1}{2}x + \tfrac{1}{16}\sin 8x + C$$

5. Let $x = 2\sin\theta$,
$dx = 2\cos\theta\, d\theta$.

$$\int \frac{dx}{x^2\sqrt{4 - x^2}} = \int \frac{2\cos\theta\, d\theta}{4\sin^2\theta\sqrt{4 - 4\sin^2\theta}}$$

$$= \frac{1}{4}\int \frac{\cos\theta\, d\theta}{\sin^2\theta\sqrt{\cos^2\theta}} = \tfrac{1}{4}\int \csc^2\theta\, d\theta$$

$$= -\tfrac{1}{4}\cot\theta + C$$

$$= -\frac{\sqrt{4 - x^2}}{4x} + C$$

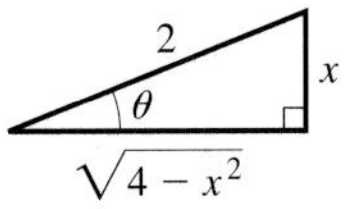

6. $\int xe^{-2x} dx;\ u = x,\ du = dx,\ dv = e^{-2x} dx,\ v = -\tfrac{1}{2}e^{-2x}$

$$\int xe^{-2x} dx = x(-\tfrac{1}{2}e^{-2x}) - \int (-\tfrac{1}{2}e^{-2x})\, dx$$

$$= -\tfrac{1}{2}xe^{-2x} - \tfrac{1}{4}e^{-2x} + C$$

7. $i = \int \frac{6t + 1}{4t^2 + 9} dt = \int \frac{6t\, dt}{4t^2 + 9} + \int \frac{dt}{4t^2 + 9}$

$$= \frac{6}{8}\int \frac{8t\, dt}{4t^2 + 9} + \frac{1}{2}\int \frac{2\, dt}{9 + (2t)^2}$$

$$= \frac{3}{4}\ln(4t^2 + 9) + \frac{1}{2}\left(\frac{1}{3}\right)\tan^{-1}\frac{2t}{3} + C$$

$i = 0$ for $t = 0$: $0 = \frac{3}{4}\ln 9 + \frac{1}{6}\tan^{-1} 0 + C$,

$$C = -\frac{3}{4}\ln 9$$

$$i = \frac{3}{4}\ln(4t^2 + 9) + \frac{1}{6}\tan^{-1}\frac{2t}{3} - \frac{3}{4}\ln 9$$

$$= \frac{3}{4}\ln\frac{4t^2 + 9}{9} + \frac{1}{6}\tan^{-1}\frac{2t}{3}$$

8. $y = \frac{1}{\sqrt{16 - x^2}};\ A = \int_0^3 y\, dx$

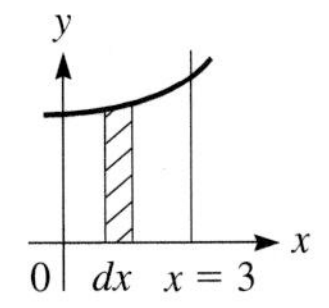

$$= \int_0^3 \frac{dx}{\sqrt{16 - x^2}}$$

$$= \sin^{-1}\frac{x}{4}\Big|_0^3$$

$$= \sin^{-1}\tfrac{3}{4} = 0.8481$$

Chapter 29

1. $f(x) = (1 + e^x)^2$ $\qquad f(0) = (1 + 1)^2 = 4$

$f'(x) = 2(1 + e^x)(e^x)$ $\qquad f'(0) = 2(1 + 1)(1) = 4$

$= 2e^x + 2e^{2x}$ $\qquad f''(0) = 2(1) + 4(1) = 6$

$f''(x) = 2e^x + 4e^{2x}$ $\qquad f'''(0) = 2(1) + 8(1) = 10$

$f'''(x) = 2e^x + 8e^{2x}$

$$(1 + e^x)^2 = 4 + 4x + \tfrac{6}{2}x^2 + \tfrac{10}{6}x^3 + \cdots$$

$$= 4 + 4x + 3x^2 + \tfrac{5}{3}x^3 + \cdots$$

2. $f(x) = \cos x$ $\qquad f\left(\frac{\pi}{3}\right) = \frac{1}{2}$

$f'(x) = -\sin x$ $\qquad f'\left(\frac{\pi}{3}\right) = -\frac{\sqrt{3}}{2}$

$f''(x) = -\cos x$ $\qquad f''\left(\frac{\pi}{3}\right) = -\frac{1}{2}$

$$\cos x = \frac{1}{2} - \frac{\sqrt{3}}{2}\left(x - \frac{\pi}{3}\right) - \frac{\frac{1}{2}\left(x - \frac{\pi}{3}\right)^2}{2} + \cdots$$

$$= \frac{1}{2}\left[1 - \sqrt{3}\left(x - \frac{\pi}{3}\right) - \frac{1}{2}\left(x - \frac{\pi}{3}\right)^2 + \cdots\right]$$

3. $\ln(1 + x) = x - \frac{x^2}{2} + \frac{x^3}{3} - \frac{x^4}{4} + \cdots$

$$\ln 0.96 = \ln(1 - 0.04)$$

$$= -0.04 - \frac{(-0.04)^2}{2} + \frac{(-0.04)^3}{3} - \frac{(-0.04)^4}{4}$$

$$= -0.040\,822\,0$$

4. $f(x) = \frac{1}{\sqrt{1 - 2x}} = (1 - 2x)^{-1/2}$

$$(1 + x)^n = 1 + nx + \frac{n(n - 1)}{2}x^2 + \cdots$$

$$(1 - 2x)^{-1/2} = 1 + \left(-\frac{1}{2}\right)(-2x)$$

$$+ \frac{-\frac{1}{2}(-\frac{3}{2})}{2}(-2x)^2 + \cdots$$

$$= 1 + x + \tfrac{3}{2}x^2 + \cdots$$

5. $\displaystyle\int_0^1 x\cos x\,dx = \int_0^1 x\left(1 - \frac{x^2}{2} + \frac{x^4}{24}\right)dx$

$\displaystyle= \int_0^1\left(x - \frac{x^3}{2} + \frac{x^5}{24}\right)dx$

$\displaystyle= \frac{1}{2}x^2 - \frac{x^4}{8} + \frac{x^6}{144}\Bigg|_0^1$

$= \frac{1}{2} - \frac{1}{8} + \frac{1}{144} - 0 = 0.3819$

6. $f(t) = 0 \qquad -\pi \le t < 0$

$f(t) = 2 \qquad 0 \le t < \pi$

$\displaystyle a_0 = \frac{1}{2\pi}\int_0^\pi 2\,dt = \frac{1}{\pi}t\Big|_0^\pi = 1$

$\displaystyle a_n = \frac{1}{\pi}\int_0^\pi 2\cos nx\,dx = \frac{2}{n\pi}\sin nx\Big|_0^\pi$

$= 0 \quad$ for all n

$\displaystyle b_n = \frac{1}{\pi}\int_0^\pi 2\sin nx\,dx = \frac{2}{n\pi}(-\cos nx)\Big|_0^\pi$

$\displaystyle= \frac{2}{n\pi}(1 - \cos n\pi)$

$\displaystyle b_1 = \frac{2}{\pi}(1+1) = \frac{4}{\pi}, \qquad b_2 = \frac{2}{2\pi}(1-1) = 0,$

$\displaystyle b_3 = \frac{2}{3\pi}(1+1) = \frac{4}{3\pi}$

$\displaystyle f(t) = 1 + \frac{4}{\pi}\sin x + \frac{4}{3\pi}\sin 3x + \cdots$

Chapter 30

1. $\displaystyle x\frac{dy}{dx} + 2y = 4$

$\displaystyle dy + \frac{2}{x}y\,dx = \frac{4}{x}dx$

$e^{\int\frac{2}{x}dx} = e^{2\ln x} = x^2$

$\displaystyle yx^2 = \int\frac{4}{x}x^2\,dx = \int 4x\,dx = 2x^2 + c$

$\displaystyle y = 2 + \frac{c}{x^2}$

2. $y'' + 2y' + 5y = 0$

$m^2 + 2m + 5 = 0$

$\displaystyle m = \frac{-2 \pm \sqrt{4 - 20}}{2} = -1 \pm 2j$

$y = e^{-x}(c_1\sin 2x + c_2\cos 2x)$

3. $x\,dx + y\,dy = x^2\,dx + y^2\,dx = (x^2 + y^2)\,dx$

$\displaystyle \frac{x\,dx + y\,dy}{x^2 + y^2} = dx$

$\frac{1}{2}\ln(x^2 + y^2) = x + \frac{1}{2}c$

$\ln(x^2 + y^2) = 2x + c$

4. $2D^2y - Dy = 2\cos x$

$2m^2 - m = 0; \qquad m = 0, \frac{1}{2}$

$y_c = c_1 + c_2e^{x/2}$

$y_p = A\sin x + B\cos x$

$Dy_p = A\cos x - B\sin x$

$D^2y_p = -A\sin x - B\cos x$

$2(-A\sin x - B\cos x) -$

$(A\cos x - B\sin x) = 2\cos x$

$-2A + B = 0, \quad -2B - A = 2, \quad A = -\frac{2}{5}, B = -\frac{4}{5}$

$y = c_1 + c_2e^{x/2} - \frac{2}{5}\sin x - \frac{4}{5}\cos x$

5. $\displaystyle \frac{d^2y}{dx^2} - 4\frac{dy}{dx} + 4y = 3x$

$m^2 - 4m + 4 = 0; \qquad m = 2, 2$

$y_c = (c_1 + c_2x)e^{2x}$

$y_p = A + Bx, \qquad Dy_p = B, \qquad D^2y_p = 0$

$0 - 4B + 4(A + Bx) = 3x$

$4A - 4B = 0, \qquad 4B = 3$

$B = \frac{3}{4}, \qquad A = \frac{3}{4}$

$y = (c_1 + c_2x)e^{2x} + \frac{3}{4} + \frac{3}{4}x$

6. $\displaystyle (xy + y)\frac{dy}{dx} = 2$

$y(x + 1)\,dy = 2\,dx$

$\displaystyle y\,dy = \frac{2\,dx}{x + 1}$

$\frac{1}{2}y^2 = 2\ln(x + 1) + c$

$y = 2 \quad$ when $\quad x = 0$

$\frac{1}{2}(4) = 2\ln(1) + c, \quad c = 2$

$\frac{1}{2}y^2 = 2\ln(x + 1) + 2$

$y^2 = 4\ln(x + 1) + 4$

7. $\displaystyle \frac{dA}{dt} = rA$

$\displaystyle \frac{dA}{A} = r\,dt$

$\ln A = rt + \ln c$

$\displaystyle \ln\frac{A}{c} = rt$

$A = ce^{rt}$

$A_0 = ce^0,$

$c = A_0$

$A = A_0e^{rt}$

8. $y'' + 9y = 9$

$y(0) = 0, \qquad y'(0) = 1$

$L(y'') + 9L(y) = L(9)$

$\displaystyle s^2L(y) - s(0) - 1 + 9L(y) = \frac{9}{s}$

$\displaystyle (s^2 + 9)L(y) = \frac{9}{s} + 1$

$\displaystyle L(y) = \frac{9}{s(s^2 + 9)} + \frac{1}{s^2 + 9}$

$y = 1 - \cos 3t + \frac{1}{3}\sin 3t$

9. $\displaystyle L\frac{d^2q}{dt^2} + R\frac{dq}{dt} = E \qquad L\frac{di}{dt} + Ri = E$

$L = 2\text{ H}, \qquad R = 8\ \Omega, \qquad E = 6\text{ V}$

$\displaystyle 2\frac{di}{dt} + 8i = 6 \qquad di + 4i\,dt = 3\,dt$

$\displaystyle e^{\int 4\,dt} = e^{4t} \qquad ie^{4t} = \int 3e^{4t}\,dt = \frac{3}{4}e^{4t} + c$

$i = 0$ for $t = 0, \qquad 0 = \frac{3}{4} + c, \qquad c = -\frac{3}{4}$

$ie^{4t} = \frac{3}{4}e^{4t} - \frac{3}{4}$

$i = \frac{3}{4}(1 - e^{-4t}) = 0.75(1 - e^{-4t})$

10. $m = 0.5$ kg, $k = 32$ N/m

$mD^2x = -kx, \qquad 0.5D^2x + 32x = 0$

$D^2x + 64x = 0$

$m^2 + 64 = 0; \qquad m = \pm 8j$

$x = c_1\sin 8t + c_2\cos 8t$

$Dx = 8c_1\cos 8t - 8c_2\sin 8t$

$x = 0.3$ m, $\quad Dx = 0, \quad$ for $\quad t = 0$

$0.3 = c_1\sin 0 + c_2\cos 0, \qquad c_2 = 0.3$

$0 = 8c_1\cos 0 - 8c_2\sin 0, \qquad c_1 = 0$

$x = 0.3\cos 8t$

Index of Applications

Biology and Medical Science

Business and Finance

Chemistry

Civil Engineering

Computer

Construction

Design

Electricity and Electronics

Energy Technology

Environmental Science

Geodesy, Geology, Seismology

Hydrodynamics and Hydrostatics

Machine Technology

Materials

Measurement

Meteorology

Optics

Petroleum

Photography

Physics

Police Science

Refrigeration and Air Conditioning

Space Technology

Statics

Strength of Materials

Surveying

Thermal

Thermodynamics

Wastewater Technology

Index

Algebra

Exponents and Radicals

$a^m \cdot a^n = a^{m+n}$

$\frac{a^m}{a^n} = a^{m-n} \quad a \neq 0$

$(a^m)^n = a^{mn}$

$(ab)^n = a^n b^n$

$\left(\frac{a}{b}\right)^n = \frac{a^n}{b^n} \quad b \neq 0$

$a^0 = 1 \quad a \neq 0$

$a^{-n} = \frac{1}{a^n} \quad a \neq 0$

$a^{m/n} = \sqrt[n]{a^m} = (\sqrt[n]{a})^m$

$\sqrt{ab} = \sqrt{a}\,\sqrt{b}$

Special Products

$a(x + y) = ax + ay$

$(x + y)(x - y) = x^2 - y^2$

$(x + y)^2 = x^2 + 2xy + y^2$

$(x - y)^2 = x^2 - 2xy + y^2$

Quadratic Equation and Formula

$ax^2 + bx + c = 0$

$$x = \frac{-b \pm \sqrt{b^2 - 4ac}}{2a}$$

Properties of Logarithms

$\log_b x + \log_b y = \log_b xy$

$\log_b x - \log_b y = \log_b\left(\frac{x}{y}\right)$

$n \log_b x = \log_b (x^n)$

Complex Numbers

$\sqrt{-a} = j\sqrt{a} \quad (a > 0)$

$x + yj = r(\cos\theta + j\sin\theta)$

$= re^{j\theta} = r\underline{/\theta}$

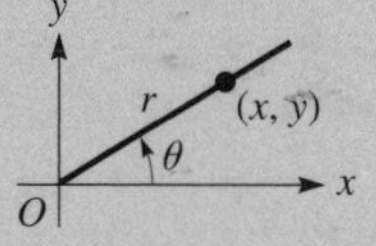

Variation

Direct variation: $y = kx$

Inverse variation: $y = k/x$

Trigonometry

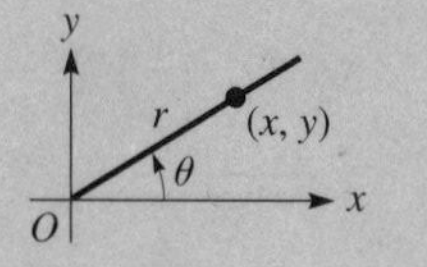

$\sin\theta = \frac{y}{r} \qquad \cos\theta = \frac{x}{r} \qquad \tan\theta = \frac{y}{x} \qquad \cot\theta = \frac{x}{y} \qquad \sec\theta = \frac{r}{x} \qquad \csc\theta = \frac{r}{y}$

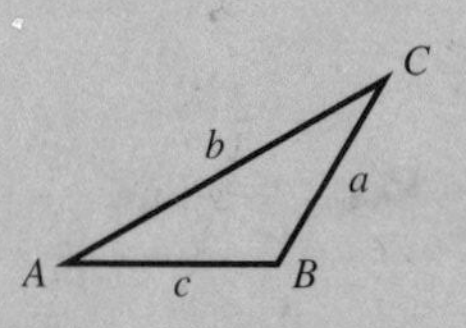

Law of Sines: $\frac{a}{\sin A} = \frac{b}{\sin B} = \frac{c}{\sin C}$

Law of Cosines: $a^2 = b^2 + c^2 - 2bc\cos A$

$\pi \text{ rad} = 180°$

Basic Identities

$\sin\theta = \frac{1}{\csc\theta} \qquad \cos\theta = \frac{1}{\sec\theta} \qquad \tan\theta = \frac{1}{\cot\theta} \qquad \tan\theta = \frac{\sin\theta}{\cos\theta} \qquad \cot\theta = \frac{\cos\theta}{\sin\theta}$

$\sin^2\theta + \cos^2\theta = 1 \qquad 1 + \tan^2\theta = \sec^2\theta \qquad 1 + \cot^2\theta = \csc^2\theta$

$\sin 2\alpha = 2\sin\alpha\cos\alpha \qquad \cos 2\alpha = \cos^2\alpha - \sin^2\alpha = 2\cos^2\alpha - 1 = 1 - 2\sin^2\alpha$

$\sin\frac{\alpha}{2} = \pm\sqrt{\frac{1 - \cos\alpha}{2}} \qquad \cos\frac{\alpha}{2} = \pm\sqrt{\frac{1 + \cos\alpha}{2}}$